Doubling Farmers Income through Horticulture

The Editors

Dr. K.L. Chadha (*Padma Shri Awardee*) is currently President of the Horticultural Society of India and also Adjunct Professor (Hort.) at IARI. He received M.Sc. (Hort.) degree from Punjab University, Chandigarh and Ph.D. from IARI, New Delhi. He had served in various capacities at IARI, New Delhi, PAU, Ludhiana and Central Mango Research Station, Lucknow. He was the Director, IIHR, Bengaluru, Horticulture Commissioner and Executive Director, National Horticulture Board, Govt. of India; DDG (Hort.), ICAR and ICAR-National Professor (Hort.) at IARI.

Dr. S.K. Singh is presently the Head, Division of Fruits and Horticultural Technology, IARI, New Delhi. He received M.Sc. and Ph.D. degrees in Horticulture from Banaras Hindu University, Varanasi and joined Agricultural Research Service in 1994. He is involved in teaching and improvement of mango & grape. He has developed three extra-early grape varieties and production technology of mango and grape. Dr. Singh is working on biotechnological application and marker-assisted breeding in fruit crops.

Dr. Pritam Kalia is the former Head, Division of Vegetable Science at IARI, New Delhi. He has significantly contributed in the field of vegetable breeding and developed 28 varieties/ hybrids of different vegetable crops. He pioneered popularizing exotic vegetable crops broccoli and leek by developing indigenous varieties for diversification and nutrition.

Dr. W.S. Dhillon is presently the Assistant Director General (Hort-I) at Indian Council of Agricultural Research, New Delhi. He did M.Sc. and Ph.D. from Punjab Agricultural University, Ludhiana. He has developed 10 varieties of fruit crops; 1 rootstock; and 35 technologies on propagation, production, protection and post-harvest management. He also served as Director, Punjab Horticultural Post-harvest Technology Centre (PHPTC), Ludhiana.

Dr. T.K. Behera is presently Professor of Vegetable Science at Indian Agricultural Research Institute, New Delhi. He received M.Sc. (Ag.) and Ph.D. in Horticulture from OUAT, Bhubaneshwar and IARI, New Delhi. He has significantly contributed in the field of vegetable breeding and developed 14 varieties/hybrids of cucurbitaceous vegetable crops. He introgressed gy-1 (gynoecious habit) gene into commercial varieties of bitter gourd and first time developed bitter gourd variety for cultivation under protected structures.

Dr. Jai Prakash is presently the Senior Scientist at the Division of Fruits and Horticultural Technology, IARI, New Delhi. He received M.Sc. (Hort.) from CSAUA&T, Kanpur and Ph.D. degree in Horticulture from Banaras Hindu University, Varanasi and joined Agricultural Research Service of ICAR in 2003. He is serving ICAR for the last 14 years and is involved in research & teaching and has developed improved papaya and pineapple varieties.

Doubling Farmers Income through Horticulture

— *Editors* —

K.L. Chadha

S.K. Singh

P. Kalia

W.S. Dhillon

T.K. Behera

Jai Prakash

2017

Daya Publishing House®

A Division of

Astral International Pvt. Ltd.

New Delhi – 110 002

ISBN 978-93-5124-828-6 (Hardbound)
ISBN 978-93-86071-67-5 (International Edition)

Published by : **Daya Publishing House®**
A Division of
Astral International Pvt. Ltd.
– ISO 9001:2008 Certified Company –
4760-61/23, Ansari Road, Darya Ganj
New Delhi-110 002
Ph. 011-43549197, 23278134
E-mail: info@astralint.com
Website: www.astralint.com

Digitally Printed at : Replika Press Pvt. Ltd.

This copy is for sale in Indian Sub-Continent.

Preface

Horticulture crop production surpassed food crop production in India for first time during 2013-14. The trend continued during 2014-15 with 283.5 MT. Share of horticulture output in agriculture GDP has increased to more than 33%. There has been highest annual growth of 9.5% in fruit production (2013-14) and 7% in vegetable production (1991-92 to 2014-15). Cut flower production has increased significantly. A virtual revolution has taken place in potato production too.Indian horticulture has been gradually penetrating in the International market with significant increase in export of different horticulture commodities. Export of horticulture produce has risen significantly by 536.42 times in quantity and 2007.80 times in value (1991-92 to 2012-13). India ranks second globally in Horticulture production. This led the Hon'ble Minister of Finance, Govt. of India, during his speech on unveiling Economic Survey 2015-16 in the Parliament to applaudethe achievements made in Horticulture sector in the country.

Though Horticulture sector has a sound technological back-up,yet marketing and retail, processing, export, value chain *etc.* are still big concerns and continue to slow down it's pace of development. These missing links need consolidation to make the Horti-business a win-win situation for all the stakeholders. Despite, ample increase in production of horticultural crops, low productivity and profitability in several crops is still a challenge and is required to be attended.

Hon'blePrime Minister of India has envisioned *"Doubling Farmers Income by 2022"*. It is felt that it can be achieved through interventions of Horticulture and allied sectors. To achieve this, factors which can improve productivity and profitability like biotechnological interventions, efficient resource management, soil and water use efficiency, precision horticulture, alternate production systems, management of abiotic stress and climate change, management of pests, diseases, physiological disorders, food safety, post-harvest management and value addition, marketing, trade and policy issues were presented in detail during the 7th Indian Horticulture Congress held at New Delhi.This compendium contains the edited full length version of 108 Invited lectures presented in detail by different scientists and experts during Congressorganized by the Horticultural Society of India, New Delhi to commemorate its Diamond Jubilee at Indian Agricultural Research Institute, New Delhi from 15-18 November 2016.

I am indebted to all contributors for responding to our request of agreeing to present lead papers on different topics and sending the manuscript well in time to be included in this publication to be released during the Inaugural Session.

I appreciate the efforts put in by the editorial team in bringing out this publication well in time. I also thank M/s Astral Publishing House, New Delhi for kindly agreeing to bringout this publication on the occasion of the Diamond Jubilee Year of the Horticultural Society of India. I am equally thankful to the Secretariat staff of the society who made tremendous effortsin providing secretarial support in this publication.

It is hoped that the publication will provide useful information on policies and technologies to double framer's income in years to come.

(K.L. Chadha)

Contents

— Reducing Post-Harvest Losses —

— Adding Value to Produce-I —

— Adding Value to Produce-II —

— Reducing Drudgery through Mechanization —

— Managing Climate Risks —

— Creating Employment Opportunities for Youth —

— Developing Horticulture in North East India —

List of Contributors

Ahmed, Nazeer
Vice-Chancellor, Sher-e-Kashmir University of Agricultural Sciences and Technology-Kashmir, Shalimar, Srinagar – 190 025, J&K
E-mail: dnak59@rediffmail.com

Ahuja, D.B.
Director (Actg.), ICAR-National Research Centre for Integrated Pest Management, LBS Building, IARI, New Delhi – 110 012
E-mail: deshbandhu4@rediffmail.com

Akshita, H.J.
ICAR-Indian Institute of Spices Research, Kozhikode – 673 012, Kerala

Alamalakala, Leela
Maharashtra Hybrid Seeds Company Private Limited, Dawalwadi, Jalna-Aurangabad Road, Jalna – 431 203, Maharashtra

Anandaraj, M.
Former Director, ICAR-Indian Institute of Spices Research, Kozhikode – 673 012, Kerala
Present Address: No. 7, S R Aptt., 8A cross, MSR Nagar, Bengaluru – 560 054, Karnataka
E-mail: arajiisr@gmail.com

Arora, N.K.
Associate Professor, Department of Fruit Science, Punjab Agricultural University, Ludhiana – 141 004, Punjab

Asrey, Ram
Principal Scientist, Division of Food Science and Post-harvest Technology, ICAR-Indian Agricultural Research Institute, New Delhi – 110 012
E-mail: ramasrey@iari.res.in

Babu, K. Nirmal
Director, ICAR-Indian Institute of Spices Research, Kozhikode – 673 012, Kerala
E-mail: nirmalbabu30@hotmail.com

Bajpai, Arpana
Senior Research Fellows, Precision Farming Development Centre, ICAR-Central Institute of Agricultural Engineering, Nabi Bagh, Berasia Road, Bhopal – 462 038, Madhya Pradesh

Ballal, Chandish R.
Director (Actg.), ICAR-National Bureau of Agricultural Insect Resources, Hebbal, Bengaluru – 560 024, Karnataka
E-mail: ballalchandish@gmail.com

Banerjee, Kaushik
Principal Scientist, ICAR-National Research Centre for Grapes, Manjri Farm, Pune – 412 307, Maharashtra
E-mail: kbgrape@yahoo.com

Banga, Km. Sheetal
Agro Produce Processing Division, ICAR-Central Institute of Agricultural Engineering, Nabi Bagh, Berasia Road, Bhopal – 462 038, Madhya Pradesh

Baranwal, V.K.
Professor and In-charge, Advance Centre for Plant Virology, Division of Plant Pathology, ICAR-Indian Agricultural Research Institute, New Delhi – 110 012
E-mail: vbaranwal2001@yahoo.com

Barman, Mandira
Scientist, Division of Soil Science and Agricultural Chemistry, ICAR-Indian Agricultural Research Institute, New Delhi – 110 012

Basu, Sudipta
Principal Scientist, Division of Seed Science and Technology, ICAR-Indian Agricultural Research Institute, New Delhi – 110 012
E-mail: bst_spu_iari@rediffmail.com

Behera, Tusar K.
Professor, Division of Vegetable Science, ICAR-Indian Agricultural Research Institute, New Delhi – 110 012
E-mail: tusar @ rediffmail.com

Bharati, Hemlata
Scientist, ICAR-Directorate of Medicinal and Aromatic Plants, Boriavi, Anand – 387 310, Gujarat

Bhattacharjee, S.
Executive Director, NERAMAC Ltd., A Govt. of India Enterprise, Guwahati – 781 005, Assam
E-mail: edfmd.niramac@gmail.com

Bhattacharya, Anjanabha
Maharashtra Hybrid Seeds Company Private Limited, Dawalwadi, Jalna-Aurangabad Road, Jalna – 431 203, Maharashtra

Chakrabarti, S.K.
Director, ICAR-Central Potato Research Institute, Shimla – 171 001, Himachal Pradesh
E-mail: skc_cpri@yahoo.co.in

Chandel, N.S.
Scientist, Agricultural Mechanization Division, ICAR-Central Institute of Agricultural Engineering, Nabi Bagh, Berasia Road, Bhopal – 462 038, Madhya Pradesh

Chandra, Pitam
Professor (Food Engg.), National Institute of Food Technology Entrepreneurship and Management, Kundli, Sonepat – 131 028, Haryana
E-mail: pc1952@gmail.com

Char, Bharat
Researcher, Maharashtra Hybrid Seeds Company Private Limited, Dawalwadi, Jalna-Aurangabad Road, Jalna – 431 203, Maharashtra

Chaturvedi, O.P.
Principal Scientist, ICAR-Central Agroforestry Research Institute, Jhansi – 284 003, Uttar Pradesh

Chaudhury, Rekha
Professor, ICAR-National Bureau of Plant Genetic Resources, Pusa Campus, New Delhi – 110 012

Chawda, Narayan
Director, VNR Seeds Private Limited, Raipur – 492 099, Chhattisgarh
E-mail: narayan.chawda @ vnrseeds.com

Chawda, Vimal
VNR Seeds Private Limited, Raipur – 492 099, Chhattisgarh

Cheriyan, Homey
Directorate of Arecanut and Spices Development, Kozhikode – 673 005, Kerala

Choudhary, Shalu
Maharashtra Hybrid Seeds Company Private Limited, Dawalwadi, Jalna-Aurangabad Road, Jalna – 431 203, Maharashtra

Dadlani, Malavika
Former Joint Director (Research), ICAR-Indian Agricultural Research Institute, New Delhi – 110 012

Damodaran, T.
Principal Scientist, ICAR-Central Soil Salinity Research Institute, RRS, Lucknow – 226 002, Uttar Pradesh
E-mail: damhort73@gmail.com

Das, Manish
Principal Scientist, Horticultural Science Division, Indian Council of Agricultural Research, Krishi Anusandhan Bhawan-II, New Delhi – 110 012

Dave, S.
Former Chairman of Codex Alimentarius Commission (a FAO/WHO body of the United Nations) and Advisor, FSSAI
E-mail: salve.codex@gmail.com

Deka, Bidyut C.
Director, ICAR-Agricultural Technology Application Research Institute (ATARI), Umroi Road, Umiam – 793 103, Meghalaya
E-mail: bidyutdeka@yahoo.com

Desai, Adavi Rao
Senior Scientist, ICAR-Central Coastal Agricultural Research Institute, Old Goa – 403 402, Goa
E-mail: adavirao.desai@icar.gov.in desaiavars@gmail.com

Deshmukh, N.A.
ICAR Research Complex for NEH Region, Umroi Road, Umiam – 793 103, Meghalaya

Devasahayam, S.
Principal Scientist, ICAR-Indian Institute of Spices Research, Kozhikode – 673 012, Kerala
E-mail: sdasahayam@gmail.com

Dhakar, Mahesh K.
Scientist (Fruit Science), ICAR-Research Complex for Eastern Region, Research Centre, Plandu, Ranchi – 834 010, Jharkhand

Dhatt, Ajmer Singh
Senior Vegetable Breeder and Head, Department of Vegetable Science, Punjab Agricultural University, Ludhiana – 141 004, Punjab
E-mail: ajmerdhatt@gmail.com

Dhillon, W.S.
Assistant Director General (Hort. Sci.), Horticultural Science Division, Indian Council of Agricultural Research, Krishi Anusandhan Bhawan-II, New Delhi – 110 012
E-mail: wasakhasingh @ yahoo.com

Dinesh, M.R.
Director, ICAR-Indian Institute of Horticultural Research, Hessaraghatta Lake Post, Bengaluru – 560 089, Karnataka
E-mail: drmrdinesh@gmail.com

Dora, Dillip K.
Dean, PGF-cum-DRI, Odisha University of Agriculture and Technology, Bhubaneswar – 751 003, Odisha
E-mail: dilipk_dora@yahoo.com

Dua, V.K.
Principal Scientist, ICAR-Central Potato Research Institute, Shimla – 171 001, Himachal Pradesh
E-mail: vkdua65@yahoo.co.in

Dwivedi, B.S.
Head, Division of Soil Science and Agricultural Chemistry, ICAR-Indian Agricultural Research Institute, New Delhi – 110 012
E-mail: bsdwivedi@yahoo.com

Eapen, Santhosh J.
ICAR-Indian Institute of Spices Research, Kozhikode – 673 012, Kerala

Gaikwad, B.
Scientist, AICRP on Farm Implements and Machinery, ICAR-Central Institute of Agricultural Engineering, Nabi Bagh, Berasia Road, Bhopal – 462 038, Madhya Pradesh

Gajanana, T.M.
Principal Scientist, Division of Social Sciences and Training, ICAR-Indian Institute of Horticultural Research, Hesaraghatta Lake Post, Bengaluru – 560 089, Karnataka
E-mail: gajanana@iihr.res.in

Ganeshamurthy, A.N.
Head, Division of Soil Science and Agricultural Chemistry, ICAR-Indian Institute of Horticultural Research, Hessaraghatta Lake Post Bengaluru – 560 089, Karnatka
E-mail: angmurthy@gmail.com

Gangwar, Suchi
Senior Research Fellows, Precision Farming Development Centre, ICAR-Central Institute of Agricultural Engineering, Nabi Bagh, Berasia Road, Bhopal – 462 038, Madhya Pradesh

George, James
Director (Acting) and Project Coordinator, ICAR-Central Tuber Crops Research Institute, Sreekariyam, Thiruvananthapuram – 695 017, Kerala
E-mail: ctcritvm@yahoo.com, ctcritvm@gmail.com

Gholap, B.S.
Research Associate, AICRP on Farm Implements and Machinery, ICAR-Central Institute of Agricultural Engineering, Nabi Bagh, Berasia Road, Bhopal – 462 038, Madhya Pradesh

Ghosh, S.P.
Ex-DDG (Horticulture), ICAR and Vice President, HSI, New Delhi – 110 012
E-mail: spghosh100@gmail.com

Gill, M.I.S
Head, Department of Fruit Science, Punjab Agricultural University, Ludhiana – 141 004, Punjab
E-mail: misgill@pau.edu

Hada, Bharat Singh
Ph.D. Student, Division of Floriculture and Landscaping, ICAR-Indian Agricultural Research Institute, New Delhi – 110 012

Haldankar, P.M.
Dean, College of Horticulture, Dr Balasaheb Sawant Konkan Krishi Vidyapeeth, Dapoli – 415 712, Maharashtra
E-mail: paraghaldankar@yahoo.in

Haldavnekar, P.C.
College of Horticulture, Dr Balasaheb Sawant Konkan Krishi Vidyapeeth, Dapoli – 415 712, Maharashtra

Hussain, K.
Sher-e-Kashmir University of Agricultural Sciences and Technology-Kashmir, Shalimar, Srinagar – 190 025, J&K

Jain, R.K.
Dean and Joint Director (Education),
ICAR-Indian Agricultural Research Institute,
New Delhi – 110 012

Jan, Aarifa
ICAR-Central Institute of Temperate Horticulture, Rangreth, Srinagar – 190 007, J&K

Jat, R.S.
Senior Scientist, ICAR-Directorate of Rapeseed-Mustard Research, Sewar, Bharatpur – 321 303, Rajasthan

Jawaharlal, M.
Dean, Horticultural College and Research Institute, Tamil Nadu Agricultural University, Coimbatore – 620 009, Tamil Nadu
E-mail: deanhort@tnau.ac.in

Jha, S.N.
Assistant Director General, (Process Engineering) Indian Council of Agricultural Research, Krishi Anusandhan Bhawan-II, New Delhi – 110 012
E-mail: snjha_ciphet@yahoo.co.in

Jindal, K.K.
Former Director Research,
Dr YS Parmar Horticulture and Forestry Nauni, Solan, Himachal Pradesh

Kadam, D.M.
Scientist, ICAR-Indian Institute of Soil and Water Conservation, Dehradun – 248 195,
Uttarakhand

Kalia, Pritam
Principal Scientist and Former Head, Division of Vegetable Science, ICAR-Indian Agricultural Research Institute, New Delhi – 110 012
E-mail: pritam.kalia@gmail.com

Kalloo, G.
Former Vice Chancellor, JNKVV, Jabalpur and Ex-Deputy Director General (Horticulture and Crop Sciences), Indian Council of Agricultural Research, New Delhi – 110 012
E-mail: gautamkallo@gmail.com

Kamal, Shwet
Former Director, ICAR-Directorate of Mushroom Research, Solan – 173 213, Himachal Pradesh

Kapoor, R.
Advance Centre for Plant Virology, Division of Plant Pathology, ICAR-Indian Agricultural Research Institute, New Delhi – 110 012

Kapoor, Tanvi
ICAR-Central Potato Research Institute, Shimla – 171 001, Himachal Pradesh

Kaur, Ratna Preeti
ICAR-Central Potato Research Station, PO Box 1, Model Town PO, Jalandhar – 144 003, Punjab

Kaushal, Rajesh
ICAR-Indian Institute of Soil and Water Conservation Dehradun – 248 195, Uttarakhand

Khanna, Manoj
Principal Scientist and In-Charge Farm Operation and Service Unit, Water Technology Centre, ICAR-Indian Agricultural Research Institute, New Delhi – 110 012
E-mail: khanna_manoj2001@yahoo.com.au

Kotwaliwale, Nachiket
Head, Agro Produce Processing Division, ICAR-Central Institute of Agricultural Engineering, Nabi Bagh, Berasia Road, Bhopal – 462 038, Madhya Pradesh
E-mail: nachiket.kotwaliwale@icar.gov.in

Kumar, Ashwani
Joint Secretary (Mechanisation and Technology), Ministry of Agriculture and Farmers Welfare, Krishi Bhavan, New Delhi – 110 001

Kumar, Chavlesh
Scientist, Division of Fruits and Horticultural Technology, ICAR-Indian Agricultural Research Institute, New Delhi – 110 012

Kumar, Dinesh
Principal Scientist and Director, ICAR-Central Citrus Research Institute, Nagpur – 440 010, Maharashtra

Kumar, Jitendra
Director, ICAR-Directorate of Medicinal and Aromatic Plants Research, Boriavi, Anand – 387 310, Gujarat
E-mail: jitendrakumar7@yahoo.com

Kumar, K.
Professor, Department of Pomology, College of Horticulture, Dr Y.S. Parmar University of Horticulture and Forestry, Nauni, Solan – 173 230, Himachal Pradesh

Kumar, Mahendra
Division of Soil Science and Agricultural Chemistry, ICAR-Indian Institute of Horticultural Research, Hessaraghatta Lake Post, Bengaluru – 560 089, Karnatka

Kumar, N.K. Krishna
Regional Director, Biodiversity International, Regional Office, NASC Complex, New Delhi – 110 012

Kumar, Pramod
Principal Scientist, Division of Agricultural Economics, ICAR-Indian Agricultural Research Institute, New Delhi – 110 012

Kumar, Rajesh
ICAR-Indian Institute of Vegetable Research, PB No. 01, P.O. Jakhini, Shahanshahpur, Varanasi – 221 305, Uttar Pradesh

Kumar, Satish
Former Director, ICAR-Directorate of Mushroom Research, Solan – 173 213, Himachal Pradesh

Kumar, Sudhir
Principal Scientist, ICAR-Central Agroforestry Research Institute, Jhansi – 284 003, Uttar Pradesh

Kumar, Susheel
Rubber Board Office, Parlakhemundi, Gajapati, Odisha – 761 200

Kumar, V.
Scientist, Agricultural Mechanization Division, ICAR-Central Institute of Agricultural Engineering, Nabi Bagh, Berasia Road, Bhopal – 462 038, Madhya Pradesh

Kumaranag, K.M.
ICAR-AICRP on Honey Bees and Pollinators, Division of Entomology, Indian Agricultural Research Institute, New Delhi – 110 012

Ladaniya, M.S.
Director, ICAR-Central Citrus Research Institute, Nagpur – 440 033, Maharashtra
E-mail: msladaniya@gmail.com

Lakhotia, Prativa
Scientist, Division of Floriculture and Landscaping, ICAR-Indian Agricultural Research Institute, New Delhi – 110 012

Lal, Barsati
ICAR-Central Institute for Sub-tropical Horticulture, Lucknow – 226 101, Uttar Pradesh

Lal, Gopal
Director (Acting), National Research Centre on Seed Spice, Tabiji, Ajmer – 305 206, Rajasthan
E-mail: glal67@yahoo.co

Lal, S.
Scientist, ICAR-Central Institute of Temperate Horticulture, Rangreth, Srinagar – 190 007, J&K

Lallan Ram
Director, Central Institute of Horticulture, Medziphema – 797 106, Nagaland
E-mail: lallanram09@gmail.com

Leela, N.K.
Division of Crop Production and Post Harvesting, ICAR-Indian Institute of Spices Research, Marikunnu, Kozhikode – 673 012, Kerala

Legha, Mohan Ram
Ph.D Student, Division of Floriculture and Landscaping, ICAR-Indian Agricultural Research Institute, New Delhi – 110 012

Mahajan, Gopal Ramdas
Natural Resource Management, ICAR-CCARI, Old Goa – 403 402, Goa

Malhotra, S.K.
Agriculture and Horticulture Commissioner, Ministry of Agriculture and Farmers Welfare, Krishi Bhawan, New Delhi – 110 001
E-mail: agricommissioner@gmail.com

Mandal, Debashis
Principal Scientist and National Fellow, ICAR-Indian Institute of Soil and Water Conservation Dehradun – 248 195, Uttarakhand
E-mail: dmandalcswcrti@gmail.com

Mangal, Manisha
Senior Scientist, Division of Vegetable Science, ICAR-Indian Agricultural Research Institute, New Delhi – 110 012

Mani, Indra
Head, Division of Agricultural Engineering, ICAR-Indian Agricultural Research Institute,
New Delhi – 110 012
E-mail: head_engg@iari.res.in

Manivannan, S.
Dean, Students' Welfare, Sikkim University, Gangtok – 737 102, Sikkim
E-mail: smanivanan@cus.ac.in

Manivel, P.
Principal Scientist, ICAR-Directorate of Medicinal and Aromatic Plants, Boriavi, Anand – 387 310, Gujarat
E-mal: manivelp@yahoo.com

Mansuri, Shekh Mukhtar
Principal Scientist, Agro Produce Processing Division, ICAR-Central Institute of Agricultural Engineering, Nabi Bagh, Berasia Road, Bhopal – 462 038, Madhya Pradesh

Masuti, Deelep
Assistant Professor, KRC College of Horticulture, Arabhavi, UHS, Bagalkot – 587 104, Karnataka

Meena, M.C.
Scientist, Division of Soil Science and Agricultural Chemistry, ICAR-Indian Agricultural Research Institute, New Delhi – 110 012

Mehta, C.R.
Project Coordinator, AICRP on Farm Implements and Machinery, ICAR-Central Institute of Agricultural Engineering, Nabi Bagh, Berasia Road, Bhopal – 462 038, Madhya Pradesh
E-mail: cr.mehta@icar.gov.in, crmehta65@yahoo.co.in

Mehta, R.S.
Principal Scientist, National Research Centre on Seed Spice, Tabiji, Ajmer – 305 206, Rajasthan

Mikkilineni, Venugopal
Maharashtra Hybrid Seeds Company Private Limited, Dawalwadi, Jalna-Aurangabad Road, Jalna – 431 203, Maharashtra

Mir, J.I.
Senior Scientist, ICAR-Central Institute of Temperate Horticulture, Rangreth, Srinagar – 190 007, J&K
E-mail: javidiqbal1234@gmail.com

Mishra, Dushyant
Principal Scientist, ICAR-Central Institute for Sub-tropical Horticulture, Lucknow – 226 101, Uttar Pradesh

Mishra, Gyan P.
Senior Scientist, ICAR-Indian Institute of Vegetable Research, PB No. 01, P.O. Jakhini, Shahanshahpur, Varanasi – 221 305, Uttar Pradesh

Misra, A.K.
Former Project Coordinator (AICRP-STF), ICAR-Central Institute for Sub-tropical Horticulture, Rehmankhera, Lucknow – 226 101, Uttar Pradesh
E-mail: misra_a_k@yahoo.co.in

Moni, M.
Professor Emeritus and Chairman, Centre for Agricultural Informatics and e-Governance Research Studies, Shobhit University, Delhi
E-mail: moni@shobhituniversity.ac.in; moni@nic.in

Murthy, D. Sreenivasa
Sr. Scientist, Division of Social Sciences and Training, ICAR-Indian Institute of Horticultural Research, Hesaraghatta Lake Post, Bengaluru – 560 089, Karnataka

Mustaffa, Mohammad
ICAR- AICRP on Honey Bees and Pollinators, Division of Entomology, Indian Agricultural Research Institute, New Delhi – 110012

Narayana, C.K.
Principal Scientist, Division of Post-harvest Technology and Agricultural Engineering, ICAR-Indian Institute of Horticultural Research, Bengaluru – 560 089, Karnataka

Nath, Vishal
Director, ICAR-National Research Centre on Litchi, Muzaffarpur – 842 002, Bihar
E-mail: vishalnath66@rediffmail.com

Nissar, Muhammad
ICAR-Indian Institute of Spices Research, Kozhikode – 673 012, Kerala

Oberoi, Harinder Singh
Head, Division of Post-harvest Technology and Agricultural Engineering, ICAR-Indian Institute of Horticultural Research, Bengaluru – 560 089, Karnataka
E-mail: hari_manu@yahoo.com

Pal, R.K.
Director, ICAR-National Research Centre on Pomegranate, Kegon, Solapur – 413 255, Maharashtra
E-mail: rkrishnapal@gmail.com

Pal, Ram
Senior Scientist, National Research Centre for Orchids, Darjeeling Campus, Darjeeling – 734 101, West Bengal

Pal, Suresh
Director, ICAR-National Institute of Agricultural Planning, Pusa, New Delhi – 110 012
E-mail: spl.econ@gmail.com

Pandey, A.K.
Dean, College of Horticulture and Forestry, Central Agricultural University, Pasighat – 791 102, Arunachal Pradesh
E-mail: pandey.ajai1@gmail.com

Pandey, G.
Principal Scientist, ICAR-Central Institute for Sub-tropical Horticulture, Lucknow – 226 101, Uttar Pradesh

Pandey, Sudhakar
Senior Scientist, ICAR-Indian Institute of Vegetable Research, PB No. 01, P.O. Jakhini, Shahanshahpur, Varanasi – 221 305, Uttar Pradesh

Pandit, A.H.
Associate Professor, Sher-e-Kashmir University of Agricultural Sciences and Technology-Kashmir, Shalimar, Srinagar – 190 025, J&K

Panduranga, G.S.
Division of Entomology, ICAR-Indian Agricultural Research Institute, New Delhi – 110 012

Pant, Vimal
Associate Professor, National Institute of Food Technology, Entrepreneurship and Management, Kundli – 131 028, Haryana

Parkhi, Vilas
Maharashtra Hybrid Seeds Company Private Limited, Dawalwadi, Jalna-Aurangabad Road, Jalna – 431 203, Maharashtra

Parulekar, Y.R.
College of Horticulture, Dr Balasaheb Sawant Konkan Krishi Vidyapeeth, Dapoli – 415 712, Maharashtra

Patel, V.B.
University Professor-cum-Chief Scientist (Hort.), Bihar Agricultural University, Sabour, Bhagalpur – 813 210, Bihar

Pathak, Prachi
Assistant Professor, Doon University, Banjarawala, Kedarpur, Dehradun – 248 001, Uttarakhand

Pathania, Radhika
ICAR-Central Potato Research Institute, Shimla – 171 001, Himachal Pradesh

Patil, Bapurayagouda
Assistant Professor, Seed Unit, University of Horticultural Sciences, Bagalkot – 587 104, Karnataka

Patil, D.R.
Special Officer (Seeds), Seed Unit, University of Horticultural Sciences, Bagalkot – 587 104, Karnataka
E-mail: adre.uhsbagalkot@gmail.com

Pongener, Alemwati
Scientist, ICAR-National Research Centre on Litchi, Muzaffarpur – 842 002, Bihar

Poornima, K.
Professor, Department of Nematology, Tamil Nadu Agricultural University, Coimbatore – 641 003, Tamil Nadu

Prakash, S.
Scientist, Agricultural Mechanization Division, ICAR-Central Institute of Agricultural Engineering, Nabi Bagh, Berasia Road, Bhopal – 462 038, Madhya Pradesh

Prasad, K.V.
Director, ICAR-Directorate of Floricultural Research, Shivajinagar, Pune – 411 005, Maharashtra
E-mail: kvprasad66@gmail.com

Prasanna, H.C.
Senior Scientist, ICAR-Indian Institute of Vegetable Research, PB No. 01, P.O. Jakhini, Shahanshahpur, Varanasi – 221 305, Uttar Pradesh

Punera, Bhoopesh
Research Scholar, Division of Agricultural Economics, ICAR-Indian Agricultural Research Institute, New Delhi – 110 012

Raghupathi, H.B.
Senior Scientist, Division of Soil Science and Agricultural Chemistry, ICAR-Indian Institute of Horticultural Research, Hessaraghatta Lake Post Bengaluru – 560 089, Karnatka

Rai, A.C.
ICAR-Indian Institute of Vegetable Research, PB-01, Jakhini (Shahanshapur), Varanasi – 221 305, Uttar Pradesh

Rai, K.K.
ICAR-Indian Institute of Vegetable Research, PB-01, Jakhini (Shahanshapur), Varanasi – 221 305, Uttar Pradesh

Rai, Nagendra
Principal Scientist, ICAR-Indian Institute of Vegetable Research, PB-01, Jakhini (Shahanshapur), Varanasi – 221 305, Uttar Pradesh

Raja, W. Hassan
Scientist, ICAR-Central Institute of Temperate Horticulture, Rangreth, Srinagar – 190 007, J&K

Rajan, Shailendra
Director, ICAR-Central Institute for Sub-tropical Horticulture, Rehmankhera, PO Kakori, Lucknow – 226 101, Uttar Pradesh
E-mail: srajanlko@gmail.com

Ram, Lallan
Director, Central Institute of Horticulture, Ministry of Agriculture and Farmers Welfare, Govt. of India, Medziphema – 797 106, Nagaland
E-mail: lallanram09@gmail.com

Rani, R. Thilaka
Scientist, Division of Plant Pathology, ICAR-Indian Institute of Horticultural Research, Hessaraghatta Lake P.O., Bengaluru – 560 089, Karnataka

Ranjan, J.K.
Senior Scientist (Vegetable Science), Division of Crop Improvement, ICAR-Indian Institute of Vegetable Research, Varanasi – 221 305, Uttar Pradesh

Rao, K.V. Ramana
Principal Scientist, Precision Farming Development Centre, ICAR-Central Institute of Agricultural Engineering, Nabi Bagh, Berasia Road, Bhopal – 462 038, Madhya Pradesh
E-mail: kondapalli.rao@icar.gov.in

Rao, Madhusudhana
Division of Biotechnology, ICAR-Indian Institute of Horticultural Research, Hessaraghatta Lake, Bengaluru – 560 089, Karnataka

Rathnamma, K.
Division of Plant Pathology, ICAR-Indian Institute of Horticultural Research, Hessaraghatta Lake P.O., Bengaluru – 560 089, Karnataka

Ravindra, V.
Principal Scientist, Division of Plant Physiology and Biochemistry ICAR-Indian Institute of Horticultural Research, Hessaraghatta Lake Post, Bengaluru – 560 089, Karnataka
E-mail: vattem@iihr.res.in; vattem_r58@yahoo.com

Ravishankar, H.
Ex-Director, ICAR-Central Institute for Sub-tropical Horticulture, Rehmankhera, Lucknow, Uttar Pradesh
Principal Scientist (Horticulture) and In-charge, IIHR-ATIC, ICAR-Indian Institute of Horticultural Research, Bengaluru – 560 089, Karnataka
E-mail: ravishankar@iihr.res.in, drhravishankar@gmail.com

Reddy, N.N.
Principal Scientist (Horticulture), ICAR-Central Research Institute for Dryland Agriculture, Hyderabad – 500 059, Telangana
E-mail: drreddynn@gmail.com

Reddy, R.S.B.
ICAR-Indian Institute of Vegetable Research, PB-01, Jakhini (Shahanshapur), Varanasi – 221 305, Uttar Pradesh

Reddy, S. Vijay Rakesh
Scientist (Horticulture), ICAR-Central Institute for Arid Horticulture, Beechhwal Bikaner – 334 006, Rajasthan

Rupa, T.R.
Division of Soil Science and Agricultural Chemistry, ICAR-Indian Institute of Horticultural Research, Hessaraghatta Lake Post Bengaluru – 560 089, Karnatka

Ryavalad, Shivayogi
Assistant Professor, Seed Unit, University of Horticultural Sciences, Bagalkot – 587 104, Karnataka

Rymbai, H.
Scientist, ICAR Research Complex for NEH Region, Umroi Road, Umiam – 793 103, Meghalaya

S.K. Malik
Principal Scientist, ICAR-National Bureau of Plant Genetic Resources, Pusa Campus, New Delhi – 110 012
E-mail: skm1909@gmail.com

Sabir, Naved
Principal Scientist, Division of Nematology, ICAR-Indian Agricultural Research Institute, New Delhi – 110 012
E-mail: n_sabir@rediffmail.com

Sadamate, V.V.
Former Adviser (Agriculture), Niti Aayog (Formerly Planning Commission), GoI, Sansad Marg, New Delhi – 110 001
E-mail: sadamatevv@gmail.com

Sagar, V.R.
Head, Division of Food Science and Post-harvest Technology ICAR-Indian Agricultural Research Institute, New Delhi – 110 012

Sahijram, Leela
Principal Scientist, Division of Biotechnology, ICAR-Indian Institute of Horticultural Research, Hessaraghatta Lake, Bengaluru – 560 089, Karnataka
E-mail: leelas@iihr.res.in

Sahni, R.
Scientist, Agricultural Mechanization Division, ICAR-Central Institute of Agricultural Engineering, Nabi Bagh, Berasia Road, Bhopal – 462 038, Madhya Pradesh

Sajeev, M.S.
Principal Scientist, Crop Utilization ICAR-Central Tuber Crops Research Institute, Sreekariyam, Thiruvananthapuram – 695 017, Kerala

Sajeev, M.V.
Sr. Scientist, ICAR-Directorate of Cashew Research, Puttur – 574 202, Karnataka

Sankaran, M.
Senior Scientist, ICAR-Indian Institute of Horticultural Research, Hessaraghatta Lake Post, Bengaluru – 560 089, Karnataka

Sarswat, B.L.
Director (C/C) and Executive Director, National Bee Board, Department of Agriculture, Cooperation and Farmers Welfare, Ministry of Agriculture and Farmers Welfare, Govt. of India, Krishi Bhawan, New Delhi – 110 001
E-mail: bl.sarswat@nic.in

Sardana, H.R.
Principal Scientist, National Research Centre for Integrated Pest Management, LBS Building, IARI, New Delhi – 110 012

Saritha, R.K.
Advance Centre for Plant Virology, Division of Plant Pathology, ICAR-Indian Agricultural Research Institute, New Delhi – 110 012

Saroj, P.L.
Director, ICAR-Central Institute for Arid Horticulture, Beechhwal, Bikaner, Rajasthan
E-mail: sarojicar@gmail.com, plsaroj@yahoo.co.in

Satisha, G.C.
Principal Scientist, Division of Soil Science and Agricultural Chemistry, ICAR-Indian Institute of Horticultural Research, Hessaraghatta Lake Post Bengaluru – 560 089, Karnatka

Sawant, S.D.
Director, ICAR-National Research Centre for Grapes, Manjri Farm Post, Pune – 412 307, Maharashtra
E-mail: ipmsawant@yahoo.co.in

Saxena, A.K.
Principal Scientist, Division of Plant Pathology, ICAR-Indian Institute of Horticultural Research, Hessaraghatta Lake P.O., Bengaluru – 560 089, Karnataka
E-mail: aks@iihr.res.in

Selvakumar, G.
Senior Scientist, Division of Soil Science and Agricultural Chemistry, ICAR-Indian Institute of Horticultural Research, Bengaluru – 560 089, Karnataka

Shakil, N.A.
Principal Scientist, Division of Agricultural Chemicals, ICAR-Indian Agricultural Research Institute, New Delhi – 110 012
E-mail: iamshakil@gmail.com

Sharma, A.
Sr. Scientist, ICAR-Central Institute of Temperate Horticulture, Rangreth, Srinagar – 190 007, J&K

Sharma, Ajay Kumar
Principal Scientist, ICAR-National Research Centre for Grapes, Manjri Farm Post, Solapur Highway, Pune – 412 307, Maharasthra
E-mail: ajay.sharma1@icar.gov.in

Sharma, Kirti
Principal Scientist, Division of Entomology, ICAR-Indian Agricultural Research Institute, New Delhi – 110 012
E-mail: kirtisharma2@yahoo.com

Sharma, Laxman
Department of Horticulture, School of Life Sciences, Sikkim University, Gangtok

Sharma, Nimisha
Scientist, Division of Fruits and Horticultural Technology, ICAR-Indian Agricultural Research Institute, New Delhi – 110 012

Sharma, O.C.
Principal Scientist, ICAR-Central Institute of Temperate Horticulture, Rangreth, Srinagar – 190 007, J&K

Sharma, R.R.
Principal Scientist, Division of Food Science and Post-harvest Technology ICAR-Indian Agricultural Research Institute, New Delhi – 110 012

Sharma, V.P.
Former Director, ICAR-Directorate of Mushroom Research, Solan – 173 213, Himachal Pradesh
E-mail: director.mushroom@icar.gov.in

Shirgure, P.S.
Principal Scientist, ICAR-Central Citrus Research Institute, Nagpur – 440 033, Maharashtra

Shirur, Mahantesh
Senior Scientist, ICAR-Directorate of Mushroom Research, Solan – 173 213, Himachal Pradesh

Shukla, S.K.
Principal Scientist, ICAR-Central Institute for Sub-tropical Horticulture, Lucknow – 226 101, Uttar Pradesh
E-mail: skshuklacish@gmail.com

Shylesha, A.N.
Principal Scientist, Entomology, ICAR-National Bureau of Agriculturally Important Insect Resources, Hebbal, Bengaluru – 560 024, Karnataka
E-mail: anshylesha@gmail.com

Singh, Awani Kumar
Senior Scientist, Centre for Protection Technology, ICAR-Indian Agricultural Research Institute, New Delhi – 110 012

Singh, B.
Director, ICAR-Indian Institute of Vegetable Research, PB No. 01, P.O. Jakhini, Shahanshahpur, Varanasi – 221 305, Uttar Pradesh
E-mail: bsinghiivr@gmail.com

Singh, B.K.
Scientist, ICAR-Indian Institute of Vegetable Research, PB No. 01, P.O. Jakhini, Shahanshahpur, Varanasi – 221 305, Uttar Pradesh

Singh, Balraj
Vice-Chancellor, Agriculture University, Jodhpur – 342 304, Rajasthan
E-mail: drbsingh2000@yahoo.com

Singh, Brahma
Founder President, Indian Society for Protected Cultivation, New Delhi – 110 012
Former Director, DRDO
E-mail: brahma88@gmail.com

Singh, D.B.
Director, ICAR-Central Institute of Temperate Horticulture, Rangreth, Srinagar – 190 007, J&K

Singh, D.R.
Director, ICAR-National Research Centre for Orchids, Pakyong – 737 106, East Sikkim
E-mail: drsinghhort66@gmail.com

Singh, I.P.
Principal Scientist, ICAR-Central Citrus Research Institute, Amaravati Road, Nagpur – 440 033, Maharashtra

Singh, K.
Senior Scientist, ICAR-Central Institute for Sub-tropical Horticulture, Rehmankhera, PO Kakori, Lucknow – 226 101, Uttar Pradesh
E-mail: patelvb7@gmail.com

Singh, Major
Project Coordinator, All India Coordinated Vegetable Research Project (Vegetable Crops), Division of Crop Improvement, ICAR-Indian Institute of Vegetable Research, Varanasi – 221 305, Uttar Pradesh
E-mail: singhvns@gmail.com

Singh, Man
Professor, Water Technology Centre, ICAR-Indian Agricultural Research Institute, New Delhi – 110 012
E-mail: mansingh61@ymail.com

Singh, Manjit
Former Director, ICAR-Directorate of Mushroom Research, Solan – 173 213, Himachal Pradesh
E-mail: manjitbhandal122@gmail.com

Singh, Narendra Pratap
Director, ICAR–National Institute for Abiotic Stress Management, Baramati – 413 115, Maharashtra
E-mail: director.niasm@icar.gov.in

Singh, P.M.
Principal Scientist, ICAR-Indian Institute of Vegetable Research, PB No. 01, P.O. Jakhini, Shahanshahpur, Varanasi – 221 305, Uttar Pradesh

Singh, Ram
Division of Entomology, ICAR-Indian Agricultural Research Institute, New Delhi – 110 012

Singh, Rashmi
Principal Scientist, Division of Agricultural Extension, ICAR-Indian Agricultural Research Institute, New Delhi – 110 012

Singh, Ravindra
National Research Centre on Seed Spice, Tabiji, Ajmer – 305 206, Rajasthan

Singh, S.K.
Head, Division of Fruits and Horticultural Technology, ICAR-Indian Agricultural Research Institute, New Delhi – 110 012
E-mail: sanjaydr2@rediffmail.com

Singh, Shrawan
Senior Scientist, Division of Vegetable Science, ICAR-Indian Agricultural Research Institute, New Delhi – 110 012

Singh, Siddhartha
Associate Professor, College of Horticulture and Forestry, Central Agricultural University, Pasighat – 791 102, Arunachal Pradesh

Singh, Sudhir
Principal Scientist, ICAR-Indian Institute of Vegetable Research, Jakhini, Shahanshapur, Varanasi – 221 305, Uttar Pradesh
E-mail: sudhiriivr@gmail.com

Singh, Sukhwinder
Senior Scientist, ICAR-Central Potato Research Station, PO Box 1, Model Town PO, Jalandhar – 144 003, Punjab
E-mail: ss.cpri@gmail.com

Soudagar, T.P.
College of Horticulture, Dr Balasaheb Sawant Konkan Krishi Vidyapeeth, Dapoli – 415 712, Maharashtra

Sridhar, V.
Principal Scientist Division of Entomology and Nematology, ICAR-Indian Institute of Horticultural Research, Hessaraghatta Lake Post, Bengaluru – 560 089, Karnataka
E-mail: vsridhar@iihr.res.in

Srinivas, K.S.
Joint Secretary (Marketing) and Agricultural Marketing Advistor to the Govt. of India, Krishi Bhavan, New Delhi – 110 001
E-mail: jsmktg_agri@gov.in

Srinivasan, V.
Principal Scientist, ICAR-Indian Institute of Spices Research, Kozhikode – 673 012, Kerala

Srivastav, Manish
Principal Scientist, Division of Fruits and Horticultural Technology, ICAR-Indian Agricultural Research Institute, New Delhi – 110 012

Srivastava, A.K.
Principal Scientist, ICAR-Central Citrus Research Institute, Nagpur – 440 033, Maharashtra
E-mail: aksrivas2007@gmail.com

Srivastava, Arpita
Scientist, Division of Vegetable Science, ICAR-Indian Agricultural Research Institute, New Delhi – 110 012

Sudha, M.
Senior Scientist, Division of Social Sciences and Training, ICAR-Indian Institute of Horticultural Research, Hesaraghatta Lake Post, Bengaluru – 560 089, Karnataka

Talang, H.D.
ICAR Research Complex for NEH Region, Umroi Road, Umiam – 793 103, Meghalaya

Tattimani, Manjunath
Technical Assistant, MHREC, University of Horticultural Sciences, Bagalkot – 587 104, Karnataka

Tendulkar, S.S.
Associate Professor, College of Horticulture, Dr Balasaheb Sawant Konkan Krishi Vidyapeeth, Dapoli – 415 712, Maharashtra

Thakur, R.K.
Project Coordinator, ICAR-AICRP on Honey Bees and Pollinators, Division of Entomology, Indian Agricultural Research Institute, New Delhi – 110 012
E-mail: rkt_apic@rediffmail.com

Thorat, S.B.
College of Horticulture, Dr Balasaheb Sawant Konkan Krishi Vidyapeeth, Dapoli – 415 712, Maharashtra

Tiwari, Jagesh K.
Senior Scientist, ICAR-Central Potato Research Institute, Shimla – 171 001, Himachal Pradesh

Tiwari, P.S.
Head, Agricultural Mechanization Division, ICAR-Central Institute of Agricultural Engineering, Nabi Bagh, Berasia Road, Bhopal – 462 038, Madhya Pradesh
E-mail: pstiwari60@yahoo.com

Tiwari, R.K.
Chief Technical Officer, AICRP on Farm Implements and Machinery, ICAR-Central Institute of Agricultural Engineering, Nabi Bagh, Berasia Road, Bhopal – 462 038, Madhya Pradesh

Tiwari, S.K.
ICAR-Indian Institute of Vegetable Research, PB No. 01, P.O. Jakhini, Shahanshahpur, Varanasi – 221 305, Uttar Pradesh

Tolety, Janakiram
Assistant Director General (Horticulture Science), ICAR-Indian Council of Agricultural Research, New Delhi – 110 012
E-mail: janakiram.kab@gmail.com

Tomar, B.S.
Head, Division of Vegetable Science, ICAR-Indian Agricultural Research Institute, New Delhi – 110 012
E-mail: bst_spu_iari@rediffmail.com

Uma, S.
Director, ICAR-National Research Center for Banana, Trichy – 620 102, Kerala
E-mail: umabinit@yahoo.co.in

Upadhyay, Ashutosh
Associate Professor (Food Sci. and Tech.), National Institute of Food Technology Entrepreneurship and Management, Kundli, Sonepat – 131 028, Haryana

Usha
Ph.D. Student, Division of Floriculture and Landscaping, ICAR-Indian Agricultural Research Institute, New Delhi – 110 012

Vani, Akella
Principal Scientist and Head, Division of Biotechnology, Indian Institute of Horticultural Research, Hessaraghatta, Bengaluru – 560 089, Karnataka
E-mail: vaniakella@gmail.com

Varshney, Richa
ICAR-National Bureau of Agricultural Insect Resources, Hebbal, Bengaluru – 560 024, Karnataka

Vijayakumari, N.
Principal Scientist, Tissue Culture Laboratory, Central Citrus Research Institute (CCRI), Amravati Road, Nagpur – 440 010, Maharashtra
E-mail: narukullav@gmail.com

Walia, R.K.
Project Coordinator and Professor, AICRP on Nematodes, ICAR-Indian Agricultural Research Institute, New Delhi – 110 012
E-mail: pcnematodes@iari.res.in

Walia, Suresh
Former Professor and Principal Scientist, Division of Agricultural Chemicals, ICAR-Indian Agricultural Research Institute, New Delhi – 110 012
E-mail: sureshwalia@yahoo.co.in

Warade, S.D.
College of Horticulture and Forestry, Central Agricultural University, Pasighat – 791 102, Arunachal Pradesh

Yalamalle, V.R.
Ph.D. Scholar, Division of Seed Science and Technology, ICAR-Indian Agricultural Research Institute, New Delhi – 110 012

Zachariah, T. John
Head and Principal Scientist, Division of Crop Production and Post Harvesting, ICAR-Indian Institute of Spices Research, Marikunnu, Kozhikode – 673 012, Kerala
E-mail: john@spices.res.in

Zehr, Borwale Usha
Director, Maharashtra Hybrid Seeds Company Private Limited, Dawalwadi, Jalna-Aurangabad Road, Jalna – 431 203, Maharashtra
E-mail: usha.zehr@mahyco.com

Improving Availability and Quality of Planting Materials and Seeds

1

Production of Quality Planting Material in Fruit Crops: A Pre-Requisite for High Productivity

M.R. Dinesh and M. Sankaran

ICAR-Indian Institute of Horticultural Research, Hessaraghatta Lake Post, Bengaluru – 560 089, Karnataka
E-mail: drmrdinesh@gmail.com

India is bestowed with varied agro-climatic conditions, which favours the cultivation of wide range fruit crops such as temperate, tropical, sub-tropical and arid zone fruits. The increasing trend in area, production, and productivity of fruit crops from 2001 to 2015 (4.01 million ha to 7.21 million ha, 43.00 to 88.97 million tonnes and 10.70 to 12.30 t/ha) (Anonymous, 2015). Out of the total production, the contribution of each fruits *viz.*, banana (33.4 per cent), mango (20.70 per cent), papaya (6.30 per cent), guava (4.1 per cent), grapes (2.9 per cent), apple (2.8 per cent), sapota (2.8 per cent), pineapple (2.0 per cent), pomegranate (1.5 per cent), litchi (0.7 per cent) and others is (11.1 per cent) (Anonymous, 2015 a). The greatest bottleneck in the expansion of area under fruits is the non-availability of genuine and quality planting material in adequate quantity from reliable government nurseries. In view of growing importance of fruit crops, the demand for quality planting material has increased manifold throughout the country in the recent past. The production and supply of genuine high quality-planting materials either seeds or grafts is of great concern due to the high demand from the farmers. The working on horticulture under Planning Commission (Niti Aayog) has projected that there has been four fold increase of demand for planting materials of horticultural crops. The significance of demand for quality planting material can be very well gauged by the fact that in last 16 years, the area under fruits nearly doubled and in the last 10 years alone there has been quantum jump of nearly 25 per cent in area under fruits. Similarly in plantation crops also there has been an increase to the tune of 13 per cent in area. In nut crops like cashew and walnut the increase in the area is of the order of nearly 11 per cent (Prakash, 2010). The modern production technologies like "High density orchards" and "Meadow orchards" have resulted in accelerated demand for quality planting material of genuine pedigree (Singh, 2009). Seeds and planting materials are the basic and vital input for a successful agriculture production. Therefore, it is very much essential to maintain its purity and quality through various stages of seed production, *i.e.* Breeder, foundation, registered and certified seed (Anonymous, 2015b). The key issues pertaining to quality planting material production are discussed below:

1. Improved Varieties in Fruit Crops

Several high yielding varieties in tropical fruits have been released for the various regions and for different purposes in fruit crops. One of the main constraints in fruit crops is the adoption of new varieties by the growers in the past. However, with the changing trends in some of the crops *viz.*, custard apple, guava *etc.*, adoption of new varieties also has been very good. However, the consumer preference in some of the crops like mango is very strong. Hence, replacing a traditional variety becomes extremely difficult.

2. Establishment of Mother Blocks

In fact, once an elite type of material has been identified by research organizations or private sectors, it should be multiplied and maintained as the original source. From this, further stock is to be maintained permanently under strict supervision so that it can be used as a primary source for all subsequent propagation. This is called a foundation block. It will serve as primary the repository of true-to-type stock, and not as a direct source of propagating material by itself. The planting stock maintained as a source for commercial propagation is referred to as a mother-block. Hence, mother-block is considered to be at the heart of the nursery, without which one cannot expect production of quality planting material in fruit crops. Therefore, for preservation of disease-free planting stock; proper isolation, sanitation, periodical inspection and testing and maintenance are very essential (Hartmann *et al.*, 2004). Even mother-block for rootstocks should be established from where regular multiplication by layering, stem cuttings, *etc.* can be done.

As far as possible, the mother-block should be in close proximity to the nursery. The scion wood should be taken only from bearing plants. To ensure quality of planting material, there is a need for quality and health management system to be in place so that increasing demand for propagules can be met. There should be proper records and regular certification of planting stock. Commercial planting stock can be referred to as certified stock if it is grown under the supervision of a legally designated agency with prescribed regulations, designed to maintain minimum standards developed for different fruit crops. These standards should be effectively implemented to ensure quality production of planting material. For improving nutrient and water use efficiency, the mother-block should have facilities like micro-irrigation and fertigation/system. Ultimately, a healthy mother-block will help produce quality-planting material in fruit crops.

Mother blocks should be established according to climate and soil conditions, topography and demand for planting materials. All the conditions required for establishment of fruit orchards should be also looked into for establishment of mother-blocks. Especially, irrigation facilities, electric supply, drainage, source of mother plants, *etc.* are of paramount significance. Various farm-operations should be performed timely, in order to produce healthy planting material. In general, 15-20 per cent less-spacing as that recommended for establishment of orchards should be followed for establishment of mother-blocks of different fruit crops. However, preservation of disease-free planting stock, proper isolation, sanitation, periodical inspection and testing and maintenance are very essential activities in maintenance of clonal stocks. To ensure quality of planting material, there is a need for quality and health management system so that increasing demand for propagules can be met.

3. Recommended Rootstocks

The rootstocks are of vital importance for successful fruit production. However, multiplication of these rootstocks in sufficient numbers with assured quality is not available for large-scale propagation in those regions affected by abiotic stresses *viz.*, as salinity, alkalinity and drought and biotic stresses such as nematodes and pests. While these rootstocks possess the resistance/ tolerance attributes, the quality of the produce by using these rootstocks may become another problem in certain situations. This has led to the situation at present wherein the plants produced by private nurserymen are procured by majority of the growers as alternate sources for raising the planting materials. The need in rootstock today is the creation of variability for rootstocks. An example that can be quoted here is the case of 'Dogridge' rootstock, which has revolutionized the grape industry in India. Out of 1.2 million ha, about 90 percent of the grape growing area is under 'Dogridge' rootstock, as it possess tolerance to drought, salinity and nematodes. Similarly, the polyembryonic mango varieties such as Vellaikolumban, Olour, 13-1 and Kurukkan are gaining the popularity in India. In guava, the Pusa Srijan and *Psidium molle* are commonly used to induce dwarf and wilt tolerance.

In order to produce true-to-the type planting materials most of the fruit crops are propagated by vegetatively except the crops like papaya, *phalsa* and mangosteen.

4. Micropropagation

Micropropagation of horticultural crops in India has been impressive with innumerable protocols developed by public and private sectors. Though several important protocols have been developed by public sector undertakings, very few commercial laboratories have been established. In India, more than 100 commercial laboratories are actively involved in micropropagation of various crops. The industry exports a sizeable quantity of micropropagated plants either in the culture stage in test tubes, or *in vitro* rooted plants/hardened plants. Besides exports, large quantities of micropropagated plants are sold in the domestic market. Among fruit crops, banana, pomegranate and strawberry are propagated commercially on a large scale. In banana, prevalence of disease problems and a need for generating clean planting-stock in large quantities has stimulated recent surge of interest in production of superior banana clones through the aseptic micropropagation technique.

5. Hi-Tech Seedling Production

The seedling production in protray/polybags has become a specialized practice and is fast catching up. The dimensions of the trays generally are 54 cm length and 27 cm breadth with cavity depth of 4 cm. These are made of polypropylene and are re-usable. Life of the tray depends on handling. Seedling trays are designed in such a way that a sapling gets pre-calculated growth media and the right amount of moisture (as the trays have pre-punched holes in each cavity for proper drainage of excess water and also have the right spacing). It is already a commercial venture to produce seedlings of papaya, banana and strawberry hybrids using seedling trays and protective structures. Many progressive farmers are coming forward to take up quality-seedling production using seedling trays. Even in fruit crops like papaya, bigger-sized trays can be used for raising seedlings. Since cost of F_1 hybrid seeds is quite high, the seedling tray method is mostly adopted for propagation.

6. Problems Encountered by the Farmers

Generally, farmers procure the planting materials from the local nurserymen on belief but very often the farmers complain about the sudden drying of grafts/ budded plants even after 2 to 3 years, non-flowering of the trees even after 6-7 years, poor fruit setting, less yield and poor quality of fruits. The only reason for all these problems could be unawareness of the farmers while procuring planting materials because they do know the techniques in identifying the correct nurseries to source the right planting materials. There are certain guidelines to be followed for selecting planting materials; nurseries should have a well maintained mother block (scion bank), proper labeling of the variety, the grafts should have proper labels of the variety and the date of grafting should be displayed. One should realize that the fruit plants will start bearing only after four or five years and if the fruits are of poor quality, faulty planting materials are responsible and whatever money and energy have been spent are a waste.

7. Problems Associated with Nurseries

The horticulture nurseries in the country have many in-built lacunae, which makes them qualitatively and quantitatively incompetent to meet the present demand (Sharma and Srivastava, 2004). Some of the problems are:

i) Poor maintenance of mother block

ii) Non-availability of skilled man-power: success of the grafting/budding is purely depends on the experience of the person who the job

iii) In most of the nurseries, health and sanitary considerations are far from desirable. Timely availability of healthy plant material should be the order of the day.

iv Even today, there are very few nurseries having their progeny orchards. Today, it is mandatory to have scion and rootstock banks as part of nursery activities. This facilitates pedigree-assessment periodically and ascertains true-to-type nature of both scion and stock material.

v) Rootstock bank of a known genotypes is becoming the order of the day. In crops like citrus, well defined rootstocks are in place where their mother plant blocks need to be maintained. It is also mandatory to periodically ascertain clonal fidelity in these. Similarly, in grapes, currently Dogridge is a widely employed rootstock in vineyards.

vi) Another important issue, which is plaguing the nursery industry today is lack of proper labelling and certification of planting material.

vii) Today, in modern nurseries, quality inputs like UV-stabilized bags, good organic substrates (media), soil solarizers, are all very essential. However, very few nurseries can boast of having these inputs, which can qualitatively improve plant health and subsequent performance in the field.

viii) Most of the modern day nurseries operate from small units around cities; with the result they do not have mother plant blocks, which require large space. They depend upon scion wood and stock plant supply either from other sources or their own facility located far away.

8. Quality Control Mechanism

Government of India has framed and brought out different legislations to protect the quality of seeds and planting materials. Seed Act (1966), Seed Rules (1968), Seed (Control) Order(1983), New Policy on Seed Development (1988), Plants, Fruits and Seeds (Regulation of import into India), 1989, The PPV and FR Act (2001), Essential Commodities Act including Seeds (1955), National Seed Policy (2002) and Seed Bill (2004) take care of seeds right from the production to marking, labeling and marketing levels to maintain the quality standards as prescribed by the Central Seed Committee. These laws are framed in order to make available quality seeds to a common farmer and equip him to approach authority for justice. It is now well established that there is a need to increase area under fruit crops, in order to step up production of fruits. In India, more than 4,409 fruit nurseries including 1,575 under government sector and 2834 under private sector are functioning, which have an annual target of producing 1,387 million fruit plants. In order to ensure the quality materials the Seeds Act and the Nursery Registration Act have been in operation since December, 1966, is in force with regard to horticulture nurseries only in the states of Punjab, Maharashtra, Himachal Pradesh, Uttar Pradesh, Uttrakhand, Jammu and Kashmir, Odisha and Tamil Nadu, whereas in other states the act is not properly implemented due to non-availability of commercial nurseries. With a view to ensure availability of genuine and quality planting material until a new regime of registration of horticulture nurseries under the provisions of a new Seeds Act, 2005, comes in force, it is essential to put a system of Recognition of Horticulture Nurseries in place to facilitate, promote and monitor production and trade of quality planting materials of horticulture crops with a special emphasis on perennial fruit crops, which are propagated vegetatively. The 'National Horticulture Board' in consultation with the Department of Agriculture and Cooperation and the ICAR has prepared the guidelines for recognition of 'Nursery'; which, in turn, not only provide for a procedure of nursery recognition but also aim at capacity building among owners of horticulture nurseries. Now a days, there is sudden increase in the demand for certain commercial plants which are being addressed by Hi-tech nurseries having greenhouse, building of glass or a plastic tunnel, designed to protect young plants from harsh weather, while allowing access to light and ventilation. Modern greenhouses allow automated control of temperature, ventilation, light, watering, fertigation and hardening chambers. State Governments must ensure supply of quality planting materials for fruit crops by enactment of 'Fruit Plant Nursery (Regulation) Act' and enforcement of its provisions through licensing of horticulture nurseries. In 2006, Government of India has also established the "National Certification System for Tissue Culture Raised Plants (NCS-TCP)" authorizing Department of Biotechnology, Ministry of Science and Technology as the Certification Agency vide the Gazette Notification dated 10th March, 2006 under the "Seeds Act, 1966" for ensuring production and distribution of quality tissue culture planting materials. The purpose of NCS-TCP is to ensure production and distribution of quality tissue culture planting materials. NCS-TCP is a unique quality management system, first of its kind in the world, which ensures recognition of Tissue Culture Production Facility for the production of quality planting material and certification of end products. Besides, the Export/Import of seeds and planting materials is governed by the Export and Import policy, 2002-07, which prescribes the rules and regulations for export and import of planting materials. Similarly, the APEDA reiterates that farmers must follow the Good Agricultural Practices (GAP) to produce the quality produce, which can fetch the premium price in the international market. Like USA, UK, Brazil *etc.*, Govt. of India also formulated and brought out a publication on IndiaGAP, which has information about crop and location specific models for production of quality planting materials. It is pertinent to mention here that there should be coordinated effort from the ICAR institutes, SAUs and State Horticulture Departments should ensure the quality planting materials supply to the farmers.

9. Conclusion

The inadequate availability/supply of quality planting material is the major constraint for realizing a sound horticulture industry in the country. Existing infrastructure could meet only 30-40 per cent of the demand for production of quality planting material in this sector. In order to supply quality seed and planting material, ICAR had undertaken crop-specific research for development of rapid diagnostic techniques. The NHB and NHM are being implementing several schemes for production of quality planting materials. However, there is an urgent need for nursery standards, diagnostics, virus indexing, production of pathogen-free clones and nursery registration with NHB, GoI for proper monitoring and ensuring the supply of quality planting materials.

10. References

Anonymous. 2015a. *Horticulture Database*. National Horticulture Board, GOI.

Anonymous. 2015b. *India GAP (Good Agricultural Practices)*. www.qcin.org/documents/GAP/**INDIAGAP**-FINAL.pdf - Cached.

Chada, K.L. 2016. Hundred years of fruit crop improvement in India. Keynote address during the *National Conference on Fruit Breeding in Tropics and Subtropics: An Indian Perspective*, April 27th–29th, held at IIHR, Bengaluru.

Hartmann, H.T., Kester, D.E., Davies, F.T. and Geneve, R.L. 2004. *Plant Propagation: Principles and Practices*. Prentice-Hall of India Private Limited, New Delhi.

Prakash, G.S. 2010. Demand and supply of quality planting material: National scenario. **In**: *Proceedings of National Conference on Production of Quality Seeds and Planting Material: Health Management in Horticultural Crops*, March 11th–14th, New Delhi, pp. 23–28.

Singh, H.P. 2009. Accelerating quality production of horticultural crops in India. *Indian Hort.*, **54**: 3–9.

Sharma, R.R. and Srivastav, M. 2004. *Plant Propagation and Nursery Management*. International Book Distributing Company, Lucknow, Uttar Pradesh.

Standards for planting material. (http://tmnehs.gov.in/writereaddata/Standards for Planting Material.pdf).

2

An Integrated Citrus Propagation Programme for Improving Production and Profitability

N. Vijayakumari

Tissue Culture Laboratory, Central Citrus Research Institute (CCRI), Amravati Road, Nagpur – 440 010, Maharashtra

E-mail: narukullav@gmail.com

Government of India has fixed a target to increase the production of fruit crops by 40 per cent from current production of 250 to 350 MT in the next 5 years. ICAR had adopted 'Farmers first policy' in Vision 2030, hence all the programs need to aim at improving the farm incomes by increasing the productivity and profitability of different horticultural crops. Citrus is the 3rd largest fruit industry of the country after banana and mango. Citrus crops are grown over an area of 9.87 lakh ha with a production of 121.81 lakh MT (Horticultural Statistics at a Glance 2015). The average productive life of citrus orchards is 20 years. The area under citrus is growing @ 5.95 per cent (0.30 lakh ha) per annum and about 5 per cent of the area (0.30 lakh ha) is replanted annually. Citrus fruits are mainly consumed as fresh, hence fruit quality parameters *viz.*, size, colour, ease of peeling, seed content, TSS, sugar-acid blend *etc.* influence the consumer acceptance significantly. While the consumer is concerned with fruit quality and cost parameters, the producer is concerned with productivity and profitability.

The problems related to productivity/profitability that are being faced by the citrus farmers presently are (i) Non availability of quality (elite, healthy) planting stock, (ii) Shorter productive life of orchards due to decline (Phytophthora, CTV, Greening *etc.*), (iii) Higher Post-harvest losses, (iv) Constrains like lack of adequate and timely irrigation, (v) Increasing cost of production due to increasing costs of inputs, manpower, (vi) Uncertainty of remunerative prices due to lack of developed marketing channels and processing facilities. Out of the above, problems listed at Sr.No i, ii and iii are related to research/technology and the remaining are concerned with non-technical issues hence ICAR has only facilitator/advisory role in solving them.

Quality of the planting material plays a major role in addressing the research/productivity related issues. The quality of planting stock refers to a set of genetic, physical, health related parameters such as (i) variety, (ii) vigor and size of stock, (iii) health of bud-graft and (iv) budding height *etc.* The citrus propagation program should aim at producing high quality citrus plants for achieving the targeted productivity of citrus orchards.

It is estimated that about 2.0 million plants are required for establishment of new orchards and replacement of old orchards. The numbers of reliable sources are very few and the quantum of planting stock supplied by them is very less. There are about 300 nurseries in Amravati district of Maharashtra alone which are supplying about 1.5 to 2.0 million plants to all the citrus growing regions of the country. The nurserymen are not paying any attention to quality parameters *viz.*, yield potential, health *etc.* of the stock and are inadvertently spreading the diseases leading to low productivity and lesser productive life. The Maharashtra Fruit Nurseries and Sale of Fruit Plants (Regulation) Act, 1969 is not being implemented properly for ensuring quality of planting stock. A technology mission for production of quality stock is started by CCRI and about 3.0 lakh seedlings are produced annually and supplied to nurserymen and farmers. The farmers are not getting adequate quantities of quality planting stock. At this rate it would never be possible to meet the total requirement. There would not be any significant improvement because the quantity supplied is not even adequate to meet the requirement of replacement of orchards. There is a need to go for large scale multiplication to meet the total requirement of plants by changing the present planting stock supply procedure of CCRI. Priority needs to be given to supply large quantity of quality nucleus mother stock only to registered nursery men and transfer the technology to approved nurseries to increase the production. It is desirable to identify 400 nurseries, each of 0.50 lakh production capacity, which are willing to implement Nursery Acts of respective States in all the major growing areas and start a sound sanitized planting stock production program for meeting the total requirement. The States Seed Development Corporations, States Horticultural nurseries, Orange Growers Associations, Progressive farmers, Farmers Producers organizations need to be encouraged and supported technically for this purpose. As regards Maharashtra, Mahabeej, PKV, private tissue culture companies, progressive Farmers have come forward for collaboration in this programme. The Nurseries of Amravati district need to be given focus to achieve quick results in 4-5 years. A technologically sound planting stock production programme integrating the best features of all the available propagation methods needs to be in place by all the stake holders jointly to meet the total requirement of quality plants.

1. Citrus Propagation

The citrus industry taps the desirable horticultural characteristics of selected scions *viz.*, precocity, high yield, uniformity and better fruit quality by vegetatively propagating them on compatible rootstocks. Since the rootstock exerts considerable influence on growth of scion, yield and quality of fruits, use of uniform rootstock seedlings for budding/grafting is advocated.

1.1 Commercial Citrus Propagation Methods

1. Seed: Acid lime, Khasi mandarin and Coorg mandarin.
2. Budding: Sweet orange, Kinnow mandarin, Nagpur mandarin, Grape fruit and Pummelo.
3. Cutting and layering: Lemons, Sweet limes and Citron.

1.2 Merits and Demerits

The merits and demerits with reference to commercial propagation are given in Table 1.

2. Advancements

has paved the way for development of newer technologies. Researchers had focused on overcoming the disadvantages of commercial propagation methods leading to refinement/development of new methods of propagation and widening the scope of application. These technologies are briefly described below:

2.1 Shoot Tip Grafting

A wide variety of infections *viz.*, virus and virus like diseases, fungus (phytophthora) and bacteria are reported from various citrus growing regions. The planting stock/trees infected with these diseases do not put on normal healthy growth and huge economic losses are caused due to increased cost of orchard maintenance, reduced production and productivity and productive life of the orchards. The commercial vegetative propagation methods have an inherent drawback of inadvertently spreading these diseases. Bud wood certification programmes were started and Nursery Registration Acts were enacted for ensuring production of quality plants/bud wood but expected results could not be achieved due to non availability of sound virus elimination and quick disease diagnostic techniques.

Shoot Tip Grafting (STG) technique developed by (Navarro *et al.*, 1975), at the University of California, Riverside and Elisa diagnostic techniques (Cambra *et al.*, 1990), had paved the way for implementing sound Citrus Sanitation and Citrus Quarantine programmes. At Central Citrus Research Institute, the protocols have been standardized for production of disease-free planting stock of Nagpur mandarin and exotic bud wood successfully. About 3.78 lakh quality plants of Nagpur mandarin had been supplied to nursery men/farmers and 23 exotic cultivars are made available to researchers. This technique is most suitable for elimination of viruses and recovery of healthy plants (Vijayakumari *et al.*, 2006; Vijayakumari and Karihaloo, 2012). *Citrus* industries of Spain and other countries were rejuvenated by

Table 1: The Merits and Demerits with Reference to Commercial Citrus Propagation

Propagation Method	*Advantages*	*Disadvantages*
Seed	☆ Uniform nucellar seedlings ☆ Cheaper ☆ Negligible disease transmission	☆ Not all species are poly-embryonic ☆ Seed availability is season dependent ☆ Long juvenility
Budding/grafting	☆ Precocious ☆ Combines the desirable characters of scion and rootstock ☆ Uniform stock	☆ Multiplication is limited by availability of suitable bud wood, season, budding skill ☆ Scope for inadvertent multiplication ofdiseases

successfully employing this technique. Nearer home (Vijayakumari *et al.*, 1994; Vijayakumari *et al.*, 2008; Vijayakumari, 2009; Vijayakumari, 2010; Vijayakumari, 2015), had succeeded in producing and supplying 3.78 lakh STG derived disease-free planting stock during 2012-16 to the citrus growers/nurserymen).

2.1.1 Double Grafting

Double grafting technique is mainly used in enhancing the survival of successful shoot tip grafted plants during lab to the land transfers. Successfully shoot tip grafted plants with 2-4 expanded leaves are transferred either by direct transfer in sterilized soil or by double grafting on vigorous rough lemon and Rangpur lime rootstock seedlings under protected greenhouse conditions. Direct transfer requires aseptic greenhouse conditions and continuous care of the plants transplanted to the sterile soil in the pots. To avoid losses by direct transfer due to contamination and poor growth, double grafting method is standardized (De Lange, 1978). In double-grafting, successful shoot tip graft is side grafted (double grafted) on one year old rough lemon or Rangpur lime rootstocks. Excellent subsequent growth is obtained by this method. Double grafting provides more base material in less time for indexing and subsequent multiplication (Vijayakumari, 2012).

2.2 Mini-Grafting

Mini-grafting can also be used as an alternate technique for the rescue of the *in vitro* regenerated plants which are difficult-to-root/weak rooting and unable to survive while transfer to the soil. Generally few *Citrus* species are recalcitrant to micropropagation due to low regeneration capacity, especially mature plant tissue. A major limitation is root regeneration rather than shoot multiplication. Mini grafting is often used where rooting capacity of micro cuttings is poor and also in regeneration of novel plants created in tissue culture that are difficult to root (Navarro *et al.*, 3; Vijayakumari, 12)

3. Micro-budding

Commercial citrus nurseries use shield/T budding method for nursery stock production. Rootstock seedlings of appropriate size/age (1-year-old in case of rough lemon and 1½ to 2-year -old in case of Rangpur lime), suitable bud wood and favorable climate/season are essential. Further bud grafts requires longer time (18-20 months) for field release. These requirements limit the rate of multiplication.

To overcome the above limitations, micro budding protocol is standardized and successfully employed for faster, economical and year-round multiplication of planting material of Nagpur mandarin and other *Citrus* species under humidity controlled screen house/low cost shade net structure and requires lesser space. Micro-budding is a miniature budding in which young citrus rootstock measuring 3 mm diameter are de-topped and scion bud is inserted in a wedge cut on the rootstock. The budded portion is immediately covered with a micro-pipette tip to modulate the climate and to protect it. The micro-pippets tips are removed after 12-14 days from the sprouted bud grafts. The bud grafts attain plantable size in 10 months. Sourcing of scion from STG derived disease free plants assures health of the plants. Further micro-budded plants are suitable for high density orchards. Researchers can use this technique for quicker biological indexing of virus, viroids, greening bacterium and other disease inoculation and expression studies.

Micro-budding was successful with different commercial scion and rootstock combinations (Skaria, 2000; Vijayakumari, 2016). Micro-budding was also used in *in vitro* rescue of genetically modified Rio Red grape fruit explants in Texas (Skaria, 2000), which were otherwise difficult-to-root and develop into complete plants. In Texas, USA, micro-budded orange, grape fruit and lemon trees produced fruits in less than 2 yrs (Skaria, 2000).

Micro-budding technique was explored for early age propagation in Nagpur mandarin, sweet orange, Kinnow mandarin, acidlime and calamondin on 6 months old two commercial rootstocks, *viz.*, rough lemon and Rangpur lime. In rough lemon rootstock all the cultivars micro-budded successfully with maximum success on Nagpur mandarin (100 per cent) followed by sweet orange (90 per cent), calamondin (70 per cent) Kinnow (60 per cent) and acid lime (40 per cent) (Vijayakumari, 2016).

4. Top Working/Tree Grafting of Citrus Tree

The top working of citrus trees is a process of changing the top of an established tree or changing the fruiting branches of a tree from one cultivar to another by budding or grafting (Williamson and Jackson, 1994). There are several reasons to change the fruiting branches of citrus trees, either to make unproductive trees and tress producing fruits of poor market value can readily be changed, for with a large healthy rootstock and to rejuvenate the declining tree for other reasons, it only takes about 3 years to have a successful top worked citrus tree in satisfactory production, sometimes even in shorter duration (Vijayakumari, 2016). About 0.30 lakh ha needs to be replanted annually due to decline. If 50 per cent of the declining orchards are rejuvenated, 6-7 years of productive life and production to the tune of 12 lakh tonnes can be increased.

5. Dovetailing of Propagation Methods

On the basis of experience gained the following recommendations are given:

i) Clean the proven elite scion using shoot tip grafting technique.

ii) Employ double grafting for faster and vigorous growth of scion wood in a protective net house.

iii) Use Micro-budding technique for faster, economical multiplication of bud grafts in a sanitized container nurseries.

iv) Introduce a mechanism for testing and certification of propagative bud wood by establishing referral labs at CCRI, State Agriculture Universities and State Seed testing labs.

v) Implement nursery acts of respective states for ensuring quality of plants produced and supplied.

6. Way Forward

There is an urgent need to establish an inter-institutional arrangement between all the stakeholders to define the roles and coordinate the efforts for boosting the Citrus Industry. ICAR institutes and SAUs need to undertake research works related to varietal improvement, sanitation, establishment and maintenance of protected foundation blocks, registered nursery men need to establish model nurseries on scientific lines and the line departments need to implement the Nursery Registration Act. The scope of existing seed testing laboratories may be widened and strengthened or referral labs may be established by the ICAR institutes and SAUs to undertake health certification.

7. References

Cambra, M., Garnsey, S.M., Permar, T.A., Henderson, C.T., Gumpf, D. and Vela, C. 1990. Detection of Citrus Tristeza Virus (CTV) with a mixture of monoclonal antibodies. *Phytopathology*, **80**: 103.

de Lange, J.H. 1978. Shoot tip grafting: A modified procedure. *Citrus Subtrop. Fruit J.*, **539**: 13–15.

Navarro, L., Roistacher, C.N. and Murashige, T. 1975. Improvement of shoot tip grafting *in vitro* for virus free citrus. *J. Amer. Hort. Sci.*, **100**: 471–479.

Skaria, M. 2000. A microbudding technique for biological indexing and ultra-high density planting of citrus. *Proc. 14th Conf.*, IOCV Riverside.

Vijayakumari, N. 2009. Superior production performance of STG derived planting stock vs conventional bud grafts in farmers fields of central India. Presented as oral paper, *Int. Conf. on Hort.*, Bengaluru, India 9th–12th November, pp. 276.

Vijayakumari, N. 2010 Citrus regeneration systems in Citrus. **In**: *Advances in Horticulture Biotechnology, Vol 1: Fruit Crops, Plantation Crops and Spices*, (Eds.) H.P. Singh, V.A. Parthasarathi, K. Nirmal Babu. Westville Publishing House, New Delhi, pp. 121–161.

Vijayakumari, N. 2015. Greenhouse transfer of commercial citrus species by mini-grafting, Central Citrus Research Institute, *Annual Report*, p. 27.

Vijayakumari, N. 2016. Micro-budding of commercial citrus scions on commercial citrus rootstock, Central Citrus Research Institute, *Annual Report*, pp. 23–24.

Vijayakumari, N. 2016. Micro-budding of commercial citrus scions on commercial citrus rootstock, Central Citrus Research Institute, *Annual Report*, pp. 22–23.

Vijayakumari, N. and Karihaloo, J. 2012. Biotechnology in citrus improvement. Souvenir and abstracts, *National Dialogue on Citrus Improvement, Production and Utilization*, Nagpur, India, 27th–29th February, pp. 33–34.

Vijayakumari, N., Dass, H.C. and Singh, A. 1994. Standardization of *in vitro* shoot tip grafting technique for elimination of virus and virus like disease from Nagpur mandarin. *Indian J. Hort.*, **51**: 311–315.

Vijayakumari, N., Ghosh, D.K., Das, A.K., Singh, S. and Singh, I.P. 2008. Healthy citrus gene bank through STG: A safe insurance policy for sustainable citrus industry. Presented at *National Symposium on Citriculture: Emerging Trends*, Nagpur, India, pp. 30–31.

Vijayakumari, N., Ghosh, D.K., Das, A.K., Singh, A. and Singh, S. 2006. Elimination of citrus tristeza virus and greening pathogens from exotic germplasm through *in vitro* shoot tip grafting in citrus. *Indian J. Agril. Sci.*, **76**(3): 209–210.

Williamson, J.G. and Jackson, L.K. 1994. *Citrus Propagation*. Florida Cooperative Extension service, fact sheet HS–**86**: 1–6.

3

Improving Availability and Quality of Vegetable Seeds for Higher Profitability

B.S. Tomar[1] and V.R. Yalamalle[2]

[1]*Division of Vegetable Science,*
[2]*Division of Seed Science and Technology,*
ICAR-Indian Agricultural Research Institute, New Delhi – 110 012
E-mail: bst_spu_iari@rediffmail.com

India ranks second in vegetable production, but productivity it lags behind in most of the vegetable crops (FAO, 2013). Apart from other factors, limited availability of quality vegetable seed and at reasonable price is one of the major cause. However, quality seed alone can lead to 15-20 per cent increase in productivity (Singh *et al.*, 2010). India grows vegetables in an area of 9.3 million ha and requires around 51,000 tonnes of seeds for sowing annually but the actual availability is around 40,000 tonnes and large quantity of seed is still being multiplied by farmers themselves (Dutta, 2004).

Indian seed industry is the 5th largest in world and is worth of 2.7 billion US $ (Dubey, 2016). Vegetable seed industry consist of agencies, *viz.*, like National Seeds Corporation, state seed corporations, SAUs, ICAR institutes and private seed companies. Unlike cereals, vegetable seed is categorised as low volume high value seed, and dominated by the private seed companies. The private sector concentrate on the production of hybrid seeds of tomato, cabbage, brinjal, chilli, okra and cucurbits, where the seed production is comparatively easy and more profitable. The government R&D institutes and production agencies are largely concentrated on development and seed production of open-pollinated (OP) varieties. Although 461 varieties have been released by AICRP-vegetable crops but more than 60 per cent varieties are open-pollinated. ICAR institutes and SAU's produce breeder seed, which is made available to the seed producing agencies/corporations and small private seed companies which don't have their own R&D setup.

Private seed companies are making available quality vegetable seeds to the farmers, but high seed cost is limiting factor to small and medium farmers who are unable to afford the high cost of seed *e.g.* capsicum and tomato seed price can be as high as 60- 90 thousand rupees/kg. The availability of quality seeds can enhanced by two approaches one by increases the production and productivity of seeds and second by increasing the quality of the seeds by enhancing the germination, seed vigour and storability *etc.* There by, reduction of seed rate could be possible. The use of improved seed production and seed enhancement technologies shall not only be helpful in increasing the profitability of the seed producers but will also helpful to vegetable growers by making available the best quality seed. The technological interventions could improve the seed productivity and increase the availability.

1. Strengthening Breeder Seed Production and Public Private Partnership

India has the large network of 63 SAUs, ICAR research institutes and many autonomous research institute working on the vegetable improvement which have large land banks, which can be utilized to produce breeder seed and the same must be provided to of 15 state seed corporations, National Seeds Corporation,

KVKs and 100's of small and medium seed companies and producers which don't have R&D setup.

2. Practicing the Generation System of Seed Production

The varieties and hybrids gain acceptance when the farmer gets genetically pure seeds of high quality. Employing the generation system comprises of breeder, foundation and certified seed could provide adequate safeguard to quality reduction by rouging frequently at farmer's field. Thus, it will cut down the labour cost and increase the net profit from seed production. The choice of a proper seed multiplication model is the key to further success of a seed programme. For self-pollinated (garden pea, *methi*, cowpea, cluster bean) and asexually (potato and garlic) reproduced crops 5 generation model, often cross-pollinated crops (okra, brinjal, broad bean, dolicus bean) 4 generation model and in cross-pollinated crops (cole crops, leafy vegetables and cucurbits) 3 generation model is suggested for multiplication to minimize the expenditure involved in rouging.

3. Seed Production of Hybrids and high Yielding Varieties

Since, the vegetables are being cultivated by resource poor farmers and largely self-saved seed is being used by them and contribution of hybrid seed is less of total seed demand and limited to selected crops. The seed production of hybrids varieties is one of the means to meet the demand of quality seed. Since hybrids/varieties have been released in number of vegetable crop by public sector. Thus, production of genetically pure and quality seeds by adopting suitable seed production techniques specifically in remote areas where market for fresh vegetables are not adequate.

Hybrid seed production on commercial scale is being done by hand emasculation and pollination. Hence, involvement of skilled labours resulting employment to rural youth and also producer fetches higher price of hybrid seed as compare to open-pollinated variety seed production and *vis-à-vis* commercial vegetable production.

4. Identification of New Seed Production Pockets

The state of Karnataka produces nearly 90 per cent of the total vegetable hybrids in India and majority of the area is concentrated around Ranebennur area, apart from it Jalna in Maharashtra and few other pockets in Telangana and Gujarat. There is a need to identify new seed production pockets which are having higher production potential, *e.g.* onion seed is traditionally produced in Maharashtra, where the seed yield is around 700-800 kg/ha, whereas the seed yield in Saurashtra region of Gujarat is 1000-1200 kg/ha (Gupta and Sharma, 9). Identification of new areas of seed production is also essential to reduce the build-up of soil borne disease, *e.g.* Sproon valley in Solan, Himachal Pradesh has become endemic to black rot disease of cauliflower and made the seed production uneconomical and forced to shifting to other crops.

5. Seed Demand Forecasting

Seed multiplication under generation system requires at least 4-5 years of advance planning. Forecasting based on production trends, consumer preference and other economic analysis is essential to ensure the adequate production and supply of seed.

6. Healthy Nursery Raising for Transplanted Crops

Quality seeds of hybrid seeds is costly and even the single seeds is valuable, but in traditional system of nursery raising the demand for seed is higher due to lower seedling establishment. In order to reduce seed demand, uniform seedling growth, seedling production under adverse condition and enhance production one should opt protected nursery raising. A plug-tray nursery raising technology by using vermiculite, cocopeat and perlite (3:1:1) as soilless media has been standardized for raising seedlings of many vegetables. This high-tech nursery raising technology is capable of vigorous root development, virus-free healthy seedling production and suitable for raising of large number of seedling at minimum space without any damage to the seedlings. Besides, the production of quality seedlings, the plug tray production under protected structures could be adopted by the agricultural graduate and farmers as a side business for enhancement of income.

7. Methods of Growing

There are four method of growing, *viz.*, flat, ridge and furrow, raised bed and hill channel method in different crops. Flat bed method reduces the crop growth, fruit set and increase susceptibility to disease and pest. Legume vegetable like garden pea should be grown in raise bed to protect the crop from frost during flowering and pod development by providing the irrigation between the raised beds, otherwise flood irrigation suppresses the growth and podding and reduces the yield. A successful demonstration of raise bed seed production of garden pea cv. Pusa Pragati has been successfully adopted by several farmers of Punjab. Onion bulb crop could be grown over raised bed for better bulb development and reduction in disease incidence, which ultimately reduces the cost of cultivation. Hill channel method in cucurbits is of economical for seed production as compared to flat bed, saved inputs.

7.1 Relay Sowing in Cucurbits for Higher Seed Yield

In north India early sowing especially in wheat belt crops like musk melon and water melon can be planted as a relay crop in hill channel method to protect the crop from low temperature and better growth. The relay planting in wheat resulting more fruiting and more time for fruit development and maturity leading to higher seed yield and better quality in musk melon and water melon.

7.2 *Stacking and Pruning*

In traditional system of growing plants are allowed to grow at the soil, hence, fruits prone to the disease infestation and leads to more number of fruit decay on the ground because fruits have to stay till maturity in seed crop. In order to enhance the seed yield and quality the intervention of stacking can play a vital role. During rainy season or under irrigated situations, the vegetable crops like tomato, chilli, capsicum and brinjal should be stacked. So that the fruits can be prevented from direct contact with the soil and water and are supported with the stick or rope for vertical growth to protect from disease and proper aeration.

7.3 *Trailing in Cucurbits*

Seed production of cucurbits by use of low cost trailing system has been done and various studies have indicated the superiority of trailing method over traditional methods. The trailing method resulting more number of fruits, seed yield and quality in case of bottle gourd seed production of Pusa Hybrid-3 and Pusa Naveen (Kalyanrao *et al.*, 2012; Sharma *et al.*, 2016). Trailing method avoids shading and allows better utilization of the sunlight. The superiority of quality attributes is due to the sound development of fruit and seed due to an early induction of male and female flower in trailing and protection of fruit from rotting which also leads to production of seeds with less fungal load as indicated from the Table 1.

7.4. *Direction of Sowing*

Seed production by adjusting the orientation of the row can influence the light interception as well as crop canopy micro climate significantly which in turn may change the crop growth parameters and pest and disease scenario and leads to increased production of fruits and higher seed yield. In bottle gourd (Sharma and Tomar, 2016) reported that sowing of seed in E-W direction (over the trailing) is beneficial as compared to other crop orientations for attaining better plant growth, fruit yield, seed setting, higher seed yield and quality traits. The comparison of direction of sowing is given in Table 2.

Table 2: Effect of Direction of Sowing on Seed Yield and Quality Traits in Bottle Gourd cv. Pusa Naveen during *Kharif* 2014

Characters	*Direction of Sowing*	
	E-W	*N-S*
No. of fruit set/vine	4.40	3.76
Seed weight/fruit (g)	107.06	89.61
Seed yield/acre (kg)	423.75	396.50
Germination (%)	97.65	97.43
EC (µmhos/cm/g)	92.20	105.05

8. Frequency of Cutting in Leafy Vegetable Seed Production

Seed production of leafy vegetable without cutting leads to lodging of crop, less reproductive phase and causes problem during monitoring and makes seed production uneconomical. Thus, leaf cutting is beneficial practically in leafy vegetables, *viz.*, palak, methi, coriander and vegetable mustard, which favoured the high penetration of light, reduces disease incidence, less crop growth and more reproductive phase and sale of cut leaves fetch additional benefit. Two leaves cutting in *palak* variety All Green has given 6 quintal seed yield per acre valued of 60,000 under Delhi conditions and makes its vegetable seed production economical in peri-urban area of Delhi.

9. Enhanced Pollinator Activity and Pollination

Pollination management is essential particularly for producing the hybrid seeds. Urbanization, pesticide application and habitat loss of pollinators have reduced the pollinator load which results in insufficient pollination of flower, less fruit and seed set and lower yield and quality in seed production plot. However, in order to increase seed set and seed yield 2-3 medium bee hives needs to be introduced in the periphery of seed production plot of cauliflower, cabbage and onion.

Methods

Hand pollination is employed for hybrid seed production in many vegetable crops being hand pollination favour to higher number of fruit set, higher number of seeds per fruit and seed yield per plant in

Table 1: Effect of Trailing on Seed Yield and Quality Traits in Bottle Gourd cv. PH-3 and Pusa Naveen

Character	*PH-3*		*Pusa Naveen*	
	Trailing	*Traditional*	*Trailing*	*Traditional*
Number of fruit set/vine	6.65	4.70	4.08	3.48
Seed weight/fruit (g)	98.14	59.78	98.33	77.64
Seed yield/acre (kg)	689.6	248.2	410.13	371.00
Germination (%)	97.85	93.11	97.54	95.21
Disease infestation (%)	44.62	87.29	45.02	83.08

bottle gourd and pumpkin (Table 3) (Vishwanath *et al.*, 2008; Tomar *et al.*, 2009) than natural pollination. Generally single pollination is recommended for higher seed set in cucurbits, but in case of tomato repeated pollination has been reported to increase fruit set and seed yield under poly-house condition.

Mulching

Black gray, bio-degradable plastic mulch enhance crop performance during winter season in north Indian condition coupled with drip irrigation lead to higher root growth and development of plant. Natural mulches like wheat or paddy straw mulching can be used in crops like onion for higher yield and quality seeds Anisuzzaman *et al.* (1) reported that synthetic plastic opaque mulches do not allow the light but control of weed is effective, it also raises soil temperature and protected from frost injury.

11. Foliar Spray of Micro-nutrients

Application of mineral nutrients is essential to fetch the higher yield and quality because Indian soils are deficient in micro nutrients. For better plant growth and development micro nutrients are needed but in small quantities however their deficiency cause a greater disturbance in the physiological and metabolic processes and ultimately reduced seed yield and quality. Micro nutrients elements, especially B, Zn, Ca and Mg avoid antagonistic effects of nutrients during uptake from soil. In onion Sanjay *et al.* (2015), reported that foliar spray in combination of B + Zn + Ca + Mg (Recommended dose at 30 and 60 DAP) is beneficial for getting higher number of productive umbels, seed yield and quality.

12. Post-harvest Ripening for Enhances Seed Yield

Fleshy fruits usually need to be stored before seed extraction for after ripening. Vinod *et al.* (2014) reported that pumpkin fruits harvested at 70 DAA and cured for 10 days had the highest seed yield and quality parameters. After-ripening facilitates the further development of the seed and removal of dormancy.

13. Use Drip-irrigation and Fertigation

The water management and fertigation studies have revealed that seed productions over low pressure drip/ pressurized drip not only saves farm inputs but also increases the seed yield and quality. Tomar *et al.* (2004) realised that seed yield and quality in onion crop was the highest over drip irrigation method as compared to surface irrigation and similar results were obtained in carrot, cauliflower, onion and turnip. The comparative performance of drip and flood irrigation is given Table 5.

14. Manipulation of Sex Expression in Cucurbits

Auxin, ethylene and gibberlic acid are the important PGRs which regulate various physiological responses in plant. Alteration of sex ratio to produce more female flowers particularly in monoecious cucurbitaceous crops like cucumber, bitter gourd and production of male lines in gynoecious cucumber lines is essential to increase the productivity. NAA at 50 ppm and cycocel 750 ppm spray

Table 3: Effect of Methods of Pollination on Seed Yield and Quality Traits in Bottle Gourd and Pumpkin Hybrid Seed Production

Character	*Method of Hybrid Seed Production*			
	Bottle Gourd		*Pumpkin*	
	Natural Pollination	*Hand Pollination*	*Natural Pollination*	*Hand Pollination*
Number of fruit set	6.70	7.40	1.78	3.27
Seed yield/plant (g)	146.01	215.60	46.00	83.66
Germination%	85.00	90.00	91.00	96.00
EC (μmhos/cm/g)	90.60	72.65	57.67	58.88

Table 4: Effect of Foliar Spray of Mineral Nutrients on Seed Yield, Quality and Disease Infection in Onion cv. Pusa Riddhi

Treatment	*Seed Setting (%)*	*Seed Yield/ umbel (g)*	*Seed Yield/ha (q/ha)*	*Germination %*	*EC (μmhos/cm/g)*	*Disease Infected Plants (%)*
RD of B at 60 DAP	83.82	3.33	7.03	90.08	1.85	15.5
RD of Zn at 60 DAP	79.69	3.69	7.14	89.42	2.09	9.5
RD of Ca at 60 DAP	81.41	3.74	7.25	89.00	1.82	12.5
RD of Mg at 60 DAP	79.29	3.67	7.25	88.50	1.75	19.0
RD of B + Zn at 30 and 60 DAP	89.05	3.64	7.48	90.40	1.68	10.0
RD of B + Zn + Ca at 30 and 60 DAP	91.37	3.93	7.89	90.83	1.50	10.0
RD of B + Zn + Ca + Mg at 30 and 60	90.18	4.18	8.35	91.84	1.65	10.5
Control (water spray)	78.13	3.17	6.00	86.83	2.32	24.0

Table 5: Comparative Performance of Seed Production under Drip and Flood Irrigation

Crop	Drip irrigation		Surface irrigation	
	Seed Yield/ Plant (g)	Germination %	Seed Yield/ Plant (g)	Germination %
Onion	94.61	92.13	44.08	79.75
Carrot	41.27	98	28.37	97
Cauliflower	34.00	99	14.93	97
Onion	15.33	98	13.20	97
Turnip	43.77	99	39.47	98

in cucurbits doubled the number of female flowers. In summer squash cv. Pusa Alankar spray of 350 ppm ethephon at 2-4 leaf stage can enhance the number of female flowers for hybrid seed production. 100 ppm GA_3 spray in chilli enhanced the seed yield by 30 per cent.

15. Bio-fertilizers and Integrated Nutrient Management

The role and importance of bio-fertilizers in sustainable crop production has been studied by several authors and found that chemical fertilizers integrated with vermi-compost and bio-fertilizers were found to be superior in improving the seed yield and seed quality. The combined application of bio-fertilizers - Rhizobium + PSB + PGPR resulted in 13 per cent increase in garden pea seed yield Mishra *et al.* (2010).

16. Use of Protected Structures for Seed Production

Grow seed crops under protected structures for higher seed yield and quality. Insect proof net house has been utilized for hybrid seed production of tomato, sweet pepper, chilli, okra, brinjal and cucurbits under north Karnataka condition. However, Kaddi *et al.* (2014) reported higher seed yield in cucumber Pant Shankar-1 under insect-proof nethouse as compared to open field conditions in Delhi (Table 6).

Table 6: Effect of Growing Condition on Seed Yield and Quality in Cucumber cv. Pant Shankar-1

Growing Conditions	Seed Yield/ 1000 m^2 (kg)	Germination %
Open	4.2	63.85
Net house	14.17	71.69
Poly house	14.06	72.30

Semi-climate controlled greenhouse are suitable for hybrid seed production of indeterminate hybrids, cherry tomato, sweet pepper and parthenocarpic cucumber varieties. Seed yield is 3-4 times higher than open filed.

17. Seed Quality Enhancement and Fruit Regulation

Seed quality and yield in vegetables can be enhanced by regulating the fruit number. One fruit per vine in pumpkin cv. Pusa Hybrid-1 (Kumar *et al.*, 2015) during summer and three fruits/vine in bottle gourd cv. Pusa Hybrid-3 during *Kharif*, (Kalyanrao *et al.*, 2012) have been reported for superior seed yield and quality under open field conditions in Delhi. Two fruits/vine in cucumber cv. Pant Shankar-1and summer squash variety and Pusa Alankar gave higher seed yield and quality under insect-proof nethouse.

Improving the Availability of the Seed by Upgrading the Seed Quality

Quality seed refers to seed which is genetically pure, free from inert matter, disease tree, higher germination per cent and seed vigour. Good quality seeds not only yield better but also helps in reducing the seed rate and following approaches shall be helpful.

a) Revising the Seed Quality Standards

A seed can be termed as quality seed if it genetically pure. Breeder seed should be 100 per cent genetically pure, foundation seed 99.5 per cent and certified seed should be 99 per cent genetically pure. Hybrids and seeds produced by hand emasculation or by use of CHAs should be 95 per cent and 90 per cent genetically pure respectively. Apart from this the seed lot should have high germination percentage and a minimum of inert, weed and other crop seeds and are free from diseases. The genetic purity of OPV and hybrids produced by use of male sterility system can be maintained by adhering to the isolation requirement, following seed village concept if the isolation distance is not adequate. IMSCS, 2013 prescribes for seed health standards at field level and health test of the seed lot is not a mandatory. Seed health testing should be included as a routine test since the seed in most case act as a primary inoculum for disease. Seed vigour determines the true planting value of the seed lot. At present, there is a need to standardize the appropriate vigour test in vegetable seed and the same needs to be incorporated as a lab test for quality test of seed lot. Germination standards of many vegetables are very low and needs to be increased, *e.g.* for cucurbits the minimum germination requirement is 60 per cent, which is very low and needs to relook.

b) Packing and Use of Dessicants

Most of the vegetable seeds are poor storers and cannot be stored beyond 8- 9 months without loss of seed vigour and viability. The seed viability can be enhanced by reducing the moisture content. Fluctuating Relative Humidity (RH) in the tropical countries like India bringing down the seed moisture to safe storage level is difficult. Khanal and Paudel (2014) reported that the moisture of onion, pea and tomato seeds can be reduce to 4-6 per cent by using zeolite beads within 3-4 days and the seeds maintained higher germination and vigour after storage.

c) Seed Coating and Pelleting

Coating of seeds with thin layer of polymers is helpful in adhesion of plant protectants, water absorption, protection against temperature stress *etc.* Polymers also increase the flowability and the ballistic properties of the seed, which help in precision planting. Hydrophobic coatings reported to reduce soaking injury (Hwang and Sung, 1991) and imbibitional chilling injury (Chachalis and Smith, 2001; Taylor and Kwiatkowski, 2001). The reduction of water uptake by hydrophobic polymer coatings, especially the absorption of water from the vapour state Henning, (10) may also have a role in improving seed storage by maintaining lower moisture content in uncontrolled storage conditions. Polymers can also be colour coded which helps in visibility, appearance and overall marketability of seeds. Brooker *et al.* (2007) reported that the polymer coating of seeds along with plant protectants like captan, metalaxyl and or thiabendazole reduced the soil borne diseases and the polymer provided extended release of active ingredients due to less leaching loss in the soil. Seed pelleting alters the shape and size of the seeds so that the direct seeding can be done in small seeded crops like onion, lettuce *etc.* Plant protectants, micronutrients, growth hormones can also be included in the peletting material. Due to improved efficacy of the applied pesticides, the polymers coating and peletting also reduces the dosage of the pesticides to be delivered with seed

d) Seed Sorting using Image Analysis and Chlorophyll Fluorescence Marker

The advances in imaging technology have been utilized in the seed processing industry. It serves two purposes i) discarding contaminating seeds of other crops or weeds and inert material, and ii) eliminating poor physiological quality seeds. In both cases, seed is conditioned using cleaning equipment which uses vibrating or rotating sieves, and air-stream separation techniques allowing a non-destructive separation of weed seeds, innert matter, undersized or immature seeds (Copeland and McDonald, 2001). The technique of chlorophyll fluorescence to enhance the germination was reported by Jalink *et al.* (1998) in white cabbage (*Brassica oleracea* L.) and in carrot (*Daucus carot* subsp. *sativus*) seeds by Sreckel *et al.* (1989).

e) Seed Stimulation Using of Electromagnetic Waves

The exposure of the seeds for short period of time to electromagnetic field has been reported to increase the seed germination as well as vigour of seeds. Podlesni *et al.* (2004) reported reduced germination time, increased seed vigour and seed yield in broad bean. Moon and Chung (2000) reported that percent germination rates of the treated tomato seed were accelerated about 1.1–2.8 times compared with that of the untreated seed. Even though the benefits of magnetic stimulation on seed have been reported by several workers its commercial exploitation is limited due to operational difficulties of the technology for treating large quantities/volume of seeds.

f) Seed Invigoration

Seed priming is defined as controlling the hydration level within seeds so that the metabolic activity necessary for germination can occur but radicle emergence is prevented. Seed priming can be effective on-farm technique to which can not only reduce the seed rate but also increases the field stand. Priming has been reported to increase germination and field emergence in onion and tomato (Singh *et al.*, 2014; Pandita and Nagarajan, 2000).

g) Monitoring and Seed Legislation

Seeds Act 1966 provides various provisions for regulation of seed quality during production and sale. Regular inspection and quality checks can check the sale of spurious/substandard vegetable seeds.

Vegetable seed industry has enormous employment generation potential. Hybrid seed production of vegetable requires lot of manual labour for emasculation and pollination. There is a need to diversify the vegetable seed production hubs to non-tradition high productive regions. Funds under various government programmes like Tribal Sub plan, MNREGA should be made available to the rural areas for training them in seed production. Village level entrepreneurship development must be encouraged. Promotion of ancillary activities and diversification of farm enterprise is one of the important strategies outlined by the government to double the farmers' income by 2022, for which seed production of vegetables can be a major player in doubling the farmers' income. India with its diverse climatic and soil types and vast pool of sklled man power can become the world leader in export of vegetable seeds and can ushering the rural prosperity.

References

Anisuzzaman, M., Ashrafuzzaman, M., Ismail, M.R., Uddin, M.K. and Rahman, M.A. 2009. Planting time and mulching effect on onion development and seed production. *African J. Biot.*, **8**: 412–416.

APEDA. 2013. http://agriexchange.apeda.gov.in/indexp/reportlist.aspx. accessed on 1st September, 2016.

Brooker, N.L., Lagalle, C.D., Zlatanic, A., Javni, I. and Petrovic, Z. 2007. Soy polyol formulations as novel seed treatments for the management of soil-borne diseases of soybean. *Comm. Agric. Appl. Biol. Sci.*, **72**: 35–43.

Chachalis, D. and Smith, M.L. 2001. Hydrophobic-polymer application reduces imbibition rate and partially improves germination and emergence of soybean seedlings. *Seed. Sci. Tech.*, **29**: 91–98.

Copeland, L.O. and McDonald, M.B. 2001. Seed conditioning and handling. **In:** *Principles of Seed Science and Technology*, (Eds.) L.O. Copeland and M.B. McDonald. Kluwer Academic Publishers, Boston, pp. 252–267.

Dubey, B. 2016. Seed export: India and world. *Seed Times*. January–June, pp. 21–22.

Dutta, O.P. 2004. Recent innovations in hybrid seed production in vegetables. *Proc. Indian Sci. Cong.*, New Delhi. November, **1**(6–9): 217–242.

FAOSTAT. 2013. http://faostat3.fao.org/.com accessed on 9th August, 2016.

Gupta, R.P. and Sharma, H.P. 2014. Onion seed production: Scope, challenges and future strategy to meet the demand of quality seed. In: *Brain Storming Session on Crop Improvement and Seed Production of Onion*, Organised at NHRDF, Nashik, 15th March, 2015, pp. 1–12.

Henning, A.A. 1990. Polymeric coatings improve the storage life of soybean seeds. *Ph.D. thesis*, University of Florida, pp. 96.

Hwang, W.D and Sung, F.J.M. 1991. Prevention of soaking injury in edible soybean seeds by ethyl cellulose coating. *Seed. Sci. Tech.*, **19**: 269–278.

Jalink, H., van der Schoor, R., Frandas, A., van Pijlen, J.G and Bino, R.J. 1998. Chlorophyll fluorescence of *Brassica oleracea* seeds as a non-destructive marker for seed maturity and seed performance. *Seed Sci. Res.*, **8**: 437–445.

Kaddi, G., Tomar, B.S., Singh, B. and Sanjay, Kumar. 2014. Effect of growing conditions on seed yield and quality of cucumber (*Cucumis sativus*) hybrid. *Indian J. Agri. Sci.*, **84**(5): 624–27.

Kaddi, G., Tomar, B.S., Singh, B. and Vidyadhar, B. 2015. Influence of seed extraction methods on seed quality in cucumber (*Cucumis sativus*) cv. Pant Shankar Khira-1. *Progr. Hort.*, **47**(1): 16–21.

Kalia, P. 2009. Hybrid seeds making available at affordable price. In: *Recent Initiatives in Horticulture*, (Eds.) Chadha *et al.*, Westville Publishing House, New Delhi, pp. 219–237.

Kalyanrao, Tomar, B.S. and Singh, B. 2012. Influence of vertical trailing on seed yield and quality during seed production of bottle gourd (*Legenaria siceraria*) cv. Pusa hybrid-3. *Seed Res.*, **40**(2): 139–144.

Khanal, A. and Paudel, P. 2014. Evaluation of zeolite beads technology for drying vegetable seeds to low moisture content prior to long-term storage in Nepal. *Int. J. Res.*, **1**(8): 140–149.

Kumar, V., Tomar, B.S, Dadlani, M., Singh, B. and Kumar, S. 2015. Effect of fruit retention and seed position on the seed yield and quality in pumpkin (*Cucurbita moschata*) cv. Pusa Hybrid 1. *Indian J. Agr. Sci.*, **85**(9): 1210–1213.

Maxwell, K. and Johnson, G.N. 2000. Chlorophyll fluorescence: A practical guide. *J. Exp. Bot.*, **51:** 659–668.

Mishra, A., Prasad, K. and Rai, G. 2010. Effect of bio-fertilizer inoculations on growth and yield of dwarf field pea (*Pisum sativum* L.) in conjunction with different doses of chemical fertilizers. *J. Agron.*, **9**: 163–168.

Moon, Jae-Duk and Hwa-Sook, Chung. 2000. Acceleration of germination of tomato seed by applying AC electric and magnetic fields. *J. Electrostat.*, **48**(2): 103–114.

Pandita, V.K. and Nagarajan, S. 2000. Osmopriming of fresh seed and its effect on accelerated ageing in Indian tomato (*Lycopersicon esculentum*) varieties. *Indian J. Agr. Sci.*, **70**(7): 479–480.

Podlesni, J., Pietruszewski, S. and Podlesna, A. 2004. Efficiency of the magnetic treatment of broad bean seeds cultivated under experimental plot conditions. *Int. Agrophys.*, **18**(1): 65–71.

Rao, R.G.S., Singh, P.M. and Rai, M. 2006. Storability of onion seeds and effects of packaging and storage conditions on viability and vigour. *Scientia Hort.*, **110**: 1–6.

Sanjay, Kumar. 2015. Optimization of quality seed production in onion (*Allium cepa* L.) cv. Pusa Riddhi. *Ph.D. Thesis*, IARI, New Delhi.

Sharma, R.K. and Tomar, B.S. 2016. Influence of growing direction on seed yield and quality in bottle gourd (*Lagenaria siceraria*). *Indian J. Agr. Sci.*, **86**(7): 895–900.

Sharma, R.K., Tomar, B.S., Singh, S.P. and Kumar, A. 2016. Effect of growing methods on seed yield and quality in bottle gourd (*Lagenaria siceraria*). *Indian J. Agr. Sci.*, **86**(3): 373–378.

Singh, P.K, Pandita, V.K., Tomar, B.S. and Seth, R. 2014. Germination and field emergence in osmotic and solid matrix priming in onion (*Allium cepa*). *Indian J. Agr. Sci.*, **84**(12): 1561–1564.

Singh, P.M., Singh, B., Pandey, A.K. and Singh, R. 2010. Vegetable seed production: A ready reckoner. *Tehc. Bull.* No. **37**, IIVR, Varanasi.

Sreckel, J.R.A., Gray, D. and Rowse, H.R. 1989. Relationships between indices of seed maturity and carrot seed quality. *Ann. App. Biol.*, **114**: 177–183.

Taylor, A.G. and Kwiatkowski, J. 2001. Polymer film coatings decrease water uptake and water vapour movement into seeds and chilling injury. *BCPC*

Symposium Proceedings: Seed Treatment: Challenges and Opportunities, **76**: 215–220.

Tomar, B.S., Pham, T.T., Singh, B. and Kalayanrao. 2009. Studies on Methods of pollination for hybrid seed production of bottle gourd (*Lagenaria scieraria*). *Seed Res.*, **37**(1 and 2): 171–174.

Tomar, B.S., Singh, B., Kumar, M., and Hasen, M. 2004. Effect of irrigation method on the yield and quality of onion seed. cv. Pusa Madhvi. *Seed Res.*, **32**(1): 45–46.

Vinod, K., Tomar, B.S., Kaddi, G. and Kumar, S. 2014. Effect of stage of harvest and post-harvest ripening of fruits on hybrid seed yield and quality in pumpkin (*Cucurbita moschata*). *Indian J. Agr. Sci.*, **84**(6): 737–741.

Vishwanath, R.Y., Tomar, B.S. and Singh, B. 2008. Studies on methods of pollination for hybrid seed production of pumpkin (*Cucurbita moschata* Poir.). *Seed Res.*, **36**(2): 214–217.

4

Plug Plant Production of Horticultural Crops through Hi-Tech Nursery Raising

Tusar K. Behera[1]* and Awani Kumar Singh[2]

[1]Division of Vegetable Science and [2]Centre for Protection Technology,
ICAR-Indian Agricultural Research Institute, New Delhi – 110 012
**E-mail: tusar @ rediffmail.com*

Protected cultivation is one of the main components of Hi-tech production technology in horticultural crops. The main objectives of the technology are to minimise biotic and abiotic stresses and to enhance quality, quantity and productivity per unit area through control microclimate in side of protected structures. With this technology the quality nursery plants can be also grown successfully in year round inside the structure, which has controlled micro-climate, irrigation, and fertigation systems. The optimal temperature, relative humidity and solar radiation and ventilation are maintained according to requirements of month and crops.

Plug plant production of horticultural crops has become an Industry in several developed countries including Israel, Japan, Spain, the Netherlands and United States. In Israel, no vegetable grower is growing his/her own nursery, but are procuring the required kind of seedlings from an established nursery owners on pre-order basis. Healthy seedlings not only reduce the amount of time of the crop but also increase the uniformity of the crop and harvesting as compared to direct sowing of the crops. Transplanting of quality seedlings also eliminates the need for thinning and provides very good opportunities for virus-free, vigorous and even off-season seedlings if grown under protected conditions in soil-less media. Hence, it becomes necessary to know the protected structures which can be efficiently used for successful raising of horticultural crops seedlings in suitable containers and growing media, method of seeding in containers in soil-less media, growing conditions and water and fertilizer requirement of seedlings along with hardening and transplanting to the field. Protected nursery raising technology is highly suitable and can be established as a small scale industry in major vegetable growing areas of our country by progressive farmers especially in peri-urban areas. By this way the vegetable growers would get virus-free healthy and off season nursery as per their requirement and it would generate extra income and employment in urban and peri-urban areas.

List of horticultural crops which are suitable for transplanting;

1. Vegetable crops: Tomato, brinjal, capsicum, chili, cole crops, cucurbits, beet root, lettuce, parsley, celery, pokchoi, onion *etc.*
2. Fruits crops: Papaya, strawberry, tree tomato, raspberry *etc.*
3. Flowers crops: Rose, gerbera, chrysanthemum, carnation, dahlias, orchids *etc.*
4. Medicinal crops: Stevia, aloe vera *etc.*

1. Plug Tray Media

Different type of media are used for nurserygrowing in plug tray

- ✰ Soil less media combination: Coco peat + perlite + vermiculite in the ratio of 3:1:1 for commercial use in urban areas in Hi-tech greenhouse.

- ✰ Soil based media combination: Compost + sand + soil in the ratio of 3:1:1 for small farmers in rural and hilly areas in simple greenhouse or tunnels.
- ✰ Soil and soilless combination: Vermicompost + coco peat in the ratio of 1:2 for commercial in peri-urban areas in hi tech greenhouse.

Mostly artificial soil-less media are used for raising healthy and vigorous seedlings of vegetable in plastic pro-trays. Mainly three ingredients, *viz.*, coco-peat, vermiculite and perlite are used as root medium for raising the nursery. These ingredients are mixed in 3:1:1 (v/v) ratio before filling in the required containers or plastic pro-trays.

1.1 Soil-less Medium

Mainly three ingredients *viz.*, coco-peat, vermiculite and perlite are being used as a medium for raising the nursery in the high-tech nursery greenhouse. These ingredients are mixed in the volume ratio of 3:1:1 before filling in the trays. The characteristics of these three ingredients are as under:

1.2 Coco-peat

It is prepared from the waste of coconut husk or coir dust. This medium has good porosity, improved drainage and air movement activity. This medium is completely free from infestation of any pest or pathogen. Coco-peat is commonly being used as a medium under protected cultivation of ornamental crops like roses and gerberas and for raising the nurseries of vegetables and other ornamental plants in the developed countries.

1.3 Perlite

It is a light rock material of volcanic origin. It is essentially heat-expanded aluminum silicate rock. The volcanic ore is heated to extreme temperatures of about 1800°F to cause the rock particles to expand and to produce the white granular product. It improves aeration and drainage. If this ingredient is used in a mix, the horticultural grade should be selected since it has larger particle size and is thus more effective. Perlite is neutral in reaction and provides almost no nutrients to the mix (except for small amounts of sodium and aluminum). A disadvantage of the use of perlite is its low weight, which makes it float when the medium is watered. During mixing it produces dust which can be eliminated by wetting the material before its use.

1.4 Vermiculite

It is heat expanded mica. This mineral is heated at a temperature of about 1,400°F (760°C) to produce the folded structure associated with the material. It is very light in weight and has minerals (magnesium and potassium) for enriching the mix, as well as good water-holding capacity. It is neutral in reaction (pH) and available in grades according to sizes. Grade 1 includes the largest particles and grades 4 and 5 are fine in texture. The most commonly used grades are 2 and 4. Its fineness, incidentally, makes it prone to being compressed easily in the mix. To reduce its potential, a mix including vermiculite should not be pressed down hard.

1.6 Advantages of Soilless Media

The major advantages of soil-less media include the following:

1. *Uniformity of mix:* The physical and chemical properties of a mix are uniform throughout the mix, a condition not found in field soil. The homogeneity of the medium makes it possible for plants to grow and develop uniformly.
2. *Ease of handling:* Mixes are light in weight and easy to transport. Ingredients are easy to scoop and move around during manual preparation.
3. *Versatility:* Mixes can be made to order (custom mixed) for specific needs. They can be used to amend soils in the field by mixing into flower beds, lawns or garden soil. They are convenient for use in container plants.
4. *Sterility:* Mixes, initially at least, are free from diseases and pests. Seed germination is less prone to diseases such as damping-off.
5. *Good drainage and moisture retention:* A mix can be custom-made to provide the appropriate degree of drainage and moisture retention.
6. *Convenience of use:* Mixes are ready-to-use when purchased.

2. Suitable Structures

Fan-pad Polyhouse, naturally ventilated polyhouse, IP-nethouse, Shade-net-house or temporary hi-tunnels are commonly used structures for protected cultivation.

2.1 Sowing of Seeds and Cutting Methods

Seeds are usually planted/sown at a shallow depth after pressing the media with finger in a gentle way into the potting plugs or cells, which are filled with artificial media. The actual depth of sowing depends on the crop and the size of the seeds. After sowing of seeds a thick layer of vermiculite is given to cover the seeds for better germination as this media is having better water holding capacity. After seed sowing the pro-trays are kept in the germination room at the optimum temperature for early and better germination. Usually one to two seed are sown per cell and after the seeds are germinated, the seedlings are thinned to one per cell and the thinned one can be gap filled in those cells where seed germination could not take place.

2.2 Seeding Trays and Covering Seed

The trays are filled with pre-moistened growing media and then the media is compressed or "dibbled" to make a uniform surface for the seed. The media should be compressed between 1/4" to 3/8" deep. If seeded too shallow, certain crops (especially onions and cole crops) will tend to push up out of the cell. The seeded trays are then covered with medium-grade vermiculite. Seed of all vegetable species must be covered after seeding. Vermiculite is preferred because it is easy to apply evenly, allows good aeration, does not support algae growth and does not allow root growth between cells.The trays are now ready for watering before being placed in the germination or "sweat" chamber. Hot water should be used to raise the temperature of the soilless mix and start the germination process as soon as possible.

2.3 Germination Chambers

Most vegetable crops will benefit from the use of a germination chamber. This is usually an insulated room in which temperature and relative humidity can be maintained at a precise level. The goal is to facilitate the germination process in a confined area to minimize the cost of heating a large greenhouse to germination temperature. Air circulation is important to ensure uniform temperature and humidity throughout the chamber. There should be a thermostat to maintain the temperature regime. If the temperature goes too high, the variability between the seeds is accentuated, resulting in both uneven germination and transplant development.

Optimum temperature and time for germination vary for different vegetable crops. Germination conditions for the major transplanted vegetables are listed in Table 1. Germination time will vary between seed lots, so growers should check the trays regularly while they are in the germination chamber. The trays should be moved to the greenhouse after the seed coat has cracked and the plummule just starts to emerge. This will prevent excessive elongation. The time in the sweat chamber may only be 2 or 3 days. If possible, warm water should be used when watering plants during early growth-stages. Water should be heated to about 21°C. Heating the volume of water required to irrigate a large greenhouse may not be feasible.

2.4 Handling Trays: Racks and Benches

Vegetable transplant growers should use a rack system for benching in the greenhouse and for moving transplants to the field. Plug trays are usually handled on racks made of either angle-iron or wood with wire-mesh tops. Trays must be elevated off the ground to prevent root growth through the bottom. Trays should be levelled to prevent water from running into low spots. The growing-on stage may be defined as the stage of growth of transplants from after emergence until the plants are hardened-off before field setting. During the growing-on stage, environmental conditions (temperature, ventilation, light), watering and fertility, all affect the growth and quality of the transplants.

3. Crop Requirements

Trying to do a good job in raising different horticultural crops within the same greenhouse environment can be a challenging process. The environmental requirements for one crop may not suit those for another. For example, cabbage transplants require relatively cool temperatures and low fertility levels, while pepper transplants require higher temperatures and more fertility. Wherever possible, crops with dissimilar requirements should be grown in different areas of the greenhouse, where they can be managed as required.

3.1 Temperature

Different vegetable crops vary in their response to temperature. The optimum day and night growing temperatures for several crops are listed in Table 1.

3.2 Height Control

Warm-season vegetable crops (tomatoes, peppers, eggplant and vine crops) are susceptible to a disorder known as chilling injury. Chilling occurs when transplants are exposed to temperatures above freezing but below 10°C for an extended period. Chilling causes stunting of growth and can have a long lasting effect on field establishment. For susceptible crops, maintain a minimum greenhouse temperature of 10°C.

The DIF (difference) Method is a method of managing greenhouse day/night temperatures to control plant height. The DIF is determined by subtracting the nighttime from the day time temperature. A higher day temperature gives a positive DIF and promotes growth while a lower day temperature gives a negative DIF which retards growth. High temperatures during the first three to four hours after sunrise can cause considerable stretching in vegetable transplants. This stretching can be reduced by keeping the greenhouse

Table 1: Optimum Temperature Ranges for Germination of Seeds and Growth of Various Vegetable Transplants

Crop	*Germination (°C)*	*Approx. No. of Days to Emergence*	*Growing Temperature Day (°C)*	*Growing Temperature Night (°C)*
Tomatoes	21-24	3-4	18-21	10-18
Sweet Pepper	26-28	4-6	18-21	12-18
Cole Crops	18-24	2-3	12-18	8-15
Vine Crops	24-30	2-3	21-24	12-18
Onion	18-24	3-4	16-18	8-15

temperature cooler during the morning hours compared to the night temperature (negative DIF). When cooling the greenhouse to a negative DIF, be careful to avoid chilling temperatures. Usually, 4-5°C negative DIF will give good height control.

3.3 Watering of Transplants

The amount and frequency of watering will vary depending on cell type, growing media, greenhouse ventilation and weather conditions. It is important to water thoroughly, and moisten the entire plug, which will promote root growth to the bottom of the plug. If the plug is not watered thoroughly, root growth will be confined to the top of the plug. Allow the plug to dry down before watering, but do not let the plant wilt severely, as this will damage roots.

Plug transplants should be watered thoroughly in the morning, but should not be watered late in the afternoon. If the plants remain wet overnight, disease problems increase.If an overhead watering boom is used, it is advisable to remove and rearrange the nozzles occasionally to avoid the "streaking" that results from variations in output from different nozzles.

3.4 Fertility

The fertility programme used in raising vegetable transplants affects the quality of the finished transplant and its ability to become established in the field. A well-grown transplant will have adequate nutrient reserves to ensure rapid establishment under a wide variety of field conditions.Vegetable transplants are usually fertilized with a soluble fertilizer which is applied in the irrigation water. Several of the fertilizer analyses that are recommended for transplant production are listed in Table 2. These materials vary in percent nitrogen (N), phosphate (P_2O_5), and potash (K_2O) and in the micronutrient content. Growers should use fertilizers that have most of the nitrogen in nitrate form. Avoid fertilizers with a high concentration of urea.

Table 2: Concentrations of N, P, K and ECs for 100 ppm Solutions of various Soluble Fertilizer Materials that are Suitable for Use in Vegetable Transplant Production

Fertiliser Analysis	*Rate for 100 ppm N (g/100 l of water)*	*Parts per Million*			*Conductivity (EC)mmho/ cm**
		N	*P*	*K*	
20-20-20	50	100	43	83	0.40
20-10-20	50	100	21	83	0.60
20-8-20	50	100	17	83	0.75
17-5-19	59	100	12	92	1.0
15-5-15	67	100	14	83	0.70
14-0-14	71	100	0	83	0.85

* Electrical Conductivity of a 100 ppm solution in micromhos.

A high concentration of phosphate (P_20_5) may promote excessive seedling elongation under certain conditions. Use a fertilizer with a low to medium phosphate concentration. An alternative is to use a fertilizer with no phosphate (such as 14-0-14) for most feedings, and apply a high-phosphate fertilizer periodically (once every four or five feedings) to promote growth. Do not withhold phosphate completely, as this will delay field establishment.Transplants should be watered as requiredand the nutrient solution concentration and application frequency should be modified to promote the desired amount of growth. Fertilizer requirements vary depending on cell size (larger cells require less fertilizer), and the nutrient charge of the growing media (less fertilizer should be used if the media have a high nutrient charge).

3.4.1 Fertility Requirements of Vegetable Crops

Different vegetable crops vary in their response to fertilizer, so the feeding programme must be modified for different crops:

Tomatoes are very responsive to fertilizer and excess fertility will reduce transplant quality. If feeding at every watering, use a fertilizer concentration of 50 to 100 ppm N, depending on the stage of plant development. There may be some advantage to feeding less often at a higher concentration. If feeding once every seven days, use a concentration of 250 to 350 ppm N. Peppers require more fertilizer than tomatoes. If feeding at every watering, use approximately 100 ppm N. and increase the concentration if feeding less often. Cole crops do not require as much fertilizer as other crops. One application per week of 100 to 150 PPM N should be sufficient under most conditions. Vine Crops have a relatively short growing cycle compared to other crops. Two to four applications of fertilizer at weekly intervals, at a 100 to 150 ppm N concentration, should be sufficient to produce good-quality vine crop transplants.

3.4.2 Calculation of Fertilizer Solutions

$$\text{Weight of fertilizer (g)} = \frac{\text{Solution concentration (ppm)} \times \text{Solution volume (litres)}}{10 \times \text{(per cent N of fertilizer material)}}$$

The concentration of fertilizer solutions is usually expressed in parts-per-million (ppm) of nitrogen. To determine how much fertilizer material is required to produce a solution of a desired concentration, use the following formula:

$$\text{Weight of fertilizer} = \frac{(100 \times 500)}{(10 \times 20)} = 250 \text{ g}$$

For example, to make a 100 ppm solution of 20-10-20 fertilizer in a 500-litre tank, the amount of fertilizer required is

Weight of fertilizer required = 250 g

If using a fertilizer injector, to determine the amount of fertilizer material required in the stock solution:

- ☆ Calculate the weight of fertilizer required to produce the desired final nutrient solution, using the above formula.
- ☆ Multiply this weight by the injection ratio of the fertilizer injector.
- ☆ Be sure to use enough water in the stock solution to fully dissolve all the fertilizer material.

The concentration of the final nutrient solution can be checked using electrical conductivity (EC). Conductivity meters are available from most greenhouse-supply companies. Subtract the background EC of the water from the fertilizer solution EC to compare with the values in Table 2.

4. Disease Prevention

The primary means of controlling diseases on vegetable transplants is by sanitation and by managing the greenhouse environment to suppress disease development.

4.1 Sanitation

- ☆ Control weeds inside and outside the greenhouse that may harbour disease organisms.
- ☆ If trays are reused between crops of transplants, they should be washed to remove any soil or plug media that may adhere to the plastic, then dipped in a solution of chlorine bleach (1 per cent) or registered disinfectant.
- ☆ Trays may need to be thoroughly rinsed after disinfecting as bleach or other residues may be toxic to young seedlings.

4.2 Damping-off Control

- ☆ Ventilation, which promotes air movement around the plants, is the best method of preventing most damping-off and foliar fungal diseases.
- ☆ If signs of damping-off diseases are noticed in the greenhouse, transplants can be treated with fungicides.

5. Hardening of Seedlings

Hardening of vegetable seedlings before transplanting in the main field is very essential for reducing transplanting shock and also to have better crop stand. Plants should be gradually hardened, or toughened by acclimating them to the anticipated growing conditions of fields, at least a week before planting them in main field. This is done by slowing down their rate of growth to prepare them to withstand such as chilling, drying winds, shortage of water, or high temperatures. Seedlings of vegetables like cabbage, lettuce, onion and many others can be hardened to withstand frost; others such as tomatoes and sweet peppers cannot. Withholding water or minimal water supply are the best ways to harden a plant. Generally, seedlings of all vegetables become ready for transplanting in 28-30 days after sowing in plastic pro-trays. Under the greenhouse conditions, cucumber seedlings become ready for planting in main field or greenhouse in 25-30 days after sowing during peak winter whereas muskmelon in 30-35 days during peak winter season but most of the seedlings of cucurbits are ready for transplanting within 20 days in summer season. In case of tomato and brinjal, the seedlings are ready for transplanting in 30-32 days if the germination room is used for initial period for seed germination during winter season whereas the optimum temperature for seed germination (*i.e.* 24-25°C) is kept by using electric heaters.

6. Advantages

i. Seedlings can be raised under adverse climatic conditions, where it is not possible under open field conditions.

ii. Healthy seedlings can be raised in short period as compared to the time taken under open field nursery raising.

iii. There is no chance of soil-borne fungus or virus infection to the seedlings and insect-vectors can not enter under the protected conditions.

iv. Drastic reduction in the mortality in transplanting of the seedlings as compared to the traditional system of nursery raising. No mortality, no transplanting shock and quick establishment of the seedlings due to perfect development of root system.

v. Early planting is accomplished by raising such nursery.

vi. It is suitable for raising the nursery of sexually and asexually propagated vegetables and ornamental crops.

vii. Management of insects and diseases under greenhouse/protected conditions is quite easy particularly the infection of viruses.

viii. Looking to the very high cost of hybrid seed of vegetables reduce the seed rate 30-40 per cent in comparison with the traditional nursery raising system, as individual seeds are sown in each cell, which produces a very healthy seedling.

ix. Easy for transportation after packing for long distances.

x. Farmers can get the nursery ready from such nursery greenhouse any time as per the requirement.

Plug Tray Nursery in India

Vegetable seedling raising in multi-celled plastic plug-trays has already developed as an industry in several developed countries and it has also become popular in India during the last 4 to 5 years. Seedlings of cucurbitaceous and solanaceous vegetable crops raised through traditional system of nursery raising on soil beds prone to damage against during their uprooting and transplanting. This kind of damage can not be avoided in the traditional system of seedling raising, therefore it was not in general practice to transplant these vegetable through seedlings. The Centre for Protected Cultivation Technology of the Indian Agricultural Research Institute, New Delhi has developed a nursery raising technology for cucurbitaceous and solanaceous vegetable crops and standardized in which seedlings are raised in multi-celled plastic plug trays containing each cell of a volume of 18-20 cc and inverted pyramid in shape by using coco-peat, vermiculite and perlite as soil-less media in 3:1:1 ratio on volume basis. The plug-tray nursery raising technology by using soilless media can be extended to the growers in various parts of northern India for growing off-season cucurbitaceous vegetables (Singh *et al.*, 2007a). This technique is not only efficient in vigorous root development but also suitable to avoid any damage to the roots and shoots of the seedlings at the time of transplanting. This technology is economical and suitable for the cucurbits growers in northern plains of India, because with the introduction of this technique, farmers can grow large number of seedlings as per requirement for off-season cultivation of cucurbits (with various varieties) for fetching high price of their off-season produce. The plug tray nursery raising technology is suitable for melons, gourds and squashes for their off season cultivation in Northern India. This kind of technology is good option for the agriculture entrepreneur by the progressive farmers or unemployed educated youths in major vegetable growing parts of the country (Singh *et al.*, 2010).

When vegetable seedlings are raised in typical containers by using soilless media, growth rate of seedlings is affected by the container cell size and volume. Tomato seedlings raised in round cells with cell volume of 68.2 cm^3 produced the higher shoot fresh weight, root fresh weight, root length, leaf area, stem diameter, true number of leaves and seedling dry weight followed by the seedlings raised in inverted pyramid shaped cells having a cell volume of 18.4 cm^2. Cost of seedling raising was minimum (Rs. 41.4/100 seedlings) for the seedlings raised in inverted pyramid shaped cells. Inverted pyramid shaped cells of 18.4 $cm^3$are the most suitable and cost effective for raising tomato seedlings in soilless media (Singh *et al.*, 2007b)

Production of Grafted Plants

The use of grafted transplants to prevent physiological disorders in vegetables is increasing. The major objectives of grafting transplants are to suppress soil-borne diseases, increase resistance to adverse environmental conditions such as high or low temperatures, and avoid physiological disorders caused by continuous cultivation. Grafting techniques are currently applied to tomato, brinjal, cucumber, watermelon, and melon. However, they will be further extended in the near future to other crops such as chili pepper. The rootstocks commonly used are pumpkin, gourd and melon. Grafting methods commonly used include inarching (cucumber), insertion grafting (watermelon), grafting of cuttings (many cucurbit transplants), inarching with only one cotyledon, and pin grafting (tomato). Graft union chambers (or graft healing chambers) were developed and put into commercial production. The air temperature, relative humidity, and air movement of the chambers can be adjusted. A cabinet type with shelves on which trays can be stacked, and a tunnel type, were both developed.

Vegetable grafting became a crucial tool to overcome soilborne diseases and other pests. In the 1990s, nearly 60 per cent of open fields and greenhouses in Japan producing muskmelon (*Cucumis melo* L.), watermelon, cucumber, tomato, and eggplant were reportedly planted with grafted seedlings and 81 per cent in Korea (Lee, 1994). Today, over 500 million grafted seedlings are produced annually in Japan (Kobayashi, 2005). The major objectives of using grafted seedlings are: i) to achieve resistances to soil borne diseases and nematodes; ii) to increase yield and quality; and iii) to improve the physiology of plants making them more adaptable to harsh environments. Consequently, many countries in Europe, the Middle East, Northern Africa, Central America, and Asia adopted the technology and the areas introducing grafted plants increased rapidly during the past two decades.

Grafting of tomato is accomplished by using an elastic plastic tube that holds the graft union cut at an angle (tube grafting method). Each plastic tube has a slit so it falls off as the stem expands in diameter. Grafting speed using this tube grafting method varies from 300 to 500 grafts per hour depending on the worker's skill. Grafting is usually performed ≈2 to 3 weeks after seeding. Some rootstock seeds exhibit slower germination and require seeding 1 week earlier than the scion to develop compatible stem sizes at the time of grafting. Propagators need to optimize the seeding schedule for each selected combination of scion and rootstock so as to maximize the grafting success. Grafting young seedlings is considered advantageous as a result of the smaller size of seedlings, which allows greater seedling density and reduces the cost. After grafting, seedlings are placed under high humidity (greater than 95 per cent), ambient temperature (27 to 28°C), and low light intensity (≈100 $\mu mol\ m^{-2}s^{-1}$ photosynthetic photon flux) for 4 to 7 d to heal the grafted union. The propagators use a humidification chamber or a simple covered structure inside a shaded greenhouse, whereas more sophisticated indoor healing units with better air temperature, relative humidity, and light intensity control capability are commercially available in other countries.

Several commercial trials using grafted watermelon and muskmelon seedlings were conducted in different countries to increase yield, quality, or to demonstrate grafting as a potential partial alternative to soil fumigants. Cucurbitaceous vegetable crops are grafted using many different methods, including approach grafting, cotyledon grafting (a type of splice grafting), cleft grafting, and hole-insertion grafting (Lee and Oda, 2003). Some propagators use unrooted cuttings as rootstocks harvested before grafting, and root the grafted cuttings while the grafted unions are healed.

Rootstocks for Grafting

Greenhouse tomato constitutes more than 90 per cent of grafted seedling production in North America (Kubota, 2008). The majority of these seedlings are currently grafted to one or two common rootstocks. There is only limited information on the use of other rootstocks, compatibility to open-field cultivars, and field performance of grafted seedlings in various climatic conditions. Resistance information of rootstocks available from seed companies may not necessarily contain all the resistance/tolerance potentially exhibited in local soil and environmental conditions. One example is the recent finding that muskmelons grafted to interspecific hybrid squash (*Cucurbita maxima* × *Cucurbita moschata*) had resistance to vine decline caused by *Monosporascus* spp. (Cohen *et al.*, 2000; Edelstein *et al.*, 1999) and tolerance to charcoal rot (*Macrophomina phaselina*). Many tomato rootstocks contain the *Mi* gene and have been incorporated into many tomato cultivars for resistance to RKN as demonstrated using 'Beaufort' rootstock by Cao *et al.* (2005). A drawback of the *Mi* gene, however, is its sensitivity to temperature, and failure of resistance can occur at high soil temperature (Williamson, 1998). Nematode isolates capable of breaking *Mi* resistance have also been identified in many areas of world (Williamson, 1998). Ioannou (2001) examined a tomato rootstock for grafting eggplant and found reduced efficacy of grafting against RKN in summer crops, presumably as a result of the high soil temperature.

Cucurbit rootstocks with RKN resistance are not commercially available, but there are a few species that are promising candidates. Igarashi *et al.* (1987) examined several wild *Cucumis* species, including Kiwano (*Cucumis metuliferus* Naud.), for resistance to RKN and grafting compatibility to muskmelon. Considering all aspects, including fruit size and yield, Igarashi *et al.* (1987) concluded that *C. metuliferus* is the most suitable rootstock exhibiting good resistance to RKN. Recently, a research group in California (Sigüenza *et al.*, 2005) reported that *C. metuliferus* can be used as a rootstock for muskmelon to prevent both plant growth reduction and nematode population increases in *M. incognita*-infested soil. Sigüenza *et al.* (2005) also found that *C. moschata* rootstock, a traditional rootstock used for cucurbits in Asia, had a high level of tolerance to *M. incognita*.

Handling Grafted Seedlings

The introduction of mechanization and automation technology will also help address large-scale production issues. Efficient labor management has been recognized as a key to success in mass production of grafted seedlings. Semi- or fully-automated grafting robots were invented by several agricultural machine industries (Kurata, 1994) and some models are available in East Asia, Europe.According to Kobayashi (2005), the first commercial model of a grafting robot (GR800 series; Iseki and Co. Ltd., Matsuyama, Japan) became available for cucurbits in 1993 and there were various semi- and fully automated grafting robots presented from nine different agricultural machine industries at an international horticultural trade show in Tokyo in 1996. The prototype model has scion and rootstock feeders, which pick, orient, and feed the scion and the rootstock shoots to the grafting processor, performing 750 grafts per hour with a 90 per cent success rate.

As a result of a short production cycle and the consistent seedling quality, use of artificial environments to produce grafted seedling plugs can be feasible, especially for mass production systems. The advantages of using an artificial environment are the predictable production schedule and standardized seedling quality as a result of the consistent environmental conditions and therefore plant growth and development. Production of scion and rootstock seedlings having a standardized quality will facilitate use of automation in the grafting process. Failure resulting from lack of uniformity among plants is a common cause of ineffective use of grafting robots (Kobayashi, 2005).

Conclusion

In the past, vegetable transplants were either grown in greenhouses using flats or ground beds, or in outdoor ground beds. Now most growers have made the transition to greenhouse-grown transplants using various types of containers, primarily plug or pro-trays. With this system, each transplant grows in an individual cell so there is less competition among plants and greater uniformity. Less labour is required for mixing and sterilizing soil, filling flats and pulling plants. Plug transplants establish better in the field because roots are not damaged in pulling.

References

Cao, Z.P., Chen, G.K., Chen, Y.F., Yang, H., Han, L.F. and Dawson, R. 2005. Comparative performance of nematode resistant rootstock and non-resistant tomato cultivars on soil biota. *Allelopathy J.*, **15**: 85–94.

Cohen, R., Pivonia, S., Berger, Y., Edelstein, M., Gamliel, A. and Katan, J. 2000. Toward integrated management of *Monosporascus* wilt of melons in Israel. *Pl. Dis.*, **84**: 496–505.

Edelstein, M., Cohen, R., Burger, Y., Shriber, S., Pivonia, S. and Shtienberg, D. 1999. Integrated management of sudden wilt in melons, caused by *Monosporascus cannonbolus*, using grafting and reduced rates of methyl bromide. *Pl. Dis.*, **83**: 1142–1145.

Igarashi, I., Kano, T. and Kawabe, T. 1987. Disease and pest resistance of wild *Cucumis* species and their compatibility as rootstock for melon, cucumber and watermelon. *Bull. Natl. Res. Inst. Veg. Orn. Plants Tea*, **A1**: 173–185.

Ioannou, N. 2001. Integrating soil solarization with grafting on resistant rootstocks for management of soil-borne pathogens of eggplant. *J. Hort. Sci. Biotech.*, **76**: 396–401.

Janice, L. 2013. Agricultural Information Contact Centre, OMAFRA, Guelph, Ontario, Canada.

Kobayashi, K. 2005. Vegetable grafting robot. *Res. J. Food Agri.*, **28**: 15–20.

Kubota, C. and McClure, M.A. 2008. Vegetable grafting: History, use and current technology status in North America. *Hort. Sci.*, **43**: 1664–1669.

Kurata, K. 1994. Development of grafting robots in Japan. 2. Cultivation of grafted vegetables. *Hort. Sci.*, **29**: 240–244.

Lee, J.M. 1994. Cultivation of grafted vegetables. 1. Current status, grafting methods, and benefits. *Hort. Sci.*, **29**: 235–239.

Lee, J.M. and Oda, M. 2003. Grafting of herbaceous vegetable and ornamental crops. *Hort. Rev.*, **28**: 61–124.

Sigüenza, C., Schochow, M., Turini, T. and Ploeg, A. 2005. Use of *Cucumis metuliferus* as a rootstock for melon to manage *Meloidogyne incognita*. *J. Nematol.*, **37**: 276–280.

Singh, B., Kumar, M., Singh, V. and Mehto, S.P. 2007a. Plug-tray nursery raising of cucurbits in soilless media: A sustainable technology for off-season crop production under northern plains of India. *Acta Hort.*, **742**: 85–87.

Singh, B., Tomar, B.S. and Hasan, M. 2010. Plug-tray nursery raising technology for off-season cucurbits cultivation. *Acta Hort.*, **871**: 279–282.

Singh, B., Yadav, H.L., Kumar, M. and Sirohi, N.P.S. 2007b. Effect of plastic plug-tray cell size and shape on quality of soilless media grown tomato seedlings. *Acta Hort.*, **742**: 57–60.

Williamson, V.M. 1998. Root-knot nematode resistance genes in tomato and their potential for future use. *Ann. Rev. Phytopathol.*, **36**: 277–293.

5

Grafting for High Productivity and Profitability in Vegetable Crops

Narayan Chawda* and Vimal Chawda

Director, VNR Seeds Private Limited, Raipur – 492 099, Chhattisgarh
**E-mail: narayan.chawda @ vnrseeds.com*

Grafting is a process of bringing together two living plant parts (root and shoot) from two different sources to produce complete plant to achieve specific purposes (Figure 1). Earlier grafting was the means of propagation of certain asexually propagated or tedious to sexually propagated fruit crops for commercial seedling production only. And grafting was confined only to fruit crops just for the multiplication purpose but not any other reason. As science started advancing this grafting took a dimension not only to improvise the yield and quality but also to address the climatic and adaphic factors like biotic and abiotic stresses. Undoubtedly grafting served its purpose in achieving its goal by producing various resistant cultivars in fruit crops

Figure 1 : Grafting Tomato Scion on Brinjal.

Grafting of vegetable seedlings is a unique horticultural technology to overcome soil-borne diseases and pests and to add extra vigour to the plants under various environmental stress conditions. The technique was practiced for many years in East Asia, and more recently it has become a common pest management technique worldwide. In 1990s, nearly 60 per cent of open fields and greenhouses for producing muskmelon, watermelon, cucumber, tomato and eggplant were reportedly planted with grafted seedlings in Japan (1998 survey data by National Research Institute of Vegetables, Ornamental Plants and Tea, 2001), and 81 per cent in Korea **(1)**. Over 500 million grafted seedlings are produced annually by specialized commercial propagators in Japan **(2)**. Today, in Spain, for example, grafting is used for almost all watermelon cultivated in Almeria region, mainly for controlling *Fusarium* diseases. Marketing effort of international seed companies and local extension services are the driving force that disseminated this labour intensive propagation technique.

In Indian context vegetable grafting is very attractive and quick solution to encounter some of important production constraints *viz.*, biotic and abiotic stresses. Vegetable grafting entered the India in recent past but gaining importance rapidly. By keeping all these things in mind we made efforts to encounter the existing production constraints through grafting and also made an effort to share our challenges and success of vegetable grafting in Chhattisgarh state through this article.

1. Need of Vegetable Grafting

A few years back, in developed countries and many big farmers were using Methyl bromide to treat the soil against the soil borne pathogens. This methyl bromide was contaminating the undergroundwater and its fumes were depleting the Ozone's sphere, which was one of the reasons of climate change. Hence, UNO banned this chemical on 1st January 2005, in the world. Then those countries and farmers were dependant on methyl bromide, diverted to vegetable grafting.

Some one can say that why to go for labour intensive technology – Vegetable grafting. Our breeder scientists can develop resistant varieties. There are limitations of breeders, to develop one variety; it takes 7-8 years. Again it takes 2-3 years for R.S.T. and O.F.D. type trials. Then this variety reaches to the farmers' field and after 5-6 years, again this variety becomes susceptible to a pathogen, because of the evolving of new races of the pathogen. In case of brinjal – every after 100 – 200 km consumers' liking is changed. There are so many varieties being grown. How many resistant varieties will be developed? Vegetable grafting provides solution to region-specific problems too. One variety showing resistance at one place could be in susceptible in other areas. This type of problem can be solved by adoption of vegetable grafting. Crops like watermelon and muskmelon are susceptible to *Fusarium* wilt and nematodes and that is why the cultivation of these crops is restricted to mostly river bed cultivation, where every year soil of the river bed is changed due to flood. There are rootstocks, which are resistant to wilt and nematode. By grafting on such rootstock, crops can be taken in wilt and nematode susceptible soil. Cucumber is eaten throughout the year in vegetable salad. But cucumber crop cannot be grown in winter due to cold climate. Below 12°C temperature, cucumber roots cannot uptake nutrients from the soil. While, in North India 5 to 10°C temperature is common during winter. This type of abiotic stresses problem can be overcome by grafting cucumber on rootstock – *Cucurbita ficifolia*, which is a cold tolerant genotype.

2. Initiative in Vegetable Grafting

In the central part of India; state like Chhattisgarh wilt is a major problem for cultivation of Solanaceous crops like tomato, eggplant *etc.* Southern parts of the state especially Bastar soil is a great host for wilt causing organisms like bacteria, fungi and nematodes. Because of this reason farmers are unable to grow Solanaceous crops like brinjal and tomato. In 2003, we got the training on the technology of vegetable grafting from AVRDC, Taiwan and many symposia **(3)**. Siorea 2011, we are producing grafted plants for farmers of Bastar and other districts too.

It took eight years to finalize the proper protocol for healing and hardening of grafted seedlings, according to our climatic conditions. We supplied the grafted plants of brinjal to Bastar farmers in 2011. We got good results. The brinjal grafted on *Solanum torvum* rootstock was found resistant to bacterial wilt, fusarium wilt and nematode. When we develop F_1 hybrid, it could increase the yield by 15-20 per cent, but by grafting the yield can be increased by 50 to 100 per cent as was in the case of brinjal in example cited above. Many of the farmer from Bastar region took the advantage of vegetable grafting in production of record yields in brinjal even in the soils infested with wilt causing pathogens (Times of India, 2015) (Figure 2). During 2015-16, we supplied 10 lakh grafted brinjal and tomato seedlings to Chhattisgarh farmers. Presently, we are also undertaking breeding of rootstocks and grafting experiment in cucurbits like musk melon, water melon, *etc.*

Growing 1,00,000kg brinjal in Bastar, this farmer earns big

Times News Network

Raipur: Adopting grafting method and other technology, a Bastar-based farmer has overcome the tribal region's airborne disease bacterial wilt, which hampered the cultivation of brinjal vegetable in this region of Chhattisgarh.

Progressive farmer Ramesh Chawda now gets an yield of 1,25,000kg of brinjal from each acre of his 2.5 acres land. To make brinjal cultivation a success in this terrain, he adopted the horticulture technique of grafting the root and stem of different varieties of brinjal plant to grow them as a single plant.

The inserted root and stem of two different varieties helps fight the bacteria and produce desired genes of healthy brinjal variety.

Head of department of vegetable science at Indira Gandhi Krishi Vishwavidyalaya, Dr Jitendra Singh said, "This is a great success story. To achieve around 1,000 quintals of brinjal in Bastar region is remarkable."

Ramesh owns 30 acres of land near Jagdalpur and is growing grafted brinjals in just 2.5 acres of land for last 10 months. For its cultivation, he bought grafted saplings from VNR seeds, learnt the technique and is also helping local farmers to adopt the method. He said, "Getting high yield was a mind-boggling experience for me. People are coming from different areas to check the reality."

"Now the crop is still standing. I am expecting around 50,000kg yield again. The local market of Jagdalpur and nearby places is getting good variety of brinjal supply from other cities. We see a promising market locally as well as in state capital Raipur and towns and cities in neighbouring Odisha and Maharashtra," he added.

Ramesh encourages youth to join the league. "A new avatar of progressive farmers can do wonders. It is a respectable profession like that of doctors and engineers."

Ramesh Chawda in his brinjal farm

No 'T' ... American

Figure 2: Success Story of Grafted Brinjal Farmer in Bastar.

4. Conclusion

After looking at above success stories we could realize the importance of vegetable grafting in India. No doubt in future vegetable farmers from all over India will adapt this technology in their package of practices for improving vegetable production, productively as well as their economy

5. References

International Symposium on Vegetable Grafting, 2011 held at Italy.

Kobayashi, K. 2005. Vegetable grafting robot. *Res. J. Food Agri.*, **28**: 15–20.

Lee, J.M. 1994. Cultivation of grafted vegetables. 1. Current status, grafting methods and benefits. *Hort. Sci.*, **29**: 235–239.

6

Micropropagation for Mass Production of Horticultural Crops

Leela Sabijram* and Madhusudhana Rao

Division of Biotechnology,
ICAR-Indian Institute of Horticultural Research, Hessaraghatta Lake,
Bengaluru – 560 089, Karnataka
**E-mail: leelas@iihr.res.in*

Micropropagation refers to the true-to-type propagation of a selected plant genotype employing *in vitro* culture techniques. Ever since George Morel first proposed application of plant tissue culture for commercial clonal propagation of virus-free plants around five decades ago, micropropagation has evolved into a competitive, worldwide industry that now produces billions of plants annually. Production has long centred on high-value ornamentals like tropical foliage plants, orchids (Kane, 1996), anthuriums, woody landscape plants, bulbous/other floriculture/ specialty crops, bamboos, *etc.* Among fruit crops, bananas occupy the centre-stage while strawberries too are produced in large numbers. It affords rapid, mass propagation of selected plants/clones/valuable hybrids *in vitro* round the year in a relatively small space under controlled conditions. Performance of tissue-culture-raised propagules is generally superior to those derived from cuttings or seedlings, provided no scope is given to 'somaclonal variation' – the undesirable scourge in this hi-tech horticulture. Undoubtedly, micropropagation has proved itself as the Cinderella among commercial biotechnologies in agriculture/ horticulture. Scientific advancements are constantly afoot to develop automated micropropagation systems as successfully demonstrated in multiplication of liliums using bioreactors. However, the basics of tissue culture have changed little since it came of age, with reliance on the sterile technique and subtle variations in media formulations as its focal points.

The number of nurseries engaged in plant propagation by tissue culture escalated rapidly in the United States during the 1970s and 1980s. This exciting technique was a result of active research which was employed by progressive horticulturists, students, specialists and hobbyists alike. Hundreds of commercial laboratories worldwide are currently practicing micropropagation, some of which produce over 20 million plants a year (Kyte and Kleyn, 1996). Statistics on commercial production are not frequently available, but it is estimated that globally 970 million plantlets were produced in 2004 and the growth has been exponential ever since. Although leafy and other ornamentals capture over three-fourths of the market share, woody perennials (recalcitrant to tissue culture) had started capturing 16 per cent of all micropropagated plants by the mid-1980s itself (George and Sherrington, 1984).

Published protocols for micropropagation in various crops are many, of which only a few have been converted into commercially viable ventures. Information generated on protocols is often available only in very expensive and specialized journals, symposium proceedings or advanced textbooks.

An Overview of Global Plant Tissue Culture Industry

As the value of globally traded micropropagated plants in terms of proportion of all the plants propagated is not too high, it is clubbed with all other categories of live plants and makes it difficult to obtain reliable statistics on current level of global trade in this particular commodity. Feedback from knowledgeable persons involved in the industry suggested an estimated global production of about a billion micropropagated plants. This has risen from 130 million in 1985-86 to 513 million in 1990-91, to 680 million in 1993-94 and 800 million in 1996-97. As per an earlier estimate on global production, major producers of micropropagated plants were the Asian countries, USA, East European countries, the Netherlands, Latin American companies, West European countries, France, Australia, New Zealand, Italy and Israel. The Netherlands is a major market and ornamental plants account for a major share. Most laboratories abroad, particularly in the European countries, are small and labour cost is high. Therefore, these countries have shifted their production base to the developing countries which offer cheaper labour.

Increases in yield of up to 150 per cent of virus-free potatoes have been reported under controlled conditions (Altaf Hussain *et al.*, 2012)

The Indian Scenario

Recognizing the potential of plant tissue culture technology in revolutionizing commercial horticulture, it was identified as a priority area and several programmes were initiated in India aimed at developing and commercializing it in an integrated manner. Besides increasing agricultural production, this industry generates rural employment. It is estimated that at present about 50 commercial, medium-to-large scale tissue culture labs are in operation successfully in India. To facilitate commercialization further, the Government of India set up a National Facility for Virus Diagnosis and Quality Control of Tissue Culture plants at New Delhi, with 5 satellite centres in various parts of the country. Commercial labs in India produce bananas, pineapple, strawberry, potato, sugarcane, turmeric, ginger, vanilla, pepper, cardamom, *Aloe vera*, geraniums, orchids, anthuriums, lilies, carnation, *etc.* Banana is the star among micropropagated fruit crops in India (Leela Sahijram *et al.*, 2003), with lion's share in the market. Bananas popular among growers are Grande Naine (abbreviated to 'G9' by growers), *Elakki*, Dwarf Cavendish, *Nendran*, *etc.* Commercial micropropagation of tropical and sub-tropical fruits, temperate fruit crops, vegetables, ornamentals, spices, plantation crops, medicinal/aromatic crops is listed by Ramesh Chandra and Maneesh Mishra (2003). Innumerable garage-level units too have cropped up lately. Maharashtra leads in terms of production at 38 per cent of the microplants. In the year 2013-14, production of tissue culture banana in India stood at 2,50,000 MT, and was projected to reach 3,00,000 MT in 2014-15 (www.desaifv.com).

As the micropropagation industry was identified as a priority area by the Government of India, over the last two decades, Ministry of Science and Technology has supported more than 150 projects for R&D and field demonstration in about 80 different universities/ Research Institutes. The strong R&D support from the Ministry of Science and Technology and favourable policies of the Ministries of Commerce, Industries and Agriculture, Government of India, encouraged entrepreneurs and technocrats to set up more than 50 commercial laboratories between 1987 and 1995, with a total installed capacity of 210 million plants per annum. It was observed that during 1986-1989, targets achieved were 50 per cent of the installed capacity. In 1991, there was a decline and only 20 per cent of target was achieved. In 1996, there was a drastic reduction in the number of plants *vis-a-vis* an increase in the number of units. The history of commercialization in India is a story of 'Rise and Fall, and Rise Again'. Percentage increase in production decreased by 50 per cent from 1991 to 1994 and, 1998 saw a negative growth, showing rapid decline. However, during 1999-2006, there was 35 per cent rise in tissue culture production per year. This trend resulted in better capacity utilization of existing facilities by 2002; additional facilities began to be set up to increase the total installed capacity in the country to 300 million plants per annum. A concerted effort is being made by the Government and industries to ensure that plant tissue culture, a technology with enormous commercial potential, becomes an important industrial activity during the 21st century (Prakash, 2006).

Several government schemes and incentives are operational to promote the micropropagation industry. Mechanisms are in place in India for accreditation of commercial tissue culture laboratories and for certification of the production facility, as stated earlier. Department of Biotechnology (DBT), Govt. of India, New Delhi, set up the National Certification Agency for Tissue Culture-raised Plants (NCS-TCP) to ensure quality through virus-free microplant production (www.dbtncstcp.nic.in). NCS-TCP is in operation since the year 2006, as per Gazette of India Notification, under the Seeds Act 1966. As of 22 September 2016, 95 commercial labs have been recognized by DBT under their NCS-TCP programme (www.dbtncstcp.nic.in/html/content/Recognized_TCPU's.pdf)It has been instrumental in building the capacity of tissue culture companies for producing quality planting material besides enhancing their market reach through a certification process (http://apeda.gov.in/apedawebsite).

From an initial estimate of commercial tissue culture units numbering over 150, India only about a third sustained and can be termed medium-to-large outfits. The rest are all small players. In view of the declining performance of commercial plant tissue culture units, *Agricultural and Processed Foods Products Exports Development Authority* (*APEDA*) commissioned *Dalal Consultants and Engineers Limited* (*DCEL*) for a study to assess the reasons for the downward trend. The major objective of the study was to identify constraints faced by the industry, to recommend remedial measures and to suggest criteria for selecting tissue culture units eligible for revival. DCEL enlisted services of subject-matter specialists and industry experts for this. A field survey was carried out during March-June 2001, wherein 60 units were contacted. Efforts were also made to enumerate all the commercial units established until then in the country for a comprehensive understanding of the Indian tissue culture industry. Pre-structured questionnaires were used. Most units were personally contacted by the DCEL team. Information on the remaining units was gathered from knowledgeable people within the industry.

Domestic Market

The major plant species micropropagated in the country for domestic sector include (http://apeda.gov.in/apedawebsite/useful_links/national_link.htm):

- ✰ ***Fruits Plants*** – Banana, pineapple, strawberry, apple, papaya, jackfruit and pomegranate.
- ✰ ***Spices*** – Cardamom, vanilla, ginger and turmeric.
- ✰ ***Beverage Plants*** – Tea, coffee.
- ✰ ***Vegetables*** – Potato, *Aspargus*, onion.
- ✰ ***Medicinal and Aromatic Plants***: 'Safed Musli', *Taxus baccata*, patchouli.
- ✰ ***Agroforestry Plants*** – Teak, eucalyptus, kadam, pinus, pauwlonia, bamboo, and poplar.
- ✰ ***Flowering Plants*** – Gerbera, carnations, chrysanthemum, anthurium, roses, orchids.
- ✰ ***Field Crops*** – Sugarcane.

Banana is the most popular plant in India, followed by sugarcane. Banana has become popular due to good results obtained by the farmers and its presence in a large number of states in the country. Sugarcane is doing well in some parts of the country like Maharashtra, Gujarat and parts of Southern India. Wherever cooperative sugar mills have adopted the strategy of promoting nurseries and selling certified seed sets to the farmers at a reasonable price, results have been good. Sugarcane has not done well in North India.

The major markets for tissue cultured plants in India are in the states of Maharashtra (38 per cent), Karnataka (17 per cent) followed by Tamil Nadu, Andhra Pradesh, Gujarat, Kerala and West Bengal. In rest of the states results are poor. One reason for this could be the large concentration of laboratories in South and Western India and area under banana cultivation is also high in these states. An acute need is to achieve higher productivity by developing superior-quality planting material. This was met when the DBT (Department of Biotechnology), Government of India, created TERI's micropropagation park for mass propagation of cash crops and forest species. Though traditional breeding helped improve the quality of crops, its tedious and time-consuming nature (especially in the case of forest trees and other perennials that take several years) prevented it from gaining wider acceptance. Micropropagation, on the other hand, has come to be seen as a far more effective way of producing high-quality planting material of various species within a short period of time. This is done by selecting elite genotypes from the natural population of a species and culturing them under germ-free conditions to produce millions of identical plants. The benefits have been multifaceted ever since – greater output, economy of time and space, freedom from seasonal constraints, clone uniformity, and disease-free nature of the regenerants. The Park, therefore, bridged the yawning gap between academic research and its application in the field (http://www.teriin.org/technology/micropropagation-technology-park).

The Park has state-of-the-art infrastructure ranging from modern laboratories and latest equipment, to greenhouses and polyhouses. By December 2003, TERI had supplied over 12.8 million plants to various state forest departments, non-governmental organizations, agro-based companies, and private growers.

Table 1: Number of Plants Dispatched from TERI's Facility between April 1991 and December 2003

Type of Plants	*Number of Plants Dispatched (in million)*
Forestry	4.14
Fruit crops	0.47
Cash crops	3.33
Medicinal and aromatic	0.47
Ornamentals and miscellaneous	5.15
Total	**13.56**

http://www.teriin.org/technology/micropropagation-technology-park.

Micropropagation protocols for mass propagation of over 70 different species are available at the MTP of TERI. MTP helped popularize strawberry cultivation in non-hilly areas by mass-multiplying day-neutral varieties obtained from overseas and made them available to growers at affordable prices. Cash crops have particularly been helped. More than 2.7 million plantlets/microtubers/minitubers of different potato varieties (low-sugar content and high dry matter) specially meant for making potato chips were supplied to the industry and farmers. Planting different varieties of tissue-cultured sugarcane has helped produce more number of tillers than conventional plants. A large number of tissue-cultured plants of vanilla produced at the MTP and hardened at TERI's satellite facility in Guwahati have been distributed through the Spices Board of India in the seven North-Eastern states to

Table 2: Plant Species for which Tissue Culture Protocols are Available at TERI's Micropropagation Technology Park (MTP)

Forest Species	*Foliage Plant*	*Flowering Plant*	*Fruit*	*Vegetable*	*Medicinal and Aromatic Plants*	*Misc.*
Anogeissus pendula	*Callistemon*	*Carnation*	*Banana*	Asparagus	*Aloe vera*	Hops
Anogeissus latifolia	*Dieffenbachia*	*Chrysanthemum*	*Strawberry*	Ginger	*Bacopa* (*Brahmi*)	Sugarcane
Bambusa bambos	*Ficus* spp.	*Dahlia*	*Citrus*	Leek	*Centella asiatica*	Turmeric
Dendrocalamus asper	*Hosta*	*Gerbera*		Potato	Geranium (scented)	
Dendrocalamus strictus	*Nandiana*	*Pelargonium*			*Pogostemon cablin* (Patchouli)	
Eucalyptus tereticornis	*Philodendron*	*Plumbago*			*Rauwolfia serpentina* (Sarpgandha)	
Eucalyptus camaldulensis	*Pogonantherum*	*Rose*			*Chlorophytum* sp. (*Safed* musli)	
Paulownia fortunei	*Syngonium*	*Thalictrum*			*Swertia chirata* (*Chirata*)	
Populus deltoides	*Yucca*	*Verbascum*			Vanilla	

popularize and increase area under acreage of this valued export earner (Table 2).

Though the cost of tissue-cultured plants is variable (it can range from Rs. 5 to 15 apiece, depending upon the species and number of plants required), they prove to be the right choice because of their high quality. MTP supplies superior-quality tissue-cultured plants with clonal uniformity, conducts contractual research/production of plants, inoculates tissue-cultured plantlets with efficient mycorrhizal biofertilizers; offers post-delivery field care; and, even helps entrepreneurs set up laboratories and greenhouses.

Commercialization of Oil Palm and Date Palm Tissue Culture Protocols

ICAR-Indian Institute of Oil Palm Research (IIOPR), Pedavegi, A.P., has successfully developed a methodology of oil palm tissue culture using explants from mature palms. The technology has been commercialized through AgrInnovate India Limited and licensed to four companies, *viz.*, M/s Bejo Sheetal Bio-science foundation, Jalna, Maharashtra; M/s Vijaya Phyto Farms Pvt. Ltd, Hyderabad; M/s Sristi Agro Biotech Private Limited, Kolkata, and M/s Sheel Biotech Ltd., Haryana (http://dopr.gov.in/Achievements.htm#Achievements).

Ace Date Palm Tissue-culture Laboratory (ADPTL), a DBT-certified company in Secunderabad, Telangana, with imported technology micropropogates high-value cultivars of date palms including production of Arabian date varieties like Mejdhool, Barhee, Khallas, Khadrawy and Sugaye (http://adptl.com/index.php).

SWOT Analysis (http://apeda.gov.in/apedawebsite)

A 'SWOT' analysis of the industry is given in Table 3.

Table 3: SWOT Analysis for Tissue Culture Industry in India

Strengths	*Weaknesses*
1. Labour-oriented industry: Wages in India considerably low when compared to that in the developed countries	1. Sustained production of quality products uncertain
2. Large pool of high caliber scientific talent available	2. Industry fragmented; a large number of small scale units present and lack of cooperation among them
3. Argroclimatic conditions in India are suitable for making round-the-year supplies	3. Over capacities and unhealthy competition among units
4. Sustainable level of large volumes can be made available	4. Inability to provide technical and marketing services
5. Good industrial infrastructure, various concessions/incentives and favorable environment for exports	5. Lack of standardization and quality certification
6. Recognition of Biotechnology as a new Sunrise Industry	6. Poor efficiency of work force
	7. For export markets dependence on contract mass multiplication

Opportunities	*Threats*
1. Growing global demand for planting material	1. Competition from countries in Asia and Africa where wage rate is low
2. Enormous potential in domestic market	2. Inconsistent process yield
3. Liberalized economic policies promising relief	3. Erratic product-demand
4. Rich Indian biodiversity: possible to develop new products	4. Fluctuation in prices particularly in export market
5. Possibilities of developing value-added products	5. High risk of contamination
	6. Risk of legal problems in case of non-performance of the planting material in the farmers' field
	7. Technological obsolescence: Synthetic seed may replace this technology

Future of the Indian Micropropagation Industry

The Indian tissue culture industry has now surpassed the teething trouble stage and reached to a stage of maturity in terms of absorbing and fine tuning of technology. Industry has realized that export market is quality sensitive and domestic market is price sensitive. While export market can be recaptured by sustained efforts the domestic market has huge potential. Even if 1 per cent acceptance level of the micropropagated plants is successfully commercialized in crops like banana, sugarcane, potato, pineapple and futuristic crops like tomato and chillies, entire capacity in the country can be utilized. There is tremendous potential for forestry trees. Medicinal/Herbal/Aromatic plants have enormous potential. Afforestation programmes can also be undertaken on the unutilized marginal and waste land available in the country. This will generate huge demand for micropropagated plants. The scope for backward integration is limited as labour and energy are the major cost components. However, companies with resources can go in for clonal orchards and packaging of branded produce. The resourceful companies could diversify to synthetic seeds, initiate breeding programmes for development of exclusive varieties, and other value-added products. Idle greenhouse capacities could be utilized for producing plugs and speedlings. A number of units have shown willingness for the horizontal transfer of technology.

References

Chandra, Ramesh and Mishra, Maneesh 2003. *Comprehensive Micropropagation of Horticultural Crops*. International Book Distribution Co., Lucknow, U.P., p. 590.

George, E.F. and Serrington, P.D. 1984. *Plant Propagation by Tissue Culture*. Eversley, Exegetics Ltd., England, UK.

http://adptl.com/index.php

http://apeda.gov.in/apedawebsite/useful_links/national_link.htm

http://dopr.gov.in/Achievements.htm#Achievements

http://www.teriin.org/technology/micropropagation-technology-park

Hussain, Altaf, Qarshi, Iqbal Ahmed, Nazir, Hummera and Ullah, Ikram 2012. Plant tissue culture: Current status and opportunities. **In**: *Recent Advances in Plant In vitro Culture,* (Eds.) Annarita Leva and Laura M.R. Rinaldi. ISBN 978-953-51-0787-3.

Kane, M.E. 1996. Foreword in "*Plants from Test Tubes: An Introduction to Micropropagation*", 3rd edn. (Eds.) L. Kyte and J. Kleyn. Timber Press, Inc., Portland, Oregon, USA, pp. 240.

Kyte, L. and Kleyn, J. 1996. *Plants from Test Tubes: An Introduction to Micropropagation*, 3rd edn. Timber Press, Inc., Portland, Oregon, USA, p. 13.

Prakash, J. 2006. Micropropagation industry in India: Biology and business. *Acta Hort.*, **725**: 293–300.

Rout, G.R., Mohapatra, A. and Jain, M.S. 2006. Tissue culture of ornamental pot plants. *Biotech. Advanc.*, **24**: 531–560.

Sahijram, Leela Soneji, Jaya R. and Bollamma, K.T. 2003. Somaclonal variation in micropropagated banana: An analysis. **In**: *Comprehensive Micropropagation of Horticultural Crops,* (Eds.) Ramesh Chandra and Maneesh Mishra. International Book Distribution Co., Lucknow, U.P., pp. 221–236.

www.dbtncstcp.nic.in

www.dbtncstcp.nic.in/html/content/Recognized_TCPU's.pdf

www.desaifv.com

7

Diagnostics for Management of Viral Diseases in Horticultural Crops

V.K. Baranwal, R.K. Jain, R.K. Saritha and R. Kapoor*

Advance Centre for Plant Virology, Division of Plant Pathology, ICAR-Indian Agricultural Research Institute, New Delhi – 110 012

**E-mail: vbaranwal2001@yahoo.com*

Horticultural crops provide wide umbrella for nutritional security in India and there has been rapid expansion of production base of horticultural crops in recent times. However, the quality and productivity of many vegetatively propagated horticultural crops like apple, citrus, banana, grapevine, potato, spices, and ornamentals are often affected by viruses or virus like pathogens. Infact, viruses are considered major constraints to many horticultural crops as once infected, complete cure is impossible. It has been estimated that plant viruses are solely responsible for 47 per cent of infectious diseases reported in plants (Anderson *et al.*, 2004). Detection of viruses is becoming more challenging as globalization and international trade of horticultural crops enhance the movement of viruses and their vectors to new geographical areas (Boonham *et al.*, 2007). Climate change further contributes to spread and establishment of viruses or their vectors at places which were previously unfavorable. Here we have discussed viral diseases and viruses known to occur in five economically important vegetatively propagated horticultural crops of India, *viz.*, apple, banana, citrus, grapevine and potato and use of diagnostics in their management.

Apple is an important temperate fruit of the northwestern Himalayan region in India, grown predominantly in hill states of Jammu and Kashmir, Himachal Pradesh, Uttaranchal, Arunachal Pradesh and Nagaland. Cultivation of apple ishindered by the incidence of a large number of viruses and viroids. In India, there are five major viruses and one viroid reported to infect apple, *viz., Apple mosaic virus* (ApMV), *Apple chlorotic leaf spot virus* (ACLSV), *Apple stem grooving virus* (ASGV), *Apple stem pitting virus* (ASPV) and *Prunus necrotic ringspot virus* (PNRSV) and *Apple scar skin viroid* (ASSVd) (Rana *et al.*, 2011). Except ApMV, all other viruses are latent viruses (Nemeth, 1986). These viruses usually do not induce visible disease symptoms in the infected trees and fruits, although the infection eventually does lead to significant reduction in fruit yield and quality (Plese *et al.*, 1975; Yanase, 1983; Tiziano *et al.*, 2003). *Apple mosaic virus* (ApMV), an Ilarvirus, is a positive-sense single stranded RNA virus with isometric virions of about 30 nm diameter.In India, the virus was first reported based on symptoms from Uttrakhand (Bhargava and Bist, 1957). Severely infected plants show yellowing of leaves, veins and surrounding leaf lamina. ApMV is transmitted through vegetative propagation and root grafting (Dhingra, 1972) and by mechanical inoculations onto herbaceous hosts (Sutic *et al.*, 1999). PNRSV, also an *Ilarvirus* is similar to ApMV in particle characteristics and host range, but is transmitted either through pollen or thrips (Alrefai *et al.*, 1995). ACLSV, type species of genus *Trichovirus* in *Betaflexi viridae* family is a filamentous virus, transmitted through grafting, mainly infecting fruit trees of *Rosaceae* family, which include apple, pear, quince, sweet and sour cherry, peach, plum and apricot. The virus is latent in most apple cultivars but in sensitive varieties, malformation as well as reduction in leaf size and chlorotic rings or line patterns are common. The severity of symptoms elicited by ACLSVdepends largely on plant species and virus strains (Németh, 1986). *Apple stem pitting virus* (ASPV) and *Apple stem grooving virus* (ASGV) belong to the genera *Fovea virus* and *Capillo virus*, respectively. Both viruses have elongated filamentous particles and are latent in most commercial applecultivars, graft transmissible, world-wide distributed in pome fruit trees, and often occur in mixed infection (Nemeth, 1986; Kummert *et al.*, 1998; Kundu, 2003). Viroids, the smallest infectious agents, are single-stranded circular RNA molecules and infect many economically important horticultural and agricultural crops (Diener, 2001). *Apple scar skin viroid* (ASSVd) type species of genus *Apsacaviroid* and family *Pospiviroidae*, is a major limitation to apple fruit quality and causes huge economic losses. It produces severe symptoms on apple fruits including colour dappling, cracking, scarring and distortion depending upon the cultivar that render the fruits unmarketable (Koganezawa *et al.*, 2003).

Banana (*Musa* spp.) is worldwide a major food crop and an important horticultural crop. This tropical and sub-tropical crop contributes significantly to many national domestic and export incomes and in some regions is the principal source of carbohydrate. As banana is a vegetatively propagated crop, a number of viruses are transmitted by the vegetative planting material (suckers, bits and corms). These viruses are also horizontally transmitted by insect vectors. There are four major viruses which naturally infect bananas *viz.*, *Banana bunchy top virus* (BBTV), *Banana streak virus* (BSV), *Banana bract mosaic virus* (BBrMV) and *Cucumber mosaic virus* (CMV). *Banana bunchy top virus* (BBTV), a member of genus *Babuvirus* and family *Nanoviridae*, is an important virus seriously affecting banana plantations in India (Selvarajan *et al.*, 2010; Banerjee *et al.*, 2014). It is associated with devastating bunchy top disease of banana throughout the world including India. BBTV is a multi-component virus with circular single stranded DNA genome, consisting of small isometric virions of 18-20 nm size and is persistently transmitted by an aphid vector, *Pentalonia nigronervosa* Coquerel (Hu *et al.*, 1996; Hooks *et al.*, 2009; Chen and Hu, 2013). BBTV induces characteristic symptoms discontinuous dark green flecks and streaks of variable length on the leaf sheath, midrib, leaf veins and petioles. New leaves emerge with great difficulty, are narrower and shorter than normal leaves and have wavy leaf lamina with yellow leaf margins (Nelson, 2004). They appear to be "bunched" at the top of the plant, the symptom for which this disease is named (Thomas *et al.*, 1994). Susceptible cultivars infected at a young stage and the suckers emerging from infected stools are severely stunted. Severely infected plants usually do not fruit and if at all fruit is produced, the hands and fingers are distorted and twisted (Nelson, 2004). There is no effective resistance to bunchy top disease in banana, thus control is mainly based on regular rouging of infected plants, production and use of virus-free banana plants. Banana streak viruses (BSVs) are mealybug transmitted plant pararetroviruses, belonging to the genus *Badnavirus*, family *Caulimoviridae*. The virus particles are bacilliform-shaped (120-150 x 30 nm), with a non-covalently closed double-stranded circular DNA genome of 7.2-7.8 kb that uses a virus-encoded reverse transcriptase (RT) to replicate. There are currently four recognized banana-infecting badnavirus (BIB) species: *Banana streak GF virus* (BSGFV), *Banana streak MY virus* (BSMYV), *Banana streak OL virus* (BSOLV), and *Banana streak VN virus* (BSVNV). BIBs exist in two forms, either the infectious episomal form resulting from cells/plant infection following transmission by mealybugs, or the endogenous form in which the viral genome is integrated into the host genome. The most common badnavirus infecting banana in India is BSMYV (Sharma *et al.*, 2014). Symptoms of banana streak disease are characterized by conspicuous leaf streaks, which initially are chlorotic and become progressively necrotic. *Banana bract mosaic virus* (BBrMV), type member of genus *Potyvirus* and family *Potyviridae*, has a positive-sense ssRNA genome (~10 kb) with flexuous filmentous virions (750 x 11 nm) and vectored non-persistently by several species of aphids (King *et al.*, 2012). The characteristic symptoms of this disease are the green or reddish-brown spindle shaped streaks along the petioles and pseudostem, dark brown streaks on the bracts of the inflorescences. The disease has been so named after its characteristic mosaic symptoms on the flower bract evident only during flowering stage. Among banana viruses, BBrMV poses a considerable quarantine risk due to its limited geographic distribution, its ability to spread easily via the use of vegetative plant parts, aphid vectors, and its inconsistent production of visible symptoms. Banana mosaic, also known as infectious chlorosis caused by *Cucumber mosaic virus* (CMV) is one of the common viral diseases affecting banana worldwide. CMV belonging to genus *Cucumovirus* and family *Bromoviridae*, has a tripartite plus sense RNA genome. It is vectored in a non-persistent manner by more than 60 aphid species. It causes variable symptoms from mild chlorosis to severe chlorotic streaks that are similar to the symptoms of BSV infection. Most isolates of CMV from banana (CMV-B) have been identified to belong to subgroup I. The majority of common CMV strains do not produce severe symptoms or cause significant damage on banana. However, some severe strains cause severe necrosis in pseudostems and even plant death (Bouhida and Lockhart, 1990).

Citrus is an important fruit crop, comprising many species of economic importance. It is the largest fruit industry in India after banana and mango. Northeastern part of India is considered to be the natural home of citrus from where a number of citrus species/varieties originated and later taken to different parts of the world (Ahlawat, 2005). The major species grown in India are sweet orange (*Citrus sinensis* L. Osbeck), madarin (*C. reticulata*) and acid lime (*C. aurantifolia* Swingle). The productivity of citrus fruit is greatly affected, owing to many biotic stresses of which fungi, bacteria, viruses, viroids and phytoplasma play a very important role. Among virus diseases, tristeza, ringspot, mosaic and citrus yellow vein clearing virus are important in India. Citrus exocortis and yellow corky vein are two most important viroid diseases of citrus. In addition, citrus greening disease caused by a bacterium *Liberibactera siaticus*is an important disease (Ahlawant and Pant, 2003). *Citrus tristeza virus* (CTV), a member of *Closterovirus*, has flexuous filamentous particles (10-12 x 2000 nm) with single stranded positive sense RNA genome of about 20 kb (Dawson, 2010). Tristeza virus, transmitted by aphids, infects nearly all citrus species, citrus relatives and hybrids expressing different symptoms, causing decline on sour orange rootstock, seeding yellow in lemon and grapefruit, stem pitting on grapefruit and/ or sweet orange. Citrus ring spot disease was first reported in India as psorosis-A disease (Ahlawat, 1989) but it was later confirmed that this disease was distinct from psorosis. *Indian citrus ringspot virus* (ICRSV) and *Citrus yellow vein clearing virus* of genus *Mandarivirus* and family *Flexiviridae* are filamentous viruses, infecting most of the commercial citrus cultivars and rootstocks used in India causing variable symptoms (Byadgi and Ahlawat, 1995; Pant and Ahlawat, 1998; Meena and Baranwal, 2016). Citrus yellow mosaic disease caused by *Citrus yellow mosaic virus* is an important disease, widely distributed in India in sweet oranges (*C. sinensis* (L.) Osbeck) and pummelo (*C. grandis* (L) Osbeck.) (Ahlawat *et al.*, 1996; Ahlawat, 1997). It is a bacilliform non-enveloped DNA virus under the genus *Badnavirus* and family *Caulimoviridae* with a genome of 7559 bp. CYMV is graft transmissible and the disease is also spread by citrus mealybug *Planococcus citri* (Huang and Hartung, 2001). Affected trees not only produce significantly less fruit but also produce less juice and ascorbic acid. The fruit also may develop irregular yellow blotches in the rind and are smaller and flatter than normal fruit. Citrus exocortis viroid of *Pospiviroid* genus is the largest citrus viroid at 371 nt and produces the most pronounced symptoms of leaf epinasty, stunting and necrosis of leaf midvein. Citrus exocortis disease is transmitted by graft and also through mechanical inoculations. In India exocortis was reported by Nariani *et al.* (1968) and Rao *et al.* (1975) and later studied by Ramachandran *et al.* (1993). Yellow corky vein disease of Citrus is widely distributed in India causing yield loss up to 89.7 per cent in Kagzi lime in Assam (Ahlawat and Pant, 2003). The disease is being disseminated inadvertently through vegetative propagation producing characteristic symptoms of yellow veins and formation of corky tissues on the lower surface of affected leaf veins. Citrus greening disease also known as 'Huanglongbing' disease is perhaps the most important citrus disease in areas where the disease and its vector are both present (Bove, 2006). The symptoms of greening disease occur either throughout the tree or is often confined to some branches of the tree. Affected sectors grow poorly, bear abnormal fruit and show a variety of chlorotic leaf symptoms of which a blotchy mottle is the most characteristic. The disease is found in all commercially grown citrus cultivars but is particularly severe in mandarins, tangelos and sweet oranges. It is mild in true lemons and rough lemons and virtually symptomless in limes and trifoliate orange and its hybrids.

Among the various diseases of grapevines virus and virus like disease are the major limiting factor to the quality and quantity of the crop (Martelli and Boudon-Padieu, 2006). Symptoms specific to leafroll disease of grapevine, *i.e.* reddening of interveinal areas of leaves and downward rolling of leaf lamina in dark fruited cultivars were observed in the vineyards of Nashik and Pune regions of India and fresh cases of infections of GLD were also reported in the vineyards of Nashik and Pune regions of India. GLD has a complex aetiologyand five phloem restricted filamentous viruses of the family *Closteroviridae*, referred to as Grapevine leafroll-associated viruses (GLRaVs) are often associated with the disease (Martelli *et al.*, 2012; Martelli, 2014; Naidu *et al.*, 2014). These viruses are GLRaV-1, GLRaV-3 and GLRaV-4 belong to the genus *Ampelovirus*, GLRaV-2 to the genus *Closterovirus* and GLRaV-7 to the genus *Velarivirus*. Strains of GLRaV-4 formerlyknown as GLRaV-5,-6,-9, GLRaV-Pr, GLRaV-De and GLRaV-Car are considered genetically divergent variants (or strains) of GLRaV-4 within subgroup II of the genus *Ampelovirus* (Martelli *et al.*, 2012). Of these, GLRaV-1, GLRaV-3 have been characterized from India (Kumar *et al.*, 2012, 2013). Recently, GLRaV-4 has also been detected in different cultivars of grapevine (Unpublished).

Potatoes are vegetatively propagated food cropknown to beinfected bymore than 36 viruses and a viroid of which eight are reported in India *viz.*, *Potato viruses Y, X, M, A, S, Potato leafroll virus, Potato apical leaf curl virus, Groundnut bud necrosis virus* and Potato spindle tuber viroid. These viruses cause significant economiclosses in potato production in India. *Potato virus Y* (PVY), of genus *Potyvirus* in the family *Potyviridae* (King *et al.*, 2012), is the most widespreadvirus infecting potatoes. It is a flexuous filamentous virus, readily spread by aphids in a non-persistent manner as well as mechanically, resulting in severeyield reductions. Symptoms of PVY infection arevariable and range from mild (foliar mottling, streaking and mosaic) to severe (leaf necrosis, leaf drop and stunting) depending upon the strain and potato variety grown (Schramm *et al.*, 2011). *Potato virus A* (PVA), also a *Potyvirus* is similar to PVY. It causes mild mosaic symptoms and is spread by aphids. *Potato virus X* (PVX), type member of the genus *Potexvirus*, is another widespread virus affecting potato. PVX is tuber-borne and is readily mechanically transmitted by human activities. PVX alone induces mild symptoms inpotato but results in severe stunting of plants when coinfected with PVY (Rochow and Ross, 1955; Paul Khurana and Singh, 1988). Synergistic interaction of PVX and PVY induces rugose mosaic, causing severe disease generally notseen in individual infection (Vance *et al.*, 1995; Pruss *et al.*, 1997). *Potato viruses S* and *M* (PVS and PVM), belong to *Carlavirus* genus respectively. Both the viruses are tuberborne, have filamentous particles and are vectored by aphids in a non-persistent manner. PVS is symptomless on the majority of cultivars, with occasional mild leaf symptoms of rugosity, vein deepening and leaf bronzing. PVM may induce symptoms including mottling, mosaic, crinkling and rolling of leaves (para-crinkle) and stunting of shoots. *Potato leafroll virus* (PLRV) of genus *Polerovirus*, family *Luteoviridae* is one of major viruses infecting potato. It consists of icoshedral virus particles and is persistently transmitted by aphids. PLRV infection is restricted to phloem-associated tissues, showing characteristic symptoms of upward leaf rolling. Primary infection typically exhibits pale upper leaves and in some cultivars, rosiness and rolling of leaves. Secondary symptoms in potatoes cultivated from infected tubers are severe with stunting of the plant and rolling of the leaves (Harrison, 1985). *Groundnut bud necrosis virus* (GBNV), of genus *Tospovirus* in the family *Bunyaviridae* is a serious viral pathogen affecting several fabaceous and solanaceous hosts such as cowpea, groundnut, mungbean, potato,soybean and tomato (Mandal *et al.*, 2012). Tospoviruses are enveloped isometric RNA viruses, transmitted by thrips vector. GBNV is responsible for stem necrosis disease in potato (Jain *et al.*, 2004). Potato spindle tuber viroid (PSTVd) of genus *Pospiviroid* is a major constraint to potato production worldwide. PSTVd has a simple structure consisting of small (241-399 nucleotides),single-stranded circular RNAs (Diener, 1987). It is highly contagious, mechanically transmissible and is carried through pollen and true seed of potato. PSTVd infection cause tubers to be spindle or oblong shaped, have deeper eyes than normal and the tuber skin is often cracked. Foliar symptoms of potato include an upright and stunted growth habit with an erect stem and shortened internodes between leaflets resulting in an overlapping appearance. The leaflets are often smaller in size with a rugose appearance.

One of the successful ways of addressing the economic impact of viral diseases in vegetative propagated crops is the use of certified planting material which involves virus indexing of mother plants and certification of plant saplings. The management strategies for the viral diseases in horticultural plants largely rely on the definite diagnosis of virus agents followed by taking preventive measures to prevent their spread in field conditions. Virus detection based on biological techniques like symptomatology are inconclusive. They are also too slow and not amenable to large number of samples. Use of electron microscopy which began in 1930's gives direct evidence of the presence of virus but needs expensive equipments and is not amenable to large scale application. Besides, viroids which consist of naked RNA as genetic material cannot be detected by this method. Two breakthroughs in virus diagnostics came about in the last four decades. Serological assay in the form of Enzyme linked immuosorbent assay (ELISA) (Clark and Adams, 1977) and nucleic acid based *in vitro* amplification of DNA commonly called as polymerase chain reaction (PCR) (Candresse *et al.*, 1998). Due to these advancements, rapid, specific and sensitive detection methods are available for most of the viral diseases.

Serological Assays

Serological assays are based on the binding of an antigen (the viral coat protein) to its corresponding antibody. Among the serological techniques, ELISA introduced into plant virology about 37 years ago, is the most robust assay for routine detection of plant viruses in plant materials especially in crude extracts, owing to its simplicity, sensitivity (detection limits of 1-10 ng/ml) and adaptability. The amount of virus present is proportional to the amount of coloured end product forms the basis of the test. Traditional method of antiserum production involves the tedious process of virus purification from plant materials. ELISA has been standardized for detection of *Indian citrus ringspot virus* (ICRSV), *Citrus trizteza virus* (CTV) and *Citrus yellow mosaic virus* (CYMV) in citrus (Ahlawat and Pant, 2003). Polyclonal antibodies were raised using purified virus particles. With the advent of recombinant DNA technology it is now possible to express the viral coat protein in bacteria and purifiy it using affinity chromatography. This is easier than the difficult and complex procedure of virus purification and has almost become the choicest method for antiserum production. Antisera to BSMYV has been produced using recombinant coat protein and was successfully employed in detection of this virus in large number of field grownand tissue cultured raised banana plants (Sharma *et al.*, 2014, 2015).

Lateral Flow Assays

This One-step lateral-flow (LF) tests or dip sticks/ quick strip have been developed for the on-site detection and identification of several plant viruses (De Boer and Lopez, 2012). This method allows disease diagnosis in the field within 5 minutes. The dipstick format of lateral flow test uses a membrane-based detection system using specific monoclonal and polyclonal antibodies. The membrane contains two capture zones, one captures the bound viral protein, the other captures color reagent. Paper strips or plastic paddles are used as support for the capture antibody that is immobilized onto a test strip in specific zone. The lateral flow test strip is dipped into the prepared sample in extraction solution and the sample migrates up the strip by capillary action. As the sample flows through the detection antibody strip and the capture antibody strip, the proteinaceous virus will accumulate and thus give a high intensity band. This technique has been applied to detection of many plant viruses mainly for viruses of vegetable, ornamental plants and fruit crops (Salomone *et al.*, 2002, 2004; Drygin *et al.*, 2009; 2012; Maejima *et al.*, 2014; Yoon *et al.*, 2014 a, b). One step LF strips available commercially include tests for *Potato virus Y*, *Plum poxvirus*, *Pepino mosaic virus*, *Tobacco mosaic virus*, *Tomato spotted wilt virus*, *Impatiens necrotic spot virus*, *Zucchini yellow mosaic virus etc.* (Agdia, Bioreba, Forsite Diagnostics). The efficiency of lateral flow assay has been found to be comparable to that of ELISA (Salomone *et al.*, 2004; Danks and Baker, 2000). The main advantage of LF is that detection can be done anywhere, with a simple device, operated by personnel with very little or no training and results are obtained within minutes. However, the major disadvantage of LF tests may be its higher cost for single test and their general unsuitability for detecting latent infections because of their relatively low sensitivity.

Nucleic Acid-based Assays

Rapid advancement in virus genome characterization has led to the development of nucleic acid based diagnostics which include conventional PCR, real time PCR, multiplex PCR, micro/macro arrays and biochips. The amplification of DNA in PCR reaction is achieved *in vitro* conditions by using 4 dNTPs and two oligonucleotides (primers) flanking the segment of DNA to be amplified. A thermostable *Taq* DNA polymerase enzyme catalyzes the reaction.

For single stranded RNA viruses, the RNA is first reverse transcribed into c-DNA using a reverse transcriptase (RT) enzyme. The specificity and sensitivity of virus detection can be improved by immunocapture PCR (IC)-PCR in which virus is trapped by antibodies coated on the wall of microfuge tube and then nucleic acid of virus is released for direct amplification for DNA viruses or for c-DNA synthesis and subsequent PCR for RNA viruses. Amplification of viral RNA in c-DNA form using reverse transcriptase enzyme and subsequent amplification of c-DNA by polymerase enzyme has been used successfully for a large number of plant viruses (Candresse *et al.*, 1998)

Since horticultural crops are often infected by more than one virus, their individual detection by PCR is not only expensive but also time consuming. Therefore, multiplex PCR has been developed where in genome of more than one virus could be amplified and detected in the same reaction mixture. The end products of RT-PCR/PCR and multiplex PCR are detected by agarose gel electrophoresis using ethidium bromide which is a neurotoxin and not suited for high throughput application. Further, this type of RT-PCR/PCR based detection of viruses is semi-quantitative and in many cases the amount of PCR product is not proportion to amount of the target DNA, making this type of PCR a qualitative tool for detection of a particular virus. Therefore a new real time PCR has been developed and it allows to actually view the increase in the amount of DNA as it is amplified dispensing the use of agarose gel electrophoresis.

Once PCR is initiated, results can be available within 20 min. TaqMan real-time reverse transcriptase (RT)-polymerase chain reaction (PCR) using purified RNA targets or coupled to tissue-print and squash procedures was developed to detect and quantify CTV RNA-targets in plant tissues and in single aphids (Bertolini *et al.*, 2008).Though PCR has been a mainstay in virus detection, yet its need for rate-limiting temperature cycling makes it less useful than isothermal methods in terms of simplicity and speed.

The Loop Mediated Isothermal Amplification (LAMP)

The loopmediated isothermal amplification (LAMP) technique is one of the most prominent of these methods explored to date which utilizes a strand displacement strategy (Notomi *et al.*, 2000). LAMP assays have been developed for many viruses. The method makes use of four specially designed primers, a pair of outer primers and a pair of inner primers, which together recognize six distinct sites flanking the target DNA sequence. LAMP reaction relies on a DNA polymerase with particularly high strand displacement activity to remove newly amplified strands in place of heat denaturation of double-stranded DNA as in PCR. The final amplified product consists of a mixture of stem-loop DNA strands with various stem lengths and structures with multiple loops (Nagamine *et al.*, 2002). Only a water bath or heating block is needed to maintain temperature in the range of 60-68°C for LAMP. LAMP assays are particularly specific because hybridization to at least six sites is required for the reaction to proceed and can also be configured in real-time quantitative formats (Aoi *et al.*, 2006). Its tolerance to template contaminantsmakes LAMP method more attractive for on-site testing. For reverse transcription-LAMP (RT-LAMP) test for peach latent mosaic viroid, the sample template simply consists of traces of plant sap eluted from a toothpick pricked into a symptomatic leaf (Boubourakas *et al.*, 2010). Similarly, RT-LAMP assays for *Tomato yellow leaf curl virus* and *Melon yellow spot virus* use toothpicks for sample template acquisition (Shiro *et al.*, 2005; Takeuchi *et al.*, 2006). Lamp has been developed and standardized for some of the viruses infecting banana, citrus, grapevine and apple (Selvarajan *et al.*, 2015; Siljo and Bhat, 2014; Zhao *et al.*, 2014; Anthony Johnson *et al.*, 2014; Walsh and Pietersen, 2013; Peng *et al.*, 2012 a,b,c). The complexity of the primers required to drive the amplification reaction is one of the downsides of LAMP assays

Recombinase Polymerase Amplification (RPA)

Another isothermal amplification technique most recently developed for detection of viruses infecting fruit and vegetable crops is Recombinase Polymerase Amplification (RPA) (Mekuria *et al.*, 2014; Zhang *et al.*, 2014; Silva *et al.*, 2015; Londono *et al.*, 2016). In RPA, the isothermal amplification of specific DNA fragments is achieved by the combination of enzymes and proteins *viz.*, recombinase, single-stranded binding proteins (SSB) and strand displacing polymerase used at a constant low temperature and yielding result in as little as 10-15 min. There is no need for initial denaturation of double stranded DNA (ds DNA) target. RPA employs two sets of primers which combines with recombinase that targets the ds DNA and facilitate strand displacement. The displaced strand is further stabilized by SSB and the polymerase initiate synthesis.

Multiplex Methods

With the increasing number of plant viruses being identified, techniques like enzyme linked immunosorbent assay (ELISA) and Polymerase Chain Reaction (PCR) become inadequate since they can be used to test for a few specific viruses only at a time.

Cocktail Antiserum

Further advancement in serological detection was made by producing immuno-cocktail of polyclonal antibodies against the recombinant fusion coat protein of two and/or three viruses infecting vegetable crops. Dual and triple cocktail of polyclonal antibodies were developed for for simultaneous detection of multiple infection of *Cucumber mosaic virus* (CMV) and *Papaya ringspot virus* (PRSV) or CMV, PRSV and *Groundnut bud necrosis virus* (GBNV) (Kapoor *et al.*, 2014a). A similar attempt was made to develop dual cocktail PAb for simultaneous detection of *Potato virus Y* (PVY) and *Potato virus X* (PVX) since their mixed infection is common in potato (Kapoor *et al.*, 2014b).

Multiplex PCR/RT-PCR

Since horticultural crops are often infected by more than one virus, their individual detection by PCR is not only expensive but also time consuming. Therefore, multiplex PCR has been developed where in genome of more than one virus could be amplified and detected in the same reaction mixture. The end products of RT-PCR/PCR and multiplex PCR are detected by agarose gel electrophoresis using ethidium bromide which is a neurotoxin and not suited for high throughput application.Simultaneous detection of four viruses and greening bacterium has been achieved in citrus by multiplex PCR (Meena and Baranwal, 2016).

Nucleic Acid-based Chips

DNA microarrays have the potential to detect thousands of viruses in parallel in a single assay and have been used for detection and typing of plant and animal viruses. The identification of viral pathogens with microarrays is based on specific probes, in most cases on oligonucleotide probes that target nucleic acid sequences that are only present solely in the organism of interest to the investigator. Arrays for plant viruses were developed aiming at detection of viruses infecting a single crop or a few families of plant viruses and viroids (Boonham *et al.*, 2003, 2007; Lee *et al.*, 2003; Agindotan *et al.*, 2007; Bystrica *et al.*, 2005; Engel *et al.*, 2010; Pasquini *et al.*, 2008). A microarray chip that has potential to detect more than 1100 plant viruses and 40 viroids has been developed on Affymetrix platform at Advanced Centre for Plant Virology. Each species is represented by a set of 8-11 unique probe sets amounting to a total of 17,292 unique probes on the chip. DNA and RNA viruses as well as viroids were detected from known and unknown samples using the same chip starting from as little as 25 ng of total plant RNA. More than 20 viruses belonging to 9 different families and 3 viroids, from 9 different crops were detected. Out of this six viruses were not known to occur previously in the region. This chip will be a valuable tool to identify emerging and exotic viruses to initiate quarantine measures as well as to prepare geographical distribution of different viruses infecting different crops.

Deep Sequencing

Deep sequencing is an in depth sequencing of individual genome a large number of times to avoid sequencing error. Next-generation sequencing has already been applied in plant virology to detect the presence of unknown viruses by metagenomic analysis of the entire plant genome along with any infecting viral genomes. The generation of sequence information for the complete virome in grapes led to identification of viruses not previously reported in South African vineyards (Coetzee *et al.*, 2010).

Application of Diagnostics

For efficient diagnosis of virus infection, following factors must be considered; sensitivity (the minimum quantity of virus that can be detected), accuracy and reproducibility among the testing labs, number of samples that can be tested at a time, initial investment for apparatus and running cost per sample, experience, training and technical knowhow required for testing the samples and suitability of test for onsite detection at the point of decision making (Hull, 2009). ELISA and PCR based assays are the most important commonly used diagnostic assays. In a recent study we have been able to eliminate four different viruses in a garlic cultivar through apical meristem culture, *Indian citrus ringspot virus* through shoot tip grafting and this could be possible because of availability of serological and PCR based detection assays of all viruses (Hoa, 2003, Unpublished). The meristem tip culture and tissue culture of explants from virus free mother plants has been used in wide scale by tissue culture industries especially banana. Virus indexing of tissue culture plants and citrus budwoods using the above said technologies has been adopted for further elimination of virus infected planting materials. For certification of tissue cultured plant for freedom from viruses, Department of Biotechnology has established a Referral Centre at IARI and accredited test laboratories in different parts of India. There is a need to develop accredited diagnostic laboratories in India for virus diagnostics to detect viruses as a part certification programme of production of planting material by nurseries.

References

Agindotan, B.O., Patrick, J.S. and Philip, H.B. 2007. Simultaneous detection of potato viruses, PLRV, PVA, PVX and PVY from dormant potato tubers by TaqMan® real-time RT-PCR. *J. Virol. Methods*, **142**: 1–9.

Ahlawat, Y.S. 1989. Psorosis: Disease of citrus in India. *Indian Phytopath.*, **42**: 21–25.

Ahlawat, Y.S. 1997. Virus, greening bacterium and Viroids associated with citrus (*Citrus* species) decline in India. *Indian J. Agric. Sci.*, **67**: 51–57.

Ahlawat, Y.S. 2005. Virus and virus-like diseases affecting citrus in India: A step forward to management of die-back complex. *Indian Phytopath.*, **58**(3): 257–268.

Ahlawat, Y.S. and Pant, R.P. 2003. Major virus and virus like diseases of citrus in India, their diagnosis and management. *Ann. Rev. Pl. Path.*, **2**: 447–474.

Ahlawat, Y.S., Pant, R.P., Lockhart, B.E.L., Srivastava, M., Chakraborty, N.K. and Varma, A. 1996. Association of a badnavirus with citrus mosaic disease in India. *Plant Dis.*, **80**: 590–592.

Alrefai, R.H., Domier, L. and Korban, S.S. 1995. The complete nucleotide sequence of *Apple mosaic virus* RNA-3. *Arch Virol.*, **140**: 1247–1256.

Anderson, P.K., Cunningham, A.A., Patel, N.G., Morales, F.J., Epstein, P.R. and Daszak, P. 2004. Emerging infectious diseases of plants: Pathogen pollution, climate change and agrotechnology drivers. *Trends Ecol. Evol.*, **10**: 535–544.

Anthony Johnson, A.M., Dasgupta, I. and Sai Gopal, D.V.R. 2014. Development of loop-mediated isothermal amplification and SYBR green real-time PCR methods for the detection of *Citrus yellow mosaic badnavirus* in citrus species. *J. Virol. Methods*, **203**: 9–14.

Aoi, Y., Hosogai, M. and Tsuneda, S. 2006. Real-time quantitative LAMP (loop-mediated isothermal amplification of DNA) as a simple method for monitoring ammonia-oxidizing bacteria. *J. Biotech.*, **125**: 484–491.

Banerjee, A., Roy, S., Behere, G.T., Roy, S.S., Dutta, S.K. and Ngachan, S.V. 2014. Identification and characterization of a distinct *Banana bunchy top virus* isolate of Pacific-Indian Oceans group from North-East India. *Virus Res.*, **183**: 41–49.

Bertolini, E., Moreno, A., Capote, N., Olmos, A., de Luis, A., Vidal, E., Pérez-Panadés, J. and Cambra, M. 2008. Quantitative detection of *Citrus tristeza virus* in plant tissues and single aphids by real-time RT-PCR. *European J. Plant Pathol.*, **120**: 177–188.

Bhargava, K.S. and Bist, N.S. 1957. Three virus diseases of hill fruits in Kumaon. *Curr. Sci.*, **26**: 324–325.

Boonham, N., Tomlinson, J. and Mumford, R. 2007. Microarrays for rapid identification of plantviruses. *Ann. Rev. Phytopathol.*, **45**: 307–328.

Boonham, N., Walsh, K., Smith, P., Madagan, K., Graham, I. and Barker, I. 2003. Detection of potato viruses using microarray technology: Towards a generic method for plant viral disease diagnosis. *J. Virol. Methods*, **108**: 181–187.

Boubourakas, I.N., Fucuta, S., Luigi, M., Faggioli, F., Barba, M. and Kyriakopoulou, P.E. 2010. Improvement of the reverse transcription loop mediated isothermal amplification (RT-LAMP) method for the detection of peach latent mosaic viroid (PLMVd). *Julius Kuhn Arch.*, **427**: 65–69.

Bouhida, M. and Lockhart, B.E.L. 1990. Increase in importance of *Cucumber mosaic virus* infection in greenhouse grown bananas in Morocco. *Phytopathology*, **80**: 981.

Bove, J.M. 2006. Huanglongbing: A destructive, newly-emerging, century-old disease of citrus. *J. Plant Pathol.*, **88**: 7–37.

Byadgi, A.S. and Ahlawat, Y.S. 1995. A new viral Ringspot disease of citrus (*Citrus* species) in India. *Indian J. Agric. Sci.*, **65**: 763–770.

Bystricka, D., Lenz, O., Mraz, I., Piherova, L., Kmoch, S. and Sip, M. 2005. Oligonucleotide-based microarray: A new improvement in microarray detection of plant viruses. *J. Virol. Methods*, **128**: 176–182.

Candresse, T., Hammond, R.W. and Hadidi, A. 1998. Detection and identification of plant viruses and viroids using polymerase chain reaction (PCR). **In**: *Control of Plant Virus Diseases*, (Eds.) A. Hadidi, R.K. Khetarpal and K. Koganezawa, APS Press, St. Paul, MN, USA, p. 399–416.

Chen, Y. and Hu, X. 2013. High-throughput detection of *Banana bunchy top virus* in banana plants and aphids using real-time TaqMan® PCR. *J. Virol. Methods*, **193**: 177–183.

Clark, M.F. and Adams, A.N. 1977. Characteristics of the microplate method of enzyme linked immunosorbent assay for the detection of plant viruses. *J. Gen. Virol.*, **34**(3): 475–483.

Coetzee, B., Freeborough, M.J., Maree, H.J. *et al.*, 2010. Deep sequencing analysis of viruses infecting grapevines: Virome of a vineyard. *Virology*, **400**: 157–163.

Danks, C. and Barker, I. 2000. On-site detection of plant pathogens using lateral-flow devices. *OEPP/EPPO Bulletin*, **30**: 421–426.

Dawson, W.O. 2010. *Citrus tristeza virus*: Complex and tristeza disease. **In**: *Molecular Genetics of Citrus tristeza Virus*, (Eds.) A.V. Karasev and M.E. Hilf, St. Paul: APS Press, pp. 53–72.

De Boer, S.H. and Lopez, M.M. 2012. New grower-friendly methods for plant pathogen monitoring. *Ann. Rev. Phytopath.*, **50**: 197–218.

Dhingra, K.L. 1972. Transmission of apple mosaic by natural root grafting. *Indian J. Hort.*, **20**: 348–350.

Diener, T.O. 1987. Potato spindle tuber. **In**: *The Viroids*, (Ed.) T.O. Diener, Plenum Press, New York, pp. 221–233.

Diener, T.O. 2001. The viroid: Biological oddity or evolutionary fossil? *Adv. Virus Res.*, **57**: 137–184.

Drygin, Y.F., Blintsov, A.N., Grigorenko, V.G., Andreeva, I.P., Osipov, A.P., Varitzev, Y.A., Uskov, A.I., Kravchenko, D.V. and Atabekov, J.G. 2012. Highly sensitive field test lateral flow immunodiagnostics of PVX infection. *App. Microb. Biot.*, **93**: 179–89.

Drygin, Y.F., Blintsov, A.N., Osipov, A.P., Grigorenko, V.G., Andreeva, I.P., Uskov, A.I., Varitsev, Y.A., Anisimov, B.V., Novikov, V.K. and Atabekov, J.G. 2009. High sensitivity express immunochromatographic method for detection of plant infection by *Tobacco mosaic virus*. *Biochemistry* (Moscow), **74**: 1212–1220.

Engel, E.A., Escobara, P.F., Rojasa, L.A., Riveraa, P.A., Fiore, N. and Valenzuelaa, P.D.T. 2010. A diagnostic oligonucleotide microarray for simultaneous detection of grapevine viruses. *J. Virol. Methods*, **163**: 445–451.

Harrison, B.D. 1985. Potato leafroll virus. **In**: *AAB Description of Plant Viruses*, no. 291.

Hoa, N.V. 2003. Molecular charecterization of *Indian ringspot virus* and its management by shoot tip culture. *Ph.D. Thesis*, Indian Agricultural Research Institute, New Delhi.

Hooks, C.R., Fukuda, S., Perez, E.A., Manandhar, R., Wang, K.H., Wright, M.G. and Almeida, R.P. 2009. Aphid transmission of *Banana bunchy top virus* to bananas after treatment with a bananacide. *J. Econ. Ent.*, **102**: 493–499.

Hu, J.S., Wang, M., Sether, D., Xie, W. and Leonhardt, K.W. 1996. Use of polymerase chain reaction (PCR) to study transmission of *Banana bunchy top virus* by the banana aphid (*Pentalonia nigronervosa*). *Ann. App. Biol.*, **128**: 55–64.

Huang, Q. and Hartung, J.S. 2001.Cloning and sequence analysis of an infectious clone of *Citrus yellow mosaic virus* that can infect sweet orange via *Agrobacterium*–mediated inoculation. *J. Gen Virol.*, **82**: 2549–2558.

Hull, R. 2009. *Comparative Plant Virology*, 2nd edn. Elsevier Academic Press, San Diego, Burlington, London.

Jain, R.K., Paul Khurana, S.M., Bhat, A.I. and Chaudhary, V., 2004. Nucleocapsid protein gene sequence studies confirm that potato stem necrosis is caused by a strain of *Groundnut bud necrosis virus*. *Indian Phytopath.*, **57**: 169–173.

Kapoor, R., Mandal, B., Paul, P.K. and Jain, R.K. 2014b. Simultaneous detection of potato viruses Y and X by DAC-ELISA using polyclonal antibodies raised against fused coat proteins expressed in *Escherichia coli*. *J. Pl. Bioch. Biot.*, **23**: 332–335.

Kapoor, R., Mandal, B., Paul, P.K., Chigurupati, P. and Jain, R.K. 2014a. Production of cocktail of polyclonal antibodies using bacterial expressed recombinant protein for multiple virus detection. *J. Virol. Methods*, **196**: 7–14.

King, A.M.Q., Adams, M.J., Carstens, E.B. and Lefkowitz, E.J. (Eds.). 2012. *Virus Taxonomy: Ninth Report of the International Committee on Taxonomy of Viruses*, Academic Press, San Diego. Oxford: Elsevier Academic Press, pp. 1327.

Koganezawa, H., Yang, X., Zhu, S.F., Hashimoto, J. and Hadidi, A. 2003. Apple scar skin viroid in apple. **In**: *Viroids*, (Eds.) A. Hadidi, R. Flores, J.W. Randles and J.S. Semancik, CSIRO Publishing, Collingwood, Australia, pp. 137–141.

Kumar, S., Baranwal, V.K., Singh, P., Jain, R.K., Sawant, S.D. and Singh, S.K. 2012. Characterization of a Grapevine leafroll-associated virus 3 from India showing incongruence in its phylogeny. *Virus Genes*, **45**(1): 195–200.

Kumar, S., Singh, L., Ferretti, L., Barba, M., Zaidi, A.A. and Hallan, V. 2013. Evidence of Grapevine leafroll associated virus-1-3, Grapevine fleck virus and Grapevine virus B occurring in Himachal Pradesh, India. *Indian J. Virol.*, **24**(1): 66–69.

Kummert, J., Marinho, V.L.A., Rufflard, G., Colinet, D. and Lepoivre, P. 1998. Sensitive detection of *Apple stem grooving* and *Apple stem pitting viruses* from infected apple trees by RT-PCR. *Acta Hort.*, 472: 97–105.

Kundu, J.K. 2003. The occurrence of Apple stem pitting virus and Apple stem grooving virus within field-grown apple cultivars evaluated by RT-PCR. *Plant Protect. Sci.*, **39**(3): 88–92.

Lee, G.P., Min, B.E., Kim, C.S., Choi, S.H., Harn, C.H., Kim, S.U. and Ryu, K.H. 2003. Plant virus cDNA chip hybridization for detection and differentiation of four cucurbit-infecting Tobamoviruses. *J. Virol. Methods*, **110**: 19–24.

Londono, M.A., Harmon, C.L. and Polston, J.E. 2016. Evaluation of recombinase polymerase amplification for detection of begomoviruses by plant diagnostic clinics. *Virol. J.*, **13**: 48.

Maejima, K., Himeno, M., Netsu, O., Ishikawa, K., Yoshida, T., Fujita, N., Hashimoto, M., Komatsu, K., Yamaji, Y. and Namba, S. 2014. Development of an on-site *Plum pox virus* detection kit based on immunochromatography. *J. Gen. Plant Pathol.*, **80**: 176–183.

Mandal, B., Jain, R.K., Krishnareddy, M., Krishna Kumar, N.K., Ravi, K.S. and Pappu, H.R. 2012. Emerging problems of tospoviruses (*Bunya viridae*) and their management in the Indian subcontinent. *Plant Dis.*, **96**: 468–479.

Martelli, G.P. 2014. Directory of virus and virus-like diseases of the grapevine and their agents. *J. Plant Pathol.*, **96**: 1–136.

Martelli, G.P. and Boudon-Padieu, E. 2006. Directory of infectious diseases of grapevines. **In**: *Options Mediterraneennes*, (Eds.) G.P. Maertelli and E. Boudon-Padieu, Serie B. **55**: 9–194.

Martelli, G.P., Abou Ghanem-Sabanadzovic, N., Agranovsky, A.A., Al Rwahnih, M., Dolja, V.V. and Dovas, C.I. *et al.*, 2012. Taxonomic revision of the family *Closteroviridae* with special reference to the grapevine leafroll-associated members of the genus *Ampelovirus* and the putative species unassigned to the family. *J Plant Pathol.*, **94**: 7–19.

Meena, R.P., and Baranwal, V.K., 2016. Development of multiplex polymerase chain reaction assay for simultaneous detection of clostero-, badna- and mandarin-viruses along with huanglongbing bacterium in citrus trees. *J. Virol. Methods*, **235**: 58–64.

Mekuria, T.A., Zhangb, S. and Eastwella, K.C. 2014. Rapid and sensitive detection of *Little cherry virus* 2 using isothermal reverse transcription-recombinase polymerase amplification. *J. Virol. Methods*, **205**: 24–30.

Nagamine, K., Hase, T. and Notomi, T. 2002. Accelerated reaction by loop-mediated isothermal amplification using loop primers. *Mol. Cell. Probes*, **16**: 223–29.

Naidu, R.A., Rowhani, A., Fuchs, M., Golino, D. and Martelli, G.P. 2014. Grapevine leafroll: A complex viral disease affecting a high-value fruit crop. *Pl. Dis.*, **98**: 1172–1185.

Nariani, T.K., Raychaudhuari, S.P. and Sharma, B.C. 1968. Exocortis in citrus in India. *Plant Dis. Reptr.*, **52**: 834.

Nelson, S.C. 2004. Banana bunchy top: Detailed signs and symptom. Cooperative extension service College of Tropical Agriculture and Human Resources, p. 22, Manoa: University of Hawaii.

Nemeth, M. 1986. *Viruses, Mycoplasmas and Ricketesia Diseases of Fruit Trees*. Norwell: Kluwer Academic Publishers Group, p. 840.

Notomi, T., Okayama, H., Masubuchi, H., Yonekawa, T., Watanabe, K., Amino. N and Hase, T. 2000. Loop-mediated isothermal amplification of DNA. *Nucleic Acids Res.*, **28**: e63.

Pant, R.P. and Ahlawat, Y.S. 1998. Partial characterization of a filamentous virus associated with ringspot disease of citrus. *Indian Phytopath.*, **51**: 225–232.

Pasquini, G., Barba, M., Hadidi, A., Faggioli, F., Negri, R., Sobol, I., Tiberini, A., Caglayan, K., Mazyad, H., Anfoka, G., Ghanim, M., Zeidan, M. and Czosnek, H. 2008. Oligonucleotide microarray–based detection and genotyping of *Plum pox virus*. *J. Virol. Methods*, **147**: 118–126.

Paul Khurana, S.M. and Singh, M.N. 1988. Yield loss potential of potato viruses X and Y in Indian potatoes. *J. Indian Potato Assoc.*, **15**: 27–29.

Peng, J., Fan, Z. and Huang, J. 2012c. Rapid Detection of *Banana streak virus* by loop-mediated isothermal amplification Assay in South China. *J. Phytopathol.*, **160**: 248–250.

Peng, J., Shi, M., Xia, Z., Huang, J. and Fan, Z. 2012b. Detection of *Cucumber mosaic virus* isolates from banana by one step reverse transcription loop-mediated isothermal amplification. *Arch. Virol.*, **157**: 2213–2217.

Peng, J., Zhang, J., Xia, Z., Li, Y., Huang, J. and Fan, Z. 2012a. Rapid and sensitive detection of *Banana bunchy top virus* by loop–mediated isothermal amplification. *J. Virol. Methods*, **185**: 254–258.

Plese, N., Hoxha, E. and Milicic, D. 1975. Pathological anatomy of trees affected with *Apple stem grooving virus*. *Phytopathol. Zeitschrift.*, **82**(4): 315–325.

Pruss, G., Ge, X., Shi, X.M., Carrington, J.C. and Bowman, V.V. 1997. Plantviral synergism: The potyviral genome encodes a broad-range pathogenicity enhancer that trans-activates replication of heterologousviruses. *Plant Cell*, **9**: 859–868.

Ramachandran, P., Pandey, P.K., Ahlawat, Y.S., Varma, A. and Kapur, S. P. 1993. Viroid diseases of citrus in India, **In**: *Proc. 12th Conf. IOCV*, *IOCV*, Riverside, pp. 438–440.

Rana, T., Negi, A., Dhir, S., Thockchom, T., Chandel, V., Walia, Y., Singh, R.M., Ram, R., Hallan, V. and Zaidi, A.A. 2011. Molecular diagnosis of apple virus and viroid pathogens from India. *Arch. Phytopathol. Plant Prot.*, **44**: 505–512.

Rao, D.G., Pandey, P.K., Chakraborty, N.K. and Capoor, S.P. 1975. Mechanically transmitted virus diseases of citrus new to India. *Indian J. Hort.*, **32**: 117–122.

Rochow, W.F. and Ross, A.F. 1955. Virus multiplication in plants doublyinfected by Potato viruses X and Y. *Virology*, **1**: 10–27.

Salomone, A., Bruzzone, C., Minuto, G., Minuto, A. and Rogerro, P. 2002. A comparison of lateral flow and ELISA for the detection of *Tomato mosaic virus* in tomato. *J. Plant Pathol.*, **84**: 193.

Salomone, A., Mongelli, M., Roggero, P. and Boscia, D. 2004. Reliability of detection of *Citrus tristeza virus* by an immunochromatographic lateral flow assay in comparison with elisa. *J. Plant Pathol.*, **86**: 43–48.

Schramm, S., Frost, K., Charkowski, A., Gray, S., Crockford, A. and Groves, R.L. 2011. Management of *Potato virus Y* (PVY) in Wisconsin Seed Potato Production (A3951). Division of Cooperative Extension of the University of Wisconsin-Extension, Madison, pp. 1–8.

Selvarajan, R., Balasubramanian, V. and Sasireka, T. 2015. A simple, rapid and solvent free nucleic acid extraction protocol for detection of *Banana bunchy top virus* by polymerase chain reaction and loop-mediated isothermal amplification. *European J. Pl. Path.*, **142**(2): 389–396.

Selvarajan, R., Sheeba, M.M., Balasubramanian, V., Rajmohan, R., Dhevi, N.L. and Sasireka, T. 2010. Molecular characterization of geographically different *Banana bunchy top virus* isolates in India. *Indian J. Virol.*, **21**: 110–116.

Sharma, S.K., Kumar, P.V. and Baranwal, V.K. 2014. Immunodiagnosis of episomal *Banana streak MY virus* using polyclonal antibodies to an expressed putative coat protein. *J. Virol. Methods*, **207**: 86–94.

Sharma, S.K., Kumar, P.V., Geetanjali, A.S., Pun, K.B. and Baranwal, V.K. 2015. Subpopulation level variation of banana streak viruses in India and common evolution of banana and sugarcane badnaviruses. *Virus Genes*, p. 450–465.

Shiro, F., Naoko, A., Masashi, K., Yukie, Y. and Masahiro, F., *et al.*, 2005. Simplification of template DNA preparation for the loop-mediated isothermal amplification (LAMP) method to detect *Tomato yellow leaf curl virus*. *Ann. Rep. Kansai Plant Prot. Soc.*, **47**: 37–41.

Siljo, A. and Bhat, A.I. 2014. Reverse transcription loop-mediated isothermal amplification assay for rapid and sensitive detection of *Banana bract mosaic virus* in cardamom (*Elettaria cardamomum*). *European J. Plant Pathol.*, **138**: 209–214.

Silva, G., Bömer, M., Nkere, C., Kumar P.L. and Seal, S.E. 2015. Rapid and specific detection of *Yam mosaic virus* by reverse-transcription recombinase polymerase amplification. *J. Virol. Methods*, **222**: 138–144.

Sutic, D.D., Ford, R.E. and Tosic, M.T. 1999. *Handbook of Plant Virus Diseases*. Virus diseases of fruit trees, Florida: CRC Press, p. 345–347.

Takeuchi, Y., Fukuta, S. and Oya, T. 2006. Detection of *Melon yellow spot virus* by reverse transcription loop-mediated isothermal amplification. *Res. Bull. Aichi-ken Agric. Res. Center*, **38**: 57–63.

Thomas, J.E., Iskra-Caruana, M.L. and Jones, D.R. 1994. Banana bunchy top disease. *Musa Disease Fact Sheet* No. 4. Montpellier, France: INIBAP, pp. 2.

Tiziano, C., Folwell, R.J., Wandschneider, P., Eastwell, K.C. and Howell, W.E. 2003. Economic implications of a virus prevention program in deciduous tree fruits in the US. *Crop Prot.*, **22**: 1149–1156.

Vance, V.B., Berger, P.H., Carrington, J.C., Hunt, A.G. and Shi, X.M. 1995. 52proximal potyviral sequences mediate *Potato virus X*/potyviral synergisticdisease in transgenic tobacco. *Virology*, **206**: 583–590.

Walsh, H.N. and Pietersen, G. 2013. Rapid detection of *Grapevine leafroll-associated virus* type 3 using a reverse transcription loop-mediated amplification method. *J. Virol. Methods*, **194**: 308–316.

Yanase, H. 1983. Back-transmission of *Apple stem grooving virus* to apple seedlings and induction of symptoms of apple top working disease in rootstocks of Mitsuba Kaido (*Malus sieboldii*) and Kobano Zumi (*Malus sieboldii* var. *arborescens*). *Acta Hort.*, **130**: 117–122.

Yoon, J-Y., Choi, G-S., Cho, I-S. and Choi, S-K. 2014a. Development of rapid immune-gold strip kit for on-site diagnosis of *Cucumber mosaic virus*. *J. Korean Soc. Int. Agric.*, **26**(1): 62–67.

Yoon, J-Y., Choi, G-S., Cho, I-S. and Choi, S-K. 2014b. Development of rapid immune-gold strip kit for on-site diagnosis of *Tomato spotted wilt virus*. *Res. Pl. Dis.*, **20**(1): 15–20.

Zhang, S., Ravelonandro, M., Russell, P., McOwen, N., Briard, P., Bohannon, S. and Vrient, A. 2014. Rapid diagnostic detection of *Plum pox virus* in Prunus plants by isothermal Amplify RP® using reverse transcription-recombinase polymerase amplification. *J. Virol. Methods*, **207**: 114–120.

Zhao, L., Feng, C.H., Li, B.Q., Hao, X.A., Liu, H., Wu, Y.F. and Wang, Q.C. 2014. Rapid detection of *Apple stem grooving virus* by reverse transcription loop-mediated isothermal amplification. *J. Pl. Pathol.*, **96**(2): 407–409.

Biotechnological Interventions

8

Role of Biotechnology for the Improvement of Production, Productivity and Resistance in Vegetable Crops

G. Kalloo[1], M. Singh[2], N. Rai[2], A.C. Rai[2], R.S.B. Reddy[2] and K.K. Rai[2]*

[1]Former Vice Chancellor, JNKVV, Jabalpur and
Ex-Deputy Director General (Horticulture and Crop Sciences),
Indian Council of Agricultural Research, Delhi – 110 001
[2]ICAR-Indian Institute of Vegetable Research,
PB-01, Jakhini (Shahanshapur), Varanasi – 221 305, Uttar Pradesh
**E-mail: gautamkallo@gmail.com*

Biotechnology is a fast emerging science, in the recent years, has created a great opportunities, not only for the manipulation of biological systems for the benefit of mankind, but also to undertake studies for better understanding of the fundamental life processes. The term biotechnology was coined in 1917 by Hungarian Engineer Karl Ereky to describe a process for large scale production of pigs. According to US National Science Foundation "Biotechnology is defined as any technique that uses living organisms or part of organism to make or modify product, to improve plants or animals, or to develop micro-organisms for specific uses". In this way, it deals accurate and cost-effective means to produce a range of novel, value-added products. It has the potential to improve food productivity, reduce the use of harmful pesticide, make low cost of raw materials and decrease the adverse environmental impacts related with traditional production methods. Along with this the researcher has improve their basic research using the advance tools of biotechnology into the nature of life and ecosystems at the molecular level can lead to improved farming practices and diagnostic tools for use in agriculture.

Biotechnology can be engaged to improve the seed quality, increase protein levels, as well as to produce the crops resistance to pathogens, insect pests and the crops tolerance for droughts, floods and extreme temperatures. Nevertheless the current use of biotechnology can make the vegetables much more healthier to improve their nutritional quality, for example the model vegetable crop tomato and other colorful vegetables containing increased levels of certain nutrients like vitamin K, C, E and beta carotene along with improved level of minerals. These vitamin and minerals may help to protect against the risk of many cardiac, chronic and cancer diseases to the human being.

In this article, we will discuss the success achieved by the researchers using this new technology and possible future developments that may have an impact on vegetable crops under the following headings.

- ✰ Molecular markers and genome mapping
- ✰ Recombinant DNA technologies and DNA delivery systems
- ✰ Tissue culture for vegetable crop improvements

1. Molecular Markers

A molecular marker is a fragment of the DNA molecule that is associated with certain traits in an organism. Markers help us in determining the location (map position) of genes that control important traits. In the last decades, a linkage map has been identified for morphological markers, resistance genes and mutants containing loci which are collectively affecting physiological functions. This type of map is usually used for understanding the genetic studies but their use in application of plant breeding programs is quite limited. Therefore recently, the research has been based on protein and DNA markers which are focused to the construction of detailed molecular map which do not affect the phenotype. Both the markers are based on polymorphisms which are due to differences in the primary sequence of genomic DNA. The polymorphisms in protein markers are usually detected as electrophoretic or antigenic peptide variants, whereas DNA markers directly reflect the variation both in a coding or non-coding region at DNA level. These markers are useful for phenotyping and genotyping, evolutionary relatedness, contamination detection, disease diagnosis forensic evidence and marker assisted breeding. In current scenario molecular markers possess several distinct advantages over morphological and other conventional markers.

i) The morphological markers can only be recognized at phenotypic level, whereas molecular markers great genotypic level and can be established at cellular, tissue and plant levels.

ii) In case of morphological markers, number of distinguishable alleles at morphological or biochemical loci is rather limited and often dependent on the application of exogenous mutagens. RFLPs, however, provide a virtually unlimited number of markers, as the polymorphism does not need to alter the phenotype or the charge of a protein to become detectable. With markers for all chromosomes at intervals of 10-15 map units, the flow of any gene of interest through segregating generations in a breeding program can be monitored, -using a linked (anonymous) RFLP-marker as 'tag'.

iii) There is no unwanted phenotypic effect with molecular markers, whereas, the morphological markers are usually associated with unwanted phenotypic effects.

iv) Most of the molecular markers are codominant, all possible genotypes can be distinguished in any segregating generation.

v) Only a few epistatic or pleiotropic effects occur with molecular markers, allowing a virtually unlimited number of segregating markers to be monitored in a single population.

Initially the isozyme molecular markers have proven their utility already in breeding programs as well as in genetic studies of various crops (Tanksley and Rick, 1980; Tanksley and Orton, 1983). In case of tomato, 41 isozymic genes corresponding to 15 unique enzyme reactions have been constructed for a isozyme linkage map. Out of forty one thirty six of these genes has been mapped to their respective chromosomes (Tanksley and Bernatzky, 1987). Thus, tomato isozymes have served many purposes like *e.g.*, verifying the purity of hybrid seed (Tanksley and Jones, 1979), tagging genes of economic importance (Rick and Fobes, 1977; Tanksley *et al.*, 1984), detecting genes and chromosomes from wild species following introgression (Rick *et al.*, 1986), identifying somatic hybrids and mapping genes underlying quantitative variation (Vallejos and Tanksley, 1983).

Nevertheless, isozyme loci suffer several limitations which are not inherent to the use of DNA markers. All the modern molecular techniques belong to the basic approaches: Southern hybridization or polymerase chain reaction (PCR). The southern hybridization based approach involves restriction digestion of genomic DNA; size fractionation of digested DNA through gel electrophoresis; transfer of fractionated DNA fragments on Nylon membranes; and finally hybridization with labeled probe sequence to visualize DNA polymorphisms. PCR based approaches involves; amplification of specific/random regions of genomic DNA using primers of different specificities, through polymerase chain reaction followed by visualization of amplified products on gel after staining with ethidium bromide or fluorescence. SSRs, CAPS and SNPs are the widely used techniques under the approach. SSRs (Simple Sequence Repeats) are tandem repeats of one to six nucleotide long DNA motifs. Polymorphism is based on the variation in the number of repeats in different genotypes (Ellegren, 2000). Cleaved Amplified Polymorphic Sequences (CAPS) polymorphisms are differences in restriction fragment lengths caused by SNPs or INDELS that create or abolish restriction endonuclease recognition sites in PCR amplicons produced by locus-specific oligonucleotide primers (www.ncbi.nl,.nih.gov). They are found in sections of DNA that contain a restriction site present in one genotype, but not in another. The CAPS markers are detected using PCR amplification and restriction analysis. Single nucleotide polymorphisms (SNPs) are DNA sequence variations that occur when a single nucleotide in the genome sequence is altered. It is the polymorphism occurring between the DNA samples for single base. Due to their abundancy nature in genome, they are highly preferred markers of choice. SNPs can be detected without gels. The Table 1 summarized the types of molecular marker linked with the particular gene.

1.1 Gene Tagging

The most interesting application of molecular markers at present time is the ability to facilitate the method of "Conventional" gene transfer. In plant breeding gene tagging is a term commonly applied where a genetic marker is used as the selection criteria for a closely linked agronomic trait. A classic example resulted from the discovery of tight linkage between the

Table 1: Identification of Markers Linked to Disease Resistance Genes in Vegetable Crops

Host	Pathogen	Gene	Type of Markers	References
Tomato	*Meloidogyne incognita*	*Mi*	CAPS	Hoogstraten *et al.* (2005)
	Cladosporium fulvum	*Cf 9*	SCAR	Truong *et al.* (2011)
		Mi3	RAPD, RFLP	Yaghoobi *et al.* (1995)
	Leveillula taurica	*Lv*	RFLP, RAPD	Chuwongse *et al.* (1994)
	Phytophthora infestans	*Ph2*	CAPS	Panthee *et al.* (2012)
	Verticillium dahliae	*Ve*	CAPS	Jung *et al.* (2015)
	Oidium lycopersicon	*Ol1*	RAPD, RFLP	Beck and Lindhout, 1994)
	Yellow leaf curl virus	*Ty1*	CAPS	Hoogstraten *et al.* (2005), Jung *et al.* (2015)
		Ty2	SCAR	Yang *et al.* (2014)
		Ty3	SCAR	Ji *et al.* (2007)
		Ty4	CAPS	Ji *et al.* (2009)
		Ty5	dCAPS	Lapidot *et al.* (2015)
	Tospovirus	*SW-5*	SCAR, SNP	Shi *et al.* (2011)
Potato	*Globodera rostochiensis*	*H1*	SCAR	Finkers Tomczak *et al.* (2011)
	Phytophthora infestans	*R1* and *R3*	SCAR	Gebhardt *et al.* (2004), Sakolova *et al.* (2010)
Lettuce	*Bremia lactuceae*	*Dm17* and *18*	RAPD, SCAR	Maisonneuve *et al.* (1994)
		Dm8 and *10*	RAPD, SCAR	Witsenboer *et al.* (1995)
	Plasmopara lactucae-radicis	*Plr*	SCAR	Witsenboer *et al.* (1995)
	Turnip mosaic virus	*Tu*	RAPD, RFLP	Robbins *et al.* (1994)
Bean	*Uromyces appendiculatus*	*Up2*	RAPD	Miklis *et al.* (1993)
	Common bean mosaic virus	*I*	CAPS, SNP	Bello *et al.* (2014)
	Pseudomonas syringae pv. *phaselicola*		RAPD and AFLP	Yaish *et al.* (2006)
Pea	Pea seed borne mosaic virus	*Sbm1*	SNP	Smykal *et al.* (2012)
	Pea common mosaic virus	*Mo*	RFLP	Dirlewanger *et al.* (1994)
	Erysiphe polygone	*Er*	SNP	Sun *et al.* (2016)
	Fusarium oxysporum	*Fw*	CAPS	Jain *et al.* (2015)
Corn	*Bipolaris maydis*	*Rhm*	SSR	Zhao *et al.* (2012)
Chineese cabbage	*Plasmodiophora brassicae*	*CRb*	SCAR and CAPS	Piao *et al.* (2004)

isozyme marker *Aps1* and *Mi*, a gene encoding root knot nematode resistance in tomato (Rick and Fobes, 1974). As a result many seed companies now use isozyme analysis as the preferred method for the incorporation of this trait into elite tomato lines.

More than 1000 markers have been mapped onto the tomato molecular linkage maps and these markers cover a total of 1,276 map units (Tanksley *et al.*, 1992). Many of the markers of the tomato map have also been mapped onto the potato and pepper molecular maps. All these solanaceous species share the same chromosome number and by having molecular maps based on the same RFLP probes, it has been possible to determine chromosomal homologies with a fairly high degree of precision (Bonierbale *et al.*, 1988; Tanksley *et al.*, 1988; Tanksley *et al.*, 1992; Prince *et al.*, 1992).

Another area where molecular markers facilitate gene transfer is the analysis of linkage drag. Linkage drag is frequently encountered when wild species are used as a trait donor source; while back crossing the trait, considerable excess linked chromosomal DNA may inadvertently be incorporated. A study by Young and Tanskley (1989) demonstrates the power of RFLP in this area. The tomato cultivar "Craigella" was found to carry a approximately 50 cM piece of DNA derived from the disease resistant doner parents following 11 backcross generations. In a single generation of selection, Young and Tanksley (1989) were able to reduce the size of this fragment to less than 7 cM. The markers linked to disease genes in vegetable crops are given in Table 1.

1.2 QTL Mapping

Two major developments during the 1980s changed the scenario: (i) the discovery of extensive, yet easily visualized, variability at the DNA level that could be as markers; and (ii) development of statistical packages that can help in analyzing variation in a quantitative trait in congruence with molecular marker data generated in a segregating population. With phenomenal improvements in molecular marker technology in the last two decades, identification and utilization of polymorphic DNA markers as a framework around which the polygenes could be located, has improved multiple-fold. It is now clear that a genetic map saturated with polymorphic codominant Mendelian markers can be generated for almost any species. Nearly saturated genetic maps have already been produced for most species of economic or scientific interest. We now refer the polygenes by a more catchy acronym, QTL (Quantitative Trait Loci), a term first coined by Gelderman (1975). A QTL is defined as "a *region* of the genome that is associated with an effect on a quantitative trait". Conceptually, a QTL can be a single gene, or it may be a cluster of linked genes that affect the trait. QTL mapping studies have been reported in most crop plants for diverse traits including yield, quality, disease and insect resistance, abiotic stress tolerance, and environmental adaptation. Here, a brief overview of the principle of QTL mapping, salient requirements for QTL mapping, common statistical tools and techniques employed in QTL analysis, and the strengths, constraints and applications of QTL mapping for crop improvement shall be provided.

The single marker approach (SMA), sometimes referred to as the single factor analysis of variance (SF-ANOVA) or single point analysis, has been used extensively, especially with isozymes (Tanksley *et al.*, 1982; Edwards *et al.*, 1987). SF-ANOVA is done for each marker locus independent of information from other loci. F-tests provide evidence whether differences between marker locus genotype classes are significant or not. Although computationally simple, this approach suffers from some major limitations: (i) the likelihood of QTL detection significantly decreases as the distance between the marker and QTL increases; (ii) the method cannot determine whether the markers are associated with one or more QTLs; (iii) the effects of QTL are likely to be underestimated because they are confounded with recombination frequencies.

Simple Interval Mapping (SIM) was first proposed by Lander and Botstein (1989) and it takes full advantage of a linkage map. The method evaluates the target association between the trait values and the genotype of a hypothetical QTL (target QTL) at multiple analysis points between pair of adjacent marker loci (the target interval). Presence of a putative QTL is estimated if the log of odds ratio (LOD) exceeds a critical threshold. Lander and Botstein (1989) developed formulae for calculating significance levels appropriate for interval mapping when the genome size, number of chromosomes, number of marker intervals, and the overall false positive rate desired are given. SIM has been the most widely approach as it can be easily accessed through statistical packages such as MAPMAKER/ QTL. By using tightly linked markers for analysis, it is possible to compensate for recombination between markers and the QTL, thereby increasing the probability of statistically detecting the QTL, and providing an unbiased estimate of QTL effect. However, when multiple QTLs are segregating in a cross (which is usually the case), SIM fails to take into account genetic variance caused by other QTLs. In such a case, SIM suffers essentially from the same shortcomings of single marker analysis.

Composite Interval Mapping (CIM) developed by Jansen and Stam (1994) combine interval mapping for a single QTL in a given interval with multiple regression analysis on marker associated with other QTL. It considers a marker interval plus a few other well-chosen single markers in each analysis, so that n-1 tests for interval-QTL associations are performed on a chromosome with n markers. The advantages of CIM are as follows: (i) mapping of multiple QTLs can be accomplished by the search in one dimension; (ii) by using linked markers as cofactors, the test is not affected by QTL outside the region, thereby increasing the precision of QTL mapping; and (iii) By eliminating much of the genetic variance by other QTL, the residual variance is reduced, thereby increasing the power of detection of QTL. CIM is more powerful than SIM, but is yet to be used extensively in QTL mapping.

Recently developed statistical packages also offer the means to analyze the QTL x environment interactions. The most commonly used statistical packages for QTL analysis are MAPMAKER/QTL, QTL Cartographer, MAPQTL, MapManager QTL, and Qgene. Most of these methods of analysis used by these packages yield essentially similar QTL locations and gene effects on a given data set, while there could be slight variation in the confidence intervals. Recent advances using nonparametric statistics or association-based approaches to identify QTLs and to calculate empirical critical (threshold) values for declaring significant QTLs will further refine QTL mapping.

Advanced backcross QTL (AB-QTL) analysis was proposed by Tanksley and Nelson (1996) for combining QTL analysis with variety development, particularly for the transfer of valuable QTLs from unadapted germplasm (such as landraces and sexually-compatible wild relatives of crop plants) into elite breeding lines. In this strategy, QTL analysis is delayed until the BC_2 or BC_3 generation and, during the development of these populations, negative selection is exercised to reduce the frequency of deleterious donor alleles. QTL-NILs can be derived from advanced backcross populations in one or two additional generations and utilized to verify QTL activity. Such lines also represent improved commercial inbreds (over the original parental line) for one or more quantitative traits. The utility of this approach has been demonstrated for quantitative trait improvement in tomato and rice.

Another facet of QTL analysis concerns the fact that QTL effects may depend on environment. Methods of analysis of QTL x E interactions are not well developed. Beavis and Keun discussed QTL x E interaction and review experiments where QTL were analysed in multiple environments. They suggested methods for combined analysis based on mixed models (Random environments and fixed genotypes) with parameters estimated by best linear unbiased prediction. Jinker and Mathur (1995) proposed a procedure of simplified composite interval mapping (SCM) for the analysis of QTL replicated in multiple environments.

Many classes of genetic markers are available to the plant geneticists and breeders (Table 2). Their efficient application in plant breeding will depend on the development of cost effective techniques that automate genetic assays in a way similar to the automation of many clinical diagnostic assays. Their use will depend on computer technology that will display resulting complex data in a way that is easily understood and accessible to the plant breeder. Considerable progress has been made in this area and a lot remains to be done. There is need for a cultural change in the plant breeding community, so as to accept the need for designing

Table 2: Gene Mapping in Vegetable Crops for various Traits

Crop Plant	*Molecular Marker(s)*	*Trait(s)*	*Reference*
Tomato	CAPS	*Fusarium oxysporum* resistance locus *I-2*	Lim *et al.* (2006)
	RFLP	Male sterile 14 gene and a hotspot for recombination	Gorman *et al.* (1996)
	RFLP	Regulation of iron-metabolism	Ling *et al.* (1996)
	RFLP	Sequencing of cDNA clone	Ganal *et al.* (1998)
	SNP	*Cf-2* gene	Zhao *et al.* (2016)
	RAPD	Integration of the classical and molecular linkage maps of chromosome-6	Weide *et al.* (1993)
	RFLP	Soluble solids, Fruit weight, fruit pH, Fruit size and shape, Fruit ripening, Yield	Tanksley *et al.* (1996)
	SCAR and CAPS	Parthenocarpy	Gorguet *et al.* (2008)
	RFLP	Ripening inhibitor and non- ripening Loci.	Giovonnoni *et al.* (1995)
	AFLP and RFLP	Root knot nematode resistance locus	Kaloshian *et al.* (1998)
	RAPD	Salt tolerance	Fooled and Chen (1997)
	RFLP	Golden nematode resistance gene (*Hero*)	Ganal *et al.* (1995)
	RFLP	Lateral suppressor(Is) locus	Sehumacher *et al.* (1995)
	RFLP	Tandemly repeated telomeric DNA sequences	Ganal *et al.* (1992)
	RFLP	Physical location of *Cf-2* gene	Dixon *et al.* (1995)
	RAPD	Integration of the classical and molecular linkage maps of tomato chromosome-6.	Weide *et al.* (1993)
	RAPD and RFLP	Close linkage between the *Cf-2/Cf-5* and *Mi* resistance Loci	Dickinson *et al.* (1993)
	RAPD	Carotenoid loci	Zhang and Stommel (1998)
	AFLP	Quantitative resistance to late blight	Meyor *et al.* (1998)
	RAPD and AFLP	High resolution map of chromosome VII harbouring the nematode resistance gene	Ballvora *et al.* (1995)
	RFLP	Mapping of simple sequence repeat loci	Micbourne (1998)
	SSR and CAPS	Senescence related QTLs	Hurtado *et al.* (2012)
	InDels	Fine mapping of *Ph3* gene conferring resistance to late blight	Zhang *et al.* (2013)
	SNP and InDels	21 SNPs and one InDel were identified in candidate genes for starch content, tuber yield and starch yield	Schonhals and Kirchhe-imbolanden (2014)
	SNP	Genetic mapping and QTL analysis of agronomic traits *viz.* (total tuber yield, number of tubers or tuber set per plant, average tuber weight, specific gravity, vigour, maturity and tuber end rot	Carpentiro *et al.* (2015)
Cabbage	RFLP	Linkage group alignment	Hu. *et al.* (1998)
	RFLP	Location of the self -incompatibility locus	Camargo *et al.* (1997)
	RFLP	Comparative genome mapping	Lagercraniz *et al.* (1996)
	RFLP	Linkage group alignment from four independent- RFLP maps	Hu (1998)
Hot pepper	RFLP, RAPD	*Xanthomonas compestris*	Kang *et al.* (1998)
	SSR	Mapping for Bacterial wilt (*Ralstonia solanacearum*)	Mimura *et al.* (2012)
	RAPD, AFLP	Genetic relationship among 34 cultivars	Paran *et al.* (1998)
Egg plant	SSR	Bacterial wilt resistance	Liu (2016)

breeding schemes that take account of the efficiency of genetic selection.

1.3 Gene Cloning

Gene cloning is the formation of multiple copies of the same gene and one of the most serious limitations to the advance of plant molecular biology and biotechnology is the difficulty in isolating genes responsible for specific characters. Yield, disease resistance, insect resistance and crop quality are just few of the important characters for which genetic variation exists within crop species, but for which the corresponding genes have not yet been cloned. The advent of genome mapping at the DNA level (especially RFLPs) has provided a method for localizing genes of economic importance to specific chromosomal positions. The ability to map any gene of economic importance to a defined chromosomal site opens the possibility of isolating genes via chromosome walking. This method is sometimes referred to as "map based cloning" or the "top down" approach since it starts with an interesting phenotype observed at the whole plant level (*e.g.*, disease resistance or high yield) and proceeds to clone the gene via genetic mapping and chromosome walking.

Chromosome walking presents serious difficulties in crop species since they tend to have relatively large genomes. For example, the "average" crop species has a nuclear genome 10-100 times larger than yeast, *Drosophila* or *Arabidopsis*. Nevertheless, the problem with chromosome walking the large genomes is rapidly being overcome by new methods for manipulating large pieces of DNA (*e.g.*, pulsed field gel electrophoresis and yeast artificial chromosome cloning). In tomato, the availability of a high density molecular linkage map and a yeast artificial chromosome library potentially provide the foundation on which to initiate map-based gene cloning for genes underlying any trait that can be genetically mapped (Martin *et al.*, 1992).

A critical prerequisite for map-based gene cloning is that the target gene must be localized definitively with respect to markers in that region of the map. With the small genetic distances being considered, analysis of very large progenies is often required for such fine mapping. New protocols for rapid DNA isolation as well as new strategies for high resolution genetic mapping should enhance our ability to meet this prerequisite for map-based cloning (Churchill *et al.*, 1992). It is anticipated that, in the future, the molecular linkage map will pay a key role in cloning genes for a variety of economically valuable and/or biologically interesting characters including disease resistance, growth habit, fruit ripening behaviour and even yield.

1.4 Transposon Tagging

Transposon genes are also called as "jumping gene." They are segments of DNA that can jump or move from one place in the genome to another. Transposons were first discovered by McClintock (1951) in case of maize when she found that a segment of DNA moved into gene coding for pigmented kernels and produced light coloured kernels. Two general approaches are currently being developed for their molecular isolation: RFLP tagging and transposon tagging.

An inherent problem associated with this RFLP-tagging approach is that it is difficult to associate the appropriate DNA sequences with the corresponding phenotype of the desired gene. A solution to this problem can be found in some cases. For example, if the cloned DNA sequences represent the wild type, such sequences might be introduced into a recessive mutant through genetic transformation. In this case, the sequence or sequences that restore wild type in the transformants can be easily associated with the corresponding mutant gene.

Transposon tagging might be a more appropriate approach for the molecular isolation of useful genes. Application of endogenous plant transposable elements has shown to be a valuable method to obtain plant mutants. The insertion of a transposable element in to a genetic locus can give rise to mutations. Transposons possess repetitive DNA, either similar or inverted, at their ends, some 5, 7 or 9-nucleotide long. Enzyme transposae separates the segment from it original cleaving the repetitive sequences at its ends. Using a transposable element from maize, the *Ac*-element, Fedoroff *et al.* (1984), were able to clone a locus which until then was only characterized genetically. Their results indicate that in principle any locus in maize, identifiable with a transposable element, can be cloned and analysed at the molecular level. This approach, called transposon tagging, has successfully been applied for the isolation of various maize genes (Wienand and Saedler, 1987).

The subsequent demonstration that the maize transposable element *Ac* can transpose in dicots has opened the possibility to use this well characterised element for gene isolation in plants other than maize (Baker *et al.*, 1986, 1987; Van Sluys *et al.*, 1987). Using the *A. tumefaciens* mediated leaf explant transformation procedure, the transposable element *Ac* has been introduced into various tomato lines and shown to transpose within these lines (Yoder *et al.*, 1988; Haring *et al.*, 1989).

In maize and also in tobacco it has been shown that *Ac* has a tendency to transpose in the vicinity rather than to far away positions (Greenblat, 1984; Dellaporta and Chomet, 1986). Therefore, to obtain a transposon insertion in a particular gene, it is advantageous to start with a plant containing the transposable element close to the position of the desired gene. Currently, the utility of this transposon tagging system for tomato gene isolation is under study in various research groups.

2. Recombinant DNA Technology

The recombinant DNA technology includes formation of recombinant DNA and introduction of recombinant DNA into an appropriate host. The cloning of the first plant nuclear encoding gene almost two decades ago signified the beginning of plant molecular biology in its own right. Since then the number of plant genes isolated and studied by molecular techniques has increased exponentially. As a result our knowledge of plant genome structure has progressed at an amazing pace in recent years. However, it is not sufficient to simply know the architectural arrangement of DNA sequences. Investigation on how the expression of particular or cluster of genes are influenced by the spatial arrangement and rearrangement of the genome are also crucial to understand how the entire genome acts as a concerted whole to produce the overall plant phenotype. Plant genome function also has considerable implications on our ability to genetically manipulate the genome to attain gene expression with the desired temporal and spatial specificity.

The various factors that affect the expression of a particular heterologous gene and subsequent production of its protein in plant cells include: (i) transcriptional determinants such as transgene copy number, site of integration of the T-DNA in the chromosome, and promoter activity; (ii) posttranscriptional factors including, 5-0-untranslated leader sequence, 3-0-polyadenylation, and mRNA stability, and (iii) translational/post-translational factors such as protein stability, modification, and trafficking.

Cis-acting promoters and enhancers play a substantial role in regulating spatial and temporal patterns of gene expression at particular stages or in response to particular surroundings. Numerous promoters and other cis-acting components including signal peptides of plant or bacterial origin exploited in heterologous expression vectors have been labeled. The *Cauliflower mosaic virus* (CaMV) 35S promoter is widely used in plant expression vectors. Manifold copies of enhancer sequences from extremelyactive promoters can be stacked to lift transgene expression.

RNA processing such as splicing, capping and polyadenylation has revealed to affect the expression stages of heterologous genes. Moreover, 3-0-untranslated sequences downstream of the stop codon playing a precarious role in processing have been encompassed along with signal sequences aimingand soothing the message for polyadenylation. Polyadenylation intensely improves the heterologous expression of bacterial genes in plants. Sequences located immediately around the translational start site, termed Kozak sequences, can also be adapted to fit the consensus initiation sequence of plants in order to accomplish advanced rates of translation.

3. Tissue Culture for Vegetable Improvement

Plant tissue culture techniques, in combination with recombinant- DNA technology are the essential requirements for development of transgenic plants. However, culture techniques like anther/pollen/ ovule culture, meristem culture can themselves be used for improvement of different traits or as an aid to conventional plant breeding. In addition, through the use of protoplast fusion and hybrid embryo rescue, hybridization between distant taxa/species has been successfully achieved. Somaclonal variation generated during culture can also be utilized for recovery of variants showing disease resistance or other desirable traits. The use of such techniques for improvement of vegetable crops is listed in Table 3.

Table 3: Achievements of Tissue Culture for Improvement of Vegetable Crops

Crop	*Technique Applied for*
Capsicum annuum	Haploid production
Lycopersicon esculentum	–do–
Solanum melongena	–do–
P. vulgaris × *P. angustissimus*	To rescue the hybrid embryo
Vigna pubescens × *V. unguiculata*	–do–
Cucumis sativus × *C. hystri*	–do–
S. melongena × *S. torvum*	–do–
Allium sativum	Elimination of *Onion yellow dwarf virus*
S. melongena	EMCV
L. esculentum	Cryopreservation
P. sativum	–do–
L. esculentum x *L. peruvianum* + *Lycopersicoides*	Fertile somatic hybrids
Amaranthus tricolor + *Celosia cristata*	Intergenic hybridization between C_3 and C_4 species
L. esculentum + *L. peruvianum*	Somatic hybrids with intermediate morphology
S. melongena	Resistance towards Little Leaf disease
L. esculentum	Tolerance to *Fusarium oxysporum* culture filtrate
	Resistance to *F. oxysporum*
	Resistance to bacterial wilt

4.1 Viral Disease Resistance

There are several approaches which have been used to engineer plants resistant to viruses. The first method aims at introducing genes encoding viral coat protein to produce transgenic plants resistant to plant virus infection (Powell *et al.*, 1986). The viral coat protein produced by the transgenic plants appears to block a receptor in the plant cell that is required for uncoating the viral nucleic acid, or to interfere with viral replication or expression of viral genes. Resistance is manifested by delay in symptom development or escape from infection altogether. A second approach to engineering virus resistance in plants has centered on satellite RNAs. It has been observed that expression of satellite RNA's in transgenic plants can reduce infection with cucumber mosaic virus (Harrison *et al.*, 1987). The third approach aims at using antisense virus RNA to govern transgenic resistance. Following the successful field testing of coat protein mediated resistance (CPMR) against TMV in tomato plants (Nelson *et al.*, 1988) additional testing of a number of plant lines has occurred. A better understanding of the cellular molecular mechanisms that confer transgene mediated resistance may lead to second generation viral genes that improve resistance beyond current levels and extend its applications to still more viruses and crop plants. The details of transgenic developed against virus resistance have given below in details (Table 4).

4.2 Fungal Disease Resistance

One of the major approaches for engineering plants resistant to fungal disease includes use of chitinase genes of plant and/or bacterial origin. Chitinase enzyme degrades chitin-rich cell walls of fungi and delays symptom development (Broglie *et al.*, 1993). Other antifungal protein genes include Osmotin, glucanase, thionin, peroxidase *etc.* Recently, an effective antifungal protein from barley seeds called ribosome inactivating protein (RIP) has been isolated. RIP acts by inhibiting protein synthesis via N-glycosidase modification of fungal 285 r DNA. Transgenic crop plants expressing RIP proteins are now being tested against many fungi (Sharma and Kumar, 1995). Scientists have found that deliberately disarmed fungal pathogens can trigger the necessary response in susceptible (*i.e.* slow responding) plants before the virulent pathogen arrives. This approach, the use of pathogen derived "bio-control" agents to induce resistance in susceptible plants has opened a new avenue in disease control.

4.3 Bacterial Disease Resistance

Using molecular biological tool, a novel approach for engineering resistance against bacteria was undertaken to isolate the tabtoxin resistance genes (*ttr*) from bacteria *Pseudomonas syringae* pv. *tabaci* and subsequently to introduce into susceptible plants (Yoneyama and Anazi, 1993) for transgenic resistance. In similar way Phaseolotoxin-insensitive *OCTase* (Ornithine carbamoyl transferase) gene was isolated from *Pseudomonas syringae* pv. *Phaseolicola* and introduced to bean to observe total resistance against the pathogen (Herrera-Estrella and Simpson, 1995). This simple and useful strategy for molecular breeding of disease resistant plants by detoxification of pathogenic toxins has a great potential to be exploited against bacterial and fungal diseases in vegetable crops.

4.4 Engineering for Insect-Pest Resistance

Today, insect resistant transgenes, whether of plant, bacterial or other origin, can be introduced into plants to increase the level of insect resistance, a technology that has dramatically extended the scope of resistance genes available to plant breeders. The first reports of transgenic insect-resistant plants were only published in 1987 (Fischnoff *et al.*, 1987; Hilder *et al.*, 1987; Vaeck *et al.*, 1987) but technological advancement has been swift since there. This new technology is seen as an additional tool for the control of crop pests and could offer certain advantages over conventional insecticides, such as more effective targeting of insects protected within plants, greater resilience to weather conditions, fast biodegradability, reduced operator exposure to toxins and financial savings (Gatehouse *et al.*, 1992).

The insect resistance genes transferred into plants to date mainly target the insect digestive system. Most have been derived from a single species of bacterium, or a range of higher plants, in addition some insect resistance genes from animals and other microorganisms have also recently been introduced into crop plants. The number of genetically engineered insect resistant transgenic vegetable crops is scanty in compared to other crops (Table 5).

Bt corn has been cleared for commercial release in the USA, Canada, Argentina, Japan and the European Community (MacKennzie, 1997).

Table 4: Examples of Successful Transgenic Vegetables against Viruses

Purpose of Genetic Manipulation	*Transgene Product*	*Origin of Transgene*	*Transformed Plant*
Tomato mosaic virus (ToMV)	Viral coat protein	ToMV	Tomato
Cucumber mosaic virus	Viral coat protein	CMV	Cucumber
Potato virus X (PVX)	Viral coat protein	PVX	Potato
Potato virus Y (PVY)	Viral coat protein	PVY	Potato
Tomato mosaic virus	Antisense RNA	ToMV	Tomato
Cucumber mosaic virus (CMV)	Satellite RNA	CMV	Tomato
Tomato spotted wilt virus (TSWV)	N-gene	TSWV	Tomato
Tomato yellow leaf curl virus (TYLCV)	Coat protein	TYLCV	Tomato
Tomato yellow leaf curl virus (TYLCV)	Truncated version of C1 gene of TYLCV	TYLCV	Tomato

Table 5: Transgenic Plants Expressing Insect Resistance Genes

Transgenic Product	*Origin of Transgene*	*Target Insects*	*Transformed Plant*
cry1AB	*Bacillus thuringiensis*	Lepidoptera	Tomato
cry1AB	–do–	Lepidoptera	Tomato
cry1H	–do–	Lepidoptera	Corn
cry3A	–do–	Coleoptera	Egg plant
cry9C	–do–	Lepidoptera	Corn
cry1AC	–do–	Lepidoptera	Brinjal
CpTI (cowpea trypsin inhibitor)	Cowpea	ColeopteraLepidoptera	Tomato
Tomato proteinase inhibitor I	Tomato	Lepidoptera	Tomato
Tomato proteinase inhibitor II	–do–	Lepidoptera	Tomato
a - Amylase inhibitors			
a AI-Pv(a - Amylase inhibitor of the common bean)	Common bean	Coleoptera	
Potato PI-II and carboxypeptidase inhibitors (PCI)	Potato	*Heliothis obsoleta* and *Liriomyza trifolii* larvae	Tomato
Snowdrop lectins (GNA)		Lepidoptera	Tomato
Wheat germ agglutinem (WGA)	–	Coleoptera Lepidoptera	Corn
Jacalin	–	Coleoptera Lepidoptera	Corn
Rice lectin	–	Coleoptera Lepidoptera	Corn
Tobacco anionic peroxidase	–	Lepidoptera	Tomato
Bovine pancreatic trypsin inhibitor (*BPT 1*)		Lepidoptera	Lettuce

4.5 Engineering for Herbicide Resistance

There is detoxification of degradation of herbicide in which a number of detoxifying enzymes have been identified in plants as well as in microbes. Some of these enzymes include (i) glutathione-S trensferase or GST, which detoxifies the herbicide atrazine; (ii) nitrilase (coded by gene *bxn* in *Klebsiella pneumoniae*, which detoxifies the herbicide bromoxynil, and (iii) phosphinothricin acetyl transferase or PAT (coded by *bar* gene in *Streptomyces* spp.), which detoxifies the herbicide PPT (L- phosphinothricin). Transgenic tomato plants using bxn gene from *Klebsiella* and *bar* gene from *Streptomyces* and transgenic plants in potato, oilseed rape (*Brassica napus*) and sugarbeet using bar gene from *Streptomyces* have been obtained and were found to be herbicide resistant. Tissue culture technique has been used to isolate herbicide tolerant mutants. A paraquat tolerant mutant has been identified in *Lycopersicon*. In this case an interspecific hybrid strain L_2 was used. The callus cells of this genotype were kept on an agar medium containing a lethal concentration of paraquat. The tolerant clone grew under these conditions. Plants produced have about 30-fold increase in paraquat tolerance as compared to strain L_2 (Thomas and Pratt, 1982). Similarly, cultures have been used to select and isolate mutants tolerant to various chemicals *e.g.* metribuzin in *Lycopersicon* (Trebst and Wietiska, 1975; Ellis, 1978) and *D. carota* (Templeton, 1981) and 2,4-D resistant cells in *D. carota* (Gressed, 1979).The later strategy has been used to produce transgenic plants with to detoxify phosphinothricin by identifying *bar* gene from *Streptomyces hygroscopicus*, by transferring to the higher plants (De Block *et al.*, 1987). Lettuce cv. Evola and Radish were successfully transformed for Bialaphos herbicide resistance by transferring this *bar* gene (Mohapatra *et al.*, 1999; Curtis *et al.*, 2001).

Table 6: Mechanism of Action of different Herbicides

Active Principle of Herbicide	*Inhibited Pathway*	*Target Product*	*Use*	*Basis of Resistance*
I. Amino acid biosynthesis inhibitor				
Glyphosate(Roundup)	Aromatic amino acid biosynthesis	EPSPS	Broad spectrum	Over expression of *EPSPS* gene; bacterial *aroA* gene
Sulphonylurea and Imidazolinones	Branched chain Amino acid	ALS	Selected crops	Mutant *ALS* gene
Phosphinothricin	Glutamine biosynthesis	GS	Broad spectrum	Gene amplification *bar* gene: detoxification
II. Photosynthesis Inhibitor				
Atrazine(Lasso)	Photosystem II	Q_B(32k Da protein)	Selected crops	Mutant *PsbA* gene: *GST* gene:; detoxification
Bromoxynill(Buctril)	Photosynthesis	–	Selected crops	*bxn* gene: detoxification

Source: Gupta (2001).

Tomato was successfully transformed for resistance to Methyviologen herbicide by transferring the Mn-superoxide dismutase transgene into it (Wang *et al.*, 2002). The mechanism of action of different herbicides and basis of achieving resistance against them in transgenic plants (Table 6).

4.6 Engineering for Parthenocarpic Plants

In plants able to develop fruits in the absence of fertilization are said to be parthenocarpic. In such fruits seeds are absent which can increase fruit acceptance by consumers. Parthenocarpy has been achieved by genetic engineering in eggplant and tomato. For the parthenocarpic development of fruit the heterologous gene def *H9-iaa*M has been used in transformation experiments to increase the concentration of IAA in the ovule. Line transformed with this gene gave normal sized parthenocarpic fruits even from the emasculated flowers. It has been observed that in a protected cultivation trial, three transgenic plant hybrids yielded more than the corresponding non transgenic control even when the latter used treated with plant growth regulators (Rotino *et al.*, 1998). The transgenic vegetables for parthenocarpic fruits are given in Table 7.

Table 7: Successful Engineered Vegetables for Parthenocarpic Fruits

Transgenic aganist	*Transgene*	*Origin of Transgene*	*Transformed Plants*
Tomato	*Rol* B gene	*A. rhizogens*	Tomato
Brinjal	*Iaah gene*	*Ps. syringae* pv. *savas*	Brinjal

4.7 Engineering for Stress Tolerance

In global agriculture, drought cause major crop loss during the past decades. It is the most significant environmental stress on world agricultural production (Tuberosa and Salvi, 2006; Cattivelli *et al.*, 2008) and sumptuous efforts are being made by plant scientists to improve crop yields during decrease in water availability. Plant response to water deficit is articulated in a complex network of morphological, physiological, and molecular changes. As plant's relative water content decreases, plant growth gets slowed, and stomatal closure occurs, which is paralleled by a decreased photosynthetic rate (Lawlor, 2002). Physiological and biological screening methods are applied to select for drought and heat tolerance/sensitivity traits, and results are used to rate crops for adaptation to extreme conditions. Park *et al.* (2005) constitutively over expressed the vacuolar H^+-pyrophosphatase in commercial tomato varieties. The resulting transgenic plants exhibited greater pyrophosphate-driven cation transport into root vacuolar fractions, increased root biomass, and enhanced survivability under soil water deficit stress. Constitutive expression of *Arabidopsis CBF1* (C repeat/dehydration responsive element-binding factor 1) in tomato resulted in enhanced plant tolerance to cold, drought, and salt stresses. However, this enhanced tolerance came at a cost, causing reduced plant growth and yield (Hsieh *et al.*, 2002a, b). When the same gene (*CBF1*) was expressed using an abcisic acid (ABA)/stress-inducible promoter from barley *HAV22* gene, the transgenic tomatoes showed enhanced tolerance to chilling, water deficit, and salt stresses as compared to untransformed plants and the use of the inducible promoter eliminated the deleterious effects of the ectopic expression of *CBF1* on plant growth and yield (Lee *et al.*, 2003).Boiling stable proteins (BSP) have been implicated in desiccation tolerance against water stress by protecting other proteins. Transgenic tomato plants transformed with a novel 66-kD BSP from *Populus tremula* were tested for drought tolerance by polyethylene glycol (PEG) test, biomass analysis, proline estimation, and electrolyte leakage measurement. These plants exhibited slightly increased tolerance to water stress compared to the parental genotype (Roy *et al.*, 2006).

Most transformation studies to improve plant drought resistance have produced transformants that display a variety of both tolerance and avoidance traits. An exception was demonstrated by (Rivero *et al.*, 2007) who manipulated a leaf senescence gene. Leaf senescence is an avoidance strategy and is accelerated in drought-sensitive plants to decrease canopy size. In crop plants, accelerated senescence is often associated with reduced yield and is thought to be the result of an inappropriately activated program cell death. Therefore, suppression of drought induced leaf senescence in tobacco plants was investigated as a tool to enhance drought resistance. Transgenic plants were developed by expressing isopentyl transferase (IPT), a key enzyme in the biosynthesis of cytokinin (a leaf senescence inhibitor) under the control of the senescence-associated receptor protein kinase promoter (PSARK). The SARK gene, which is induced during late maturation and drought and decreased during senescence development, encodes a maturation/- senescence-dependent protein kinase. Although transgenic plants wilted under a 15-day glasshouse drought stress treatment, senescence did not occur and a reduced photosynthetic capacity was maintained. Following rewatering, transgenic plants recovered full leaf turgor and resumed growth and maximum photosynthetic capacity, while the control plants were unable to recover from the drought stress (Rivero *et al.*, 2007). Water use efficiency (WUE) of the transgenic plants was also markedly higher than wild-type (WT) plants, and was two to three times higher after re-watering than before the drought treatment. An experiment to assess whether the transgenic plants could produce significant yields under water-limiting conditions determined that biomass and seed yield were not as affected in transgenic plants than in WT controls, although this result was not significant (Rivero *et al.*, 2007). Cuticular biosynthesis was investigated by transgenic expression of AtMYB41, which encodes an R2R3-MYB transcription factor (TF) in *Arabidopsis* AtMYB41 is expressed at high levels in response to drought, abscisic acid and salt treatments, and was demonstrated to have a role in cell expansion and cuticle deposition. The transformation of *Arabidopsis* with AtMYB41 was associated with undesirable pleiotropic phenotypes including dwarfism, enhanced sensitivity to desiccation, and enhanced permeability of leaf surfaces (Cominelli *et al.*, 2008).

Environmental stress tolerance is a complex trait and involves many genes (Wang *et al.*, 2003). In response to stresses, both RNA and protein expression profiles change. Approximately 130 drought-responsive genes have been identified using microarrays (Seki *et al.*, 2001, Reymond *et al.*, 2000). These genes are involved with transcription modulation, ion transport, transpiration control and carbohydrate metabolism. *DREB1A* and *CBF*, and *HSF* genes are transcription factors implicated in drought and heat response, respectively (Sung *et al.*, 2003, Sakuma *et al.*, 2002). Cell wall invertase (*INV*) and sucrose synthase (*SUSY*) play key roles in carbohydrate partitioning in plants (Déjardin *et al.*, 1999) and this regulation of carbohydrate metabolism in leaves may represent part of the general cellular response to acclimation and contribute to osmotic adjustment under stress. The *ERECTA* gene regulates plant transpiration efficiency in *Arabidopsis thaliana* (Masle *et al.*, 2005), and the *NHX* and *AVP1* genes are associated with ion transport (Zhang and Blumwald, 2001). There are many more genes implicated with stress response and the current challenge is to identify the ones that confer a tolerant phenotype in the crop of interest. Although the function of these genes has been elucidated, particularly in *A. thaliana*, only a few genes have contributed to a tolerant phenotype when over-expressed in vegetables (Zhang *et al.*, 2004). Expression of *AVP1*, a vacuolar H+ pyrophosphatase from *A. thaliana*, in tomato resulted in enhanced performance under soil water deficit (Park *et al.*, 2005). The engineered tomato has a stronger, larger root system that allows the roots to make better use of limited water. The control plants suffered irreversible damage after five days without water as opposed to transgenic tomatoes which began to show water-stress damage only after 13 days but recovered completely as soon as water was supplied. The *CBF/DREB1* genes have been used successfully to engineer drought tolerance in tomato and other crops (Hsieh *et al.*, 2002). Orthologous genes of *CBF* have been found in most crop plants and functional tests indicate conservation of the pathway in these plant species. Constitutive over-expression of *CBF* genes results in salt, cold, or drought tolerance in several plant species.

The ability of an organism to acclimate to its environment is a key determinant in its global distribution and capacity to compete with other organisms (Suzuki *et al.*, 2008). The heat stress response is a highly conserved response involving multiple pathways, regulatory networks and cellular compartments (Kotak *et al.*, 2007). Short duration heat stress or even a few days can have a negative effect on yield production globally causing devastating economic and social impacts (Shah *et al.*, 2013; Shah and Nahakpam, 2011). Tomato is an important vegetable crop whose productivity is severely affected when exposed to high temperature (Bar- Tsur *et al.*, 1985). The phenomenon of formation of reactive oxygen species (ROS) by a stepwise one-electron transfer to molecular oxygen is common in normal metabolism in plants (Halliwell and Gutteridge, 1989), however ROS level is known to increase under heat and drought stress-conditions (Shah and Nahakpam, 2011). To contest the implications of enhanced ROS and to restore and regulate it at the cellular-steady state levels, the plants have developed constitutive or inducible protective systems (Inze and Van Montagu, 1995; Rai *et al.*, 2012a,b 2013; Shah *et al.*, 2013; Shah and Dubey, 2005; Shah and Nahakpam, 2011, 2012). Tissue degradation, loss of membrane semi-permeability, altered gene expression, accumulation of transcripts followed by synthesis of stress-related proteins, also constitute important adaptive strategies in plants subjected to stress (Rai *et al.*, 2013; Shah *et al.*, 2013; Shah and Dubey, 1998).

A mild heat-shock (HS) is sufficient to induce the synthesis of small heat-shock proteins (sHsp), which in turn leads to gaining of thermo-tolerance in organisms (Almoguera *et al.*, 1998; Nover, 1991; Wehmeyer *et al.*, 1996). At least 20 types of sHsps are reported in plants (Vierling, 1991) which are located in cytosol, chloroplast, endoplasmic reticulum, mitochondria and membranes (Waters *et al.*, 1996). Molecular pathways which lead to the gene-expression for synthesis of heat-shock proteins (Hsps) begin with temperature sensing. This acts as signal and stimulates the heat-shock factors (Hsfs). These Hsfs then activate Hsp-gene expression by regulating heat-shock element (HSE) present on DNA (Larkindale *et al.*, 2005; Schoffl *et al.*, 1998). In tomato, the cytosolic Hsp17.4 and chloroplastic Hsp21 are reported to be induced under heat treatment and play a role in Acquired Thermotolerance (AT) (Larkindale *et al.*, 2005; Schoffl *et al.*, 1998). The tomato chloroplastic Hsp21 (Lawrence *et al.*, 1997) is reported to be induced under heat treatment in leaves, flowers and fruits which protects photosystemII in transgenic tomato from temperature-dependent oxidative stress and during fruit reddening at maturation (Neta-Sharir *et al.*, 2005).

Over-expression of transcription factor *ZAT12* enhanced the expression of ROS-scavenging genes and tolerance to salt, drought or osmotic stresses (Sakamoto *et al.*, 2004; Davletova *et al.*, 2005; Wu *et al.*, 2008; Rai *et al.*, 2013; Shah *et al.*, 2013) respectively. As *ZAT12* gene is known to respond to multiple abiotic stress tolerance there is a possibility that *BcZAT12*-transformants of tomato may also be tolerant towards heat-shock induced oxidative stress (Shah *et al.*, 2013). Further studies established that plants, respond to Heat shock by synthesizing a new set of proteins called heat shock proteins (Hsp). Hsps have been extensively-analyzed for their physiological, biochemical, cellular and molecular properties (Vierling, 1991; Agarwal *et al.*, 2001; Katiyar-Agarwal *et al.*, 2001; Scharf *et al.*, 2001; Agarwal *et al.*, 2003; Shah and Dubey, 2005; Shah *et al.*, 2013). It has been shown that Hsp are highly-conserved proteins across different species (Katiyar-Agarwal *et al.*, 2001). Over the years, large number of *hsp* genes has been isolated, sequenced and cloned from varied plant species representing diverse taxonomic classes (Agarwal *et al.*, 2003; Grover *et al.*, 2003)

Schramm *et al.* (2007) provide evidence that HsfA3 is transcriptionally induced during heat shock by DREB2A a heat stress- induced regulator of the heat stress response of *Arabidopsis*. Tomato *Lehsp* (mitochondrial) gene over-expressed in tobacco showed that transgenics were more thermotolerant at 48°C than the transgenics produced with the antisense construct of the same gene (Sanmiya *et al.*, 2004). The low molecular weight compounds such as amino acids (*e.g.* proline), polyamines (*e.g.* putrescine), quaternary ammonium compounds (*e.g.* glycine betaine), sugars (*e.g.* mannitol, fructans, sorbitol, trehalose) and sugar alcohols (*e.g.* polyols) commonly referred to as osmolytes help plants to acclimatize largely against the osmotic stresses (Gepstein *et al.*, 2005). Glycine betaine is proposed to protect the photosynthesis machinery by stabilizing O_2 evolving photosystem II complex (Papageorgiou and Murata, 1995).

However, in addition to increased stress tolerance, the transgenic plants were dark-green and were stunted, with higher levels of soluble sugars and proline (Liu *et al*; 1998). The use of stress-inducible promoters that have a low background expression of *CBF* under normal growth conditions can achieve increased stress tolerance without growth retardation (Lee *et al.*, 2003). Maintaining a low cytosolic Na^+ concentration is essential to achieve salt tolerance and can be achieved by restricting inflow, increasing outflow, or increasing vacuole sequestration of Na+ (Zhang *et al.*, 2004). Increased expression of the *A. thaliana* tonoplast membrane Na^+/H^+ antiporter, *AtNHX1*, under a strong constitutive promoter, was reported to result in salt-tolerant tomatoes (Zhang and Blumwald, 2001). The transgenic tomato plants grown in the presence of 200 mM NaCl were able to flower and set fruit. While the leaves accumulated high concentrations of sodium, the tomato fruits continued to contain only low concentrations of sodium. The *NHX1* system seems to be highly conserved between many different plant species and manipulation of this system in crop species is likely to result in improved salt tolerance. Research on the physiology of stress tolerance has demonstrated that tolerance to a specific stress is determined by several component traits and controlled by corresponding genes. A combination of a genome-wide scan of expression, using DNA arrays, and QTL analysis could provide important information in identifying the major genes association with stress tolerance.

The damaging effects of salt accumulation in agricultural soils have severely affected agricultural productivity in large swathes of arable land throughout the world (Kumar *et al.*, 2000). About 20 per cent of irrigated agricultural land is adversely affected by salinity (Flower and Yeo, 1995). Problem of salinity is further increasing because of the use of poor quality of water for irrigation and poor also the poor drainage. It is distributed largely amongst coastal salt marshes or inland desert sands. About one-fifth of the irrigated, agriculture land is impacted by salinity and it is rapidly expanding due to the increased use of undergroundwater (Chinnusamy *et al.*, 2005).

A putative osmosensory histidine kinase (AtHK1) - MAPK (mitogen activated protein kinase) cascade is one signaling pathway hypothesized to regulate osmotic homeostasis and reactive oxygen species (ROS) scavenging. Several genes with a potential role in imparting tolerance to salinity stress have been identified. These include calcium sensor protein, SOS3, which binds to and activates a Ser/Thr protein kinase SOS2 that regulates activities of a plasma membrane Na^+/H^+ antiporter SOS1, and a tonoplast Na^+/H^+ antiporter NHX1 (Sahi *et al.*, 2006). It is expected that understanding of ion transporters and their regulators, and of the CBF (C-repeat-binding factor) regulons, will help in the future development of salt-tolerant crops (Chinnusamy *et al.*, 2005). Throughout its ontogeny, tomato, like most other crop plants, is sensitive to salinity (Foolad, 2004). Zhang and Blumwald (2001) over expressed the *Arabidopsis NHX1* vacuolar Na^+/H^+ antiporter in tomato plants. These transgenic plants were able to grow, flower, and produce fruit in the presence of 200 mM sodium chloride. Interestingly, their leaves accumulated high concentrations of sodium, but the transgenic fruits contained very low levels of sodium.

Salt and other osmotic stresses have been shown to increase levels of glycine betaine, which is an important quaternary ammonium compound in many organisms. Mohanty *et al.* (2002) demonstrated that *Agrobacterium*-mediated transformation of an elite indica rice cultivar to increase GlyBet synthesis through the incorporation of the choline oxidase encoding codA gene, was an effective way to improve salinity tolerance. Challenge studies performed with R1 plants by exposure to salt stress for one week followed by a recovery period, revealed survival of more than 50 per cent of the transgenic plants under salt stress and could set seed whereas WT plants failed to recover. Over expression of betaine aldehyde dehydrogenase encoded by *BADH* gene catalyzes conversion of betaine aldehyde into glycine betaine from sorghum contributed to maintenance of osmotic potential under salt stress in tomato (Moghaieb *et al.*, 2000). The tomato hairy root lines transformed with *BADH-1* gene using root inducing plasmid (pRi) plasmid accumulated betaine. Constitutive expression of *Atriplex hortensis BADH* gene in tomato led to enhanced tolerance of transgenic tomato to salt stress (Jia *et al.*, 2002) (Table 8).

4.8 Engineering for Post-harvest Traits

The transgene used for improved self-life and post-harvest have listed in Table 9. Polygalacturonase (PG) is an endo-acting polygalacturonic acid hydrolase that is synthesized specifically during ripening of some fruit (for example, tomato) and it degrades the pectin fraction of fruit cell walls, and thus plays a role in softening. It has been shown that expression of tomato *pg* gene during ripening could be inhibited by a single antisense gene (Sheehy *et al.*, 1988, Smith *et al.*, 1988). One of the first transgenic products released for commercial cultivation in USA is 'Flavr-Savr' tomato with increased 'shelf-life'. Ethylene production was inhibited by 93 per cent ripening transgenic tomatoes expressing antisense gene (Hamilton *et al.*, 1990).

Table 8: Genes Conferring Abiotic Stress Tolerance and their Mode of Action

Gene(s)	*Function*	*Mechanism of Action*	*References*
Drought			
DREBs/CBFs; ABF3	Stress induced transcription factors	Enhanced expression of downstream stress related genes confers drought/cold/salt tolerance. Constitutively overexpression can lead to stunting growth	Oh *et al.* (2005), Ito *et al.* (2006), Rai *et al.* (2012c)
SNAC1	Stress induced transcription factor	SNAC1 expression reduces water loss increasing stomatal sensitivity to ABA	Hu *et al.* (2006)
ZAT12	Stress induced transcription factors	Enhanced expression of downstream stres srelated genes confers drought/heat/salt tolerance.	Rai *et al.* (2013)
OsCDPK7	Stress induced Ca-dependent protein kinase	Enhanced expression of stress responsive genes	Saijo *et al.* (2000)
Farnesyl-transferase (*ERA1*)	Negative-regulator of ABA sensing	Down-regulation of farnesyltransferase enhances the plant's response to ABA and drought tolerance reducing stomatal conductance	Wang *et al.* (2005)
Mn-SOD	Mn-superoxide dismutase	Overexpression improves stress tolerance also in field conditions	McKersie *et al.* (1999)
AVP1	Vacuolar H^+ pyrophosphatase	Overexpression facilitate auxin fluxes leading to increased root growth	Gaxiola *et al.* (2001), Park *et al.* (2005)
HVA1; OsLEA3	Stress induced LEA proteins	Over-accumulation of LEA increases drought tolerance also in field conditions	Bahieldin *et al.* (2005), Xiao *et al.* (2007)
ERECTA	A putative leucine-rich repeat receptor like kinase is a major contributor to a locus for D on *Arabidopsis* chromosome 2	ERECTA acts as a regulator of transpiration efficiency with effects on stomatal density, epidermal cell expansion, mesophyll cell proliferation and cell-cell contact	Masle *et al.* (2005)
otsA and *otsB*	Escherichia coli trehalose biosynthetic genes	Increased trehalose accumulation correlates with higher soluble carbohydrate levels, elevated photosynthetic capacity and increased toleranceto photo-oxidative damage	Garg *et al.* (2002)
P5CS	d-Pyrroline-5-carboxylate synthetase	Enhanced accumulation of proline leads to increased osmotolerance	Kavi Kishor *et al.* (1995), Zhu *et al.* (1998)
mtlD	Mannitol-1-phosphate dehydrogenase	Mannitol accumulation leads to increased osmotolerance	Abebe *et al.* (2003)
GF14l	14-3-3 protein	Lines overexpressing GF14l have a "staygreen" phenotype, improved water stress tolerance and higher photosynthetic rates underwater deficit conditions	Yan *et al.* (2004)
NADP-Me	NADP-malic enzyme	The overexpression decreased stomatal conductance and improves WUE	Laporte *et al.* (2002)
AREB cupida dehydrine	bZIP transcription factor in tomato Leaf necrosis in tomato Increased distances from dehydrins activated by abscisic acid in tomato	Overexpression increasing dehydrin expression. Overwilting or stomatal defect. Probably the dehydrins protect membranes during stresses. Dehydrins are upregulated by abscisic acid.	Hsieh *et al.* (2010) Anonymous (2006) Weiss and Cortines (2010)
Chloroplast drought induced stress protein	Thiol-disulfide exchange intermediateactivity in potato	Preservation of the thiol: disulfide redox potential of chloroplastic proteins during water deficit.	Anonymous (2006)
CDSP 32	Thiol-disulfide exchange intermediate activity in potato	Preserve chloroplastic structures against oxidative injury upon drought.	Anonymous (2006)

Contd...

Table 8–*Contd...*

Gene(s)	*Function*	*Mechanism of Action*	*References*
CDSP 34	Increases in CDSP 34 transcript and protein abundances were also observed inpotato plants subjected to high illumination.	The CDSP 34 protein is proposed to play a structural role in stabilizing stromal lamella ethylakoids upon osmotic or oxidative stress.	Beyly *et al.* (1998)
Wilty.	Dominant TGRC gene in tomato Leaves overwilt when drought stressed	Wilting under field or greenhouse conditions; marginal leaf narcrosis	Anonymous (2006)
Wilty dwarf	Recessive TGRC gene in tomato	Grayish-green, droopy leaves; stunted plants; leaves droop when drought stressed.	Anonymous (2006)
Water stress induced *ER5* protein	Stress induced CaLEA6 (for *Capsicum annuum* LEA) is 709 bp long with anopen reading frame encoding 164 amino acids	Predicted to produce a highly hydrophobic, butcytoplasmic, protein	Kim *et al.* (2005)
Abscicic acid stress ripening 2	Putative DNA binding and chaperon like activity	A member of the Asr gene family. It is inducedby abiotic stress such as water and is expressedin the leaf phloem companion cells.	Giombini *et al.* (2009)
OsWRKY45	Transcription factor	Drought tolerance	Qiu and Yu (2009)
SodERF3	Ethylene responsive factor	Increased tolerance to drought and osmotic stresses	Trujillo *et al.* (2008)
CAP 160 andCAP 85	Dehydrin and Lea-like	Slightly enhanced freezing tolerance	Kaye *et al.* (1998)
ZAT12	Stress induced transcription factors	Enhanced expression of downstream stress related genes confers drought/heat/salt tolerance.	Rai *et al.* (2013)
Heat			
HsfA1	Transcription factor	Heat shock tolerant/Heat shock sensitive	Mishra *et al.* (2002)
Hsf3 (2002)	Transcription factor	Heat shock tolerant	Panchuk *et al.*
Hsp70	Heat shock protein	High temperature tolerant/sensitive	Lee *et al.* (1999)
Hsp101	Heat shock protein	Acquired thermo tolerance	Nieto-Sotelo *et al.* (2002)
Hsp17.7	Heat shock protein	High temperature tolerant/sensitive	Malik *et al.* (1999)
Fad7	Fatty acid denaturation	High temperature tolerance	Murakami *et al.* (2000)
Hvapx1	Active oxygen species metabolism	Heat shock tolerant	Saijo *et al.* (2000)
OsHSFA2e	Transcription factor	*Arabidopsis* plants showed enhanced thermotolerance	Yokotani *et al.* (2007)
AtHsfA2	*A. thaliana*	Thermo tolerance as well as oxidative stress tolerance	Li *et al.* (2005)
Salt stress			
hal1	Protein involved in regulation of K+ transport	Salt stress tolerance	Gisbert *et al.* (2000)
BADH-1	Over expression of betaine aldehyde dehydrogenase encoded by *BADH* gene catalyzes conversion of betaine aldehyde into glycine betaine	Salt stress tolerance	Moghaieb *et al.* (2000)
NHX1	Over expressed the *NHX1* vacuolar Na^+/H^+ antiporter in tomato plants	Salt stress tolerance	Zhang and Blumwald (2001)
MAPK gene (*CsNMAPK*)	ROS scavenger and osmotic adjustment Transcription factor	Salt stress tolerance	Xu *et al.* (2011)

The "ripening related " genes encode a range of enzymes thought to be required to bring about changes leading to ripe fruit. Tomato, being a climacteric fruit, shows a dramatic increase in respiration at the onset of ripening, accompanied by increased synthesis of phytohormone, ethylene production of which appears

Table 9: Examples of Successful Transgenic Vegetables for Post-harvest Traits

Trait	*Transgene Product*	*Origin of Transgene*	*Transformed Plant*
Improved storage/shelf-life	Antisense polygalacturonase	Tomato	Tomato
	α-mannosidase (α-Man) and β-D-N-acetyl-hexosaminidase (β-Hex) RNAi	Tomato	Tomato
Ripening	Antisense ACC oxidase	Tomato	Tomato
	Antisense ACC synthase	Tomato	Tomato
	Endo-1,4-beta-D-glucanase (EGase) CaCel1	Pepper	Tomato
	Antisense ACC oxidase	Tomato	Melon
Fruit pigmentation	Phytoene synthase gene	Tomato	Tomato
Improved flavour and aroma	*LeCCD1*	Tomato	Tomato
Modify asour taste into a sweet taste	Miraculin	Richadelladulcifica	Tomato

to be involved in the initiation, modulation and co-ordination of expression of many of the genes required for the ripening process (Picton *et al.*, 1994). The anti-sense RNA approach has been successfully utilized to bring about reduced ethylene synthesis in fruit by expression of an antisense RNA for ACC Oxidase (Hamilton *et al.*, 1990, 1991) or ACC synthase RNA (Oeller *et al.*, 1991) and to disrupt fruit pigmentation by down-regulation of the phytoene synthane gene (Bird *et al.*, 1991).

Carotenoids, such as β-carotene and lycopene, give the fruit its characteristic color. An effort to increase the provitamin A content in tomato was made by engineering a bacterial gene encoding for the phytoene-desaturase enzyme that converts phytoene to lycopene into transgenic tomato. Another group of metabolites exploited for its antioxidant property are the flavonoids and they have been correlated with a decreased risk of cardiovascular disease in humans. Since natural variation for flavonoid content of tomato fruit is limited and most of it is present in the peel, a transgenic approach has been used to increase the flavonoid content by over-expression of either the enzymes involved in flavonoid biosynthesis or transcription factors that regulate the genes of this pathway. Transgenic tomato plants expressing petunia CHI-A gene encoding chalcone isomerase showed significant increase in flavonoid content (Muir *et al.*, 2001).

4.9 Engineering for Male Sterility and Fertility Restoration

During 1990-1992, an exciting example of producing transgenic plants with male sterility and fertility restoration genes has become available in *Brassica napus*. This faciltates hybrid seed without manual emasculation and controlled pollination. In 1990, C. Mariani and others from Belgium, successfully used a gene construct having an other specific promoter (from*TA29* gene of tobacco) and bacterial coding sequence for a ribonuclease (*barnase* gene from *Bacillus amyloliquefaciens*) for production of transgenic plants in *B. napus*. The same group of workers (1992) used another gene construct later involving the same anther specific promoter, *i.e. TA29* and the *barstar* gene from *B. amyloliquefaciens*, for production of transgenic plants in *B. napus*. The product of *barstar* gene is a ribinuclease- inhibitor. It forms a complex with ribonuclease and neutralizes its cytotoxic properties. In *Bacillus*, the ribonuclease is active extracellularly and the *bacterium* itself is produced by a ribonuclease inhibitor protein coded by *barstar* gene. When transgenic male sterile plants (with *barnase* gene) were crossed with transgenic male fertile plants (with *barstar* gene), the F_1 plants expressed both genes so that male fertility was restored due to suppression of cytotoxic ribonuclease activity in the anther by the formation of cell specific Rnase/Rnase – inhibitor complexes. This system known as "*Barnase-Bar Star*" systems has also been proved successful in tomato and corn.

5. Transgenic Achievements

China was the first country to commercialize transgenic in the early nineties with the introduction of virus resistant tobacco, which was later followed by a virus resistant tomato (James, 1998). Subsequently, the global area of transgenic increased from 170.7million ha in 2006 to 549.9 million ha in 2015 (Table 10).

6. Future Strategies

Though remarkable work has been done for improvement of production, productivity and resistance in vegetable crops through biotechnology, however, there are some suggestions that could be used as future strategies for improvement of vegetable production in India.

- Recombinant DNA technology based use of genetic material from animals, micro-organisms and genetically distant plant species can be efficiently applied to crop improvement. Moreover, new strategies, based on genetic manipulation of native or homologous gene from related species, can be introduced.
- Genetic engineering of extremely nutritional food crops entails both gene silencing and counter-

Table 10: Global Area of Transgenic Crops in 2006 and 2015 (Country-wise in descending order)

Country	*2006*	*2015*	*Increase in Area (%)*	*Crops (2015)*
	Area (m ha)	*Area (m ha)*		
USA	54.6	70.09	22.1	Soybean, maize, cotton, canola, sugarbeet, alfalfa, papaya, squash
Brazil	11.5	44.2	73.9	Soybean, maize, cotton
Argentina	18.0	24.5	53.0	Soybean, maize, cotton
India	3.8	11.6	67.2	Cotton
Canada	6.1	11	44.5	Canola, maize, soybean, sugarbeet
China	3.5	3.7	5.40	Cotton, papaya, poplar, tomato, sweet pepper
Paraguay	2.5	3.6	30.5	Soybean
Pakistan	–	2.9	–	Cotton
South Africa	1.4	2.3	39.1	Soybean, maize, cotton
Uruguay	0.4	1.4	71.4	Soybean, maize
Bolivia	–	1.1	–	Soybean
Australia	0.2	0.7	71.4	Cotton, canola
Philippines	0.2	0.7	71.4	Maize
Myanmar	–	0.3	–	Cotton
Burkina Faso	–	0.4	–	Cotton
Mexico	0.1	0.1	0	Cotton, soybean
Spain	0.1	0.1	0	Maize
Colombia	<0.1	<0.1	0	Cotton
Chile	–	<0.1	–	Maize, soybean, canola
Honduras	<0.1	<0.1	0	Maize
Portugal	<0.1	<0.1	0	Maize
Czech Republic	<0.1	<0.1	0	Maize
Poland	–	<0.1	–	Maize
Egypt	–	<0.1	–	Maize
Slovakia	<0.1	<0.1	0	Maize
Romania	0.1	<0.1	0	Maize
Sweden	–	<0.1	–	Potato
Costa Rica	–	<0.1	–	Cotton, soybean
Germany	<0.1	<0.1	0	Potato
Total		**170.70**	**549.9**	**100**

Source: James 2015; International Service for the Acquisition of Agro-Biotech Applications (ISAAA), 2015.

silencing tools. Constructing vectors that can overturn the RNAi pathway and overexpressing these engineered trans genes will revolutionize this field.

- ✰ Genetic engineering can then be exploited to excluded detrimental traits while augmenting positive traits. Transgenic and xenogenic approaches will primarily be useful in introducing prevailing new traits into commodity crops, whereas intragenic and cis-genic approach can deliver additional cost-effective and adequate means for the advancement of specialty crops.
- ✰ Merger of precise genetic maps and DNA sequences of genomes can expedite the discovery of desired genetic traits and simultaneously can effectively link it with specific DNA sequences by using molecular probes.
- ✰ Model plant *Arabidopsis thaliana;* provides a complete understanding of entire process occurring at transcriptome and proteome level with in the plant genome and this information can offers a best chance for a parallel studies in other economically important agricultural crops.
- ✰ The exploration and sequencing of tomato gene that confers resistance to bacterial disease has offers scientists first impression of biotechnology in the development of disease resistance crops in decades.
- ✰ Soil micro-organisms that are in association with specific clan of crops also comprehend

disease resistance genes and little is known about these propitious micro-organisms. Hence, by deploying these micro-organisms and their genes in developing disease resistance crops will undoubtedly help in increased crop production.

- Crops without suitable regeneration techniques can also be enhanced by mutagenesis through TILLING. It is imperative to deliver breeders with the widest variety of biotechnology tools since each stress has its own discriminations so there is need of one or more combination of specific biotechnological approaches to tackle them efficiently.
- The focus should be the development of infrastructure tools for integration of high throughput data for effects of abiotic stress factors and genomics, associated with climate change and generation of functional genomics, proteomics, metabolomics, and comparative genomics data.
- Multidisciplinary approach to use current biological knowledge to relate genetic change to phenotypic outcome and incorporation of prediction of responses of phenology, plant architecture, and molecular and metabolic events related to climate change into conventional plant breeding.

References

Abebe, T., Guenzi, A.C., Martin, B. and Cushman, J.C. 2003. Tolerance of mannitol-accumulating transgenic wheat to water stress and salinity. *Plant Physiol.*, **131**: 1748–1755.

Agarwal, M., Katiyar-Agarwal, S., Sahi, C., Gallie, D.R. and Grover, A. 2001. *Arabidopsis thaliana* Hsp100 protein: Kith and kin. *Cell Stress Chap.*, **6**: 219–224.

Agarwal, M., Sarkar, N. and Grover, A. 2003. Low molecular weight heat shock proteins in plants. *J. Pl. Biol.*, **30**: 141–149.

Almoguera, C., Prieto-Dapena, P. and Jordano, J. 1998. Dual regulation of a heat shock promoter during embryogenesis: Stage dependent role of heat shock elements. *Plant J.*, **13**: 437–446.

Bahadur, A., Rai., N., Kumar, Rajesh, Tiwari, S.K., Singh, A.K., Rai, A.K., Singh, Umesh, Patel, P.K., Tiwari, Vivek, Rai, A.B., Singh, Major, and Singh, B. 2015. Grafting tomato on eggplant as a potential tool to improve waterlogging tolerance in hybrid tomato. *Veg. Sci.*, 42.

Bahieldin, A., Hesham, H.T., Eissa, H.F., Saleh, O.M., Ramadan, AM., Ahmed, I.A., Dyer, W.E., El-Itriby, H.A. and Madkour, M.A. 2005. Field evaluation of transgenic wheat plants stably expressing the HVA1 gene for drought tolerance. *Physiol Plant.*, **123**: 421–427.

Baker, B., Coupland, G., Fedoroff, N., Starlinger, P. and Schell, J. 1987. Phenotypic assay for excision of the maize controlling element Ac in tobacco. *EMBO J.*, **6**(6): 1547.

Baker, B., Schell, J., Lörz, H., and Fedoroff, N. 1986. Transposition of the maize controlling element "Activator" in tobacco. **In**: *Proc. National Acad. Sci.*, **83**(13): 4844–4848.

Ballvora, A., Hesselbach, J., Niewöhner, J., Leister, D., Salamini, F. and Gebhardt, C. 1995. Marker enrichment and high-resolution map of the segment of potato chromosome VII harbouring the nematode resistance gene Gro1. *Mol. Gen. Genet. MGG*, **249**(1): 82–90.

Bar-Tsur, A., Rudich, J. and Bravdo, B. 1985. High temperature effects on CO_2 gas exchange in heat tolerant and sensitive tomatoes. *J. Amer. Soc. Hort. Sci.*, **110**: 582–586.

Bello, M.H., Moghaddam, S.M., Massoudi, M., McClean, P.E., Cregan, P.B. and Miklas, P.N. 2014. Application of in silico bulked segregant analysis for rapid development of markers linked to Bean common mosaic virus resistance in common bean. *BMC Genomics*, **15**(1): 1.

Beyly Gillet, B., Peltier, A. and Rey, G.P. 1998. Molecular characterization of *CDSP-34*: A chloroplastic protein induced by water deficit in *Solanum tuberosum* L. plants, and regulation of *CDSP-34* expression by ABA and high illumination. *Plant J. Cell. Mol. Biol.*, **16**(2): 257–262.

Bonierbale, M.W., Plaisted, R.L. and Tanksley, S.D. 1988. RFLP maps based on a common set of clones reveal modes of chromosomal evolution in potato and tomato. *Genetics*, **120**(4): 1095–1103.

Camargo, L.E.A., Savides, L., Jung, G., Nienhuis, J. and Osborn, T.C. 1997. Location of the self-incompatibility locus in an RFLP and RAPD map of *Brassica oleracea*. *J. Hered.*, **88**(1): 57–60.

Cattivelli, L., Rizza, F., Badeck, F.W., Mazzucotelli, E., Mastrangelo, A.M., Francia, E., Mare, C., Tondelli, A. and Stanca, A.M. 2008. Drought tolerance improvement in crop plants: An integrated view from breeding to genomics. *Field Crops Res.*, **105**: 1–14.

Chawda, N. and Chawda, B. 2015. Vegetable grafting: A case study. *Veg. Sci.*, **42**(1): 126–127.

Chinnusamy, V., Jagendorf, A., and Zhu, J.K. 2005. Understanding and improving salt tolerance in plants. *Crop Sci.*, **45**: 437–448.

Churchill, G.A., Giovannoni, J.J. and Tanksley, S.D. 1993. Pooled-sampling makes high-resolution mapping practical with DNA markers. *Proc. Nat. Acad. Sci.*, **90**(1), 16–20.

Davletova, S., Schlauch, K., Coutu, J. and Mittler, R. 2005. The zinc-finger protein Zat12 plays a central role in reactive oxygen and abiotic stress signaling in *Arabidopsis*. *Pl. Physiol.*, **139**: 847–856.

Déjardin, A., Sokolov, L.N. and Kleczkowski, L.A. 1999. Sugar/osmoticum levels modulate differential ABA-independent expression of two stress responsive sucrose synthase genes in *Arabidopsis*. *Biochem. J.*, **344**: 503–509.

Dellaporta, S.L. and Chomet, P.S. 1985. The activation of maize controlling elements. **In**: *Genetic Flux in Plants*, Springer Vienna, pp. 169–216.

Dickinson, M.J., Jones, D.A., and Jones, J.D.G. 1993. Close linkage between the Cf-2/Cf-5 and Mi resistance loci in tomato. *Mol. Pl. Micr. Int.*, **6**: 341–341.

Dirlewanger, E., Isaac, P.G., Ranade, S., Belajouza, M., Cousin, R. and De Vienne, D. 1994. Restriction fragment length polymorphism analysis of loci associated with disease resistance genes and developmental traits in *Pisum sativum* L. *Theor. Appl. Genet.*, **88**(1): 17–27.

Dixon, M.S., Jones, D.A., Hatzixanthis, K., Ganal, M.W., Tanksley, S.D., and Jones, J.D.G. 1995. High-resolution mapping of the physical location of the tomato Cf-2 gene. *MPMI–Mol. Pl. Micr. Interac.*, **8**(2): 200–206.

Edelstein, M. 2004. Grafting vegetable-crop plants: Pros and Cons. *Acta Horticulturae*, p. 65.

Ellegren, H. 2000. Microsatellite mutations in the germline: Implications for evolutionary inference. *Trends Genet.*, **16**(12): 551–558.

Finkers-Tomczak, A., Bakker, E., de Boer, J., van der Vossen, E., Achenbach, U., Golas, T. and Goverse, A. 2011. Comparative sequence analysis of the potato cyst nematode resistance locus H1 reveals a major lack of co-linearity between three haplotypes in potato (*Solanum tuberosum* ssp.). *Theor. Appl. Genet.*, **122**(3): 595–608.

Flowers, T.J. and Yeo, A.R. 1995. Breeding for salinity resistance in of crop plants. Where next? *Australian J. Pl. Physiol.*, **22**: 875–884.

Foolad, M.R. and Chen, F.Q. 1998. RAPD markers associated with salt tolerance in an interspecific cross of tomato (*Lycopersicon esculentum* × *L. pennellii*). *Plant Cell Reports*, **17**(4): 306–312.

Foolad, M.R. 2004. Recent advances in genetics of slat tolerance in tomato. *Plant Cell Tiss. Org. Cult.*, **76**: 101–119.

Ganal, M.W., Broun, P., and Tanksley, S.D. 1992. Genetic mapping of tandemly repeated telomeric DNA sequences in tomato (*Lycopersicon esculentum*). *Genomics*, **14**(2): 444–448.

Ganal, M.W., Czihal, R., Hannappel, U., Kloos, D.U., Polley, A. and Ling, H.Q. 1998. Sequencing of cDNA clones from the genetic map of tomato (*Lycopersicon esculentum*). *Genome Res.*, **8**(8): 842–847.

Ganal, M.W., Simon, R., Brommonschenkel, S., Arndt, M., Phillips, M.S., Tanksley, S.D., and Kumar, A. 1995. Genetic mapping of a wide spectrum nematode resistance gene (Hero) against *Globodera rostochiensis* in tomato. *MPMI–Mol. Pl. Micr. Interac.*, **8**(6): 886–891.

Garg, A., Kim, J., Owens, T., Ranwala, A., Choi, Y., Kochian, L. and Wu, R. 2002. Trehalose accumulation in rice plants confers high tolerance levels to different abiotic stresses. *Proc. Nat. Acad. Sci., USA*, **99**: 15898–15903.

Gaxiola, R.A., Li, J., Undurraga, Dang, L.M., Allen, G.J., Alper, S.L. and Fink, G.R. 2001. Drought- and salt-tolerant plants result from over expression of the AVP1 H+ – pump. *Proc. Nat. Acad. Sci., USA*, **98**: 11444–11449.

Gepstein, S., Grover, A. and Blumwald, E. 2005. Producing biopharmaceuticals in the desert: Building an abiotic stress tolerance in plants for salt, heat and drought. **In**: *Modern Biopharmaceuticals*, (Eds.) J. Knablein and R.H. Muller, Wiley-VCH Verlag GmbH and Co., Weinhaum, pp. 967–994.

Giombini, M.I., Frankel, N., Iusem, N.D. and Hasson, E. 2009. Nucleotide polymorphism in the drought responsive gene *Asr-2* in wild populations of tomato. *Genetica*, **136**(1): 13–25.

Giovannoni, J.J., Noensie, E.N., Ruezinsky, D.M., Lu, X., Tracy, S.L., Ganal, M.W. and Tankslev, S.D. 1995. Molecular genetic analysis of the ripening-inhibitor and non-ripening loci of tomato: A first step in genetic map-based cloning of fruit ripening genes. *Mol. Gen. Genet. MGG*, **248**(2): 195–206.

Gisbert, C., Rus, A.M., Bolarin, M.C., Lopez-Coronado, M., Arrillaga, I., Montesinos, C., Caro, M., Serrano, R. and Moreno, V. 2000. The yeast HAL1 gene improves salt tolerance of transgenic tomato. *Plant Physiol.*, **123**: 393–402.

Gorguet, B., Eggink, P. M., Ocana, J., Tiwari, A., Schipper, D., Finkers, R., van Heusden, A.W. 2008. Mapping and characterization of novel parthenocarpy QTLs in tomato. *Thero. Appl. Genet.*, **116**(6): 755–767.

Gorman, S.W., Banasiak, D., Fairley, C. and McCormick, S. 1996. A 610 kb YAC clone harbors 7 cM of tomato (*Lycopersicon esculentum*) DNA that includes the male sterile 14 gene and a hotspot for recombination. *Mol. Gen. Genet. MGG*, **251**(1): 52–59.

Greenblatt, I.M. 1984. A chromosome replication pattern deduced from pericarp phenotypes resulting from movements of the transposable element, Modulator, in maize. *Genetics*, **108**(2): 471–485.

Grover, A., Aggarwal, P.K., Kapoor, A., Katiyar-Agarwal, S., Agarwal, M., Chandramouli, A. 2003. Addressing abiotic stresses in agriculture through transgenic technology. *Curr. Sci.*, **84**: 355–367.

Halliwell, B. and Gutteridge, J.M.C. 1989. Oxygen is poisonous: An introduction to oxygen toxicity and free radicals. **In**: *Free Radicals in Biology and Medicine*, 2nd edn, Clarendron Press, Oxford, UK.

Haring, M.A., Gao, J., Volbeda, T., Rommens, C.M., Nijkamp, H.J.J. and Hille, J. 1989. A comparative study of Tam3 and Ac transposition in transgenic tobacco and petunia plants. *Pl. Mol. Biol.*, **13**(2): 189–201.

Hoogstraten, J.G.J., and Braun III, C.J. 2014. U.S. Patent No. 8,785,720. Washington, DC: U.S. Patent and Trademark Office.

Hsieh, T.H., Lee, J.T., Charng, Y.Y. and Chan, M.T. 2002a. Tomato plants ectopically expressing Arabidopsis CBF1 show enhanced resistance to water deficit stress. *Pl. Physiol.*, **130**: 618–626.

Hsieh, T.H., Lee, J.T., Yang, P.T., Chiu, L.H., Charng, Y.Y., Wang, Y.C. and Chan, M.T. 2002b. Heterology expression of the *Arabidopsis* C-repeat/dehydration response element binding factor 1 gene confers elevated tolerance to chilling and oxidative stresses in transgenic tomato. *Pl. Physiol.*, **129**: 1086–1094.

Hsieh, T.H., Li, C.W., Su, R.C., Cheng, C.P., Sanjaya, Tsai, Y.C. and Chan, M.T. 2010. A tomato *bZIP* transcription factor, *SlAREB*, is involved in water deficit and salt stress response. *Planta*, **231**(6): 1459–1473.

Hu, H., Dai, M., Yao, J., Xiao, B., Li, X., Zhang, Q. and Xiong, L. 2006. Overexpressing a NAM, ATAF, and CUC (NAC) transcription factor enhances drought resistance and salt tolerance in rice. *Proc. Nat. Acad. Sci., USA*, **103**: 12987–12992.

Hu, J., Sadowski, J., Osborn, T.C., Landry, B.S. and Quiros, C.F. 1998. Linkage group alignment from four independent *Brassica oleracea* RFLP maps. *Genome*, **41**(2): 226–235.

Inze, D. and Van Montagu, M. 1995. Oxidative stress in plants. *Curr. Opin. Biot.*, **6**: 153–158.

Ito, Y., Katsura, K., Maruyama, K., Taji, T., Kobayashi, M., Seki, M., Shinozaki, K. and Yamaguchi-Shinozaki, K. 2006. Functional analysis of rice DREB1/CBF-type transcription factors involved in cold-responsive gene expression in transgenic rice. *Plant Cell Physiol.*, 47: 1–13.

Jansen, R.C. and Stam, P. 1994. High resolution of quantitative traits into multiple loci via interval mapping. *Genetics*, **136**(4): 1447–1455.

Ji, Y., Schuster, D.J. and Scott, J.W. 2007a. Ty-3, a begomovirus resistance locus near the Tomato yellow leaf curl virus resistance locus Ty-1 on chromosome 6 of tomato. *Mol. Breed.*, **20**: 271–284.

Ji, Y., Scott, J.W. and Schuster, D.J. 2009a. Toward fine mapping of the Tomato yellow leaf curl virus resistance gene Ty-2 on chromosome 11 of tomato. *Hort. Sci.*, **44**(3): 614–618.

Jia, G.X., Zhu, Z.Q., Chang, F.Q. and Li, Y.X. 2002. Transformation of tomato with the BADH gene from Atriplex improves salt tolerance. *Plant Cell Rep.*, **21**: 141–146.

Jung, J., Kim, H.J., Lee, J.M., Oh, C.S., Lee, H.J., and Yeam, I. 2015. Gene-based molecular marker system for multiple disease resistances in tomato against Tomato yellow leaf curl virus, late blight, and verticillium wilt. *Euphytica*, **205**(2): 599–613.

Kaloshian, I., Yaghoobi, J., Liharska, T., Hontelez, J., Hanson, D., Hogan, P. and Zabel, P. 1998. Genetic and physical localization of the root-knot nematode resistance locus Mi in tomato. *Mol. Gen. Genet. MGG*, **257**(3): 376–385.

Kang, B.C., Nahm, S.H., Huh, J.H., Yoo, H.S. and Kim, B.D. 1998. Construction of a molecular linkage map in hot pepper. **In**: *XXV International Horticultural Congress, Part II: Application of Biotechnology and Molecular Biology and Breeding–Gene*, pp. 165–172.

Katiyar-Agarwal, S., Agarwal, M., Gallie, D. and Grover, A. 2001. Search for the cellular functions of plant Hsp100/Clp family proteins. *Crit. Rev. Plant Sci.*, **20**: 277–295.

Kavi Kishor, P.B., Hong, Z., Miao, G.H., Hu, C.A.A. and Verma, D.P.S. 1995. Over-expression of d-pyrroline-5-carboxylate synthetase increases proline production and confers osmotolerance in transgenic plants. *Pl. Physiol.*, **25**: 1387–1394.

Kaye, C., Neven, L., Hofig, A., Li, Q.B., Haskell, D. and Guy, C. 1998. Characterization of a gene for spinach CAP160 and expression of two spinach cold-acclimation proteins in tobacco. *Pl. Physiol.*, **116**: 1367–1377.

Kim, Hyung Lee, Kim, J.H., Kim, J.J., Sung Chang, Jun, and Young, Hong. 2005. Molecular and functional characterization of *CaLEA6*, the gene for a hydrophobic *LEA* protein from *Capsicum annuum*. *Gene*, **344**: 115–123.

Kotak, S., Larkindale, J., Lee, U., Koskull-Doring, P., Vierling, E. and Scharf, K.D. 2007. Complexity of the heat stress response in plants. *Curr. Opin. Pl. Biol.*, 10: 310–316.

Kumar, R.G., Shah, K. and Dubey, R.S. 2000. Salinity induced behavioural changes in malate dehydrogenase and glutamate dehydrogenase activities in rice seedlings of differing salt tolerance. *Plant Sci.*, **156**: 23–34.

Lagercrantz, U., and Lydiate, D.J. 1996. Comparative genome mapping in *Brassica*. *Genetics*, **144**(4): 1903–1910.

Laporte, M.M., Shen, B. and Tarczynski, M.C. 2002. Engineering for drought avoidance: expression of maize NADP-malic enzyme in tobacco results in altered stomatal function. *J. Exp. Bot.*, **53**: 699–705.

Larkindale, J., Hall, J.D., Knight, M.R. and Vierling, E. 2005. Heat stress phenotypes of *Arabidopsis* mutants implicate multiple signaling pathways in the acquisition of thermotolerance. *Pl. Physiol.*, **138**: 82–89.

Lawlor, D.W. 2002. Limitation to photosynthesis in water-stressed leaves: Stomatal metabolism and the role of ATP. *Annals Bot.*, **89**: 871–885.

Lawrence, S.D., Cline, K. and Moore, G.A. 1997. Chromoplast development in ripening tomato fruit: Identification of cDNAs for chromoplast-targeted proteins and characterization of a cDNA encoding a plastid-localized low-molecular-weight heat shock protein. *Plant Mol. Biol.*, **33**: 483–492.

Lee, H.J., Xiong, L., Ishitani, M., Stevenson, B. and Zhu, J.K. 1999. Cold-regulated gene expression and freezing tolerance in an *Arabidopsis thaliana* mutant. *Plant J.*, **17**: 301–308.

Lee, J.T., Prasad, V., Yang, P.T., Wu, J.F., Ho, T.H.D., Charng, Y.Y. and Chan, M.T. 2003. Expression of Arabidopsis CBF1 regulated by an ABA/stress promoter in transgenic tomato confers stress tolerance without affecting yield. *Plant Cell Env.*, **26**: 1181–1190.

Li, C., Chen, Q., Gao, X., Chen, N., Xu, S., Chen, J. and Wang, X. 2005. AtHsfA2 modulates expression of stress responsive genes and enhances tolerance to heat and oxidative stress in *Arabidopsis*. *Sci. China C. Life Sci.*, **48**(6): 540–550.

Lim, G.T.T., Wang, G.P., Hemming, M.N., Basuki, S., McGrath, D.J., Carroll, B.J., and Jones, D.A. 2006. Mapping the I-3 gene for resistance to *Fusarium* wilt in tomato: Application of an I-3 marker in tomato improvement and progress towards the cloning of I-3. *Australasian Pl. Path.*, **35**(6): 671–680.

Ling, H.Q., Pich, A., Scholz, G., and Ganal, M.W. 1996. Genetic analysis of two tomato mutants affected in the regulation of iron metabolism. *Mol. Gen. Genet., MGG*, **252**(1–2): 87–92.

Liu, Q., Kasuga, M., Sakuma, Y., Abe, H., Miura, S., Yamaguchi-Shinozaki, K. and Shinozaki, K. 1998. Two transcription factors, DREB1 and DREB2, with an EREBP/AP2 DNA binding domain separate two cellular signal transduction pathways in drought- and low-temperature-responsive gene expression, respectively, in *Arabidopsis*. *Plant Cell*, **10**: 1391–1406.

Liu, T., Yu, Y., Cai, X., Tu, W., Xie, C. and Liu, J. 2016. Introgression of bacterial wilt resistance from *Solanum melongena* to *S. tuberosum* through asymmetric protoplast fusion. *Pl. Cell Tiss. Org. Cult.*, **125**(3): 433–443.

Maisonneuve, B., Bellec, Y., Anderson, P., and Michelmore, R.W. 1994. Rapid mapping of two genes for resistance to downy mildew from *Lactuca serriola* to existing clusters of resistance genes. *Theor. Appl. Genet.*, **89**(1): 96–104.

Malik, M.K., Slovin, J.P., Hwang, C.H. and Zimmerman, J.L. 1999. Modified expression of a carrot small heat shock protein gene, hsp17.7, results in increased or decreased thermotolerance. *Plant J.*, **20**: 89–99.

Martin, G.B., Ganal, M.W., and Tanksley, S.D. 1992. Construction of a yeast artificial chromosome library of tomato and identification of cloned segments linked to two disease resistance loci. *Mol. Gen. Genet., MGG*, **233**(1–2), 25–32.

Masle, J., Gilmore, S.R. and Farquhar, G.D. 2005. The ERECTA gene regulates plant transpiration efficiency in *Arabidopsis*. *Nature*, **436**: 866–870.

Matsubara, S. 1989. Studies on salt tolerance of vegetables-3. Salt tolerance of rootstocks. *Agric. Bull.*, Okayama Univ., **73**: 17–25.

McKersie, B.D., Bowley, S.R. and Jones, K.S. 1999. Winter survival of transgenic *alfalfa* over expressing superoxide dismustase. *Pl. Physiol.*, **119**: 839–847.

Meyer, R.C., Milbourne, D., Hackett, C.A., Bradshaw, J.E., McNichol, J.W. and Waugh, R. 1998. Linkage analysis in tetraploid potato and association of markers with quantitative resistance to late blight (*Phytophthora infestans*). *Mol. Gen. Genet., MGG*, **259**(2): 150–160.

Midmore, D.J., Roan, Y.C. and Wu, D.L. 1997. Management practices to improve lowland sub-tropical summer tomato production: yields, economic returns and risk. *Exptl Agric.*, **33**: 125–137.

Milbourne, D., Meyer, R.C., Collins, A.J., Ramsay, L.D., Gebhardt, C. and Waugh, R. 1998. Isolation, characterisation and mapping of simple sequence repeat loci in potato. *Mol. Gen. Genet., MGG*, **259**(3): 233–245.

Mimura, Y., Inoue, T., Minamiyama, Y. and Kubo, N. 2012. An SSR-based genetic map of pepper (Capsicum annuum L.) serves as an anchor for the alignment of major pepper maps. *Breed. Sci.*, **62**(1): 93–98.

Mishra, S.K., Tripp, J., Winkelhaus, S., Tschiersch, B., Theres, K., Nover, L. and Scharf, K.D. 2002. In the complex family of heat stress transcription factors, HsfA1 has a unique role as master regulator of thermotolerance in tomato. *Genes Dev.*, **16**: 1555–1567.

Moghaieb, R.E.A., Tanaka, N., Saneoka, H., Hussein, H.A., Yousef, S.S., Ewada, M.A.F., Aly, M.A.M. and Fujita, K. 2000. Expression of betaine aldehyde dehydrogenase gene in transgenic tomato hairy roots leads to the accumulation of glycine betaine and contributes to the maintenance of the osmotic potential under salt stress. *Soil Sci. Pl. Nutr.*, **46**: 873–883.

Mohanty, A., Kathuria, H., Ferjani, A., Sakamoto, A., Mohanty, P., Murata, N. and Tyagi, A.K. 2002. Transgenics of an elite indica rice variety Pusa Basmati 1 harbouring the coda gene are highly tolerant to salt stress. *Theor. Appl. Genet.*, **106**: 51–57.

Murakami, Y., Tsuyama, M., Kobayashi, Y., Kodama, H. and Iba, K. 2000. Trienoic fatty acids and plant tolerance of high temperature. *Sci.*, **287**: 476–479.

Neta-Sharir, I., Isaacson, T., Lurie, S. and Weiss, D. 2005. Dual role for tomato heat shock protein 21: Protecting photosystem II from oxidative stress and promoting color changes during fruit maturation. *Pl. Cell*, **17**: 1829–1838.

Nieto-Sotelo, J., Martinez, L.M., Ponce, G., Cassab, G.I., Alagon, A., Meeley, R.B., Ribaut, J.M. and Yang, R. 2002. Maize HSP101 plays important roles in both induced and basal thermotolerance and primary root growth. *Pl. Cell*, **14**: 1621–1633.

Nover, L. 1991. Inducers of HSP synthesis: heat shock and chemical stressors. In: *Heat Shock Responses*, (Ed.) L. Nover, CRC press, Boca Raton, FL, pp. 41–127.

Oh, S.J., Song, S.I., Kim, Y.S., Jang, H.J., Kim, S.Y., Kim, M., Kim, Y.K., Nahm, B.H. and Kim, J.K. 2005. *Arabidopsis* CBF3/DREB1A and ABF3 in transgenic rice increased tolerance to abiotic stress without stunting growth. *Pl. Physiol.*, **138**: 341–351.

Okimura, M., Matsou, S., Arai, K. and Okitso, S. 1986. Influence of soil temperature on the growth of fruit vegetable grafted on different stocks. *Bull. Veg. Ornam. Crops Res. Stn., Japan*, **C9**: 3–58.

Panchuk, I.I., Volkov, R.A. and Schoffl, F. 2002. Heat stress and heat shock transcription factor-dependent expression and activity of ascorbate peroxidase in *Arabidopsis*. *Pl. Physiol.*, **129**: 838–853.

Panthee, D.R., and Foolad, M.R. 2012. RETRACTED ARTICLE: A reexamination of molecular markers for use in marker-assisted breeding in tomato. *Euphytica*, **184**(2): 165–179.

Papageorgiou, G.C. and Murata, N. 1995. The unusually strong stabilizing effects of glycine betaine on the structure and function of the oxygen-evolving Photosystem II complex. *Photosynth. Res.*, **44**: 243–252.

Paran, I., Aftergoot, E., and Shifriss, C. 1998. Variation in *Capsicum annuum* revealed by RAPD and AFLP markers. *Euphytica*, **99**(3): 167–173.

Park, B.J., Liu, Z., Kanno, A. and Kameya, T. 2005. Genetic improvement of Chinese cabbage forsalt and drought tolerance by constitutive expression of a *B. napus* LEA gene. *Pl. Sci.*, **169**: 553–558.

Park, S., Li, J., Pittman, J.K., Berkowitz, G.A., Yang, H., Undurraga, S., Morris, J., Hirschi, K.D. and Gaxiola, R.A. 2005. Up-regulation of a H+-pyrophosphatase (H^+-PPase) as a strategy to engineer drought-resistant crop plants. *PNAS*, **102**: 18830–18835.

Piao, Z.Y., Deng, Y.Q., Choi, S.R., Park, Y.J. and Lim, Y.P. 2004. SCAR and CAPS mapping of CRb, a gene conferring resistance to *Plasmodiophora brassicae* in Chinese cabbage (*Brassica rapa* ssp. *pekinensis*). *Theor. Appl. Genet.*, **108**(8): 1458–1465.

Prince, J.P., Loaiza-Figueroa, F. and Tanksley, S.D. 1992. Restriction fragment length polymorphism and genetic distance among Mexican accessions of capsicum. *Genome*, **35**(5): 726–732.

Qiu, Y. and Yu, D. 2009. Over-expression of the stress-induced OsWRKY45 enhances disease resistance and drought tolerance in *Arabidopsis*. *Env. Exp. Bot.*, **65**: 35–47.

Rai, A.C., Singh, M. and Shah, K. 2012a. Environmental stresses and Transgenics: Role of ZFP (ZAT) gene in multiple stress tolerance in plants. **In**: *Advances in Plant Physiology, Vol. 13*, (Ed.) A. Hemantranjan, Scientific Publishers, Jodhpur, India, pp. 197–232.

Rai, A.C., Singh, M. and Shah, K. 2012b. Effect of water withdrawal on formation of free radical, proline accumulation and activities of antioxidant enzymes in ZAT12-transformed transgenic tomato plants. *Pl. Physiol. Biochem.*, **61**: 108–114.

Rai, A.C., Singh, M. and Shah, K. 2013. Engineering drought tolerant tomato plants over-expressing *BcZAT12* gene encoding a C_2H_2 zinc finger transcription factor. *Phytochem.*, **85**: 44–50.

Rai, G.K., Rai, N.P., Kumar, S., Yadav, A., Rathaur, S. and Singh. M. 2012. Effects of explant age, germination medium, pre-culture parameters, inoculation medium, pH, washing medium, and selection regime on *Agrobacterium*-mediated transformation of tomato. *In vitro Cell Dev. Biol. Plant*, **48**: 565–578.

Reymond, P., Weber, H., Damond, M., Farmer, E.E. 2000. Differential gene expression in response to mechanical wounding and insect feeding in *Arabidopsis*. *Pl. Cell*, **12**: 707–720.

Rick, C.M., and Fobes, J.F. 1974. Association of an allozyme with nematode resistance. *Tomato Genet. Coop. Rep.*, 24, 25.

Rick, C.M., DeVerna, J.W., Chetelat, R.T., and Stevens, M.A. 1986. Potential contributions of wide crosses to improvement of processing tomatoes. **In**: *2nd International Symposium on Processing Tomatoes, XXII IHC*, pp. 45–56.

Rick, C.M., Fobes, J.F., and Holle, M. 1977. Genetic variation in *Lycopersicon pimpinellifolium*: Evidence of evolutionary change in mating systems. *Pl. Syst. Evol.*, **127**(2–3): 139–170.

Rivero, R.M., Kojima, M., Gepstein, A., Sakakibara, H., Mittler, R., Gepstein, S. and Blumwald, E. 2007. Delayed leaf senescence induces extreme drought tolerance in a flowering plant. *Proc. Natl. Acad. Sci., USA*, **104**: 19631–19636.

Robbins, M.A., Witsenboer, H., Michelmore, R.W., Laliberte, J.F., and Fortin, M.G. 1994. Genetic mapping of turnip mosaic virus resistance in *Lactuca sativa*. *Theor. Appl. Genet.*, **89**(5): 583–589.

Romero, L., Belakbir, A., Ragala, L. and Ruiz, M.J. 1997. Response of plant yield and leaf pigments to saline conditions: effectiveness of different rootstocks in melon plant (*Cucumis melo* L.). *Soil Sci. Pl. Nutr.*, **3**: 855–862.

Roy, R., Purty, R.S., Agrawal, V. and Gupta, S.C. 2006. Transformation of tomato cultivar 'Pusa Ruby' with bspA gene from *Populus tremula* for drought tolerance. *Pl. Cell Tiss. Org. Cult.*, **84**: 49–55.

Sahi, C., Singh, A., Blumwald, E. and Grover, A. 2006. Beyond osmolytes and transporters: Novel plant salt-stress tolerance-related genes from transcriptional profiling data. *Physiol. Pl.*, **127**(1): 1–9.

Saijo, Y., Hata, S., Sheen, J. and Izui, K. 2000. Overexpression of a single Ca2þ-dependent protein kinase confers both cold and salt/drought tolerance on rice plants. *Pl. J.*, **23**: 319–327.

Sakamoto, H., Maruyama, K., Sakuma, Y., Meshi, T., Iwabuchi, M., Shinozaki, K. and Yamaguchi-Shinozaki, K. 2004. *Arabidopsis* Cys2/His2-type zinc-finger proteins function as transcription repressors under drought, cold and high-salinity stress conditions. *Pl. Physiol.*, **136**: 2734–2746.

Sakuma, Y., Liu, Q., Dubouzet, J.G., Abe, H., Shinozaki, K. and Yamaguchi-Shinozaki, K. 2002. DNA-binding specificity of the ERF/AP2 domain of *Arabidopsis* DREBs, transcription factors involved dehydration- and cold-inducible gene expression. *Biochem. Biophys. Res. Comm.*, **290**: 998–1009.

Sanmiya, K., Suzuki, K., Egawa, Y., Shono, M. 2004. Mitochondrial small heat-shock protein enhances thermotolerance in tobacco plants. *FEBS Lett.*, **557**: 265–268.

Scharf, K.D., Siddique, M. and Vierling, E. 2001. The expanding family of *Arabidopsis thaliana* small heat stress proteins and a new family of proteins containing alpha-crystallin domains (ACD proteins). *Cell Stress*, **6**: 225–237.

Schoffl, F., Prandl, R. and Reindl A. 1998. Regulation of the heat shock response. *Pl. Physiol.*, **17**: 1135–1141.

Schramm, F., Larkindale, J., Kiehlmann, E., Ganguli, A., Englich, G., Vierling, E. and Koskull-Doring, P. 2007. A cascade of transcription factor DREB2A and heat stress transcription factor HsfA3 regulates the heat stress response of *Arabidopsis*. *Pl. J.*, **53**(1): 264–274.

Schumacher, K., Ganal, M., and Theres, K. 1995. Genetic and physical mapping of the lateral suppressor (ls) locus in tomato. *Mol. Gen. Genet., MGG*, **246**(6): 761–766.

Seki, M., Narusaka, M., Abe, H., Kasuga, M., Yamaguchi-Shinozaki, K., Carninci, P., Hayashizaki, Y. and Shinozaki, K. 2001. Monitoring the expression pattern of 1300 Arabidopsis genes under drought and cold stresses by using a full-length cDNA microarray. *Pl. Cell*, **13**: 61–72.

Shah, K. and Dubey, R.S. 1998. Cadmium elevates level of protein, amino acids and alters activity of proteolytic enzymes in germinating rice seedlings. *Acta Physiol. Pl.*, **20**: 189–196.

Shah, K. and Nahakpam, S. 2011. Heat stress and cadmium toxicity in higher plants: An overview. In: *Advances in Plant Physiology, Vol. 12*, (Ed.) A. Hemantranjan, Scientific Publishers, Jodhpur, India, pp. 243–280.

Shah, K. and Nahakpam, S. 2012. Heat exposure alters the expression of SOD, POD, APX and CAT isoenzymes and mitigates low cadmium toxicity in seedlings of sensitive and tolerant rice cultivars. *Pl. Physiol. Biochem.*, **57**: 106–113.

Shah, K. and Dubey, R.S. 2005. Plant metabolism under temperature stress. In: *Physiology of Abiotic Stress in Plants*, (Eds.) P. Dwivedi, R.S. Dwivedi, Agrobios Publishers, Jodhpur, India, pp. 243–274.

Shah, K., Singh, M. and Rai, A.C. 2013. Effect of heat-shock induced oxidative stress is suppressed in BcZAT12 expressing drought tolerant tomato. *Phytochemistry*, **95**: 109–117.

Shi, A., Vierling, R., Grazzini, R., Chen, P., Caton, H. and Panthee, D. 2011. Identification of molecular markers for Sw–5 gene of tomato spotted wilt virus resistance. *Amer. J. Biot. Mol. Sci.*, **1**: 1.

Smýkal, P., Aubert, G., Burstin, J., Coyne, C.J., Ellis, N.T., Flavell, A.J. and McPhee, K.E. 2012. Pea (*Pisum sativum* L.) in the genomic era. *Agronomy*, **2**(2), 74–115.

Sokolova, E.A., Beketova, M.P. and Khavkin, E.E. 2010. DNA markers of the R1 and R3 genes as predictors of potato late blight resistance. *Russ. Agril. Sci.*, **36**(5): 334–337.

Stommel, J.R. and Zhang, Y. 1998. Molecular markers linked to quantitative trait loci for anthracnose resistance in tomato. *Hort. Sci.*, **33**(3): 515–515.

Sun, S., Fu, H., Wang, Z., Duan, C., Zong, X. and Zhu, Z. 2016. Discovery of a novel er1 allele conferring powdery mildew resistance in Chinese pea (*Pisum sativum* L.) landraces. *PloS One*, **11**(1): e0147624.

Sung, D.Y., Kaplan, F., Lee, K.J. and Guy, C.L. 2003. Acquired tolerance to temperature extremes. *Trends Plant Sci.*, **8**: 179–187.

Suzuki, N., Bajad, S., Shuman, J., Shulaev, V. and Mittler, R. 2008. The transcriptional coactivator MBF1c is a key regulator of thermotolerance in *Arabidopsis thaliana*. *J. Biol. Chem.*, **283**: 9269–9275.

Tanksley, S.D. and Nelson, J.C. 1996. Advanced backcross QTL analysis: A method for the simultaneous discovery and transfer of valuable QTLs from unadapted germplasm into elite breeding lines. *Theor. Appl. Genet.*, **92**(2): 191–203.

Tanksley, S.D. 1984. Linkage relationships and chromosomal locations of enzyme-coding genes in pepper, *Capsicum annuum*. *Chromosoma*, **89**(5): 352–360.

Tanksley, S.D. 1988. Resolution of quantitative traits into Mendelian factors by using a complete linkage map of restriction fragment length polymorphisms. *Nature*, **335**(721726), 6170Schneeberger.

Tanksley, S.D. and Orton, T.J. (Eds.) 1983. *Isozymes in Plant Genetics and Breeding: Part A*. Amsterdam New York: Elsevier, pp. 289–302.

Tanksley, S.D., Ganal, M.W., Prince, J.P., De Vicente, M.C., Bonierbale, M.W., Broun, P. and Martin, G.B. 1992. High density molecular linkage maps of the tomato and potato genomes. *Genetics*, **132**(4): 1141–1160.

Trujillo, L.E., Sotolongo, M., Menendez, C., Ochogava, M.E., Coll, Y., Hernandez, I., Borras-Hidalgo, O., Thomma, B.P.H.J. Vera, P. and Hernandez, L. 2008. SodERF3, a novel sugarcane ethylene responsive factor (ERF), enhances salt and drought tolerance when over-expressed in tobacco plants. *Pl. Cell Physiol.*, **49**: 512–515.

Truong, H.T.H., Choi, H., Cho, M.C., Lee, H.E. and Kim, J.H. 2011. Use of Cf-9 gene-based markers in marker-assisted selection to screen tomato cultivars with resistance to *Cladosporium fulvum*. *Hort. Env. Biot.*, **52**(2): 204–210.

Tuberosa, R. and Salvi, S. 2006. Genomics-based approaches to improve drought tolerance of crops. *Tr. Pl. Sci.*, **11**: 405–412.

Vallejos, C.E. and Tanksley, S.D. 1983. Segregation of isozyme markers and cold tolerance in an interspecific backcross of tomato. *Theor. Appl. Genet.*, **66**(3–4): 241–247.

Van Sluys, M.A., Tempe, J. and Fedoroff, N. 1987. Studies on the introduction and mobility of the maize Activator element in Arabidopsis thaliana and *Daucus carota*. *EMBO J.*, **6**(13), 3881.

Vierling, E. 1991. The roles of heat shock proteins in plants. *Ann. Rev. Pl. Physiol. Pl. Mol. Biol.*, **42**: 579–620.

Wang, W.X., Vinocur, B. and Altman, A. 2003.3 Plant responses to drought, salinity and extreme temperatures: Towards genetic engineering for stress tolerance. *Planta*, **218**: 1–14.

Wang, Y., Ying, J., Kuzma, M., Chalifoux, M., Sample, A., McArthur, C., Uchacz, T., Sarvas, C., Wan, J., Tennis, D.T., McCourt, P. and Huang, Y. 2005. Molecular tailoring of farnesylation for plant drought tolerance and yield protection. *Plant J.*, **43**: 413–424.

Waters, E.R., Lee, G.J. and Vierling, E. 1996. Evolution, structure and function of the small heat shock proteins in plants. *J. Exp. Bot.*, **47**: 325–328.

Wehmeyer, N., Hernandez, L.D. Finkelstein, R.R. and Vierling, E. 1996. Synthesis of small heat-shock proteins is part of the developmental program of late seed maturation. *Pl. Physiol.*, **112**: 747–757.

Weide, R., Van Wordragen, M.F., Lankhorst, R.K., Verkerk, R., Hanhart, C., Liharska, T. and Koornneef, M. 1993. Integration of the classical and molecular linkage maps of tomato chromosome 6. *Genetics*, **135**(4): 1175–1186.

Weiss, J. and Egea Cortines, M. 2010. Transcriptomic analysis of cold response in tomato fruits identifies dehydrin as a marker of cold stress. *J. Appl. Genet.*, **50**(4): 311–319.

Wienand, U., and Saedler, H. 1987. Plant transposable elements: Unique structures for gene tagging and gene cloning. **In**: *Plant DNA Infectious Agents*, pp. 205–227, Springer, Vienna.

Wu, L., Zhang, Z., Zhang, H., Wang, X.C and Huang, R. 2008. Transcriptional modulation of ethylene response factor protein JERF3 in the oxidative stress response enhances tolerance of tobacco seedlings to salt, drought, and freezing. *Pl. Physiol.*, **148**: 1953–1963.

Xiao, B., Huang, Y., Tang, N. and Xiong, L. 2007. Over-expression of a *LEA* gene in rice improves drought resistance under the field conditions. *Theor. Appl. Genet.*, **115**: 35–46.

Xu, H., Sun, X., Wang, X., Shi, Q., Yang, X. and Yang, F. 2011. Involvement of a cucumber MAPK gene (CsNMAPK) in positive regulation of ROS scavengence and osmotic adjustment under salt stress. *Scientia Hort.*, **127**: 488–493.

Yaghoobi, J., Kaloshian, I., Wen, Y. and Williamson, V.M. 1995. Mapping a new nematode resistance locus in *Lycopersicon peruvianum*. *Theor. Appl. Genet.*, **91**(3): 457–464.

Yaish, M.W., Sosa, D., Vences, F.J. and Vaquero, F. 2006. Genetic mapping of quantitative resistance to race 5 of *Pseudomonas syringae* pv. *phaseolicola* in common bean. *Euphytica*, **152**(3): 397–404.

Yan, J., He, C., Wang, J., Mao, Z., Holaday, S.A., Allen, R.D. and Zhang, H. 2004. Over-expression of the *Arabidopsis 14-3-3* protein *GF14* lambda in cotton leads to a "stay-green" phenotype and improves stress tolerance under moderate drought conditions. *Pl. Cell Physiol.*, **45**: 1007–1014.

Yoder, J.I., Palys, J., Alpert, K. and Lassner, M. 1988. Ac transposition in transgenic tomato plants. *Mol. Gen. Genet., MGG*, **213**(2–3): 291–296.

Yokotani, N., Ichikawa, T., Kondou, Y., Matsui, M., Hirochika, H., Iwabuchi, M. and Oda, K. 2007. Expression of rice heat stress transcription factor OshsfA2e enhances tolerance to environmental stresses in transgenic *Arabidopsis*. *Planta*, **227**(5): 957–967.

Young, N.D., and Tanksley, S.D. 1989. RFLP analysis of the size of chromosomal segments retained around the Tm-2 locus of tomato during backcross breeding. *Theor. Appl. Genet.*, **77**(3): 353–359.

Zhang, H.X. and Blumwald, E. 2001. Transgenic salt-tolerant tomato plants accumulate salt in foliage but not in fruit. *Nat. Biot.*, **19**: 765–768.

Zhang, J.Z., Creelman, R.A. and Zhu, J.K. 2004. From laboratory to field: Using information from Arabidopsis to engineer salt, cold, and drought tolerance in crops. *Pl. Physiol.*, **135**: 615–621. An Open Access Journal published by ICRISAT

Zhao, T., Jiang, J., Liu, G., He, S.H., Chen, X. and Xu, X. 2016. Mapping and candidate gene screening of tomato *Cladosporium fulvum*-resistant gene Cf-19, based on high-throughput sequencing technology. *BMC Pl. Biol.*, **16**(1): 1.

Zhao, Y., Lu, X., Liu, C., Guan, H., Zhang, M., Li, Z. and Lai, J. 2012.Identification and fine mapping of rhm1 locus for resistance to southern corn leaf blight in maize F. *J. Integ. Pl. Biol.*, **54**(5): 321–329.

Zhu, B.C., Su, J., Chan, M.C., Verma, D.P.S., Fan, Y.L. and Wu, R. 1998. Over-expression of a d-pyrroline-5-carboxylate synthetase gene and analysis of tolerance to water-stress and salt stress in transgenic rice. *Pl. Sci.*, **139**: 41–48.

9

Role and Status of Genome Sequencing and Post-genomic Research in Improving Productivity

S.K. Chakrabarti* and Jagesh K. Tiwari

ICAR-Central Potato Research Institute,
Shimla – 171 001, Himachal Pradesh
****E-mail: skc_cpri@yahoo.co.in***

The Indian horticulture sector has become a key driver for economic development in India and it contributes 30.40 per cent to agricultural Gross Domestic Product. Of the total horticulture production (283.47 million tonnes) in 2014-15, vegetables contribute (168.30 mt; 59.40 per cent) maximum followed by fruits (88.80 mt; 31.30 per cent), plantation crops (17.10 mt; 6 per cent), spices (5.90 mt; 2.10 per cent) and flowers and aromatics (2.30 mt; 1.10 per cent). Horticulture crops are the important sources of food and also provide nutritional security. In the 21st Century, with the burgeoning population and limited cultivable lands, availability of food would be a serious challenge. Although, much has been achieved in terms of production of these crops through conventional methods, intervention of biotechnology, especially genomics tools, is inevitable to enhance crop productivity for sustainable horticulture development. Information on the entire genetic blueprint (genome) of the particular crop plant is necessary for harnessing the benefits of modern biotechnology in improving that crop. Sequencing the *Arabidopsis* model plant genome in 2000 (AGI, 2000) was a major milestone not only for plant research but also for genome sequencing. The approach chosen for sequencing the *Arabidopsis* genome relied on overlapping bacterial artificial chromosomes (BAC) clones that represent a minimal tiling path to cover each chromosome arm. The BAC sequences were individually assembled and arranged according to the physical map, creating a very high quality genome sequence. The prohibitively high effort and time associated with this approach limited its applicability to a few genomes. Nevertheless, only two years later, the approximately three times larger genome of the first crop plant, rice (Yu *et al.*, 2002, Goff *et al.*, 2002), was elucidated using the same BAC based approach. Poplar was the next crop genome sequenced and employed the emerging whole genome shotgun (WGS) strategy (Tuskan *et al.*, 2006). Here, the genome is randomly broken down into smaller pieces which are then sequenced and subsequently assembled. Whilst this strategy significantly lowers preparation time and cost, the data are more difficult to assemble and typically results in a more fragmented genome sequence. The analysis of further crop genomes employed both strategies, balancing the various disadvantages and advantages. In 2007, the grapevine genome sequence based on the WGS strategy was published (Jaillon *et al.*, 2007). Although WGS reduced the time and effort required, genome sequence generation was still expensive and time-consuming due to the high cost of Sanger sequencing. The adaptation of Next Generation Sequencing (NGS) improved the output/cost ratio of genome sequencing dramatically. NGS technology covers a broad range of 'post Sanger' approaches, which sequence multiple DNA fragments in parallel to yield a much larger number of sequence reads, but generally of shorter length and lower quality. The first NGS technology, 454, was based on pyro sequencing and was initially suited mainly to bacterial-sized genomes due to low data yield and relatively short reads. Subsequent improvements in the technology enabled 454 sequencing to become applicable also to more complex genomes in conjunction with Sanger sequencing or alone. Another early NGS technology was the sequencing by synthesis based approach offered by Illumina. This approach generated much greater data yields than 454, albeit at the cost of even shorter reads. Since its introduction both read count and in particular read length improved dramatically. Illumina sequencing was adopted to sequence the cucumber (Huang *et al.*, 2009), where it was combined with conventional Sanger sequencing in a hybrid approach, showing that this technology is a feasible approach to plant genome sequencing. The application of NGS to plant genomes then became an increasingly strong trend. With the advent of NGS technologies, several horticultural crops have been sequenced namely, fruits: sweet orange, papaya, oil palm, strawberry, apple, banana, mango, date palm, Chinese plum, Japanese apricot, peach, pear and grape; vegetables: sugarbeet, Chinese cabbage, capsicum, watermelon, melon, cucumber, cassava, common bean, tomato and potato; flowers and others: capsella, rubber tree, lotus and cocoa; and a few are in pipeline of sequencing. Moreover, several protein coding genes have been identified in these crops for breeding applications. Now, in the 21st century the next step is to employ the myriad of genomics data in the real improvement of horticultural crops to enhance yield and meet food and nutritional security of the world.

To achieve this, integrated approach of characterization of a huge germplasm lines based on precise phenotyping for different agro-physio-biochemical traits, genotyping by NGS-based tools like genome wide selection (GWS) and/or genomic selection (GS) through high-throughput molecular markers such as SNPs and identification of marker-trait association for further genomics/markers-assisted breeding of the horticultural crops. Although, marker-assisted selection (MAS) has been demonstrated in many horticultural crops, breeding methods as to be decided based on the crop type. In doing so, in the recent years, for example, GS and/or GWS could be an ideal approach for annual and well as perennial crops. GWS allows genotyping of a group of genotypes through method, genome-wide markers 1000s of SNPs panel, unlike linked markers of trait of interest in MAS. While, GS is a very novel method that calculates genomic estimated breeding values based on molecular genetic information from the whole genome, and shows promise for breeding for complex traits. Taken together, NGS-based tools would definitely strengthen the horticultural crops breeding in development of varieties suitable to the climate change scenario to meet the food and nutritional security of the country for sustainable agriculture development.

Genome of Horticultural Crops

Here we update in brief about the genome sequencing of horticultural crops and use of new breeding tools to hasten the same as a post-genomics research. The genome sequence of horticultural crops provides a platform for genetic improvement.

I. Fruit Crops

Apple

Genome sequence of the domesticated apple (*Malus × domestica*) was reported in the year 2010 having the genome size of 742 mb and 57,386 protein-coding genes. Study has revealed that a relatively recent genome-wide duplication results in the transition from nine ancestral chromosomes to 17 chromosomes in the Pyreae. Gene families further explain the fruit development of the 'pome'- a Pyreae-specific false fruit that develops by proliferation of the basal part of the sepals, the receptacle. Where *MADS-box* genes are normally involved in flower and fruit development involving 15 gene families members of Rosaceae-specific sugar transport and assimilation metabolism (Velasco *et al.*, 2010).

Banana

Banana (*Musa* spp.) genome sequence was deciphered in 2012 of the 523 mb genome with 36,542 genes of a *Musa acuminata* doubled-haploid genotype. Bananas including dessert and cooking types, are giant perennial monocotyledonous herbs of the order Zingiberales. Banana is important for food security in many tropical and sub-tropical countries and the most popular fruit in industrialized countries. The *Musa* domestication process started some 7,000 years ago in Southeast Asia that involved hybridizations between diverse species and subspecies. This first monocotyledon high-continuity whole-genome sequence reported represents an essential bridge for comparative genome analysis in plants (D'Hont *et al.*, 2012).

Cacao

Cacao (*Theobroma cacao*) is an economically important tropical-fruit tree crop that is the source of chocolate. The genome sequencing was completed in 2011 with genome size of 430 Mb having 28,798 protein-coding genes. The assembly corresponds to 76 per cent of the estimated genome size and contains almost all previously described genes, with 82 per cent of these genes anchored on the ten chromosomes. Analysis of sequence data highlighted specific expansion of some gene families during evolution, for example, flavonoid-related genes and also provides a major source of candidate genes for improvement of this crop (Argout *et al.*, 2011).

Citrus

Cultivated citrus (*Citrus* spp.) are selections from, or hybrids of wild progenitor species and so citrus domestication remain controversial. Genome sequencing of citrus genomes–a high-quality reference haploid clementine genome and mandarin, pummelo, sweet-orange and sour-orange genomes–revealed that cultivated types derived from two progenitor species. Understanding citrus phylogeny through genome analysis distinct taxonomic relationships and thus facilitates sequence-directed genetic improvement of citrus species (Wu *et al.*, 2014). The most widely cultivated citrus, sweet orange (*Citrus sinensis*) is an important nutritional source for human health and have immense economic value. Study evidenced that sweet orange originated from a backcross hybrid between pummelo and mandarin. Sweet orange was sequenced in 2013 having 367 Mb genome size with 29,445 protein-coding genes. Focused analysis on genes involved in vitamin C metabolism showed that *GalUR*, encoding the rate-limiting enzyme of the galacturonate pathway, is significantly upregulated in orange fruit, and the recent expansion of this gene family may provide a genomic basis. This draft genome represents a valuable resource for understanding and improving many important citrus traits in the future (Xu *et al.*, 2013).

Date Palm

Date palm is one of the most economically important woody crops and is a good source for improving agricultural yields in arid environments. Genome sequencing of date palm was completed in 2011 having genome size of 658 Mb with 28,890 protein-coding genes. Nevertheless, long generation times (5–8 years) and dioecious (separate male and female trees) nature have complicated its cultivation and genetic analysis. Genome sequencing of Khalas variety (female date palm) and eight other varieties (including females of the Deglet Noor and Medjool) enables identification of the genome linked to gender (male and female plants). Further, study found evidence that date palm employs an XY system of gender inheritance (Al-Dous *et al.*, 2011).

Grapevine

Genome sequencing of grapevine (*Vitis vinifera*) of a highly homozygous genotype was completed in 2007 having genome size of 505 Mb with 29,585 genes. The draft sequence of the grapevine genome is the fourth one produced so far for flowering plants, the second for a woody species and the first for a fruit crop (cultivated for both fruit and beverage). Grapevine was selected because of its important place in the history of human evolution beginning during the Neolithic period. The detailed genome analysis provides insight into polyploidization events from a truly diploids that may have importance in plant evolution. Several large expansions of gene families with roles in aromatic features are also observed (Velasco *et al.*, 2007).

Mango

Mango (*Mangifera indica* L.) is the most popular fruit in India and called 'King of fruits' belong to the family Anacardiaceae and is an allotetraploid. The recent whole genome sequencing by Indian scientists revealed genome size of 450 Mb with 63,130 genes (Singh *et al.*, 2014).

Oil Palm

Palm (*Elaeis guineensis*) is the predominant source of oil production worldwide. The genome sequencing of oil palm revealed 1800 Mb genome size with 34,800 genes. Transcriptomes data from 30 tissue types were used to predict the genes, including oil biosynthesis genes and homologues of WRINKLED1 (WRI1), and other transcriptional regulators, which are highly expressed in the kernel. The oil palm sequence enables the discovery of genes for important traits as well as somaclonal epigenetic alterations that restrict the use of clones in commercial plantings (Singh *et al.*, 2013).

Papaya

Papaya (*Carica papaya*) is an important fruit crop cultivated in tropical and sub-tropical regions. It is known for its nutritional (vitamin A) benefits and medicinal applications. The genome sequencing of papaya in 2008 revealed genome size of 372 Mb with 28,629 genes. This was the first genome sequence the first commercial virus-resistant transgenic fruit tree (cv. 'SunUp') to be sequenced. The papaya genome is three times the size of the *Arabidopsis* genome, but contains fewer genes, including significantly fewer disease resistance gene analogues. Papaya offers numerous advantages as a system for fruit-tree functional genomics, and this genome sequence provides the foundation for revealing the basis of *Carica*'s distinguishing morpho-physiological, medicinal and nutritional properties (Ming *et al.*, 2008).

Peach

Rosaceae is the most important fruit-producing clade, and its key commercially relevant genera (*Fragaria, Rosa, Rubus* and *Prunus*) show broadly diverse growth habits, fruit types and compact diploid genomes. Peach (*Prunus persica*), is one of the best genetically characterized deciduous trees. The whole genome sequencing of a completely homozygous genotype in 2013 revealed 265 Mb genome size with 27,852 genes. The path of peach domestication through whole-genome resequencing of 14 *Prunus* accessions was investigated. The analyses suggest major genetic bottlenecks that have substantially shaped peach genome diversity (Verde *et al.*, 2013).

Pear

The genome sequencing of pear (*Pyrus bretschneideri*) in 2013 revealed total genome size of 527 Mb with 42,812 genes. Evolutionary and phylogenetic data analysis with other temperate fruits shows that pear and apple diverged from each other ~5.4–21.5 million years ago. Further, genome size differences between the apple and pear genomes are confirmed mainly due to the presence of repetitive sequences predominantly contributed by transposable elements, while genic regions are similar in both species. Genes critical for self-incompatibility, lignified stone cells (a unique feature of pear fruit), sorbitol metabolism, and volatile compounds of fruit were also identified. Moreover, alpha-linolenic acid metabolism is a key pathway for aroma in pear fruit (Wu *et al.*, 2013; Chagné *et al.*, 2014).

Plum

Chinese plum or Japanese apricot (*Prunus mume*) was domesticated in China more than 3,000 years ago as ornamental plant and fruit, is one of the first genomes among *Prunus* subfamilies of Rosaceae been sequenced in 2012. The total genome size is 280 Mb with 31,390 genes. *P. mume* genome data enables reconstruction of nine ancestral chromosomes of Rosaceae family, as well as depicting chromosome fusion, fission and duplication history in three major subfamilies. Transcriptome analysis revealed regulation of early blooming in endodormancy, immune response against bacterial infection and biosynthesis of flower scent. The *P. mume* genome sequence adds to our understanding of Rosaceae evolution and provides important data for improvement of fruit trees (Zhang *et al.*, 2012).

Strawberry

Strawberry (*Fragaria vesca*) is a versatile experimental plant system. The whole genome sequencing of strawberry in 2011 revealed a total genome size of 240 Mb with 34,809 genes. This crop is amenable to genetic transformation and shares substantial sequence identity with the cultivated strawberry (*Fragaria* × *ananassa*) and other economically important rosaceous plants. This diploid strawberry sequence lacks the large genome duplications seen in other rosids. Genes critical to valuable horticultural traits including flavor, nutritional value and flowering time wereidentified (Shulaev *et al.*, 2011).

II. Vegetables

Sugar Beet

Sugar beet (*Beta vulgaris* ssp. *vulgaris*) is an important temperate crop and sources of sugar, bioethanol and animal feed. The genome sequence of sugar beet was deciphered in 2014 having 714-758 Mb genome size and 27,421 genes. The assembly covers a large proportion of the repetitive sequence content that was estimated to be 63 per cent. Intraspecific genomic variation was analysed based on the genome sequences of sea beet (*Beta vulgaris* ssp. *maritima*; progenitor of all beet crops) and four additional sugar beet accessions. Further, identification of seven million variant positions in the reference genome, and also large regions of low variability, indicating artificial selection. The sugar beet genome sequence enables the identification of genes affecting agronomically relevant traits, supports molecular breeding and maximizes the plant's potential in energy biotechnology (Dohm *et al.*, 2014).

Common Bean

Common bean (*Phaseolus vulgaris*) is the most important legume for human consumption and sustainable agriculture.The genome sequencing of common bean was deciphered in 2014 as 520 Mb genome size and 31,638 genes distributed on 11 chromosomes. A set of genes were identified linked with increased leaf and seed size and combined these results with quantitative trait locus data from Mesoamerican cultivars. Further, resequencing of several hundred of landraces from the Mesoamerican and Andean gene pools showed two independent domestications before human colonization. Genes affected by domestication may be useful for genomics-enabled crop improvement (Schmutz *et al.*, 2014).

Capsicum

Hot pepper (*Capsicum annuum*) is the most widely grown spice crop in the world. The whole genome sequencing and assembly of the hot pepper (Mexican landrace of *Capsicum annuum* cv. CM334 and cv. Zunla-1) was completed in 2014 having 3480 Mb genome size with 34,903 and 35,336 protein-coding genes, respectively. Integrated genomic and transcriptomic analyses suggested that change in gene expression for capsaicin biosynthesis and also differential molecular patterns of ripening regulators and ethylene synthesis in hot pepper and tomato. The reference genome will serve as a platform for improving the nutritional and medicinal values of *Capsicum* sp. and its improvement (Kim *et al.*, 2014).

Carrot

Carrot (*Daucus carota*) is an important vegetable crop and good source of vitamin A. The genome sequencing of carrot was completed very recently in 2016 having 473 Mb genome size with 30,824 protein-coding genes. A high-quality chromosome-scale assembly and a comparative evolutionary analysis of the carrot genome have been reported among members of the euasterid II clade. Further, identification of a candidate gene (DCAR_032551) for carotenoid accumulation (*Y*) in carrot taproot and is coexpressed with several isoprenoid biosynthetic genes pathways. It has been hypothesized that DCAR_032551 regulates upstream photosystem development and functional processes, including photomorphogenesis and root de-etiolation (Iorizzo *et al.*, 2016).

Cassava

Cassava (*Manihot esculenta*) is a major tropical food crop of the world that has high carbohydrate production potential and adaptability to diverse environments. The genome sequencing of cassava was completed in 2012 having 760 Mb genome size with 30,666 protein-coding genes. Analyses of a wild ancestor and a domesticated variety of cassava identified 1,584 and 1,678 gene models specific to the wild and domesticated varieties, respectively. Moreover, analyses revealed involvement of genes in photosynthesis, starch accumulation and abiotic stresses are positively selected, whereas those involved in cell wall biosynthesis and secondary metabolism, including cyanogenic glucoside formation, are negatively selected in the cultivated varieties, indicting the result of natural selection and domestication. The genome sequencing results can contribute to genetic improvement of cassava through better understanding of its biology (Wang *et al.*, 2014).

Cucumber

Cucumber (*Cucumis sativus*) is an economically important salad crop. The genome sequencing of cucumber was completed in 2009 having 367 Mb genome size with 26,682 protein-coding genes. The sequenced cucumber genome affords insight into traits such as its sex determination, disease resistance, biosynthesis of cucurbitacin and 'fresh green' odor. The cucumber genome provides a valuable resource for developing elite cultivars and for studying the evolution and function of the plant vascular system. Researchers have identified 686 gene clusters related to phloem function. Moreover, study established that five of the seven chromosomes arose from fusions of ten ancestral chromosomes after divergence from *Cucumis melo* (Huang *et al.*, 2009).

Eggplant

Eggplant (*Solanum melongena* L.) or brinjal is an indigenous to India. Unlike other important Solanaceae crops such as tomato, potato, chili pepper, and tobacco, all of which originated in South America and are cultivated worldwide and in this respect it is phylogenetically unique. The genome sequencing of brinjal revealed genome size of 833 Mb with 85,446 genes. The detailed comparative analysis of the brinjal and tomato genomes will facilitate our understanding of the genomic architecture of solanaceous crops, which will contribute to cultivation and further utilization of these crops (Hirakawa *et al.*, 2014).

Melon

Melon (*Cucumis melo*) is an important horticultural crop worldwide. The genome sequencing of melon was deciphered in 2012 having 450 Mb genome size with 27,427 genes. The absence of recent whole-genome duplications in the melon lineage was observed since the ancient eudicot triplication, and the sequence data suggest that transposon amplification may in part explain the increased size of the melon genome compared with the close relative cucumber. Further, a low number of nucleotide-binding site–leucine rich repeat disease resistance genes were annotated, suggesting the existence of specific defense mechanisms in this species. The use of the genome sequence in future investigations will facilitate the understanding of evolution of cucurbits and the improvement of breeding strategies (Garcia-Mas *et al.*, 2012).

Radish

Genome sequencing of radish (*Raphanus raphanistrum*) revealed genome size of 254 Mb with 21,226 genes. It is reported that radish experienced a whole-genome triplication event prior to diverging from *Brassica rapa*. Genome sequence data analysis showed with most of the gene losses occurring prior to their divergence. Overall, study provides new insights into the processes of plant duplicate loss, retention, and functional divergence and highlights the need for further understanding factors controlling duplicate gene fate (Moghe *et al.*, 2014).

Potato

Potato (*Solanum tuberosum* L.) is the third world's most important food crop after rice and wheat. It is clonally propagated, highly heterozygous, autotetraploid, and suffers acute inbreeding depression. The whole genome sequencing of potato (a homozygous doubled-monoploid potato clone) revealed 844 Mb genome and predicted 39,031 protein-coding genes. Further, sequencing of a heterozygous diploid clone shows that genic variants that are a likely cause of inbreeding depression. Gene family expansion, tissue-specific expression and recruitment of genes to new pathways contributed to the evolution of tuber development (Xu *et al.*, 2011; Aversano *et al.*, 2015).

Tomato

Tomato (*Solanum lycopersicum*) is a major crop plant and a model system for fruit development. *Solanum* is one of the largest angiosperm genera and includes annual and perennial plants from diverse habitats. A high-quality genome sequence of domesticated tomato revealed 900 Mb genome size with 34,727 genes. A comparative analysis among the domesticated tomato, its closest wild relative *Solanum pimpinellifolium* (Current tomato), and potato genomes showed only 0.6 per cent nucleotide divergence between tomatoes and signs of recent admixture, but >8 per cent divergence from potato. The *Solanum* lineage has experienced two consecutive genome triplications: one that is ancient and shared with rosids, and a more recent one. These triplications set the stage for the neofunctionalization of genes controlling fruit characteristics, such as colour and fleshiness (Sato *et al.*, 2012). *Solanum pennellii* is a wild tomato species endemic to Andean regions in South America and important for its extreme stress tolerance. Genome sequencing and mapping study identified candidate genes for stress tolerance and provide evidence that transposable elements had a role in the evolution of these traits. This work paves a path toward further tomato improvement and for deciphering the mechanisms underlying the myriad other agronomic traits that can be improved with *S. pennellii* germplasm (Bolger *et al.*, 2014).

Watermelon

Watermelon (*Citrullus lanatus*) is an important cucurbit crop grown throughout the world. Genome sequence of watermelon revealed 425 Mb genome size with 23,400 genes. Comparative genomics analysis provided an evolutionary scenario for the origin of the watermelon chromosomes and the extent of genetic diversity and population structure of watermelon germplasm. In addition to loss of many disease-resistance genes during domestication, integrative genomic and transcriptomic analyses yielded important insights into aspects of phloem-based vascular signaling in common between watermelon and cucumber. Study identified genes crucial to valuable fruit-quality traits, including sugar accumulation and citrulline metabolism (Guo *et al.*, 2013).

III. Flowers

Capsella

Capsella (*Capsella rubella*) is an important flower crop which shows an example of shift from outcrossing (progenitor: *Capsella grandiflora*) to selfing (*C. rubella*) in flowering plants. The genome sequencing of *C. rubella* deciphered in 2013 having 219 Mb genome size and 26,521 protein coding genes.This transition was evidenced by genome sequencing of a *C. rubella* and comparison of RNA expression and polymorphism patterns with its outcrossing *C. grandiflora*. A similar shift in the expression of genes associated with flowering behaviours, similar to that seen in *Arabidopsis*, where self-fertilization evolved about 1 million years ago (Slotte *et al.*, 2013).

Sacred Lotus

Sacred lotus (*Nelumbo nucifera*) is a basal eudicot with agricultural, medicinal, cultural and religious importance. It is cultivated for its rhizomes, seeds as a food crop and notably known for its 1,300-year seed longevity. The genome sequencing of lotus revealed genome size of 929 Mbwith 4,223 genes. The sacred lotus has 16 COG2132 multi-copper oxidase family proteins with root-specific expression; these are involved in root meristem phosphate starvation, reflecting adaptation to limited nutrient availability in an aquatic environment (Ming *et al.*, 2013).

Way Forward

Marker-Assisted Selection (MAS)

Once the molecular markers linked to QTLs or candidate genes are identified, their deployment in breeding program is required for developing varieties. With the advancement in genome sequencing technologies, identifying accurate genetic polymorphisms have accelerated the discovery of molecular markers. As a result, there has been much development on the use of computational methods for the development of SNPs markers from large collection of DNA sequences. Over the decades, marker-assisted selection has been used by potato breeders for traits of interest, such as late blight resistance/allele mining (Tiwari *et al.*, 2013a, 2013b) and virus resistance (Tiwari *et al.*, 2012). Moreover, markers for complex potato traits such as tuber quality, chip colour or starch content have been developed but with limited success of their deployment in breeding. As the costs of genotyping/sequencing decrease, new methods for screening potatoes are being developed. Thousands of SNPs, SSRs, AFLPs, and DArT markers were mapped to the published potato genome, which would likely to develop markers for MAS. When the QTLs/genes would be identified for trait with higher phenotypic variation could be validated in desired germplasm.

Genome-Wide Selection (GWS)

Currently, with the advancement of genome sequencing, the Next-Generation Sequencing-based breeding tools like genome wide selection (GWS) and/or genomic selection (GS) are emerging methods to hasten crop breeding processes. First, in Genome-wide selection (GWS) allows genotyping of a group of genotypes/cultivars/lines through method, genome-wide markers like 1000s of SNPs panel, unlike selected markers of trait of interest in MAS. A novel method of genome-wide association mapping has been used in potato population using 8.3 K SolCAP SNPs panel. Several groups have recently started exploring the GWS approach in many crop species. The success of the GWS approach is dependent on the availability of a diverse and representative training population and their precise phenotyping. This approach has been used recently to improve durable stem rust resistance in wheat and eventually could be systematically explored to bring different components of multigenic drought tolerance using the GWS approach.

Genomic-Selection (GS)

Genomic selection is a very novel method that calculates genomic estimated breeding value-based on molecular genetic information from the whole genome, and shows promise for breeding for complex traits, but it has yet to be applied in many horticulture crops including potato. GS is a novel approach to develop elite lines with overall excellent performance in a target environment and it reduces the frequency of phenotyping and similarly also increases annual gains from selection by reducing cycle time. Taken together, next generation breeding methods based on the sequencing technologies promise to revolutionize screening and further breeding for desirable traits of crops.

Genotyping-by-Sequencing (GBS)

Another recent approach of developing SNPs by genotyping-by-sequencing (GBS) that reduces complexity using selective hybridization prior to next-generation sequencing has been followed in crops to enhance breeding. Advances in next generation technologies have driven the costs of DNA sequencing down to the point that genotyping-by sequencing (GBS) is now feasible for high diversity, large genome species. As this is based on the constructing GBS libraries based on reducing genome complexity with restriction enzymes (REs). This approach is simple, quick, extremely specific, highly reproducible, and may reach important regions of the genome that are inaccessible to sequence capture approaches. Future application of GBS to breeding, conservation, and global species and population surveys may allow plant breeders to conduct genomic selection on a novel germplasm or species without first having to develop any prior molecular tools, or conservation biologists to determine population structure without prior knowledge of the genome or diversity in the species.

Genome Editing, a new breeding tool

With the advent of new techniques for manipulating genes and genomes that are applicable now to plants also. The use of gene editing tools such as zinc finger nucleases, mega-nucleases, hybrid DNA/RNA oligo-nucleotides, TAL effect or nucleases and more recent one is CRISPR/Cas9. These tools have ability to precisely target one specific DNA sequence molecule within a genome to modify a double-stranded DNA. DNA repair to such breaks sometimes leads to gene knockouts or gene replacement by homologous recombination if exogenously supplied homologous DNA fragments are made available. Genome rearrangements are also possible to engineer. Creation and use of such genome rearrangements, gene knockouts and gene replacements by the plant science community is gaining significant momentum. To document some of this progress and to explore the technology's longer term potential, the future uses of designer nucleases has potential to greatly expedite research with model plant systems and to engineer genes and genomes in major and minor crop species for enhanced food production.

Conclusions

The horticulture germplasm collection provides excellent genetic resources for the utilization of wild relatives in addressing global food security in the era of climate change and the emergence of new races of pests and pathogens. Crop genetic resources including wild relatives have largely untapped genetic variation for abiotic and biotic stress tolerances, and could greatly expand the available domesticated gene pools in the predicted extremes of climate change. Crop genebanks collection is critical in order to collect, conserve, characterize, and utilize the germplasm in a systematic and integrated way to meet future food needs. Current genomic strategies can assist in the introgression of these valuable traits into the domesticated crop gene pools, where they can be better evaluated for crop improvement. Knowledge and tools that can be developed through the use of next-generation sequencing technologies, SNP discovery, transcriptomics, other–Omics approaches and genetic modification will pave the way forward for exploitation of an amazing germplasm diversity for greater use of genomic research in crop improvement. Nevertheless, genomics-assisted breeding will play key role, since it is not only cost-effective and efficient but also amenable to high throughput automation. Indeed, to break yield barriers in many horticultural crops will essentially need novel breeding efforts like novel discovery of SNPs/allele/genes from germplasm resources and their deployment broad marker/genomics-assisted selection, enrichment of the narrow genetic base of crop varieties, stability QTL analysis and genetic modifications that regulate the phenotypic variation for yield associated traitswill certainly augment crop improvement programmes.

References

AGI Initiative, 2000. Analysis of the genome sequence of theflowering plant Arabidopsis thaliana. *Nature*, **408**: 796–815.

Al-Dous, E.K. *et al.*, 2011. *De novo* genome sequencing and comparative genomics of date palm (*Phoenix dactylifera*). *Nature Biotech.*, **29**(6): 521–527.

Argout, X. *et al.*, 2011. The genome of *Theobroma cacao. Nature Genet.*, **43**(2): 101–107.

Aversano, R. *et al.*, 2015. The *Solanum commersonii* genome sequence provides insights into adaptation to stress conditions and genome evolution of wild potato relatives. *The Plant Cell* **27**: 954–968.

Bolger, A. *et al.*, 2014. The genome of the stress-tolerant wild tomato species *Solanum pennellii. Nature Genet.*, **46**(9): 1034–1038.

Chagne´ D. *et al.*, 2014. The draft genome sequence of European pear (*Pyrus communis* L. 'Bartlett'). *PLoS One*, **9**(4): e92644

D'Hont, A. *et al.*, 2012. The banana (*Musa acuminata*) genome and the evolution of monocotyledonous plants. *Nature*, **488**: 213–217.

Dohm, J.C. *et al.*, 2014. The genome of the recently domesticated crop plant sugar beet (*Beta vulgaris*). *Nature*, **505**(23): 546–549.

Garcia-Mas, J. *et al.*, 2012. The genome of melon (*Cucumis melo* L.). *PNAS*, **109**(29): 11872–11877.

Goff, S.A., Ricke, D., Lan, T.H., Presting, G., Wang, R., Dunn, M., Glazebrook, J., Sessions, A., Oeller, P. and Varma, H. *et al.*, 2002. A draft sequence of the rice genome (*Oryza sativa* L. ssp. *japonica*). *Sci.*, **296**: 92–100.

Guo, S. *et al.*, 2013. The draft genome of watermelon (*Citrullus lanatus*) and re-sequencing of 20 diverse accessions. *Nature Genet.*, **45**(1): 51–58.

Hirakawa, H. *et al.*, 2014. Draft genome sequence of eggplant (*Solanum melongena* L.): The representative *Solanum* species indigenous to the old world. *DNA Res.*, p. 1–12.

Huang, S., Li, R., Zhang, Z., Li, L., Gu, X., Fan, W., Lucas, W.J., Wang, X., Xie, B., Ni, P. *et al.*, 2009. The genome of the cucumber, *Cucumis sativus* L. *Nat. Genet.*, **41**: 1275–1281.

Iorizzo, M. *et al.*, 2016. A high-quality carrot genome assembly provides new insights into carotenoid accumulation and asterid genome evolution. *Nature Genet.*, **48**(6): 657–666.

Jaillon, O., Aury, J.M., Noel, B., Policriti, A., Clepet, C., Casagrande, A., Choisne, N., Aubourg, S., Vitulo, N. and Jubin, C. *et al.*, 2007. The grapevine genome sequence suggests ancestral hexaploidization in major angiosperm phyla. *Nature*, **449**: 463–467.

Kim, S. *et al.*, 2014. Genome sequence of the hot pepper provides insights into the evolution of pungency in *Capsicum* species. *Nature Genet.*, **46**(3): 270–278.

Ming *et al.*, 2013. Genome of the long-living sacred lotus (*Nelumbo nucifera* Gaertn.). *Genome Biology*, **14**: R41.

Ming, R. *et al.*, 2008. The draft genome of the transgenic tropical fruit tree papaya (*Carica papaya* Linnaeus). *Nature*, **452**(7190): 991–996.

Moghe, G.D. *et al.*, 2014. Consequences of whole-genome triplication as revealed by comparative genomic analyses of the wild radish *Raphanusra phanistrum* and three other Brassicaceae species. *The Plant Cell*, **26**: 1925–1937.

Sato, S. *et al.*, 2012. The tomato genome sequence provides insights into fleshy fruit evolution. *Nature* **485**: 635–641

Schmutz, J. *et al.*, 2014. A reference genome for common bean and genome-wide analysis of dual domestications. *Nature Genetics*, **46**(7): 707–713.

Shulaev, V. *et al.*, 2011. The genome of woodland strawberry (*Fragaria vesca*). *Nature Genetics*, **43**(2): 109–116.

Singh, N.K. *et al.*, 2014. A draft genome of the king of fruit, mango (*Mangifera indica* L.). *Plant and Animal Genome XXII Conference*, San Diego, CA, USA.

Singh, R. *et al.*, 2013. Oil palm genome sequence reveals divergence of interfertile species in old and new worlds. *Nature*, **500**: 335–339

Slotte, T. *et al.*, 2013. The *Capsella rubella* genome and the genomic consequences of rapid mating system evolution. *Nature Genetics*, **45**(7): 831–835.

Tiwari, J.K. *et al.*, 2012. Marker-assisted selection for virus resistance in potato: Options and challenges. *Potato J.*, **39**: 101–117.

Tiwari, J.K. *et al.*, 2013a. Molecular markers for late blight resistance breeding of potato: An update. *Plant Breed.*, **132**: 237–245.

Tiwari, J.K. *et al.*, 2013b. Identification of host gene conferring resistance to Potato virus Y using *Ry* gene-based molecular markers. *Indian J. Hort.*, **70**: 373–377.

Tuskan, G.A, Difazio, S., Jansson, S., Bohlmann, J., Grigoriev, I., Hellsten, U., Putnam, N., Ralph, S., Rombauts, S., Salamov, A. *et al.*, 2006. The genome of black cottonwood, *Populus trichocarpa* (Torr. and Gray). *Science*, **313**: 1596–1604.

Velasco, R. *et al.*, 2007. A high quality draft consensus sequence of the genome of a heterozygous grapevine variety. *PLoS One*, **2**(12): e1326.

Velasco, R. *et al.*, 2010. The genome of the domesticated apple (*Malus* × *domestica* Borkh.). *Nature Genetics*, **42**(10): 833–840

Verde, I. *et al.*, 2013. The high-quality draft genome of peach (*Prunus persica*) identifies unique patterns of genetic diversity, domestication and genome evolution. *Nature Genetics*, **45**(5): 487–494.

Wang, W. *et al.*, 2014. Cassava genome from a wild ancestor to cultivated varieties. *Nature Comm.*, **5**: 5110.

Wu, G.A. *et al.*, 2014. Sequencing of diverse mandarin, pummelo and orange genomes reveals complex history of admixture during citrus domestication. *Nature Biotech.*, **32**(7): 656–662.

Wu, J. *et al.*, 2013. The genome of the pear (*Pyrus bretschneideri* Rehd.). *Genome Res.*, **23**: 396–408.

Xu, Q. *et al.*, 2013. The draft genome of sweet orange (*Citrus sinensis*). *Nature Genetics*, **45**(1): 59–66.

Xu, X. *et al.*, 2011. Genome sequence and analysis of the tuber crop potato. *Nature*, **475**: 189–195.

Yu, J., Hu, S., Wang, J., Wong, G.K.S., Li, S., Liu, B., Deng, Y., Dai, L., Zhou, Y., Zhang, X. *et al.*, 2002. A draft sequence of the rice genome (*Oryza sativa* L. ssp. *indica*). *Science*, **296**: 79–92.

Zhang, Q. *et al.*, 2012. The genome of *Prunus mume*. *Nature Comm.*, **3**: 1318.

10

Vegetable Breeding through New Techniques for Higher Productivity

Leela Alamalakala, Shalu Choudhary, Anjanabha Bhattacharya, Vilas Parkhi, Venugopal Mikkilineni, Bharat Char and Usha Borwale Zehr*

Maharashtra Hybrid Seeds Company Private Limited,
Dawalwadi, Jalna-Aurangabad Road, Jalna – 431 203, Maharashtra
**E-mail: usha.zehr@mahyco.com*

Historically, conventional plant breeding based on natural genetic variation, selection and hybridization has had an enormous impact on agricultural productivity. With the advent of biotechnological tools, plant breeders have adopted the latest scientific and technological innovations available to them to enhance crop productivity. The application of biotechnological tools that include genomics, mutagenesis, tissue culture, genetic transformation, embryo rescue and doubled haploids has resulted in germplasm diversification and genetic improvement. The collective impact of this is observed in improvement of vegetable production, sensory properties and shelf life of vegetables. The introduction of genetic engineering technology in the early 1990s has changed the face of commercial agriculture. Herein, plant breeding transformed from conventional intra-species breeding to transgenic approaches for developing plant traits (Sprink *et al.*, 2016). Genes from another species could be introduced into plant species through recombinant DNA technology and new genetic combinations were made possible that otherwise were not possible through conventional breeding. In many crop plants, limitations with transformation efficiency, and regulatory or acceptance hurdles have slowed the adoption of genetic engineering technology although solutions may exist. More recently, researchers have found ways to precisely edit genes within species using genome editing technologies. Genome editing has the potential to overcome these limitations and has emerged as a complement to conventional breeding, DNA marker technology and genetic engineering approaches for developing new traits and improved cultivars.

In the last decade, in addition to genome editing, a number of new breeding technologies (NBTs) that include cisgenesis, intragenesis, grafting on GM rootstock, and RNAi have provided multiple options for innovations in agriculture (Tait and Barker, 2011; Lusser *et al.*, 2011). Targeted genome engineering technology is ushering in a new revolution in the agriculture landscape by increasing our ability to use the complete genetic repertoire of the targeted crop species for improving yields, nutritional value and other traits of commercial value (Mahfouz *et al.*, 2016). With shrinking cultivable land coupled with limited available water resources and increasing human population and consumption patterns, vegetable crops can ensure food and nutrition security because of their high productivity per unit area in the shortest possible time. Therefore, vegetable crops are good candidates for applying genome editing technologies. Genome editing technologies utilize engineered endonucleases (EENs) that cleave DNA in a sequence-specific manner because of the presence of a sequence-specific DNA-binding domain (DBDs) or RNA sequence. Through the recognition of specific target DNA sequence, these nucleases can efficiently and precisely introduce double-strand breaks (DSBs), stimulating the cell's endogenous DNA repair mechanisms. DSBs are repaired either by homology-directed repair (HDR) or the error-prone non-homologous end joining breaks (NHEJ) leading to gene modification at the target sites (Sung *et al.*, 2013). The error-prone NHEJ repair process introduces errors such as the insertion or deletion of nucleotides due to imperfect ligation at the DSB break points. The HDR process uses a template for repairing DSBs and is therefore more precise than the NHEJ. Gene editing technologies can thus be used for understanding gene function and for developing commercial pipeline of products. The complete loss-of function of genes is a prerequisite for reverse genetics which is a popular approach for gene functional studies in plants (Alonso and Ecker, 2006). The discovery of Zinc-finger nucleases (ZFNs), transcription activator-like effector nucleases (TALENs) and CRISPR/Cas9 systems have facilitated gene functional studies in a variety of organisms, including plants (Gaj *et al.*, 2013; Doudna and Charpentier, 2014). In this paper, we will explore the opportunities and challenges presented by gene editing in vegetable breeding, and the advantages this technology offers to achieve breeding goals.

Zinc Finger Nucleases (ZFNs) are synthetic proteins consisting of two protein domains, the engineered zinc finger (ZF) DNA-binding domain and an endonuclease domain (most frequently FokI). The α-helices in the ZF DNA-binding domain insert into the major groove of the double helix, by recognizing three contiguous nucleotide bases within the DNA substrate (Pavletich and Pabo, 1991). In nature, there are >1000 known zinc finger domains. Typical ZFNs consist of three to six ZF domains each recognizing a triplet within the DNA substrate. All ZFN work as pairs, since FokI only introduces DSBs as a dimer. The targeted DNA sequence varies between 18 and 36 bp in length. Gene corrections with ZFNs have been demonstrated in plants such as *Arabidopsis* (Osakabe *et al.*, 2010), tobacco (Townsend *et al.*, 2009), and maize (Shukla *et al.*, 2009). However, the time-consuming screening process involved in obtaining functional ZFNs and other practical caveats, such as off-target effects or host-cell toxicity, limit the application of ZFNs in plant genome editing (Cathomen and Joung, 2008; Hsu and Zhang, 2012).

Transcription Activator-Like Effector Nucleases (TALEN), have emerged as an alternative to ZFNs for genome editing. TALE nucleases are a combination of an endonuclease (most frequently FokI) and a subset of DNA-binding protein domains derived from the plant pathogenic bacteria *Xanthomonas* (Boch and Bonas, 2010). The DNA recognition property of the TALE nucleases is mediated by tandem repeats of 33-35 amino acids each binding to a certain single base pair. However, the specificity is defined by two hyper-variable amino acids known as repeat-variable di-residues (RVDs) located at the 12th and 13th position in each repeat (Mak *et al.*, 2012; Deng *et al.*, 2012). TALENS are most commonly used as pairs to introduce double-strand breaks (DSBs) in the DNA, as FokI only introduces DSBs as a dimer. TALEN-mediated gene modifications have been successfully carried out in rice (Li *et al.*, 2012), wheat (Wang *et al.*, 2014), potato (Sawai *et al.*, 2014) and tomato (Lor *et al.*, 2014).

The CRISPR (Clustered regularly interspaced short palindromic repeats)/Cas (CRISPR associated) system is the most recent addition of the gene editing technologies and is emerging as an leading tool for crop improvement due to its simplicity, versatility and efficiency. The basic principle of this system was derived from a type II prokaryotic organism adaptive immune system and involves two components, the CRISPR-associated Cas9 endonuclease and a single RNA chimera (consisting of a short CRISPR RNA (crRNA) and a transactivating crRNA (tracrRNA)) termed as the guide RNA (gRNA). The DNA recognition and specificity are achieved by a designable guide RNA and the presence of the NGG (or NAG) protospacer adjacent motif (PAM) downstream of the target sequence (Jinek *et al.*, 2012). The DNA targeted by CRISPR/CAS9 nuclease is typically 16-20 bp long. The Cas9-single guide RNA (sgRNA) system has been successfully used for genome editing in both dicot and monocot crop species (Jiang *et al.*, 2013). However, a major concern when using an RNA-guided Cas9 is the introduction of unwanted off-target mutations (Pattanayak *et al.*, 2013). The potential off-target effects could be minimized through approaches such as replacing the *Streptococcus pyogenes* Cas9 protein with *Neisseria meningitides* Cas9 protein. *N. meningitides* Cas9 recognizes a longer target sequence than the *S. pyogenes* Cas9 thereby improving the target specificity of the CRISPR/Cas9 complex (Hou *et al.*, 2013). Other strategies such as the utilization of Cas9 nickase (Cho *et al.*, 2014) and manipulation of the length of the recognition sequence in guide RNA were also evaluated to reduce the off-target effects (Fu *et al.*, 2014).

Potential Applications of Genome Editing Technologies in Vegetable Breeding

The impact of biotechnology in vegetable breeding has demonstrated the continuous need to adopt technological interventions that can accelerate crop improvement programmes. With the advent of new technologies such as the gene-editing, the possibility of achieving precise gene mutations is achievable, thereby providing an alternative way to add value to breeding and crop improvement. Through the development of high efficiency and designer nucleases, creating heritable genome changes including transgene insertion, deletion or replacement in the targeted genes is a reality. Proof-of-concept experiments have demonstrated the disruption in the *ARGONAUTE7* gene, which plays a mediating role in RNA silencing (Brooks *et al.*, 2014); *PROCERA* gene involved in auxin signaling pathway, controls flower morphology, cell division and expansion in tomato (Lor *et al.*, 2014); *StIAA2* an auxin coding gene in potato (Wang *et al.*, 2015) and *BolC.GA4.a* gene in *B. oleracea* resulting in a dwarf phenotype (Lawrenson *et al.*, 2015) when edited using this approach. This further demonstrates the potential of the editing system to generate multiple targeted mutations simultaneously for different vegetable breeding traits. Here, we discuss several examples of how genome editing tools can contribute in bringing novel improved vegetable crops to the market.

Fruit Ripening in Vegetables

Quality in vegetable crops is one of the more important traits in addition to yield. Vegetable quality, appearance, shelf life, taste, and nutritional value drive consumer's preferences. Fruit ripening is of key importance in vegetables because excessive softening caused by ripening limits the shelf life of the fruit, thus making the fruits prone to pathogen infection. Tomato is often used as model crop for fruit ripening studies. Molecular analysis of fruit ripening process in tomato revealed that silencing of cell-wall related genes (Uluisik *et al.*, 2016), ethylene biosynthesis and ethylene signal transduction pathway genes (Yokotani *et al.*, 2009) can enhance shelf life. Ito *et al.*, 2015 generated mutations in the targeted tomato *RIN* gene, confirming the efficacy of CRISPR/Cas9 gene editing system and the possible role of *RIN* in regulating fruit ripening. Therefore, the key ripening associated pathway genes can be targeted for precise gene editing in various vegetable crops thus improving the shelf life of vegetable crops.

Acrylamide Free Non-browning Potatoes

Acrylamide in potato chips and French fries has raised global food safety concerns, and tremendous efforts have been undertaken to limit acrylamide levels in fried potato products (Vinci *et al.*, 2012). An effective approach is to decrease reducing sugars (glucose and fructose) and subsequently acrylamide in cold-stored potato tubers. However, due to the complexity of the autotetraploid genome and the high degree of heterozygosity, efforts on improving this trait have met with challenges. Technologies such as RNAi, TILLING have been deployed but desired allele combinations/trait could not be developed. Recently, through TALEN-mediated gene editing, new potato lines were generated with knock out mutations in all alleles of *VASCULAR INVERTASE* (*Vinv*) gene (Clasen *et al.*, 2016). Thus, without using foreign DNA, with high efficient genome editing system, Clasen group were able to mutate multiple alleles of *Vinv* gene in a single generation, thus illustrating that in the coming years these technologies are likely to be key driving forces for genetic engineering of plants for trait improvement. Further, other genes like Asparagine synthetase-1 gene (*Asn1*) and Polyphenol oxidase (PPO) could be targeted through these approaches for lowering the acrylamide level and browning in potatoes.

Enhancement of Capsaicin and Oleoresin Contents in Pepper

Chilli fruit is known for its pungency, rich flavour, aroma, and nutritional value that are beneficial to human health. Capsaicin (pungency) and oleoresin (colour) are the most unique metabolites of peppers, used widely in food products and in diverse pharmaceutical applications. Candidate genes *for* capsaicin biosynthesis pathway have been identified as *Pun1*, *CCR*, *KAS* and *HCT* (Han *et al.*, 2013). In addition, genetic variations have been identified at these genomic regions that are controlling capsacinoids accumulation (Reddy *et al.*, 2014). Therefore, gene editing technologies can help to engineer capsaicin and oleoresin pathway genes to enhance capsaicin and oleoresin.

Introgression of Disease Resistance in Vegetable Plants

Breeding for disease resistance is a major objective of vegetable breeding. Several plant genes conferring disease resistance have been identified and introgressed through backcross breeding approach. Genome editing in plants permits simultaneous targeting of several genomic loci (multiplexing) through designing multiple sgRNAs, thus making it feasible to develop plants that can confer resistance to multiple viruses. Edited plants exhibiting broad spectrum resistance to *Cucumber vein yellowing virus* and other potyviruses have been developed in cucumber by editing of the *eIF4E* and *eIF(iso)4E* genes (Chandrasekaran *et al.*, 2016) through CRISPR/Cas9 system.

Gene editing technology can be used to knock out host susceptibility genes that are serving as binding sites of pathogen TAL effectors (Kim *et al.*, 2015). Blanvillain-Baufumé *et al.* (2016) generated an allele library of OsSWEET14 gene promoter in rice through TALENs approach and obtained edited plants showing resistance to Xanthomonas oryzae bacterial strains. The broad spectrum resistance to powdery mildew was obtained in wheat when three homeo-alleles of the susceptibility gene *Mlo* were edited through this system (Wang *et al.*, 2014). Gene editing technologies, therefore offer possibilities for modifying disease susceptibility endogenous gene expression to impart disease resistance against TAL-effectors.

Often, the source germplasm is not available to the researcher to introgress the favourable allele or the hybridization between the species is not possible due to species incompatibility. In such a scenario, gene editing technologies enable us to edit the genes based on the information available in other species. Therefore, desired disease resistant allele can be created in any species without having source germplasm or by wide hybridization.

Doubled Haploid (DH) Inducer Line Development

Doubled haploids have been successfully developed in vegetable crops like onion, pepper, melon and crucifer vegetables that helped to accelerate breeding. However, the feasibility and efficiency of DH production is highly dependent on the crop and the genotype. Recent studies have shown that through modifications of the centomere specific H3-histones (CEN-H3) gene (Ravi and Chan 2010), haploid inducer lines can be generated for a given species that could potentially make DH production, genotype independent. The utilization of this approach has been demonstrated in *Arabidopsis* (Britt and Kuppu 2016) and maize (Kelliher *et al.*, 2016). This objective can be achieved in majority of the other crops as CEN-H3 is present in almost all plants (Britt and Kuppu, 2016). Gene editing technology thus provides an opportunity to create doubled haploid in crops or genotypes that are recalcitrant to tissue culture techniques.

Challenges

Although genome editing and other NBTs offer unlimited possibilities for crop genetic improvement, their viability and use in the commercial development of products depends on overcoming technical hurdles, addressing safety issues (food, feed and environmental safety), creating a positive perception among consumers towards the technology, intellectual property (IP) and the regulatory pathway for approval.

Technical Challenges

For gene editing to be successful, knowing the functional gene/allele information is important. As genome sequence information of many species is not yet known, we may not be able to ascertain the off-target effects caused due to gene editing. This is a limitation in many crop species. Phenotype information of the gene/allele should also be known and measurable. Herein is another limitation, as phenotyping for many important traits is slow, cumbersome and environmentally sensitive. Limitations in tissue culture and transformation processes in certain crops like hot pepper also add to the challenges of gene editing. With the development of cost-effective high-throughput sequencing technologies, genomic information for vegetable crop species has been dramatically increased. This will certainly accelerate our understanding of many important crop traits and will boost applications of genome edited technologies in vegetable breeding.

GE-Intellectual Property and Product Development

As the genome editing technologies are being increasingly used in plant research and product development, an understanding of the intellectual property (IP) and license implications of the chosen technology is needed for reliable launch of products into the desired market (Schinkel and Schillberg 2016). All of the genome editing methods commonly used seem to be patented, hence, the IP situation is rather ambiguous for some methods but less so for others. The IP for ZFN belongs to Sangamo BioSciences, however, for the development of commercial products concerning agriculture crops, the technology was licensed to Dow AgroSciences in 2008 (Scott 2005). ZFN use for plant-derived products was sublicensed by Dow AgroSciences under the trademark EXZACT™ Precision technology (http://www.exzactprecisiontechnology.com). No plant product developed with ZFN has entered the market yet. The IP for TALEN is complex and patents based on one patent family (Voytas *et al.*, 2010) owned by the University of Minnesota and the Iowa State University Research Foundation have been granted in the USA and in the European Community and exclusive rights for their use was given to Calyxt Inc. (erstwhile Cellectis Plant Sciences Inc.) in January 2011. There is another patent (and its family), however, for protecting the specificity of the TALEN through DNA-binding domain design (Bonas *et al.*, 2010) that is owned by the Two blades foundation (http://2blades.org), for developing and deploying durable resistance in crop plants. The two blades foundation has a cross-license agreement with Cellectis and uses the TALEN technology for non-profit humanitarian efforts while the latter uses the technology for commercial product development, making Calyxt Inc. the key player for TALEN IP. Field trials of plants modified using TALEN technology *viz.* high oleic soybean and cold storable potato (Clasen *et al.*, 2016) were completed in 2015. The CRISPR technology is at the center of a major legal dispute over who owns the rights to exploit it commercially.

Regulations

The future application of genome editing technologies to agriculture depends on how the regulatory framework evolves for commercializing products. The regulation of genome-edited crops has been widely discussed and reviewed by regulators and scientific community. It has been proposed that products developed using genome editing technology should be subjected to a more nuanced regulatory process and governance since the EENs used for editing can be crossed out of the target crop leaving no "transgenic" residue in the product. Technology developers are currently focusing on consumer traits and emphasizing the consumer benefits to ensure the social acceptance of crops produced via genome editing.

References

Alonso, J.M. and Ecker, J.R. 2006. Moving forward in reverse: Genetic technologies to enable genome-wide phenomic screens in *Arabidopsis*. *Nat. Rev. Genet.*, **7**: 524–536.

Blanvillain-Baufumé, S., Reschke, M., Solé, M., Auguy, F., Doucoure, H., Szurek, B., Meynard, D., Portefaix,

M., Cunnac, S., Guiderdoni, E., Boch, J. and Koebnik, R. 2016. Targeted promoter editing for rice resistance to *Xanthomonas oryzae* pv. *oryzae* reveals differential activities for SWEET14-inducing TAL effectors. *Pl. Biot. J.*, doi: 10.1111/pbi.12613.

Boch, J. and Bonas, U. 2010. *Xanthomonas AvrBs3* family-type III effectors: Discovery and function. *Ann. Rev. Phytopathol.*, **48**: 419–436.

Bonas, U., Boch, J., Schornack, S. and Lahaye, T. 2010. Modular DNA-binding domains and methods of use. *Int. Pat. App.*, WO2010079430, 12 January.

Britt, A.B. and Kuppu, S. 2016. Cenh3: An emerging player in haploid induction technology. *Front. Pl. Sci.*, doi: 10.3389/fpls.2016.00357.

Brooks, C., Nekrasov, V., Lippman, Z.B. and Van, E.J. 2014. Efficient gene editing in tomato in the first generation using the clustered regularly interspaced short palindromic repeats/CRISPR–associated9 system. *Pl. Physiol.*, **166**: 1292–1297.

Cathomen, T. and Joung, J.K. 2008. Zinc-finger nucleases: The next generation emerges. *Mol. Ther.*, **16**: 1200–1207.

Chandrasekaran, J., Brumin, M., Wolf, D., Leibman, D., Klap, C., Pearlsman, M., Sherman, A., Arazi, T. and Gal-On, A. 2016. Development of broad virus resistance in non-transgenic cucumber using CRISPR/Cas9 technology. *Mol. Pl. Pathol.*, **17**: 1140–1153.

Clasen, B.M., Stoddard, T.J., Luo, S., Demorest, Z.L., Li, J., Cedrone, F., Tibebu, R., Davison, S., Ray, E.E., Daulhac, A., Coffman, A., Yabandith, A., Retterath, A., Haun, W., Baltes, N.J., Mathis, L., Voytas, D.F. and Zhang, F. 2016. Improving cold storage and processing traits in potato through targeted gene knockout. *Pl. Biot. J.*, **14**: 169–176.

Cho, S.W., Kim, S., Kim, Y., Kweon, J., Kim, H.S. and Bae, S. *et al.*, 2014. Analysis of off-target effects of CRISPR/C as-derived RNA-guided endonucleases and nickases. *Genome Res.*, **24**: 132–141.

Deng, D., Yan, C., Pan, X., Mahfouz, M., Wang, J. and Zhu, J.K. *et al.*, 2012. Structural basis for sequence-specific recognition of DNA by TAL effectors. *Science*, **335**: 720–723.

Doudna, J.A. and Charpentier, E. 2014. Genome editing: The new frontier of genome engineering with CRISPR-Cas9. *Science*, **346**(6213): 1258096

Fu, Y.F., Sander, J.D, Reyon, D., Cascio, V.M. and Joung, J.K. 2014. Improving CRISPR-C as nuclease specificity using truncated guide RNAs. *Nat. Biotech.*, **32**: 279–284.

Gaj, T., Gersbach, C.A. and Barbas, C.F. 2013. ZFN, TALEN, and CRISPR/C as based methods for genome engineering. *Trends Biotech.*, **31**: 397–405.

Han, K., Jeong, H.J., Sung, J., Keum, Y. and Cho, M.C. *et al.*, 2013. Biosynthesis of capsinoid is controlled by the Pun1 locus in pepper. *Mol. Breed.*, **31**: 537–548

Hou, Z., Zhang, Y., Propson, N.E., Howden, S.E., Chu, L-F. and Sontheimer, E.J. *et al.*, 2013. Efficient genome engineering in human pluripotent stem cells using Cas9 from *Neisseria meningitidis*. *Proc. Nat. Acad. Sci.*, USA, **110**: 15644–15649.

Hsu, P.D and Zhang, F. 2012. Dissecting neural function using targeted genome engineering technologies. *ACS Chem. Neurosci.*, **3**: 603–610.

Ito, Y., Nishizawa-Yokoi, A., Endo, M., Mikami, M. and Toki, S. 2015. CRISPR/Cas9-mediated mutagenesis of the *RIN* locus that regulates tomato fruit ripening. *Biochem. Biophys. Res. Comm.*, **6**: 76–82.

Jiang, W., Zhou, H., Bi, H., Fromm, M., Yang, B. and Weeks, D.P. 2013. Demonstration of CRISPR/Cas9/sgRNA-mediated targeted gene modification in *Arabidopsis*, tobacco, sorghum and rice. *Nucleic Acids Res.*, **41:** e188.

Jinek, M., Chylinski, K., Fonfara, I., Hauer, M., Doudna, J.A. and Charpentier, E. 2012. A programmable dual-RNA-guided DNA endonuclease in adaptive bacterial immunity. *Science*, **337**: 816–821.

Kim, H., Kim, S.T., Kim, S.G. and Kim, J.S. 2015. Targeted genome editing for crop improvement. *Pl. Breed. Biot.*, **3**: 283–290.

Kelliher, T., Starr, D., Wang, W., McCuiston, J., Zhong, H. and Nuccio, M.L. *et al.*, 2016. Maternal haploids are preferentially induced by CENH3-tailswap transgenic complementation in maize. *Front. Pl. Sci.*, **7**: 414. doi: 10.3389/fpls.2016.00414.

Li, T., Liu, B., Spalding, M.H., Weeks, D.P. and Yang, B. 2012. High-efficiency TALEN-based gene editing produces disease-resistant rice. *Nat. Biot.*, **30**: 390–392.

Lawrenson, T., Shorinola, O., Stacey, N., Li, C., Stergaard, L., Patron, N., Uauy, C. and Harwood, W. 2015. Induction of targeted, heritable mutations in barley and *Brassica oleracea* using RNA-guided Cas9 nuclease. *Genome Biol.*, **16**: 258. doi: 10.1186/s13059-015-0826-7.

Lor, V.S., Starker, C.G., Voytas, D.F., Weiss, D. and Olszewski, N.E. 2014. Targeted mutagenesis of the tomato *PROCERA* gene using transcription activator-like effector nucleases. *Pl. Physiol.*, **166**: 1288–1291

Lusser, M., Parisi, C., Plan, D. and Rodriguez-Cerezo, E. 2011. New plant breeding techniques: State of the art and prospects for commercial development. Joint Research Centre Technical Report EUR 24760. European Commission Joint Research Centre, Brussels.

Mahfouz, M.M., Cardi, T. and Neal Stewart, C. 2016. Next-generation precision genome engineering and plant biotechnology. *Plant Cell Rep.*, **35**: 1397–1399.

Mak, A.N., Bradley, P., Cernadas, R.A., Bogdanove, A.J. and Stoddard, B.L. 2012. The crystal structure of TAL effector PthXo1 bound to its DNA target. *Science*, **335**: 716–719.

Osakabe, K., Osakabe, Y. and Toki, S. 2010. Site-directed mutagenesis in *Arabidopsis* using custom-designed zinc finger nucleases. *Proc. Nat. Acad. Sci., USA*, **107**: 12034–12039.

Pattanayak, V., Lin, S., Guilinger, J.P., Ma, E., Doudna, J.A. and Liu, D.R. 2013. High-throughput profiling of off-target DNA cleavage reveals RNA-programmed Cas9 nuclease specificity. *Nat. Biotech*, **31**: 839–843

Pavletich, N.P. and Pabo, C.O. 1991. Zinc finger-DNA recognition: Crystal structure of a Zif268-DNA complex at 2.1 A. *Sci.*, **252**: 809–817.

Ravi, M. and Chan, S.W. 2010. Haploid plants produced by centromere-mediated genome elimination. *Nature*, **464**: 615–618.

Reddy, U.K., Almeida, A., Abburi, V.L., Alaparthi, S.B., Unselt, D., Hankins, G., Park, M., Choi, D. and Nimmakayala, P. 2014. Identification of gene-specific polymorphisms and association with capsaicin pathway metabolites in *Capsicum annuum* L. collections. *PLoS One*, **9**: e86393. doi: 10.1371/journal.pone.0086393.

Sawai, S., Ohyama, K., Yasumoto, S., Seki, H., Sakuma, T. and Yamamoto, T. *et al.*, 2014. Sterol side chain reductase 2 is a key enzyme in the biosynthesis of cholesterol, the common precursor of toxic steroidal glycoalkaloids in potato. *Plant Cell*, **26**: 3763–3774.

Schnell, J., Steele, M., Bean, J., Neuspiel, M., Girard, C., Dormann, N., Pearson, C., Savoie, A., Bourbonnière, L. and Macdonald, P. 2015. A comparative analysis of insertional effects in genetically engineered plants: Considerations for pre-market assessments. *Trans. Res.*, **24**: 1–17.

Schinkel, H. and Schillberg, S. 2016. Genome editing: Intellectual property and product development in plant biotechnology. *Plant Cell Rep.*, **35**: 1487–1491.

Scott, C.T. 2005. The zinc finger nuclease monopoly. *Nat. Biotech*, **23**: 915–918.

Shukla, V.K., Doyon, Y., Miller, J.C., DeKelver, R.C., Moehle, E.A. and Worden, S.E., *et al.*, 2009. Precise genome modification in the crop species *Zea mays* using zinc-finger nucleases. *Science*, **459**: 437–441.

Sprink, T., Eriksson, D., Schiemann, J. and Hartung, F. 2016. Regulatory hurdles for genome editing: Process-vs. product-based approaches in different regulatory contexts. *Plant Cell Rep.*, **3**: 1–4.

Sung, Y.H., Baek, I.J., Kim, D.H., Jeon, J., Lee, J., Lee, K., Jeong, D., Kim, J.S. and Lee, H.W. 2013. Knockout mice created by TALEN-mediated gene targeting. *Nat. Biotech*, **31**: 23–24.

Tait, J. and Barker, G.C. 2011. Global food security and the governance of modern biotechnologies. *EMBO Rep.*, **12**: 763–768.

Townsend, J.A., Wright, D.A., Winfrey, R.J., Fu, F., Maeder, M.L. and Joung, J.K. *et al.*, 2009. High frequency modification of plant genes using engineered zinc-finger nucleases. *Nature*, **459**: 442–445.

Uluisik, S., Chapman, N.H., Smith, R., Poole, M., Adams, G., Gillis, R.B., Besong, T.M.D., Sheldon, J., Stiegelmeyer, S., Perez, L., Samsulrizal, N., Wang, D., Fisk, I.D., Yang, N., Baxter, C., Rickett, D., Fray, R., Blanco-Ulate, B. and Powell, A.L.T. *et al.*, 2016. Genetic improvement of tomato by targeted control of fruit softening. *Nat. Biotech*, **34**: 950–952.

Yokotani, N., Nakano, R., Imanishi, S., Nagata, M., Inaba, A. and Kubo, Y. 2009. Ripening-associated ethylene biosynthesis in tomato fruit is autocatalytically and developmentally regulated. *J. Exp. Bot.*, **60**: 3433–3442.

Vinci, R.M., Mestdagh, F. And Meulenaer, B.D. 2012. Acrylamide formation in fried potato products: Present and future, a critical review on mitigation strategies. *Food Chem.*, **133**: 1138–1154.

Voytas, D.F., Bogdanove, A., Zhang, F., Christian, M., Cermak, T., Schmidt, C.L., Doyle, E. and Wang, L. 2010. Tal effector-mediated DNA modification. *Int. Pat. App.*, WO2011072246, 10 Dec 2010.

Wang, S., Zhang, S., Wang, W., Xiong, X., Meng, F. and Cui, F. 2015. Efficient targeted mutagenesis in potato by the CRISPR/Cas9 system. *Plant Cell Rep.*, **34:** 1473–1476.

Wang, Y., Cheng, X., Shan, Q., Zhang, Y., Liu, J. and Gao, C. *et al.*, 2014. Simultaneous editing of three homoeoalleles in hexaploid bread wheat confers heritable resistance to powdery mildew. *Nat. Biotech*, **32**: 947–951.

11

Molecular Breeding for Hastening Varietal Development in Horticultural Crops

Pritam Kalia*, Shrawan Singh, Arpita Srivastava and Manisha Mangal

Division of Vegetable Science, ICAR-Indian Agricultural Research Institute, New Delhi – 110 012
**E-mail: pritam.kalia@gmail.com*

Horticultural crops are comprised of heterogeneous groups of fruits, vegetables, flowers, medicinal plants and plantation crops with greater diversity within each group (Bliss, 2010). The breeders dealing with these crops need to have insight about various crop species that provide genetic diversity as well as different reproductive systems, for instance inbreeding, out breeding and mixed breeding. These factors collectively influence breeding methods and tools employed to develop improved cultivars. In spite of their high value per unit area and role in health and well-being of humans, there are limited breeding programmes in horticultural crops in the public sector nationally and internationally. Therefore, there is often less genetic diversity among accessible breeding materials which in turn impedes progress towards developing molecular markers linked to key loci controlling economically important traits. However, the possibility of using genetic markers to improve breeding efficiency is particularly attractive for perennial horticultural crops due to their long breeding cycles, large plant size and difficulty in selection for traits of importance.

In India, low productivity in horticultural crops is partly due to the occurrence of various biotic and abiotic stresses, which has been a concern for breeders. Pesticide use is common in commercial horticulture in India and in vegetables; samples investigated are contaminated with pesticide residues. Breeding efforts have resulted in the release of several disease resistant varieties/ F_1 hybrids in the past few decades. A recent advance in molecular marker technology is playinga key role in enhancing breeding efficiency. With a long history of development, especially since the fundamental principles of inheritance were established in the late 19th and early 20th centuries, plant breeding became an important component of agricultural sciences and now genomics is set to revolutionize plant breeding in 21th century. Conventional breeding methodologies have extensively proven successful in development of improved cultivars and germplasm. However, subjective evaluation and empirical selection play a considerable role in crop improvement as the availability of genomic tools and resources will facilitate the study of the genotype and its relationship with the phenotype, in particular for complex traits. Molecular breeding (MB), more often, implies molecular marker-assisted breeding (MAB), is the application of molecular biotechnologies, specifically molecular markers, in combination with linkage maps and genomics, to alter and improve plant or animal traits on the basis of genotypic assays. This term is used to describe several modern breeding strategies, including marker-assisted selection (MAS), marker-assisted backcrossing (MABC), marker-assisted recurrent selection (MARS), and genome-wide selection (GWS) or genomic selection (GS). These tools and techniques allow the identification of markers linked to genes and QTLs, using a diversity of techniques like bulked segregant analysis (BSA), fine genetic mapping, or association mapping. These new methodologies are being used for 'breeding by design'. Advances in genomics are providing breeders with new tools and methodologies that allow a great leap forward in plant breeding, including the 'super domestication' of crops and the genetic dissection and breeding for complex traits and represents a paradigm shift, by facilitating the direct study of the genotype and its relationship with the phenotype. Molecular breeding is playing a great role in supplementing conventional breeding in making it targeted saving the time thereby and handling tens of thousands of plants in search of few desirable ones. However, high costs and technical or equipment demands of MAB will continue to be a major obstacle for its large-scale use in the near future, especially in the developing countries (Collard and Mackill, 2008; Ribaut *et al.*, 2010). Therefore, integration of MAB into conventional breeding programme will be an optimistic strategy for crop improvement in the future. It can be expected that the drawbacks of MAB will be gradually overcome, as its theory, technology and application are further developed and improved.

1. Components of Molecular Breeding

Molecular breeding is the integration of molecular biology techniques into classical breeding. It ensures precision and high speed to obtain desired results in crop breeding programme in a short period.

1.1 Molecular Markers

Molecular markers are sequence of nucleic acid which are located near or within the desired gene and tend to stay together generation after generation. These provide opportunity for marker assisted breeding not only to address the issues that previously have been difficult using classical approaches but also to improve efficiency of routine breeding. Molecular markers of choice can be used in a breeding programme for identifying desirable parents, enhancing selection of elite alleles at loci controlling important traits, pyramiding favourable alleles at multiple loci affecting either a single trait or multiple traits and fingerprinting cultivars for identification and intellectual property protection. Marker locus-trait combinations can be used as selection criteria to complement parent selection and that of desirable plants in segregating populations. DNA markers can be used to improve breeding efficiency for multigene selection by allowing early discard of undesirable plants. Before discussing how molecular tools can be used in programmes of population improvement through recurrent selection, we will first describe the different types of markers, presenting their respective qualities and limitations. The range of molecular markers available has expanded rapidly. The commonest are restricted fragment length polymorphisms (RFLPs), amplified fragment length polymorphisms (AFLPs), random amplified polymorphic DNA (RAPD) and simple sequence repeats (SSRs), also known as microsatellites. Single-nucleotide polymorphisms (SNPs) are the newest type. No absolute hierarchy exists between the marker types-all having advantages and disadvantages (Table 1). The choice of a type of marker, therefore, will depend on the user's objectives and thus be determined case by case.

1.2 Mapping Populations

Mapping population is important for mapping the gene(s) or QTLs for use in molecular breeding. It is usually obtained from controlled crosses between selected contrasting parents for the trait of interest(s). The mapping populations are F_2, $F_{2:3}$, doubled haploids, backcross population (BC), Recombinant Inbreed Lines (RILs) and Near Isogenic Lines (NILs). The F_2 mapping population is the simplest, best for preliminary mapping and require less time. For this, two contrasting parents for trait of interest are crossed to get F_1individuals which are selfed to generate an F_2 population. However, it has limitations in use for QTLs and fine mapping of gene(s), because F_2 population is the outcome of only one meiotic cycle and they are not immortal. Selfing of F_2 individuals for single generation results in F 2:3 populations, which are suitable for mapping of quantitative trait and mapping of recessive genes. Doubledhaploids (DH) are produced from colchicine induced chromosome doubling of haploid plants derived from pollen or ovule culture from F_1 plants. Wide crossing is also used for doubled haploids. The DH population is immortal and useful for mapping of both gene(s) and QTLs particularly in cross-pollinated crops where inbred development is still a challenge. However, it requires high cost and there are chances of appearance of somaclonal variation. Backcross population (BC) is also used in mapping genes and it is developed by crossing a F_1 hybrid with one of its parents (usually recessive parent) for genetic analysis and mapping of traits. It requires less time and preferred for recessive traits but has limitations for QTLmapping as these are not immortal. Recombinant Inbred Lines (RILs) are developed by continuous selfing or sib mating members of F_2 population until complete homozygous state is attained. Single seed descent method is best for development of RILs and bulk or pedigree method without selection can also be used. It is immortal population and important for QTL mapping besides identifying tightly linked markers. However, it requires many seasons to develop RILs and difficult in crops having high inbreeding depression. Near Isogenic Lines (NILs) are lines of genetic codes that are identical except for differences at a few specific locations or genetic loci. These are developed by repeated selfing or backcrossing of F_1 with recurrent parent. The NILs are immortal population and used for tagging traits, in functional genomics and fine mapping. Since it is time consuming and costly, therefore it has limitation for its use in linkage mapping. Besides, Advanced Intercross Lines (AILs) and Advanced Backcross lines (ABLs) or Backcross Inbred Lines (BILs) are also useful in understanding the genetics of target traits. The TILLING (T_{arget} I_{nduced} L_{ocal} L_{esions} I_n G_{enomes}) populations are also common as an exploratory tool for trait improvement. However, these are not strictly a mapping population.

Table 1: Main Characteristic Features of Major Molecular Markers

Characteristic	*Marker System*							
	RAPD	*RFLP*	*ISSR*	*SSR*	*AFLP*	*SCAR*	*CAPS*	*SNP*
Type of visualization	Multi-loci	Single locus	Multi-loci	Single locus	Multi-loci	Single locus	Single locus	Single locus
Allelism	Dominant	Co-dominant	Dominant	Co-dominant	Dominant	Co-dominant	Co-dominant	Co-dominant
Type of polymorphism	Sequence	Sequence	Sequence	No. of repeats	Sequence	Sequence	Sequence	Sequence
Level of polymorphism	Good	Good	Good	Excellent	Excellent	Excellent	Excellent	Excellent
Number of polymorphic loci analyzed	1.5-50	1-3	Presence/ absence	1-3	20-100	Presence/ absence	Presence/ absence	1.0
Quantity of DNA Small needed	Small (0.02 µg) Small (0.05 µg)		Large (10 µg)	Small	Small (0.05 µg)		Small (0.5-1.0 µg)	Small
Quality of DNA needed	High	Good	Good	Moderate	Moderate	Good	Good	High
Reproducibility	Low	Good	Good	Good	Good	Good	Good	Good
Time once markers developed	Fast	Long	Fast	Fast, once markers developed		Fast	Fast	Fast Fast,
Amenable to automation	Moderate	Low	Moderate	High	Moderate	Moderate	Moderate	High
Cost	Average	Expensive	Average	Average	Average	Average	Average	Expensive
Technical difficulty	Medium	High	Medium	Low	Medium	Medium	Medium	High
Limitations	Sensitive to reaction conditions (reproducibility issues)	Pre-screen for single copy sequences to be used as probes	Sensitive to reaction conditions for reproducibility	Requires sequence data, therefore expensive to develop primers	Multi-step, therefore high technical requirements	Requires sequence data, therefore expensive to develop primers	Requires restriction enzyme digestion of PCR product, enzymes can be expensive	Less alleles, Multiplexes don't actually work yet
	Assumption that when two samples share a fragment, it is the same locus	Assumption that when samples share a fragment, they share flanking cleavage sites	Assumption that when two samples share a fragment, it is the same locus				Requires sequence data, therefore expensive to develop primers	Requires sequence information Technically challenging

TILLING populations are mutagenized populations that allow for the identification of point mutations within genes of interest following a screening procedure of Henikoff and Comai (2006).

1.3 Gene Tagging

Linked molecular markers have been identified in different horticultural crops (Table 2). Molecular markers linked with reproductive traits such as sex forms in cucurbits, asparagus, and papaya. Adhikari *et al.* (2012) identified two sex-linked markers, namely OPS_{250}-11 and OPS_{600}-11 for male and bisexual individuals, respectively in cucumber for their use to understand the developmental biology and stable sex expression. In asparagus (*Asparagus officinalis* L.), all-male cultivars are desired for higher yields. To distinguish male and female individuals, Li *et al.* (2012) identified a 1600 bp amplicon with T35R54 (5′-TTCACGGTGG-3′) analysis in both super male and all-male lines indicating its usefulness as a sex-related marker to increase the reliability of sex determination in asparagus. Kumar *et al.* (2008) reported two RAPD markers, male-specific marker $OPC05_{1000}$ and female-specific marker $OPC14_{400}$, which together can reliably differentiate between male and female plants of pointed gourd (*Trichosanthes dioica*). Male sterility has been reported in various vegetable crops such as radish, carrot, peppers, brinjal, cole crops, beet root *etc.* Although, conventional programmes for introgression of male sterility trait into elite genetic backgrounds has accomplished satisfactory results in different vegetable crops but use of molecular markers enhanced the speed of conversion process. In case of papaya, more number of female and hermaphrodite plants are required for higher productivity which can be done by detection of

Table 2: Linked Molecular Markers to Trait of Interest in Horticultural Crops

Crop	*Trait*	*Gene*	*Marker*	*Reference*
Vegetable crops				
Cauliflower	High carotene	Or		Li and Garvin (2003)
	High anthocyanin	Pr	*BoMYB3BoMYB4*	Chiu *et al.* (2010)
	Black rot	*Xca1bo(B. oleracea)*	RAPD, ISSR	Saha *et al.* (2014)
		Xca1bc(B. carinata)	SSR, ILP	Sharma *et al.* (2016)
	Downy mildew	*Ppa3*	RAPD, ISSR	Singh *et al.* (2012)
Cabbage	Black rot	BRQTL-C3, BRQTL-C6	SNP	Lee *et al.* (2015)
	Head shape	QTLs (*Htd* 3.1, *Htd* 8.1)	SSR, InDel	Pang *et al.* (2015)
	Yellow-green leaf	ygl-1	InDel	Lv *et al.* (2014)
	Plant height (Ph3.1), leaf length (Ll3.2) and head transverse diameter (Htd 3.2)	QTLs Ph 3.1, Ll 3.2, Htd 3.2	InDel	Lv *et al.* (2016)
	Resistance to head splitting	QTL SPL-2-1	SSR, InDel	Pang *et al.* (2015)
	Fusarium wilt resistance	Gene FOC1	InDel	Lv *et al.* (2014)
Pepper	Tomato spotted wilt virus	Tsw	RAPD	Jahn *et al.* (2000)
	Xanthomonas vesicatoria	Bs2	AFLP	Tai *et al.* (1999)
	Temperature-sensitive gene	Seychelles-2('sy-2')	SNPs	Liu *et al.* (2016)
	Male sterility	ms10	SSR	Aulakh *et al.* (2016)
	Thrips resistance	QTLs (L1 and L2 (1st and 2nd instar larval stages)	SSR, 109 AFLP, and 5 SNP	Maharijaya *et al.* (2015)
	Anthracnose resistance	QTLs (AnRGO3, AnRGO5, AnRGO7, AnRGO10, AnRGO12)	SSRs, InDel	Sun *et al.* (2015)
	Pungency (*Pun1)*	*Pun1, CCR, KAS, HCT*	SNP	Reddy *et al.* (2014)
Bean	Common bean mosaic virus	*I*	RAPD	Melotto *et al.* (1996)
	Common bacterial blight	*QTL*	SNP	Shi *et al.* (2011)
	Anthracnose	*Co-5, Co-7*	STS	Sousa *et al.* (2014)
	Bean rust	*Ur-3*	SSR, SNP	Hurtado-Gonzales *et al.* (2016)
	Cowpea golden mosaic virus	*2 genes*	AFLP	Rodrigues *et al.* (2012)
	Golden mosaic virus	*bgm-1*	SCAR	Blair *et al.* (2007)
Garden Pea	Pea common mosaic virus	*Mo*	RFLP	Dirlewanger *et al.* (1994)
	Powdery mildew	*er1*	SCAR	Sun *et al.* (2015)
		er2	ISSR, RAPD	Katoch *et al.* (2009)
		Er3	SCAR, RAPD	Fondevilla *et al.* (2007)
Cucumber	*Fusarium oxysporum* f. sp.*melonis*	*Fom2*	SSP	Wechter *et al.* (1998)
Melon	*Fusarium oxysporum* f. sp.*melonis*	*Fom2*	RAPD	Wechter *et al.* (1995)
Tomato	*Meloidogyne incognita*	*Mi*	RAPD	Williamson *et al.* (1994)
	Meloidogyne javanica	*Mi3*	RAPD	Yaghoobi *et al.* (1995)
	Cladosporium fulvum	*Cf2*	RFLP	Dixon *et al.* (1995)
	Verticillium dahaliae	*Ve*	RFLP	Diwan *et al.* (1999)
	Fusarium oxysporum f. sp. Radicislyco persici	*Fr2*	RAPD	Fazio *et al.* (1999)
	Cucumber leaf curl virus	*Cmr*	RFLP	Stamova and Chetalat (2000)
	Tomato mosaic virus	*Tm2*	SCAR	Sobir *et al.* (2000)
	Yellow leaf curl virus	*Ty2*	RFLP	Hanson *et al.* (2000)
	Leaf mold	*Cf-19*	EST	Zhao *et al.* (2016)
	Fruit size and contents	*QTLs*	EST	Causse *et al.* (2004)
	Fusarium wilt	*I-3*		Lim *et al.* (2006)

Contd...

Table 2–Contd...

Crop	Trait	Gene	Marker	Reference
Carrot	Alternative oxidase (AOX) enzyme	*DcAOX1* gene	SNP, InDel	Nogales *et al.* (2016)
	Carotenoid accumulation	Gene DCAR_032551	DCAR_032551	Iorizzo *et al.* (2016)
	Abiotic stress responses like heat stress	*DcPHD* genes like *DcPHD-I-15*	RT-PCR	Wu *et al.* (2016)
	Root and leaf anthocyanin pigmentation	QTLs (Cy3XG, Cy3XGG, Cy3XSGG, Cy3XFGG)		Cavagnaro *et al.* (2014)
	Root knot nematode	Mj-1	SSR	Ali *et al.* (2003)
		Mj-2	SSR	Ali (2013)
Potato	Cyst nematode resistance	Hi		Heller *et al.* (1996)
Onion	Downy mildew	Pd	DMR1	Kim *et al.* (2016)
	Fructon	QTLs	SSR	McCallum *et al.* (2006)
Fruit crops				
Mango	NBS–LRR resistance gene analogues	NBS-LRR	SLAF-seq	Luo *et al.* (2016)
Banana	*Fusarium* wilt	*pd1*	DaRT	Ssali *et al.* (2000)
		Defense genes associated with CEBiP, BAK1, NB-LRR, PR proteins, transcription factor	Sequence	Tin *et al.* (2013)
Papaya	Sex determination	*Sex1*	SNP, EST	Wanga *et al.* (2012)
Ornamental and flower plants				
Rose	Powdery mildew	Rpp1	SCAR	Linde *et al.* (2004)
Chrysanthemum	Aphid feeding	CK-VS-Y, CK-VS-Z and Z-VS-Y	RNA-Seq technology	Xia *et al.* (2014)
Marigold	Genic male sterility trait	*Tems*	AFLP	He *et al.* (2010)
Carnation	Bacterial wilt	QTL	STS-WG44	Yagi *et al.* (2013)

such plants at early growth stage. For this, researchers have reported RAPD and microsatellite markers linked to sex form and their conversion to SCAR (sequence characterized amplified region) (Urasaki *et al.*, 2002). Ejaz *et al.* (2015) investigated that out of 105 plants (49 male, 56 female), all male plants exhibited amplification of sex-linked alleles with markers T12 and W11, whereas, 96 and 95 per cent of female plants showed the absence of sex-linked allele with these markers, respectively. Markers SDP, PKBT5 and Napf-76 showed the presence of sex-linked alleles in 98, 96 and 93 per cent of male plants, respectively, whereas the same markers showed the absence of sex-linked alleles in 100, 96 and 94 per cent of female plants. Hsu *et al.* (2012) performed multiplex loop-mediated isothermal amplification (mLAMP) assay consisted of two sex-LAMP reactions for amplifying two male-specific markers (T12 and Cpsm90) in one reaction for rapid sex identification of papaya. The established method can be easily performed in the field by visual inspection and facilitates the selection of all hermaphroditic individuals in papaya production. Gynoecism is a trait of interest in cucurbits for use in hybrid breeding programme. In bitter gourd, it is controlled by a recessive gene (*gy-1*) and Mishra *et al.* (2014) identified a RAPD marker (OPZ 13_{700}) which could produce polymorphism between the monoecious and gynoecious bulks. Feng *et al.* (2009) reported C1A/C1a (TI L- 5′ AACGGATGAAGAAGGAAAACG 3′; TI R- 5′ ATATTGGGCAGTGTCCACACA 3′) a co-dominant molecular marker linked to monoecy in melon and could be used in marker assisted selection in melon breeding. Fu-zhong *et al.* (2009) found that parthenocarpic gene in brinjal is controlled by single dominant nucleus gene (*Pat*) and reported one AFLP marker E75/M53-70 linked at 15.38 cM distance with *Pat* gene suggesting fine mapping. Mayatake *et al.* (2012) analysed two QTLs on chromosomes 3 (*Controlling parthenocarpy3.1, Cop3.1*) and 8 (*Cop8.1*) with confirmed effect of *Cop8.1*. Two flanking SSR markers (emf21H22, emh11J10) to *Cop8.1* were also reportedfor use inMAS and in revealing the genomic mechanism of the trait. Budahn *et al.* (2014) investigated genes governing floral architecture and pollen development in carrot using SSR markers. He *et al.* (2010) analysed the novel male sterile type resulting from spontaneous floral organ homeotic conversion in marigold (*Tagetes erecta* L).

Yagi *et al.* (2013) highlighted that the use of STS-WG44 marker, linked to a QTL with a large effect for resistant to bacterial wilt in carnation helped in narrowing down the tested population by half. Repeated crossing and selection *via* both conventional disease screening and MAS led to successful development of the first CBW-resistant carnation cultivar, Karen Rouge.

Regulation of flowering and vernalization requirement infruits and biennial winter vegetables can be explored by studying regulatory pathway mechanisms using molecular markers. In carrot, European types require vernalization to induce flowering and one major gene controls floral initiation. Alessandro *et al.* (2013) developed the first linkage map of carrot locating the genomic regions that control vernalization response and fertility restoration. The vernalization gene *Vrn1* mapped to chromosome 2 with flanking markers at 0.70 and 0.46 cM, and fertility restore gene (*Rf1*) mapped to chromosome 9 with flanking markers at 4.38 and 1.12 cM. *Vrn1* and *Rf1* were the first two reproductive traits mapped in carrot genome along with flanking markers forreproductive biologystudy and to facilitate carrot breeding.

The linked molecular markers to genes for resistancein tomato (leaf curl, bacterial wilt, late blight, root knot nematode), cauliflower (black rot, down mildew), cucumber (papaya ring spot virus, *Fusarium* wilt, downy mildew), banana (*Fusarium* wilt, browning nematode), potato (cyst and golden nematodes), grape (powdery mildew), apple (powdery mildew, scab), rose (black spot) and coconut (mites and lethal yellowing), and quality traits such as carotenoids and anthocyanins (carrot and cauliflower); lycopene (papaya and watermelon); pungency and sulphur assimilation (onion), glucosinolates(cole crops) and reduction of antinuritional factors such as tannins (broad bean); fruit acidity (citrus); and seedlessness (grape and cucumber) will have great promise in marker assisted selection (Table 2). For abiotic stresses like salinity, drought, temperature extremes and heavy metals, breeding resistant genotypes continue to be a challenge due to genetic complexities of the traits and difficulties in sequential phenotyping conventionally.

1.4 Process of Molecular Breeding

Marker-Assisted Backcrossing

There are three levels of selection in which markers can be applied in backcross breeding. In the first level, these can screen for the target trait, which may be useful for traits that have laborious phenotypic screening procedures such as drought and salinity tolerance. The second level of selection involves selecting backcross progeny with the target gene and tightly-linked flanking markers in order to minimize linkage drag. This is referred to as recombinant selection. The third level of MAB involves selecting backcross progeny, that have already been selected for the target trait, with background markers. In other words, markers can be used to select against the donor genome, which may accelerate the recovery of the recurrent parent genome. With conventional backcrossing, it takes a minimum of five to six generations to recover the recurrent parent. Data from simulation studies suggests that at least two, but possibly three or even four backcross generations can be saved by using markers.

Zhang *et al.* (2008) found SCAR markers linked to recessive "*or*" gene inducing beta-carotene accumulation in Chinese cabbage. Ripley and Roslinsky (2005) identified an ISSR Marker for 2-propenyl glucosinolate content in *Brassica*. The eye-appealing orange cauliflower was first discovered in Bradford Marsh, Ontario, Canada in 1970. The orange cauliflower curds results from a spontaneous mutation of a single dominant gene designated as '*Or*' (Dickson *et al.*, 1988). This *Or* mutant was originally found in white curded autumn crop cv. Extra Early snowball. Since cauliflower grown and consumed extensivelyin almost every household for most part of the year, therefore in order to tackle widespread malnutrition due to betacarotene deficiency, a rigorous breeding programme was initiated on genetic biofortification of Indian cauliflower about half a decade ago at IARI with β-carotene enhancing native '*Or*' gene following marker assisted breeding, which led to development of betacarotene (8-10 ppm) rich variety Pusa KesarivitA-1 recommended for cultivation in mid-late group (December-January) (Kalia, 2016).

The marker assisted breeding is also being used to reduce anti-nutritional factors in vegetable crops as reported in faba bean (*Vicia faba*). Faba bean contains condensed tannins that reduce the value of the inherently high protein levels of the crop. Gutierrez *et al.* (2008) reported that these tannins can be removed by the activity of two genes, *zt-1* and *zt-2*, which are pleiotropic for white-flowered plants. A sequence characterized amplified repeat (SCAR) marker linked to the *zt-2* gene was identified, one of the genes controlling absence of tannins, that is associated with increased protein levels and reduced fibre content of faba bean seeds, which would facilitate the development of tannin-free faba bean cultivars.

Marker-Assisted Pyramiding

Pyramiding is the process of simultaneously combining of multiple genes/QTLs together into a single genotype. In this, assembly of many genes and alleles with similar effect from different loci is practically difficult through phenotypic analysis. The QTLs dispersed in parental lines can be recombined through molecular screening. Using conventional phenotypic selection, individual plants must be phenotypically screened for all the traits tested which is extremely difficult or impossible at early generations. Therefore, it may be very difficult to assess plants from certain population (segregants) or for root trait bioassays in which the plants are to be uprooted for the study. The DNA markers may facilitate selection because DNA marker assays are non-destructive and markers for multiple specific genes/QTLs can be tested using a single DNA sample without phenotyping. The most widespread application for pyramiding has been for combining multiple disease resistance genes in order to develop durable disease resistance. Gene pyramiding for disease resistance in vegetable crops particularly in tomato for leaf curl resistance is successful. The genes for different characters can also be assembled into one genotype with the help of markers (Chahal and Gosal, 2002).

Early Generation Marker-Assisted Selection

One of the most intuitive stages to use markers to select plants is at an early generation (especially F_2 or F_3). The main advantage of this is that many plants with unwanted gene combinations, especially those that lack essential disease resistance traits and plant height, can be simply discarded. This has important consequences in the later stages of the breeding programme because the evaluation for other traits can be more efficiently and cheaply designed for fewer breeding lines (especially in terms of field space). Ribaut and Betran (1999) proposed early generation selection which is useful to discard many lines in an early generation permitting evaluation of fewer promising lines in later generations thereby.

1.5 Gene Introgression

The molecular marker-assisted breeding has become a routine step in breeding of most crops where the gene and the markers for a specific trait are known. This technique is being used in the efficient introgression of important genes in to various horticultural crops including TLCV and late blight resistance in tomato, black rot and downy mildew resistance in cauliflower and cabbage, powdery mildew resistance in garden pea, increased beta-carotene content in cauliflower, cassava, Chinese cabbage and banana, to name a few. Molecular markers are also used to determine the genetic profile of a line or variety.

Efforts to bring marker assisted breeding into wide adoption for horticultural crops improvement will benefit from targeted utilization of synteny among related crops and information from model plants to develop markers for use in standard breeding methods like backcrossing and in new approaches such as genomic selection. Advanced molecular genetics techniques and information from *Arabidopsis* studies has also been used to identify a gene encoding the leucoanthocyanidin dioxygenase (PpLDoX) enzyme as potentially responsible for a QTL controlling enzymatic browning of peach fruit mesocarp (Ogundiwin *et al.*, 2008) which is a major system for chilling injury. Combined with insight from biometrical analysis of chilling injury components, the molecular information can be used to breed for tolerance to chilling injury in peach (Peace *et al.*, 2005; Ogundiwin *et al.*, 2007) and other stone fruits affected similarly. To fully realize the potential of markers in genetic improvement of horticultural crops, advances in genomics of model species such as rice and *Arabidopsis* should be integrated with DNA marker technology.

The varietal development traditionally can take up 8 to 12 years or more depending on the crop as the breeders select plants based on their visible or measurable traits, the process which is difficult, slow, influenced by the environment besides being costly. With marker assisted selection, the process has been made easier and faster shortening the time considerably for the new crop varieties to be brought to market.

1.6 Enhancing Selection Efficiency

Fruit ripening and firmness are important quality traits of apple that are the focus selection by breeders in apple improvement programme.The phenotypes associated with fruit quality of seedling progenies cannot be measured for at least 4-5 years or until apple trees are old enough to produce sufficient fruit for evaluation at the time of harvest and usually after a period of cold storage similar to that encountered in a commercial post-harvest regime. Besides, long juvenile period for seedling trees which consumes time and considerable resources, the low heritability of fruit firmness and storability contributes to predicted small gains from selection. The importance of climacteric ripening for fruit firmness and storability make genes in the ethylene biosynthetic pathway prime candidates for developing markerlocus-trait associations that can be used as selection criteria for improved fruit texture in apple. Alternative alleles at the Md-ACS1 and Md-ACO1 loci were identified (Zhu and Barritt, 2008). Combinations of favourable alleles were generally associated with fruit firmness in 347 existing cultivars suggesting that MAS can be used either to eliminate clearly undesirable genotypes or facilitate phenotypic selection through improved trait heritability. By using a bi-allelic marker at the ACS1 locus associated with ethylene production levels, substantial savings can be realized using MAS to discard seedlings prior to costly nursery and field operations (Edge Garza *et al.*, 2010).

2. Challenges and Perspectives in MAB and Conventional Breeding

Marker-assisted breeding became a new member in the family of plant breeding as various types of molecular markers in crop plants were developed during the 1980s and 1990s. The extensive use of molecular markers in various fields of plant science, *e.g.* germplasm evaluation, genetic mapping, map-based gene discovery, characterization of traits and crop improvement, has proven that molecular technology is a powerful and reliable tool in genetic manipulation of agronomically important traits in crop plants. Comparatively, MAB has advantages over conventional breeding approaches:

i. MAB can allow selection for all kinds of traits to be concentrated upon at seedling stage and thus reduce the time and cost required to raise the crop to maturity to assess individual plants phenotypically. However, for the traits that are expressed at later developmental stages, undesirable genotypes can be eliminated at early stages using MAS saving time, expenditure and cumbersome handling of unwanted material. This feature is particularly important and useful for some breeding schemes such as backcrossing and recurrent breeding, in which introgression of some useful trait (s) is to be carried out.

ii. Genetic markers are environment independent, thus allow selection to be performed under any environmental conditions. This is very helpful for improvement of traits like disease/insect pest and abiotic stress tolerance which are expressed only underfavourable environmental conditions. For the low-heritability traits that are easily affected by the environments, MAS-based on reliable markers tightly linked to the QTLs can be more effective for quickening the process than phenotypic selection.

iii. The co-dominant markers such as SSRs can allow effective selection of recessive alleles of desired traits in the heterozygous state. No selfing or test crossing is needed to detect the traits controlled by recessive alleles, thus saving time and accelerating breeding progress.

iv. For the traits controlled by multiple genes/QTLs, individual genes/QTLs can be identified and selected using MAB at the same time and in the same individuals revealing its suitability for gene pyramiding.

v. Genotypic assays based on molecular markers may be faster, cheaper, accurate and reliable than conventional phenotypic assays, depending on the traits and conditions, and thus MAB may result in higher effectiveness and higher efficiency in terms of time, resources and efforts saved.

The research and applicability of MAB is continuously increasing in crop plants in the public and private sectors since 2000s. However, MAS and MABC were primarily focused on simply-inherited traits, such as monogenic or oligogenic resistance to diseases/insect pests, although quantitative traits were also involved (Collard and Mackill, 2008; Segmagn *et al.*, 2006; Wang and Chee, 2010). The application of molecular markers in plant breeding has not achieved the results as expected previously in terms of extent and success (*e.g.* release of commercial cultivars). Collard and Mackill (2008) listed ten reasons for the low impact of MAS and MAB in general, which are:

- The DNA marker technology is still at the early stages of development.
- The results of MAS may not be published, particularly by private seed sector.
- Reliability and accuracy of QTLs mapping studies is critical for more complex traits.
- Insufficient linkage between marker and gene/QTLs.
- Limited markers and limited polymorphism of markers in breeding material.
- Effectiveness of QTLs influenced by genetic background due to interactions with other loci or epistasis.
- QTLs × environment interaction influence magnitude of effect of QTLs.
- Initial high cost of marker-assisted selection.
- Application gap between research laboratories and plant breeding institutes due to difficulties in the transfer of markers and relevant information to breeders.
- Knowledge gap among molecular biologists, plant breeders and other disciplines.

Improvement of most agronomic traits,which have complex inheritance and economic importance such as yield and quality, is still a great challenge in spite of Hi-Tech technology availability.The limitations with Hi-Tech Technologies are non-availability of cost-effective and high-throughput genotyping systems, poor infrastructure, inadequate capacity and operational support, lack of breeder-friendly markers and an application gap between research laboratories and public plant breeding institutes (Xu and Crouch 2008, Ribaut *et al.*, 2010, Xu *et al.*, 2012, Jiang 2013b).

3. Drawbacks in Application of Molecular Technologies

- All the markers are not breeder-friendly. This problem can be overcome by converting of non-breeder-friendly markers to other types of breeder-friendly markers (*e.g.* RFLP to STS, sequence tagged site, and RAPD to SCAR, sequence characterized amplified region).
- All the markers cannot be applicable across populations due to lack of marker polymorphism or reliable marker-trait association. Multiple mapping populations are helpful in understanding marker allelic diversity and genetic background effects. In addition, QTL positions and effects also need to be validated and re-estimated by breeders in their specific germplasm (Heffner *et al.*, 2009).
- False selection may occur due to recombination between the markers and the genes/QTLs of interest. Use of flanking markers or more markers for the target gene/QTL can help.
- Imprecise estimates of QTL locations and effects result in slower progress than expected. The efficiency of QTL detection is attributed to

multiple factors, such as algorithms, mapping methods, number of polymorphic markers, and population type and size (Wang *et al.*, 2012). High marker density fine mapping with large populations and well-designed phenotyping across multiple environments may provide more accurate estimates of QTL location and effects.

- A large number of breeding programmes have not been equipped with adequate facilities and conditions for a large-scale routine adoption of MAB.
- The methods and schemes of MAB must be easily understandable, acceptable and implementable by plant breeders, unless they are not designed for a large scale use in practical breeding programmes.
- Higher startup expenses and labour costs.
- Private sector has got successful examples but they are not publishing successful MAS-applications and by putting intellectual property rights into effect, private companies usually prevent breeders from public institutions to have free access to the private-sector marker technologies.

5. Genome Wide Selection

There has been a striking decrease in the cost and effort needed to read a plant's DNA sequence, besides qualitative improvement in the infrastructure needed to store and analyze these data during the past two decades. As a result there is reduction in the number of cycles in horticultural breeding programmes and increase in precision and efficiency of new cultivar development. Among the most rapidly developing approaches, genome-wide selection (GWS) (Nockerand Gardiner, 2014) and genomic estimated breeding values (GEBVs) are used as selection parameters rather than estimated breeding values (EBVs) traditionally used by fruit breeders. GEBVs are derived for individuals in a phenotyped training population using dense genome-wide single-nucleotide polymorphism (SNP) markersto establish marker effects on complex phenotypes controlled by a large number of genetic loci. Individuals in breeders' selection populations are then screened and GEBVs of individuals calculated based on genetic marker information to identify outstanding *'elite'* individuals. These may then be employed to advance generations and/or evaluated in the field as potential cultivars. In kiwifruit, *Actinidia chinensis*, GBS was employed to construct a genetic map that was used to anchor a draft genome assembly and also experimented as speedy means of identifying genetic markers for resistance to *Pseudomonas syringae* pv. *actinidiae.* It was concluded that use of this system could at least halve the time taken for future genetic marker development in kiwifruit, by enabling the simultaneous screening of several thousand markers over population phenotype extremes, compared with the previous requirement to screen genome-wide markers individually. However, improvements need to be made in the area of genomic data analysis, as reduced-representation sequencing has a high degree of technical uncertainty which can result in uneven sequencing of samples and consequently missing data issues in GBS analysis.For the establishment of GWS in pear (*Pyrus communis*), a training population was developed using germplasm from an interspecific breeding programme and genotyping completed by GBS. A population of 1000 plants from a raspberry (*Rubus idaeus*) breeding population already phenotyped for yield and quality components and development of GWS is inprogress.

6. Achievements of Molecular Breeding

Around 136 MAS-derived varieties released by public sector breeders in 22 countries have been found mentioned in literature/institutional websites *etc.* (Greenpeace, 2014). This search was necessarily restricted by budget and confined to information in English language, therefore it is assumed that there are more MAS-derived varieties released from public sector breeding programmes.

One of the most successful applications of MAS in the public sector has been for introgression and/or pyramiding of major genes. This approach has led to the commercial release of several biotic stress resistant varieties in cassava, barley, bean, chilli, pearl millet, peanut, rice, sorghum, soybean, tomato and wheat. About two-third of the identified varieties have been bred using MAS for biotic stress tolerance. Further successful MASapplication resulted in the release of varieties with improved quality, such as a high-quality protein maize, lowamylose rice, as well as low-cadmium and high-protein wheat. In addition, public sector molecular breeding also achieved the improvement of complex traits such as drought- and salt-tolerance in rice and yields in tomato and rice.

Tomato leaf curl virus (TLCV) is a major challenge for tomato production as it devastates tomato in the tropics, subtropics causing 100 per cent yield loss from early infection. Sources of resistance for this malady have been found only in the wild species like *Solanum chilense*, *S. habrochaites* and *S. peruvianum.* Transfer of these genes into cultivated background was a major challenge. Markers linked to genes controlling resistance to this disease namely Ty1, Ty2, Ty3, Ty4, ty5 and Ty6 have been found to be located on chromosome 6, 11, 6, 3, 11and 10, respectively. In tomato, molecular markers linked to Tomato leaf curl virus (ToLCV), bacterial wilt, late blight and root-knot nematode resistance available in public domain have been validated at ICAR institutions, namely Indian Agriculture Research Institute, New Delhi; Indian Institute of Horticultural Research, Bengaluru and Indian Institute of Vegetable Research, Varanasi and are being used for pyramiding and stacking of resistance genes. New sources of resistance to ToLCV, Groundnut bud necrosis virus (GBNV), high temperature and drought have also been identified and mapping populations have been generated to identify molecular markers linked to these menaces. At Indian Institute of Vegetable Research, marker-assisted pyramiding of Ty-2 (*S. habrochaites*) and Ty-3 (*S. chilense*) has been accomplished and at IARI, New Delhi the work is in progress. The pyramided line VRT6-1-4 from IIVR, Varanasi showed better performance over all other lines in agronomic evaluation trials conducted over two seasons. Allele specific primers associated with pvr1 in chili were used to identify potyvirus resistant lines. Candidate gene analysis of *Fom-1* and *Fom-2* genes in melon for lines resistant to a local race of Fusarium wilt revealed that the local race is different from those reported earlier (Races 0, 1 and 2), and *Fom-1* and *Fom-2* based resistance cannot be deployed in India. RIL and backcross introgression line (BIL) populations were developed in watermelon using wild species *viz., Citrulluslanatus* var. *citroides* (resistant to *Fusarium* wilt) and *Citrullus colocynthis* (resistant to WBNV) for mapping of these resistance traits. Bulk segregant 14 AVRDC - The World Vegetable Center analysis was employed to identify a random amplified polymorphic DNA (RAPD) marker linked to Mungbean yellow mosaic virus (MYMV) resistance in French bean (*Phaseolus vulgaris* L.). Allele mining for acyltransferase gene linked to pungency in an array of chili germplasm revealed several SNPs linked to the trait. In chilli, two cytoplasmic male sterility specific sequenced characterized amplified region (SCAR) markers, *CoxII* and *Atp6*, were validated in an array of genotypes. Mitochondria specific markers linked to petaloid male sterility in carrot were identified. In okra, two RAPD markers were found to be tightly linked to genic male sterility. Candidate gene analysis for sex expression in melon resulted in the development of high frequency female lines.

Brassica genus includes many important vegetables like *B. rapa* (Chinese cabbage, pokchoi, and turnip) and *B. oleracea* (broccoli, cabbage and cauliflower), and oilseed crops (*Brassica napus, B. rapa, B. juncea* and *B. carinata*), which collectively provide 12 per cent of the world's edible vegetable oils. The *B. rapa* Genome Sequencing Project Consortium reported the annotation and analysis of the draft genome sequence of *B. rapa* accession Chiifu-401-42, a Chinese cabbage. Anchoring of genome assembly to the *B. rapa* genome linkage includes two *B. rapa* genome sequences (JWF3p and VCS_DH, http://www.brassica-rapa.org) available in public domaingenerated from RDA. Genome-wide variable genetic markers are developed from expressed sequence tags (ESTs), tandem repeat sequences, and bacterial artificial chromosome (BAC) sequences. A variable set marker includes restriction fragment length polymorphisms (RFLPs), simple sequence repeats (SSRs), cleaved amplified polymorphic sequences (CAPs) and intron-based amplicon polymorphisms (IBAPs), respectively. Launching a Chinese cabbage design for kimchi could produce the same cultivars in four seasons in the near future.

Bacterial wilt caused by *Ralstonia solanacearum* is another devastating disease of solanaceous vegetables. Tomato cultivar 'Hawaii 7996' has been shown to have stable resistance against different strains of the pathogen under different environments. In order to locate the quantitative trait loci (QTLs) associated with stable resistance, 188 recombinant inbred lines (RILs) derived from 'Hawaii 7996' and 'West Virginia 700' were developed. The population was evaluated against phylotype I and II strains at seedling stage and in the field in Indonesia, Philippines, Taiwan, Thailand, and Reunion. Two major QTLs were identified to be associated with stable resistance: Bwr-12 on chromosome 12 controlled 17.9 to 56.1 per cent of total resistance variation, and Bwr-6 on chromosome 6 explained 11.5 to 22.2 per cent of the phenotypic variation. Bwr-12, delimited by simple sequence repeat (SSR) markers SLM12-2 and SLM12-10, is important for resistance to phylotype I strain. Two populations of near isogenic lines (NILs) CLN3125A-23 and CLN3125L were evaluated to confirm the usefulness of SLM12-2 and SLM12-10 for marker-assisted selection. The NILs group with the resistant allele on Bwr-12 wilted less than those with the susceptible allele by 27.9 per cent and 24.2 per cent individually. Research on fine-mapping Bwr-6 is continuing. Markers of Bwr-12 and Bwr-6 used for assisted selection should increase the efficiency of pyramiding resistance QTLs, and help in selection of novel sources for bacterial wilt resistance.

7. Future Prospects of Molecular Breeding in Horticultural Crops

- Micronutrient malnutrition is the most common dietary problem affecting women and children worldwide and fruits and vegetables are important sources of these dietary elements. Their role in food based approaches to counter micronutrient malnutrition is a well accepted particularly in vulnerable communities. Successful examples of molecular approaches for transfer of gene(s) or QTLs for nutritive traits in various food crops including vegetables and fruits show the future path for further attempts.
- The molecular marker-assisted breeding is picking up fast for its wider use among horticultural crops and is likely to accelerate as advanced technology becomes available in more crops. This technique is being used in the efficient introgression of important genes in to various horticultural crops including TLCV and late blight resistance in tomato, black rot and downy mildew resistance in cauliflower and cabbage, powdery mildew resistance in garden pea, increased betacarotene content in cauliflower,

cassava, Chinese cabbage and banana, to name a few. Molecular markers are also used to determine the genetic profile of a line or variety. The information on genetic diversity of the lines is utilised in selecting for extremely unrelated parents useful for hybrid seed technology.

- Efforts to bring marker assisted breeding in to wide adoption for horticultural crops improvement will benefit from targeted utilization of synteny among related crops and information from model plants to develop markers for use in standard breeding methods like backcrossing and in new approaches such as genomic selection. Advanced molecular genetics techniques and information from *Arabidopsis* studies has also been used (Ogundiwin *et al.*, 2008) to identify a gene encoding the leucoanthocyanidin dioxygenase (PpLDoX) enzyme as potentially responsible for a QTL controlling enzymatic browning of peach fruit mesocarp which is a major system for chilling injury. Combined with insight from biometrical analysis of chilling injury components, the molecular information can be used to breed for tolerance to chilling injury in peach (Peace *et al.*, 2005b; Ogundiwin *et al.*, 2007) and other stone fruits affected similarly.
- Molecular markers linked with different traits of interest suggest possibilities for their use in speeding up the process. Shrinking land resources and for doubling the yield, dwarf plant architectures would be more preferable for high density planting in fruit crops. Breeding for such traits finds some of the supportive studies wherein molecular markers have been identified in papaya, apple and other non-horticultural crops like rice, *Arabidopsis*, barley *etc.* Regulation of flowering and vernalization requirement in fruits and biennial winter vegetables is possible but need systematic breeding programme using DNA markers.
- Education and training of breeders and exposure to the science and practicalities of MAB and collaborations among genomicists and breeders will hasten the use of markers for innovative futuristic cultivar development.
- The marker assisted breeding can be made effective based on functional breeding programme in place and it needs to be understood that DNA-based genotyping is not a stand-alone process; it is rather another tool available for cultivar development in a time frame.

8. References

Adhikari, S., Bandyopadhyay T.K. and Ghosh, P. 2012. Hormonal control of sex expression of cucumber (*Cucumis sativus* L.) with the identification of sex linked molecular marker. *Nucleus*, **55**(2). DOI: 10.1007/s13237–012–0061–5.

Alessandro, M.S., Claudio R. Galmarini, Massimo Iorizzo and Simon, Philipp W. 2013. Molecular mapping of vernalization requirement and fertility restoration genes in carrot. *Theor. Appl. Genet.*, **126**: 415–423.

Ali, A., Matthews, W.C., Cavagnaro, P.F., Iorizzo, M., Roberts, P.A. and Simon, P.W. 2013. Inheritance and mapping of *Mj-2*, a new source of root-knot nematode (*Meloidogyne javanica*) resistance in carrot. *J. Heredity*, doi: 10.1093/jhered/est090.

Blair, M.W., Rodriguez, L.M., Pedraza, F. and Morales, F. 2007. Genetic mapping of the bean golden yellow mosaic gemini virus resistance gene bgm-1 and linkage with potyvirus resistance in common bean (*Phaseolus vulgaris* L.). *Theor. Appl. Genet.*, **114**: 261–271.

Bliss, F.A. 2010. Marker-assisted breeding in horticultural crops. **In**: *Proc. IS on Molecular Markers in Horticulture, ISHS 2010*, (Eds.) N.V. Bassil and R. Martin. *Acta Hort.*, **859**: 339–350.

Budahn, H., Barañski, R., Grzebelus, D., Kie³kowska, A., Straka, P., Metge, K., Linke, B. and Nothnagel, T. 2014. Mapping genes governing flower architecture and pollen development in a double mutant population of carrot. *Front. Plant Sci.*, **5**: 504.

Chiu, L.W., Zhou, X., Burke, S., Wu, X., Prior R. L., and Li, L. 2010. The purple cauliflower arises from activation of a myb transcription factor. *Pl. Physiol.*, **154**(3): 1470–1480.

Collard, B.C.Y. and Mackill, D.J. 2008. Marker-assisted selection: An approach for precision plant breeding in the twenty-first century. *Philos. Trans R. Soc. Lond. B. Biol. Sci.*, **363**(1491): 557–572.

Edge-Garza, D.A., Peace, C.P. and Yanmin Zhu. 2010. Enabling marker-assisted seedling selection in the Washington apple breeding program. *Acta Hort.*, **859**: 369–373.

Feng, H., Li, X., Liu, Z., Wei, P. and Ji R. 2009. A co-dominant molecular marker linked to the monoecious gene CmACS-7 derived from gene sequence in *Cucumis melo* L. *African J. Biot.*, **8**(14): 3168–3174.

Fondevilla, S., Rubiales, D., Moreno, M.T. and Torres, A.M. 2008. Identification and validation of RAPD and SCAR markers linked to the gene *Er3* conferring resistance to *Erysiphepisi* DC in pea. *Mol. Breed.*, **22**: 193–200.

Fu-zhong, L., Xiang, W., Yu-hui, C., Yong, L. and Ming, S. 2009. Inheritance of the eggplant parthenocarpy and AFLP molecular marker. *Acta Hort. Sinica*, 2008–09.

Greenpeace. 2014. Smart breeding: The next generation marker assisted selection: A biotechnology for plant breeding without genetic engineering. Greenpeace International, Amsterdam, The Netherlands.

Gutierrez, N., Avila, C.M., Moreno, M.T. and Torres, A.M. 2008. Development of SCAR markers linked to *zt-2*, one of the genes controlling absence of tannins in faba bean. *Australian J. Agric. Res.*, **59**: 62–68.

He, Y., Ning, G. and Sun, Y. 2010. Cytological and mapping analysis of a novel male sterile type resulting from spontaneous floral organ homeotic conversion in marigold (*Tagetes erecta* L.). *Mol. Breed.*, **2619**(1): 19–29.

Henikoff, S. and Comai, L. 2003. Single-nucleotide mutations for plant functional genomics. *Ann. Rev. Pl. Biol.*, **54**: 375–401.

Hsu, T.H., Gwo, J.C. and Lin, K.H. 2012. Rapid sex identification of papaya (*Carica papaya*) using multiplex loop-mediated isothermal amplification (mLAMP). *Planta*, DOI 10.1007/s00425–012–1681–3.

Katoch, V., Sharma, S., Pathania, S., Banayal, D.K., Sharma, S.K. and Rathour, R. 2009. Molecular mapping of pea powdery mildew resistance gene *er2* to pea linkage group III. *Mol. Breed.*, **25**(2): 229–237.

Kim, S., Kim, C. and Choi, M. 2016. Development of a simple PCR marker tagging the *Allium roylei* fragment harboring resistance to downy mildew (*Peronospora destructor*) in onion (*Allium cepa* L.). *Euphytica*, **208**(3): 561–569.

Kumar, S., Singh, B.D., Sinha, D.P. and Rai, M. 2008. Sex expression-associated RAPD markers in pointed gourd (*Trichosanthes dioica*). **In**: *Proceedings of the 9th EUCARPIA meeting on Genetics and Breeding of Cucurbitaceae*, (Ed.) M. Pitrat, INRA, Avignon (France), Cucurbitaceae.

Lee, J., Izzah, N.K., Jayakodi, M. and Perumal, S. 2015. Genome-wide SNP identification and QTL mapping for black rot resistance in cabbage. *BMC Plant Biol.*, doi: 10.1186/s12870–015–0424–6.

Li, L. and Garvin, D.F. 2003. Molecular mapping of *Or*, a gene inducing beta-carotene accumulation in cauliflower (*Brassica oleracea* L. var. *botrytis*). *Genome*, **46**(4): 588–594.

Li, Y., Uragami, A., Uno, Y., Kanechi, M. and Inagaki, N. 2012. RAPD-based analysis of differences between male and female genotypes of *Asparagus officinalis*. *Hort. Sci.*, **39**(1): 33–37.

Lim, G.T.T., Wang, G. and Hemming, M.N. 2006. Mapping the *I-3* gene for resistance to Fusarium wilt in tomato: application of an *I-3* marker in tomato improvement and progress towards the cloning of *I-3*. *Australasian Pl. Path.*, **35**: 671–680.

Linde, M., Mattiesch, L. and Debener, T. 2004. Rpp1, a dominant gene providing race-specific resistance to rose powdery mildew (*Podosphaera pannosa*): molecular mapping, SCAR development and confirmation of disease resistance data. *Theor. Appl. Genet.*, **109**(6): 1261–1266.

Mahwish Ejaz, M., Iqbal, M., Naeemullah, M., Ahmed I., Shahzad A., Masood, M.S. and Ali, G.M. 2015. Validation and use of DNA markers for sex determination in papaya (*Carica papya*). *Pakistan J. Bot.*, **47**(3): 1051–1059.

McCallum, J., Clarke, A., Pither-Joyce, M., Shaw, M. and Butler, R. 2006. Genetic mapping of a major gene affecting onion bulb fructan content. *Theor. Appl. Genet.*, **12**(5): 958–967.

Miyatake, M., Saito, T., Negoro, S., Yamaguchi, H., Nunome, T., Ohyama, A. and Fukuoka, H. 2012. Development of selective markers linked to a major QTL for parthenocarpy in eggplant (*Solanum melongena* L.). *Theor. Appl. Genet.*, **124**(8): 1403–1413.

Nocker, S. van and Gardiner, S.E. 2014. Breeding better cultivars, faster: Applications of new technologies for the rapid deployment of superior horticultural tree crops. *Hort. Res.*, **1**: 14022.doi: 10.1038/hortres.2014.22.

Ogundiwin, E.A., Peace, C.P., Gradziel, T.M., Dandekar, A.M., Bliss, F.A. and Crisosto, C.H. 2007. Molecular genetic dissection of chilling injury in peach fruit. *Acta Hort.*, **738**: 633–638.

Ogundiwin, E.A., Peace, C.P., Nicolet, C.M., Rashbrook, V.K., Gradziel, T.M., Bliss, F.A., Parfitt, D. and Crisosto, C.H. 2008. Leucoanthocyanidin dioxygenase gene (PpLDOX): A potential functional marker for cold storage browning in peach. *Tree Genet. Genom.*, **4**: 543–554.

Peace, C.P., Ahmad, R., Gradziel, T.M., Dandekar, A.M. and Crisosto, C.H. 2005b. The use of molecular genetics to improve peach and nectarine post-storage quality. *Acta Hort.*, **682**: 403–410.

Saha, P., Kalia, P., Sonah, H. and Sharma, T.R. 2014. Molecular mapping of black rot resistance locus *Xca1bo* on chromosome 3 in Indian cauliflower (*Brassica oleracea* var. *botrytis* L.). *Plant Breed.*, **133**(2): 268–274.

Sharma, B.B., Kalia, P., Yadava, D.K., Singh, D. and Sharma. T.R. 2016 Genetics and molecular mapping of black rot resistance locus *Xca1bc* on chromosome B-7 in Ethiopian mustard (*Brassica carinata* A. Braun). *PLoS ONE*, **11**(3): e0152290. doi: 10.1371/journal.pone.0152290.

Singh, S., Sharma, S.R., Kalia, P., Deshmukh, R., Vinod, Sharma, P. and Sharma, T.R. 2012. Molecular mapping of the downy mildew resistance gene *Ppa3* in cauliflower (*Brassica oleracea* var. *botrytis* L.). *J. Hort. Sci. Biotech.*, **87**(2): 137–143.

Sousa, L.L., Cruz, A.S., Filho P.S.V., Vallejo, V.A., Kelly, J.D. and Gonçalves-Vidigal, M.C. 2014. Genetic mapping of the resistance allele Co-5 2 to *Colletotrichum lindemuthianum* in the common bean MSU 7-1 line. *Australian J. Crop Sci.*, **8**(2): 317–323.

Syun, S., Wang, Z., Fu, H., Duan, C., Wang, X. and Zhu, Z. 2015. Resistance to powdery mildew in the pea cultivar Xucai 1 is conferred by the gene *er1*. *The Crop J.*, **3**(6): 489–499.

Wanga, J., Naa, J.K., Yub, Q., Gschwenda, A.R. and Hana, J. 2012. Sequencing papaya X and Yh chromosomes reveals molecular basis of incipient sex chromosome evolution. *Proc. Natl. Acad. Sci. USA*, **109**(34): 13710–13715.

Xia, X., Shao, Y., Jiang, J. and Ren, L. 2015. Gene expression profiles responses to aphid feeding in chrysanthemum (*Chrysanthemum morifolium*). *BMC Genom.* DOI: 10.1186/1471–2164–15–1050.

Yagi, M. 2013. Application of DNA markers for breeding carnations resistant to bacterial wilt. *JARQ*, **47**(1): 29–35.

Zhao, T., Jiang, J., Liu, G. and He, S. 2016. Mapping and candidate gene screening of tomato *Cladosporium fulvum*-resistant gene Cf-19, based on high-throughput sequencing technology. *BMC Plant Biol.*, **16**: 51, DOI 10.1186/s12870–016–0737–0.

Zhu, Y. and Barritt, B. 2008. Md-ACS1 and Md-ACO1 genotyping of apple (*Malus x domestica* Borkh.) breeding parents and suitability for marker-assisted selection. *Tree Genet. Genom.*, **4**: 555–562.

12

Biotechnological Interventions in Canopy Architecture of Tree Fruits

S.K. Singh*, Chavlesh Kumar, Manish Srivastav and Nimisha Sharma

Division of Fruits and Horticultural Technology,
ICAR-Indian Agricultural Research Institute, New Delhi – 110 012
**E-mail: sanjaydr2@rediffmail.com*

Trees are perennial woody plants with a distinct primary shoot, or trunk, from which lateral branches followed by leaves emerge. These physical compositions (stem, branches and leaves) of tree is called canopy and its orientation infers architecture. Thus, canopy architecture can be defined as orientation or organization of physical composition of above ground parts of the plant. This canopy architecture has determined the shapes of the trees. The canopy architecture development in perennial tree crops has a seasonal and a lifetime developmental pattern.The sum of development over individual seasons results in the final canopy dimensions and form (Robinson *et al.*, 1991). Throughout the evolutionary history various forms of tree has been evolved and do not constitute a single clade or group of plants (Groover, 2005). Canopy architectures have different functions in the plant system and affect the overall physiology of the plants including reproductive biology. Light interception can be influenced intrinsic canopy architectures in a tree (Sansavini and Corelli-Grappadelli, 1992). In these ways canopy is responsible for maximum potential assimilation of photosynthates while absorbing the incoming light (Wagenmakers, 1995)and subsequently the photo- inception is positively correlated with the fruit productivity of a tree. If the light inception by a canopy is lacking or insufficient due some reasons then it may reduce the fruit quality like fruit size, colour and TSS (Robinson *et al.*, 1991; Palmer *et al.*, 1992). Canopy architectures also influence the pests and diseases paradigm in trees. Thus modification of canopy architecture to improve global light interception is a major objective of all planting systems (Lakso and Corelli-Grappadelli, 1992; Palmer, 1980; Sansavini and Corelli-Grappadelli, 1992; Tustin *et al.*, 1988, 1998; Wagenmakers *et al.*, 1991; Willaume *et al.*, 2004). Thus, the canopy architecture manipulation programmesglobally are giving emphasis on (i) optimizing the light distribution at both canopy level of individual tree and at orchard level, (ii) Improved biomass partition among the fruiting shoots, and (iii) avoiding competition with vegetative sinks (Fumey *et al.*, 2011; Miller *et al.*, 2001).

Tree canopy architectures can be manipulated different ways by adopting pruning methods, training methods to control tree form, dwarf scion cultivars selection to control tree size, and closer tree spacing to control orchard density (Robinson *et al.*, 1991). Rootstocks also influence the growth habit, size-shape of the trees and tolerance of various adverse situations to the composite plant (Tubbs, 1974; Hirst and Ferree, 1996; Lauri *et al.*, 2006; Tworkoski and Miller, 2007). The canopy architecture modification through the physical methods like pruning, training involved labour and time, which are the major concern in tree fruit culture. The importance of modified plant architecture has already revealed greater production and productivity in rice and wheat during the 'Green Revolution' (Peng *et al.*, 1999; Hedden, 2003) and now it is also well realized in fruit trees also. Breeding for architectural modified tree species would revolutionize the tree fruits industries. The ideotype architecture of fruit trees are dwarf and semi-dwarf types, which have ability to give high productivity and quality with optimum use of natural recourses (Scorza *et al.*, 1999; Scorza, 2005). Further, availability of rootstocks and scion cultivars with desired traits are much limited in fruit trees. Hence, development of new scion and rootstockgenotypes with desired traits, using conventional breeding tools is laborious and also time consuming due to long juvenility, highly heterozygous natureof several fruit tree species (Grosser and Gmitter, 1990).

Under such situation biotechnological tools and methods are important to make the tree fruits breeding more sophisticated. The new biotechnological tools are helping in speedy generation of new genomic resources in tree species. Biotechnological tools are now allowing us to select the desired plant types with speed and efficiency (Merkle and Dean, 2000). Biotechnological tools like molecular markers are widely using for identification of quantitative trait loci (QTLs) associated with desired canopy architecture trait in tree species (Knäbel *et al.*, 2015). Now many QTLs has been detected which allowed identification of genomic regions involved in controlling the tree architectures (Costes and Gion, 2015). The markers tightly linked with traits of interest will further help in marker-assisted breeding in tree species. Therefore, genetic engineering is one of the important approaches to manipulate the tree canopy architecture, while other traits remain the same in the otherwise popular cultivars or rootstocks. While modifying the different growth regulating hormones pathways and incorporation of different genes related to plant growth traits can be successfully manipulated the canopy architecture. The different advanced biotechnological tools and methods which are directly and/or indirectly helping canopy architecture manipulation in tree species are discussing hereunder.

1. QTL Mapping for Canopy Architecture

The trait of interest governed by many genes or polygenes, controlled by locus within the genome of individual crop species referred to as quantitative trait loci (QTL). QTL analysis through the help of various molecular markers would help in identification of genomic regions controlling the canopy architecture traits (Costes and Gion, 2015). The molecular markers tightly linked to the trait of interest will help in marker assisted breeding in the tree crop species for desired canopy architecture. Using growth traits in different tree fruits species, QTL have been identified which control shape of the trees, for example in apple, traits related to canopy architecture (Segura *et al.*, 2007; 2008; 2009), columnar growth habit (Kenis and Keulemans, 2007; Tian *et al.*, 2005). Foster *et al.* (2015) were successful in identifying two QTL, *Dw1* and *Dw2* for inducing dwarng in apple rootstock. Other QTL were also detected in *Populus* (Bradshaw and Stettler, 1995; Wu, 1998) and *Eucalyptus* (Verhaegen *et al.*, 1997), spruce (Prunier *et al.*, 2013) and rubber (Souza *et al.*, 2013) for the canopy architecture. The important QTLs were listed in Table 1. The identification of these QTL will provide new molecular markers for the early screening of progenies in early stage of the desired canopy architecture in tree fruit species. Further, QTLs are not solely the genetic background of the genotypes from where they are originated but are often specific to their species of origin. Thus, tree genome can be critically fully explored and exploited for candidate gene identification for canopy architecture related traits as well as their characterization. The generated information would thus help in conventional as well as non-conventional breeding of tree species for specific tree canopy related traits.

Table 1: List of Quantitative Trait Loci Related to Tree Architecture Identified in Fruit Trees

Trait(s)	*Tree Species*	*Materials*	*Linkage Group*	*References*
Tree height	Apple	Fiesta × Discovery F_1	3,5,8	Liebhard *et al.* (2003)
Trunk basal diameter	Apple	Starkrimson × Granny S.F_1	1,7	Segura *et al.* (2009)
Tree height and circumference	Eucalyptus	*Eucalyptus urophylla* × *E. grandis* full-sibs	5,6,8,10	Bartholomé *et al.* (2013)
Number of laterals	Apple	Starkrimson × Granny S.F_1	4,13	Segura *et al.* (2009)
Number of laterals	Apricot	Harostar × Rougede MauveF_1	6	Socquet-Juglard *et al.* (2013)
Number of latent buds	Apple	Starkrimson × Granny S.F1	17	Segura *et al.* (2009)
% Branching nodes	Apple	Starkrimson × GrannyS.F1	4,13,17	Segura *et al.* (2009)
Sylleptic branching	Apple	Starkrimson × Granny S.F1	1,9,13,16	Segura *et al.* (2009)
Columnar growth habit	Apple	F_1s from *Malus* × *domestica*Borkh.	10	Tian *et al.* (2005) and Moriya *et al.* (2012)
Stem growth and form	Eucalyptus	*E. urophylla* × *E. grandis*full-sibs	1,2,3,5,8,11	Verhaegen *et al.* (1997)
Internodal length	Apple	Starkrimson × Granny S.F1	3,6,7,10,14,15,16	Segura *et al.* (2007, 2009)
Internodal length	Apple	Telamon × Braeburn F_1	3,14	Kenisand and Keulemans (2007)
Vigour and precocity	Pear	Old Home × Louise Bonne de Jersey	5,6	Knäbel *et al.* (2015)
Dwarfing	Apple	'M9' × 'Robusta5'	5, 6, 9, 10, 11, 12	Foster *et al.* (2015)

Partly taken Costes and Gion, 2015.

2. Development of Genomic Recourses in Tree Fruits Species

In the present era, plant science research is moving form plant genetic recourses to genomic recourses, while realizing its importance in tree improvement programme. The new biotechnological tools likenext generation sequencing (NGS) technologies like FLX-454, Illumina, SOLiD and Helicoseare helping to rapid accessibility of genomics resources. As the consequences, several genome sequencing programmes have been initiated in different tree species. The different types of genomic recourses like expressed sequence tags genomic survey sequences) and high throughput genome sequences have generated in different fruit crop species. The consequences of genome and transcriptome sequence project, 20 tree genome has been published and others important tree fruit crops the same is under construction in construction stage, particularly mango, olive, avocado*etc.* (Costes and Gion, 2015). Further these genome sequence data can be used for distinguishing the underlying pathways and genes responsible for canopy architectures in tree species (Table 2) (Hollender and Dardick, 2015). The RNA sequence and genome sequence can help in identification of alleles related to canopy architecture. These genomic recourses can be used for generation of new molecular markers, mapping and cloning of the genes, identification of QTL and marker- assisted selection in their precision genetic enhancement (Sonah *et al.*, 2011). The application of these will further make the tree improvement more speedy and efficient.

Table 2: The Published Genome Sequences of different Fruit Crops and Tree Species

Species	Botanical Name	References
African oil palm	*Elaeis guineensis*	Singh *et al.* (2013)
Apple	*Malusx domestic a*Borkh.	Velasco *et al.* (2010)
Banana	*Musa acuminata*	D'Hont *et al.* (2012)
Black cottonwood	*Populus tricocarpa*	Tuskan *et al.* (2006)
Cacao	*Theobroma cacao*	Argout *et al.* (2011)
Date palm	*Phoenix dactylifera* L.	Al-Mssallem *et al.* (2013)
Dwarf birch	*Betula nana*	Wang *et al.* (2013)
Japanese apricot	*Prunus mume*	Zhang *et al.* (2012)
Loblolly pine	*Pinus taeda*	Zimin *et al.* (2014)
Mulberry	*Morus notabilis*	He *et al.* (2013)
Norway spruce	*Picea abies*	Nystedt *et al.* (2013)
Papaya	*Carica papaya* Linnaeus	Ming *et al.* (2008)
Peach	*Prunus persica*	Verde *et al.* (2013)
Pear	*Pyrus bretschneideri* Rehd	Wu *et al.* (2013)
Rubber tree	*Heceabra siliensis*	Rahman *et al.* (2013)
Silver birch	*Betula pendula*	birchgenome.org
South American oil palm	*Elaeis oleifera*	Singh *et al.* (2013)
Sweet orange	*Citrus sinensis*	Xu *et al.* (2013)
White spruce tree	*Picea glauca*	Birol *et al.* (2013)
Eucalyptus	*Eucalyptus grandis*	Myburg *et al.* (2014)

3. Genetic Engineering for Tailoring Canopy Architecture

The identification of candidate genes related to different growth, flowering for different canopy architectures and their functional characterization has initiated. Genetic engineering has pave the way to incorporate the desirable gene(s)into the desired scion and rootstock cultivars through which desired manipulated canopy architecture is possible.Keeping this in view, genetic modification of different tree species have been done with different genes. Efforts related to manipulation of canopy architecture are enumerated here:

3.1 Hormone Biosynthesis Genes

The various plant hormones are responsible for growth and development of the tree species. The alteration of these genes may alter the growth pattern in tree species. Realizing this fact recently, there have been attempts to alter hormone biosynthetic genes to manipulate tree size, morphology and performance. Plant growth regulators play an important role in growth and development and they also affect formation of wood (Little and Savidge, 1987). Thus, the altering plant hormone synthesis in an important approach to manipulate different bio-synthetic pathways. Some advances have been made using TDNA auxin and cytokinin biosynthetic genes from *A. tumefaciens* and *A. rhizogenes*. It has been demonstrated that the manipulation Indole-acetic-acid (IAA) levels by the over expression of *iaa*M and *iaa*HA. *tumefaciens* T-DNA in *Populus*are expected to change growth, development and wood formation (Tuominen *et al.*, 1995). Over expression of these genes in the transgenic plants brings about alterations in growth pattern (Tzfira *et al.*, 1998). Leaf explants were transformed with *Agrobacterium* containing *rol*A gene and transgenic tissues developed into transgenic plants in apple rootstock M26 (Holefors *et al.*, 1998). It would be of commercial interest to introduce genes to make M26 dwarf without affecting rooting ability. The over expression of *GA-20 oxidase* gene from *Arabidopsis* in hybrid aspen has shown altered phenotype with faster growth in height, girth, larger leaves, larger xylem fibre and increased biomass (Eriksson *et al.*, 2000). This approach may have a direct application in fruit crops. Down regulation of *GA-20-oxidase* in transgenic fruit trees could generate plant types with reduced height. This will allow higher planting density, mechanical fruit harvesting and easier management with reduced labour costs.

3.2 Incorporation of Floral Induction Genes and Growth Cessation

Different genes have been identified, which were responsible for flowering and other growth cessation event in the plant species especially in the terminal flowering onetypes. *FT*-like genes, which are known to promoting flowering, are likely to be also involved in seasonal growth and dormancy in trees (Bohlenius, 2006; Foster *et al.*, 2013; Hsu *et al.*, 2011). These genes have also been shown to regulate branching (Srinivasan *et al.*, 2012). These studies have suggested the coordination of hormonal and nutritional controls of shoot elongation and growth cessation (Wingler, 2015) with genes previously known to be involved in flowering control, particularly *FT2*in Norway spruce (Karlgren *et al.*, 2013) and hybrid aspen (Ericksonn *et al.*, 2015). This is consistent with an environmental control of meristem activity, especially during the period of growth cessation prior to dormancy as emphasized by van der Schoot *et al.* (2014), but also suggests a combination with unfavourable nutritional status of buds within the plant during this period as emphasized by physiological studies on the role of carbohydrates in floral induction (Bangerth, 2009; Wahl *et al.*, 2013) or axillary shoot control (Mason *et al.*, 2014).

3.3 Genetic Engineering for Dwarfism

Leaf explants were transformed with *Agrobacterium* containing *rol*A gene and transgenic tissues developed into transgenic plants in apple rootstock M26 (Holefors *et al.*, 1998). It would be of commercial interest to introduce genes to make M26 dwarf without affecting rooting ability. This study examines if the introduction of the *rol*A gene in the apple rootstock M26 could reduce the shoot growth without changing rooting performance. Mercuri *et al.* (2001) developed an *Agrobacterium*-mediated transformation system for *Limonium*. leaf explants of a sterile hybrid L116 (*Limoniumotolepis* Kuntze× *L. latifolium* Kuntze) inoculated with *A. tumefaciens* LBA4404 harbouring the binary vector pBin19 containing a T-DNA fragment encompassing *rol A, B* and C genes of *A. rhizogenes*Ri plasmid (pRi1855). Three transgenic clones were selected and based on their phenotypic characteristics were named 'super compact', 'compact' and 'semi-compact' types.

The genes for height control that have received the most attention are those from the class of hormones known as gibberellic acids (GAs). The active forms of GA promote the elongation of plant cells. When active GA synthesis, or signaling caused by GA, is disrupted, cells do not fully elongate and plants remain short and stocky. When the activities of genes that degrade active GA are elevated, plants similarly remain short (Salamini, 2003). Research in the last several years has demonstrated that dwarfism is commonly associated with deficiencies in GA levels or signaling (Pena *et al.*, 1999; Spielmeyer *et al.*, 2002). The level of bioactive GAs is precisely controlled by several mechanisms, including transcriptional regulation of the genes encoding enzymes from both biosynthetic and catabolic pathways (Olszewski *et al.*, 2002). By modifying regulation of genes controlling GA flux, it is possible to modify processes regulated by GA and, thus, plant form. GA biosynthetic genes have received much more study than catabolic genes, but recently identification and functional analysis of the genes encoding the major catabolic enzyme, GA 2-oxidase (GA2ox), have revealed that the catabolic genes are also important in control of GA levels (Reid *et al.*, 1992; Thomas *et al.*, 1999; Sakamoto *et al.*, 2001). *GA2ox*s are encoded by small gene families in Arabidopsis and pea (Hedden and Phillips, 2000a; Schomburg *et al.*, 2002), but only one *GA2ox* gene has been identified in rice (*Oryza sativa;* Sakamoto *et al.*, 2001). Little is known about the developmental role of *GA2ox* genes in plants. Over expression of the rice *GA2ox* gene (*OsGA2ox1*) causes a dwarf phenotype and delay in reproductive development (Sakamoto *et al.*, 2001). Similarly, in *Arabidopsis*, overexpression of two *GA2ox* genes (*AtGA2ox7*and *AtGA2ox8*) causes dwarfing and delayed flowering under non-inductive short days (Schomburg *et al.*, 2002), and loss-of-function mutations of the same genes led to increased hypocotyl elongation and higher levels of active GAs. RNA *in situ* localizations of *OsGA2ox1* in rice revealed a ring-like expression pattern in the basal region of leaf primordia, near the vegetative shoot apical meristem (Sakamoto *et al.*, 2001). Very low mRNA levels were detected in the inflorescence meristem. This expression pattern suggests that *OsGA2ox1* may prevent GA flux into the shoot apical meristem, but *OsGA2ox1* is not expressed during flower development when high GA levels are needed. No *GA2ox* genes have yet been cloned from a tree, and their developmental function and regulation during woody plant development is completely unknown.

The over expression of *GA-20 oxidase* gene from *Arabidopsis* in hybrid aspen has shown altered phenotype such as faster growth in height, girth, larger leaves, larger xylem fibbers and increased biomass. These findings not only open up tree architecture but also wood quality opportunities (Eriksson *et al.*, 2000). Salamini (2003) inserted GA-related genes into a variety of hybrid poplar (*Populus tremula*x *P. alba*). Most of the genes were obtained from the model plant *Arabidopsis*, but one also originated from pea and from poplar itself. Most of the genes were strongly active throughout the plant because a generic kind of a gene expression controlling element was used (the 35S promoter). The products of these genes metabolized active GA faster than usual (*2-oxidase*), or inhibited the cellular signal needed for GA to activate the metabolic pathways required for cell growth (*gai, rgl2*). Seven distinct kinds of GA-inhibiting genes were used and gener-ated more than 160 different types of insertions. Busov *et al.* (2003) identified a dwarf transgenic hybrid poplar (*Populus tremula* × *P. alba*) after screening of 627 independent activation-tagged transgenic lines in tissue culture, greenhouse, and field environments. The cause of the phenotype was a hyper-activated gene encoding *GA 2-oxidase* (*GA2ox*), the major GA catabolic enzyme in plants. The mutation resulted from insertion of a strong transcriptional enhancer near the transcription start site. Over expression of the poplar *GA2ox* gene (*PtaGA2ox1*) caused hyper-accumulation of mRNA transcripts, quantitative shifts in the spectrum of GAs, and similarity in phenotype to transgenic poplars that overexpress a bean (*Phaseolus coccineus*) *GA2ox* gene. This poplar *PtaGA2ox1* sequence was most closely related to *PsGA2ox2* from pea (*Pisum sativum*) and two poorly known *GA2oxs* from *A. thaliana* (*AtGA2ox4* and *AtGA2ox5*). The dwarf phenotype was reversible through gibberellic acid application to the shoot apex. *MdDRB1* gene controlled tree architecture adventitious rooting and leaf curvature in apple transformed plant by modulating the accumulation of miRNAs and the transcript levels of miRNA target genes (You *et al.*, 2014). Transgenic approaches to producing semi-dwarf trees for use in arboriculture, horticulture, and forestry could have significant economic and environmental benefits, including altered fibre and fruit production, greater ease of management, and reduced risk of spread in wild populations.

This approach may have a direct application in tree fruit crops genetic enhancement. Down regulation of *GA-20-oxidase* in transgenic trees fruit could generate planttypes with reduced height. This will thus allow higher planting density, mechanical fruit harvesting and easier management with reduced labour costs above all higher productivity.

4. Conclusion

The tree fruit cropsbeing perennial in nature are difficult to improve solely through conventional breeding. Now biotechnological tools and methods are playing important role in making tree breeding efficient and easy. In this connection several QTLs have been identified in relation to canopy architecture, genomic regions have been identified for specific traits. The molecular markers linked to traits of interest will further helping in molecular-assisted breeding. These techniques would reduce the breeding periods in tree fruit species. There are several candidate genes identified and well characterized for the canopy architectural traits, which can be incorporated in the desired scion and rootstock genotypes without any changes in the recipient genome except for the genes, which was transformed. Further advancement of biotechnological tools like cisgenesis and genome editing technology can make more precise manipulation of traits,*i.e.* canopy architectures. The manipulation in canopy architecture traits will facilitates the tree fruits species to be more efficient and productive.

5. References

Al-Mssallem, I.S., Hu, S., Zhang, X., Lin, Q., Liu, W., Tan, J., Yu, X., Liu, J., Pan, L., Zhang, T. *et al.*, 2013. Genome sequence of the date palm *Phoenix dactylifera* L. *Nature Communications*, **4**: 2272.

Argout, X., Salse, J., Aury, J.M., Guiltinan, M.J., Droc, G., Gouzy, J., Allegre, M., Chaparro, C., Legavre, T., Maximova, S.N. *et al.*, 2011. The genome of *Theobroma cacao*. *Nature Genetics*, **43**: 101–108.

Bangerth, K.F. 2009. Floral induction in mature, perennial angiosperm fruit trees: similarities and discrepancies with annual/biennial plants and the involvement of plant hormones. *Scientia Horticulturae*. **122**(2): 153–163. http://doi.org/10.1016/j.scienta. 2009.06.014.

Bartholomé, J., Salmon, F., Vigneron, P., Bouvet, J.M., Plomion, C. and Gion, J.M. 2013. Plasticity of primary and secondary growth dynamics in *Eucalyptus* hybrids: A quantitative genetics and QTL mapping perspective. *BMC Plant Biology*, **13**(1): 120.

Birol, I., Raymond, A., Jackman, S.D., Pleasance, S., Coope, R., Taylor, G.A., Yuen, M.M.S., Keeling, C.I., Brand, D., Vandervalk, B.P. *et al.*, 2013. Assembling the 20 Gb white spruce (*Picea glauca*) genome from whole-genome shotgun sequencing data. *Bioinformatics*, **29**: 1492–1497.

Bohlenius, H. 2006. CO/FT regulatory module controls timing of flowering and seasonal growth cessation in trees. *Science*, **312**(5776): 1040–1043. http://doi.org/10.1126/science.1126038.

Borkh.)by QTL analysis of growth traits. *Molecular Breeding*, **19**(3): 193–208. http://

Busov, V.B., Meilan, Richard, Pearce, D.W., Ma, C., Rood, S.B. and Strauss, S.H. 2003. Activation tagging of a dominant gibberellin catabolism gene (*GA 2-oxidase*) from poplar that regulates tree stature. *Plant Physiol.*, **132**: 1283–1291.

Costes, E. and Gion, J.M. 2015. Genetics and genomics of tree architecture. *Adv. Bot. Res.*, **74**. http://dx.doi.org/10.1016/bs.abr.2015.05.001.

D'Hont, A., Denoeud, F., Aury, J.M., Baurens, F.C., Carreel, F., Garsmeur, O., Noel, B., Bocs, S., Droc, G. and Rouard, M. *et al.*, 2012. The banana (*Musa acuminata*) genome and the evolution of monocotyledonous plants. *Nature*, **488**: 213–217.

Davies, P.J. 1995. *Plant Hormones: Physiology, Biochemistry and Molecular Biology*, Kluwer Academic Publishers, London, pp. 6–7. doi.org/10.1007/s11032-006-9022-5.

Eriksson, M.E., Hoffman, D., Kaduk, M., Mauriat, M. and Moritz, T. 2015. Transgenic hybrid aspen trees with increased gibberellin (GA) concentrations suggest that GA acts. *New Phytologist*, **205**(3): 1288–1295. http://doi.org/10.1111/nph.13144.

Eriksson, M.E., Israelsson, M., Olsson, O. and Mortiz, T. 2000. Increased gibberellin biosynthesis in transgenic trees promotes growth, biomass production and xylem fiber length. *Nat. Biotech.*, **18**: 784–788.

Foster, T.M., Celton, J.M., Chagné, D., Tustin, D.S. and Gardiner, S.E. 2015. Two quantitative trait loci, Dw1 and Dw2, are primarily responsible for rootstock-induced dwarfing in apple. *Hort. Res.*, doi:10.1038/hortres.2015.1031.

Foster, T.M., Watson, A.E., van Hooijdonk, B.M. and Schaffer, R.J. 2013. Key flowering genes including FT-like genes are upregulated in the vasculature of apple dwarng rootstocks. *Tree Genetics and Genomes*, **10**(1): 189–202. http://doi.org/10.1007/s11295013-0675-z.

Fumey, D., Lauri, P.E., Guédon, Y., Godin, C. and Costes, E. (2011). How young trees cope with removal of whole or parts of shoots: an analysis of local and distant responses to pruning in 1-year-old apple (*Malus* x *domestica*; Rosaceae) trees. *Amer. J. Bot.*, **98**(11): 1737–1751. doi: 10.3732/ajb.1000231

Groover, A.T. 2005. What genes make a tree a tree? *Tr. Pl. Sci.*, **10**: 210– 214.

Grosser, J.W. and Gmitter, F.G.Jr. 1990. Protoplast fusion and citrus improvement. *Pl. Breed. Rev.*, **8**: 339–374.

He, N., Zhang, C., Qi, X., Zhao, S., Tao, Y., Yang, G., Lee, T.H., Wang, X., Cai, Q., Li, D. *et al.* (2013). Draft genome sequence of the mulberry tree *Morusnotabilis*. *Nature Comm.*, **4**: 1–9.

Hedden, P. 2003. The genes of the Green Revolution. *Trends in Genet.*, **19**: 5–9.

Hedden, P. and Phillips, A.L. 2000a. Gibberellin metabolism: new insights revealed by the genes. *Trends Pl. Sci.*, **5**: 523–530.

Hirst, P.M., and Feree, D.C. 1996. Rootstock effects on the bud development and floral formation of "Starkspur Supreme Delicious" apple. *Fruit Var. J.*, **50**(1): 25–34.

Holefors, A., Xue, Z.T. and Welander, M. 1998. Transformation of the apple rootstock M26 with the *rolA* gene and its influence on growth. *Plant Sci.*, **136**: 69–78.

Hollender, C.A. and Dardick, C. 2015. Molecular basis of angiosperm tree architecture. *New Phytologist*, **206**: 541–556. doi: 10.1111/nph.13204.

Hsu, C.Y., Adams, J.P., Kim, H., No, K., Ma, C., Strauss, S.H., Drnevich, J., Vandervelde, L., Ellis, J.D., Rice, B.M., Wickett, N., Gunter, L.E., Tuskan, G.A., Brunner, A.M., Page, G.P., Barakat, A., Carlson, J.E., DePamphilis, C.W., Luthe, D.S. and Yuceer, C. 2011. FLOWERING LOCUS T duplication coordinates reproductive and vegetative growth. *Proc. Nat. Acad. Sci. USA*, **108**(26): 10756–10761.

Knäbel, M., Friend, A.P., Palmer, J.W., Diack, R., Wiedow, C., Deng, C., Gardiner, S.E., Tustin, D.S., Schaffer, R., Foster, T. and Alspach, P. 2015. Genetic control of pear rootstock-induced dwarfing and precocity is linked to a chromosomal region syntenic to the apple Dw1 loci. *BMC Pl. Biol.*, **15**: 230.

Lakso, A.N. and Corelli-Grappadelli, L. 1992. Implications of pruning and training practices to carbon partitioning and fruit development in apple. *Acta Hort.*, **322**: 231–239.

Lauri, P.E., Maguylo, K. and Trottier, C. 2006. Architecture and size relations: An essay on the apple (*Malus x domestica*, Rosaceae) tree. *Amer. J. Bot.*, **93**(3): 357–368.

Liebhard, R., Kellerhals, M., Pfammatter, W., Jertmini, M. and Gessler, C. 2003. Mapping quantitative physiological traits in apple (*Malus* x *domestica* Borkh.). *Pl. Mol. Biol.*, **52**(3): 511–526.

Little, C.H.A. and Savidge, R.A. 1987. The role of plant growth regulators in forest tree cambial growth. *Plant Growth Regul.*, **6**: 137–169.

Mason, M.G., Ross, J.J., Babst, B.A., Wienclaw, B.N. and Beveridge, C.A. 2014. Sugar demand, not auxin, is the initial regulator of apical dominance. *Proc. Nat. Acad. Sci. USA*, **111**(16): 6092–6097.

Mercuri, Antonio, Bruna, Simona, Benedetti, Laura De, Burchi, Gianluca and Schiva, Tito 2001. Modification of plant architecture in *Limonium* spp. induced by *rol* genes. *Pl. Cell, Tiss. Org. Cult.*, **65**: 247–253.

Merkle, S.A. and Dean, J.F.D. 2000. Forest tree biotechnology. *Curr. Opp. Biotech.*, **11**: 298–302.

Miller, S.A., Broom, F.D., Thorp, T.G. and Barnett, A.M. 2001. Effects of leader pruning on vine architecture, productivity and fruit quality in kiwifruit (*Actinidia deliciosa* cv. Hayward). *Scientia Hort.*, **91**(3): 189–199.

Ming, R., Hou, S., Feng, Y., Yu, Q., Dionne, L. A., Saw, J.H., Senin, P., Wang., W., Ly, B.V., Lewis, K.L.T. *et al.*, 2008. The draft genome of the transgenic tropical fruit tree Papaya (*Carica papaya* Linnaeus). *Nature*, **452**: 991–996.

Moriya, S., Okada, K., Haji, T., Yamamoto, T. and Abe, K. 2012. Fine mapping of *Co*, a gene controlling columnar growth habit located on apple (*Malus* x *domestica* Borkh.) linkage group 10: Fine mapping of Co in apple (*Malus* x *domestica* Borkh). *Pl. Breed.*, **131**(5): 641–647.

Myburg, A.A., Grattapaglia, D., Tuskan, G.A., Hellsten, U., Hayes, R.D., Grimwood, J., Jenkins, J., Lindquist, E., Tice, H., Bauer, D., *et al.*, 2014. The genome of *Eucalyptus grandis*. *Nature*, **510**(7505), 356–362. doi: 10.1038/nature13308.

Nystedt, B., Street, N.R., Wetterbom, A., Zuccolo, A., Lin, Y.C., Scoeld, D.G., Vezzi, F., Delhomme, N., Giacomello, S., Alexeyenko, A., *et al.*, 2013. The

Norway spruce genome sequence and conifer genome evolution. *Nature*, **497**: 579–584.

Olszewski, N., Sun, T.P. and Gubler, F. 2002. Gibberellin signaling: Biosynthesis, catabolism, and response pathways. *Plant Cell*, **14**: S61–S80.

Palmer, J.W. 1980. Computed effects of spacing on light interception and distribution within hedgerow trees in relation to productivity. *Acta Hort.*, **114**: 80–88.

Palmer, J.W., Avery, D.J., Wertheim, S.J. 1992. Effect of apple tree spacing and summer pruning on leaf area distribution and light interception. *Sci. Hort.*, **52**: 114–139.

Pena, L., Martin-Trillo, M., Juarez, J., Pina, J.A., Navarro, L. and Martinez-Zapater, J.M. 2001. Constitutive expression of *Arabidopsis LEAFY* or *APETELA1* genes in citrus reduces their generation time. *Nat. Biotech.*, **19**: 263–267.

Peng, J., Richards, D.E., Hartley, N.M., Murphy, G.P., Devos, K.M., Flintham, J.E., Beales, J, Fish, L.J., Worland, A.J., Pelica, F. *et al.*, 1999. 'Green revolution' genes encodemutant gibberellin response modulators. *Nature*, **400**: 256–261.

Prunier, J., Pelgas, B., Gagnon, F., Desponts, M., Isabel, N., Beaulieu, J. and Bousquet, J. 2013. The genomic architecture and association genetics of adaptive characters using a candidate SNP approach in boreal black spruce. *BMC Genomics*, **14**: 368.

Rahman, A.Y.A., Usharraj, A.O., Misra, B.B., Thottathil, G.P., Jayasekaran, K., Feng, Y., Hou, S., Ong, S.Y., Ng, F.L., Lee, L.S. *et al.*, 2013. Draft genome sequence of the rubber tree *Hevea brasiliensis*. *BMC Genomics*, **14**: 75.

Reid, J.B., Ross, J.J. and Swain, S.M. 1992. Internode length in *Pisum*: A new, slender mutant with elevated levels of C19 gibberellins. *Planta*, **188**: 462–467.

Robinson, T.L., Lakso, A.N. and Ren, Z. 1991. Modifying apple tree canopies for improved production efficiency. *Hort. Sci.*, **26**(8): 1005–1012.

Sakamoto, T., Kobayashi, M., Itoh, H., Tagiri, A., Kayano, T., Tanaka, H., Iwahori, S. and Matsuoka, M. 2001. Expression of a gibberellin 2-oxidase gene around the shoot apex is related to phase transition in rice. *Pl. Physiol.*, **125**: 1508–1516.

Salamini, F. 2003. Hormones and the green revolution. *Sci.*, **302**: 71–72.

Sansavini, S., and Corelli-Grappadelli, L. 1992. Canopy efficiency of apple as affected by microclimatic factors and tree structure. *Acta Hort.*, **322**: 69–77.

Schomburg, F.M., Bizzell, C.M., Lee, D.J., Zeevaart, J.A.D. and Amasino, R.M. 2002. Over expression of a novel class of gibberellin 2-oxidases decreases gibberellin levels and creates dwarf plants. *Plant Cell*, **15**: 1–13.

Scorza, R. 2005. Theory and practice of genetically manipulating peach tree architecture. *New York Fruit Quart.*, **13**: 27–31.

Scorza, R., Bassi, D., Dima, A. and Rizzo, M. 1999. Developing new peach tree growth habits for higher density plantings. Presented at the 42nd Annual IDFTA Conference, 20–24 February. International Dwarf Fruit Tree Association, Hamilton, Ontario, Canada, pp. 18–20.

Segura, V., Cilas, C. and Costes, E. 2008. Dissecting apple tree architecture into genetic, ontogenetic and environmental effects: mixed linear modeling of repeated spatial and temporal measures. *New Phytologist*, **178**(2): 302–314. *http://doi.org/10.1111/j.14698137.2007.02374.x.*

Segura, V., Denancé, C., Durel, C.E., and Costes, E. 2007. Wide range QTL analysis for complex architectural traits in a 1-year-old apple progeny. *Genome*, **50**(2): 159–171. http://doi.org/10.1139/G07–002.

Segura, V., Durel, C.E., and Costes, E. 2009. Dissecting apple tree architecture into genetic, ontogenetic and environmental effects: QTL mapping. *Tree Genetics and Genomes*, **5**(1): 165–179.

Singh, R., Ong-Abdullah, M., Low, E.T.L., Manaf, M.A.A., Rosli, R., Nookiah, R., Ooi, L.C.L., Ooi, S.E., Chan, K.L., Halim, M.A. *et al.*, 2013. Oil palm genome sequencereveals divergence of interfertile species in Old and New worlds. *Nature*, **500**: 335–339.

Socquet-Juglard, D., Christen, D., Devenes, G., Gessler, C., Duffy, B. and Patocchi, A. (2013). Mapping architectural, phenological, and fruit quality QTLs in apricot. *Pl. Mol. Biol. Rep.*, **31**(2): 387–397.

Sonah, H., Deshmukh, R.K., Singh, V.P., Gupta, D.K., Singh, N.K. and Sharma, T.R. 2011. Genomic resources in horticultural crops: Status, utility and challenges. *Biotech., Adv.*, **29**: 199–209.

Souza, L.M., Gazaf, R., Mantello, C.C., Silva, C.C., Garcia, D., Le Guen, V., Cardoso, S.E., Garcia, A.A and Souza, A.P. 2013. QTL mapping of growth-related traits in a full-sib family of rubber tree (*Hevea brasiliensis*) evaluated in a sub-tropical climate. *PLoSONE* **8**: e61238.

Spielmeyer, W., Ellis, M.H. and Chandler, P.M. 2002. Semidwarf (*sd-1*), "green revolution" rice, contains a defective gibberellin 20-oxidase gene. *Proc. Nat. Acad. Sci. USA*, **99**: 9043–9048.

Srinivasan, C., Dardick, C., Callahan, A. and Scorza, R. 2012. Plum (*Prunus domestica*) trees transformed with poplar FT1 result in altered architecture, dormancy requirement, and continuous flowering. *PLoS One*, **7**(7): e40715. http://doi.org/10.1371/journal.pone. 0040715.

Thomas, S.G., Phillips, A.L. and Hedden, P. 1999. Molecular cloning and functional expression of gibberellin 2-oxidases, multifunctional enzymes involved in gibberellin deactivation. *Proc. Nat. Acad. Sci. USA*, **96**: 4698–4703.

Tian, Y.K., Wang, C.H., Zhang, J.S., James, C. and Dai, H.Y. 2005. Mapping Co, a gene controlling the columnar phenotype of apple, with molecular markers. *Euphytica*, **145**(1–2): 181–188. http://doi.org/10.1007/s10681–005–1163–9.

Tubbs, F.R. 1974. Rootstock/scion relations in horticultural crop physiology. *Scentia Hort.*, **2**(3): 221–230.

Tuominen, H., Sibon, F., Jacobsson, C., Sandberg, G., Olsson, O. and Sundberg, B. 1995. Altered growth and wood characteristics in transgenic hybrid aspen expressing *Agrobacterium tumefaciens* T-DNA indoleacetic acid-biosynthetic genes. *Plant Physiol.*, **109**: 1179–1189.

Tuskan, G.A., DiFazio, S., Jansson, S., Bohlmann, J., Grigoriev, I., Hellsten, U., Putnam, N., Ralph, S., Rombauts, S., Salamov, A. *et al.*, 2006. The genome of black cottonwood, *Populus trichocarpa* (Torr. and Gray). *Sci.*, **313**: 1596–1604.

Tustin, D.S., Cashmore, W.M. and Bensley, R.B. 1998. The influence of orchard row canopy discontinuity on irradiance and leaf area distribution in apple trees. *J. Hort. Sci. Biotech.*, **73**: 289–297.

Tustin, D.S., Hirst, P.M. and Warrington, I.J. 1988. Influence of orientation and position of fruiting laterals on canopy light penetration, yield, and fruit quality of 'Granny Smith' apple. *J. Amer. Soc. Hortic. Sci.*, **113**: 693–699.

Tworkoski, T. and Miller, S. 2007. Rootstock effect on growth of apple scions with different growth habits. *Sci Hortic.*, **111**(4): 335–343.

Tzfira, T., Zuker, A. and Altmann, A. 1998. Forestry biotechnology: Genetic transformation and its application to future forest. *Trends Biotech.*, **16**: 439–446.

van der Schoot, C., Paul, L.K., and Rinne, P.L.H. 2014. The embryonic shoot: A lifeline through winter. *J. Expt. Bot.*, **65**(7): 1699e1712. http://doi.org/10.1093/jxb/ert413.

Velasco, R., Zharkikh, A., Affourtit, J., Dhingra, A., Cestaro, A., Kalyanaraman, A., Fontana, P., Bhatnagar, S.K., Troggio, M., Pruss, D. *et al.*, 2010. The genome of the domesticated apple (*Malus* x *domestica* Borkh.). *Nature Genet.*, **42**: 833–839.

Verde, I., Abbott, A.G., Scalabrin. S., Jung, S., Shu, S., Marroni, F., Zhebentyayeva, T., Dettori, M.T., Grimwood, J., Cattonaro, F. *et al.*, 2013. The high-quality draft genome of peach. *Nature Genet.*, **45**: 487–494.

Verhaegen, D., Plomion, C., Gion, J.-M., Poitel, M., Costa, P. and Kremer, A. 1997. Quantitative trait dissection analysis in Eucalyptus using RAPD markers: 1. Detection of QTL in interspecic hybrid progeny, stability of QTL expression across different ages. *Theor. Appl. Genet.*, **95**(4): 597–608.

Wagenmakers, P.S. 1995. Light relations in orchard systems. Thesis Wageningen. CIP– gegerens Koninklijke Bibliotheek, Den Haag, the Netherlands, p. 149

Wagenmakers, P.S., Nijsse, F. and De Gendt, C.M.E. 1991. Planting systems and light climate. Research Station for Fruit Growing, Wilhelminadorp, The Netherlands. *Ann Rep*. p. 39–40.

Wahl, V., Ponnu, J., Schlereth, A., Arrivault, S., Langenecker, T., Franke, A., Feil, R., Lunn, J.E., Stitt, M. and Schmid, M. 2013. Regulation of flowering by trehalose-6-phosphate signaling in Arabidopsis thaliana. *Sci.*, **339**(6120): 704–707. http://doi.org/10.1126/science.1230406.

Willaume, M., Lauri, P-É. and Sinoquet, H. 2004. Light interception in apple trees influenced by canopy architecture manipulation. *Trees*, **18**: 705–713.

Wingler, A. 2015. Comparison of signaling interactions determining annual and perennial plant growth in response to low temperature. *Front. Pl. Sci*, **5**. http://doi.org/10.3389/fpls.2014.00794.

Wu, J., Wang, Z., Shi, Z., Zhang, S., Ming, R., Zhu, S., Khan. M.A., Tao, S., Korban, S.S., Wang, H. *et al.*, 2013. The genome of the pear (*Pyrusbrets chneideri* Rehd.). *Genome Research*, **23**: 396–408.

Wu, R. and Stettler, R.F. 1998. Quantitative genetics of growth and development in Populus. III. Phenotypic plasticity of crown structure and function. *Heredity*, **81**(3): 299–310. http://doi.org/10.1046/j.1365–2540.1998.00397.x.

Xu, Q., Chen, L.L., Ruan, X., Chen, D., Zhu, A., Chen, C., Bertrand, D., Jiao, W.B., Hao, B.H., Lyon, M.P. *et al.*, 2013. The draft genome of sweet orange (*Citrus sinensis*). *Nature Genet.*, **45**: 59–66.

You, C.X., Zhao, Q., Wang, X.F., Xie, X.B., Feng, X.M., Zhao, L.L., Shu, H.R. and Hao, Y.J. 2014. A dsRNA-binding protein MdDRB1 associated with miRNA biogenesis modies adventitious rooting and tree architecture in apple. *Pl. Biotech. J.*, **12**: 183–192.

Zhang, Q., Chen, W., Sun, L., Zhao, F., Huang, B., Yang, W., Tao, Y., Wang, J., Yuan, Z., Fan, G. *et al.*, 2012. Th genome of *Prunus mume*. *Nat. Comm.*, **3**: 1318. doi: 10.1038/ncomms2290.

Zimin, A., Stevens, K.A., Crepeau, M.W., Holtz, M.A., Koriabine, M., Marcais, G., Puiu, D., Roberts, M., Wegrzyn, J.L., de Jong, P.J. *et al.*, 2014. Sequencing and assembly of the 22-gb loblolly pine genome. *Genet.*, **196**: 875–890.

13

Silencing the Pathogen in a Non-transgenic Scion through SiRNA

Akella Vani

Principal Scientist and Head,
Division of Biotechnology, Indian Institute of Horticultural Research,
Hessaraghatta, Bengaluru – 560 089, Karnataka
E-mail: vaniakella@gmail.com

Almost 3 decades ago, Beachy's lab demonstrated that the virus attacking a plant can itself be a source of resistance. Here it was shown that the coat protein gene of tobacco mosaic virus (TMV), when expressed constitutively in the plant, conferred resistance to the virus. This method called as pathogen derived resistance became one of the most versatile approaches to confer resistance. Several other viral genes, other than the coat protein, also conferred the same level of resistance. While early successes were obtained by expressing the protein product of the viral genes, later it became evident that resistance can also be conferred by the RNA itself without having it to be translated in to a protein product.

Post transcriptional gene silencing has been studied extensively and has been shown to operate in all classes of living organisms from the simplest single celled to the most complex eukaryote. It is one of the major ways in which an expression of a gene is regulated. Here the transcript, which is an mRNA molecule, is degraded into precisely sized fragments, at predictable sites, by a concerted effort of several proteins produced by the organism. It is thought to have arisen primarily to combat invading nucleic acid molecules such as viruses and transposons *etc.* and has subsequently evolved to be a part of a complex mechanism to (down) regulate the expression of protein from a transcript during various developmental stages. It is an important fundamental metabolic process required at the cellular level for cell development, maturation, and is also required to respond to changes in the environment. Additionally, the small interfering RNA can also methylate DNA of viral genomes or DNA of the host genome, causing them to be inactivated

Genes control the expression of a phenotype by coding for a protein. For doing so, it makes a copy of itself in the form of an RNA molecule which is then transported to the cytoplasm, and is used as a template to code for a protein. The fact that this straight forward gene-to-phenotype has its own twists and turns was learnt, when it was realized that the RNA copy of the gene undergoes several modifications before it is translated. Further, it became evident that not all RNA gets translated to a protein product, and that the RNA itself can be silenced such that no protein product comes out.

RNA Silencing

RNA is degraded by cellular machinery in a sequence specific manner involving several players such as a protein moiety called Dicer, whose role is to recognize double stranded RNA and dice it into smaller segments of 21 nucleotides each, RNA Induced Silencing Complex (RISC) whose role is to grab the 21mer and search for other RNA molecules with a close homology to this 21mer and dice them. The 21mer itself is an important player because it base pairs with homologous RNA and causes it to be replicated by RNA dependent RNA polymerase (RDRP) which then forms a substrate for Dicer to act upon and cleave the larger double stranded molecule in to smaller 21mers. Thus the population of small 21mer RNAs increases in the cell. Owning to their small size, they are able to move from cell to cell or through the conducting vessels, throughout the plant and initiate a systemic response.

How do Viruses Establish themselves Despite RNA Degrading Mechanisms in the Host?

Viruses have evolved mechanisms to suppress host silencing components. This has been achieved by viral proteins involved in movement, replication, or coating *etc.* or in other words, no new proteins evolved, but existing proteins of the virus doubled up to carry out additional function of suppressing the host silencing mechanism. Several viral proteins carry out multiple functions and one such protein found in all members of the potyvirus group is the helper component protein (Hc-Pro). The main function of this protein is to proteolytically cleave the long polyprotein of the virus. It is also involved in aphid transmission as well as cell to cell movement of the virus. The most important role however, is to bind to SiRNA as well as double stranded RNA and prevent both the RISC and the DICER from carrying out their functions, thereby suppressing silencing. HC-Pro is capable of acting in a manner where even after silencing has set in the host, it can block its progression. Several viruses are known to produce proteins whose role is to sequester and bind to the double stranded 21-mer SiRNA with the result that the single stranded SiRNA is not available for the RISC to bind. Examples of this include the protein 2b of cucumber mosaic virus, P22 of tomato chlorosis virus and NSs gene of tomato tospo virus. The2b protein of CMV virus has been shown to initiate silencing at the growing points of the plant. Some viruses code for more than one silencing suppressor. A case in point here is that of the Citrus Tristeza Virus (CTV) which is known to code for three different proteins *viz.*, p20, p23 and the coat protein (CP) that exhibit suppressor activity.

Having understood the myriad mechanisms of how a virus is able to establish and sustain an active infection in a host, scientists have aimed their strategies to beat the virus in its own game. Viral genes involved in silencing the host have been successfully targeted through dsRNA (double stranded RNA) constructs in transgenic plants. This approach, variously called as RNAi mechanism or SiRNA strategy, involves knocking out key proteins of the virus in addition to viral genes involved in suppression of host silencing mechanism. The transitivity property of the small RNA molecules, namely amplification of the original SiRNA by the host RDRP, and their movement systemically throughout the plant can be further exploited in a transgenic system. Here the transgenic plant can be used as a rootstock to produce the necessary silencing signals and the non-transgenic scion of a commercial hybrid can be grafted on to it which then gets protected against the invading virus because of the pre-made SiRNA molecules provided to the scion by the rootstock. Automated machines to do the grafting are available commercially which can be used by a nursery man to sell the grafted seedling. The non-transgenic scion actually helps address a number of issues simultaneously. The transgenic event need not have commercial qualities, further it can even be a male sterile line (with a restorer). This way, selecting an event may be based only on its ability to produce the SiRNA molecules. Each commercial company can graft their specific hybrid on to the transgenic event thereby obviating the need to develop different events or the need to introgress the event into their own hybrid background. The scion being non-transgenic, biosafety issues regarding spread of the transgene through pollen into wild relatives or into other cultivated varieties does not arise. The scion will not contain novel proteins but may contain SiRNA molecules derived from the transgene located in the rootstock. However these molecules are no different from those present in an infected plant. This technology can be extended to include resistance to multiple viruses simultaneously.

Existing orchards with viral infections may also be cleansed of their virus by grafting a non-flowering transgenic shoot producing the necessary SiRNA signals which would then move throughout the plant and effectively halt viral replication. Rejuvenating an existing orchard is definitely advantageous than replanting one. The SiRNA approach thus holds a lot of promise for the future for doubling farmers income by 2022.

Technology Innovations

14

Technology Driven Horticulture Development: Suggested Model for Small and Marginal Farmers

S.P. Ghosh

Ex-DDG (Horticulture),
ICAR and Vice President, HSI, New Delhi – 110 012
E-mail: spghosh100@gmail.com

Although Indian horticulture has attained gigantic size with production 283.5 million tonnes during 2014-15, it is still dominated by unorganised sector. Low productivity and profitability in a number of crops is a challenge and post-harvest loss of perishable fruits and vegetables is still very high. The diversification towards horticulture crops and demand projection of fruits and vegetables by 2020 show that diversification in consumption pattern has becomes more pronounced with income growth.

Horticulture development in India received institutional support since VIII FY Plan, when policy changes in terms of financial incentives started with a financial outlay of Rs. 1,000 crores. This pace was continuously upgraded to 50 per cent and 100 per cent increase for IX and X Plans, respectively. Major Indian policy initiatives impacted commercial horticulture promotion in the country. The benefits of number of fiscal reliefs and initiatives to encourage commercialization and value-addition to horticulture products, however, could not be accessed by small and marginal farmers very effectively, although small holders constitute more than 80 per cent of the total farming households in the country. As a result total area under horticulture crops is stagnated and remained low.

Small and marginal farmers, having small land holdings for cultivation dominate in many Asian countries. China stood first in concentration of small holdings (47 per cent of small holders of the world in 1997), followed by India, Indonesia, Bangladesh and Vietnam. High dependency on agriculture for livelihood by large number of small holders is going to become a major concern for policy makers. A time has come to discuss and deliberate on public policy and governmental interventions for small holder horticulture.

In India technological development in horticulture crops is reasonably good and technology interventions can definitely influence faster growth. Land productivity of fruits and vegetables being much higher compared to that in cereals (rice and wheat in particular), proper transfer and adoption of advanced technologies should be able to improve profitability of horticultural crops significantly. Small farmers access to technology and provision funding support for enabling them to procure purchased inputs in time are essential production supports for which institutional arrangements must be in place.

In the past, institutional support in this form of Horticulture Missions (TMNE/HTM for North East and western Hill states and NHM for the rest of the states) in India have helped in increasing horticultural crops production significantly. Similarly, in the neighbouring Bangladesh two Mega projects (Northwest crop diversification project - NCDP add Second crop diversification project- SCDP, under ADB funding support) have increased production of HVCs and helped in integrating small holder horticulture with commercial horticulture. An analysis of the concepts, designs and impacts of these four flagship projects has been attempted for designing a suitable model for small holder farmers in the sub-region. Group extension (farmer mobilization as SHGs), capacity building (training and technology demonstrations), technology transfer and micro financing (micro credit) will be in the key words for working out viable models applicable for small holder horticulture. Capacity building for productivity enhancement should be possible under the suggested model.

1. Need for Promoting Small Holder Horticulture

In India, the average size of land holding declined to 1.32 ha in 2000-01 from 2.30 ha in 1970-71. It is estimated that if the trend continues, the average size of holding would be about 0.68 ha in 2020 and 0.32 ha in 2030 (source: Vision 2030, ICAR, 2011). This trend needs to be halted through crop diversification towards high value crops (HVCs), fruits and vegetables in particular.

Small farm holders prefer cereals for consideration of (i) food security; (ii) low risk and (iii) easy market access. Resource poor farmers generally cannot access modern technology and integrate the perishable horticulture crops in value chains: Small holder farmers are however, as competitive as large farmers where institutional support and assistance are available. Generally, small farmers have higher land productivity in short term but over the long term land productivity tends to decline.

In order to increase larger participation of small holders in raising fruits and vegetable crops in their land holdings, public policy support and governmental interventions in terms of production and value-addition supports are essential. It is imperative to ensure small holders effective exposure to new technologies through periodic training and re training at a time when economy is diversifying and per capita income is rising. Institutional innovations for achieving the objectives of inclusive growth of small farm holders have become more important in Indian agriculture. Crop diversification through expansion of area under HVCs (fruits and vegetables) can only ensure increased income and employment generation, Horticulture led agrarian prosperity is a distinct opportunity and should receive due support.

2. Approach for Sustainable Small Holding Horticulture

Growth rate of horticultural crops after 2005-06 was almost double the growth rate of non-horticultural crops (Chand, Raju and Chauhan, 2014). In the growth of horticulture crops, small holders' contribution is reasonably good. Short gestation crops (*e.g.* vegetables, potato) fitted well in cropping patterns in many states of India. Optimum utilization of water, seed fertilizers, pesticides are important for small - marginal land holdings. Precision farming technology adoption for crop cultivation with funding support of RKVY in Tamil Nadu have made positive productivity gains in horticultural crops. There is a scope of large yield improvements in small holder farmers through focused R&D initiatives. Release of wide range of improved crop varieties suitable for different agro-climatic zones of the country and greater availability of seeds, quality planting material of improved varieties have influenced in promotion of high value crops (HVCs). Technology interventions and production support are therefore, very crucial. Group extension of precision low input technologies will be more appropriate for resource poor small farmers. To overcome production constrains such as drought, pest and disease problems and inadequate supply of quality planting materials, development of project for small-marginal farmers will be essential. For desired improvement in productivity and economic and ecological security a 'business as usual' approach will not work.

3. Experience Sharing: Horticulture Development Projects in India and Bangladesh

a) Indian Experience

In the recent past two mega projects, namely (i) Technology Mission in North east (TMNE) covering seven states of North East, subsequently renamed as Horticulture Technology Mission (HTM) including additional states of Sikkim, Jammu and Kashmir, Himachal Pradesh and Uttarakhand promotion of hill horticulture, and (ii) National Horticulture Mission (NHM) covering all remaining 18 states to promote holistic growth of horticulture through an area based regionally differentiated development strategies/ approach. Both the missions were central sector schemes with adequate funding support.

The National Horticulture Mission (NHM) launched in 2005 in 18 states and 3 Union Territories (UTs) aimed to double the national horticulture production to 300 million by 2011-12. In is reported that upto 2009-10, 16.57 lakh ha additional area has brought under different horticultural crops under NHM alone. Similarly, under HTM in NE states alone area expansion was 5.75 lakh ha in fruits and 6.15 lakh ha in vegetables. The perennial fruit species have a long productive life and full potential in production will be realised in coming years. In spite of good funding support and area expansion under these two flagship projects, total horticulture production of 283.5 million tonnes could only be achieved (falling short of 2012 target of 300 million tonnes). However, the benefits in terms of market arrivals became visible. Fruits cultivation is considered as a game changer in any parts of rural India but yet fruits import in terms of value exceeds that of export in recent years.

b) Bangladesh Experience

Two joint venture projects of Ministry of Agriculture, Government of Bangladesh and the Asian Development Bank (ADB) of recent times (2000-2016), namely (a) Northwest Crop Diversification Project (NCDP) and (b) Second Crop Diversification Project (SCDP) with project area spread over 43 districts in the Northwest and southwest Bangladesh were instrumental in boosting horticulture crops (fruits, vegetables, spices, flowers, medicinal plants, potato-maize relay cropping *etc.*) production towards commercialization and making country self sufficient in HVC production and supply. Big production gains in vegetable crops including potato, and some tropical fruits (bananas, pineapple, papaya, mango, jack fruit *etc.*) already experienced. According to ADB (source: Project Completion Report, Bangladesh: Northwest Crop Diversification Project, November, 2011), project interventions increased annual production of HVCs by 2.7 million tonnes of fresh produce mainly through increasing the area under HVCs and adoption of modern high yield technologies in project area. In both the crop diversification projects (project beneficiaries) are small-marginal farmers, who spared part of their farm land for horticulture crop production. Against the credit linked area expansion target of 50,000 ha in the NCDP project area, with the participation of non-project farmers the additional area of 1,53,330 ha could be recorded under HVCs. The demonstration impacts of high productivity and profitability encouraged even the non-project farmers to get involved in the HVC production.

Also the loanee farmers repaid the full money disbursed to them by the NGOs. In the SCDP, also the trends in area expansion and production gains are in encouraging. Knowledge and skill development through farmers trainings, technology demonstrations in farmers field and provision of micro-credit made small holder horticulture a successful venture in Bangladesh, where pressure on land is very high and poverty alleviations in rural area is still a priority economic agenda.

4. The Suggested Model for Small and Marginal Farmers

The National Commission on Farmers (NCF) of MoA, Government of India, developed the concept of organizing "Small Farmers Horticulture Estate (SFHE)", covering an area of 200 to 400 hectare each to capture the economies of scale for involving small farmers in the organized horticulture production and marketing systems. The involvement of SAUs and ICAR institutions in identifying transferable technologies to fit in the 'regionally differentiated developmental programmes on cluster basis' under NHM; Mini Mission- I In for generation of need based location specific technology in HTM/TMNE and inbuilt mechanism of involving Technical Assistance (TA) Team comprising of International and National experts/specialists in NCDP and SCDP in Bangladesh are appropriate options for technology interventions to promote Technology Driven horticulture.

4.1 Brief Design of the Model

The different activities are as follows:

a. Group Formation

Each SHG with 15-25 members formed and trained with technical knowledge to undertake horticulture production profitably. State department officials, reputed NGOs will be responsible for the formation of groups based on local needs and agro- economic consideration.

b. Group Capacity Building

To promote group based extensions support through training and demonstrations of technologies involving ZRS of SAUs, KVKs, ATMA *etc.* Development of skills to produce quality planting materials vermi-compost, rain water harvesting, moisture conservation and others and post-harvest handling and primary processing of perishables commodities should be aimed. Woman friendly technology developments and extension are some focused areas of training under production support.

c. Group Marketing

In the pattern of "Farmers Producer Companies" in Tamil Nadu and similar others for linking farmers with markets may be ventured. For post-harvest handling operations, pack-house facilities need to be created.

d. Micro-credit Support

Micro-finance institutions (MFIs) as project partners to provide micro credit support with easy loan terms (low interest rates) and repayment in 10-12 instalments are possible options. The arrangements made in ADB funded crop diversification projects in Bangladesh maybe as possible arrangements be considered in India. NABARD may play the role of ADB (in Bangladesh). Logical Frame Work (Log Frame) of the developmental project should include only deliverable items and may vary from place. Details of operational modalities and loan agreements can be worked out once the concept is accepted and policy support provided.

15

Crop Regulation in Fruit Crops for Improving Quality and Income of Farmers

W.S. Dhillon* and Manish Das

Horticultural Science Division,
Indian Council of Agricultural Research,
Krishi Anusandhan Bhawan-II, New Delhi – 110 012
**E-mail: wasakhasingh @ yahoo.com*

Crop regulation is the basis for obtaining regular and quality fruits. A range of methods are used to increase production with enhanced fruit quality by crop regulation. It can be achieved through manual thinning, chemical thinning, selective harvesting, training, summer and winter pruning, prevention of pre-harvest fruit drop *etc.* It is well known that competition between fruits is one of the principal factors affecting fruit growth, ripening and quality. Fruit competes with each other for assimilates according to their relative sink strength and the availability of assimilates. When crop density is high owing to high fruit set conditions, competition for assimilates is intense, which results not only in decreased fruit size and quality but also promotes alternate bearing. Crop density, a measure of fruit crop, influences fruit growth and ultimately fruit size at harvest. It is largely dependent upon flowering and fruit set.

Under ideal conditions, most fruits often set heavier crop that cannot be adequately sized to meet market requirement for realization of higher price. This necessitates retention of well spaced fruits after selective removal of some on the trees. Leaf to fruit ratio fairly guides about the extent of thinning to be carried out in a particular orchard. Usually 30-40 leaves are sufficient to produce a good quality fruit. Though, thinning may reduce the total yield, but net return increases as larger fruits fetch much higher price in the market.The regulation of crop load begins at pruning. The management of irrigation is also vital, especially during the critical period of fruit development, where the most rapid fruit growth occurs. Tree nutrition also affects the eventual fruit size. However, thinning has a much greater influence on fruit size by adjusting optimum crop load on the tree during heavy cropping year. It can be achieved manually by selective removal of young fruitlets at pea size stage. In Red-Haven peach cultivar, retention of 4-5 fruits per shoot at the time when fruits are of pea size (14-15 mm in dia) assure optimum leaf to fruit ratio and higher production of large size fruits. However, manual thinning is labour intensive, expensive and only small number of trees in an orchard can be thinned at the optimum time.

Use of chemicals including plant growth regulators (PGRs) as chemical thinning agents is gradually replacing hand and mechanical thinning in stone fruits. Various considerations like tree age and vigour, irrigation, pruning and fertilizer practices followed, chemical used, its concentration and time of application and local climatic conditions should be taken into account, while adopting chemical thinning. Fruit set in stone fruits is often limited by frost injury to flowers in spring frost affected areas. Evaporative cooling of flowering buds in spring has a delaying effect on bloom in stone fruits. Trees are sprayed with water during day time at frequent intervals from bud swell stage until bloom. Frequency of tree misting depends upon the evaporation rate, which is done as and when water film on the aerial parts of the tree dries up. Evaporation of water from the shoots cools the buds and therefore delays their break. Fall (mid-October) application of growth regulators like GA_3 and ethrel delay flowering in spring by 3-7 days depending upon species and thus markedly increase fruit set.

Principles

The basic principle of crop regulation is to manipulate the natural flowering and fruiting is fruit trees in a desired season of the year that contribute to increased fruit yield, quality, profitability and sustainability of the environment by reducing the use of excessive pesticides. Many fleshy fruit species like strawberry, pineapple, mulberry, and pome fruits (apple, pear) important to humans additionally develop mature fruit tissues, including extra-carpellary floral components, in which the receptacle, bract, calyx, and floral tube (the fused base of floral organs), respectively, constitute the majority of mature fruit tissue. Even species with fruit derived from carpel tissue exclusively can display a range of developmental programmes, spanning the relatively uniform single expanded carpel or drupe of stone fruit to the differentiated carpel tissues giving rise to the peel (flavedo) and multi-carpel flesh of citrus and banana.

Molecular Basis

Fruit development and ripening are unique to plants and represent an important component of living beings. Recent discoveries have shed light on the molecular basis of developmental ripening control, suggested common regulators of climacteric and non-climacteric ripening physiology. Analyses of fruit-ripening mutants and ripening-related gene expression suggest higher levels of a developmental regulatory cascade that remain to be defined. The role of light signalling in fruit carotenoid accumulation is being examined and may represent a target for practical manipulation of fruit pigmentation and nutrient content. The continuing development of genomics tools, including ESTs and cDNA microarrays, for important fruit crops should foster accelerated discovery in fruit development and ripening research (Giovannoni, 2004).

Crop Regulators

Crop regulators are very helpful for growth and development of plant. The fruiting is regulated at different times of flower emergence and at different intensities to maintain the productivity of the plants. For example, to regulate the guava crop, it is essential to reduce the fruit set during the rainy season and subsequently increase the fruit set during winter season by the use of different chemicals like NAA, ethrel and urea (Boora *et al.*, 2016) *etc.* In guava, the profitability of crop regulation through chemical spray was maximum in case of 2,4-D followed by NAA. The different timings of crop regulation with different chemicals have produced different effects on fruit yield and quality. They have the potential of increasing plant productivity and quality through influence on various metabolic processes.

Pruning

Pruning is also a very effective method to regulate fruit yield through manipulations appropriately which is aptly proven in number of studies in different fruit crops. Plant growth regulators are known to improve fruit size, appearance and aril quality by direct effect on fruit growth and development or indirectly by regulating crops load, tree vigour and canopy architecture. The exogenous application of growth regulators has been found very effective in improving fruit size and quality of several fruit crops. Now-a-days commercial formulations of different hormones are available in the market and their use is relatively cheap and economical to the farmers. Many such studies were undertaken and evidences are noteworthy to take research further to prove that this aspect of crop regulation is of practical significance to bring economy to the doorstep of farmers to boost their economy and improve livelihood. However, studies are required to be carried out in many such important fruit crops in the light of climate change.

Pruning and Training Systems

In apple, many modern tree fruit orchards are planted high-density using dwarfing rootstocks and training systems designed for maximum sunlight interception, higher fruit yields and quality, and easier worker access. As a result, growers are seeing increased profitability and greater potential for enhanced mechanization of operations. Choosing the correct scion and rootstock combination is essential to ensure that trees will have the appropriate vigour and perform properly for the training methods used. The soil type, irrigation system, tree spacing, and management experience will also contribute to the success of a rootstock and training system combination. Researchers are conducting trials on new fruit tree training systems and rootstocks, to find advanced, economically feasible ways to grow highly productive trees with excellent fruit quality.

Fruit trees are pruned for a number of reasons including:

- ☆ **Architecture and size**: Fruit trees are pruned to establish the basic tree architecture and to limit the tree size.
- ☆ **Sunlight interception**: Pruning opens space in the tree canopy for sunlight to penetrate and support leaf photosynthesis and production of carbohydrates and energy that drive the growth of roots, shoots and fruits. Without adequate sunlight penetration during the growing season, flower bud production can be reduced with detrimental effects on cropping in the future.

- **Air circulation**: Pruning can reduce branch crowding, reduce the amount of foliage inside the canopy and enable better air circulation that reduces humidity and allows for better coverage of protectant products (*e.g.*, pesticides).
- **Strength**: Branch angles between a vertical trunk and lateral shoots are stronger as they approach a 90 degree angle. Narrow branch angles are typically weak and they may break under the weight of developing fruits.
- **Control vegetative growth**: Dormant pruning is an invigourating process that stimulates vegetative growth, while summer pruning is a devigorating process that reduces growth because it involves removing part of the functional leaf area of the tree.
- **Production**: Pruning is done to maintain the right balance between vegetative growth, flower production and fruiting. Dormant pruning is a reliable crop load management tool.
- **Tree health**: Trees are pruned to prevent the spread of disease or insect damage. This includes the removal and destruction of dead and/or infested wood.

Summer Pruning

Sometimes light pruning is done in the summer to let light penetration into the tree canopy, or to thin branches for training. Timing and amount of shoot removal is important since it can impact carbohydrate supply to developing fruit and also direct sun exposure of fruit that could lead to sunburn. Summer pruning to remove fire blight requires care to make cuts 8-10 inches below the canker and it requires disinfestation of tools between cuts to prevent disease transmission.

Mechanical Pruning

Hand pruning is laboured intensive and expensive for growers. Concerns about both the cost and the future availability of labour have been an incentive for research on mechanical pruning. Tall, narrow tree shapes and tree wall canopies are better suited for mechanical pruning. Research programmes should focus investigating techniques and machines to speed up the process.

Root Pruning

Root pruning is common in Europe to restrict root growth and canopy size in high-density orchards. This type of pruning promotes return bloom and fruit set in the next year. It is most effective when it occurs on both sides of the tree.

Flower Thinning

The decision on strategy for flower thinning should be done depending upon different factors such as site and in general weather conditions during blossom, cultivar, the number of blossom clusters and occurrence of pollinators such as wild bees, honey-bees *etc.* in the orchard, available equipment, availability of seasonal workers for thinning by hand and financial aspects. Depending on the flower set and the weather conditions especially during bloom, the time for thinning by hand may vary (Sinatsch *et al.*, 2014).

Tree Training

Training fruit trees starts when trees are planted, as setting the structure early helps young trees produce early. It's more efficient to train trees early by placing limbs in a position, rather than pruning them later. Growth in fruit trees is regulated by plant hormones. Cytokinins produced in the roots move up to the highest points in the tree to break dormancy. When the buds on the tips of limbs start to grow, auxin is produced, and moves down the limb through gravity, and inhibits the growth of lateral buds below the highest points, an occurrence known as apical dominance. The best way to overcome apical dominance is to train branches early to grow more horizontal and remove unwanted upright shoots. The key to producing a well-trained tree in the shortest amount of time is to follow a training programme consistently.

Training Systems

Training helps develop a strong tree architecture (framework) that can support crop loads. Training also helps in bringing young trees into production early. Basic training systems include:

- **Central Leader training**: Semi-dwarf and standard size rootstocks can be trained as freestanding central leader trees. A vigorous shoot near the centre is chosen at planting time, and competing upright shoots around it are removed. Limbs are spread so as not to compete with the central leader. This system is considered to be outdated. However, many modern systems have their origin from this method.
- **Open-Centre training**: Three or four shoots are chosen to form the main architecture during the first year, and competing limbs are removed. Other systems have developed from this basic tree architectural system.
- **Espalier training**: The tree structure is created in two dimensions by using trellis systems. The main trunk is tied to the wires. Buds are selected to form branches, and shoots are fastened to training wires. Other styles that built on this method include bush and palmette systems.

Advances in Crop Regulation

Prohexadione-Ca use in Apple and Pear

Prohexadione-Ca, as BAS 125 10W (a.i. 10 per cent), was tested as a growth retardant and fire blight control agent on young and mature apple (*Malus × domestica* Borkh.) and pear (*Pyrus communis* L.) orchards (Buban *et al.*, 2004). Single or repeated treatments significantly reduced shoot growth by 16-43 per cent. The decrease in internode length rather than in number of nodes was responsible for the retarded shoot growth. Treatments did not influence the trunk circumference increase or the return bloom, fruit set or yield the year after application. Prohexadione-Ca reduced the incidence of shoot blight in the apple and pear orchards significantly (Buban *et al.*, 2004).

Leaf Senescence of Fruiting Apple Trees

Key processes of autumnal leaf senescence in apple trees after commercial harvest date were studied by Tartachnyk and Blanke (2004). The study on changes in source-sink relationships and environmental effects on leaf senescence were evaluated. Results revealed that the photosynthetic potential of the senescing apple leaves remained high at 8-14 µmol CO_2 m^{-2} s^{-1} after the commercial harvest date. Delayed harvest enhanced autumn leaf photosynthesis due to larger *in vivo* RUBISCO activity in comparison with trees harvested at the commercial harvest date. Therein, transpiration was not affected by fruit harvest and stomata did not limit leaf photosynthesis. Nitrogen content declined more rapidly in the leaves of apple trees harvested at the commercial date, an indication of earlier onset of senescence and more rapid remobilization to the woody, perennial parts of the tree. The study concluded that the combined environmental factors and the time of fruit harvest determined the rate of autumnal leaf senescence in terms of photosynthetic activity, chlorophyll breakdown and nitrogen translocation (Tartachnyk and Blanke, 2004).

Root Pruning and Root IBA Application in Apple

The effects of timing of root pruning (RP) and application of IBA to roots of 15-year-old 'Cox's Orange Pippin'/M.9 were investigated by Retamales *et al.* (2004). Trees were maintained in a weed-free strip and received overhead irrigation. Treatments applied were: 1 control; 2 RP on 9 May (RP1); 3 RP on 29 June (RP2); 4 RP1 + 5,000 ppm IBA, and 5 RP2 + 5,000 ppm IBA. Root pruning was carried out by trenching and cutting (35 cm deep), 70 cm from the tree; trenches were refilled within 6 h. IBA was sprayed immediately after RP with a handheld sprayer. RP1 (9 May) significantly reduced leaf N, K, and Cu concentrations measured on 4 June with respect to control treatment, while IBA had no significant effect. On 3 August, the main effect of both RP treatments was to reduce significantly leaf N, P and K concentrations. Trees that received RP2 treatment had significantly lower leaf Mg and Ca concentrations than those that received RP1. The mean area of mid-season growth leaves was reduced significantly by RP. For this variable, a significant interaction occurred between IBA application and RP timing; thus, IBA application decreased mean leaf area at RP1, but it had the opposite effect when applied at RP2. In both vertical and horizontal shoots, RP1 had the greatest reduction in growth either with or without IBA. The effect of treatments on canopy development were indirectly established by measuring non-intercepted radiation through two visible radiation sensors placed on 29 June under each tree at 30 cm above ground level and 30 cm from tree trunk. Non-intercepted radiation was not affected by RP treatments when measured before RP2. Afterwards, both RP treatments, as well as IBA application reduced canopy development. When radiation interception was measured again, before harvest, only RP timing significantly affected this variable (Retamales *et al.*, 2004).

Floral Biology and Pollinizers for Walnut

In walnut, ninety genotypes were evaluated for floral biology and pollinizers selection in ICAR-CITH, Srinagar. Among these, 38 genotypes were found protogynous and 52 were protoandrous. Male flower blooming ranged from 7 to 23 days in different genotypes, whereas, female flower bloom ranged from 6 to 34 days. The highest blooming duration was recorded with CITH-W- 23, whereas, BP-3 recorded maximum male flower blooming period. Maximum synchronization of male and female bloom was recorded 13-16 days in CITH-W-26, CITH-W-40, CITH-W-37 and Nugget, whereas minimum of two days synchronization was recorded in Wussan-2. No synchronization of male and female flowers was observed with CITH-W-10, CITH-W-19, CITH-W-24, CITH-W- 48, Opex- culchery, Turtle, Wussan-8 and FPTBr-1 indicating that these varieties need other genotypes for pollination and fruit set.

Crop Regulation in Grapevine

Maintaining the most cost-effective balance between vegetative and reproductive growth is one of the most testing problems in modern viticulture. Grapevines which exhibit excessive vegetative vigour are likely to produce less fruit of reduced quality, and vines with inadequate vigour may be compromised in terms of their yield potential. The requirement for techniques to better manage excess vigour has become more acute in recent years with the increased use of irrigation, adoption of vigour-imparting rootstocks and the expansion of vineyards into cooler geographic regions. A number of strategies may be used to control vine vigour. Chemical growth regulators, although capable of reducing shoot vigour, have never received acceptance due to undesirable side effects and concerns over chemical residues. Devigorating rootstocks, likewise, may have the potential to control vigour but none are in wide commercial use. Restriction of the effective root volume, achieved through manipulation of planting densities, competition by cover crops, regulation of the soil volume wetted by drip irrigation or regulation of water availability can all achieve a degree of devigoration but often at the expense of fruit yield. Manipulation of vines through pruning and trellis design are probably the most commonly used methods for the control of shoot vigour. A high number of nodes retained at pruning combined with trellises, which allow open canopies have proved successful. Advances in the understanding of the physiological factors influencing shoot growth and transpiration have allowed the development of novel irrigation methods for the control of vine vigour. These techniques exploit the fact that chemical signals originating in the roots are primarily responsible for the control of shoot growth and transpiration. Stimulation of the production of these signals through partial drying of the root system results in a significant reduction in shoot growth and water-use while maintaining crop yield and improving fruit quality. These new techniques, in combination with appropriate pruning and trellising methods, are providing new viticultural tools for controlling vine vigour and water-use efficiency (Dry and Loveys, 1998).

Crop Regulation in Guava

Crop regulation has manipulated the natural flowering and fruiting of guava plant in desired season of the year that contributed to increased fruit yield, quality, profitability and sustainability of the environment by reducing the use of the frequency of the pesticides (Boora *et al.*, 2016). This concept is based on the fact that guava flowers are borne only on new, succulent, vigorously emerging vegetative growths. These new growth flushes can be either new emergences of lateral budon older stems or extensions of already established terminals of various size and vigour. The crop regulation could be achieved by the adoption of the various practices like withholding irrigation after harvesting during the months of April-May in Northern Indian plains. This results in the shedding of flowers and the tree goes to rest. The basin of the tree is dug up, manure and irrigated in June. After about 30-35 days the tree put forth profuse flowering and fruit matures in winter. Terminal portion of the shoots up to 20 or 30 cm length should be pruned between 20 to 30 April. Severe pruning in guava should always be avoided. Apply the recommended dose of fertilizers during the month of June to encourage vegetative growth in July-August for getting maximum flowering during August-September for winter season crop. To regulate the guava crop, it is essential to reduce the fruit set during the rainy season and subsequently increase the fruit set during winter season by the use of different chemicals like NAA, ethereal, urea *etc.*

The rainy season crop of guava is poor in quality and crop is affected by many biotic and abiotic stresses as compared to winter season crop. The winter season crops which ripen from second fortnight of October to first fortnight of January are superior in quality, free from diseases and pests and fetch higher income. This requires regulation of flowering to obtain most profitable crop by withholding irrigation, root exposure, pruning and thinning of flowers. In different regions various methods of crop regulations are followed depending on climatic factors, cropping pattern, cultivar *etc.* The flowering is more in summer season due to the break of winter stress that lead to more fruit production in rainy season. In this season due to high temperature and rainfall during fruit harvesting the duration of harvesting is reduced to 30 days that causes glut in the market, so the monetary return is low due to poor price and less demand in the market. Winter season crop is superior in quality which fetches higher prices than rainy season crop. In rainy season there is a serious attack of fruit flies which deteriorates its quality and fruits become unfit for human consumption. So far getting the quality fruits in guava only winter season crop should be taken and rainy season crop should be avoided.

Effect on Floral and Yield Parameters

In general, guava tree flowers twice a year, *i.e.* in April-May and August-September, of which fruits ripen in rainy and winter seasons, respectively (Gupta and Nijjar, 1978). Sometimes, a third flowering occurs in October-November (Singh and Kumar, 1993), particularly in Maharashtra and Tamil Nadu (Hayes, 1974). In guava, floral buds appear soon after the break of winter stress and new growth starts in the axil of the leaves. Floral buds appear soon after first pair of leaves mature, but there is no direct association between leaf appearance and flower production (Menzel and Paxton,1986). Flowers occur either singly or in cymose of 2-3 at leaf axils of current (Braganza, 1990). The bearing twig normally grows a few centimetres putting forth 4-5 pairs of leaves and thereafter, either flower buds start developing or twigs cease to grow till the next season. This is quite evident from the heaviest flowering that has always been obtained insummer season. Because food reserved is exhausted in flowering and vegetative growth during summer, the rainy season flowering for winter cropping is always less (Rathore, 1975).

Quality Improvement in Pomegranate

To regulate flowering and fruiting in pomegranate during *Ambe-bahar* under Punjab conditions, unopened flower buds were removed manually on and after 7, 15 and 22 April and 15, 22 and 30 April for two years in 'Ganesh-1' and 'Kandhari' cultivars of pomegranate, respectively. Control trees were allowed to flower continuously up to the end of the flowering period during both the years. Trees de-blossomed on earlier dates produced a substantially higher number of flower buds/flowers; percent fruit set, fruit weight with uniform fruit size; and the lowest flower buds/flowers drop over the control. Uninterrupted flowering in control trees for about two months produced the maximum number of fruits in both the cultivars. Physico-chemical attributes, *viz.*, aril juice content, TSS/acid ratio and sugars percent were also considerably improved in trees subjected to various de-blossoming treatments. Fruit drop, fruit cracking and incidence of diseases were not affected notably with various de-blossoming treatments including the control. However, the first peak of fruit drop (per cent) was recorded between the second week of April to the first week of May and the second started from the first week of July onwards. Maximum fruit cracking was observed between the fourth week of May to the third week of July, which coincided with a rapid fruit development period. Incidence of black spot disease in both cultivars depended upon climatic conditions like relative humidity, rainfall and temperature. To get fruits of good size and quality, the flower buds in pomegranate cultivars 'Ganesh-I' and 'Kandhari' should be removed manually from 15 April and 22 April onwards, respectively under north Indian conditions (Singh *et al.*, 2011).

Developmental Regulation of Fruit Ripening (Molecular Basis)

The molecular basis of ethylene synthesis and regulation has been a focal point of ripening analysis in the past, with more recent emphasis shifting to characterization of ethylene signal transduction. Antisense repression of tomato *ACS* (Oeller*et al.*, 1991) and *ACO* (Hamilton *et al.*, 1990) genes and similar experiments in melon (Ben-Amor *et al.*, 1999) clarified the role of these genes in regulating climacteric ethylene synthesis. It is well known, however, that ethylene alone is not sufficient for ripening and that a developmental "competence" to respond to ethylene must be achieved (Wilkinson *et al.*, 1995a; *Lelievre et al.*, 1997; Giovannoni, 2001). Consequently, immature fruit typically does not ripen in response to exogenous ethylene. Careful examination of multiple ripening phenotypes, including the production of volatile aroma compounds in ethylene-repressed transgenic melons, demonstrated that aspects of climacteric fruit ripening are regulated by developmental factors that must be properly coordinated with ethylene synthesis (Bauchot *et al.*, 1998).

However, developmental mutations that affect all aspects were studied in tomato, of which the fruit-ripening process have been available for decades and include the *rin* mutation. A second mutation, termed *non-ripening* (*nor*), is phenotypically similar to *rin* in that *nor* fruit fail to produce climacteric ethylene or ripen yet show responsiveness to ethylene at the molecular level while similarly failing to ripen in response to ethylene (Lincoln and Fischer, 1988). Both the *rin* and *nor* loci were positioned on the tomato genetic map (Figure 1) as first steps in positional cloning (Giovannoni *et al.*, 1995). Isolation of the *rin* locus revealed tandem MADS box genes separated by 2.6 kb of intervening genomic DNA (Vrebalov*et al.*, 2002). The *rin* lesion resulted in a deletion starting in the last intron of *LeMADS-RIN*, removing the final exon, and extending into the region separating *LeMADS-RIN* from the adjacent MADS box gene (termed *LeMADS-MC*). *LeMADS-MC* was shown to be the tomato ortholog of the Arabidopsis *AP1* gene and to be responsible for the *macrocalyx* (*mc*) large-sepal phenotype of *rin*.

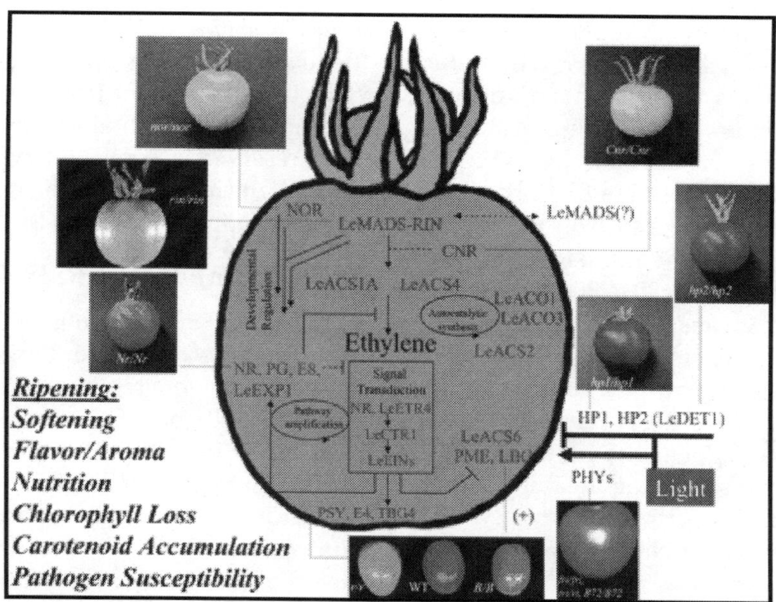

Figure 1: Model for the Molecular Regulation of Tomato Fruit Ripening.

(Fruit harbouring homozygous mutations for the indicated genes or loci are shown. The *nor*, *rin*, *Nr*, *Cnr*, *hp 1* and *hp 2* mutants are all nearly isogenic with cv Ailsa Craig. The *r* and *B* mutants are from reported introgression lines nearly isogenic with cv. M82 (WT). A triple phytochrome (PHY) mutant deficient in *PHYA* (*fri*), *PHYB1* (*tri*), and *PHYB2* (*B72*) with associated modification of carotenoid accumulation also is shown. Source: Eshed and Zamir, 1994).

Genomics in Fruit Species

The emergence of genomics technologies holds the promise of more rapid and comprehensive strides in elucidating ripening phenomena in coming years. Using a strawberry cDNA microarray, Aharoni (2000) identified an acyl-transferase that contributes to flavour development in one of the first demonstrations of large-scale expression analysis in fruit. Although the genome of a major fruit crop remains to be sequenced, review of the public EST collections is an indicator of where developments are likely to occur in the near future. More than 150,000 tomato ESTs have been developed and deposited in public databases, making this the largest EST collection of any fruit crop species (Table 1). Represented are 27 different cDNA libraries defining at least eight tissue types and including >40,000 ESTs from fruit at various stages of development (Hoeven *et al.*, 2002). A similar number of total ESTs are available from grape. In fact, grape is currently the species for which the greatest number of fruit ESTs are publicly available (nearly 50 per cent more fruit ESTs than for tomato). Interestingly, public EST resources are sparse to non-existent for many of the most important fruit species in terms of worldwide consumption (Table 1). ESTs derived from non-normalized and non-subtracted libraries can be informative in their own right in that the numbers and origins of ESTs within a contig can serve as a reflection of relative gene expression (Ewing *et al.*, 1999; Ogihara *et al.*, 2003). As such data become available for additional fruit species, it will become possible to compare the expression of homologous genes across species boundaries on a genomics scale, serve as a reflection of a relative gene expression facilitating, for example, comparisons of genome activity between climacteric and non-climacteric species to identify both conserved and unique ripening functions.

Although digital expression analysis can be useful for characterizing expression in tissues from which ESTs have been developed, this is not a feasible approach for targeted gene expression analysis. A publicly available tomato cDNA array has been developed with National Science Foundation funding that contains >13,000 elements representing 8,700 independent gene sequences (http://bti.cornell.edu/CGEP/CGEP.html). A companion database (http://ted.bti.cornell.edu/) has been developed to allow public deposition and retrieval of raw microarray data resulting from the use of the public tomato array according to MIAME guidelines (Brazma, 2001). Researchers across the world have already begun to use these resources for comprehensive analysis of fruit development and ripening (http://ted.bti.cornell.edu/). The recent development of public genomics resources for tomato will promote continued and expanded discovery in this model of fruit development and ripening.

Table 1: Total and Fruit ESTs Available from Fruit Species

Species	*Common Name*	*Total ESTs*	*Fruit ESTs*
Ananas comosus	Pineapple	0	0
Citrus x paradisi	Grapefruit	312	312
Citrus sinensis	Orange	16,180	5,623
Citrus unshiu Satsuma	Orange	2,561	2,561
Citrullus lanatus	Watermelon	593	593
Coffea arabica	Coffee	453	0
Cucumis melo	Melon	50	50
Cucumis sativus	Cucumber	358	0
Fragaria x ananassa	Strawberry	58	0
Lycopersicon esculentum	Tomato	150,228	41,186
Malus x domestica	Apple	945	903
Musa acuminata	Banana	27	27
Persea americana	Avocado	0	0
Prunus armeniaca	Apricot	4,535	4,535
Prunus avium	Sweet cherry	21	21
Prunus domestica	Plum	0	0
Prunus dulcis	Almond	3,845	0
Prunus persica	Peach	10,185	10,185
Pyrus communis	Pear	212	212
Vitis spp.	Grape	144,483	59,735
Total		**357,483**	**134,565**
Total excluding tomato and grape		**62,726**	**33,644**

Source: National Center for Biotechnology Information, www.ncbi.nlm.nih.gov/dbEST/index.html, October 2003.

Conclusion

Crop regulators are very helpful for growth and development of plant. In particular, over a period of time, plant growth regulators are known to improve fruit size, appearance and aril quality by direct effect on fruit growth and development or indirectly by regulating crops load, tree vigour and canopy architecture. The exogenous application of growth regulators has been found very effective in improving fruit size and quality of many fruit crops. Now a days commercial formulations of different hormones are available in the market and their use is relatively cheap and economical to the farmers. Many such studies were undertaken and evidences are noteworthy to take research further to prove that this aspect of crop regulation is of practical significance to bring economy to the doorstep of farmers to boost their economy and improve livelihood. However, such studies are required to be carried out in many such important fruit crops in the light of climate change and in relation to direct benefits to the doorstep of farmers and stakeholders.

References

Aharoni, A. 2000. Identification of the SAAT gene involved in strawberry flavor biogenesis by use of DNA microarrays. *Plant Cell*, **12**: 647–661.

Bauchot, A., Mottram, D., Dodson, A. and John, P. 1998. Effect of aminocyclopropane-1-carboxylic acid oxidase antisense gene on the formation of volatile esters in cantaloupe *Charentais melon* (cv. Vedrandais). *J. Agric. Food Chem.*, **46**: 4787–4792.

Ben-Amor, M., Flores, B., Latche, A., Bouzayen, M. and Pech, J. 1999. Inhibition of ethylene biosynthesis by antisense ACC oxidase RNA prevents chilling injury in Charentais cantaloupe melons. *Plant Cell Environ.*, **22**: 1579–1586.

Boora, R.S., Dhaliwal, H.S. and Arora, N.K. 2016. Crop regulation in guava: A review. *Agril. Rev.*, **37**(1): 1–9.

Braganza, M.A. 1990. Floral biology and varietal evaluation in genus *Psidium*. *M.Sc.(Agric.) Thesis*, University ofAgricultural Sciences, Bengaluru.

Brazma, A. 2001. Minimum information about a microarray experiment (MIAME): Toward standards for microarray data. *Nat. Genet.*, **29**: 365–371.

Bubán, T., Csiszár, L., Sallai, P. and Varga, A. 2004. Experiences with the bioregulator prohexadione-ca used in apple and pear orchards, DOI: 10.17660/ *ActaHorticulturae*,636.7.

Dry, P.R. and Loveys, B.R. 1998. Factors influencing grapevine vigour and the potential for control with partial rootzone drying. *Australian J. Grape Wine Res.*, **4**: 140–148. doi: 10.1111/j.1755–0238.1998. tb00143.x.

Eshed, Y. and Zamir, D. 1994. A genomic library of *Lycopersicon pennellii* in *L. esculentum*: A tool for fine mapping of genes. *Euphytica*, **79**: 175–179.

Ewing, R.M., Kahla, A.B., Poirot, O., Lopez, F., Audic, S. and Claverie, J.M. 1999. Large-scale statistical analyses of rice ESTs reveal correlated patterns of gene expression. *Genome Res.*, **9**: 950–959.

Giovannoni, J. 2001. Molecular regulation of fruit ripening. *Ann. Rev. Pl. Physiol. Pl Mol. Biol.*, **52**: 725–749.

Giovannoni, J. 2004. Genetic regulation of fruit development and ripening. *The Plant Cell*, **16**: 170–180.

Giovannoni, J., Noensie, E., Ruezinsky, D., Lu, X., Tracy, S., Ganal, M., Martin, G., Pillen, K., Alpert, K. and Tanksley, S., 1995. Molecular genetic analysis of the ripening-inhibitor and non-ripening loci of tomato: A first step in genetic map-based cloning of fruit ripening genes. *Mol. Gen. Genet.*, **248**: 195–206.

Gupta, M.R and Nijjar, G.S. 1978. Crop regulation in guava. *Indian J. Hort.*, **35**: 23–27.

Hamilton, A., Lycett, G. and Grierson, D. 1990. Antisense gene that inhibits synthesis of the hormone ethylene in transgenic plants. *Nature*, **346**: 284–287.

Hayes, W.B. 1974. *Fruit Growing in India*. Kitabistan, Allahabad.

Hoeven Van der, R., Ronning, C., Martin, G., Giovannoni, J. and Tanksley, S. 2002. Deductions about the number, organization, and evolution of genes in the tomato genome based on analysis of a large expressed sequence tag collection and selective genomic sequencing. *Plant Cell*, **14**: 1441–1456.

Lelievre, J.M., Latche, A., Jones, B., Bouzayen, M. and Pech, J.C. 1997. Ethylene and fruit ripening. *Physiol. Plant*, **101**: 727–739.

Lincoln, J. and Fischer, R. 1988. Regulation of gene expression by ethylene in wild-type and rin tomato (*Lycopersicon esculentum*) fruit. *Plant Physiol.*, **88**: 370–374.

Menzel, C.M. and Paxton, B.F. 1986. The pattern of growth, flowering and fruiting of guava varieties in sub-tropical Queensland. *Aust. J. Exp. Agric.*, **26**: 123–28.

Oeller, P.W., Wong, L.M., Taylor, L.P., Pike, D.A. and Theologis, A. 1991. Reversible inhibition of tomato fruit senescence by antisense 1-aminocyclopropane-1-carboxylate synthase. *Sci.*, **254**: 427–439.

Ogihara, Y., Mochida, K., Nemoto, Y., Murai, K., Yamazaki, Y., Shin, I.T. and Kohara, Y. 2003. Correlated clustering and virtual display of gene expression patterns in the wheat life cycle by large-scale statistical analyses of expressed sequence tags. *Plant J.*, **33**: 1001–1011.

Rathore, D.S. 1975. Deblossoming of rainy season crop of guava by NAA. *Prog. Hort.*, **7**: 63–65.

Retamales, J.B., Hipps, N.A. and Atkinson, C.J. 2004. Effect of timing of root pruning and root IBA application on leaf mineral concentration and canopy light interception of 'cox's' orange pippin' apple trees. DOI: 10.17660/*Acta Horticulturae*,636.5.

Sinatsch, S., Pfeiffer, B., Schult, T., Zimmer, J. and Brockamp, L. 2014. Crop regulation in organic grown apples: Results of different trials on three different sites. www.ecofruit.net/2014/20RP–Sinatsch, *Crop Regulation*, 142–152Pp.

Singh, U.R. and Kumar, R. 1993. Crop regulation in guava. **In**: *Advances in Horticulture*, Vol. 3, (Eds.) K.L. Chadha and O.P. Pareek. Malhotra Publishing House, New Delhi, pp. 1197–1204.

Singh, N.P., Dhillon, W.S. and Gill, P.P.S. 2011. Quality improvement studies in pomegranate under sub-tropics of India. *Acta Hort.*, **890**: 363–369.

Tartachnyk, I. and Blanke, M. 2004. Key processes in leaf senescence of fruiting apple trees. DOI: 10.17660/ *Acta Horticulturae*, 636.6.

Vrebalov, J., Ruezinsky, D., Padmanabhan, V., White, R., Medrano, D., Drake, R., Schuch, W. and Giovannoni, J. 2002. A MADS–box gene necessary for fruit ripening at the tomato ripening-inhibitor (*rin*) locus. *Sci.*, **296**: 343–346.

Wilkinson, J., Lanahan, M., Yen, H., Giovannoni, J. and Klee, H. 1995a. An ethylene-inducible component of signal transduction encoded by Never-ripe. *Sci.*, **270**: 1807–1809.

16

Innovative Technologies for Improving Production and Economic Security of Banana Farmers

S. Uma

Director, ICAR-National Research Center for Banana, Trichy – 620 102, Kerala
E-mail: umabinit@yahoo.co.in

India has made significant achievements in food production, which is marked by *Green Revolution, White Revolution, Blue Revolution, Yellow Revolution, and now Golden Revolution.* Emphasis on horticulture during the last two plan periods has provided a viable alternative for diversification of agricultural land use, employment generation and for ensuring food, nutritional and livelihood security. However, more than 300 million people below the poverty line are suffering from malnutrition without nutritional security. In the present context of dwindling land and water resources, availability, and accessibility to resources is a challenge, receiving worldwide attention of researchers. Bananas and plantains are known for its antiquity and also interwoven with Indian heritage and culture. The plants are considered as the symbol of 'prosperity and fertility'. Owing to its greater socio-economic significance and multifaceted uses, they are referred as 'Kalpatharu' (plant of Virtues). Banana and plantain fruit is a wonder berry which forms the staple food of millions of people across the globe, providing more balanced food than any other fruit or vegetable. As a dessert, banana is more filling, easy to digest, fat free, rich source of carbohydrate with a calorific value of 90 Kcal per 100 g fruit and is free from sodium, making it a salt free food fruit. It contains various vitamins and has therapeutic values for the treatment of many diseases.

1. Banana Composition and their Uses

Banana is an important source of CHO and dietary fibre. Green banana fruit contain higher Banana produces easily available energy (90-120 kcal/100g), more carbohydrate and low fat. The protein level is slightly lower than cereals. Due to low sodium and high potassium contents in banana and plantain fruits, it is recommended for heart and high blood pressure patients. Banana contains a considerable amount of mineral elements, and could serve as a source of minerals in human and animal diets. Banana is a major source of macro-elements, especially potassium, and contains health-beneficial ingredients such as resistant starch, total dietary fibers, rapidly digestible starch, and slowly digestible starch. Oligosaccharides (fructo oligosaccharides and inulin) and polyphenols, catechin, epicatechin, epigallocatechin, and gallic acid) are other ingredients present in bananas that have found application in the prevention of muscular contractions, regulation of blood pressure, prevention of colon cancer and diabetes, and in the cure of intestinal disorders when unripe.

The fruit is often called as fruit tonic due to high level of potassium which makes a person very alert. The protein Serotonin present in banana helps to get rid of depression and thus called happy hormone. Presence of vitamin B6, B12 and C along with potassium and magnesium help the body to recover from nicotine withdrawal and helps to give up smoking. Morning sickness, a common suffering for pregnant women could be overcome by eating banana between meals. Banana peel proved to be good against mosquito bite and the ripe banana has moisturizer effect and nourishing the dry skin. It also helps influence the proper functioning of the digestive tract, microbial flora, blood cholesterol level, and the glycemic index. It provides fullness to the stomach. Plantain banana is uses as baby food due to its easy digestibility. Bananas are low on the glycemic index and release their energy into the bloodstream slowly. According to the International GI Database, fully ripe banana has a glycemic index of 51. Steryl esters and free sterols are the major lipophilic component found in unripe banana peel, while free fatty acids and sterols dominate banana pulp, banana is rich in fatty acids, phytosterols and steryl glucosides. Plantains are rich in Vitamin C providing approximately 20 mg/100 g fresh which is higher than banana (10 mg/100 mg). Because of the low lipid and high energy value they are recommended for obese and geriatric patients (Gaster, 1963). Banana contains lectins sugar binding proteins and can identify foreign invaders like a virus, attach themselves to the pathogen, thus blocking its entry into the human body. Team has developed a way to isolate BanLec from banana.

2. Climate and Soils

Bananas and plantains are cultivated in wide array of soils under different climatic conditions and the area under its cultivation also expands rapidly. The soil depth and drainage are the two soil factors important for successful banana cultivation. Since banana is more sensitive to water logging, rich loamy soil with good drainage and water table below one meter depth is the most ideal soil for the crop. A soil pH of 5.5 to 8.0 is found to be optimum for bananas and plantain.

3. Varieties

Banana is one of the few crops where the man-made hybrids failed to excel the natural hybrids. Most of the present day bananas grown both nationally and internationally are the natural hybrids, which have evolved and acclimatized to various agro-ecological conditions. Popular varieties can be categorized as high value varieties and traditional varieties. High value varieties include Gros Michel until 1980's, which was wiped out due to Fusarium wilt and substituted by Grande Naine and other Cavendish ecotypes in the last few decades. Presently, Grande Naine is the ruling variety in the global banana industry. Situation is almost the same in India where Cavendish clones like Robusta, Dwarf Cavendish, Shrimanthi, Giant Governor,

Table 1: The State-wise Distribution of Local Varieties

State	*Varieties of Commerce with Genomic Groups*
Andhra Pradesh	Amrithpani (Silk) (AAB), Karpura chakkarakeli (AAB), Monthan (ABB), Yenagu Bontha (ABB), Nukala Bontha (ABB)
Assam	Amritsagar (AAA), Chenichampa (Mysore) (ABB), Jatikol (Mysore) (ABB), Digjowa (Silk) (ABB), Kulpait (Silk) (ABB), Malbhog (ABB), Amritman (ABB), Alpon (ABB), Garomoina (ABB), Dasaman (ABB), Manohar (ABB), Kanchkol (ABB), Bharatmoni (ABB), Ankur (ABB), Geda (ABB), Gera (ABB), Bankel (ABB), Sabji Kela (ABB), Nute pong (ABB), Kait (ABB), Khullung (ABB), Kait Shejeng (ABB), Singhalaji (ABB) *Balbisiana clones*: Bhimkol, Attikol, Bhimkel, Athiakol, Rissue, Manohar, Manguthamng *Bacharia Malbhog*
Bihar	Chenichampa (AAB), Alpon (Mysor) (AAB), Chinia (Mysor) (AAB), Malbhog (Silk) (AAB), Muthia (Bluggoe) (ABB), Kothia (Bluggoe) (ABB), Gauria (Bluggoe) (ABB), Kanthali (ABB)
Gujarat	Mortman (Silk) (AAB), Soneri (Mysore) (AAB), Muthia (Peyan) (ABB), Kachkel (ABB)
Karnataka	Elakki Bale (AB), Poovan (Mysor) (AAB), Rasabale (Silk) (AAB), Jwaribale (Pome) (AAB), Karibale (ABB), (Monthan) (ABB)
Kerala	Njali poovan (AB), Sevvazhai (AAA), Nendran (Plantain) (AAB), Palayankodan (AAB), (Mysore), Poovan (Silk) (AAB), Monthan (ABB)
Maharashtra	Safed velchi (AB), Lalvelchi (Mysor) (AAB), Rajeli (Plantain) (AAB)
Tamil Nadu	Ney Poovan (AB), Sevvazhai (AAA), Poovan (Mysore) (AAB), Rasthali (Silk) (AAB), Virupakshi (AAB), Pachanadan (Pome) (AAB) Nendran (Plantain) (AAB), Karpuravalli (Pisang Awak) (ABB), Sakkai (Bluggoe) (ABB), Monthan (ABB), Peyan (ABB)
West Bengal	Amrit Sagar (AAA), Mortman (Silk) (AAB), Champa (Mysore) (AAB), Monthan (ABB), Kanthali (ABB), (Pisang Awak) (ABB),
Tripura	Chompa (AAB), Sabri (AAB), Somai (AAB), Banria (ABB) *Balbisiana clone* Rissue
Mizoram	Somai (AB), Amritsagar (AAA), Champa (AAB), Sabri (AAB), Banria (ABB), Monthan (ABB), Sagarkol (ABB),
Manipur	Somai (AB), Amritsagar (AAA), Champa (AAB), Sabri (AAB), Banria (ABB), Monthan (ABB), Sagarkol (ABB),
Nagaland	Somai (AB), AmritSagar (AAA), Champa (AAB), Sabri (AAB), Banria (ABB), Monthan (ABB), Sagarkol (ABB),
Meghalaya	Champa (AAB), Banria (ABB), Monthan (ABB), Sagarkol (ABB),
Arunachal Pradesh	Champa (AAB), Banria (ABB), Monthan (ABB), Sagarkol (ABB),Pakte (ABB)

Mahalakshmi *etc.*, are being widely cultivated. Present survey indicates commercialization of very few selected varieties. The distribution of Cavendish clones was only 40-45 per cent until 1990's, which has enhanced to more than 60 per cent in 2015-16. This increase has been at the cost of other traditional and semi-commercial varieties with local importance. Recently, some states have made efforts to conserve, popularise, multiply and make access to quality planting material for local landraces. The Table 1 indicates the state wise distribution of local varieties.

4. Collections

India has polyclonal situation with respect to banana and plantains with 8-10 popular commercial clones and more than 20 traditional cultivars. Hence improvement with one cultivar fails to suffice the needs of banana farming community. In the process of selection, the individuals are selected either from the natural variants or from those artificially created population using tissue culture or mutation techniques. Natural mutation, selection and perpetuation have been the basis of evolution in vegetatively propagated crops like *Musa*. Some of the classical examples are Gandevi is a tall mutant of Giant Cavendish has been selected for its high yield potential upto 70 kgs. Though it has a long crop cycle, it produces a good mass for processing. Udhayam is a single plant selection of PisangAwak with an yield potential of 70-75 kgs which is 35 per cent higher than local Karpuravalli and field tolerant to Fusarium wilt and nematodes. Kozhikodu and Bangladesh Malbhog are selections of Silk and Pisang Ceylan is a selection of Mysore subgroup. At TNAU, Coimbatore a high yielding diploid accession of Sannachenkadali, with field tolerance to nematodes, infectious chlorosis, bract mosaic virus, shade tolerant, resistant to Eumusae leaf spot and bacterial head rot diseases has been developed. Ambala Kadali, Erachivazhai, Thattilla Kunnan, Veneetu Kunnan and Rasa Kadali are the selection from natural mutant which showed significant variations from their parental genotypes.

5. High Yielding Banana Varieties

ICAR-NRC on Banana has identified two cultivars namely NRCB-7 and NRCB-8 and have been recommended for release as national varieties.

NRCB-7

It is a selection from local culinary landrace, Bangrier, having genomic group ABB and subgroup Bontha. In case of NRCB-7, the bunch is heavy with cylindrical in shape, with 10-12 hands of fruits in a bunch and are tightly packed. Elongated fruits of 20-22 cm and has a pointed tip. Bunch weight is ranges from 28- 30 kgs compared to the local check (Monthan) ranging from 22- 25kgs. It is tolerant to Fusarium wilt (Foc race-2) and moderately resistant to root-lesion nematode. As a culinary variety it can replace the local cultivar with 15 per cent higher yield.

NRCB-8

It is an exotic introduction of cv. Saba belongs to ABB genomic group. This is a hardy cultivar tolerant moisture deficit stress and moderately resistant to root-lesion nematode. It can be cultivated in marginal lands without much yield compensation. The yield potential is 30 kg/bunch with dark green fruits, flattened with a blunt tip. Fruit pulp is starchy in nature. Fruits can be used for both culinary and dessert purpose. It performed well both in plains and higher altitudes.

Udhayam (ABB)

This variety released by NRC Banana, Trichy. It belongs to Pisang Awak subgroup. The average bunch weight is37 kg and has a potential to yield up to 45 kg under good management practices which is 40 per cent higher than local Karpuravalli. Crop duration is 13 months and yield remains stable even in ratoons. Has field tolerance to Sigatoka leaf spot and nematodes. The bunch is cylindrical in shape with 13-15 hands and 14-15 fruits per hand. Fruits are elongated with 14-16 cm in length and 10-13 cm in circumference. The cylindrical bunch and well spaced hands makes it suitable for long distance transportation and less prone to fruit damage during transit. It has high sugar content with 31°B with harmonious blend of acidity and sweetness and hence suitable for processing industry for products like juice, jam, banana fig *etc.*,

6. Exotic Introductions and Hybrids

To supplement the growing demand for improved varieties, exotic accessions introduced from various countries through ITC, Belgium and NBPGR, New Delhi have been evaluated. Promising ones are tested under multi location trials. Some of the accessions have been doing exceedingly well including the global hybrids. The details of exotic introductions is provided below:

a) BITA-2 (TMB x 1378)

This is a hybrid developed from IITA, Nigeria from the cross of Pisang Awak cv. Fougamou x M. balbisiana 1-63. It is a tetraploid hybrid for cooking purpose. The promising features are, high yield and resistant to black sigatoka. But the resistance to *Fusarium* wilt needs to be ascertained. This grows to a height of 3.2-3.3 m with sturdy stature.

b) BITA-3 (TMB x 5295)

This hybrid is developed the cross between AAB Laknau, cv. Laknau x Tjau lagada from IITA, Nigeria. Tetraploid hybrids of this fruits are suitable for cooking purpose especially for frying. It has high yield potential, big fingers resistance to black sigatoka and low BSV symptom incidence. Height of the plant is 3.00 m.

c) CRBP – 39

This hybrid is a plantain derived the cross of AAB – Plantain cv. French clair x M53 (4X) from CRBP,

Camaroon. It is a 4x hybrid type with good flavour and cooking character. Medium to tall, plant height of average 3.05 m and has highly resistance to black sigatoka with normal short cycles of leaf habit. This cultivar has good characteristic feature of cooking qualities like boiling, fried as chips and pounded and flour.

d) FHIA – 01

This is a hybrid derivative of AAB – Pome cv.Prata Ana X Prata Ana (Dwarf Prata) SH 3142 from FHIA, Honduras. This hybrid has good apple flavour and resistance to black sigatoka, burrowing nematodes, and race 1 and 4 of Fusarium Wilt with the plant height of 2.58 m. The plants are strong and semi – dwarf in nature and tolerance to drought and cold temperatures. It has received good reviews in Uganda as a green cooking banana and as a beer banana. Harvested fruit has long green life.

e) FHIA – 02

FHIA – 02 hybrid is developed the cross of AAA – Cavendish cv. Williams x SH 3393 from, Honduras. It has sweet in taste similar to that of cavendish but short post-harvest green life and good attributes to local markets. Undesirable character of this hybrid is susceptible to Fusarium Wilt.

f) FHIA – 03

This is a hybrid developed the cross of (Gaddatu x BB) x SH 2471- SH 3386 SH 3386 x SH 3320 – FHIA 03, from FHIA, Honduras. Tetraploid nature of this hybrid having good cooking quality and dessert flavour when ripe. It has drought resistant and cold tolerant. Very vigorous strong plant with the height of 2.76 m. which also resists high wind. Tolerant to marginal soil fertility, rainfall and good resistance to black sigatoka. Performs well at both sea level and high aititutes. Bunches are having long green life about 6 weeks after harvest. So, this hybrid is highly suitable for homestead gardens.

g) FHIA – I7

Cultivar AAA – Gros Michel Highgate x SH 3362 are the parents of this hybrid FHIA 17 developed from FHIA, Honduras. The desirable features are resistant to race 1 of the Fusarium Wilt for comparing with Gros Michel and good resistance to black sigatoka. Plant height of 3.03 m. Average bunch weight and number of hands per bunches are 25 and 11 respectively.

g) FHIA –18

Parentage of this hybrid is AAB – Pome cv. Prata Ana x SH 3142 developed from FHIA, Honduras. All characters are similar to that of FHIA – 01. It is a dessert, sweet acid banana, apple like flavour that has received very good reviews in Cuba; and resistance to black sigatoka.

h) FHIA – 21

Plantain hybrid FHIA – 21 developed the cross between AAB – Plantain cv. AVP – 67 x SH 3142 from FHIA, Honduras. Genomic status of this hybrid is AAAB. Tetraploid nature of this French plantain has resistant to black sigatoka and excellent plantain flavour which is twice as productive as False Horn. It is recommended for home gardens but relatively short green life.

i) FHIA – 23

This is a another important hybrid developed from cv.AAA – Gros Michel Highgate x SH 3362 from FHIA, Honduras. It is a tetraploid hybrid in the genomic status of AAAA. The promising features are tolerance to black sigatoka and resistant to race 1 of Fusarium Wilt for comparing with Gros Michel. It is being cultivated extensively in Cuba where it is twice as productive as Cavendish because of tolerant to black sigatoka.

j) FHIA – 25

Development of this hybrid from FHIA, Honduras obtained the cross of SH 3648 (4X) x SH 3142 (2X). Important character of this hybrid is producing very large bunch of weight 47 kgs. It has compactness typical of the EA varieties. other desirable features are dwarf and strong plant, high level of resistance to black sigatoka, long green life of fruit after harvest, easy to peel, fast to cook and excellent texture and flavour both boiled and fried. It could also be useful in West Africa and other areas where bananas and Plantains are eaten when boiled green. It is not yet known if it is suitable as a beer banana, but this possibility should be investigated.

7. Development of Hybrids

The use of conventional breeding in the development of pest and disease-resistant banana hybrids/improving the fruit quality has received very little attentions most of the banana cultivars are female sterile. Success of banana breeding programme depends on the selection of potential parents. This required the basic knowledge on the male and female fertility parents to be included in the breeding programmes. Hence, germplasm was evaluated to identify polleniferous male parent and to identify the compatible parental combinations based on seed setting ability and production of progenies. Despite their development of progenies compatible is tricky due to low germination percentage, long time span required from one generation to the next (almost two years from seed to seed), and a large area required for field testing. Some of these hurdles have been overcome through non-conventional biotechnological methods such as *in vitro* germination of seeds and rapid micropropagation

Diploid x diploid, triploid x diploid and tetraploid x diploid, breeding results in the development of commercial triploids. TNAU, Coimbatore attempted to improve the Pome cv. KallarLadan and has successfully developed a hybrid CO 1, which is a promising hybrid resultant of a three way sequential crosses *i.e.*, Kallar Ladan (AAB) x *M.balbisiana* clone Sawai (BB). F1 (AB) x Kadali (AA). This is very much similar to Virupakshi (AAB). Plants are medium tall (2.7m). The bunch weighs 10.5 kg with 7 hands and 80-85 fruits. TSS of fruits ranged from 22-24 Brix. Crop duration is 14 – 15 months. Fruits are acidic in taste but become sweet at full ripening.

Subsequent triploid x diploid breeding attempts at TNAU, Coimbatore like Karpuravalli x H.201 lead to the development of a promising ABB hybrid designated as 96/7. The fruit colour is bright yellow without any ashy coating. Likewise Karpuravalli x Pisang Lilin also resulted in a parthenocarpic triploid AAB (H-212). It is female sterile and poor in male fertility, possessed resistance to nematodes and Sigatoka leaf spot. The bunch weight was 13.52 kg with high TSS of 32.00 per cent.

Triploid x diploid breeding undertaken at KAU lead to the successful release of two triploid hybrids namely BRS 1 (Agnishwar x Pisang Lilin) and BRS 2 (Vannan x Pisang Lilin). Evaluation of these hybrids showed that they are superior to the parents with respect to bunch weight and quality characters. BRS 1 (AAB) is 100 days earlier than Rasthali with significant difference in bunch weight. It is resistant to Sigatoka leaf spot and suitable for homestead cultivation in Kerala. BRS 2 (AAB) is a medium statured hybrid which produces 14 kg bunches with 8 hands and 118 fruits in a crop duration of 314 days. It is also tolerant to leaf spot and Panama diseases.

Tetraploid and diploids used in the 4X x 2X schemes should be initially improved through recurrent selection to accumulate the favorable alleles. Attempts at TNAU, Coimbatore to cross H-02-02 (AABB) with Pisang Lilin has resulted in the development of a primary triploid H-03-15 (AAB) with good bunch weight (14.5 kgs) and resistant to wilt and nematodes (Damodaran, 2004). This gives an indication that tetraploid x diploid cross has to be done in large numbers to develop triploids with novel genomic constitution.

8. Production and Management

Banana being a tropical crop, has acclimatized to sub-tropical, semi-arid and to some extent arid conditions. Depending on the longevity and the method of production system, it has been classified an annual, biennual and perennial cultivation. Plantains and Cavendish clones are replanted annually, but recent results have indicated the suitability of Cavendish clones for one plant crop followed by two ratoons which has resulted in substantial reduction in the production cost of banana.

Banana is conventionally propagated through suckers and use of tissue culture has made a mark in the last decade. There are two types of suckers, sword suckers – with a well developed base, pointed tip and narrow leaf blades, and water suckers, which are small, less vigorous, broad leaved and emerge in clumps. A plant produces only 5-20 suckers during its life time of 12-14 months. For accelerating the propagation rate, suckers with growing buds or cut rhizomes called 'bits' and 'peepers' are used. Several good bits, each with a centrally placed germinating eye can be cut from an unbunched rhizome after trimming the roots. Selection of appropriate mother plant for raising new propagules either through *in vivo* or *in vitro* methods is important.

On an average, the requirement of fresh planting material works out to be 25 billion/annum, at a planting density of 2500 plants/ha which is an enormous demand. 99 per cent of the planting material requirement is met by suckers and the rest by tissue cultured planting material. Presently 15-20 viable companies are involved in the production of tissue cultured banana with an annual production of 2 - 5 million plantlets.

Despite the fact that tissue culture plants have several advantages over conventional suckers, its application is hindered by high initial cost of planting material. This calls for adoption of innovative low cost tissue culture technologies to cut down the cost without compromising the quality. Mechanization in media dispensing, dish washing, cheaper alternatives for media components like agar and sucrose and harnessing natural light over fluorescent light have been tried with success. But up-scaling of these technologies and their commercialization is yet to be achieved.

Somatic embryogenesis with the adoption of bioreactor for up-scaling can also be standardized as an alternative to shoot tip culture technology. Following validation for genetic fidelity and stability across well known commercial varieties, somatic embryogenesis technology can be exploited commercially. One ml of settled cell volume (SCV) of the suspension is expected to produce 4000-5000 plants enough to plant 1 ha of land. With a potential to produce very large number of plantlets, the cost of somatic embryogenesis based TC plants is expected to come down drastically to the level affordable even by small farmers.

Quality planting material is the key for successful production of banana. To ensure the quality of tissue culture multiplied plants, certification standards have been developed by DBT for virus indexing and genetic fidelity. Recently DBT has recognized NRCB, Trichy as the Accredited lab for fidelity testing and virus indexing. The strict certification standards will help the banana growers in getting quality planting materials. On the other hand, the planting material as suckers, has serious implications on the profitable production of bananas due to the lack of quality certification.

9. High Density Planting (HDP)

Banana is an important fruit crop grown in India and the choice of plant spacing depends upon the cultivar and further varies from region to region depending upon the cultural practices of the area. In recent years, there has been considerable emphasis on high density planting wherein the yield of an individual plant cannot be increased beyond certain limit. The total yield and net returns could be increased per unit area by adopting closer spacing. Besides, closer planting also reduces weed growth and provides protection against wind damage. However, the major limiting factor will be sunlight, which affects flowering, crop duration and subsequent performance of the ratoon crops. There can be single plant system, double row system or hill planting.

In single plant system, planting distance varies from 1.5 x 1.5 m, 1.5 x 1.8 m, 1.8 x 1.8 m, 1.8 x 2.1 m, 2.1 x 2.1 m, 2.1 x 2.4 m and 2.4 x 2.4 m. Accordingly, plant population also varies at different plant spacing and in general closer planting is followed for dwarf cultivars like Dwarf Cavendish and Robusta while wider spacing is advocated for very tall cultivars like Red Banana, Karpuravalli and Udhayam.

In double row planting system, 1.2 x 1.2 x 2.0 m is advocated for dwarf cultivars, 1.2 x 1.5 x 2.1 m is advocated for tall cultivars. In three suckers per hill planting system, the suckers are planted at 1.8 x 3.6 m distance in one pit with 4630 plants per hectare whereas; only 3086 plants are accommodated by planting two plants per pit.

The ICAR-National Research Centre for Banana, Tiruchirappalli has developed many innovative technologies to address the major concerns of banana growers and effectively disseminated the technologies. This is highly evident from the fact that the area and production of banana has rapidly increased from 16.4 million tonnes of banana from 0.62 million hectare to the current banana production of about 30.0 million tonnes from 0.83 million hectares.

The high-density planting and fertigation technologies developed by ICAR-NRCB have made a remarkable impact in promoting banana production and enhancing productivity. High density planting is a system in which a higher number of plants are accommodated per unit area when compared to conventional planting.

In case of banana, optimum planting density is very important for bridging the gap between the actual and potential yield of banana per unit area. The advantages of the technology are : Accommodation of more number of plants/unit area, Reduction in the cost of cultivation through minimal efforts for clump management, Saving of water (30-40 per cent), fertilizer (25 per cent) and plant protection chemicals are ensured in this HDP. In addition, there is significant reduction in staking cost (3 plants are tied using a single rope), Less initial investment on drip system because of wider spacing and easy fertigation, 40-50 per cent increase in productivity with enhanced fruit quality provision for inter cropping at ease.

10. Organic Production

Organic bananas have created a niche market both at the national and international levels. Although, organic cultivation is not profitable in the initial years, it has been found promising in the long run. In India, organic farming gaining importance and is now successfully practiced by thousands of organic loving farmers of Tamil Nadu, Karnataka, Andhra Pradesh, Madhya Pradesh, Gujarat and Maharashtra states. The success stories of such organic farming practices motivated many farmers to adopt the natural way of farming, which has been they found economically viable too. Since, banana is mostly eaten raw consumers prefer organic bananas instead of inorganic ones.

Organic farming is a method of farming system which avoids or largely excludes the use of compound chemicals and encourage the use of natural resources such as organic matter, minerals and microbes to maintain clean environment, ecological balance and to provide stability to the production level without polluting soil, water and air. Organic farming system relies on large-scale application of animal wastes or farmyard manure, compost, crop rotation, crop residues, green manures, vermicompost, bio-fertilizers, bio-pesticides and biological control. The constituents of organic farming are:

i) Integrated soil and water management

ii) Integrated plant nutrient supply management

iii) Integrated insect pest and disease management

It is essential to maintain proper co-ordination between natural resources like soil, organic matter, biotic life and water for plant nutrient supply management for higher production. This can be achieved by using farmyard manure, oil cakes, composts, green manure, green leaf manure, coir pith compost, vermicompost and bio-fertilizers. The organic material improves the soil physical conditions and there by increases the plant growth. Organic materials in soil promotes microbial and soil enzyme activity and often the level of soil enzyme activity increases with increasing soil organic matter.

11. Scope of Availability of Organic Manures

India, being an agricultural country, the availability of agricultural wastes including the cattle manures, pig manure and poultry manures is estimated to be plenty. It is estimated that the total availability of agricultural wastes in the country is about 1000 million tonnes, which includes 350 million tonnes from crop wastes and remaining 650 million tonnes from animal wastes. If used effectively, these agricultural wastes can contribute a major share to the total nutritional requirement of organic farming. Moreover, the animal manures also contain considerably higher dry matter content and thereby high in nutrient contents also.

The organic manures base is composed of farm yard manure, compost, chicken manure, oil cakes and also through bio-fertilizers like *Azotobacter*, *Azospirillum*, *Phosphobacteria*, VAM and green manuring with sunnhemp, daincha, cowpea *etc.* Nitrogen fertilization is through composted farmyard manure, chicken manure supplemented with Chilean nitrogen and blood meal. Supplementing the phosphorus need is through rock phosphate, blood meal, green manure, green leaf manure and FYM. Potassium is provided by FYM, chicken manure; cement factory kiln dust (CKFD) *etc.* Additional nutrient is provided by gypsum, zinc sulphate dissolved in plant extracts and composts.

12. Use of Green Manures

In general, it is a commonly recommended to grow the green manures/legumes at the time of planting of suckers or tissue cultured plants and incorporate into the soil just before or immediately after flowering when the plants are still tender for better decomposition. Growing and incorporation of green manures such as sunnhemp, daincha, cowpea, green gram, black gram *etc.* is one of the principal supplements for the supply of plant nutrients in organic farming. In addition to supply of nitrogen, the addition of green manures also supply large quantum of biomass and also improves the physicochemical properties of soil.

13. Biofertilizers

Use of mycorrhizal fungi colonies assist in the uptake of both macro as well as micro nutrients. The commonly used bio fertilizers are VAM (*Glomus mossae* and *Glomus fasciculatum*) and PSB (*Bacillus polymixa* and *Pseudomonas striata*). In banana cv. Ney Poovan, soil inoculation with VAM and PSB significantly increased the available phosphorus in the soil while inoculation with Azosprillum has enhanced the available nitrogen.

14. Mulches

Mulching has the advantages of protection of the soil, disease control and improving soil organic matter. Banana plants produce a considerable amount of biomass (pseudostem residues, shed leaves and leaves cut at harvesting) that can be used as mulch. Polythene mulch was generally superior to hoeing and weeding in its effect on plant growth and yield.

15. Precision Farming

Increasing efficiency in management system is the key aspect in Precision farming especially in small holdings by better management of farm resources. It is basically information and technology based management system, which includes Geographical Information System (GIS), Geographical Positioning System (GPS), yield monitoring devices, soil, plant and pest sensors, remote sensing *etc.* This technology allows the farmers to identify, analyse and manage the spatial and temporal variability of soil and plants for optimum profitability, sustainability and environmental protection. Precision farming in banana should include soil fertility status, weather data, forecasting models for yield, pest and disease outbreak, controlled irrigation and fertigation for a profitable and sustained production.

16. Nutrient Management

Being a high feeder, banana prefers high amounts of mineral nutrients, augmented externally. Generally recommended dose for a sucker derived banana plant is 200-225:30-50:300-400g NPK plant^{-1} depending on the soil and variety, while, the tissue culture raised banana is supplied with 300:50:400g NPK plant^{-1}. The well developed roots at establishment and fibrous root system ensured good uptake eventually leading to a good bunch. The fertilizers are applied in basins dug around the plants at 45cm, 60cm and 75cm ias split doses during 3rd, 5th and 7th months of plant growth and development. It is established that application of 1/4th of N and 1/3rd of K_2O at reproductive phase *i.e.*, after the bunch emergence delayed the leaf senescence, significantly improved finger filling and also enhanced the bunch weight and fruit quality.

17. Fertigation

Fertigation, a technique where the fertilizers (N and K) are applied daily to the roots of individual plants in small quantities through irrigation water. HDP coupled with fertigation increased the yield by 2-2$^1/_2$ times. In Robusta, 1.8 x 1.8m spacing with weekly fertigation of 75 per cent of recommended dose of N and K produced highest yields with a mean bunch weight of 35.5 kgs/plant. Similarly paired row planting recorded 23.3 per cent higher yield than conventional planting. Soluble nitrogen and potassium fertilizers feeding at the rate of 150 g/plant/crop can take care of nitrogen requirement of the crop. Application of N in the form of urea and K as Muriate of Potash (MOP) through the system could be advantageous. The MOP may be dissolved in the water and kept overnight to facilitate complete dissolving while urea can be dissolved instantly. These fertilizers may be supplied into the system either on daily basis or weekly basis and it may be stopped 10-15 days prior to harvest of bunches. Apart from the straight fertilizers, there are several formulations of water-soluble fertilizers available in the market. The commercially available fertigation grade water soluble fertilizers are found highly effective in banana and even in places where the quality of irrigation water is fully not suitable for drip system. In addition, with précised application of nutrients and higher nutrient use efficiency under fertigation, the system has the added advantage of saving of 25 per cent of the recommended dose of fertilizers and thus significantly reduces the cost of cultivation.

Application of 25 per cent N FYM + 50 per cent N as neem cake + 25 per cent N as inorganic were found optimum and recorded maximum growth parameters, shorter crop cycle, T.S.S. and highest yield in in most dessert cultivars, whereas, application of 25 per cent N FYM + 25 per cent N as neem cake + 50 per cent N as inorganic (urea) recorded the highest bunch weight in banana cultivars cooking varieties and long duration cultivar like Karpuravalli. Application of organic sources reduced the time taken for flowering, maturity and total crop duration in all cultivars. In addition, organic nutrition reduced the soil pH, bulk density and increased the porosity and thereby improved the physico-chemical properties of the soil. Significantly higher organic carbon (0.68 per cent) and K/Na ratio (0.47) were found under organics whereas, 100 per cent inorganic nutrition recorded the least organic carbon (0.28 per cent) and K/Na ratio (0.21). Integrated management of nutrients, inorganic sources along with organic fertilizers and biofertilizers have been successfully demonstrated to improve yield and other soil parameters. The highest and optimum leaf nutrient concentrations were observed in the treatment 80 per cent recommended NPK + phosphobacteria + rice husk ash or vermicompost along with 30 per cent enhanced bunch weight. Irrespective of organic manures, the performanceof phosphobacteria was superior to azospirillum and VAM, integration of phosphobacteria and rice husk ash in the NPK fertilization is reported to accrue a net additional profit of Rs. 50,000/- per ha.

Natural Resource Management for Problem Soil

Banana is being cultivated in problem soils, which are having more than PH 8.5 Due to the high salt accumulation in the soil and availability of Na in the soil the absorption of potassium is affected and the bunch weight and yields are reduced. To overcome this problem, use of available natural resources and industrial water like pro manure. Gypsum, rice husk ash and cement kiln flue dust were used for reclamation of the problem soils. The results indicated that use of 15kg poultry manure or 15kg Rice husk ash + 80 per cent of recommended dose of NPK produced higher yield and higher benefit cost ratio as compared with 100 per cent NPK along application of 0.5kg CKFD + 60 per cent of recommended K recorded the highest bunch weight.

Alleviation of Soil Moisture Deficit Stress through Chemicals

Banana, being a long duration crop, requires assured irrigation throughout the year and soil moisture stress at floral primordial initiation and flowering stages may significantly reduce the yield. In the event of non-availability of sufficient water at critical stages NRCB has come out with foliar chemical sprays to alleviate the negative effects of soil moisture stress to get assured banana yield. This foliar priming helps the farmers to irrigate with a gap of two-three weeks and may. Foliar priming of cv. Grand Naine plants at flowering stage with acetyl salicylic acid (0.1mM) combined with butylated hydroxy toluene (100 ppm) before the imposition of soil moisture stress recorded bunch weight (18.78 kg), which is comparable to irrigated control (19.72 kg) and higher than other chemicals such as glycine betaine (20mM), beta aminobutyric acid (100 ppm) and 2 per cent urea individually.

Alleviation of Salt Stress

Exogenous foliar application of acetyl salicylic acid (0.1mM) prior to NaCl (50, 100, 150 and 200 mM) treatment of Grand Naine banana plants, increased an antioxidant enzyme, superoxidetive dismutase compared over negative control (NaCl stressed plants without salicylic acid foliar spray). There was dramatic increase (43.26 per cent) of SOD after 21 days of acetyl salicylic acid treated plants. Similarly the ascorbic acid (0.1 mM) and ABA (100 ppm) treated NaCl stressed (150 mM) plants recorded higher activities of antioxidative enzyme (15.42) than negative control (NaCl treated plants without, ASA, ABA and Ascorbic acid treatment).

18. Quality Improvement

Bunch Spraying for Enhancing the Bunch Grade

For enhancing the bunch grade in terms finger size and bunch weight as a supplement to the soil applied fertilizers it is recommended to spray the bunch with nutrient solutions. Banana responds well for the post emergence spraying of nutrients on bunches.

Studies have proved that spraying of bunches with potassium sulphate has significantly enhanced the fruit characteristics including bunch grade and total yield as well as fruit quality in many commercial cultivars. It is recommended to spray the bunches with 2 per cent Potassium Sulphate (20g/l of water) with surfactant by thoroughly drenching the bunch first one week after the opening of the last hand and a second spray one month later. The spray has reported to increase the length, girth and fruit weight as well as bunch weight by about 10-15 per cent.

Bunch Covering/Bunch Sleeving

Bunch covering is an important operation to improve the bunch quality by overcoming the ill effects of direct exposure to Sun light, pests and diseases. Direct Sun light, scorches the peduncle region which becomes a predisposing factor for disease innoculum. It has been proved that bagging after the emergence of the last hand in the bunch with 100 gauge thick white or blue polythene sleeves with 6 per cent ventilation or non-woven polypropylene sleeves was effective to avoid any insect attack.

Use of Non-Woven Fabric as Bunch Sleeves for Enhancing Bunch Characteristics

Covering of banana bunches shortly after the bunch emergence using polypropylene non-woven bunch sleeves (both blue and white) has been found to significantly advance the fruit maturity by 4.4 - 9.3 days, increased the bunch weight by 1.6 - 2.8 kg over the control with a benefit cost ratio of 2.00 to 3.50 and also increased the finger characteristics and fruit quality parameters as compared to control. The bunches were fully blemish free without any mechanical damage and injuries caused by insects such as thrips and fruit scarring beetles and had significant reduction in the incidence of anthracnose disease and the fruits developed attractive green colour thereby, fetching about 10-15 per cent higher price.

19. Plant Health Management

Pests

Corm weevil (*Cosmopolites sordidus*), pseudostem borer (*Odoiporus longicollis*) and nematodes are the major pests threatening banana production and productivity in the recent years.

Corm Weevil

This is globally important, infesting most of the commercial varieties. In India, it is more prevalent under perennial system of cultivation across the country and causeing great yield loss. Chemicals have been successfully used to control corm weevil infestation but they are expensive and not eco-friendly. Hence, bio control and ecofriendly control methods have been identified and standardized in banana. CTCRI, segregated the insecticidal compound from bio-waste that subsequently helped in the formulation of the bio-pesticide'Nanma' was useful against a spectrum banana pests, including rhizome weevil. CTCRI has also designed a new applicator (stem injection syringe) for banana plants to control pseudo stem weevil. A dose of 10 to 15 ml has to be injected 5 cm below the infested region (the region where the exudates come due to infestation by weevil) for the effective control of the weevil. This has been commercialized in Kerala among the growers.

Pseudostem Borer

Of late this pest becoming a very serious in all commercial varieties and Cavendish bananas are being the most susceptible varieties followed by plantain. Chemical control through pseudostem swabbing using Chloropyrifos has been commonly used. To monitor and control the banana weevil, a simple, eco-friendly technique using longitudinal split banana stem as a trap and as a delivery system for the entomopathogenic fungi (*Beauveria bassiana*) and nematodes (*Heterorhabditis indica* and *Steinernema* sp.) for weevil management has been developed. Entomopathogenic microbials such as *Beauveria bassiana, Metarhizium anisopliae* and *Lecanicillium lecanii* were isolated from rhizosphere soil and also from different parts of banana plants were evaluated for the control of pseudostem weevil. of these, *Metarhizium anisopliae* (M. a Endo 66) was found be more effective with 76 per cent mortality of stem weevil. Evaluation of endophytic fungi against banana weevils by leaf sheath method indicated that *Metarhizium anisopliae* strains Endo *M.a*-66 was most effective with maximum mortality of 76 per cent. Semio chemical based Kairomones have also been identified. It is being tested under laboratory conditions as an effective trap. Controlling of pseudostem borer using pseudostem trap has been successfully demonstrated under highland production system as an ecofriendly approach.

Banana Rust Thrips (Chaetanophothrips signipennis)

This is becoming a serious concern in the context of export banana production. They causes blemishes and reduced the cosmetic value of the fruit. Polythene sleeves impregnated with 0.1 per cent Chloropyrifos + paraffin oil + adjuvant controlled the thrips attack by 25-30 per cent. Bell injection (flower bud injection) immediately after shooting with Imidacloprid 0.001 per cent has controlled the rust infestation to a greater extent. Scarring beetle (*Besilepta subcostatum*) is serious only in North-Eastern and Eastern states. The fruits affected by this pest have poor market acceptability. It can be controlled using 0.05 per cent Monochrotophos sprays. Survey and identification of effective new biocontrol agents and their effective utilization in controlling banana pests needs immediate attention.

Nematodes

More than 36 species of nematodes are affecting banana, of which four species namely *Radopholus similis, Pratylenchus coffeae, H. multicinctus* and Meloidogyne incognita are of global importance. Nematodes cause yield losses ranging from 30 to 50 per cent. Most nematodes are inhibit nematodes activity only for a shortwhile as most of the compounds are not toxic to the nematodes.

The repetitive application of non fumigant nematicides in long run leads to worldwide biodegradation syndrome. Nematodes are usually controlled by using nematicide like Carbofuran, which has high residual toxicity. So, focus is now on the development of eco-friendly management strategies like use of neem based formulations like Nimbicidin and Econeem @ 30ml/plant. Biocontrol agents like *Pseudomonas flurescens. Paecilomyces lilacinus, Verticillium chlamydosporum, V.lecanii, Baccillus subtilis, Trichoderma harzianum and T.viride* have been effectively used for the management of banana nematodes. Use of VAM (*Glomus fasiculatum* and *Glomus mosseae*) at planting and after 3 months effectively reduced the infestation of nematodes and increased the growth of the plants. Use of AMF (Arbuscular Mycorrhizal Fungi) are obligate symbions colonizing in roots and know to reduce the effect of impact of soil borne pathogens. Though mechanism of action is still debated but induced resistance is demonstrated. Some non-pathogenic Fusarium oxysporum endophyles have shown the ability to reduce pathogen population both in controlled and field conditions.

Fungal Diseases

Among the fungal diseases, *Fusarium* wilt and Sigatoka leaf spot complex are quite serious and *Erwinia* rot is gaining significant importance in tissue culture plants of Cavendish group.

Fusarium Wilt

In India, wilt disease caused by *Fusarium oxysporum* f. sp. *cubense* (*Foc*) is becoming a major constraint for banana production. Recent survey conducted in different parts of the country revealed that the disease is rampant in Tamil Nadu, Andhra Pradesh, Kerala, Bihar, Odhisa, West Bengal and Northeastern hilly states. The banana cultivars severely affected by this disease are: Rasthali –Silk AAB (Amritapani, Mortaman, Malbhog, Sabri, Pattakupra, Nanjangod Rasabale); Ney poovan –AB, Monthan and Bluggoe (ABB); Hill banana (Pome AAB) and Karpuravalli (Kanthali-ABB). The wilt incidence ranged from 1-100 per cent in Rasthali, 6-40 per cent in Karpuravalli (Pisang Awak) and 0-50 per cent in Monthan banana and the maximum incidence of 100 per cent was recorded in Rasthali in different banana growing districts in Tamil Nadu. The disease has caused extinction of a famous local cv. Nanjangod Rasabale (an eco type of cv. Rasthali) in Mysore district of Karnataka. Similarly in Andhra Pradesh, the Fusarium wilt disease has forced the farmers to abandon the cultivation of famous Amritapani (Silk-AAB) for more than 15 years. However, the Fusarium wilt incidence was not observed in Red Banana (AAA) and Nendran (Plantain-AAB) in any of the banana growing states in India. Because of heavy losses being incurred, farmers have already started growing other crops such as jasmine, turmeric, sugarcane etc, in many parts of the country. Besides, the choice of commercial varieties like Rasthali, Virupakshi etc, have been brought to the verge of extinction, due to this lethal disease

Fusarium Wilt Disease in Grand Naine in Theni district of Tamil Nadu

For the first time, in 2008, an intensive survey conducted by pathologists of ICAR-NRCB in Tamil Nadu revealed the incidence of Fusarium wilt disease in Grand Naine and Robusta of Cavendish group (AAA) of banana which was not reported in India earlier. Even the TC plants of one month old found killed due to this disease as against expression of symptoms mostly during the flowering stage of the crop. The incidence was ranged from 5 to 25 per cent and the maximum incidence was observed in the Chinnamanur block. In certain field at K.K Patti of Cumbum block, in the main crop, 2300 out of 9100 plants died due to Fusarium wilt and in the ratoon, 4300 plants succumbed to Fusarium wilt. Similarly in another field in the same village, which composed of virgin soil without any history of Fusarium wilt incidence and given an average bunch weight of 60-70 kg, about 500 out of 32000 plants succumbed to Fusarium wilt. The characterization of the pathogen revealed that this Fusarium wilt pathogen belongs to VCG 0124 of race 1 and found to infect almost all the varieties of banana except Red Banana and Nendran cultivars. The *Foc* isolated has been characterized as VCG 0124 of race 1 which is very unique as they are not supposed to infect the Cavendish group of bananas.

Though chemicals and other cultural methods have been reported to control the disease, current emphasis is on biocontrol agents and botanicals. Fusaric acid detoxifying strain of *P.flurescens*, *T.viride* and *Bacillus subtilis* @ 10g/plant each at the time of planting, 3rd and 5th MAP reduced the wilt incidence. Fungicides such as Propiconazole, Mancozeb and Bavistin have been found effective in controlling the leaf spot diseases during winter period. Biocontrol agents like *T.viride* and *P.flurescens* and botanicals such as Solanum, Emblica and Jatropa effectively controlled the pathogen. Options available for the control of Fusarium wilt are limited and developing resistant cultivars is the sustainable options. The recent release of genomic sequence of *Fusarium oxysporum* fsp. *lycoperscii* is expected to aid in gene discovery and functional analysis.

Sigatoka Leaf Spot Complex

Mycosphaerella leaf spot complex is the biggest constraint to commercial banana production throughout the world. The three main causal agents are Mycosphaerella musicola (Sigatoka disease, also known as yellow Sigatoka), Mycosphaerella fijiensis (black leaf steak disease, BLSD, also known as black Sigatoka) and Mycosphaerella eumusae (eumusae leaf spot disease). These species are closely related and have very similar morphology. These three major species attack all commercial varieties of banana with species specificity. At the global level, Black Sigatoka or referred as Black Streak Disease is very serious on Cavendish clones and in the export oriented western farms, 40-45 sprays are undertaken to control the disease. In India it is serious in major banana growing states like Tamil Nadu, Kerala, Karnataka, Andhra Pradesh, Maharastra (Jalgaon district), Gujarat, West Bengal and North-eastern hilly states of India and inflicts severe yield losses up to 50 per cent in almost all the commercial cultivars. Besides, the disease also found to cause reduction in quality of the fruit, delayed flowering, less number of hands and fingers, poor filling of the fruit, Peel splitting and premature ripening. The disease also predisposed the plant to another lethal disease called 'peduncle rot' which occurs during bunch maturing phase during summer months.

Petroleum oil, dithiocarbamates, chlorothalonils, strobilurin are the major chemicals used to control Sigatoka along with Benomyl. But strategic resources like some of the chemical forecasting models, logistics, inoculum management curative effects using carrier and adjuvant and post pathogen sensitivity studies will help to control the Sigatoka in a more sustainable way. Spraying of Paraffinic oil 1 per cent + half the dose of any one of the fungicides *viz.*, Propiconazole 0.05 per cent or Companion 0.05 per cent or Carbendazim 0.05 per cent in banana 4-6 times at 20-25 days interval controlled the Sigatoka leaf spot disease of banana effectively in both Robusta (AAA) and Nendran (AAB) and also increased the yieldupto 20 per cent.

But more sustainable way to contain the disease will be developing partial but durable resistance over complete resistance. Though there is no confirmatory evidence for the occurrence of Black Sigatoka studies like micro-array assisted comparisons along with DArT markers would be able to identify the different species of *Mycosphaerella*, facilitating their effective control.

Bacterial Rot

This disease is caused by soil borne bacteria *Erwinia carotovora* that gains entry into the young banana plants through physical injuries. It is known by different names like Erwinia attack, Head rot, Soft rot and Tip Over. Tissue culture raised Banana plants are prone to this bacterial attack. Underground rhizome of young plants get infected, leading to extensive rotting and emitting of foul odour. In older plants mostly the collar region gets affected become so weak that the plant breaks off easily. In severe cases, pseudostem base becomes swollen and splits. Affected portion looks yellowish brown with a definite brown margin. Disease also spreads to the daughter suckers of the clump. The disease is mainly observed during hot summer months and during rainy season after 3rd to 5th months of planting up to bunching in the field. The pathogen is soil borne and spreads through water. Rotting of collar region leads to drying of outer and inner leaves. If affected plants are pulled out it comes out from the collar leaving the corm that is why called as 'Snap off' or 'Tip over'.

The pathogen of rhizome rot (*Erwinia* spp.) was isolated and confirmed by electron microscopy and biochemical studies. The pathogenicity studies revealed that only 60 per cent of the plants inoculated showed corm rot symptoms indicating difficulties in carrying out other management studies under pot culture condition. Field trial conducted for the management of Erwinia rot disease in banana indicated that the planting of apparently healthy suckers + drenching with *Pseudomonas fluorescens* 1-2 litres per plant (50g/lit of water at 0 + 2 +4 +6 month after planting + growing sun hemp in the interspaces for 2-3times till 5months after planting or planting of apparently healthy suckers and application bleaching powder 4g/plant at 0 + 1 +2 +3 + 4 month after planting + growing sun hemp in the interspaces for 2- 3 times till 5 months after planting.

Viral Diseases

In the recent years knowledge on these viruses has expanded but some of the fundamental epidemiological question remains unanswered. Five viral diseases have been reported in India and among them BBTV, BBMV, CMV and BSV are considerably important. Occurrence of BBTV has wiped out large area under hill banana cultivation. It has been recorded even in wild bananas of Arunachal Pradesh. BBTV isolates of North Eastern states and lower Pulney hills exhibited more than 90 per cent similarity which indicated that it is quite prevalent in the hilly tracts also. Now, it is possible to do indexing of hill bananas against BBTV for large scale sucker multiplication and supply to the hill banana growers.

Although management of BBTV is effective thrust a strategy of reintroduction of planting material and internal quarantine. But internal quarantine has some of the hurdles like constitution of quarantine protecting individual liberties has hindered the introduction and enforcement of quarantine regulations. Earlier cv. Gros Michel was thought to be tolerant but later proved to be due to slow expression of symptoms. In nature there is no immunity for BBTV. Development of resistant plant through genetic engineering seems to be the only the possible option.

BBMV is another devastating viral diseases affecting plantain and many important commercial varieties. In Robusta and Ney Poovan, application of higher dose of fertilizers (150 per cent) increased the yield in BBMV infected plants. Often Bract mosaic virus is confused with sugarcane mosaic strain of Abaca. It might have alternate hosts in its epidemiology.

CMV occurs in almost all banana growing states. Control can be achieved only through effective legislative measures and quarantine procedures.

Banana Streak Virus (BSV) is present in every banana-producing region worldwide. BSV infections can lead to severe decline in yield especially in Poovan varieties. BSV is the main constraint for germplasm movement, breeding and tissue culture of *Musa acuminata* x *M. balbisiana* interspecific hybrids and species due to the presence of infectious BSV endogenous pararetrovirus (EPRV) sequences in the genome of *M. balbisiana*. Tissue culture techniques trigger the activation of infectious BSV EPRV in such interspecific hybrids and species, giving rise to infectious viral particles. Therefore meristem tip culture, which was widely adopted for generating virus free germplasm from infected plant material cannot be used safely for generating virus free hybrids. Furthermore, the presence of infectious BSV EPRVs in the B genome limits the use of B-rich genotypes for creating much needed new banana hybrids although they confer resistance to major pests and diseases, highly female fertile and contribute to cross compatibility with other cultivated varieties. Therefore, screening *M. balbisiana* resources for germplasm free from BSV EPRVs has become a pre-requisite for any classical *Musa* breeding programme. This endogenous badna virus has long association for more than million years and sequence has integrated with both A and B genomes. This is normally latent but activable only in B genome. *Musa balbisiana* is homozygous for endogenous obino l'ewai strain between *Musa* genotypes.

Molecular Diagnostics of Banana Viruses

Coat protein gene of BBTV, BBrMV, CMV and viral associated protein gene of BSMYV were cloned and over-expressed in the bacterial system and produced polyclonal antisera for diagnosis. Antisera are being supplied to TC laboratories accredited by DBT and also to the SAU's for virus indexing. PCR, RT-PCR, Multiplex-PCR/RT-PCR, Nucleic acid spot hybridization (NASH) and ELISA based detection techniques developed for the detection of four banana viruses. Realtime PCR techniques for detection of BBTVand BBrMV using TaqMan and SYBR chemistry have been standardized. Loop Mediated Isothermal Amplification (LAMP) has been standardized for detection of BBTV. Lateral Flow Immuno- Assay (LFIA) or dipstick technique has been standardized for the detection of CMV. RCA based detection of BBTV and BSMYV has been standardized. Developed a highly sensitive, cost effective simple extraction protocol for template preparation for detection of banana bunchy top virus by PCR.

Two new virus Banana Mild Mosaic Virus (Ban MMV) and banana virus X (BVX) have been reported from the west. There is indirect evidence that Ban MMV is transmitted between plants in the field but no vector is identified. But Ban MMV is expected to have synergistic effects with other viruses.

Currently, protocols have been developed to detect viruses even in vectors. This would be helpful to study the epidemiology of banana viruses.

20. Post-harvest Management

The post-harvest life of fruit starts from the time it is cut from the mother plant. Banana being a climacteric fruit is highly perishable and has got a short shelf-life after harvesting. The post-harvest losses are very high (to the tune of 22-30 per cent) due to faulty handling adopted by farmers, traders and retailers. In terms of value, it is more than Rs. 3,000 crores annually. The post-harvest losses are estimated to be highest at retailers level (20-25 per cent) followed by whole-sellers (15 per cent). The loss at farmers' level is due to mishandling during harvesting and loading which is estimated to be about 8-9 per cent. Besides these losses, there is an invisible loss due to the physiological loss in weight during ripening phase. The mishandling of fruit not only leads to post-harvest losses but also it reduces the appearance of the fruits and its cosmetic value. Therefore, in order to reduce the post-harvest losses and to enhance the shelf-life, NRCB has developed scientific technologies of post-harvest handling of fruits. The quality of bananas is influenced by many factors. The stage of maturity is of prime importance, which depends on whether the fruit is meant for long distance transport or local consumption. Bananas to be marketed locally can be harvested at a higher maturity stage than those to be transported for long distance markets. Generally fully mature bunches (90-95 per cent) are harvested for domestic market, where fruits are ripened and consumed within a week. For regional markets, where it takes about 2 weeks, the fruits have to be harvested at 85 per cent maturity with fairly defined angles. For long distance transportation and export, the harvesting is done at 75 per cent maturity, where the fruits have well defined angles. Several parameters have been suggested for determining the maturity of banana fruits on the basis of their external appearance and chemical constituents at the time of harvest.

Maturity and Harvesting

Generally, development and bunch maturity is accomplished normally between 110-120 days and harvested at various maturity levels depending upon the market they are aimed at. For export market, bunches at 70-75 per cent maturity, for distant markets 75-80 and for local markets it is normally 80-90 per cent maturity are preferred.

Storage studies of green banana fruits indicated that the sugar content of pulp significantly increased at 13.5 C than the fruits stored at 10 C and 12 C after 21 days of storage owing to higher b-amylase and antioxidative enzymes like SOD, catalase and ascorbate peroxidase. The critical temperature for storage of bananas was 13.5 C for Cavendish varieties like Robusta and Grand Naine, Rasthali (Silk), NeyPoovan (AB diploid) and Poovan (AAB), while it was 12 C for Nendran (Plantain, AAB) and 15 C for Pachanadan (Pome). These fruits ripened normally without any chilling injury when the fruits are sprayed with Ethrel @ 500 ppm. Ethrel @ 500 ppm at 22±1 C as dip for 5 min significantly improved the ripening with excellent colour development. The fruits ripened within 72 h had better quality as compared to control.

Integration of all post-harvest treatments like dehanding, washing, Bavistin dip at 500 ppm concentration, removing the surface moisture, packing using 400 gauge polyfilm, use of ethylene scrubber and storage at 13.5 C extended the green life of Robusta for 110 days, Grand Naine up to 90 days and Rasthali up to 120 days. Ethylene aided ripening could ripen the fruit normally after removal from cold room and had a yellow life of 5 days.

Hydrocooling, use of ethylene absorbant, waxol treatment and packing with polythene wrapping increased the green shelf-life. Combination of heat shock, modified atmosphere packaging, storing in CFB box in cold storage at 13.5°C extended the storage life upto 126 days in Rasthali and 136 days in Robusta. The cold stored fruits ripened normally after treatment with Ethrel (500ppm).

Green Life Enhancement of Pre-climacteric Bananas using Biochemicals

In order to minimize the post-harvest loss of bananas during the season of high harvest and natural calamities, a technology to store and increase the green life of cvs. Grand Nain and Poovan bananas was developed with treatment of 1-methylcyclopropene, which enhanced the green life of pre-climacteric Grand Nain and Poovan bananas for four months at 14°C and for 15 days at ambient temperature. The treated bananas can be ripened normally without any change in the physiological, biochemical and qualitative parameters during storage and ripening.

Value-addition in Banana

Value-addition is the only viable option to convert the excess banana production into value-added products, which can be stored for long time and it may fetch premium price in the market. In banana and plantain, the value-addition through processing have resulted in development of various value-added products such as Fig, Juice, Bar, Jam and Sweet *Chutney* from pulp of ripe banana; Flour, Baby Food, Health Drink, Sauce, Pickle and Chips from pulp of unripe banana; Pickle from flower (male bud) of banana; Candy and Pickle from centre core stem Procedures for making value-added products from banana fruit *viz.*, Jam, fruit bar/leather, and sweet *chutney* have been developed. Besides, ready to-eat/serve products like '*Pazhampori*' and 'Banana pulp based beverage' were also developed.

Banana can be processed into different value-added products. The NRCB, Trichy has developed protocols for more than 14 value-added products and 4-5 products have been commercialized already. Apart from Nendran banana chips, aseptically packed pulp has high demand for export. Banana puree has maximum application in dairy products, bakery, beverage and baby food. Blending of banana (70 per cent) with pineapple (30 per cent) juice recorded maximum organoleptic value. Banana flour is also becoming popular. Banana flour based functional foods like Chapathies, bread and health drink are found to control the blood sugar level on par with other diabetic food available in the market.

Banana can be processed into different value-added products. The NRCB, Trichy has developed protocols for more than 14 value-added products and 4-5 products have been commercialized already. Aseptically packed pulp has high demand for export. Banana puree has maximum application in dairy products, bakery, beverage and baby food. Blending of banana (70 per cent) with pineapple (30 per cent) juice recorded maximum organoleptic value. Banana flour is also becoming popular. Banana flour based functional foods like Chapathies, bread and health drink are found to control the blood sugar level on par with other diabetic food available in the market.

Plantains are usually processed into chips. For every tonne of chips produced, 3-4 quintals of raw banana peel is generated as waste. A technology has been developed at NRCB to convert the waste into pickle of acceptable quality. Preparation of pickle from male bud of banana has been standardized. This technology has been transferred to several entrepreneurs for commercialization.

Waste Utilization

Conversion of waste into a wealth is very much possible in banana. The banana male bud is a waste material produced during crop production with less economic value, while central core (true stem), a component of pseudostem after harvesting of bunch and peel, a component of fruit also go as wastages. For the utilization of both male bud and peel, recipes were developed for the preparation of pickles (*thokku*). About 5-7 tonnes of central cores can be extracted from one hectare. A process was also standardized for preparation of stem pickle, juice and candy from central core stem. Biochemical constituents in banana peel were analyzed for commercially cultivated varieties

On comparative evaluation of four banana based pickles (fruit, flower, central core stem and peel) stored upto six months, both peel and flower pickles (*thokku*) were accepted as fresh as well as even after six months of storage. About 30-40 per cent of banana fruits go as peel waste. A process for production of commercial grade ethanol from waste banana fruit peel was standardized.

Processed Functional Products

Diabetic, antioxidant and micronutrient rich functional products could be prepared from various parts of banana. Many kinds of instant mixes, soup powders and baby foods can be prepared from the banana flour. Unripe banana have a higher starch concentration than ripened fruits and the modern food industries growing demand for starches has created interest in identifying new sources of this polysaccharide. Modified food starches generally show better taste and stability and increased resistance to Retrogradation, and Freeze-thaw stability. The green banana process requires 8–10 kg of raw green bananas to produce 1 kg of banana flour. Banana flour is used not only as a replacement for wheat flour in baking, but also as a thickener for sauces and soups and as a breading for meats and vegetables. Banana flour is used as animal feed in various parts of the world.

The development of nutraceutical ingredients from the banana is current interest in all food industries. The fibre rich powder could be prepared from the green banana flour using many physical, chemical and enzymatic modification processes. Attempts are made to use rejected bananas in animal feed and products such as chips, flakes and powders, but they are used only to a limited extent for these purposes due to the low value of such products. Banana flour traditionally made of culinary banana is often used as a gluten free replacement to wheat flours or as a source of resistant starch. The functional food market is innovation driven and dominated by mostly Multinational companies.

21. Conclusion

The status of banana has been uplifted from a mere backyard fruit to a high value crop both in the national and global scenario. Consumption has proportionately increased with the magnitude of production. India being the largest producer has witnessed 120 per cent increase in production and 70 per cent in productivity due to the adoption of improved production technologies. Simultaneously, the knowledge on pests, diseases has witnessed but the sustainable solutions are limited in view of the constraints in developing improved hybrids through conventional breeding, biotechnological approaches seem to play a major role. Availablity of clean planting material, either as suckers or as tissue cultured plant, should be given priority. Development of certification standards for the tissue culture companies by the DBT, New Delhi is a positive approach in that line. Some of the emerging opportunities for organic banana production for local and export market, varietal diversification for export have to be considered seriously to surge ahead in the global market. During this process, development of feasible environment and market policies, understanding of innovative systems, improved capacity building, adoption of knowledge.

17

Technological Innovations for Bridging the Yield Gap in Spices

K. Nirmal Babu[1], Homey Cheriyan[2], V. Srinivasan[1], H.J. Akshita[1] and Muhammad Nissar[1]*

[1]ICAR-Indian Institute of Spices Research, Kozhikode – 673 012, Kerala
[2]Directorate of Arecanut and Spices Development, Kozhikode – 673 005, Kerala
**E-mail: nirmalbabu30@hotmail.com*

Spices are integral to human life especially in Indian society – in tradition, food, aroma, health, and economy. Black pepper, ginger, turmeric, cardamom and tree spices such as nutmeg, cinnamon, garcinia and tamarind are the tropical spices of importance in Indian context. Coriander, cumin, fennel and fenugreek are important seed spices. Mintis a herbal spice of great importance in world mrket. *Garcinia*, black cumin, ajowain, saffron, mint, oregano, lavender, star anise *etc.* are the future crops.

They are low volume high value commodities used as condiments, functional foods and more importantly as sources of high value medicinal compounds.Because of their high value spices increase the farm income where ever they are grown. All over the world, the fast growing food industry depends largely on spices as taste and flavour makers. Health conscious consumers in developed countries prefer natural colours and flavours of plant origin to cheap synthetic ones. Thus, spices are the basic building blocks of flavour in food preparations in addition to their use as functional foods, nutraceuticals and sources of many high value phytochemicals. The major phytochemical classes associated with herbs and spices include a diverse array of compounds such as terpenes and terpene derivatives. Both *in vitro* and *in vivo* studies have suggested that dietary spices and herbs maintain human health by their antioxidative, chemo preventive, antimutagenic, anti-inflammatory and immune modulatory effects.

The consumption of spices is growing in the country with increase in purchasing power. It is envisaged that everyone in India would be consuming one spice or the other with a high per capita consumption. It is estimated that we may have a population of about 1.69 billion people during 2050 and approximately the per capita consumption of black pepper, cardamom, turmeric, and ginger is expected to be about 148 g, 54 g, 1.6 kg and 1.2 kg, respectively. This may increase further due to rapid urbanization which needs spices as natural food preservatives.

Production, Export and Import

India is the largest producer of spiceswith an annual Production of 6.20 Million MT in an area of 3.30 Million Hectares during 2014-15 is.After consuming more than 70 per cent of the spices produced, India still make up 48 per cent of the total world trade in quantity and 43 per cent of the value. In spite of a global economic slowdown the growth of the Indian spice industry and its exports have continued its increasing trend both in volume and value, registering an increase of 9 per cent in volume and in value 7 per cent in dollar terms. With an estimated growth rate 3.19 per cent, the world spice trade is expected to touch 18 billion and that of India is US$ 5 Billion by 2020

Irrespective of this India still imports large quantities of spices.The total import of spices in India during 2012-13 is about 1.31 lakh tonnes valued Rs. 2102.3 crores. The import value has increased by 18 per cent as compared to previous years. The major spices imported are ginger, black pepper, poppy seed, clove, star anise and cassia. Out of the total import of black pepper (15800 t), more than 60 per cent is light (immature) pepper for oleoresin extraction and re-export. Fresh ginger is imported mainly from Nepal (30 per cent of import volume) for industrial consumption. We must minimise the imports by increasing the productivity, production, sustainability and quality of spices from India to meet the global requirements in the coming years.

Challenges

There are challenges that threaten the spices marketing and industry. Declining productivity, high cost of cultivation, reduction of area under spice cultivation, labour shortage, problems in supply chain, rising demand v/s raw material storage, market instability, reducing quality and active ingredient content, increasing incidence of contaminants (aflatoxin, pesticides, illegal dyes, microbials *etc.*) stringent food safety standards,overwhelming regulations and ecological sustainability has affected spice industry and trade resulting in lower prices and which in turn leads to loss of interest in spices cultivation resulting in lower productivity. This is further complicated by noncompetitive production in comparison to other countries such as Vietnam, Guatemala, China Indonesia *etc.* where the productivity and production is high and the labour is cheap.

High Production Technologies

The involvement of Indian Institute of Spices Research (IISR), National Research Center for Seed Spices (NRCSS) with support from sister agencies like All India Coordinated Research Project on Spices (AICRPS), Spices Board, Central Food Technological Research Institute (CFTRI) and Directorate of Arecanut and Spices Development (DASD) has resulted insubstantial increase in production of spices, their mechanization, post-harvest processing, value-addition and product development.

Research activities on spices has resulted in establishment of world collection of spices germplasam, development of over 250 improved varieties of spices, integrated nutrient, disease and pest management, farm mechanization, post-harvest technologies for most of the important spices.

Bridging the Yield Gap

The main causes of low production and productivity in India are unstable prices, cultivation of low yielding low quality varieties, limited availability of quality disease free planting materials of high yielding varieties, degradation of soil health, climate change and changes in rain fall pattern, prevalence of diseases and pests, labour shortage and deficiencies in information dissemination and transfer of technology and lack of efficient market intelligence. Though over 200 improved high quality varieties, some of them with diseases and pest resistance, sustainable high production technologies, efficient techniques for production of disease free planting materials a most important component in spices, integrated IPM, IDM and INM are available for most of these their adoption is hampered by non-remunerative prices. Good agricultural practices (GAPs) are also available in some spices and are getting ready for other spices.

Further Strategies to Fill the Gaps

a) Greater efforts focused on sustainable increase in production through high production technologies integrating improved varieties, disease, pest and nutrient management, cropping system approach and organic farming and GAPs are needed.

b) Many facets of spices cultivation, harvesting, storage and processing can be mechanized with available technologies.

c) Improving and harmonization of Standards for production, post-harvest technology, processing, marketing and bio-safety issues of spices.

d) To create a final product that is acceptable as 'food safe', one needs to be able to control all the elements on the supply chain, and with the support of all the stakeholders.

e) Product development, diversification and value-addition need a greater thrust and innovation to achieve greater commercial importance.

f) The intrinsic quality and flavour and other components of spices are dependent of the agro climatic conditions in which they are grown. Hence development of Geographical Indication (GI) for spices important for getting the best spices and for better pricing.

g) These in turn gives better remuneration higher income hence higher care and involvement in spics technology adoption resulting in higher production and productivity.

h) Effective flow of information to all stake holders for quick implementation the technologies available.

To achieve the targeted production and export in spices in 2050, sustained research, development and industrial initiatives to overcome the constraints becomes pivotal. Increasing the productivity per unit area through spice based farming systems, development of varieties with high degree of resistance to biotic and abiotic stresses, development of agro technology towards low input management, precision farming, developing eco-friendly IPM strategies, post-harvest technologies with value-addition and exploiting its medicinal properties, and popularization of proven technologies though extension network are the major areas.

Besides, the new areas on nanotechnology, bioinformatics, carbon (C) and water foot prints and knowledge management would further strengthen the research programmes on climate resilient agriculture. These technological advancements will bring out the surge in productivity of spices, and 2-5 fold increase in all major spices, to meet the consumption and export demand. The forecasted spices production by 2050 is expected to be achieved with an annual compound growth rate (ACGR) of about 3 per cent. Awareness programs for producers to follow hygienic practices to develop clean spices will help partially but better pricing will automatically help.

The advances made in spices improvement, production, protection, microbial and post-harvest technology must be popularized among all stake holders through the NGOs, SHGs through participatory technology development. In addition to visual and print media, mobile technology can also be used for quick transmission of information and to give alerts.

Harmonization of global standards on bio-safety, ensuring sustainability of supplies and to bridge the demand and supply imbalances without adversely affecting the environment involving producers, traders, industry and consumers and adjusting to emerging Market Trends.

Nanotechnology has been identified as an important component in future in maximizing the productivity of agriculture. Though there are many toxicology concerns this technology has great potential in developing new tools for rapid disease diagnosis and to combat pathogens, enhancing nutrient absorption, developing efficient pesticides, separation, identification and quantification of biological molecules, soil and water management, bio processing, post-harvest technology, monitoring the identity and quality of agricultural products and precision agriculture.

Web based interactive data bases providing expert systems and instant information on, for example, weekly analysis for spices, future predictions on demand, supply and pricing, directories of products, buyers, sellers, agents, and statistics *etc.* are available for traders but more are needed especially for supporting farming community also.

Product Development

Diversification and value-addition need a greater thrust and innovation to achieve greater commercial importance. In a world with many lifestyle diseases spices with its chemo preventive, antioxidant, anti-inflammatory, antimicrobial and insecticidal properties, bioenhancer potential are gaining importance as Functional foods and are used as dietary supplements. With greater amounts of biomolecules with important health notes the spices need to be exploited towards problem solving approaches to enhance the human health.

The awareness about spices is increasing. The demand for organic spices is growing @ 20 per cent annually. Spices are finding newer applications in food, cosmetics and pharmaceuticals. Sustainability and Food Safety are of crucial importance today and need serious deliberation, proactive action and an active coming-together of all the stakeholders. Global initiatives need to be intensified by bringing together the global spice industry to define and set an implementable pathway to secure a sustainable future for the industry and the agriculture system which supports the industry, while also examining the industry's ecological and social impact. The rise of e-commerce must help in alternate price discovery mechanisms. If we utilize these opportunities the spices industry and marketing is poised for exponential growth.

Conclusion

The mission is to capture India's preeminent position as *'spice bowl of the world'* by producing and exporting the best spices and spice products. This can be achieved by obtaining high productivity of clean spices through varietal deployment, plant and soil health, optimal and safe technologies. Most technologies needed are either already available or in advanced stage of development. They just need to be fine-tuned and implemented with the urgency needed. The industry along with developmental and trade agencies also need to get involved in this efforts by funding contractual research as well as contract cultivation as 2 of the important initiatives among others.The most important thing is the farmer need to get supported for better remuneration for his efforts. Thus up gradation of technology across the entire production and supply chain – a key driver of growth.

This cannot be achieved without effective linkage, convergence and cooperation between all stake holders in the supply chain from farmers to the industry and optimal and equitable sharing of efforts, knowledge and profits. The coming together of all stake holders in a PPP mode under one umbrella is very much needed to meet the demands of industry, regulatory bodies and consumers. The vision of the Indian Spice Industry is to become the processing hub and premier supplier of value-added spices and herbs in the industrial, retail and food services segments of the global spices market by meeting the quality requirements and thus to achieve US$10 billion worth Spices Exports by 2030 from India.

Finally the Governmental involvement is also required in providing incentives to agriculture, industry and trade, to remove trade barriers in the form of unfair regulations simplification of policies and procedures and in export-import trade agreements with emerging markets is an area that requires governmental involvement. If we utilize these opportunities the spices industry and marketing is poised for exponential growth

References

AISEF. 2012. *A Report on the Indian Spice Industry*. All India Spices Exporters Forum, www.aisef.org

Krishnakumar, N.K. 2015. Spices: Way forward. **In**: Souvenir and Abstracts, *National Seminar on Spices and Aromatic Crops (SYMSAC VIII), Towards 2050: Strategies for Sustainable Spices Production*, (Eds.) K.S. Krishnamurthy *et al.*, Indian Society for Spices, Kozhikode, p. 3–14.

Spices Board of India, 2015. *Spices Industry in India*.

18

Technology Options for Doubling Farmers' Income in Cashew Cultivation

P.L. Saroj[1]* and M.V. Sajeev[2]

[1]*ICAR-Central Institute for Arid Horticulture, Beechhwal, Bikaner, Rajasthan*

[2]*ICAR-Directorate of Cashew Research, Puttur – 574 202, Karnataka*

**E-mail: sarojicar@gmail.com, plsaroj@yahoo.co.in*

The cashew (*Anacardium occidentale* L.), a native of Brazil, was introduced in India by Portuguese travelers in 16th Century. However, the export potential of cashew kernels in the international trade was first realized by India in the early 1900s'. Presently, India has the largest area under cashew and also largest producer of cashew in the world. Cashew plantation occupies an area 1.027 million ha with production of 0.725 million tonnes but the productivity is only 706 kg/ha (2014-15). Most of the cashew plantations are along the coastal regions mainly in Maharashtra, Goa, Karnataka and Kerala in the West coast and Tamil Nadu, Andhra Pradesh, Odisha and West Bengal in the East coast. It is also spreading in non-traditional areas such as Bastar region of Chhattisgarh and Kolar (Plains) region of Karnataka, Gujarat, Jharkhand and in North East hill (NEH) region. Maharashtra ranks first in area (0.186 million ha), production (0.235 million tonnes) and productivity (1262kg/ha) of cashew in the country. The area, production and productivity of cashew in different states are given in Table 1. It is also pertinent to mention that Andhra Pradesh and Maharashtra both are having individually more than 18 per cent area under cashew but Maharashtra alone contributing 31 per cent of the total raw cashew nut production. This is only because of adoption of new varieties and advanced production technologies. The productivity level of traditional

Table 1: Area, Production and Productivity of Cashew in India: 2014-15

State	*Area ('000 ha)*	*Production ('000 tonnes)*	*Productivity (kg/ha)*
Kerala	84.53	80.00	946
Karnataka	124.71	80.50	645
Goa	58.17	32.00	550
Maharashtra	186.2	235.00	1262
Tamil Nadu	140.42	67.00	478
Andhra Pradesh	185.45	100.00	539
Odisha	180.41	85.50	474
West Bengal	11.36	13.00	1096
Jharkhand	14.83	4.50	303
Chhattisgarh	13.7	8.50	620
Gujarat	7.22	8.50	1177
Pondicherry	5.00	3.00	600
Assam	1.05	0.57	543
Tripura	4.25	2.50	588
Meghalaya	8.5	4.50	529
Manipur	0.90	0.15	167
Nagaland	0.50	0.20	400
Total	**1027.20**	**725.42**	**706**

DCCD – 2015.

cashew growing states like Tamil Nadu, Andhra Pradesh and Karnataka is very low. The percentage share of area, production and productivity of cashew in different states is given in Figures 1–3. The domestic production of cashew nut is insufficient to fulfill the processing requirement of cashew industries in the country (2.0

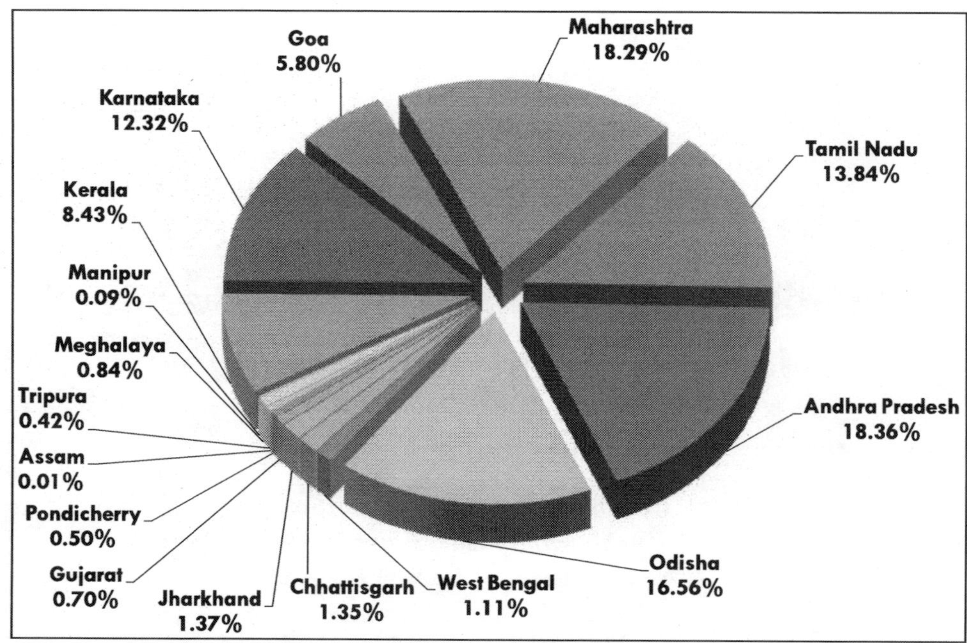

Figure 1: Percentage Area under Cashew in different States: 2013-14.

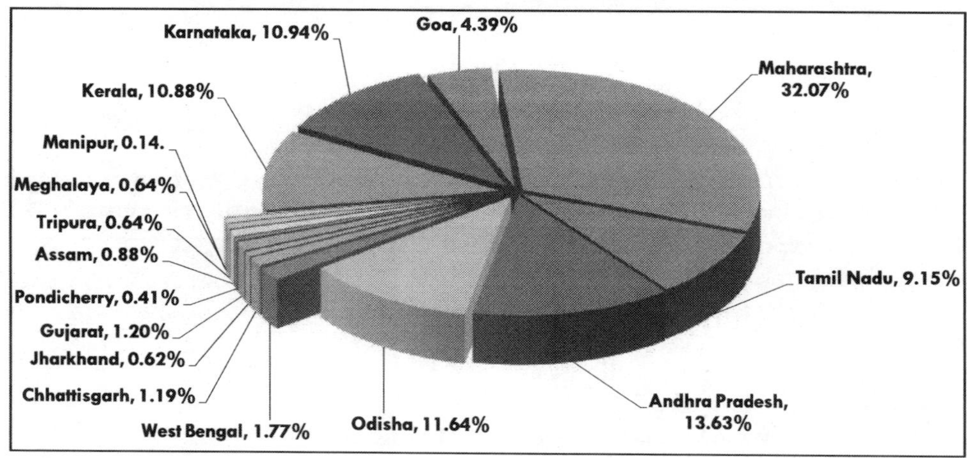

Figure 2: Percentage of Cashew Nut Production in different States: 2013-14.

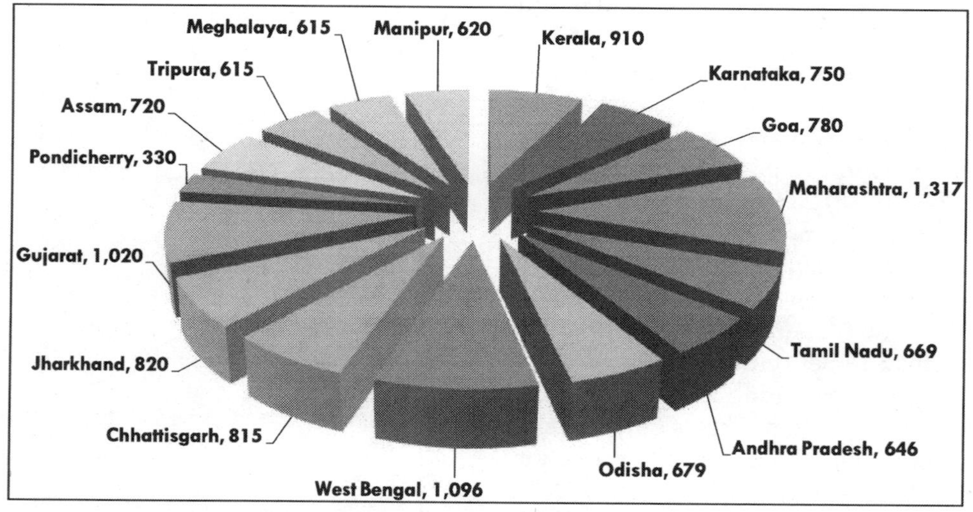

Figure 3: Productivity (kg/ha) of Cashew in different States: 2013-14.

million tonnes), which calls for immediate action to increase our domestic production of raw cashew nut, in order to reduce dependency on import of raw nut.

The area and production of cashew nut is increasing continuously but from the last fifteen years, productivity varied in between 600-800 kg/ha (Figure 4) At the present level of productivity, farmers are getting an income of Rs. < 50,000/ha. Poor productivity and low income of cashew farmers is a serious concern. At this juncture, when country is looking towards doubling farmers' income, major constraints in cashew production are; (i) most cashew plantations being seedling in origin of non-descript types and senile in condition, (ii) planted under degraded lands and are in neglected conditions, (iii) often heavy yield loss due to outbreak of tea mosquito bug (TMB) and severe attack of cashew stem and root borer (CSRB), (iv) absence of dwarf and compact high yielding varieties suitable for high density planting and (v) inadequate transfer of technology resulting in poor adoption of recommended package of practices.

On the other hand, ICAR-Directorate of Cashew Research, Puttur (Karnataka) and All India Coordinated Research on Cashew with its fourteen centres located across the cashew growing states as well as several developmental organizations are involved in India for research and development of cashew. The Directorate of Cashewnut and Cocoa Development (DCCD), Kochi, under the Union Ministry of Agriculture and State Department of Horticulture/Agriculture of several cashew growing states are involved in technology dissemination through Central Sector Schemes including National Horticulture Mission (NHM). The Cashew Development/Plantation Corporations and Forest Department in the states of Kerala, Karnataka, Goa, Tamil Nadu, Andhra Pradesh and Odisha are also involved in the developmental activities of cashew. Whereas, export of cashew kernel is being looked after by Cashew Export Promotion Council of India (CEPCI). Owing to the efforts of all these organizations both in research and development, the status of cashew has now been changed from a crop for soil conservation and afforestation into a horticultural crop and have developed various varieties and technologies for improving productivity of cashew. Keeping these facts in view, the major technology options available to enhance the productivity of cashew nut and thereby increasing income of cashew farmers' have been discussed in this paper.

1. Technologies for Productivity Enhancement

1.1 Edaphoclimatic Delineation

Most of the cashew plantations are on loamy red and lateritic soils, mixed red and black soils, coastal and deltaic alluvium derived soils. The distribution of cashew is restricted to altitude below 1000 m where the temperature does not fall below 20°C for prolonged periods. However, best production was noticed up to the altitude of 700 m, mean annual temperature range of 22.5-27.5°C with at least 8-9 hrs of sunlight/day from December-May. Cashew grows well at reasonably high temperatures and does not tolerate prolonged periods of cold and frost especially during the juvenile period. However, temperature above 36°C during the flowering and fruiting period could adversely affect the fruit set and retention. Cashew can adapt very well to dry conditions and its performance is better in the areas which receive fairly distributed rainfall of between 600-1500 mm per annum. Prolonged dry spells, frost, foggy weather and heavy rains during flowering and initial

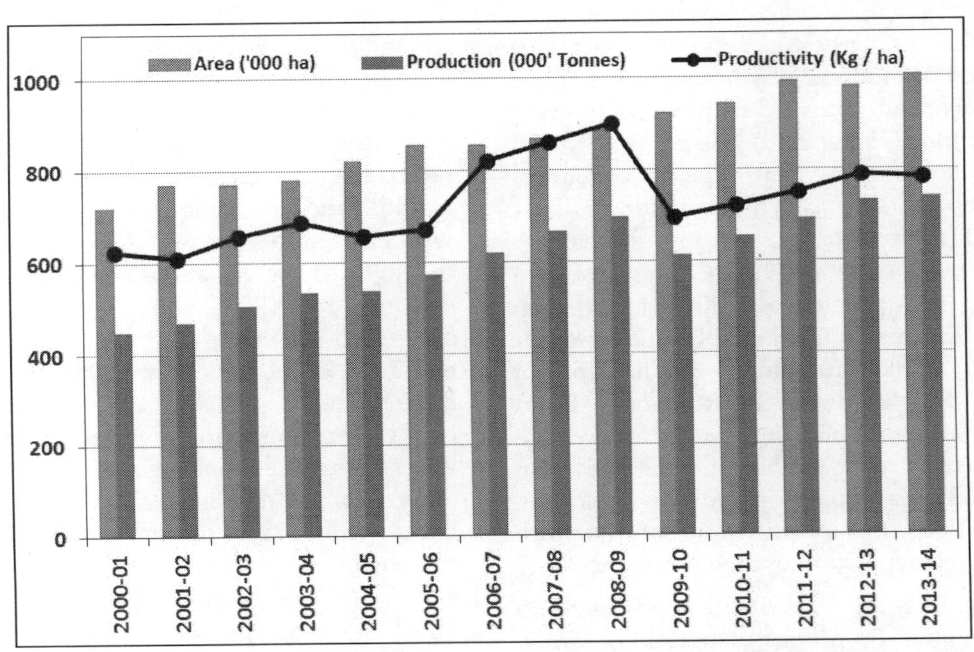

Figure 4: Trends in Area, Production and Productivity of Cashew in India.

fruit setting, adversely affect fruit set and production. Cashew is categorized as a climate smart crop and tolerates wide range of ecological factors. Therefore, even at some degree of variations in climatic parameters like rainfall, temperature, humidity, sun shine hours, dew and wind velocity, there is no significant difference in the cashew yield at national level. But this does not mean that cashew is not affected by climatic aberrations. Since, most of the cashew plantations are in ecologically sensitive areas such as coastal belts, hilly areas and areas with high rainfall and humidity as well as rainfed conditions, weather aberrations particularly during reproductive phase adversely affects the cashew productivity.

1.2 Rejuvenation of Senile Plantations/ Replanting

In India, a sizeable area under cashew plantation is seedling in origin and left either unmanaged or least managed, owing to huge size trees with intermingling branches, poor light interception, acting as a shelter for various pests and thereby very little peripheral fruiting is often observed. Such plantations are more than 30-years-old and need either replacement by fresh plantation with superior varieties or severe heading back followed by top working with desired varieties. The technique for rejuvenation of unproductive trees has been standardized by beheading the healthy trees at 1 m height in the month of March. Thereafter, large number of new sprouts will emerge but retain only 4-6 healthy branches considering all directions, so as to develop a good canopy structure and rest should be thinned out. These new sprouts after attaining appropriate size should be top worked by softwood grafting of desired scion shoots. Whereas, frame working can also be done by heading back of major branches from near the site of its origin leaving the main tree frame as such followed by the similar grafting operation. In case of 10-15 years old cashew plantations, highest success rate (80 per cent) in the month of April followed by June grafting (77.08 per cent). In case of 5-10 years old plantations, grafting in the month of June (70 per cent) was better followed by grafting in May (55.3 per cent). The uneconomical cashew trees top worked with high yielding Ullal-1 variety resulted in four-fold increase in nut yield per tree within five years (Kumar, 1990). Cashew growing regions in which the population of CSRB is high, the top working technology may not be suitable unless regular follow-up action is taken to manage the incidence of CSRB (Swamy, 1995). Besides CSRB, attack of TMB and other pests are also common on new sprouts which should be managed properly. The cut end should also be pasted with Bordeaux paste in order to avoid any secondary infections. Proper nutrition and moisture management of rejuvenated trees are also essential.

If rejuvenation is not possible, it is advisable to go far new plantation by proper preparation of area. In most of the cases, old plantations are subjected to heavy weed growth of perennial plants, which compete for moisture, nutrients and light. Sometimes, these weeds act as host plant for various pest and diseases. Therefore, new plantations should be taken up by using high yielding varieties. Proper pit digging at desired spacing, filling by appropriate pit filling mixture and subsequent planting of grafts with care. Being a new plantation, proper training can be given to plants, since beginning, to develop into a desired shape and thereafter need based regular pruning to promote better fruiting. If plants are trained well and spaced properly, intercultural operations are also feasible which intern minimize the cost of cultivation.

1.3 Soil and Water Conservation

In India, cashew is commonly grown in degraded sloppy and undulating lands both in the West and East coasts. Soil erosion and leaching of plant nutrients are generally expected under such situations. In general, cashew experiences moisture stress from January to May, which adversely affects its flowering and fruit set. To overcome this problem, *in situ* rainwater harvesting and making it available to the cashew plant during critical periods is very important. For conserving soil moisture, preparing terraces around the plant and opening of catch pits have been found beneficial. Before the onset of south-west monsoon (May-June), terraces of 1.5m radius should be made during the second year of planting and this should be widened up to 2.0m during the third year. Terraces are prepared by removing soil from the elevated portion of slope and it is spread to the lower side and a flat basin of 1.5m to 2.0m radius is made depending upon the age of the plant. Terraces may be crescent-shaped with inwardly sloping, so that the top soil which is washed off from the upper side due to rain water is deposited in the basin of the plant. A catch pit (200cm long x 30cm wide x 45cm deep) across the slope at the peripheral end of the terrace is made for withholding water during pre-monsoon and post-monsoon showers in sloppy areas. A small channel connecting catch pit sideways is made to drain out the excess water during rainy season. Further, modified crescent bund (at 2 m radius having a crescent shaped bund of 6 m length, 1 m width and 0.5 m height on the upstream of the plant) was also found very useful. Formation of trenches (5 m length, 1m width and 0.5 m depth in the middle of 4 plants) and filling with coconut husk burial is also suggested for improving yield of cashew grown on degraded lands. Besides, mulching with coconut husk burial in trenches conserves more moisture and coconut husks serves as a source of potassium too. The available water during pre-monsoon, monsoon and post monsoon season can be efficiently harvested and utilized for higher yields (25-30 per cent more than control) using proper soil and water conservation techniques (Rejani and Yadukumar, 2010). Using coconut coir pith as soil mulch in cashew plantations resulted in 14.15 per cent more water retention and suppression of weeds to an extent of 73.52 per cent (Kumar *et al.*, 1989). Black polythene mulch was helpful to conserve soil moisture (Nawale *et al.*, 1985).

1.4 Planting of Recommended Varieties

With the concerted efforts of ICAR-Directorate of Cashew Research, Puttur and Centres of AICRP on Cashew located in different SAUs/ICAR institutes, a total of 42 varieties (29 selections and 13 hybrids) have been developed and released in the country for cultivation in varied agro-ecological situations. Twenty-one varieties having export grade kernels (W 180 to W 240). Cashew varieties recommended for different states are given in Table 2. The varieties released by DCR *viz.*, NRCC Selection-2 and Bhaskara are medium nut types with high yield and recommended for cultivation in Karnataka. These varieties exhibited better yield performance in other cashew growing regions of the country as well. Among the hybrid seedlings under evaluation, hybrids namely H-43, H- 66, H-68 (cross combinations of NRCC Sel-2 × Bhutnath-II), H-125 and H-126 (cross combinations of NRCC Sel-2 × Bhedasi) have been identified as promising. Cashew varieties suitable for drought, high density planting (Salam, 1999) and nutrient deficient soils have been reported (Latha, 2000; Latha and Salam, 2001).

Table 2: Cashew Varieties Recommended for different States

State	*Recommended Varieties*
Andhra Pradesh	BPP-4, BPP-6 and BPP-8
Chhattisgarh	BPP-4, BPP-6, BPP-8 and Indira Caju
Goa	Goa-1, Goa-2, Vengurla-4 and Vengurla-7.
Gujarat	Vengurla-4 and Vengurla-7
Karnataka	NRCC Sel-2, Bhaskara, Ullal-I, Ullal-3, Ullal-4, Vengurla-4, Vengurla-7, Madakkathara-2,VRI-3, Chintamani-1 and Chintamani-2.
Kerala	Madakkathara-1, Madakkathara-2, K-22-1, Kanaka, Dhana, Priyanka, Amrutha and VRI-3.
Maharashtra	Vengurla-I, Vengurla-3, Vengurla-4, Vengurla-7 and Vengurla-8.
Odisha	Bhubaneswar-1, Jagannath, Balabhadra, BPP-8 and Dhana.
Tamil Nadu	VRI-1, VRI-3 and VRI (Cw) H-1
West Bengal	Jhargram-1, Bidhan Jhargram-2 and BPP-8
North Eastern states	Ullal-3, Ullal-4, NRCC Sel. 2, Vengurla-1, BPP-8 and Jhargram-1

1.5 Use of Quality Planting Material

Cashew being a cross-pollinated crop shows a wide range of variation with respect to plant growth, yield, nut size, apple colour *etc.*, if orchard is raised by seedlings. Whereas, vegetatively propagated clonal progenies are genetically identical to the mother plant and give relatively more uniform in plant vigour and yield and come to fruiting earlier. Moreover, cashew rootstocks are still being raised by seed nuts only and neither rootstocks have been identified nor its clonal multiplication technique has been standardized.

In fact, production of quality planting materials of high yielding varieties through softwood grafting technique has been a great success story in cashew and has made tremendous impact in improving cashew productivity especially in states like Maharashtra and Odisha. India produces over 150 lakh cashew grafts annually by adopting soft wood grafting technique which has revolutionalized the availability of planting material in the country. There are over 100 Regional Cashew Nurseries, including both public and private sector. Now, the plants are produced as per standard. To maintain quality standard in cashew nurseries, accreditation is being done by National Horticulture Board, Gurgaon (Haryana) and Directorate of Cashewnut and Cocoa Development, Kochi (Kerala).

1.6 Canopy Management

Training of young plants during first 3-4 years is essential for providing proper shape and to promote better fruiting in subsequent years. The sprouts coming from the rootstock portion of the graft (below the graft joint) should be removed and proper staking of plants is required to avoid lodging due to wind during the initial years. The trees are shaped by removing lower branches and water sprouts coming from the base during first 2-3 years. The plants should be allowed to grow by maintaining a single stem up to 0.75 – 1.00 m from the ground level. After that the main stem may be de-topped at a height of 2.5m from the ground level and allow the branches in different directions. Proper shape of plants facilitates terrace making, weeding, fertilizer application, nut collection and plant protection operations. The flower panicles emerging from the graft during first year of planting should also be removed (de-blossoming).

In bearing trees, removal of dried/dead wood, crisscross branches and water shoots, should be done at least alternate years. This allows proper growth of the canopy and adequate exposure to sunlight. Pruning of cashew plants should be done after harvesting of nuts. Pruning intensity depends on variety, plant vigour, growing sites and planting systems. It has been shown that nut yield of cashew increased from 1.44 kg/tree before pruning to 2.36, 4.39, 5.53 and 6.08 kg/tree during the 1st, 2nd, 3rd and 4th years, respectively (Khan *et al.*, 1987). Pruning of leader shoots in July enhanced the production of productive laterals and number of bisexual flowers per panicle (Chattopadhyay and Ghosh, 1994). It has been shown that considerable improvement in nut yield in pruned plants of selected cashew varieties *viz.*, VTH 30/4, NRCC Sel-1, VRI-1 and Ullal-l from third year onwards after planting (Nayak *et al.*, 2008). The cut surfaces are smeared with Bordeaux paste. Sometimes after pruning of branches, there is attack of pests on new flushes and thus spraying of carbaryl (0.1 per cent) should be done to control the pests.

1.7 High Density Planting

So far very little database has been generated on high density planting system in cashew. There is a significant role of rootstock in high density planting but even today the rootstock for scion cultivars of cashew has not been standardized. The canopy management for different varieties and resource situations has not been attempted precisely. Though, cashew plants respond to need based pruning. In fact, high density planting system not only provides higher yield and economic returns per unit area in the initial years, but also input use efficient if managed properly. Therefore, plants are required to be trained in initial years for desired and strong framework. In general, recommended spacing for cashew is 7.5 m × 7.5 m or 8 m x 8 m. Maintaining a tree density of 625 plants/ha (4 m × 4 m) for the first 11 years and diagonal thinning thereafter to reduce the population to 50 per cent doubled the nut yield during the first 10 years of planting (Yadukumar *et al.*, 2001). Salam (1999) reported that the varieties 44/3, Anakkayam-1 and H-1608 were suitable for high density planting system. The precocious, less vigorous and pruning responsive varieties such as NRCC Sel-2, K 22-1, VRI 3, V-4 and Ullal-3 have shown good response to high density planting (Saroj and Nayak, 2013). Shape pruning is needed to maintain the canopy. Also, high density planting system with spacing of 6.5 m × 4 m (384 plants/ha), 5 m × 5 m (400 plants/ha) and 5 m × 4 m (500 plants/ha) were found superior over normal spacing.

High density planting system is more suitable for soils with low fertility and it helps to utilize the space available between cashew plants during the initial years of orchard thereby increases the yield per unit area. There is a also need to standardize high density planting on terraces/along contours on sloppy lands. Further, high density planting reduces the weed growth by early canopy coverage, reduces soil temperature thereby increasing the soil moisture content especially during peak summer season and provides mulching effect. The aspects like canopy area, photosysnthetic efficiency, root development pattern, pest and disease interaction, longevity of orchard, operational feasibility *etc.* need to be worked out. The nutrition and irrigation management of high density planting is altogether different than normal spacing as we maintain more number of plants per unit area with low canopy volume.

1.8 Integrated Nutrient Management

Most of the cashew plantations are on the wastelands, where inherent fertility status is poor. In high rainfall and sloppy lands, erosion of soil and loss of nutrients through runoff and leaching are other constraints. A large chunk of cashew growing soils is reported to be acidic. Deficiencies of Mg, Zn, B and Mo are more likely common in acid soils. Fe and Al toxicity is a problem. Micronutrient deficiencies in soil limit the cashew crop production to some extent; hence micronutrient management options need to be worked out to enhance yield and quality of cashew.

In general, integrated nutrient management in cashew has not been taken up seriously by the farmers. There is a perception among the farmers that cashew does not require nutrition like other crops. However, researches clearly shown that cashew respond to nutrition. Application of fertilizer dosage and time schedule under different agroclimatic zones has been standardized. The dose of N, P and K recommended to cashew for different states is shown in Table 3. Studies conducted at Directorate of Cashew Research, Puttur indicated that application of 500 g N and 125 g each of P and K and 10 kg poultry manure per tree per year under normal density planting system (200 trees/ha) and 250 g N and 50 g each of P and K and 10 kg poultry manure per tree per year under high density planting system (625 trees/ha) is found superior in terms of higher nut yield for rainfed cashew. Based on the initial fertility status of soil, nutrient dose may vary from location to location. About 10 to 15 kg farmyard manure/plant/year is recommended in addition to primary nutrients (N, P and K). The recommended dose of fertilizers during first year after planting is $1/5^{th}$ of the full dose, 2nd year after planting is $2/5^{th}$, 3rd year after planting is $3/5^{th}$, 4th year after planting is $4/5^{th}$ and 5th year onwards is full dose should be applied.

Table 3: Recommended Dose of Fertilizers to Cashew

State	Nutrient Dose for Mature Cashew Plantations (5th year of planting) (g/tree/year)		
	N	P_2O_5	K_2O
Kerala	500	125	125
	750	325	750
Karnataka	500	250	250
	500	125	125
Tamil Nadu	500	200	300
Andhra Pradesh	500	125	125
	1000	125	125
Maharashtra	1000	250	250
Odisha	500	250	250
West Bengal	1000	250	250

During fertilizer application, it should be ensured that adequate moisture is available in soil profile. The fertilizers should be applied in 2 split doses during pre-monsoon (May-June) and post-monsoon (September-October) periods. However, if single application is to be done, it should be during the post-monsoon season (September-October). In red loamy soils and in low rainfall areas (east coast), the fertilizers should be applied in circular bands at a distance of 0.5m, 0.75m, 1.0m and 1.5m away from the tree trunk during first, second, third and fourth year onwards of planting respectively. The fertilizer is raked into the soil. In laterite soils and soils of slopy land in heavy rainfall areas (west coast), the fertilizers should be applied in circular trenches of 25cm width and 15cm depth at above distances. The trenches should be closed immediately after the fertilizer application.

Foliar sprays of nutrients (urea 2 to 4 per cent; DAP 1 per cent; orthophosphoric acid; $ZnSO_4$ 4 per cent; Cu 0.3 to 0.6 per cent) at the emergence of the flush, panicle initiation and fruit set stages ensure better fruit set and also enhance nut yield in cashew (Ankaiah and Rao, 1987; Sapkal, 2000). Yellow leaf spot in low soil pH (4.5-5.0) could be corrected by foliar sprays of Mo salts (Subbaiah *et al.*, 1986). Foliar spray of growth regulators Planofix, Nutron, IAA, IBA, NAA, 2,4-D and ethrel were effective in increasing the total number of flowers, hermaphrodite flowers, sex ratio, fruit and yield per panicle, besides improving physico-chemical composition of apples and nuts (Ghosh, 1988; Singh *et al.*, 1992; Kumar *et al.*, 1994).

Cashew plantations have large amount of organic biomass available for recycling. The availability of leaf litter from cashew plantations of 10 to 40-years-old varied from 1.38 to 5.20 t/ha (Guruprasad *et al.*, 2009). About 5.5 t of available cashew biomass waste per ha can be converted into 3.5 t of compost or vermicompost and helps in meeting nutrient requirement to cashew by 50 per cent (Yadukumar and Nandan, 2005; Yadukumar, 2007). *Ex-situ* green manuring by using biomass of *Gliricidia sepium* is also beneficial. There is an also need to standardize organic production of cashew nuts especially for international market.

1.9 Water Management

Cashew is generally grown as a rainfed crop but it responds well to soil moisture management. The largest area under cashew cultivation is along the steep hillocks of west coast region of India where the mean annual rainfall ranges from 3000 to 3500 mm and 80 per cent of the total rainfall confined during June to September. Due to the non-uniform distribution of rainfall, cashew experiences severe moisture stress from January to May which adversely affects its flowering and fruit set, resulting in immature nut drop and lower productivity of cashew. During fruiting season of cashew (February to May), a mean rainfall of around 67 to 415 mm is received. The water deficit is highest during March to May (112 to 183 mm). Cashew yields can be enhanced by providing protective irrigation with 200 L of water per tree once in 15 days from January to March during the summer season. Irrigation can be started after the commencement of flowering for better nut set, filling and yield. The frequency and quantity of irrigation varies with soil conditions and age of trees. The stagnation of water near cashew tree hampers the plant growth. Therefore, a proper drainage in cashew orchards is essential.

Fertigation of water soluble fertilizers such as urea, DAP, and MOP can be done once in a week from December to March. With fertigation, the quantity of nutrients to be applied can be reduced to half the quantity of recommended nutrients. Richards (1993), Yadukumar and Mandal (1994) and Mishra *et al.* (2008) reported that fertigation can save 50 per cent fertilizer requirement and doubled the cashew yield. The experiments conducted at Directorate of Cashew Research (DCR), Puttur on fertigation in cashew indicated that under normal density planting system (7 m x 7 m), the effective canopy coverage per tree is 12.56 m^2. The quantity of irrigation water calculated based on the effective canopy area was 12.56 l/tree/day from December to January (daily open pan water evaporation is 5 mm) and 20 l/tree/day from February to March (daily open pan water evaporation is 6.5 mm) to meet 20 per cent of the cumulative pan evaporation (CPE). Similarly, for 40 per cent CPE and 60 per cent CPE, the irrigation rate was 24 l/tree/day and 38 l/tree/day from December to January and 36 l/tree/day and 58 l/tree/day from February to March, respectively. In order to meet 20 per cent CPE, four drippers of 2 l/hour discharge rate can be fitted at two equidistant points 1 m away from the base of the tree. Similarly, to meet 40 per cent and 60 per cent CPE, four drippers and six drippers of 4 l/h discharge rate can be fixed. Drip irrigation can be given for 1 h 30 min during December and January and 2 h in February and March (Yadukumar and Rejani, 2008).

1.10 Weed Management

Weeds are serious problem in cashew cultivation particularly under humid regions. Weeds compete for moisture, nutrients and light with cashew plantations owing to heavy yield loss, if not managed properly. In many cases, cashew orchards are neglected, thus weeds act as host for the pests also. The major weeds of cashew plantations are *Eupatorium odoratum, Tridex procumbent, Mimosa pudica, Lantana, camara, Brachiaria* spp., *Ischaemum indicum, Chromolaena odorata* L, *Syndrella nodiflora* Gaertn, *Elephantopus scabber* L. *Desmodium triflorum* DC, *Glycosmis arborea* DC, *Hemedesmus indicus* R. Br., *Ichnocarpus frutescens, Passiflora foetida* L., *Cyclea peltata* Cooke, *Tragia involucrate* L., *Triumfetta rhomboidea* Jacq, *Zizyphus* spp., *Narengamia elata* W and A., *Macaranga peltata etc.* In most of the cases, weeding is done manually. Moreover, weeds must be slashed or uprooted before seed setting in order to minimize the weed intensity. The slashed biomass can be used as mulch material or after decomposition may be applied to the plants. Application of glyphosate at 0.8 kg/ha (single spray) or glyphosate at 0.4 kg per ha + 2, 4-D 1.0 kg per ha (single spray) or paraquat 0.4 kg per ha + 2, 4-D 1.0 kg per ha (single spray) during August can effectively control the weed growth in cashew plantations (Abdul Salam *et al.*, 1993).

1.1 Cashew Based Cropping Systems

Cashew canopy takes 8 to 10 years to cover entire area under normal spacing; thereby there is good scope for growing intercrops. However, cashew is a light demanding plant, thus integration of any tall growing trees as over storey component is not advisable. Also, in coastal hilly regions there is a possibility of soil erosion, nutrient losses and weed growth during initial years of planting. Intercropping is the best option to minimize erosion, nutrient losses and conserve soil and moisture and to realize higher returns from unit area. Pineapple can be grown as a suitable intercrop between two rows of cashew for the first seven years. The spacing to be maintained for cashew is 8 x 8 m (156 trees/ha) or 7.5 m x 7.5 m (175 trees/ha) or 10 m x 5 m (200 trees/ha). Growing pineapple in trenches across the slope between two rows of cashew helped to conserve the soil moisture, which in turn increased the yield of cashew (main crop) by 1.5 times compared to cashew alone. Pineapple can be grown as a profitable intercrop under irrigated as well as rainfed conditions in west coast region (Yadukumar *et al.*, 2003).

In some homestead gardens of coastal region, cashew is integrated with banana, jackfruit, turmeric, pine apple, beans, amaranthus *etc.* Trailing pepper on the stem and branches of grown up cashew trees (more than 6 years) is also adopted. Other crops such as tapioca, turmeric, ginger, cucurbits, colocasia, elephant foot yam have been found to be suitable intercrops in West coast region while colocasia, brinjal, groundnut, blackgram were found to be good intercrops in East coast region. Leguminous cover like *Peuraria javanica*, *Calapagonium mucunoides* and *Centrosema pubescens* as well as medicinal plant like *Aloe vera* can be successfully intercropped with cashew. In Gujarat, vegetables like okra and pulses like cowpea and pigeon pea are highly profitable under cashew based cropping system. Growing seed spices like fenugreek and dill as intercrops are profitable under the Eastern humid region. In Baptala areas of Andhra Pradesh, inter cropping with flower crop like marigold with cashew recorded the highest net profit of Rs. 65,967/- ha. Whereas in Thane and Nasik areas of Maharastra, *Mogra is* grown by the farmers.

1.12 Use of Plant Growth Regulators

Plant growth regulators play important role in promoting, inhibiting and otherwise modifying physiological processes in plant system but its usage in cashew has not been exploited well. Cashew produces abundant flowers but only less than 10 per cent of which are hermaphrodite, about 85 per cent of the hermaphrodite flowers are fertilized under standard conditions and only 4-6 per cent of them reach maturity to give fruits. The remaining flowers shed away at different stages of development. Maiappan *et al.* (1995) reported that best production of hermaphrodite flowers and highest nut count and nut weight were found with application of NAA (200 ppm) followed by ethephon (100 ppm). Lakshmipathi *et al.* (2014) reported that application of 50 ppm Ethrel and 25 ppm NAA were found to be beneficial for increasing the nut yield through improvement in sex ratio, fruit set and fruit retention in cashew var. Bhaskara. For increasing the yield of cashewnut, two sprays of 10 ppm ethrel *i.e.* first at flushing and second at flowering stage are recommended from AICRP on Cashew centre, Vengurla. Under field plantations of cashew variety Ullal-3 at spacing of 3 m apart, soil application of Paclobutrazol (2.0 g alternate year) at pre-flushing stage is effective in reducing plant height, canopy spread, intermodal length but increased nut yield to the tune of 51.78 per cent over control (Meena *et al.*, 2014).

2. Plant Protection

2.1 Tea Mosquito Bug (TMB)

Under West Coast condition, three species of TMB *Helopeltis* (*H. antonii* S., *H. theivora* W. and *H. bradyi* W.) infest cashew, whereas, in other regions, only one species (*H. antonii*) exists on cashew. *H. antonii* is the dominant species infesting cashew causing severe damage of tender shoots, inflorescences, immature nuts and cashew apples and yield losses are more than 40 per cent. Both nymphs and adults of *Helopeltis* suck sap from tender shoots and leaves, floral branches and from developing nuts and apples by making a number of feeding lesions. During outbreak situation, the entire flush dries up and the trees present a scorched appearance.

Integrated Pest Management strategy involves proper management of orchard, planting of tolerant varieties and adopting need based spraying schedules of insecticides to manage TMB. Mid season/late season flowering cashew varieties such as Bhaskara, are able to escape from the severity of the pest to a certain extent. Even though, all groups of insecticides and plant products (botanicals) were evaluated against this pest, none of them exhibited any ovicidal action (Raviprasad *et al.*, 2005). However, λ-cyhalothrin (0.003 per cent) and carbaryl (0.1 per cent) had shown longest residual action against nymphs and adults (Sundararaju *et al.*, 1993). Need based sprays are recommended during most vulnerable periods of crops such as flushing, flowering and fruiting stage of the crop. Though, cashew is an insect pollinated crop, spraying of these insecticides during flowering season did not influence the fruit set (Pillai *et al.*, 1984; Rai, 1984, Sundararaju *et al.*, 1993).

Erythmelus helopeltidis Gahan. (Hymenoptera: Mymaridae), *Telenomus* sp. (Laricis group) (Scelionidae), *Chaetostricha* sp. (Trichogrammatidae) and *Gonatocerus* sp. nr. *bialbifuniculatus* were the egg parasitoids reported on this pest from West Coast regions while, *Ufens* sp. was the egg parasitoid reported East Coast (Vridhachalam). The build of this pest is naturally regulated through these egg parasitoids (Sundararaju,1993a). *Crematogaster wroughtonii* Forel (Formicidae) has been recorded as a predator on nymphs of the pest. Several species of spiders, *Hyllus* sp., *Oxyopes sehireta*, *Phidippus* patch and *Matidia* sp. five species of reduviid bugs (*Sycanus collaris* (Fab), *Sphedanolestes signatus* Dist. and *Endochus inornatus* Stal., *Irantha armipes* Stal. and *Occamus typicus* Dist. have also been recorded as predators of the TMB. *Aspergillus flavus* and *A. tamarii* were confirmed to be pathogenic to *H. antoni* (Ambika and Abraham, 1979; Devasahayam and Nair, 1986; Sundararaju, 1984).

2.2 Cashew Stems and Root borer (CSRB)

CSRB (*Plocaederus ferrugineus* L.) (Coleoptera: Cerambycidae) is the primary species, infesting cashew in all parts of India. The symptoms of damage include the, extrusion of frass, occurrence of gummosis, premature yellowing and shedding of leaves, drying of twigs and finally death of the tree. Integrated Pest Management approaches include examination of trunk portion for any symptoms, collection of grubs by careful removal of the bark and swabbing of the affected portion with either carbaryl or lindane or chlorpyripos (0.2 per cent). The kernels obtained after treatment of these insecticides were free from residues at 0.2 and 0.4 per cent concentrations, thus cashew kernels are free from pesticides contamination (Bhat and Raviprasad, 2008).

Swabbing the pest infested portion after removal and drenching the root zone with chlorpyriphos (0.2 per cent) as post extraction prophylaxis (PEP) was found effective in reducing reinfestation upto 70.7 per cent. There exists large population of CSRB grubs in dead cashew trees and also in CSRB infested cashew trees. The trees having more than 50 per cent of bark circumference damaged/having yellow canopy are considered as trees beyond recovery. It is essential to adopt phytosanitation by removing the dead trees and trees beyond recovery by uprooting during monsoon season, prior to emergence of adult beetles and dispose off suitably without delay to reduce build-up of pest inoculum in a given locality. It has been seen that through phytosanitation, the spread of CSRB to healthy trees can be effectively prevented (Sundararaju and Bakthavatsalam, 1990 and Raviprasad *et al.*, 2009). Mycopathogens like *Metarhizium anisopliae* and *Beauveria bassiana* were known to cause mycosis of grubs of CSRB (Bhat and Raviprasad, 1996). The spores of *M. anisopliae* survive for three months under field condition. Mixing the spawn with organic matter like FYM, neem cake and cashew apple can enhance the spore load under the field conditions.

2.3 Disease Management

Less information is available on occurrence and management of diseases in cashew. However, a large number of diseases are associated with different growing stages of cashew. Some diseases infecting cashew plants are given in Table 4. The die back disease occurs due to severe infestation by tea mosquito bug, and is more pronounced under cloudy weather. The causal organisms for this disease are (*Gloeosporium mangiferae* and *Phomopsis anacardii* in causing blossom blight and *Colletotrichum gleosporioides* and *Bortryodiplodia theobromae*. For management of this disease, the infected twigs need to be pruned off up to the portion wherein the inner xylem is clear and not pinkish/brownish. The cut ends need to be smeared with Bordeaux paste. In case, the infected twigs show fresh sprouting they can be cut above the flushed portion and a spray with Bordeaux solution may be given to prevent any spread of the pathogen. The occurrence of blossom blight is reported from Northern Kerala, wherein the flowering panicles develop black spots and the flower buds dry off. In severe cases development of fungal mycelia is also noticed. For management of this disease, spraying with Dithane M-45 (2 g/l) is recommended along with the insecticidal sprays (Bhat *et al.*, 2002). Under heavy rainfall situations, occurrence of collar rot and root rot due to several fungal pathogens may be noticed. Under such conditions, it is recommended to remove the poly bags having infected seedlings/grafts and spray the remaining seedlings/grafts with 1 per cent Bordeaux mixture when there is a short dry spell of at least 4-6 h. This will effectively take care of the spread of the infection to healthy seedling/grafts.

Table 4: Diseases of Cashew

Disease	*Causal Organism*
Nursery diseases	
Damping-off	*Phytopthora palmivora* Butler
Seedling Blight	*Cylindrocladium scoparium*
Root-rot disease	*Pythium ultimum* Tron
Inflorescence diseases	
Anthracnosis	*Colletotrichum gloesporioides*
Inflorescence Blight	*Gloeosporium mangiferae* and *Phomopsis anacardii*
Foliage diseases	
Anthracnosis	*Phytophthora nicotinae* and *Asterina carbonacea*
Diplodia rot	*Diplodium anacardiacearim*
Leaf blight	*Pestalotia paeoniae*
Nut Diseases	*Aspergillus tamari, Penicillium citrium, Lasiodiplodia theobromae, Gloeosporium* spp.
Other diseases	
Dieback	*Pellicularia salmonicolor*
Gummosis	*Diplodia natalensis*
Powdery mildew	*Odium* spp.

3. Post-Harvest Technology

The cashew fruit has two distinct parts, *viz.*, the swollen and pear shaped peduncle (cashew apple) and a kidney-shaped nut attached to the lower end of the apple. The nut contains the outer hard shell and the inside seed; the seed is consisting of the outer testa and the edible kernel. Cashew kernel is a rich source of protein (21 per cent), carbohydrate (22 per cent) and fat (47 per cent).

Besides these, it contains minerals such as Ca, P, Na, K, Mg, Fe, Cu, Zn and Mn. Cashew kernel lipids are rich in unsaturated fatty acids such as oleic acid (73.7 per cent), linoleic acid (7.67 per cent) and stearic acid (11.2 per cent). Cashew kernels are free from cholesterol as it contains sizeable quantity of monounsaturated fatty acid (oleic acid) which is probably helpful in lowering down the blood cholesterol. Cashew kernel contains few vitamins like thiamine, niacin and vitamin E. Another by-product known as cashew nut shell liquid (CNSL) finds extensive industrial use like furnace lining, textile industry, preparation of varnish, paints, resins, timber protection *etc.* Other valuable products from cashew industry are cashew testa and tannin.

Cashew nut is either collected with fallen cashew apple or harvested at ripe stage of apple and then the nuts are to be separated from the cashew apple and sun dried for 2 to 3 days. Raw nuts at the time of harvest will have moisture content ranging between 16 to 25 per cent. Drying is important in order to prevent spoilage of nuts during storage due to fungal attack. Safe permissible moisture level of raw nuts is 8 per cent and well dried nuts can be stored at ambient temperature for 12 months without any quality deterioration. In India, cashew processing is mostly manual and it consists of steaming or roasting, shelling, kernel drying, peeling, grading and packing. Conditioning makes the raw cashew nuts hard and loosens the kernel within the shell. Various method such as Drum roasting, Steam boiling and Oil bath roasting have been employed for commercial processing of raw cashewnuts. In recent times Drum roasting and Steam roasting methods are being employed for processing. Commercial processing by steam roasting involves drying of raw nuts, steam roasting for 18-20 min., shelling of the raw nuts, peeling, grading and packing the kernels under vacuum (Ohler, 1979).

The thin-skinned edible cashew apple (pseudo fruit) has a light yellow spongy flesh, which is very juicy, pleasantly acidic and slightly astringent when eaten raw and highly astringent when green. Cashew apple contains sugars, tannins, phenols, amino acids, ascorbic acid, minerals and fibre. Besides these, it also contains riboflavin and fairly good amount of ascorbic acid (240 mg/100 g). In case the cashew apple is processed either for producing syrup/liquor or for bio-fuel or for extracting nutraceuticals, the income of farmers can increase from cashew crop. Cashew apple pomace (the residue left after removal of juice), which is rich in fibre could be blended with cereals (ragi, rice, wheat) and pulses (green gram) up to 10 per cent without affecting the quality in terms of *in vitro* digestibility of both proteins and carbohydrates.

The Indian Cashew industry is almost export oriented. Apart from earning valuable foreign exchange from the industry, the major substantial contribution is in the field of providing sustainable employment opportunities to about 10 lakh people and especially for women folk. Cashew is processed and marketed on an industrial basis. Even though organized marketing structures are not available but cashew does not face any serious marketing problem in India. There are about 3799 registered cashewnut processing units spread in the cashew growing regions of India of which 51 per cent industries are grouped under organized sector. The processing capacity of Indian cashew industries is to the tune of 20 lakh tonnes. Although, Maharashtra state stands first in the total number of processing units in this country, most of the processing units are categorized under cottage scale units having processing utilization up to 40 tonnes/annum. Kerala and Tamil Nadu together processed around 70 per cent of the total quantity in the country and serves as nucleus centre of cashew trade.

4. Enhancing Farmers Income

As an estimate, with the present level of cashew nut productivity (706 kg/ha), growers are hardly getting Rs. 40,000-50,000/ha income by selling raw cashew nut. If scientific production technologies are adopted, the productivity can be increased at three times, particularly by using soft wood graft of high yielding varieties, adoption of high density planting, integrated nutrient and water management as well as need based plant protection measures. At the level of about two tonnes productivity, a net profit of 1-1.5 lakhs/ha can be realized easily. If nuts are graded before selling, one can fetch more income from the same level of productivity.

There is no systematic marketing system of raw cashew nut in India. Industries has established their access directly to growers through middle man who are collecting at a common point and supplying to the processing industries. Therefore, it is suggested that farmers have not only to adopt proper growing technologies but also to form an association of cashew growers for selling their produce directly to the industries. This way profit margin can be increased by avoiding middle man. Through the registered association, purchase of various inputs like fertilizers, pesticide, tools and implements *etc.* can also be done as subsidized rates.

India is also importing raw nuts of about 7-8 lakh tonnes annually particularly from African countries for processing without any import duty. Now, government has imposed import duty on import of raw nuts, thereby per unit cost of raw nut and kernel both have increased, which impacted increase in raw nut cost in domestic market also. Now, Indian farmers are getting more price for sale of raw nut compared to previous years.

Besides cashew nut production, cashew apple which is about 8 times that of nut is not used except in Goa where it is fermented to an alcoholic beverage called 'Feni'. Thus, processing of cashew apple into various value-added products like jam, jelly, pickle, candy, toffee, syrup, juice is another way of increasing income by the farmers. Now, some small scale units are started in cashew growing areas for processing of cashew apple, where farmers can supply apples which otherwise wasted only. Moreover, care should be taken than cashew apple must be harvested at right stage of maturity.

Cashew based cropping system is another area, where profit margin from the unit area can be increased by integrating compatible intercrops, as cashew plants takes 8-10 years to develop full cover. Cashew also respond to pruning, hence proper canopy management can provide better opportunity for growing intercrops for longer period of time. Integrating crops like pineapple, turmeric, cowpea, bottle gourd *etc.*, more than Rs. 50,000/ha additional income can be obtained without affecting yield of cashewnut.

The maximum labour input in cashew industry goes for the processing of nut, particularly for shelling, peeling, grading. Development of such small scale mechanized processing units or on-farm processing units will be a boon to the cashew growers to reduce the share of middle man in chain of cashew nut industry and increase the return from their produce. Thus, farmers can sell their produce in the form of cashew kernel instead of cashew nuts. On-farm level cashew nut processing is advantages for getting better return from the cashew nut produced by the farmers. Cashew farmers normally sell their cashew nuts to traders who in turn sell the nuts to the cashew processors. The profit which otherwise would have gone to middle men can be earned by the farmers themselves through processing the cashew nuts by the adoption at on-farm level.

5. References

Abdul Salam, M., Pushpalatha, P.B., Suma, A. and Abraham, C.T. 1993. Efficacy of chemical weed control in cashew plantations. *J. Plant. Crops*, **21**(1): 54–56.

Ambika, B. and Abraham, C.C. 1979. Bioecology of *Helopletis antonii* Sign. (*Miridae Hemiptera*) infesting cashew trees. *Entomon.*, **4**: 335–342.

Ankaiah, S. and Rao, P.V. 1987. Studies on the age of leaf, time of absorption and concentration of urea spray on cashew. *Indian Cashew J.*, **15**(3): 9–11.

Bhat, P.S. and Raviprasad, T.N. 2008. Studies on lindane and chlorpyriphos residues in cashew kernels. *J. Plant. Crops*, **36**(2): 142–144.

Bhat, P.S. and Raviprasad, T.N. 2006. Studies on endosulfan and carbaryl residues in cashew kernels. *J. Plant. Crops*, **34**: 373–376.

Chattopadhyay, N. and Ghosh, S.N. 1994. Studies on the effect of time and extent of pruning in increasing the yield of cashew. *J. Plant. Crops*, **22**(2): 111–114.

Devasahayam, S. and Nair, C.P.R. 1986. The mosquito bug, *Helopeltis antonii* Sign. on cashew in India. *J. Plant. Crops*, **14**: 1–10.

Ghosh, S.N. 1988. Effect of nitrogen, phosphorus and potassium on flowering duration, yield and shelling percentage of cashew (*Anacardium occidentale* L.). *Indian Cashew J.*, **19**(1): 19–23.

Guruprasad, T.R., Hanumanthappa, M., Sudhir, K.V. and Dhananjaya, B. 2009. Technologies for improvement of senile cashew gardens. *The Cashew and Cocoa J.*, **1**(3): 18–22.

Kumar, D.P., Subbarayappa, A., Hiremath, I.G., Khan, M. M. and Sadashivaiah. 1989. Use of coconut coir-pith: A biowaste as soil mulch in cashew plantations. *The Cashew*, **3**(3): 23–24.

Kumar, D.P. 1990. The growth and yield performance of top-worked cashew trees. *Cashew Bull.*, **27**(9): 7–10.

Kumar, D.P., Khan, M.M. and Melanta, K.R. 1994. Effect of growth regulators on sex expression, fruit set, fruit retention and yield of cashew. **In**: *Proc. of the XI Symposium on Plantation Crops*, 30th November–3 December, Calicut, Kerala, India, pp. 610–627.

Latha, A. 2000. Cashew varieties tolerant to N deficient soils. *The Cashew*, **14**(2): 29–39.

Latha, A. and Salam, M.A. 2001. Screening of cashew varieties for drought tolerance. *J. Plant. Crops*, **29**(2): 27–30.

Lakshmipathi, Adiga, J.D. and Kalaivannan, D. 2014. Influence of growth regulators on certain reproductive parameters of cashew (*Anacardium occidentale* L.) variety Bhaskara. *J. Plant. Crops*, **42**(1): 113–116

Maiappan, S., Prabakaran, J. and Sambandamoorthy, S. 1995. Effect of growth regulators on sex expression and fruit set in cashew (*Anacardium occidentale* L.). *Cashew*, **9**(1): 11–13.

Meena, R.K., Adiga, J. D., Nayak, M.G., Saroj, P.L. and Kalaivanan, D. 2014. Effect of paclobutrazol on growth and yield of cashew (*Anacardium occidentale* L.). *Vegetos*, **27**(1): 11–16.

Mishra, J.N., Paul, J.C. and Pradhan, P.C. 2008. Response of cashew to drip irrigation and mulching in coastal Odisha. *J. Soil Water Cons.*, **7**(3): 36–40.

Nawale, R.N., Sawke, D.P. and Salvi, M.J. 1985. Effect of black polyethylene mulch and supplemental irrigation on fruit retention in cashewnut. *Cashew Causerie*, **7**(3): 8–9.

Nayak, M.G., Swamy, K.R.M., Yadukumar, N. and Raviprasad, T.N. 2008. Effect of limb pruning on growth and yield of four cashews (*Anacardium occidentale* L.) varieties. *J. Plant. Crops*, **36**(3): 300–303.

Ohler, J.G. 1979. Cashew processing. *Tro. Abs.*, **21**(9): 1792–2007.

Pillai, G.B., Singh, Vijay, Dubey, O.P. and Abraham, V.A. 1984. Seasonal abundance of tea mosquito bug, *Helopeltis antonii* on cashew in relation to meteorological factors. In: *Cashew Research and Development*, (Eds.) E.V.V.B. Rao and H.H. Khan. Indian Society for Plantation Crops, CPCRI, Kasaragod, India, pp.103–110.

Rai, P.S. 1984. *Handbook of Cashew Pests*. Researchco Publications, Delhi.

Raviprasad, T.N., Sundararaju, D. and Bhat, P.S. 2005. Efficacy of botanicals against *Helopeltis antonii* Sig. infesting cashew. *The Cashew*, **19**(4): 9–14.

Rejani, R. and Yadukumar, N. 2010. Soil and water conservation techniques in cashew grown along steep hill slopes. *Scientia Hort.*, **126**: 371–378.

Richards, N.K. 1993. Cashew Research in Northern Territory, Australia, 1987–1991. *Tech. Bull.* **202**.

Salam, M.A. 1999. Cashew varieties suitable for high density planting. *Cashew Bull.*, **37**(6): 3–5.

Sapkal, B.B. 2000. Integration of foliar application of urea and insecticidal spray for yield maximization in cashew. *The Cashew*, **14**(1): 41–44.

Saroj, P.L. and Nayak, M.G. 2013. Potential of high density plantation for enhancing quality production of cashew. **In**: *Canopy Management and High Density Planting in Sub-tropical Fruit Crops*, CISH, Lucknow, pp. 55–63.

Singh, S.K., Syamal, M.M. and Maurya, A.N. 1992. Effect of NAA and Ethrel on vegetative growth, flowering, fruiting and yield of cashewnut. *The Cashew*, **6**(2): 11–12.

Sundararaju, D. 1993. Studies on the parasitoids of the mosquito bug, *Helopeltis antonii* Sign. (Hymenoptera: mymaridae) on cashew with special reference to Telenomous sp. (Hymenoptera: Scelionidae). *J. Biol. Cont.*, **7**: 68.

Sundararaju, D. and Bakthavatsalam, N. 1994. Pests of cashew. **In**: *Advances in Horticulture*, Vol. 10, (Eds.) K.L. Chadha and P. Rethinam. Malhotra Publishing House, New Delhi, India, pp. 759–785.

Swamy, K.R.M. 1995. Top working of cashew in Goa and Maharashtra States: A case study. *The Cashew*, **9**(2): 12–17.

Yadukumar, N. and Mandal, R.C. 1994. Effect of supplementary irrigation on cashewnut yield. *Water Management for Plantation Crops: Problems and Prospects*. Centre for Water Resource Development and Management, Calicut, Kerala, pp. 79–84.

Yadukumar, N. and Rejani, R. 2008. Soil and water management in cashew plantations. *Tech. Bull.*, 10. NRC – Cashew, Puttur, Karnataka, pp. 20.

Yadukumar, N., Rao, E.V.V.B. and Mohan, E. 2001. High density planting of cashew. *Trop. Agri.*, **78**: 19–28.

Yadukumar, N., Raviprasad, T. N., Nagaraja, K.V., Haldankar, P. M., Godase, S. K., Susanamma, K., Gajendran, G., Mahalingam, T., Lenka, P.C., Mohapatra, R.N. and Bandyopadhyay, B. 2003. National Agricultural Technology Project. Final Report on developing integrated production packages for enhancing productivity of cashew. National Research Centre for Cashew, Puttur, D.K., Karnataka, pp.1–95.

19

Development of Technologies for Increasing Productivity and Availability of Onion

Ajmer Singh Dhatt

Department of Vegetable Science, Punjab Agricultural University, Ludhiana – 141 004, Punjab
E-mail: ajmerdhatt@gmail.com

Onion (*Allium cepa* L.) belongs to family Alliaceae and thought to have ancestors in Central-Asia. Physiologically onion is a long day plant, but due to suitability for cultivation to specific photoperiod and temperature, it is classified into temperate long, intermediate and short day and, tropical short day bulb, shallot and multiplier groups. Short day varieties, includes not only populations developed within the tropics, but also developed well out side the tropics as over winter crop to provide early fresh bulb. Such varieties can also be grown as over winter crop under sub-tropical conditions due to adaptation for bulbing at approximately equinoctial day lengths.

Onion is mainly grown for bulb, which is used almost daily in every home. Its main use is due to its aromatic, volatile oil ally-propyl disulfide that impart cherish flavor to the food. Worldwide onion is grown on 43.64 lakh ha with 863. 44 lakh MT production and 19.79 MT ha^{-1} productivity. In last decade 53.49 per cent increase in area, 72.87 per cent in production and 12.63 per cent in productivity are the indicators of onion growth on the globe. India ranked first in area (12.03 lakh ha) and second in production (194.02 lakh MT) of onion in the world. Though, onion production in India has increased by 300 per cent in last two decades, but country requires 211 lakh MT onion by 2050 with present rate of consumption (6.7 kg per capita), export (9 per cent), processing (6.75 per cent) and losses (20 per cent). If export and processing is increased by 25 and 9 per cent, respectivelythan 286 lakh MT onion would be required.Which can be achieved by increase into 28 MTha^{-1} or by expansion of area under onion. Hence, for increasing productivity and making availability of, have to focus ondevelopment of high yielding varieties/ hybrids, improved production technologies, efficient crop management practices and better post-harvest handling and proficient marketing systems.

1. Crop Improvement

1.1 Objectives

The crop improvement of onion is dexterous and time consuming due to its biennial and cross- pollinated nature. Thousand years of conscious and unconscious selection made it adaptable to variable latitudes and climatic conditions. Bulb, made up of swollen leaf sheaths is an economical part of onion and all objectives of breeder revolve around for its improvement. The bulb traits include size, shape, colour of skin and flesh, single centers, skin retention, firmness, dormancy, pungency and concentration of soluble solids. Production of commercially acceptable bulbs is dependent on resistance to diseases, pests and bolting. Production of high-quality seed requires vigorous bulbs, uniform flowering, straight seed stalks and high sucrose in the nectar. Bulbing in onion is initiated by combination of day length and temperature and, onion breeders must understand that cultivars adapted to one region may not produce desirable bulbs at the another location. As unadapted material, may reach the critical day length for bulbing too early and produce small bulbs. Conversely, genotypes requiring long days to bulb may never mature properly under the short days. Well-matured bulbs with tight necks store better and are less likely to develop storage rots. Tight outer dry skins protect the bulb and reduce the losses during storage. Therefore, for maintenance of true to type the breeders and seed producers always do selection of desired bulbs. Without selection, bulbs have the tendency to develop multiple centers and become softer and flatter.

1.2 Varietal Development

In population improvement, farmers made selections as per growing season and location for desirable traits and yield, and cultivatingthose old land races since long. Important selections are *'Pune Fursungi'* in Maharashtra, *'Pili Patti'* in Gujarat, *'Bellary Red'* in Karnataka, 'Sukhsagar' in West Bengal, *'K.P'*. in Andhra Pradesh, *'Nirmal Local'* in Madhya Pradesh and *'Balwan'* in Haryana. Several public sector institutes improved the local populations of onionand released more than 60 varieties of bulb and six of multiplier onion for cultivation in the country. Wherein, N-53 (*Kharif*), N-2-4-1 (Late *Kharif* and *Rabi*), N-257-9-1 (*Rabi*) were released by Department of Agriculture, Maharashtra; Agrifound Dark Red (*Kharif*), Agrifound Red-multiplier (*Kharif* and *Rabi*), Agrifound Light Red (Late *Kharif* and *Rabi*), Agrifound White, L-28 and L-355 (*Rabi*) by NHRDF, Nashik; Baswant 780 (*Kharif*), Phule Samrath (Late *Kharif*), PhuleSafed and Phule Suvama (Late *Kharif* and *Rabi*) by MPKV, Rahuri; PKV White (*Rabi*) by PDKV, Akola; Bhima Dark red (*Kharif*), Bhima Red and Bhima Shubra (*Kharif* and Late *Kharif*), Bhima Raj and Bhima Shweta (*Kharif* and *Rabi*), Bhima Shakti (Late *Kharif* and *Rabi*), Bhima Super (*Kharif*, Late *Kharif* and *Rabi*), BhimaKiran (*Rabi*) by DOGR, Rajgurunagar; Co-1(multiplier), Co-2, Co-3, Co-4 and Co-5 (*Kharif* and *Rabi*), MDU (*Rabi*) by TNAU, Coimbatore; ArkaKalyan (*Kharif*), Arka Pragati and Arka Viswas (*Kharif* and *Rabi*), Arka Niketan (*Late Kharif* and *Rabi*), Arka Pitamber, Arka Ujjawal-multiplier, Arka Swadista, Arka Sona, Arka Bheem-Syn. and Arka Akshay-Syn. (Rabi), Arka Bindu (*Kharif*, Late *Kharif* and *Rabi*) by IIHR, Bangaluru; GWO-1 (*Rabi*) by GAU, Junagarh; Udaipur-101, Udaipur-102 and Udaipur-103 (*Rabi*) by RAU, Udaipur; RO-59 and RO-252 (*Rabi*) by RARS, Durgapura; Kalyanpur Red Round (*Rabi*) by CSAUAT, Kanpur; Pusa White Flat, Pusa White Round, Pusa Ratnar and Pusa Madhvi, Sel-126 (*Rabi*), Early Grano and Pusa Red (Late *Kharif* and *Rabi*), and Brown Spanish (Long day) by IARI, New Delhi; Hisar-1, Hisar-2 and Hisar-3 (*Rabi*) by HAU, Hisar; Punjab Selection, Punjab-48, Punjab Red Round, Punjab White, Punjab Naroya and PRO-6 (Rabi) by PAU, Ludhiana; RO-1, VL-67 and VL-3 (Long day) by VPKAS, Almora. For further strengthening of varietal development programme inbred lines are to be developed by selfing individual plants from OP cultivars or by recombining with previous inbreds. The large GCA effects indicate that superior inbred lines are more likely to be selected from populations improved by recurrent selection or synthesized by recycling inbreds with good combining ability. Over years, locations and seasons evaluation before recycling into new populations is important for stability of the populations. The day length, temperature and season adaptability is very important for suitability of the variety under diverse growing conditions of India.

1.3 Hybrid Development

The hybrid development in onion began with the discovery of cytoplasmic male sterility by Jones and Emsweller in 1936. A high degree of heterosis has been reported in onion and being exploited world-wide. The market share of F_1 hybrid in USA (81 per cent), Japan (71 per cent) and European countries (>50 per cent) is very high compared to India (<1 per cent). So far, there are only two hybrids Arka Lalima and Arka Kirtiman from public sector and their adaptability is also confined to Karnataka state. Some short day exotic hybrids introduced by private companies are very good in yield, but very poor in storability under tropical conditions of India. Therefore, efforts are needed to develop male sterile lines as per adaptability for different parts of the country.The aid of molecular markers can be useful in exploiting this phenomenon. PAU isolated CMS lines from the adapted populations with the help ofmolecular markers and named as D-97A and B (Light red from variety Punjab Naroya), D-121 A and B (Red from variety Punjab Selection), D-26A and B (Dark red from population P-266) and D-30 A and B (White from variety Punjab White). *A. galanthum* cytoplasmic lines can also be developed for having diversityin cytoplasm for male sterility. Punjab Agricultural University also took lead in the country by developing three CMS lines *viz.*, D-G422A and B, D-G408A and B and D-G407A and B of*A. galanthum* cytoplasm following the backcross breeding approach. The identification of pollinators or C-lines from distant group can give potential F_1's for good yield and storability. The synthesis of heterotic groups and their further use in development of inbred lines should be a long term planning for strengthening the hybrid programme in country. The di-haploid technology can further help in developing inbred lines at faster rate for use in heterosis breeding, synthesis of synthetic varieties and genomic studies.

1.4 Resistance Breeding

It is desirable due to environmental pollution, development of resistance in pests and social opposition against pesticides. Sources of onion germplasm showing resistance to purple blotch, downy mildew, *Botrylis* neck rot and white rot have been reported. PBR-287, PBR-5, Arka Kalyan and MS-65-268 reported resistant to purple-blotch; Welsh onion lines TA104, TA108, TA198, AF468 and TA204 moderately resistant to *Stemphylium* leaf-blight; Bhima Raj and Red Creole tolerant to anthracnose; IC 48045, IC 32149, IC 49371, DOP 2 and *A. roylei* resistant to the downy mildew diseases. Resistance to pink root is inherited as a recessive allele at a single locus and has been incorporated into many commercial cultivars, although stable expression has been problematic in some environments. Cultivars showing adequate levels of resistance to *Fusarium* basal rot have also been developed and a single dominant gene controls the resistance. Whereas, two dominant genes are controlling downy mildew resistance in onion. Thrips routinely cause economically important losses by damaging the leaves during bulb production and also for virus transmission. Genotypes with wide angled leaves and glossy foliage are less prone to attack of thrips in onion. Protease inhibitor showed significant effect on inhibiting the oviposition to restrict thrip population. The resistance to thrips in cultivar 'White Persian' appears to be due to reduced waxiness (glossy) of the foliage. Two recessive genes are controlling wax production; one locus condition glossy foliage and second glossy scapes. The expression of a glossy scape occurs only when glossy foliage is present. The USDA has released four glossy inbredsB9885, B9897, B11278, and BI137T. Tolerance to thrips have also been reported in Texas Grano 502, Brownville, Daquesa, Chiltan 89, Phulkara, Sariab red, Swal 1544, Kalyanpur Red Round, Udaipur 103, N 53, N-2-4-1, Sello 04, Ratnar, Sel 71, Hisar Brown, Sel 202, Pusa Red, Giza 6, Giza 20, Valeouro, IPA11, Roxa IPA3, Belem IPA9, Meshkat, Sefid-e-Kurdistan, Sefid-e-Qom, Eghlid, Safid-e-Kashan and Safed-e-Khomein. Cultivated and wild alliums possess disease resistances potential and can be exploited in genetic improvement of the bulb onion. *A. ampeloprasum* provides resistance to white-rot, *A. roylei* to downy mildew and *A. fistulosum* to pink rootsmut and onion maggot. Fertile hybrids between *A. roylei* and *A. cepa* and successful transfer of downy mildew resistance achieved in The Netherlands. Hybrids between *A. cepa* and *A. fistulosum* have long been known, but the F_1 always shows extremely low fertility due to poor chromosomal pairing. The low degree of fertility exhibited by the hybrids between *A. cepa* and other *Allium* species restricts that successful introgression of disease resistances.

1.5 Quality Improvement

Quality in onion is defined by the appearance (size, shape and color), shelf-life and availability of biochemical compounds (pungency and soluble solids). The consumer preference for shapes and colours varied all over the world. Bulbs may be white, red, brown or yellow; round, pyriform or flat; and soft or firm. Dormancy is important in onion for storage. Higher soluble solids are important for dehydrating industry to produce onion chips or powder. The amounts of s-alkyl cysteine sulfoxide precursors and the enzyme alliinase contribute to the yield of volatile sulfur compounds that constitute the pungency of the onion bulb. Consumers in India often prefer strongly pungent onion, whereas in USA and many European countries low pungency is preferred. The bulb quality characteristics have a genetic base and can be manipulated by onion breeders. However, population under study influence the heritability. Generally, diameter and weight of bulbs show low heritability. Bulb height and shape index (polar: equator diameter) showed relatively large genetic components. Therefore, breeders can expect response to selection for bulb height, but not for bulb size. These estimates are supported by practical observations; 'Italian Red Torpedo', a piriform onion, was selected from the flat cultivar 'California Early Red'. It may be that the onion bulb will reach a certain height and continue to increase in diameter. The extent to which the environment is conducive to diameter increase would determine the shape of the bulb. If there is less time for the bulb diameter to increase, the bulb will be pyriform; if the diameter of the bulb increases for a longer time, the bulb will appear flat. Onion populations show marked differences in storage ability, firmness, pungency and moisture content. The environment can have a large effect on pungency and storage ability. The bulbs subjected to stress, over watering during the growing season are more likely to sprout and rot during storage. Nevertheless, these traits also have a genetic base. Phenotypic correlations between large size, softness, low pungency and poor storage ability have been widely recognized. Short-day onions having large, soft, non-pungent bulbs and low solids are poor in storage ability, however, it is reverse for small, firm, pungent and high solid bulbs. Similarly, long-day onions having small, firm, pungent and high solids bulbs show better storage. The genotypic correlation is high between high solids and small bulb size. The phenotypic correlations could be explained by water retention of onion at maturity. A large, less pungent, soft onion may retain more water at maturity, diluting the alkyl sulfides responsible for pungency. The softer onion sprouts more quickly during storage. The *rabi* season onions stored for longer period have less water, firm flesh, high solids and pungency compounds per unit dry weight, which is not true for the *kharif* season crop. However, phenotypic correlations do not necessarily mean that the same genes control the traits. An interesting observation is that 'Sweet Sandwich', a three-way hybrid developed by the USDA, is a firm, well-storing and significantly less pungent onion.

1.6 Biotechnological Interventions

Use of biotechnological tools are more essential due to two-year breeding cycle of onion. Successful micropropagation, callus culture and embryo rescue of *Allium* species have been reported and are potentially useful tools in onion breeding. Embryo rescue offers the possibility of increasing the number of interspecific hybrids, but introgression of beneficial genes may be severely restricted by low fertility. Meristem tip culture can be used to get virus free plants. *In vitro* evaluations for resistance to pink root or *Fusarium* basal rot of onion have been proposed by treating cultured cells with toxins produced by these fungi. Culture of the inflorescence or basal plate can be used to propagate male sterile plants without the maintainer line. Vegetatively propagated female parents of the hybrid can be transplanted directly in the field for pollination. The use of this technology will depend directly on the cost: benefit ratio of propagating the female and income generated from the hybrid seed. The development of haploid plants by ovule culture is another potentially useful application of tissue culture. The doubled haploid plant can give homozygous inbred in 2-3 years than 10-12 years (5-6 generations) with conventional approach. Exploitation of this technology can fasten the inbred development lines for use in hybrids and synthetics developments. Advances in the identification and mapping of biochemical markers are expanding the genetic map of onion; it will provide basic information on phylogeny, genetics and breeding. Identification of genetic markers linked with desirable traits would help identification of plants at very early stage. This is more important in onion due to its two year breeding cycle and bulky handling of the produce. The markers linked to the genes conditioning traits such as high solids, low pungency, bulb colour or long dormancy would reduce the number of evaluation completed on each bulb or family. A breeder would evaluate for the genotypes at the marker loci instead of completing numerous independent evaluations. The marker assisted selection made easy identification of male sterile, maintainer and restorer locus. The genetic transformation offers great potential in onion breeding and its application depends upon the availability of cloned genes. The development of 'Tearless' and 'Round Up Ready' onions are the successful examples of this technology.

2. Production Technologies

Production technologies have key role in enhancing the production and productivity of onion. All cultural requirements depend upon soil type, season and geographic conditions for cultivation of the onion. Yield determinant of onion are the bulb size and bulb numbers produced from per unit area. Its potentials can be realized from well-drained soils, where water is not limited and nutrients are readily available. Affect of high temperature, drought, floods and insect-pests at critical stages significantly harm the growth and lower the productivity. Plant density and spacing are essential for determining bulb size and numbers. Yields are higher, where early leaf cover is produced and then maintained for a long period prior to bulb formation.

2.1 Sowing and Transplanting

Under Indian conditions onion cultivation begin with sowing of seed in the nursery followed by transplanting of the seedlings in field. Onion seed is highly perishable in nature for emergence and storage life. Therefore, seed must be genetically true to type, high in germination and vigour. Onion seed may be infected with seed-borne pathogens like *Aspergillus niger*, *Botrytis aclada* and *Fusarium oxysporum* f. sp. *cepae*, which are known as causal agents of black mould, neck rot and basal rot diseases, respectively. These pathogens can be transmitted from infected seeds to seedlings, sets or bulbs. Soil and seed treatment with copper fungicides or Carbendazim can be effective in preventing such diseases. Use of *Trichoderma viride* for seed and soil application found beneficial in controlling the damping off in nursery. Solarization of nursery beds with clear plastic sheet also found effective for checking the pathogens and weeds. Seedling growth can also be promoted with applications of soluble fertilizers, vermin-wash, vermin-compost and neem-cake. Depending upon season, variety and germination, 6 - 8 kg seed is required to raise nursery for transplanting one hectare. Seed for *rabi* should be sown in Oct-Nov., *kharif* in May-June and late *kharif* in Aug.-Sept on the raised seed beds in lines at 5-7 cm spacing. It takes 8-10 weeks to attain 0.8-1.0 cm thickness and 20-25 cm height, which are considered ideal for transplanting. Over aged seedlings are prone to bolting and establish late, whereas, under aged do not establish well. Crop geometry of 65-70 plants/m^2 is considered ideal for getting optimum yield in onion.

2.2 Fertilizers and Manures

Onion is a shallow rooted (25-30 cm) crop and a balance inflow and out flow of the nutrients in the soil must be maintained for getting good yield, otherwise major share of the nutrients lost through leaching. The diagnosis of soil and plant nutrient status is helpful in determining the nutrient status required for healthy growth. To produce one tonne of onion 2.10-2.16 kg N, 0.7-0.8 kg P, 2.00-2.69kg K are removed under tropical conditions of India. The nutrient uptake starts after establishment of plants and increases upto 60 days of transplanting. The application of nutrients through external sources must match with the crop nutrient demand for higher nutrient use efficiency. N, P, K and S in mature healthy leaves should be 3-4, 0.3-0.4, 3.5-5.5 and 0.3-0.5 per cent at pre-bulbing and 2, 0.1-0.3, 2-4 and 0.2-0.3 per cent at bulbing stage of onion, respectively. It is considered deficient below 1 per cent of N and K each, and 0.1 per cent ofP concentration. The use N from 150-200 kgha^{-1}increased the yield andexcessive use causes thick stem, thick neck, delay maturity, soften bulbs, poor storage and high incidence of foliar diseases. To have higher efficiency, N should be applied in split dozes, incorporated in soil to avoid volatization in granular forms or with organic manures. Phosphorus is essential for root development and boost establishment in early phase of the plant growth.The recommended doze of P is 50-80 kgha^{-1} and given as basal doze at the time of transplanting. P get fix in the soil complexes and remains in available equilibrium also. Onion root development is severely affected below 10ppm level of P in the soil. Phosphorus Solubilizing Bacteria (PSP) can further enhance P availability along with application of organic manure. Potassium accumulates high dry matter and sugars and provides diseases resistance. It is important under conditions of high nitrogen application. K requirement in onion is more or less equal to N. Generally, 120-200 kg ha^{-1} K is recommended and Potassium Chloride (muriate of potash) is a main source. It is highly mobile and 0.5 per cent spray of sulphate of potash gives good results. Sulphur is critical for taste and pungency of the bulb. It should be applied @ 15 kg in sulphur sufficient and 25 kg ha^{-1} in deficient soils. Low level of S also affects the availability of P. The application of potassium sulphate is considered better than potassium chloride.

Micronutrients like Zn, Fe, Cu, Mn, B and Mo are found 48, 12, 4, 5, 33 and 13 per cent deficient in Indian soils, respectively. Among them availability of Zn and B is critical. Soil pH, organic matter, calcium carbonate, sodium, calcium and magnesium governed the availability of micronutrients. Soil application of micronutrients should be given only after the diagnosis. The indiscriminate use of micronutrients causes toxicity to the plants, therefore, critical appraisal through soil and plant tissue test is must before application. Soil application generally, affects its availability; hence foliar sprays are preferred. Regular application of FYM @ 10-15 $mTha^{-1}$ keeps the check on deficiency of micronutrients in onion.

Continuous use of organic manures enhances organic matter and physical properties like hydraulic conductivity, porosity, aggregate stability and soil biological activity along with lower bulk density. Soil microorganisms like *Azospirillium* sp., *Enterobacter* sp., *Azotobacter* sp. and *Pseudomonas* sp. encourages plant growth by promoting secondary roots, protectors against microorganism by releasing plant hormones and siderophores. Therefore, for increasing onion production at sustainable manner we have to opt combination of chemical fertilizers, organic manures andmicrobial inoculants at optimum time of application.Further, nutrient use efficiency can be enhanced with a focus on uptake pattern of the plant, losses management, microbial inoculation and developing efficient ideotype.

2.3 Irrigation

Onion is a shallow rooted crop and demand frequent and light irrigation for growth and bulb development. Onion needs to be irrigated at transplanting, three days after transplanting, and subsequent irrigation at 7-10 days interval depending upon soil type and moisture content. In general *kharif* crop needs 5-8 irrigation, late *kharif* 10-12 and *rabi* 12-15 irrigations.In general frequent irrigations are required at early growth, more frequent at bulb development and none at maturity of the crop. Bulb development stage is considered crucial for availability of moisture. In onion, water use efficiency was considered higher at IW/CPE of 1.2. Heavy irrigation at harvest time causes immature thick-skinned bulbs, delay maturity, become cause of *Fusarium* infection. Onion gives high response to drip irrigation and fertigation due to its shallow root system. Yield with this approach increased by 30 per cent, water saved 25 per cent, fertilizer 20 per cent along with better storage than the flood irrigation system. Results of micro-sprinkler also give encouraging results for saving the water and improving the onion yield.

2.4 Weed Control

Onion is a poor competitor of weeds due to short stature, non-branching habit, sparse foliage, shallow root system and frequent irrigation requirement. Manual, mechanical and chemical methods are followed for weed control in onion. Close spacing makes the manual and mechanical method of weed control tedious, expensive and time consuming. Some times non-availability of labour at optimum time delay the operation and resulted in poor yield. Therefore, use of pre and post emergence weedicides commonly usedfor controlling the weeds. Pre-emergence herbicides are Mertibuzin (Sencor Lexone) @ 0.35, Propachlor (Ramrod) @ 4.5, Oxyfluorfen (Goal) @ 0.15-0.25, Fluchloralin (Basalin) @ 1.5-2.0, Pendimethalin (Stomp) @ 2.5-3.5, Oxadiazon (Ronstar) @ 1.25, Methabenzthiazuron (Tribunal) @ 1.4, Nirtofen (Tok) @ 2.0, Linuron (Lorox) @ 0.94, Trifluralin (Treflan) @ 1.0 and Oxydiargyl (Raft) @ 0.25 kg or liter ha^{-1}. Whereas, pre-emergence and post planting herbicides are Loxynil (Toxynil) @ 0.5, Pendimethalin (Stomp) @ 2.5-3.5, Butrlin (Tamex, Sector) @ 3.5, Oxyfluorfen (Goal) @ 0.15-0.25, Fluchloralin (Basalin) @ 1.5-2.0, Alachlor(Lasso) @ 2.0, Trifluralin (Treflan) @ 1.0 and Oxydiargyl (Raft) @ 0.25 kg or liter ha^{-1}. These herbicides are applied at recommended doze under optimum moisture conditions. Manual weeding for 2-3 times during the crop season also keeps the weed under control.

3. Mechanization

In onion, all operations staring from nursery sowing, transplanting, weeding, harvesting, grading and storage are carried out manually, consequently cost of cultivation is increased and net-returns are lowered. Involvement of manual labour in high number and lack of mechanization is also one of the reasons for limited area under onion in north-westernplains of India. Various institutes made sporadic efforts for development of machinery required in onion cultivation.As nursery seed drill have been developed by PAU, seed drill for direct sowing in field by PAU and IIHR; Pneumatic seeder imported from Italy by CIAE and PAU; seedling transplanter by IIHR; seedling transplant marker by PAU, bulb digger by IARI and PAU; seed extraction machine by IIHR and bulb grader by DOGR and IIHR. All these machines should be tested thoroughly and for comprehensive use as a package for onion cultivation in the country.

4. Seed Production

Seed is the basic input and contributes one-third in total productivity of the crop. Thegenetic purity is of more concern in this highly cross-pollinated and entomophilous crop for maintaining the uniformity of colour, size, shape and maturity; tolerance to bolting, splitting, non-bulbing and insect-pests; longer storability. Genetic purity begin with choice of true to type variety from reliable source, healthy bulb production practices, proper storage of mother bulbs, maintenance of more than 1000 m isolation distance, ideal soil and climatic conditions for seed production, proper harvesting, threshing, cleaning and storage practices for onion seed. Onion seed production requires dry andmoderate sunny days (18-25°C) with cool nights (4-14°C) at flowering and seed setting stage. Warm weather and high humidity leads towards perpetuation of fungal diseases. Some of the traditional areas considered suitable for onion seed production are Saurashtra in Gujarat; Nashik, Ahmednagar, Satara and Marathwada in Maharashtra; Khargaon, Indore and Dhar in Madhya Pradesh; Jaipur, Chittorgarh, Udaipur and Sriganganagar in Rajasthan; Kurnool in Andhra Pradesh and northern Karnataka. Thebulbs used in seed production should be produced in recommended regions of the variety for maintaining the genetic stability. Onion seed is produced through annual and biennial methods. In annual method, seed-to-seed approach can be followed by advance sowing of seed and transplanting of nursery by two months, but seed produced through this approach shouldbe used for commercial bulb crop only, not again in seed cycle. The *kharif* season varieties are of short duration with low vernalization requirement and can complete the seed-bulb-seed cycle in one year also. However for *rabi* season varieties, biennial approach has to be opted due to more low temperature requirement for induction of flowering. About 25 quintals of medium sized (4-6 cm) bulbs are enough for planting one-hectare seed crop in month of Oct-Nov. All healthy production practices should be followed to get good seed yield. Seed setting percentage can further be enhanced buy putting 4-6 honeybees boxes in one hectare. After harvesting seed should be cleaned, graded and dried to 6 per cent moisture for storage at 16-20°C and 30-40 per cent RH. Onion seed yield depends upon variety and weather conditions, but on an average 6-8 q ha^{-1}seed can be harvested.

The annual onion seed requirement of the country is about 9,600 tonnes @ 8.0 kgha^{-1}, besides 20 per cent additional stock required to cover poor germination, storage losses and as buffer stock.In this,8 per cent is supplied by the public sector organizations, 9 per cent by the private seed companies, 13 per cent by private traders and rest 70 per cent by the farmers from their own saved seed. Thus, a big gap is there between the availability of improved varietiesseed and the requirement. To cope this gap holistic planning is needed for production of breeder, foundation, certified and truthfully labeled seed by identifying areas for bulb and seed production of recommended varieties for different regions and for export. The objective should be to replace at least 50 per cent area under improved varieties in the next ten years.

5. Inset-Pest and Diseases Management

World wide, 40 fungal, 14 bacterial, six nematodes, three viral, one phytoplasamal and one prarasitic plant have been reported, however, in India 29 fungal, three viral and four bacterial pathogens have observed. These diseases can cause up to 50 per cent losses during crop production and storage. Among them, purple-blotch, Stemphyllium-blight and Anthracnose causes severe losses, whereas, downy mildew prevalent in temperate regions and, pink-rot and *Fusarium* basal-rot also cause damage in onion growing areas. The severity of the diseases is influenced by the seasonal effect, variety and growing conditions.

Purple-blotch (*Alternaria porri* (Ellis) Cif.) requires 22-25°C temperature and 90-100 per cent RH for sporulation and infection. It forms small, water-soaked areas on leaf or seed stalk and turns brown. The enlarged spot becomes zonate and more or less purplish. The margin is a shade of purple or red and surrounded by yellow halo that extends upwards and downwards for some distance. It can cause upto 80 per cent yield losses under Indian conditions. *Stemphylium* leaf-blight (*Stemphylium vesicarium* Wallr.) occurs at same time and even on same plant as of purple blotch and forms the diseases complex. Its symptoms start with small, yellow to brown purple, water soaked lesions. These lesions elongate, become spindle shaped to ovate-elongate and turn into diffusive spot, often extending to leaf tips. The spots frequently coalesce into patches blighting leaves. As conidiophores and conidia develop on the lesions, turn light brown to tan purple at the center, and later dark olive-brown to black. Similar symptoms also occur on the scape of onion. Perithecia may appear in blighted areas of leaves and scape as small, black, pin-head like raised bodies. Anthracnose (*Colletotrichum gloeosporiodes* (Penz.) Penz. and Sacc.) mainly shows white sunken-oval lesions, which turns into pale yellow water soaked spots and cover whole leaf in later stage. In severe cases pseudo-stem twists and show abnormal elongation. The favorableconditions for conidia germination and infection are high humidity and 23-30°C temperature. It is more prevalent in *Kharif* crop. Downy mildew (*Pernospora destructor* (Berk.) Casp.) is highly destructive under cooler and humid regions anddevelops white to purplish downy growth first on entire surface of the older leaves.The affected parts become pale-green, yellow and collapse. The fungus requires cool moist nights and moderately warm days for best development. The conidia produced in humid atmosphere between 4-25°C with optimum at 13°C. Cloudy days favour most, as 8 hour of light kills the conidia.

Above fungal diseases can be managed by following clean cultivation, crop rotation, raised bed planting and proper drainage. Seed treatment with Bavistin (0.1 per cent) andhot water (50°C for 20 min), high rates of calcium, phosphorus and potassium also reduce diseases infection. Application of *Trichoderma viride, Bacillus subtilis, Saccharomyces cerevisiae* or Salicylic acid helps in managing these diseases. Foliar sprays of Mancozeb @ 0.25 per cent, Tricyclazole @ 0.1 per cent and Hexaconazole @ 0.1 per cent effectively controls the diseases also.

Viral diseases are becoming major constraint in onion production. However, number of viruses like Iris yellow spot, Groundnut bud necrosis, Tobacco streak, Onion yellow dwarf, Allexi and Rymo viruses infect onion, but Iris yellow spot virus (IYSV) and onion yellow dwarf virus (OYDV) are more common. IYSV is Tospovirus, which develops spindle/diamond shaped chlorotic lesions on leaves and scapes and ultimately wither and twists the infected parts. This is sap transmitted and thrips are the main vector. OYDV is Potyvirus, which develops yellow streaks, flattened leaves, dwarf plants, twisted flower stalk and under developed bulbs. It is transmitted by aphid vector. To control these viruses use virus free bulbs, control insect-vector and follow seed-seed approach.

Large number of insects including thrips, leaf miner, leafhopper, psyllid, beet army worm, curt worm, gram borer, red spider mite, onion maggot and nematodes attack onion. Among these all, thrip (*Thrips tabasci*) is wide spread and cause damage not only by feeding and oviposition, but as a vector of tospoviruses. Thrip infestation is directly associated with spread of fungal diseases also. Thrips can be managed with regular monitoring of inner-most whorl of the leaves. Thrip nymphs and adults are visible on plant, eggs laid inside the plant, while pre-pupal and pupal stages occur in the soil and debris. Use yellow and blue sticky traps, lure with ethyl isonicotinate and methyl isonicotinate. Time of planting, crop rotation, barrier crops, irrigation, fertilization, intercropping, deep ploughing, weed control, reflective mulches andclean management followed for cultural control of thrip. So far, non of the genotype with complete and stable resistance have been reported, but some tolerate thrips. Biological control of thrip with mired bug, *Orius* sp. and predatory mite *Amblyseius;* and entomopathogens, fungi, *Verticillium, Metarrhizium*and *Beauveria;* nematodes, *Steinernema feltiae, Thripinema* and *Heterorhabditis* have been reported under the greenhouses. Chemical insecticides belonging to organophosphate, carbamate and pyrethroid groups are widely used for controlling the trip in tropical onion. Seedling root-dip with Carbosulfan (0.025 per cent) or Imidacloprid (0.04 per cent) solution for 2 h protect the plants up to 30 days. Foliar sprays of Lambda Cyhalothrin, Imidacloprid, Abamectin, Spinosad, Thiamethoxam, Fipronil, Acetamiprid, Acibenzolar-S-Methyl and Clothianidin lower the thrip population in onion.

6. Post-harvest Management

A huge loss (40-50 per cent) occurs during post-harvest handling and storage of onion, which includes weight loss (25-30 per cent), rotting (10-15 per cent) and sprouting (8-10 per cent).Post-harvest losses depends upon variety, cropping season, cultural practices, crop maturity and post-harvest handling. Pre-harvest application of Maleic Hydrazide (MH), ethephon, calcium nitrate (0.5 per cent), zinc sulphate (0.5 per cent), abscisic acid (10 µmol l^{-1}) and CIPC (2 per cent) suppress the sprouting, particularly in *Kharif* onion. Exogenous application of calcium nitrate and calcium chelate improves the cell wall structure integrity and reduce microbial decay caused by black rot, neck rot and basal rot. Abalance use of nitrogenous fertilizer and controlled irrigation before harvesting keep check on physiological weight loss of onion. Among post-harvest management practices, curing is the most important, which allows formation of strong intact outer protective skin and closure of onion neck. Curing significantly decrease the surface area and volume and, increase the bulk density and hardiness of the freshly harvested bulbs. Curing under field conditions for 3-4 days with wind-row method and at 30°C in controlled chamber found effective for storage of bulbs. Exogenous application of ethylene and ethylene binding inhibitor, 1-methylcyclopropene (1-MCP) suppress sprouts during storage in onion. However, application of CIP at 20°C found effective than the hot fogging. Gamma-irradiation (0.15 kGy) inhibit the sprouting, but its effect depends upon pre-harvest growing conditions, seasonal variation, curing period, bulb dormancy as well as storage environment/structures. Pre-cooling at <4°C for 4-6 hours reported effective for maintaining the quality of onion produce. This can be done with hydro, forced air and vacuum cooling.

Storage environment has significant effect on dormancy and storage life of onion. Two distinct temperature ranges *i.e.* 0-2° and 25-30°C found suitable for storage of onion, whereas, 10-20°C is favorable for sprouting. Under, Indian conditions Directorate of Onion and Garlic recommended different ventilated storage structures and needs promotion to install in whole country. The basic is provision of ventilation to the stored bulbs from all sides with 1-1.2 m loading height and protection from rains with waterproof roof. To reduce the storage losses forced air ventilation flow has been suggested @ 114 m^3hr^{-1} MT^{-1} of onion bulbs. Keeping all precautions under ambient conditions, about 30 per cent losses are inevitable, which can be reduced with cold storage at 1-2°C and 55-70 per cent RH. To get maximum storage benefit, bulbs should be fully mature and properly cured with tight neck. The air circulation in the store should be proper maintainrequired temperature and humidity without condensation of water on the bulb surface. Excessive humidity promotes root development and, higher temperature and humidity favor sprouting and pathological disorders. The controlled atmosphere (CA) storage having 2-4 per cent O_2, 0-5 per cent CO_2, 86-92 per cent RH and 2±1°C temperature can store onion bulbs for eight months with minimum losses. The onion availability to the consumers can further be enhanced by processing and value-addition. It can be in form of pre-sliced fresh onion, dehydrated, powdered, paste, pickle, vinegar, oil and wine.

7. Marketing and Export

Onion showed 8.28 per cent per annum growth in production, 6.01 per cent in area and 2.48 per cent in productivity during the last two decades. However, its growth was exponential for production (14.24 per cent), area (8.84 per cent) and productivity (4.97 per cent) during the last ten years. This made the increase in per capita availability of onion from 4.56 to 13.97 kg in the country. The per capita consumption growth of 12 per cent indicated remarkable preference of onion by the Indian consumers. No other food crop in India has shown this type of growth in recent past. The increase in income, change of life-style and dietary pattern increased the onion consumption by 53 per cent in urban area. Large expansion of eating joints, out side eateries, snack corners and restaurants enhanced the onion consumption considerably due to main ingredient, particularly in spicy food. There is a high rigidity in consumer demand and is difficult to imagine acceptance of any food without the onion. It has replaced its substitute spices due to theirhigh cost. The low price elasticity of demand for onion implies from the fact that 10 per cent increase in price resulted in 1 per cent decline in demand, whereas, 10 per cent shortfall in supply increase 100 per cent price of onion. Such, situation is exploited by the traders to hike the price. Presently, about 70 per cent of onion production is from Maharashtra, Karnataka, Madhya Pradesh, Gujarat and Bihar. Maharashtra contributes about 30 per cent of the total andin March 2013, due to abnormal weather 4 per cent low production was predicted in the state. Understanding this situation private traders procured onion in a big way at higher price from April to June (24 per cent higher than normal), as a result supply from the farmer dried up after June. Market arrival in July lowered by 17 per cent and in August by 22 per cent than the previous year. Then traders control the supply and price of onion, which was 186 and 293 per cent higher in July and August than the previous year price, respectively. So, decrease in production of onion responsible to a small extent, but predominant part is withdrawl of stock from the market. Such shock in onion prices occurs frequently and we usually attribute it to unfavorable weather, rather than to cartelization, hoarding *etc.* by the traders, which needs solution to avoid their recurrence.

The world trade of onion exports is about 6.77million tonnes of worth US$ 2,856. India ranked first in export (1.82 million MT), followed by China (0.74), The Netherlands (0.6), Egypt (0.49), Mexico (0.37), USA (0.35), Spain (0.25), Argentina (0.21), Pakistan (0.17) and Turkey (0.12). The onion export in India has risen with 9 per cent growth rate during last decade. Malaysia, Sri Lanka, Bangladesh and UAE are the major and traditional markets for export due to high demand of Indian red onion. Further, penetration in European, American and Japanese markets can be made with the development of big size yellow onions.

Though, horizontal and vertical growth of onion occurred due to increase in demand, high returns, better varieties and improved production technologies, but hike in prices many times lead to sensitive social, political and economical unrest in the country. Such situation should be handled with strict vigil on the distribution system specially hoarding and black marketing; exploring way to withheld stock with the farmers; extending cultivation to alternative geographic regions than the traditional; developing mechanism for forecasting; strengthening existing storage system; proper distribution system; augmentation with import; developing high yielding varieties/hybrids; adoption of improved production technologies and reducing post-harvest losses.

8. References

Bang, H., Kim, S, Park, S.O., Yoo, K.S. and Patil, B.S. 2013. Development of a codominant CAPS marker linked to the *Ms* locus controlling fertility restoration in onion (*Allium cepa* L.). *Sci. Hort.*, **153**: 42–49.

Brewster, J.L. 1994. *Onions and other Vegetable Alliums.* CAB International, Wallingford.

Bose, T.K. and Som, M.G. 1986. *Vegetable Crops in India.* Naya Prokash, Calcutta.

Chadha, K.L. 1990. Onion and garlic research in India. In: *Proceedings of the National Symposium on Onion and Garlic*, pp. 47–60.

Coureel, A.G.L. de, Veder, F. and Boussac, J.M. 1989. DNA polymorphism in *Allium cepa*cy to plasms and its implications concerning the origin of onions. *Theor. Appl. Genetic*, **77**: 793.

Currah, L. and Proctor, F.J. 1990. *Onions in Tropical Regions*, National Resources Institute Bulletin, No. **35**, Kent, U.K.

Dhatt, A.S. and Havey, M.J. 2011. Marker assisted selection for identification of male sterile cytoplasm in onion. *Crop Improv.*, p. 235–36.

Eady, C.C., Weld, R.J. and Lister, C.E. 2000. *Agrobacterium tumefaciens* mediated transformation and transgenic plant regeneration of onion (*Allium cepa* L). *Plant Cell Report.*, **19**: 376–381.

Engelke, T., Terefe, D. and Tatlioglu, T. 2003. A PCR–based marker system monitoringCMS (S), (T) and N cytoplasm in the onion (*Allium cepa* L). *Theor. Appl. Genet.*, **107**: 162–167.

Gokce, A.F. and Havey, M.J. 2002. Linkage equilibrium among tightly linked RFLPs and the Ms locus in open-pollinated onion populations. *J. Amer. Soc. Hort. Sci.*, **127**: 944–946.

Gowda, V.R., Rao, E.S., Pathak, C.S. and Singh, T.H. 2002. Development and commercialization of F_1 hybrids in short day onion: Indian perspective. **In**: *Proc. Intl. Conf. Vegetables*, Bengaluru, 11th–14th November.

Hagler, J.R., Cohen, A.C. and Loper, G.M. 1990. Production and composition of onion nectar and honey bee foraging activity in Arizona. *Env. Eneomol.*, **19**: 327.

Hanelt, P. 1990. Taxonomy, evolution, and history. In: *Onions and Allied Crops*, (Eds.) H.D. Rabinowitch and J.L. Brewster, CRC Press, Boca Raton, FL, USA.

Havey, M.J. 1991. Phylogenetic relationships among cultivated *Allium* species from restriction enzyme analysis of chloroplast genome. *Theor. Appl. Genet.*, **81**: 752–757.

Havey, M.J. 1993. A putative donar of S-cytoplasm and its distribution among open-pollinated populations of onion. *Theor. Appl. Genet.*, **86**: 128–134.

Havey, M.J. 1993. Onion (*Allium cepa* L.). **In**: *Genetic Improvement of Vegetable Crops*, (Eds.) G. Kalloo and B.O Bergh, Pergamon Press, Oxford.

Havey, M.J. 2000. Diversity among male-sterility-inducing and male-fertile cytoplasms of onion. *Theor. Appl. Genet.*, **81**: 752–757.

Havey, M.J. 2000. Transgenic onions are on the way. *Onion World*, **16**(6): 8–10.

Havey, M.J. and Bark, O.H. 1994. Molecular confirmation that sterile cytoplasm has been introduced into open-pollinated Grano onion cultivars. *J. Amer. Soc. Hort. Sci.*, **119:** 90–93.

Jones, H.A. and Clark, A.E. 1943. Inheritance of male sterility in onion and the production of hybrid seed. *Proc. Amer. Soc. Hort. Sci.*, **43**: 189–194.

Kohli, U.K. and Chauhan, V.S. 1990. Improvement onion through interspecific hybridization with allies species. **In**: *Proceedings of the National Symposium on Onion and Garlic*, pp. 6–19.

Kumar, N.K.K., Gopal, J. and Parthasarthy. 2015. *The Onion*. ICAR, New Delhi.

Lawande, K.E. 2005. Onion improvement: Present status and future thrusts. **In**: *Proc. SYMSAC–II: National Symposium on Current Trends in Onion, Garlic, Chillies and Seed and Spices Production, Marketing and Utilization*, NRCOG, Rajguru Nagar, 25th–27th November.

Melgar, S.A. 2009. Genetic, genomic and proteomic studies of male-fertility restoration in onion. *Ph.D Dissertation*, University of Wisconsin, USA.

Molenaar, N.D. 1984. Genetics, thrips resistance, and epicuticular wax characteristics on nonglossy and glossy onions. *Ph.D. Thesis*, University of Wisconsin, Madison, USA.

Pathak, C.S., Blck, L.L., Ko, S.S., Cherng, S.J. and Wang, T.C. 1996. Genetic improvement of onion for the tropics. *Onion Newsletter for the Tropics*, **7**: 12–15.

Pathak, C.S. and Gowda, V.R. 1993. *Improvement of onion*. **In**: *Advances in Horticulture Vol. 5: Vegetable Crops*, (Eds.) K.L. Chadha and G. Kallo, Malohtra Publishing House, New Delhi.

Pike, L.M. 1986. Onion breeding. **In**: *Breeding Vegetable Crops*, (Ed.) M.J. Bassett, AVI, Westpon, CT.

Sato, Y. 1998. PCR amplification of CMS-specific mitochondrial nucleotide sequences to identify cytoplasmic genotypes in onion (*Allium cepa* L.). *Theor. Appl. Genet.*, **96**: 367–70.

Singh, R.S. 1999. *Diseases of Vegetable Crops*. Oxford and IBH Publishing Co. Pvt. Ltd., New Delhi.

Rabinowitch, H. and Brewster, J. 1990. *Onions and Allied Crops*. CRC Press, Boca Raton, FL, USA.

Van Heusden, A.W., Ooijen, J.W. van, Ginkel, R.V., Verbeek, W.H.J., Wietsma, W.A. and Kik, C. 2000. A genetic map of inter-specific cross in *Allium* based on amplified fragment length polymorphisim (AFLPTM) markers. *Theor. Appl. Genet.*, **100**: 118–126.

Wein, H.C. 1997. *The Physiology of Vegetable Crops*. CAB International, Wallingford.

Yang, Y.Y., Huo, Y.M., Miao, J., Liu, B.J., Kong, S.P., Gao, L.M., Liu, C., Wang, Z.B., Tahara, Y., Kitano, H. and Wu, X. 2013. Identification of two SCAR markers co-segregated with the dominant *Ms* and recessive *ms* alleles in onion (*Allium cepa* L.). *Euphytica*, **190**: 267–277.

20

Development of Orchid Biodiversity-based Microenterprises for Conservation and Income Generation

D.R. Singh

ICAR-National Research Centre for Orchids,
Pakyong – 737 106, East Sikkim
E-mail: drsinghhort66@gmail.com

The orchids are admired world over for their bewitching beauty, bewildering shape and size, breadth taking colors, therapeutic value and more importantly keeping quality of flowers that surpasses with other flowering plants of the world. Owing to these qualities, orchids are now ruling over the world floriculture market. However, the wild genetic resources under this family are declining very fast due to various regions including overexploitation and clandestine trade for various purposes. The collection of orchids from their natural habitats is still cheaper option for the traders. Potential of orchid species in floriculture and medicine and food has been debated at various forums and concluded that a large number of species hold good promise for exploitation. However, a very few systematic efforts have been made to this direction. The biodiversity based enterprise has the potential to lift individuals and communities out of poverty while conserving biodiversity. The joint forest management policies of the government are directed to provide livelihood to forest dwellers by harvesting minor produce from the natural habitats. One can see, forest dwelling communities selling live plants including orchids in Shillong, Meghalaya. There is big demand for Indian orchids in foreign markets. A large number of orchid species are exported annually to the destination like USA, Germany, Japan and Canada. Apart from export, the wild orchids are used as potted plant, cut flower and medicine. At presently, such uses of orchids are localized and primarily based on the plants collected from the natural habitats. Here is an opportunity to expand the market based on plants produced in the tissue culture laboratories. The twin strategy development of orchid biodiversity based enterprises and dealing firmly with unscrupulous collectors would be helpful in establishing genuine trade orchid species.

1. Orchid Distribution, Habit and Habitat

There are 1331 species known to occur within the political boundaries of India. North Eastern Himalaya is very rich in orchid biodiversity and nearly 860 species are known to found in this region. The largest biodiversity found in Arunachal Pradesh followed by in Sikkim and Meghalaya. The least orchid biodiversity has been recorded from Tripura (Figure 1).

Orchids can be found growing in open grass lands, marshy land or under the shady forest floors (terrestrials), clasping the tree trunks or branches in open or hiding under the tree canopies (epiphytes), clinging with the stones or rocks (lithophytes) and surviving on dead or decayed organic matter (saprophytes). The orchids that grow on ground have varying requirements requirement for substrate, water requirement (soil) to which they are adapted. For example, *Paphiopedilum* prefer to grow on lime stone or calcareous soil while *Dactylorhiza hatagirea* grows best on acidic soils; *Spiranthes sinensis* and *Satyrium nepalensis* grows in stony open grassy lands having very low moisture content whereas *Goodyera procera* prefers marshy soil; *Anthgonium gracile* grows

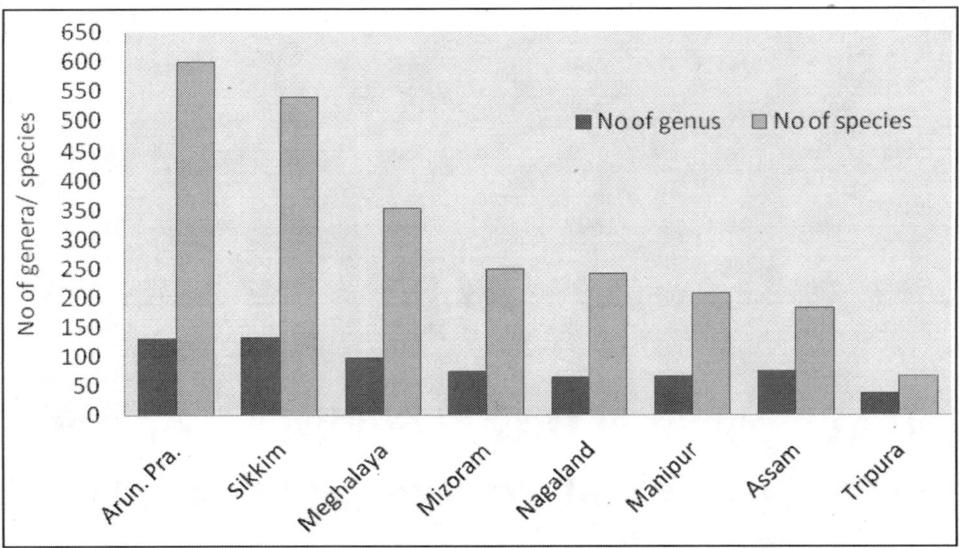

Figure 1: Orchid Biodiversity Distribution in North Eastern States of India.

best on sun facing steep mud embankments while species of *Calanthe, Odotochilus* prefer to grow on humus rich soils under the dense forest cover. In some orchids, the growing point (apical bud) grows ceaselessly and flowers are borne on the stem (monopodial), *e.g. Vanda*, whereas in others the growing point ceases the growth and is replaced by a new auxiliary bud (Sympodial) *e.g. Dendrobium, Cymbidium, Coelogyne* etc. In general, new shoot buds emerge just before or after the flowering.

The orchid rich habitats of the country lie in alpine and sub-alpine zone (3500–5000 m), temperate and sub temperate zone (1850–3500 m), sub-tropical zone (850–1850 m), tropical zone (250–850 m) and plains (sea level–250 m). Each zone has its own characteristic features with regard to climatic and microclimatic conditions. Alpine and sub alpine zone is characterized by open grassy meadows exposed rock surfaces, scrubby vegetation, very high light intensity and very low temperature during winters. in this zone orchid species like *Didicea cunnighmaii, Orchis puberula* and *Cypripedium* species grow on rocky banks along the streams; *Satyrium nepalensis, Orchis* species and *Herminium* species in open grassy lands; *Aphyllorchis* species and *Epipactis* on the floors of forests. The temperate to sub temperate zone is characterized by presence of fog and mist almost round the year and high light intensity coupled with low temperature during winters. Here orchids like *Pleione humilis, P. praecox, Cymbidium hookerianum, C. devonianum, C. elegans, C. gammieanum, Dendrobium hookeriana* can seen growing on tree trunks or on the rocks while *Satyrium nepalense, Herminium* species and species *Habaneria* grow on grassy slopes. The sub-tropical zone is characterized by heavy summer rains and mild day temperature between 25 -30°C along with slightly cooler nights (20°C) and dry and cold winters. The orchids like *Anoectochilus brevilabris, Odontochilus lancelatus, Eulophia brevilabris*, grow on forest floors rich in humus; the species of *Coelogyne, Cymbidium* can be seen growing on trees whereas species like *Anthogonium gracile* grow on sun facing mud embankments. The tropical zone is characterized by hot summers (30-35°C) and cooler nights (20-25°C) accompanied with a very high humidity. Species of *Dendrobium, Cymbidium alofolim, C. pendulum, Coelogyne favida*, etc. grow on trees whereas species like *Anoectochilus roxburghii, Goodyera hispida* on shady embankments. The plains are characterized by very high summer temperature with relatively cooler nights, lower humidity. The species which inhabit in this zone include *Aerides multiflorum, Dendrobium aphyllum, Vanda tesselata, Eulophia comprestis, Phalaenopsis speciosa* etc.

2. Economic Value

Orchids represent royalty and aristocracy in medicine and floriculture. The flowers of orchids are well known for their uniqueness of shape, breath taking colors, and exceptionally long shelf-life that surpasses many other floriculture crops. The life of flowers on the plant varies with the species. It has been found that life of the flowers on the plants of *Zygopetalum intermedium* (18 days), *Phaius tankervillae* (24 days), *Aerides odoratum, A. multiflorum, Cymbidium iridioides, Dendrobium nobile* and *Renanthera imschootiana* (28-56 days), *Paphiopedilum hirtutissimum, P. wardianum* (56 days) and *Vanda coerulea, Vanda teres* (28-42 days). These qualities have made orchid growing a highly profitable industry all-over the world. Today, more than 1, 25,000 hybrids are known and cultivated for cut flowers and potted plants and more and more new ones are being registered every month. A large number of species bear attractive flowers and have capacity for interspecific as well as intergeneric hybridization that has generated tremendous possibilities producing hybrids of diverse floral characteristics. There are about 70 species of Indian orchids that have been used for producing several primary and other generation hybrids in and outside country. Thus, wild native orchids are likely to play a unique role in the development of new cultivars/ hybrids and also restructuring of existing ones with one or other attributes. The orchids also have inherent capacity for drought resistance, nutrient conservation, long shelf-life *etc.* which are not fully understood today. For these reasons, it is believed that orchids can contribute many valuable genes to other crops.

The Indian orchids have also been used in various indigenous systems of medicines since time immemorial. In the Ayurvedic system of medicine, a group of eight drugs known as *'Ashtavergha'* is used in the preparation of various rejuvenating formulations and the tonics such as *'Chyavanprash'*. Among eight drugs of *ashtawarga* group, *Jeevak* (*Malaxis muscifera*), *Rishbhak* (*Malaxis acuminata*), *Ridhi* (*Habenaria intermedia*) and *Vridhi* (*Habenaria edgeworthii*) come from orchids. The ethnobotanical studies carried out in the country have shown that various orchids are used for treating ailments like *Dactylorhiza hatagirea*, *Cypripedium parviflora*, *Vanda roxburghii*, *Orchis latifoia*, used for aphrodisiac and nervine tonic; *Habenaria pectinata* for snake bite; *Eria nuda* for worm infection; *Dendrobium nobile* for pulmonary tuberculosis; *Anoectochilus formosus* for Hepatitis, hypertension and cancer are mention a few. There are many more orchids that are used as medicine. Though most important commercial product isolated from orchids is Vanillin from cured beans of *Vanilla planifolia*, which is used for flavouring chocolates, pudding ice cream and also used as deodorants in soaps, perfumes and powders. Till recent past little attention was paid for isolation and identification of therapeutically important phytochemical constituents of orchids but recently the utility of *Cymbidium hybrids*, *Epipactis hellebonie* and *Liparis ovata* in anti-AIDS medicine has been identified. Orchids have now come under ambitious chemical study for extraction of alkaloids, polyphenols, terpenoids, steroids and other molecules of biological interest. The future application of such studies may eventually turn commercial to create new forms of drugs.

3. Orchid Biodiversity Based Enterprise

Orchids are one of the most threatened plants on this earth. The vulnerability stems from two sources: the first being their highly specialized lifecycle - dependency on pollinators for pollination, lack of reserved food material in the seeds, reliance on mycorrhizal fungi for seed germination and second ornamental and therapeutic value which they possess has made them so sought after the man. Each species is adapted to life in a specialized environment because of their specialized requirements and many species are very restricted in distribution. Any destruction, degradation or defragmentation of natural habitat beyond a tolerable limit causes threat for their survival. But such disturbances in orchid rich belts are unabatedly continuing causing serious damage to orchid biodiversity. Apart from above mentioned factors a large number of orchid species traded clandestinely, and also used indigenously for various purposes like medicine, research etc. In result, the population of these species has declined and have been pushed to rarity. The number of species facing various kinds of threats in north eastern states is given in Figure 2. There is an urgent need to develop the orchid bio diversity based enterprises that can reduce the pressure on natural habitats. The prospective orchid biodiversity based enterprises are given as below:

3.1 Export of Live Plants

The orchids collected from natural habitat are not allowed for export, however, orchids propagated in nurseries are allowed for export as per law. A large number of plants are being exported from India to several destinations like Germany, Japan, USA, UK etc.

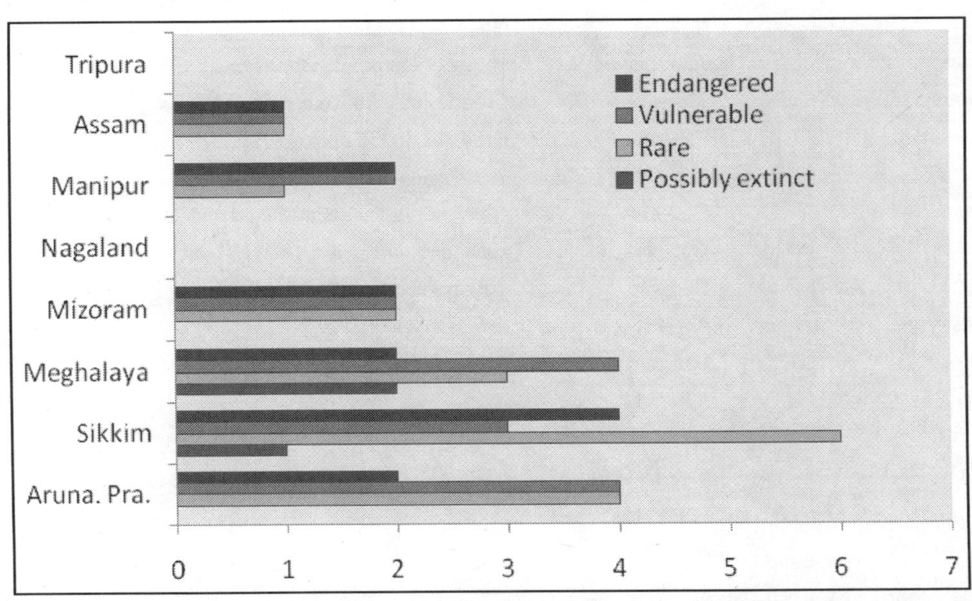

Figure 2: The Number of Orchid Species under Various Categories of Threats in North Eastern States of India.

It is unfortunate that most of the trade is met through the plants collected from the wild despite of law of the land, which prohibits the collection of orchids from the natural habitat for commercial purpose. There is great scope for development of enterprise meant for export of live orchids. The entrepreneurs of these enterprises should have good knowledge about the species found in the countries and also knowledge about growing of such plants especially rare ones.

3.2 Potted Plant and Cut Flower Production

A large numbers of species having bizarre shape, exquisite colors size and long lasting flowers for which some people even might have not thought in their dreams, are well suited as potted plants that could be utilize for harnessing the markets for potted plants. The species suitable floriculture purposes are listed in Table 1. Apart from floriculture importance, a large number of plants are required by research institutions as well as private and government garden, which can also can be a part of such enterprise.

3.3 Herbal Medicine Industry

A large number orchid species are used in traditional system of medicine. Largely, at present, the demand of orchids for therapeutic purpose is met through the plants collected from the natural habitats. There is a large scope for cultivation of medicinally important species in the country. Some medicinally important species are listed in Table 2.

Table 1: Some Floriculturally Important Orchids

Orchid Species	*Flowering Season*	*Flower Colour*
Aerides multiflorum	April	White flushed with purple
Aerides odoratum	May–June	White with purple blotch
A.fieldingii	May	Pink to dark pink purple
Phalaenopsis cornu cervi	Varies	Yellow green barred and botched with red brown
Phalaenopsis mannii	May	Sepal and petal golden yellow barred and blotched with chestnut brown
Phalaenopsis lobii	Mar–Apr	White rusty brown lip
Rhynchostylis retusa	June	Pink fragrant
Holcoglossum amesiana	Feb–Mar	White lip pink
Vanda coerulea	Oct–Nov	Blue also known as blue vanda
Vanda coerulescens	Apr–May	Blue also known as shy blue vanda
Vanda tessellata	Mar–Jun	Yellow dotted purple
Vanda stangeana	Mar–Apr	Yellow lined with purple
Renanthera imshootina	June	Red also known as red vanda
Cymbidium aloifolium	Apr–Jul	Yellowish purple
Cymbidium devonianum	May–Jun	Greenish brown with purple blotch on lip
Cymbidium elegans	Sep–Oct	Tubular yellow
Cymbidium erythraeum	Sep–Oct	Pale green veined purple lip white blotched purple
Cymbidium cyperifolium	Nov–Dec	Green lip blotched purple
Cymbidium tigrinum	Apr–may	Yellow green purple stripes on lip
Cymbidium lowianum	Mar–Apr	Green with lip red in v shape at apex
Cymbidium hookerianum	Feb–Mar	Green lip dotted purple
Cymbidium mastersii	Oct–Nov	White with yellow tinged lip
Cymbidium tracyanum	Sep–Oct	Green sepal and petal veined and lip blotched purple
C. iridioides	Oct–Nov	sepal and petal veined and lip blotched purple
Calanthe biloba	Sep–Oct	Purple with pink lip
Calanthe herbacea	Apr–May	Green with white lip
Calanthe plantaginea	Mar–May	Lilac
Calanthe puberula	Apr–May	Lilac
Calanthe brevicornu	Mar–Apr	Purple
Calanthe triplicata	Sep–Oct	White with red
Calanthe masuca	Apr–May	Violet
Calanthe chloroleuca	Mar–Apr	Yellow
Dendrobium jenkinsii	May	Yellow

Contd...

Table 1–*Contd...*

Orchid Species	*Flowering Season*	*Flower Colour*
Dendrobium lindleyi	Feb–Mar	Yellow
Dendrobium densiflorum	Apr–May	Yellow with deep orange yellow throat
Dendrobium chrysotoxum	Apr–May	Bright yellow
Dendrobium farmeri	Apr–May	White with orange yellow lip
Dendrobium aphyllum	Apr–May	Pink
Dendrobium thyrsiflorum	Apr–May	White with bright golden yellow lip
Dendrobium capillipes	Mar–Apr	Bright golden yellow
Dendrobium nobile	Apr–May	White tipped deep pink lip dark purple throat
Dendrobium devonianum	Apr–May	White tipped purple two orange blotch in throat of lip
Dendrobium draconis	Jul–Aug	Shiny white orange blotch in lip throat
Coelogyne nitida	Mar–Apr	White lip blotched yellow
Coelogyne cristata	Mar–Apr	Pure white
Coelogyne flaccida	Mar–May	Dull yellow
Coelogyne viscosa	Jan–Feb	White
Coelogyne barbata	Sep–Dec	White with brown lip
Coelogyne corymbosa	Mar–Apr	White with 4 orange yellow blotch on lip
Coelogyne fuscescenns	Oct–Dec	White lip flushed with purlpe
Coelogyne occultata	Apr–May	White with 4 orange yellow blotch on lip
Peione humilis	Feb–Mar	White with purple spotted lip
Peione maculata	Oct–Nov	White with yellow and purple botched lip
Peione praecox	Oct–Nov	Rosy pink with purple blotched lip
Paphipedilum fairrieanum	Oct–Dec	Dorsal sepal white veined purple, green
Paphipedilum hirsutissimum	Feb–Apr	Pale green
Paphipedilum insigne	Oct–Dec	Pale green with white band on dorsal sepal
Paphipedilum spicerianum	Oct–Feb	Green and white streaked of purple
Paphipedilum venustum	Feb–Mar	Whitish green, lip pale yellow, tinged with purple
Paphipedilum villosum	Feb–Mar	Green and yellow
Anoectochilus brevilabris	Jan–Mar	Yellow with dark purple nerves
Liparis cordifolia	Nov–Dec	Green
Neogyne gardenerana	Dec	White
Pholidota imbricata	May–Aug	Dirty white
Pholidota rubra	Nov–Dec	Yellowish green

3.4 Education/Botanical Tours

The tourism industry is gaining importance in north eastern states of the country. The linking of this enterprise with tourist operators for educational/ botanical tours would benefit both travel operators as well as entrepreneurs.

3.5 Micropropagation

Tissue culture is a tool by which a large number planting material could be produced in a limited period of time. For the functioning of above mentioned enterprises would require continuous supply of planting material. This could be met through tissue culturing the desired plants in tissue culture laboratories. Thus, there is scope for development of enterprises for supply of planting material to the above mentioned enterprises.

3.6 Entrepreneurship Model

The commercial cultivation of orchids is not very developed. A very few species of orchids are cultivated by a very few person as a hobby. Initially, the institutions like Ministry of Forest or North Eastern Hill Council or both may fund some projects to demonstrate the profitability and sustainability of such enterprises. These enterprises could be set up under the institutions engage in development of orchids in the region. The NERMAC may help these enterprises for the marketing and finding out the buyers of the produce. The schematic diagram involving various prospective institutions for achieving the given objective is given in Figure 3.

Table 2: Some Therapeutically Important Orchids of North Eastern Himalayas

Botanical Name	*Part(s) Used*	*Therapeutic Use(s)*
Acampe papillosa	Roots	Rheumatism, sciatica, neuralgia, syphilis and uterine diseases.
Aerides odorata	Ground fruits	Healing wounds. Juice of leaves to heal boils in ears and nose.
Anoectochilus regalis	Stem and leaves	Ingredients in certain medicinal oils
Calanthe sylvatica	Flowers	to stop nose bleeding
Calanthe triplicata		Swollen, diarrhea
Cephalanthera longifolia	Roots and Rhizome	As a tonic
Corymborkis veratrifolia	fresh leaf juice	to cure fever in children
Cremastra appendiculata	Roots paste	toothache and as emollient
Cymbidium aloifolium	Whole pant with ginger	Vomiting and diarrhea, to cure chronic illness, weakness of the eyes, vertigo and paralysis
Dendrobium nobile	Fresh and dried stems	Preparation of Chinese drug Shih-hu for longevity and as an aphrodisiac, stomachic, analgesic.
Dendrobium jenkinsii	Fresh and dried stems	Preparation of Chinese drug Shih-hu for longevity and as an aphrodisiac, stomachic, analgesic
Echioglossum williamsoni	Leaf juice	Cure swellings of hands and legs and bone fractures
Eria pannea	Roots and leaves decoction	Used in bathing in cases of ague
Goodyera repens reptile bites	Roots and leaves	Female disorders, stomach and bladder diseases,chewed leaves applied to
Goodyera schlechtendaliana	Tincture of the plant in rice wine	Used as a tonic for internal injuries and to improve circulation
Gymnadenia orchidis	Salep made from digitate tubers	Used as aphrodisiac, tonic
Phaius tankervilliae	Paste of Pseudobulbs	Used to heal swellings of hands and legs, poultice
Pholidota imbricata	Pseudobulbs finely macerated in mustard oil	Applied on joints forrheumatic pains
Pleione maculata	Pseudobulbs	Liver complaints and stomachache
Ponerorchis chusua	Tubers	Diarrhea, dysentery and chronic fever
Rhynchostylis retusa	Roots (*Rasna*)	Plant used against Asthma, tuberculosis nervous twitching, cramps, infantile epilepsy, vertigo, palpitation, kidney stone, and menstrual disorders.
Satyrium nepalense	Tubers	Malaria, dysentery, also aphrodisiac
Spathoglottis plicata	Decoction of the boiled plant	Rheumatism and used in hot asa foment
Tropidia curculigoides	Decoction of plant	Cold stage of malaria, decoction of boiledroots given for diarrhea
Vanda coerulea	Leaf juice	Diarrhea, dysentery, external application for skin diseases
Arundina graminifolia	Scrapped bulbous stem	Applied on heels to treat cracks
Dendrobium densiflorum	Leaves crushed to paste with salt	Applied on fractured area to setbone
Cymbidium hookerianum	Seeds	Applied on cuts and injuries as haemostatic

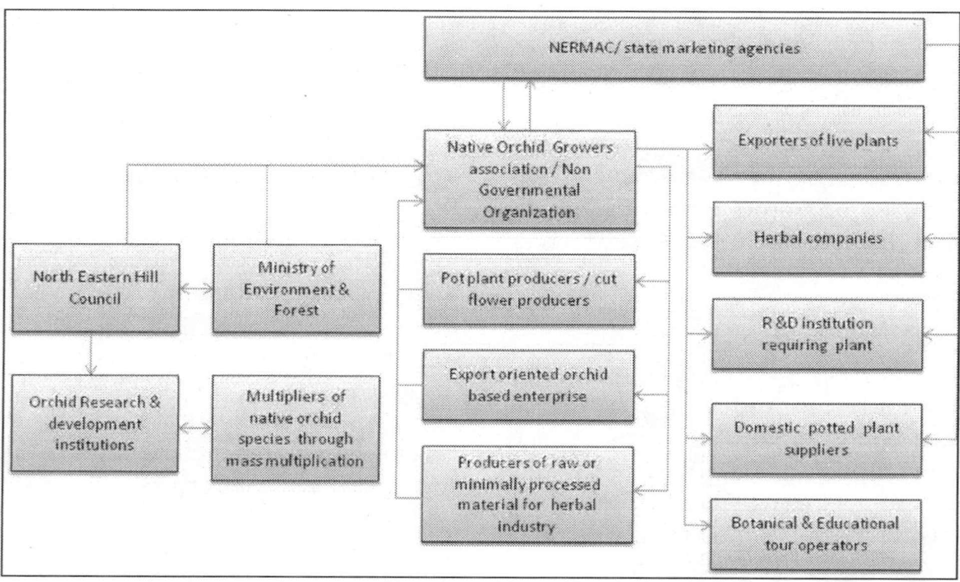

Figure 3: Schematic Diagram for Development of Orchid Biodiversity based Enterprise in North Eastern States of India.

4. Conclusion

The genetic resources of orchids in their natural habitats are declining very fast due to various regions including clandestine trade for various purposes. The orchid biodiversity based enterprises would be helpful in income generation as well as reducing the pressure on natural habitat. As the collection of orchids from their natural habitats still cheaper option for the clandestine traders. The unscrupulous collectors would require to be dealt firmly so that a space for the genuine trades is created.

HDP and Canopy Management

21

Canopy Architecture and HDP in Temperate Fruits: An Innovative Approach for Doubling Farmers' Income

N. Ahmed[1]*, A.H. Pandit[1], J.I. Mir[2] and K. Hussain[1]

[1]Sher-e-Kashmir University of Agricultural Sciences and Technology-Kashmir, Shalimar, Srinagar – 190 025, J&K
[2]ICAR-Central Institute of Temperate Horticulture, Srinagar – 190 007, J&K
E-mail: dnak59@rediffmail.com

The agro climatic conditions in entire Himalayan region which ranges from subtemperate and temperate to cold arid ideal conditions for growing variety of temperate fruit crops like apple, pear, plum, apricot, almond, walnut, pecan nut, kiwi fruit, strawberry, hazelnuts, pistachio nuts and other small fruits and berries. This region which extends from Jammu and Kashmir in north to Arunachal Pradesh in east has a unique and fragile eco-system, where more than 65 percent of people are dependent on horticulture or agriculture and draws about 60 percent of its Gross Domestic Product (GDP) from the surrounding ecological resources.

Before independence, the temperate horticultural activities in the region were insignificant. However, the activities especially in apple and other temperate fruits were initiated after 1950's with the introduction of exotic varieties leading to significant improvement (Tables 1 and 2). In 1960-61 the area under temperate fruits in the country which was just 0.8 lakh hectares increased to 8.35 lakh hectares 2014-15 and production from 3.03 to 39.38 lakh tonnes. The productivity however increased marginally from 3.7 to 4.32 t/ha. As per the dietary recommendations, the modest requirement of temperate fruits as on 2015 is about 50 lakh tonnes which is expected to increase to 200 lakh metric tonnes by the end of 2050 for meeting growing domestic consumption, processing and export market. Among fruits, apple represent major crop accounting for 65 per cent of temperate fruit production while rest of the production comes from other fruits like pear, peach, plum, almond, walnut, apricot, cherries *etc.* which too contribute towards region's economy. The trend in growth of different fruits indicated that apple showed

Table 1: Area (Lakh ha), Production (Lakh tonnes) and Productivity (t/ha) in Fruit and Nut Crops in different States of India during 2014-15

State	*Area*		*Production*		*Productivity*	
	1960-61	*2014-15*	*1960-61*	*2014-15*	*1960-61*	*2014-15*
Jammu and Kashmir	–	3.55	–	20.73	–	5.8
Himachal Pradesh	–	2.20	–	8.66	–	3.9
Uttarakhand	–	1.716	–	6.78	–	4.0
Arunachal Pradesh	–	0.89	–	3.21	–	3.6
Total	**0.82**	**8.356**	**3.03**	**39.38**	**3.7**	**4.32**

reasonable growth both in area and production from 1960-61 to 2014-15 with slow growth in productivity. When we compare our productivity of apple (6.08 t/ha) with world average (15 t/ha) and of other developed countries (50 t/ha), it is very low. Among apple growing states of India, there is a large variation in productivity ranging from just 2.2 t/ha in Arunachal Pradesh to 10.2 t/ha in J&K. Due to increasing demand, apple is now being imported especially during off-season and is increasing every year. This needs to be reversed by increasing total production, productivity and quality innovative approaches of canopy architecture and high density planting with suitable rootstock varieties and effective technologies.

Table 2: Area (000 ha), Production (000 MT) and Productivity (MT/ha) in Apple in different States of India during 2014-15

State	*Apple*		
	Area	*Production*	*Productivity*
Jammu and Kashmir	160.9	1647.7	10.2
Himachal Pradesh	107.7	738.7	6.9
Uttarakhand	30.0	77.5	2.6
Arunachal Pradesh	14.3	31.9	2.2
Others	0.2	2.0	8.5
Total	**313.0**	**2497.7**	**6.08**

In India, there is increase in area consistent under the temperate fruit crop but the production, productivity and quality has not shown a proportional trend which is very low in comparison to other advanced countries due to many factors especially the non availability of quality planting material; traditional orcharding system involving less yielding varieties, monocropping and low density plantations; poor canopy, nutrient and water management; high incidence of insect-pests and diseases and heavy pre- and post-harvest losses.

Among several factors, the poor canopy management and traditional low density orcharding are the important factors leading to low productivity in temperate fruit crops. In the absence of full canopy and high density planting, the full potential of the trees cannot be exploited as they have direct bearing on productivity and quality.

Canopy management is the manipulation of tree canopy to optimize its production potential with excellent quality fruits. In fruiting tree, the fruiting potential is largely governed by its architecture, canopy density and photosynthetic efficiency (Kallow *et al.*, 2005). Where, the term canopy refers to the physical composition of plant comprising of mainly stem, branches, shoot and leaves, while the canopy density is determined by considering the number and size of leaves. Its architecture is determined by the number, length and orientation of stem, branches and shoots while canopy volume is determined by spread and height of canopy (Pathak, 2009). It is one of the predominant technologies by which huge and unmanageable trees are properly managed to make them more productive. Canopy management is based on the choice of tree size, number of tree per hectare and pruning technique, so as to optimize the production of assimilation and its conversion into economic yield. This is only possible with the management of canopy to intercept maximum light in the innermost parts of trees. The canopy management has direct correlation with dry matter production, flower bud initiation and fruit quality.

Essential Features of an Ideal Canopy

Some of the features are:

- Strong frame of primary branches.
- Wider crotches in scaffold branches.
- Healthy and well distributed secondary and tertiary branches.
- Sufficient fruiting terminals in most productive areas.
- Healthy foliage with high photosynthetic efficiency to maximize the solar radiation use efficiency.
- Enough space for air circulation inside canopy.
- Should support adequate to nurture and protect the fruits from sun burn.

Objective of Canopy Management

The basic objective of canopy management is to maximize light interception, to optimize light distribution within canopy and to maintain proper airflow. Canopy management enhances productivity, improves fruit quality, facilitates cultural practices and help in management of pest and disease. In new plantations initial training and pruning is given to develop strong frame work of the tree whereas in old plantation the aim of canopy management is to reduce tree height and make provision of solar radiation inside the canopy by thinning excessive bio mass. The canopy management is also useful in rejuvenating old orchards, increasing input use efficiency, reduced cost of production and facilitates mechanization.

Principle of Canopy Management

The basic principle of canopy management of the fruit trees is to make the best use of the land and the solar radiations for increased productivity. Light is fundamental aspect of fruit production. The canopy management optimizes the plant model in such a form that it intercepts maximum light. In many fruit crops, improved production and fruit quality has come from producing more fruit from smaller trees as small trees are more efficient and better at capturing and converting sunlight into fruit than larger trees and thus need to be considered when developing new planting system.

Fruit production involves the capture and conversion of light energy into fruit biomass (dry matter). The main controlling factors are (i) the amount of incoming radiation, (ii) the percentage of that radiation which is intercepted by the tree, and (iii) how efficiently the tree converts that energy in to fruits. Both climate and length of growing season determine the amount of incoming radiation. The grower can control neither of these, except the initial choice of orchard site. The length of growing season is also important in deciduous plants which shed their leaves in autumn and thus have a finite growing season. The percentage of incoming radiation intercepted by the tree is where growers have most control so decisions relate to block size and planting density is important. In small orchard blocks the surrounding shelter trees shade a greater proportion of the trees than they do in larger blocks. The situation is similar with high density plantings where shading from adjacent trees results in large volumes of shaded canopy and leaves operating well below their photosynthesis potential. The high density is thus determined more by tree size than by planting distance. In low density plantings high percentage of incoming radiation is generally captured by the grass growing between the tree rows and not by the fruit trees. Interception of light by total orchard canopy is therefore very important and is correlated to total biomass production. In tree canopy development two aspects are important firstly the interception of available light by the total orchard canopy which is correlated to total biomass production of the crop system (Monteith, 1977). Secondly, penetration and/or distribution of light within a single tree canopy which must be optimized for maximum tree performance and crop development, while Shikhamany (2001) indicated that principles of canopy management lies in (i) maximum utilization of light, (ii) avoidance of the build-up of micro-climate congenial for the disease and pest, (iii) convenience in carrying out the cultural operations, (iv) maximizing the productivity and quality; and (v) economy in obtaining the required canopy architecture.

Light Interception

Light interception is the amount of available light intercepted by the tree canopy and not striking the orchard floor. Interception then is a function of plant density, canopy shape and size, and leaf area index (LAI, m^2 leaf area/m^2 ground area) within the canopy. Increase in biological yield has been attributed to enhancement in the amount of seasonal total light interception with improved cultural practices that affect leaf growth, leaf duration and/or canopy display (Gifford and Jenkins, 1982). The total amount of light intercepted by a fruit orchard system depends mainly on orchard design factors such as planting system, tree spacing, tree shape, tree height, alley width, row orientation aswell as leaf area index (ratio between the total area of all leaves per tree to the land area allocated to the tree) and the length of the growing season (Jackson, 1980; Lakso, 1994; Palmer, 1989; Wagenmakers, 1991).

Light Interception Influenced by Orchard Canopies and Plant Density

Orchard canopies, unlike those of other crops, due to the alleyways maintained between rows, a large proportion of the land area not covered by trees (Jackson, 1980). This distribution results in low values for LAI and total light interception for orchard canopies. Light interception is low when the orchard is just planted and increases as the orchard develops in relation to total LAI. Cain (1970) introduced the idea that orchard density should be considered over the lifetime of the orchard. Since significant land and light resources are wasted in the early life of an orchard the planting of higher and higher tree densities is therefore encouraged in 'Golden Delicious' low-density hedge row plots (1100 trees ha^{-1} on M.9 and 660 trees ha^{-1} on M.2) intercepted roughly half of the incident light at maturity and yields peaked at 40 t ha^{-1} Whereas, the high-density hedge row plots (3300 trees ha^{-1} on M.9 and 2260 trees ha^{-1} on M.2, intercepted two-thirds and three-quarters of available light, respectively and yielded more than 70 t ha^{-1} in their 6th and 7th years (Verheij and Verwer, 1973). Likewise, in apple cv. 'Golden Delicious' with plant densities ranging from 853 to 3746 trees ha^{-1}, yield and light interception were approximately linearly related and light interception is related to LAI, not just planting density (Palmer and Jackson, 1974). In discontinuous orchard canopies LAI is the largest single factor influencing light interception (Jackson *et al.*, 1977). Increasing tree planting density has been the most important means of increasing the early yield and early light interception of young orchards (Wertheim *et al.*, 1986). In the same way, Barritt (1989) reported that tree density was more important than training system or rootstock for improving light interception and yield of 'Granny Smith' in the 3rd year. Pruning generally reduces total light interception because of reduced growth and canopy size. Summer pruning has been shown to improve light penetration within tree canopies (Marini and Barden, 1987). Light interception varies during a season due to leaf surface area development and solar angle, and during orchard establishment due to tree size increase. Light interception during the season closely follows leaf area development (Palmer and Jackson, 1977). Total light interception by north-south rows did not vary after complete canopy formation (Jackson and Palmer, 1972). From the above findings it can be concluded that in early life of the orchard, light interception is a function of tree planting density whereas in mature orchard, light interception is a function of the tree height alley width ratio and the density of the canopy. Thus light interception in orchards can be raised by increasing the number of trees ha^{-1}, increasing the height of the trees relative to the clear alley width, increasing the density of foliage in the canopy (Corelli and Sansavini, 1989), orienting rows in north-south direction (Wunsche and Lakso, 2000) and by summer pruning.

Light Distribution

In an ideal orchard, light should be evenly distributed under tree rows and alley's and should be around 60-70 per cent. Light penetration into the tree is usually determined by the size and shape of the tree.

Tree Form and Light Distribution

Two approaches have been used to improve the light distribution in tree canopies. One is to use relatively natural tree forms that allow light penetration through the canopy by providing many small openings in the foliage such as in the multiple leader, central leader, vertical axis, or slender spindle forms (Wertheim, 1968; Lespinasse and Delort, 1986). A second approach is to provide fewer large, permanent openings for light penetration into canopies restricted into geometric forms. Examples are thin restricted planes of foliage such as narrow hedgerows, tree walls, and A, V, or T forms (Chalmers and van de Ende, 1975).

Effect of Light Interception and Distribution on Fruit Plants

The productivity and quality of fruits largely depends upon tree vigour and weather parameter particularly light interception, temperature, relative humidity and air flow. The proper sunlight exposure of foliage and fruits and prevailing temperature play vital role in vegetative growth, photosyntheyic efficiency, flower initiation, fruit set, fruit bud differentiation, ripening, fruit colour, fruit quality *etc.*, while increase air flow reduce disease pressure. Fruit trees where there is a higher proportion of 'shade' to 'sun' their leaves were found photosynthetically inactive under deeper shade and act as unproductive sink. Shaded leaves are thinner and contain less palisade layers compared to control leaves. For the induction of flowering, certain level of irradiance and photosynthates accumulation are needed.

The size and colour of the fruit is associated with light level of more than 70 per cent of full radiation. Large and best coloured fruits are borne near the outer periphery of the canopy where irradiance is high, whereas smaller poor colour fruits are borne in the shady part of the tree (Awasthi and Mehta, 2004). Tree canopy receiving less than 30 per cent light produced virtually no flower. Reduction in light from 100 to 70 per cent in tree canopy had a small reduction in flowering, whereas increased shading increased fruit shedding and reduced flower set in the following season. Jackson and Palmer (1977) observed 36 per cent reduction in the number of flowering spurs when light to Cox's Orange Pippin tree was reduced to 37 per cent by shading in early June to leaf fall. The increased shading led to progressive reduction in flower bud production and further shading decreased, fruit weight and produced symptom of fruit immaturity, *e.g.* decrease in fruit colour, fruit dry matter, fruit soluble solids but increased fruit firmness (Lakso, 1994). The decline in fruit quality with shading within tree canopy has also been reported in peach and cherry (Flor and Kesner, 1982). Proietti and Tombesi (1996) opined that shading reduced flower by about 50 per cent and starch content of root tissues. Low irradiance reduced fruit set in open-pollinated flower of persimmon (George *et al.*, 1993).

Canopy Management Techniques

Canopy management reflects an inter-relation of the physiology underlying the relationship between growth and production. Orchard architecture largely depends upon the orchard production system, which is a combination of variety, rootstock, tree spacing, training and pruning. To optimize the utilization of light for increased yield of quality fruits, canopy management deserves greater attention by exploiting the various available techniques like training, pruning (dormant, summer and root pruning), braches orientation (bending), scoring, girdling, bark inversion, selection of proper rootstock, use of plant growth regulators, appropriate use of fertilizer, deficit irrigation, use of genetically engineered plants with altered architectural characters would help in maintaining the ideal canopies of trees.

Training

Basic objectives of training is to control vegetative vigour, increase light exposure for fruit/foliage, increase air flow to prevent disease infestation and facilitate pruning and harvest methods and mechanization. Whatever system is chosen, the objective of training and pruning is to achieve maximum tree life and productivity. In training we should consider (i) ground clearance, (ii) shoot orientation, and (iii) ventilation and light interception into canopy. In order to harvest the advantage of light and proper air circulation, training is performed in the initial years by maintaining desired shapes and strong framework. While deploying the training systems, utilization of vertical and horizontal space should also be taken into consideration. The choice of training system is part of the whole orchard management influencing planting distances, light interception and finally the success of the orchard in terms of obtained fruit quantity and quality (Palmer, 1999).

Pruning

The pruning like training is an important operation performed every year in a tree to improve both the fruit crop and tree appearance but it is a practice frequently neglected and misunderstood. Pruning is defined as the judicious removal of parts to obtain good and qualitative yield. The ultimate aim of pruning is to develop a canopy, which intercepts maximum light, restore root to shoot ratio, develop specific geometry within the plant and its overall topology, maintain optimum balance between vegetative and reproductive parts and maintains dwarfing and promote maximum shoots/ fruiting points. Pruning is affected both during dormant season as well as in summers as dormant pruning and summer pruning.

Dormant Pruning

All pruning's have a dwarfing effect, but dormant pruning produces the most new growth, it reduces trunk size and canopy volume as compared to the same tree that is not pruned. Miller and Byers (2000) reported 21 per cent reduction in canopy volume when peach trees were severely pruned. Dormant hedging or shearing has shown success in controlling tree size, but often results in a proliferation of shoots over the periphery of the tree resulting in very dense tree canopies (Forshey *et al.*, 1992). The effect of dormant pruning may seem to have stimulated growth, but the total dry matter produced by the pruned tree is less than the non-pruned tree (Myers and Saville, 1996). Therefore, to avoid localized and invigorating effects, it is recommended that pruning consist primarily of thinning cuts and few heading cuts be used (Marini *et al.*, 1993).

Summer Pruning

Summer pruning in combination with dormant pruning reduce light interception by 15 per cent (Wertheim *et al.*, 1986) but summer pruning increase photosynthetic photon flux density (PPFD) within the tree canopy (Morgon *et al.*, 1984). According to Marini and Barden, (1987) summer pruning increased light penetration within the tree canopies, but may decrease levels during the following year (Taylor and Ferree, 1984). Summer pruning improves fruit color for coloring varieties grown in climates where color development is less than ideal and also increases size and weight of fruits along with soluble solids. Trimming of apricot shoots is an interesting approach to avoid having the productive area shift to the canopy periphery. The early pruning intervention increased the number of flowers per tree from 2,212 (June 6th) to 2,524 (June 20th) compared to the un-pruned control that exhibited only 1,938 flowers. Pruning in early July (4th) induced a 0.8 kg per tree increase in yield compared with the unpruned control. Despite the same number of flowers/tree, fruit mass of the early July pruning was 10 g less than in fruits of the unpruned control trees (Musacchi *et al.*, 2014). In peach summer pruning is useful in traditional open centre trees to increase fruit size, colour and soluble solids and to reduce pre-harvest rots. However, in the perpendicular V, summer pruning is essential to maintain fruiting wood and fruit quality along the scaffolds.

Use of Dwarf and Semi-dwarf Rootstocks

In temperate fruits species, two main factors that modulate tree architecture and overall tree size. The first one is the genotype, which affects branching density, proportion of short vs. long shoots and flowering pattern (*i.e.*, lateral vs. terminal flowering). A second factor is the root system, which has been used as an efficient although empirical means to control tree size, with variable results on flowering (Lauri *et al.*, 2006). The root system, from "dwarf" to "vigorous," strongly influences overall tree size, with own-rooted trees generally equivalent to or larger than trees of the same genotype grafted on the most vigorous clonal rootstocks (Quamme and Brownlee, 1993) or on seedling rootstocks (Zimmerman and Miller, 1991). Trunk cross-sectional area (TCSA) and branch cross-sectional area (BCSA) are the main traits used to characterize the influence of the root system on tree growth (Moore, 1978). Dwarfing rootstock reduced tree size and increase the partitioning of dry matter to fruit production and also results in precocious flowering. Rootstock control tree size through a direct effect on growth and indirectly through enhanced crop load. With the advent of more dwarfing rootstocks, the temperate fruit industry went through a revolution in training systems especially in apple. The most dwarfing *i.e.* M.27, V.3, G *etc.*, dwarf, *i.e.* M.9, P.22, P.2, B.9, B.146, B.49, Mark, MAC.39, G.16, G.65, OAR1, Bemani, 0.3, V.2, J.9, F *etc.* and semi-dwarf M.7, MM.106, B.490, G.13, G.30, 0.8, H *etc.* are available for commercial high density planting in apple. However, among different dwarf rootstocks M.9, B.9 and Mark are most popular rootstocks. Most of dwarfing and Semi-dwarfing clonal rootstocks have poor anchorage due to brittle roots and shallow root system, requiring staking or trellis wire support. Use of dwarfing interstock, provides best means of overcoming this problem. Generally dwarfing rootstocks, M.8, M.9, M.27, C.6, B.9 and Mark –are used as interstocks to control tree size. It allows the use of well-anchored and strong root system of seedings or vigorous clonal rootstock and dwarfing effect of interstock. The extent of dwarfing effect and precocious bearing depends upon length of interstock, according to Kanwar (1987) a minimum of 15 cm interstock is necessary to produce some effect. Tustin *et al.* (2001) conducted field trail in apple cv. Fuji trained on Slender Pyramid system utilizing medium vigour (MM. 106, 5.0 × 3.0 m), semi-dwarf (M.26, 4.0 × 2.0 m), or dwarf (Mark, 4.0 × 2.0 m) rootstocks. They concluded that light interception was up to 50 per cent less in the Mark dwarf tree system. Further, continued canopy growth in the systems on MM.106 and M.26 resulted in higher yields than from the Mark planting system from the fourth year onwards. Peak yields within 6 or 7 years reached 110 t ha^{-1}.

The most common dwarfing rootstocks for pear are Old Home x Farmingdale (OH x F) and quince (*Cydonia oblonga* L.) selections that result in trees about 50 to 70 per cent of a standard size pear tree (Alonso *et al.*, 2011).Trees on quince rootstocks are reduced in vigor, precocious and productive in cropping and, in most cases, bear fruits of good size and quality. Pear growers in Western Europe use trees propagated on Quince C, Adams, A or BA.29 clone of quince rootstock (Webster, 2002). On the basis of TCSA, canopy volume and pruning weight Massi *et al.* (2008) ranked different pear rootstock in order of decreasing vigor: Kirschensaller > Farold 282 = FOX 11 > Farold 87 = FOX 16 > Farold 40 > Sydo = EM A > BA 29 > Adams > EM C.

Cherry rootstock Mahaleb is about 75-90 per cent of Mazzard size and is often called "semi-dwarf", simply because it is smaller than Mazzard. The introduction of the interspecific hybrid Gisela (Gi) rootstocks, largely based on crosses between *P. canescens* and *P. cerasus*, provided new important traits of precocious flowering, reduced vigor, and high productivity (Lang, 2000). Gisela rootstocks now available commercially include Gi3 (very dwarfing), Gi5 (dwarfing), Gi6 (semi-vigorous), and Gi12 (semi-vigorous) (Lang *et al.*, 2014). There are really no commercially utilized dwarf or semi-dwarf rootstocks for peach, plums, prunes and apricots.

Use of Plant Growth Regulators

Phytohormones play a major role in controlling many physiological processes of plant growth and development. They regulate division, expansion and differentiation of cells thus determining plant morphological traits. It was shown, that different amounts of phytohormones are found in the shoots and leaves of apple trees depending on their canopy architecture (Watanabe *et al.*, 2008).

Induction of Feathers (Lateral Branching)

The development of branched trees in the deciduous tree-fruit nursery has become more important as availability of dwarf rootstock has made higher-density plantings commercially feasible. Planting branched (feathered) apple, pear and sweet cherry trees from the nursery improves total production earlier in the life of orchard. Lateral branch development in vigorously growing nursery fruit trees is under the control of a physiological system termed "apical dominance". Apical dominance is driven by inhibition of lateral bud development by auxin, while cytokinin promotes growth of axillary buds, and these hormones influence each other (Leyser, 2009). The relative amounts of each type of hormone, and hence the relative degree of inhibition or stimulation of lateral-bud growth, can be altered by environmental factors as well as by various cultural practices, such as exogenous applications of bioregulators or pinching techniques such as leaf removal and notching (Ono *et al.*, 2002; Sazo and Robinson, 2011). Unfortunately, pinching and notching are inconsistent and most of the nursery plant requires more than one "leaf pinching" to maintain low auxin content due to rapid stem elongation. The use of plant growth regulator chemicals has been seen as a more consistent and less expensive method to induce lateral branching.

Among the many bioregulator products tested over the past half century, cytokinin especially, 6-benzyladenine (BA) and gibberellic acid especially (GA_{4+7}) and combination of these two- (a mixture of 50 per cent BA and 50 per cent GA_{4+7}) are the most common plant growth regulators that promote formation of lateral branches. They are successfully applied for different fruit crops. BA is responsiblefor overcoming apical dominance (Muller and Leyser, 2011) and enhancing feathering in apple nursery trees (Sazo and Robinson, 2011), while the main role of GA is to elongate feathers (Hrotko *et al.*, 2000). BA is also used in young trees newly planted in orchards to induce proleptic shoot formation and to weak acrotony (Buban, 2000). Recently, Cyclanilide has been tested for growth-related effects on nursery apple, pear, sweet cherry and plum trees and found promising (Magyar and Hrotkao, 2002; Elfving, 2010; Sazo and Robinson, 2011; Robinson and Sazo, 2014).

Chemical Control of Vegetative Growth

Typically, tree growth management is performed by physical measures, particularly by dormant and summer pruning. In extreme cases, root pruning, girdling or even stem sawing is being used. Any of these methods is cost-intensive and/or bears a high risk of failure. Furthermore, part of the trees assimilates or potential assimilates are lost. Alternatively, an improved balance between vegetative growth and fruit formation may also be achieved by employing plant growth retardants. Historically, daminozide has been most successfully applied in apples for growth control. However, relatively high levels of residues with potentially critical toxicological features have led to a general ban of this compound for use in edible crops. The registration of chlormequat chloride, particularly value in pears, has recently been terminated in many countries because of non-tolerable residues resulting from too intensive use (Frank Maas, 2006). Paclobutrazol is registered only in some countries for growth control in fruittrees. The problems of this compound are particularly related to its extreme persistency and its strict acropetal movement in treated plants, which make adequate treatments relatively difficult. Residues in fruits and flattened fruit shapes are additional obstacles to a wider use of this compound (Rademacher, 2004). Ethylene, released from ethephon and compounds inhibiting the formation of growth-active gibberellins (GAs) would be candidates for shoot growth control. Ethephon is required at dosages, which, in addition to growth reduction, mayinduce fruit drop, precocious ripening or reduce alternate bearing.

Recently prohexadione-calcium (calcium 3-hydroxy-5-oxo-4-propionyl-cyclohex-3-enecarboxylate, Pro-Ca) has been introduced (trade names: *Apogee*® and *Regalis*®) in a number of countries as a plant bioregulator for pome fruit trees may offer an attractive new solution for the chemical control of excessive shoot growth. Several reports are available dealing primarily with the effect of Pro-Ca on vegetative growth and fruit formation (Byers *et al.*, 2004; Costa *et al.*, 2004; Rademacher *et al.*, 2004). It is relatively short-lived and only low, if any, residues are detected in fruits from treated trees. Pruning costs in treated trees are reduced by approximately 30 per cent. Often, higher yields are obtained as a result of increased fruit set and the availability of additional assimilates for fruit growth. Likewise, improved fruit quality and fruit storing quality are observed in many instances (Rademacher, 2004). Surprisingly, trees treated with Pro-Ca turned out to be less infected by fire blight (*Erwinia amylovora*) (Norelli and Miller, 2004) and apple scab (*Venturia inaequalis*) (Costa *et al.*, 2004).

5. Alteration in Canopy Architecture through Genetic Engineering

Engineered trees with altered architectural characteristics should enable substantial advancements in productivity, and sustainability. By altering gene in temperate fruit we can alter their growth habit and impart dwarfness. Over-expression of rol (root loci) genes in plant has demonstrated a drastic effect on the plant architecture. The *rol* genes from *Agrobacterium rhizogenes* have been popular transgenes for rootstock. It reduce plant height, shortened internode and increase rooting number and percentage in apple rootstock (Sedira *et al.*, 2001; Igarashi *et al.*, 2002) and kiwifruit (Rugini *et al.*, 1999). Cytokinin-over-expressing transgenic peach plants containing the *Agrobacterium tumefaciens* are dwarf in stature and produce more branches (Hammerschlag and Smigochi, 1998). Correspondingly, introduction of tumour-inducing (T)-DNA from *Agrobacterium rhizogenes* in Colt rootstock of cherry produced trees with significant reductions in plant height, internode length, leaf area and petiole length (Gutierrez-Pesce *et al.*, 1998). Likewise, Negri *et al.* (1998) reported that the rice *phytochrome*A (*phy*A) gene had dwarfing effect on Colt rootstock when over-expressed in transgenic. Srinivasan *et al.* (2012) introduced FT gene form *Populus trichocarpa* in European plum (*Prunus domestica* L.) cv. Blue Byrd. They reported that FT plum lines expressed pleiotropic phenotypes that rendered them different from normal plum trees. They did not grow with a typical upright tree growth habit, but instead became highly branched resulting in a shrub-type growth habit. Further, they found that the architectural phenotypes of increased lateral branching (*i.e.* bushy) were not correlated with

Table 3: Some Important Dwarfing Rootstocks and Interstocks Used in Temperate Fruits

Fruit(s)	*Rootstock(s)*	*Salient Features*
Apple	EMLA 111/MM111	Suit to drought prune areas
	EML-7	Suit to sloppy, virgin lands, semi vigorous
	EMLA.106/MM-106	Suit to sloppy, and less clay soils, semi vigorous
	EMLA-9/M9	For high density planting with assured irrigation and deep fertile soils, very dwarf
	M779	For hilly areas of Uttarakhand and HP
Pear	Quince -A	Standard rootstocks, semi vigorous
	BA 29C	Standard rootstocks, semi vigorous
	Quince C	Very dwarf
	Quince-B	Semi vigorous
Apricot	Apricot seedling	Vigorous, drought tolerant and compatible
	Peach seedling	Suitable for dry and light soils
Peach	Peach seedling	Vigorous and compatible
	GF-557 and GF-677	Drought tolerance
	Siberian-C, Rubia	Dwarf
Plum	Wild Peach X Apricot seedling	Seedling are vigorous
	Myrobalan-B,Myrobalan-29C, Myrobalan-GF-31, Myrobalan-2621, GF-8/1	Clonal and semi vigorous, compatible rootstocks
	PIXY X ST. Julian-K	Dwarfing rootstocks
Cherry	Peach X Apricot seedling	Semi vigorous
	COLT	Semi dwarf
	F-12/1	Vigorous
	Mazzard and Mahaleb	Vigorous
Almond	Apricot, peach and almond seedling	Vigorous
	Peach and almond hybrids GF-557 X GF-677	Semi dwarfing, good for high density
Walnut	Walnut seedling	Vigorous

higher PtFT1 expression but these traits require only low levels of PtFT1 expression. Shrub-like growth habit of FT plum potentially makes them suitable for novel production systems.

6. Influence of Rootstock on Canopy

Rootstocks have been extensively employed as a means to regulate canopy vigour. It influences the size and shape of canopy of scion grafted on them and maintains shallow canopy or thin hedgerow which can be adequately illuminated throughout its value. In temperate fruit a wide range of vigour controlling rootstock are available by which large number of plants per unit area can be accommodated, which intercept maximum solar energy, heavy yield and quality.

Orchard Planting System on Light Interception and Distribution

Interception and distribution of light within tree canopies be determined by various inter-related factors. One factor is genotype of cultivar or architectural pattern ranging from the non-branching, columnar type to the spreading, non spur type. Further, a given cultivar rootstock also play an important role *via* the canopy volume. A second factor is the planting system which included tree arrangement in orchard (planting spacing, and orientation of row) and training of canopy by pruning as well as limb orientation (height and shape of tree). Training partly defines the proportion and distribution in space of the various shoot categories, long versus short shoots and vegetative versus fruiting shoots (Robinson *et al.*, 1991). Since whole-tree photosynthesis is primarily light limited, thus manipulation of canopy architecture to enhance global light interception is a major objective of all planting systems

On the basis of tree canopy shape Robinson (2003) categorized orchard planting systems in four basic shapes: spherical canopy shapes, conic canopy shapes, flat-fan canopy shapes and Y or V canopy shapes.

Spherical-Shaped Canopy Systems

Traditionally, in this form of canopy shape, the trees are globe shaped with 6-8 m height and 2 m trunk length. According to (Wertheim, 1981) after development of clonal rootstocks in apple in the 1930s, a smaller version of the large globe-shaped tree on semi-dwarf rootstock developed in Holland and England, named the bush tree (the bush-tree system) and on dwarf in Germany named Spindle-bush system. The main advantage of spherical shapes is that they are generally the natural shape, however, the canopies generally have a large shaded centre core that is unproductive and produces poor fruit quality (Robinson, 2003).

Conic-Shaped Canopy Systems

As large globe shaped tree result in less light distribution within the canopy, many researchers proposed conic or pyramidal canopy. In this tree form several planting systems developed like Central-leader system, Mini-central-leader system, Palmette-leader system, Slender-spindle system, North Holland spindle, Slender-spindle multi-row or bed system, Vertical-axis system, SolAxe system, Slender-pyramid system, HYTEC system, Super-spindle system *etc.* and now it is a dominant tree form in commercial orchards in most part of the world. The primary advantage of this tree from are that it is a natural form with high light interception and distribution in initial age of orchard establishment. However, when mature, the upper branches of the tree begin to outgrow the lower branches, causing excessive shade on the lower canopy. Keeping this in view later many new systems have been developed to increase light interception and distribution in conic shape tree form.

The conic shaped systems have chosen by majority of apple growers all over the world. In the 1960s and 1970s, the central-leader system was dominant system in the world while in the 1970s and 1980s, the slender spindle and the vertical axis were the most common systems in Europe. At the end of the 20th century, the most common systems are the slender spindle and its variants and the vertical axis and its variants. Over the last decade, the height of most slender-spindle trees has increased to close to 2.75 m and thus the trees are more similar to vertical-axis trees. This has allowed slender-spindle orchards to intercept more light and achieve higher yields, while reducing the amount of pruning required to contain the tree to a short stature of 2 m. Similarly over the last decade, the planting density of vertical axis orchards has increased from 1,500 to more than 2,000 trees ha^{-1}, which is similar to the density of slender-spindle orchards (Robinson, 2003).

Flat Planar Canopy Systems

From very early, European apple growers have utilized trellises to confine tree canopies to a two dimensional plant and developed many geometric shape of apple trees (Huggard, 1980) and among them espaliers was most commonly grown along walls and fences. Further, in the mid-1950s Italian fruit growers developed first commercial apple orchard with two dimensional plane canopy to improve labour efficiency and named as palmette training system. After that this system widespread in Italy and further, significant technical innovation in pruning of this system lead to hasten fruit bearing and reduce pruning costs and increased planting densities to secure the benefits of earlier and greater returns (Sansavini, 1983). With the time, a number of variations of this system have been developed by using dwarf rootstocks, *i.e.* 'Regular' palmette system, 'Free' palmette system, Penn State thin-wall trellis system, Lincoln canopy system, Ebro trellis system, Solen system *etc.* The major advantages of flat planar systems are that it improve labour efficiency and allow good light distribution to all parts of the canopy.

V-Shaped Canopy Systems

V-shaped fruit trees were popularized for commercial orchards by Chalmers and Van Den Ende (1975) with the growing system known as the Tatura trellis. This system was originally developed for peaches to increase yields which allowed mechanical harvesting but was later adapted for apples (Chalmers *et al.*, 1978). With the development of dwarf rootstock and to improve light interception and penetration to the centre of the canopy, a number of variation of the Tatura trells have been developed *i.e.* Mini-Tatura trellis system, Geneva Y-trellis system, Mikado and Drilling systems, MIA trellis system (A-shaped trellis), Mini-V-trellis system, Gutingen V slender-spindle system, *etc.* The primary advantage of V system has high light interception that result in high yield (Robinson and Lakso, 1991).

Training Systems and Planting Densities for Rootstocks

There is single planting system to suit all situations. Factors such as soil, cultivar, rootstock, management regimes and socio-economic conditions will determine the optimal combination for each orchard. Here are some most common systems with plant density and rootstock which are being used in India as well as across the temperate regions of world is presented in Table 4.

Effect of High Density Plantation and Canopy Architecture on Fruit Yield and Quality

High density planting system is first pioneered for temperate fruits in Europe in 1960's in which higher number of plants are accommodated within a unit area in comparison to the conventional planting density. The high density planting envisages the planting of small trees densely, restricted their vegetative growth by use of dwarfing rootstocks, growth retarding chemicals, pruning and thereby diverting much of the plant energy to the economical parts. Close planting of the tree and development of dense canopies may alter the, micro-climate around the tree canopy. The best principle advantage of high density planting is better utilization of incident solar radiation and increase in bearing surface

Table 4

Training System	*Plant Density (trees/ha)*	*Commonly Used Rootstock(s)*
Central Leader System	300-700	MM.106, MM.111, M.7
Modified Central Leader System	625-1100	–do-
Open Centre System	400-625	–do-
Mini-central Leader System	500-1000	–do-
Dwarf pyramid	4,444-5,000	M.9
Head and spread	625	MM106
Palmette-Leader System	1,500-2,000	M.9
Spindle Bush	2,000, 2,222	MM.106, M9
Vertical Axis	1,000-2,500/4,444	M.9, M.26M.9
Bi-axis	2500-3,300	M.9
SolAxe system	2,250-5,500	M.9
Slender Spindle	1500-4000	M.9, B.9, G.16
Slender-pyramid system	600	MM.106
	2000	M.9
Super Spindle	4,000	M.9
Tall Spindle Planting System	2,500-3,300	M.9
Cone system	625	MM.106
Trellis system	2,222	M.27, Bud.9, M.9, or M.26, pears on Quince C rootstock
Palmette system	7,00-1,500	MM.106, M.26, MM.111.In pear Quince A is suitable
V or Y shaped canopy systems	2,976-7143	M.9
V hedge	2,057	M.9
Open Tatura Trellis (OTT)	2,000-5,000	M.9
Cordon system	2,000	M.27, M.9, M.26 Pears are best grafted on Quince C or, for very poor soils, Quince A.

per unit land area. Tree canopy architecture however plays an important role in adjusting trees under HDP. Traditional system can easily accommodate trees trained as open centre system, central leader system or modified central leader system. But in modern orcharding system involving high density plantations more number of trees need to be accommodated per unit area and therefore need architectural engineering of plants *viz.*, Espalier system, v-trellis, Y-trellis, vertical axis, biaxis cordon, solaxe, spindle bush, slender spindle,tall spindle, head and spread *etc.* These systems can accommodate more number of plants and allow sufficient light to pass and diffuse for production of quality apple.

Apple

In apple, Palmer (1980) worked out LAI for three different training systems, namely, slender spindle, palmette and full field, capable to ensure adequate distribution of intercepted light within 90 per cent of the canopy. The LAI value for three systems were 1.6, 1.6 and 1.9 respectively, and the fraction of total intercepted light was 53 per cent for both the slender spindle, palmette and 66 per cent for full field. Light sharply decreased inside the first 0.5 to 1 m from the canopy edge in Delicious apple trained on central leader (Barritt *et al.*, 1987). Ferree (1980) observed that cumulative production in the sixth year for central leader was 41 per cent of the field for the slender spindle in Golden Delicious. Dallabetta *et al.* (2014) carried an experiment at Italy to compare Spindle training system of training, with Bi-axis, on two main cultivars *i.e.* 'Galaxy Evolution' and 'Pink Lady'. Where, 'Galaxy Evolution' orchard was established in 2005 at a planting distance of 3.45 x 1 m for Spindle (2,900 trees per ha) and 3.45 x 1.2 m for Bi-axis (2,400 trees per ha). While, the 'Pink Lady' orchard was established in 2008 at a planting distance of 3.5 x 0.8 m (3,570 trees per ha) for both training systems. They found that "Bi-axis Galaxy Evolution fruit" showed higher size in the lower level as compared to Spindle ones. Fruit size in 'Pink Lady'was significantly higher in Spindle as compared to Bi-axis while fruit colour was not affected. In general, they concluded that fruit harvested from "Bi-axis" resulted more homogeneous than "Spindle" training system. In apple the Tall Spindle system is more profitable than either higher (3,300 trees per ha) or lower density systems (2,500 trees per ha), early yield has exceeded 150 per ha over the first 5 years (Robinson *et al.*, 2014).

Apple trees 'Royal Gala'/M.9 were trained in five ways-Slender spindle (SS), Geneva Y-trellis (GY), a modified Solen training- 'Solen Y-trellis' (SY), or V-trellis(LDV), all at the same spacing (1.2m × 2.8 m), giving a planting density of 2,976 trees per ha. In addition, a higher density (7,143 trees per ha) version of the V-trellis (HDV). After 8 years, the two Y-shaped training systems had 11-14 per cent greater cumulative yield per ha than the SS. Relative to the LDV, the HDV yielded less per tree, but far more per ha, particularly in the first 3 years. After 8 years, the cumulative yield per ha was still 65 per cent greater than with LDV, further, the HDV intercepted more light (73 per cent) than SS (53 per cent) (Hampson *et al.*, 2002). Higher and better quality yields (160.6 t per ha) obtained from 7 year old non spur cultivar Jonagold/MM-106 planted at 4.5 x 2.5 m in cone training system compared to the yields (111.6 t per ha) harvested from trees trained to the free growing spindle type bush system (Gandev, 2009). At CITH, Srinagar a productivity of 55-66 t per ha was recorded in apple with vertical axis, espalier and cordon system of training system. High yield obtained in these training systems because of high levels of light interception and its transmission through the canopy. Open tree canopies with the fruit bearing spurs being well exposed throughout the growing season provides not only the basis for high fruit numbers and large fruit size but also good fruit color and quality.

Table 5: Cumulative Yield under different Training Systems at CITH and SKUAST Srinagar

Canopy Management System	*Number of Plants/ha (Spacing)*	*Recommended Rootstock*	*Average Yield (t/ha) (Age of Plants)*
Modified Central	625 (4 x 4 m)	Seedling	25-40 (10-12 yrs)
Leader System	1600 (2.5 x 2.5 m)	M-106	40-45 (8-10yrs)
	2222 (3.0 x 1.5 m)	M-9	30-35 (7-8 yrs)
Spindle Bush	2222 (1.5 x 3.0 m)	M-9	35-45 (7-8 yrs)
Head and Spread	2222 (1.5 x 3.0 m)	M-9	40-50 (7-8 yrs)
Cordon system	8888 (1.5 x 0.75 m)	M-9	55-60 (4 yr)
Espalier	2222 (1.5 x 3.0 m)	M-9	62-66 (5 yrs)
Vertical Axis	8888 (1.5 x 0.75 m)	M-9	58-62 (5 yr)
Tall spindle	2222 (3.0 x 1.5 m)	M-9	30-40 (4 yr)

The data validated over the years for HDP in apple at ICAR-CITH and SKUAST-Kashmir showed that showed that yields are related to tree density with the highest tree density producing the highest cumulative yield. Both fruit yield and quality can be obtained under high density plantation with suitable training system. In apple with modern high density orchards, a small but significant yield is expected during the second growing season of the orchard. Substantial yields are expected in the third year and mature yields are expected by year 5. In contrast, traditional low density systems on vigorous rootstocks begin production around year 6 or 7 and do not reach mature yields until year 10-12. In the early years the relationship of tree density and cumulative yield is essentially linear but rapidly becomes curvilinear in later years. By year 5 the relationship of tree density and cumulative yield is only slightly curvilinear but as the orchard age increases the relationship of cumulative yield and tree density becomes strongly curvilinear by the end of year 9. Light management within horticultural canopies has been an invariable rationale of fruit tree architecture strategy during the development of training systems. ICAR-CITH, Srinagar attempts to compare fruit quality characteristics of three apple cultivars Coe Red Fuji, Granny's Smith and Spartan trained on three canopy architectural engineering system *viz.*, Espalier, vertical axis and cordon were grafted on M9 rootstock. These systems provide the yield in the range of 50-60 t/ha with superior quality fruits. Among training systems, maximum fruit weight, highest yield per tree and maximum yield efficiency was observed in espalier training system. Light interception demonstrated maximum photon flux density in espalier system with minimum LAI. Fruit size, TSS and colour parameters of fruits in all varieties were found to be influenced by light intensity. Higher the light intensity, higher was the TSS and colour development in coloured varieties like Spartan and Coe Red Fuji. Therefore espalier training system was found the best canopy management system allowing maximum PAR penetration and diffusion leading better fruit quality and productivity.

Pear

Duckers and Schoofs (2001) have been suggested the V-hedge system (3.5 x 1.25 m and 3.2 x 0.8 m, 2,057 and 3,516 trees per ha respectively), Tiense hedge system (3.5 x 1.75 m, 1,632 trees per ha) and free spindle system (3.75 x 1.5 m, 1,777 trees per ha) for high density planting of pear. Vercammen (2014) compared bush spindle shape (3.5 x 1.5 m), hedge of Tienen (3.5 x 1.5 m), long pruning (3.5 x 1.5 m), candlestick system (3.5 x 1.5 m), V-system (3.2 x 1.0 m), spindle trees (3.2 x 0.50 m) and Drapeau system (3.5 x 1.75 m-tree were planted under an angle of 45°). He found that long pruning system and hedge of Tienen had the lowest total costs after 11 years of growth. The highest yields were obtained with long pruning, followed by "Drapeau" system and V-system in 'Conference' pear on quince Adams rootstock.

Peach and Nectarine

In peach new training system two-scaffold perpendicular V system have been suggest by many researcher. Research trails indicate that the V system is easier to prune and train than traditional systems if simple pruning and training rules are followed. The primary advantage of this system are early and sustained fruit cropping, high fruit quality and ease to management. In peaches among different training system light interception was 74 per cent in Y-shape, 71 per cent in open vase and 89 per cent in slender spindle. However, sunlight distribution in Y shape canopy was 35 per cent higher than the slender spindle canopy (DeSalvador and Dejong, 1989). At CITH, Srinagar a productivity 50 tonnes per ha was obtained in peach from V shape Tatura trellis system while the traditional vase trained trees planted at low densities of less than 400 trees per ha yielded only 15-20 t per ha.

Plum

In plum four different single row planting systems (vertical axis, free spindle, hedgerow and Y-trellis) and three planting densities (0.5, 1.0, 1.5 x/4.0 m for the Y-trellis and hedge trees and 2 x 4 m for the central leader trees) were evaluated by Meland (2005) on semi-dwarf rootstock St. Julien. He workout that trunk cross sectional area of the trees increased with increasing tree spacing. During the first four cropping years, the yield ha^{-1}were positively correlated with tree density. After nine years, cumulative yield ha^{-1} was highest on the Y-trellisand hedge trees with the highest density (5000 trees ha^{-1}). Cook (2007) suggested hedge row (4.0 x 1.5 m- 1,667 trees ha^{-1}) and V system (4 x 0.75 m- 3,333 trees ha^{-1}) for high density planting of plum.

Sweet Cherry

Sweet cherry performed well oncentral leader, multiple leader bush, palmette, and Y-trellis system of training (Whiting *et al.*, 2005). Training systems that incurred theleast amount of pruning (*e.g.*, a trellised palmette system, some central leader/spindle-type systems,) generally have had the highest yields but the smallest fruit size. Training systems with a higher proportion of pruning, particularly heading-type cuts (*e.g.*, multiple leader/bush architectures), generally have had lower yields but larger, sweeter fruit (Lang and Ophardt, 2000). Lang *et al.* (2014) compared Kym Green Bush (KGB),Tall Spindle Axe (TSA), Upright Fruiting Offshoots (UFO) and Super Slender Axe (SSA). They found that KGB, TSA, SSA and UFO is likely to be most suitable for higher vigor, moderate vigor, lower vigor and moderate to higher vigor sites and/or rootstock; respectively. Further, among the different canopy system the SSA is most precocious.

Apricot

Apricot has undergone a great renewal in the orchard training system and planting density have been substantially increased. The principal training systems used are vase, palmette, spindle and V-system. The V-training system is suitable for orchards with over 1200 trees ha^{-1} (De Salvador *et al.*, 2000) while, Stanica and Eremia (2014) suggested Parallel Trident System (4 x 2 m) for high density planting.

Kiwifruit

Commercially grown kiwifruit plants are trained to a permanent framework, either a T-bar (like a clothesline with 4-5 wires) or pergola (solid overhead canopy with wire going in both directions). However, a T-bar system has yielded well, is easier to prune and may be better suited for small-scale production. For a T-bar trellis, vines commonly are planted 4.5m apart in rows 4.5 m apart,while, for pergola plants spaced at 4.5 to 6.0 m apart. There are other systems like Geneva double curtain, free spindle and Tatura (V-trellis)- which allow

Figure 1: Apple Flowering and Fruiting on Espalier System at ICAR-CITH, Srinagar.

much higher densities. Costa *et al.* (1991) compared. Geneva double curtain- 1000 vines ha^{-1}, Free spindle- 1,500 vines ha^{-1}, Tatura trellis- 1800 vines ha^{-1} and T-bar- 700 vines ha^{-1} and they found that both GDC and T-bar produced higher fruit yields (54.9 kg $tree^{-1}$ or 40.6 t ha^{-1}) than the free spindle and V-trellis system.

Conclusion

Optimum light interception and adequate distribution of light holds the key to attain higher production and maintain superior quality of fruits. Canopy management practices would control the tree vigour, provide proper framework for ease of cultural operations and mechanical harvesting and also help in maintaining the orchard health in terms of reduced microbial and insect pest load. Irrespective of land holding, it is the need of the hour to adopt scientific canopy management practices to achieve higher productivity. This system can accommodate more number of plants and increase the productivity of the crop in addition to the quality of the fruit. By adopting HDP and proper canopy management strategies productivity can be doubled with enhancement in the fruit quality.

The experimentations made at ICAR-CITH, Srinagar and SKUAST-K, Srinagar on HDP using advanced canopy management systems promise to double the productivity of apple in the region. Productivity to the tune of 50 t/ha under HDP on modern training system can improve the livelihood of the orchardists due to high returns. Through this system not only the productivity but also the quality of the produce is increasing due to proper light interception and diffusion.

References

Awasthi, R.P. and Mehta, K. 2004. Strategies for canopy management in tree fruit crops. **In:** *Proceedings of the First Indian Horticultural Congress*, 2004, pp. 320–330.

Barritt, B.H. 1989. Influence of orchard system on canopy development, light interception and production of third year 'Granny Smith' apple trees. *Acta Hort.*, **243:** 121–130.

Barritt, B.H., Rom, C.R., Guelich, K.R., Drake, S.R. and Dilley, M.A. 1987. Canopy position and light effects on spur, leaf and fruit characteristics of 'Delicious' apple. *Hort. Sci.*, **22:** 402–405.

Buban T. 2000. The use of benzyladenine in orchard fruit growing: a mini review. *Plant Growth Regulation*, **32:** 381–390.

Byers, R.E., Carbaugh, D.H. and Combs, L.D. 2004. The influence of prohexadione calcium sprays on apple tree growth, chemical fruit thinning, and return bloom., **58**: 111–117.

Cain, J.C. 1970. Optimum tree density for apple orchards. *Hort. Sci.*, **5**: 232–234.

Chalmers, D.J. and van den Ende, B. 1975. The 'Tatura Trellis', a new design for high yielding orchards. *J. Agri. Vict.*, **73**: 473–476.

Chalmers, D.J., van den Ende, B. and Van Heek, L. 1978. Productivity and mechanization of the Tatura trellis orchard. *Hort. Sci.*, **13**: 517–521.

Cook, N.C. 2007. An overview of plum training system in South Africa. *Acta Horticulturae*, **732**: 435–437.

Corelli, L. and Sansavini, S. 1989. Light management and photosynthesis related to planting density and canopy management in apple. *Acta Horticulturae*, **243**: 159–174.

Costa, G., Biasi, R. Giuliani, R. and Succi, F. 1991. Comparison of kiwifruit training systems. *Acta Horticulturae*, **297**: 427–434.

Costa, G., Sabatini, E., Spinelli, F., Andreotti, C., Spada, G. and Mazzini, F. 2004. Prohexadione-Ca controls vegetative growth and cropping performance in pear. *Acta Horticulturae*, **653**: 127–132.

Dallabetta, N., Costa, F., Pasqualini, J., Wehrens, R., Noferini, M. and Costa, G. 2014. The Influence of training system on apple fruit quality. *Acta Horticulturae*, **1058**: 55–62.

De Salvador, F.R. and Dejong, T.M. 1989. Observation of sunlight interception and penetration into the canopies of peach trees in different planting densities and pruning configuration. *Acta Horticulturae*, **254**: 341–346.

De Salvador, F.R., Nuzzo, V. and Mattatelli, B. 2000. Gestionedell'albero e sistemid' impiantodell' albicocco. *Rivista di Frutticoltura e Ortofloricoltura*, **4**: 19–26.

Elfving, D.C. 2010. Plant bioregulators in the deciduous fruit tree nursery. *Acta Horticulturae*, **884**: 159–166.

Ferree, D.C. 1980. Canopy development and yield efficiency of 'Golden Delicious' apple trees in four orchard management systems. *J. Amer. Soc. Hort. Sci.*, **105**: 376–380.

Flor, J.A. and Kesner, C. 1982. Orchard design for stone fruit based on light interception. *Compact Fruit Tree*, **5**: 159–165.

Forshey, C.G., Elfving, D.C. and Stebbins, R.L. 1992. *Training and Pruning Apple Trees*. American Society for Horticultural Science, Alexandria, Virginia, p. 166.

Frank Maas, 2006. Carry-over Effects of CCC-applications in pear orchards. *Acta Horticulturae*, **727**: 125–132.

Gandev, S. 2009. Training and pruning of apple tree according to the new system 'Cone'. *Acta Horticulturae*, **825**: 465–470.

George, A.P., Nissen, R.J., Morley–Bunker, M.J. and Collins, R.J. 1993. Effects of pollination and irradiance on fruiting of persimmon (*Diospyros kaki* L.) in sub-tropical Australia. *J. Hort. Sci.*, **68**: 447–454.

Gifford, R.M. and Jenkins, C.L.D. 1982. Prospects of applying knowledge of photosynthesis toward improving crop production. In: *Photosynthesis*, (Ed.) Govindjee, p. 419–457. Academic Press, New York.

Gutierrez-Pesce, P., Taylor, K., Muleo, R. and Rugini, E. 1998. Somatic embryogenesis and shoot regeneration from transgenic roots of the cherry rootstock Colt (*P. avium* x *P. pseudocerasus*) mediated by pRi 1855 T-DNA of *Agrobacterium rhizogenes*. *Plant Cell Reports*, **17**: 574–580.

Hammerschlag, F.A. and Smigocki, A.C. 1998. Growth and *in vitro* propagation of peach plant transformed with the shoot mutant strain of *Agrobacterium tumefaciens*. *Hort. Sci.*, **33**: 897–899.

Hampson, C.R., Quamme, H.A. and Brownlee, R.T. 2002. Canopy growth, yield, and fruit quality of 'Royal Gala' apple trees grown for eight years in five tree training systems. *Hort Sci.*, **37**: 627–631.

Hrotko, K., Magyar, L. and Ronay, Z. 2000. Improved featheringon apple nursery trees by BA application. *Acta Horticulturae*, **514**: 113–119.

Huggard, J. 1980. High density planting in French orchards, developments and current achievements. *Acta Horticulturae*, **114**: 300–308.

Igarashi, M. Ogasawara, H., Hatsuyama, Y. Saito, A. and Suzuki, M. 2002. Introduction of *rol*C into Marubakaidou (*Malus prunifolia* Borkh. var. Ringo Asami Mo 84-A) apple rootstock via *Agrobacterium tumefaciens*. *Pl. Sci.*, **163**: 463–473.

Jackson, J.E. 1980. Light interception and utilization by orchard systems. *Hort. Rev.*, **2**: 208–267.

Jackson, J.E. and Palmer, J.W. 1972. Interception of light by model hedgerow orchards in relation to latitude, time of the year and hedgerow configuration and orientation. *J. Appl. Ecol.*, **9**: 341–358.

Jackson, J.E., Palmer, J.W. Perring, M.A. and Sharples, R.O. 1977. Effects of shade on the growth and cropping of apple trees. III Effects on fruit growth, chemical composition and quality at harvest and after storage. *J. Hort. Sci.*, **52**: 267–282.

Kallow, G., Reddy, B.M.C., Singh, G. and Lal, B. 2005. *Rejuvenation of Old and Senile Orchard*. CISH, Lucknow, p. 40.

Kanwar, S.M. 1987. *Apples: Production Technology and Economics*. Tata McGraw Publishing Company Limited, New Delhi, pp. 88–98.

Lakso, A.N. 1994. Apple. In: *Environmental Physiology of Fruit Crops, Vol. 1: Temperate Crops*, (Eds.) B. Schaffer and P.C. Andersen. CRC Press, Boca Raton, FL, pp. 3–42.

Lang, G.A. and Ophardt, D.R. 2000. Intensive crop regulation strategies in sweet cherries. *Acta Horticulturae*, **514**: 227–234.

Lang, G.A., Blatt, S.S., Embree, C., Grant, J., Hoying, S., Ingels, C., Neilsen, D., Neilsen, G. and Robinson, T. 2014. Developing and evaluating intensive sweet cherry orchard systems: The NC140 regional research trial. *Acta Horticulturae*, **1058**: 113–120.

Lespinasse, J.M. 1980. La conduite du pommier. II.– L'axe vertical. La rénovationdeverges. INVUFLEC (Ed.), Paris. 120p.

Lespinasse, J.M. and Delort, J.F. 1986. Apple tree management in vertical axis, appraisal after ten years of experiments. *Acta Horticulturae*, **160**: 139–155.

Magyar, L. and Hrotko, K. 2002. Effect of 6-benzyladenine (BA) and gibberellic acid (GA_{4+7}) application on feathering of plum cultivars in nursery. *Acta Horticulturae*, **577**: 345–349.

Marini, R.P. and Barden, J.A. 1987. Summer pruning of apple and peach trees. *Hort. Rev.*, **9**: 351–375.

Marini, R.P., Barden, J.A. and Sowers, D. 1993. Growth and fruiting responses of 'Redchief Delicious' apple trees to heading cut and scaffold limb removal. *J. Amer. Soc. Hort. Sci.*, **118**: 446–449.

Meland, M. 2005. High density planting systems of European plums-the effect of growth and productivity of three cultivars after nine years. *Acta Agriculturae Scandinavica, Section B–Soil and Plant*, **55**: 51–57.

Miller, S.S. and Byers R.E. 2000. Response of winter-injured peach trees to pruning. *Hort. Tech.*, **10**: 757–765.

Monteith, J.L. 1977. Climate and the efficiency of crop production in Britain. *Philosophical Transactions of the Royal Society of London. Series B, Biological Sciences*, **281**: 277–294.

Muller, D. and Leyser, O. 2011. Auxin, cytokinin and the control of shoot branching. *Ann. Bot.*, **107(7)**: 1203–1212.

Myers, S.C. and Savelle, A.T. 1996. **In**: *Tree Fruit Physiology: Growth and Development*, (Ed.) K.M. Maib Good Fruit Grower, Yakima, WA.

Negri, P., Magnanini, E., Cantoni, L., Berardi, G. and Sansavini, S. 1998. Gene transfer in woody fruit species: initial trails in growth habit control. **In**: *Proceedings of AGRO-BIO-FRUIT*, Cesena, Italy, p. 13.

Norelli, J.L. and Miller, S.S. 2004. Effect of prohexadione-calcium dose level on shoot growth and fire blight in young apple trees. *Pl. Dis.*, **88**: 1099–1106.

Ono, T., Koike, H., Tamai, H. and Sigehara, I. 2002. Effects of leaf removal and benzyladenine application on branch development of two-year-old 'Fuji' apple nursery trees grown by pruning and bud removal for one-year-old whip trees. *Hort. Res.* (Japan), **1**: 31–34.

Palmer, J.W. 1989. Canopy manipulation for optimum utilization of light. **In**: *Manipulation of Fruiting*, (Ed.) C.J. Wright. Butterworths, London, p. 245–262.

Palmer, J.W. 1999. Light, canopies, fruit and dollars. *Compact Fruit Tree*, **32**: 119–122.

Palmer, J.W. and Jackson, J.E. 1974. Effects of tree population and variations in spacing within and between rows of Golden Delicious on M.9. In: *Report of the East Malling Research Station for 1973*, East Malling Research Station, Maidstone, UK, pp. 66–68.

Palmer, J.W. and Jackson, J.E. 1977. Seasonal light interception and canopy development in hedgerow and bed system apple orchards. *J. Appl. Ecol.*, **14**: 539–549.

Pathak, R.K. 2009. Canopy management in minor fruit crops. In: Summer School on *Management of Canopy Architecture for Higher Productivity in Sub-tropical Fruits*. Pub. CIAH, Lucknow, India, pp. 43–48.

Proietti, P. and Tombesi, A. 1996. Translocation of assimilates and source-sink influences on productive characteristic of the olive tree. *Adv. Hort. Sci.*, **10**: 11–14.

Quamme, H.A. and Brownlee, R.T. 1993. Early performance of micropropagated trees of several *Malus* and *Prunus* cultivars on their own roots. *Can. J. Pl. Sci.*, **73**: 847–855.

Rademacher, W. 2004. Chemical regulation of shoot growth in fruit trees. *Acta Horticulturae*, **653**: 29–32.

Robinson, T.L. 2003. Apple orchard systems. In: *Apples: Physiology, Production and Uses*, (Eds.) D.C. Ferree and I.J. Warrington. CABI Publishing, Wallingford, United Kingdom, pp. 345–407.

Robinson, T.L. and Lakso, A.N. 1991. Bases of yield and production efficiency in apple orchard systems. *J. Amer. Soc. Hort. Sci.*, **116**: 188–194.

Robinson, T.L. and Sazo, M.M. 2014. Effect of promalin, benzyladenine and cyclanilide on lateral branching of apple trees in the nursery. *Acta Horticulturae*, **1042**: 293–302.

Rugini, E., Muganu, M., Pilotti, M., Balestra, G.M., Varvaro, L., Magro, P., Bressan, R. and Taratufolo, C. 1999. Genetic stability, transgene hereditability and agronomic evaluation of transgenic kiwi (*Actinidia deliciosa* A. Chev.) plants for *rol*ABC, *rol*B and osmotin gene. In: *Fourth International Symposium on Kiwifruit*, Santiago de Chile, 11th–14th February, p. 26 (Abstract).

Sansavini, S. 1983. High density orchards of various crops. In: *Proceedings of the 21st International Horticultural Congress*, Vol. 1. International Society for Horticultural Science, Leuven, Belgium, pp. 182–197.

Sazo, M.M. and Robinson, T.L. 2011. The use of plant growth regulators for branching of nursery trees in NY State. *New York Fruit Quarterly*, **19**(2): 5–9.

Sedira, M., Holefors, A. and Welander, M., 2001. Protocol for transformation of the apple rootstock Jork 9 with *rol*B gene and its influence on rooting. *Plant Cell Reports*, **20**: 517–524.

Shikhamany, S.D. 2001. Canopy management of the tropical and sub-tropical fruit crops. *Indian J. Hort.*, **58**: 28–32.

Srinivasan, C., Dardick, C., Callahan, A. and Scorza, R. 2012. Plum (*Prunus domestica*) trees transformed with poplar FT1 result in altered architecture, dormancy requirement, and continuous flowering. *PLoS ONE*, **7**(7)e 40715. doi: 10.1371/journal.pone.0040715.

Stanica, F. and Eremia, A. 2014. Behaviour of some new apricot cultivars under the parallel trident planting system. *Acta Horticulturae*, **1058**: 129–136.

Taylor, B.H. and Ferree, D.C. 1984. The influence of summer pruning and cropping on growth and fruiting of apple. *J. Amer. Soc. Hort. Sci.*, **109**: 19–24.

Tustin, D.S. 2000. The slender pyramid tree management system: In pursuit of higher standards of apple fruit quality. *Acta Horticulturae*, **513**: 311–319.

Vercammen, J. 2014. Comparison of different planting systems for 'Conference'. *Acta Horticulturae*, **1058**: 37–43.

Wagenmakers, P.S. 1991. Planting systems for fruit trees in temperate climates. *Crit. Rev. Pl. Sci.*, **10:** 369–385.

Watanabe M., Bessho H., Suzuki A., Komori S. 2008. Seasonal changes of IAA and cytokinin in shoots of columnar type apple trees. *Acta Horticulturae*, **774**: 75–80.

Wertheim, S.J. 1968. The training of the slender spindle. Pub. No. 7, Proefstation Fruitteelt, Wilhelminadorp, The Netherlands (37p.).

Wertheim, S.J. 1981. High density planting, development and current achievements in the Netherlands, Belgium and Germany. *Acta Horticulturae*, **114**: 318–327.

Whiting, M.D., Lang, G. and Ophardt, D. 2005. Rootstock and training system affect sweet cherry growth, yield and fruit quality. *Hort. Sci.*, **40**(3): 582–586.

Wunsche, J.N. and Lakso, A.N. 2000. Apple tree physiology: Implications for orchard and tree management. *The Compact Fruit Tree*, **33**(3): 82–88.

Zimmerman, R.H. and Miller, S.S. 1991. Orchard growth and fruiting of micropropagated apple trees. *J. Amer. Soc. Hort. Sci.*, **116**: 780–785.

22

Canopy Architecture Management for Doubling India's Fruit Production

Vishal Nath* and Alemwati Pongener

ICAR-National Research Centre on Litchi,
Muzaffarpur – 842 002, Bihar
****E-mail: vishalnath66@rediffmail.com***

Sustainable production in perennial fruit crops is one of the biggest challenges in India under changing environmental conditions. Erratic behavior of trees with respect to flushing and flowering poses a serious threat to the fruit industry for sustained supply and price trend. Hence, there is need to develop state of the art technologies for harnessing the naturally available resources in a sustainable manner to meet the growing demands of the society with respect to nutritive and healthy food. Canopy architecture and its management is one of the important operations in perennial fruit crops to harnessing solar energy and utilizing available space, soil nutrients and moisture in new as well as existing productive and old senile fruit orchards. It has been noticed that lack of focus upon regulating growth of tree, and appropriate canopy architecture right from the establishment stage are the main reasons for such problems. As a result of casual approach and negligence, the trees acquire their natural shape, which very often is not ideal for optimum quality production. A majority of the trees attains tall, upright and overlapping growth and the canopy is marked with criss-cross branches leading to highly dense vegetative mass with very poor penetration of photosynthetically active radiation (PAR). Such conditions not only affect the photosynthetic rate but also facilitate proliferation of pests and diseases. Consequently, with pressure of pest abundance and poor photosynthetic efficiency over the years for want of due care to canopy, the trees turn senile, unproductive and uneconomical. Canopy management therefore can be termed as, "Designing the plant as per need, using inherent plant characteristics in accordance with given set of conditions and resources to perform the plant maximum" (Nath *et al*, 2014).

Canopy management and tree architectural engineering in fruit trees are very important operations, which should be practiced in the orchards right from the planting stage. Even during selection of plants in the nursery, the aim of canopy architecture must be kept in mind so that right plant is selected and established. Every individual tree in the orchard should be subjected to proper management for development of ideal architecture with outward spread of canopy to facilitate greater penetration of active solar radiation in the canopy and thus enhancing the photosynthetic efficiency. Besides, trees regulated to a manageable height and canopy size make it amenable for available machinery for pesticide spray, training and pruning of branches, intercultural operations and harvesting of fruits. A sound architecture of the trees with open and outward spread canopy is critical for health of plant and sustainable production. Based on numerous research investigations, it can be inferred that fruiting potential of the tree is largely governed by its architecture, canopy density and photosynthetic efficiency of leaves (Nath, 2013).

The productivity of fruit crops depends on several crop husbandry practices in which management of canopy is the most important one. The tree architecture and canopy management, especially size control, has become a priority for reducing production cost and increasing yield and quality. Canopy design and shape influences light interception with assured higher monetary returns to fruit growers. Therefore, early height control and tree canopy management should be practiced in fruit crops for doubling orchard income. Some of these factors are being enumerated.

Plant Characters Affecting Architectural Design

The architectural design of the individual fruit plant depends upon inherent plant characters like fruit species and variety because every plant has inbuilt inherent characteristics and growth behavior which includes angling of branches and pattern of branching, flushing pattern, dormancy pattern and bearing behavior, level of endogenous growth hormones, *etc.* Based on these factors, the principles and action for canopy architectural design are planned and decided. Some plants have shoot growth with extension of flushes (mango) while others have continuous flushing during active growth phase (litchi). Few species have inherent dormancy for certain period (ber, guava, grapes, temperate fruits) and produce current shoots or fruiting terminals/spurs after pruning which finally bear fruits. Therefore, the architecture of each plant is decided according to their inherent growth characteristics.

External Factors Affecting Plant Canopy Architecture

Besides internal factor, several other factors like planting system, spacing, solar radiation, wind velocity, rainfall pattern and light distribution, *etc.* also play very crucial role in plant canopy development in fruit crops. Hedgerow planting system with a principle of single canopy and multiple root system is ideal for those fruit species which are responsive to pruning and bear on either current season shoots or spurs (guava, peach etc), whereas open centre system may be more useful for plants requiring more solar radiation for fruiting and colour development (litchi). Bearing area alignment in canopy, depth requirement in canopy, rootstocks, soil depth, *etc.* are also important external factors, which influence canopy architecture and management in a species.

Architectural Behaviour of Plants

Prior to canopy architecture management, understanding of nature and behavior of plant is essential because it depends on two very important factors. Firstly, the nature and relative arrangement of each branch, and secondly, the expression of an equilibrium between endogenous growth process and exogenous constraints exerted by the environment. In view of above points,canopy architecture concept for perennial woody fruit species has been (Figure 1) proposed by Nath *et al.* (2012) which includes suggested series of operations (cutting, bending, angling) on the above ground portion of the plant for maximizing production of quality fruits per unit canopy area. The crown structure of a tree wholly depends on branching pattern and crotch angles. For example, the litchi tree normally has a thick, straight and short trunk and the branches often have 'V'-shaped crotches which are easily broken by strong winds. Some trees have naturally narrow crotch angles (40° or less from the vertical). Narrow crotches result in greater loss of limbs from windstorm and heavy fruit load than crotches with larger angles. The strongest crotches are those in which branches grow up from the trunk at angles ranging from 40-90° and scaffold branches evenly spaced around the tree, with each main branch at least 90° from the next one. An appropriate canopy architecture management

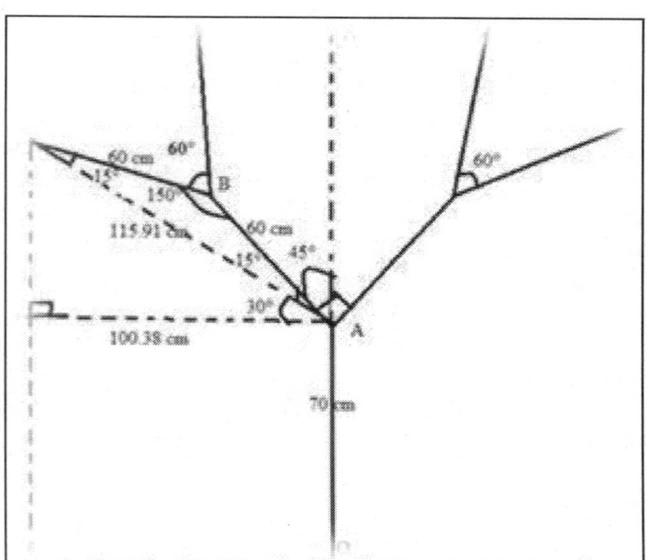

Figure 1: Scheme of Canopy Skeleton in Fruit Trees (Left); Frame Development and Distribution of Primary Branches (Right).

in fruit trees encompasses the development of following tree features.

- Strong frame of primary branches
- Wider crotches in scaffold branches
- Healthy and well distributed secondary and tertiary branches
- Sufficient fruiting terminals in most productive areas
- Healthy foliage with high photosynthetic efficiency to maximize the solar radiation use efficiency
- Enough space for air circulation inside canopy

Relevance of Plant Canopy Architecture in Doubling Fruit Production

Maintaining Tree Size

Certain principles of canopy architecture management have been in focus for most fruit trees. These are

- Facilitating aeration inside the canopy
- Maximizing light penetration in and out of canopy
- Facilitating conversion of sunlight into chemical energy by plant leaves
- Controlling plant structure including stature
- Improving fruit colour, quality and yield

Nature does not care how many good sized, high quality fruits are to be borne in a tree but is certainly affected by the capacity of plant to sustain these. For economical fruit growing, the natural form and shape of fruit trees has to be modified through the practices of pruning to achieve targeted yield with better quality fruits. Fruit growers desire to regulate the growth and development of their trees in a specific manner for production of maximum fruit, because it is not always wise to allow a plant to develop naturally, since unwanted portions may develop at the expense of those which are essential from the cultivator's point of view to increase production. Canopy architecture management is thus one of the most important production factors confronting the rapidly increasing fruit industry in India. In many fruit crops, 15-20 per cent increase in production with enhanced fruit quality has been witnessed by managing appropriate and efficient canopies particularly in short-statured trees. Small trees capture and convert sunlight into fruit production in a better way than larger ones, hence, it has direct impact on orchard efficiency.

Improving Light Interception

Light is critical for plants as well as for fruit growth and development. The green leaves harvest sunlight to produce carbohydrates and sugars, which are transported to the sites/sinks where they are most needed like buds, flowers and fruits. Improving light penetration into the tree canopy improves tree growth, productivity, yield and fruit quality. High density planting and orientation of planting often pose a problem for light penetration in an orchard. Generally, in close planting, quicker shading becomes a problem. An East-West row orientation results in more shading compared to Western and Southern orientation of trees. Strong bearing branches tend to produce larger fruits. They also transport nutrients and water more efficiently throughout the trees. Pruning aims to encourage new and strong growth of all branches. Some of the basic principles of the canopy management are maximum utilization of light, avoidance of build-up of microclimate congenial for disease and pest infestation, convenience in carrying out cultural operations, maximizing the productivity with quality fruit production, economy in obtaining the required canopy architecture (Nath and Kumar, 2013).

Fruit production involves capturing and conversion of sunlight into production of fruit biomass (dry-matter content). The main controlling factors in this context are the amount of incoming radiation and percentage of total radiation received by photosynthetically active leaves (PAL) for conversion of solar energy into fruits yield.

Light Environment

Light limits the photosynthetic productivity of crops, and is considered to be the most important variable affecting productivity and quality. The transpiration rate of any crop is the function of ambient temperature, humidity and light of which light is usually out of our control as it is received from the sun. Supplementary lighting does offer opportunity to increase yield during low light periods, but is generally considered commercially unprofitable. The other means for manipulating light are limited to screening or shading and are employed when light intensities are too high. However, there are also general strategies to help maximize the crop's access to available light.

In order to understand how to control the environment to make maximum use of available light, it is important to know about the properties of light and how light is measured. Light has both wave properties and properties of particles. Depending on how light is considered, the measurement of light can reflect either its wave or particle properties in terms of photons. Different types of light sensors are available for use with computerized environmental control systems. As long as the sensors measure the light available to plants, it is for growers to be able to relate these measurements to how the crop is performing. Light is a form of electromagnetic radiation produced by the sun. A narrow range of this electromagnetic radiation falls within the range of 400 to 700 nanometers (nm) of wavelength. One nanometer being equal to 0.000000001 meter. The portion of the electromagnetic spectrum which falls between 400 to 700 nm is referred to as the spectrum of visible light; this is essentially the range of the electromagnetic spectrum that can be seen. Plants respond to light in the visible spectrum and uses this light to drive photosynthesis. Photosynthetically active radiation (PAR) is defined as radiation in the 400 to 700 nm waveband. PAR is the general term which covers both photon and energy terms (LI-COR Inc.). The rate of flow of radiant (light) energy in the form of an electromagnetic wave is called the radiant flux, and the unit used to measure this is the Watt (W). The unit of Watts per square meter (W/m^2) is used by some light meters and is an example of an "instantaneous" measurement of PAR (LI-COR Inc.). Other meters commonly seen in greenhouses take "integrated" measurements reporting in units of joules per square centimeter (j/cm^2) (LI-COR Inc.). Photosynthetic Photon Flux Density (PPFD) is another term associated with PAR, but refers to the measurement of light in terms of photons or particles. It is also sometimes referred to as Quantum Flux Density (LI-COR Inc.). Photosynthetic Photon Flux Density is defined as the number of photons in the 400 - 700 nm wavebands reaching a unit surface per unit of time (LI-COR Inc.). The units of PPFD are micromoles per second per square meter ($micromol/m^2$).

Light exposure is a factor that varies with the position within the canopy of the fruit-bearing branch and of the fruit itself. The effect of light on photosynthesis includes both a direct effect of the photosynthetic photon flux on the rate of electron flow and an indirect effect of light on leaf photosynthetic capacity, since plants allocate nitrogen resources within the canopy to enhance photosynthetic capacity in portions of the tree receiving high irradiances.

In a mango tree, photosynthetic acclimation to light is mainly determined by the mass-to-area ratio, M_a, and, to a lesser extent at low irradiance, by changes in the allocation of total leaf nitrogen, whereas changes in leaf nitrogen on a mass basis, N_m, play a minor role. In fact, on mango, leaf nitrogen on an area basis, N_a(calculated as the N_m-to-M_a ratio), which reflects photosynthetic capacity, was linearly related to the fraction of intercepted light (Figure 2). The lowest mass-to-area ratio and N_a were observed in the shaded leaves (Table 1). These effects of light exposure lead to a lower carbon assimilation in shaded leaves. Fruit growth in dry mass depends on the partitioning of carbohydrates between maintenance of the system, *i.e.* the fruit-bearing branch, fruit growth and storage in leaves and stems. If the carbon supply decreases, fruit growth in terms of dry mass is reduced as well. It has been shown that fruit size and dry matter content decrease in 'Kensington' fruit from upper to lower positions in the canopy. Another quality criterion, soluble solids content and total sugars, which can be related to dry matter content, were determined to be lower in mango fruit from the lower portion of the canopy.

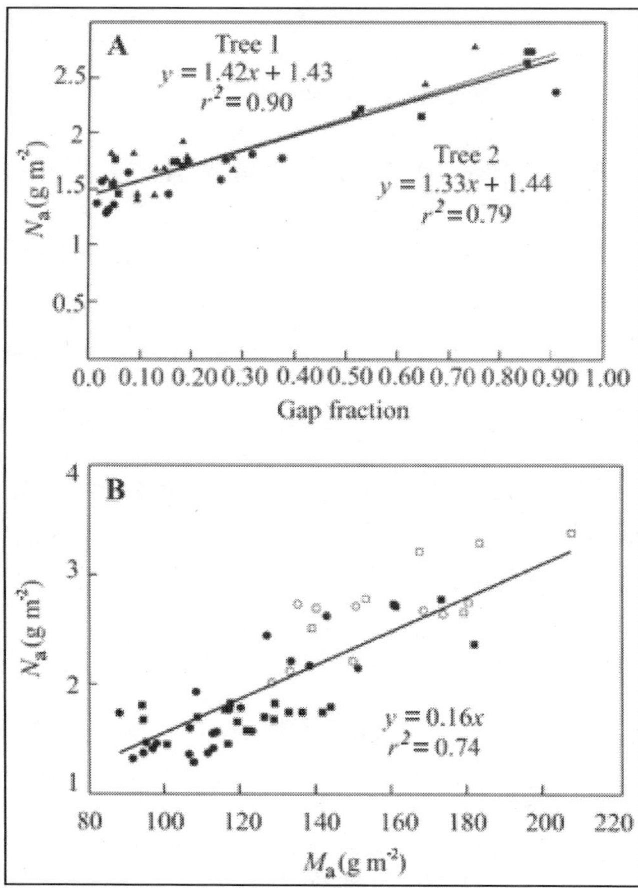

Figure 2: Relationship between Leaf Nitrogen Concentration per unit Leaf Area (Na) and either (A) the Gap Fraction for Leaves in the Crowns of 3 Year-Old Tress.

[Measurements were performed on leaves less than 2 month old (•), 8 month old (■) 12-to 14 –month old (▲) and 17 to 20 month old (▲)]. or (B) the lead mass to area ratio M_a (measurements were performed on leaves of 3- year-old tress [Tree (•) and Tree 2 (■)] leaves far from (standard leaves. o) and close to (o) developing fruits of 11-year-old trees] [adapted from Ubran *et al.* (2003)].

Table 1: Effect of Fruit Position on the Canopy on Quality of Kensington Pride Mango Fruit. Fruit were harvested on the same day and ripened at 22ºC [adapted from Honfman *et al.*]

Fruit Position	*Fruit Mass (g)*	*Dry Matter (%)*	*Days to Eating Soft*	*Total Sugars (ºBrix)*
Upper	360	13.1	13.4	11.6
Middle	370	13.0	14.5	11.6
Lower	346	11.6	15.9	10.5

Another effect of light exposure on mango quality deals with its attractive trait, especially the red pigmentation of skin through the influence of light on anthocyanin production. Mangoes inside the canopy retain a greener skin colour due to the decrease of fruit exposure to sunlight.

Fruit plants respond to light quantity, quality, direction, and periodicity. There are numerous photoreceptors in plants, including chlorophylls, phtyochromes, cryptochromes, phototropins, and ones that react to green light. Light, along with other environmental cues, enables plants to adapt to environmental conditions. Efforts to manipulate plant morphology and physiology using photo-selective filters have been ongoing for decades, especially in greenhouse environments. More recently, there are some tools/practices, which can manipulate the light quality specifically for plant development and growth as well as quality production.

The fruit crops grown in different climatic zones (tropical, subtropical and temperate zones) are required to be trained and pruned differently depending upon the type of plant and specific objectives of the fruit growers. Canopy architecture depends on the choice of trees size, their population per unit area and pruning needs to get optimum yield. Pruning should be done when trees are young to develop their basic shapes and frame, and continued as tree matures to provide sufficient fruiting area. In commercial orchards, pruning operations are normal and routine activity to be done once/twice in a year as the case may be but in the orchards where trees have outgrown their allotted space, pruning operation only can bring them within their allotted limit and increase their bearing surface through rebuilding the effective canopy area. Trees that grow too close together shade out the lower canopy portions and in return decrease their fruit setting ability. Here, also pruning plays a crucial role to bring them back in fruiting condition through midterm architecture corrections like Centre opening, side branch chopping.

Objectives of Canopy Architecture

The basic objective of canopy architecture is the of tree canopies to optimize the production of quality fruits through various techniques. They are objectives are as follows:

- ✰ To restore proper root/shoot ratio for balance growth and source-sink relationship.
- ✰ To obtain specific form of the plant as per geometry and production aim.
- ✰ To develop specific geometry within the orchard and its overall topology for commercial production.
- ✰ To maintain optimum balance between vegetative and reproductive parts for quality assurance.
- ✰ Plants respond differentially in different environment hence the architecture will be available as per location.
- ✰ Development of shoot and root system in the plant has symbiotic relationship, which should be maintained.

Canopy Architecture Management

For a farmer, it is very important to decide the strategies and operations for canopy architecture and management. Some of the broader points of concern and considerations have been suggested as follows:

- ✰ Analysis of architectural engineering aspects such as angling of branches and pattern of branching in plant.
- ✰ Understanding the endogenous growth process like flushing pattern and action of growth hormones.
- ✰ Understanding the level of influence of environment on architecture at a point of time/ location.
- ✰ Objectives for which canopy is being managed.

Major Considerations

Certain considerations have to be in mind before initiating canopy architecture and its further furthermanagement. These points are:

- ✰ Plants respond differently in different environment.
- ✰ Development of shoot and root system in the plant has symbiotic relationship.
- ✰ Weak crotches in the branches and direction of the branches can be corrected by unequal cuttings and need based training.
- ✰ Pruning aims for dwarfing and promotion of maximum shoots/fruiting points
- ✰ Vegetative growth and fruiting/yield have antagonistic effects
- ✰ Apical dominance is a natural phenomenon in tress, which appears time and aging even after its removal.
- ✰ Vigour of shoots varies inversely with number of fruits and crop load in previous crop.
- ✰ Fruit quality is affected by heavy cropping.
- ✰ Pruning effects are localized.
- ✰ Young shoots afford the best material for blossom bud formation.
- ✰ Genetically dwarf cultivars produce excellent canopy.

Steps

To make canopy architecture more relevant and effective in present day context of fruit production with excellent quality produce, a complex system of modern orchard management including those of canopy management has to be followed. Certain steps for canopy management have been suggested:

- ✰ Selecting dwarf cultivars
- ✰ Selecting appropriate planting system preferably the medium density
- ✰ Use of dwarfing rootstock and inter-stocks as much as possible
- ✰ Initial frame development in plants is must, which should be decided as per the space provided
- ✰ Training the plants with open centre to harness maximum solar radiation
- ✰ Allow the canopy to grow horizontally for ease of operations
- ✰ Centre opening of grown up trees to induce stereo fruiting and minimizing pest and disease load
- ✰ Use of spacers for positioning the branches to make wide crotch and strong frame.
- ✰ Pruning of the shoots and excessive growth to make a balance

Strategy for Plant Architecture Management

Based on major principles, few important strategies have to be thought of for canopy architecture management in fruits trees. These strategies revolves around management of light, air, radiation, inherent nutrients, bearing surface, ease of operation, cost involved in maintenance, *etc.* Light management is quite crucial, hence there should be enough light exposure to all parts of the canopy to maximize bearing area in minimum time, and sustain maximum bearing area for maximum time. As the light intensity decreases with the depth of canopy, an ideal situation would have been 3 m canopy depth with 50 per cent light penetration. Tree canopy should be well within the reach to perform orchard operations, which can be managed both manually and mechanically. Manual operation requires skilled workers and is time consuming, involving high cost of operation. However, mechanical pruning is less costly and less time consuming. Therefore, large orchards are always profitable when pruned mechanically including other orchard operations like sod culture, fertigation, other spray treatment to control pest (Kumar, 2011).

Plant Architectural Engineering and Management in Selected Fruits

Mango

Mango is a large tree fruit crop, which can grow up to 10-12 m under natural conditions. For commercial orchard, an area of 50-60 m^2 is allowed for individual tree. To obtain desired form and effective canopy, an initial branching height of 60-70 cm is considered appropriate. Train the plants and allow 3-4 horizontal branches to grow. Remove the upright growing branches and selected horizontal shoots form the main scaffold limbs of the trees (Figure 2). Allow the branches to grow and during the second flush, the tips should be pinched for maturation of the shoots. This procedure is to continue during the 2nd-3rd year also. If selected shoots grow vertically this should be removed. Thus, the horizontal scaffold limbs are stronger which flower and fruit earlier. Short branches within tree canopy produce a complex, compact and strong structure for fruiting shoots after 3rd year. Each plant should be allowed to grow 2-2.5 m in each direction to have uniform, semispherical canopy because mango bears on 2/3 lower middle canopy orienting outside.

Figure 2: Canopy Development in Mango Plant during Initial Stage.

For rejuvenation, first objective is to identify senile orchards and study its bearing behavior. Mark the branches, which are to be removed and mark those branches that are to be kept for development of canopy in future and heading back operation to be done during November-December. All the cut branches and twigs to be removed from the plot and clean surface for making the basin for irrigation, manures and fertilizer application. Profuse emergence of new shoots starts after few months (March-April) on pruned branches, which require selective light thinning of excessive shoots. About 8-10 healthy shoots with outwardly growth are retained per branches and remaining to be removed (Singh, 2011). Adopt the recommended intercultural operation and for control of insects and diseases. The properly managed trees start bearing from third year and onwards with quality fruits.

Litchi

In general, litchi trees grow unstopped and develop into larger canopy. However, for systematic orcharding and promotion of large bearing area vis-s-vis accommodation of large number of plants/per unit area, tree architecture management particularly the tree height, the canopy having vigorous shoots concentration to provide more fruiting twigs and penetration of light to inner portion become more relevant. Thus, giving a proper shape to the litchi plants from initial year of growth and pruning of branches after harvesting in bearing trees is more essential. Since more sprouting takes place in young plants, proper shape is quite convenient. Single stem air-layered plants (40-50 cm) should be transplanted in field and allowed to grow straight up to 60-70 cm, allowed 3-4 branches in all direction from 60-75 cm onwards (Figures 3A and B) and remove all other branches from initial stage. It is necessary to continue shaping of plants by removing those branches forming narrow crotches with main branches. The litchi plant has the tendency to bear fruits on outer surface of the canopy receiving maximum solar light and the economical yield is confined to 2/3rd of the lower middle canopy. Therefore, the tree frame, II, and III order branches should be so spaced and arranged on tree frame which can provide a comprehensive and uniform canopy. To develop good and compact canopy, 15-20 cm in current season fruit bearing shoots are removed at the time of harvesting. In this way, 2-3 new terminals develop which consequently develop into fruiting branches for the next season.

Unproductive trees are pruned heavily to develop new fruitful shoots. In such cases heavy reiterative pruning, usually up to limbs level at a height of 2-2.5 m has been suggested is commonly adopted followed by supplementation with application of nutrients. These shoots start fruiting 2-3 years after pruning. Thereafter, general pruning is followed to maintain ideal vigour and productivity of trees in case of rejuvenated orchards after 4th year. Rejuvenation of old senile unproductive orchard through severe pruning is recommended in the month of August-September at a height of 2-2.5 m depending upon the tree structure (Figure 4). After topping, 3-4 main branches with outward growth are kept and rest are be removed. All cut branches to be removed from the field for making the basin and application of manures and fertilizer. This will also avoid the pest buildup in the field after pruning. The new shoots start to emerge

(A) (B)

Figure 3: (A) Development of Frame in Litchi Plant with Two Primary and Two Secondary (Plants of cv. Shahi in flowering stage below); (B) Development Frame with Three Primary and Two Secondary in Each Primary (Plants of cv. China in flowering stage below).

Figure 4: Rejuvenation in 70 Years old Litchi Plant and Developing Canopy in Rejuvenated Plants.

within 30 days of topping. A large number of new shoots emerge and cover the entire branches. Thinning of the new shoots are important and allow only 5-6 outward growing shoots to mature and form the canopy in each branches as semi-circular or open umbrella type canopy (Singh *et al.*, 2011). There should be access of light and air in all parts of rejuvenated trees.

Advances in Canopy Architecture Management

Planting Systems

The major advances in planting systems are square system with high density planting (Figure 5), rectangular system with high density planting, double hedge row planting and single hedge row planting (Figure 6). These systems (Table 2) are adopted in different fruit trees to obtain the maximum yield and quality fruits per unit area.

Training Systems

The canopy management includes tree training and pruning, which decides the quantity of light intercepted by trees, as tree shape determines the presentation of leaf area to incoming radiation. An ideal training strategy revolves around the arrangement of plant parts especially to develop a structure that optimizes the utilization of sunlight and promotes productivity. Different system of training have been standardized fruit trees (Table 3) with better fruiting area.

Table 2: Planting System Adopted in different Fruit Crop with Canopy Architecture

Crop	*Planting System*
Litchi	Square system with medium density planting (6 × 6 m, 6 × 4 m) for Shahi and China, Single hedge-row system (9 × 4.5 m, 8 × 4 m)
Mango	Square system with medium density planting for Dashehari (6 × 6 m) and Amrapali (5 × 5 m)
Guava	Single hedge row system with ultra-high density planting (2 × 1 m), ultra density rectangular system (1 × 1 m)

Table 3: Latest Methods of Training in Fruit Plants for Canopy Architecture

Crop	*Training System*
Litchi	Open centre or modified centre leader system
Mango	Open centre, modified centre leader system, Y- shape trellis system
Pomegranate	Multi-stem training
Peach	*Tatura* trellis system

Figure 5: Square Planting System in Litchi.

Figure 6: Single Row Planting System in Litchi.

Rejuvenation

Old and senile tree of litchi, mango, aonla, guava, apple, peach and pear have been successfully rejuvenated through pruning followed by canopy management at various places. By adopting crop specific recommendations and time rejuvenation more than 95 per cent success with improved fruit quality and orchard operations have been experiences.

Regulation of Flushing and Flowering

Canopy management through pruning is an integral part in many fruit crops including litchi to regulate flushing, flowering and fruiting. In deciduous fruit trees, pruning induces new growth and reproductive phases for better fruiting and crop production. Mechanical and chemical methods of flush regulation have been found effective.

Improvement in Yield and Fruit Quality

Canopy architecture and its management has been key factor for improving fruit yield and quality in most of the fruits. In litchi, apple, peach and certain other fruits, development of red colours in the peel of the fruit depends mainly on availability of sunlight, which can be manipulated with ideal canopy architecture. Fruit yield in many crops can be increase with maintaining appropriate canopy size especially in temperate fruits, grapes, guava *etc.*

Designs for different Densities

Orchards are planted at different spacing in various planting system. Among the prevalent planting systems, square and rectangular systems are most commonly adopted whereas hexagonal and contour systems are need-based depending upon the availability of area, land topography and purpose thereof. The modern planting systems like single hedge-row and double hedge row are the derived form of rectangular system where concept of single canopy with multiple root system is followed. These methods are adopted to harness maximum solar radiation and potential use of mechanization.

The designing of canopy architecture in different density depends on the space provided to each plant in each direction. Under square system of planting the plants are designed in such way that the canopy grow uniformly in each direction where as in rectangular system the canopy is allowed to grow wider in two directions and shorter in two directions where space is less. While developing design for canopy, the frame of skeleton, the angle of branches, the axis of growing branches, the spacing between the growing branches, the orientation of branches *etc.* are essentially kept in mind (Figure 7).

As a matter of fact, 60 per cent space in one direction should be allotted for skeleton, 25 per cent for fruiting terminals and 15 per cent for gallery space in square system. The situation is almost similar in rectangular system except the side where distance between two plants is narrow. A concept of canopy design in different densities has been given in Table 4.

Most of the perennial fruit plants especially the mango and litchi bears on 2/3rd lower and middle canopy and fruiting in top region is generally prone to biotic and abiotic stress. Well-designed canopy facilitates many orchard operations like bunch/fruit bagging, spraying, netting *etc.* (Figure 8) The result has been so

Figure 7: Canopy Designing Concept in Litchi under Square System.

Table 4: Canopy Design Concept for different Densities

Plant Spacing (m)	*Canopy Design Concepts*				
	I Order Branch (m)	*II Order Branch (m)*	*III Order Branch (m)*	*Fruiting Terminals (m)*	*Gallery Space (m)*
8 × 8	1.0	0.80	0.70	1.00	0.50
6 × 6	0.8	0.60	0.50	0.70	0.40
5 × 5	0.7	0.50	0.40	0.60	0.30
4 × 4	0.6	0.40	0.30	0.50	0.20
8 × 4	1.0	0.80	0.70	1.00	0.50
6 × 3	0.6	0.40	0.30	0.70	0.00
	0.8	0.60	0.50	0.70	0.40
"Y" Shaped trellis (5 × 3)	0.5	0.40	0.30	0.30	0.00
	0.6	Parallel training of I, II and III branches			0.80
	–	Fruiting surface on II and III branches			0.00

far indicated that hedge row system in Shahi and square system in China produces higher yield (Anon 2004) square system in Shahi (TSS : acid ratio, 91.74) and single hedge row China (TSS : acid ratio 89.04) produced best fruit quality in litchi. Therefore, it has been concluded that row planting is the most suitable planting system for cultivars Shahi and China at Ranchi, Pantnagar and Pusa, Samastipur, Bihar (Anon., 2004).

Figure 8: Bagging of Litchi Fruit Bunches.

Figure 9: Dwarf Canopy Facilitate Spraying Operation.

Canopy Zonation

Effective canopy through designing based on canopy zonation (Figure 11), the effective canopy designing has been calculated for litchi and effective volume has been calculated in Table 5.

Future Thrusts

A lot of work has been done on training and pruning in different fruit crops but the effect has not been viewed in totality. With the advancement of knowledge in area of tree architectural design, now the effect is being realized in larger prospective and almost all crops are coming under the umbrella of tree architecture. Hence, furtherance is needed in this area of research for establishing basic facts and logical explanation of its impact on production quantum, productivity and quality of fruits. Some specific future thrust areas of research are suggested.

- ✰ Standardization of tree architectural design and frame for range of fruit crop is needed.
- ✰ Optimum canopy for sustainable yield needs to be worked out for commercial fruit crops.
- ✰ Influence of canopy architecture on fruit quality needs to be quantified.

Figure 10: Individual Plant/Row Netted in Well Designing Canopy in Litchi.

Figure 11: Canopy Zonation in Evergreen Fruit of Litchi.

Figure 12: Modern Canopy Desing in Fruit Plants to Produce Heavy Yield and Quality Fruits; Mango in Hedge Row.

Figure 13: Tree Design for Square System at Normal Density.

Table 5: Canopy Design for various Planting System

Spacing (m)	*No. of Plants/ha*	*Maximum Permitted Plant Height (m)*	*Maximum Permitted Plant Spread (m)*	*Canopy Volume (m^3/ha)*	*Effective Canopy Volume (m^3/ha)**
4 × 4	625	3.0	3.0x3.0	79,545	78,750
5 × 5	400	3.5	4.0x4.0	1,87,712	1,83,958
6 × 6	277	4.0	5.0x5.0	3,62,697	3,51,816
6 × 4	415	3.0	5.0x4.0	2,60,828	2,55,612

Canopy volume m^3 has been calculated by the formula 4/3ð a^2b where a = half of spread and b= half of the height (Bestwood, 1963).

* Reduction in canopy by 10, 20, 30, 10%, respectively from top non yielding areas.

Figure 14: Litchi in Y-shaped Trellis.

Figure 15: Litchi Flowering and Fruiting under Hedgerow System.

- ✰ Canopy architecture v/s light interception study needs to be conducted for each crop.
- ✰ Canopy architecture and its management for different planting systems needs to be worked out.
- ✰ Tree architectural plan against natural calamities needs to be developed/postulated.
- ✰ Canopy architecture in context to climate resilient option also needs proper attention.

Conclusion

Canopy architecture and its management is one of the important operations in fruit crops for harnessing natural resources through utilization of proper space, nutrients, moisture, solar energy and aeration in the new as well as existing productive and old senile fruit orchards. The tree architectural management, for strong frame and size control, has become a priority for reducing production cost and increasing input use efficiency, yield and quality in various perennial tree fruit species. Canopy design and shape influence light interception with assured higher monetary returns to fruit growers. This management process is an essential tree management operation that starts from the first year of plant establishment which of-course varies with species and environments but enables the plant to produce high quality and quantity by providing proper framework, more fruiting area and utilization of available natural resources. Therefore, early height control, open centre canopy, strong limbs with wide crotch angle and more fruiting area and sustained progressive tree canopy architecture management are the important techniques, which should be practiced in fruit crops for higher returns to growers.

References

Kumar, R. 2011. Orchard canopy management for enhanced productivity of fruit crops. **In**: Souvenir National seminar on *Development of Horticulture in Bihar: Issues and Strategies* organized by Bihar Horticultural Society, RAU, Pusa on 28th–29th January, Patna, Bihar.

Nath, V. 2013. Canopy architecture and management in litchi under high density. **In**: *National Seminar on Plant Canopy Management in Fruit Trees under High Density Plantation*, 21st–24th September, CISH, Rehmankheda, Lucknow.

Nath, V. and Kumar, A. 2013. Canopy architecture management for precision farming in fruits. **In**: *Precision Farming in Horticulture*, (Eds.) Singh *et al.*, New India Publishing Agency, New Delhi, p. 87–93.

Nath, V., Pandey, S. and Purbey, S.K. 2015. *Compendium of Lectures on GAP in Litchi*. Publication of ICAR–NRC on Litchi, Muzaffarpur, pp. 1–114.

Nath, Vishal. 2014. Canopy architecture. **In**: *6th Horticulture Congress*, TNAU, Coimbatore, 6th–9th November.

Purbey, S.K. and Kumar, A. 2015. Effect of Pre–harvest bagging on quality and yield of litchi (*Litchi chinensis* Sonn) fruits. *The Ecoscan* **7** (Special issue): 197–201.

Salisbury, F.B. and Ross, C.W. 1985. *Plant Physiology*. Wadsworth Publishing Company.

Shahak, Y. 2008. Photo-selective netting for improved performance of horticultural crops: A review of ornamental and vegetable studies carried out in Israel. *Acta Hort.*, **770**: 161–168.

Shahak, Y., Gussakovsky, E.E., Cohen, Y., Lurie, S., Stern, R., Kfir, S., Naor, A., Atzmon, I., Doron, I. and Greenblat-Avron, Y. 2004. Color Nets: A new approach for light manipulation in fruit trees. *Acta Hort.*, **636**: 609–616.

Singh, G. 2011. Innovative production technologies: Canopy architecture. **In**: *Horticulture to Horti-Business*, (Eds). K.L. Chadha, A.K. Singh and V.B. Patel, Westville Publisher, New Delhi.

Singh, G., Kumar, R. and Nath, V. 2011. *Rejuvenation in Litchi*. FAO bulleting NRC on Litchi, Muzaffarpur, Bihar, India, pp. 1–14.

Wang, Z. and Stutte, G. 1992. The role of carbohydrates in active osmotic adjustment in apple under water stress. *J. Amer. Soc. Hort. Sci.*, **117**: 816–823.

Wilkinson, B.G. 1968. Mineral composition of apples. IX. Uptake of calcium by the fruit. *J. Sci. Food Agri.*, **19**: 646–647.

Wilson, R.N., Heckman, J.W. and Somerville, C.R. 1992. Gibberellin is required for flowering in *Arabidopsis thaliana* under short days. *Pl. Physiol.*, **100**(1): 403–408.

23

Canopy Management, High Density Planting and Advanced Production Systems in Citrus

M.S. Ladaniya

ICAR-Central Citrus Research Institute,
Nagpur – 440 033, Maharashtra
E-mail: msladaniya@gmail.com

Global annual citrus production is 121.27 million tonnes and India stands third in citrus production in the world with 11.14 million tonnes followed by China with 29.56 million tonnes and Brazil 18.96 million tonnes. Indian production comes from vast area under citrus, *i.e.* 1.07 million hectares. The average citrus productivity is stagnant at 9-10 tonnes per ha in India since many years although area has expanded considerably (NHB, 2011). It is estimated that from the level of 121 crores (2010-11) it will increase to 130 crores in next 6-8 years (2017-19) and to 140 crores in further ten years (by 2027-28). To meet the growing demand and the nutritional security of the growing population, it is imperative to increase production by raising productivity levels. At present, per capita per year citrus fruit availability is 7 kg, which is quite low considering world average of 11.62 kg. With increasing population, this availability of fruit is going to drop further below if production is not increased. Citrus production needs to be increased from present level of 111 lakh (11.1 million tonnes) to 125 lakh (12.5 million) tonnes in next 5-6 years to maintain citrus fruit availability at 7-8 kg/capita/year. This increase of 10-12 lakh tonnes in production gradually during next 5-6 years is possible only with innovative approach like new canopy architecture development, intensive management practices and higher plant density to harness solar radiation for more fruit production because further area expansion has limited scope. Unfortunately there is not much research work on this aspect in the country. High density planting and canopy architecture management are complementary and supplementary to each other.

Canopy architecture engineering and canopy management have not received desired attention considering evergreen nature of the citrus plants. In deciduous crops like cherry, apple, pear and peach small canopy management using dwarfing rootstocks and training and pruning is a commercial practice. In citrus, small canopy can be obtained by using dwarfing rootstocks. Other approaches can be utilization of techniques like training, pruning, shaping tree structure using trellising so as to accommodate maximum number of plants in a unit area without compromising tree and soil health, fruit quality and at the same time increasing productivity. Citrus trees including mandarins (Nagpur, Khasi, Coorg and Kinnow), sweet oranges (Mosambi and Sathgudi), acid limes and lemons are perennial and evergreen and if neglected, they tend to overgrow and produce dense unmanageable canopy. In few years, canopy becomes so dense that interior branches become unproductive due to lack of sunlight and diseases and pests. After few years, trees start bearing fruit at the surface of the canopy only and become ultimately unproductive. Training and pruning to develop proper tree architecture in young plants promote photosynthesis by exposing leaves to light and air besides keeping trees in right size and shape for efficient orchard management and desired yields. This also controls pests and diseases by providing ample light, aeration and reduced relative humidity inside the canopy. Citrus growers, in general do not follow training and pruning and if they do at all any, it is arbitrary and it is to remove dry twigs only. Systematic training and pruning right from young age of the citrus plants is generally neglected by growers. Citrus orchards have traditionally been spaced widely at 6 x 6 m or 7 x 7 m, with little or no subsequent canopy manipulation apart from maintenance pruning once overcrowding occurs after 15-20 years.

Training and pruning affect tree canopy growth and it has direct bearing on fruiting wood right from the beginning. On one hand improperly trained and overgrown trees lack sunlight and hence produce less while on other hand trees with insufficient canopy also produce less due to insufficient leaves. Proper tree shape/canopy form has direct bearing on land area utilization, harnessing of solar energy/radiation, yield, fruit quality and economic returns. Vegetative growth and floral flushes are affected in response to pruning and use of plant bio-regulators especially growth retardants. If training and pruning is applied judiciously with growth retardants, tree vigour and thereby fruiting behaviour of commercial citrus cultivars can be modified and kept in control.

Mandarin, sweet orange and acid lime are three important citrus species grown commercially in India presently. Only twig blight affected dry branches and few weak twigs are removed by the growers. Structure, shape, flowering, fruiting and several other physiological and horticultural characteristic of citrus trees have direct bearing on orchard management and productivity. Citrus trees put forth considerable growth every year and canopies grow large within 10-12 years even with 5 x 5 m or 6 x 6 m spacing. Commercial crop is taken from 4th or 5th year onwards. High density planting may allow early cropping- right from 3 rd year. In this paper, issues related to citrus flowering and fruiting behaviour, canopy management in young and overgrown orchards, high density planting and some advanced systems of production are discussed.

Flowering Flushes, Bearing Habit and Shoot Growth

Most cultivated citrus species are evergreen and they do not have a period of true dormancy. Trifoliate orange, *i.e. Poncirus trifoliata* Raf. is deciduous. In India with tropical and sub-tropical climate, new shoot growth occurs in spring, summer (monsoon), autumn (October –November), and sometimes even in winter in central and southern parts where winter is mild. The spring and summer (monsoon) shoots are the most important ones. They should develop properly, and not be too vigorous. In mature citrus trees, the bearing wood develops mainly from the spring and summer shoots. The spring shoots which sprout from the bearing woods are the most productive ones. Single flowers or inflorescences may develop from the shoot apex or from auxilary buds. During early phase of monsoon, trees have active growth and put forth vegetative and reproductive shoots with lots of flowering. In places where winters are severe, the new growth and flowering takes place during spring.

Basic Concept and Principles of Canopy Management

In a perennial tree like citrus, canopy is managed to make best use of light and resources (land and other inputs) in the given climatic condition to get maximum crop and returns per canopy volume. Principles of canopy management involve manipulating tree vigour to make maximum use of solar radiation for higher photosynthetic efficiency. Canopy management is also aimed at reducing micro-climate that is congenial for diseases and pests. The important principle behind canopy management is to make space for convenient cultural operations with economic feasibility. Maximum possible yield and best quality fruit from the given unit area is also an important principle of canopy management.

Tools for Canopy Management and High Density

i. Training and pruning
ii. Plant growth regulators
iii. Dwarfing rootstocks
iv. Root growth control

1. Training and Pruning

Citrus growth and development processes are governed by following factors, *viz.*, i) genetic or hormonal factors, ii) environmental factors – such as rainfall and temperature, iii) cultural factors like irrigation, pruning, training and growth regulators and nutrient/fertilizer application. Under ideal growth conditions, proper training and pruning can alter physiological processes for maximum benefit of growers through utilization of maximum potential of the tree. After full growth of the tree, selective pruning remains a management practice and has significant effect on citrus yield. In South Africa, selective pruning showed that benefits are cost–effective, sustainable and achieved with minimal environmental impact.

There could be three main objectives of training and pruning *viz.*,

a) To increase the total effective leaf area index (LAI), and promote photosynthesis by exposing maximum canopy to light and air. If branches are well spaced and properly oriented, they have plenty of space and light. This improves the efficiency of water use, and also conversion of available nutrients into metabolites. In turn, this increases the yield and quality of the fruit.
b) To maintain right size for easy management and inter-culture. It also increases the vigor of the

tree, enhances its tolerance to various stresses, and helps maintain the most efficient balance between vegetative and reproductive growth.

c) To reduce incidence of pests and diseases in the orchard.

2. Plant Growth Regulators

Most plant growth retardants inhibit the formation of growth-active gibberellins and can, thus be used to reduce unwanted shoot elongation. To date several compounds have been developed which include - CCC, Mapiquat chloride, Alar, Amo-1618, paclobutrazole and uniconazole. These can be used to reduce the height of the plant. Growth retardants have several beneficial side effects such as better colour to fruit, resistance to abiotic stresses and better quality (Rademacher, 1998). Growth retardants such as cycocel (CCC), Amo 1618 and paclobutrazol were also found to reduce shoot growth in citrus (Harty and van Staden, 1988). Application of CCC produced early flowering, early and better fruit set and improved yield in lemon and acid lime.

Training and pruning combined with growth retardants can be effectively used for managing vegetative growth and fruiting behaviour. Some work has been done to control growth with pruning in overcrowded trees for better sunlight penetration. Growth retardant such as CCC has been used to check the growth and for inducing flowering to get particular crop (bahar) in acid lime. Training and pruning from early stage could restrict vegetative growth and encourage compact fruitful canopy in citrus. Higher yield with increased fruit weight, juice and TSS content were reported as a result of pruning and crop regulation with CCC in acid lime (Ladaniya and Huchche, 2012a). In young 'Nagpur' mandarin, canopy size could be effectively reduced by 25-30 per cent using CCC (3000 ppm) and pruning. Trees were trained at young age to provide 4-5 scaffold branches for a strong framework.

3. Dwarfing Rootstocks

In citrus, *Poncirus trifoliata* has been extensively used in breeding programmes for dwarfing. Several such hybrids involving Trifoliate orange as one parent and pummel, mandarin or sweet orange as another parent have been developed and commercially used. *Poncirus trifoliata* var. Monstrosa (Flying Dragon) has been successful in several countries but not in India particularly at CCRI (erstwhile NRCC), Nagpur. Citranges like Troyer and Carrizo and citrumelo like Swingle citrumelo have been very promising for high density planting in USA, Spain, Italy and many other countries.

4. Root Growth Control

It is common practice to control root growth of plants in a given container or restrict root spread so that canopy can be controlled. Bonsai is one well known example. Such control of root growth can be possible technique in large scale commercial citrus plantations.

Canopy Management

Keeping in view the different parts like limbs, scaffold branches and secondary and then tertiary branches of the citrus tree, it should be determined how these different parts can be distributed in given spacing. Fruit should not be allowed to grow on the plants until the tree is mature for commercial crop (minimum 2-3 years of juvenile phase). If the tree sets fruit during its second and third year, the fruit should be removed while they are small and green. Fruit thinning in young trees will promote the growth of the canopy. Fruit thinning in grown up but young trees improves the fruit quality. Trees which bear heavy crops tend to produce small fruit. Thinning will also correct the tendency found in citrus trees to bear a heavy crop of fruit only in alternate years, with few or no fruit during the year in between.

Canopy management in citrus trees with the help of training and pruning can be divided in to three categories according to purpose. 1). Canopy architecture to develop tree frame and increase plant density, 2). Maintaining productivity of overgrown orchard, and 3). Invigoration of old and declining orchard.

1. Canopy Architecture to Develop Tree Frame and Increase Plant Density

Citrus trees become unmanageable by the age of 10-12 years at normal/standard (6 x 6m) spacing if not trained properly right from young age. Citrus trees are evergreen in growth habit and their canopies grow exceptionally large and most of the inside area in canopy remain unutilized. With training and pruning since beginning, trees can be kept in bound and shorter canopy can be managed effectively for cultural practices. Proper leaf area can be maintained for fruiting without superfluous unproductive growth. This helps in proper aeration, light interception, easy harvesting, less incidence of insect–pests and diseases. The other major advantages of canopy management are that more number of trees can be planted per unit area (increased plant density) occupying the orchard space early with early returns.

Generally large mandarin trees grow with main limbs and branches occupying lot of space in the centre of the tree and this space without productive shoots and flushes is almost wasted. Bearing is mostly in 1 metre outer periphery of the canopy. In such trees, wood area index is relatively high. The strategy should be to create balanced growth of the plants in all parts/direction inside the tree and also outside on the periphery so that leaves are distributed in all parts of the canopy.

i) Procedures for Training and Pruning

a) After planting, side shoots and water sprouts need to be removed regularly. The strategy should be to create strong, balanced tree framework and to maintain these through the orchard's life by selective pruning. As new "internal" growth arises and develops, fruiting takes place in sheltered but well-lit and ventilated canopy interiors.

b) In general, care should be taken never to remove more than 10-15 per cent of the total shoots at a time in small trees. Four to five scaffolds at wide angel in different directions are retained. The structure of a tree should be carefully seen before they are pruned. Pruning should be done from outside and top twigs and branches. The top scaffold branch should be pruned first, followed by the second and finally the first. For each scaffold, pruning should start from the secondary branch, and then the side shoots.

c) Diseased as well as heavily insect- infested branches should be removed. Branches and shoots which are in the wrong position, or growing in the wrong direction (undesirable branches and shoots which are in cross or overlapping manner or at close angle) needs to be removed from the young age. Pruning secateurs can be used on young trees for heading back and thinning out cuts. Heading back promotes the growth of lower buds and increases branching. Thinning removes the entire branch or shoot, to reduce the total number of the side shoots.

d) Correct time for pruning is important. The metabolism of the tree is less active in the period after fruit harvest generally in dry season and this is the best time for pruning. In tropical parts of India with mild winter, the metabolism is less active before the spring flush, because temperatures are lower and less water is available to the tree. It is this period of reduced metabolic activity that is suitable for pruning. Light pruning can be conducted to remove unwanted and overcrowded shoots. In summer months (March- April) also trees can be pruned. Never prune during wet climate/rainy season.

ii) Systems of Training

There are several systems of training young trees and important one are a) Central leader, b) Open centre and c) Modified leader system. Each system has its effect on plant growth, flowering and productivity. These systems are basic for evergreen and deciduous fruit trees that are to be trained without any support or trellising.

In central leader system of training, the main central shoot is allowed to grow tall and trees get the pyramid shape or oval to round canopy. In this method of training, apical dominance is not checked. The canopy is compact in this system.

In open centre system, the main central shoot/stem is headed back at about four feet height and lateral branches are allowed to grow. Thus the centre part of the tree is open and sunlight interception is better. The tree canopy is quite open and aeration is better in this system. This system is not used for citrus.

In modified leader system, the centre shoot is cut back relatively late (after 7-8 feet height) so that canopy is opened up to some extent. This is a mid-way between central leader system and open centre system. The tree shape is open from top and not pyramid or round. The modified leader system can be most suitable for citrus in India depending on species and plant growth habit. This system can be useful in case of Kinnow in which canopy becomes very dense after certain age. There are several advantages of this system as the light interception is better, trees are easy to manage, including spraying and harvesting the fruit. This system could also be useful in acid lime under high density planting system.

Attempts were made to combine training and pruning with growth retardants to develop smaller canopy architecture and determine fruiting behaviour. Findings at CCRI (erstwhile NRCC), Nagpur have revealed that Paclobutrazole (1000 ppm) reduced inter-nodal distance in young acid lime trees of 3-4 year age and thereby significantly reduced shoot length and canopy spread. Chlormequat chloride (1000 ppm) treatment was less effective as compared to paclobutrazole in containing growth. Canopy spread and canopy volume were less when growth retardants (paclobutrazole and CCC) were combined with training and pruning (NRCC, 2013). Least canopy volume (1.98 m^3), plant height (1.65 m), canopy height (1.30 m), canopy spread (1.30 m) and trunk girth (22.16 cm) were recorded in trees under combined treatment of paclobutrazole (1000 ppm) + training and pruning. Maximum canopy volume and plant height were recorded in untreated trees. At the age of 4 years, canopy height and canopy volume were 25-30 per cent lower in paclobutrazole (1000 ppm) treatment than control. When paclobutrazole (1000 ppm) was combined with training and pruning, canopy volume was 40 per cent less as compared with untreated trees. Maximum vegetative shoot length, inter-nodal distance and leaves per shoot were recorded in untreated trees. Floral shoot length was maximum in control but number of flowers per shoot were maximum in paclobutrazole (1000 ppm).

Pruning along with application of chlormequat chloride (CCC) contained/reduced tree growth and canopy of 'Nagpur' mandarin (Ladaniya and Huchche, 2012, b). The least canopy volume (2.72 m^3) and spread (1.64 m) were recorded in trees under treatment of CCC (3000 ppm) + pruning and training. CCC treatment reduced inter-nodal distance. Pruning + CCC had significant positive effect on flowering and fruiting of 'Nagpur' mandarin.

Training and pruning from early stage could restrict vegetative growth and encourage compact fruitful canopy in citrus (Goswami *et al.*, 1993). Higher yield with increased fruit weight, juice and TSS content were reported as a result of pruning in sweet lime (Arora and Yamdagni 1985) and Assam lemon (Nath, 1994).

2. Maintaining Productivity of Overgrown Healthy Citrus Orchards

In an overgrown citrus orchard, the trees may be healthy but lose vigor due to crowding and poor light interception. Their yields decline and their canopies become too large. This is the time when they need pruning. The trees which are showing low productivity due to overcrowding of branches but are otherwise healthy should be pruned gradually over the period of 2-3 years. Drastic pruning would reduce the yield considerably. Trees get round canopy shape after the pruning. Optimal light interception for the entire orchard is the first thing to be considered while pruning. The individual tree form and tree vigor should be taken in to account. Tree's response to pruning varies according to its growth or age. Twenty-year-old trees which are totally neglected may need 25-40 per cent canopy reduction initially while 15 year old tree may need less pruning.

In overgrown acid lime trees, *Hasta* crop (Autumn bloom), which is very profitable, can be induced during October-November and it can be harvested in summer (April–May). Chlormequat chloride can be effectively used in combination with pruning to get *Hasta* crop. In pruned + CCC treated trees, average yield was nearly 50 kg per tree which was about 20 tonnes per ha (extrapolated to per ha basis with 400 trees at 5 × 5 m distance as in the existing plantation). In non-pruned and non-treated trees, yield was significantly low (30-35 kg/tree or 12-14 tonnes/ha). Yield efficiency of non-treated trees was also low as compared to pruned + CCC treated trees. Higher yield efficiency in pruned trees indicated higher fruiting per cu m^3canopy volume (Ladaniya and Huchche, 2012, a). These findings consecutively for three years revealed that canopy of overgrown acid lime trees of 20 year age can be managed very efficiently by regular pruning and desired canopy volume around 45-50 cu m can be maintained and *Hasta* crop can be taken regularly with CCC application. With application of recommended plant protection and cultural practices (including nutritional requirements), mature acid lime trees produce very good crop every year with higher yield efficiency as compared to non-pruned trees. It is very difficult to carry out normal cultural operations in overgrown and non-pruned trees as branches touch to ground and practically no space to move around due to very dense canopy (Ladaniya and Huchche, 2013). In pruned trees, re-growth occurs rapidly and it was observed that around 15 m^3 of canopy grows again in a period of one year.

In grown up plants, pruning stimulates production of healthy and strong branches directly subtended by scaffold branches in which sap flow is abundant. These strong branches bear large crop and unblemished fruit of uniform high quality.

3. Invigoration of Old and Declining Orchard

In citrus production, decline due to old age and also due to various diseases and pests is a big problem and dry branches need to be pruned to arrest the disease spread. This is must to invigorate the trees. There are thousands of citrus trees which are in various stages of decline in the country. The pruning required is slight to severe in these orchards depending on stage of decline. Tree twigs or branches need to be pruned 1-2 cm below the dry part with suitable tool depending on thickness of the shoot/branch. It should be done in summer or winter when weather is dry. Applications of copper oxychloride (COC) or Bordeaux paste on thick branches followed by COC (0.3 per cent) sprays with carbendazim (0.1 per cent) fungicide to drench entire canopy immediately after pruning are must. Normal horticultural practices need to be followed for tree nutrition, irrigation and to maintain orchard health.

High Density Planting in Citrus

Basic Concept

High density planting is the technique of growing plants at short or very short distance as compared to standard (traditional) planting distance and thus accommodating large number of plants per unit area to get maximum yield and possible profit per unit of tree volume without impairing fruit quality. High density planting should also be economically and environmentally sustainable.

Advantages

There are several advantages which are as follows:

i) Efficient utilization of land
ii) Early and higher yields per unit area
iii) Higher returns and early economic profit from investments.
iv) Higher water and nutrient efficiency.
v) Lower harvesting costs.
vi) Increased possibility of mechanized operations

Ways to Achieve High Density

i) Dwarfing rootstock
ii) Moderate growing scion with upright growth
iii) Training and pruning
iv) Growth retardants
v) Root restrictions
vi) Use of Exocortis viroid

The ability to control tree size is central to the philosophy of high density plantings (HDP). Tree size can be controlled with dwarfing rootstock or less vigorous rootstocks or by training and pruning.

The effects of mild strains of exocortis viroid has also been encouraging and being utilized commercially in many countries especially in Australia. The viroid method has been used successfully to control Valencia orange [*Citrus sinensis* (L) Osb.] tree size propagated on *P. trifoliata* or on citrange (*P. trifoliata* x orange hybrid) rootstock. Trees were deliberately inoculated (young trees) with budline tissue taken from mildly dwarfing citrus exocortis viroid (CEV) infected trees. The cropping efficiency (Yield/canopy surface area (CSA) in terms of Kg/m^2) of trees so grown was not significantly depressed and yields were proportional to canopy surface area (CSA).

The yield potential of a citrus orchards is related to the CSA per hectare. A substantial increase in orchard productivity was demonstrated by utilizing HDP of dwarfed "Valencia" orange trees and no detrimental effects on fruit quality were identified. Greater productivity can be achieved with close within-row spacings and row spacings. In many countries, on semi-dwarfing rootstocks like Carrizo citrange and Trifoliate, mandarin trees are grown at spacing of 6 x 2 m and after 4-5 years of bearing, one tree within the row is removed to make it 6 x 4 m.

Normal recommended spacing for citrus is 6 x 6 m with 277 plants/ha. More number of plants per ha can be planted at 5 x 5 m spacing (400 plants/ha). Spacing can be further reduced as 6 x 3 m (555 plants/ha). The per ha yield of Nagpur mandarin during seven year period (1994-95 to 2000-01) was maximum at spacing of 6 x 3 m due to better light intervention and higher number of plants per ha than normal recommended spacing of 6 x 6m (Ingle and Athavle, 2005). Similar results were obtained in Kinnow mandarin at 6 x 3 m spacing as compared with 7.6 x 7.6 m spacing (Sharma *et al.*, 1992). In Kinnow, spacing of 6 x 3 m and 6 x 4 m is being followed presently by some farmers in Punjab.

If Rough lemon or Rangpur lime is used as rootstock, then pruning of trees and plant protection measures should be the regular practice in high density plantations. Pruning of trees is must after certain age whether rootstock is dwarfing, semi-dwarfing or vigorous. Goswami *et al.* (1993), studied root activity and root distribution pattern of rootstocks and reported that Troyer and Karna khatta are suitable for Kinnow. With Troyer citrange as rootstock, Kinnow could be planted at 2 x 2 m spacing with 2990 plant density per ha. With Karna khatta, 8 x 8 m and 10 x 10 m was ideal. Plants were trained to multiple shoot system. This provided adequate area for fruiting and it was easy for crop management. This training system also checked terminal growth. Fruiting started after 3 rd year and 50-60 fruits per plant were produced in initial year resulting in yield of 26-27 tonnes per ha. Production was quite good up to 12 years. Thereafter regular pruning was necessary every year or removal of alternate plant in alternate row was suggested for reducing density. It was observed that fruit maturity was early (in November) on Troyer citrange rootstock. On Karna khatta, harvesting was staggered up to mid –January due to late fruit maturity.

In Brazil, spacings of 1, 1.5, 2, 2.5 and 3 m between trees and 6 m between rows have been evaluated for tree growth and fruit quality of Tahiti lime (*Citrus latifolia*) cv. IAC-5 grafted on Flying Dragon rootstock. The productivity was higher per hectare in 6 x 1.5 m spacing than that of other spacings. Effect of spacing was minor on fruit quality, however, spacing of 6 x 3 m provided greater fruit juice, vitamin 'C' and TSS contents. Findings suggested that high density plantings are possible in future using Flying Dragon rootstock (Machado, *et al.*, 2012).

In Florida, Wheaton *et al.* (1990) have tried high density planting of sweet orange, grapefruit and Murcott tangor. Spacing tried was 1.5 x 3.3 m with density of 800 trees/acre (2000 trees/ha). Trees reached near containment size after 7 years. It was observed that variety and rootstock combination of moderate vigour and good fruiting were most desirable. Hamlin and Valencia sweet oranges, Red blush grapefruit and Murcott tangor were budded on six rootstocks namely orange, Swingle citrumelo, Carrizo citrange, Koethen Rubidox, Morton and Rusk citrange. Canopy volume achieved after 7 years at 1.5 x 3.3 m spacing varied from 7.22–8.40 m^3. These all stionic combinations produced moderate vigour and good fruit quality. Trifoliate rootstock did not perform well. Murcott tangor appeared particularly suitable for high densities because of smaller tree size and upright growth habit. Both sweet orange varieties also adapted well to high density but grapefruit had higher vigour. The production was 20.5 tonnes per acre (51 tonnes/ha) in 5th year and even reached 23-25 tonnes per acre (57.5-62.5 tonnes/ha) in 7 th year up to 15 years.

Tribulato *et al.* (1992) reported that, in double row system (2.5 x 2.5 m plant to plant in two rows and 4.5m between the rows) with density of 1200 trees/ha, drastic pruning was required to contain canopy volumes of Navelina orange on sour orange rootstock. The cumulative per ha yield for 13years was 493 t/ha and 289 t/ha in case of density of 1200 plants/ha (high density) and 416 plants/ha (standard density). With increase in canopy volume per ha over the years, yield increased proportionately. Higher density of 800 plants/ha (4.3 x 2.8m) in single row, produced better results where fast hedge formation provided better space utilization and higher production. Double rows are valid where plants have more compact canopy such as 'Navelina' orange. After 11 years, one row was removed to get density of 600 plants/ha. Working with standard vigour cultivar and rootstock, high density sooner or later requires thinning which must be taken in to consideration right from the start of the orchard.

In 'Valencia' orange budded on sour orange and Troyer citrange rootstocks, closer planting distance resulted in higher yields per unit area. Planting at the distance of 6 x 4 x 2 m and 5 x 2.5 gave 123 per cent higher yields compared to standard 6 x4 m. Highest values for mean fruit weight were obtained in the 5.5 x 3 m spacing which also produced highest cumulative yields per tree together with greater canopy volume. A 5.5 m between row spacing can be sufficient for mechanized cultural operations (Introgliolo, *et al.*, 1992).

The trend of initial results at Central Citrus Research Institute (CCRI), Nagpur has indicated that close planting of Nagpur mandarin (in 2012) on conventional vigorous rootstock like Rangpur lime resulted in vertical lanky growth of trees at 2 x 2 m spacing probably due to competition for sunlight. During spring blossom crop (*Ambia* season) in 2015, first fruiting was allowed during third year after planting that resulted in 12.36 kg yield per tree which was extrapolated to 30.89 tonnes/ha at 2 x 2m spacing (2,500 plants/ha). At 6 x 6m, 6 x 3 m, 4 x 4m, and 4 x 2 m, the per plant yield was 4.30, 6.49, 2.59 and 5.12 kg/tree, respectively (CCRI, 2016). In acid lime 'Kagzi' budded on Rangpur lime rootstock, first fruiting was allowed during third year after planting (in 2012) and per plant yield was 4.17 kg at 3 x 1 m spacing which was extrapolated to 13.88 tonnes. At 4 x 1 m, 6 x 3 m and 6 x 6m spacing per plant yield was 4.56, 6.53 and 8.50 kg, respectively. At 6 x 6m spacing (277 plants/ha), per hectare yield was 2.36 tonnes. Next year (4th year after planting) at 3 x 1m spacing, yield was 57.66 tonnes per hectare while it was 6.73 tonnes/ha at 6 x 6m spacing.

Challenges of High Density Planting

In high density planting, generally problems arise when allotted space to the trees is filled with canopy and pruning becomes must. The cost of investment on irrigation could be higher besides cost of trees. The cost of planting/layout and material cost of fertilizers pesticide in initial years could be higher. The cost of operations of equipments (small trucks and tractor drawn implements) could also be higher due to close planting and higher travel cost between the close rows. The costs which are independent of tree density include cost of land and its preparation. Higher and early returns offset the high initial costs.

Advanced Citrus Production System (ACPS) or Open Hydroponic System (OHS)

Advanced production system or open hydroponic system (OHS) was designed for consistently good crops of right sized fruits and the system was developed by Prof. Raphael Martinez of Spain. Since then it has been adopted in many countries and modified to local requirements.

The hydroponic system is soil-less culture (in fact it is in water with required nutrients and other chemicals). Here in open hydroponics, nutrients are not reused or re-circulated as in a conventional closed hydroponic system. In open hydroponic system it is in open field, mostly in sandy or gravel soils with less/very less fertility and application of water and nutrients is quite frequent. The normal role of soil is diminished and relegated to providing anchoring to trees only. The wind breaks are required as the root system is fairly shallow in such plants.

Generally open hydroponic system as followed in South Africa is in soils with 80 per cent sand and 20 per cent clay and silt. The ridges are formed (18 – 30 inches high and 3 – 5 feet wide). Oranges or lemons or mandarins on Carrizo citrange rootstock are planted 5 feet apart on ridge. The row to row distance in plants is 16 feet (540 trees/acre). The drip irrigation system with single line of drippers spaced at 32 inches apart and delivering 0.4 gallons/h is installed with filtration and fertilizer injection system. The controllers are linked to soil water sensors and weather system. Once the orchard is established, the trees are fertigated daily throughout the year with 30-min. pulses every hour from 6 a.m. to 6 p.m. Nutrients are applied by dividing the total yearly dose in to day-wise and also considering active and relatively inactive growth phases. The application of nitrogen when active growth taking place and phosphorus to support root growth when shoot growth is not occurring. Girdling is followed to induce flowering after 2 years. Upright branches are pruned to keep growth under control, canopy open for light penetration.

The yield in such a system is generally 500 boxes per acre (one box contains roughly 40 kg fruit). The yield of 345 boxes/acre of Murcott mandarin/tangor on Carrizo rootstock has also been reported.

The primary objective of open hydroponic system is to manipulate tree growth and get the crop to favourable ends. In open hydroponics we have to manage feeder root system and supply required nutrients and water continuously for vegetative and reproductive growth as the case may be. The advantages of open hydroponic or advanced production system is reliable high returns and managing diseases. Furthermore this system can also help researchers of crop improvement. The new varieties can be evaluated faster and delivering them as early as possible to growers.

In Florida micro-sprinklers are common instead of drip. Micro-sprinklers provide protection from frost/severe cold. In Florida, recent research results have indicated that OHS/ACPS is quite promising. The main goals of ACPS are early, high production, early economic returns on investment and disease avoidance.

Main Components of OHS

i) Intensive fertigation using sensors and controllers guided by meteorological data.

ii) Balanced and complete nutrition
iii) Use of dwarfing rootstock
iv) High density planting
v) Effective psyllid and Huaglongbing management

The objectives accomplished in OHS are:

i) Increased water and nutrient use efficiency (partial root zone drying) and reduced environmental impacts.
ii) Good protection of new flush (leaves and blossom) during establishment.
iii) It is synonymous with computerized fertigation using pulsed drip emitters, frequent monitoring and remote control.

Field Layout, Spacing and Yield

In ACPS, dwarfing rootstocks like C-35 citrange (hybrid of Ruby orange x *Poncirus trifoliata*), Swingle citrumelo, X639 and Carrizo citrange were commonly used. The distance between two rows was 15 feet and varied up to 18 feet. Plant to plant distance could be 5 – 8 feet. Thus plant densities could be 363 trees/acre (8 x 15 feet) or 303 trees/acre (18 x 8 feet). Trees could be planted in single row or double rows in ACPS.

In private commercial plantations in Florida, high density blocks resulted in sustained production in Valencia orange on Swingle citrumelo rootstock and such plantations were reported since 1992. Production was as high as 615 boxes per acre at density 247 trees/acre (9.37 x 18.75 feet). The trees required hedging twice per year. Schumann *et al.* (2012) reported grove designs for higher efficiencies in ACPS. In "tramline" or zigzag planting APS experiment, Valencia orange planted at a density of 538 trees/acre. In this system, actually 20 per cent more row length and 30 per cent larger canopy was achieved in this triangular planting pattern as compared to straight row. Plant to plant distance was 9 feet. In between two plants, one additional plant was adjusted thus twin plants were 5.4 feet. The distance between the rows was 18 feet.

In Hamlin orange on C-35 rootstock planted in 2008 December, the crop was harvested in December 2010 after 24 months. The spacing was 8 x 15 feet (363 trees/acre). Per tree yield of first crop was 10 lbs/tree (1650 kg/acre) (Gapway Grove Corpn, Citrus Res. and Dev. Foundation Inc. Florida). In ACPS, depth of root system of plants was not much and two drippers 15 inch apart were placed. The water and electrical conductivity sensors were placed at the depth of 4 inch and 18 inches, respectively.

Huanglongbing (HLB) or Citrus Greening Management

In Florida, objective behind ACPS is to avoid HLB and get maximum returns before economical losses start occurring due to HLB. With high densities, HLB infected tree removal do not have much economic impact. There is possibility that HLB infected trees produce little lower yield and that is compensated by many trees in high density plantation. Eventually when enough diseased trees have been removed to make orchard unprofitable, the whole orchard is re-planted with disease-free trees. Slightly higher costs of pest control are compensated by higher production and cost-saving with early yields so economic pay-back is reached in fewer years to offset some losses due to disease. Early economic fruit production and net profits should be reached in as little as 3 years thus achieving early returns on investment.

Citrus on Trellising System

Can we grow citrus on trellising? The answer could be yes. In South Africa, W. Murcott trees on X639 rootstock planted at 6.5 X 15 feet spacing have been tried and trained to form a fruit wall on a 4- wire trellis.

Containerized Production under Protected Conditions

In fresh fruit production, research indicated that very high density of 2000 trees/acre (at 3.0 x 7.5 feet spacing) can yield up to 1680 boxes/acre (fruit weight 90 pound or 40.90 kg per box) at 4 year age in mandarin. The central irrigation system was used in combination with fertilizer injection. The plants were planted in containers and it was a sort of Hydroponic System so that water and nutrients were controlled with computer aided system. The first crop was harvested at 14 months age. The plantation was taken under shed cloth (insect-free screen to avoid insect vector citrus psyllids). In United states, several growers (with 10-40 acre plantation) of mandarin and sweet orange varieties for fresh fruit market have come forward for under cover production. Under HLB-endemic environment, integrated sustainable fresh citrus fruit production systems have been reported and called as Citrus Undercover Production Systems (CUPS). Different growing medias like Canadian peat and perlite have been used in pots of 5, 7 and 10 gallon (1 gallon = approx. 4 litre) capacities to grow disease-free plants undercover. High value citrus varieties for fresh fruit market can be grown for higher economic returns (Schumann *et al.*, 2015).

Future Thrust Areas

i) More dwarfing rootstocks need to be tried for high density plantation in India. *Poncirus trifoliata* and its hybrids are successful in cool climate and in little acidic soils. For central and south Indian agroclimatic conditions, dwarfing rootstock need to be bred/identified.
ii) Effect of canopy shape and size (architecture) management on tree yield and fruit quality needs to be studied in all commercial cultivars in India.

iii) It is to be determined whether high density can be maintained through canopy architecture engineering on sustained basis in standard (vigorous or semi-vigorous) rootstocks and if so, how long in terms of duration/life of that orchard ?.

iv) The economic viability of various canopy architecture and high density systems needs to be worked out.

v) Containerized advanced production system with dwarfing rootstock, sensors and fertigation under cover (protected condition) need to be evaluated and its economics and environmental impact to be determined for citrus cultivars in India.

References

Arbizo, H., Ferenczi, A., Gambetta, G., Rivel, F. and Graviva, A. 2000. Physiological and productive response to pruning in three citrus cultivars. Abstract. *Int. Soc. Citrus Congress.* Dec. 3–7, Florida, p. 132.

Arora, R.K. and Yamadagni, R. 1985. Effects of different levels of pruning on flowering, fruit set, final retention and fruit quality in sweet lime (*C. limettioides* Tanaka). *Prog. Hort.*, **17**(1): 1–4.

CCRI, 2016. Annual report 2015–16. Central citrus research institute, Nagpur, India.

Goswami, A.M., Saxena, S.K. and Kurien, S. 1993. High density planting in citrus. **In**: *Advances in Horticulture, Vol. 2: Fruit Crops*, Part 2, pp. 645–648.

Harty, A.R. and van Staden, J. 1988. Paclobutrazol and temperature effects on lemon. *Proc. 6th Int. Citrus Cong.*, **1**: 343– 353.

Ingle, H.V. and Athavale, R.B. 2001. Studies on high density planting in Nagpur mandarin. *South Indian Hort.*, **49**(Special Issue): 16–17.

Introgliolo, F., Reforgiato-Recupero, G. and Giuffrida, A. 1992. Effect of planting density and rootstock on performance of 'Valencia' orange. *Proc. Int. Soc. Citriculture. VII Proc. Intn. Citrus Cong.*, Italy, **2**: 705–708.

Ladaniya, M.S. and Huchche, A.D. 2012a. Effect of pruning and CCC on overgrown acid lime (cv. Kagzi) trees to contain growth, regulate cropping and increase productivity. Paper presented in *"National Dialogue on Citrus Improvement, Production and Utilization"* organized at NRCC, Nagpur on 27–29th February.

Ladaniya, M.S. and Huchche, A.D. 2012b. Canopy growth and flowering in young Nagpur mandarin plants as influenced by training, pruning and chlormequat chloride (CCC). Paper presented in *5th Indian Hort. Cong.*, Punjab Agriculture University, Ludhiana.

Ladaniya, M.S. and Huchche, A.D. 2013. Pruning and Management of overgrown and senile acid lime orchards for sustained productivity. Extension Bulletin No. 45, National Research Centre for Citrus, Nagpur, pp. 16.

Machado, D.L.M., Alves, R.R., Siqueira, D.L. Salomao, L.C.C. and Silva, S.D.R. 2012. Development and production of Tahiti lime IAC–5 grafted on 'Flying Dragon' (*Poncirus trifoliata* var. *monstrasa*) grown in high planting densities. Paper presented in *12th Intn. Citrus Cong.*, Valencia, Spain, 18–23 November, pp. 181.

Nath, J.C. 1994. Effect of pruning intensity on growth, yield and quality of Assam lemon (*C. limon* Brun). *Haryana J. Hort. Sci.*, **23**(4): 281–285.

NHB, 2011. *Horticulture Database*. National Horticulture Board, Gurgaon, Haryana.

NRCC, 2013. Plant growth, canopy and productivity of acid lime (Kagzi) as affected by training, pruning and growth retardants. *Annual Report* (2012–13), National Research Centre for Citrus, Nagpur.

Rademacher, W. 1998. Growth retardants: Bio-chemical features and applications in horticulture. *Acta Hort.*, **394**: 345–349.

Schumann, A., Hostler, K. and Waldo, L. 2012. Advanced citrus production systems: Grove designs for higher efficiencies. *Citrus Industry*. September, pp.6–9.

Schumann A., Rogers M., Brlansky, R., Grosser, J., Gmitter, F and Vashisth, T. 2015. Advanced citrus production systems. Field Day Handout., October 29, UF, IFAS,CREC, Lake Alfred, Florida.

Sharma, J.N., Chohan, G.S., Vij, V.K. and Monga, P.K. 1992. Effect of spacing on growth, yield and quality of Kinnow mandarin. *Indian J. Hort.*, **49**: 158–164.

Tribulato, E., Continella, G. and La Rosa, G. 1992. Research on higher density planting for orange and lemon. *Proc. Int. Soc. Citriculture. VII Int. Citrus Cong.*, Italy. **2**: 702–704.

Wheaton, T.A., Castle,W.S, Witney, J.D., Tucker, J.D.H. and Muraro. R.P. 1990. A high density planting. *Proc. Flo. State Hort. Soc.* **103**: 55–59.

24

Role of Pollinizers and Pollinators in Enhancing Horticulture Crop Productivity

R.K. Thakur[1]*, K. Kumar[2], K.M. Kumaranag[1] and Mohammad Mustaffa[1]

[1]ICAR- AICRP on Honey Bees and Pollinators, Division of Entomology, Indian Agricultural Research Institute, New Delhi – 110012
[2]Department of Pomology, College of Horticulture, Dr. Y.S. Parmar University of Horticulture and Forestry, Nauni, Solan – 173 230, Himachal Pradesh
_*E-mail: rkt_apic@rediffmail.com_

Pollination is crucial for plant reproduction, and animal pollination (primarily bees) is responsible for 35 percent of the world's cropproduction. In agro-ecosystems, pollinators are essential for orchard, horticultural and forage production, as well as the production of seed formany root and fibre crops (FAO 2011). The total horticulture produce reached 277.4 million metric tonnes in 2013, making India the second largest producer of horticultural products after China. Many horticultural crops are actually self-sterile and require cross-pollination to produce seeds and fruit (McGregor, 1976; Free, 1993). But it is not only self-sterile varieties that benefit from cross-pollination, but self-fertile varieties also produce more and better quality seeds and fruits if they are cross-pollinated (Free, 1993). Therefore, it is necessary to explore all possible ways of increasing the sustainable productivity and carrying capacity of the farming systems in order to improve the livelihoods of marginal households (Partap, 1998). In view of the importance of bees in increasing the yield of cross-pollinated crops, different species of honey bees and solitary bees are being utilized in North India. Himachal Pradesh, Uttar Pradesh and Kashmir are the principle temperate fruit growing regions of the country. In Himachal Pradesh, more than 75,000 hectares of land is under temperate fruit cultivation and it requires more than 2,00,000 colonies of honey bees only for pollination. Amongst several factors attributing to increase productivity, the most important of which include the number of bee pollinators. Thus, keeping in view the importance of bee as an agro-based industry and its role in increasing crop yields through bee pollination this chapter will review some literature regarding the importance of pollination in improving horticultural crop productivity.

Pollination and its Types

The act of "pollination" occurs when pollen grains are moved between two flowers of the same species by wind or animals. Successful pollination results in the production of healthy fruit and fertile seeds, allowing plants to reproduce. Almost 90 per cent of all flowering plants rely on animal pollinators for fertilization, and about 200,000 species of animals act as pollinators. Of those, 1,000 are hummingbirds, bats, and small mammals such as mice. The rest are insects like beetles, bees, ants, wasps, butterflies and moths. There are two types of pollination called self-pollination and cross-pollination. Self-pollination is the more basic type of pollination because it only involves one flower. This type of pollination occurs when pollen grains from the anther fall directly onto the stigma of the same flower.

Although this type of pollination is simple and quick, it does result in a reduction in genetic diversity because the sperm and egg cells of the same flower share genetic information. Cross-pollination is a more complex type of pollination that involves the transfer of pollen from the anther of one flower to the stigma of a different flower. This type of pollination results in an increase in genetic diversity because the different flowers are sharing and mixing their genetic information to create unique offspring (Thakur and Gupta, 2016).

Pollinizer

Is a source plant for compatible pollen that normally blooms at the same time as the target plant. For good yield, *e.g.*, apples require a different variety of tree (for example Golden Delicious for Royal Delicious) as a pollinizer. It's very important to select a pollinizer variety that blooms profusely and has a flowering period that overlaps with that of the commercial varieties. In India, only 2 per cent of farmers in the Shimla Hills, and Kullu Valley and none in Kinnaur met the standard requirement of 30 per cent pollinizer trees in their orchard. About one quarter of farmers in Himachal Pradesh had between 20 and 25 per cent pollinizer trees. Anyhow there is a lack of knowledge about the importance of pollinizer-pollination systems, farmers have ignored the need to plant appropriate ratios of pollinizer apple varieties.

Pollinator

Is the agent that moves pollen from the anthers to the stigmas of flowers, thus enabling pollination. The most common pollinators are bees.Over 75 per cent of the world crops and 80 per cent of all flowering plant species rely on animal pollinators (Kevan *et al.*, 2002). Of Hundred or so animal-pollinated crops which make up most of the world's food supply 15 per cent are pollinated by domestic bees, while at least 80 per cent are pollinated by wild bee species and other wildlife (Ingram *et al.*, 1996). Decline in the diversity and abundance of natural insect pollinators is the second most important factor affecting crops pollination, and thus productivity. In recent years the diversity and population of these natural insect pollinator have been declining for several reasons: including loss of food and nesting habitats due to clearance of forest and grassland for agricultural purposes and indiscrimination use of pesticides. Further, changes in climate might be affecting insect numbers (Verma *et al.*, 1990).

Pollination in Important Temperate Fruits

Apple

The flowers are fragrant and borne in groups of six. Each flower consists of five sepals, five pinkish- white petals, and 20-25 stamens surrounding the carpel having a single ovary, a style and five stigmas. The ovary is divided into five chambers, each having 1-4 ovules. Although fertilization of every ovule in the ovary is not necessary for fruit development, for a larger perfect fruit a larger number of ovules should be fertilized. Inadequate pollination results in low number of seed, which may result in lop- sided or symmetrical fruits. Moreover, fruits with few seeds are more likely to drop. Almost all commercial varieties are self-incompatible and require pollen from compatible polliniser varieties. Moreover, the pollen is sticky and so wind pollination is not effective. Pollination largely depends on insects, especially honeybees. The flower produces plenty of nectar and pollen, which helps to increase the strength of honeybee colonies. Strong colonies also collect surplus honey from its flow. The blooming period of apple, pear, cherry, apricot, plum is very short and fifty percent of flowering occur within 3-4 days, honeybee colony should be moved to orchard as soon as tree start blooming. Because the shape and size of apple fruit depend upon the number of ovule fertilized, there should be plenty of bees/insect pollinators in the orchard. In case attractive crop like mustard or rubus are in flowering in the vicinity of orchard, increase the number of recommended colonies for pollination of crop.

Apricot

The apricot blooms February-March for 2-3 weeks, depending on the weather. The flower is usually white and occurs either singly or doubly. It has five sepals, five petals, and occurs either singly or doubly. It has five sepals, five petals and about 30 stamens surrounding a carpel having a single ovary containing two ovules, and one stigma. It produces plenty of nectar and pollen. Some varieties are self-compatible and some are completely self incompatible and require pollen from a compatible polliniser. Cross-pollination is essential for self incompatible varieties and is beneficial to self compatible varieties. Honeybees are its primary pollinator.

Cherry

The cherry blooms during February-March for 7-10 days. The flower is pinkish-white and produces plenty of nectar and pollen. It has five sepals, five petals, 20-25 stamens, and one pistil consisting of and ovary having one or two ovules, a style, and a stigma. While cross-pollination are essentials for self incompatible varieties. It is also beneficial to self compatible varieties. Honeybees are the primary pollinators. Pollination is simple. Since its blooming period is short and 50 per cent of the flowering occurs within 3-4 days, place honeybee colonies as soon as flowering start.

Pear

The pear blooms during February-March for about 7-12 days. The flowers are white and produced in clusters of 7-8. The flower has five sepals, five petals, 20-25 stamens, and one pistil consisting of an ovary a style, and a stigma. The stigma is receptive before its anthers release pollen. Some varieties are self incompatible and some are self compatible. Cross-pollination is essential and some are self compatible varieties and beneficial to self compatible varieties.Flowers produce plenty of nectar and pollen. Honey bees visit mainly for its highly attractive pollen. Pear produce a large number of flowers and for a satisfactory crop, only five per cent are required to set fruit. Commercial varieties are self incompatible and the blooming period is short with 50 per cent of the flowering occurring within 3-4 days. Therefore, for sufficient pollination, place recommended number of colonies as soon as flowering starts.

Plum

The plum blooms during February for 1-2 weeks depending on the weather. It produces white flowers in clusters of 2-3. The flower consists of five sepals, five petals, 25-30 stamens and a single pistil that has an ovary with a single ovule, a style and a stigma. Varieties vary from completely self compatible to completely self incompatible; however, the major varieties are self incompatible. Flower produces good amount of nectar and pollen and are visited by many species of insects. Honeybees are the primary pollinator. The blooming period is short with 50 per cent of the flowering occurring within 3-4 days. Therefore for sufficient pollination, place recommended number of colonies as soon as flowering starts.

Impact of Pollination on Fruit Set

Experiment in apple orchard was conducted at Kullu, where one colony each of *A. mellifera* and *A. cerana* were kept (Sharma *et al.*, 2001) to record the fruit set at different distances from the bee colonies. Per cent fruit set in apple on spur basis at 25, 50 and 75 m away from placement of bee colonies was 27.31, 29.22 and 29.52 per cent, respectively which was identical however, it differed statistically from the fruit set recorded at 100 m (18.35) from the bee colonies, indicating more effectiveness of the colonies within 75 m. Another experiment was carried out in two locations at Katrain (Kullu) *viz.*, Kamalda and Ghurdaur, where the orchards were having optimum pollinizers (Sharma and Gupta, 2001). At Kamalda, bees were kept in groups of four and at Ghurdaur in groups of two. The results showed that the per cent fruit set did not vary upto 100 m distance from the colonies when kept in a group of four, however, per cent fruit set was statistically less at 100 m when the group consisted of two colonies. This indicated that if the honeybee colonies are kept in a group of four their pollination efficacy in respect to the distances increased. Gupta *et al.* (1993) also studied the effect of placement of varying number of *A. mellifera* colonies on fruit set in apple orchards having different proportion of pollinizers. Their findings suggested that in orchards having about 20 per cent pollinizers, for maximum fruit set (49.31 per cent after June), bee colonies should be placed in groups of four and each group separated by a distance of 50 metres. In orchards having only 3 per cent pollinizer, greater fruit set (3.33 to 14.30 per cent) can be obtained by keeping 3 colonies in a group than by distributing them singly. They further reported that at Kullu, where the orchard had an ideal pollinizer ratio (30 per cent), a uniformly good crop of apple was recorded by keeping only 4 colonies (fruit set of 29.28 to 43.08 per cent).

Rana *et al.* (1998) conducted experiment in apple orchards at Kotkhai where effect of honeybee pollination on yield parameters of apple in relation to pollinizer proportion was studied. A significant increase in per cent fruit set from 32.48 to 53.55 with increase in *A. mellifera* colonies (0, 2 and 4 colonies) was noticed in apple orchards having adequate pollinizer proportion. Similar increasing trend in fruit set was observed by placing 0, 2, 4 and 8 colonies per hectare in orchards with low pollinizer proportion. Fruit set was also significantly higher on the plants closer to the hives (placement at 25 or 50 m) than when 75 m away.

Studies on the effect of placement of *A. cerana* colonies in apple orchard showed that there was about 3-4 times increase in *A. cerana* visits in such orchards along with significant increase in fruit set and fruit yield (Sharma *et al.*, 2002 a). However, increase significantly higher in the orchards with sufficient pollinizers as compared to the pollinizer deficient orchards. Demonstration given in Kullu valley in the apple orchards of the orchardists showed that there was upto 100 per cent increase in the crop yield (Sharma *et al.*, 2002b).

Similarly significant increase in apple yield had also been found where bees were kept (Sharma *et al.*, 2003). At Larankelo, an increase of 108 and 53.8 per cent in apple yield, respectively was recorded during 2000 in the orchards having sufficient and insufficient pollinizer proportion due to placement of *A. cerana* colonies. In 2001, an increase of 39.50 and 24.40 per cent was recorded, respectively in apple orchards with sufficient and insufficient pollinizers.

Studies conducted on role of honey bees in apple production in one of the orchard in Kashmir valley revealed that as a result of introduction of *Apismellifera* colonies in apple orchards, there was an increase in fruit set (9-12 per cent). The maximum fruit set was observed in Red Gold (57.74 per cent) followed by Golden Delicious (33.63 per cent). However, the minimum fruit set was observed in Vista Bella (0.10 per cent).

The data presented in Table 1 revealed that percentage fruit set varied in different treatments. In case of bee pollination, fruit set was significantly higher (84.14 per cent) followed by hand pollination (76 per cent) followed by open pollination (74.76 per cent). All these treatments were superior over self pollination where fruit set was only 22.5 per cent. The data clearly shows that bee pollination resulted in higher fruit set followed by hand pollination and open pollination as compare to treatments with insecticides and self-pollination. Similarly, the percentage increased over control was much higher in bee pollination as compare to other treatments.

Table 1: Impact of Pollination Treatments on Quality and Quantity of Fruit Set in Cucumber

Treatments	*Fruit Set (%)*	*Percent Increases Over Control (Self pollination)*
Hand Pollination	76.00	237.3
Self Pollination	22.53	–
Bee Pollination	84.14	273.4
Open Pollination	74.76	231.8

Pollination by Honey Bees

Nearly 70 percent of the cultivated crops all over the world are cross-fertile and depend on insects like honey bees for pollination. Insects are the most commonly occurring pollinators of many agricultural and horticultural crops. Different kinds of insect pollinators such as bees, flies, beetles, butterflies, moths and wasps are important pollinators of many crops. Of all the pollinating insects, honey bees are considered as the most efficient and reliable crop pollinators. The beekeeping industry has rarely put itself forward as a Key Factor in agricultural production. The value of honey bees and bumble bees as pollinators of major selected UK crops for which market statistics are available has been estimated to be £172 million for outdoor crops (rape, beans, tree and soft fruit) and £30 million for glasshouse crops like tomatoes and sweet peppers A range of studies have shown that pollination makes a very significant contribution to the agricultural production of a broad range of crops, in particular fruits, vegetables, fibre crops and nuts (Levin, 1984; Costanza *et al.*, 1997). Dependency of some horticultural crops on insects for pollination is given in the Table 2.

Table 2: Per cent Dependence of Horticultural Crops on Insects for Pollination

Crops					
Oil Seeds	*% Dependence*	*Fruit*	*% Dependence*	*Vegetable Seeds*	*% Dependence*
Sunflower	100	Grape Fruit	80	Pumpkin	90
Safflower	100	Lemon	20	Water Melon	70
Rape Seed	100	Lime	30	Vegetable Seed	100
Niger	100	Strawberry	40		

Promoting use of beekeeping for pollination of horticultural crops will be of benefit to both the beekeeper who will receive money for the pollination services of his honeybees and harvest honey and to the farmer whose income will be increased through boosting crop productivity as a result of pollination services of bees.

Table 3: Impact of Honeybee (*Apis cerana*) Pollination on Vegetable Seed Production

Crop	*Increase in Pod Setting (%)*	*Increase in Seed Setting (%)*	*Increase in Seed Weight (%)*
Cabbage	28	35	40
Cauliflower	24	34	37
Radish	23	24	34
Broad Leaf Mustard	11	14	17
Lettuce	12	21	9

Source: Verma and Partap (1993; 1994).

This will help ensure food security and enhance the livelihoods of both the farmers and the beekeepers. This system of hiring and renting honeybee colonies for apple pollination is being practiced in Himachal Pradesh in northwest Indian Himalayas.

Effect of Pollination in Improving Horticultural Crop Productivity

Pollination of crops by honey bees is one of the most practical and most promising methods of increasing crop production. Honeybees have flowers fidelity/constancy have a potential for long working hours and can be managed in sufficient numbers where and when required. Bees collect pollen and nectar to supply to the next generation, whereas the other insects collect for their individual needs only. Under AICRP project on honey bees and pollination observations were made on role of honey bees in increasing yield of cross-pollinated crops given in Figure 1.

From the Figure 1 it is evident that bee pollination significantly increases productivity of the various crops *e.g.*, banana, mango, apple, citrus, litchi, papaya *etc.*, compared to that of open-pollinated. The bee pollinated crops also resulted in better quality fruits, with enhanced colour, reduced misshapen fruits, which helped in fetching good market and price. Similar studies were made by (Verma and Pratap, 1993) to confirm the usefulness of bee pollination and its role in increasing crop productivity and increasing quality of fruits.

Qualitative and Quantitative Changes in Crop Plants due to Bee Pollination

As a result of cross-pollination by bees,somatic, reproductive andadaptive heterosis or hybridseffect occurs in plant progeny. Such hybrid effect brings the following qualitative andquantitative changes in plants:

- ✰ Stimulate germination of pollen on stigma.
- ✰ Increase viability of seeds, embryos and plants.
- ✰ More nutritive and aromaticfruits.

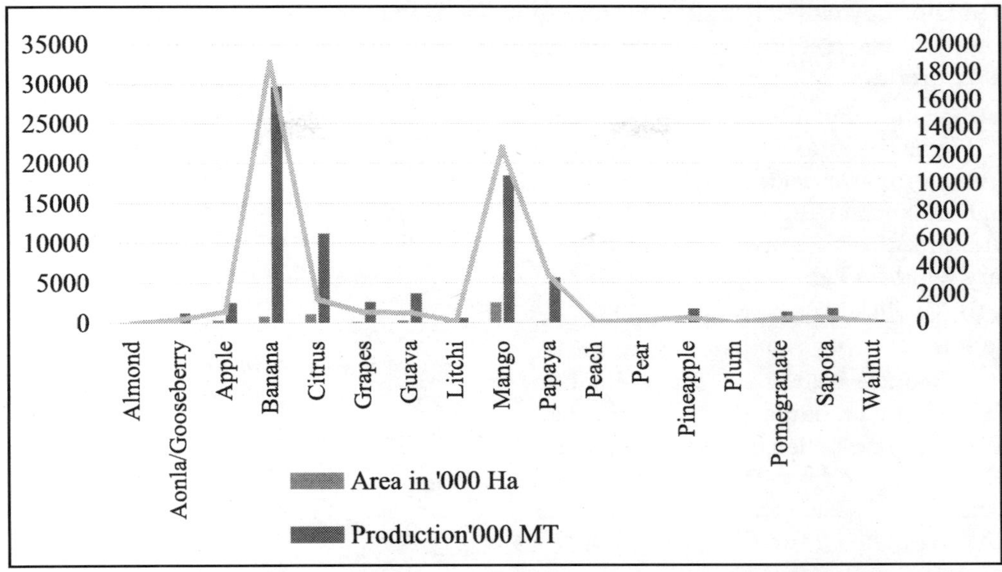

Figure 1: Increase in Production of Important Horticultural Crops.

- ✰ Stimulate faster growth of plants.
- ✰ Increases number and sizes of seeds and yield of crops.
- ✰ Increase nectar production in the nectarines.
- ✰ Increases fruit set and reduces fruit drop.
- ✰ Enhances resistance todiseases and other adverseclimatic conditions.
- ✰ Increases the oil content in oil seed crop (Mohapatra, 2010).

Use of Hive Bees for Pollination

A simple method of crop pollination is to place honeybee colonies in a field when the crop starts flowering. However, for effective pollination, a farmer should follow the points given below

a) Large strong colonies are better pollinators than small colonies because larger colonies have more forager bees. Also, good honey yielding colonies are more efficient pollinators. A bee colony with 60,000 worker bees produces one- and a- half-times more honey than for colonies with 15000 workers each. The same is true for pollination, thus, as far as possible, farmers should use strong colonies. The colonies should be well settled disease free and have young laying queens. The strength of a honeybee colony depends above all on the season. In mountain areas, colony strength is poor during winter because of low temperatures and a dearth of bee flora. Thus, when required in early spring for pollination of temperature fruit crops these colonies do not have enough strength to act effectively. To help colonies maintain their strength, mountain farmers and commercial beekeepers should move colonies to low hill plain areas during winter, because it is warmer and floral sources are available, and return them to the hills in spring when temperate fruit and vegetable crops are blooming, such a migration of bee colonies is the practice in Himachal Pradesh in India.

b) The number of colonies required for pollination varies for different crops. This depends on the total number of plants; the total number of flowers per plant; attractiveness of flowers to bees; duration of flowering; amount of nectar and pollen. Colony condition with respect to strength of colonies; number of pollen foragers, presence of young and prolific queen and amount of unsealed brood in the colonies is also very important. The number of colonies required also varies from crop to crop to crop and from season to season for the same crop and from season to season for the same crop. In general, two to three strong honeybee colonies having 6-8 frame strength with 3-4 frames of brood per acre are recommended for orchards having recommended number of polliniser proportion. Honey bees prefer to visit sources of nectar and pollen that are near to their colonies (200-300 m). At over 500 m pollination activity diminishes greatly. For effective pollination, two colonies should be placed at 200 m distance and distributed evenly in the field.

c) The time of placement of bee colonies is important for high yields and good quality produce. It is important to synchronies flower opening and foraging activities of bees. Freshly migrated colonies are more likely to visit a crop than those in place for a long time. Colonies should be brought to the field when 5-10 per cent of the crop is in bloom. If colonies are placed early, bees will forage on flowers of wild plants nearby, becoming conditioned to these and ignoring the target crop when it blooms. If bees are moved late, they will only pollinate late and less vigorous flower, resulting in poor yields and low quality produce.

d) Cool weather and wind affect foraging activities of bees. Place colonies in sunny, sheltered locations giving protection from wind to encourage maximum flights in spring where no natural windbreak is available, provide a temporary wind shelter.

e) Bees that forage for pollen (pollen collectors) are known to be better pollinators than nectar collectors. Colonies being used for pollination should have large amounts of unsealed brood. This will increase the pollen forgers. The amount of unsealed brood in a colony can be increased by adding frames of unsealed brood from another colony that is not being used for pollination.

f) For crops that are poor nectar producers and relatively unattractive to honeybees, *e.g.* kiwi fruit, bees should be fed sugar syrup in which a few fresh flowers and pollen from the crop have been soaked for some hours. Sugar syrup should be fed in late evening when all foragers have come back to hive. Sugar feeding increases pollen collection and thus pollination.

g) Sometime the crop to be pollinated is ignored. Bees may forage on other more attractive plants, *e.g.*, weeds nearby. In such case, remove the weeds to avoid competition in attracting bees and used good agronomic practices for a healthy crop such as proper manuring, irrigation, and pesticide application. Weeds and other competing crops are useful to sustain insect pollinators as they provide additional nectar and pollen sources. Therefore, it is advised to condition the honeybee to the target crop and increase the number of recommended colonies required for pollination.

h) Do not spray pesticides while the crop is flowering and bees are visiting it. Remove colonies immediately after petal fall and before pesticides are applied. However, if it is required urgently to control the pests, these should be applied only in the evening or early morning when pollinators are not visiting the bloom. The colonies kept in the orchards should be covered with burlap cloth before the spraying (Thakur and Gupta, 2016).

Colony Required

The number of colonies required for pollination of different cultivated crops depends upon the several factors like: density of plant stand, number of flowers perinflorescence, duration of flowering, strength of bee colonies and number of flowers per unit area. In general three colonies of *Apisceran indica* and two colonies of *Apis mellifera* are required per hectare. Requirement of honey bee's colonies for optimum pollination in apple has been assessed on the basis of abundance of bees on the marked bloom at different distances from the hive(s) and fruit set at three distances. Rana *et al.* (1998) have suggested that in order to ensure a good commercial crop, at least eight *A. mellifera* colonies are needed per hectare of apple for pollen distribution where pollinizer proportion is low. In general, 2-3 colonies per hectare are recommended in apple orchards with optimum pollinizer (Goyal, 1993). According to Rana *et al.* (1998), in order to ensure good crop, at least 8 *A. mellifera* colonies are required/ha of apple for pollen distribution in orchards having low pollinizer proportions.

Pear

At Katrain, Kullu five types of insect pollinators namely *Apis cerana*, *Apis mellifera*, syrphids, other dipteran and lepidopteran were recorded visiting bloom of the pear (Sharma *et al.*, 2001). *A. cerana* and *A. mellifera* worker bees carried 1124 and 11.76 loose pollen grains on their bodies, respectively. The number of flowers visited per minute for a single bee was 7.99 (*A. cerana*) and 8.57 (*A. mellifera*). The time spent for flower was 5.52 and 4.70 for *A. cerana* and *A. mellifera*, respectively. Episyrphus sp. visited 1.52 flowers/minute and spent 51.19 seconds per flower.

Cherry

According to Sharma and Rana (2000), *A. cerana* was most dominant (40.66 per cent) among all the insect pollinators. More number of loose pollen grains was found on the body of *A.mellifera* (1962.22) than on *A. cerana* (1865.55) and *Episyrphus* sp. 144.44. The bees were active upto 75 m distance from the placement of the colonies and the fruit set obtained was also higher upto this distance. The fruit set was found to vary between 16 to 33.66 per cent.

Apricot

Five types of insect pollinators were found visiting on the bloom of *Prunus armeniaca* L. var. New Castle (Rana *et al.*, 1996). *A. mellifera* and *A. cerana* bees collected nectar and pollen from the bloom. Data collected on fruit set did not show any specific trend and it varied between 10.98 and 24.85 per cent at different distances from the colonies, indicating that pollinators other than honeybees are also playing important role in the pollination of this crop.

Plum

Six types of pollinators visited the bloom of plum, Santa Rosa cultivar. No specific trend in the distribution of *A. mellifera* and *A. cerana* on the bloom was found at different distances from the colonies. Nectar and pollen collectors of *A. mellifera* spent more time/flower than that of *A. cerana* foragers. The fruit set at different distances from the colonies did not decline with the increase in distance and it varied between 2.9 and 13.08 per cent. Thus there seems to be no need to keep bee colonies for self-fruitful Santa Rosa cultivar of plum. The pulp to stone ratio varied between 10.96:1 to 23.3:1 at different distances.

Kiwi

Six types of insect pollinators were found visiting on the bloom of kiwi (*Actinidia chinensis*) and out of them, *A. dorsata* was the most frequent visitor (63.50 per cent). Insect pollinators get only pollen from the bloom. No specific trends in population of honeybee pollinators on the bloom and fruit set were observed with regard to distance from the placement of honeybee colonies. The fruit set varied between 66.67 and 100.00 per cent in open-pollinated crop, while it was cent per cent in hand pollinated crop. Heavier fruits were produced from hand pollinated flowers (56-75 g) than from open-pollinated ones (40-58 g). However, in both the cases, almost similar numbers of seeds were counted.

Almond

Pollination studies were conducted on Almond orchards located in Tral and Pulwama districts of Kashmir valley. The foraging behaviour and the pollinating efficiency of native pollinators (*Osmia* spp., *Lassio glossum* spp. and *Xylocopa* spp.) and *Apis mellifera* were studied for this purpose. The flowers that received visits with *Apis mellifera* bee was having fruit set of 25.5 per cent and 32.17 per cent. The results revealed after placing 2-3 *Apis mellifera* colonies/ha in orchards in addition to the native pollinators already present, there was 20 times increase in the yield

Table 3: Placement and Distribution of Colonies in the Field

Distance from Colony (m)	No.of Honeybees (per m^2 branch in 10 min)	Fruit Set (in 100 spurs)
20	2.12	107.40
40	1.47	88.89
60	1.06	81.36
80	0.81	81.36
100	0.66	78.89
150	0.50	76.72
200	0.31	52.76

Time, placement and Distribution of Colonies

For betterpollination, bee colonies shouldbe placed in the field ororchard at night times when 10 to 20 per cent of the crop is inbloom. Honeybees as a rule visitprimarily those sources of nectar which are within 0.3 to 0.5 km radius from the apiary. The pollination activitydiminishes significantly at adistance more than 0.5 km. Thenumber of honey bees on the crop decreases with increased distance of the crop from the colonies and there is corresponding decrease in fruitset, production and productivity. For effective pollination, *Apis ceranaindica* hive should be placed singly (Thakur and Gupta, 2016).

Attracting Bee Pollinators

Bees should be fed sugar syrup flavoured by theflowers required to beepollinated in order to attractlarge number of bees for effective pollination.

Planting Plan/Placement of Pollinizers

To achieve optimum fruit set in fruit crops, there is an absolute need to have suitable pollinizers which possess the following distinguishing characteristics *e.g.*, prolonged flowering season, viable pollen in abundance, synchronization of bloom period, cross-compatibility with main cultivar (s), and produce fruit with acceptable quality. Not only that the pollinizer should have the above qualities but there is an equal importance of having the pollinizers in good proportion and planted systematically. It has been seen that though there is a provision of pollinizers in an orchard but their inter-planting is haphazard, which results into low availability of pollen at the time it is required. In case of wind pollinated (anemophilous) species such as walnut and pecan, the proporation of pollinizers should not be less than 10 per cent and in insect pollinated (entomophilous) species such as apple, pear, cherry, plum, apricot, almond *etc.*,the proportion of pollinizers should not be less than 20 percent. The ideal proportion is 25 to 30 percent in an orchard. For effective pollination of a self sterile variety *e.g.* apple, the minimum requirement is to plant a pollinizer. The pollinizer should be well distributed in the orchard. Several planting plans are being adopted for fruit crops. The important ones provide 11, 15, 25 and 33 per cent pollinizers (Thakur and Gupta, 2016).

The planting plans for pollinizers shown in Figure 2.

Economic Scenario under Pollinated Conditions

In the below given table above given observations were collected from the work of various authors (Abrol 2012; Mahopatra *et al.*, 2010). Based on their observations expected increase in production was calculated and for the other crops standard of 20 per cent increase in yield was used. It is evident from the above table there is an increase in 57 per cent, 66 per cent, 15 per cent and 20 per cent increase in production of fruits, vegetables, flowers, spices respectively. There is an overall increase of 62 per cent increase in total production of horticultural crops.

Effect of Pollination on Productivity

Increasing the productivity in horticultural crops under conventional agriculture system is possible mainly by two ways firstly by increasing agriculture inputs such as water, land, fertilizers, agricultural chemicals *etc.* Increasing agricultural inputs is not

i) 11% Planting Plan						
o	o	o	o	o	o	o
o	x	o	x	o	x	o
o	o	o	o	o	o	o
o	o	o	o	o	o	o
o	x	o	x	o	x	o
o	o	o	o	o	o	o
o	o	o	o	o	o	o
o	x	o	x	o	x	o
o	o	o	o	o	o	o

O - Commercial cultivar x ş Pollinizer

In this plan every third plant of every third row should be a pollinizer. This plan is preferable for self-fruitful apple, plum and kiwifruit.

ii) 15 % Planting Plan								
o	o	o	o	o	o	o	o	o
o	x	o	o	x	o	o	x	o
o	o	o	o	o	o	o	o	o
o	x	o	o	x	o	o	x	o
o	o	o	o	o	o	o	o	o
o	x	o	o	x	o	o	x	o
o	o	o	o	o	o	o	o	o
o	x	o	o	x	o	o	x	o
o	o	o	o	o	o	o	o	o

O - Commercial cultivar x ş Pollinizer

In this plan, every second plant in third row should be a pollinizer. This plan is modified form of 11% planting plan and is applicable for kiwifruit only.

iii) 25 % Planting Plan						
o	o	o	o	o	o	o
o	x	o	x	o	x	o
o	o	o	o	o	o	o
o	x	o	x	o	x	o
o	o	o	o	o	o	o
o	x	o	x	o	x	o
o	o	o	o	o	o	o
o	x	o	x	o	x	o
o	o	o	o	o	o	o

o- Commercial cultivar x ş Pollinizer

In this plan every second plant in alternate row should be a pollinizer. Plan is suitable for apple and pear. This plan is applicable for those areas where the climate fluctuation is dominant.

iv) 33 % Planting Plan								
o	x	o	o	x	o	o	x	o
o	x	o	o	x	o	o	x	o
o	x	o	o	x	o	o	x	o
o	x	o	o	x	o	o	x	o
o	x	o	o	x	o	o	x	o
o	x	o	o	x	o	o	x	o
o	x	o	o	x	o	o	x	o
o	x	o	o	x	o	o	x	o
o	x	o	o	x	o	o	x	o

o- Commercial cultivar x ş Pollinizer

In this plan, the number of plants of main cultivar is reduced to 67 per cent. This plan is suitable for apple and cherry. The planting plan differs from crop to crop.

Figure 2: The Planting Plans for Pollinizers.

possible in every scenario and is not desirable under limited availability of water and land. Furthermore, adding more chemicals to the field has adverse effects on environment. In Organic agriculture it is not possible at all. The second method includes the use of biotechnological techniques, such as manipulating rate of photosynthesis and biological nitrogen fixation, *etc.* These conventional techniques ensure healthy growth of crop plants, but work up to a limit. At some stage crop productivity becomes stagnant or declines with additional inputs for the known agronomic potentials of crop will have been harnessed (Partap and Partap, 1997). Pollination of crops by honeybees is one of the most practical and most promising methods of increasing crop production. Honey bees have flowers fidelity/ constancy have potential for long working hours and can be managed in sufficient numbers where and when required. Bees collect pollen and nectar to supply to the next generation, whereas the other insects collect for their individual needs only. Bees, therefore, make more flower visits. Their long tongues, coats of long collecting hairs, and the ability to warm them and to work in cool

Table 4: Production and Expected Increase in Production of Important Horticultural Crops (Area in '000 Ha and Production in '000 MT)

Crop	*2013-14*		*Expected% Increase in Production*	*Increase in Production Under Bee Pollination*	*Total Production*
	Area	*Production*			
			Fruits		
Almond	21	13	20	2.6	15.6
Aonla/Gooseberry	104	1225	20	245	1470
Apple	313	2498	24.40-39.50	799.36	3297.36
Banana	803	29725	63	18726.75	48451.75
Citrus	1078	11147	15	1672.05	12819.05
Grapes	119	2585	29	749.65	3334.65
Guava	268	3668	20	733.6	4401.6
Litchi	84	585	20	117	702
Mango	2516	18431	68	12533.08	30964.08
Papaya	133	5639	60	3383.4	9022.4
Peach	18	94	50	47	141
Pear	42	317	50	158.5	475.5
Pineapple	110	1737	20	347.4	2084.4
Plum	23	76	20	15.2	91.2
Pomegranate	131	1346	20	269.2	1615.2
Sapota	177	1744	20	348.8	2092.8
Walnut	122	241	20	48.2	289.2
Total fruits	4984	69924		41683.04	122754.04
			Vegetables		
Beans	138	1370	128.57	1761.40	3131.40
Bittergourd	79	807	20	161.4	968.4
Bottlegourd	103	1819	20	363.8	2182.8
Brinjal	711	13558	31.25	4236.87	17794.87
Cabbage	400	9039	20	1807.8	10846.8
Capsicum	30	167	227.05	379.17	546.17
Carrot	62	1074	20	214.8	1288.8
Cauliflower	434	8573	20	1714.6	10287.6
Cucumber	43	678	5-15	67.8	745.8
Chillies (Green)	140	1687	20	337.4	2024.4
Muskmelon	37	761	20	152.2	913.2
Mushroom		17	20	3.4	20.4
Okra/Ladyfinger	533	6346	20	1269.2	7615.2
Onion	1204	19402	93	18043.86	37445.86
Peas	434	3869	20	773.8	4642.8
Potato	1973	41555	20	8311	49866
Radish	173	2485	20	497	2982
Sweet Potato	106	1088	20	217.6	1305.6
Tapioca	228	8139	20	1627.8	9766.8
Tomato	882	18736	160.61	30091.88	48827.88
Total Vegetable	7710	141170		71890.43	213060.43
Flowers Cut		543	79-30	266.07	809.07
Flowers Loose	255	1754	79-30	859.46	2613.46
Total Flowers	255	2297		1263.35	3560.35
Coconut	2140	14911	40	5964.4	20875.4
Spices	3163	5908	20	1181.6	7089.6
Total	**28038**	**441693**		**236819.64**	**706714.64**

weather make them generally more efficient pollinators than most other insects. Under AICRP project on honey bees and pollination observations were made on role of honey bees in increasing yield of cross-pollinated crops given in Figure 2.

From the above chart it is evident that bee pollination significantly increases productivity of the various crops over open-pollinated and without bee pollinated. The bee pollinated crops also resulted in better quality fruits, with enhanced colour, reduced misshapen fruits, which helped in fetching good market and price. Similar studies were made by Verma and Pratap (1993, 1994) to confirm the usefulness of bee pollination and its role in increasing crop productivity and increasing quality of fruits and vegetables. Figure 3 sums up the findings of experiments carried out on vegetable crops in Kathmandu region.

Conclusion

Pollinators can be used as an important resource to improved variety, quantity and quality of fruits. Applied pollination, pollinator management, and managed pollination are the common efforts recently being practiced for maximization of production in cross-pollinated crops and to bring the pollinator to the target crop. In view of great role of bees in pollination of various crops, they need tobe conserved, augmented and supplemented with domesticated honey bees. Honey bee is one of the best methods to enhance the productivity which is expected not only to be a tool of 2nd green revolution. However, more works needs to be performed in order to elucidate certain aspects for improving the crops productivity. To effectively address this issue it is necessary to bring pollination concerns

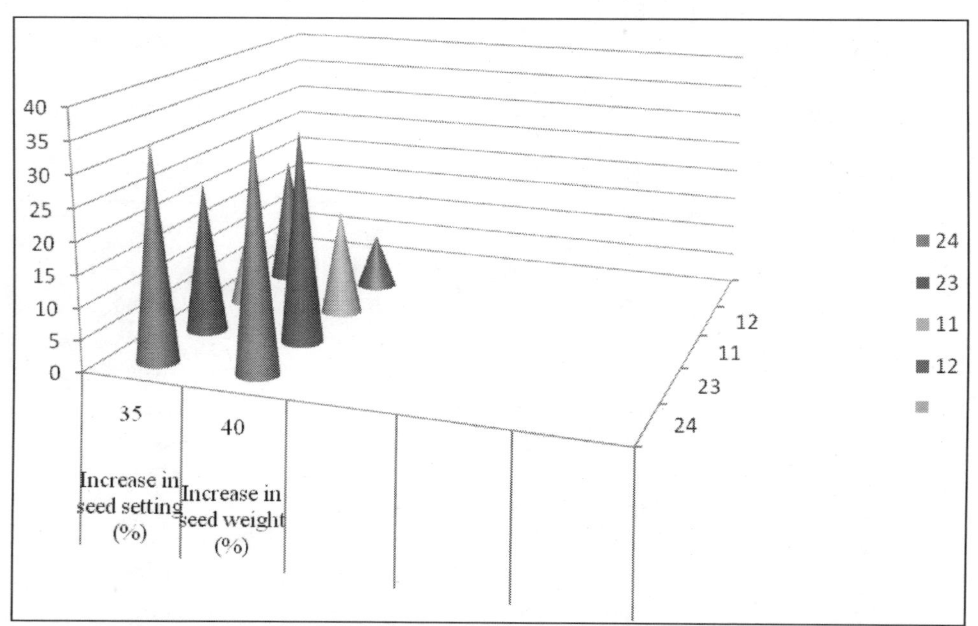

Figure 2: Role of Honey Bees and Other Pollinators in Increasing the Yield of Various Crops.

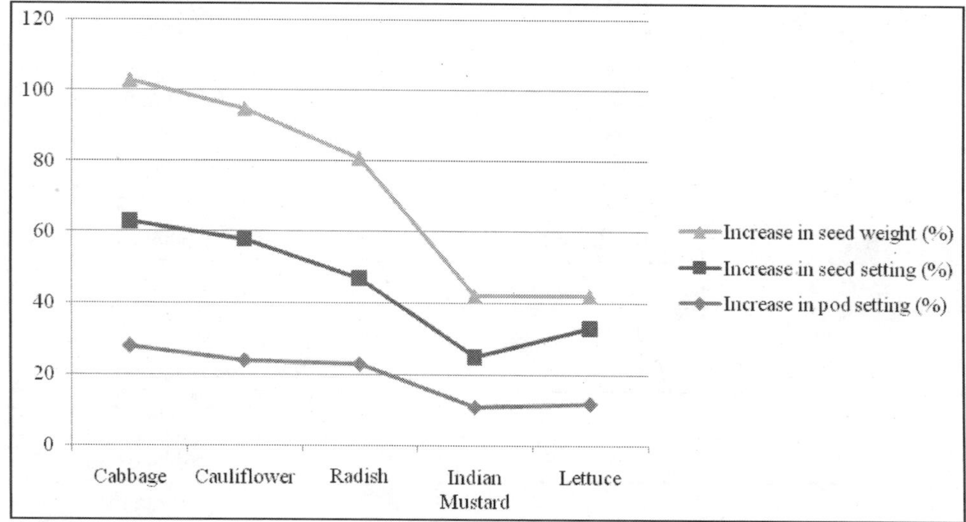

Figure 3: Honey Bee (*Apiscerana*) Pollination Impacts on Vegetable Seed Productivity.

e.g. (protection, conservation, and restoration) into the policy, research and development mainstream.

References

Abrol, D.P. 2012. Value of bee pollination. **In**: *Pollination Biology: Biodiversity Conservation and Agricultural Production*. Springer Netherlands, New York, London, pp. 185–222.

Costanza *et al.*, 1997a. The value of the world's ecosystem services and natural capital. *Nature*, 387: 253–260.

Free, J.B. 1993. *Insect Pollination of Crops*. Academic Press, London, pp. 684.

Goyal, N.P. 1993. Role of honeybees in improving agricultural productivity. **In**: *First National Conference on Beekeeping*, (Eds.) N.P. Goyal and R.K. Sharma. National Horticulture Board at Chandigarh, 29th June.

Gupta, J.K., Goyal, N.P., Sharma, J.P. and Gautam, D.R. 1993. The effect of placement of varying number of *Apis mellifera* colonies on the fruit set in apple orchards having different proportion of pollinizers. Pollination in tropics, *Proc. Int. Symp. Polln. Trop.* (Eds.) G.K. Veeresh, R. Uma Shankar and K.N. Ganeshaiah. IUSSI, Indian Chapter, Bengaluru.

Ingram, M., Nabhan, G.C. and Buchmann, S.L., 1996. Impending pollination crisis threatens biodiversity and agriculture. *Tropinet*, **7**: 1–10.

Kevan, P.G., Imperatriz-Fonseca, V., Frankie, G.W., O'Toole, C., Jones, R. and Vergara, C.H. (Eds.) 2002. Pollinating bees: The conservation link between agriculture and nature. **In**: *Proceedings of the Workshop on the Conservation and Sustainable Use of Pollinators in Agriculture, with Emphasis on Bees*.

Levin, M.D. 1984. Value of honeybee pollination to United States agriculture. *Amer. Bee J.*, **124**: 184–186.

McGregor, S.E. 1976. *Insect Pollinations of Cultivated Crop Plants*. US Department of Agriculture, Agriculture Handbook, pp. 496.

Mohapatra, L.N., Sontakke, B.K. and Ranasingh, N. 2010. Enhancement of crop production through bee pollination. *Odisha Review*, pp. 44–47.

Mohapatra, S., Deepa, M., Jagdish, G.K., Rashmi, N., Kumar, S. and Prakash, G.S. 2010. Fate of fipronil and its metabolites in/on grape leaves, berries and soil under semi arid tropical climatic conditions. *Bull. Env. Contam. Toxicol.*, **84**: 587–591.

Partap, U. 1998. Successful pollination of apples in Himachal Pradesh. *Beekeeping and Development*, **48**: 6–7.

Partap, U. and Partap, T. 1997. *Managed Crop Pollination: The Missing Dimension of Mountain Agricultural Productivity*. Mountain Farming Systems' Discussion Paper Series No. MFS 97/1. Kathmandu, ICIMOD.

Rana, B.S., Gautam, D.R., Goyal, N.P. and Sharma, H.K. 1998. Effect of honeybee pollination on yield parameters of apple in relation to pollinizer proportion. *Indian Bee J.*, **60**: 9–11.

Sharma, H.K, Gupta, J.K. and Thakur, J.R. 2001. Pollination studies on apple and pear. In: *Proceedings of the 7th International Conference on Tropical Bees: Management and Diversity and 5th AAA Conference*, March 19–25. Chiang Mai, Thailand, pp. 265–270.

Sharma, H.K, Gupta, J.K., Khajuria, D.R. and Thakur, J.R. 2002a. Factors influencing natural crop pollination in apple and effect of placement of *Apis cerana* colonies in increasing apple productivity. **In**: "*6th AAA International Conference and World Apiexpo*, February 24–March 1, Bengaluru, India, pp. 63.

Sharma, H.K. and Gupta, J.K. 2001. Impact of bee pollination on sustainable production in apple orchards. *Pest Mgt. Econ. Zool.*, **9**(1): 45–48.

Sharma, H.K., Gupta, J.K. and Thakur, J.R. 2003. Effect of bee pollination and pollinizer proportion on apple productivity. **In**: *7th International Symposium on Temperate Zone fruits in the Tropics and Sub-tropics*, October 14–18, Solan India, pp. 315.

Sharma, H.K., Gupta, J.K. Thakur, J.R. and Sharma, J.P. 2002b. Impact of farmer's participatory research in pollination of temperate fruits and beekeeping training cum educating programme in Kullu valley of Himachal Pradesh, India. **In**: *6th AAA International Conference and World Apiexpo*, February 24–March 1, Bengaluru, India, pp. 205.

Thakur, R.K. and Gupta, S. 2016. Apiculture to improve productivity in horticultural crops. **In**: *Innovations in Horticultural Sciences*, (Ed.) K.V. Peter. New India Publishing Agency, New Delhi, p. 391–451.

Verma, L.R. and Pratap, U. 1993. The Asian Hive Bee, *Apis cerena* as the Pollinator in Vegetable Seed Production: An awareness handbook (ICIMOD, Kathmandu, Nepal).

Verma, L.R. and Partap, U., 1994. Foraging behaviour of *Apis cerana* on cabbage and cauliflower and its impact on seed production. *J. Apicultural Res.*, **33**: 231–236.

25

Rejuvenation of Fruit Plants as a Means of Improving Productivity and Profitability

S.K. Shukla*, G. Pandey, Barsati Lal and Dushyant Mishra

ICAR-Central Institute for Sub-tropical Horticulture,
Lucknow – 226 101, Uttar Pradesh
**E-mail: skshuklacish@gmail.com*

India is a leading fruit producer with a total fruit production of 89 million MT from 7.2 m ha, next only to China, which produces 137 m MT from 11.83 m ha area. Though the fruit productivity in India (12.3 MT/ha) is far lower than that of USA (23.3 t/ha), Indonesia (22.3 t/ha) and Brazil (16.5 t/ha) (Saxena and Gandhi, 2015). Fruit trees, invariably after a certain period of successful fruit production, have a tendency to lose their productivity and quality depending upon the species, making the orchards unproductive and non-remunerative. Fruit orchards have different production phases like pre-commercial bearing stage (new plantations), commercial bearing orchards and old senile declining orchards even up to the stage of unproductive ones (Jahangeer *et al.*, 2011). In our country, 30-40 per cent fruit orchards have become old and unproductive (Jahangeer *et al.*, 2011; Kumar, 2015). Replacing the old and senile orchards through replanting is a long term strategy, requiring long time to reach the desired level of productivity. Rejuvenation technology for such unproductive trees offers many benefits such as early and quality production from old unproductive senile trees through better light penetration and efficient utilization of available space and inputs (Kumar, 2015). Manifold increase (2-6 times) in fruit productivity has been recorded in various fruit crops with the use of this technology. Efforts have been made to record the experiences of various workers especially in tropical and sub-tropical fruit crops as far as the technology is concerned. Thinning of overcrowded branches to facilitate light penetration and air circulation improves photosynthetic efficiency, fruit yield and fruit quality especially size. Favorable effects of pruning in mango have been reported on light interception and chlorophyll content of leaves, growth parameters (Lal *et al.*, 2001) and fruit yield (Lal and Mishra, 2007, Rao and Shanmugavelu, 1976; Shinde *et al.*, 2002) and regularity in bearing (Rao, 1971).

1. Fruit Production Constraints

India is the second largest producer of fruits next only to China, yet the per capita consumption of fruits and vegetables in our country is lower than recommended doses which are 100 g of fruits and 300 g of vegetables/day/person. Poor productivity of existing orchards, burgeoning pressure of population and the perishable nature of fruits resulting in around 30-35 per cent of post-harvest losses are the major causes of concern. There exists a wide gap between the actual yield and the potential yield in all the tropical and sub-tropical fruits which needs immediate attention. The area, production and productivity of various fruits have been given in the Table 1.

There is an overall increase in fruit production in the country but the productivity is still very low. Besides, the quality of produce is also not very good. The reasons responsible of poor productivity and quality of fruits in the country are dominance of seedling orchards, prominence of old, dense and seedling

Table 1: Area and Production of Major Fruits

Fruit Crop	*Area (000 ha)*	*Production (000 ha)*	*Productivity (MT/ha)*
Mango	2516	18431.3	7.3
Guava	268.2	3667.9	13.7
Citrus	1077.73	11147.06	10.34
Litchi	84	585	7
Cashew nut	1011	753	0.7
Pomegranate	131	1346	10.3
Ber	80	900	11.25
Aonla	108	1266	11.7

Source: Saxena and Gandhi, 2015; Singh *et al.*, 2015.

plantations, dominance of malnourished orchards, poor management and high incidence of insect pest and diseases. Rejuvenation technology in crops like mango, guava, *aonla*, litchi, cashew nut, *ber*, *etc.* offers unique opportunity to combat the problem of unproductive seedling, old and senile orchards. Thickly shaded mango orchards in Malihabad areas of Uttar Pradesh (Kalloo *et al.*, 2005; Mishra, 2013), lanky as well as sick tall unproductive litchi orchards in Muzaffarpur and its adjoining areas in the state of Bihar (Kumar, 2008), seedling orchards of guava and citrus fruits throughout the country, including disease affected *aonla* and other perennial fruit trees are commonly seen in large tracts in different parts of country (Pandey *et al.*, 2013; Singh, 2007). In pomegranate, old orchards become unproductive due to neglect and incidence of insect pests and diseases (Hiwale, 2009). Other production constraints are lack of information regarding canopy management, non-adoption of appropriate agro-techniques and timely management practices leading to large number of old and senile orchards turning them unproductive (Hare Krishna, 2012). Singh (2007) enumerated several other causes which make the orchards uneconomical, *viz.*, growth of wild shrubs and grasses, unsystematic planting at initial stages, inferior varieties, heavy incidence of diseases, pests, rodents and other enemies, development of hollows in older tree trunks and main branches with advanced age.

2. Why to Rejuvenate Trees?

In order to get rid of old unproductive orchards existing in abundance, large scale uprooting and replanting is a long term and expensive strategy, while such trees can be made productive and profitable through rejuvenation technique which, in turn, restricts the further deterioration of the health of plant health, extending its life and improving the yield of quality of produce within minimum period of time as compared to establishment of new orchard which attains the commercial bearing stage after 8-10 years (Kalloo *et al.*, 2005; Lal and Mishra, 2008; Singh, 2007; Hiwale, 2009; Pandey *et al.*, 2013).

Rejuvenation technology is a unique integration of invasive exhaustive operations and skillful management for transforming the trees from unproductive to productive ones by restoration of youthful vigor. It deals with many issues like restoration vigor, enhancement of quantum and quality production, sustainability and economic feasibility *etc.* All these aspects have been proved with sound experimentation, implementation and success stories (Kalloo *et al.*, 2005; Kumar, 2008; Lal *et al.*, 2000; Singh and Mishra, 2007; Pandey *et al.*, 2013). Reiterative pruning and canopy rebuilding for rejuvenation of perennial fruit trees is medium strategy for making the senile trees productive in a sustainable way within a time gap of three years for mango, litchi and cashew nut (Lal *et al.*, 2000; Kumar, 2008), while two years for guava, *aonla*, *ber* and pomegranate (Singh, 2007; Hiwale, 2009; Ghosh and Mathew, 2002; Singh *et al.*, 2015).

3. Physiology of Rejuvenated Trees

Perennial fruit trees build their woody architecture with some building blocks, *i.e.* modular units, known as metamer or phytomer (White, 1979; Room *et al.*, 1994; Hare Krishna, 2012), which consists of meristems, initially an apex. When a meristem faces periods of dormancy during low temperature or pruning, it appears as bud, whenever active, which produces a growing point with sequence of meristematic tissues simultaneously called a sprout, *i.e.* extension unit or growth unit (Halle *et al.*, 1978; Bell, 1991). The driving mechanism behind the pattern appears to be auxin production when the new shoot and leaves developed and are exposed to light (Sterck, 2005). Other features of trees are to produce more xylems relative to phloem, accumulate xylem over time and may thus produce thick woody branches and stems. The studies have also suggested that attaining reproductive phase and reproduction processes compete with vegetative functions, depleting resources necessary for maintenance and growth (Fakhri *et al.*, 1987). The concept of source and sink helps to explain allocation patterns at both the physiological and evolutionary levels. The allocation to vegetative structures has compounding interest for plant growth, as plant life cycles begin with pure vegetative growth (Abrahamson, 1979). Photosynthesis either directly or indirectly provides the building blocks for all biomass production and growth but the plant growth and productivity is determined more by the way in which the product of photosynthesis are allocated and utilized during plant growth and development.

The decline in productivity of orchards is largely due to poor photosynthetic efficiency besides several other factors including age of orchard, dense and intermingling branches, neglect and poor management, bearing poor quality fruits, attaining unproductive stage. This requires formative or corrective pruning to reinforce the tree to solve the bearing problem of old orchards, *i.e.* making them productive again. This is what is done by rejuvenation technology. After severe pruning, the left over thick branches of the tree are the primary source of reserve carbohydrates in the absence shoots and leaves (Hare Krishna, 2012). The appearance of sprout, giving rise shoot and young leaves arising from the pruned branches of the old trees depend on imported carbohydrate from the sieve tubes of the phloem and other tissues to provide metabolism and carbon skeletons for metabolic inter conversions and bio-synthetic process. Vegetative re-growth as sprout emergence is a sink for resources, and so allocation to it must be under ecological controls similar to those that govern sexual reproductive allocation. There is existence of meristematic tissues in the left over trunk portion after heavy pruning, which are capable of active cell division and differentiation into specialized tissues as shoots and leaves at initial stage importing stored carbohydrates (Singh, 2003; Hare Krishna, 2012). New shoots emerge after pruning but they don't possess flower buds for one to two years. The developing new shoots of current season continue growth taking enough time to mature to for phase change. It experiences dormancy during exposure to winter or low temperature for one or two seasons thereby changing to floriferous condition as its vegetative meristem has differentiated into inflorescence (Singh, 2003; Hare Krishna, 2012; Kumar, 2012; Kumar, 2015).

It is necessary to develop a framework strong enough to support the load of fruit yield. The consequences of severe pruning, the method and time gap of rejuvenation coupled with cultural practices are required to support the reinvigoration process (Kumar, 2012). Thinning of emerged shoots after six months of growth and selection of primary scaffolds are done to properly train them outward to grow fast and outward. Scaffolds should be headed during the dormant season of first two years to promote lateral branching to be stiffened and strengthened as well as for maturing the scaffolds for shaping desired framework (canopy) (Kumar, 2015). Floral induction in some perennial fruit crops like mango, litchi, guava, citurs, *etc.* occur in response to cool temperatures perceived by mature leaves largely regulated on physiologically mature shoots (Menzel and Simpson, 1995; Menzel *et al.*, 2000; Hare Krishna, 2012; Kumar, 2012). Contribution of only some percentage of primary blushes in blooming rather than whole still needs further investigation which may involve biochemical and hormonal assays of shoots (Hare Krishna, 2012; Kumar 2012). Periodic monitoring of pruning operation, emergence of re-growth and its maintenance avoid complexities and disappointments due to undesired framework development or chances of mortality (Kumar, 2008).

4. Selection of Orchards for Rejuvenation

The exact reasons for the decline or poor productivity of orchards should be determined and corrected, if possible, otherwise response to rejuvenation will be temporary. If the trunk and basic scaffold limbs of trees are not structurally sound due to disease, heart rot and cold injuries, the rejuvenation is not justified. A limited root system due to high water table, a hard pan or poor subsoil can result in early tree decline with trees being more readily stressed by drought. When this occurs, the top becomes out of balance with the more limited root system with consequent insufficient uptake of water and nutrients by the roots. Topping such trees will temporarily restore them to a more favorable top to root ratio, but not alleviate the long term problem. The severity of pruning for rejuvenation will depend on the cause and degree of decline (Kumar, 2015).

It has been observed that orchards of mango, litchi, cashew and *aonla* planted at 10X10m, after attaining the age of 30-40 years turn dense with compact top canopy cover and long barren (leafless) branches and bearing confined only to the top periphery. It is often uneconomical to maintain old orchards after 30-40 years of age due to production of poor quality fruits on high tops which poses problem in harvesting and other horticultural operations, increased incidence of diseases, insect pests and physiological disorders, alternate bearing habit. In case of guava, citrus, ber and pomegranate, the orchards after attaining the age of 10-20 years depending upon the tree spacing, become dense, compact top canopy cover with long lanky branches devoid of leaves and with little bearing. Such orchards need to be rejuvenated for further continued production of quality fruits (Kumar, 2015). It is always important to look for health, location and value of the fruit trees, which should be determined before going for rejuvenation (Kalloo *et al.*, 2005; Lal *et al.*, 2000; Kumar, 2008; Singh and Mishra, 2007). Sometimes the condition of orchard is so poor that it is not at all profitable to rejuvenate. It is advisable to for new plantation in such cases instead of rejuvenation or the efforts should be made for satisfactory growth maintenance by other required cultural and management practices (Singh, 2007).

5. Rejuvenation different Crops

Mango

Mango trees generally begin commercial production in three to four years after planting and continue to produce progressively increasing yields as canopies enlarge until shading by adjacent trees forces the growth upwards and out of reach of harvests. Lower branches supporting previous year's productive stems become unproductive due to shading by branches located higher. Flowering and fruit production in mango occurs on stem terminals; therefore, as canopy size increases, production moves to the top as competition for available light continues.

Like other fruit trees, mango trees fail to sustain their productivity and economic viability after a certain age.

Sometimes it may be natural or a number of factors are responsible for deterioration in the productivity. It has been observed that after about 45-50 years of planting, the branches of one tree start intermingling with those of the others. As results, the orchards become dense and are devoid of sunlight affecting the photosynthesis of the leaves. Such a condition leads to poor and sickly growth of shoots, reducing their potential to bear normal fruit load. Whatever new shoots emerge, they become weak and are unsuitable for flowering and fruiting. The average national productivity of mango in country is very poor (7 t/ha). Dominance of such orchards is one of the major reasons for this low level of productivity. In our country, about 35-40 per cent mango orchards are old and senile.

By the age the trees reach 45 years, they have grown tall with less canopy area and that too only on the top. Such trees harbour insect-pests and diseases. Further, it becomes difficult to control them easily. Pruning of trees in old orchards helps develop new shoots which grow into a desired canopy, increases light interception by the foliage. It also helps in an effective management of insect pests and diseases, restoring their vigour and productivity.

Steps of Rejuvenation

i) Identification of unproductive mango orchards by yield records of last five years.

ii) Heading back of 3-5 wide angled branches during December at approximately 2.5 to 3.0 m height from ground and thinning out of remaining branches from their base.

iii) Heading back should be done with power/petrol operated chain saw for getting sharp and smooth cuts.

iv) Fertilizer application (2.5 kg urea, 3.0 kg single super phosphate and 1.5 kg muriate of potash with 50 kg FYM) in February followed by light irrigation.

v) Thinning of excessive shoots during April and June to allow only 6-8 healthy shoots per branch for ideal canopy development.

vi) Monitoring and management of stem borer and other leaf eating insect pests.

vii) After three years of heading back, the trees become rejuvenated with development of healthy and productive canopy and start-bearing fruits.

viii) Orchardist can get additional income from sale of pruned wood and intercropping of short duration crops such as vegetables and flowers in open space of orchard during the initial years.

ix) Fruit yield is enhanced up to 4-5 times and quality is also improved with bigger fruit size. Higher yield is due to better harvesting of solar radiation by the tree canopy.

Lal *et al.* (2000) reported that the growth of emerging shoots on pruned branches was influenced by severity of pruning. Length and girth of emerging shoots were more in first, second and third order pruned trees than in fourth and fifth order treatments, whereas it was minimum in the control. Fruit yield was higher in fourth and fifth order pruned trees than the control. No fruit yield was obtained in first, and second order pruned trees during first and second year of fruiting, but yield was obtained in an increasing trend in subsequent years. Cumulative fruit yield of 6 years indicated that fourth order pruned trees gave the highest yield, whereas it was lowest in the control. Further, Lal and Mishra (2008) studied the continuous effect of different pruning treatments on the growth parameters, bearing behaviour and quality of fruits of aforesaid rejuvenated Dasherari trees. After twelve years of pruning, highest increase in tree height was recorded in first order pruned trees. Highest shoot length was recorded in fourth order pruned trees, while shoot girth was highest in first order pruned trees. Canopy spread was the highest in control. There was loss of fruiting for initial two years in severe pruning treatments (first and second order pruned trees), while from third year onwards, there was a gradual increase in fruit yield in these treatments. In moderate and light pruning treatments (third to fifth order pruned trees), there was continuous fruiting from first year onwards with an increasing trend for initial 3-4 years, there after inconsistent yields were obtained.

The mango rejuvenation is rather critical and somewhat difficult and under high rainfall conditions it is not recommended, since the tree may not recover after pruning and thus it requires much care and attention. The technique is common in sub humid and semi-arid areas. However, reports suggest that in warm humid tropical areas this technique is not widely used because of subsequent loss of fruit production and excessive re-growth (3-4 vegetative flushes per year) that occurs (Crane *et al.*, 2009). Rejuvenation pruning studies conducted in old non-bearing trees of mango cv. Alphonso revealed that heading back operation along with paclobutrazol @ 7.5 g a.i./tree and canopy rebuilding has been found beneficial for earliness of flowering and getting higher fruit yield (Mistry and Patel, 2009; Kumar and Kavino, 2012). Banik and Sen (2002) successfully rejuvenated the unproductive trees of mango cv Fazli aged 30-50 years with the rejuvenation pruning comprising leader intact, removal of first order branch and pruning of secondary branches and application of 1kg each of N and K_2O and 0.5 kg P_2O_5/plant/year along with 3 per cent spray of urea in September, resulting in maximum growth and yield (388.42 kg/plant) of older trees above 50 years of age. Depending upon the extent of pruning, there are four model approaches that have been in practice at various locations to take up rejuvenation in mango (Hackett, 1985; Kumar and Kavino, 2012; Kumar, 2015).

Model-1

In this case the tall central trunks are cut back to about 3-4 m height from the ground level. It is a partial rejuvenation where one half of the tree (all around) is cut back and the remaining trunks and leaves will help protect the stump from sunburn (Crane *et al.*, 2009), Stassen (1999) compared the different systems of pruning with un-pruned controls under high density orchards of mango and concluded that about closed vase system (informal pyramid) performed the best.

Model-2

This is the form of skeletonizing the tree, *i.e.* cutting back the branches of the tree till only basic frame is left. Particularly large unthrifty trees would benefit from more severe skeletonizing.

Model-3

In this case, the extent of pruning is decided on the basis of first, second, third and fourth order branches. It was found that for rejuvenation, the pruning treatment with paclobutrazol application recorded significantly higher and quality fruit yield in fourth and fifth order pruned trees than the control and succeeded in restoring the health, vigor and sustainable fruit yield in declining mango orchard (Burondkar *et al.*, 2000; Lal *et al.*, 2000; Shinde *et al.*, 2002). Reddy and Kurian (2011) recorded significantly higher and almost double mean cumulative yield with third order pruning in mango cv Alphonso when compared with un-pruned control.

Model-4

In this model, reiterative pruning is done in phased manner up to height of 3-4 m from the ground level of senile trees and later canopy rebuilding in open umbrella shape with intensive care and management transforming successfully from unproductive to productive phase with enhanced quantum of quality fruit production within a gap of three years onwards (Kalloo *et al.*, 2005; Kumar and Kavino, 2012).

Guava

Heavy pruning in guava involved the heading back of exhausted trees, showing marked decline in annual production, to the extent of 1.0 to 1.5 m height from the ground level during May with the objective of facilitating production of new shoots below the cut point during the monsoon months have been found to give promising effect. The skillful pruning operation and time of pruning (April) have been found to give significant effect on fruit set, fruit yield and quality and proved successful in rejuvenating old guava orchards (Kalloo *et al.*, 2005; Singh *et al.*, 2005; Basu *et al.*, 2007), while the beneficial effect of pruning for rejuvenation with the integration of IPNM and AM fungi have been recorded up to a significant level on shoot emergence, fruit yield and fruit quality in guava (Chandra *et al.*, 2012; Kumar, 2015)

Guava orchards often lose their productivity after 20-25 years due to various biotic and abiotic factors. CISH technology on rejuvenation involves heading back of trees to a level of 1.0-1.5 m from ground level. Emerging shoots are allowed to grow (40-50 cm) for 4-5 months after heading back. Shoots are further pruned to 50 per cent of their length to allow emergence of multiple shoots. These shoots offer fruit bearing surface in the following season leading to healthy and productive canopy development, profuse fruiting and yield gets enhanced to the tune of 70-90 per cent over non-rejuvenated trees as observed in the very first year of operation. Yield, improved fruit size and better fruit quality are attributed to better light penetration into the canopy.

Singh *et al.* (2007) restored yield and quality production of exhausted trees of guava cvs Allahabad Safeda and Sardar. Topping and hedging increased new shoots (below the cut portion) over the un-pruned control trees. Topping and hedging operations particularly performed in May and October, reduced tree size and consequently improved the fruiting of trees. As a result, tree height was reduced by 34 to 43 per cent over the control trees in both the varieties. Similarly, May and October pruning restricted tree canopy by 29.89 to 50.03 per cent (different topped trees) over the control trees without adversely affecting yield. An enhancement of yield, *i.e.*, 82.39 kg plant in Allahabad Safeda and 53.10 kg plant in Sardar as against un-pruned trees after the first year of topping and hedging was recorded. Further, yield enhancement in the range of 104-112 to 74.90-79.20 kg/plant over the control trees was recorded in Allahabad Safeda and Sardar after second year of pruning as result of topping and hedging and vegetative to reproductive balance began to stabilize. The fruit produced from pruned trees had better quality as compared to control trees in both the varieties.

Basu *et al.* (2007) conducted an experiment to study the response of pruning on rejuvenation of old guava orchard (cv. L-49) in West Bengal. Eleven-year old guava plants were pruned drastically leaving only four scaffold branches per tree at monthly interval from March 2003 to July 2003. The height and spread of plants increased significantly after pruning as compared to control. Only rainy season crop was obtained one year after pruning and pruned plants flowered one month ahead of un-pruned (control) plants. Time of pruning had significant effect on fruit set. Fruit yield and physicochemical characters of mature fruits were found to be best in fruits obtained from plants pruned in April. Pruning proved to be successful in rejuvenating an old guava orchard of cv. L-49 (Sardar).

Singh *et al.* (2012) conducted a field experiment to study the response of varying rejuvenation periods in an old guava orchards (cv. Allahabad Safeda). Fifteen-year-old guava plants were severely pruned leaving only four scaffold branches per tree at monthly interval from

April to June. The tree height, tree spread and canopy volume were affected significantly by different periods of rejuvenation. Only winter season crop was obtained in first year after pruning. Time of pruning had significant effect on fruit set. Fruit yield and physico-chemical characters of mature fruits were found to be best in fruits obtained from trees pruned during May. Maximum number of fruits and yield were obtained from May rejuvenated trees (272 fruits and 46.22 kg/tree) and minimum in control (104 fruits and 27.12 kg yield/ tree). Pruning proved to be successful in rejuvenating old guava orchards and the economics also suggested gain in returns compared to senile orchards.

Litchi

In case of litchi, heavy reiterative pruning of litchi trees is done either with manual saw or power operated saw at an effective height of 2.0 to 2.5 m depending upon the nature and growth as well as girth and type of main trunk (Kumar, 2008; Singh *et al.*, 2011). Care is taken to avoid bark splitting or debarking at the cut end due to falling of heavy branches at the time of pruning.

For the success of rejuvenation technology in old senile orchards of litchi, the sequential operational steps should be followed in scientific manner. About 3 to 5 main branches with outward growth from the base are marked for pruning at recommended height, with the plan of developing umbrella like or semi circular frame work of tree canopy to enhance bearing surface area (Singh, 2006; Kumar, 2008). To avoid any external infection at the cut portion, it should be pasted with Bordeaux mixture or copper-oxy-chloride (Blitox) immediately after pruning (Bal, 2006). The best time to go for reiterative pruning for rejuvenation have been found August-September under Bihar conditions. There should be integrated nutrient management for maintenance of soil fertility and plant nutrient supply to an optimum level for sustaining the desired growth and further productivity level. As such, it is necessary that fertilizer applications are made using both organic and inorganic in pruned orchard with the main intention to boost initiation of new shoots with vigor at an early date. After the reiterative pruning, the tree should be fed with optimum fertilizer doses, followed by irrigation in the root zone, just like the bearing orchards. The manures and fertilizers should be applied through ring/trench method and the dose per tree as recommended for an adult bearing tree. The nutrition should be applied after one month of rejuvenation and thereafter it should be preferably in two split doses one during the end of February-March and remaining half dose in August and in both the time manure and fertilizer should be spread around the main trunk and mixed well in the surface soil followed by light irrigation.

Rejuvenated trees require immediate irrigation if any dry spell occurs just after the reiterative pruning operation is over, as to avoid drying out of the trees. Irrigation is must after the manure and fertilizer application. In litchi, depending upon the temperature, soil type and soil moisture status, pruned trees must be irrigated at regular interval of 8-10 days during hot summer season and 15-16 days during winter season helps the fast vegetative growth and good canopy development. Among the various irrigation systems, basin system is well suited but drip irrigation has also given better response. It is advisable to do mulching in the basins around the pruned tree trunk for conserving soil moisture in the basins during April to June and November to January. Various mulches in rejuvenated plants in the orchards have been found better with paddy straw, sugarcane thrash, dried grasses, mango/ litchi/banana dried leaves *etc.* applied thick enough (10-15 cm) or black polythene (200 gauge) sheet might also be used for mulching. Mulching at the base immediately surrounding the main trunk of the pruned trees during the months of April-May and September-October have been found beneficial in conserving soil moisture beneath the tree canopy, reducing the frequency of irrigation and enhancing growth rate of vegetative flush (Kumar, 2008).

A number of branches arise from the reiteratively pruned tree. At this juncture, thinning operation is done carefully with the ultimate aim of developing solid, semi-circular or open umbrella type canopy (Kumar, 2008; Singh, 2010). In this attempt, it is required to thin out the excessive branches as well and keeping only 3-4 numbers nearly top-side portion on each pruned limbs of the trunk projecting all the directions and further training for acquiring desired shape (Srivastava, 2007; Singh, 2010). The air and light must have access to all the parts of the tree. The study indicated that the overlapping of more branches and shading effect will reduce the flowering, fruiting, fruit yield and quality production. It requires shoot thinning followed by proper pruning operations to shape the canopy of the trees and with proper management of new shoots and flower initiation, the young shoots can be forced again to flower and fruit in the next season. The success of this technique depends upon the proper management of shoots through precise and timely pruning, when tree starts bearing after rejuvenation. During initial 2-3 years after rejuvenation, there is an excellent opportunity for economic utilization of inter-space in the rejuvenated orchards through intercropping with profitable kharif and rabi crops. Apart from the significant income obtained by the intercrops in rejuvenated orchards, the added advantages like improvement in the soil tilth (health), almost complete check over weed population/ growth as well as less incidence of pests-diseases are also received.

Intensive care to control the infestation of important pests like stem borer, bark eating caterpillar, mite, leaf roller and leaf cutting weevil *etc.* and the diseases like microbial infestations are required. Control methods may be biological, mechanical or chemical or the combination depending upon the growers' preference and specific circumstances.

Fruit yield and physicochemical characters of mature fruits were better in fruits obtained from rejuvenated trees. Maturity period is found to be slightly delayed in rejuvenated plants. Fruit yield in the 3rd year was obtained with enhanced quality production and the categorization of fruits under different quality classes revealed high percentage of fruits under extra class and class I as compared to un-pruned trees (Kumar, 2008). The quantum of fruit yield with enhanced quality increased significantly every year after the start of bearing in rejuvenated trees. It has been found that the quality fruit yield increased fruits almost three times in 10 years after rejuvenation as compared to old senile trees.

Cashew Nut

The productivity of existing cashew plantations in India appears to be quite low (700 kg/ha) against the target of 2000 kg/ha. The probable reasons for the poor productivity are cultivation in poor soils/ under neglected condition, poor maintenance of the existing cashew plantations, lack of awareness about latest production technology and presence of senile plantations. Hence, necessary corrective measures need to be taken to elevate the productivity and production of cashew.

Most of the cashew plantations taken up during fifties of the 20th century have almost become senile and unproductive which should either be replaced or rejuvenated. Rejuvenation would yield early result and rejuvenated plants also continue to be in productive state for some more years. An attempt was made to suggest a suitable strategy for rejuvenation of senile cashew plantations. Rejuvenation of old declining trees in cashew is done by heading back (mostly limb pruning of overlapping branches) of the exhausted tree at 1.0 - 1.50 m height from the ground level before the onset of monsoon (Kalloo *et al.*, 2005; Suharto *et al.*, 2012). The best season for heading back is May-June, *i.e.* immediately after harvesting of annual crop. The cut ends of branches should be smeared with 10 per cent Bordeaux paste to prevent the gummosis and entry of pathogens. Remaining portion of the trunk and exposed part of roots is to be swabbed with 0.2 per cent chlorpyriphos at least 2-3 times after beheading till new canopy is developed. The new flushes arise from the dormant buds of the trunk within a period of 6-7 months depending on the variety. The yield during the first season after pruning is slightly lower, while from second season onwards it is better than the un-pruned plants. The limb pruned trees should be pruned or trimmed as and when required in subsequent years. It may be noted that yield of limb pruned trees is much better than freshly planted cashew grafts due to better root system (Singh, 2010).

Aonla

Old and unproductive trees of *aonla* can be subjected to hardcore pruning for rejuvenation. The plants are beaded back during the dormant season to the extent of 2.0-3.0 m above the ground level with the sole objective of top working with superior varieties (Kalloo *et al.*, 2005; Basu *et al.*, 2007; Singh *et al.*, 2007; Singh and Mishra, 2007). Rejuvenation pruning to restore canopy and to enhance fruiting in old and senile *aonla* orchard has also been found successful at Lucknow through top working with superior cultivars (Mishra *et al.*, 2007) and Ranchi (Nath *et al.*, 2008).

Pomegranate

In case of pomegranate, 15-20 year old orchards showing the signs of decline mainly due to drying of old branches resulting in sparse flowering and fruiting, non-selective pruning up to a height of 30 cm to 120 cm from the ground level are headed back but pruning plants up to 30 cm was the best to rejuvenate the old pomegranate orchard (Hiwale, 2006, 2009). The pomegranate should be pruned to a multiple-trunk system so that if one scaffold is damaged the tree can be reshaped and white washing is preferred to protect against frost/freezing injury (Stebbins, 1985)

Ber

Ber trees also lose their productivity after 20-25 years of age. Sometimes, ber plantations if not managed properly with annual pruning in a scientific manner often attain huge size with the fruiting confined only to the outer periphery. Such trees can be successfully rejuvenated. Singh *et al.* (2015) successfully rejuvenated the 33 year old ber trees, heading back from the ground level and top working with different cultivars, with Gola giving the maximum fruit yield (36.4 kg/tree) in the third year. Ber trees were severely pruned, heading back the branches to 50-75 per cent and rejuvenated successfully after proper after-care by Bal *et al.* (2013). Besides, seedlings of *Ziziphus rotundifolia* and *Z. nummularia* have been success rejuvenated through top working with superior ber cultivars in arid and semi-arid parts of our country (Deb Roy, 1994; Deb Roy and Rai, 1993; Singh *et al.*, 2006). Ghosh *et al.* (2009) compared fruiting in different ber varieties after top working, with Jogia variety giving the highest fruit yield.

Citrus

Fruiting branches decline in productivity after 2-3 years and pruning of declining old fruiting wood, removal of dead, dried and diseased wood is necessary to check the further spread of the diseased wood is necessary to check the further spread of the diseased and mechanical injury of fruits. Heavy hand pruning called skeletonization is given to old and neglected citrus trees to rejuvenate them (Naik, 1963, Reddy and Patil, 2004). Fruit grade was improved (*i.e.* substantially larger), thus increased net cash returns. In case of citrus, severe pruning for rejuvenation is yet to give significant and consistent results (Lewis *et al.*, 1963).

6. Time of Heading Back or Reiterative Pruning

In principle, perennial trees are to be pruned when they are in dormant phase more particularly practiced in winter season or after the monsoon season is over (Lonsdale, 1999). Time of rejuvenation pruning may vary from crop to crop depending upon the nature of growth and time of dormancy (Table 2). Rejuvenation is labour intensive operation and it is extremely necessary to complete the operations timely for achieving the desired results. The imbalance in root:shoot ratio can be corrected by judicious pruning. Dead, cris-cross, broken branches or the water sprouts should be removed at any time but not before flowering period. If the exposed wounds are big, they should be smeared with the anti-fungal paste.

7. Nutrition of Rejuvenated Trees

There should be integrated nutrient management for maintenance of soil fertility and plant nutrient supply to an optimum level for sustaining the desired growth and further productivity level. It is necessary that fertilizer applications are made using both organic and inorganic forms in pruned orchard with the sole purpose of initiating new shoots with vigor at an early date. After the reiterative pruning, trees should be fed with optimum fertilizer doses, followed by irrigation near the root zone, just like the commercial bearing orchards. The dose per tree as an adult bearing stage has been presented in the Table 2. Based on soil and leaf nutrient status, micronutrient deficiencies can be corrected by

Table 2: Recommended Height and Time for Rejuvenation of Old Senile Orchards of some Perennial Fruit Trees

Crop	*Recommended Height for Pruning (m)*	*Time of Pruning*	*References*
Mango	3.0-5.0	November-December	Kalloo *et al.* (2005)
Guava	1.0-1.5	May-June	Kalloo *et al.* (2005)
Litchi	2.0-2.5	August-September	Kumar (2008)
Cashew	1.0-1.5	May-June	Kalloo *et al.* (2005)
Aonla	2.0-2.5	December-January	Kalloo *et al.* (2005)
Ber	2.0-2.5	May -June	Bal *et al.* (2013)
Pomegranate	1.0	May-June	Hiwale (2009)
Citrus	1.0-1.5	May-June	Kalloo *et al.* (2005)

Table 2: Doses of Manure and Fertilizers for different Fruit Crops (Kumar, 2015)

Crop	*Doses of Manures and Fertilizers*	*References*
Mango	120 kg well rotten FYM, 4 kg neem cake, 2.5 kg urea, 3.0 kg single super phosphate and 1.5 kg muriate of potash per tree (apply preferably in two split doses one during the end of February and remaining half dose in June)	Kumar and Adak (2013)
Litchi	75-80 kg well rotten FYM, 2 kg neem/castor cake, 1.00 kg urea, 1.5 kg single super phosphate and 0.5 kg muriate of potash per tree (apply preferably in two split doses one during August-September and another February-March). The micronutrients like zinc ($ZnSO_4$ @ 2-3 g/l) and boron (borax @ 2 g/l) may be applied through foliar application form one year after the reiterative pruning.	Kumar (2008, 2012)
Guava	50 kg well rotten FYM, 2 kg neem cake, 1.5 kg urea, 2.0 kg single super phosphate and 1.0 kg muriate of potash per tree (apply preferably in two split doses one during the end of May-June and remaining half dose in September)	Singh and Mishra (2007)
Citrus	50 kg well rotten FYM, 2 kg neem cake, 1.5 kg urea, 2.0 kg single super phosphate and 1.0 kg muriate of potash per tree (apply preferably in two split doses one during the end of May-June and remaining half dose in September)	Srivastava and Singh (2004), Srivastava and Singh (2007)
Aonla	50 kg well rotten FYM, 1.5 kg urea, 2.0 kg single super phosphate and 1.0 kg muriate of potash per tree (apply preferably in two split doses one during the February-March and remaining half dose in August, while another recommendation is apply during July-August and December-January)	Singh and Mishra (2007), Maity and Das (2004)
Cashew nut	50 kg well rotten FYM, 0.75 kg N, 0.325 kg P_2O_5 and 0.75 kg K_2O per tree	Janik and Paull (2008), Veeraraghavan *et al.* (1985)
Pomegranate	20 kg FYM, 0.25 kg N, 0.325 kg P_2O_5 and 0.25 kg K_2O/tree/year at the onset of monsoon	Hiwale (2009)

soil application of micronutrients containing fertilizer, micronutrient based formulations or through foliar nutrition (Kumar, 2015).

8. Thinning of Shoots and Canopy Rebuilding

After pruning to improve productivity of old and unproductive perennial fruit orchards, rebuilding canopy is very vital (Srivastava, 2007). The reiterative pruning brings invigorating effect and subsequent emergence of numerous shoots on the entire body of trunk requires proper thinning. Thinning of shoots and canopy should be skillfully done by just removing the extra, overlapping down growing, weak shoots after a certain gap period. The ideal gap period is at least 6 months for mango, litchi, 3 months for *aonla* and cashew, while one month for guava, citrus and pomegranate (Kalloo *et al.*, 2005; Kumar, 2008, Hiwale, 2009; Singh and Mishra, 2007). The operation is repeated 3-4 times with the same time interval but just before the onset of rainy and winter season. Enormous shoots arise and cover the entire top of the pruned plant. There is a need to thin out the excessive branches and allowing only 3-4 numbers nearly top side portion on each pruned limbs of the trunk projecting towards all the directions and further training for acquiring desired shape (Kumar, 2015). The thinning operation is done carefully with the ultimate aim of developing solid, semi-circular or open umbrella type canopy. A sort of tree training is required for canopy rebuilding to make a semi-circular/open umbrella shape canopy with enhanced surface bearing area (Kumar, 2008). Strategic thinning and pruning help in controlling tree size but it must be carried out at the right time to ensure shoot maturity, panicle emergence, flowering and cropping (Olesan *et al.*, 2002; Kalloo *et al.*, 2005; Kumar, 2013).

9. Plant Protection

Controlling the insect pests, the restoration of vigor and keeping the tree more healthy help in enhanced and sustained quality fruit production (Kalloo *et al.*, 2005; Kumar, 2008; Singh and Mishra, 2007; Singh, 2010). Intensive care to control the infestation of important pests like stem borer, bark eating caterpillar, mite, leaf roller and leaf cutting weevil, *etc.* and the diseases like microbial infections are required (Dodd *et al.*, 1997; Masarrat *et al.*, 2000; Ploetz and Prakash, 1997; Ploetz, 2003). Control methods may be biological, mechanical or chemical or the combination depending upon the growers preference and specific circumstances.

10. Fruit Yield and Quality

Fruit yield and physiological characters of mature fruits were found to be better in fruits obtained from rejuvenated trees. Maturity period is found to be slightly delayed in rejuvenated plants. Fruit yield in the 2nd year in case of guava, citrus and *aonla*, while third year in the rejuvenated trees of mango, litchi, cashew and ber obtained with enhanced quality production and the categorization of fruit under different quality class revealed high percentage under extra class and class I as compared to un-pruned ones. The quantum of fruit yield with enhanced quality increased significantly every year onwards after the start of bearing in rejuvenated trees.

In case of mango, fruit production starts from the third year and the productivity increases up to about four times after proper canopy development of about 7-8 years of the operation. In this fashion, productivity is restored and with progressive increase in fruit yield making the orchard productive and remunerative (Kalloo *et al.*, 2005; Lal *et al.*, 2000, 2001; Lal and Mishra, 2007, 2008; Reddy and Kurian, 2011; Kumar and Kavino, 2012; Mishra, 2013). An enhancement of about 22.35 per cent to 45.65 per cent in fruit yield and that too of superior quality in different cases have been recorded from the rejuvenated trees as compared to un-pruned ones in litchi (Kumar, 2008) and that of guava cv. Allahabad Safeda and Sardar after the first year of pruning (Nimbalkar *et al.*, 2010; Singh, 2010; Jahangeer *et al.*, 2011; Singh *et al.*, 2012; Mishra, 2013). In case of *aonla*, all the old senile trees responded well to rejuvenation and the technology helped in securing quality yields in the range of 35 to 40 kg/tree in demonstration plots after second year of rejuvenation (Singh and Mishra, 2007).

11. Economics of Rejuvenation

In fact, various operations of rejuvenation are performed over a period of 2-3 years and thereafter the management continues. In case of litchi, economics has been worked out for five consecutive years after reiterative pruning and it was found that old senile trees could be rejuvenated and made productive just like young bearing orchards within a time gap of 2-3 years without any income loss (Kumar, 2013). All the factors and particulars like sale of fruits before rejuvenation, cost of reiterative pruning, sale of wood, cost and returns from intercropping, sale fruits from rejuvenated trees and finally gross income were taken into account. Rejuvenation of one hectare of senile litchi orchard gave net returns of Rs 3,83, 700 (in first year) mainly through sale of wood and intercrops, Rs 42,500 in second year, Rs 41,00 in the third year mainly through intercrops, Rs 91,500 in fourth year and Rs 1,15,500 in fifth year through both sale of fruits after rejuvenation and growing of intercrops. Thus gains from fruits started from fourth year (Kumar, 2013). Economics has been worked out and significant gains have been recorded after rejuvenation of unproductive trees in mango (Lal and Padaria, 2001; Hasan *et al.*, 2009), guava (Singh *et al.*, 2005; Singh *et al.*, 2012).

12. Constraints

It is proven that rejuvenation technology has several advantages and is the need of hour for a number of fruit crops. But slow level of adoption of the technology clearly indicates that there are some inhibitions from the farmers' side. The technology involves a series of steps and processes like hard pruning followed by nutritional, irrigation, mulching, thinning of shoots, training for canopy rebuilding in desired shape, intensive plant protection measures for insect pest especially stem borer and intercropping to ensure income during initial years of rejuvenation. Getting permission from the respective Govt. State Departments for rejuvenation is also a problem in some areas. The chief hindrances in significant adoption are lack of implementing team equipped with required machinery (Singh *et al.*, 2011; Kumar, 2012).

13. Conclusion

Rejuvenation of old, senile orchards is the viable option in number of fruit crops such as mango, litchi, guava, cashew, *ber*, citrus, pomegranate to regain the tree vigor and make the tree productive within a short time gap instead of replacement with new plantation which is a long term strategy. Critical assessment of all the factors is a must before going for rejuvenation. A series of operations such as reiterative pruning followed by aftercare, intercropping, nutrition, irrigation over a time period are to be performed with sincerity in order to achieve success in rejuvenation. Significant gains from the rejuvenated orchards are proven fact but the constraints like lack of trained people, tools/ machinery are still the constraints in faster adoption of the technology. The sensitization among farmers, government departments through demonstrations, speedy process for granting permission to the farmers for rejuvenation are some of the issues which need attention.

14. References

Abrahamson, 1979. Patterns of resource allocation in wildflower populations of fields and woods. *American J. Bot.*, **66**: 71–79.

Bal, J.S. 2006. *Fruit Growing: Fruit Plant Management Practices.* Kalyani Publishers, pp. 54–66.

Bal, J.S., Sharma, D.R. and Singh, P. 2013. Historic ber (jujube) trees in Golden Temple Amritsar (Punjab): Twenty years of consistent care. *Acta Hort.*, **993**: 37–41.

Banik, B.C. and Sen, S.K. 2002. Effect of pruning and fertilization on growth and yield of mango cv. Fazli. *The Odisha J. Hort.*, **30**(1): 13–18.

Basu, J., Das, B., Sarkar, S., Mandal, K.K., Banik, B.C., Kundu, S., Hasan, M.A., Jha, S. and Ray, S.K. 2007. Studies on the response of pruning for rejuvenation of old guava orchard. *Acta Hort.*, **735**: 303–309.

Bell, A.D. 1991. *Plant Form.* Oxford University Press, New York.

Burondkar, M.M., Gunjate, R.T., Magdum, M.B. and Govekar, M.A. 2000. Rejuvenation of old and overcrowded Alphonso mango orchard with pruning and use of paclobutrazol. *Acta Hort.*, **509**: 681–686.

Chandra, K.K., Pandey, S.K. and Singh, A.K. 2012. Influence of tree rejuvenation, IPNM and VAM fungi on shoot emergence, yield and fruit quality of Psidium guajava under farmers' field condition. *Int. J. Biosci.*, **2**(11): 9–17.

Crane, J.H., Garcia, L. Salazor, Lin T.S., Pinto A.C.de Queiroz and Shu, Z.H. 2009. Crop production management. **In**: *The Mango: Botany, Production and Uses*, 2nd edn, pp 432–83, (Ed.) Richard E. Litz. CABI Publication, UK.

Deb Roy, R. 1994. Agroforestry: Principles and practices. **In**: *Dryland Agriculture*, (Ed.) R.P. Singh. Scientific Publishers, Jodhpur, p. 554.

Deb Roy, R. and Rai, P. 1993. More crops for Jharberi through *in situ* budding. *Wasteland News*, **10**(1): 77.

Dodd, J.C., Prusky, D. and Jeffries, P. 1997. Fruit diseases. **In**: *The Mango: Botany, Production and Uses*, pp. 281–325, (Ed.) R.E. Litz. CABI Publication, Wellington, UK.

Fakhri, A., Bazzaz, Nona, R., Chiariello, Phyllis D. Coley and Louis F. Pitelka. 1987. How plants cope: Plant physiological ecology. *Biosci.*, **37**(1): 58–67.

Ghosh, S.N. and Mathew, V. 2002. Performance of nine ber (*Ziziphus mauritiana* Lamk) on top-working in semi-arid region of West Bengal. *J. Appl. Hort.*, **4**(1): 49–51.

Ghosh, S.N., Tarai, R.K., Bera, B. and Kundu, A. 2009. Performance of ber cultivars raised through top-working in close spacing. *Acta Hort.*, **840**: 181–188.

Hackett, W.P. 1985. Juvenility, maturation and rejuvenation in woody plants. *Hort. Rev.*, **7**: 109–155.

Halle, F., Oldeman R.A.A. and Tomlinson, P.B. 1978. *Tropical Trees and Forests: An Architectural Analysis.* Springer Verlag, Berlin-Hiedelberg.

Hare Krishna, 2012. *Physiology of Fruit Production.* Studium Press (India) Pvt. Ltd., p. 101.

Hasan, M.A., Singh, B. and Sarkar, S., Jha, S and Ray, S.K. 2009. Canopy management of unproductive mango (*Mangifera indica* L.) orchards. *Acta Hort.*, **820**: 339–346.

Hiwale, S.S. 2006. Rejuvenation of pomegranate orchards through non-selective pruning. **In**: *Proceedings of National Symposium on Production, Utilization and Export of Underutilized Fruits with Commercial Potentialities*, Kalyani, WB, pp. 174–176.

Hiwale, S.S. 2009. Orchard rejuvenation. **In**: *The Pomegranate*. New India Publishing Agency, New Delhi, pp. 67–72.

Jahangeer, A.B., Akbar, I. and Kumar, V. 2011. Rejuvenation of old and senile orchards: A review. *Annals Hort.*, **4**(1): 37–44.

Janik, J. and Paull, R.E. 2008. Anacardiaceae–Cashewnut (*Anacardium occidentale* L.). **In**: *The Encyclopedia of Fruits and Nuts*, pp. 8–14. CABI publication, Wellington, UK.

Kalloo, G., Reddy, B.M.C., Singh, G. and Lal, B. 2005. Rejuvenation of old and senile orchards. *Tech. Bull.*, CISH, Lucknow, 40 p.

Kumar, K. and Adak, T. 2013. Integrated nutrient and water management. **In**: *Recent Advances in Sub-tropical Fruit Crop Production*, pp. 40–50, (Eds.) V.K. Singh and H. Ravishankar. CISH, Lucknow.

Kumar, N. and Kavino, M. 2012. Canopy management in mango. **In**: *Mango: Production and Processing Technology*, Vol. 1: 366–383, (Eds.) G. Valavi Sudha, K. Rajamohan, J.N. Govil, K.V. Peter and George Thottappilly. CABI Publication, UK.

Kumar, R. 2012. Adoption of improved production technologies by litchi growers for quality production. *J. Comm. Mobiliz. Sust. Dev.*, **7**(2): 210–216.

Kumar, R. 2013. Reiterative pruning and canopy rebuilding in old senile litchi trees for rejuvenation and enhanced quality fruit production. **In**: *Canopy Management and High Density Planting in Sub-Tropical Fruit Crops*, pp. 217–224, (Eds.) V.K. Singh and H. Ravishankar. Proceedings of National Seminar-cum-Workshop, CISH, Lucknow.

Kumar, R. 2013. Rejuvenating old senile litchi trees: A remunerative venture. **In**: *Canopy Management and High Density Planting in Sub-Tropical Fruit Crops*, pp. 264–269, (Eds.) V.K. Singh and H. Ravishankar. Proceedings of National Seminar-cum-Workshop, CISH, Lucknow.

Kumar, Rajesh 2008. Rejuvenation of old senile litchi orchard. Tech. Bull. No. 5. NRC for Litchi, Muzaffarpur, pp. 1–8.

Kumar, Rajesh 2015. Rejuvenating old, senile orchards of tropical and sub-tropical fruits for enhanced production and improved quality: A review. *Indian J. Agril. Sci.*, **85**(3): 295–313 .

Lal, B. and Mishra, D. 2007. Effect of pruning on growth and bearing behavior of mango cv Chausa. *India J. Hort.*, **64**(3): 268–270.

Lal, B. and Mishra, D. 2008. Studies on pruning in mango for rejuvenation. *Indian J. Hort.*, **65**(4): 405–408.

Lal, B. and Padaria, R.N. 2001. Rejuvenation of mango orchards. Ext. Bull.–9. CISH, Lucknow, 20 p.

Lal, B., Rajput, M.S., Rajan, S. and Rathore, D.S. 2000. Effect of pruning on rejuvenation of old mango trees. *Indian J. Hort.*, **57**(3): 240–242.

Lal, B., Rajput, M.S., Rajan, S. and Rathore, D.S. 2001. Rejuvenation of old and unproductive mango orchards. *Indian Hort.*, **57**(3): 240–242.

Lewis, L.N., McCarty, C.D. and Moore, P.W. 1963. Effects of skeletonization on the rejuvenation and production of declining orange trees. *Proc. Amer. Soc. Hort. Sci.*, **83**: 295–299.

Lonsdale, D. 1999. *Principles of Tree Hazard Assessment and Management*, 6th impression 2008, Forestry Commission, Great Britain.

Maity, R.G. and Das, P.C. 2004. *Fruits Crops of India.* Kalyani Publishers, New Delhi.

Masarrat, H., Abbas, S.R., Srivastava, R.P. and Sharma, S. 2000. Studies on insect pests of *Aonla* (*Emblica Officinalis* Gaertn). *Ann. Pl. Prot. Sci.*, **8**: 85–88.

Menzel, C.M. and Simpson, D.R. 1995. Nutritional studies on lychee trees in sub-tropical Australia. *Acta Hort.*, **273**: 581–585.

Menzel, C.M., Olesan, T., McChonchie C., Wiltshire, N., Diczabilis, Y. and Wicks, C. 2000. Lychee, longan and rambutan: Optimizing canopy management. A report for the Rural Industries Research and Development Corporation (RIRDC), RIRDC Publication No. 2000/29, Canberra.

Mishra, D. 2013. Improving orchard productivity through rejuvenation and canopy management. **In**: *Recent Advances in Sub-tropical Fruit Crop Production*, pp. 35–39, (Eds.) V.K. Singh and H. Ravishankar, CISH, Lucknow.

Mishra, D., Pandey, D., Mishra, R. and Pathak, R.K. 2007. Performance of improved *aonla* cultivars during top-working of senile trees. *Indian J. Hort.*, **64**(4): 396–398.

Mistry, P.M. and Patel, B.N. 2009. Impact of heading back plus paclobutrazol on rejuvenation of old and overcrowded Alphonso orchards. *Indian J. Hort.*, **66**(4): 520–521.

Naik 1963. *South Indian Fruits and their Cultivars*. P. Vasadachari and Co., Chennai.

Nath, V., Kumar, D. and Pandey, V. 2008. *Fruits for the Future: Well Versed Arid and Semi-arid Fruits*, Vol. 1, SSPH, Delhi.

Nimbalkar, C.A., Idate, G.M., Gaikwad, S.P. and Desai, U.T. 2010. Rejuvenation technology for improvement of guava (cv Sardar) orchard. *J. Mh. Agril. Univ.*, **35**(3): 486–487.

Olesan, T., Menzel, C.M., Wiltshire, N. and McChonchie, C.A. 2002. Flowering and shoot elongation of lychee in Eastern Australia. *Aust. J. Agril. Res.*, **53**: 977–983.

Pandey, D., Pandey, A. and Yadav, S. 2013. Horticultural interventions for increasing productivity of *aonla* and bael. **In**: *Recent Advances in Sub-tropical Fruit*

Crop Production, pp. 35–39, (Eds.) V.K. Singh and H. Ravishankar. CISH, Lucknow.

Ploetz, R.C. 2003. Diseases of mango. **In**: *Diseases of Tropical Fruit Crops*, pp. 327–363, (Ed.) R.C. Ploetz. CABI Publication, Wellington, UK.

Ploetz, R.C. and Prakash, O. 1997. Foliar, floral and soil born diseases. **In**: *The Mango: Botany, Production and Uses*, pp. 281–325, (Ed.) R.E. Litz. CABI Publication, Wellington, UK.

Rao, V.N.M. 1971. A note on pruning as a remedy for irregular bearing mango. *Andhra Agril. J.*, **18**: 242–245.

Rao, V.N.M. and Shanmugavelu, K.G. 1976. Studies on the effect of pruning in mango. *Progr. Hort.*, **8**: 21–28.

Reddy, B.M.C. and Patil, Prakash. 2004. Training and pruning. **In**: *Advances in Citriculture*, pp. 206–219, (Eds.) Shyam, Singh, V.J. Srivastava, A.K. Srivastava and I.P. Singh. Jagminder Book Agency publication, New Delhi.

Reddy, Y.T.N. and Kurian, R.M. 2011. Studies on rejuvenation of old, unproductive Alphonso mango trees in orchards. *J. Hort. Sci.*, **6**(2): 145–147.

Room, P.M., Maillette, L. and Hanan, J.S. 1994. Modular and metamer dynamics and virtual plants. *Adv. Ecol. Res.*, **25**: 105–157.

Saxena, M. and Gandhi, C.P. 2014. *Indian Horticulture Database–2014*. Ministry of Agriculture, National Horticulture Board, Gurgaon, Haryana.

Shinde, A.K., Waghmare, G.M., Godse, S.K. and Patil, B.P. 2002. Pruning for rejuvenation of overcrowded old Alphonso mango (*Mangifera indica* L) gardens in Konkan. *Indian J. Agril. Sci.*, **72**(2): 90–92.

Singh, A. 2003. *Fruit Physiology and Production*. Kalyani Publishers, New Delhi, pp. 488.

Singh, A. 2006. *Horticulture at a Glance: Fruits*. Kalyani Publishers, pp. 76–99.

Singh, A., Meghwal, P.R., Saxena, A. and Morwal, B.R. 2015. Rejuvevation of old and uneconomical ber tres and its effect on growth, yield and fruit quality under rainfed conditions of western India. *Indian J. Hort.*, **72**(4): 472–478.

Singh, D., Shukla, A.K., Singh, S.P., Meena, S.R. and Kumar, R.N. 2006. Improvement in natural plantation of underutilized fruits through top-working. *Proc. National Symp.* Nov 22–24, BCKV, West Bengal, p. 120–122.

Singh, G. 2010. *Practical Manual on Canopy Management in Fruit Crops*. Department of Agriculture and Cooperation, MOA, GOI, New Delhi.

Singh, G. and Mishra, R. 2007. *Aonla* Rejuvenation. PFDC, CISH, Lucknow, 16 p.

Singh, G. Kumar, R. and Nath, V. 2011. *Rejuvenation in Litchi*. NRC on Litchi, Muzaffarpur, Bihar, pp. 1–14.

Singh, G., Mishra, R. and Gupta, S. 2007. Modifying existing guava tree canopies for increased production efficiency. *Acta Hort.*, **735**: 243–248.

Singh, G., Mishra, R. and Singh, G.P. 2005. Guava rejuvenation. Ext. Bull.–28, CISH, Lucknow, 20 p.

Singh, N.K., Shrivastava, D.C. and Bhandarkar, A.P. 2012. Growth, yield and quality of guava as influenced by varying rejuvenation periods. *Indian J. Hort.*, **69**(2): 181–184.

Srivastava, A.K. and Singh, S. 2004. Citrus nutrition. In: *Advances in Citriculture*, pp. 305–362, (Eds.) Shyam, Singh, V.J. Srivastava, A.K. Srivastava and I.P. Singh, Jagminder Book Agency Publication, New Delhi.

Srivastava, A.K. and Singh, S. 2007. Nutrient Management in citrus orchards of northeast India. *Indian Farm.*, **59**(6): 27–32.

Srivastava, K.K. 2007. *Canopy Management of Fruit Crops*. IBDC, Lucknow.

Stassen, P.J.C., Grove, H.G. and Davie, S.J. 1999. Tree shaping strategies for higher density mango orchards. *J. Appl. Hort.*, **1**(1): 1–4.

Stebbins, R.L. 1985. *Pruning: How to Guide for Gardeners*. HP Books, Price Stern Sloan, Los Angeles, California.

Sterck, Frank 2005. Woody Tree Architecture. **In**: *Plant Architecture and its Manipulation, Annual Plant Reviews*, pp. 209–237, (Ed.) G.N. Turnbell Colin. Blackwell Publishing, CRC Press.

Suharto, I., Ambarawati, I.G.A.A., Agung, I.G.A.M.S. and Nurjaya, I.G.M.O. 2012. The number of grafted scions and remaining productive branches after new shoot growth and flowering of side grafted cashew (*Anacardium occidentale* L.). *J. Int. Soc. South East Asian Agril. Sci.*, **18**(1): 160–172.

Veeraraghavan, P.G., Celine, V.A. and Balakrishnan, S. 1985. Study on the fertilizer requirements of cashew (*Anacardium occidentale* L.). *Cashew Causerie*, **7**(2): 6–8.

White, J. 1979. The plant as a meta-population. *Ann. Rev. Ecol. Syst.*, **10**: 109–145.

26

Effect of Cycocel on Suppression of Post-monsoon Vegetative Growth of Mango cv. Alphonso

P.M. Haldankar*, Y.R. Parulekar, T P. Soudagar, S.S. Tendulkar, P.C. Haldavnekar and S.B. Thorat

[1]College of Horticulture,
Dr. Balasaheb Sawant Konkan Krishi Vidyapeeth, Dapoli – 415 712, Maharashtra
**E-mail: paraghaldankar@yahoo.in*

Abiotic factors play important role in flowering and fruiting of fruit crops. Mango is greatly influenced by abiotic stresses. The effect of abiotic factors vary from variety to variety. Alphonso is a premium mango variety well known for its unique flavour both for table as well as processing. It is planted on very large scale in Konkan region of Maharashtra which is a coastal belt receiving assured rainfall and moderate climatic conditions with respect to temperature and humidity (Anon., 2015). The flowering in Alphonso commences in month of October – November and fruits are harvested from March onwards. The early harvested fruits fetch premium rates in the market. In recent years, the climatic abberations especially with respect to rainfall at the onset of flowering adversely affect the Alphonso mango production in Konkan region. The rainfall which ceases in the month of September under normal conditions is many times extended upto October – November. The delayed rainfall induces vegetative shoot instead of flowering panicles in mango in spite of use of paclobutrazol which is well known growth retardant for induction of flowering. The flowering is thus late which further delays the harvesting. The fruits fetch relatively lower rates in the market causing remarkable economic loss. Hence, it is necessary to restrict the vegetative growth immediately after withdrawl of monsoon and induce flowering in mango cv. Alphonso. Growth retardants play important role in suppression of vegetative growth (Mansuroglu *et al.*, 2009). The use of cycocel in suppression of vegetative growth and induction of flowering is well emphasized by Rademacher (2000). An attempt was therefore made to study the effect of cycocel for suppression of vegetative growth in mango cv. Alphonso.

1. Materials and Methods

The experiment was conducted at Department of Horticulture, College of Agriculture, Dr. B.S. Konkan Krishi Vidyapeeth, Dapoli (MS) during 2015 -16. Forty years old mango plants of cv. Alphonso were selected on which post monsoon vegetative flush occurred instead of flowering. All recommended package of practices were followed alongwith the application of paclobutrazol @ 0.75 g a.i. was applied per meter of canopy spread in second fortnight of July as per the recommendation. The experiment was conducted in randomized block design with five treatments *viz.*, T_1 =Cycocel 1250 ppm (1 spray), T_2 = Cycocel 1250 ppm (2 spray), T_3 = Cycocel 2500 ppm (1 spray), T_4 = Cycocel 2500 ppm (2 spray), T_5 =Control (No spraying), replicated four times. The foliar application of required concentration of cycocel was done on experimental plants in month of October as per the treatment details. The second spray of cycocel as per the treatment details was given 15 days after first spray. The various observation *viz.*, emergence of panicle, length and breadth of panicle, sex ratio, fruit set, fruit retension, days required for flowering and yield (kg/plant) were recorded along with physico – chemical properties. The data obtained was analysed statistically as per the method suggested by Panse and Sukhatme (1995).

2. Results and Discussion

The foliar spray of cycocel had non-significant effect on the days required for flower emergence (Table 1). However, it took 92.5 days which were shortest for the panicle emergence. Further, panicle produced in T3 had significantly shortest length 15.02 cm and width of 9.46 cm. The fruit set recorded by all cycocel treatments were greater than that of control but the fruit retention recorded non-significant deviations among all treatments. The higher concentration of cycocel reduced the length of panicle might be due to blocking of GA synthesis which curtailed the length and width of panicle. Similar results were recorded in mango by Suryanaryana (1986) and Sarkar *et al.* (1997). The higher fruit set in mango by application of cycocel was also recorded by Daulta *et al.* (1981) and Gautam (1998).

It took 89.63 days from flowering to first harvesting in T_1 which was at par with T_4 (Table 2). The yield obtained was maximum in T_2 (120.26 fruits and 60.13 Kg) which was significantly superior over rest of treatments. The fruit length, diameter and weight exhibited non-significant variations. Choudhari *et al.* (2014) also reported increase in total fruit yield when the trees were sprayed with cycocel. Similar increase in yield of Kesar mango was also reported by Sarkar *et al.* (1998). Also regarding the physical parameters the similar results were observed by Kulkarni (1988), Desai (1993) and Padhiar (1999) in mango when trees were sprayed with various growth retardants. The non-

Table 1: Effect of Foliar Spray of Cycocel on Days Required for Emergence of Panicle, Panicle Length and Width, Fruit Set and Retention in Mango cv. Alphonso

Treatment	*No. of Days for Panicle Emergence*	*Panicle Length (cm)*	*Panicle Width (cm)*	*Fruit Set (%)*	*Fruit Retention (%)*
T_1- Cycocel 1250 ppm 1 spray	98.00 98.00±2.44	26.82 26.82±8.03	16.25 16.25±3.65	3.11 3.11±0.13	0.77 0.77±0.33
T_2-Cycocel 1250 ppm 2 sprays	93.00 93.00±4.54	24.29 24.29±5.28	15.73 15.73±3.95	3.44 3.44±0.29	0.75 0.75±0.19
T_3-Cycocel 2500 ppm 1 spray	92.50 92.5±6.35	15.02 15.02±0.41	9.46 9.46±1.79	1.67 1.67±0.40	0.69 0.69±0.20
T_4-Cycocel 2500 ppm 2 sprays	98.75 98.75±1.5	18.92 18.92±7.022	10.38 10.38±2.67	4.78 4.78±0.55	0.67 0.67±0.26
T_5–Control	95.25 95.25±0.96	26.38 26.38±2.09	13.80 13.80±4.81	1.37 1.37±0.1	0.65 0.65±0.20
Range	92.50–98.75	15.02–26.82	9.46–16.25	1.37–4.78	0.65–0.77
Mean	95.50	22.29	13.13	2.87	0.70
S.E.±	1.99	2.46	1.68	0.19	0.13
C.D.at 5%	NS	7.57	5.16	0.58	NS

Table 2: Effect of Foliar Spray of Cycocel on Days Required for Harvesting, Yield and Physical Parameters of Fruit

Treatment	*Days Required from Flowering to 1st Harvesting*	*No. of Fruits per plant*	*Yield Kg/plant*	*Fruit Length (cm)*	*Fruit Diameter (cm)*	*Fresh Fruit Weight (g)*
T_1- Cycocel 1250 ppm 1 spray	89.63 89.63±1.79	83.65 83.65±17.89	46.34 46.34±7.41	9.17 9.17±0.52	7.23 7.23±0.19	232.70 232.70±11.89
T_2-Cycocel 1250 ppm 2 sprays	96.00 96.00±5.35	120.26 120.26±20.49	60.13 60.13±7.11	9.08 9.08±0.35	7.46 7.46±0.28	255.13 255.13±18.74
T_3-Cycocel 2500 ppm 1 spray	95.75 95.75±1.5	79.63 79.63±10.17	39.8 139.81±7.08	9.68 9.68±0.50	7.85 7.85±0.43	266.55 266.55±21.69
T_4-Cycocel 2500 ppm 2 sprays	90.00 90.00±4.08	68.94 68.94±9.13	34.47 34.47±3.46	9.52 9.52±0.79	7.35 7.35±0.36	252,48 252.48±14.96
T_5–Control	95.00 95.00±4.08	74.73 74.73±13.48	38.85 38.85±8.21	8.77 8.77±0.03	7.65 7.65±0.66	267.40 267.40±7.94
Range	89.63–96.00	68.94–120.26	34.47–60.13	8.77–9.68	7.23–7.85	232.70–267.40
Mean	93.28	77.44	43.92	9.04	7.51	254.05
S.E.±	1.66	6.92	3.03	00.35	0.22	8.70
C.D.at 5%	5.11	21.31	9.35	NS	NS	NS

Table 3: Effect of Foliar Spray of Cycocel on Chemical Composition of Fruits of Mango cv. Alphonso

Treatment	Acidity (%)		Ascorbic Acid (mg/100g)		Reducing Sugars (%)		Total Sugar (%)		TSS (° B)	
	At Harvest	Ripe Stage	At Harvest	Ripe Stage	At Harvest	Ripe Stage	At Harvest	Ripe Stage	At Harvest	Ripe Stage
T_1- Cycocel 1250 ppm 1 spray	4.51 4.51±0.21	0.40 0.40±0.11	72.66 72.66±3.45	49.31 49.31±2.71	1.41 1.41±0.06	3.01 3.01±0.70	1.43 1.43±0.09	13.67 13.67±0.35	8.58 8.58±0.36	17.38 17.38±1.49
T_2-Cycocel 1250 ppm 2 sprays	4.75 4.75±0.53	0.38 0.38±0.09	76.89 76.89±1.73	47.97 47.97±1.75	1.37 1.37±0.21	2.66 2.66±0.32	1.34 1.34±0.18	13.97 13.97±0.24	8.30 8.30±0.29	18.59 18.59±0.62
T_3-Cycocel 2500 ppm 1 spray	4.46 4.46±0.43	0.48 0.48±0.05	76.99 76.99±3.29	48.71 48.71±3.30	1.43 1.43±0.05	3.18 3.18±0.62	1.44 1.44±0.12	13.76 13.76±0.35	8.38 8.38±0.21	18.43 18.43±0.75
T_4-Cycocel 2500 ppm 2 sprays	4.74 4.74±0.43	0.52 0.52±0.09	76.26 76.26±2.55	50.89 50.89±5.46	1.33 1.33±0.19	3.37 3.37±0.60	1.47 1.47±0.21	13.81 13.81±0.50	8.45 8.45±0.14	19.19 19.19±1.14
T_5-Control	4.24 4.24±0.23	0.41 0.41±0.02	78.97 78.97±1.94	47.22 47.22±3.06	1.35 1.35±0.22	3.71 3.71±0.78	2.10 2.10±0.20	13.86 13.86±0.35	8.35 8.35±0.38	19.72 19.72±1.73
Range	4.24–4.75	0.38–0.52	72.66–78.97	47.22–50.89	1.33–1.43	2.66–3.71	1.34–2.10	13.67–13.97	8.30–8.58	17.38–19.72
Mean	4.54	0.43	76.36	48.82	1.38	3.19	1.56	13.81	8.41	18.66
S.E.±	0.17	0.04	1.11	1.80	0.07	0.30	0.07	0.19	0.14	0.55
C.D.at 5%	NS	NS	3.42	NS	NS	NS	0.22	NS	NS	NS

significant differences in various physical and chemical parameters due to spraying of various growth retardants were also reported by Vahora (2010) in Alphonso and Kesar varieties of mango.

Foliar sprays of cycocel had non-significant effect on acidity, reducing sugars and TSS at harvest as well as ripe stage (Table 3). Further, the ascorbic acid content as well as total sugars also remained non-significant due to cycocel treatments. The similar findings are also cited by Kulkarni (1988), Desai (1993) and Padhiar (1999) in mango when trees were sprayed with various growth retardants.

3. References

Anonymous. 2015. Mango production in Konkan shows downtrend. *The Hindu Business Line.*

Chaudhari, Darshana, Patel, N.L., Patel, N.B., Tandel, Y.N. and Gurjar, Tulsi 2014. Effect of foliar spray of growth retardants on flowering, fruiting and yield of mango (*Mangifera indica* L.) cv. Kesar. *Tr. Biosci.*, **7**(22): 3765–3767.

Daulta, B.S., Singh, H.K. and Chauhan, K.S., 1981. Effect of Zink and CCC sprays on flowering, fruiting and physico-chemical composition of mango (*Mangifera indica* L.) cv. Dashehari. *Har. J. Hort. Sci.*, **10**(3/4): 161–165.

Desai, M.M. 1993. Regulation of flowering in mango (*Mangifera indica* L.). cv. Alphonso. *Ph.D. Thesis*, G.A.U., S.K. Nagar.

Gautam, J.N. 1998. Variability studies in mango (*Mangifera indica* L.) for growth, flowering yield and quality. *Ph.D. (Agri.) Thesis*, Dr. B.S. Konkan Krishi Vidyapeeth, Dapoli

Kulkarni, V.J. 1988. Chemical control of tree vigour and the promotion of flowering and fruiting in mango (*Mangifera indica* L.) using paclobutrazol. *J. Hort. Sci.*, **63**(3): 557–566.

Mansuroglu, S., Karaguzel, O., Ortacesme, V. and Sayan, M.S. 2009. Effect of paclobutrazol on flowering, leaf and flower colour of *Consolida orientalis*. *Pak. J. Bot.*, **41**: 2323–2332.

Padhiar, B.V. 1999. Use of paclobutrazol (cultar) in mango (*Mangifera indica* L.). in South Gujarat condition. *Ph.D. Thesis*, G.A.U., S.K. Nagar.

Panse, V.G. and Sukhatme, P.V. 1995. *Statistical Methods for Agricultural Workers*. ICAR Rev. Ed. By P.V. Sukhatme and V.N. Amble, pp. 97–156.

Rademacher, W. 2000. Growth retardants: Effects on gibberellin biosynthesis and other metabolic pathways. *Ann. Rev. Pl. Physiol. Mol. Biol.*, **51**: 501–531.

Sarkar, S.K., Gautham, B., Srihari, D. and Seethambaram, Y. 1998. Regulation of tree vigour in mango. *Indian J. Hort.*, **55**(1): 37–41.

Suryanarayana, V. 1986. Effect of foliar sprays of growth regulators on flower production, sex expression, fruit set and retension in mango. *Hort. Adv.*, **12**: 82 – 87.

Vahora, A.M. 2010. Induction of early flowering, yield and quality of mango cv. Alphonso and Kesar through various chemicals. *Ph.D. Thesis*, Navsari Agriculture University, Navsari, Gujarat.

Crop Management

27

Increasing Farmers Income from Mango and Guava in Sub-tropics

Shailendra Rajan* and K. Singh

[1]ICAR-Central Institute for Sub-tropical Horticulture,
Rehmankhera, PO Kakori, Lucknow – 226 101, Uttar Pradesh
**E-mail: srajanlko@gmail.com*

In India, sub-tropics contribute about 35 per cent of the fruit production with a major share of mango and guava sharing more than 50 per cent of the total production. Across the sub-tropical region, substantial yield gaps in mango and guava provide ample opportunities for leveraging and adopting newer technologies, getting timely crop advisories and marketing information for filling the yield gap. Increasing farmers' income, interventions are required to increase productivity, enhanced quality, reduced farmer level post-harvest losses and intelligent marketing. Several varieties of mango and guava have potential for increasing farmers' income through better yield, regular bearing and improved fruit quality. Well established hybrids like Amrapali and Mallika have become popular among consumers and fetching good price in the markets because of late availability in the season. Mango hybrids, Ambika, Arunika developed at ICAR-CISH are highly attractive in fruit peel colour with regularity in cropping and may get premium price in local as well as distant markets. Guava varieties Lalit, Shweta and Dhawal are heavy yielder, responsive to pruning and suitable for high density plantation.

Plant density manipulation has become an important feature of mango and guava production system by accommodating a more number of plants which starts yielding economic yield in initial years of the planting. Canopy architecture management can convert trees into high yielding canopies with good fruit quality. Technologies developed for pruning can be useful in keeping the plants of small stature, requiring less plant protection measures and enabling easy and safe fruit harvesting. Sub-tropical zone is traditionally mango growing region where majority of the commercial orchards are old and have become unproductive and senile. Rejuvenation technology can be utilized for making them productive. Interspace in young and rejuvenated orchards can be used for producing vegetables for income enhancement. Diseases and pest problems can be addressed with the effective management techniques using integrated pest management.

Post-harvest interventions required for reducing losses during transportation and storage are available. Technologies for production of value-added products are helpful in utilization of surplus produce and glut reduction in the market. In addition to these opportunities, there are several challenges also. New disease like mango wilt is becoming threat to mango industry. Several new diseases and pests are becoming alarming in some agro-ecologies. Guava wilt still remains an unresolved problem. The wilt resistant interspecific rootstock has been developed but in absence of mass multiplication technology sufficient number of plants could not be produced. Foolproof protocol for increasing shelf-life during sea transportation is required for enhance export possibilities of mangoes from this region. Enhancing farmers' income of mango and guava farmers through various interventions is possible, which can significantly improve the economy and livelihood of a large number of the fruit growers of the region.

Mango and guava are the crops of significance having special significance in terms of economy farmers and important for processing and export. There is wide gap in productivity of these crops from state to state even when there is no varietal difference. This indicates that we should incorporate production and plant protection technology which enables us bridge the gap between potential achievable productivity and actual productivity.

1. Challenges and Constraints

Though, sub-tropical region is a major contributor towards fruit production and fruit growers income is low because of low average yield as compared to other parts of the world. Majority of mango plantations in the region are old, traditional, seedling origin planted at wider spacing of >10 m and orchards lack homogeneity. Alternate bearing tendencies of choice varieties and widespread prevalence of nutrient deficiencies and their unsatisfactory management, lack of irrigation during critical stages of fruit growth and development and inadequate canopy management are some of the other factors contributing to low productivity. Low factor productivity of mango orchards prevailing threatens sustainability. A large number of pests and diseases attack mango causing serious damage to fruit yield and quality. Indiscriminate use of chemicals for pest management restricts the population of beneficial pollinators and predators. Mango suffers from many physiological disorders like malformation, soft nose, jelly seed, black tip, clustering of fruitlets, internal necrosis *etc.* of varying severity influencing sustainability. Satisfactory management strategies are not fully worked out for many of them.

The facilities of post-harvest handling, transport, marketing and storage are either inadequate or non-functional or seriously lacking in several production belts. However, mango transport and marketing is expanding fast from the traditional mango producing regions to non-traditional countries. In sub-tropical region, low productivity is among the major causes of poor income of the farmer. Several constraints are responsible for the low productivity of the orchards. Orchards are of irregular and low yielding varieties. Old orchards are over-crowded, unproductive and have become less remunerative. Further due to inadequate supply of genuine and quality planting material, lack of adoption of improved package of practices of cultivation by orchardists, production, productivity and fruit quality has remained low. Inadequate pest management practices, emergence of new pests and diseases due to emerging environmental challenges and indiscriminate use of pesticides are also cause of concern. Efficient harvesting, grading, ripening, product standardization and packaging to minimize post-harvest losses are also not much in use. A number of opportunities, mitigating diverse challenges and constraints addressed through scientific innovations and technologies portrayed.

2. Opportunities

Mango and guava are the major crops of Indian sub-tropics and substantial yield gaps provide ample opportunities for adopting newer technologies, getting timely crop advisories and marketing information for filling the gap. Increase productivity, enhanced quality, reduced farmer level post-harvest losses and intelligent marketing are important aspect for increasing farmers' income. Adoption of technologies like high yielding regular bearing varieties with better fruit quality, plant density optimization, canopy architecture manipulation, rejuvenation of old and senile orchards, effective management of diseases and pest, post-harvest interventions required for reducing losses, value-addition, marketing network are the interventions, which can improve the mango and guava farmers' income leading to improved the economy and livelihood of a large number of the fruit growers of the region.

3. Wedge Grafting for Large Scale Multiplication

Fruit planting nursery has become an industry and livelihood and number of farmers are dependent on the supply of planting material. Wedge grafting a method developed for mango and guava has been successfully adopted by the nurserymen. It gives good grafting success and is contributing to the mass scale production of the new varieties like Amrapali, Mallika, Ambika and Arunika. The method has helped in production of grafts for newly developed high density plantings where a large numbers are required. Farmers have developed their own nursery and they are using wedge grafting for guava multiplication and earning good profits. Wedge grafting is a relatively easy method of propagation having higher success percentage. The plants grafted with wedge method takes less time (9-12 days) for sprouting and plant gets ready for sale within five months from day of grafting operation. The grafted plants showed very high field establishment on account of undisturbed tap root system. Even higher success rate can be achieved using polythene cap even at low temperature (during winter season).

During last twenty century, mango was grown using both seedlings and grafted plants, however, now a days, orchards are planted using plants prepared asexually (Singh and Rajan, 2007). Asexual methods of mango propagation being followed are inarching, Veneer grafting, epicotyl/stone grafting, cleft or wedge grafting, soft-wood grafting, splice or whip grafting, side grafting, notch grafting, ring or crown grafting, budding, rootage methods. However, veneer grafting, cleft grafting (variants : wedge, epicotyl, soft wood, splice, whip grafting) are generally followed for the propagation. Rootage methods are now losing the commercial arena (Ram and Rajan, 2003).

In sub-tropics, ICAR Institutions and State Agricultural Universities are producing quality planting material at a large scale. Several entrepreneurs have established nurseries producing genuine planting material using mother plants from reliable sources. Newer methods like wedge grafting have been adapted by nursery industry and a good number of planting material is being produced. Through accreditation by National Horticulture Board, GoI these nurseries are star rated for supply of quality planting material.

4. New Varieties for Better Profits

A good number of mango and guava varieties have been developed and evaluated under sub-tropical conditions. However, only few of these have been adopted commercially by the farmers. In mango, due to high fruit yield, regular bearing, good fruit quality and red colour demand of new varieties has increased. Whereas, guava varieties have been accepted because of their high yield, increasing consumer demands, attractive fruit and suitability for processing. The limitation in the expansion of new varieties seems to be availability of quality planting material. Guava varieties like Lalit and Shweta have been found suitable for high density planting and are responsive to pruning. As commercialization of varieties in tree crops takes considerable time, the varieties released (Amrapali and Mallika) in later part of 20th century are now becoming popular. Ambika and Arunika are attaining popularity due to coloured fruit and regularity in cropping. These mango varieties are also getting better sale price due to late ripening when there is not much competition.

4.1 Mango

Amrapali

This is cross between Dashehari and Neelum developed at IARI, New Delhi being commercially grown in eastern India and several other parts of the sub-tropics. It is highly successful in the areas where irrigation facilities are available or tree do not face water stress during fruit development. Because of good quality fruits and late maturity, it is fetching a good price in the market. Gradually, popularity of new varieties is increasing because of availability of planting material and in adequate quantity fruits in the market. These are suitable for high density planting due to regular bearing habit, semi-vigorous nature and responsiveness to pruning.

Mallika

This hybrids is a cross between Neelum and Dashehari, developed at IARI, New Delhi. It is high yielder with regular cropping, good shelf-life and medium to large size fruits with good quality. The variety has potential for export as well as domestic market. The taste and appearance of the variety has been liked by foreigners because of good sugar and acid blend. Variety is well adapted in all parts of the country does extremely well in southern states, especially Karnataka, Andhra Pradesh and Tamil Nadu.

Ambika

This is a variety developed from a cross between Amrapali and Janardan Pasand for commercial cultivation and export. The variety is regular bearing with red peel colour. It is a regular bearing variety and matures late in the season. Average fruit yield is 80 kg per plant at the age of 10 years. The fruit is oblong oval in shape weighing 250-350 g. Fruits are purple green at maturity and turn bright yellow with dark red blush upon ripening. The ripe fruit is firm with scanty fibre, pulp colour is orange yellow and TSS develops up to 21°Brix. The variety has potential for export and internal market because of its attractive fruit colour.

Arunika

This is a hybrid variety developed from a cross between Amrapali x Vanraj. Plants are dwarf than Amrapali. Fruits weigh 180-200 g, cordate in shape with smooth and tough skin, orange yellow with dark crimson blush; flesh dark yellow, firm with scanty fibres; TSS 23°Brix. Late ripening with potential for export and internal market because of attractive fruit colour and good quality.

4.2 Guava

Lalit

The guava variety Lalit released by the Institute is a high yielder with pink pulp colour. Lalit is responsive to pruning and is suitable for high density planting. The fruit yield is 65-100 kg/plant at the age of 6 years. Fruits are of saffron yellow colour with red blush and weigh 185-200 g; pulp firm and pink due to lycopene with good blend of sugar and acid. The content of vitamin C in fruit is 250 mg per 100 g. It is suitable for both table and processing purposes. The pink colour in the beverage remains stable for more than a year during storage.

Shweta

It is a selection from half-sib population of Apple Colour. Tree is semi vigorous with medium height and prolific bearer. The fruit yield is 90 kg/plant at the age of 6 years. It is a variety with medium size globose fruits, weighing 225g, creamy white epicarp having red spots or blush, snow-white pulp, TSS 12.5-13.2°Brix, vitamin C content 300 mg/100 g pulp and good keeping quality. The variety is suitable for cultivation in guava growing areas of sub-tropics and is a high yielder. Fruits are very attractive and have good nutritive value.

Dhawal

It is a heavy bearer with round, smooth, medium to large fruits (200-250 g). It is a half sib selection from Allahabad Safeda. Mature fruits develop whitish light yellow colour on ripening. Pulp is white, sweet in taste having muskiness, TSS 13.4°Brix, acidity 0.42 per cent and vitamin-C more than 250 mg/100 g fruit. Seeds are soft, moderate in number.

Lalima

It is a selection from half-sib population from Apple guava, having attractive crimson colour fruits, higher proportion of coloured fruit of the tree. Fruits have longer shelf-life, weight 190 g and TSS 13.7°Brix.

Several mango hybrid varieties have been developed under south Indian conditions and their evaluation under sub-tropics has shown limited possibility of adaption. Many mango hybrids have been released by several public funded institution but area under these varieties will expand with the availability of the planting material. Guava varieties like Pant, Prabhat, Hissar Surkha and Hissar Safed are being grown in some pockets. The limiting factor is the availability of large scale planting.

5. Planting Density Manipulation

Poor yields experienced in mango industry are partly due to the wide tree spacings of conventional orchards with spacing ranging from 10 to 12 m between trees in the row and between rows. Canopies of these trees often take more than 10 years to fill the allocated space in the orchard row. This problem can be resolved with higher density plantings. Although, the optimal spacing for commercial cultivars still has to be determined, it appears that there is tremendous scope for increasing orchard productivity by increasing planting density (Rajan and Singh, 2002; Singh, 2010).

At CISH, Lucknow, a trial to standardize medium/high density orcharding in mango cv Dashehari was conducted at CISH, Lucknow with densities *viz.*, 1600, 800, 400, 266, 178 and 100 plants ha^{-1} with spacing of 2.5 × 2.5 m, 5.0 × 2.5 m, 5.0 × 5.0 m, 7.5 × 5.0 m, 7.5 × 7.5 m and 10.0 × 10.0 m respectively. This trial was conducted without using dwarfing rootstock or any growth retardant. At 19 years of establishment highest fruit yield (16.00 t ha^{-1}) was recorded in medium density planting (400 plants ha^{-1}) followed by 10.00 t ha^{-1} in density of 266 plants ha^{-1} and the least yield (5.50 t ha^{-1}) in 100 plants ha^{-1}. High density planting at GBPUA&T, Pantnagar representing Tarai region of Uttarakhand, in alternate bearer Dashehari and regular bearer Amrapali without the use of dwarfing rootstock planted at 3.0 × 2.5 m (1,333/ha) was successfully demonstrated. Regular bearing and tree growth was controlled by annual pruning, soon after harvest, and soil application of paclobutrazol 2.0 to 2.5 months after pruning. Pruning and use of paclobutrazol also decreased floral malformation to a very low level (0.2 per cent). However, use of paclobutrazol was not required for regular bearing Amrapali. Recent observations also have indicated that even in Amrapali, pruning after harvest is critical to sustain its regular bearing feature. Concerted efforts are necessary to adopt HDP systems for different varieties of mango with complementary package of technologies to break the yield barriers (Ram *et al.*, 2001; Ram and Rajan, 2003; Nath *et al.*, 2007; Carlos *et al.*, 2012).

The traditional system of guava cultivation often does not attain desired levels of productivity due to less number of plants and large tree canopies. Hence, high density planting involving planting of fruit trees closely, accommodating maximum possible number of the plants per unit area to maximize possible profit per unit of the tree volume. Under HDP, guava plant can be grown at a spacing of 2.0 × 1.0 m (5,000 plants/ha) under Ultra High Density Planting (UHDP) and 3 × 3 m (1,111 plants/ha), 3 × 6 m (555 plants/ha) under HDP as compared to traditional system 6 × 6 m (277 plants/ha). The plants are trained to dwarf canopy structure by pruning for better light interception and distribution resulting in increase and earliness in yield. HDP and UHDP orcharding gives higher yield as well as returns/unit area due to increasing the number of trees/unit area (Singh, 2010). Overall production is increased (30-50 t ha^{-1}) as compared to traditional system (12-20 t ha^{-1}).

In guava, plants are topped 2 months of planting to a uniform height of 50 to 60 cm in October for emergence of new shoots below cut ends. 50 per cent length of each new shoot, pruned again in December-January for induction of more shoots and good spread is attained by May. The flower buds differentiate after second pruning. Recurrent pruning of shoots is carried out in the month of May and October for getting the fruiting in the next winter or rainy season. Under UHDP pruning is done three times in a year in February, May and September so that the plants are maintained at height of 1.0-1.5 m in form of hedge. On an average 12 t ha^{-1} fruits are obtained in first year. The yield increases up to 55 t ha^{-1} after 5 years of planting. Among different cultivars, Lalit performs very well in UHDP system.

In HDP, about 70-90 per cent increase in yield was obtained after five years of planting. The planting density can be modified as per the soil, climate, skill and resources of the farmer. After five years of planting, if the canopy becomes too dense, back pruning to a height of 50-60 cm above ground level can be adopted to develop a new canopy as mentioned above.

6. Drip Irrigation and Mulching

Young mango plants should be watered at 2-3 day interval during dry season till they are well established. Drip irrigation system having 2 emitters per plant of 8 lph capacity placed equidistant in at 50 per cent distance of canopy radius. In older trees the number of emitters per plant is increased to 4. Drip irrigation 60 per cent

PE level results in significantly high yield and quality. Use of 400 gauge (100 micron) UV stabilized thick black polyethene mulch in mango and guava orchards conserves water and helps in controlling weeds (Singh *et al.*, 2015; Ramniwas *et al.*, 2012).

Irrigation efficiency is very low 40-45 per cent under traditional flood or basin irrigation, the modern irrigation methods especially drip irrigation is used in only 1 per cent of the total irrigated area. Drip irrigation offers unique agronomical, agro-technical, and economical advantages for the efficient use of water. It is one of the most efficient methods and the system usually makes use of saving water supply in order to apply the precise amount of water to the root zone. It is planned to deliver frequent light applications of water to wet only root zone. Drip irrigation method saves upto 60 per cent of water and gives 25-40 per cent more yield as compared to that of furrow irrigation method. Irrigation is applied on a daily basis to deliver the require quantity of water for irrigation namely 80 per cent replenishment of crop ET, calculated every day. The discharge rates are achieved through appropriated combination of 2 and 4 emitters with 8 lph to deliver the desired volume of water as per plant spacing. The irrigation period is decided based upon the previous day's crop ET. The amount of water varies based on the current soil moisture content in the field. The irrigation is applied through drip on the basis of replenishment of daily ET. Saving of water has been recorded to the tune of 45 to 50 per cent in mango and guava and 40 to 45 per cent in *aonla* with drip along with plastic mulching besides efficient weed control.

The high initial cost of installing drip system can be minimized by installing the gravitational drip system through overhead tank and low pressure emitters. The capacity of the tank to create sufficient and even discharge should be according to the crop and area irrigated.

7. Intercropping in Young and Rejuvenated Orchards

Normally, grafted trees are spaced at 8 x 10 m or 10 x 12 m. Planting at 5 x 5 m distance is becoming common with proper training and pruning operations starting from the first year. Without which, within 10 years, tree canopy intermingle with each other and fruit yield become very low. Intercrops of fruits such as papaya, guava and vegetables could be grown for economical utilization of land in widely spaced young orchards (Rajan, 2016). Resource conserving technologies like intercropping vegetables in fruit orchards improves resource use efficiency and farm profitability by year round production, reduce costs, increase food and nutritional security and sustainable crop productivity. In young and rejuvenated orchards, intercrops may be raised for utilizing the vacant interspaces. This practice, besides providing income to the grower, suppresses weed growth, improves soil fertility and prevents leaching of nutrients and erosion of soil. Vegetables that have their roots within 25 cm depth of the soil are considered good for intercropping in mango. Great care should be taken in selecting the right type of intercrops at the centre, between and within rows of fruit trees. The crops like cowpea, bean, cabbage, cauliflower, chillies, okra and partial shade- loving plants (ginger, turmeric and elephant- foot yam *etc.*), as intercrop in the orchard provide sustainable return from the initial stage of canopy development. Brinjal intercrop was found effective in sustaining income in mango and guava especially during non bearing phase. The highest monetary return (Rs. 1,60,300/ha) with brinjal in mango orchard was obtained as compared to Rs. 4,9920/ha in mango sole crop. Whereas, intercropping of brinjal in guava gave monetory returns of Rs. 1,96,400/ha as compared to Rs. 73,500/ha in guava sole crop.

8. Efficient Canopy Management

Tree canopy management, specially size control is of priority for reducing production costs and increasing yield and fruit quality. Tree canopy management techniques specifically for mangoes have been developed and being used in different parts of the world, which can be adopted after certain modifications in sub-tropics. Early height control and tree canopy management are important techniques and should be practiced in India. Similarly, the problem of large tree size in fruit crops can be tackled by using topping and hedging because large and crowded trees pose many disadvantages. Appropriate height, topping and hedging, cutting angles, as well as the time and frequency of hedging determined for mango, which are common practices in Israel, USA, Australia and South Africa are useful for increasing working efficiency and production (Rajan and Singh, 2002).

Proper control of vegetative growth is a prerequisite for high density planting and without it there is overcrowding and shading, which reduces flower bud formation, fruit retention, fruit size and fruit colour. Control of apical growth must begin within the first year of planting and continue each year in high density planting system. If management techniques are not applied early in the life of the tree, encouraging lateral canopy growth and fruiting, the resulting tree canopy will be difficult and expensive to control in the future.

Shaping the mango tree immediately after planting should be kept in mind and first branching should be allowed at desired height so that proper clearance for equipment, *etc.* is possible. An initial branching height between 60-70 cm is appropriate. A single cut is made on the main stem of the graft at this point, which gives rice to multiple, leafy branches will grow from the buds just below the initial cut (Rajan and Singh, 2002). Of these, three or four horizontal shoots, spaced equally around the stem are allowed to grow. Upright growing shoots should be removed. These selected horizontal shoots will form the main scaffold limbs of the tree. These shoots should be allowed to grow to about 40-50 cm long, this length is generally reached during the second growth flush of leaves. The flush at this time is pinched off with the fingers or left to mature and then pruned. Multiple shoots emerge from the buds below the cut. These are allowed to extend about 40-50 cm then pruned. This procedure is continued in second year also. If selected shoots grow vertically, these can be pull down by attaching some weight with the help of string. These weights are left for about three months. These horizontal scaffold limbs are stronger and flower and fruit earlier. Short branches within the tree canopy will produce a complex, compact, strong structure after three years.

Fruiting is expected in these plants after second or third year after planting depending on the cultivar. Fruiting should be allowed because it is one of the best tools for maintaining tree size. After harvest, trees are pruned and heading the small branches to a length of 50 cm and favours horizontal shoots. In vigorous cultivars, proper pruning may lead trees to develop large, structural limbs with less energy fruit production. Annually some portion of this wood is remove by removing an entire limb. These cuts affect the vigor of the tree less and allow to maintain tree dwarf without excessive vegetative growth. Therefore, efficient mango tree canopy management starts early in the life of the tree and continues indefinitely (Ram and Rajan, 2003; Rajan and Singh, 2002).

In absence of dwarfing rootstock, application of mechanical tree size control has increased in high-density orchards. Use of growth retardants like paclobutrazol can be used to keep canopy small and induce regularity in cropping. This may be the only practical method of tree size control of inherently large trees in HDP (Ram and Rajan, 2003; Rajan and Singh, 2002). Tree height control is accomplished by removal of the multiple central leaders within the canopy of the young tree. The dominant, upward shoot or shoot from each growth umbel (whorl of new shoots) is removed and when there are parallel branches within the canopy, the more dominant upright shoots are removed. Experimental evidence indicates that topping and hedging is an adequate approach to control the tree size.

Overcrowding of mango tree in mature orchards growing is common feature in sub-tropics. In such cases, pruning and topping should be practiced, mostly on an annual or biennial cycle. In addition, one or more major limbs are sometimes removed from center of each tree (opening).

High-density planting of Dashehari with regular pruning after harvesting followed by application of paclobutrazol in tarai region of Uttarakhand and Uttar Pradesh has been recommended. Amrapali being a regular bearer, high density planting (1600 plants/ha) can be recommended provided that the pruning after harvesting is performed for sustainable high production (Ram and Rajan, 2003).

9. Management of Irregular Bearing in Mango

Majority of the choice mango varieties in the Indian sub-continent are beset with the problem of alternate bearing tendency, with rhythms of 'on' (year of high yields) and 'off' (years of low or no yield). The intensity and degree of such phenomenon may vary with the varieties and agro-ecologies leading to surpluses and deficits in production that influence crop management and sustainable livelihoods of farming community. In this background paclobutrazol (PP_{333}) as a broad spectrum growth retardant tested at different location in different alternate bearing commercial cultivars of mango proved to be a very potential chemical which control the alternate bearing problem in mango (Singh and Singh, 2003; Burondkar and Gunjate, 1993). Several reports indicate that paclobutrazol exerts its antagonistic effect due to lowering of endogenous gibbrellins levels which in turn causes reduced vegetative growth but profuse flowering in mango (Voon *et al.*, 1991).

Persistence of paclobutrazol and its subsequent impact on the yield of different commercial cultivars of mango, *viz.*, Chausa, Dashehari and Langra have been tested at CISH, Lucknow. Results based on five year averages indicate that trees treated with 6 g a.i./tree of paclobutrazol, recorded maximum yield in Chausa and Langra, whereas, only 4 g a.i./tree was most effective in Dashehari. The application of paclobutrazol, at half the above doses, was effective in inducing flowering as well as fruiting in the 3rd year, but only in cv. Dashehari; still substantially higher yields were recorded over control, but not so in cvs. Chausa and Langra. The trees which were not given paclobutrazol treatment in the third year, showed residual effect only in Dashehari, while Chausa and Langra did not show any residual response. The paclobutrazol residue in the soil collected during the third year from the root zone of trees was in the range of 0.4898 to 1.0005 µg/g (Singh and Bhattacherjee, 2005). The residue of paclobutrazol in leaves and inflorescence was noticed after long period of its application (Bhattacherjee and Singh, 2015), therefore for its effectiveness it should be applied in much advance to the initiation of flowering.

Long term experiment on different cultivars of mango indicates that its effect is fruit load/cultivars dependent even its residue remain in soil for longer period. Heavy fruit load trees receiving the same amount of paclobutrazol in the same season scarcely flowered as fruit to some extent nullified the effect of paclobutrazol. Application of paclobutrazol (PBZ) was found successful in producing off-season mangoes in cultivars Totapari, Rumani, Dashehari in Andhra Pradesh and Alphonso in Konkan region on a limited commercial scale. Extended mango harvest season from Ist week of February to IInd week of March under Konkan conditions in Maharashtra with PBZ application was also established to enhance the total yield and economic status of farmers (Rajan *et al.*, 2014). Few pprecautions for use the use of paclobutrazol are: (i) application should be started from the age of 10 years tree age and onwards; (ii) the treated orchards should be adequately fertilized; (iii) unhealthy and disease trees should be avoided for the use of apclobutrazol; (iv) soil moisture should be maintained under the tree canopy for about 20-25 days after its application

10. Crop Regulation in Guava for Higher Profits

Guava being a crop with bearing on new shoots, fruit harvesting time can be manipulated as per the time of market demand for increased income of the farmer by inducing new growth as per the schedule. The regulation of flowering time is successful in areas where extreme temperatures are not prevailing. Compared to monsoon crop, winter crop is much superior in fruit quality and fetch premium price. This may be true in regions where winter is not very mild and summers are too hot. Therefore, farmers can reduce monsoon crop by deblossoming/shoot pruning/chemical application for inducing flowering resulting crop maturity when higher sale price is prevailing in the market. At many places it can be achieved by withholding the irrigations during summers. Fifty percent flower removal in May can be a good alternative for higher income during winters.

The rainy season crop is eliminated by spraying fertilizer grade urea (10 per cent) twice at the interval of 10 days, in cv. Allahabad Safeda and at 15 per cent in cv. Sardar during bloom (April-May). This results in increase of yield of good quality fruits during winter. Pruning of shoots to their half length (50 per cent) on entire tree in the month of May resulted in shortening of shoot growth and production of maximum yield with better quality fruits during winter. One leaf pair shoot pruning (retaining one leaf pair at the base of the new shoot growth) should be done in the first week of May. Bending of branches is one of the methods being used for off season production in West Bengal. Branches of guava plants are bent down about 45-60 days before the expected date of flowering and to produce fruits in the off season. Leaves, small shoots, flowers and fruits from branch are removed before bending of branches or cut off, keeping 10-12 inch of terminal twig intact.

11. Rejuvenation of Senile Mango Orchards

Many of the mango orchards especially in the traditional mango producing regions are quite old, majority of them more than 60 years; their canopies have become dense, overlapping with low sunlight penetration rendering them unproductive. ICAR-CISH, Lucknow has standardized technology for rejuvenation of old and unproductive mango orchards. It involves pruning/heading back of branches during December. About three to four main branches, with outward growth, are retained. The undesirable branches are completely removed from their base. The main branches (3-4 only) are allowed to produce shoots and develop canopy. The tree attains umbrella like open canopy of healthy shoots with good bearing potential within three years of pruning. The technology has made a good impact and is in demand for rejuvenating unproductive orchards. It not only generates income through sale of cut wooden branches but also through intercropping during the initial years. It has immense potential to revolutionize mango productivity from old and senile orchards.

There exist great possibilities to rejuvenate such orchards, centre open, top work with newer improved trait specific varieties, adopt package of technologies developed under NARS for improving the productivity potential. For this, appropriate machineries for pruning and optimized adoption of triazoles across varieties and agro-ecologies though paclobutrazol is ubiquitously found useful.

12. Rejuvenation of Unproductive Guava Orchards

Over the years, productivity of guava crop continues to be low. Guava orchards lose their productivity after 25 years due to ageing and biotic stresses. Thickly shaded guava orchards are common in large numbers in different parts of country that have significantly reduced the average productivity. The problem of large tree size and crowded guava tree has poor distribution of light throughout the canopy followed by poor yield and quality of fruits. Since, guava trees respond well to canopy modification by pruning and training, canopy modification through topping and hedging is a step for enhancing orchard production efficiency. The old orchards which have turned unproductive and produce low grade fruits should be rejuvenated through heavy and systematic pruning, followed by fertilization and plant protection measures. Heading back of unproductive guava trees at the height of 1-1.5 m from the ground level is done in the month of May. The newly emerging shoots are allowed to grow up to a length of about 40 to 50 cm, which could be attained in 4 to 5 months after cutting the tree. These shoots are further pruned to about 50 per cent of the total length for emergence of multiple shoots below the cut ends. Multiple shoots developed as a result of second pruning lead to flowering and fruiting in the following season.

Yield enhancement in the range of 70 to 90 per cent over the un-pruned orchard can be recorded. Additional income from the sale of pruned wood and intercrops can also be obtained. Higher yields are realized due to better harvesting of solar radiation by the tree canopy, removal of dead and diseased branches and judicious intercultural operations.

13. Effective Insect-Pest and Disease Management

Efficient management of Insect pests and diseases is important because they affect essentially every phase of mango tree growth and development. Many a times they cause a great loss to the farmer's income due to epidemics. Most of the insect pests and diseases require annual control measures. Fortunately, except fungal and bacterial diseases, the attack of other pathogens like virus, viroid, phytoplasma on mango is not common. Control measures for most of the major diseases have been developed. If some of the insect pests, like mango mealy bug, stem borer and shoot gall maker, are controlled continuously for couple of years, can be eradicated for a long period of time. However, fruit flies, seed weevils, shoot borer, mango hoppers *etc.* are required to control every year. Some of the diseases and pests such as anthracnose, powdery mildew, mango hopper, midges *etc.* cause severe damage and are present throughout the mango growing areas of the world Therefore, the future of mango insect pests and disease control has a great promise with the development of IPM technology in view of large number of insect pests and diseases attacking mango tree and crop throughout the year. A large number of chemicals are recommended to control them. It may be difficult to produce mango crop in the absence of some of these control measures. Following of IPM technology, reduce chemical residues in mango.

Many of the major insects and pests have developed resistance due to rapid change in agro-ecosystem, advancement in orchard management practices and indiscriminate use of chemicals, insecticides and pesticides. This has led orchardist to use high dose of toxic insecticides, thereby causing imbalance in population dynamics of pollinators and other useful fauna and incorporating high toxic residue in the fruits. Therefore, it is necessary to bring out modern concepts of IPM for important mango pests.

Major pests identified are for guava are fruit-fly, bark-eating caterpillar, fruit-borers, scale insects and mealy bugs. Guava fruit fly can be managed by using methyl eugenol-baited trap. Root-knot nematode (*Meloidogyne*) has become a serious threat to guava in different part of the country causing reduction in plant vigor; yellowing and wilting of plants. Vegetables, susceptible for root-knot nematode should not be grown in guava field. Major fungal diseases of guava are guava wilt, anthracnose, dieback, canker and fruit-rot.

14. Managing Physiological Disorders

Few manageable disorders like black tip and internal necrosis are important constraints in some part of sub-tropics and can be managed using recommended package of practices.

Black Tip

This disorder is mainly noticed in Punjab, Uttar Pradesh, Bihar and West Bengal where low height chimney brick-kilns are operative. The distal-end of the affected fruits turns black and becomes hard. These fruits ripen prematurely, unevenly and become unmarketable. Brick-kilns located within a distance of 600 m impact the crop because it is caused by the smoke containing gases like carbon monoxide and carbon dioxide, sulphur dioxide and acetylene. Symptoms are caused because of impaired physiology of the fruit due exposure to these gases. It can be controlled by raising the chimney height of the of the brick-kilns. Spraying borax (0.6 per cent) at 10–14 days intervals starting from fruit set also controls it.

Internal Necrosis

This disorder of mango fruits is caused by boron deficiency and use of boron was recommended to control the disorder. The disorder results drop of immature fruits. The internal fruit necrosis is characteristically different from black tip (Ram *et al.*, 1989). The first noticeable symptom of internal necrosis is appearance of dark green colour in lower half of the fruit. Subsequently, browning of seed and mesocarpic tissues in a small isolated area of indefinite outline from first to third week of May before endocarp hardening is observed. The brown tissues then turn into brownish black necrotic lesions which later on extends towards the epicarp. At this stage, water soaked isolated necrotic areas appear on the surface of the fruit exuding gummy substances below the green tip. These brown areas develop dark brown gummy cavities in the mesocarp before external symptoms appear on the fruit surface. At advanced stages, the affected tissues of the entire lower half of the fruit turnes necrotic, brown-black and results into longitudinal cracking of the fruit through the necrotic region exposing the seed. An inverse linear relationship of internal fruit necrosis to boron levels in leaf and fruits was found in cultivar Dashehari. In cultivar Dashehari, spray of 0.6-0.8 per cent borax solution is effective (Ram and Rajan, 2003).

Clustering (*Jhumka*)

This malady is characterized by a cluster of small fruits at the tip of the panicle giving an appearance of bunch tip called *jhumka*. These fruitlets are dark green with a deeper curve in the sinus beak region compared with normally developing fruitlets. These fruitlets may grow to marble size after which their growth ceases and become yellow and drop. Young fruit-lets are devoid of seed, the centre of synthesis of endogenous hormones. One of the main reasons for clustering is the adverse climate during flowering and fruit set (February–March), particularly the low temperature. Sunny days with temperature dip during night during fertilization may increase the incidence. In most of the fruits embryo are aborted or shriveled which do not develop further, signifying the role of normal embryo growth in the growth and development of fruits. Almost all commercial varieties are affected with the malady with a varying degree of incidence.

Jelly Seed in Mango

Jelly seed is one of the serious problems in mango particularly in Northern belt is characterised by loosening of pulp, jelly formation, off-flavoured and dull colour around stone, resulting reduction in shelf-life of fruit. Dashehari, the choicest variety of north India is highly prone to this disorder. The affected fruits looks like normal from the outside but after slicing give an unpleasant appearance and become unfit for consumption. In jelly seed besides high ethylene evolution, pectin hydrolysing enzymes was found higher as compared to healthy fruits. Affected tissue had low Ca, K with higher N. Extensive disintegration of cell walls along with large air spaces and interspersed starch grain was noticed in affected fruits whereas in healthy tissue the cell walls and starch idioblasts were found intact even at advance stage of ripening. Maximum incidence (> 60 per cent) of jelly seed was noticed in the fruits harvested after second/third week of June or after a spell of heavy rains. For the management of this disorder an integrated approach was found very effective which involve application of black plastic mulch (100 μ thick) and foliar spray of dihydrated calcium chloride (2.0 per cent), potassium sulphate (1.0 per cent) one month before harvesting of fruit along with application of 150 g borax in soil.

15. Efficient Post-harvest Management

Poor infrastructure for storage, processing and marketing of mango and guava in many parts of sub-tropics contributes to a high proportion of waste, which average between 10 and 30 per cent. Improvement in major infrastructural facilities can reduces losses due to severe constraints in domestic distribution of these crops. Farmer income can be significantly increased by reducing the waste which occurs owing to the fact that small farmers lack resources and are unable to implement suitable post-harvest handling practices. Post-harvest management and processing of produce are of considerable significance in light of increasing demand for these fruits. Post-harvest handling of mango comprise of all activities concerning bringing fruit from tree to table. These activities are important because shelf-life of the fruit after harvest depends upon the health and quality of the fruit at the time of harvest and post-harvest handling including fruit maturity, colour, shape, size, sweetness, position on tree, occurrence of pest and diseases, weather conditions, soil moisture, nutrient availability *etc.* These factors vary from tree to tree, orchard to orchard and season to season. Hence, precise management keeping view of factors is the key for getting better returns. Oozing of latex after detachment of stem trickles over skin and imparts shabby appearance to the fruit by blackening the skin. Thus, picking of fruit is followed by washing, desapping, precooling, hot water dipping and fungicide application, grading, waxing, packaging, ripening, transport and marketing. Fruit maturity is very important at the time of harvest among all the considerations which affect the quality of the fruit after harvest.

16. Safe Harvesting

Majority of mango orchards are large in stature thus, efficient harvesting operations pose problems which are also compounded by scarce labour availability. This has necessitated canopy management and mechanization of harvesting so as to obtain fruits without bruises and improved efficiency. Besides, aspects like de-sapping, sorting, grading, pre-treatment, pre-cooling, ethylene treatment, packing, palettisation and cold transport all have considerable impact on minimizing the post-harvest losses and improve the quality and market appeal of mangoes. Protocols developed by CISH and APEDA could be harmonized for the purpose of both domestic as well as export trade in this regard. Widespread use of Calcium carbide, a banned chemical and potential health hazard as ripening aid in the marketing channel is a matter of great concern. This should be substituted by use of ethrel (ethephon)/ethylene gas by popularizing low-cost ripening chambers and techniques through appropriate sensitization of stakeholders and policy interventions.

The mango fruits should be harvested at mature green stage in the morning hours. The maturity stage is determined on the basis of skin and pulp colour, specific gravity and number of days from fruit set. Cultivars 'Dashehari' and 'Langra' require 12 weeks after fruit set for maturity, while 'Chausa' and 'Mallika' require 15 weeks. Fruits should be handpicked or plucked with a harvester. Shaken of branches to drop them should not be followed for better shelf-life. Simple, low cost and portable mango harvesters, designed and developed at different centers in the country can be used. With the harvester, fruits are harvested with stalks, which appear better on ripening as undesired spots on skin caused by sap burn are prevented. Fruits are also less prone to stem end rot disease during storage when harvested in this manner.

17. Efficient Packaging and Transport

During post-harvest handling and distribution, there is a great loss of produce. Timber and bamboo have been used for making packaging containers in many parts of the sub-tropics. However, alternative packaging material such as corrugated card board packages have offered themselves as suitable alternative to timber and bamboo. Wrapping in tissue paper and polythene foam paper can be used for high value fresh mango. Establishment of cold chain in the post-harvest management system is crucial for enhanced sustainability and profitability.

18. Increased Income by Value-addition

Endless numbers of processed mango products are made from mango and variations exist from region to region. Peeled, sliced and chopped mangoes are preserved by dehydration, canning, freezing, and pickling. Raw fruits are utilized for products like chutney, pickle, amchoor, green mango beverage, *etc.* Besides conventional pickling using oil, oilless pickle, fat free pickle with storage life of 9 months are also made. Beverage from raw mango called *pana* can be prepared and stored for about 9 months. Mango puree production is a good example of a highly mechanized process which requires less labor and offers other benefits. Mango puree is one of the products produced in large volume used for developing several mango based products. Selected varieties of mango when fully matured are harvested, quickly transported to fruit processing plant and ripened in controlled ripening chambers. Fully ripened fruits are washed, blanched, pulped, deseeded, homogenized, and concentrated when required, thermally processed and aseptically filled in containers for maintaining commercial sterility (Rajan, 2016). Numerous usage of the puree are for making products *viz.*, fruit breads, cakes, tarts, muffins, pie-fillings, icings, donuts, fruit drinks, nectars, Ice-creams, fruit bars, milk shakes, yogurts, puddings, toppings, deserts, cereals, juices, strained fruit, fruit deserts, spray dried freeze-dried powders *etc.* Such product diversity expands the market potential for mango puree (Ram and Rajan, 2003). Within the past few years the largest increase in consumption of mango products has been for the mango juice and mango puree. The large increase in consumption of mango juice is part of the world-wide increase in fruit beverages. The increase in consumption reflects the changes composition preference, which has prompted a large number of baby food manufacturer to promote tropical fruit ingredients in its food line (Nanjundswamy, 1991).

19. Intelligent Market Information

Intelligent marketing is a powerful source of economic growth of mango and guava farmers especially in sub-tropics where farmers are have no organized marketing system through marketing or cooperative societies. Smart marketing of fresh mango and guava fruits is helping in increased productivity and better exposure of farmer to new technologies. To ensure that benefit of mango and guava trade percolate down to the grass root stakeholders, complementary domestic policies need to be in place. Synergy between trade and farmers economic development thus requires proper international trade policies that are complemented by appropriate domestic policies.

Fruit production system is characterized by variety diversity, diverse/alternate land use systems, regional preferences and alternate bearing tendency of choice varieties. Further climatic aberrations such as high temperatures during flowering and fruit set, hail storms, outbreak of pests and diseases greatly impact the quantum of marketable out puts in any given agro-ecology. This situation impacts considerably regional market arrivals. There is a need to develop market intelligence network based on region-wise yield forecasting models so that farmers could be empowered with appropriate market access in addition to strengthening of local markets.

For better income from these crops, the farmers may come together by either forming cooperatives or or self-help groups. This could also help in adopting the latest production technology practices over a contiguous and wider production area. This in turn would facilitate production and export of quality fruits in required quantities as per the international quality standards on a sustainable basis. The direct purchases and trading by already established private commercial houses, *e.g.* M/s TCI, Reliance, Godrej, Bharti Delmonte, Essar *etc.*, has also emerged as another alternative. Here, the commercial houses purchase the produce directly form the producer, grade and pack, brand or process it. Thereafter, the produce is sold through their own retail chains (Ravishankar *et al.*, 2012).

In conclusion, farmers' income can be significantly increased though high production and productivity of mango and guava through efficient and judicious use of inputs like water, nutrients, need-based eco friendly pesticides expanding area in problematic soils through developing/selecting suitable cultivars/rootstocks; high-density plantation, canopy, pest, and disease management and proper post-harvest management. The guava is an ideal fruit for increasing farmers income because of its hardiness, high yield, long supply season and increasing demands due to high nutritive value.

20. References

Bhattacherjee, A.K. and Singh, V.K. 2015. Uptake of soil applied paclobutrazol in mango cv. Dashehari and its persistence in soil, leaves and fruits. *Indian J. Pl. Physiol.*, **20:** 39–42.

Burondkar, M.M. and Gunjate, R.T. 1993. Control on vegetative growth and regulation of regular and early cropping in 'Alphonso' mango with paclobutrazol. *Acta Hort.*, **341:** 155–160.

Nath, V., Das, B. and Rai, M. 2007. Standardization of high-density planting in mango (*Mangifera indica*) under sub-humid Alfisols of Eastern India. *Indian J. Agril. Sci.*, **77:** 3–7.

Osthuyse, S.A. 2009. Management of a 'Tommy Atkins', ultra-high density orchard and recognized benefits associated with small tree mango orchards. *Acta Hortic.*, **820**: 335–338.

Rajan, S. 2016. Cultivation practices of mango in India. In: *Mango Production and Protection from Fruit Flies*, (Eds.) Mujeebur Rahman Khan, Fayaz Ahmad Mohiddin and Ziaul Haque. Write and Print Publications, New Delhi.

Rajan, S. and Singh, G. 2002. Approaches and strategies for precision farming in mango. In: *Precision Farming in Horticulture*, (Eds.) Singh, H.P., Singh, Gorakh, Samuel, J.C. and Pathak, R.K. NCPAH, DAC, MOA, PFDC, CISH, Lucknow, pp. 124–144.

Rajan, S., Singh, V.K., Reddy, Y.T.N., Upreti, K.K., Kennedy, R., Burondkar, M.M. and Bhagwan, A. 2014. Understanding the mechanism of off-season flowering and fruiting in mango under different environmental conditions. NAIP Subproject Report submitted to PIU, New Delhi, pp. 48.

Ram, S. and Rajan, S. 2003. Status report on genetic resources of mango in Asia-Pacific region. p. 196. International Plant Genetic Resource Institute, New Delhi.

Ramniwas, Kaushik, R.A., Sarolia, D.K., Pareek, S. and Singh, V. 2012. Effect of irrigation and fertigation scheduling on growth and yield of guava (*Psidium guajava* L.) under meadow orcharding. *African J. Agri. Res.*, **7**: 6350–6356.

Ravishankar, H., Rajan, S., Singh, V.K., Pandey, B.K., Singh, Achal and Verma, Ajay 2012. Status of mango production and its appraisal for improving sustainability of mango industry in India. *Agriculure Yearbook*, pp. 73–78.

Singh, G. 2008. High density and meadow orcharding of guava. Central Institute for Sub–tropical Horticulture, Rehmankhera, P.O.Kakori, Lucknow, pp. 20.

Singh, G. 2010. Development of meadow orchard in Guava for higher production. *Progr. Hort.*, **42**(2): 129–133.

Singh, Ram, S. and Shukla, C.P. 2001. Effect of different planting densities on growth and yield of mango. *Indian J. Hort.*, **58**(3): 191–195.

Singh, V.K. and Bhattacherjee, A.K. 2005. Genotypic response of mango to persistent of paclobutrazol (PP_{333}) in soil. *Scientia Hort.*, Elsevier, **106**: 53–59.

Singh, V.K. and Singh, A. 2003. Effect of paclobutrazol on regularity of bearing in mango (*Mangifera indica* L.) *Physiol. Mol. Biol. Plants*, **9**(2): 239–248.

Singh, V.K., Singh, Gorakh and Bhriguvanshi, S.R. 2009. Effect of polyethylene mulch on soil nutrient level and root, leaf and fruiting characteristics of mango (*Mangifera indica* L.). *Indian J. Agril. Sci.*, **79**(6): 411–417.

Singh, V.K., Soni, M.K. and Singh, Anurag. 2015. Effect of drip irrigation and polyethylene mulching on fruit yield and quality of guava cv. Allahabad safeda under meadow orcharding. *Indian J. Hort.*, **72**(4): 479–484.

Voon, C.H., Pitakpavian, C. and Tan, S.J. 1991. Mango cropping regulation with cultar. *Acta. Hortic.* **291:** 219–228.

28

Improving Profitability through Enhanced Nutrient Use Efficiency in Horticultural Crops

V.B. Patel[1]* and Mahesh K. Dhakar[2]

[1]University Professor-cum-Chief Scientist (Hort.),
Bihar Agricultural University, Sabour, Bhagalpur – 813 210, Bihar
[2]Scientist (Fruit Science),
ICAR-Research Complex for Eastern Region, Research Centre, Plandu, Ranchi – 834 010
**E-mail: patelvb7@gmail.com*

The global population is projected to reach 9 billion people by 2050. The continued population growth and the increasing per capita real income will further increase a total food demand, with changing dietary patterns toward higher proportions of meat, dairy, and fish as well as horticulture. Horticulture sector in India is one the fastest growing sector within agriculture. India has a wide variety of climate and soil on which a large range of horticultural crops are grown. Horticulture sector needs promotion for attaining sustainability of small farm holdings, crop diversification, increasing employment, providing an enormous export potential, improving environment, and more than anything for achieving nutritional security. As a result horticulture has received due emphasis during the last few decades by policy makers. A due importance to nutrients is essential as these affect the productivity, quality and profitability of horticultural crops. Improving nutrient use efficiency (NUE) is a worthy goal and fundamental challenge faced by the fertiliser industry and agriculture in general. Nutrient management for horticultural crops is one of the most challenging operation. There is an enormous need to increase the nutrient use efficiency of horticultural crops, to reduce emissions of ammonia and nitrous oxides related to fertiliser use. An increase in nutrient use efficiency is needed because fertilisers are becoming scarce: *e.g.* phosphate rock and also energy to produce nitrogen fertilisers. World fertilizer

Table 1: Nutrient Removal by different Horticultural Crops

Crop	*Yield (t/ha)*	*Nutrient Removal (kg/ha)*		
		N	*P_2O_5*	*K_2O*
Fruit				
Mango	15.0	100.0	25.0	110.0
Banana (Robusta)	57.5	322.0	73.0	1180.0
Citrus	20.0	22.0	12.0	57.0
Apple	29.0	18.0	2.0	40.0
Pineapple	84.0	150.0	45.0	530.0
Papaya	80.0	225.0	60.0	180.0
Grape	20.0	160.0	40.0	180.0
Vegetable				
Potato	28.0	202.0	50.0	225.0
Brinjal	60.0	175.0	40.0	300.0
Tomato	37.0	104.0	22.0	141.0
Cauliflower	50.0	250.0	100.0	350.0
Cabbage	37.0	112.0	28.0	112.0
Beans	35.0	130.0	40.0	160.0
Green peas	25.0	55.0	20.0	40.0
Lettuce	30.0	107.0	30.0	234.0
Spinach	25.0	120.0	45.0	200.0
Celery	20.0	140.0	55.0	220.0
Onion	30.0	73.0	36.0	68.0

Source: Ganeshamurthy *et al.* (2011).

use has increased more than fivefold and it is estimated that a1 per cent increase in NUE could save $1.1 billion annually. There is a need of integrated, balanced and effective nutrient management in the growth of horticultural crops. Finally an increase in fertilizer efficiency is also important to reduce production costs and increase profits for farmers.

Nutrient Use Efficiency

The main objective of increasing the nutrient use efficiency is to minimize the losses of nutrients by different measures for quality production and clean environment under local biophysical conditions, its feasibility and cost effective for farmers. An integrated approach is needed to make a coherent solution that fulfils the objectives of both farmers and society as best as possible. This system approach should focus on the farm level as the farmer is responsible to select the measures that will be applied in the field.

Various measures are available to improve nutrient use efficiency in horticulture crops. Foliar fertilization, bio-fertilizers, nutrient solubilizing microorganisms (mycorrhiza) and fertigation are already established technologies for enhancing NUE in horticulture crops. Many more new measures are coming to solve the nutrient issue in crops. Some of the important methods are described below:

Soil Management Practices

Both chemical amendments such as lime and gypsum and physical management involving tillage are important for increasing crop yields and in doing so, they improve NUE. Liming in acid soils maintain the soil pH and help in availability of nutrient to the plants. Crop yields are low on sodic soils and can be largely increased by gypsum application. Weeds compete with crop plants for nutrients, water, and sunlight and as a consequence reduce crop yield and NUE. These beneficial management practices help in maintaining the soil fertility.

4R Nutrient Stewardship Approach

International Plant Nutrition Institute (IPNI), The Fertilizer Institute (TFI), the Canadian Fertilizer Institute (CFI) and the International Fertilizer Industry Association (IFA), along with other organizations, described "4R Nutrient Stewardship", programmes. It provides a framework for using the right nutrient source, applied at the right rate, at the right time, in the right place, to achieve improved sustainability. These approaches consider economic, social, and environmental dimensions essential to sustainable agricultural systems and therefore provide an appropriate context for specific NUE indicators.

1.3 Nutrient Recycling

Soil fertility can easily be maintained through nutrient recycling from soil-plant-animal-environmental system. The plant residues and manure from animals returned to the soil and act as pool of carbon compounds which act as food for bacteria, fungi and other beneficial microorganisms. After decomposition of residue, nutrients are released in the soil and remain available for uptake. This cycle run continuously and maintained the soil fertility. Thus integration of horticulture with other enterprises like livestock, fishery, and apiculture is essential to promote the holistic growth of the sector.

Foliar Fertilisation

Timely application of nutrients through foliar spray is the most effective methods to correct specific nutrient disorder and to improve plant growth and NUE. Foliar application of water-soluble nutrients especially micronutrient fertilizers certainly ensures many fold higher use efficiency than soil application as it is directly spayed on crop foliage and skips irreversible retention in the soil or chemical transformation. The use of foliar cyanobacterial biofertilizer also showed higher percentage of fertilizer recovery and NUE compared to the composted manure in many horticultural crops. Humic substances (HS) are formed by chemical and biological transformations of plant and animal matter and from microbial metabolism. Humic substances comprise more than 60 per cent of the soil organic matter and are the major component of organic fertilizers and areknown to contain significant amounts of nutrients. HS bioactivity can help to reduce fertilizer application rates, enhance efficiency of nutrient use, replace synthetic plant regulators, improve fruit quality, increase water stress tolerance, decrease disease incidence, enhance

Table 2: Effect of Humic Substances (HS) on Vegetables, Fruit Crops and Ornamental Plants

Crop	*Effects*
Apricot	Increased of yield which ranged from 16 to 33 per cent
Chrysanthemum	Dose-effect response curve for shoot and root fresh and dry weight; increased flower diameter by 33%
Common beans	Increased yield by 25–35%
Eggplant	Enhanced fruit yield from 23 to 63% and N use efficiency
Faba bean	Increased growth and mineral content, 100-seed weight (by 26%) and decreased the damage by chocolate spot and rust diseases
Garlic	Enhanced efficiency of mineral nutrition, and increased from 1.96 to 2.28 fold the pyruvic acid concentration as an indicator of pungency
Onion	Increased total yield by 5–6%
Peach	Increased fruit yield by 80%
Potato	Enhanced nutrient use efficiency, and tuber weight (by 13%) and decreased incidence of hollow heart
Strawberry	Enhanced yield (33%), fruit firmness and total soluble solid percent

Source: Canellas *et al.* (2015).

early growth and flowering (Canellas *et al.*, 2015). Table 2 summarises reports of the bio-stimulant effects of HS on horticultural crops.

Biofertilizers

Biofertilizer is a natural product carrying living microorganisms derived from the root or cultivated soil. Hence, they don't have any ill effect on soil health and environment. Besides their role in atmospheric nitrogen fixation and phosphorous solubilisation, these also help in stimulating the plant growth hormones providing better nutrient uptake and increased tolerance towards drought and moisture stress. Role of *Rhizobium*, *Azotobacter*, *Azospirillum* in vegetable production is given in Table 3.

Table 3: Response of Vegetable Crops to Bio-fertilizer Inoculations (for nitrogen)

Bio-fertilizer	*Crop*	*Increase in Yield (%)*	*Nitrogen Economy (%)*
Rhizobium	Pea	13.38	–
	Pea	5.10	–
Azotobacter	Cabbage	24.30	25
	Garlic	14.23	25
	Knol khol	9.60	25
Azospirillum	Cabbage	11.87	25
	Onion	21.68	25
	Radish	9.00	–
	Sweet potato	8.50	–

Source: Rao *et al.* (2014).

Nutrient Solubilizing Microorganisms (Mycorrizha)

Mycorrhizal activity is particularly important under limiting soil fertility so that they can release the necessary nutrients to plants. Arbuscular mycorrhizal fungi (AMF) are essential components of soil biota which maintain ecological interactions with plant community and soil. The establishment of AMF in root allows the improving of nutrimental uptake by plants. Mycorrhizal activity influences the soil pH modifications and also the beneficial bacteria populations, which further contribute in plant growth promotion. AMF also help in reducing number and activities of nematodes and pathogens, increasing absorption of low mobility nutrients like phosphorus, zinc and copper, increasing plant vigour and reducing transplant injury, increasing photosynthetic rate to host plants and helping formation of water-stable aggregates through the action of mycelium and glomalin-related soil proteins.

Drip Irrigation and Fertigation

Irrigation and fertilizers are the most important inputs which directly affect the plant growth, development, yield and quality of produce. Application of irrigation water and fertilizers through drip are the most effective ways of supplying water and nutrients to the plant roots. These inputs are effectively utilized by plants as these are placed near crop root zone. Fertigation is an appropriate answer, particularly for horticultural and cash crops as it permit the irrigator to limit the watering as per water requirement of plants and optimum application of fertilizers through drip irrigation system which will enhance the production and productivity per unit area to meet the growing demand of the population. Drip irrigation technology has become potential alternative for fruit crop survival and growth in arid and semi-arid conditions. Apart from water saving (30 to 50 per cent) salt concentration is maintained within permissible limit near the root zone due to frequent irrigation. Dhakar *et al.* (2013) observed that the interaction between fertigation levels and irrigation methods were found to be quite superior in pomegranate (Table 4). When water is applied through drip irrigation, plant may benefit most from frequent fertigation as opposed to conventional basal dose.

Table 4: Effect of Interaction of Fertigation Levels and Irrigation Methods on Vegetative Growth Characteristics of Newly Planted Pomegranate

Treatment	*RGR of Shoot (%)*		*Plant Spread (m)*		*Plant Height (m)*	*Leaf Area (cm^2)*
	90th day	*120th day*	*E-W*	*N-S*		
75% RDF + HW	11.99 (20.26)	6.30 (14.54)	1.19	1.29	1.43	46.83
100% RDF + HW	14.00 (21.97)	7.5 (15.90)	1.22	1.53	1.47	53.75
125% RDF + HW	13.00 (21.13)	6.96 (15.13)	1.43	1.33	1.50	53.23
150% RDF + HW	14.00 (21.97)	8.00 (16.43)	1.45	1.36	1.51	49.73
75% RDF + DI	17.99 (25.09)	9.69 (18.14)	1.63	1.43	1.83	56.76
100% RDF + DI	17.86 (25.00)	11.90 (20.18)	1.69	1.46	1.82	57.13
125% RDF + DI	18.34 (25.36)	12.06 (20.32)	1.62	1.54	1.81	57.53
150% RDF + DI	18.14 (25.20)	10.57 (18.97)	1.68	1.55	1.80	58.09
CD P=0.05	0.52	0.18	0.09	0.13	0.03	2.563

HW: Hand watering; DI: Drip irrigation; RDF: Recommended dose of fertilizers.

Partial Root-zone Drying

Partial root-zone drying (PRD) is a new irrigation technique that improves the water and nutrient use efficiency of fruit crop production without significant crop reduction. Partial root-zone drying (PRD) is an irrigation technique where half of the root zone is irrigated while the other half is allowed to dry out. The treatment is then reversed, allowing the previously well watered side of the system to dry down while fully irrigating the previously dry side. Fully irrigated plants usually have widely opened stomata. In PRD part of the root system in drying soil can respond to drying by sending a root-sourced signal to the leaves where stomata may be closed to reduce water loss. A small narrowing of the stomatal opening may reduce water loss substantially with little effect on photosynthesis. Partial root-zone drying leads to preferential root growth and increase root hydraulic conductivity by cell-to-cell channels known as aquaporins which helps in enhancing the nutrient uptake from this soil zone. PRD irrigation may have benefits on WUE and fruit quality. There is saving of water approximately up to 30-50 per cent in PRD with no significant yield reduction. PRD advanced 5-7 days ahead of scheduled harvest which means better prices for farmers in the market. Dodd (2009) compared the crop yield ratio of PRD to drip irrigation at similar irrigation volumes (a ratio of 1 indicates that yield with both techniques is equivalent) and suggested that the ratio was more than one in most of the cases (Figure 1).

Nitrification Inhibitors

To control gaseous emissions from fertilizer N, nitrate leaching and to obtain high N use efficiency by crop plants, nitrification inhibition is thus desirable. Nitrification is the biochemical oxidation of ammonium to nitrate. *Nitrosomonas* and *Nitrobacter* bacterial species are involved in nitrification in soil. *Nitrosomonas* converts ammonium to nitrite by the action of ammonia mono-oxygenase enzyme, while *Nitrobacter* oxidizes nitrite to nitrate. Nitrification inhibition is the inactivation of ammonia mono oxygenase enzyme under the chemical action of some nitrification inhibitor. A number of chemicals have been used to inhibit nitrification in soil. These include 2-chloro-6 (trichloromethyl) pyridine (nitrapyrin), sulfathiazole, dicyandiamide, 2-amino-4-chloro-6-methyl pyrimidine, 2-mercaptobenzothiazole, thiourea and 5-ethoxy-3-trichloromethyl-1,2,4-thiadiazole (terrazole), 3,4-dimethyl pyrazol phosphate (DMPP) *etc.* Nitrapyrin, dicyandiamide (DCD) and acetylene are most commonly tested in laboratory and field experiments. There are several nitrification inhibitors of plant origin and are highly cost effective such as *neem* (*Azadirachta indica*), *karanj* (*Pongamia glabra*), pyrethrum, Norway spurce (*Picea abies* L.) *etc.* The use of a small quantity of neem oil can serve the purpose and may be used successfully for the coating of urea. However, not all the chemical components (group of compounds) of neem oil have nitrification-inhibiting properties. The major components in neem oil are

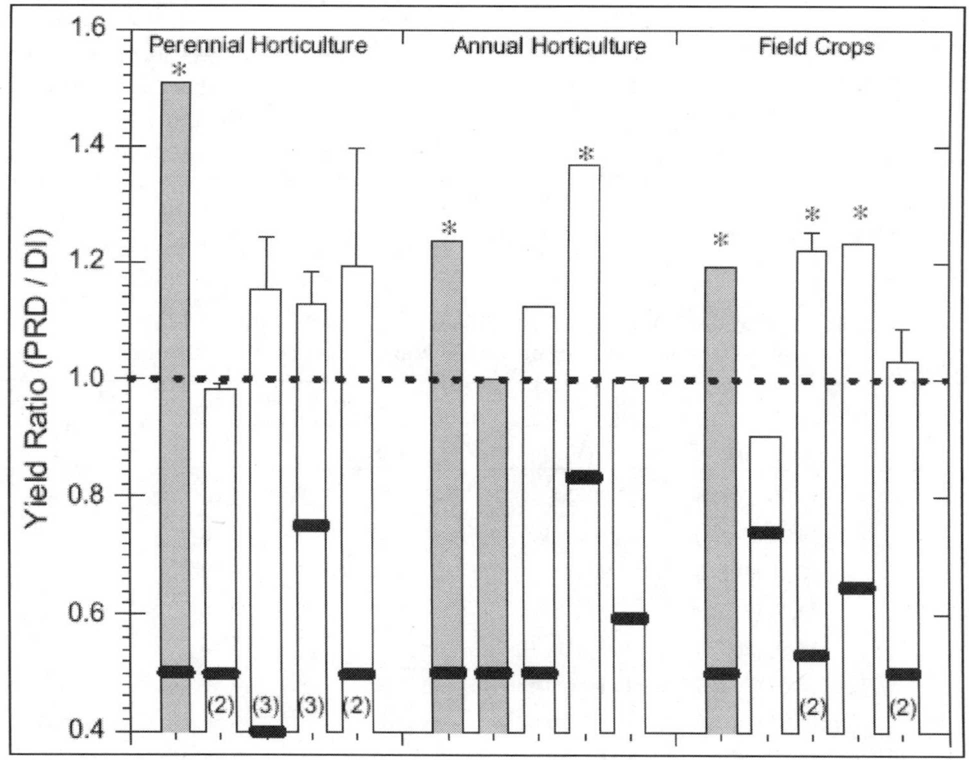

Figure 1: Crop Yield Ratio of PRD to Drip Irrigation at Similar Irrigation in different Crops.
***Source*: Dodd (2009).**

free fatty acid (FFA), pure oil, meliacins, saturated and unsaturated fractions. Kumar *et al.* (2007) found in a soil incubation experiment that the meliacins content in neem oil directly affected the nitrification inhibition. Dhakar *et al.* (2015) also applied DCD and meliacins along with different forms of N-fertilizers in the Kinnow field and found that the inhibitory effect of DCD and meliacins on the nitrification process resulted in more NH_4^+ accumulation in soil. Due to nitrification inhibitors, NO_3^--N availability remains high on surface level, reduced NO_3^- leaching and increased N fertilizer utilization efficiency in Kinnow production.

Nano Technology Applications

Fertilizers play a pivotal role in improving the productivity of agricultural crops. According to Subramanian *et al.* (2015) the nutrient use efficiencies of conventional fertilizers hardly exceed 30–35, 18–20, and 35–40 per cent for N, P, and K, respectively which remained constant for the past several decades. Nano-fertilizers intended to improve the nutrient use efficiencies by exploiting unique properties of nanoparticles. Nanotechnologybased products and their applications in agriculture may include nanonutrients, nano pesticides, nanoscale carriers, nanosensors, nano chips, nano cellulose, nano barcode, quantum dots *etc.* Realizing that the effective use of modern technology is needed to improve the nutrient use efficiency, nanotechnology has a potential to revolutionize the agricultural systems. The nano-nutrients may be applied as foliar and aerosol sprays and enters into the plants through leaf hole, stomata orifices and also through root hairs. In general, nanoparticle size less than 20 nm is the best for penetration through foliar application. Controlled foliar application of nanoparticle can stimulate different enzyme release by the plant to the rhizosphere resulted in 12-243 per cent more release of beneficial enzymes such as esterase, acid phosphatase, alkaline phosphatase, phytase, aryl sulphatase, nitrate reductase, urease, cellulase, hemi cellulase, lignase, after triggering different co-factors inside the enzymes (Table 5).

Table 5: Increase in Activity of Beneficial Enzymes with Nano-nutrients Application

Beneficial Enzyme	*% Increase in Activity*
Dehydrogenanse	25-68
Esterase	23-90
Acid phosphatase	21-72
Alkaline phosphatase	18-136
Phytase	23-83
Nitrate reductase	12-47
Aryl sulphatase	19-68
Cellulase	48-243
Hemi-cellulase	37-115
Lignase	19-47

Source: Tarafdar *et al.* (2015).

The nano-fertilizers are synthesized by fortifying nutrients singly or in combinations onto the adsorbents with nano-dimension (Table 6). Both physical and chemical approaches are used to produce nano-materials, and the targeted nutrients are loaded as it is for cationic nutrients (NH_4^-, K^+, Ca^{2+}, Mg^{2+}) and after

Table 6: Synthesis, Characteristics and Nutrient Release from Nano Fertilizers/Formulations

Nutrient	*Adsorbent*	*Approach*	*Size*	*Nutrient Release (h)*
N	Zeolite	Physical	25–30 nm	1,200
	Montmorillonite	Physical	35– 40 nm	400
	Zeolite	Chemical	200 nm	–
	Surface crosslinked super absorbents (hydrogels)	Chemical	40–80 nm	672
	Zeolite	Physical	420 μm	16
	Hydroxyapatite nanoparticles + *Gliricidia sepium*	Biological	19–25 nm	1,440
P	Zeolite	Physical	25–30 nm	1,104
	Montmorillonite, bentonite	Physical	35–40 nm	284
K	Zeolite	Physical	25–30 nm	1,176
	Montmorillonite, bentonite	Physical	35– 40 nm	216
NPK	Nano-coating of sulfur layer Chitosan	Chemical	100 nm78 nm	–
Nanocomposite	Kaolinite	Chemical	30–80 nm	–
Zn	Zeolite	Physical	25–30	1,176
	Montmorillonite, bentonite	Physical	35–40 nm	312
	Nano-Zn	Chemical	35 nm	–
	Nano-ZnO	Chemical	20 nm	–
B	Zeolite	Physical	60 nm	1,500

Source: Subramanian *et al.* (2015).

surface modification for anionic nutrients (NO_3^-, PO_4^{2-}, SO_4^{2-}). Nano-fertilizers are known to release nutrients slowly and steadily for more than 30 days which may assist in improving the nutrient use efficiency without any associated ill-effects. Since the nano-fertilizers are designed to deliver slowly over a long period of time, the loss of nutrients is substantially reduced *vis-à-vis* environmental safety.

NUE and a Look Forward

Manipulation of key genes controlling nutrient use efficiency would be an ideal approach and considerable effort has been expended in identifying and testing possible key steps limiting efficiency of utilization of mineral nutrients. Breeding for high nutrient use efficiency in horticultural crops is to be performed and exploit variation in existing germplasm. Screening of crops not just for yield performance but for individual components that contribute to efficient nutrient use is required, not just for modern commercial varieties, but for older varieties, land races and wild relatives.

Conclusion

There is need to adopt fertigation recommendations for every horticultural crop which including the basic nutrients (NPK) and micro nutrients according to the local soil type, climate, crops and their physiological stages. New irrigation strategies must be established to use the limited water resource more efficiently as PRD approximately save 50 per cent irrigation water without significant yield loss, while may improve the quality. Nano fertilisers can enhance the nutrient use efficiency and improve soil fertility in an ecofriendly manner. The addition of the effective nitrification inhibitor can substantially reduced NO_3^- leaching and increase N fertilizer utilization efficiency in horticulture production.

References

Canellas, L.P., Olivares, F.L., Aguiar, N.O., Jones, D.L., Nebbioso, A., Mazzei, P. and Piccolo, A. 2015. Humic and fulvic acids as biostimulants in horticulture. *Scientia Hort.*, **196**: 15–27.

Dhakar, M.K., Kaushik, R.A., Sarolia, D.K. and Bana, M.L. 2013. Effect of fertigation through low cost drip irrigation system on vegetative growth characteristics in pomegranate cv. Bhagwa. *Green Farm.*, **4**(1): 33–36.

Dhakar, M.K., Singh, A.K., Patel, V.B., Singh, S.K., Datta, S.P., Kumar, R. and Khanna, M. 2015. Effect of different nitrogen sources and nitrification inhibitors on soil nitrogen distribution in Kinnow orchard. *Indian J. Hort.*, **72**(2): 178–182.

Dodd, I.C. 2009. Rhizosphere manipulations to maximize 'crop per drop' during deficit irrigation. *J. Exp. Bot.*, **60**: 2454–2459.

Ganeshamurthy, A.N., Satisha, G.C. and Patil, Prakash 2011. Potassium nutrition on yield and quality of fruit crops with special emphasis on banana and grapes. *Karnataka J. Agric. Sci.*, **24**(1): 29–38.

Kumar, R., Devakumar, C., Sharma, V., Kakkar, G., Kumar, D. and Panneerselvam, P. 2007. Influence of physicochemical parameters of neem (*Azadirachta indica* A. Juss.) oils on nitrification inhibition in soil. *J. Agric. Food Chem.*, **55**: 1389–93.

Rao, K.M., Singh, P.K., Ryingkhun, H.B.K. and Maying, B. 2014. Use of bio-fertilizers in vegetable production. *Indian Hort. J.*, **4**(1): 73–76.

Subramanian, K.S., Manikandan, A., Thirunavukkarasu, M. and Rahale, C.S. 2015. Nano-fertilizers for balanced crop nutrition. **In**: *Nanotechnologies in Food and Agriculture*, (Eds.) M. Rai *et al.*, Springer International Publishing Switzerland, pp. 69–80.

Tarafdar, J.C., Rathore, I. and Thomas, E. 2015. Enhancing nutrient use efficiency through nano technological interventions. *Indian J. Fert.*, **11**(12): 46–51.

29

Harnessing Heterosis in Vegetable Crops: A Means for Achieving High Productivity and Profitability

B. Singh, H.C. Prasanna, P.M. Singh, Sudhakar Pandey, Rajesh Kumar, Gyan P. Mishra, S.K. Tiwari and B.K. Singh*

ICAR-Indian Institute of Vegetable Research, PB No. 01, P.O. Jakhini, Shahanshahpur, Varanasi – 221 305, Uttar Pradesh

**E-mail: bsinghiivr@gmail.com*

Commercial and strategic exploitation of heterosis is of vital importance to increase production, productivity and thereby to meet the challenge of providing food and nutritional security for an ever-increasing human population particularly in the Indian sub-continent. Hybrid technology is one of the potential and pragmatic options to meet the huge food, fodder and fiber requirements that are the need of hour. The effective use of hybrid technology has accelerated the development and expansion of seed-industry worldwide. Rapid advances in plant breeding and associated seed production technologies have served to enhance the competitiveness of hybrids in terms of increased yield and improving seed quality and thereby increasing profitability to the vegetable growers. The phenomenon of heterosis, also referred to as hybrid vigour has been the basis of much of the yield improvements in crop plants including vegetables during last century. Heterosis in tomato was reported a century ago by Hedrick and Booth (1907). Later on many researchers have reported heterosis for yield and earliness. The hybrid advantage up to the extent of 300 per cent has been reported over the best inbred. Following the success of F_1 hybrid sweet corn breeding in the USA in the 1940s and 1950s, other F_1 hybrid vegetables were bred. The year 1945 more or less marked the beginning of the global hybrid vegetable seed industry and during this time the hybrid seed production technologies were spread from the US and Japan to other countries. In India, vegetable production is expanding at a rapid rate with an increase of three folds during the last 50 years (Tay, 2002).

In India, hybrid breeding in vegetables got a fillip when the Govt. of India, Ministry of Agriculture and Cooperation announced a new policy on import of seed/planting material on Sep. 16, 1988 and the New Seed Policy came into effect from 1 October, 1988, which prompted a large number of private sector seed companies to enter into vegetable seed business on a large scale. This trend got further boost when Protection of Plant Varieties and Farmers' Rights (PPV and FR) Act-2001 came on the statute book in India (Ram, 2005). Accordingly, several private sector seed companies in India have strong R&D, production and marketing programmes in various hybrid vegetables. A large area is now planted with F_1 hybrid cultivars and thus hybrid vegetable seed production is growing at a rapid pace. The production of hybrid seeds is primarily concentrated in Karnataka and Andhra Pradesh due to the regional climatic suitability.

In vegetables, manual emasculation followed by manual pollination in case of hermaphrodite crops and manual pollination on pistillate flowers in case of monoecious crops is the widely used method in seed production of commercial hybrids. This method has been economically feasible in tomato, eggplant, chillies, cucurbits and few other vegetables, in which a large number of seeds are obtained from one manual cross. Ever since the discovery of male sterility in onion during 1930s and self-incompatibility in cabbage, these mechanisms have been used for development of commercial hybrids in onion and brassicas. Further, other mechanisms such as gynoecism, use of sex-regulators and chemical hybridizing agents have been used in development of hybrids in selected vegetable crops. To upscale the benefits of heterosis, greater efforts are required on broadening genetic base of inbred lines, developing improved germplasm for biotic and abiotic stress tolerance, exploring diverse sources for cytoplasmic genetic male sterility, use of innovative techniques such as doubled haploid lines as parents and use of transgenic approaches to develop male sterility.

Cultivation of F_1 hybrids in vegetable crops is quite expensive, due to very high cost of seeds. Male sterility (MS) and or self-incompatibility (SI) based F_1s are gaining popularity in various vegetable crops, since it significantly reduces the cost of hybrid seed. Male sterility systems based F_1s are commercially available in chilli, radish, cauliflower and muskmelon and SI based F_1s are commercially available in many cross-pollinated crops like cabbage, cauliflower, radish, *etc.* However, there is ample possibility to use such system even in self or often cross-pollinated vegetable crops. In cucurbits, hybrid seed production cost can be minimized by utilizing gynoecious line in cucumber and bitter gourd. Use of PGR in cucurbits is also an effective way for economic hybrid seed production. Although hybrid technology is most extensively used to sustain crop productivity in general, the phenomenon of heterosis that underlies hybrid performance is inadequately understood. Vegetables are one of those many agricultural crops that have benefitted from this mysterious phenomenon. In the recent past, efforts are underway in different crops to understand this key phenomenon. In tomato, both genetic and epigenetic causes of heterosis have been investigated. Numerous genetic factors for superior yield, chemical composition, morphology, and extreme heterosis have been defined. The recessive mutant locus called single flower truss (SFT), capable of increasing yield by up to 60 per cent was recently identified in tomato. These developments in the area of heterosis have reinforced the possibility of improved understanding of heterosis. Translation of this knowledge would further enhance productivity and profitability.

The aim of this article is to understand the advantages of hybrids and the powerful impact of the heterosis and to find out as to how advances in hybrid seed production technologies are affecting the productivity and quality of selected vegetables, namely, tomato, brinjal, chilli, okra, cole crops, root crops, bottle gourd, bitter gourd, smooth gourd, ridge gourd, ash gourd, cucumber, pumpkin, watermelon and muskmelon owing to increased emphasis on R&D programmes in India. It is further hoped that this article will act as a catalyst for further research and development in this field and also shall contribute towards better public-private-partnership mode in order to leverage advantages of both the sectors.

1. Hybrids for Enhancing Productivity and Profitability

It is presumed that by 2025, India will have 1.4 billion people, majority of whom will have relatively higher living standards. The consumption of vegetables has attained tremendous importance with these changing dimensions of food and nutritional securities. Due to increasing pressure on land through urbanization and industrialization, it is not feasible to increase the area under vegetables commensurate to our requirements; hence, the preciousness of increasing the yield per unit area becomes much more significant than it has ever been. This is the point at which hybrids carve a niche for themselves over non-hybrids. Development in vegetable production particularly through hybrids, contributes not only to food and nutritional security but also towards poverty alleviation and income generation since it is labour intensive, earns higher returns and involve extra skills.

The seed holds the key for increased productivity. Coupled with biotechnology and other crop improvement technologies, it offer tremendous opportunity for improving the productivity of Indian vegetable sector. High productive seeds and the involvement of private sector are the keys for achieving higher production. During past 5 years the Indian Seed Industry has been growing at a CAGR of 12 per cent compared to global growth of 6-7 per cent. In value terms the major growth has come from the increased adoption of hybrid vegetables apart from *Bt* cotton hybrids and single cross corn hybrids.

India is the second highest producer of vegetables in the world. However, productivity level of different vegetables is low in relation to other developed countries of the world. For example, the average productivity of tomato in India is 19 tonnes/ha. Comparatively, it is higher by several folds in developed countries like USA and Spain. The low productivity levels of different vegetables is due to several reasons such as production of vegetables in open field conditions rather than in protected conditions which exposes vegetables to weather adversities and lack of adoption of advanced production technologies. This is evident due to the fact that the maximum share in vegetable production is contributed by open-pollinated varieties compared to hybrids in majority of vegetables. Presently, considerable numbers of vegetable farmers cultivate local varieties and use farm saved seeds in vegetable production. If the share of hybrids in the vegetable production is enhanced, the productivity of vegetables is also likely to be increased along with the seed replacement rate. The adoption of hybrid technology should also be complemented with appropriate production and protection technologies to achieve higher levels of productivity. In the past several decades in India, there has been increasing trend in adoption rate of hybrid technology in vegetables such as tomato, cabbage, brinjal and okra (Sudha *et al.*, 2006). This trend needs be extended to other vegetables and sustained for achieving higher levels of productivity.

Farm level profitability can be increased from vegetable hybrids in two different ways, (i) through enhanced productivity, and (ii) by undertaking hybrid seed production. The hybrid seed production of vegetables is very lucrative and can be used as an opportunity to double the income of farmers. In comparison with the open-pollinated varieties, hybrid seeds are expensive as the cost of hybrid seed production is more due to the involvement of skilled labour in emasculation and pollination and also due to their higher productivity levels. Hybrid seed production of sweet pepper has potential to generate an income of Rs. 1,36,000 in an area of 0.75 acre followed by hot pepper with an income of Rs. 41,500 in 0.25 acre (Sharma, 2011). The hybrid seed production of tomato is highly remunerative with a benefit-cost ratio of 2.77. Similarly, okra hybrid seed production has a benefit-cost ratio of 2.02 (Sudha *et al.*, 2006).

There is great potential for earning revenues through export of vegetable hybrid seeds. India is one of the major exporters of fruit and vegetable seeds in the world. Fruit and vegetable seed exports constitutes considerable portion of total horticultural exports from India. There is a considerable trade imbalance. In the year 2015-16, 10684.30 MT of fruit and vegetable seeds was exported and earned export revenue of Rs. 48,502.86 lacs. However, India imported 14,328.09 MT of fruit and vegetable seeds worth Rs. 70,303.49 lacs (Source: APEDA). Therefore, efforts should be made to reduce the import and increase the export. This is likely to benefit the farmers engaged in hybrid vegetable seed production in specific and seed producers in general.

Since hybrid seed production is a labour intensive farm operation, it has great potential to create employment opportunities in rural India. In tropical vegetable seed sector, available reports indicate that a total of 10,394 farm families are engaged in hybrid seed production and annually generate a net income of Rs. 373 million (Kalia, 2009). Manual emasculation and pollination require skilled labour force and it is being performed mainly by women and young girls. The skilled work force engaged in hybrid seed production is paid higher than work force engaged in other farm activities (Sudha *et al.*, 2006). The highest share in employment generation is contributed by hybrid seed production of Solanaceous vegetables followed by cucurbits and okra (Hazra *et al.*, 2005; Kalia, 2009; Sharma, 2011). Mostly hybrid seed production is done in farmers' fields by private seed industry on buy-back basis. These private firms provide farm inputs to the farmers and provide technical know-how to the farmers. This reduces farmers risk and uncertainty in farming. According to the estimates available, 0.17 million farmers are engaged in such contract seed production (Hazra *et al.*, 2005).

2. Seed Market Scenario of Vegetables

The vegetable seed improvement programme is backed up by a strong crop improvement programme in both the public and private sectors. The requirement and market scenario of vegetable seeds impacting the hybrid technology on production-productivity advancement in vegetable crops is given in Table 1.

From the above Table it can be inferred that even if the share of F_1 is less, it carries more value as compared to the high OPV share of the same crop. It is a clear indication of the manifold profitability that can be harnessed by way of utilizing heterosis in vegetable crops.

3. Manifestation of Heterosis

The popularity of F_1 hybrid cultivars is due to their vigour, uniformity, disease resistance, stress tolerance and good horticultural traits including earliness and long shelf-life expressed and therefore giving consistent stable high yield. A range of manifestation of heterosis for various traits in different vegetable crops has been observed for various traits including yield and its contributing traits in different vegetable crops (Table 2). From the breeder point of view, it is a fast and convenient way to combine desirable characters of a vegetable together, for example fruit size and colour, plant type and disease resistance, and as a mean to control intellectual property rights through control and protection of the parental lines by the breeders.

Hybrids also provide opportunities for the rapid deployment of dominant genes for resistance to diseases and pests. However, it must be well understood and kept in mind that the advantages of F_1 hybrids in terms of yield superiority may be difficult to quantify. Yield *per se* is often of little significance, since many vegetable crops are harvested before maturity and may be graded before marketing. Thus, marketable yield is variable proportion of the total yield at harvest, which may not in any case represent the biological potential of the crop.

4. Status of Developed Vegetable Hybrids

The first vegetable hybrid in India, Pusa Meghdoot was developed and released in bottle gourd in 1971 from IARI, followed by the development and release of F_1 hybrids of summer squash (Pusa Alankar) and cucumber (Pusa Sanjog) in 1973. Out of these, however, only Pusa Meghdoot spread among the growers due to non-availability of hybrid seeds of others in sufficient quantity. A large portion of vegetable hybrid seed demand in the country is met by the private seed companies. In private sector, M/s Indo American Hybrid Seed Company, Bengaluru was the pioneer,

Table 1: Vegetables Seed Market Scenario in India (2014-15)

Crop	*Organised Sector (%)*	*Area (000.ha)*	*Total Seeds (OPV +F_1)*	*% Share (OPV)*	*% Share (F_1)*	*OPV Valuel (Mln) Rs.*	*F_1 Value (Mln Rs.)*
Brinjal	30.0	680.0	204.00	70.00	30.00	142.80	214.20
Cabbage	90.0	401.0	120.30	10.00	90.00	18.05	1082.70
Capsicum	80.0	30.0	19.50	40.00	60.00	31.20	292.50
Carrot	70.0	63.0	194.04	81.82	18.18	39.69	52.92
Cauliflower	70.0	436.0	130.80	50.00	50.00	98.10	850.20
Chilli	74.0	925.0	277.50	70.00	30.00	233.10	1248.75
Cucumber	60.0	82.0	98.40	20.00	80.00	19.68	629.76
Bottle Gourd	35.0	200.0	390.00	50.00	50.00	58.50	487.50
Bitter gourd	50.0	120.0	312.00	70.00	30.00	109.20	234.00
Ridge gourd	50.0	40.0	104.00	40.00	60.00	20.80	156.00
Sponge gourd	40.0	42.0	109.20	60.00	40.00	32.76	96.09
Pumpkin	35.0	50.0	140.00	60.00	40.00	67.20	224.00
Muskmelon	70.0	37.0	22.20	60.00	40.00	26.64	79.92
Water melon	75.0	140.0	336.00	30.00	70.00	90.72	823.20
Okra	50.0	524.0	5030.40	65.00	35.00	980.93	2200.80
Onion	70.0	1192.0	8344.00	99.00	1.00	12390.84	208.60
Peas	60.0	433.0	18186.00	100.00	0.00	3637.20	0
Radish	80.0	174.0	678.60	90.00	10.00	183.22	81.43
Tomato	75.0	846.0	296.10	60.00	40.00	355.32	1776.60
Others	155.0	1728.0	18186.00	100.00	0.00	11607.44	2033.20
Total						32535.64	12788.13

Source: Seed Association of India.

Table 2: Range of Heterosis (Per cent) for Yield Traits in Important Vegetable Crops

Crop	*Fruit Weight*	*No. of Fruits*	*Yield*	*Reference*
Tomato	17	83	62	Ahmad *et al.* (2011)
Brinjal	86	80	45	Patel *et al.* (2013)
Chilli	34	227	448	Patel *et al.* (2010)
Capsicum	(-) 4	173	70	Kamble *et al.* (2009)
Okra	8	8	9	Reddy *et al.* (2013a)
Cucumber	135	63	162	Singh *et al.* (2016)
Bottle gourd	19	21	37	Jha (2015)
Bitter gourd	49	38	60	Kandasamy (2015)
Pumpkin	14	17	29	Pandey *et al.* (2010)
Watermelon	114	87	205	Omran *et al.* (2012)
Cauliflower	–	–	37-63(curd)	Dey *et al.* (2011)
Cabbage	–	–	62 (head)	Kibar *et al.* (2015)
Radish	52 (avg. leaf length)	5 (plant weight)	186 (root weight)	Nur-E-Afrin Jahan (2010)

which, in 1973, launched its tomato hybrid 'Karnataka' and capsicum hybrid 'Bharath'. Thereafter, tremendous progress has been made by public and private sectors in the development of hybrids in several vegetable crops. Though the credit for popularizing hybrids in vegetables goes mainly to private seed companies, it is imperative to underline that the public sector hybrids are as good as, or sometimes even better, than the hybrids of private seed companies. The main reasons for non- spreading of public sector hybrids are lack of publicity, unavailability of seeds of parental lines in large quantities and inadequate, marketing network.

A large number of vegetable hybrids identified through AICRP (Vegetable Crops) from both public and private organizations. Till date, 151 vegetable hybrids have been recommended for cultivation in different parts of the country based on multi location testing, which includes 73 from private sector (Table 3) accounting for about 48 per cent contribution. In addition to recommendation of these hybrids through AICRP (VC), the private sector companies develop and release hybrids at their own level driven by market demand.

5. Hybrids Developed by IIVR

The significance of any technology is determined by consumer (farmers) demand. The need for farmers to enhance incomes through high value crops is demand of the time. The Institute has developed 10 vegetable varieties/hybrids in 5 different crops. The detail of the varieties/hybrids with salient features is given in Table 4.

6. Development of F_1 Hybrids as per Market Demand

It is a well known fact that private sector operates based on the profitability of an enterprise and this adds to the fact that heterosis harnessed through these hybrids is cost-effective, which help sustain private seed companies. Further, both private-sector and

Table 3: Vegetable Hybrids Recommended through AICRP-VC Developed by Private and Public Sectors

Crops	*Hybrids Developed (No.)*			*Share of Public Sector (%)*
	Total	*Private*	*Public*	
Tomato	39	16	23	41.03
Brinjal	40	27	13	32.50
Chilli	15	10	05	33.33
Sweet pepper	05	01	04	80.00
Okra	13	05	08	61.54
Cabbage	09	06	03	33.33
Cauliflower	07	03	04	57.14
Bottle gourd	05	02	03	60.00
Bitter gourd	04	02	02	50.00
Ridge gourd	02	01	01	50.00
Ash gourd	02	00	02	100.00
Cucumber	05	03	02	40.00
Watermelon	02	01	01	50.00
Muskmelon	02	00	02	100.00
Carrot	01	00	01	100.00
Total	**151**	**84**	**67**	**44.37**

Source: Anonymous 1975-2016.

Table 4: List of Released Variety/Hybrid, Salient Feature and their Adoptability

Crop/Hybrid	*Salient Features*	*Adaptability*
Tomato		
Kashi Abhimaan	Determinate, moderate resistance to tomato leaf curl virus, fruits colour deep red, firm, yield 8000-900 q/ha	J&K, H.P., Uttarakhand, Punjab, U.P., Bihar and Jharkand
Brinjal		
Kashi Sandesh	Fruits shape round, medium size, light purple, yield 750-800 q/ha.	Uttar Pradesh, Bihar and Jharkhand.
Kashi Komal	Plant 80-90 cm tall, erect growth habit, picking starts 65-70 days after transplanting, fruits small, long (6-7 cm long), colour purple, each fruit weighing 60-65 g, suitable for summer and autumn season cultivation yield 500-550 q/ha.	Recommended for growing in all the brinjal growing region of the country.
Chilli		
Kashi Surkh	Fruits colour light green, suitable for green as well as red fruit production, green fruit yield 240 q/ha, red fruit yield 140 q/ha.	Punjab, U.P., Bihar and Jharkhand.
Kashi Early	Indeterminate (80-100 cm in height), fruits colour green, suitable for green chilli production, tolerant to thrips, mites and viruses, green fruit yield 300-350 q/ha.	Punjab, Uttar Pradesh, Jharkhand, Bihar, Chhattisgarh, Odisha, A.P., Karnataka, Tamil Nadu and Kerala.
Okra		
ShitlaUphar	Flowering starts 38-40 days after sowing, resistant to yellow vein mosaic virus and OLCV fruits colour green, length 11-13 cm at marketable stage, yield 150-170 q/ha.	Punjab, U.P., Bihar, M.P. and Maharashtra.
ShitlaJyoti	Flowering starts 30-40 days after sowing, resistant to YVMV and OLCV, fruit colour green, length 12-14 cm at marketable stage and yield 20-22 t/ha.	Rajasthan, Gujarat, Haryana, Delhi, Chhattisgarh, Odisha and A.P.
Kashi Bhairav	Fruit colour dark green, resistant to YVMV and OLCV under field condition, length 10-12 cm at marketable stage, yield 160-180 q/ha.	Recommended for growing in all the okra growing region of the country.
Kashi Mahima	Flowering starts 36-40 days after sowing, resistance against YVMV and OLCV, fruit colour green, length 12-14 cm at marketable stage, yield 200-220 q/ha.	Punjab, U.P., Bihar, Jharkhand, Chhattisgarh, Odisha, A.P., M.P. and Maharashtra.
Bottle Gourd		
Kashi Bahar	Long fruit, single plant bears 12 light green straight fruits of 30-32 cm length and 780-850g weight, suitable for both season cultivation (rainy and summer), yield 500-550 q/ha.	U.P., Punjab, Bihar and Jharkhand.

Table 5: Market Demand and Consumer Preferences for different Segments in Vegetables

Crop	*Segment*	*Specific Traits*
Tomato	Fresh market	Uniform maturity, oblong to round shape, red colour, firm fruits, average 3-4 locules
	Processing	High sugar: acid ratio, high total soluble solids, large size, thick pericarp, fleshy
	Cherry type	Small size (20-25 g), oblong shape, highly clustered, acidic
Chilli	Fresh market	Colour, pungency, shape, size, flavour, exocarp thickness, endocarp: seed ratio, vitamin A and C
	Fresh processing	Colour, pungency, size, peduncle length, pericarp thickness, endocarp: seed ratio
	Dried fruits/powder	Colour, pungency, shape, size, dry weight, low crude fibre, endocarp: seed ratio
	Oleoresin extraction	Colour (paprika oleoresin), pungency (capsicum oleoresin)
Sweet pepper	Fresh market	Colour, shape, size, flavour, exocarp thickness, endocarp: seed ratio, vitamin A and C
Brinjal	Long	Fruit colour (dark purple/light purple/green/striped), size (small for kalaungi to large for curry), fleshy, soft pulp, less seeded (<400 seeds), 125-150 g.
	Round	Fruit colour (dark purple/light purple/green/striped), medium large for bharta, fleshy, soft pulp (manual pressure applied by hand), less seeded (< 400 seeds), average fruit weight (influenced by local demand) 200-250 g
	Oblong	Fruit colour (dark purple/light purple/green/striped), size (small for kalaungi to large for curry), fleshy, earliness, soft pulp, less seeded (< 400 seeds), average fruit weight (influenced by local demand) 125-150 g
Okra	Dark Green	Dark green fruit color, no seed protuberance on the immature fruits, 5-6 ridges
	Reddish-purple	Reddish-purple fruit color, no seed protuberance on the immature fruits, 5-6 ridges,
	Light green	Light green fruit color, no seed protuberance on the immature fruits, 5-6 ridges
Cucumber	Greenish white/Pune Kheera	Greenish white, cylindrical fruits, 16-18 cm long, 150-160 g, crispy texture, non-bitter
	Green, long	Light green to medium green fruits with specks, cylindrical fruits, 18-22 cm long, 180-220 g, non-bitter
	Uniform dark green	Dark green, cylindrical fruits, 6-20 cm long, mostly parthenocarpic and suitable for greenhouse cultivation, non-bitter
Watermelon	Jubilee type	Oval to oblong, 8-12 kg fruits with light green rind with distinctive dark green stripes, deep crimson flesh, granular texture, 11-12% TSS, excellent transportability and better shelf-life
	Sugar Baby type	Round, dark bluish green rind, 6-10 kg fruits, deep crimson flesh, 12% TSS, excellent transportability, better shelf-life, 75-80 days maturity
	Ice-Box type	Oblong fruit, dark green rind with light stripes, 3-4 kg, dark red, firmly textured flesh, 12% TSS, sweet and juicy, suitable for storage in fridge and to be consumed in one go by small families
Muskmelon	Non-netted	Sweet, juicy fruit, 13-14% TSS, oval-round fruit, 1 kg fruit weight, small seed cavity,attractive deep orange flesh, longer shelf-life and better transportability,
	Netted	Highly netted skin, non-sutured, sweet, juicy fruit, 13-14% TSS, oval-round fruit, 1 kg fruit weight, small seed cavity, attractive deep orange flesh, longer shelf-life and better transportability
Pumpkin	Small, medium, large, very large	Flat round mottled green, oval to round mottle green, oval to oblong mottle green, dark green and light greenSize- small (< 3kg), medium (3-6 kg), large (6-12 kg), very large (>12 kg)
Bottle gourd	Medium long, cylindrical	Perfect cylindrical, 25-30 cm long, shiny green fruits, 700-800 g fruit weight
	Round with neck (pear shape)	High round oval, green fruits with pronounced neck, 700-800 g fruit weight
	Dark green with specks	Bottle shaped, dark green with specks, 1000-1200 g fruit weight
Bitter gourd	Green, short, spindle	Dark green, 5-6 cm long, oval shape, 30-40 g weight, strong spines, suitable for stuffing
	Green, medium long, spindle	Glossy dark green, spindle shape, 16-18 cm long, 120-150 g weight, strong spines
	Green, long, spindle	Uniform dark green, spindle shape, 20-25 cm long, 175-200 g weight, strong spines, resistant to downy mildew and bitter gourd virus
	White, long, spindle	White, thick, spindle shape, 20-25 cm long, 200-250 g weight, strong spines
	Green, long, smooth (Jhalari type)	Shiny light green, largely smooth with continuous ribs, 30-35 cm long, 150-200 g weight
Ridge gourd	Light green fruit type	Almost cylindrical fruits, 30-40 cm long, 200-250 g, non-bitter
	Dark green fruit type	Almost cylindrical fruits, 40-45 cm long, 200-250 g, non-bitter
Sponge gourd	Light green/white fruit type	Cylindrical fruits, 30-35 cm long, 180-200 g, 45-50 days maturity
	Dark green fruit type	Dark green, cylindrical fruits, 25-30 cm long, 200-250 g, 45-50 days maturity
Carrot	Orange	Nantes type root shape, uniform external root colour, self-coloured and small core, smooth root, fewer secondary roots, lesser root scars, 20-23 cm long root.
	Red	Danvers type root shape, uniform external root colour, self-coloured and small core, smooth root, fewer secondary roots, lesser root scars, 20-23 cm long root.
Cauliflower	–	Retentive white, compact and hemispherical curd; marketable curd weight 800-1200 g.
Cabbage	–	Compact and round head, green coloured smooth leaves, small core, head weight 700-1000 g
Radish	–	Iciclical root shape, attractive white coloured smooth root, fewer secondary roots, 22-25 cm long root, non-pithiness

public-sector researchers are now focusing on consumer preference in the vegetable cops besides yield, so that the hybrids can not only yield more but also can fetch premium price of it produce. The demand of F_1 hybrids in market depend upon the consumers liking. The developed F_1 hybrids should be eye catching, easy to carry and fit to home (refrigerated) storage and. The traders also influence the marketing and major value traits for traders are transportability and shelf-life. Therefore, yield becomes of lesser importance in case of vegetables compared to consumer preference. Accordingly research personnel's have to deliver the desired product. The consumer preferences for different vegetables are mentioned in Table 5.

7. Cost-effective Systems to Minimize the Hybrid Seed Cost

The vegetable hybrid seeds production methods can be divided into two groups: hand-pollinated and the gene-control system. Manual emasculation pollination method involves lots of labour and cost. The genetic control system can be either due to the self-incompatible (SI) system where pollen of the same plant or flower cannot pollinate itself or due to the male-sterility (MS) systems where a female plant has no functional pollen for self-pollination. When no such genetic control system is available, tedious hand-emasculation and pollination are used to produce F_1 seed. The methods of hybrid seed production exploited in different vegetable crops is listed (Table 6).

7.1 Hand-Pollination

Most of the F_1 vegetables seed are produced by hand-pollination and the method involves the manual emasculation of the anthers, followed by hand-pollination using pollen of male parent and then preventing other pollen from contaminating the pollinated flowers. However, it is labor intensive, requires a team of skillful growers, gentle hands, a lot of patience and commitment. Most hybrids in tomato are produced by hand emasculation and hand pollination. Hand emasculation results in good number of seeds as high as 90 per cent pollination depending on the year and location. However, it is possible to produce seeds economically using male sterility or exerted stigma.

Most of the hybrids currently under seed chain in India are often produced using manual emasculation and pollination method. In solanaceous vegetables, hybrids are developed using hand emasculation and pollination procedures. Most of the inter-varietal hybrids of eggplant are reported to have exhibited considerable vigor in economic characters, particularly the yield. For this reason, most commercial cultivars of eggplant are F_1 hybrids. In hybrid seed production of eggplant, hand emasculation is still used. In cucurbits, hybrids are produced by hand pollination of emasculated flowers on the female parent using staminate flowers from the male parent.In okra, reports are available for yield improvement from 50 to 70 per cent through heterosis breeding (Jagan *et al.*, 2013; Reddy *et al.*, 2013b). In okra, number of fruits has been found to be the most effective factor contributing to yield, followed by plant height, fruit weight and size, and earliness. Improving yield and ensuring its sustainability under adverse conditions through resistant hybrids is the major objective of heterosis breeding. Through intensive research efforts more than 50 improved varieties and hybrids have been released so far. Some of these varieties have already made significant impact/contribution in revolutionizing the production of okra in the country. In okra, till now 13 hybrids have been released by AICRP of which 8 were from public sector and 5 from private seed companies (Table 3).

7.2 Male-sterility Systems

Hybrid seed production of chillies, sweet corn, carrot and onion are based on MS system and the genetic control can be either just clear-cut MS genes (genetic male sterility) or the interaction of a MS gene with a cytoplasmic factor (cytoplasmic-genetic male sterility).

Table 6: F_1 Seed Production Methods Mainly Used in Various Vegetables

Vegetable	*F_1 Seed Production Method*	*Vegetable*	*F_1 Seed Production Method*
Bitter gourd	HP	Broccoli	SI
Cabbage	SI	Carrot	CMS
Cauliflower	SI	Chilli	HP, GMS, CMS
Cucumber	HP	Eggplant	HP
Gourds	HP	Luffa	HP
Melons	HP	Okra	HP; GMS reported
Onion	CMS	Pumpkin	HP
Radish	SI	Sweet corn	HP and CMS
Tomato	HP and CMS	Turnip	SI
Watermelon	HP	Zucchini	HP

HP: Hand pollination; GMS: Genetic male sterility; CMS: Cytoplasmic male sterility; SI: Self incompatibility.

Source: Kumar and Singh, 2005.

Genetic male sterility system has also been attempted in other vegetable crops such as tomato, okra *etc.*, however, it still remained at the experimental level. In recent years, brassica breeders are trying to use MS system instead of the SI system. Utilization of male sterility systems in major vegetable crops as been discussed hereunder:

7.3 Solanaceous Vegetables

Genetic male sterility (GMS) is most common in cultivated tomato, but cytoplasmic male sterility (CMS) is not natural in the genus *Lycopersicon*. Rick (1945) used male sterility for the first time in tomato hybrid seed production. Different types of male sterility found in tomato are classified as structural, sporogenous and functional. The structural male sterility mutants are formed due to aberrations in genes that determine the stamen structure. The *stamenless* mutant is considered to be useful in tomato breeding. The fertility in this mutant can be induced temporarily by changing environment conditions. For example, *stamenless 2* can be altered by changing temperature conditions; under 28°/23°C day/night temperature carpel-like structures are produced, whereas under 18°/15°C day/night temperature normal stamens are produced with viable pollen (Sawhney, 1983).

The sporogenous male sterile mutants are characterized by the absence of fertile pollen where break down might occur at premeiotic, meiotic, tetrad or microspore stage. Of the numerous male sterile mutants, only *ms10* and *ms32* are considered as useful in hybrid seed production (Atanassova and Georgiev, 1986). The gene *ms10* is linked to gene *anthocyanin absent* (*aa*). Therefore, *anthocyanin absent* (*aa*) phenotype can be used as phenotypic marker in maintenance of male sterile line. Functional male sterile mutants produce viable pollen but show sterility because of mechanical reasons. This type of sterility have been described in *tomato* and *positional sterile* (*ps*), *positional sterile-2*, (*ps-2*), and the *exserted* mutants have so far been reported (Larson and Paur, 1948; Rick and Robinson, 1951). The *positional sterile-ps* tomato mutant sterility is due to petal coalescence with the corolla especially at the apex of the flower. Significantly higher hybrid seed yield was reported on the *ps 2* seed parents compared to the fertile isogenic line.

Among the genetic emasculation tools, both genetic male sterility (GMS) and cytoplasmic-genetic male sterility (CMS) have been employed in hybrid seed production of chillies. In India, MS-12 (*ms-509/ms-10*) line has been utilized to develop first commercial male sterile based chilli hybrid, namely, CH-1 and CH-3 and farmers in Punjab are producing seeds of these hybrids. The *ms-10* gene is linked with taller plant height, erect growth and dark purple anthers (Das *et al.*, 2001). Male sterile lines possessing *ms-3* gene are being utilized for hybrid seed production in Hungary. The first cms *Capsicum* plant was identified in an Indian *C. annuum* population (PI-164835) in USA. Subsequently, two more male sterile cytoplasms were isolated, but all the three independently isolated male sterile cytoplasms were found to be genetically identical. Yet another male sterile cytoplasm was located in India during 1990's, which has been transferred in several desirable genotypes and utilized for hybrid development at IIHR, Bengaluru. The Peterson's sterile cytoplasm has been commercially exploited in China, Korea and India, although at limited scale due to the temperature sensitivity of male sterility expression (Kumar *et al.*, 2007). At IIVR, Varanasi and IIHR, Bengaluru, cms lines have been successfully utilized to develop commercial chilli hybrids (Table 2).

Male sterility in eggplant has been described in several reports, however, not much has been attained for commercial use. Functional male sterility in a population of Blackie eggplant was developed from the cross Black Beauty by Black Nagasaki, which was due to failure of the anthers to dehisce normally. Cases of cytoplasmic male sterility have been reported in backcross progenies derived from the interspecific cross between *S. gilo* (*i.e., S. aethiopicum*, Gilo group) and *S. melongena*. None of these lines have been utilized for the practical hybrid seed production of eggplant. Functional male sterility was obtained in the Ottawa eggplant mutant. From a test of three F_2 populations of different pedigrees showed that this trait inherited as a single recessive gene. UGA 1-MS line derived from a spontaneous male sterile mutant of the cultivar Florida High-bush had the functional male sterility of which was due to a failure of anther dehiscence. Genic male sterility is also caused by malfunctioning of the tapetum and the sterility controlled by the interaction of two recessive nuclear genes designated as *ms1ms1ms2 ms2*. Although male sterility controlled by nuclear genes found in eggplant, they have not been utilized for the practical hybrid seed production of eggplant.

7.4 Cole Crops

CMS is not apparently found in cauliflower or other cole crops but has been introduced from several sources such as *Raphanus sativus*, *B. nigra*, *B. napus*, *B. juncea* and *B. tournefortii*. The CMS from various sources have been transferred in broccoli, cauliflower, cabbage and other cole crops through repeated backcross. The most extensively investigated CMS system among cole crops started with the discovery of male sterility in Japanese radish by Ogura (1968). After introgression through conventional breeding to *B. oleracea*, several developmental and floral abnormalities were commonly observed among CMS plants that affected their agronomic and breeding value, but these abnormalities have been overcome *via* protoplast fusion. In India, Ogura type CMS are being used to transfer the male sterility system in tropical and Snowball cauliflower. IARI have developed few hybrids such as Pusa Kartik Sankar, Pusa Hybrid-2 and Pusa SB cauliflower Hybrid-1 and Pusa Cabbage Hybrid-1 and Pusa Cabbage

Hybrid-81 of cabbage by utilizing SI and CMS system. Watts (1965) observed sufficient heterosis for earliness and curd size in cauliflower. Kumaran (1971) made a study of three maturity groups of Indian cauliflower and recorded appreciable amount of heterosis for curd weight, yield, curd size index, maturity, stalk length and leaf size index. Appreciable amount of heterosis for various morpho-horticultural traits have also been reported by Sharma *et al.* (2004), Kucera *et al.* (2006), Dey *et al.* (2011) and Jiafu *et al.* (2012). Additionally, greater amount of heterosis for ascorbic acid, anthocyanin and carotenoids in cauliflower has been observed by Dey *et al.* (2014).

Appreciable amount of heterosis has been reported and realized in cabbage for high yield, early maturity and bigger head by various researchers (Xu *et al.*, 2004; Kibar *et al.*, 2015). Further, desirable hybrid effects for ascorbic acid, carotenoid, earliness, frame size, net head weight and head compactness have been estimated by Singh *et al.* (2009a).

7.5 Root Crops

The CMS system was first identified in a Japanese radish cultivar by Ogura (1968) popularly known as 'Ogura CMS', and thereafter the Ogura CMS has been transferred into radish as well as various *Brassica* crops too. Thereafter, several types of CMS system have been characterized and reported, namely, 77-01A CMS (He *et al.*, 1981, Kosena CMS (Ikegaya, 1986), NWB CMS (Nahm *et al.*, 2005), DCGMS (Lee *et al.*, 2008) and 805A CMS (Wang *et al.*, 2012). Significant heterosis for higher root yield, earliness and long sized roots has been observed in radish hybrids. So far, ICAR-IARI Regional Station, Katrain, HP; and ICAR-IARI, Pusa, New Delhi has developed two carrot hybrids, *i.e.*, Pusa Nayanjyoti (orange root) and Pusa Vasuda (red root), respectively by using CMS system.

The heterosis breeding in carrot is generally based on system of CMS with different genetic backgrounds. The 'brown anther' type (*ba*) is present in all cultivated orange coloured open-pollinated varieties, which is results from an interaction of the 'Sa'-cytoplasm with two independent nuclear genes, *i.e.* homozygous recessive *aa* or dominant *B-* (Banga *et al.*, 1964); and the phenotype is characterized by deformed, brown-coloured anthers without functional pollen, caused by a genetic block in meiosis. On the other hand, the cytoplasm of the 'petaloid' type (pt) is derived from a wild form of *Daucus carota* L.; inherited by an interaction of Sp-cytoplasm and two independent dominant genes, *viz.*, *Ml* and *M2* (Morelock, 1974); and is characterized by a transformation of the anthers into petals or petal-like structures, which do not produce functional pollen (spoon-type, strap-type or carpeloid-type). A third system of CMS, gu-type, derived from the cross of orange coloured carrot and wild carrot *D. carotagummifer* Hook. (Nothnagel 1992); governed by an interaction of the '*gummifer*' cytoplasm with a recessive nuclear gene, gugu (Stein and Nothnagel, 1995); and this can potentially be used as a new source of CMS for carrot breeding because of the simple mode of selection and phenotypic stability.

8. Self-incompatibility System

The vegetables with highly developed SI system are present in the Crucifereae family which includes *Brassica oleracea*, *B. rapa* and *Raphanus sativus* (Table 5). The genetics of the SI system in the cruciferous crops are well developed and a series of genes (loci) and alleles has been identified. Vegetable breeders have been successfully using them for F_1 hybrid seed development. However, in India, the enthusiastic development on this part is lagging. Use of SI system in some vegetables is being discussed hereunder:

8.1 Cole Crops

In cauliflower, heterosis was first reported by Jones (1932). Likewise, the possibility of using self-incompatibility (SI) to produce hybrids in cole crops was suggested by Pearson (1932), but first hybrid appeared in Japan during 1950 and in USA during 1954, but later on CMS system dominates over SI for hybrid breeding.

8.2 Root Crops

Sporophytic SI was first observed in radish (Stout, 1920), and its inheritance pattern was first demonstrated by Bateman (1955). Unlike cabbage, SI system in most of the radish genotypes is very weak and unstable; hence there is always chance to get sib-seeds, while producing the hybrid seeds. Facing the problem with SI system; globally, CMS system is generally used to harness the heterotic vigour for root yield, earliness, non-pithiness, adaptation and uniformity. The first hybrid radish was developed by Frost (1923) who observed that the crosses between selfed lines were very vigorous and usually exceeded the better parent in root size and all plant characters.

8.3 Gynoecism

With the availability stable pistillate seed parents/ gynoecious lines are being used in hybrid combination for yield and earliness. Development of gynoecious lines suitable for the tropical conditions holds great promise in giving a fillip to the evolution of hybrids. True breeding tropical gynoecious lines of cucumber, *viz.*, 87-304-6, 87-316, 87-319-12 and 87-338-15 have been isolated by More and Budgujar (2002). Heterosis over better parent with 77.6 per cent was evidenced in tropical gynoecious hybrid 304 x R.K.S. 296. Results also revealed the usefulness of tropical gynoecious lines in production of hybrid seeds compared to temperate gynoecious lines or monoecious parents. For hybrid seed production gynoecious lines are grown under open field condition, simply by isolated planting in alternate with pollinated line in the ratio of 4 female: 1 male.

The most noticeable effect of sex expression on yield is not in total yield over multiple harvests, but on early yield. Gynoecious x gynoecious and gynoecious x monoecious hybrids produce significantly higher yields in the first harvest than monoecious x monoecious hybrids, but there is no significant difference among these hybrids for total yield over multiple harvests (Wehner and Miller, 1985). Because of their early, concentrated fruit set, gynoecious hybrids were instrumental in establishing a system for once-over mechanical harvesting of processing cucumber (Wehner, 1989). Silver nitrate and gibbrellic acid will promote male blooms on gynoecious plants (Peterson and Anhder, 1960). Recently, gynoecism developed in bitter gourd at IIVR, Varanasi is also being utilized for hybrid development, however, it has yet to be reached to the farmers.

8.4 Chemical Hybridizing Agents

There are various chemicals available which can artificially induce MS, are called gametocides. This method of inducing male sterility is fast and inexpensive compared to backcross method and there is no necessity to use maintainer lines. The effective chemicals used in various crop plants are sodium methyl arsenate, zinc methyl arsenate, naphthalene acetic acid (NAA), gibberellins, ethrel, maleic hydroxide and FW 450 (Santhy *et al.*, 2008). Ethephon [2-chloroethyl [hosphonic acid] also called as ethrel (or) camposan, its gametocidal potential has been extensively studied in many vegetable crops such as brinjal, baby corn, lettuce and sugar beet (Sharma and Sharma, 2005). Summer squash has been shown to have heterosis of 11-84 per cent for yield in zucchini types and 0-82 per cent yield advantage over open-pollinated types in yellow types. In hybrid seed production of summer squash, one of the inbreds is rendered as female for first two to three weeks of flowering by spraying the plants with ethephon at the two and four- leaf stage. Ethephon does not work properly to suppress induction of male flowers in *C. maxima* and *C. moschata*.

9. Biotechnological Interventions

9.1 Hybrid Purity Testing using Molecular Tools

Molecular markers are constantly being used to for both fingerprinting and genetic purity testing of various hybrid vegetables like tomato, beans, pepper and potato (Ansari, 2015). Genetic purity can also be tested using technique as reported by Mongkolporn *et al.* (2004). Purity of F_1 chilli hybrids was determined using two molecular techniques RAPD and ISSR. Four molecular marker systems—RAPD, ISSR, SRAP and SSR were used to evaluate seed genetic purity of a hybrid cabbage cultivar 'Zaoxia 16' by Liu *et al.* (2007). Markers for identification of hybrids and parental lines have been identified for six hybrids (Table 7) developed by ICAR-IIVR, Varanasi.

Besides working on molecular tools, a range of morphological markers (Table 8) are available in vegetables which can be utilized for testing the purity of F_1 hybrid produced.

9.2 Double-Haploid (DH)

ADH is aparental line formed when haploid cells (gamete) undergo chromosome doubling. In order to propagate them through seeds and to include them in breeding programs, their fertility has to be restored with chromosome doubling. The obtained DHs are homozygous at all loci and can represent a new variety (self pollinated crops) or parental line for the production of hybrids (cross-pollinated crops). The production of pure lines using doubled haploids has several advantages over conventional methods. The time saving is substantial, particularly in biennial crops and in crops with along juvenile period. For self-incompatible species,

Table 7: Molecular Markers for Purity Testing of Hybrids Developed at IIVR

Crop	*Hybrid*	*Marker for Hybrid Testing/Identification*		
		Parent 1	*Hybrid*	*Parent 2*
Brinjal	Kashi Komal	UBC 840_300 bp	UBC 840_300 bp	–
		–	UBC 840_1200 bp	UBC 840_1200 bp
	Kashi Sandesh	UBC 811_1000 bp	UBC 811_1000 bp	–
		–	UBC 809_1300 bp	UBC 809_1300 bp
Tomato	Kashi Abhiman	–	UBC 811_1000 bp	UBC 811_1000 bp
Chilli	Kashi Surkh	UBC 840_400 bp	UBC 840_400 bp	–
		–	UBC 811_300 bp	UBC 811_300 bp
	Kashi Early	UBC 815_1000 bp	UBC 815_1000 bp	–
		–	UBC 890_400 bp	UBC 890_400 bp
Bottle gourd	Kashi Bahar	UBC 835_600 bp	UBC 835_600 bp	–
		–	UBC 825_1100 bp	UBC 825_1100 bp

Table 8: Utilization of Morphological Markers in Hybrid Development

Vegetable	Marker	Advantage
Tomato	Brown seed: *bs, bs2, bs3, bs4* genes	Reduced cost on hybrid seed production; only 10% brown seeds (non-hybrid = selfed), which can be eliminated easily.
	Anthocynin less (*aa*) linked with ms gene	Planting of 100% male sterile plants in hybrid seed production field.
	Enzyme marker (*prx-2*) linked with ms-10 gene	Proposed indirect use of marker in male sterile line development
	Anthocynin less (*aa*) and exerted stigma	Reduction in hybrid seed production cost and identification of selfed seeds at seedling satge.
Chilli	The ms-10 gene is linked with taller plant height, erect growth and dark purple anther.	Identification of male sterile plants at comparatively early stage in the hybrid seed production field.
Watermelon	Linkage of ms gene with delayed-green (*dg*) seedling marker gene	Planting of 100% male sterile plants in hybrid seed production field.
Onion	Brown seed	Eliminate necessity of specific planting arrangement of female (*cms*) and male parents

Source: Kumar and Singh (2005).

dioecious species and species that suffer from inbreeding depression due to self-pollination, haploidy may be the only way to develop in bred lines. DH technique can be used to transfer sterile cytoplasm of CMS lines into a desirable background in a single step avoiding many generations of backcrossing. Several vegetable seed companies like Monsanto, Syngenta, Nunhems (Now Bayer), Namdhari, HyVeg, *etc.* have already made significant progress in this regard, especially in crops like cauliflower, cucumber, pepper, and melons (Sadashiva *et al.*, 2014).

9.3 QTL Mapping

In tomato, both genetic and epigenetic causes of heterosis have been investigated. *S. pennellii* × *S. lycopersicum* ILs have been used to define numerous quantitative trait loci (QTLs) for superior yield, chemical composition, morphology, abiotic stress tolerance and extreme heterosis (Kamenetzky *et al.*, 2010). The recessive mutant locus called *SINGLE FLOWER TRUSS* (*SFT*), identified by Professor Dani Zamir's group in Israel is capable of increasing yield by up to 60 per cent (Krieger *et al.*, 2010). This was achieved while working under a hypothesis that mutations have the potential to be over-dominant. This remarkable outcome reinforced the possibility of identification of loci for heterosis. With the availability of genomic resources, genomic and genome editing tools powerful platform to identify underlying causes of heterosis would be available. These advances are likely to provide ways to efficiently exploit heterosis in vegetables and agricultural crops in general.

10. Strategic Use of Dominant Genes to Develop Multiple Resistant Hybrids

Hybrids offer advantages in utilizing the genes conferring resistance to diseases and the pests as the F_1 hybrids manifest the resistance immediately. Fortunately, resistances against many diseases in various vegetables are dominant in nature. This permits development of hybrid cultivars resistant to a disease when only one of the parents is resistant. It also permits development of hybrids with multiple disease resistance when a line with dominant resistance to one disease is crossed with another line dominant for resistance to other disease. Considering the importance of biotic stresses, there is a need to develop a variety which would be resistant to major diseases and pests. The major biotic stresses of different vegetables are described in Table 9.

Table 9: Major Biotic Stresses in different Vegetable Crops

Vegetable		Major Biotic Stress
Tomato	:	Tomato leaf curl virus, early blight, late blight, *Fusarium* wilt, *Phytopthora* root rot, bacterial wilt, root not nematode
Brinjal	:	Phomopsis, bacterial wilt, little leaf, fruit and shoot borers
Chilli and Capsicum	:	Leaf curl, anthracnose, *Phytopthora* root rot, powdery mildew, thrips, mites, bacterial leaf spot
Okra	:	Yellow vein mosaic virus, enation leaf curl virus, jassid, mites, fruit borer
Cucumber	:	Downy mildew, powdery mildew, *Fusarium* wilt, mosaic virus, sucking pest
Muskmelon	:	Powdery mildew, downy mildew, anthracnose, Fusarium wilt, sucking pests
Watermelon	:	Powdery mildew, downy mildew, anthracnose, bud necrosis virus, sucking pests
Cabbage	:	Black rot, *Sclerotinia* rot, *Alternaria* leaf spot, Diamondback moth
Cauliflower	:	Black rot, *Sclerotinia* rot, *Alternaria* leaf spot, Diamondback moth
Radish	:	*Alternaria* leaf spot, aphids
Onion	:	*Stemphyllium*, purple blotch, thrips

Several dominant gene controlled traits that can be potentially be utilized in tomato hybrids include resistance to *Fusarium* wilt, *Verticillium* wilt, root-knot nematode, tobacco mosaic virus, tomato spotted wilt and tomato leaf curl virus disease (*Ty-2*). Nonetheless the traits controlled by recessive genes or partially dominant genes can also exploited if these genes are introgressed into both seed parent and pollen parent. Single dominant gene has been reported to confer resistance in cucumber, water melons and melons for various diseases caused by fungi and viruses. Dominant genes have also been reported for resistance/tolerance to insects in squash, melon, and bottle gourd.

11. Hybrid Production under Protected Conditions

Vegetable cultivation faces various constraints in the production of disease free, healthy and genetically pure seed for commercial cultivation. Seed production under open fields often faces problems such as lack of proper isolation, insects-pests, diseases and virus free environment. Compared to open field practice, hybrid seed production under protected cultivation can deliver higher seed yield along with better quality. Insects and viruses are the most devastating problems for quality seed production in most of the vegetable crops in open fields, and if the insect vectors are checked by using protected structures this problem could be managed quite well. This concept has shown significant results not only in quality seed production but also in increasing the per unit seed yield in different vegetables. The protected cultivation models has proved highly economical and profitable with limited resources and man power to provide best quality seeds of numerous vegetables cultivated in India and across the globe (Singh and Tomar *et al.*, 2015).

In parts of the country especially in Northern plains, although the soils are very fertile but extremes of temperature ranging from 0-48°C do not allow year round outdoor vegetable cultivation. Similarly, many other parts of the country biotic stresses do not allow successful production of vegetables like tomato, chilli, okra, cauliflower *etc.* in the fields. In trans-Himalayan cold-arid desert conditions during winter months, the temperature is extremely low (-5 to -30°C) and it is very difficult to grow any vegetables in such climate under open field conditions (Mishra *et al.*, 2010). In many European countries, USA, Japan, China, Israel, Morocco, Turkey *etc.*, where climate limits the choices for year round outdoor production, vegetables are being produced under protected structures. Greenhouses being the most efficient means to overcome climatic diversity, it is being used for vegetable production especially hybrid vegetables. Tomato hybrids when grown under both open-field conditions and under protected-conditions, gave significantly higher yield under protected conditions under cold-arid desert conditions of Ladakh (Singh *et al.*, 2009b). India is also entering into the greenhouse vegetables cultivation and total area under protected vegetable production is nearly 10,000 hectares. India being a vast country with diverse and extreme agro-climatic conditions, the protected vegetables cultivation technology can be effectively and economically used for year round and off-season production of high-value, low-volume vegetables, virus-free quality seedlings and quality hybrid-seed production (Wani *et al.*, 2011).

12. Approaches for Realization of Yield Potential of Hybrids

The potential yield of hybrids of vegetables could only be achieved if the appropriate seedling raising/ crop production practices including recommended level of inputs at proper time is applied. For harnessing maximum yield from the hybrids the following practices needs to be attended:

- ✰ Protect tomato, chilli and capsicum seedlings from whitefly to manage insect transmitted viruses. Nursery bed covering with cheese-cloth or fine nylon nets is advocated.
- ✰ Protect all seedlings in nursery by damping-off. Solarization of nursery bed should be done in month of May-June every year.
- ✰ Tomato Mosaic Virus in tomato can be prevented by seed treatment with 10 per cent trisodium phosphate solution for 45 min. followed by rinsing in running water and drying.
- ✰ Raise cucurbits seedlings in low cost poly tunnels to advance early fruiting.
- ✰ Root treatment of seedlings of with imidacloprid (3 ml/10 l of water) before transplanting prevents attack of sucking pests up to 40 days.
- ✰ Careful uprooting of seedlings with maximum number of intact roots helps in rapid establishment of plants after transplanting.
- ✰ Raised bed method of transplanting and placement of fertilizers in root zone should be adopted to minimize the input requirements and get the maximum productivity,
- ✰ Mulching should be used as it moderate soil temperature and conserve moisture. During winter season, polythene sheets (transparent and black), paper bags and straw mulch and during summer and rainy seasons straw mulch have been found effective.
- ✰ Use of plant growth regulator (PGR) for earliness, quality and yield in cucurbits. Similarly, spray of PCPA (50 ppm) and boric acid (5 ppm) improves the fruit set under low and high temperatures in tomato and reduce fruit drop of chilli.
- ✰ Staking of indeterminate tomato and cucurbits is essentially required for good quality fruit yield in rainy season.

Pruning and training for quality and yield is required for tomato, sweet pepper and cucurbits. Pruning in indeterminate tomato (single stem), cucumber (single stem allowed to grow with 2-3 fruiting branches) and muskmelon (lateral branches removed up to 6-9 nodes leaving 2-3 fruiting branches) should be followed.

13. Future Thrusts

After the introduction of new seed policy in1989, there has been a spurt of activity in the Indian vegetable seed scenario with the hybrid acreage soaring and a number of vegetable seed companies coming into existence as the private sector got a huge boost in growth with the removal of restrictions and liberalization of seed imports through open general license and removing tariff barriers. This move resulted in import of hybrid seeds by private seed companies in cabbage, cauliflower, capsicum *etc.* besides large quantities of seeds of carrot and beetroot. In the recent past, there is an increased interest on heterosis and efforts are underway in different crops to understand this key phenomenon.

In general, there are insignificant inbreeding depression and heterosis in cucurbits except few specific combinations where improvement in earliness and quality has been noted. Hybrid breeding in cucurbits has been in place mostly on account of strategic deployment of dominant genes conferring resistance to diseases, insect-pests and a few other key characters related to sex expression and fruit development and for product designing as per demand of consumers and market. Hybrid development in cucurbits provides an opportunity for PPP (public-private-partnership) mode of project implementation where public funded institutions can devote their resources and manpower to develop desirable inbreds/parental lines and develop hybrids as per demand of consumers/market and once these are found to be acceptable by the seed industry, the seed industry can enter into some kind of mutually beneficial agreement for seed production and marketing of the hybrids using their vast resources and net-work in this segment. This arrangement shall provide leveraging of synergy between the two and open-up vast possibility of meaningful interaction between the two organizations which at the moment appear to be working in isolation. The hybrid breeding programme shall have to travel in reverse gear from market/consumer feedback to the product designing and value-addition by R&D team, which is happening in the private sector but perhaps not to the desired level in the public sector. Deployment of dominant genes conditioning resistance to disease and pests needs to be implemented on war footing to reduce cost on extensive use of fungicides and insecticides and ultimately to reduce the cost of cultivation and to minimize the toxic residues on the fresh vegetables.

F_1 hybrids have almost entirely replaced open-pollinated varieties of tomato, cabbage, cucumber, squash, melons, onion and sweet pepper in Japan, the Netherlands, Denmark, France, Canada, Australia, UK and USA. Though India has also made a revolutionary progress particularly in case of cabbage, cauliflower, tomato, cucumber, bottle gourd and bitter gourd and accelerating its pace in the direction, still the path is long to move neck on with global vegetable giants. The future research strategies for the exploitation of heterosis in various vegetables include, diversification of sterile cytoplasm, identification of stable sterile and SI lines across environments, identification of linked markers with restorers, understanding the role of metabolic inhibitors, intensifying efforts on TGMS and PGMS, targeting mitochondrial genes using genetic engineering to introduce MS and understanding the reasons for expression of male sterility and SI. In those vegetables where no MS of SI system naturally exist, mutagenesis can be used to get the success. The future requisite is in harnessing heterosis in vegetables is stable and economic hybrid seed production along with incorporation of biotic- and abiotic-stress tolerance and nutraceutical properties.

14. References

Ahmad, S., Quamruzzaman, A.K.M. and Islam, M.R. 2011. Estimate of heterosis in tomato (*Solanum lycopersicum* L.). *Bangladesh J. Agril. Res.*, **36**(3): 521–527.

Ansari, A.M. 2015. Molecular markers in vegetable improvement. *Horticultural Biotechnol. Res.* (1): 5–10. doi: 10.5455/jhcbt.2015–07–05.

Atanassova, B. and Georgiev, Hr. 1986. Investigation of tomato male sterile lines in relation to hybrid seed production. *Acta Hort.*, **190**: 553–557.

Banga, O., Petiet, J. and van Bennekom, J.L. 1964. Genetical analysis of male sterility in carrots *Daucus carota* L. *Euphytica*, **19**: 263–269.

Bateman, A.J. 1955. Self-incompatibility systems in angiosperms. III. Cruciferae. *Heredity*, **9**: 52–68.

Das, M., Das, D., Mukherjee, A. and Nandy, M. 2014. Growing interest in doubled haploidy for improvement of horticultural crops in India. *Curr. Sci.*, **107**(1): 16–17.

Dey, S.S., Sharma, S.R., Bhatia, R., Parkash, C. and Barwal, R.N. 2011. Superior Ogura based CMS lines with better combining ability improve yield and earliness in cauliflower (*Brassica oleracea* var. *botrytis* L.). *Euphytica*, **182**: 187–197.

Dey, S.S., Singh, N., Bhatia, R., Parkash, C. and Chandel, C. 2014. Genetic combining ability and heterosis for important vitamins and antioxidant pigments in cauliflower (*Brassica oleracea* var. *botrytis* L.). *Euphytica*, **195**: 169–181.

Frost, H.B. 1923. Heterosis and dominance of size factors in *Raphanus*. *Genetics*, **8**: 116–153.

Hazra, P., Singh, M. and Kumar, S. 2005. *Vegetable Hybrid Research and Development at Indian Public Sector*. Souvenir, 23 Annual Group Meeting, AICRP Vegetable Crops, BCKV, Mohanpur, pp. 97–106.

He, Q.W., Shi, H. and Liu, E. 1981. The development of 77–01A CMS in Chinese radish. *Shangdong Agril. Sci.*, **1**: 13–16.

Hedrick, U.P., and Booth, N.O. 1907. Mendelian characters in tomato. *Proc. Amer. Soc. Hort. Sci.*, **5**: 19–24.

Ikegaya, Y. 1986. Frequent appearance of cytoplasmic male sterile plants in a radish cultivar Kosena. *Japanese J. Breed.*, **36**(Suppl 2): 106–107.

Jagan, K., Reddy, R.K., Sujatha, M., Sravanthi, V. and Reddy, M.S. 2013.Heterosis for yield and yield components in okra (*Abelmoschus esculentus* L.). *IOSR J. Pharmacy Biol. Sci.*, **7**(4): 69–70.

Jha, A. 2015.Genetic diversity in bottle gourd (*Lagenaria siceraria* Mol.) (Standl.) inbred using molecular markers and their relation with heterosis. *Ph.D. Thesis*, MGCGV, Chitrakoot, Satna.

Jiafu, Y., Libing, R. and Honghui, G. 2012. Analysis of heterosis for agronomic traits in cauliflower at different environments. *Acta Agr. Zhejiangensis*, **24**(3): 415–420.

Jones, H.A. 1932. Vegetable breeding at the University of California. *Proc. Amer. Soc. Hort. Sci.*, **29**: 572–581.

Kalia, P. 2009. Hybrid seeds making available at affordable price. In: *Recent Initiatives in Horticulture*, (Eds.) Chadha *et al.*, Westville Publishing house, New Delhi, pp. 219–237.

Kamble, C., Mulge, M.B., Madalageri, M.B. and Jadeesha, R.C. 2009. Studies on heterosis in capsicum (*Capsicum annuum* L.) for yield and yield traits. *Karnataka J. Agril. Sci.*, **22**(1): 155–157.

Kamenetzky, L., Asís, R., Bassi, S., de Godoy, F., Bermúdez, L., Fernie, A. R., Van Sluys, M.A., Vrebalov, J., Giovannoni, J.J., Rossi, M. and Carrari, F. 2010. Genomic analysis of wild tomato introgressions determining metabolism- and yield-associated traits. *Plant Physiol.*, **152**: 1772–1786.

Kandasamy, R. 2015. Heterosis in bitter gourd (*Momordica charantia* L.). *Asian J. Hort.*, **10**(1): 158–160.

Kibar, B., Karaaðaç, O. and Kar, H. 2015. Heterosis for yield contributing head traits in cabbage (*Brassica oleracea* var. *capitata*). *Cien. Inv. Agr.*, 42(2): 205–216.

Krieger, U., Lippman, Z.B. and Zamir, D. 2010. The flowering gene single flower truss drives heterosis for yield in tomato. *Nat. Genet.*, 42: 459–463.

Kucera, V., Chytilova, V., Vyvadilova, M. and Klima, M. 2006. Hybrid breeding of cauliflower using self-incompatibility and cytoplasmic male sterility. *Hort. Sci. (Prague)*, 33(4): 148–152.

Kumar, S, Singh, P.K. 2005. Mechanisms for hybrid development in vegetables. *J. New Seeds*, DOI: 10.1300/J153v06n04_05.

Kumar, S., Singh, V., Singh, M., Rai, S.K., Kumar, S., Rai, M. and Kalloo, G. 2007. Genetics and distribution of fertility restoration associated RAPD markers in pepper (*Capsicum annuum* L.). *Scientia Hort.*, **111**: 197–202.

Kumaran, N.M. 1971. Studies on combining ability, gene effects and heterosis in cauliflower. *Ph.D. Thesis*, PG School, IARI, New Delhi.

Larson, R.E. and Paur, S. 1948. The description and the inheritance of a functional sterile flower mutant in tomato and its probable value in hybrid tomato seed production. *J. Amer. Soc. Hort. Sci.*, pp. 355–364.

Lee, Y.P., Park, S., Lim, C., Kim, H., Lim, H., Ahn, Y.S., Sung, S.K., Yoon, M. K. and Kim, S. 2008. Discovery of a novel cytoplasmic male-sterility and its restorer lines in radish (*Raphanus sativus* L.). *Theor. Appl. Genet.*, **117**: 905–913.

Liu, L., Liu, G., Gong, Y., Dai, W., Wang, Y., Yu, F. and Ren, Y. 2007. Evaluation of genetic purity of F_1 hybrid seeds in cabbage with RAPD, ISSR, SRAP, and SSR markers. *Hort. Sci.* 42(3): 724–727.

Mishra, G.P., Singh, N., Kumar, H. and Singh, S.B. 2010. Protected cultivation for food and nutritional security at Ladakh (Review article). *Def. Sci. J.*, **61**(2): 219–225.

Mongkolporn, O., Dokmaihom, Y., Kanchana-Udomkan, C. and Pakdeevaraporn, P. 2004. Genetic purity test of F_1 hybrid Capsicum using molecular analysis. *J. Hort. Sci. Biotech.*, **79**(3): 449–451.

More, T.A. and Budgujar, C.D. 2002. Isolation of parthenocarpic tropical gynoecious lines in cucumber (*Cucumis sativus* L.). *Acta Hort.*, **588**: 255–260.

Morelock, T.E. 1974. Influence of protoplasm source expression of male sterility in carrot. *Ph.D. Thesis*, Wisconsin University, USA.

Nahm, S.H., Lee, H.J., Lee, S.W., Joo, G.Y., Harn, C.H., Yang, S.G. and Min, B.W. 2005. Development of a molecular marker specific to a novel CMS line in radish (*Raphanus sativus* L.). *Theor. Appl. Genet.*, **111**: 1191–1200.

Nothnagel, T. 1992. Results in the development of alloplasmic carrots (*Daucus carota* var. *sativus* Hoffm.). *Plant Breed.*, **109**: 67–74.

Nur-E-Afrin, Jahan. 2010. Heterosis for yield related traits in radish, *M.Sc. Thesis*, Bangbandhu Sheikh Mujibur Rahman Agricultural University, pp. 63.

Ogura, H. 1968. Studies on the new male-sterility in Japanese radish, with special reference to the utilization of this sterility towards the practical raising of hybrid seeds. *Memoirs of the Faculty of Agriculture, Kagoshima University*, **6**: 36–78.

Omran, S.A., Ramadan, W.A.E. and Mostafa, Y.A.M. 2012. Heterosis and combining ability in watermelon hybrids. *J. Plant Production*, **3**(12): 3139–3148.

Pandey, S., Jha, A., Kumar, S. and Rai, M. 2010. Genetics and heterosis of quality and yield of pumpkin. *Indian J. Hort.*, **67**(3): 333–338.

Patel, J.P., Singh, U., Kashyap, S.P., Tiwari, S.K., Singh, D.K. and Singh, M. 2013. Heterosis for productivity traits in brinjal (*Solanum melongena* L.). *Veg. Sci.*, **40**(1): 80–82.

Patel, M.P., Patel, A.R., Patel, J.B. and Patel, J.A. 2010. Heterosis for green fruit yield and its components in chilli (*Capsicum annuum* var. *longicum* (D.G.) Sendt) over environments. *Electron. J. Plant Breed.*, **1**(6): 1443–1453.

Pearson, O.H. 1932. Breeding plants of the cabbage group. *California Agriculture Experiment Station Bulletin*, **532**: 3–22.

Peterson, C.E. and Anhder, L.D. 1960. Induction of staminate flowers on gynoecious cucumbers with gibberellic acid. *Science*, **131**: 1673–1674.

Ram, H.H. 2005. *Vegetable Breeding: Principles and Practices*. Kalyani Publishers, Ludhiana, 514 p.

Reddy, M.A., Sridevi, O., Salimath, P.M., Nadaf, H.L., Patil, M.S. and Hosamani, R.M. 2013b. Heterosis for yield and yield components in okra. *Int. J. Adv. Res.*, **1**(8): 287–302.

Reddy, M.T., Haribabu, K., Ganesh, M. and Begum, H. 2013a. Exploitation of heterosis in Okra (*Abelmoschus esculentus* (L.) Moench). *Int. J. Agric. Food Res.*, **2**(4): 25–40.

Rick, C.M. 1945. Field identification of genetically male sterile tomato plants to use in producing F_1 tomato seeds. *Proc. Amer. Soc. Hort. Sci.*, **46**: 277–283.

Rick, C.M. and Robinson, J. 1951. Inherited defects of floral structure affecting fruitfulness in *Lycopersicon esculentum*. *Amer. J. Bot.* **38**: 639–652.

Sadashiva, A.T., Aghora, T.S., Rddy, M.K., Mohan, N. and Rao, E.S. 2014. Growing interest indouble haploidy for improvement of horticultural crops in India. *Curr. Sci.*, **107**: 16–17.

Santhy, V., Khadi, B.M., Singh, P., Vijaya Kumari, P.R., Deshmukh, R.K. and Vishwanathan A. 2008. Hybrid seed production in cotton. In: *CICR Technical Bulletin No.* **35**.

Sawhney, V.K. 1983. Temperature control of male sterility in a tomato mutant. *J. Heredity*, **74**: 51–54.

Sharma, J P. 2011. Indian vegetable seed Industry: Potential and problems. In: *Quality Seed Production of Vegetable Crops Technological Interventions, Vol. 1: General Aspects*, (Ed.) Sharma, J.P., Kalyani Publishers, pp. 60–78.

Sharma, S.R., Singh, P.K., Chable, V. and Tripathy, S.K. 2004. A review in hybrid cauliflower development. In: *Hybrid Vegetable Development*, (Eds.) Singh, P.K., Dasgupta, S.K. and Tripathi, S.K. The Haworth Press, New York, pp. 217–221.

Sharma, Y. and Sharma, S.N. 2005. Chemical Hybridizing Agents (cha): A tool for hybrid seed production – A review. *Agric. Rev.*, **26**(2): 114– 23.

Singh, B. and Tomar, B.S. 2015. Vegetable seed production under protected and open field conditions in India: A review. *Indian J. Agr. Sci.*, **85**(10): 1251–1259.

Singh, B.K., Sharma, S.R. and Singh, B. 2009b. Heterosis for antioxidants and horticultural traits in single cross hybrids of cabbage (*Brassica oleracea* var. *capitata* L). *Indian J. Agr. Sci.*, 79(9): 703–708.

Singh, R., Mishra, G.P., Ahmed, Z. and Singh, S.B. 2009a. Effect of various greenhouse structures on yield and quality of different table tomato varieties under cold arid desert of trans-Himalayas. *Indian J. Agril. Sci.*, **79**(4): 243–47.

Singh, S.K., Singh, S.V. and Srivastava, J.P. 2016. Studies on heterosis and inbreeding depression in cucumber. *Agriways*, **3**(2): 107–111.

Stein, M. and Nothnagel, T. 1995. Some remarks on carrot breeding (*Daucus carota* var. *sativus* Hoffm.). *Plant Breed.*, **114**: 1–11.

Stout, A.B. 1920. Further experimental studies on self-incompatibility in hermaphroditic plants. *J. Genetics*, **9**: 85–129.

Sudha, M., Gajanana, T.M. and Murthy, D.S. 2006. Economic impact of commercial hybrid seed production in vegetables on farm income and farm welfare: A case of tomato and okra in Karnataka. *Agri. Econ. Res. Rev.*, **19**: 251–268.

Tay, D. 2002. Vegetable hybrid seed production. In: *Proceedings of the International Seed Seminar: Trade, Production and Technology*, (Eds.) McDonald, M. and Contreras, S. Pontificia Universidad Católica de Chile, Facultad de Agronomía e IngenieríaForestal, Departamentode Ciencias Vegetales, Santiago, 15–16 October, pp. 128–139.

Wang, Z.W., Gao, L., Liu, H.Z., Mei, S.Y., Zhou, Y., Xiang, C.P. and Wang, T. 2012. Genetic and cytological analysis of a new spontaneous male sterility in radish (*Raphanus sativus* L.). *Euphytica*, **186**: 313–320.

Wani, K.P., Singh, P.K., Amin, A., Mushtaq, F. and Dar, Z.A. 2011. Protected cultivation of tomato, capsicum and cucumber under Kashmir valley conditions. *Asian J. Sci. Technol.*, **1**(4): 056–061.

Watts, L.E. 1965. The inheritance of curding periods in early summer and autumn cauliflower. *Euphytica*, **14**: 83–90.

Wehner, T.C. 1989. Breeding for improved yield in cucumber. *Plant Breeding Review*, **6**: 323–359.

Wehner, T.C. 1999. Heterosis in vegetable crops. **In**: *Genetics and Exploitation of Heterosis in Vegetable Crops*, (Eds.) J. C. Coors and S. Pandey. Amer. Soc. Agron., Crop Sci. Soc. Amer., Soil Sci. Soc. Amer., Wisconsin, pp. 387–397.

Wehner, T.C. and Miller, C.H. 1985. Effect of gynoecious expression on yield and earliness of a fresh-market cucumber hybrid. *J. Amer. Soc. Hort. Sci.*, **110**: 464–466.

Xu, Z.M., Zhang, E.H. and Cheng, Y.A. 2004. A new cabbage F_1 hybrid 'Qingan-70'. *China Veg.*, **1**: 20–21.

30

Utilizing Indigenous Flowers for Livelihood Security of Farmers

M. Jawaharlal

Horticultural College and Research Institute,
Tamil Nadu Agricultural University, Coimbatore – 620 009, Tamil Nadu
E-mail: deanhort@tnau.ac.in

Use of flowers in the social life of India has a long tradition. The origin of flower and ornamental plant growing is contemporary with agricultural crops. In the beginning their use was exclusively for esthetic and religious purposes. References regarding the use of flowers during ancient time exist in the old Hindu classics such as *Veda*, *Puran* and *Ramayana*. Mention of the lotus was made by Kalidasa in his play Shakuntala. The poet Asvaghosa (A.D.100) also mentions the lotus in his *Buddha Charita*. Many paintings and carvings on the walls of old structures provide strong evidences that flowers were an integral part of daily life and adornment. Motivation for cultivating flowers arose due to their extensive use for religious purposes, particularly for worship. A large number of festivals and rituals are celebrated, using flowers in various ways. Such usages along with the steady increase of population virtually have created huge domestic demand for these so that floriculture has become an increasingly important industry in India.

India's Diversity of Native Flowers and Ornamental Crops

India is bestowed with a wide variety of agroclimatic conditions suitable for growing all kinds of flowers and ornamental plants. High rise hills to vast coastal areas together with the fertile Indo-Gangetic plains provide a unique combination of climate and soil, the envy of many countries. India has been designated as one of the twelve mega-diversity states of the world. The flora of the Indian sub-continent is very rich in diversity with many endemic species, largely due to its varied agro-climatic conditions. Ten bio-geographic zones, namely Trans-Himalayan, Himalayan, North-east India, Indian Desert, Semiarid Zone, Indo-Gangetic Plain, Western Ghats, Deccan Peninsula, Coastal Zone and Islands, depict diversity of climate and flora. Each zone has its own unique and characteristic natural vegetation. Some of the zones like the North East (Assam, Arunachal Pradesh, Meghalaya, Sikkim) harbour indigenous economically important ornamentals such as orchids, zingibers and *Nepenthes*. Due to congenial agro-climatic conditions, an enormous opportunity exists throughout India for cultivation of various flowers and ornamental plants on a commercial scale.

Among the flower crops which are under cultivation in different parts of the world, those that are native of India include orchids, rhododendrons, musk rose (*Rosa moschata*), begonia (*Begonia* spp.), balsam (*Impatiens balsamina*), globe amaranth (*Gomphrena globosa*), gloriosa lily (*Gloriosa superba*), foxtail lily (*Eremerus himalicus*), primula (*Primuladenticulata rosea*), blue poppy (*Meconopsis*), lotus (*Nelumbo nucifera*), water lily (*Nymphae* spp.), clematis (*Clematis montana*) and the wild tulip of the Himalayas (*Tulipa stellata* and *T. aitchisonii*).

Among the native shrubs and climbers, the most important ones are the jasmines (*Jasminum* spp.). Jasmines are native to tropical and warm temperate regions of Europe, Asia and Africa. The centres of diversity of jasmine are South Asia and Southeast Asia. India is one of the centers of origin of jasmine. Of the 200 species of jasmine reported, 40 inhabit the Indian sub-continent (Veluswamy *et al.*, 1975). The other indigenous species which are under commercial cultivation are *Ixora* spp. (*I. coccinea, I. parviflora, I. barbata, I. undulata*), *Crossandra infundibuliformis, Glorosia superba* and *Tabernaemontana coronaria.* The other native shrubs popularly used in gardening are *Bauhinia acuminata, Mussaenda frondosa, Hamiltonia sauveolens, Holmskioldia sanguinea, Clerodendron inerme, Plumbago rosea, Plumbago zeylancia, Trachelospermum fragrans, Osmanthus fragrans, Passiflora leschenaulti, Clitoria ternatea, Porana paniculata,* and *Clematis montana.*

Among the ornamental trees which are native to India, the important ones include *Saraca indica, Terminalia arjuna, Mesua ferrea, Ficus benghalensis, F. religiosa, F. infectora, Michelia champaka, Butea monosperma, Butea frondosa, Anthocephalus indicus, Bauhinia variegata, Cassia fistula, Cassia nodosa, Erythrina blakei, Lagerstroemia flos-reginae, L. thorelli, Cordia sebestena, Cochlospermum gossypium, Pongamia glabra, Thespesia populnea, Crataeva roxburghii, Sterculia colorata* and *Rhododendron* spp.

Many flowers including primula, orchids, aconitum, delphinium, erigeron, anemone, aquilegia, aster, bergenia, campanula, corydalis, gentiana, geum, saxifraga, allium, fritillaria, lilium, iris, paeonia, clematis, prunus, rhododendrons, sorbus, viburnum and several others were introduced from their wild habitats in India into England and other European countries. Some of these, like the Blue poppy (*Meconopsis*), *Clematis montana,* many species of orchids, rhododendrons, primula, balsam, begonia, foxtail lily (*Eremurus himalaicus*), glorios a lily (*Gloriosa superba,* musk rose (*Rosa moschata*) etc, are now widely grown in gardens in several parts of the world. Several species of orchids and rhododendrons, which are native of India, have been extensively used in breeding new varieties and hybrids. Most of the plants species introduced from India into Great Britain are being maintained in the Kew and Edinburgh Botanic Gardens.

Avenues in Utilizing Indigenous Flowers

The various avenues existing in the floriculture based on India's indigenous flower and ornamental crops for securing farmers' livelihood include cultivation of loose flowers and cut flowers, ornamental pot plant culture, value-addition with fresh and dry flowers, establishment of ornamental plant nurseries for landscaping and the rapidly expanding export potentials.

Cultivation of Loose Flowers

The most popular native flower crops cultivated commercially in different parts of the country as loose flowers are jasmine (*J. sambac, J. grandiflorum, J. auriculatum, J. pubescens*), crossandra (*Crossandra infundibuliformis*), star flower (*Tabernaemontana coronaria*), Ixora (*Ixora coccinea, I. parviflora, I. barbata, I. undulata*) and globe amaranth (*Gomphrena globosa*). Among the native aquatic ornamental plants, lotus (*Nelumbo nucifera*) and water lily (*Nymphae* spp.) are the most popular both as garden plants as well as loose flowers. The lotus has the added merit of being the National Flower of India and it symbolizes divinity, fertility, wealth, knowledge and enlightenment.

In India, Tamil Nadu is the leading producer of jasmine in the country and the flowers produced in the state are being exported to the neighbouring countries, *viz.,* Singapore, Malaysia, Sri Lanka, Dubai, Bahrain and Muscat and also to the United States. Musk rose (*Rosa moschata*) is cultivated for its loose flower and for value-added products such as essential oil, attar, *etc. Crossandra infundibuliformis* is a popular plant grown for its loose flowers, which have no perfume but stay fresh for several days on the bush. It is also grown as a houseplant in containers and also in beds

Cultivation of Cut Flowers

The northeastern states (Assam, Arunachal Pradesh, Meghalaya and Sikkim) are the natural habitat of hundreds of orchid types. Many of them are cultivated for cut flowers and potted plants. Private entrepreneurs have started large scale micropropagation mainly for export of plantlets. *Cattleya, Cymbidium, Dendrobium, Phalaenopsis* and *Vanda* are the commercially important genera marketed in India.

Plants for Gardening

Clematis, begonia (*Begonia* spp.), balsam (*Impatiens balsamina*), foxtail lily (*Eremerus himalicus*), primula (*Primula denticulata rosea*), blue poppy (*Meconopsis*), *etc.* are some of the native ornamentals used in gardening and for pot plants. *Tabernaemontana coronaria* is a popular evergreen ornamental plant and also has importance in temple garden. It flowers throughout the year. *Clerodendron inerme is* commonly planted in the gardens for making hedges. It is also well adapted to the topiary art for cutting and making various designs. The juice of the leaves has much reputation to bring down fever.

Saraca indica is considered as a historically important tree since it is associated with ancient literature. It produces attractive flowers. Many of the ornamental trees possess medicinal value apart from their aesthetic values. *Terminalia arjuna* is an excellent medicine for heart disease. *Ficus benghalensis* is used in indigenous system of medicines namely Ayurveda, Siddha, Unani and Homeopathy. *Cassia fistula* which produces attractive yellow flowers is a popular tree for gardening. The flowers are of ritual importance in the *Vishu* festival of Kerala.

Value-addition with Fresh and Dry Flowers

Value-added products from flowers including

floral ornaments, garlands, essential oils, dry flowers and export packaging of loose flowers such as jasmine, crosandra, gomphrena have proved to be potential activities for women employment. Many of India's native ornamental species have valuable medicinal importance. *Gloriosa superba* is a herbaceous ornamental climbing perennial plant. Its rhizomes, seeds and leaves possess several medicinal values. India's wide diversity of indigenous ornamental plant species are a rich source of raw materials for the dry flower industry.

Potential Unutilized Flowers

India is also native to a wide array of ornamental plant species which are considered 'wild' since they remain unutilized. Many of these have potential as ornamentals for use as loose flower, cut flower, pot plant as well as medicinal uses. If strategies are formulated to utilize them optimally, keeping in view the essential norms of biodiversity conservation, these potential unutilized species can yield great economic benefits to the farming community. Some of the potential unutilized ornamental plant species are listed in Table 1 with brief details.

Success Stories in India

Floriculture has proved to be a reliable livelihood securing venture for many farming communities in India. A few of the success stories that have taken place in the recent past are quoted here to emphasize this fact.

Experiences under the ICAR-NAIP project 'Value chain on flowers for domestic and export markets' implemented at TNAU during 2008-2014 have indicated that floriculture ventures are playing a key role in the livelihood of the farming communities of Tamil Nadu involved in cultivation of jasmine, marigold, carnation and those entrepreneurs associated with the dry flower industry (Jawaharlal *et al.*, 2014). Similarly, an analysis of the cultivation of jasmine flower in Tamil Nadu indicated that it generated impressive returns to the farmers and good employment opportunities for farm family as well as agricultural labourers especially for female workers (Prawin Arya, 2015). Over 500 tribal families in Thane district of Maharashtra are now engaged in floriculture since it has proved to be a highly remunerative farming venture besides, providing them livelihood security.

An economic analysis of commercial flower cultivation in Sirmaur district of Himachal Pradesh has clearly indicated that flower cultivation was a profitable venture for the farmers of the region (Sharma, 2014).

Women Empowerment

Floriculture is a highly woman-friendly venture. Floriculture farming and value-addition activities have proved to be a reliable source of employment and income generation to womenfolk. Value-addition in flowers is a predominantly woman-friendly sector of the floriculture industry, since it involves more of creativity and artistic skills. This area offers tremendous scope for women empowerment through women entrepreneurship promotion. Owing to the potential of value-added floral products in generating higher levels of income, employment opportunities, greater scope for women empowerment and export promotion, value-addition in floriculture is growing rapidly as a profitable and successful industry.

Table 1: Potential Unutilized Ornamental Plant Species

Unutilized *Jasminum* spp.	*Jasminum nitidum* - It is a species with year-round flowering potential. The flowers buds resemble those of *J. grandiflorum* and are attractive and mildly fragrant. This species can yield flowers during off-season (Sep to Feb) when the commercial jasmine species don not produce flowers. The plant architecture is attractive making it an ideal ornamental plant for gardening also.
Rosa brunanii (Himalayan musk rose)	Flowers are fragrant and ornamental. Ideal for preparation of value-added products namely, attar and marmalade. A soothing cough syrup is made out of the hips which have a high vitamin Ccontent.
Hedychium spicatum (Sweet snow)	A perfume prepared from its roots is used for perfuming tobacco. The roots are used medicinally as a carminative and stimulant. The leaves are woven into mats.
Echinops cornigerus (Globe thistle)	The flower heads are almost perfect spheres and made up of small closely packed blue flowers. They are quite outstanding in flower beds.
Delphinium denudatum (Wild delphinium)	The flowers are attractive and are produced in shades of pink, white and blue.
Viola canescens (Dog violet)	It produces attractive purple fragrant fowers. It has medicinal value and a very effective cough mixture is made from it. In the Ayurvedic and Unani systems of medicine, violet flowers are used for treating several diseases. Homeopathicremedies for skin, eye and ear troubles also use the violet plant.
Vinca minor (Blue periwinkle)	It is an attractive garden plant producing purple-blue flowers. It is used in herbal medicines to treat diarrhea and dysentery
Commelina forkalii	This is an unusually pretty blue wild flower. The petals are abeautiful sky-blue and have an orchid-like appearance. It is used in herbal medicine for remedies for inflammation of the skin.
Gentiana cochemirica (Kashmir gentian)	It is an ornamental garden plant. It is used in herbal medicine.
Tulipa stellata (Wild Indian tulip)	It is an ornamental plant.

Future Prospects

India's share in global floriculture trade may not be significant but the country has, of late, shown enough potential to eventually turn itself as a favourite destination of flower importers in near future. Surprisingly, the small land-holding pattern, considered a handicap for the country's agricultural production, comes as an advantage in floriculture due to its 'low volume high value nature. Since the sector has huge export potential, a number of small and marginal farmers have started turning towards flower production.

The country is bestowed with ideal temperature conditions for commercial floriculture throughout the year in some or other part. This has helped entrepreneurs and growers in recognizing diversification into floriculture as of a commercial value. Even non-arable land can be put to agricultural use since protected floriculture involves specific growing media or amendment of the existing soil,

References

Arya, Prawin, 2015. An economic analysis of jasmine cultivation in Tamil Nadu. OnlineAuthors@gmail.com.

Jawaharlal, M., Ganga, M. and Karthikeyan, S. 2014. Value chain on flowers: Production to consumption system in Jasmine, Marigold, Carnation and Dry flowers. Sreedevi Lithographics, Coimbatore. ISBN No. 7881923306553.

Sharma, Manoj Kumar, 2014. Economic analysis of commercial flower cultivation in Sirmaur district of Himachal Pradesh. *Ph.D. Thesis*, CSK Himachal Pradesh Krishi Vishvavidyalaya, Palampur.

31

Cultural Practices and Disease Management in Mushroom for Improving Profitability

V.P. Sharma*, Satish Kumar and Shwet Kamal

ICAR-Directorate of Mushroom Research,
Solan – 173 213, Himachal Pradesh
**E-mail: director.mushroom@icar.gov.in*

Mushrooms have been cultivated since ancient times for their nutritional value and flavour especially in the far eastern countries. Mushrooms are considered as a potential substitute of muscle protein on account of their (i) high digestibility (Digestibility coefficient around 89 per cent), (ii) good amino acid content and (iii) about 1,000 times higher production of mushroom protein per unit area. They have low fat content, high fibre and all essential amino acids and with the exception of iron, contain all important minerals too. On exposure

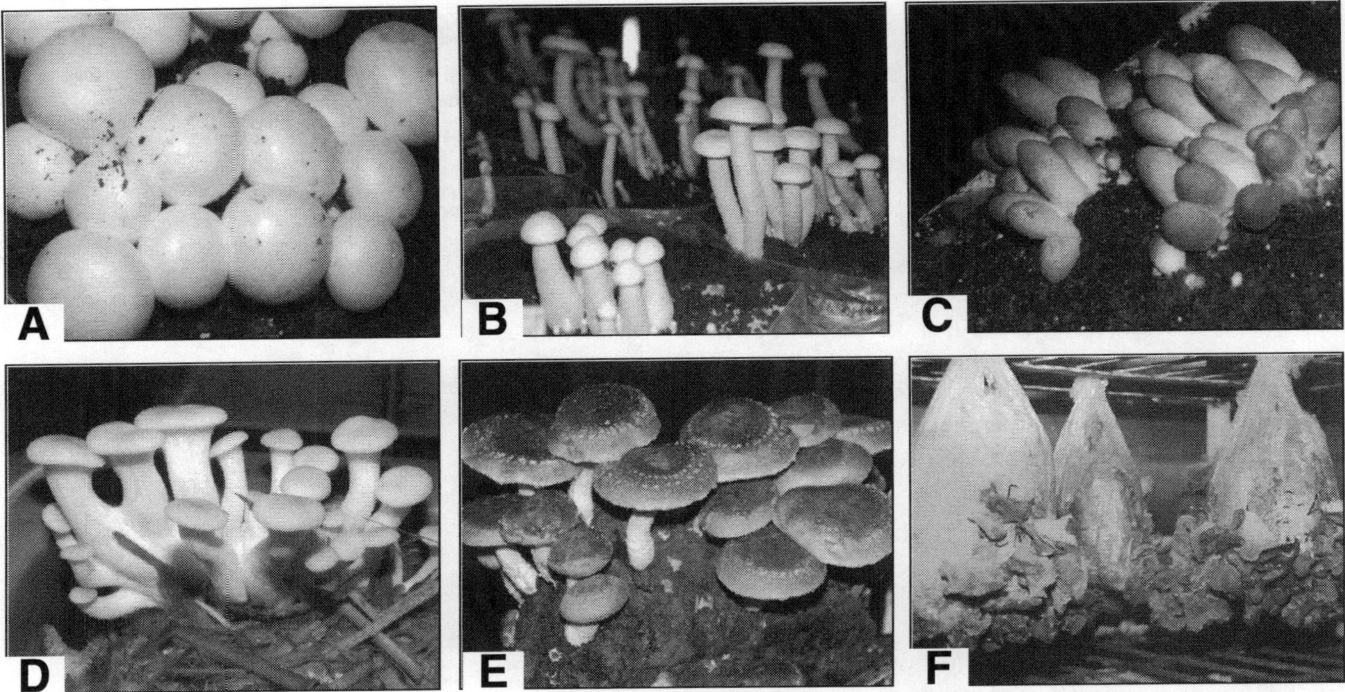

Figure 1: Commonly Cultivated Mushrooms in India.

A: *Agaricus bisporus*; B: *Calocybe indica*; C: *Volvariella volvacea*; D: *Pleurotus* spp.; E: *Lentinula edodes*; F: *Auricularia* spp.

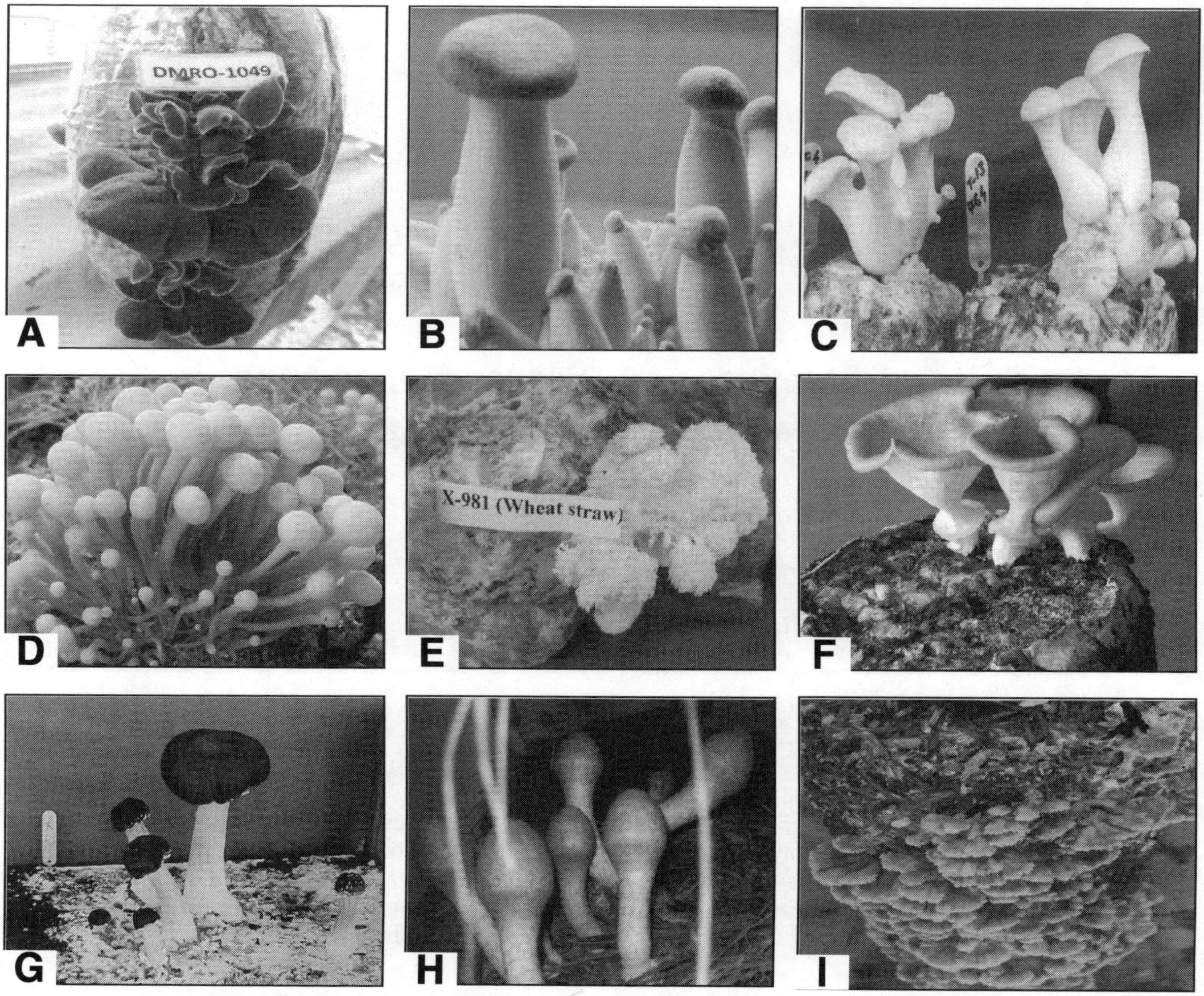

Figure 2: Some of the Newly Cultivated Mushroom at ICAR-DMR, Solan.

A: *Auricularia polytricha*; B: *Pleurotus eringii*; C: *Pleurotus fossulatus*; D: *Flammulina velutipes*; E: *Hericium erinaceus*; F: *Lentinus sajor-caju*; G: *Stropharia rugosa-annulata*; H: *Volvariella bombycina*; I: *Schizophyllum commune*.

to UV-light, mushrooms also produce large amounts of vitamin D, which is normally difficult to obtain from a regular diet intake. Among the possible direct medicinal value of mushrooms, the most important ones are anti-cancer, hypolipidemic, hypocholesterolemic, and anti-hypertensive. There is tremendous potential and appeal for growing a highly nutritious food with excellent taste from substrates that are plentiful and not very expensive. Further, it is very environmental friendly, capable of converting the lignocellulosic waste materials into food, feed and fertilizers.

The major efforts to promote mushroom cultivation took off in the early 1980s in India. The ICAR-Directorate of Mushroom Research, Solan (established as National Centre for Mushroom Research and Training in 1983 and renamed as National Research Centre on Mushroom in 1997) with its efforts in research and training has contributed in making considerable progress in

Figure 3: Artificial Cultivation *Macrocybe gigantea* in Trays.

mushroom cultivation in the country. The technological interventions and approaches in mushroom in the recent past together hold the potential to increase the production as well as productivity in different mushrooms.

Diversification is essential for sustainability of any farming system. Mushroom component in a farming system augurs well to impart the diversification as it makes use of agro-residues of the farm and also recycle the spent mushroom compost after harvesting the mushrooms. Some of the most important refinements in technologies of mushroom production are discussed in the communication.

Continuous efforts to cultivate/domesticate many wild mushrooms are underway at the Directorate of Mushroom Research, Solan. Much progress has been achieved in standardizing the technologies by using different agro-residues to cultivate different mushrooms

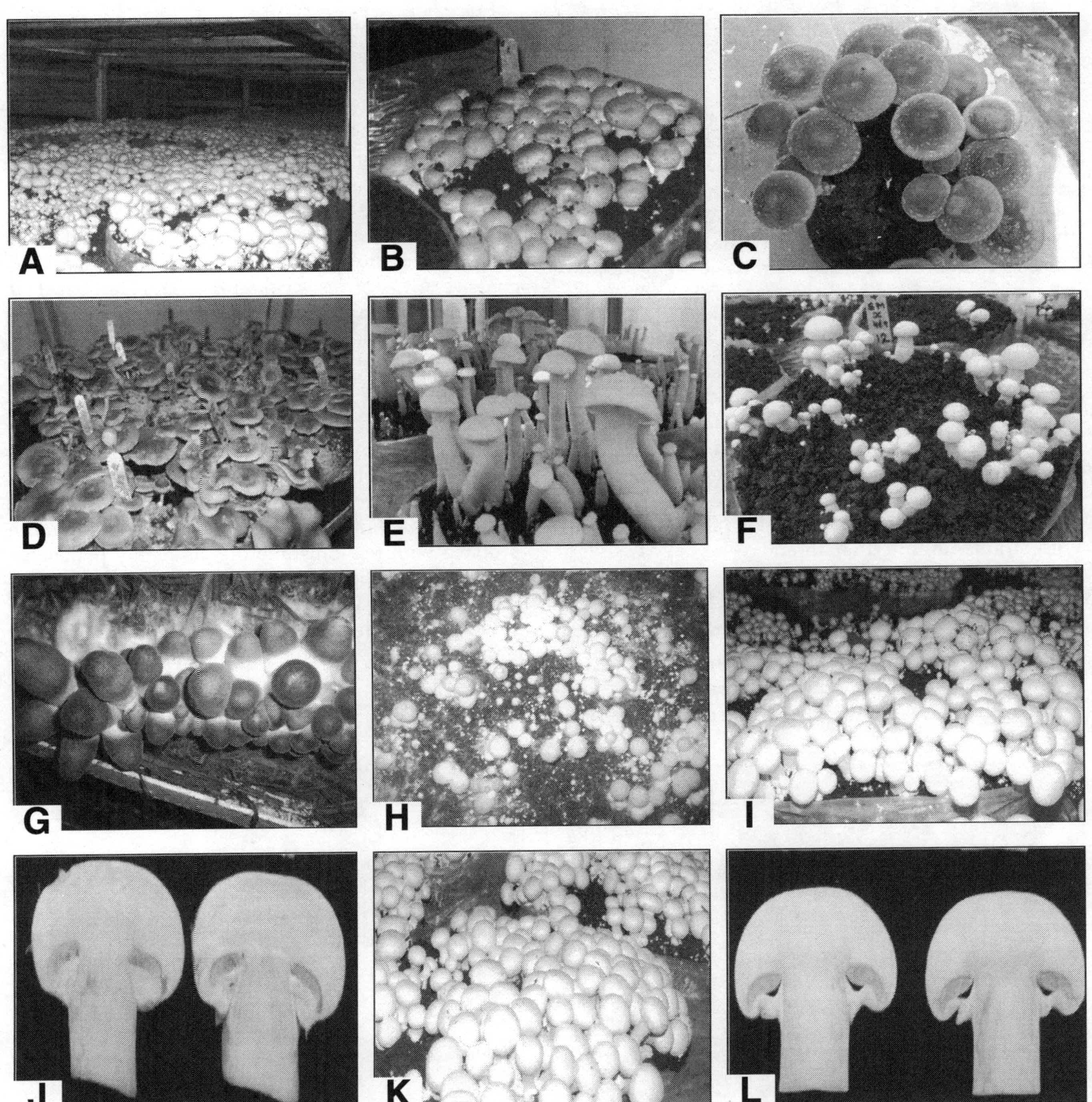

Figure 4: New Improved Variety of different Mushrooms Released by ICAR-DMR, Solan.

A: DMR-Button-03; B: DMR-Button-06; C: DMR-Shiitake-38; D: DMR-Shiitake-388; E: DMR-Milky-334; F: DMR-Macrocybe-01; G: DMRO-247; H: DMRO-484; I: NBS-1; J: Longitudinal section of NBS-1 showing no browning; K: NBS-5; L: Longitudinal section of NBS-5 showing no browning.

such as *Flammulina velutipes, Auricularia polytricha, Auricularia rosea, Hericium coralloides, Lentinus sajor-caju, Lentinus squarrosulus, Macroletpiota procera, Pleurotus eryngii* (King Oyster), *Volvariella bombycina, Macrocybe gigantea, etc.*

Cultivation technology of a new mushroom *Macrocybe gigantea* has been standardized. This new tropical edible mushroom is rich in protein, vit-D, potassium, iron and zinc. This mushroom resembling white button mushroom and milky mushroom makes its marketing easy. Further, the pungency associated with milky mushroom is not found in this mushroom and its suitability for tropical climate can be a helpful for growing in most parts of India.

For enhancing productivity as well as quality, it is of paramount importance to combine characters from different sources by breeding new and improved strains. In the recent past ICAR-DMR has developed and released eight new varieties of button (2), paddy straw (2), shiitake (2), milky (1) and Macrocybe (1) mushrooms. More recently, ICAR-DMR, Solan has developed and released two new browning resistant hybrids; NBS-1 and NBS-5 in *Agaricus bisporus*, which have shown promising results in research trials as well as farmers' unit on commercial scale. The two hybrids besides possessing browning resistance, gave higher yields compared to other varieties and hybrids under testing. NBS-1 and NBS-5 yielded 23 and 25 per cent yield, respectively at M/s Tirupati Balaji Mushroom at Baramati, Pune while it was 22 and 23 per cent respectively at M/s Weikfield Mushrooms, Pune. The yield level of NBS-5 was at par with strains imported (A-15 and S-465) at M/s Flex Foods, Dehradun.

The agriculture residues are either composted or pasteurized or autoclaved to prepare suitable medium to grow different mushrooms. The compost prepared by short method or total indoor composting method to grow *Agaricus* mushroom have proven to be superior over long method of composting in terms of productivity, environment safety and resource use efficiency. Around 20 per cent reduction in the water requirement in the composting process through the advanced composting technology envisages the sustainable practice of "more crop per drop".

The compost prepared in the pasteurization tunnels with air circulation by active pressure though yields better than the long method of composting, the cost involved in machinery and the infrastructure is beyond the reach of resource poor farmers. As an alternative, ICAR-DMR has evolved a novel approach where in the high density polyethylene (HDPE) pipes with holes at regular intervals are inserted through the compost pile. It was demonstrated that passive aeration effectively controlled the composting process, shortened the compost period, enhance the substrate quality and *Agaricus* yield as of pile composting. The parallel arrangement (10 per cent perforations) of pipes was found best among different passive aeration treatments for achieving the quality substrate in just 16 day and enhance the *A. bisporus* yield up to 27.6 per cent. This new intervention also opens the possibility to cultivate *A. bisporus* on a ligno-cellulosic, non-pasteurized, non-conditioned, aerated substrate. The shortened composting period from 30 to 16 days saves considerable amount of energy, labour and cost on infrastructure.

Paddy straw mushroom is one of the most cultivated mushrooms in Indian coastal areas of Odisha, Andra Pradesh, *etc.* The temperature required for the mushroom is normally 35-40°C with high humidity. The problem with this mushroom is its highly perishable nature. ICAR-DMR has selected a new variety of this mushroom *V. bombaciana*, which can grow at a lower temperature and the keeping qualities are very good. Cultivation technology of this new variety is standardized for commercial production. Further, in the coastal area paddy straw mushroom are cultivated

Figure 5: Design of Zero-Energy Polytunnel for Substrate Preparation for White Button Mushroom.

Figure 6: Indoor Cultivation Method of Paddy Straw Mushroom.

outdoors below the shade of the trees with low productivity of 8-10 kg per 100 kg straw. ICAR-DMR has come up with an indoor technology for the mushroom, where the production level can be as high as 4-45 kg mushroom per 100 kg composted substrate.

Shiitake is world second most cultivated mushrooms and has high nutritive and medicinal attributes. Traditionally shiitake mushroom has been grown on natural logs of various trees species with yields up to 33 per cent. This traditional method takes 8-18 months. Another shiitake cultivation technology is synthetic log production system where sawdust is used to prepare synthetic logs. This method takes around 80-90 days time with biological efficiency is 75 to 100 per cent. Recently a new strain DMRO-Shiitake-338S has been developed, which along with some modified package of practice, can fruit within a period of 42-45 days with an average biological efficiency of 125-130 per cent in a total cropping period of four months. Cultivation was carried out on sawdust of selected trees (100 kg), wheat bran (20 kg), $CaCO_3$ (1 kg) and $CaSO_4$ (0.5 kg). Mixed substrate was filled in Poly Propylene bags (1.2-1.5 kg wet/bag). Bags sterilized in an autoclave at 22 psi for 90 min. Grain spawn was inoculated @ 3 per cent (wet wt basis, top spawning) and bags incubated at 4 h/20 h light/dark cycles at 23-25°C. Spawn run took 36-40 days including bump formation and browning. Cold water treatment by immersing spawn run synthetic logs in cold water (6-8°C) for 10-15 min. after removing PP bag. A room temperature of 20-22°C, RH (85-90 per cent) and light for 10-12 h were maintained. After 2-3 days of the cold water treatment small primordia developed which mature into full grown fruit bodies in next 3-4 days.

The control of mycopathogens/competitor moulds in mushroom cultivation usage very limited chemical substances, cultural protection practices and sanitation precautions. The lack of registered chemicals is a problem in the practice. Integrated disease management has been used in mushroom cultivation but there is a need for novel selective fungicides and cheap reliable disinfectants. Among the several factors, occurrence of pests and diseases significantly affect the mushroom production and also the interest of growers in this enterprise. Many growers cultivate mushroom under natural climatic conditions and thereafter abandon it after 2-3 crops because of pests and diseases problems. Moreover, mushroom growing is a kind of mono-cropping, which provides an easy access to food for insect-pests and medium for disease development. Due to short duration of the crop, deleterious effect of pesticides and residue problems limit the scope of liberal pesticide application in mushrooms. ICAR-DMR has tried to identify cultural practices and natural products, which can be used effectively in mushroom

Figure 7: Short Duration Cultivation Technology for Shiitake Mushroom.

cultivation against fungal/bacterial diseases/competitor moulds as preventive and/or curative methods. The trials on cultural practices conducted showed that use of unpasteurized substrate/compost for raising crop is one of the major reasons of inoculums build up in the area of mushroom farm. Further, lower moisture contents of casing soil for cultivation of *A. bisporus* at the time of pasteurization favours the survival of pathogens and ultimately the disease development. In case of wet bubble, the most dreaded disease of white button mushroom, pre spawning of casing soil 20 days prior to pasteurization with grain spawn, liquid spawn of button mushroom, spawn run compost and grinded mushrooms resulted in reduced incidence of wet bubble disease. Maximum reduction was recorded with spawn run compost, followed by grain spawn. Liquid spawn was least effective. Out of five fungicides and two other chemicals tried, chlorothalonil and carbendazim proved equally effective in managing wet bubble disease among all the fungicides. Different plant extract *viz.*, *Dhatura strumonium*, *Tagetus erecta*, *Nyctanthes arortristis*, *Parthenium* spp, *Callistomon lanceolatus*, *Cannabis sativa*, *Gadenia* spp., *Thooja campacta*, *Ricinus cummunis* and *Eucalyptus* spp. were evaluated (both dry, hot water and alcoholic extract) against *Cladobotryum* and *Verticillium*. The maximum inhibition in mycelium growth was recorded in *Cannabis sativa* (32.02 per cent) and the least inhibition was recorded in case of *Parthenium* spp. (16.52 per cent) in all the extracts. Further research is necessary to determine correct application and concentrations of different botanicals and fungicides to reduce the incidence of diseases and pests in mushroom.

The technologies and strategies discussed above are only few among the many potential ones in mushroom cultivation. No doubts, these refined technologies will help farmers to increase the production and productivity of the mushroom.

32

Production of Seed Spices through Hi-Tech Interventions for Enhancing Farmers Income

Gopal Lal, R.S. Mehta and Ravindra Singh*

National Research Centre on Seed Spice,
Tabiji, Ajmer – 305 206, Rajasthan
**E-mail: glal67@yahoo.co*

Seed spices are annual herbs, whose dried seeds or fruits are used as spices. They are nature's gift to humankind as they add flavour to our food. In addition, they also have preservative and medicinal values. India is consistent source of seed spices for importing countries world-wide. There has been ever increasing demand of seed spices and importing countries look at India for quality produce of seed spices. Out of the 109 spices listed by the International Organization for Standardization (ISO), India produces as many as 63 of them at diverse agro-climatic regions of the country. Almost all the States and Union Territories (UTs) of the country grow one or the other spice. It is a source of livelihood and employment for large number of people in the country, both for rural population, who grows them, and the urban population, who process and trade them. Out of the total 63 spices grown in India, 20 are classified as seed spices. A sizable area lies under 10 important seed spices contributing nearly 41.70 per cent share in area and 19.72 per cent share in production of total spices in the country. India is the largest producer, consumer and exporter of seed spices and their products. It produces a wide variety of seed spices like coriander, cumin, fennel, fenugreek, dill, nigella, *ajwain*, anise, celery, caraway, parsley, black cumin, *etc.* The trend of seed spices with respect to area, production and productivity is given in Table 1.

Table 1: Trend of Area, Production and Productivity of Seed Spices in India (Area in lakh ha, production in lakh metric tonne and productivity q/ha)

Crop	*2004-05*			*2007-08*			*2015-16 (Est)*		
	AR	*PR*	*PRTY*	*AR*	*PR*	*PRTY*	*AR*	*PR*	*PRTY*
Coriander	3.90	2.96	7.47	4.50	2.86	6.25	5.85	5.57	9.50
Cumin	3.68	1.76	4.79	4.80	1.64	5.54	7.01	3.72	5.30
Fennel	0.40	0.51	12.81	0.54	1.31	15.58	0.47	0.78	16.60
Fenugreek	0.50	0.63	12.45	0.55	0.67	12.31	1.24	1.34	10.80
Ajwain	0.15	0.095	8.19	0.20	0.11	5.72	0.24	0.17	7.10
Celery	0.012	0.046	38.24	0.032	0.042	13.42	0.041	0.055	13.40
Total	8.85	6.10	6.89	11.14	7.84	7.04	14.85	11.63	7.83

Note: AR: Area; PR: Production; PRTY: Productivity.

Agriculture has evolved across the world with time, from traditional farming system now we have moved to precision farming system, *i.e.*, technology package for precise application of inputs for harvesting high yield with better quality and also safe for human and environmental. In the present era of science and technology, innovation and intervention of precision faming technologies have shown high impact in agriculture and more specifically in horticulture crops. Focusing the horticultural crops, numerous hi-tech and precision farming techniques have been developed looking to the need and demand. The basic concept of precision fanning is to minimize the inputs with high efficacy. Seed spices are the golden wealth of the arid and semi-arid regions of the country, among the group cumin and coriander are the leading crops, followed by fennel, fenugreek, *ajwain*, dill, nigella, celery, caraway and anise. Most of these crops are belonging to the Apiaceae family; the seed size is also small with low vigour. They all are grown in *rabi* season,

Export of Seed Spices

India exports raw spices as well as value-added products to nearly 135 countries in the world meeting 50 per cent of the global demand. The share of seed spices export to total spices export is only 30.22 per cent in terms of quantity and 18.90 per cent in terms of value. Spices exports have registered substantial growth during the last five years, registering an annual average growth rate of 21 per cent in value and 8 per cent in volume. In 2014-15. the export of seed spices from India is estimated to be 2,70,150 tonnes valued Rs. 2,815.64 crores. The share of different seed spices in export is given in Figure 1. The leading countries which import Indian seed spices are given in Table 2.

However, in recent years area and production of seed spices is increasing but there is no perceptible increase in productivity of seed spices. Presently productivity of seed spices is only 7.83 q/ha which need to be enhanced 15.20 q/ha by the year 2050 for meeting the international and domestic demand. The efforts in different directions are made for enhancing production and productivity of seed spices. Hi-tech interventions played remarkable role in enhancing productivity of fruits, vegetables and flowers. The recent research done at ICAR-NRCSS, SAUs and other organisations on various hi-tech interventions in seed spices have proved that yield of seed spices can be increased remarkably over existing methods of cultivation. Precision farming models fitting to seed spice cultivation needs to be intervened and popularized for maximum utilization of available resources to produce more with better quality. The various hi-tech interventions are described in this chapter.

Table 2: Leading Countries which Import Indian Seed Spices

Crop	*Major Importing Countries with Decreasing Order of % Share*
Coriander	Pakistan, Malaysia, U.A.E., Saudi Arabia, U.K., South Africa, Nepal, Singapore, Egypt (A.R.E.), Sri Lanka, U.S.A., Kuwait, Oman, Baharain
Cumin	U.A.E., Brazil, U.S.A., Nepal, U.K., Malaysia, Pakistan, Peru, Ecuador, Y.A.R., Mexico, Indonesia, South Africa, Columbia, Japan, Australia, Canada, Venezuela
Celery	U.S.A., South Africa, Netherland, Canada, Germany, Japan, U.K., Australia
Fennel	Malaysia, U.S.A., U.A.E., Saudi Arabia, U.K., Pakistan, Nepal, South Africa, Canada, Singapore, Sri Lanka, Egypt (A.R.E.), Australia, Japan, Morocco
Fenugreek	Y.A.R., U.A.E., Egypt (A.R.E.), Japan, Saudi Arabia, South Africa, U.S.A., Nepal, U.K., Malaysia, Bangladesh, Netherlands, Israel, France, Iran, Canada, Sudan.
Other seed spices	Nepal, Sri Lanka, U.A.E., Saudi Arabia, U.K., U.S.A., Malaysia, Canada, A.R.E.

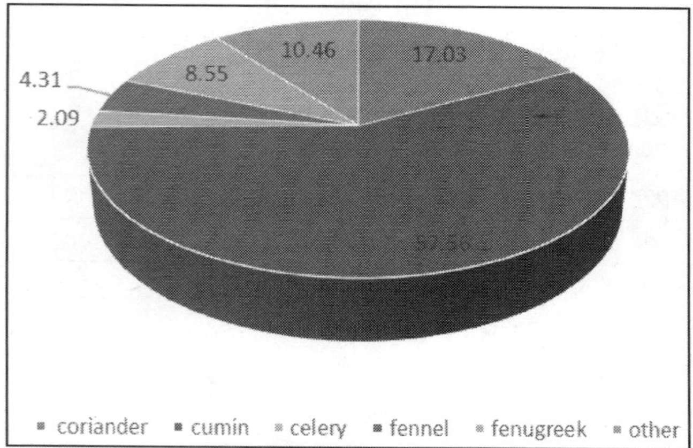

% Share in Quantity of Seed Spices in Total Seed Spice Export in 2014-15.

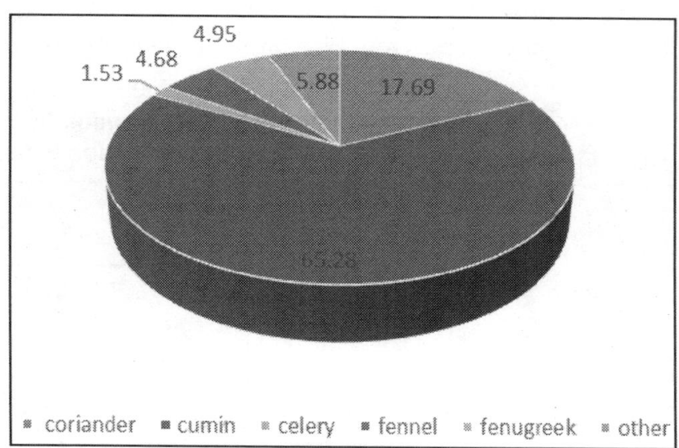

% Share in Value of each Seed Spices in Total Seed Spice Export in 2014-15.

Figure 1: The Share of different Seed Spices in Export in Quantity and Value Term.

Soil Solarization

Soil health is having importance in agriculture. Sanitation of fields is very much required periodically. Seed spices are cultivated in the western part of country where the intensity of sun light is very high, in summers the temperature goes beyond 45°C. The solar power can be converted into a lethal agent for sanitizing the fields under see spices. Soil solarization with the help of transparent plastic sheet of 50-100μ thickness spread over moist fied during peak summer months over a period of 3-4 weeks will raise the temperature underneath the soil surface by 10-12°C than the ambient. This will kill the entire living microorganism in the soil and will also desiccate the weed seeds making a significant reduction in the viable weed seed bank. This is a very effective method to control soil born fungus and nematodes.

Seed Priming and Pelleting

In the Apiaceae crops, there is an inherent problem of delayed germination and low vigour. The seed size is also small that does not permit mechanize sowing with precise seed rate, sowing depth and spacing per unit area. The traditional practice of bio-priming in crops like cumin, coriander *etc.*, is prevalent but the advanced techniques of seed priming and pelleting is having a better option over the traditional one. Priming advances the germination duration and pelleting increases the size of the seed. Which permits mechanized sowing with modified seed drills? This not only provides better crop geometry per unit area but simultaneously reduces the seed rate required in unit area with uniform germination.

Pneumatic Seeding

Traditional practice of sowing seeds of seed spices like cumin is by broadcasting, available seed drills can't manage the depth and seed rate with high precision. Now a days pneumatic seed drills are available that can be used effectively in crops like cumin, coriander, *ajwain*, dill, celery, nigella *etc.* In this era of civilization man power availability is also becoming a big problems in rural areas due to high migration of population in urban or peri-urban zones. Hence, mechanization is the future to deal with these challenges and to increase the seed input efficiency by précising the rate and depth of sowing.

Drip Fertigation

Water is going to be the most limiting factor in coming times, and these crops are cultivated in the arid and semi-arid pockets of the country where availability of water is very less. The technique of drip irrigation has advanced upon by fertigation methods *i.e.*, application of fertilizers with water as per need of the plant. The method of fertigation has to be standardized with respect to each crop for large scale application. Sensor based technology is available now a days, to deliver the commanded dose of fertilizer and water. These technologies are commercially available with pressurized drip system. Other than pressurized drip system, low pressure drip is also having immense scope. A simple water tank of 1,000 l capacity elevated to a height of 1.5 to 2.0 m can irrigate an area of nearly 1,000 sq m. Low pressure drip is an effective technology not only for hilly areas, but even for arid and semi-arid plains where poor or marginal farmers having low cultivable lands with limited water resources. Since seed spice crops are prone to air borne diseases which generally infect host plants due to high humidity and temperature, drip irrigation reduces the humidity occurring over the canopy of the crop compared with that what is there by flood and sprinkler irrigation methods. This will reduce the chance of disease incidence and in-turn will reduce the fungicide spray load over the farmers and crops. The seed spices are *rabi* season crops, they can be grown only when assured supply of irrigation is available, hence water management is the key issue to sustain these crops in the adapted pocket of the country, therefore, the drip irrigation systems is highly suitable for seed spice cultivation.

Raised Bed Technology

The traditional package of cultivation was not based on efficient land use management and other resources as the prices of agricultural commodities are not fixed taking into account each and every input from land to water as followed in deciding the MRP of finished goods produced under industries. Land degradation has been a major event that has happened in past due to excessive surface irrigation in flat bed system. Ridge and furrow method of cultivation has come up as a scientific method for proper growth and development of plants as it offers a pulverized loose zone for root growth in the raised beds prepared by mechanized implements available in market. Seed spices can be effectively grown over raised beds, it not only increase the water use efficiency but also provides aerated environment for proper root growth thus reducing chances of infection by soil borne pathogens. Further this technology is also helpful for reduction of fungal and bacterial diseases when coupled with drip fertigation and mulching.

Plug Tray Seedling Production

In seed spices, the process of seed germination is very slow which may take several days to germinate after sowing. On the other hand, most of the seed spices are sown in dry fields by broadcasting followed by light surface irrigation. Low vigour and slow germination rate of these crops permit the weeds to proliferate faster in the initial 30 days, which creates tough plant-weed competition. Therefore, at initial stage of 50-60 days of sowing high labour input is required for weeding and hoeing. Moreover, for maintaining the soil moisture status the upper layer of sandy or sandy loam soil, a minimum of two to three irrigations are inevitable for enhancing germination and growth in the initial period of 30 days. This practice is common in seed spices growing areas of Rajasthan and Gujarat. In addition, some times when the farmers grows cluster bean during *kharif*, the crop is mostly harvested by the end of October to mid-November, whereas, the recommended optimum time for sowing of some seed spice crops like fennel, dill, *ajwain*, celery is by mid-September to mid October and therefore, it may not be possible for the farmer to take such crops in continuation to cluster bean. Looking to these problems, an effective solution to have an alternative technology rather than direct sowing of seeds after harvest of the *kharif* crop is nursery raising. The seedlings of crops, *viz*., fennel, dill, *ajwain* and celery were raised in soilless media (cocopeat) in plastic beds and also in plug trays and 50-60 days old seedlings were directly transplanted in the field. Both the systems of nursery raising showed 95-100 per cent success rate in seedling establishment. Therefore, plug tray nursery raising technology was found suitable for all these crops, *viz*., fennel, dill, *ajwain* and celery, whereas, the plastic made and cocopeat filled bed technology was only suitable for fennel, dill and celery except, *ajwain*. The basic advantages of this technology are:

- ✰ We can save 50-60 days as the seedlings can be directly transplanted to the field *i.e.*, it gives an advantage of 50-60 days over traditional method.
- ✰ Lot of manpower can be saved in terms of weeding and hoeing at initial stage during which the crop-weed competition is high.
- ✰ Crops like cluster bean which is harvested by the end of October to mid-November is also suitable for the cropping system following fennel, dill, *ajwain* or celery crops in *rabi*, as the advantage of 50 to 60 days permits the *kharif* crop to mature in field and simultaneously take care of the successive crop to be transplanted in *rabi*.
- ✰ It also offers savings of resources in terms of nutrients and irrigation water which is applied in very high amount during early stage to protect young seedlings from high temperature and other environmental stresses.

Plastic Mulching

The technique is not new among the farmers and been used primarily in vegetables and fruit crops. As compared to the normal seed sown crop, as discussed above plug tray or plastic beds filled with cocopeat or traditional nursery raising practice offers a control over the crop-weed competition in the first phase, *i.e.* for 30 days. To increase the crop potential, a complete control over the weed population from the very first day of crop establishment is highly essential. The technique of mulching is a fool-proof method to suppress the weed growth in the commercial crop sown area. Mulching layer also reduces the evaporation loss that occurs in open fields. Seed spices being grown in arid and semi-arid areas, *rabi* crop faces tremendous pressure of water management. The best combination of the mulching practice is with plug tray nursery raised seedling and drip irrigation for better resource management. Crops like celery are sown in the North Western plains of Punjab where the cold period and water availability status are high compared to that of the areas lying in the western arid and semi-arid zones of Rajasthan and Gujarat. It has been recorded by experimentation at ICAR-NRCSS, Ajmer that, celery grows well when sown as transplanted crop under mulching with drip fertigation. The yield levels realized in the experiment shown tremendous potential of the technology in introducing the celery crop as a future seed spice for few tracts of western hot/humid zone of the country. Similarly, this technology is also equally suitable for fennel, dill and *ajwain* in arid and semi-arid regions of the country.

Temporary Plastic Walls for Protection against Frost

The main problem in the cultivation of seed spices is their susceptibility to frost. Due to frost, significant damage has been observed almost every year and during severe conditions total loss of crop has also been recorded. Frost occurs mostly during end of the year or at start of the year, when cold period is at its peak (mid-December to mid-January) and the crop is at its flowering stage. The umbel stalks are very tender and they hardly survive the cold breeze causing cold injury and death of the flower buds resulting in heavy yield losses. To manage this condition, a barrier is needed to block the cold wave over the crop canopy. Rather, low cost technologies like raising of plastic walls of 1-1.5 m height depending upon crop height from the field surface in the northern direction efficiently blocks the flow of cold waves and thus reduce the loss caused by frost significantly. A 150-200 μm thick transparent plastic sheet is suitable for creating artificial temporary walls in the field. These walls can be laid parallel if the field is large leaving a distance of 10-15 m. A single wall in the north-south direction will reduce the chances of cold wave movement along the wall laid in north-west direction. The investment in raising the structure seems to be costly, but it is not, as the plastic is laid only for 20-25 days during peak winters and they are leaned along GI pipes or bamboo sticks which is durable and also can be reused for 3-4 years and in multiple activities like soil solarization *etc.* For crops like cumin and coriander, it can be advantageous, considering acreage of the crops and significant damage by frost is reported yearly in one or the other parts of the country. In addition, this can also be a boon in the vegetable crops, where the crops get suppressed and damaged due to low temperature or frost in several northern plains of India.

Transparent Plastic Covered Walk-in-Tunnels

Transparent plastic covered temporary walk-in-tunnels are also suitable and effective for seed spice cultivation. Plastic sheet of 150-200 µm thickness is laid in a semi-circle shape with 1/2 inch GI pipes fitted on the iron made plates fixed in the field at a distance of 2.5-3.0 m. The height of these walk-in-tunnels is seven feet at the center. This type of structure provides protection not only against frost but against unseasonal rains and hailstorms. Since plastic is used, the internal temperature rises and results in vernalization, as it induces early flowering compared to normal open field crop. A crop advancement of 20-25 days was observed for flowering as well as maturity. The investment cost is slightly higher but the inputs required are of long lasting nature. GI pipes can be used for 20-25 years and plastic is having a life span of 3-5 years depending upon the quality (if it is UV stabilized, it will last more). This technique is actually very low of cost and feasible for crops like coriander, *ajwain* and cumin, where heavy losses are reported due to abiotic factors. The additional advantage of these barriers are, it reduces the movement of insect pests and even wind-borne pathogens in the cropping zone, thus reducing the inputs required for insect and disease management.

Transparent Plastic Covered High-Tunnels

Transparent plastic covered temporary high-tunnels are also suitable and effective for seed spice cultivation limited to tall crops like fennel and dill. The height of the structure at the center from the ground is normally nine feet. This type of structure provides protection not only against frost but against unseasonal rains and hailstorms. It also helps in vernalization and other advantages are as those stated in transparent plastic covered walk-in-tunnel technology.

Insect Proof Net Covered Walk-in-Tunnels

The technique of insect proof net covered temporary walk-in-tunnels are as same as above, the only difference is that the basic covering material used is UV stabilized nylon net of 40 mesh. This provides a more open type micro-climate compared to plastic covered as the temperature difference between the outer and inner structure is not much due to free air movement through the holes of the net mesh. The basic advantage of the technology is prevention against hailstorms and insects. The size of the mesh is smaller than the size of hailstorm and insects, which does not permit the entry of these abiotic and biotic elements. It is an excellent technology for raising virus-/insect-free seedlings in vegetables. An intervention of this technique can limit the use of insecticides in these high value low volume crops facing tremendous pressure from western world of high pesticide toxicity in Indian produce. Crop like cumin, when it is to be grown in organic system, efficient pest management strategy is by implementing insect proof net covered walk-in-tunnel completely closed from all sides for complete restriction of aphids and other insects.

Shade Net Covered Walk-in-Tunnel Technology for Off-season Cultivation

The present technology is a very simple approach to make the crops to be raised during off-season for green leafy purpose. Seed spices, specifically coriander and fenugreek are grown during *rabi*, are also having importance for their green leaves. Coriander leaves are an integral part of Indian cooking system and are needed round the year for adding taste to the cuisines. Leaves being the most perishable part have a limitation of transportation from one corner to other. During the peak summer months (May to July) the availability of coriander leaves is highly influenced and there is a sharp rise in the rates per kg leaves sold during summer. The present technology of shade net covered walk-in-tunnel is an intervention found to be most suitable for the farmers of arid and semi-arid regions of the country where the summer temperature goes beyond 45 °C coupled with hot waves. Mostly 50-60 per cent shading intensity shade net covered walk-in-tunnel can be coupled with low pressure drip irrigation system to obtain good quality coriander leaves. During summer months, problem of early bolting occur in open field, even genotypic differences are noticed for bolting under open and shade net conditions. Presently, large areas are sown by temperate type coriander by farmers for leaves. However, this temperate type attains good vegetative growth providing huge biomass of leaves for sale, but does not possess the aroma as the temperature required for metabolism of flavouring biochemical constituents is not as per the temperate conditions in plains. These temperate type late bolters due to high leaf yield have gained popularity among farmers. But the other side of the coin the consumers are not happy, as the leaves of temperate types do not possess the desired aroma for adding the flavour we want. A study was conducted at ICAR-NRCSS, Ajmer to screen the genotypes for adaptability under shade net walk in tunnels for leaf yield and with high aroma. Huge variation was observed for both bolting time and aromatic quality. Compared to the temperate type of coriander Indian genotypes had high aroma. Variety ACr-1, which is a dual type bolted in 40-42 days during peak summer months, whereas, one genotype Ajmer Green Coriander-1 bolted in 52-55 days and had the high aroma in leaves. It was also found to be free from powdery mildew which occurs in the humid micro-microclimate of shade net equipped with drip. The said genotype AGCr-1 is under pipeline for release in the near future for off-season cultivation. The intervention of low pressure drip in the technology is very much feasible for the arid and semi-arid areas as during peak summer months water is the most limiting factor, a plastic tank of 1000 l is sufficient to irrigate an area of 1000 m^2. The technology is a boom for seed spice growers and other farmers to earn more from selling coriander leaves in peak summer months at high price.

Conclusion

The introduction of hi-tech interventions for seed spices cultivation is an important aspect for managing various factors of production for enhancing yield and productivity of seed spices. Most of the seed spice growers are small and marginal farmers and their income level is very low. By growing seed spices under hi-tech interventions one can have 25-30 per cent more yield of almost all the seed spices and 3-4 times more green leaves of coriander in off-season under shade nets consequently get the higher income.

Nutrient and Water Management

33

Soil Health Management for Raising Profitable Horticultural Crops

B.S. Dwivedi*, Mandira Barman and M.C. Meena

Division of Soil Science and Agricultural Chemistry,
ICAR-Indian Agricultural Research Institute, New Delhi – 110 012
****E-mail: bsdwivedi@yahoo.com***

Horticulture is a fast emerging vital sector in Indian agriculture, covering fruit crops, vegetables, flowers and other ornamentals, and aromatics.Area and production of horticultural crops have been consistently increasing over the years. In 2001-02, area and production were 10.4 mha and 132.3 MT, respectively, which was increased to 16.7 mha and 260.4 mha, respectively during 2014-15 (ICAR, 2016). Fruits and vegetable crops have a very significant role in human nutrition, and therefore, their production levels as well as quality need greater emphasis. These issues can be addressed through improvement in soil fertility, and adoption of judicious nutrient management options.

Soil, the most valuable natural resource, has been over exploited, misused and even abused, partially for meeting our need and more often due to our greed. Majority of Indian soils have low organic matter content, and are inherently poor in nutrient supply. Incidence and expansion of multi-nutrient deficiencies in soils owing to inadequate and unbalanced nutrient input through fertilizers is considered one of the major reasons for decline in productivity of crops. Indiscriminate use of fertilizers not only aggravates the extent of soil fertility depletion, but also proves ultimately harmful in terms of low nutrient use efficiency, poor quality of produce, enhanced susceptibility of crops to biotic and abiotic stresses, and potential threats of groundwater pollution and other environmental hazards. Extensive deficiencies of six nutrients *i.e.*, N, P, K, S, Zn and B have been recorded in soils, indicating that continued neglect of these nutrient-starved soils would render them completely unproductive. Unless all deficient nutrients are addressed in fertilizer prescriptions and adequate input of organic manures is ensured, it would be difficult to restore soil health and sustain high productivity of crops including horticultural crops (Dwivedi and Dwivedi, 2007).

Some Important Issues

Soil Fertility and Productivity

It is the inherent capacity of soil to supply plant nutrients in right amount and right proportion. Thus it is more related to nutrient availability to the plants. On the other hand, soil productivity is a broader term that indicates capacity of soil to produce biomass depending on soil fertility and other factors like soil texture, soil-borne diseases, waterlogging (drainage), soil depth, erodibility *etc.*

Soil Health and Quality

It could be defined as 'the capacity of soil to function within ecosystem boundaries, to sustain biological productivity, maintain environmental quality and promote plant and animal health' (Doran *et al.*, 1996). Many a times, the state of soil is also explained loosely in terms of soil health, soil productivity, soil environment *etc.* On the other hand, soil quality connotes soil's usefulness for a particular purpose over a long time scale (Goswami, 2006). Nonetheless, scientific literature as well as popular press often uses the terms 'soil health' and 'soil quality' interchangeably, with farmers preferring the former and scientists the later. As the World recognizes the vulnerability of earth's soil, interest has been increased to protect its health through different management practices to sustain the civilization.

Major Soil Ailments Affecting Horticultural Crops

Soils may have very low pH (acid soil) affecting horticultural crops as in humid, hilly regions or coastal areas, or high pH (alkali or sodic soil) as experienced in arid and semi-arid climatic conditions. Acid soils need to be limed in order to bring soil pH near neutrality. Saline soils found in arid and semi-arid climatic conditions have excess of soluble salts, and leaching (or washing out) the salts by repeated flooding with good quality watercan convert saline soils into normal soil. Alkali (or sodic) soils on the other hand, need to be amended with gypsum followed by repeated leaching with a known good quality water. It is noteworthy that only normal soils or reclaimed soils respond well to inputs for expected production. Saline soils pose the problem of poor absorption of nutrients by plant roots. In extreme conditions, soil solution becomes too concentrate and even seed germination may not take place, or young saplings may not survive. In fact, almost all nutrients may become limiting, though in mildly saline soils K, Ca, and Mg may not be deficient as saline soils have salts of these nutrients. With or without excess of salts, the alkali soils contain high proportion of Na and consequently low of K, Ca and Mg. These soils have pH sufficiently high to retard plant growth. Most horticultural crops badly suffer in such soils. The acid soils due to low pH have toxic levels of certain (metallic) nutrients. Only between a pH range 6.0 and 8.0, the nutrient availability remains at optimum. Obviously due to restricted plant growth and consequently limited residue recycling soil organic matter content in these soils is low.

Accumulation or build-up of metals in soil due irrigation with sewage or industrial effluents is another alarming problem. Toxic levels of heavy metals not only adversely affect production but also substantially reduce the quality of all crops in general and those of fruits and vegetables in particular. High concentration of metals such as lead, chromium, cadmium, nickel *etc.* ultimately contaminate the food chain. Such soils must be reclaimed before cultivation of edible crops. There are techniques available for this. For example, phytoremediation is a process under which specific plant species known to be hyper-accumulators of metals are grown. Besides phytoremediation, chemical remediation using lime is also very effective technique to reclaim metal-contaminated soils. Also, non-edible plantation species can be raised.

Water-logging is another serious problem in many areas. In such situation, poor aeration in soil is the obvious consequence, leading to adverse effect on root respiration. Most horticultural crops require well-drained soils. Sub-soil drainage may also be improved by mechanically breaking hard impervious layers, if any. Often this can be achieved while digging the pits for orchard plantation.

Soil Testing and Health Card

Soil testing as a means of judicious fertilizer use works on the principle of probability, meaning that if all other factors of productivity are at optimum, there is all probability to obtain profitable response to applied nutrients based on soil testing than to those applied on ad-hoc basis (Sharma and Meena 2012; Dwivedi and Meena, 2015). In soil testing, a part of total nutrient content which is accessible and present in the plant-utilizable form is determined. Thus, it is the index of availability of nutrients that serves as a guide for fertilizer use. Soil testing includes analysis of soils for detection of soil problems also along with available nutrient status. Major parameters included in soil testing are:

- ✰ *Problem-related parameters*: Salinity, acidity, alkalinity/sodicity
- ✰ *Nutrient availability*: Available N, P, K, S, Zn, Fe, Cu, Mn, B
- ✰ *Auxiliary tests*: Calcium carbonate, texture
- ✰ *Special tests*: Heavy metals, BOD, COD *etc.* (need-based)

Majority of soil testing laboratories (STLs) follow a 3-category soil fertility rating system (Table 1). These ratings proposed in 1950s have become redundant in several farming situations owing to changed nutrient uptake pattern of modern cultivars and also due to ever-increasing multi-nutrient deficiencies. At IARI and in some of the states, however, the fertility ratings are revised within broad framework of the earlier categories, and the revised ones are used for fertility evaluation.

Table 1: Soil Fertility Ratings Adopted in the STLs

Parameter	*Low*	*Medium*	*High*
Organic carbon (%)	<0.5	0.5-0.75	>0.75
Available nitrogen (N, kg/ha)	<280	280-560	>560
Available phosphorus (P, kg/ha)	<10	10-25	>25
Available potassium (K, kg/ha)	<110	110-280	>280

For secondary and micronutrients, critical limits are used to differentiate deficient soils from non-deficient ones. A deficiency critical limit is a value beyond which sufficiency of the nutrient in question sets-in, with lesser predictability of crop response to applied nutrients. It is thus apparent that the critical limits would vary according to soils and crops, and also with productivity levels. Nonetheless, on the basis of analysis of large number of soil samples as well as crop response studies, some critical levels are suggested for use in generalized fertilizer prescriptions and for soil fertility delineation (Table 2).

Table 2: Critical Limits of Secondary and Micronutrients

Nutrient	*Extractant*	*Critical Limit (mg/kg of soil)*
Sulphur (S)	0.15% $CaCl_2$	10
Zinc (Zn)	DTPA	0.6
Iron (Fe)	DTPA	4.5
Copper (Cu)	DTPA	0.2
Manganese (Mn)	DTPA	2.0
Boron (B)	Hot water	0.5

Soil Testing Service in India

With continued emphasis laid by the Government, the STL network involving static and mobile STLs expanded consistently (Figure 1). At present, a total of 1244 STLs (1048 static + 196 mobile STLs) are functioning in different states. Of these, 84 STLs (*i.e.*, about 7 per cent) belong to fertilizer industry and remaining STLs are owned mostly by the State Departments of Agriculture. The growth of STL network was relatively faster in the North and West zones. The annual analyzing capacity of the STLs also got increased over the years (Figure 2). Highest annual analyzing capacity of 6.72 million samples exists in South zone, followed by 5.44 million samples in North, 4.70 million samples in West, 0.65 million samples in East and 0.31 million samples in North-East zone. Out of the total installed capacity of 17.83 million samples, 13.52 million samples were analyzed during 2014-15, indicating an average capacity utilization of 75.8 per cent. Also, vast inter-zonal disparities exist in the capacity utilization.

In addition, soil testing facilities have been created at more than 300 *Krishi Vigyan Kendras* (KVKs). The SAUs and some of the ICAR institutes also offer soil testing service to the farmers on limited scale. In recent years, a few private entrepreneurs have established STLs. These STLs are neither listed in the STL network, nor are the samples analyzed therein included in the statistics.

Collection and Processing of Soil Samples

The biggest source of error in soil testing is a poorly-collected sample itself. The chemical analysis becomes meaningless if there is an error in collection and preparation of soil samples. Hence, the collection and preparation of soil samples should be done with utmost perfection. The steps involved are:

Collection of Soil Samples

The soil sample collected should be representative of the area sampled. A field can be treated as a single sampling unit if it is uniform in all respects. Variation in slope, texture, color, crop growth and management should be taken into account for soil sampling. Separate sets of composite samples need to be collected from areas differing in these characteristics. Recently fertilized plots, bunds, channels, marshy tracts, areas near trees, farm ways, buildings, wells, compost piles or other non-representative locations must be avoided during sampling. When crops are grown in rows, samples can be taken in between the rows. Soil samples should be taken in *zig zag* pattern or randomly.

Depth of Sampling

The plant root penetration is very important for deciding the sampling depth. The surface/plough layer

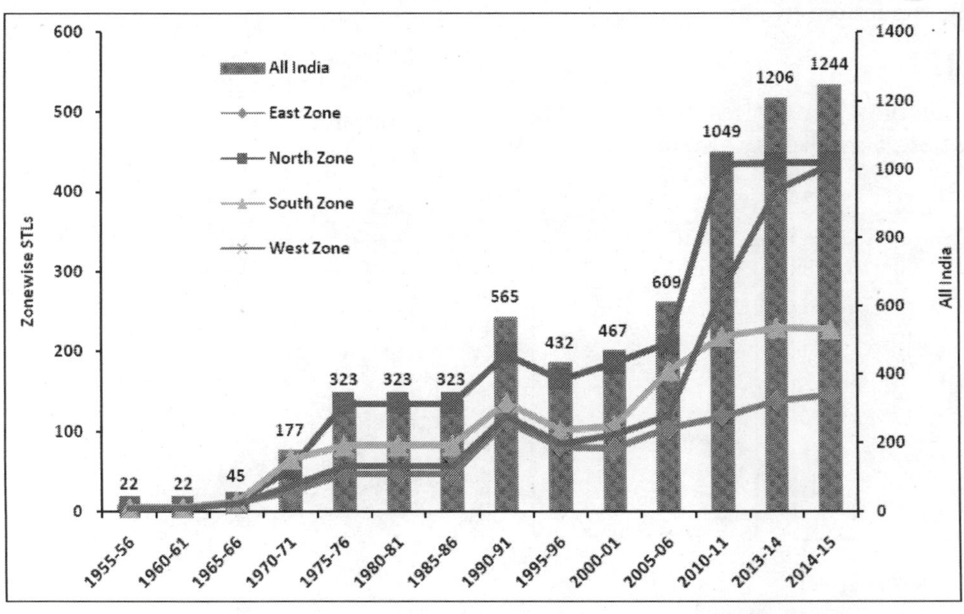

Figure 1: Expansion of STL Network during Past Six Decades.

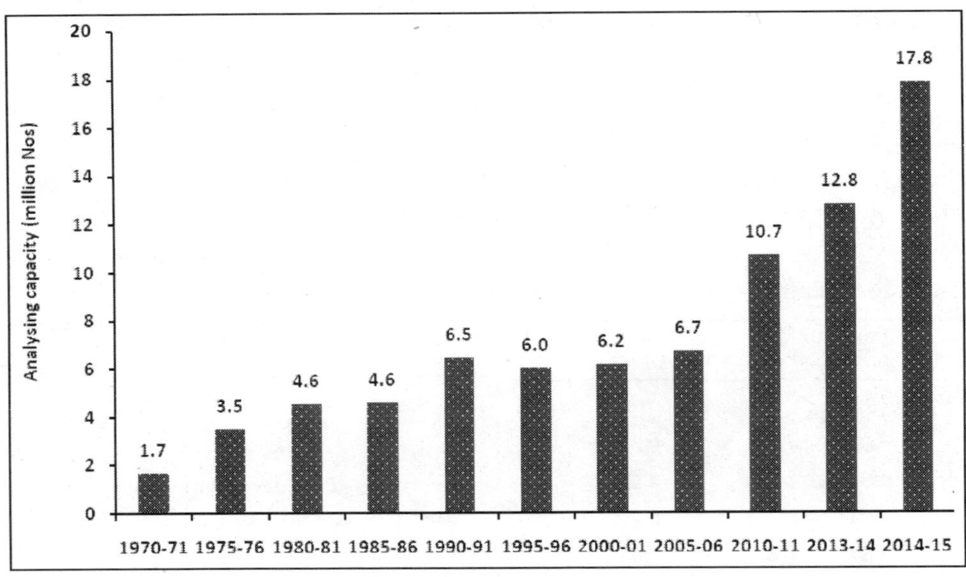

Figure 2: Annual Analyzing Capacity of STLs in India.

(0-15 cm) is appropriate depth for field crops, whereas deep samples are advised for orchards or plantations. For this, pits dug for planting of saplings can be used to collect samples from 0-30, 30-60 and 60-90 cm with *khurpi* or spade (*kassi/phawda*). Such samples can also be taken with the help of long-handle augers from 5 to 6 places from one acre to make a composite sample following quartering process.The three depths mentioned above are not to be mixed together. Thus from each point/ pit, there will be one sample each for the three depths.

Preparation of Composite Sample

For making composite sample, collected soil samples should be thoroughly mixed by hand on a clean thick paper or cloth or polythene sheet. Amount of the bulk sample should be reduced to about 500 g by quartering process in which entire soil mass is spread, divided into four quarters, two opposite samples are discarded and the remaining two are remixed.

Sampling Tools

A *khurpi* is the most convenient and simple tool available everywhere. Samples can also be drawn with the help of different types of augers shown in Figure 3. While sample from a soft soil can be taken with the help of tube auger, a post-hole auger is used for excessively-wet fields, and a screw auger for a hard and dry soil.

For sampling soft and moist soil, a tube auger, spade or khurpi is quite satisfactory. A screw type auger is more convenient on hard or dry soil, while the post-hole auger is useful for sampling in excessively wet areas, *viz.*, rice fields. Tube Auger is convenient for sampling from lower depths. If a spade or *khurpi* is used, a V-shaped cut may be first made up to the plough layer and a uniform 2 cm thick slice is taken out.

Processing of Soil Samples

Processing of soil samples involves drying, grinding, sieving. After collection, soil samples should be dried in shade at room temperature. Dried samples should be ground with the help of wooden pestle and mortar and sieved using 2 mm stainless steel sieve. Plant residues, gravel and organic material retained on the sieve should be removed. For specific type of analysis (*e.g.* organic C), soil sample should be ground further to pass through a 0.2 to 0.5 mm sieve. Before a sample

Khurpi

Spade

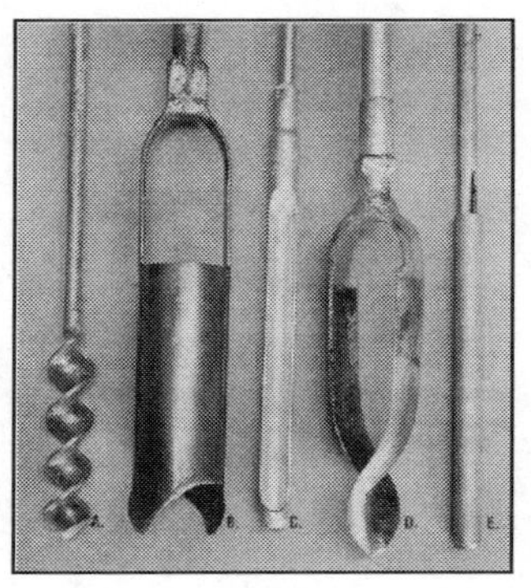

Augurs

Motorized Augur

Figure 3: Soil Sampling Tools.

is to be weighed for analysis, entire quantity of sieved soil should be remixed properly.

Precautions

In order to avoid contamination, following precautions should be taken to minimize error:

- Keep away the sample from chemicals, fertilizers or manure.
- For micronutrient analysis, collect samples using auger made up of stainless steel instead of rusted *khurpi* or spade.
- Do not use bags previously used for storing fertilizers or any chemical.
- Store soil samples in clean cloth or polythene bags.
- Use glass or polythene jars for storage of soil samples for a longer duration.

Plant Analysis for Nutritional Status

In case of horticultural crops, especially perennial crops, soil testing has significance before establishing or at very beginning of any orchard. After establishment, plant/leaf analysis has been recognized as a better indicator of the nutritional status than soil analysis. Concentrations of various nutrients in leaf tissues as well as their ratios are often taken into considerations for fertilizer application. Generally accepted nutrient concentrations in mango for example are given in Table 3.

Table 3: Ratings of Nutrient Concentration in Mango Tissues

Nutrient	*Concentration*		
	Deficient	*Sufficient*	*Excessive/Toxic*
Primary nutrients (per cent)			
N	0.70-0.99	1.00-1.50	>1.50
P	0.05-0.07	0.08-0.25	>0.25
K	0.25-0.39	0.40-0.90	>0.90
Secondary nutrients (per cent)			
Mg	0.15-0.19	0.20-0.50	>0.50
S	0.05-0.19	0.20-0.60	>0.60
Micronutrients (μg/g)			
Zn	15-19	20-200	>200
Fe and Mn	25-49	50-250	>250
Cu	5.0-6.0	7.0-15.0	>15
B	20-49	50-100	>100
Mo	0.01-0.04	0.05-1.00	>1.00

Source: Tandon and Tiwari (2008).

Tissue Sampling

The most appropriate part index tissue of a fruit tree is specific to its nutritional status should be sampled at right stage. For a meaningful plant analysis, collection of right plant tissue and stage of growth is very important. It would be wrong and wasteful to pluck any of the growing plant part at any time and send to laboratory for analysis. Thus, depending on the purpose of analysis, plant sampling should be planned. Plant parts recommended to be sampled for analysis are given in Table 4.

After collection, plant samples should be freed from dust and other foreign material by washing with detergent solution followed by 0.1NHCl and deionized water. Liquid detergent solution (0.2 per cent) would remove waxy coating on the leaf surface, 0.1 NHCl removes metallic contaminants and deionized water washes the previous two solutions. The washed samples should be placed on filter paper sheets and air-dried for 24 h. Then plant samples should be kept in new paper bags and dried in a forced air oven at 65°C ± 2°C for 48 h. The samples so dried should be ground in a stainless steel grinder using 0.5 mm sieve, oven-dried again for few hours for constant weight, andstored in paper bags for further analysis.

Recommendations for Horticultural Crops

Diagnosis and Recommendation Integrated System (DRIS)

One of the most scientific approaches using nutrient ratio in tissues is Diagnosis and Recommendation Integrated System (DRIS) proposed by Beaufils (1973). This approach has got several advantages:

i) Importance of nutritional balance is taken into account in deriving the norms and making diagnosis.

ii) The norms for the elemental content in leaf tissues can be universally applied to a particular crop, regardless of farming situations.

iii) Diagnosis can be made over a wide range in stages of crop development, irrespective of cultivars.

iv) The nutrient limiting yield either through excess or insufficiency can be readily identified and arranged in order of their limiting importance on yield.

It is relatively new approach for interpretation of leaf or plant analysis which identifies all nutritional factors limiting crop production, and increases the chances of obtaining high crop yields by improving fertilizer recommendations.

State Fertilizer Recommendations

State fertilizer recommendations for fruits, vegetables and flowers being adopted are summarized by Tandon and Tiwari (2008), from which it can be visualized that the recommendations vary widely from one state to the other, and of course from one species to the other. Development of recommendations for

Table 4: Plant Parts of Horticultural Crops Recommended to be Sampled for Analysis

Category	*Crop*	*Part to be Sampled with Stage/Age*
Commercial crops	Tea	3rd leaf from tip of young shoots
	Coffee	3rd or 4th pair of leaf from apex of lateral shoots, at bloom
Vegetables	Potato	Most recent, fully developed leaf (half-grown)
	Tomato	Leaves adjacent to influorescence (mid-bloom)
	Onion	Top non-white portion (1/3 to 1/2 grown)
	Brinjal	Blade of most recent fully developed leaves
	Beans	Uppermost, fully developed leaves
	Cauliflower	Most recent, fully mature leaf, at heading
	Cabbage	Wrapper leaf at 2-3 months age
	Pea	Leaflets from most recent, fully developed leaves at first bloom
	Carrot	Most recent, fully mature leaf, at mid-growth
	Radish	Most recent, fully developed leaf
	Turnip	Most recent, fully developed leaf
	Beetroot	At 30-50 days age
	Spinach	5th leaf from tip (omit unfurled) at the stage of bud starting to small fruits
	Cucumber	Most recent, fully developed leaf
Ornamental plants	Bougainvillea, jasmine, croton, fern, ficus, geranium, gerbera, gladiolus, lilly, orchid, hibiscus and rose	Most recent, fully developed leaves at the stage offlower bud of pea size
Fruit crops	Almond	3rd leaf from top, at the beginning of bloom
	Apple, pear	Leaves from middle of terminal shoot growth, 8-12th weeks after full bloom, 2-4 weeks after formation of terminal buds in bearing trees
	Blackberry	Latest matured leaf from non-tipped canes,4-6th weeks after peak bloom
	Cherry	Fully expanded leaves, mid shoot current growth in July-August
	Peach	Mid-shoot leaves, fruiting or non-fruiting spurs, mid-summer leaves, fruiting
	Strawberry	Youngest, fully expanded mature leaf without petiole, atpeak or harvest period
	Plum	Leaves from middle of current season's extension growth,in January-February
	Banana	Petiole of 3rd open leaf from apex, 4 months after planting
	Cashew	4th leaf from tip of mature branches, at beginning of flowering
	Grapes	5th petiole from base at bud differentiation for yield, and petiole opposite to bloom for quality
	Citrus fruits	3-5th month old leaves from new flush, 1st leaf of the shoot, in June
	Guava	3rd pair of recently matured leaves, at bloom (August or December)
	Mango	Leaf with petiole (4-7 month-old) from middle of shoot
	Papaya	6th petiole from apex, six months after planting
		PineappleMiddle 1/3rd portion of white basal portion of 4th leaf from apex, at 4-6 months stage
	Pomegranate	8th leaf from apex at bud differentiation in April and August
	Phalsa	4th leaf from apex, one month after pruning
	Ber	6th leaf from apex from secondary or tertiary shoot, 2 months after pruning

horticultural crops not as simple as for general field crops, and continues to be an important researchable area.

Fruit Crops

Fertilizer use techniques for fruit/plantation crops have not been detailed as much as field crops because:

(a) Fruit crops have a long pre-bearing period.

(b) Nutrient demand increases with age of plant, depending on the extent of growth and development.

(c) Stems and branches accumulate large amount of nutrients often supplying these to young growing parts, not allowing deficiency symptoms to appear even if soil is deficient.

(d) With a deep and extensive root system, plants may extract nutrients from deeper soil layer and hence surface soil tests become less-relevant, and

(e) Nutrient demands different for early and later growth stages.

Soil testing seems to be of hardly any use for fully grown trees. More reliable and factual assessment of nutritional status of fruit crops is analysis of leaf (or other indicator parts) through which malnutrition of plants can be detected even if definite deficiency symptoms are absent. In some laboratories, facilities for plant (leaf/ petiole *etc.*) analysis are available. Portable tissue testing kits can also be used on a limited scale but interpretation of the generated data is an expert's job. Nevertheless for a regular care, the trees may be sprayed with the micronutrient solutions as suggested in Table 5.

Table 5: Suggested Micronutrient Solution Concentrations for Foliar Spray

Element	*Solution Concentration (%)*	*Salt to be Used*
Fe	1.0-3.0	Ferrous sulphate
Mn	0.5-2.0	Manganese sulphate
Zn	0.5	Zinc sulphate
Cu	0.025	Copper sulphate
B	0.2-0.5	Solubor
Mo	0.1-0.3	Ammonium molybdate

For normal soilshaving medium initial soil fertility level, fertilizer recommendations are available which vary widely from one state to the other (Tandon, 1991). These are evolved from research institutes and state agricultural universities. These are given in brief for selected crops in Table 6. The doses are recommended along with varying quantities of organic manure per tree. A generally debated point in these recommendations is that the same dose is recommended for all the soils in a given state, irrespective of the type of soil, depth of soil, profile characteristics, climate conditions, and variety of crop and soil fertility level. Nevertheless, these recommendations serve as a general guideline and necessary corrections or improvements can be made under specific conditions.

Besides primary nutrients, use of secondary and micronutrients is necessary in deficient soils. There is narrow range between minimum critical level and toxic levels of some of the micronutrients (like B) and therefore, only need-based use of micronutrient fertilizers is advised, preferably as foliar spray.

Vegetable Crops

Fertilizer schedules for vegetable crops have been at a low key. In all states, the recommended doses are not as well defined as for cereals and other field crops due to more frequently conducted response studies in the latter case. The recommended doses vary from

Table 6: State-wise Fertilizer NPK Recommendations for Important Fruit Crops

Fruit	*Age (Years)*	*N*	*P_2O_5 (g/tree)*	*K_2O*
		Andhra Pradesh		
Mango	1st yr	200	100	100
	6th yr and above	1000	2000	500
Sapota	1-3 yrs	50	20	75
	11 yrs and above	400	160	450
		Bihar		
Mango	1st yr	50	160	Woodash (10 kg)
	Above 12 yrs	500	800	Woodash (22 kg)
Litchi	1 yr	50	40	Woodash (10 kg)
	Above 12 yrs	200	320	500
		Delhi		
Mango	1st yr	0.5 kg*	–	–
	Above 6 yrs	2.0-3.5 kg*	* mixture of 1:3:1 of AS, SSP and SOP	
		Gujarat		
Banana	3rd to 5th month	67	33	67
Sapota	1 yr	90	45	45
	10 yrs	900	450	450
		Himachal Pradesh		
Apple	Fully grown	700	350	700
Citrus	Fully grown	800	500	600
		Karnataka		
Mango	1 yr	73	18	68
	10 yrs and above	730	180	680
Sapota	1-3 yrs	50	20	75
	11 yrs and above	400	160	450
		Kerala		
Mango	1 yr	20	18	50
	Above 10 yrs	500	360	750
Sapota		500	360	750
		Maharashtra		
Mango	1 yr	50	50	50
	10 yrs and above	500	500	500
Orange	2 yrs	240	120	0
Vidarbha)	10 yrs and above	1000	500	0
Grape		200	120	180
		Punjab		
Mango	1-3 yrs	50-100	40-100	100-200
	10 yrs and above	250	160	600
Citrus	1-3 yrs	50-150	0	0
	10 yrs and above	400-800	0	0
		Uttar Pradesh		
Mango	1 yr	100	50	100
	10 yrs and above	1000	500	1000
Guava	1 yr	50	40	50
	7 yrs and above	350	280	350
		West Bengal		
Mango	1 yr	100	50	50
	10 yrs and above	1000	750	750
Banana (Robusta)	Yearly	140	35	90

state to state and with the productivity level (Tandon, 1991). Certain general/adhoc NPK doses recommended for major vegetables crops are given in Table 7, which are subject to necessary changes for specific situations. The general practice followed in the STLs is to increase and decrease the recommended doses by 20-30 per cent for soils testing low and high, respectively, in respect to a particular nutrient. However, the soil type involved should also be considered while making this adjustment. No definite recommendations are available for secondary and micronutrients which are equally essential. Therefore, need-based application of these nutrients as per soil testing is advised.

Table 7: General NPK Recommendations (kg/ha) for some Vegetable Crops

Crop	N	P_2O_5	K_2O
Cauliflower	150	60	60
Cabbage	100	60	60
Knolkhol	100	50	50
Radish	100	50	80
Carrot	60	30	75
Turnip	50	30	30
Brinjal	100-120	50-60	50-60
Tomato	100-120	60-80	80-100
Okra	100-150	50-100	50-100
Capsicum	100	50	50
Chilly	100	50	50
Cowpea	25-30	60	50
Beans	30	40	50
French bean	100-120	50-60	50-60
Pea	20-40	60	30-60
Cucurbits	40-60	40-60	30-60
Potato	150-250	80-150	100-250
Onion and garlic	75-125	25-75	25-125

Source: Tiwari (2001).

Use of B fertilizer has been found to benefit many vegetable crops including tomato for which a 0.3 per cent borax solution is sprayed. Similarly, a 0.2 per cent solution of zinc sulphate is used in case of Zn deficiency. However, the effect of soil applied zinc sulphate leaves a long-lasting effect. Needless to mention that a healthy soil provides sustainable yield, high quality produce and higher profits.

Significance of Manuring

Irrespective of the type of crops, manures have well known favourable effects on soil health in terms of soil physical, chemical and biological environment. Although the total amount of nutrients supplied through manure may be substantial due to large quantity added, manures are considered as a soil conditioner rather than a nutrient source. In fact, the advantage in terms of soil quality improvement overshadows their nutrient supplying potential.

Early nutrient requirement of most of the horticultural species is fulfilled by soil available nutrients as a result of initial manuring. However, the subsequent nutritional needs of fruit trees have to be addressed mainly through judicious fertilizer application, because (i) nutrient demand becomes greater with the plant age, (ii) the per cent content of nutrients in manure being very low compared to fertilizers, huge quantities of manure are required, (iii) availability of manure/compost is limited, and (iv) it takes time for nutrients in manures to get released (mineralized) for plant use.Nevertheless, manureshave to be applied along with fertilizers for improving soil health, and also for meeting secondary and micronutrient demands to a great extent. Table 8 shows the average composition of two commonly used organic manures and the plant nutrients supplied through their application.

Table 8: Average composition of FYM and poultry manure and nutrients supplied through them

Nutrient	General Composition (Dry wt. basis)		Nutrient (kg/ha) Supplied through Manures @10 t/ha	
	FYM	Poultry Manure	FYM	Poultry Manure
N (%)	0.75	2.9	75	290
P (%)	0.08	1.3	8	130
K (%)	0.50	2.0	50	200
Ca (%)	0.91	4.48	91	448
Mg (%)	0.19	0.60	19	60
S (%)	0.49	0.99	49	99
Zn (ppm)	14.5	50.0	0.145	0.50
Fe (ppm)	1465	1075	14.65	10.75
Mn ppm)	69.0	190	0.69	1.90
Cu (ppm)	2.8	6.9	0.028	0.07
B (ppm)	5.0	15	0.050	0.15
Mo (ppm)	21.0	65	0.21	0.65

Source: Tandon (2005).

Besides, organic manures help in improving soil organic matter (SOM) levels. SOM is epicenter of soil health and majority of Indian soils are inherently low in SOM, improvement in its content has several advantages beyond plant nutrition. In fruit trees/orchards, substantial litter fall results in SOM aggradation, although quantitative data on the effect of leaf litter recycling on SOM build-up for different crop species and growing conditions is not available.

References

Beaufils, E.R. 1973. *Soil Sci. Bull.* **1**: University of Naetal, SA.

Doran, J.W., Sarrantonio, M. and Liebig, M.A. 1996. *Adv. Agron.*, **56**: 1–54.

Dwivedi, B.S. and Dwivedi, V. 2007. *Indian J. Fert.*, **3**(4): 11–23.

Dwivedi, B.S and Meena, M.C. 2015. *Indian J. Fert.*, **11**(10): 110–122.

Goswami, N.N. 2006. **In**: *Proc. Int. Cong. Soil, Water Environ. Quality: Issues and Strategies*, ISSS, New Delhi, pp. 43–58.

ICAR. 2016. *Agricultural Research Data Book 2016*. ICAR–IASRI, New Delhi.

Sharma, B.M. and Meena, M.C. 2012. Soil test based integrated nutrient management under cropping systems. In: *System Based Integrated Nutrient Management*, (Eds.) B. Gangwar and V.K. Singh, New India Publishing Agency, New Delhi, pp. 155–167.

Tandon, H.L.S. 1991. *Fertilizer Recommendations for Horticultural Crops*. FDCO, New Delhi

Tandon, H.L.S. 2005. *Methods of Analysis of Soils, Plants, Waters, Fertilizers and Organic Manures* (edited), FDCO, New Delhi.

Tandon, H.L.S. and Tiwari, K.N. 2008. *Nutrient Management in Horticultural Crops*, FDCO, New Delhi.

Tiwari, K.N. 2001. *Balanced Fertilizer Use in Vegetable Crops*. PPIC–India Programme, Gurgaon, Haryana.

34

Soil Fertility Management for Better Productivity of Fruit Crops

A.K. Srivastava* and P.S. Shirgure

[1]ICAR-Central Citrus Research Institute,
Nagpur – 440 033, Maharashtra
**E-mail: aksrivas2007@gmail.com*

Perennial fruit crops represent hardly 1 per cent of the global agricultural land area, but Mediterranean region covers maximum of 11 per cent area, which are of great economic importance in world trade and tariff (FAO, 2011). On the other hand, Indian fruit industry occupies 6.82 million ha with a total production of 80.96 million tonnes (IHB, 2013). Furthermore, at least 60 per cent of the world's arable lands have mineral deficiencies or elemental toxicity problems, and on such soils fertilizers and lime amendments are essential for achieving improved crop yields (Pathak and Nedwell, 2011).

Fruit crops by the virtue of their perennial nature of woody framework (Nutrients locked therein), extended physiological stages of growth, differential root distribution pattern (root volume distribution), growth stages from the point of view of nutrient requirement and preferential requirement of some nutrients by specific fruit crop, collectively make them nutritionally more efficient than the annual crops (Srivastava and Singh, 2008a; Srivastava *et al.*, 2015, Scholberg and Morgan, 2012). There will be an increasing importance of nutrient efficient cultivars that are higher producers. Nutrient efficient plants are defined as those plants, which produce higher yields per unit of nutrient, applied or absorbed than other plants (standards) under similar agroecological conditions. Horticultural crops occupy 10 per cent of cultivated area. Share of fertilizer use in horticulture has risen from 2 per cent in 1990s to 8 per cent in 2012-13 (Av. application rate: 159 kg/ha with N:P:K use ratio 1.6:0.9:1.0), with banana and grapes being two most heavily fertilized fruit crops (Chanda, 28). According to Malhotra and Srivastava (2015), estimated fertilizer requirement of fruit industry comes out as 7.56 million tonnes, out of which 75 and 60 per cent of total fertilizers used in horticultural sector, are consumed by potato and grapes, respectively. Ghosh *et al.* (2015), while analysing the carrying capacity of Indian horticulture observed, annual growth rate in domestic demands for fruits is estimated at 3.34 per cent with nearly 76 per cent of fruits still consumed as fresh.

Fruit crops by the virtue of their perennial nature of woody framework (Nutrients locked therein), extended physiological stages of growth, differential root distribution pattern (root volume distribution), growth stages from the point of view of nutrient requirement and preferential requirement of some nutrients by specific fruit crop, collectively make them different than the annual crops (Srivastava *et al.*, 2008a; Srivastava, 2013a, 2013b).

Microbes are considered gateway to improved use efficiency of applied fertilizers. A still bigger question emerges, whether rhizosphere competent microbes could collectively contribute towards improved resilience of plant's rhizosphere (Wang *et al.*, 2014). And if those microbes are so successful in promoting growth response, addition of starter nutrients in such combination may further magnify the magnitude of response called nutrient-microbe synergy. Our earlier studies have shown that rhizosphere effective microbes have the tendency to play multiple roles (Rao and Dass, 1989; Srivastava and Singh, 2005; Keditsu and Srivastava, 2014) to overcome various biotic and abiotic stresses while interacting with an environment. A sound understanding of nutrient- microbe synergy could possibly lay a solid foundation in unlocking the productivity potential of perennial fruit crops, besides safeguarding the soil health, both physic-chemically as well as biologically. In this background, incise efforts have been made to analyse various aspects of an effective nutrient management in giving the desired fillip to the productivity of fruit crops.

1. Soil Fertility and Functions: Some Facts

Soil function fertility refers to the ability of soil to support and sustain plant growth, which relates to making all the essential nutrients available for root uptake. This is facilitated by their storage in soil organic matter, nutrient recycling from organic to plant available mineral forms, and physicochemical processes that control their fixation and release (Srivastava and Kohli, 1997). On the other hand, managed soils are highly dynamic system that makes the soil work and supply ecosystem services to humans. Overall, the fertility and functioning of soils strongly depend on interactions between soil mineral matrix, plants and microbes. These are responsible for both building and decomposing soil organic matter, and therefore for the preservation and availability of nutrients in soils, cycling of nutrients in soils must be preserved (Srivastava *et al.*, 2008).

Soil health considers the physical, chemical/ biological properties of the soil and the disturbance and ameliorative responses of land managers. Soil health also describes the capacity of a soil to meet performance standards relating to nutrient and water storage and supply, biological diversity and function and resistance to degradation. The most important of these manageable services include BNF (Biological nitrogen fixation), other symbiotic and beneficial organisms, nutrient and moisture supply, carbon storage and protection from erosion (Srivastava *et al.*, 2014a). Let us look at soil nutrient imbalances, nutrient mining and sustainability of nutrient management practices.

Managing soil carbon for multiple benefits address to enhance a range of ecological services. Increasing the soil organic matter of degraded soils can boost crop productivity, sequester CO_2, enhance soil microbial growth and activities and improve water capture and retention. Soil carbon stocks, highly vulnerable to human activities, decrease significantly in response to changes in land capability and land use such as deforestation and increased tillage continues (Srivastava and Singh, 2009b; Tagaliavini *et al.*, 2007).

Promotion of proper management practices to protect and enhance soil organic matter as an essential element of good soil and environmental quality. Promotion of sources of plant nutrients (*e.g.* cover crops, legumes, crop diversification that enhance soil organic carbon stocks. Integration of several crops in a field at the same time to increase soil organic matter, soil biodiversity and soil health. Decline in soil fertility is the major constraint limiting the productivity of fruit crops. Continuous reduction in nutrient density of different fruit crops is an indication of nutrient mining induced decline in fruit crop productivity (Srivastava and Ngullie, 2009).

1.1 Improving Soil Fertility

Nutrient mining (extracting more nutrients than are returned) is a major factor impoverishing the soil. Massive mobilisation programmes have been proposed, especially for phosphorus, but their economic benefit and long term sustainability need case-by-case evaluation. Mineral fertilisers have had greater impact, particularly when subsidised, but continuously using them alone is not sustainable (Srivastava *et al.*, 2002; Srivastava and Malhotra, 2014).

Integrated soil-fertility management is an ecological approach that uses locally available amendments and inorganic fertilisers in an integrated way. Standard technology packages, delivered in a top-down manner, are inappropriate; rather, approaches must help farmers make decisions based on a combination of their own knowledge and research based options. Synergistic fertility management packages are being tested that include soil amendments like organic matter and rock phosphate, and wise fertiliser applications, aiming at sustainable and profitable crop production (Mengel and Kirby, 2000). Programmes to promote fertiliser use alongside other improved practices have had considerable impact for some marketable crops in high potential areas. But remote areas and those with marginal agro-ecology are riskier, so often require different approaches.

Organic alternatives to mineral fertilisers, such as manure and compost, are bulky and often scarce and of poor quality. Using them needs a lot of labour, which is in short supply in many places. Improved transport and research to develop labour- saving techniques may overcome this problem. The small but growing demand for organic fruits, and the higher prices it fetches, have stimulated organic production, often by smallholders. Certified organic systems prohibit the use of mineral fertilisers, but farmers may be confused, if fertilisers are promoted in other circumstances (Srivastava *et al.*, 2002; Srivastava and Singh, 2004c).

2. Nutrient Removal Pattern

Fruit crops by the virtue of their perenniality longer growth period, and developing fruits acting as major sink, are considered nutrient extracting in nature, and hence so nutrient responsive. Interestingly, nutrient uptake pattern (Table 1) by major fruit crops display the fact that K – removal is far higher than N or K, however P-removal is nearly half of N- removal. Crops like banana, citrus, grape, kiwifruit, mango *etc.* are considered highly nutrient exhausive crops, warranting their replenishment on long term basis sustainability in production is to be ensured (Tandon and Muralidharadu, 2010; Srivastava and Singh, 2004a, 2004b).

Table 1: Nutrient Removal Pattern by Major Fruit Crops

Crop	*Nutrients Removal (kg/ha)*			
	N	*P_2O_5*	*K_2O*	*Yield (t/ha)*
Apple	100.0	45.0	180.0	25.0
Avocado	11.3	3.9	23.4	10.0
Banana	250.0	60.0	1000.0	40.0
Citrus spp.				
Nagpur mandarin	3.4	0.37	3.72	1.0
Khasi mandarin	1.81	0.16	2.18	1.0
Coorg mandarin	3.04	0.27	3.31	1.0
Kinnow mandarin	2.40	0.57	2.35	1.0
Acid lime	1.22	0.21	4.06	1.0
Durian	16.1	6.6	33.5	6.7
Grapes	170.0	60.0	220.0	20.0
Guava	120.0	50.0	150.0	20.0
Ber	80.0	35.0	125.0	20.0
Kiwifruit	165.0	50.0	273.0	40.0
Mango	100.0	25.0	110.0	15.0
Papaya	90.0	25.0	130.0	50.0
Pecan	9.7	5.3	5.4	1.2
Pistachio	30.0	12.0	15.0	0.057
Passion fruit	60.0	15.0	75.0	15.0
Pineapple	185.0	55.0	350.0	50.0
Sapota	130.0	50.0	170.0	80.0
Strawberry	200.0	100.0	900.0	0.75
Walnut	37.9	10.7	21.2	2.7

Note: In case of citrus, figures are expressed as g/tonne fruits.

Source: Tandon and Muralidharadu (2010), Srivastava *et al.* (2008), Chadha and Bhargava (1997), Kemmeler and Hobt (1985), Chadha (2007), Sparks (2005).

3. Nutrient Diagnosis for Early Warning

Plant nutritionists across the globe are on their toes to find ways and means to identify nutrient constraints as early in standing crop season as possible while dealing with perennial crops (Srivastava *et al.*, 2014a; Srivastava and Singh, 2005). Exciting progress has been made over the years, and accordingly, the basis of nutrient management strategy has experienced many paradigm shifts (Srivastava, 2013a). While doing so, it is being increasingly felt to have some diagnostic tool to identify nutrient constraint as and when it originates by capturing the signals released at subcellular level (Srivastava and Singh, 2003b; 2004a). On the other hand, conventionally used diagnostic tools of identifying nutrient constraints such as leaf analysis, soil analysis, juice analysis, and to some extent, metalloenzyme-based biochemical analysis, all have been under continuous use and refinement (Srivastava and Singh, 2001a; 2001c; 2007; Srivastava, 2010a; 2013a). But despite so much of genuine efforts worldwide, no one of these alone provides complete information, except the combined use of leaf and soil analysis, which are used on a comparatively wider scale (Srivastava *et al.*, 2006). Establishment of absolute figures of normal, deficient or excess nutrient level is not real, unless the dynamic aspect of leaf nutrient concentration is considered, especially when various nutrients interactions produce resonance within close space of tissue composition (Srivastava, 2013c; Srivastava and Singh, 2008b). Tertiary diagrams and nutrient ratios are early representation of interacting nutrients in the tissue compositional space (Srivastava, 2013c).

Productivity of the plant depends essentially on the nutrient balance and the biological activity (Srivastava *et al.*, 2007). There are definite limitations with the leaf analysis application which is largely dependent upon composition of index leaves or any other plant parts. On the other hand, overlapping phenotypic symptoms of plants deficient in N, S or Fe accompanied by lowered chlorophyll concentration makes the distinction between nutrients most often very difficult (Srivastava and Singh, 2006; 2009a). Other example of Fe-deficiency showing typical chlorosis pattern due to interrupted chlorophyll synthesis may be cited, even though chlorophyll chelates Mg rather than Fe. Nutrient deficiency, thus, involves degradative changes in chloroplast components and additional cellular compartments (Srivastava, 2013d). In the light of this information, an integrative physiological approach was suggested. For example, the use of the peroxidase in the diagnosis of Fe- and Mn- deficiencies prompted checking the utility of the method for citrus cultivars grown on differentially fertile soils. Parallel to what was observed with peroxidase, catalase, and aconitase reduced their levels of activity with Fe-deficiency and increased with Mn-deficiency, facilitated to establish the possibility of using the latter enzyme as an alternative mean of diagnosing Fe- and Mn- deficiencies. In the early diagnosis of mineral deficiencies in lemon trees, aconitase has proved as precise as peroxidase, a specific Fe-metalloenzyme or even more so (Srivastava and Singh, 2008a; Srivastava and Singh, 2006; Aseri *et al.*, 2008).

Studies have shown that nutritional status of the responsive tissues transmits signals as a regulator of gene expression and at times, that can become a limiting factor in the process of plant development. There are other interesting improvements such as the determination of the nutrient evolution along the vegetative cycle, the substitution of the critical levels by the critical zone, fractionating the nutrient contents (especially the biologically active ones), and finally implementing the biochemical diagnosis. For the latter, the use of activities of specific enzymatic systems and also of metabolites concerned with photosynthesis, has a good potential to improve the accuracy of nutrient constraint diagnosis over other conventional diagnostics. Many studies suggested that the levels of enzymatic activity could be effectively used as an alternative diagnostic tool to leaf analysis. Since functional analysis of the nutrients is, thus, based on the examination of certain molecular compounds linked with their functional activity (Srivastava and Singh, 2006; Aseri *et al.*, 2008).

Not surprisingly, proximal sensing through spectral signatures of crop canopies in the field are more complex and often quite dissimilar from those of single green leaves measured under carefully controlled conditions (Van Maarschalkerweerd and Husted, 2015; Das *et al.*, 2015). Even when leaf spectral properties remain relatively constant throughout the season, canopy spectra change dynamically depending upon variation in soil type, vegetation, and architectural arrangement of plant components. Vegetation indices provide a very simple yet elegant method for extracting the green plant quantity signal from complex canopy spectra (Adams *et al.*, 2000). Narrower band indices such as the photochemical reflectance index, water band index, and normalized pigment chlorophyll ratio index are examples of reflectance indices that are correlated with certain physiological plant responses, and have promise for diagnosing water and nutrient stress. Such studies hold promise for nutrient like nitrogen. Ironically, micronutrient deficiencies are diagnosed through specific pattern of chlorosis, *e.g.*, Fe versus Mn or Fe/Mn versus Zn backed up by nutrient concentration, capturing symptomatic pattern of chlorosis via spectral norms (signatures) irrespective of crop species further limit this concept towards more wider application (Adams *et al.*, 2000). Spectral reflectance method show significant promise as an alternative to traditional wet chemistry anahbes with regard to ease, costs and speed with wider range of applicalhous including natural resources (Mylavarapu, 2010).

Specific nutrient-signalling pathways have made it feasible to enhance the uptake and use efficiency of applied nutrient. Nutrient mobility in the phloem from the leaves to the fruits and from the older to the younger fruits is reported in perennial crop like banana. Considering the thumping success of trunk nutrition, won't it be more advisable to analyse the xylem sap or phloem tissue for chemical and microbial constituents since the signal transduction for various nutrients functioning mediate through these tissues only (Srivastava and Singh, 2006). Such attempts could provide some meaningful clues about the presence or absence of those signals to be later utilized in understanding the underlying principles of nutrient stress induced warning mechanism. These studies could lay the solid foundation for developing some probe linked to transpiration stream of plant to act as early warning system for identifying deficiencies of various nutrients. In this regard, functioning of nutrients linked through metalloenzymes and enzyme antioxidants would be the target areas to uphold such novel possibilities as research and development issues related to perennial trees (Srivastava, 2013c; Srivastava *et al.*, 2014a).

Potential response of nutrient application is largely jeopardized in the absence of precision-based nutrient diagnostics capable of diagnosing nutrient deficiency early in the season (independent of crop stage) or early signaling towards origin of nutrient deficiency so that strategic changes in fertilizer schedule could be effectively exercised within the same crop season. In this background, it is strongly advocated to develop early warning system for nutrient deficiency by developing metalloenzyme-based probe or trapping the functioning of regulator(s) associated with nutrient signaling at cellular level using genomics. Once such probes are in contact with trunk sap flow, signals of nutrient deficiency can be befittingly used in more effective nutrient management in years to come.

4. Inorganic Fertilizers and Fertigation

4.1 Fertilizer Use

Optimum soil fertility induced plant nutrition has a key role to maximise yield and quality production of perennial fruit crops. The fertilizer requirement of perennial fruit crops is determined by many approaches, including surveys, growers' experience, following the fertilization program of high yielding orchards, replacing the amount of nutrients removed by fruits, deficiency symptoms, applying results from sand or soil culture and field experiments and leaf/soil analysis, with each one of these having distinct advantages and limitations.

Soil provides nearly all the nutrients essential to complete the life cycle of a plant. Different soil properties primarily determine the extent of a fertilizer response (Bronick and Lal, 2005) and the crop rotation on some recently published review articles changes in physico-chemical (Lehoczky *et al.*, 2005) and biological properties of soil (Manna *et al.*, 2005; Srivastava, 2013a). Since the subject of fertilization in horticultural crops is so vast and complex, the readers may refer to the few review articles exclusively devoted to different issues of nutrient management in perennial fruit crops (Lipecki and Barbeæ, 2005; Zahir *et al.*, 265; Hazarika and Ansari, 2007; Srivastava, 2009, 2013a; Malhotra and Srivastava, 2009; Berg, 2009; Srivastava and Ngullie 2009; Maity *et al.*, 2012; Singh *et al.*, 2012). However, the subject dealing with multipronged action to rhizocompetent microbes in combination of various nutrient sources in fruit crops, is distinctly missing.

Three approaches to fertilizer recommendations that are widely used: the deficiency correction philosophy (originates from nutrient constraints based crop response through nutrient additions to the point of maximum economic yield), maintenance concept (aims to maintain soil fertility level slightly above the point of maximum economic yield), and nutrient removal or balanced philosophy (emphasizes the return to the soil what is removed by the crop to maintain productivity, but often over recommends nutrient need, since it does not take into account for the soil's ability to supply available nutrients to the plants over time) (Srivastava *et al.*, 2008a). According to summarized work of Tandon (2002), different fruit crops have been observed remove nutrients (kg/ha) as : 100 N-45 P-180 K (1.00: 0.45:1.80) for 25 tonnes yield of apple, 250 N-60 P-100 K (1.00: 0.24:4.00) for 40 tonnes yield of banana, 100 N-60 P-35 K (1.00: 0.60: 3.50) for 30 tonnes yield of citrus, 170 N-60 P-200 K (1.00 :0.35: 1.30) for 20 tonnes yield of grape, 100 N-25 P-110 K (1.00: 0.25: 1.10) for 15 tonnes yield of mango, 90 N-25 P-130 K (1.00:0.28:1.40) for 50 tonnes yield of papaya and 185 N-55 P-350 K (1.00:0.30:1.90) for 50 tonnes yield of pineapple. Hence, optimum fertilizer requirement is governed by expected yield level. A soil fertility level optimum for a given yield level, will become sub-optimum at higher yield level.

Optimum Macronutrients

There are varied fertilization and doses schedules followed across a variety of crops (Table 2). Some fertilization plans recommend N application since the beginning of bud break until 6 weeks after full bloom for bearing pear trees (Neto *et al.*, 120), whereas others defend that N must be applied during the whole growth cycle considering that after harvest, trees can still improve their reserves through N uptake from soil. Some authors have studied the fertilizer N use efficiency in pears and apples (Cheng *et al.*, 2001; Neilsen *et al.*, 2001; Chen *et al.*, 2004). Most studies were performed in pots in sand culture, comprising its applicability to field conditions. The re-cycling of N as a result of the decomposition of senescent leaves in soils was only addressed in one study with apple trees (Tagaliavini *et al.*, 2007; 2007). Fertilizer N use efficiency by trees increased from the first to the third year, but was generally small (6, 14, and 33 per cent), and estimated N losses were large (89, 46, and 53 per cent, respectively, in the first, second, and third year). Irrigation water and soil provided more N to the trees than fertilizer N (Neto *et al.*, 2008).

For many years, several authors have tested the response of different crops to application of different nutrients, especially K, with respect to yield and quality, in many crops like coffee, *Coffea arabica* L. (Silva *et al.*, 2001); almond, *Amygdalus communis* L. (Reídel *et al.*, 2004); pistachio, *Pistacio vera* L. (Zeng *et al.*, 2001); pecan, *Carya illinoinensis* Koch (Worley, 1994); olive, *Olea europaea* (Jasrotia *et al.*, 1999) *etc.* However, such basal fertilizer application technique is greatly conditioned by different soil properties, particularly soil moisture, which affects the mobility of the supplied nutrients (Mengel and Kirkby, 2000). This is mainly attributed to large variation in fertilizer doses to be really effective in different crops, annual versus perennial fruit crops. Such variation in optimum doses is dictated by climate, soil types, crops, and farming practices in such a way that the correct balance of nutrients necessary for one farm, may be quite different from that necessary for a farm somewhere else in the world. Therefore, determining the appropriate balance of nutrients to increase crop yield and soil fertility will require localized research.

Table 2: Optimum Nutrient Requirement (g/tree) for different Fruit Crops

Fruit Crop	*N*	*P_2O_5*	*K_2O*	*Source*
Mango	800	200	300	Sharma *et al.* (2000)
Acid lime	800	200	100	Huchche *et al.* (1996)
Guava	500	250	250	Singh and Singh (2007)
Grape*	300	500	1000	Patil *et al.* (2008)
Pomegranate	400	100	300	Ghosh *et al.* (2012)
Ber	500	200	300	Lal *et al.* (2003)
*Aonla**	212	55	234	Biswas *et al.* (1995)
Sapota	400	100	300	Ghosh *et al.* (2012)
Date palm	460	500	500	Munir *et al.* (1992)
Fig**	430	200	430	Irget *et al.* (2008)
Phalsa	200	75	100	Sharma *et al.* (2008)
Apple	1065	650	1500	Singh *et al.* (2011)
Litchi***	600	350	140	Pathak and Mitra (2008)
Pear	1000	2000	1500	Arora and Singh (2006)

*: Figures in kg/ha; **: Addition of 280 g/tree Ca; ***: Addition of B at 7 g/plant.

Optimum Micronutrients

Soil application of micronutrients, especially inorganic salts, is often not so effective due to immediate reaction of added micronutrient cations with the mineral portion of soil through various processes like adsorption, fixation, chemical precipitation *etc.* (Srivastava and Singh, 2008b) irrespective of crop, annual or perennial in nature. This issue in the past has been addressed in depth. Researchers even today are not unanimous about the efficacy of soil versus foliar fertilization with reference to micronutrients (Srivastava and Singh, 2002, 2003b). Elevating Zn concentration only in the tops of Zn-deficient sour orange plants with foliar sprays partially restored normal root growth but clearly was not as effective as the roots absorbing Zn directly from high Zn concentration solutions (Swietlik and Zhang, 1994). Duxbury *et al.* (2006) suggested that micronutrient-enriched seed successfully addressed Zn and Mo deficiencies in rice and wheat, and increased yields beyond those achieved by soil fertilization due to difference in root health activating early seedling emergence.

Interestingly, some recommendations have advocated soil application of micronutrients as one of the means to realize good yield of a crop, *e.g.*, $ZnSO_4$ (300 g/tree)-$FeSO_4$ (300 g/tree)-600 N-200 P_2O_5-100 K (g/tree) in citrus (Srivastava *et al.*, 2006; Srivastava, 2011; Srivastava and Singh, 2015). The micronutrient-based Zn chelater complexes on the other hand are poorly or not at all absorbed by plant roots, as demonstrated through water culture studies (Chaney, 1988; Swietlik and Zhang, 1994). Under field conditions, however, the addition of Zn micronutrient-chelate elevated the amount of exchangeable nutrients in the soil solution due to adsorption and exchange properties of minerals present in soil (Chaney, 1988). Soil application of a micronutrient, *e.g.*, Zn from $ZnSO_4$ is fixed in the surface soil, while the chelated-Zn remains soluble and becomes distributed evenly throughout the soil, as evident from 46-times higher uptake of Zn by a perennial fruit crop like citrus from Zn-EDTA as $ZnSO_4$ on sandy soils (Parker *et al.*, 1995). In non-citrus crops like wheat (Modaihsha, 1997), banana (Mostafa *et al.*, 2007), pear, apple, grapevine (Sohlegel *et al.*, 2006) *etc.* similar results have been reported. Of late, micronutrient seed treatment including seed priming and seed coating has offered an attractive and easy alternative (Farooq *et al.*, 2012). One of the major obstacles of conventional practices of addressing nutritional requirements of perennial fruit crops, either through soil fertilization or through foliar feeding, is the precise diagnosis if the nutrient constraint type, their doses as per crop age and soil type, with the result more often such practices have not been able to facilitate the realisation of potential productivity of fruit crops. Neither any due consideration is given to exploit the nutrient reserve of the plant's rhizosphere (native nutrient supplying capacity of soil) while formulating the fertilizer doses. Most importantly in perennial fruit crops, nutrient doses need to be recommended in tandem with level of fruit yield targeted, a nutrient dose optimum for one fruit yield target, will become suboptimum for higher targetted fruit yield level in couple in subsequent years. Where is such nutrient monitoring tool to keep vigil on nutrient input and output relationship, a kind of nutrient budgeting (Srivastava *et al.*, 2008; 2013b).

4.2 Fertigation

Intensive growing of perennial fruit crops under high density and ultra-high density plant population has further put an additional pressure on soil fertility conservation. But, concept of fertigation in perennial crops has given a definite edge over conventionally used basin method of irrigation coupled with basal or top dressing of fertilizers within the perimeter of trees (Shirgure and Srivastava, 2013a; Shirgure *et al.*, 2014). Fertigation is considered synonymous to nutrient use efficiency which can further be fine tuned with nitrification inhibitors (restrict the microbial conversion of ammonium to nitrate that it is mobile in soils) or plant growth-promoting bio-effectors (microorganisms and active natural compounds involved in plant growth). Thus, in addition to all these, precise soil sampling, whether to take samples from below drippers or in between drippers or mixing soil samples from both the sites and finally, drawing a representative soil samples, find a greater intervention while evaluating nutrient-water interaction in citrus (Srivastava *et al.*, 2003; Srivastava and Singh, 2009a, 2009b). Another prominent concern is often raised with regard to threshold values of leaf nutrient diagnostics. Whether or not, optimum leaf nutrient values hold some application efficacy when compared with basin irrigation coupled with basal fertilizer application. Under two contrasting fertilizer application techniques, due to difference in nutrient use efficiency, optimum values could warrant minor adjustments (correction factor), depending upon mode of fertilizer application (Schumann *et al.*, 2006).

Agriculture has advanced at a rate that should complement the net food requirement for ever growing population either at regional level or global level. Perennial fruits of tropical nature (citrus, banana, mango, papaya and guava) have emerged as an alternative source of nutrients to lessen the burden on per capita consumption of cereals crops. Intensive growing of perennial fruit crops under high density and ultra-high density plant population has further put an additional pressure on soil fertility conservation. But, concept of fertigation in perennial crops has given a definite edge over conventionally used basin method of irrigation coupled with basal or top dressing of fertilizers within the perimeter of trees (Shirgure *et al.*, 2001a). Reports are candidly visible on fruit yield, quality indices soil fertility improvements coupled water use efficiency (WUE) and fertilizer use efficiency (FUE) covering a variety of fruit crops (Shrigure *et al.*, 2001c; 2004b; Srivastava *et al.*, 2003). Hence, fertigation has proved beyond doubt about its utility to fruit culture (Solaimali *et al.*, 2005).

Techniques and managements of perennial fruits production are nowadays directed towards the need to conserve resources, energy and a commitment to the environment. In this sense, fertigation has risen as a valuable tool in recent years spreading around the world in major fruit belts. The million dollar question strikes why does fertigation out-perform conventional basin method of irrigation carrying fertilizers. A plant exposed to uniform regime of moisture and nutrient flow within rhizosphere zone has to spend much less energy than the growing conditions constantly changing over time. Fertigation, is thus, energy efficient as well (Shirgure *et al.*, 2001d). This has led to an increase in both fertilizer use efficiency and water use efficiency. In the future, fertigation would continue to replace traditional flood irrigation so extensively adopted in water surplus irrigated fruit growing areas (Srivastava *et al.*, 2008b).

Open hydroponics (OH), a concept synonymous to fertigation is a management practice to address low fertility gravel base soils and saline water. The nutrient uptake is maximised, if the ratio of ions in the solution matches with scion/stock requirements. In Spain, the performance of 'Nova', 'Marisol', and 'Dalite' mandarins at density of 1000 plants/ha under OH system was evaluated (Krugger *et al.*, 2000). The average yield in sixth year was 60-75 tonnes/ha, which is much higher than many conventionally managed orchards. In Israel the response of 'Shamouti' orange under restricted root zone practices, a sprinkler versus drip irrigation treatment with three fertilizer rates, maintained at high moisture status (8-12 Kpa) as a part of intensive OH program. A significant increase in yield in the restricted root zone drip irrigation compared treatment was observed with the highest rate of fertilizer application, 400 kg N/ha. Similar promising results were obtained in South Africa where OH system increased the yield of 'Valencia' orange and 'Clementine' orange by 19 per cent and 25 per cent, respectively, using 16 per cent less water with 25-31 per cent higher returns compared to micro-irrigation with broadcast method of fertilizer application as control (Martinez and Fernandez, 2004). More information on critical issues like capability of manipulate the soil solution as a restricted root zone versus conventional drip irrigation root zone, buffering capacity of soil manipulating specific nutrient ratios at different physiological stages, evaluating orchard productivity-energy relationship through ionic balanced nutrient solution, planting density *etc.* are further required before OH system under citrus is adopted on a commercially large scale. Therefore, OH and IFP use a more intensive nutrition program that may push the trees into a higher level of vigor and productivity requiring higher nutrient application rates to maintain production, besides higher nutrient use efficiency (Shirgure *et al.*, 2000; 2001b; 2013b).

Possibility of integrating liquid biofertilizers (broth using native isolates) as a starter with nutrients in order to further hasten the bioavailablity of nutrients within plant rhizosphere needs to be looked afresh (Ngullie *et al.*, 1996; Srivastava *et al.*, 2002). Concept of rhizosphere hybridization triggering the nutrient dynamics across crop phenophases then finds its candid way into the curriculum of fertigation oriented research (Srivastava *et al.*, 2007; 2010). However, it remains to be seen that net impact on soil solute fluctuations in both soluble and exchangeable phase of soil if nutrient use efficiency has to be targeted, besides an eye on changes in soil salinity buildup. While addressing these concerns, however, the major concern emerges as orchard efficiency vis-a-vis water use efficiency/nutrient use efficiency by adopting variable rate fertilizer application taking into consideration the spatial variability in soil properties in a time domain manner known as automated fertigation or pulse-based fertigation (Shirgure, 2012a).

Large number of studies have, however, demonstrated much better nutrient-use-efficiency with fertigation in crops like guava (Ramniwas *et al.*, 2012), banana (Reddy *et al.*, 2002), apple (Banyal and Sharma, 2011), kiwifruit (Chauhan and Chandel, 2008), sweet cherry (Ahmed *et al.*, 2010), litchi (Dey *et al.*, 2010), sapota (Khot *et al.*, 2012), mango (Singh *et al.*, 2009), Banana (Pawar and Dingre, 2013), citrus (Shirgure *et al.*, 2002; 2003b; Srivastava *et al.*, 2012; 2015b), papaya (Jeyakumar *et al.*, 2010), pomegranate (Haneef *et al.*, 2014; Shanmugasundaram and Balakrishnamurthy, 2015), peach (Baldi *et al.*, 2010) *etc.* using various bases of drip irrigation scheduling and NPK-based fertilizers, but without much success with micronutrient fertigation. These studies have provided a wealth of information with unanimous result that fertigation reduced both irrigation and nutrient requirement by 30-50 per cent compared to conventional split application within plant basin. Open hydroponics (Krugger *et al.*, 2000; Martinez and Fernandez, 2004) and variable rate fertilizer linked site specific nutrient management (Zaman and Schumann, 264; Johnston *et al.*, 2009; Srivastava and Singh 2015; Srivastava *et al.*, 2006) in fruit crops like citrus, olive, avocado, coconut *etc.* have also started showing their utility in improving fertilizers use efficiency to various dimensions.

Several experiments were carried out in Egypt and Germany to study whether fruit shrubs and trees can be fertilized by injection through the trunk. Results showed that dicotyledonous vascular trees (mango and grapevine) can be fully fertilized by injection through xylem. Only 5-10 per cent of the levels used in soil fertilization were sufficient for good growth and high yield. Growth of the injection-fertilized mango (var. 'Sukkary White') trees was 20-25 per cent higher than soil fertilized plants while in grapevine.

'White Riesling' and 'Spacet Burgunder' fruit yield increases were 32-49 per cent higher compared to soil fertilization Fruits quality of grapevine clusters assessed (juice Brix, pH, reduced sugars, total acidity, grape vinegar, apple vinegar, ethanol and glycerine content) of the plants fertilized through injection was better that those fertilized through soil. Grapevine fresh juice content of the reduced sugars and ethanol increased by 7.5-11.9 and 41.4-50 per cent, respectively while the total acidity decreased by 6.2-19.7 per cent. Using injection fertilization, there was no need to control weeds because they never competed with tree roots for nutrient absorption. Since there is no soil fertilization, there was no use of herbicides or pesticides, nor leaching of these compounds to undergroundwater, which was expected to be clean enough to be used as drinking water with no or less health hazards (Shaaban, 2009).

Table 3: Fertigation Requirement of major Fruit Crops

Crop	*Fertigation Requirement*	*Crop Response*
Mango		
Variety : Dashenari 10 x 10 m spacing	75% RDF (380 g N-380 g P_2O_5-380 g K_2O/tree) + mulch + drip irrigation in V-area of hasin in 16 year old orchard	Increase in growth yield and quality, in addition to soil fertility
Variety : Alphonso 2.5 x 2.0 m spacing	100% RDF (120 g N-100 g P_2O_5-100 g K_2O/tree) + 24 trees water/day in 3 year old orchard	Increase in fruit yield and quality
Variety : Dashenari 10 x 10 m spacing	75% RDF (750 g N-375 g P_2O_5-750 g K_2O/tree) + 100% irrigation through drip based 0.60W/C PE ratio in 30 year old orchard	Improvement in nutrition distribution nutrient use efficiency coupled with better soil moisture distribution and yield
Guava		
Variety : Shweta 2 x 1 m spacing	75% RDF (30 g N-10 g P_2O_5-10 g K_2O/tree) + 75% irrigation requirement based on CPE in 3 year old meadow orchard	Increase in flowering leaf nutrient status, growth, yield and quality
Variety : Shweta 2 x 1 m spacing	100 RDF + surface drip irrigation at 1.0 W/CPE ratio	Increase in growth and fruit yield response
Pomogranate Variety : Bhgwa 2 x 2 m spacing	100% RDF (200 g N-200 g P_2O_5-500 g K_2O/tree using water soluble 19:19:19 fertilizer) + 100% irrigation at alternate daythrough drip irrigation in 4 year old orchard	Better soil moisture distribution, fruit yield and quality
Variety : Bhgwa 2 x 2 m spacing	50% RDF using water soluble fertilizer + drip irrigation at 100% CPE in 2 year old orchard	Improvement in flowering soil available nutrients and fruit yield
Almond		
Variety : Waris	75% RDF (330 g N-45 g P_2O_5-455 g K_2O/plant through water soluble fertilizers) + drip irrigation based on pan evaporation in 8 year old orchard	Increase in growth response and yield in addition to soil fertility
Banana		
Variety : Robusta	Fertigation along with microbial consortium	Increase in uptake of secondary nutrients and micro nutrients besides fruit yields
Variety : Robusta	100% RDF (13 g N-40 g P_2O_5-13 g K_2O/plant using urea, DAP and KNO_3 + drip irrigation at 1.0 IW/CPE	Improvement in growth, yield and quality
Apple		
Variety : Gala on M-26 rootstock 2 x 4 m spacing	40 g N-17.5 g P_2O_5 using ammonium nitrate and ammonium pyrophosphate (10 N – 15 P – 0 K) in 5 year old orchard	Improved root distribution and alleviated K-deficiency
Variety : Gala on M-9 rootstock 3.5 x 1.25 m spacing	60% RDF + irrigation based on pan evaporation	Improved growth and yield initially with N-fertigation and later NPK fertigation proved better
Strawberry		
Variety : Chandler 25 x 50 m spacing	75% RDF (112 kg N-75 kg P_2O_5-90 kg K_2O/ha + 100% ETC	Increase in growth yield and quality
Papaya		
Variety : Taiwan 2.5 x 2.0 m spacing	100% RDF (250 g N-250 g P_2O_5-500 g K_2O/plant) + 50 mm CPE as 100% CPE	Improvement in growth, yield and quality
Acid lime		
Variety : Kagzi lime 6 x 6 m spacing	75% RDF (450 g N-150 g P_2O_5-175 g K_2O/tree) + irrigation at 30% AWC depletion	Improvement in growth, yield, quality and leaf nutrient composition
Nagpur mandarin		
Variety : Nagpur mandarin 6 x 6 m spacing	80% RDF (48 g N-160 g P_2O_5-240 g K_2O/tree) + irrigation at 20% AWC depletion	Increase in growth, yield, quality, leaf nutrient composition and soil fertility

1. Panwar *et al.* (2007), 2. Prakash *et al.* (2015), 3. Adak *et al.* (2014), 4. Ramniwas *et al.* (2013), 5. Sharma *et al.* (2013), 6. Mohd Haneef *et al.* (2014), 7. Shanmugasundaram *et al.* (2013), 8. Dinesh Kumar and Ahmed (2014), 9. Senthilkumar *et al.* (2014), 10. Mahendran *et al.* (2013), 11. Neilsen *et al.* (1996), 12. Treder (2006), 13. Kachwaha *et al.* (2014), 14. Deshmukh Hardaha (2014), 15. Shirgure *et al.* (2016b), 16. Srivastava *et al.* (2003).

Nutrient and Irrigation Partitioning *vis-a-vis* Crop Phenology

Tailoring the nutrient and water application as per crop phonological stages is a pre-requisite to an elevated nutrient use efficiency (Shirgure, 2013b; Shirgure *et al.*, 2014; 176; Srivastava *et al.*, 2009; 2002). And occurrence of nutrient constraint or water stress at any - phenological growth stage in any perennial fruit crop, *e.g.* citrus, could jeopardize the incentive accruing through other management practices. Studies on dynamics of different nutrients and water application (ER-based) across all six growth stages (January-February as Stage I, March-April as Stage II, May-June as Stage III, July-August as Stage IV, September-October as Stage V and November - December as Stage VI) of Nagpur mandarin trees grown on alkaline calcareous Haplustert, have shown some interesting observations (Table 4).

Different treatments having varying schedules of fertilization as well as irrigation based on evaporation rate influenced both the production oriented parameters *viz.*, canopy volume, fruit yield and fruit quality parameters (Table 4) through treatments T_2 and T_4, respectively. These studies revealed application of 30 per cent N, 40 per cent P and 10 per cent K out of total requirement at fruit set stage; 30 per cent N, 35 per cent P and 10 per cent K at marble size stage; 30 per cent N, 25 per cent P and 55 per cent K during fruit development stage and remaining 10 per cent N and 25 per cent K during fruit maturity colour break stage. While, irrigation requirement across crop phenophases showed 9, 18, 45, 20 and 8 per cent out of total water requirement needs to be given to corresponding stages of flowering, fruit set, fruit development, harvesting and colour break stage. Thus, such studies on partitioning nutrient and water requirement across crop phonological stages would lay a sound research and development strategy for scheduling fertigation for any perennial fruit crop with improved demand-driven nutrient-use-efficiency (Srivastava, 2009; 2011). Such attempt on nutrient and water partitioning has been attempted in other fruit crops like apple, banana, grapes, mango crops but distinctly lacking in other perennial fruit crops to develop crop demand oriented fertigation scheduling including variable rate fertigation technique. Reports are candidly visible on fruit yield, quality indices soil fertility improvements coupled water use efficiency and nutrient-use-efficiency covering a variety of fruit crops (Panigrahi *et al.*, 2012a; 2012b; Panigrahi and Srivastava, 2016). Hence, fertigation has proved beyond doubt about its utility to fruit culture.

Emitters Arrangements of Drip Irrigation

Soil moisture deficit stress-induced reduction in fruit yield and quality in citrus is most common across major commercial citrus cultivars, irrespective of soil

Table 4: Response of Stagewise Nutrient and Water Treatment on Tree Volume, Fruit Yield and Fruit Quality Parameters of Nagpur Mandarin (Pooled data 2006-12)

Treatment	*Tree Volume (m^3)*	*Fruit Yield (kg/tree)*	*Fruit Quality (%)*		
			Juice	*TSS*	*Acidity*
Stagewise nutrient requirement					
T_1 (100% N and 50% K upto stage IV; 100% P upto stage III; 50% K divided at V and VI)	19.62* (16.47)	42.77** (11.85)	45.79	8.58	0.81
T_2(80% N, 100% P and 50% K upto stage IV; 20% N and 50% K divided at V and VI)	19.11 (15.91)	56.52 (15.66)	48.01	9.40	0.71
T_3(100% N, 100% P and 30% K upto IV; 70% K divided at V and VI)	19.31 (13.51)	45.41 (12.58)	46.03	8.66	0.76
T_4(67% N and 100% P upto IV; 33% N and 100% K at V)	19.47 (16.51)	46.24 (12.81)	46.99	8.54	0.80
CD (*p=0.05*)	0.19	1.95	0.86	0.26	0.05
Stagewise water requirement					
T_1 (30% ER in stage I, III and V + 40% ER in stage II, IV and VI)	63.22* (61.06)	34.62** (9.59)	36.10	9.46	0.81
T_2 (40% ER in stage I, III and V + 60% ER in stage II, IV and VI)	69.53 (67.28)	55.09 (15.26)	36.80	9.52	0.80
T_3 (60% ER in stage I, III and V + 80% ER in stage II, IV and VI)	71.18 (68.01)	65.45 (18.13)	37.42	10.04	0.77
T_4 (80% ER in stage I, III and V + 80% ER in stage II, IV and VI)	73.75 (70.05)	74.11 (20.53)	39.00	10.09	0.78
CD (*p=0.05*)	0.30	1.94	1.03	0.38	0.07

*: Figures in parenthesis indicate initial values; ** Figures in parenthesis indicated fruit yield in tonnes/ha.

Source: Shirgure *et al.* (2016b), Srivastava *et al.* (2014b).

and climate (Shirgure and Srivastava, 2012a). At the same time, it is equally important to rationalise the water use synchronising right amount of water across different critical growth stages, collectively to enhance the growth, yield and fruit quality (Shirgure and Srivastava, 2012b). The common methods of applying irrigation water to the citrus orchards include: basin, micro-jet irrigation and drip irrigation (Cevik *et al.*, 1987). Surface basin irrigation using the circular ring generally followed in early establishment phase of trees (Shirgure, 2012b; 2013a). Drip irrigation have the distinct advantage over surface irrigation methods, due to more constant, uniform and complete wetting of effective rootzone of the plant (Dasberg, 1995) in addition to better water use efficiency at lesser expense of energy (Kanber *et al.*, 1996). Uneven water distribution and associated complexity involved in application of water, the importance of surface irrigation in perennial crop like citrus is fast diminishing very fast (Peng and Rabe, 1999; Shirgure *et al.*, 2001a; 2003a). While, under tree drip irrigation with more number of emitters, better option having more uniform distribution of soil moisture and accurate amount of irrigation water has made definite inroads to exploit for better crop responses. With the drip irrigation systems, water savings may be obtained because water is applied only to the active rootzone, leaving the inter-plant area un-irrigated (Shirgure and Srivastava, 2015).

Of late, there is a gradual shift from surface irrigation involving furrow irrigation to under-tree surface drip irrigation systems (Smajstrala, 1993; Dasberg, 1995). Various developments in surface drip irrigation systems have been reported from time-to-time with regard to improvements in water-use-efficiency vis-a-vis quality citrus production worldwide (Grieve, 1988; Germana, 1994; Cevik *et al.*, 1987). However, some studies in citrus comparing irrigation methods and irrigation schedules showed higher yield and quality with better performance using drip irrigation systems in India (Kumar and Bhojappa, 1994; Shirgure *et al.*, 2004a; 2004c) our studies showed maximum plant canopy volume was observed with T_3 (84.96 m^3) followed by T_1 (81.87 m^3) due to better soil moisture distribution coupled with higher nutrient availability-induced changes in leaf nutrient concentration in the former treatment. While maximum fruit yield was observed with treatment T_3 (28.78 tonnes/ha) followed by T_4 (27.26 tonnes/ha). The TSS/acid ratio as the major fruit quality parameter followed the same pattern of response. Similarly, leaf analysis displayed maximum concentration of nutrients in treatment T_3 compared to other treatments. These observations suggested T_3 as the best emitter's arrangement for harnessing maximum water-use-efficiency vis-a-vis quality production of Nagpur mandarin (Table 5). The conventional drip-irrigation method does not cover even 60-70 per cent soil water regime of the basin area of Nagpur mandarin compared to requirement of higher soil moisture to be constantly maintained throughout the period of plant growth and fruit development (Shirgure *et al.*, 2001a). With expanding scarcity of available irrigation water, intervention of drip irrigation is gradually gaining momentum in citrus orchards (Koo and Smjstrala, 1984).

Interactive Effect of Irrigation Schedules and Fertigation Levels

There are two major production constraints, *viz.*, irrigation and fertilization warranting right amount of water and nutrients to be added across different growth stages to enhance the growth, yield and fruit quality of Nagpur mandarin (Bielorai *et al.*, 1984; Shirgure *et al.*, 2016a; 2016b). The water need for 6-year-old *Kinnow* mandarin, varied from 539-1276 mm depending upon the level of irrigation with average consumptive use from 66.7-132.5 cm (Mageed *et al.*, 1988). In contrast, the water requirement of young, middle age and mature mandarin trees was 651.9, 849.0 and 997.3 mm/year, respectively, under clean cultivation (Ghadekar *et al.*, 1989). Past research work on irrigation scheduling and fertigation in perennial fruit crop like citrus indicated the irrigation at 65 per cent field capacity caused drought

Table 5: Plant Canopy Volume, Fruit Yield and Fruit Quality of Nagpur Mandarin in Response to different Treatments (Pooled data: 2010-13)

Treatment	*Plant Growth*		*Fruit Yield*			*Fruit Quality*			
	Plant Height (m)	*Canopy Volume (m³)*	*No. of Fruits*	*Av. Fruit Weight*	*Yield (tonnes/ha) (g)*	*Juice Content*	*TSS (%) (%)*	*Acidity (%)*	*TSS/Acid Ratio*
T_1	5.2	69.89	580	139.5	24.02	41.95	9.84	0.80	12.3
T_2	5.4	79.98	583	143.4	23.14	40.91	8.87	0.85	10.4
T_3	5.8	84.96	724	142.7	28.78	42.90	10.16	0.72	14.1
T_4	5.0	81.87	676	147.3	27.26	39.10	10.43	0.75	13.9
LSD (*p=0.05*)	0.32	1.68	32	2.1	1.32	1.01	0.40	0.06	–

T_1 (4 lph six emitters per plant in hexagonal arrangement), T_2 (4 lph three emitters on double lateral arrangement at 1 meter spacing), T_3 (4 lph eight emitters on single lateral in octagonal arrangement), T_4 (3 lph inline emitters at 0.75 meter spacing on the lateral and double lateral arrangement 3 feet apart from the trunk). – Note: lph stands for litre per hour.

Source: Shrigure and Srivastava (2014), Shirgure *et al.* (2016a).

injury symptoms, excessive defoliation and less water consumption compared to irrigation at 85 per cent field capacity in citrus. It was observed that partial fertigation of N and K resulted in lower N content in leaves with higher total soluble solids and acid concentration in juice and the yield remained unaffected in Valencia orange (Koo and Smjstrala, 1984; Smajstrla *et al.*, 1986; Ferguson, 1990; Smajstrla, 1993; Panigrahi *et al.*, 2008; 2012c). Fertigation has shown good responses on growth, yield, quality and uniform distribution pattern of applied water as well as nutrients within the active rootzone compared to band placement involving localized fertilization (Shirgure *et al.*, 2000; Shirgure and Srivastava, 2013b). Interestingly yield efficiency computed as fruit yield per unit canopy volume was also observed maximum (0.26) with treatment carrying irrigation at 80 per cent ER and fertigation with 80 per cent RDF. The sustained productivity of Nagpur mandarin can be achieved with irrigation scheduled at 80 per cent ER along with fertigation technology at 80 per cent RDF without any potential nutrient mining (Table 6).

4.2 Development of Rationale of Fertilizer Use

Of late, there have some distinct headways towards development of rationale of fertilizer use. In this regard, exploiting spatial variability has huge practical significance. Characterizing spatial variability of soil physico-chemical properties is a fundamental element of i. soil quality assessment, ii. modeling non-point source pollutants in soil and iii. Site-specific crop management. The heterogeneity of soil physico-chemical properties has been known since the classic study of (Nielsen *et al.*, 1973), which characterized the spatial variability of soil-water properties for a 150 ha field at the University of California's West Side Field Station in the San Joaquin Valley. The protocols developed by (Corwin and Lesch, 2005) for site specific evaluation of soil properties (EC appraisal) comprised of eight general steps: i. site description, and EC_a, survey design; ii. EC_a data collection with mobile GPS-based equipment; iii. soil sampling design; iv. soil core sampling; v. laboratory analysis; vi. calibration of EC_a to EC_e; vii. spatial statistical analysis and viii. GIS database development and graphic display. Of late, (Siqueira *et al.*, 2010) suggested that relief may be considered an integrating factor that expresses the interaction of various soil and plant attributes. The study further analyzed the potential use of landforms to predict the variability of soil and orange attributes, with large spatial variability in soil and temporal variability in orange quality.

4.3 Soil Fertility-Based Spatial Variogram as a Decision Support Tool

Soil fertility - based spatial variograms were developed based on georeferenced grid -based soil sampling. As many 56 grid based plants identified within the Nagpur mandarin orchard using two layered variograms (spatial variogram for fruit yield and other soil test values for various nutrients and delineating different management zones using DRIS-based analysis and formulated variable rates of fertilizer doses depending upon the soil test value of various grids (Srivastava, 2013b) These variable rates of fertilizer doses comprised of:

- ✰ 200 g N – 100 g P_2O_5 – 200 g K_2O – 100 g $FeSO_4$ – 100 g $MnSO_4$ – 100 g $ZnSO_4$

Table 6: Effect of Irrigation and Fertigation Schedules on Growth, Yield, Leaf Nutrient Composition and Plant available Nutrients in Nagpur Mandarin (Pooled data : 2010-13)

Treatment		*Canopy Volume (m^3)*	*Yield (t/ha)*	*Yield Defficiency (t/m^3)*	*Leaf Nutrient Composition (%)*			*Plant available Nutrients (mg/kg)*		
Irrigation	*Fertigation*				N	P	K	N	P	K
I_1	F_1	71.8	8.4	0.12	2.03	0.08	0.92	115.8	11.8	241.3
	F_2	71.3	9.9	0.14	2.07	0.08	0.97	115.8	11.1	243.3
	F_3	74.2	10.9	0.15	2.10	0.09	0.98	116.8	11.8	246.3
I_2	F_1	72.8	11.0	0.15	2.15	0.09	1.00	117.1	11.3	251.8
	F_2	72.3	14.8	0.20	2.16	0.10	1.07	119.3	13.0	231.8
	F_3	74.9	19.2	0.26	2.21	0.10	1.17	123.8	14.8	261.2
I_3	F_1	74.1	12.6	0.17	2.16	0.09	0.98	121.1	14.1	248.1
	F_2	73.8	15.4	0.21	2.16	0.09	1.03	120.2	14.0	241.1
	F_3	74.4	17.6	0.24	2.17	0.08	1.08	122.3	14.2	252.3
LSD	(I)	0.95	1.2	–	0.04	0.006	0.04	1.2	0.70	1.2
(*P* = *0.05*)	(F)	0.24	0.71	–	0.03	NS	0.02	0.8	0.64	1.8
	(I x F)	0.14	0.23	–	0.06	NS	0.07	2.1	1.40	3.2

I_1, Irrigation at 70% ER; I_2, Irrigation at 80% ER; I_3, Irrigation at 90% ER; F_1, Fertigation at 60% RDF; F_2, Fertigation at 70% RDF and F_3, Fertigation at 80% RDF.

Source: Shirgure *et al.* (2012a).

Table 7: Highlighting the Magnitude of different Types of Responses

Response Variability	*Yield ($q\ ha^{-1}$)*	*N ($mg\ kg^{-1}$)*	*P ($mg\ kg^{-1}$)*	*K ($mg\ kg^{-1}$)*	*Zn ($mg\ kg^{-1}$)*	*Fe ($mg\ kg^{-1}$)*	*Mn ($mg\ kg^{-1}$)*
Fertilizer dose : 200 g N – 100 g P_2O_5 – 200 g K_2O – 100 g $FeSO_4$ – 100 g $MnSO_4$ – 100 g $ZnSO_4$							
Initial	85.79	114.0	12.00	147.0	0.82	11.64	8.72
Final	163.74	116.8	12.78	152.3	0.88	12.42	9.68
Difference	37.45	2.8	0.78	5.3	0.06	0.77	0.96
Fertilizer dose : 400 g N – 150 g P_2O_5 – 300 g K_2O – 200 g $FeSO_4$ – 200 g $MnSO_4$ – 200 g $ZnSO_4$							
Initial	90.63	106.8	9.95	120.5	0.66	8.65	5.97
Final	102.30	113.3	11.36	128.9	0.78	10.35	8.12
Difference	11.68	6.63	1.40	8.41	0.12	1.70	2.15
Fertilizer dose : 600 g N – 200 g P_2O_5 – 400 g K_2O – 300 g $FeSO_4$ – 300 g $MnSO_4$ – 300 g $ZnSO_4$							
Initial	47.82	98.5	8.83	110.3	0.51	8.58	6.11
Final	85.79	117.3	11.07	138.9	0.87	11.29	9.24
Difference	37.95	19.01	2.23	28.9	0.35	2.70	3.30

Source: Srivastava *et al.* (2014a; 2014b), Srivastava *et al.* (2015).

- ✰ 400 g N – 150 g P_2O_5 – 300 g K_2O – 200 g $FeSO_4$ – 200 g $MnSO_4$ – 300 g $ZnSO_4$
- ✰ 600 g N – 200 g P_2O_5 – 400 g K_2O – 300 g $FeSO_4$ – 300 g $MnSO_4$ – 300 g $ZnSO_4$

After the application of variable rates of fertilizers, the difference in yield responses over initial values, and likewise differences in soil test values with respect to $KMnO_4$-N, Olsen-P, NH_4OAc-K, DTPA-Fe, DTPA-Mn, DTPA-Cu and DTPA-Zn were computed and summarized as below highlighting the magnitude of different types of responses are given in Table 7.

The changes in area distribution either with regard to fruit yield or with regard to different nutrients ($KMnO_4$-N, Olsen-P, NH_4OAc-K, DTPA-Fe, DTPA-Mn and DTPA-Zn)as a result of variable doses of fertilizer application (Table 8) further revealed some distinct reorientation. Out of different grid sizes, the grid size upto 40 x 40 m, 42.42 – 43.02 per cent area of the identified orchard (Total area of 11,184 sq.m) was observed to possess optimum fruit yield. While, sampling with both grid size of 60 x 60 m, as high as 68.33 per cent area was observed having optimum yield, suggesting further the optimum grid size of 40 x 40 m, using all the grid sizes (20 x 20 m to 60 x 60 m) no significant area under any category was observed to be significantly influence since most of the orchard area 99.77 to as much as 10 per cent area was observed within the optimum range. Across all the three grid size sampling from 20 x 20 m to 60 x 60 m, an area ranging from 99.83 to as high as 100 per cent area was observed to fall within optimum range. On the other hand, zoning of $NH\text{-}_4OAc$-K based on variograms developed through soil sampling via three grid sizes of 20 x 20 m, 40 x 40 m and 60 x 60 m, 73.93 per cent, 72.40 per cent and 73.14 per cent area were observed within low range. While rest of the 26.07 per cent, 27.60 per cent and 26.86 per cent area, respectively showed $NH\text{-}_4OAc$-K within optimum range at grid size of 20 x 20 m, 40 x 40 m and 60 x 60 m. Under grid size of 20 x 20 m, 40 x 40 m and 60 x 60 m, respectively, 29.54 per cent, 21.50 per cent and 25.67 per cent area of the orchard was observed displaying optimum range of DTPA-Fe. While other 70.47 per cent, 78.50 per cent and 74.33 per cent area of the orchard showed low level of DTPA-Fe under grid size sampling of 20 x 20 m, 40 x 40 m and 60 x 60 m. DTPA-Mn showed a distant variation under differing zone when compared across three grid size sampling. Under the grid size sampling 20 x 20 m, 6.34 per cent, 44.05 and 49.60 per cent area of the orchard was registered to fall within deficient, low and optimum range, respectively. Likewise under grid size of 40 x 40 m, 3.5 per cent, 39.56 per cent and 56.86 per cent area of the orchard, respectively, showed low, optimum and optimum range of DTPA-Mn. These changes were well corroborated with changes in fruit yield zones (Figure 1).

4.4 Development of Fertilizer Prediction Models

Basic data developed for targeted yield using prediction model

Nutrient	*Parameter*		
	NR	*CS*	*CF*
Nitrogen	4.49	81.5	34.3
Phosphorous	0.27	65.0	5.55
Potassium	1.10	21.1	64.9
Zinc	0.01	30.4	1.23
Iron	0.04	8.69	2.06
Manganese	0.01	1.98	1.30

Table 8: Distribution of Area Statistics (sq. m) of Nutrients under different Grid Size Sampling at Village Ladgaon, Katol, Nagpur (Total area: 11,184 sq.m)

Nutrient Indices	*Grid Size (m)*					
	20 x 20		*40 x 40*		*60 x 60*	
	Area (m^2)	*Percent-age*	*Area (m^2)*	*Percent-age*	*Area (m^2)*	*Percent-age*
Iron						
Deficient	0.0	0.0	0.0	0.0	0.0	0.0
Low	7881.5	70.47	8780.0	78.50	8313.0	74.33
Optimum	3302.5	29.53	2404.0	21.50	2871.0	25.67
High	0.0	0.0	0.0	0.0	0.0	0.0
Excess	0.0	0.0	0.0	0.0	0.0	0.0
Manganese						
Deficient	711.8	6.34	400.3	3.58	108.5	0.97
Low	4926.7	44.05	4424.5	39.56	6561.0	58.67
Optimum	5545.0	49.60	6359.3	56.86	4514.5	40.36
High	0.0	0.0	0.0	0.0	0.0	0.0
Excess	0.0	0.0	0.0	0.0	0.0	0.0
Zinc						
Deficient	0.0	0.0	0.0	0.0	0.0	0.0
Low	1456.5	13.02	494.0	4.42	564.3	5.04
Optimum	9727.5	86.98	10690.0	95.58	10619.8	94.96
High	0.0	0.0	0.0	0.0	0.0	0.0
Excess	0.0	0.0	0.0	0.0	0.0	0.0
$KMNO_4$-N						
Deficient	0.0	0.0	0.0	0.0	0.0	0.0
Low	25.5	0.23	0.0	0.0	15.8	0.14
Optimum	11158.5	99.77	11184.0	100	11168.3	99.86
High	0.0	0.0	0.0	0.0	0.0	0.0
Excess	0.0	0.0	0.0	0.0	0.0	0.0
Olsen-P						
Deficient	0.0	0.0	0.0	0.0	0.0	0.0
Low	0.0	0.0	0.0	0.0	0.0	0.0
Optimum	11177.0	99.94	11164.5	99.83	11184.0	100
High	7.0	0.06	19.5	0.17	0.0	0.0
Excess	0.0	0.0	0.0	0.0	0.0	0.0
NH_4OAc-K						
Deficient	0.0	0.0	0.0	0.0	0.0	0.0
Low	8268.8	73.93	8097.0	72.40	8180.3	73.14
Optimum	2915.3	26.07	3087.0	27.60	3003.8	26.86
High	0.0	0.0	0.0	0.0	0.0	0.0
Excess	0.0	0.0	0.0	0.0	0.0	0.0
Fruit yield						
Deficient	62.3	0.56	3.3	0.03	3.0	0.02
Low	6377.3	57.02	6369.3	56.95	7663.5	68.53
Optimum	4744.5	42.42	4811.5	43.02	3517.5	31.45
High	0.0	0.0	0.0	0.0	0.0	0.0
Excess	0.0	0.0	0.0	0.0	0.0	0.0

Source: Srivastava *et al.* (2014b; 2015).

Where, NR (Nutrient requirement) computed as the amount of nutrient required (kg) to produce one quintal of fruits, using formula: NR = Uptake of nutrient (kg ha^{-1})/Fruit yield (q ha^{-1}) CS (Percent contribution from soil to total nutrient uptake) computed as the amount of per cent contribution of soil nutrient, using formula: CS = Uptake of nutrient (kg ha^{-1}) by fruits x 100/Available soil test value for nutrient (kg ha^{-1}) from control treatment CF (Percent contribution from fertilizer to total uptake) computed as the amount of per cent contribution of fertilizer nutrient, using the formula: CF = Uptake of nutrient (kg ha^{-1}) from fruit crop)-(Avail. STV from nutrient (kg ha^{-1}) x CS/100 x 100/Fertilizer nutrient applied (kg ha^{-1}).The Following predication equations for computing fertilizer doses were developed:

i) Fertilizer Nitrogen = 13.09 (Targeted fruit yield) – 2.37 (Soil test value for $KMnO_4$-N)

ii) Fertilizer Phosphorous = 4.08 (Targeted fruit yield) – 26.83 (Soil test value for Olsen-P)

iii) Fertilizer Potassium = 1.69 (Targeted fruit yield) – 0.39 (Soil test value for NH_4OAc-k)

iv) Fertilizer Zinc = 0.98 (Targeted fruit yield) - 24.73 (Soil test value for DTPA-Zn)

v) Fertilizer Iron = 1.94 (Targeted fruit yield)-4.21 (Soil test value for DTPA-Fe)

vi) Fertilizer Manganese=0.77 (Targeted fruit yield)-1.52 (Soil test value for DTPA-Mn)

Using these fertilizer prediction equation, ready reckoner (Table 9) developed for different soil test values-based fertilizer recommendation in relation to different target fruit levels, have shown good promise to rationalized the soil test-crop response based fertilizer recommendations (Srivastava *et al.*, 2014b; 2015).

4.5 Tailoring Fertilizer Requirement

Experiments have shown that soils initially rich in fertility developed a variety of nutrient constraints following the non-synchronisation in demand and supply of nutrients with advancing orchard age (Tiwari, 2007; Srivastava and Singh, 2015). For example when exchangeable K is not rapidly replenished, crops start drawing on the non-exchangeable K, resulting in soil mining and depletion in soil K reserve (Srivastava *et al.*, 2006; Srivastava, 2011). A nutrient level sufficient for one particular productivity level may not hold sufficient for other higher productivity levels. An extensive survey of 18,000 ha area of Nagpur mandarin orchards of central India showed large scale deficiency of three nutrients, *viz.*, N, P and Zn, with most of the sites expressing multiple nutrient deficiencies (Srivastava and Singh, 2015). Reduced longevity of commercial citrus orchards due to varying nutritional constraints is one of the major production related constraint heavily responsible for poor orchard efficiency (Srivastava *et al.*, 2006; 2009). On the other hand, conventionally designed long term fertilizer trials revealed that: i. omission of limiting

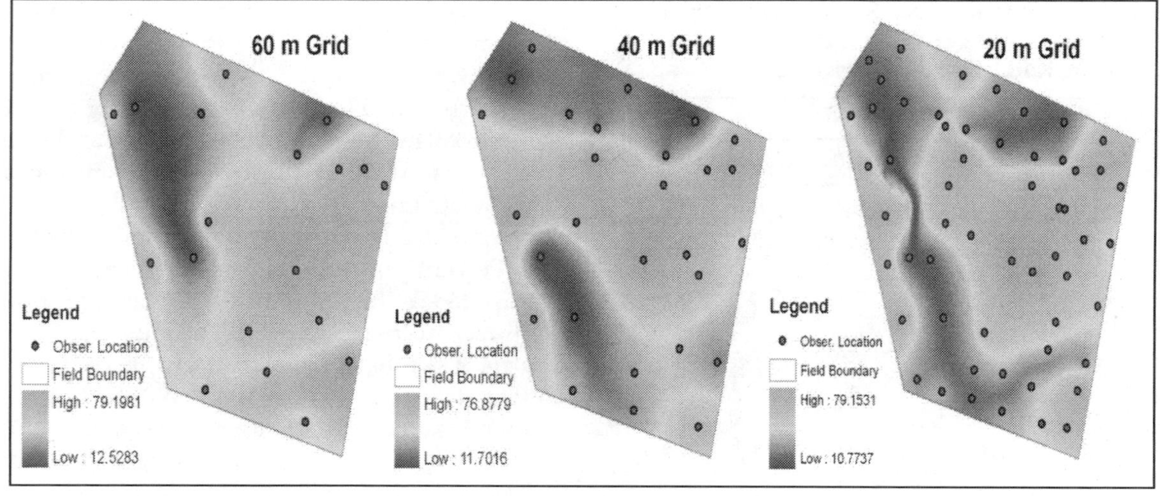

a. Spatial variation in FRUIT YIELD based on actual FRUIT YIELD data at different grid sampling.

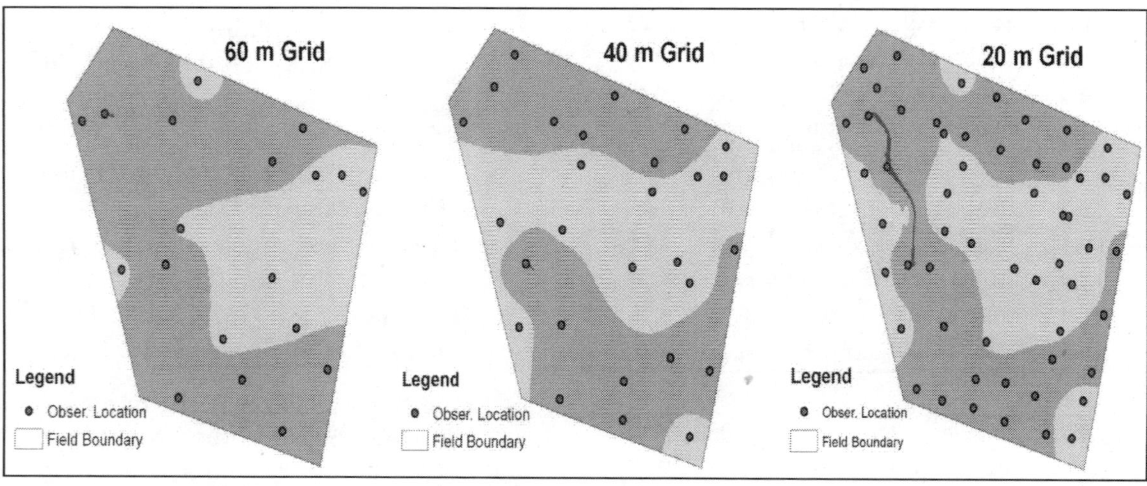

b. Spatial variation in FRUIT YIELD based on DRIS-based different FRUIT YIELD classes at different grid sampling.

Figure 1: Spatial Variation in FRUIT YIELD Based on Actual FRUIT YIELD Data at different Grid Samplings Sizes, (20 x 20 m, 40 x 40 m and 60 x 60 m) in the Selected Nagpur Mandarin Orchard at Ladgaon, Katol, Nagpur. ***Source*: Srivastava *et al.* (2014a; 2015).**

macro- or micronutrients led to their progressive deficiencies due to heavy removals; ii. sites initially well supplied with P, K or S become deficient when continuously cropped using N alone and iii. fertilizer rates considered optimum still resulted in nutrient depletion at higher productivity levels, if continued, become sub-optimum rates (Tiwari, 2002).

Knowing the required nutrients for all stages of growth, and understanding the soil's ability to supply those needed nutrients are critical to profitable crop production. The recommendations on fertilizer application may not, however, produce the same magnitude of results when practiced on an orchard of large area, because of its inability to accommodate variation in soil fertility status (Srivastava and Singh, 2001a). Slight changes in the nature of soil, local climate, and agronomic practices *etc.* may seriously affect the nutrient utilization capacity of the plant. However, application of a single rate of nutrients may result in over-application of nutrients at some sites and under-application at others often lead to reduced fertilizer use efficiency. Under such circumstances, site specific nutrient management (SSNM) is adopted in orchards requiring different treatment in patches as per the nature of surface and sub-surface soil properties so as to improve the orchard efficiency (Average yield of specified trees in relation to average orchard yield) in ultimate terms (Srivastava *et al.*, 2007).Therefore, fertilizer application should be tailored according to the crop's need keeping in view, the capacity of these soils to fulfill various demands (Srivastava *et al.*, 2006; 2009). To achieve this, it is necessary to keep an overall nutrient balance in relation to total crop load. This may indicate the need for the application of different nutrients at specific times, in a particular order to derive the maximum benefit from the application of a given quantity of nutrients keeping in mind the spatial variability in availability of different nutrients within

Table 8: Soil Test-based Fertilizer Recommendation in Relation to different Targeted Fruit Yield of Nagpur Mandarin as a Ready Reckoner

Soil Test Value (kg/ha)	*Targeted Fruit Yield (q/ha)*				
	100	*150*	*200*	*250*	*270*
Soil nitrogen test-based fertilizer recommendation					
100	107.2	172.65	238.10	303.55	369.0
200	101.27	166.72	232.17	297.62	363.07
300	95.35	160.80	226.25	291.70	357.15
400	89.42	154.87	220.32	285.77	345.30
500	83.50	148.95	214.4	279.85	315.22
Soil phosphorous test-based fertilizer recommendation					
5	345.85	585.85	825.85	106.5	130.5
10	211.7	451.7	691.7	931.7	117.1
15	77.55	317.55	557.55	797.55	103.7
20	–56.6	183.4	423.4	663.4	903.4
25	–190.8	49.25	289.25	529.25	769.25
Soil potassium test-based fertilizer recommendation					
100	130.10	214.70	299.30	383.90	468.50
200	91.00	175.60	260.20	344.80	429.40
300	51.90	136.50	221.10	305.70	390.30
400	12.80	97.40	182.00	266.60	351.20
500	–26.30	58.30	142.90	227.50	312.10
Soil zinc test-based fertilizer recommendation					
0.89	75.84	124.84	173.84	222.84	242.44
1.79	53.68	102.68	151.68	200.68	220.28
2.68	31.53	80.53	129.53	178.53	198.13
3.58	9.37	58.37	107.37	156.37	175.97
4.48	–12.79	36.21	85.21	134.21	153.81
Soil iron test-based fertilizer recommendation					
8.96	156.28	253.28	350.28	427.88	486.08
13.4	137.42	234.42	331.42	409.02	467.22
17.9	118.56	215.56	312.56	390.16	448.36
22.4	99.70	196.70	293.70	371.30	429.50
26.8	80.84	177.84	274.84	352.44	410.64
Soil manganese test-based fertilizer recommendation					
11.2	59.98	98.48	136.98	175.48	190.88
20.1	46.36	84.86	123.36	161.86	177.26
29.1	32.74	71.24	109.74	148.24	163.64
38.0	19.12	57.62	96.12	134.62	150.02
44.8	8.90	47.40	85.90	124.40	139.80

Source: Srivastava *et al.* (2014b; 2015).

an orchard. In this background, the present experiment was carried out to determine fertilizer doses of Nagpur mandarin grown on an orchard having large spatial variability in soil fertility under sub-humid tropical climate of central India.

Conventional nutrient management practices in perennial crop like citrus hardly exploit the spatial variability in soil fertility, with the result; fertilizer recommendation is always inclined to reduce fertilizer-use-efficiency. Efforts were made to develop soil fertility variogram-based site specific nutrient management (SSNM) practices in Nagpur mandarin grown on smectite rich black clay soil (Typic Ustorthent and Typic Haplustert) of central India. Different SSNM treatments were observed more responsive on shallow (Typic Ustorthent) than deep soil (Typic Haplustert) due to difference in the type and magnitude of multiple nutrient deficiencies (Figure 2). Best SSNM treatment in terms of response on canopy growth, yield, fruit quality and soil-plant nutrient buildup were observed to be T_9 ($N_{1200} + P_{600} + K_{600} + M_1S_1$) and T_6 ($N_{600} + P_{400} + K_{300} + M_1S_1$) on shallow soil (Typic Ustorthent) and deep soil (Typic Haplustert), respectively, within the same orchard. And fertilizer use needs to be tailored as per these doses instead of using recommended doses of fertilizers (RDF) (T_{16}) as a basis. Higher application of K at the rate of 900 g/tree alongwith 600 g N + 400g P(T_8) produced much higher acidity which imparted greener colour to the fruits, thereby, took comparatively longer time for colour break and attain fruit maturity on both shallow as well as deep soil types. Increasing application of K from 0 to 900 g/tree with $N_{600} + P_{400}$ or from 600 to 1500 g/tree with $N_{1200} + P_{600}$ maintained concurrently higher DTPA-Zn in both the soil types which aided in higher Zn accumulation in index leaves. Effect of K was more pronounced at $N_{600} + P_{400}$ (T_5–T_8) than at $N_{1200} + P_{600}$ ($T_9 - T_{12}$), irrespective of soil types. The synergism between Zn and K can be befittingly exploited with the objectivity of fertilizer-use-efficiency vis-à-vis quality citrus production, besides dictating the time of fruit maturity (Table 10).

5. Nutrient-Microbe-Crop Synergy

The use of inorganic fertilizers continuously and excessively without using organic fertilizers as a balancer may result in soil becoming barren and decrease productivity. "Microbial consortia" is a beneficial microbial mixture that has been developed to improve soil quality and crop yield while simultaneously dramatically reducing inorganic chemical application (Maiyappan *et al.*, 2010) Rhizospheric microbe plant interactions have a great influence on plant health and soil nutritional and soil borne pathogen stress conditions (Lareen *et al.*, 2016). Plant growth promoting rhizobacteria (PGPR) can be considered among rhizosphere beneficial microorganisms (Poszytek *et al.*, 2016). Rectifying the disadvantages of the carrier based biofertilizers, liquid biofertilizers have been developed which would be the only alternative for the cost effective sustainable production (Pindi and Satyanarayana, 2012). These PGPRs (*Bacillus pumilus* INR7), *Bacillus subtillis* (GB03) and *Curtobacterium flaccumfaciens* (ME1) have capacity to act as antagonist against certain diseases in crop like cucumber (Raupach and Kloepper, 1998).

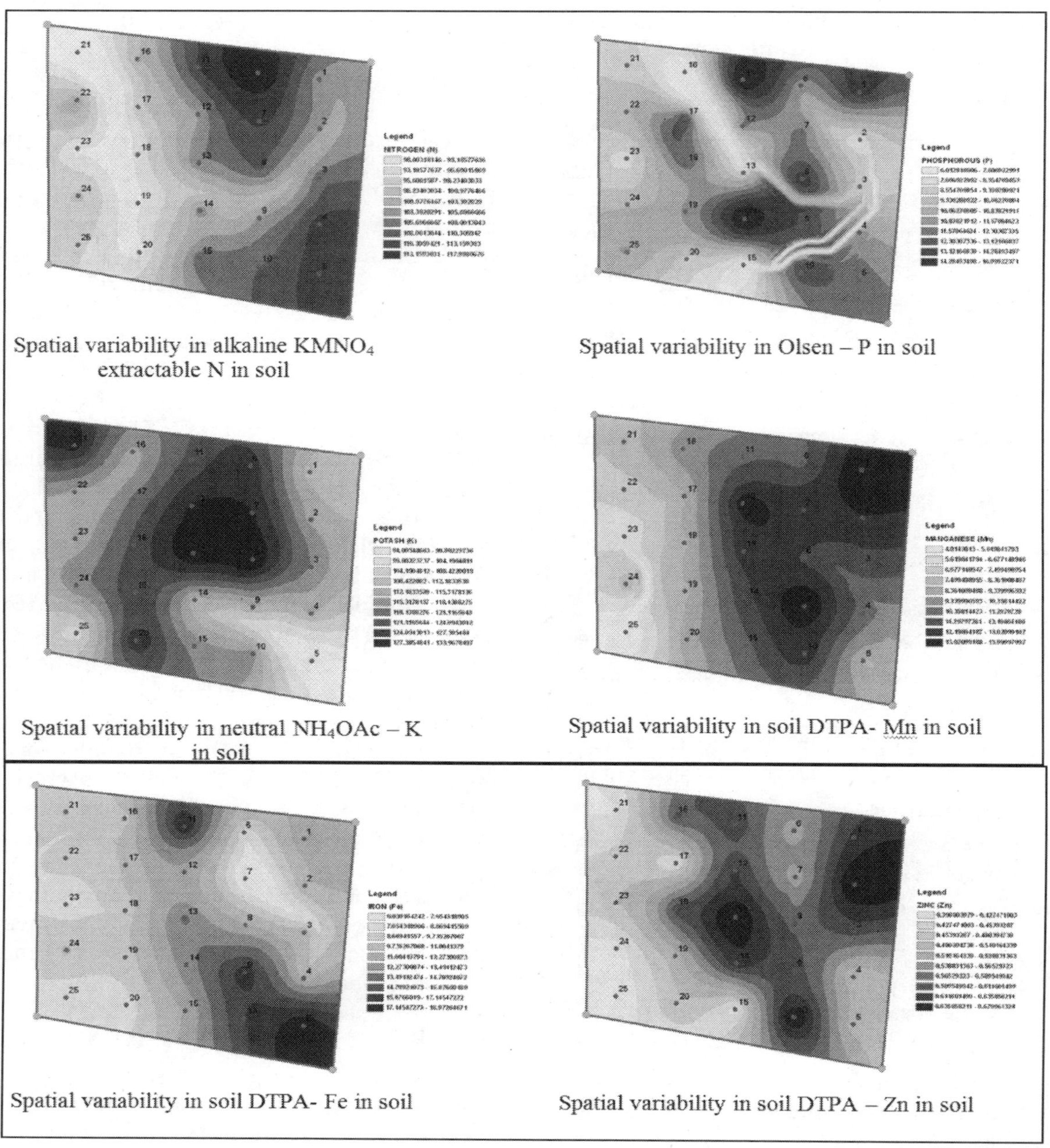

Spatial variability in alkaline $KMNO_4$ extractable N in soil

Spatial variability in Olsen – P in soil

Spatial variability in neutral NH_4OAc – K in soil

Spatial variability in soil DTPA- Mn in soil

Spatial variability in soil DTPA- Fe in soil

Spatial variability in soil DTPA – Zn in soil

Figure 2: Spatial Variability in Soil Fertility Across the Grid Sampling within the Identified Orchard. ***Source*: Srivastava and Singh (2015).**

5.1 Microbial Consortium and Plant Response

Coinoculation or combined inoculation of different microbe types is another area which can be gainfully exploited in formulating the microbially-rich substrate, provided that information on the synergism between different microbes is known (Marschner *et al.*, 2004). In the past, a number of studies have suggested the coinoculation of different microbes, which can be summarized as: *Bacillus subtillis*, *B. licheniformis*, *B. megatericum*, *B. polymyxa*, *B. macerams*, *Pseubomonas fluorescens*, *Pseubomonas putida*, *Nocardia corallina*, *Sacharomyces cerivisiae* and *Thichoderma viride* (20 g dissolved in 4 litres water with 200 ml/pot to have 10^6 cfu/ml (Schoebitz and Vidal, 2016); combination of *Funelifirnus mosseae* and *Bacillus sonorensis* in chilly (Thilagar *et al.*, 251); C2 and C25 strains as *Bacillus* spp. and C32 strain as *Streptomyces* sp. in chilli (Datta *et al.*, 1995); combination of PGPR strains Sp7 and UPMB10 as crop-enhancer and bio-fertilizer in banana (Mia *et al.*, 2010); consortium of *Bacillus* in banana (Jaizme-Vega *et al.*, 2003) in banana. for vigour seedling and production of banana. Many studies on coinoculation of microbes involving AM fungi and bacteria havae also been suggested for improvement in both yield and quality. These include: *A. brasilense- G. fasciculatum* in wheat (Gori and Favilli, 1995), strawberry (Bellone and de Bellone 1995); *A. brasilense-Pantoea dispersa* in sweetpepper (Amor *et al.*, 2008); and *A. chroococcum - G. mosseae* in pomegranate (Aseri *et al.*, 2008).

Table 10: Response of different Site Specific Nutrient Management-based Treatments on Canopy Growth, Fruit Yield and Quality of Nagpur Mandarin on Typic Ustorthent (S_1) and Typic Haplustert Soil (S_2) Soil Types (pooled data)

Treatment	*Canopy Volume (m^3)**		*Fruit Yield (kg/tree)*		*Fruit Quality Parameters (%)*					
					Juice Content		*TSS*		*Acidity*	
	Tu	*Th*	S_1	S_2	S_1	S_2	S_1	S_2	S_1	S_2
T_1 ($N_{600}+P_{400}+K_{600}+M_1S_1$)	3.9 (25.7)	3.5 (19.3)	37.4	58.9	45.7	45.5	8.2	8.1	0.56	0.77
T_2 ($N_{600}+P_{200}+K_{600}+M_1S_1$)	3.7 (27.3)	2.7 (21.3)	30.6	57.2	44.5	41.6	8.5	7.6	0.64	0.68
T_3 ($N_{600}+P_0+K_{600}+M_1S_1$)	3.4 (29.5)	3.1 (23.0)	27.9	57.1	44.1	42.4	9.1	8.4	0.60	0.75
T_4 ($N_{600}+P_{600}+K_{600}+M_1S_1$)	4.6 (28.7)	2.9 (20.6)	39.2	58.0	44.7	43.7	8.8	7.9	0.63	0.68
T_5 ($N_{600}+P_{400}+K_0+M_1S_1$)	4.2 (31.7)	2.4 (21.4)	33.4	55.3	41.9	42.4	9.6	8.7	0.56	0.64
T_6 ($N_{600}+P_{400}+K_{300}+M_1S_1$)	4.7 (32.6)	5.4 (21.7)	33.9	68.5	44.9	49.8	9.3	8.6	0.58	0.67
T_7 ($N_{600}+P_{400}+K_{600}+M_0S_0$)	3.8 (27.0)	2.6 (19.6)	25.1	39.2	45.2	46.5	8.6	7.8	0.62	0.81
T_8 ($N_{600}+P_{400}+K_{900}+M_1S_1$)	5.7 (28.5)	4.3 (20.5)	49.9	48.7	48.6	48.2	8.2	7.9	0.75	0.82
T_9 ($N_{1200}+P_{600}+K_{600}+M_1S_1$)	6.6 (31.2)	3.7 (23.4)	52.7	60.9	45.4	42.7	8.9	8.8	0.55	0.62
T_{10} ($N_{1200}+P_{600}+K_{900}+M_1S_1$)	6.6 (29.1)	3.3 (23.4)	41.8	50.4	42.6	43.6	8.6	8.2	0.59	0.71
T_{11} ($N_{1200}+P_{600}+K_{1200}+M_1S_1$)	5.8 (30.0)	3.9 (20.5)	39.3	56.1	44.9	44.8	8.5	8.1	0.62	0.76
T_{12} ($N_{1200}+P_{600}+K_{1500}+M_1S_1$)	4.6 (28.4)	4.3 (19.7)	36.3	56.3	48.4	46.3	8.2	7.6	0.80	0.86
T_{13} ($N_{1200}+P_{600}+K_0+M_1S_1$)	3.8 (26.6)	3.3 (19.5)	33.3	46.5	41.6	42.5	8.2	9.1	0.51	0.66
T_{14} ($N_{1200}+P_{600}+K_{1200}+M_0S_1$)	4.5 (28.1)	2.9 (21.2)	33.9	46.3	43.2	43.7	9.5	8.5	0.64	0.77
T_{15} ($N_{1200}+P_{600}+K_{1200}+M_0S_0$)	3.9 (27.2)	2.9 (23.2)	30.0	45.5	44.6	43.8	9.6	8.2	0.63	0.74
T_{16} ($N_{600}+P_{200}+K_{100}$) – RDF	3.5 (24.5)	3.0 (23.7)	31.5	43.7	44.0	43.1	8.6	8.5	0.57	0.68
LSD (*p=0.05*)	1.2	0.8	8.0	8.1	3.1	2.2	0.5	0.6	0.09	0.08

M_1 stands for application of 300 g each of $ZnSO_4$, $FeSO_4$, $MnSO_4$ and 100 g borax/tree; Mo stands for no application of micronutrient fertilizers; S_1 stand for application of 400 g $MgSO_4$/tree and 100 g elemental S/tree; Tu and Th stand for Typic Ustorthent and Typic Haplustert, respectively; So stands for no application of Mg and S; RDF stands for recommended doses of fertilizers; *expressed as increase over initial value as given in parenthesis; Source: Srivastava and Singh (2015; 2015).

Differential response of inoculation of different microbes of microbial consortium on seed germination, at various dates of observation. The germination was observed to increase from 1.0 per cent at 10 days to 35.2 per cent at 100 days of germination with treatment T_0, 0.83 per cent to 46.4 per cent with T_1, 10.8 per cent at 20 days of germination (First 10 days, no germination was observed) with treatment T_2, 10.8 per cent at 20 days of germination to 23.9 per cent with T_3, 1.9 per cent to 73.9 per cent with T_4 and 1.7 per cent to 76.4 per cent with T_5 during 100 days of observations. But first 30 days of observations showed almost maximum germination in all the treatments (Figure 4). The maximum seed germination of 76.4 per cent, was observed with treatment T_5 (*Th + Bm + Pp + Ac + Pf*), while treatment T_3 (*Th + Bm + Pp*) observed minimum seed germination of 23.95 per cent, significantly lower to even control treatment T_0 (35.20 per cent). The maximum seed treatment T_3 (*Th+ Bm + Pp*). The maximum reduction in germination percentage of 76.0 per cent was observed with treatment T_3 followed by T_2, T_0, T_1, T_4 and T_5 in decreasing order (Table 11).

Seed inoculation of microbial consortium influenced the seed germination to varying proportions, (Table 11) and at various dates of observation. The maximum seed germination of 72.29 per cent, was observed with treatment T_5 (*Th + Bm + Pp + Ac + Pf*), while treatment T_1 (*Th*) showed minimum seed germination of 61.66 per cent, statistically inferior to control treatment T_0 (63.33 per cent) at all the stages of observation. The maximum seed viability index of 2.76 was observed with treatment T_5 (*Th + Bm + Pp + Ac +Pf*) followed by 2.68 with T_4 (*Th*

Table 11: Growth Response of Acid Lime Seedlings in Response to different Microbial Inoculations (Period: 120 days)

Treatment	*Shoot Parameter*				*Root Parameter*	
	Shoot Length (cm)	*Shoot Wt. (g)*	*No. of Leaves/ Plant*	*Girth (mm)*	*Root Length (cm)*	*Root Wt. (g)*
T_0 (Control)	18.6	1.60	14	1.87	10.5	0.40
T_1 (*Th*)	21.5	1.72	16	2.03	11.1	0.43
T_2 (*Th+Bm*)	22.7	2.53	19	2.32	13.6	0.67
T_3 (*Th+Bm+Pp*)	23.9	2.64	19	2.38	8.6	0.71
T_4 (*Th+Bm+Pp+Ac*)	24.9	2.60	22	2.21	11.1	0.78
T_5 (*Th+Bm+Pp+Ac+Pf*)	23.4	2.81	26	3.09	19.4	0.89
CD (*P=0.05*)	1.2	0.10	1.8	0.11	0.40	0.06

Th, Bm, Pp, Ac and *Pf* stand for *Trichoderma harzianum, Bacillus mycoides, Paenibacillus polymyxa, Azotobacter chroococcum* and *Pseudomonas fluorescens* respectively.

Source: Srivastava *et al.* (2015).

+ *Bm* + *Pp* + *Ac*), 2.57 with T_3 (*Th* + *Bm* + *Pp*), 2.53 with T_2 (*Th* + *Bm*), 1.73 with T_1 (*Th*) and 1.65 with T_0 (Control). The changes in reduction in germination percentage over time in response to different microbial inoculants likewise showed a similar response (Table 12), with a 27.77 per cent reduction with T_5 (*Th* + *Bm* + *Pp* + *Ac* + *Pf*), 30.37 per cent reduction with T_4 (*Th* + *Bm* + *Pp* + *Ac*), 33.33 per cent reduction with T_3 (*Th* + *Bm* + *Pp*), 32.91 per cent reduction with T_2 (*Th* + *Bm*), 38.33 per cent reduction with T_1 (*Th*) and 36.66 per cent reduction with T_0 (Control). These observations while comparing the response of different microbial inoculants in soil, showed much higher germination percentage of seeds in soil inoculated with broths of different microbes constituting microbial consortium.The maximum reduction in germination percentage of 38.33 per cent was observed with treatment T_1 (*Th*), while treatment T_5 (*Th* + *Bm* + *Pp* + *Ac* + *Pf*) observed minimum reduction in germination percentage of 27.70 per cent.

Table 12: Growth Response of Acid Lime Seedlings in Response to different Microbial Inoculants (Period: 120 days)

Treatment	*Shoot Parameter*				*Root Parameter*	
	Shoot Length (cm)	*Shoot Wt. (g)*	*No. of Leaves/ Plant*	*Girth (mm)*	*Root Length (cm)*	*Root Wt. (g)*
T_0 (Control)	16.6	1.61	19	1.96	9.6	0.38
T_1 (*Th*)	17.5	2.12	21	1.99	10.7	0.43
T_2(*Th+Bm*)	22.5	2.26	28	2.16	15.3	0.51
T_3(*Th+Bm+Pp*)	20.1	2.56	32	2.46	15.5	0.64
T_4(*Th+Bm+Pp+Ac*)	21.4	3.25	33	2.92	17.1	0.77
T_5(*Th+Bm+Pp+Ac+Pf*)	23.3	3.48	38	3.17	19.1	1.01
CD (p = *0.05*)	0.50	0.20	3	0.18	0.82	0.22

Th, Bm, Pp, Ac and *Pf* stand for *Trichoderma harzianum, Bacillus mycoides, Paenibacillus polymyxa, Azotobacter chroococcum* and *Pseudomonas fluorescens*, respectively.

Source: Srivastava *et al.* (2015)

5.2 INM and Nutrient Dynamics

In recent years, nutrient additions have been exclusively in favor of inorganic fertilizers (IF) due to demographic pressure and demands related to life styles and trade involvement. While the quick and substantial response to fruit yield due to IF eclipsed the use of organic manures (OM), the inadequate supply of the latter sources exacerbated this change (Srivastava, 2009). Although, differential efficacy of two conventional methods of fertilization (soil versus foliar application) has helped in improving the quality production of citrus (Srivastava *et al.*, 2008a; 2008b) and other fruits (Singh *et al.*, 2011; 2012).

In recent years, continuous fertilization has failed to sustain the yield expectancy on a long term basis due to depletion of soil carbon and consequently, multiple nutrient deficiencies have emerged irrespective of soil type (Srivastava and Singh, 2009b). The menace of multiple nutrient deficiencies has further been triggered through increase in air temperature via changes in microbial communities and activities within the rhizosphere in the light of climate change (Wu and Srivastava, 2012). Such changes will adversely dictate on the orchard's productive life in long run. Gradual shift from purely IF to OM started gaining wide scale application for enhanced nutrient cycling (Srivastava *et al.*, 2002).

Integrated nutrient management (INM) as a dynamic concept of nutrientmanagement considers the economic yield in terms of fruit yield coupled with quality on one hand, and soil physico-chemical and microbial prospects on other hand as a marker of resistance against the nutrient mining (arises because of failure to strike a balance between annual nutrient demand and quantity of nutrients applied). Soils under citrus differ from other cultivated soils, with respect to fallow period of 3-6 months every year forcing depletion of soil organic matter in latter case (Bhargava, 2002). In contrast, biological oxidation of existing carbon (C) continues in soil covered under citrus (Srivastava *et al.*, 2002). Multiple nutrient deficiencies are considered to have a profound effect on potential source of atmospheric carbon dioxide (CO_2). Soil carbon stock is, hence, considered as an important criterion to determine the impact of INM in the longer version of impact assessment (He *et al.*, 1997). The amount of accumulated C within the rhizosphere soil does not continue to increase with time with increasing C outputs. An understanding of the mechanism involved in C stabilization in soils is needed for controlling and enhancing soil C sequestration (Goh, 2004) under different modes of nutrient management. In this background, studies were carried out with three major objectives: i. response of INM on canopy growth, fruit yield and quality indices, ii. soil carbon dynamics related fertility changes in response to INM, and iii. INM-induced changes in plant nutrition. These objectives are anticipated to collectively provide a strong database support for suitability of INM module versus sole application either Inorganic fertilizers or green manure.

Different treatments displayed significant response on fruit yield (Table 13). The results on response of various treatments revealed diverse effect on the yield/tree at maturity. The maximum fruit yield of 80.95 kg/tree was observed with treatment T_5 which was better than 72.66 kg tree^{-1} with T_4 or 65.68 kg/tree with T_3 than 60.23 kg/tree with treatment T_2 However, all these treatments were far superior in magnitude of response when compared with 100 per cent RDF as T_1 (58.07 kg/tree). Various fruit quality parameters including fruit weight, juice content, peel colour, disease infestation on trees, fruit firmness, shape at maturity, uniformity in yield/tree, uniformity in fruit size, peel smoothness,

Table 13: Response of Microbial Consortium Loaded INM - Treatments on Canopy Volume, Fruit Yield and Soil Fertility Changes on Vertic Ustochrept (Pooled data 2010-16)

Treatment	*Canopy Volume (m^3)*	*Fruit Yield (kg/tree)*	*Plant Available Nutrients in Soil (mg/kg)*						
			Macronutrient			*DTPA-Micronutrients*			
			$KMnO_4$-N	*Olsen-P*	*NH_4OAc-K*	*Fe*	*Mn*	*Cu*	*Zn*
T_1 (100% RDF)	14.69 (13.04)*	58.07	145.6	9.22	184.5	10.40	10.54	2.13	0.96
T_2 (75% RDF + 25% Vm)	21.36 (18.51)	60.23	148.2	9.31	193.2	12.91	11.38	2.39	1.21
T_3 (75% RDF + 25% Vm +MC)	23.12 (20.28)	65.68	153.7	9.85	204.6	15.81	12.74	2.80	1.37
T_4 (50% RDF + 50% Vm)	25.20 (22.42)	72.66	160.6	10.38	221.5	17.22	11.93	2.88	1.47
T_5 (50% RDF + 50% Vm+ MC)	21.45 (18.33)	80.95	175.0	11.42	233.6	19.11	13.13	2.95	1.57
CD (p=0.05)	–	3.20	1.4	0.41	5.3	0.90	NS	NS	0.10

* Initial observations.

RDF, Vm and MC stand for recommended doses of fertilizers, vermicompost and microbial consortium (*Paenibacillus polymyxa, Trichoderma harzianum, Bacillus mycoides, Azotobacter chroococcum* and *Pseudomonas fluorescens*), respectively.

Source: Srivastava *et al.* (2009; 242).

TSS, Acidity, fruit diameter, number of seeds/fruit, *etc.* were considered in such comprehensive evaluation of fruit quality. Based on accumulated index value of 100, highest index value was observed with treatment T_5 (69/100) which was significantly higher than other treatments such as 66/100 with T_3, 65/100 with T_4, 57/100 with T_1 and 55/100 with T_2 in decreasing order.

5.3 Nutrient Dynamics

The changes in concentration of crunch nutrients like N, K, and Zn were started across crop critical growth stages in response to different treatments (Figures 3–5).

Leaf nitrogen concentration was observed maximum as 1.01 per cent at stage I,1.56 per cent at stage II, 1.61 per cent at stage III, 2.30 per cent at stage IV, 2.38 per cent at stage V and 2.20 per cent at stage VI, which significantly superior to T_4 (0.98 per cent at stage I,1.16 per cent at stage II,1.40 per cent stage III,1.61 per cent at stage IV, 2.18 per cent at stage V and 1.98 per cent at stage VI),T_3 (0.94 per cent at stage I,1.06 per cent at stage II, 1.28 per cent at stage III, 1.38 per cent at stage IV, 2.01 per cent at stage V and 1.90 per cent at stage VI),T_2(0.92 per cent

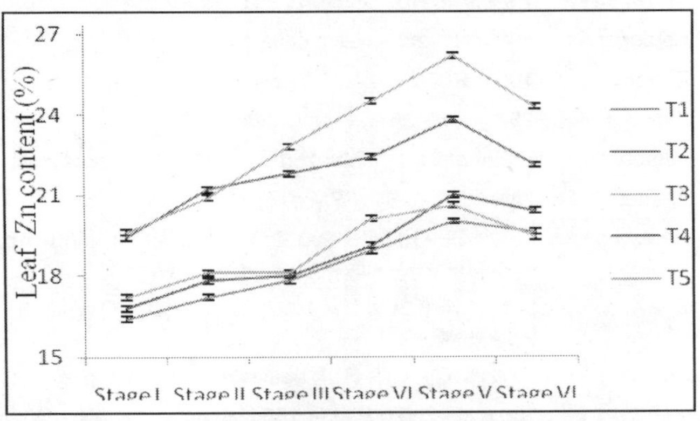

Figure 4: Response of different Treatments on K Accumulation Across Critical Growth Stages of Nagpur Mandarin.

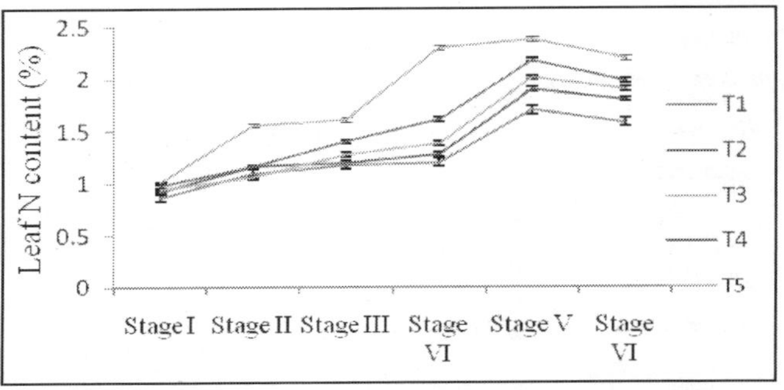

Figure 3: Response of different Treatments on the N Accumulation Across Critical Growth Stages of Nagpur Mandarin.

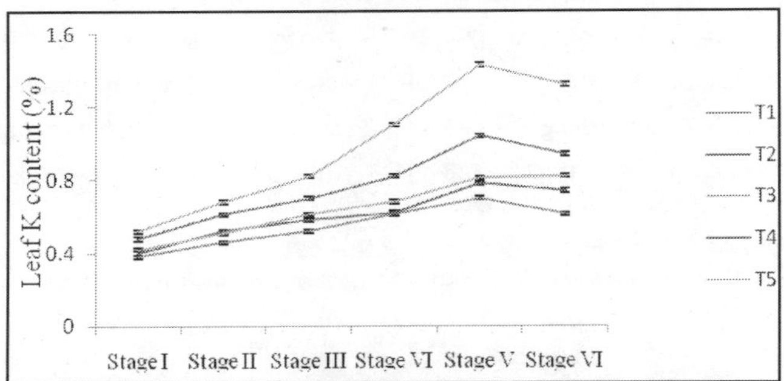

Figure 5: Response of different Treatments on Zn Accumulation Across Critical Growth Stages of Nagpur Mandarin.

***Source*: Srivastava *et al.* (2015).**

at stage I, 1.16 per cent at stage II, 1.20 per cent at stage III, 1.28 per cent at stage IV, 1.90 per cent at stage V and 1.80 per cent at stage VI) and T_1 (0.87 per cent at stage I,1.10 per cent at stage II, 1.18 per cent at stage III,1.20 per cent at stage IV, 1.70 per cent at stage V and 1.58 per cent at stage VI) in decreasing order of their effectiveness (Figure 3).

Likewise the treatment T_5 maintained almost optimum K-concentration from 0.68 per cent at stage II to as high as 1.43 per cent at stage V and 1.32 per cent at stage VI, significantly superior to any of the other treatments including T_4 (Figure 4).

Like other two nutrient (N and K), dynamics of Zn was also observed to follow the same physiological pattern of accumulation at various crop phenophases (Figure 5). These observations suggested that better nutrient dynamics is maintained across all critical growth stages by mobilising the both the nutrient fractions, applied nutrients as well as the native source of nutrients through treatment T_5. And this is the reason for better effectiveness of treatments, comparatively higher nutrient concentration is maintained across all important growth stages, since our earlier findings advocated, it is not possible to forego the nutrient application at any growth stage considered critical to that crop.

Table 14: Different Components of Integrated Nutrient Management Recommended for different Fruit Crops

Crop	*Nutrient-Microbe Combination*	*References*
Guava	FYM 50 kg/plant – *Azotobacter* sp. 50 g/plant - *Azospirillum* sp. 50 g/plant – *Sesbania* sp. as green manure	Ram and Rajput (144)
Pomegranate	400 g N- 100 g P_2O_5 – 300 g K_2O/plant – FYM 20 kg/plant,	Ghosh *et al.* (54)
Papaya	Vermicompost 20 kg/plant – rhizosphere culture 50 g/plant- 150 N – 200 P_2O_5 – 200 K_2O g/plant (75% RDF)	Kirad *et al.* (88)
Banana	FYM 12 kg/plant – *Azospirillum* sp. 50 g/plant - Phosphate Solubilising Bacteria 50 g/plant *T. harzianum* 50 g/plant	Hazarika and Ansari (64)
Banana	50% RDF- FYM 20 kg/plant – *Azotobacter* sp. 50 g/plant – Phosphate solubilising bacteria 50 g/plant –VAM 250 g/plant	Patil and Shinde (136)
Guava	488 g N – 244 g P_2O_5 - 281 g K_2O/plant – FYM 50 kg/plant – *Azotobacter* 250 g/plant – phosphate solubilising bacteria 25 g/plant	Barne *et al.* (9)
Strawberry	75% N as RDF – 25% N as FYM – *Azotobacter* sp.	Umer *et al.* (255)
Pomegranate	300 g N/plant – neem cake 1 kg/plant	Ray *et al.* (149)
Banana	100% RDF – 40% Wellgrow organic manure	Kuttimani *et al.* (93)
Peach	75% RDF - 25% N equivalent FYM	Shah *et al.* (159)
Lemon	N 525 g/plant – FYM 150 kg/plant – *Azotobacter* sp. 18 g/plant	Khehra and Bal (85)
Apricot	75% RDF – 25% FYM	Shah *et al.* (159)
Papaya	50% RDF (100 N – 100 P_2O_5 – 125 K_2O g/plant)- *Azotobacter* sp. 50 g/plant – Phosphate solubilising bacteria 2.5 g/m^2	Singh and Varu (197)
Guava	50% RDF (250 g N – 100 g P_2O_5 - 250 K_2O g/plant) - FYM 25 kg/plant – vermicompost 5 kg/plant	Dwivedi (45)
Sapota	75% RDF + 25% RDF equivalent vermicompost	Hebbarai *et al.* (66)
Mango	500 g N - 250 g P_2O_5 – 250 K_2O g/plant –50 kg FYM/plant – *Azospirillum* sp. 250 g/plant	Singh and Banik(195)
Mango	250 N – 425 P_2O_5 – 1000 K_2O – *Azospirillum* sp. 250 g/plant – PSB – 250 g/plant – $ZnSO_4$ 100 g/plant – Borax 100 g/plant	Hasan *et al.* (62)
Banana	100% RDF – FYM 10 kg/plant – *Azospirillum* sp. 25 g/pant Phosphate solubilising bacteria 250 g/plant	Bhalerao *et al.* (14)
Guava	236 g N – 66 g P_2O_5 – *Azospirillum* sp.30 g/plant – VAM 30 g/plant	Dutta *et al.* (43)
Mosambi	300 g N – 250 g P_2O_5 – 300 g K_2O – AMF 10 g/plant - *Azospirillum sp* 25 g/plant	Patel *et al.* (133)
Guava	250 g N – 100 g P – 250 g K_2O/plant – *Azotobacter* sp. 250 g/plant	Shukla *et al.* (193)
Litchi	500 g N – 250 g P_2O_5 – 500 g K_2O/plant – FYM 50 kg/plant – *Azotobacter* sp. 150 g/plant – VAM 100 g/plant	Dutta *et al.* (42)
Aonla	50% NPKS (105 kg N – 7.20 kg P_2O_5 – 125.25 kg K_2O/ha) – Biofertilizers (*Azotobacter* sp.– *Azospirillum* sp. – Phosphate solubilising bacteria) – FYM (2 tonnes/ha)	Yadav *et al.* (263)
Aonla	100 g N – 25 g P_2O_5 – 150 g K_2O/plant – FYM 10 kg/plant – Phosphate solubilising bacteria 50 g/plant	Mandal *et al.* (105)
Sapota	1500 g N – 1000 P_2O_5 – 500 g K_2O/plant – 75 kg FYM – 12.5 g/plant PSB	Dalal *et al.* (35)
Guava	50% RDF (225 g N – 195 g P_2O_5 – 150 g K_2O/plant)- FYM 50 kg/plant – *Azospirillum* 250 g/plant	Goswami *et al.* (58)

RDF and PSB stand for recommended doses of fertilizers and phosphate solubilising bacteria predominantly (*Pseudomonas fluorescens*), respectively; FYM and AMF stand for farmyard manure, and arbuscular mycorrhizal fungi, respectively.

An array of fruit crops have been reported to respond to the synergies originated through combination of organic nutrient-microbe-inorganics (Table 14). And such associations have invariably witnessed substantially higher productivity than any single component alone. However, there is a greater need to expand such plant response advantages using more rhizocompetent microbes preferably in consortium mode, plant response as well as soil health response both have to be sustained on a long term basis.

6. Way Forward

Despite many cutting edge technologies, addressing a variety of core issues on role of soil type-based nutrient management (Lipecki and Berbec, 98) in raising the productivity of perennial fruits is the major research and developmental issue. Fruit crops, despite some distinct developments, are confronted with number of soil fertility challenges. These challenges could be convincingly elaborated as: sustaining the sustainable soil fertility management, we need to define role of rhizosphere in ecological service (CO_2 sink), integrating soil nutrient and microbial pool over different spatial and temporal scales cropwise, soil fertility resilience and human health, plant-soil-microbial interface (For example, crop microbiome), coupled reaction processes in soil to add new frontiers of soil fertility management, besides robust and dynamic crop-based nutrient modelling duly tested spatio- temporally.

Until recently, research has focused on those organisms that are culturable. However a wealth of information is now being collected from both culturable and, as yet, unculturable organisms. Functions of the soil microbial population impact many processes and, therefore, productivity (Kennedy and Gewin, 83), if mechanisms involved in the plant-microbe interaction are better understood, since, a plant manufactures microbial communities according to its metabolic requirements. The microbial biomass is one of the biological properties of soil that undergoes immediate change in response to fertilizer like input. Studies, therefore, need to be undertaken with a view to explore the possibility of which soil microbial property could be used as a potential tool for finding out soil health related constraint instead of concentration of available nutrients in soil using some indicator fruit crop(s). While the genetic, functional and metabolic diversity of soil microorganisms within the rhizosphere of wide range of fruit crops is important, the capacity of soil microbial communities to maintain functional diversity of those critical soil processes could ultimately be more important to ecosystem productivity and stability than mere taxonomic diversity (Caldwell, 22).

In this context, it remains to be assessed how nutrient-microbe synergism is associated with productivity of perennial fruits. New research methods involving molecular techniques will extend our understanding of taxonomic and functional diversity in soil systems. With the availability of more technical know-how on combined use of organic manures, prolonged shelf-life of microbial bio-fertilizers, and inorganic chemical fertilizers, an understanding on nutrient acquisition and regulating the water relations would help switch orchards to CO_2 sink (expanding carbon capturing capacity of rhizosphere) so that a more sustainable fruit-based integrated crop production system under biotic and abiotic stress could be evolved. The molecular approach to breeding of mineral deficiency resistance and mineral efficiency would facilitate to produce nutritionally efficient biotypes in order to maximise the quality production of fruit crops on sustained basis. The work related to microbial inoculants for mass production, formulation coupled with innovative marketing, interaction and signalling with the plant and soil environment need further redressal to reorient fruit nutrition research. The further efforts on homologous microbes as rhizosphere microbes versus plant endophytes would further after way forward approaches in identifying more crop endosphere competent microbes for elevated crop response. It remains to be investigated, how nutrient-microbe association could bring better dividends to accurate estimation of orchard C budget vis-a-vis time scale and feedback mechanisms of changes in soil carbon pool and steady state level under specific fruit crop in order to expand potential of C credits through perennial fruit crops.

The above nutrient management-based issues and strategies would find their worth only when backed up with sound methodologies of nutrient constraints diagnosis. The currently available leaf nutrient standards for different fruit crops have certain distinct limitations in terms of: i). Sampling index plant part(s) leaving very short time for remediation of identified constraints, ii). application of diverse interpretation tools bring out varying limits of nutrient concentration, and iii). do we need to apply some correction factor when such nutrient standards are applied in fruits under fertigation. A better rationale of nutrient constraints diagnosis as early as possible in a standing crop would surely pave the way for precision-based nutrient response accruing in elevated productivity. Proximal sensing, preferably hyper-spectral analysis would add a new dimension in this regard, since the current basis of nutrient constraints diagnosis is skewed more towards next season crop than current season crop. An early warning system exploiting the metallo-enzyme sensor could be another viable option in years to come. Delineation of production management zones linked with variable rate fertilizer application as per the crop phenology is expected to tailor the fertilizer requirement without altering the fertilizer requirement of a crop soil fertility based spatial variogram would further act as a decision support tool for precise fertilizer recommendation. Evaluating nutrient response at a cellular, sub-cellular, tissue level and plant part level instead of whole plant basis, would lay a solid foundation of nutrient management strategy in fruit crops. Eventually, such attempts warrant for developing Nutrient Experts based on 4R Stewardship Concept advocated by International Plant Nutrition Institute, Gurgaon which have displayed some definite yield advantages in cereal crops, but such serious efforts are direly needed in fruit crops, if nutrient management is to be linked with nutrient use efficiency. On the line of 4R in plant nutrition, there is also a need to look 4W in terms of irrigation management, so that 4R and 4W operate complimenting each other to facilitate not only the sustained quality production, but aid in moderating the climate change – related issues as well.

7. References

Adak, T., Kailash Kumar and Singh, Vinod Kumar. 2014. Fertigation regime impacting productivity, moisture, and nutrient distribution in mango under sub-tropical condition. *Indian J. Soil Consev.* **42**: 289– 292.

Adams, M.L., Norvell, W.A., Philpot, W.D. and Peverly, J.H. 2000. Spectral detection of micronutrient deficiency in 'Bragg' soybean. *Agron. J.*, **92**: 261–268.

Ahmed, Feza M, Abhijit, Samanta and Abida Jabeen. 2010. Response of sweet cherry (*Prunus avium*) to fertigation of nitrogen, phosphorus and potassium under Kerawa land of Kashmir valley. *Indian J. Agric. Sci.*, **80**: 512–16.

Amor, del F.M., Serrano-Martinez, A., Fortea, M.I., Legua, P. and Nunez-Delicadom, E. 2008. The effect of plant-associative bacteria (*Azospirillum* and *Pantoea*) on the fruit quality of sweet pepper under limited nitrogen supply. *Scientia Hort.* **117**: 191–196.

Arora, N.K. and Singh, R. 2006. Effect of time of supplemental nitrogen application on canopy volume, flowering and yield of semi–soft pear cv. Punjab Beauty. *Indian J. Hort.* **63**: 202–204.

Aseri, G.K., Jain, N., Panwar, J., Rao, A.V. and Meghwal, P.R. 2008. Biofertilizers improve plant growth, fruit yield, nutrition, metabolism and rhizosphere enzyme activities of pomegranate (*Punica granatum* L.) in Indian Thar desert. *Sci. Hort.* **117**: 130–135.

Baldi, E., Toselli, M., Marcolini, G., Quartieri, M., Cirillo, E., Innocenti, A. and Marangoni, B. 2010. Compost can successfully replace mineral fertilizers in the nutrient management of commercial peach orchard. *Soil Use Mgmt.* **26**: 346–353.

Banyal, S.K. and Sharma, S.K. 2011. Effect of fertigation and rootstock on yield and quality of apple. *Indian J. Hort.* **68**: 419–24.

Barne, Varsha G., Bharad, S.G., Dod, V.N. and Baviskar, M.N. 2011. Effect of integrated nutrient management on yield and quality of guava. *Asian J. Hort.* **6**: 546–548.

Barua, Pankaj. 2013. Yield, fruit quality and water productivity of drip fertigated AssamLemon (*Citrus limon*). *Internat. J. Agric. Engg.*, **6**(2): 339–344.

Bellone, C.H. and Bellone, de S.C. 1995. Morphogenesis of strawberry roots infected by *Azospirilium brasilense* and VA mycorrhiza. *North Atlantic Treaty Organisation, Advance Study Institute, Series G* (Ecological Science) **37**: 251–255.

Berg, Gabriele. 2009. Plant–microbe interactions promoting plant growth and health: perspectives for controlled use of microorganisms in agriculture. *Appld. Microbiol. Biotechl.* **84**: 11–18.

Bernardi, Alberto Carlos de Campos, Pedro Luiz Oliveira de Almeida Machdo, Beata Emoke Madari, Silvio Roberto de Lucena Tavares, David Vilas Boas de Campos and Lindbergue de Araújo Crisóstomo. 2007. Carbon and nitrogen stocks of an Arenosol under irrigated fruit orchards in semiarid Brazil. *Sci. Agricola* (Piracicaba, Braz) **64**: http://dx.doi.org/10.1590/S0103–90162007000200010

Bhalerao, N.M., Patil, N.M., Badgujar, C.O. and Patil, D.R. 2009. Studies on integrated nutrient management for tissue cultured Grand Naine banana. *Indian J. Agric. Res.* **43**: 107–112.

Bhargava, B. S. 2002. Leaf analysis for nutrient diagnosis recommendation and management in fruit crops. *J. Indian Soc. Soil Sci.* **50**: 352–373.

Bhatnagar, P., Singh, J., Chauhan, P.S., Sharma, M.K., Meena, C.B. and Jain, M.C. 2016. carbon assimilation potential of Nagpur mandarin (*Citrus reticulata* Blanco). *Internat. J. Sci., Env. Tech.* **5**: 1402–1409.

Bielorai, H., Deshberg, E. and Brum, M. 1984. The effect of fertigation and partial wetting of the rootzone on production of Shamouti orange. *Proc.Int. Soc. Citriculture*, **1**: 118–120.

Bindi, M., Fibbi, L., Gozzini, B., Orlandini S., Seghi, L. and Poni, S. 1997. The effect of elevated CO_2 concentration on grapevine growth under field conditions. *Acat Hort.* **42**: 325–330.

Biswas, H., Dev Narajan and Lakaria, Brij Lal. 2012. Effect of integrated nutrient management on soil properties, performance of Aonla (*Emblica officinais* Gaertn) based on agri–horti system in Bundelkhand region. *Indian J. Soil Conserv.* **40**: 141–146.

Brady, N.C. and Weil, R.R. 2004. Elements of the Nature and Properties of Soils (2nd edn.), Pearson Prentice Hall, NJ.

Bronick, C.J. and Lal, R. 2005. Manuring and rotation effects on soil organic carbon concentration aggregate size fractions on two soils in northeastern Ohio, USA. *Soil Tillage Res.* **81**: 239–252.

Caldwell, Bruce, C. 2005. Enzyme activities as a component of soil diversity: A review. *Pedobiologia* **49**: 637–644.

Centritto, M., Magnani, F., Lee, H.S.J. and Jarvis, P.G. 1999a. Interactive effects of elevated [CO_2] and drought on cherry (*Prunus avium*) seedlings. II. Photosynthetic capacity and water relations. *New Phytol.* **141**: 141–153.

Cevik, B., Kaplankiran, M. and Yurdakul, O. 1987. Studies for determining the most efficient irrigation method for growing lemons under Cukurova conditions. *Doga, Tarum ve, Ormaniciuk* **11**: 42–43.

Chadha, K.L. (Ed) 2007. *Handbook of Horticulture*, ICAR, New Delhi, 1031 p.

Chadha, K.L. and Bhargava, B.S. 1997. Plant nutrient supply, needs, efficiency and policy issues for fruit crops in Indian agriculture from 2000 to 2025 AD. In: Kanwar, J.S. and Katyal, J.C. (Eds). *Plant Nutrient Needs, Supply, Efficiency and Policy Issues: 2000–2025*, NAAS, New Delhi, pp. 1134–132.

Chanda, T.K. 2014. A critical analysis of fertilizer use by crops in India. *Indian J. Fert.* **10**(3): 14–20

Chaney, R.L. 1988. Metal speciation and interaction among elements affect trace element transfer in agricultural and environmental food chains. In: Kramer J.R. and Allen H.E. (Eds.), *Metal Speciation: Theory, Analysis, and Application*, Lewis Publishers, Chelsa, Madison, USA, pp. 319–360.

Chauhan, Neena and Chandel J.S. 2008. Effect of fertigation on growth, yield, fruit quality and fertilizer use efficiency on kiwi fruit (*Actinidia deliciosa* Chev.). *Indian J. Agric. Sci.* **78**: 389–93.

Chen, G.S., Yang, Y.S., Xie, J.S., Li, L. and Gao, R. 2004. Soil biological changes for a natural forest and two plantation in sub-tropical China. *Pedosphere* **14**: 297–304.

Cheng L, Dong, S., Guak, S. and Fuchigami, L.H. 2001. Effects of nitrogen fertigation on reserve nitrogen and carbohydrate status and regrowth performance of pear nursery plants. *Acta Hort.* **564**: 51–62.

Conroy, J.1992. Influence of elevated atmospheric CO_2 on concentrations of plant nutrients. *Austral. J. Bot.* **40**: 445–456.

Corwin, D.L. and Lesch, S.M. 2005.Characterizing soil spatial variability with apparent soil electrical conductivity I., Survey protocols. *Computers and Electronics in Agric.* **46**: 103–33.

Dalal, S.R., Gonge, V.S., Jogdande, N.O. and Moharia, Anjali 2004. Response of different levels of nutrients and PSB on fruit yield and economics of sapota. *PKV Res. J.* **28**: 126–128.

Das, B.S., Sarathjith, M.C., Santra, P., Sahoo, R.N., Srivastava, R., Routray, A. and Ray, S.S. 2015. Hyperspectral remote sensing: opportunities, status and challenges for rapid soil assessment in India. *Curr. Sci.* **108**(5): 860–868.

Dasberg, S. 1995. Drip and spray irrigation of citrus orchards in Israel. In: *Micro–irrigation for a Changing World: Conserving Resources/Preserving the Environment. Proceedings of Fifth International Micro–irrigation Congress*, April 2–6, Orlando, Florida, U.S.A. pp. 281–287.

Datta, M., Palit, R., Sengupta, C., Pandit, M.K. and Banerjee, S. 2011. Plant growth promoting rhizobacteria enhance growth and yield of chilli (*Capsicum annuum* L.) under field conditions. *Australian J. Crop Sci.* **5**(5): 531–536.

Deshmukh, G. and Hardha, M. K. 2014.Effect of irrigation and fertigation scheduling under drip irrigationin papaya. *J. Agrisearch* **1**: 216–220.

Dey, P., Mathura Rai, Gangopadhyay, K.K., Biaksh, Das, Vishal Nath and Reddy, N. 2010. Effect of phosphorous on growth yield and nutrient use efficiency of litchi grown on Alfisol. *Indian J. Hort.* **67**: 394–95.

Dinesh Kumar and Ahmed, N. 2014. Response of nitrogen and potassium fertigation to Waris almond (*Prunnus dnlcis*) under northwestern himalayam region of India. *Sci. World J.* Aticle ID 1413286 pages.

Dutta, P., Kundu, S. and Biswas, S. 2010. Integrated nutrient management in litchi cv Bombai in new alluvial zone of West Bengal. *Indian J. Hort.* **67**: 181–184.

Dutta, P., Maji, S.B. and Das, B.C. 2009. Studies on the response of bio–fertilizer on growth and productivity of guava. *Indian J. Hort.* **66**: 39–42.

Duxbury, J.M., Borduzzaman, M., Johnson, S.E., Lauren, J.G., Meisner, C.A. and Welch, R.M. 2006. Opportunities and constraints for addressing human mineral micronutrient malnutrition through soil management. In: Abstract **105**: 4.2B 18th World Congress of Soil Science July 9–15, 2006, Philadelphia, Pennsylvania, USA, p. 82.

Dwivedi, Vandana. 2013. Effect of integrated nutrient management on yield, quality and economics of Guava. *Ann Plant Soil Res.* **15**: 149–151.

Ehleringer, J. and Cerling, T.E. 1995. Atmospheric CO_2 and the ratio of intercellular to ambient CO_2 concentrations in plants. *Tree Physiol.* **15**: 105–111.

Fageria, N.K., Baligar, V.C. and Li, Y.C. 2008. The role of nutrient efficient plants in improving crop yields in the twenty first century. *J. Plant Nutri.* **31**: 1121–1157.

Farooq, M., Wahid, A. and Siddique K.H.M. 2012. Micronutrient application through seed treatments–a review. *J. Soil Sci. Plant Nutri.* **12**: 125–142.

FAO. 2011. *Statistical Year Book of Food and Agricultural Organization FAO State Div. 2011.* Metalink: P3.Reu. FAO, ESS.FRD.AH Sci. p. 108–113. www.fao.org. docrep/017/i3138e/i3138e05.pdf.

Ferguson, J.J., Davies, F.S. and Bulger, J.M. 1990. Fertigation and growth of young 'Sunburst' tangerine trees. *Proceedings of Florida State. Hort. Sci.* **103**: 8–9.

Fischer, G., Almanza–Marchan, P.S and Ramirez, F. 2012. Source–Sink relationship in fruit species: A review. *Revista Colombiana de Ciencias Horticolas*, **6**: 238–253.

Germana, C. 1994. Increasing water use efficiency through irrigation management. *Proc. Int. Soc. Citriculture.* **2**: 638–642.

Ghadekar, S.R., Dixit, S.V. and Patil, V.P.1989. Climatological water requirement of the citrus under Nagpur agroclimatic conditions. *PKV Res.J.* **13**(2): 143–148.

Ghosh, S.N., Roy, S.N. and Bora, B. 2012. Nitrogen and potassium nutrition in sapota grown in lateriate soil. *J. Crop Weed.* **8**: 152–154.

Goodfellow, J., Eamus D. and Duff, G. 1997. Diurnal and seasonal changes in the impact of CO_2 enrichment on assimilation, stomatal conductance and growth in long term study of *Mangifera indica* in the wet–dry tropics of Australia. *Tree Physiol.* **17**: 291–299.

Goh, K. M. 2004. Carbon sequestration and stabilization in soils: Implications for soil productivity and climate change. *Soil Sci. Plant Nutr.* **50:** 467–476.

Gori, A. and Favilli, F.1995. First results on individual and dual inoculation with *Azospirillum–Glomus* on wheat. *North Atlantic Treaty Organisation, Advance Study Institute, Series G (Ecological Science)* **37**: 245–249.

Goswami, Amit Kumar, Lal, Shant and Misra, K.K. 2012. Integrated nutrient management improves growth and leaf nutrient status of guava cv Pant Prabhat. *Indian J. Hort. Sci.* **69**: 168–172.

Grieve, A.M. 1988. Water use efficiency of micro–irrigated citrus. (*In*) *Proceedings of Fourth International Micro–irrigation Congress, 23–28 October 1988, Albury–Nodonga, Australia.* 2 (6) *Tropic. Agric.* (Trinidad), **80** (2): 2003.

Guimarães, V., Danielle, Maria, I.S. Gonzaga. and José de O. Melo Neto 2014. Management of soil organic matter and carbon storage in tropical fruit crops. *Revista Brasileira de Engenharia Agrícola e Ambiental.* **18**: 301–306.

Haneef, Mohd, Kaushik, R.A., Sarolia, D.K., Mordia, A. and Mahesh Dhakar. 2014. Irrigation scheduling and fergitaion in pomegranate cv Bhagwa under high density planting system. *Indian J. Hort.* **71**: 45–48.

Hasan, M.A., Manna, M., Dutta, P., Bhattacharya, K, Mandal, S., Banerjee, H., Ray, S.K. and Jha, S. 2012. Integrated nutrient management in improving fruit quality of Mango 'Himsagar'. *ISHS Acta Hort.* 992, IX International Mango Symposium.

Hazarika, B.N. and Ansari, S. 2007. Biofertilizers in fruit crops–A review. *Agric. Rev.* **28**: 69–74.

Hazarika, B.N. and Ansari, S. 2010. Effect of integrated nutrient management on growth and yield of banana cv. Jahaji. *Indian J. Hort.* **67**: 270–273.

He, Z.L., Yao, H., Chen, G., Zhu, J. and Huang, C.Y. 1997. Relationship of crop yield to microbial biomass in highly weathered soils of China. In: *Plant Nutrition for Sustainable Food Production and Environment*, eds. T. Ando, K. Fujita, T. Mae, H. Matsumoto, S. Mori, and J. Sekiya, pp. 745–746. Tokyo: Kluwer Academic Publishers.

Hebbarai, M., Ganiger, V.M., Masthana Reddy B.G. and Joshi V.R. 2006. Integrated nutrient management in sapota (*Manikara* zapota) using vermicompost to increase yield and quality. *Indian J. Agric. Sci.* **76:** 587–590.

Hocking, P.J. and Meyer, C.P. 1991a. Effects of CO_{2-} enrichment and nitrogen stress on growth, and partitioning of dry matter and nitrogen in wheat and maize. *Austral. J. Plant Physiol.* **18**: 339–356.

Hocking, P.J. and Meyer, C.P. 1991b. Carbon dioxide enrichment decreases critical nitrate and nitrogen concentration in wheat. *J. Plant Nutr.* **14**: 571–584.

Houghton, R.A. and Skole, D.L. 1990.Carbon, p. 393–408. In: B.L. Turner, II, W.C. Clark, R.W. Kates, J.F. Richards, J.T. Mathews, and W.B. Meyer (eds.). *The earth as transformed by human action, global and regional changes in the biosphere over the past 300 years.* Cambridge Univ. Press, Cambridge.

Huchche, A.D., Dass H.C., Lallan Ram, Srivastava, A.K. and Kohli, R.R.1996. Response of acid lime (*Citrus aurantifolia* Linn.) to nitrogen fertilisation. *Indian J. Hort.* **53**: 14–18.

Idso, S.B. and Kimball, B.A. 2001. CO_2 enrichment of sour orange trees: 13 years and counting. *Environ. Exp.Bot.* **46**: 147–153.

Indian Horticulture Database 2013. Area and production trends in horticulture crops. Rajendra Kumar Tiwari, Mistry, N.C., Brajendra Singh and Chander P. Gandhi (Ed.). National Horticulture Board New Delhi, pp. 42–67.

Irget, M.E., Aksoy, U., Okur, B., Ongun, A.R. and Tepecik, M. 2008. Effect of calcium based fertilization on dried fig (*Ficus carica* L. cv Sarilop) yield and quality. *Scien. Hortic.* **118**: 303–313.

Ito, J., Hasegawa, S., Fujita K., Ogasawara S. and Fujiwara T. 1999. Effect of CO_2 enrichment on fruit growth and quality in Japanese pear (*Pyrus serotina* Reheder cv Kousui). *Soil Sci. Plant Nutr.* **45**: 385–393.

Jaizme–Vega, M. Rodriguez–Romero, A.S. and Pinero Guerra, M.S. 2003. Potential use of rhizobacteria from *Bacillus* genus to stimulate the plant growth of micropropagated banana. *Fruits* **59**: 83–90

Jasrotia, A., Singh, R.P., Singh, J.M. and Bhutani, V.P. 1999. Response of olive trees to varying levels of N and K fertilizers. *Acta Hort.* **474**: 337–339.

Jeyakumar, P., Amutha, R., Balamohan, T.N., Auxxilia J. and Nalina, L. 2010. Fertigation improve fruit yield and quality of papaya. *Acta Hort.* **851**: 369–373.

Johnston, A.M., Khurana, H.S., Majumdar, K.and Satyanarayana T. 2009. Site specific nutrient management–Concept, current research and future challenges in Indian agriculture. *J. Indian Soc. Soil Sci.* **57**: 1–10.

Kachwaha, D.S. and Chandel, J.S. 2015. Effect of fertigation on growth yield, fruit quality and leaf nutrient contents of strawberry (*Fragariax ananassa*) cv. Chandler. *Indian J. Agric. Sci.* **85**: 1319–1323.

Kanber, R., Koksal, H., Ouder, S. and Eyten, M. 1996. Effect of different irrigation methods of yield, evapo–transpiration and root development of young orange trees. *Turkish J. Agric. Forestry*, **20**: 163–172.

Keditsu, Rokolhuii and Srivastava, A.K.2014. Substrate dynamics: Development and issues. *Ann. Pl Soil Res.* **16**(1): 1–8.

Kemmler, G. and Hobt, H.1985. *Potassium a product of nature.* Kali and Salz, Kassel, Germany.

Kennedy, Ann C. and Gewin Virginia L.1997. Soil microbial diversity: present and future considerations. *Soil Sci.* **162**: 607–617.

Keutgen, N. and Chen, K. 2001. Responses of citrus leaf photosynthesis, chlorophyll fluorescence, macronutrient and carbohydrate contents to elevated CO_2. *J. Plant Physiol.* **158**: 1307–1316.

Khehra, Savreet and Bal, J.S. 2014. Influence of organic and inorganic nutrient sources on growth of lemon (*Citrus limon* (L.) Burm.) cv. Baramasi. *J. Expt. Biol. Agric. Sci.* **2**: 126–129.

Khot, A.B., Ashoka, P., Neelkanth, J.K., Rajkumar, S. and Gundlur, S.S. 2012. Performance of early growth period of sapota (*Achras sapota* L.) to drip method of irrigation and fertigation of Vertisol. *Environ. Ecol.* **30**: 1513–516.

Kimmins, J.P., 1997. *Forest Ecology*, Pearson Prentice Hall, NJ.

Kirad, K.S., Barche, S. and Singh, D.B. 2010. Integrated nutrient management in papaya cv. Surya. *Acta Hort.* **1**: 377–380.

Koo, R.C.J. and Smjstrala, A.G. 1984. Effect of trickle irrigation and fertigation on fruit production and fruit quality of Valencia orange. *Proc. Florida Sta. Hort. Sci.* **97**: 8–10.

Kruger, J.A., Britz, K., Tolmay, C.D. and Du Plessis, S.F. 2000. Evaluation of an open hydroponic system (OHS) for citrus in South Africa: Preliminary Results. In: *Proc. Intl. Soc.Citriculture*, **2**: 239–242.

Kumar, A.P. A. and Bojappa, K.M. 1994. Studies on the effect of drip irrigation on yield and quality of fruit in sweet oranges and economy in water use. *Mysore J. Agric. Sci.* **28**: 338–344.

Kuttimani, R., Velayudham, Somasundram, E. and Muthukrishnan, P.2013. Effect of integrated nutrient management on yield and economics of banana. *Global J. Biol. Agric. Helth. Sci.* **2**: 191–195

Lal, G., Pareek, C.S., Sen, J.L. and Soni, A.K. 2003. Effect of N, P and K on growth, yield and quality of ber cv Umran. *Indian J. Hort.* **60**: 158–162.

Lakso, Alan L. 2010. Esimating the environmental footprint of New York apple orchards. *New York Fruit Qrtly.* **181**: 26–28

Lareen, A., Burton, F. and Schafer, P. 2016. Plant root–microbe communication in shaping root microbiomes. *Pl. Mol. Biol.* **90**(6): 575–587.

Lehoczky, E., Debreczeni, K. and Szalai, T. 2005. Available micronutrient contents of soils in long–term fertilization experiments in Hungary. *Commun. Soil Sci. Plant Anal.* **36**: 423–430.

Lipecki, J. and Berbeæ, S. 1997. Soil management in perennial crops: orchards and hop gardens. *Soil Till. Res.* **43**: 169–184.

Lobell, D.B., Field, C.B., Chahill, K.N. and Bonfils, C. 2005. Impacts of future climate changes on California perennial crop yields: Model projections with climate and crop uncertainties. *Agril. Forest Meteorol.* **141**: 208–218.

Mageed, K.J.A., Sharma, B.B. and Sinha, A.K. 1988. Influence of irrigation and nitrogen on water use and growth of Kinnow mandarin. *Indian J. Agric. Sci.* **58**(6): 284–286.

Maity, A., Jadhav, V.T. and Ram Chandra.2012. Exploration of microbial wealth for sustainable horticultural production. *Int. J. Bioresour. Stress Mgmt.* **3**: 489–500.

Maiyappan, S. Amalraj, Leo Daniel E., Santosh, A. and Peter John A. 2010. Isolation, evaluation and formulation of selected microbial consortia for sustainable agriculture. *J. Biofertil Biopest.* **2**: doi: 10.4172/2155–6202.1000109.

Mahendran, P.P., Yuraj, M., Paremeswari, C., Gurusamy, A. and Krishnasamy, S. 2013. Enchaining growth, yield and quality of Banana through sursurface drip fertigation. *Int. J. Chem., Envroni. Biol. Sci.* **1**(2): 391–394 ISSN 2320–4087.

Malhotra, S.K. and Srivastava, A.K. 2015. Fertilizer requirement of Indian horticulture: An analysis. *Indian J. Fert.* **11**: 16–25.

Mandal, K.K., Rajak, A., Debnath, S. and Hasan, M.A. 2013. Integrated nutrient management in Aonla cv A–7 in the red lateritic region of West Bengal. *J. Crop Weed* **9**: 121–123.

Manna, M.C., Swarup, A., Wanjari, R.H., Singh, Y.V., Ghosh, P.K., Singh, K.N., Tripathi A.K., and Saha, M.N. 2005. Soil organic matter in a West Bengal Inceptisol after 30 years of multiple cropping and fertilization. *Soil Sci. Soc. Amer. J.* **70**: 121–129.

Marshner, P., Crowley, D. and Yang, C.H. 2004 Development of specific rhizosphere bacterial communities in relation to plants species, nutrition and soil type. *Plant and Soil* **261**: 199–208.

Martinez–Valero R., and Fernandez C. 2004. Preliminary results in citrus groves grown under MOHT system. In: Proceedings *Int. Soc. of Citriculture* Vol.1, p.103.

Mengel, K. and Kirkby, E.A. 2000. *Principles of Plant Nutrition*. Kluwer Academic Publishers, London, UK, pp. 169–212.

Mia, M.A., Shamsuddin, Z.H. Wahab, Z. and Marziah, M. 2010. Effect of plant growth promoting rhizobacterial (PGPR) inoculation on growth and nitrogen incorporation of tissue–cultured Musa plantlets under nitrogen–free hydroponics condition. *Austral. J. Crop Sci.* **4**: 85–90.

Modaihsha, S. 1997. Foliar application of chelated and non–chelated metals for supplying micronutrients to wheat grown on calcareous soil. *Exptl. Agri.* **33**: 237–245.

Mohd Haneef, Kaushik, R.A., Sarrolia, D.K, Mordia, A. and Mahesh Dhakar 2014. Irrigation scheduling and fertigation in pomegranate cv. Bhagwa under high density planting system. *Indian J. Hort.* **7** : 45–48.

Montanaro, G., Celano, G., Dicho, B. and Xiloyannis, C. 2010. Effects of soil–protecting agricultural practices on soil organic carbon and productivity in fruit tree orchards. *Land Degrad Dev.* **21**: 132–138.

Montanaro, G., Dicho, B., Briccoli Bati, C. and Xiloyannis, C. 2012. Soil management affects carbon dynamics and yield in a Mediterranean peach orchard. *Agr. Ecosyt. Environ.* **161**: 46–54.

Mostafa, E.A.M., Sakeg, M.M.S. and El–Migeed Abd M.M.M. 2007. Response of banana plants to soil and foliar applications of magnesium. *American–Eurasian J. Agri. Environ. Sci.* **2**: 141–146.

Munir, Muhammad, Baloch, Jalal–ud–Din, Alizai, Atiq Ahmed and Ahmad, Zia 1992. Resposne of date palm cultivar Dhakki to NPK fertilizers. *Paksitan J. Agric. Res.* **13**: 347–349.

Mwamba Bwalya Jackson 2013. Estimation of net carbon sequestration potential of citrus under different management systems using the life cycle approach. *Research Thesis*. University of Zambia Research Repository Online. http://dspace.unza.zm: 8080/xmlui/handle/123456789/2202

Mylavarapu, R.S. 2010. Diagnostic nutrient testing. *HorlTech.***20**(1): 19–22.

Negi, J.D.S., Manhas, R.K. and Chauhan, P.S. 2003. Carbon allocation in different components of some tree species of India: A new approach of carbon estimation. *Curr. Sci.* 85:

Neto, C., Carranca, C., Clemente, J. and Varennes de A. 2008. Nitrogen distribution, remobilization and re–cycling in young orchard of non–bearing 'Rocha' pear trees. *Scien. Hortic.* **118**: 299–307.

Nielsen, D.R., Biggar, J.W. and Erh, K.T. 1973 Spatial variability of field–measured soil–water properties. *Hilgardia*, **42**(7): 215–59.

Neilsen, D., Millard, P., Neilsen, G.H., Hogue, E.J., Parchomchuck, P. and Zebarth, B.J. 2001. Remobilization and uptake of N by newly planted apple (*Malus domestica*) trees in response to irrigation method and timing of N application. *Tree Physiol.* **21**: 513–521.

Neilson, G.H., Parchoonchuk, P., Neilsen, D. and B.J. Zebarth 1996. Drip–fertigation of apple trees affects 2001 distribution and development of K deficiency. *Canadian J. Soil Sci.*

Ngullie, E., Singh, A.K., Sema Akali and A.K. Srivastava, 2015. Citrus growth and rhizosphere properties. *Commun. Soil Sci. and Pl. Anal.* **45**: 1540–550.

Pan, Q., Wang, Z. and Quebedeaux, B. 1998. Responses of the apple plant to CO_2 enrichment: changes in photosynthesis, sorbitol, other soluble sugars, and starch. *Austral J. Plant Physiol.* **25**: 293–297.

Panigrahi, P., Huchche, A.D., Srivastava, A.K. and Shyam Singh. 2008. Effect of drip irrigation and plastic mulch on performance of Nagpur mandarin (*Citrus reticulata*) grown in central India. *Indian J. Agricult. Sci.* **78**(12): 911–115.

Panigrahi, P. and Srivastava, A.K. 2016. Effective management of irrigation water in citrus orchards under a water scarce hot sub–humid region. *Scientia Hort.* **210**: 6–13.

Panigrahi, P., Srivastava, A.K. and Huchche, A.D. 2012a. Effects of drip irrigation regimes and basin irriation on Nagpur mandarin agronomical and physiological performance. *Agric. Water Mgmt.* **104**: 79–88.

Panigrahi P., Srivastava A.K., Huchche, A.D. and Shyam Singh. 2012b. Plant nutrition in response to drip versus basin irrigation in young 'Nagpur' mandarin on inceptisol. *J. Pl. Nutri.* **35**(2): 215–24.

Panigrahi, P., Srivastava, A.K., Huchche, A.D. and Shyam Singh. 2012c. Plant nutrition in response to drip versus basin irrigation in young Nagpur mandarin on Inceptisol. *J. Pl. Nutri.* **35**: 215–24.

Panwar, Rashmi., Singh, S.K., Singh, C.P., Singh P.K. 2007. Mango fruit and quality improvement through fertigation along with mulch, *Indian J. Agrl. Sci.* **77** (10): 680–4.

Parker, D.R., Norvell, W.A. and Chaney, R.L. 1995. GEOCHEM–PC: A chemical speciation program for IBM and compatible personal computers. In: Loeppert R.H., Schwab A.P. and Goldberg S. (Eds.), *Chemical Equilibrium and Reaction Models*, Soil Science Soc. Amer. Special Publication, Soil Science Soc. Amer., Amer. Soc. Agron., Madison, Wisconsin USA, 42, pp. 253–269.

Patel, V.B., Singh,S.K., Asrey Ram, Nain Lata, Singh, A.K. and Laxman Singh 2009. Microbial and inorganic fertilizers application influenced vegetative growth, yield, leaf nutrient status and soil microbial biomass in sweet orange cv Mosambi. *Indian J. Hort.* **66**: 163–168.

Pathak, P.K. and Mitra, S.K. 2008. Effect of phosphorus, potassium, sulphur and boron on litchi. *Indian J. Hort.* **65**: 137–140.

Pathak, H. and Nedwell, D.B. 2011. Nitrous oxide emission from soil with different fertilizer, water levels and nitrification inhibitors. *Water, Air Soil Pollu.* **129**: 217–228.

Patil, V.K. and Shinde, B.N. 2013. Studies on integrated nutrient management on growth and yield of banana cv.Ardhapuri (Musa AAA). *J. Hort. Forest.* **5**: 130–138.

Patil, D.R., Sulikeri, G.S., Patil, H.B. and Balikar, R.A. 2008. Studies on integrated nutrient management in Thompson seedless grapes. *Acta Hort.* 785.

Pawar, D.D. and Dingre, S.K. 2013. Influence of fertigation scheduling through drip on growth and yield of banana in western Maharashtra. *Indian J. Hort.* **70**: 200–05.

Peng, L., Wang, C., He, S., Guo, C. and Yan, C. 2000. Effects of elevation and climatic factors on the fruit quality of Navel orange. *South China Fruits* **29**: 3–4.

Peng, Y.H. and Rabe, E. 1999. Effect of irrigation methods and ground cover on the fruit quality, yield and light levels in the canopy of microwave Satsuma. *J. Fruit Sci.* **15**: 128–132.

Pindi, P.K., and Satyanarayana, S.D.V. 2012. Liquid Microbial consortium–A potential tool for sustainable soil health. *J. Biofertil Biopestici* http://dx.doi.org./10.4172/2155–6202.1000124

Prakash, K., Vijaykumar, R.M., Balamohan, T.N. and Sundhar Singh, S.D. 2015. Effect of drip irrigation regime and fertigation regime on yield and quality of mango cultivar Alphonso under ultra high density planting. Doi: 10.17660/*Acta Hort.* 2015. 1066.17

Poszytek, K., Ciezkowska, M., Sklodowska, A. and Drewniak, L. 2016. Microbial consortium with high cellulolytic activity (MCHCA) for enhanced biogas production. Front. Microbiol. http://dx.doi.org/10.3389/fmicb.2016.00324

Ram, R.A.and Rajput, M.S. 2000. Role of biofertilizers and manures on production of guava (Psidium *guajava* L.) cv. Allahabad Safeda. *Haryana J. Hort. Sci.* **29**: 193–194.

Ramniwas, Kaushik, R.A., Sunil Pareek, Sarolia, D.K. and Vivendra Singh. 2013. Effect of drip irrigation on fertilizer use efficiency, leaf nutrient status, yield and quality of shweta guava (*Psidium guajava* L.) under medow orcharding. *Nat. Acad. Sci.* Lett. **36**: 483–488.

Ramniwas, Kaushik, R.A., Sarolia, D.K., Pareek Sunil and Sunil V. 2012. Effect of irrigation and fertigation scheduling on growth and yield of guava (*Psidiuim guajava* L.) under meadow orcharding. *African J. Agri. Res.* **7**: 6350–356.

Rao, A.V. and Dass, H.C. 1989. Growth of fruit plants as influenced by nitrogen fixing bacteria. *Ann. Arid Zone* **28**: 142–147.

Raupach, G.S. and Kolepper, J. W. 1998. Mixture of plant growth promoting rhizobacteria enhance biological control of multiple cucumber pathogens. *Phytopathol.* **88**: 1158–1164.

Ray Dutta, S.K., Takawale, P.V., Chatterjee, R. and Hnamte, V. 2014. Yield and quality of pomegranate as influenced by organic and inorganic nutrients. *The Bioscan*, **9**: 317–620.

Reddy, B.M.C., Srinivas, K., Padma, P. and Raghupati, H.B. 2002. Response of Robusta banana to N and K fertigation. *Indian J. Hort.* **59**: 342–48.

Reídel, E.J., Brown, P.H., Duncu, R.A., Heerema, R.J. and Weinbaum, S.A. 2004. Sensitivity of yield determinants to potassium deficiency in 'Nonpareil' almond. *J. Hort. Sci. Biotech.* **79**: 906–910.

Schaffer, B., Whiley, A.W., Searle, C. and Nissen, R.J. 1997. Leaf gas exchange, dry matter partitioning, and mineral element concentrations in mango as influenced by elevated atmospheric carbon dioxide and root restriction. *J. American Soc. Hort. Sci.* **122**: 849–855.

Schoebitz, M. and Vidal, G. 2016. Microbial consortium and pig slurry top improve chemical properties of degraded soil and nutrient uptake. *J. Soil Sci.*

Pl. Nutrient. http://dx.doi.org/10.4067150718-95162016005000018

Scholberg, J. and Morgan, K.T. 2012. Nutrient use efficiency in citrus. In : Advances in Citrus Nutrition. Srivastava, A.K. (Ed.). Springer Verlag, the Netherlands pp, 205–229.

Schumann, A.W., Miller,W.M., Zaman,Q.U., Hostler, K.H., Buchanon, S. and Cugati, S. 2006. Variable rate granular fertilization of citrus groves: Spreader performance with single–tree prescription zones. *Appld. Engg. Agric.* **22**: 19–24.

Sekikawa, S.sup, Kibe, T.sup, Koizumi, H., and Mariko, S. 2003. Soil carbon sequestration in a grape orchard ecosystem in Japan. *J. Japanese Agri. Systems Soc.* **19**: 141–150.

Senthilkumar, M.,Ganesh, S., Srinivas, K. and P.Panneerselvam 2014. Enhancing uptake of secondary and micronutrients in banana cv. Robusta through intervention of fertigation and consortium of biofertilizers. *Scholars Acad. J. Biosci.* **2**: 472–478.

Shaaban, M.M. 2009. Injection fertilization: A full nutritional technique for fruit trees saves 90–95 per cent of fertilizers and maintains a clean environment. *Fruits, Vegetables and Cereal Sci. Biotech.*3(1): 22–27.

Shah Azam S., Mohammad, W., Shah Mahmood S., Elahi Rizwan, Ali, and Azaz Abdul Basir Haroon 2014. Integrated effect of organic and inorganic nitrogen on peach fruit yield and orchard fertility. *Agric. Sci. Res. J.* **4**: 78–82.

Shah, Mahmood S., Mohammad, Wisal, Shah, Azam S. and Nawaz Haq. 2006. Integrated nitrogen management of young deciduous apricot orchard. *Soil Environ.* **25**: 59–63.

Shanmugasundaram, T. and Balakrishnamurthy. G. 2013. Effect of fertigation on flowering and yield of tissue culture pomegranate (*Punica granatum* L.) cv. Mridula grown under ultra high density planting. i–scholar **8**: 601–604.

Shanmugasundaram T and Balakrishnamurthy, G. 2015. Effect of fertigation on yield and economics of tissue culture pomegranate (*Punica granatum* L.) *J. Sci. Industr. Res.***74**: 457–460.

Sharma, R.C., Mahajan, B.V.C., Dhillon, B.S. and Azad, C.S. 2000. Studies on the fertilizer requirement of mango cv. Dashehari on sub–montaneous region of Punjab. *Indian J. Agric. Res.* **34**: 309–310.

Sharma, J.R., Kaushik, R.A. and Panwar, R.D. 2008. Influence of nitrogen, phosphorus and potassium on yield and physico–chemical properties of phalsa. *Indian J. Hort.* **65**: 326–327.

Sharma, S., Patra Sanmay, K.R., Roy, G.B. and Bera, S. 2013. Influence of drip irrigation and nitrogen fertigation on yield and water productivity of guava. *The Bioscan* **8**(3): 783–786.

Shirgure, P.S. 2012a. Effect of pulse irrigation scheduling with hybrid station controller on fruit yield and quality of Nagpur mandarin (*Citrus reticulate* Blanco). *Sci. J. Crop Sci.*, **1**(5): 76–82.

Shirgure, P. S. 2013a. Yield and fruit quality of Nagpur mandarin (*Citrus reticulata* Blanco) as influenced by evaporation based drip irrigation schedules. *Sci. J. Crop Sci.*, **2** (2): 28–35.

Shirgure, P. S. 2012b. Micro–irrigation systems, automation and fertigation in Citrus. *Sci. J. Rev.*, **1**(5): 156–169.

Shirgure, P. S. 2013b. Citrus fertigation–a technology of water and fertilizers saving. *Sci. J. Crop Sci.*, **2** (5): 56–66.

Shirgure, P.S. and Srivastava, A.K. 2012a. The effect of four under tree micro–jet irrigation (180°–300°) systems on fruit yield and quality of Nagpur mandarin in Central India. *Sci. J. Agr.*, **1**(7): 177–186.

Shirgure, P.S. and Srivastava, A.K. 2012b. Effect of automatic micro–irrigation scheduling on productivity and quality of Nagpur mandarin (*Citrus reticulata* Blanco). *Indian J. Hort.* **71**(1): 112–116.

Shirgure, P. S and Srivastava A. K. 2013a. Plant growth, leaf nutrient status, fruit yield and quality of Nagpur mandarin (Citrus reticulate Blanco) as influenced by potassium (K) fertigation with four potash fertilizer sources. *Sci. J. Crop Sci.* **2**(3): 36–42.

Shirgure, P.S. and Srivastava A.K. 2013b. Nutrient–water interaction in citrus: recent developments. *Agri. Adv.* **2** (8): 224–36.

Shirgure, P.S. and Srivastava, A.K. 2015. Evaluation of drip irrigation emitters arrangement and its effect on soil moisture, leaf nutrients, yield and quality of Nagpur mandarin (*Citrus reticulata*). *Indian J. Agri. Sci.* **85**(12): 1586–1591.

Shirgure, P.S., Srivastava, A.K. and Huchche, A.D. 2014. Water requirements in growth stages and effects of deficit irrigation on fruit productivity of drip irrigated Nagpur mandarin (*Citrus reticulata* Blanco). *Indian J. Agri. Sci.* **84**(3): 317–322.

Shirgure, P.S., Srivastava, A. K. and Huchche, A. D. 2016a. Effect of drip irrigation scheduling on yield and quality of Nagpur mandarin (*Citrus reticulata* Blanco) fruits. *Indian J. of Hort.* **73** (1): 30–35.

Shirgure, P. S., Srivastava, A. K., Huchche, A. D. and Prakash Patil. 2016b. Interactive effect of irrigation schedules and fertigation levels on fruit yield, quality and plant nutrition of Nagpur mandarin. *Indian J.of Agri. Sci.* **86** (11): In press

Shirgure, P.S., Srivastava, A.K. and Singh, S. 2000. Water management in citrus–A review. *Agri. Rev.*, **21**(4): 223–230.

Shirgure, P.S., Srivastava, A.K. and Singh, S. 2001a. Effect of pan evaporation based irrigation scheduling on yield and quality of drip irrigated Nagpur mandarin. *Indian J. Agri. Sci.* **71**(4): 264–266.

Shirgure, P.S., Srivastava, A.K. and Singh, S. 2001b. Growth, yield and quality of Nagpur mandarin (*Citrus reticulata* Blanco) in relation to irrigation and fertigation. *Indian J. Agri. Sci.*, **71** (8): 547–550.

Shirgure, P.S., Srivastava, A.K. and Singh, S. 2001c. Fertigation and Drip Irrigation in Nagpur mandarin (*Citrus reticulata* Blanco). *South Indian Hort.*, **49**: 95–97.

Shirgure, P.S., Srivastava, A.K. and Singh, S. 2001d. Effect of nitrogen fertigation and band placement fertilizer application on soil–leaf nutrient build–up and incremental growth of acid lime. *J. Soil Water Conserv.***45** (3 and 4) **:** 176–181.

Shirgure, P.S., Srivastava, A.K. and Singh, S. 2002. Economics of drip and fertigation in acid lime orchards. *J. Soil Water Conserv.*, **46** (1): 56–60.

Shirgure, P.S., Srivastava, A.K. and Singh, S. 2003a. Evaluating micro–irrigation systems in Nagpur mandarin under sub–humid tropical climate. *Trop. Agr.*, **80** (2): 91–96.

Shirgure, P.S., Srivastava, A.K. and Singh, S. 2003b. Irrigation scheduling and fertigation in acid lime (*Citrus aurantifolia* Swingle). *Indian J. Agri. Sci.*, **73** (7)**:** 363–367.

Shirgure, P.S., Srivastava, A.K. and Singh, S. 2004a. Growth, yield and quality of acid lime under pan evaporation based drip irrigation scheduling. *Indian J. Soil Conserv.*, **32** (1): 32–35.

Shirgure, P.S., Srivastava, A.K. and Singh,S. 2004b. Integrated water and nutrient management in acid lime. *Indian J. Soil Conserv.*, **32** (2): 148–151.

Shirgure, P.S., Srivastava, A.K., Singh, S. and Pimpale, A.R. 2004c. Drip irrigation scheduling growth, yield and quality of acid lime (*Citrus aurantifolia* Swingle). *Indian J. Agri. Sci.*, **74**(2): 92–94.

Singh, S.R. and Banik, B.C. 2011. Response of integrated nutrient management on flowering, fruit setting, yield and fruit quality in mango cv. Himsagar (*Mangifera indica* L.). *Asian J. Hort.* **6**: 151–154.

Singh, Sandeep Kumar, Thakur Nidhika and Sharma Yamani.2012. Effective nutrient management in fruit crops. *Asian J. Hort.* **7**: 606–609.

Singh, Sanjay K., Singh, C.P. and Rashmi Panwar. 2009. Response of fertigation and plastic mulch on growth characteristics of young Dashehri mango. *Indian J. Hort.* **66**: 390–92.

Siqueira, D.S., Marques, Jr J. and Pereira, G.T. 2010. The use of landforms to predict the variability of soil and orange attributes. *Geoderma.***155** (1–2): 55–66.

Shukla, A.K., Sarolia, D.K., Bhavana Kumari, Kaushik, R.A., Mahawer, L.N. and Bairwa, H.L. 2009. Evaluation of substrate dynamics for integrated nutrient management under high density planting of guava cv. Sardar. *Indian J. Hort.* **66**: 461–464.

Silva,E.D., Nogueira, F.D., Guimaraes, P.T.G. and Neto, A.E.F. 2001. Coffee tree response to potassium fertilization in low and high yields. *Pesquisa Agropecuaria Bisileira* **36**: 1331–1337.

Singh, S.R. and Banik, B.C. 2011. Response of integrated nutrient management on flowering, fruit 151–154.

Singh, H.P. and Singh, G. 2007. Nutrient and water management in guava. *Acta Hort.* **735**: 389–398.

Singh, J.K. and Varu, D.K. 2013. Effect of integrated nutrient management in Papaya (*Carica papaya* L.) cv Madhubindu. *Asian J. Hort.* **8**: 667–670.

Smajstrla A. G., Parsons, L. R., Aribi, K. and Yelledis, G. 1986. Response of young citrus trees to irrigation. *Proc. Florida State Hort. Soc.* **98**: P 25.

Smajstrala, A. G. 1993. Micro–irrigation for citrus production in Florida. *Hort. Sci.* **28**: 295–298.

Sofo, A., Nuzzo, V., Palese, A.M., Xyloyannis, C. and Cellano, G. 2005. Net CO_2 storage in Mediterranean Olive and Peach orchards. *Sci. Hortic.***107**: 17–24.

Sohlegel, T.K., Schonherr J. and Schreiber L. 2006. Rates of foliar penetration of chelated Fe (II): Role of light, stomata, species and leaf age. *J.Agric. Food Chem.* **2**: 141–146.

Solaimalai, A., Baskar, M., Sadasakthi, A. and K. Subburamu. 2005. Fertigation in high value crops: A review. *Agric. Rev.* **26**: 1–13.

Sparks, D. 1989. Pecan nutrition–a review. *Proc. Southeastern Pecan Growers Assoc.***82**: 101–122.

Srivastava, A.K. 2009. Integrated nutrient management: Concept and application in citrus. In: Tennant P. Beakebhia N (eds.) Citrus II. *Tree Forestry Sci. and Biotechnol.* **3**(1): 32–58.

Srivastava, A.K. 2010a. Development of INM to harmonize with improved citrus production: Changing scenario. *Adv. Pl. Physiol.* **12**: 294–268.

Srivastava, A.K. 2010b. Integrated nutrient management in citrus: Frontier developments. *Indian J. Fert.* **6**(11): 34–44.

Srivastava, A.K. 2011. Site specific potassium management for quality production of citrus. *Karnataka J. Agric. Sci.* **24**(1): 60–66.

Srivastava, A.K. 2013a. Nutrient management in Nagpur mandarin: Frontier developments. *Sci. J. Agri.* **2**(1): 1–14.

Srivastava, A.K. 2013b. Site specific nutrient management in citrus. *Sci. J. Agri.* **2**(2): 1

Srivastava, A.K. 2013c. Recent developments in diagnosis and management of nutrient constraints in acid lime. *Sci. J.Agri.* **2**(3): 86–96.

Srivastava, A.K. 2013d. Nutrient deficiency symptomology in citrus: An effective diagnostic tool or just an aid for post–mortem analysis. *Sci.J.Agri.* **2**(6): 177–194.

Srivastava, A.K., Das S.N., Malhotra S.K. and Kaushik Majumdar. 2014a. SSNM–based rationale of fertilizer use in perennial crops: A review. *Indian J. Agric. Sci.* **84**(1): 3–17.

Srivastava, A.K., Huchche, A.D. and Dinesh Kumar. 2014b. Integrated nutrient management for sustained productivity of Nagpur mandarin. Annual Report 2014–15. ICAR–Central citrus Research Institute, Nagpur, Maharashtra, pp. 53–59.

Srivastava, A.K., Huchche, A.D. and Dinesh Kumar.2015. Integrated nutrient management for sustained productivity of Nagpur mandarin. Annual Report 2015–16. ICAR–Central citrus Research Institute, Nagpur, Maharashtra, pp. 56–59.

Srivastava, A.K., Huchche, A.D., Lallan Ram and Shyam Singh. 2007. Yield prediction in intercropped versus monocropped citrus orchards. *Sci. Hort.* **114**: 67–70.

Srivastava, A.K. and Kohli R.R. 1997. Soil suitability criteria for citrus–An appraisal. *Agric. Rev.* **18**(3): 134–46.

Srivastava, A.K. and Malhotra, S.K. 2014. Nutrient management in fruit crops : Issues and strategies. *Indian J. Fert.* **10**(12): 72–88.

Srivastava, A.K., Malhotra, S.K. and Krishna Kumar, N.K. 2015a. Exploiting nutrient–microbe synergy in unlocking productivity potential of perennial fruits: A review. *Indian J. Agric.Sci.* **85**(4): 459–481.

Srivastava, A.K. and Ngullie, E. 2009. Integrated nutrient management: Theory and practice. *Dynamic Soil, Dynamic Pl.* **3**(1): 1–30

Srivastava, A.K., Shirgure, P.S. and Shyam Singh. 2003. Differential fertigation response of Nagpur mandarin (*Citrus reticulata* Blanco) on an alkaline Inceptisol under sub–humid tropical climate. *Trop. Agri.* **80**(2): 91–96.

Srivastava, A.K. and Shyam Singh. 2001a. Soil properties influencing yield and quality of Nagpur mandarin (*Citrus reticulata* Blanco). *J. Indian Soc. Soil. Sci.* **49**(1): 226–229.

Srivastava, A.K. and Shyam Singh. 2001b. Soil fertility limit in relation to optimum yield of Nagpur mandarin (*Citrus reticulata* Blanco). *J. Indian. Soc. Soil Sci.* **49**(4): 758–762.

Srivastava, A.K. and Shyam Singh. 2001c. Development of optimum soil property limits in relation to fruit yield and quality of *Citrus reticulata* Blanco cv. Nagpur mandarin. *Trop. Agri.* **78**(3): 174–181.

Srivastava, A.K. and Shyam Singh. 2002. Soil analysis based diagnostic norms for Indian citrus cultivar. *Commun. Soil Sci. Pl. Anal.* **33**(11): 1689–1706.

Srivastava, A.K. and Shyam Singh. 2003a. Soil–plant nutrient limits in relation to optimum fruit yield of sweet orange (*Citrus sinensis* Osbeck) cultivar Mosambi. *Indian. J. Agric. Sci.* **73**(4): 209–211.

Srivastava, A.K. and Shyam Singh. 2003b. Plant and soil diagnostic norms for optimum productivity of Nagpur mandarin (*Citrus reticulata* Blanco). *Fert. News* **48**(2): 47–63.

Srivastava, A.K. and Shyam Singh. 2004a. Zinc nutrition, a global concern for sustainable citrus production. *J. Sustain. Agri.* **25**(3): 5–42

Srivastava, A.K. and Shyam Singh. 2004b. Soil and plant nutritional constraints contributing to citrus decline in Marathawada region, India. *Commun. Soil Sci. and Pl. Anal.* **35**(17/18): 2537–2550.

Srivastava, A.K. and Shyam Singh. 2004c. Zinc nutrition and citrus decline–A review. *Agric. Rev.* **25**(3): 173–188.

Srivastava, A.K. and Shyam Singh. 2005. Diagnosis of nutrient constraints in citrus orchards of humid tropical India. *J. Pl. Nutri.* **29**(6): 1061–1076.

Srivastava, A.K. and Shyam Singh. 2006. Biochemical markers and nutrient constraints diagnosis in citrus: A perspective. *J. Pl. Nutri.* **29**(5): 827–855.

Srivastava, A.K. and Shyam Singh. 2007. DRIS–based nutrient norms for Nagpur mandarin (*Citrus reticulata Blanco*). *Indian J. Agric. Sci.* **77**(6): 363–365.

Srivastava, A.K. and Shyam Singh. 2008a. Citrus nutrition research in India: Problems and prospects. *Indian J. Agri. Sci.* **78**: 3–16.

Srivastava, A.K. and Shyam Singh. 2008b. DRIS norms and their field validation in Nagpur mandarin (*Citrus reticulata* Blanco). *J. Pl. Nutri.* **31**(6): 1091–107.

Srivastava, A.K. and Shyam Singh. 2009a. Zinc nutrition in Nagpur mandarin on haplustert. *J. Pl. Nutri.* **32(7):** 1065–1081.

Srivastava, A.K. and Shyam Singh. 2009b. Citrus decline: Soil fertility and plant nutrition. *J. Pl. Nutri.* **32(2):** 197–145.

Srivastava, A.K and Shyam Singh. 2015. Site–Specific nutrient management in Nagpur mandarin (*Citrus reticulata* Blanco) raised on contrasting soil types. *Commun. Soil Sci. and Pl. Anal.* **47**(3): 447–456.

Srivastava, A.K., Shyam Singh. and Albrigo, L.G. 2008. Diagnosis and remediation of nutrient constraints in Citrus. *Hort. Rev.* **34**: 277–64

Srivastava, A.K., Shyam Singh, Das, S.N. and Harmandeep Singh. 2010. Delineation of productivity zones in mandarin orchards using DRIS and GIS. *Better Crops*–South Asia. **4**(1): 13–15.

Srivastava, A.K., Shyam Singh, Diware, V.S.and Haramandeep Singh. 2009. Site–specific nutrient management in 'Mosambi' sweet orange. *Better Crops–India* **3**(1): 10–11.

Srivastava, A.K., Shyam Singh and Huchche, A.D. 2012. Evaluation of INM in citrus (*Citrus reticulata* Blanco): Biometric response, soil carbon and nutrient dynamics. *Int. J. Innov. Hort.* **1**(2): 126–134.

Srivastava, A.K., Shyam Singh and Huchche, A.D. 2015b. Evaluation of INM in citrus on Vertic Ustochrept: Biometric response and soil health. *J. Pl. Nutri.* **38**(5): 1–15.

Srivastava, A.K., Singh, Shyam, Marathe, R.A. 2002. Organic citrus: Soil fertility and plant nutrition. *J. Sustain. Agri.* **19**(3): 5–29.

Srivastava, A.K., Singh, Shyam and Tiwari, K.N. 2006. Site specific nutrient management for Nagpur mandarin (*Citrus reticulata* Blanco). *Better Crops.* **88**(2): 22–25.

Srivastava, A.K., Shyam Singh and Tiwari, K.N. 2007. Diagnostic tools for citrus: Their use and implications in India. *Better Crops–India* **1**(1): 26–29.

Sugiura T., Kuroda, H. and Sugiura H. 2007. Influence of the current state of global warming on fruit tree growth in Japan. *Hortic. Res.* Japan. **6**: 257–263.

Swietlik, D. and Zhang, L. 1994. Critical Zn^{2+} activities for sour orange determined with chelator buffered nutrient solution. *J. Am. Soc. Hort. Sci.* **119**: 693–701.

Tagaliavini, M., Tonon, G., Scandellari, F., Quinones, A., Palmieri, S., Menarbin, G., Gioacchini, P. and Masia, A. 2007. Nutrient recycling during the decomposition of apple leaves (*Malus domestica*) and mowed grasses in an orchard. *Agri. Eco. Environ.* **118**: 191–200.

Tagaliavini, M., Tonon, G., Solimando, D., Gioacchini, P., Toselli, M., Boldreghini, P. and Ciavatta, C. 2004. Nitrogen uptake by ryegrass (*Lolilum perenne*) as affected by the decomposition of apple leaves ad pruning wood in soil. In: Hatch D.J. (Ed.), Proceedings of the 12th N Workshop: Controlling N Flows and Losses, Wageningen Academic Publishers, Wageningen, the Netherlands, pp. 239–241.

Tandon,H.L.S. and Muralidharadu,Y. 2010. Nutrient uptake, removal and recycling by crops. Fertilizer Development and Consultation Organization, New Delhi. pp. 1–140.

Thilagar, G., Bagyaraj, D.J. and Rao, M.S. 2016. Selected microbial consortia developed for chilly reduces application of chemical fertilizers by 50 per cent under field conditions. *Sci. Hort.* **198**: 27–35

Tiwari, K. N. 2002. Site specific nutrient management for high yield and quality of fruit crops. (*In*) *Proceedings of Workshop: Nutrient Status, Needs and Recommendations for Major Fruit Crops*, pp. 105–111, Lucknow, Uttar Pradesh, India.

Tiwari, K.N. 2007. Breaking yield barriers and stagnation through site–specific nutrient management. *J. Indian Soc. Soil Sci.* **55:** 444–54.

Treder, W. 2006. Influence of fertigation. With nitrogen and a complete fertilizer on growth and yielding Gala apple trees. *J. Fruit Ornamental evaporation–based P. Res.* **14**: 143–146.

Umer Iqbal, Wali, Vinod Kumar, Ravi Kher and Mahital Jamwal. 2009. Effect of FYM, Urea and Azotobacter on growth, yield and quality of strawberry cv. Chandler. *Not. Bot. Hort. Agrobot.*Cluj **37**: 139–143.

van Maarschalkerweerd. and Husted, S. 2015. Recent developments in fast spectroscopy for plant mineral analysis. *Front.Plant Sci.* **6**: 169.doi: 10.3389/fpls.2015.00169.

Wang Shuang, Srivastava, A.K., Qiang–Sheng, Wu, and Fokom, R. 2014. The effect of mycorrhizal inoculation on the rhizosphere properties of trifoliate orange (*Poncirus trifoliata* L. Raf.). *Scientia. Hort.* **170**: 137–142.

World Health Organisation 2014. Promoting fruit and vegetable consumption around the world. Global Strategy on Diet, Physical Activity and Health, pp. 14–18.

Worley, R.E. 1994. Long term performance of pecan trees when potassiumapplication is based on prescribed threshold concentration in leaf tissue. *J. Am. Soc. Hort. Sci.* **119:** 434–434.

Wu, Ting, Wang Yi, Yu Changjilang, Rawee Chiarawipa, Zhang Xinzhoing, Han Zhenhai and Lianhai Wu 2012. Carbon sequestration by fruit trees–Chinese apple orchards as an example.PLOS ONE **7**: e38883, doi: 10.1371/journal.pone.0038883./http://www.plosone.org/article/info per cent 3Adoi per cent 2F 10.1371 per cent 2Fjournal.pone.0038883.

Wu, Qiang–Sheng and Srivastava, A.K. 2015. Effects of mycorrhizal symbiosis on growth behaviour and carbohydrate metabolism of trifoliate orange under different substrate P levels. *J. Pl. Growth Regul.* **34**: 499–508.

Xiloyannis, C., Montanaro, G., Sofo, A. 2002. Proposte per contenere i danni da siccita' alle piante da frutto. *Frutticoltura* 7–8, 19–27.

Yadav, Rajesh, Baksh, Hari Singh, Singh, H.K. and Yadav, A.L. 2007. Effect of integrated nutrient management on productivity and quality of aonla (*Emblica officinalis* Gaetm.) fruits. *Plant Arch.* (1 and 2): 881–883.

Zaman, Q.U. and Schumann, A.W. 2006. Nutrient management zones for citrus based on variation in soil properties and tree performance. *Precis. Agri.* **7**: 45–63.

Zahir, A., Zahir, Mohmmad, Arshad, William T. Frankenberger Jr. 2003. Plant growth promoting rhizobacteria: Applications and perspectives in agriculture. *Adv. Agron*. **31**: 97–166.

Zeng, Q., Brown, P.H. and Holtz, B.A. 2001. Potassium fertilization affects soil K, leaf concentration, and nut yield and quality of mature pistachio trees. *Hort. Sci.* **36:** 85–89.

35

Microbial Consortia in Bio-capsule for Enhancing the Farmers' Income in Spices

M. Anandaraj

Former Director, ICAR-Indian Institute of Spices Research, Kozhikode – 673 012, Kerala
Present Address: No. 7, S R Aptt., 8A cross, MSR Nagar, Bengaluru – 560 054, Karnatka
E-mail: arajiisr@gmail.com

Plant growth-promoting rhizobacteria (PGPR) are root associated bacteria, when introduced onto the seeds, roots or soil colonize the rhizosphere and improve plant growth in a variety of agricultural and horticultural crops. It has been estimated that more than 100 million tonnes of nitrogen, potash and phosphate-chemical fertilizers have been used annually in order to increase plant yield. The negative effects of chemical fertilizers on the global environment and the cost associated with their production have given impetus to the research on bioinoculants. The beneficial free-living bacteria referred to as PGPR are found in the rhizosphere of plants. The rhizosphere is the narrow zone of soil surrounding the root that is under the influence of the root system. The rhizosphere, influenced chemically, physically and biologically by the plant root, is a favourable habitat for the proliferation of microorganisms and exerts a potential impact on plant health and soil fertility. Root exudates rich in amino acids, monosaccharides and organic acids, serve as the source of nutrients, and support the dynamic growth and activities of various microorganisms within the vicinity of the roots. Among PGPR the fluorescent Pseudomonads (*Pseudomonas putida, P. fluorescens*) a genus of Gram-negative, rod-shaped bacteria with polar flagella reported to often predominate among the rhizosphere bacteria have attracted considerable attention due to the beneficial effects on plants through direct and indirect mechanisms of plant growth promotion. The common phenotypic trait that is shared by the various Pseudomonas species is the production of fluorescent, iron-binding siderophores when cells are cultured under low-iron conditions.

Mechanisms of Plant Growth Promotion by PGPR

PGPR promotes plant growth and yield either directly or indirectly. The direct growth-promoting mechanisms involve nitrogen fixation, solubilization of phosphorus, sequestering of iron by production of siderophores, production of phytohormones such as auxins, cytokinins, gibberellins and lowering of ethylene concentration. The indirect mechanisms of plant growth promotion by PGPR include antibiotic production, depletion of iron from the rhizosphere, synthesis of antifungal metabolites, production of fungal cell wall lysing enzymes, competition for sites on roots and induced systemic resistance. Many PGPR possess multiple plant growth-promoting attributes which influence plant growth at different developmental stages. PGPR may stimulate cell division through the early plant development by producing phytohormones. Further plant growth may be enhanced by PGPR by providing sufficient amount of iron and phosphorus from the soil.

Phosphate Solubilization

Phosphorus (P) is an essential mineral nutrient for plant growth as it affects plant structure at cellular level, and stimulates growth and hastens maturity. Plants with

P deficiency exhibit stunted growth, wilting of leaves, delayed maturity and reduced yield. Most of the soils are poor in available P. A large portion of soluble inorganic phosphates applied as chemical fertilizers to the soil gets rapidly fixed into the forms plants are unable to utilize. The fixation and precipitation of P is largely dependent on pH and soil type. P becomes fixed in the acidic soils by free oxides and hydroxides of aluminium and iron, while in the alkaline soils it gets fixed by calcium. Phosphorus in the labile organic compounds is slowly mineralized as available inorganic P or immobilized into more stable organic materials as a part of the soil organic matter.

Phosphate-Solubilizing Microorganisms

Microorganisms play a critical role as inhabitants of the rhizosphere and bulk soil in mediating the availability of soil P to the plants. Phosphate-solubilizing microorganisms belonging to diverse actinomycetes, bacteria and fungi have been isolated from rock phosphate deposits and rhizosphere of several crops. Several bacterial species belonging to *Achromobacter, Aerobacter, Agrobacterium, Bacillus, Burkholderia, Erwinia, Flavobacterium, Gluconoacetobacter, Micrococcus, Pseudomonas, Ralstonia, Rhizobium, Serratia* and others have been reported for the conversion of insoluble inorganic phosphates into soluble forms.

Phytohormone Production

Auxins constitute an important class of phytohormones which influence many cellular functions regulating plant growth and development. Auxins play an important role in the orientation of root and shoot growth in response to light and gravity, differentiation of vascular tissues, apical dominance, initiation of lateral and adventitious roots, cell division and elongation of stems and roots. Auxins particularly indole-3-acetic acid (IAA) has also been reported to be involved in host-parasite interactions. The biocontrol action of IAA either due to the inhibition of spore germination and mycelium growth of pathogenic fungi or due to its involvement together with glutathione S-transferase in defense-related plant reactions. The auxin concentration is critical to the physiological response by the plant. The production of auxin is widespread among plant associated bacteria. PGPR predominantly synthesize IAA from tryptophan mainly through the indole-3-pyruvic acid (IPA) and indole-3-acetamide (IAM) pathways. Tryptophan exuding from roots serves as the primary precursor of IAA biosynthesis by microorganisms

ACC-deaminase Activity

Alcaligenes, Bacillus spp., *Burkholderia, Enterobacter cloacae, Pseudomonas* spp., *Ralstonia solanacearum, Rhizobium, Rhodococcus, Sinorhizobium meliloti* and *Variovorax paradoxus* produce the enzyme ACC deaminase which cleaves the plant ethylene precursor ACC to -ketobutyrate and ammonia, thereby lowering the level of ethylene in the rhizosphere. IAA produced by plant growth-promoting bacteria can stimulate plant cell proliferation, plant cell elongation or induce the transcription of ACC synthase catalyzing indole-3-cetaldehyde, Indole-3-pyruvic acid, indole-3-acetonitrile, indole-3-acetic acid, indole-3-acetamide, tryptamine, tryptophan, indole-3-acetaldoxime the formation of ACC. The cleavage of ACC by ACC deaminase leads to more ACC exudation from inside the plant to maintain the equilibrium, thus reducing ACC and ethylene levels evolved by the plant. The plants grown in association of these bacteria have longer roots and shoots. ACC-deaminase producing bacteria have also been found to provide resistance to plants against pathogens, tolerance against salinity, drought, flooding, metal and organic contamination.

Siderophore Production

Iron is an essential plant micronutrient as it serves as a co-factor of many enzymes with redox activity. A large portion of the soil iron is in highly insoluble form of ferric hydroxide which acts as a limiting factor for plant growth even in iron rich soils. Several microorganisms including *Alcaligenes, Azotobacter, Azospirillum, Bacillus, Enterobacter, Pseudomonas, Rhizobium* produce low-molecular weight iron-binding molecules usually less than 1 kDa called siderophores under low-iron condition. Studies on inoculation with siderophore producing bacteria have provided evidence for the absorption of bacterial iron-siderophore complexes by the plants especially in the calcareous soils. Siderophore-producing rhizobacteria suppress fungal pathogens by making iron unavailable for fungal growth. Siderophore-mediated competition for iron is also a major factor in determining the interaction among bacterial strains during rhizosphere competence by PGPR.

Production of Anti-fungal Metabolites

Pathogenic microorganisms affecting plant health are a major threat to crop production.Biological control of plant pathogens using antagonistic rhizobacteria is an effective and eco-friendly alternative to the use of synthetic chemicals. PGPR produce several metabolites including siderophores, antibiotics such as 2, 4-diacetylphloroglucinol, hydrogen cyanide (HCN), phenazine, pyoluteorin, pyrrolnitrin and microorganisms depends not only on the phosphate form but also on the soil type and its buffering capacity, plant species and PGPR strain. Other mechanisms including H+ excretion, production of inorganic acids, carbon dioxide and H_2S have also been implicated in phosphate solubilization. Proton-excretion accompanying ammonium ion assimilation has been reported as the most plausible explanation for microbial solubilization without organic acid production.

Effect of PGPR on Plant Growth

Inoculations with a target microorganism at a much higher concentration than that found in the soil is necessary to take advantage of PGPR for plant growth and yield enhancement. The benefits of PGPR application depend upon plant species/variety, growth conditions, and PGPR strains. Increased yields have been reported with the application of PGPR selected for various plant growth promoting activities. *Acinetobacter rhizosphaerae, Bacillus megaterium, B. subtilis, Bradyrhizobium* sp., *Burkholderia* sp., *Enterobacter* sp., *Pantoea* sp., *Pseudomonas fragi, P. putida, P. striata, Pseudomonas* spp., *Serratia marcescens, Serratia* sp. and *Rhizobium leguminosarum* solubilizing phosphate have been reported to increase growth and yield in several crops. The potential use of phosphate-solubilizing bacteria along with rock phosphates has been studied intensively for increasing P availability to plants.

Stress Tolerance in PGPR

The abiotic and biotic stimuli beyond the optimal range for survival induce stress, causing metabolic dysfunction that can lead to death in microorganisms. The performance of microorganisms is severely affected by the stress generated by environmental factors like moisture status, temperature, alkalinity/acidity and salinity in the soil. The ability of the microorganisms to adapt to fluctuations in the osmolarity and temperature of their surrounding is of fundamental importance for their survival.

Rhizosphere Competence of PGPR

Successful colonization and persistence in the rhizosphere are necessary for PGPR to exert their beneficial effect on plants. The knowledge about the impact of introduced PGPR strains on the native microflora is important as the microbial inoculants may affect the dynamics of resident microbial population.

Different approaches and techniques have been used to quantify and identify the inoculated strains on the host plant. The most common method to study the rhizosphere competence is based on comparing the colonization ability of mutants with their parental strains. Spontaneous antibiotic resistance has provided a simple and effective method to genetically mark the bacterial strains for monitoring after introduction into the plant rhizosphere. The chromosomal nature of the mutation affords greater stability than that occurs with plasmid-borne markers and is also advantageous as the mutation is not easily transferable. Because of the greater stability and non-transferable nature of the mutation, rifampicin-resistant mutants have been used widely to study the root colonization by PGPR strains-fluorescent *Pseudomonas* strains in potato and tomato.

In spice crops the fate of the introduced organism has been monitored using antibiotic resistance and the introduced organisms have survived in the rhizospehere. It also indicated the positive influence of the host roots in maintaining the populations.

Trichoderma as Growth Promoting Organism

Initially PGPR were credited with all the beneficial effects of growth promotion and disease suppression. Antagonistic fungi-like *Trichoderma* spp. were thought to be as only biological control agents and even termed as "bio-pesticides" and brought under the purview of Central Insecticide Board (CIB). Latest research reveals that species of *Trichoderma* also possess all the functions of nutrient mobilization, growth promotion, bio-control properties besides growing as endophytes and triggering internal defences. Latest technologies like proteomics and meta- genomics also indicate that inoculation of efficient strain of *T. harzianum* in black pepper rhizosphere promotes enhanced colonization of roots by arbuscular mycorrhiza and promote native microbial communities to grow and assist the plant in better uptake of nutrients thereby enhancing the growth.

Bio-inoculants for Spices

The Research conducted at ICAR- Indian Institute of Spices Research, Kozhikode and the All India Coordinated Research Project on Spices has lead to the development of several bio-inculants for spice crops. This includes a consortia of bacterial inoculants for black pepper, pathogen suppressing antogonistic fungi like *Trichoderma harzianum* and *Pochonia chlamydosporium, Bacillus amiloliquefaciens* for ginger both for enhancing growth and suppressing diseases. The seed coating technology for delivering beneficial organisms on seeds and delivery of beneficial organisms in capsule form that reduces the use of bulk quantities of carrier material has been the novel technologies.

Delivery of Beneficial Microbes

The majority of plant-associated bacteria are from the soil environment. They may migrate to the rhizosphere and subsequently the rhizoplane of their hosts before they are able to show beneficial effects. Some rhizoplane colonizing bacteria can also penetrate plant roots and live as endophytes without causing any visible damage to the host plants. To be successful, the beneficial organisms must possess competence to colonize the root surface or enter the roots to reside as endophytes. Scientists have been successful in isolating microorganisms that posses useful agronomic traits like nutrition mobilizing, synthesising growth hormones and also producing metabolites that prevent disease causing pathogens. Some of them possess multiple traits of growth promotion and pathogen suppression. Identifying an efficient strain is only part of the story. To deliver the same in the field is a formidable task as there are several challenges. The first and foremost is the retaining viability of the organism during storage (shelf-life) and transportation.

Traditionally, the beneficial organisms are cultured in the laboratory, mixed with a carrier medium and delivered to the field by making formulations with inert materials. Various forms of PGPR formulations are available namely liquid formulations and dry powder based formulations. Liquid formulations include cell suspension in water, buffer, concentrated and diluted liquid formulations with nutrients *etc.* Dry formulations include talc based formulations, dry carrier based formulations like peat, coffee husk, rice bran, sorghum *etc.* The main drawback of this formulation is short lifespan or short term viability in formulations. And also dry powders has bulk weight and size, hence it needs more space for storage and difficult to transport to long distances. Lot of efforts are required to carry the material within the plot also. Often the application becomes cumbersome due to the bulkiness of the carrier media. Secondly, the population of bio-inoculants during storage gets reduced making the product less effective. It is also a fact that quality of suffers because of reduced populations during storage and long distance transport.

Novel Delivery Method for Beneficial Microbes

A novel method of delivery of beneficial organisms was developed at ICAR- Indian Institute of Spices Research and a patent for the process has been filed (Appl. No.3594/CHE/2013 dt 13/08/2013). The invention of PGPR formulation as Biocapsule is the easy and reliable technology of storing and delivering PGPR bioagents in hard gelatin capsule termed as biocapsule. It is preparation of viable microbial agents in a capsule form. Initially, this technology was developed using efficient plant growth promoting rhizobacteria (PGPR) for ginger namely *Bacillus amyloliquefaciens* GRB35. Based on extensive field studies and monitoring the shelf-life for over 18 months patent application was filed.

In this method, the shelf-life of the bacteria are maintained and for bacterial inoculants, before application the cells can be activated for twelve hours, diluted and applied. The method is available for delivering both bacteria and antagonistic fungus like *Trichoderma harzianum.*

Advantages of Capsule Method of Delivery

The viable propagules delivered through biocapsule technology is easy to handle and the volume gets reduced by one hundred times. To apply one hectare of ginger 20 kg talc formulations are required. Whereas, with biocapsule technology a mere 200 capsules (1g each)weighing 200g will suffice). Since the bulk is reduced, only the organism along with minimum ingredients are stored. The organism gets activated once it comes in contact in water and thus the propagule number is enhanced in case of bacterial bioinoculants. Transportation and handling costs are reduced considerably with increased efficiency.

Antagonistic fungi like *Trichoderma* can also be delivered by this method. Since conidia are present it is ideally suited to deliver it at the root zone along with irrigation water. Thus here is a method of delivering microorganism in the field with ease and increased efficiency.

The technology has already been licensed to entrepreneurs for commercial production by ICAR-Indian Institute of Spices Research, Kozhikode.

Conclusion

Though the benefits of bioinoculants is well known, the adoption of the technology is fraught with several glitches. The first and foremost is the logistic problem in handling bulk of the quantity. A lot of effort has to be made to deliver in the field. The second one is the shelf-life of the product. Often, the number of propagules required for the effective control is not maintained in the product. The capsule technology ensures easy handling of the bio-inoculant as the bulk is reduced by one hundred per cent. The shelf-life is also addressed as there is large number of viable units and the number in case of bacterial inoculants also increases during the incubation period. Since these days farmers follow drip irrigation and fertigation the technology is quite easy to adopt and the cost in handling is reduced considerably.

Another issue that remains to be sorted out is the inclusion of bio-inoculants such as *Trochoderma* and PGPR under the purview of Central Insecticide Board (CIB) along with chemical pesticides. Since most of the bio-inoculants have multifarious activities like growth promotion, pathogen suppression and modulating defense responses against pathogens it should be considered as bio augmentation in the crop production and never be considered as "bio-pesticide" so that the technology spreads to all the producers and the farmers are benefited. There should be strict guidelines to maintain the quality of products and accredited laboratories from among the National Agricultural Research System (NARS) across the country for quality management. Since we do not have any dearth of laboratories and expertise in the country this needs to be regulated through this system rather than by CIB.

Acknowledgments

My sincere thanks my all colleagues at ICAR- Indian Institute of Spices Research, Kozhikode and several reviewers whose material I have been used liberally in this compilation.

References

Amer, G.A. and Utkhede, R.S. 2000. Development of formulations of biological agents for management of root rot of lettuce and cucumber. *Can. J. Microbiol.*, **46**: 809–816.

Anelise, Beneduzi, Adriana, Ambrosini and Luciane, M.P. Passaglia 2014. Plant growth-promoting rhizobacteria (PGPR): Their potential as antagonists and biocontrol agents. *Genet. Mol. Biol.*, **35(**4)(suppl): 1044–1051.

Bora, T., Ozaktan, H., Gore, E., and Aslan, E. 2004, Biological control of *Fusarium oxysporum* f. sp. *melonis* by wettable powder formulations of the two strains of *Pseudomonas putida*. *J. Phytopath.*, **152**: 471–475.

Callan, N.W., Mathre, D.E. and Miller, J.B. 1990. Biopriming seed treatment for biological control of *Pythium ultimum* pre-emergence damping-off in *sh2* sweet corn. *Plant Dis.*, **74**: 368–372.

Charpentier, C.A., Gadille, P. and Benoit, J.P. 1999, Rhizobacteria microencapsulation: Properties of microparticles obtained by spray drying. *J. Microencapsulation*, **16**: 215–229.

Dieter Haas and Geneviève Défago, 2005. Biological control of soil-borne pathogens by fluorescent pseudomonads. *Nat. Rev. Microb.*, AOP, published online 10 March; doi:10.1038/nrmicro1129

Kloepper, J.W., Leong, J., Teintze, M. and Schroth, M.N. 1980a. Enhancing plant growth by siderophores produced by plant growth-promoting rhizobacteria. *Nature*, **286**: 885–886.

Kloepper, J.W., and Schroth, M.N., 1981, Development of powder formulation of rhizobacteria for inoculation of potato seed pieces. *Phytopathology*, **71**: 590–592.

Kloepper, J.W., Leong, J, Teintze M and Schroth, M.N. 1980b. *Pseudomonas* siderophores: A mechanism explaining diseasesuppressive soils. *Curr. Microbiol.*, **4**: 317–320.

Kloepper, J.W., Lifshitz, R. and Zablotowicz, R.M. 1989. Free-living bacterial inocula for enhancing crop productivity. *Trends Biotechnol.*, **7**: 39–43.

Kloepper, J.W., Ryu, C-M and Zhang, S.A., 2004. Induced systemic resistance and promotion of plant growth by *Bacillus* spp. *Phytopathology*, **94**: 1259–1266.

Loon, L.C., Bakker, P.A.H.M. and Pieterse, C.M.J. 1998. Systemic resistance induced by rhizosphere bacteria. *Ann. Rev. Phytopathol.*, **36**: 453–483.

Lugtenberg, B. and Kamilova, F., 2009. Plant growth promoting rhizobacteria. *Annu. Rev. Microbiol.*, **63**: 541–556.

Manjula, K., and Podile, A.R., 2001, Chitin supplemented formulations improve biocontrol and plant growth promoting efficiency of *Bacillus subtilis* AF1. *Can. J. Microbiol.*, **47**: 618–625.

Munees, Aa. and Mohammad, S.K., 2011. Functional aspects of plant growth promoting Rhizobacteria: Recent advancements. *Insight Microbiology*, **1**: 39–54.

Nakkeeran, S., Dilantha, W.G. Fernando and Zaki, A.S. 2005. Plant growth promoting rhizobacteria formulations and its scope in commercialization for the management of pests and diseases, (Ed.) Z.A. Siddiqui. *PGPR: Biocontrol and Biofertilization*, pp. 257–296.

Rathore, Pragya, 2014. A review of N approaches to develop plant growth promoting rhizobacteria. *Int. J. Rec. Sci. Res.*, **5**: 403–407.

Verma, J.P., Yadav, J., Tiwari, K.N., Lavakush and Singh, Vimal 2010. Impact of plant growth promoting rhizobacteria on crop production. *Intl. J. Agric. Res.*, **5**: 954–983.

36

Improving Productivity and Income of Horticultural Crops from Degraded Lands

T. Damodaran[1]* and S. Rajan[2]

[1]ICAR-Central Soil Salinity Research Institute,
RRS, Lucknow – 226 002, Uttar Pradesh
[2]ICAR-Central Institute for Sub-tropical Horticulture,
Lucknow – 226 101, Uttar Pradesh
**E-mail: damhort73@gmail.com*

Current scenario and future projections suggest that to produce food for the world's expanding population; there need to be an increase in use of marginal-quality water and land resources. India, which constitutes only 2 per cent of the world's total land, provides shelter to the 16 per cent of the world population (Bower, 2000). Decades of anthropogenic activities and inadequate agricultural practices have resulted in the pollution of soil, degradation, loss of soil fertility, and desertification (McLauchlan, 2006). In addition, arid and semi-arid regions are particularly prone to salinization, as a consequence of mismanaged irrigation (Szabolcs, 1994). Major emphasis has been given on technologies for irrigated and resource rich soils while marginal and problem soils which harboured a majority of marginal farmers had been ignored. A total of 120.20 million ha are classified under degraded lands among, which majority of the land (70 per cent) is under area exposed to wind and water erosion. About 6.73 million ha are covered with sodic and saline soils while 17.93 million ha has been covered with acidic soils (Qadir and Schubert, 2002). Sodicity and salinity are serious environmental issues that cause soil congestion and osmotic stress, which reduces crop productivity in irrigated areas of arid and semiarid regions (Cicek and Cakirlar, 2002). Conversion of waste and degraded lands in to productive commercial horticulture necessitates innovative approaches for increasing the biological dynamism of the soils. With this scenario the current paper highlights on various innovative initiatives for increasing the productivity of horticultural crops in the marginal and salt affected soils.

1. Rhizosphere Diversity

Plants have always been susceptible to various abiotic stresses in their immediate environment due to their sessile nature. Survival depends on their ability to adjust to their physiology, development, and growth to mitigate the impacts of stress (Hirt, 2009). The rhizosphere is the zone of soil around roots that is the interface between plants and their environment which is essential for the acquisition of water and nutrients for beneficial interactions with soil-borne microorganisms. This region is more prone to biotic and abiotic stresses (Neumann and Römheld, 2007). Exploitation of rhizosphere microbial diversity for plant growth promoting rhizobacteria (PGPR) and mycorhizae enhances the possibility of increasing the productivity of the horticultural crops in sodic soils of pH 9.2 (Damodaran *et al.*, 2016). Engineering or modification of rhizosphere involves release of inorganic and organic substances from the plant roots (rhizodeposition). The acidification increases the availability of Fe^{3+} and phosphorus through secretion of acids like gluconic acids, which are often fixed to the surface of minerals that are present as insoluble complexes. Iron uptake in plants involves the release of organic compounds called siderophores by bacteria that chelate Fe^{3+} and facilitate its reduction to Fe^{2+} prior to uptake (Neumann and Römheld, 2007).

Salinity enhances ethylene production in plants which is a stress harmone resulting in early senescence. Certain PGPR are known to produce ACC (1-aminocyclopropane, 1-carboxylate) de-aminase which lowers the synthesize of ethylene by cleaving the ACC to form ammonia and α-ketobutyrate instead of ethylene (Glick *et al.*, 1998). This enzyme hydrolyzes ACC, the immediate biosynthesis precursor of ethylene in plants. The products of this hydrolysis, ammonia and α-ketobutyrate are used by the bacterium as a source of nitrogen and carbon for growth (Klee *et al.*, 1991). The bacterium acts as a sink for ACC preventing some of the potentially deleterious consequences of high ethylene concentrations like early senescence of plants under stress (Steenhooudt and Vanderleyden, 2000). PGP functions by synthesizing specific compounds which enhance the nutrient mobilization and delay senescence under saline environment and increases plant growth.

1.1 Microbial Consortia and Substrate Dynamism

In the rhizosphere, a single root often carries a multiple symbiosis like mycorhizzal and bacterial. Net gain obtained by plant is different from a simple one to one gain obtained from application of single microbe, which necessitates the requirement of microbial group or consortia. Substrate for delivery of the efficient microbial consortia plays a vital role in establishment of the microbe of interest in the soil system for performing the role of growth promotion, stress tolerance and disease control (Damodaran *et al.*, 2016). The origin of a substrate and its pH are considered two most important guiding principles in developing a substrate dynamic to plant's rhizosphere (Altland, 2006). Microbial consortia comprising of microbes for nutrient mobilization under stress, growth promotion, siderophore and antibiotic activity initiation in a single formulation using a dynamic media and substrate prone to increase the productivity of the soils and alleviate the crops from stress (Damodaran *et al.*, 2013).

1.2 Plant Growth Promoting (PGP) Inducing Stress Tolerance

Under salt stress, PGPR (Plant growth promoting rhizobacteria) have shown positive effects in plants on parameters like germination rate, tolerance to drought, weight of shoots and roots, yield and plant growth. Treating of tomato plants with ACC de-aminase producing PGPR reduced significantly the damage caused due to early senescence under water stress (Grichko and Glick, 2001). *Archomobacter piechondii* ARV 8 significantly increased the ability of the plant to tolerate drought in pepper and tomato. Salt tolerant *Pseudomonas chlororaphis* isolate TSAU 13 promoted growth and yield of cucumber and tomato (Egamberdieva, 2012). A12ag a P-solubilzing salt tolerant bacterium promoted growth of tomato under poor and marginal lands with lower fertility (Chookietwattare and Maneewan, 2012). Treating the seeds and soil with *Bacillus megaterium* and *Glomus* sp. increased the tolerance of clove to drought. *Pseudomonas mendocinia* and *Glomus intraradices* mitigate the impact of soil drought in lettuce crop through production of exopolysaccharates, induction of resistance genes and increased circulation of water in the plant (Grover *et al.*, 2010). Bio-inoculation (*Bacillus amylolequifaciens*) leads to the production of polysaccharates that facilitates the formation of macroaggregates of soil which increases plant resistance to stress and drought. Two salt tolerant bacterial epiphyte CSR-B-2 (*Bacillus pumilus*), CSR-B-3 (*Bacillus thuringiensis*) and one fungal isolate CSR-T-1 (*Trichoderma harzianum*) identified from 36 isolates isolated from sodic soils alleviate the salt stress in the tomato and *bhendi* grown under sodic conditons (Damodaran *et al.*, 2013). Bio-priming of seeds, soil application and foliar spray during the flowering with CSR-BIO a bio-growth enhancer using consortia of microbes on dynamic media resulted in 20 -22 per cent increase in the average yield of commercial vegetables like tomato, capsicum and *bhendi* under surface reclaimed sodic soils with soil pH 9.00 to 9.20 and SAR 15.43. Experiments with *bhendi* hybrid No.10 under sodic conditions showed that the product facilitates the uptake of potassium (K) to about 2.246 meq/l in indicator leaves while it was 0.606 meq/l, which was significantly lower than the treated plant (Damodaran, 2015). Beans and clover inoculated with AM fungi improved their osmoregulation and increased proline accumulation which resulted in salinity resistance (Feng *et al.*, 2002). Treatment of coriander and fenugreek with 1 per cent CSR-BIO a bio-growth enhancer for salt affected soils as seed treatment and soil application along with FYM increased the yield to 14 per cent and reduced the incidence of blight. Application of CSR-BIO in tomato and *bhendi* increased the plants inherent potential to produce reserve oxygen scavenger (ROS) enzymes like superoxy dimuatase and proline attributing for the induced systemic tolerance of the plants to abiotic stresses (Damodaran *et al.*, 2014).

1.3 Plant Growth Promoting (PGP) inducing Stress Tolerance

Stress limits crop growth and productivity of fruit crops, especially in arid and semi-arid regions. Some microbial species and strains that inhabit plant rhizosphere use different mechanisms to mitigate negative effects of drought on plants. Due to perennial growth habit the productivity of fruit crops are poor in marginal and salt affected soils. Microbial-based alleviation technologies can become successful only when package of application (POA) on integration with media and substrate dynamism of the bio-formulations were attempted and a simple supplementation of bio-formulation as seed treatment or one time soil application results in reversion of the crop to the stress. Application of bio-products targeted at critical stages of growth aid in incorporating the stress tolerance in perennial fruit crops. Soil application of 25 GR gypsum in the pits along with CSR-B-3 strain of *Bacillus thuringienesis* during planting, 5th, 7th and 9th month after planting using vermicompost as substrate reclaimed the rhizosphere soil and alleviated the salt stress in the rhizosphere of the crop resulting in higher bunch weight of 22.7 kg/plant (Damodaran *et al.*, 2013a). Owing to the AAA genome the Grand Naine tissue culture plantlets failed to establish in sodic soils even with the use of salt tolerant bioformulations. Substrate management using enriched green manuring with adoption of patented *in vitro* four-tier biopriming technology increased the productivity of Grand Naine banana from 8.5 to 28 kg/bunch in salt affected soils of pH 9.1 (Damodaran *et al.*, 2016). Bio-hardened cormlets of gladioulus using CSR-B-3 strain using CSR-BIO patented media resulted in higher productivity of marketable spikes of length 75-77 cm in sodic soils of 9.2 (Damodaran *et al.*, 2014: Damodaran *et al.*, 2014a). The bacterium *Burkholderia phytofirmans* PSJN colonizes grapevine residues and protects the plant against heat and frost through increase in the levels of starch, and proline and phenols (Saint-Etienne *et al.*, 2006). Interactions between Vesicular-Arbuscular Mycorrhizal Fungi (AMF) and horticultural crops (Menge, 1983) are, in incorporating tolerance to abiotic stress is well-documented. Association of AM fungi with N fixing bacteria enhanced the nutrient mobilization and growth of the Mosambi seedlings and grafts by increasing the efficacy of nitrogen fixation and availability of the phosphorous (Manjunath *et al.*, 1983). VAM fungi also aid in assimilation of NH_4^+ (Barea *et al.*, 1987) under stress conditions apart from increasing the soil phosphorous availability to ferns and palms.

1.4 Impact of Bio-formulaitons in Increasing the Profitability and Productivity

The profitability of the CSR-BIO formulation in commercial crops like tomato and banana with its impact on reducing use of chemical fungicides toxic to the environment was assessed with adopters and non adopters. An overall increase in yield up to 22.43 and 15.62 per cent in the adopters of tomato and banana, which simultaneously increased the gross profitability to 20.11 and 17.39 per cent in banana and tomato respectively (Tables 1 and 2). The use of plant protection chemicals was 47.33 and 33.36 per cent lower than the non-adopters who didn't practiced the technology in tomato and banana respectively (Damodaran *et al.*, 2013).

Table 1: Impact of CSR BIO on Profitability and Environment among the Adopters and Non-adopters in Tomato (N=100)

Parameter	*Group**	*Mean*	*SD***	*Significance (P = 0.05)*	*SED****
Yield (kg)	1	19.640	4.079	0.004	1.137
	2	16.053	1.663	0.005	1.137
Expenditure (Rs)	1	0.378	4.224	0.026	1.414
	2	0.412	3.489	0.026	1.414
Income (Rs)	1	2.216	0.448	0.008	0.130
	2	1.845	0.229	0.010	0.130
Pesticide	1	6.533	1.567	0.001	0.569
	2	12.400	1.549	0.001	0.569

**SD: Standard Deviation; ***SED: Standard Error Difference.

Source: Damodaran *et al.*, 2013.

Table 2: Impact of CSR BIO on Profitability and Environment among the Adopters and Non-adopters in Banana (N=100)

Parameter	*Group**	*Mean*	*SD***	*Significance (P = 0.05)*	*SED****
Yield (kg)	1	31.533	4.612	0.030	1.867
	2	27.266	5.574	0.030	1.867
Expenditure (Rs)	1	0.647	5.522	0.006	2.026
	2	0.707	5.574	0.006	2.026
Income (Rs)	1	2.782	0.393	0.013	0.157
	2	2.366	0.464	0.013	0.157
Pesticide	1	7.333	1.195	0.004	0.658
	2	11.067	2.250	0.005	0.658

**SD: Standard Deviation; ***SED: Standard Error Difference.

Source: Damodaran *et al.*, 2013.

1.5 Stress Tolerant Rootstock

Abiotic stress is a major productivity constraint which limits cultivation of several fruit crops. Development of true-to-type rootstock tolerant to drought, salinity and high temperature is essential for mitigating the abiotic stress through climate resilent agriculture. The mango rootstock '13-1' is a promising salt tolerant polyembryonic type developed from Israel in 1984. Post tsunami selections and screening from inundated orchards in South Andaman Islands resulted in development of polyembryonic salt tolerant rootstock ML-2 and ML-6, which have sustained high pH of 9.42 without any mortality and reduction in growth (Damodaran *et al.*, 2013b). Rootstocks attributing tolerance to drought and salinity were developed in grapes (Dogridge, Salt Creek, 1616), citrus (Cleopatra mandarin, Karna Katta) and apples (M9, M26, M27) that changed the productivity status of the horticultural crops (Chadha, 1998). Productivity of guava in salt affected soils increased with the use of hybrid of *P. mole* and *P. guajava* as rootstock.

2. Conclusion

With the demand for increasing the productivity of horticultural crops in the degraded lands which are subjected to drought, salinity and erosion, microbial based alleviation using substrate dynamism suggest to be a eco-friendly strategy for adoption and implementation. The bio-enhancers and bio-formulations modifies the plant inherent potential for tolerance through expression of stress tolerant genes like SOD1, higher stress tolerant enzyme activity that plays a key role in increasing the productivity of commercial horticultural crops. Use of stress tolerant rootstocks in fruit crops with the bio-inoculant technology tends to increase the productivity of important horticultural crops.

3. References

Barea, J.M., Azcbn-Anuilar, C. and An. R. 1987. Vesicular-arbuscular mycorrhiza improve both symbiotic N-fixation and N uptake from soil as assessed with a "N technique under field conditions. *New Phytol.*, **106**: 717–725.

Bouwer, H. 2000. Integrated water management: Emerging issues and challenges. *Ag. Water Mgmt.*, **45**: 217–228.

Chadha, K.L. 1998. Improvement of tree and plantation crops. *Ind. J. Hort.*, **55**: 265–296.

Chookietwattana, K. and Maneewan, K. 2012. Screening of efficient halotolerant phosphate solubilizing bacterium and its effect on promoting plant growth under saline conditions. *World Appl. Sci. J.*, **16**: 1110–1117.

Cicek, N. and Cakirlar, H. 2002. The effect of salinity on some physiological parameters in two maize cultivars. *Bulg. J. Pl. Physiol.*, **28**: 66–74.

Damodaran, T., Rai, R.B., Jha S.K., Dhama, K., Mishra V.K., Sharma, D.K., Singh, A.K. and Dixit, Himanshu 2013. Impact of CSR-BIO: An eco-friendly bio-growth enhancer on increasing the profitability of horticultural crops to small and marginal land holders. *Int. J. Curr. Res.*, **5**(9): 2682–2685.

Damodaran, T., Mishra, V.K., Sharma, D.K., Jha, S.K., Verma, C.L., Rai, R.B., Kannan, R., Nayak, A.K. and Dhama, K. 2013a. Management of sub-soil sodicity for sustainable banana production in sodic soil: An approach. *Inr. J. Curr. Res.*, **5**: 1930–1934.

Damodaran, T., Rajan, S., Ramkumar, S., Sharma, D.K., Mishra, V.K., Jha, S.K. and Rai, R.B. 2013b. Post-tsunami collection of polyembryonic mango diversity from Andaman islands and their *ex situ* reaction to high sodium in sodic soil. *J. Appl. Hort.* 15(1): 21–25.

Damodaran, T., Rai, R.B., Kannan, R., Pandey, B.K., Sharma, D.K., Misra, V.K., Vijayalaxmi, S. and Jha, S.K. 2014. Rhizosphere and endophytic bacteria for induction of salt tolerance in gladioulus grown in sodic soils. *J. Pl. Interact.*, **9**: 577–584.

Damodaran, T., Rai, R.B., Sharma, D.K., Misra, V.K., Pandey, B.K., Prabat Kumar, Himanshu Dixit and Kannan, R. 2014a. Utilization of native bioagents as bioregulators to improve growth, flower quality and vase life of gladiolus. *Ind. J. Hort.*, **71**(2): 242–246.

Damodaran, T., Rai, R.B, Pandey, B.K., Sharma, D.K., Mishra, V.K., Jha, S.K. and Prabhat Kumar 2015. CSR-BIO: A boon for production of horticultural crops in sodic lands. *Ind. Hort.*, pp. 16–18.

Damodaran, T. and Mishra, V.K., 2016. Rhizosphere engineering: An innovative approach for sustainable crop production in sodic soils. **In**: *Innovative Saline Agriculture*, (Eds.) J. Dagar, P. Sharma, D. Sharma and A. Singh. Springer Publications pp. 224.

Egamberdieva, D. 2012. *Pseudomonas chlororaphis*: A salt-tolerant bacterial inoculant for plant growth stimulation under saline soil conditions. *Acta Physiol. Plant.*, **34**: 71–56.

Feng, G., Zhang, F., Li, X., Tian, C., Tang, C and Rengel, Z., 2002. Improved tolerance of maize plants to salt stress by arbuscular mycorrhiza is related to higher accumulation of soluble sugars in roots. *Mycorrhiza*, **12**: 185–190.

Glick, B.R., Penrose, D.M. and Li, J., 1998. A model for lowering plant ethylene concentration by plant growth promoting rhizobacteria. *J. Theo. Biol.*, 190: 63–68.

Grichko, V.P. and Glick, B.R., 2001. Flooding tolerance of transgenic tomato plants expressing the bacterial enzyme ACC deaminase controlled by the 35*S*, *rolD* or PRB–1*b*promoter. *Plant Physiol. Biochem.*, 39: 19–25.

Grover, M., Ali, S.Z., Sandhya, A., Rasul, B. and Venkateswarlu, B., 2010. Role of micro-organisms in adaptation of agriculture crops to abiotic stresses. *World J. Micr. And Biotech.*, **27**(5): 1231–1240.

Hirt, H., 2009. *Plant Stress Biology: From Genomics to Systems Biology*. Wiley, West Sussex.

Klee, H.J., Hayford, M.B., Kretzmer, K.A., Barry, G.F. and Kishore, G.M., 1991. Control of ethylene synthesis by expression of a bacterial enzyme in transgenic tomato plants. *Plant Cell.*, **3**: 1187–1193.

Manjunath, A., Mohan, R. and Bagyaraj, D.J., 1983. Responses of citrus to VAM inoculation in unsterile soils. *Canadian J. Bot.*, **61**: 2779–2732.

McLauchlan, 2006. The nature and longevity of agricultural impacts on soil carbon and nutrients: A review. *Ecosystems*, **9**(8): 1364–1382.

Menge, J.A. 1983. Utilization of vesicular-arbuscular mycorrhizal fungi in agriculture. *Canadian J. Bot.*, **61**: 1015–1024.

Neumann, G. and Romheld, V. 2007. The release of root exudates as affected by the root physiological status. **In**: *The Rhizosphere: Biochemistry and Organic Substances at the Soil-Plant Interface*, (Eds.) R. Pinton, Z. Varanini, P. Nannipieri. CRC, Boca Raton, pp. 23–72.

Qadir, M. and Schubert, S., 2002. Degradation processes and nutrient constraints in sodic soils. *Land Degrad. and Develop.*, **13**: 275–294.

Saint-Etienne, L., Paul, S., Imbert, D., Dulormne, M., Muller, F., Toribio, A., Plenchette, C. and Bâ, A.M. 2006. Arbuscular mycorrhizal soil infectivity in a stand of the wetland tree *Pterocarpus officinalis* along a salinity gradient. *Forest Ecol. and Mgt.*, **232**: 86–89.

Steenhoudt, O. and Vanderleyden, J. 2000. *Azospirillum*: A free-living nitrogen-fixing bacterium closely associated with grasses–Genetic, biochemical and ecological aspects. *FEMS Microbiol. Rev.*, **24**: 487–506.

Szabolcs, I., 1994. Soils and salinization. **In**: *Handbook of Plant and Crop Stress*, (Ed.) M. Pessarakali. Marcel Dekker, New York, pp. 3–11.

37

Improving Productivity and Profitability of Horticultural Crops through Micro-Irrigation Technologies

Man Singh

Water Technology Centre,
ICAR-Indian Agricultural Research Institute, New Delhi – 110 012
E-mail: mansingh61@ymail.com

Micro-irrigation has been one of the technical means to improve water use efficiency. Improved water and fertilizer management through fertigation can reduce associated costs and compensate for the higher irrigation system costs. One of the major concerns raised by farmers about micro-irrigation system is its economic viability. Most of the horticultural crops are highly sensitive to environmental factors, thus, it is mostly cultivated under glasshouses in Western Europe and in greenhouses in the USA, Canada, East European countries and Israel. In India, the yield levels of horticultural crops are generally very low and are attributed to several environmental factors mainly because of viruses and all kinds of abiotic stresses, as it is mostly cultivated under open field conditions.

Fertigation offers precise control on fertilizer application and could be adjusted to the rate of plant nutrient uptake. Nitrogen over-application is a serious problem in intensive agricultural production areas. N and K applications have been neglected in many parts of India. This has resulted in poor productivity of horticultural crops in terms of quantity and quality. Nitrogen and potassium interaction is currently a topic of great interest for those who are engaged in enhancing the productivity of an array of horticultural crops. A better understanding of the mechanism of N and K interaction could be a useful guide to best nutrient management practice in order to achieve high yields with higher nutrient use efficiency while practising fertigation at large.

The trend in recent years has been towards conversion of all kinds of methods of surface irrigation like basin, border, strip or furrow into trickle (drip) irrigation which is often considered to be the most efficient method of irrigation that suits to the horticultural crops. High-frequency irrigation by the pressurised irrigation system minimizes soil depth requirement as a storage reservoir for water, provides at least daily requirements of water to a portion of the root zone of each plant, and maintains a high soil metric potential in the rhizosphere to reduce plant water stress (Nakayama and Bucks, 1986).

Fertigation Supported by Micro-irrigation System

Fertigation supported by micro-irrigation ensures that every plant that is irrigated receives its requirement of nutrients as and when needed by the plant. Adequate supplies of water and nutrients are essential for enhanced production of all horticultural crops. Effects of application rate on the water distribution pattern demonstrated that increasing the water application rate allows more water to distribute in horizontal direction, while decreasing the rate allows more water to distribute in vertical direction for a given volume applied. An extremely high ammonium concentration existed in the proximity of the point source because ammonium is adsorbed in soil peds. Accumulation of nitrate towards the boundary of the wetted volume was observed for each fertigation strategy (Li *et al.*, 2003), which suggests that flushing of the remaining fertilizer solution in the drip pipeline system should be as short as possible after fertilizer application is finished to avoid the potential loss of nitrate from the root zone.

The strategy of first applying water for one-fourth of the total irrigation time, then applying fertilizer solution for one-half of the total irrigation time, followed by applying water for the remaining one-fourth of the total irrigation time left the most nitrate close to the source and was therefore recommended (Li *et al.*, 2004).

Nitrogen and Potassium Interaction on Quality of Crops

Potassium is acknowledged as a nutrient element improving crop quality and protecting plants against abiotic and biotic stress, while excessive N dressings often cause lower quality and higher susceptibility of crops to disease. An important aspect of the N and K interaction, therefore, is its relationship with crop quality and stress tolerance. It was established that K application could alleviate the N pollution problem by inducing a high uptake rate of N by crops (Ardjasa *et al.*, 1 and Yang *et al.*, 11). The positive interaction of N and K may offer the opportunity for considerable savings in the cost of N fertilizer and food security for the rapidly expanding human population (Nijamudeen *et al.*, 7). Therefore, N and K fertilizers should be applied with optimal ratios at the right time and right rate according to the nutrient uptake pattern of the crops, soil nutrient status, soil texture and climate changes, in order to reach the target yields with good quality and minimize K and N losses to the environment (Nijamudeen *et al.*, 2012; Bowen and Frey, 2002).

The N and K interaction was related to soil nutrient status, cultivation practices, crop species, yield level *etc.* When the exchangeable K in the soil is below the critical value, excessive N fertilization becomes much more critical because the nutrient cannot be used efficiently and the N loss becomes much greater. These principles could be used in practice as a guide to the scheduling of fertigation through trickle (drip) irrigation to most of the horticultural crops should one is aiming to intensive cultivation.

Field Experiment in Low Cost Greenhouse

The experiment was carried out in the naturally ventilated greenhouse coupled with low head drip irrigation system. The irrigation tank with 500 litres capacity was kept at one meter height from the surface (Figure 1). The low cost greenhouse was made of polythene cover on the roof (Figure 2). Side of the greenhouse and the part of roof were made of 40 mesh insect proof white net for the ventilation. There was an additional layer of polythene outside of the side net. The polythene layer was kept rolled during the summer in order to provide ventilation and released to cover the side net of greenhouse during the winter to raise temperature inside. The black color shade net was fixed on top side below roof to use during the summer season to reduce penetration of solar radiation and temperature inside. Field experiment was conducted for the two consecutive seasons September 2010 to May 2011 and September 2011 to May 2012. Cropping duration in the

Figure 1: Low Cost Low Head (LCLH) Water Tank to be Connected to Drip Line.

Figure 2: Structure of Low Cost Naturally Ventilated Greenhouse.

field was 8 months from 1st week of October in both years, 2010 and 2011 (Nijamudeen, 2012).

The sweet pepper (*Capsicum annum*) variety of Indra (Red colour fruit) was selected for the investigation. First, virus free healthy seedlings of the sweet pepper variety was raised in a most modern nursery greenhouse in soil less media in one month prior to the date of transplanting

Procedure Adopted

The investigation was conducted jointly at Water Technology Centre (WTC) with the collaboration of Centre for Protected Cultivation Technology (CPCT) at Indian Agricultural Research Institute, New Delhi. The experiment was carried out with 11 treatments replicated three times in a randomized block design. First, virus free healthy seedlings of the sweet pepper were raised under the most modern nursery greenhouse in one month prior to the date of transplanting. The soil less media for the nursery was prepared with the mixture of cocopeat, perlite and vermiculite in 3:1:1 ratio, respectively. The seedlings were transplanted in the field with the spacing of 45 x 30 cm in each dripping point where the same emitter spacing was under low head drip irrigation system. Eleven N and K fertilizer combinations, including nine combinations of urea and MOP, one commercially available ready mix dry liquid (water soluble) fertilizer with a ratio of 20:20:20 and a control, were applied through low head drip irrigation system weekly in 32 equal split doses. Uniform P (through surface) was applied in all the treatments. Greenhouse sweet pepper cultivars generally have an indeterminate pattern of growth. Therefore, the activity of pruning, training and trellising were required for the sweet pepper cultivation.

Pruning, Training and Trellising

Greenhouse sweet pepper cultivars generally have an indeterminate pattern of growth. Because the plants can grow up to 150 cm feet tall during a growing season of 250 days, they need to be supported vertically. Sweet pepper plants can be trellised to the Dutch "V" system or to the "Spanish" system. The "V" trellis system is used by Dutch and Canadian growers. The "V" trellis system was practice for the experiment with sweet pepper (Fig. 3 A-B). Trellising plants with the "V" system consists of forming a plant with two main stems by removing one of the two shoots developed on each node and leaving one or more adjacent leaves per node. The pairs of stems were kept vertically by the use of hanging twines that were wound around the stems as they grow (Elio, 2009). Two main stems were maintained on each plant after pruning and pinching on each internode. These two main stems were trained upon strings to the main over head wire running on the parallel length of the rows at a height of 8-9 feet. The stems were clipped with the strings using plastic clips (Figure 3).

Irrigation Scheduling

The irrigation water was applied to the experimental plots through drip irrigation method twice a week as per actual crop ET needed. The volume of water applied was estimated based on pan evaporation, pan coefficient, crop coefficient of the sweet pepper, cropped area, percentage wetted area and application efficiency of the drip system. Also the value of coefficient (factor) of evapotranspiration under greenhouse was considered as 0.45 (Hasan *et al.*, 2010) in the estimation of irrigation requirement. Actual crop ET under greenhouse with drip irrigation system was estimated as per the established standard procedure.

Fertigation Schedule

Fertilizer was applied through fertigation as per the treatment in every week. The N, P and K requirement for sweet pepper is 320, 150 and 360 kg ha^{-1} under fertigated greenhouse, respectively (Hasan *et al.*, 4).

A. Pruning and training operation

B. Finished plant with "V" trellis system

Figure 3: Demonstration of Pruning, Training and Trellising Skills.

A. Mature green food at 35 to 40 days after flowering.

B. Development of red color at 35 days after green mature stage.

Figure 4: Different Stages of Sweet Pepper Fruit Development in Greenhouse.

These amounts of N and K were considered as 100 per cent of the fertilizer application in calculation of the amount of fertilizer in deferent treatments. As conventional fertilizer application (T1 to T9), the source for N and K was urea and muriate of potash (MOP), respectively. For dry liquid fertilizer treatment (T11), the fertilizer of 20:20:20 was applied. One treatment was included as no N and K (T10). In all the treatments, P was maintains equal. Di-ammonium phosphate (DAP) was applied as P source except T10. To avoid the addition of N to treatment T10, the single supper phosphate (SSP) was applied. While calculating the urea as N source in conventional fertilizer treatments, the amount of N added through DAP was considered. Also, to maintain the 100 per cent N, P and K level in dry liquid fertilizer treatment, the urea and MOP were supplied as N and K supplementary.

In addition, as usual practice, multi micronutrients liquid fertilizer (Multiplex) was provided to the crop as foliar application uniformly in the rate of 200 litres mixture for 1000 m2. The fertilizer mixture was prepared by 2.5 ml Multiplex in one litre of water. This fertilizer contains zinc-3 per cent, iron-1.5 per cent, lead-0.003 per cent, copper-0.5 per cent and boron-0.5 per cent. The micronutrients liquid fertilizer was applied 3 times during the cropping period in both the seasons.

Sweet Pepper Fruiting in Low Cost Greenhouse

Harvesting of sweet pepper fruits were started from 75th day after transplanting by using sharp pruning knives. The sweet pepper fruits were harvested at red colour stage, except first 2 harvests of green fruits, as the red colour fruits have high market value of about Rs. 100 kg^{-1} where the green fruits were around Rs. 40 kg^{-1}. The green harvests were made in the month of December. If it is retained for red colour change, the crop growth would be affected as it was facing winter season in which the plant growth is very poor and long days required for fruit colour change into red. The time period for colour change ranges from 20-40 days and depended on the light and heat prevailed in greenhouse. There were total 8 harvests during the crop growth season of 8 months (Figures 4. A-B).

Statistical Analysis

The statistical analysis of the data of each parameter was done by the statistical method known as "Analysis of variance" appropriate for randomized block design. The coefficient of variance (C.V.) was worked out. Wherever the results were significant, the least significant difference (LSD) at 5 per cent level of significance was estimated. The statistical software of MSTAT was used for the analysis. The data are suitably illustrated with graphs and figures in the following sections.

Estimation of Irrigation Requirement

The irrigation requirement was computed based on daily pan evaporation and the field was irrigated twice a week. The seasonal depth of water required was of 249 and 254 mm in 2010/11 and 2011/12, respectively. However, considering irrigation efficiency, water applied was 277 and 282 mm in year 2010/11 and 2011/12 respectively. All the treatments were received the same amount of water through drip irrigation throughout the cropping season. The above applied amount of irrigation water was 45 per cent of the outdoor estimation. Hence, the irrigation water could be saved in greenhouse condition more than 50 per cent due to the reduction of evapotranspiration inside the naturally ventilated greenhouse.

Plant Growth

Plant growth pattern representing in plant height throughout the cropping period in both the seasons was graphically shown in Figures 5 and 6. The rate of increase in plant height was rapid at the early stages of the growth till the month of December in both the

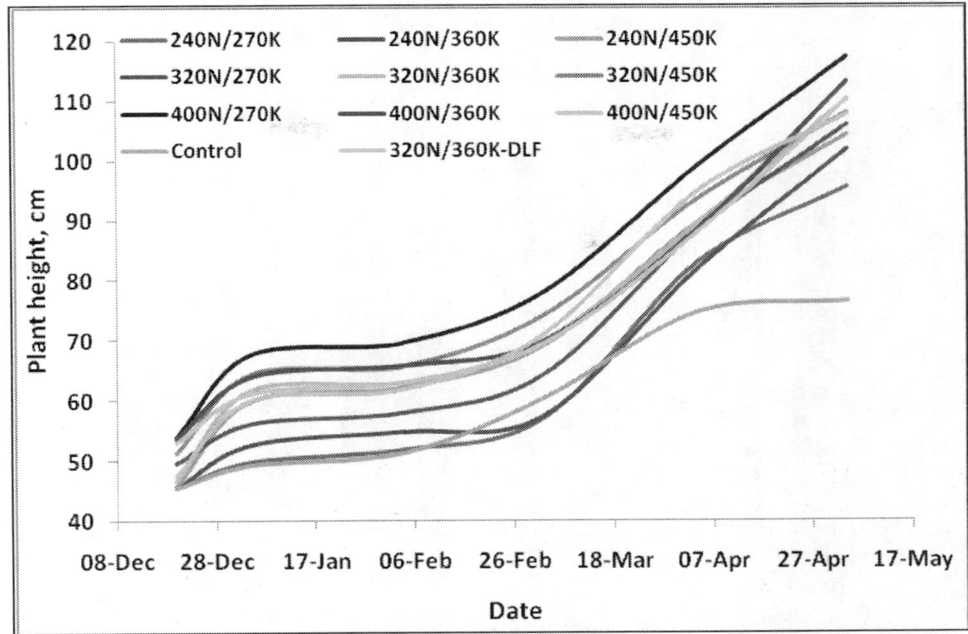

Figure 5: Plant Height during Crop Growth Period in Year 2010/11.

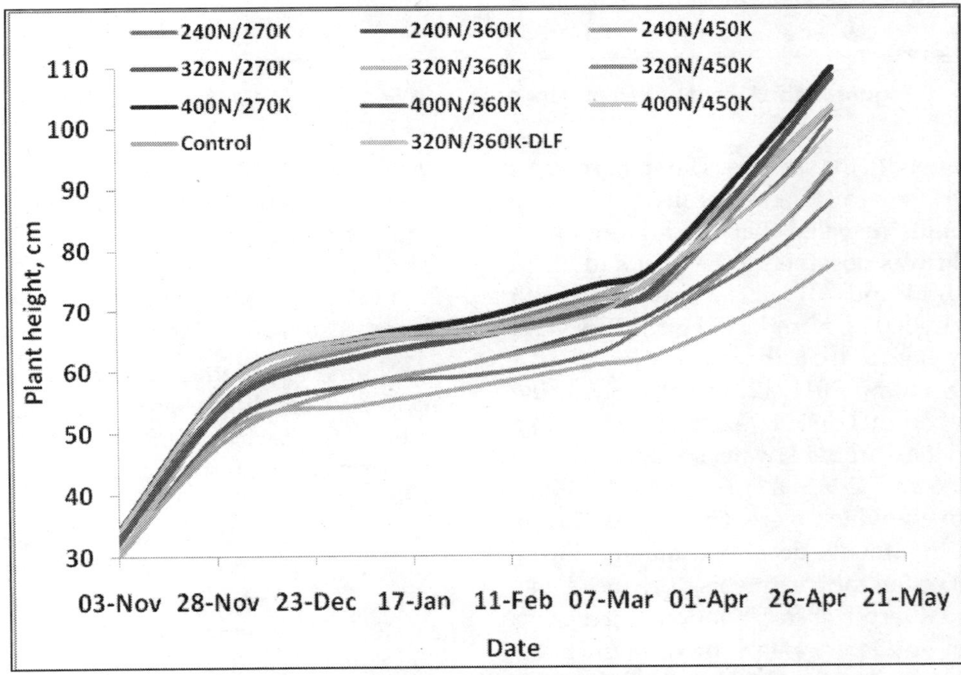

Figure 6: Plant Height during Crop Growth Period in Year 2011/12.

years and was poor during January and February. Average growth rates during the period of 2 months in January and February was 9 and 7 cm in year 2010/11 and 2011/12, respectively. It could be due to the effect of low temperature during that period. The minimum temperature during January to February observed was in the range of 4-11 and 4-9 °C in year 2010/11 and 2011/12, respectively. several studies reported that the optimum temperature for vegetative growth in sweet peppers is between 21 and 23°C. Plant growth in terms of plant height was observed to be 7 per cent lower in year 2011/12 compared with year 2010/11. It was attributed to the variation in weather condition prevailed such as the minimum temperature that became near zero for 4 consecutive days in December 2011 and the minimum temperature in 2010 was above 5 °C during cropping period.

Yield

There were 9 and 7 harvests of sweet pepper fruits in year 2010/11 and 2011/12, respectively. The first two harvests were made in green colour stage and remaining

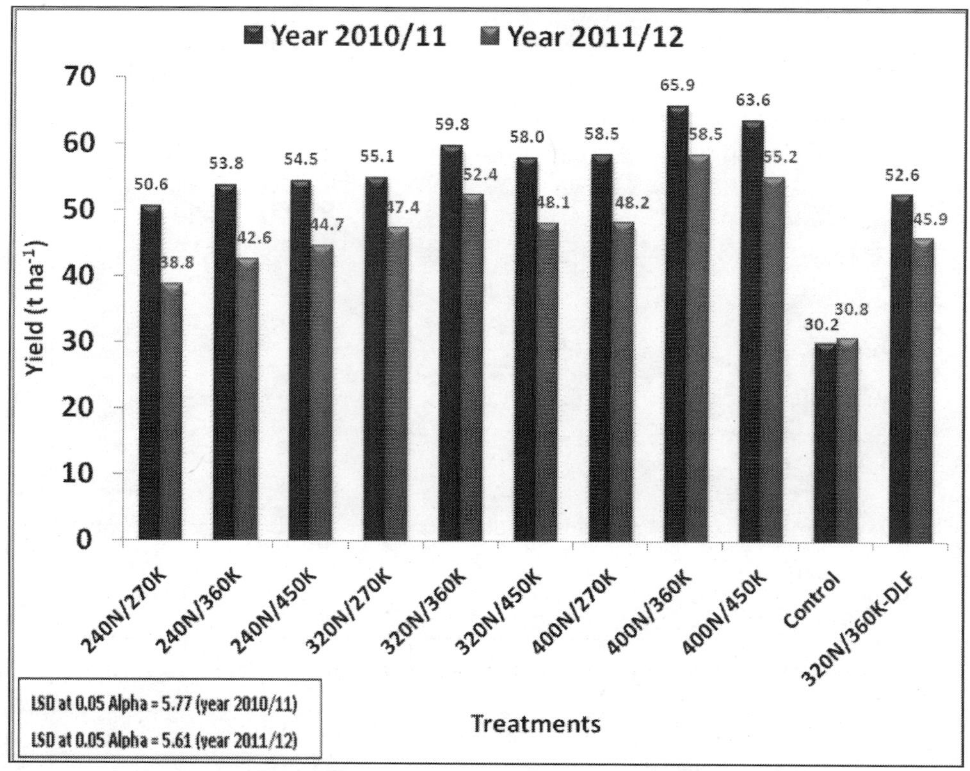

Figure 7: Total Yield of Sweet Pepper in both Seasons (2010/11 and 2011/12).

was red in colour in both the seasons. The status of total fruit yield in both the years is graphically depicted in Figure 7. The results revealed that the highest sweet pepper fruit yield was observed to be 65.9 and 58.5 t ha^{-1} in year 2010/11 and 2011/12, respectively with the treatment of T_8 (400 kg N and 360 kg K). This yield was significantly higher than that of 59.8 and 52.4 t ha^{-1} in year 2010/11 and 2011/12, respectively at the application rate of normal dose T_5 (320 kg N and 360 kg K). The yield from the dry liquid (water soluble) fertilizer treatment was 52.6 and 45.9 t ha^{-1} in year 2010/11 and 2011/12, respectively which were significantly lower than that of the treatments described above. Lower yields were observed in the treatments of reduced and imbalanced N and K fertilizer combinations. Increasing the nitrogen level by 25 per cent of the recommended dose (320 kg ha^{-1}) promoted the yield level significantly. However, additional amount of potassium leads to a reduction in the yield. It may be due to the antagonism effect between NH_4^+ and K^+ in which the plant uptake of these nutrients are controlled. The experimental result is in conformity with the following past research reports which evidenced that the K uptake rate was reduced with the increase of NH_4^+-N concentration.

Beneficial effect of NH_4^+-N on sweet pepper yield was more significant at lower rate of application than that of the higher application of K. Further, observed yields recorded in year 2011/12 was average of 15 per cent lower than that of yield recorded in 2010/11 due to the variation in weather condition prevailed such as the minimum temperature that became near zero for 4 consecutive days in December 2011 and the minimum temperature in 2010 was above 5 °C during cropping period (Figure 7). Also during the 2nd cropping season (2011/12), rain fall was not experienced till end of December 2011 after planting in 1st week of October 2011.

Normal application of nitrogen for sweet peppers in greenhouse condition has been 320 kg ha^{-1}, which may not fulfill the development of yield potential under low head drip fertigation practice fixed into the naturally ventilated greenhouse. The nitrogen rate increased from 320 to 400 kg ha^{-1} in accordance to the increased crop duration *i.e.* 8 months, with the normal dose of 360 kg ha^{-1} potassium enhanced the average yield from 56.1 to 62.2 t ha^{-1}. This average yield was 11, 26 and 104 per cent higher than the fertilizer application of normal dose, dry liquid fertilizer and the control, respectively. However, it was also observed that an increase in the amount of potassium from 360 to 450 kg ha^{-1} with the N level of 400 kg ha^{-1} could not improve the yield anymore. The maximum water productivity of sweet pepper in the naturally ventilated greenhouse was found to be 23.5 to 25.9 kg m^{-3} with the application of 400 kg ha^{-1} N and 360 kg ha^{-1} K. Two years average maximum yield of 62.2 t ha^{-1} was achieved on fertilizer application rate of 400 kg N ha^{-1} and 360 kg K ha^{-1}. This yield was 11 per cent higher as compared to the normal dose of fertilizer. The information obtained from the study is useful in the design, operation and management of fertigation system.

Visionary Models for Horticultural Crops

New Paragon 2030

Paragon refers to a model of excellence that could be in vogue for 30 years. Singh (2016) proposed the concept of the model first time in January, 2016 with an aim to find generic and sustainable solution to achieve the Sustainable Development Goal (SDG postulated by the United Nations in September, 2015. The 5 Es in the model refers to Economy, Environment, Energy, Empathy and Equity which were announced by the PM in the Science Congress. These 5 Es are very important in today's context for initiating massive development in integrated irrigation and drainage sector. This model assumes that *Pradhan Mantri Krishi Sinchai Yojana* (PMKSY) would be able to create large scale installations of irrigation and infrastructure in next 30 years with the declared massive investments and would make poorest of the poor farmers independent of monsoon rainfall as well as climate change by ensuring irrigation water supply to each and every farm spread over in 5 to 6 lakh villages (Figure 8).

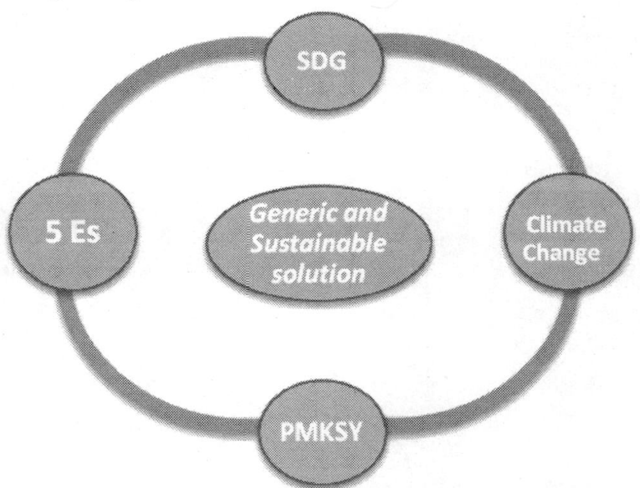

Figure 8: Concept of Paragon 2030 (Singh, Man. 9)

Once large scale irrigation drainage infrastructure are created by integrating important factors outlined in the periphery of the proposed model and earlier proposed A-V Model (Singh, 2015) farmers of India at large and the marginal farmers in particular may be able to take diversified horticultural crops and earn money like any salaried person round the year. In A-V model, an analogy of human body was drawn equivalent to the cultivable land of any X, Y or Z ha area. The human artery, vein, tissues and cells system corresponds to irrigation, drainage, village and individual small farms system, respectively. This pragmatic model fits very well into the entire cultivable land that is covered through extensive nodes and links of irrigation networks in water deficit areas and network of drainage in water excess areas.

Drainage Not to be Forgotten alongwith Irrigation Network

Surface and sub-surface drainage safeguard the investments made in irrigation infrastructure, promotes agricultural intensification and diversification of horticultural crops. Land drainage is essential for permanent intensive, high density plantations and finally conserves land resources and prevents the land from retirement. The absence of land drainage infrastructure is fully felt by mango and grape grower. It is right time to consider the importance of drainage facilities in garden crops.

Productivity Enhancement of Horticultural Crops in Hilly Region

Hilly regions invariably are electricity scarce and supply of electricity may not exceed on an average 6 to 8 hours daily. In these circumstances, an innovative idea of overhead irrigation tank (OIT) is mooted for storage of water and its distribution through the network of pipelines of micro-irrigation for small and marginal farmers without consumption of electrical energy. It is further clarified that if the OITs are put up at an elevation of 2 to 14 meter from the reference of the last point to be irrigated, is sufficient to operate most of the low head and low discharge micro-irrigation emission devices. In this scheme network of OITs may be filled up from the water source when supply is available. If this scheme is put in place it is likely to create a revolution in the field of horticultural crops.

References

Ardjasa, W.S., Tomonori, A., Ho, A., Ken-lchi, K. and Makoto, K. 2002. Fate of basal N and growth of crops cultivated under cassava-based intercropping system with reference to K application rate. *Soil Sci. Plant Nutr.*, **48**: 365–370.

Bowen, P. and Frey, B. 2002. Response of plasticultured bell pepper to staking, irrigation frequency, and fertigated nitrogen rate. *Hort. Sci.*, **37**: 95–100.

Elio, J., Daniel, J.C., Steven, A. S., Lance, S. and Osborne 2004. Production of greenhouse-grown peppers in Florida, Dept. of Hort. Sciences, IFAS, Univ. of Florida. Publication #HS979, Available from http://edis.ifas.ufl.edu.

Hasan, M., Singh, B., Singh, M.C., Singh, A.K., Kaore, S.V., Tarunendu, Sabir, N. and Tomar, B.S. 2010. Fertigation scheduling for horticultural crops. Tech. Bull. No. TB–1CN: 80/2010, IARI, New Delhi, pp. 1–44.

Nakayama, F.S. and Bucks, D.A. 1986. Trickle Irrigation for crop production design, operation and management. *Developments in Agricultural Engineering*, Elsevier, Amsterdam, **9**: 383.

Nijamudeen, M.S. 2012. Evaluation of conventional and dry liquid nitrogen and potassium fertilizer under drip irrigation system for sweet pepper (*Capsicum annum* L.). *Ph.D. Thesis*, Water Technology Centre, P.G. School, Indian Agricultural Research Institute, New Delhi, 113 p.

Nijamudeen, M.S., Singh, M., Khanna, M., Singh, B., Singh, R., Pandey, R. and Hasan, M. 2013. N and K fertilizer application rate under drip-fertigation for greenhouse grown sweet pepper, *Indian J. Hort.*, **70**(1): 54–59.

Nijamudeen, M.S., Singh, M., Khanna, M., Singh, B., Singh, R., Pandey, R. and Hasan, M. 2013. Nitrogen and potassium distribution in the soil profile of sweet pepper crop with different drip fertigation methods under greenhouse condition, *Ann. Sri Lanka Dept. Agri.*, **15**: 234–245.

Singh, Man 2015. Large scale creation of irrigation and drainage infrastructure through A–V model. Invited Lecture, *Proc. Workshop on Waterlogging and Salinity Control*, 3–4 September, Central Board of Irrigation and Power, Chandigarh.

Singh, Man 2016. Trickle irrigation and vegetable crops' production technology: Examples of integration. Special Lecture, *Proc. XXXIV Group Meeting, All India Coordinated Research Project (Vegetable Crops)*, 10–13 May, Indian Agricultural Research Institute, New Delhi.

Yang, S.M., Malhi, S.S., Song, J.J., Xiong, Y.C., Yue, W.Y., Lu, L.L., Wang, J.G. and Guo, T.W. 2006. Crop yield, nitrogen uptake and nitrate-nitrogen accumulation in soil as affected by 23 annual applications of fertilizer and manure in the rainfed region of Northwestern China. *Nutr Cycl Agroecosys.*, **76**(1): 81–94.

38

Deficiencies and Management of Micronutrients in Horticultural Crops

A.N. Ganeshamurthy, T.R. Rupa, H.B. Raghupathi, G.C. Satisha and Mahendra Kumar*

Division of Soil Science and Agricultural Chemistry,
ICAR-Indian Institute of Horticultural Research, Hessaraghatta Lake Post
Bengaluru – 560 089, Karnatka
E-mail: angmurthy@gmail.com

India being a subcontinent enjoys a variety of climate and soil on which a large range of perennial horticultural crops can be grown. Perennial horticultural crops include fruits, palms, spices and cashew, cocoa, medicinal and aromatic plants. Importance of horticulture in providing nutritional security and livelihood to small farmers was felt by the planners in recent times. As a result horticulture received due emphasis during the last few decades. The green revolution of the 1960s and 1970s ended chronic food deficits and while cereals still command the attention of policy makers, fruit production has surged impressively, making India the second largest global producer behind China. In achieving nutritional security fruits play an important role and the current requirement of fruits to achieve nutritional security is 92 million tonnes. Although India has achieved fairly good production levels of fruits (Figure 1), however, the present production of 82.63 m tonnes falls short of the need of the country (Times of India Jan 2016). Hence we have to go a long way in achieving this target (Ganeshamurthy 2016). Challenges to be faced and fulfill the needs of ever increasing population in a sustainable way would therefore require a tremendous effort to increase the production of fruits both horizontally and vertically through a rational and balanced use of production

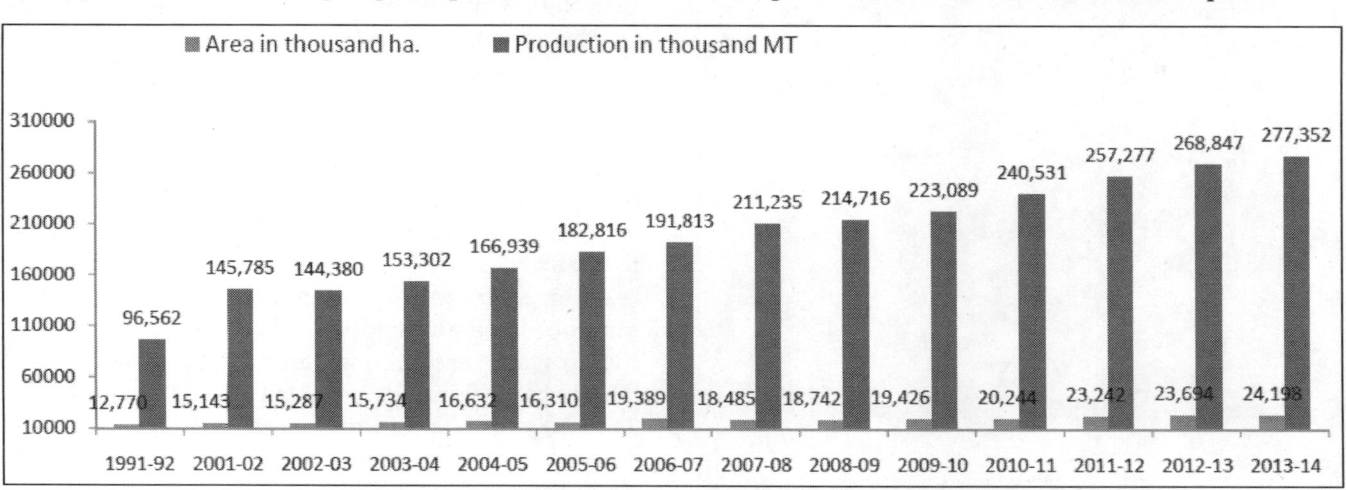

Figure 1: Area and Production Growth Trends for Horticulture Crops.

inputs, specially the mineral fertilizers. This would require a better and more comprehensive insight into ecologically sound production processes as the fruit crops production is concentrated in the most fragile environment and with resource poor farmers where unscientific and over exploitation of basic resources has already produced a large scale degradation and ground water depletion.

Extent of Micronutrient Deficiencies

Most soil in India is lacking in vital nutrients. A majority of the soils in India are deficient in secondary nutrients and micronutrients vital for both plant and human health (Singh 2008; Patel *et al.*, 2016). The soils of as many as 174 districts across 13 states are deficient in secondary nutrients like sulphur and micronutrients like zinc, boron, iron, manganese and copper. The chemical analysis of 70,759 soil samples collected from Andhra Pradesh, Assam, Bihar, Gujarat, Haryana, Madhya Pradesh, Maharashtra, Odisha, Punjab, Tamil Nadu, Uttar Pradesh, Uttarakhand and West Bengal found zinc to be deficient by 39.9 per cent and sulphur by 27.8 per cent. The boron deficit was found to be 20 per cent while iron, manganese and copper were deficient by 12.9, 6.0 and 4.3 per cent, respectively. The consistently low yield of almost all crops in states like Odisha, West Bengal, Gujarat and Bihar despite application of nitrogen, phosphorus, potassium and zinc has been attributed to the deficiency of boron. The removal of secondary nutrients and micronutrients was found to be the highest in the trans-Gangetic part of the Indo-Gangetic plains. Tamil Nadu recorded the maximum zinc deficit of 62.2 per cent, while West Bengal is the lowest at 8.5 per cent.

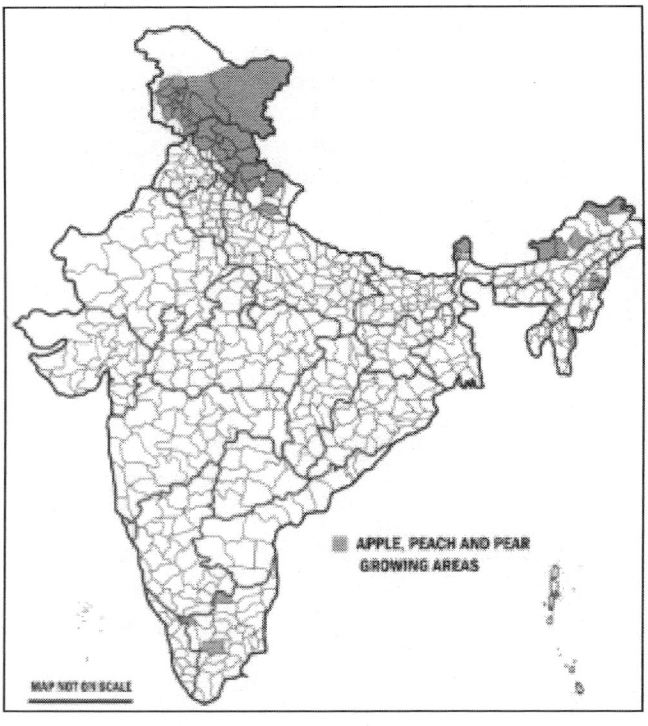

Figure 2: Major Apple Growing Areas in India.

Micronutrient deficiency maps prepared by the ISSS, Bhopal is presented in Figure 1. Crop boundaries of different fruit crops were prepared and superimposed over these micronutrient deficiency maps. From these maps district wise deficiencies of micronutrients in different fruit crops were delineated. Fruit crop-wise the deficiency maps were prepared. The status of deficiency of micronutrients and their management is discussed for the most important fruit crops below:

Apple, Peach and Plum

Eight states were identified as prime temperate fruits producing states of India. Based on categorization with respect to area and production per unit area, 48 districts were identified as prime temperate fruits growing districts. Boundary map of temperate fruits was prepared using district maps of respective leading producing states (Figure 2).

Micronutrient Deficiencies

Zinc

The zinc fertility map of apple, peach and plum (Figure 3A) prepared using the DTPA zinc status of Indian soils shows that out of 49 districts growing these crops, only seven districts, *viz.*, Leh, Kargil, Srinagar, Pulwama, Shopian, in J and K, and Krishnagiri, Nilgiris and Dindigul, T.N. showed deficiency in soil available zinc. Soils in 31 districts are reported to be well supplied with zinc. However, 23 per cent of the apple growing districts (11 districts) do nothave information on soil available zinc status.

Management

Zinc is not readily mobile within the tree and applications must be thorough and timely for optimal response. Zinc applied to the soil is usually ineffective for fruit trees because it is not mobile in the soil and does not reach the roots at the rates required by trees. Soil applications need to be applied or banded around the root zone, but uptake is limited and doesn't occur until 4/5 leaves are unfolded in the spring. Zinc deficiency is usually corrected by applying annual sprays of zinc sulphate heptahydrate to the bark when the trees are dormant. Care is needed in choosing the right form of foliar-applied zinc so as to minimize damage to young leaves and the fruit. Both inorganic forms such as $ZnSO_4$ and EDTA chelates are effective. The dormant spray is a safe treatment for bearing trees as foliar zinc sprays applied in spring may cause russeting of the fruit. If the deficiency is mild, a zinc-based fungicide normally used for scab control like -zineb, ziram, mancozeb or metiram generally applied up to three times during the growing season, may correct the deficiency. As a late-dormant Sprays farmers may give zinc sulphate at 2 to 2.5 kg per 500 l of water as a dilute spray. This may supply adequate zinc for growth during bud break to silver tip stage.

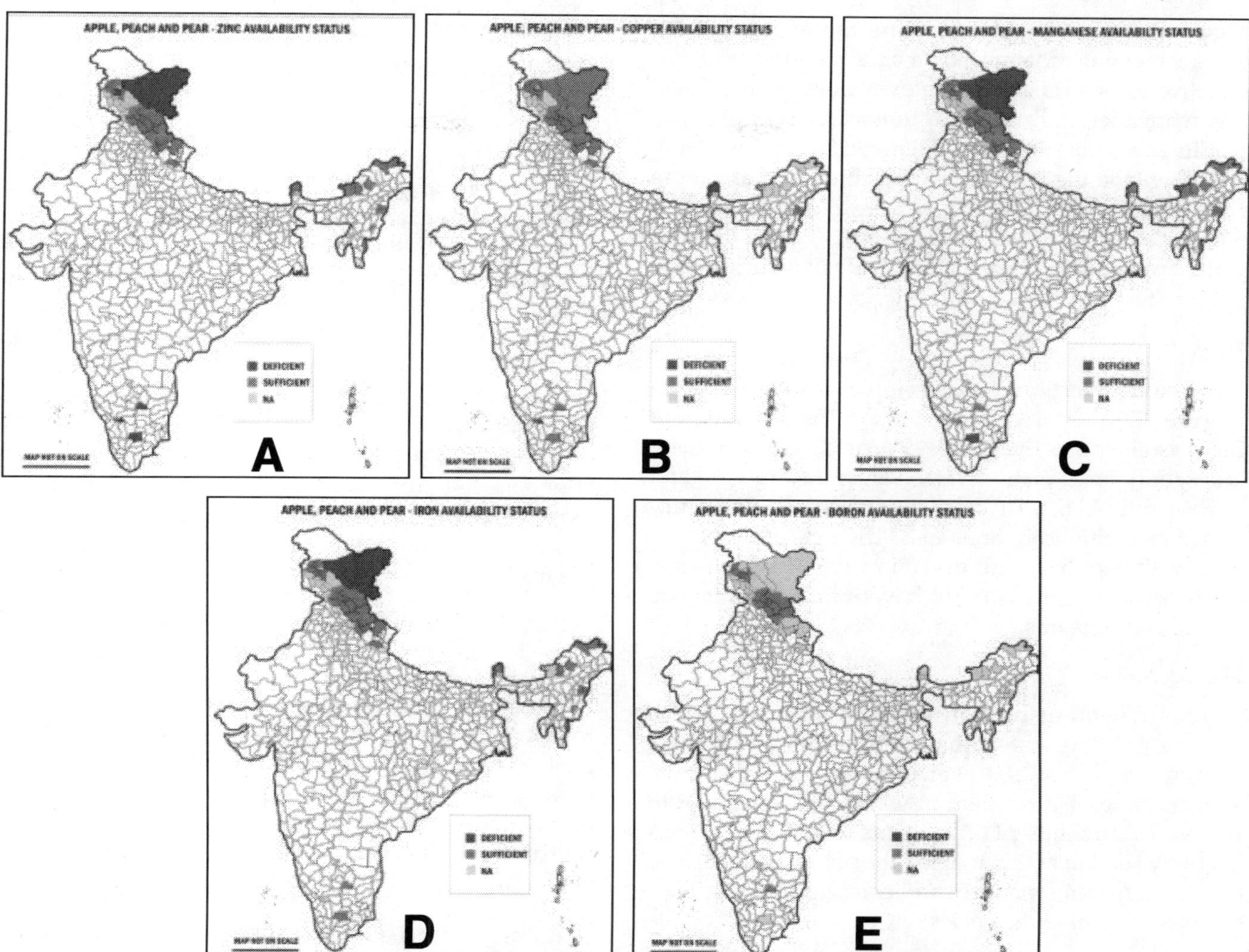

Figure 3A-E: Micronutrient Fertility Status of Apple Growing Regions of India.

Copper

The copper fertility map of apple, peach and plum (Figure 3B) prepared using the DTPA copper status of Indian soils shows that out of 49 districts growing these crops, only three districts, *viz.*, North Sikkim and West Sikkim, in Sikkim and Nilgiris in Tamil Nadu showed deficiency in soil available copper. Soils in 29 districts are reported to be well supplied with copper. However, 35 per cent of the apple growing districts (17 districts) do not have information on soil available copper status.

Management

Deficiency is rare in apples as soil supplies are normally sufficient. Copper deficiency causes 'wither tip' symptoms, leading to the death of young shoots. Copper deficiency can occur on young trees on very deep, sandy soil. With fewer copper fungicides being applied growers should be more aware of copper deficiency. The best way to correct copper deficiency is to apply a soil dressing to each tree every 2-3 years, @ 100 g of copper sulphate for young trees or 250 g for mature trees. Foliar applications quickly correct the supply problem. Copper sulphate at 500 g/100 l can be mixed with a winter zinc spray. A nutritional copper spray should only be needed every 2-3 years. Copper sprays can russet apples if applied post-bloom, or cause injury if applied under freezing conditions.

Manganese

The manganese fertility map of apple, peach and plum (Figure 3C) prepared using the DTPA manganese status of Indian soils shows that out of 49 districts growing these crops, only three districts, *viz.*, Leh and Kargil in J&K state and Dundigal in Tamil Nadu showed deficiency in soil available manganese. Soils in 36 districts are reported to be well supplied with manganese. However, 11 districts in apple growing states do not have information on soil available manganese status.

Management

Deficiency of manganese is not common in apple orchards. However, toxicities are reported in orchards on acid soils. Although manganese can be effectively supplied through applications to the soil, this method is considered to be too expensive for general recommendation. Foliar sprays are the most cost effective means for supplying manganese to orchards. The manganese in mancozeb fungicides is apparently readily available to the trees. Hence, if farmers are using this fungicide then separate application of manganese is not necessary. Usually, a single application of manganese sulphate at the rate of 0.5 kg per 100 l of water applied 7 to 10 days after petal fall is sufficient to prevent the appearance of deficiency symptoms.

Iron

The Iron fertility map of apple, peach and plum (Figure 3D) prepared using the DTPA iron status of Indian soils shows that out of 49 districts growing these crops, only three districts, *viz.*, Leh and Kargil in J and K state and Nilgiris in Tamil Nadu showed deficiency in soil available iron. Soils in 33 districts are reported to be well supplied with iron. However, 13 districts in apple growing states do not have information on soil available iron status.

Management

Management of iron in apple depends upon the soil pH and water quality. Too much carbonate accumulation in soils prevents the absorption of iron. This condition is prevalent in water logged soils or in regions having high pH calcareous soils. Keeping soils properly aerated and managing the pH will control iron chlorosis. Adding manures and compost to the soil can help correct iron deficiencies.

Soil applications of Fe are not recommended due to the extensive soil fixation of any applied Fe. Little information exists on the effectiveness of foliar Fe on apple, peach and plums or other tree fruit. Fortunately iron deficiency is not a serious problem in the main pome fruit districts. Corrective treatments are either ineffective or too expensive. If solution pH is maintained between 6.2 to 6.7 then foliar application of ferrous sulphate at 0.5 grams per litre of water corrects iron deficiency.

Boron

The boron fertility map of apple, peach and plum (Figure 3E) prepared using the hot water soluble boron status of Indian soils shows that out of 49 districts growing these crops, none of the districts for which data is available showed deficiency in soil available boron. However, boron deficiency is very common in orchards indicating that the expression of the deficiency may be more due to imbalance among potassium, calcium and boron rather than per se boron content in the soil. Further excessive use of nitrogen, potassium, and magnesium, and prevailing deficient levels of calcium, and zinc predisposes the trees to boron deficiency. Soils in 15 districts are reported to be well supplied with boron. However, 34 districts in apple growing states do not have information on soil available boron status.

Management

A complete boron program frequently includes both a soil application to meet the basic need of the crop, plus one or more foliar applications to supply additional boron at critical stages of crop development. In general, avoid excessive levels of nitrogen, potassium, and magnesium, and avoid deficient levels of calcium, and zinc, since these conditions may contribute to boron deficiency. Boron foliar sprays during Pre-bloom are more beneficial in improving bud development, pollination, fruit set, early season leaf and shoot growth. This further assists in improved calcium uptake. To avoid fruit disorders, post-bloom sprays of boron are frequently needed. These applications may be made at 7-10 days after petal fall and at approximately 30 days after petal fall. Applications of boron sprays later in the season should not be made because of the possibility of stimulating abnormal ripening and breakdown of the fruit.

Soil application at the rate of 100 g boric acid or borax per tree in the beginning of the monsoon season may be done. However, regularly monitor the soil boron levels after soil application to avoid the risk of toxicity. It is advisable to go for foliar sprays only if leaf analysis results indicate a need as the difference between deficient and excessive B levels is small, and B toxicity is very damaging. Foliar spray can be undertaken at the rate of 50 ppm boron concentration either through boric acid or through borax.

Micronutrient Deficiencies in Banana

Eleven states were identified as prime banana producing states of India. Based on categorization with respect to area and production per unit area, 76 districts

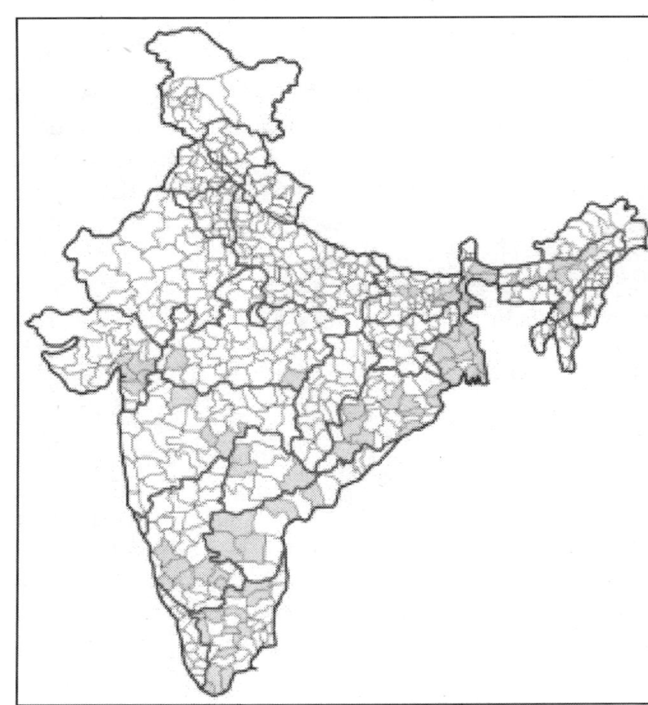

Figure 4: Major Banana Growing Areas in India.

were identified as prime banana growing districts. Boundary map of banana was prepared using district maps of respective leading producing states (Figure 4).

Zinc

Sufficient data base is available on soil zinc availability in India (Figure 5A). Of the 76 banana growing districts of India there is no information available on five districts, *viz.*, Goalpara in Assam, Bengaluru Urban District in Karnataka and Birbum, Malda and Midnapur in West Bengal. Twenty two districts have deficiency of zinc in the soils and 49 districts showed sufficient levels of soil available zinc (Figure 5A). Keeping this in mind and the pattern of fertilizer use in banana and the crop ability to absorb nutrients it is advisable to go for foliar application of zinc rather than soil application as a precaution to avoid unnecessary accumulation in the soils leading to nutrient imbalances.

Management

Zinc deficiency has often been confused with virus infection. But a uniform spread of the symptoms in the field rather than in patches confirms zinc deficiency. This is the most widely reported micronutrient element deficiency in bananas. Deficiency is more common in calcareous and organic soils. Since Zn is leached under acid conditions and fixed under alkaline conditions it is easy to correct its deficiency with a spray of 0.5 per cent $ZnSO_4$. But the pH of the spray solution is very important. Maintain a pH in the range of 6.2 to 6.6 so that the scorching on leaves and fruits may be avoided. The Zn requirement of the banana is relatively small. On the basis of whole plant analysis application of 1.0 kg Zn/ha/y is sufficient to meet the zinc requirement of a reasonably high yielding crop. Farmers are advised not to make it a habit to Zn applications and are advised to correct deficiencies as it occurs.

Manganese

Out of 76 banana growing districts only Vishali district in Bihar showed low manganese availability in soils. For seven districts, *viz.*, Goalpara in Assam, Bengaluru urban, Dakshina Kannada and Shimoga in Karnataka and Birbhum, Malda and Midnapur in West Bengal the status of soil Mn is not available. Soils in the remaining 69 districts are well supplied with Mn for banana cultivation (Figure 5B). Because of ability of

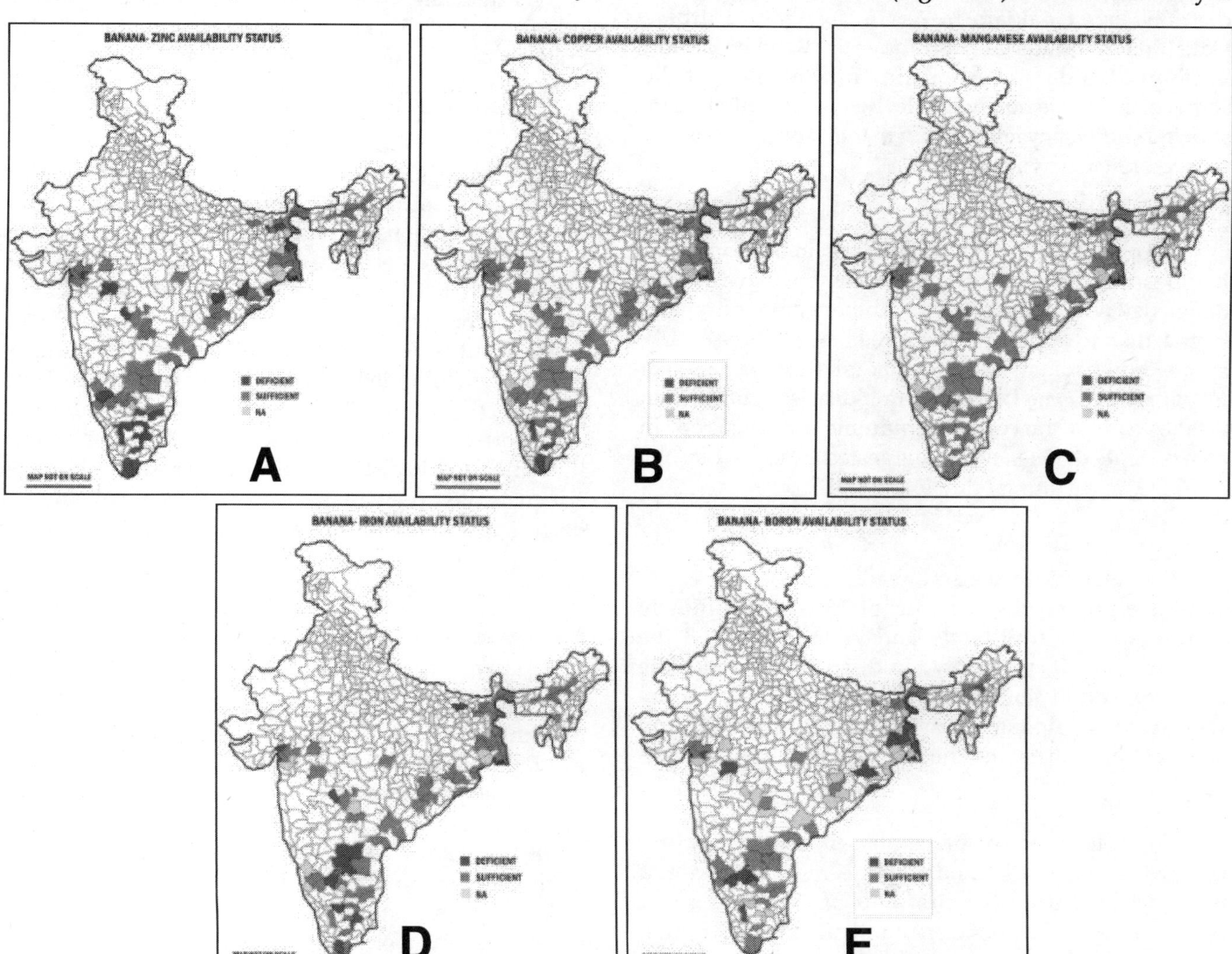

Figure 5A-E: Micronutrient Fertility Status of Banana Growing Regions of India.

banana to acidify its rhizosphere and because of good availability of Mn in most of the banana growing regions it is advised not to render to soil application of Mn. Only if need arises by way of expression of deficiency symptoms growers may undertake foliar application of Mn.

Management

As already stated toxicity of manganese is more important than its deficiency. Banana is generally tolerant to high concentrations of Mn in the soil solution. But high Mn could be a problem in acid soils with marginal amounts of available Ca, Mg and Zn. The Mn toxicity observed in the field could therefore be attributed to the indirect effects of other elements rather than to a high concentration of Mn. Application of dolomite corrects manganese toxicities. In case of deficiency, correction can be achieved by 0.3-0.4 per cent foliar sprays of $MnSO_4$ or soil application. Application of ten kg/ha of Mn is suggested as a remedial measure.

Copper

Banana soils have adequate supply of copper (Figure 5C). Out of 76 banana growing districts seven districts, *viz.*, Goalpara in Assam, Bengaluru urban, Dakshina Kannada and Shimoga in Karnataka and Birbhum, Malda and Midnapur in West Bengal the status of soil Cu is not available. Remaining all districts showed sufficiency levels of Cu in their soils. No district showed low Cu availability.

Management

Copper deficiency is not common in banana and if it appears in the field then it is preferable to correct Cu deficiency with a leaf spray of 0.5 per cent neutralized $CuSO_4$ than to apply it to the soil. If copper is regularly sprayed on bananas as Bordeaux mixture for Sigatoka (*Mycosphaerell amusicola*) and other disease control, then it takes care of the copper requirement of the crop. In case of acute deficiency it is suggested to apply 1 kg/ha of $CuSO_4$ as an annual rate to the soil.

Iron

Information on the soil availability of iron is not available for ten districts out of 76 districts mostly in Karnataka, Tamil Nadu and West Bengal. Of the remaining 66 districts only ten districts fell under the category of deficient rating (Figure 5D). Foliar spray with ferrous sulphate or with chelated forms of iron is effective in correcting the iron deficiency in banana.

Management

The total amount of Fe absorbed by a healthy plant is only about I g and 80 per cent of this amount is absorbed during the first half of the plant's life. Appearance of iron deficiency in banana is more due to adverse soil conditions rather than low Fe content per se in the soil. Before application of external source of iron farmers must first adjust the soil pH to near neutral or slightly acidic conditions through addition of amendments and organics and other means. In case the irrigation water containing high bicarbonates then this water need to be passed through gypsum column or alternate with good quality water, if available. If soil is alkaline or in calcareous soils, go for an inter crop with green manure crops and incorporate into the soil. In such soils farmers must use ammonium sulphate as source of nitrogen and it is also recommended to apply iron pyrites if available along with ammonium sulphate @ 10 kg/ha. As a field level corrective measure farmers may undertake repeated spray of 0.3-0.4 per cent spray of ferrous sulphate adjusting the pH to about 6.4-6.4 at 10-15 days interval untill the deficiency is corrected.

Boron

Boron deficiency has emerged as an important micronutrient problem in banana growing soils of India. Boron deficiencies occur on a wide variety of soils, however, boron availability reduces as pH increases. B uptake rate in field is constant from sucker to harvest – 40 mg/plant/month.

Its deficiency has been widely found in highly calcareous of Bihar, Tamil Nadu, eastern Uttar Pradesh and Saurashtra, sandy soils of Haryana and Rajasthan, hill and sub mountainous soils of north Himalayan and north eastern states and in red and lateritic soils of Odisha, Karnataka, Andhra Pradesh and Kokan region. Extent of boron deficiencies varied widely with soil types, cropping systems and agro-ecological situations. Intensive cultivation of banana has aggravated the emergence of boron deficiency in several soils which were earlier not testing low in available boron. Information on soil boron is available for 49 districts out of 76 districts. Map showing the status of boron availability in banana growing regions has been prepared (Figure 5E). There are only nine districts, which are reported to be deficient in soil available boron and for the remaining 40 districts the hot water soluble boron is adequate. Because banana is a heavy feeder, the rate of boron supply and the boron demand by the crop do not match. Hence in most cases boron application results in very good crop response.

Management

The demand for boron in banana is fairly uniform throughout the growth period. Hence continuous supply of B must be maintained. A soil concentration of 0.1 to 1.0 ppm B should be maintained for proper growth and development of banana. About 10 kg/ha of boric acid is suggested as soil application to take care of the boron problems. Care must be taken to see that a proper calcium, potassium and boron ratio is maintained in the soil. Foliar application of boric acid at 0.2 per cent may be undertaken at fortnightly interval at 4 - 5 month after planting until the deficiency is corrected.

Molybdinum

Symptom of Mo deficiency, in banana have not been recorded in sand culture or in the field. However Mo increases the root growth and development of banana plants in sand culture with Mo spray @ 4 ppm. But no deficiency symptoms were recorded. Data on Mo content of banana growing soils is not available and hence the map could not be generated for Mo.

Micronutrient Deficiency in Citrus

Eleven states were identified as prime citrus producing states of India. Based on categorization with respect to area and production per unit area, 131districts were identified as prime citrus growing districts. Boundary map of citrus was prepared using district maps of respective leading producing states (Figure 6). Although 131 districts are producing citrus, the species of citrus vary from region to region. For example Vidarbha region of Central India (Maharashtra and Madhya Pradesh) grow mostly Nagpur mandarins and acid limes, while Punjab grows Kinnow mandarins. Similarly, Coorg and Neelgiris grow Coorg mandarins while North Eastern states grow Khasi mandarins. Andhra Pradesh, Telangana and Karnataka produce most of the sweet oranges and limes. Irrespective of these differences in crops species the citrus map is prepared as one unit combining all these cultivated species.

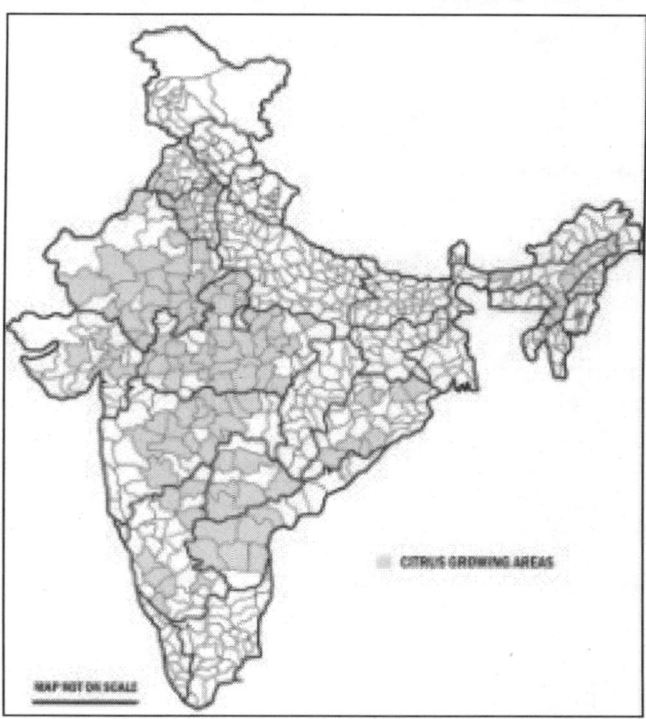

Figure 6: Major Citrus Growing Areas in India.

India ranks fifth in citrus production in the world contributing 5.63 per cent of the total world production of citrus. United Andhra Pradesh (Andhra and Telangana) is the leading producer of citrus occupying 12.75 per cent of the total area under citrus in 2013-14 with a production of 3800.6 thousand MT. Each of the species of citrus has specific problems leading many times to low yields or no yield, major among them is the general citrus decline caused both due to disease and nutrition.

Zinc

The citrus zinc map (Figure 7A) prepared using the soil DTPA zinc status of Indian soils shows that out of 131 citrus districts, 44 districts are reported to be low in available zinc and 63 districts were reported to be well supplied with soil zinc. Twenty-four districts of the citrus growing region (18 per cent) do not have information soil zinc status. Keeping this in mind and the pattern of fertilizer use in citrus and the crop ability to absorb nutrients it is advisable to go for foliar application of zinc rather than soil application as a precaution to avoid unnecessary accumulation in the soils leading to nutrient imbalances.

Management

Zinc deficiency in citrus, can be ameliorated by foliar application of zinc sulphate. Spray the plants with 0.25-0.5 per cent zinc sulphate solution (250-500 g zinc sulphate per 100 l of water depending upon severity of deficiency) during April-May on new flushes. However, in some places where new flush emerges late, the spray should be done in June and if further delayed flush appears then zinc sulphate should be sprayed during August-September. Apart from this in citrus, zinc deficiency becomes acute immediately after first fruit set on plants; therefore one maintenance spray of 0.3 per cent zinc sulphate solution may be given in pre-monsoon from the third year of planting.

Copper

Out of 131 citrus growing districts only nine districts showed low copper availability in soils. For 24 districts mostly in Assam and Rajasthan the status of soil copper is not available. Soils in the remaining 98 districts are well supplied with copper for citrus cultivation (Figure 7B) Generally copper is not recommended for soil application. Foliar sprays or soil applications of Cu fertilizer can prevent or cure Cu deficiency. Spraying a solution containing 2 to 3 kg per hectare of elemental Cu applied during flowering commonly results in near-immediate recovery followed by a normal fruit set. Copper deficiency can be a controlling factor in fruit production, and acute Cu deficiency may put trees entirely out of production. Foliar sprays are useful emergency treatments when symptoms of Cu deficiency are first observed.

Management

Copper has an important role in plant nutrition through operating in trace amounts as an essential micronutrient, but copper deficiency in citrus is uncommon and hence is not externally applied as a plant nutrient. Instead copper containing compounds have a

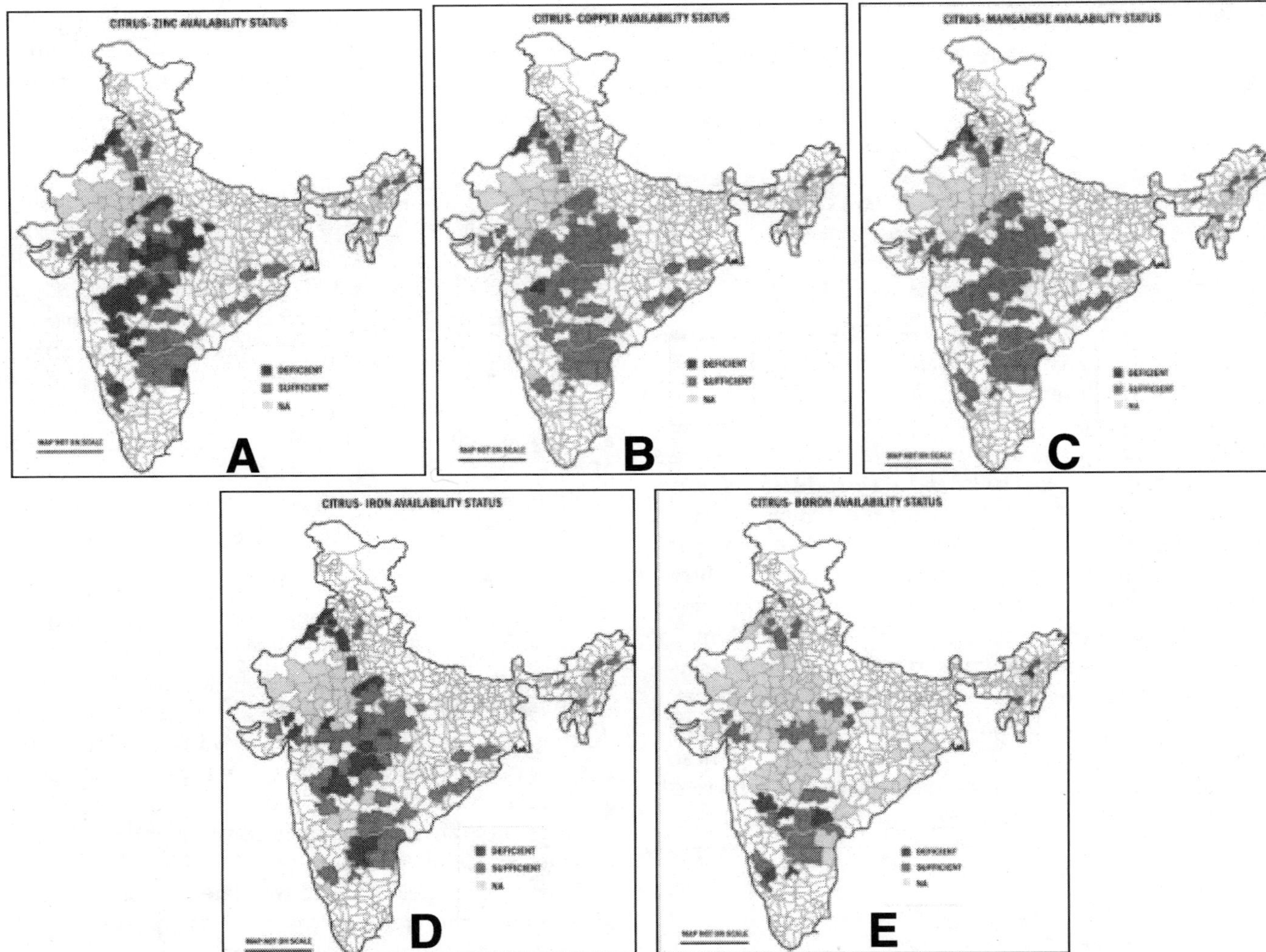

Figure 7A-E: Micronutrient Fertility Status of Citrus Growing Regions of India.

unique place in citrus rehabilitation due to their broad spectrum activity and longevity of their deposits. The copper ion (Cu^{2+}) is best known as a fungicide but it is also an effective bactericide and molluscicide, as well as suppressing algae, lichens and even mosses, ferns and other plants, that grow on citrus trees in the hot wet tropics. The quantity of copper supplied to plant through the feungicides is adequate, rather even excess, as plant nutrient.

Manganese

Out of 131 citrus growing districts only nine districts showed low manganese availability in soils. For 24 districts mostly in Assam and Rajasthan the status of soil Mn is not available. Soils in the remaining 98 districts are well supplied with Mn for citrus cultivation (Figure 7C). Generally Mn is not recommended for soil application. Only if need arises by way of expression of deficiency symptoms growers may undertake foliar application of Mn.

Management

Manganese is best applied as a foliar spray during late winter or beginning of summer in central India and post monsoon dry season in other parts when new growth is being made. Where severe, persistent deficiency exists, use 100 g manganese sulphate in 100 l water. Do not spray if rain is expected within 48 h. Citrus Special developed by IIHR, Bengaluru takes care of the manganese deficiency in citrus.

Iron

Information on the soil availability of iron is not available for six districts out of 131 districts, *viz.*, Golaghat in Assam, Hassan, Gulbarga, Bengaluru (Rural),in Karnataka, Solapur in Maharashtra and Nalgonda in Telangana. Of the remaining 125 districts 70 districts fell under the category of deficient rating (Figure 7D). The most reliable means of correcting citrus Fe chlorosis is by soil application of iron chelates. However, Fe chelates are not equally effective. Foliar application of Fe is not effective and a risk of leaf and fruit burn is involved. At best, foliar sprays of Fe produce a speckled greening of the leaves rather than an overall greening.

Management

Iron chlorosis is a major production concern for trees grown on calcareous soils. Application of iron to such soils may not be fruitful unless a chelated form of iron is applied. Foliar application of iron will be necessary to grow a healthy tree on these soil types. Iron fertilizer formulations are available that can correct chlorosis; however, the required application rate and frequency make the treatment expensive. Inorganic sources of Fe such as ferrous sulfate ($FeSO_4$) are not effective unless applied at extremely high rates; these sources should not be used on calcareous soils. The most popular synthetic organically chelated forms of Fe include Fe-EDTA, Fe-HEDTA, Fe-DTPA, and Fe-EDDHA. The effectiveness of these fertilizers varies greatly, depending on soil pH. Fe-DTPA may be used on mildly alkaline soils (with pH values of 7.5 or less), whereas Fe-EDDHA is the chelate of choice for use on highly calcareous soils (with a pH value greater than 7.5).

Boron

Deficiency of boron has been widely found in all citrus growing areas particularly in highly calcareous of Bihar, Tamil Nadu, eastern Uttar Pradesh and Saurashtra, sandy soils of Haryana and Rajasthan, hill and sub mountaneous soils of north Himalayan and north eastern states and in red and lateritic soils of Odisha, Karnataka, Andhra Pradesh and Konkan region. Extent of boron deficiencies varied widely with soil types and agro-ecological situations. Intensive cultivation of citrus has aggravated the emergence of boron deficiency in several soils which were earlier not testing low in available boron. Information on soil boron is available only for 51 districts out of 131 districts. Map showing the status of boron availability in citrus districts has been prepared (Figure 7E). There are only six districts which are reported to be deficient in soil available boron and for the remaining 45 districts the hot water soluble boron is adequate. Because citrus is a heavy feeder, the rate of boron supply and the boron demand by the crop do not match. Hence in most cases boron application results in very good crop response.

Management

Applying B to the soil can provide unsatisfactory results during dry seasons and may result in toxicity problems Basal application of boron through broadcast on surface soil is better than top dressing in boron deficient soils. Caution should be exercised in its soil application as continuous soil application leads to its accumulation and reaches toxic levels in short periods. Since foliar sprays of 0.2 per cent borax or boric acid solution were found more efficient in citrus it is better to supply boron to citrus through foliage. Because of its greater solubility, boric acid might be preferred over borax for foliar spray application. The spray may be applied either during the dormant period or post-bloom. The applicator should be careful not to apply more than the recommended amount because any excess amount would result in B toxicity.

Molybdenum

Application of ammonium molybdate at the rate of 25 g tree^{-1} as soil application ('Sathgudi' sweet orange) and foliar spray at 0.08-1.0 per cent ('Sathgudi' sweet orange and 'Kinnow' mandarin) have shown promising results as effective corrective measures.

Micronutrient Deficiency in Grapes

Crop Boundary

Ten states were identified as prime grapes producing states of India. Based on categorization with respect to area and production per unit area, 43 districts were identified as prime grapes growing districts. Boundary map of grapes was prepared using district maps of respective leading producing states and is presented in Figure 8.

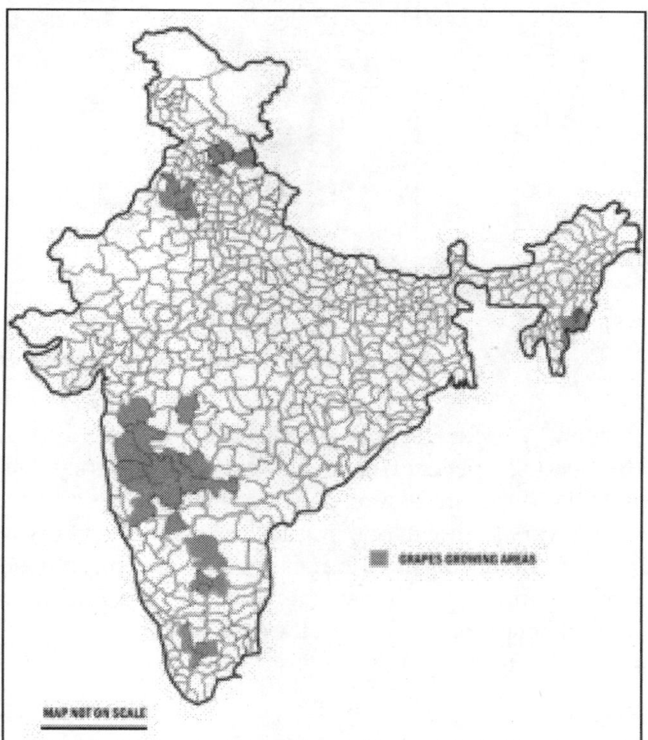

Figure 8: Major Grapes Growing Areas in India.

Zinc

Using the zinc fertility status map of India developed by AICRP MSNPE and adding additional information available in published papers the grapes growing soils area was delineated for zinc deficiency on the basis of the percentage of samples found deficient in each district and is shown in Figure 9A. It is found that of the 43 grapes growing districts soils in 29 districts (67.4

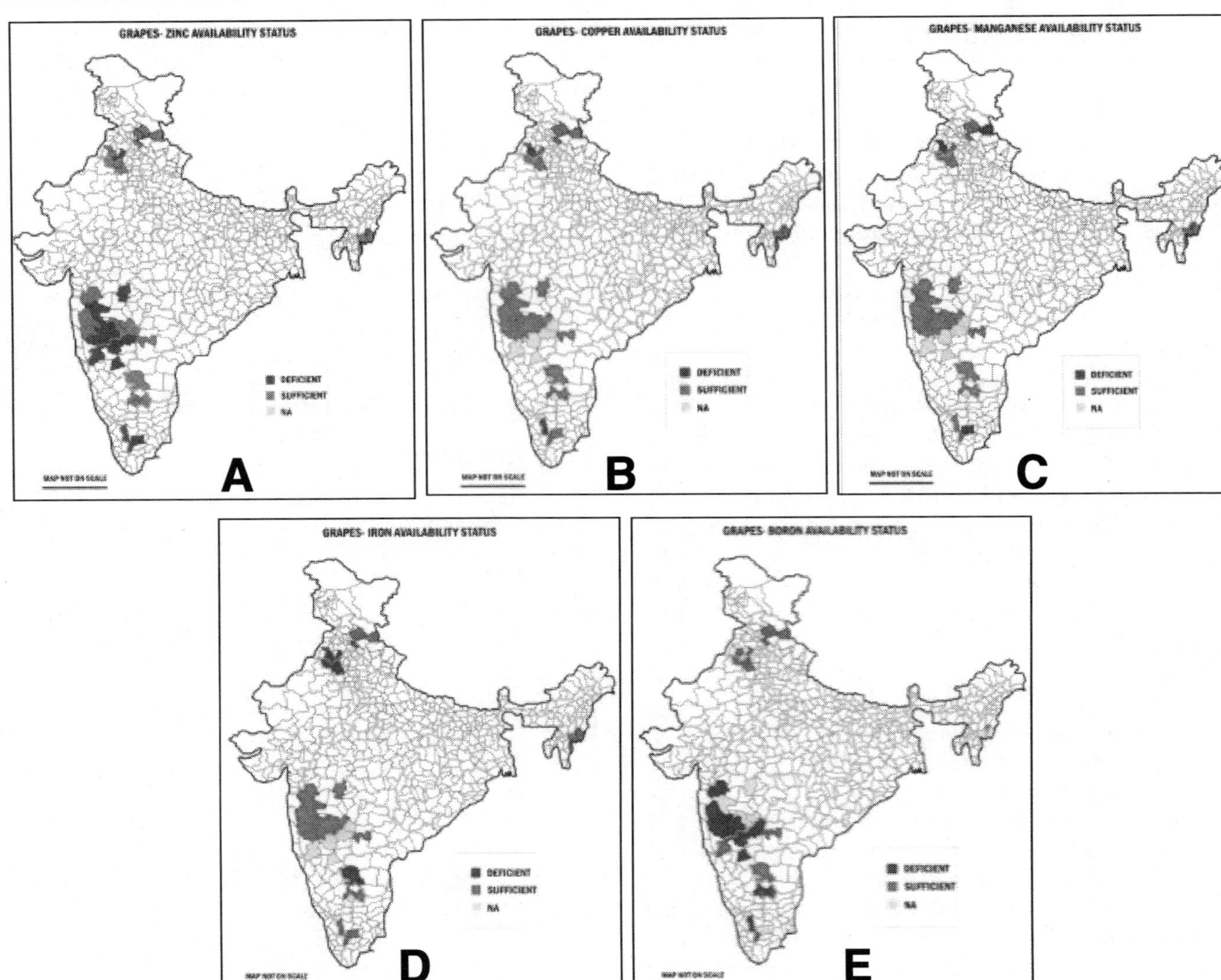

Figure 9A-E: Micronutrient Fertility Status of Grapes Growing Regions of India.

per cent) were well supplied with available zinc and 14 districts (32.6 per cent) soils showed zinc deficiency. The data on the extent of zinc deficiency in grapes growing soils is very high, but one must exercise caution in using this information as most grapes growers are enlightened farmers and regularly use zinc as both soil application and foliar sprays. As a result the soil status of individual vineyards may be high.

Management

Response to applied zinc in grapes is impressive compared to other fruit crops, if the soil status is low. Hence, zinc application should be made to grapes on the basis of soil test. A blanket application of zinc to soils may be avoided as build up would cause more damage than doing good. Zinc concentrations in seeded grapes petioles range from 25-94 ppm and in seeded grapes from 30-88. If the concentration of zinc in the petioles fall below these levels then corrective measures are required.

Foliar spray of zinc is the most common because it is the most widely deficient micronutrient. Treatment can also be quite effective if the correct material and methods of application are used. Neutral solution of zinc sulphate heptahydrate and zinc oxide are the most economical and effective on a recommended label basis. There is no advantage in using chelated zinc products in sprays. They were originally intended for soil application, are more expensive, and less effective than zinc sulphate and zinc oxide. The deficiency of zinc is readily corrected with zinc sulphate heptahydrate and zinc oxide spray at 0.25 to 0.3 per cent depending upon the severity of the deficiency. Spray of lower concentrations is better than higher concentrations because the uptake is greatest with dilute application. The optimum timing to influence fruit set is three weeks before bloom up to bloom.

Copper

Most Indian soils contain adequate amounts of copper for optimum crop yields. Organic or peat soils are exceptions but grapes is not grown in such soils. Out of 43 grapes growing districts only two districts, one in Punjab (Bathinda) and one in Tamil Nadu (Coimbatore) are reported to have copper deficiency (Figure 9B). No data is available for seven districts, all in Karnataka and the soils of those districts are not likely to have copper deficiency.

Management

Copper concentration in the petioles of seeded grapes varies from 10-30 ppm and in seedless grapes from 5-10 ppm. In most of the Indian vineyards rarely it falls below this range to show any deficiency of copper on the vines. Grapes is one fruit crop which receives huge quantity of copper through fungicides. Its requirement being small, the copper deficiency in grapes is rarest of the rare situation. Instead we may expect copper toxicity in this crop. Hence as a thumb rule copper application or a spray of copper is not recommended for grapes in India.

Manganese

Manganese availability in grapes soil is generally not a production constraint. Availability status of manganese in grapes soils (Figure 9C) shows that only five districts out of 43 districts are reported to be low in DTPA extractable Mn. Data for seven districts is not available. The soils in the remaining 31 districts are well supplied with Mn.

Management

In the soil It is present as free Mn^{2+}, which is readily available to plants. The proportion of manganese in various forms in the soil is dependent both on chemical reactions and on microbial activity. Very high pH soils greatly reduce the solubility of soil manganese, and therefore its availability to roots. Thus, manganese deficiency is most likely to occur in alkaline soils or soils after liming. Such situations may arise in some alkaline soils of Punjab and western Uttar Pradesh and after liming in acid soils of Karnataka and Tamil Nadu. Application of bio-fertilizers containing nutrient mobilizers can mobilize soil manganese. Further the quantity required being very small the requirement is met mostly from applied organics, pesticides containing manganese and contaminations from fertilizers. Manganese concentrations in seeded grapes petioles range from 18-86 ppm and in seedless grapes from 42-209. If the concentration of manganese in the petioles fall below these levels then corrective measures are required. It is readily corrected with manganese sulfate spray at 0.25 to 0.3 per cent depending upon the severity of the deficiency. There are no advantages in using chelated manganese in a foliar spray.

Iron

The iron status map (Figure 9D) show that out of 43 districts mostly in Punjab, Haryana and Ananthpur in Telangana are reported deficient in available iron. Data for equal number of districts, all in Karnataka, is not available. Remaining 29 district soils showed sufficiency in available Fe in soils.

Management

Iron concentration in the petioles of seeded grapes varies from 38-107 ppm and in seedless grapes from 54-80 ppm. If the concentration in the petioles fall below these ranges then it is necessary to take corrective measures to avoid any crop loss. Although it is easy to see the visual symptoms of Fe deficiency, it is relatively hard to alleviate the problem in calcareous soils where it normally occurs. Foliar sprays are generally ineffective as the ion is immobile in the plant. A drenching of diluted sulphuric acid (@ 10 to 15 ml in 100 l of water) in calcareous soils/vineyards irrigated with bicarbonate rich waters can correct iron deficiency problems. If it is decided to apply Fe to grapes, foliar applications are the most viable option. The grapes special developed by Indian Institute of Horticultural Research, Bengaluru can correct the iron deficiency problems in most situations. This can be used at 4 grams per litre of solution. Ferrous sulphate neutral solution at 0.2 to 0.3 per cent can also be sprayed to overcome iron deficiency.

Boron

Information on available boron status of Indian soils is very poor. We have tried to compile available information and attempted to prepare a boron status map of grape growing area (Figure 9E). Of the 43 grape growing districts we have data for 29 districts (Figure 9E). Out of this 14 districts showed boron deficiency and 15 districts are well supplied with boron. No data is available for the remaining 14 districts.

Management

Boron deficiency may be observed on alkaline soils (high pH) or sandy soils. Soil tests are recommended to establish base levels but should be confirmed with tissue analyses before resorting to boron application. Boron levels in tissue fluctuate during the growing season. Hence boron deficiency may be temporary and may be noticed during unusually dry spells. Recommendations for B vary widely across the country. We have chosen to avoid recommending soil application and recommend foliar application of B at rates that are conservative, in order to avoid potential toxicity. Growers should monitor their crops with leaf or petiole analysis to determine the correct rate in their specific situation. If deficiency is noticed then a foliar spray of boric acid at the rate of 2 g per l of water corrects the deficiency. But the number and frequency is to be decided based on petiole analysis. Grapes nutrient formulation developed by IIHR caters to the need for boron in grapes across the country and varieties. This can be used at the rate of 4 g per l and the number of sprays may be adjusted to suit the individual grape vineyards.

Molybdenum

Molybdenum is involved in enzymatic reactions essential for growth and reproduction in grape vines. It is involved in the conversion of nitrate nitrogen, taken up by the roots, into a form that the vine can use. It is also involved in fruit set; seed formation; berry formation and development; and bunch yield. Molybdenum deficiency may be exhibited as small leaves (size of a two rupee coin or smaller); the edge of leaves burn, poor leaf colour, wood fails to mature; rubbery feel to shoots, papery feel to leaves; distorted or zig- zag growth habit of shoots. Vines with these symptoms accumulate excessive nitrate-nitrogen in the petiole."Hen and chickens" - a form of berry asynchrony in seedless berries, where the bunch at harvest consists of a mixture of a few large, normal berries (hens) and many small berries (chickens) of uneven ripeness are typical symptoms of molybdenum deficiency in grapes. Data for molybdenum status of grape growing soils is not available. Hence the status maps could not be prepared.

Management

Molybdenum deficiency in Indian grapes is rare. Hence application of molybdenum may be avoided unless tissue testing is done. If petiole analysis shows excessive nitrate levels (about 10,000 ppm) and molybdenum deficiency symptoms are exhibited then only molybdenum should be applied. However, molybdenum deficiency that is only temporary during flowering may be a major cause of poor fruit set, without any vegetative signs of deficiency. Care should be taken to see that fruit set problems caused due to other factors, such as periods of cold, wet conditions between bud burst and fruit set, and zinc or boron deficiency must be cross checked and confirmed before taking up corrective measures for molybdenum deficiency. Vines with petiole Mo concentrations at peak flowering less than 0.09 ppm or nitrate nitrogen more than 10000 ppm needs molybdenum application. Molybdenum, as a soluble fertiliser is available as ammonium molybdate (54 per cent Mo) or sodium molybdate (39 per cent Mo). As molybdenum is only required in a small quantity, one annual foliar spray of 500 g ammonium or sodium molybdate per 1000 l of water should be adequate to overcome a vegetative growth deficiency where vegetative symptoms have been observed or petiole tests indicate high nitrate-nitrogen concentrations

Micronutrient Deficiency in Mango

Crop Boundary

Mango is grown throughout the country on all types of soils and in all the agro ecological sub-regions except in very cold or extremely dry desert areas. Based on area and production, ten states were identified as prime mango producing states of India (Figure 10). Similarly within these states, based on categorization with respect to area and production per unit area, one hundred eleven districts were identified as prime mango growing districts. The boundary map of mango was prepared using district maps of respective leading mango producing states.

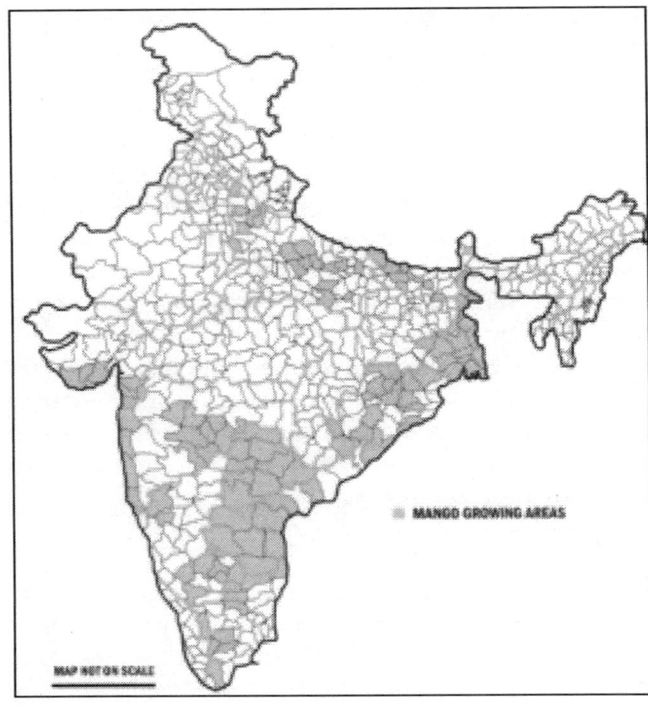

Figure 10: Major Mango Growing Areas in India.

Zinc

Soil status of zinc and appearance of deficiency in mango donot match closely. Out of 111 mango districts, soils in 60 per cent of area (67 districts) are well supplied with zinc. Only 33 per cent area (33 districts) shows deficient levels of soil zinc (Figure 11A). However, no information is available for seven districts, one each in UP (J.P. Nagar) and Karnataka (Bengaluru-urban) and five in West Bengal (Malda, West Midnapore and East Midnapore, Purulia, Birbhum). Zinc deficiency occurs during peak growth period or fruit development stage when the demand for zinc exceeds supply rate from the soil. Being immobile retranslocation from older leaves or woody part is slow. As a result the deficiency is exhibited both as hidden hunger and expressed as leaf symptoms. Foliar application of zinc corrects such deficiencies.

Management

Continuous soil application of zinc leads to its accumulation in the soil leading to nutrient imbalance in the soil. Further the uptake of foliarapplied zinc in mango is more rapid than that of soil applied zinc. Hence foliar application is preferred over soil application. Zinc deficiency in mango, can be ameliorated by foliar application of zinc sulphate. Spray the plants with 0.25-0.5 per cent zinc sulphate solution (250-500 g zinc sulphate per 100 l of water depending upon severity of deficiency) soon after cessation of monsoon and before October flush and repeat the spray before panicle emergence. However, in some places where new flush emerges late, the spray should be done in June and if further delayed flush appears then zinc sulphate should be sprayed during August-September. Apart from this

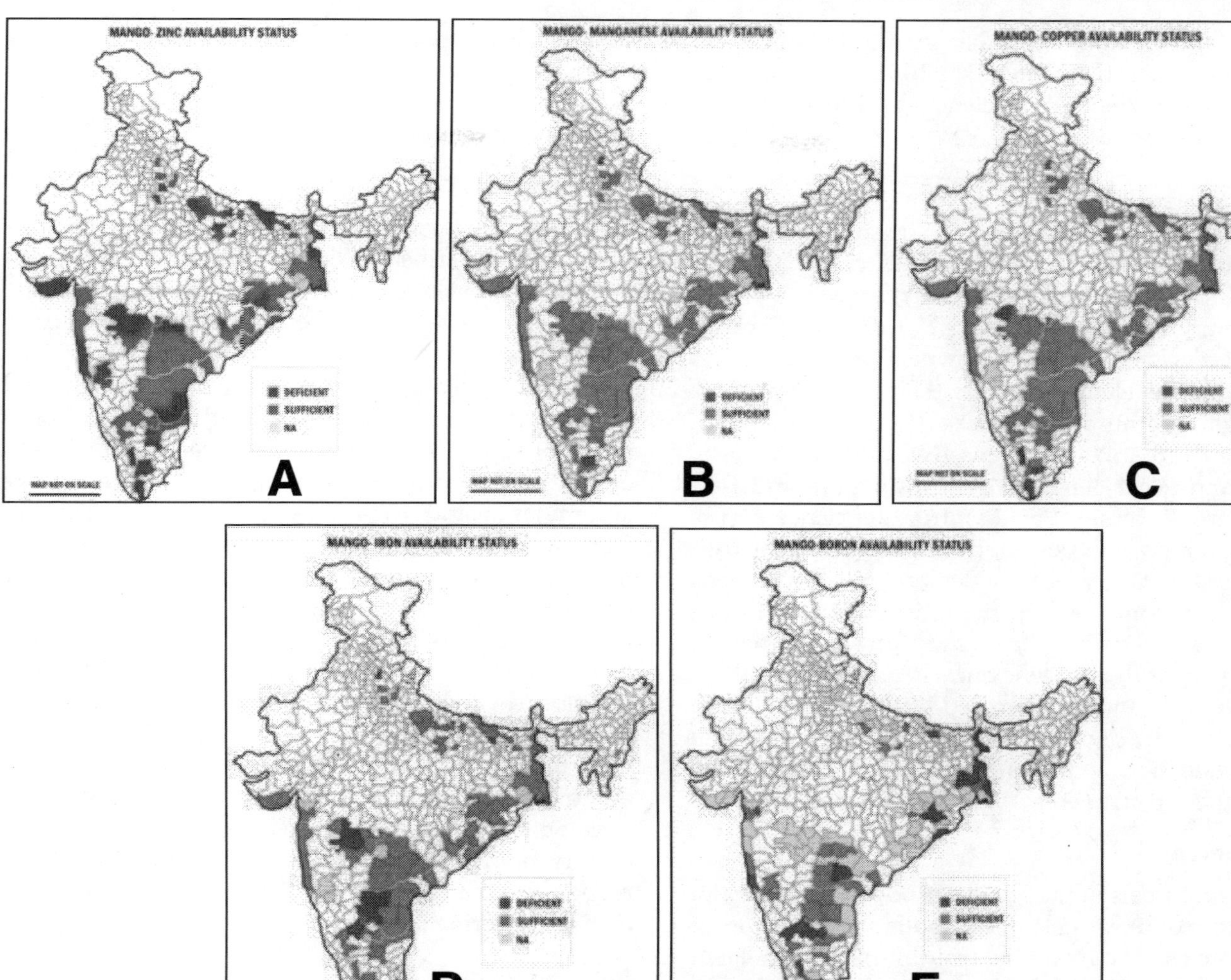

Figure 11A-E: Micronutrient Fertility Status of Mango Growing Regions of India.

in mango, zinc deficiency becomes acute immediately after first fruit set on plants; therefore one maintenance spray of 0.3 per cent zinc sulphate solution may be given in pre-monsoon from the third year of planting. Arka mango Special developed by IIHR may be used at 5g per litre to correct zinc deficiency in addition to other micronutrient deficiencies. Those growers who wish to apply Zn to soil may apply one kg zinc sulphate per tree once in two to three years.

Manganese

Soil status of available manganese is presented in Figure 11B Soils in only three districts, *viz.*, Saharanpur in UP, Vaishali in Bihar, Dindigul in Tamil Nadu showed manganese deficiency. The remaining 108 districts showed adequate supply of soil manganese.

Management

Manganese deficiency is not common in mangos. However, hidden hunger might be there in sandy and light textured neutral to alkaline soils. Many fungicides and insecticides used in mango contain traces of manganese. This partially meets the manganese requirement of the trees. In case of deficiency foliar application of 0.2 per cent manganese sulphate may be adopted rather than soil application. Arka Mango Special developed by IIHR, Bengaluru may be used at 5g per litre to correct Mn deficiency in addition to other micronutrient deficiencies.

Copper

Similar to manganese, soil available copper is sufficient in 90 districts out of 111 districts. Soils in only five districts were reported deficient in available copper. However, no information is available for 16 districts, mostly spread in Karnataka, Tamil Nadu and West Bengal (Figure 11C).

Management

Copper deficiency is not common in mangos. However, hidden hunger might be there in high organic matter soils and sandy and light textured neutral to alkaline soils. As already stated many fungicides and insecticides used in mango contain traces of copper. This partially meets the copper requirement of the trees. In case of deficiency foliar application of 0.2 per cent copper sulphate may be adopted rather than soil application. Arka Mango special developed by IIHR may be used at 5 g per litre to correct Cu deficiency in addition to other micronutrient deficiencies.

Iron

Soil status of iron and appearance of deficiency in mango do not match closely. Out of 111 mango districts, soils in 82 per cent of area (91 districts) are well supplied with iron. Only 8 per cent area (9 districts) shows deficient levels of soil iron. However, no information is available for eleven districts (Figure 11D), in U.P. (J.P. Nagar, Mathura), Karnataka (Belgaum, Dharwad, Bengaluru (urban)), Telangana (Nizamabad), Gujarat (Surat) and West Bengal (West Midnapore and East Midnapore, Birbhum, Purulia). Iron deficiency occurs during peak growth period, during first rains or fruit development stage when the demand for iron exceeds supply rate from the soil. Being immobile in tissue, retranslocation from older leaves or woody part is slow. As a result the deficiency is exhibited both as hidden hunger and as expressed leaf symptoms. High soil bicarbonate, calcium carbonate and irrigating with bicarbonate rich water causes iron deficiency even though the soil status may be high in such cases.

Management

Iron chlorosis in mango is a production constraint for trees grown on calcareous soils. Application of iron to such soils may not be fruitful unless a chelated form of iron is applied. Foliar application of iron will be necessary to grow a healthy tree on these soil types. Iron fertilizer formulations are available that can correct chlorosis; however, the required application rate and frequency make the treatment expensive. Inorganic sources of Fe such as ferrous sulfate ($FeSO_4$) are not effective unless applied at extremely high rates; these sources should not be used on calcareous soils. The most popular synthetic organically chelated forms of Fe include Fe-EDTA, Fe-HEDTA, Fe-DTPA, and Fe-EDDHA. The effectiveness of these fertilizers varies greatly, depending on soil pH. Fe-DTPA may be used on mildly alkaline soils (with pH values of 7.5 or less), whereas Fe-EDDHA is the chelate of choice for use on highly calcareous soils (with a pH value greater than 7.5). The mango special developed by IIHR, Bengaluru is a proven product that corrects both hidden and expressed iron deficiency in mango. This can be used as a foliar spray @5 g per litre during flowering season.

Boron

Soil availability status of boron indicated that out of 111 mango districts, no information is available for 58 per cent of the area (64 districts). Data from 19 districts showed deficient levels of hot water soluble boron. However, only 28 district soils showed sufficiency in boron supply (Figure 11E). Deficiency of boron can be easily corrected through foliar feeding. Many a times an imbalance in potassium and calcium supply relative to boron supply may also cause boron deficiency even under adequate boron supply. Hence, before undertaking remedies, leaf analysis is essential.

Management

Though appearance of boron deficiency in mango is very common, farmers seldom take corrective measures. If properly applied correction of boron deficiency is very easy. Applying B to the soil can provide unsatisfactory results during dry seasons and may result in toxicity problems. Basal application of boron through broadcast on surface soil is better than top dressing in boron deficient soils. Caution should be exercised in its soil application as continuous soil application leads to its accumulation and reaches toxic levels in short periods. It is better to supply boron to mango trees through foliage as foliar sprays of 0.2 per cent borax or boric acid solution were found efficient in mango. Because of its greater solubility, boric acid might be preferred over borax for foliar spray application. The spray may be applied either on recently matured flush orduring flowering. The applicator should be careful not to apply more than the recommended amount because any excess amount would result in B toxicity. The Mango Special developed by IIHR, Bengaluru is a proven product that corrects both hidden and expressed boron deficiency in mango. This can be used as a foliar spray @5 g per litre during flowering season.

Molybdenum

Molybdenum as an essential element is necessary for the activator of enzymes in respiration. It also has a role in the protein synthesis. Mo is also essential for auxin synthesis and can act as electron carriers in enzyme systems bringing about oxidation-reduction reactions.

Generally Mo is not a limiting factor in mango production. No reported Mo deficiency is reported from anywhere in the country. However, fully matured leaf should contain Mo in the range of 0.05 to 0.10 ppm. Unless leaf analysis shows deficiency it is advisable to avoid application or spray of Mo in mango.

Micronutrient Deficiency in Papaya

Crop Boundary

Eleven states were identified as prime papaya producing states of India. Based on categorization with respect to area and production per unit area, 86 districts were identified as prime papaya growing districts. Boundary map of papaya was prepared using district maps of respective leading producing states and presented in Figure 12.

Zinc

Sufficient data base is available on soil zinc availability in India (Figure 13A). Of the 86 papaya

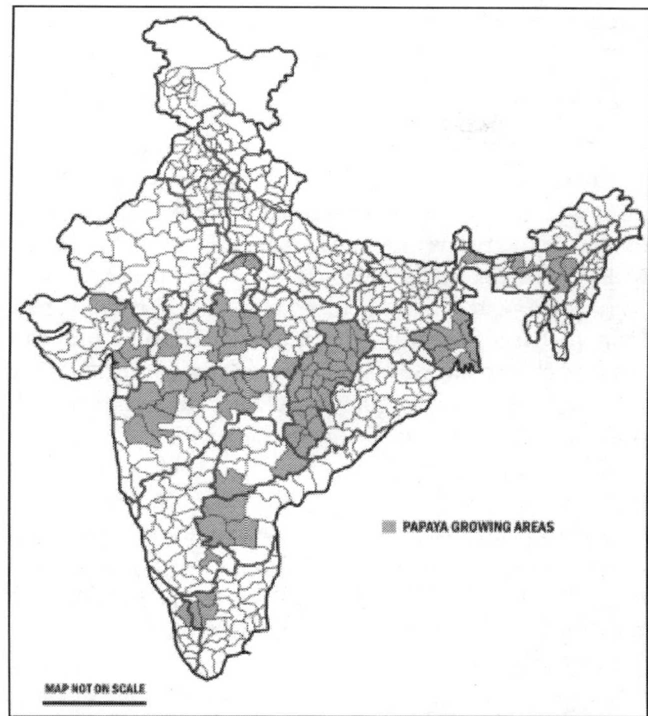

Figure 12: Major Papaya Growing Areas in India.

growing districts there is no information available for21 districts. 20 districts have deficiency of zinc and 45 districts showed sufficient levels of soil available zinc.

Management

Fertilizing papaya with zinc containing chemicals is uncommon. However, those farmers who are applying zinc as fertilizers to soils have continued to apply every year leading to accumulation of high levels of zinc in such soils. This has caused imbalance in plant nutrition leading to induced phosphorus and other deficiencies. Keeping this in mind and the pattern of fertilizer use in papaya and the crop ability to absorb nutrients it is advisable to go for foliar application of zinc rather than soil application as a precaution to avoid unnecessary accumulation in the soils leading to nutrient imbalances. A general spray of 0.25 to 0.40 per cent solution of $ZnSO_4.7.H_2O$ would take care of the zinc requirement of papaya. Use of Arka papaya micronutrient formulation developed by IIHR, Bengaluru @ 2 g per litremeets the micronutrient requirement of papaya including zinc.

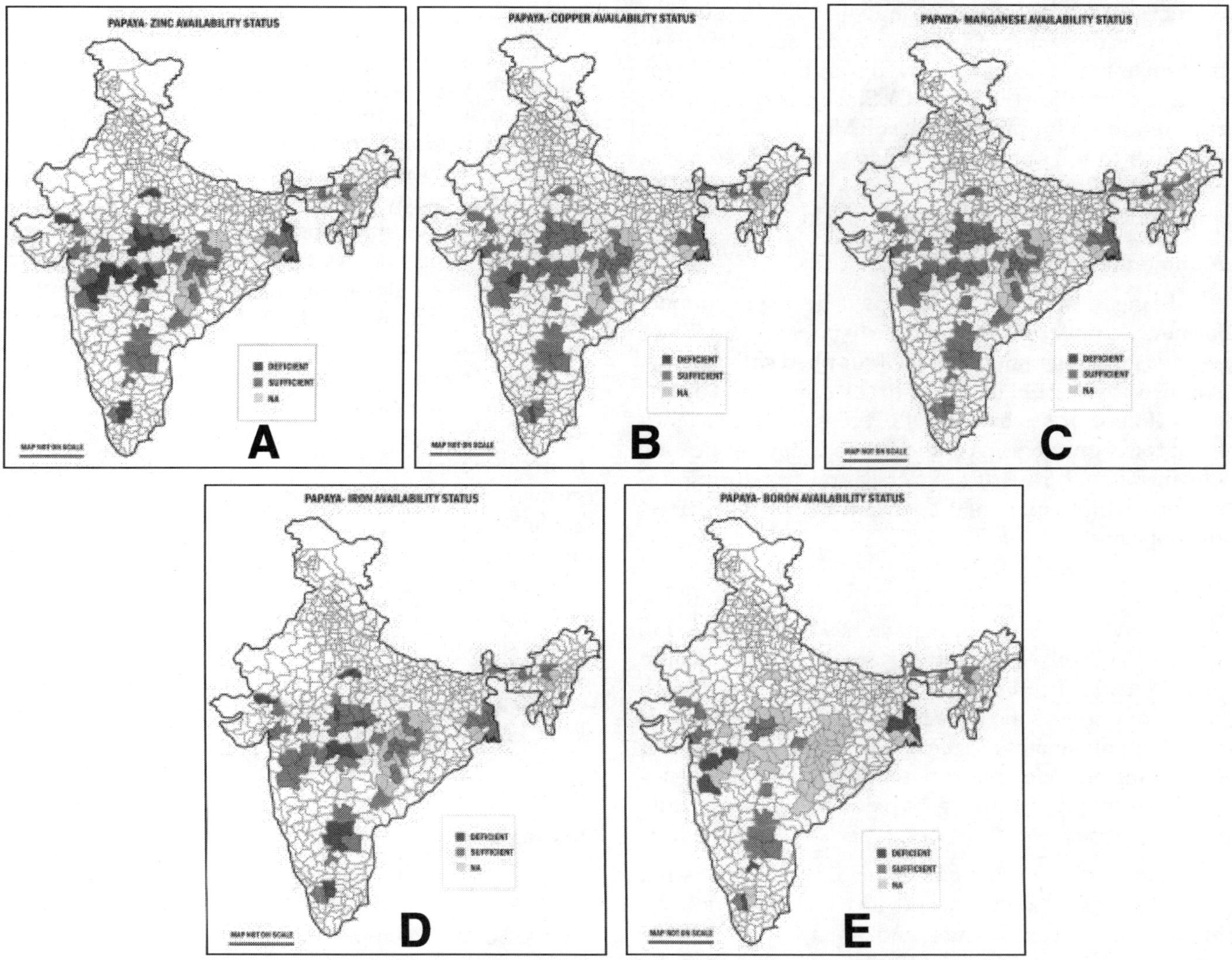

Figure 13A-E: Micronutrient Fertility Status of Papaya Growing Regions of India.

Copper

Papaya soils have adequate supply of copper (Figure 11B). Out of 86 papaya growing districts only two districts, *viz.*, Aurangabad in Maharashtra and Coimbatore in Tamil Nadu showed low DTPA copper content in their soils. For 19 districts the status of soil Cu is not available. Remaining all districts showed sufficiency levels of Cu in their soils.

Management

Copper is not reported as deficient in any of the papaya growing regions. Hence copper application to papaya may be avoided. Sufficient copper is supplied to crop through the addition of organic matter and also through the copper containing fungicides and as a contaminant with other agrochemicals. Hence, soil application of copper to papaya may be avoided.

Manganese

Manganese deficiency does occur in sandy soils which witness frequent flooding and drying. Under such situation soluble Mn gets leached out of root zone. Papaya in such areas shows Mn deficiency during dry periods. Out of 86 papaya growing districts not a single district showed low manganese availability in soils. For 21 districts the status of soil Mn is not available. Soils in the remaining 65 districts are well supplied with Mn for papaya cultivation (Figure 11C). Since papaya growing regions are well supplied with soil Mn, it is advised not to render to soil application of Mn. Only if need arises by way of expression of deficiency symptoms growers may undertake foliar application of Mn.

Management

Manganese deficiency in papaya being uncommon, farmers do not apply manganese to papaya crop. Since papaya growing regions are well supplied with soil Mn, it is also advised not to render to soil application of Mn. Only if need arises by way of expression of deficiency symptoms growers may undertake foliar application of Mn. Though rare, if deficiency appears during dry period a simple spray of 0.2 per cent $MnSO_4$ will correct the deficiency

Iron

Deficiency of iron in papaya has been reported in the field in Tamil Nadu, Bihar, Karnataka, Maharashtra and Madhya Pradesh, where it is associated with calcareous soils, soils with high water tables and soils with high manganese. If papaya is irrigated with water containing high bicarbonate content then iron deficiency occurs in plants as iron gets oxidised to biologically inactive ferric form.

Information on the soil availability of iron is not available for 22 districts out of 86 districts mostly in Karnataka, Tamil Nadu and West Bengal. Of the remaining 64 districts only 12 districts fell under the category of deficient rating (Figure 11D) and 42 districts are well supplied with DTPA iron.

Management

Iron chlorosis is a major production concern for papaya production on calcareous soils and soils irrigated with waters containing high bicarbonates. Application of iron to such soils may not be fruitful unless a chelated form of iron is applied. Foliar application of iron will be necessary to grow a healthy tree on these soil types. Iron fertilizer formulations are available that can correct chlorosis; however, the required application rate and frequency make the treatment expensive. If plant exhibits iron deficiency then foliar application of 0.25-0.4 per cent ferrous sulfate ($FeSO_4$) corrects the deficiency but care should be taken to see that the spray solution pH must be in the range of 6.25 to 6.75. Synthetic organically chelated forms of Fe like Fe-EDTA, Fe-EDTA, Fe-DTPA and Fe-EDDHA are available in the market. The effectiveness of these fertilizers varies greatly, depending on soil pH. Fe-DTPA may be used on mildly alkaline soils (with pH values of 7.5 or less), whereas Fe-EDDHA is the chelate of choice for use on highly calcareous soils (with a pH value greater than 7.5).

Boron

Boron deficiencies occur on a wide variety of soils. The extent of deficiencies varies widely with soil types, cropping systems and agro ecological situations. The deficiency has been widely found in highly calcareous soils of Bihar, Tamil Nadu, eastern Uttar Pradesh and Saurashtra, sandy soils of Haryana and Rajasthan, hill and sub-mountainous soils of north Himalayan and north eastern states and in red and lateritic soils of Odisha, Karnataka, Andhra Pradesh and Konkan region and Andaman and Nicobar Islands. Intensive cultivation of papaya has aggravated the emergence of boron deficiency in several soils which were earlier not testing low in available boron. Information on soil boron is available for 30 districts out of 86 districts. Map showing the status of boron availability in papaya growing regions has been prepared (Figure 11E). There are only nine districts which are reported to be deficient in soil available boron and for the remaining 21 districts the hot water soluble boron is adequate. Since, papaya is a heavy feeder, the rate of boron supply and the boron demand by the crop do not match. Hence boron application, in most cases, results in very good crop response. It has been observed that a plant well supplied with boron can tolerate virus diseases and produce reasonably good fruits and extend the productive life of the orchards.

Management

Correction of boron deficiency in papaya is easier than that of iron. To correct a boron deficiency care should be taken to apply the right amount of boron to the soil or through foliage. Excess boron severely damage or kill a plant. At the same time too little boron will not resolve the deficiency. If sufficient boron is supplied through foliar spray, then do not re-apply boron for at least three months, as this much time is required to see the effects of a treatment. While using the commercial products of boron care must be taken to follow the following points:

- Feed papaya at least twice per year with a fertilizer designed for papaya like Arka papaya special. It should contain all of the minor elements, including boron. Even if papaya fertilizers containing boron are applied regularly, however, boron-deficiency symptoms may still develop.
- If orchards exhibit boron deficiency then spray foliage or drench the soil with a 0.25 per cent borax or boric acid solution. The effect can be seen after three months of treatment.
- If the farmers intend to apply to soil then apply 0.5 g boric acid around the base of a papaya plant. If this amount fails to solve the deficiency, up to 5 g borax per plant may be applied. For symptomatic plants a drench to the root systems of using boric acid @ 2 g per litre may be applied followed by irrigation or application of 10 to 15 litres of water per plant so that the roots can take up the boron-containing water.
- Wherever, soil application is done farmers must get the soil analysed before next application of boron to any crop in the system

Molybdenum

Molybdenum is involved in enzymatic reactions essential for growth and reproduction in papaya. It is involved in the conversion of nitrate nitrogen, taken up by the roots, into a form that the tree can use. It is also involved in fruit set; seed formation, fruit development and fruit yield. Molybdenum deficiency may be exhibited as small leaves; the edge of leaves burn, poor leaf colour, papery feel to leaves; distorted or zigzag growth habit of shoots. Trees with these symptoms accumulate excessive nitrate-nitrogen in the petiole. Data for molybdenum status of papaya growing soils is not available. Hence, the status maps could not be prepared.

Management

Molybdenum deficiency in Indian papaya is rare. Hence, application of molybdenum may be avoided unless tissue testing is done. If petiole/leaf analysis shows excessive nitrate levels (about 10,000 ppm) and molybdenum deficiency symptoms are exhibited then only molybdenum should be applied. However, molybdenum deficiency may be a major cause of poor fruit set, without any vegetative signs of deficiency. Care should be taken to see that fruit set problems caused due to other factors, such as periods of cold, wet conditions and zinc or boron deficiency must be cross checked and confirmed before taking up corrective measures for molybdenum deficiency. Trees with petiole Mo concentrations at less than 0.09 ppm or nitrate nitrogen more than 10000 ppm needs molybdenum application. Molybdenum, as a soluble fertiliser is available as ammonium molybdate (54 per cent Mo) or sodium molybdate (39 per cent Mo). As molybdenum is only required in a small quantity, one annual foliar spray of 500 g ammonium or sodium molybdate per 1,000 l of water should be adequate to overcome a vegetative growth deficiency where vegetative symptoms have been observed or petiole tests indicate high nitrate-nitrogen concentrations

References

Ganeshamurthy, A.N. 2016. Fertilizer systems of key horticultural crops in India. *NPK 2016*, New Delhi.

Patel, K.P., Chiklekhar, S., Ramtek, S., Sahu, B.L., Daharia, N.S., Sharma, R. 2015. Micronutrient status in soils of central India. *Am. J. Pl. Sci.*, **6**. 3025–3037.

Singh, M.V. 2008. Micro and secondary nutrient mapping for Indian soils. AICRP of Micro and secondary nutrients and pollutant elements. IISS, Bhopal.

Times of India 2016. India 2nd largest fruit producer in world Vishwa Mohan | TNN | Jan 18, 2016,http://timesofindia.indiatimes.com/india/India-2nd-largest-fruit-producer-in-world/articleshow/50618234.cms

39

Fertigation and Innovative Approach for Fertilizer Use Efficiency in Horticultural Crops

Manoj Khanna

Principal Scientist and In-Charge Farm Operation and Service Unit, Water Technology Centre, ICAR-Indian Agricultural Research Institute, New Delhi – 110 012
E-mail: khanna_manoj2001@yahoo.com.au

Micro-irrigation is an efficient method of water application for vegetable and horticultural crops. This method is widely used because it not only conserves water but also allows efficient management of water and fertilizer. It consists of network of small diameter pipes and emitters placed on the soil surface or buried, which discharges water and fertilizer at a controlled rate. In this method water infiltrates in the region surrounded by emitter, which is very small as compared to the total cropped area.

Though the area under micro-irrigation in the world is less than 1 per cent but it is increasing rapidly. Micro-irrigation system has gained wide popularity in areas of acute water scarcity and in areas where horticultural crops are grown. In the world scenario, almost 54 per cent of the micro-irrigation is applied to fruit orchard and vegetable crops. This technology was introduced in early seventies in India but large adoption of it was in eighties particularly in fruit, vegetable crops. Now, micro-irrigation is becoming increasingly popular in areas of water scarcity. In India almost 1.5 lakh hectares of fruit crops and 2.5 lakh hectares of vegetable crops are being added annually under micro-irrigation. As per the estimate of Ministry of Agriculture and Farmers Welfare, GoI about 27 million ha in the country may be brought under micro-irrigation.

Another advantage of micro-irrigation is that it is capable of delivering precise amount of nutrients to the soil in measured quantities at any time at no additional cost. This referred as fertigation. Application of water and fertilizer through micro-irrigation improves water and nutrient use efficiency. Fertigation management is aimed at maximizing farmer's income and minimizing environmental pollution. Micro-irrigation fertigation offers various advantages such as: easy application of amount and concentration of nutrients suited to the crop according to its stage of development and its climatic conditions; reduces the salinisation and groundwater pollution; reduces fluctuation in nutrient concentration in soil during the crop growing season; easy use of solid as well as balanced liquid fertilizer and micronutrients (Bar-Yosef, 1999); no wetting of crop foliage thus controlling the attack of pathogens (Yarwood, 1978); precise application of nutrient thus minimizing the leaching of fertilizer out of the root zone; application of water and nutrient in the root zone enhance water and fertilizer application efficiency and reduce leaching of nutrient to groundwater; fertilizer application is unaffected by wind and no surface runoff loss of fertilizer; easy availability of nutrients to the plant roots. Under micro-irrigation fertigation the spatial distribution of soil moisture and consequently crop root zone is restricted to small volume of soil directly below emitter (Haynes, 1990). Studies carried out worldwide indicate that fertilizer savings through drip fertigation can be in the rage of 25 to 30 per cent.

Materials for Fertigation

All chemicals applied through micro-irrigation system must avoid corrosion, softening of plastic pipe and tubing, or clogging any component of the system. It must be safe for field use, must increase or at least not decrease crop yield, must be soluble or emulisifiable in water; and it must not reach adversely to salts or other chemicals in the irrigation water. In addition, the chemicals or fertilizers must be distributed uniformly throughout the field. Uniformity of distribution requires efficient mixing, uniform water application and knowledge of the flow characteristics of water and fertilizer in the distribution lines. To avoid clogging, chemicals are applied through micro-irrigation systems, to dissolve the deposits in drip lines. Currently, urea ammonium nitrate (UAN) which is 100 per cent soluble is also available, which could be easily used for micro-irrigation fertigation.

Design and Management of Fertigation System

Micro-irrigation fertigation is an efficient for irrigation and fertilizer application in a variety of crops and farmers using this method, because of its higher water and N use efficiency and higher yield. Efficiency of this method can further be increased with proper design and operation. This can be achieved by increase knowledge of crop nutrient requirement, soil nutrient supply, fertigation technique, irrigation scheduling and nutrient leaching pattern under various conditions in soil. A properly managed micro-irrigation fertigation system reduces overall fertilizer application rate, better nutrient supply to the plants and minimizes the adverse effect of fertilizer application on soil, water and environment. Better control mechanism available in this system also results in timely application and uniform distribution of nutrients in the soil. Fertilizer application through micro-irrigation fertigation system reduces fertilizer use and minimize ground water pollution due to the fertilizer leaching from rain or excess irrigation (Hochmuth, 2000; Hartz, 1993). The review of literature suggests that soil water, solute propagation and soil profile characteristic are often not adequately incorporated in the design and operational of micro-irrigation fertigation system.

Appropriate design of micro-irrigation fertigation system requires detailed knowledge of water and nutrient distribution pattern in the root zone, nutrient availability in the vicinity of roots and nutrient leaching below the root zone, which is the function of discharge of emitter and soil hydraulic and physical properties. Major objectives of the design of micro-irrigation system are to control the quantity of water passing though the root zone by regulating the emitter discharge or selecting appropriate size of the emitter, minimizing deep percolation, leaching of nutrients beyond the root zone and keeping the water and nutrients distribution near to the roots. Hydraulic properties of the soils play a critical role in water and nutrient distribution in the soil.

Development of design and management options for micro-irrigation fertigation systems involves study of water and nutrient dynamics using an appropriate model under different boundary conditions including flux, no-flux and free drainage boundary. Many studies have been carried out to investigate the water and nutrient distribution under drip irrigation with or without crops. These studies need to be extended to include field experimental data and modelling to develop different operational scenarios leading to appropriate operation and management options. Before using model for simulating water and nutrient distribution, it need to be calibrated with observed field data. In Indian conditions sufficient studies have not been conducted to develop crop-specific design and operation of fertigation system to minimize the leaching considering the N distribution in the soil and uptake by plant in particular with horticultural crops. Studies conducted elsewhere also suggest the lack of operational guidelines on this aspect (Gardenas *et al.*, 2005).

Water and Nitrogen Transport in Micro-irrigation Fertigation

The water and nitrogen transport need to be determined under different fertigation levels in horticultural crops. To achieve best results from fruit crops water content in the root zone is to be maintained in the range of field capacity of the soil. In all the types of micro-irrigation fertigation, nitrogen availability in the active root zone should be such that adequate nitrogen availability should be there in the active root zone, even 72 h after fertigation. Nitrogen balance studies indicated that leaching of nitrogen beyond the root zone increased with the level of nitrogen applied in micro-irrigation. All these results or recommendations are available for vegetable crops however, similar results needs to be developed for horticultural crops.

Development of Design and Operation Guidelines for Micro-irrigation Fertigation System

The calibrated and validated HYDRUS 2D model could be used for simulation of hypothetical scenarios generated through combinations of various emitter discharge rates, soil types and fertigation levels in horticultural crops. The criteria for best selecting guidelines would be have least vertical drainage or leaching of nutrients under micro-irrigation fertigation. For example, similar guidelines were developed by Sivasakti (2010) with garlic crop. The study simulated leaching of nitrogen under different scenarios in micro-irrigation fertigation. The highest leaching of nitrogen was observed in sand with 4 lph discharge rate and 120 kg N ha^{-1} of fertigation level followed by loamy sand, sandy loam, sandy clay loam, loam, silt silty loam, clay loam, silty clay loam, silty clay and clay. The negligible amount of leaching was observed in clay soil with 1 Lph discharge rate and 40 kg N ha^{-1} fertigation level followed by silty clay, silty clay loam and clay loam. From the leaching information under different emitter and discharge rate and fertigation level, it was concluded that irrespective of soil type, 1-2.5 lph emitter with 80 kg N ha^{-1} is best to reduce the leaching of nitrogen and free drainage of water in all the types of soil. This strategy will help to reduce the effect of nitrogen leaching in garlic crop. The results obtained in the study were used for development of design and operational guidelines for micro-irrigation fertigation in garlic crop. These guidelines will be useful for the farmers using micro-irrigation fertigation in garlic crop. Similar guidelines needs to be developed for horticultural crops.

Fertigation Research and Modeling in Horticultural Crops

Water and nutrient transport needs to be studied in field experiments in various horticultural crops under varying fertigation levels. Conducting field experiments in large number of soils with varying emitter discharge rates to investigate water and nutrient distribution for evolving appropriate design and operational guidelines is a costly and time consuming affair. A properly calibrated and validated flow and solute transport model can reduce time and cost required for studying the water and nutrient dynamics under micro-irrigation fertigation system. HYDRUS 2D model is an established model which needs to be calibrated and validated with the observed field data collected from field experiments and applied to simulate the water and nitrogen distributions from other soils. The information obtained from the HYDRUS 2D simulations is helpful in the design, operation, and management of micro-irrigation fertigation. Design and operation of micro-irrigation fertigation system requires more understanding of nutrient dynamics in case of horticultural crops, which are deep rooted and having more scope for nitrogen leaching. Although some information on micro-irrigation fertigation are being developed for vegetable crops however, very little information is available in horticultural crops. It is therefore essential to study and model the nitrate transport dynamics under micro-irrigation fertigation to develop design and operation guidelines.

Conclusion

In India, the growth of adoption of micro-irrigation system has taken place during last two decades in horticultural crops this technology to save irrigation water and enhancing the water use efficiency. Although, the micro-irrigation fertigation offers numerous advantages but it is not being used widely due to many reasons. There is lack of research information in respect to fertigation strategies such as its rate of application, amount to be applied and frequency to be adopted. However, the research efforts are being focused on this aspect in vegetable crops but there is a lack of information in respect to varied agro climatic conditions, soils and horticultural crops. There is need for developing detailed design and operational guidelines for micro-irrigation fertigation in horticultural crops. Also as a policy issue, In India there is a subsidy policy for normal NPK fertilizers in the specified grades. However, for fertigation the requirement of fertilizer is in different grades and it should be 100 per cent water soluble for its effective application. The fertigation material is either not available in desired form or it is available at higher price, than the conventional fertilizers. Once the micro-irrigation fertigation practice is being followed as per detailed operational guidelines the benefits of the technology and adoption could be enhanced.

References

Bar-Yosef, B. 1999. Advances in fertigation. *Adv. Agron.*, **65**: 1–75.

Gardenas, A.I., Hopman, J.W., Hanson, B.R. and Simunek, J. 2005. Two-dimensional modeling of nitrate leaching for various fertigation scenarios under micro-irrigation. *Agr. Wat. Mgt.*, **74**(3): 219–242.

Hartz., T.K., Strange, M. Le and May, D.M. 1993. Nitrogen requirement of drip irrigated pepper. *Hort. Sci.*, **28**: 1097–1099.

Haynes, R.J. 1990. Movement and transformations of fertigated nitrogen below trickle emitters and their effects on pH in the wetted soil volume. *Fert. Res.*, **23**: 105–112.

Hochmuth, G.J. 2000. Nitrogen management practices for vegetable production in Florida. Circular 1222. University of Florida. http://edies.Ifas. ufl.edu/BODY_CV237.

Sivasakti, S. 2012. Modeling of nitrate transport in micro-irrigation fertigation in garlic (*Allium sativum* L.). Unpublished *Ph.D. Thesis*, PG School, IARI, New Delhi

Yarwood, C.E., 1978. Water and the infection process. In: *Water Decit and Plant Growth*, Vol. 5, (Ed.) T.T. Kozolowski, Academic Press, New York, pp. 141–165.

40

Micro-Irrigation Systems for Enhancing Input Use Efficiency in Horticultural Crops

K.V. Ramana Rao*, Arpana Bajpai and Suchi Gangwar

ICAR-Central Institute of Agricultural Engineering,
Nabi Bagh, Berasia Road, Bhopal – 462 038, Madhya Pradesh
**E-mail: kondapalli.rao@icar.gov.in*

The annual horticultural production in India is about 149 Million tonnes. Vegetable cultivation is an important constituent of Indian agriculture and nutritional security due to their short duration, high yield, nutritional richness, economic viability and ability to generate on-farm and off-farm employment. With the diversified agro-climatic conditions coupled with distinct seasons, a wide varieties of vegetables can be grown in India. Today, India is the second largest producer of the vegetables (90.8 million tonnes) in the world, contributing 14.45 per cent of the total world production (NHB, 2013). Conventionally the vegetable growers adopt furrow irrigation practices for irrigating their crops, which has very low application efficiency and ranges between 40 to 60 per cent.

Micro-irrigation has emerged as an appropriate water application technique for horticultural cropsas well as agricultural crops. Drip irrigation is one such method receiving better acceptance and adoption, particularly in areas of water scarcity. Therefore, the efforts are now warranted to harness the available quantities of water and put them to efficient use to realize higher productivity per drop.This irrigation provides most effective way to supply water and nutrients to the plant, which not only saves water but also increases yield of vegetable crops (Nadiya *et al.*, 2013). Water saving in drip irrigation is due to the fact that maximum amount of water is stored in the root zone andsurface ponding, evaporation, seepage and deep percolation losses are minimized.Moisture is also maintained in the rootzone with low and more frequent application of water. Maintaining available soil moisture at low water tension and almost constant during entire growth period through micro-irrigation with considerable water saving upto 50 per cent was possible with micro-irrigation (Patel *et al.*, 2006). Uniform moisture content maintenance always in the root zone of the crop is possible by way of frequent irrigation at shorter intervals.In drip irrigation system, only a fraction of the soil surface generally between 15 to 60 per cent is wetted. Several experiments have shown positive responses in most of the crops to high frequency drip irrigation.Because of high frequency water application, concentrations of salts remain manageable in the rooting zone.

Fertigation is an effective crop cultural practice, by which fertilizers are applied along with irrigation water through drip system to get higher fertilizer use efficiency besides increasing the crop yields. Small applications of soluble nutrients save labour, reduces compaction in the field, thereby enhances crop productivity. Fertigation allows nutrient placement directly to the plant root zone during critical periods in the required quantities (Jat *et al.*, 2011). Application of high dose of fertilizers not only causes economic loss but also leads to chemical changes in the soil and reduces the yield. Fertilizer requirement can be reduced by 15-25 percent with fertigation through drip without affecting the yield.

Drip irrigation along with plastic mulching is effective in increasing the yield when compared either with surface irrigation or with drip irrigation standalone (Tiwari *et al.*, 2003). Mulching is used to cover soil surface around the plants to create congenial condition for the plant growth.Polyethylene mulches are widely used in the cultivation of vegetables, due to their ease in transport, availability and installation. With the adoption of mulching the desirable effects that can be expected are weed control, temperature moderation, salinity reduction *etc.* It also exerts decisive effects on earliness, yield and quality of the crop.Apart from weed control, application of mulch films along with drip irrigation enhanced the nutrient use efficiency of crops (Bharadwaj, 2013).

Water Use Efficiency

A well designed drip irrigation system benefits the environment by conserving water and can save 80 to 90 per cent of the water normally used in other types of irrigation systems. Another advantage of drip irrigation is that the water is applied more evenly throughout the field, thus eliminating the need to run the irrigation longer to wet the whole field (Carter and John Howell, 2000). Several research studies reported the enhanced crop yields with adoption of drip irrigation system. In capsicum (45.60 t/ha), tomato (67.0 t/ha), potato (28.95 t/ha), okra (23.9 t/ha), cucumber (44.70t/ha), chilli (2.74 t/ha) and pea (77.47 t/ha) better yields were reported over conventional practices (Table 1). A close perusal of this table indicates that about 10 to 60 percent increase in yield is possible with drip irrigation when compared with conventional method of irrigation.It can also be seen from the same table, compared to conventional, drip irrigation produced higher water use efficiency in capsicum (60.80 kg/ha-mm), tomato (80 kg/ha-mm), potato (41.36 kg/ha-mm), okra (29.89 kg/ha-mm), cucumber (89.40 kg/ha-mm), chilli (5.48 kg/ha-mm) and pea (193.50 kg/ha-mm) (Table 1).

From the literature reviewed it was found that under the conventional irrigation technique, additional amounts of water is being applied in capsicum (4508m^3/ha), tomato (2820 m^3/ha), potato (2520 m^3/ha), okra (3680 m^3/ha), cucumber (850 m^3/ha), chilli (550 m^3/ha) and pea (173 m^3/ha) as compared to drip irrigation.

Energy Use Efficiency

Water-energy productivity is the amount of yield that is obtained from a unit of energy (KWh) consumed during the irrigation. Irrigation operation is the major user of energy in agricultural production (Khan *et al.*, 2009). Energy for water pumping alone may be several times more than that of all other agricultural field operations. Energy requirements for agricultural production increaseswhen the water conveyance and application becomes inefficient; Furthermore, achievable energy conversion efficiency for lifting water is up to about 20 per cent in developed parts of the world but elsewhere it is to be nearer to 12.5 per cent.Therefore, inorder to have more sustainable energy utilization application of water through is the best way. More the energy utilized for lifting water, more would be the operational cost such as diesel or electricity. Rafiee *et al.* (2010) reported that Energy equivalents of water for

Table 1: Irrigation Parameters under Conventional Irrigation and Drip Irrigation for different Vegetable Crops

Crop	*Practices*	*Yield (kg/ha)*	*Water Requirement (mm)*	*Quantity of Water Applied (m³/ha)*	*The Total Energy Input (MJ/ha)*	*The Total Energy Output (MJ/ha)*	*Water Energy Productivity (kg/ KW h)*	*Water Saving over Conventional*	*References*
Capsicum	Conventional	18200	750	7500	7650	1456000	13.5	60.1%	Muhammad *et al.* (2016)
	Drip	45600	750	2992	3051.84	3648000	84.6		
Tomato	Conventional	20000	600	6000	6120	1600000	18.5	58.33	Makau *et al.* (2014)
	Drip	48000	600	3180	3243.6	3840000	83.85		
Potato	Conventional	18600	700	7000	7140	1488000	14.8	36.00	Fathia *et al.* (2014)
	Drip	28950	700	4480	4569.6	2316000	35.9		
Okra	Conventional	12860	800	8000	8160	1028800	8.9	46.00	Puneet *et al.* (2015)
	Drip	23910	800	4320	4406.4	1912800	30.7		
Cucumber	Conventional	37100	500	5000	5100	2968000	41.2	17.00	Yaghi *et al.* (2013)
	Drip	44700	500	4150	4233	3576000	59.8		
Chilli	Conventional	1990	500	5000	5100	159200	2.21	27.37	Gulshan *et al.* (2007)
	Drip	2740	500	4450	4539	219200	3.42		
Pea	Conventional	58700	400	4000	4080	4696000	81.5	24.00	Narayanamoorthy *et al.*
	Drip	77400	400	3040	3100.8	6192000	141.4		(2003)

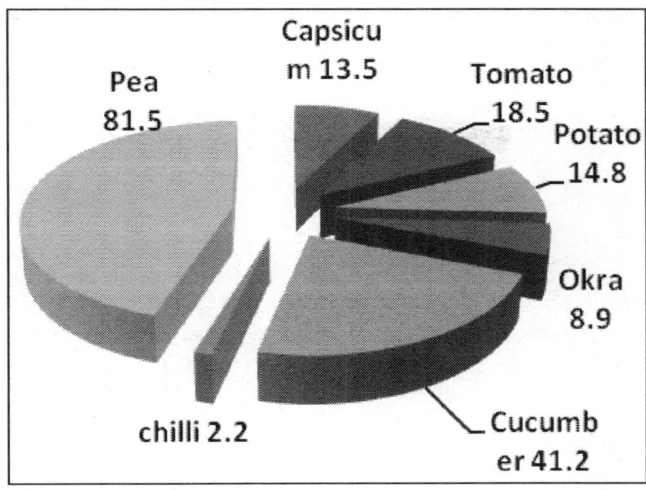

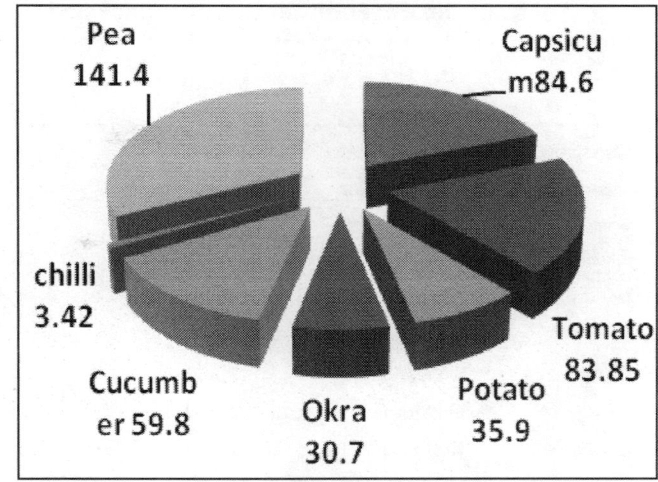

Figure 1: Water Energy Productivity (kg/KWh) under Conventional (a) and Drip Irrigation (b).

irrigation (m^3) is 1.02 MJ/ha and outputs in vegetable production was 0.80 MJ/ha (Ibrahim 2011). The total energy input were calculated as for capsicum (7,650 and 3,051 MJ/ha), tomato (6,120 and 3,243 MJ/ha), potato (7,140 and 4,569 MJ/ha), okra (8,160 and 4,406 MJ/ha), cucumber (5,100 and 4,233 MJ/ha), chilli (5,100, 4,539 MJ/ha), pea (4,080 and 3,100 MJ/ha) conventional and drip irrigation respectively. The total energy output were calculated as for capsicum (14,56,000 and 36,48,000 MJ/ha), tomato (16,00,000 and 38,40,000 MJ/ha), potato (14,88,000 and 23,16,000 MJ/ha), okra (10,28,800 and 19,12,800 MJ/ha) cucumber (29,68,000 and 35,76,000 MJ/ha), chilli (1,59,200, and 2,19,200 MJ/ha), pea (46,96,000 and 61,92,000 MJ/ha) conventional and drip irrigation respectively. Figure 1 shows that water-energy productivity was found to be highest under drip irrigation system for all vegetable crops.

Fertilizer Use Efficiency

Drip fertigation optimize the use of water and fertilizer enabling to harness high crop yield, simultaneously ensuring a healthy soil and environment. Crop growth and yields under drip irrigation can be lower than those achieved under conventional irrigation methods if fertilizer placement is not modified to meet the needs of drip irrigated crops. Fertigation can be affected by using single or multiple nutrient fertilizers in their solid or liquid form. Some of the desirable characteristics of the fertilizer material for use in fertigation are full solubility, quick dissolution in water, fine grained product, high nutrient content in the saturated solution, compatibility with other fertilizers, absence of chemical interaction with irrigation water and minimum content of conditioning agents. Fertilizer applied to the soil have to be dose the water source (emitter) in order to be used effectively by the crops. Fertigation enhances the overall root activity, improves the mobility of nutritive elements and their uptake, as well as reducing the contamination of surface and groundwater. Hebbar *et al.* (2004) suggested that for fertigation, use of 100 per cent water soluble fertilizer is recognized to safe guard the drip system in a long run.A properly designed drip fertigation system delivers water and nutrient at a rate, duration andfrequency, so as to maximize crop water and nutrient uptake, while minimizing leaching of nutrients and chemicals from

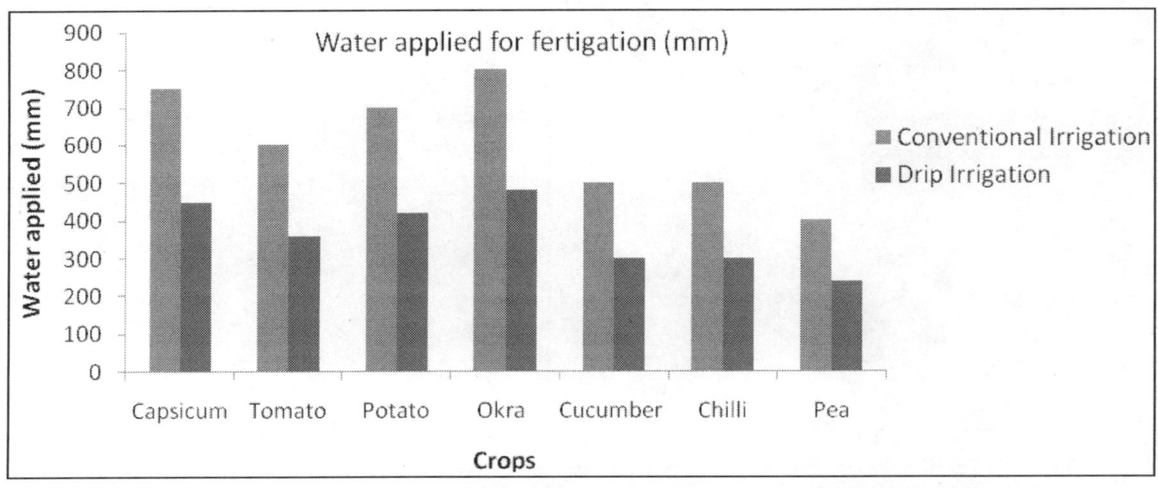

Figure 2: Water Applied for Fertigation Drip over Conventional Method of Irrigation.

Table 2: Fertilizer Use Efficiency under Conventional Method and with Drip Irrigation

Crops	*Practices*	*Yield (kg/ha)*	*Total Quantity of Nutrient Applied (kg/ha)*			*FUE (%)*
			N	*P*	*K*	
Capsicum	Conventional	18,200	250	150	150	33.09
	Drip Irrigation	45,600				82.91
Tomato	Conventional	20,000	250	150	250	30.76
	Drip Irrigation	48,000				103.08
Potato	Conventional	18,600	120	240	120	38.75
	Drip Irrigation	28,950				60.31
Okra	Conventional	12,860	200	100	100	32.15
	Drip Irrigation	23,910				59.78
Cucumber	Conventional	37,100	150	75	75	123.67
	Drip Irrigation	44,700				149.00
Chilli	Conventional	1,990	120	80	80	6.78
	Drip Irrigation	2,740				9.78
Pea	Conventional	58,700	60	80	70	279.52
	Drip Irrigation	77,400				368.57

the root of agricultural field. Application of NPK from liquid fertilizer significantly increased the concentration and total uptake of N, P and K in vegetable crop over solid fertilizer. Study conducted by Patel *et al.* (2001) indicated that drip fertigation resulted in approximate 40 per cent saving of water (Figure 2).

Analysis of reviewed data indicated that the drip irrigation exhibited higher fertilizer use efficiency for all vegetable crops compared to conventional irrigation (Figure 3). Stewart 2005 concluded that the higher fertilizer use efficiency (FUE) in drip irrigation might be due to the reason that fertilizer fixation was delayed when fertilizer applied with drip irrigation and plants absorbed nutrient more quickly and directly from the soil solution. He also concluded that the positive effect of fertigation was enhanced due to optimum soil moisture content in the soil, which facilitated maximum utilization of applied nutrients to crops.

Labour Use Efficiency

Irrigation systems are often designed to maximize efficiencies and minimize labour and capital requirements. The most effective management practices are dependent on the type of irrigation system and its design. Drip system has high capital and operating costs but may utilize minimal labour and conserve water. Their use tends toward high value cropping patterns. Conventional systems are relatively less expensive to construct and operate but have high labour requirements. The major capital expense of the surface system is generally associated with land grading. Recent developments in surface irrigation technology have largely overcome the irrigation efficiency advantage of micro-irrigation systems. An array of automating devices roughly equates labour requirements. Comparison of labour costs for drip and conventional irrigation method

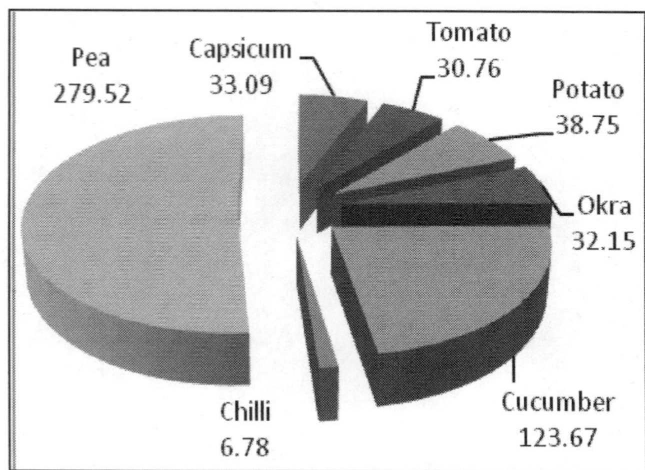

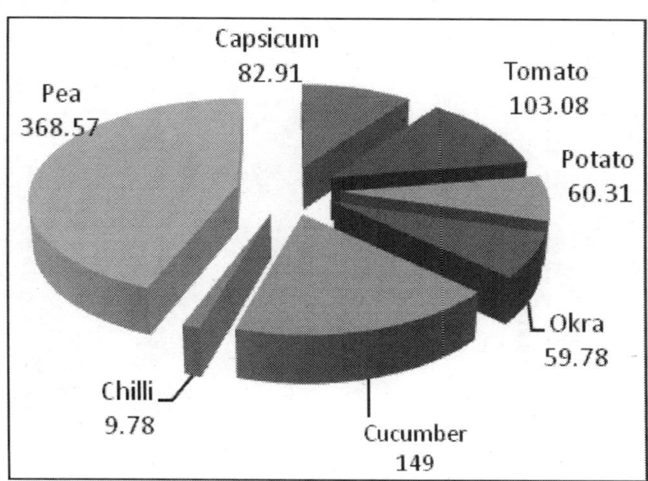

Figure 3: Fertilizer Use Efficiency (per cent) under Conventional (a) and Drip Irrigation (b).

(Table 3), the maximum cost savings are possible in drip irrigation, i.e capsicum (25 per cent), tomato (25 per cent), Potato (19 per cent), okra (13.8 per cent), cucumber (15.7 per cent), chilli (21.8 per cent) and pea (19.23 per cent) over conventional irrigation.

Table 3: Labour Requirement in Conventional and Drip Irrigation Practice

Vegetable	*Conventional (Days)*	*Drip Irrigation (Days)*	*Labour Saving over Conventional*	*Reference*
Capsicum	185	160	25%	Khatri *et al.* (2011)
Tomato	194	169	25%	
Potato	124	105	19%	Kaur *et al.* (2013)
Okra	185	160	13.8%	Esmatet al. (2011)
Cucumber	121	102	15.7%	Everaarts *et al.* (2015)
Chilli	128	100	21.8%	Jagtap (2012)
Pea	130	105	19.23	Everaarts *et al.* (2015)

Conclusion

Agricultural sector is the largest consumer of water. The overall efficiency of the flood irrigation system ranged between 25 to 40 per cent. To assure food security of ever increasing population under depleting water resources, adoption of micro-irrigation systems in almost all the horticultural crops and some of the agricultural crops is the viable solution. As Micro-irrigation Systems helps in saving significant amounts of water and increases the quality and quantity of produce to achieve 'more crop per drop'. In this system, irrigation events are fine-tuned to spoon feed water and nutrients just in time to avoid plant stress for enhanced crop productivity. These systems proved to be effective not only in water saving but also in fertilizer, labour and energy saving. Therefore, wider promotion this technology amongst the farming community in India, would help in maximizing the input (water, fertilizer, labour and energy) use efficiency.

References

Bharadwaj, R.L. 2013. Effect of mulching on crop production under rainfed condition: A review. *Agril. Rev.*, **34**: 188–197.

Esmat, A.B., NozrulIslam, M.D., Alam, Q.M. and Byazid Hossain, S.M. 2011. Profitability of some *bari* released crop varieties in some locations of Bangladesh. *Bangladesh J. Agril. Res.*, **36**(1): 111–122.

Everaarts, H. and Maerere, A.P. 2015. Profitability, labour input, fertilizer application and crop protection in vegetable production in the Arusha region, Tanzania A.P. Wageningen, p. 1–40.

Fathia E.M., Kamel, N., Mohamed, M.M. and Netij, B.M. 2014. Effects of surface and subsurface drip irrigation regimes with saline water on yield and water use efficiency of potato in arid conditions of Tunisia. *J. Agri. Env. Int. Dev.*, **108**(2): 227–246.

Gulshan, M., Singh, K.G., Sharda, R. and Siag, M. 2007. Response of red hot pepper to water and nitrogen under drip and check basin method of irrigation. *As. J. Pl. Sci.*, **6**(5): 815–820.

Hebbar, S.S., Ramachandrappa, B., Nanjappa, H.V. and Prabhakar, M. 2004. Studies on NPK drip fertigation in field grown tomato (*Lycopersicon esculentum* Mill.). *European J. Agron.*, **21**: 117–127.

Ibrahim, H.Y. 2011. Energy use pattern in vegetable production under fadama in north central Nigeria. *Tro. Subtro. Agroecol.*, **14** : 1019–1024.

Jagtap, P.P., Wagh, K.K., Shingane, U.S. and Kulkarni, K.P. 2012. Economics of chilli production in India. *Afri. J. Bas. Appl. Sci.*, **4**(5): 161–164.

Jat, R.A., Wani, S.P., Sahrawat, K.L., Singh, P. and Dhaka, B.L. 2011. Fertigation in vegetable crops for higher productivity and resource use efficiency. *Indian J. Fert.*, **7**: 22–37.

Kaur, J., Quadri, J., Ahmad, P., Nafees, A., Chesti, M.H., Ahmad, H.S. and Bhat, B.A. 2013. Study on economics of potato growing towards livelihood security. *African J. Agril. Res.*, 8(45): 5639–5644.

Khan, S., Khan, M.A., Hanjra, M.A. and Mu, J. 2009. Pathways to reduce the environmental footprints of water and energy inputs in food production. *Food Policy*, **34**(2): 141–149.

Khatri, R.T., Mistry, H. and Patel, K.S. 2011. Comparative economics of production of important vegetables in Surat district international research. *J. Agril. Econ. Stat.*, 2(1): 58–62.

Makau, J.M., Masinde, P.W., Home, G.P., Njoroge, C.K. and Mugai, E.N. 2014. Growth and yield of tomato under alternate furrow irrigation. *African J. Hort. Sci.*, **8**: 24–37.

Muhammad, A., Malik Muhammad, A., Rana Muhammad, A. and Muhammad Asim, R. 2016. Impact of drip and furrow irrigation methods on yield, water productivity and fertilizer use efficiency of sweet pepper (*Capsicum annuum* L.) grown under plastic tunnel. *Sci. Lett.*, **4**(2): 118–123.

Nadiya, N., Kurien, E.K., Mathew, E.K. and Varughese A. 2013. Impact of fertigation and drip system layout in performance of chilli (*Capsicum Annum*). *Int. J. Engg. Res. Dev.*, **7**: 85–88.

Narayanamoorthy, A. and Deshpande, R.S. 2003. Irrigation development and agricultural wages: An analysis across states. *Econ. Polt. Week.*, **38**(35): 3716–3722.

Patel, J.C. and Patel, B.K. 2001. Response of potato to nitrogen under drip and furrow methods of irrigation. *J. Indian Potato Assoc.*, **28**(2–4): 293–295.

Patel, R.S., Patel, P.G., Patel, J.C. and Patel, M.M. 2006. Effect of irrigation and nitrogen on growth and yield of *brinjal* under drip system. *J. Maharashtra Agri. Univ.*, **31**: 173–175.

Rafiee, S., Mousavi Avval, S.H. and Mohammadi, A. 2010. Modeling and sensitivity analysis of energy inputs for apple production in Iran. *Energy*, **35**: 3301–3306.

Stewart, W.M., Reiter, J.S. and Krieg, D.R. 2005. Cotton response to multiple applications of phosphorus fertilizer. *Better Crops*, **89**: 18– 20.

Tiwari, K.N., Singh, A., Singh, A. and Mal, P.K. 2003. Effect of drip irrigation on yield of cabbage under mulch and non-mulch conditions. *Agril. Wat. Mgt.*, **58**: 19–28.

Managing Emerging Insect Pests

41

Biological Control of Pests in Horticultural Crops to Improve Profitability and Quality

Chandish R. Ballal* and Richa Varshney

ICAR-National Bureau of Agricultural Insect Resources,
Hebbal, Bengaluru – 560 024, Karnataka
**E-mail: ballalchandish@gmail.com*

An arthropod pest or weed exists in nature along with a pool of natural enemies, which can comprise of vertebrates, invertebrates, and microorganisms. Plant disease antagonists also exist in nature, which can be used for control of fungal or bacterial diseases. The fundamental focus in applied biological control should be to select an appropriate species or combination of species from this pool and to work on a strategy to bring out the desired level of pest suppression with minimal impact on non-target species. More than one-and-half million insect species occur in this world, out of which only about 1.0 per cent have attained the status of pests. Many species have pestilent potential, yet remain at low levels because of the perpetual regulatory action exerted on them by their natural enemies. This in itself reflects the great potential of biological control, which can be exploited for management of some of our major pests, diseases and weeds by restoring the natural balance through purposeful human interventions (Bale *et al.*, 2008).

In India, several parasitoids, predators and microbials, *viz.*, trichogrammatids, bethylids, chrysopids, ladybird beetles, nuclear polyhedrosis viruses, entomofungal pathogens, *Bt*, *etc.* have been identified, mass produced, evaluated and recommended for field releases against pests of fruit crops like grapevine, citrus, papaya and apple and vegetable crops like tomato, *bhendi*, cabbage and brinjal. To support such bio-control programmes, mass-production of natural enemies followed by a strategy for field utilization are necessary. ICAR-National Bureau of Agricultural Insect Resources, Bengaluru, has standardized several promising technologies, which can be used for bio-control based management of crop pests and diseases. Several bio-intensive pest management modules developed by NBAIR for management of pests lay emphasis on release of bio-control agents *viz.*, parasitoids, predators and pathogens and reducing chemical pesticide applications to the possible extent. The pre-requisite of any bio-control programme is to have an ensured source for large-scale supply of beneficial agents. There are several success stories in the field of bio-suppression of crop pests in our country. Biological control has gained maximum acceptance in India through use of *Trichogramma* species against several lepidopteran pests *viz.*, *Helicoverpa armigera*, *Leucinodes orbonalis*, *Cydia pomonella* and *Plutella xylostella* attacking horticultural crops. Augmentative releases of the Australian lady bird beetle, *Cryptolaemus montrouzieri* have led to successful suppression of mealybugs on grapevine, guava and custard apple.

Potential Bio-agents for Pest Management in Horticultural Crops

Use of *Trichogramma* spp. and other Parasitoids

The egg parastioids of the genus *Trichogramma* have been widely used against many lepidopteran pests for several years. Since their widespread use, these are considered the *Drosophila* of the parasitoid world as they have been used for inundative releases. Several exotic *Trichogramma* spp which have proved effective through augmentation include, *T. pretiosum* (thelytokous and arhenotokous forms) (Origin: USA) against *Helicoverpa armigera* infesting tomatoes; *T. embryophagum* (Origin: Russia) against *Cydia pomonella* infesting apple and *Trichogrammatoidea bactrae* (Origin: Taiwan) against *Plutella xylostella* infesting cruciferous crops. Other pests that are controlled by this egg parasitoid include *Papilio* spp., *Virachola isocrates, Leucinodes orbonalis, etc. Trichogramma* species are mass produced in Indian insectaries on eggs of rice meal moth, *Corcyra cephalonica*, which in turn is multiplied on sorghum/bajra/maize grains following standard protocol developed by NBAIR. The parasitoids are released @ 50,000 to 100,000/ release/ha and are released 3-8 times on different crops.

Recent studies indicate that the production of *T. chilonis* on eri silkworm *Samia cynthia ricini* eggs is a farmer friendly system (Lalitha *et al.*, 2013) and it could potentially yield trichogrammatids with superior biological attributes. The quality of *T. chilonis* mass produced on eri silkworm eggs was assessed in the field against early shoot and internode borers

Research focus at NBAIR has also been on identification/development of parasitoids and predators tolerant to high temperature and resistance to pesticides. A strain of an egg parasitoid *T. chilonis* was developed for tolerance to multiple insecticides belonging to six different groups, *viz.*, organochlorine, organophosphate (monocrotophos), pyrethroids (λ cyhalothrin), oxadiazine (indoxacarb), neonicotinoids (imidacloprid) and spinosyn (spinosad) (Jalali and Singh, 1993, Jalali *et al.*, 2006). The strain developed can be integrated with farmers' practice, particularly on vegetable crops where there is an indiscriminate use of insecticides. The resistance factor of this strain ranged from 3.5 to 175-fold over susceptible populations for different insecticides. Field demonstrations of this multiple insecticides tolerant strain of *T. chilonis* were carried out in Patiala (Punjab) on tomato and in Dharmapuri (Tamil Nadu) on tomato and brinjal. The results of the trial in Patiala indicated that insecticide sprays could be reduced by 25 per cent, per cent bored fruits by 35 per cent and an increase in yield of 8.16 q/ acre was recorded resulting in an additional revenue of Rs. 13,872/acre. At Dharmapuri, 40 and 65 per cent reduction of insecticide sprays was recorded on tomato and brinjal, respectively, and after the 4th release of parasitoids, no larval population was recorded in the parasitoid released fields. The increase in the yield was 2400 kg/acre, resulting in additional revenue of Rs. 20,000/acre. In Srinagar, releases of *Trichogramma embryophagum* (@ 2000 parasitised eggs or 1/8 of a tricho-card per tree) in combination with chlorpyriphos significantly reduced the infestation by codling moth on apple trees.

Encyrtid nymphal parasitoid, *Leptomastix dactylopii* introduced from West Indies in 1983 successfully established on the common mealybug, *Planococcus citri* infesting citrus in south India, thus almost completely eliminating sprays of chemical insecticides.

The papaya mealybug *Paracoccus marginatus* W and G was first recorded on papaya plants from Coimbatore in 2008 and later spread to different states, *viz.*, Kerala, Karnataka, Maharashtra and Tripura. Chemical pesticides and indigenous natural enemies could not keep the papaya mealy bug population under check. NBAIR imported three species of parasitoids *Acerophagus papaya, Pseudoleptomastix mexicana* and *Anagyrus loecki* (from USDA-APHIS at Puerto Rico), which are known to effectively suppress the papaya mealy bug in its native range (Shylesha *et al.*, 2010). The parasitoids could be successfully multiplied and supplied to stake-holders all over the country. Inoculative releases of the parasitoids, especially *A. papaya* were made in farmers' fields in different villages. The parasitoids could successfully establish in all the areas of release and suppress the papaya mealybug infestation on papaya and other horticultural crops. *A. Acerophagus* was the most successful parasitoid. Recent surveys have indicated that *P. mexicana* has also established well in the released areas and is spreading to nearby areas too.

The spiraling whitefly, *Aleurodicus dispersus*, was first reported in 1993 from Kerala and later from other parts of peninsular India and the Lakshadweep islands. The pest is highly polyphagous and has been recorded on 253 host plants in India, including several polyhouse crops. Two aphelinid parasitoids, *Encarsia guadeloupae* and *E.* sp. nr. *meritoria*, have been fortuitously introduced together with the host into India. With the accidental introduction of both species of *Encarsia* into India, there has been a perceptible reduction in the population of *A. dispersus* (Ramani *et al.*, 2002). These parasitoids can be effectively used for management of the spiralling whitefly.

Indigenous parasitoids play a major role in the management of the coconut black-headed caterpillar in the coconut ecosystem. Field release of the three stage specific *Opisina arenosella* parasitoids *viz., Goniozus nephantidis, Elasmus nephantidis* and *Brachymeria nosatoi* at fixed norms and intervals in a heavily infested coconut garden for a period of five years resulted in highly significant reduction in *Opisina* population. Follow up observations revealed that even after three years no build up of the pest was noticed in the released site.

Recent outbreaks of the litchi stink bug, *Tessaratoma javanica* (Thunberg) (Hemiptera: Tessaratomidae), in Jharkhand has generated a pressing need for developing suitable mass rearing protocols for the parasitoids associated with this pest. *Anastatus acherontiae* Narayanan, Subba Rao and Ramachandra Rao (Hymenoptera: Eupelmidae) and *A. Bengalurunsis* Mani and Kurian, egg parasitoids of *T. javanica*, were reared successfully on eggs of eri silk worm (Gupta *et al.*, 2013). This is a technology which can be adopted for rearing and utilising the indigenous parasitoids for biological control of the litchi stink bug.

Use of Predators

Cryptolaemus montrouzieri

Cryptolaemus montrouzieri is being effectively used against mealybugs infesting vegetables and ornamentals grown in polyhouses. In fruit and plantation crops, the beetles are released @ 5-50 per plant, depending upon the severity of infestation and crop canopy. On each mealy bug infested plant of coorg mandarin, robusta coffee, *arabica* coffee and san ramon coffee release of 10, 5, 3 and 2 beetles per plant resulted in significant reduction of mealy bug population and by 5th week the pest population reduced to negligible level. Field releases of *C. montrouzieri* @ 20 adults per tree gave excellent control of different species of mealy bugs *viz.*, *Ferrisia virgata*, *Maconellicoccus hirsutus* and *Planococcus lilacinus* on guava within 50 days in the presence of other local natural enemies (Mani *et al.*, 1990). It was also found to be highly effective in suppressing the populations of *M. hirsutus* in grapes within 75 days. This predator was found effective in suppressing the mealy bug populations on citrus, guava, grapes, mulberry, mango, pomegranate, custard apple, ber *etc.* and green shield scale on sapota, mango, guava, brinjal and crotons in Karnataka. It did not seriously impair the efficiency of local biocontrol agents.

Anthocorid Bugs

Thrips and mites are a serious threat to polyhouse crops and chemical insecticides have proved to be ineffective, also leading to problems of pest resurgence and development of resistant populations. Field releases of anthocorid predators in several western countries have indicated their effectiveness against several sucking pests on different crop ecosystems, especially on polyhouse crops. In India, anthocorid bugs have gained prominence recently. Several anthocorid bugs have been recorded for the first time as potential predatos of sucking pests. Their conservation and application in different crop ecosystems will be useful for the protection of crops from lepidopteran pests, aphids, thrips, mites, *etc.* (Ballal and Gupta, 2011). Rearing of anthocorids is a challenge if we have to rear them on thrips or mites. However, methods heve been developed to rear *Orius maxidentex* Ghauri, *Orius tantillus* (Motsch.) and *Blaptostethus pallescens* Poppius (Ballal *et al.*, 2003b) on alternate laboratory host eggs for *e.g. C. cephalonica* and *Sitotroga cerealella*. *B. pallescens* is highly amenable to rearing. Trials conducted at NBAIR and AICRP biocontrol centres clearly indicate that these anthocorids are potential bio-agents, which can be used for management of sucking pests in polyhouses. NBAIR conducted trials on evaluation of *B. pallescens* against *Frankliniella schultzei* infesting chilli in a commercial polyhouse. There was a 50 per cent reduction in the thrips population in the biocontrol plot in comparison to the chemical control plot which was subjected to 14 chemical applications. The yield and quality parameters of the harvested chillies from the biocontrol plot were comparable with those of the chemical control plot.

In Pune, Maharashtra, four releases of anthocorid predator *B. pallescens* on carnation at weekly intervals @ 20 per plant was found effective in reducing mite *T. urticae* population by 64 per cent in comparison to control. Similar releases against *T. urticae* infesting rose, resulted in 77 to 83 per cent reduction in the mite population compared to control. In Kerala, three releases of the anthocorid resulted in 97 per cent reduction in mite population on rose. In Srinagar, five releases of the anthocorid resulted in 88 per cent reduction in mite population on rose.

The anthocorid predator *Cardiastethus exiguus* has been observed to be highly amenable to mass production and it has also proved to be highly effective against the egg and neonate larval stages of *O. arenosella* as indicated in the field trials conducted at Kerala and Karnataka (Ballal *et al.*, 2003a). *Amphiareus constrictus* (Stål), a litter inhabiting anthocorid, biology of which has been less studied, was successfully reared in the laboratory on the eggs of *C. cephalonica* without any plant substrate. The mass reared *A. constrictus* was evaluated against the tomato pinworm *Tuta absoluta* (Meyrick) and the brown plant hopper *Nilaparvata lugens* Stål and was found to be a promising bioagent.

Use of Microbials

Bt and Entomopathogenic Fungi

Bio-control technologies have proved to be effective in management of pests attacking vegetable crops. *Bacillus thuringiensis* (@ 2 kg/ha) is effective for the management of *Earias* sp. on okra, brinjal fruit borer and diamondback moth on cabbage in Punjab. In Himachal Pradesh, *Bt* subsp. *kurstaki* at 1 kg/ha with EcoNeem Plus at 2 ml/l was effective in suppressing early instar larvae of *Pieris brassicae* on cole crops. The module comprising of two sprays of *Bt* and 6 releases of egg parasitoid *Trichogramma brassicae* could significantly reduce DBM population on cabbage, with an increase in marketable yield in Karnataka. Use of bio-agents – *Bt* @ 0.5 ml and 1 ml/l or *Metarhizium anisopliae* @ (1x10^9 spores/ml) at weekly intervals - resulted in reduction in larval population and damage by *Diaphania indica* on gherkins in Karnataka.

Fungal pathogens have also been evaluated against insect pests in polyhouse conditions (Ramanujam, 2007). Spraying of *Verticillium lecanii* (10^{10} conidia/l) was effective against thrips on chilli in polyhouses in Pune with a 71 per cent reduction in thrips population in comparison to control. In another trial in Pune, *Metarhizium anisopliae, V. lecanii, Beauveria bassiana* and *Hirsutella thompsoni* were found to be equally effective and they could bring about a 55 to 63 per cent reduction in thrips population in comparison to control. In Kerala, after three sprays, *V. lecanii* (10^{10} conidia/l), *B. bassiana* (10^{10} conidia/l) and *H. thomsoni* (10 g/l) there was 88 to 97 per cent reduction in aphid population. After three sprays of *M. anisopliae, V. lecanii, B. bassiana* and *H. thompsoni* there was 85 to 98 per cent reduction in population of thrips (*Frankliniella* sp.) on Gerebera in Kerala. In Pune, *H. thompsonii* @ 10 g formulation/lit was on par with three sprays of abamectin 0.5 ml/lit in suppressing mite population (*Tetranychus urticae* Koch.) infesting polyhouse rose. In Tamil Nadu, *V. lecanii* and *B. bassiana* could reduce aphid incidence significantly (82 to 86 per cent reduction) on carnation. Recently, some of the isolates of entomofungal pathogens are being tested for their ability to establish as endophytes in crop plants and to exploit their endophytism as a novel delivery mechanism for managing the insect pests especially borer pests. In cabbage and cauliflower, isolates of *B. bassiana* (Bb-5a, Bb-45 and Bb-68) and *M. anisopliae* (Ma-4, Ma-6 and Ma-35) have been established as endophytes and are being tested against *Plutella xylostella* (B. Ramanujam, Personal Comm.).

Entomopathogenic Nematodes and Antagonists

Utilisation of entomopathogenic nematodes (EPNs) to manage white grub incidence has proved to be successful and NBAIR has developed wettable powder formulations for EPNs with a shelf-life of 8-10 months. Methods are available to incorporate or deliver appropriate doses at appropriate time, which would be suitable to specific crop and pest. These technologies were validated in white-grub endemic areca farms in Sulya, Bandkal, Shringeri areas and transferred to eight DSIR- recognized companies across the country on non-exclusive basis. Through front line demonstrations, 1500 acres of whitegrub-infested areca and cardamom farms in Karnataka and Kerala were covered. Solid-state and diphasic fermentation and post-production technologies for contamination-free antagonistic fungi for the management of root-knot and cyst nematodes have been standardised (Nagesh *et al.*, 2007, Nagesh and Hussaini, 2013). NBAIR has filed applications for grant of Indian patent rights for WP formulations of EPN and *Paecilomyces lilacinus* (Indian Patent application No. CHE/3490/2010, CHE/2664/2010 and CHE/1310/2012) and Indian patent right has been granted to NBAIR on "A simple and novel design for small-scale solid state mass production unit for antagonistic fungi" (Indian Patent No. 275009).

Use of Predatory Mites

Phytophagous mites are serious pests of field and polyhouse crops causing crop losses in terms of quantity and quality of produce. Spider mites and broad mites can cause serious problems to polyhouse crops. Development of pesticide resistance has aggravated the problem of mites. Predatory mites are used efficiently to manage phytophagous mites in polyhouses, especially in the Mediterranean regions.

The exotic predatory mite *Phytoseiulus persimilis* and indigenous ones *Amblyseius tetranychivorus, Neoseiulus longispinosus* and *Neoseiulus barkeri* are efficient predatory mites for targeting phytophagous mites and these indigenous mites are amenable to mass rearing. Novel and farmer friendly techniques have been developed at NBAIR to mass rear some of the potential predatory mites. This technology, which is ready for commercialization can be a safe and solution for management of spider mites infesting polyhouse crops. In Solan, three releases of *N. longispinosus* proved to be the best for management of *T. urticae* on capsicum. This treatment was as effective as three sprays of Profenophos. Evaluation trials conducted in other countries like Israel and Netherlands indicate that predatory mite *Amblyseius swirski* is an effective bioagent of whiteflies. In India, there is a need to search for predatory mites, which can be used for targeting thrips and whiteflies.

Reports have clearly indicated the compatibility of some of the anthocorid predators and predatory mites in IPM programmes. Preliminary trials have indicated that indigenous anthocorid predator *B. pallescens* can also target pupal stages of thrips in soil. Hence, it would be worthwhile to investigate an integrated approach by utilising predatory mites and anthocorid predators for management of thrips and mites in polyhouses.

Disease Management through Antagonists

The management of plant diseases using biocontrol agents has been emphasized during the last two decades. Among the fungal antagonists, formulations of *Trichoderma* spp. play an important role. For the seed borne diseases of horticultural crops especially damping off, caused by *Pythium, Phytophthora, Rhizoctonia etc.* the use of seed treatment with *Trichoderma* formulations as well as use of *Trichoderma* cocopeat in nursery have proved to be very useful. Soil borne diseases in the main field crops can be managed through application of *Trichoderma* enriched farm yard manure (1:100 w/w) for one acre (Ramanujam and Sriram, 2007). The use of *Trichoderma* or *Pseudomonas* has been found very effective in reducing the quick wilt in pepper caused by *Phytophthora capsici*. The isolates of *Trichoderma* species that produce more hydrolytic enzymes like chitinase and glucanase have been identified. Similarly, induced systemic resistance is another developing area and at

NBAIR, *Trichoderma* isolates that induce the systemic resistance in chilli and tomato have been identified. *Trichoderma virens* isolates with more expression of Sm1 protein that elicits ISR have been identified to reduce infection by *Phytophthora infestans* in tomato and *P. capsici* in chilli.

The shelf-life of *Trichoderma* formulations is of paramount importance considering the gap between the production and end use by farmers. Research efforts at NBAIR have led to interventions like addition of glycerol at 3 per cent, colloidal chitin at 0.2 per cent in production medium or chitin in formulation at 2 per cent, heat shock at the end of log phase for 30 min at 40°C to enhance the shelf-life of *Trichoderma harzianum* formulation. The solid state derived formulations of *T. harzianum* have been developed, which are more desiccation tolerant and with a shelf-life of 12 months. The salinity tolerant (upto 1.5 M NaCl concentration) and carbendazim tolerant *T. harzianum* (200 ppm) can be effectively employed in chemical pesticide free farming. The liquid formulation of *P. fluorescens* and invert-emulsion formulation of *Trichoderma* can be applied for the successful management of foliar diseases infecting several crops.

A promising plant growth promoting strain of *Bacillus megaterium* (NBAII EXB53) has been identified which can promote growth in different chilli varieties and other vegetable crops like capsicum, tomato, egg plant, cauliflower and cabbage. Seed treatment with talc based formulation of *B. megaterium* NBAII EXB53 (2×10^8 cfu g^{-1}) at the rate of 10g/kg seeds increased the seedling vigor index from 20 -22 per cent in these crops. Liquid and oil formulations for promising antagonists, *T. harzianum* (Th-3) and *B. megaterium* (EXB-5) with a shelf-life of 6-9 months were also developed for biological control of chilli anthracnose disease.

Proven, well-tested biocontrol technologies are now available for management of several of the pests and diseases infesting horticultural crops. However, commercial availability of quality bioagents is a problem in many of the states in our country. Producers of biological control agents, researchers, state departments and NGOs should sincerely facilitate the spread of these farmer-friendly technologies and work towards making the bioagents available to farmers.

References

Bale, J.S., Van Lanteren, J.C. and Bigler, F. 2008. Biological control and sustainable food production. *Phil. Trans. R. Soc. B.*, **363**: 761–776.

Ballal, C.R. and Gupta, T. 2011. Anthocorids as potential predators. pp. 92–111. **In**: *Plant Protection through Eco-friendly Techniques*, (Eds.) S.C. Bhardwaj, S. Singh and A. Bhatnagar. Pointer Publishers, Jaipur, 272 pp.

Ballal, C.R. Singh, S.P., Poorani, J. and Gupta, T. 2003a. Feasibility of mass multiplication and utilization of *Cardiastethus exiguus* Poppius, a potential anthocorid predator of *Opisina arenosella* Walker (Lepidoptera : Oecophoridae). pp. 29–33. **In**: *Biological Control of Lepidopteran Pests*, 354 pp.

Ballal, C.R., Singh, S.P., Poorani, J. and Gupta, T. 2003b. Biology and rearing requirements of an anthocorid predator, *Blaptostethus pallescens* Poppius (Heteroptera : Anthocoridae). *J. Biol. Cont.*, **17**(1): 29–33.

Gupta, A., Lalitha, Y. and Ballal, C.R. 2013. Production protocol for *Anastatus* (Anastatus) *acherontiae* Narayanan, Subba Rao and Ramachandra Rao (Hymenoptera: Euplemidae), a potential egg parasitoid of litchi stink bug, *Tessaratoma javanica* (Thunberg) (Hemiptera: Tessaratomidae). p. 65. **In**: Abstracts of the IOBC *MRQA 13th workshop on "Emerging Opportunities for the Mass Production and Quality Assurance of Invertebrates"*, Bengaluru, India, 94 pp.

Jalali, S.K. and Singh, S.P. 1993. Superior strain selection of the egg parasitoid *Trichogramma chilonis* Ishii: Biological parameters. *J. Biol. Cont.*, **7**: 57–60.

Jalali, S.K., Singh, S.P., Venkatesan, T. Murthy, K.S. and Lalitha, Y. 2006. Development of endosulfan tolerant strain of an egg parasitoid *Trichogramma chilonis* Ishii (Hymenoptera: Trichogrammatidae). *Indian J. Exp. Biol.*, **44**: 584–590.

Lalitha, Y., Ballal, C.R. and Patel, V.N. 2013. Quality assessment of mass reared *Trichogramma chilonis* Ishii (Hymenoptera: Trichogrammatidae) based on field performance. p. 60. **In**: Abstracts of the IOBC *MRQA 13th Workshop on "Emerging Opportunities for the Mass Production and Quality Assurance of Invertebrates"*, Bengaluru, India, 94 pp.

Mani, M., Krishnamoorthy, A. and Singh, S.P. 1990. The impact of the predator, *Cryptolaemus Montrouzieri* Mulsant on pesticide–resistant populations of the striped mealybug, *Ferrisia virgata* (Ckll.) on guava in India. *Int. J. Trop. Sci.*, **11**(2): 167–170.

Nagesh, M. and Hussaini, S.S. 2013. Development of fungal bionematicides: Scale–up, PH processing, storage stability, toxicology and field evaluation. DBT Project Report (2009–13).

Nagesh, M., Hussaini, S.S. Chindandaswamy, B.S. Shubha, M.R. Ruby, K.M. 2007 Relationship between initial water content of the substrate and mycelial growth and sporulation of the nematophagous fungi, *Paecilomyces lilacinus* and *Pochonia chlamydosporia*. *Nematologia Mediterranea*, **35**, 57–60.

Ramani, S., Poorani, J., and Bhumannavar, B.S. 2002. Spiralling whitefly, *Aleurodicus dispersus*, in India. *Biocontrol News and Information*, **23**(2): 55N–62N.

Ramanujam, B. 2007. Microbial control of crop pests using entomopathogenic fungi. pp. 231–260. **In**: *Biological and Biotechnological Approaches to Insect Pest and Disease Control, Proceedings of the Winter School*, PDBC, Bengaluru, 400 pp.

Ramanujam, B. and Sriram, S. 2007. Biocontrol of plant diseases with fungal antagonists. pp. 305–324. **In**: *Biological and Biotechnological Approaches to Insect Pest and Disease Control, Proceedings of the Winter School*, PDBC, Bengaluru, 400 pp.

Shylesha, A.N., Joshi, S., Rabindra, R.J. and Bhumannavar, B.S. 2010. Classical biological control of the papaya mealybug. Folder, National Bureau of Agriculturally Important Insects, Bengaluru, 3 p.

42

Role of IPM Protocols in Horticultural Crops for Improving Profitability

D.B. Ahuja* and H.R. Sardana

ICAR-National Research Centre for Integrated Pest Management, LBS Building, IARI, New Delhi – 110 012
E-mail: deshbandhu4@rediffmail.com

Adoption of frontier farm technologies has led to increase in agricultural production ensuring food and nutritional security of the country. Food grains and horticultural production was reported as 269 and 283 mt, respectively during 2015-2016. While nation achieved a commendable position in food and horticultural production, farming itself has reached to non profitable level due to rising cost of production and small and fragmented holdings. This has shaken the farmers' confidence in the adoption and implementation of advanced farm technologies which is of great concern for policy makers and scientists. Unless farmers' income increases sustainably, the farmers distress cannot be tackled (Chand, 2016). Honourable Prime Minister in his recent public addresses exhorted to double the farmer's income by 2022. Recent estimates show that, the average income of farm house holds is growing at 5.24 per cent and doubling of income would take 14 years (Satyasai and Mehrotra, 2016). In general, larger farmers' double their income in lesser number of years as compared to small and marginal farmers. However, country has more number of smaller farmers than larger farmers more so in horticultural sector, thus challenge assumes the bigger status in context of horticultural sector. There are many options for addressing the issue of increasing the farmers' income, a few may give results soon and rated as short term options but in many cases, the results may require more time and rated as long term options. Development of high yielding planting material/seeds requiring low cost inputs are considered as long term options. Reducing cost of production through better soil health management, balanced fertilizer and water use efficiency, and preventing damage from pests through IPM practices can give a boost to horticultural production in shorter duration. Bridging yield gaps and raising farmers' income through implementation of low cost IPM practices focusing on adoption of bio-pesticides and good agricultural practices restoring natural mechanism of reducing pest densities is one of the short term options that can bring additional farm income. It is in this context IPM adoption has become more relevant and can play important role in enhancing farmers' income, reducing the farm distress as well as safeguarding the environment, human health and sustain productivity. Over the years constant efforts put forth by plant protection ICAR institutes like NCIPM and crop based institutes, AICRPs, SAUs, KVK's and NGOs, in developing IPM options, which when practiced together in farmers participatory modes have demonstrated their worth and has improved the planners confidence level in the technology.

Several studies on development and validation of IPM modules in vegetable and fruit crops in different climatic zones and geographical regions of India indicated sizeable gaps between the attainable yield and actual yield at farmers' fields/orchards in spite of over/repeated application of pesticides by farmers. Perhaps low cost plant protection options embedded in IPM modules may step up yield levels and bring higher profits to farmers enhancing their income. Projects aimed at increasing production by the development and implementation of low cost IPM strategies/package replacing the high cost pesticide reliant practices have been executed by ICAR NCIPM, New Delhi in both vegetables and fruits throughout the country during the last one and half decade.

The changing pest complex in the spatial and temporal scale and different methods of cultivation in different parts of the country desires the development of IPM packages suited for local needs. Sustained efforts are needed to integrate locally-adapted, farmer friendly, conventional and bioregional methods of pest management. A major aim of implementing IPM technologies is to increase productivity, lowering the cost of plant protection and restore the population of natural enemies, establishing natural and sustainable pest control and finally enhancement in farmers' income.

IPM focus includes choice of cultural practices, crop management options that don't favour the insect pests and diseases development, cost effective choices of environmentally friendly pesticides that favour the conservation of beneficial fauna and flora, harmonising community and policy issues with IPM awareness for catering to improvement in farming system. IPM protocols incorporated seed treatment; nursery application and soil application in main field with bio agents such as *Trichoderma viride* and/or *Pseudomonas fluorescens* for disease management; *Metarhzium anisoplae*, *Beauveria bassiana* and *Verticillium lecanii* against various type of insect pests; application of neem cake; selection of good and virus disease-free seedlings for planting material.

Over a period of about fifteen years (2000-2015) several farmer-centred and farmer-driven participatory programmes were put in operation for the development and validation of IPM protocols in major fruit and vegetable crops *viz*., citrus, mango, apple, tomato, cabbage, brinjal, chillies, bell pepper, cauliflower and cucurbits, in a total area of over 1,000 hectares covering more than 1,500 farming families in the states of Uttar Pradesh, Karnataka, Haryana, Rajasthan, Gujarat, Punjab, Andhra Pradesh, Himachal Pradesh, Maharashtra, Assam and Jharkhand by NCIPM with the support of University of Agricultural Sciences, Raichur (Karnataka), Sardar Vallabh Bhai Patel University of Agricultural Sciences and Technology, Modipuram (Uttar Pradesh), Rajasthan Agricultural University, Durgapura, Jaipur (Rajasthan), Gujarat Agricultural University, Surat, Indian Institute of Horticultural Research, Bengaluru; Indian Institute of Vegetable Research, Varanasi and Rama Krishan Mission, Krishi Vigyan Kendra, Ranchi, Punjab Agricultural University, Ludhiana, Assam Agricultural University, Guwahati, Directorate of onion and garlic research (DOGR), Pune *etc*.

Socio-economic Status of IPM Adopted Villages/Farmers

The socio-economic status and base line information of adopted IPM farming families was studied which formed the basis for systemic synthesis and validation of crop-specific IPM protocols. Any indigenous technical knowhow was also recorded for incorporation in the IPM programme. Certain farmers were practising old age rural based techniques to suppress these pests problems, like the use of dung ash (for cut worms, thrips, and red pumpkin beetle), multiplication of predatory ants by providing wheat floor (for controlling termites), sun drying, attracting the predatory birds for caterpillars.

Agro-ecosystem analysis or habitat analysis for management is another most important activity in the IPM. This is to know the field situations with regard to pests, soil conditions, plant health, climatic factors *etc*. This can be employed in making decisions on pest management. Like knowledge on climatic factors impact on pest incidence. Some of the plant ideotypes like intensity of branching, compensatory mechanism of the plant, presence of hairs, and presence of different chemicals in the plant body are used to check the increase in pest populations. Alteration in cropping system and crops sequence to avoid peak activity period of pest is helpful to check the mode of carryover of the pests. Simultaneously, collateral hosts are major factor to provide shelter to pests and diseases during the crop period as well as in the off season. Crop hygiene also can be an important approach in the pest management. All the flora and fauna existing in the ecosystem are not harmful but some of them are farmer-friendly, *viz*., spiders, ants and other parasitoids, predators and pollinators. Therefore, conservation of these beneficial insects is very important and needs attention. Pollinators, *viz*., honey bee are playing important role in increasing crop production. The population of bees must be protected.

The detailed IPM protocols tested and validated for different fruits and vegetable crops in different regions of the country are described along with the yield and other economic parameters in both IPM and non-IPM fields.

IPM Protocols for Fruit Crops, their Economic Viability and Impact

Mango

During the period 2001 - 2004, the IPM protocol for mango crop was developed and validated in about 58 ha area spread in 11 villages *viz*., Gulabkhera, Habibpur, Budhadia, Pathakganj, Rehmankhera and Kanar in Malihabad and Kakori belt of mango near Lucknow in five districts of Uttar Pradesh covering 112 farming families in Dashehari variety in a sustained manner for four years. Despite 7-9 sprays of pesticides given for diseases and other insect pests control, farmers were unable to raise the productivity of mango from 35-70 q/ha. The IPM protocol was formulated and successfully validated and adopted by mango farmers, comprised of sprays of neem seed kernel extract (NSKE) 5 per cent, neem soap (obtained from IIHR, Bengaluru) and *Verticillium lecani* for mango hopper, *Beauveria bassiana* for mealy bug, use of alkthene bands for mealy bugs, installation of methyl eugenol pheromone traps for fruit fly, collection and destruction of fruit fly affected fruits from time to time, collection of die-back and anthracnose affected twigs, mechanical collection of leaf webber, deep ploughing of orchards during October-November *etc.* and need based application of insecticides based on monitoring of insect pests population and diseases severity. For die-back, pruning of infected and dried branches, 10 cm below the dried portion and pasting of copper oxy chloride and spray of 0.3 per cent copper oxy chloride (3 g/l) (repeated in November if required) after pruning were adopted. The mango fruit fly had been very effectively managed in mango orchards using methyl eugenol plastic bottle traps having small wooden blocks. This was highly economical and environmentally safe. Wooden blocks were very cheap and efficient in capturing fruit flies @ 2,500 flies/trap. These traps could even be prepared by farmers themselves using waste plastic bottles. The fruit fly traps had been further modified and mixture containing ethanol were replaced with some other material, which prolonged the efficacy of mixture due to slow release of methyl eugenol in fruit fly traps. Ethanol is not freely available in open market and is sold under government supervision. Wooden block size has also been recently reduced (IIHR) resulting in more number of traps required per acre. Early harvesting of mature fruits also avoids fruit fly infestation and anthracnose. Regular removal of weeds in orchards which harbour insects and diseases was also recommended. Mango mealy bug was kept under check utilizing non chemical method *i.e*: alkathene (25 cm wide, 400 gauge) banding during 3rd week of December and drenching of 5 per cent Neem Seed Kernel Extract and *Beauveria bassiana* around the trunk in December-January. This method prevented migration of early instars of mealy bugs on tree. Deep ploughing of orchards during November which is also a common practice in mango, exposed the eggs of mealy bugs, pupae of fruit fly, leaf webber and midge to birds and removal of weeds helped in reducing the population of these insects harbouring them. Heavy irrigation of orchards in October helped in destruction of eggs of mealy bugs, diapauses pupae of midge and fruit fly. Mango hoppers, inflorescence midge and mealy bug which ascended on tree by chance were effectively controlled by spraying Neem Seed Kernel Extract (5 per cent), monocrotophos and chlorpyriphos (0.05 per cent) spray. These two insecticides did not persist for long in mango orchards and were found to be below maximum residual limits (MRL) fixed by WHO. In fields where this package was not followed (non - IPM) higher residue of these insecticides, *i.e.* 0.01 and 0.005 ppm were found. Looking into the slightly higher residues of monocrotophos insecticide, farmers were advised to use chlorpyriphos and desist from spraying monocrotophos. Mango leaf webber was effectively controlled by leaf webber removal device which is locally available and very cheap. Webs collected were later burned in August-September. Prunning of overcrowded and overlapping branches for control of leaf webber dried portion; twig infected with anthracnose disease, malformed panicles and powdery mildew affected leaves on plant was followed by all mango growers. To reduce the use of synthetic insecticides vigorous campaigning was done which was readily accepted by the farmers. Sooty mold was easily controlled by using non-chemical starch @ 2 per cent. Powdery mildew disease was easily controlled by giving a single spray of sulfex fungicide @ 2 g/l along with removal of infected leaves and malformed panicles. For bacterial canker, spray of streptocyclin @ 200 mg/l is recommended with repeated spray if the pathogen persists (Ahuja *et al.*, 2015a).

Figure 1: Polythene Banding.

The detailed yield and economics in both IPM and farmers fields in different regions studied are presented in Table 2 which clearly implied that IPM was superior to the farmers practices.

Apple

During the period 2001-2004, integrated pest management (IPM) protocol for apple crop was developed and validated in about 30 ha of orchards in villages like Kotkhai, Jhubbal, Theneder and Rohroo, the apple growing belt of Himachal Pradesh. The farmers of these villages have been raising apple crop since several decades and was the main source of their income and livelihood. Because of the old orchards and several years old apple trees the pests and disease problems have also multiplied and rather new pests and disease have emerged in these orchards. Despite 8 -9 sprays of pesticides given for the control of these pests and diseases, farmers were not able to raise the production and productivity of apple from 7.9 tonnes/

Table 1: IPM Protocols for Apple Crop for Implementation during different Seasons

Stage	*Target Pest*	*IPM Activities*
	Dormant season *i.e.* October–November	
Dormant (after natural leaf fall)	Sun burn, Lichen and Green Moss	✰ Bordeaux mixture paint + linseed oil in 1:1 ratio @ 1 kg/9 l of water to stem portion only.
	Canker	✰ Pruning of disease shoots/twigs ✰ Apply Chaubattia paste (copper carbonate + linseed oil + red lead oxide) in the ratio of (8:8:15) on stem. This is very important after pruning.
Leaf fall to bud swell stage	Canker	✰ Pruning of diseased shoots/twigs. ✰ Application of Bordeaux mixture after pruning to prevent development of canker.
Green tip stage	Apple scab	✰ Apply captan @ 600 g/200 l of water.
Tight cluster stage	San Jose scale and over-wintering eggs of mite	✰ Apply dormant oil @ 4 litres and is very important for San jose scale and over wintering eggs of mites.
	Spring season module, *i.e.* April–June	
Pink bud, Petal fall	San Jose scale, Wooly apple aphid, Blossom thrips, Mites, Powdery mildew, Apple scales.	✰ Dormant oil spray in the first week of May for San Jose scale and European mite when 1st stage crawlers emerge. ✰ Release of Coccinelid, *Chilocorus bijugus* for San Jose scale. ✰ Release of *Chrysoperla scelestes* and *C. carnea* for wooly aphids. ✰ Predatory mites, *Amblysius fallacis* and *Zitzellia mali* for mites ✰ Spray of carbendazim @ 100 g/200 l of water for powdery mildew and apple scab. ✰ Spray of chlorpyriphos @ 400 ml/200 l of water. ✰ Remove all the suckers and water sprouts at the base of tree trunk and an major scaffold limbs and paint all the cuts with commercial paint.
	Summer season module, *i.e.* July–September	
Fruit set (pea seed size)	Marssonina blotch, Apple scab	✰ Spray of Mancozeb @ 600 g/200 l of water for premature leaf fall and Apple scab.
Fruit development (Walnut size)	European red spider mite, Caterpillar	✰ Spray of Horticulture mineral oil for mites when ETL < 6-8/leaf and temperature is < 24° C. encourage predatory mites. Spray of endosulfan for caterpillar @ 300 ml/200 l of water.
Fruit development	Marssonina blotch, apple scab and mite	✰ Mancozeb spray @ 600 g/200 l of water for premature leaf fall and apple scab if fruit set spray is not given. ✰ Encourage predatory mites. ✰ Spay of horticultural mineral oil for mites.
Fruit development (20 days after earlier spray given)	Marssonina, apple scab, mites	✰ Encourage predatory mites. ✰ Spray acarcide after monitoring the population ETL. ✰ Spray of mancozeb for scab and premature leaf fall. ✰ Spray of horticulture mineral oil for mites.
Fruit development (20 days after earlier spray given)	Marssonina blotch, apple scab, sooty blotch and fly speck.	✰ Monitor the disease and give the spray of fungicide mancozeb @ 500 g/200 l of water. This spray is important for all summer diseases and quality fruit finish.
Pre-harvest stage and harvesting stage	Marssonina blotch, apple scab, sooty blotch and fly speck.	✰ Monitor the disease and give the spray of fungicide mancozeb @ 500 g/200 l of water. This spray is important for all summer diseases and quality fruit finish.
Post harvest	Pre- mature leaf fall, apple scab and stain injury.	✰ Collection and destruction of fallen leaves. ✰ Sprays carbendezion @100 g or COC @ 600 g/200 l of water. This is very important to delay the leaf fall till late autumn or to recover skin injury and sun burns.
Fruit Development (Rainy season)	White root rot	✰ Sowing of cabbage and garlic and soil solarisation. ✰ Apply carbendezion @100 g/100 l of water or Bordeaux mixture in 1-2 feet deep holes around the tree basin. This gives good control in small trees at an early stage of infection when soil is wet up to root zone ✰ Use of strains of antagonists, *Enterobacter aerogenes and B. subtillis.* ✰ Addition of *deodar* needles and neem cake in soil and around the stem.
Early rainy season	Collar rot	✰ Apply ridomil or Bordeaux mixture 1:1 around tree trunk as soil drenching ✰ Use of strains of *Enterobacter aerogenes* and *Bacillus subtills* ✰ Sowing of cabbage and garlic
Leaf yellowing stage	Premature leaf fall complex and apple scab	✰ Urea @ 10 kg/ha very important for early decomposition of diseased leaves on orchard floor/tree basin

ha. Rather decline in productivity apple by 50 per cent was recorded in some of the old orchards.

IPM module could be adopted in apple crop during different stages and periods of the year which can be more effective and can be broadly divided in three seasons as described in Table 1.

The detailed yield and economics in both IPM and farmers fields in different regions studied are presented in Table 2 (Trivedi *et al.*, 2004a,b, 2011).

Citrus

Protocols for IPM for citrus crops *i.e.* mandarin in three different agro-climatic regions of India were developed, validated during 2013-16 in 150 ha orchards covering 150 farming families and are being implemented at farmers orchards in collaboration with PAU, Ludhiana (*kinow* mandarin); Assam Agricultural University, Assam (for *Khasi* mandarin) and NRC Citrus, Nagpur. Report of the joint inspection team of the National Horticulture Mission (*http://nhm.nic.in/JIT_Reports/JIT-Punjab-FEB per cent 202013.pdf*) revealed the use of highly toxic chemical pesticides in Kinnow and the need to practice IPM in Punjab. Chemical pesticide sprays numbering 15-20 were given in a span of ten months between February and December. Kinnow growers of the region were also anxious to know about nonchemical management practices to prevent the pest damage in an eco-friendly way. Hence, an IPM for validation was initiated in Panjkosi village of Abohar, Fazilika. Simultaneously, IPM practices for *Khasi* mandarin cultivated in North Eastern Region of India were taken up with a view to validate combination of different management components for a sustainable and effective integrated pest management. Only the actionable IPM practices, which were economically viable, were validated. Base line information collected at Panjkosi village about the Kinnow mandarin by the farmers indicated gummosis to be the major disease prevalent in the area. It is known by various names as *Phytophthora* foot rot, crown-rot, root-rot, fruit rot and leaf fall (Ahuja *et al.*, 2016).

IPM Protocol for Kinnow

Spring Season (February–March)

- ✰ Occurrence of citrus psylla should be checked on the border trees. It starts after harvesting of fruit at bud swelling stage towards the end of February or beginning of March. If found infested, spraying of *Lecanicillium lecanii* (1 x 10^7 conidia/ml) or dimethoate 30 EC @ 25 ml/10 l of water needs to be done, for the management of early (1st and 2nd) instar nymphs of citrus psylla.
- ✰ In the month of March, if the occurrence of the pest is still observed above ETL, spray of imidacloprid 17.8 SL (4 ml/10 l water), for the management of citrus psylla, leaf miner and aphids, may be considered. If required, spray of thiamethoxam 25 WG @ 3 g/10 l water, 10 days after application of Imidacloprid should be done.
- ✰ *Phytophthora* affected bark portions of the tree trunk, branches and limbs along with some healthy green bark should be scrapped with the help of sharp knife or *khurpa*. Proper destruction of the diseased bark is advocated by burning. Apply the fungicidal paint of Metalaxyl (2 g in 100 ml linseed oil) or *Trichoderma harzianum* (1 x 10^8 cfu/g @ 100 g/l) and paint linseed oil after 5 days with the help of painting brush. Drenching the root zone area of the infected tree with Metalaxyl (25 g/tree in 10 l of water) should also be done.
- ✰ Farmers adopted the trunk method of irrigation by flooding of orchards through use of tractor driven raised bed planter which created a depression between the row of trees, for preventing the passage of water around the bud scion.
- ✰ Prune the trees and spray Bordeaux mixture.

Summer Season (April–June)

- ✰ Erection of yellow sticky traps @ 10 per ha could be useful. If whitefly population is more than 5-10/trap/week (ETL), spray of Triazophos 40 EC @ 25 ml/10 l of water coinciding with 50 per cent of egg hatching or first nymphal stage could be done, which would also take care of leaf miner and/or psylla. Spray of Ethion 50 EC @ 20 ml/10 l water, 8-10 days after Triazophos application needs consideration if whitefly reappears beyond ETL. Spray mixture of Aureofungin/carbendazim + 2,4-D @ 0.4 g/10 g + 0.1 g/10 l

Table 2: Economics of IPM in different Horticultural Crops

Parameter	Mango (Lucknow)		Apple		Kinnow Mandarin (Punjab)		Khasi Mandarin (Assam)	
	IPM	Non-IPM	IPM	Non-IPM	IPM	Non-IPM	IPM	Non-IPM
No. of Chemicals sprays	3	7	6	9	11	17	–	–
Cost of Plant Protection (Rs./ha)	10444.4	21000	14444.4	23000	26595	34350	–	–
Yield (q/ha)	50.86	40.21	49.9	41.2	290	270	320	244
Gross Income (Rs/ha)	158622.2	107666.6	178622.2	147666.7	290000	270000	510000	396000

water for the management offruit drop. Use GA_3 instead of 2, 4-D when cotton or other broad leaved crop is cultivated in oraround the orchard could be useful. If infestation of *Phytophthora* is observed, spray fosetyl-Al 80 WP @ 25 g/10 l water could be undertaken.

- ✰ Foliar application of fosetyl- Al 80 WP (2.5 g/l) may be made during April for the effective *Phytophthora* management. The spray application will check the lesions on the tree parts as well as regeneration of feeder roots. Fosetyl- Al spray can be combined with drenching of root zone area with Metalaxyl.
- ✰ If whitefly still persists beyone ETL, spray Triazophos 40 EC @ 25 ml/10 l water. Carbendazim 50 WP could be applied @ 10 ml/10 l for the management of fruit drop, if needed.

Monsoon Season (July–September)

- ✰ Scrap the affected bark portions of the tree trunk, branches and limbs along with some healthy green bark with the help of sharp knife or khurpa. Properly collect and destroy the diseased bark by burning and do not allow it to fall on the ground. Apply the fungicidal paint of Metalaxyl (2 g in 100 ml Linseed oil) or *Trichoderma harzianum* @ 100 g/l and paint linseed oil after 5 days with the help of painting brush. Drenching the root zone area of the infected tree with Metalaxyl (25 g/ tree in 10 litres of water).
- ✰ Spray Fosetyl- Al 80 WP (2.5 g/l) for the effective *Phytophthora* control. The spray application will check the lesions on the tree parts as well as regeneration of feeder roots. Fosetyl- Al spray can be combined with drenching of root zone area with Metalaxyl.
- ✰ During this period, psylla or leaf miner or white fly starts reinfesting the trees. Spray Thiamethoxam 25 WG @ 3 g/10 l of water or Triazophos 40 EC @ 25 ml/10 l of water.
- ✰ Spray mixture of 2,4-D at 0.1g + Tilt Propiconazole 25 EC 10 ml/10 l water for the management of fruit drop. Spray carbendazim 50 WP @ 10 g/10 l water, 2 weeks after application of 2,4-D, if needed.
- ✰ If the whitefly damage still persists beyond ETL and sooty mould is observed, spray mixture of Triazophos 40 EC at 25 ml + Ziram 80 WP at 25 g/10 l water could be useful.
- ✰ Clearing the infested branches of the fross, faeces and injection of 5-10 ml of Dichlorvos (0.1 per cent) in to the tunnel and covering it with cotton swap reduces infestation of bark eating caterpillar. Spray Fosetyl-Al 80 WP @ 25 g/10 l water to manage *Phytophthora*. Need-based spray of Thiophanate methyl 70 WP @ 10 g/10 l water for the management of fruit drop at end of September.

Post-monsoon Season (October–December)

- ✰ If the spraying of triazophos was done in September, citrus psylla was still observed beyond ETL, spray of thiamethoxam 25 WG @ 3 g/10 l water could be useful. There was need to destroy the fallen fruits by burying.
- ✰ Generation of smoke in the late evening hours and foliar application of neem oil (1 per cent) or malathion 50 EC @ 2 ml/l reduces infestation of fruit sucking moth.
- ✰ For the effective management of canker/scab, spray mixture of streptocycline + Copper oxy chloride @ 1 g + 25 g/10 l water.

Need-based implementation of IPM options in North Western region resulted in reduced pest incidence and increased *Coccinellids* and spiders in the IPM as compared to non-IPM orchards. Fruit yield (t/ha) was higher from IPM orchards. Number of applications and amount of chemical pesticides was also reduced as a result of implementation of IPM along with the replacement of hazardous molecules by greener pesticides as given in Table 2 (Ahuja *et al.*, 2015b; 2016).

IPM Protocol for *Khasi* Mandarin

Spring Season (February–March)

- ✰ Application of petroleum spray oil (2 per cent) for citrus butterfly and citrus psylla.
- ✰ Application of Bordeaux mixture (1 per cent) for most of the diseases.
- ✰ Soil application of *Paecilomyces lilacinus* (2 x 10^7 spores/g) infested grain @ 30 g/tree for managing citrus nematodes (*www.nrccitrus.nic.in*).
- ✰ Smearing the tree trunk up to 1m from the ground level by the mixture of 50 ml dimethoate + 2 kg lime in 10 litres of water along with gum in March to prevent early infestation of trunk borer and bark eating caterpillars.
- ✰ Soil drenching and spraying of tree trunk with Metalaxyl 8 per cent + Mancozeb 64 per cent @ 0.2 per cent in February to manage *Phytophthora* foot rot.

Summer Season (April–June)

- ✰ Installation of yellow sticky traps @ 10/ha for soft bodied insect-pests and methyl eugenol trap @ 10/ha for fruit fly.
- ✰ Application of biorational insecticide, spinosad 45 SC @ 1 per cent for managing leaf miner.
- ✰ Spraying of Aliette 80 WP (2 g/l) during May for management of *Phytophthora.*

- ✰ Application of Bactronol 1,000 ppm for management of citrus greening.
- ✰ ITK (paddy straw is tied at a height of 1m around the tree trunk during April to prevent insect from crawling upward).

Monsoon Season (July–September)

- ✰ Smearing the tree trunk up to 1 m from the ground level by the mixture of 50 ml Dimethoate + 2 kg lime in 10 l of water along with gum in March to prevent early infestation of trunk borer and bark eating caterpillar Soil application of *Paecilomyces lilacinus* infested grain 30 g/tree for managing citrus nematodes.
- ✰ Need-based spray of Thiophanate methyl 70 WP @ 10 g/10 l water for the management of fruit drop at end of September.

Post-monsoon Season (October–December)

- ✰ For the effective management of canker/scab, spray mixture of Streptocycline + copper oxychloride @ 1 g + 25 g/10 l water.
- ✰ Application of petroleum spray oil (2 per cent) for citrus psylla.

At North Eastern Region of India, with different pests scenario, *viz.* citrus trunk borer, bark eating caterpillar, citrus psylla, leaf miner, lemon butterfly and *Phytophthora* foot rot, twig blight, citrus greening, citrus scab, pre-harvest fruit drop, the location specific IPM strategy also resulted in decreased incidence of the pests as well as higher economic benefits.

IPM Protocols for Vegetable Crops, their Economic Viability and Impact

Scientifically tested individual components, including insecticides, resistant varieties reported effective by the researchers as well as vegetable growers including opinion of farmers were systematically synthesized in the form of IPM modules in order of crop growing stages for important vegetable crops for different locations, resulting into precise IPM protocols/modules and packages. The critical inputs like seeds of disease resistant hybrids, bio-pesticides, pheromone traps, biopesticides like *Ha*NPV, *Sl*NPV, Spinosad and quality chemical pesticides were provided to IPM farmers. The initiatives were also taken to strengthen the bioagents production at the project site for timely availability to the farmers. The regular monitoring of insect pests and diseases and finally need-based application of pesticides were some of the unique features of the programme.

The major objective of the IPM protocols/technologies development and validation programmes in vegetables was to create a visibility of IPM at fields level keeping in mind the economic, environmental and social impact factors. The data on number of chemical sprays, total cost of cultivation, expenditure on plant protection, yield and cost-benefit ratio was worked out for IPM and non-IPM farmers to find out the economic viability and sustainability of the programme at all the locations for all the vegetables. The structured and unstructured impact analysis as increase in yield, cost of production, nature of plant protection measures, profitability level and increase of bio-diversity has been studied.

Tomato

Initially small scale trials on IPM protocols validation were initiated during 2001-2002 at three locations. Encouraged with the success of these trials, IPM technologies were validated on a large scale covering more farmers, area and villages. IPM protocols for tomato crop were validated in more than 125 ha area covering 33 villages in Karnataka, UP, Jharkhand and Haryana during 2001-2014.

The integrated pest management protocol\module successfully validated and adopted by farmers for tomato mainly consisted of raising of healthy nursery using seed treatment with *Trichoderma viride*, soil solarisation, use of nylon net for avoidance of insect vector, wide spacing of 60 x 45 cm, installation of pheromone traps, release of *Trichogramma chilonis*, based on monitoring of fruit borer using pheromone traps, sowing of marigold as trap crop, spray of good quality NPV on tomato and spray of NSKE @ 5 per cent against leaf miner and need based application of insecticides based on monitoring of insect pests and diseases.

In addition to the egg laying by fruit borer, growing marigold in between tomato rows had the twin advantages of attracting lead miner adults as well as repelling the parasitic nematodes. The various components of IPM for tomato aimed at mainly fruit borer management also favourably regulated the population of larval parasite *Campoletes chloridae*. The maintenance of higher row to row and plant to plant spacing of 60 x 50 cm and 90 x 60 cm was followed by almost all the farmers and reduced the incidence of early blight and powdery mildew in tomato. Removal of leaf curl affected plants in tomato from time to time was followed by all the farmers and easily adopted by them. The economic impact on the implementation of IPM technology in different locations is presented in Table 4 (Sardana *et al.*, 2013).

Cabbage/Cauliflower

IPM protocols for cabbage/cauliflower crop were validated in more than 125 ha area covering 33 villages in Karnataka, UP, Jharkhand and Haryana during 2001-2014.

The integrated pest management module successfully validated and adopted by farmers for cabbage at three locations is presented in Table 5.

Table 3: IPM Protocols for Tomato Developed and Validated at different Locations

Bengaluru, Karnataka	*Varanasi, U.P.*	*Ranchi, Jharkhand*	*Meerut, U.P.*	*Sonepat, Haryana*
		Nursery		
☆ Raised seed bed ☆ Soil solarization for 3 weeks. ☆ Grow leaf curl resistant hybrid/varieties Avinash. ☆ Seed treatment with *T. viride* @ 4g/kg of seed. ☆ Use of nylon nets for white fly *etc.*	☆ Raised seed bed of 10 cm height ☆ Soil solarization using transparent polythene sheets. ☆ Soil treatment with *Trichoderma harzianum* @ 4 g/kg of seed. ☆ Spray of copper oxy chloride.	☆ Raised seed bed of 15 cm height ☆ Soil solarization using transparent polythene sheets. ☆ Grow leaf curl resistant hybrid/varieties Avinash ☆ Soil treatment with *T. harzianum* @ 100 g/kg of FYM 1 kg/t.	☆ Raised seed bed of 15 cm height ☆ Soil solarization. ☆ Grow leaf curl resistant hybrid/varieties Hemsona. ☆ Soil treatment with *T. harzianum* @ 100 g/kg of FYM 1 kg/t.	☆ Raised seed bed of 15 cm height ☆ Grow leaf curl resistant hybrid/varieties Hemsona or Krishna. ☆ Soil treatment with *T. harzianum* @ 100 g/kg of FYM 1 kg/t. ☆ Cover nursery with nylon net
		Main Field		
☆ Apply neem cake @ 250 kg/ha while planting or at 20 DAP ☆ Wide spacing @ 90 x 60 cm ☆ Dip seedling roots for 15 mts in imidacloprid @ 0.5 ml/l. ☆ Transplant marigold per 14 rows of tomato. ☆ Spray imidacloprid/ @0.5 ml/l at 15 DAP for whitefly or 5 per cent NSKE against leaf-miner. ☆ Pheromone traps @ 5/ha for monitoring ☆ Monitor top three leaves for borer eggs at flowering. ☆ Release of *T. pretiosum* @ 1.0 lakh/ha six times ☆ Spray of *Ha*NPV 250 LE/ha. ☆ Spray indoxacarb for fruit borer. ☆ Roguing of damaged fruits, leaf-curl and wilt affected plants. ☆ Spray 0.02% mancozeb/captan for the control of early and blight.	☆ Spray imidacloprid at 15 DAP ☆ Installation of pheromones traps @ 5/ha for monitoring of *H. armigera*. ☆ Spray of *Ha*NPV 250 LE/ha ☆ Release of *T. bactrae* @ 1.0 lakh/ha six times. ☆ Regular collection and destruction of damaged fruits. ☆ Need base application of pesticides like thiomethoxam and mancozeb @ 0.02 %	☆ Spray imidacloprid @ 0.5 ml/l at 20 DAP ☆ Installation of phero-mone traps @ 5/ha for monitoring of *H. armigera* population ☆ Monitor top three leaves for *Helicoverpa* eggs. ☆ Release of *T. chilonis* @ 1.0 lakh/ha after appearance of adults ☆ Spray of *Sl*NPV @ 250 LE/ha twice at 15 day interval or *Ha*NPV for *H. armigera* ☆ Regular collection and destruction of damaged fruits from time to time. ☆ Need based application of pesticide rynaxpyr 20 EC	☆ Spray imidacloprid @ 0.5 ml/l at 20 DAP ☆ Pheromone traps @ 5/ha for monitoring of borer population ☆ Monitor top three leaves for *Helicoverpa* eggs. ☆ Spray of *Sl*NPV @ 250 LE/ha twice at 15 day interval or *Ha*NPV for *H. armigera* ☆ Regular collection and destruction of damaged fruits ☆ Need based application of pesticide like endo-sulfan @ 0.07% or emmamectin benzoate.	☆ Spray imidacloprid @ 0.4 ml/l in March ☆ Pheromone traps @ 5/ ha for monitoring of borer population ☆ Monitor top three leaves for *Helicoverpa* eggs. ☆ Spray of *Ha*NPV @ 250 LE/ha twice at 15 day interval or *Ha*NPV for *H. armigera* ☆ Regular collection and destruction of damaged fruits ☆ Need based application of pesticide like rynxpyr 20 EC

Table 4: Yield and Economics of IPM in Tomato at different Locations

Parameter	*Bengaluru (Karnataka)*		*Varanasi (UP)*		*Ranchi (Jharkhand)*		*Meerut (UP)*			
	2001-04		*2001-04*		*2001-04*		*2007-08*		*2008-09*	
	IPM	*Non-IPM*	*IPM*	*Non-IPM*	*IPM*	*Non-IPM*	*IPM*	*Non-IPM*	*IPM*	*Non-IPM*
No. of chemical sprays	5	28	3.83	14.17	1	5	6.6	11.2	7.0	11.2
Plant protection cost (Rs/ha)	3048	27668	3667	8917	3665	36000	4425	7437	4650	5075
Production Cost (Rs/ha)	51844	114678	12083	18833	30238	31455	44488	46500	52689	54481
Yield (q/ha)	740.38	450.56	142.5	130.0	222.93	187.72	190.4	171.2	451.3	420.4
Gross returns (Rs/ha)	301565	184382	52000	57000	86943	73211	85680	77040	203228	189180
Net returns (Rs/ha)	249721	69704	39917	38167	56705	41756	41193	30541	150539	134700
C:B ratio	1: 4.82	1: 0.61	1: 3.3	1: 2.02	1: 1.87	1: 1.32	1:0.92	1:0.65	1:4.85	1:4.47

Table 5: IPM Protocols Developed for Cabbage/Cauliflower

Bengaluru, Karnataka	Varanasi, U.P.	Ranchi, Jharkhand	Sonepat, Haryana
Nursery			
☆ Raised seed bed upto 10 cm height ☆ Soil drenching with COC ☆ Seed treatment with *Trichoderma viride* @ 4 g/kg ☆ Use of nylon net to avoid entry of sucking pests ☆ Spray of *Bt* @ 0.5 ml/l one day before transplanting ☆ Spray of Dithane M-45/ Ridomyl @ 0.2% against downy mildew	☆ Raised seed bed of 15 cm height ☆ Soil solarization for three weeks with 45 um polythene sheet ☆ Seed treatment with *Tricho-derma harzianum* @ 4 g/kg ☆ Use of nylon net for sucking insects	☆ Raised seed bed of 10 cm height ☆ Soil solarization for three weeks ☆ Soil treatment with *Tricho-derma harzianum* mixed with FYM @ 1 kg/t.	☆ Raised seed bed of 15 cm height ☆ Soil solarization for three weeks with 45 um polythene sheet ☆ Seed treatment with *Tricho-derma harzianum* @ 4g/kg ☆ Seed treatment with imidaclo-prid @3-5 g/kg
Main Field			
☆ Growing of two rows of mustard after every 25 rows of cabbage as a trap crop at the time of planting. ☆ Wider spacing of 60 x 45 cm ☆ Use light traps @ 5/ha for trapping adult DBM ☆ Spray of *B. t* (1 g/litre) if DBM 1/plant is noticed early or spray of NSKE 5% or Pongamia soap 1% at primordial stage. Repeat 3-4 sprays. ☆ Rogue out disease affected bottom leaves from time to time. ☆ Need based sprays of chloro-thalonil/mancozeb @ 0.2 per cent for *Alternaria* and Blitox + streptomycin @ 0.2% for black rot.	☆ Spray *B. t.* (1 g/l) if DBM 1.0/ plant is noticed early ☆ Spray of NSKE 5% at primordial stage, *i.e.* 15-20DAS. Repeat 3-4 sprays. ☆ Release of *T. bactrae* @ 0.75 lakh/ha at weekly interval ☆ Remove disease affected heads and bottom leaves from time to time. ☆ Need based application of pesticides like mancozeb @ 0.2%	☆ Spray of Neemarin @ 5 per cent when DBM appeared early ☆ Spray *B. t* (1 g/l) if DBM 1.0/ plant is noticed ☆ Spray of pongamia soap @ 5% at 15-20 days after planting ☆ Installation of pheromone traps @ 5/ha for *Spodoptera litura* monitoring ☆ Need based application of fungicide like chlorothalonil for *Alternaria* ☆ Spray of insecticides like Triazophos @ 0.05%.	☆ Seedling dip in *T. harzianum* solution for 10-15 mts ☆ Installation of pheromone traps @ 5/ha for *S. litura* monitoring ☆ Collection of egg masses of *S. litura* and plucking of leaves with neonate larvae ☆ Spray *SL*NPV for *S. litura* ☆ Spray of NSKE 5% for *S. litura* ☆ Need-based spray of spinosad or novaluron. ☆ Spary of mancozeb @ 2 g/l against *Alternaria* leaf spot ☆ Release of *T. bactrae* @ 0.75 lakh/ha at weekly interval ☆ Remove disease affected heads and bottom leaves from time to time ☆ Need based application of pesticides like mancozeb @ 0.2%

Mustard attracted 80 per cent of diamond back moth for colonization. Besides, it also attracted entire population of leaf webber and stem borer. Depending upon the severity of diamond back moth and other insects, NSKE as it was was safe to *Cotesia plutella*, the dominant parasite of diamond back moth. Release of *Diadegma semiclausam* and *Cotesia plutella* can also suitably be incorporated in the IPM programme as one of components and gave encouraging outcome. In cauliflower in Sonepat, Haryana IPM protocol was vailated mainly in rainy season cauliflower and rabi season cauliflower in more than 25 ha area. The implementation of IPM by farmers resulted in increased bio-diversity major natural enemies recorded were egg/ larval parasitoid (*Telenomus* sp.) and larval parasitoid (*Cotesia glomeratus*) of *S. litura*, *Chrysoperla carnea* (Stephans) predating neonate larvae of *S. litura*. During rainy season, extent of parasitization by *C. glomeratus* was seen in both IPM and farmers' field, but was higher in the former (Ahuja *et al.*, 2011).

The economics of implementation of IPM in cabbage and cauliflower at different locations was presented in Table 6. The results in the present study established that IPM had the economic potential to substitute chemical pesticides without demanding any enhancement in cost of cultivation. Over and above it also ensured higher economic returns as well as higher curd yield with added advantage of no adverse effect on environment, natural enemies and human health. Implementation of IPM in irrigated cauliflower (*Brassica oleracea* L var *botrytis* subvar *cauliflora*) at Anantpura, Rajasthan also led to reduction in number of chemical pesticide sprays by 50-60 per cent and their replacement with safer pesticides. The reduction in cost of plant protection has taken place due to replacement of Cyclodiene, Organophosphates, Carbamates and synthetic pyrethroid, with newly introduced insecticides such as Spinosad, Indoxacarb, *Sl*NPV, Rimon with proven efficacy against *S. litura* and low residual effect with shorter waiting period for harvest of the produce (Ahuja *et al.*, 2009, 2012, 2013 and 2015c).

Bell Pepper

Capsicum (*Capsicum annuum*) also known as bell pepper, once restricted to mid hills of the country, is now conspicuous in non-traditional region and area under

Table 6: Yield and Economics of IPM in Cabbage and Cauliflower at different Locations

Parameter	Cabbage						Cauliflower	
	Bengaluru, Karnataka		Varanasi, UP		Ranchi, Jharkhand		Sonepat, Haryana	
	IPM	Non-IPM	IPM	Non-IPM	IPM	Non-IPM	IPM	Non-IPM
Number of chemical sprays	6	16	3.50	11.17	0	2	4	11
Plant protection cost (Rs/ha)	5259	21806	3000	8000	2094	5090	6510	12105
Cost of production (Rs/ha)	26414	66196	9000	16667	81942	85140	38321	51709
Yield (q/ha)	620.13	580.13	240.8	211.7	347.50	287.60	74	62
Gross returns (Rs/ha)	107348	106857	42333	48167	312750	263340	199800	167400
Net returns (Rs/ha)	80934	40661	33333	37500	231386	156900	161479	115691
C:B ratio	1: 3.04	1: 0.61	1:3.70	1: 1.89	1 : 2.8	1:1.84	1:5.02	1:3.02

its cultivation is on the rise. In Karnal (Haryana) alone, bell pepper has witnessed a sharp increase in area in recent past and it is grown on more than 1000 acres with much of the fresh harvest transported to Delhi. In this region, damping off, thrips, aphids (nursery), *Fusarium* wilt, *Rhizoctonia* root rot, *Pythium* fruit rot (sporadic), virus complex, fruit borer *Helicoverpa armigera* put together cause about 20-25 per cent losses depending on the occurrence and severity of above pests. To manage these pests, farmers resort to 14-15 chemical sprays in a season without appreciable increase in yield. It is in this context, that an eco-friendly IPM protocol comprising raising healthy nursery using soil solarisation, mixing of *Trichoderma*, erection of pheromone traps, sprays of *Ha*NPV, collection and destruction of borer affected fruits, periodical rouging of diseased damaged plants, spraying biopesticide like spinosad, release of *T. chilonis* based on pheromone monitoring was synthesised, validated, refined and implemented over a large area in Daha-Bajidan-Sirsi, a hamlet of villages of Karnal district of Haryana during 2007-16.

Hot Pepper/Chillies

It is a crop of tropical and sub-tropical regions and can be grown successfully even as a rain-fed crop. Hot pepper occupies an area of one million ha with an annual production of one million tonnes, its productivity however, still remains low owing to damage by several pests *viz.* thrips (*Scirtothrips dorsalis* Hood), white fly (*Bemisia tabaci* Gennadius), aphids (*Myzus persicae* Sulzer), gall midge (*Asphondylia capsici* Barnes), broad mites (*Polyphagotarsonemus latus* Banks), fruit borer (*Helicoverpa armigera* Hubner), tobacco caterpillar (*Spodoptera litura* Fabricius), *Cercospora* leaf spot, Die-back and anthracnose, powdery mildew, *Fusarium* wilt and virus mosaic complex. The detailed IPM modules for bell pepper and chilly is presented in Table 7 and their economics in Table 8.

Brinjal

Egg plant (*Solanum melongena* L.) is one of the important vegetable crops grown extensively in different parts of India. Attack by several insect pests and diseases

Table 7: Validated IPM Modules for Bell Pepper and Hot Pepper

Bell Pepper (Sardana et al., 2012)	Hot Pepper (Naik et al., 2015)
Nursery	
✩ Mixed fungal antagonist *T. harzianum* (c.f.u. 2 x 10^9/gm) in soil with FYM. ✩ Spray dithane M-45 and imidacloprid in nursery for collar rot and aphids, respectively	✩ Seed treatment with imidacloprid 70 WS and *T. viride* ✩ Spray of imidacloprid for sucking pest ✩ Neem cake application @ 1 q/acre for nematodes
Main Crop	
✩ Dipping of seedlings in *Pseudomonas fluorescens* solution @ 5 ml/litre and imidacloprid for ten minutes before transplanting ✩ 1-2 sprays of neem (neem ban) against thrips and aphids in initial stages. ✩ Spray spinosad @ 1ml/litre if population of thrips is high ✩ Erect pheromone traps @ 5/ha for borer monitoring. ✩ Two sprays of *Ha*NPV 250 LE/ha in initial stages in evening. ✩ One spray of green labeled rynaxpyr 20EC for fruit borer. ✩ Periodically remove and destroy borer damaged fruits collar rot damaged plants ✩ One or two application of carbendazim for Fusarium wilt. ✩ Use of Phenyl tubes for blue bull	✩ Vermicompost application @ 1 q/acre ✩ Pegasus spray @1 g/litre for thrips and mites ✩ Application of thiophenate methyl @ 1g/L for leaf spot ✩ General spray of *Pseudomonas fluorescence* to induce systemic resistance and as a biocontrol agent ✩ Another spray of Pegasus for thrips ✩ One spray of Wettable Sulphur 75 WP for powdery mildew ✩ Spray of spinosad for fruit borers ✩ Drenching with carbendazim 50 WP @ 2g/litre for wilt or rot management ✩ Crop rotation for wilt management

Table 8: Impact of IPM on Productivity of Bell Pepper and Chilli

Parameter	Bell Pepper (Karnal, Haryana) (Sardana et al., 2012)				Chilli (Raichur, Karnataka) (Naik et al., 2015)			
	2010-11		2011-12		2010-11		2011-12	
	FP	IPM	FP	IPM	FP	IPM	FP	IPM
No. of chemical sprays	6.2	14.0	6.5	16	17.0 (27)	10.0	22.0	11.0
Number of biopesticide sprays	–	–	–	–	–	5.0	3.0	5.0
Cost of plant protection (Rs/ha)	10977	15525	11251	16585	–	–	–	–
Culti. Cost (Rs/ha)	78296.5	79622	62936	64124	44869.0*	38389.0*	41150*	36280*
Yield (q/ha)	310.5	245	163.755	124.8	25	30.7	10	15
Gross return (Rs/ha)	205077	161700	212275	175240	235825	289593	64600	96900
Net return (Rs/ha)	126781	82077	149338	111116	190,956	251204	23450	60620
C:B Ratio	1:2.61	1:2.03	1:3.37	1:2.37	1:5.25	1:7.54	1:1.57	1:2.68

Figures in parentheses indicate the number of insecticides used in mixtures.

* Cost of cultivation includes the cost of plant protection (Rs/ha).

viz; leaf folder, *Eublema olivacea* (Walker), hadda beetle *Epilachna dodecastigma* Mulsant, leaf hopper, *Amrasca bigutulla bigutulla* Ishida, aphid *Aphis gossypii* Glover, shoot and fruit borer *Leucinodes orbonalis* Guenee, little leaf, phomopsis blight and fruit rot, *Phomopsis vexan,* wilt *Sclerotinia sclerotiorum* and bacterial wilt, *Pseudomonas solanacearum etc.* at different stages of crop growth hampers the production potential of the brinjal crop.

Multifaceted larger area validation of adaptable IPM technology in egg plant were carried out in a participatory approach at farmers' fields in village Raispur of Ghaziabad in an area of 80 acrea covering 50 farming families during 2004-05 and 2005-06.

IPM Protocol for Brinjal Crop

Nursery

- ✰ Raised seed bed of 10-15 cm height to avoid flooding of bed
- ✰ Soil solarisation for three weeks during June using polythene sheets of 45 mm width
- ✰ Soil enrichment with *T. viride* @ 100 g/kg FYM for three weeks before mixing in soil
- ✰ Seed treatment with hot water, carbendazim @ 0.25 per cent/captan

Main Field (After Transplanting)

- ✰ Installation of Delta traps and yellow sticky traps @ 5/ha) for hoppers, mites and white fly
- ✰ Three sprays of NSKE @ 5 per cent
- ✰ Need-based spray of imidacloprid @ 0.5 ml/l
- ✰ Seedling dip in rynaxypyr @ 0.5 ml/l solution for three hours
- ✰ Pheromone traps installed @ 100/ha for *L. orbonalis* mass trapping as well as monitoring
- ✰ Soil application of neem cake @ 250 kg/ha along the plant rows at planting and 30 DAP
- ✰ Clipping of borer damaged shoots and early infested fruits at weekly interval
- ✰ 3- 4 releases of egg parasite, *T. chilonis* @1.0 lakh/ha
- ✰ Need based spray (ETL > 5 per cent) of rynaxypyr and Emmamectin benzoate 5 SG @ 100 g/acre formulated and spinosad 45 SC and alternatively give second round of all three insecticides (All these three have different mode of action at flowering initiation and fruiting
- ✰ Collection and destruction of borer affected fruit
- ✰ Collection and destruction of egg masses, larvae and adults of hadda beetle
- ✰ Rougeing of little leaf affected plants
- ✰ Add bleaching powder @ 12 kg/ha for bacterial wilt and follow crop rotation with non-solanaceous crop

IPM package consisted of healthy nursery raising using raised seed bed, mixing of *Trichoderma* in FYM in soil, soil solarisation, installation of pheromone traps, application of neem cake, release of *Trichogramma,* spray of NSKE 5 per cent, imidacloprid and cypermethrin insecticides, collection and destruction of borer damages fruits from time to time and need based spray of rynaxypyr and Emmamectin benzoate and spinosad and alternatively give second round of all three insecticides (All these three have different mode of action at flowering initiation and fruiting).There was a high reduction in number of sprays from 6.6 and 7.20 in non-IPM trials to 2.0 and 2.70 in IPM trial in brinjal crop. A slightly higher cost of cultivation *i.e.* Rs. 27,310 and 24,340.38/ha was observed in IPM fields as against Rs. 26,781 and 25,863/ha in non-IPM fields; but this higher cost was compensated by the higher yields obtained in IPM fields (450.05 and 454.14 q/ha) than non-IPM fields (274.12 and 354.62 q/ha). This resulted in higher cost: benefit ratio of 1: 3.27 and 1:3.73 in IPM than non-IPM fields (1: 2.02 and 1:2.74). IPM farmers earned a net income of Rs. 62700 and 66487/ha as compared to farmers who did not adopt IPM (Rs. 28,043 and 45,061/ha) (Sardana *et al.*, 2008).

However, the adoption and promotion of IPM in Fruit and vegetable crops still remains a challenge in view of several constraints faced by farmers because of small holdings, their raw consumption, harvest at short intervals and less time available for reaching from harvest to table. Wider area adoption of IPM is further challenging because the vegetable cultivation is often fragmented in terms of area and space and IPM being knowledge intensive, dependent on quality critical inputs like bio-pesticides and often requiring community approach, thereby making the adoption of larger areas extremely difficult proposition. Use of curative tools like pesticides is based on ETL and monitoring of pest population is nominal and complex to adopt in spite of use of ICT. Extreme climatic conditions, regional variations in crops and pests problems warrants frequent orientation in IPM formats for development of season and location specific IPM practices. Establishment of national IPM network through using ICT technology for major pests across crops and agro-ecological regions for real time pests scenario building and the implementation of IPM practices is the need of time. Establishment of national hub for decision support system for crop protection services of data is the future thrust area NCIPM is the key institution that shall play a key role in development and spread of eco- friendly IPM practices across the country in cooperation with crop institutes (ICAR) AICRPs, SAUs, NGOs, ZPDs and KVKs in a phase manner. Effectiveness of IPM should be made obvious to research managers/policy makers to draw financial assistance to support future IPM programmes.

References

Ahuja, D.B, Ahuja, Usha Rani, Singh, S.K. and Singh, Niranjan. 2015c. Comparison of integrated pest management approaches and conventional (non-IPM) practices in late-winter-season cauliflower in Northern India. *Crop. Prot.*, **78**: 232–238.

Ahuja, D.B, Ahuja, Usha Rani, Srinivas, P., Singh, R.V., Malik, Meenakshi, Sharma, Pratibha and Bambawale, O.M. 2012. Development of farmer-led integrated management of major pests of cauliflower cultivated in rainy season in India. *J. Agril. Sci.*, **4**(2): 79–90.

Ahuja, D.B., Rao, C.N., Das, A.K., Arora, P., Deka, S., Sardana, H.R. and Chattopadhyay, C. 2015b. IPM in citrus. **In**: *National Symposium on Sustainable Citrus Production: Way Forward*, 28th November, ICAR-CCIR, Nagpur.

Ahuja, D.B., Sabir, N., Singh, Jitender, Arora, Purshotam, Arora, Jitender, Rao, C.N., Das, A.K., Ladaniya, M.S., Bairwa, B.S., Deka, Shikha and Ahuja, Usha Rani. 2016. IPM in Kinnow and Khasi mandarins. **In**: *Success Stories of IPM in India*, (Eds.) S. Vennila, A. Birah, Kanwar and C. Chattopadhyay, NCIPM, New Delhi.

Ahuja, D.B., Saxena, S.K., Kalaria, G.B., Sharma, H., Singh N. and Singh, R.V. 2015a. Pest management in 'Alphonso' mango orchards of Southern Gujarat region of India using elements of cultural and mechanical methods (under publication).

Ahuja, D.B., Sharma, Pratibha, Sardana, H.R., Singh, Saroj, Singh, S.K., Sabir, Naved, Singh, R.V. and Singh, S.P. 2009. *Barsati Phul Gobhi avm Patta Gobhi Mein Samekit nashijiv Prabandhan* (In Hindi). National Centre for Integrated Pest Management, New Delhi.

Ahuja, D.B., Singh, S., Singh, R.V., Ahuja, Usha Rani, Malik, Meenakshi and Bambawale, O.M. 2011. Farmers led development and validation of IPM technology in cabbage. *Ind. J. Hort.,*. **68**(4): 578–580.

Ahuja, D.B., Singh, Swaroop, Sharma, Pratibha, Malik, Meenakshi, Singh, R.V. and Kalra, Bharti. 2013. Development and on-farm validation of location specific IPM technology for rainy season cauliflower. *Ind. J. Agril. Sci.*, **83**: 78–82.

Chand, R. 2016. Addressing agrarian distress: Sops versus development. 23rd B.P. Pal memorial lecture, 26th May, IARI, New Delhi.

Naik, M.N. Sardana, H.R., Arun Kumar, Sehgal, Mukesh, Bheemenan, M and Bhat, M.N. 2015. Area-wide Farmers driven approach for validation and economic analysis of sustainable and adaptable IPM technology for hot pepper, *Capsicum annum*. *Ind. J. Pl. Prot.*, **43**(1): 54–59.

Sardana, H.R, Kanwar, Vikas, Bhat, M.N. and Singh, R.V. 2013. Farmers' driven approach for field implementation and economic analysis of IPM technology in tomato. *Ind. J. Ent.*, **75**(3): 185–188.

Sardana, H.R., Bambawale, O.M. and Batra, Poonam 2008. Wider area validation and economic analysis of adaptable IPM technology in egg plant *Solanum melongena* L. in a farmers' participatory approach. *Pest. Res. J.*, **20**(2): 204–209.

Sardana, H.R., Bhat, M.N. and Sehgal, Mukesh. 2012. Wider area validation and economic analysis of adaptable IPM technology in bell pepper (*Capsicum annum* var. frutescens)*. *Ind. J. Agril. Sci.*, **82**(2): 186–9.

Satyasai and Mehrotra. 2016. Enhansing farmers income. NABARD foundation day seminar, Vigyan Bhawan, 12th July, New Delhi.

Trivedi, T.P. and Ahuja, D.B. 2011. Integrated pest management: Approaches and implementation. *Ind. J. Agril. Sci.*, **81**(11): 81–93.

Trivedi, T.P., Sardana, H.R. and Thakur, V.S. 2004a. Apple. **In**: *Validated IPM Technologies for Selected Crops*, (Eds.) Amerika Singh, H.R. Sardana and Naved Sabir, pp. 163–180.

Trivedi, T.P., Sardana, H.R., Shukla, R.P. and Misra, A.K. 2004b. Mango. **In**: *Validated IPM Technologies for Selected Crops*, (Eds.) Amerika Singh, H.R. Sardana and Naved Sabir, pp. 149–162.

43

Integrated Management of Fruit Flies in Horticultural Crops

Kirti Sharma*, G.S. Panduranga and Ram Singh

Division of Entomology,
ICAR-Indian Agricultural Research Institute, New Delhi – 110 012
****E-mail: kirtisharma2@yahoo.com***

Improving productivity and quality of fruits and vegetables enhances food security, employment and trade opportunities. The Asian region (East, Southeast, and South) is among the top three regions worldwide for both exporters and importers of fruits and vegetables. Among the various sub-regions in Asia, the Mekong river basin countries are major producers of a range of economically important fruits and vegetables for domestic consumption, regional and international trade. Almost a decade before in 2004, Asian countries produced 178 million tonnes of tropical fruits which amounted to 66 per cent of the total global production and earned US$ 2.5 billion. However, several factors constrain fruit and vegetable production such as tephritid fruit flies (*Bactrocera dorsalis* and *B. cucurbitae*, *B. zonata* and possibly others like *B. correcta*). These fruit flies cause direct damage to fruits and vegetables which can lead to up to 90-100 per cent yield loss depending on fruit fly population, locality, variety and season. In addition, to the direct losses, fruit fly infestation can result in serious losses in trade value and export opportunity due to strict quarantine regulations imposed by most importing countries. Fruit flies are major pests of several fruit and vegetable crops throughout the tropical and subtropical world. There are about 325 species of fruit flies occurring in the Indian subcontinent, of which 205 are from India alone. The major fruit fly pests in India belong to the genus *Bactrocera*: *B. Cucurbitae* (Coquillett), *B. dorsalis* (Hendel) and *B. zonata* (Saunders). Other species of *Bactrocera*, such as *B. correcta* (Bezzi), *B. diversa* (Coquillett) and *B. Latifrons* (Hendel), although moderate pests, are localized in their distribution. *B. correcta* is occasionally reported to dominate *B. zonata* and *B. dorsalis* on mango. To date, five or six species of the *B. dorsalis* complex have been recorded in India, and at least 10 species may occur there, as well as three or four species of the *B. zonata* complex. *B. tau* (Walker) and *B. scutellaris* (Bezzi) have not been recognized even as moderate pests, whereas *B. Caudate* (Walker) is still not fully confirmed in India (V.C. Kapoor, 2005). Nearly 35 per cent of the known fruit fly species attack soft fruits of which mango, guava, citrus, ber, peach and several cucurbitaceous vegetables are important.

India is the second largest producer of fruits in the world with an annual production of 88 million tonnes from an area of 7.2 million hectares and contributes to more than twelve per cent of world's fruit production (Indian Horticulture Database, 2014-15). Major species occurring in India are *B. dorsalis*, *B. cucurbitae*, *B. zonata*, *B. tau*, *B. correcta*, *B. diversa*, *B. latifrons*, *B. versicolor*, *B. nigrofemoralis*, *B. minax*, *D. ciliatus* and *Carpomya vesuviana*.

Among various tropical fruits, guava and mango fruits are nutritive and attractive delicious commodities to eat for man. Guava (*Psidium guajava* Linn.) also called the apple of tropics is the fourth most important fruit crop in India, with an area of 2.6 lakh hectares and production of 36.68 lakh tonnes (Indian Horticulture Database, 2014-15). Delayed harvesting makes it vulnerable to the attack of fruit flies. Both guava and mango are equally attractive to fruit flies. Peach accounts for 1.8 lakh hectares with production of 9.35 lakh tonnes (Indian Horticulture Database, 2014-15) and is equally attractive to Dorsalis group of fruit flies. Mango, the king of fruits, has a great export potential due to exquisite taste and flavour. Our country ranks first in area and production of Mango with 2.31 million ha and 12.75 MT, respectively. But infestation with oriental fruit fly (*Bactocera dorsalis*) has been a major impediment to mango export. It accounts for about 27 per cent of the harvesting loss.

Fruit flies belong to the family Tephritidaes which is one of the largest, most diversified and fascinating acalypterate families of Diptera. Tephritidae includes about 4000 species arranged in 500 genera (White and Elson-Harris, 1992). These are commonly called "fruit flies" as these dipterans have their close association with fruits and fruiting vegetables. The severity of infestation is recorded upto 100 per cent in guava, mango, bitter gourd, chilli and carambola (ICMPFF). Female fruit flies lay eggs in fruit and single female can ruin more than 400 different fruits and vegetables in its life. They are considered as high priority quarantine pests in many countries worldwide. Because of their infestations, India has been included in the list of those countries from where fruit import to developed countries is banned. Most common species attacking mango and guava are *Bactrocera dorsalis* Hendel, *B. correcta* Bezzi and *B. zonata* (Saunders) (Verghese and Sudhadevi, 1998; Rajitha and Viraktamath, 2005). For the large number of small fruit and vegetable growers with limited resources no control is undertaken, fruit is often picked while green, avoiding loss but reducing value, or if left to mature damaged fruit is simply culled.

Fruit Flies in Asia

The tephritid fruit flies of genus, with more than 500 species currently described constitutes important pests of reproductive stages of a number of fruits and vegetable crops in Asian countries. A number of economically important and widely prevalent species of this genus such as *B. dorsalis* (Oriental fruit fly), *B. correcta* (Guava fruit fly), *B. cucurbitae* (Melon fly), *B. papayae* (Asian papaya fruit fly), *B. carambolae* (Star fruit fly) are commonly found in South east and South Asia. *B. dorsalis* is one of the most common species and it is similar to the closely related species *B. carambolae*, *B. papayae*, *B. occipitalis*, *B. philippinensis* and *B. invadens* in colour pattern and referred as a species complex.

Distribution, Abundance and Host Preference

Bactrocera dorsalis Complex

The Indian species of this complex are: *B. dorsalis* (Hendel), *B. caryeae* (Kapoor), *B. poonensis* (Kapoor), *B. carambolae* Drew and Hancock, *B. vishnu* Drew and Hancock, *B. occipitalis* (Bezzi) and *B. kandiensis* Drew and Hancock. *B. dorsalis* (Hendel) has been known as *Dacus ferrugineus* (Fabricius) in the Indian literature. It is also called the "Oriental fruit fly" due to its wide distribution in the oriental region. It is a serious pest of mango and also attacks guava but mostly in association. *Bactrocera carambolae* Drew and Hancock species is similar to *B. dorsalis* and *B. occipitalis*. The species is presently known from Andaman Islands in India. *Bactrocera (Bactrocera) caryeae* (Kapoor) described from the many unidentified fruit fly specimens present in the collection of the Division of Entomology, Indian Agricultural Research Institute, New Delhi. The species is currently widely distributed in southern India as well as Sri Lanka (Drew and Hancock, 1994). It attacks guava and mango. The species seems to be confined for the present to the southern parts of the Indian subcontinent. *B. vishnu* is another species of the *B. dorsalis* complex described by Drew and Hancock (1994) from Kodaikanal, a famous hill resort of southern India. The host is not known as it was collected using cue lure as attractant.

Bactrocera (Zeugodacus) cucurbitae (Coquillett)

The melon fly, *B. cucurbitae* is a serious pest of many cucurbits and is considered the foremost fruit fly pest of India. It is the only species that is uniformly widespread, attacking a large array of cucurbit fruits. It is easily recognized while at rest by its predominant orange-brown color, postsutural yellow stripes, facial spots, and a characteristic wing pattern. The species is extremely variable morphologically especially in body coloration, chaetotaxy and wing markings. *B. cucurbitae* is commonly known as the "melon fly" due to its preference for melons. In India it destroys either partially or completely, over 60 per cent of the cucurbit crops if left untreated (Kapoor and Agarwal, 1983). During the hot summer months (*i.e.* April-June) attacks on cucurbits by *B. cucurbitae* often occur together with *Dacus (Didacus) ciliatus* Loew. During July-August infestations of *B. cucurbitae* are so severe that almost all cucurbits are attacked if left unprotected resulting in heavy damage. In some cases the entire crop is lost. The most vulnerable hosts are bitter gourd, ribbed gourd, bottle gourd, long melons, squash melons, snap melons, ash gourd and cucumber. The infestation continues till the appearance of fruit continues in October but Intensity of attack diminishes considerably after October.

Bactrocera (Bactrocera) zonata Saunders

The peach fruit fly, *B. zonata* is prevalent in almost all of south-east Asia. It is one of the main species of the *zonata* group. The other species are *B. affinis* Hardy, *B. versicolor* Bezzi, and *B. correcta*. It is red-brown in colour with incomplete costal band and without anal streak. *B. zonata* is similar in distribution to *B. cucurbitae* and attacks more than 50 host plants, including guava, peach and mango. Since it prefers peach and guava, it is commonly referred to as "peach fruit fly". It attacks soft and tender fruit. During April to June the attack on peaches is sometimes so severe that it destroys the entire crop. With the beginning of monsoon, *i.e.*, from July onwards, it prefers guava. Mango is also attacked at this time, but in India this is mixed with attack by *B. dorsalis* too.

Bactrocera (*Bactrocera*) *correcta* (Bezzi)

The guava fruit fly, *B. correcta* is distributed throughout the Indian subcontinent and Thailand but has also invaded places distant from Asia. In India it usually occurs in association with *B. zonata* and *B. dorsalis* (Kapoor, 1989), and has similar hosts. Its distribution in India is peculiar: it is found in isolated pockets, due to the dominancy of ecological homologues such as *B. zonata* and *B. dorsalis*. In recent years it has increased its intensity of attack in some parts of India, dominating the other two species on guava and mango. Its numbers start increasing by late June or early July, when other fruit fly species are still low in number. In Tamil Nadu this species is showing signs of dominancy over *B. zonata* and *B. dorsalis* on guava for the first time, indicating that *B. correcta* may become a major threat to guava in south India.

Bactrocera (*Bactrocera*) *latifrons* (Hendel)

It is a serious pest of solanaceous crops including chilli (*Capsicum* sp.). In the Indian subcontinent it has been reared from eggplant, *Solanum melongena* and *S. sisymbriifolium* (Narayanan and Batra, 1960). Udayagiri (1987) reared it from berries of *Solanum vivarum* in Karnataka. Recent collections have shown its predominance in southern India and its potential threat for other parts of India.

Bactrocera (*Zeugodacus*) *tau* (Walker)

This species is widely distributed in the Oriental region and has been known as *Chaetodacus tau* (Walker). It attacks a wide range of cucurbitaceous plants, tomatoes and fleshy fruits such as jack fruit, guava, mango, sapodilla and wax apple. The species is fairly well distributed across the whole of the Indian subcontinent and commonly occurs in the hilly tracts of northern India, attacking all kinds of cucurbits and tomatoes. The species is easily recognized by the orange-brown scutum marked with lateral and medial yellow stripes, facial spots not interconnected, and wing with a costal band expanded into a large apical spot.

Bactrocera (*Zeugodacus*) *caudata* (Fabricius)

This species has also been known as *Bactrocera maculipennis* Doleschall, *Chaetodacus caudatus* Fabricius and *Zeugodacus caudatus* (Fabricius). It is easily recognized by being predominantly black with lateral and medial yellow scutal stripes; face with a black line across mouth opening or with the black spots in the antennal furrows extended laterally and almost forming a line across the mouth opening; and wing with costal band expanded into an apical spot. The species is distributed in Indonesia (Java, Sumatra), Malaysia, Myanmar, Taiwan, Thailand, Vietnam and Brunei. The species was also recorded from India and Nepal (Kapoor *et al.*, 1976; White and Elson-Harris, 1992).

Bactrocera (*Zeugodacus*) *scutellaris* (Bezzi)

This is distributed throughout almost the whole of the Indian subcontinent, Myanmar and Thailand. It is similar to *B. cucurbitae* in having 3 frontal setae. It attacks various cucurbits, especially pumpkin and gourds. It is noteworthy that the species is quite active in cool climates and is becoming an important fruit fly pest in some areas of Himachal Pradesh.

Bactrocera (*Daculus*) *oleae* Gmelin

The olive fruit fly, *B. oleae* is distributed throughout Europe, Africa, Middle East, Northern India and North-West Pakistan. It is a serious pest of olives and involves millions of dollars in its control every year. It is easily recognized by the predominantly black scutum (White and Elson-Harris, 1992); no yellow stripes on thorax; facial spots present. Fletcher (1917) was the first to report it from India and Pakistan.

Table 1: List of Category 'A' Pest of *Bactrocera* Based on Pest Severity, Host Range, Invasiveness and Frequency of Infestation (Vargas *et al.*, 2015)

Species	*Hosts*	*Distribution*
B. carambolae Drew and Hancock	Polyphagous fruit pest	Vietnam to Indonesia. Introduced into South America
B. correcta (Bezzi)	Polyphagous fruit pest	Pakistan to Vietnam.
B. cucurbitae (Coquillett)	Primarily Cucurbitaceae (fruit and flower)	Tropical Asia (widespread). Introduced into Africa and Oceania.
B. dorsalis (Hendel)	Polyphagous fruit pest	Tropical Asia (widespread). Introduced into Africa and Oceania.
B. latifrons (Hendel)	Mainly Solanaceae	Pakistan to Taiwan; south to Sulawesi. Introduced into Hawaii and Africa.
B. neohumeralis (Hardy)	Polyphagous fruit pest	Australia, New Guinea.
B. oleae (Gmelin)	Olive	Africa. Introduced into southern Europe, the Middle East and California.
B. tryoni (Froggatt)	Polyphagous fruit pest	Australia. Introduced in Oceania.
B. zonata (Saunders)	Polyphagous fruit pest	India to Vietnam

***Bactrocera* (*Paratridacus*) *diversa* Coquillett**

B. diversa has invaded almost the whole of the Indian subcontinent and Thailand. It is easily recognized by its predominantly black scutum with bright yellow stripes. It has also been seen congregating on undersides of leaves of guava and mango alone or in association with *B. dorsalis* or *B. zonata*.

Life Cycle and Biology of Fruit Flies

Fruit flies deposit their eggs on host fruitswith the help of ovipositor by puncturing the fruit when they are physiologically ripe. Each female lays 8-15 eggs in each puncture that encourages bacterial/mold growth on the fruits. On hatching, maggots feed on fruit pulp for 6-13 days and make tunnels inside the fruit. Damage causes premature falling of fruits. Mature larvae jump out of the fruit, make tunnel and pupate inside the soil. Pupation takes 7-11 days. Fruit fly lives for more than 75 days and lays thousands of eggs intermittently in its whole life. Being polyphagous pests with high reproductive potential, wide host range, adaptability to climate and overlapping of generations, their management is rather difficult.

Fruit Fly Damage

The damage starts when the female fruit fly punctures the fruit with its long and sharp ovipositor. The fruit skin is breached, and bacteria enter and the fruit starts to decay. The larvae that hatch from the eggs feed on the decaying fruit tissue, and on the yeasts and bacteria that multiply in it. It is believed that fruit fly females carry bacteria with them that they inject into the fruit at oviposition so that the fruit decays faster (making it more nutritious for the larvae). Fruits with fruit fly larvae in them decay quickly. It is sometimes possible to cut out the damaged part and use the remaining part of the fruit for home consumption, but infested fruits are generally unmarketable, and can certainly not be exported. Crop losses can vary from a few per cent up to 100 per cent, and losses of 90 per cent or over are common. In some cases losses can be reduced by other treatments applied by the farmers in an area, *e.g.* against another orchard pest, or in another crop intercropped in the orchard.

Strategies for Area-Wide Management

Wide area management is not a unitary concept, but incorporates a number of related but distinct methods including local area management such as bagging of fruits, field sanitation, protein baits and lure traps, host plant resistance, biological control, and soft insecticides, can be employed to keep the pest population below economic threshold in a particular crop over a period of time to avoid the crop losses without health and environmental hazards. The methods used for a wide area management includes:

1. Insecticide-based Suppression Tools including Cover Sprays, Protein Bait Sprays, and Soil Drenches

Addition of protein food baits to insecticide sprays reduced the amount of pesticide needed for fruit fly control and has been used successfully in many eradication programs (Steiner *et al.*, 1961). Female flies after emergence from the pupae feed on nitrogenous substances for maturation of gonads and in need of protein for full ovarian development and egg production, readily feed on a protein source containing a toxicant. Enzymatic protein hydrolysate baits were first used in Hawaii for control of *B. dorsalis*, and malathion became the organophosphate insecticide additive of choice due to its low mammalian toxicity, affordable price, and low levels of fruit fly resistance (Steiner, 1952). In the late 1990s, a new bait spray formulation containing the reduced-risk biopesticide spinosad was shown to be as effective as Nulure-malathion bait sprays for use in Central America and the USA against *Ceratitis capitata* (Wiedemann) (Vargas *et al.*, 2002). Spinosad, a bio-pesticide derived from a soil-dwelling actinomycete bacterium (*Saccharopolyspora spinosa* Mertz and Yao), has low mammalian toxicity and reduced environmental impact on natural enemies (Stark *et al.*, 2004). A spinosad-based hydrolysed protein bait that attracted, induced feeding and killed fruit flies was initially developed by Moreno and Mangan. Later it became known as GF-120 Fruit Fly Bait (Dow AgroSciences, Indianapolis, IN).

Fruit fly larvae after maturation in the fruit, leaves the fruit any jump down on the soil underneath the tree. The make their way to soil and pupate inside the soil. Therefore, another component in many tephritid fly area-wide IPM programs has been the application of insecticidal soil drenches under host trees where fruit flies have been detected (Ekesi *et al.*, 2007). Florida uses diazinon as a soil drench underneath host plants where fruit fly larvae or gravid adult females are detected during eradication programs.

2. Male Annihilation

Male Annihilation Technique (MAT) involves the use of a high density of bait stations consisting of a male lure combined with an insecticide (usually technical malathion, and more recently fipronil), to reduce the male population of fruit flies to such a low level that mating does not occur. This technique may be used for the control of those fruit fly species of the genera *Bactrocera* and *Dacus* that are attracted to male lures (cuelure or methyl eugenol). The commercial product methyl eugenol is used for trapping of Dorsalis fruit flies complex, whereas cue-lure is used to trap the male of melon fruit fly complex, and Trimedlure captures Mediterranean fruit fly complex. These lures can be also used as monitoring traps, to convey the presence of fruit fly. When traps are used in the large numbers they disturbs the male: female ratio and reduces the insect's chances of mating and these females produce few progeny. Accordingly, the insect population at the target area declines and the insects could be eradicated at the end (Zaheeruddin, 2007). Stonehouse *et al.* (2008) described that this technique can also be used for rapid detection of accidental introductions of species of Dacinae (comprised of the two major genera, Bactrocera Macquart and Dacus Fabricius).

At least 324 species of the male *Bactrocera* are attracted to cue-lure (C-L) and 123 species to methyl eugenol (ME) (Vargas *et al.*, 2014). Recently, Vargas *et al.* (*2015*) summarized the history of Male Annihilation Technique (MAT) and the many successful programs worldwide. In all cases, eradication was attempted or achieved through repeated area-wide applications, by airdrops and/or manual ground applications, of bait stations (fiberboard or coconut husk blocks, cotton string or wick, or molded paper pulp) impregnated with male lures mixed with a toxicant (naled, malathion, or fipronil). The first fruit fly eradication using ME-MAT was achieved with *B. dorsalis* in the Mariana Islands (Steiner *et al.*, 1965). From 1960 to 2012, nine different *Bactrocera* species have been detected in California, with *B. dorsalis* detected most frequently (126 times). Altogether, there have been 140 eradication programmes with 25 quarantines (Vargas *et al.*, 2014). STATIC™ Spinosad-ME has been tested and registered for use in California and Florida and offers a novel and convenient ready-to-use MAT formulation that is safer than the current use of organophosphates. SPLAT-MAT-C-L formulation has also been tested successfully in Hawaii (Vargas *et al.*, 2008) for use against C-L responding flies. MAT is well suited to large areas of fruits with lower levels of attack (for example mango plantings over several hectares) in which the blocks can be applied and replaced from the beginning of the fruiting season. Thus the male annihilation technique (MAT) is a chemical control method designed to deplete the males available for mating in a population and thus breaks the reproductive cycle of fruit flies. The growers are advised to implement MAT in their orchards as a supplement to baiting.

3. Sterile Insect Releases

It is the technique where the sterile males are released in the fields for mating with the wild females. Sterilization is accomplished through irradiation, chemo-sterilization, or by genetic manipulation. In sterile insect programs the terms 'sterility' or sterile insect' refer to the transmission of dominant lethal mutations that kill the progeny. The females either do not lay eggs or lay sterile eggs. Ultimately, the pest population can be eradicated by maintaining a barrier of sterile flies. A sterile insect program is species specific, and considered to be an ecologically safe procedure and has been successfully used in area-wide approaches to suppress or eradicate insect pest. This technique also known as the sterile insect technique (SIT) (Knipling, 1979), and is well suited for the suppression or eradication of *Bactrocera* flies that display elaborate courtship behavior that can be mediated by male lures (Benelli, 2015). Scientists from Hawaii and Australia carried out the original pilot SIT tests to eradicate *B. cucurbitae* from Rota in the Northern Mariana Islands (Steiner *et al.*, 1965), suppress *B. tryoni* in Australia (Andreawartha *et al.*, 1967), eradicate *B. dorsalis* in Micronesia (Steiner *et al.*, 1970) and suppress or eradicate *C. capitata* in Hawaii (Harris *et al.*, 1986). In Japan, *B. cucurbitae* was eradicated by SIT (Koyama, 1996) using the approach developed in Hawaii. The development of a sexing strain for *B. cucurbitae* (McInnis *et al.*, 2004) allowed for the application of SIT to small-farm situations in selected areas and reduced the local *B. cucurbitae* population to near extinction (McInnis *et al.*, 2007).

4. Biological Control

Eggs and larvae are main stage. Hymenopteran parasitoids are commonly employed; biological control alone does not provide high degree of control on sustainable basis. Srinivasan (1994) reported *Opius fletcheri* to be a dominant parasitoid of *B. cucurbitae*, but the efficacy of this parasitoid has not been tested under field conditions in India. Mexican strain of the nematode, *Steinernema carpocapsae* Weiser (*Neoaplectana carpocapsae*), has been reported to cause 0 to 86 per cent mortality to melon fruit fly. Sinha (1997) reported that culture filterate of the fungus, *Rhizoctonia solani* Kuhn, to be an effective bio-agent against *B. cucurbitae* larvae. The efficacy of most of these bio-agents is unclear under field conditions. Another biopesticide with activity against larvae and pupae of major *Bactrocera* (*B. dorsalis* and *B. cucurbitae*) and *Ceratitis* (*C. capitata*, *C. cosyra* (Walker)) species in soil is the fungal pathogen *Metarhizium anisopliae* (Metchnikoff) Sorokin (Ekesi *et al.*, 2007).

5. Cultural Controls

Cultural controls involve a variety of approaches ranging from use of fruit fly resistant varieties, to early harvesting of fruit, bagging of fruit and field sanitation (Allwood, 1997; Ekesi and Billah, 2007). In the case of tephritid fruit flies, cultural practices can support synergies with other components of IPM, such as protein bait sprays and biological control. Field sanitation is a technique that either prevents fruit fly larvae from developing or sequesters young emerging adult flies so that they cannot return to the crop to reproduce (Vargas *et al.*, 2008). To break the reproduction cycle and population increase, growers need to remove all unharvested fruits or vegetables from a field by completely burying them deep into the soil. Burying damaged fruits 0.46 m deep in the soil prevents adult fly eclosion and reduces population increase (Klungness *et al.*, 2005). Bagging of fruits on the tree (3 to 4 cm long) with 2 layers of paper bags at 2 to 3 day intervals minimizes fruit fly infestation and increases the net returns by 40 to 58 per cent (Jaiswal *et al.*, 1997). Wrapping materials can be newspaper, paper bags, or polythene sleeves in the case of long/thin fruits (Sarker *et al.*, 2009). This system also provides physical protection from mechanical injuries (scars and scratches) and, in some cases, reduces fungal spots on the fruits.

6. Quarantine

The import and export of infested plant material from one area or country to other non-infested places is the major mode of the spread of insect-pests. When exporting tropical fresh fruit, such as mango and papaya, the fruit needs to go through strict plant quarantine inspections in the importing country to ensure that quarantine pests such as eggs and larvae of Melon flies and oriental fruit flies (which originate in tropical regions) and are present in fruits grown in these regions are sterilised and incapable of growing and multiplying in the importing country. The spread of fruit fly can be blocked through tight quarantine and treatment of fruits at the import/export ports.

7. Vapour Heat treatment (VHT)

Vapour Heat Treatment is a perfect method for such a treatment and the process complies with strictest international standards. The Vapour Heat Treatment System achieves this objective with the use of hot saturated water vapour, and without the use of any chemicals. At the same time, during the process, the temperature and humidity are meticulously controlled so that the fruit is not damaged and natural freshness is preserved. The Vapour Heat Treatment System is a safe way to treat fruit so that it passes the quarantine inspection, enabling the fruit to be exported. Importing countries where such pests do not exist insist that such fruit is treated by suitable process or processes to ensure extinction and sterilisation of such pests as a means to protect the flora of the importing country. The range of temperature and humidity in VHT is varies from 10 to 60°C and 55 to 95 per cent RH, respectively depending upon the perishability of fruit and vegetables.

Vapour Heat treatment (VHT) of avocado fruits infested with eggs and larvae of *B. cucurbitae* for 40° C for 24 h reduced the estimated surviving population by 99.5 to 100 per cent (Yang 1994). Import controls carried out in airports in France since 1993 on tropical fruits have revealed the presence of 12 non-European and one European species of Tephritidae, (Bayart *et al.*, 1997).

Fruit flies are modern-day pestilences which ravage production of horticulture. These flies, moreover, are good fliers and therefore their spread is extensive. Hence, area wide adoption of management strategy is recommended. Male Annihilation Technique (MAT), which uses sexual lures to capture males of fruit flies and kill them, is an effective to control growth and spread of fruit flies. This however, requires a community approach for implementation. Because of less initiative taken on community basis/administrative initiative farm level management is prevalent in India. The following are three general scenarios where these technologies can be most effectively applied in India.

Farm-Level Management of Fruit Flies

(i) For Cucurbits

Fruit fly management by bait sprays is roughly equivalent in effect to management by cover sprays, so the advantage of bait control lies in its greatly reduced costs and workload. It should be emphasized that bait control does not immediately reduce fly populations to very low levels, and so the advantage of its ease of deployment should be maximised by frequent applications.

The most suitable baits in any situation may depend on local conditions and the preferences of local flies, so an initial check on the effects of individual candidate baits is advisable before wide-scale use. Farmers prefer baits in traps, but these are generally less effective and convenient than liquid baits applied to foliage as "squirts".

Traps or squirts should be used just below the level of the pandal (supporting framework), and with emphasis on the field borders, which are frequented by flies that roost in hedges and surrounding vegetation. In some areas (*e.g.* Hawaii, USA) BAT against *B. cucurbitae* is more effective when applied on border plants (such as maize) which form particularly attractive feeding and resting places for the melon fly (CIRAD, pers. comm., 2006).

(ii) For Orchards

In order to be effective, management of orchard fruit flies by methyl-eugenol MAT appears to need doses substantially greater than those used in India at the moment, and much more powerful than those obtainable using holy basil or "tulsi" (*Ocimum sanctum*, a known methyl eugenol analogue that is used in parts of India)

Farmers should be made aware of the dangers of "psychological control" whereby the appearance of dead flies in traps is taken as evidence of "success" – farmers should be aware that these are males, and that a dozen or so dead in a trap in a week is having little effect on ovipositionthe catch in a trap baited with a powerful methyl-eugenol block, which can catch several hundred flies in few days, is a graphic and effective illustration of how many flies are left alive when a weak trap is killing small numbers.

Methyl-eugenol male annihilation technique while effective at crop protection in most conditions and most seasons even on quite small plots, may not be relied on to provide reliable control under all conditions, and that BAT as a back-up should always be remembered. In most cases, the likelihood is that an economic return in mango protection would be obtained by using both soaked-block MAT and banana-pulp liquid bait application on a weekly basis.

Village-Level Management of Fruit Flies

The requisites for successful and sustainable village-level fruit fly management were identified in the social studies accompanying the experimental evaluation. Initiatives should be taken by either an existing cooperative or by an energetic, popular and effective local mayor. A cooperative which buys produce or otherwise assists in marketing, and already has a role in the distribution of inputs, is the most promising starting point.

Farms should either be quite large and technology-oriented, or quite small and intensively cultivated. Fruit flies should be perceived as particularly serious – as one of the three or four critically limiting factors which come to farmers' minds when first asked. The local community should possess some aspects of social cohesion and, in particular, institutions in which all participate.

Mass Trapping at Indian Agricultural Research Institute, Pusa Campus

Keeping these observations in view, male annihilation traps were installed in and surroundings of IARI campus including the various fields, residential area and different orchards and NASC complex for mass trapping of fruit flies. In these studies effects of attractant, trap density and deployment on the efficacy of the mass trapping methods for the control of the Dorsalis fruit fly complex were reported. The population monitoring was also conducted to study the effect of mass trapping on population reduction.

The mass trapping activity was planned for Pusa campus and the area under mass trapping was approximately 1,200 acres, where 460 traps with methyl eugenol as lure were installed to catch the flies of Dorsalis fruit fly complex. Mass trapping was conducted from October 2007 to May 2009. Traps were prepared by dipping porous and light wooden blocks (5.0 x 4.0 x 0.6 cm^3) in lure and killing agent. The methyl eugenol impregnated blocks were used in transparent plastic jars having three holes of 20 mm diameter on the side walls. The lids of these traps had a hole to pass the wire for hanging. The traps were serviced at one month interval at each and every location.

The populations of native male adult of Dorsalis fruit fly complex dominated by peach fruit fly *Bactrocera zonata* (F.) and oriental fruit fly *Bactrocera dorsalis* (Hendel) were estimated weekly by the capture technique using live traps baited with methyl eugenol, at Indian Agricultural Research Institute orchards where many varieties of fruit trees were grown. Trapping of fruit flies was done with the help of lure (methyl eugenol) combined with an insecticide, to trap the male population of fruit flies infesting mango, guava, sapota, peach *etc*. Fourteen locations were selected and traps were installed. Weekly population was monitored for the whole year. The average fruit fly population of all the fourteen traps per day per trap is shown in the graph below (Figures 1 and 2).

The monitoring was mainly done to know about the fluctuation in population over the period of time, whereas mass trapping was done to reduce the male population. The flies per trap per day is a population index that estimates the average number of flies captured in one trap in one day by the trap exposed in the field. The function of this population index is to have a relative measure of the size of the adult pest population in a given space and time. It is used as base-line information to compare the size of the population before, during

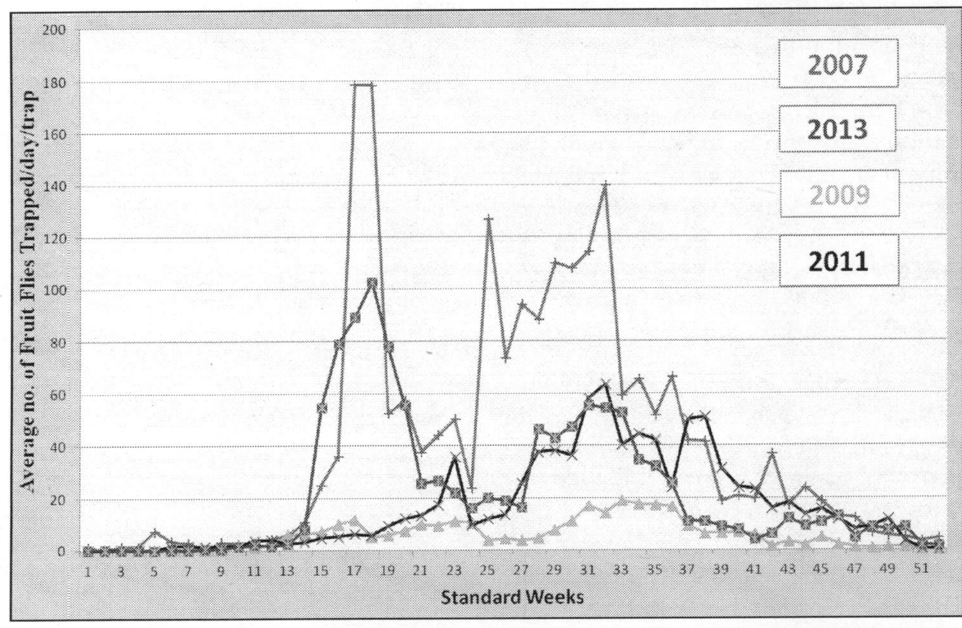

Figure 1: Monitoring of Dorsalis Fruit Fly Complex during 2007, 2009, 2011, and 2013 at Pusa Campus, IARI, New Delhi.

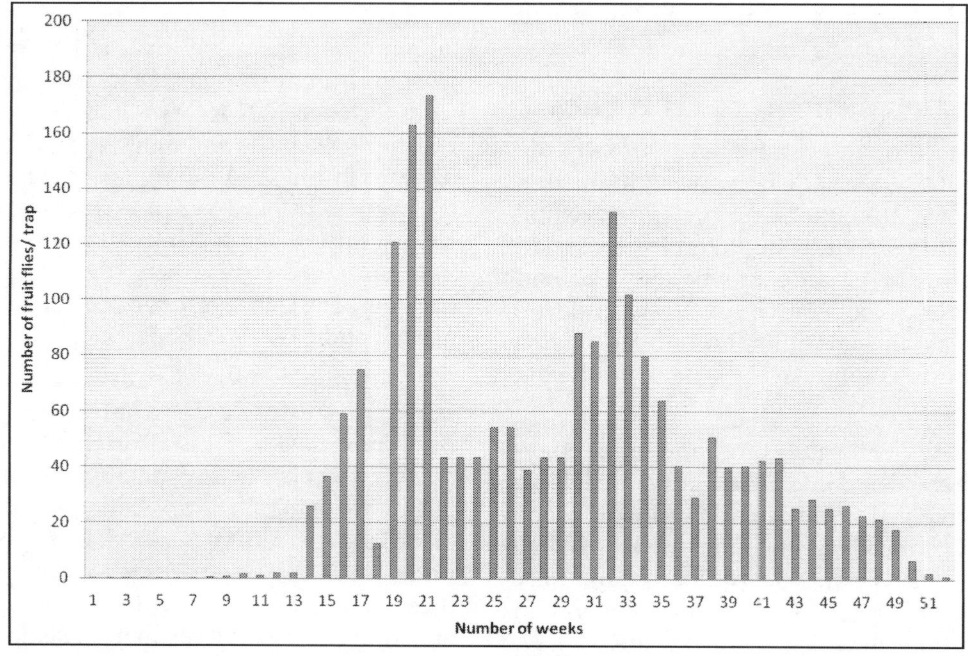

Figure 2: Monitoring of Dorsalis Fruit Fly Complex during 2015 at Pusa Campus, IARI, New Delhi.

and after the application of a fruit fly mass trapping programme. The index was calculated by dividing the total number of captured flies by the product obtained from multiplying the total number of traps by the average number of days the traps were exposed. The formula followed was:

$$\frac{F}{T \times D}$$

where,

F = Total number of flies

T = No. of traps

D = Avg. No. of days traps were exposed.

The experiment on monitoring of Dorsalis fruit fly complex was conducted from 2007 to till date at fourteen locations by hanging one trap at each location and data on population was recorded at weekly interval by servicing the traps at weekly. The mass trapping was done during October, 2007 to May 2009 at 460 stations and population were recorded at weekly interval by servicing the traps at every month throughout the year.

Experimental results on monitoring of Dorsalis fruit fly complex at PUSA IARI, New Delhi found that highest number of flies (178 flies/trap/day) was trapped during 15th to 17th standard weeks of the year 2007. Again in rainy season during 31st standard week of the year 2007, about 140 flies were trapped and it is followed by population (130 flies/trap/day) recorded during 25th to 27th week of 2007 (Figure 1). Maximum flies caught in single trap during 27.04.2007-04.05.2007 were 5030. Population build up started from 11th standard week and reached its peak in 17th standard week. Total three peaks were observed. Population declined after 33rd standard week.

Table 2: Active Methyl Eugenol Traps Installed at Eight different Locations for Mass Trapping of Dorsalis Fruit Fly Complex at PUSA, IARI, New Delhi

Month	*L1*	*L2*	*L3*	*L4*	*L5*	*L6*	*L7*	*L8*
October	18	42	73	32	25	53	73	13
November	25	45	72	31	30	60	80	12
December	17	52	55	30	27	54	60	7
January	11	23	55	15	7	45	52	14
February	11	24	44	17	22	42	49	9
March	25	47	48	33	28	57	68	10
April	26	53	48	35	30	49	68	8
May	18	33	28	27	18	40	70	13
June	25	36	5	6	27	47	72	4
July	25	54	19	27	22	42	27	15
August	25	18	88	32	30	56	77	16
September	25	51	87	29	30	44	71	15
October	13	45	74	26	20	40	59	16
November	20	40	89	25	22	50	97	18
December	23	55	94	26	30	52	73	14
January	21	49	84	32	25	49	86	12
February	20	33	56	10	11	34	55	5
March	16	41	67	24	26	47	73	14
April	27	57	95	32	32	61	81	18
May	20	52	94	22	27	59	69	16

The mass trapping was done during 2007 and found that highest number of flies (8,682 flies/53 active traps) were trapped @164/trap at L6 followed by L1(2302

Table 3: Mass Trapping of Fruit Flies at Eight different Locations of Pusa Campus, IARI, New Delhi during 2007-09

Month		*1L*	*2L*	*3L*	*4L*	*5L*	*6L*	*7L*	*8L*
October 2007	TFT	2302	3851	6133	3018	1781	8682	2100	137
	AFT	127.8	91.69	84.01	94.31	71.24	163.81	28.76	10.53
November 2007	TFT	434	1831	1583	1953	959	3379	738	34
	AFT	17.36	40.68	21.98	63.00	31.96	56.31	9.22	2.83
December 2007	TFT	624	1536	520	593	367	1002	467	9
	AFT	36.70	29.53	9.45	19.76	13.59	18.55	7.78	1.28
January 2008	TFT	55	129	852	87	42	212	251	82
	AFT	5.00	5.60	15.49	5.80	6.00	4.71	4.82	5.85
February 2008	TFT	69	152	368	71	96	239	292	36
	AFT	6.27	6.33	8.36	4.17	4.36	5.69	5.95	4.00
March 2008	TFT	201	566	534	521	331	921	617	115
	AFT	8.04	12.04	11.12	15.78	11.82	16.15	9.07	11.50
April 2008	TFT	1045	2530	679	1587	2022	5175	1172	193
	AFT	40.19	47.73	14.14	45.34	67.40	105.61	17.23	24.12
May 2008	TFT	498	2723	382	2112	1232	2015	1081	183
	AFT	27.66	82.51	13.64	78.22	68.44	50.37	15.44	14.07
June 2008	TFT	338	1520	202	80	993	2317	1750	41
	AFT	13.52	42.22	40.40	13.13	36.77	49.29	24.30	10.25
July 2008	TFT	915	3665	479	2031	397	2345	628	431
	AFT	36.60	67.87	25.21	75.22	18.04	55.83	23.25	28.73
August 2008	TFT	1280	2157	2205	1902	1315	3599	3886	702
	AFT	51.20	119.83	25.05	59.43	43.83	64.26	50.46	43.87
September 2008	TFT	1384	4811	4471	893	2338	5649	7374	2291
	AFT	55.36	94.33	51.39	30.79	77.93	128.38	103.85	152.73
October 2008	TFT	72	3208	3758	2493	2007	3719	2569	1821
	AFT	5.53	71.28	50.78	95.88	100.35	92.97	43.54	113.81
November 2008	TFT	750	2970	3080	760	985	3624	4834	1119
	AFT	37.5	74.25	34.60	30.40	44.77	72.48	49.83	62.16
December 2008	TFT	319	1283	1424	391	308	987	2247	336
	AFT	13.86	23.32	15.14	15.03	10.26	18.98	30.78	24.0
January 2009	TFT	157	323	645	186	143	339	675	168
	AFT	7.47	6.59	7.67	5.81	5.72	6.91	7.84	14.00
February 2009	TFT	96	131	242	24	50	120	263	30
	AFT	4.80	3.96	4.32	2.40	4.54	3.52	4.78	6.00
March 2009	TFT	84	210	286	117	187	446	642	115
	AFT	5.25	5.12	4.26	4.87	7.19	9.48	8.79	8.21
April 2009	TFT	1125	3445	1515	2342	1326	1677	1355	256
	AFT	41.66	60.43	15.94	73.18	40.18	27.49	16.72	14.22
May 2009	TFT	844	3127	3086	734	857	2504	1925	186
	AFT	42.20	60.13	32.82	33.36	31.74	42.44	27.89	11.62

TFT: Total number of flies in active traps; AFT: Average number of flies/trap.

flies/18 active traps) during the month of October, 2007 (Table 3). Average number of Fruit fly population trapped in the month of November and December, 2007 were ranged from 2.83 to 63.00 and 1.28 to 36.70/trap, respectively. In the year 2009, highest population during mass trapping was recorded during April and May with 3,445 flies/57 active traps and 3,217/52 active traps, respectively. Whereas, in the winter months of the year, lowest population was recorded from 2.40 to 14.00 flies/trap; while the trap catches reached above 60 flies per trap from the month of April and May. Similarly, mass trapping was carried out during the year 2008 and results found that highest population (2291 flies/15 active traps) was recorded @ 152.93/trap during the month of September, 2008 at L8 followed by L2 with average trapping of 119.83 flies/trap during August, 2008 (Table 3).

In the year 2009, it was found during monitoring of fruit flies that population has drastically declined (Figure 1) due to mass trapping in the previous years upto 18.1 flies per trap per day. After the mass trapping was suspended the monitoring of population was going on. In next two years population increased nearly 60 flies/trap in two years as compared to 18 flies/trap when mass trapping was going on. Again in 2013 monitoring of population showed that the population was increased to 100 flies/trap/day at peak (Figure 1), because there was no mass trapping was carried out during 2010 and 2015. After 2013, monitoring for the incidence of fruit fly population was carried out in the year 2015 and it was observed that the population keeps on increasing from the beginning (13th standard week) up to 21st Standard week of the year with highest population of nearly 175 flies/trap. The population density was fluctuating throughout the year depending on the season.

The fruit fly population was usually high during 13th to 23rd standard weeks of all the year and population was in a decline phase after every mass trapping with methyl eugenol traps as it trapping the males of dorsalis fruit fly complex and leaving females unmated, which are unable to lay viable eggs. In case it is not possible to continue the mass trapping throughout the year, mass trapping during summer months (13th to 23rd standard weeks) of every year is best time to trap the flies and to reduce the population, in north Indian conditions. These results gain support from Makhmoor and Singh (1998) found that the peak population of *B. dorsalis* from the last fortnight of July to the first fortnight of August, with highest trap catches of 170.66 in the last week of July at Jammu. Further, Jalaluddin *et al.* (1999) have also recorded three peaks of *B. zonata*, *B. correcta* and *B. dorsalis* in Tamil Nadu. The first peak occurred during 2nd fortnight of June, 2nd peak during 2nd fortnight of July and 3rd peak during 2nd fortnight of August. Similar results were reported in guava by Chaudhary and Jamal (2000) who observed the peak activity of these fruit flies from August to October which coincided with the maturity period of guava. Our results were also supported by Ravikumar (2005), as he observed the peak incidence of *B. correcta* during first fortnight of July. *B. dorsalis* had two major peaks during 1st week of July and 3rd week of November (132.67 fruit flies/trap week) in 2004 at Dharwad. *B. zonata* had the major peak of 37.67 fruit flies/week during last week of May in mango at kambapur and Dharwad regions of Karnataka. Hussain *et al.* (2010) evaluated the efficacy of different control methods including cultural control, Male Annihilation Technique (MAT) and Bait Application Technique (BAT) for controlling oriental fruit fly in Peshawar district, Pakistan and found that maximum control was achieved in MAT followed by BAT. Rodrigo *et al.* (2014) tested the trap-lure efficacy and trap density in minimizing the fruit infestation in orchards and opined that handmade plastic bottle traps showed significantly higher capture rate in trapping *Anastrepha ludens*. The male annihilation technique (MAT) is a chemical control method designed to deplete the males available for mating in a population and thus breaks the reproductive cycle of fruit flies. The growers are advised to implement MAT in their orchards as a supplement to baiting.

References

Ali, Hussain, Ahmad, Sajjad, Jan, Salim and Safiullah. 2010. Efficacy of different control methods against Oriental fruit fly *Bactrocera zonata* (Saunders). *ARNP J. Agri. Biol. Sci.*, **5**(2): 1–3.

Allwood, A.J. 1997. Control strategies for fruit flies (Diptera: Tephritidae) in the South Pacific. **In**: *Management of Fruit Flies in the Pacific: A Regional Symposium*, Nadi, Fiji 28th–31st October, (Eds.) A.J. Allwood and R.A.I. Drew, Australian Centre for International Agricultural Research: Canberra, Australia, pp. 171–178.

Andreawartha, H.G., Monro, J. and Richardson, N.L. 1967. The use of sterile males to control populations of Queensland fruit fly, *Dacus tryoni* (Frogg.) (Diptera: Tephritidae). II. Filed experiments in New South Wales. *Aust. J. Zool.*, **15**: 461–473.

Bayart, J.D., Phalip, M., Lemonnier, R. and Gueudre, F. 1997. Fruit flies. Results of four years of import control on fruits in France. *Phytoma*, **49**: 20–25.

Benelli, G. 2015. Aggression in Tephritidae flies: Where, when, why? Future directions for research in integrated pest management. *Insects*, **6**: 38–53.

Chaudhary, M.M.K. and Jamal, Q. 2000. Effect of abiotic factors on population fluctuation of oriental and peach fruit flies at Rawalpindi, Pakistan. *Pakistan J. Zool.*, **32**: 182–185.

Dow Elanco. 1994. *Spinosad Technical Guide;* DowElanco: Indianapolis, IN, USA.

Drew, R.A.I. and Hancock, D.L. 1994. The *Bactrocera dorsalis* complex of fruit flies (Diptera: Tephritidae: Dacinae) in Asia. *Bull. Ent. Res.*, **2**: 1–68.

Ekesi, S. and Billah, M.K. 2007. *A Field Guide to the Management of Economically Important Tephritid Fruit Flies in Africa.* ICIPE Science Press: Nairobi, Kenya.

Ekesi, S., Dimbi, S. and Maniania, N.K. 2007. The role of entomopathogenic fungi in the integrated management of tephritid fruit flies (Diptera: Tephritidae) with emphasis on species occurring in Africa. **In**: *Use of Entomopathogenic Fungi in Biological Pest Management*, (Eds.) S. Ekesi and N.K. Maniania. Research Sign Post: Kerala, India. pp. 239–274.

Fletcher, T.B. 1917. Annotated list of Indian crop pests. **In**: *Proceedings of the Second Entom. Meeting*, Pusa, **2**: 307.

Harris, E.J., Cunningham, R.T., Tanaka, N., Ohinata, K. and Schroeder, W.J. 1986. Development of the sterile-insect technique on the Island of Lanai, Hawaii for suppression of the Mediterranean fruit fly. *Proc. Hawaii. Ent. Soc.*, **26**: 77–88.

Indian Horticulture Database. 2014–15.Website: www.nhb.gov.in

Jaiswal, J.P., Gurung, T.B. and Pandey, R.R. 1997. *Findings of melon fruit fly control survey and its integrated management 1996/97*, Kashi, Nepal. Lumle Agriculture Research Centre Working Paper No. 97/53, pp 1–12.

Jalaluddin, S.M., Natarajan, K., Sadakathulla, S. and Balasubramanian, S. 1999. Discovery of the guava fruit fly *Bactrocera correcta* (Bezzi). *Ent.*, **24**: 195–196.

Kapoor, V.C. 1989. Indian sub-continent. **In**: *Fruit Flies: Their Biology, Natural Enemies and Control*, (Eds.) A.S. Robinson and G. Hooper, W. Helle, World Crop Pests, **3**(A). Elsevier, Amsterdam, pp. 59–62, Xii :372 pp.

Kapoor, V.C. 2005. Taxonomy and biology of economically important fruit flies of India. *Israel J. Ent.*, **459**: 35–36.

Kapoor, V.C. and Agarwal, M.L. 1983. Fruit flies and their natural enemies in India.104–105. **In**: *Proceedings of the CEC/IOBC International Symposium on Fruit Flies of Economic Importance*, (Ed.) R. Cavalloro, Athens, Greece, 1982. Balkema Publications, Rotterdam.

Kapoor, V.C., Agarwal, M.L. and Grewal, J.S. 1976. Identified Indian fruit flies (Diptera: Tephritidae) in National Pusa Collection, India. *Bull. Ent.*, **17**: 1–15.

Klungness, L.M., Jang, E.B., Mau, R.F.L., Vargas, R.I., Sugano, J.S. and Fujitani, E. 2005. New approaches to sanitation in a cropping system susceptible to tephritid fruit flies (Diptera : Tephritidae) in Hawaii. *J. Appl. Sci. Env. Mgt.*, **9**: 5–15.

Knipling, E.F. 1979. *The Basic Principles of Insect Population Suppression and Management*; U.S. Department of Agriculture: Washington, DC, USA. pp. 1–659.

Koyama, J. 1996. Eradication of the melon fly, Bactrocera cucurbitae by the sterile insect technique in Japan. **In**: *Proceedings of IAEA Training Course on the Use of Sterile Insect and Related Techniques for the Area-Wide Management of Insect Pests*, Gainesville, FL, USA.

Makhmoor, H.D. and Singh, S.T. 1998. Effective concentration of methyl eugenol for the control of guava fruit fly *Dacus dorsalis* Hendel in guava orchard. *Ann. Pl. Prot. Sci.*, **6**: 165–169.

McInnis, D., Leblanc, L. and Mau, R. 2007. Melon fly (Diptera: Tephritidae) genetic sexing: All-male sterile fly releases in Hawaii. *Proc. Hawaiian Ent. Soc.*, **39**: 105–110.

McInnis, D.O., Tam, S., Lim, R., Komatsu, J. and Albrecht, C. 2004. Development of a pupal-colour based genetic sexing strain of the melon fly, *Bactrocera cucurbitae* (Diptera: Tephritidae). *Ann. Ent. Soc. Amer.*, **97**: 1026–1033.

Moreno, D.S. and Mangan, R.L. 1995. Responses of the Mexican fruit fly (Diptera: Tephritidae) to two hydrolyzed proteins and incorporation of phloxine B to kill adults. **In**: *Light Activated Pest Control*, (Eds.) J.R. Heitz and K.R. Downum, ACS Symposium Series: Washington, DC, USA, pp. 257–279.

Narayanan, E.S. and Batra, H.K. 1960. *Fruit Flies and their Control*. Indian Council of Agricultural Research Publication, New Delhi, 1–68 pp.

Rahman, O., Rahman, S. and Agarwal, M.L., 1995. Guarding against oriental fruit fly, *Bactrocera dorsalis* by the use of biotechnical control. *Indian Hort.*, **39**: 13–15.

Rajitha, A.R. and Viraktamath, S. 2005. Efficiency of different types of traps in attracting fruit flies in guava orchard at Dharwad, Karnataka. *Pest Mgt. Econ. Zool.*, **13**: 111–120.

Ravikumar, C.H. 2005. Studies on fruit flies in guava and mango orchards with special reference to their management through mass trapping. *M.Sc. (Agri.) Thesis*, University of Agricultural Sciences, Dharwad.

Rodrigo Lasa, Toxtega Yeni, Herrera Fortino, Cruz Alejandar, Navarrete Manuel A and Antonio Sofia. 2014. Inexpensive traps for use in mass trapping *Anastrepha ludens* (Dipteral : Tephritidae). *Florida Entomologist*, **97**(3): 1123–1130.

Sarker, D., Rahman, M.M. and Barman, J.C. 2009. Efficacy of different bagging materials for the control of mango fruit fly. *Bangladesh J. Agric. Res.*, **34**: 165–168.

Sinha, P. 1997. Effects of culture filtrates of fungi on mortality of larvae of *Dacus cucurbitae*. *J. Env. Biol.*, **18**: 245–248.

Srinivasan, K. 1994. Recent trends in insect pest management in vegetable crops. **In**: *Trends in Agricultural Insect Pest Management*, (Eds.) G.S. Dhaliwal and R. Arora, 345–372. Commonwealth Publishers, New Delhi, India.

Stark, J.D., Vargas, R.I. and Miller, N.W. 2004. Toxicity of spinosad in protein bait to three economically important tephritid fruit fly species (Diptera: Tephritidae) and their parasitoids Hymenoptera: Braconidae). *J. Econ. Ent.*, **97**: 911–915.

Steiner, L.F. 1952. Methyl eugenol as an attractant for oriental fruit fly. *J. Econ. Ent.*, **45**: 241–248.

Steiner, L.F., Harris, E.J., Mitchell, W.C., Fujimoto, M.S. and Christenson, L.D. 1965. Melon fly eradication by over flooding with sterile flies. *J. Econ. Ent.*, **58**: 519–521.

Steiner, L.F., Hart, W.G., Harris, E.J., Cunningham, R.T., Ohinata, K. and Kamakahi, D.C. 1970. Eradication of the oriental fruit fly from the Mariana Islands by the methods of male annihilation and sterile insect release. *J. Econ. Ent.*, **63**: 131–135.

Steiner, L.F., Mitchell, W.C., Harris, E.J., Kozuma, T.T. and Fujimoto, M.S. 1965. Oriental fruit fly eradication by male annihilation. *J. Econ. Ent.*, **58**: 961–964.

Steiner, L.F., Rohwer, G.G., Ayers, E.L. and Christenson, L.D. 1961. The role of attractants in the recent Mediterranean fruit fly eradication program in Florida. *J. Econ. Ent.*, **54**: 30–35.

Stonehouse, J.M., Mumford, J.D., Verghese, A., Shukla, R.P., Stapathy, S., Singh, H.S., Jiji, T., Thomas, J., Patel, Z.P., Jhala, R.C., Patel, R.K., Manzar, A., Shivalingaswamy, T.M., Mohantha, A.K., Nair, B., Vidya, C.V., Jagadale, V.S., Sisodiya, D.B. and Joshi, B.K. 2008. Village-level area-wide fruit fly suppression in India: Bait application and male annihilation at village level and farm level. *Crop Protec.*, **26**: 788–793.

Suresh Babu, K. and Viraktamath, S. 2003. Species diversity and population dynamics of fruit flies (Diptera : Tephritidae) on mango in Northern Karnataka. *Pest Mgt. Econ. Zool.*, **11**: 103–110.

Udayagiri, S. 1987. New records of natural enemies of the Malaysian fruit fly: *Dacus latifrons* (Diptera: Tephritidae). *J. Pl. Protec. Tropics*, **4**: 69–70.

Vargas, R.I. 2004. Area-wide integrated pest management for exotic fruit flies in Hawaii. **In**: FLC Awards Program. The FLC–TPWG National Meeting, 3rd–6th May, San Diego, California, p. 12.

Vargas, R.I., Jang, E.B. and Klungness, L.M. 2003. Area-wide pest management of fruit flies in Hawaiian fruits and vegetables. **In**: *Recent Trends on Sterile Insect Technique and Area–wide Integrated Pest Management*, Research Institute of Subtropics, Okinawa, Japan, p. 37–46.

Vargas, R.I., Leblanc, L., Piñero, J.C. and Hoffman, K.M. 2014. Male annihilation, past, present, and future. **In**: *Trapping Tephritid Fruit Flies. Lures, Area-Wide Programs, and Trade Implications*, (Eds.) T. Shelly, N. Epsky, E.B. Jang, J. Reyes-Flores and R. Vargas, Springer: Berlin, Germany, pp. 493–511.

Vargas, R.I., Mau, R.F.L., Jang, E.B., Faust, R.M. and Wong, L. 2008. The Hawaii fruit fly area-wide pest management programme. **In**: *Area-wide Pest Management: Theory and Implementation*, (Eds.) O. Koul and G. Cuperus, CABI Books: London, UK, pp. 300–325.

Vargas, R.I., Miller, N.W. and Prokopy, R.J. 2002. Attraction and feeding responses of Mediterranean fruit fly and a natural enemy to protein baits laced with two novel toxins, Phloxine B and Spinosad. *Ent. Exp. Appl.*, **102**: 273–282.

Vargas, R.I., Piñero, Jaime C. and Luc Leblanc. 2015. An overview of pest species of *Bactrocera* fruit flies (Diptera : Tephritidae) and the integration of biopesticides with other biological approaches for their management with a focus on the Pacific region. *Insects*, **6**: 297–318.

Vargas, R.I., Stark, J.D., Hertlein, M., Mafra-Neto, A., Coler, R. and Piñero, J.C. 2008. Evaluation of SPLAT with spinosad and methyl eugenol or cue-lure for "attract-and-kill" of oriental and melon fruit flies (Diptera: Tephritidae) in Hawaii. *J. Econ. Ent.*, **101**: 750–768.

Verghese, A. and Sudhadevi, K. 1998. Relation between trap catch of *Bactrocera dorsalis* Hendel and abiotic factors. **In**: *Proceedings of First National Symposium on Pest Management in Horticultural Crops*, Bengaluru, pp.15–18.

Verghese, A., Kamala Jayanthi, P.D., Shivananda, T.N, Soumya, C.B, and Mumford, J.D. 2013. From chemical ecology to application: A flashback of olfactory response leading to area-wide management of the fruit fly, *Bactrocera dorsalis* in India. *International Chemical Ecology Conference*, Melbourne, August 19th–23rd, 152 pp.

Verghese, A., Madhura, H.S., Jayanthi, K.P.D. and Stonehouse, J.M. 2002. Fruit flies of economic significance in India with special reference to *Bactrocera dorsalis*. **In**: *Proceedings of 6th International Fruit fly Symposium*, 6th–10th May, South Africa. Pp. 317–324.

White, I. and Elson-Harris, M.M. 1992. *Fruit flies of Economic Significance: Their Identification and Bionomics*. CAB International with Australian Centers for International Agricultural Research, 601p.

Yang, P.J., Carey, J.R. and Dowell, R.V. 1994. Tephritid fruit flies in China: Historical background and current status. *Pan-Pacific Entomologist*, **70**: 159–167.

Zaheeruddin, M. 2007. Study of diffusion and adoption of male annihilation technique. *Int. J. Educ. Dev. Inf. Comm. Tech.*, **3**: 89–99.

44

Strategies for Managing Sap Sucking Pests in Horticultural Crops

V. Sridhar

Division of Entomology and Nematology,
ICAR-Indian Institute of Horticultural Research, Hessaraghatta Lake Post,
Bengaluru – 560 089, Karnataka
E-mail: vsridhar@iihr.res.in

Among the agriculturally important insect pests, sap sucking ones are assuming greater importance in post *Bacillus thuringiensis* (*Bt*) phase and impending climate change. In this regard, sap sucking pests like thrips, aphids, mealy bugs, leafhoppers, whiteflies and mites inflict damage by direct feeding and as a vector of many plant viruses. Several authors reviewed the emerging insect pest problems in horticultural crops and their management in India, *i.e.*, vegetables (Rai *et al.*, 2014), fruits and ornamentals.

These pests suck nutrients from the plants they feed on. Some sucking pests (*e.g.* broad mite and silver leaf whitefly (SLW) also inject toxic saliva, while feeding. Sucking pests like hoppers, whiteflies and aphids excrete sticky honey dew on which sooty mould develops affecting the plant growth and quality of the economic parts. Their management also relies on understanding the pest's life cycle and environmental conditions that favour population increase of pests and their natural enemies.

Sustainable management of sucking pests starts with regular monitoring and their correct identification. Though an array of chemical options are available for individual pests, there is a need to focus on Integrated Pest Management (IPM) for minimising reliance on chemicals keeping in view their negative effects like residues, resistance and resurgence. Using sticky colour traps like blue traps for chilli thrips; yellow traps for whiteflies and leaf miners helps in monitoring of pests. IPM components includes cultural practices like trap and mixed cropping, field sanitation and avoidance of peak activity period of pest by altering sowing and harvesting time of the crop. Biological control agents *viz.*, predators, parasites, entomo-pathogens; use of sex pheromones, botanicals are other important tools of IPM. Thrips populations are effectively checked by certain predatory anthocorids, lygaid bugs, few species of mites and need based sprays of neonicotinoids, neem products or spinosad or fipronil. Botanicals like NSPE @ 5 per cent manage leaf hoppers on okra, aphids on cole crops, leaf miner in tomato. Plant oils of *neem* and *Pongamia* at 0.2 per cent acts as good synergists when used with different insecticide sprays for managing sucking pests. Rotation of chemicals with different modes of action as per the IRAC classification is to be followed for effective Insecticide Resistance Management (IRM).

Transfer of pest resistance genes through molecular techniques, use of semio chemicals, identification of new insect target sites for development of novel insecticides, RNA interference (RNAi)/specific and strong silencing of various genes, exploitation of gut microbes, optical manipulation, use of plant natural products, nanotechnology for development of potent insecticides are other emerging strategies for managing sap sucking pests. In this paper various management options against major sucking pests in horticultural ecosystems are discussed.

Integrated pest management is an important principle on which sustainable crop protection can be based. IPM allows farmers to manage pests in cost effective, environmentally sound, and socially acceptable way. IPM defined as "A pest management system that in the context of the associated environment and the population dynamics of the pest species utilizes all suitable techniques and methods, in compatible manner as possible and maintains the pest populations at levels below those causing economic injury.

Bio-intensive Integrated Pest Management (BIPM)

It incorporates ecological and economic factors into agricultural system design and decision making, and addresses public concerned about environmental quality and food safety. The benefits of implementing BIPM can include reduced chemical inputs costs, reduced on farm and off farm environmental impacts, and more effective and sustainable manner.

Sucking pests include thrips, whiteflies, aphids, mites and bugs. The feeding activity of sucking pests can damage plant buds, leaves, vegetable and fruit directly. Calendar-based spray programmes to control these pests are expensive, and often ineffective due to insecticide resistance amongst the pest populations. Integrated Crop Protection, ICP (also referred to as Integrated Pest Management (IPM) considers the production system as a whole, including all pests, soil and plant health.

Important Tips for IPM of Sap Sucking Pests: Basics for Protecting Crops

i) Know the potential threats and the target pests
ii) Know which beneficial insects (natural and introduced) may be relevant to your cropping system
iii) Know the impact of potential treatments on beneficial
iv) Maintain thorough field sanitation – remove and destroy weeds, infested plants and crop debris
v) Use clean transplants : clean planting material, compost or growing media
vi) Monitor regularly and track changes in pest and beneficial populations
vii) Follow Crop-free periods; avoid crops of different ages in close proximity
viii) Understand 'soft' treatments options and how to achieve maximum coverage
ix) Use chemical insecticides only when necessary and do not rely on them
x) Understand resistance management and rotating chemical groups

Various components of IPM followed for sap sucking pests in various horticultural crops are discussed below.

Host Plant Resistance

Resistant/tolerant varieties are the most important component in IPM. Wherever resistant sources for different sucking pests are available even from wild germplasm, efforts are to be made for incorporating the genes resistant into commercial varieties. Various tolerant varieties of some vegetable/fruit crops are given in Tables 1 and 2.

Newer Molecules against Sucking Pests

- ✰ **Miticides/Acaricides:** Abamectin, milbemectin, bifenthrin, bifenazate, flufenoxuron, fenpyroximate, spiromesifen, propargite, chlofentizine, diafenthuron and petroleum oils.
- ✰ **Insecticides:** Spinosad, fipronil, imidacloprid, buprofezin, acetamiprid, danitol, thiamethoxam, cyantriniprole *etc.*

Use of Biorational Pesticides

The terms "biorational pesticide" and "biopesticides" are gaining popularity in the current

Table 1: Tolerant Varieties of some Vegetable/Fruit Crops

Crop	*Pest*	*Varieties/Lines*
Brinjal	Aphid, jassid, thrips, whitefly	SM 17-4, PBr 129-5, Punjab Barsati, ARV 2-C, Pusa Purple Round, Punjab Neelam, Kalyanpur-2, Punjab Chamkila, Gote-2, PBR-91, GB-1, GB-6
Capsicum	Whitefly, *Bemisia tabaci*	California wonder, Yellow Wonder, Koral
Cabbage	Aphid (*Brevicoryne brassicae*)	All season, Red Drum Head, Sure Head, Express Mail
Okra	Jassid (*Amrasca biguttula biguttula*)	IC-7194, IC-13999 New Selection, Punjab Padmini
Onion	Thrips (*Thrips tabaci*)	PBR-2, PBR-6, Arka Niketan, Pusa Ratnar, PBR-4, PBR-5, PBR-6
Citrus	California red scale, *Aonidiellaa urantii*	Mandarin, clementine, Santra Baladi and rough lemon
	Citrus psylla, *Phyllocnistis citrella*	Cleopatra and Rubidoux
Mango	Hopper, *Amritodus atkinsoni*	Rajmanu,Vanraj
Apple	Apple woolly aphid, *Eriosoma lanigerum*	Malling-Merton-9 (rootstock)

Table 2: Source of Resistance in Vegetables/Fruit Crops

Crop	Pest	Source of Resistance
Tomato	White fly	*Solanum hirsutum* f. *glabratum*
Cucurbits	aphids	*Cucumis cantaloupe*
Potato	Green peach aphid	*S. brachistotrichum*
	Potato aphid	*S. hjertingii, S. bulbocastanum*
Citrus	California red scale, *Aonidiella aurantii*	Mandarin, clementine, Santra Baladi and rough lemon
	Citrus psylla, *Phyllocnistis citrella*	Cleopatra and Rubidoux

climate of environmental awareness and public concern. Both terms are derived from two words, "biological" and "rational", referring to pesticides that have limited or no adverse effects on the environment, non- target organisms including humans. Khater (2012) reviewed about the current biorational insecticides and their mode of actions, uses, commercial products, and safety concerns.

Biorational insecticides include:

i) **Biochemical insecticides** (botanicals like pyrethrum, neem, essential oils as fumigants and repellents), insect growth regulators (*e.g.* chitininhibitors-benzyl phenylureas, triazines, pyrimidine derivatives; juvenile hormone analogues, ecdysteroid agonists and antijuvenile hormones), insect pheromones, photosanitizers and organic and inorganics);

ii) **Biological control** (using of natural enemies such as parasitoids, predators, nematodes, and pathogens (virus, bacteria, fungi, or protozoa); Natural enemies play an important role in limiting potential pest populations and they are more likely to survive in case of application of eco-friendly biopesticides (Khater, 2012). Botanicals and mycopathogens were found effective against okra leafhopper, *Amrasca biguttula biguttula* and aphid, *Aphis gossypii*. Among botanicals and mycopathogens, neem oil @ 2 per cent was found effective in controlling the leafhopper population while NSKE @ 5 per cent was found effective against aphid. Among biorationals, NSKE proved to be the most economical treatment as realized by highest net returns (Anitha and Nandihalli, 2008). Similarly, *Lecanicillum* (*Verticillium*) *lecanii* was found effective against green scale, Coccus *viridis* on coffee and citrus. *L. lecanii* and *Isaria* (*Paecilo myces*) *fumoroseus* fungi mainly attacks sucking pests such as aphids and whiteflies. *I. fumoroseus* has strong epizootic potential against *Bemisia* and *Trialeurodes* sp. In both greenhouse and open field environments. Entomopathogenic fungi, *P. tenuipes* were documented to control chemical resistant whiteflies, on greenhouse vegetables (Ratnakumari *et al.*, 2014; Ramanujam *et al.*, 2014).

iii) **Transgenic insecticides:** Genetically modified crops or organisms could be engineered to augment pest control.

Current and Future Development: Biological Methods

Sucking pests management by biological methods has become the greatest successes reveals that biological management is merely the beginning of a long series of

Table 3: Insecticides for Use in Vegetable

Common Name	Crop	Target Pest(s)	Dose/ha (g a.i.)
Imidacloprid 17.8% SL	Chilli	Jassid, aphid, thrips	25-20
Thiamethoxam 25% WG	Okra	Jassid, aphid, thrips	20
	Tomato	Whitefly	50
	Brinjal	Whitefly	50
Thiacloprid 21.7% SC	Chilli	Thrips, aphid	40-50
Fipronil 5% SC	Chilli	Thrips, aphid	40-50
Emmamectin benzoate	Chilli	thrips, mite	10
Chlorfenapyre 10% SC	Chilli	Mites	75-100
Spiromesifen 22.9% SC	Brinjal	Red spider mite	96
	Chilli	Yellow mite	96
Diafenthurion 50% WP	Chilli	Mite	300
	Brinjal	Whitefly	300
Buprofezin 25% SC	Chillies	Yellow mite	75-100

new and safer technologies to augment productivity to bring sustainable crop productivity in an eco-friendly manner. For sucking pests' control, bio pesticides are increasingly used in agricultural and horticultural production. In comparison to chemical pesticides, the potential benefit of using bio-pesticides continues to stimulate the search for new biocontrol agents.

For aphid management in cole crops, coriander is recommended as borer crop. Spray of entomopathogenic fungi, *L. lecanii* formulation @ 500 g/ha at 10 days interval is very effective.

Against papaya mealybug, *Paracoccus marginatus* an invasive pest recently introduced into India, three exotic hymenopteran parasitoids were tried and among them *Acerophagous papayae* was found very effective (Myrick *et al.*, 2014). Mani and Shivaraj (2016) reviewed the management of mealybugs on various horticultural crops.

Geetha *et al.* (2013) reviewed the various new bioagents and their effect on sucking pests in various cropping systems based on the number of patents filed/obtained. They include different methods of biological control of sucking pests of agricultural crops and strains of *Bacillus thuringiensis*, and fungal strain *Entomophthora thaxteriana* for bio-preparations, herbal formulation for the control of aphid, white fly; *Bacillus thuringiensis* isolates active against sucking insects; environment friendly sticky traps using colour contrast. Ramanujam *et al.* (2014) reviewed the management of insect pests by micro-organisms.

Insecticide Resistance Management (IRM)

Sucking pests develops resistance to various synthetic pesticides, when the pesticides are used indiscriminately. IRM has to be followed as a component of IPM for delaying or management of insecticide resistance.

Myzus persicae and *Aphis gossypii* are among the most economically important insect pests in the world due to direct feeding damage and transmission of over 100 plant viruses to their large range of host crops. These pests have developed resistance to most classes of insecticides including neonicotinoids. Isoclast is an active insecticide to manage neonicotinoids target-site resistant in *M. persicae* and *A. gossypii.*

Resistance Management Guideline for Green Peach Aphid *Myzus persicae*

Major mechanisms of insecticide resistance in green peach aphid *M. persicae* resistance include enhanced expression of esterases, modified acetyl choline esterase (MACE) for OP and carbamates, nAchR target site resistance against neonicotinoids and knock down resistance against pyrethoids. Effective IRM as a component of IPM should include: (i) Compounds should be used according to the label recommendations; (ii) rotating compounds from different mode of action groups; (iii) Inclusion of non-chemical control measures in IPM. For easy understanding of IRM approach an example of green peach aphid is presented in Table 5.

Table 4: Important Pests Showing Resistance/Resurgence against Insecticides

Name of the Pest	*Resistance/Resurgence*	*Insecticide/Acaricide*
Whitefly (*Bemisia tabaci*)	Resistance	Cyfluthrin, cypermethrin, monocrotophos, quinalphos, methamidophos
Okra jassid (*A. biguttula biguttula*)	Resistance	Malathion, monocrotophos
Yellow mite of chilli (*P. latus*)	Resistance	Acephate, cypermethrin, deltamethrin, fenvalerate, monocrotophos, imidacloprid
Bitter gourd aphid (*Aphis malvae*)	Resurgence	Deltamethrin, permethrin, malathion
Red spider mite of brinjal (*Tetranychus urticae*)	Resurgence	Deltamethrin, fenvalerate
Brinjal aphid (*Myzus persicae*)	Resurgence	Cypermethrin, deltamethrin
Red spider mite of okra (*T. urticae*)	Resurgence	Ethion

Source: www.pesticideresistance.org.

Table 5: Resistance Management Guideline for Green Peach Aphid *Myzus persicae*

IRAC Group	*Mode of Action*	*Subgroup*	*Chemical Class*
1	Acetyl cholin esterase inhibitors	AB	Carbamates, Organophosphates
3	Sodium channel modulators	A	Pyrethroids
4	nAChR agonists	ACD	Neonicotinoids, Sulfoxaflor, Flupyradifurone
9	Effectors of chordotonalorgans	BC	Pymetrozine, Flonicamid
23	Inhibitors ofacetyl—CoA carboxylase	None	Spirotetramat
28	Ryanodine receptormodulators	None	Cyantraniliprole

Source: http://www.irac-online.org/documents/myzus-persicae-irm-poster/?ext=pdf.

Table 6: Insect Pests Scenario under Protected Environment in India

Group	*Insect and Mite Pest*	*Host(s)*	*Distribution*
Aphids	*Aphis gossypii*	Capsicum	Punjab, Delhi
	Macrosiphoniellasanborni	Chrysanthemum	Karnataka, HP
	Macrosiphum luteum	Orchid	Sikkim
	Myzus escalonicus	Strawberry	New Delhi
	Myzus persicae	Capsicum, gerbera	Punjab, Maharashtra
	Toxoptera aurantii	Orchid	Sikkim
Mites	*Polyphagotarsonemuslatus* (yellow mite)	Capsicum	Karnataka, Punjab,Delhi, H.P.
	Stenotarsonemusfragariae	Strawberry	New Delhi
	Tetranychuscinnabarinus	Carnation	Maharashtra
	Tetranychusneocalidonicus	Cucumber	New Delhi
	Tetranychus urticae (Spider mite)	Tomato, capsicum, cucumber, carnation, gerbera	H.P., Maharashtra
Thrips	*Scritothrips dorsalis*	Rose	Karnataka
	Thrips palmi	Gerbera	Karnataka
	Thrips tabaci	Gerbera	Maharashtra
Whiteflies	*Bemisia tabaci*	Gerbera, capsicum	Karnataka, Punjab
	Trialeurodes vaporariorum	Tomato, cucumber, capsicum, beans, gerbera	H.P. and Nilgiri hills (TN)

Sucking Pest Management in Greenhouses/ Polyhouse

Various strategies for management of thrips under greenhouses is presented here as a case study, similar measures are to be followed for other sucking pests like mites, whiteflies *etc.* Since greenhouse conditions allow rapid development of pest populations, successful control of insect pests on greenhouse vegetables and ornamentals depends on several factors. The IPM programme for protected cultivation can be described as a pyramid constructed having three key components namely, Avoidance of problem; Sampling and early detection; and Curative measures. With "Avoidance" as the foundation, virtually all the noncurative preventive management components (*e.g.* physical, mechanical and cultural means) can be fitted (Table 6).

Thrips Management Options

I. Eliminate the Alternate Hosts

Weed control is essential for a successful thrips control program. Weeds can serve as a refuge both for thrips and tospo virus. Eliminate weeds inside the greenhouse and also from areas immediately outside, particularly near vents and doors. Black plastic mulch covered with coarse gravel can be used for weed control and to reduce thrips habitat around the greenhouse and under the greenhouse benches. Coarse gravel on plastic is an unfavourable site for prepupal and pupal stages of thrips.

II. Exclusion/Avoidance

i) Use of physical barriers (insect proof screens, provision double door, UV radiation absorbent sheets)

ii) Sanitation and cultural practices (pre-season clean-up, inspection on arrival, balanced use of fertilizers, pinching and pruning)

iii) Plant quarantine

iv) Trap crop/indicator plants

The best control is to prevent thrips from coming into the greenhouse. Where a thrips invasion from outside is likely, micro screening may be more cost effective than frequent insecticide applications. Micro screening with 100 holes per square inch is effective in screening out all life stages of thrips. One concern with the use of micro-screening is the reduction of air movement into the greenhouse. Suggested screen mesh sizes for various pests are presented in Table 7.

Table 7: Screen Mesh Sizes Needed to Exclude Major Greenhouse Pest Species

Insect Pest	*Hole Size (Microns)*	*Mesh No. of Threads/Linear inch*
Melon aphid, *Myzus persicae*	340	52
Whitefly, *Bemisia tabaci*	462	42
Greenhouse whitefly, *Trialuerodes varopariarum*	290	58
Thrips, *Thrips* sp.	192	76

III. Early Detection through Scouting and Monitoring

Early detection of a thrips infestation is critical because the symptoms of thrips feeding are not often noticed until after damage or virus transmission. Blue sticky cards catch more thrips than yellow ones. For monitoring other pests like winged aphids, whiteflies yellow sticky cards are more preferred.

IV. Curative Measures (Biological Control and Chemical Measures)

a) Biological Control

The predatory phytoseid mites, *Amblyseius cucumeris*, *Iphiseius degenerans*, and *A. swirskiietc* are well suited for immature thrips control on greenhouse crops. Use biological control for other pests or be selective in pesticides used, selecting insect growth regulators or using biorational materials such as *Beauveria bassiana* or azadirachtin that have minimal impact on predators. Three out of about 70 species of predatory true bugs in the genus *Orius* (minute pirate bugs) are generally commercially available for thrips control. Pirate bugs are voracious, reproduce well in greenhouses, and may provide better thrips control because they attack all stages of thrips. In floriculture crops apply 2-6 *Orius* per square yard of production area. The entomopathogenic fungi, *Beauveria bassiana* and *Metarhizium anisopliae*, effectively manages thrips in greenhouses.

b) Chemical Control

Effective chemical control is complicated by the development of insecticide resistance. Since thrips often feed in protected areas of the plant, apply insecticides with equipment that produces very small spray particles of 100 microns or less.

If insecticides must be used, rotate insecticides among different chemical classes to help delay the development of insecticide resistance. Maintain the use of an effective insecticide for more than one generation of a pest before rotating to another class of insecticide.

Emerging Strategies for the Management of Sucking Pests

I. Understanding the Role of Gut Microbes

The volatiles produced by the honeydew associated bacteria in aphid, *A. pisum* enhanced the efficacy of the hoverfly, *Episyrphus balteatus*, by driving prey location and ovipositional preferences (Leroy *et al.*, 2011). Similarly, several reports are available, regarding the role of endosymbionts/microbes in influencing efficacy of parasites, predators and susceptibility to pesticides in various sap sucking pests like aphids, whiteflies, mealybugs *etc.*

II. Nanoparticles for Pest Management

The potential uses and benefits of nanotechnology are enormous. These include enhancement involving nanocapsules for vector and pest management and nanosensors for pest detection. Nanoparticles are 1-100 nm in diameter, whereas the size of a virus is roughly 100 nm. Nanotechnology deals with the targeted nanoparticles which exhibit different physical strength, chemical reactivity, electrical conductance, and magnetic properties. Nanoparticles help to produce new pesticides, insecticides and insect repellents (Owolade *et al.*, 2008). Nanoencapsulation is a process through which a chemical (ex. an insecticide) is slowly but efficiently released to a particular host for insect pest control. Recently, nanotechnology is widely acceptable publicly because it is not yet linked to any toxicological and ecotoxicological risks. Research on nanoparticles and insect control should be directed toward production of faster and eco-friendly pesticides to deliver into the target host tissue through nanoencapsulation. This will control pests efficiently and accelerate the green revolution (Khater, 2012).

III. Optical Manipulation for Reducing Sucking Pests

Many insects are attracted to plants following a series of events, initiating from far-orientation, and ending with landing on plants to feed and lay eggs for which insects use mostly optical or vision cues. Thus, manipulation of these cues in insects can interfere with their orientation and impede insect-plant communication (Antignus, 2010). Sucking insect pests, such as aphids, white flies and thrips have receptors for UV light (peak at 360 nm) and green-yellow light (peak at 520–540 nm), and preventing or diminishing of these spectra can hamper insect's various activities. Optical manipulations can be performed by repelling, attracting and camouflaging of vision cues, and they can be achieved by incorporating optical additives to mulches (below plants), to cladding materials (plastic sheets, nets and screens above plants) or to other objects. This technology would not give sufficient protection in crop production but optical manipulation should be integrated along with other IPM tactics with the hope that their use will reduce the use of insecticides. Colour responses of four cell sap sucking insect pests are recorded indicated that spiralling whitefly, *Aleurodicus dispersus* and psyllid, *Heteropsylla cubana* were attracted to yellow colour while, *Coccidohystris insolita* and aphid, *Aphis craccivora* attracted to red colour mostly (Sathe *et al.*, 2015). Of the different colour sticky traps evaluated for monitoring chilli thrips, *Scirtothrips dorsalis* on rose both under open and protected conditions, blue colour attracted highest number of *S. dorsalis* adults followed by yellow and pink (Sridhar *et al.*, 2015).

IV. Molecular Approaches for Sucking Pest Management

RNA interference involves the down regulation of gene expression by exogenous double stranded RNA molecules (dsRNA). RNAi has facilitated significant technological advances in diverse fields from functional genomics to agricultural biotechnology and pest management. Recent studies have demonstrated that RNAi responses in insects can be achieved by exposure to dsRNA in artificial diet or expressed in transgenic host plants, resulting in mortality of targeted species. There is clear evidence for potential applications of RNAi for the control of insect pests. However, the environmental and ecological risks of RNAi including potential toxicity to non-target organisms, environmental fate and the risk of resistance evolution have yet to be determined (Ana Maria and Siegfried, 2014). In the absence of effective *Bt* toxins against sap-sucking homopteran pests such as aphids, leafhoppers *etc.* feeding of siRNAs through an artificial diet can be exploited for the screening of siRNAs for insect pests control and functional genomic studies in both sucking and chewing insect pests (Francis, 2010; Ana Maria and Siegfried, 2014).

Conclusion

IPM programmes have demonstrated that current levels of pesticide use in many circumstances are not necessary and, frequently, are even counter-productive. Excessive and otherwise inappropriate pesticide use is an unnecessary burden on farmers' health and income, on public health, and on the environment. Under this scenario, bio-rational insecticides and emerging molecular tools like use of RNAi have to be incorporated into the IPM of various sap sucking pests. For the management of disease vectors, research should focus on alternative control strategies, including commercialization of attractant/repellent strategies, disruption of vector lifecycles with RNA interference, and other tactics. Wherever use of chemicals is to be inevitable, IRM should form an integrated component of IPM in various fruit, vegetable and ornamentals crops.

References

Ana María, V. and Siegfried, B.D. 2014. RNA interference in insect pest management: Assessing potential benefits and environmental risk. *ACS Symposium Series*, **1172**: 59–70.

Anitha, K.R. and Nandihalli, B.S. 2008. Utilization of botanicals and mycopathogens in the management of sucking pests of okra. *Karnataka J. Agric. Sci.*, **21**: 231–233.

Antignus, Y. 2010. Optical manipulations block the spread of *Bemisia tabaci* in greenhouses and the open field. **In**: *Bemisia: Bionomics and Management of a Global Pest*, (Eds.) P.A. Stansly and S.E. Naranjo. Springer, Dordrecht, pp. 349–356.

Francis, B.J. 2010. RNAi mediated gene knockdown in sucking and chewing insect pests. *J. Biopest.* **3**: 386–393.

Geetha, G.T., Baby, Nesil Liz, Venkatesan, T. and Verghese, Abraham. 2013. Analysis of opportunities and challenges in patenting of management of sucking pests like aphids, hoppers, whiteflies and thrips in agriculture and horticulture fields. *Int. J. Curr. Microb. Appl. Sci.*, **2**: 164–173.

Khater, H.F. 2012. Ecosmart biorational insecticides: Alternative insect control strategies. **In**: *Insecticides: Advances in Integrated Pest Management*, (Ed.) Farzana Perveen. Intech Europe, Croatia, pp. 17–60.

Leroy, P.D., Sabri, A., Heuskin, S., Thonart, P., Lognay, G., Verheggen, F.J., Francis, F., Brostaux, Y., Felton, G.W. and Haubruge, E. 2011. Microorganisms from aphid honeydew attract and enhance the efficacy of natural enemies. *Nat. Comm.*, **2**: 348.

Mani, M. and Shivaraju, C. 2016. *Mealybugs and their Management in Agricultural and Horticultural Crops*. Springer, India, 655 p.

Myrick, S., Norton, G.W., Selvaraj, K.N., Natarajan, K., and Muniappan, R. 2014. Economic impact of classical biological control of papaya mealybug in India. *Crop Prot.*, **56**: 82–86.

Owolade, O.F., Ogunleti, D.O. and Adenekan, M.O. 2008. Titanium dioxide affects disease development and yield of edible cowpea. *Elec. J. Env. Agri. Food Chem.*, **7**: 2942–2947.

Rai, A.B., Halder, Jaydeep and Kodandaram, M.H. 2014. Emerging insect pest problems in vegetable crops and their management in India: An appraisal. *Pest Mgt. Hort. Ecosys.*, **20**: 113–122.

Ramanujam, B., Rangeshwaran, R., Sivakumar and Mohan, M. 2014. Management of insect pests by microorganisms. *Proc. Indian Nat. Sci. Acad.*, **80**: 455–471.

Ratnakumari, B., Vijayabharathi, R., Srinivas, V. and Gopalakrishnan, S. 2014. Microbes as interesting source of novel insecticides: A review. *Afr. J. Biotech.*, **13**: 2582–2592.

Sathe, T.V., Gophane, Anna and Shendage, Nilam 2015. Colour attractively and occurrence of some cell sap sucking pests on crop plants. *Biolife*, **3**: 540–546.

Sridhar, V. and Onkar Naik, S. 2015. Efficacy of colour sticky traps for monitoring chilli thrips, *Scirtothrips dorsalis* Hood (Thysanoptera: Thripidae) on rose. *Pest Mgt. Hort. Ecosys.* **21**: 101–103.

45

Nematode Management in Horticultural Crops for Improving Productivity

R.K. Walia[1]* and K. Poornima[2]

[1]AICRP on Nematodes,
ICAR-Indian Agricultural Research Institute, New Delhi – 110 012
[2]Department of Nematology,
Tamil Nadu Agricultural University, Coimbatore – 641 003, Tamil Nadu
**E-mail: pcnematodes@iari.res.in*

Plant parasitic nematodes are ubiquitous and associated with all types of crops. It is the kind of nematode (ectoparasite/endoparasite or migratory/ sedentary) and their population density that matters for the extent of losses. The susceptibility of host crop and their cultivars, cropping sequences also regulate nematode population build-up. Besides, the edaphic factors like soil texture, moisture also play significant roles in determining the extent of crop losses. However, in nature nematodes are often involved in complex disease syndromes, where they play the role of predisposers/vectors for other soil-borne fungal, bacterial and viral pathogens.

Horticultural cropping systems are invariably more liable to damage by nematodes. The intensive cultivation systems of these crops, shorter durations, high density planting, drip system of irrigation and many times monoculture of the same crop are highly condusive for rapid nematode multiplication. Protected cultivation systems are the most appropriate examples where everything is ideal for nematode population explosion – the most susceptible crops (tomato/ cucumber), favourable temperature for most part of the year, optimum moisture, well textured soils *etc.* leading to heavy crop losses.

The major plant parasitic nematodes attacking horticultural crops include – root-knot nematodes (*Meloidogyne* spp.), reniform nematode (*Rotylenchulus reniformis*), potato cyst nematode (*Globodera rostochiensis/G. pallida*) on vegetable crops; root-knot, reniform, lesion (*Pratylenchus* spp.) on fruit tress, citrus nematode (*Tylenchulus semipenetrans*) specifically on citrus; root-knot and burrowing (*Radopholus similis*) on banana and other plantation crops, spices, condiments; and root-knot and foliar nematodes (*Aphelenchoides* spp.) on ornamental plants.There are some nematode problems *e.g.*, stem and bulb nematode (*Ditylenchus dipsaci*) on vegetables and bulbous ornamentals, sugarbeet cyst nematode (*Heterodera schachtii*) on vegetable crops, *etc.*; these are widespread in Europe and other temperate parts of the world, but have not been intercepted in India so far.

On a global basis, average annual losses in 20 horticultural crops have been estimated up to 14 per cent; while in India avoidable losses estimated in individual fields in various high value horticultural crops varies from 10-69 per cent in fruits, 13-99 per cent in vegetables, 13.8-70 per cent in ornamentals, 24-64 per cent in medicinal plants, 18.6-76 per cent in spices and condiments, and 35-100 per cent in mushroom.

In this article we have attempted to identify the major nematode pests of different horticultural crops, and the losses caused by them with special reference to India. The effective management practices recommended on the basis of multi-location field trialsconducted under the aegis of All India Coordinated Research Project on Nematodes are provided.

Table 1: Estimated Annual Losses due to Nematodes for Selected Horticultural Crops - Global

Crop	*Loss (%)*	*Crop*	*Loss (%)*
Banana	19.7	Ornamentals	11.1
Brinjal	16.9	Papaya	15.1
Cacao	10.5	Pepper	12.2
Citrus	14.2	Pineapple	14.9
Coffee	15.0	Potato	12.2
Cowpea	15.1	Sweet potato	10.2
Grape	12.5	Tea	8.2
Guava	10.8	Tomato	20.6
Melons	13.8	Yam	17.7
Okra	20.4	Misc. others	17.3

Pathogenic Potential and Symptoms of Nematode Infestations

In Nurseries

Vegetable nurseries often suffer poor germination, seedling vigour is adversely affected, their growth remains moribund. Infected seedlings do not establish well in the main field; heavily infected ones may die eventually. Root galling due to *Meloidogyne* spp. at nursery stage is generally inconspicuous and can easily escape the notice; these are primary galls and very minute. Fruit nurseries such as pomegranate, guava *etc.* show bronzing of leaves at nursery stage, defoliation is common, growth is stagnant.

In Fields

In general, above ground symptoms by most of the nematodes (except foliar nematode) are not diagnostic. Stunted growth, yellowing of leaves typical of nitrogen

Table 2: Yield Losses by Economically Important Nematodes Reported by Cooperating Centers of AICRP (Nematodes) (2005-15)

Crop	*Nematode*	*State*	*Yield Loss in Hot Spot Areas (% range)*
Banana	*Meloidogyne incognita*	Gujarat	07.90-34.60
		Karnataka	06.00-11.50
		Tamil Nadu	05.00-09.50
		West Bengal	15.00-25.70
	Radopholus similis	Karnataka	05.00-26.00
		Kerala	12.00-45.00
		Maharashtra	15.00-18.46
		Tamil Nadu	11.70-24.60
Betelvine	*Meloidogyne incognita* *Rotylenchulus reniformis*	Kerala Tamil Nadu	10.00-54.00
Bitter gourd	*Meloidogyne incognita*	Kerala	13.00-15.00
		Odisha	12.90-14.30
Bottle gourd	*Meloidogyne incognita*	Rajasthan	19.25-23.21
		Uttar Pradesh	22.50-42.20
Brinjal	*Meloidogyne incognita*	Karnataka	08.00-18.00
		Maharashtra	08.00-21.05
		Rajasthan	18.10-23.67
		Tamil Nadu	10.40-24.70
		Uttar Pradesh	18.00-23.70
		West Bengal	20.00-27.90
	Rotylenchulus reniformis	Rajasthan	16.74-22.5
Carrot	*Meloidogyne incognita* *Meloidogyne hapla* *Meloidogyne nassi*	Tamil Nadu	50-65
Chilli	*Meloidogyne incognita*	Assam	05.00-07.00
		Karnataka	12.00-17.00
		Odisha	11.6-15.80
		Rajasthan	10.00-17.45
		Tamil Nadu	18.40-17.69
		Uttar Pradesh	23.30-40.40

Contd...

Table 2–*Contd...*

Crop	*Nematode*	*State*	*Yield Loss in Hot Spot Areas (% range)*
Citrus	*Tylenchulus semipenetrans*	Haryana	06.87-47.70
		Karnataka	10.00-17.50
Cowpea	*Meloidogyne incognita*	Tamil Nadu	9.45-11.6
Cucurbits	*Meloidogyne incognita*	Karnataka	11.00-23.50
Cucumber	*Meloidogyne incognita*	Assam	06.00-08.00
		Odisha	15.00-18.10
		Uttar Pradesh	06.00-07.00
Grapes	*Meloidogyne incognita*	Haryana	11.16-57.12
		Maharashtra	22.31-28.57
		Tamil Nadu	15.30-20.60
	Meloidogyne incognita *Rotylenchulus reniformis* *Pratylenchus vulnus*	Tamil Nadu	15.6
Guava	*Meloidogyne enterolobii*	Tamil Nadu	27.70
Mulberry	*Meloidogyne incognita*	Karnataka	28.00
Okra	*Meloidogyne incognita*	Assam	09.00-17.00
		Karnataka	10.00-13.50
		Kerala	05.00-10.00
		Odisha	16.18-17.46
		Tamil Nadu	17.90-22.10
		Uttar Pradesh	15.00-49.30
		West Bengal	10.00-25.00
Pepper	*Meloidogyne incognita*	Kerala	20.00-30.00
Pointed gourd	*Meloidogyne incognita* *Meloidogyne javanica*	West Bengal	39.80-100.00
Pomegranate	*Meloidogyne incognita*	Karnataka	08.00-10.00
		Maharashtra	19.65-33.33
Potato	*Globodera rostochiensis* *Globodera pallida*	Tamil Nadu	26.1
	Meloidogyne incognita	Gujarat	10.00
Snake gourd	*Meloidogyne incognita*	Tamil Nadu	17.20-19.80
Tomato	*Meloidogyne incognita*	Assam	07.00-20.00
		Haryana	27.27-40.48
		Himachal Pradesh	18.00-33.33
		Karnataka	10.00-21.50
		Odisha	10.00-19.79
		Rajasthan	18.88-22.13
		Tamil Nadu	11.00-25.80
		Uttar Pradesh	32.30-49.20
		West Bengal	18.57-31.30
	Rotylenchulus reniformis	Haryana	17.00-22.00

deficiency, symptoms of water stress (epinasty) in the beginning on hot days and recovery by evening, ultimately leading to permanent wilting, shedding of leaves, poor fruit setting, undersized fruits are the common symptoms. Nematode infected trees depict dieback symptoms, that accentuates year after year, rendering trees uneconomical prematurely.

However, specific symptoms are discernible on roots, and are described below nematode-wise.

Root-knot Nematodes (*Meloidogyne* spp.)

This nematode is widespread, polyphagous and is considered economically most important for being a sedentary parasite of vascular tissues of roots. Starting with minute primary galls at the nursery stage, conspicuous root galling appears after 1-2 months due to compounding of galls. The galling pattern is highly variable according to plant species and nematode species. Tomato, brinjal and okra are considered to be most susceptible crops with heavy galling. Compared to this, cucurbits exhibit profuse galling, so much so that sometimes the whole root system looks swollen along the length. In case of leguminous crops, the root galls can be distinguished from the rhizobium nodules easily. The bacterial nodules are side appendages and can easily be detached from the roots and are soft in texture; while nematode induced galls are axial swellings of root itself, are not detachable and hard in consistency. Certain crops like carrots, radish and turnip exhibit minute galls on roots, but the edible tap root may show forking leading to poor quality. Tuber crops like potato, sweet potato, besides showing galling on the roots, also develop pimple-like outgrowths on the tuber surface. Such tubers harbour nematode females in the outer layers of the tubers, leading to significant drop in consumer acceptability. In perennial crops, like most of the fruits trees, galls may not appear very prominent due to woody texture, but fresh roots exhibit prominent galling. Papaya with succulent roots shows very big galls, while crops like grapes have minute galls.

Nematode induced galled roots initially look creamish in colour, but as the crops growth reached flowering stage, such galled roots become prone to secondary infection by rotting and wilt pathogens, leading to root decay and ultimately plant mortality. Such situations are more common towards crop maturity in annuals. Similar disease complexes in perennial crops result in premature death of the established plants.

Reniform Nematode (*Rotylenchulus reniformis*)

Reniform nematode is also fairly wide spread in tropical and sub-tropical regions, but prefers heavy to medium textured soils. This sedentary semiendoparasite of roots can attack a wide range of plants and draws its nutrition from stelar region of roots, hence it is considered as highly pathogenicThe incidence of this nematode often goes unnoticed for want of clear symptoms on roots as well.

The above-ground symptoms are again indicative of fertilizerdeficiency, water stress. However, this particular nematode does not induce any galls on the roots, therefore, very difficult to diagnose under field conditions. The roots show malformation and necrosis. The nematode is potentially equally damaging as root-knot nematode, many times occurring concomitantly with the latter.

Golden Nematode of Potato (*Globodera rostochiensis/G. pallida*)

Initially confined to Nilgiri hills of Tamil Nadu and adjoining areas of Kerala and Karnataka in about 4000 ha, it has recently been recorded on potato in many parts of Himachal Pradesh and high altitudes of Jammu region in J&K. Two species, namely *Globodera rostochiensis* and *G. pallida* occur as mixed populations; however, the latter is more common at lower altitudes. The nematode forms cysts at the end of the life cycle that remain in soil and harbour viable eggs for many years. The nematode is sedentary endoparasite, feed on vascular tissues of roots and is highly pathogenic to potato. Heavily infected plants mayshow only a tuft of green leaves at the top, while the remaining leaves defoliate. Such plants can be easily uprooted as the roots are shallow and poorly developed. The only confirmatory diagnosis is the appearance of white, pale, golden and brown coloured cysts attached to roots and can be seen with naked eyes.The number and size of the daughter tubers are drastically reduced.

Lesion Nematodes (*Pratylenchus* spp.)

Many species are polyphagous and attack several horticultural crops. Being a migratory endoparasite, all stages of this nematode cause extensive damage to the cortical parenchyma tissues while migrating, besides feeding. The nematodes can leave the roots and come out into the soil at the end of crop season, to infect the new crop. Symptoms on roots initiate as discrete elliptical water-soaked lesions that intensify as the crop grows. The lesions merge and necrotic tissues girdle the roots, eventually the cortical tissues slough off.

Burrowing Nematode (*Radopholus similis*)

Only one species is predominant and wide spread in tropical areas. This is a migratory endoparasite and attacks a wide range of crops that include plantations, spices, condiments, aromatic and medicinal plants. Because of its migratory habit, the nematode causes extensive necrosis on the roots and other underground parts of plants, particularly the banana rhizomes. The infected banana rhizomes facilitate its dissemination to newer areas. The necrotic tissues induced by nematode often involve other pathogens leading to disease complexes like Panama wilt or toppling disease of banana. Corms of crops like turmeric, ginger ultimately show rotting.

Citrus Nematode (*Tylenchulus semipenetrans*)

Invariably associated worldover on all citrus types, this sedentaqry semiendoparasite is a cortical root feeder and causes "Slow decline" of citrus. During the initial 5-7 years, no symptoms are evident. Consequently, the nematode population attains pathogenic levels and the trees start showing dieback symptoms that go on aggravating year after year. The feeder roots, however, show diagnostic symptoms; the infected roots are brownish, having loose cortical layer that tends to slough off in case of heavy infection. In contrast, the healthy rootsappear creamish.

Foliar Nematodes (*Aphelenchoides* spp.)

They belong to a different group of plant parasitic nematodes (*Aphelenchoides* spp.), that specialize in attacking above ground plant parts only, mainly the leaves, stems and buds. They are more active during humid climate. The symptoms of angular spots on leaves, malformation of inflorescence and devitalized buds are diagnostic. These nematodes possess the ability to survive moisture stress conditions by undergoing anhydrobiosis, and in that stage can remain viable for several years. They revive once the moisture is available.

Disease Complexes

Nematodes are often involved in disease complexes, wherein they play the role of an incitant, predisposer, aggravator and vector of other plant pathogens, besides being pathogens themselves. Contrary to the earlier perception that plant parasitic nematodes provide only entry avenues for other pathogens by their stylet insertions and mechanical wounding on the roots, it is now established that the damage caused by most nematodes on plants is systemic. Consequently, the qualitative and quantitative changes in the root exudates of nematode infected plants leads to activation of otherwise dormant plant pathogens in rhizosphere. Nematodes as vectors of other plant pathogens externally (fungal, bacterial) and internal (viruses) is already well known.

Potential Nematode Management Strategies

Selection of Nematode-free Propagating Materials

Vegetable and fruit nurseries are a perpetual source of nematode origin and dissemination. Unlike other pests and pathogens, nematode damage starts from the very beginning *i.e.*, sowing/planting. It is, therefore, utmost essential to propagate healthy planting material. Sanitization of nursery beds in case of vegetable crops, hardening of air layers/ground layers/grafted material in nematode-free soil must be ensured. There are numerous examples that provide ample evidence of wide spread dissemination of nematodes along

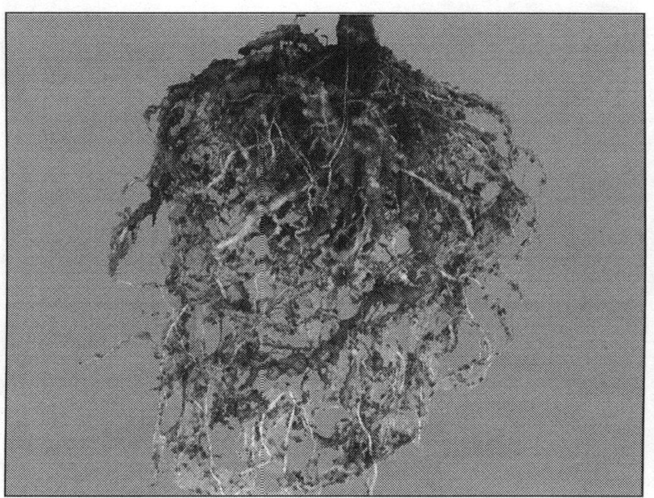

Figure 1: Tomato Roots Heavily Galled by Root-knot Nematode.

Figure 2: Root-knot Nematode Sick Okra Field Showing Patchiness and Stunting of Plants.

Figure 3: Large Sized Compound Galls Caused by Root-knot Nematode in Ash Gourd.

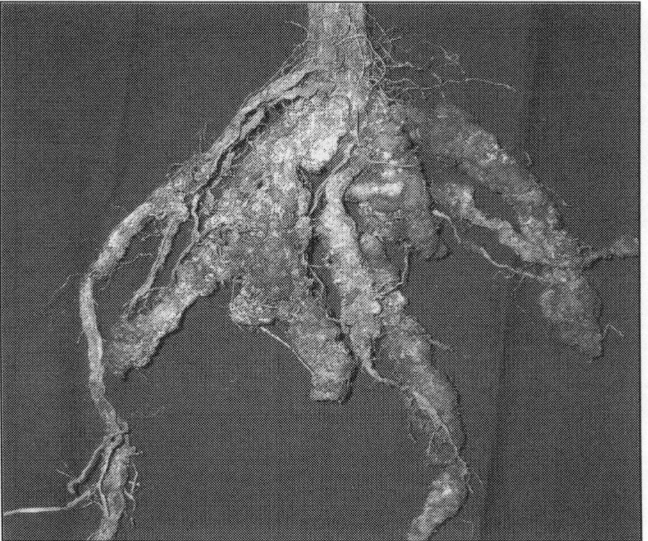

Figure 4: Snake Gourd Roots Showing Galling and Rotting due to Secondary Infection.

Figure 5: Root-knot Nematode Infected Carrots showing Forking.

Figure 6: Potato Tubers Showing Root-knot Nematode Infection.

Figure 7: Potato Field showing Yellowing of Plants due to Golden Nematode.

Figure 8: Cysts of Golden Nematode Attached to Potato Roots.

Figure 9: Pomegranate Tree showing Yellowing Symptoms due to Root-knot Nematode.

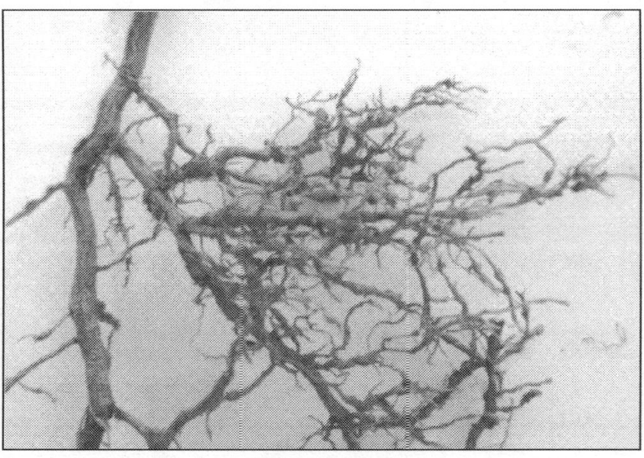

Figure 10: Root-knot Nematode Infested Pomegranate Roots

with planting materials. Particularly in fruit crops, the infected saplings do not grow properly, the plants remain moribund and succumb after a few months, causing huge losses to growers because of enormous expenditure involved in procuring elite fruit cultivars and establishing new orchards. Similarly, the bulbs and corms of banana and ornamental plants like tuberose carry nematode infections. The fast emerging new root-

Figure 11: Root-knot Nematode, *Meloidogyne enterolobii* Infestation in Guava Plants.

Figure 12: Guava Roots Showing Galls of Various Sizes Caused by Root-knot Nematode, *Meloidogyne enterolobii.*

Figure 13: Toppling of Banana due to Burrowing Nematode.

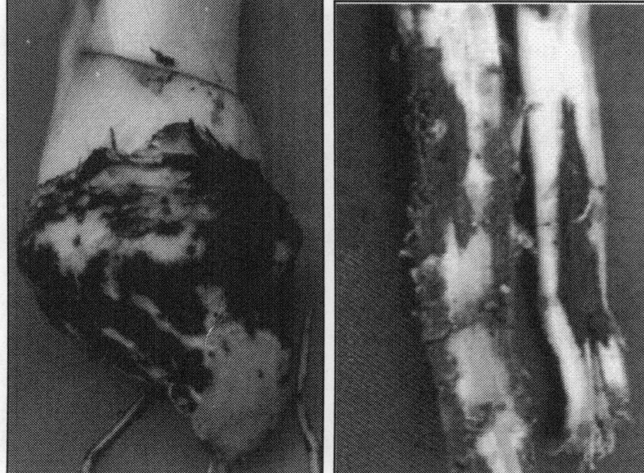

Figure 14: Banana Corm and Roots showing Necrosis.

Figure 15: Citrus showing Slow Decline due to Citrus Nematode.

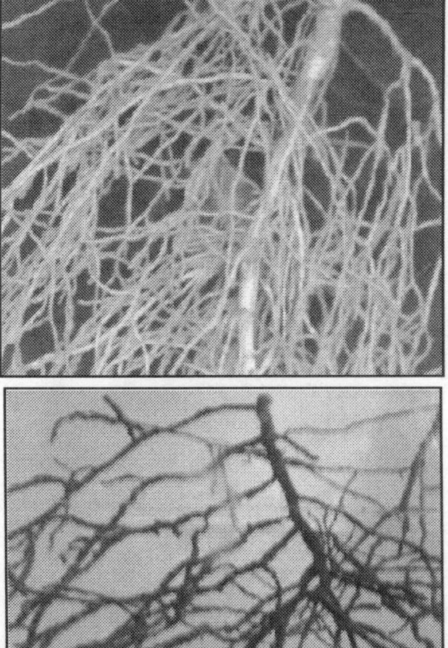

Figure 16: Healthy (Top) and Nematode Infected (Below) Roots of Citrus.

Figure 17: Heavy Infection of Root-knot Nematode in Cucumber in Polyhouse.

Figure 18: Tuberose – Helathy (Lleft) and Infected (Right) with Foliar Nematode.

knot nematode, *Meloidogyne enterolobii* in guava and *M.incognita* in pomegranate causing major devastations in many parts of the country are being spread through nematode-infested planting materials such as ground layers, grafts and rootstocks used for grafting.

Soil Solarisation

Solarisation of nursery beds in open fields of vegetable crops, and beds in polyhouses has given very good results in mitigating the nematode infestations. This practice alone, if done meticulously, has the potential of managing not only nematodes, but other soil-borne pathogens as well,besides weeds. Use of thin transparent polythene sheets for about 15-20 days during peak summer has been found to be very effective and may be used as a routine good horticultural practice.

Enforcement of Domestic Quarantine Regulations

The utter failure of domestic quarantine regulations in checking the spread of dreaded golden nematode of potato has come to limelight recently. Hitherto confined to Nilgiri hills in Tamil Nadu and some adjacent areas in about 4000 ha only, wide occurrence of this nematode has been recorded in Himachal Pradesh and some of areas of Jammu division of J&K. Whether it is a case of new introduction from Nilgiris or it was already prevalent in that area, needs to be ascertained.

Plant Growth-Promoting Rhizobacteria (PGPR

PGPR colonize the roots of plants following inoculation onto seed before planting and enhance plant growth and/or reduce disease, nematode or insect damage. There has been much research interest in PGPR and there is now an increasing number of PGPR being commercialized for crops. Organic growers may have been promoting these bacteria without knowing it. The addition of compost promotes existing PGPR and may introduce additional helpful bacteria to the field. The absence of pesticides and the more complex organic rotations are likely to promote existing populations of these beneficial bacteria, *e.g.*, *Pseudomonas* spp. and *Bacillus* spp. There is a need to search for bacteria that would be active in the rhizosphere and produce compounds lethal to nematodes, inhibitory to nematode movement or development and repellent to nematodes on roots.

Bio-agents

Commercial formulations of nematode egg parasitic fungi, *Purpureocillium lilacinum*, *Pochonia chlamydosporia* are now available. Additionally, broad spectrum bio-agents like *Trichoderma viride*, *T. harzianum*, *Pseudomonas fluorescens* can be applied through organic material, like FYM fortified with bio-agents are very effective in long-term, particularly in perennial crops.

Botanicals

Control of plant parasitic nematodes through antagonistic plants has some advantages like biodegradability, selective toxicity to target by utilizing plant products. Use of botanicals as nematicidal and nematostatic products are economical and eco-friendly. Application of botanicals or botanical-based byproducts to the soil leaves no residues in the field and are economically viable. Concerning the environmental safety, now-a-days botanicals are gaining much importance in the integrated nematode management practices. Many botanical-based products are available in the market for the control of pathogens and nematodes. More understanding of the active nematicidal principles of the botanicals are likely to aid in the development of next generation bionematicides with nematode antagonism and improved plant growth and yield attributes.

Biofumigation

Cultivation and incorporation of cover/rotation crops, especially Brassicaceae plants, occasionally suppress soil-borne diseases, including nematodes. Utilization of volatile toxic compounds, such as isothiocyanates generated from the glucosinolates in

Table 3

Crop	Nematodes	Recommendations
Tomato	*Meloidogyne incognita*	✰ Soil solarization through tarping with 100 gauge (25 μm) LLDPE clear plastic film for 15 days during May. ✰ Application of Poultry manure @ 2 t/ha. ✰ Application of carbofuran 3G @ 10 g/m^2 at sowing and 1 kg a.i./ha in the main field one week after transplanting. Treat the seeds with antagonistic fungi *Trichoderma viride* @ 4 g/kg seed along with press mud at 5 kg/m^2 for nematode disease complex. ✰ Grow root-knot nematode resistant variety - Hisar Lalit. ✰ Nursery bed treatment with *Pasteuria penetrans* @ 10 g culture (1x10^8 spores/g)/m^2 before sowing. ✰ Nursery bed treatment with *Glomus fasciculatum* @ 50 g culture (200 spores/g)/m^{2}before sowing.
Okra	*Meloidogyne incognita*	✰ Follow okra-rice-fallow or okra-cucumber-mustard cropping sequence. ✰ Soil application of egg parasitic fungus, *Purpureocillium lilacinum* (*Paecilomyces lilacinus*) @ 2.5 kg/ha in 2.5-tonne FYM. ✰ Seed dressing with carbosulfan 25 ST @ 3% a.i. (w/w) with summer ploughing. ✰ Application of saw dust/paddy husk @ 500 g per plant or neem leaves or eupatorium leaves @ 250 g/plant in the basins 3 weeks prior to planting and watering daily. ✰ Seed treatment with *Bacillus macerans* @ 3% w/w of seed.
Brinjal	*Meloidogyne incognita*	✰ Two to three deep ploughings during May/June at 10-15 days' interval. ✰ Application of carbofuran @ 1 kg a.i./ha at nursery sowing. ✰ Use of non-host crops like mustard, sesame and cluster bean at least for 2-3 years in vegetable-based cropping system. ✰ Application of carbofuran 3G @ 10G or phorate 10G @ 3 g/m^2 at seeding in the nursery beds and spot application of carbofuran 3G @ 33 kg or phorate 10G @ 10 kg + neem cake @ 400 kg/ha at the base of plant 10 days after transplanting. ✰ Nursery bed treatment with *Pasteruia penetrans* @ 10 g culture (1x10^{8}spores/g)/m^{2}before sowing. ✰ Nursery bed treatment with *Glomus fasciculatum* @ 50 g culture (200 spores/g)/m^2 before sowing. ✰ Intercropping with sweet potato cv. Shree Bhadra. ✰ Nursery treatment with *B. macerans* @ 25 g/m^2 + drenching with 2% solution of formulated product of *B. macerans* 7 days after sowing. ✰ Bare root dip of brinjal seedlings in neem leaf extract.
Tomato/Brinjal/ Capsicum	*Meloidogyne incognita*	✰ Application of *P. lilacinum* (cfu 2 x 10^6) @ 50 g/m^2 in nursery bed (0.5 tonnes/ha) + *P. lilacinum* (cfu 2 x 10^6) @ 5 kg along with 2.5 tonnes of FYM/ha in the main field prior to planting. ✰ Seed treatment with *P. lilacinum* (cfu 2 x 10^6) @ 20 g/kg seed in combination with its application @ 2.5 kg along with 2.5 tonnes of FYM/ha prior to sowing.
Cowpea	*Meloidogyne incognita* *Rotylenchulus reniformis*	✰ Application of neem or eupatorium leaves @ 15 t/ha two weeks before sowing. ✰ Soil application of neem and calotropis leaf extract.
Bottlegourd/ Bittergourd/ Pointed gourd	*Meloidogyne incognita*	✰ Application of neem cake @ 30 g/spot + seed treatment with *Gluconoacetobacter* diazotrophicus strain 35-47 @ 50 ml/2 kg seed. ✰ Seed treatment of carbosulfan 25 DS or carbofuran 25 ST @ 3% a.i. w/w (120 g/kg seed). ✰ Dipping of planting material of pointed gourd in 0.05% carbosulfan solution for 6 hours. ✰ Application of neem/karanj oilcakes @ 1 tonne/ha in the main field. ✰ Seed dressing of bittergourd/cucumber with carbosulfan DS @ 3% a.i w/w
Carrot	*Meloidogyne incognita* *Meloidogyne hapla*	✰ Application of neem cake/mustard cake @ 300-400 kg/ha
Potato	*Globodera* spp.	✰ Rotation with radish, cruciferous vegetables or garlic. ✰ Planting resistant cultivar, Kufri Swarna ✰ Application of FYM at 20 t/ha. ✰ Application of eucalyptus leaf waste from oil distillation plants at 2.5 t/ha. ✰ Application of carbofuran 3G at 33 kg/ha or *P. fluorescens* @ 2.5 kg/ha.
Citrus	*Tylenchulus semipenetrans* *Meloidogyne indica*	✰ Application of carbofuran 3G @ 13 g/m^2 (about 9 m^2 around a plant) just before flowering; pulverize the soil in the basin area and mix the chemical thoroughly followed by flood irrigation. ✰ Application of carbofuran 3G @ 7 g/m^2 + 1 kg neem cake (about 9 m^2 around a plant) just before flowering; pulverize the soil in the basin area and mix the chemical thoroughly followed by flood irrigation. ✰ Soil application of *Pseudomonas fluorescens* (cfu 2 x 10^6/g) @ 20 kg/ha through 100 kg FYM. ✰ Application of *Purpureocillium lilacinum* @ 20 kg/ha or *P. fluorescens* @ 20 kg/ha having a spore load of 10^6 cfu/g of talc based formulated product or neem cake100 g/m^2 ✰ Application of *P. lilacinus* @ 20 kg/ha enriched with FYM at planting for the management of *M. indica*.

Contd...

Table 3–*Contd...*

Crop	Nematodes	Recommendations
Banana	*Meloidogyne incognita* *Radopholus similis* *Pratylenchus coffeae* *Helicotylenchus multicinctus*	✰ Banana sucker paring and pralinage with carbofuran 3G @ 10 g/sucker before planting. ✰ In heavily nematode infested area paring banana suckers and dipping in hot water @ 55°C for 20 minutes and application of neem cake 1 kg/plant (two split doses first at planting and second at 3 months after planting) or carbofuran 0.5 g a.i./ha for the control of nematode complex in banana. ✰ *Radopholus similis* and *M. incognita* infestation in banana can be managed by the application of *P. fluorescens* @ 20 g/m^2 or *T. viride* @ 20 g/m^2. ✰ Bio-fumigation with crop residues of cabbage and cauliflower.
Papaya	*Meloidogyne incognita* *Rotylenchulus reniformis*	✰ Cultural method of control of the nematode as mentioned under other crops. ✰ Application of 25 ml of copper oxychloride (1g/litre) and 1 g of carbofuran 3G/polybag, 15 days after sowing, containing one seedling. ✰ Application of carbofuran 3G at 40 g/tree before flowering.
Grape	*Meloidogyne incognita* *Rotylenchulus reniformis*	✰ Appl. of carbofuran3 G @ 7 g/m^2 in combination with garlic as intercrop in between vine rows. ✰ Application of carbofuran 3G at 60 g/vine or phorate 10G at 20 g/vine. ✰ Soil application of *Pseudomonas fluorescens* (cfu 2 x 10^6/g) @ 20 kg/ha through 100 kg moist FYM at October pruning. ✰ Application of neem cake at 200 g/vine. ✰ Application of *Pseudomonas fluorescens* at 100 g/vine for the management of root knot nematodes and disease complex involving *Fusarium moniliformae.*
Guava	*Meloidogyne enterolobii*	✰ Application of carbofuran 3G @ 33 kg/ha or phorate 10G @ 10 kg per ha at the time of planting. ✰ One tonne of FYM/biocompost can be enriched with each 1 kg of *Pseudomonas fluorescens, Trichoderma viride/T.harzianum* and *P. lilacinum*, kept for a week or two with enough moisture in shade for multiplication and applied @ 500 g/pit around the rhizosphere region. This may be repeated every six months.
Pomegranate	*Meloidogyne incognita*	✰ Soil application of *Pseudomonas fluorescens* or *Trichoderma viride* (cfu 2 x 10^6/g) @ 20 kg/ha through 100 kg FYM at 'Bahar' (Blossom). ✰ Application of *Trichoderma* plus @ 20 kg/ha or *P. lilacinum* @ 20 kg/ha + castor cake @ 2.0 ton per ha.
Tuberose	*Meloidogyne incognita* *Aphelenchoides besseyi*	✰ Use of healthy bulbs, free from galls. ✰ Dipping of cleaned bulbs in 0.05% carbosulfan solution for 4-6 hours. ✰ Field application of carbofuran/cartap @ 1 kg a.i./ha. ✰ Foliar spray of carbosulfan/monocrotophos @ 2 ml/l at 2 months after planting (2 sprays at 15 days' interval).
Crossandra	*Meloidogyne incognita* *Pratylenchus delattrei*	✰ Deep ploughing and fallowing of the selected field for about a month. ✰ Avoid planting crossandra after a graminaceous crop. ✰ Application of phorate 10G at 10 kg/ha, 15 to 30 days after transplanting. ✰ Application of neemcake @ 500 kg/ha
Ginger	*Meloidogyne incognita*	✰ Dipping of ginger rhizomes in hot water at 45° C for three hours. ✰ Summer ploughing/solarization (covering the ploughed area with transparent polythene sheet from 15 to 30th May). ✰ Combination of hot water treatment of ginger with either solarization or summer ploughing gave better result as compared to individual treatment. ✰ Application of carbofuran @ 1 kg a.i./ha at planting + neem cake @ 2.5 tonne/ha at 45 days after planting.
Pepper	*Meloidogyne incognita*	✰ Application of *Bacillusmacerans* (1.2 x 10^8 cfu/g) per pit just before monsoon. ✰ Application of either *P. fluorescens* or *B. macerans* @ 20g/plant.
Turmeric	*Meloidogyne incognita* *Radopholussimilis* *Pratylenchuscoffeae*	✰ Washing turmeric seed material free of soil and shade drying before planting. ✰ Rotation of turmeric with rice in wetland. ✰ Rotation of turmeric with cereal crops in gardenland. ✰ Application of carbofuran 3G at 12 kg/acre at 3 and 5 months after planting.
Betel vine	*Meloidogyne incognita*	✰ Appl. of FYM at 25 t/ha or neem cake at 1 t/ha or calotropis leaves 2.5 t/ha before planning. ✰ Rotation of betelvine with rice. ✰ Chopped green leaves of *Calotropis* and *Azadiracta indica* at the time of planting. ✰ Panal leaf (*Glycosmis pentaphylla*) as green leaf mulch in existing betelvine plantation. ✰ Oil cakes of neem or mustard @ 500 kg per ha. ✰ Carbofuran 1.5 kg a.i. per ha in fresh planting in endemic areas two weeks prior to planting.
Protected cultivation (in pipe line)	*Meloidogyne incognita*	✰ Raising of nurseries in plug trays. ✰ Mixing of carbofuran @ 1 g per plant at the time of transplanting, if nematode infestation is high. ✰ Sterilization of severely infested polyhouses with dazomet (@ 40 g/m^2) or metham sodium (@ 40 ml/m^2). ✰ Addition of neem cake in (40 g per pit) about 15 days before the transplantation of the crop. ✰ *Purpureocillium lilacinum* @ 10 g/kg seed + its soil application with 50 g/m^2. ✰ Combined application of neem cake @ 200 g/m^2 + *Purpureocillium lilacinum* @ 50 g/m^2 15 days before transplanting along with FYM.

such crops, for soil-borne disease control is generally termed biofumigation. Isothiocyanates are known to possess broad pesticidal activity against weeds, bacteria, fungi and nematodes and are similar or relative to compounds found in commercial soil fumigants, such as metham sodium and dazomet, which release methyl isothiocyanate into the soil.

Green Chemical Nematicides

Certain situations like very high initial nematode populations, often encountered in polyhouses, warrant knock-down treatment, that is possible only with chemical treatments. Selective use of green chemicals is indispensable in such situations. New molecules safe to the environment, and having fungicidal properties as well will preferably be the best choice for therapeutic remedies of disease complex quite often encountered in fruits crops like guava, pomegranate, as well as cucurbits and other vegetable crops.

Recommended Nematode Management Practices

The concerted efforts made by the cooperating centers of AICRP on Nematodes through on-farm trials, have led to the development of nematode management technologies in different horticultural crops; and these have been included in the package of practices of participating centers. These are reproduced in the Table 3.

Acknowledgements

Most of the photographs on nematode symptoms used in this article have originated from different cooperating centers of AICRP on Nematodes.

References

Khan, M.R., Jain, R.K., Ghule, T.M. and Pal, S. 2014. *Root Knot Nematodes in India: A Comprehensive Monograph.* All India Coordinated Research Project on Plant Parasitic nematodes with Integrated approach for their Control, Indian Agricultural Research Institute, New Delhi, 78 p.

Khan, M.R., Jain, R.K., Singh, R.V. and Pramanik, A. 2010. *Economically Important Plant Parasitic Nematodes Distribution Atlas.* Directorate of Information and Publications of Agriculture, Indian Council of Agricultural Research, New Delhi, 145 p.

Manju, P. and Sankari Meena, K. 2015. Antinemic properties of the botanicals (Review article). *Int. J. Sci. Nat.*, **6**: 125-134.

Parvatha Reddy, P. 2013. *Biointensive Integrated Pest Management in Horticultural Ecosystems.* Springer India, 277 p.

Parvatha Reddy, P. 2013. *Recent Advances in Plant Protection.* Springer India. 268 p.

Parvatha Reddy, P. 2013. *Recent Advances in Plant Protection*, Springer India, 281 p.

Perry, R.N. and Moens, M. 2006. *Plant Nematology.* CABI Publishing, 463 p.

Ravichandra, N.G. 2014. *Nematode Diseases of Horticultural Crops.* Springer India, 547 p.

Sasser, J.N. 1989. *Plant Parasitic Nematodes: The Farmer's Hidden Enemy.* A Cooperative publication of the Department of Plant Pathology and the Consortium for International Crop Protection, 115 p.

Singh, A.U. 2015. *Yield Losses in Crops Due to Phytonematodes.* Project Coordinating Cell, All India Coordinated Research Project, Division of Nematology, ICAR-IARI, New Delhi, 33 p.

Managing Emerging Diseases and Disorders

46

Online Advisory for Disease Management in Horticultural Crops

S.D. Sawant

ICAR-National Research Centre for Grapes,
Manjri Farm Post, Pune – 412 307, Maharashtra
E-mail: ipmsawant@yahoo.co.in

Recent times due to global warming, climate is changing and we do not experience similar weather every year. This demands that our plant protection practices should be climate smart. In most cases plant protection practices are carried out based on time bond schedules. Such schedules are prepared considering annual crop cycle and known climate of the area where crop is grown. Wherever, the crop is cash crop and high investment is done for intensive cultivation, such plant protection schedules usually include protection practices such as sprays of pesticide more than required to ensure zero risk of damage by diseases. This not only increase the cost of production, but many times do not provide expected protection when applications are not given when risk of diseases were high due to unexpected weather. Plant protection in vineyards in Maharashtra and Karnataka is aclassic example. Downy mildew in grapes is very important disease and can cause very high yield damage if climate smart protection is not provided. New shoots developed after fruit pruning are highly sensitive to infection till first 50 days after pruning. During this period well timed about 4-6 preventive sprays can provide enough protection when weather is normal. The number of sprays could increase to 10-11 when weather is favourable for the disease. However, there are many situations when farmer give more than 11 sprays and do not get the desired disease control. This situation justifies need for climate smart advisory for disease management.

The development of a disease is a result of interaction between host, pathogen and environment. We may assume that inoculum of the pathogen is present in problem area, but to prepare accurate advisory information on growth stage of the host and weather over specific area and time interval in needed. The different types of advisory systems collect this information through different methods. HORTSAP program implemented in 23 districts of Maharashtra for preparing advisory for various important diseases and insect pests of banana, mongo, pomegranate, orange and sweet orange, and sapota (Vennila *et al.*, 2016) covering more than three lakh hectares. In this program surveillance is actually conducted to collect information on crop stages and incidence of diseases and insect pests by the experts located at different locations. The information is sent to National Research Centre for Integrated Pest Management New Delhi, were the information is compiled validated and advisory is prepared and released to department staff involved in extension. The project is being implemented for last few years and is successful in rationalizing plant protection practices in different fruit crops in Maharashtra. However, the system needs consistent performance from crop experts conducting surveillance. ICAR-CPRI, Shimla, in their Pan India Blightcast system collect weather information from different locations and estimate weather based risk of the late blight of potato to release the advisory (Singh *et al.*, 2000). The crop in specific area has mostly similar growth stage hence it is relatively easy to give advisory on weather based disease risk. The situation in grapes is different. Shoot growth is triggered by pruning in vines. The time of pruning could be different in different grape growing areas, while even at same location different plots owned by same growerwill have different time of pruning. This means at every location there will be different growth stages present. Sensitivity to damage by downy mildew at different growth stages is different and need for protection practices at different stages also will be different. Therefore, for effective disease management advisory in grapes should be plot specific. It will be difficult to manage such advisory system if we depend on manually collected information. The system thus has interactive web page which collects plot specific information from the user and auto generates advisory. The scheme of collection of information, interpretations and generation of advisory in grapes is indicated below.

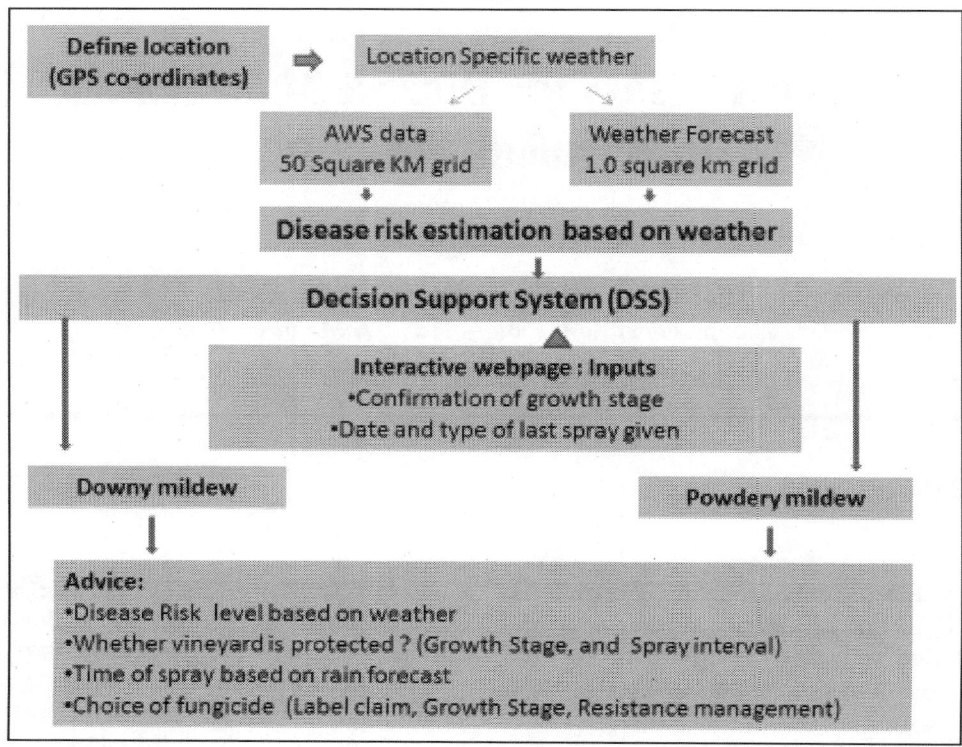

Installation of Automatic Weather Stations (AWS)

AWS are installed in vineyards 2.0 ft above the canopyto record the weather data. Only a sensor for leaf wetness is installed within canopy. Data at hourly interval is recorded on temperature, RH, precipitation, wind speed and light intensity and period of leaf wetness, and sent to central server using data card from local mobile service operators. There are about 100 weather stations to cover most of the grape growing area in Nasik and Sangli districts. One weather station approximately covers 50 square kilometers in 2.5 km radius. Weather forecast for each location of AWS is generated.

Estimation of risk of the Disease Based on Weather

Rule based models developed at NRCG, Pune are used for estimation of risk of the disease.

Downy Mildew

The downy mildew becomes active only after the onset of monsoon in June or July. It is considered active when following conditions are fulfilled

- ✰ Maximum temperature drops below 30°C coinciding rains.
- ✰ There are atleast 3 rainy days (above 4.0 mm rain per day) in a week and RH remains 90 per cent continuously for 3 days.

Once these conditions are fulfilled it is assumed that downy mildew is active. Downy mildew risk is estimated only after the downy mildew is declared active. Seven days AWS data and 7 days weather forecast data is used for estimation of disease risk. Whenever, possibility of rain is forecasted or leaf wetness more than 2.5 h between 6.0 to 9.0 am is recorded high risk of downy mildew is declared. Thereafter, "Temp–RH" model is used for estimation of day to day disease risk. Spray intervals at different growth stages is decided based on risk estimated as per "Temp-RH" model.

Powdery Mildew

Powdery mildew is believed to be active in the vineyards throughout the year. "Temp-RH" model for powdery mildew is thus directly used for estimation of day to day risk of powdery mildew. Spray intervals and type of fungicide are decided based on estimated risk.

Temp-RH Model for Estimation of Disease Risk

The daily minimum and maximum temperature and RH are used for estimation of risk of the disease. Model includes the range of these parameters in three classes in which the disease cannot develop, can develop and can develop luxuriously. When actual recorded parameter for the day falls in the range in which disease cannot grow gets 0 score, and when it falls in range in which disease can grow or luxuriantly grow gets 1 and 2 score respectively. Separate score is given to four parameters, *i.e.* Minimum and maximum temperatures and RH and daily summation is calculated, which will be in the range of 0 to 8. Summation of daily scores is calculated up to 8 days and is called cumulative score. The risk of the disease is estimated on the basis of days required to reach threshold cumulative value of 24 in case of downy mildew and 32 in case of powdery mildew as detailed below.

Days Needed to Reach Threshold Value	*Estimated Disease Risk*
3	High
4-5	Medium
6-7	Low
>= 8	Nil

Growth Related Risk of Damage by Diseases

The growth in grapevine is triggered by pruning, hence growth stage in grapes is defined on the basis of days after pruning. Other factors such as temperatures, application of growth regulators also influence rate of growth. Thus often it is difficult to correctly identify growth stage in vineyard. However, one can consider growth stages standard for the area and most common period of pruning. Recently, ICAR-NRCG in collaboration with IACR-IARI had developed mathematical model based on weather data which can indicate correct growth stage. Yet this model has not been incorporated in the advisory system for grapes. Presently, in advisory system growth stage is defined on the basis of days after pruning. 15th October has been considered as standard pruning date. Further, in an interactive web page photographs of different growth stages have been included and user first selects the growth stage before getting the advice. Growth stage risk have been classified as low, medium and high based on likely economical damage due to diseases. Spray intervals in advisory have been decided based on growth stages alone, while action on plant protection has been decided considering growth stage and weather based disease risk. All the advices on disease management are preventive advices based on estimated risk. Risk related to growth is estimated as detailed below.

Days after Pruning	*Stage*	*Scale of Risk*
0 to 8	Pre sprouting	Nil
9 to 25	Ponga to 5 leaf	Very high
26 to 50	Bunch elongation to berry set	High
51 to 90	Berry growth to veraison	Medium
90 to harvest	Veraison to harvest	Low

Two way table is used to estimate risk ratings and to decide the action based of growth stage and weather based disease risk as detailed below:

		Weather Risk			
		No (0)	*Low (1)*	*Medium (2)*	*High/ Very High (3)*
Growth	No (0)	0	1	2	3
Stage	Low (1)	1	2	3	4
Risk	Medium (2)	2	3	4	5
	High (3)	3	4	5	6

Based on risk rating guidelines on action is decided

Risk Rating	*Guideline*
1 to 2	Avoid using fungicides
3 to 4	Prefer Non-systemic and/or low profile compound
5 to 6	Preventive application of systemic and bio-control agents

Many supportive parameters can improve the quality of advice and can be included easily in the system. List of approved pesticides for use in grapes is prepared by NRCG every year, which includes doses of pesticides alongwith Maximum Residue Limits (MRL), and Pre Harvest Intervals (PHI). Choice of fungicides at different growth stages can be based on MRL and PHI. Development of resistance in pathogen against some group of fungicides is more common than others. Fungicide Resistance Action Committee (FRAC) of "Crop-Life" classify the fungicides on the basis of risk of development of resistance and suggest preventive actions to avoid such resistance development. While suggesting fungicide applications, guidelines of FRAC can be considered.

Experience of the crop expert can give value to the advice. High Powdery mildew incidence was often associated with presence of potash deficiency symptoms. While, untimely winter rains often reduce or slow down powdery mildew for few days. These observations have been introduced as correction factors in the advisory to fine tune the advisory. It is often experienced that if rains are followed immediately after fungicide applications, effective control of the disease is less likely. The system has weather forecast on rains, and it used very effectively to give advice on time of spray and has resulted in reduction of number of sprays. The rule based automatic generation of advisory eliminates the bias of individual expert and thus more reliable.Similar advisory system can be developed in any horticultural crop.

Use of grape advisory developed at ICAR-NRCG for management of downy mildew and powdery mildew was demonstrated at farmers vineyards in Nasik and Sangli districts of Maharashtra in more than 125 locations over the period of three years. Presently, the the application is provided to growers through licensed service provider through android based app on smart phone. Currently about 10 thousand subscribers are using the advisory. All user data submitted on interactive web page is saved on the server and is available to subscribers. This data can also be used for plotting disease incidence, and intensity of pesticide applications in different areas. Data on date of pruning can be used to forecast market arrivals of grapes during harvesting season.

Future Innovations

Remote sensing using satellite imagery or using different sensors in drones or tractor mounted cameras can be used to generate relevant information on vineyards to provide appropriate advisory. Leaf wetness is an important parameter for disease development and it is difficult to know about it based on few sensors. Remote sensing can observe leaf wetness over larger area and give better idea for disease forecasting. Image analysis softwares even for simple RGB image can accurately distinguish different diseases on leaf surface. Estimating disease intensity and identifying disease present on plants, is thus possible by remote sensing. Hyper spectral image can give good indication of presence of pesticide residue on berry. The information can be used for generating advise on residue management. At present location specific advisory is based on weather information. Application of remote sensing can provide more detailed information on plants and pathogen in given location and can make the advisory more relevant.

References

Annonymous. 2015. HORTSAP guidelines. *State Department of Horticulture, Commissionerate of Agriculture, Pune, MS.* Websites: www.ncipm.org.in, www.mahaagri.gov.in

Singh, B.P., Islam, Ahammed, Sharma, V.C. and Shekhawat, G.S. 2000. Jhulsacast: A computerized forecast of potato late blight in western Uttar Pradesh. *J. Indian Potato Assoc.*, **27**: 25–34.

Vennila, S., Lokare, R., Singh, N., Ghadge, S.M. and Chattopadhyay, C. 2016. *Crop Pest Surveillance and Advisory Project of Maharashtra: A Role Model for an e-pest Management in India.* ICAR-National Research Centre for Integrated Pest Management, New Delhi, India, 56 p.

47

Minimizing Pesticide Residues in Horticulture Produce for Ensuring Higher Returns in Exports

Kaushik Banerjee

ICAR-National Research Centre for Grapes,
Manjri Farm, Pune – 412 307, Maharashtra
E-mail: kbgrape@yahoo.com

Horticultural commodities contribute a major share in the agricultural trade of India. India is a major food basket of the world. But in spite of being a recognized producer and exporter of the horticultural commodities, its share in the global food market is still < 2 per cent only. The global food trade has an ever rising trend, which has got more than tripled since 1990 to almost US$ 1 trillion of worth annually. A major challenge that the Indian exporters face while competing in international trade is to meet the challenge of differing standards across different countries, often with little clarity on how to meet those standards. Significant costs are usually involved in testing and certifications for compliance to such regulations and harmonization often seems like an unattainable goal.

The international trade of food is governed by three international agreements as outlined below:

i) Article 20 of the General Agreement on Tariffs and Trade (GATT) allows the governments to act on trade in order to protect human, animal or plant life or health, provided they do not discriminate or use this as disguised protectionism.

ii) Sanitary and Phytosanitary Measures Agreement (SPS) allows countries to set their own standards with an understanding that the regulations are based on science, with the provision that "precautionary principle" may be applied.

iii) The Technical Barriers to Trade Agreement (TBT) tries to ensure that regulations, standards, testing and certification procedures do not create unnecessary obstacles in the trade flow of food items. The TBT agreement encourages countries to recognize each other's procedure for assessing whether a product conforms to the regulatory requirements.

Who takes Care of Food Safety?

i) Academia- organizes research, generates science-based conclusions that in turn helps in developing food quality standards and also the package of practices for meeting those standards.

ii) Government bodies- do surveillance, develop regulations related to various policies, coordinate inspections, and implement various food safety traceability and monitoring programs. The research institutions under the state and central governments also conduct need-based fundamental, strategic, and applied research in priority areas of regional and national interests.

iii) Industry- Implements food safety practices, conducts applied research, sponsor projects to research institutes/universities/laboratories. In many cases the industries have their own R&D wing, which conducts practical oriented strategic and applied research.

iv) Consumers- play a driving role and facilitate the initiation, evolution and implementation of the food safety quality systems through expression of complaints, worries and trust.

In any case, an individual organization with the perspective from only one sector cannot successfully address the issue of food safety. And hence, "Achieving food safety is a shared responsibility and different types of stakeholders – including Government, the food industry, consumers and their organizations, academic and scientific institutions, *etc.* contribute to this capacity"[*FAO – strengthening National Food Control Systems: A Quick Guide to Assess Capacity Building Needs, 2007*] (FAO, 1).

Internationally, Codex Alimentarius Commission (Codex) is responsible for implementing the Joint FAO/WHO Food Standards Programme. The Codex international food standards, guidelines and codes of practice, have become the global reference points for consumers, food producers and processors, national food control agencies and the international food trade (Codex, 2).

Although the Codex standards are, in general, acceptable across the UN member countries, individual countries still may continue to have their own standards, because the food safety concerns could be different across different countries based on the differences in how food is grown, prepared and consumed across different countries and populations.

The monitoring of agro-chemical residues is becoming increasingly important in India and certain other Asian countries because of frequent non-compliances reported for the food commodities originated from these geographical regions. An examination of the rapid alert notifications of the European Union between the years 2000 to 2015 (RASFF, 3) clearly reflects the maximum number of food safety non-compliances from the Asian countries. And across the various food types, the most significant non-compliances are reported for the fruits and vegetables. A major reason for such non-compliances originate from the fact that the cultivation of horticultural crops often involves unregulated applications of agro-chemicals leading to their detections at harvest.

The major contaminants in fruits and vegetables include the residues of pesticides, mycotoxins (aflatoxins, ochratoxins, *etc.*), microbial contaminations (bacteria, fungus, virus, *etc.*), other chemical residues (dioxins, Furans, PAHs, *etc.*), antibiotics, disinfectants, colouring agents, food additives, allergens, *etc.* As per the EU-RASFF alerts, the major food safety concern relates to various pathogenic microorganisms, and this is sequentially followed by mycotoxins, pesticide residues and heavy metals.

Organic Contaminants: Major Concerns in Food Safety

- ✰ Frequent alerts issued by importing countries (due to presence of pesticide residues, mycotoxins, *etc.*)
- ✰ Growing concern over the food safety (growing consciousness among consumers for safe food in the country), andgrowing recognition of the potential impact of various food contaminants on public health, food security, and trade competitiveness.

The agro-chemical residues may appear as a major trade barrier. Very often there is lack of a strategic approach to manage these contaminants, although strict food safety regulations are in place in the domestic (FSSAI) and international (EU, USA, Japan, *etc.*) trade sectors. The "trade barriers" are usually those restrictions induced by Government or public authorities that render the imported commodities less competitive *vis-à-vis* locally produced items. In general, different MRLs in different countries create such trade barriers.

Expectations of Trade from Horticultural Food Producers

- ✰ The items should have adequate traceability in the entire chain from production to the consumption level.
- ✰ Implementation of Good Agricultural Practices for production. This includes appropriate documentation of the crop production process, implementation of officially recommended package of practices, strict adherence to the list of authorized chemicals for pest and disease managements, and appropriate testing for multi residue monitoring of agro-chemicals to establish food safety compliance for consumption.
- ✰ Compliance of the produce to Maximum Residue Limits (MRLs).

Pesticide residue analysis is thus considered as an important aspect in food safety monitoring and management systems. The national and international regulatory bodies, trade partners, and food industry stakeholders demand analysis of products at the place of origin. There are demands for residue testing for a large number of compounds, the residues of which might enter the food chain through direct applications and indirect sources (*e.g.* drift from adjoining agricultural fields, contaminated agro-inputs *etc.*), and enable comprehensive risk assessment. The methods of residue analysis should therefore be on par with the internationally accepted methods of analysis so that the results are mutually agreeable.

Maximum Residue Limit (MRL)

To guarantee the use of pesticides in the field according to good agricultural practices and to protect consumer health, regulatory agencies establish tolerances, or maximum residue limits (MRL), in crops. Maximum Residue Limits (MRLs) are the highest permitted concentrations of pesticide residues that can be present in or on food or feed following good agricultural practices and maintain the presence of toxic contaminants at lowest possible level. MRL is established taking into account the toxicological data of the pesticide, food consumption pattern as well as the trials on crops under good agricultural practices. Some of the heavy metals also have regulatory limits fixed by authorized bodies. Where MRL is not established on the basis of risk assessment, a default value applies (also called as extraneous MRL) set at the limit of analytical determination (0.01 mg/kg) (EU-MRL, 4).

Pre-harvest Interval (PHI) for Food Safety

Pre-harvest interval (PHI) is the most important in field level management of pesticide residues. PHI is also termed as the safe waiting period, which refers directly to the number of days that must pass between the time of the last application of a pesticide and when the crop is harvested. The residues from products used for pest control are only metabolized and broken down by living plants, not by dead/cut plants in the field. PHI can vary depending on the crop being sprayed and the product being used. The same product may be registered for use on multiple crops but the PHI could be different for each of those crops. On the basis of the MRL, the PHIs of pesticides are calculated. Estimating PHI of a pesticide ideally involves multi-location field trails wherein the pesticides are applied following the guidelines of GAP. With the help of dissipation studies, the crop-specific PHI is established. Strictly following PHIs, the pesticide contamination in harvested produce can be minimised below MRL. This also can remove the pesticide burden on environment efficiently.

Management System for Contaminants

Good Agricultural Practice (GAP)

To ensure safety, establishment of quality control and inspection systems is essential. The intention of monitoring programmes concerning the hazardous contaminants is to bring out the safe food and ensure that the contamination level in fruits and vegetables do not exceed the limits prescribed by the regulatory bodies. It also deters misuse of pesticides that could result in presence of unexpected contaminants in food and facilitates implementation of good agricultural practices (GAP).

Agriculture is expected to assure food security in a range of settings, now and in the future, and is increasingly called upon to produce positive environmental, social and economic benefits. While agriculture can be a key contributor to sustainable development and to meeting these challenges, the paradigm is dramatically shifting for its many primary producers in the context of a rapidly changing food economy and globalization.

These challenges can be tackled in part through a Good Agricultural Practice (GAP) - an approach that improves environmental, economic and social sustainability of on-farm production and results in safe and quality food and non-food agricultural products. A GAP approach can contribute concretely to implementing sustainable agriculture and rural development while addressing the demand-side priorities of consumers and retailers, the supply-side priorities of producers and labourers, and those institutions and services that are bridging supply and demand. While a GAP approach may respond to the growing demands of increasingly globalized and integrated agricultural sectors, it is also very important for local and national markets.

Residue Monitoring Programme (GrapeNet/HortiNet)

The pesticide residue monitoring programme (RMP) in table grapes for export to the EU was started in 2003-04 season in response to the rapid alerts issued against Indian table grapes in EU in 2003. The programme was initiated to establish a complete traceability system for export of table grapes to the EU countries through a web based residue monitoring system *viz.*, Grapenet (APEDA-Grapenet, 5), declaring ICAR-NRC for Grapes, Pune as the NRL (first crop based Referral Laboratory in the country). Over the last ten years, ICAR-NRCG has successfully managed the agro-chemical residues (pesticides, plant growth regulators, heavy metals, mycotoxins, *etc.*) in Indian grapes for both export and domestic markets. This has significantly improved the export of table grapes from India to EU and Middle East countries with simultaneous overall improvement in quality. The internal alerts for pesticide residues in grapes got significantly reduced to around 3.5-7 per cent over the last few grape seasons from the initial 23.7 per cent in 2003-04. The detection of persistent, old-generation, highly toxic pesticides (*e.g.* organophosphates, carbamates, *etc.*) has been substantially reduced. Similarly, detection of non-recommended, non-label claim pesticides has also been minimized significantly. For example, in table grapes, the residues of chlormequat chloride (CCC), a plant growth retardant, was detected at above-MRL (0.05 g/kg) in around 40 per cent of the export samples in 2010, and this has been minimized to only 0.7 per cent in the recently concluded 2016 grape season. In domestic market too, the grapes are now found much safer than earlier, which is a direct influence of implementation of the RMP that has brought an overall quality consciousness among all the stake holders of the grape industry regarding agro-chemical usage and has emerged as a model system for all agricultural crop in the country.

Considering this success story several initiatives are currently in progress to establish food safety traceability system in other horticultural commodities through an integrated system called HortiNet involving a comprehensive network of the stakeholders, *e.g.* State Horticulture Departments, farmers' organizations, exporters, quality evaluation laboratories, Agmark, *etc.*

Conclusion

Consumers are now more concerned about what they eat than ever before. In recent years, food safety has received extensive media exposure also. Comprehensive and effective food legislation, and through the enforcement mechanism to give it effect, it is possible to ensure that the production and handling of food is hygienic, free from contamination and properly described. Sincere adherence to the recommendations of the GAP and PHI can definitely minimize the residue load of agro-chemicals in fruits and vegetables at field level, which in turn, will ensure safety to the health of the consumers and ecosystem as a whole. It is important that the crop protection scientists should keep in mind the consequence of residue issues while issuing recommendation for a chemical at any particular crop growth stage. Integration of bio-efficacy and residue dissipation experiments is the need of the hour to bring out effective GAP.

Risk assessments methodologies associated with different contaminants need to be scientifically executed in an organised way. A well structured residue monitoring system with an institutional frame work can possibly ensure field-level traceability and management of contaminant residues in fruits and vegetables through implementation of good agricultural practices. As mentioned previously, well structured residue monitoring system in grapes initiated by APEDA and implemented through NRC Grapes has created a significant impact in improving the quality of the table grapes. Considering the success of Residue Monitoring Plan (RMP) in grapes, the traceability network has to be extended in other horticultural crops considering the typical crop-specific requirements.

References

APEDA-Grapenet. Agricultural and Processed Food Products Export Development Authority. apeda.gov.in/apedawebsite/Grapenet/GrapeNet_new.htm

Codex. Joint FAO/WHO Food Standards Programme. Codex Alimentarius Commission. fssai.gov.in/CCASIA/pdf/ALINORM_05_28_15.pdf

EU-MRL. EU pesticides database. http://ec.europa.eu/food/plant/pesticides/eu-pesticides-database-redirect/index_en.htm

FAO. Strengthening national food control systems A quick guide to assess capacity building needs. http://www.fao.org/3/a-a1142e.pdf

http://ec.europa.eu/food/safety/docs/rasff_annual_report_2015_preliminary.pdf

RASFF. Rapid Alert System for Food and Feed.

48

Management of Guava Wilt for Improving Productivity

A.K. Misra

ICAR-Central Institute for Sub-tropical Horticulture, Rehmankhera, Lucknow – 226 101, Uttar Pradesh
E-mail: misra_a_k@yahoo.co.in

Guava (*Psidium guajava* Linn.) is one of the important fruits of our country. In India it is grown in an area of 268 thousand ha with a total production of 3,668 thousand MT and average productivity of 13.7 MT/ha. It is an easily grown crop and is grown almost in all the states of India with highest area in UP, *i.e.* 45.0 thousand ha, followed by Maharashtra, *i.e.* 40.0 thousand ha. It is a hardy crop and is cultivated even in neglected soils or near river bed *etc.* Highest productivity is reported from Madhya Pradesh, *i.e.* 37.6 MT/ha. Wilt of guava is the most destructive disease, which is a limiting factor for guava cultivation in India. The average productivity can at least be doubled, if wilt disease is managed effectively. Wilt affected plants show varied symptoms, but two major symptoms are described *i.e.*, slow and quick wilt. Quick wilt can be termed as true wilt. In this, plant take 16 days to two months period for the complete wilting of plant after the appearance of first visible symptom. In slow wilt, plants take one to two years for complete wilting. Here, sometimes partial wilting also takes place. In one year some portion or few branches show symptoms and later other parts are affected. *Fusarium oxysporum* f. sp. *psidii*, *F. solani*, *F. chlamydosporum*, *Gliocladium roseum* are identified to cause wilt in guava, though *F. oxysporum* f. sp. *psidii*, *F. solani* were identified as the two major common pathogens. Efforts to manage wilt by chemical means generally fail. Work in last two decades at CISH, Lucknow revealed that the disease can now be effectively managed, if complete package of integrated disease management practices are followed, which include resistant rootstock, biocontrol and use of intercrop with cultural practices.

Historical Background and Geographical Distribution

In India, guava (*Psidium guajava* Linn.) is grown almost in all the states. Wilt of guava was first reported in 1935 from Allahabad. To understand the disease initial work was done at Allahabad Agricultural Institute, Naini under the leadership of Dr J.C. Edward and at Bidhan Chandra Agriculture University, Kalyani under the leadership of Prof S.B. Chattopadhyay. In the recent past efforts were made to understand the disease at CISH, Lucknow and HARP, Ranchi. The work on this problem was also reviewed (Misra and Prakash, 1990; Misra, 1995, 2001, 2003, 2005; Misra and Pandey, 1996; Negi *et al.*, 2001 and Gupta, *et al.*, 2010). After the first report of guava wilt in 1935 from Babakkarpur, Allahabad, it was reported from different parts of U.P. by several workers (Das Gupta and Rai, 1947; Dey, 1948; Anonymous, 1949, 1950; Prasad *et al.*, 1952; Mathur, 1956; Edward and Srivastava, 1957; Singh and Lal, 1953; Pandey and Dwivedi, 1985; Misra and Prakash, 1986; Misra, 1987). After the first few reports from U.P., in West Bengal the disease was reported from several places (Chattopadhyay and Sen Gupta, 1955; Chattopadhyay and Bhattacharjya, 1968a, b). Later, the disease was reported from Haryana (Suhag, 1976; Mehta, 1987), Punjab (Chandra Mohan *et al.*, 1986), Rajasthan (Katyal, 1972; Bhargava *et al.*, 2003), Delhi State (Anonymous, 1953), Jhakhand (Srivastava *et al.*, 2001), Andhra Pradesh (Jhooty *et al.*, 1984) and M.P. (Personal observation), Odisha and Tamil Nadu (Personal observation).

Wilt in guava plants have also been reported from other countries, *viz.*, Florida, USA (Webber, 1928), Taiwan (Hsieh *et al.*, 1976; Leu and Kao, 1979), Cuba (Rodrigvez and Landa, 1977), South Africa (Grech, 1985; Vos *et al.*, 2000), Brazil (Tokeshi *et al.*, 1980; Rodriguez *et al.*, 1987; Junqueira *et al.*, 2001), Pakistan (Ansar *et al.*, 1994), Bangladesh (Hamiduzzaman *et al.*, 1997), Canberra and Australia (Lim and Manicom 2003).

Losses

Loss assessments due to wilt disease in guava were estimated in different terms by different workers. Singh and Lal (1953) estimated 5-15 per cent loss amounting to almost 1 million Rupees due to guava wilt every year in 12 districts of U.P. In West Bengal, the disease reduced the yield by 80 per cent, *i.e.*, from 113.5 q/ha in healthy plantations to about 18.16-22.7 q/ha in affected orchards (Chattopadhyay and Sen Gupta, 1955). Seven thousand acres of land in Andhra Pradesh, reduced the land value to half by the presence of the disease (Jhooty *et al.*, 1984). About 150 and 300 acres of guava orchards in Punjab and Haryana, respectively were uprooted during 1978-81 (Jhooty *et al.*, 1984). Around Lucknow loss due to wilt in guava varied from 5-60 per cent (Misra and Shukla, 2002).

Symptoms

Symptoms of wilt in guava is generally confusing. Any guava plant dying can not be termed as wilt. It is necessary to discriminate the two *i.e.* wilt and death of plant. Death of the plant may be due to heavy infestation of insects like bark eating caterpillar, termite or root white grubs *etc.*, it can also be due to root rotting fungi like *Pythium*, *Phytophthora*, *Rhizoctonia* or *Macrophomina etc.* or it can be due to cultural or physiological reasons or soil reasons. It is also to be understood that guava wilt is a disease of grown up plants and not the disease of guava seedlings or young guava plants. True wilt is when the xylem vessels are affected and movement of water and nutrition is restricted and finally plant die. Xylem vessels may be blocked with the physical presence of pathogen mycelium or spores, gums or formation of tylose *etc.* or effect of toxin production by the pathogens. Hence, it is essential to understand the basic concept before studying guava wilt.

When we describe the symptom of true wilt, the affected plants show sudden drooping of leaves with yellow colouration. At the later stage subsequently shedding of leaves take place. Twigs become bare and fail to bring forth new leaves or flowers and eventually dry up. Fruits of all the affected branches remain underdeveloped, become hard, black and stony and remain on branches for some time. Later the entire plant becomes defoliated and eventually dies (Figure 1). Some trees show only partial wilting and only half or some branches wilt and other half remain healthy, which die next year. This type of partial wilting is common in guava wilt. It was found in our studies that in quick wilt, it requires only 16 days for complete wilting of plant after the appearance of first visible symptom. Some affected trees linger on even up to 252 days and then die (Misra and Pandey, 2000b). We also studied variations in the symptoms during different months of the year and noticed yellowing of the leaves with inter-venial chlorosis during the month of August. At this stage the leaves drop even with the slight shaking of the plants. During September, general drooping of the leaves takes place. During October complete wilting of plants are seen with almost dried leaves and small dried black fruits hanging on the branch. Few plants also show partial wilting, which is a very common symptom of wilt in guava. It is also recorded that some plants show wilting of variable degree (leaf yellowing, drooping of leaves, drying of terminal branches or partial wilting) during different months but later escape/resist wilting. These plants start recovering from December onward. It was recorded that out of total wilting plants, around 17 per cent plants, which initially show some symptoms of wilting, ultimately escape/resist wilting (Misra and Pandey, 2000b).

Figure 1: Symptoms of Complete Quick Wilting and Partial Wilting.

The finer roots show black streaks, which become prominent on removing the bark (Das Gupta and Rai, 1947). The roots also show rotting at the basal region and the bark is easily detachable from the cortex. The cortical regions of the stem and root show distinct discoloration and damage. Light brown discoloration is noticed in vascular tissues (Chattopadhyay and Bhattacharjya, 1968a,b). Wilted plants later show bark splitting. The pathogen attacks young as well as old fruit bearing trees but older trees are more prone to the disease (Misra and Shukla, 2002).

Causal Organism

Various pathogens involved with the affected plants, *viz.*, *Fusarium oxysporum* f. sp. *psidii*, *F. solani*, *Macrophomina phaeseoli*, *Rhizoctonia bataticola*, *Cephlosporium* sp., *Gliocladium vermoesenii*; *Gliocladium roseum* and *Verticillium albo-atrum*, *Acremonium* sp. *etc.* (Vestal, 1941; Dey, 1948; Das Gupta and Rai, 1947; Prasad *et al.*, 1952; Edward and Srivastava, 1957; Pandey and Dwivedi, 1985; Chattopadhyay and Bhattacharjya, 1968a,b; Chattopadhyay and Sengupta, 1955; Chandra Mohan, 1985; Pandey and Dwivedi, 1985 and Sohi, 1983a,b; Misra and Pandey, 1997; Misra and Pandey, 2000a).

The reports from other parts of the world are different. *Clitocybe tabescens*, *Myxosporium psidii*, *Septofusidium* sp., *Penicillium vermoensenii*, *Colletotrichum gloeosporiodes*, *Fusarium oxysporum* f. sp. *psidii*, *Meloidogyne* sp., *Helicotylenchus* sp., *Pratylenchus* sp., *Pseudomonas* sp. and *Erwinia psidii* are reported by various workers (Webber, 1928; Rodriguez and Landa, 1977; Hsieh *et al.*, 1976; Leu *et al.*, 1979; Tokeshi *et al.*, 1980; Rodrigues *et al.*, 1987; Jungueira *et al.*, 2001; Grech, 1985; Ansar *et al.*, 1994; Hamiduzzaman *et al.*, 1997; Vos *et al.*, 2000; Lim and Manicom, 2003).

As earlier advocated that the symptom of wilt disease of guava is generally confused and generally co-related with the death of plant, the various causal organisms are reported. These organisms may be either only associated or responsible for the death of plants either alone or in combination or may help as cofactor or predisposing factor for the wilt syndrome. Gomes *et al.* (2011) advocated that the guava decline/wilt can be characterize as a complex disease, caused by the synergistic effect of these organisms, in which parasitism by the nematode predisposes the plants to root decay or entry of pathogen.

The recent studies at CISH, Lucknow (Misra and Pandey, 1997; Misra and Pandey, 2000a) revealed *Gliocladium roseum* as one of the pathogen of guava wilt, which reproduces symptom of wilt on artificial inoculation. The author succeeded in reproducing wilt by artificial inoculation in grown up guava plants of 6-7 years by *Gliocladium roseum* by stem hole inoculation technique (Figure 2).

Figure 2: Guava Plants (5-6 years) Inoculated in Field by Stem Inoculation with *Gliocladium roseum* Reproduced the Symptom.

They earlier also reported *Gliocladium virens*, *G. penicilloides*, *Fusarium oxysporum* f. sp. *psidii*, *F. solani* and *Acremonium* sp. associated with guava wilt symptoms, indicating it's complex nature (Misra and Pandey, 1999c). They standardized the inoculation technique, *i.e.* stem hole inoculation technique for reproducing wilt symptoms quickly. Compared to soil inoculation and root inoculation techniques, their stem inoculation method proved superior in producing quick wilting symptoms within a month.

At CISH Lucknow *Fusarium oxysporum* f. sp. *psidii*, *F. solani*, *F. chlamydosporum*, *Gliocladium roseum*, *G. virens*, *G. penicilloides* and *Acremonium* sp. caused wilting on inoculation test and were consistently re-isolated from affected tissue also, hence, fulfilled Koch's postulates. Later *Gliocladium virens* and *Acremonium* sp. inoculated plants regenerated after showing initial symptoms. Hence, only *Fusarium oxysporum* f. sp. *psidii*, *F. solani*, *F. chlamydosporum*, *Gliocladium roseum* and *G. penicilloides* may be considered as the causal organism for wilt in guava. It was further found that the occurrence of *Fusarium oxysporum* f. sp. *psidii* and the *F. solani* are most frequently isolated, and can be considered the two most important pathogens of wilt disease of guava, which can be considered as the real pathogen of guava wilt. The other pathogens are also responsible, but with lesser frequency. It was also found that out of two pathogens, the frequency of *F. solani* was more compared to *Fusarium oxysporum* f. sp. *psidii* (Misra, 2004, 2008). It was also found that the isolates of *Fusarium oxysporum* f. sp. *psidii* and *F. solani* collected from different places showed great variability. In an extensive studies, morphological variations were recorded in the isolates of *F. oxysporum* f. sp. *psidii* and *F. solani* and these pathogens were grouped on the basis of colony colour, texture, colour of metabolite released, frequency of their occurrence on culture, sporulation and macro-conidia production and type/percentage of wilting they caused. High degree of variations was recorded among the different isolates. These are the probable reasons for the variable incidence and severity of the disease in different places.

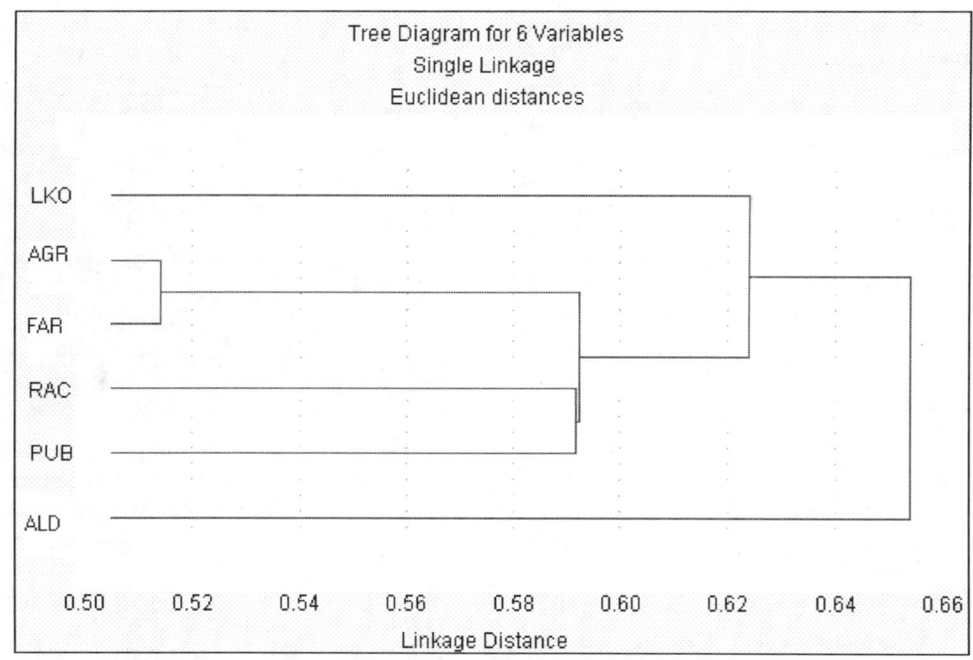

Figure 3: Dendogram Derived from Random Amplified Polymorphic DNA Analysis of Six Isolates of *Fusarium solani* of Six different Regions. The bottom scale is the percentage of similarity. (LKO: Lucknow; AGR: Agra; FAR: Farukhabad; RAC: Ranchi; PUB: Punjab; ALD: Allahabad).

Partial sequencing of *F. oxysporum* f. sp. *psidii* isolate VKGF01, *F. solani* isolate VKGFS1 and *F. chlamydosporum* isolate VKGFC01 were also done. Total 36 representative isolates of *F. oxysporum* f. sp. *psidii* (6 from one location) and 36 representative isolates of *F. solani* (6 from one location) collected from 6 different agro-climatic zones/ guava growing areas of India were subjected to estimate the molecular characterization. Out of 20 RAPD primer screened, OPM 11(RBM3) was found best to produce good allelic band with all isolates of *F. oxysporum* f. sp. *psidii* and *F. solani*. Product size ranged from 200 to 2000 and 600 bp were obtained almost in all the isolates of *Fusarium* spp. For *F. solani* RAPD data were subjected to construct a dendogram using UPGMA software. It showed that isolates from Agra and Farukhabad regions (UP) were almost similar and form one group. Punjab and Ranchi region isolates formed second group. *F. solani*, isolates of Allahabad region were distinctly related to isolates of other regions (Figure 3).

Similarly, for *F. oxysporum*, RAPD data were subjected to construct a dendogram using UPGMA software. It showed that isolates from Agra and Farukhabad regions were almost similar and form a group. Punjab and Ranchi region formed second group Figure 4.

The association of nematode *Heliocotylenchus dihystera* was found higher in the field of affected orchards and is considered to enhance the incidence of wilt (Khan and Misra, 2003). It was interpreted that *Heliocotylenchus dihystera* make the entry point in guava root for the easy entry of pathogen. Recently, *Meliodogyne* spp. was also found associated with the disease in some localities of Badayu (UP). It was found that this area had

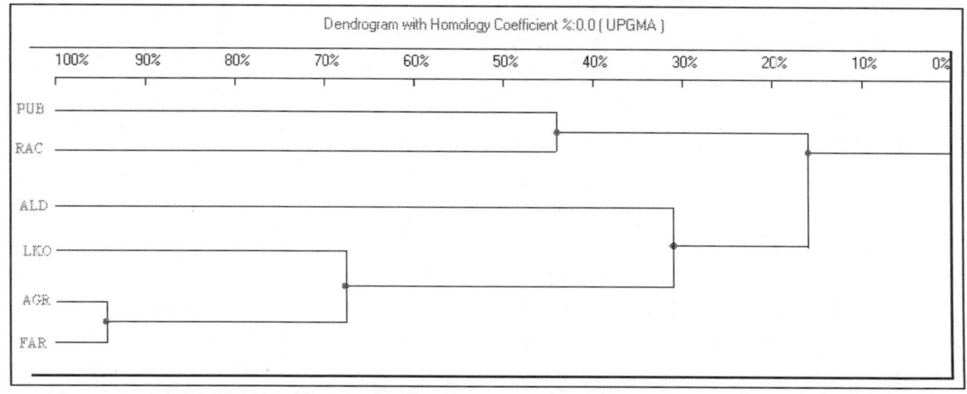

Figure 4: Dendogram Derived from Random Amplified Polymorphic DNA Analysis of Six Isolates of *F. oxysporum* of Six different Regions.

history of growing vegetables, which had problem of root gall nematodes. The soil has developed very high build up of *Meliodogyne* spp. Under the heavy pressure of this nematode population slow death of plants take place. Hence, it can be said that *Meliodogyne* spp. cause death of the plant and not wilt.

Epidemiology

There are variable reports about the severity of disease at different pH levels and variable fertilizer levels. However, in general it was found that disease is common in different types of soil and at different pH or different levels of fertilizers. Hence, it cannot be correlated either with soil types or pH or different levels of fertilizers *etc.* However, there is clear picture on the period of higher disease incidence during the year. Extensive studies on the progress of natural wilting of guava plants during different months have been made. Misra and Pandey (1999a, d, 2000b) at Lucknow found that wilting generally start after rains during August, September and maximum wilting occur during October month. Some plants, which show slight yellowing that start recovering from December onwards. On analyzing the weather data, it was found that higher rainfall during July-September with maximum temperature ranging from 31.3 to 33.5°C and minimum temperature ranging from 23 to 25°C and humidity around 76 per cent favour the wilt incidence. It was also found that in general two months are required for the complete wilting of plants (from appearance of 1st visible symptom to complete wilting). However, minimum period was found only 16 days.

Histopathology

Histopathological studies of the root samples revealed breaking and opening in the epidermis, through which pathogens are supposed to enter in the host tissue (Figure 5). Necrosis in the internal tissue and vascular bundle were recorded, which restricts the movement of water and nutrient in the plant and thus, results wilting of guava plants.

Physical presence of the pathogen and tylose formation was also recorded in the root tissues, which restrict the movement of water and nutrition in the plant (Figure 6).

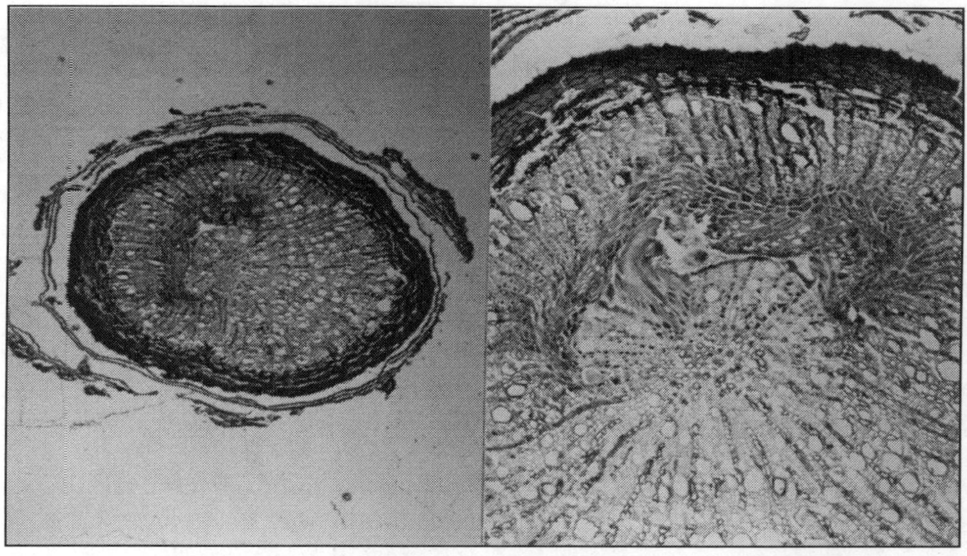

Figure 5: T.S. of Diseased Root Showing Disintegration of Epidermis and Cortex and Necrosis of Vascular Tissue.

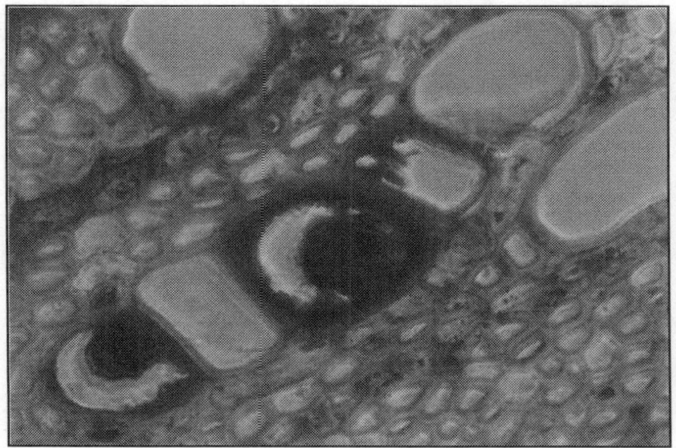

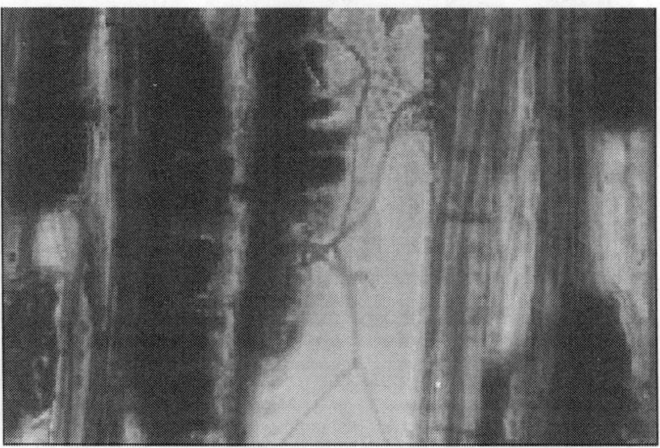

Figure 6: TS and LS of Stem Showing Tylose and Fungal Mycelium Causing Blockage of Vascular Vessels.

Disease Management

Time to time recommendations for the control of guava wilt has been suggested by different workers. These are summarized below.

Disease Management through Chemicals

Different chemical managements were suggested by different previous workers. Chaubatia paste, water-soluble 8-quinolinol sulphate, benlate or bavistin, metasystox and zinc sulphate, thiophanate methyl, captafol and thiabendazole were suggested (Anonymous, 1949; Jain, 1956; Suhag, 1976; Bhargava *et al.*, 2003; Leu *et al.*, 1979). In South Africa tebuconazole, propiconazole, prochoraz, triforine and carbendazim + flusilazole were found effective *in vitro* evaluation (Joubert and Frean, 1993). Misra and Pandey (1999b) reported that though different fungicides *viz.*, bavistin, topsin M, indofil M-45, thiram, blitox to check the various wilt pathogens in laboratory effectively, but these pathogens increases it's aggressiveness with profuse spore mass production in the soil, once the effect of these fungicides diminishes in soil (Misra and Pandey, 1999). These chemicals are costly and repeated application is not economical also. Moreover, seeing the soil mass, increasing aggressiveness of the pathogen, profuse spore production and economics, chemical management seems not practical.

Besides fungicides some soil amendment chemicals/ cakes/fertilizers were also evaluated for control of wilt. Mathur *et al.* (1964) found wilt control by soil treatment with 1.82 kg lime or gypsum/tree, although the control mechanism was not well understood. Oil cakes like neem, *mahua* and *kusum* supplemented with urea @ 10 kg and 1 kg respectively also check the disease (Das Gupta and Ghoshal, 1977). At CISH, Lucknow also it was found that wilt was checked by application of 6 kg. neem cake + 2 kg gypsum per plant (Misra and Pandey, 1994-95). These are useful and can be integrated in the integrated disease management practices as one of the component.

Disease Management through Cultural Practices

Mathur (1956) advocated that wilt could be controlled by proper sanitation in the orchard. Wilted trees should be uprooted, burnt and trench should be dug around the tree trunk. Edward (1960a) suggested that while transplanting, roots of plants should not be severely damaged. Maintenance of proper tree vigour by timely and adequately manuring, inter-culture and irrigation enable them to withstand infection. The pits may be treated with formalin and kept covered for about 3 days and then transplanting should be done after two weeks. Prasad *et al.* (2003), Khan and Misra (2003) and Misra *et al.* (2004b) reported intercropping with turmeric or marigold to restrict the wilting of guava. Misra (2016) also observed that the orchards, which were having frequent tillage had more incidence of wilt compared to less tilled orchards. Tillage during monsoon enhance wilt incidence. Hence, tillage should at least be avoided during monsoon. As the disease is soil borne, flood irrigation spreads the disease. Hence, separate basin irrigation or drip irrigation should be encouraged for the management of disease. These cultural practices are useful and should be adopted for integrated management practices.

Disease Management through Varietal Resistance

Edward (1961) reported cultivar Chittidar, Hafsi, Safeda Riverside, Rolf and Stone acid susceptible and guava species *Psidium cattleianum* var. *lucidium* and other genera *Syzigium cumini* (*Jamun*) resistant to wilt. Edward and Gaurishanker, (1964) further reported that *Syzigium cuminii*, *Lagerstraemia indica*, *Psidium cattleianum* (*P. molle*), *P. guianense*, Chinese guava (*P. friedrichsthalianum*) and Philippine guava are resistant to wilt. Stock-scion compatibility was also evaluated. Cultivar Safeda as scion and the above resistant material as rootstock *Lagerstraemia indica* proved incompatible, *Syzigium cumini* (Jamun) as partially compatible and other compatible. They suggested use of resistant rootstock. At CISH, Lucknow Misra (1998-99) studied the relative field tolerance of 20 guava cultivars and categorized them into different groups on the basis of their natural susceptibility. The cultivars Allahabad Safeda, Florida Seedling, Guinenees, Hafsi, Karela, Mirzapuri Seedling, Nasik, Pear Shaped, Sindh, Superior and White Fleshed proved highly susceptible; Behat Coconut and Pourtgal as susceptible; Apple Colour, Chittidar, Seedless and Spear Acid, Superior Sour Lucidium, Red Flesh and Smooth Green as tolerant. The cv. Chittidar was graded as reasonably good cultivar, which has reasonable level of resistance and quality and directly be reported as resistant material. Later Misra *et al.* (2003c) identified F1 population of *Psidium molle* x *Psidium guajava* free from wilt, when grown in wilt sick plot and inoculated repeatedly with *Gliocladium roseum*, *Fusarium solani* and *Fusarium oxysporum*. This material is now used at CISH, Lucknow as resistant rootstock for multiplying plants of choicest variety of guava. The strawberry guava (*Psidium cattleianum*) has been reported relatively hardy species from Reunion (Normand, 1994).

Disease Management through Bio-control Agents

Being the soil borne nature of wilt pathogens, it is desired that any management material, which itself multiples in the soil is more useful for the management of the disease. The chemicals do not fulfill this requirement. The effects of chemicals are also hazardous for the soil and environment, moreover when the effect of chemicals diminishes, the pathogen become more virulent and aggressive (Misra and Pandey 1999b). Hence, considering the above facts, it was considered more desirable to use the bio-agents for the management of the wilt disease.

At CISH, Lucknow three bio-agents were found effective for the management of the wilt disease, *i.e.* *Aspergillus niger* strain AN 17, *Trichderma harzianum* and *Penicillium citrinum* (Misra *et al.*, 2000, 2004; Prakash *et al.*, 2002). When these fungi were tested for the management of wilt pathogen in laboratory conditions, they were found quite effective (Misra *et al.*, 2004). When relative growth of the three bio-agents was studied, it was found that *Aspergillus niger* was fastest growing and most effective in controlling the wilt disease in field (Misra and Prasad, 2003). Besides this quality, it is also growth enhancer and the plants treated with *Aspergillus niger* developed faster with more height, more thickness and higher numbers of leaves (Misra *et al.*, 2000). These can be grown easily on any substrate like maize/*bajra* seeds *etc.* and can also be multiplied on cheap substrates like *Sacchrum* sp. (grass) and dry and green leaves of *Psidium guajava* (Shukla *et al.*, 2003). It was also found that at village level, these bioagent can be multiplied even in earthen pots (Misra and Prasad, 2004) (Figure 7).

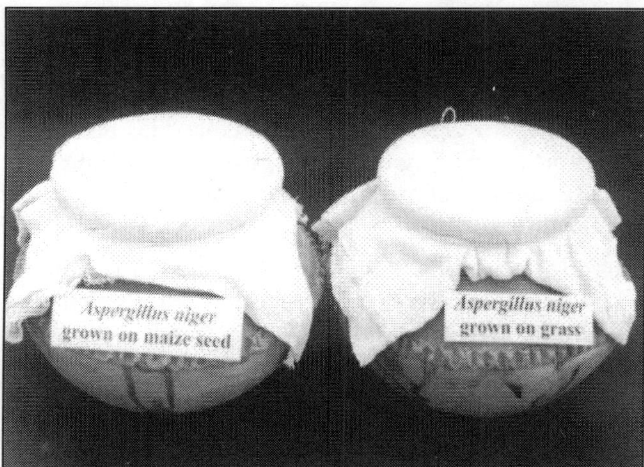

Figure 7: Multiplication of *Aspergillus niger* in Earthen Pots..

Technique of multiplication of bio-agents and the application in field were standardized (Misra *et al.*, 2003 b,d; Misra and Singh, 2005) and is summarized below.

Method of Multiplication of *Aspergillus niger*

- ☆ First the culture of *Aspergillus* is multiplied on maize seed.
- ☆ Then it is further multiplied on FYM.
- ☆ FYM bed is prepared of known quantity (1 m x 2 m x 30 cm).
- ☆ Multiplied culture is added in bed @ 0.1 per cent *i.e.* 1g/kg of FYM (Figure 8).
- ☆ Light amount of water is sprayed over the bed to increase moisture and covered with polythene sheet.
- ☆ After 4-5 days, it is reopened and again light amount of water is sprayed and FYM is turned properly for mixing it and again covered with polythene sheet.
- ☆ This process is repeated 3-4 times to allow *A. niger* to multiply in FYM properly.

Figure 8: Mass Multiplication of Bio-agents in FYM.

Method of Application

- ✰ While planting the new guava plants, each pit is filled with 5 kg of this enriched FYM with *A. niger* and then the new guava plant is planted in this pit.
- ✰ In grown up plants 10 kg of this enriched FYM is added near the root zone of guava plants and properly mixed in soil by turning the soil and mulched with organic mulch.
- ✰ This practice is done every year in the month of June end as preventive measure.

Similarly the other bio-agents *viz., Trichoderma harzianum, T. viride* and *Penicillium citrinum* can also be multiplied and applied.

Integrated Eco-friendly Approach

Considering the complexity of the problem integrated eco-friendly approach for the control of guava wilt was suggested by Misra *et al.* (2003c, 2004b) and Misra (2005).

- ✰ Use resistant rootstock (*P. molle* x *P. guajava*).
- ✰ Apply Bioagent (*Aspergillus niger, Trichoderma* spp. or *Penicillium citrinum*) at the time of planting and regularly once every year enriched in FYM before monsoon.
- ✰ Intercropping with marigold or turmeric.
- ✰ Application of neem cake and gypsum.
- ✰ Minimum tillage. Avoid tillage at least during monsoon.
- ✰ Separate basin ring irrigation/drip irrigation.
- ✰ Maintain ideal plant population.
- ✰ Maintain sanitation in orchard.

It is very necessary to implement complete management practice so that effective management is achieved. Following only one or two component do not give desired result. If disease is managed, the production of guava can easily be doubled.

Conclusion

Follwoing conclusion emerge out of the above:

- ✰ Disease is complex.
- ✰ Several pathogens may cause wilt of guava, However, *Fusarium oxysporum, F. solani, F. clamydosporum* and *Gliocladium roseum* are important pathogens.
- ✰ Role of nematode is in aggravating the disease.
- ✰ Frequency of *F. solani* was found maximum.
- ✰ *G. roseum* was found as an aggressive pathogen.
- ✰ Minimum tillage should be done in guava field especially during monsoon/just after rain.
- ✰ Flood irrigation to be avoided.
- ✰ Integrated disease management practices to be adopted for its management.
- ✰ Bioagents (fresh on-farm prepared) need to be applied every year with FYM.
- ✰ Practical and long lasting solution lies with adoption resistant rootstock/cultivar.
- ✰ Maintenance of plant population is a practical solution and the gap made by wilted plant immediately be filled.
- ✰ In West Bengal crop rotation has proved effective, where plant growth is fast and 5-6 year old plants gives good yield.

References

Anonymous. 1949. *Annual Administrative Report of the Department of Agriculture*, United Provinces for the year 1947-48, 89 pp.

Anonymous. 1950. *Annual Administrative Report of the Department of Agriculture*, United Provinces for the year ending 30th June, 1949, 125 pp.

Anonymous. 1953. *Report of the Directorate of Plant Protection*, Quarantine and Storage for the period 1946-1951. Govt. of India Press, Delhi, 44 pp.

Ansar, M., Saleem, A. and Iqbal, A. 1994. Cause and control of guava decline in Punjab (Pakistan). *Pakistan J. Phytopath.*, **6**: 41–44.

Bhargava, A.K., Sobti, A.K. and Ghasolia, R.P. 2003. Studies on guava (*Psidium guajava* L.) drying/wilt disease in orchards of Pushkar Valley. *J. Phytological Res.* **16**: 81 –84.

Chandra Mohan, 1985. Studies on guava decline in Punjab with special reference to wilt. *Ph.D. Thesis*, Punjab Agric. Univ., Ludhiana, India, p. 90.

Chandra Mohan, Jhooty, J.S. and Chand, T. 1986. Prevalence of guava decline in Punjab. *Plant Dis. Res.*, **1**: 77–78.

Chattapadhyay, S.B. and Bhattacharjya, S.K. 1968a. Investigations on the wilt disease of guava (*Psidium guajava* L.) in West Bengal I. *Indian J. Agric. Sci.*, **38**: 65 –72.

Chattapadhyay, S.B. and Bhattacharjya, S.K. 1968b. Investigations on the wilt disease of guava (*Psidium guajava* L.) in West Bengal II. *Indian J. Agric. Sci.*, **38**: 176–183.

Chattopadhyay, S.B. and Sengupta, S.K. 1955. Studies on wilt of *Psidium guajava* L. in West Bengal. *Indian J. Hort.*, 12: 76–79.

Das Gupta, M.K. and Ghoshal, B.K. 1977. Is it possible to control guava wilt through oil cake amendments? *Sci. Cult.* **43**: 131–133.

Das Gupta, S.N. and Rai, J.N. 1947. Wilt disease of guava (*P. guajava*). *Curr. Sci.*, **16**: 256–258.

Dey, P.K. 1948. Plant Pathology, Adm. Rep., Agric. Dept., U.P., 1945-46, 43–46 pp.

Edward, J.C. 1960a. Wilt disease of guava. *The Allahabad Farmer*, **34**: 289–293.

Edward, J.C. 1961. Rootstock trials for guava wilt control. *The Allahabad Farmer*, **35**: 5–9.

Edward, J.C. and Gaurishanker. 1964. Rootstock trial for guava (*Psidium guajava* L.). *The Allahabad Farmer*, **38**: 249–250.

Edward, J.C. and Srivastava, R.N. 1957. RE 502: Studies on guava wilt. *The Allahabad Farmer*, **31**: 144–146.

Grech, N.M. 1985. First report of guava rapid death syndrome caused by *Septofusidium* sp. in South Africa. *Pl. Dis.*, **69**: 726.

Gomes, V.M., Souza, R.M., Mussi-Dias, V., Silveira, S.F.D. and Dolinski, C. 2011. Guava decline: A complex disease involving *Meloidogyne mayaguensis* and *Fusarium solani*. *J. Phytopath.*, **159**: 45–50.

Gupta, V.K., Misra, A.K., Gaur, R.K., Jain, P.K., Gaur, D. and Sharma, S. 2010. Current status of Fusarium wilt disease of guava (*Psidium guajava* L.) in India. *Biotech.*, **9**: 176–192.

Hamiduzzaman, M.M., Meat, M.B. and Ahmad, M.U. 1997. Effect of *Fusarium oxysporum* and nematode interaction on guava wilt. *Bangladesh J. Pl. Pathol.*, **13**: 9 –11.

Hsieh, S.P.Y., Liang, W.J., Kao, C.W. and Lau, L.S. 1976. Morphological and physiological characters of *Myxosporium psidii*, the causal organism of guava wilt. *Plant Prot. Bull.* (Taiwan), **18**: 309–317.

Jain, S.S. 1956. A preliminary note on the inactivation of *Fusarium oxysporum* f. *psidii* in guava plants by chemotherapeutic treatment. *Indian J. Hort.*, **13**: 102–104.

Jhooty, J.S., Chand, J.N. and Krishnamurthy, V. 1984. Report of committee constituted by ICAR on guava decline in Punjab and Haryana. Submitted to ICAR, New Delhi.

Joubert, M.H. and Frean, R.T. 1993. An *in vitro* evaluation of fungicides against guava wilt. Inligtings bulletin Instituut Vir Tropiese en Subtropiese Gewasse, (246): 3.

Junqueira, N.T.V., Andrade, L.R.M. de, Pereira, M., Lima, M.M. and Chaves, R.da.C. 2001. Diseases of guava (*Psidium guajava* L.) cultivated in Brazilian Cerrdos. *Circular Tecnica Embrapa Cerrados* (15): 31.

Katyal, S.L. 1972. Twenty five years of research on fruit crops. *Indian Fmg.*, **22**: 14–16.

Khan. R.M. and Misra, A.K. 2003. Influence of co-cultivation of marigold, garlic and turmeric on nemic population in guava cropping system. Souvenir and Abstract, Zonal Conference (East Zone), Indian Society of Mycology and Plant Pathology and Seminar on Plant Diseases of National Importance with Special Reference to Guava Wilt and Mango Malformation. 4–5th April, CISH and IISR, Lucknow, 24–25 pp.

Leu, L.S. and Kao, C.W. 1979. Artificial inoculation of guava with *Myxosporium psidii*. *Plant Dis. Report.*, **63**: 1077–1079.

Leu, L.S., Kao, C.W., Wang, C.C., Liang, W.J. and Hsieh, S.P.Y. 1979. *Myxosporium* wilt of guava and its control. *Plant Dis. Report*, **63**: 1075–1077.

Lim, T.K. and Manicom, B.Q. 2003. Diseases of guava. **In**: *Diseases of Tropical Fruit Crops*, pp. 275–289, (Ed.) R.C. Ploetz. CABI, Wallingford, U.K.

Mathur, R.S. 1956. Guava diseases in India. *Indian J. Hort.*, **13**: 26–29.

Mathur, R.S., Jain, S.S. and Swarup, J. 1964. Chemical treatment for guava wilt. *Proc. Nat. Acad. Sci.*, India. Sect B., **34**: 33–36.

Mehta, N. 1987. Distribution of guava wilt in relation to age, soil type, management practices and varieties grown in Haryana. *Plant Dis. Res.*, **2**: 116–119.

Misra, A.K. 1987. Studies on diseases of fruits crops. *Annual Report*, CIHNP, Lucknow, pp. 124–125.

Misra, A.K. 1995. Guava Wilt. **In**: *Advances in Diseases of Fruits in India*, (Ed.) S.J. Singh, Kalyani Publishers, Ludhiana, pp. 183–190.

Misra, A.K. 1998-99. Screening against wilt. Annual Report, CISH, Lucknow, 10 pp.

Misra, A.K. 2001. Diseases of Guava and their Management. **In**: *Diseases of Fruits and Vegetables and their Management*, (Ed.) T.S. Thind, Kalyani Publishers, Ludhiana, pp. 128–138.

Misra, A.K. 2003. Guava diseases: Their symptoms, cause and Management. **In**: *Diseases of Fruits and Vegetables: Diagnosis and Management*, Vol. 2, (Ed.) S.A.M.H. Naqvi, Kluwer Academic Publishers, The Netherlands, pp. 81–119.

Misra, A.K. 2005. Important diseases of guava in India with special reference to wilt. Souvenir. *1st International Guava Symposium*, December 5–8, CISH, Lucknow, pp. 75–90.

Misra, A.K. 2008. Important diseases of guava and their management. **In**: Souvenir, *National Guava Symposium: Improvement, Production and Utilization*, 24th–26th November. Guava Grower Association, Shiridi, AMPUS, Pune, CISH, Lucknow and MPKV, Rahuri. pp. 8–14.

Misra, A.K. 2016. Progressive steps in understanding and solving guava wilt: A national problem. Abstracts: *6th International Conference on Plant, Pathogens and People*, 23rd February–27th February. Indian Phytopathological Society, New Delhi, pp. 8.

Misra, A.K. and Pandey, B.K. 1994-95. Studies on guava wilt. Annual Report, CIHNP, Lucknow, 28 pp.

Misra, A.K. and Pandey, B.K. 1996. Present status of wilt disease of guava. **In**: *Disease Scenario in Crop Plants, Vol. 1: Fruits and Vegetables*, (Eds.(V.P. Agnihotri, Om Prakash, Ram Kishun and A.K. Misra. International Books and Periodical Supply Service, New Delhi, pp. 61–70.

Misra, A.K. and Pandey, B.K. 1997. Pathogenicity and symptom production of wilt disease of guava by a new potent pathogen *Gliocladium roseum*. Indian Phytopathological Society–*Golden Jubilee International Conference on Integrated Plant Disease Management for Sustainable Agriculture*, IARI, New Delhi, November 10–15, 319 pp.

Misra, A.K. and Pandey, B.K. 1999. Guava wilt disease: A challenge for the coming millennium. *Proc. Nat. Symp. Challenges and Prospects of Plant Pathology in the Coming Millennium*. December 9–11, NBRI, Lucknow, pp. 22.

Misra, A.K. and Pandey, B.K. 1999a. Natural wilting of guava plants during different months. *51st Annual Meeting and Nat. Symp. on Seed Healthcare and Phytosanitation for Sustainable Agriculture*, 17–19 February, IISR, Lucknow, 64–65 pp.

Misra, A.K. and Pandey, B.K. 1999b. Pathogenicity and evaluation of fungicides against guava wilt pathogens. *J. Mycol. Pl. Pathol.* **29**: 274–275.

Misra A.K. and Pandey, B.K. 1999d. Natural wilting of guava plants during different months. *Indian Phytopathol.*, **52** (3): 312.

Misra, A.K. and Pandey, B.K. 2000a. Pathogenicity and symptom production of wilt disease of guava by a new potent pathogen *Gliocladium roseum*. Proceedings, Indian Phytopathological Society–Golden Jubilee, International Conference on Integrated Disease Management for Sustainable Agriculture Vol. II Pub. Indian Phytopathological Society, New Delhi, 749–750 pp.

Misra, A.K. and Pandey, B.K. 2000b. Progressive natural wilting of guava plants during different months. *Indian Phytopathol.*, **53:** 423–427.

Misra, A.K. and Prakash, Om. 1986. Studies on diseases of fruit crops. *Annual Report*, CIHNP, Lucknow, 67–68 pp.

Misra, A.K. and Prakash, Om. 1990. *Guava Diseases: An Annotated Bibliography 1907–1990.* Bishen Singh Mahendra Pal Singh, Dehradun, 132 p.

Misra, A.K. and Prasad, B. 2003. Relative efficacy of different bio-agents for the control of guava wilt. *J. Mycol. Pl. Pathol.* **33**(3): 494.

Misra, A.K. and Prasad, D. 2004. *Aspergillus niger* strain AN 17 potent bioagent to control wilt disease and its easy multiplication. *Symposium on Recent Advances in Fungal Bioagents and their Social Benefits*, 10th September, NBRI, Lucknow, 12 pp.

Misra, A.K. and Shukla, S.K. 2002. Assessment of loss due to guava wilt around Lucknow. *Nat. Seminar on Production and Post-Harvest Technology of Guava.* Dept. Hort. CSAUA and T, Kanpur, 9–10 January, 34–35 pp.

Misra, A.K., Singh, V.K. 2005. The CISH, Lucknow, marches towards solving guava wilt and mango malformation. *Indian Hort.* (April–June), **50**(1): 25–27.

Misra, A.K., Prakash Om and Sen, B. 2000. Biological control of guava wilt by *Aspergillus niger* strain AN17 (Pusa Mrida). *Nat. Seminar on Hi-tech Horticulture*, 26–28th June, Bengaluru, 149 pp.

Misra, A.K., Pandey, B.K. and Prakash Om. 2003b. Wilt of guava, a disease of national importance and possible solutions. Souvenir and Abstract, Zonal Conference (East Zone), Indian Society of Mycology and Plant Pathology and *Seminar on Plant Diseases of National Importance with Special Reference to Guava Wilt and Mango Malformation*. 4–5th April, CISH and IISR, Lucknow, 23 pp.

Misra, A.K., Rajan, S., Babita Prasad, Shukla, S.K. and Prasad, D. 2003c. Resistant source of wilt disease of guava. Souvenir and Abstract, Zonal Conference (East Zone), Indian Society of Mycology and Plant Pathology and *Seminar on Plant Diseases of National Importance with Special Reference to Guava Wilt and Mango Malformation*. 4–5th April, CISH and IISR, Lucknow, 51–52 pp.

Misra, A.K., Om Prakash, Pandey, B.K., Babita Prasad, Shukla, S.K. and Prasad, D. 2003d. Biological control of guava wilt. Souvenir and Abstract, *National Symposium on Organic Farming in Horticulture for Sustainable Production*. 29–30th August, CISH, Lucknow, 65 pp.

Misra, A.K., Om Prakash, Pandey, B.K., Babita Prasad, Shukla, S.K. and Prasad, D. 2004. Biological control of guava wilt. **In**: *Organic Farming in Horticulture*, (Eds.) R.K. Pathak, Ram Kishun, R.M. Khan and R.A. Ram, CISH, Lucknow, 302–305 pp.

Misra, A.K., Prasad, D., Babita Prasad and Shukla, S.K. 2004b. Effective management of wilt disease of guava. Souvenir and Abstract. *Nat. Symp. on Crop Surveillance: Disease Forecasting and Management*, Indian Phytopathological Society, February 19–21, Div. of Plant Pathology, IARI, New Delhi, 92–93 pp.

Negi, S.S., Misra, A.K. and Rajan, S. 2001. Guava wilt. Proc. National Seminar on New Horizon in Production and Post-harvest Management of Tropical and Sub-tropical Fruits, IARI, New Delhi, held on December 8–9, 1998, Special issue, *Indian J. Hort.*, **54** (1 and 2), March and June: 145–151.

Normand, F. 1994. Strawberry guava, relevance for Reunion. *Fruits–Paris*, **49**: 217–227.

Pandey, R.R. and Dwivedi, R.S. 1985. *Fusarium oxysporum* f. sp. *psidii* as a pathogen causing wilt of guava in Varanasi district in India. *Phytopath. Z.*, **114**: 243–248.

Prasad, D., Shukla, S.K., Babita Prasad and Misra, A.K. 2003. Effect of intercrop and different doses of NPK on the incidence of guava wilt. *Souvenir and Abstract, Zonal Conference (East Zone), Indian Society of Mycology and Plant Pathology and Seminar on Plant Diseases of National Importance with Special Reference to Guava Wilt and Mango Malformation*. 4–5th April, CISH and IISR, Lucknow, pp. 53.

Prasad, N., Mehta, P.R. and Lal, S.B. 1952. *Fusarium* wilt of guava (*Psidium guajava* L.) in Uttar Pradesh, India. *Nature*, 169: 753.

Prakash, Om, Misra, A.K. and Shukla, S.K. 2002. *Penicillium citrinum*: A potent pathogen against wilt disease of guava. *Asian Congress of Mycology and Plant Pathology*, October 1–4, 2002, Univ. Mysore, Manasgangotri, Mysore, pp. 180.

Rodriguez, F.M.E. and Landa, J.B. 1977. Chemical soil disinfection against parasitic nematode in guava nurseries. *Centrol Agricola de la Facultod de Cieneias Agricolos*, **4**: 57–77.

Rodrigues, N.J., Robbs, C.F. and Yamashino, T. 1987. A bacterial disease of guava (*Psidium guajava*) caused by *Erwinia psidii* sp. nov. *Fitopathologia Brasileira*, **12**: 345–350.

Shukla, S.K., Babita Prasad, Prasad, D. and Misra, A.K. 2003. A cheap substrate for the mass multiplication of bio–agents of guava wilt. *Souvenir and Abstract, Zonal Conference (East Zone), Indian Society of Mycology and Plant Pathology and Seminar on Plant Diseases of National Importance with Special Reference to Guava Wilt and Mango Malformation.* 4–5th April, CISH and IISR, Lucknow, 54–55 pp.

Singh, B. and Lal, S.B. 1953. Wilt of guava. *Agril. and An. Hus.*, III: 78–79.

Sohi, H.S. 1983a. Diseases of tropical and sub-tropical fruits and their control. **In**: *Recent Advances in Plant Pathology*, (Eds.) A. Husain, K. Singh, B.P. Singh and V.P. Agnihotri. pp. 73–86. Print House, Lucknow, 521 p.

Sohi, H.S. 1983b. Studies on wilt disease of guava. *Annual Report*, Indian Inst. Hort. Res., Hessaraghatta Lake Post, Bengaluru, 102 pp.

Srivastava, A.K. Ahmad, R., Kumar, S. and Sukhada Mohands. 2001. Role of VA-Mycorrhiza in the management of wilt disease of guava in the alfivols of Chotanagpur. *Indian Phytopathol.*, **54**: 78 –81.

Suhag, L.S. 1976. Observations in guava decline in Haryana and its control. *Pesticides*, **10**: 42–44.

Tokeshi, H., Valdebenito, R.M. and Dias, A.S. 1980. Occurrence of a bacterial disease of guava in Sao Paulo State. *Summa Phytopathologica*, **6**: 85–87.

Vestal, E.F. 1941. *Textbook of Plant Pathology*. Kitabistan Allahabad, 406 p.

Vos, J.E., Schoeman, M.H., Berjak, P., Watt, M.P. and Toerien, A.J. 2000. *Acta Horticultural* (513): 69–79.

Webber, G.F. 1928. *Plant Pathology*. Annual Rep., Florida Agric. Univ. Exp. Stat. June 30, 1928.

49

Major Physiological Disorders in Horticulture Crops Focus Spongy Tissue in Mango

V. Ravindra

Division of Plant Physiology and Biochemistry
ICAR-Indian Institute of Horticultural Research, Hessaraghatta Lake Post,
Bengaluru – 560 089, Karnataka
E-mail: vattem@iihr.res.in; vattem_r58@yahoo.com

India's diverse agro-climatic regions facilitate production of a wide variety of fruits and vegetables. According to FAO statistics, India produces nearly 160 million tonnes of fruits and vegetables, and ranks second in the global map next only to Brazil and China. The horticulture sector has, in past few years, steered the growth in agriculture that has led to improvement in productivity per unit area. Horticulture sector has also acted as a source of generating employment, improving the economic well being of the farming community and the entrepreneurs involved in agriculture and horticulture by enhancing exports and earning considerable foreign exchange.

The productivity as well as the quality of horticultural crops is affected to a greater extent due to various physiological disorders. Physiological disorders refer to the breakdown of tissue that is not caused by either invasion by pathogens *i.e.* disease causing organisms or by mechanical damage or may be referred to as non-parasitic or inanimate diseases of fruit and vegetable crops. They may develop in response to an adverse environment, especially temperature, or to a nutritional deficiency during growth and development. Adequate supply of nutrients, pollution-free soil environment and optimum temperature and moisture favour normal growth of the plants. Any deviation from these conditions result in expression of disorders of various magnitudes. The deficiency or excess of any of the nutrient element, heavy metals, soluble toxic salts in the irrigation water, toxic gaseous pollution in the air, unsuitable prevalent temperature and moisture and soil pH have direct effect on plant growth. Temperature affects horticultural crops in many ways, including influencing timing and reliability of plant growth, flowering, fruit set and growth, ripening and product quality.

The extremes of environmental variables like temperature, moisture, light, aeration and nutritional imbalances result in disturbances in the plant metabolic activities leading to these disorders. While the symptoms may appear disease-like, they can usually be prevented by altering environmental conditions. In fruit crops, the deficiency of micronutrients causes many more disorders than that of macronutrients. These disorders have become widespread with diminishing use of organic manures, adoption of high density planting, use of rootstocks for dwarfing, disease and salt tolerance, unbalanced NPK fertilizer application, extension of horticulture to marginal lands and climate change. To get high quality fruit and yields, micronutrient deficiencies have to be detected before visual symptoms are expressed. The critical analysis of the nature, origin and causative factors of these non-pathogenic disorders will help in formulation of management strategies, reducing the losses to a significant level. This article presents a critical review on cause and characteristics of physiological disorders in important horticultural crops including vegetables and fruit species with emphasis on mango besides providing an insight into the gaps and researchable issues. The insights from the successful management of spongy tissue disorder in 'Alphonso' mango and its relevance in the management of physiological disorders in other horticultural crops particularly fruits will be the focus of this review.

Physiological Disorders of Vegetable Crops

Vegetables form an important part of our diet. They contain a number of nutritionally important compounds such as vitamins, minerals, carbohydrates and proteins. They often attract the consumer for their aesthetic qualities like flavour, colour and texture and contain a high percentage of water. Consequently, they exhibit relatively high metabolic activity when compared to other plant derived foods as seeds. Vegetable production increased more rapidly than other crop production. This was due to improved facilities for production, processing and distribution of educational and promotional programmes dealing with the importance of vegetables in diet and to the rising purchase power, changing food habits and life styles.

The quality of vegetables that we consume is influenced by deficiency or excess of mineral elements, hormonal imbalance, improper pollination or fertilization, injuries due to adverse climatic and growing conditions and some physiological factors. Sometimes more than one factor may be responsible for physiological disorders. Almost all major vegetable crops are prone to various types of disorders affecting different plant organs thus rendering them unfit for human consumption. Therefore, control of physiological disorders is essential for profitable production of the crop. Physiological disorders may develop during growth, storage or in transit thus rendering the produce unmarketable. Examples of physiological problems in vegetable crops are blossom-end-rot of fruit (a leather like decay of the blossom-end of several vegetable fruits associated with calcium deficiency). Secondary growth of potatoes and bolting of many biennial and perennial crops (a pre-mature flowering that occurs when the crops are exposed to inappropriate day lengths or temperatures). The production of premature flower stalk (bolting) in spinach, and ensuing seed production can render the produce unmarketable. This can occur when spinach is grown under long days and warm conditions. Other disorders that effect the economical production of the crops are sun scald, puffiness, cat face, growth cracks, blotchy ripening (tomato), chlorosis and cracked stem (celery), hollow heart, black heart and greening of potato, russet spotting (lettuce and sweet potato), buttoning, whiptail and browning (cauliflower), splitting and doubling of bulbs in onion, albinism (strawberry), red heart (lettuce), pithiness in radish, forking or root splitting in radish and carrot and cavity spot in carrot, to name a few. High and low temperature during developmental phase and after harvest also render the produce unfit for market. Therefore, increased production of vegetables will have significance provided they reach the consumer in good condition, that is without any blemish or misshapen. This can be possible only when they are grown under adequate and balanced nutrition essential for sustaining crop productivity. For this reason, the soil critical level of a nutrient defined for one vegetable may not apply to another crop species. An attempt has been made to provide scientific knowledge on the role of nutrients, adverse climatic and other related factors in the development of physiological disorders of vegetable crops, making them unfit for human consumption. In this endeavour many sources such as books, bulletins, scientific papers, journals, web sites *etc.*, related with the topic have been consulted, to give most relevant information as possible regarding symptoms, causes, and role of elements and temperature associated in the development of these disorders.

Blossom-End-Rot (BER)

Blossom end rot is a disorder that affects tomato, pepper, eggplant, squashes and melons. Symptoms of BER start out as a small spot, which grows larger and darker with fruit growth. BER can most easily be identified by a discoloured, sunken spot at the blossom end of the fruit (Vanderlinden, 2009). BER has water soaked spots, which enlarge and can cover up to half the fruit or even whole fruit. The tissue turns light to dark brown. The spot then dries out and becomes leathery. It may be infected by secondary pathogens. BER occurs when fruit is one third to one half fully grown. Fruit affected by BER ripens prematurely resulting in inedible product. The disorder usually affects the first few fruits of tomatoes growing in cold soils. This may be a result of low Ca levels in the soil, quite often it is the result of erratic watering (Vanderlinden, 2009). BER is caused by Ca deficiency in the fruit. When a plant does receive inadequate water or when there is excess water in the soil, a plant's capabilities of absorbing Ca are diminished. The disorder is usually most severe following extremes in soil moisture (either too dry or too wet). These conditions result in a deficiency of Ca available to the maturing fruit, at the spot, where damage becomes apparent (McLaurin, 1998). Cause of

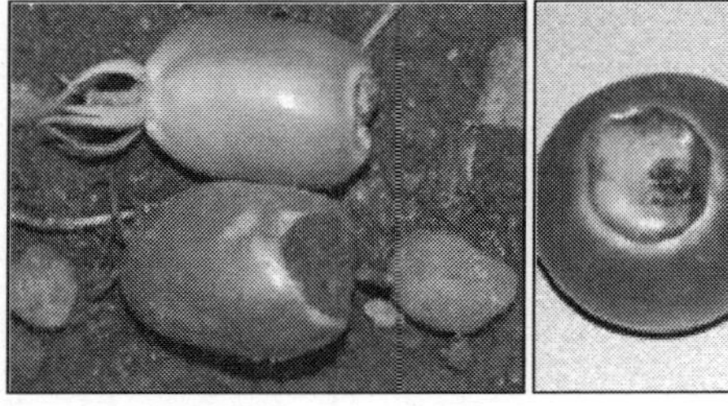

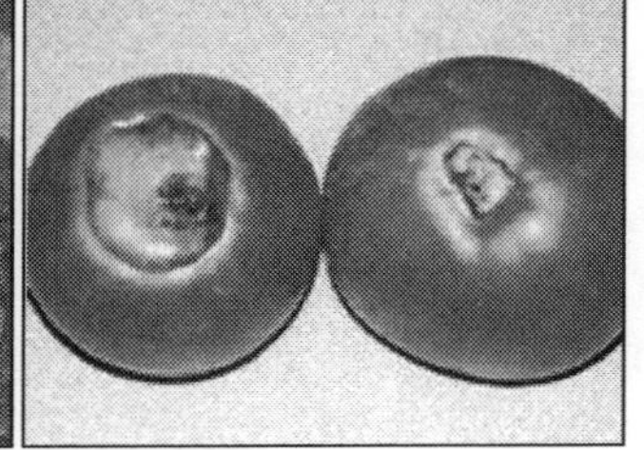

Figure 1: Blossom-end-Rot in Tomato (M. Grabowski).

injury is due to localized Ca deficiency, which occurs during a critical phase of fruit growth. Waterlogging, excess ammonium, K, Mg ions interfere with Ca uptake. Development of BER was observed to be positively correlated to the leaf K:Ca ratio but weakly to the K:Ca ratio in mature fruits. The incidence of external and internal BER has been reported to strongly depend on ion activity ratios between K and Ca+Mg (ak/√[aCa+aMg]) and between Mg and Ca [aMg/aCa] in the root zone at both moderate (3 to 6 mS cm^{-1}) and high (6 to 12 mS cm^{-1}) salinity levels (Willumsen *et al.*, 1996). There is a positive correlation between higher risk of BER and higher activity ratios due to lower Ca uptake and its reduced availability in the fruit tissue, which may be caused by increased concentration of organic acids in fruit juice. Environmental conditions that promote rapid transpiration worsen the problem like drastic fluctuations in soil moisture.

Catface

Catface is a disorder characterized by gross deformity of tomato fruit, which usually renders them unmarketable. The defect is usually located on the blossom end of the fruit. The deformity is caused by something (internal and external) that occurs during the formation of the flower that results in the fruit not developing normally (Olson, 2004; Peet, 2009). There is dearth of published information on the exact cause of this disorder. Catface is commonly observed in first harvest fruit. The symptoms include enlarged scars and holes in the blossom end.

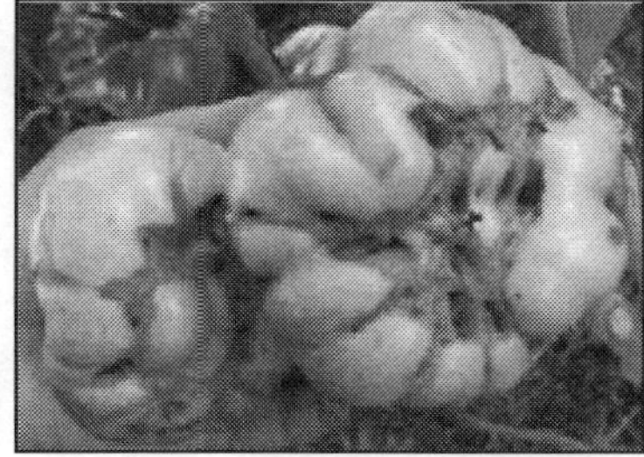

Figure 2: Catface in Tomato (M. Grabowski).

Internal White Tissue

Fruit affected by this disorder usually show no apparent symptoms. When ripe fruits are cut, white hard areas especially in the vascular region are present in the outer walls (Olson, 2004). This disorder is more of a problem on fruits, which ripen on the vine. Under severe conditions fruit may also show white tissue in cross-wall and center of fruit. The problem may be reduced by adequate K fertilization (Imas, 1999; Olson, 2004; Peet, 2009) or use of less susceptible cultivars especially the highly colored ones.

Watery Fruit

Watery tomato is a physiological disorder resulting from an imbalance between plant water absorption and ambient climatic conditions. Excessive root pressure results in the massive influx of water into the fruit and increases the volume of cells and sometimes damages the cells. This disorder develops after over-irrigation under a situation, where there is development of a strong root system, which triggers an elevated root pressure. Shelf-life of such fruit is reduced and organoleptic quality is negatively affected. Maintaining plant leaf area index at a reasonable level during summertime helps reduce root pressure and minimize the incidence of this physiological disorder (Dorais *et al.*, 2001).

Sunscald

This disorder can cause serious damage to fruit especially during the very hot South African (Sakata, 2009) and Swaziland summers. Symptoms of this disorder include bleached patches on green tomato fruit accompanied by rapid desiccation, which leads to sunken areas. Tomato fruit exposed to direct sun were found to be affected by this disorder. The situation is aggravated when there is insufficient leaf cover, trellising and susceptible varieties (due to certain genetic predisposition). This disorder may be controlled by avoiding any factor that causes leaf loss, such as diseases, insects and abiotic factors like hail damage may also be responsible for the problem. Control measures should be aimed at avoiding leaf loss by whatever environmentally friendly means possible.

Figure 3: Sunscald in Tomato (M. Grabowski).

Fruit Cracking

Several types of fruit cracking injury have been reported: radial cracking (star shaped and originating from the peduncle), fruit bursting, concentric cracking (circular cracks originating from the peduncle) and cuticle cracking also known as russeting (Peet, 1992; Peet and Willits 1995; Dorais *et al.*, 2001; Olson, 2004; Kennelly, 2009). Fruit cracking occurs due to rapid growth when there is abundant water and high temperatures, especially when these conditions prevail following periods of stress. Fruit cracking is generally associated with rapid movement of water and sugars towards the fruit when cuticle elasticity and resistance are weak. Bakker (1988) reported that this disorder generally occurred six to seven weeks after fruit set. An imbalance between water supply (influx) and water loss (efflux) has been implicated in the fruit cracking phenomena. Fruit cracking disorder can be controlled by using the cultivars that are less prone to cracking, provision of even water and balanced nutrition. The cause of this disorder is under genetic control.

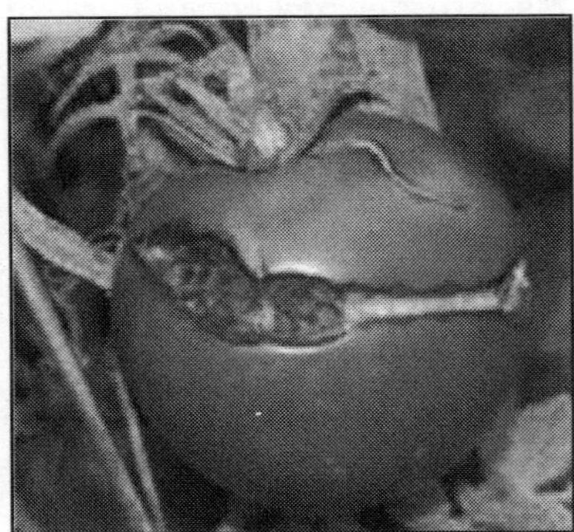

Figure 4: Fruit Crack in Tomato (M. Grabowski).

Physiological Disorders of Fruit Crops

The physiological disorders in fruits include but not limited to bitter pit, lenticels breakdown, Jonathan spot, internal breakdown, water core and stem-end splitting in apple, 'Black End' and 'Cork' in pear, leaf tip burn in strawberry, fruit cracking in citrus, peaches, plums, litchi, pomegranate *etc.*, granulation in citrus, jellyness in papaya, corky tissue in sapota, aril browning in pomegranate, spongy tissue in Alphonso mango, jelly seed in certain mango varieties like Dasehari, Amrapali, Mallika *etc.*, and stem-end-cavity in Tommy Atkins variety of mango. Fruit splitting or cracking has been reported for almost every horticultural crop of economic importance. This disorder occurs most notably in apple (Visai *et al.*, 1989), apricot (Benson 1994), cherry (Belmans and Keulemans, 1996), grape (Considine and Kriedmann, 1972), nectarine (Gibert *et al.*, 2007) and prune (Milad and Shackel, 1992). Splitting of a fruit is a consequence of cracks developing in the epidermis or in the rind of a developing fruit. In most crops that are prone to cracking or splitting the spatial distribution of cracks varies. Cracks could develop at the stylar- and/or stem-end, cheeks and shoulders of the fruit. Although tendency to crack is genetically controlled with certain cultivars being crack-resistant (Belmans and Keulemans, 1996; Gül°en *et al.*, 1995; Sperry *et al.*, 1996), the intensity of the disorder in crack-prone cultivars is episodic, with fluctuations in severity between seasons. In terms of mineral nutrient deficiency, cracks develop due to boron (B) deficiency in apricots (Benson, 1994) and calcium (Ca) deficiency in cherry (Meheriuk *et al.*, 1991) and tomato (Huang and Snapp, 2004).

Fruit physiological disorders have often been attributed to fruit calcium levels irrespective of temperate or tropical in origin. Fruits which are soft on arrival, or which ripen too rapidly create marketing problems, while those with internal physiological disorders are probably even more damaging, as the buyer becomes aware of the problem only after purchase. Bangerth (1979) indicated that transpiration (and by implication tree water relations) may play a vital role in Ca uptake and distribution. Shear (1975) indicated that excessive soil water may also impede uptake. There are indications that irrigation regimes can affect fruit Ca concentrations at a critical period in fruit development. The importance of Ca in the nutrition of plants has long been recognized. Inadequate concentration of Ca has been associated with the development of physiological disorders in many types of fruit crops.

Physiological Basis of Effects of Ca on Quality

Physiological disorders that occur in a wide range of fruits and vegetables are associated with cell death in tissues where localized deficiency of Ca exists. Loss of membrane integrity with resultant loss of cell compartmentalization is commonly considered as the primary cause of cell death (Christiansen and Foy, 1979; Poovaiah, 1979, White and Broadley, 2003). Loss of quality due to structural failure of tissues may be associated primarily with the lack of Ca in cell walls. During ripening there are major changes in the pectin-rich middle lamella region of cells where Ca ions have a role in linking adjacent acidic pectin polymers (Seymour *et al.*, 1993). In addition to its structural and functional role in plant cell walls and membranes, Ca is important as an intracellular 'second messenger' (Zocchi and Mignani, 1995). More recent research has concentrated on the role of Ca at the cellular level and in particular its role in biochemistry and cell signaling (White and Broadley, 2003). However, it is more difficult to directly associate the metabolic aspects of Ca with specific physiological disorders or changes in quality.

In the fruit, the absence or deficiency of Ca inflow is related to physiological disorders (Saure, 2005). In specific organs, increases in Ca content delay "softening" in papaya (Qiu *et al.*, 1995) and maintain quality in apples (Coway and Sams, 1987), tomatoes (Rigney and Wills, 1981), avocados (Eaks, 1985) and mangoes (Schaffer and Andersen, 1994). The relationship between Ca absorption and climatic factors was reported by Clarkson (1984) and discussed by Saure (2005). In mangoes, several reports have suggested that edaphic and environmental factors are related to the development of physiological disorders (Schaffer and Andersen, 1994). In addition to edaphic factors, environmental variables such as relative humidity and air temperature have been involved in the development of physiological disorders in mangoes and may be indirectly one of the factors of Ca deficiency in the fruit.

According to Saure (2005) the mechanisms of Ca uptake are related to a phyto-hormonal control by gibberellin action. Active gibberellins inhibit Ca translocation in the fruit (Cohen and Greene, 1989) and high auxin/gibberellin ratios promote integrated differentiation of the xylem in relation to the phloem that may contribute to optimizing Ca arrival in the fruit, because this mineral is moved to the canopy exclusively in the xylem (Aloni, 2001). Reduction in gibberellin levels, either by means of inhibitors such as paclobutrazol or by restriction in N supply, may be an important action in the control of physiological disorders in fruit (Cohen and Greene, 1989; Aloni, 2001). A preliminary study on papaya pulp jellification found that the mass of Ca extracted by the seeds with pulp jellification was higher than from the healthy fruit (Ali and Lazan, 1998).

Mineral Nutrition, Fruit Position and Crop Load

Calcium is the nutrient most commonly associated with post-harvest disorders. In bitter pit in apple fruit, almost all pre-harvest factors, which influence incidence of the disorder can be directly or indirectly related to Ca nutrition of the fruit with the exception of maturity. The reason for higher incidence of bitter pit in less mature fruit is not clear, but is perhaps related to subsequent rates of ripening (Ferguson and Watkins, 1989). A number of pre-harvest factors are associated with movement of Ca into the developing fruit. Apart from pollination, these are associated largely with relationships between fruit development, position on the tree, and the dynamics of nutrient flow in relation to leaves on the fruiting wood.

Pre-harvest factors which may predispose fruit for subsequent disorder development are dominated by position of the fruit on the tree, characteristics of the fruiting site, crop load, mineral and carbohydrate nutrition of the developing fruit, water relations, and response to temperatures. Expression of disorders which are the direct result of storage conditions, such as low temperature and high CO_2, will still be modied by similar factors. An understanding of these factors allows modication of fruit development to optimize storage quality, and development of methods for predicting disorder risk.

Ca has received considerable attention in recent years not only because of its relationship to physiological disorders but also because of other desirable effects, particularly in fruits where it can reduce respiration (Bangerth *et al.*, 1972; Bramlage *et al.*, 1974), delay ripening, extend storage life (Sharples and Johnson, 1977), increased firmness (Bangerth *et al.*, 1972; Cooper and Bangerth, 1976), vitamin C content (Bangerth *et al.*, 1972), and reduced storage rot (Sharples and Johnson, 1977).

Physiological Disorders of Mango

Mango (*Mangifera indica* L.) is the choicest fruit of India in terms of production, economy and acceptability by the consumers. In India, it covers around 36 per cent of total fruits growing area. Owing to delicious taste, succulence type and exotic flavour it enjoys the same popularity in the tropics as apple in the temperate region. Mango orchards in India are facing declining growth in recent times leading to decrease in yield and productivity. Although a number of factors are responsible for mango yield decline, yet physiological disorders play more devastating and vital role. As a result, mango growers are losing their faith and interest in mango orchards and trying to shift their cultivation practices from mango to other crops. Among the different factors responsible for low productivity and low export potentiality, physiological disorders associated with mango is one of the main factors which affect mango cultivation at all stages, right from the plants in the nursery to the fruits in storage or transit. These physiological disorders not only reduce the production efficiency of the tree but also hamper fruit quality which is the key factor for export, resulting in huge economic losses to the growers every year. Therefore, to obtain higher mango production with better quality, utmost attention must be given to address the physiological disorders. Different physiological disorders along with their management strategies are detailed below.

Mango Malformation

During last few decades, farmers, particularly in northern India are facing huge problem due to this disorder which is now showing its face in southern India. Although, it was first observed in 1891 in Bihar it is now an alarming threat in Punjab, Delhi and Uttar Pradesh and to some extent in Gujarat, Maharashtra, Bihar, West Bengal and Odisha. It is mainly of two types: vegetative and floral. Vegetative malformation is more common in nursery seedling and in young plants. Malformed vegetative shoots assume the appearance of bunchy top with narrow scaly leaves. Vegetative buds growing in cluster, swells and losses the apical dominance. But the most destructive one is the floral malformation which appears at the bearing stage and produce short, thicker and hypertrophied panicles as condensed clusters. Ratio of sterile male to hermaphrodite flower increases in the malformed panicles, resulting in failure of fruit set in those panicles. Almost all the commercial mango cultivars like 'Bombay Green', 'Dashehari', 'Lucknow Safeda' and 'Chausa' show huge susceptibility to this malady, however the cultivar 'Baramasi' is less affected. Researchers have reported that *Fusarium moniliformae* var. subglutinans, a fungus, is strongly associated with this disorder. Temperature range of 21-27°C (max.) and 8°C (min.) along with relative humidity of 85 per cent during late winter or early spring is conducive for the growth of the fungi, resulting in severe incidence of malformation on newly emerged panicles.

Management Strategies

- Application of NAA/Planofix® (200 ppm) during October, prior to fruit bud differentiation followed by debloosming of newly emerged panicles or bud during January-February is beneficial to control the disorder.
- Spraying of different phenolic compounds like catechol, cinnamic acid and tannic acid during 1st week of October at 2000 ppm is equally effective in reducing floral malformation.
- Several anti-malformins have been suggested for beneficial results when sprayed on panicles just after emergence (4-6 cm). Sprays of glutathione at 2,250 ppm, ascorbic acid at 2,110 ppm, $AgNO_3$ at 600 ppm have also been found effective to control the malformation.
- Application of nutrients likes P and K and micronutrients especially Zn and B just after harvesting are quite effective in reducing the incidence of malformation. Application of moderate amount of N helped to minimize the vegetative malformation while Zn and B helped in biosynthesis of auxins which ultimately helps to reduce the incidence of mango malformation.
- Removal of just emerged malformed panicles or vegetative shoots at a distance of 15-20 cm below the point of occurrence is beneficial to minimize the spread of the disorder further.

Black Tip

This malady is widely prevalent in Punjab, Uttar Pradesh, Bihar and West Bengal. It causes considerable economic losses to the growers. Among the commercial cultivars, Dashehari is the highly susceptible one, while Lucknow Safeda is the least. This disorder has been detected mainly in orchards located in the vicinity of brick kilns. Gases like carbon dioxide, sulphur dioxide and ethylene constituting the fumes of brick kiln, damage growing tips of fruits and give rise to the symptoms of black tip. Etiolation of distal end of the fruit leaving mesocarp and seed unaffected followed by appearance of grey spots of indefinite outline, turns brown in colour, coalesces and the entire fruit tip turns brown black. In severe cases, the affected fruits are left over with a thin layer of collapsed tissues.

Management Strategies

- Establishment of the orchards away from brick kiln by at least 1.6 km in east to west and 0.8 km in north to south direction and increasing the chimney height to at least 15-18 m can be preventive measures for this malady.
- Spray of 1 per cent borax (10 g/lit) thrice - first before flowering, second during flowering and third at fruit set stage is very effective. Moreover, spraying of other alkaline solutions like caustic soda (0.8 per cent) or washing soda (0.5 per cent) at pea stage followed by two more sprays at 15-day intervals may neutralize the acidic fumes of brick kilns, precipitating on the tip of the fruits.

Fruit Drop

Despite initial high fruit set, the ultimate retention is quite low in mango (only 0.1 per cent perfect flowers develop fruit to maturity). The intensity of fruit drop varies with variety. Among the commercially grown varieties, 'Langra' is more susceptible to drop, while 'Dashehari' is the least. Fruit drop in mango is classified into three groups: (i) Pinhead drop, (ii) Post-setting drop, and (iii) pre-harvest drop (May drop). The first two drops result in less economic losses to the growers but the third group is most destructive and cause huge economic losses to the growers as the fruits totally drop down at pre-harvesting stage. A good deal of fruit drop is attributed to disintegration of embryos, ovular abnormalities and competition among fruits. Other factors like climatic conditions, disturbed water relations, nutrition deficiency, and pests and diseases also play an important role in fruit drop.

Management Strategies

- The extent of fruit drop in mango can be significantly reduced by regular and frequent irrigations during the entire fruit developmental period.
- Orchards should be protected from desiccating winds by planting wind breaks.
- Timely and effective control measures against major pests and diseases
- Growth regulators like NAA and 2,4-D in varying concentrations depending upon the variety and time of application can be sprayed. The optimum concentration lies between 10-15 ppm during the month of April-May.

Clustering (*Jhumka*)

Development of fruits in clusters at the tip of the panicles is called as clustering or *Jhumka*. Such fruits do not grow beyond pea or marble stages and drop after a month of fruit set. Mainly it is due to lack of pollination/ fertilization, which may be attributed to many reasons such as the absence of sufficient population of pollinators in the orchards, indiscriminate spray of pesticides during flowering, spraying of synthetic pyrethroids, monoculture and bad weather during flowering.

Management Strategies

- Spraying of pesticides during flowering should be avoided. Apart from this, monoculture in the orchard should be avoided by planting at least 5-6 per cent of other cultivars in new plantations. In old orchards, a few branches should be top worked with pollinizing varieties.

☆ Introduction of beehives in the orchards during flowering season for increasing the number of pollinators.

Internal Breakdown Disorders in Mango

Past Perspective

Wainright and Burbage (1989) reviewed many internal breakdown disorders like spongy tissue (ST) in 'Alphonso', jelly seed in 'Dasehari' and stem-end-cavity in 'Tommy Atkins' mango. 'Alphonso' a premium cultivar of mango, constituting more than 60 per cent of the mango being exported from India is in great demand globally for its unique flavour, texture, aroma and taste. Unfortunately, the export trade of 'Alphonso' suffers a major setback from ST, a physiological internal breakdown disorder. ST is characterized by unripe, acidic, pale yellow/white, corky tissue with or without air pockets associated with an unacceptable off flavour in mesocarp tissue adjacent to endocarp (Cheema and Dhani, 1934). This is reported to be caused by inactivation of ripening enzymes during fruit maturation stage due to high temperature, convective heat and post-harvest exposure of fruit to sunlight. In extreme cases, the whole fleshy portion becomes too soft resembling bacterial rot. Fruits affected by this disorder do not show any external symptoms and the malady is detected only after cutting the fruits open, posing a challenge for quality control. ST is higher in the coastal Konkan region of Maharashtra than in other inland regions of India (Subramanyam *et al.*, 1971). Extensive investigations have been carried out over the past several decades to understand the cause of this disorder, but with little success (Katrodia, 1979; Rane *et al.*, 1976). Factors as diverse as ecological, nutritional, environmental, microbial, physiological and biochemical were attributed to ST. Katrodia (1988) while ruling out Ca deficiency attributed ST to accumulation of soil convective heat at the endocarp. However, the nature of this disorder could not be proved conclusively as both healthy and ST-affected fruits were traced to the same panicle (Desai, 1966; Ravindra and Shivashankar, 2004) and the problem remained an enigma as the studies were confined to measuring effects rather than the cause. About 30 to 70 per cent fruits were reported to be affected annually by ST incidence resulting in losses to farmers, traders, exporters and consumers in different dimensions.

Fruit Maturity

Spongy tissue was found to vary directly in proportion with maturity status of the fruit. Early picked fruits might escape ST, while late harvested and tree ripe fruits invariably showed higher incidence (Katrodia, 1979; Ravindra and Shivashankar, 2004). Therefore, harvesting fruits before they attain 80 per cent maturity compromising on quality has become a practice in the Konkan. Harvested fruits are sorted into three maturity grades based on their density in plain and salt water. Fruits floated on plain water were classified as 12-annas (75 per cent mature) while fruits floated on 2.5 per cent salt water as 14-annas (80 per cent mature) and the remaining as 16-annas (85 per cent mature). Fruits in the 16-anna category had > 90 per cent, 14-anna category ≤ 50 per cent and the 12-anna category ≤ 30 per cent ST-affected fruits. Farmers and contract harvesters use a thumb rule of selling lighter fruits (those which float in water) at premium markets as they are presumed to be free from ST and the rest in low value local markets.

Nutritional Factors

ST was attributed to deficiency of Ca and B (Selvaraj *et al.*, 2000; Subramanyam *et al.*, 1971; Shanta, 1981) but was later ruled out as trees well supplied with these nutrients also developed ST (Katrodia, 1988). ST is localized and is observed invariably in mesocarp tissue on surface of the stone only. Logically, any nutrient deficiency arising out of lack of its supply affects growth of the whole fruit while its remobilization results in localized deficiency, as in ST. Moisture and Ca contents were significantly higher in STS (seed from ST affected fruit) than HS (seed from healthy fruit) indicating their mobilization from the pulp. Nutrient requirements of seed are normally met directly from parent tree through the funiculus. When direct supplies from the tree are severed during ontogeny, the seed sustains on mesocarp by drawing required resources randomly from tissue close to the endocarp (stone) leaving it dry and devoid of such nutrients (Ravindra and Shivashankar, 2004).

Convective Heat Accumulation Near Endocarp

Katrodia (1988) attributed ST to accumulation of convective heat at endocarp. If convective heat were to be considered, the effect would be more pronounced at epicarp than endocarp, as high water content of the mesocarp would have dissipated it before accumulating at endocarp. Further, ST cannot be attributed to external factors like temperature, convective heat, *etc.* as these do not vary, logically, between healthy and ST affected fruits on the same panicle. However, accumulation of heat of respiration of seed at endocarp makes sense (Ravindra and Shivishankar, 2004).

Spongy Tissue in Alphonso Mango

Presence of ST invariably in mesocarp tissue on the surface of stone, occurrence of vivipary and faster and higher rate of germination of stones from ST-affected fruits provided clues that ST in 'Alphonso' mango might be associated with physiological status of the seed (Ravindra, 1998, unpublished). Flesh breakdown in mango cultivar 'Tommy Atkins', similar to ST in 'Alphonso', was attributed to disconnection of funiculus between peduncle and endocarp even while fruit is still attached to the tree, rendering the seed physiologically independent of tree for its supplies (Wainwright and Burbage, 1989), prompting further, to pursue the pivotal role of seed in ST, thus making a paradigm shift from the earlier thinking.

Inter-fruit Competition and Vascular Disconnection

Panicle-wise analysis revealed that ST was >30 per cent in panicles with more than one fruit as compared to only 5.7 per cent in 1-fruit panicles indicating influence of inter-fruit competition (Ravindra and Shivashankar, 2004). Individual sinks on a tree change their competitive ability leading to diversion of assimilates towards the stronger sinks (Wright, 1989) and developing seeds have the highest priority (Cannell, 1985). Inter-fruit competition influences fruit size and early fruit thinning results in increased fruit size in many crops. Partial thinning of fruits on a tree was not as effective as selective whole tree thinning and the best results were obtained by thinning within fruit clusters suggesting that competition effects are rather localized (Knight, 1980). Genetically weak seeds cannot sustain competition and gets separated from the main tree anatomically and physiologically (Wainwright and Burbage, 1989). Apparently, the vascular disconnection takes place at some point during ontogeny (Greenwood and Bewley, 1982) and inter-fruit competition triggers dissociation of such weaker sinks (seed) to prevent reverse flow of resources resulting also in abortion of fruits.

Seed Recalcitrance and Shift to Germination Mode

In contrast to dormant species viviparous and recalcitrant species enter germination phase directly as they lack maturation drying and the consequent metabolic quiescence (Fransworth, 2000). Loss of vasculature between seed and the parent tree triggers a cascade of developmentally important events as the programme for embryonic growth is overruled by that of germination, prompting the embryo to shift into germination mode (Dure III, 1975). Radicle emergence in mango during cold storage and its possible association with internal breakdown was reported (Acosta *et al.*, 2000; Vergos Ramos and Durigan, 1995). Recalcitrant seeds, as in mango, initiate pre-germination events during ontogeny and consistently demand moisture to remain viable for perpetuation (Berjak *et al.*, 1990; Farrant *et al.*, 1988). Being a high-priority sink and capable of draining resources from fleshy parts of fruit the metabolically active and growing seed, can, therefore, be expected to induce ST.

Environmental Factors

Higher ST in the Konkan region appears due to prevalence of persistent warm weather (30-35°C) coupled with high RH (> 85 per cent), that favour faster growth and development of fruits resulting in shift of embryo in to germination mode and the consequent dependence on mesocarp for its sustenance. However, in inland regions where temperatures vary from 35 to 48°C with a lower RH (30 to 60 per cent), ST incidence remains lower as the recalcitrant seed loses viability during ontogeny and MSW infestation further compounds the problem. Therefore, ST in the inland regions stems mainly from interplay of seed viability and MSW infestation.

Significance of the Physiological Role of Seed

Physiological and biochemical studies (Ravindra and Shivashankar, 2004) have provided ample evidences on the causative role of seed and vindicated the direct relation between ST and shift of recalcitrant embryo in to germination mode. The various biochemical constituents of seed like contents of soluble carbohydrates, DNA, RNA, soluble protein, FAA, oil, FFA, and activities of amylase and lipase have confirmed the shift of embryo in to germination mode and corroborated well with seed germination under field. Duplication of DNA is initiated only after hydration of embryo leading to *de novo* synthesis of enzymes associated with onset and rapid progress of germination processes (Chen and Osborne, 1970). Radiotracer tritium counts from preharvest PGR treated fruits clearly established that an increased mobilization of water from mesocarp to seed precedes ST formation and is closely associated with metabolic status of the seed (Ravindra *et al.*, 2010).

Natural Evidence for the Role of Seed

All but a few mango stone weevil (MSW) infested fruits were found to be free from ST, in which the embryonic axis and/or the funiculus attachment at hilum were damaged by the random feeding activity of the weevil larvae confirming the direct role of seed in ST development. A few per cent (< 5) of fruits, though infested with MSW, exhibited spongy tissue, which was traced to the presence of intact embryonic axis and funiculus attachment at hilum that escaped infestation by MSW (Ravindra and Shivashankar, 2006).

Radiotracer Tritium Studies

It was observed that a significant loss of water from ST-affected region of mesocarp was accompanied by a concurrent and proportionate gain in seed and the extent of water lost determined the degree of sponginess (Ravindra and Shivashankar, 2004, 2006). However, the questions remained as to:

i) Whether the mobilization of water from mesocarp to seed was occurring randomly or from a pre-destined region; and

ii) How the exogenously applied PGRs (GA_3 and PBZ) were able to regulate amylase activity in seed without affecting the same in mesocarp in spite of diffusing through the same.

Experiments conducted on developing 'Alphonso' mango fruits using tritiated water ($1H_3$) and PGRs provided conclusive evidence for increased mobilization of water into seed from a pre-destined region of mesocarp during ST formation in 'Alphonso' mango (Ravindra *et al.*, 2010). Results on tritium counts from preharvest PGR treated fruits clearly established that an increased mobilization of water from mesocarp to seed precedes ST formation and is closely associated with physiological status *vis-à-vis* metabolic activity of the seed. The fact that applied PGRs did not directly affect amylase activity in mesocarp irrespective of whether the seed was physiologically active or damaged due to mango stone weevil (MSW) infestation, further substantiated the causative role of seed in ST formation of 'Alphonso' mango. The findings of radiotracer studies besides resolving the enigma of ST in 'Alphonso' mango have also opened up new vistas in understanding and management of similar internal breakdown disorders of tropical and/or temperate fruit species endowed with recalcitrant seeds. The key to manage such internal fruit disorders clearly lies in down-regulation of the metabolism of recalcitrant seed. The causative role of seed in ST formation of 'Alphonso' mango is further elucidated by the following observations (Ravindra *et al.*, 2010):

i) The tritium counts in the mesocarp tissue of fruits in which the seed was damaged by MSW infestation was similar to that of healthy fruits.

ii) PGRs could not directly affect amylase activity in mesocarp tissue in spite of diffusing through it but could regulate the same significantly in seed.

iii) Amylase activity in mesocarp was not affected by PGRs in fruits where the seed was damaged by MSW infestation.

iv) Manipulation of seed metabolic activity by PGRs significantly influenced ST incidence levels in mesocarp.

Seed Dormancy and Viability

The endogenous abscissic acid (ABA) and jasmonic acid (JA) levels of seed from healthy and PBZ treated fruits were three times higher over those from STS and GA_3 treated fruits indicating their dormancy status. Mango seed, being recalcitrant, lacks the desired level of dormancy to keep the seed metabolically quiescent. Higher levels of ABA and JA in seed ensure metabolic quiescence and therefore prevention of ST. Seed viability is as important as dormancy and absence of ST due to loss of embryo viability or its abortion during ontogeny further substantiated the significant role of seed (Ravindra, 2010, unpublished).

Creation of Anerobiosis in the Mesocarp

Climacteric respiration of mesocarp tissue results in production of sugars and CO_2. Rate of O_2 diffusion from mesocarp into seed enhances to meet the respiratory demands of the hydrating STS (Ravindra and Shivashankar, 2004; Shivashankar *et al.*, 2007). Consequently, diffusion of CO_2 from the actively respiring seed into mesocarp creates conditions akin to anaerobiosis in mesocarp favouring synthesis and accumulation of a variety of organic acids that lowers pH. As a result, cell wall degrading enzymes like cellulose, polygalacturonase *etc.* are activated but amylase activity is inhibited resulting in unhydrolized starch in the pulp. Accumulation of CO_2 also leads to development of off flavour, internal pitting (Chaplin, 1984) and inhibition of ripening (Thompson, 1971; Selvaraj *et al.*, 2000). Increased mobilization of moisture into seed disrupts homeostasis of the mesocarp tissue along endocarp resulting in loss of membrane integrity and death of cells seen as necrotic spots. Studies on preharvest treatment of fruits with GA_3 and PBZ indicated that amylase activity in pulp and seed are tightly but inversely inter-linked (Shivashankar *et al.*, 2007). The higher osmotic potential achieved through enhanced hydrolytic activities enable STS to act as a metabolic pump and meet the desired demand for water from adjoining mesocarp resulting in ST (Ravindra and Shivashankar, 2004; Ravindra *et al.*, 2010).

Cold Storage Studies

It was thought that fruits harvested before 80 per cent maturity would escape ST under cold storage during export (Cheema and Dani, 1934). However, fruits harvested prior to 80 per cent maturity and stored at 15°C for 4 weeks also showed a higher degree of ST except PBZ treated fruits. Thus, maintaining seed dormancy *vis-a-vis* metabolic quiescence is critical to prevent ST even under cold storage (Ravindra, 2010, unpublished).

Failure to induce ST in 'Totapuri', a ST free cultivar, with GA_3 provided clinching evidence on the role of seed in ST of 'Alphonso'. The non-response of 'Totapuri' to applied GA_3 was traced to the disconnected funiculus at hilum unlike 'Alphonso' mango in which it was intact throughout ontogeny (Ravindra and Shivashankar, 2006). Thus ST in 'Alphonso' mango is a phenomenon arising out of seed recalcitrance and an *in situ* germination signal leading to the following sequence of events holds the key:

i) Disconnection of vascular strands (funiculus) between peduncle and endocarp during fruit/seed maturation phase triggered by inter fruit competition leading to shift of embryo into germination mode due to its recalcitrant nature;

ii) Mobilization of water, O_2 and solutes from mesocarp near the endocarp into seed and the consequential progress of germination associated events;

iii) Creation of anaerobiosis and metabolic disturbances in mesocarp tissue near endocarp due to accumulation of CO_2 and heat of respiration leading to loss of membrane integrity, accumulation of organic acids, lowering of pH, activation of cell wall degrading enzymes, decrease in amylase activity and dissociation of Ca from pectin complex; and

iv) Spongy tissue development near the endocarp.

Thus, down-regulation of seed metabolism appears to be a viable strategy to prevent ST in 'Alphonso'. Use of PBZ provided valuable academic insights, but had limited practical application due to export restrictions and associated health hazards. A few, eco-friendly formulations have therefore been developed and tested for their efficacy in preventing ST at the experimental orchard of IIHR and farmer's orchards in the Konkan region of Maharashtra, with highly promising results (Ravindra, 2011, unpublished). This paradigm shift in unravelling the cause of the mysterious spongy tissue problem in 'Alphonso' mango paved the way to work out environmentally-safe preventive strategies in managing physiological fruit disorders not only in mango but also in other fruit crops endowed with recalcitrant seeds.

Non-destructive Detection Techniques of Spongy Tissue

Non-destructive detection through X-ray diffraction instruments though available these days, are only tools limited to sort the ST-incident fruits but not a solution to prevent its occurrence altogether. The non-destructive detection is not only cumbersome but also very expensive and uneconomical. Though fruits can be sorted out before exporting the consignment the probability of ST development in the sorted fruits is quite high during cold storage transport chains as the recalcitrant seed continues to depend on the surrounding mesocarp for its sustenance enabling it drain water and nutrients leading to higher incidence of ST unlike when the fruit is attached to the tree.

Management Strategies

All observations reported so far, on the phenomenon of ST/internal breakdown in 'Alphonso' mango could be explained by the systematic work done while elucidating the causative role of seed (Ravindra, 2015; Ravindra and Shivashankar, 2004, 2007; Ravindra *et al.*, 2010 and Shivashankar *et al.*, 2007). The cause of ST has thus been traced to shift of recalcitrant seed in to germination mode and the consequent drawl of water and other resources from mesocarp by the physiologically active recalcitrant embryo with funiculus connection intact at hilum. The paradigm shift in unraveling the cause of ST in 'Alphonso' mango, helped in working out the right preventive strategy for the first time in the world. The key to manage such internal fruit disorders clearly lies in inducing metabolic quiescence of the recalcitrant seed through induction of appropriate levels of dormancy while maintaining the fruit quality through environmentally safe approaches. Physiological fruit-disorders akin to spongy tissue, like jelly seed in 'Mallika', 'Dashehari' and 'Amrapali' varieties bearing recalcitrant seeds might also be prevented in a similar manner.

Environmentally-Safe Management Strategy

Physiological studies indicated that inducing seed dormancy and/or down regulation of seed metabolism without affecting the fruit growth and development was found to be the right strategy to prevent/arrest ST formation as it cannot be cured once developed. Use of paclobutrazol provided valuable academic insights, but had limited practical application due to export-restrictions and health hazards associated with this carcinogenic chemical. A few eco-friendly formulations were therefore developed and field evaluated successfully for prevention of ST incidence by more than 98 per cent including tree ripe fruits. This is for the first time in the world that not only the exact cause of ST in 'Alphonso' mango was unravelled but also an environmentally-safe formulation was developed for its successful prevention.

"Arka Saka Nivarak" – An Innovative Technology

It is obvious that physiological fruit disorders like ST, jelly-seed, stem-end-cavity *etc.*, can only be prevented/arrested successfully but not cured once developed. Therefore, down-regulation of seed metabolism without affecting fruit growth was resorted to as the right preventive-strategy to address internal breakdown disorders like ST in 'Alphonso' mango successfully. The product, "Arka Saka Nivarak" (pending patent award), is novel and an environmentally-safe formulation developed by ICAR-IIHR (V. Ravindra) for successful prevention of internal breakdown disorders in mango including spongy tissue in Alphonso. The uniqueness of the formulation lies in its action that delays/prevents the shift of recalcitrant seed in to germination mode through induction of dormancy and/or down-regulation of its metabolism. Lack of dependable technology for prevention/control of ST has been affecting the farmers and export agencies in different dimensions. Whenever export consignments were rejected for fruit disorders the agencies and the government were clueless as to what GAP practices were needed to send assured quality fruits, posing a serious challenge for certification by export promoting agencies. The product is a breakthrough and novel innovation and is unique as no technology is in place till date anywhere in the world for prevention of physiological fruit disorders. Prevention of physiological disorders boosts the export of mango in general and 'Alphonso' in particular by many folds resulting in higher foreign exchange earnings and improves the socio-economic status of the fruit growers.

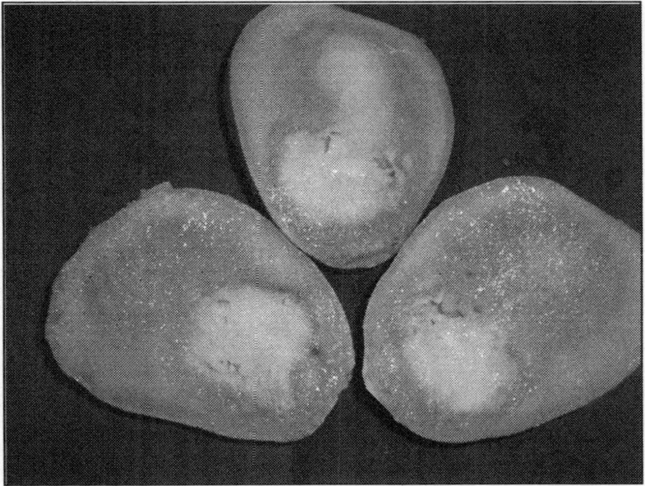

Figure 5: Spongy Tissue in Alphonso Mango.

Co-benefits of Treatment with "Arka Saka Nivarak"

Fruits treated with "Arka Saka Nivarak" were found to be not only free from spongy tissue but also of any harmful residues. "Arka Saka Nivarak", was also found to improve many quality aspects of fruit including organoleptic as follows

- Increased fruit weight (>230 g)
- Attractive appearance
- Uniform external and internal colour development
- Improved fruit firmness
- Excellent pulp texture
- Improved flavour, aroma and TSS
- Extended shelf-life of fruits (up to 3 weeks under ambient conditions)
- Enhanced antioxidant capacity of the fruit pulp (Figure 6).

Insights

Physiological disorders afflicting storage organs such as fruits, some vegetables and root crops in horticulture, irrespective of being temperate or tropical in origin, are associated with cell-death in tissues where localized calcium (Ca) deficiency remains an undisputed observation. Over 30 physiological disorders due to Ca deficiency are reported in various horticultural crops, and this list will probably extend as research progresses. Poor Ca distribution within a plant can also explain frequent appearance of Ca deficiency disorders (even in Ca-rich soils) and failure to respond to the Ca applied. Success at improving the plant's Ca content mitigated the disorder only partially, as, Ca application at best is palliative in action. In several physiological disorders, distinction between cause and effect is not easy to make due to variable symptoms of the malady in crops, compounded by complex inter-relationships among different factors. A history of investigations on physiological disorders has repeatedly demonstrated that knowledge on the cause of a disorder is generally imperfect until a single, primary cause is assigned. As physiological disorders can only be prevented (and not cured *post facto*), a remedy takes long to develop, years or decades after the cause is established as is the case in 'Alphonso' mango. Systematic studies in 'Alphonso' mango have helped separate the cause from effect. We traced the cause of spongy tissue to demand-driven mobilization of Ca into the recalcitrant seed, along with water from the mesocarp, rather than poor uptake of Ca. Therefore, we could successfully evolve preventive management strategy for spongy tissue in 'Alphonso' mango by pre-harvest induction of dormancy and/ or down-regulation of seed metabolism. This novel strategy has opened up new vistas to address many more physiological disorders attributed to inefficient Ca uptake/distribution in several fruit species having recalcitrant seeds. Thus, we have shown that an in-depth understanding of factors and their interaction with underlying physiological processes is vitally important

Figure 6: Alphonso Mango Fruits Free from Spongy Tissue after Treatment with "Arka Saka Nivarak".

to generate innovative, science-based management strategies for export-quality horticultural production free of physiological disorders.

Acknowledgements

The author is thankful to the Director of ICAR-IIHR for providing all facilities.

References

Acosta, R.M., Neito, A.D., Mena, N.G., Vaquera, H.H., Teliz, O.D. and Nieto, A.R. 2000. Effect of post-harvest temperatures on the development of internal darkening in mango fruits (*Mangifera indica* L.,) cv. Haden and their quality. *Acta Hort.*, **509**: 401–412.

Ali, M. and Lazan, H. 1998. Papaya. In: *Tropical and Sub-tropical Fruits*, (Eds.) P.E. Shaw, H.J. Jr, Chan and S. Nagy, Ag. Science, Inc. Florida, USA, p. 401–445.

Aloni, R. 2001. Foliar e axial aspects of vascular differentiation: hypotheses and evidence. *J. Pl. Gr. Reg.*, **20**: 22–34.

Bakker, J.C., 1988. Russeting (circular cracking) in glasshouse tomatoes in relation to fruit growth. *J. Hort. Sci.*, **63**: 459–463.

Bangerth, F. 1979. Calcium related physiological disorders of plants. *Ann. Rev. Phytol.*, **17**: 97–102.

Bangerth, F., Dilley, D. R. and Dewey, D.H. 1972. Effect of post-harvest calcium treatments on internal breakdown and respiration of apple fruits. *J. Amer. Soc. Hort. Sci.*, **97**: 679–682.

Belmans, K. and Keulemans, J. 1996. A study of some fruit skin characteristics in relation to the susceptibility of cherry fruit to cracking. *Proc. Intl. Cherry Sym. Acta Hort.*, **410**: 547–550.

Benson, N.R. 1994. *Nutrient Disorders in Tree Fruits*. Pacific Northwest Extension Publications. http://cru.cahe.wsu.edu/cepublications/pnw0121e/pnw0121e.pdf

Berjak, P., Farrant, J.M., Mycock, D.J. and Pammenter, N.W. 1990. Recalcitrant (homoiohydrous) seeds: The enigma of their desiccation sensitivity. *Acta Hort.*, **18**: 297–310.

Bramlage, W.J., Drake, M. and Baker, J.H. 1974. Relationships of calcium content to respiration and post-harvest condition of apples. *J. Amer. Soc. Hort. Sci.*, **99**: 376–378.

Cannell, M.G.R. 1985. Dry matter partitioning in tree crops. In: *Attributes of Trees as Crop Plants*, (Eds.) M.G.R. Cannell and J.E. Jackson, Natural Environment Research Council.

Chaplin, G.R. 1984. Post-harvest physiology of mango fruit. **In:** *First Australian Mango Research Workshop*, Proceedings, CSIRO, Melbourne, Australia, November, p. 261–270.

Cheema, G.S. and Dani, P.G. 1934. Report on the export of mangoes to Europe in 1932 and 1933. Department of Agriculture, Bombay, Bulletin, **170**: 1–31.

Chen, D. And Osborne, D.J. 1970. Hormones in the translational control of early germination in wheat embryos. *Nature* (London), **1**(226): 1157–1160.

Christiansen, M.N. and Foy, C.D. 1979. Fate and function of calcium in tissue. *Com. in Soil Sci. Pl. Anal.*, **10**(1 and 2): 427–442.

Clarkson, D.T. 1984. Calcium transport between tissues and its distribution in the plant. *Plant Cell Env.*, **7**: 449–452.

Cohen, R.A. and Greene, D.W. 1989. A possible relationship between fruit calcium and gibberellin A4 response on 'Delicious' apple. (Abstract). *Hort. Sci.*, **24**: 223.

Considine, J.A. and Kriedmann, P.E. 1972. Fruit splitting in grapes: Determination of the critical turgor pressure. *Australian J. Agr. Res.*, **23**: 17–24.

Cooper, T. and Bangerth, F. 1976. The effect of Ca and Mg treatments on the physiology, chemical composition and bitter pit development of "Cox's Orange" apples. *Scientia Hort.*, **5**: 49–57.

Coway, W.S. and Sams, C.E. 1987. The effects of post-harvest infiltration of calcium, magnesium, or strontium on decay, firmness, respiration and ethylene production in apples. *J. Amer. Soc. Hort. Sci.*, **112**: 300–303.

Desai, M.C. 1966. Sponginess in mango fruits. *Junagadh Agril. Coll. Mag.*, **5**: 4–6.

Dure III, L.S. 1975. Seed formation. *Ann. Rev. Pl. Physiol.*, **26**: 259–278.

Dorais, M., Papadopoulos, A.P. and Gosselin, A. 2001. Greenhouse tomato fruit quality, (Ed.) J. Janick, *Hort. Rev.*, **5**: 239–319.

Eaks, I.L. 1985. Effects of calcium on ripening, respiration rate, ethylene production and quality of avocado fruit. *J. Amer. Soc. Hort. Sci.*, **110**: 145–148.

Ferguson, I.B. and Watkins, C.B. 1989. Bitter pit in apple fruit. *Hort. Rev.*, **11**: 289–355.

Farnsworth, E. 2000. The ecology and physiology of viviparous and recalcitrant seeds. *Ann. Rev. Ecol. Syst.*, **31**: 107–138.

Farrant, J.M., Pammenter, N.W. and Berjak, P. 1988. Recalcitrance: A current assessment. *Seed Sci. Tech.*, **16**: 155–166.

Greenwood, J.S. and Bewley, J.D. 1982. Seed development in Ricinus communis (Castor bean). I. Descriptive morphology. *Canadian J. Bot.*, **60**: 1751–1760.

Gibert, G., Chadoeuf, J., Vercambre, G., Génard, M. and Lescourret, F. 2007. Cuticular cracking on nectarine fruit surface: Spatial distribution and development in relation to irrigation and thinning. *J. Amer. Soc. Hort. Sci.*, **132**: 583–591.

Gülsen, Y., Dumanoðlu, H. and Kunter, B. 1995. Fruit cracking in some Turkish apricot cultivars. *Acta Hort.*, **384**: 277–282.

Huang, J. and Snapp, S.S. 2004. The effect of boron, calcium, and surface moisture on shoulder check, a quality defect in fresh-market tomato. *J. Amer. Soc. Hort. Sci.*, **129**: 599–607.

Imas, P. 1999. Quality aspects of K nutrition in horticultural crops. **In**: Paper Presented at IPI–PRII–KKV Worksop on: *Recent Trends in Nutrition Management in Horticultural Crops*, 11th–12th February, Dapoli, Maharashtra, India.

Katrodia, J.S. 1979. The study in to the causes of the development of spongy tissue in mango (*Mangifera indica* L.) fruit of cultivar Alphonso. *Ph.D. Thesis*, Marathwada Agricultural University, Parbhani, India.

Katrodia, J.S. 1988. Spongy tissue in mango-causes and control measures. *Acta. Hort.*, **231**: 815–826.

Kennelly, M., 2009. *Tomato Leaf and Fruit Diseases and Disorders*. Kansas State University Agricultural Experiment Station and Cooperative Extension Service.www.ksre.ksu.edu.

Knight, J.N. 1980. Fruit thinning of the apple cultivar Cox's Orange Pippin. *J. Hort. Sci.*, **55**: 267–273.

McLaurin, W.J., 1998. Blossom-end rot. *Hort., Fact Sheet*, H-98-036: 1–4.

Meheriuk, M., Neilsen, G.H. and McKenzie, D.L. 1991. Incidence of rain splitting in sweet cherries treated with calcium or coating materials. *Canadian J. Pl. Sci.*, **71**: 231–234.

Milad, R.E. and Shackel, K.A. 1992. Water relations of fruit end cracking in French prune (*Prunus domestica* L. cv. French). *J. Amer. Soc. Hort. Sci.*, **117**: 824–828.

Olson, S.M., 2004. *Physiological, Nutritional and other Disorders of Tomato Fruit*, University of Florida IFAS Extension HS-954. EDIS. website at http://edis.ifas.ufl.edu

Peet, M.M., 1992. Fruit cracking in tomato. *Hort. Tech.*, **2**: 216–223.

Peet, M.M., 2009. Physiological disorders in tomato fruit development. *Acta Hort.* (*ISHS*), **821**: 151–160.

Peet, M.M. and Willits, D.H. 1995. Role of excess water in tomato cracking. *Hort. Sci.*, **30**: 65–68.

Poovaiah, B.W. 1979. Role of calcium in ripening and senescence. *Comm. Soil Sci. Pl. Anal.*, **10**(1 and 2): 83–88.

Qiu, Y., Nishina, M.S. and Paull, R.E. 1995. Papaya fruit growth, calcium uptake, and fruit ripening. *J. Amer. Soc. Hort. Sci.*, **120**: 246–253.

Rane, D.A., Katrodia, J.S. and Kulkarni, D.N. 1976. Problem of spongy tissue development in mango (*Mangifera indica* L) cv. Alphonso: A review. *J. Agric. Univ.*, **1**: 89–94.

Ravindra, V. 2015. Recent understanding of spongy tissue problem in 'Alphonso' mango: A review. **In**: *Proc. of the Global Conference on Augmenting Production and Utilization of Mango: Biotic and Abiotic Stresses. Acta. Hort.*, **1006**: 169–175.

Ravindra, V. and Shivashankar, S. 2004. Spongy tissue in Alphonso mango-significance of *in situ* germination events. *Curr. Sci.*, **87**(8): 1045–1049.

Ravindra, V. and Shivashankar, S. 2006. Spongy tissue in Alphonso mango II. A key evidence for the causative role of seed. *Curr. Sci.*, **91**(12): 1712–1714.

Ravindra, V., Shivashankar, S., Muraleedhara Reddy, T. and Kotur, S.C. 2010. Spongy tissue in Alphonso mango. III. Radiotracer evidence for increased mobilization of water from mesocarp to seed. *Curr. Sci.*, **99**(5): 571–574.

Rigney, C.J. and Wills, R.B.H. 1981. Calcium movement, a regulating factor in the initiation of tomato fruit ripening. *Hort. Sci.*, **16**: 550–551.

Sakata, 2009. *Fortress, F_1 Hybrid Determinate Salad Tomato.* Technical Bulletin, www.sakata.com

Saure, M.C. 2005. Calcium translocation to fleshy fruit: Its mechanism and endogenous control. *Scientia Hort.*, **105**: 65–89.

Schaffer, B. and Andersen, P.C. 1994. Introduction. **In**: *Handbook of Environmental Physiology of Fruit Crop*, (Eds.) B. Schaffer and P.C. Andersen. New York, Boca Raton, Florida, p. 1–2.

Selvaraj, Y., Raja, E.M. and Rawal, R.D. 2000. Biochemical studies on internal breakdown: A ripening disorder in mango fruits. *Indian J. Hort.*, **57**: 183–188.

Seymour, G., Taylor, J. and Tucker, G. 1993. Introduction. **In**: *Biochemistry of Fruit Ripening*. Chapman and Hall, London.

Shanta, K. 1981. Chemical studies on internal breakdown in Alphonso mango. *J. Hort. Sci.*, **56**: 247–250.

Sharples, R.O. and Johnson, D.S. 1977. The influence of calcium on senescence changes in apple. *Ann. Appl. Biol.*, **85**: 450–453.

Shear, C.B. 1975. Calcium related disorders of fruits and vegetables. *Hort. Sci.*, **10**: 361–365.

Shivashankar, S., Ravindra, V. and Louis, L. 2007. Biochemical changes in seed and mesocarp of mango (*Mangifera indica L.*) cv. Alphonso and their significance during the development of spongy tissue. *J. Hort. Sci. Biotech.*, **82**: 35–40.

Sperry, W.J., Davis, J.M. and Douglas, D.C. 1996. Soil moisture and cultivar influence cracking, blossom-end rot, zippers, and yield of staked fresh-market tomatoes. *Hort. Tech.*, **1**: 21–24.

Subrahmanyam, H., Krishnamurthy, S., Subhadra, N.V., Dalal, V.B., Randhawa, G.S. and Chacko, E.K. 1971. Studies on internal breakdown: A physiological ripening disorder in Alphonso mango (*Mangifera indica* L). *Trop. Sci.*, **13**: 203–210.

Thompson, A.K. 1971. The storage of mango fruits. *Tro. Agri.*, **48**: 63–70.

Vanderlinden, C. 2009. *Identifying and Controlling Blossom End Rot*. http://organic gardening.about.com/od/disease/p/blossomendrot.htm.

Vergas Ramos, V.H. and Durigan, F. 1995. Occurrence of "Radicel Sprout" and "Breakdown" in mangoes type 'Parvin', 'Keitt' and 'Tommy Atkins', in the region of Jaboticabal–SP. *Acta Hort.*, **370**: 173–176.

Visai, C., Failla, O. and Eccher, T. 1989. Effects of promalin and paclobutrazol on cracking and quality of Neipling Stayman apple. *Acta Hort.*, **239**: 451–454.

Wainwright, H. and Burbage, M.B. 1989. Physiological disorders in mango (*Mangifera indica* L.) fruits: A review. *J. Hort. Sci.*, **64**: 125–135.

White, P.J. and Broadley, M.R. 2003. Calcium in plants. *Ann. Bot.*, **92**: 487–511.

Willumsen, J., Petersen, K.K. and Kaack, K. 1996. Yield and blossom-end rot of tomato as affected by salinity and cation activity ratios in the root zone. *J. Hort. Sci.*, **71**: 81–98.

Wright, C.J. 1989. Interactions between vegetative and reproductive growth. **In**: *Manipulation of Fruiting*, (Ed.) C.J. Wright. Butterworth's, Essex, p. 15–27.

Zocchi, G. and Mignani, I. 1995. Calcium physiology and metabolism in fruit trees. *Acta Hort.*, **383**: 15–23.

50

Slow Release Nano-formulations of Bio-active Molecules from Horticultural Crops

N.A. Shakil

Division of Agricultural Chemicals,
ICAR-Indian Agricultural Research Institute, New Delhi – 110 012
E-mail: iamshakil@gmail.com

The importance of polymers in medical applications has been well recognized. The most widely used area is their use as carriers for pharmaceuticals in drug delivery and a considerable amount of research has been directed towards the use of natural and synthetic polymers as polymeric drugs and drug delivery systems. The low molecular weight pharmaceutical surfactants, which are currently used, have low toxicity and high solubilization power. Further, these suffer from the drawback that they have relatively high critical micelle concentration (CMC) and are unstable upon large dilution. However, by using amphiphilic copolymers, these drawbacks can be eliminated as they have very high solubilization power and rather low CMC values, thus making them more stable in vivo. Taking this into account, studies the world over have largely been confined to two different areas of research. One utilizes the insoluble polymers for the development of controlled release systems in which the drug is physically adsorbed or entrapped within an insoluble polymeric matrix, while the other involves the development of soluble prodrugs or drug carriers where the drug is either chemically bound to the polymer backbone or is the part of the backbone itself.

Use of PEG-based Copolymers

Poly(ethylene glycol) (PEG)-based (co)polymers have been utilized as one of the primary components of DDS, not only because PEGs have been shown to be nontoxic, biocompatible and durable enough for biological systems, but also because of their good solubility, low cost, easy availability, and ease with which these can be chemically modified or linked with other polymers. A key property of PEG is its attachment to other molecules and surfaces to provide a biocompatible, protective coating. This protective coating slows the rejection of materials in biological systems (such as human body) and greatly reduces bacterial adsorption. Also, it can be attached to other molecules without effecting their chemistry, but can control their solubility and increase their sizes. These novel properties can be further enhanced by chemical modification of the molecular structure.

The presence of just two terminal groups in PEG limits its utility to attach drugs which in turn effects the drug loading capacity. In order to make the drug delivery systems more versatile, researchers the world over have attempted to synthesise copolymers based on PEGs by the synthesis of PEG-based polymers/copolymers by taking the PEGs of different molecular weights and polymerizing them with a moiety having a hydrophobic segment so that the polymer formed is amphiphilic in nature. It is desirable that in order to design a micelle that is able to physically encapsulate small active reagents (*i.e.* the drugs, pesticides), the polymers should have amphiphilic structures. By attaching the hydrophilic PEG segment with one or more hydrophobic ligand(s) or segment(s), an amphiphilic polymer can be prepared. The amphiphilic polymer should be able to self-assemble into micelles in a specific solvent system, such as water.

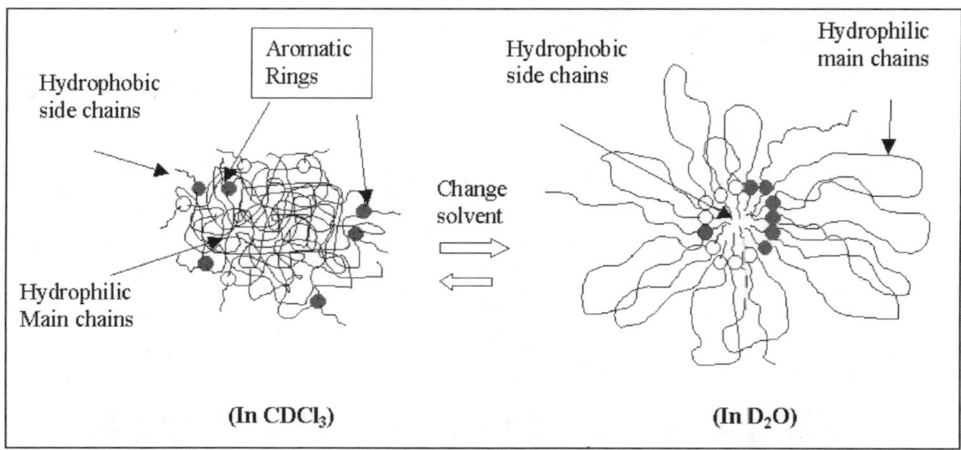

Figure 1: Formation of Micelle by Amphiphilic Copolymers.

Amphiphilic Copolymers

Amphiphilic block copolymers with hydrophilic and hydrophobic segments have been investigated extensively not only because of their unique self-organization characteristics but also for their wide range of potential applications, such as in drug delivery and separation technology systems. The micellar characteristics of amphiphilic block copolymers depend on the nature of each block and surface properties of self-organized micelles are highly dependent on the structures of the hydrophilic blocks. The formation of micelles by the self assembly of amphiphilic copolymers in aqueous media as shown in Figure 1.

As shown above, by changing the type of solvents, *i.e.* from organic to an aqueous system, the polymer does self-assemble into a stable aggregation or micelle. The hydrophobic side chains were most likely in the core of the micelle. The radius of gyration of these micelles fall within the nano range and hence it is possible to develop slow release nano formulations of bioactive molecules based on PEG-based polymers.

Nanotechnology

Nanotechnology has emerged as a highly attractive tool for the formulation and delivery of active ingredients as well as enhancing their application efficiency. It has the potential to revolutionize the agricultural and food industry with new tools for the molecular treatment of diseases, rapid disease detection, enhancing the ability of plants to absorb nutrients *etc.* Smart sensors and smart delivery systems will help the agricultural industry combat various crop pathogens. In the near future nanostructured catalysts will be available which will increase the efficiency of pesticides and herbicides, allowing lower doses to be used.

Our group has been actively involved in the synthesis of environmental friendly amphiphilic polymers having the tendency to aggregate in aqueous medium into nanospheres. These polymers have been utilized to developed slow release formulations of various pesticides analogous to drug delivery.

Using these copolymers of poly(ethylene glycols) and various dimethyl esters, which self assemble into nano micellar aggregates in aqueous media, controlled release (CR) formulations of azadirachtin-A, a bioactive constituent derived from the seed of *Azadirachta indica* A. Juss (Meliaceae), imidacloprid, thiamethoxam, thiram, β-cyfluthrin, carbendazim, azoxystrobin, mancozeb, carbofuran, *etc.* have been prepared. It was observed that the release from the commercial formulation was faster than with the developed CR formulations. The rate of release of encapsulated bioactives from nano micellar aggregates reduced on increasing the molecular weight of PEG.

We have extended this concept of developing controlled release formulations for encapsulation of bioactive compounds such as β-carotene and curcumin, have also been synthesized by attachment of saturated and unsaturated long chain acid chlorides to the copolymer of PEG and dimethyl-5-aminoisophthalate via an amide linkage. Slow release formulations of β-carotene have been developed employing poly(ethylene glycols) (PEGs) based functionalized amphiphilic polymers. Encapsulation efficiency and loading capacity of the developed formulations were determined which ranged from 22.60 to 28.08 per cent and 2.2 to 2.8 per cent respectively. The release kinetics of β-carotene from developed formulations in water revealed increased solubility and prolonged stability of β-carotene.

Several controlled release/slow-release formulations of pesticides and bioactives from plant species have been developed and their behavior has been recorded. The studies show that the developed controlled release formulations provide improvement in their pest control behavior as well as increased bioavailability of nutritional products.

The lecture will focus on the synthesis of poly(ethylene glycol) based polymeric materials utilizing different linker molecules to be used for encapsulation of different bioactive products. The details of the synthesis of amphiphilic polymers and development of slow release formulations based on these polymers will be presented during the symposium.

References

Adak, T., Kumar J., Shakil, N.A., Walia, S., Kumar, A. and Watterson, A.C. 2011. Synthesis and characterization of novel surfactant molecules based on amphiphilic polymers. *J. Macromol. Sci. Pure Appl. Chem., Part A*, **48**(10): 767–775.

Danprasert, K., Kumar, R., Chen, M.H., Gupta, P., Shakil, N.A., Prasad, A.K., Parmar, V.S., Kumar, J., Samuelson, L.A. and Watterson, A.C. 2003. Synthesis of novel poly(ethylene glycol) based amphiphilic polymers. *Eur. Polymer J.*, **39**: 1983–1990.

Kaushik, P., Shakil, N.A., Kumar, J. and Watterson, A.C. 2012. Synthesis and characterisation of novel poly(ethylene glycol) based amphiphilic polymers. *J. Macromol. Sci. Pure Appl. Chem., Part A*, **49**: 111–115.

Kaushik, P., Shakil, N.A., Kumar, J., Singh, M.K. and Yadav, S.K. 2013. Development of controlled release formulations of thiram employing amphiphilic polymers and their bioefficacy evaluation in seed quality enhancement studies. *J. Evn. Sci. Hlth., Part B*, **48**(8): 675–685.

Koli, Pushpendra, Shakil, N.A., Kumar, Jitendra, Singh, Braj B. and Watterson, A.C. 2014. Synthesis and characterization of novel encapsulating materials based on functionalized amphiphilic block copolymers. *J. Macromol. Sci., Part A*, **51**(9): 729–736.

Koli, Pushpendra, Singh, Braj B., Shakil, Najam A., Kumar, Jitendra and Kamil, Deeba. 2015. Development of controlled release nanoformulations of carbendazim employing amphiphilic polymers and their bio efficacy evaluation against *Rhizoctonia solani*. *J. Env. Sci. Hlth., Part B: Pesticides, Food Contaminants, and Agricultural Wastes*, **50**(9): 674–681.

Krishnani, K.K., Kathiravan, V., Shakil, N.A., Singh, Manish Kumar, Brahmane, M.P., Meena, K.K., Sarkar, Biplab, Choudhary, K., Singh, Mukesh Kumar and Kumar, Jitendra. 2015. Bactericidal activity of nanopolymers against shrimp pathogenic bacterium *Vibrio harveyi*. **In**: *Proceedings of the National Academy of Sciences, India Section B: Biological Sciences*, **85**(4): 1079–1086.

Kumar, R., Shakil, N.A., Chen, M.H., Parmar, V.S., Samuelson, L.A., Kumar, J. and Watterson, A.C. 2002. Chemo-enzymatic synthesis and characterization of novel functionalised amphiphilic polymers, *J. Macromol. Sci., Pure and Appl. Chem. Part A*, **39**: 1137–1149.

Kumar, R., Tyagi, R., Shakil, N.A., Parmar, V.S., Kumar, J. and Watterson, A.C. 2005. Self-assembly of PEG and diester copolymers: Effect of PEG length, linker, concentration and temperature. *J. Macromol. Sci. Pure and Appl. Chem., Part A*, **42**(11): 1523–1528.

Kumari, Anu, Kumar, Jitendra, Shakil, Najam A., Singh, Braj B. and Watterson, Arthur C. 2015. Synthesis and characterization of amphiphilic PEG based polymers and their self assembling behaviour. *J. Macromol. Sci.*, Part A, **52**(6): 417–424.

Loha, K.M., Shakil, N.A., Kumar, J., Singh, M.K. and Srivastava, C. 2012. Bio-efficacy evaluation of nanoformulations of β-cyfluthrin against *Callosobruchus maculatus* (Coleoptera: Bruchidae). *J. Env. Sci. Hlth., Part B*, **47**(7): 687–691.

Loha, K.M., Shakil, N.A., Kumar, J., Singh, M.K., Adak, T. and Jain, S. 2011. Release kinetics of β-cyfluthrin from its encapsulated formulations in water. *J. Env. Sci. Hlth., Part B*, **46**(3): 201–206.

Majumder, Sujan, Shakil, Najam A., Kumar, Jitendra, Banerjee, Tirthankar, Sinha, Parimal, Braj, Singh, B. and Garg, Parul. 2016. Ecofriendly PEG-based controlled release nanoformulations of Mancozeb: Synthesis and bioefficacy evaluation against phytopathogenic fungi *Alternaria solani* and *Sclerotium rolfsii*. *J. Env. Sci. Hlth.*, Part B.

Majumder, Sujan, Shakil, Najam A., Singh, Braj B. and Watterson, Arthur C. 2016. Synthesis and characterization of functionalized amphiphilic polymers for utilization as surfactants. *J. Macromol. Sci., Part A: Pure Appl. Chem.*, **53**(2): 75–81.

Sarkar, D.J., Kumar J., Shakil, N.A., Adak, T. and Watterson, A.C. 2012. Synthesis and characterization of amphiphilic PEG based aliphatic and aromatic polymers and their self assembling behavior. *J. Macromol. Sci. Pure Appl. Chem., Part A*, **49**(6): 455–465.

Shakil, N.A., Singh, M.K., Pandey, A., Kumar, J., Pankaj, Parmar, V.S. and Watterson, A.C. 2010. Development of poly(ethylene glycol) based amphiphilic copolymers for controlled release delivery of carbofuran. *J. Macromol. Sci. Pure Appl. Chem., Part A*, **47**(3): 241–247.

Singh, Braj Bhushan, Shakil, Najam A., Kumar, Jitendra, Walia, Suresh and Kar, Abhijit. 2015. Development of slow release formulations of β-carotene employing amphiphilic polymers and their release kinetics study in water and different pH conditions. *J. Food Sci. Tech.*, DOI 10.1007/s13197-015-1925-x.

Ensuring Higher Price of Farm Produce and Products

51

Role of Credit and Price Policy in Development of Horticulture

Bhoopesh Punera*[1], *Pramod Kumar*[1] *and Suresh Pal*[2]

[1]*Division of Agricultural Economics,*
ICAR-Indian Agricultural Research Institute, New Delhi – 110 012
[2]*ICAR-National Institute of Agricultural Planning, Pusa, New Delhi - 110 012*
**E-mail: spl.econ@gmail.com*

India, with its wide variability of climate and soil, is highly favourable for growing a large number of horticultural crops. However, only in mid 80's did the Govt. of India identify horticulture crops as a means of diversification for making agriculture more profitable through efficient land use and creating skilled employment for rural masses especially women folk. The changing scenario encourages private investment, to go for Hi-tech horticulture with micropropagation, protected cultivation, drip irrigation, fertigation, and integrated nutrient and pest management, besides making use of latest post-harvest measures particularly in the case of perishable commodities. India's horticulture production currently nearing 280 million tonnes a year has not only brought prosperity to small and marginal farmers but also provided food and nutritional security. Horticulture in India has today emerged as one of the most important sectors for diversification in agriculture. India is the second largest producer of horticultural products in the world after China with the production and productivity of 283.47 million tonnes and 12.10 t/ha, respectively. Horticultural crops covered an area of 23.42 million ha in 2014-15 as compared to 20.2 m ha in 2007-08, registering an increase of about 15.94 per cent. Despite this impressive growth, the production is much less than the requirement. The target can be achieved by bringing additional area under horticultural crops by using hybrid seeds, improved agro-techniques and promotion of protected cultivation (Singh *et al.*, 1999). Precision farming in horticulture has been found very useful for achieving higher level of productivity and product quality. Precision farming and protected cultivation technologies are capital intensive in nature so credit availability plays an important role in horticultural development. Due to lack of credit and institutional rigidities India lags behind many countries. The very nature of horticultural crops *i.e.*, long gestation period, and the need for costly input requirement especially of seeds/planting material, *etc.* demand heavy investment. Such a huge investment can come by only through the credit support from financial institutions.

Importance of Horticulture in the Economy

Plan wise share of horticulture in agriculture showed an increasing trend for IX to XI five year plans (Table 1). For XIIth Five Year Plan it has increased in absolute term. This showed high investment in horticulture subsequently followed by investment in other agricultural activities. National Horticulture Mission (NHM) and National Mission on Micro-irrigation (NMMI) are the major schemes of horticulture sector for XI and XII Five Year Plans.

Focused attention to horticultural research and development has paid dividend by way of increased production and productivity and enhanced exports. Large areas have been brought under improved cultivars as a result of considerable increase in production of quality planting material and seeds. A large number of farmers were trained in innovative technologies like drip irrigation, greenhouse cultivation, micro-grafting *etc.* The increased investment in horticulture

Table 1: Plan-wise Share of Horticulture in Agriculture (Rs. Crores)

	Agriculture and Allied	*Horticulture*
IX Plan (1997-2002)	*37546*	*1453 (3.9 per cent)*
X Plan (2002-07)	*58933*	*5025 (8.5 per cent)*
XI Plan (2007-12)	*136381*	*15800 (11.6 per cent)*
XII Plan (2012-2017)	*363273*	*16840 (4.6 per cent)*
Annual Plan (2012-13)	54748	3062 (5.6%)
Annual Plan (2013-14)	18781	4128 (21.9%)
Annual Plan (2014-15)	11531	2922 (25.3%)

Source: GoI (2015).

Table 2: Production of Horticulture *vis-à-vis* Foodgrains (in million tonnes)

Year	*Horticulture*	*Foodgrains*	*% of Horticulture to Foodgrains*
2001-02	145.8	212.9	68.5
2005-06	182.8	208.6	87.6
2010-11	240.5	244.5	98.4
2014-15	283.5	257.1	110.3

Sources: (1) NHB (Various issues), and (2) GoI (Various issues).

sector brought about improvement in production and productivity of horticultural crops. The increase in horticultural production was so much that it surpassed the production of foodgrains by the year 2014-15 (Table 2). The per capita net availability of fruits rose from 114 in 2001-02 to 172 grams per day in 2011-12, while that of vegetables increased from 236 to 350 g per day over the same period (Table 3).

The value of output from horticulture sector has also shown a consistent increase from 0.16 lakh crores in 2008-09 to 0.21 lakh crores during 2012-13 (Table 4). It accounts for 34.4 per cent of the total agricultural output of the country. Among the horticultural crops the contribution of fruits and vegetables was the highest. This has immense potential for meeting the household nutritional security of the country.

The export of horticulture commodities recorded a four fold increase from 10 thousand crores to Rs 45 thousand crores during the period 2004-05 to 2013-14 (Table 5). The sector thus recorded a compound annual growth of 18 per cent during the similar period. It contributes to 17 per cent of the total agriculture sector exports.

Hi-tech Horticulture

Hi-tech horticulture includes use of modern technologies such as greenhouses, plastic tunnels, walk-

Table 3: Per Capita Availability and Production of Fruits and Vegetables in India

Year	*Per Capita Availability (gram/per person/day)*			*Production (million tonnes)*		
	Fruit	*Vegetables*	*Total*	*Fruit*	*Vegetables*	*Total*
2001-02	114	236	350	43	89	132
2007-08	158	309	468	66	128	194
2008-09	163	306	469	68	129	197
2009-10	167	313	480	71	134	205
2010-11	170	332	502	75	147	222
2011-12	172	350	522	76	156	232

Table 4: Value of Output of Horticultural Crops (at constant prices) (Rs '00 crores)

Particulars	*2008-09*	*2009-10*	*2010-11*	*2011-12*	*2012-13*
Agricultural crops	5250	5241	5761	6094	6091
Fruits and vegetables	1288.6	1286.2	1292.0	1558.2	1639.4
Condiments and spices	155	170	185	224	202
Floriculture	70	75	80	118	106
Plantation crops	122	131	140	145	148
(i) Coconut	*76*	*82*	*88*	*94*	*94*
(ii) Cashew nut	*17*	*20*	*14*	*20*	*27*
(iii) Cocoa	*0.2*	*0.19*	*0.2*	*0.3*	*0.2*
(iv) Areca nut	*28.5*	*29.07*	*37.6*	*30.4*	*26.2*
Total horticulture	1635 (35.2)	1662(31.7)	1697(29.5)	2045(33.6)	2095(34.4)

Note: The figures in parenthesis are% to the value of output of agricultural crops.

Source: CSO (2013).

Table 5: Total Horticulture Exports during 2004-05 to 2013-14 (Rs '000 crores)

Particulars	2004-05	2009-10	2013-14	CAGR (%)
Tea	1.84	2.94	4.83	13.5
Coffee	1.07	2.03	4.80	17.5
Floriculture products	0.22	0.30	0.46	3.2
Spices	1.88	6.16	15.98	29.1
Cashewnut shell liquid	0.01	0.03	0.04	17.6
Cashewnut	2.48	2.80	5.10	8.6
Fruits and vegetable seeds	0.07	0.15	0.40	20.5
Fresh fruits	0.86	2.27	4.36	17.4
Fresh vegetables	0.86	2.94	5.12	19.9
Processed vegetables	0.36	0.75	1.32	13.2
Processed fruit juices	0.37	1.16	2.56	20.6
Total horticulture	10.03 (24)	21.53 (24)	44.96 (17)	18.4
Total agriculture	41.60	89.34	268.47	23.2
Total national export	375.34	845.53	1894.18	19.9

Note: Figures in parenthesis are percent to the total agriculture

Source: GoI (2014).

in tunnels and micro-irrigation, to raise the productivity and quality of produce. Hi-tech practices are useful for production of various horticultural commodities like fruits, vegetables and flowers. Among fruit crops, grapes, orange and pomegranate are cultivated using hi-tech technologies, *viz.*, micro-irrigation and fertilizer management. Trends for these commodities show high growth rates. Area and production of grapes have increased from 47.5 to 118.7 thousand ha and 1,184.2 to 2,585.3 thousand metric tonnes, respectively, during the period 2001-02 to 2013-14 with compound annual growth rate (CAGR) 8.79 and 4.4 per cent, respectively. During the period 2001-02 to 2013-14, area under orange has increased from 198.9 to 330 thousand ha with CAGR of 6.69 per cent and production has increased from 1660.1 to 3431.4 thousand metric tonnes with CAGR 9.66 per cent. Production of grape decreased in year 2009-10 due to unseasonal rains. The growth and production of both grapes and oranges was particularly high after 2009-10. This was attributed due to high investment in horticulture by various government programmes. The production of pomegranate has almost doubled within last three years. Now the country produces about 1345 thousand metric tonnes of pomegranate with an area of 130 thousand ha (2013-14).

The National Committee on Plasticulture Applications in Horticulture assessed the area under protected cultivation at the national level in 2010 (Table 6). China is observed to be having the highest area under polyhouse cultivation followed by South Korea (Hickman, 2011). Among European nations Spain is the leading country followed by Italy and the Netherlands.

Table 6: Area under Protected Cultivation in different States

State	Greenhouse including Polytunnels (ha)	Shade Net (ha)	Protected Cultivation Achieved under NHM (ha)
Andhra Pradesh	5.16	118.19	3,639.23
Bihar	22.27	33	2077
Chhattisgarh	–	22,598	24,988.25
Gujarat	166.92	213.71	15,546.11
Haryana	109.12	36.26	2,991.03
Himachal Pradesh	141.54	6.8	–
Jammu and Kashmir	22.02	0.89	–
Karnataka	35.44	778.32	7,043.35
Kerala	13.07	37.58	2,116.8
Madhya Pradesh	1.4	1,790.02	8,228.5
Maharashtra	103.27	2,112.18	12,806.65
Odisha	–	–	16,545.23
Mizoram	45.62	108.5	–
Punjab	–	–	1,140.86
Rajasthan	–	–	1,435.78
Tamil Nadu	162.82	18.45	2,841.25
Uttar Pradesh	978.66	670.56	7,135.78
Uttarakhand	23.16	0.57	–
West Bengal	7.6	61.48	1,311.39
Others	70.17	33.2	4,173.09
Total	**1,908.24**	**28,617.71**	**11,6892.4**

Source: National Committee on Plasticulture Applications in Horticulture (2010).

Chang (2011) estimated that there were more than 4 million ha of protected cultivation of vegetables in China (Table 7). Various methods of protected cultivation like plastic mulches, low tunnels and walk-in-tunnels are used on more than 80 per cent of the total area under protected cultivation in China. Netherlands is also known as global leader in greenhouse production of horticultural products.

Table 7: Area under Greenhouse at Global Level

Country	Greenhouse Area ('000 ha)	Country	Greenhouse Area ('000 ha)
China	2760.0	Italy	26.5
Korea	57.4	Mexico	11.8
Spain	52.2	Netherlands	10.4
Japan	49.1	France	9.6
Turkey	33.5	United States	8.4

Source: Hickman (2011).

The importance of subsidy to farmers for adoption of protected cultivation was analysed using data from 30 farmers collected during 2015 from Solan district of

Himachal Pradesh. Economic feasibility of carnation cultivation is calculated with and without subsidy. Subsidy was provided to farmers on the cost of infrastructure comprising both structure and irrigation supply. The variable cost constituted 69 per cent of the total costs, whereas fixed cost constituted 31 per cent of the total costs if no subsidy is provided (Table 8). If subsidy is provided to farmers, variable costs accounted for 81 per cent of the total cost and fixed cost formed only 19 per cent of the total cost of cultivation. Among the variable cost, cost of labour was found to be the major cost followed by cost of planting material. Among fixed costs interest on fixed capital and cost of infrastructure were the major costs. Cost of infrastructure and interest on fixed capital is lower in case if subsidy is provided to farmer because additional cost is paid as subsidy.

It was found that carnation cultivation is a profitable venture with and without subsidy. Benefit-cost ratio was estimated to be 1.20 if no subsidy was provided and 1.43 if subsidy was provided by the government (Table 8). Average amount of subsidy provided to the farmer was around 50 per cent of the cost of infrastructure. Net present worth of the project was positive even without subsidy. High internal rate of return (28 per cent) ensured profitability of the project. Pay-back period was found to be fifth year if no subsidy was provided and if subsidy was provided it was estimated to be third year. So farmer can obtain nearly 55 per cent more income if subsidy is provided. However, cultivation of capsicum with carnation gives low returns if subsidy is withdrawn (IRR 13 per cent). Therefore, sustainability of this technology will require financial incentives from the government in case low value crop are taken.

Table 8: Cost of Cultivation under Polyhouse with and without Subsidy (Lakh Rs. per 1000 sq. m)

Particulars	*Carnation Crop*	*Export-oriented*	*Diversified Pattern (Carnation and coloured capsicum)*
Total fixed cost	20.65	17.53	22.11
Total variable cost	48.54	59.15	47.94
Total costs	69.18	76.68	70.05
Total returns	89.98	114.69	73.92
Net returns	20.80	38.02	3.86
Total discounted costs	55.55	60.84	52.77
Discounted revenue	66.68	84.85	54.92
Benefit cost ratio	1.20 (1.43)	1.39 (1.60)	1.04 (1.21)
Net present worth	11.13	24.01	2.15
Pay-back period	5th year	3rd year	7th year
Internal rate of return (%)	28 (73)	58 (117)	13 (31)

Notes: (1) Figures in parenthesis are with subsidy; (2) Horticulture: Mission for Integrated Development of Horticulture (MIDH), DAC and FW.

The sensitivity analysis for export-oriented famers revealed that cultivation without subsidy is also profitable as financial ratios like benefit-cost ratio is greater than one and net present worth and internal rate of return are high. Even without subsidy export-oriented farmer can earn positive net income which is around 75 per cent of income obtained with subsidy and IRR is as high as 58 per cent. Thus improved practices and better prices in export market can make this technology financially viable for growers. Evidences of sensitivity analysis showed that practicing protected cultivation of high value flowers is profitable even if no subsidy was provided. This is also supported by the fact that some of the farmers construct their own polyhouses without availing any subsidy from the government.

Enabling Factors for Growth of Horticulture Sector

Maintaining sustainable growth in horticulture requires continual research in developing innovative technologies for conservation and management of limited natural resources. The ICAR is responsible for providing leadership for agricultural research in the country. The financial outlay has kept pace with time and the money allocated has increased from Rs 1389 crores in 2001 to Rs. 3,818 crores in 2010. It is also observed that about 8 per cent of the total budget is allocated to horticulture sector. Looking at the contribution of the horticulture sector in total agriculture GDP it deserves a higher share of total budgetary allocation.

The investment in research and development has led the ICAR and SAUs to develop new horticultural crop varieties with specific traits that improve yield and nutritional quality along with tolerance/resistance to various biotic and abiotic stresses. The number of varieties released during the period 1951-2009 is 1086 (Table 9). The varieties embody the technology having potential to provide higher yield. This is the cheapest and most effective among all the horticulture technologies. The maximum number of varieties are developed in vegetable crops followed by plantation and

Table 9: Number of Horticultural Crops Varieties Released during 1951-2009

Particulars	*Selection*	*Hybrid*	*Mutant*	*Others*	*Total*
Vegetable crops	392	97	5	113	607
Ornamental crops	14	9	8	37	68
Plantation and spice crops	175	43	7	11	236
Fruit crops	50	70	4	10	134
Medicinal and aromatic plants	24		2	15	41
Total	655	219	26	186	1086

Source: IASRI (Various issues), Agricultural Research Data Book, Indian Agricultural Statistical Research Institute, ICAR, New Delhi.

spice crops and fruits crops. However, the popularity of the vegetable and flower varieties is a concern resulting in the dominance of imported seed/planting material.

The emphasis received by the horticulture sector by the government during the X and XI Five Year Plans boosted the sectors performance. This was also revealed from the number of programmes, which were implemented and the consistent deployment of funds towards achieving the intended objectives of these programmes. The total fund allocation towards the major programmes increased from Rs. 1,941 crores in 2007-08 to Rs. 2,922 crores during 2014-15 (Table 10). National Horticulture Mission followed by National Mission on Micro-irrigation and Horticulture Mission for North East and Himalayan States (HMNEH) account for the highest share in the total budgetary outlay.

The major programme, *i.e.*, NHM and HMNEM have benefitted 33.41 lakh ha of land (Table 12). The production of horticulture crops is intended through bringing area under protected cultivation (1.51 lakh hectare), rejuvenation of existing orchards (6.39 lakh ha). The programmes also aims at improving the inputs and outputs through development of nurseries (4,524 no.), water resources (87,612 no.), organic farming (2.67 lakh ha) and markets (2,045 no.).

In India, the horticultural commodities are faced with a high level of losses in the supply chain. The available cold storage facility is sufficient for only 10 per cent of fruits and vegetables produced in India (Planning Commission 2011). A study conducted by the Central Institute of Post-Harvest Engineering and Technology (CIPHET) in 2010 put the losses in the range of 0.8 per cent to 18 per cent and attributed them to several factors including non-availability of facilities for aggregation, packaging, storage, transportation, and cold chain and low level of processing of horticultural produce. To overcome the problem the government is supporting creation of modern enabling infrastructure and efficient processing facilities. These initiatives include setting up of (i) mega food parks, and (ii) cold chain, value-addition and preservation infrastructure. The Ministry of Food Processing Industries (MoFPI) has appointed NHB to play the role of Project Management Implementation Agency (PMIA) in respect of Integrated Cold Chain Infrastructure for horticulture projects under the scheme of cold chain, value-addition and preservation infrastructure since beginning of 2012 in concurrence

Table 10: Trend in Plan Expenditure under Horticulture Sector (Rs in cores)

Name of the Scheme	*2007-08*	*2008-09*	*2009-10*	*2010-11*	*2011-12*	*2012-13*	*2013-14*	*2014-15*
National Horticulture Board (NHB)	122.5 (6.3)	124.4 (6.1)	143.0 (7.6)	151.8 (5.6)	144.0 (4.7)	140.7 (4.6)	202.8 (4.9)	206.4 (7.1)
Coconut Development Board (CDB) including Technology Mission on Coconut	52.5 (2.7)	65.4 (3.2)	64.9 (3.5)	53.3 (2.0)	50.0 (1.6)	55.0 (1.8)	70.0 (1.7)	123.0 (4.2)
Horticulture Mission for North East and Himalayan States (HMNEH)	321.8 (16.6)	291.4 (14.2)	325.7 (17.4)	399.9 (14.8)	493 (16.0)	442.3 (14.4)	532.9 (12.9)	402.0 (13.8)
National Mission on Bamboo Technology and Trade Development (NBM)	114.4 (5.9)	84.7 (4.1)	50.9 (2.7)	89.7 (3.3)	85.7 (2.8)	87.0 (2.8)	128.0 (3.1)	80.1 (2.7)
National Horticulture Mission (NHM)	917.3 (47.3)	1010.5 (49.3)	800.0 (42.6)	970.9 (36.0)	1049.9 (34.0)	1089.3 (35.6)	1809.6 (43.8)	1142.8 (39.1)
National Mission on Micro Irrigation (NMMI)	411.3 (21.2)	470.0 (22.9)	480.0 (25.6)	997.3 (36.9)	1227.0 (39.7)	1203.3 (39.3)	1271.5 (30.8)	963.6 (33.0)
Central Institute of Horticulture in Nagaland (CIH)	1.4 (0.1)	2.2 (0.1)	1.4 (0.1)	2.1 (0.1)	2.3 (0.1)	1.7 (0.1)	2.9 (0.1)	4.1 (0.1)
Grant for Replanting and Rejuvenation of Coconut Gardens			10.0 (0.5)	34.0 (1.3)	35.0 (1.4)	42.6 (1.4)	110.0 (2.7)	–
Grants to coconut development board for coconut palm insurance scheme			0.1	0.9	1.0	0.5	0.5 (0.1)	–
Total Horticulture	1941.0	2048.5	1876.5	2699.9	3088.1	3062.2	4128.3	2922.0

Notes: (1) From 1 April 2014, the ongoing schemes of NHM, HMNEH, NBM, CDB, NHB, CIH, Replanting and Rejuvenation of Coconut Gardens and Coconut palm insurance have been subsumed under Mission for Integrated Development of Horticulture (MIDH); (2) From 1 April 2014, NMMI has been subsumed under the National Mission on Sustainable Agriculture and is now being implemented as "On Farm Water Management"(OFWM). During 2014-15; (3) Figures in parenthesis are percent to the total.

Source: GoI (2015).

with Department of Agriculture and Cooperation, Ministry of Agriculture, Govt. of India. The NHB has released grants to the number of MoFPI projects implementing the integrated cold chain. A National Centre for Cold-Chain Development (NCCD) has been set up and is expected to provide the necessary boost for adding capacity and creating a cold chain network in the country.

The availability of credit plays a major role in the development of any enterprise or sector. The financial institutions have played a major role in facilitating the flow of credit to horticulture sector in alignment with the national goals and objectives and have a crucial role in the growth of the sector. This is revealed from the allocation of the term loan towards the plantation and horticulture sector and the Hi-tech agriculture (Table 11). Together these two components have availed highest amount of term loan credit from the financial instititions. The other sub-components like minor irrigation and that of land development also reveal major chunk of the term loan being used for the horticulture sector development. Even among the crop loans, *i.e.* the short term loans of which major share is allocated for the cultivation of vegetables, fruits and flowers, which are highly capital intensive and is not possible without the credit facility made available by the financial institutions. For improving the agricultural credit flow and bringing down the rate of interest on farm loans the following steps have been taken by the government (i) Farmers could avail of crop loans up to a principal amount of Rs 3.0 lakhs at 7 per cent rate of interest. The effective rate of interest for farmers who promptly repay their loans was 4 per cent per annum during 2013-14; and (ii) Farmers were granted post-harvest loans against negotiable warehouse recepits (NWRs) at commercial rates. To encourage storage of produce in warehouses against NWRs, the benefit of interest subvention was extended to small and marginal farmers with Kisan Credit Cards (KCC) for a further period of up to siz months post-harvest on the same rate as crop loans.

In order the facilitate the larger resource flow to the horticulture sector, the National Bank for Agriculture and Rural Development (NABARD) is providing the refinance facility to the financial institutions like cooperative and regional rural banks. Sector wise distribution of refinance showed that plantation and horticulture is the one of the major sector of refinancing by NABARD (Table 12).

Subsidy plays a major role in promotion of capital intensive technologies. Protected cultivation and other hi-tech horticultural activities are capital intensive in nature. Most of the farmers in India are small and marginal with fewer resources, and therefore, subsidy provided by the government for construction of polyhouse or other infrastructure is an important factor which influenced the adoption of this technology. There are various schemes implemented by the central and state governments. Subsidy provided under the central scheme like National Horticulture Mission and Horticulture Mission for North-East and Himalayan states was 50 per cent of the cost and subsidy in the state schemes like Pt. Deen Dayal Kisan Bagwan Samriddhi Yojna of Himachal Pradesh, which was 80 per cent to male and 90 per cent to woman farmer.

The horticulture crops are risky enterprises especially the vegetables, fruits and flowers due to their perishability and bulky nature of the produce. They also are prone to price risk due to glut in the market. However the most important risk is the production risk in which the farmers suffer loss of crop due to pest and disease attack, vagaries of weather also brings in

Table 11: Sector-wise Groundlevel Credit Flow to Agriculture and Allied Sector (Rs '000 crores)

Sector/Sub-sector	*2003-04*	*2004-05*	*2005-06*	*2006-07*	*2007-08*	*2008-09*	*2009-10*	*2010-11*	*2011-12*	*2012-13*	*2013-14*	*2014-15*
Short term-production credit												
I. Crop loan	54.9	76.1	105.4	138.5	181.4	210.5	276.7	335.6	396.2	473.5	573.0	636.1
Medium term and Long term investment credit												
II. Term loan	32.0	49.2	75.1	90.9	73.3	91.5	107.9	132.7	114.9*	133.9*	138.6*	204.5*
Minor irrigation	2.7	4.2	8.6	8.6	2.8	3.2	5.2	4.4	3.8	4.4	4.6	6.7
Land development	0.6	0.8	1.7	2.3	2.6	2.9	3.7	3.6	3.1	3.6	3.8	5.6
Farm mechanization	3.9	4.6	9.7	10.1	8.3	8.3	10.2	12.8	11.1	12.9	13.4	19.7
Plantation and Horticulture	1.4	1.7	4.5	5.3	5.9	6.0	6.4	6.6	5.7	6.7	6.9	10.2
Animal Husbandry	2.9	3.1	7.3	8.0	9.0	10.4	10.3	12.8	11.1	12.9	13.3	19.7
Fisheries	1.1	1.3	1.0	1.4	1.2	1.3	1.9	1.9	1.7	1.9	2.0	2.9
Hi-tech Agriculture	4.0	6.6	9.7	21.5	33.3	41.7	50.8	82.8	71.6	83.5	86.4	127.5
Others	15.2	26.9	32.5	33.7	10.1	17.6	19.5	7.9	6.8	7.9	8.2	12.1
Total (I +II)	**86.9**	**125.3**	**180.5**	**229.4**	**254.7**	**301.9**	**384.5**	**468.3**	**511.0**	**607.4**	**711.6**	**840.6**

Note: *The sub-sectoral allocation is derived based on the proportionate share of the year 2010-11.

Source: NABARD (Various issues).

Table 12: Sector-wise Disbursement of Refinance by NABARD (Rs. crore)

Purpose	*2012-13*		*2013-14*		*2014-15*	
	Amount in Rs. Crore	*% of Total*	*Amount in Rs. Crore*	*% of Total*	*Amount in Rs. Crore*	*% of Total*
Minor irrigation	739.27	4.18	517.86	2.41	1147.17	3.65
Land development	817.69	4.63	834.05	3.88	1300.16	4.14
Farm mechanization	2282.79	12.92	3062.4	14.25	2383.6	7.59
Plantation and horticulture	1361.92	7.71	690.59	3.21	2033.45	6.47
PF/SGP/AH-Other	411.29	2.33	620.59	2.89	672.52	2.14
Fisheries	37.85	0.21	21.81	0.10	64.15	0.2
Dairy development	872.87	4.94	931.89	4.34	1770.63	5.63
Forestry	7.8	0.04	6.39	0.03	8.29	0.03
Storage godown and market yard	295.3	1.67	379.32	1.77	294.21	0.94
SGSY	111.72	0.63	54.96	0.26	0.81	0.01
Non-farm sector	5150.88	29.14	7806.46	36.3	11117.83	35.38
SC/ST-AP	19.35	0.11	14.70	0.07	2.47	0.07
SHG	3916.64	22.16	3745.83	17.43	4493.67	14.29
Others	1648.92	9.33	2799.32	13.03	6138.34	19.54
Total	17674.29		21486.17		31427.30	

Source: NABARD (Various issues).

Table 13: Crop-wise Area Insured under all Insurance Schemes

Crops	*Gross Area Sown (lakh ha)*	*Area Insured (lakh ha)*		
		2012-13	*2013-14*	*2014-15*
Paddy	427.6	105.0 (24.6)	93.9 (21.9)	100.5 (23.5)
Wheat	304.9	64.6 (21.2)	79.7 (26.1)	77.6 (25.5)
Coarse grains	251.5	51.2 (20.4)	44.4 (17.7)	5953.0 (23.7)
Sugarcane	54.4	3.2 (5.9)	3.1 (5.6)	1.5 (2.7)
Cotton	118.8	13.5 (11.3)	12.7 (10.7)	15.4 (12.9)
Jute and mesta	8.6	0.1 (0.6)	0.04 (0.5)	0.7 (8.2)
Oilseeds	290.9	98.4 (33.8)	112.1 (38.5)	104.6 (35.9)
Pulses	219.6	66.1 (30.1)	64.7 (29.5)	57.7 (26.3)
Vegetables	55.1	7.3 (13.3)	10.2 (18.6)	20.9 (37.9)
Fruits	37.7	5.6 (14.9)	1.3 (3.3)	15.1 (40.1)
TA	1943.9	444.0 (22.8)	427.2 (21.9)	437.6 (22.5)

Source: GoI (2014).

miseries to the farmers. There have been a number of steps to bring horticultural crops under the purview of the crop insurance programmes. The recent programme of Pradhan Mantri Fasal Bima Yojna has a number of positive features and is a boon to the horticulture sector. The result is that the almost 40 per cent of the fruits and vegetables area has been brought under the purview of crop insurance (Table 13). The Agricultural Insurance Company (AIC) implemented MNAIS and WBCIS in many districts and also developed crop insurance products for risk mitigation of various crops namely, coffee, rubber plantation, bio-fuel plants, grapes, mango and potato. Hi-tech capital intensive infrastructure like poly-houses, net houses *etc.* may be covered by a suitable insurance scheme. Orchards of perennial crops like mangoes, dates, apples, *etc.* may be covered by insurance scheme on line with coconut palm insurance scheme (CPIS).

Conclusion

Advanced horticulture is a capital intensive activity. With advancement of technology it is becoming more capital intensive and efficient. To cater the need of domestic as well as of international market and to make farming profitable business farmer has to take up these technologies. Here government plays an important role for adoption and development of these hi-tech practices by providing subsidy and price support. For a country like India where nearly 86 per cent of the farmers are in small and marginal category with limited resources role of public institutions become more crucial. Role of credit in horticulture in present era is as important as it was during green revolution. To meet world standard and production Indian horticulture has to advance itself and this can be done with support of credit and favourable price policies.

References

Agricultural Research Data Book, Indian Agricultural Statistical Research Institute, ICAR, New Delhi.

Agricultural Statistics at a Glance, 2014. Directorate of Economics and Statistics, Department of Agriculture and Cooperation, Ministry of Agriculture, Government of India, Published by Oxford University Press, New Delhi.

Annual Report. National Bank for Agriculture and Rural Development, Mumbai, India.

Annual Report. National Horticulture Board, Ministry of Agriculture, Government of India.

Chang, J., Wu, Xu, Liu, A., Wang, Y., Xu, B., Yang, W., Meyerson, A.L., Gu, B., Peng, C. and Ge, Y. 2011. Assessment of net ecosystem services of plastic greenhouse vegetable cultivation in China. *Ecological Economics*, **70**: 740–748.

Hickman, G.W. 2011. *A Review of Current Data on International Production of Vegetables in Greenhouses*. www.cuestaroble.com, p. 73.

Horticulture Statistics at a Glance, 2015. Ministry of Agriculture and Farmer's Welfare, Government of India, Published by Oxford University Press, New Delhi

Horticulture Statistics at a Glance, Horticulture Statistics Division, DAC and FW, Government of India.

ICAR Data Book, Indian Agricultural Statistical Research Institute, ICAR, New Delhi.

Indian Horticulture Database. National Horticulture Board, Ministry of Agriculture, Government of India

Singh, Narender, Diwedi, S.K. and Paljor, Elli. 1999. *Ladakh Mein Sabjion Kei Sanrakshit Kheti*. Regional Research Laboratory of DRDO, Leh.

State-wise Estimates of Output from Agriculture and Allied Activities, 2013, CSO, MoSPI.

52

National Agriculture Market: Status and Future Prospects

K.S. Srinivas

Joint Secretary (Marketing) and Agricultural Marketing Advistor to the Govt. of India, Krishi Bhavan, New Delhi – 110 001
E-mail: jsmktg_agri@gov.in

While self-sufficiency in agriculture has been a priority for the Government and several policy initiatives weave around this objective, the post-harvest management including agricultural marketing has not kept pace with the changes in economy, particularly relating to setting up of an efficient supply chain. We are, therefore, faced with new challenges in terms of finding an efficient market for the marketable surplus. There is enormous concern over several aspects of our agricultural marketing system. For one agriculture marketing has so far been administered by the states as per their Agricultural Produce Marketing and Regulation Acts. As per the provisions of the State Acts, each State is divided into several market areas, each of which is administered by a separate Agricultural Produce Marketing Committee (APMC), which imposes its own marketing regulation (including fees). Any agriculture produce, which transits from one market area to another, therefore, becomes liable to pay the market fees again to the APMC of the next market area, if traded or sometimes when presumed to be traded. Not only traders are required to obtain separate licenses for each market area, even within a market area, produce transacted anywhere outside the market yard is liable to pay market fees. The very concept of market areas is, therefore, against the free-flow of agri-commodities from one market area to another and multiple levies of *mandi* charges end up escalating the prices for the consumers without commensurate benefit to the farmer. The need to unify market both at state and national level is, therefore, clearly the requirement of time, in order to provide better price to farmers, improved supply chain, reduce wastages and create a unified national market.

National Agriculture Market (NAM) is envisaged as a pan-India electronic trading portal which seeks to network the existing APMC and other market yards to create a unified national market for agricultural commodities. NAM is a "virtual" market but it has a physical market (*mandi*) at the back end. The NAM Portal will provide a single window service for all APMC related information and services. This will include commodity arrivals and prices, buy and sell trade offers, provision to respond to trade offers, among other services. While material flow (agriculture produce) shall continue to happen through *mandis*, an online market would reduce transaction costs and information asymmetry.

Scheme Design

It was with the objective to usher in reform of the agri-marketing system and to provide farmers/ producers with access to markets across the country that the NAM was conceived and approved by the Cabinet Committee on Economic Affairs (CCEA) on 1st July, 2015 with a budget allocation of Rs. 200 crores. The scheme envisages implementation of the NAM by setting up of an appropriate common e-market platform that would be deployable in selected 585 regulated wholesale markets in states/UTs desirous of joining the e-platform. Small Farmers Agribusiness Consortium (SFAC) will implement the national e-platform and will cover 400 *mandis* upto March 2017 and remaining 185 upto March 2018. Department of Agriculture, Cooperation and Farmers Welfare (DAC and FW) will meet expenses on software and its customisation for the States and provide it free of cost. DAC and FW will also give grant as one time fixed cost subject to the ceiling of Rs. 30.00 lakhs per *Mandi* (other than to the private *mandis*) for related equipment/infrastructure in 585 regulated *mandis*, for installation of the e-market platform. State Governments will suggest names of APMCs, where this project would be initiated.

National Agriculture Market has the following components:

- ✰ A national e-market platform for transparent sale transactions and price discovery initially in regulated markets and to be followed in *kisan mandis*, warehouses and private markets. Willing states to accordingly enact provision for e-trading in their APMC Act.
- ✰ Liberal licensing of traders/buyers and commission agents by state authorities without any pre-condition of physical presence or possession of shop/premises in the market yard.
- ✰ One license for a trader valid across all markets in the State.
- ✰ Harmonisation of quality standards of agricultural produce and provision for assaying (quality testing) infrastructure in every market to enable informed bidding by buyers.
- ✰ Single point levy of market fees, *i.e.* on the first wholesale purchase from the farmer.

Ministry of Agriculture, Department of Agriculture, Cooperation and Farmers Welfare (DAC and FW) have mandated Small Farmers' Agribusiness Consortium (SFAC) to act as the Lead Promoter of NAM. SFAC will operate the NAM with technical support from a Strategic Partner (SP) selected for a period of five years to develop, operate and maintain the NAM platform (Table 1).

M/s Nagarjuna Fertilizers and Chemicals Ltd. in consortium with Techno Brain Global FZE has been selected as the Strategic Partner (SP). The broad role of the Strategic Partner is comprehensive and includes

- ✰ Making available the NAM software to the SFAC and its customization according to the Legal framework of states willing to integrate with NAM.
- ✰ Customizing the software to meet the specific requirements of the *mandis* in the states willing to integrate with NAM.
- ✰ Hosting the software in a data centre (NIC), including operation and maintenance of associated hardware.
- ✰ Maintaining the software application and continuously developing the same to cater to the requirements that may arise from time to time as per the provisions of the contract.
- ✰ Training all market participants (like commission agents, traders, staff of the implementing agency, *etc.*) in using the NAM software for participating in the market.
- ✰ Migrating operations of selected markets to the NAM.
- ✰ Information dissemination to all stakeholders, for efficient price discovery.
- ✰ Marketing and promotional activities for ensuring acceptability and increasing usages (transactions).

As per the Activity-wise Timeline for roll out of NAM project (Annexure I) the software solution will be initially rolled out in 585 selected *mandis* in States ready with the necessary reforms for NAM. Of these, 200 *mandis* are to be rolled out by September 30, 2016 and next 200 *mandis* shall be rolled out by March 31, 2017. Remaining 185 *mandis* shall be rolled out by March 31, 2018. The *mandis* would be identified by the States.

The approved schedule for provision of budgetary grant support for this initiative by DAC and FW is as follows:

Table 1: Budgetary Support for Establishing NAM

Implementation Period	*Physical Target (Nos.) of Mandis*	*Government of India Grant (Rs. in Crores) @ Rs. 30 lakh approx. per Mandi*
2015-2016	250	75.00
2016-2017	200	60.00
2017-2018	135	40.00
Grand Total	**585**	**175.00**

The number of *Mandis* may increase depending upon the demand of the states. In addition to the above, funds have been provided for writing the software for common e-platform and its customization to meet States' needs, its annual maintenance, data centres, servers, training costs and administrative expenses.

Benefits of NAM

The Central Government has embarked on the idea of creating a unified market through an online platform. NAM would offer a common market place by providing real-time prices on a national level to the participants of the agri-produce eco-system. It would create an efficient trading system for agri-produce, which would enable transaction between buyers and sellers from their existing location, thus expanding the existing market and also facilitating transactions in places where markets physically do not exist, over a period of time.

A new distribution channel for procurement and sale will come into existence, which would result in better price discovery for Agri-Produce, largely benefiting the farmers. The prices reflected would be determined by demand and supply prevailing from time to time and shall be captured on real time basis.

Once this system is implemented, it will provide a number of benefits to various stakeholders of the system.

- For farmers NAM promises more options for selling their produce and competitive returns.
- For local traders, NAM will provide access to larger national market for secondary trading.
- For bulk buyers, processer, exporters, NAM will enable direct participation in the local *mandi* trade reducing intermediation cost.
- Stable prices and availability to consumer.
- For *mandis*, the benefits will be in terms of the followings:
  - Reduction in book keeping and reporting system – This system eliminates the need of preparing the number of reports (*viz.* daily minimum, maximum and modal prices and arrivals of commodities) at APMC. The statements would be generated automatically.
  - Monitoring and regulation of traders and commission agents – The APMC staff officers can easily monitor the activities of the traders and commission agents and regulate more effectively.
  - Transparency – It will be completely transparent system and will not provide any scope of intentional/un-intentional manipulation of tendering/auctioning process.
  - Improvement in market fee collection – It helps the APMC level officers in improving the market fee collection by means of accounting all the transactions that are taking place in the market.
  - Reduction in manpower requirements – As tendering/auctioning process takes place through the system, there is no need to have the manpower requirements in bringing, opening the tender box, processing of the quotes and declaring the tender results.
  - Faster tender declaration - Presently, tender declaration runs till late in night as there are thousands of LOTs. This system helps to declare the tenders for the LOTs within few seconds even for thousands of LOTs in the market.
  - Analysis and forecasting of the arrivals and prices – This helps to analyze the incoming and outgoing commodities, prices, arrivals and quality of the commodities and generate the useful reports for decision-making and forecasting.
  - APMC progress on the web-site – It will help to post the activities of each APMC on the web-site directly.

Progress of Implementation

As far as integration of markets across States is concerned, the States will undertake prior reforms in their APMC Acts on 3 items, *viz.* (i) a single trade license to be valid across the State, (ii) single point levy of market fee and (iii) provision for electronic auction as a mode for price discovery. The status of requisite reforms is at Annexure II.

Table 2: Status of in Principle Approved Proposals

State/UT	*No. of Mandis*	*Proposal Cost (Rs. in crore)*	*Maximum Admissible Assistance (Rs. in crore)*	*Funding from State/UT Own Resources (Rs. in crore)*	*Sanction/Release Status*
Gujarat	40	27.86	12.00	15.86	Released Rs.12.00 crore on 20.11.15.
Maharashtra	30	16.74	9.0	7.74	Approved in principle on 18.11.15. Certain inputs awaited to enable release.
Telangana	44	13.264	12.165	1.099	Released Rs. 12.165 crore on 30.12.2015.
Jharkhand	19	5.92	5.70	0.22	Released Rs. 5.70 crore on 18.01.16
Chhattisgarh	05	2.58	01.50	1.08	Released Rs. 1.50 crore on 09.03.16
Madhya Pradesh	50	35.21	'15.00	20.21	Approved in principle on 31.12.2015.
Rajasthan	25	18.80	7.50	11.30	Released Rs. 7.50 crore on 05.02.16
Chandigarh	01	2.01	0.30	1.71	Approved in principle on 31.12.2015
Haryana	54	38.62	16.20	22.42	Approved in principle on 19.02.2016.
Himachal Pradesh	19	5.70	5.70	5.70	Approved in principle on 19.02.2016
Uttar Pradesh	66	32.94	19.80	13.149	Approved in principle on 19.02.2016
Andhra Pradesh	12	3.88	3.60	0.283	Recast proposal Approved in principle on 19.02.2016

A Project Appraisal Committee (PAC) has been set up in DAC and FW under the Chairmanship of Secretary (AC and FW) to examine the projects from the states and sanction grant of upto Rs.30.00 lakhs per *mandi*. So far, projects from the following States have been appraised (Table 2).

Simultaneously to facilitate assaying of commodities for trading on NAM, common tradable parameters have been developed for 22 commodities. These are at Annexure III.

The NAM is to be lauched shortly on 14th of April, 2016 in 20 markets in 8 states. The detail list is at Annexure IV.

Annexure I: Activity-wise Timeline for Roll Out of NAM project as on 18.03.2016

Description of Deliverables	*Date of Completion of Work*	*Status*
Selection of Transaction Advisor	03.08.2015	Completed
Signing of agreement with Strategic Partner	23.12.2015	Completed
Submission of draft SRS, FRS and PDD	22.02.2016	Completed
Development of electronic trading portal	30.03.2016	
Hosting of NAM portal on NIC cloud	By 01.04.2016	
Actual Testing of Software	1st week of April, 2016	
Fixing of Bugs in the Software	By 10.04.2016	
Mock and Actual Testing in the States	11.04.2016	
Launch of NAM in 20 pilot *mandis*	14.04.2016	
Roll out of software in next 180 *mandis*	30.09.2016	
Roll out of software in next 200 *mandis*	31.03.2017	
Roll out of software in remaining 185 *mandis*	**31.03.2018**	

Annexure II: Status of Agricultural Marketing Reforms (as on 18.03.2016) in Respect of

(a) a single license to be valid across the State,

(b) single point levy of market fee and

(c) provision for electronic auction as a mode for price discovery.

Sl.No.	*State*	*E-trading*	*Single Point Levy (SPL) of Market Fee*	*Unified Single License (USL)*	*Remarks*
1.	Andhra Pradesh	Yes	Yes	Yes	—
2.	Arunachal Pradesh	No	No	No	—
3.	Assam	No	No	No	—
4.	Chhattisgarh	Only e-auction is provided	Yes	Yes	Under Section 36(3)(ii) only e-auction is provided, which is sufficient for NAM State has not yet notified detailed rules thereunder.
5.	Delhi	No	No	No	—
6.	Goa	Yes	Yes	Yes	—
7.	Gujarat	Yes	Yes	Yes	—
8.	Haryana	Yes	No	Yes	State needs to notify the rules for e-trading and USL and also make for provision for SPL in the APMC Act and simultaneously notify rules.
9.	Himachal Pradesh	Yes	Yes	Yes	While under Section 39 (3)(iii)- Powers and Functions of APMC, e-trading is provided, however there is a need for detailed rules
10.	Jharkhand	Yes	Yes	No	Rule No. 146 clearly provides for Private e-trading referring Section-5 (Private market) of the APMC Act, while e-trading has not appeared in the Act. However, provision is sufficient. Provisions for USL is not provided.
11.	Karnataka	Yes	Yes	Yes	—
12.	Madhya Pradesh	Yes	Yes	Yes	Under Section 32 (A) read with Section 79, detailed rules are notified in 2009 for issue of Special License to traders desiring to trade in more than one market area by establishing purchase centres and commodity exchanges for e-trading

Contd...

Annexure II–*Contd...*

Sl.No.	*State*	*E-trading*	*Single Point Levy (SPL) of Market Fee*	*Unified Single License (USL)*	*Remarks*
13.	Maharashtra	No	No	Yes	Though no specific provision for e-trading exists, however, State has issued licenses to commodity exchanges under direct marketing clause. Further, SPL for general trading does not exist while for specific purposes (export and processing), it exists.
14.	Meghalaya	No	No	No	—
15.	Mizoram	Yes	Yes	Yes	Since under Section 26 (5)- Powers and Functions of APMC, e-trading is provided, therefore, there is a need for detailed rules. Further, rules for all amended provisions need to be notified.
16.	Nagaland	No	Yes	Yes	Rules for all amended provisions need to be notified.
17.	Odisha	No	No	No	Licenses to commodity exchanges have been issued under private market clause. State needs to make detailed rules for e-trading. Further, State needs to make provision for SPL and USL also.
18.	Punjab	No	Yes	No	—
19.	Rajasthan	Yes	Yes	Yes	—
20.	Tamil Nadu	No	No	No	Provisions for e-trading, SPL and USL need to be made and simultaneously rules be notified.
21.	Telangana	Yes	Yes	Yes	—
22.	Tripura	No	No	No	Provisions for e-trading, SPL and USL need to be made and simultaneously rules be notified.
23.	Uttarakhand	Yes	Yes	NO	While e-trading is provided under Section 26 (2)(ix)- Powers and Functions of APMC, however, detailed rules need to be notified. Further, rules for all amended provisions need to be notified. Provision for USL in the APMC Act does not exist.
24.	Uttar Pradesh	Yes (?)	Yes	Yes (?)	E- Trading is mentioned in " Definition" of the Act only but has not appeared in body of the Act. Single unified license is for primary trade not for secondary trade. Rules are also not notified.
25.	West Bengal	No	No	No	—
26.	Chandigarh	No	Yes	No	—
27.	Andaman and Nicobar Islands				No AMPC Act
28.	Dadar and Nagar Haveli				NO APMC Act
29.	Daman and Diu				NO APMC Act
30.	Lakshdweep				NO APMC Act
31.	Puducherry				APMC Act not implemented
32.	Bihar				APMC Act repealed in 2006.
33.	Jammu and Kashmir				APMC Act not implemented.
34.	Kerala				No APMC Act.
35.	Manipur				No APMC Act.
36.	Sikkim				APMC Act not implemented

Annexure III: List of Commodities to be Tradee in NAM

Sl.No.	*Commodity*	*Sl.No.*	*Commodity*	*Sl.No.*	*Commodity*
1.	Wheat	2.	Maize	3.	Chana
4.	Castor seed	5.	Jwar	6.	Bajra
7.	Barley	8.	Groundnut	9.	Soyaean
10.	Mustard seed	11.	Paddy	12	*Arhar*
13.	Urad	14.	Masoor	15.	*Moong*
16.	Red chilli	17.	Cumin	18	Cotton
19.	Potato	20.	Onion	21.	Apple
22.	Turmeric	23.	Shelling peas	24.	Tamarind

Annexure IV: State-wise List of *Mandis* for Pilot Launch

Sl.No.	*State*	*APMC*	*District*	*Commodity Proposed*
1.	Gujarat	Patan*	Patan*	Castor Seed
2.		Botad	Bhavnagar	*Chana* (Black gram)
3.		Himmatnagar	Sabarkantha	Wheat
4.	Telangana	Tirumalgiri	Nalgonda	Paddy
5.		Nizamabad*	Nizamabad*	Turmeric
6.		Badepally	Mahboobnagar	Maize
7.		Hyderabad	Hyderabad	Onion
8.		Warangal	Warangal	Maize
9.	Rajasthan	Ramganj *mandi**	Kota*	*Chana*(Black gram)
10	Madhya Pradesh	Karond, Bhopal*	Bhopal*	*Chana*(Black gram)
11.	Uttar Pradesh	Sultanpur*	Sultanpur*	Wheat
12.		Lakhimpur	Lakhimpur Kheri	Wheat
13.		Lalitpur	Lalitpur	Wheat
14.		Bahraich	Bahraich	Wheat
15.		Saharanpur	Saharanpur	Wheat
16.		Mathura	Mathura	Wheat
17.	Haryana	Ellanabad	Sirsa	Mustard
18.		Karnal*	Karnal*	Wheat
19.	Jharkhand	Pandra, Ranchi	Ranchi	Tamarind, *Mahua* flower
20.	Himachal Pradesh	Fruit Mandi, Solan	Solan	Shelling Peas
21.		Dhalli, Shimla	Shimla	Shelling Peas

* 6 *Mandis* will have live interactions with Hon'ble Prime Minister.

53

Re-Orienting Horticultural Extension Services for Doubling Farmers Income

V.V. Sadamate

Former Adviser (Agriculture),
Niti Aayog (Formerly Planning Commission), GoI, Sansad Marg, New Delhi – 110 001
E-mail: sadamatevv@gmail.com

1. Agricultural and Horticultural Sector Growth Scenario

1.1 Agricultural Growth

India with 1.25 billion people can boast of a robust Agriculture sector that accounts for 14 per cent of GDP and having a share of about 11 per cent in total exports. Agriculture is the principal source of livelihood for more than 58 per cent of the population providing employment to 57.4 per cent of work force in rural areas. It remains the main source of raw materials for a large number of industries and thus fundamental to the economy. Agricultural growth in the XIth plan has accelerated by 3.3 per cent compared to 2.4 per cent in the Xth Plan. However, its share in overall GDP has declined from 23.4 per cent in IXth Plan to 19 per cent in the Xth and to 14.2 per cent at the end of XIth Plan without commensurate decrease of number of people dependent on agriculture. The crop sector has recorded an average growth of about 1.8 to 2 per cent, recording increase in farm diversification from cereals. However, allied sectors such as Horticulture, Animal Husbandry and Fisheries have registered output growth of 4.5 to 6 per cent. State wise growth trends (CSO) showed that despite usual explanations for low growth like changing climate, soil degradation, stress on water resources, technology slowdown and policy constraints, a few States seemed to have done exceedingly well, recording average annual growth of above 4 per cent during the XIth Plan, though there are medium and slow performers too. Rapid growth of agriculture is critical to reduction of poverty and hunger and inclusiveness. Therefore, enhancing productivity, resource use efficiency, technology and innovations, sustainability and total factor productivity require serious attention. Other important issues like investments, market infrastructure, price incentives, convergence and programme delivery, agricultural diversification from cereals towards high value agricultural commodities requite adequate focus to achieve targeted agricultural growth of 4 per cent for the XII th Plan period.

1.2 Crops[1]/Horticulture[2]

Horticulture accounts for about 30 per cent of India's agricultural GDP from 13.08 per cent of cropped area. It also provides about 37 per cent of the total exports of agricultural commodities. India produces nearly 11 per cent of the vegetables and 15 per cent of the fruits of global output, yet its share in global exports of vegetables is only 1.7 per cent and the fruits of meagre 0.5 per cent. Thus, there is a vast scope for Indian Horticulture sector in WTO regime.

2. Agriculture and Allied Sectors: Current Concerns

Agriculture sector faces formidable challenges of shrinking land and water, adverse impact of climate change, skewed development, crisis in farm labour, stagnant productivity, increasing production costs, dwindling natural resources, market uncertainties and weak technology delivery systems across states and sectors. The major concerns revolve around the following.

2.1 Shrinking Land Holdings and Natural Resource Base[3]

Predominance of small holders in Indian agriculture continues to rise. Average size of land in India has reduced to 1.30 hectare per farm holding[4]. Small and marginal farmers now comprise of 83 per cent of total land holdings and they occupy about 41 per cent of area under all operational holdings[5]. Natural Resource base that sustains farming is under increasingly serious strain. Land, water, soil health and biodiversity resources are shrinking and deteriorating. Due to overexploitation, groundwater levels are going down even in irrigated areas in several parts of the country. Continuing imbalance in use of fertilizer has adversely affected soil health. Thus, long term sustainability requires that land and water are used judiciously and efficiently.

2.2 Regional Imbalance[6]

Present level of productivity of most of the crops and other sectors is significantly low as compared to what is attainable and also are being attained by some other countries. This production potential is believed to be very large, for example, in the eastern and central regions. Regional imbalances in growth and productivity are surprisingly high in rainfed conditions. Dryland areas that are home to a large section of marginalized areas and communities continue to be in disadvantageous position in terms of desired development interventions. Recent focus on agricultural investment and infrastructure in some of the backward states and regions have shown significant promise in improving growth performance.

2.3 Sustaining Food and Nutritional Security

India ranks at 67th as per Global Hunger Index (IFPRI) amongst 171 nations. Half of the children (51 per cent) are malnourished, over 55 per cent married women suffer from anaemia and 30 per cent adults suffer from protein –energy malnutrition. Food availability to BPL families and supply and distribution apparatus in disadvantaged areas along with food quality standards are basic concerns. Access to food and affordability of the impoverished families to harness the social entitlement are also issues that need greater attention to expand the income and livelihood options for such category of people. Horticulture, milk, poultry and nutri-cereals may provide an excellent opportunity on nutritional improvement; however, the awareness and reach are still limited. Malnutrition particularly with reference to essential minerals is a major problem, which can be addressed by inclusion of coarse cereals and millets in the NFSM and PDS, food fortification and propagation of nutrient-dense food. Farm women could play an important role in managing and sustaining house hold food security and extension services need that kind of orientation.

2.4 Rainfed Farming and its Sustainability[7] and Agricultural Distress

Nearly 60 per cent cultivable area would continue to be rain fed - typically characterized by vulnerable natural resource base, inhabiting 40 per cent population, giving 40 per cent food grain output and sustaining about 60 per cent of livestock. Watershed has been taken up as a key intervention for development of such areas. As against the target of 36.6 million ha the area covered under watershed is about 19 million ha so far, that indicates greater scope for resource conservation through WSD and thereby improving the productivity of dryland agriculture. Climate change and adaptation are dominant issues in farm sustainability. Better convergence and programme delivery below of the sub-district levels are urgently called for. Recent innovative programmes like NICRA and NMSA have demonstrated good potential that need to be mainstreamed into the future strategies. Technology transfer models adopted so far have not been able to make any perceptible dent in rain fed areas. It calls for alternative thinking in policy framework and delivery mechanism involving farm women.

The agricultural distress reflected in some parts of the country needs to be addressed by alternate strategies as the normal programmatic formulations are yet to impact there. The focus would need to be on the conservation technologies, alternate crops and cropping patterns, crop diversification and alternate livelihood options and training of farm women in alternate livelihoods.

2.5 Convergence with other GoI Programmes

Agriculture development programmes and related programmes of other departments like rural development, Panchayati Raj, Social Development *etc.* generally operate in vertical silos, with very little or no horizontal convergence, especially at the block level and below. Hence, field level convergence is one of the major challenges existing today across the departments. The major task therefore would be to reorient the district and the block functionaries for the task and provide them opportunities for experience sharing across the departmental boundaries so as to obtain required coherence. The farmers and the farm women need required support for implementing programmes on convergence mode, especially below of the blocks and at the cluster level.

2.6 Re-orientating Farm Extension Services for Women in Agriculture[8]

Foregoing text has adequately focused on growth

and development of agriculture and allied sectors wherein farm women have been playing important role practically in all the field operations. Also, the challenges in the system have also been pointed out. Serious attention needs to improve the access of farm women to the extension services, both public and private.

Women's participation in agriculture has been going up due to a variety of socio-economic factors. Therefore, there is need for intensive capacity building and empowerment efforts for women and women groups at every level that includes re-orienting farm services to improve their access to credit, technology, inputs, markets and gender friendly mechanization. It may call for feminization of farm extension services enabling effective women extension workers for better participation and extension delivery. Farm women including women from weaker sections need to be given adequate representation in various programme management fora at all the levels. The ongoing programmes and schemes should ensure greater gender coverage through improved strategies, financial allocation and well defined extension strategies.

3. Challenges in Agricultural/Horticulture Extension and Solutions

Present Extension Models and their Performance

Several Extension models are in operation for dissemination of information such as ATMA Model, KVK model, PPP model, NGO model, Extension services provided by input support services providers, Extension services extended by co-operatives and Farmer Organizations, ICT Extension Modes, *etc.* There is need to improve performance of extension models in operation to their optimum levels with focus on R&D Linkages, especially research backstopping, widening the sectoral and area coverage. Resource sharing, training coordination, convergence and programme delivery, improving penetration to the small and marginal producers would also need to be addressed by the respective models in their area of jurisdiction on supplementary and complementary mode.

Good or bad there is Agricultural Extension System exists for cops sector. However, in allied areas like horticulture, animal husbandry, dairy, poultry, fisheries etc, its either weak or missing, though the allied sectors are contributing significantly in agricultural growth. Hence, Extension needs to be re-oriented thoroughly (additional support from KVKs/ATMAs), promoting Grower's Associations(in horticultural sector), use of milk cooperatives, fisher co-operatives, use of entrepreneurs, progressive farmers, *etc.* could play an important role to cater to the extension needs of allied sectors.

Empowerment of Farmers

Farmers could be empowered by various methods like organizing them into SHGs, FOs, FPCs, Coops, etc, by providing training and capacity building and by making them partners in decision making and by providing role space in extension operations. There is need for re-orienting extension services to address extension needs of farm women and farm youths. It has been observed that the Farm Youths are taking interest in secondary agriculture (processing, value-addition etc) and the extension may need to address this dimension too.

Future Extension Strategies must focus on organizing producer aggregates at various levels. This would provide strong backward and forward linkages including market led extension strategies for enhanced income opportunities for the farmers. Such Farmer Aggregates would also provide adequate pressure on the R&D agencies making extension operations demand driven and participatory.

Farmer to Farmer Technology Dissemination

This is one of the most effective methods of technology dissemination wherein the innovators and progressive farmers share the technological information with the fellow farmers; however, they need to be oriented on their roles and responsibilities. The Farmers Field Schools are being organized on the fields of such farmers where they become the host and the teacher for the technology in which they have already excelled or have had an unique experience and achievements. Now, the issue is how to promote, encourage and fund it under various programmes? One of the methods tried by the ATMAs and KVKs was to involve such farmer's innovations for validation by the research system and using these innovators as training resources and disseminators. FOs and FFSs are few such successful experiments wherein innovative farmers have been used to promote farmer to farmer extension.

There is a strong need for promoting informal sources of technology delivery like farmer friends, model farmers, contact farmers, Kisan Sahayyaks(farmer facilitators), *etc.* across the states/sectors, to enable faster technology transfer at the cutting edge levels. There are different models promoting farmer facilitators at the community levels. NGOs like Ramakrishna Mission have trained young boys and girls from the rural areas, built their capacities and used them as local farm volunteers. Every three months these young volunteers are brought back to the local KVKs for training and retraining. The Department of Agriculture, Andhra Pradesh have selected local practicing farmers as Model farmers (Adarsh Raytu) for technology transfer. Kisan Sahayyaks (Farmer facilitators) are used in the States of Uttar Pradesh, Bihar *etc.* In fisheries and in the animal husbandry sectors the states like West Bengal are promoting Matsya Mitras (Fish friends) and Prani Bandhus (Animal brothers).

These facilitators would need to be properly selected and trained for the specific tasks in horticulture sector. They should be innovative, practicing farmers and should have acceptability of the community. So the selection process would need to be far fairer in selecting the farm facilitators. He/she needs to be oriented to handle community level extension advisories and therefore need periodic updating of their knowledge and skills. The field extension agencies like ATMAs, KVKs, NGOs, farmer organizations would need to promote the Farm Facilitators and use them at community levels. This sort of intervention is necessary especially if extension has to be location specific at the field, cluster and below of the block.

The states could take a few pilots in selected blocks and promote farmer facilitators (FFs) and demonstrate how effectively they could take on extension and technology transfer functions at the community level. This would largely fill up the extension gaps at the community level especially in the remote areas, hill and mountain eco-systems and tribal areas where the public extension services are practically non- existing or remained very weak.

Farmer Facilitators (FFs) is an effective extension tool, practiced widely, needs to be up-scaled as it provides an opportunity for farmer to farmer extension which has been found to be quite effective. The host farmer needs to be properly oriented about his roles and responsibilities and the technology that is being demonstrated on his field. The fellow farmers would have to have a faith and access to the host farmer, his competence and results that are being obtained on the farmers field schools.FFS differs from demonstration since it has more emphasis on experimentation, science based learning, sharing of knowledge and skills in adoption of technologies. Up-scaling of FFSs is a challenging task.

Management of Field Extension Activities

Field formations, by themselves are not able to penetrate down the line, especially to the small and the marginal farmers, and across the sectors and areas. This has resulted due mainly to inadequate numbers, their quality and limited outreach. Therefore, it would be better if extension services promote chains of extension agents, formal/informal for enhanced and appropriate coverage. Change agents in extension chains may include progressive farmers, entrepreneurs, the FOs, FPCs, Cooperatives, local self government agents, *etc.*

Field formations of various extension streams would need to supplement and complement in selecting programme sites, identification of beneficiaries, joint diagnosis, problem solving, information support, training and capacity building of FFs, BTMs, ABTMs. Sharing of training resources, providing mentoring support to the block level formations (*e.g.*, RSKs of Karnataka and Krishi Bhavan's of Kerala) in the specialized areas, jointly promoting farmer to farmer extension *etc.* may also be thought of. As above is far more applicable in case of KVK and ATMA in operations.

Flexibilities are required to be promoted to the field formations to respond to the emerging local extension challenges. Thus, making extension operations accountable to the local situations and farming communities and making extension to address allied sector extension needs effectively. Operational flexibilities may facilitate entering local MOU based PPPs, which are essential to make the extension interventions effective at the local level.

Programmatic convergence down the line is essential for wider coverage to benefit from each other's strengths. Policies and instructions are necessary to drive it. Convergence appears to be most effective at the Block/Cluster level. Matrix mode/MOU approach is recommended to make it happen at the field level. Likewise, the programme delivery is crucial in disadvantaged areas, far flung areas, tribal areas, hill and mountainous areas, desert areas, *etc.* Hence, Extension Strategies need to be re-oriented and strengthened accordingly for convergence and programme delivery.

There are series of district level planning instruments available as follows:

- ✰ Comprehensive District Agricultural Plans (C-DAPs) under RKVY,
- ✰ Strategic Research and Extension Plans (SREPs) under ATMA
- ✰ District Profiles (DPs) prepared under KVK Programme
- ✰ Detailed Project Reports (DPRs) developed under Watershed Projects
- ✰ Potential Liked Credit Plans (PLPs) developed by NABARD for the districts
- ✰ District Irrigation Plans (DIPs) developed under PMKSY.

The District/Block extension functionaries may scan these instruments for capturing the horticultural extension priorities and integrating in the convergence and PD strategies.

Farmer distress and challenged areas need clear extension strategies/advisories for climate adaptation strategies. It would need clear technology options along with credit - market linkages. Appropriate enterprise combinations would need to be worked out for the blocks, clusters and the micro level agro-situations for such areas to be promoted by the extension operations. This would need re-orientation of the extension machinery and improved programme delivery. Similarly, the tribal areas, the hill and mountain areas, the rainfed areas, the desert areas and the border areas would perhaps need regionally differentiated extension strategies.

Performance standards need to be set for Extension Performance, in a given cluster, block and district. It would make the extension services accountable to the local area and farming communities, enabling extension services to respond to the emerging extension needs. Such performance benchmarks would need to be set for agriculture and allied areas too.

Technology Back Stopping and R&D Linkages

Enhanced research backstopping to the field formations is required so as to facilitate strong technological input in the extension processes. The scientists interface with Field Functionaries and with the farmers would need to be augmented. Technology options would need to be generated for the micro level farm situations. Also, validation of farmers own innovations needs to be prioritized and mainstreaming them. Farmer Entrepreneurs/innovators need to be promoted in extending extension services by using them in training and CB activities.

Feedback collation from the farmers, field functionaries is generally weak in the extension system. So also, its systematic documentation and analysis. Systematic documentation and analysis would not only improve the relevance of the research, but also, it would make the extension operations flexible and far more relevant/productive suiting to the various categories of farmers.

Technology options for agro-climatic zones and sub-zones needs to be systematically worked out and integrated into programme delivery mechanism, as per specificity of each micro agro- eco situation. Socio-economic dimensions and technology options would need to be matched and addressed in extension operations accordingly, would bring in enhanced focus in extension processes as opposed to mere TOT approach.

Field extension functionaries from agriculture and line departments would need to be adequately oriented in promotion of crucial technologies like soil health management, IPM, INM, microbial applications, climate change, especially adaptation strategies, farm mechanization, secondary agriculture, IUE and WUE, project based extension at various levels, *etc.* The approach would benefit the farmers in judicious input use enhancing farm incomes.

There are various R&D linkage mechanisms available at various levels as follows:

- ✰ National : Pre-*kharif* and pre-*rabi* interfaces between DAC and FW and ICAR
- ✰ Regional : ICAR Regional Committee Meetings
- ✰ State level : Pre-Seasonal Meetings, Agro - Climate Zone Wise
- ✰ District Level: SAC of KVKs, GB of ATMA and District level ATMA FACs
- ✰ Block level : Block Level FACs of ATMAs

Experiences in various States indicate that the performance of these forums need lots of improvement so that they function efficiently. It is also necessary that both research and development sides participate pro-actively in improving R&D agenda, making these forums dynamic and responding to the needs of horticultural farmers as well. ATMA was proposed to be a platform for broad based and intergraded extension operations, the states and the districts would need to think on this proposition seriously.

Training and Capacity Building

Farmers training responsibility down the line in agricultural and allied sectors is shared by the large number of agencies and organizations. For example at the district level the farmers are trained by the agencies like KVKs, ATMAs, SAU Outposts, FTCs, Input Support Agencies, NGOs, *etc.* Can we think of focussed and segregated training responsibility? Could Director Extension, ATARI and SAMETI take a call and address this challenge? MANAGE and Extension Division of ICAR may take a overview and guide the process. It would streamline the whole training function

It is estimated that over 1.25 lakh field functionaries are operating for the field extension work in agriculture and allied departments in all the Sates. This would include about 7,000 block level officers, about 700 District Agri/Horticulture Officers, about 21,000 SMS (3 per block). Further, it is assumed that there would be about 3.5 lakh Farmer Friends (one for two villages) are engaged in the field extension services. The training and capacity building of such a huge manpower is a gigantic task which could be taken up by the SAUs and ICAR institutes in a systematic way. This issue needs to be focussed adequately. The scientific institutions like SAUs and ICAR institutes have a great scope for updating technical competence of the SMSs and middle level functionaries. Modalities may have to be worked out jointly by the ATARIs and SAMETIs under the guidance of Extension Division of ICAR and MANAGE. Similarly, the KVKs and FTCs of the State Departments of Agriculture could take on the training and capacity building task of the huge number of Farmer Friends.

Gender Issues

Farm women play major roles in farming and their presence is is increasing in all the farm operations (agriculture, horticulture, animal husbandry, dairy, poultry, fisheries, sericulture, *etc.*). Male migration to towns and cities is one of the main reasons. Further, the extension services are manned by male workers, barring a few States and the development services are not reaching the women farmers at the desired level. Hence there is need to orient these services in favour of favour of women farmers training and capacity building and customized training and capacity building modules need to be operated as per the needs of the farm women at a micro level. Farm women need to be empowered by various means like grouping them into SHGs, Farm Women Groups, Farm Women Organization, *etc.* Also, there is need to provide representation to the farm women on various forums.

PPPs in Horticultural Extension

PPPs are required down the line to promote participatory extension arrangements between the public and private extension service providers. It is expected that large number of MoU based PPPs must come up at various levels to make the extension services respond to the local situations. However, PPP operations need policy directions (otherwise provisions do not get implemented as it has been seen in ATMA). ATMAs can be a good platform for working out operational flexibilities and resource sharing, keeping in view large number of government, private and NGO extension players in the field. The input support agencies, the companies and corporate, the NGOs, the Agri Clinics and Agri Business Centres (ACABCs) *etc.*, can also play an important role in promoting and operating PPPs in Extension in Agri/Horti and allied areas.

Farm Entrepreneurs could play a major role in agricultural extension. It has been observed through various studies that the. He being highly motivated person could take on activities like promotion of aggregated models, providing better forward linkages, driving market led extension strategies, better price realization, *etc.*

Promote ICTs and Farmer Knowledge Groups

ICT has lots of scope in future extension strategies, especially at the awareness and interest stages of adoption process. However, content development and delivery in respect of print and electronic media is crucial, Kisan Call Centre (KCC) : 1800-180-1551 has becoming very popular day by day, however, the pertinent question is how to increase its access and reach to the large number of farmers. Mobile applications in extension need to be expanded (*e.g.* Agri Apps Bengaluru).Internet access needs to be improved by organizing Farmer Knowledge Groups (FKGs) in the villages facilitating/improving arrangements for knowledge interpretation and knowledge sharing.

Special Focus Areas in Future Extension Strategies

Future extension strategies must address areas having high yield potentials but performing at lower levels. Extension strategies must follow sequential extension process equilibrium as: Technology Validation (TV), working out technology Options (TOs), Technology Dissemination, Participatory Processes and obtaining feedback for further improvement. There is need to organize large number of FFSs/FOs and FPCs (may be one per Block) in potential areas. Extension functionaries must identify potential areas for export opportunities (for progressive farmers) and involve processing and export industries in such assessments. Promotion of forward linkages through appropriate extension interventions and enhancing outreach to small and marginal farmers would be essential. Hence, focus on social dynamics would be required in future strategies.

The future extension strategies must draw strengths from international extension models and private sector experiences(learning from companies, corporate sectors, multi-nationals, Non- Government Agencies: Entrepreneurs, Farmer Professionals and Para professionals and Outreach Programmes of Research Organizations). This would enable extension contents and delivery move on improved and dynamic mode as per need of the situation.

In broad terms, farm journalism is the science of conceptualizing, developing and operationalizing information activities through various media supportive to agricultural extension. Farm journalism is an integral part of the agricultural extension system and growth of farm journalism is directly proportional to the performance of extension processes, coverage and impact/adoption. Access to information and improved communication is a crucial requirement for sustainable agricultural development. Modern communication technologies when applied to conditions at grassroots level can help improve communication, increased participation and disseminate information and share knowledge and skills. However, it is observed that the rural population, especially farm women still have difficulty in accessing crucial information. The challenge is not only to improve the accessibility of information and communication technologies but also to make it available to the fellow women farmers.

Media Categories

Information support to Technology Dissemination in agriculture and allied sectors broadly falls in three categories namely print media, the electronic media and the other media. Each having large scope to influence women farmers by depicting success stories, entrepreneurial experiences, *etc.*

Media alternatives given below would need to be considered in future extension strategies.

Social media is a powerful communication tool that widens lines of communication, engages a large number of people quickly and has a potential to reach out farmers/farm women and stakeholders. Face book and Twitter are two well known social media sites whose usage in agriculture field would witness radical changes in near future.

Emails have come up as fast moving communication channel across the globe. Its use in scientific communications is growing manifold. Young agri-preneurs and the farmers have started using internet information for accessing technologies and credit-market information. This medium has a great scope in high tech agriculture development.

Farm portals are also available for dissemination of information, e-commerce and distance learning. Information given includes crop production and protection technologies, inputs, prices, e-commerce, *etc.* Portals vary widely in their contents, updates, user friendliness and use of visuals.

Kisan Call Centres (KCC)

the Ministry of Agriculture, Government of India took a new initiative by launching the scheme Kisan Call Centres in January,2004, aimed at answering farmers queries on telephone calls in farmers own dialect. These call centres are now working at 14 locations covering all the States/locations. A country wide common eleven digit Toll free number 1800-180- 1551 has been allotted for Kisan Call Centre. This number is accessible through all mobile and land line telephone networks including private service providers. Replies to the farmers' queries are given in 22 local languages and the calls are attended from 6AM-10 PM on all the seven days of the week at each KCC location. Farm Tele Advisers (FTAs who are post graduates in agriculture and allied areas), called as Level-1 respond to the farmers queries instantly. The queries which are not answered by the level-1 are transferred to higher level experts Level-II (SMSs of the State Department of Agriculture or scientists of State Agricultural Universities or ICAR institutes) in call conferencing mode, there are 4-7 identified experts in every state for answering KCC calls. If the calls are not answered even at level-II, then the calls are escalated to level-III, *i.e.* identified nodal institutions those look after the working of KCC in the concerned state. Access of women farmers to KCC needs to be improved.

Smart Mobile phones featured with talk, text and special features have emerged as widely available option providing multiple opportunities for communication. They can access internet and can be used for wide range of applications such as SMS transmission, voice to voice specific queries, and voice to machine for obtaining automated response, video screenings, text to pre- packaged response and text to call centres for individualized text or recorded voice response. Communication of Extension workers and of scientists with the farmers/farm women and reverse has improved greatly by application of mobile phones. Most of the KVKs, ATMAs, SAUs and the ICAR institutes have taken up M-Extension in a big way.

Research in Extension is a crucial area but not attended to adequately to look into the efficacy of different Extension approaches being followed in various states. Strong extension research input is required to strengthen the basic knowledge in extension discipline. MANAGE, SAMETIs, ATARIs and Directorates of Extension of the SAUs would need to have a strong extension research window. They may deliberate and define broader Extension research areas annually. The specificities should be left with the local institutions. Synthesis of selected MSc and Ph.D gender dissertations be considered for academic and application purpose, and the Directorates of Extension of SAUs and ATARIs may initiate this process.

Doubling Farm Income: A Few Strategies

Improved Backward Linkages Especially Technology Interventions

Strong technological interventions would be required in terms of seeds and planting material to get quality production. The production scenario should be open to both public and private sources including technologies and innovations from international institutions.

Market-led Horticultural Extension Strategies

There is need for continuous assessment of Market demands and horticultural production, both in short and long term, enabling farmers to obtain better prices for their produce. Access to e-NAM could offer better price scenario in various markets across the country.

Empowered Horti Groups

Both public and private sector extension service providers may promote commodity groups and empower them through proper training and capacity building. Empowered groups provide adequate demands on the R&D mechanism to respond to the emerging needs.

Horti Farmer Producer Aggregates

Aggregation of the produce at field or cluster level could give an advantage to the small producers in realizing better prices to their marketable produce. This would also reduce the transport and handling costs and make the net returns remunerative, especially for the perishables like fruits and vegetables. Aggregated sizable produce could also get a better place in operations like storage.

Promoting Involvement of Horti Entrepreneurs and PPPs

Promoting Entrepreneur farmers could enhance the investments in horticultural production operations. Besides investments, they could play an important role in technology and market linkages and in aggregating the produce. The storage and supply lines would also be streamlined by entrepreneurial interventions enabling horticultural producers better price realizations for their produce. It has been observed that properly oriented entrepreneur would be a better technology disseminator as compared to the grass roots level field extension functionaries.

Crop Diversification, Low Volume and High Value Crops

Crop diversification especially in horticultural crops, could lead to a better economic proposition to the farmers. For example, producers going for floriculture, intensive vegetable production would get a better price per unit area, if the technologies and the market negotiations are handled properly. Therefore, the horti extension functionaries would need to be oriented on this aspect too.

Promoting Secondary Agri/Horticulture

Processing and value-addition activities have a tremendous potential for better economic returns provided strategic production and linkages with the processing units are streamlined, ensuring sustained supply. Good number of horticultural crops could be used as raw material for industrial use enabling farmers better income regime for their produce.

Strengthening Horticultural advisories by reorienting both public and private extension service providers(contract buyers and input suppliers)

Extension services operated both by the public and private agencies would need to strengthen horticultural advisories. The subject matter specialists of KVKs and the field functionaries of ATMAs who operate at the cutting age would need to be thoroughly oriented on latest production technologies and the market linkages. The Farmer friends would need to be transformed into horti entrepreneurs in potential areas. There is need for carrying out specific HRD programmes for extension functionaries, horti market managers, cold chain operators, nursery managers, *etc.*

Large number of private input support providers and buyers who enter into contract production of horti crops for domestic or export markets provide extension services including technological backup in horticultural clusters and offer better price deals.

Outreach Programmes and Training Modules

The outreach programmes of the concerned ICAR institutes and of the SAUs would need to focus on the demonstrating the best possible technological options at the clusters having potentials for a selected crops. Doubling the farm income through horticultural interventions should be carefully worked out and modules developed by the training organizations. The SAMETIs and the ATARIs could play an important role in providing better training modules incorporating better technological options (through research agencies) for different macro and micro situations.

Export Opportunities for Innovators and Investors

Horticultural Sector provides great opportunities for export operations. Exporters and producers interface could be promoted for this purpose for better price realization to the growers. Horticultural R&D agencies could play an important role in mapping such areas.

Growers Associations

Promoting Horticultural Growers associations for various commodities could be anther strategy for realizations of better prices for the produce. Such associations have come up in good number of States. They have better access to the technologies and market linkages resulting in better price realization.

Recommendations

Contribution of agriculture in GDP is sliding down; however, still > 60 per cent population depends on agriculture and allied sectors. Horticulture contributes 30 per cent to the agriculture GDP. There are over 135 million farm families, average size of farm holdings being 1.33 ha. Around 60 per cent cropped area would continue to be rainfed, which demands innovative development strategies. Agricultural Extension is one of the crucial inputs to address these issues with focus on improving performance of existing extension models in operation.

Good or bad, there exists an agricultural extension system for crops sector, however, in allied areas like horticulture, animal husbandry, dairy, poultry, fisheries, *etc.* it is either weak or missing. Horticultural extension has been integral part of agricultural extension, however, needs special focus considering skilled operations in horti sector.

Horti Extension strategies need augmentation in the areas of enhanced researches backstopping, feedback based extension strategies, working out technology options for Agro-Climatic Zones, Sub-Zones and micro situations, focus on integrated into programme delivery mechanism, prioritization of focus areas, empowerment of women farmers, *etc.* Future extension strategies should focus on technology mapping and activating existing linkage forums. Follow-up by the field formations by themselves may not able to penetrate down the line, especially to the small and the marginal farmers, across the sectors, due mainly to inadequate numbers, their quality and limited outreach. Therefore, it would be better if the chains of extension agents (formal/ informal) like farmer groups, farm entrepreneurs are systematically promoted and used. Flexibilities to the field formations are required enabling them to respond to the emerging local extension challenges.

Horti Extension Strategies must focus on producer aggregates at various levels and provide strong forward linkages. Farmer to Farmer Extension has been one of the most effective mechanisms, the question is how to promote, encourage and fund it. Farm women's presence is seen increasing in all the farm operations because of male migration to towns and cities, hence extension services need to be re-oriented to their technology dissemination needs. ICT has lots of scope for future extension strategies, however, content development and delivery for print and electronic media is crucial. Internet access needs to be improved by organizing Farmer Knowledge Groups (FKGs) in the villages facilitating improved arrangements for knowledge interpretation and sharing. Training co-ordination (segregated responsibility), convergence and programme delivery, extension priority setting, extension for farmer distress areas and PPPs in extension management need to be streamlined as pertained to the local extension needs.

Future dissemination strategies must address yield potentials/gaps and organize large number of FFSs/FOs and FPCs in potential areas. Identification of potential areas for export opportunities (for entrepreneur farmers), involving processing and export industries in such assessments, drawing strengths from international extension experiences and setting quality standards for extension services may form essential parameters of extension performance. Further, there is an urgent need for intensive research in horti extension to look into efficacy of different approaches.

Farm Journalism is an effective instrument for augmentation of information support to horti/agricultural extension. This could be achieved through strategic and location specific media combination focussed on print and electronic, and keeping the targeted audience in view. Important considerations like content development, presentation and delivery of the same are pre-requisites for effective communication strategy in horti sector.

The agricultural/horticultural universities and the ICAR institutes would need to take a lead in promoting training and capacity development of field functionaries. This has to be done in collaboration with the user departments at various levels and as per felt needs of the clientele. Specific modules are required to be designed for this purpose through specialized institutes. Perhaps Zonal Agricultural Technology Application and Research Institutes of the ICAR, the State Agricultural Management and Extension Training Institutes (SAMETIs) are required to develop systematic training modules on the subject under the guidance of Extension Division of Council and MANAGE. A few such modules are necessary to be offered to KVK scientists and ATMA functionaries. Also, there is urgent need to collate, analyze and internalize farm information augmentation experiences across the globe.

We may have the technologies, the investments, but if the dissemination arrangements are not properly defined/streamlined, then development efforts would be far more insufficient. Agri/Horti extension has to deal with the applied social sciences when dealing with the different socio-economic groups and categories of farmers. Hence, information augmentation and dissemination are important for effective horti extension performance.

Horticultural Sector offers better price realization opportunities provided the production and the market linkages are streamlined. Specific strategies like processing and value-addition, sustained supplies to processing units, crop diversification, empowered horticultural groups, re-oriented extension services, producer aggregates, growers associations,etc would certainly enable the producers to double their farm incomes.

References

Chand, Ramesh *et al.*, 2012. *Instability and Regional Variation in Indian Agriculture*, Policy Paper, June, NCAP, New Delhi.

Department of Agriculture and Co-operation. 2011. *Agricultural Statistics at a Glance*. Department of Agriculture and Co-operation, Ministry of Agriculture, Government of India, October, New Delhi.

Department of Agriculture and Co-operation. 2011. *Status of Indian Agriculture*, Department of Agriculture and Co-operation, Ministry of Agriculture, Government of India, October, New Delhi

Department of Agriculture and Co-operation. 2012. Report of the Committee of the Governors to Study and Recommend Measures to Enhance Productivity, Profitability and Sustainability of Rainfed/Dryland Farming with Special Reference to Farmer-Industry Partnership, Ministry of Agriculture, Government of India, July, New Delhi.

Planning Commission. 2011. Report of the Working Group on Crop Husbandry, Agricultural Inputs, Demand and Supply Projections and Agricultural Statistics for the Twelfth Five Year Plan (2012–17), Planning Commission, Government of India, October, New Delhi.

Planning Commission. 2011. Report of the Working Group on Horticulture and Plantation Crops, Planning Commission, Planning Commission, Government of India, November, New Delhi.

Planning Commission. 2011. Report of the XIIth Plan Working Group on Natural Resource Management and Rainfed Farming, Planning Commission, Government of India, October, New Delhi.

Planning Commission. 2011. Twelfth Plan Working Group on Disadvantaged Farmers, including Women, Final Report, Planning Commission, Government of India, November, New Delhi.

54

Innovations in Horticulture Markets and Value Chains Development

T.M. Gajanana[1]*, M. Sudha and D. Sreenivasa Murthy

Division of Social Sciences and Training,
ICAR-Indian Institute of Horticultural Research, Hesaraghatta Lake Post,
Bengaluru – 560 089, Karnataka
**E-mail: gajanana@iihr.res.in*

The importance of horticultural crops, especially fruits and vegetables, in improving the nutritional status and economy needs no emphasis. The varied agro-climatic conditions prevailing in India are conducive for the cultivation of fruits and vegetables in different stages round the year. The importance of this sector has been realized by the planners and policy makers only recently. Establishment of National Horticulture Board (NHB) in 1984, creation of APEDA in 1986 for encouraging exports, formation of a separate Ministry of Food Processing (MFPI) in 1988 for improving processing of horticultural crops and adoption of liberalized seed import policy in 1988 to overcome the constraints of non-availability of quality planting material are some of the important steps taken by the Government of India for encouraging the horticulture sector. The major boost to horticultural development was given during VIII Plan by an allocation of Rs.10,000 millions which coincided with the liberalization policies initiated by the GoI. A special provision of Rs.975 millions was also made for export of horticultural crops. During the liberalization period many policies were initiated for the overall development of horticultural sector. The important ones are: according 100 per cent EOU status to floriculture, 'Extreme Focus Area' for fruit and vegetable processing industries and floriculture, abolition of excise duty on processed food, subsidy on export of fruit, vegetables and flowers by APEDA *etc.* It is heartening to note that in order to improve the level of processing to reduce wastage, increase marketability and to help the farmers to enjoy higher value realization, APEDA has developed a concept of Agri-Export Zones (AEZ). It takes a comprehensive look at a particular produce/product located in a contiguous area for the purpose of developing and sourcing raw materials, their processing/packaging leading to final exports. Thus, cluster approach is followed to address marketing problems of certain produce/products in a specified area. In fact, there are 60 AEZ for horticultural produce/products. The plan allocation has further increased to Rs. 1200 crores during IX Plan and to Rs. 5,580 crores during 10th FYP. Further, since 2005, horticulture sector is being supported by the government under National Horticulture Mission (NHM) and the projected budget allocation for 11th FYP was Rs. 49,640 crores (Chadha, 2008) and Rs. 69,500 crores during the XII Plan (Sudha *et al.*, 2011).

The steps taken and policies initiated by the GoI have started giving results in terms of increased production. The introduction of hybrids in case of vegetables like tomato, cabbage, cauliflower, chillies *etc.* during the late eighties has increased the productivity. However, concurrent improvements were not made in production and marketing of fruits and vegetables which has resulted in gluts in the market and violent fluctuations in prices.

Production of Fruits and Vegetables

With the GoI initiatives having started bearing fruits, India is now the second largest producer of fruit and vegetables contributing about 12.40 and 13.30 per cent to the total world fruit and vegetable production, respectively. Among fruits, India ranks first in the production of mangoes (41 per cent), banana (28 per cent), and papaya (30 per cent). Among vegetables, India is the largest producer of peas (30 per cent); second largest in brinjal (29 per cent), cauliflower (29 per cent), onion (18 per cent) and cabbage (8 per cent).

Growth in Production of Fruits and Vegetables

In India, fruits are grown on an area of 6.405 million hectares with a production of 91.44 million tonnes (NHB, 2016). The important fruits are mango, citrus, banana, apple, litchi, guava, pomegranate and sapota. Vegetables are grown on an area of 9.57 million hectares with a production of 166.61 million tonnes. The important vegetables grown are potato, brinjal, tomato, onion, okra, cauliflower, cabbage and green peas. Production of fruits and vegetables has increased from about 87.13 million tonnes in 1991-92 to about 258.05 million tonnes in 2013-14 [NHB, 2016].

Fruits and vegetables together registered a growth of 4.65 per cent in production, 3.54 per cent in area and a meager 1.07 per cent in productivity. As regards fruits, production registered a growth 4.49 per cent which was mostly area led (4.37 per cent) rather than productivity led (0.12 per cent). In case of vegetables, production grew at 4.73 per cent and here also area led growth was noticed as the area grew at a rate of 2.99 per cent and yield registered a growth of only 1.69 per cent. However, it may be noted from Table 3 that during the latest period (2001-2014), production growth was much faster (>6 per cent) compared to the period 1991-2001(<5 per cent).

Keeping in view the rising production of horticultural crops, efforts were also made to introduce innovations in marketing of the increased production of horticultural crops. The paper discusses some such important innovations which made a dent in creating enabling environment for improved marketing of horticultural crops.

Innovations in Horticulture Markets

Model Market Act, 2003

Regulated markets have been the major marketing avenues for farmers. However, due to anomalies in trade, the farmers were not getting the benefits of regulation. Realizing this, amendments to APMC Act (Model Market Act, 2003) were made in September 2003 which paved the way for establishment of private markets/yards, contract farming, direct purchase centres and a provision of separate constitution for Special Markets for perishable commodities such as fruits and vegetables. The role of the Agricultural Produce Market Committee is to promote an alternative marketing system, contract farming, direct marketing, and farmer/consumer markets, as well as redefine the role of State Agricultural Marketing Boards to promote grading, quality certification, and standardization. These market reforms were done to promote pledge financing and e-trading. Accordingly, efforts were made to institutionalize contract farming in horticultural crops like fruits, vegetables and flowers both for sourcing raw material for processing, export and also for seed production.

Distant Market Sale Model

Distant market model to realize the higher price advantage has been in vogue for a long time. However, of late, emphasis was laid on organized distant market sale for some of the produce like tomato, onion, oranges, banana, pomegranate, guava, grapes *etc.* on the lines of spatial equilibrium concept.

Pink Flesh Guava in Tamil Nadu and Kerala

Guava produced in Karnataka is also transported to distant markets like Cochin in Kerala and Chennai in Tamil Nadu. Transit loss ranged from 3.6 to 4.62 per cent and retail level losses were 4.59-6.09 per cent. However, the wholesaler's margin ranged from 26-30 per cent and retailers' margin was 54-84 per cent [table 1] which more than offset the marketing cost and also the losses during marketing of pink flesh guava [Gajanana *et al.*, 2014]. Considering these facts, the farmers may take initiatives to transport their guava to distant market places and take advantage of the prevalent higher price.

Table 1: Marketing Cost, Price Realized, Intermediaries Margin and Producer's Share in Guava (Pink flesh) in Distant Markets (Rs/kg)

Sl.No.	*Particulars*	*Kerala*	*Tamil Nadu*
1.	Selling price of PHC/Purchase price of wholesaler	20.83	24.00
	Marketing cost of WS	2.88	2.75
	Margin of WS	6.21 (26.19%)	7.96 (29.76%)
	Selling price of WS/Purchase price of retailer	29.92	34.71
	Marketing cost of retailer	0.27	1.40
	Margin of retailer	16.35 (54.16%)	30.42 (84.24%)
2.	Selling price of retailer/ consumer's price	46.54	66.53

Note: Figures in parentheses are percentage to retailer's selling price.

The introduction of "Banana Special" train from Nimbhora in Raver taluk, Jalgaon in Maharashtra to Delhi should be emulated in other parts of the country for exclusive transport of fruits and vegetables to remunerative distant market. It is interesting to note that the state governments are thinking about introducing 'Green Track' for onion transport.

The sale of Nagpur mandarins during 2016 in Bengaluru is another case of taking advantage of the higher price in the distant market. The Maharashtra Orange Growers' Association (MOGA) managed to sell oranges @ Rs. 25-30/kg which fetched only Rs.7-8/kg in the production area of Amravati.

Roadside Stall for Marketing of Indigenous Mangoes

There is no separate market for indigenous mangoes and are normally brought to APMC along with other varieties. The price realized in the APMC is less as it is treated like other varieties (Other than Choice varieties). Further, value of indigenous varieties such as Early bearing, Late bearing, Good keeping quality; Multipurpose (Fresh, processing and pickle); High temperature tolerant and their use in ecosystem services like Pollination are not yet known to farmers.

Considering the importance of indigenous mangoes and also realizing the inadequate market for them, under the UNEP-GEF TFTGR project on "Conservation and use of cultivated and wild tropical fruit diversity; promoting sustainable livelihoods, food security and ecosystem services", efforts were made to create an alternate markets (Lamers *et al.*, 2016) in the form of roadside stalls along the Bengaluru-Tirupathi highway with proper packaging material (CFB boxes). A small market intervention in the form of erection of roadside stalls in Chittoor in Andhra Pradesh resulted in higher prices to indigenous mangoes [Table 2].

Table 2: Road Side Stall for Marketing of Indigenous Mangoes-Experience of Chittoor Farmers (Price: Rs./kg)

Indigenous Varieties	*Mandi Sale*	*Stall Sale*	*Difference*
Kudadut	8.99	47.61	38.62
Naati	8.04	11.91	3.87
Reddy Pasand	6.14	7.61	1.47
Kalepadu	13.03	24.41	11.38
Green Baneshan	–	12.61	12.61
Atimadhuram	–	22.61	22.61
Dil Pasand	7.57	11.20	3.63

Emerging Markets for Safe and Nutritious Horticultural Produce

Domestic Markets

Health and quality conscious consumers are now looking for safe and nutritious horticultural produce and in order to make available such produce, markets, though on a small scale, have emerged in the form of outlets for organic produce, fresh and dried horticultural produce markets, carbide free produce etc,. The recent experience of HOPCOMS selling carbide free mangoes is a case in point. Emphasis on organically produced mangoes in Chittoor being supplied and market for such mangoes is another example of making available the safe and nutritious mangoes to the consumers. Enterprises like Era Organic, Phalada Agro Research Foundation, ISKON, Bengaluru, Organic Junction *etc.* are involved in production and marketing of organically grown commodities in their own retail outlets in metropolitan cities.

International Markets

Since the overseas importers of horticultural commodities like fruits, vegetables and medicinal crops are insisting on Global GAP, SPS and MRL, the exporters of these commodities procure them after meeting the required standards which ensures safe horticultural produce in the international markets. Further, organic products with 'Certified Organic' label have an overseas market. International importers like the Netherlands based Eosta, Thailand based Swift are showing keen interest in organically grown fruits and vegetables.

FPO and Producer Companies Model of Marketing for Horticultural Produce

Co-operatives were considered the only option available to get organised themselves and move up. But, these have been infected by various inadequacies. The committee headed by eminent economist Dr Y. K. Alagh conducted a study and recommended for the formation of Producer Company in 2002. Producer Company is a form of co-operative entity that has been introduced in the Indian economy by virtue of Amendment IX A of the Companies Act, 2002 which became effective from February 6, 2003. A producer company is a hybrid between a private limited company and a cooperative society. It combines the benefits of a cooperative enterprise and the vibrancy and efficiency of a company.

Indian Organic Farmers Producers' Company Ltd. (IOFPCL), Kerala

The first producer company, IOFPCL has been providing higher price to the organically certified spices, cocoa *etc.* grown by the member farmers in the face of competition from MNCs like Cadbury's. The producer company has successfully demonstrated its power by challenging the MNC Cadbury's and the MNC was forced to increase the procurement price of cocoa from Rs. 70 to Rs.100/kg. With organic certification and Fair Trade Certification of spices, the member of the PC could get 10-20 per cent premium at the time of procurement as about 90 per cent of the produce entered the international markets.

Small and Marginal Agriculture (India) Producers' Company (SAMAGRI)

This is a case of a PC engaged in establishment of their own market (Thottam market in Chennai) on par with any organized retail outlet for fruit and vegetables and also to meet the daily needs of the member farmers. The farmer members are benefited by their own market.

In order to make farming profitable and to have market access, in 2014-15, the Government set the goal of supporting the creation and/or registration of 2000 FPOs across India over the next few years. In fact, the year 2014 was labelled as the 'Year of FPO'. In order to achieve this goal, multiple financial support packages were created. The main qualifying criterion for an FPO to attract benefits under various schemes and programmes of the Central and State Governments is that it must be registered and administered by farmers focused on activities in the agricultural sector. As part of these available packages the SFAC provides credit guarantees to farmers to mitigate risks and provides grants worth around $15,000 dollars. NABARD made available funding support of around 30 million dollars for the creation and financial support of up to 2,000 FPOs and provides credit support for business operations and capacity building. Through the Government of India, FPOs are also eligible to receive higher subsidies.

Contract Farming and Price Contract

Contract farming has been in existence for export oriented produce like gherkins, baby corn, chillies, seed production of vegetables and flowers.

Contract Price

Ooty is also called as 'Carnation capital' of India and besides carnation, gerbera and liliums are also grown on a large scale due to the favourable climate for these flowers. However, marketing was the main problem for these Ooty grown flowers. Price contract between the traders of Bengaluru, Kolkata, Delhi and Coimbatore and the flower producers eased this problem to some extent.

e-marketing/Online Marketing of Flowers

In order to bring in transparency and competition, e-marketing was introduced in marketing of flowers in the form of 'dew drops', Coimbatore during the last decade. International Flower Auction, Bengaluru (IFAB) is a perfect case of an institution created based on the principle of PPP for online trading of flowers for both domestic market and export.

Direct Linking of Farmers to Market

Start-ups, Ninja Cart, Apna Haat, Helpchat, Fair trade, Honey Bee solutions are slowly making their entry in to marketing of horticultural crops by establishing a direct linkage between the producers and the market.

Value Chains Development in Fruits and Flowers

Value-addition by self help groups (SHG) in fruits, especially underutilized fruits like *amla, kokum* (*G. indica*) and Uppage (*G. gummigutta*) by NGOs like BAIF/ DHAN/Life trust was seen as a case of value-addition by way of creating value chains for underutilized fruits. Further, converting aromatic mangoes (*Appemidi*) in to pickle by women SHGs to capture the niche markets is another case of value chain development for fruits.

Value-addition to *kokum* fruit by processing it in to butter using the seed and Kokum juice/concentrate and HCA from the rind are examples of value-chain development.

Underutilized fruit products like Amla pickle, candy, mouth freshener, squash, Karonda pickle, Jack fruit papad, chips, pulp, tamarind chutney, *pani puri* masala, sauce prepared by SHG women promoted by BAIF in Gujarat, Maharashtra and Karnataka are cases of value chains developed for underutilized fruits (Gajanana *et al.*, 2010).

Tomato Crush, Puri

In the event of glut situation, it is advisable to convert excess tomato in to crush or puri so that the

Figure 1: Kokum Fruit to Kokum Concentrate/Squash.

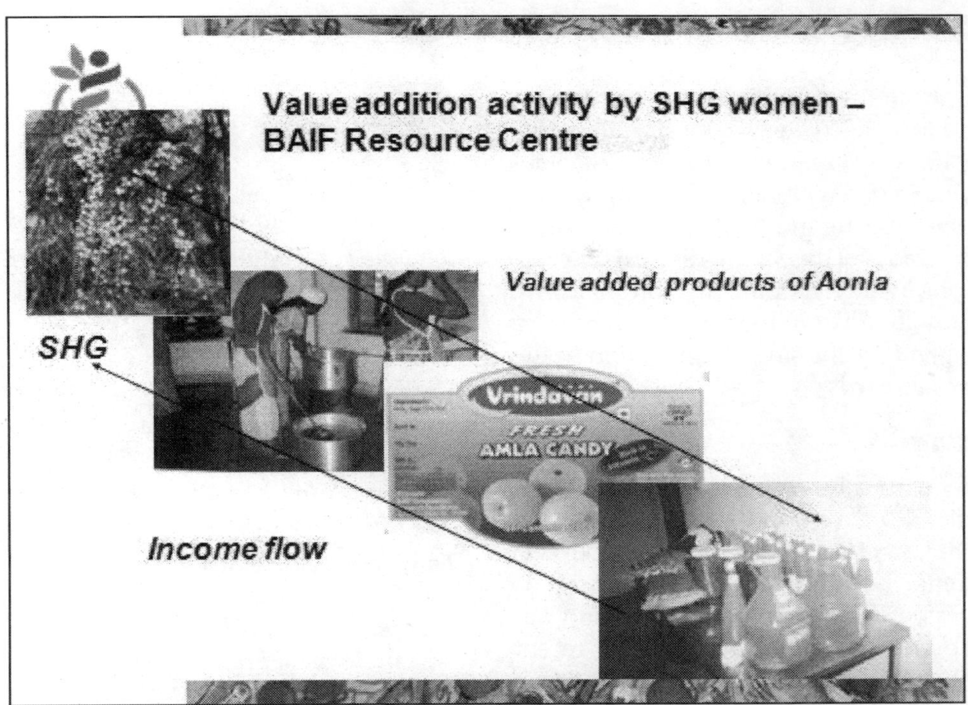

Figure 2: Value Chain in Aonla.

price fluctuations due to glut can be minimized. Tomato crush can be prepared even at the farmers' group level. One tonne of tomato is processed in to 300 kg of crush and the present trend of formation of FPOs suits this concept of preparation of tomato crush.

Cut Flowers as Garlands, Bouquets, Office Arrangements, Wreath *etc.*

The margin associated with flower arrangements was found much higher (50-100 per cent) compared to fresh flower sale (28-40 per cent).

Figure 3: Value-addition to Flowers.

Unified National Agriculture Market

The Agricultural Produce Market Committee Act has been restrictive regarding trading of produce within the state. Under APMC Act, states are divided and local markets are created (*mandis*), with farmers selling their produce through auction in the mandi. To operate in the mandi, the trader needs a license. Wholesalers, traders and food processing companies cannot buy the produce directly from the farmers and they have to go through the mandi.

Recently, Department of Agriculture and Cooperation under the Ministry of Agriculture emphasized on creating a Unified Market that is well-integrated across the nation to increase the net returns of the farmer. The budget speech of Finance Minister for 2015 announced the establishment of a "Unified National Agriculture Market." This prompted Department of Agriculture and Cooperation (DAC) to formulate the Central Sector Scheme for Promotion of National Agriculture Market through Agri-Tech Infrastructure fund (ATIF) *via* provision of common e-platform. This e-platform will be deployed in 585 regulated wholesale markets in States/Union Territories (UTs) which will be implemented by Small Farmers Agribusiness Consortium (SFAC), and a budgetary provision of approximately 31.5 million US$ will be spent over the next three years (2015-16 to 2017-18).

The states/UTs need to undertake intra-state reforms: (i) a single license to be valid across the state, (ii) single-point market fee, and (iii) electronic auction for price discovery, to be eligible for being part of this scheme. Successful formulation and implementation of this market will require active participation of the State Governments. If this happens, it is expected to serve the interest of the farmers and the agricultural sector by creating greater direct participation at the farmer level as middle agent involvement will be eliminated from the supply chains.

Karnataka is the first state which has taken steps in this direction and around 51 market yards/sub yards (mandi), among 155 main market yards and 354 sub-yards, have been integrated with single-licensing system. It offers modern facilities for grading, dissemination of prices for different grades of the commodities in different market yards along with electronic auction platform. There is further possibility of reducing intra-state price differences by expanding the single registration to the remaining yards/sub-yards.

The Way Forward

Many innovations have been introduced in marketing of horticultural crops. While, Model Market Act, 2003 paved the way for creation of private markets and contract farming, the policy intervention to support FPO and formation of producer companies stand out which may make a dent in this direction. The recent announcement by the GoI for the unified national agriculture market, though looks attractive, it is highly ambitious as the unification of all the agricultural markets is a herculean task considering the number of produces sold, varieties and the grades of produce brought to the market. If implemented in true letter and spirit, e-NAM may be an ideal thing to happen. However, whether all the perishable horticultural commodities will benefit from this is yet to be ascertained. The start-ups can make a difference, if they are really serious about the task before them. Organization of SHGs is still a potential venture to group the gullible, small and marginal farmers and empower them to engage themselves in value-addition activities.

References

Gajanana, T.M., Gowda. I.N.D. and Reddy, B.M.C. 2010. Exploring market potential and developing linkages: A case of underutilized fruit products in India. *Agril. Econ. Res. Rev.*, **23** (Conference Number): 437–443.

Gajanana T.M., Sreenivasa Murthy, D. Saxena, A.K. Sudhakar. Rao, D.V. Sudha, M. and Dakshinamoorthy, V. 2014. Marketing and PHL assessment in pink flesh guava: An economic analysis (Kannada). *Krishi Pete*, **40**(6): 25–26.

Lamers, H.A.H., Kruijssen, F., Sthapit, B. and Ramanatha Rao, V. 2016. How can markets contribute to the conservation of agricultural biodiversity on farms? In: *Tropical Fruit Tree Diversity: Good Practices for in situ and on-farm Conservation*, (Eds.) Bhuwon Sthapit, Hugo A.H. Lamers, V. Ramanatha Rao and Arwen Bailey. Earthscan Books, Routledge, N.Y.

Sudha Mysore, Malhotra, S.K. and Gajanana, T.M. 2011. Assessing the growth trend of horticulture sector, future projections and needed interventions. Presented in *The Third Meeting of Sub-group 2: Assessing the Growth Rates of Horticulture Sector and XII Five Year Plan Outlay*, 29.08.2011, NASC, New Delhi.

Technology-led Organic Horticulture and Safe Food

55

Food Safety begets Food Security

S. Dave

Former Chairman of Codex Alimentarius Commission (a FAO/WHO body of the United Nations) and Advisor, FSSAI
E-mail: salve.codex@gmail.com

The other day, I happened to be home a bit early in the evening and was chatting with my wife over a cup of tea, when she said, "You know several food retail chains will also sell fresh fruits and vegetables. We can buy these from there when we go for our weekly grocery shopping."

"But why from them? All these years, we have been buying fresh from the street vendors or from the nearby weekly markets. Why should we go all the way to a distant place and then store vegetables for the whole week?", I retorted.

"Because they maintain quality", she said.

"But, you will still get the same stuff from these retailers that you buy from the street vendors next door or from the weekly markets; after all, the farmers are the same", I responded.

I could understand what she was trying to say, but I continued to extend the conversation to assess consumer behaviour.

She replied, "You don't understand, the retailers have a name for quality and safety of food, and maintain high standards. They might be selecting their farmers. If necessary, they will import safe produce. At least, we will not get pesticides in our food."

I wasn't quite sure of what she was saying. Most retailers buy from traders and on many occasions they do not even know which farmer produced the fruit or vegetable they are selling. How many retailers, leave alone the roadside vendors, are checking the produce for pesticide residues, is doubtful. Even these small vendors have no idea of the adverse health effects of pesticide residues. On many occasions, even sweetening injections are, reportedly, inserted in the fruits or colours smeared on the vegetables. Thinking from the angle of farmers, it is important for them to make sure that the fruit or vegetable produced in their farm does not get damaged due to pests and so they sell the produce as early as possible without thinking of the residual effect. In many areas, the farmers lease out their farm to traders and take advance money for the crop season; resultantly, the farmers or the traders spray pesticides and insecticides as guided by the salesmen of pesticide manufacturers. On most occasions, they do not know nor are they told that there is a pre-harvest time gap prescribed for each pesticide. Most farmers have also not received the required training or knowledge in this direction.

The Food Safety and Standards Authority of India (FSSAI) has framed several standards for quality and safety. The Maximum Residue Limits for pesticides in fruits and vegetables are also being established. But, what is our expectation from these food laws? Are we in a position to ensure that setting standards will provide safe food to all of us? Have we made any assessment about the need and practices required to be followed to ensure compliance, who will give this training to them, and what is the technical manpower required to meet and monitor the intended objectives? FSSAI's mandate under the Act does not extend to the farm level. Whether we like it or not, the fact is that if we want safe fruits and vegetables and processed food products, safety will have to begin from the farm. Then, are we aware of the growing disease burden on our Govt. and the tax payer due to lack of compliance with food safety requirements? Is it not appropriate to ensure prevention of diseases than trying to cure them. Now that there are a growing number of consumers ready to pay an extra buck for safer food, why can't we prepare our farmers with the practices they need to follow to ensure that the pesticide residues are within limits? These were some of the questions that traversed my mind while I was looking for answers to the points raised by my wife.

Thus, in order that farmers in India are able to market safe produce, it is important to promote the use 'best practices', possibly in a certifiable form among the farmers. It would be prudent to direct resources in this direction. Farmers need to be encouraged to implement Integrated Pest Management by using bio-pesticides. It is also important that the manufacturers of pesticides and insecticides ensure that their products are appropriately registered with the Ministry of Agriculture and that they abide by a Code of Conduct in selling their products to the farmers. Proper and complete labeling of pre-harvest diffusion period of pesticides in a legible form should, essentially, be followed by all engaged in pesticide business. This is in the interest of farmers, environment and the consumers. There should also be an effective mechanism to monitor adulteration with banned pesticides.

Good Agricultural Practices (GAP) and registration of pesticides is practiced by several countries to ensure availability of safe fruits and vegetables to the consumers. The retail chains insist on compliance through third party verification. Even in India, the exporters and their constituent farmers implement GAP and comply with the importers' requirements. Of course, there is a cost to it. In spite of the best efforts by the exporters, safeguard measures were taken by several countries against our okra (bhindi), curry leaves, green chilly, table grapes, pomegranate, groundnuts, *etc.*

All said and done, if safety can be ensured for exports, why not for our own consumers? What should the farmer do if the lady of house would decide to buy from a store that guarantees safe food even if it means a little extra price? The answer is implementation of GAP and appropriate registration of all pesticides in the market. While the national food laws intend to promise availability of safe food, implementation of appropriate practices at the farm level will continue to remain the 'X' factor to make a difference. Else, our consumers will continue to buy unsafe fruits and vegetables and fall sick of unknown diseases. If the retailers implement GAP and are able track the source of produce and get their practices certified, they should be able to display it on the packages by stating that the produce is GAP certified. This will create positive ripples among the consumers with the result that our consumers will benefit. The consumers will start demanding safe and GAP certified produce. At the same time, our farmers will also benefit. A win-win.!!

So, how do we go through the process of enabling the farmers meet the challenges before him? He is only too eager to sell safe produce and take pride in doing so. How can we empower even the smallest farmer to stand up and say, "Yes, I can do it." ? Can we bring about that change for his sake and for the sake of our consumers? Since most of the farmers in the country are small and marginal players, the ideal step forward would be to encourage farmers to form groups and implement an Internal Control System so that the cost gets divided and also generates marketable volumes. This should be followed by a third party certification process.

The development of GAP certified farms would need to be taken up on a project mode with the help of state governments and the state agricultural universities/national research centers under ICAR. This work requires a cluster development approach by setting up model GAP compliant farms with train-the-trainer initiatives. Thereafter, all we need to collectively do is to motivate the farmer and educate the consumer to demand GAP certified produce.

Implementation of GAP requires the development of packages of practice (POP) for each horticulture or agricultural crop. A generic POP document with season and location specific annex attached to it with the list of pesticides having a label claim. Where ever, there is a requirement of pesticide that does not have a label claim, best option is to use bio-pesticides. The Government should facilitate its registration, if it should be necessary.

Implementation of this programme would increase awareness among the farmers as well as the consumers about the need for consumption of good quality and safe produce. This would facilitate compliance with better standards for our citizens. It would meet the objectives of the government to provide safe food to the consumers. As a result, it would lead to import of produce of comparable quality and safety.

Implementation of GAP would give a boost to contract farming, whereby, processors would have an added advantage both in the domestic and international market in terms of better quality produce for processing. The programme would also contribute to improvement in the environment and soil fertility. Needless to add, this would also lead to employment generation. And one of the most important outcome of implementation of GAP is support to food security for our country through reduction in wastage.

The short point is to make it possible for even the smallest farmer and retailer of fresh and processed food in our country sell safe food to the consumers and contribute to food security. Let us empower them to take pride in their business of little means !

Together, we can make a difference !!

56

Income-Assisted Safe Food Production for Remunerative Organic Horticulture

S. Manivannan

Students' Welfare, Sikkim University, Gangtok – 737 102, Sikkim
E-mail: smanivanan@cus.ac.in

Chemicals occupy human food consumption like never before. The advent of human civilization has made many chemicals as a part of food chain. Though many of the technologies involving chemicals were developed with good intentions, the continuous use of these technologies have made the bio ecosystem more vulnerable for many stress factors. It is reported that about 97 percent of sprayed agro-chemicals reach non-target destinations creating a huge environmental risk.

Even today the impact assessment of agrochemicals mostly stopped at the residual compounds in soil and food chain. It conveniently oversteps the impact of heavy metals and other elemental toxicity. Many of the elements are introduced into soils from various sources. Some of them are sewage sledge, phosphate fertilizer, pig slurry and pesticides. The risk associated with such polluted soils is contamination of food chain. Since, first compartment of the terrestrial food chain is plants the effect is more prominent in the human beings, who depend on settled modern agriculture. It is further more magnified when the plants freshly consumed without any cooking or processing. As most of the horticultural crops especially fruits and salads are consumed freshly, they stand in the high risk end.

It is in this context, organic production system gained importance by the turn of this century. Though organic horticulture was practiced in one form or other in different parts of the world, a systemic approach to that end started just two decades ago. Even if horticulture is practiced with complete non chemical use, the agro chemical residual compounds might be avoided, but trace elements and heavy metals will always pose a threat as they might be indigenously available in the soil due to soil withering process or irrigation through contaminated groundwater. Problem of arsenic contamination in groundwater is reported from seven states in India (Shamugapriya *et al.*, 2015) is a testimony to that.

Such a contamination apart from creating health issues, it also has the adverse effect on the remuneration to all those involved in the cultivation business. According to International Trade in Agriculture and Agro-based Industries (CITA), around 30 consignments from India were rejected by the EU in May and June, 2013 for the presence of heavy metals. These included grapes (eight consignments), egg powder (four consignments), chilli powder, honey, soft drink concentrate and prawns. The US rejected around 50 consignments of food articles and health supplements in April, 2013 alone.

Hence, it becomes imperative to analyze the soil, irrigation water and horticulture produces for their nutritional quality as well as for the presence of heavy metals and trace elements which pose health and market risks. For analyzing all the inputs and produce, the one stop solution would be ionome profiling. Ionomic profiling reveals the entire profile of ion status of the soil, water and plant in the given environment. In plants, the relatively high throughput and low cost of ionomic analysis means that it has the potential to provide a powerful approach to not only the functional analysis of the genes and gene networks that directly control the ionome, but also to the more extended gene networks that control developmental and physiological processes that affect the ionome indirectly. In addition it reveals the interaction effects of different elements in different entities like soil, water and plants.

Sikkim being one of the declared organic state, the safety parameters and remunerations of the organically produced horticultural produce as food is of the paramount importance. The Plant Nutrition and Ionome Laboratory, Department of Horticulture, Sikkim University is pioneering itself in the scientific validation of safety of growing environment and the farmers' products. Some of the major results pertaining to mediated analysis of growing environment and horticultural crops are discussed here.

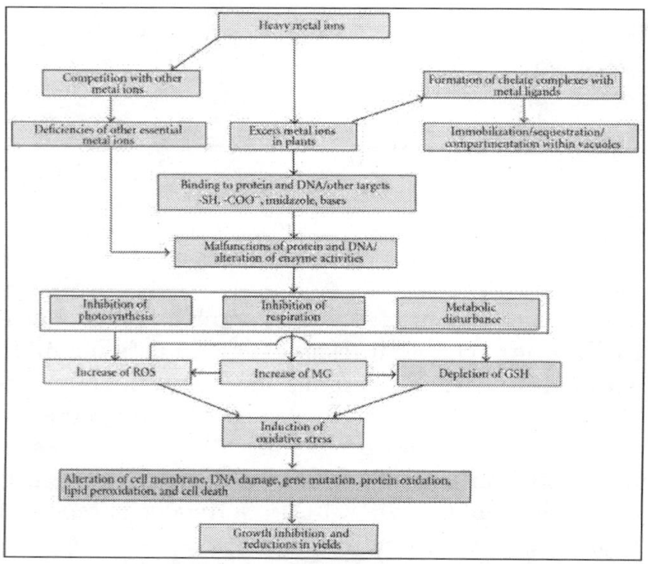

Figure 1: Mechanism of Heavy Metal Toxicity in Plants.

Irrigation Water Quality Assessment

Irrigation water samples collected from different streams of Sikkim were analyzed for the presence of different elements (Table 1). Samples collected from North Sikkim had least amount of heavy metals, in fact one sample had lesser than the safest drinking water. However, samples collected from other parts especially down streams from East and South Sikkim had high amount of arsenic and mercury. The arsenic (As) content of the samples varied from trace amount to 39222.94 ± 20 ng l^{-1} and mercury (Hg) content found up to 3381.59 ± 25.2 ng l^{-1}.

The high amount of these elements could be correlated with high anthropogenic activities in addition to natural source like parent rocks. Such contaminated irrigation water when used for agriculture, pollutes both soils and crops grown on it, which in turn creates heath issues and reduces the marketability of the produce.

Soil Analysis under Organic Production System

As the irrigation water was containing high amount of As and Hg, it was foregone conclusion that soils also will have high amount of these elements. Contrary to that expectation as content was low in the analyzed soil samples. However, higher amounts of Hg content could be observed (Table 2). This may be due to Hg has the strong affinity to organic matter (OM) and it increases with increasing soil acidity (Lodenius, 1989). Sikkim soils are highly acidic and has high amount of OM. In addition, by organic farming practices more OM is enriched at each cycle of farming. However, the Hg bound by OM is leached out from the soil profiles in an acid environment in a form bound organic compounds. Humus-Hg is more easily released through evaporation and leaching (Mou *et al.*, 1999). Hence, presence of Hg need not be a concern and organic production practices aids in their complexation and leaching.

Basically soil crust contains on an average of 1.8 mg kg^{-1} of As. Major source of As in soil are other than natural sources, industrial activities (metal processing, chemical works based on S and P minerals, coal combustion, and geothermal power plants) and use of arsenical agro chemicals sprays, particularly in orchards. Even though, As content in soil may be high, but its bioavailability to the plants depend upon various

Table 1: Quality of Irrigation Water (Streams) in Sikkim

Sample	*Concentration (ng/l)*						
	As	*Al*	*Cd*	*Pb*	*Ni*	*Hg*	*Fe*
Water 1	864.22	776.88	824.4	591.4	851.31	959.28	65.78
Water 2	trace	61.74	2.58	trace	1.84	trace	trace
Water 3	16201.35	trace	trace	trace	trace	840.28	trace
Water 4	39222.94	trace	trace	trace	trace	2726.4	trace
Water 5	11371.52	trace	trace	trace	trace	3381.59	36257.57
Water 6	12309.53	trace	trace	trace	trace	2083.44	trace
Water 7	16540.58	trace	trace	trace	trace	2665.36	trace
Water 8	2627.76	trace	trace	trace	trace	3204.63	13800.52

(Water 1 and 2 are from North Sikkim; 3 and 4 from East Sikkim; 5 and 6 from South Sikkim and 7 and 8 from West Sikkim).

Table 2: Multi-elemental Profiling of Soil Samples of Sikkim

Sample	Concentration (mg/kg)																
	As	*Al*	*Cd*	*Pb*	*Ni*	*Hg*	*Fe*	*P*	*K*	*Ca*	*Mg*	*S*	*Cu*	*B*	*Mo*	*Mn*	*Zn*
Soil (1)	trace	1008.584	0.10907	0.15318	4.67786	0.10213	652.1266	trace	23.1185	11.61582	47.15253	12.11152	9.63023	11.25254	0.28417	87.53217	3.09837
Soil (2)	trace	900.5486	0.1098	0.08868	2.66174	0.09543	565.4209	trace	29.52018	11.1092	43.79048	12.81158	6.68831	12.24963	0.08981	66.68281	3.37522
Soil (3)	trace	1050.728	0.08885	0.09524	3.54524	0.15738	659.0829	trace	34.29961	13.16034	50.72769	12.16962	7.44912	12.53527	0.20635	83.18227	3.52783
Soil (4)	trace	1667.096	0.08451	0.24865	5.38551	0.10548	1083.495	trace	42.25119	12.99891	66.42027	11.9934	12.52243	11.98469	0.30485	145.6474	4.17005
Soil (5)	trace	2130.257	0.13363	0.23056	7.05769	0.16575	1213.039	trace	53.3018	20.47513	89.85911	13.12355	19.10536	11.45297	0.38309	177.8292	6.5126
Soil (6)	trace	1194.634	0.09607	0.17763	3.75193	0.11385	758.3136	trace	29.10926	12.22535	54.51588	12.42007	11.72779	12.22821	0.1848	107.4781	4.32574
Soil (7)	trace	892.3364	0.1033	0.29217	4.81724	0.34491	647.2778	trace	21.16152	11.40865	48.11368	12.19402	13.57554	12.32003	0.28055	72.21191	4.67351
Soil (8)	trace	803.1296	0.08668	0.33453	5.25327	1.64757	562.1408	trace	19.81966	10.52009	43.08465	12.70547	16.86526	11.56191	1.07147	58.73918	4.22161
Soil (9)	trace	966.4984	0.11413	0.26353	6.11923	0.74507	689.5941	trace	24.47289	10.46676	45.46543	14.28037	12.8666	11.8868	0.82255	72.49569	3.86156
Soil (10)	trace	864.753	0.14302	0.17024	4.81424	0.17245	558.7078	trace	21.29703	9.71584	40.11471	13.45665	9.35161	11.96808	0.4376	60.07922	4.19961

factors. According to Lazo *et al.* (2007), As enrichment in soils is mainly associated with Fe–S, Cu–S, and Ni–S mineralization.

Bioavailability of As from soils of an abandoned mining area was relatively low, and the highest was by earthworms (Gal *et al.*, 2007). However, analysis of vermicompost at Sikkim University for the presence of As could not prove this observation. In our study, the water and FYM with high As was used in vermicomposting but the As content in vermicompost was as low as 2.75 ng kg^{-1}.

Even though, we could find higher amount of As in irrigation water, only a limited fraction of the total amount of As is easily mobile, and over 80 per cent is strongly associated with Iron (Fe) and aluminum (Al) oxides and is not available for plants (Thornton and Farago, 1997). As the soils of Sikkim are acidic and rich in Fe and Al due to continuous leaching of calcium (Ca) and magnesium (Mg), arsenic complexes with these elements and are not available to the plants. Hence, the horticultural crops cultivated in the marshy lands only will have the effect of the As pollution.

Assessment of Germplasm Tolerance for Al Toxicity

Apart from As and Hg, being acidic, Sikkim soil were containing toxic level of Aluminum (Al). Following any amendment practices will run contrary to the accepted organic production norms. Hence, major effort was diverted towards the finding tolerant species to grow in the Al toxicity. Two major crops, *viz.*, citrus (14 germplasm) and cow pea (20 germplasm) were tested for tolerance against Al toxicity under hydroponic system. They were also analyzed for growth parameters and dietary nutrients.

Among citrus germplasm based on the growth parameters and ionome profiling Indian mandarin and rough lemon were found to be highly tolerant followed by Sikkim mandarin. In Sikkim mandarin, the dietary nutrients were also unaffected in high levels of Al concentrations. Among the cow pea accessions IC-559386 obtained from ICAR-IIVR, Varanasi found to be tolerant. These varieties/accessions could be effectively adopted in the organic production system where Al toxicity prevails.

Yield and Quality under Organic Production System

One of the important components of the organic production system is the usage of PGPRs. At Sikkim University three different PGPR capsule, *viz.*, FL-18, GRB-35 and TR-23 obtained from ICAR-IISR, Calicut were tested in two tomato varieties under organic production system. Highest yield and ascorbic acid could be obtained with TR-23. In addition, increased amount of N, P, K, Mg and I in tomatoes was observed.

However, TSS, Ca and Zn contents were found to be less. Additionally, it was observed that the incidence of fruit borer was completely absent in TR23 treated plants either alone or in combinations with other organic inputs. The pest incidence had a negative correlation with K and TSS.

Certification of Organic Horticulture Produce for Higher Marketability

The marketability of any product will be increasing on two counts, one with its brand value another by display of its nutritional content. Horticultural produce are known for their minerals. Hence, certification of presence of dietary minerals in the organically produced horticulture produce was carried out at Sikkim University. Several organically produced horticulture crops, *viz.*, mandarin, jack fruit, edible ferns, water cress, yellow raspberry, Malero (indigenous fruit), *Nakima* (indigenous vegetable) *etc.* were profiled for dietary minerals as well as for presence/absence of heavy metals to increase its marketability. Based on such profiling some of the crops especially edible ferns and *Nakima* can be even designated as crops for the future as they will serve the human beings in the changed environmental scenario. In addition, Plant Nutrition and Ionome Laboratory of Sikkim University analyzed plant samples namely ginger, turmeric, buck wheat *etc.* for ionome profile to certify for the SIMFED (Sikkim State Cooperative Supply Marketing Federation Ltd.), which markets farmers' products. Further, a survey also was conducted in Nepal to ascertain the income levels of the organic vegetable growers and it has given the encouraging results. The income levels have increased up to one and half times in some of the crops.

Conclusion

Organic production of horticultural crops starts at soil, irrigation water and culminates in end product. So all the entities involved in the organic production system were analyzed in a continuum through ionome profiling for their safety and dietary nutrients which eventually reduced the health risks and increased the marketability. The model could easily be replicated elsewhere for safeguarding the interest of all the stake holders as well as the environment as whole.

References

Gal, J., Hursthouse, A. and Cuthbert, S. 2007. Bioavailability of arsenic and antimony in soils from an abandoned mining area, Glendinning (SW Scotland). *J. Environ. Sci. Health, Part A-Toxic/Hazard. Subst. Environ. Eng.* **42**: 1263–1274.

Lazo, P., Cullaj, A., Arapi, A. and Deda, T. 2007. Arsenic in soil environments in Albania. **In:** *Arsenic in Soils and Groundwater Environment*, (Eds.) B. Bhattacharya, A.B. Mukherjee and J. Bundschuh. Elsevier, Amsterdam, pp. 237–256.

Lodenius, M. 1987. Factors affecting the mobilization of mercury from soil. **In:** *Heavy Metals in the Environment*, Vol. 2, (Eds.) S.E. Lindberg and T.C. Hutchinson. CEP Consult., Edinburgh, pp. 417.

Lodenius, M. 1989. Heavy metals in the soil: interactions and long-term changes. *Proc. Joint Symp. Ecological Impact of Acidification*, Szabolcs, I., Budapest, pp. 131.

Mou, S., Qing, C. and Yao, A. 1999. Transfer of soil mercury. *Proc. 5th Int. Conf. Mercury as a Global Pollutant*, Rio de Janeiro, pp. 562.

Shanmugapriya, S.P., Rohan, J. and Alagiyameenal, D. 2015. Arsenic pollution in India: An overview. *J. Chem. Pharmac. Res.*, **7**(10): 174–177.

Thornton, I. and Farago, M. 1997. The geochemistry of arsenic. **In:** *Arsenic Exposure and Health Effects*, (Eds.) C.O. Abernathy, R.L. Calderon and W.R. Chappell. Chapman and Hall, London, p. 1.

57

Organic Cultivation of Medicinal and Aromatic Plants

P. Manivel, Hemlata Bharati and Jitendra Kumar*

ICAR-Directorate of Medicinal and Aromatic Plants,
Boriavi, Anand – 387 310, Gujarat
**E-mal: manivelp@yahoo.com*

Though, organic farming is getting popularity in recent times, but it was initiated in 10000 years back to 'Neolithi age, practiced by ancient civilization depending on natural sources only. There is brief mention of several organic inputs in our ancient literatures like Rigveda, Ramayana, Mahabharata, Kautilya Arthasastthra but the concept of organic farming is not clear to many concerns. Many people think that traditional agriculture, sustainable agriculture, *Jaivik Krishi etc.*, are organic farming. Some people are of the idea that the use of organic manures and natural methods of plant protection instead of using synthetic fertilizers/pesticides are organic farming. The organic farming in real sense envisages a comprehensive management approach to improve the health of under lying productivity of the soil. Earlier Lampkin *et al.* (1999) mentioned that organic agriculture is a production system which avoids or largely excludes the use of synthetic compounded fertilizers, pesticides, growth regulators and livestock feed additives. It relies on crop rotation, crop residues, animal manure, legumes, green manure, off farming organic waste and aspects of biological pest control (Bhattacharya, 2004).

Organic agriculture is developing rapidly; it share in agricultural land and farm continues to grow in many countries. According to the FiBL-IFOAM Survey (Willer and Kilcher, 2012) on certified organic agriculture worldwide, 37 million hectares of land is under organic agricultural (including in-conversion areas). The largest areas of organic agricultural land are in Oceania (12.1 million hectares), Europe (10 million hectares), and Latin America (8.46 million hectares). In total, 80 million hectares (agricultural and non-agricultural areas) are organic in the world and 1.6 million organic producers (Figure 1). India is leading country (0.8 million hectares) in Asia (2.8 million hectares) followed by China (1.4 million hectares). There were almost 0.4 million organic producers in India. Currently, India ranks 10th among the top ten countries in terms of cultivable land under organic certification. The certified area under organic certification is 4.72 million hectare in India which includes 15 per cent cultivable area with 0.72 million hectares and rest 85 per cent (3.99 million hectares) is forest and wild area for collection of minor forest produces. India produced around 1.24 million MT of certified organic products which includes all varieties of food products namely sugarcane, cotton, oil seeds, Basmati rice, pulses, spices, tea, fruits, dry fruits, vegetables, coffee and their value-added products. The

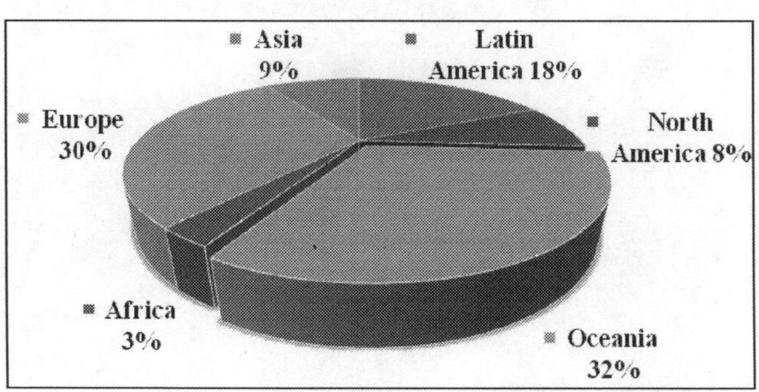

Figure 1: Worldwide Distribution of Organically Managed Lands by Regions (2014). ***Source*: FiBL-IFOAM Survey 2016.**

production is not limited to the edible sector but also produces organic cotton fibre, functional food products *etc.* among all the states, Madhya Pradesh has covered largest area under organic certification followed by Himachal Pradesh and Rajasthan. India exported 135 products in the year 2013-14 with the total volume of 194088 MT including 16,322 MT organic textiles. The organic-agri export realization was around 403 million US $ including organic textiles registering a 7.73 per cent growth over the previous year. India is exporting organic products to US, European Union, Canada, Switzerland, Australia, New Zealand, South East Asian countries, Middle East, South Africa *etc.* (APEDA, 2013-14).

Emergence of Organic Agriculture in India

With the phenomenal growth in area under organic management and growing demand for wild harvest products India has emerged as the single largest country with highest arable cultivated land under organic management (Figure 2). India has also achieved the status of single largest country in terms of total area under certified organic wild harvest collection. The growth of organic agriculture in India has three dimensions and is being adopted by farmers for different reasons. First category of organic farmers are those which are situated in no-input or low-input use zones, for them organic is a way of life and they are doing it as a tradition (may be under compulsion in the absence of resources needed for conventional high input intensive agriculture). Second category of farmers are those which have recently adopted the organic in the wake of ill effects of conventional agriculture, may be in the form of reduced soil fertility, food toxicity or increasing cost and diminishing returns. The third category comprised of farmers and enterprises which have systematically adopted the commercial organic agriculture to capture emerging market opportunities and premium prices. While majority of farmers in first category are traditional (or by default) organic they are not certified, second category farmers comprised of both certified and un-certified but majority of third category farmers are certified. These are the third category commercial farmers which are attracting most attention. The entire data available on organic agriculture today, relates to these commercial organic farmers. This method is now widely used to grow medicinal plants. The annual global market for medicinal herbs is estimated at around US$ 65 billion. Several methods have been adopted to carry the process of organic farming. These Medicinal plants are not only a major resource base for the traditional medicine and herbal industry but also provide livelihood and health security to a large segment of Indian population. The domestic trade of the AYUSH industry is of the order of Rs. 80 to 90 billion. As per study commissioned by the Associated Chamber of Commerce and Industry (ASSOCHAM), the Indian herbal industry is projected to double in the next 10 years. The supreme goal of organic production is to develop enterprises that are sustainable and amicable with the environment.

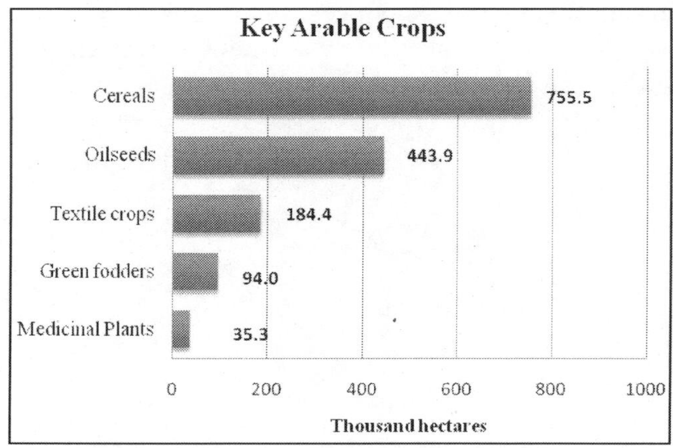

Figure 2 Asia: Use of Agricultural Organic Land 2014. *Source*: FiBL survey 2016; based on information from the private sector, certifiers and governments.

Organic Medicinal Plants

Medicinal Plants, 'the plant is said that all of its components are either fresh, dried or processed for diagnosis, treatment, prevention, helping to maintain the health of humans or animals and other physiological functions of the plant used process'. Some studies have reported the application of organic fertilizers can enhance antimicrobial and anticancer activities of secondary metabolites like phenolic in medicinal plant (Luo, 2008). The increase in production of secondary metabolites under organic fertilization might be due to high in micronutrients content in the organic fertilizer used. Plants grown under organic agricultural conditions are reported to have higher micronutrient content in more cases than conventionally grown plants (Weibel *et al.*, 2000; Ibrahim and Jaafar, 2011). Considering the fact that some of chemical reactions in cells involve minor elements, either directly or indirectly, this could explain why organic-fertilized plants exhibited higher production of secondary metabolites (Bimova and Pokluda, 2009). It was observed that the use of chicken dung (organic fertilizer) resulted in higher production of secondary metabolites and increased antioxidant activity compared to the use of NPK (inorganic fertilizer). It was also observed that higher rates of fertilizer enhanced the production of nitrate. The nitrate content was found to be the lowest with chicken dung fertilization. The results indicated that the use of chicken dung was superior compared to the NPK fertilizer in producing high quality herbal plants (*L. pumila*) for pharmaceutical use (Ibrahim *et al.*, 2013). Nitrate has been attributed to negative effects to human health. Toxicity of nitrate to human can be manifested by headaches, syncope, vertigo and discolouration that manifest in fingers or lips (Winchester *et al.*, 2009). The results indicate that the usage of organic fertilizer can minimize the nitrate content in plants. Organic fertilizer contains nitrogen bound to organic material that is slowly released (Benbrook *et al.*, 2008).

Inorganic fertilizer which is absorbed rapidly into the plant usually has higher nitrosamines which have been associated with chronic diseases such as leukemia and gastrointestinal cancer (Rembialkowska, 2007). The nitrate levels that acceptable to human are 10–100 ppm in drinking water. Intake more than 100 ppm would induce the toxic effects (Van and Loch, 1983). Usually, the processing of food (*e.g.*, chopping, grinding and heating) would reduce the nitrate content and thus reduce the negative effects (Gatseva and Dimitrov, 1997). This implies the application of organic fertilizer plays an important role in the production of healthier herbal products (Table 1). The work research work on organic cultivation of many medicinal plants has been started at ICAR-DMAPR, Anand and its All India Coordinated Project on Medicinal and Aromatic Plants centers at various locations of India (Figure 3).

Table 1: Response of Medicinal and Aromatic Plants to Farmyard Manure (FYM)

Crop	*Rate of Application (t/ha)*	*Increase in Yield (%)*	*References*
Ambrette	20 FYM	308.7	Rajeswara Rao *et al.* (2002)
Bacopa	20 FYM	33.8	Prakasa Rao *et al.* (2003)
Periwinkle	15 FYM	NS	Prakasa Rao *et al.* (1988)
Kalmegh	10FYM	150.0	Rajeswara Rao *et al.* (2004)
Ashwagandha	7.5 FYM	53.8	Rajeswara Rao (2001)
Velvet bean	5VC	30.4	Rajeswara Rao (2001)
Davana	30 FYM	36.7	Prakasa Rao *et al.* (1997)
Jamrosa (*Cymbopogon*)	24 FYM	22.8	Chand *et al.* (1996)
Menthol mint	20 FYM	22.1–37.7	Ram and Kumar (1997) Patra *et al.* (2000)
Palmarosa	15 FYM	10.3	Rajeswara Rao (2001)

Organic Essential Oils

In the flavors and fragrances market, the market for essential oils and extracts is predicted to grow the fastest, at more than 5 per cent per year (CBI, 2009b). Fruity flavors in food and beverages are currently very popular, and are believed to be driven by consumers wanting healthy and natural products. This should contribute to an increased industrial demand for fruity essential oils (Table 2). An increasing demand for natural and organic flavors is requiring manufacturers to use compatible production processes. Apart from an increasing demand for essential oils produced from organically-grown plants, there is also a growing need for fractionated oils. Fractionation is a separation process that allows the

Figure 3: Organic Cultivation of *Ashwagandha* and *Tulsi* at ICAR-DMAPR, Anand.

Table 2: Global Production of Essential Oils, Examples of Top Essential Oils in each Production Category (Compiled from table in CBI, 2009a)

Production above 100 tonnes	*Production 50–100 tonnes*	*Production 1–50 tonnes*
Eucalyptus (*Eucalyptus globulus*)	Spike lavender (*Lavandula latifolia*)	Rose (*Rosa damscena, R. centifolia*)
Peppermint (*Mentha piperita*)	Basil (*Ocimum basilicum*)	Hyssop (*Hyssopus officinalis*)
Citronella (*Cymbopogon winterianus*)	Palmarosa (*Cymbopogon martinii*)	Calamus (*Acorus calamus*)
Patchouli (*Pogostemon cablin*)	Rosewood (*Aniba rosaeodora*)	Carnation (*Dianthus caryophyllus*)
Scotch spearmint (*Mentha spicata*)	Sandalwood (*Santalum album*)	Mimosa (*Acasia dealbata, A. mearnsii*)
Eucalyptus (*E. citriodora*)	Chamomile, Roman and blue (*Anthemis nobilis*)	Jasmine (*Jasminum grandiflorum*)
Vetiver (*Vetiveria zizanoides*)		Tuberose (*Polianthes tuberosa*)
		Violet leaf (*Viola odorata*)
		Cumin (*Cuminum cyminum*)

isolation of individual or groups of compounds from a complex essential oil mixture. Applying fractionation to organic essential oils does not change their status as organic, and in this way individual compounds responsible for a particular fragrance can be obtained and combined with others to obtain a wide range of new organic flavors and fragrances (CBI, 2009b).

Production of Herbal Medicines by Organic Farming Practices

Ayurvedic practitioners have not given much emphasis to the evaluation of adverse side reactions. The efficacy of some herbal products is beyond doubt. The toxic effect of herbal drugs may be because of adulteration, contamination by pesticides and their diseased state. Toxic materials like arsenic, mercury, lead, *etc.* having been detected from time to time in some herbs. The Food and Drug Administration, USA has banned utilization of many plants by human beings on this background. In order to promote Indian herb preparations in the global market, care should be taken during drug standardization. Synergy in organic cultivation of herbal plants may lead to improved products and it would be better if new and improved drug formulations are prepared with some of the globally demanded drugs of Indian medicinal plants so that India may occupy a leading position in the world market of herbal drugs. The government and drug industry must come forward in funding and undertaking this task, so that good quality and safe herbal preparations are a reality.

Certification

This means that a crop was produced without input of synthetic agrochemicals, and that the farmland used has been free of such chemicals for a certain number of years, that the growing site was inspected and the whole growing process audited. Different countries have different organizations dealing with organic certification, while the International Federation of Organic Agriculture Movements (IFOAM), an NGO, is a main international organization (www.ifoam.org). Social certification may be relevant to the cultivation of medicinal and aromatic plants (MAPs), depending on where it occurs. The aim of social certification is to ensure that trade is based on fair and ethical standards. For example if plants are cultivated in a developing country for a company in the EU this certification ensures states that producers and workers are given fair terms of trade and fair prices and wages. Examples of social certification systems include those of the Fair Trade Labeling Organization International (FLO) (www.fairtrade.net), International Fair Trade Association (www.ifat.org), and of Social Accountability International (SAI) (http://www.sa-intl.org). The International Organization for Standardization (ISO) is an international body that also establishes quality standards. Two standards are relevant to the production of MAPs. ISO 9001:2001 for quality management systems was implemented to ensure quality requirements are satisfied and to enhance satisfaction in supplier-consumer relationships. An updated new standard (ISO 9001:2008) is now gradually being implemented for quality management systems. A second standard, ISO 14001:2004 is for environmental management systems, and is relevant for organizations that wish to operate in an environmentally sustainable way (www.iso.org).

Chemical vs. Organic Cultivation

In medicinal plants, the quality of end product is very important; the secondary metabolites from medicinal and aromatic plants are value in commerce. Chemical fertilizers resulted in overall growth of crop plants and enhancement of active constituents of many medicinal plants but their overuse and unbalanced application without prior analysis of soil nutrient status coupled with intensive farming operations, inadequate land management and cultivation of high yielding crop varieties often lead to non-sustainability in production and also pose a serious threat to soil health in terms of decline in soil organic matter (Zia *et al.*, 1998; Saha *et al.*, 2007; Qazi *et al.*, 2009). A number of studies reported the effects of agricultural practices on the secondary metabolites in medicinal and aromatic plants (Ozguven *et al.*, 2002; Ashraf *et al.*, 2006; Sekeroglu and Ozguven 2006; Ozguven *et al.*, 2008). Fertilizer applications generally affect oil yield by enhancing the amount of biomass yield per unit area. Rao *et al.* (1983) reported that *C. winterianus* responds very well to applied nitrogen with cultivars differing by up to 42 per cent, with respect to herbage content and up to 36 per cent, with respect to oil yields per ha at the same nitrogen levels (Singh *et al.*, 1980) S however, the crop response, in terms of higher herbage or oil yield with more nitrogen application, may diminish in succeeding years (Rao and Singh, 1991). Potassium has little influence on leaf oil content and may exert a depressing effect at high levels. Nitrogen directly affects total herbage yield in geranium, but has little or no effect on oil content or oil characteristics. In pot trials, magnesium frequently increases the geranium herb yield with no significant effect on oil concentration or composition. It is scientifically well established that active ingradients present in the medicinal plants, generally, are low. Therefore, it is most appropriate and need of the hour to develop an appropriate technology to increase the active ingrediants in medicinal plants under cultivation by the intervention of biofertilizers. Application of biofertilizers and compost significantly improved most vegetative growth characters as plant height, fresh and dry weight of leaves, stems and roots, highest accumulation of essential oil of leaves and flowers, rutin and coumarin percentage as well as nutrieant content in *Ruta graveolens* (El-Sherbeny *et al.*, 2007b). The irregular and low germination is the main problem in the propagation of many medicinal plants (Koppad and Umarbhadsha, 2006) that can be rectified by using the cocktail of biofertilizers to maximize the seed germination and vigour of the medicinal plants.

Ways to Promote Organic Farming

- ☆ Researches should be done to improve knowledge and relevant methods that will enhance productivity in Ayurvedic pharmaceutical sector.
- ☆ Common standards in organic agriculture in Ayurvedic pharmaceuticals should be developed.
- ☆ Local markets and exports should be strengthened.
- ☆ Organic products should be processed at home and marketed by local Ayurvedic pharmacies.
- ☆ Local NGOs and self-help groups must be formed by government to promote and provide safe herbal medicines cultivation, which can give quality to patients.
- ☆ Soil fertility should be enhanced by Ayurvedic and indigenous traditional methods as using cow urine preparations with ferments of turmeric, rotten fruits, nettles, vegetables, cow dung, and essential microorganisms.
- ☆ *In situ* and *ex situ* conservation of medicinal plants, including rare species should be promoted.
- ☆ Technical assistance in establishing herbal gardens should be provided.
- ☆ Training programs in cultivation of medicinal plants should be conducted.
- ☆ Awareness on the importance of medicinal plants within the society should be created.

Conclusion

India has a vast and rich resource of herbal raw materials and it can create a front position for itself in the global herbal market if the domestic industry produced quality products of international standards. Adopting Organic farming in medicinal plants cultivation is the need of the day. Whereas on one hand pollution due to advancement in technology is increasing at a great pace, on the other population is also increasing. It is best if the amount of crops grown is increased and the time taken to grow them is reduced but this should not be done at the price of posing health hazards to the public. In this direction, the organic cultivation of medicinal and aromatic plants is one of the best viable options readily available to us, although more better and profitable techniques can be introduced by carrying out further research in this regard. The medicinal plants are the basic source of raw-material for preparation of Ayurvedic medicines. By implementing good agro-technique and organic farming practices in medicinal plants cultivation, the safety and quality of plant materials and finished products could be assured. Authors believe that integration of internationally accepted agro-techniques with traditional Indian system of harvesting would help in following Good Agriculture and safe Practices as well as getting certificates. In this direction, encouraging results are coming from public and private sectors towards the organic cultivation of medicinal and aromatic plants and the future is very bright in this field.

References

Ashraf, M., Qasim, A. and Zafar, I. 2006. Effect of nitrogen application rate on the content and composition of oil and minerals in black cumin (*Nigella sativa* L.) seeds. *J. Sci. Food Agric.* **86**(6): 871-876.

Benbrook, C., Zhao, X., Yáñez, J., Davies, N. and Andrews, P. 2008. *New Evidence Confirms the Nutritional Superiority of Plant-Based Organic Foods. State of Science Review*; the Organic Center: Boulder, CO, USA.

Bhattacharya, A.K. 2004. Integrated nutrient management in king of bitters. Proceedings of Symposium on Organic Farming-Prospects and Challenges in the Millennium. Society of Agronomists of Acharya N.G. Ranga Agricultural University, Hyderabad, pp. 149-152.

Bimova, P. and Pokluda, R. 2009. Impact of organic fertilizers on total antioxidant capacity of head cabbage. *Hortic. Sci. (Prague)*, **36**: 21–25.

CBI. 2009a. Natural Ingredients for cosmetics: the EU market for essential oils for cosmetics. Centre for the Promotion of Imports from Developing Countries.

CBI. 2009b. The Natural colours, flavours and thickeners market in the EU. Centre for the Promotion of Imports from Developing Countries.

Chand, S., Kaul, P.N., Bhattacharya, A.K., Singh, K., Singh, C.P. and Rajeswara Rao, B.R. 1996. Response of jamrosa (Cymbopogon species) to spacings, FYM, nitrogen, phosphorus and zinc. Indian Perfum., **40**: 41-46.

El-Sherbeny, S.E., Khalil, M.Y. and Hussein, M.S. 2007b. Growth and productivity of rue (*Ruta graveolens*) under different foliar fertilizers application. *J. Appl. Sci. Res.*, **3**(5): 399-407.

Frick and Bonn. 2016. FiBL and IFOAM-Organics Intervention: The World of Organic Agriculture. pp-182-188.

Gangolli, S.D. 1994. Assessment: Nitrate, nitrite and N-nitroso compounds. *Eur. J. Pharmacol.*, **292**: 1–38.

Gatseva, P. Dimitrov, I. 1997. Population morbidity in a community with nitrate contamination of drinking water. *Folia Medica*, **39**: 65–71.

Ibrahim M.H. and Jaafar, H.Z.E. (2011). Photosynthetic capacity, photochemical efficiency and chlorophyll content of three varieties of *Labisia pumila* Benth exposed to open field and greenhouse growing conditions. *Acta Physiol. Plant.*, **33:** 2179–2185.

Ibrahim, M.H., Jaafar, H.Z.E., Karimi, E. and Ghasemzadeh, A. 2013. Impact of organic and inorganic fertilizers application on the Phytochemical and antioxidant activity of Kacipfatimah (*Labisia pumila* Benth). Molecules, **18**: 10973-10988.

Lampkin, N. 2007. Organic farming's contribution to climate change and agricultural sustainability. Welsh organic producer conference.

Luo, Y.L. 2008. Inhibition of cell growth and VEGF expression in ovarian cancer cells by flavonoids. *Nut. Cancer*, **60**: 800–809.

Ozguven, M., Kirpik, M. and Sekeroglu, N. 2002. Determination of the optimal sowing time and nitrogen fertilization for lavender (*Lavendula angustifolia* Mill.) in the Cukurova conditions, **In**: Proceedings of the Workshop on Agricultural and Quality Aspects of Medicinal and Aromatic Plants, Adana, Turkey, May 29-June 1, pp.217-223.

Ozguven, M., Sener, B., Orhan, I., Sekerglu, N., Kirpik, M., Kartal, M., Pesin, I. and Kaya, Z. 2008. Effects of varying nitrogen doses on yield, yield componentsand artimisinin content of *Artimisia annua* L. *Ind. Crops. Prod*, **27**: 60-64.

Patra, D.D., Anwar, M. and Chand, S. 2000. Integrated nutrient management and waste recycling for restoring soil fertility and productivity in Japanese mint and mustard sequence in Utter Pradesh, India. *Agric. Ecosys. Environ.*, **80**: 267-275.

Prakasa Rao, E.V.S. 1992. Fertilizer management in medicinal and aromatic crops. In: NonTraditional Sectors for Fertilizer Use. Ed. Tandon, H.L.S. Fertilizer Development and Consultation Organization, New Delhi, India. pp. 73-93.

Prakasa Rao, E.V.S. and Puttanna, K. 2000. Role of leaf mulch and farmyard manure in N economy of Java citronella (Cymbopogon winterianus). Proceedings of National Seminar on Developments in Soil Science, Bengaluru, pp. 97.

Prakasa Rao, E.V.S., Ganesha Rao, R.S. and Puttanna, K. 2001. Studies on in situ soil moisture conservation and additions of phosphorus and potassium in rainfed palmarosa (*Cymbopogon martinii* var. motia) in a semi-arid tropical region of India. *European J. Agron.*, **167**: 172. 45

Prakasa Rao, E.V.s., Narayana, M.R. and Rajeswara Rao, B.R. 1997. The effect of nitrogen and farm yard manure on yield and nutrient uptake in davana (*Artemisia pattens* Wall. ex D.C.). J. *Herbs Spices Med. Plants*, **5**: 39-48.

Prakasa Rao, E.V.S., Puttanna, K. and Singh, M. 1988. Effect of nitrogen, phosphorus and farmyard manure on yield and alkaloid concentration in *Catharanthus roseus* (L.) G. Don. *J. agric. Sci. Camb.*, **110**: 205-206.

Prakasa Rao, E.V.S., Puttanna, K., Ganesha Rao, R.S. and Chandravamshi, P. 2003. Response of Bacopa monnierito inorganic and organic sources of nutrients. Proceedings of National Seminar on New Perspectives in Spices, Medicinal and Aromatic Plants, 27-29 November 2003, Calicut, Kerala, pp.'166-168.

Prasad, R., Rajale, G.B. and Lakhdive, B.A. 1971. Nitrification retarders and slow-release nitrogen fertilizers. *Adv. Agron.* **23**: 337-383.

Puttanna, K. and Prakasa Rao, E.V.S.1997. Effect of incorporation of citronella spent grass in a soil on nitrogen economy and yield of citronella (*Cymbopogon winterianus* Jowitt.). *J. Indian Soc. Soil Sci.*, **45**: 835-837.

Qazi, M.A., Akram, M., Ahmad, N., Artiola, J., Tuller, M. 2009. Economical and environmental implications of solid waste compost applications to agricultural fields in Punjab, Pakistan. *Waste management*, **29**(9): 2437-2445.

Rajeswara Rao, B.R. 2001. Biomass and essential oil yields of rainfed palmarosa (*Cymbopogon martinii* (Roxb.) Wats. var. motia Burk) supplied with different levels of organic manure and fertilizer nitrogen in semi-arid tropical climate. *lndus. Crops Prod.*, **14**: 17 1-178.

Rajeswara Rao, B.R. 2002. Biomass yield, essential oil yield and essential oil composition of rose-scented geranium (*Pelargonium* species) as influenced by row spacings and intercropprng with cornmint (*Mentha arvensis* Lt. piperascens Malinv. ex Holmes). *lndus. Crops Prod.*, **16**: 133-144.

Rajeswara Rao,8.R., Bhattacharya, A.K., Chand, S. and Kaul, P.N. 1993. Cultural and Chemical methods of weed control for citronella java (*Cymbopogon winterianus* Jowitt.). *lndian Perfum*. **37**: 270 – 274

Ram, M. and Kumar, S. 1997. Yield improvement in the regenerated and transplanted mint *Mentha arvensis* by recycling the organic wastes and manures. *Bioresource Technol*. **59**: 141-149.

Rao, E.V.S.P., Narayana, M.R., Singh, M. and Putatanna, K. 1983. Effect of N, P, and K fertilizers on growth, yield and nutrient uptake in *Java citronella* (*C. winterianus* Jowitt). *Z Acker Pflanzenbau*, **152**: 279-283.

Rembialkowska, E. 2007. Quality of plant products from organic agriculture. *J. Sci. Food Agric.*, **87**: 2757–2762.

Saha, P.K., Ishaque, M., Saleque, M. A., Miah, M.A.M., Panaullah, G.M. and Bhuiyan, N.I. 2007. long-term integrated nutrient management for rice-based cropping pattern: Effect on growth, yield, nutrient uptake, nutrient balance sheet, and soil fertility. *Commun. Soil Sci. Plant Anal.*, **38**(5-6): 579-610.

Sekeroglu, N. and Ozguven, M. 2006. Effects of different nitrogen doses and plant densities on yield and quality of *Oenothera biennis* L. grown in irrigated lowland and un-irrigated dryland conditions. *Turk J Agric For.*, **30**: 125-135.

Sharma, S.N., Singh, A. and Tripathy, A. 1980. Response of Palmarosa to nitrogen, phosphorous, potassium and zinc. *Indian J. Agro.*, **25**(4): 719-723.

Van, D.W. and Loch, J.P.G. 1983. Nitrate in the Netherlands: A serious threat to groundwater. *Aqua*, **2**: 59–60.

Weibel, F.P., Bickel, R., Leuthold, S. and Alfoldi, T. 2000. Are organically grown apples tastier and healthier? A comparative field study using conventional and alternative methods to measure fruit quality. *Acta Hort.*, **517**: 417–426.

Willer, H. and Kilcher, L. (Eds.). 2012. The World of Organic Agriculture: Statistics and Emerging Trends 2012: Bonn: International Federation of Organic Agriculture Movements (IFOAM); Frick, Switzerland: Research Institute of Organic Agriculture (FiBL).

Winchester, P.D., Huskins, J. and Ying, J. 2009. Agrichemicals in surface water and birth defects in the United States. *Acta Paediat.*, **98**: 664–669.

Zia, M.S., Baig, M.B. and Tahir, M.B. 1998. Soil environment issues and their impact on agricultural productivity of high potential areas of Pakistan. *Science Vision*, **4**(2): 56-61.

58

Role of Weather Factors in the Progression of Leaf Blight in Mango

A.K. Saxena*, K. Rathnamma, and R. Thilaka Rani

Division of Plant Pathology,
ICAR-Indian Institute of Horticultural Research,
Hessaraghatta Lake P.O., Bengaluru – 560 089, Karnataka
**E-mail: aks@iihr.res.in*

Mango (*Mangiferae indica* L.) is known as the king of fruits, is indigenous to India. It is grown in area of about 2516 million ha in different states like Uttar Pradesh, Telangana, Karnataka,Gujarat, Maharashtra *etc.* with an annual production of 18431million MT (NHB data base, 13). There are more than 1000 varieties of mango. However only few varieties are commercially cultivated throughout India that include Alphonso, Totapuri, Mallika, Dasheri, Banishan *etc.* India is a major exporter of fresh mango and mango pulp. Mango crop is experiencing the impact of climate change. Variability in climatic factors namely temperature, relative humidity, total rainfall and its distribution not only influencing the flowering pattern and fruit set but also disease dynamics in mango and some of the diseases that were considered minor earlier have become serious. Leaf blight caused by *Pestalotiaopsis mangiferae* (*Henn.*) Steyaert in mango has attained serious proportion in recent days in India. Disease has been reported from Malaysia (Lim and Khoo, 7); Taiwan (Kho *et al.*, 6); UK (Mordue, 11); US (Poeltz *et al.*, 17); Italy (Ismail *et al.*, 5) and in India it occurred in Gujarat (Vala *et al.*, 27); Punjab (Verma and Singh, 29); M.P. (Rai, 18) However, though Karnataka is one of the major mango producing state in the country, yet the status of leaf blight disease in mango orchards located in the state is not known. Determining the effect of temperature, rainfall, leaf wetness and relative humidity on the formation, release and germination of inoculums in different pathosystems have been focused by many researchers worldwide (Rowe and Beute, 20; Sutton, 26; Pinkerton *et al.*, 15; MacHardy *et al.*, 8; Mondal and Timmer, 10). Based on understanding the disease epidemiology, effective control measures could be developed and implemented (Hopkins and McQuilken, 4). Prevailing weather conditions influence the onset, initiation and progression of disease, besides others (Saxena, *et al.*, 21). It is worthwhile to determine the role of weather factors in relation to disease initiation and development aimed at developing disease prediction models that in will turn lead to need based, judicious and effective disease management programme through fungicidal application (Venugopalan *et al.*, 31). Therefore, efforts were made to determine the status of leaf blight in the state of Karnataka and study the role of weather factors on the incidence and development of leaf blight in mango and statistical correlation were drawn to determine the relationship between weather parameters with disease progression so as to develop prediction models and apply suitable management strategy.

Materials and Methods

A roving survey was conducted for the status of leaf blight in different mango growing areas of the Karnataka state during 2015. Percentage disease incidence and severity was recorded. Leaves were scored for the severity of leaf blight following a 0-5 disease rating scale and data thus obtained were converted Percent Disease Severity Index (PDI) following McKinney (9).

Isolations were made from the disease leaf samples collected during survey and identity of the associated pathogen was confirmed.

Influence of weather factors on the initiation and progression of leaf blight was studied during the cropping season 2015. Thirty year old selected mango trees (var. Totapuri) located in the experimental orchard of ICAR-IIHR, Bengaluru were regularly visited and disease severity was recorded following McKinney (9) as explained above. Data on prevailing weather factors were also recorded and correlated with disease severity. Selected fungicidal molecules were evaluated for their efficacy against the pathogen *in vitro* following 'food poison technique'.

Results and Discussion

Roving survey conducted for the status of diseases in the mango orchards in the state of Karnataka revealed that the severity of leaf blight was least (8.00 – 18.00 per cent) in the mango orchards located in Mysore and Tumkur districts whereas it was maximum (20.00 – 24.00 per cent) in orchards located in Bengaluru district. Mango orchards in Kolar and Chitrdurga districts recorded disease severity in the range of 14.00 -20.00 per cent and 16.00 – 22.50 per cent respectively. Isolations from infected leaves samples and subsequent re-isolations made from the inoculated leaves constantly yielded a fungus that was identified as *Pestalotiopsis mangiferae* based on cultural characteristics and the identity of same was confirmed following (Mordue, 11) as *Pestalotiopsis mangiferae* (Henn.) Steyaert. The pathogen confirmed the Koch's postulates. Increased incidence and severity of the leaf blight in different localities was noticed by Vala *et al.*, 27; Verma and Singh, 28 and Rai, 18) and the same pathogen was isolated from leaf blight sample by them. However the present status of leaf blight in the state of Karnataka was document for the first time. Occurrence of the disease at these levels may be due non application of required and effective management strategy and unawareness among the growers about the emerging threat in view of climate change. In mango variety 'Totapuri' disease severity (PDI) was 16.00 during (January 2015) Ist standard meteorological week (SMW) when meteorological factors conditioning the weather were: maximum temperature = 28.6°C; minimum temperature = 20.4°C, RH at 7.00h at and 14.00 h were 71 per cent and 53 per cent, respectively whereas bright sunshine h were 7.00h; wind speed = 2.83 km/h; and pan evaporation was 2.64 mm. without any rain fall. Severity of leaf blight recorded the peak value of 29.50 per cent during 47th SMW (November, 2015) when meteorological factors conditioning the weather were: maximum temperature = 28.1°C; minimum temperature = 17.6°C, RH at 7.00 h and 14.00 h were 84 and 52 per cent, respectively with total rain fall of 26.00 mm received in four rainy days whereas bright sunshine h were 2.20; wind speed = 2.89 km/h; and pan evaporation was 0.86 mm. Subsequently, disease progression was not steady, recording increasing and decreasing trend during the following met. weeks and finally during 9th SMW, 2016 disease severity was 23.50 per cent when meteorological factors conditioning the weather were: maximum temperature = 33.57°C; minimum temperature = 19.14°C, RH at 7.00 h and 14.00 h were 71 per cent and 51 per cent, respectively whereas bright sunshine h were 7.7; wind speed = 4.90km/h; and pan evaporation was 6.0 mm. without any rain fall (Table 1). Correlation analysis revealed that disease severity was positively correlated with RH at 7.00 h (r = 0.3133); RH at 14.00 h (r = 0.0943) and clouds at 7.00 h (r = 0.0454); whereas negative correlation was recorded with maximum temperature (r = - 0.2969); minimum temperature. (r = - 0.2394); total rainfall (r = - 0.0328) and no. of rainy days (r = - 0.0937); bright sunshine (r = - 0.1369) and Clouds (Octas) at 14.00 h (r = - 0.2004); wind speed (r = -0.4219); and pan evaporation (r = - 0.3319) (Table 2). Fluctuations in the progression of disease severity and similar correlation pattern were recorded by Vala *et al.* (27). According to Verma and Singh, (30) also, the incidence and development of grey blight was negatively correlated with maximim and minimum temperature and positively correlated with maximum and minimum RH and total rainfall. All these factors played an important role in the development of mango grey leaf blight disease (Patil, 16). Temperature in the range of 25 – 35°C was suitable for the germination and subsequent infection by the conidia of the pathogen (Vala *et al.*, 27., Patil, 16) whereas high relative humidity helped in faster development (Derso, 2; Okigbo, 14) of such diseases. For developing a reliable disease prediction model, long term data on the influence and relationship between meteorological factors on the disease initiation and progression for several disease cycles is utmost required. Climate change will change patterns of disease through changes in host distribution and phenology, changes in plant-associated microflora and direct biological effects on rapidly evolving pathogens. Short-term forecast models coupled with weather generated from climate simulations may be a basis for projection; however, they will often fail to capture long-term trends effectively (Shaw, 25). If we know the basic factors limiting the abundance of a pathogen and understand quantitatively their relation to weather, It may then be appropriate to use a weather-based prediction system (Seem, 23; Semenov, 24) to predict future abundance. Garrett *et al.* (3) argue that this reductive, process-based approach is practical, though challenging, for many pathosystems. Therefore, the problem of preparing for change in plant disease due to changing climate is inevitably complex, and specific prediction with any useful degree of skill more than a short time ahead is unlikely to be possible This is because the uncertainties in predicting all the factors influencing disease are likely to spread out the initial distribution of uncertainty in climate prediction until the distribution is almost uninformative (Scherm, 22). If we wish to forecast

Table 1: Inluence of Weather Factors on the Progression Laf blight (*P. mangiferae*) in Mango during 2015-16

Std Met Week	PDI	Temperature (ºC)		Relative Humidity (%)		Wind Speed	Bright Sun Shine	Cloud Amt (Octas)		Rainy Days	Rainfall	Pan Evap.
	(%)	Max	Min	(730h)	(1430 h)	(km/h)	(h)	(730h)	(1430 h)	(No)	(mm)	(mm)
(2015)												
1	16.00	28.57	20.43	71.43	52.57	2.83	7.30	2.00	2.00	0.00	0.00	2.64
2	15.33	28.14	18.43	69.57	41.14	2.74	8.30	1.00	2.00	0.00	0.00	2.93
3	14.67	27.29	17.00	70.71	42.29	3.13	8.60	1.00	2.00	0.00	0.00	3.01
4	14.67	27.86	14.57	72.29	45.71	4.79	9.20	0.00	1.00	0.00	0.00	3.74
5	18.17	28.29	13.71	68.29	37.29	4.90	8.50	1.71	1.71	0.00	0.00	3.53
6	14.50	28.86	13.43	63.86	41.71	4.57	9.90	0.00	1.10	0.00	0.00	4.26
7	**14.17**	29.43	14.71	63.29	40.86	2.90	10.20	0.00	1.10	0.00	0.00	4.21
8	17.67	29.86	15.00	45.29	34.29	4.59	10.30	0.00	1.40	0.00	0.00	5.09
9	15.00	30.86	18.86	67.00	41.29	3.94	7.57	0.71	1.71	2.00	1.24	5.03
10	18.50	30.71	18.86	66.86	51.86	4.37	8.40	0.00	1.00	0.00	0.00	4.91
11	14.17	31.14	19.43	63.43	45.71	4.95	9.40	1.00	1.00	0.00	0.00	6.00
12	19.50	32.00	18.86	64.14	47.57	4.10	9.20	0.00	1.00	0.00	0.00	6.46
13	16.17	33.29	19.57	66.86	52.29	4.67	9.40	1.00	2.00	0.00	0.00	6.23
14	16.17	33.71	21.71	74.71	56.29	4.90	8.10	0.30	1.10	0.00	0.00	6.57
15	14.83	31.43	20.00	69.00	58.29	4.13	6.70	1.00	2.00	3.00	3.00	4.06
16	17.00	31.71	19.43	76.86	55.71	3.69	9.30	1.00	1.00	2.00	2.54	5.50
17	18.17	32.29	20.43	73.86	51.71	3.70	8.20	1.00	2.00	2.00	3.14	4.64
18	15.00	32.86	20.86	75.00	52.71	2.69	8.20	3.00	1.00	4.00	6.89	3.30
19	14.00	33.00	21.29	72.43	54.57	4.64	5.10	2.00	2.00	2.00	2.14	5.34
20	14.50	30.00	21.43	80.43	54.71	3.77	5.60	1.00	1.00	4.00	12.57	4.03
21	14.33	33.14	22.71	76.57	50.71	5.83	8.90	1.00	2.00	0.00	0.00	5.61
22	14.67	31.86	20.57	78.29	56.71	4.37	5.90	1.00	2.00	3.00	8.57	3.19
23	12.83	31.14	21.29	79.43	52.43	4.64	7.90	1.00	1.00	2.00	1.29	4.66
24	17.50	30.14	20.86	77.71	56.29	5.54	4.30	2.00	2.00	2.00	6.86	3.50
25	14.33	31.00	22.14	80.14	55.00	9.80	2.60	2.00	3.00	2.00	0.71	2.71
26	17.33	30.29	21.00	75.71	53.00	7.44	6.20	2.00	2.00	1.00	1.43	3.96
27	20.83	30.00	21.00	77.43	54.86	6.21	7.60	2.00	2.00	1.00	1.57	4.43
28	20.83	30.14	21.43	78.57	51.43	5.66	7.20	2.00	2.00	3.00	2.17	3.23
29	17.33	30.86	21.86	75.43	52.86	6.27	4.30	3.00	2.00	3.00	3.57	3.57
30	20.17	29.86	20.14	79.71	49.43	5.76	4.40	3.00	3.00	2.00	1.57	2.97
31	18.74	30.14	20.57	79.71	49.29	6.03	4.90	2.00	2.00	1.00	1.00	4.21
32	17.33	30.00	20.43	79.86	50.86	5.79	6.80	2.60	2.10	2.00	4.06	3.50
33	15.17	30.29	20.57	81.14	51.71	3.69	4.60	2.60	2.10	3.00	5.57	2.93
34	16.50	30.00	20.29	78.71	51.14	4.31	6.70	2.60	2.30	4.00	8.43	2.64
35	11.92	29.86	19.43	79.43	48.43	4.39	4.10	3.00	2.20	2.00	1.00	3.79
36	18.33	30.43	20.43	80.29	51.14	4.29	6.00	2.00	2.00	2.00	13.43	3.00
37	13.67	30.00	20.00	80.71	53.71	5.19	3.80	2.00	2.00	2.00	1.71	3.00
38	17.00	30.29	20.29	77.71	50.14	4.06	5.30	2.00	1.00	3.00	0.89	3.33
39	18.00	30.00	19.43	82.00	51.00	2.54	5.90	1.00	2.00	4.00	19.86	2.57
40	21.00	30.00	19.29	83.14	54.14	2.96	2.50	2.00	2.00	4.00	19.64	2.21
41	14.00	29.86	19.00	79.71	51.86	2.59	5.20	1.00	1.00	1.00	0.71	2.54
42	21.17	30.57	20.71	75.71	51.14	2.13	9.20	0.00	0.00	0.00	0.00	4.57
43	21.00	30.86	20.57	77.86	53.43	2.37	8.80	0.00	0.00	0.00	0.00	4.43
44	22.00	30.00	20.14	81.14	53.29	2.39	4.90	2.00	2.00	4.00	6.43	2.79

Contd...

Table 1–*Contd...*

Std Met Week	PDI	Temperature (ºC)		Relative Humidity (%)		Wind Speed	Bright Sun Shine	Cloud Amt (Octas)		Rainy Days	Rainfall	Pan Evap.
	(%)	Max	Min	(730h)	(1430 h)	(km/h)	(h)	(730h)	(1430 h)	(No)	(mm)	(mm)
45	24.67	28.86	18.57	85.00	54.71	3.46	4.80	3.00	2.00	4.00	11.86	2.07
46	26.50	26.29	16.43	85.43	53.14	2.77	0.10	5.00	4.00	3.00	3.14	0.79
47	**29.50**	28.14	17.57	83.86	51.86	2.89	2.20	4.00	3.00	4.00	3.71	0.86
48	27.00	28.29	18.43	83.86	53.43	2.56	7.50	3.00	2.00	0.00	0.00	1.29
49	23.83	27.71	19.29	79.43	52.00	3.03	2.80	0.00	0.00	0.00	0.00	1.50
50	25.67	26.14	16.43	81.71	56.14	1.56	6.10	0.00	0.00	0.00	0.00	1.29
51	21.41	28.86	17.14	82.29	52.57	2.11	9.30	0.00	0.00	1.00	0.07	1.89
52	25.00	29.38	18.13	73.38	49.88	3.05	9.10	0.00	0.00	0.00	0.00	2.16
(2016)												
1	21.33	28.00	14.71	75.71	46.57	2.74	9.70	0.00	0.00	0.00	0.00	4.20
2	23.17	29.00	15.71	72.14	43.00	1.47	8.70	0.00	0.00	0.00	0.00	4.50
3	22.67	27.00	15.14	78.71	42.43	2.29	5.80	2.00	2.00	0.00	0.00	3.90
4	20.00	26.57	15.14	82.43	51.14	2.70	5.40	0.00	1.00	0.00	0.00	4.10
5	18.50	30.00	17.57	82.00	44.71	2.63	9.31	0.00	0.00	0.00	0.00	4.51
6	22.67	31.57	18.57	73.86	47.86	2.61	8.90	0.00	0.00	0.00	0.00	4.90
7	22.33	32.14	18.86	75.57	41.71	4.17	9.00	0.00	0.00	0.00	0.00	5.10
8	27.20	33.29	20.00	74.14	52.57	2.37	9.00	0.00	0.00	0.00	0.00	5.70
9	23.50	33.57	19.14	71.71	51.29	4.90	7.70	0.00	0.00	0.00	0.00	6.00

Table 2: Inter- and Intra-class Correlation among Progression of Leaf Blight (*P. mangiferae*) in Relation to Weather Parameters

Parameters	PDI	Max. Temp. ºC	Min. Temp. °C	RH % (7.00 h)	RH % (14.00h)	Ws (km/h)	BSS (h)	Clouds (Octas) (7.00h)	Clouds (Octas) (14.00 h)	Rainy Days (No.)	Rainfall (mm)	PEV (mm)
PDI	1											
Max. Temp °C	–0.2969	1										
Min. Temp °C	–0.2394	0.6292	1									
RH% (7.00 h)	0.3133	–0.2315	0.3034	1								
RH% (14.00 h)	0.0943	0.2529	0.6498	0.5986	1							
Ws (km/h)	–0.4219	0.3408	0.4277	–0.1105	0.1226	1						
BSS (h)	–0.1369	0.2694	–0.2733	–0.6173	–0.4602	–0.1778	1					
Clouds (Octas) (7.00 h)	0.0454	–0.1543	0.2937	0.4552	0.2973	0.2530	–0.6940	1				
Clouds (Octas) (14.00 h)	–0.2004	–0.1201	0.2285	0.1877	0.1649	0.4360	–0.5969	0.8158	1			
Rainy days (no)	–0.0937	0.0828	0.4148	0.4572	0.4562	0.1122	–0.6070	0.6524	0.5331	1		
Rain fall (mm)	–0.0328	0.0463	0.2646	0.3796	0.3358	–0.0546	–0.4124	0.3453	0.3240	0.7488	1	
PEV (mm)	–0.3319	0.6838	0.1291	–0.5636	–0.2109	0.2294	0.5993	–0.5048	-0.3384	–0.4113	–0.3242	1

changes in plant disease in relation to climate, we need to test our forecasts.

For the management of leaf blight, among the evaluated molecules Azoxystrobin was most effective as it recorded 100 per cent inhibition of the pathogen *in vitro* at 100 µg/ml whereas Chitosan and Hexaconazole could record 69.15 per cent and 97.20 per cent inhibition of pathogen growth at the same concentration. Azoxystrobin was effective against *Pestalotia* leaf spot in guava (Moustafa *et al.*, 12). In view if systemic nature of the molecule, its efficacy and acceptance for use in mango, need based pre harvest applications with Azoxystrobin will be effective for the management of leaf blight disease in mango. Spraying strategy will be more effective and useful, if it is adopted based on proven effective disease forecasting model.

Acknowledgement

Financial support under project 'NICRA' and facilities provided by Director, ICAR-IIHR, Bengaluru for the work are gratefully acknowledged.

References

Bourgeois, G., Bourque A. and Deaudelin G. 2004. Modelling the impact of climate change on disease incidence: A bioclimatic challenge. *Can. J. Plant Path.*, **26**: 284–290.

Derso, E. 1999. Occurrence, prevalence and control methods of phaeoramularia leaf and fruit spot disease of citrus in Ethiopia. *Fruits*, **54**: 225–232.

Garrett, K.A., Dendy S.P., Frank E.E., Rouse M.N. and Travers S.E. 2006. Climate change effects on plant disease: genomes to ecosystems. *Ann. Rev. Phytopath.*, **44**: 489–509.

Hopkins, K.E., and McQuilken, M.P. 2000. Characteristics of *Pestalotiopsis* associated with hardy ornamental plants in the UK. *Europ. J. Plant Pathol.*, **106**: 77–85.

Ismail, A.M., Cirvilleri, G. and Polizzi, G. 2013. Characterisation and pathogenicity of *Pestalotiopsis uvicola* and *Pestalotiopsis clavispora* causing grey leaf spot of mango (*Mangifera indica* L.) in Italy. *Europ. J. Plant Pathol.* **135**(4): 619–625.

Ko, Y., Yao, K.S., Chen, Y.C. and Lin, H. 2007. First report of Gray leaf spot of mango (*Mangifera indica*) caused by *Pestalotiopsis mangiferae* in Taiwan. *Plant Dis.*, **91**(12): 1684.

Lim, T.K. and Khoo, K.C. 1985. *Diseases and Disorders of Mango in Malaysia*. Kuala Lumpur: Tropical Press.

MacHardy, W.E., Gadoury, D.M. and Gessler, C. 2001. Parasitic and biological fitness of *Venturia inaequalis*: relationship to disease management strategies. *Plant Dis.*, **85**: 1036–1051.

Mckinney, H.H., 1923. A new system of grading plant diseases. *J. Agric. Res.*, **26**: 195–217.

Mondal, S.N. and Timmer, L. W. 2002. Environmental factors affecting pseudothecial development and ascospore production of *Mycosphaerella citri*, the causal of citrus greasy spot. *Phytopathology*, 92: 1267–1275.

Mordue J.E.M. 1980. *Pestalotiopsis mangiferae*. CMI descriptions of pathogenic fungi and bacteria, **No. 676.** Kew, Surrey: Commonwealth Mycological Institute.

Moustafa M.S.H., El-Dakar Hala, A.M. and Alkolaly, Asmaa M.A. 2015 *Pestalotia* leaf spot a new disease affect guava trees in Egypt. *Int. J. Sci. and Eng. Res.*, **6**(10): 1306–1312.

NHB, 2015. *Indian Horticuture Database*. Agric. Res. Data Book, ICAR, New Delhi pp. 90–99.

Okigbo, R.N. 2001. Occurrence, pathogenicity and survival of *Macrophoma mangiferae* in leaves, branches and stems of mango (*Mangifera indica* L.). *Plant Protect. Sci.*, **37**: 138–144.

Pinkerton, J.N., Johnson, K.B., Stone, J.K. and Ivors, K.L. 1998. Factors affecting the release of ascospores of *Anisogramma anomala*. *Phytopath.* **88**: 122–128.

Patil, V.A. 2012. Investigation on grey leaf blight disease (*Pestalotia anacardii*) of Mango (*Mangifera indica* L.) under south Gujarat conditions. *Ph.D. Thesis*. Navsari Agricultural University, Navsari.

Ploetz, R.C., Zentmyer, G.A., Nishijima, W.T., Pohrbach, K.G. and Ohr, H.D. 1998. *Compendium of Tropical Fruit Diseases*. APS Press, The American Phytopathological Society.

Rai, M.K. 1996. *In vitro* evaluation of medicinal plant extracts against *Pestalotiopsis mangiferae Hind. Anti. Bull.*, **38**(1/4): 53–56.

Rizwana, H., Siddiqui, I. and Bukhary N. 2012. A post-harvest disease of *Mangifera indica* fruit caused by *Pestalotiopsis mangiferae*, in Saudi Arabia.*African J. Micro. Res.*, **6**(27): 5723–5724.

Rowe, R.C. and Beute, M.K. 1975. Ascospore formation and discharge by *Calonectria crotalariae*. *Phytopathol.*, **65**: 393–398.

Saxena, A.K., Rathnamma, K. and Thilaka Rani, R. 2016. Influence of weather factors on Black Banded Disease in Mango in view of climate change. A paper presented in 5th International conference on Plant Pathogens and people with the mission "*Challenges in Plant Pathology to Benefit Humankind*" held at New Delhi during February 23–27.

Scherm, H. 2004. Climate change: can we predict the impacts on plant pathology and pest management? *Can. J. Plant Pathol.*, **26**: 267–273.

Seem, R.C. 2004. Forecasting plant disease in a changing climate: A question of scale. *Can. J. Plant Pathol.*, **26**: 274–283.

Semenov, M.A. 2007. Simulation of extreme weather events by a stochastic weather generator. *Climate Res.*, **35**: 203–212.

Shaw, M.W. 2009. Preparing for changes in plant disease due to climate change. *Plant Protect*, **45,** (Special issue S3–S10).

Sutton, T.B. 1981. Production and dispersal of ascospores and conidia by *Physalospora obtuse* and *Botryosphaeria dothidea* in apple orchards. *Phytopathol.*, **71**: 584–589.

Vala, D.G., Solanky, K.U., Desai, V.D. and Joshi, H.U. 1989. Disease of mango occurring in Gujarat state. *Acta Hortic.* **231**: 544–550.

Verma K.S. and Singh, J. 1993. Identity and pathogenic potential of *Pestalotiopsis mangiferae* caused grey blight of mango. *Plant. Dis. Res.*, **8**: 194.

Verma, K.S. and Singh, T. 1996. Prevalence and control of grey blight of mango caused by *Pestalotiopsis mangiferae*. *Plant Dis. Res.*, **11**(1): 69–71.

Verma, K.S. and Singh, T. 1996d. Survival and epidemiology of *Pestalotiopsis mangiferae*: The cause of grey blight of mango. *Plant Dis. Res.*, **11**: 25–30.

Venugopalan, R., Rawal, R.D. and Saxena, A.K., 2005. Statistical model for ascertaining the influence and reliability of weather parameters on incidence of blossom blight in Mango (*Mangifera indica*). *J. Hort. Sci.*, **1**(1); 64–67.

59

Green Technologies for Production of Clean and Safe Food Spices

Santhosh J. Eapen, V. Srinivasan and S. Devasahayam*

ICAR-Indian Institute of Spices Research,
Kozhikode – 673 012, Kerala
E-mail: sdasahayam@gmail.com

Excessive use of chemical fertilizers and plant protection chemicals affects soil health and non-target organisms, result in pesticide residues in the produce and cause human and environmental hazards. In recent years, organic agriculture that favours lower chemical inputs, conserves non-renewable resources, preserves the environment and results in production of clean and safe foods including spices has gained considerable importance. Spices are high-value products obtained from diverse crops and are extensively used all over the world to add flavour and taste to human food besides being used in indigenous medicine. Spices are of considerable economic importance for the country, being major sources for foreign exchange among agriculture commodities and play a crucial role in the economy of various states were they are produced. India produces about 6.11 million tonnes of spices from 3.31 million ha (2014-15) and exports around 180 spice products to more than 150 countries. The total export of spices from India crossed 8.4 lakh tonnes worth Rs. 16,238 crores during 2015-16. Export of organic spices has been growing rapidly with 25 per cent increase in quantity and 11 per cent increase in value. The ICAR-IISR, Kozhikode has developed green technologies for organic production of clean and food-safe spices like black pepper, cardamom, ginger and turmeric. These practices include application of PGPR as nutrient supplements to promote rooting and growth of plants in the nursery and field, cultivation of pest and disease resistant varieties and use of biocontrol agents and natural products for pest and disease management. Post-harvest practices laying stress on hygiene to avoid contamination with harmful organisms during processing, packaging and storage have also been developed.

Cropping Systems

Spice crops are highly amenable for mixed and intercropping. Crops like black pepper and cardamom are ideally suited for cultivation along with other plantation and horticultural crops like coconut, arecanut, cocoa, coffee, tea, citrus, *etc.* Rhizomatous crops such as ginger and turmeric are highly remunerative as intercrops in coconut and other horticultural tree crops under partial shade. Ginger and turmeric can also be intercropped with a wide variety of other crops including vegetables, pulses, cereals, oil seeds, fodder and medicinal plants. Green manure crops like *daincha* (*Sesbania aculeata*) can be raised successfully in inter-spaces of ginger and turmeric beds in a row, which adds to 50 per cent of the green leaves required for mulching and suppresses weed growth, reducing the cost of cultivation. In mixed cropping systems addition of organics due to litter fall result in organic matter build and enhanced nutrient availability in soils. There is also great scope of covering all the shade trees with black pepper vines in tea and coffee estates in South and North East India. The coffee and tea plantations in South India can accommodate about 3.8 and 2.0 million standards for black pepper, respectively, and can result in additional black pepper yield up to 20,500 tonnes per year.

Plantation crops being perennial in nature contribute large quantities of waste by-products, which by composting and recycling will meet the nitrogen requirement of the crop and partly other nutrients. In South India where coconut is a major plantation crop, abundance of waste materials such as coir-dust (about 7.5 million tonnes of coir dust annually) from coir based industries provides a good scope for recycling the organic materials into the system. Composts made out of leaf and sheath wastes and husks of these plantation crops are also valuable sources of plant nutrients (Table 1). Leaf and bunch wastes of coconut, arecanut and oil palm, coffee husk and spent tea leaf have been found to give very high bioconversion efficiency. Application of dried coconut leaves as mulch in ginger beds after removing the petiole at the time of planting reduced weed growth and enhanced the yield by 10 per cent compared to green leaf mulch application. Application of paddy straw and green leaf mulch in ginger beds at the time of planting recorded 36 per cent increased yield compared to green leaf mulch application alone. Multiple cropping systems also results in continuous uptake, addition of biomass and higher level of nutrient recycling that would lead to a positive influence on the physico-chemical and biological properties of soil.

Table 1: Biomass Availability in Plantations and its Nutritive Values

Crops/Manures	Biomass Available	Nutritive Value (%)		
	t/ha/yr	N	P	K
Coconut: Coir pith compost	8-12	1.2	0.60	1.20
Vermicompost	–	2.0	0.80	1.20
Arecanut: Leaf/sheath compost	5-8	2.1	0.12	0.90
Coffee: Pulp compost	3-4	3.2	0.16	2.30
Shade trees: Leaf compost	8-10	1.8	0.84	0.78

The microbial biomass, organic content, total N, P and K is higher in the root region soils of multistoried cropping systems when compared to monocropping systems. The beneficial effects are reflected through enhanced soil fertility status, microbial activity and better utilization of natural resources for the benefit of plant growth and sustainable crop productivity. The organic addition brought through leaf litter recycling and microbial activity would lead to integrated nutrient management for enhanced system productivity.

Crop Management

The ICAR-IISR, Kozhikode, has developed package of practices for organic production of major spices like black pepper, ginger and turmeric. For organic production of black pepper, traditional varieties adapted to the local soil and climatic conditions that are resistant to pests and diseases, are preferable. All crop residues and farm wastes like green loppings, crop residues, grasses, cow dung slurry, poultry droppings, *etc.* available on the farm can be recycled through composting, including vermicomposting so that soil fertility is maintained at high level. Farmyard manure (FYM) may be applied @ 5-10 kg/vine along with vermi/leaf compost @ 5-10 kg/vine. Further, supplementation of oil cakes like neem cake (1 kg/vine), composted coir pith (2.5 kg/vine) or composted coffee pulp rich in potassium and *Azospirillum* sp. and phosphate solubilizing bacteria will improve the fertility. Based on soil test, application of lime/dolomite to correct the pH, rock phosphate/bone meal and wood ash or sulphate of potash (mineral potassium) may be done to get required quantity of phosphorus and potassium. Application of composts namely, leaf and vermicomposts, neem cake and FYM improved the soil physico-chemical properties, nutrient availability, yield and quality significantly in black pepper and cardamom. Leaf compost and vermicompost treatments supported significantly higher population of free-living N-fixing bacteria, phosphate solublizing bacteria populations. Trials conducted through AICRP on Spices at various locations revealed that sustainable yield up to 6.5 kg/vine could be achieved in black pepper by application of FYM (10 kg/vine) + rock phosphate (40 g/vine) + wood ash (2 kg/vine) along with *Azospirillum* sp. which was on par with the conventional practice in nutrient supplementation. Recently a talc based formulation (IISR Biomix) consisting of a consortium of PGPR [*Micrococcus luteus* (BRB 3)] + [*Enterobacter aerogenes* (BRB 13)] + [*Micrococcus* sp.(BRB 23)] has been developed by ICAR-IISR for application in the nursery and main field for enhanced growth and yield.

In ginger and turmeric, transplanting technology using single bud sprouts (about 5 g) raised in nursery medium of decomposed coir pith and vermicompost mixture (75:25) enriched with PGPR and *Trichoderma* sp. (10 g/kg) for planting in pro-trays has been developed. This will reduce the seed rhizome requirement by 1/4th and also helps in weeding out infected plantlets at the initial stage itself. Application of 20 t FYM + 2 t neem cake + 1 t ash + 4 t vermicompost/ha, *Azospirillum* sp. and P-solubilising bacteria (50 g/bed) for nutrient supplement, *Bacillus amyloliquefaciens* (GRB 35) as seed treatment and drenching of *B. amyloliquefaciens* (GRB 35) are helpful in sustaining the yield and keeping the rhizome rot disease incidence under check. Based on soil test, application of lime/dolomite, rock phosphate and wood ash may be done to get required quantity of phosphorus and potassium. Combined application of different organic sources like FYM + *Pongamia* oil cake + Neem oil cake + Stera meal + Rock phosphate + Wood ash yielded on par with the conventional practice in addition to high quality of the produce. The quality parameters of ginger (high oleoresin and low fibre) and turmeric (curcumin and starch) were found to be high under organic farming.

The population of bacteria and fungi and dehydrogenase enzyme activity were higher under organic practice. Exclusive use of chemical fertilizers led to a significant reduction in microbial biomass C, which was on an average 52 per cent and 28 per cent lower compared to organic and integrated nutrient managment systems, respectively. The organic carbon, nitrogen, phosphorus and potassium contents were highest in integrated system which was on par with organic management and significantly higher availability of calcium, magnesium and zinc under organic management systems. Organic management system showed highest population buildup of total bacteria, *Azospirillum* sp., *Pseudomonas* sp., Actinomycetes and *Trichoderma* sp. with less incidence of rhizome rot. With respect to yield, reduction was observed during the initial period of organic cultivation as in turmeric where, the yield was higher under integrated management system followed by organic system which was on par with conventional (chemical) system over years.

Pest and Disease Management

Spice crops are prone to attack by many pests and diseases for which several eco-friendly green technologies have been developed utilizing cultural methods, use of resistant varieties, natural products and bio-control agents (Tables 2 and 3). Improved varieties of black pepper, cardamom and ginger have been developed that are resistant to major pathogens in these crops. Soil solarization and fortifying the nursery mixture with *Trichoderma* sp. and *Pochonia chlamydosporia* in black pepper and cardamom gives protection from *Phytophthora capsici*, *P. meadii* and *Pythium* sp. and the nematodes *Radopholus similis* and *Meloidogyne incognita*. PGPR consortiums have also been developed to promote rooting and growth of plants and controlling diseases in the nursery. Effective control of foot rot disease under field condition can be achieved by adopting cultural practices like shade regulation, phyto-sanitation, minimum tillage, adequate drainage and application of *T. harzianum* @ 50 g/vine mixed with 1 kg of neem cake or 5 kg of FYM. In cardamom, shade regulation, phytosanitation, adequate drainage and application of *Trichoderma* sp. @ 50 g/clump, multiplied in decomposed coffee compost and mixed with cow dung (2.5 kg/plant) gives good control of capsule rot caused by *P. meadii* and rhizome rot caused by *Pythium vexans* and *Rhizoctonia solani*. Liquid formulation of *Trichoderma* sp. has been developed containing minimum population of 10^8 fungal spores per ml that can be stored up to 1 year without significant reduction in viable cells. In ginger, *Bacillus amyloliquefaciens* (PGPR) as seed treatment and drenching are helpful in managing rhizome rot disease incidence in the field. Organic amendments such as coconut oil cake, ground nut cake and sesamum cake when applied to the soil to give nitrogen requirement of 50 kg/ha were also found to reduce rhizome rot incidence of ginger. ICAR-IISR has made a significant breakthrough in encapsulation and delivery of PGPR for growth promotion and disease control in ginger. The encapsulation process is simple and can be used to deliver all kinds agriculturally important microorganisms such as N fixers, nutrient solubilizers and mobilizers, PGPR, *Trichoderma* sp. *etc.*

Table 2: Resistant Varieties Developed against Pests and Diseases

Crop	Variety	Resistant against
Black pepper	IISR Shakthi	*Phytophthora* foot rot
	IISR Pournami	*Melodogyne* spp.
	IISR Thevam	*Phytophthora* foot rot
Cardamom	IISR Vijetha	CdMV virus
	IISR Avinash	Rhizome rot
	Appangala -2	CdMV virus
Ginger and	IISR Mahima	*Meloidogyne* spp.
Turmeric	IISR Kedaram	Leaf blotch

Table 3: Promising Biocontrol Agents Identified against Pests and Pathogens

Crop	Biocontrol Agent	Pathogen/Pest
Black pepper	*Trichoderma* spp.	*Phytophthora capsici*
	Pochonia chlamydosporia	*Radopholus similis, Meloidogyne incognita*
	Chilocorus spp.	*Lepidosaphes piperis, Aspidiotus destructor*
Cardamom	*Trichoderma* spp.	*Phytophthora meadii, Pythium vexans, Rhizoctonia solani*
	Lecanicillium psalliotae	*Sciothrips cardamomi*
	Metarhizium anisopliae	*Basilepta fulvicorne*
	Beauvaeria bassiana	*Basilepta fulvicorne*
	Heterorhabitidis indica	*Basilepta fulvicorne*
Ginger and	*Bacillus amyloliquefaciens*	*Pythium* spp.
Turmeric	*Trichoderma* spp.	*Pythium* spp.
	Steinernema sp.	*Conogethes punctiferalis*
	Oschieus gingeri	*Conogethes punctiferalis*

Cultural methods like shade regulation are important for reducing the population of pollu beetle on black pepper and cardamom thrips in the field. Pruning and destruction of infested shoots during initial stages of infestation can also be adopted against shoot borer infestations in cardamom and ginger. Various plant products like neem oil, fish oil rosin, tobacco extract and custard apple seed extract are promising against pollu beetle, scale insects and root mealybugs infesting black pepper. Spinosad (derived from the Actinomycetes, *Saccharopolyspora spinosa*) can substitute synthetic insecticides for thrips control in cardamom, when sprayed three times (March, May and August) under Karnataka conditions. Coccinellid predators such as *Chilorus* spp., anthocorid bugs, hymnopterous parasitoids and mermithid nematodes that are widely distributed in black pepper, cardamom, ginger and turmeric fields are to be conserved. Entomopathogens such as *Metarhizium anisopliae* and *Beauvaeria bassiana* are effective against root grubs in cardamom and white grubs in ginger. Recently, the entomopathogen *Lecanicillium psalliotae*, was found effective against cardamom thrips when applied 3-4 times during (May–September) and is ideal for organic cultivation. The entomophagous nematode *Heterorhabitidis indica* was found promising in controlling root grub infestation in cardamom and *Oschieus gingeri* and *Steinernema* sp. against shoot borer infestation in ginger and turmeric.

Post-harvest Management

Post-harvest operations like harvesting, processing and packaging play a major role in maintaining quality of spices to the specifications of international trade. Many of these operations have been mechanized which in addition to reducing the labour, helps in maintaining the quality and food safety standards. Improvements in hygiene, packaging and storage facilities will not only help in keeping quality of spices but also play a major role in reducing aflatoxin and salmonella contaminations which are major non-tariff trade barriers. Post-harvest practices lay stress on hygiene to avoid contamination with dirt, contaminants and harmful organisms during processing, packaging and storage. Drying is generally done in the sun on clean drying platforms preferably concrete or concrete lined with polythene sheets raised above the ground level. Solar and mechanical dryers are recommended for quick and hygienic drying and to protect from dust and other contaminants. Drying is to be undertaken to the appropriate level to prevent spoilage by microorganisms, especially aflatoxin-producing fungi during storage. A number of machineries have been developed for threshing, cleaning and grading of black pepper, peeling and polishing of ginger and boiling and polishing of turmeric. Hygiene of personnel involved in the processing, cleaning and packaging processes is also very important. Workers should follow good hygienic practices to protect against contamination of the product and should regularly receive training in handling techniques, personal hygiene and sanitary practices. Storage is preferably done in suitable polythene-lined bags to prevent absorption of moisture and infestation by storage pests. The processing, drying and storage areas should be well-ventilated and kept clean and hygienic and free from all contaminants.

Future Thrusts

India is blessed with having several regions where use of agro-chemicals is low as in north eastern states and many tribal areas in various states, which facilitates conversion to organic production. The share of organic products is negligible compared to conventional export. Organic horticultural products export constituted 6 per cent -8 per cent of total organic export of India. However, the global demand for organic products is also growing fast at the rate of 20 per cent per annum. Organic spices export from India have been growing rapidly with 25 per cent increase in quantity and 11 per cent increase in value. The main organic spices traded by volume are black pepper (39 per cent), ginger (19 per cent), turmeric (11 per cent), cardamom (8 per cent) and chilli (7 per cent). Though the current market size is small the demand is expected to grow at a rate of 15 per cent in the coming years. Given the ever-growing shift towards chemical-free organic food items in western world the potential for export of organic commodities is high, and India is placed at an advantageous position being a major exporter in the global market. With increase in per capita income and accelerated growth of health conscious population, the demand for organic produce is on increase which is expected to further accelerate within the country also. As spices are major sources of nutraceutical compounds and with global demand for functional foods picking up, India could emerge as a major player by exploiting its own strength in self sustaining organic systems with the adoption of these green technologies.

The main strategies for organic farming in the future should be on:

- ✰ Promoting adoption of biofertilizers, biocontrol agents, and other eco-friendly inputs
- ✰ Strengthening the research base on the
- ✰ economic viability of organic production systems
- ✰ Market promotion and publicity for increasing the demand for organic spices in the local market.

References

Anandaraj, M. and Bhai, R.S. 2011. Diseases. **In**: *Turmeric: Production and Utilization*, (Eds.) H.P. Singh, V.A. Parthasarathy, V. Srinivasan and K.N. Shiva. Westville Publishing House, New Delhi, pp. 139–161.

Anandaraj, M. and Dinesh, R. 2007. Use of microbes for spices production **In**: *Organic Spices*, (Eds.) V.A. Parthasarathy, K. Kandiannan and V. Srinivasan. New India Publishing Agency, New Delhi, pp. 101–132.

Anandaraj, M., Bhai, R.S. and Bhat, A.I. 2012. Diseases. **In**: *Piperaceae Crops: Production and Utilization*, (Eds.) H.P. Singh, V.A. Parthasarathy, V. Srinivasan, V. and K. Saji. Westville Publishing House, New Delhi, pp. 122–143.

Bhai, R.S. 2007. Biological control of fungal diseases. **In**: *Organic Spices*, (Eds.) V.A. Parthasarathy, K. Kandiannan and V. Srinivasan. New India Publishing Agency, New Delhi, pp. 153–192.

Devasahayam, S. 2007. Biological control of insect pests. **In**: *Organic Spices*, (Eds.) V.A. Parthasarathy,

K. Kandiannan and V. Srinivasan. New India Publishing Agency, New Delhi, pp. 133–152.

Devasahayam, S. and Eapen, S.J. 2011. Insect and nematode pests. In: *Turmeric: Production and Utilization*, (Eds.) H.P. Singh, V.A. Parthasarathy, V. Srinivasan and K.N. Shiva. Westville Publishing House, New Delhi, pp. 118–138.

Devasahayam, S. and Koya, K. M. A. 1999. Integrated management of insect pests of spices. *Indian J. Arecanut Spices Med. Plants*, **1**: 19–23.

Devasahayam, S., Eapen, S.J., Jacob, T.K. and Pervez, R. 2012. Insect and nematode pests. **In**: *Piperaceae Crops: Production and Utilization*, (Eds.) H.P. Singh, V.A. Parthasarathy, V. Srinivasan and K. Saji. Westville Publishing House, New Delhi, pp. 144–163.

Devasahayam, S., Eapen, S.J., Jacob, T.K. and Pervez, R. 2012. Pests. **In**: *Zingiberaceae Crops: Present and Future*, (Eds.) H.P. Singh, V.A. Parthasarathy, K. Kandiannan and K.S. Krishnamurthy. Westville Publishing House, New Delhi, pp. 332–347.

Dinesh, R., Srinivasan V., and Hamza S. 2012. Crop nutrition. **In**: *Zingiberaceae Crops: Present and Future*, (Eds.) H.P. Singh, V.A. Parthasarathy, K. Kandiannan, and K.S. Krishnamurthy. Westville Publishing House, New Delhi, pp. 255–287.

Eapen, S.J. 2008. Management of plant parasitic nematodes. **In**: *Organic Spices*, (Eds.) V.A. Parthasarathy, K. Kandiannan and V. Srinivasan. New India Publishing Agency, New Delhi, pp. 153–192.

International Pepper Community. 2013. *Good Manufacturing Practices (Code of Hygienic Practices for Pepper and Other Similar Spices)*. International Pepper Community, Jakarta, 21 pp.

Kumar, A., Biju, C.N. and Praveena, R. 2012. Diseases. **In**: *Zingiberaceae Crops: Present and Future*, (Eds.) H.P. Singh, V.A. Parthasarathy, K. Kandiannan and K.S. Krishnamurthy. Westville Publishing House, New Delhi, pp. 288–331.

Srinivasan, V., Krishnamurthy, K.S., Dinesh, R., Hamza, S. and Kandiannan, K. 2011. Input use efficiency. **In**: *Turmeric: Production and Utilization*, (Eds.) H.P. Singh, V.A. Parthasarathy, V. Srinivasan and K.N. Shiva. Westville Publishing House, New Delhi, pp. 103–117.

Srinivasan, V., Thankamani, C.K., Dinesh, R., Kandiannan, K., Zachariah, T.J., Leela, N.K., Hamza, S., Shajina, O. and Ansha, O. 2016. Nutrient management system in turmeric: Effect on soil quality, rhizome yield and quality. *Industrial Crops Products*, **85**: 241–250.

Thankamani, C.K., Chempakam, B., Jayashree, E., Agalodia, A.V., Chitra, R., Lal, G., Singh, D., Pethe U.B. and Babu, K.N. 2013. Post-harvest handling and value-addition in spices. **In**: *Souvenir and Abstracts, National Symposium on Spices and Aromatic Crops (SYMSAC VII): Post-Harvest Processing of Spices and Fruit Crops*, 27th–29th November, Madikeri, Indian Society for Spices, Kozhikode, pp. 44–56.

Promoting Innovative Production Systems

60

Innovative Protected Horticulture and its Potential for Doubling Farmers' Income

Brahma Singh

Founder President,
Indian Society for Protected Cultivation, New Delhi – 110 012
Former Director, DRDO
E-mail: brahma88@gmail.com

Protected cultivation has an important aspect of cultivation that is minimization of abiotic and biotic stress during crop production.Protected cultivation has emerged as a way to protect crops from adverse weather conditions allowing year-round production and the application of an integrated crop production and protection management approach for better control over pests and diseases.Protected horticulture practices, which are defined as a cropping techniques wherein the micro climate surrounding the plant is modified/ controlled partially or fully as per the requirements of the plant species grown during their period of growth, further enhance return to the farmers with added advantage of assured remuneration on account of quality produce and in some cases off season production, early production and prolonged production-a desirable trait in fruits, vegetables and flowers production.

Protected cultivation has two major components of technology.

a. Structure and Engineering

It deals with structures involving frames, cladding materials, irrigation system, tools, implements, robots and other engineering inputs which ensure optimal light, air temperature, water and plant growth requirements leading to yield higher and quality produce during on and off-season or round the year. This aspect takes care mainly of abiotic stress on the crop inside structures.

b. Improved Crop Production Technology

It includes production package of practices with or without hi-tech involving development of high-yielding varieties and hybrids of crops suitable for protected cultivation conditions grown for longer period in comparison to open field cultivation.

Major Advantages of Protected Cultivation

There are several advantages of adopting protected cultivation in horticulture in different agro-climatic regions. Some of the major advantages of protected cultivation are mentioned hereafter:

i) There is large increase in yield, produce quality and revenue under protected cultivation of vegetables/flowers and certain fruits.

ii) Since, generally, drip irrigation and/or fertigation practices are followed under large scale production under protected cultivation, water evaporation rate is reduced and weed population is minimal, high water use efficiency is there. There is saving in significant amounts of water under protected cultivation. Advantage of water harvesting from roof tops of structure is also there.

iii) Use of pesticides, generally, is less under protected cultivation and there is significant reduction in pesticide use leading to reduction in production costs which results in healthier produce.

iv) Off- season and year-round production of certain crops is possible under protected cultivation. This allows farmers to take advantage of market seasonality and off season higher prices. It is well established advantage of protected cultivation.

v) Protected cultivation is an attraction to educated unemployed youths in villages and pre-urban areas which is helpful in attraction of them to farming.Protected cultivation adds pride to the farming profession which is otherwise very low.

vi) Affordable low-cost poly-houses or required structures such as poly tunnel, rain shelter *etc.* are fabricated locally by small-scale entrepreneurs leading to local employment generation.

vii) Hi-tech greenhouse cultivation methods such as hydroponics, vertical farming, aeroponics, aquaponics, hi-tech plant nursery production in soilless medium, balcony gardens and others are pre-urban and unban farming of tomorrow which are to be promoted through protected cultivation in the country. These technologies are known as new age protected cultivation technologies in horticulture crops particularly for vegetables and flowers.

Limitations

- ☆ Structure cost initially looks high and unaffordable by farmers. Zero risk affordability by majority of farmers does not allow them to venture protected cultivation.
- ☆ Claddings/covering materials are expensive coupled with short life and short supply.
- ☆ Lack of repair/tools and expertise for maintaining protection structures.

Protected Structures

In different regions of the country several types of structures made of local material are being used for raising horticulture crops. Such structures are called low cost structures to be used in the country on large scale. A brief description of these structures is as under:

Plastic Mulches

Plastic mulching is covering the soil around the plant with plastic film (10-15 micron thick) to conserve the soil moisture that prevents weed growth and regulate soil temperature. There are different colour plastic films used as mulches such as black, silver-black, red, yellow, while-black *etc.* Advantages of plastic mulching are *in situ* moisture conservation, prevention of weed growth and action as barrier to soil pathogens. Plastic mulching accelerates uptake of micro nutrients from the soil by the active root zone, conserves soil moisture thereby reduces the irrigation water requirement of the crop and enhances quality of the produce with cleaner crop. Development ofcommercial grade photo or biodegradable plastic suitable for mulching is awaited. The use of plastic mulch in cultivation of horticulture crops (fruits, vegetables, flower crops, medicinal, aromatic and spices crops) is becoming remuneratively (20 per cent more yield) popular in several parts of India.

Plastic Low Tunnels

These are miniature structures producing greenhouse like effect. Plastic low tunnel facilitates entrapment of carbon dioxide thereby enhancing the photosynthetic activity. It protects plants from harsh climatic conditions such as rain, wind, hail, snow, low temperature, *etc.* These are mainly used for raising nursery for early germination of seeds and production of healthy seedlings. Besides protecting plants from hostile climate these are helpful in round the year cultivation of certain crops. These tunnels are helpful in raising early cucurbits in north India by protecting them from low temperature and frost. Like plastic mulch the cost of low tunnels is affordable by most of the farmers. Besides, they can be used to initially cover the crops for growth to save from abiotic and biotic stress and later expose to natural environment for fruiting by removing cover. Plastic mulch under poly tunnels are beneficial to check weeds and conserve water better.

Walk-in Tunnels

Walk-in tunnels are structures covered with UV grade film. These are designed to withstand wind up to 120 km/h, and trellising loads up to 25 kg/sq. m. Structure in gable configuration can be 8 or 10 meters wide. The height of the tunnels is 2 or 3 meter. Besides raising plant nursery these are suitable for production of tomato, brinjal, capsicum, parthenocarpic cucumbers, leafy vegetables, early production of certain vegetables and flower crops.

High Roof Tunnels with Ventilation

These tunnels are similar to walk-in tunnels with suitable size ventilation at the top to facilitate escaping the hot air. These are suitable for cucumber, capsicum, tomato, gerbera and carnation production.

Insect Proof Net House

Insect proof net house is a closed crop production system with 6 -8 feet height that excludes insect pests via physical barrier of synthetic insect proof 40 mesh, 20 per cent shade net. Double layer net houses are also available which lowers the temperature in summer months. Open top insect proof net houses with more than eight feet height have been found suitable in protecting crops from insects and insect vectors.

Shade Net Houses

Shade net houses are considered as one of the major technologies to provide development of healthy grafts/

Figure 1: Virus Free Kinnow Saplings in Insect Proof Net House in Haryana.

seedlings, potted foliage and flower plants, medicinal and spices crops and hardening for various horticultural crops irrespective of climatic conditions. Shade nets are available with different shade percent, 20, 40, 70 *etc.* and color. Shade net houses are made keeping in view the requirements of crop to be raised generally with 6 to 8 feet height.Shade houses are popular among horticulture nurseries throughout the country.

Naturally Ventilated Poly Houses

These are naturally ventilated green/poly houses with single span and total central height of 5 m, side ventilation of 3 m with roll-able poly-cover with or without roof ventilation and double door entry. Such poly houses are found suitable for round the year production of crops depending on the geographical area. There can/should be multi-span naturally ventilated poly house with a central height of 6.5 m, gutter height of 4.25 m, side ventilation of 1.5 m and resistant to wind storm. Naturally ventilated polyhouses have been found suitable for cultivation of large number of vegetables, flowers and strawberries besides raising their nurseries in many parts of the country. Large acreage at farms can be brought under such polyhouses for large scale production.

Polyhouse with Pad and Fan System

These poly houses unlike naturally ventilated poly houses has fan and pad system to regulate temperature and humidity. They have generally height of 4.5 m with common side and top ventilation. These structures are clubbed with exhaust fans and cellulose cooling pads of 1.8 m in height and 150 mm thickness. The cost of construction, running and maintenance is high. Lack of assured power supply becomes a limiting factor for such structures in our country.

Hi-tech or Climate Controlled Greenhouses

Hi-tech greenhouses are climate controlled structures. Greenhouses have a variety of applications, the majority being, off-season growing of vegetables, foliage and flower plants, planting material multiplication and acclimatization, crop growing for export market, seed production, plant breeding and new varieties and hybrids development. These are available in different sizes and constructed as per requirement. The sizes vary from as small as 100 sq. m to 10,000 sq.m and more. Here most of the parameters temperature, light, humidity, fertilizer, irrigation are sensed and corrected as per program through auto-control systems. These can be fully or partially automated. These are considered, based on experience, very good and essential structures for planting material multiplication of horticulture crops in the country. Farmers possessing such structures can multiply their income manifold.

Figure 2: Multiplication of Rootstock and Seedlings in Soilless Medium in Haryana and Tomato Crop in Uttar Pradesh under Hi-tech Greenhouse.

Innovative Structures

There can be several different innovative structures suiting to local environment/climate. Two of such structures i.e soil trench/underground poly house and Ladakhi polyhouse ideally suited for vegetable cultivation in cold desert of Ladakh, J&K have made cultivation of all sort of vegetables possible which was considered in 1980s as impossible due to around six months sub-zero temperature prevailing atmosphere there. This contribution of DRDO laboratory named Defence Institute of High Altitude Research (DIHAR), Leh is unique globally where author had the privilege to serve as Director and is still associated as chairman, DIHAR, research council.

Soil Trench

Since cropping period is small in Ladakh and other arid high altitudes, it was difficult to raise vegetables requiring transplanting, such as cauliflower, cabbage, knolkhol, broccoli, kale, tomato, capsicum, chili,brinjal, *etc.* Defence Research and Development Organization (DRDO) has standardized soil trench cultivation method of raising early nursery of vegetables. Seedlings are raised in polythene covered trenches of appropriate sizes, generally 15 x 5 x 3 ft during March and April and transplanted as soon as cropping season starts. These trenches during summer are used for cultivation of cucurbits and solanaceous vegetable crops. Trench can be used to raise off season potato crop (March-May/ June). During severe winter months these trenches are used for raising several vegetable crops.

Figure 3: Underground Polyhouse/Soil Trench in Cold Desert of Ladakh – Effective under Sub-zero Conditions.

Ladakhi Polyhouse

Another structure, Ladakhi polyhouse made of unbaked bricks on three sides and polythene on southern sidemeasuring 12 x 4 x 2 m size is made using local poplar and willow wood. It is covered with ultraviolet stabilized white polyethylene film of 200 micron thickness or other films like rigidex. Crop inside polyhouse is covered over night with 150 gauge black polythene. During frozen winters this structure like soil trench produces leafy vegetables (palak, mustard, fenugreek, celery, parsley, mint, coriander, chard, chenopodium, kale, sorrel, spinach, buck wheatetc). Ladakhis used to laugh on the idea of producing vegetables in long sub-zero winters. Now they themselves are practicing it to meet their kitchen requirement and that of local defence forces. Such polyhouses are also popular in arid region of China.

Figure 4: Ladakhi Polyhouse in Leh Valley.

Rain Shelters

Rain shelter, a low cost naturally ventilated poly house is found ideal for nursery raising and off-season vegetable and flower cultivation in Kerala, Andaman and Nicobar Islands, north east region and other high rainfall areas of the country. Off-season cultivation of vegetables by utilizing low cost protected structures like rain shelter can be considered as a profitable enterprise besides protecting crop nurseries from high rain during monsoon months. Such structures are boon for horticulture production in high rainfall areas of the country to multiply farmers' income.

Suitable Varieties/Hybrids for Protected Cultivation

Suitable varieties/hybrids for protected cultivation/greenhouses are generally different than those grown under open field conditions.In vegetable crops indeterminate tomatoes and cherry tomatoes, parthenocarpic cucumbers, colou red capsicums and among flowers certain varieties/hybrids of gerbera, carnation and rose have been found suitable for polyhouse cultivation. Cucurbit varieties other than parthenocarpic cucumber are to be pollinated manually or by putting honey bee colonies under polyhouses. Bumble bees are considered better pollinators in green/poly house crops but their colonies in market are not available in India. Some of the suitable varieties/hybrids of certain vegetables and flowers are given below which have been found to do well in different parts of the country under different structures.

Vegetables

- ☆ **Tomato:** GS-600, Pant Polyhouse Hybrid Tomato-1 and 2, Rakshita, Naveen 2000, Avinash-2, Snehlata, Heem Sona, VL Tomato-4, Tropic, Dombello, Tuckcross-520, Avtar, Indam Hybrid, All Rounder, Arka Meghali, Kalash 137, Pusa Hybrid 4 and 8, Arka Samarat, Snehlata, VNR F1 (Sakura, Cobra) and several others.
- ☆ **Cherry tomato:** VR 124, Pusa Cherry tomato-1, Ole, Sairan, N S Cherry-1, Pant Cherry tomato, Paireso, BSS-834, Rosa, Sheeja (yellow) and Nag Moti.
- ☆ **Capsicum:** Bharat (green), Mahbharat (green), Natasha (red), Indra (green),Swarna (yellow), Orobelle (yellow), Tanvi (yellow), US26 (yellow), DARL 71 (yellow), Mekong(green), KSP-1070 (green), US 81 (red), Cab-120, Tarquino,1DARL-72 (violet), California Wonder (green), baby capsicum orange-9967422, yellow- 9956434, and red -995559. Capsicum green varieties become red on maturity/ripening.
- ☆ **Brinjal:** Saropawa
- ☆ **Cucumber** (parthenocarpic): Isatis, Kian, Green Star2535 (Rizwan), Pant Parthenocarpic Cucumber-2 and 3, Multi Star (Rizwan), Hilton, Kuk-9, Cucumber Hybrid-1(KPCH-1) from KAAU, KSP-1734, Pusa Parthenocorpic Cucmber-6, Sandhya, Defender.
- ☆ **Bitter gourd:** Abhisek, Pali, Pusa Rasdar
- ☆ **Leafy Vegetables (off season)**: Kale, kailon, pok-choi, basil, lettuce, spinach, komatsuna, vegetable mustard (suitable for vertical production)
- ☆ **Beans:** Pole type varieties of different beans
- ☆ **Squash:** Australian Green, Pusa Alankar, Pusa Pasand, Dhoom-2, Zuccini
- ☆ **Bottle gourd**: Pusa Naveen and several hybrids
- ☆ **Watermelon**: Sugar Baby and Arka Manik
- ☆ **Muskmelon**: Pusa Sarda.

Flowers

- ☆ **Gerbera**: Goliath, Pink Elegance, Iton, Red Monarch, Rosalin, Salvador, Sangria, Tiramissu, Tropic Blend, Savannah, Rising Sun, Lion, Rich, RCGH-22, 22, 114 and 117, Piton, Pink Elegance, Goliath, Dalma
- ☆ **Carnation**: Yellow Firato, Sonsara, Dakar, Solar, Isaac and Master
- ☆ **Rose:** Tropical Amazon, Grand Gala, Kiss, First Red, Golden Gates, Mercedez, Konfetti, Starlite, Vivaldi, Ravel and Noblesse
- ☆ Orchids: Some of native genera like *Cymbidium, Paphiopedilum, Vanda, Arachnis* and *Dendrobium* are cultivated on a large scale for cut flower production under polyhouses.
- ☆ Sonia-17, White Daisy, Copper Queen, Thanachi Gold, ThoshikaInabu, F1 NRCO-42 (Dendrobium 'Emma White' x Dendrobium 'Pampadour'), F_1 Cymbidium ('Red Beauty' x Golden Elf')
- ☆ **Asiatic Lilies**: Connecticut King, Elite, Prieto, Pollyanna, Gran Paradiso, *etc.*
- ☆ **Chrysanthemum**: Yellow Gold, Arka Ravi and Ravi Kiran
- ☆ **Anthurium**: Esmeralda, Ivory, Chichas, Aymara, Flame Jewel
- ☆ **Marigold**: Pusa Narangi Gainda, Pusa Basanti Gainda, African hybrids and French hybrids
- ☆ **Aster**: Bonita (Blue, Pink, Rose, Scarlet and White)
- ☆ **Alstroemeria**: Aladdin, Amor, Capri, Cinderella, Pluto, Rosita and Serena.
- ☆ **Bird of paradise**: *Strelitzia* and *Caesalpinia* species

Fruits

- ☆ **Strawberry**: Sweet Charlie, Chandler, Elista, Fair Fox, Camaros, Festival, and Seascape.
- ☆ **Pomegranate**: Bhagava (under net house)
- ☆ **Papaya**: Pusa Nanha, Taiwan, Honey Dew

Innovative Protected Horticulture

There are several evolving innovative greenhouse production technologies which have proven potential of doubling Indian farmers' income. Some of them are briefly mentioned below.

Soilless Cultivation/*Beena Kheti ke Kheti*/ Hydroponics/Aeroponics

One of the innovations in protected horticulture is soilless horticulture which is becoming popular not only in raising plant nurseries but in production of micro-greens, vegetables, flowers and certain fruits like strawberry. This innovation is known as hydroponics where plants are fed through water and fertilizer mix (fertigation) in different type of containers, pipes, local material like bamboo and containers/pots *etc.* having substrates/growing media other than soil. This innovation though is very old but has recently become popular world over including India wherein plant is fed with required nutrients through irrigation water and is supported by media other than soil. Multiplication of horticulture nursery plants even of fruit crops like coconut, mango, several vegetables and flowers in soilless medium is picking up in the country. A number of soil less media (called artificial media or mixtures) are used as substrates for soilless cultivation, of which the most popular are: rock wool, coco peat, perlite, vermiculite, sawdust, bark chips, sand, gravel, pumice, polyurethane mats, water, polyester soil and mixtures of the these. Commercial production of nursery plants, fruits, vegetables, flowers, medicinal plants and others has become popular both under protected as well as open conditions resulting in added income to its practitioners. This innovation has great future. Button mushroom, tomato, cucumber, leafy vegetables and orchid's production under covers are examples of soilless cultivation. It is most suited for vertical gardens under protected conditions or balconies. This technique can make best use of waste land or sick soils roof tops and others. Water is a major and essential resource in soilless cultivation. Difficulty and cost of controlling soil born pests and diseases, soil salinity, lack of fertile soil, water and land shortage *etc.*, have led to the development of soilless substrates and technology. Of late grow bags with soilless medium are becoming popular for raising several vegetable and other horticulture crops. These grow bags are replacement of earthen pots or other containers filled with soil substrates. Grow bags have lighter substrate other than soil.

Another innovation in soilless cultivation is use of frequent spray on system of plant roots with water and fertilizer mixture as per requirement of plants with automation with the help of sensors. This is called popularly as aeroponics. Micro-tubers for breeder seeds of potato are multiplied using this innovation with better returns. Progressive farmers do take advantage of this innovation.This conserves water, space and is less cumbesome. It demands constant supervision.

Some of the advantages of soilless cultivation are listed below:

i) Soilless cultivation can be conducted within protected environment virtually anywhere because it does not rely on soil suitable for agriculture.
ii) Soilless cultivation conserves land and water, not affected by environmental changes and requires lesser pesticides.
iii) There is an increased yield per unit of water, space and energy.
iv) It facilitates maximizing space utilization through vertical growing.
v) Soilless cultivation is the future for an ever growing world population. One cannot afford to continuously plant and re-plant in the soil or risk their yields to soil deterioration and climate change.
vi) It would be common practice in next few years to have vegetable and other crop nurseries in plug trays multiplied in soilless medium (grow bags *etc.*) under protected structures.

Verticalization of Horticulture Production System

Certain vegetables particularly leafy vegetables, fruit crops like strawberry and some of the annual flowers have become commercial innovations in vertical farms grown in several tiers depending on the height of the plants and protective structure where plants are fed hydroponically and natural or LED light helps in photosynthesis. Utilization of vertical space by growing indeterminate tomatoes, capsicums, cucurbits (cucumber, bottle gourd, bitter gourd and others) by training them has given several fold yield in comparison to their conventional methods of production. This innovation results more produce per unit area leading to multiply farmers' income. Enterprising educated youths in India have adopted this innovation by linking supply to star hotels in metro cities which is potential to give better returns.

Figure 5: Vertical Poly House Production of Leafy Vegetables in Chennai by Young Engineer Mr. Sai Krishna.

Plasticulture

Use of plastics in agriculture is known as plasticulture. Use of plastics in horticulture for irrigation, micro-irrigation, fertigation, mulching, claddings to protective structures, nets, containers- innovative plastics-plastics with shelf-life enhancers, crop covers *etc.* economize cost of production of horticulture crops leading to better returns to farmers. Use of 10-15 micron plastic mulch along with micro-irrigation and fertigation in horticulture crops have more than doubled the return in comparison to conventional horticulture practices by reducing expenditure on water and weed control in several crops. There is National Committee on Plasticulture Applications in Horticulture (NCPAH), Ministry of Agriculture and Farmer Welfare, Govt. of India to promote plasticulture application in horticulture. Research on reuse, recycle and bio as well as photo degradation of plastics is in progress. Protected cultivation sector in agriculture is one of the major users of plastics.

Tissue Culture of Horticultural Plants

China is leading in production of tissue culture plants. Three billion micro propagated horticulture plants are raised annually in China. This is 90 per cent of Asia's tissue culture plants production. Asia produces 80 per cent of worlds total tissue culture plants production. There is increasing trend of use of LEDs (light – emitting diodes) for production of tissue culture plants as there is energy saving to the extent of 40 per cent. Banana, carnation, gerbera, orchids, chrysanthemum and other herbaceous crops multiplication by micropropagation is an innovation which adds to the income of nurserymen and farmers. Scion and rootstock for horticulture crops grafts are multiplied better through micropropagation particularly the vegetable grafts. Modern genetic manipulation of plants is assisted by tissue culture and hardening under protected structures. Protected cultivation and micropropagation are inter dependent. Farmers practicing micropropagation of horticulture crops are earning several times more than conventional nurserymen.

Vegetable Grafts

Vegetable grafts with resistant rootstock and high quality as well as yield scion are becoming popular to manage biotic and abiotic stress. Even though grafting originated in East Asia (Japan, 1920), could not get attention of South Asia but became popular in European countries where greenhouse cultivation is not uncommon. Grafted plants are produced to a tune of several million tonnes in Spain, France and Italy. In Japan 93 per cent watermelon and 73 per cent cucumber area of total area covered are grafts. AVRDC- Asian Vegetable Research and Development Centre, Taiwan recommends growing tomato on eggplant accessions EG 195 and EG 203 and sweet pepper on chili accessions PPO273-7502 and PPO 242-62 and Lee B for flood tolerance. In China 20 per cent area under watermelon and cucumber is covered with their grafts. In India, sporadic attempts have made and resistant/tolerant rootstocks have been identified. Most common bio-stress in vegetables such as root-knot nematodes (solanaceae and cucurbitaceae vegetables), *Fusarium* wilt (melons and tomato), bacterial wilt (tomato and eggplant), *Verticilium* wilt (tomato), *Phytophthora* (sweet pepper) are being solved by using resistant root-stocks. In vegetable crops following rootstocks have been developed in India.

Figure 6: Tomato Grafted on Root-knot Nematode Resistant Brinjal Wild Species.

- ✫ Wilt and root-knot nematode resistant in solanaceous vegetables
- ✫ *Phytophthora* resistant for chilli and sweet pepper
- ✫ *Fusarium* wilt resistant for cucurbits
- ✫ Salt tolerant for cucurbits
- ✫ Cold tolerant for cucumber, muskmelon and bitter gourd

Vegetable Seedling Plugs

Most of the vegetables can be transplanted by raising their seedling plugs in soilless medium under small protected area well in advance (one month or so). The seedlings can be grafts also. Seedling plugs ensure better stand in the field and help in regulating transplanting date reducing their time in main field/structures to avoid competition with weeds and protection from adverse weather. Fruit crop nurseries are adopting raising cuttings/saplings in soilless medium.

Figure 7: Vegetable Plugs Production at Gharaunda-Haryana.

Aquaponics

It involves production of fish and vegetables with same input resource of water and nutrients. Feed is given to the fishes in the pond not to the vegetables or plants grown as hydroponics. Fish pond or tank water is used as hydroponics or areoponics with or without additional fertilizers. This results in effective utilization of fish pond water and its nutrients for production of vegatables/plants simultaneously with added returns on input resources. Ammonia from the fish waste in the pond gets converted into nitrites and nitrates through bacterial action. Nitrates aid in plant development. Only certain kinds of plants can be grown this way. Leafy vegetables like broccoli, lettuce, spinach, *palak, pudina* are the best, apart from tomato, lady's finger and brinjal. Tilapia and carp fish are best suited for aquaponics.

Micro-greens

Microgreens are tiny edible plants, rich in flavor and nutrition, that are older than a sprout, but younger than a full-grown plant. Microgreens are harvested after the first "true" leaves have developed. They are the smallest of the salad greens, and can be grown from almost any plant. They are rich source of vitamins and minerals. They can be grown in every home around the year. Commonly grown microgreens belong to plant familyBrassicaceae (arugula, radish, mizuna, pepper cress, knol-khol, mustard, red cabbage, pakchoi, tatsoi, wasabi, broccoli); Fabaceae (lentil, pea, Bengal gram, alfalfa, methi; Amaranthaceae (amaranth); Chenopodiaceae (Swiss chard, *palak*, quinoa, orach, spinach, beet root); Apiaceae (carrot, coriander, celery); Poaceae (popcorn); Lamiaceae (basil), Convolvulaceae (water spinach) and Polygonaceae (sorrel, buckwheat). Useful bio-active compounds present in microgreens are mostly phenolics, glucosinolates, betalians, carotenoids, polyacetylene, folic acids, anthocyanins and minerals.

Figure 8: Micro-greens in Polyhouse.

Suggestions to Strengthen Protected Horticulture

i) Develop appropriate cropping package for different agro-climatic regions of the country for utilization of the protected structures round the year.

ii) Large-scale production of horticulture crops nursery under protected structures using soil-less medium (coco peat) be encouraged throughout the country on priority.

iii) There is need for development of innovative protected structures for different agro-climatic regions of the country

iv) Horticulture crops with potential of doubling farmer's income for protected cultivation need to be identified and promoted.

v) There is an immediate need for developing suitable varieties and hybrids of horticulture crops for protected cultivation with their package of production.

vi) Pollination protocols under protected structures needs to be developed including development of parthenocarpic varieties and rearing of insect pollinators.

vii) There is requirement of research on production of fruit crops like banana, pomegranate, papaya, strawberry and others under protected structures.

viii) Micro-irrigation, fertigation and plastic mulch research for protected cultivation is to be given due importance.

ix) Promote human resource development on protected cultivation at national, state level and block level throughout the country.

x) There is need to have a national research center or project directorate exclusively on protected cultivation under aegis of ICAR which may also take up research on soil-less cultivation (vegetable grafts, microgreens, plug seedlings, hydroponics, aeroponics) and organic production.

xi) Promote protected cultivation in horticulture clusters.

In conclusion innovative protected horticulture practices mentioned above have potential to more than double farmers' income in India.

References

Bala, B. 2013. Upliftment of rural economy through protected cultivation. *Global Research Analysis*, **2**(2): 42–43.

Development of protected cultivation in Haryana. Working Group Report, Haryana Kisa nAyog, Panchkula, Haryana, 2014, 68 p.

Parvatha Reddy, P. 2016. *Sustainable Crop Protection under Protected Cultivation.* Spinger e- book, 433 p.

Singh, Brahma. 2016. *New Age Protected Cultivation,* e-magazine. ISPC, New Delhi Publication, **2**(1): 41.

Spehia, R.S. 2015. Effect of poly house on productivity of vegetable crops as compared to open cultivation. *Curr. Sci.,* **108** (2254): 2254–2257.

Yadav, R.K., Kalia, P., Choudhary, H., Zakir Husain and Brhama Dev. 2014. Low-cost polyhouse technologies for higher income and nutritional security. *Int. J. Agril. Food Sci. Tech.,* **5**(3): 191–196.

61

Technologies for Hybrid Seed Production of Vegetable Crops under Protected Conditions for Improving Farmer's Income

Balraj Singh[1]* and B.S. Tomar[2]

[1]Agriculture University, Jodhpur – 342 304, Rajasthan
[2]Division of Vegetable Science,
ICAR-Indian Agricultural Research Institute, New Delhi – 110 012
E-mail: drbsingh2000@yahoo.com

Quality seeds and healthy planting material including hybrid seeds contribute nearly 30 per cent enhancement in overall productivity thereby increased seed yield, reduce pesticide use by inbuilt disease resistance, and produce slandered products with improved keeping quality. Despite the enormous potential for hybrid seeds, development and seed production of open-pollinated varieties continued to provide affordable planting material to the poor farmers and for home gardening. Attention need to be raised on popularizing the use of hybrids and reduction in hybrid seed production cost. It has been indicated that the development of vegetable industry is linked to the direct availability of quality seed in the country. Experience has shown that OP or F_1 hybrids of vegetable and flower crops developed locally for the domestic market have a better success and acceptability among the farmers.

Healthy Seedling Raising under Protected Conditions

Majority of vegetable and flower crops are grown through transplanting method and traditionally the seedlings are raised in flat bed, raised bed or different kinds of pans in open field condition makes the seedling prone to the attack of insects, diseases specially the damping off and viruses which hampers the performance of seed production programme and quality of the seed.

The healthy and virus free seedlings can successfully be produced under low cost insect proof nursery structures. The same insect proof net house can be utilized during winter by covering with UV stabilized and transparent polythene for growing of cucurbits, tomato, chily, and zinnia, cosmos, gaillardia. To mitigate the effect of high temperature the structure can be covered with various color shade nets. Large scale production of disease free seedlings of vegetables, ornamentals and papaya could be organized under high tech nursery structures with the help of soilless growing media (3 parts cocopeat; 1 part vermiculite; 1 part perlite) in multi-celled plug trays. Even the nursery of cucurbits can also be raised under one system (Singh *et al.*, 2010). The cost of media could be brought down with the intervention of vermicompost/leafcompost *etc.*

Grafting of Vegetable Seedlings for Healthy Seed Production

Numerous vegetable crops face the problem of soil born fungus like (*Fusarium* wilt, *Verticillium* wilt, damping off, nematodes), which causes severe damage to the seed crop and reduce the seed production level. Therefore, the grafting of seedlings on resistant rootstocks can play a vital role to overcome these soil borne fungus and nematodes. Grafted seedlings are very popular and widely used among farmers in some Asian and some European countries and it has been increasing in the US, especially with tomatoes. In Japan, almost 95 per cent of watermelon, oriental melon, egg plant, and greenhouse cucumber and tomato plants are grafted before being transplanted to the field or greenhouse. In South Korea, the use of grafted seedlings has been increasingly popular in the production of many fruit bearing vegetables. The number of vegetables being grafted annually is estimated to be more than 540 million seedlings in Korea and 750 million in Japan. An increase in plant purchases and shortage of skilled workers has made it necessary to automate grafting in Japan. Development of a grafting robot targeted efficient mass production of seedlings especially cucumbers. Grafting robot produced 815 plants per hour (about three times the manual grafting) at a success rate of 97.1 per cent.

Protected Structures

Recently there has been a technical shift towards the production of high quality hybrid seed of important high-value vegetables under protected conditions. Selection criteria of protected structures to be used for vegetable seed production certainty depends upon few important factors, *viz.*, climatic conditions of the area selected for seed production, vegetable crop, type of seed crop, season of seed production and targeted quantity of seed production. Generally, semi-climate controlled greenhouses are only suitable for hybrid seed production of determinate type varieties/hybrids of standard tomato, cherry tomato, sweet pepper and parthenocarpic cucumber varieties, which are suitable for protected cultivation. Seed yield of such crops can be 3-4 times more compared to their open field cultivation along with high quality of seed. Similarly, zero energy naturally ventilated greenhouses are also equally suitable for hybrid seed production of these high value vegetable crops, where the seed yield is usually 2-3 times more than open fields, but the cost of seed production is only one third (1/3) of the seed produced under semi-climate controlled greenhouse conditions.

Potential Use of Greenhouses

Generally semi-climate controlled greenhouses are suitable for hybrid seed production of indeterminate type varieties/hybrids of standard tomato, cherry tomato, sweet pepper and parthenocarpic cucumber varieties with long duration and higher seed yields, moreover these varieties are only suitable for protected cultivation. Seed yield of such crops can be 3-4 times more compared to their open field cultivation along with high quality of seed. Similarly, zero energy naturally ventilated greenhouses are more suitable for hybrid seed production of these high value vegetable crops, where the seed yield is usually 2-3 times more over open fields, but the cost of seed production is only one third (1/3) of the seed produced under semi-climate controlled greenhouse conditions due to less basic cost of fabrication and running cost in terms of power supply *etc.* for cooling or heating of the structures.

Potential Use of Insect Proof Nethouses

Insect proof net houses are the most suitable for large number of areas in India and low cost protected structures suitable for quality hybrid or seed production of either OP varieties or hybrids in large number of vegetables, *viz.*, tomato, sweet pepper, chilli, okra, brinjal and important cucurbits. The major objectives of seed production under insect proof net houses are to grow virus free seed crops, and also to protect the seed crops against other major insects/pests like shoot and fruit borer in brinjal, fruit borer in okra and tomato, fruit fly in cucurbits, leaf minor in cucurbits and solanaceous vegetables and red pumpkin beetle in cucurbits. Not only this, insect proof net house also provides protection to the seed crops against frost conditions mostly prevailing in northern plain of India. The design of insect proof net houses can be in shape of walk-in-tunnels with double door entry as temporary structures or a permanent structure in shape and design of a greenhouse where maximum vertical space can be used and the seed crops can be grown for longer duration with proper training, pruning and with optimum pollination management. The cost of fabrication on such insect proof net houses may be Rs. 200 to 600/m^2 depending upon the shape and designs. Whereas, some other low cost protected structure like walk in tunnels, plastic low tunnels and temporary plastic walls can also be used for off season seed production or advancing the season of seed production, but these structures are technically suitable only for northern plains of India during winter season with a basic objective to protect the crops against frost or extreme low temperature.

Research work conducted on hybrid seed production of vegetables, *viz.*, pumpkin, bitter gourd, bottle gourd, summer squash, cucumber, tomato, cherry tomato, brinjal and capsicum has clearly showed that there is a significant and wide gap not only in the hybrid seed yield but also in seed quality in the seed crops produced under various protected structures, *viz.*, insect proof net houses, poly-houses, naturally ventilated greenhouses and semi-climate controlled greenhouses as compared to their seed production under open field conditions. Not only this during various studies carried out on the above mentioned crops during different seasons it was also clearly emerged out that under open field conditions the hybrid seed crops are severely damaged by several viruses, insects and pests, birds, rodents and abiotic stresses like severe frost, hail storms and unseasonal rains but no or very minor problems were recorded to be encountered under various protected conditions. On the basis of the above research work done on various vegetables there are lot of advantages of hybrid seed production of vegetable crops under protected conditions in comparison to open field conditions.

Raised Bed Technology with Drip Fertigation and Plastic Mulching

Mostly the seed production programs of bulbous and root vegetable crops are taken up under flatbeds with surface irrigation system. Under this system high incidence of fungal diseases like stemphylium blight, purple blotch, powdery mildew, *etc.* have been recorded through field experimentations and also under several seed production programs in states of Haryana, Punjab, Delhi, U.P *etc.* It was observed that this is happening mainly due to high status of humidity on the soil surface which invites the high incidence of such fungal diseases but with simple intervention with raised bed technology equipped with drip fertigation alone or coupled with plastic mulching can really solve these problems very easily. The basic reason of no or low incidence of such fungal diseases in such vegetable seed crops is mainly due to very low status of humidity on the soil surface. And this has been further established and proved through comparison of the above vegetable seed crops under both the systems at IARI, New Delhi. Therefore, in these vegetables high quality seed production programmes can be taken up under raised bed technology equipped with alone drip fertigation or coupled with plastic mulching.

The major advantages of vegetables seed production under protected conditions are:

- ✰ Seed yield is generally 2-3 times more compared to open fields in major vegetables. On the other hand quality of seed is always high compared to open field produced seed. The problem of high isolation distance in cross-pollinated vegetables can be minimized under protected conditions.
- ✰ Problem of synchronization of flowering in parental lines can be minimized.
- ✰ Maximum plant population can be maintained along with appropriate ratio of male and female parents for higher hybrid seed yield.
- ✰ Seed production under adverse climatic conditions is possible otherwise it is not possible under open field conditions.
- ✰ Duration of seed crops can be extended, which is not possible under open field conditions.
- ✰ Field standards can be well enforced under protected conditions and virus free and disease free vegetable seed crops can be grown, which is otherwise very difficult under open field conditions.
- ✰ Training, pruning and hand pollination practices are very easily manageable under protected conditions compared with open field seed crop.
- ✰ More number of seed crops can be grown under protected conditions.
- ✰ By using soil less media seed crops can also be grown even under saline and acidic soil conditions.
- ✰ Protected conditions provide the best opportunity for organic seed production in vegetables.
- ✰ Judicious use of natural resources like soil and water is possible for seed production under protected conditions.
- ✰ Virus free and healthy seedling production of the parental lines is possible under protected conditions.
- ✰ Grafting technology in seedling can easily be applied to over come the soil borne problems in solanaceous and cucurbitaceous vegetables.
- ✰ Quality seed production of vegetables can certainly be adopted as an agri- entrepreneur in various regions of the country and it can generate lot of employment in several areas.
- ✰ Seed production cost in vegetables can be reduced by growing seed crop under protected conditions.
- ✰ In hybrid seed production the emasculation practices on the female parents are not required as there are no insect pollinators inside the protected structures.
- ✰ Complete protection of the crop from rodents and birds compared to open field seed crops.
- ✰ Seed crops can not be damaged by un-seasonal rains at the time of their maturity alike open field seed crops.
- ✰ Off-type plants, objectionable weeds or plants affected by designated diseases can not pose any problem in the seed crops raised under protected conditions.
- ✰ Seed viability and seed vigour can be extended through integrated nutrient management in seed crops under protected conditions.

Conclusion

In India, similar strategy can be adopted like China to have huge area under protected cultivation for boosting horticulture production. India, can not only provide fresh quality food/vegetables from protected cultivation but can also be an industrial hub for production of quality hybrid seed and healthy planting material by undertaking production programmes in large number of vegetables and other horticultural crops under protected conditions. These protected cultivation can be undertaken in various structures like insect proof net houses, zero energy naturally ventilated greenhouses, walk-in-tunnels, high tunnels, *etc.* as per region and crop specific requirements. China has placed more than 3.5 million hectare area under various categories of protected cultivation of horticultural crops out of which 96 per cent of the area is only being used for protected cultivation of vegetables and further out of the 96 per cent area under protected vegetables a major portion of which is being used only for hybrid seed production in case of vegetable crops. On the same pattern and pace like China, India is also required to replicate in all efforts to shift its vegetable seed production programs from conventional open fields to protected conditions looking to large number of problems in terms of several viruses, insects and pests and also due to several abiotic stresses encountered under open field conditions. Work on quality seed production of vegetables has been carried out under different protected conditions on brinjal, tomato, cherry tomato, pumpkin, bitter gourd, cucumber, summer squash and on other vegetables, which clearly established that higher seed yield and higher seed quality can be achieved under protected conditions as compared to open field seed production of these crops. Hybrid seed production by using insect proof net houses can be highly profitable venture not only for seed companies but also for progressive farmers for increasing their income in various regions of the country.

62

Role of Urban and Peri-urban Horticulture (UPH) for Meeting Aesthetics, Nutritional and Livelihood Security

Janakiram Tolety[1] and S. Vijay Rakesh Reddy[2]*

[1]ICAR-Indian Council of Agricultural Research, New Delhi – 110 012

[2]ICAR-Central Institute for Arid Horticulture, Beechhwal Bikaner – 334 006, Rajasthan

**E-mail: janakiram.kab@gmail.com*

Horticulture is the science and art of growing fruits, vegetables and flowers or ornamental plants. Urban horticulture, can be defined as a plant production activity conducted in a city or suburb to grow horticultural plants that are wholly or partially edible, economically viable and has the potential to reduce CO_2 emissions. It focuses on the functional use of horticulture so as to maintain and improve the surrounding urban area. Urban and Peri-urban Horticulture (UPH) is the growing of horticultural plants and/or the raising of animals for food and other uses (*e.g.* recreation, environment improvement) within urban and peri-urban areas; as well as the related input production and the processing and marketing of horticultural products.

Urban and peri-urban horticulture (UPH) includes all horticultural crops grown for human consumption and ornamental use within and in the immediate surroundings of cities. Although crops have always been grown inside the city, the practice is expanding and gaining more attention. The products of UPH include a large variety of vegetables, flowers, ornamental trees, aromatic vegetables and mushrooms. Table 1 presents the main species cultivated in urban and peri-urban horticultural systems. Although most of these species are not specific to UPH systems and can also be grown elsewhere. Horticulture in urban areas minimizes the transportation time for the supply of fresh produce to city dwellers. Generally, the types of crops cultivated vary according to the area, influenced by culture and tradition. In cities, short-cycle crops are preferred; while in the surroundings of the city, crops with longer cycles are cultivated.

With the expansion of cities and rapid urbanization, UPH is a new term gained momentum very recently due to its large field of study. It has an irrefutable relationship to production horticulture in which the fruits, vegetables and other plants are grown for harvest, aesthetic, architectural, recreational and psychological purposes; but it extends far beyond these benets. The value of plants in the urban environment has yet to be thoroughly researched or quantied.

UPH, like urban agriculture (UA) in general, has multiple functions. The main function is supplying fresh food, but emerging functions that are becoming more and more essential are economic (income generation), social (labour), cultural, living environment (open spaces and greening), environmental (recycling) and security (food and natural risks). Under UPH system, the horticultural crops are grown in small gardens or larger fields, using traditional or high-tech and innovative practices. These include production on built-up land using various types of substrates (*e.g.* roof top, organic production and hydroponic production); water saving

Table 1: Horticultural Plants Cultivated in the Urban and Peri-urban Areas

Vegetables
Amaranth (*Amaranthus* spp.)
Beans (*Vigna radiata* and *Phaseolus vulgaris*)
Broccoli (*Brassica oleracea* var. *italica*)
Cabbage (*Brassica oleracea* var. *capitata*)
Cassava leaves (*Manihot esculenta*)
Cauliflower (*Brassica oleracea*)
Chinese cabbage (*Brassica rapa* var. *pekinensis*)
Cucumber (*Cucumis sativus*)
Eggplant (*Solanum melongena*)
French bean (*Phaseolus aureus*)
Gourds (*Cucurbita* spp.)
Indian mustard (*Brassica juncea*)
Leek (*Allium ampeloprasum*)
Lettuce (*Lactuca sativa*)
Lotus (*Nelumbo nucifera*)
Mung bean (*Phaseolus aureus*)
Okra (*Hibiscus esculentus*)
Onion (*Allium cepa*)
Palak (*Beta vulgaris*)
Pea (*Pisum sativum*)
Potato (*Solanum tuberosum*)
Squash (*Cucurbita maxima*)
Sweet pea (*Lathyrus odoratus*)
Sweet pepper (*Capsicum annuum*)
Snow pea (*Pisum sativum*)
Tomato (*Lycopersicon esculentum*)
Water morning glory (*Ipomea aquatica*)
Yardlong bean (*Vigna unguiculata sesquipedalis*)
Water chestnut (*Trapa* spp.)
Fruits
Banana (*Musa* spp.)
Melon (*Cucumis melo*)
Orange (*Citrus sinensis*)
Papaya (*Carica papaya*)
Peach (*Prunus persica*)
Pineapple (*Ananas comosus*)
Strawberry (*Fragaria* spp.)
Ornamentals
Bougainvillea spp.
Chrysanthemum spp.
Kumquat (*Fortunella* spp.)
Rose (*Rosa* spp.)
Petunia spp.
Nerium spp. and other annuals

in highly populated areas; the production of pesticide-free vegetables year-round with a low content of heavy metals and human pathogens; and control of wastes and leaching (fertilizers, pesticides, organic matter, water) in the urban environment. Urban and peri-urban cultivation systems differ from rural systems by their proximity to cities and by the constraints of space, which often lead to greater intensification of production. Through the large variety of crops that are produced, urban horticulture makes a major contribution to food and economic security of the urban population. It also contributes to strengthening social sustainability and increasing ecological sustainability by reducing pollution, conserving natural resources, transforming urban wastes, and preventing soil erosion.

1. History, Scope and Importance of UPH

Horticulture and the integration of nature into our civilization has been a major part in the establishment of our cities. When nomadic civilizations began settling down, their major trading centers were the market gardens and farms. Urban horticulture rapidly progressed with the birth of cities and the increase in experimentation and exchange of ideas. These insights led to the ûeld being dispersed to farmers in the hinterlands.

For centuries, the built environment such as homes, public buildings, *etc.* were integrated with cultivation in the form of gardens, farms, and grazing lands, kitchen gardens, farms, common grazing land, *etc.* Therefore, horticulture was a regular part of everyday life in the city. However, the industrial revolution and population outbreak has rapidly changed the outlook of the city landscape and replaced green spaces with brick and asphalt. After the nineteenth century, Horticulture was then selectively restored in some urban spaces as a response to the unhealthy conditions of factory neighborhoods and thus the cities began seeing the development of parks.

Even if the consumption of vegetables per person is relatively low, consumer demand remains the major driving force behind UPH. In developing countries, the consumption of fruits and vegetables is generally lower than the FAO recommendation of 400 g/capita/day. The importance of fruit and vegetable consumption depends on the population group. Average household consumption of fruits and vegetable (149-152 g/capita/day) during this decade indicate that the annual per capita consumption, though far below the recommended intake but is slightly better from the previous survey (120-140 g/capita/day). Annual consumption (kg/person/annum) of fruits was 9.6 (rural), 15.6 (urban) and 11.8 (India), while it was 74.3 (rural), 79.1 (urban), 76.1 (India) for vegetables. Other recent studies on average consumption of fruits and vegetables amongst urban residents of Chennai and Jaipur reported as 265 g/day and less than three servings by 72.6 per cent respondents respectively (Sachdeva *et al.*, 2013). However, the urban consumption is related to the size of households, income and socio-cultural characteristics (Bicas, 1998).

The proximity to urban markets often defines the production of specific fruits or vegetables, while there are also seasonal differences between rural and urban areas in terms of supply to the urban market. Factors such as climate, soil, access to water, insects and diseases, costs of production and, most importantly, the shelf-life of the crop itself, influences the location of vegetable production. The last factor explains why, for most urban markets, leafy vegetables are produced in urban and peri-urban areas. Some leafy vegetables are well adapted to a hot wet season. The very short shelf-life of cut flowers such as roses and chrysanthemums explains the development of these horticultural crops around cities. Urban horticultural areas may also supply the urban market more regularly than the rural areas. The tendency to crop year-round is increased by the UPH producers' need to derive an income from various high-value crops throughout the year. This bias towards UPH may also be due to production constraints and access to transportation infrastructure during the rainy seasons or to socio-economic causes.

2. Factors Influencing UPH

The development of horticultural systems in and around the urban areas is determined by specific opportunities and constraints in the city. The major constraints being the resource scarcity (water, land, labour and access to other inputs) and pollution.

2.1 Resource Constraints

In the current scenario of rapid urbanization, there is great pressure on the land resources and the prices are rising very high. The land ownership and tenure arrangements are becoming very important. Under such circumstances, the access to land by the urban and peri-urban producers was difficult and pose a major constraint for their activities. Being the migrants from rural areas, these urban dwellers are not usually the land owners and they are obliged to rent from others or to squat on public land in order to cultivate. Thus the land tenure uncertainty has a strong influence on the land-use strategy and its maintenance. Hence, the producers usually select fast growing crops (usually leafy vegetables) rather than the perennial crops (such as fruit trees); also they select the swampy places which are unsuitable for dwellings.

Due to insecurity of the land tenure, the urban growers generally choose strong inputs with quick effects such as chemical fertilizers and pesticides, rather than improving the soil using organic and integrated production techniques. Some producers have even turned to soil-less production systems based on diverse substrata.

Though different sources of water are available in the urban and peri-urban areas *viz.*, potable water, wastewater, rivers, lakes and ponds, the horticultural systems have specificity in getting adapted to these sources, in particular with the use of wastewater. In all cases this scarce resource should be used efficiently with precaution. Though drip/micro-irrigation is possible, the use of watering tank is quite popular in most of the UPH systems. However, the use of wastewater has a special advantage of cutting the cost of fertilizer and its labour as it provides nutrients along with the water.

2.2 Pollution constraints

The soil, water and atmosphere of the urban areas are highly polluted by industries, traffic, and the high population density. The major challenge for the urban agriculture, especially horticulture is to supply safe products in this highly polluted environment. The major pollutants of the UPH systems include heavy metals, pesticide residues, and biological contaminants. These pollutants present risk not only to the consumers but also to the producers coming into contact with the contaminated materials such as waste/sewage water. This pollutant problem arises mainly in the areas close to the active/old industrial sites, urban waste disposal sites; when the crops are irrigated with water contaminated by heavy metals, fertilizers, or organic manures; or when contaminated soils are used for cultivation.

2.3 Heavy Metals

The major sources of heavy metal (Pb, Cd, Cr, Zn, Cu, Ni, Hg, Mn, Se and As) pollution are irrigation water from the streams and wastewater contaminated by industrial effluents. Heavy metal pollution may also occur with the application of contaminated solid wastes and through use of former industrial lands contaminated with oil spil and other industrial wastes. The heavy metal toxicity can directly affect the growth and physiology of the plants. The soils in major cities of developing countries have very high heavy metal content and with increase of their concentration in human food, they may cause toxic symptoms and damage human health (carcinogenic and mutagenic effects). For example, Jorgensen *et al.* (2005) show that intensive horticultural systems (particularly in greenhouses) in urban areas may be threatened by soil toxicity through trace elements such as Zn, Cu, As and Pb. Puschenreiter *et al.* (1999) have concluded that there are several pathways available to reduce the transfer of heavy metals to the human food chain with proper precautions. However, Birley and Lock (2000) argue that little research was done in studying the chronic health effects of consuming tiny amounts of heavy metals over long periods of time and that further research is indispensable.

The heavy metal accumulation varies depending on the species and the type of plant part. Leaves can accumulate high levels while seeds are often less affected. It is possible to adapt the choice of crops in relation to the degree and type of contamination. Some horticultural crops such as beans, peas, melons, tomatoes and peppers show very low uptake of heavy metals and these crops can be recommended for the polluted areas. The risk of pollution and rate of absorption depends on the location of the fields and their levels in the soil. The heavy metal lead (Pb) is taken up by the plant roots and transported to the leaves. However, lead from the traffic fumes in the air also gets settled on the leaf surface. This heavy metal can be washed away by watering the leaves, especially when they are waxy (cruciferous plants and alliums). Conversely, cadmium (Cd) can be taken up by the plants through roots as well as leaves. These two heavy metals are very toxic with no positive biological functions in the human body. Hence, the location for urban and peri-urban vegetable production should be chosen very carefully with regard to roads and polluting industries. Bio-remediation of soils by plants and installation of soil mycorrhizae that limit heavy metal uptake are the long-term projects for heavy metal management in the future.

In addition to the heavy metals, air pollution too contributes to the crop toxicity in the urban and peri-urban areas. For instance, Agrawal *et al.* (2003) observed significant effect of SO_2, NO_2, and O_3 on the physiological characteristics of bean, *palak*, wheat and mustard in the polluted environment of Varanasi, India. These gases are very common in the large cities of the developing countries like India, especially with the fast growth of personal transport system.

2.4 Chemical Residues

Like agriculture, horticulture is confronted with chemical residues in the plants as well as exportation to the environment which leads to the major health problems of the producers and consumers. The major source of chemical residues being the agricultural inputs, the cultivation in contaminated lands and irrigation with contaminated wastewater also contributes to the increasing residue levels in the horticultural commodities. The pesticide and fertilizer contaminants are characterized by a very long half-life and most of the producers use the banned chemicals due to lack of proper awareness. Among various vegetables cultivated in the UPH systems, the chemical residues were found more in the tubers and root vegetables.

2.5 Biological Contaminants

In urban horticultural systems, solid wastes are mainly used to improve the soil (household wastes, market refuse, sewerage, night soil, manure, fish wastes and agro-industrial wastes). These organic solid wastes along with the recycled urban wastewater are the main sources of human pathogenic organisms that contaminate the produce. These pathogens not only affect the consumers who eat these contaminated fruits and vegetables, but also the producers who handle the contaminated material.

To reduce the risk of pathogen contamination, the urban organic wastes should be composted properly and this process could significantly reduce the health risks. If the organic wastes are not composted properly, they can still contain pathogens (bacteria, helminth eggs *etc.*). The risk can be very high if organic materials are mixed with night soils from latrines, manure or hospital waste, causing pathogens to breed. The use of domestic sewage for irrigating and fertilizing field crops, perennials and trees is widespread in the UPH systems. A large part of the wastewater used is untreated or poorly treated and contains various bacteria, protozoan parasites, enteric viruses and helminths. Coliform bacteria are mainly transmitted to humans from wastewater *via* the contamination of crops irrigated with wastewater or through consumption of contaminated meat from domestic animals that have ingested tapeworm eggs from faeces in untreated sewage.

3. Pollution Caused by UPH Systems

The urban horticultural systems may pose certain risk to their environment because of its proximity to the people. There's always been conflict between the urban gardeners and the city dwellers in the fear of pollution caused by the intensive use of agro-chemicals (fertilizers, pesticides and fungicides) which might leave crop residues, contaminate surface/groundwater, and cause negative effect to the health of workers as well as consumers. Hence, the agro-chemicals affecting the human and environmental health should be used with utmost care.

Fruits and vegetables containing pesticide residues above the maximum residue limit have been identified in certain urban markets of developing countries (Moustier, 2000; Midmore, 1998; Fatou Diop Gueye and Sy, 2001). This occurs very often, in spite of the fact that regulations for the use of pesticides and recommendations for health safety are in place. The application of pesticides on crops also endangers workers if little information is available on how to use them and when no protective measures are taken. This mainly affects low-income gardeners who cannot afford to buy proper protective clothing and equipment or are not aware of the importance of doing so. Awareness of the risks caused by excessive use of chemical pesticides should be raised among all the stakeholders, ranging from producers, consumers and public authorities to agrochemical companies. At this point in time, the penalties are not high enough to drastically reduce the over-use of pesticides. More negotiation between all players in the commodity chain might be one solution. In any case, there will be a cost, implying that the consumer must be ready to pay more to have a better-quality product and a safer environment. The development of new technologies such as integrated pest management and biological control can help in reducing pesticide use.

Nitrates deserve a special mention in the urban pollution as they cause eutrophication of water bodies in combination with phosphorus. In UPH systems, nitrates stem from fertilization and from irrigation water. If the source of pollution is close to the water resource, as is often the case with UPH, the risk of pollution of water by nitrates is enhanced. This is particularly true in developing countries that do not have a good network of water supply and where many people depend on the local water resources for their supply.

4. Recommendations for Safe UPH Systems

A number of preventive and control measures were suggested to reduce the risk of pollution by heavy metals, agrochemical residues, pathogens and diseases. The urban farmers should be educated to respect the rules to limit/stop contamination of the horticultural commodities, which necessitates good communication with the health sector.

4.1 Horticultural Practices

Horticulture in urban areas will continue to be adapted to specific circumstances, as and when required including combinations of practices from traditional horticulture and more modern, innovative practices. In tropical areas, the distribution of rainfall often varies greatly between the dry season, which is usually colder, and the wet season, which is usually warmer.

In the wet season, heavy rains may stop horticultural activities even though the consumer demand is high. In solving this problem, producers in some areas, use shelters as "umbrellas" to prevent an excess of water for the crops. In some areas, despite the tropical location, heated shelters have to be used during winter when the temperatures are low. In some other cases, an insect-proof greenhouse has to be used to protect the crops (at least in its early stage of growth) from a virus frequently transmitted by insects. This is the case of tomatoes which can be infected by PYMV (*Potato yellow mosaic virus*) and TYLCV (*Tomato yellow leaf curl virus*) through the white fly (*Bemisia tabaci*). These shelters help increase yields but require significant investment and may lead to side effects, such as the soil becoming too poor to further sustain production. Producers, whether rural or urban, are always willing to adapt and improve their practices based on their own experiences and new information. Most of the new techniques however require access to capital for investments and access to specific knowledge.

4.2 Irrigation Methods

Water is a highly essential input for plant growth and its requirement is variable depending on the climatic conditions and plant species. In most capital cities of developing countries located in tropical and sub-tropical areas, the quantities needed vary from 0.1 to 1 l/m^2/day in very hot and dry weather. For a crop of 30 days, the quantity of water needed by a leafy vegetable during the dry season is around 15 l/m^2. In general, the water availability in the cities was showing a decreasing trend and production of 1 kg of tomato requires approximately 60-140 l of water based on the climate and yield.

Vegetables, especially leafy ones such as lettuce and cabbage, need to be watered twice a day, every day or at least every other day to obtain a good quality (freshness, tenderness) for marketing. For UPH in developing countries, the watering can is the most commonly used system. Each can holds 8–15 l and one worker can usually carries two cans. This manual system is efficient because, most of the time, the gardener applies exactly the quantity of water needed by the crop. This is highly labour intensive, and this operation takes 60 percent of the total labour requirement in the urban areas for vegetable production.

Drip or trickle irrigation is another popular irrigation technique that has been promoted for nearly twenty years. It saves water by 10-20 percent compared to overhead irrigation, but requires clean water in order to avoid blocking of the emitters. The full-fledged system includes filters, pumps and a pressure regulator, which low-income vegetable growers cannot usually afford. The advantage of this technique is that, water do not come in direct contact with the fruits and leaves thus eliminates the option of contamination with pathogenic microbes. However, this system cannot avoid contamination of the soil and roots of vegetables with biological pathogens. Underground irrigation provides water to plants through capillary action. Such an underground system can limit the transmission of pathogens to the vegetables due to the filtrating effect of the soil. But installation (flat soil) and operation (control of the flow to the plants) are rather difficult.

4.3 Fertilization

All crops require nutrients especially macro (N, P, K, Ca, Mg, S) and micro-nutrients (Fe, Cu, Zn, B, S, Mg, Ni *etc.*) for their proper growth and development. In the intensive cropping systems such as UPH systems, the crop nutrient demand is not met optimally since the urban growers mostly use only the solid and liquid urban wastes. In general two main groups of fertilizers are used; (i) organic and (ii) inorganic/chemical fertilizers. There has always been a heavy use of organic fertilizers in the intensive production of vegetables and ornamental crops. The quantity may vary from few tonnes/ha to 50 or even 100 tonnes per year. The most commonly used organic manures include manure from livestock or poultry, compost from vegetable wastes or wastes from urban activities: sewage sludge, night soils, household wastes *etc.* that provide most of the micro-nutrients and in addition improve the soil structure. Over many centuries, peri-urban and urban farmers have managed and recycled urban wastes.

Solid organic fertilizers have the limitation of slow release of the nutrients especially nitrogen. In contrast, the liquid fertilizers act very quickly and hence liquid organic fertilizers are often used in the production of short-cycle leafy vegetables such as amaranth and mustard. In certain countries, pig urine is used as a liquid organic fertilizer to supply nitrogen during crop growth. However, the current research is focused in utilizing the combination of organic and inorganic fertilizers to enhance their efficiency. AVRDC (World Vegetable Center), Taiwan is working on producing an organic liquid fertilizer that does not endanger consumer health (AVRDC, 2000).

Tough inorganic fertilizers are easy to use and allow application of the right dosage, there are risks of over application and contamination of soil and water by nitrates and phosphates which led to eutrophication of the neighbouring water bodies. In addition, they can also be a source of heavy metals. Urea is the main inorganic fertilizer used in the horticulture systems especially for vegetables and this along with lack of phosphorus and potash creates an imbalance of nutrient proportions in the urban soils.

4.4 Crop Protection

The use of chemical pesticides in the UPH systems have increased significantly because of their easy access *via* national and international companies, retailers and wholesalers along with technical information. However, this intensive use of chemicals have also increased the negative perception of horticultural production in and around the cities with the involvement of three major risks; i) health risks for consumers; ii) risks of polluting the environment (mainly water sources); and iii) risks for producers/agricultural workers. Pesticides in the city's surface and wastewater does not necessary come from urban horticultural production. Still when this water is used for urban crop irrigation, it constitutes a high risk.

5. Modern Urban Horticulture Systems

Urban horticulture systems have been evolved from the rural horticulture systems and differ in terms of the crops produced and the input sources utilization based on the economic strategies adopted by the urban farmers. Mixed vegetable farming with watering cans and/or pumps to cultivate short- and long-cycle vegetables like lettuce, cabbage, carrots, onions, *etc.* is a typical example. Farming systems in the peri-urban areas of Hubli-Dharwad (India) comprise vegetable production, agroforestry systems, Napier grass (fodder) production and small-scale livestock production (Bradford *et al.*, 2002). In Hyderabad (India), the predominant system is paragrass production, which is used as fodder. Green leafy vegetables are grown here on small sections for subsistence needs and for sale. Other crops include rice, fruit trees and flowers. There is also coconut and banana as well as livestock keeping (Buechler *et al.*, 2002). The urban horticulture production can be either for home consumption or for market sale. Peri-urban horticulture is often dominated by irrigated vegetable production, as is the case in Vietnam or Malaysia. Other systems that can be counted are commercial and domestic livestock production, flowers and seldom agroforestry (Potutan *et al.*, 2000). In china, the major UPH systems are for cereals, vegetable and livestock production (Yi-Zhang and Zhangen, 2000).

New and novel horticultural practices have been developed to maximize the use of urban spaces, to optimize the use of inputs, and to minimize the impact of horticulture on the health of mankind as well as on the environment. Some of the new techniques described here are: growing horticultural crops on urban built-up land with various types of urban substrates (*e.g.* on roof tops, organic farming and hydroponic production); to save water in highly populated areas; to produce residue-free vegetables year-round with a low content of heavy metals and human pathogens; and to control wastage and leaching (fertilizers, pesticides, organic matter, water) into the urban environment. These techniques are developed to take into account the specific constraints of UPH systems, but are more demanding than traditional or conventional techniques in terms of new knowledge and/or investments.

5.1 Hydroponics

Hydroponics is an innovative technology with soil less culture and is often considered as a subset of hydroculture. It requires less space, labour, external inputs and time but often demands proper management and higher investments. Since, it is difficult to control or quantify the nutrient availability in the soil, the hydroponic system provides a convenient system to control nutrient uptake by the plants. Additional advantages of this soil less culture include reduction in accumulation of soil toxins and cross contamination with soil borne pathogens.

In hydroponics, the plant roots are suspended in the flowing nutrient solutions so that they can rapidly absorb food and oxygen. The major concern of the growers is to balance the combination of water, nutrients and oxygen with the plants requirement in order to maximize the yield and quality. Thus the use of inputs especially water and nutrients are optimized. At times, some inert substrates (sponges, artificial mineral marbles, rock wool *etc.*) can also be used in order to fix the crop roots in the required position.

The quantity and quality of water is a key factor in the hydroponic system. The urban growers generally use water from different sources *viz.*, surface water (lakes, natural and artificial ponds), groundwater (wells), municipal tap water, rainwater and combinations of these. Rain water if available, can be treated as a better option since it has very low ionic strength and possess very low microbial and algal densities. However, the rainwater is not always available for irrigation because of the technical problems in collection and storage. Even if collected from greenhouse roofs into ponds, they are highly susceptible to environmental changes *i.e.* eutrophication and acidification since they are fed by the atmospheric precipitation. Hence, the growers should rely on other water sources such as rivers and ponds but in most cities these are polluted (Schwarz *et al.*, 2005).

Hydroponics is a perfect technology for urban or peri-urban areas where the soil is poor or polluted. It allows production in abundance of healthy fresh vegetables, ornamentals, medicinal and aromatic plants thus suiting the requirements of poor urban farmers. When adopted technically the productivity generated by the hydroponic systems is greater than that from the traditional gardening systems (Tabares, 2003; Rios, 2003). In many countries of South America, this technology is gaining importance now-a-days.

Small hydroponic units which can be operated by urban families shall help them in meeting their food requirement and also in getting some additional income. Some special hydroponic techniques have been developed to suit people living in the developing countries with limited urban spaces. These simplified hydroponic systems use recycled materials often and are easier to understand, learn and implement, which are well suited to fresh fruits (with a high water content) and vegetables such as lettuce, tomato, bell pepper, basil, celery and radish (Caldeyro-Stajano, 2004).

5.2 Aeroponics

Aeroponics is the technique of growing crop plants in an air or mist environment without the use of soil or an aggregate medium. Aeroponic culture differs from both conventional hydroponics, aquaponics, and *in vitro* (plant tissue culture) growing. Unlike hydroponics, which uses a liquid nutrient solution as a growing medium and essential minerals to sustain plant growth; or aquaponics which uses water and fish waste, aeroponics is conducted without a growing medium. Since water is used in aeroponics to transmit nutrients, it is sometimes considered a sort of hydroponics.

The basic principle of aeroponic growing is to grow plants suspended in a closed or semi-closed environment by spraying the plant's dangling roots and lower stem with an atomized or sprayed, nutrient-rich water solution. The leaves and crown, often called the canopy, extend above. The roots of the plant are separated by the plant support structure. Ideally, the growing environment is kept free of pests and disease so that the plants may grow healthier and more quickly than their counter parts grown in media/water. However, since most aeroponic environments are not perfectly closed off to the outside, pests and disease may still cause a threat. Controlled environments advance plant development, health, growth, flowering and fruiting for any given plant species and cultivars. Due to the sensitivity of the root systems, aeroponics is often combined with conventional hydroponics, which is used as an emergency "crop saver" – backup nutrition and water supply – if the aeroponic apparatus fails.

5.3 Organoponics

Organoponic systems grow crops on organic substrata to replace unavailable chemical inputs. The crops are sown in holes or furrows filled with an organic substratum, the concentration allows maximization of its effect. The organic substratum used can be of diverse origin, including compost or organic residues from other sources (faeces, wastewater residues *etc.*). This technique is widely used in the urban areas of countries such as Cuba and Venezuela. This system is particularly suited to low fertile soils as it helps in their restoration with increase of soil organic matter content. Since it maximizes the use of space and water, it is well suited for vegetable production in urban and peri-urban areas. However, in an urban context, the supply of organic substrata can be limited. Based on the origin of the substratum, certain problems relating to pollution and sanitation may be raised, especially when solid waste compost with heavy metals is used. Linking horticultural organoponic systems with ecological sanitation or use of manure, could increase the productivity of organoponic systems.

5.4 Permaculture

Permaculture (permanent agriculture) is an integrated system of cultivation developed to meet the constraints of limited space in the urban areas. The main aim of this permaculture is to improve the efficiency of the agricultural activities with minimal impact on the environment. They combine growing fruits, vegetables or grains along with livestock keeping by creating a symbiotic ecosystem, with an ethical foundation in sustainability and for copying nature with scientific basis in ecology. This permaculture is highly relevant to the urban and peri-urban areas because of its flexibility and suitability to the city conditions such as local recycling of energy and resources. The variety of production limits the risk and gives financial security to some extent. This system is well suited to the developing countries since the external inputs (synthetic fertilizers and pesticides) are limited or absent. Thus, permaculture can be considered as one ultimate cropping system concept that uses a wide range of techniques and concepts: rainwater collection, excrement composting, reuse and recycling of resources, energy saving, green building and planning, and development of the local economy.

5.5 Microgardens

Micro-gardening is the intensive cultivation of a wide range of vegetables, roots and tubers, and herbs in small spaces, such as balconies, patios and rooftops. While urban residents have long grown vegetables in backyard plots, modern micro-gardening makes use of containers such as plastic-lined wooden crates, custom-built tables and even old car tyres. It integrates horticulture production techniques with environmentally friendly technologies suited to cities, such as rainwater harvesting and household waste management. Micro-gardens allow low-income families to meet their needs for vitamins, minerals and plant protein by providing direct access to fresh, nutritious vegetables every day. They also offer a source of extra income from the sale of small surpluses.

5.6 Community/Allotment Gardens

These are a special type of garden established in highly urbanized areas of developing and developed countries, particularly to benefit urban poor families. These gardens are characterized by several small land parcels assigned to individual families and are concentrated at one place. At times these families may also organize into associations. The main aim of this community garden movement is to cleanup neighbor hoods; restoration of nature to industrial areas; growing and preserving food for consumption; and bringing the rural farming traditions to the urban and peri-urban areas. The main essence of the community garden lies in creation of bond between the people and place through social and physical engagement. Most of the urban garden are created on vacant lands of variable sizes and are generally gardened as individual plots by the community members and they help in supporting social, cultural and artistic events by contributing to the rebuilding of local community spirit. However, the modern community garden movement is initiated by neighborhoods with the support of government as well as the non-profit organizations. Some of these community gardens can be linked to the public housing projects, schools, temples and social agencies. This movement helped to create and sustain relationships between city dwellers and the soil while contributing to a diûerent kind of urban environmentalism that did not have any characteristics of reform charity. A more recent survey showed that more community gardens are being created in cities as opposed to being lost to economic development resulting in the community development.

5.7 Green Buildings

In the highly developed countries, green building strategy (covering building with plants) is a part of the ecological design of the urban landscapes and is becoming popular now a days (Calkins, 2004). This rooftop growing systems should be profitable with nutrient capture system, organic waste recycling and employment generation. In developing countries, horticulture on buildings, mainly on rooftops, is gaining importance and allows production of various fruits, vegetables, or flowers. For instance, in Senegal rooftop gardening, based on bricks or wooden box beds filled with compost, allows growing a wide variety of crops, including: fibrous root crops, tomato, hot pepper, eggplant, *etc.* (Deesohu Saydee and Ujereh, 2003). The green buildings are characterized by high level of intensity due to trivial spaces available on the building rooftops. Here, they might use hydroponics or organoponics (often in containers, boxes, pots or cells). There is a great positive effect of rooftop gardening on the thermal performance of the building since these plants block solar radiation and reduce daily temperature variation as well as thermal changes between winter and summer (Eumorphopoulos and Aravantinos, 1998).

A green project was developed in Singapore for planting of trees, shrubs and grasses in the city for upkeep of pleasant living environment. Though not a new concept, the roof top gardens increase the percentage of greenery in the urban built-up areas for bringing back the vanishing urban green spaces. Wong *et al.* (2003) observed that the roof top gardens reduced the energy consumption of the commercial buildings by about 14.5 per cent in Singapore. By sprucing up the underutilized portions of the commercial buildings, we can create a network of vegetation linking roofs in order to increase the ratio of greenery to the ever increasing population. The major limitation in this UPH system is that they face natural attack by insects and birds as they are located outdoors and hence may need protection to some extent.

5.8 Vertical Farming

Vertical farming refers to any kind of construction and support structure for growing plants in a vertical way; and thereby making efficient use of the limited space for food or other kind of plant production. In the UPH systems, space is one of the major limiting factor and the vertical farming concept is highly relevant which needs to be promoted. In vertical gardening, a huge variety of concepts and designs are available including greywater towers, green walls or living walls. However, The design of vertical garden (sometimes also called green walls) depends on the available material, space and local preferences as well as on the creativity and imagination of the users. In vertical gardening, reuse of different composts (*e.g.* compost, vermi compost or terra preta) from household wastes and reclaimed water (*e.g.* Grey water) or urine can be considered to enhance food production. In vertical farming, the cultivable crops include food crops (fruits, vegetables, herbs *etc.*) and non-food crops (ornamentals, medicinal and aromatic crops). The suitable growing media include soil, compost, vermicompost, terra preta compost, acrylic material as well as aquaponic and aeroponic solutions. The crops can also be grown in sacks, bags, flowerpots and all kinds of available receptacles like bins, cans, tins, bottles, tanks or boxes. They can be even placed on yet unused places like on the roof of houses, balconies, on the top of walls or just hung up where ever space is available. Thus this technology has great potential for growing food in the urban and peri-urban areas where the space is limited, particularly in the dense populated urban areas.

The diverse vertical garden technologies are an effective, simple and sustainable method of enhancing food security for urban communities, slum dwellers and other communities where horticultural production areas are limited. Most models can be implemented easily with locally available material at low cost while offering livelihood opportunity at the same time and contributing to the local food security situation.

6. Emergence and Development of Urban Aesthetics

This concept of urban aesthetics has become increasingly imperative since the urban settlements are becoming more and more crowded due to work pressure and stressful life they offer. Thus there is a great need for development of urban aesthetics under integrated and systemic approaches. Parks and green spaces are the backbone for sustainable and high quality urban environment. Urban aesthetics is primarily concerned with environment enhancement, air and noise pollution, and modification of the micro climates. High quality green spaces with a wide collection of trees and other plants have huge educational potential in addition to improving the physical and mental health of the urban dwellers. Public parks in the urban areas are an important recreational facility in the developed as well as developing countries. People derive quantifiable benefits from viewing of green trees which have psychological as well as physiological positive effects.

In developing countries like India, the urban green spaces such as parks and gardens are also valued as a wild life habitat since the tree populations sustain a huge biodiversity of birds, animals and other urban fauna. In general, the people appreciate the urban green space for their social, psychological health, aesthetic, ecological and economic functions but most of the times they take it for granted. Studies in the Indian cities suggest that the visitors are willing to pay entry fee for these green spaces and gardens if managed sustainably. Air in urban areas of many developing countries suffer from serious level of air pollution and may cause health problem. Development of urban green spaces have the potential to make marked improvement on the urban air quality by absorbing CO_2 and other pollutants (Sinha, 2013).

In order to counter the negative effects of rapid urbanization in the last few decades, special attention must be given to future urban planning, so that cities, regardless of their size, stop developing chaotically, and have not only functional attributes but also, aesthetic ones, and to harmoniously combine the built surfaces and the ones destined to recreation and green spaces.

8. Policy Implications and Recommendations

The importance of urban and peri-urban horticulture in improving the economic security of the urban poor and nutritional security of the urban dwellers is recognized by the international policy makers and put on a high political agenda. If managed well, urban horticulture can play an important role in reducing socio-economic and environmental problems of the urban areas. Planners and policymakers should develop and support community-wide plans, in improving the poor people's incomes and in using urban organic waste, to improve urban food safety and to create sustainable food systems.

Policymakers around the world are showing an increased interest in urban horticulture, although their major focus is still on the temporary use of peri-urban lands. UPH is encouraged in poor countries, mainly because it improves food security of poor households and the urban population's nutritional status (freshness of products and better access to fruit and vegetables, considered as a major source of vitamins and micronutrients), especially in view of the inefficient transportation and storage facilities in these countries. Policymakers also encourage UPH because it provides jobs and incomes to poor and landless urban dwellers and because it is well adapted to the urban environment where water and land are scarce.

9. Conclusion

In many expanding cities of the developing countries, the UPH systems act as a major contributor in supply of fresh produce to the urban markets and is expected to retain its contribution in the near future. Currently we have two major challenges (i) reduction in the available urban lands for cultivation and (ii) increased demand for fresh fruits, vegetables and flowers with raising standards of living and ever growing urban population. As the urban areas develop, the UPH system units shall evolve and adapt to new environments while the urban vegetable production remains as an essential source of high income and healthy food for the growing cities.

Worldwide, about one quarter to two thirds of urban and peri-urban households are involved in horticulture. In the coming decades, fruit and vegetable production will continue to play a key role in feeding cities and providing income to farmers. Many specific techniques have been developed or adapted specifically for urban areas but there is still some research needed in order to better understand these complex anthropised agro-systems. Systems such as permaculture that combine various forms of production can be very complex ecologically. It is therefore important to undertake agro-technical studies that could provide more in-depth on the conditions required for obtaining good-quality vegetables.

As a part of UPH, all stakeholders in the commodity chains have to be involved in developing better conditions for emerging production technologies of integrating fruits, vegetables and flowers. During urban planning process, certain key points such as supply of inputs and materials, management of crop residues and linkage between activities should be considered strictly.

The food supply function remains the most important, even though economic, social (labour), cultural, living environment, environmental (recycling) and security (food and natural risks) functions appear to be essential too. Of all the agricultural production systems, UPH was found to have a multifunctional role which should be taken into account by researchers and policy makers. For well-balanced urban development, implementation of an urban planning policy including the sustainability of UPH systems is highly essential as they play a substantial role in the development of local (micro) enterprises, including input supply, processing and marketing sectors. It also reduces the distance that fresh food needs to travel from producer to consumer.

In addition to the socio, cultural, nutritional and environmental benefits, there are various economic benets from urban and peri-urban gardening such as money savings on purchase of food and even on the utility bills. The developing countries spend approximately up to 60- 80 percent of income on buying food alone. Having green roofs can reduce the cost of heating in the winter and help stay cool in the summer. They also lower the cost of roof replacement. While green roofs are an addition to urban horticulture people are eating healthy while also improving the value of their property. Other benets include increased employment from non-commercial jobs where producers include reductions on the cost of food. Thus urban and peri-urban horticulture (UPH) helps to grow greener cities by contributing to food security, employment, waste management and community well-being.

References

Agrawal, M., Singh, B., Rajput, M., Marshall, F. and Bell, J. 2003. Effect of air pollution on peri-urban agriculture: a case study. *Env. Poll.*, **126**: 323–329.

AVRDC. 2000. *AVRDC Report 1999*. Asian Vegetable Research and Development Center, Shanhua, Tainan, Taiwan, vii+152, p. 47–48.

Birley, M. and Lock, K. 1999. A review of health impacts of peri-urban natural resource development. International Health Impact Assessment Research Group, University School of Medicine.

Bradford, A., Brook, R. and Hunshal, C. 2002. Risk reduction in sewage irrigated farming systems in Hubli, Dharwad, India. **In**: *Urban Agriculture Magazine*, **6**, April. RUAF, Leusden, The Netherlands.

Bricas, N. 1998. Cadre conceptuel et méthodologique pour l'analyse de la consommation alimentaire urbaine en Afrique. Montpellier, France, série Urbanisation, alimentation des filières vivrières, n°1, 48 p.

Buechler, S., Hertog, W. and Veenhuizen, R.van 2003. Editorial. **In**: *Urban Agriculture Magazine*, **8**. December, 2002. Wastewater use for Urban Agriculture. RUAF, Leusden, the Netherlands.

Caldeyro-Stajano, M. 2004. Simplified hydroponics as an appropriate technology to implement food security in urban agriculture. *Practical Hydroponics and Greenhouses*, **76**: 1–6.

Calkins, M. 2004. Strategy use and challenges of ecological design in landscape architecture. *Landscape and Urban Planning*.

Deesohu Saydee, G. and Ujereh, S., 2003. Rooftop gargening in Senegal. *Urban Agr. Mag.*, **10**: 16–17.

Eumorphopoulos, E. and Aravantinos, D. 1998. The contribution of a planted roof to the thermal protection of buildings in Greece. *Energy and Buildings*, **27**: 20–36.

Fatou Diop Gueye, N. and Sy, M. 2001. The use of wastewater for urban agriculture. *Urban Agr. Mag.*, **1**(3): 2.

Jorgensen, N., Laursen, J., Viksna, A., Pind, N. and Holm, P. 2005. Multi-elemental EDXRF mapping of polluted soil from former horticultural land. *Env. Int.*, **31**: 43–52.

Midmore, D.J. 1998. Importance of peri-urban vegetables to Asian cities. **In**: Paper presented at meeting on *Peri-urban Vegetable Production in the Asia-pacific Region for the 21st Century*, 29th September–1st October, Kasetsart University, Bangkok, Thailand, 14 p.

Moustier, P. 2000. Urban and peri-urban agriculture in West and Central Africa: An overview. Provisional paper (30/10/00) for SIUPA stakeholder meeting and strategic workshop, Sub-Saharan region, 1–4th November, Nairobi, Kenya.

Potutan, G.E., Schnitzler, W.H., Arnado, J.M., Janubas, L.G. and Holmer, R.J. 2000. Urban agriculture in Cagayan de Oro: A favourable response of city government and NGOs. **In**: *Growing Cities Growing Food; Urban Agriculture on the Policy Agenda*, (Eds.) M. Bakker, M. Dubbeling, U. Sabel-Koschella and H. de Zeeuw. DSE, Feldafing, Germany, pp. 413–428.

Puschenreiter, Markus, Hartl, Wilfried and Horak, Othmar. 1999. Urban agriculture on heavy metal contaminated soils in Eastern Europe. Vienna: Ludwig Boltzmann Institute for Organic Agriculture and Applied Ecology.

Rios, J.A. 2003. Hydroponics technology in urban Lima, Peru. *Urban Agr. Mag.*, **10**: 9–10.

Sachdeva, S., Sachdev, T.R. and Sachdeva, R. 2013. Increasing fruit and vegetable consumption: Challenges and opportunities. *Indian J. Comm. Med.*, **38**(4): 192–197.

Schwarz, D., Grosch, R., Gross, W. and Hoffmann-Hergärten, S. 2005. Water quality assessment of different reservoir types in relation to nutrient solution use in hydroponics. *Agril. Wat. Mgt.*, **71**: 145–166.

Sinha, R.S. 2013. *Urban Forestry: Urbanisation and Greening of Indian Cities– Efforts for Green Delhi*. Assignment: MCT Phase IV: June–August.

Tabares, C.M. 2003. Hydroponics in Latin America. *Urban Agr. Mag.*, **10**: 8.

Wong, N., Cheong, D., Yan, H., Soh, J., Ong, C. and Sia, A. 2003. The effect of rooftop garden on energy consumption of a commercial building in Singapore. *Energy and Buildings*, **35**: 353–364.

Yi-Zhang, C., and Zhangen, Z., 2000. Shanghai: Trends towards specialised and capital-intensive urban agriculture. **In**: *Growing Cities Growing Food; Urban Agriculture on the Policy Agenda*, (Eds.) M. Bakker, M. Dubbeling, U. Sabel-Koschella and H. de Zeeuw. DSE, Feldafing, Germany, pp. 467–476.

63

Agroforestry Interventions in Fruit Orchards for Improving Productivity, Profitability and Sustainability

Sudhir Kumar and O.P. Chaturvedi*

ICAR-Central Agroforestry Research Institute,
Jhansi – 284 003, Uttar Pradesh
**E-mail: chaturvediopc@gmail.com*

Agroforestry is an age old practice of growing trees on farms for benefit of farm family. India has a long history of tree based farming system under diverse ecological conditions. Harappan civilization has indicated that the inhabitants were familiar with the species of date palm, pomegranate, lemon, melon, coconut *etc.* The agri-horticulture plantation was fostered by emperor Ashoka (274-237 BCE) with mango, jackfruit and grapes. The earliest evidences of agroforestry from India can also be found in the travelogue of Ibn Battuta (1325-1354 CE, Kumar *et al.*, 2012). Although, agroforestry is a very old practice, but it was not in planned way; as the agricultural research advanced and scientists engaged in agroforestry research, they came out with the conclusion that by incorporating agroforestry in planned way farmers/planters can improve productivity, profitability and sustainability from their existing agriculture/orchards and could be a best solution to solve the problems of food, fuel, fodder, timber, unemployment, malnutrition, GHGs, environment pollution, soil health and so on. The intervention of scientific community to harness full benefits of agroforestry land use and address global issues of land degradation and agricultural sustainability has led to horizontal (expansion of area) and vertical (innovative research) growth of agroforestry (Prasad *et al.*, 2016).

Indian agriculture is facing diverse challenges and constraints due to growing demographic pressure, increasing needs of food, feed, pulp, fodder, fruits and timbers, degradation of natural resources and climate change (Pandey, 2007; NRCAF, 2013 Vision 2050). It is assumed that diversification of land use system with agroforestry can address some or most of these challenges. Therefore, agroforestry has been receiving greater attention by researchers', policy makers and others for its ability to contribute significantly in economic growth, poverty alleviation and environmental quality. Agroforestry has been recognized as an important part of 'evergreen revolution' movement in the country. India launched the National Agroforestry Policy 2014 with focus on improving productivity of small and marginal farmers and providing them sustainable livelihoods besides helping in natural resource management and improving forest covers (Dhyani, 2014).

India is blessed with different types of agro-climatic conditions, so there are huge variation in agroforestry systems in their structural complexity and species diversity, their protective and productive attributes and their socioeconomic dimensions. They range from simple form of shifting cultivation to complex form of home- gardens; from sparse stands of trees on farmlands as in arid region of Rajasthan to high density complex multi-storied homesteads of Kerala; from systems in which trees play a predominantly 'service' role (shelter belts) to those in which they provide main saleable products (intercropping with plantation crops) (Dhyani *et al.*, 2005 and 2009).

There are number of agro-forestry systems based on agricultural components used in the orchard and almost all the systems are being used in one or other part of the country. These systems not only provide fruits but also help in fulfilling the requirement of timber, food, fuel, fodder, employment generation and provide economic stability.

Why Intervention is Required?

There are number of reasons to prove the intervention of Agroforestry in fruit orchards and few of them are as under:

- ✰ Land limitation
- ✰ Increasing human population
- ✰ Increasing cattle population
- ✰ Industrialization
- ✰ Livelihood support
- ✰ Gap between supply and demand
- ✰ Depleting soil fertility
- ✰ Conservation of natural resources
- ✰ Prevention or reduction in environmental pollution.

Agriculture is of fundamental importance in India's economy contributing about 20-21 per cent of gross domestic product (GDP) and generates two third of the employment. Total food grain production, which was only about 51 million t during 1950's has gone up to 257.07 million t in 2014-15. However, the population of India which is expected to rise to 1389 million by 2025. For meeting the requirements of the burgeoning population, the country will require to produce 320 m t of food grains from a current level of 257 million t. At presently, the net sown area in the country is only 142 m ha although the total cultivable land is 183.2 m ha. The agricultural land under cultivation, culturable waste, and barren and unculturable land are 155.56, 13.12 and 17.7 M ha, respectively. Currently, about 55 per cent of the cultivated area is under rainfed agriculture contributing 45 per cent of the total production. It is estimated that even after achieving the full irrigation potential (barring for successful completion of river linking project) nearly 45-50 per cent of the total cultivated area will remain dependent on rain. At present, about 40 per cent of population and 60 per cent of the total livestock inhabit these lands. The human and animal population exceeds the supporting capacity of many fragile ecosystems in the country.

With the rapid growth of urbanization and economic growth in the country, farming community have witnessed unprecedented opportunities for moving beyond subsistence farming to supplying products needed by urban dwellers. Agroforestry products such as timber, fruit, food, fibre, fodder, medicine and others are progressively meeting the subsistence needs of households and providing the platform for greater and sustained productivity. Now due to change in dietary habits and consumption patterns, it is estimated that there will be less demand for cereals and many times more demand for animal products like milk, meat, pulses and edible oils by 2025. This requires a matching reallocation of land use in favour of livestock production, pulses and oilseeds. Therefore, research strategy has to match this reality.It is in this context agroforestry systems offer opportunities to farmers for diversifying their income and to increase farm production, allowing them to increase the resources they devote to purchasing food and countering diseases. Agroforestry can improve soil fertility, provide fodder, produce fruits, expand fuel wood supplies, and produce a variety of wood products for farmers' home use and sale. Research results from different agro-climatic regions of the country show that financial returns generated from agroforestry systems vary greatly but are generally much higher than returns from continuous unfertilized food crops. The higher returns associated with agroforestry can translate into improved household nutrition and health, particularly when women control the income.

Timber Tree vs Fruit Trees

It is estimated that shisham trees produced 25 q of timber and 15 q of fire wood in 50 years, whereas the trees of mango and jamun produced 20 and 30 q of timber; 18 and 15 q fire wood in addition to 50 and 35 q of fruits, respectively in the same period. Leaves of some of the trees and most of the fruit trees are used as fodder for cattle, camel, sheep and goat.

Models developed through Agroforestry Intervention in Fruit Orchards

There are number of ways to intervene agroforestry in fruit orchards as per the local climate and need of the dwellers. On the basis of either planting of trees, pastures, arable crops *etc.* in standing fruit orchards or by planting fruit trees in existing pasture or timber/forest trees different agroforestry models have been developed in almost all the region of India. The evolved models are as under:

Agri-Horti System

Horticulture based production system is considered effective strategy for improving productivity, employment opportunities, economic condition and nutritional security. Several drought hardy fruit crops like *Cappirs decidua, Salvadora oleoides, Cordia dichotoma, Cordia gharaf, Zizyphus nummularia* var. *rotund ifolia, Z. mauritiana* are suitable for the area receiving rainfall <300 mm. Besides providing fruit these plant produce moisture laded nutritious leaves for animal. Several other fruit such as aonla, pomegranate, *bael* (*Aegle marmelos*), date palm, *imli* (*Tamarindus indica*) can be grown in the area having irrigation facilities. Agri-horti system comprising *Zizyphus* + mungbean provided fruit, fuel wood and round year employment even in below average rainfall year. Intercropping of bottle guard and during *kharif* season and pea (arkel) and kasurimethi with ber plantation did not cause adverse effect on three year old ber. Pomegranate has been found compatible with pearl millet, mungbean, *isabgol*, sorghum and cumin in Jallore district of Rajasthan. Studies revealed that the yields of both jujube and the intercrops were higher than in monoculture (Singh 1997). Intercropping of guar, green gram and sesame with jujube (cv. Seb) increased the fruit yield to 14.8 from 5.2 kg tree^{-1}. Cultivation of guar in this system gave additional advantage of 782 kg seed yield, which was higher than that obtained from green gram and cowpea in drought years. Agri-horti system under rainfed and irrigated conditions provide assured sustainability due to additional gains from under storey crops, better utilization of natural resources like water and soil, recycling of on-farm wastes, improving soil fertility, preventing soil erosion and land degradation. The system was found more remunerative particularly during juvenile stages of the horticultural crops. However, for long term sustainability, choice of crops for intercropping is more important, which are photo insensitive, less exhaustive and economical. During juvenile phase of fruit tree, there are ample opportunities for raising annual, biennial and perennial crops which can meet diversified need of farmers.

Agri-Horti-Silvi System

This system is developed to avoid the soil erosion and evaporative water losses through speedy winds from field; it has been scientifically advised to adopt the planting of perennials as shelter belts in or around the field. Such systems were termed as agri-horti-silvi system to provide more security, a symbiotic affinity which leads to the higher production of fruit, fodder, fuel and timber wood per unit area. Drought hardy fruit crops can survive and provide income to the farmers even under severe drought. Silvicultural plantations would check the drift of sand, provide forage, fuel and timber wood and would help in creating favorable micro-climatic conditions. To manage the arid lands in judicious way, a combination of horticultural crops, silvicultural plantations and fodder crops has been suggested.

A study was conducted under extreme arid condition (Yadav *et al.*, 2013) to get maximum biomass and production per unit area from agri-horti-silvi production system. The green gram and guar were grown as intercrops under this system with 9-year-old plantation of citrus (*Citrus aurantifolia*), mopane (*Colophos permummopane*) and shisham (*Dalbergia sissoo*). Highest biological and seed yield of green gram (1412.5 and 471.3 kg ha^{-1}) and guar (1352 and 419.8 kg ha^{-1}) was recorded with citrus over intercropping with shisham and mopane, respectively. The study revealed that with use of minimum irrigation through sprinkler system, intercropping under horti-silvi system can be successful with established citrus plantations without any reduction in productivity.

In an another study having wheat as winter season intercrop, irrigation and fertilizer for wheat promoted the growth of horti and silvi species compared to sole plantation (Singh *et al.*, 2012). *C. mopane* with *C. myxa* combination was found more competitive because of vigorous growth of these species by utilizing above and below ground resources extending competition with the wheat reducing yield by 25 per cent. However, a combination of P. *cineraria* and Z. *mauritiana* appeared to be the best where yield reduction in wheat was bridged down to 6 per cent. This combination needs validation and promotion for greater benefits in terms of fodder, fruit and food production in the arid environment. Further, other tree species and suitable arable crops should also be explored for better sustainability.

Horti-Pastoral System

Under horti-pastoral system grasses are growing in the fruit orchard for rearing cattle. It isa traditional farming system mainly adopted in dry areas. This system is also adopted as an alternative land use system in northern regions of India. Farmers in these areas grow fruit trees like guava, mango, citrus, pomegranate, jujube, Indian gooseberry, *etc.* and rear small cattle that graze on pastures. Now adays, hybrid varieties of fruit trees are being grown and people inhabiting these areas earn money from selling fruits, milk and other different products. In horti-pastoral system, the inter spaces between fruit trees are utilized for the cultivation of grasses and grass legume mixtures. In this system, fruit trees usually form the first tier, whereas grasses are grown as ground storey crop. Only during dormant season of the fruit tree, the livestock are allowed to graze on the available pasture for a period of 3-4 months in a year depending on stocking rate. This system is more suitable for rainfed conditions where animal husbandry is also a source of livelihood. In the integrated farming system model with a land holding of 7-8 ha, 10–15 per cent of land is allocated for this system. Z. *nummularia* (*jharber*) is an important top feed species commonly found in farmers' fields. During drought years when crops fail, this plant comes to the rescue of the farmers and provide them fodder, fuel wood and fruits. A study conducted on Z. *nummularia* and Z. *rotundifolia* in an established buffel grass pasture concluded that jujube species can be successfully grown in buffel grass pasture provided pasture is established after two years of the establishment of the jujube plants (Sharma and Vashishtha, 1985). Economics of the system over a period of 9 years was worked out by method of discounting (Sharma and Diwakar 1989) and found thatthe cost was positively related with the yield of forage and its seed production throughout the life of the system except the year of establishment.

In an experiment, where *C. ciliaris* was introduced in an 18 year old jujube orchard no adverse effects were observed on jujube yield but additional 2.5 t ha^{-1} dry forage and 25–30 kg grass seed yield was achieved, besides average jujube fruit yield of 7.5 t ha^{-1} (Vashistha and Prasad 1997). In semi-arid conditions also, *C. ciliaris* and *Stylosanthus hamata* in various combinations were grown in the interspace of jujube (cv. Gola). Introduction of pasture was reported to have no adverse effect on the growth parameters of jujube during the establishment phase of first five years (Kumar *et al.*, 2002). In Indian gooseberry based system, marvel grass (*Dicanthium annulatum*) as under storey forage species with different doses of nitrogen was assessed for ten consecutive years (Kumar *et al.*, 2009). The pasture production was marginally higher with association of trees as compared to pure pasture. Further, high dose of nitrogen (60 kg ha–1) gave significantly higher dry pasture as well as fruit yield compared to sole pasture. The mean B:C ratio over ten years was 1.85 for pure pasture and 3.7 with this system. The employment generation was 2.07 and 4.74 man days per month in pure pasture and in horti-pasture system, respectively. The study suggested that adoption of horti-pastoral system in rainfed agriculture is economically viable, ecologically feasible and socially accepted to minimize the migration from rural areas for the hunt of jobs.

Horti-Silvi-Pastoral System

The concept of horti-silvi-pastoral production system is based on both biological and socio-economic considerations. The former includes all the advantages of fruit and trees on the soil and environment, such as close and efficient nutrient cycle, maintenance of organic matter, prevention of run off and soil erosion, regulation of microclimate, and above all adaptability of trees to soils that are incapable of sustaining annual crops. The socio-economic factors that substantiate the potential value of this production system are the burgeoning human and animal population pressure and lack of resources (Bhandari *et al.*, 2014). The consumption of fruits in the daily diet of rural mass is almost negligible. Animal husbandry being major source of income, the availability of fodder in adequate quantity is important. Therefore, this system has great potential to provide a sustainable land use system. In this system, the selection of fruit trees would depend upon the agro-climatic conditions and availability of irrigation facilities. Recognizing the importance of trees as top feed during lean periods, it is useful to know the potential feed species along with their palatability and nutritive value. As far as possible, multipurpose species would be an ideal choice under the given conditions.

Some identified feed species based on palatability ratings for leaves are *A. tortilis, P. cineraria, S. oleoides, Z. nummularia, Acacia senegal, Albizia lebbeck, Anogeissus rotundifolia, A. pendula, Calligonum polygonoides, Azadirachta india, Grewia tenax, Gymnosporia spinosa, P. juliflora and T. undulata* (Ganguli *et al.*, 1964). These species occur in distinct habitats and remain green even during hot summer months. To provide balanced nutritive fodder to the animals, both grasses and legumes in the pasture are advocated. The selection of the species would, however, depend upon the soil and climatic conditions. Based on the edaphic factors suitable grasses are classified, which include *L. sindicus, Panicum antidotale, Saccharum bengalensis, C. ciliaris* for sand dunes and sandy plains; *D. annulatum, Heteropogon contortus* for sandy loam and sandy clay loam soils; *Sehima nervosum, D. annulatum* for sand stones and rocky sites; *C. ciliaris* and *C. setigerus* for well drained sandy alluvial soils; and *Sporobolus marginatus, Chloris virgata* for low lying saline soils (Bhati, 1984), but the selection of grass species will also depend on the rainfall. The management of trees, fruit crops and pasture grasses will involve planting all the three components in such a way so as to avoid competition for light, nutrition and water. The tree species should be planted on the periphery of the orchard in south-west direction, which would act as shelterbelt and wind break. While planting trees/shrubs for shelterbelt, selection of species should be on the basis of height. The tall species should be planted on the boundary, the middle row should consist of shrub, having low height and again the third row should be of tall species. Fruit plants require more care, therefore, selected cultivars of fruit plants should be planted at recommended distance, which would enable the pasture to grow in between rows of the fruit trees. In horti-silvi-pasture system, silvicultural plantation (trees/bushes) should be planted as shelterbelt and fruit trees can be planted simultaneously. The sowing/planting of grass should be taken up 1 or 2 years after planting the trees/shrubs and fruit plants. The horti-silvi-pastoral system can produce more biomass per unit area, which also meets the basic requirement of fruit, fuel and fodder for the rural masses of the region. Not much work has been done on this multidimensional aspect and hence there is a need to conduct intensive research on multi-location to generate information. This system will not only provide the judicious use of available land resources but will also generate additional employment opportunities for rural people.

Impact of Agro Forestry Intervention in Fruit Orchards

Based on anticipated basic needs of the country like food, fruit, fuel, timber and fodder besides enhancing export prospects in medicinal and aromatic crop sectors, Agroforestry has the potential to contribute and fulfill the basic needs through increased production and decrease the requirement. In terms of specific empirical evidence, the area under pastures/grazing lands has moderately decreased from 3.7 to 3.5 per cent of reported geographical area of 306 m ha of the country. On the other hand, the area under fallow lands is gradually increasing from about 8 per cent to over 10

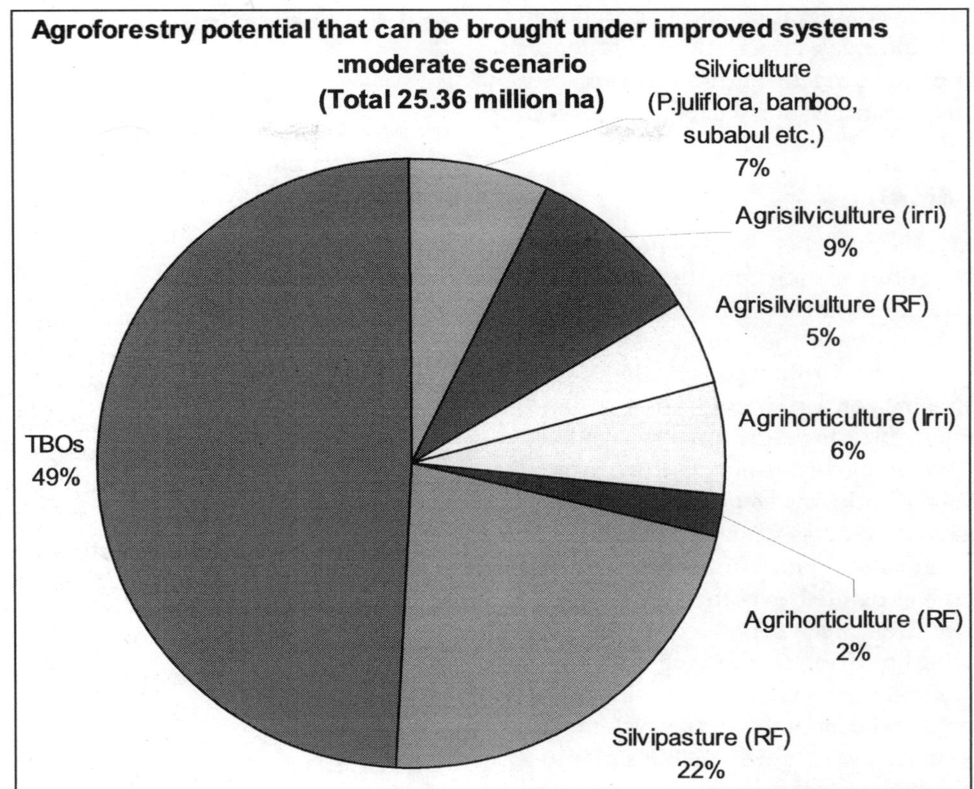

Figure 1: Agroforestry Potential under Moderate Scenario.
***Source:* NRCAF Perspective Plan Vision 2025.**

per cent. Thus, the potential for agroforestry exists in such lands, which has been taken into account in the potential area and investment. The total potential for agroforestry area by 2025 has been estimated at 25.36 million ha (moderate scenario) with almost half of it under TBOs followed by hortipasture and others (Figure 1). However, agroforestry promotion for realizing the true potential will require efforts from all the concerned including forest departments, panchayats, NGOs and farmers. The policy thrust must be on creating infrastructure, ensuring quality in planting material, plugging the vacuum of extension network, favorable pricing mechanism including support price system, encouraging public-private partnerships in all spheres of agroforestry among others.

The tangible contributions of agroforestry under three scenarios *viz.*, Business as usual (BAU), Moderate (Mod) and Optimistic (Opt) both for traditional and improved systems was worked out, by keeping the potential area and total demand as constant (Table 1).

Table 1: Contribution of AF Produce to National Demand by 2025

Product	*Total Demand*	*Contribution from AF in different Scenario*			*% Contribution from AF*		
		BAU	*Mod*	*Opt*	*BAU*	*Mod*	*Opt*
Timber (million cum)	116	80	100	120	69	86	103
Fuel wood (million tonnes)	330	50	73	95	15	22	29
Dry fodder (million tonnes)	653	89	112	125	9	11	12
Fruits (million tonnes)	43	4.5	5.9	9	10	14	21
TBO's (million lit biodiesel)	10	4.4	5.9	11	40	59	100
Food* (million tonnes)	308	15	22	29	5	7	9
Bioenergy (MW)	16000	3600	5000	6000	2	3	4
Forest cover (million ha)	100	18	25	10	18	25	30

*Food grain production from agrisilviculture/agrihorticulture only considered

BAU: Business as usual; Mod: Moderate scenario; Opt: Optimistic scenario.

Source: NRCAF Perspective Plan Vision 2025.

Efficient Land Use

The horticulture based production systems can be practiced both in rainfed and irrigated conditions for enhanced productivity, profitability and sustainability. These systems are also useful for those marginal and degraded lands where arable farming is either not possible or uneconomical. Adopting horti-pastoral or horti-silvi- pastoral system can ameliorate these lands in a long term due to soil conservation and improvement in fertility levels. Windbreak effect due to live fences and tree plantation leading to better soil moisture retention and reduced transpiration will be an additional advantage.

Improves Productivity

In agroforestry, the potentially higher productivity could be due to the capture of more growth resources *e.g.* light or water or due to improved soil fertility. Several studies in different parts of the country suggested that agroforestry is more profitable to farmers than agriculture or forestry for a particular area of land. National Research Centre for Agroforestry has been working since 1989 on agri-silvi-horticulture system which included 4 varieties of *aonla viz.*, Chakaiya, Kanchan, Krishna and NA 7 as fruit trees, *Leucaena* as multi-purpose tree and blackgram as intercrop in rainfed areas. The *Leucaena* was planted on both sides of the fruit trees at 2 m distance. The *aonla* was planted at 10 m x 6 m and 5 m x 6 m spacing but 10 m x 6 m spacing was proved an ideal spacing among these and it was considered for calculating the yield and economics of the system. The cost of cultivation in first year which includes planting of fruit trees and *Leucaena* as multipurpose trees and cultivation of crop was Rs. 8,666/ha but during next year the cost of cultivation was reduced and it was again increased with subsequent increase in the cost of input during different years. The gross income from the system was less during initial yearbut when fruiting started in aonla, the gross income was increased and it went up to Rs. 60,712/ha at age of 13 years. Similarly, the net income was positive in all the years except at first, third and six years when *aonla* had no fruiting/ or less fruiting. In age of 13 years, the B: C ratio from the system was 3.28 and on discounted rate it was 2.61, which indicated that *aonla*-based agroforestry system is a profitable enterprise in marginal lands under rainfed conditions (Ram Newaj and Rai, 2005).

Increase Soil Fertility

The primary objective of soil conservation is to improve/maintain soil fertility. To achieve this, control of erosion, maintenance of organic matter and physical properties, organic matter addition, maintenance of nutrient is necessary. In this way agroforestry system constitute sustainable land use and helps to improve soils in the number of ways. Maintenance and enhancement of soil fertility is vital for global food security and environmental sustainability. Ecologically sound agroforestry systems such as intercropping and mixed arable-livestock systems can increase the sustainability of agricultural production while reducing on-site and off-site consequences and lead to sustainable agriculture. Alternate land-use systems such as agro-forestry, agro-horticultural, agro-pastoral and agro-silvipasture are more effective for soil organic matter restoration (Manna *et al.*, 2003).

Soil Conservation

Suitable alternate land use systems involving agriculture, horticulture, forestry and agroforestry has been designed with the support of local natural resources for almost identical hydrological behavior as under the natural system. The model land use suggests utilizing slopes below 50 per cent towards lower foothills and valley lands for agricultural crops, middle slopes between 50 and 100 per cent for horticulture and top slopes over 100 per cent for forestry/silvi-pastoral establishment. Under agri- horti-silvi- pastoral systems, the reduction in runoff was 99 per cent and in soil loss 98 per cent (Singh, 1988). Silvi-pastoral system comprising *Alnus nepalensis*, pineapple and forage crops like *Panicum maximum* or *Setarias phacelata* coupled with *S. guyanensis* in 1:1 ratio was found to be a sustainable agroforestry practice in soils having 30-60 per cent slopes. Forage yield of 13.5 t/ha was obtained from the combination of *Stylo* and *Setaria*. In addition, 2.3 t/ha litter from *Alnus nepalensis* and 4,000 fruits/ha from pineapple were obtained. This system also restored the fertility of these soils (Chauhan *et al.*, 1993).

Fodder Supply

The fodder scarcity (Table 2) could be reduced through fodder cultivation on farm lands and through silvipasture/hortipasture on farm/community wastelands. Assuming availability of another 10 percent of community wastelands for silvipasture/hortipasture, the additional fodder production (@ 2.0 t/ha/yr, *i.e.* at 40 per cent of research yield) would be 11.99 million tonnes in the immediate future. At a modest price of Rs. 800/t, the contribution would work out to Rs. 11,800 million.

Table 2: Projected Supply and Demand Scenario (million tonnes) of Forage in the Country

Year	Supply (mt)		Demand (mt)		Deficits% of Demand	
	Green	Dry	Green	Dry	Green	Dry
2010	395.2	452.7	1057	588.2	62.63	22.91
2015	400.5	464.0	1095	609.2	63.44	23.7
2020	406.0	475.7	1135	630.9	64.26	24.57
2025	411.5	487.6	1174	653.3	65.10	25.44

Socio-economic Issues

Adoption of horticulture based production systems can improve the socio-economic conditions of resource deficient farming community. These will provide an assured source of livelihood as perennial component will generate farm produce even during low rainfall or drought conditions. Many of these underutilized fruit species play an important role in the social economy and livelihoods of tribal, small, marginal and landless farmers. Produce of trees provides additional income to these farmers and substantial livelihood support in addition to the nutritional security to the children and women. Organized production and processing for value-addition of products would further enhance income of small and marginal farmers and also help in on-farm conservation of valuable germplasm (Malik *et al.*, 2013). Increased biomass availability will help in integration of livestock in to the farming system. Most of the perennial components also provide fodder for huge animal population under adverse conditions.

Employment Opportunity

These production systems will certainly open an opportunity for additional employment for rural youth during most of the months compared to arable farming, which is exclusively dependent on rains and confined to a limited period. These systems involve collection of fodder from grass and top feed tree component, raising nurseries, grafting of fruit trees, training, application of agronomical and plant protection measures, irrigation, grading, marketing, pruning, and managing farm produce for compost preparation or recycling in soil. There is a tremendous potential for employment generation with improved agroforestry systems including fruit trees to the tune of 943 million person days per annum (Table 3).

Table 3: Employment Generation Potential through Improved Agroforestry

AF System	*Area ('000 ha)*	*Employment/ ha/yr (man days)*	*Total Employment year ('000 man days)*
Silviculture	1775.2	30	5325.6
Agri-silviculture (Irri)	2282.4	40	9129.6
Agri-silviculture (RF)	126.8	30	3804.0
Agri-horticulture (Irri)	1521.6	50	76080.0
Agri-horticulture (RF)	507.2	40	202880.0
Silvi-pasture	5579.2	30	167376.0
TBOs	12426.4	40	497056.0
Total	25360.0		943392.0

Source: NRCAF Perspective Plan Vision 2025.

Nutritional Security

The rural masses do not have adequate fruits and vegetables in their diet. With large scale plantation of horticultural crops in their farm under integrated system, the inhabitants of region will certainly get nutrition in the form of vegetables and fruits.

Biodiversity Conservation

Over exploitation of natural resources is a major challenge for sustainable production and livelihood security. Agroforestry with components like timber trees,fruit trees, agricultural crops, grasses, livestock *etc.* provides all kinds of life support.Trees in agroforestry system act as a refuse to biodiversity after catastrophic events such as fire (Griffith, 2000). The traditional society of coastal belts and tropics of the country practicing homegardens and sacred groves help in bio- diversity conservation.

Carbon Sequestration

Intervention of agroforestry in fruit orchard can be significant sink of atmospheric carbon (C) due to their fast growth and high productivity. By including fruit trees along with timber trees in agricultural production systems, canincrease the amount of C stored in lands devoted to agriculture, while still allowing for the growing of food crops (Kursten, 2000). In fruit based agroforestry system, tree components are managed, often intensively by pruning of minimizing competition and maximize complementarity. The pruned materials are mostly non- timber products. Such materials are often returned to soil. Besides, the amount of biomass and therefore C that is harvested and exported from the system is relatively low in relation to the productivity of the tree. Therefore, unlike in tree plantations and other mono culture systems, an agroforestry model seems to have unique advantage in terms of C sequestration. The C storage capacity varied from region to region and also depends upon the growth and nature of tree species involved in the system.

Conclusion

In present scenario when the population (human and animal), unemployment, malnutrition is increasing, land is shrinking and soil fertility is going down and down, climate is changing, environment is polluting and food security is threatening- agroforestry intervention in existing and new fruit orchards could be a best solution to fight against such problems as different fruit based agroforestry models not only provide food, fruit, fodder but also provides employment, income, conserve the soil and its fertility, conserve biodiversity and helps in carbon sequestration and mitigation.

References

AICRPAF. 2006. *All India Coordinated Research Project on Agroforestry*, Report, NRCAF, Jhansi.

Bhati, T.K. 1984. Grass component in silvipastoral system with special reference to Indian arid zone. **In**: *Agro-forestry in Arid and Semi-arid Zone*, (Ed.) K.A. Shankarnarayan. CAZRI, Jodhpur, pp. 150–155.

Bhandari, D.C., Lodha, S. and Meghwal, P.R. 2014. Horticulture based production systems in Indian arid regions. **In**: *Sustainable Horticultural Systems, Sustainable Development and Biodiversity* 2, (Ed.) D. Nandwani. DOI 10.1007/978-3-319-06904-3-2.©Springer International Publishing Switzerland, pp. 19–47.

Chauhan, D.S., Dhyani, S.K. and Desai, A.R. 1993. Productivity potential of grasses in association with *Alnus nepalensis* and pineapple under silvi-horti-pastoral system of agroforestry in Meghalaya. *Indian J. Dryl. Agril. Res. Dev.*, **8**: 60–64.

Dhyani, S.K. 2014. National Agroforestry Policy 2014 and the need for area estimation under agroforestry. *Curr. Sci.*, **107**(1): 9–10

Dhyani, S.K., Newaj, R. and Sharma, A.R. 2009. Agroforestry: Its relation with agronomy, challenges and opportunities. *Indian J. Agron.*, **54**(3): 70–87

Dhyani, S.K., Sharda, V.N. and Samra, J.S. 2005. Agroforestry for sustainable management for soil, water and environment quality: looking back to think ahead. *Range Mgmt. Agroforestry*, **26**(1): 71–83

Ganguli, B.N., Kaul, R.N. and Nambiar, T.N. 1964. Preliminary studies on a few top feed species. *Ann. Arid Zone*, **3**(2): 33–37

Griffith, D.M. 2000. Agroforestry: A refuge for tropical biodiversity after fire. *Cons. Biol.*, **14**: 325–326.

Kaur, B., Gupta, S.R. and Singh, G. 2002. Carbon storage and nitrogen cycling in silvopastoral systems on a sodic soil in northwestern India. *Agroforestry Syst.*, **54**: 21–29.

Kumar, B.M., Singh, A.K. and Dhyani, S.K. 2012. South Asian Agroforestry: Traditions, transformations and prospects. In: *Agroforestry: The Future of Global Land Use, Advances in Agroforestry* 9, (Eds.) P.K.R. Nair and D. Garrity. DOI 10.1007/978–94–007–4676–3_19, © Springer Science+Business Media Dordrecht.

Kumar, S., Kumar, S., Baig, M.J., Choubey, B.K., Sharma, R.K. 2002. Effect of nitrogen on the productivity of aonla based hortipastoral system. *Indian J. Agroforest.*, **4**: 94–97.

Kumar, S., Kumar, S. and Choubey, B.K. 2009. Aonla-based hortipastoral system for soil nutrient build-up and profitability. *Ann. Arid Zone*, **48**(2): 153–157.

Manna, M.C., Ghosh, P.K. and Acharya, C.L. 2003. Sustainable crop production through management of soil organic carbon in semiarid and tropical India. *J. Sustain. Agric.*, **21**: 87– 116.

NRCAF 2005. *Annual Report*, NRCAF, Jhansi, Uttar Pradesh.

NRCAF 2013. *Vision 2050*, NRCAF, Jhansi, UP, p. 30.

Prasad, R., Dhyani, S.K., Newaj, R., Kumar, S. and Tripathy, V.D. 2016. Contribution of advanced agroforestry research in sustaining soil quality for increased food production and food security. *J. Soil and Wat. Cons.*, **15**(1): 31–39

Pandey, D.N. 2007. Multifunctional agroforestry systems in India. *Curr. Sci.*, **92**(4): 455–463.

Rai, P., Yadav, R.S., Solanki, K.R., Rao, G.R. and Singh, R. 2001. Growth and pruned biomass production of multipurpose tree species in silvi-pastoral system on degraded lands in semi-arid region of Uttar Pradesh, India. *Forest Tree and Livelihood*, **11**: 347–364.

Ram Newaj and Rai, P. 2005. *Aonla*-based agroforestry system: A source of higher income under rainfed conditions. *Indian Farm.*, **55**(9): 24–27.

Sharma, S.K. and Diwakar, G.D. 1989. Economic evaluation of horti-pastoral system in arid region of western Rajasthan. *Indian J. Range Mgt.*, **10**(2): 119–122.

Sharma, S.K. and Vashishtha, B.B. 1985. Evaluation of jujube-buffel grass hortipastoral system under arid environment. *Ann. Arid Zone*, **24**(3): 303–309.

Singh, A. 1988. Hydrological behaviour of experimental watersheds. *Annual Report*, ICARRC for NEH Region, Barapani, Meghalaya, pp. 169–171.

Singh, G. 2005. Carbon sequestration under an agrisilvicultural system in the arid region. *Indian Forester*, **131**(4): 543–552.

Singh, R.S. 1997. Note on the effect of intercropping on growth and yield of ber (*Z. mauritiana*) in semi-arid region. *Curr. Agric.*, **21**(1–2): 117–118.

Singh, B., Bishnoi, M. and Baloch, M.R. 2012. Tree growth and wheat yield in agri-horti-silvi system in the arid region of Rajasthan. *Indian Forester*, **138**: 726–732.

Swamy, S.L., Puri, S. and Singh, A.K. 2003. Growth, biomass, carbon storage and nutrient distribution in *Gmelina arborea* Roxb. stands on red lateritic soils in central India. *Bioresource Tech.*, **90**: 109–126.

Vashistha, B.B. and Prasad, R.N. 1997. Horti-pastoral systems for arid zone. **In**: Silvipastoral Systems in Arid and Semi-arid Ecosystems, (Eds.) M.S. Yadav, M. Singh, S.K. Sharma, J.C. Tewari and U. Burman. CAZRI, Jodhpur, pp. 277–288.

Yadava, N.D., Soni, M.L., Nathawat, N.S. and Birbal, 2013. Productivity and growth indices of different intercrops in agri-horti-silvi system in arid western Rajasthan. *Ann. Arid Zone*, **52**(1): 61–66.

64

Horticulture Based Integrated Farming Systems: A Way Forward in Doubling Farm Income

Narendra Pratap Singh[1]* and Gopal Ramdas Mahajan[2]

[1] ICAR–National Institute for Abiotic Stress Management, Baramati – 413 115, Maharashtra

[2]Natural Resource Management, ICAR-CCARI, Old Goa – 403 402

**E-mail: director.niasm@icar.gov.in*

The Indian economy is strongly linked to the advancement of agriculture and that too of resource poor marginaland small farmers who constitute 80 per cent of >115 million farm families cultivating only 29 per cent of the consolidated and scattered arable land (Gill *et al.*, 2009). Sustaining house-hold food security has been an issue of prime importance to majority of the farmers. Farmers under small and marginal category are economically poor and work in diverse, location specific and risk prone environment invite attention to develop technologies for interdisciplinary enterprises as farming system research. During last five decades, agricultural research emphasized mainly component and commodity based research involving development of animal breed, crop varieties, farm implements, and machinery, fertilizer use, and other production and protection technologies mostly conducted in isolation and at the Institute level which enabled the farmers to grow more but at the same time over exploited the resources. It resulted in decreasingfactor productivity, resource use efficiency and ultimately less farm productivity and profitability. It furthercoupled with the national problems like environmental degradation, groundwater contamination and entry of toxic substances in to the food chain has become a significant problem. To tackle such problems, farming system approach has been widely recognized and advocated as one of the tool for harmonious use of inputs and their compounded response to make the production system sustainable.

The horticulture sector includes fruit crops, vegetable crops, medicinal and aromatic crops and spices and plantation crops. It contributes in poverty alleviation and nutritional security. Presently, horticulture contributes 28 per cent of agriculture gross domestic product. The national goal of achieving 4.0 per cent growth in agriculture can be achieved through major contribution from horticulture growth. Introduction of improved technology of vegetable production became an important component in overall livelihood improvement of the respective region. Vegetable crops are highly income intensive if improved management practices are adopted along with development of appropriate market linkages. They are also rich source of nutritional security. Horticulture is not only an integral part of food and nutritional security, but also an essential ingredient of economic security. Vegetable cultivation is considered one of the major sources of food security and income generation among the rural community. The role of horticulture is changing rapidly from traditional to high income generating activity. The process of transition from low-input largely subsistence horticulture to a more intensive market-oriented version present many challenges. (Kashyap *et al.*, 2015). India's share in world fruits and vegetables production is 10 and 13.28 per cent, respectively. India is the second largest producer of horticultural crops in the world after China, but the productivity is very low. Since, fruits and vegetables are perishable in nature the losses are very high. However, despite all adversities such as falling farmland and scarcity of water the country is still able to produce major share of food crops to feed its population. Marginal and small categories of all farmers,representing more than 80 per cent of Indian farm families with holding size below 1.2 ha are living in risk prone diverse production conditions. Small and fragmented land holdings do not allow farmers to have independent farm resources like drought animals, tractors, bore wells or tube wells and other sophisticated farm machineries for various cultural operations. In the past, the focus had been on maximization of crop yields only and that to for well- endowed resource rich farmers. To fulfil the basic needs of households including food for humans, feed, fodder, fuel and fibre, a well-focused attention towards Integrated Farming System Research is warranted (Kashyap *et al.*, 2015). Horticultural and vegetable crops can provide2-3 times more energy production than cereal crops on thesame piece of land and will ensure the nutritional securityon their inclusion in the existing system (Gill *et al.*, 2009).

The paper reviews critically the horticulture based farming systems to improve the income of the farmers. The aim of the paper is to identify key region or location specific horticultural interventions for the integrated farming systems for small and marginal farmers keeping in mind improvement in the farm income.

The studies on the horticulture based integrated farming systems for different regions have critically reviewed and discussed here.

1. Multi-Storied Horticulture Crop based Farming Systems of Kerala (Parthasaraty *et al.*, 2004)

India is considered as the land of spices and the glory of Indian spices are known world over. Spices used in foods impart flavour, pungency and colour. In Kerala, each home is having small land, in which coconut, banana, jack fruit, mango, pepper, nutmeg, cloves, cinnamon, ginger, turmeric are commonly cultivated for their requirement in a multi-storey or multi-tier cropping. To achieve high resource use efficiency, multi-storey or multi-tier system are adopted where crops ranging from tall trees to low growing annual form different canopy layers are grown together that normally have different rooting pattern with complementarity to each other. Results of a studyconducted in Kerala clearly advocate the supremacy ofmulti-storied cropping where coconut + ginger and coconut+ clove system gave 3.52 and 3.70 folds more netreturn over pure crop of coconut. Likewise, developedsubstantial additional employment (Table 1).

Table 1: Economics and Employment Generation in Coconut + Spice System

System	*Net Return (Rs./ha/year)*	*Additional Employment (Man days/ha/year)*
Coconut alone	10,400	–
Coconut + Ginger	36,590	500
Coconut + Clove	38,500	NA

Source: Parthasaraty *et al.* (2004).

2. Vegetable-based Farming System (Kumar *et al.*, 2013)

For Urban and peri-urban areas, where the land is available and facility for truck gardening isavailable, vegetable based farming system will be the best option for better economic returns from thesystem. Keeping in view ICAR- RCER, Patna has also developed one acre vegetable based farmingsystem model in which cereals were also integrated with seasonal vegetables, fodder crop, goatry and vermicomposting. Crop enterprises have been established on 80 per cent, having cereal and pulses in26 per cent and seasonal vegetables in 54 per cent area where as 20 per cent area was allocated forother enterprises such as fodder crop (12.5 per cent), goatry (5 per cent) and vermin composting pits (2.5 per cent) (Figure 1).

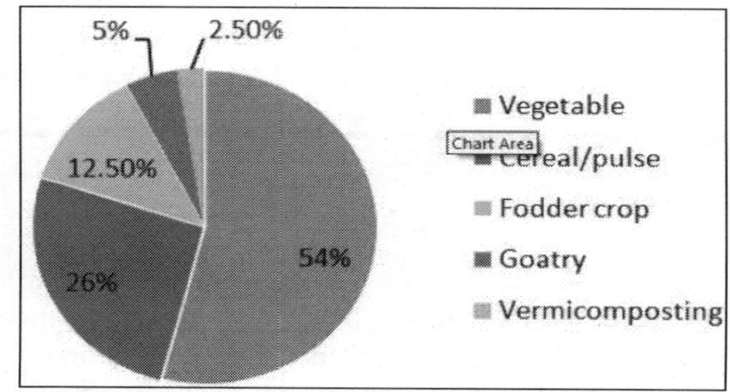

Figure 1: Area Allocation of different Enterprises of the Vegetable Based Integrated Farming System.

Rice-wheat-green gram cropping system is one of the prevalent cropping system in the eastern region of India. Diversification of the existing system with vegetables was experimented at ICAR Research Complex for Eastern Region, Patna. In the vegetable based integrated farming system, *i.e.* rice-potato-green gram system washigher than rice-wheat-green gram cropping system. Based on potato yield equivalence the yieldof rice-potato-green gram was found nearly doubled than rice-wheat-green gram.The net income from this one acre model was estimated to be Rs. 87,060/. In this model, all components areinterdependent on one another and by product of one component act as input for other component thus reduces the cost of cultivation and fetches more income than growing crops only.

3. Coconut – Forage Crops – Livestock Based Integrated Farming Systems for Midlands and Valleys (Korikanthimath and Manjunath, 2009)

Cultivation of coconut alone in the west coast region of India is very prevalent. Thus, there is ample scope to integrate different interventions to utilize the space and improve the farm income. Field investigations were carried out during 1998-2001 to identify a suitable forage grass with legume combination for intercropping in coconut. Further, dairy integration with the forage produced from coconut garden and recycling of wastes from coconut and dairy for betterment of soil was attempted.

Growing 'Benaulim' coconut alone resulted in a net return of Rs. 6,225/ha. However, the net returns from a high yielding forage grass (PBN-16) with legume (*Centrosema*) intercropped in coconut was Rs. 19,350/ha, totalling the net returns to Rs. 25,575/ha for coconut intercropping with forage grass. The mean net returns from two milch cows was worked out to Rs. 17,710/year. Thus, the overall net returns of the integrated system was Rs. 32,335/year. High yielding nature of hybrid napiers used for intercropping coupled with on-farm generation of inputs in terms of feed for cattle and manure for the crop resulted in higher net returns. The benefit:cost ratio was found higher for intercropping of forage crops (1.15) compared to all the others as the hybrid napiers used were multi-cut (perennial) high yielding ones with least maintenance costs. Christopher Lourduraj *et al.* (1992) also noted lesser cost of cultivation with hybrid napiers due to lesser cost of seed and fertilizers and no involvement of inter cultural operations like weeding or plant protection.

Income Generation and Nutritional Security through Intercropping in Juvenile Dwarf Coconut Plantation

A field experiment was conducted during the initial growth phase of dwarf coconut at ICAR-CPCRI, Regional Station, Kayamkulam, India (9°8′ N and 76°30′ E) from 2009 to 2012 to study the performance of selected horticultural crops for income generation and nutritional security. The cost economics of intercropping various fruit/vegetable/spice and tuber crops in the initial growth phase of dwarf coconut plants (in one ha area for one year) indicated a total cost of cultivation (labour, Planting material and others in the case of intercropping) Rs 3.34 lakhs with the total production and income as 36.8 t and Rs 5.70 lakhs, respectively, giving a BC ratio of 1.71.The Study revealed that various short duration crops such as vegetables, spices, fruit crops and tuber crops can be successfully grown in the interspaces of dwarf coconut plantations during initial years without any adverse effects of growth of coconut plants. These crops will also be able to provide sufficient economic support and nutritional security through vegetables, fruit crops and spices.

Performance of Coconut on Intercropping with *Heliconia* (Nihad *et al.*, 2016)

Field experiments on coconut - heliconia intercropping were conducted at ICAR-CPCRI, Regional Station, Kayamkulam, Kerala India.The yield performance of coconut in the intercropped helicona showed improvement in both DxT and WCT varieties of coconut. The palms recorded an improvement in yield parameters such as total number of bunches (12 to 14; 5 to 9), average number of nuts per bunch *i.e.*, below fist size (10.9 to 21.4; 14.1 to 27.9) and above fist size (4.3 to 7.8; 4.6 to 7.9) as well as 20 per cent reduction in the number of bunches with barren nuts and percentage of fallen nuts (51.2 to 37.4, 73.9 to 46.2) in D x T and WCT, respectively. Intercropping of *Heliconia* was found to increase the yield of palms in both D x T (40 to 59 nuts/palm) and WCT (29 to 45 nuts/palm) by the end of the study period. The yield improvement in coconut could be attributed to the favourable soil condition (maintaining adequate soil moisture in the intercropped area as well as application of organic manures at bimonthly interval) in the interspaces of coconut. Intercropping *Heliconias*which require intensive cultivation can enhance the yield performance of coconut varieties by improving the microclimatic conditions of agroecosystem.

Coconut Based Tuber Intercropping System (Mhaskarn *et al.*, 2015)

In India, coconut is predominantly cultivated in small and marginal holding. Coconut as monocrop, does not fully utilize the basic resources such as soil and sunlight available in the garden Tropical tuber crops such as yams, elephant foot yam, tannia and colocasia can be cultivated successfully and economically in the cropping system. There is no reduction in yield of elephant foot yam and yam compared to open field condition under 25 per cent shades. In 25-30 year-old coconut garden, under partially shaded conditions, cassava and sweet potato can also be grown under such diffused sunlight to generate additional income and employment in Odisha. African white yam variety SreePriya was found suitable as a profitable intercrop in coconut garden under Kerala conditions. Planting of arrowroot under partial shade at spacing 30x15 cm produced higher rhizome yield and economical returns. Several field experimented conducted over the years at ICAR-Central Plantation Crops Research Institute, Kasargod revealed that among the tuber crops evaluated in coconut garden, tapioca, elephant foot yam, greater yam, lesser yam, taro, tannia, sweet potato, Chinese potato were found to be suitable intercrops to enhance the productivity and profitability from the coconut gardens At the AICRP on Palms Centre, Veppankulam in Tamil Nadu State Different Models of coconut based cropping systems were evaluated. The results revealed that the maximum nut yield increment of 11.4 per cent was recorded in model IV (coconut+black paper +banana+ elephant foot yam). Coconut based intercropping with sweet potato, greater yam, lesser yam and Chinese potato resulted in net returns of Rs. 16,420, Rs. 18,562-, Rs. 13125, and Rs. 12,700, respectively under Kerala condition.

4. Plantation Crop Based Farming System for West Coast Region of India (Singh and Mahajan *et al.*, 2015)

A model of area 0.8 ha comprising of enterprises – plantation crops Crops: (Cashew (variety Bhaskara) + Pine apple (variety Giant Kew), Coconut + Pineapple + Papaya + Noni, Arecanut + Banana), piggery, poultry, vermi-compost unit, compost unit was evaluated for upland situations of Goa. The total cost of cultivation (inclusive of the farm labor employed) of the system was around Rs. 1.26 lakhs, whereas the net profit was Rs. 0.92. This net return was significantly higher based to the sole cashew crop cultivation. The highest contribution to net profit was from the cashewnut-pineapple (39 per cent) system as both the components started yielding. The contribution of the piggery component to the net profit was 22 per cent.

5. Cashew Based Intercropping System under Northern Transition Zone (zone-8) of Karnataka (Hegade and Jagadeesh, 2016)

The field trials conducted for three successive years to study Cashewnut based integrated farming system at Horticultural Research Station (HRS), Kanburgi, Belgaum (Dist.), Karnataka. Different intercrops included for the study were *Coleus* (*Plectranthus forskohlii*), cowpea, ginger, ground nut, sweet potato and turmeric.Mean yield of intercrops (mean of three years) varied from 10.55 q per ha in cowpea to 126.81 q per ha in sweet potato. Net profit calculated based on mean yield of three years was maximum in coleus (Rs. 68,740/ha) followed by turmeric (Rs. 57,650/ha) compared to the minimum in cowpea (Rs. 18,150/ ha). Mean Cashew yield was also increased due to intercropping. It was maximum in cashew +ground nut (6.10 kg/tree) followed by cashew +turmeric (6.03 kg/ tree) and cashew + cowpea (5.90 kg/tree) compared to the minimum in cashew alone (5.17 kg/tree). Among the cashew based intercropping systems the highest net profit was recorded by cashew +coleus (Rs. 1,48,755/ ha) followed by Cashew +turmeric (Rs. 1,44,035/ha) compared to the lowest in cashew alone (Rs. 71,175/ ha). Benefit-cost ratio ranged between 3.65 (in cashew alone) and 1.04 (in Cashew + ginger).

6. Horticulture based Integrated Farming System for Western Plain Zone of Uttar Pradesh (Kashyap *et al.*, 2015)

The farm holding size of India has been declining over years as a result of indiscriminate fragmentation of land holding, industrialization und urbanization. Small farm holders including the small and marginal farmers constitute more than 86 per cent of Indian farm families. Under Such situation horticulture plays a significant role in contributing to food and livelihood security along with employment generation. Three horticulture based modules *viz.*, fruit tree based (CS-1, 0.3 ha), vegetable crop based (CS-2, 0.22ha) and field crop based (CS-3, 0.4 ha) were evaluated. In CS-1 mango, guava and banana were grown as the main crops whereas cucumber, radish, carrot and onion as intercrops in mango, brinjal, vegetable pea and okra as intercrop in guava and turmeric as intercrop in banana. In Cs-2, turmeric, bottlegourd-cauliflower-tomato and brinjal-potato were grown while under in CS-3 system Rice-wheat and sugarcane-sugarcane ratoon-wheat were grown. Result obtained from the study revealed that among the three modules, vegetable based system has been found most effective in terms of net returns for the first two years. The net returns ha^{-1} from the vegetable based system was Rs. 1,51,640 followed by CS-3 (Rs. 68,765) followed by CS-1 being Rs. 46,660. Vegetable based system also resulted in higher employment generation. Fruit based system gave lowest net returns $hectare^{-1}$ as low yield was recorded from the fruit trees during the initial years. Initial results of the study revealed that through diversification of existing farming system livelihood of poor farmers can be improved significantly.

7. Intercropping of Tuber and Spices Crop in Plantation Crops (Dademal *et al.*, 2016)

An experiment was conducted at Arecanut Research Station, Shrivardhan, Maharashtra in the year 2013-14 to 2015-16 to study performance of tuber and Spices crop as intercrop in arecanut plantation. Elephant foot yam (var. Gajendra), Xanthosoma (var. Konkan Haritparni), Colocasia (var. Sanjivani), Lesser Yam (var. Konkan Kanchan), Greater yam (var. Sree Kartika), Arrow root (var. Trivendram), Turmeric (var. Salem), Ginger (var. Mahima), were intercropped in arecanut (var. Shrivardhini), at recommended planting distance. The growth and yield parameters were recorded during course of investigation. Significantly maximum arecanut equivalent yield was observed in elephant foot yam 56.47 q/ha followed by ginger (55.8 q/ha) and turmeric (52.67 q/ha). Maximum net returns of Rs. 2,42,671 was also recorded in elephant foot yam followed by ginger (Rs 2,38,359) and turmeric (Rs 2,18,658) under arecanut plantation. It is concluded that elephant foot yam, ginger and turmeric can be grown as intercrop in arecanut plantation. To fetch maximum yield per unit area.

8. Horticulture-based Integrated Farming Systems for the Arid Region of India (Bhandari *et al.*, 2014)

Arable cropping enterprise in dry lands of arid and semiarid regions of India is often un-remunerative due to aberrations of monsoon like late onset, prolonged dry spell, early withdrawal and unequal distribution of rainfall (Gill *et al.*, 1998). Hence an integrated approach of land management is essential for efficient utilization of natural resources to meet the requirements of farmers without deteriorating the land productivity and also to stabilize the income (Gill and Deb Roy, 1998). Even the traditional cropping system has multi species character example pearl millet, green gram, guar, moth bean, sesame between Indian mesquite and between fruit trees like Indian cherry, *jharber, kair, pilu* (*Salvadorao leoides*). Leaves of these fruit trees are lopped and fed to livestock during the lean period. The prevalence of this system reveals its potential advantage in production, stability, resiliencies and ecological stability. Farmers are often seen mixing the seeds of above arable crops altogether for sowing in their fields as mixed cropping with an optimism to harvest at least some produce even when the conditions are not congenial for one or another crop. Under such a situation, horticultural based production systems are considered an ideal strategy for the overall development and wellbeing of its farming community of the region.

Intercropping of agricultural crops with woody species is an age old practice in traditional farming systems in arid region of India. Growing of arable crops in association with horticultural plants is one of the viable approaches to sustain the agricultural production and stabilize the rural economy in the region. Studies revealed that the yields of both *jujube* and the intercrops were higher than in monoculture (Singh, 1995). Intercropping of guar, green gram and sesame with jujube (cv. Seb) increased the fruit yield to 14.8 from 5.2 kg tree^{-1}. Cultivation of guar in this system gave additional advantage of 782 kg seed yield, which was higher than that obtained from green gram and cowpea in drought years. However, Gupta *et al.* (2000) reported that 3-year old plantation of jujube at a density of 400 plant ha^{-1} in association with green gram performed well with seasonal rainfall of 210 mm. Intercropped green gram yielded only 160 kg ha^{-1} as against 620 kg ha^{-1} from pure crop, but the fruit yield from intercropped system increased the net profit to Rs. 2,886. The economic analysis of cost and returns ofjujube and Indian goose berry based agri-horti systems revealed that highest net profit of Rs. 13,487 from one hectare was realized from jujube with guar followed by Rs. 11,700/- from jujube with green gram (Meghwal and Henry, 2006). Jujube, being a multipurpose species provided better income and interspaces were best utilized for growing of leguminous crops. Least gross return and negative profits were obtained from Indian gooseberry with guar, since fruit plants were just 3-year-old without bearing of fruits; nevertheless an additional income of Rs. 3,720 from a hectare was obtained just by growing guar, which met out 79 per cent of the expenditure on maintaining this system. The highest B:C ratio (2.15) was achieved in *jujube* and guar followed by sole plantation of jujube (2.10) and *jujube* with green gram (2.02),which was higher than the B:C ratio obtained by either of sole crops. This shows that agri-horti system minimizes the risk and helps in imparting economic stability. Cover cropping with lobia (*Dolichos biflorus*) was found to increase water holding capacity of light soils as a result of increased organic carbon content in these regions (Pareek and Vishal Nath, 1996). The legume, Caribbean stylo (*Stylosanthes hamata*) is a good cover crop in the semi-arid regions (Raturi and Chadha, 1993). Studies onstandardization of jujube based production system revealed that 5 × 5 m spacing (400 plants ha^{-1}) for cultivation of jujube (cv. *Gola*) was ideal for legume intercropping.

9. Success Story on Sapota/Mango–Teak-Based Agroforestry System for Peninsular India (DARE/ICAR Annual Report, 2011-12)

A multi-component agroforestry system with sapota as the base crop, teak in the sapota line and agricultural crop (horsegram, *jowar* and *bajra*) in the interspace was developed for high rainfall areas having irrigation facilities of District Dharwad, Karnataka. Sapota (or mango, as modified by farmer) crop served as an insurance against failure of field crops. Two demonstrations at two different sites revealed the sapota is yielding 30 to 40 kg/plant which fetches for Rs. 22,000 to 25,000/ha. The fruit yield from mango is 30-50 kg/ plant and that fetches Rs. 36,000 to 60,000/ha.

The income generated from field crops in both the cases is about Rs. 2,500 to 3,500/ha. The value estimation of teak reveals that each teak pole costs about Rs. 120. With age, the crown size of perennial component (sapota/mango and teak) increased and consequently of field crops was not in cultivation from 2007 onwards. The system generated employment to an extent of 180 man days per year. The socio-economic status of the farmers improved as farmer is earning on an average of Rs. 23,500/ha/year with sapota and Rs. 48,000/ha/year with mango based system as against Rs. 3,000/ha/year only during initial period from the same land.

While designing the horticulture based integrated farming system, following concerns may be kept in mind,

i) Soil and climate suitability,
ii) Market demand,
iii) Family requirements – food and nutrition demand (as per the ICMR recommendations),
iv) Farm resources,
v) Dependence on external inputs may be minimized to make the system more profitable.

Conclusion

Based on the review made in the present paper, horticultural crops as an enterprise in the integrated farming system triggers the profitability of particular farming system. The selection of horticultural component for the integrated farming systems should be based on the location, soil and climate suitability, market demand, family requirement, available farm resources, *etc.* Besides the yield and income, it also addresses partially the nutritional security and the issue of unemployment and has a potential to produce round the year income.

References

All India Co-ordinate Research Project-Integrated Farming System Annual Progress Report. 2013-14. Development of region specific on- station IFS models. Karmana (Kerala). p. 29.

Bhandari, D.C., Meghwal, P.R. and Lodha, J. 2014. Horticulture based production systems in Indian arid regions. Chapter Sustainable Horticultural Systems Vol. 2 of the series *Sustainable Development and Biodiversity*, pp. 19-49.

Christopher, Lourduraj, Geethalakshmi, V., Rajamanickam, K. and John Joel, A. 1992. Performance of forage crops in coconut shade. *Indian Coconut J.* September, pp. 7-8.

Dademal, A.A., Mhaskar N.V., Khandekar, R.G., Chavan, L.S. Kasture M.C. and Mahadkar. U.V. 2016. Intrercropping of tuber and spices crop in arecanut plantation. National seminar on plantation based cropping system for improving livelihood security. Page No. 2.

DARE/ICAR Annual Report, 2011-12. http://www.icar.org.in/files/reports/icar-dare-annual-reports/2011-12/farming-cropping-system-AR-2011-12.pdf. Last accessed on 27.9.2016.

Gill, A.S. and Roy, Deb. R. 1998. Tree/crop interface competition studies under semiarid conditions-relative yield of wheat. *Range Manage. Agroforest.*, **19**(2): 154–164.

Gill, A.S., Roy, Deb. R. and Chaudhary, B.L. 1998. Effect of multipurpose tree species on pulse production under semi-arid conditions. *Indian J. Agron.*, **43**(3): 650–565.

Gill, M.S., Singh, J.P. and Gangwar, K.S. 2009. Integrated farming system and agriculture sustainability. *Indian J. Agron.*, **54**(2): 128-139.

Gupta, J.P. 2000. Technology approach for greening degraded arid lands. *CAZRI*, Jodhpur, p. 51.

Hegde, N.K. and Jagadeesh, R.C. 2016. Cashew based intercropping system under northern transition zone (zone-8) of Karnataka. *National Seminar on Plantation Based Cropping System for Improving Livelihood Security*, p. 31.

Kashyap, P., Kansal, A., Prushty, A.K. and Singh J.P. 2015. Evaluation of horticulture based integrated farming based model for providing nutritional security to small and marginal farmers of western plain zone of Uttar Pradesh, India. *Int. J. Econ. Pl.*, **2**(1): 015-017.

Korikanthimath, V.S. and Manjunath, B.L. 2009. Integrated farming system for sustainability in agricultural production. *Indian J. Agron.*, **54**(2): 140-148.

Krishnakumar, V., Nihad, K. and Sudhanandan, K.K. 2016. Income generation and nutritional security through intercropping in juvenile dwarf coconut plantation. *National Seminar on Plantation Based Cropping System for Improving Livelihood Security*, p. 7.

Kumar, S. 2013. Integrated farming system models for food and nutritional security. Lecture Delivered during *Model Training Course on Gender Perspective in Integrated Farming System* w.e.f 17-24 January 2013 at ICAR Research Complex for Eastern Region, Patna, Bihar.

Meghwal, P.R. and Henry, A. 2006. Economic evaluation of different agro-horticulture systems under rainfed condition of arid zone. In: *Legumes in Dry Areas*, (Eds.) D. Kumar, A. Henry and K.P.R. Vittal. Indian Arid Legume Society and Scientific Publishers, Jodhpur, pp. 299–303.

Mhaskaran, V*., Shankar, Deo., George, James, and Nedunchezhian, M. 2016. Coconut based tuber intercropping system: A review. *National Seminar on Plantation Based Cropping System for Improving Livelihood Security*, p. 6.

Nihad, K., Krishnakumar, V., Haris Abdul and Jeena, Mathew 2016. Yield performance of coconut on intercropping with *heliconia*. *National Seminar on Plantation Based Cropping System for Improving Livelihood Security*, p. 11.

Pareek, O.P. and Nath, V. 1996. Coordinated fruit research in Indian arid zone: A two decade profile. NRC for Arid Horticulture, Bikaner.

Parthasarathy, V.A., Chempakam, B., Kandiannam, K. and Srinivasan, V. 2004. Spices in farming systems. In: *Proceeding of Symposium on Alternative Farming Systems: Enhanced Income and Employment Generation Options for Small and Marginal Farmers*, pp. 144–153.

Raturi, G.B. and Chadha, K.L. 1993. Horti-silvi-pastoral system. In: *Advances in Horticulture.* Vol. 2, No. 2, (Eds.) K.L. Chadha and O.P. Pareek. Malhotra Publishing House, New Delhi, pp. 805–811.

Singh, N.P. and Mahajan, G.R. 2015. Research findings of AICRP on IFS. *Annual Report*, ICAR – CCARI, Old Goa.

Singh, R.S. and Vishwanath, S. 1995. Performance of pea varieties as intercrop in *ber* orchard in semi-arid region. *Indian J. Hort.*, **52**(2): 137–140.

65

Aeroponics: A Soilless Method of Potato Minitubers

Sukhwinder Singh* and Ratna Preeti Kaur

ICAR-Central Potato Research Station,
PO Box 1, Model Town PO, Jalandhar – 144 003, Punjab
**E-mail: ss.cpri@gmail.com*

Potato being vegetatively propagated crop tends to accumulate and transmit various seed borne diseases to subsequent generations including most prominent viral diseases (Hide and Lapwood, 1992). This leads to significant degeneration of seed stocks over the years which significantly affect production. For sustaining the good production levels frequent replacement of quality seed material is of paramount importance. For enhancing the availability of seed potatoes tissue culture technique have been widely adopted throughout the world. In this process potato minitubers, also called G-0 (generation zero), are produced under 40 mesh insect proof net from *in vitro* plants (Singh *et al.*, 2011). Microplants are planted in fertile soil of net houses and minitubers thus produced are harvested after the stipulated time, giving ample time for skin curing after haulm cutting. Net covering protects potato crop from viral infestation by creating a physical barrier for vectors like aphidand white fly.

Minituber production through direct transplanting of *in vitro* plantlets, involves a risk of infection from pathogens present in soil, such as *Erwinia* sp. (bacterial soft rot), *Spnogospora subterranea* (powdery scab), *Streptomyces scabies* (common scab) and nematodes (Hide and Lapwood, 1992). Though soil solarization (Arora *et al.*, 2008) is performed in the hot summer months, still infection cannot be ruled out. Looking into factors like soil health, risk of soil borne diseases and temperature profile, only one crop in a season is taken. Further, in this method of producing minitubers, multiplication rate, though depends on plant density and geometry, but is generally low from 5-8 mini-tubers per plant (Sharma and Pandey, 2013). Drawback of low multiplication rate is more cost of production and loweravailability and affordability of quality seed potatoes.

Looking into the gargantuan requirement of 30-35 q/ha of seed potatoes, new quicker, safer and cost effective methods of seed stock multiplication have always been a serious concern. For this many new techniques like aeroponics, nutrient film technique, deep culture hydroponics sand culture *etc.* (Correa *et al.*, 2008; Kang *et al.*, 1996; Muro *et al.*, 1997; Nicolas, 2005; Rolot, 1999), have been explored. Out of these, aeroponics have been found most suitable considering high multiplication rate as compared to traditional system. Also aeroponic method is capable of addressing issues like crop environment control, avoidance of soil borne diseases and simultaneously takingup to twocrop cycles in a season. Lot of work have been done for understanding aeroponics for potato seed production and for comparing this technique with other methods already in use (Correa *et al.*, 2009; Farren and Mingo, 2006; Kim *et al.*, 1997; Otazu, 2010).

1. Aeroponics

Aeroponics is a unique type of hydroponic system where the plant roots are suspended in air and are sprayed with a fine mist of nutrient solution either continually or intermittently (Singh *et al.*, 2010; Christie and Nicholas, 2004). Aeroponic systems are mainly of two types, close and open system. In close system, nutrientsolution sprayed over the roots is collected in tanks and is re-circulated. In open system solution once misted is discarded and fresh solution is sprayed every time. Advantage of open system is that there are minimum chances of any infection from solution but is not environment friendly. Close system is considered environment friendly as chemicals are not dumped unnecessarily but caution is required to avoid infection through repeated use of the same solution. Schematic of aeroponic system is presented in Figure 1.

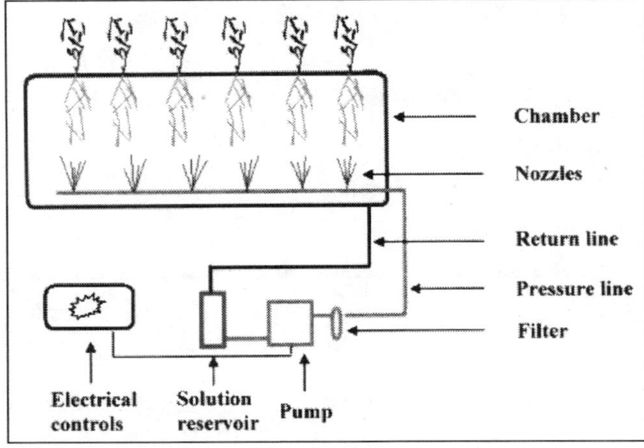

Figure 1: Schematics of Aeroponic System.

While we tend to think of aeroponics as a recent development in the hydroponics field, it has actually been in use since the 1940's, although largely as a research tool rather than as an economically feasible method of crop production. Though, it was seldom thought an appropriate technology for raising root crops or crops like potato, but recently a lot of interest has been shown in aeroponics for high quality disease free potatoes seed production.

1.1 Advantages

Aeroponics offers the potential to increase production in terms of number of minitubers per plant from four to six times and thus reduces cost as compared to conventional method. Considerably less number of *in vitro* plants is required. Its other advantages are:

- Avoids exposure to unfavorable soil conditions and thus freedom from soil borne diseases.
- Lot of space available for aeration, growth and development of roots and progeny tubers.
- Easy inspection of tuber zone and picking of minitubers of desired size by sequential harvesting.
- Aeroponic growing is safe and ecologically friendly for producing natural, healthy plants and crops. About 90-95 per cent less water and 50-60 per cent nutrients are saved as compared to traditional cultivation.
- Presence of oxygen in the rhizosphere (root zone) is essential for healthy plant growth. As aeroponics is conducted in air combined with micro-droplets of water, almost any plant can grow to maturity in air with a plentiful supply of oxygen, water and nutrients.
- At most of the places two crops in a year can be taken as system is installed under fan pad controlled greenhouse.
- In terms of weight, 20-25 times less aeroponic minitubers as compared to traditional potato seed are required for producing same amount of subsequent generation of seed.
- Minitubers produced through aeroponics are usually of size 2-3 gm and logistics is not a problem which is always an issue in traditional seed potato movement to distant places.

1.2 Components of Aeroponic System

Comprehensive aeroponic based potato minituber production system consists of following broad sub systems:

Greenhouse

Aeroponic growing system is generally housed in fan and pad cooled greenhouse so that environment can be controlled and suitable conditions in terms of temperature, RH, light *etc.* can be maintained as per requirement of crop. Greenhouse with climate control facilities helps in extending useful period to have two crops in a season. If temperature profile of the location ranges between 14-28°C (maximum) and 4-16°C (minimum), aeroponic system can be installed in simple insect proof netswithout fan and pad cooling system. But polycarbonate or polythene sheet roofing is must to avoid rain water entering the system.

Aeroponic Unit

Aeroponic system is most complex amongst all the hydroponic systems and involves lot of components and sub units. Following are the major components of aeroponic system:

- *Hydraulic pumps*: Supply nutrient solution to the nozzles through pipes at preset pressure.
- *Solution tank*: Stores nutrient solution which is sprayed over the roots through nozzles.
- *Growth chambers*: For housing roots,stolons and tubers in dark environment.
- *Nozzles or misters*: Spray a fine mist of nutrient solution over the roots in grow boxes.
- *Filters*: Filter impurities in the nutrientsolution to avoid choking of nozzles.
- *Piping*: Deliver nutrient solution to nozzles and return solution to tanks.

- ☆ *Plant holders*: Hold plants upright and lift upwards plants for tuber harvesting/picking.
- ☆ *DM unit*: Removal of salts from water before preparing nutrient solution.

Electrical Controls, Sensors and Instrumentation

For successful running of an aeroponic system, selection and installation of instruments and sensors at right place is of great importance. All the instruments should have provision of easy calibration and should be installed at appropriate location to avoid errors in measurement and operation. Major instruments and sensors used in aeroponic system are mentioned in Table 1.

Table 1: List of Instruments Used in Aeroponic System

Instrument/ Component	*Function*
Timers	For operation of pumps, solenoid valves, nozzles and foggers.
Temperature controllers	Senses temperature of greenhouse and nutrient solution reservoirs and operate cooling/heating system and chilling system whichever is required.
pH sensing and dozing instrument	Automatically senses the pH level of the nutrient solution and operates dozing unit for adding acid or base solution for bringing pH to desired level.
EC sensing and dozing system	Senses the EC level and operates dozing pump for adding water or stock solution for bringing EC to desired level.
Solution level indicators	Indicate the level of nutrient solution in the tank.
Pressure indicators	Pressure gauges are installed at various points for monitoring nutrient supply pressure in supply lines.

Chilling Unit

Hundred percent relative humidity is maintained in the root zone in aeroponic grow boxes. High temperature can be detrimental to the crop and bacterial infection can spread affecting roots, stolons and mini-tubers. To avoid this, chilling unit is installed for maintaining desired temperature of nutrient solution tank and thereby of grow chambers.

Storage Unit

Freshly picked minitubers from highRH environment, are highly juicy and have immature skin with open lenticels. These tubers cannot be directly sent to storage. For curing and final storage of minitubers two separate cold chambers with RH control are required. Harvested minitubers are first washed and treated with fungicide and then stored at 14-16°C for skin curing for 8-10 days. After curingthese are shifted to 2-4°C compartment for next season use. In both these chambers RH is maintained at 95 ± 2 per cent to avoid moisture loss from the tubers.

1.3 Design Consideration in Aeroponics

Application of principal of aeroponics for raising potato or any other crop requires in depth application of engineering fundamentals. A poorly designed and fabricated unit can be a nightmare for the producer. Following points are of importance for having a successful aeroponic system for potato mini-tuber production:

I. Aeroponic facility should be designed considering local weather conditions for efficient control of temperature, RH, light and aeration.

II. Grow boxes in which roots, stolons and tubers will grow should be designed in such a way that:
   a) Boxes are black from inside, light proof and there are no leakage issues.
   b) There is enough space for roots and stolons to grow in well aerated environment.
   c) Picking of tubers is convenient, damage free and should not involve drudgery to the workers.
   d) There is provision for temperature regulation in root zone during warmers/cooler periods.

III. Nozzles' mist pattern should cover all the roots during whole crop period and there should be minimum blockage of nozzles.

IV. Pumps, valves, nozzles, filters, pipes *etc.* should have matching capacity.

V. System should be designed in such a way that it is convenient to operate, repair and maintain.

VI. System should be equipped with canopy management system and should have convenient walkways for picking and movement of workers.

VII. Aeroponic unit, hardening chamber, storage room and control room should be connected internally through pathways so that workers don't need to go out frequently.

1.4 Operation and Crop Raising

Sequence of activities of seed production through aeroponics is depicted in Figure 2. Details of major activities is as follows:

Hardening and Planting

Planting material such as rooted cuttings and tuber sprouts can be used for planting in aeroponics. However, *in vitro* plants are preferred for planting in aeroponics. *In vitro* plants can be transplanted directly in the aeroponic system after acclimatizing them for 5-7 days or the *in vitro* plants can be planted in sand, peat moss, vermiculite or sand for rooting and then after hardening for 15 days at 24-26°C can be transferred to grow boxes in aeroponic system. *In vitro* plants of

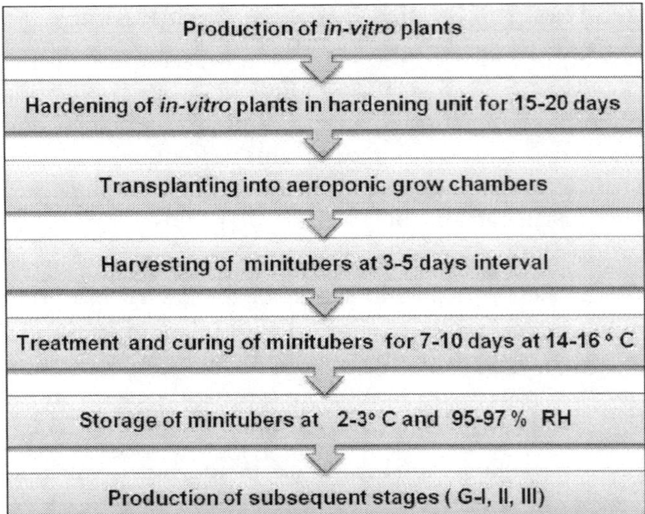

Figure 2: Sequence of Activities for Production Seed Potatoes.

15-21 days are considered appropriate for planting in greenhouse. Very old and yellowish plants are not suitable for aeroponics. It should be ensured that plants are free from any disease and plants with any symptoms should be discarded. These hardened plantlets of 10-15 cm height are planted in the growth box holes. Before starting planting system should be well sanitized and switched on for creating mist in the grow boxes.

Management of Nutrient Solution

In aeroponic system nutrient solution management is a key factor. All the essential nutrient elements nitrogen (N), phosphorus (P), potassium (K), calcium (Ca), magnesium (Mg), iron (Fe), sulfur (S), manganese (Mn), copper (Cu), zinc (Zn), boron (B) and molybdenum (Mo) required for normal plant growth are dissolved in demineralized water so that plants can absorb them through their root system. Solution composition and concentration is varied from time to time according to the crop age. Maintaining pH is must. For potato pH is maintained in the range of 5.8-6.4. With the passage of time some of the elements are taken up by plants and exudates are released in the solution. pH and EC will either go up or will go down. For maintaining pH acid or base solution is added manually or with auto dozing system daily. Similarly EC is adjusted every other day by adding water or concentrated nutrient solution.

Crop Growth and Harvesting

Roots, stolons and tubers develop inside the chamber and leaves are exposed to light allowing them to carry out photosynthesis. After a week, root system starts developing inside the growth chambers. As in the soil system, stolon and tuber formations are initiated at different intervals depending upon the variety. After about 25-30 days, lower leaves are removed with a dissecting blade following strict aseptic measures and the plant are lowered by 4-6 cm. The process of lowering the stems is important and is equivalent to earthing up in the field. If stolons start to form in the upper part of the root system, lowering of plants lead to better stolon formation. Night temperatures lower than 4°C are too cold for aeroponic plants. Day temperatures higher than 30°C are too warm. When tuberization begins, it is desirable to have maximum night temperatures of 10 - 15°C and day temperatures around 20°C. Picking of the tubers starts after 45-65 days when some of the tubers attain 14-15 mm diameter. Once the first flush is harvested, it triggers formation of more minitubers/ plant. In this system, harvesting is done at regular interval when the tubers attain the desired size. On an average 35-60 mini-tubers can be harvested from a single plant, depending on variety. Tubers are harvested sequentially as they attain the desired size. In suitable environments two crop cycles can be taken in a year. For this, first crop is terminated and the system is sanitized thoroughlybefore planting second crop. Selection of cultivarsis important in second crop, as all the cultivarsdon'tperform well in second crop.

1.5 Minituber Utilization

Aeroponically produced minitubers are planted in open fields for production of generation I, II and III. Generation III seed is generally given to the farmers for their use. In some case farmers buy G-0 or other early generations to produce quality seed under their own supervision. On the other hand some of the farmers prefer to have G-IV or G-V considering high cost of earlier generations. Selected farmers plant G-0 under insect proof nets for maintaining quality of seed to extended generations.

1.6 Progress of Aeroponics in India

For potato minituber production through aeroponics, research was initiated during 2009 at ICAR-Central Potato Research Institute Campus, Modipuram. Emphasis was given on engineering design suited to Indian conditions and material selection along with standardizing production protocols (Figures 1–3). Till now seven aeroponic units of capacity of 1-2 lakh minitubers/year, have been established at various stations of Central Potato Research Institute (CPRI) for production and research. Aeroponic technology has also been transferred to 12 companies for production and fabrication. More companies are coming forward and production of minitubers through aeroponics is expected to touch 30 million/year by the year 2020. With this availability of G-4, potato seed will be around 1.5 million metric tonnes per year.

2. Future Thrusts

Though, potato seed production, significant dent has already been made by this technology till now, yet following areas will need to be focused in future:

Figure 3: Initiation and Bulking of Minitubers in Aeroponic System.

Figure 4: Size Variation in Traditionally and Aeroponically Produced Minitubers.

Figure 5: G-I Crop from Aeroponic Minitubers.

I. Development of cost effective models which are affordable even to small seed companies/ farmers.

II. Enhancing production per unit area and per unit time to reduce the cost of production.

III. Focusing on developing low energy requiring and solar power operated units for remote/ power shortage areas.

IV. Location specific designs especially focusing NE and Bengal conditions.

V. Developing multi-crop all season systems.

Thus, aeroponics can be the technology for future when shrinking land and other resources would help in intensive frustration. This technique can be adopted in other crops after thorough research.

3. References

Arora, R.K., Sunaina, V. and Somani, A.K. 2008. Eco-friendly management of soil and tuber borne diseases. **In**: *Twenty Steps towards Hidden Treasure*, (Eds.) S.K. Pandey and S.K. Chakarabarti. Central Potato Research Institute, Shimla, India, pp. 137–150.

Christie, C.B. and Nichols, M.A. 2004. Aeroponics: A production system and research tool. *Acta Hort.*, **648**: 185–190.

Correa, R.M., Pinto, J.E.B., Pinto, C.A.B.P., Faquin, V., Reis, E.S., Monteiro, A.B. and Dyer, W.E. 2008. A comparison of potato seed tuber yields in beds, pots and hydroponic systems. *Scientia Hort.*, **116**: 17–20.

Correa, R.M., Pinto, J.E.B., Pinto, C.A.B.P., Faquin, V. and Reis, E.S. 2009. The production of seed potatoes by hydroponic methods in Brazil. *Fruit, Veg. Cer. Sci. Biotech.*, **3**(1): 133–139.

Farran, I. and Mingo-Castel, A.M. 2006. Potato minituber production using aeroponics: Effects of plant density and harvesting intervals. *Amer. J. Pot. Res.*, **83**: 47–53.

Hide, J.A. and Lapwood, D.H. 1992. Disease aspects of potato production. **In**: *The Potato Crop*, (Ed.) Paul Haris. Chapman and Hall, London, UK, pp. 403–437.

Kang, J.G., Kim, S.Y., Kim, H.J., Om, Y.H. and Kim, J.K. 1996. Growth and tuberization of potato (*Solanum tuberosum* L.) cultivars in aeroponics, deep flow technique and nutrient film technique culture systems. *J. Korean Soc. Hort. Sci.*, **37**: 24–27.

Kim, K.T., Kim, S.B., Ko, S.B. and Park, Y.B. 1997. Effect of minituber picking intervals on the yield and tuber weight of potato grown in aeroponics. *RDA J. Hort. Sci.*, **39**: 65–69.

Muro, J., Diaz, V., Goni, J.L. and Lamsfus, C.1997. Comparison of hydroponic culture and culture in peat/sand mixture and the influence of nutrient solution and plant density on seed potato yields. *Potato Res.*, **40**: 431–438.

Nichols, M.A. 2005. Aeroponics and potatoes. *Acta Hort.*, **670**: 201–206.

Otazú, V. 2010. *Manual on Quality Seed Potato Production using Aeroponics*. Lima: International Potato Centre (CIP).

Rolot, J.L., and Seutin, H. 1999. Soilless production of potato mini-tubers using hydroponic technique. *Potato Res.*, **42**: 457–469.

Sharma, Ashwani K. and Pandey, K.K. 2013. Potato minituber production through direct transplanting of *in vitro* plantlets in green or screen house: A review. *Potato J.*, **40**(2): 95–103.

Singh S., Singh V., Singh, B.P. and Pandey, S.K. 2010. Aeroponics for potato minitubers production. *ICAR News*, Indian Council of Agricultural Research, India, New Delhi, **16**(2): 1–2.

Singh, B.P., Pandey, K.K. and Venkatasalam, E.P. 2011. Potato seed production system in India. **In**: *Production of Disease Free Quality Planting Material Propagated through Tubers and Rhizomes*, (Eds.) B.P. Singh, V.K. Dua and Brajesh Singh. Central Potato Research Institue, Shimla, India, pp. 1–16.

66

IPM and GAP in Protected Cultivation

Naved Sabir

Division of Nematology,
ICAR-Indian Agricultural Research Institute, New Delhi – 110 012
E-mail: n_sabir@rediffmail.com

Ever since intensive cultivation technologies have dawned on the agricultural scenario, horticultural crops are continually replacing traditional crops due to higher productivity, professionalism, entrepreneurship and export potentials. In view of shrinking land resources and the recent spurt in high technology and entrepreneurial ventures in agriculture, the protected cultivation is fast picking up in the country. Higher productivity under protected cultivation is attainable mainly on account of reduced competition, healthy environment, better nutrient management, increased surface area and longer crop duration besides biotic management. Apart from higher productivity, there are several other reasons of importance of the protected cultivation such as- shrinking per-capita land resources- constantly pushing large populace, huge potential of expansion in future, pesticide free/pesticide-checked produce leading to direct consumer health gain, great export potential, environmental value-addition, purity of even cross-pollinated crops such as cucurbits, self-employment and attraction of educated youth – entrepreneurship, farmer empowerment, reduced drudgery, honourable worksmanship and last but importantly Govt. subsidy on protected cultivation is fairly encouraging for entrepreneurs. However, losses due to pests especially endemic build-up of pests is alarming, where, once the pests set in, the losses caused may be total. While greenhouses are established with much hope and hype, they are often rendered and relegated to apathy of non-guidance, unavailability of inputs, repair and maintenance issues apart from serious pest and disease problems.

For effective implementation of any IPM strategy, it is imperative for organizations affiliated to state to lend a helping hand to the growers, who have had large amounts of money invested in this sector. Lack of organized support from the institutions is slowly driving such hapless and distraught growers and indeed maligning technology in its infancy rather unduly. It is therefore, a high time that policy makers and administrators laid due emphasis and garnered diligent efforts to converge a helping hand to growers for pest management, and allied support for maintenance and quality inputs *etc.* so that the protected cultivation technology meets its deserving seat of higher productivity, quality and entrepreneurship.

The major factors inducing these pests in the greenhouse and protected structures are ideal or homogeneous conditions, high humidity, supple nature of crops grown, mono-culture, pest endemicity over the years, reduced competition *etc.* Most of these factors contribute towards the creation of microclimatic conditions favourable to the pests often ironically more than the crops.

To tackle these pests, integrated pest management (IPM) strategy is the most advisable environment friendly option. *"IPM is a systems approach involving integration of three or more plant protection strategies to minimize the pest population below economic injury level keeping in view the environmental health and economic feasibility"*. The use of cultural practices and biological control agents are the corner stones of any IPM strategy, however, the judicious use of pesticides under pest contingencies is also adoptable. IPM indeed is a part of larger goals of healthy crop production to processing chain within in the ambit of Good Agricultural Practices (GAP). Actually the four major pillars, *viz.*, environmental sustainability, economic viability and social acceptability besides health and food safety are the spirit and philosophy of GAP, IPM and Food Safety issues.

Greenhouse IPM and GAP are all the more crucial because of higher initial and operational costs, and rightly so for expecting higher returns. Higher productivity linked with quality and catering to upper niche markets and eventually export, make IPM and GAP imperative for greenhouse production systems. Implementation of these demands a holistic approach starting from baseline studies involving identification of key pests and culminates in all preparations for quality production including issues of certification.

Consideration for IPM Strategies in Greenhouses

The protected nature and microclimatic conditions in the greenhouses make it rather imperative to suitably alter the IPM strategies as per requirements. There is a lack of specific chemicals suited for greenhouse conditions because most of the persistent pesticides used under field conditions cannot be used in greenhouses. Containment of toxic fumes from pesticides are very hazardous to the greenhouse workers. Due to frequent harvesting and direct table value of the produce, pesticides with lower waiting periods only can be used. Use large sprayers due to limited space, cannot be made. Therefore, the grower must be extra careful in the long range planning, selection, use and application of pesticides in greenhouse vegetable production system.

IPM: its Sub-systems and Tools

a. Exclusion, avoidance and repellence
b. Habitat management and sanitary measures
c. Cultural control
d. Host plant resistance
e. Mechanical and physical measures
f. Chemical control
g. Biological control
h. Regulatory measures
i. Other methods/Tools in IPM

Major Pests given in Table 1 are soil-borne pathogens and sucking arthropod pests.

Exclusion, Avoidance and Repellence

Avoiding sick plots for the construction of greenhouses is essential. Insect proof screening is available for vents and other openings. Therefore, the openings including those in cooling pads and ventilation units should be screened to prevent entry of even the smallest insects. Use of ultraviolet absorbing plastics can reduce insect problems. Mesh sizes above 50 do not allow the entry of pests such as thrips. The UV-free light inside the greenhouse alters insects landing and feeding behaviour and can greatly reduce the spread of insect-vectors of viruses as well as discourage the establishment of aphids, whiteflies and thrips. New crops should not be planted near greenhouses. Use of allele-chemicals or repellent plants is one of the important tools in integrated pest management. Intercropping of marigold is also helpful as a trap crop in case of cabbage, while as a repellent against plant parasitic nematodes. Use of disease free planting material is very important. Some important points are as following:

i) Domestic quarantine is an important issue. Once any new pest enters in any area, its spread has to be checked in the initial stages only else measures have to be taken at far footing. Lately, two important pests have been intercepted in India, *viz.*, South American tomato pinworm, *Tuta absoluta* from Pune and later from Kolar and Bengaluru districts of Karnataka (Anon, 2015). Almost concomitantly another report of western flower thrip, *Frankliniella occidentalis* occurrence in India has been confirmed, though earlier reports were not true (Tyagi and Kumar, 2015; Anon, 2015). Similarly, other crops like seed tubers of potato from Nilgiri Hills of Tamil Nadu should not be used due to domestic quarantine against golden cyst nematode. Otherwise also only nematode-free planting material should be used. Now another suspected by exotic nematode pest been intercepted in Dindigul area of Tamil Nadu *viz.*, *Meloidogyne enterolobii* infesting guava orchards with an advice to destroy such orchards (Anon. 2015). Risk of its spread to newer areas/crops is currently imminent and has to be countered on priority.

ii) Workers should avoid wearing yellow clothing which attracts insects such as white flies.

iii) Follow long-term crop rotation, general rule being the replacement of dicot (broad-leaved) with monocot (thin-leaved) crops. Rotation specifically with crops like cruciferous vegetables, flowers like marigold or intercrop like sunhemp may be helpful in reducing problems of nematodes in the greenhouses.

iv) Avoid excessive watering or overdoses of fertilizers that may lead to plant root or shoot damage.

v) Adjacent areas free from wild or cultivated plants, which act as reservoirs of the pests or pathogens. Hedges around the greenhouses should be avoided as they harbour lot of thrips which are serious pests of many greenhouse crops like capsicum and roses.

Table 1: Key Pests under Protected Agriculture in India

Pests	Common Name	Scientific Name	Host
Aphids	Melon aphid	*Aphis gossypii*	Tomato, capsicum, cucumber, lettuce
	Green peach aphid	*Myzus persicae*	Capsicum, tomato, lettuce
	Potato aphid	*Macrosiphum euphorbiae*	Lettuce, tomato
Leaf-miner	American serpentine leafminer	*Liriomyza trifolii*	Tomato, cucumber,
Mites	Broad or yellow mite	*Polyphagotarsonemus latus*	Capsicum, gerbera
	Carmine Spider mite	*Tetranychus cinnabarinus*	Tomato, gerbera
	Vegetable mite	*Tetranychus neocalidonicus*	Cucumber
	Two spotted Spider mite	*Tetranychus urticae*	Tomato, capsicum, cucumber, gerbera
Thrips	Chilli thrips	*Scritothrips dorsalis*	Capsicum
	Western flower thrips	*Frankliniella occidentalis*	Reported in rose
	Others–melon, onion thrips	*Thrips palmi, T. tabaci*	Many vegetables
Whiteflies	Sweetpotato whitefly	*Bemisia tabaci*	Capsicum, lettuce
	Greenhouse whitefly	*Trialeurodes vaporariorum*	Tomato, cucumber, capsicum, beans, lettuce
Mealy bugs	Papaya mealybug, Citrus mealybug *etc.*	*Planococcus* sp., *Pseudococcus* sp., *Pseudococcus* sp., *etc.*	Gerbera, tomato *etc.*
Caterpillars	Cotton boll worm	*Helicoverpa armigera*	Capsicum, tomato,
	Asian cotton leafworm or tobacco cutworm	*Spodoptera litura*	Tomato, capsicum, cucumber
	Cabbage looper	*Trichopulsiani*	Lettuce
Powdery mildew		*Erysiphe cichoracearum, Podosphaera xanthii, Leveillula taurica*	Cucumber, lettuce, capsicum
Gray mould		*Botrytis lactucae*	Lettuce
		Botrytis cineria	Tomato,
Fusarium wilt		*Fusarium oxysporum* f. sp. *lycopersici*	Tomato
		Fusarium oxysporum f. sp. *cucumerinum*	Cucumber
Fusarium Crown and Root rot		*Fusarium oxysporum* f. sp. *radicis lycopersici*	Tomato
Damping off/Root rot		*Pythium ultimum, Pythium aphanidermatum*	Cucumber, lettuce
		Phytophthora spp., *Rhizoctonia solani*	Capsicum
Downy mildew		*Pseudoperenospora cubensis*	Cucumber
		Bremia lactucae	Lettuce
Root-*Knot nematode*		*Meloidogyne incognita*	Tomato, capsicum, cucumber

http://www.entomology.umn.edu/cues/inter/inmine/Thripc.html; http://dsp-psd.pwgsc.gc.ca/collection_2009/agr/A118-10-19-2006E.pdf.

vi) Proper drainage helps minimizing *Fusarium* wilt, root/stem rot and other soil-borne fungal and bacterial diseases.

vii) Avoidance of viral infection in crops like tomato and cucumber in greenhouses is done best by rouging out the infected plants and destroying them away from the greenhouse.

Habitat Management and Sanitary Measures

It is closely related to exclusion and should be practiced to manage insects and diseases. The following practices should be used to keep the greenhouse tidy.

i) House preparation: Basically involves sanitation, summer ploughing weeding during off-season.

ii) Media preparation/hygiene/sterilisation *etc.*

Sterilization of media, troughs, trays, tools *etc.* Clean tools are helpful in minimizing the risk of mechanical transmission of plant diseases especially those caused by the viruses.

Use pathogen-free clean water of deep wells (60-75 feet) to minimize pest propagules. Filters should be used for the irrigation water, mainly to avoid pathogenic bacteria.

The production of seedlings should utilize only new seed and new sterile media in a separate greenhouse away from on-going crop production. The greenhouse irrigation tanks should be cleaned as frequently as possible. Apart from this, the pH and EC of the water have to be kept in check.

Production Sanitation

A number of recommended sanitation measures have been adopted by successful growers. Some of the more common ones are given below:

- ✰ Use of an air-lock type entrance to each production house so that, wind-carried insects, soil, *etc.* do not enter the production area directly from the outside or along with the workers.
- ✰ Use of foot baths to prevent entry of unwanted soil.
- ✰ Restricted access by visitors.
- ✰ Raising transplants at a height of at least one foot above the ground on raised platforms.
- ✰ Prohibition of cigarette, chewing tobacco, *etc.* used by workers so as to minimize contamination.

Damping Off in Cucumber

Powdery Mildew in Cucumber

Fusarium Wilt in Cucumber

Wilt Infested Capsicum

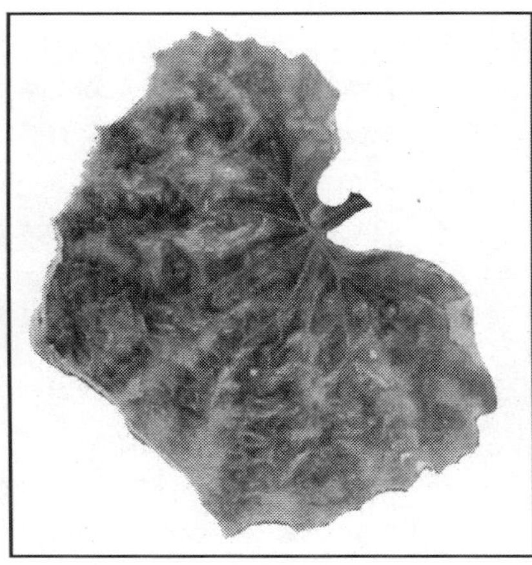

Cucumber Mosaic Virus in Cucumber

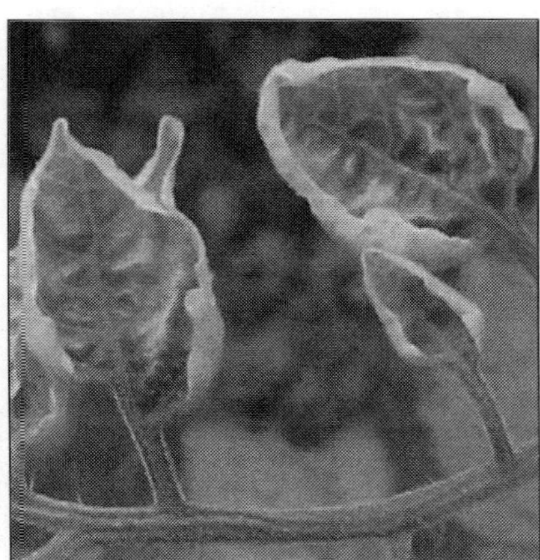

Tomato Yellow Leaf Curl Virus (TYLCV)

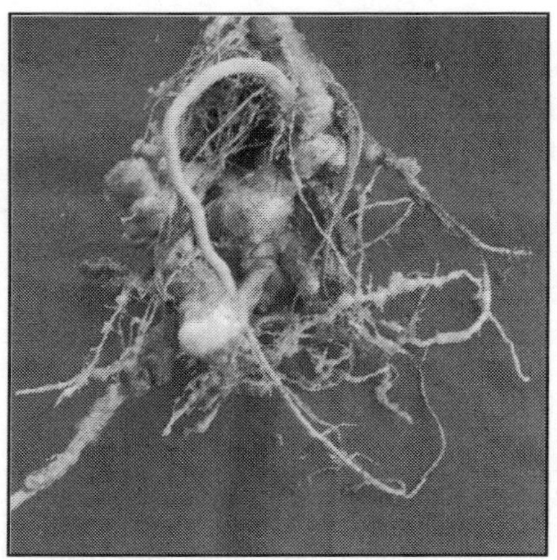

Root-knot Nematode in Cucumber

Nematode Wilt Complex in Cucumber

Whitefly in Gerbera

Figure 1a: Key Pests of the Greenhouse.

Leaf Miner in Tomato

Two-Spotted Spider Mite

Thrips in Rose

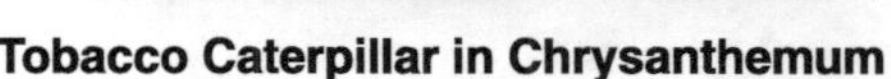

Tobacco Caterpillar in Chrysanthemum

Mealy Bugs in Gerbera

Figure 1b: Key Pests of the Greenhouse.

- Rigid hand-scrubbing rules for personnel involved in pruning, pollination, tying and harvesting activities.
- Minimize condensation of moisture in greenhouse, which provides leaf wetness for infection.
- Filters on all air intakes to restrict air-blown soil and insect vectors entry.
- Clean before planting to remove crop residues and debris containing any eggs, larvae, nymphs, pupae or adults of the pests.
- Clean or sterilize to make the greenhouse insect-free. Wash the walls, stem cracks and crevices, and fumigate with appropriate broad-spectrum insecticides.
- Strict weed management during off-season especially summers helps in reducing mites.
- Continue the sanitation practices throughout the growing period as well. The pruned material and overripe fruits should be immediately disposed off.
- Avoid planting outdoor crops near greenhouses because they serve as host for insect pests and diseases. Follow weed control measures within and around the greenhouse (10-30 feet area).

Cultural Control

Summer Ploughing/Solarization/Mulching: Soil disinfestations can be done by steam and by fumigation or application of chemicals. It is practiced in hot countries to control soil-borne pathogens and the elimination of weeds. In greenhouses, solarization is carried out in summers in the absence of a crop. Mulching of the soil with polythene or polyvinyl is done inside closed greenhouses to intensify soil heating. Formaldehyde is one of the commonly used chemical to sterilize the soil. Drenching the soil with formaldehyde @ 10-25 ml/l of water and then covering it for a week with plastic film. It is found very effective to check the attack of damping-off disease. Better fumigants like methyl bromide, metham sodium or dazomet granules are not available in India and even if available they are mostly unregistered. Albeit very serious damage of greenhouses by the root-knot nematodes and heavy demand of fumigants for highly sick polyhouses, their registration in India is indeed required for restricted use considering heavy amount of incurrence by the farmers on these accounts and hence to be considered at priority for a policy level decision.

Deep summer ploughing keeping greenhouse doors closed and formaldehyde mulching (@ 10-25 ml/l) using polythene sheets, for 15 days and removing the same 15 days before transplantation helps against several soil-borne pathogens including nematodes.

Figure 2: Solarization (with fumigation) during Summers is Crucially Important (Second drip is under soil).

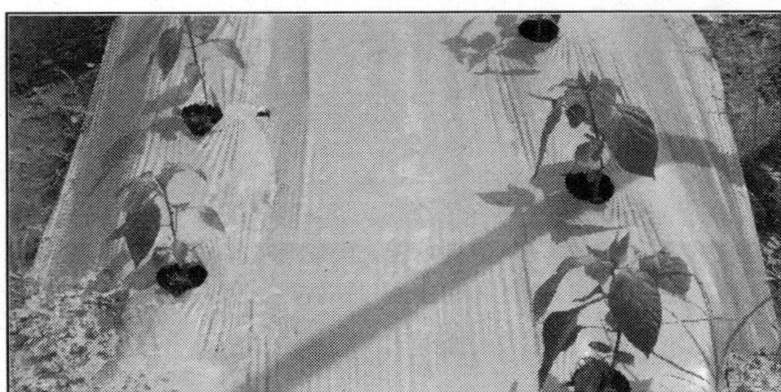

Figure 3: Mulching has Multifarious Uses– Weed Control, Nematode Management, Thrip Repellence *etc.*

Figure 4: Plastic Low Tunnels can be Used for Effective Avoidance of Winter Frost and Aphid Attacks.

Similarly, low tunnels can help in controlling the winter frost and aphid damage in several crops by early planting that gives additional advantage of:

Eradication

Eradication involves (i) removal or destruction of infected plant parts or plant as a whole, and their dispose, (ii) uprooting of alternate hosts within and outside a greenhouse which harbour the pathogens to complete their life-cycle and abridge the two seasons, (iii) crop rotation is effective against soil-borne pathogens. Roguing out of viral affected plants at the first sight of the symptoms is always advisable before the disease took its toll. Viruses affect many greenhouse crops particularly; cucumber and tomato, often transmitted by whiteflies and aphids, hence such pests have to be regularly monitored in addition.

Tillage

Deep summer ploughing is an effective practice to mitigate the soil inhabiting stages of insects. Deep ploughing of the soil after harvesting and then closing the door of greenhouse kills the pupae of fruit fly, red pumpkin beetle and cutworms as these remain in the soil as cocoons to complete the dormant stage of their life cycle. Besides, this helps in removal of weeds and their perennation in the crop.

Intercropping/Trap Cropping

Intercropping with diverse plant geometry and insect-pests breaks the standard mono-cropping and limits the infestation of insects. Diverse cropping system besides frustrating adult insects from egg laying also deters the adult insects from damaging the crops due to emission of volatile allelo-chemicals by certain crops. These effects do not let the development of micro-climatic conditions favouring the multiplication of any particular pests. Such planting combinations also enhance the activity of predators and parasites. The important crop combinations which can be used in greenhouse are broccoli-faba bean, which is effective against flea beetle, cabbage-carrot and cabbage-tomato against diamond back moth. Intercropping of marigold is also helpful as a trap crop in case of cabbage, for selective control of *Spodoptera* sp. that lays egg, while it is a repellent crop against plant parasitic nematodes.

Organic Amendment/Microbial Supplementation

Application of organic amendments like neem or *mahua* cake @ 500 kg/ha, or at lower doses to the soil in case of row application is good for nematode management. Apart from this addition of phosphate solubilizing bacteria enriches nutritional availability of the plants increasing the general health and tolerance against the pests. Neem seed kernel powder (preferably from seeds without extracting oil) gives very good results as amendments, controlling nematodes, ovicidal to insects, adds organic carbon to soil besides suppressing the denitrifying bacteria (Prasad *et al.*, 1971, 1999).

Figure 5

Environmental Control

Environmental control of diseases can minimize the need to use chemicals. The factors which affect development of diseases in the greenhouse are temperature of both soil and air, vapour pressure deficit, dew, soil water content and light (quality, day length and intensity). All these factors can be controlled to a certain extent depending on the facilities available. Interplay among these factors affects sporulation, dispersal, germination of conidia, penetration of the germ tubes and lesion development.

Figure 6

Soilless Cultures

The use of soil-less media eliminates the need for regular soil solarization, thus avoiding a very costly and laborious task. The plant roots are no longer exposed automatically to the pathogens which almost inevitably build up in intensively cropped greenhouse soil. The major advantage of soil-less is non-conventional means of disease control. Soil-less culture without a soil substrate, can easily be made pathogen-free before planting, albeit the possibility of re-infestation later. Major methods of soil less culture include the hydroponics or aeroponics, rockwool, sandbag culture, or cocopeat filled grow bags/strips *etc.*, wherein good degree steam or chemical sterilization can be attained.

Host Plant Resistance and Grafting

Use of resistant varieties holds good in pest management that provides an inbuilt ability to resist pests. Managing pests with host plant resistance is not only economic but safer to environment as well. Unlike cereals, few resistant/tolerant varieties have been developed in vegetables. Varieties such as Nun 3020, Orobellee and Indira are tolerant to mites in sweet pepper. Similarly, varieties GS-600, R-144, Delphi and Astona of greenhouse tomato and Satis and Nun 9729 of parthenocarpic cucumber are also tolerant to mites under greenhouse cultivation. However, care should be taken as for example tomato cv. R-144 and cucumber cv. Satis are tolerant to mites but susceptible to root-knot nematodes. Some nematode resistant tomato varieties include Pusa-120, PNR-7 and Hisar Lalit though at higher level of nematode incidence and higher temperature their resistance tends to fail badly. Several hybrid seed companies also claim resistance against nematodes.

Similarly, several diseases, mainly, *Fusarium* and *Verticillium* wilts of tomato, *Fusarium* wilt of cucumber and melons; *Phytophthora capsici* and tobacco mosaic virus (TMV) on pepper and leaf mould and TMV on tomato and bacterial blight are effectively prevented either by employing resistant varieties, by grafting

Figure 7: Grafting is One of the Best Methods for Pest Avoidance, Especially Nematodes, if Proper Rootstocks are available.

desirable varieties on resistant rootstocks or by both techniques.

Mechanical and Physical Measures

In a country like India, where cheap labour is available, mechanical collection of pest larvae/ propagules and their destruction (or some time use in preparing biocontrol agents such as NPV) is very useful. Daily misting to maintain an RH of >90 per cent to reduce thrips and mites. Sprinkler irrigation system reduces mites build up. Give hot water treatment to planting material or steam-sterilization of the planting beds.

Chemical Control

Due to the easy accessibility, sure and quick action against the target pests, chemical control still remains first choice of farmers. There are large number of reports of successful chemical control of vegetable pests, *i.e.* insecticide resistance development in diamond- back moth, fruit borer, tobacco caterpillar, resurgence of white fly and red spidermite have now threatened the wide spread use of synthetic insecticides. However, these insecticides can be judiciously used considering certain aspects like waiting period of pesticides, economic threshold level and initiation of controlled measures through pheromone/light-trap catches.

The following steps are suggested when using any pesticides:

a. Insecticides are sometimes insect-specific (or stage-specific) and are not effective against their close related ones. To avoid the persistent pesticide residues, choice of pesticide plays a crucial role. Insecticides should be chosen considering the picking-interval and waiting period. These with less picking intervals should be sprayed with insecticides having less waiting period. Further, the insecticides should be quickly alternated with their counterparts from different group, so as to avoid development of resistance in target pests. Grower should be aware of the insect pest biology and appropriate sprayer and nozzle.

b. Judicious use of pesticides is essential in terms of dosage, frequency, time and schedule.

c. Proper application is one of the most important steps in pest control efforts. There are many factors and components of spray methods that add up to proper application of insecticides.

Important Approaches

- ☆ Seed treatment Thiram/Vitavax/Bavistin @ 2-3 g/kg seed is good against many seed borne diseases or spray with Dithane M-45 as in case of early *Alternaria* blight. Drench the soil with formaldehyde before sowing of seeds helps against damping off. Drenching with metalaxyl and carbendazim can also help against soil borne pathogens.
- ☆ Spray imidacloprid (0.01 per cent) or acetamiprid 0.01 per cent against (aphids, whiteflies).
- ☆ Spray of dimethoate (0.05 per cent) or imidacloprid (0.04 per cent) or fipronil (0.0075 per cent) (against thrips). Spray of dichlorvos (0.05 per cent) against tobacco caterpillar *Spodoptera litura*. Spray of Coragen @ 0.3 ml/l can take control of most of the caterpillars and beetles.
- ☆ Spray of dicofol @ 1 ml/l or spinosad 0.4 ml/l is helpful against mites, vertimec @ 0.4 ml/l is good against mites, although expensive.
- ☆ For nematodes and other soil-borne pathogens fumigate soil with formaldehyde @ 1 ml/l or if better fumigants like metham sodium/methyl bromide/dazomet are available and registered (used under expert supervision taking stock of health and safety hazards) and or apply carbofuran @ 1-2 kg a.i./ha in the soil before planting against nematodes. Fumigation is better done with polythene mulching and the transplantation has to be avoided upto 10 after removal of the polythene mulches. Nematode infested patches can also be treated using the liquid preparation of triazophos/cadusaphos *etc.*

Biological Control

Biological control in greenhouse environment means releasing of insect predators, parasitoids, to control insect-pests, nematodes and disease causing organisms. They are eco-friendly and compatible to natural enemies of exiting the insect pests in the system. However, high insect pest populations will be difficult to control biologically. Some predators and parasites are better adapted to particular temperature and humidity conditions than others, and some perform better on specific crops than others. The natural enemies are under-exploited for pest management in vegetables even today. Among the egg parasites *Trichogramma* spp. has been utilized to some extent for the control of tomato fruit borer. Inundative release of 2,50,000 parasitized *Corcyra* eggs has been found to control the fruit borer effectively. *Chrysoperla carnea* is an effective predator for control of aphids, jassids and eggs of some lepidopterous insects with the release of first instar larvae @ 50,000 per hectare.

Unfortunately, only a few of them are available commercially in India. Some of the other useful bioagents are mentioned below:

1. The most common fungal bio-agent used are for biological suppression of soil-borne pathogens are *Trichoderma harzianum* and *T. viride* but the formeris considered superior.
2. Microbial biocontrol is so far limited to a few insects only, among them, *Bacillus thuringiensis*

Table 3: Key Biocontrol Agents for Application in Greenhouses (Sabir and Singh, 2012; Sood, 2011)

Pest	Beneficial Insect(s)
Spidermites (*Tetranychus urticae*)	✰ Predatory mites (*Phytoseiulus persimilis*) (*Amblyseius longispinosus, A. swirskii*) ✰ Anthocorid beetle (*Blaptostethus pallescens*)
Whitefly (*Bemisia tabaci, Trialeurodes vaporarium*)	Predatory mite (*Amblyseius swirskii*)
Leafminer (*Liriomyza trifolii*)	Parasitic wasps (*Diglyphus isaea, Dacnusa sibiriica*)
Aphids (*Aphis gossypii, Macrosiphum* spp.)	Parasitic wasp (*Aphidius colemani*)
Thrips (*Thrip tabaci, Thrip palmi, Scirtothrips dorsalis, Frankliniella occidentalis*)	Predatory bug (*Orius laevigatus*) – effective against *F. occidentalis* only *O. insidiosus*is the other predatory bug, Generalist predatory mites *Amblyseius swirskii, A. mckenziei*
Mealy bug (*Phenacoccus solenopsis*)	Predatory beetle (*Cryptolaemus montrouzieri*)
Soft-bodies/sucking arthropod pests	Predatory lacewings (*Chrysoperla carnea*)

is the most extensively used bio-control agent in vegetables. It has been widely applied against *Plutella xylostella, Helicoverpa armigera* and *Earias vittella* @ 500 g a.i./ha. Spray of *Bt* (*Bacillus thuringiensis*) @ 500 g a.i./ha at 10 days interval against *Helicoverpa armigera*. Bacterial antagonists, *Pseudomonas fluorescens* is the widely studied agent in India and is used to control the diseases caused by the fungal pathogens such as *Fusarium oxysporum, Pythium* spp., *Rhizoctonia solani, Phytophthora* spp. and *Sclerotium rolfsii.*

3. Because of the host specificity, the use of nuclear polyhedrosis virus (NPV) is limited to *Helicoverpa armigera, Spodoptera litura* and *Plutella xylostella* only. The lack of commercial production has limited its popularity even in potential areas. Application of *Ha*NPV and *Sl*NPV @ 250 larval equivalents in the evening hour with some UV protectants like charcoal powder and adjuvants like molasses reduce the populations of *H. armigera* and *S. litura*, respectively. Proper adjuvants and time of application need to be regulated for its success. The success of biological control depends on the quality, timely availability and appropriate release timing and methodology of bio-agents.
4. Use of parasitic wasp (*Encarsia formosa*) and bio-agents like *Verticillium lecanii* is quite effective against whiteflies.
5. Release of predacious mite (*Neoseiulus cucumeris*) is effecting in controlling thrips.
6. Predacious mites *Amblyseius longispinosus* is effective in suppressing spider mite.

Regulatory and Quarantine Control

✰ Quarantine plants coming from other greenhouses strictly. Workers should avoid the wearing of yellow clothes as insects get attracted to it. and may land and ride along into the greenhouse or from one greenhouse to another.

✰ Seed tubers of potato from Nilgiri hills of Tamil Nadu should not be used due to domestic quarantine against golden cyst nematode. Otherwise also only nematode-free planting material should be used.

Other Methods/IPM Strategies

a. Scouting and Monitoring

Monitoring of the pests such as *Helicoverpa/Spodoptera* with the help of pheromone traps helps to schedule the spray of insecticide/HaNPV/SlNPV, one week after > 10 moths are caught in the trap. However, for greenhouses, yellow and blue sticky traps are better.

b. Trapping/Baiting

Various traps such as yellow sticky traps (for whiteflies), light traps (for lepidopterous moths) and lures such as 'Qlure' are useful for controlling the pests as well as in monitoring of the pests. Pheromone traps and lures are very helpful in controlling the pests. Older larvae of *Spodoptera* have a habit of hiding under clods/fallen leaves on the soil and climbing on to the top plants during night. To kill the hidden larvae, food bait can be effectively used. The bait can be prepared by mixing 10 kg rice bran, 1 kg jaggery and 150 ml monocrotophos with a small quantity of water. Small pallets of this bait are distributed on the soil throughout the greenhouse.

Good Agricultural Practices (GAP)

Actually, GAP are practices that address environmental, economic and social sustainability for on-farm processes, and result in safe and quality food and non-food agricultural products [FAO]. Thus, GAP actually encompass:

✰ Best management practices for every crop or animal

✰ Production and post-production (harvest, storage, processing) system

✰ Result in safe and nutritious food/security

- ✰ Promote economic, social and environmental protection
- ✰ Promote natural resources (water and soil) sustainability
- ✰ Driven by market and consumer-demanded

Indeed integrated pest management is part and parcel of good agricultural practices.

Four Pillars of GAP

- ✰ Economic Viability
- ✰ Environmental Sustainability
- ✰ Social Acceptability
- ✰ Food Quality and safety

Major Principles of GAP

- ✰ Traceability
- ✰ Record keeping and self-inspection
- ✰ Varieties and rootstocks
- ✰ Site history and site management
- ✰ Soil and substrate management
- ✰ Fertilizer use
- ✰ Irrigation/Fertigation
- ✰ Crop protection
- ✰ Harvesting
- ✰ Produce handling
- ✰ Waste and pollution management, recycling and re-use
- ✰ Worker health, safety and welfare
- ✰ Environment issues
- ✰ Complaint form

Sabir *et al.* (2010) have enlisted the importance of IPM and GAP and their relevance to Indian Horticulture. Now that IndiaGAP protocols have been published, their adoption by horticultural entrepreneurs would certainly position Indian products in International markets with much greater acceptability. Producer, retailer and consumer relationships have to be facilitated by platforms and markets by the govt. through promotional schemes so that such relationships fructify in future and ultimately establish in some times ahead in near future.

Conclusions

- ✰ Greenhouse technology has spread to very large area in the country presently to estimates of more than 50,000 ha. There are many success stories in the country in the area of protected cultivation. However, there are many issues that need to be addressed, failing which the technology stands on cross roads of success and liability to farmers having invested huge amounts of money, indeed most of which is without subsidy.
- ✰ Construction of greenhouses should preferably be done in clusters with proper feasibility studies. Individual farmers may be encouraged to take up only when ample avenues of trainings, inputs, repairs and markets are available to them.
- ✰ Quality control of greenhouse structure, insect-proof nets, ventilation arrangements, soil and bed preparation besides these technical as well as market support should be facilitated in organized manner.
- ✰ Diagnostic supports for soil, water and pests and their natural enemies should be ensured as far as possible, preferably with proper pest monitoring and advisory services.
- ✰ Suitable seeds and healthy planting material availability may be mobilized for the greenhouse growers.
- ✰ Pesticide residues should be managed through proper sampling, advisory and local level management.
- ✰ Critical inputs like biocontrol agents, pollinators, soil-health and other bio-rationals should be encouraged. Currently, mass production of biocontrol agents for greenhouse industry in particular is still in its infancy and abysmally low to cater to their demand. Tangible solution to their present availability through import alone should be found sooner before the technology finds it difficult to survive, just as the case of pollinators.
- ✰ Strategies to rope-in commercial industry with scientists for collaborative production and promotion of local biocontrol agents with commercial production expertise from abroad can also be explored.
- ✰ Quality parameters, food-safety and bio-safety issues need to be looked into seriously
- ✰ Good Agricultural Practices need to be brought into practice through encouragement and preferably linked with niche markets or exports.
- ✰ Export promotion needs to be the priority area for greenhouse growers.

Protected cultivation technology holds a great future in the country considering small landholdings, better economic viability, higher profitability, entrepreneurship potential and engagement of youth in agriculture which has lately proven otherwise. It is high point to remember, that erection of larger number of greenhouses is definitely an achievement, but their maintenance and perseverance as profitable venture needs very serious engagement for managing them through adoption and facilitation of proper technology, better pest management and good agricultural practices *etc.*

References

Anon. 2015. *Tuta absoluta*: A New Invasive Pest Alert. http://www.icar.org.in/en/node/8600.

Anon. 2015. Search for invasive species. CABI compendium. http://www.cabi.org/isc/datasheet/24426

Anon. 2015. Farmers advised to destroy nematode-infested guava trees http://timesofindia.indiatimes.com/city/coimbatore/Farmers-advised-to-destroy-nematode-infested-guava-trees/articleshow/49170062.cms

Prasad, R., Rajale, G.B. and Lakhdive, B.A. 1971. Nitrification retarders and slow release nitrogenous fertilizers. *Adv. Agron.*, **23**: 337–405.

Prasad, R., Singh, S., Saxena, V.S. and Devkumar, C. 1999. Coating of prilled urea with Neem (*Azadirachta indica* Juss.) oil for efficient nitrogen use in rice. *Naturwissenschaften*, **86**: 538–539.

Sabir, N., Singh, B., Hasan, M., Sumitha, R., Deka, S., Tanwar, R.K., Ahuja, D.B., Tomar, B.S., Bambawale, O.M. and Khah, E.M. 2010. Good Agricultural Practices (GAP) for IPM in Protected Cultivation, Technical Bulletin No. 23, July 2010, National Centre for Integrated Pest Management, New Delhi, p. 16.

Sabir, N. and Singh, B. 2012. Protected cultivation of vegetables in global area: A review. *Indian J. Agril. Sci.*, **83**(2): 123–135.

Sood, A. 2011. Integrated pest management under protected environment: principles and practices. http://agropedia.iitk.ac.in/content/management-insect-pests-protected-environment

Tyagi, K. and Kumar, V. 2015. First report of western flower thrips, *Frankliniella occidentalis* (Pergande) (Thripidae: Thysanoptera) from India: a potential havoc to Indian Agriculture. *HALTERES*, **6**: 1–3.

Reducing Post-Harvest Losses

67

Infrastructure Requirements in Post-harvest Management for Horticultural Crops

S.K. Malhotra

Agriculture and Horticulture Commissioner,
Ministry of Agriculture and Farmers Welfare, Krishi Bhawan, New Delhi – 110 001
E-mail: agricommissioner@gmail.com

Contribution of horticulture to agricultural growth in India has been commendable during the XIth and XIIth Plan period. The country witnessed an unprecedented growth in production of fruits and vegetables and is the world leader in several crops. This has been possible due to technology and favourable policy environment for the development of horticulture. In nutritional security of the people of India, fruits and vegetables play a significant role and it has been observed that any decrease in the availability of fruits and vegetable due to weather vagaries drastically trigger the inflation. However, incomes to the farm are volatile because of number of unforeseen factors, like weather, pest and disease outbreak and post-harvest losses which necessitate the effective management. Most of horticultural produces, being perishable, and seasonal in nature, are prone to post-harvest losses, which ranged between 4.58 to 15.88 per cent (Jha *et al.*, 2015). The orientations of farmers have mostly been towards production in the past. The negligent attitude towards post-harvest losses, lack of quality consciousness and inadequate number of food processing units and cold storages were found responsible for post-harvest losses. It can be inferred that there is immediate need to augment cold chain infrastructure, apart from increasing processing of horticulture produces, to minimize post-harvest losses.

Status of Production and Post-harvest Losses

Area, production, productivity, availability and export of horticulture have increased manifolds, which provided ample opportunities for utilization of waste lands, employment generation and effective land use planning. Diversification, recognised as one of the options for improving land use planning has dramatic impact. The technology-led horticulture has shown impressive impact on production, productivity and profitability of all horticultural crops. The production of horticultural produce has jumped more than 11 times from the level of 25 million tonnes production during 1950-51. Currently, the total production of horticultural produce is 283.3 million tonnes (2015-16) from the 9 per cent of the total cultivated area in India, which contributes 30 per cent in value term to GDP of Agriculture and the trend of development, has been marked as "Golden Revolution". But there is a growing demand for horticultural produce both in domestic and overseas markets, and at the same time competition is also increasing (Saxena *et al.*, 2015).

During the last two decades, the production of horticultural produce has increased manifold but the gap in demand and supply continues. At present there is considerable gap between the gross production and net availability of fruits and vegetables where post-harvest losses have been identified as one of the reasons. Constraints to proper post-harvest management in developing countries are: inadequate adoption of harvesting and post-harvest handling and management. It was, however, evident that post-harvest and value-addition are integral components of strategies to improve agricultural productivity, improving need-based post-harvest infrastructure, food quality and safety standards, supply chain, storage, processing and waste utilization. It is also important for effective exploitation of the export potential of fruits and vegetables.

Largely, with the warm climate in India, sub-tropical and tropical horticulture is prevalent throughout the year; therefore, low temperature storage and refrigerated transportation system are mandatory for extending shelf-life. The transportation of produce from farm to pack house, than to pre-coolers and cold storage facilities to market place are very crucial cold chain links which are lacking in India. Most of the transportation is done in un-refrigerated open trucks without proper packaging leads to losses.

Post-harvest Losses

Ministry of Food Processing Industries commissioned a study in 2012 on "Assessment of Quantitative Harvest and Post-harvest Losses of Major Crops/Commodities in India" which was conducted by Central Institute of Post-harvest Engineering and Technology (CIPHET), Indian Council of Agricultural Research (ICAR). The report of the study was published in 2015. As per this report, post-harvest losses including losses during storage were in the range, 6.70 to 15.88 per cent (fruits) and 4.58 to 12.44 per cent (vegetables). The details of harvest and post-harvest losses of major horticultural crops and commodities estimated by this study are at Table 1. The CIPHET study assessed the losses in most channels of the supply chain but did not cover the losses incurred in long haulage transport from farms to markets. The report was unable to differentiate between losses in and without the cold-chain from farm to market.

According to estimates from CIPHET, quantitative post-harvest losses in fruits and vegetables is about 4.58-15.88 per cent of total production, valued at Rs. 22,000 crores. These losses are mainly due to improper methods of harvest, handling, transportation and storage. However, inclusive of the qualitative deterioration like weight loss, spoilage and loss of other sensory characters, the actual loss is reported to be between 6-29 per cent in different horticultural commodities. There is a constant change in these figures as post-harvest losses are dependent on several factors like soil and climate of the region, variety, prevailing marketing practice of the region and other biotic stresses. This is not only loss of produce but also wastage of labour, energy and inputs involved in production. Establishment of an integrated cold chain system supported by processing facilities is the need of the hour to prevent this colossal loss. Therefore, efficient post-harvest management has become an absolute necessity and needs all infrastructure related to it.

Table 1: Harvest and Post-harvest Losses of Major Horticultural Crops and Commodities

Name of Crop/ Commodity	*% Average Loss*	*Name of Crop/ Commodity*	*% Average Loss*
Apple	10.39	Banana	7.76
Citrus	9.69	Grapes	8.63
Guava	15.88	Mango	9.16
Papaya	6.70	Sapota	9.73
Onion	8.20	Tomato	12.44
Cabbage	9.36	Cauliflower	9.56
Green pea	7.45	Potato	7.32
Tapioca	4.58	Arecanut	4.91
Black pepper	1.18	Cashew	4.17
Chilli	6.51	Coconut	4.77
Coriander	5.87	Turmeric	4.44

Source: CIPHET Report (Jha *et al*., 2015)

Post-harvest Handling and Pre-conditioning

In agriculture, post-harvest handling is the preliminary stage in a crop's lifecycle, which immediately follows harvest. Post-harvest handling includes aggregation, pre-conditioning (cleaning, sorting, packaging, pre-cooling *etc*.) of a crop and is important to extend the marketable life of any produce. Post-harvest treatment does not change the essential characteristics of the agricultural produce and determines the quality and value realized from a crop. In case of fruits and vegetables, the most important aspects of post-harvest management are reducing the field heat of the freshly harvested crop, keeping the produce at appropriate temperature and humidity and avoiding physical damage such as bruising while keeping the produce physiologically active or alive. Farm-level traceability and sanitation of the fresh produce is also an important factor in scientific handling, to safe-guard the consumers' health.

The preconditioning of fresh produce occurs through a series of activities that are commenced at a pack-house. This can be a simple shed, providing shade and running water, or a larger scale, sophisticated, modern facility, with sorting, grading, waxing and packing stations, pre-coolers and dispatch cold rooms. The activities covered under preconditioning are exempted from service tax under Finance Act. Once the produce undergoes precooling, the next phase in its cold-chain is the refrigerated transport, which is used to consolidate the daily volumes at a market linked cold store. This cold store would typically serve as a distribution platform for delivery to the front end of the supply chain. Alternately, certain fruits and vegetables, may follow a drastically different sequence of activities as per requirements. For *e.g.*, potato is not pre-cooled or package but stored in bulk format in refrigerated warehouses, warmed up when exiting storage and dispatched to market in ordinary transport systems. However, in all cases of post-harvest handling, including the cold-chain, the post-harvest care and parameters maintained are critical to keeping marketability and quality of the produce. Each crop has an optimal range for storage temperature and humidity. Also, certain crops cannot be effectively stored together, as unwanted cross contamination and tainting can result.

Regardless of the scale of harvest, from domestic garden to industrialized farm, the basic principles of post-harvest handling for most crops are the same: careful handling to avoid damage (cutting, crushing and bruising), immediate cooling and maintaining in cool conditions, and culling (remove damaged items). India over the years witnessed a marked increase in production of perishable high nutrition products like fruits, vegetables, meat and poultry products *etc.* but development of supply chain infrastructure was not strategically directed, for safe handling and to convey these perishable products to markets, except in the dairy sector. A resultant demand supply mismatch emerged across these agricultural commodities, frequently contributing to wide spread price fluctuations and inflation. The inadequacy of technology aided farm-to-market logistics, also contributes to high food losses in case of perishable foods, further adding to inflationary pressures.

Cold Chain: Status and Requirement of Infrastructure

Cold chain refers to the process used to maintain optimal conditions during the transport, storage, and handling of perishables from the point of origin to the point of consumption in the food supply chain. A well-organized cold chain reduces spoilage, retains the quality of the harvested produce and guarantees a cost efficient delivery system to the consumer. A cold chain in horticulture needs to start at the farm level (*e.g.* harvest, pre-cooling) and cover up to the consumer level or at least to the retail level. The main feature of the chain is that if any of the links is missing or is weak, the whole system fails. Cold chain logistics infrastructure primarily includes, *in situ* or *ex-situ* pack house; pre-cooling facilities; packaging; cold storages; refrigerated carriers; multi-modal transportation; information management system; warehouse management system; traceability; financial and insurance institutions.

Currently, there are approximately 6,300 cold storages in India, and they are designed originally for single commodity storage and can store only less than 11 per cent of the country's total horticultural produce. More than 75 per cent of these cold storages currently available are suitable for potato storage and only 0.4 per cent are used for other fruits and vegetables. While about 105 million MT of perishable produce is transported across India annually, only about 4 million MT is transported *via* reefers. Food industry experts believe that controlled atmosphere storage and other cold storage facilities with the technology for storing and handling different types of fruits and vegetables at variant temperatures have a very good potential market in India. Another major constraint is the lack of refrigerated vehicles for movement of perishable produce (with the exception of milk). Although there are currently more than 25,000 refrigerated vehicles and 250 operators involved in refrigerated transport, 80 per cent of this capacity is dedicated to transporting milk. When compared with world standards for cargo movement through cold chain, India is still far behind. The percentage of movement of fruits and vegetables through cold chain in U.S. is around 80 to 85 percent,

Table 2: Commodity-wise Distribution of Cold Storage in India

Commodity	*Capacity (Million MT)*	*% Share*	*No. of Cold Storages*
Potatoes	18.43	75.4	2,862
Multi-purpose	5.64	23.1	1,584
Fruits and vegetables	0.10	0.4	160
Meat and fish	0.19	0.8	497
Milk/Milk products	0.07	0.3	191
Others	0.03	0.1	87
Total	**24.45**	**100.0**	**5381**
Institution-wise cold stores			
Private	23,406.7	95.7	4,885
Cooperative	936.9	3.8	356
Public	107.0	0.4	140
	Total	**24450.7**	**5381**

Source: Ministry of Agriculture, Government of India.

Thailand is 30 to 40 percent and India is negligible. Currently, most of the refrigerated transport in India is operated by small firms that do not make use of state–of–the–art technology or management practices.

Cold storage in India has been largely adopted for long-term storage of potatoes, and other high value crops like apples, grapes and imported fruits (Fiji apples, pear, plum and American grapes *etc.*). Cold storages for meat, fish and dairy items and for other items such as chilies and spices account for only 1 per cent of total cold storage capacity (Table 2). The percentages of multi-purpose cold storages available in the country are only 23.1 per cent of the total. More than 90 per cent of the cold stores in the country are in private sector with 95.7 per cent of share in capacity, 6.6 per cent in cooperative sector with 3.8 per cent share of capacity (Table 2). In 2010, the cold storage gap of about 37 million MT was worked out on the basis of peak season production and highest arrival/harvesting of storable fruits and vegetables in a month (Table 3). Against this, normally 50 per cent capacity is required for storable surplus of the identified fruits and vegetables. In addition, India has about 250 reefer transport operators (this includes independent firms) that transport perishable products. Of the estimated 25,000 vehicles in use, 80 per cent transport dairy products (wet milk); only 5,000 refrigerated transport vehicles are available for all other commodities.

For calculations, the estimated cost taken is Rs. 2000 crores per million tonne cold store. The commercial rates for establishment of cold stores under different schemes of GoI presently are Rs. 0.20 lakhs/tonne capacity. Estimated cost of one refrigerated vehicle of 5 MT capacity was Rs. 0.35 crore. It was estimated that fruits and vegetables worth Rs. 13,259 crores can be saved annually by handling the produce through cold chain logistics.

Table 4: Cold Storage Gap Assessed by National Spot Exchange in 2010

State	*Cold Storage Requirement in '000 MT*	*Present Capacity* in '000 MT*	*Gap in '000 MT*
Andhra Pradesh	2,324	901	1,423
Assam	919	88	831
Bihar	4,241	1,147	3,094
Chhattisgarh	543	342	201
Gujarat	2,748	1,267	1,481
Haryana	804	393	411
Himachal Pradesh	487	20	467
Jammu and Kashmir	737	43	694
Jharkhand	796	170	626
Karnataka	2,404	407	1,997
Kerala	2,771	58	2,713
Maharashtra	6,273	547	5,726
Manipur	80	0	80
Meghalaya	239	3	236
Mizoram	74	0	74
Madhya Pradesh	1,213	808	405
Nagaland	70	6	64
Odisha	1,835	291	1,544
Punjab	1,318	1345	–27
Rajasthan	391	324	67
Tamil Nadu	7,906	239	7,667
Tripura	163	30	133
UP and Uttarakhand	12,228	10,187	2,041
West Bengal	10,566	5,682	4,884
Total	61,130	24,298	36,832

* Present capacity estimates as of 2009

Source: NSE and DMI (present capacity in Delhi–126,158 MT, Goa-7,705 MT, A and N-210 MT, Puducherry–85 MT).

Table 5: Additional Requirement of Infrastructure and Financial Estimates for Establishment of Cold Chain

Commodity	*Production in 2012-13 (mT)*	*% to be Placed in Cold Chain*	*Additional Capacity Required (mT)*	*Financial Requirement (Rs. Crores)*		
				Cold Store	*Refrigerated Transport*	*Total*
Potato	45.3	20	9.00	18,000	2,989	20989
Onion	19.0	20	3.80	7,600	1,330	8930
Tomato	18.2	25	4.55	9,100	1,593	10693
Apples	1.9	40	0.76	1,520	266	1786
Citrus	10.1	15	1.52	3,040	530	3570
Banana	26.5	25	6.63	13,260	2,319	15579
Grape	2.5	20	0.50	1,000	88	1088
Mango	18.0	40	7.20	14,400	2,520	16920
Total			**33.95**	**67,920**	**11,844**	**79764**

Gap in Cold-chain Infrastructure in the Country

The Planning Commission (now Niti Aayog) constituted a Committee on Encouraging Investments in Supply Chain. The Committee made a key recommendation that a cold-chain system would expressly not be a pure price arbitrage business model, but have objective to smoothen episodic output with regular demand and result in greater price stabilization and market connectivity. The report estimated additional 37 million tonnes cold storage capacity requirement. This estimate was based on a study carried out by NSEL.

Hansa Report

As per Base line Survey conducted during December, 2013 by M/s Hansa Research Group, 5,367 cold storages with a capacity of 26.85 million MT were in operation. It was reported that around 1,200 were permanently closed, 95 per cent of cold storage are owned and operated by private sector, 3 per cent are with cooperatives and remaining 2 per cent are under PSUs. Overall average capacity utilization was as under:

Table 6: Cold Storage Capacity Utilization for different Products

Product Type	Capacity Utilization (%)
Horticulture	75
Processed food	71
Animal husbandry	74
Pharma products	70
Others	65

NCCD-NABCONS Report

As per a recent study on All India Cold-chain Infrastructure Capacity (AICIC-2015) conducted by NCCD-NABCONS. This study assessed demand and reported the gaps in end-to-end connectivity. Overall status of cold chain infrastructure created and required is tabulated as follows:

Table 7: Status of Cold Chain Infrastructure Created

Component	Existing Capacity (2014) (No.)	Approximate Requirement (No.)
Integrated pack houses	250	70,000
Reefer trucks	<10,000	62,000
Cold stores (bulk and distribution hubs)	32 million	35 million
Ripening chambers	800	9,000

Future focus should be for creation of pack houses, refrigerated transport, ripening chambers instead of standalone large cold storage/CA store projects. The Study conducted by NSEL/Planning Commission was mostly based on cold storages while as study conducted by NABCONS has enumerated cold chain requirement. The assessment of requirement and existing capacity of cold chain including cold storages is a continuous process and needs to be studied further. Statewise and agencywise distribution of Cold Storages in the country as on 31st March, 2016 is given in Table 7.

Status of Indian Cold Chain industry

Indian cold chain industry is expected to grow at a CAGR of more than 28 per cent over the next 4 years and reach a market size of more than US$ 13 billion in 2017 and is largely unorganised in nature. Thereby, presenting a great opportunity for foreign companies through increased investments, modernization of existing facilities, and establishment of new ventures *via* private and government partnerships. Features of Indian cold chain industry are:

i) India's cold chain industry is still evolving, not well organized and operating below capacity
ii) The Indian cold chain market is highly fragmented with more than 3,500 companies in the whole value system
iii) Organized players contribute only ~8–10 per cent of the cold chain industry market
iv) Most equipment in use is outdated and single commodity based

Cold stores are the major revenue contributors of the Indian Cold Chain industry and are majorly used for storing potatoes. However, the market is gradually getting organised and focus towards multipurpose storages is rising. There are more than 6,300 number of total cold chain storages in India with 95 per cent of total storage capacity under private players. More than 36 per cent these cold storages in India have capacity below 1,000 MT. More than 65 per cent of India's cold chain storage capacity is contributed by the states of Uttar Pradesh and West Bengal. With the current capacity only less than 11 per cent of what is produced can be stored. Cold storage capacity is expected to grow at more than 13 per cent per annum on a sustained basis over the next 4 years, with the organized market growing at a faster pace of more than 20 per cent. Key growth drivers of cold chain include:

i) Growth in organized retail and food service industry, government's initiatives, rising export demand for processed and frozen food.
ii) Huge variation in the share of potatoes storage capacity and revenue generation.
iii) Share of multipurpose cold storages is low in storage capacity but high in revenue generation.
iv) There is growing focus towards multi-purpose cold storages.
v) India offers great market potential for cold chain logistic solution providers, including refrigerated transport.

vi) According to industry estimates, approximately 104 million metric tonnes of perishable produce is transported between cities each year

vii) Of this figure, about 100 million metric tonnes moves *via* non–reefer mode and only four million metric tonnes is transported by reefer.

viii) Even though, India has about 250 reefer transport operators (mostly small and non integrated firms) that transport perishable products and >30,000 refrigerated vehicles currently ply in India.

ix) Majority of the refrigerated vehicles (~80 per cent) are utilized for milk and milk products transportation

x) The high cost of transportation is a major challenge for refrigerated vehicles market.

Mission Programme for Promotion of Cold Chain

Post-harvest management includes activities like handling, grading, pre-conditioning, packaging, transient storage, transportation, distribution, curing and ripening and where possible long term storage can be taken up. These facilities are essential for increasing the marketability of the horticulture produce, adding value to the produce, increasing profitability and reducing losses. It is necessary to strengthen a network of infrastructural facilities for horticulture storages, transportation, marketing, packaging and grading, and export. The existing schemes of the National Horticulture Board (NHB), Directorate of Marketing and Inspection (DMI) and National Cooperative Development Corporation (NCDC) have played a significant role in this context. MIDH would include projects relating to establishment of pre-cooling units, 'on-farm' pack houses, mobile pre-cooling units, staging cold rooms, cold storage units with and without controlled atmosphere capability, integrated cold chain system, supply of refrigerated vans, refrigerated containers, primary/mobile processing units, ripening chambers, evaporative/low energy cool chambers, preservation units, onion storage units and zero energy cool chambers. These projects are entrepreneur driven and provide facility for credit linked back-ended subsidy. PSUs/Government agencies/cooperatives/ growers' association recognized/registered by the DMCs, having at least 25 members, are entitled to avail assistance for such activities. They may avail back ended subsidy without credit link, subject to condition that they are able to meet their share of the project cost. Assistance for market intelligence will be project based.

Infrastructure Developed under NHM and NHB

Cold Chain

Assistance for setting up of new cold storage infrastructure will be available only to multi-chamber cold storage units with technologies which are energy efficient with provision for thermal insulation, humidity control, advanced cooling systems, automation, *etc.*, having specifications and standards approved by the Ministry. While cold storages (Long term storage and distribution hubs) upto 5,000 MT capacity will be promoted under NHM/HMNEH sub-schemes, capacity above 5,000 MT up to 10,000 MT will be promoted under NHB sub-scheme. In this context, for cold storages, 3.4 cubic metres (cum.) (120 cubic feet (cft) of chamber volume shall be considered equivalent to one MT of storage capacity. Similarly, for refrigerated transport, 3 cum (106 cft) of chamber volume shall be equivalent to one MT of storage capacity and for ripening chambers, 11 cum of chamber volume shall be equivalent to one MT of storage capacity. Assistance for pre-cooling unit will be linked to pack-houses and cold rooms (staging). Similarly assistance for staging cold room will be linked to existing and new pre-coolers.

Details of infrastructure established during the XIth Plan under NHM are 4,611 pack houses, 510 cold storage units, 12 CA storage, 39 refrigerated vans, 696 mobile/primary processing, 1,083 low cost onion storage structures, 87 ripening chambers, 244 evaporative/low energy cool chambers and 830 pusa zero energy cool chambers were established. About 25 lakh MT capacity of cold warehouses were developed during the period bringing in private investments of approx Rs. 450 crore. Through NHB schemes around 982 cold storage projects were sanctioned during the XIth Plan period for creating an additional capacity of 48.65 lakh MT with a back ended capital investment subsidy of Rs. 178.91 crore, which has attracted private investment of more than Rs. 750 crores in this sector. The physical target to be achieved during XIIth Plan for integrated post-harvest management is for creation of 7,700 pack houses, 95 pre-cooling units, 60 cold room staging, 29 mobile pre-cooling units, 446 single temperature zone, 70 multiple temperature zone and 19 CA type cold storages, 135 reefer vans, 160 ripening chambers and 288 low cost onion storages.

Assistance can also be availed for a combination of PHM infrastructure components by a beneficiary, within the prescribed norms of individual items. Assistance will be available to individuals, Group of farmers/growers/ consumers, Partnership/Proprietary firms, Self Help Groups (SHGs), Farmers Producer Organization (FPOs), Companies, Corporations, Cooperatives, Cooperative Marketing Federations, Local bodies, Agricultural Produce Market Committees (APMC) and Marketing Boards and State Governments. Assistance will also be available for taking up cold chain components so as to integrate the activities into a single project, as a new component during XII Plan. Proposals on long distance transport solutions on project basis shall be supported under NHB sub-scheme.

Strategies

- ☆ Cold storages are extremely energy intensive and the supply of which is erratic in rural areas. In fact, energy consumption typically represents the second-highest operating expense in these facilities, trailing only labour costs. Use of alternate and renewable energy to supplement hydro-electric power should be thought of. Solar concentrators hybridized with biomass gasifier can be used to power cold storage facilities during periods of power shortage. This type of cold storage design boosts overall efficiency of the agricultural produce value chain by reducing wastage and improving storage conditions for agricultural produce in rural areas. In addition to the solar technology, there are various other sub-systems required in a solar cooling project including power block, water treatment, balance of plant, plant EPC, O and M, *etc.* Further research efforts based on solar technology for low cost cold storage can help to develop the cold chain status in India.
- ☆ Technical Standards for storing fresh fruits and vegetables had been notified for precooling, cold storages and fruit ripening units by National Horticulture Board for implementation are available on website http://nhb.gov.in/guideline. Extant specifications standards and protocols on cold storage and cold-chain components are required to be adhered to while approving cold storage projects. Revisions to technical standards and adherence protocols are now updated by National Centre for Cold-chain Development (NCCD), New Delhi as necessary when improved technologies and efficiencies are introduced/understood/approved.
- ☆ Awareness on food safety and entrepreneurship are the other needs to be attended with urgency to establish a strong cold chain system in the country and reduce post-harvest losses of all the perishables. Creation of awareness on usefulness of cold storages to farmers, encouraging small and medium entrepreneurs to enter into cold storage, ripening and allied businesses and development of human resource through training and mentoring on scientific methods of post-harvest handling and processing operation are immediate essential initiative required.
- ☆ Taking advantage of low-cost opportunities in combination with establishing a strategy for continuous energy efficiency improvement can yield impressive cost and energy savings. R&D is required for value-addition of produce stored in the cold storage to achieve complete reduction in post-harvest losses. The produce deteriorates in quality with a slight change in cold storage environment in the event of power failure or even during prolonged storage. The cold stored produce from such environment may not be suitable for fresh marketing but can be utilized in development of processed products. Simple processing technologies with minimum sophistication in terms of plant and machinery should be developed to supplement and support cold chains in India.

References

Jha, S.N., Vishwakarma, R.K., Ahmad, T., Rai, A. and Dixit, T.K. 2015. Report on assessment of quantitative harvest and post-harvest losses in major crops and commodities in India. ICAR-CIPHET, Ludhiana, p. 1–162.

Malhotra, S.K. 2014. Future requirements of infrastructure in post-harvest management of horticulture crops. **In**: *Horticulture for Inclusive Growth*, (Eds.) K.L. Chadha, Pritam Kalia and S.K. Singh. The Horticulture Society of India and Westville Publishing House New Delhi, pp. 596–604.

Narayana, C.K., Malhotra, S.K. and Pandey, V. 2013. *Post-harvest Losses in Selected Fruits and Vegetables Crops*. IIHR, Bengaluru, p. 1–30.

Saxena, Mamta, Swapna, Bhatacharya and Malhotra, S.K. 2015. *Horticulture Statistics 2015*. Oxford University, p. 1–200.

68

Post-harvest Losses and their Mitigation to Improve Farmers' Income

S.N. Jha

Indian Council of Agricultural Research,
Krishi Anusandhan Bhawan-II, New Delhi – 110 012
E-mail: snjha_ciphet@yahoo.co.in

Agriculture is one of the important sectors of the Indian economy. It contributes about 14 per cent to the GDP of India and about 11 per cent of its total exports. About 50 per cent Indian population still relies on agriculture as its principal source of income and serves as a source of raw material for a large number of industries. India accounts for only about 2.4 per cent of the world's geographical area and 4 per cent of its water resources, but it supports about 17 per cent of the world's human population and 15 per cent of the livestock. Accelerated growth of agriculture production is therefore necessary, not only to achieve higher contribution towards GDP and meet the rising demand for food, but also to increase farmers' income to ensure their inclusiveness.

Indian agriculture had shown tremendous evolution since independence and converted India as exporting country from importer even after four fold increase in population. India produced 235.85 million tonnes horticultural produce in 2012-13 (Table 1; DoAC, 2013) and is increasing rapidly. India ranks second in the world in production of fruits and vegetables. Thus the boom in production brought the country in a position to provide food for about 1.25 billion people. Ever-increasing population however poses serious challenge to prolong this scenario and ways has to be thought of to provide safe and quality food to the masses.

Increasing agricultural production is one aspect of fulfilling food demand. Delivering food to the consumers by saving produced commodities from losses in fields, transport, storage, retailing, processing *etc.* without straining our fields, water and environment seems much better option. After production, agricultural produce undergo series of post-harvest unit operations, handling stages and storage before they reach to the consumers. Each operation and handling stage results into some losses. These post-harvest losses result into decrease in food availability. A recent study showed that 5.8-18.1 per cent fruits, and 6.9-13.0 per cent vegetables are lost during harvest, post-harvest operations, handling and storage in India (Nanda *et al.*, 2012). Thus a huge quantity of agricultural production is reduced from the food chain. Therefore it becomes inevitable to identify the operations and channels where losses are considerable. Improvement in technology in future for these operations and channels will lead towards more availability of produce. The farmer can save his

Table 1: Production of different Commodities in India

Name of Crop/ Commodity	*Production (million tonnes)*		
	2010-11	*2011-12*	*2012-13*
Fruits	71.52	76.42	79.40
Vegetables	134.10	156.33	156.45
Plantation crops	11.93	16.36	16.39
Spices	4.02	5.95	5.79

Source: Directorate of Economics and Statistics, Department of Agriculture and Cooperation, 2013.

valuable produce and get more prices in the market. The reduction in losses in different channels will help in providing the quality produce for the consumers and hence all stakeholders including farmers, marketing persons and consumers will be benefited. Reduction in post-harvest losses of horticulture produce will also be helpful in ensuring nutritional security of the country.

The Ministry of Food Processing Industries (MoFPI) therefore sponsored a project to conduct the nation-wide concurrent repeat study to assess the post-harvest losses of crops and commodities. In order to assess the change in magnitude of losses and to identify the initiatives to be taken in reducing losses, it was imperative to conduct a survey in continuation of the previous study. This paper is therefore a part of accepted report by MoFPI, Govt. of India for major horticultural produce.

1. Concepts, Definition and Assumptions of Losses

Reduction in weight of available amount for human consumption was defined as the quantitative loss. Losses such as quality deterioration, food value, kitchen loss, plate/table loss, loss of goodwill or reputation, seed vigor loss, *etc.* are difficult to quantify, hence were not considered under quantitative loss. Further, it was decided to estimate the post-harvest losses both by enquiring various stakeholders and actual observations in the field. Major assumptions and considerations taken for this study are:

i) The data for harvest and post-harvest losses was collected for one full crop cycle (one crop year) of the selected commodities.

ii) Initial point to start data collection in farm operations was harvesting operation.

iii) In case the crops are grown more than once in a year, the data of farm operations were collected for each harvest in that year.

iv) Multiple picking is common practice in perishables, cottonseed, plantation crops and spices. The data on losses in farm operations for such crops were collected for at least three harvests/pickings.

v) No intervention in the farm practice should occur during data collection.

vi) Actual farm practices should be followed for collection of data by observation.

vii) Data collection for losses during storage was started immediately after the selection of respondents.

viii) Selection of respondents, fields, districts etcwere standard using statistical methods.

ix) Data were collected by trained manpower actually appointed for the purpose.

2. Commodities, their Channels and Unit Operations

Altogether 20 major horticultural cropsfor estimating the quantitative harvest and post-harvest losses in different operations and channels were selected. The crops are listed in different groups as per farm operations, crops such as fruits, vegetables, plantation crops and spices *etc.* Selections of crops/commodities were based on their national production. The farm operations and channels for selected crops/commodities and their extent of coverage in the study are summarized in Table 2.

3. Sampling Design and District Selection

Sampling is a process of selecting a subset of number of respondents of population for a study. This study was planned to estimate the harvest and post-harvest losses of crops/livestock produce at national level. Therefore stratification of the country was carried out on the basis of climatic conditions, agricultural practices and crops grown. Stratification done by Planning Commission of India in form of agro-climatic zones was found to be the most appropriate for this study. The whole country is divided into 15 agro-climatic zones. The island region was not included in the survey as the total contribution in Indian agricultural production from this zone is negligible.

Districts were selected as the sampling unit in the sampling design for further selection of respondents. To estimate the post-harvest losses accurately using sample survey, it is essential to cover at least 10 per cent units of first stage sampling. Hence, total 120 districts were selected from 14 agro-climatic zones (about 20 per cent of the total districts in India, excluding the urban districts where cultivation is not practiced). The number of districts in each agro-climatic zone was proportionately taken after rounded off to the nearest integer.

For selection of respondents to collect the data for assessment of harvest and post-harvest losses, the stratified multistage random sampling method was used. The group of agro-climatic zones was considered as a stratum. Districts, blocks, villages and farmers were taken as first, second, third and fourth stage units, respectively in each stratum.

The blocks and villages in each district were selected randomly. Two additional districts (Sikkim south and Sikkim west) were added for the survey. The selected districts, blocks and villages were allocated to the nearest centres of AICRP on PHET for data collection.

4. Survey Schedule, Sampling Size and Data Collection

Structured schedules developed after numerous deliberations in workshops, meetings and using expert advice were used for in farmers' fields, villages, markets,

Table 2: Farm Operations/Channels and Extent of their Coverage for Selected Crops/Livestock Produce

Operation/Channel	*Extent of Coverage of the Operation*	*Crop(s) Covered*
Harvesting	Plucking of fruits/bunch from tree/plant/vines	Apple, Banana, Mango, Papaya, Sapota, Grapes, Black Pepper, Citrus, Guava, Chili, Arecanut, Coconut, Cashew, Tomato, Cauliflower, Cabbage, Mushroom
	Digging/uprooting of the tubers from soil	Onion, Potato, Tapioca, Turmeric
Collection	Stacking, filling in baskets/bags, and transportation to sorting/grading place	Apple, Banana, Mango, Papaya, Sapota, Grapes, Onion, Citrus, Guava, Arecanut, Coconut, Cashew, Chili, Cauliflower, Cabbage, Potato, Tapioca, Green Pea, Turmeric, Tomato, Cottonseed, Mushroom
Sorting/grading	Separation of material not fit for human consumption due to damage and injury, unripe harvest, removal of soiled portion of mushroom, removal of first layer of cabbage leaves	Apple, Banana, Mango, Papaya, Sapota, Grapes, Citrus, Guava, Chili, Cauliflower, Cabbage, Onion, Potato, Green Pea, Turmeric, Tomato, Mushroom
	Trimming of tubers	Tapioca
Threshing/Dehusking	Separation of seed from plant/pods, removal of husk from nuts	Arecanut, Coriander, Black Pepper, Coconut, Cashew
Winnowing/cleaning	Collection of threshed material, winnowing to remove chaff, dust *etc.*	Black Pepper, Arecanut, Chili, Turmeric
Drying	Collection of material after cleaning, spreading for drying, heaping after drying	Black Pepper, Arecanut, Coconut, Cashew, Turmeric, Chili
Packaging	Collection after winnowing/cleaning/drying/sorting/grading/threshing (in case of use of thresher having blower), filling in the bags/baskets/other packaging material	Black Pepper, Arecanut, Chili, Turmeric, Apple, Banana, Mango, Papaya, Sapota, Grapes, Citrus, Guava, Coconut, Cashew, Cauliflower, Cabbage, Onion, Potato, Tapioca, Green Pea, Tomato, Mushroom
Transporta-tion	Loading of packed material in threshing yard/sorting/grading place, transportation to farmers, store, unloading, transportation from threshing yard/sorting/grading place/store to market yard and unloading at market yard	Black Pepper, Arecanut, Coconut, Cashew, Cottonseed, Turmeric, Chili, Apple, Banana, Mango, Papaya, Sapota, Grapes, Citrus, Guava, Cauliflower, Cabbage, Onion, Potato, Tapioca, Green Pea, Tomato, Mushroom
Storage at farm/ household level	During storage, cleaning/grading before sending to market for sale or own consumption	Black Pepper, Arecanut, Coconut, Cashew, Cottonseed, Turmeric, Chili, Apple, Banana, Mango, Papaya, Sapota, Grapes, Citrus, Guava, Cauliflower, Cabbage, Onion, Potato, Tapioca, Green Pea, Tomato
Storage at godown/ warehouse/cold stores	Unloading, during storage, loading for further sale/disposal	Onion, Black Pepper, Arecanut, Coconut, Cashew
	Unloading, during storage, loading for further sale/disposal (in cold stores)	Chili, Apple, Banana, Papaya, Citrus, Cauliflower, Cabbage, Potato, Green Pea, Tomato
Storage at wholesale level	Unloading, during storage, loading for further sale/disposal	Black Pepper, Arecanut, Coconut, Cashew, Turmeric, Chilly, Apple, Banana, Mango, Papaya, Sapota, Grapes, Citrus, Guava, Cauliflower, Cabbage, Onion, Potato, Tapioca, Green Pea, Tomato
Storage at retailer level	Unloading and loading, during storage, sorting/grading for sale	Black Pepper, Arecanut, Coconut, Cashew, Turmeric, Chilly, Apple, Banana, Mango, Papaya, Sapota, Grapes, Citrus, Guava, Cauliflower, Cabbage, Onion, Potato, Tapioca, Green Pea, Tomato. Mushroom
Storage at processing units	Unloading and loading, during storage of raw material.	Black Pepper, Arecanut, Coconut, Cashew, Chili, Apple, Banana, Mango, Papaya, Grapes, Citrus, Guava, Cabbage, Onion, Potato, Tapioca, Tomato

public and private agencies, godowns, cold storages, and processing units. The sample size for data collection was decided on the basis of standard statistical sampling procedures. Selections of farmers, and respondents in market channels were performed using random sampling method. The sample size for each operation and channel and sampling procedure are described here under different sub-sections.

Farm Operations

Two blocks, in which survey were conducted in previous study, were taken from each selected district (Table 2). Two blocks were selected randomly from every newly added district (Sikkim West, Sikkim South, Ramgarh and Ranchi districts). Then five villages were randomly taken from each block according to the random number allocated to each of them. A random sample of ten farmers was drawn from each village for data collection by enquiry at farm level and two farmers from the list of 10 selected ones for data collection by actual observations.

Storage at Producer Level

Same sample of farmers (as taken for data collection at farm level) was taken for data collection by enquiry and observation at this level.

Storage at Market Level

Two units of each channel such as wholesaler, retailer, godown, and processing unit for each crop/livestock produce were taken randomly from the list of the respondents prepared after complete enumeration of units for each channel of each selected district. In case a particular channel was not available in the selected district then nearby districts for data collection by enquiry/actual observation were considered. The data by enquiry as well as by observation were collected from all selected respondents.

Sampling Procedure

Selection of sampling units was based on random sampling technique without replacement for each crops. The sampling procedure followed for each stage is as follows:

(a) **Selection of blocks in the district:** A list of all blocks of the district was prepared and two of them were selected randomly as laid down in standard statistical survey method.

(b) **Selection of villages:** List of villages falling in the selected block was prepared. The villages, which were not growing the selected crops of the region, were removed from the list and five villages were selected randomly from the remaining list. In some of the cases where villages were big with more than 1500 households, one segment of the village was enumerated and farmers were selected from that list only.

(c) **Selection of farmers**: After complete enumeration of each selected village, the households not related to the identified commodities of the agro-climatic zone were discarded and list of farmers growing or expected to grow the identified crops/commodities in the current survey period were prepared. The farmers were sub-stratified into two categories *i.e.* the farmers growing more than 70 per cent of the selected commodities available in the village (nearest integer number) and farmers growing less than 70 per cent of selected commodities. Random samples of 6 farmers were selected from the first list. Remaining four farmers were randomly selected from the second list. In case the number of farmers in the first list was less than 6, all these farmers were selected and rest of the farmers were taken from the second list.

(d) **Selection of field and plot**: The orchard (A cluster of minimum 12 fruit bearing trees of particular crop on a single piece of land) was demarcated for assessment of losses by observation. Four fruit bearing trees were selected randomly from this demarcated area for harvesting.

(e) **Selection of wholesalers:** A list of market yards/mandies at the district headquarter was prepared and one fruits/vegetables mandi were selected randomly. The market yard/mandi was enumerated and two wholesalers for each commodity were selected randomly from the list. Priority was given to the wholesalers handling more than one crop/commodity.

(f) **Selection of retailers:** A list of main retail markets at district headquarters including the retail fruit and vegetable markets was prepare, and one market for fruits and vegetables were randomly selected and enumerated. Two retailers were selected randomly for each allocated crop giving priority to the retailer handling more than one crop.

(g) **Selection of processing units**: A list of processing units for the identified crops was prepared for each district and two units were selected randomly for each crop. In case the processing unit was not available in the identified districts, units located in neighboring district were taken.

Data Collection

Data were collected both by enquiry and observations using the developed schedule for each farm operations (Jha *et al.*, 2016) from 120 districts. The data received from centers were scrutinized for any discrepancies and errors during data collection and entry. Wherever there were inconsistencies, possible corrections were made after referring records in the filled schedules of concerned centers and revalidating the same by resending to field. If it was not possible to rectify the errors/discrepancies, data were discarded.

The scrutinized data of enquiry method were analysed using Statistical Analysis Software (SAS) whereas data obtained by observation method were analysed using advance version of Microsoft Excel at district level and results were pooled by assigning appropriate weights at agro-climatic zone levels. In this analysis, sampling weights were obtained for each record according to sampling design implemented for data collection at district level (*i.e.* weightage of sample, no. of farmers, villages and blocks to their actual number).

For estimating the losses at agro-climatic zone level, weightage was assigned based on the production of the specific crop/commodity in all the sampling districts, obtained separately from the state report. Similarly, post-harvest losses at the national level were estimated by assigning weightage on the basis of the production of a specific crop/commodity in all the agro-climatic zones.

The post-harvest losses of major fruits, vegetables, plantation crops and spices are presented in Tables 3–5.

Survey for assessment of losses in apple was conducted in only one agro-climatic zone *i.e.* western Himalayan regions comprising of Jammu and Kashmir, Himachal Pradesh and Uttarakhand, the only apple

Table 3: Harvest and Post-harvest Losses in Percentage of Fruits at National Level

Crop	Operations					Total Loss in Farm Operations	Storage Channels					Total Loss in Storage	Overall Total Loss
	Harvesting	Collection	Sorting/ Grading	Packaging	Transport		Farm	Godown/ Cold Store	Wholesaler	Retailer	Processing Unit		
Apple	4.33±0.35	0.29±0.22	3.94±0.44	0.11±0.11	0.42±0.19	9.08#±0.30	0.02±0.00	0.13±0.02	0.57±0.08	0.34±0.29	0.25±0.11	1.31±0.11	10.39#±0.24 (12.26)
Banana	1.62±0.35	0.26±0.14	2.06*±0.37	0.19±0.31	1.91±0.40	6.04±0.33	0.03±0.01	0.08#±0.04	1.16±0.35	0.45±0.07	0.00±0.00	1.72±0.17	7.76±0.29 (6.60)
Citrus	1.68±0.48	0.33±0.10	3.71±0.59	0.18±0.13	1.65±0.24	7.55*±0.38	0.04±0.01	0.02*±0.01	0.91±0.13	1.12±0.20	0.06±0.07	2.14*±0.11	9.69*±0.29 (6.38)
Grapes	1.77±0.2	0.30±0.05	3.36±0.42	0.10±0.06	0.98±0.18	6.52±0.26	0.01#±0.00	–	0.78±0.14	1.24±0.17	0.09±0.03	2.11±0.15	8.63±0.22 (8.30)
Guava	5.33±1.78	0.31±0.28	4.95±1.86	0.09±0.20	1.21#±0.64	11.90±1.24	0.23±0.06	–	1.62±0.52	2.08±0.73	0.04#±0.03	3.98±0.55	15.88±1.11 (18.04)
Mango	2.09±0.71	0.30±0.44	3.26±0.77	0.23±0.07	1.04±0.14	6.92#±0.54	0.11±0.01	0.01±0.01	0.69±0.39	1.18±0.42	0.25±0.09	2.24±0.28	9.16#±0.50 (12.74)
Papaya	0.98±0.45	0.42±0.15	1.46±0.47	0.34±0.06	0.92±0.39	4.12±0.37	0.05±0.02	0.01*±0.00	0.79±0.17	1.71±0.25	0.03*±0.01	2.58±0.12	6.70±0.26 (7.34)
Sapota	2.53±1.18	0.35±0.07	2.55±0.56	0.28±0.17	1.70±0.48	7.41*±0.67	0.01±0.01	0.25*±0.05	0.89±0.20	1.13±0.13	0.03*±0.02	2.31*±0.09	9.73*±0.46 (5.77)

* Estimated losses are significantly higher in comparison to losses observed in 2005-07 (previous study);

Estimated losses are significantly lower in comparison to losses observed in 2005-07(previous study).

Figures in parenthesis represent the losses of previous study 2005-07.

Table 4: Harvest and Post-harvest Losses in Percentage of Vegetables at National Level

Crop	*Operations*					*Total Loss in Farm Operations*	*Storage Channels*					*Total Loss in Storage*	*Overall Total Loss*
	Harvesting	*Collection*	*Sorting/ Grading*	*Packaging*	*Transport*		*Farm*	*Godown/*	*Wholesaler*	*Retailer*	*Processing*		
Cabbage	1.74±0.55	0.38±0.16	3.32±0.46	0.36±0.06	1.02±0.49	6.81±0.42	0.16±0.06	0.08±0.02	0.89±0.15	1.42±0.35	0.02±0.01	2.56±0.21	9.37±0.36 (6.94)
Cauliflower	2.21±0.79	0.26±0.38	3.78±0.48	0.38±0.13	0.92±0.49	7.55*±0.52	0.09±0.07	0.07±0.05	0.83±0.25	1.00±0.38	0.00±0.00	2.00±0.22	9.56*±0.45 (6.88)
Green pea	2.25±0.46	0.32±0.12	2.41±0.47	0.13±0.09	0.61±0.08	5.72±0.32	0.05±0.02	0.00*±0.00	1.09±0.27	0.55±0.34	0.03*±0.03	1.73±0.21	7.45±0.28 (10.28)
Mushroom	0.99±0.14	0.04±0.00	5.34±0.33	0.18±0.06	0.77±0.06	7.32±0.19	0.66±1.22	–	–	1.52±1.66	–	2.19±1.45	9.51#±0.65 (12.54)
Onion	2.62±0.25	0.44±0.32	2.35±0.72	0.12±0.27	0.51±0.15	6.05±0.40	0.35±0.19	0.30±0.08	0.77±0.16	0.72±0.14	0.01*±0.01	2.16±0.17	8.20±0.28 (7.51)
Potato	2.58±0.84	0.25±0.30	2.93±0.99	0.06±0.07	0.72±0.18	6.54±0.63	0.15±0.06	0.17±0.18	0.34#±0.08	0.11±0.05	0.02±0.01	0.78#±0.06	7.32±0.44 (8.99)
Tomato	3.16±0.53	0.52±0.22	3.74±0.48	0.24±0.14	1.75±0.29	9.41±0.39	0.12±0.04	–	1.26±0.20	1.63±0.33	0.02±0.00	3.03±0.23	12.44±0.35 (12.47)
Tapioca	1.23±0.39	0.30±0.07	0.99±0.16	0.09±0.05	0.61±0.18	3.22#±0.24	0.28±0.23	–	0.31±0.06	0.59±0.20	0.17±0.02	1.36±0.20	4.58#±0.23 (9.77)

* Estimated losses are significantly higher in comparison to losses observed in 2005-07 (previous study);

Estimated losses are significantly lower in comparison to losses observed in 2005-07(previous study).

Figures in parenthesis represent the losses of previous study 2005-07.

growing region in India. Total loss in farm operations was 9.08 per cent owing to harvesting, sorting/grading and transportation operations. The practices of farm operations are different in different states. In Jammu and Kashmir, the farmers harvest the fruits and pile them in temporary structure made in the orchards. The vehicle for transport is called thereafter which takes 3-10 days. In mean time sorting and packaging of fruits are carried out. Long duration keeping in fields caused higher loss in sorting operation. In Himachal Pradesh and Uttarakhand the farmers usually harvest the apple after confirming vehicle availability for transport. In some cases unripe fruits are also plucked, which leads to higher loss thereafter. There is a need to synchronize various operations properly to curtail these losses. It is important to note that the loss during transportation of apple has been reduced significantly in comparison to previous study. This probably happened due to improved roads and transport systems in the past few years. Total loss during storage was 1.31 per cent, which was slightly higher than that of previous study (1.21 per cent). The increase in storage loss was mainly due to higher losses at retailer level storage. Overall total

Table 5: Harvest and Post-harvest Losses in Percentage of Plantation Crops and Spices at Bational Level

Crop	Operations								Total Loss in Farm Operations	Storage Channels					Total Loss in Storage	Overall Total Loss
	Harvesting	Collection	Sorting/ Grading	Thre-shing	Winnowing/ Cleaning	Drying	Packaging	Transport		Farm	Godown	Wholesaller	Retailer	Processing Unit		
Arecanut	1.24±0.35	0.39±0.10	–	0.71#±0.27	1.19*±0.35	0.19±0.15	0.05±0.01	0.17±0.06	3.94#±0.23	0.02±0.00	–	0.48±0.31	0.10#±0.06	0.36±0.5	0.97#± 0.15	4.91#±0.22 (7.87)
Black pepper	0.47±0.19	0.21±0.10	–	0.23±0.11	0.02±0.10	0.04±0.04	0.01±0.00	0.00±0.00	0.99#±0.11	0.01±0.00	–	0.00#±0.00	0.18±0.14	–	0.20±0.07	1.18#±0.11 (3.86)
Cashew	1.45±1.59	0.57±2.23	–	1.34*±0.53	0.30±0.76	0.07±0.08	0.00±0.00	0.07±0.04	3.82*±1.12	0.00±0.00	0.00±0.00	0.14±0.11	0.03±0.03	0.17±0.07	0.35*±0.07	4.17*±0.84 (1.12)
Chili	1.60±0.34	0.84±0.09	2.18±0.59	–	–	0.02±0.07	0.15±0.05	0.30±0.36	5.11±0.36	0.03±0.00	–	0.99±0.24	0.31#±0.12	0.06±0.02	1.40±0.14	6.51±0.28 (5.60)
Coconut	1.37±0.27	0.20±0.14	–	1.02*±0.32	0.37*±0.10	0.36±0.13	0.08±0.05	0.05±0.13	3.45±0.23	0.08±0.01	–	0.61±0.23	0.25±0.10	0.38±0.05	1.32±0.07	4.77±0.20 (5.36)
Turmeric	2.41±0.28	0.10±0.19	0.79±0.17	–	–	0.16±0.13	0.09±0.04	0.04±0.29	3.60±0.21	0.09±0.03	–	0.62±0.05	0.06±0.05	0.06*±0.01	0.84±0.04	4.44±0.15 (7.37)

* Estimated losses are significantly higher in comparison to losses observed in 2005-07 (previous study).

Estimated losses are significantly lower in comparison to losses observed in 2005-07 (previous study).

Figures in parenthesis represent the losses of previous study 2005-07.

Table 6: Estimate of the Monitory Value of Harvest and Post-harvest Losses in India at Production of Year 2012-13 and Prices of 2014

Crop/ Commodity	Production (million tonnes)	Price (Rs./tonne)	Overall Total Loss (%)	Monitory Value of the Losses (Rs. crores)
Apple	1.90	68078	10.39	1341
Banana	27.06	18601	7.76	3903
Citrus	11.47	14011	9.69	1557
Grapes	2.52	44564	8.63	969
Guava	2.62	20628	15.88	858
Mango	17.29	45355	9.16	7186
Papaya	5.19	16023	6.70	557
Sapota	1.50	18770	9.73	273
Cabbage	8.53	10928	9.37	874
Cauliflower	7.79	16321	9.56	1214
Green Pea	3.87	33698	7.45	971
Mushroom	0.04	119049	9.51	46
Onion	16.66	16920	8.20	2312
Potato	41.09	16649	7.32	5008
Tomato	17.85	16510	12.44	3666
Tapioca	7.32	22436	4.58	751
Arecanut	0.53	182865	4.91	475
Black pepper	0.05	570547	1.18	35
Cashew	0.75	76026	4.17	239
Chilli	1.31	64411	6.51	547
Coconut	15.09	28587	4.77	2058
Turmeric	0.98	24845	4.44	108
Grand Total				**34948**

Table 7

Operation/Action points	*Description and Strategies*
Pre-harvest planning	1. Decide purpose of harvesting and marketing place before actual harvest. 2. Accordingly maturity stage suitable for the purpose of harvest should be determined using various developed technique.
Safe harvesting	1. Highest portion of loss is during or due to faulty harvesting time, methods and means. 2. As far as possible horticultural crops should be harvested in early morning or if it is late should be pre-cooled immediately. 3. Harvesting tools, gadgets and fruit savers developed should be used. 4. Awareness programme/trainings are must for farmers for safe harvesting
Handling and Transportation	1. Awareness programmes for farmers that how they should handle fruits and vegetables should be conducted. 2. Handling in bags must be stopped, instead crates, CFB boxes *etc.* should be encouraged. 3. Ventilated/cooled containers should be used for long distance transport. 4. Good road and rails facilities should be developed to avoid damage of fruits/fruits due to excessive vibration.
Storage/retailing	1. Before storing and fruits/vegetables they should be cleaned thoroughly and sorted. All injured/cut/rotten *etc.* must be removed before storing. Primary processing is must. 2. Storage of fruits and vegetables should be done in cool and high humid conditions. 3. Storage facility should be built-up at farm. 4. The developed formulae should be used to know the shelf-life of produce in the store in advance so that they can make strategies for selling before rotting. 5. Cold chain of each commodity should be complete from Farm to Fork. 6. Shelling high perishable products should legally banned.
Processing-Conversion to high self-life products	1. Intensive training should be conducted on secondary and tertiary processing of surplus produce in production catchments. 2. Establishment of pilot and commercial plants for high value and low volume products should be supported.
Development of new varieties	1. Horticulturists are usually concerned for development of variety for increasing production. They should focus on development of climate and injury resistant cultivars. 2. Focus should be on development of cultivars which are machine harvestable. Dwarf varieties of uniform and homogeneous canopy will helpful in purposeful mechanical harvesting. 3. Boundary conditions for purposeful varietal development with the help of process engineers should be fixed before start of breeding/selection process.

loss in apple was found to be 10.39 per cent, which was lower than that of previous study (12.26 per cent) but the difference was found to be statistically non-significant.

Economic Value of Harvest and Post-Harvest Losses

The monitory value of the losses was estimated at national level. Base year of this study was 2013-14 hence the all India production of year 2012-13 for selected crops and commodities were taken to estimate the quantity lost. Average wholesale prices of each crop at national level for the year 2013-14 was taken to calculate monitory loss. The calculated economic values of the loss for each selected commodity have been presented in Table 6.

The economic value of quantitative loss of 22horticultural crops was found to be in the tune of Rs. 34948crores at average annual prices of 2014. The economic loss from all 45 commodities (not shown here) is Rs. 92651 croreagainst the value of Rs. 44143 crore at 2008-2009 prices. The highest contribution (34 per cent) towards economic loss was from horticulture produce sector (fruits and vegetables). Soft texture and high water content which make these commodities highly perishable and thus make them difficult to handle and store. Thus, fruits and vegetables need immediate attention from farm to fork chain.

Loss Mitigation Strategies

It is evident from our data and many other studies that post-harvest losses are huge and altogether hampering the economy, natural resources are being exploited to produce more and causing environmental impact. It is therefore necessary to prepare a matrix to reduce post-harvest loss to bare minimum by keeping zero tolerance towards post-harvest losses. A mitigation strategy is presented in Table 7.

References

Jha, S.N., Viswakarma, R.K., Ahmed, T., Rai, A. and Dixit, A.K. 2015. *Report on Assessment of Quantitative Harvest and Post-harvest Losses of Major Crops and Commodities in India.* ICAR-All India Coordinated Research Project on Post-harvest Technology, CIPHET, Ludhiana.

Nanda, S.K., Vishwakarma, R.K., Bathla, H.V.L., Rai, A. and Chandra, P. 2012. *Harvest and Post-harvest Losses of Major Crops and Livestock Produce in India.* All India Coordinated Research Project on Post-Harvest Technology, (ICAR), Ludhiana.

69

Harnessing Profit through New Innovations in Storage of Horticultural Produce

R.K. Pal

ICAR-National Research Centre on Pomegranate,
Kegon, Solapur – 413 255, Maharashtra
E-mail: rkrishnapal@gmail.com

With the rapid increase in production of horticultural crops surpassing the total cereal production in India, harnessing profit of farmers growing horticultural crops is an emerging challenge due to their high degree of perishability, glut in the market and availability of poor post-harvest infrastructure for handling, packaging and storage in India. Farmers face several problems with high production and productivity particularly of horticultural crops and recent severe fall in the price of onion in retail market is a glaring example. In order to cope up with this situation more and more innovation are needed. First of all the growers should be empowered with the technical knowledge on very basic aspects of the crop physiology after harvest. As a layman many of them do not understand the gravity of dealing with a live commodity after harvest. Therefore, just putting the harvested produce in the store or keeping those in adverse conditions is not at all desirable even the expenses incurred in producing the crop. Therefore, growers engaged in horticulture as their livelihood should be made smart enough with the appropriate directives/advisory and skills for harnessing profit from the harvested produce. Post-harvest management to many people is poorly understood with the idea of taking the harvested produce to the nearby wholesale market and sell those as per the wishes of the middlemen or traders. Instead cooperative movement of creation of community pack houses should be encouraged for better return of the harvested produce.

Key Issues Need to be Understood

1. Type of the Harvested Produce

It is very important to know your produce

i) Whether it has thick skin or thin skin (peel)
ii) Whether it is vulnerable to slightest rough handling
iii) Whether it needs any crucial post-harvest operations before marketing
iv) Whether it needs any special treatment to fetch better price
v) Whether the produce quality will drastically be reduced if not marketed immediately

All the above questions need to be thoroughly analyzed before taking any hasty decision. Although our farmers are not so enlightened and many of them cannot wait long due to their sole dependence on the earnings from the sale of the harvested produce. At this juncture appropriate community/cooperative support shall be highly beneficial to these farmers.

2. Availability of Infrastructure Facilities

The availability of post-harvest infrastructure facilities in India is disproportionate with respect to locality and regions. Connecting road from farm yard to nearest market yard is one of the major constraints. Next important is the availability of storage facilities. There are tremendous scope of innovation in creation of low cost infrastructure facilities. For example, most of the horticultural commodities after harvest carry out high physiological and metabolic activities: some are desirable and most of them are undesirable. High respiratory activity of these produce cause rapid deterioration and invite pathogen invasion. However, it is a proven fact that keeping the harvested produce under cool condition can reduce the respiratory activity. Cool condition around the produce could be maintained by simple use of evaporative cooling principle. Even simple well ventilated sheds equipped with few desert coolers could be sufficient to withhold the produce for short term before marketing.

Again if we take the example of another important crop *viz.*, Onion and Garlic. Drying of outer scales is the most important factor that contributes to the maximum extent in minimizing post-harvest loss. A simple technique of stacking the harvested onion and garlic in perforated plastic crates and allowing cross flow of dehumidified air through the produce can very quickly dry the outer skin, thicken the neck and make the produce more tolerant to invasion by pathogen.

Simple grading of harvested produce as per the market need can fetch better income of the farmers. Old government concept of community canning can be modified similar to creation of new mega food parks or utilization of existing mega food parks. A consolidated effort of both scientific and extension personnel is the need of the hour to create mass awareness. Establishing FPO (Farmers' Producers Organization) by the farmers engaged in growing of horticulturalcrops is one step forward in harnessing the profit. However, implementation of total utilization concept for end-to-end solution would be the key mantra for success of FPO. For instance a progressive farmer can produce approximately 60 per cent of his/her produce that match the premium quality either for export or for high end domestic market. However, he/she will be always bothered by selling the B or C grade fruits at a throw away price. Here lies the importance of total utilization. For example in case of pomegranate a large number of high value products can be made using B or C grade fruits for production of wine, seed oil, bio-colour and organic mouth wash instead of selling those at a throwaway prices.

Now coming to innovation in storage with respect to availability of infrastructure, most of the refrigerated storages (cold stores) are having old technology and does not cater to the need for commodity-specific storage requirement. Hence, using the modern refrigeration system it is possible for multi-chamber and multi-temperature cold stores in the same premises. Further, with the advancement on the recent development of Controlled Atmosphere Storage units it is possible to develop low cost CA/MA units using appropriate scrubbing system for ethylene and CO_2. The role of ethylene should be properly understood for smart post-harvest management. It is both beneficial and detrimental to horticultural produce in many ways.

Use of pre-treatments are also very important for fetching better return from the stored produce. Although several recommendation are available on the use of edible surface coating on extending the marketability of horticultural produce, yet in most of the cases use of these surface coatings on removal from storage before marketing can certainly fetch better price and farmers can earn more profit. This practice is commercially followed all over the world in case of apples and pears.

In the end some of the innovative means are described below that can harness the profit for storage of horticultural produce.

A. Use of Eco-friendly Packaging Material

Biodegradable plastics are those, which can be decomposed into CO_2 and H_2O by naturally occurring microorganisms. Different types of biodegradable plastics are (i) starch based plastic; (ii) polymers synthesized by microorganisms ex. Polysacchrides (Pollulan, Curdlan, Biocellulose, Chitosan); (iii) Polyesters (Polyhydroxyalkanote, Polyhydroxybutyrates and their co-polymer compounds) and (iv) Polyamino acids (Polyester, polyvinyl chloride). The biodegradable laminate was considered suitable as a packaging material for storage of fresh produce.

Use of a biodegradable laminate of a chitosan-cellulose and polycaprolactone as a film for modified atmosphere packaging (MAP) of fresh produce (shredded lettuce and shredded cabbage, cut broccoli, whole broccoli, tomatoes, sweetcorn and blueberries) have been standardized where the gas composition in each biodegradable package including the fresh produce was simulated to be close to the optimal composition.

Photodegradable Plastics

Plastics obtained by giving photo-degradation to main chains of polymers by adding additives that are light sensitive. Other group contains those obtained by introducing carbonyl base as light sensitive base to the main chain of polymer.

Edible Films and Coating

Edible films and coating are traditionally used to improve the appearance and conservation of fruits and vegetables. Polysaccharides, proteins and lipids can be used as edible films or coatings.

"Net Film" is another new innovation to provide quality packaging using LLDPE. The designed holes on its surface in various dimensions make it ideal for packaging of fruits, vegetables, and flowers. Because of its sticky property, the Net Film has a smaller volume on unwrapping than other kinds of similar products and also it does not tangle in the wheel since it changes into a ball, as it is unwrapped. It is also elastic with a tendency to contract and, therefore, hold of the goods on the pallet during their transport and storage. It is very strong, prevents condensation of the packaged product because of the ready-made holes, the humidity is allowed to leave thus allowing the packaged products to breathe.

B. CA Storage

Technology has been standardized for controlled atmosphere storage of several fruits. The technique involves the utilization of specific composition of O_2 and CO_2 in the storage chambers. Usually O_2 below 8 per cent and CO_2 above 1 per cent are used. Exposure of fresh fruits and vegetables to low O_2 and elevated CO_2 reduces the respiration/ethylene evolution rate of the commodity and thereby increases the storage or shelf-life. For example, apples can be stored in CA storage for about 8 months without significant loss in texture and other quality attributes.

C. Hypobaric Storage

In hypobaric or low pressure (LP) storage systems, the produce is kept under a reduced pressure (200 mm Hg) as O_2 concentration is reduced proportionately to the atmospheric pressure. Research has shown that LP has been very useful in the storage of mango, avocado, banana, and papaya. The system was commercially used for the transportation of several commodities but is comparatively costly.

D. Mobile Cool Chamber for Retail Vending

Thermal management system using phase changing materials could be very well used for development of mobile cool chamber. For short distance transportation of fruits and vegetables, a mobile cool chamber consisting of an insulated box with TMS could be used. This box can hold 8 plastic crates in two layers each. About 100 kg vegetables or fruits can be transported in this mobile cool chamber.

70

Ethylene Management for Effective Post-harvest Handling of Horticultural Produce

Ram Asrey*, V.R. Sagar and R.R. Sharma

Division of Food Science and Post-harvest Technology
ICAR-Indian Agricultural Research Institute, New Delhi – 110 012
****E-mail: ramasrey@iari.res.in***

Fruit ripening is an important natural physiological process which leads to all sensory and biochemical changes, these phenomena are place required by consumers to reach them eating soft stage. The quality, safety and shelf-life after ripening depend on pre harvest factors (Madani *et al.*, 2014; Ferreyra *et al.*, 2016), ripening agents (Ram *et al.*, 2009; Das and Balamohan, 2013), methods and storage environment (Ahmed *et al.*, 2001). On the basis of ripening behaviour, fruits are classified as climacteric and non-climacteric fruits. Fruits undergo a pronounced increase in respiration coincident with ripening. Such an increase in respiration is known as respiratory climacteric, and group of fruits is classed as climacteric *e.g.* mango, banana, papaya, apple, plum, apricot *etc.*, whereas, non-climacteric does not exhibit respiratory climacteric *e.g.* citrus, pineapple, strawberry *etc.* The ripening process is the result of complex of changes, many of them probably occurring independently of one another (Wills *et al.*, 2007). These changes that together make fruit ripening are seed maturation, colour change, abscission, change in respiration rate, change in rate of ethylene production, change in tissue permeability and cellular compartmentation, softening, organic acid change, production of flavour, development of wax on skin (Pratt, 1975). Induction of ripening of fruits is old age practice and traditionally various methods were used like smoke, use of paddy straw, newspaper *etc.* In all these methods, the rate of ripening was not uniform and others problems were also related to this. Then traders/ growers shifted towards the use of various chemicals which has become questionable because of various health risks due to heavy metal toxicity and impact on fruit quality as well.

1. Traditional Methods of Ripening

1.1 Use of Smoking Chambers

This method is very common for ripening of banana and mango (Ram *et al.*, 1979). Smokes are produced inside the chambers which released ethylene, acetylene and other gases resulting trigger the ripening process and changes colour of surface. But this method is not suitable for commercial level and sometimes it may hazardous and produces uneven ripening and off flavored fruits due to evolution of toxic gases like CO, acetylene *etc.*

1.2 Use of Paddy Straw and Banana Leaves

This type of practices is mainly followed in rural areas where commercial facilities of ripening are not available. Mango and banana are common fruits which are ripened by this method. In this method cushioning material and wrapping material help to conserve heat and there is release of ethylene by fruits. By the autocatalytic reaction of ethylene fruit ripen. But this method takes long time and uneven ripening.

1.3 Use of Newspaper and Jute Bags

Particular commodity wrapped with newspaper, jute bags or may be other paper based air tight material and place in a chamber. Existing ethylene evolution rate become faster due to heat at room temperature and produce get ripened.

2. Use of Chemicals

2.1 Use of Calcium Carbide

Calcium carbide is a organic chemical compound which is used for ripening the fruits like mango, papaya, apple avocado, peach, pear, avocado, sapota and banana and widely used with name of *"masala"*. On the hydrolysis it produces acetylene which provides same effect as ethylene (Siddiqui and Dhua, 2010). It is used in steel industries and in automobiles for welding purpose. It is very cheap source and easily available. Reaction of calcium carbide with water produces calcium and ethylene

$$CaC_2 + 2H_2O \rightarrow Ca\,(OH)_2 + C_2H_2$$

2.2 Effects of CaC_2 on Fruit Quality

Fruits ripened with CaC_2 are overly soft and less tasty. They also have a shorter shelf-life. An artificially ripened fruit would present a yellow outer skin, but the tissue inside would not be ripe or itself remains green and raw. Thought mangoes ripen quickly (two days), they cannot be stored for more than two days. When CaC_2 is used on raw fruit, the amount of the chemical needed to ripen the fruit has to be increased. This results in the fruit becoming even more tasteless and possibly toxic.

Figure 1: Making CaC_2 Sachets.

It was observed that papaya fruit showed poor flavour when ripened with calcium carbide compare to ethrel and ethylene treatment (Jayawickrama *et al.*, 2001). Calcium carbide has many health hazards and having carciogenic properties due to traces of arsenic and phosphorous hydride (Rahman *et al.*, 2008; Jayawickrama *et al.*, 2001).

2.3 Ban on use of CaC_2

As per food safety and standard regulations, 2011, 2.3.5 of food safety and standard regulations (Prohibition and restriction on sales)Regulations 2011- Prohibition of use of carbide gas in ripening of fruits. No person can sell and offer or expose for sale or have in his premises for the purpose of sell under any description, fruits which has artificially ripened by the use of acetylene gas, commonly known as carbide gas.

Figure 2: CaC_2 Sachet inside Mango Box.

Those convicted under this Act could face imprisonment for three years and a fine of Rs. 1000. But there are hardly any cases where the traders or retailers have been booked for accelerating ripening by the use of harmful chemicals.

2.4 Harmful Effect of Calcium Carbide

Use of calcium carbide as ripening agent is hazardous because it contains traces of arsenic and phosphorous which may affect the different body organs and causes various health problems like headache, dizziness, mood disturbances, sleepiness, mental confusion, memory loss, cerebral edema, seizures and prolonged hypoxia (Ashraf-Ur-Rahman *et al.*, 2008; Siddqui and Dua, 2010; Asif, 2012). Temi-Sogo *et al.* (2014) reported the highest quantity of metals, *i.e.* Pb, Cu, Zn and Mn in calcium carbide ripened banana and the lowest in the control. The use of calcium carbide is harmful not only to the mankind but researchers proved that it is dangerous to other flora and fauna also. Harmful effect of calcium carbide was well documented in heterotrophic microorganisms (Peck, S.B., 1969; Lavoie, K.H., 1980). Calcium carbide waste causes a decrease in the population of nitrifying, phosphate solubilizing, heterotrophic, cellulolytic and lipolytic bacteria and also decrease the activities of dehydrogenase, acid

Table 1: Effect of Calcium Carbide on various Organisms

Organism	*Consequences*	*Reference*
Bacteria	Decrease in the population of nitrifying bacteria, phosphate solubilizing bacteria, total heterotrophic, cellulolytic and lipolytic bacteria	Ibe *et al.* (2014)
Fungus	Act as fungicidal leading to cellular physiology, growth retardation, death, general loss of biodiversity and reduction of ecosystem processes	Ihejirika *et al.* (2014)
Fish	Carbide reacts with liquid in the gut to form a large quantity of acetylene gas. The inflation of the gut leads to death of the fish.	Huston (1955)
Enzymes	Decrease in the activities of dehydrogenase, acid phosphatase, alkaline phosphatase, urease, cellulase and lipase enzymes	Ibe *et al.* (2014)
Human	Headache, dizziness, mood disturbances, sleepiness, mental confusion, memory loss, cerebral edema, seizures and prolonged hypoxia	Ashraf-Ur-Rahman *et al.* (2008); Siddqui and Dua (2010); Asif (2012)

phosphatase, alkaline phosphatase, urease, cellulase and lipase enzymes (Ibe *et al.*, 2014).

2.5 Precautions for Minimizing CaC_2 Risks

i) While purchasing fruits like mango, papaya, banana *etc.*, one should not select those that are homogenously ripened and with eye catching bright colours.

ii) Banana can be identified if the stem is dark green whereas the fruits are all yellow, it indicates application of CaC_2.

iii) Wash and peel the fruits before eating, it minimizes CaC_2 ill effect.

iv) Before eating mangoes and apples, it is better to cut the fruit into pieces, rather than consuming them directly. This facilitates escaping of harmful gases.

v) Do not buy fruits when they arrive in the market before the due period (before season) because these are forcefully ripened with higher doses of CaC_2.

3. Use of Acetylene Gas

Acetylene gas is not a natural plant hormone but it is produced by hydrolyzing of calcium carbide. It contains many toxic chemicals and causes many health hazards. It is also banned by the PFA rules. But its limited use by the farmers and vendors is allowed in India, However, the commercial application of acetylene by the corporate and retail chains is banned.

4. Healthy Practices of Fruit Ripening

4.1 Application of Ethylene and Ethylene Based Safe Chemicals

Ethylene is a fruit ripening phytohormone. The amount of endogenous ethylene produced by plant tissue generally increases auto catalytically at a specific stage of growth or development to initiate a physiological response. Application of exogenous or supplemental ethylene to a plant tissue prior to the autocatalytic increase will generally initiate the physiological response as well as the increase in ethylene production (Watada., 1986). Now a day's ethylene is widely used in fruit ripening at commercial level. Mahajan *et al.* (2010) reported that banana fruits treated with ethylene gas @ 100 ppm or ethephon @ 500 ppm resulted in uniform ripening with uniform colour, pleasant flavour, desirable firmness and acceptable quality and better shelf-life. Ethylene has numerous effects on the growth, development and storage life of many fruits, vegetables and ornamental crops at very low concentrations ml µl/lit. concentrations, (Table 2). However, important differences can be found in ethylene sensitivity among the species of fruit and vegetables (Saltveit, 1999).

Table 2: Ethylene Production and Sensitivity of different Horticultural Commodities

Commodity	*Ethylene Production*	*Ethylene Sensitivity*
Climacteric fruit		
Apple, kiwifruit, pear, cherimoya	***	*** (0.03-0.1 µl l^{-1})
Avocado, cantaloupe melon, passion fruit	***	**(>0.4 µl l^{-1})
Apricot, banana, mango	**	***(0.03-0.1 µl l^{-1})
Nectarine, papaya, peach, plum, tomato	**	**(>0.4 µl l^{-1})
Vegetables and non-climacteric fruit		
Broccoli, brussels sprouts, cabbage, carrot	*	***(0.01-0.02 µl l^{-1})
Cauliflower, cucumber, lettuce, persimmon	*	***
Potato, spinach, strawberry	*	***
Asparagus, bean, celery, citrus, eggplant	*	** (0.04-0.2 µl l^{-1})
Artichoke, berries, cherry, grape, pineapple	*	*(>0.2 µl l^{-1})
Pepper		*

* Low, ** Medium and *** High ethylene production or sensitivity.

Adapted from Martinez-Romero *et al.*, 2007)

4.2 Ethylene based Temporary Structure

In one of the simple and harmless techniques, 10 ml of ethrel and 2 g of sodium hydroxide pellets are mixed in five litres of water taken in a wide mouthed vessel.

This vessel is placed inside the ripening chamber near the fruits and the room is sealed air tight. About a third of the room is filled with fruits leaving the remaining area for air circulation. Ripening of fruits takes place about 12 to 24 hours. Fruits are further allowed to store for about 12 hours without any treatment in order to get brighter color on peel surface.

4.3 Ethylene Based Chambers

Artificial ripening of climacteric fruits generally carried out in the ethylene based chambers having different capacity.

For better and uniform ripening of fruits by ethylene require some optimum conditions like:

- Temperature- 18 to 25°C
- Relative Humidity- 90-95 per cent
- Ethylene concentration 10-100 ppm
- Duration of treatments- 24-72 h depends upon commodity
- Air circulation
- Ventilation

4.4 Ripening Chambers in India

Fruit ripening chamber may consist different ripening chambers with or without pack house. In many chamber, 2-3 ripening rooms are sufficient for no of cycles of ripening because single room create problems if raw material in huge and market share is high. According to protocol standards for ripening chambers by National Horticulture Board (NHB, 2011) there are two types of ripening chambers are more prevalent in India. In first type structure of ripening chamber, ethylene equipment added with an insulated cold chamber. For maintaining desire temperature and relative humidity inside the chamber ceiling mounted fin coil evaporators will be connected to the condensing unit outside. In this system air circulation is maintained with modifying air flow through the arranging of fruit crates. For ventilation, a proper ventilation system is operated either manually or automatically. In manual system, door will be opened for 20 min at every 12 h after 24 hrs of ripening. In automatic ventilation, fans are provided for exchange of fresh air.

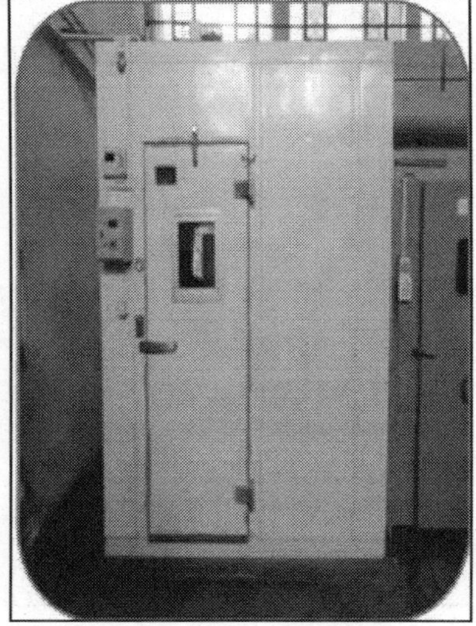

The second type ripening chamber is recommended by committee on standards and protocols of ripening chambers in India. This type of ripening chambers consist a special type of air flow system which creates optimum static pressure inside the chamber. The insulated cold rooms have a system of false ceiling, separated and sealed annular space between wall and palletized crates/CFB boxes with or without air inlet locking system to isolate designated pallets *etc.* Cool air is routed through false ceiling in to boxes with perforated holes of crates/CFB boxes for air circulation) which, in turn are stacked in single/multi-tier system. In case of multi-tier stacking, fork lift operation is necessary to handle the produce.

Table 3L Ideal Conditions for Ripening of Horticultural Commodities

Produce Details	*Ethylene Conc. (ppm)*	*Ethylene Exposure Time (h)*	*Ripening Temp. (ºC)*	*Storage Temp. After Ripening (ºC)*
Mango	100	24	20-22	10-13
Banana	100-150	24-48	15-18	13-14
Papaya	100	24-48	20-25	≈7
Pear	100-150	24-72	18-22	≈ 0
Tomato	100-150	24-48	18-20	12.5

Source: Abdel A. Kader, Department of Plant Sciences, University of California, Davis, CA http://post-harvest.ucdavis.edu.

5. Method of Application

Before the application of ethylene for fruit ripening, room should well ventilate. If the room is well sealed than open the doors and about 30-40 minutes for proper ventilation. In large commercial room ventilators are fitted to exchange the air. Mainly two methods are more common for applying ethylene inside the chamber (Kader. 1992).

5.1 Shot System

In this system a shot of measured quantity of ethylene is introduced into the chamber at regular interval. The shots may be applied by weight or by flow using a gauge that registers the discharge of ethylene in cubic feet per minute. In this method optimum temperature maintains and air circulation is important considerations for better ripening of fruits.

5.2 Trickle or Flow through System

In this method ethylene is continuously introducing inside the chamber. Care should be taken during regulation by this method because flow of ethylene is very small. This is done by lower down the pressure using a two stage regulator and passing gas into room through a metering valve and flow meter. Air circulation is important inside the room to prevent the higher CO_2 and C_2H_4. For this exhaust port and fan are attached with wall of room.

Ethylene response could be considered as detrimental for a wide range of fruit and vegetables, including acceleration of physiological disorders, abscission, bitterness, toughness, sprouting, colour changes, softening, over ripening, and off-flavours (Table 4).

Table 4: Summary of the Detrimental Effects of Ethylene Related to Quality

Ethylene effect	*Symptom or Affected Organ*	*Commodity*
Physiological disorders	Chilling injury	Persimmon, avocado
	Russet spotting	Lettuce
	Superficial scald	Pear, apple
	Internal browning	Pear, peach
Abscission	Bunch	Cherry tomato
	Stalk	Muskmelon
	Calyx	Persimmon
Bitterness	Isocoumarin	Carrot, lettuce
Toughness	Lignification	Asparagus
Off-flavors	Volatiles	Banana
Sprouting	Tubercle, Bulb	Potato, onion
Colour	Yellowing	Broccoli
	Stem browning	Sweet cherry
Discoloration	Mesocarp	Avocado
Softening	Firmness	Avocado, mango, apple, strawberry, kiwifruit, melon

Adapted from Martinez-Romero *et al.* (2007).

6. Sources of Ethylene

6.1 Ethylene Generators

Ethylene generators contains ethylene in liquid form which produced by heating in the presence of catalyst. These type of generators are widely used for ripening. The generator delivers about 14liters of ethylene gas per hour.

6.2 Ethephon (Commercially known as Ethrel/ CEPA)

Ethephon is a strongly acidic in nature and chemically known as 2-Chloroethane phosphonic acid. Any solution of more than 5 pH hydrolyzes the ethephon molecules resulting a ethylene production (Silva *et al.*, 2011). Ethrel is applied in the form of liquid, *i.e.* form of spray or dip in solution. Many researchers applied ethephon for improving colour and ripening of fruits. Latundan banana showed uniform and rapid ripening with attractive colour when treated with 2.5 per cent ethephon solution (Quoc *et al.*, 2014). Jayawickrama *et al.* (2001) reported that overall acceptability was higher in papaya fruits treated with ethrel solution rather than ethylene gas and calcium carbide. Dhillon and Mahazan, (2011) also concluded that pear fruit treated with 1000 ppm ethephon and fogging with 100 ppm ethephon showed best quality at 20°C temperature.

7. Ethylene Management

7.1 Terminating C_2H_4 Biosynthesis through ACC Synthase Inhibition

Aminoethoxyvinylglycine (AVG) is known to be a strong inhibitor of ACC synthase. The commercial product ReTainR (Valent BioSciences Corp., Libertyville, Illinois, USA), which contains 15 per cent w/w AVG, is registered for use on apple in USA and several other countries. Pre-storage applications of AVG has delayed fruit softening, starch degradation, accumulation of soluble sugars, the decrease in acidity, and generation of ester volatiles related to flavour during storage. The ethylene biosynthesis inhibition by AVG in both pre and post-harvest treatments could induce an increase in fruit shelf-life.

7.2 Terminating C_2H_4 Biosynthesis through ACC Oxidase Inhibition

Post-harvest application of acetaldehyde and ethanol to several types of fruit has shown a reduction and/or delay of softening, chlorophyll breakdown and lycopene synthesis by suppression of ethylene biosynthesis through the inhibition of ACO activity. However, acetaldehydehas been recently added to the list of known carcinogenic materials, and thus its use at commercial post-harvest application is limited. Therefore, as an alternative, more studies with ethanol treatments (a GRAS compound) are being carried out. Thus, for practical purposes, formulations based on an alcohol powder that gradually releases ethanol vapour are currently being tested. Appropriate ethanol vapour treatment has been shown to be effective delaying yellowing in broccoli florets through the retardation of the changes from chloroplasts to chromoplasts in sepal parenchyma cells, leading to 2-fold increase in shelf-life. In addition, ethanol application was able to inhibit superficial scald on apple, which is a maturity-related disorder. However, ethanol application failed on delaying banana ripening since it did not penetrate into the pulp. Thus, no general conclusion can be withdrawn from ethanol treatments, since fruit response depends on a wide number of variables (species, cultivar, ripening stage, concentration, and duration of treatment, among others), and would limit its commercial use. Other inhibitors, such as cobalt (Co^{2+}) are known to block the autocatalytic production of ethylene by inhibition of ACO activity. Thus, some experiments have been conducted in several fruit crops resulting in effective delays of the ripening process, *e.g.* bananas, apples, and tomatoes. However, these compounds are toxic heavy metals and are not allowed to be used in the food industry. Nevertheless, Ag+, formulated as silver thiosulfate, is usually the active component of preservative solutions for cut flowers, for which an extension of flower longevity by ethylene inhibition is achieved.

7.3 Application of Polyamines

Other means of altering ethylene production have been the use of exogenous treatments with polyamines (PAs) like putrescine (Put), spermidine (Spd), and spermine (Spm). Unlike the inhibitors described above, PAs are naturally-occurring compounds in most living organisms (both plant and animals) where they fulfil a wide array of physiological roles. Since there is a demonstrated competition between PAs and ethylene, through their common precursor SAM, the balance between these two opposite growth regulators is critical in retarding or accelerating the ripening process.

7.4 Inhibition of Ethylene Action trough 1-MCP Application

The discovery and commercial application of 1-methylcyclopropene (1-MCP) has revolutionized many post-harvest industries across the world. 1-MCP is a gaseous ethylene action inhibitor, which is thought to bind irreversibly to ethylene receptors and thereby prevent ethylene-dependent responses. The affinity of 1-MCP for the receptor is approximately 10 times greater than that of ethylene although there is debate on whether 1-MCP binds similarly to all ethylene binding sites. 1-MCP also influences ethylene biosynthesis in some species through feedback inhibition. 1-MCP has been used as a supplement to molecular approaches for identifying and understanding the spectrum of senescence and ripening processes under the direct control of ethylene perception. The gas has non-toxic mode of action, negligible residue and is effective at very low concentration (usually ≤ 0.6 ppm).

The major beneficial effects of 1-MCP in fresh fruits include suppression of ethylene production and respiration rates resulting in delayed fruit ripening and senescence, retardation of changes in fruit softening and skin colour, and alleviation of certain physiological disorders such as chilling injury and superficial scald.

Table 5: Summary of Effects of 1-MCP on some Fruits

Fruit	*Beneficial Effects of 1-MCP*	*Detrimental Effects of 1-MCP*
Apple	Retention of fruit texture, reduction in physiological disorders (senescent breakdown, core browning, coreline browning, watercore and superficial scald)	Susceptibility to external CO_2 injury increases in 1-MCP-treated fruit.
Apricot	Delayed fruit softening, reduced respiration and ethylene production rates	Reduced the evolution of aroma volatiles
Avocado	Delayed fruit ripening, reduced flesh browning during cold storage	
Banana	Delayed skin colouration and ethylene production	Uneven skin colouration, response is highly maturity dependent.
Citrus	Inhibition of rind pitting, inhibition of decay at low concentration	
Guava	Reduction in chilling injury and decay, retardation of fruit softening	Response may vary among climacteric and suppressed-climacteric cultivars
Kiwifruit	Retention of flesh firmness, reduction in ethylene production and respiration rate	
Litchi	Delayed pericarp browning	
Mango	Inhibited fruit softening, increased shelf-life	Increased stem-end rot
Nectarine	Inhibited fruit softening	Increase in chilling injury, flesh browning, woolliness and flesh reddening
Papaya	Inhibition of fruit softening, delayed fruit ripening, if 1-MCP is applied at >25% skin colouration stage.	Failure to ripen, leathery and rubbery fruit, if 1-MCP is applied at <25% skin colouration stage
Peach	Retention of flesh firmness and acidity	Chilling injury symptoms such as flesh browning, woolliness and flesh reddening are aggravated
	Retention of fruit texture, reduction in superficial scald, senescent breakdown, core browning, and watercore	Scald appears when ripening is resumed

Adapted from Singh and Terry (2013).

The formulation of 1-MCP as a stable powder is reliant on it being complex with γ-cyclodextrin. 1-MCP gas is easily released when the powder is dissolved in water. In 1999, the Environmental Protection Agency (EPA) approved the use of 1-MCP on ornamentals, marketed as EthylBloc® by Floralife, Inc. (Waterboro, SC, USA). SmartFresh™ as the trade name was subsequently developed by AgroFresh, Inc., a subsidiary of Rohm and Haas (Springhouse, PA, USA) for edible horticultural products. 1-MCP is registered for application in several fruits, especially climacteric ones, in many countries. The registration of 1-MCP is commodity and country specific. 1-MCP is internationally used in the apple industry to improve retention of textural and taste attributes of fruit during long-term cold storage. The application of 1-MCP in other fruits is also expanding. The beneficial or detrimental effects of 1-MCP in major fruit crops are described in Table 5.

8. Ethylene Control in Storage Structures/ Packages

8.1 Ventilation

Ventilation is the simplest method of ethylene removal from storage, packing house and retail sections. However, it suffers from several drawbacks such as loss of cool air, additional defrosting requirement for evaporative coils and impracticality in CA/MA rooms.

8.2 Controlled/Modified Atmospheres Storage

Controlled atmosphere (CA) storage involves the precise modification of O_2 and CO_2 concentrations in the storage atmospheres to retard fruit metabolism, resulting in extended storage life and better fruit quality. Various researchers and technologists have endeavoured to reduce the concentration of O_2 to the lowest possible levels and enhance the concentration of CO_2 to the highest level so that the maximum benefits from inhibition of fruit metabolism are gained with uncompromised fruit quality. Ethylene biosynthesis and action are inhibited under CA storage which is one of the major contributing factors responsible for enhanced storage life and better quality. However, very low levels of ethylene tend to accumulate in CA storage over the extended storage period which requires removal using appropriate (C_2H_4 scrubbing) technologies.

8.3 Ethylene Scrubbers

Potassium permanganate ($KMnO_4$)-based ethylene scrubbers are the most widely used technology. They are especially adapted to small volume storage rooms, since the potassium permanganate is rapidly consumed and must be replaced with a fresh supply several times during a conservation season. Potassium permanganate readily stoichiometrically oxidizes ethylene at ambient temperature to carbon dioxide and water, simultaneously forming manganese dioxide and releasing KOH. A range of commercial potassium permanganate products like Purafil, Circul-Aire, Ethylsorb, Bloomfresh or supported on zeolitic materials (ethylene control) are available in market. These products are mainly used to remove organic contaminants from the atmosphere inducted air-conditioning systems. The main disadvantages of using $KMnO_4$-based scrubbers are related to the recuperative nature of the technology and to the release of non-negligible amounts of by products such as KOH. $KMnO_4$ also requires frequent replacement (*e.g.*, for every journey in cargo transport, or several times during the storage season), and necessary amount depends upon the ethylene production.

A palladium (Pd) promoted powdered material that has significant ethylene adsorption capacity (4162 μl g^{-1} material) at 20°C and approximately 100 per cent RH was identified and was shown to be superior to potassium permanganate-based scavengers when used in low amounts and in conditions of high relative humidity. Optimum ethylene adsorption capacity was calculated as approximately 10,000 μL g^{-1}. Accordingly, corresponding inhibition of ethylene-induced ripening was observed. When removed, Pd-material did not disrupt subsequent ripening.

8.4 Biofilters–Microbial Ethylene Removal

An interesting alternative for the ethylene removal is the use of biological filters for purifying air suggested as method for removing ethylene from horticulture storage structures. Biofiltration is a complex technological process which involves many physical, chemical and biological phenomena. The air flows through a packed bed, and the pollutant transfers into a thin biofilm at the surface of the packing material. A prerequisite for successful biofiltration is the existence of ethylene-degrading microorganisms (bacteria, fungi *etc.*) with sufficient activity at low temperatures. In addition to the choice of microorganisms, the selection of an appropriate filter medium is another key factor in biofiltration efficiency. The major drawbacks of biofilters are their small surface area, low permeability, and a limited sorption capacity resulting in poor performance.

8.5 Ozone

Ozone is a powerful oxidant/disinfectant which remains technically more difficult to use, especially due to its toxicity. It can be easily generated using light generators or corona discharge systems. Benefits of adding ozone to air in packing houses and storage rooms include the control of post-harvest diseases on fruit, the delay of spore production from decaying fruits, the sanitation of surfaces, and ethylene removal. Ethylene oxidation with Ozone: $2O_3 + C_2H_4 = 2CO_2 + 2H_2O$. The current exposure limits established by the U.S. Occupational Safety and Health Administration (US-OSHA) for a 8 h work day are 0.1 ppm ozone; the short-term exposure limit (15 min.) for ozone is 0.3 ppm, while long-term exposure to concentrations above 1 ppm causes irreparable damage and even death.

8.6 Catalytic Oxidation

Among the complementary ethylene removal techniques used for storing fruits and vegetables, compared to the use of potassium permanganate, based on ethylene chemisorption, the catalytic oxidation of ethylene at high temperature (200-250°C) appeared as an economically viable alternative, especially for treating large volumes storage rooms. As an example of this technique, we can cite the catalytic gas purification process called "Swingthern" adapted to ethylene removal. Scrubbing units based on the catalytic oxidation of ethylene are available on market and are used to a limited extent in some commercial storage facilities.The claimed advantage of the system is that the units are easy to install and to operate, and that they maintain a constant low ethylene concentration due to stable conversion efficiency of 90-98 per cent. The disadvantage of the system is that the catalyst bed is very expensive and is only warranted by the manufacture for 2-3 years. Furthermore, the scrubber, even with an efficient heat exchanger, puts a significant heat load on the refrigeration system, which results in some dehumidification and additional energy use.

8.7 Ethylene Adsorbents

Adsorption phenomena can be efficiently used to achieve a very selective separation process for removing a small quantity of an adsorbate from a fluid stream. Zeolites are aluminosilicate minerals displaying a three-dimensional framework structure with interconnected cages and channels. They have a high ability for adsorption of VOCs at concentrations below parts per million. The hydrophobic property of zeolite is an important factor for the removal of nonpolar VOCs, such as ethylene, because such VOCs are usually accompanied by water vapor at higher concentration (over 5,000 ppmv) in air. There are several other types of adsorbents such as granulated activated charcoal, carbon nano-sized balls, clays *etc.* which have been used for ethylene adsorption.

9. Conclusion

The ethylene management decisions are usually dependent on the type and value of produce, handling and storage conditions and destination market. Return on investment guides the adoption of any post-harvest technology including ethylene management practices. Maintaining very low ethylene level in the post-harvest systems is beneficial to extend the storage stability and controlling some physiological disorders in fresh horticultural produce. The efficacy of post-harvest system can be enhanced several fold by following hybrid approaches to combination of technologies aimed at regulating vital physiological processes *viz.*, respiration, ethylene and transpiration. The best-fit approach for each commodity needs to be worked out in consultations with all major stakeholders in the post-harvest supply chain.

References

Ahmad, S., Thompson, A.K., Hafiz, I.A. and Asi, A.A. 2001. Effect of temperature on the ripening and quality of banana fruit. *Int. J. Agri. Biol.*, **3**(2): 224–227.

Ashraf-Ur-Rahman, Chowdhury, F.R. and Alam, M.B. 2008. Artificial ripening: what we are eating. *J. Med.*, **9**: 42–44.

Asif, M. 2012. Physico-chemical properties and toxic effect of fruit ripening agent calcium carbide. *Ann. Tro. Med. Pub. Hlth.*, **5**(3): 150–156.

Das, S.C. and Balamohan, T.N. 2013. Early and uniform ripening of mango with ethrel treatment cv. Sendura. *The As. J. Hort.*, **8**(1): 336–341.

Dhillon, W.S. and Mahajan, B.V.C., 2011. Ethylene and ethephon induced fruit ripening in pear. *J. Stored Prod. Post-harvest Res.*, **2**(3): 45–51

Ferreyra, R., Sellés, G., Saavedra, J., Ortiz, J., Zúñiga, C., Troncosoa, C., Rivera, S.A., González-Agüero, M. and Defilippi, B.G. 2016. Identification of pre-harvest factors that affect fatty acid profiles of avocado fruit (*Persea americana* Mill) cv. 'Hass' at harvest. *South African J. Bot.*, **104**: 15–20.

Huston, J.E. 1955. Selective poisoning of carp with calcium carbide. *Quart. Rep. Colo. Fish. Res. Un.*, 2(1 and 2): 17–21.

Ibe, I.J., Orji, J., Nwabueze, R.N., Nwanze, P.I., Ukagwu, N., Emeh, A.A. and Okechi, R.N. 2014. Effects of calcium carbide waste on soil bacteria and enzymes. *Int. J. Pharm. Med. Res.*, **2**(4): 1–3.

Ihejirika, C.E., Nwachukwu, M.I., Njoku-Tony, R.F., Ihejirika, O.C., Enwereuzoh, U.O., Imo, E.O. and Obenade, M. 2014. Impact of calcium carbide waste dumpsites on soil chemical and microbial characteristics. *Int. J. Env. Sci. Toxic. Res.*, **2**(6): 124–129.

Jayawickrama, F., Wijeratnam, R.S.W. and Perera, S. 2001. The effect of selected ripening agents on organoleptic and physico-chemical properties of papaya. *Acta Hort.*, **553**: 175–178.

Kader, A.A. 2002. *Post-harvest Technology of Horticultural Crops*, 2nd edn. University of California ANR Publications, California, 3311, 535

Lavoie, K.H., 1980. Toxicity of carbide waste to heterotrophic microorganisms in caves. *Microb. Ecol.*, **6**(2): 173–179.

Madani, B., Mohamed, M.T.M., Watkins, C.B., Kadir, J., Awang, Y. and Shojaei, T.R. 2014. Preharvest calcium chloride sprays affect ripening of Eksotika

II papaya fruits during cold storage. *Scientia Horticulturae*, **171**: 6–13.

Mahajan, B.V.C., Kaur Tajender, Gill, M.I.S., Dhaliwal, H.S., Ghuman, B.S. and Chahil, B.S. 2010. Studies on optimization of ripening techniques for banana. *J. Food Sci. Tech.*, **47**(3): 315–319.

Martinez-Romero, D., Bailen, G., Serrano, M., Guillen, F., Vaverde, J.M., Zapata, P., Castillo, S. and Valero, D. 2007. Tools to maintain post-harvest fruit and vegetable quality through the inhibition of ethylene action: A review. *Crit. Rev. Food Sci. Nut.*, **47**: 543–560.

Peck, S.B., 1969. Spent carbide: A poison to cave fauna. *Bull. Nat. Speleological Soc.*, **31**(2): 53–54.

Pratt, H.K. 1975. Role of ethylene in fruit ripening. *Coll. Int. Cent. Nat. Rech. Sci.*, **238**: 153.

Quoc, Le Pham, Tan, An Huynh, Thi Truong, Lam, Le Thanh, Ly Le Thi Hong and Suong, Nguyen Thi Tuyet. 2014. The effects of ethephon on the ripening of Vietnamese Latundan bananas (*Musa sapientum*). *Emir. J. Food Agric.*, **26**(3): 229–237.

Rahman, A., Chowdhury, F.R. and Alam, M.B. 2008. Artificial ripening: what we are eating. *J. Med.*, **9**: 42–44.

Ram, H.B., Singh, S.K., Singh, R.V. and Singh, Surjeet. 1979. Effect of ethrel and smoking treatment on ripening and storage of banana cultivar. *Himachal Prog. Hort.*, **11**(3): 69–75.

Saltveit, M.E. 1999. Effect of ethylene on quality of fresh fruits and vegetables. *Post-harvest Biol. Tech.*, **15**: 279–292.

Semikolennykh, A.A. Rahleeva, A.A. and Poputnikova, T.B. 2012. Spent carbide waste retains toxicity long term after disposal in caves and mines. *Acta Carsologica*, **41**(1): 129–137.

Siddiqui, M.W. and Dhua, R.S. 2010. Eating artificially ripened fruits is harmful. *Curr. Sci.*, **99**(12): 1664–1668.

Singh, S.P. and Terry, L.A. 2013. Recent advances in storage technologies for fresh fruits. In: *Advances in Food Science and Nutrition*, Vol. 3, (Eds.) P.M. Visakh, Laura B. Iturriaga and Pablo Daniel Ribotta. Scrivener Publishing, USA, pp. 391–411.

Technical Standards and Protocol for the Fruit Ripening Chamber in India, 2011 published by National Horticulture Board (MoAC, Govt. of India) available at www.nhb.com

Temi-Sogo, C.M., Idowu, O.A. and Idowu, E. 2014. Effect of biological and chemical ripening agents on the nutritional and metal composition of banana (*Musa* spp). *J. Appl. Sci. Env. Mgt.*, **18**(2): 243–246.

Watada, A.E., 1986. Effects of ethylene on the quality of fruits and vegetable. *Food Tech.*, U.C. Davis Bulletin, pp. 82–85.

Wills, R.B.H., Warton, M.A., Mussa, D.M.D.N. and Chew, L.P. 2001. Ripening of climacteric fruits initiated at low ethylene levels. *Aust. J. Exp. Agri.*, **41**: 89–92.

71

Nanotechnology Interventions in Post-harvest Management of Horticultural Crops

*Dinesh Kumar[1], Milind S. Ladaniya[1] and Lallan Ram[2]**

[1]ICAR-Central Citrus Research Institute, Nagpur – 440 010, Maharashtra
[2]Central Institute of Horticulture, Medziphema – 797 106, Nagaland
E-mail: lallanram09@gmail.com

India has been bestowed with wide range of climate and physiogeographical conditions and as such is most suitable for growing various kinds of horticultural crops. Total production of fruits during 2014-15 was 86.60 million tonnes while that of vegetables was 169.47 million tonnes. Production share of major fruit and vegetable crops in India during 2014-15 is shown in Figure 1. Nanotechnologies concerns with the study of phenomena and manipulation of materials at atomic, molecular and macromolecular scales, where material properties differ significantly from those at a larger scale (Ram *et al.*, 2011). A nanometer (nm) is one billionth of a meter. Nanotechnology using of particular characteristics of nano-particles can be a very useful technology in all science and industry branches (Yadollahi *et al.*, 2010).

The application of nanotechnology to the agriculture and food industry was first addressed by United States Department of Agriculture in 2003. An estimate by market research and industry analysis shows that, the market for the nanotechnology was 7.6 billion US$ in

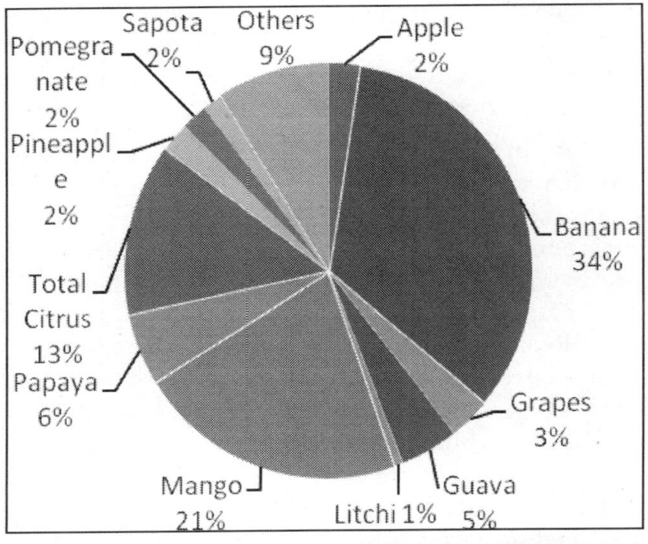

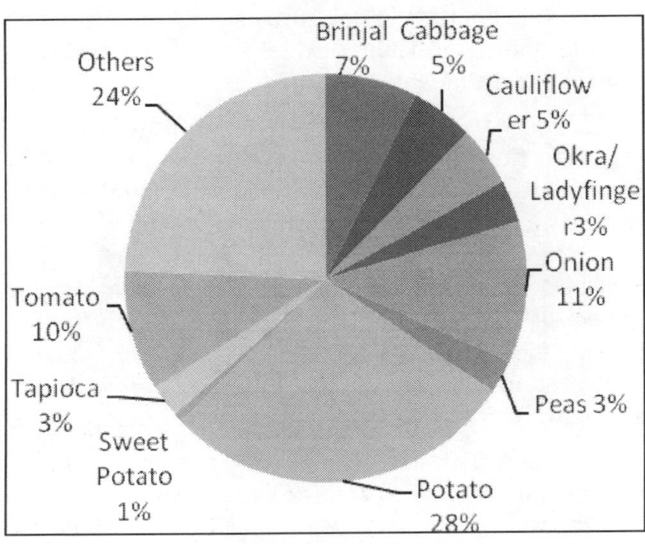

Figure 1: Production Share of Major Fruit and Vegetable Crops in India (2014-15).
***Source*: NHB 2014-15, India.**

2003 and increased upto US$ 1 trillion in 2011 (Misra *et al.*, 2013). The total nano-enabled food and beverage packaging market in the year 2008 was US$ 4.13 billion, US$ 4.21 million in 2009 and increased upto US$ 7.30 billion in 2014, at an annual growth rate of about 11.65 per cent. In relation with extension of horticultural products shelf-life, nanotechnology can help us in many ways, for *e.g.* controlling growth and development of micro-organisms, introducing a new generation of packaging coverage's (films) and controlling influence of gases and the harmful rays (UV), increasing strength, quality and packaging using the multiple chips (Nanobiosensors) for labeling products that considered as fundamental step to automated control of storages (Yadollahi *et al.*, 2010).

1. Interventions

Nanosciences and nanotechnologies potential application in food packaging includes modifying the perme-ation behaviour of foils, increasing barrier properties, improving mechanical and heat-resistant properties, developing active anti-microbic and antifungal surfaces, sensing and signaling microbiological and biochemical changes as well as anti-microbial packaging materials. Materials showing anti-microbial properties caused by nanoparticles silver or other substances have already entered the market. The most prominent products in R&D in food industry include new polymer nanocomposites for packaging and wrapping. Foils or membranes-based on nanocomposites offer adjustable gas permeability in food packaging, which can help to better protect food.

Another development is packaging with self-cleaning surfaces. Dirt-repellent coatings at nanoscale can prevent the invasion of microorganisms and ensure food safety. In smart packaging, food packagers are developing materials with properties that can change depending on external or internal condition such as temperature. Thus in the future, food packaging that can adjust its properties to pH, pressure, temperature and light or show self-healing properties when perforated may become available.

Other nanotechnological intervention has made the detection of pathogen in food products easier. The microscopic biological sensor is developed to detect the *Salmonella* bacteria at laboratory level in which biosensors like fluorescent organic dye particles attached to *Salmonella* antibodies which hook into the *Salmonella* bacteria and dye lights up like a beacon making the bacteria easier to see. The nano-technological intervention is required to use in longer period during storage of fruits and its product in Indian perspective.

Packaging and monitoring are major focus of horticultural produce industry. Packaging that incorporates nanomaterials can be "Smart" which can respond to environmental conditions or repair itself or alert a consumer to contamination and presence of pathogens. Smartness in packaging is a broad term which covers a number of functionalities depending on product being packaged. The packages must retain integrity and actively prevent fruit product spoilage and enhance the product attributes like appearance, tastes, flavour, aroma *etc.* It should also respond actively to changes in product or package environment and also communicate product information and condition to users. Smart packaging involves intelligent and active packaging of products. Hence, nano packaging might provide an attractive alternative to preserve quality of fruits during extended storage. Blending the polyethylene with nano-powder (nano – Ag, Kaolin, anatase TiO_2) to enhance shelf-life and preserve quality of fruits and its products.

Development of smart packaging to optimize product shelf-life to have antimicrobic and antifungal surfaces and sensing as well as signaling microbiological and biochemical changes (Ram *et al.*, 2011).

1.1 Nanotechnology in Preservation and Packaging of Fruits/Fruit Products

Fruits have always been the essential dietary supplement being the natural sources of different vitamins, minerals, fibers, and phytochemicals like flavonoids, limonoids, carotenoids, vitamins, *etc.* The usefulness of nanotechnological approach in fruit preservation and packaging has been considered in its ability to retain the original physical, biochemical, and organoleptic properties of different fruits and to provide the physiological and health benefits of fruit consumption (Kalia and Parshad, 2015).

1.2 Controlling Growth and Development of Micro-organisms

Gray mold disease caused by *Botrytis cinerea* is considered an important post-harvest pathogen around the world. Moreover, public awareness regarding the negative impact of synthetic fungicide residues on human health and environment, there is growing emphasis on environmentally friendly technologies in the fresh fruit and vegetables industry, and the search for safer alternatives to chemical fungicides.

Coatings can be made from polysaccharides, proteins, lipids or a combination of these materials. Coatings can be used as carriers of useful ingredients such as antimicrobial compounds, color or aroma additives, anti-oxidants, or anti-ripening compounds. Advantages of using coatings include reduction of water loss, reduction of chilling and mechanical injury, reduced decay, and added shine or gloss to the coated products (Yadollahi *et al.*, 2010). Chitin, together with its derivative chitosan, has been reported as a promising alternative to control post-harvest diseases. Chitin and chitosan are polysaccharides by inhibiting spore germination, germ tube elongation, mycelial growth of fungal phytopathogens, enhancing the efficacy of

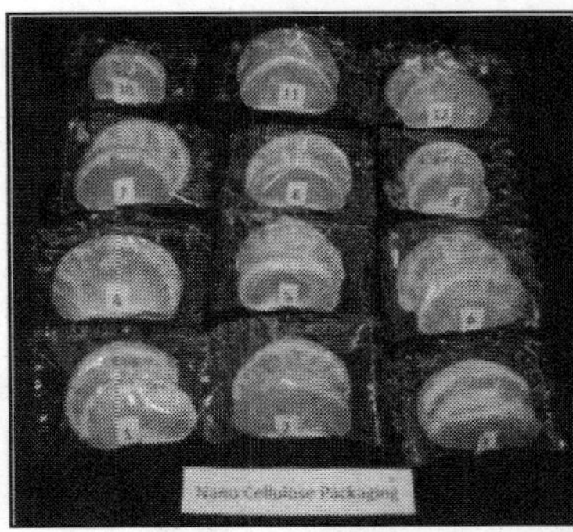

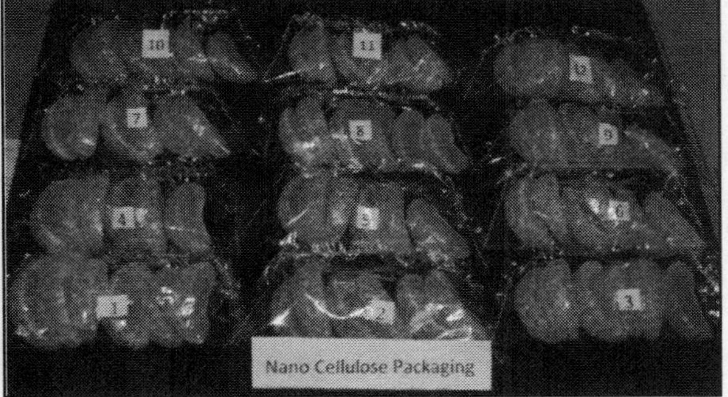

Figure 2: Nagpur Mandarin and Pummelo Segments Packed in Nanocellulose Packaging.

antagonistic yeasts, and boosting the activity of defense-related enzymes or pertinent substances (Zhang *et al.*, 2011).

1.3 Nanocomposites for Packaging

Nanocomposite materials include one-dimensional, two dimensional, three-dimensional, and amorphous materials made of distinctly dissimilar components that are mixed at the nanometer scale. Compared to the conventional packaging materials, nanocomposites have several additional benefits like enhanced strength or elasticity, improved biodegradability and better control over gaseous molecules which are required for developing better-performing packaging materials. There are many reports, which suggest the higher organoleptic and fruit safety qualities of the minimally processed fruits by using the nanopacking materials over the conventional packaging (Kalia and Parshad, 2015). Li *et al.* (2009) reported better physicochemical and sensory qualities of Chinese *jujube* fruit packed using nanocomposites over the conventional fruit packaging. Fuji apples with Nano-SiOx/chitosan preservation agents had good quality according to the research (Yadollahi *et al.*, 2010).

1.3.1 ZnO Composed Nano Packaging for Extended Shelf-life of Citrus Juice

Nano packaging provides an attractive alternative to improve preservation quality of the fruits during extended storage. Blending polyethylene with nano-powder like nano-Ag, kaolin, anatase TiO_2 has been reported to expand shelf-life and improve preservation quality of fruit and its value-added products. The effect of nano-packaging materials (Figure 2) on preservation quality of citrus species (fruit and value-added products) is been investigated (Ram and Kumar, 2013). A study was made at ICAR-CCRI, Nagpur on Nano zinc oxide composed packaging and LDPE packaging and on Nanocellulose and PVA packaging on shelf-life and quality of Nagpur mandarin and pummelo segments during refrigerated storage. The physico-chemical and microbiological observations were recorded after 30 days (Anon, 2015-16).

1.3.1.a Physico-chemical study

The physicochemical parameters, *viz.*, TSS, acidity, vitamin C, juice colour, limonin and browning were studied under refrigerated storage conditions. ZnO nanopackaging and nano cellulose packaging recorded superiority over fresh segment colour indicating the attractive colour of segments maintained in packagings than that of LDPE and PVA packaging. Study also revealed that Nanopackaging maintained the physico-chemical quality than the LDPE, PVA packaging during the storage (Ram *et al.*, 2014).

1.3.1.b Microbiological Study

The microbiological evaluation of Nagpur mandarin and pummelo segments packed in different packaging was done. Total aerobic count during the storage was enumerated on nutrient agar and total yeast and mould

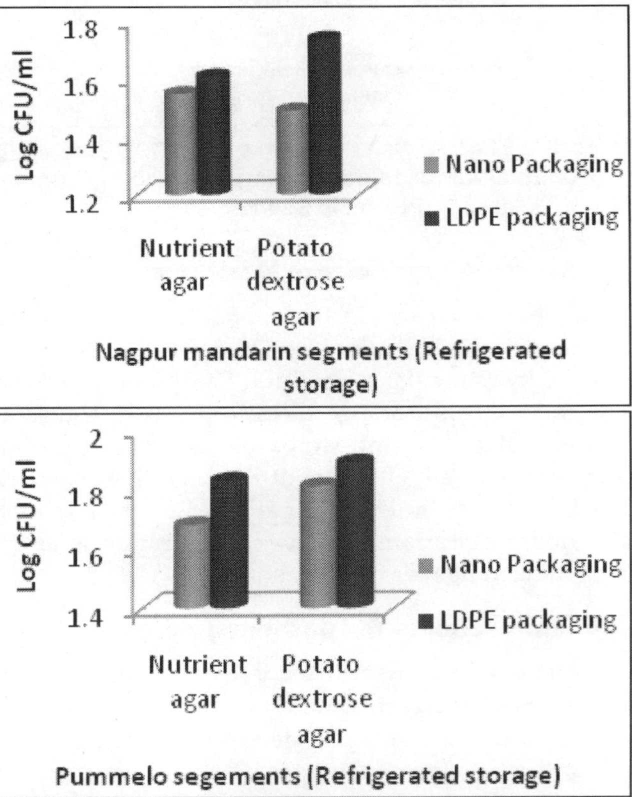

Figures 3: Effect of Nano Packaging and LDPE Packaging on Total Aerobic Count of Nagpur Mandarin and Pummelo Segments.

count on potato dextrose agar. Study revealed that the segments packed in Nano ZnO packaging (Figure 3) recorded lower total aerobic count as compared to that of the LDPE packaging. Similar trend was observed for Nano cellulose (Figure 4) and PVA packaging (Ram *et al.*, 2013).

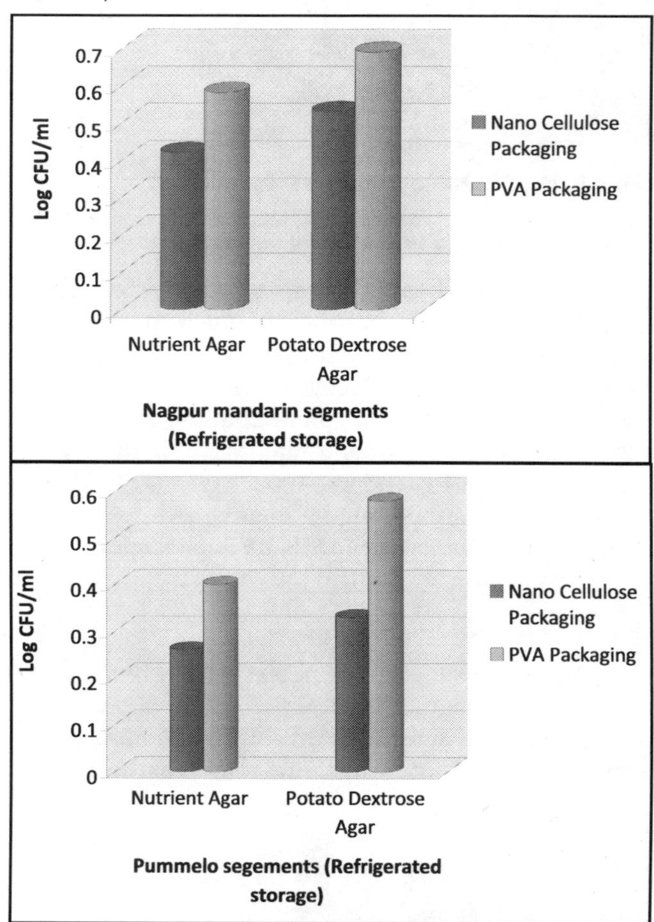

Figure 4: Effect of Nano Cellulose and LDPE Packaging on Total Aerobic Count of Nagpur Mandarin and Pummelo Segments.

1.3.1.c Metal Ions Release Measurement

The residual activity of nanoparticles of silver and zinc ions in orange juice after 28 days of storage was studied by Emamifar *et al.* (2010). The quantity of silver ions migrating into orange juice is less than its allowable concentration (10 ppm). Moreover, the quantity of Zinc ions indicated a higher rate of Zn migration than that of silver but as zinc is proved to be a GRAS compound for food applications; its low concentration is in the acceptable range for food consumers.

1.4 Nanosensors for pathogen detection

Nano sensors would serve as 'electronic tongue' or 'noses' by detecting chemicals released during food spoilage. Sensors can detect substances in parts per trillion and would trigger a colour change in the packaging to alert the consumer if a food has become contaminated or if it has begun to spoil (Ram and Kumar, 2013). Nanosensors in plastic packaging can detect gases

Figure 5: Nano-silica Coated High Oxygen Barrier Nylon (OPA) Films for Food.

given off by food when it spoils and the packaging itself changes color to alert. These films (Figure 5) are packed with "silicate nanoparticles" to reduce the flow of oxygen into the package and the leaking of moisture out of the package to stay food fresh (Wesley *et al.*, 2014).

1.5 Ripeness Indicator Label

RipeSense (Figure 6), the world's first ripeness indicator label developed in New Zealand. The intelligent ripeness indicator responds to the aroma released as fruits ripens, giving consumers a better way to determine the shelf-life of the fruits before they take it home. Help shoppers to reduce wastage.

Figure 6: RipeSense (Ripeness indicator label).

1.6 Printed Oxygen Indicator

Many foods are packed under modified atmosphere packaging (MAP). An oxygen-sensitive ink can be used to show if the MAP remains intact, throughout the whole distribution chain. "Intelligent ink" was created using light-sensitive nanoparticles such as titania nanoparticles.

1.7 Nano Bar Codes

The Nano barcodes (Figure 7) fluoresce under ultraviolet light when target compounds are detected. It improves packaging's security, assure brand

Figure 7: Nano Barcode.

and authenticity. This technology has a number of competitive advantages (Ram and Kumar, 2013):

1. Unlimited numbers of unique, machine-readable codes
2. Low cost of manufacture and implementation
3. Extraordinary compatibility/durability
4. Difficult to counterfeit

2. Nanotechnology in Shelf-life Extension of Vegetables

India is the second largest producer of vegetables with annual production of 141 million tonnes. The nanotechnology may provide an attractive alternative in improving the preservation qualities of horticultural vegetables crops during extended storage. Ouzounidou and Gaitis (2011) reported the use of colloidal copper (Nano-copper particles suspended in water) in shelf-life extension of green vegetables. Post- harvest decaying of vegetables is attributed to both oxidative reactions as well as microbial deterioration (mainly by fungi). Nano- Cu spray can be used as an alternative method in vegetable preservation and maintenance of good quality after harvest. Nanotechnology (use of nano bio fungicides) could also impart in developing eco-friendly, sustainable methodology to mitigate post-harvest losses due to fungal diseases in onion (Kumar *et al.*, 2015).

3. Nanotechnology and Post-harvest Management of Flowers

Advances in post-harvest science and technology aim to provide information for the floriculture industry to enable them to supply attractive and long-lived flowers to consumers. Nanotechnology has potential applications for cut flower vase life improvement, focusing on ethylene control strategies (Figure 8). Nano-scale systems could be applied to cut flowers for ethylene detection in the store environment (nano-metals based sensors) and along the distribution chain (nano-chip labels). They could also be used for ethylene removal (nano-metals for photocatalitic degradation in the warehouse or nanocomposites for scrubbing in active packaging). Titanium dioxide (TiO_2) has been the focus for light-activated photocatalytic degradation under ultraviolet (UV) irradiation, either from natural (sun) or artificial (lamps), because of its physical and chemical stability, low cost, availability and non toxicity. Silver (Ag) ions also show photoactivity, semiconductor photocatalysis, and antibacterial activity: nano-Ag absorbs and decomposes ethylene and can have more effective antibacterial activity than Ag. Thus, packaging films incorporating nano- Ag or TiO_2 (*e.g.* nanocomposite polyethylene film) contribute to preserve quality of fruits and vegetables, retarding senescence and decreasing microbial growth (Scariot *et al.*, 2014).

4. Commercialization of Nanotechnology

Nanotechnology has been described as the new industrial revolution. Both developed and developing

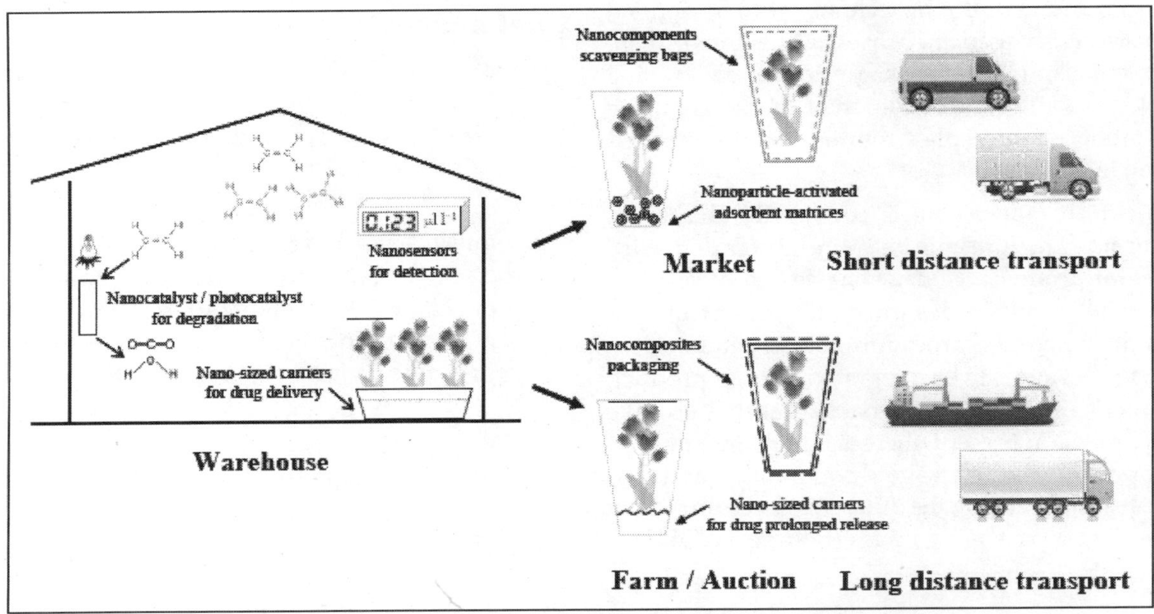

Figure 8: Example of Futuristic Nanotechnology-Based System for Ethylene Control in Ethylene-Sensitive Cut Flowers.

countries are investing in this technology to secure a market share. At present the Government of India has launched mission mode R&D during 2007 with a budgetary provision of Rs. 10 billion over five years under nanomission council guided by two advisory groups, *viz.*, Nanoscience Advisory Group and Nanosciences Applications and Technology Advisory Group, with a mission for funding research projects to foster and promote the R&D in this area in the country. However, developed countries like USA is leading with investment followed by Japan and the European Union, which have both committed substantial funds annually. While US $ 3 million have been allocated to foods and agricultural application through USDA Cooperative State Research, Education and Extension Service.

5. Prospects

India has significant potential for growth of nanotechnology in post-harvest management for nanopackaging of fruit and its products, nanomaterials and formulations that can be realized through application-oriented R&D. Keeping nanotechnology potential in a view, the Government of India encourage implementation of industry-centric and application-driven projects in Nano Science and Technology particularly in post-harvest management of fruits and preservation of juice as well as these value-added products. They are: basic understanding of fruits, value-added products and animal feed for intelligent innovation, systems biology in PHT research, biological renewal in fruits sector/biological production, technology development, nutrigenomics and consumer needs-driven innovation and food communication.

Much focus on these areas will create a holistic and an interdisciplinary approach in food research and development in India. That will help in production of nanomaterials with functional properties, along with nanosensors and nanofluidic technology to be applied in post-harvest management of fruits. Other interests are development of intelligent packaging materials, making it possible to monitor the condition of products during transportation or in display counters, and bio based packaging techniques.

The role of packaging is constantly changing, in future; its function will be more integrated with the image of products. Concurrent design of product and package involves design management of two components where the procedure model can be used. Thus, if the package is to be integrated into the product, development of product and packaging has to take place in parallel. At present, there is limited inte-gration between product and package development, and also limited integration among the different players in supply chain, regarding packaging issues (Ram *et al.*, 2011).

6. Conclusion

The impacts of nanotechnology on food industry and products entering the market, the safety of foods will remain the prime concern. This need will strengthen the adoption of nanotechnology in sensing applications, which will ensure food safety and security, as well as technology which alerts customers and shopkeepers when a food is nearing the end of its shelf-life. New antimicrobial coatings and dirt repellent plastic bags are a remarkable improvement in ensuring the safety and security of packaged foods. Already some research groups are exploring the infusion of nanoparticles in foods but still from a top-down approach, using cells rather than molecules. Although practical application of such technology is far into the future, it is expected that this could allow a more efficient and sustainable food production process to be developed where less raw materials are consumed and food of a higher nutritional quality is obtained.

Despite the considerable work done on related aspects of nanotechnology for food and agricultural industry in India, elucidation of nanosciences and nanomaterials in for post-harvest technology and value-addition is meagerly available. The new techniques like, improving mechanical and heat-resistant properties, developing active anti-microbic and antifungal surfaces, sensing and signaling microbiological and biochemical changes and active polymer sheets for debittering citrus juice as well as anti-microbial packaging materials. Success of nanotechnology for post-harvest management including preservation and value-addition will not only enhance the shelf-life of produce and products but also provide an opportunity to establish the processing industry in country on sound footing basis. The farmer will also get appropriate remuneration of their produce during the glut in the market.

References

Anonymous. 2015–16. Study on Nanocellulose and PVA packaging on shelf-life and quality of Nagpur mandarin and pummelo segments during refrigerated storage, ICAR–CCRI *Annual Report*, 2015–16, pp. 83–85.

Emamifar, A., Kadivar, M., Sahedi, M. and Soleimanian-Zad, S. 2010. Evaluation of nanocomposite packaging containing Ag and ZnO on shelf-life of fresh orange juice. *Innovative Food Science and Emerging Technologies*, **11**: 742–748.

Kalia, A. and Parshad, V.R. 2015. Novel trends to revolutionize preservation and packaging of fruits/fruit products: Microbiological and nanotechnological perspectives. *Food Sci. Nut.*, **55**: 159–182.

Kumar, V., Neeraj, Sharma, S. and Sagar, N.A. 2015. Post-harvest management of fungal diseases in onion: A review. *Int. J. Curr. Microb., App. Sci.*, **4**(6): 737–752.

Li, H., Feng, L., Wang, L., Sheng, G., Xin, Z., Zhao, L., Xiao, H., Zheng, Y. and Qiuhui, H. 2009. Effect of nano-packing on preservation quality of Chinese jujube (*Zizyphus jujuba* Mill. var. *inermis* (Bunge) Rehd). *Food Chem.*, **114**: 547–552.

Misra, A.N., Misra, M. and Singh, R. 2013. Nanotechnology in agriculture and food industry. *Int. J. Pure Appl. Sci. Technol.*, **16**(2): 1–9.

Ouzounidou, G. and Gaitis, F. 2011. The use of nanotechnology in shelf-life extension of green vegetables. *J. Innov. Econ. Mgt.*, **8**: 163–171.

Ram, L., Kumar, D. and Shivankar, V.J. 2011. Nanotechnology for post-harvest Management in citrus fruits. Section VII. **In**: *Nanotechnology in Agriculture: An Indian Perspective*, pp. 244–251.

Ram, L. and Kumar, D. 2013. Current trends in value-addition, packaging and labeling of citrus fruit products. **In**: *Horticulture for Economic Prosperity and Nutritional Security in 21st Century*, (Eds.) T.K. Hazarika and B.P. Nautiyal, Westville Publishing House, New Delhi, pp. 173–190.

Ram, L. and Kumar, D., Vigneshwaran, N. and Khewle, A. 2013. Effect of ZnO nano particle containing packaging on shelf-life of fresh Nagpur mandarin (*Citrus reticulata* Blanco) segments. *J. Biol. Chem. Res.*, **30**(2): 381–386.

Ram, L. and Kumar, D., Vigneshwaran, N. and Khewle, A. 2014. Comparative performance of nanocomposite film and LDPE packaging for extending shelf-life of fresh Nagpur mandarin segments. *Indian J. Hort.*, **71**(1): 87–91.

Scariot, V., Paradiso, R., Rogers, H. and De Pascale, S. 2014. Ethylene control in cut flowers: Classical and innovative approaches. *Post-harvest Biol. Tech.*, **97**: 83–92.

Wesley, S J., Raja, P., Raj, A.A. and Tiroutchelvamae, D. 2014. Review on–Nanotechnology applications in food packaging and safety. *Int. J. Engg. Res.*, **3**(11): 645–651.

Yadollahi, A., Arzani, K. and Khoshghalb, H. 2010. The role of nanotechnology in horticultural crops post-harvest management. *Acta Hort.* **875**: 49–56.

Zhang, H., Li, R. and Liu, W. 2011. Effects of chitin and its derivative chitosan on post-harvest decay of fruits: A review. *Int. J. Mol. Sci.*, **12**: 917–934.

Adding Value to Produce-I

72

Value-addition in Floriculture for Nutraceuticals

K.V. Prasad*[1]**, Mohan Ram Legha*[2]*, Bharat Singh Hada*[2]*,
***Usha*[2] *and Prativa Lakhotia*[2]

[1]ICAR-Directorate of Floricultural Research, Shivajinagar, Pune – 411 005, Maharashtra
[2]Division of Floriculture and Landscaping, ICAR-Indian Agricultural Research Institute,
New Delhi – 110 012
**E-mail: kvprasad66@gmail.com*

Healthy living has become a norm in the present day world. Increased incidence of diseases associated with changing lifestyle attained a threatening proposition. Although life expectancy has increased over the past decades due to better medicare, inadequate nutrition due to the current lifestyle choices people make today has brought about the need to develop functional foods that are rich in nutraceuticals. Lifestyle diseases that are attributed to nutrition are on the rise with nearly 40 per cent of the population is in the high risk zone in the developing nations like India. A number of such ailments can be tackled through the nutraceuticals available in nature. Nutraceutical in broad term is defined as a food or part of a food that purportedly provides medicinal or health benefits, including the prevention and treatment of disease. A nutraceutical may be a naturally nutrient-rich or medicinally active food or it may be a specific component of a food.

The global nutraceutical industry is valued at US$ 142.1 billion in 2011 and is expected to be around US$ 204.8 billion by 2017, with an impressive growth rate of around 6.0 per cent. Out of this quantum of business the functional food and beverage market contributed US$ 93 billion with more than 60 per cent of contribution (US$ 56.4 billion) coming from North America in 2011. The Indian nutraceutical industry during 2015 is worth about US$ 5.00 billion that mainly comprises of functional foods and beverages (67 per cent) and dietary supplements comprise of the rest 33 per cent (FICCI, 2016). Indian Nutraceutical industry is valued at US$ 2 billion which is 1.5 per cent of the global nutraceutical industry. The Indian industry is characterised by marketing of traditional herbal ingredients (usually ayurvedic) into the nutraceutical portfolio. Key example is the chyawanprash supplements market in India, which stood at US$ 74.5 million in 2010. The existence of alternative medicine in India, and the Indian consumer's belief in them are the major factors that drive this industry.

Natural pigments including anthocyanins, carotenoids, betalains, shikonin, and other pigmented phytochemicals are conspicuously found in most of the plants. The anthocyanins, carotenoids, and betalains play important ecological functions in the plants besides possessing important nutraceutical functions. Among the different group of pigments available in the plants, anthocyanins, carotenoids and betalains are more frequently exploited as the source of nutraceuticals to address a number of human ailments. It is now established that the Anthocyanins play a vital role in curing vision disorders, neuroprotection, cardiovascular diseases (CVD) and certain cancers. Whereas carotenoids are found to cure cancer, photooxidative damage and CVD besides enhance Immune system. The betalains and chlorophylls are known for their definite role in alleviating certain types of cancers.

The range of plant pigments, their class, compounds and the colour they impart are summerised in Table 1. They play important ecological and metabolic functions in the plants. They are also important for humans as they provide very essential pharmaceutical and nutraceutical principles. Realizing the ill effects of artificial colour on human health, there is a worldwide interest in use of food colorants from natural sources like fruits, vegetables and flowers.

Very often the food we consume is laced with attractive colours that potentially contain highly carcinogenic agents including heavy metals like mercury, lead and cadmium *etc.* The recent controversy on detection of Sudan Red colour in the chilli powder exported from India to EU is a classical case to contemplate. Synthetic pigments dominate the trade worldwide but due to increased awareness about the ill effects of synthetic pigments on human health and environment, nowadays research is oriented towards identification and isolation of pigments from natural sources as these natural pigments are safe for human consumption and are eco friendly. In the present era the synthetic coal tar based artificial food colourants are being extensively replaced with the natural colourants from fruits, vegetables and flowers. Natural pigments like anthocyanins, carotenoids, betalains, shikonin, and other pigmented phytochemicals are found in most of the plants.

Over the last few decades, researchers have also explored the idea of generating plant cell cultures for the production of natural colourants. This offers the advantages of consistent supply and quality as well as the opportunity to control the type of pigment produced. In spite of the amount of interest in this field no commercially viable system has been developed in case of flowers. With an increasing consumer preference for healthy foods there is now considerable demand for the use of anthocyanins as natural colour. As a direct result, the production of anthocyanin pigments is one of the fastest growing segments of the food colourant industry in recent times.

However, there are several limitations to the current supply of anthocyanin pigments. The number of different plant sources is very limited. In addition, the supply can be subject to long cultivation times, seasonal effects, climate variations, pest/disease attack and the increasing cost of intensive agriculture coupled with the decreasing availability of low-cost arable. All of these factors affect quality and quantity of supply. As the demand for more prepared and processed food increases and the expectation of high health products also rises, food manufactures require a wider range of colourants and anti-oxidative additives. One alternative source for the production of anthocyanins is through the use of plant cell cultures. The potential of utilizing anthocyanin pigments for different purposes motivated use of plant cell or callus cultures to produce several of these vital secondary metabolites irrespective of seasonal barriers.

1. Plant Pigments

All biological pigments selectively absorb certain wavelengths of light while reflecting others. The light that is absorbed may be used by the plant to power plant reactions, while the reflected wavelengths of light determine the colour the pigment will appear to the eye. Pigments also serve to attract pollinators.

Chlorophyll is the primary pigment in plants; it absorbs yellow and blue wavelengths of light while reflecting green. It is the presence and relative abundance of chlorophyll that gives plants their green color. Most of the plant parts that appear green possess this pigment. This pigment is involved in photosynthesis, a chemical reaction that produces the food material for the plant.

Anthocyanins (literally "flower blue") are water soluble flavonoid pigment that appear red to blue, according to pH. They occur in all tissues of higher plants, providing color in leaves, stem, roots, flowers, and fruits, though not always in sufficient quantities

Table 1: Plant Pigments, their Class, Compounds and the Colour they Impart

Pigment Class	*Compound Type*	*Compound Example(s)*	*Typical Colour*
Porphyrins	Chlorophyll	Chlorophyll 'a' and 'b'	Green
Flavonoids	Anthoxanthins	Pelargonidin, cyanidin, delphinidin, peonidin	Red, blue, violet
	Flavonols	Kaempferol, quercetin, fisetin, kaempferide, morin, myricetin, myricitrin, rutin	Yellow
	Isoflavonones	Apigenin, biacalein, chrysin, diosmetin, flavone, luteolin	Yellow
	Flavonones	Diadzin, genistein, enterodiol, coumestrol, biochanin	
	Flavans	Biflavan, catechin, epicatechin,	Colour less co-pigments
Carotenoids	Carotenes	Lycopene, α-carotene, β-carotene, γ-carotene	Colour less co-pigments
	Xanthophylls	Lutein, cryptoxanthin, zeaxanthin, neoxanthin, rhodoxanthin, violaxanthin, canthaxanthin, astaxanthin	Yellow, orange, red
Betalains		Betaxanthin, betacyanin	Red, magenta, deep pink

to be noticeable. Anthocyanins are most visible in the petals of flowers, where they may make up as much as 30 per cent of the dry weight of the tissue.

Carotenoids are red, orange, or yellow in colour. They function as accessory pigments in plants, helping to fuel photosynthesis by gathering wavelengths of light not readily absorbed by chlorophyll. The most familiar carotenoids are carotene (an orange pigment found in carrots), lutein (a yellow pigment found in fruits and vegetables), and lycopene (the red pigment responsible for the color of tomatoes). Carotenoids have been shown to act as antioxidants and to promote healthy eyesight in humans.

Betalains are red or yellow pigments. Like anthocyanins they are water-soluble, but unlike anthocyanins they are indole-derived compounds synthesized from tyrosine. This class of pigments is found only in the Caryophyllales (including cactus and amaranth), and never co-occur in plants with anthocyanins. Betalains are responsible for the deep red color of beet, and are used commercially as food-coloring agents.

2. Nutraceutical Function of Pigments

Flowers are the major source of a wide range of natural pigments amongst horticultural crops. Flowers consist of various bioactive compounds such as phenolic acids, flavonoids, pigments, *etc.* which act as natural anti-oxidants (Youwei *et al.*, 2008). Among different group of pigments available in flowers, anthocyanins, carotenoids and betalains are frequently used as source of nutraceuticals and for other purposes such natural colorants or dyes. Among flowers, rose is a rich source of anthocyanins, marigold in carotenoids and bougainvillea in betalains. In roses, anthocyanins are widely distributed in flowers, leaves and stems and flower colour is structurally composed of simple cyaniding 3, 5-diglucosides (Ogata *et al.*, 2005). Marigold flowers are rich source of carotenoids especially yellow carotenoids such as xanthophylls (lutein and zeaxanthin) and β carotene, which are used to improve egg yolk colour (Hasin *et al.*, 2006). The pharmacological and nutraceutical activities of marigold are related to the content of several classes of secondary metabolites such as flavonoids, sterols, carotenoids, tannins, saponins, triterpene alcohols, polysaccharides, a bitter principle, mucilage and resin (Piccagliy *et al.*, 1998). Bougainvillea is edible flower which is frequently used in most of the parts of the world for the preparation of salads and flower teas. The floral bracts are the rich source of betalain pigments particularly betacyanin (Moreno *et al.*, 2008) which can be used as a natural colourant in food industries. Nutraceutical pigments that are found in various flower crops are summerised in Table 2.

Table 2: Nutraceutical Pigments found in various Flower Crops

Crop	*Pigments*
Anthurium	Anthocyanin [cyanidin 3-rhamnosylglucoside and pelargonidin 3-rhamnosylglucoside]
Calendula	Lutein (80%), zeaxanthin (5%), β-carotene
Chrysanthemum	Carotenoids, anthocyanins [cyanidin 3-O-(6-O-malyonyl β-D-glucopyanoside)], flavonols, other glucosides
Gerbera	Anthocyanidin [42-hydroxylation (pelargonidin) over 32,42-hydroxylation (cyanidin)], carotenoids
Gladiolus	Anthocyanins (violet to red shades of lavender, pink and rose), anthoxanthins (ivory, cream to intense yellow)
Lilium	Anthocyanins, carotenoids (beta-carotene, cryptoxanthin, zeaxanthin)
Marigold	Lutein and zeaxanthin
Orchids	Anthocyanins, anthoxanthin, plastid pigments
Rose	Anthocyanidin (Cyanidin 3,5-diglucoside), pelargonidin, cyanidin, delphinidin, peonidin

2.1 Chlorophylls

Chlorophyll is the most abundant plant pigment in the plant world. It is essential to drive the life force on earth. It is not a single compound, but made up of two classical components—the bluish black chlorophyll-a and dark green chlorophyll-b. Chlorophyll is photoreceptive, absorbs blue and red spectrum of the visible light (400-700 nm) and reflects the unabsorbed spectrum in the green region. It is the green reflection that makes plant parts look green. The spectacular show of green turning to brilliant shades of yellow and red in autumn before plants shed leaves is nothing but manifestation of pigment change, from one to another. Chlorophyll is present in the membrane of disc-like structures called 'chloroplasts', which are like chemical factories powered by the sun to synthesize carbohydrates. It is one of the greatest miracles of nature that chlorophyll and its companions, sparked by sunlight, produce simple sugar (carbohydrate). The process is called 'photosynthesis', which beyond doubt is the most fundamental of all biochemical reactions and one that helps plants to grow, flower, produce seed and sustain life on earth. The chlorophylls produce basic carbohydrates which get further processed in the plants in to various other metabolites to form a complex network of biomolecules. Chlorophylls play a vital role in human health as well.

2.1.1 Health Benefits

Chlorophylls are typically consumed in much higher doses in a diet through green and leafy vegetables. Chlorophylls in green foods are credited with numerous interventions in human health maintenance including reduction of blood pressure, blood sugar control, brain activation, antimutagenic and anticlastogenic effects. In Asian countries, the dietary sources for chlorophyll-rich foods are far more extensive than in Western diets. The dark green vegetables are simultaneously excellent sources of carotenoids, in particular lutein, which contribute with the chlorophylls to the health-protective properties.

2.1.2 Protection against Cancer

The anti-mutagenic properties of chlorophylls have been demonstrated in various assays, and clearly, intake of chlorophyll has potential to act as a chemopreventive compound in humans. However, much of the research is equivocal. Chlorophyllin has shown both anticancer properties and cancer-promoting activity. Dietary consumption of green vegetables has been associated with the protection against mutagenic and clastogenic activity of genotoxicants. The crude leaf extract or the chlorophyllin proved to reduce clastogenic effects. Research supporting chlorophyll's roles in cancer chemoprevention indicates that these pigments are capable of deactivating potential carcinogens in the digestive tract. Chlorophyllin significantly reduces the incidence and multiplicity of skin tumors in mice.

2.2 Anthocyanins

Anthocyanins are water-soluble vacuolar pigments in the diverse flavonoid group, which are responsible for imparting blue, purple, red and orange colour in higher plants. The natural colour imparted by anthocyanins can be used as food and beverage additive to obtain attractive natural colour. Besides they can also be used as nutraceuticals as they possess strong anti-oxidant properties. The positive role of anthocyanins in mitigating vision disorders, neuroprotection, cancer, cardio vascular diseases is well documented.

More than 500 different structures have been identified in family of anthocyanin pigments. Anthocyanins are present in different plant organs, such as fruits, flowers, stems, leaves and roots. These pigments are normally found dissolved uniformly in the vacuolar solution of epidermal cells. However, in certain species, the anthocyanins are localized in discrete regions of the cell vacuole, called anthocyanoplasts. The main sources of anthocyanin (Table 3) are red fruits, mainly berries and red grapes, cereals (purple maize), vegetables and red wine among drinks. The glycosides of the three non-metylated anthocyanidins (delphinidin, cyanidins, and pelargonidin) are the most abundant in nature, which represent 80 per cent of leaf pigments, 69 per cent in fruits and 50 per cent in flowers. The distribution of the six most common anthocyanidins in the edible parts of plants is cyanidin (50 per cent), pelargonidin (12 per cent), peonidin (12 per cent), delphinidin (12 per cent), petunidin (7 per cent) and malvidin (7 per cent). The most widespread anthocyanin in most fruit is cyanidin-3-glucoside. Anthocyanins are interesting for two reasons. The first being technological, namely, their impact on the organoleptic characteristics of food which may influence their technological behaviour during food processes and the second reason being anthocyanin implication in human health through different pathways. Anthocyanins are relatively unstable and easily oxidised. They are sensitive to many factors, apart from pH, that may affect their stability and colour such as temperature and UV radiation. Many studies in recent years have focused on the study of the health properties of anthocyanins from different perspectives.

Table 3: Concentration of Anthocyanins in some of the Horticultural Crops

Crop	*Anthocyanin (mg 100 g^{-1})*
Pomegranate (juice)	600–765
Bilberry	300–698
Red cabbage	322
Red grape	30–750
Red radish	100–154
Cranberry	67–140
Strawberry	19–55
Red onion	23.3–48.5
Eggplant	8–85
Peach	4.2
Red apple	1.3–12

Many beneficial activities have been attributed to anthocyanins and the majority of work investigating these activities focuses on the antioxidant characteristics of anthocyanin rich diets and the enhancement of cardiovascular health.

2.2.1 Combating Free Radicals

The free-radical scavenging and antioxidant capabilities of anthocyanin pigments are the most highly publicized benefits that are in vogue. Free radical damage contributes to the aetiology of many chronic diseases and thus antioxidants may have beneficial effects on human health at different levels. Anthocyanins from roses have shown to be potent antioxidant and free radical scavengers in different *in vitro* models. Within the assayed anthocyanins the 3 glucoside, which is the only anthocyanin containing the 30, 40 and 50-hydroxy group (gallocatechol structure) in the B-ring, has shown the highest antioxidant capacity.

2.2.2 Cardiovascular Health

Anthocyanins are possibly the most widely studied for their effect on cardiovascular health and significant number of epidemiological evidence exists. Anthocyanins are capable of acting on different cells involved in the development of atherosclerosis, one of the leading causes to cardiovascular dysfunction. Vascular endothelial growth factor (VEGF) is a major pro-angiogenic and proatherosclerotic factor but anthocyanins have been shown to prevent the expression of VEGF in vascular smooth muscle cells. Anthocyanin extracts from chokeberry, bilberry, and elderberry have shown endothelium-dependent relaxation capacity in porcine coronary arteries. Intake of flavonoids, including tannins and anthocyanins, is associated with reduced incidence of coronary heart disease. A typical daily dietary intake (25–215 mg) without supplementation is sufficient to provide pharmacological benefits. The rich anthocyanin pigment and 'resveratrol' content of red wines, as well as associated flavonoids, are purported to be responsible for the well-publicized correlation between red wine consumption and reduced cardiovascular mortality (the French paradox). Because anthocyanins are potent inhibitors of nitrated tyrosine formation *in vitro*, they may have a role in prevention of atherosclerotic lesions in human coronary arteries. Bioflavonoid mixtures derived from fruit sources (*Rubus occidentalis*, and bilberry) have demonstrated ability to accelerate ethanol metabolism, reduce inflammatory and edematic symptoms, and improve permeability and strength of capillaries. Anthocyanins in red wines and grape juice contributed to the ability of these dietary sources to inhibit heart attacks, by virtue of their antithrombotic potential, and to inhibit platelet function and enhance nitric oxide release. Moreover, anthocyanins have an effect on cholesterol distribution, protecting endothelial cells.

2.2.3 Cancer Mitigation

It has been suggested that anthocyanins possess anti-inflammatory as well as chemopreventive properties. It has been shown that the anthocyanin-containing fraction of red wines is capable of suppressing proliferation of human colon cancer cells and gastric adenocarcinoma (AGS) cells. The proliferation of human hepatocellular liver carcinoma cells is inhibited by raspberry extracts, containing among other flavonoids anthocyanins. Some studies reported an inhibitory effect of the anthocyanins on cell proliferation in human vulva carcinoma cells and human colon carcinoma cells. Angiogenesis plays a critical role in the development of cancer. The treatment of human fibroblasts with cyanidin-3-glucoside prevents ochratoxin A-induced DNA damage. Anthocyanins were noted to inhibit cycloxygenase enzymes, which can be one marker for the initiation stage of carcinogenesis. Recently, anthocyanins from tart cherries reduced cell growth of human colon cancer cell lines. Anthocyanins and related flavonoids from berries including cranberry, lingonberry, bilberry and blueberry have demonstrated the ability to inhibit not only the initiation stages of chemically induced carcinogenesis but also the later promotion and proliferation stages. The anti oxidant ability of anthocyanins is the major factor that helps in mitigating some of the cancers.

2.2.4 Diabetes Management

Anthocyanins play a beneficial effect on diabetes at two levels. On one hand they can interfere with glucose absorption and on the other hand they may have a protective effect on pancreatic cells. It has been shown that the simultaneous and daily administration to rats of anthocyanins from red wine together with streptozotocin, a diabetogenic compound, decreases glucose concentration in urine and serum. Additionally, anthocyanins from red wine prevent the generation of free oxygen radicals, and decreases lipid peroxidation. Anthocyanins are also able to stimulate insulin secretion from rodent pancreatic beta-cells.

2.2.5 Liver Disorders

Anthocyanins are believed to possess the ability to protect the hepatic tissues from the free radicals thereby preventing the possible damage to the liver cells. Studies in animal models suggest that anthocyanins protect the liver cells when external toxin are induced for instance, anthocyanins from *Hibiscus sabdariffa* L. have shown to protect rat liver from hepatotoxicity induced by tertbutyl hydroperoxide by lowering the serum levels of alanine and aspartate amino-transferase and reducing oxidative liver damage. In rats in which hepatic injury was induced by the administration of D-galactosamine red potato extracts had a protective effect.

2.2.6 Gastric Effects

The protective effect of anthocyanins on gastric damage is closely related with the capacity of this group of flavonoids to prevent or ameliorate the inflammatory process. Inflammation is implicated, at least initially, in processes of gastric injury and in this regard, it has been shown that cyanidin protects gastric mucosa from the damage caused by aspirin.

2.2.7 Eye Care

Anthocyanins are well known for their ability in improving eyesight, including night vision. Black currant (*Ribes nigrum* L.) anthocyanins concentrated juice, when administered to human subjects, markedly improved visual adaptation to darkness. Anthocyanin pigments have been administered to remedy vision disorders, enhance visual acuity, and increase capillary resistance. After oral administration of three anthocyanin species to rats, intact pigments can be found in plasma; similar results were achieved subsequently in human subjects. Retinopathy and cataracts, serious consequences of diabetes mellitus, can be combated using plant-derived anthocyanin pigments. Anthocyanin extracts from a wide range of plant species including *Ipomoea batatas* and *Brassica oleracea* have demonstrated ability to inhibit AGH (a-glucosidase) enzyme.

2.2.8 Antibiotic and Antiviral Role

The effect of anthocyanins on microbial pathogens has not been studied in depth up to now. Few studies exist, however, on the effect of berry extracts on pathogens. Anthocyanins exhibit significant antimicrobial properties, and (in association with other flavonoids) have demonstrated quite effective inhibition of aflatoxin biosynthesis. Extracts from different berries had an inhibitory effect on gram-negative bacterial growth. The blackcurrant extracts inhibited the growth of *Escherichia coli*. Recent studies have demonstrated that blackcurrant concentrates inhibit the growth of different strains of *Staphylococcus aureus* and *E. coli*. Different cyanidin glycosides present in extracts of *Ribes nigrum* L. have shown antiviral properties towards influenza A and B viruses and Herpes-1 virus.

2.2.9 Neurological Impairments

Recent research has revealed a role for anthocyanin pigments in protection against neurological disorders, especially declines related to aging. *In vivo* evidence demonstrating that rats administered an anthocyanin-rich diet showed significant improvements in protection against age-associated impairments. Some of the researchers demonstrated not only protection but also reversal of age-related decrements when rats were fed a diet that comprised of 2 per cent blueberry extract. Anthocyanins were found to confer significant protection from oxidative stress and to be highly bio available in endothelial cells, which has direct relevance to neurodegenerative disorders. These polyphenolic compound was found to enhance red blood cell resistance to oxidative stress. Anthocyanin-rich fruits may be beneficial in reversing the course of neuronal and behavioral aging. Moreover, some of the researchers also reported that the blueberry extracts have the beneficial effect in Alzheimer's disease (a neurodegenerative illness) in transgenic mice.

2.3 Carotenoids

Carotenoids are the second most abundant pigments in nature, and consist of more than 700 members. Carotenoids, a large group of fat-soluble compounds, are responsible for many of the yellow, orange and red colors in plant and animal. Carotenoids make up a diverse class of natural pigments that are of interest for pharmaceutical use, as natural food colourants, nutrient supplements and animal feed. Carotenoids play an important role in plant reproduction, through their role in attracting pollinators and in seed dispersal, and are essential components of humans' diets. Several carotenoids protect humans by hindering free radical chain reactions caused by aging, smoking, pollution and other stresses. β-carotene is a precursor of vitamin A, and lutein and zeaxanthin have been implicated in the reduction of age-related macular disease (AMD). In terms of nutraceutical contribution, carotenoids are best recognized for their antioxidant capacity. Carotenoids are considered the most potent of biological quenchers of singlet oxygen. About 50 carotenoids have been identified in the human diet. β-carotene, lycopene, and lutein are the most prevalent carotenoids in the diet.

2.3.1 Combating Free Radicals

Carotenoids provide protection to vision and eye function, and against macular degeneration and cataracts. Macular degeneration, caused by blue light-mediated free radical damage to the retina, is the leading cause of irreversible blindness in aging adults. Lutein has been loosely referred to as 'the eye-protective nutrient' and its intake has been shown to be inversely associated with ocular diseases such as AMD and cataracts. Carotenoids have a photo-protective function in human metabolism, and hence have been used in therapies for photosensitivity diseases. Mechanisms suspected to be involved in the protective effects of these macula carotenoids are filtering blue light, and scavenging reactive intermediates generated in photo-oxidation.

2.3.2 Immune Fortification

Carotenoids are credited with biological promotion of immune system response. A broad range of studies point to multiple potential mechanisms by which carotenoids may boost the immune system by protecting immune cells from the reactive bactericidal species, increasing natural killer cell activity, and increasing total white blood cells.

2.3.3 Cardiovascular Ailments

A series of studies have now demonstrated a correlation between high intakes or high serum concentrations of carotenes and lowered risk of coronary artery disease. Inhibition of low-density lipoprotein (LDL) oxidation has been demonstrated, especially with low levels of carotene. Two-week dosage of tomato juice (rich in carotenoids) effectively lowers LDL in male subjects. Another potential mechanism for carotenoids to inhibit cardiovascular disease may be through a feedback mechanism to inhibit HMG-CoA reductase in humans.

2.3.4 Combating Cancer

Carotenoids enhance cell-to-cell communication because they can up-regulate gap junctions. Carotenoids are associated with inhibition of several types of cancers including cervical, esophageal, lung, prostate, colorectal and stomach. In particular, the carotenoid lycopene has been extensively researched for its role in cancer inhibition. Lutein and zeaxanthin intake has been linked to reduced risk of prostate cancer. Both β-carotene and vitamin A supplementation inhibits cell proliferation in carcinogenesis. Lycopene's inhibitory effect on prostate cancer cells has been linked to antioxidant mechanisms. Diets that incorporate diverse mixtures of different carotenoids are recommended due to this wide range of activities associated with different individual carotenoid compounds.

2.4 Betalains

Betalains have been widely used as natural colorants for many centuries, but their attractiveness for use as colorants of foods (or drugs and cosmetics) has increased recently due to their reportedly high anti-oxidative, free radical scavenging activities and concerns about the use of various synthetic alternatives. The main commercial sources of betalains are powders and concentrates of red beet (*Beta vulgaris*) or cactus pear (*Opuntia ficus indica*) extracts. Betalains comprise a class of nitrogen-containing plant pigments found in the cell sap of plants. In the ice plant (*Mesembryanthemum crystallinum* L.) betacyanin accumulation was discussed to screen tissue against damaging UV radiation. Little has been published with respect to the physiological role of betalains in mammals.

2.4.1 Health Benefits

Betalains have not been as extensively researched in terms of health benefits as the other major plant-derived pigments; however, antiviral and antimicrobial properties are well documented. Recent findings rank beet among the 10 most potent vegetables with respect to their antioxidant capacities. Commercially, betaxanthin pigments have been introduced as a food supplement in order to fortify processed food products with a nutraceutical natural colourant. Recent reports suggested that betalain-containing plant extracts have high antioxidant capacities. Betalains reportedly have anti-inflammatory, hepatoprotective activities, and protect low density lipoproteins (LDLs) from oxidation. A striking characteristic of betalain pigments is that they can be naturally absorbed by the body, that means a minimal transformation occurs during metabolism. The antioxidant potential was found to be related to structural features of the respective betalains.

2.4.2 Combating Cancer

Preliminary cancer-chemopreventive studies with crude beetroot extract have shown that the protection against skin and lung tumors in animal models is slightly higher than that anthocyanin pigments and carotenoids. Betanin appears to be a potent anticancer compound. Beetroot extract has demonstrated chemopreventive properties against both lung and skin cancers. Recent reports showed a significant inhibitory effect of beetroot towards skin and lung cancer in mice. Future studies will have to ascertain the alleged promising physiological and pharmacological effects of betalains on humans.

3. *In vitro* Approaches for Pigment Induction

Conventionally the flowers and leaves are used as sources for the extraction of nutraceutical pigments by solvent extraction method. However, the plant pigments are localized in the flowers and other young vegetative parts thereby restricting their availability only during the cropping season. On the other hand, natural pigments obtained from the fresh plant materials involve the limitations such as low metabolite yield, seasonal availability of raw material, fresh material losses, inconsistent product quality, and pigment degradation caused by storage and the extraction process. Therefore, there is a need to develop a system so as to freely express such nutraceutical pigments at will in controlled conditions so as to harvest them without the seasonal boundaries. Plant pigments can be produced by growing plant cells in tissue culture. A considerable amount of research efforts has been directed towards this goal, ranging from strain improvement, medium optimization, and optimization of culture conditions, elicitation, and intelligent process strategies. To achieve the commercialization of plant cell suspension technology for pigments production, the major task of related research and development is the achievement of large-scale cultivation of high yielding cell lines, high product level and productivity, and the stability of production over extended cultivation time. Different strategies are used to enhance pigment production, which may contribute to the scale up and commercialization of plant cell cultures for the production of pigment based food colourants and nutraceutical compounds. The protocols, thus developed would be helpful to upgrade the technology to pilot scale bioreactors to manufacture the pigments in *in vitro* conditions throughout the year in a highly controlled manner. There are distinct advantages of producing plant pigments in *in vitro* culture, rather than *in vivo*. These include the; Independent of seasonal boundaries, *in vitro* cultures are a superb model to test elicitation, Production can be more reliable, simpler, and more predictable, Interfering compounds that occur in the field-grown plant can be avoided, *in vitro* cultures can yield a source of defined standard pigments in large volumes, Compounds produced in *in vitro* cultures can be directly used, Isolation of pigments can be rapid and efficient, as compared to extraction from complex whole plant.

Being highly perishable the flowers are subjected to the vagaries of fluctuating market with steep fall in prices very often. To insulate the farmers from such price crash the excess produce could be used for the in vivo extraction of pigments. Similarly *in vitro* approaches could be developed to express the pigments independent of the season.

Keeping the unique advantage the flowers offer attempts have been made to express the high value nutraceutical pigments in *in vitro* conditions to harvest specific pigments. Systematic studies were made to identify the location of pigments in the cellular layers of roses to identify suitable varieties for *in vitro* expression. The paper discusses the beneficial effects of various pigments and the success achieved in expressing the pigments from rose, chrysanthemum (anthocyanins), calendula (carotenoids) and bougainvillea (betalins) by introducing precursors, enhancers, elicitors, besides manipulating the culture conditions.

3.1 Strategies Employed

Researchers across the world have adopted various strategies to induce *in vitro* pigment induction and enhancement (Figure 1).

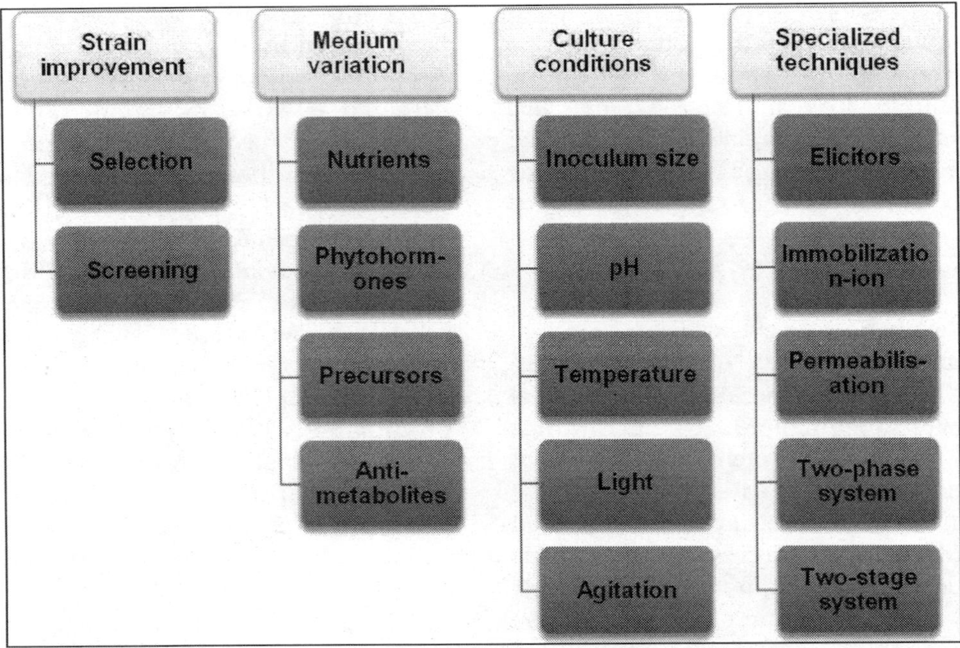

Figure 1: Strategies Evolved for *in vitro* Induction of Nutraceutical Pigments.

Selection of suitable varieties that are inherently rich in pigments and screening such varieties in search of specific plant part or layers of cells that are rich in pigments formed the first strategy.

- ✰ Manipulating the *in vitro* media constituents in terms of nutrients, plant growth regulators, precursors and anti metabolites formed the second strategy.
- ✰ Altering culture conditions in terms of manipulating the inoculums size, media pH, temperature, light and agitation of cultures became another strategy.
- ✰ Specialized techniques involving use of elicitors, immobilization ion technique, permibialization, two phase system of media were some of the recent strategies adopted in various labs.

3.1.1 Selection of Varieties

Flowers are known to contain a wide range of pigments of myriad colours. However, we have chosen those flowers that are edible in nature to address the major issue of food safety. Rose for instance is consumed in India as *gulkand*, rose petals are used in confectionery and traditional sweets, rose water is used as a food flavour. Calendula is a medicinal plant widely used in Ayurveda and homeopathy. Chrysanthemum ray florets are widely used as salad and herbal tea in China, Japan and Korea.

The varieties are chosen based on the initial investigations by estimating the total antocyanin or carotenoids or betalain pigments. Histological investigations were conducted to understand the distribution of pigments in different cell layers. Investigations in our lab at Indian Agricultural Research Institute, New Delhi revealed that significant differences exist in the amount of antocyanin pigments. Similarly, the distribution of pigments among cell layers in petals also vary from variety to variety. For instance the pigment distribution in rose cv. Dulhan is distributed uniformly in adaxial and abaxial epidermal cells besides sub epidermal cells and mesophyll cells (Figure 2).

In some of the varieties like Raja Surendra Singh of Nalagarh the distribution of pigments was restricted to the upper and lower epidermis without any presence in the intermediary cell layers (Figure 3).

3.1.2 Culture Medium Modifications

3.1.2.1 Nutrient Variations

In vitro anthocyanin production in cell lines is well documented in several crops such as grape, oxalis, carrots, radish, strawberry and sweet potato. Earlier reports on roses indicated that *in vitro* anthocyanin production was attained in cell cultures in suspension cultures but not in solid calli. The Division of Floriculture and Landscaping, IARI, New Delhi achieved a significant breakthrough by standardizing reliable protocol for *in vitro* induction and expression of anthocyanin pigments from callus cultures of rose cv. Pusa Ajay (Prasad *et al.*, 2012).

Early pigment initiation, enhancement and maximum anthocyanin production from calluses were recorded when leaf discs were cultured on *Euphorbia millii* (EM) medium supplemented with 7 per cent sucrose compared with calluses cultured at 4 per cent sucrose concentration under 16/8 h (light/dark) photoperiod regime. Reducing the concentration of

a. Rose cv. Dulhan

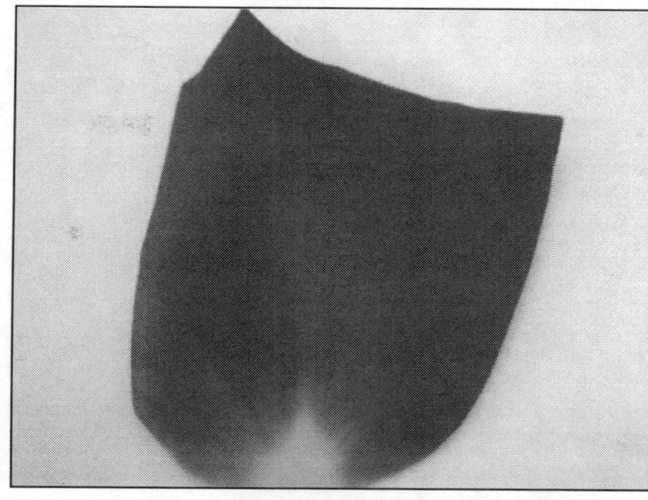

b. Pink Pigmentation Uniform in Petals

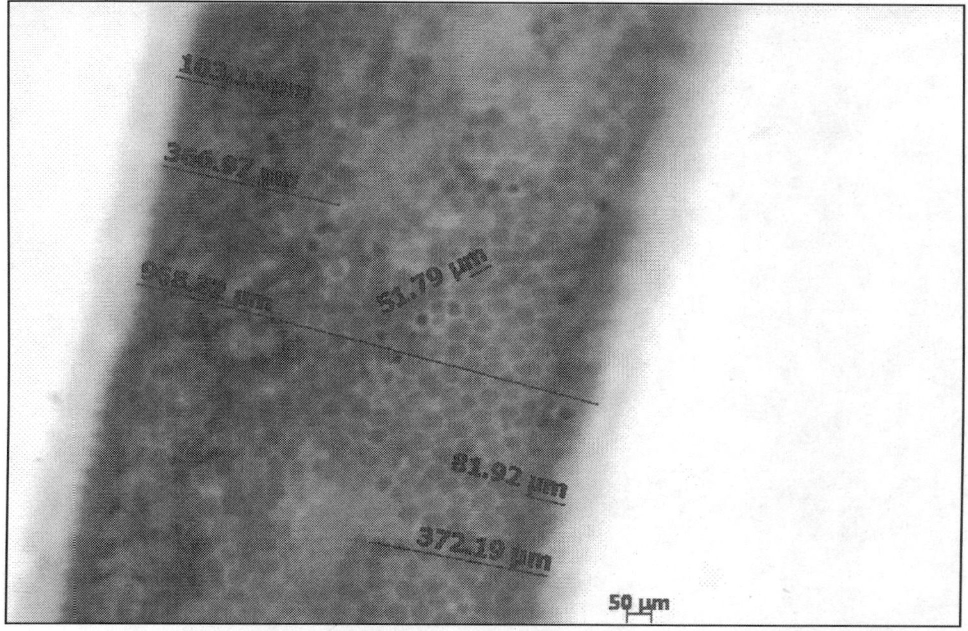

c. Both Abaxial, Adaxial Epidermal Cell and Mesophyll Cell contain Pigments

Figure 2: Distribution of Pigments in different Cell Layers in Rose cv. Dulhan.

NH^{4+} nitrogen in the solid MS medium led to slight improvement in anthocyanin production in rose leaf calluses (Mohan Ram, 2012).

Nishiyama *et al.* (2014) studied the anthocyanin accumulation in hairy root culture of *Ipomoea batatas* cv. Ayamurasaki. The amount of accumulated anthocyanins in hairy roots increased as a result of choosing the medium and increasing the sucrose concentration in the medium. The maximum anthocyanin was reported when these cultures were cultured on PRL- 4C medium containing 3 per cent sucrose in the continuous light irradiation (25-35 µE $m^{-2}s^{-1}$).

High sucrose (30 g l^{-1}) and nitrogen (100 per cent N of standard MS) concentrations and low temperature (15°C) significantly promoted anthocyanins accumulation in cultured shoots of *Clematis pitcheri*. Reduction of nitrogen compounds level in the medium to 50 per cent and lowering sucrose concentration to 10 g l^{-1} leads to decrease of anthocyanins accumulation in the shoots cultured at 15 and 20°C (Kawa-Miszczak *et al.*, 2009).

Colojn *et al.* (1981) studied the synthesis of malvidin and petunidin in pigmented tissue cultures of *Petunia hybrida* by culturing leaf explants on culture medium of Murashige and Skoog 1962 (MS medium) containing α-naphthalene acetic acid (NAA) at a concentration of 1 µg/ml. They observed that sucrose concentration in the culture medium higher than 3 per cent decreased pigment synthesis strongly. A decrease of pigment synthesis was also observed when the auxin (2,4-D or NAA) concentration in the medium exceeded a concentration of 0.5 µg/ml.

a. Rose cv. Raja Surendra Singh of Nalagarh

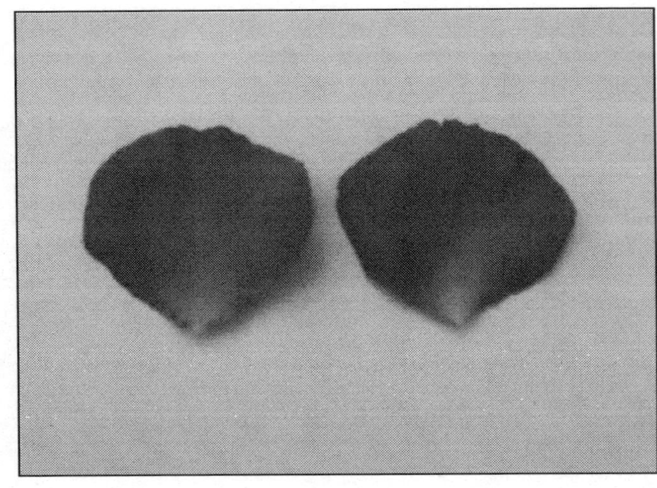

b. Uniform Orange Pigmentation in Petals

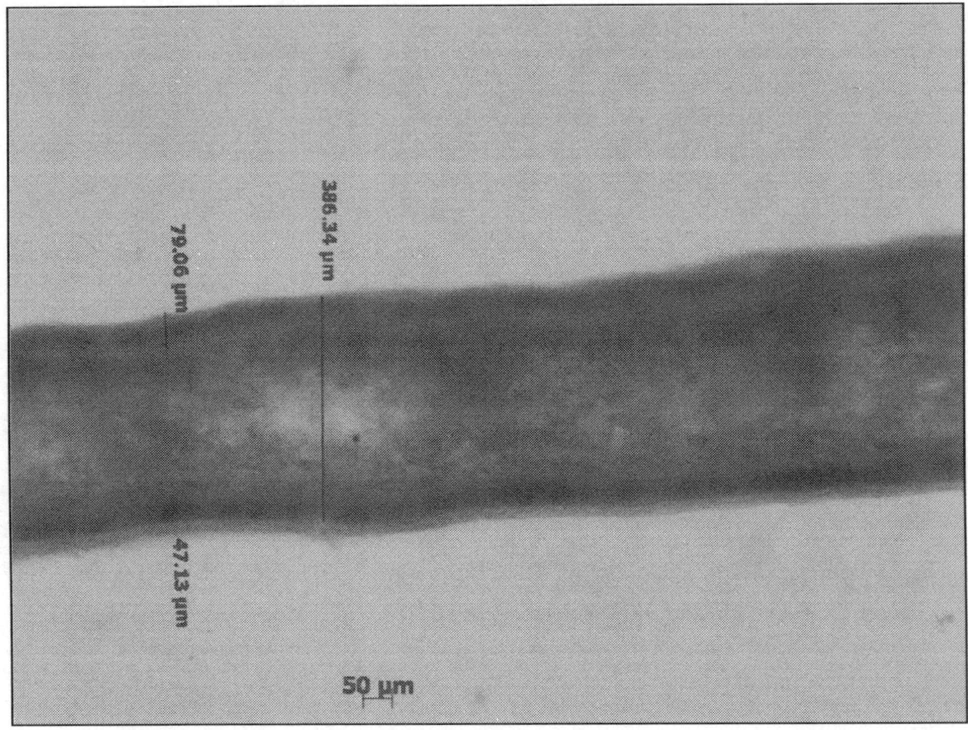

c. Only Abaxial, Adaxial Epidermal Cell contain Orange Pigments

Figure 3: Distribution of Pigments in different Cell Layers in Rose cv. Raja Surendra Singh of Nalagah.

The alternative iron source FeEDDHA enhances anthocyanin production, perhaps through the greater availability of iron in a high light environment or through the irons ability to inhibit anthocyanin degradation (Fang *et al.*, 1999).

3.1.2.2 Phytohormone Variations

In order to regulate anthocyanin and lignin synthesis in *Petunia hybrida* cell suspensions, Hagendoorn *et al.* (71991 induced anthocyanin in batch cultures by dilution stress method in which MS medium containing 30 g-sucrose and 2 mg/l NAA was mixed with 1 volume of medium from a six-day old culture using hypocotyl as explant. The cells were grown in the dark at 28°C on a gyratory shaker at 125 rpm. Dilution stress method led to an increased production of anthocyanin.

Callus tissues derived from cotyledonary leaf pieces from *Hyocyamus muticus* and hypocotyl pieces grown on a revised medium for growth and bioassays with tobacco tissue cultures. Medium supplemented with 0.5 mg/l BAP and 0.5 or 1.0 mg/l 2,4-D, or NAA, showed pigment accumulation after 15–17 days of inoculation (Basu and Chand, 1996).

Callus cultures of *Hyoscyamus muticus* L. cv. HMI-80 I was showed anthocyanin pigment accumulation after 15–17 days of inoculation when the cultures were grown on MS medium supplemented with 0.5 mg l^{-1} BAP and 0.5 or 1.0 mg l^{-1} 2,4-D. Cyanidin was found as a major compound by paper chromatography and spectrophotometric analysis (Basu and Chand 1996).

3.1.2.3 Carbon Source Variations

Mohan Ram *et al.* (2011) observed earliest expression (5.0 days) in maximum number of cultures (82.5 per cent) when EM medium was supplemented with 7 per cent sucrose. The red pigmentation became more intense as the concentration of sucrose in the culture medium was increased to 7 per cent.

In chrysanthemum pigment content (PC, 2.50±0.09 CV g^{-1} FCW) and pigment production (PP, 1.11±0.04CV test-tube^{-1}) were maximum in the calluses cultured on MS medium fortified with 204.5 mM sucrose (Bharat, 20154).

Individual florets (4–5 mm long) of a purple cultivar (*Fandango*) of the horticultural chrysanthemum (*Chrysanthemum morifolium* Ramat) were taken from flower buds just prior to opening and cultured in a sterile liquid medium (containing inorganic salts and sucrose) at 15°C under a 12-h day. For the first 14 days increase in wet weight was exponential. Anthocyanin appeared on the third day and was then synthesized rapidly. Chlorophyll and carotenoid were present initially: carotenoid levels rose quickly while chlorophyll remained almost constant. Highest pigment content and most growth were found when the florets were grown on 3 per cent sucrose. However, the highest anthocyanin concentration was found with 4 per cent sucrose, the highest carotenoid concentration with 0.6 per cent sucrose. No anthocyanin was produced when the florets were grown at 6 or 30°C; maximum yield was at 15 °C. Most carotenoid was formed at 30 °C and most chlorophyll was found at 20–5°C (Stickland, 1974).

Nagira and Ozeki (2004) developed a system in which anthocyanin synthesis can be induced under defined conditions in regenerated *Torenia fournieri* shoots. The various sugars (glucose, fructose, galactose, or maltose) were induces anthocyanin biosynthesis, but in medium containing more than 7 per cent glucose, fructose, or galactose, the growth of shoots was strongly suppressed. Solfanelli *et al.* (2000) reported that the sugar acts as a signalling molecule and the whole genome transcript profiling reveals that the flavonoid and anthocyanin biosynthetic pathways are strongly up-regulated by sucrose treatment.

Yamamoto *et al.* (1989) investigated the optimal conditions for anthocyanin production in suspension cultures of *Euphorbia millii* cell strain obtained by successive cell-aggregate selections. A modified Gamborg's solution containing 5 to 7 per cent sucrose (w/v), a 1:16 ratio of NH^+_4 : NO^-_3, 30 mM of nitrogen source, 10^{-6} M 2,4-D, 10^{-8} M BA, and low concentrations of Fe^{2+} and SO_4^{2-} greatly increased anthocyanin production. The optimum concentration of sugar was 5 and 7 per cent (w/v) for the cell growth and anthocyanin content, respectively.

Zwayyed *et al.* (1991) found that glucose gives a high growth rate and fructose gives higher anthocyanin content. However, a combination of both sources reduced both factors.

Mohan Ram *et al.* (2011) observed earliest expression (5.0 days) in maximum number of cultures (82.5 per cent) when EM medium was supplemented with 7 per cent sucrose. The red pigmentation became more intense as the concentration of sucrose in the culture medium was increased to 7 per cent.

Nagira and Ozeki (13) developed a system in which anthocyanin synthesis can be induced under defined conditions in regenerated *Torenia fournieri* shoots. The various sugars (glucose, fructose, galactose, or maltose) were useful in inducing anthocyanin biosynthesis, but in medium containing more than 7 per cent glucose, fructose, or galactose, the growth of shoots was strongly suppressed.

In pigmented cells of *Vitis vinifera* suspension cultures, best accumulation of anthocyanins was obtained when nitrate concentration was reduced from 25 mM to 6.25 mM and when sucrose concentration was increased from 88 to 132 mM. The high sucrose and low nitrate concentrations can be one of the important culture factors in controlling of anthocyanin production by cell cultures (Chi and Cormier, 4).

3.1.3 Culture Conditions

Chan *et al.* (3) found that the cultures exposed to 10 days continuous darkness showed the lowest pigment content, while the cultures exposed to 10 days continuous irradiance (301–600 lux) showed the highest pigment content in the cells of *Melastoma malabathricum*. The cell cultures incubated at a lower temperature range (20 ± 2°C) grew better and had higher pigment content than those grown at 26 and 29 ± 2°C.

3.1.4 Specialized Techniques: Elicitor Supplementation

Increased production of secondary metabolites under nutrient stress conditions in cell culture systems has been reported in many species. Under in vivo conditions anthocyanin was widely used as a marker of stress. The callus subjected to phosphate and nitrate stress produced 7.2 per cent and 8.5 per cent anthocyanin respectively (Rajendran *et al.*, 19). Pirie and Mullins (17) reported that the 20 µM ABA stimulated anthocyanin accumulation in the culture of both old and young leaf-discs of grapevine (*Vitis vinifera*) in the presence of 0.04 to 0.12 M sucrose and under intermittent light. Pirie and Mullins (1976) also found that the 20 µM ABA stimulated anthocyanin accumulation in the culture of both old and young leaf-discs of grapevine (*Vitis vinifera*) in the presence of 0.04 to 0.12 M sucrose and under intermittent light.

Sudha and Ravishankar indicated that the methyl jasmonate (MeJA) and salicylic acid (SA) were both found to stimulate the anthocyanin production in the callus cultures of *Daucus carota* by increasing the activity of Ca^{2+}ATPase. Treatment with the calcium ionophore A23187 resulted in the enhancement of both growth and anthocyanin production.

Calluses of chrysanthemum cultured on MS medium fortified with $CoCl_2$ exhibited maximum induction coefficient percentage (85.35±2.09) and took minimum days (7.50±0.65) for anthocyanin biosynthesis with maximum colour value index (PC= 2.48±0.09 CVg^{-1}FCW; PP= 0.74±0.02 CV test $tube^{-1}$). Similarly MS medium fortified with $NiCl_2$ medium exhibited maximum induction coefficient percentage (65.42±1.85) with highest colour value index (PC= 1.68±0.11 CVg^{-1}FCW; PP= 0.39±0.03 CV test $tube^{-1}$) but cultures on nickel chloride did not exhibit satisfactory results as compare to $CoCl_2$.

Amongst the elicitors tested [salicylic acid (SA), methyl jasmonate (MeJA), abscisic acid (ABA)], MeJA was most effective in inducing anthocyanin biosynthesis in calluses. MS media fortified with MeJA resulted in maximum response coefficient (92.42±1.12) per cent, with highest PC (3.20±0.11 CV g^{-1} FCW) and PP (1.45±0.05 CV $test\text{-}tube^{-1}$). Incorporation of in MS medium also resulted in good response coefficient percentage (82.84±1.87) with higher colour value index (PC= 2.68±0.10CV g^{-1}FCW; PP= 1.19±0.04CV $test\text{-}tube^{-1}$). However, no additive effect for response coefficient, PC and PP was observed in any treatment which had combined SA and MeJA. MS mediumfortified with ABA exhibited (88.28±0.68)induction coefficient percentage with pigment content (2.73±0.05 CVg^{-1}FCW) and pigment production (1.24±0.02 CV test $tube^{-1}$) values (Bharat, 2015).

4. Conclusions

Plants ant their parts are traditionally part of the human diet containing bioactive components that may exert physiological effects beyond nutrition promoting human health and well-being. The combined evidence of numerous research trials indicates that variety of different pigments in a well-balanced diet is key to prevention and therapy for a range of chronic human disease conditions. Regular consumption of fruits and vegetables is associated with reduced risks of chronic diseases such as cancer, cardiovascular disease, stroke, Alzheimer's disease, cataract and age-related functional decline. Thus, in addition to their immediate function ameliorating the visual quality, colouring foodstuffs such as juice, their respective concentrates but also fruit, vegetable and flower powders incorporating the pigments may improve the nutritional value of the target food. These potential nutraceutical properties are even more important at a time, where human diet is increasingly based on processed foods. Because the natural colours market is forecasted to prosper in the future, further activities aim at finding forgotten or even new pigment sources. The selective activation of those metabolic pathways conducting to the synthesis of more stable or bioactive final products will open a promising perspective for the design of functional foods or new foods with health related claims. The physicochemical principles that have evolved in plants to enhance and modify colour have been mostly neglected and should be considered in the future by plant breeders, horticulturists, food technologists and dieticians for selection, growing, processing, utilization of the respective colour sources as well as the nutritional benefit offered by the ingested compounds.

5. References

Basu, P. and Chand, S. 1996. Anthocyanin accumulation in *Hyoscyamus muticus* L. tissue cultures. *J. Biotech.*, **52**: 151–159.

Chan, L.K., Koay, S.S., Boey, P.L. and Bhatt, A., 2010. Effects of abiotic stress on biomass and anthocyanin production in cell cultures of *Melastoma malabathricum*. *Biol. Res.*, **43**: 127–135.

Chi, B.D. and Cormier, F. 1991. Effects of low nitrate and high sugar concentrations on anthocyanin content and composition of grape (*Vitis vinifera* L.) cell suspension. *Plant Cell Rep.*, **9**: 500–504.

Colijn, C.M., Jonsson, L.M.V., Schram, A.W. and Kool, A.J. 1981. Synthesis of malvidin and petunidin in pigmented tissue cultures of *Petunia hybrida*. *Protoplasma*, **107**: 63–68.

Fang, Y., Smith, M.A.L. and Pepin, M.F. 1999. Effects of exogenous methyl jasmonate in elicited anthocyanin-producing cell cultures of ohelo (*Vaccinium pahalae*). *In vitro Cell. Dev. Biol. Plant*, **35**: 106–113.

Hada, Bharat Singh 2015. *In vitro* Induction and characterization of anthocyanin pigments in chrysanthemum. *Ph.D. Thesis*, PG School, IARI, New Delhi.

Hagendoorn, M.J.M., Zethof, J.L.M., Hunnik E. and Plas, L.H.W. 1991. Regulation of anthocyanin and lignin synthesis in *Petunia hybrida* cell suspensions. *Pl. Cell, Tiss. and Org. Cult.*, **27**: 141–147.

Hasin, B.M., Ferdaus, A.J.M., Islam, M.A., Uddin, M.J., and Uddin Islam, M.S. 2006. Marigold and orange skin as egg yolk colour promoting agent. *Int. J. Poul. Sci.*, **5**, 979–987.

Kawa-Miszczak, L., Wegrzynowicz-Lesiak, E., Gabryszewska, E. and Saniewski, M. 2009. Effect of different sucrose and nitrogen levels in the medium on chlorophyll and anthocyanin content in *Clematis pitcher* shoots cultured *in vitro* at different temperatures. *J. Fruit Orn. Pl. Res.*, **17**(1): 113–121.

Mohan Ram 2012. Induction and characterization of anthocyanin pigments from callus cultures of rose (*Rosa hybrida* L.). *Ph.D Thesis*, PG School, IARI, New Delhi.

Mohan Ram, Prasad, K.V., Kaur, C., Singh, S.K., Arora, Ajay and Kumar, S. 2011. Induction of anthocyanin pigments in callus cultures of *Rosa hybrida* L. in response to sucrose and ammonical nitrogen levels. *Pl. Cell Tiss. Org. Cult.*, **104**: 171–179.

Moreno, D.A., Garcý´a-Viguera, C., Gil, J.I., Gil-Izquierdo, A., 2008. Betalains in the era of global agri-food science, technology and nutritional health. *Phytochem. Rev.*, **7**: 261–280.

Nagira, Y. and Ozeki, Y. 2004. A system in which anthocyanin synthesis is induced in regenerated torenia shoots. *J. Pl. Res.*, **117**: 377–383.

Nishiyama, Y. and Yamakawa, T. 2004. Effect of medium composition on the production of anthocyanins by hairy root cultures of *Ipomoea batatas*. *Plant Biotech.*, **21**(5): 411–414.

Ogata, J., Kanno, Y., Itoh, Y., Tsugawa, H. and Suzuki, M. 2005. Plant biochemistry: Anthocyanin biosynthesis in roses. *Nature*, **435**: 757–758.

Piccagliy, R., Marotti, M., and Grandi, S. 1998. Lutein and lutein ester content in different types of *Tagetes patula* and *T. erecta*. *Industrial Crops and Products*, **8**, 45–51.

Pirie, A. and Mullins, M.G. 1976. Change in anthocyanin and phenolics content of grapevine leaf and fruit tissues treated with sucrose, nitrate and abscissic acid. *Plant Physiol.*, **58**: 468–472.

Prasad, K.V., Mohan Ram Legha and Janakiram, T. 2012. *In vitro* anthocyanin production from rose ICAR–News Volume **17**(4), October–December 2011.

Rajendran, L., Ravishankar, G.A., Venkataraman, L.V. and Prathiba, K.R. 1992. Anthocyanin production in callus cultures of *Daucus carota* as influenced by nutrient stress and osmoticum. *Biotech. Lett.*, **14**(8): 707–712.

Solfanelli, C., Poggi, A., Loreti, E., Alpi, A. and Perata, P. 2000. Sucrose specific induction of the anthocyanin biosynthetic pathway in *Arabidopsis*. *Plant Physiol.*, **140**: 637–646.

Stickland, R.G. 1974. Pigment production by cultured florets of *Chrysanthemum morifolium*. *Ann. Bot.*, **38**(1): 1–6.

Sudha, G. and Ravishankar, G.A. 2003. Influence of putrescine on anthocyanin production in callus cultures of *Daucus carota* mediated through calcium ATPase. *Acta Physiol. Plantarum*, **25**: 69–75.

Yamamoto, Y., Kinoshita, Y., Watanabe, S., and Yamada, Y. 1989. Anthocyanin production in suspension cultures of high producing cells of *Euphorbia millii*. *Agric. Biol. Chem.*, **53**: 417–423.

Youwei, Z., Jinlian, Z. and Yonghong, P. 2008. A comparative study on the free radical scavenging activities of some fresh flowers in southern China. *LWT–Food Sci. and Tech.*, **41**: 1586–1591.

Zwayyed, S.K., Frazier, G.C. and Dougall, D.K. 1991. Growth and anthocyanin accumulation in carrot cell suspension cultures growing on fructose, glucose or their mixtures. *Biotech. Prog.*, **7**: 288–290.

73

Tuber Crop Based Novel Products and Functional Foods for Improving Profitability

James George and M.S. Sajeev*

ICAR-Central Tuber Crops Research Institute,
Sreekariyam, Thiruvananthapuram – 695 017, Kerala
**E-mail: ctcritvm@yahoo.com, ctcritvm@gmail.com*

Tropical tuber crops, which include cassava (tapioca), sweet potato, elephant foot yam, yams, taro, coleus *etc.*, form an important group of crops which are produced with very low inputs, are consumed by tribals and economically backward sections of the population, contribute significantly to food and nutritional security and are also used for animal feed or as raw material for processing industries. Many of these crops are considered as household food bank because of their ability for staggered harvesting.

It is predicted that climate change will reduce the yield of most of the grain crops such as rice and wheat, to the extent of 2-6 per cent by 2030 and 5-11 per cent by 2050. At the same time, studies conducted at CTCRI clearly show that tuber crops such as cassava, elephant foot yam and yams are climate resilient crops as they are little affected by drought or surge in temperature. Also, there will be huge demand for food grains in the energy sector. More than half of cereal produced in 2006-07 was used for non-food purpose including bioenergy and this may be aggravated by 2050 when fossil fuel is predicted to be exhausted. The likely demand-supply gap of major food grains can be effectively bridged by tuber crops that are concentrated source of carbohydrate. It is also predicted that cereal consumption in India will decline from 60 per cent of total calorie uptake at present to 48 per cent by 2050. On the contrary, there will be considerable increase in vegetable consumption including tuber crops which are rich in minerals, vitamins, antioxidants and dietary fibre.Tuber crops are mostly boiled and eaten as food in Kerala, north eastern States and in most of the tribal regions of India, however there are various ethnic preparations and dishes in different parts of the country favoured by a large section of the society. Some tuber crops like taro, tannia, elephant foot yam are very important vegetables in most part of the country. However, there is a great potential for value-addition in these crops which probably would rewrite the destiny of these 'future crops'. It is expected that value-addition in these crops would expand the area of cultivation as well as benefit the farmers in fetching higher returns.

1. Cassava Based Food Products

1.1 Cassava Rava and Porridge

Cassava or tapioca (*Manihot esculenta* Crantz) is one of the important tuber crops valued for its high starch content (20-35 per cent). Cassava rava is a pre-gelatinized granular product similar to wheat semolina and finds use as a breakfast recipe product. For the preparation of cassava rava, the tubers are peeled and sliced into round chips. It is then partially cooked by boiling in water for 5 min., decanting the steep water is sun-drying the parboiled pieces and powdering coarsely in a hammer mill. Then, sieved through fine sieve to separate out the finest fraction which can be converted to porridge powder by flavoring with cardamom and fried powdered cashew nuts. The residue is sieved through larger mesh size sieve to obtain rava. The uneven large pieces are again powdered to recover the rava.

1.2 Fried Cassava Chips

Fried cassava chips presently available in the market are often too hard to bite and bear no comparison with potato chips. This leads to poor acceptability of the product and lower price. Research at ICAR-CTCRI has shown that excellent quality fried chips can be made from cassava tubers, by soaking the chips in acetic acid-brine solution for 1 h, parboiling for 5 min., surface drying and deep frying in oil. This facilitates in the removal of excess starch and sugars from the cassava slices, with the result that light yellow crispy chips can be obtained, having soft mouth feel and good texture.

1.3 Cassava Based Extruded Products

The demand for extruded snack products is expanding at a phenomenal rate in developed and developing countries. Extrusion cooking is a high temperature short time cooking process designed for processing of starchy as well as proteinaceous materials. The use of extrusion cooking has distinct advantages like versatility, high productivity, low cost, product shapes, high product quality and production of new products. Several cereal based extruded products are available commercially. But tuber crop based extruded snack product has yet to appear in the market as we lack both technologies for value-addition and products acceptable to elite/urban populace. Being the treasure house of starches with complex physico-chemical properties, cassava can be extruded to obtain a variety of nutritionally enriched, ready to eat/cook products. Cassava tubers after washing, peeling and slicing into chips are dried and powdered in a hammer mill. The dry flour after conditioning to 12-15 per cent moisture content is extruded by maintaining appropriate temperatures at different sections of the barrel and dieof the food extruder. Cassava being rich in carbohydrates and lacking in protein content, addition of low cost protein sources like wheat, finger millet, soy flour *etc.* gave more nutritional and market value products (Sajeev and Sheriff, 2013).

1.4 Protein and Fibre Enriched Functional Foods

The cost of making value-added snack foods from cassava could be considerably reduced, if wet cassava paste is used instead of cassava flour. Such an innovation was made in making a highly acceptable crisp snack food *viz.*, chitchore from cassava. The wet cassava tuber paste is mixed with ingredients like *maida*, cheese, salt, sugar, baking powder and white pepper. The dough after proofing for 1h is spread into sheets and cut into small discs of 1cm diameter. These are then deep fried in oil.

Keeping an eye on the health conscious consumers, mini- papads are developed from cassava flour by adding fibre sources like wheat bran, oat meal, rice bran and cassava fibrous residue. The fibre sources are added to gelatinized cassava slurry and mixed thoroughly (Padmaja *et al.*, 2013). The spicy condiments are also added and spread on plastic sheets which are then dried in the sun for 36h. The papads are peeled off from the sheets and packed. The deep fried products have soft and crisp texture (Mini- papads with high protein content (7-15 per cent) could be made from cassava flour by adding protein sources like cheese, defatted soy flour, prawn powder and whey protein concentrate along with other spicy condiment. The papads are allowed to dry for 36 h, after which they are separated from the sheets and packed. This could be deep fried in oil before use.

1.5 Fried Snack Foods from Cassava

Cassava Pakkavada

This is a hot snack food having good texture and taste made out of cassava flour. The other ingredients include *maida*, bengal gram flour, salt, chilli powder, asafoetida, baking soda and oil. The ingredients are thoroughly mixed and made into dough with hot water (50°C), proofed for 1h and then extruded through hand extruder having flat rectangular holes, into hot oil.

Cassava Sweet Fries

This is a sweet snack food made out of cassava flour, *maida*, baking soda and oil. The ingredients are mixed well and made into dough with hot water (50°C). The dough after proofing for 1h is hand extruded through die having round holes, into hot oil. The fried product is then coated with sugar by dipping for a few minutes in sugar syrup having thick consistency.

Cassava Nutrichips

This is a high protein snack food made out of cassava flour by mixing with other ingredients like *maida*, groundnut paste, egg, salt, sugar, *sesame*, coconut milk, baking soda and oil. After mixing the ingredients, hot water is added and mixed to form smooth dough. The dough after proofing is made into small balls which are then spread into sheets of 0.2 cm thickness. This is then cut into dimon shape using a sharp knife and deep fried in oil.

Cassava Crisps

This is a soft and good textured crispy snack food made from cassava flour, *maida*, rice flour, bengal gram flour, salt, baking soda, turmeric powder and oil. The dough made with hot water is proofed for 1h and then extruded through the small pore size die having round holes. The deep fried material is mixed with fried nuts, curry leaves *etc.* before packing (Sajeev *et al.*, 2013).

Other Products

Cassava nutrichips (without egg), cassava salty dimons, cassava hot sticks, cassava salty fries, cassava sweet dimons *etc.* for which also formulations are available.

1.6 Cassava Starch, Sago and Wafers

The process of extraction of starch consists of peeling, rasping, screening, settling and drying. Peeled roots of cassava are disintegrated into pulp by a rasper which releases the starch granules from the fibrous matrix. The resulting slurry is pumped onto a series of vibratory screens and the fibrous waste (thippi) is retained on them and the starch milk passing through the sieves are channeled into sedimentation tanks. After at least 8 hours of settling, the supernatant liquor is run off and the starch cake settled at the bottom is scooped up for sun drying on a cement floor. Sago (Saboodana) is manufactured from the partially dehydrated (35-40 per cent m.c) starch cake. The lumps are broken in a spike mill and then globulated on a gyratory shaker. The globules are graded according to size and then partially gelatinized by roasting on shallow metal pans. Finally the sago pearls are dried in the sun on cement flour. The agglomerates are separated by means of a spike beater and polished before bagging (Sajeev *et al.*, 2012).

Wafers are made by arranging the wet granules in suitable dies and steaming. The steamed granule take the shape of the die and after drying, it can be separated out from the dies and packed.

1.7 Tea Time Snack Foods

Tasty cutlets could be made from cooked and mashed cassava tubers by mixing with other ingredients like onion, ginger, green chillies, green peas (cooked), beet root spice mix *etc.* and preparing cutlets in the conventional manner. Cooked cassava tubers could be made into tasty samosas and bondas by onion, ginger, green chillies, green peas and spices and this is used as a filler to thin chapattis, shaping to the samosas by folding and fixing the edges. This is then deep fried in oil. Bondas are made by shaping the mix into balls and coating with thick batter of basan and deep frying.

2. Functional Pasta from Sweet Potato and Cassava

Cassava is known to be a high starch, high glycaemic food, while sweet potato is a low glycaemic health food. Whilst the global prevalence of diabetes is projected to increase from 4 per cent in 1995 to 5.4 per cent by 2025, approximately 170 per cent increase in diabetic population has been predicted in developing countries, with India topping the list, followed by China. FAO-WHO Expert Consultation recommends the increased consumption of low glycaemic foods rich in resistant starch, non-starch polysaccharides and oligosaccharides. Foods with low glycaemic response are reported to be of use in the treatment of obesity, type 2 diabetes mellitus and in weight management. Considering the projected rise in diabetic population in India to 80 million by 2030 and in an attempt to diversify the use of these root crops, studies were made at ICAR-CTCRI to develop an array of pasta products having high functional value coupled with low starch digestibility. Besides, pasta getting wide popularity among the young Indians and in the metros as a convenient food, transformation to new health and wellness foods is essential to add value to cassava and sweet potato and sustain their cultivation of in India (Padmaja *et al.*, 2015).

2.1 Protein-enriched Pasta

As protein content and gluten strength are critical factors deciding the cooking quality of pasta and since sweet potato lacks gluten, fortification with protein sources like whey protein concentrate (WPC), defatted soy flour (DSF) and fish powder (FP) was attempted to understand their impact on starch-protein network formation. The study showed that slowly digestible functional pasta with good quality could be developed form sweet potato, which also has high protein content.

Lack of protein in cassava flour is a major impediment to the development of pasta type of products from it. Nevertheless, technology for making protein rich pasta from cassava through fortification with other ingredients was standardized at ICAR-CTCRI.

2.2 Fibre Enriched Pasta

The rapid increase in lifestyle diseases have led to an increasing awareness among the consumers about the health benefits of dietary fiber, and several reports indicate that diets rich in fiber could reduce the risk from coronary heart disease, cancer, obesity, and diabetes. The physiological effects of dietary fiber depend on the type of fiber, its chemical and physical composition, solubility, *etc.*, and have reported effects like prevention of colon cancer, constipation, *etc.* Water-soluble dietary fibers increase the viscosity of food in the intestines and thereby slow down glucose and sterol absorption. Low glycaemic diets rich in dietary fiber have been reported to reduce the insulinemic responses to food and exert hypocholesterolemic action as well as protective effects against certain types of cancer (Björck *et al.*, 200). Several studies have been carried out to improve the nutritional properties of pasta using protein and dietary fiber sources, isoflavones, ω-fatty acid, β-glucans, carotene sources, resistant starches, *etc.* Cereal bran sources have been used to fortify wheat pasta to enhance their dietary fiber content. The type and amount of added fiber have been reported to influence the cooking and starch degradation in durum wheat semolina pasta.

Sweet potato flour based pasta rich in dietary fiber was made at ICAR-CTCRI using dietary fiber sources like oat bran, wheat bran, and rice bran, with the objective of enhancing the functional value of sweet potato pasta as a prophylactic/therapeutic diet in the prevention or management of conditions like diabetes, cardiovascular conditions, and cancer. Fortification of sweet potato flour with dietary fiber sources resulted in products having slow and progressive starch digestibility and very high percentage of starch remained undigested (RS) after 2 h of digestion under *in vitro* conditions (Bouwkamp, 1985). The slow digestibility of the fiber enriched sweet potato pasta coupled with the high level of residual undigested starch make these pastas ideal foods for diabetic and obese people.

2.3 Hydrocolloid Fortified Pasta from Tuber Crops

Hydrocolloids (gums) have been extensively used in the food industry to modify the dough quality and have been reported to affect the pasting properties of starch, retrogradation *etc.* As the cooking quality of pasta dried at high temperature is reported to be superior to that dried at low temperature, it was thought worthwhile to study how the three gums such as guar gum, xanthan gum and locust bean gum modify the cooking quality of sweet potato pasta and its starch digestibility and texture, when dried at two temperatures *viz.*, 55°C (low) and 70°C (high). Edible gums such as guar gum (GG), xanthan gum (XG) and locust bean gum (LBG) were also incorporated at 1 per cent level in cassava-maida blends and their cooking as well as digestibility characteristics were studied. *In vitro* starch digestibility was slow and progressive over a period of 2 h for the control as well as the gum fortified pastas, with a high retention of resistant starch after the digestion

2.4 Low Glycaemic Spaghetti from Tuber Crops

The high incidence of metabolic diseases such as type 2 diabetes, cardiovascular conditions and obesity, among people consuming foods rich in carbohydrates has led to increased research efforts in the development of low glycaemic foods. The effect of fortification of sweet potato flour with banana and legume starches as well as sweet potato starch itself in producing low glycaemic spaghetti (flour based noodles) from sweet potato was investigated. Also the effect of three gum sources such as guar gum, xanthan gum locust bean gum in improving the cooking characteristics and reducing the starch digestibility of sweet potato spaghetti was studied. Sweet potato: maida flour blend was fortified with commercial (edible) gum sources such as guar gum, xanthan gum and locust bean gum at three levels *viz.*, 0.5 per cent, 1.0 per cent and 1.5 per cent. The formulation contained 10 per cent whey protein concentrate which by virtue of its ability to mimic gluten, helped to give excellent binding with starch and also elevated the crude protein content (Padmaja *et al.*, 2015).

Several researchers have reported the benefits of pulses in controlling metabolic diseases such as diabetes mellitus and cardiac diseases as well as their effect in weight management due to their low glycaemic indices. The effect of fortification of sweet potato with three common legume flours such as black gram (Urad dhal), Chickpea (basan) and green gram (mung bean) on the *in vitro* starch digestibility and glycaemic index of sweet potato spaghetti was also studied to develop low glycaemic spaghetti.

Resistant starch, is gaining lot of importance recently as a food additive due to its number of physiological effects such as prevention of colonic cancer, hypoglycaemic action, reduction in gall stone formation, hypocholesterolaemic effect, control of obesity *etc.* Hence low glycaemic spaghetti was also developed successfully using the resistant starch source, NUTRIOSE (Lefranc-Millot, 2008).

2.5 Starch Noodles from Sweet Potato

Starch noodles, presently known by different names such as glass noodles, cellophane noodles, vermicelli, *bihon* noodles *etc.* have been a favorite food in China since 1400 years. They are now produced from purified starches of various plant sources and have become popular in several Asian countries as well. Starch noodles differ in their quality and texture from Asian or Italian pasta/spaghetti, as the latter are made from flour or semolina, although many fortified pasta and noodle products have been attempted by subsequent researchers with a view to improving the quality. Starch noodles have been classified according to the type of raw materials, size of the noodle strands, manufacturing method *etc.* Traditionally, starch noodle made from mung bean starch is considered as the best owing to the transparent appearance, fine threads, high tensile strength and low cooking loss.

Although starch noodles have been developed form sweet potato starch and are very popular in China, Korea and the Philippines, the product has a bland taste and low nutritive value as starch is the only ingredient. Hence, it was thought worthwhile to enhance the protein content through fortification and to study how this affects the cooking and starch digestibility characteristics. Further, the effect of blending sweet potato starch with banana, green gram (mung bean) and cassava starches native form as well as with resistant starch enhanced annealed cassava starch in improving the quality and reducing the starch digestibility of sweet potato starch noodles was also studied. Sweet potato starch noodles fortified with NUTRIOSE were also developed and its low gklycaemic nature studied.

2.6 Gluten-free Spaghetti from Sweet Potato

Coeliac disease is an immune-mediated disease affecting approximately 15 million people round the globe. The disease has assumed alarming proportions in the last decade with one in 133 people suffering from coeliac disease in the USA and Canada, one in 100 in UK, one in 200 in Germany and one in 300 in Europe. Coeliac disease patients have intolerance to gluten, the proteins in wheat, rye and barley. Consumption of gluten leads to inflammation of the small intestine ultimately damaging the villi and affecting the absorption of iron, vitamins, minerals *etc.* (Green and Jabri, 2003). A strict gluten-free diet as a lifelong diet strategy is the only treatment for coeliac disease. With the rapid rise in coeliac disease patients and the advent of new techniques to diagnose this chronic health condition, the demand for gluten-free diets has also increased worldwide.

Gluten-forming proteins are essential for proper dough development and texture of a large number of baked foods and pasta products. The typical viscoelastic behaviour exhibited by the durumwheat proteins allows good networking of the matrix and provides the quality attributes of pasta. Nevertheless, as gluten containing substances have to be totally avoided in the development of gluten-free spaghetti, the networking ability of gluten has to be imparted to the dough using other substances that mimic gluten.

Higher intake of refined sugar by celiac patients than normal individuals has been reported increasing their chances of becoming diabetic patients as well, which stresses the need to develop low glycaemic gluten-free spaghetti for such people. Several researchers have attempted the development of gluten-free pasta and spaghetti products using materials such as buckwheat, maize, quinoa, sorghum *etc.* However, there was only limited information on the production of gluten-free pasta from sweet potato (*Ipomoea batatas* Lam.). Hence studies were made to develop gluten-free pasta from sweet potato flour-rice flour blends along with additives such as WPC and guar gum.

3. Sweet Potato Based Products

Sweet potato is cultivated throughout the tropics and warm temperate regions of the world for its starchy roots, which can provide nutrition, besides energy. A number of novel food products with functional value are being developed worldwide. Sweet potato tubers with their low glycaemic index have additional value as a food for diabetics There are a range of primary food products that could be made from sweet potato like chips, flakes, frozen products, French fries, puree *etc.*, while it is also the raw materials for a host of secondary products like noodles, sugar syrups, alcohol, pasta *etc.*

Sweet potato based composite flours have been used in many countries for making small baked goods like cakes, cookies, biscuits, doughnuts *etc.* Sweet potatoes are consumed at home level, mainly after cooking, baking or converting into fried chips. The roots are often converted to canned or pureed form, to enhance the shelf-life. Sweet potato based baby foods are preferred in many countries as the first solid food for infants.Canning of sweet potato is widely practised in the United States, to enhance the storage life and ensure round- the year availability of the product (Woolfe, 1992).

Sweet potato roots are transformed into more stable edible products like fried chips, crisps, French fries *etc.* Sweet potato puree, is a primary processed product from the roots, which is used directly as a baby food or used for mixing various food items like patties, flakes, reconstituted chips *etc.* High quality puree can be made from white, cream or orange- fleshed sweet potatoes and also from tubers of any size or shape.

Sweet potato roots can be termed as a '3-in-1' product, as it integrates the qualities of cereals (high starch), fruits (high content of vitamins, pectins *etc.*) and vegetables (high content of vitamins, minerals *etc.*). The beneficial effects of these ingredients have been appropriately put to use by converting the roots into a number of intermediary food products like jam, jelly, soft drinks, pickles, sauce, candies etc

4. Sweeteners from Starch

Starch is an important raw material for the production of sweeteners and functional oligosaccharides. Liquid glucose and high fructose syrup are made from cassava starch by liquefying the starch and saccharifying it using enzymes or acids. After saccharification, decolourisation is done using charcoal to produce glucose syrup. Fructose syrup is obtained from glucose syrup using glucose isomerase at 60-62°C for 2h. This is then decolourised and concentrated. HFS is a highly valued product for the confectionery industries, due to its specific properties like non-crystallising nature, extra sweetness *etc.* Maltose syrup is another product which is commercially prepared from starch, which finds application in food and confectionery industries. The essential composition includes maltose, glucose, maltotrioses, and higher oligosaccharides. The process consists in hydrolysing the starch using α-amylases from native sources like the rice seedling extracts. CTCRI has also standardized techniques for the production of maltodextrins with different dextrose equivalents (DE), which can be used as fat replacers in low calorie foods (Blanchard and Katz, 2006). The process involves the treatment of starch with heat-stable bacterial amylase and purification of the product.

5. Bioethanol from Starch

Fresh cassava tubers, dry chips/flour or starch can be used for the production of ethanol. The process consists of three steps *viz.*, liquefaction, saccharification and fermentation. During liquefaction, the cooked starch slurry is hydrolysed to maltose and low molecular weight dextrins using either acids or α-amylase. In the saccharification step, hydrolysis to glucose is achieved using acids or glucoamylases. During fermentation, the glucose formed is fermented to ethyl alcohol using yeast, *Saccharomyces cerevisiae*. The optimum concentration of sugars for ethanol fermentation is 12-18 per cent and optimum pH and temperature are 4.0 to 4.5 and 28-32°C respectively. After 48-72 h of fermentation, alcohol is recovered through distillation.

This process is highly energy-intensive and involves use of costly enzymes and hence the overall energy balance is negative. However, presently, improved enzymes are available and a global shift has already occurred. Research under a Department of Biotechnology (DBT) funded project led to the development of new technology for ethanol production from cassava. This new process has saving of time and energy over the earlier process patented in 1983, as the process could be done at room temperature (30°C) and completed in 48h. The yield of alcohol in the new process is 680 l/t of starch against 450 l/t of starch in the old process. The process has to be upscaled before transfer: establishment of Pilot Plant.

6. Starch-based Adhesives

Liquid adhesives and gum pastes can be made from cassava starch using simple low cost technologies. Starch is cooked with water, cooled and preservatives like formaldehyde or copper sulphate are added. The shelf-life of the gums can be improved by adding borax, urea, glycerol, carboxymethyl cellulose *etc.* These chemicals help in improving and stabilising the paste viscosity (Jeon *et al.*, 1999). Another potential area is the development of corrugating adhesives from starch.

7. Biodegradable Plastics

One of the innovations in the potential use of cassava starch is in the production of biodegradable plastics. In the context of increasing risk of pollution from plastics, use of biodegradable plastics can help in a big way to provide a healthy and pollution less environment for the future generations. Research has shown that cassava starch incorporated into polypropylene granules can yield products having good strength and enhanced biodegradability. The starch incorporated plastic films (up to 25-40 per cent) possess adequate mechanical strength and flexibility and can be processed just like normal plastics, *i.e.*, heat-sealed, printed, coloured *etc.* The granules and finished products can be stored almost like synthetic plastics and biodegradable under soil burial conditions. From the outdoor weathering and soil-burial tests, the biodegradation time has been reported to vary from 6 months to 5 years depending on its composition and soil conditions.

8. Starch-Graft-Copolymers and Super-absorbent Polymers

Among the various modifications of starch, graft copolymerization with vinyl monomers is a fascinating field for research with unlimited possibilities for improving starch properties. A number of graft-copolymers have been prepared by grafting different monomers onto starch which are gaining increasing importance in the manufacture of natural biodegradable polymer based plastics, ion-exchange resins and cosmetics.

Grafted starches possess very high water absorption capacity and thermal stability. Superabsorbent polymers (SAP) can absorb a large amount of water ranging from hundreds to thousands times of that of the polymer. These have been extensively used as absorbents in personal care products and incontinence products. Starch can be modified as superabsorbent polymers through radical graft polymerization with vinyl monomers and proper cross-linking. Due to the biocompatibility, biodegradability and abundant nature, natural biopolymer based superabsorbents can act as potential replacers for petroleum based products (Stuerga and Delmotte, 2002). Starch is a low cost and easily available biopolymer which can be effectively utilized in the development of such products. These polymers can be utilized in soil moisture retention as soil conditioners for water conservation.

9. Starch Based Edible Films and Composite Films

The high viscosity of cassava starch enables its use as binding material in pelleted fish feeds and is being practised in Latin American countries. Economical fish feeds leading to remunerative and sustainable aquaculture systems, can be made using cassava flour as the energy component and cassava starch as the binder. Cold water miscible starch prepared from cassava starch is a new innovation of CTCRI. The technology has been perfected and process optimised through a series of trials. Cassava and sweet potato starch based edible films are developed by incorporating various hydrocolloids and glycerol in starch (Alves *et al.*, 2007). These films can be utilized as edible coatings for various food products. Presently we are working on the development of starch based nanocomposite films for specialized applications such as edible coatings for food, packaging films and slow release matrices for tablets.

10. Small Scale Processing Machineries

10.1 Chipping Machines

The simplest and also the most common mode of processing cassava is conversion of tubers into chips. Dried cassava chips/flour are used in animal feed formulations, as a raw material for manufacturing starch, dextrin, glucose and ethyl alcohol.Under the conventional practice, cassava tubers are sliced with the help of hand-knives with or without peeling the outer skin and rind.

Conventional method of chipping cassava tubers by hand-knives produces about 10 to 40 kg/h for chip thickness varied from 2.7 to 12.5 mm. The chipping machines has the following advantages: i) higher output, ii) low operational cost, iii)moderate initial cost, iv) accommodates all sizes of tubers, v) easy to operate,

vi) requires no special skill to operate, vii) production of uniform chips, viii) adjustable chip thickness, ix) convenience of feeding the tubers into the machine.

The basic parts of these machines are two concentric mild steel drums, the annular space between which is divided into compartments for feeding the tubers. A rotating disc at the bottom of the drum carries the knives assembly. Thickness of chips can be changed by introducing spacing washers between the disc and the blade. Tubers are fed into the compartments from the top and the chips are collected at the bottom (Nanda and Kurup, 1994).

In the hand operated chipping machine, a pair of H.S.S. bevel gears is provided to operate the machine manually with a crank arm. The pedal operated chipping machine is a modified version of the earlier prototype with additional provision of a pivoted pedal for transmitting the power to the cutting disc through suitable belt and pulley drive mechanism. The motorized chipper is run with a 0.5 hp single phase motor through suitable belt drive

The outturn of the hand-operated cassava chipping machine is from 40 to 120 kgh^{-1} for chip thickness varied form 2.3 to 6.9 mm. The capacity of the pedal operated machine ranged from 80 to 770 kgh^{-1} for increase in chip thickness from 0.9 to 6.9 mm. The output of the motorized machine ranges from 290-1090 kgh^{-1} for chip thicknesses from 2.5 to 9.9 mm, respectively.

10.2 Starch Extraction Machineries

The starch content in the tubers varies according to cassava varieties Commercially available machines for starch extraction are of high capacity and the cost is prohibitive for small farmers. Farmers are often exploited by middlemen who are the suppliers of raw materials to large-scale starch extraction units. Extraction of starch from the tubers at the production site itself will be a boon to the farmers as it avoids the middlemen to exploit the peasant. The high return will also ensure promotion of cottage and small-scale industries (Nanda *et al.*, 2005). In order to facilitate *in situ* starch extraction in the villages for value-addition, different types of starch extraction equipment, including a peeling knife of novel design, were developed at CTCRI.

Mobile Starch Extraction Plant

The method of preparation of starch is as follows. The tubers are washed by hand and peeled with handknives. These are then crushed or rasped by adding enough water to get a pulp or mash and sieved with excess water to remove the fibrous residues. The starch milk passing through the sieve is channeled for settling in settling tanks. When starch granules settle down, the supernatant water is decanted and the moist starch is crumbled and sun dried.

The major components of the machine are a hopper to feed the tubers, crushing disc or cylinder with nail-punched protrusions rotating inside the crushing chamber to crush the tubers, sieving tray to remove the fibrous and other cellulosic materials, stainless steel or plastic tanks to collect the sieved starch suspension, tuber storage chamber, handle and wheels for easy transportation from place to place and a frame to support these components. Addition of water during the processing can be controlled through a water pipe with holes fixed inside the hopper along its length and during sieving by a shower attachment connected to the water line. An electric motor (¾ hp) or a generator (kerosene–petrol) attached to the frame can be used as the energy source to operate the machine Sajeev and Balagopalan, 2005). The extraction efficiency ranged from 75-85 per cent with overall crushing capacity of the machine varied from 400-500 kg/h

Cassava Rasper

The rasper consists of a crushing drum made up of a mild steel pipe with power hacksaw blades fixed on its circumference. The crushing drum is fixed on a shaft, power to which is provided by anelectric motor with belt and pulley. The drum is rotated inside the crushing chamber which is made up of two halves, the upper being act as hopper and the bottom half portion acts as outlet for the crushed mash to flow. Gap between the blade set and crushing chamber is adjusted to minimum. A changeable sieve plate is provided in the bottom half to filter the starch pulp with out any bigger pieces. Capacity machine ranged from 800 to 1000 kg h^{-1}

10.3 Sago Manufacturing

The sago (saboodana) marketed in India is manufactured using cassava starch. The initial steps of the process are similar to starch production *i.e.*, peeling and washing, disintegration and settling. The partially dried starch cake is globulated on power-driven gyratory shakers made of wooden trays with gunny cloth flooring. The oscillatory movement enables the starch granules to adhere together and form into spherical beads. The globules are graded by size. The next step is partial gelatinisation of the globules by roasting on shallow metal pans which are smeared with small quantities of oil and heated by fire. The granules are stirred continuously throughout the operation for around 15 min. The granules are then dried in the sun, the dried mass is passed through a spike beater to separate the large clumps, and finally polished before packing in gunny bags. The yield of sago is around 25 per cent of the weight of fresh tubers (Sajeev and Sheriff, 2013). Sago is used mainly as infant and invalid food, and in preparation of puddings and sago wafers.

11. References

Alves, V.D., Mali, S., Beleia, A. and Grossmann, M.V.E. 2007. Effect of glycerol and amylase enrichment on cassava starch film properties. *J. Food Engg.*, **78**: 941–946.

Björck, H., Liljeberg, H. and Ostman, R. 2000. Low glycaemic index foods. *Br. J. Nutr.*, **83**: S149–S155.

Blanchard, P.H. and Katz, F.R. 2006. Starch hydrolysates. In: *Food Polysaccharides and their Applications*, 2nd edn. (Eds.) A.M. Stephen, G.O. Phillips and P.A. Williams. Taylor and Francis, Boca Raton, FL., pp. 119–145.

Bouwkamp, J.C. 1985. *Sweet Potato Products: A Natural Resource for the Tropics*. CRC Press, Boca Raton, Florida, pp. 271.

Green, P.H.R. and Jabri, B. 2003. Coeliac disease. *Lancet*, **362**: 383–391.

Jeon, Y.S., Viswanathan, A. and Gross, R.A. 1999. Studies of starch esterification: Reactions with alkenyl-succinates in aqueous slurry systems. *Starch/Stärke*, **51**(2–3): 90–93.

Lefranc-Millot, C. 2008. NUTRIOSE®06: A useful soluble dietary fibre for added nutritional value. *Nutr. Bull.*, **33**: 234–239.

Nanda, S.K. and Kurup, G.T. 1994. Processing and process equipments for tuber crops. In: *Advances in Horticulture, Vol. 8: Tuber Crops*, (Eds.) K.L. Chadha and G.G. Nayar. Malhotra Publishing House, New Delhi India, pp. 703–714.

Nanda, S.K., Sajeev, M.S., Sheriff, J.T. and Hemasankari, P. 2005. Starch extraction machinery for tuber crops. *Technical Bulletin* **40**. Central Tuber Crops Research Institute, Sreekariyam, Trivandrum.

Padmaja, G., Renjusha, Menon, Jyothi, G. Krishnan and Sajeev, M.S. 2015. Pasta and noodles form tuber crops as novel health foods. *Technical Bulletin* Series, ICAR-CTCRI, **62**: 204.

Padmaja, G., Sheriff, J.T. and Sajeev, M.S. 2013. Healthy food from tubers. *Ind. Hort.*, **58**(3): 18–21.

Sajeev, M.S. and Balagopalan, C. 2005. Performance evaluation of multi-purpose mobile starch extraction plant for small scale processing of tuber crops. *J. Root Crops*, **31**(2): 106–110.

Sajeev, M.S. and Sheriff, J.T. 2013. Mechanization add value in cassava product. *Ind. Hort.*, **58**(3): 48–51.

Sajeev, M.S. Padmaja, G. and Sheriff, J.T. 2012. Value-addition and product diversification on tuber crops. State level work shop on *"Agro Processing: Strategies and Action Plan for 12th Five Years Plan"* organized by Small Farmers Agribusiness Consortium, 29–30th June 2012 at KAU, Vellanikara, Trichur.

Sajeev, M.S., Padmaja, G. and Sheriff, J.T. 2013. Novel value-addition technologies to change the status of traditional crops into high value commodities: Experience in tuber crops. **In**: *National Conference on Tuber Crops for Sustainable Agriculture and Livelihood Security in the Climate Change Scenario*, 29th April, Assam Agricultural University, Jorhat, Assam.

Stuerga, D. and Delmotte, M. 2002. Wave-material interactions: Microwave technology and equipment. **In**: *Microwaves in Organic Synthesis*, (Ed.) A. Loupy Wiley V.C.E., Weinheim.

Woolfe, J. 1992. *Sweet Potato: An Untapped Food Resource*. Cambridge Univ. Press, Cambridge, England.

74

Waste Utilisation in Horticultural Crops for Supplementing Farmers Income

Harinder Singh Oberoi*[1]*, C.K. Narayana[1] and G. Selvakumar[2]

[1]Division of Post-harvest Technology and Agricultural Engineering,
[2]Division of Soil Science and Agricultural Chemistry,
ICAR- Indian Institute of Horticultural Research, Bengaluru – 560 089, Karnataka
**E-mail: hari_manu@yahoo.com*

India accounts for nearly 10 per cent of the total crop residues in the world. Overall, India produces 686 MT gross crop residue biomass on annual basis, of which 234 MT (34 per cent of gross) are estimated as surplus for bioenergy generation (Hiliodhari *et al.*, 2014). As per the estimates of Ministry of New and Renewable Energy (MNRE) in 2009, 500 million tonnes of crop residues is generated every year in India which is exclusive of the residue/waste generated during processing. Processing of horticultural commodities leave behind a substantial amount of residues/by-products in the form of peel, pods, kernel, seeds, *etc.* which are not put to any significant commercial use (Babbar and Oberoi, 2013). Although these residues/by-products are considered as wastes, they have immense nutritional and therapeutic potential (Babbar *et al.*, 2015). Assuming the total annual production of fruits and vegetables at 250 million and 10 per cent as residues or by-products generated during processing, India alone generates about 25 million tonnes of horticultural waste/residues every year. The mass of by-products obtained as a result of processing of tropical crops may approach or even exceed that of the corresponding valuable product affecting the economics of growing tropical exotic crops (Miljkovic and Bignami, 2002). Conversion of abundant fruit and vegetable processing waste (FVPW) into value-added products presents a viable and sustainable option for reducing environmental pollution, improving energy security and reducing greenhouse emissions. Horticultural wastes can be differentiated from the agricultural wastes largely by their higher moisture content, significant nutritional value, low lignin content facilitating easy extraction of high value compounds and therapeutic potential. Hence, it is felt that processing waste utilization is a necessity and a challenge for the food industries and needs first and foremost consideration to make units economically viable through the recovery of the value-added products. Therefore, the exploitation of fruit and vegetable processing wastes for the development of value-added products is a promising area of research. This huge quantity of waste has tremendous potential if it is put to a significant use.

Horticultural Wastes as Source for Enzyme and Bioethanol Production

Conversion of abundant agricultural waste biomass to biofuels, oligosaccharides and other value-added products presents a viable option for reducing environmental pollution, improving energy security and reducing greenhouse emissions. Hence, it is felt that processing waste utilization is a necessity and a challenge to the agro-food industry and needs first and foremost consideration to make units economically viable through the recovery of the value-added by-products (Babbar and Oberoi, 2013). Horticultural wastes are rich in cellulose, hemicellulose, pectin, minerals and have very low concentrations of lignin in comparison to the crop residues, making them more suitable as substrates for enzyme and ethanol production. Without proper storage and transport facilities, the wastes from fruit and vegetable processing industries generally contain large amounts of solid suspensions and a high biochemical oxygen demand (BOD). Some other parameters usually of interest to the waste treatment are pH, chemical oxygen demand (COD), dissolved oxygen (DO) and total solids (Thassitou and Arvanitoyannis, 2001). The horticultural wastes are rich in carbohydrates, fibre and proteins. Such nutritional characteristics are essentially required for the production of enzymes through fermentation. Microorganisms need a carbon, nitrogen and mineral sources for their growth and production of secondary metabolites. In addition, many studies have reported that cellulose, hemicellulose, pectin, *etc.* help in induction of cellulases, hemicellulases (xylanases and other associated enzymes) and pectinases (Dhillon *et al.*, 2011; Thangaratham and Manimegalai, 2014).

Various horticultural wastes, such as citrus wastes, apple peel and pomace, grape pomace, pineapple peel, banana peel, mango peel, jackfruit waste, tomato waste, carrot pomace, onion hulls, olive processing waste and potato peels have been used for production of various enzymes, such as cellulases, pectinases, hemicellulases, lipases and amylases (Sharma *et al.*, 2016). As mentioned previously that processing wastes generated during processing of horticultural crops contain significant amounts of carbohydrates and fibre, they are considered as potent sources for bioethanol production. Ethanol from citrus peels, banana peels, pineapple peel, mango peel have been reported previously.

Horticultural Wastes as Sources of Phytochemicals

Phytochemicals are pronounced as "fight-o-chemicals" *i.e.* they fight to protect health. Phytochemicals naturally occur in fruits and vegetables that work together with vitamins, minerals and fiber to promote health benefits in many different ways. Phytochemicals that are present in the diet, and have been associated with health benefits include glucosinolates, sulphur containing compounds of the alliaceae, terpenoids (carotenoids, monoterpenes) and various groups of polyphenols (anthocyanins, flavones, flavan-3- ols, isoflavones, stilbenoids, ellagic acid, *etc.*). Bioactivity of these phytochemicals has been, to some extent, associated to their antioxidant properties *i.e.* capacity to scavenge free-radicals which are involved in the onset of many of chronic degenerative diseases (LDL oxidation in atheroma plaque development, DNA oxidation and cancer, oxidation and ageing, inflammation, *etc.*). Phenolics have been considered powerful antioxidants *in vitro* (Frankel et *al.*, 1995) and have been proved to be more potent antioxidants than vitamins E, C and carotenoids (Rice-Evans *et al.*, 1997). Carotenoids, flavonoids, cinnamic acids and their derivatives, anthocyanins, *etc.* are classified as phenolic compounds

Table 1: Composition of different Fruit and Vegetable Wastes (per 100 g)

Waste	*Moisture (g)*	*Protein (g)*	*Fat (g)*	*Minerals (g)*	*Fibre (g)*	*Carbohydrates (g)*
Carrot pomace	4.61	10.06	1.75	7.29	24.73	27-60
Green pea peels	4.28	13.27	1.34	7.18	19.82	–
Apple pomace	3.97	4.45	3.49	2.13	48.70	48
Olive cake	–	–	–	–	15.5	18.4
Peach pomace	–	–	–	–	19.1	–
Mausami peel	5.3	5.4	–	5.8	–	14.2
Pineapple peel	9.4	8.7	–	3.9	–	29.1
Mango peel	7.3	9.5	–	6.4	27.95	20.7
Banana peel	10.5	6.02	–	4.2	–	17.8
Orange bagasse	4.15	7.59	1.57	10.03	39.19	35.56
Orange peel	4.23	5.97	1.21	7.12	28.56	25.93
Banana peel	–	5.62	–	–	46.79	36.52
Potato solid waste	85-87	3-5	2.25	6-12	19.86	27-35
Tomato solid waste	85-90	17-22	–	3-5	–	35-45
Onion peelings and whole bulbs	82-92	–	–	4.7	–	–
Sugar beet pulp, silage and leaves	85	–	–	3.8-8.8	–	–
Cauliflower leaves	86	16.1	–	14.0	28.0	24.0
Cabbage leaves	82.8	20.4	–	17.2	34.0	22.5
Pea pods, peel, shell waste	91.5	20.2	–	8.5	57.0	46.5
Pea vines	91.2	12.2	–	8.8	60.0	47.5
Green oats	90.8	10.9	–	9.2	67.0	57.5

Adapted from Sharma *et al.* (2016).

with significant antioxidant activity while ascorbates and tocopherols are non-phenolic compounds with established antioxidant activity.

The concentration of bioactive compounds of the wastes are enormous (Table 2) and they have the potential to serve as ingredients for developing functional foods. However, cost effective and efficient extraction methods for bioactive compounds have to be optimized in order to envisage the development of functional foods using bioactive compounds from horticultural wastes.

Table 2: Composition and Content of Bioactive Compounds in Methanolic Extracts of Various Plant Materials as Determined by HPLC

Material	*Compound(s)*	*Concentration (mg/g)*
Tomato peel	Cis-lycopene	22.02
	Beta- carotene	6.87
	Trans-lycopene	36.49
	Lutein	1.08
	Ascorbic acid	12.27
	Quercetin	2.89
	Kaempferol	7.2
Cucumber peel	Chlorophyll	3.46
	Pheophytin	1.95
	Phellandrene	1.21
	Caryophellene	1.49
Water melon peel	Chlorophyll	5.28
	Diosmetin	1.57
	Pheophytin	1.27
	Malvidin 3,5 diglycoside	1.23
Potato peel	Gallic acid	0.16
	Protocatecheic	1.84
	p-Hydroxybenzoic	0.26
	Caffeic acid	0.19
	Vanillic acid	0.04
	Chlorogenic acid	0.28
Orange peel	p-coumaric	1.02
	Ferulic acid	0.91
	Syringic acid	7.71
	Naritutin	1.21
	Nazingin	3.83
	Ascorbic acid	14.9
Olive leaves	Oleuropein	71.61
	Apigenin 7-glucoside	4.1
	Rutin	0.15
	Vanillin	0.15
	Vanillic acid	1.87
	Caffeic acid	1.02
	Hydroxytyrosel	3.29

Adapted from Zeyada *et al.* (2008).

Horticultural Wastes for Composting

Composting is the natural decomposition of organic material (from plants and animals) by microorganisms, into a dark brown crumbly organic substance called compost. Compost mixed with soil is beneficial for growing plants as the compost retains soil moisture, improves soil structure and provides nutrients. There are different types of composting methods such as using a typical compost heap (outdoor or in a compost bin), using earthworms in vermicomposting, or using effective microorganisms in Bokashi composting. Factors influencing the rate at which composting proceeds include raw material particle size, aeration, moisture content, and the carbon -to- nitrogen (C/N) ratio. The predominant bacteria in composting are aerobic (needing oxygen). The particle size of raw materials determines the porosity of the pile which in turn affects aeration. Material which is too finely ground compacts so densely that air cannot penetrate the pile resulting in anaerobic (without oxygen) or undesirable conditions. One of the major advantages that horticultural wastes have over the typical agricultural wastes is low lignin content and lower C:N ratio. Composting can be a feasible treatment to stabilise horticultural wastes and, thus, to improve their properties for use as organic fertilizers (Gavilens *et al.*, 2016). Authors conducted a composting study using hen manure, saw dust and broccoli/tomato wastes using window tunneling system and concluded that the type of vegetable waste used influenced organic matter and nitrogen mineralisation and the final properties of the composts, showing the mixture with tomato waste with higher fertilising capacity and less environmental problems. Composting of horticultural wastes does not require high capital and huge investments and thus could be a useful technology for farmers interested in augmenting their income through compost preparation. Microbial cultures generally used for composting are available with either the ICAR Institutes/State Agricultural Universities/ National Microbial collection centres, such as MTCC, IMTECHH, Chandigarh, NBAIM, Mau Nath Bhanjan and NCIM, NCL, Pune. Three composting methods are most commonly employed: windrows, aerated static pile, and bins or aerated chambers. For high-volume composting of raw materials, such as animal manure, poultry litter, sewage sludge, or municipal solid waste, the windrow or aerated static pile method is most often used. Windrows 3-5 feet tall and 10-15 feet wide at the base are formed by combining the proper raw materials along with the microbial inoculum, if desired on open earth surfaces. Windrows are turned periodically, and the frequency of turning is largely dependent on temperature and the equipment available at the farm for such purpose. The turning machinery, however, has a high initial cost. Minimum composting time is one month followed by 1-2 months of curing before the material is ready to be used or marketed.

Raw Coir Pith Fermented Cocopeat Peat

Figure 1: Raw Coirpith and Arka Fermented Cocopeat.

Arka Fermented Cocopeat

Coirpith a by-product of the coir industry is a mass of tiny, brown, irregularly shaped particles, that has posed a problem of disposal until its utility as a horticultural growth media was known. It contains less than one per cent of any of the major nutrients and thus cannot be used as manure. Besides it has a carbon : nitrogen ratio in the range of 80-100:1 *i.e.* it contains 80 parts of carbon for every part of nitrogen making it difficult to decompose and utilize as manure for soil application. It is primarily composed of the structural molecule lignin that comprises nearly 40-50 per cent percent of its composition and roughly around 10- 12 percent of pentosans. This low pentosan- to-lignin ratio is an indicator of its extremely slow decomposability and utilization (Tejano, 1985). Another issue with coir-pith is its extremely high content of tannins and phenols, which are largely dependent on the age of the material and the extent to which they are weathered by natural elements. These tannins and phenols retard a wide range of natural biological processes including decomposition and the germination of seeds that are sown onto raw coir pith. In order to overcome these deficiencies, the leaching of the raw coir pith with good quality water has been recommended in order to remove the water extractable tannins and phenols, thereby making the material amenable for use as a plant growth substrate. The end product of this process is called as compost and to qualify as compost the carbon to nitrogen (C: N) ratio of the finished has to be ideally below 20: 1. The methodology for the conversion of coir pith into compost has been developed and is usually achieved by the utilization of lignolytic fungi such as *Pleurotus sajor-caju* and is usually a 90-120 day process. The end product of this process is called as the coirpith compost and is usually recommended for large scale soil application. However, for utilization as a horticultural growth media, the process of composting is not required, and the process of cocopeat production is entirely different from coir pith composting. The end products of the two products *viz.*, coir pith compost and cocopeat are two entirely different entities that ought not to be confused.

Arka Fermented Cocopeat production process utilizes the principle of solid state fermentation to produce fermented cocopeat, using raw coir pith as a substrate and a consortium of tannase producing filamentous fungal strains *Aspergillus heteromorphus* HSGLF-1 (MTCC 11212) and *Aspergillus heteromorphus* PATHLF-1 (MTCC 11211). The essential process that takes place during this process is the breakdown of the seed germination and plant growth inhibiting tannins/ phenols that are present in raw coir pith to threshold levels that no longer affect seed germination and plant growth. Therefore, it is very clear that this is essentially a process of fermentation and not composting since the C: N ratio of the fermented cocopeat is usually in the range of 30-40: 1 and the structural molecule *viz.*, lignin is not subjected to extensive chemical alteration.The term "Fermented" is used in this technology to clearly differentiate this product from compost and the earlier methods of cocopeat production.

Technology for Arka fermented cocopeat has immensely benefitted the farmers involved in raising the vegetable seedling nurseries from Karnataka, Tamil

Figure 2: (A) Cauliflower seedling production on Arka Fermented Cocopeat at Ekalavya Nursery, Bengaluru. (B) satisfied nursery entrepreneur with seedlings produced using Arka Fermented Cocopeat.

Nadu and other states. This technology has real potential to augment the farmer's income as is evident from the Table 3 and Figure 2.

Other Uses of Horticultural Wastes

The major waste resulting from onion bulb peeling is the peels and the fleshy leaves which have the potential for use as pectic-oligosaccharides (POS) which have a considerable prebiotics potential (Babbar *et al.*, 2016) and also for the development of bio-adsorbent for wastewater treatment (Narayana *et al.*, 2013). Banana peels account for about 20-40 per cent of the fresh fruit weight of banana and are rich in phenolic compounds and dietary fibre. It has been reported that 1000 banana plants yield about 20-25 tonnes of the pseudostem which comprise of 5 per cent edible starch (Narayana *et al.*, 2013). Lignocellulose constitutes about 60–85 per cent dry weight of the banana pseudostem sheath, in which cellulose accounts for about 50 per cent. Therefore, pseudostems from banana are used for extraction of fibre, which has diverse industrial applications, such as in preparation of ropes, papers and handicrafts. Edible dietary fibre derived from horticultural wastes can also be used for supplementation of fibre in bakery products, such as muffins, bread, *etc.* Biomethanation is another strategy that can be thought of as a possible potential application of horticulture waste. As the horticulture waste is rich in organic carbon, it could be used as a substrate for production of biogas which can be used by the farmer for carrying out basic household activities. One of the important transformations that could be thought of in the mushroom production is the use of dried horticultural waste as substrate for production of wide variety of mushrooms. Presently, only agricultural residues like wheat and rice straw are only being used as substrates for the production of mushrooms. The most suitable candidates for use as substrates for mushroom production are the low moisture and carbon rich residues, such as grape prunings, jackfruit rind and pineapple peel.

Conclusion

The horticultural wastes generated in huge quantities during primary and secondary processing, are actually a source of wealth. Though, there are relatively difficult to handle because of high moisture content, they are nutritionally rich and are good sources of phytochemicals which have not only nutritional but therapeutic potential. Horticultural wastes or by-products/residues because of their good nutritional

Table 3: Economics of Production of Arka Fermented Cocopeat as practiced by Mr. Prakash, M/s Ekalavya Nursery, Bengaluru, Karnataka, India

Substrate	*Approx. Requirement of Substrate per month*	*Unit rate of Substrate (per kg)*	*Cost of the Media per Month*
Commercially procured cocopeat	30 tonnes	Rs. 4.00	Rs. 1,20,000
Cost of production of Arka fermented cocopeat	30 tonnes	Rs. 2.00 (1.00 raw material cost including transportation from nearest available point *i.e.* Krishnagiri district of T.N.) + 0.50 for other inputs)	Rs. 60,000
Net saving in cost of cocopeat	Rs. 60,000 per month approximately.		

content could be used as substrates for enzyme and ethanol production. Extraction of bioactive compounds from horticultural wastes and their use in health food formulations or their use as substrates for production of value-added products, such as enzymes and ethanol are largely industrial processes which may not benefit farmer directly. However, if the farmer is able to collect the huge residual biomass and dry it by using solar tunnel dryers for which Government of India provides subsidy, sale of such biomass is likely to provide him extra remuneration. Composting or production of fermented cocopeat for vegetable seedling production could add to the farmer's income directly through its use in his own field and also sale of additional material. Biomethanation of horticultural residues and use of such residues for production of mushrooms are other farmer-friendly technologies which can help in augmenting his income.

References

Babbar, N. and Oberoi, H.S. 2013. Potential of agro-residues as sources of bioactive compounds. **In**: *Biotransformation of Waste Biomass: Fine Chemicals*, (Eds.) G.S. Dhillon and S.K. Brar. Springer, USA, pp. 261–298.

Babbar, N., Oberoi, H.S. and Sandhu, S.K. 2015. Therapeutic and nutraceutical potential of bioactive compounds extracted from fruit residues. *Crit. Rev. Food Sci. Tech.*, **55**: 319–337.

Babbar, N., Roy, S.V., Wijnants, M., Dejonghe, W., Caligiani, A., Sforza, S. and Elst, K. 2016. Effect of extraction conditions on the saccharide (neutral and acidic) composition of the crude pectic extract from various agro-Industrial residues. *J. Agril. Food Chem.*, 64: 268–276.

Dhillon, G.S., Oberoi, H.S., Kaur, S., Bansal, S. and Brar, S.K. 2011. Value-addition of agricultural wastes for augmented cellulase and xylanase production through solid-state tray fermentation employing mixed-culture of fungi. *Ind. Crops Prod.*, **34**: 1160–1167.

Frankel, E.N., Waterhouse, A.L. and Teissedre, P.L. Principal phenolic phytochemicals in selected California wines and their antioxidant activity in inhibiting oxidation of human low-density lipoproteins. *J. Agri. Food Chem.*, **43**: 890–894.

Gavilens, I., Jara-Saminiago, J., Idrovo-Novillo J. and Paredes, C. 2016. Windrow composting as horticultural waste management strategy: A case study in Ecuador. *Waste Mgt.*, **48**: 127–134.

Hiliodhari, M., Das, D. and Barauha, D.C. 2014. Bioenergy potential from crop residue biomass in India. *Renew Sustainable Energy Revs.*, **32**: 504–512.

Miljkovic, D. and Bignami, G.S. 2002. Nutraceuticals and methods of obtaining nutraceuticals from tropical crops. USA. Application number: 10/992.502 Published in Google Patent No 10/992.502.

Narayana, C.K., Naik, P. and Nagappa. 2013. Novel products from peels and seeds. *Keralaprakashan e-j.*, 10–13.

Rice-Evans, C.J. and Miller, N.J. 1998. Structure-antioxidant activity relationships of flavonoids and isoflavonoids. **In**: *Flavonoids in Health and Disease*, (Eds.) C.A. Rice-Evans, and L. Packers, Marcel Dekker, Inc, New York, pp. 199 –219.

Sharma, R., Oberoi, H.S. and Dhillon, G.S. 2016. Fruit and vegetable processing waste: Renewable feed stocks for enzyme production. **In**: *Agro-Industrial Wastes as Feedstock for Enzyme Production: Apply and Exploit the Emerging and Valuable Use Options of Waste Biomass*, (Eds.) G.S. Dhillon and S. Kaur, Academic Press, pp. 23–59.

Tejano, E.A. 1985. State of the art of coconut coir dust and husk utilization. Paper presented during the *National Workshop on Waste Utilization, Coconut Husk* at the Philippine Coconut Authority, Diliman, Quezon City, Philippines.

Thangaratham, T. and Manimegalai, G., 2014. Optimization and production of pectinase using agro waste by solid state and submerged fermentation. *Int. J. Curr. Micro. Appl. Sci.*, **3**: 357–365.

Thassitou, P.K. and Arvanitoyannis, I.S. 2001. Bioremediation: A novel approach to food waste management a review. *Tr. Food Sci. Tech.*, **12**: 185–196.

Zeyada, N.N., Zeitoum, M.A.M. and Barbary, O.M. 2008. Utilization of some vegetables and fruit waste as natural antioxidants. *Alex J. Food Sci. Tech.*, **5**: 1–11.

75

Value-addition in Vegetables for Increasing Farmers' Income

Sudhir Singh* and B. Singh

ICAR-Indian Institute of Vegetable Research,
Jakhini, Shahanshapur, Varanasi – 221 305, Uttar Pradesh
**E-mail: sudhiriivr@gmail.com*

Vegetables play a significant role in nutritional and food security for our growing population for not only fulfilling the appetite but also supplying the essential nutrients of vitamins and minerals along with non-nutritive health promoting attributes such as phytochemicals, antioxidants, essential fatty acids and dietary fibre. These health promoting attributes significantly control the life style diseases such as controlling cardiovascular diseases, high blood pressure, diabetes, cancer, arthritis, gastro-intestinal diseases, *etc.*

India is second largest producer of vegetables in the world with the total production of 169.5 million metric tonnes of vegetables on an area of 9.542 million hectare with the productivity of 17.6 million tonnes/hectare (NHB DataBook, 2014-15). The area in horticultural sector grew by about 2.7 per cent per annum and annual production increased by 7 per cent. A number of vegetables such as onion, okra, bitter gourd, green chillies, mushrooms and potatoes contribute largely to the vegetable export basket. There has been increasingly demand of fruits and vegetables in our country due to growing awareness of health benefits. Still India's share in global market is nearly 1 per cent only. Apart from huge investment in private sector, public sector has also taken initiative with the APEDA's assistance for creating infrastructure facilities towards post-harvest handling of vegetable crops. Several centres for preserving the quality of perishable vegetables and integrated post-harvest handling facilities have been created across the country.

However, the integrated efforts of handling of fresh harvested vegetables with good post-harvest management practices are confined to few big cities in India as large quantities of vegetables produced in our villages perish due to poor post-harvest management such as lack of washing, grading, packaging and cold chain storage facilities. It is estimated that around 15-20 per cent vegetables produced in our country perish and around 4.6 per cent of total vegetable produced in our country is being commercially processed. Vegetables after harvest spoil rapidly due to high moisture content as well as high respiratory activity during storage at room temperature. The changes are faster in leafy vegetables as compared to tuber and root crops as a result wilting take place in large quantities of leafy vegetables during storage.

The shelf-life extension of fresh vegetables is of greater importance in the present context for increasing the nutritional and food security for growing population. Due to recent impact of global warming, there has been gradual increase of the temperature of the environment. The increase in shelf-life of perishable vegetables at ambient temperature is of greater importance.

In many Agricultural Universities and ICAR institutes, concerted efforts have been carried out to increase the shelf-life of vegetables. The cold chain management is best way to increase the shelf-life of the harvested horticultural crops after primary processing but as per reality the generation of the facilities at larger level of operation is very difficult due to bulk handling of vegetables.

1. Value-addition of Fresh Vegetables

1.1 Packaging in Plastic Pouches

Vegetables packaged in plastic pouches after harvest retain colour and texture for longer period as compared to those stored in open plastic crates. Freshly harvested vegetables after cooling, grading and washing are packaged in low density polyethylene or polypropylene pouches and are stored at low cost zero energy cool chamber. The shelf-life and quality attributes of vegetables are significantly increased during storage.

1.2 Edible Coating

Edible coating provide partial barrier to moisture and gas especially carbon dioxide and oxygen exchange. Coating on the surface of vegetables improve mechanical handling properties thus handling the mechanical handling properties retain volatile flavour compounds and carry food additives containing antimicrobial agents, antioxidants, *etc.* The purpose of edible coatings in vegetables is basically to enhance the natural barrier properties or replace with new barrier where handling and washing has partially removed or altered it. Various commercial wax based edible formulation in the form of carnauba wax and shellac based emulsions are available for increasing the shelf-life of vegetables.

The shelf-life of carnauba wax coated carrot sample at 5.3 per cent wax level in polypropylene pouches was increased to 15 days as compared to 5-6 days in fully control of carrot samples during storage at 26-32°C, while the acceptable overall acceptability was 25 days at 15°C as compared to 8-10 days in fully control carrot samples during storage. Similarly, the shelf-life of carnauba wax coated capsicum followed by packaging in polypropylene pouches was extended to 20-22 days at 15°C as compared to 5-6 days in fully control capsicum. The decrease in ascorbic acid of carnauba wax treated capsicum was 22-24 per cent after 20 days of refrigerated storage as compared to 33-35 per cent after decrease after 5 days of storage in fully control capsicum samples.

The shelf-life of shellac coated capsicum is increased to 30-35 days at refrigerated storage temperature of 10°C as against 5-6 days of fully control capsicum during refrigerated storage. There had been 28-30 per cent decrease in ascorbic acid, 33-35 per cent decrease in green colour and 28-30 per cent decrease in hardness during shellac coated capsicum after 30-35 days of refrigerated storage. Similarly the shelf-life of shellac coated pointed gourd is increased to 15 days at refrigerated storage as against 6-8 days of control pointed gourd during refrigerated storage. There has also been significant increase in shellac coated eggplant during ambient storage. The shelf-life of shellac coated eggplant is increased to 7 days as against 2-3 days of control eggplant during ambient storage. There has been increase in physiological loss in weight from 6.0-25.7 per cent and increase in total solids from 8.5-11.9 per cent after 7 days of ambient storage. There had been 35-40 per cent decrease in texture and 50-55 per cent decrease in antioxidant activity.

2. Potential Vegetable Processing Technologies for Commercial Usage

The use of appropriate post-harvest technology reduces the post-harvest and storage losses, adds value to the product, generate employment in village and promote establishment of agro-industries in rural sector. Presently, the farmers sell their products without processing. If they do primary processing and value-addition in the villages, it will generate more income and employment in rural sector.

2.1 Dehydration

The techniques of dehydration offer a highly effective and practical means of preserving horticultural produce to reduce post-harvest losses. Still the maximum trade of processed vegetables are in dried form. However, the energy consumption in conventional dehydration techniques is quite high and judicious use of heating is required for rapid and slow drying rate in vegetables to attain desirable product quality in dried vegetables. Another dehydration technique, *viz.*, osmo-air drying is a simple process and has potential advantages for preservation of vegetables for longer duration. The inclusion of osmotic process in conventional dehydration has major objectives of quality improvement and energy savings. This process also results in quality improvement in terms of colour, flavour, texture, product stability, nutrient retention and prevention of microbial spoilage during storage. Different factors such as pre-treatment, nature and concentration of osmotic solution, raw material characteristics, stage of maturity, size of slices, duration of osmosis, ratio of slice to brine concentration and agitation influence the product quality. At the terminal stage, the drying can be achieved by using cabinet dryers or solar driers. Osmo air-drying has greater potential for drying of large quantities of vegetables with good sensory properties. This drying can be adopted as a rural based simple technology by small entrepreneurs, home-scale industry and also by self-help group in close association with NGOs. Small entrepreneurs can adopt this process on large scale. There is ample scope for cost reduction through the use of solar energy for brine concentration and dehydration process. Osmo-air drying of bitter gourd slices, cauliflower, carrot, okra, broccoli, cabbage, cow pea and ivy gourd slices during osmotic diffusion treatment in 5-10 per cent salt solution at 40 – 60°C osmotic diffusion treatment for 2-3 h followed by drying at 65-70°C for 6-8 hrs. The effectiveness of drying technology requires the treatment of vegetables with permissible additives to ensure good reheological, sensory and rehydration properties.

2.2 Hurdle Technology

Innovative technologies based on hurdle techniques have been developed to give stable high moisture foods for extended shelf-life without refrigeration. Hurdle technology based foods are stabilized by a combination of two or more factors as none of which alone is sufficient to prevent microbial spoilage. Methods have been standardized to preserve high moisture (50-60 per cent moisture) fruit slices by reduction in water activity, lowering down the pH and mild heat treatment in controlled osmotic conditions. The process can be adapted to several ranges of tropical and temperate fruits. Hurdle technology treated fruits were found microbiologically safe with extended shelf-life under ambient conditions in flexible pouches. The product can be eaten as such and also have the utility in preparation of salads or can be used in other food formulations. The preservation of onion bulb and paste was standardized using hurdle technology with 1-2 per cent brine concentration, 1-7 days of acetic acid steeping at pH 2.5-2.75, potassium metabisulphite treatment at 100-500 ppm and heat treatment at 100°C for 2-5 min for reduction of microbial load. The good sensory quality of onion bulb and paste was maintained with hurdle technology.

2.3 Minimal Processing

Minimal processing is an emerging technological concept, which has gained increased popularity in recent past. The technique enables global marketing of pre-cut fruits and vegetables in pre-packaged form and the products are meant for specific end uses, *viz.*, curry, salads, pies, stuffings, toppings and garnishing. Apart from offering fresh/fresh like vegetable products minimal processing also results in elimination of kitchen drudgery with sufficient convenience and restriction of packaging and transportation costs. Minimal processing allows consumers to have fresh like quality fruits and vegetables with convenience. The minimal processing operations like cutting, slicing, chopping, peeling *etc.* cause injuries to the plant tissues and initiate enzymatic changes. As the tissues are alive, these operations trigger respiratory activity, ethylene production, respiration, accumulation of secondary metabolites and water loss from the tissues and subsequently make them vulnerable to microbial spoilage and toxin production. A number of preservative methods have been used for minimally processed fruits and vegetables such as addition of texture improvers, antibrowning agents, acidulants, antimicrobial agents, reduction in a_w and mild heat treatment coupled with modified atmospheric storage. The technology is less intensive with low energy consumption and requirements of manpower making it suitable for rural based industries. Minimal processing technology for bitter gourd, okra, carrot and cauliflower samples has been standardized utilizing antimicrobial agents, *viz.*, sodium hypochlorite, hydrogen peroxide and citric acid. Sodium hypochlorite and hydrogen peroxide treatment at 50-100 ppm for 1-2 min reduced the microbial count 10 -100 times in cauliflower and bitter gourd slices without affecting the sensory properties during refrigerated storage for 10-15 days.

2.4 Steeping Preservation

Vegetable production in India is seasonal. The production of vegetables exceeds the demand during the peak season of production; hence he producer is bound to dispose the perishable horticultural crops in no time of storage. Large quantities of vegetables during peak season of production can be preserved in steeping solution consisting of permissible chemical preservatives and other food additives which is non- thermal and alternate to processing technology with considerable scope for adoption at rural sector by women. Consumers are more conscious on health benefits and are demanding additive free processed food products.

Food Safety Standards Authority of India (FSSAI) is regularly monitoring the level of the permitted food additives and some of the permitted additives are either deleted or the level is reduced depending upon nature of health hazards associated with additives. However, the steeping treatment involves the use of permitted additives at a low level of maximum permissible limit and at the same time it should not affect the sensory and nutritional properties of the product and should also be convenient and economical to use. The steeping preservation technology has been standardized at pilot scale at our institute for steeping preservation of cauliflower and carrot. Blanched cauliflower at 100°C for 1 min and steeped in the chemicals consisting of 2-4 per cent sodium chloride, 1-1.5 per cent acetic acid and 350 ppm sulfur di oxide remained acceptable upto 75 days of storage at room temperature. Similarly carrot slices steeped in 25 per cent sugar syrup, 0.4 per cent citric acid, 400 ppm benzoic acid and 350 ppm sulfur di oxide remained acceptable organoleptically for flavour, texture, colour and appearance and overall acceptability.

2.5 Ready-to-Eat Convenience Vegetables

Convenience foods refer to the category in foods which have undergone major processing by the manufacturers such that these type of products require or no secondary processing and cooking before consumption Convenience foods – the choice of the day can be designed to suit all segments of population including armies, airways, railways and even patients with suitable supplement. The demand for convenience food is growing at a faster pace due to changes in social and economic patterns, as well as increase in urbanization, buying power and awareness about health foods, changes in meal pattern and existing food habit, desire to taste new products, *etc.*

The consumer preference is focused towards convenience form of food products which are easy to procure and consume instantly. The vegetables are not only rich in vitamins and minerals but these vegetables also contain health promoting properties such as neutraceutical compounds. These neutraceutical compounds exhibit anticancer and other disease preventing properties. Hence, the vegetables can be suitably be processed in ready-to-eat convenience form such as processing the vegetables for *chips* production

2.6 Easy-to-Cook Leafy Vegetables

Leafy vegetables are store house of nutrients of vitamins, minerals and bioactive components contributing health benefits. However, large quantities of vegetables spoil due to rapid loss of moisture. It is recommended that around one third quantities of our daily requirement of vegetables should be fulfilled by leafy vegetables. Seasonal variability also affects the availability of leafy vegetables throughout the year. Leafy vegetables such as amaranth, spinach, fenugreek, *bathua* and *basela* can be developed in the form of easy-cook-leafy vegetables. Leafy vegetables after sorting, cleaning and washing are blanched with permitted additives followed by osmotic diffusion treatment and subsequent drying at 50-55°C for 7-8 h to reduce the final moisture to less than 1 per cent. These dried leafy vegetables can be effectively preserved in polyethylene pouches for 6-87-8 months of ambient storage.

2.7 Bio-preservation of Vegetables

The bio-preservation of vegetables in common salt or in vinegar is known as pickling. It is one of the most ancient methods of preserving vegetables. Pickles are good appetizer and add to the palatability of meal. Pickles help in stimulation of gastric juice and thus help in digestion. Several kinds of pickles are sold in the Indian market. Mango pickles rank first followed by cauliflower, onion, turnip and lime pickles. Pickles are either manufactured at cottage or home scale as well as commercially manufactured and exported. Pickling is the result of fermentation by lactic acid bacteria, which are generally present in large numbers on the surface of fresh vegetables and fruits. These bacteria can grow in acid medium and in the presence of 8-10 per cent salt solution, whereas, the growth of majority of undesirable organisms is inhibited. Lactic acid bacteria are most active at 30°C, so this temperature must be maintained as far as possible in the early stage of pickle manufacturing process. When vegetables are placed in brine, it penetrates into the tissues of vegetables and soluble material present in vegetable tissues diffuses into brine solution by osmosis. The soluble material basically includes fermentable sugars and minerals. The fermentable sugar is broken down by lactic acid bacteria, which convert them into lactic and other acids. The acid brine thus formed acts upon vegetable tissues to produce characteristics taste and aroma of pickle.

2.8 Instant Bottle Gourd *kheer* Mix

Sweets are very popular in our Indian culture. Sweets are of great demand in every functions and festivals. It is liked by all sections of society such as children, adult and old. It is a symbol to welcome guests by offering sweets in our culture. The demand of vegetable based sweets is increasing because of many nutritional and functional attributes from vegetables in sweets. Traditionally bottle gourd *kheer* mix is very popular and is relished during festivals. However, the process for the manufacture of bottle gourd *kheer* mix is cumbersome and time consuming and shelf-life is also limiting factor. Instant bottle gourd *kheer* mix can be developed by mixing dried bottle gourd pieces with sugar, thickener and milk powder. Bottle gourd *kheer* can be prepared by reconstitution of mix in known quantity of water followed by boiling the reconstituted mix for 2 min.

3. Conclusion

The processed fruit and vegetable industry has great potential to realize the agricultural resources. There has been a huge challenge to our industry to increase the processing of the perishables to provide sufficient growth to the global trade by adding increased value to the end products. The Ministry of Food Processing, Govt. of India, also intends to increase the processing of horticultural crops from 2 per cent to 20 per cent, value-addition from 7 per cent to 35 per cent and share of global trade from 1 to 3 per cent. Farmers particularly women in rural sector can play a crucial role to give a boost in the processing of perishable products at small sector. There is an urgent need to support farmer-managed rural production and marketing ventures in horticulture, and post-harvest processing and to provide technology training and input support to farmers to take advantage of emerging high-value agribusiness sector. Agro-processing particularly vegetable processing can generate employment to whole family of farmer so that farmers income can be increased.

76

Utilisation of Winery By-products: Initiatives and Future Strategies

Ajay Kumar Sharma*, Kaushik Banerjee and S.D. Sawant

ICAR-National Research Centre for Grapes, Manjri Farm Post, Solapur Highway, Pune – 412 307, Maharasthra
**E-mail: ajay.sharma1@icar.gov.in*

India has now moved from table grapes to wine grape production. Systematic wine making in India was started about 30 year back. After adoption of wine policy by Maharashtra state and followed by Karnataka state, wine production was boost up. At present, 93 registered wineries exist in India spread in Maharashtra and Karnataka. Wine making is centuries old practice of grape processing. From grape harvesting to bottling,various byproducts like pomace, marc, lees *etc.* are generated. These by-products are used in different countries for various purposes. Wine producing countries have their own policy to deal with winery by-products. The winery waste contains 61 per cent moisture and 25 per cent crude fiber with a pH of about 3.6. Grape marc can account for 22 per cent of the fruit processed at small wineries. Stalks are a significant waste for both red and white wines accounting for up to 5.5 per cent of the crush. After wine and juice are recovered from yeast lees in both red and white ferments, there are still 0.5-0.7 tonnes dry lees per 100 tonnes fruit. However, not much work has been done on these aspects worldwide; some countries are using/ recycling about 30–40 per cent of winery waste. These wineries are generating about 3 lakh tonnes of grape waste. Its disposal is a major environmental problem. Presently, waste material is disposed by burning in open fields, which results in loss of potential source of organic matter and valuable plant nutrients and functional materials. However, one commercial winery has started the use of a small part of grape pomace which is generated during the process of wine making, as manure in the grape vineyards and using some quantity of grape seeds to extract seed oil for supplying in the domestic market. Winery by-products, even after going through fermentation process for winemaking, have a huge amount of phenolic compounds, depending on the type of grape, the part of the tissue as well as processing conditions. Theyare good sources of flavanols, flavonoids, anthocyanins and phenolic acids (Lafka *et al.*, 2007). Moreover, grape pomace and stem are valuable sources of soluble dietary fiber (SDF) and Klasonlignin(Llobera and Canellas, 2007). The skin of wine grape pomace is ideal source of dietary fiber having good content of bioactive compounds (Deng *et al.*, 2011). Red wine grape pomace shows higher values in total phenolic content in comparison to white wine grape pomace powder. The phenolics have potential beneficial impact on human health as these phenolics have antioxidant, antimicrobial, antiviral, and anti-inflammatory properties. Dietary fiber is another predominate functional component in wine grape pomace, and has benefits of reducing the risk of cardiovascular diseases, cancer, and diabetes (Lizarraga *et al.*, 2011).Wine lees is not utilized in any form by any industry in our country. The fine wine lees is a wastage product collected after 2nd racking and doesn't attract attention of wine industry to make it more feasible by use of this product apart from grape seeds. Unexpectedly, it also shows significant antibacterial properties as well as antioxidant properties from wine lees could include spreads, none caloric thickeners, flavour enhancers, and functional food additive, such as β-glucans. The fine wine lees collected from fermentation of red grapes is source of natural colour also. Negro *et al.* (2003) demonstrated that GWL (grape wine lees) exhibited strong antioxidant activity due to its high phenolic contents, and proanthocyanidins in grape seed also had strong antioxidant ability. The lees of red wine which consist of tannins and plant pigments precipitated around crystals of potassium tartrate can advantageously be used directly as a tonic or demulcent (Edward 2011). Goni *et al.* (2005) confirmed that the wine lees is abundant of fibers (3 per cent) and proteins (6 per cent). The value-added conversion of the bio-products from winemaking can help in reducing the negative costs and demonstrating sustainability in winemaking. In our country, no policy is available for utilization of winery waste. While, winery waste can further utilize in various forms to enrich the food materials.Strategic utilization of winery waste not only solves the environmental issues, it will provide opportunities for generation of food, fuel and employment.

Possible Uses of Winery by-products

Composting

Modest amounts of pomace (less than 25 per cent) can be fed to cattle without any adverse effects. Pomace cannot be added directly back to the vineyards because of potential pathogens associated with it and also because it cannot be ploughed into the soil without damaging the shallow roots of grape vines. On the other hand, it is possible that pomace could be easily composted to eliminate potential pathogen problems and then can be used as a soil amendment, perhaps in a potting mixture in nursery operations. The composting of winery waste is an alternative to the traditional disposal of residues, and also involves a commitment to reducing the production of waste products. The resulting compost will have a high agronomic value and is particularly suitable for the soils of the vineyards which have very low organic matter content. The compost can be reintroduced into the production system, thereby closing the residual material cycle.

Fodder

Pomace can, and has, been used as animal feed, especially in the dry season. Its use is limited due to its very low nutritional value and its antinutritional factors. Low grade winery waste having a low economic value can be upgraded to have a higher economic value by converting the wine waste into high grade animal fodder which can be used as animal fodder in times of drought to maintain cattle healthy.

Fuel and Biofuel

Fuel ethanol production from grapes is already a reality in Europe. This situation arises as a result of surplus wine production and is a continuing problem. Bioethanol is an octane enhancer. It was estimated that a net energy output of 69 GJ/ha is possible (with an energy output: input ratio of 1.8), and production costs (after deduction of revenue for co-products) could be reduced to $0.80 per liter. Ethanol could be produced by fermenting the majority of the carbohydrates found in grape marc. This production was found to yield up to 270 liters per ton of grape marc. The ethanol yield can be increased by pre-treatment with acid and enzymes up to 400 liters a ton. Biodiesel production is also possible. After mechanical separation, the seeds are pressed to yield (at approx. 12 per cent by weight) poly-unsaturated oils. While this oil can be burned directly in internal combustion engines (possibly resulting in some fouling or glazing over time), splitting the tri-glyceride molecule can result into cleaner-burning single-stranded methyl or ethyl esters. Because of the high production cost seed derived oils are more attractive as friction-reducing fuel additives that improve delivered horsepower and extend engine life by reducing internal heating and wear.

The "press cake," from the seed press, also has energy potential. It can be pelletized and burned, yielding – per initial calorimetric studies – approximately 50 per cent of the heating value of wood pellets. Many of the organic components of the press-cake pellets can also be volatilized to form a biogas for subsequent heat and power production. Grape seed with the oil removed also has uses as fuel-pellets which have 133 percent the heating value of wood pellets. This strategy of producing biofuels can be initiated/planned in India in years when there will be excess wine production to recover the cost from unsold wines.

Wine Lees

Wine lees is the residue that forms at the bottom of recipients containing wine, after fermentation, during storage or after authorized treatments, as well as the residue obtained following the filtration or centrifugation of this product. Unexpectedly it also shows significant antibacterial properties as well as antioxidant properties from wine lees could include spreads, non-caloric thickeners, flavor enhancers, and functional food additive, such as β-glucans. The fine wine lees obtained from red grapes is source of natural color also. The lees of red wines which consist of tannins and plant pigments precipitated around crystals of potassium tartrate can advantageously be used directly as a tonic or demulcent. Due to nutritive value, rheological properties, attractive flavour and colour, fine wine lees may be used as a suitable source of natural additive in ice cream making to enhance nutritional values, sensory properties and improve physicochemical properties.

Grape Seed Flour

Grape seed flour has 3,000 to 5,000-fold more antioxidant value than regular flour. Grape skin and seed contain flavonoids (catechin, epicatechin, procyanidins and anthocyanins), phenolic acids (gallic acid and ellagic acid) and stilbenes (resveratrol and piceid). Grape seed procyanidin extract has in vivo antioxidant activity and could be as important as vitamin E in preventing oxidative damage in tissues by reducing the lipid oxidation and/or inhibit the production of free radicals.

Grape seed oil

Grape seed oil is a vegetable oil pressed from the grape seeds. It has a relatively high smoke point, approximately 216°C, so it can be safely used for heating. In addition, it has a clean, light taste that has been described as 'nutty' and is safe for cooking food. Less grape seed oil is needed for cooking purposes, compared with other oils. Grape seed oil is reputed to contain plentiful antioxidants, as well as substances which lower serum cholesterol levels. It also contains vitamin E (0.8 to 1.2 g/kg), vitamin C and β-carotene. The availability of grape seed oil in domestic market will also help in managing prices of cooking oils.

Grape Seed Extract (GSE)

GSE is 50 times more potent than vitamin C and 20 times more potent than vitamin E. Apart from it, GSE retains for longer time in blood than either of the antioxidants (up to 72 hours), giving benefit of sustained antioxidant power against free radicals. Many topical face creams have GSE in them but an oral dose in supplement form can raise the level of GSE (and vitamin C) in human body in hours, working from the inside out. GSE may also slow down skin damage from sun and pollution exposure. When treated with mega doses of GSE via a feeding tube, the tumors shrank 44 per cent compared to controls but more importantly, an increase in the protein that helps cancer cells self-destruct increased as well.

Nanoparticles

Use of nanoparticles for various purposes is new era. Metal nanoparticles of Au, Ag, Pd, and Pt have been synthesized in aqueous media using red grape pomace as a reducing agent and a capping agent. The opportunities for synthesizing nanoparticles from winery waste can be exploited.

Research Initiatives at ICAR-NRC for Grapes

Utilization of Fine Wine Less in Ice Cream

The fine wine lees was collected from 2nd racking of Cabernet Sauvignon wine. Standard methodology of ice cream making was followed and 5, 10, 20, 40 g/kg fine wine lees was added at homogenization stage. The physico-chemical parameters of ice cream were significantly affected over the control. The antioxidant properties comprising DPPH, phenols and anthocyanins content were also significantly increased in comparison to control. The rheological activities *viz.;* melting and specific gravity also influenced in positive way. With the increased doses of fine lees, colour intensity was also increased. As per results based on sensory evaluation, the ice cream prepared by adding 20 g/kg (T4) was recorded with higher values of acceptance for flavour, texture, taste, body and overall acceptability while in case of colour, addition of 40 g/kg (T5) had more acceptability than other treatments.The addition of processed Fine Wine Lees improved nutritive value, antioxidant activities, rheological and sensory properties. The ice creams preparedby addition of FWL were having natural attractive flavor and colour. Fine wine lees may be used as a suitable sourceof natural additive in ice cream production to improvephysicochemical properties, enhance nutritional values andsensory properties (Sharma *et al.*, 2015).

Enrichment of Low Sugar Ice Cream by Addition of Fine Wine Less

Five treatments namely T1, T2, T3, T4 and T5 with the fine wine lees concentration of 0 g/kg, 5 g/kg, 15 g/kg, 25 g/kg and 35 g/kg ice cream, respectively were prepared by replacing the sugar with the sodium saccharin to make available low calorie product. The results showed that the addition of FWL improved physical, functional and rheological properties of ice cream. Treatment T5 was recorded with maximum content of antioxidants, phenols, flavonoids, flavonols, anthocyanins and tannins as compared to other treatments. Also, with the increasing wine lees, the values of pH, specific gravity and overrun were decreased; however increment in viscosity, acidity, colour intensity and fat destabilization index was reported. Due to adding of wine lees viscosity increased thereby reducing the incorporation of air in the ice cream and thus the overrun decreased. The products added with carbohydrate generally exhibit higher viscosity of ice cream and FWL contains more carbohydrates. The results of sensory evaluation showed that T4 having fine wine lees concentration of 25 g/kgice cream was most acceptable based on qualitative parameters *viz.;* flavor, texture, body and taste; however T5 scored maximum in case of colour. On the basis of results obtained from the study it may be concluded that addition of FWL improved physical, functional and rheological properties of low sugar ice cream. Higher score from the sample of FWL added ice cream during sensory evaluation is positive sign and specific consumers who require low sugar ice cream may have additional benefits by taking antioxidant rich ice cream(Pundhir *et al.*, 2016).

Use of Wine Grape Pomace Powder for Enrichment of Cookies

Wine grape pomace has been recorded with nutritional and antioxidant properties. A study was carried out by replacing part of wheat flour with wine grape pomace powder (WGPP) at levels of 5, 10, 15 and 20 per cent. While in case of control cookies were made without adding WGPP. Cookies were prepared using two stage method of cookies making. Wine grape pomace powder was added to dry ingredients. The addition of wine grape pomace powder increased antioxidant properties comprising ferric reducing antioxidant power, total phenol content, flavonoid and anthocyanin. Wine grape pomace powder imparts brown colour to cookies as compared to control. The colour intensity was increased with increasing concentration of wine grape pomace powder in cookies. Maximum colour intensity was observed in cookies obtained from T5. Addition of 5 per cent of wine grape pomace powder (T2) was registered with maximum score in theorganoleptic testing. However, the availability of various constituents in pomace is affected by many factors like variety, harvesting at particular stage, maceration duration, yeast strains or culture used for winemaking, winemaking conditions, pressure used for separation of wine from pomace and pomace drying temperature *etc.* Focused and right efforts are needed to increase the acceptance of cookies having more wine grape pomace powder as consumers will get more antioxidants with cookies(Maner *et al.*, 2015). The change in baking temperature and duration, the acceptability of cookies was increased and the cookies made by addition of 15 per cent WGPP were also well accepted during.

Strategies

- Adoption of zero waste concept by wineries
- Encourage and award to Green Wine production
- Initiation of intensified research approaches for utilization of winery by-products
- Encourage the food industries to adopt the winery by-products for value-addition
- Treat these specific products in the market as high value products
- Supply of these high value food products for a particular sector
- Promote the use of winery by-products in other industries also like cosmetic, nutraceuticals *etc.*

Conclusion

As many number of wastes generated in wineries and no attempts have been made to utilize those wastes in productive way, it is essential to understand the potential benefits of these wastes so that they can be properly processed. Effective utilization of such wastes will reduce the negative cost and thus increase the revenue to the persons engaged in wine grape cultivation and wine making. The Indian wine industry is facing financial problems. The adoption of strategies for better utilization of winery waste will help them to sustain and provide additional employment opportunities in this sector.

References

Deng, Q., Pemmer, M.H. and Zhao, Y. 2011. Chemical composition of dietary fiber and polyphenols of five different varieties of wine grape pomace skins. *Food Res. Int.*, **44**(9): 2712–2720.

Edward, S. 2011. Antimicrobial lees. http://www.patentgenius.com/patent/6821533.html Accessed on 20 June 2011

Goni, I., Martin, N. and Saura Calixto, F. 2005. *In vitro* digestability and intestinal fermentation of grape seed and peel. *Food Chem.*, **90**: 281–286.

Lafka, T.I., Sinanoglou, V. and Lazos, E.S. 2007. On the extraction and antioxidant activity of phenolic compounds from winery waste. *Food Chem.*, **104**: 1206–1214.

Lizarraga, D., Vinardell, M.P., Noé, V., van Delft, J.H., Alcarraz-Vizán, G., van Breda, S.G., Stall, Y., Günther, U.L., Reed, M.A., Ciudaad, C.J., Torres, J.L. and Cascante, M. 2011. A lyophilized red grape pomace containing proanthocyanidin-rich dietary fiber induces genetic and metabolic alterations in colon mucosa of female C57BL/6 J mice. *J. Nutr.* **141**: 1597–1604.

Llobera, A. and Canellas, J. 2007. Dietary fibre content and antioxidant activity of Manto Negro red grape (*Vitis vinifera*): Pomace and stem. *Food Chem.*, **101**: 659–666.

Maner, S., Sharma, A.K. and Banerjee, K. 2015. Wheat flour replacement by wine grape pomace powder positively affects physical, functional and sensory properties of cookies. **In**: *Proceedings of the National Academy of Sciences, India Section B: Biological Sciences*, DOI 10.1007/s40011-015-0570-5.

Negro, C., Tommasi, L. and Miceli, A. 2003. Phenolic compounds and antioxidant activity from red grape marc extracts. *Biores. Tech.*, **87**: 41–44.

Pundhir, A., Sharma, A.K., Banerjee, K., Jogaiah, S. and Somkuwar, R.G. 2016. Improvement in functional, rheological and sensory properties of low sugar ice cream by adding fine wine less. *Prog. Hort.* (Communicated)

Sharma, A.K., Kumar, R., Azad, Z.R.A.A. and Adsule, P.G. 2015. Use of fine wine lees for value-addition in ice cream. *J. Food Sci. Tech.*, **52**(1): 592–596. DOI: 10.1007/s13197-013-1007-x.

77

Probiotic Micro-organisms: Avenue for Processing and Value-addition in Horticultural Produce

Dillip K. Dora

Dean, PGF-cum-DRI,
Odisha University of Agriculture and Technology,
Bhubaneswar – 751 003, Odisha
E-mail: dilipk_dora@yahoo.com

Recently, the Food and Agricultural Organization of the United Nations (FAO) predicted that the world population would surpass eight billion by the year 2030. Therefore, the demand for food would increase quite rapidly. As stated in the FAO report, "Agriculture: Towards 2015/30", remarkable progress has been made over the last three decades towards feeding the growing population of the world. While global population has increased over 70 percent, per capita food consumption has been almost 20 per cent higher. In developing countries, despite a doubling of population, the proportion of those living in chronic states of under nourishment was cut in half, falling to 18 per cent in 1995/97. According to the report, crop output is projected to be 70 per cent higher in 2030 than the current output. Fruits and vegetables would play an important role in providing essential vitamins, minerals, and dietary fiber to the world, feeding populations in both developed and developing countries. Fruits and vegetables contribute approximately 91 per cent of vitamin C, 48 per cent of vitamin A, 27 per cent of vitamin B6, 17 per cent of niacin, 16 per cent of magnesium, 19 per cent of iron and 9 per cent of calories to the human diet. The trend towards consumption of fresh fruits and vegetables in developing countries is one indication of the population's standard of living. Urban population is exploding in developing countries, having risen from 35 per cent of the total population in 1990, and projected to rise 54 per cent in 2020. The effect of income growth on consumption is more pronounced in developing countries, compared to developed countries, they are expected to spend larger shares of extra income on food items like meat, fruit and vegetable products. With the continuous elevation in the socio-economic standards, there has been an increased demand for the functional foods and beverages. Functional foods can be considered to be those whole, fortified, enriched or enhanced foods that provide health benefits beyond the provision of essential nutrients (*e.g.*, vitamins and minerals), when they are consumed at efficacious levels as part of a varied diet on a regular basis. That foods might provide therapeutic benefits is clearly not a new concept. The tenet, "Let food be thy medicine and medicine be thy food" was embraced ~2500 years ago by Hippocrates, the father of medicine. However, this "food as medicine" philosophy fell into relative obscurity in the 19th century with the advent of modern drug therapy. In the 1900s, the important role of diet in disease prevention and health promotion came to the forefront once again. India at present stands in a much comfortable position in respect of production of horticultural crops. Currently nearing 280 million tonnes of horticultural produce a year has not only brought prosperity to small and marginal farmers but also provided food and nutritional security to the Nation. Ranked as the second largest producer of Fruits and Vegetables in the world, horticulture in India has today emerged as one of the most important sectors for diversification in agriculture. However, it is unfortunate to note that a major chunk of the horticultural produce is lost only because of improper post harvest management. A March 2015 report of the Indian Council of Agricultural Research (ICAR) – Central Institute of Post-Harvest Engineering and Technology (CIPHET) – in Ludhiana showed that the cumulative percentage of post-harvest losses of cereals was low in the range of 4.65–5.99 per cent, while that of pulses was between 6.36–8.41 per cent and oil seeds 3.08–9.96 per cent. As per the report, the maximum losses were in fruits and vegetables (4.58–15.88 per cent) and fisheries–marine (10.52 per cent). However, an earlier paper put out by the Department of Industrial Policy and Promotion (or DIPP which formulates the FDI policy) had quoted data from agriculture ministry to say that 25-30 per cent of fruits and vegetables and 5-7 per cent of food grains in India are wasted. This loss in quantitative terms is equivalent to the annual consumption of UK and in financial terms it comes around Rs. 65,000 crores per annum. In a developing country like India a major chunk of the processed produce is lost due to lack of post harvest management, which has been a subject of concern. Here under such a predicament, it becomes the responsibility of the scientific community working in the post harvest sector to get into new ides relating to processing and value addition in horticultural crops for the development of functional foods and beverages for the world population. Irrespective of many approaches for development of functional foods, the intervention of microbiological approach appears to be an excellent avenue for translating aforesaid concept into action. Though microbes can play a major role in value addition in horticultural produce and products in many different ways, their role as probiotic supplementation is quite a new concept and a lot needs to be done for creating market mobilization for the same. Probiotics are "live microorganisms that, administered in adequate amounts, confer a health benefit on the host." as defined by the FAO/WHO. Immense attention is presently addressed to probiotics, prebiotics, or their combined use as synbiotics, to improve human health in natural ways.

Microorganisms as Probiotic Agents

The most tried and tested manner in which the gut microbiota composition may be influenced is through the use of live microbial dietary additions, as probiotics. In fact, the concept dates back as far as prebiblical ages. The first records of ingestion of live bacteria by humans are over 2,000 years old. However, at the beginning of this century probiotics were first put onto a scientific basis by the work of Metchnikoff (1908). He hypothesised that the normal gut microflora could exert adverse effects on the host and that consumption of 'soured milks' reversed this effect. The word "probiotics" was initially used as an anonym of the word "antibiotic". It is derived from Greek words 'pro' and 'biotos' and translated as "for life". The origin of the first use can be traced back to Kollath (1953), who used it to describe the restoration of the health of malnourished patients by different organic and inorganic supplements. Later, Vergin (1954) proposed that the microbial imbalance in the body caused by antibiotic treatment could have been restored by a probiotic rich diet; a suggestion cited by many as the first reference to probiotics as they are defined nowadays. Similarly, Kolb recognized detrimental effects of antibiotic therapy and proposed the prevention by probiotics (Vasiljevic and Shah, 2008) Later on, Lilly and Stillwell (1965) defined probiotics as "microorganisms promoting the growth of other microorganisms". Following recommendations of a FAO/WHO (2002) working group on the evaluation of probiotics in food, probiotics, are live microorganisms that, when administered in adequate amounts, confer a health benefit on the host Sanders, (2008). The idea of health-promoting effects of LAB is by no means new, as Metchnikoff proposed that lactobacilli may fight against intestinal putrefaction and contribute to long life. Such microorganisms may not necessarily be constant inhabitants of the gut, but they should have a "beneficial effect on the general and health status of man and animal."

Other definitions advanced through the years have been restrictive by specification of mechanisms, site of action, delivery format, method, or host. Probiotics have been shown to exert a wide range of effects. The mechanism of action of probiotics (*e.g*, having an impact on the intestinal microbiota or enhancing immune function) was dropped from the definition to encompass health effects due to novel mechanisms and to allow application of the term before the mechanism is confirmed. Physiologic benefits have been attributed to dead microorganisms. Furthermore, certain mechanisms of action (such as delivery of certain enzymes to the intestine) may not require live cells. However, regardless of functionality, dead microbes are not probiotics (Sanders, 2008). In relation to food, probiotics are considered as "viable preparations in foods or dietary supplements to improve the health of humans and animals". According to these definitions, an impressive number of microbial species are considered as probiotics (Holzapfel *et al.*, 2001). For gastrointestinal ecosystem, however, the most important microbial species that are used as probiotics are lactic acid bacteria (LAB) (Table 1).

Selection Criteria of Probiotic Microbes

Many *in vitro* tests are performed when screening for potential probiotic strains. The first step in the selection of a probiotic LAB strain is the determination of its taxonomic classification, which may give an indication of the origin, habitat and physiology of the strain. All these characteristics have important consequences on the selection of the novel strains (Morelli, 2007). An FAO/WHO (2002) expert panel suggested that the specificity of probiotic action is more important than the source of microorganism. This conclusion was

Table 1: Selected Micro-organisms that are Used as Probiotic Agents (Gareau *et al.*, 2010)

Probiotic	*Human Disease in which Benefit is Shown*	*Animal Model in which Benefit is Shown*
Yeast		
Saccharomyces boulardii	*Clostridium difficile* infection	*Citrobacter rodentium*-induced colitis
Gram-negative bacteria		
Escherichia coli Nissle 1917	NA	DSS-induced colitis
Gram-positive bacteria		
Bifidobacteria bifidum	NA	Rat model of necrotizing enterocolitis
Bifidobacteria infantis	IBS29	NA
Lactobacillus rhamnosus GG(used with lactoferrin)	Sepsis in very low birth weight infants	NA
Lactococcus lactis (engineered toproduce IL-10 or trefoil factors)	Crohn's disease	DSS-induced colitis and IL-$10^{-/-}$ mice (spontaneous IBD)
*Lactobacillus plantarum*299v	Antibiotic-associated diarrhea	IL-$10^{-/-}$ mice (spontaneous IBD)
Lactobacillus acidophilus	NA	Visceral hyperalgesia 40 and *C. rodentium*-induced colitis
Lactobacillus rhamnosus	Pediatric antibiotic-associated diarrhea	–
Lactobacillus casei	NA	DNBS-induced colitis
Bacillus polyfermenticus	NA	DSS-induced colitis and TNBS-induced colitis
Combination regimens		
Lactobacillus rhamnosus GG combined with *Bifidobacterium lactis*	Bacterial infections	NA
Lactobacillus rhamnosus combined with *Lactobacillus helveticus*	NA	*C.rodentium*-induced colitis, chronic stress, and early life stress
VSL#3 (*Lactobacillus casei, Lactobacillus plantarum, Lactobacillus acidophilus, Lactobacillus bulgaricus, Bifidobacterium longum, Bifidobacterium breve, Bifidocacterium infantis and Streptococcus thermophilus*)	Pouchitis and pediatric ulcerative colitis	DSS-induced colitis, IL-$10^{-/-}$mice (spontaneous IBD; DNA only), and SAMP mouse model of spontaneous IBD

brought forward due to uncertainty of the origin of the human intestinal microflora since the infants are borne with virtually sterile intestine. However, the panel also underlined a need for improvement of *in vitro* tests to predict the performance of probiotics in humans. While many probiotics meet criteria such as acid and bile resistance and survival during gastrointestinal transit, an ideal probiotic strain remains to be identified for any given indication. Furthermore, it seems unlikely that a single probiotic will be equally suited to all indications; selection of strains for disease-specific indications will be required (Shanahan, 2003). The initial screening and selection of probiotics includes testing of the following important criteria: phenotype and genotype stability, including plasmid stability; carbohydrate and protein utilization patterns; acid and bile tolerance and survival and growth; intestinal epithelial adhesion properties; production of antimicrobial substances; antibiotic resistance patterns; ability to inhibit known pathogens, spoilage organisms, or both; and immunogenicity. The ability to adhere to the intestinal mucosa is one of the more important selection criteria for probiotics because adhesion to the intestinal mucosa is considered to be a prerequisite for colonization (Tuomola *et al.*, 2001). The table below (Table 2) indicates key creteria for selecting

Table 2: Key and Desirable Criteria for the Selection of Probiotics in Commercial Applications (Vasiljevic and Shah, 2008)

General	*Property*
Safety criteria	Origin Pathogenicity and infectivity Virulence factors–toxicity, metabolic activity and intrinsic properties, *i.e.*, antibiotic resistance
Technological criteria	Genetically stable strains Desired viability during processing and storage Good sensory properties Phage resistance Large-scale production
Functional criteria	Tolerance to gastric acid and juices Bile tolerance Adhesion to mucosal surface Validated and documented health effects
Desirable physiological criteria	Immunomodulation Antagonistic activity towards gastrointestinal pathogens, *i.e., Helicobacter pylori, Candida albicans* Cholesterol metabolism Lactose metabolism Antimutagenic and anticarcinogenic properties

probiotic candidate for commercial application, and Figure 1 presents major and cardinal steps for selecting probiotic candidates. It is of high importance that the probiotic strain can survive the location where it is presumed to be active. For a longer and perhaps higher activity, it is necessary that the strain can proliferate and colonise at this specific location. Probably only host-specific microbial strains are able to compete with the indigenous microflora and to colonise the niches. Besides, the probiotic strain must be tolerated by the immune system and not provoke the formation of antibodies against the probiotic strain. Hence, the host must be immuno-tolerant to the probiotic. On the other hand, the probiotic strain can act as an adjuvant and stimulate the immune system against pathogenic microorganisms. It goes without saying that a probiotic has to be harmless to the host: there must be no local or general pathogenic, allergic or mutagenic/carcinogenic reactions provoked by the microorganism itself, its fermentation products or its cell components after decrease of the bacteria (Desai, 2008).

Potential Mechanisms of Action

A wide variety of potential beneficial health effects have been attributed to probiotics. Claimed effects range from the alleviation of constipation to the prevention of major life-threatening diseases such as inflammatory bowel disease, cancer, and cardiovascular incidents. Some of these claims, such as the effects of probiotics on the shortening of intestinal transit time or the relief from lactose maldigestion, are considered well-established, while others, such as cancer prevention or the effect on blood cholesterol levels, need further scientific backup (Leroy *et al.*, 2008). The mechanisms of action may vary from one probiotic strain to another and are, in most cases, probably a combination of activities, thus making the investigation of the responsible mechanisms a very difficult and complex task. In general, three levels of action can be distinguished.

Probiotic organisms can provide a beneficial effect on intestinal epithelial cells in numerous ways. **a**: Some strains can block pathogen entry into the epithelial cell by providing a physical barrier, referred to as colonization resistance or **b**: create a mucus barrier by causing the release of mucus from goblet cells. **c**: Other probiotics maintain intestinal permeability by increasing the intercellular integrity of apical tight junctions, for example, by upregulating the expression of zona-occludens 1 (a tight junction protein), or by preventing tight junction protein redistribution thereby stopping the passage of molecules into the lamina propria. **d**: Some probiotic strains have been shown to produce antimicrobial factors. **e**: Still other strains stimulate the innate immune system by signaling dendritic cells, which then travel to mesenteric lymph nodes and lead to the induction of TREG cells and the production of anti-inflammatory cytokines, including IL-10 and TGF-β. **f**: Some probiotics (or their products) may also prevent (left-hand side) or trigger (right-hand side) an innate immune response by initiating TNF production by epithelial cells and inhibiting (or activitating) NFκB in Mφ and dampening (or priming) the host immune response by influencing the production of IL-8 and subsequent recruitment of Nφ to sites of intestinal injury.

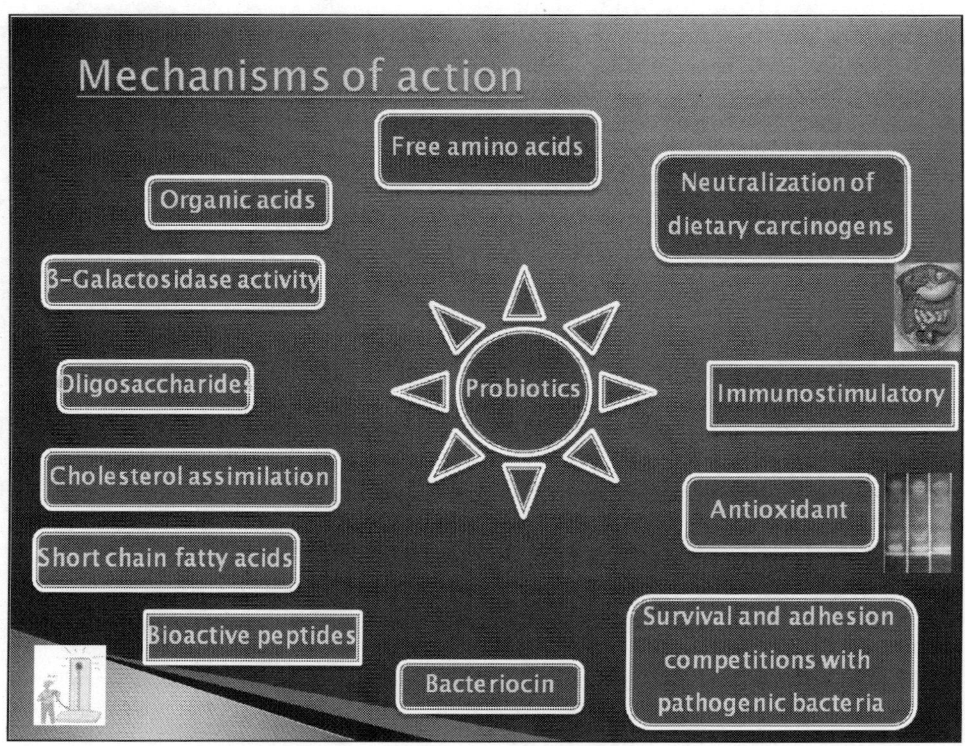

Figure 1: Different Mechanisms of Action of Probiotics Bacteria.

Abbreviations: Mφ, macrophage; Nφ, neutrophil; TREG cell, regulatory T cell. Reproduced from, *Gareau M. G., P. M. Sherman and W. A. Walker (2010) Nature Reviews Gastroenterology and Hepatology 7, 503-514.*

Probiotics and Human Health

Probiotic research is moving forward on two stages: laboratory studies and clinical trials to evaluate the safety and efficacy of probiotics in treatment and prevention of various medical conditions. Evidences of probiotics in various diseases by clinical trials and meta-analysis are discussed below.

Antibiotic-Associated Diarrhea

The incidence of antibiotic-associated diarrhea (AAD) ranges between 5 per cent and 30 per cent. The risk is greatest with aminopenicillin therapies (Ampicillin or Amoxicillin), aminopenicillin combined with clavulanic acid, cephalosporins, and clindamycin. Probiotics given in conjunction with antibiotics have been extensively studied for the prevention of AAD in both adults and children. The major changes in the microbiota of the gut with antibiotics are decrease in total number and species diversity of Bacteroides and Bifiobacteria associated with decreased amylolytic activity with increase in facultative anaerobes such as Fusobacteria, Clostridia, and Eubacteria species. Decreased short chain fatty acid production and increased proteolytic activity was also noted in elderly patients treated with antibiotics. Several clinical trials have been conducted using *Saccharomyces boulardii* for the prevention of AAD. All but one concluded that *S. boulardii* was an effective agent for prevention of AAD. With increasing number of trials over the last several years on the role of probiotics in preventing AAD, new single-strain meta-analysis are now being published. A meta-analysis of several randomized controlled trials testing the efficacy of *S. boulardii* in preventing AAD in adults showed *S. boulardii* was significantly protective for AAD with an overall pooled relative risk of 0.47 (95 per cent Confidence Interval = 0.35, 0.63; p

Clostridium difficile Infections

Clostridium difficile is a spore-forming, anaerobic, Gram-positive bacterium that causes gastrointestinal infection with diarrhea and colitis. There has been a marked increase in the incidence and severity of *Clostridium difficile* infection (CDI) during the past decade. The clinical outcomes of CDI range from asymptomatic carriage to mild diarrhea to fulminant, often fatal, pseudomembranous colitis. Recurrent CDI is one of the most challenging aspects of the disease. Approximately 25 per cent of patients treated for CDI with metronidazole or vancomycin experience recurrent symptoms, typically within 4 weeks of completing antibiotic therapy. Owing to increasing incidence, rising death rates, and frequent recurrences, there is a substantial need for more effective approaches to CDI prevention and therapy. Castagliuolo *et al.*, found a 54 kDa serine protease produced by *S. boulardii* which directly degrades *C. difficile* toxin A and B and also produces a protease capable of degrading the colonic receptor site for C. difficile. S. boulardii may cause an increase in anti-toxin secretory IgA levels in the intestine. Probiotics have been studied in prevention, and treatment of *Clostridium difficile* infections (CDI) and recurrent CDI.

Irritable Bowel Syndrome (IBS)

Epidemiological, physiological and clinical studies have suggested the role of intestinal bacteria in the pathogenesis of IBS. Many previous studies indicate that gastroenteritis is a trigger for IBS. In Canada following an outbreak of gastroenteritis, a cohort analysis revealed an increased odd of IBS within 2 years (OR 4.8) and continued for 8 years. In another study, incidence of gastroenteritis in the previous 2 years was associated with almost a four-fold increase in the risk of developing IBS. Physiological studies on animals and humans showed a profound effect of alterations in the composition of the intestinal microbiota (dysbiosis) on the intestinal physiological functions and IBS. A review of several case-control studies revealed abnormal breath test results in patients with IBS following a sugar challenge when compared with controls. The increased risk of developing IBS following gastroenteritis, dysbiosis, and elevated luminal gas production and immune activation, indicate that the gastrointestinal microbiota may be a therapeutic target in IBS.

Allergy and Immune Response

Recent research in mucosal immunology demonstrated interactions between microbes and host at an early age even when mucosal barrier and immune system are still immature. Probiotics have been found to enhance the innate immunity and modulate pathogen induced inflammation via toll-like receptor-regulated signaling pathways. The mode of delivery has a great impact on the acquisition of the intestinal bacteria, also beyond the immediate neonatal period. Vaginally born infants and infants born by cesarean section show major differences in cultural microbiota up to 6 months of age. Infants harboring *Bacteroides fragilis* and *Bifidobacterium* species had more circulating immunoglobulin (Ig) A-secreting and IgM-secreting cells. Bacteria in breast milk and microbes potentially present in the amniotic fluid may affect the composition of gut microbiota. Gut microbiota stimulates the TH1, TH3, and T regulatory cells, which can balance the IL-4, IL-5, and IL-13 secreted by TH2 cells in atopic diseases like allergic rhinoconjunctivitis, asthma and atopic eczema. In a randomized double blind placebo-controlled studies of probiotic use, *Lactobacillus* GG or placebo when given to pregnant mothers with a strong family history of eczema, allergic rhinitis or asthma, and to their infants for the first six months after delivery. The frequency of developing atopic dermatitis in the offspring's of pregnant mothers who received *Lactobacillus* GG was significantly reduced by 2, 4, and 7 years, by 50, 44 and 36 per cent respectively. Lactobacillus acidophilus strain was not able to produce same result in a different study suggesting the strain specificity. Lactobacillus GG in combination with *B. lactis* during pregnancy and breastfeeding reduced the risk of atopic eczema and allergic sensitization in child, whereas a mixture of probiotics (*Lactobacillus GG, L. rhamnosus LC705, B. breve, and Propionibacterium freudenreichii* ssp. *Shermanii* JS) failed to reduce the risk atopic eczema, indicates the strain differences and interactions.

Lactose Intolerance

It is a genetically determined beta-galactosidase deficiency resulting in the inability to hydrolyse lactose into the monosaccharides glucose and galactose. Upon reaching the large bowel the undigested lactose is degraded by bacterial enzymes leading to osmotic diarrhoea. Acquired, usually reversible, causes of beta-galactosidase deficiency include pelvic radiotherapy which damages the mucosa, as well as infection with rotavirus which infects lactase producing cells, and short bowel syndrome. Lactose intolerant individuals develop diarrhoea, abdominal discomfort and flatulence after consumption of milk or milk products. Although conventional yoghurt preparations, using *S. thermophilus* and *L. delbrueckii* ssp. *bulgaricus*, are even more effective in this direction, partly because of higher betagalactosidase activity, improvement of lactose metabolism is a claimed health benefit attributed to probiotics and seems to involve certain strains more than others and in specific concentrations. Therefore and as certain individuals have responded positively to probiotic supplementation, clinicians should consider it as a therapeutic alternative.

Non-dairy Based Probiotic Products

Despite potential sensory challenges, there is a genuine interest in the development of fruit-juice based functional beverages, fortified with the probiotic and prebiotic ingredients. The fruit juices have been suggested as an ideal medium for the functional health ingredients because they inherently contain beneficial nutrients, they have taste profiles that are pleasing to all the age groups, and because they are perceived as being healthy and refreshing. The fruits and vegetables are rich in the functional food components such as minerals, vitamins, dietary fibers, antioxidants, and do not contain any dairy allergens that might prevent usage by certain segments of the population. Current industrial probiotic foods are basically dairy products, which may represent inconveniences due to their lactose and cholesterol content. Technological advances have made possible to alter some structural characteristics of fruit and vegetables matrices by modifying food components in a controlled way. This could make them ideal substrates for the culture of probiotics, since they already contain beneficial nutrients such as minerals, vitamins, dietary fibers, and antioxidants, while lacking the dairy allergens that might prevent consumption by certain segments of the population. There is a genuine interest in the development of fruit juice based functional beverages with probiotics because they have taste profiles that are appealing to all age groups and because they are perceived as healthy and refreshing food. There is every reason to believe that beverages would be the next food category where the healthy bacteria will make their mark. Likely candidates are chilled fruit juices or fermented vegetable juices. The probiotic microorganisms also have been directly incorporated into the drinks. The key to the development of this second generation of the probiotic products is a special Direct Liquid Inoculation system. It allows food producers to add the probiotic bacteria directly to the finished food product. The technology uses Tetra Pak's aseptic dosing machine Flex Dos that allows the bacteria to be added to liquids just before they are filled into the cartons. The innovation is expected to significantly boost the market for the probiotic beverages, which have so far been restricted by the delicate nature of the ingredient and concerns over the contamination. The probiotic bacteria must live to exert their health benefits but they can be destroyed by a number of processing situations. In beverages, for example, the heat treatment in a standard production run would kill the live bacteria. Adding probiotics to the juices is more complex than formulating in the diary products because the bacteria need protection from the acidic conditions in the fruit juice. However, with microencapsulation technologies, the probiotics can become an important ingredient in the functional foods, expanding the probiotic application outside the pharmaceutical and supplement industries. Microencapsulation technologies have been developed and successfully applied using various matrices to protect the bacterial cells from the damage caused by the external environment (Del Piano *et al.*, 2006). It is the process by which small particles or droplets are surrounded by a coating to produce the capsules in the micrometer to millimeter range known as microcapsules. The microencapsulation allows the probiotic bacteria to be separated from its environment by a protective coating. Several studies have reported the technique of the microencapsulation by using gelatin, or vegetable gum to provide protection to acid-sensitive Bifidobacterium and Lactobacillus (Chandramouli *et al.*, 2004; Lee *et al.*, 2004; O'Riordan *et al.*, 2001; Sultana *et al.*, 2000). The commercial success of the probiotic products ultimately depends on taste and appeal to the consumer (Heenan *et al.*, 2004; Yoon *et al.*, 2006). The consumer needs to receive a comprehensible and reasonable message about probiotics, without appearing to be exaggerated. The "health claims" must be defensible when placed under the scrutiny by the controlling authorities. In the coming years of the new Millennium,

Table 3: Non-Dairy (Plant-based) Probiotic Formulations

Probiotic Foods	Examples	Probiotic Strains
Cereal based	Wheat, Malt, maize, oat based beverages	*L. plantarum, L. fermentum, L. acidophilus and L. reuteri*
Fruit juices	Pineapple, apple, orange, grape,watermelon juice, cashew apple juice and blackcurrant juice	*L. plantarum, L. delbruekii, L. casei,L. paracasei. L. fermentum, L. plantarum andL. acidophilus*
Vegetable juices	Beets-based drink, tomato-based drink, cabbage juice, carrot juice, ginger juice, cucumber juice and banana puree	*Latobacillus acidophilus LA39,Lactobacillus plantarumC3, Lactobacillus casei A4, Lactobacillus delbrueckii D7* and *Lactobacillus plantarum 299v (LP299V®)*

changes would occur in the interface between the scientific studies and the acceptance by the consumers (Gorbach, 2002). According to Sheehan *et al.* (2007), when adding *Lactobacillus* and *Bifidobacterium* to orange, pineapple and cranberry juice, extensive differences regarding their acid resistance were observed. All of the strains screened survived for longer in orange and pineapple juice compared to cranberry. *Lactobacillus casei, Lactobacillus rhamnosus, Lactobacillus paracasei* display a great robustness surviving at levels above 7.0 log cfu/ml in orange juice and above 6.0 log CFU/ml in pineapple juice for at least 12 weeks. However, after thermal pasteurization at 76°C for 30 s and 90°C for 1 min in addition to a high-pressure treatment of 400MPa for 5 min, *L. casei, L. rhamnosus* and *L. paracasei* were not capable of withstanding the treatments required to achieve a stable juice at levels >6.0 log CFU/ml (Sheehan *et al.*, 2007). Yoon *et al.* (2004) determined the suitability of the tomato juice as a raw material for the production of probiotic juice by *L. acidophilus* LA39, *L. plantarum* C3, *L. casei* A4 and *Lb. delbrueckii* D7. The tomato juice was inoculated with a 24 h-old culture and incubated at 30°C. The lactic acid cultures reduced the pH to 4.1 and the viable cell counts reached nearly (1.0-9.0) x 10^9 CFU/mL after 72 h fermentation. The viable cell counts of the four lactic acid bacteria in the fermented tomato juice ranged from 10^6 to 10^8 CFU/ml after 4 weeks of cold storage at 4°C.

Yoon *et al.* (2005) also evaluated the potential of red beets as the substrate for the production of probiotic beet juice by the above four species of lactic acid bacteria. All the lactic cultures are capable of rapidly utilizing the beet juice for the cell synthesis and lactic acid production. *L. acidophilus* and *L. plantarum* produced higher amount of lactic acid than other cultures and reduced the pH of the fermented beet juice from an initial value of 6.3 to below 4.5 after 48 h of fermentation at 30°C. Although the lactic cultures in fermented beet juice gradually lost their viability during the cold storage, the viable cell counts of these bacteria, except for *L. acidophilus*, in the fermented beet juice still remained at 10^6–10^8 CFU/ml after 4 weeks of cold storage at 4°C. Mousavi *et al.*, in 2011 have studied the suitability of fermented pomegranate juice using probiotic strains such as *L. plantarum, L. delbruekii, L. paracasei, L. acidophilu* and found to be a suitable probiotic drink with desired results. Optimized growth conditions of *L. casei* in cashew apple juice were studied by Pereiraa *et al.* (2011). *L. casei* has shown suitable survival ability in cashew apple juice during 42 days of refrigerated storage. It was observed that *L. casei* grew during the refrigerated storage and cashew apple juice showed to be suitable probiotic product.

Cereal-based Products

Cereals are good nutrient substrates for the growth of probiotic strains and their fermentation with probiotic microorganisms could be beneficial for health due to their availability of the vitamin B, the decrease of non-digestible carbohydrates and the improvement of the quality and level of lysine Fermented probiotic cereal products commercially available worldwide are followings; Rice-based yogurt, Anarshe, Aya-bisbaya, Oat-based drink, Malt-based drink, Oat-bran pudding (Yosa), Tempeh Wheat fermented probiotic beverages, Adai, Injera, Togwa, ProViva, Fermented maize beverage (Mahewu), Millet or sorghum flour fermented probiotic beverage, Sorghurt (Sorghum based), Burukutu, Bhatura.

Fruits and Vegetable-based Probiotics

Fruits and vegetables are rich in several nutrients such as minerals, vitamins, dietary fibers, antioxidants and do not contain any allergens. This advantage of fruits has encouraged several food industries to manufacture healthy probiotic fruit juices like Pineapple, apple, orange, grape, watermelon juice, ginger, cashew apple juice and blackcurrant juice. Few examples of vegetable juices are Tomato-based drink, Cabbage juice, carrot juice, Ginger, cucumber based drink Vege puree and pulp, powdered fruits, non-fermented fruit juice, Beverages

Conclusion

In spite of all our achievements, the microbiological interventions for value addition as probiotics in horticultural produce are in its infancy. There are some challenges to overcome, *i.e.*, the survival of probiotics, and the effects on the sensory traits; however, there are some possible solutions that show that there is a promising way. Some probiotic juices are already on the market, but many other products can be launched. Thus, perhaps, the combination of probiotics with some new methods (encapsulation, fortification with other ingredients, using non conventional juices or non conventional probiotics) shall be able to catch up with consumers and became the winning strategy.

References

Chandramouli, V., Kailasapathy, K., Peiris, P., and Jones, M. 2004. An improved method of microencapsulation and its evaluation to protect Lactobacillus spp. in simulated gastric conditions. *J. Microb. Meth.*, **56**: 27–35.

Del Piano, M., Morelli, L., Strozzi, G.P., Allesina, S., Barba, M. and Deidda, F., *et al.*, 2006. Probiotics: From research to consumer. *Dig. Liv. Dis.*, **38**(Suppl. 2): S248–S255.

Desai, A. 2008. Strain identification: Viability and probiotics properties of *Lactobacillus casei. Ph.D. Thesis*, Victoria University, Victoria, Australia 228 pp.

FAO/WHO. 2002. Guidelines for the evaluation of probiotics in food. Joint Food and Agricultural Organization of the United Nations and World Health Organization. Working Group Meeting Report, London Ontario, Canada.

Gareau, M.G., Sherman, P.M. and Walker, W.A. 2010. Probiotics and the gut microbiota in intestinal health and disease. *Nat. rev. Gastr. Hepat.*, **7**: 503–514.

Gorbach, S.L. 2002. Probiotics in the third millennium. *Dig. Liv. Dis.*, **34**(Suppl. 2): S2–S7.

Heenan, C.N., Adams, M.C., Hosken, R.W., and Fleet, G.H. 2004. Survival and sensory acceptability of probiotic microorganisms in a non-fermented frozen vegetarian dessert. *Lebensmittel-Wissenschaft und-Technology*, **37**: 461–466.

Holzapfel, W.H., Haberer, P., Geisen, R., Björkroth, J. and Schillinger, U. 2001. Taxonomy and important features of probiotic microorganisms in food nutrition. *Amer. J. Clin. Nutr.*, **73**: 365S–373S.

Kollath, W. 1953 Ernährung und Zahnsystem. *Deutsche Zahnärztliche Zeitschrift*, **8**: 7–16.

Lee, J.S., Cha, D.S., and Park, H.J. 2004. Survival of freeze-dried Lactobacillus bulgaricus KFRI 673 in chitosan-coated calcium alginate microparticles. *J. Agr. Food Chem.*, **52**: 7300–7305.

Leroy, F., Falony, G. and de Vuyst, L. 2008. Latest developments in probiotics. **In**: *Meat Biotechnology*, (Ed.) F. Toldrà, Springer Science+Business Media, LLC, 217–229 pp.

Metchnikoff, E. 1908. *The Prolongation of Life*. G.P. Putnam's Sons, New York, 376 p.

Morelli, L. 2007. *In vitro* assessment of probiotic bacteria: From survival to functionality. *Int. Dairy J.*, **17**: 1278–1283.

Mousavi, Z.E., Mousavi, S.M., Razavi, S.H., and Kiani, H. 2011. Fermentation of pomegranate juice by probiotic lactic acid bacteria. *World J. Microb. Biot.*, **27**: 123–128.

O'Riordan, K., Andrews, D., Buckle, K., and Conway, P. 2001. Evaluation of microencapsulation of a Bifidobacterium strain with starch as an approach to prolonging viability during storage. *J. Appl. Microb.*, **91**: 1059–1066.

Pereira, D.I.A. and Gibson, G.R. 2002. Effects of consumption of probiotics and prebiotics on serum lipid levels in humans. *Crit. Rev. Biochem Mol. Biol.*, **37**(4): 259–281.

Sanders, M.E. 2008. Probiotics: definition, sources, selection and uses. *Clin. Inf. Dis.*, **46**: S58–61.

Sanders, M.E. 2008. Probiotics: Definition, sources, selection, and uses. *Clin. Inf. Dis.*, **46**: S58–61.

Shanahan, F. 2003. Probiotics: A perspective on problems and pitfalls. *Scand. J. Gastroenterol.*, **38**(S): 34–36.

Sheehan, V.M., Ross, P. and Fitzgerald, G.F. 2007. Assessing the acid tolerance and the technological robustness of probiotic cultures for fortification in fruit juices. *Innov. Food Sci. Emerg. Tech.*, **8**: 279–284.

Sultana, K., Godward, G., Reynolds, N., Arumugaswamy, R., Peiris, P., and Kailasapathy, K. 2000. Encapsulation of probiotic bacteria with alginate-starch and the evaluation of survival in simulated gastrointestinal conditions and in yoghurt. *Int. J. Food Microb.*, **62**: 47–55.

Tuomola, E., Crittenden, R., Playne, M., Isolauri, E. and Salminen, S. 2001. Quality assurance criteria for probiotic bacteria. *Amer. J. Clin. Nutr.*, **73**: 393S–398S.

Vasiljevic and Shah, N.P. 2008. Review: Probiotics - from Metchnikoff to bioactives. *Int. Dairy J.*, **18**L 714–728.

Vergin F. 1954. Anti- and probiotika. *Hippokrates*, **25**: 16–19.

Yoon, K.Y., Woodams, E.E. and Hang, Y.D. 2004. Probiotication of tomato juice by lactic acid bacteria. *J. Microb.*, **42**: 315–318.

Yoon, K.Y., Woodams, E.E. and Hang, Y.D. 2005. Fermentation of beet juice by beneficial lactic acid bacteria. *Lebensmittel-Wissenschaft und-Technologie*, **38**: 73–75.

Yoon, K.Y., Woodams, E.E., and Hang, Y.D. 2006. Production of probiotic cabbage juice by lactic acid bacteria. *Biores. Tech.*, **97**: 1427–1430.

Adding Value to Produce-II

78

Exploitation of Natural Colours and Dyes from Horticultural Crops

Suresh Walia

Former Professor and Principal Scientist, Division of Agricultural Chemicals,
ICAR-Indian Agricultural Research Institute, New Delhi – 110 012
E-mail: sureshwalia@yahoo.co.in

Food colour is any substance that is added to food or beverage to change or enhance its colour, make it more attractive, acceptable and sometimes restore colour loss as a result of exposure to light, air, temperature, moisture or long storage. They provide distinct visual identity to the food and increase consumer's preference towards a particular food product. Most of the functional foods available in the market are usually coloured with synthetic colorants which can be harmful to humans. Because of their harmful effects, they are being replaced world over by natural colorants. Natural food colours are generally extracted from vegetables, fruits, seeds, leaves, flowers, algae, fungi, insects, or mineral resources. Since natural food colorants have several health benefitting effects, they also find application as nutraceutical and phyto-medicine.

The global food colorant market is growing at 3-4 per cent CAGR and is projected to reach $2.3 billion by 2019 (www.marketsandmarkets.com). In 2014, North America and Europe accounted for the largest share in the total natural food colours and flavours market. At present Indian share in domestic and international trade is very small. In the next five years, Asia-Pacific region particularly India and China is expected to experience highest growth in consumption. Major drivers of growth being, increasing demand for natural foods, ever increasing health concerns, and ever growing health conscious consumers that demand natural foods and food colorants for various use applications.

1. Natural Food Colourants

Natural food colorants can make a variety of different hues ranging from yellow, orange, red, blue, purple, green *etc.* They are capable of imparting desired color to food, drug, cosmetic, fabric, or any produce/ article of human interest, making such products more attractive, appealing, and informative. In view of their natural origin and perceived safety, they find wide application in food, nutraceutical, pharmaceutical, cosmetic and textile industry. They are being used in a variety of non-food applications, for example dyeing of hairs, textile fibres and in home decorations. Natural colorants require stringent certification before use in food industry. Regulatory agencies in USA and the European Union have a list of approved food grade colorants which are tested and certified to meet the legal specifications. In the United States, colour additives are regulated by the FDA as **"Certified" (synthetic or artificial colours),** and **those "Exempt from Certification" (natural colours).** Certified colours are synthetically produced, while natural colours that are exempt from certification are derived from natural sources or may be synthetic duplicate of natural ones. FD and C number is given to each FDA approved synthetic food dye that do not exist in nature. In European Union (EU), E numbers are given to both synthetic and natural additives that are approved in food applications. FDA regulations (21 CFR, section 70.3) do not consider any colour added to a food product as natural. The only exception is when the colour is 'natural' to the food product itself. For example colouring strawberry ice cream with strawberry juice. If natural beet juice is added to colour strawberry ice cream, it is then considered as colour additive and not a natural colour. Food Safety and Standards Authority of India (FSSAI) has added some natural colorants to the permitted list of colorants in regulation 3.1.2 (Sub-clause 2) for use in various food products. These include anthocyanins, copper cholorophyllin, natural carotenes, paprika, carthamus and lutein. They are however, not permitted to be used in infant foods. List of some of the authorized natural food colorants in EU and US is given in Table 1.

Majority of the food colourants are categorized into five broad segments namely caramels, anthocyanins, betalains, carotenoids and chlorophylls. Caramel color is produced by heating carbohydrates consisting of glucose, fructose, sucrose, sugar syrups, mollasses *etc.*, alone or in the presence of acids, alkalis, and/ or salts. Natural colourants from other four segments are extracted from various horticultural crops/plants and are the subject of discussion in this paper. Under the ICAR-NAIP sub-project on nutraceuticals, know-how has been developed for the production of several food grade natural oxidants and food colorants (Walia, 2013). Recent developments in production and use of anthocyanin, betalain, carotenoid and chlorophyll group of natural colorants derived from horticultural crops are highlighted below.

1.1 Anthocyanin Group of Natural Pigments

Anthocyanins imparting blue-violet and red hues may be extracted from food crops like black carrot (*Daucus carota*), purple cabbage (*Brassica oleracea*), grape (*Vitis vinifera*), pomegranate (*Punica granatum*), black rice cultivar (*Oryza sativa*), under-utilized fruits like jamun (*Syzygium cumini*), kokum (*Garcinia indica*), *karonda* (*Carissa carandus* L.) and purple-red coloured flowering plants like gulmohar (*Delonix regia*).

1.1.1 Anthocyanins from Black Carrot

Black or purple carrot (*Daucus carota*) is one of the richest sources of anthocyanins which impart black carrots the dark violet colour. It is widely grown in India, Turkey, Afghanistan, Egypt, and Pakistan. In India, it remains an underutilized crop and does not find much consumer acceptance as a vegetable due to its colour. Phytochemical investigation of black carrot revealed anthocynanin content of 17.4 to 45.4 g kg^{-1} dry matter (Kammerer *et al.*). Most of the black carrot anthocyanins are glycosides of cyanidin, but some contain peonidin or pelargonidin as aglycone

Table 1: List of some Authorized Natural Food Colorants

Natural Colourant	*Source(s)*	*Colour*	*E number*
Curcumin	*Curcuma longa* (turmeric)	Yellow	E 100
Riboflavin (Vit. B2)	Nuts, broccoli, wild rice, whole grain, mushrooms, soybean	Yellow orange	E 101
Indigo dye (Indigotine)	*Indigofera tinctoria*	Blue	E 132
Chlorophyll	Plant leaves (spinach, alfalfa)	Green	E 140
Copper complexes of chlorophyll (Chlorophyllin)	Plant leaves	Green	E 141
Caramel, caramelized sugar syrup	Heating of sugars	Brown	E 150a-d
α-, β- et γ-carotene	Carrot	Red	E 160a
Bixin, nor-bixin, annatto extract	Annato seeds (*Bixa oleracea*)	Red	E 160b
Paprika oleoresin, capsanthin, capsorubin	Paprika, Chilli	Red	E 160c
Lycopene	Tomato, watermelon	Red	E 160d
β-apo-8'-carotenal	Spinach, citrus fruits	Orange-yellow	E160e
Ethyl ester of beta-apo-8'-carotenic acid	Variety of plants Apocarotenal	Orange-yellow	E160f
Flavoxanthin	Variety of plants	Golden yellow	E161a
Lutein	*Tagetes* sp. (marigold)	Yellow, orange	E 161b
Cryptoxanthin	Orange rind, papaya, peppers, pumpkins	Orange red	E 161c
Canthaxanthin	Mushrooms or shrimp	Violet	161g
Zeaxanthin	Paprika, corn spinach, lettuce	Orange red	E 161h
Astaxanthin	*Microalgae* (*Haematococcus pluvialis, Chlorella zofingiensis*)	Red	E 161j
Betanin (Betalain)	Beet root	Red purple	E 162
Anthocyanin	Coloured fruits and vegetables (strawberry, cherry, grapes, pomegranate, *jamun*, black carrot, black rice, red cabbage *etc.*)	Purple	E 163
Saffron oleoresin	Saffron	Saffron	E 164

moiety. Bench scale know-how has been developed for the extraction, purification, and analysis of the natural colorant from black carrot. The developed extraction technology employs heat treatment followed by treatment with cell wall degrading enzyme(s) for recovering higher amounts of anthocyanins in the juice Various steps involved in the extraction process include i) extraction, ii) filtration, iii) purification by passing through adsorbent resin, iv) concentrating, and v) spray drying with suitable substrates to obtain free flowing powder. The LC-DAD-ESI-MS of anthocyanin powder concentrate extracted from Pusa Asita variety resulted in the characterization of eight anthocyanins, which included two non-acylated and four acylated anthocyanins of cyanidin aglycone, and one acylated and one non-acylated anthocyanin of peonidin aglycone (Walia, 2013; Singh *et al.*, 2016). As evident from FRAP and DPPH assays, the anthocyanin powder concentrate exhibited significant antioxidant activity. It was found to be a potent inhibitor of α-glucosidase (IC_{50} of 0.692 mg ml^{-1}) and showed moderate antibacterial activity. It also inhibited proliferation of human cancer cells (HT-29 colorectal adenocarcinoma and HL-60 promyelocytic leukemia) (Netzel *et al.*, 2007). Spanish black carrot anthocyanin has been reported to exhibit significant antioxidant activity (Algarra *et al.*, 2014).

Black carrot extract and concentrate is water soluble and unlike other natural colourants, it is more stable under heat, light, and broad pH range. It is particularly most stable in acidic pH as it generates a red shade in pH 3, a magenta shade in pH 6, and a purple shade in pH 7. The black carrot natural colorant is often used as a natural alternative to synthetic azo dye Allura-Red (FD and C Red #40, or E-129). Its stability can be vouched from the fact that it is being used to enhance colour stability of jams, juices and other functional foods. Since discoloration of strawberry products due to the rapid degradation of pigments is considered as the quality defect, addition of purple carrot juice up to 20 per cent in strawberry jam processing can stabilize the colour of jam without any effect on its acceptance (Abdel-Hady *et al.*, 2014). 'Kanji' and 'Salgam' are two popular fermented drinks made from black carrot, turnip, and/or beet juice. While, *kanji* is popular in north India, *salgam suyu* (turnip juice) is a popular street drink in south Turkey. During fermentation, a lactic acid bacterium is added to give the fermented drink typical taste, flavour and attractive purple hue. Typical taste and flavor of the drink is due to lactic acid, ethanol and other organic compounds formed during fermentation (Huseyin *et al.*, 2008; Bige Sncedayi *et al.*, 2008).

1.1.2 Anthocyanins from *Jamun*

Jamun tree (*Syzygium cuminii*) is a tropical and sub-tropical evergreen tree found widely in the Asian sub-continent and other tropical regions of the world. The pulp of the *jamun* berry contains anthocyanins, which impart bright purple colour to the fruit (DeBrito *et al.*, 2007; Veigas *et al.*, 2007; Ayyanar *et al.*, 2012). In traditional Indian medicine, it is extensively used against diabetes and other ailments. This underutilized tropical fruit is also valued for its use as natural colourant. Know-how has been developed for the preparation of anthocyanin powder and liquid concentrate from *jamun* fruit skin, pulp or juice. The process involves i) preparing extract, ii) filtering and passing the extract through a column filled with a suitable adsorbent(s), iii) eluting the colorant from the absorbent with a suitable solvent, and iv) concentrating the eluate to provide anthocyanins concentrate. The concentrate is later purified to obtain food grade anthocyanin powder concentrate (Walia, 2013). On the basis of LC-MS-MS, three major anthocyanins have been identified as i) myricetin 3-O-(4-acetyl)-α-L-rhamnopyranosides, ii) delphinidin-3-gentiobiside, and iii) malvidine-3-laminariboside. Other two minor anthocyanins include: myricetin 3-O-(4-acetyl)-α-rhamnopyranosides and delphinidin-3-gentiobiside. These anthocyanins in mixture form exhibit excellent antioxidant activity and demonstrated anti-cancer properties against the early stage HCT-116 human colon cancer cells. It also induced apoptosis and inhibited self-renewal ability in colon CSCs (Charepalli *et al.*, 2016). In an earlier study, antiproliferative activity of *Jamun* extracts against human lung cancer A549 cells has also been reported (Aqil *et al.*, 2012). For diabetic patients, *jamun* fruit juice is considered as healthy as the grape juice for non-diabetics.

Natural colorants from jamun can be used to provide light pink to dark purple/violet hue to various functional foods. It can be used for colouring value-added products like ice creams, sauces, jams, jelly, syrup, squash, energy drinks, soy drinks, *aonla* candy *etc.* Purple wine and vinegar has also been made from the fermented fruit juice.

1.1.3 Anthocyanins from Purple Grapes

Pigmented grapes (yellow, green, crimson, black, and pink) are the storehouse of numerous health promoting anthocyanins, proanthocyanidins, stilbenes (resveratrol), poly-phenolics (catechins), and phenolic acids (gallic acid and ellagic acid). While dark purple varieties are rich source of anthocyanins, catechins are more abundant in green varieties. Grape juice obtained from crushing and blending into a liquid may further be fermented to obtain wine, alcoholic drinks or vinegar. Anthocyanins can be traditionally extracted with methanol, ethanol, or their aqueous mixtures acidified with hydrochloric acid. Ultra-sound assisted extraction and pressurised fluid extraction (PFE) techniques can be used for rapid extraction. In PFE extraction, optimum yield can be obtained by extraction with aqueous ethanol at 70°C for 60-75 min. Heat and light had little effect on the pigment stability (Zou *et al.*, 2014).

Grape peel extract and anthocyanin powder concentrates has been successfully used as natural food colourants (Polovka *et al.*, 2010). The dark reddish or purple grape peel powder concentrate is soluble in water, alcohol but insoluble in oil and absolute alcohol. The pH of the product influences its hue and stability. For example in the pH range of 2.0-6.0, the colour is red; at pH 7.0, it is blue, and at pH above 7.0, it is unstable green.

Grapes colourants can be used for colouring spirits, carbonated drinks, juices, energy drinks, soy drinks, jams, jellies, ice creams, yogurts and candy. Red wine containing anthocyanins offer more health benefits than other wines. Grape anthocyanins also have many beneficial effects on human health, *e.g.*, reduction of coronary heart disease incidence, and anti-carcinogenic and antioxidant properties.

1.1.4 Anthocyanins from Red Cabbage

Red or purple cabbage (*Brassica oleracea*) is a colorful variety from the brassica family. Due to the anthocyanin pigment content, cabbage grows more reddish in acidic soils, more purple in neutral soils, and greenish-yellow in alkaline soils. Purple cabbage leaves have a relatively high content of natural colorant, nutrients and vitamins. The extraction process involves various steps of washing, slicing, extraction, filtration, purification, vacuum concentration, and spray-drying. It contains several anthocyanins having different substituents and functional groups. Cabbage anthocyanins concentrate has been partially purified (Coutinho *et al.*, 2004) to obtain food grade colourant. HPLC-MS/MS of the extract concentrate has led to the identification of twenty derivatives of cyanidin glucosides (Wiczkowski *et al.*, 2013), wherein the glucoside residues were non-acylated, monoacylated and/or diacylated with sinapic, ferulic, caffeic and/or *p*-coumaric acids. The predominant anthocyanins include non-acylated cyanidin-3-diglucoside-5-glucoside, cyanidin-3-(sinapoyl) (sinapoyl)-diglucoside-5-glucoside, cyanidin-3-(*p*-coumaroyl)-diglucoside-5-glucoside, cyanidin-3,5-diglucoside, and cyanidin-3-monoglucoside. Among these, cyanidin-3-diglucoside-5-glucoside diacylated with sinapic acid exhibited the highest radical-scavenging activity.

Good stability of red cabbage anthocyanins is attributed to the presence of acyl groups in the glycone moiety. The water soluble food colourant is noted for its excellent heat stability and light fastness. It has greater stability at pH 3.8 or lower and at higher pH, the colorant becomes less stable. It generates a reddish-magenta solution at pH 3; a pinkish solution at pH 4-6; a purple solution in a pH of 7; a bluish-green solution in pH 8-9; and yellowish solution in pH > 9. As a result of the typical changes at different pH, red cabbage powder is used as a pH indicator. Purple colour of red cabbage pigment can be changed to rarely occurring sapphire blue by addition of baking soda.

Red cabbage can be used in wine, beverages, fruit sauce, candy, cake, sauerkraut *etc.* Since red cabbage anthocyanins are more stable than those from other fruits and vegetables, it can have many applications in functional foods, beverages, food supplements, and pharmaceutical industry.

1.2 Carotenoid Pigments

Carotenoids are yellow, orange, or red pigments that are widely distributed in animal and plant kingdom. These are of two types. Those with a hydrocarbon chain are called carotenes, while the other type with oxygenated (alcoholic) chain is referred to as xanthophylls. Some prominent carotenoids from food and non-food crops include lycopene from tomato, capsanthin from chillies and lutein from marigold (Saha *et al.*, 2015a, 2015b).

1.2.1 Lycopene from Tomato

Lycopene is a naturally occurring red plant pigment found in tomatoes (*Lycopersicon esculentum* M.), watermelon, grapefruit, papaya, apricot, guava and other fruits. Due to its strong colour and non-toxicity, lycopene is considered as the useful coloring agent. This acyclic carotenoid contains 40 carbons and 11 conjugated double bonds, which impart deep orange to red colour to lycopene. It may be used to colour food products at concentration ranging from 5-500 mg/kg food. Besides food colourant, it also possesses high antioxidant and other healthy benefitting effects. Bench scale know-how has been developed for the extraction of lycopene concentrate from fresh tomato fruits or tomato paste. Protocols have been standardized for optimum extraction of lycopene from raw tomato (Walia, 2013). Pure lycopene powder can crystallized from the de-fatted extract. Lycopene in pure form consists of 94-96 per cent all *trans*, 3-5 per cent 5-*cis*, 1 per cent 13-*cis*, and less than 1 per cent other *cis* isomers.

As compared to β-carotene, lycopene is two times more efficacious free radical scavenger, and its presence in the diet positively correlates with reduced cancer incidence. Lycopene is well known specifically to help prevent many forms of cancer as well as the prevention and treatments of heart diseases, infertility, diabetes, age-related macular degeneration, cataracts and ageing of skin.

Lycopene is used as natural colorant in preserves, confectionery, flavoured dairy products, jams, jellies, snacks, soup and soup mixes, liquid and solid food supplements, meat and fish coatings, non-alcoholic beverages and wines. Tomat-O-Red®, a product of PTL Limited is approved as a food and beverage colorant in the US, Europe and Japan. In India, many companies manufacture lycopene for domestic use and export as natural colorant and as ingredient of medicines for prevention of prostate and other forms of cancer.

1.2.2 Capsanthin from Red Chilli

Red chilli (*Capsicum annum* L. and *C. frutenscens* L.) is primarily used as a colouring and/or flavouring agent in various food products. Red chilli powder and paprika powder is made from grinding the dried pods of hot and mild red varieties of the pepper respectively. Extraction is done by solvent percolation employing hexane, acetone, trichloroethylene, 2-propanol, methanol, and/or ethanol. Extraction may also be done by super-criticial fluid extraction with or without the use of moderator solvent. The oleoresin is oil-soluble, but when emulsified becomes water dispersible. India is the largest producer and exporter of chillies and chilli products. Colour and pungency are the two main attributes in chillies that determine its quality. While the red color of chillies is due to the presence of the carotenoid pigments like capsanthin (major), capsorubin, zeaxanthin, violaxanthin, cryptoxanthin, β-carotene *etc.*, the pungency (hot flavour) is attributed to capsaicinoids, namely, capsaicin and dihydrocapsaicin. An improved process has been developed for extraction of paprika oleoresin from paprika powder (Walia, 2013). The process is simple, cost effective and use food grade solvents for extraction. The process involves i) cleaning to remove any extraneous matter in red chilli, ii) drying to bring down the moisture to nearly 5 per cent, iii) grinding and sieving, and iv) extraction with GRAS solvents with or without enzymes, and v) desolventization and further processing to get deep coloured oleoresin and powder concentrates. Paprika oleoresin and technical capsanthin concentrate is used as natural colorant in beverages, pickles, cheese, sauces, snack foods, meat and poultry products, salad dressing, seasoning and confectionery industry. Capsaicinoid concentrate on the other hand find application as flavouring agent that cause burning (hot) sensation.

1.2.3 Lutein from Marigold (*Tagetes erecta*)

Lutein is a carotenoid group of plant pigments naturally found in a number of fruits, vegetables and flowers especially those with deep orange and yellow color. Marigold petals are a rich source of lutein in which it is primarily found esterified with saturated fatty acids like lauric, myristic, palmitic and stearic acid. Lutein is one of two major carotenoids found as a colour pigment in the human eye (macula and retina). It is thought to function as a light filter, protecting the eye tissues from sunlight damage. Marigold petals contains approx. 0.03 per cent lutein per dry weight (Sowbhagya *et al.*, 2004; Zuorro and Lavecchia, 2010; Murillo *et al.*, 2010). Some preferred methods of extraction include pressurized solvent extraction, and supercritical fluid extraction (Gao *et al.*, 2010; Hazare *et al.*, 2013; Kamalambigeswari *et al.*, 2016. This pigment has acquired greater significance because of its antioxidant property and for its potential in eye health protection. Marigold flower extract has also been used in veterinary and poultry feeds. Compared with beta-carotene and 'annatto', marigold colour is stronger yellow colour, which dissolves easily in oil and fat. It is acid resistant, heat resistant, light resistant and is relatively stable. The effect of pH is negligible. A wide array of products is available ranging from oil soluble oleoresin, water dispersing emulsion, and water soluble liquid. Lutein is used in alternative medicine for eye diseases such as cataracts and age-related macular degeneration and to protect the eye from injury induced by free radicals. Lutein supplements can also help prevent colon cancer, breast cancer, diabetes and heart disease.

Unlike other natural colorants, lutein has good stability to light and heat. It can therefore be used as a food colouring agent and nutrient supplement in a wide range of baked goods and baking mixes, beverages, breakfast cereals, chewing gums, dairy products, egg products, fats and oils, frozen dairy desserts and mixes, sauces, soft and hard candies, infant and toddler foods, milk products, processed fruits and juices, soups and soup mixes in levels ranging from 2 to 330 mg/kg. Marigold flower extract can be used as a source of textile dyeing. The various colour shades can be obtained using ecofriendly mordants. The dyeing property is comparable to other synthetic dyes (Jha *et al.*, 2015). Lutein dye has also been successfully used for improving colour strength and fastness characteristics. Good colour strength was obtained when irradiated cotton (RC, 30 kGy) was dyed with extract of radiated marigold flower powder. 7 per cent of tannic acid as pre-mordant and 5 per cent of Cu as post-mordant were the best treatments to further improve colour strength (Adeel *et al.*, 2017).

2. Betalain from Beetroot

Beetroot (*Beta vulgaris*) is a rich source of betalain group of natural colourants. The name "betalain" comes from the Latin name of the common beetroot (*Beta vulgaris*), from which betalains were first extracted. There are two categories of betalains - betacyanins, which appear reddish to violet, and betaxanthins which appear yellow to orange in colour. They mainly comprise of red–violet-colored betacyanins (betanin, isobetanin, probetanin and neobetanin) and yellow–orange-coloured betaxanthins. Among these betanin is the most predominant betalain in beetroot. Fresh beets, beet powder or extracted pigments are often added to improve the red colour of tomato pastes, sauces, soups, desserts, jams, jellies, ice creams, sweets and breakfast cereals *etc.* They also contribute to consumers' health and wellbeing because betalains are known to have antioxidant properties (Singh and Hathan, 2014).

Beet root colourant is comparatively unstable and prone to degradation when exposed to light, heat, and air. It also degrades during cooking if cooking time is increased. Due to its poor heat stability it is ideally used in products which are designed for a short shelf-life, *e.g.* frozen foods. Beet juice formulations containing metal cations such as iron or copper degrade faster and need to be stabilized with antioxidants/stabilizers to extend their shelf-life. Beet extract concentrate is pH dependent and generates a bluish-red shade at pH 4-5, a brighter red as the pH rises, and yellowish-brown if the pH becomes too alkaline. The bluish-red colour is attributed to betanin which is stable at a higher pH. In the dry powder state, it is more stable than in the liquid state. It finds wide use in soup, ice cream, yogurts powdered beverages, confectionary products, and frozen foods of various types. Beetroot extract is also a good source of nitrates. Recent studies have suggested that proper intake of nitrates in the food may be beneficial to human body. Thus intake of beetroot extract containing desired nitrate content is more beneficial as it can regulate vasodilation, improve blood flow, and boost human stamina, vigour and athletic performance.

3. Chlorophylls from Plant Leaves

Chlorophyll is the green photosynthetic pigment present in chloroplasts which provide the energy necessary for photosynthesis. It is found in almost all plants, algae, and cyanobacteria. It is extracted as green pigment from dark green leafy vegetables like spinach, lettuce, parsley, alfalfa; microalgae like spirulina and chlorella; and super grasses like wheatgrass and alfalfa. The intense green colour of chlorophyll is due to its strong absorbencies in the red and blue regions of the electromagnetic spectrum. It is structurally similar to haemoglobin, a porphyrin pigment in blood where iron is replaced with magnesium at the centre of the molecule. The major chlorophyll a and chlorophyll b are common in all plant tissues and are present in approximate ratio of 3:1. Chlorophylls contain a 20-carbon hydrophobic "tail," referred to as phytyl group. Since chlorophyll pigment is not very stable, freezing storage is an effective method of preserving the green color of vegetables.

Chlorophyllin or copper chlorophyllin is a chlorophyll-derived food colorant formed by displacing magnesium from the core of the molecule with copper. It is commercially available as a complex mixture of different molecules. The improved stability and brightness of copper chlorophyllin leads to its much wider use than chlorophyll as a food colour. Besides food colouring, this compound exhibit antimutagenic, anticarcinogenic and antioxidant activities, however, its mechanism of action is not yet well understood. Unlike natural chlorophyll, copper chlorophyllin is more stable to heat, light and pH variations.

Chlorophylls and chlorophyllins are used to impart green shades to a wide range of food applications. They are used in sugar confectioneries, mint based products, pasta, sauces, pickles, jams and dry mixes *etc.* They are available as an oil soluble or water soluble liquid or powder formulations and provide green hues to beverages, baked goods, soups, milk based ice cream, pet products *etc.* Sodium copper chlorophyllin is a water soluble powder approved for use in citrus-based dry beverage mixes. It has got good heat stability, good light fastness, and good stability in a range of pHs. It is used as a commercial green dye in food industry, cosmetics industry, washing soaps and detergents industry, and tooth care industry.

4. Storage Stability of Natural Colorants

Natural colourants have less stability than the synthetic ones. Their stability is impacted by temperature, light, pH, and oxidative degradation with air during long storage. High temperature affects the stability of most of the natural colorants. For example betalain in beet juice and carotenoids like bixin and lycopene get degraded under the impact of high temperature. The longer the period of heat application, the more severe is the degradation. Most of the natural colourants are sensitive to light and fade with time during storage. Because of the high degree of unsaturation and conjugation, they strongly absorb light and get degraded. Exposure to light, especially sunlight, hastens this process in foods containing colouring agents. This is a major consideration when clear transparent packaging is used and the product is displayed on the shelf for longer duration. Among the natural colours, the yellow pigment curcumin from turmeric, betanin from beetroot, and anthocyanins from black carrot and cabbage are more sensitive to light and get degraded or turn brownish even under mild heating and light exposure. The pH is another factor, which influences the stability of natural colouring agents. A pH over 7 causes turmeric to appear red and to fade rapidly. Some natural colourants are more prone to irreversible oxidative degradation. For example carotenoids are very good oxygen scavengers and under the impact of oxygen/air get readily oxidized causing oxidative rancidity. Such processes can be thwarted by the use of oxygen barrier in the packaging.

Microencapsulation could be one way to enhance the stability of natural colorant. In this process the active ingredient is enveloped in a coating or capsule, and is protected from the surrounding environment. In addition, microencapsulation and complexation can also be used to modify or regulate solubility and other physico-chemical properties. Complexation of colorants with β-cyclodextrin may also promote intensification of colour and increase water solubility. Application of curcumin-β-CD in cheese and yogurt and bixin-β-CD in the curd make the products more stable, attractive and well accepted without altering basic characteristics (Marcolino *et al.*, 2011).

5. Application of Natural Colorants in Textile Dyeing

Dyes derived from horticultural crops and other natural sources have emerged as an important alternative to synthetic dyes. They are renewable, carbon neutral and unlike many synthetic dyes, are not dependent on petroleum or coal tar products. Being of natural origin, they provide environmentally sound alternative to the synthetic dyes. Historically, natural dyes were used to colour fabrics (cotton, wool, silk), garments, clothing and other textiles. Their use was subsequently decreased following introduction of more stable and cheaper synthetic dyes. Because of the various environmental problems and undesired toxic effects, interest in synthetic dyes has faded drastically. Lately, interest in natural dyes has increased substantially as they cut down on the amount of toxic effluent resulting from the synthetic dyeing process. Among the various natural colorants, blueberries, blackberries, black carrot, red cabbage, marigold and gulmohar flowers create lovely blue, violet and pink and orange colorations. Raspberries and beetroot create red shades, turmeric and lemon peels make light yellow dye, spinach creates a nice green, and carrots and onion skins provide orange shade. Many such natural dyes show resistance to fading due to colour fastness and can make textile industry more competitive by reducing production costs and eliminating the huge expenses of chemical imports. In spite of its various advantages, they have limitations too as they are costly, may not ensure colour fastness and guarantee sustainability. However, in view of their safety to humans and the environment, natural dyes produced from horticultural crops can provide ecologically sound alternative to the complex chemical dyes.

6. Conclusion

Due to hazardous impact of synthetic colourants, many countries have banned their use and are now focusing more on natural colourants. As compared to the developed countries where natural colorants are increasingly used, their availability is limited in India. Farmers are not aware of the market potential of food colourants and horticulture crops producing them. Since sufficient scope exists for value-addition, horticultural crops rich in natural colourants can be used as raw material for food, cosmetics, pharmaceutical and textile industry. In recent years, a number of bioactive natural molecules like carotenes, capsanthin, astaxanthin, lycopene, anthocyanins, curcuminoids, phenolics, flavonoids, betalains *etc.* have attracted a great deal of scientific interest as they offer versatile applications in food and beverage, nutraceutical, phytopharmaceutical, bakery, confectionery, and dairy industry. In addition to food applications, natural colorants may also be used to colour medicines (tablets, pills, capsules, liquid formulations), cosmetics like lipsticks, nail enamels, facials, and as natural dye to colour hairs, cotton, wool, silk fabrics, clothing and other textile materials. There is a need to develop better solid–liquid extraction, processing and formulation techniques so that farmer's themselves can adopt such technologies and become entrepreneurs and produce natural food colorants for domestic use and export. Proper training and financial support to the farmers and other stakeholders will enable them not only undertake large scale cultivation of such crops but also use the produce as raw material for the industry. Guaranteed buy-back of the raw material and subsequent help in marketing of the finished products will ensure commercial viability of such start-ups. Such steps will go a long way for improving farmer's income and prosperity.

7. References

Abdel-Hady, M.M., Gamila, Y., Attia and Ali, Afaf M. 2014. Color stability of strawberry jam fortified by purple carrot puree. *Egypt. J. Agric. Res.*, **92**(1): 323–336.

Adeel, S., Gulzar, T., Azeem, M., Fazal-ur-Rehman, M., Hanif, I. and Iqbal, N. 2017. Appraisal of marigold flower based lutein as natural colorant for textile dyeing under the influence of gamma radiations. *Radi. Phy. and Chem*, **130**: 35–39.

Algarra, M., Fernandes, A., Mateus, N., Victor de Freitas, Joaquim, C.G., Esteves, da Silva and Casado, J. 2014. Anthocyanin profile and antioxidant capacity of black carrots (*Daucus carota* L. ssp. *sativus* var. *atrorubens* Alef.) from Cuevas Bajas. *Spain J. Food Comp. Anal*, **33**: 71–76.

Aqil, F., Gupta, A., Munagala, R., Jeyabalan, J., Kausar, H., Sharma, R.J., Singh, I.P. and Gupta, R.C. 2012. Antioxidant and antiproliferative activities of anthocyanin/ellagitannin-enriched extracts from *Syzygium cumini* L. (*Jamun*, the Indian blackberry). *Nutr. Cancer*, **64**(3): 428–38.

Ayyanar, M. and Subash-Babu, P. 2012. *Syzygium cumini* S.: A review of its phytochemical constituents and traditional uses. *Asian Pac. J. Trop. Biomed.*, **2**: 240–246.

Bige sncedayi, Vildan Uylaíer, and Ömer Utku Çopur 2008. A traditional Turkish beverage shalgam: Manufacturing technique and nutritional value. *J. Food, Agr. Env.*, **6**(3–4): 31–34.

Jha, C.K., Kumar, R., Venkat Kumar, S. and Devi Rajeswari, V. 2015. Extraction of natural dye from marigold (*Tagetes erecta*) and dyeing of fabric and yarns: A focus on colorimetric analysis and fastness properties. *Der Pharmacia Lettre*, **7**(1):185–195.

Charepalli, V., Reddivari, L., Vadde, R., Walia, S., Radhakrishnan, S. and Jairam, K.P. Vanamala 2016. *Eugenia jambolana* (Java plum) fruit extract exhibits anti-cancer activity against early stage HCT–116 colon cancer cells and colon cancer stem cells Stem cells *Cancers*, **8**(3): 29; doi:10.3390/cancers 8030029.

Coutinho, M.R., Quadri, M.B., Moreira, R.F.P.M. and Quadri, M.G.N. 2004. Partial purification of anthocyanins from *Brassica oleracea* (Red cabbage). *Sep. Sci. Tech.*, **39**: 3769–3782.

DeBrito, E.S., De Araujo, M.C.P., Alves, R.E., Carkeet, C., Clevidence, B.A. and Novotny, J.A. 2007. Anthocyanins present in selected tropical fruits: Acerola, jambolão, jussara, and guajiru. *J. Agric. Food Chem.*, **55**: 9389–9394.

Gao, Y., Liu, X., Xu, H., Zhao, J., Wang, Q., Liu G. and Hao, Q. 2010. Optimization of supercritical carbon dioxide extraction of lutein esters from marigold (*Tagetes erecta* L.) with vegetable oils as continuous co-solvents. *Sep. Puri. Tech*, **71**: 214–219.

Hajare, R., Ray, A., Shreya and Selvaraj, C.I. 2013. Extraction and quantification of lutein from various plant sources. *Int. J. Pharm. Sci. Rev. Res.* **29**: 152–157.

Huseyin, E., Hasan, T. and Ahmet, C. 2008. A traditional Turkish lactic acid fermented beverage: Shalgam (Salgam). *Food Reviews Int.*, **24**(3): 352–359.

Kamalambigeswari, R. and Jeyanthi Rebecca, L., 2016. Extraction of antioxidant lutein from various flowers. *Int. J. Pharm. Sci. Rev. Res.*, **39**(1): 122–124.

Kammerer, D., Carl, R. and Schieber, A. 2004. Quantification of anthocyanins in black carrot extracts (*Daucus carota* ssp. *sativus* var. *atrorubens* Alef.) and evaluation of their color properties. *European Food Res. and Tech.*, **219**: 479–486.

Marcolino, V.A., Zanin, G.M., Durrant, L.R., Benassi Mde, T. and Matioli, G. 2011. Interaction of curcumin and bixin with β-cyclodextrin: Complexation methods, stability, and applications in food. *J. Agric. Food Chem.*, **59**(7): 3348–3357.

Murillo, E., Meléndez-Martínez, A.J. and Portugal, F. 2010. Screening of vegetables and fruits from Panama for rich sources of lutein and zeaxanthin. *Food Chem.*, **122**: 167–172.

Netzel, M., Netzel, G., Kammerer, D.R., Schieber, A., Carle, R., Simons, L., Bitsch, I., Bitsch, R. and Konczak, I. 2007. Cancer cell anti-proliferation activity and metabolism of black carrot anthocyanins. *Innov. Food Sci. Food Emerg. Tech.*, **8**: 365–372.

Polovka, M., St'avíková, L., Hohnová, B., Karásek, P. and Roth, M., 2010. Off-line combination of pressurized fluid extraction and electron paramagnetic resonance spectroscopy for antioxidant activity of grape skin extracts assessment. *J. Chromat.*, **A 1217**: 7990–8000.

Saha, S., Walia, S., Kundu, A., Sharma, K. and Paul, R.K. 2015a. Optimal extraction and fingerprinting of carotenoids by accelerated solvent extraction and liquid chromatography with tandem mass spectrometry. *Food Chem.*, **177**: 369–337.

Saha, S., Walia, S., Kundu, A., Kaur, C., Singh, J. and Sisodia, R. 2015b. Capsaicinoids, tocopherol, and sterols content in chili (*Capsicum* sp.) by gas chromatographic-mass spectrometric determination. *Int. J. Food Prop.* **18**(7): 1535–1545.

Singh, B. and Hathan, B.S. 2014.Chemical composition, functional properties and processing of beetroot – review. *Int. J. Sci. Eng. Res.*, **5**(1): 679.

Singh, J., Walia, S., Shukla, R.K., Saha, S., Sharma, K., Kaur, C. and Kalia, P. 2016. Characterization of anthocyanins in Indian black carrot (*Daucus carota* var. Pusa Asita) and their evaluation for antioxidant, antidiabetic and antibacterial activity. *Food Res. Int.* (in press).

Sowbhagya, H.B., Sampthu, S.R. and Krishnamurthy, N. 2004. Natural colorant from marigold: Chemistry and technology. *Food Rev. Int.*, **20**: 1, 33–50.

Veigas, J.M., Narayan, M.S., Laxman, P.M. and Neelwarne, B. 2007. Chemical nature, stability and bioefficacies of anthocyanins from fruit peel of *Syzygium cumini S. Food Chem.*, **105**: 619–627.

Zou, Y., Liu, Y., Zhao Y., and Wen Y. 2014. Extraction and stability of red pigment from Grape skin. *Adv. J. Food Sci. Tech.*, **6**(8): 1005–1007.

Walia, S. 2013. Final Report – NAIP Sub–project on *"Value chain on production of food grade nutraceuticals for use as natural antioxidants and food colorants* NAIP–PIU, (ICAR), p. 71.

Wiczkowski, W., Szawara-Nowak, D. and Topolska, J. 2013. Red cabbage anthocyanins: Profile, isolation, identification, and antioxidant activity. *Food Res. Int.*, **51**(1): 303–309.

Zuorro, A. and Lavecchia, R. 2010. New functional food products containing lutein and zeaxanthin from marigold (*Tagetes erecta* L.) flowers. *J. Biotech.*, **150**: 296–296.

79

Phytomedicines from Horticultural Crops for Human Health and Prosperity

Jitendra Kumar[1]* *and R.S. Jat*[2]

[1]*ICAR-Directorate of Medicinal and Aromatic Plants Research, Boriavi, Anand – 387 310, Gujarat*

[2]*ICAR-Directorate of Rapeseed-Mustard Research, Sewar, Bharatpur – 321 303, Rajasthan*

**E-mail: jitendrakumar7@yahoo.com*

About 25 per cent of modern drugs derived from plants and more than 120 active compounds isolated from higher plants are widely used in modern allopathic medicines today. Besides this, at least 7,000 medicinal compounds derived from plants are included in the modern pharmacopoeia of drugs. In the face of the increasing use of herbal medicines and fast-growing market in both developing and developed countries of the world, policy-makers, health professionals and the public are increasingly expressing concerns about the safety, efficacy, quality, availability, preservation, and further development problems of these herbal products. In order to allay these concerns and to meet public demands, extensive research on herbal medicines is to be undertaken not only for their great healthcare value but also for the commercial benefits. In the recent past, extensive researches on phytochemical and pharmacological uses of medicinal plants and herbal medicines are already in place to isolate and identify their active chemical constituents and to substantiate the claims of their efficacy and safety. Besides this, medicinal and aromatic plants (MAP) production either through collection or cultivation, also requires a strategic plan, which takes a holistic view of the entire state of affairs from sustainable utilization to conservation of vital biodiversity, crop diversification of the existing harvest systems and value-addition and marketing. Herbal medicines may emerge as good substitutes or better alternatives for synthetic chemicals-based allopathic drugs or may even replace them.

Phytomedicines from Traditional to *Bona Fide* Drugs

The use of medicinal and aromatic plants has long history and most likely began with the first afflictions and the recognition that smell-ing, chewing, and/ or eating some plant materials could provide relief from nausea, pain, and/or other infirmities. Based on scientific evidences, it is proved that these plants contain unique chemical profiles that offer pain relief, pleasant aromas, and enhanced food flavours and fragrance. MAPs have benefits of immediately available at doorstep for use in medical treatment of man and animals, cost effective and lesser side effects. Various human migration and settlements in the history led to discover new plants of medicinal values and there distribution over wider areas. However, technological innovations and political and social forces at the beginning of the 20th century caused a rapid decline in the use of plant based medicines with the introduction of sulfa drugs in the 1930s and synthesis of organic chemicals in the 1940s. Later on in the mid of 20th Century, it became more tough as the medical practitioners and consumers were demanding scientific proof of efficacy of medicinal plants through double-blind clinical trials in contrast to the traditional use and mythological practices. As mentioned that the human migration and settlement played role in distribution of plants as well as knowledge across the geographical boundaries, again came to the rescue through rejuvenated interest in plant based products in later part of the same century. Further, increased knowledgeable and consciousness among the consumers about health during the 1980s and 1990s, aroused the interest in organic and natural foods using medicinal and aromatic plants. Many of the developed countries used to considered herbal medicines as an attractive alternative to the comparatively high costs of conventional medicines by the end of the 20th century. For example the use of herbal medicines in the US by regular and occasional users increased from 2 per cent of the population in 1990 to 37 per cent in 2000 (WWF, 2000). In most areas of Africa and Asia, the traditional healer continues to be the main source of medical care into the 21st century, primarily due to the relatively high cost of conventional medicines and the lack of trained physicians (Craker and Gardner, 2006). As demand for medicinal species grown, the domes-tic and foreign growers increased production which posed major concerns about the conservation of genetic stock of medicinal plants. Similarly, quality standards for plant material increased with processors and consumers demanding clean (no physical nor chemical contaminating adulterants), consistent (dependable production and bioactive levels), and certifiable (identifiable for origin and history) products (Khan *et al.*, 2005). To address these issues, research on medicinal and aromatic plants initiated in the 1980s and 1990s internationally led to improve the production, extraction of bioactive constituents, and confirmation of medicinal applications (Khan *et al.*, 2005).

Growers in each production center world-wide have been affected by changing market places. In the late 1990s and early 2000s, a number of new medicinal and aromatic plant products were brought to market. The development, marketing, and sales of health and wellness drinks (herbal and flavored teas and energy, health, functional, and sports drinks that often contain herbal extracts) exploded after 2000, with total sales of several billion dollars in current markets. Besides this, consumer choice spending has lead to interest in natural products for animal care (feed and non-feed items labeled natural or organic), significantly increasing demand during the past years. While the global market for medicinal and aromatic plants can be estimated to be at least US$ 60 billion (WHO, 2003), exact market figures and market trends are difficult to ascertain due to herbal materials in a vast array of products being sold through a large number outlets, ranging from entrepreneurial sales over the internet to mass market sales in supermarkets and natural product stores. Over the near future, continued globalization of trade and markets along with ethno botanical exploration can be projected to continue to bring awareness of new plant materials for home, medicinal, and industrial use. In addition to the demand created by population diversification, the relatively high cost of medical treatment in the US (Craker and Gardner, 2006; Schippman *et al.*, 2005, 2006) and the failure of conventional medicine to have satisfac-tory treatments for aliments, and currently incur-able afflictions, such as HIV, cardiovascular and degen-erative diseases, and cancer (JTF, 1999), have stimulated many healthcare consumers to test alternative medicines (AARP, 2007). The fact example, *Artemisia annua* for treatment of malaria, responsible for almost three million deaths annually over the past 30 years, as synthesized drugs became ineffective (IPBO, 2004; WAC, 2004). Continued loss of habitat in the future due to deforestation and development can be expected to remain a threat to many medicinal and aromatic species in both developing and industrialized countries (Shanley and Luz, 2003; Schippmann *et al.*, 2006). Which frequently creating environments unfavorable to growing native medicinal and aromatic species and posing detrimental effects on traditional healthcare. Such destruction in natural ecosystems and the resultant losses in medicinal and aromatic species will surely increase pressure for preservation and cultivation of endangered flora. Shortages of available plant material for collection in the natural environment of medicinal and aromatic plants can be expected to lead to high costs for plant material until cultivation systems are in place.

The global trade in medicinal plants industry is increasing at the rate of 15-25 per cent annually, while the total value of traditional medicinal products worldwide is estimated to be as large as that of modern medicines (about US$ 80-90 billions) and the demand is expected to grow to the tune of 5-6 trillion US$ by the year 2050. At present there is huge consumption of herbal drugs in India, for some plants it is more than 200 tonnes annually (Table 1). The MAPs industry is plagued by a lack of standards in formulations and finished product ingredients. This is one of the major hindrances for sustainable growth of this sector and can partly be attributed to the absence of any recognized central or regional authority with this responsibility.

Table 1: Medicinal Plants of Annual Consumption over 200 tonnes

Name	*Botanical Name*	*Part Used*	*Consumption in tonnes*
Amrita	*Tinospora cordifolia*	Whole plant	727.90
Satavari	*Asparagus racemosus*	Root	630.08
Vasa	*Adhatoda vasica*	Aerial part	321.23
Pippali	*Piper longum*	Spike	296.76
Prisniparn	*Desmodium gangeticum*	Root	252.06
Shaliparni	*Pseudarthria viscida*	Root	252.61
Chitraka	*Plumbago indica*	Root	245.96
Gokshura	*Tribulus terrestris*	Fruit	313.21

The human population alongwith animals increasing much faster rate than growth in the sector. The world healthcare situation in 2050 will be most awful due to the factors of population, food habit, climate change, life style, genetic pollution. In this scenario the most dangerous health problems like cancer, hypertension, diabetes, malarial, viral dengue, chikengunea, Alzheimer's and others will be at alarming place. Parasitic diseases are among the most prevalent infections worldwide. Humans have learned to use plants to treat parasitic diseases, and in the last century modern medicine has developed pure compounds from plants into pharmaceutical drugs. Ethnopharmacological research has brought to light many plants with anti-parasitic properties, and often these plants contain several compounds which contribute together to this activity (Table 2). Some plants also have insecticidal or molluscicidal properties, or interfere in some way with the vectors or intermediate hosts of parasitic diseases and can be used to control them or reduce transmission to humans.

Table 2: Diseases, Medicinal Plants and their Therapeutic Activities

Activity	*Plant Species*
Antimalarial	*Cinchona officinalis, Artemisia annua*
Anticancer	Catharanthus roseus, *Taxus baccata, Podophyllum emodi, Juniperus communis, Mappia foetida*
Antiulcer	*Azadirachta indica, Glycyrrhiza glabra*
Antidiabetic	*Catharanthus roseus, Momordica charantia*
Anticholesterol	*Allium sativum*
Antiinflammatory	*Curcuma domestica, Desmodium gangeticum*
Antiviral	*Acacia catechu*
Antibacterial	*Plumbago indica*
Antifungal	*Allium sativum*
Hepatoprotective	*Silybum marianum, Andrographis paniculata*
Antiallergic	*Nandina domestica, Scutellaria baicalensis*
Cardiotonic	*Digitalis* sp., *Thevetia* sp.
Antiprotozoal	*Ailanthus* sp., *Cephaelis ipecacuanha*
Anthelmentic	*Quisqualis indica, Punica granatum*
Astringent	*Piper betle, Abrus precatorius*
Antidiarrhoeal	*Curcuma domestica, Psidium gujava*
Diuretic	*Phyllanthus niruri, Centella asiatica*
Spasmolytic	*Atropa belladona, Hyoscyamus niger*
Anaesthetic	*Erythroxylum coca*
Hypotensive	*Coleus forskohlii, Alium sativum*
Tranquilizing	*Rauvolfia serpentina*

Herbal Production and Trade in India

The international market of medicinal plants is over US $ 60 billion per year, which is growing at the rate of 7 per centper annum (Planning Commission, Govt. of India, 2000). The present export of herbal rawmaterials and medicines from India is about US $ 100-114 million

Table 3: Important Medicinal Plants with Reference to Trade

Botanical Name	*Part Used*	*Trade tonnes/ Year*
Aconitum heterophyllum	Root	20
Acorus calamus	Rhizome	150
Aloe vera	Leaf	200
Anacyclus pyrethrum	Fruit	50
Andrographis paniculata	Aerial part	250
Asparagus recemosus	Root	500
Berberis Aristata	Root	500
Cedrus deodara	Heart Wood	200
Chlorophytum borivilianum	Root	25
Cinnamomum zeylanicum	Bark	200-300
Commipphora wrightii	Gum resin	500
Crocus sativus	Stigma	5
Cyprus rotundus	Rhizome	150
Eclipta alba	Aerial part	500
Elettaria cardamomum	Seed	60
Embelia ribes	Fruit	200
Glycyrrhiza glabra	Root	5000
Hedychium spicatum	Rhizome	400
Hemidesmus indicus	Root	200
Holarrhena pubescens	Bark	150
Justicia adhatoda	Leaf	500
Mucuna pruriens	Seed	200
Myristica fragrans	Fruit	500
Nardostachy gradiflora	Root	200
Embelica officinalis	Fruit	10000
Picrorhiza kurroa	Root	200
Piper cubeba	Fruit	150
Piper longum	Fruit	200
Piper nigrum	Fruit	150
Plumbago zeylanica	Root	500
Pueraria tuberose	Root	200
Saraca indica	Bark	1200
Senna Alexandrian	Leaf and pod	1000
Strychnosnux vomica	Seed	1000
Swertiachirayta	Whole plant	300
Syzygium aromaticum	Flower bud	150
Syzygium cumini	Seed	300
Trachyspermum ammi	Fruit	200
Terminalia bellrica	Fruit	500
Terminalia chebula	Fruit	500
Tinospora cardifolia	Stem	1000
Valeriana jatamansi	Root and Rhizome	150
Withania somnifera	Root	500
Zingiber officinalis	Rhizome	500

Source: Kalia, 2005.

per year mainly to USA, Germany, France,Switzerland, U.K. and Japan.India has tremendous potential to expand this sector because of its unique bio-geographical positions having all known types of Eco-systems ranging from coldest place, dry cold desert to temperate, alpine and sub-tropical regions of north-west and trans-Himalayas; rain forests with high rainfall; wet evergreen humid tropics of western ghats and arid and semi-arid regions of peninsular India; dry desert conditions of Rajasthan and Gujarat to the tidal mangroves of Sunder ban whereall type of vegetation exists.Some of the highly consumed medicinal plants are presented in the Table 3 with reference to their turnover.

About 70 per cent of families spend 60 per cent of their annual income on health care out of which 93 per cent amount spent on curative and emergency care. In this context, MAP and their derivatives can prove their significance in prevention as well as curing of human health problems.Herbal medicines are; a)cheaper, accessible and affordable,b) societal acceptance and preference, c) established fact of effective and safe as conventional medicines. Due to this, herbal medicines have undergone a major revival in the last twenty years and now exists side-by-side with western forms of medicine.Because of this current trend of increasing use of herbal medicines and their growing popularity all over the world, the search for drugs and dietary supplements derived from plants have accelerated. Pharmacologists, Pharmacognosists, Microbiologists, Botanists and natural-products chemists are combing the earth for phytochemicals and leads from herbs and plants that could be developed for treatment of various diseases.

Production Constraints and Approaches

Medicinal and aromatic plants are also gradually recognized as source of significant livelihood opportunities for many rural communities, especially, primitive forest-dependent tribes, landless poor and marginalized farmers. However, the every stakeholder playing role in MAPs sector is facing problems; farmers are unable to cultivate medicinal plants in large scale due to lack of knowledge, unavailability of quality planting material and proper market; pharmaceutical companies are unable to maintain the quality in their drugs due to variation in raw drug materials; traders facing the problem due to fluctuation in demand and supply and policy support. Overall the sector is facing the problems of conservation of spp. and habitats; lack of novel, high yield, quality and resistant to biotic and abiotic stresses varieties; technology for economic and efficient processes for extraction and isolation of bioactive molecules; development of GAP, GCP, GLP and GMP; emerging biotic and abiotic stresses and changing climatic conditions; lack of availability of reference and high value innovative products; and lack of coordination among different stakeholders of MAPs. Under these situations, MAP sector require a strategic plan, which takes a holistic view of the entire situation from sustainable utilization to economic development, conservation of vital biodiversity, crop diversification of the existing cropping systems and value-addition and marketing with the advancement of technologies.

More than 90 per cent of the plant material used in pharma and aromatic industries is presently collected as non-descriptive uncharacterized land races from forest and other wild resources. More often than not, such collections are usually made using destructive harvest from the wild. This is neither sustainable nor acceptable in a globally competitive scenario. In order to arrest extinction and rapid decline in the diversity of medicinal plants and to maintain the quality of the raw material to be used for the production of herbal formulations, there is a pressing need for a change in the mind-set of Indian pharma industry to shift from collection to cultivation of MAPs. The agri economics of such efforts have also been worked out to attract adoption of such approaches by the farmers. Initial testing results are promising in terms of Land Use Efficiency (LUE), Area Time Equivalency Ratio (ATER), efficient water and fertilizer usage, lesser pesticidal applications that may ultimately add to improved per unit earning of the farmers. The advantage of medicinal and aromatic crops is that most of them require low-inputs, and can withstand stresses of salinity, alkalinity, drought, temperature, heavy metal toxicities, pest infestations *etc.* and, hence can be very good candidates for the utilization of *hitherto* under-utilized poor soils.

Quality Standard Drugs

In order to bring this shift in the paradigm, farmers and traders need to be sensitized, and provided with material of improved MAPs varieties with quality tags to initiate cultivation in the country. Itshould be produced by following good agricultural practices from cultivation to harvest and post-harvest operations to produce quality raw drugs and finished products of international standard.It also requires proactive initiatives in establishing linkages among farmers, end-user industry, funding institutions and quality accreditation agencies for creating a complete value-chain for the MAP sector. It also needs to identify appropriate locations within the country where strong cultivation and/or marketing hubs of in-demand MAPs can be established to strengthen these linkages. In addition to advancing R&D domains of genetic up-gradation, varietal improvement and development of suitable agro technologies for cultivation of industrially important MAPs in different agro-climatic zones of the country, we also need to generate bench-level data on potential inter- or mix-cropping systems of medicinal and aromatic crops like mints, aromatic grasses, scented rose, geranium, garlic, coleus, ashwagandha, centella, sarpagandha, ocimums, senna, aloe vera, kalmegh, brahmi, satawar *etc.* within the cycles of cultivation of food crops such as wheat, sugarcane, lentil, arhar, rice, vegetables and brassicas. Information generated by CSIR-CIMAP and ICAR-DMAPR on integration of MAP crops in agro-forestry, agro-horticulture and agri-orchards systems is also very encouraging. Since useful secondary metabolites present in MAPs generally make them unpalatable to animals, pests and microbes, they can also be inter-cropped with food crops as protection shields.

References

American Associated Retired People (AARP) and National Center for Complimentary and Alternative Medi-cine (NCCAM). 2007. *Complimentary and Alternative Medicine*. Research report by AARP Knowledge Management and National Center for Complimentary and Alternative Medicine. Online www.globalaging.org/health/us/2007/cam.pdf.

Craker, L.E. and Gardner, Z.E. 2006. Medicinal plants and tomorrow's pharmacy. p. 29–41. **In**: *Medicinal and Aromatic Plants*, (Eds.) R.J. Bogers, L.E. Craker, and D. Lange, *Proc. Frontier Workshop on Medicinal and Aromatic Plants*, Wageningen, The Netherlands, 17th–20th April 2005. Nucleus for Strategic Expertise Wageningen University and Research Centre, Wageningen.

Institute of Plant Biotechnology for Developing Countries (IPBO). 2004. Artemisia research (Work in progress). www.worldagroforestrycentre.org/.

Joint Taskforce on Older People (JTF). 1999. Report of the healthcare and aging population panels. www.foresight.gov.uk.

Kalia, A.N., 2005. *Textbook of Industrial Pharmacognosy*. Oscar Publication.

Khan, I.A., Smillie, T.J., Craker, L.E. and Gardner, Z.E. (Eds.) 2005. Quality and safety issues related to botanicals. *Acta Hort*., p. 720.

Nucleus for Strategic Expertise Wageningen University and Research Centre, Wageningen.

Schippmann, U., Leaman, D. and Cunningham, A.B., 2006. A comparison of cultivation and wild collection of medicinal and aromatic plants under sustainability aspects. p. 75–95. **In**: *Medicinal and Aromatic Plants*, (Eds.) R.J. Bogers, L.E. Craker, and D. Lange, *Proc. Frontier Workshop on Medicinal and Aromatic Plants*, Wageningen, The Netherlands, 17th–20th April 2005. Nucleus for Strategic Expertise Wageningen University and Research Centre, Wageningen.

Schippmann, U., Leaman, D.J., Cunningham, A.B. and Walter, S. 2005. Impact of cultivation and collection on the conservation of medicinal plants: Global trends and issues. *Acta Hort*., **676**: 31–44.

Shanley P. and Luz, L. 2003. The impacts of forest degradation on medicinal plant use and implications for health. *BioScience*, **53**(6):573–584.

World Agroforestry Centre (WAC). 2004. Africa malaria day on 25 April: Africa resolves to promote use of anti-malaria herbs. News release. www.worldagroforestrycentre.org/.

World Health Organization (WHO). 2003. *WHO Guidelines for Good Agricultural and Collection Practices (GACP) for Medicinal Plants*. whqlibdoc.who.int/publications/2003/9241546271.pdf.

World Wildlife Foundation (WWF). 2000. *Medicinal Plant Trade*. Wildlife Trade, FAQs. www.worldwildlife.org/tade/faqs_medicinal.cfm.

80

Intrinsic Quality and Phytochemical Profiling of Major Horticultural Crops

T. John Zachariah* and N.K. Leela

Division of Crop Production and Post Harvesting, ICAR-Indian Institute of Spices Research, Marikunnu, Kozhikode – 673 012, Kerala

E-mail: john@spices.res.in

Plants produce various phytochemicals that are of nutritional and medicinal value to humans. Phytochemicals having antioxidant capacity are drawing increased interest from consumers. Improving intake of fruits, vegetables and spices has been a major public health effort for many years with minimal success. Development of fresh produce containing a greater concentration of phytochemicals is known to improve health.

The human body uses various anti oxidants for defense, some of which are dietary derived. The plant-based dietary antioxidants are believed to have an important role in the maintenance of human health because our endogenous antioxidants provide insufficient protection against the constant and unavoidable challenge of reactive oxygen species (ROS). ROS are constantly produced and the production is enhanced by various environmental and biological factors. Generally, ROS induce oxidative damage to key biological sites such as DNA and the lipid membranes of cells, resulting in the accumulation of oxidative damage with age, which eventually contributes to the senescence and development of age-related disease.

Plants produce various biologically active compounds that can promote anti oxidation in the human body. Such compounds in fruit and vegetables include vitamin C, vitamin E, oloy phenols, and carotenoids among others. The concentrations and profile of these components vary with cultivation practices and environmental conditions well as genotypes (species and cultivars). For example, concentration of lycopene in tomato was reportedly altered by light quantity and quality, air temperature and level of salinity in the nutrient solution.

Value-added Greenhouse Crop Production

Controlled environments provide a unique opportunity to modify the concentrations of selected phytochemicals in fruit and vegetables, yet practical information is limited regarding methods effective in optimizing antioxidant capacity. Research at the University of Arizona

Controlled Environment Agriculture Programme has shown that application of moderate salt stress to tomato plants can enhance lycopene and potentially other antioxidant concentration in fruit. The increase in lycopene in response to salt stress in to tomato fruits was shown to be cultivar specific, varying from 34 per cent to 85 per cent. Another experiment showed that low temperature during postharvest increased antioxidant capacity of tomato fruit while it maintained the lycopene concentration. More detailed study in this area is needed including accumulation of antioxidant phytochemicals as affected by environmental conditions during cultivation and the postharvest (Kubota *et al.*, 2006).

It also has been well established that diets high in fruit and vegetables, including tomatoes rich in lycopene, can reduce disease incidence. In fact, diet-based epidemiological studies, from case control to intervention trials, have demonstrated that diets rich in fruit and vegetables can protect against cardiovascular disease, cancer, diabetes, rheumatoid arthritis, hypertension and obesity. High intake of fruit and vegetables, particularly those low in caloric content and sufficient in fiber, such as tomatoes, can also increase satiety and contribute to an overall reduced energy-density diet. Although limited data are available to describe the health benefits of tomatoes specifically, growing evidence indicates that lycopene, the key anti oxidant carotenoid in tomato, may have a unique role in overall health, particularly when it comes to the prevention of chronic disease. More importantly, dietary intake of tomatoes and processed tomato products has been inversely associated with risk of several cancers in case-control and prospective cohort studies.

Presently, producers are striving to create products which contain a value-added factor, such as dietary fibre or in more recent times, phytochemicals. The production and addition of such nutrients can be quite costly for the producer. In the fruit and vegetable industry, the preparation and processing procedures can lead to one third of the product being discarded. This can be costly for the manufacturer and also may have a negative impact on the environment. Research has shown that these by-products can have a high nutritional value. It has also been suggested, that they could be used as a food ingredient due to their functional abilities such as gelling and water binding (Norah *et al.*, 2012).

Fruits and Vegetables and Phytochemicals

Consumption of fruit and vegetables is an important part of a healthy diet and is associated with reduced risk of chronic diseases, including cardiovascular disease, stroke, and certain types of cancer. Fruit and vegetables contain a variety of vitamins, minerals, fiber and phytochemicals. Phytochemicals are compounds produced by plants that are believed to affect health, but are not traditional 'essential' nutrients. That is, if not consumed, clinical deficiency symptoms do not occur. They are, however, thought to be protective against disease, likely working synergistically with other compounds, including vitamins and minerals. More than 5,000 individual phytochemicals have been identified in fruit, vegetables, and grains, but a large number are still unidentified.

Fruits have become increasingly important in human nutrition because of their nutrient composition and potential beneficial health effects. The nutritional value and health related properties of fruits depend not only on the concentration of nutrients and phytochemicals but also on the daily intake and bioavailability. *Annona cherimola* Mill., commonly known as cherimoya, is a tropical fruit well known due to its tasty flavor. Moreover, nutritional composition (proximates and vitamins) and bioactive compounds content were determined. The peel showed the highest antioxidant capacity, with an EC50 of 0.97 mg/mL, and total flavonoids (44.7 epicatechin equivalents/100 g). The most abundant carotenoid was lutein, with values ranging from 129 to 232 lg/100 g. The highest L-ascorbic acid content (4.41 mg/100 g) was found in the peel of Perry Vidal cultivar. These results highlight *A. cherimola* Mill. antioxidant properties, especially in its by-products and encourage their application in cosmetic, pharmaceutical and food processing industries, as added value natural extracts (Tânia *et al.*, 2016). Consumers always have doubts regarding consumption of phytochemicals. Some of FAQs are as follows

1. How many phytochemicals are enough?

Many research reports suggest that half of what you eat each day should be fruit and vegetables. Currently there are no national recommendations for the amounts of phytochemicals needed to prevent disease.

Fruit and Vegetables are essential components of a healthy diet. The World Health Organization (WHO) recommends an intake of five to eight portions (400-600 g) daily of fruits and vegetables to reduce risk of cardiovascular disease, cancer, poor cognitive performance, and other diet-related diseases, as well as for the prevention of micronutrient deficiencies. Their consumption as supplements must be carried out with care, because it may exceed the recommended nutritional intake. Nonetheless, there is a general consensus that whole fruit and vegetables intake is more important in providing health benefits than that of only one of their constituent, because of additive and synergistic effects. Importantly, since some phytochemicals regulate the same genes and pathways targeted by drugs, diets rich in fruits and vegetables in combination with medical therapies are being considered as novel approaches to treatment. Therefore, phytochemicals in fruits and vegetable might be a promising tool for the prevention and/or amelioration of a wide range of disease (Arantxa, 2016)

2. Does the colour of a fruit or vegetable indicate what phytochemicals it contains?

Colour is a good, but not exclusive indicator of phytochemical content. Many, but not all, phytochemicals are pigments that give plants color. Some phytochemicals are colorless, and some are responsible for taste. Sometimes one color may mask another color. Products in a color group will have varying amounts of the same phytochemicals.

3. What affects the amount of phytochemicals in foods?

Both genes and the environment affect the amount of phytochemicals in foods. Some varieties have higher amounts of phytochemicals than others. Other factors in the environment, like soil, altitude, climate, temperature, plant maturity, the presence of predators, and processing and preparation methods can all influence phytochemicals content.

4. What part of a fruit or vegetable contains the phytochemicals?

Phytochemicals are found in all edible portions of a fruit or vegetable, although they are frequently concentrated in the skin, so eating them with their peels on is a plus.

5. Do fresh fruit and vegetables have more phytochemicals than canned, frozen, or dried?

Not necessarily, For example, the carotenoids in vegetables and fruit can be absorbed by the body more efficiently once cooked. The process of removing water during dehydration can also help concentrate phytochemicals in dried fruit and vegetables.

6. Are phytochemicals stored in the body?

Carotenoids are fat-soluble compounds, like vitamins A, D, E and K, and they can be stored in body fat the same way. Also like vitamins, one is much less likely to accumulate toxic levels from whole foods than from concentrated supplements.

Flavonoids are more water-soluble, like B-vitamins. They get washed out of the body and are rarely stored, so food sources of flavonoids may have to be eaten more frequently than foods containing fat-soluble carotenoids in order to get their benefits.

7. How do phytochemicals work in the body?

Many phytochemicals are strong antioxidants that help moderate the damage to cells resulting from oxidation. Phytochemicals are also involved in many of the metabolic pathways that regulate the body's functions. Also, some phytochemicals reduce inflammation that occurs in the walls of arteries. Phytochemicals may also enhance the body's ability to detoxify chemicals, slow or stop the growth of cancer cells and even kill cancer cells. Features of common phytochemicals in fruits and vegetables are illustrated

Carotenoids

There are more than 600 carotenoids identified in nature and these compounds are the source of the yellow, orange, and red colors of many plants (though the chlorophyll in some green plants hides these yellow-orange-red pigments). Carotenoids can be broadly classified in to two classes: carotenes (α-carotene, β-carotene, and lycopene) and xanthophylls (β-cryptoxanthin, lutein, and zeaxanthin). These six cartenoids are the most studied because of their abundance in the food supply. B-cryptoxanthin, α-carotene, and β-carotene are provitamin A carotenoids that can be converted by the body to vitamin A. Lutein, lycopene, and zeaxanthin do not have vitamin A activity. Chopping, pureeing, and cooking carotenoid-containing vegetables in oil generally increases the bioavailability of carotenoids. Lycopene gives tomatoes pink, grapefruit, watermelon, and guava their red color. α-carotene and β-carotene offer the yellow-orange pigments; β-cryptoxanthin offers red/orange.

Chlorophyll

Chlorophyll is used by plants to trap light needed for photosynthesis and also gives plants their green color. It is fat soluble and insoluble in water, though little is known about its bioavailability and metabolism, chlorophylls are the most abundant pigments in plants, with the richest sources being dark green leafy vegetables.

Polyphenols

Flavonoids are the most abundant of the polyphenols. Several subclasses include: anthocyanins, flavanols, flavanones, flavones, and isoflavones. It is estimated that flavonoids account for approximately two thirds of the polyphenols in our diet and the bulk of the remaining are from phenolic acids. Many of the biological effects of flavonoids appear to be related to their ability to modulate cell-signaling pathways, rather than their antioxidant activity.

Tropical Tuber Crops

Tropical root and tuber crops are considered as the third important crops after cereals and legumes. These mainly include cassava (*Manihot esculenta* Crantz), sweet potato (*Ipomoea batatas*), taro (*Colocasia esculenta*), yams (*Dioscorea* sp.), elephant foot yam (*Amorphophallus paeoniifolius*), arrowroot (*Maranta arundinacea*) and other minor root crops. Tuber crops offer immense scope as nutraceuticals and biocolours. They contain an array of bioactive phytochemicals which are actively being researched for their medicinal and culinary potentials. These include anthocyanins, carotenoids, tocopherols, sterols, and soluble fibres. Anthocyanins are one of the best known natural pigments used in food due to their bright and attractive colours, non-toxicity, and water solubility, which allows their incorporation into aqueous food systems. Violet or purple fleshed varieties of greater yam (*D. alata*) and sweet potato, as well as purple leaves of sweet potato were found to contain significant amount of anthocyanins and other flavonoids which possess tremendous antioxidant potential. These anthocyanins were extracted and purified. The purified anthocyanins (0.1 per cent w/v) were encapsulated with maltodextrin (30 per cent w/v) using a spray dryer, since the pure anthocyanins are highly sensitive to temperature, pH and sunlight. The encapsulation process increased the stability of these purified anthocyanins in aqueous solution. The anthocyanins isolated from sweet potato tuber showed more stability in solution based on the color values compared to sweet potato leaf anthocyanins and greater yam anthocyanins. The *in vitro* antioxidant properties of encapsulated anthocyanins were also studied and found that greater yam anthocyanins exhibited more antioxidant potential than sweet potato leaf and tuber anthocyanins evident from the IC_{50} values. In brief, sweet potato tuber anthocyanins have better value as natural colorant whereas greater yam anthocyanins have better nutraceutical value.

Cultivation of roots and tubers in addition to cereals provides greater food security and a more varied and interesting diet. Besides, processing of root crops could also provide cottage industry employment for rural women

Processing and Value-addition in Spices

During 2014-15, India produced 61,08210 tonnes of spices form an area of 33,17,210 hectares. Spice export registered substantial growth during the last one-decade. During 2014-15, a total of 8,93,920 tonnes of spices and spice products valued at 14,899.68 crores (US$ 2,432.85 million) were exported from India, registering a nine percent increase in volume and eight percent in rupee terms and seven percent in dollar terms of value as compared to 8,17,250 tonnes valued at Rs. 13,735.39 crores (US$ 2,267.67 million) exported in 2013-14.

Chilli, mint products, cumin, spice oils and oleoresins, pepper, turmeric, coriander, small cardamom, curry powder/paste and nutmeg and mace contributed substantially to the spice export basket as the demand for Indian spices scaled up phenomenally at the global level.

Chilli continued to propel the growth story as India's largest exported spice item both in volume and value, accounting for Rs. 3,47,000 tonnes in quantity and 3,517.10 crores in value during 2014-15. The export grew by 11.04 percent in quantity and 29.20 percent in value as compared to 2013-14.

Turmeric, ginger, nutmeg and mace, fennel, fenugreek, garlic, celery and other seed spices (mustard, aniseed, bishops weed, dill seed, *etc.*) also chipped in significantly in bolstering spice exports from India and earning precious forex for the country.

In 2014-15, chilli (39 per cent), cumin (17 per cent), turmeric (10 per cent), coriander (5 per cent) and ginger (5 per cent) accounted for more than 75 percent of the total volume of spice exports whereas chilli (24 per cent), mint products (18 per cent), spice oils and oleoresins (13 per cent), cumin (12 per cent) and pepper (8 per cent) accounted for around 75 (per cent) of the total export earnings.

Value-addition in spices is yet another area of activity in which India is moving forward. The consistent effort by various agencies during the last one decade has improved the share of the value-added products in the export basket to more than 53 per cent. India can now boast as the monopoly supplier of spice oils and oleoresins the world over. During the year 2014-15, India exported 24650 tonnes of curry powder/paste worth Rs. 476.26 crores, 25750 tonnes of mint products worth Rs. 2689.25 crores and 11475 tonnes of spice oils and oleoresins worth Rs. 1910.10 crores.

Spices and Functional Foods

Functional food products have been defined as providing an added health benefit over and above the food product's traditional nutritional value. The main consumer motive for purchasing functional foods is the growing desire to use foods either to help prevent chronic illnesses such as cardiovascular disease, Alzheimer's disease and osteoporosis or to optimize health, for example by increasing energy, boosting the immune system and generation of wellbeing. This need has led to one of the fastest growing food sectors, with a compound annual growth rate of >8.6 per cent in the last decade. The emergence of a new market segment called 'Health and Wellness' reached a global value of US$ 625 billion in 2012. This segment incorporates fortified/functional foods, but also including organic foods, "better for you" food and beverages (BFY), "naturally healthy" (NH) foods, products catering to food intolerance, vitamins and dietary supplements, traditional herbal products, slimming products and sports nutrition. Of this market, the functional foods part alone was valued at US$ 168 billion in a global market that is 2.5 times the size that of vitamins and dietary supplements.

Spices and Health

Spices are an important part of human nutrition and have a place in all the traditional cultures of the world. One of the changes occurred in food industry is from the concept of 'adequate nutrition' to 'optimal nutrition' which includes potentially of food to promote health, improve well being and reduce the risk of developing disease. That is how new terminologies like functional foods, designed or therapeutic foods, super foods or medicinal foods emerged.

The positive effect of a functional food may include the maintenance of health or well being or a reduction in the risk of suffering a given illness. Functional food may be obtained by modifying one or more of the ingredient or by eliminating the same.

Culinary Herb and Spice

The leaves of a plant used in cooking are denominated as culinary herbs while any other part of the plant is known as a spice. Spices can be leaf (*e.g.* bay leaf), buds (clove), bark (cinnamon), root (ginger), berries (grain of black pepper), seeds (cumin) or even stigma of flower (saffron). Both spices and herbs can be used fresh, dried, whole, chopped or ground and due to their colour, aroma and/or flavour characteristics are used in the preparation of foods and drinks. Spices and herbs contain proteins, fiber, sugars, essential oils, minerals, pigments besides bioactive compounds such as phenolic acids, sterols and cumarins. Many of the functional properties attributed to spices associate with presence, type and concentration of phenolic compounds although exact composition depend on several factors such as the plant part used, vegetative state, environmental conditions, harvesting techniques *etc.*

Other compounds present in spices are the essential oils which due to their content in terpenes, monoterpenes and susquiterpenes (as hydrocarbons, alcohols, ketones *etc.*) are responsible for many functional properties. These are composed of more than 70 components some of which may represent more than 85 per cent of total content.

Phenolic Compounds

Phenolic compounds present in spices and herbs include phenoles, flavons, flavanones, flavanols, anthocyanins *etc.* which are responsible for their functional properties. Properties of spices such as antioxidant, antibacterial, antiviral, anti-inflammatory capacities, cardio protective and anti carcinogenic effects and their ability to act as inhibitors of platelet aggregation are significant. Phenolic compounds contribute to colour and taste. Simplest phenols are liquid or solid with a low fusion point and high boiling point as they form hydrogen bonds. They are easily oxidized which is why they frequently appeared to be coloured.

Terpenes

These group of chemicals are the fundamental components of essential oils. They all result from the condensation of isoprenic units. Among their different form are monoterpenes, diterpenes and sesquiterpenes. They are extremely volatile and have been demonstrated to possess multiple functional properties including antioxidant, antimicrobial and antiviral capacities.

Antioxidant Properties

Spices have shown their ability to slow down process of lipidic oxidation. This process is one of the principal causes of food spoilage and may occur during the storage of raw materials, processing or at the storage of final product. To avoid such spoilage food industry has resorted to synthetic antioxidant formulations, some of the most commonly used being butylated hydroxy toluene (BHT) butylated hydroxyl anisole (BHA) and propyl gallate. However, because of their synthetic origin their safety and efficacy are frequently questioned. The result has been a growing interest in compounds of vegetable origin which show antioxidant potential.

Possible mechanism of antioxidant action by spices may be by scavenging free radicals, hydrogen donation, the chelating of metallic ions or the capacity to act as substrate of radicals such as superoxide or hydroxyl.

Free radicals, produced as a result of normal biochemical reactions in the body, are implicated in contributing to cancer, atherosclerosis, aging, immune suppression, inflammation, ischemic heart disease, diabetes, hair loss, and neurodegenerative disorders such as Alzheimer's disease and Parkinson's disease. The human body possesses innate defense mechanisms to counter free radicals in the form of enzymes such as superoxide dismutase, catalase, and glutathione peroxidase. Vitamin C, vitamin E, selenium, β-carotene, lycopene, lutein and other carotenoids have been used as supplementary antioxidants.

Apart from these, plant secondary metabolites such as flavonoids and terpenoids play an important role in the defense against free radicals.

Table 1 illustrate some of the medicinal properties known to be imparted by spices

Table 1: Medicinal Properties of Spices Recognized for a Long Time

Spices	*Medicinal Properties Recognized for a Long Time*
Coriander (*Coriandrum sativum*)	Antidyspeptic, flavorant
Cumin (*Cuminum cymminum*)	Antispasmodic, carminative, digestive stimulant
Fenugreek (*Trigonella foenumgraecum*)	Diuretic, emmenagogue, emollient, useful in heart diseases
Garlic (*Allium sativum*)	Antidyspeptic, antiflatulent, for ear infection, duodenal ulcers, as rubefacient in skin diseases
Ginger (*Zingiber officinale*)	Sialogogue, useful in diseases of heart and blood
Onion (*Allium cepa*)	Diuretic, emmenagogue, expectorant, for bleeding piles
Peppr (*Piper nigrum*)	Antipyretic; Rubefacient
Red pepper (*Capsicum annuum*)	Anti-inflammatory, for pain relief (rheumatism/ neuralgia)Useful in indigestion, rubefacient
Turmeric (*Curcuma longa*)	Anti-inflammatory, diuretic, laxative, good for affections of the liver, jaundice, diseases of blood

The spice trade, probably, is the most ancient trade practiced by man. The affluence generated by the spice trade was responsible for several historic voyages and discoveries of new lands. Today, the annual global spice trade is estimated to be in the order of $2000 million involving a quantity of 500,000 tonnes. India is the largest producer, consumer, and exporter of spices. India's annual spice exports amount to 2.3 lakh tonnes and of a value of Rs. 16,000 million. Black pepper, also called "king of spices", is the major spice commodity exported from India followed by red pepper, turmeric, ginger, cardamom, seed spices, curry powders, spice oils, and oleoresins.

Spices are not only used individually, but also in the form of spice mixtures, known as 'curry powder," to suit different tastes and dishes. Although spices have never been considered to contribute anything to human nutrition, this group of food adjuncts has been used in human diets for centuries as flavor modifiers to make food more palatable. Interestingly, the protein content in spices varies from 4.5 per cent in rosemary leaves to 31.5 per cent in mustard, and the fat level varies from 0.6 per cent in garlic to 42.6 per cent in mustard. The ash content can be anywhere from 2.3 per cent in marjoram to 16.7 per cent in basil leaves reflecting high mineral levels in them. Some spices contain significant levels of vitamins and minerals, which cannot be ignored. A few spices are also rich sources of dietary fiber. Amongst common spices consumed, the dietary fiber is highest in red pepper, as high as 43.3 per cent, although black pepper (27.8 per cent), coriander (36.2 per cent), cumin (23.0 per cent), fennel (28.7 per cent), and fenugreek (33.5 per cent)also are rich sources of dietary fiber, both soluble as well as insoluble. However, due to the low levels of spices consumption, their impact on nutrient make up may not be as dramatic as that of other food ingredients. The components of spices responsible for the quality attributes have been designated as "active principles" and in many instances, they are also responsible for the beneficial physiological effects of spices.

Extensive animal studies carried out to evaluate the safety aspects of spices have indicated that even at much higher dietary levels (up to 100 times the normal intake), red pepper, black pepper and turmeric have no adverse effects on growth, organ weights, Feed Efficiency Ratio, nitrogen balance, and blood constituents. In the past two to three decades, many more beneficial physiological effects of spices have been experimentally documented, which suggest that the use of these food adjuncts extend beyond taste and flavor. Among the health problems that effect mankind, diabetes, cardiovascular disease, and inflammatory disorders including arthritis, and cancer have received considerable attention. During recent years, spices and their active principles have been studied as possible ameliorative or preventive agents.

Hypolipidemic/Hypocholesterolemic Effect

The spices fenugreek, red pepper, turmeric, garlic, onion and ginger were found to be effective as hypocholesterolemic agents under various conditions of experimentally induced hypercholesterolemic/ hyperlipemia. Further, fenugreek, onion, and garlic are effective in human with hyperlipemic condition. Curcumin and capsaicin, the active principles of turmeric and red pepper, respectively, are also efficacious at doses comparable to calculated human daily intake. Turmeric and curcumin showed excellent hypocholesteremic effect in experimental animals. However, endogenous cholesterol synthesis was not affected. Extracts of garlic and onions inhibit platelet aggregation and lower cholesterol levels. The raw form is more effective than the cooked form. About 50g of onion and garlic corresponding to 5-6 cloves per day may be adequate to bring these beneficial effects.

Other work also documents hypocholesterolemic effects of spices. Fenugreek seeds were hypocholesterolemic in rats with hyperlipidemia induced by either high fat or a high cholesterol diet. Defatted fenugreek seed was effective in diabetic hypercholesterolemia in dogs and in humans. The hypolipidemic effectiveness of turmeric/curcumin, red pepper/capsaicin and of onion and garlic has been periodically reviewed in recent years by different authors.

With the introduction of dehydrated garlic powder containing a standardized level of the parent sulfur compound, alliin, effective clinical work could be undertaken with the relatively low and acceptable daily dosage of 300-900 mg (= 1 clove of garlic). This level produced a consistent 10-13 per cent reduction in blood cholesterol and triglycerides.

In a recent study, dietary supplementation with aged garlic extract showed better beneficial effects, relative to fresh garlic, on the lipid profile and blood pressure of moderately hypercholesterolemic subjects. In other study, garlic supplementation significantly decreased both total and LDL cholesterol in hypercholesterolemic subjects. Co-administration of garlic with fish oil had a better beneficial effect on serum lipid and lipoprotein concentrations by providing a combined lowering of total cholesterol, LDL cholesterol, and tiglyceride concendration as well as the rations of total cholesterol to HDL cholesterol and LDL cholesterol to HDL cholesterol. The anti-platelet aggregation, the anti platelet adhesion and the antiproliferation properties of aged garlic extracts appear to contribute more to cardiovascular protection than do the hypolipidemic properties. Apart from the hypocholesterolemic effect of capsaicin, its beneficial effect on overall lipid metabolism under different conditions of lipemia has also been reported. In a further study, capsaicin treatment is also shown to enhance energy metabolism in rats.

These reports establish the fact that value-addition based on phytocemical composition has great scope and farmers or group of farmers should explore this possibility for entrepreneurship. Table 2 depicts some of the value-added products from different spices.

Narasinga Rao (2003) has illustrated bioactive phytochemicals in Indian foods and their potential in health promotion and disease prevention. He has reported that foods, particularly plant foods, are a rich source of bioactive phytochemicals or bionutrients. Studies carried out during the past 2–3 decades have shown that these phytochemicals have an important role in preventing chronic diseases like cancer, diabetes, coronary heart disease and hyper-cholesterolaemia. The major classes of phytochemicals with disease-preventing functions are dietary fibre, antioxidants, detoxifying agents, immunity-potentiating agents and neuro-pharmacological agents. Each class of these functional agents consists of a wide range of chemicals with differing potency. Foods rich in these chemicals and exhibiting disease protecting potential are called functional foods. Indian habitual diets, which are based predominantly on plant foods like cereals, pulses, oils

Table 2: Value-added Products from Spices

Spice	*Value-added Product*
Black pepper	
Green pepper based products	Canned green pepper, green pepper in brine, cured green pepper, frozen green pepper, freeze dried green pepper, dehydrated green pepper, green pepper oil, green pepper pickle, mixed green pepper pickle, green pepper sauce and green pepper-flavoured products.
Black pepper based products	Whole black pepper, sterilized black pepper, ground black pepper, cryoground black pepper powder, pepper oil and oleoresin.
White pepper based products	Whole white pepper, white pepper powder.
Other products of black pepper	Pepper mayonnaise, pepper cookies and pepper tofu. Pepper extract is a valuable adjunct in the flavouring of sausages, canned meat, soups, table sauces and certain beverages and liquor. Pepper oil is used in perfumery and also for manufacturing soaps. Products like lemon pepper, garlic pepper, sauces and marinades have pepper as major ingredient. Curry powders and spice blends have pepper as one of the major ingredient.
Cardamom	Bleached cardamom, decorticated seeds and seed powder, cardamom volatile oil, cardamom oleoresin. Other products include encapsulated cardamom which is free flowing and having uniform flavor, cardamom tea, cardamom coffee, cardamom soft drink mix.
Ginger	Ginger powder, salted ginger, ginger oil, ginger oleoresin, ginger-based beverages like ginger beer and ginger ale, ginger candy, ginger paste, salted ginger, salted ginger, crystallized ginger.
Turmeric	Ground turmeric, turmeric oil, turmeric oleoresin, curcumin
Chillies	Paprika oleoresin, chilli colour, chilli pungency, dehydrated chilli, canned chilli, brined/pickled chill fermented chilli, brined/pickled chilli, fermented chillies.
Nutmeg	Nutmeg powder, nutmeg oleoresin, nutmegs butter, mace oleoresin, mace oil
Cinnamon	Cinnamon bark oil, cinnamon oleoresin, cinnamon leaf oil, cinnamon powder, cinnamon root bark oil
Clove	Clove powder, clove oil

Table 3: Disease-Preventing Phytochemicals in Plant Foods

Specific Foods/Food Groups	*Main Health Promoting Chemical*	*Useful in the Prevention of*
Whole cereal, grains	Dietary fibre, tocopherols†	Cancer, diabetes, cardiovascular disease, hypercholesterolemia
Vegetables, fat	α-linoleic acid, linoleic acid,tocopherols,† sterols	Cancer, diabetes, cardiovasculardisease, hypercholesterolemia
Palm oil diseases,muscle injury	Tocopherols,†tocotrienols,†cartenoids	Cancer, heart diseases, atherosclerosis,cataract, pulmonary
Yellow or green leafy vegetables, yellow fruits	Carotenes,† ascorbic acid,† dietary fibre, omega-3 fatty acids, pectins	Cancer, heart diseases,atherosclerosis, cataract, pulmonary diseases, muscle injury
Rice bran oil	Sterols, PUFA	Hypercholesterolemia, diabetes, cardiovascular disease
Linseed oil, fish oil	Omega-3 fatty acid	Hypercholesterolemia, diabetes, cardiovascular disease
Spices, fenugreek seeds, turmeric	Gums, curcumin, eugenol, capsaicin	Cancer, cardiovascular disease, detoxification of drugs and toxins

†: Antioxidant; PUFA: Polyunsaturated fatty acid.

Table 4: Bioactive Phytochemicals in Foods

Classification	*Main Groups of Compounds*	*Biological Function*
NSA	Cellulose, hemicellulose, gums, mucilages, pectins, lignins	Water holding capacity, delay in nutrient absorption, binding toxins and bile acids
Antioxidants	Polyphenolic compounds, flavonoids, carotenoids, tocopherols, ascorbic acid, anthocyanins, phenolic indoles	Oxygen free radical quenching, inhibition of lipid per-oxidation
Detoxifying agents	Reductive acids, tocopherols, phenols, indoles, aromatic isothiocyanates, coumarins, flavones, diterpenes, carotenoids, retinoids, cyanates, phytosterols, methyl xanthines, protease inhibitors	Inhibitors of procarcinogen activation, inducers of drug-metabolising enzymes, binding of carcinogens, inhibitors of tumourogenesis
Others	Alkaloids, volatile flavour compounds, biogenic amines, terpinoids and other isoprenoid compounds	Neuropharmacological agents, anti-oxidants, cancer chemoprevention

NSA: Non-starch polysaccharides.

and spices, are all good sources of these classes of phytochemicals, particularly dietary fibre, vitamin E, carotenoids and phenolic compounds. There is, however, much scope for further systematic research in screening Indian foods and diets for these phytochemicals and assessing their potential in protecting against chronic diseases. Tables 3-13 illustrate the health potential of Indian foods and the components involved. This under lines the fact that the type of items used in each food item and the processing or cooking methods have great role in the upkeep of these nutritional benefits. Excellent phytochemical rich foods are also value-added products.

Table 5: Components of Dietary Fibre

Major Class	*Property*	*Chemical Nature*
Cellulose	Water insoluble	Linear 1–4 β-glycans
Non-cellulose polysaccharides		
Pectin	Water soluble	Galacturonic acid, neutral sugars
Hemicellulose	Water insoluble	Xylose, arabinose, galactose, mannose
Gums	Water soluble	Xylose, arabinose, rhaminose
Mucilages	Water soluble	Galactose, galacturonic acid, rhaminose
Lignin	Water insoluble	Polymer of hydroxyphenyl-propane derivatives

Table 6: Some Properties of Dietary Fibre and their Health Consequences

Function	*Health Consequence*
Water absorbing and bulking property	Energy diluent to formulate low calorie diets
Increased transit time of food in the gut	Reduced risk of inflammatory bowel disease
Bile acid and steroid binding	Bile acid and steroid binding Hypo-cholesterolemic activity and reducing the risk of cardiovascular diseases
Retardation of carbohydrate absorption and impaired glucose tolerance	Management of certain type diabetes
Binding of toxins	as a detoxifying agent
Binding of divalent cations	Reduced bioavailability of Ca, Mg, Zn, Fe

Table 7: Total Antioxidant Activity of some Indian Foods

Food Item	*µg α-tocopherol Equivalent/ 100 g Fresh Weight*
Rice (*Oryza sativa*)	80–102
Whole wheat (*Tritium aestivum*)	220–500
Spinach (*Spinacia oleracea*)	750–890
Amaranth (*Amaranthus gangeticus*)	620–810
Coriander leaves (*Coriander sativum*)	610–750
Fenugreek leaves (*Trigonella sativum*)	520–690
Gogu (*Hibiscus cannabinus*)	700–810
Mint (*Mentha spicata*)	630–720
Ponnuganti (*Alternanthera sessitis*)	500–630
Ambati chukka (*Rumex vesicarius*)	530–610

Table 8: β-Carotene and Ascorbic Acid-Rich Foods Available in India

Food	*β-Carotene (µg/100g)*	*Ascorbic Acid (Mg/100g)*
Amaranth	5520	99
Agathi	5400	169
Coriander leaves	6918	135
Curry leaves	7560	4
Mint	1620	27
Mustard leaves	2620	33
Spinach (*Bachali*)	5580	28
Rape leaves	1330	65
Ponnaganti	1926	17
Fenugreek leaves	2340	52
Drumstick leaves	6700	220
Cluster beans	198	49
Chillis (green)	1007	111
Carrots	6460	168
Amla	–	600
Lime	–	63
Tomato (ripe)	590	27
Papaya (ripe)	880	57
Mango (ripe)	1990	16
Guava (country) lemon	–	21239

Table 9: Average Tocopherol Content of Foods and Average Daily Intake of Tocopherol in India

Food	*Average Content (mg/100 g)*	*Daily Intake (mg/CU*)*	*Contribution of Food (%)*
Cereals and millets	1.2	5.57	37.7
Pulses and legumes	3.8	1.29	8.7
Leafy vegetables	2.0	0.34	2.3
Other vegetables	1.0	0.48	3.3
Roots and tubers	0.1	0.05	0.3
Fruits	0.5	0.06	0.4
Nuts and oilseeds	16.0	1.35	9.5
Oils and fats	47.5	4.75	32.2
Milk	0.1	0.72	4.9
Animal foods	0.1	0.16	1.1
Total	–	14.77	100.0

* Consumption unit of an adult reference man.

Table 10: Daily Intake Polyphenolic Compounds (per CU*) in Indian Rural Diets

Regional Diet	*Catechin Equivalents (mg)*	
	Low Income	*Very Low Income*
Rice based	2230	1495
Wheat based	2474	2148
Sorghum based	2318	1507

* Consumption unit of an adult reference man.

Table 11: Hypocholesterolaemic and Antioxidant Potentials of some Edible Oils

Edible Oils	*Hypocholesterolemic Agents (%)*				*Antioxidant Agents (%)*		
	Phytosterols	*Triterpene Alcohols*	*Sesaminol*	*Tocopherols*	*Tocotrienols*	*Sesmolin†*	*Carotene‡*
Palmolein	0.14	0.09	–	0.64	0.49	–	0.50
Rice bran oil	1.66	1.08	–	0.18	0.37	–	–
Corn oil	1.70	0.02	–	0.78	–	–	–
Sesame oil	0.56	0.17	0.85	0.07	–	0.4	–

† Hydrolysed to sesamol; ‡ In red palm oil.

Table 12: Inhibitor Compounds of Carcinogenesis Occurring in Plant Foods

Category	*Inhibitory Compounds*
Compounds preventing formation of carcinogens from precursor compounds	Ascorbic acid, α- and r-tocopherols, caffeic acid, ferrulic acid Phenols (*e.g.*, ellagic acid, caffeic acid, ferrulic acid, P-hydroxy cinamic acid)
	Indoles (*e.g.*, indole-3-acetonitrile, indole-3-carbinol, 3-3' Indolymethane)
	Aromatic isothiocyanates (*e.g.*, benzyl isothiocyanates, phenthyl Isothiocyanates)
	Coumarins (*e.g.*, courmarin, limethin)
	Flavones (*e.g.*, quercetin pent methyl ether)
	Diterpenes (*e.g.*, kahweel palmitate)
	Retinoids and carotenoids (*e.g.*, retinyl palmitate, retinyl acetate, β-carotene
Suppressing agents	Soybean protease inhibitor, benzyl-isothiocyanate, B- Sitosterol, caffeine, fumeric acid, selenium

Table 13: Thiocyanate and Phenolic Contents of Indian Foods

Food Group	*SCN (mg/100 g)*	*Phenolic (g/100 g)*
Cereals and millets	0.2–0.9	0.1–2.9
Pulses	0.3–0.8	0.6–5.6
Vegetables (non-leafy)	0.5–5.7	0.2–3.8
Leafy vegetables	0.3–10.1†	1.4–3.7
Roots and tubers	0.2–18.6‡	0.6–2.7
Oilseeds and other seeds	0.8–16.2§	0.7–5.6
Spices and condiments	0.3–8.5	0.2–3.7
Fruits	0.2–1.6	0.9–4.4

Conclusion

Oxidative stress is clearly implicated in a wide range of diseases, including cardiovascular diseases, cancer, inflammatory diseases, neurodegenerative diseases, cataract, *etc.* Epidemiological data generally suggest a benefit of consuming diets rich in antioxidant nutrients and phytochemicals. Phytochemical rich items like fruits, vegetables,spices *etc.*, that are normal ingredients of our diet are now known to exert health beneficial antioxidant effects in the mammalian system and in several model systems through their bioactive compounds. Antioxidant properties of spices are of particular interest in view of the impact of oxidative modification of LDL cholesterol in the development of atherosclerosis, and suppression of oxidative stress and inflammation by spices is important in their cancer preventive role, since both oxidative stress and inflammation are risk factors for cancer initiation and promotion.

The antioxidant activity of phytochemicals in the mammalian system involve one or more of the following: (i) free radical scavenging, (ii) suppressing lipid peroxidation, (iii) enhancing the antioxidant molecules in tissues, (iv) stimulating the activities of endogenous antioxidant enzymes, (v) inhibiting LDL oxidation and (vi) inhibiting 5-LO and COX-2 enzymes. It is invariably extrapolated by different independent investigators that by virtue of their antioxidant activity, these bioactivecompounds can be anti-inflammatory, antimutagenic and cancer preventive, antiatherogenic and cardioprotective, hepatoprotective, neuroprotective, anticataractogenic, *etc.* The challenge lies in integrating this knowledge to ascertain whether any effects can be observed in humans.

References

Araceli, M.V., José, L.C., José, C.C.R. and Mercedes, G.L. 2011. Phytochemical evaluation of wild and cultivated pepper (*Capsicum annuum* L. and *C. pubescens* Ruiz and Pav.) from Oaxaca, Mexico. *Chilean J. Agril. Res.*, **71**(4): 578–584.

Arantxa, R.C. 2016. The health potential of fruits and vegetables phytochemicals: Notable examples. *Crit. Rev. Food Sci. Nutr.*, **56**: 1097–1107.

Chwan-Li Shena, Brenda J. Smithd, Di-Fan Loa, Ming-Chien Chyua, Dale M. Dunna, Chung-Hwan Chenf and In-Sook Kwung. 2012. Dietary polyphenols and mechanisms of osteoarthritis. *J. Nut. Biochem.*, **23**: 1367–1377.

Juan Carlos Espý´n, Marý´a Teresa Garcý´a-Conesa, Francisco A. and Toma´s-Barbera´n. 2007. Nutraceuticals: Facts and fiction. *Phytochemistry*, **68**: 2986–3008.

Khan, Rao Sanaullah, Grigor, John, Winger, Ray and Win, Alan. 2013. Functional food product development: Opportunities and challenges for food manufacturers. *Tr. Food. Sci. Tech.*, **30**(1): 27–37.

Kubota, C., Cynthia A.T., Wu, M. and Jamal, J. 2006. Controlled environments for production of value-added food crops with high phytochemicals concentrations: Lycopene in tomato as an example. *Hort. Sci.*, **41**(3): 522–525.

Narasinga Rao, B.S. 2003. Bioactive phytochemicals in Indian foods and their potential in health promotion and disease prevention. *Asia Pacific J. Clin. Nutr.*, **12**(1): 9–22

Norah, O., Elke, K.A. and Eimear, G. 2012. Dietary fibre and phytochemical characteristics of fruit and vegetable by-products and their recent applications as novel ingredients in food products. *Innov. Food Sci. Em. Tech.*, **16**: 1–10.

Srinivasan, K. 2005. Role of spices beyond food flavoring: Nutraceuticals with multiple health effects. *Food Rev. Int.*, **21**(2): 167–188

Tânia, G.A., Filipa, S., Ana, S.S., Beatriz, M.O., Ana, C.B. and Helena, S.C. 2016. Nutritional and phytochemical composition of *Annona cherimola* Mill. fruits and by-products: Potential health benefits. *Food Chem.*, **193**: 187–195.

Reducing Drudgery through Mechanization

81

Initiatives on Reduction of Drudgery through Mechanisation in Horticulture

Ashwani Kumar

Joint Secretary (Mechanisation and Technology), Ministry of Agriculture and Farmers Welfare, Krishi Bhavan, New Delhi – 110 001

India is the second largest producer of fruits and vegetables in the world. It produces nearly 10.7 percent of world's vegetables and 12.2 per cent of fruits (year 2013). During the last decade, there has been a tremendous growth in the area under horticultural crops. Similarly, the production of horticultural crops has also nearly tripled from 96.53 million tonnes in 1991-92 to 283.47 million tonnes in 2014-15 (Table 1). The percentage share of production of various horticulture crops in total horticulture crop is 31.3 per cent fruits, 59.4 per cent vegetables, 1.1 per cent in flower and aromatics, 6 per cent plantation crop and 2.1 per cent spices. Horticultural crops present a great diversity in terms of morphology, agronomy, plant protection, nutrition, reproduction, microclimate, and utilization. There is a shortage of appropriate machinery for field preparation, plant protection, canopy management, weeding, inter-culture, *etc.* Critical operations such as transplanting, earthing up, pruning, picking/harvesting *etc.* are generally carried out manually and are not only time consuming but also involve a lot of drudgery. In India Large area is comes under horticulture, 95.41 lakh ha under vegetables, 63.58 lakh ha under fruits, 35.38 lakh ha under plantation crop, 31.63 lakh under spices, 4.99 lakh ha under aromatic and 3.17 lakh ha under flowers. As a result of a number of thoughtful research, technological and policy initiatives and inputs, horticulture in India, today, has become a sustainable and viable venture for the small and marginal farmers. It is a matter of satisfaction that their food consumption levels and household income have increased. Besides, this sector has also started attracting entrepreneurs for taking up horticulture as a commercial venture. Therefore, there is a great scope for the horticulture industry to grow and flourish contribution of horticultural and floricultural crops to the total agricultural production in the country is quite significant due to highly favorable and varied agro-ecological diversities.

For Horticulture preferable tractor is high-clearance, small, tall, narrow tires work well in vegetable crops and provide good traction and flotation as well as clearance. Major field operations for horticultural crops include nursery/seedling preparation, post hole digging for planting, inter-culture, aeration, earthing, irrigation, plant protection, harvesting, handling, packaging transport. The cultivation of horticultural crops is predominantly dependent upon human labour, since commercial cultivation is only on a limited scale.

Machinery for Land Preparation

Animal/power tiller or tractor-drawn mould board ploughs, disc ploughs, harrows, cultivators and rotavators, are available and used for land preparation. The fruits plants are generally perennial plants planted in pits. The tractor or power tiller operated posthole diggers are available in the country up to 30 cm diameter. These pits are filled with good soil and manure. Inter-row spaces of the pits can be tilled with power tiller/ tractor drawn tillage machinery to destroy the weeds,

Table 1: Area and Production of different Horticultural Crops in India

Crop	*Area Lakh (ha)*	*Production* 2014-15 (Lakh tonne)*	*Crop*	*Area Lakh (ha)*	*Production* 2014-15 (Lakh tonne)*
			Vegetables		
Potato	20.60	448.93	Onion	11.81	189.24
Brinjal	6.80	127.06	Tomato	7.91	173.98
Okra	5.07	58.53	Peas	4.59	43.29
Cabbage	3.79	85.97	Cauliflower	4.14	78.97
Tapioca	2.53	85.42	Sweet potato	1.11	13.38
			Fruits		
Mango	22.17	185.06	Citrus	9.87	121.81
Banana	8.80	300.08	Apple	3.2	18.85
Guava	2.51	40.83	Sapota	1.21	14.57
Pomegranate	1.43	17.74	Pineapple	1.13	18.92
Papaya	1.26	55.08	Grape	0.88	25.54
Litchi	0.8	5.9			
Spice					
Chilies	7.75	14.92	Garlic	2.31	12.52
Turmeric	2.33	11.90	Ginger	1.33	6.55
Cumin	8.59	5.14	Coriander	4.47	3.14
Plantation crop					
Coconut	19.79	157.03	Areca nut	4.52	6.23
Cashew	10.29	7.89	Cocoa	0.78	0.16

Tractor Operated Laser Guided Land Leveller

Different Farm Machineries for Use in Horticultural Crops.

Different Farm Machineries for Use in Horticultural Crops.

increase aeration and conserve moisture. Small power tillers can be used in terraced fields.

Plastic mulch and raised beds work together to provide an ideal environment for vegetable growth. The plastic warms the soil, prevents weed growth, and controls moisture, while the raised bed provides for a deeper soil profile and better drainage around the root system. The easiest way is the use of a raised bed mulch layer; an "all-in-one" machine that creates the bed, lays the plastic and lays the drip line in one pass.

Fertilizer and Farm Yard Manure Spreader

For spreading fertilizer manually as well as tractor mounted centrifugal spreader is commonly used, while farm yard manure is spreading manually.

Nursery Preparation, Grafting, Sowing, Planting and Transplanting Machinery

The rootstock are sown manually for raising the nursery seedlings for grafting. For planting bold seeds, manual dibblers have been used. The inclined plate and pneumatic planters have been successfully used for direct sowing of small vegetable seeds. Vegetable trans-planters are now available for planting vegetable seedlings.

Hand tools such as budding and grafting knife, pruning knife and secateurs are used for plant propagation and pruning. Telescopic handles can be used for the pruning of the twigs and branches, which are not in the reach of human hand.

Different Farm Implements and Equipments for Use in Horticultural Crops.

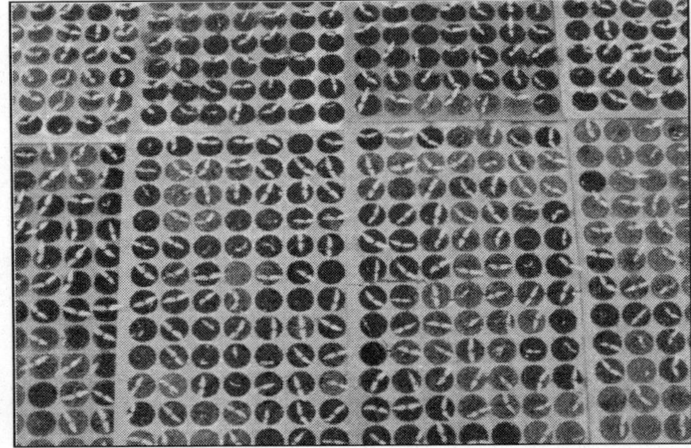

Tractor Operated 6-row Inclined Plate Planter

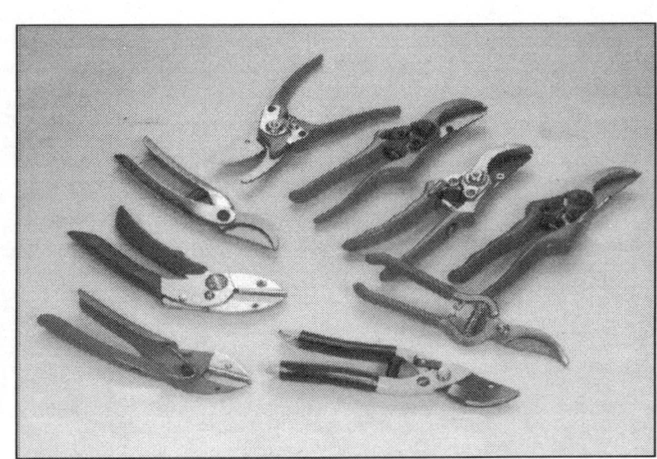

Weeding, Inter-culture and Plant Protection Equipment

Hand tools or power weeders are used for weeding and inter-culture.

Tractor Operated Orchard Sprayer

Machinery for different Inter-cultural Operations in Annual and Perennial Crops.

Different types of sprayers and dusters are available from manually to power operated such as knapsack sprayers foot operated sprayers, power operated mist blowers and dusters for plant protection. Fogging machines and mist blowers are available, which can be used in the greenouses and for covered crop cultivation.

Picking/Harvesting Machineries

Harvesting is the most labour intensive operation due to delicate nature of fruits. Mechanical harvesters for fruits can be used where fruits are sent immediately to

Machinery for Harvesting and Grading of Horticultural Produce.

Mechanical Harvest of Grapes.

Preparation of Raised Bed with Heavy Duty Machine.

processing industries for extraction of juices or making other value-added products. Grading of the fruits is done according to the size and colour, which is presently being done manually.

Government Efforts to Promote Horticultural Tools

The operational guidelines of Sub-Mission of Agricultural Mechanization (SMAM) have been recently revised, includes the various Horticulture equipment along with Post-Harvest Technology equipment for oilseed, food grain and horticulture crops for financial assistance to the extent of 50 per cent of the cost of equipment for SC, ST, small and marginal farmers, women, and NE states beneficiaries and 40 per cent for other category beneficiaries. The details of Financial Assistance for Procurement of Agriculture Machinery and Equipment for various Horticulture equipments is available on operational guideline on Sub-Mission of Agricultural Mechanization (SMAM) which is available on website http://farmech.gov.in

Sensor Based Fruit Grader.

Irradiation and Post-harvest Treatment of Fruit Produce before Export.

Future Requirements

The unrelenting **globalization** of the horticultural industry will see a rapid transfer and adoption of knowledge. This will comprise the transfer of major enterprises from developed to developing countries as they would have to meet the needs of year-round production and in response to securing lower-cost land and labour. In India horticulture machinery industry is still at beginning edge there are lot of machinery needs to be developed for reduction of drudgery and scarcity of labour at peak time. Though few tractor companies manufactures small horticulture tractors, but we need very specific tractors for the special horticultural needs having less height, hinge at centre for less turning radius, work in both direction, tractor with front, side and rear PTO, tractor with front and rear three point linkage, tractor with the powerful hydraulic system to attach appropriate horticulture equipment with the external hydraulic circuit. Further we also need good quality of trencher, step seeders, automatic pro-tray de-stacking, filling, dibbling, sowing machine, high capacity grafting machines, offset weeder with obstruction sensor, air-assisted multi-head sprayers, canopy sprayers, electrostatic sprayers, tractor mounted trimmers for vineyard, tree pruner, washers for fruit and vegetables, graders for potato tomato, sapota, onion garlic *etc.* Under the given situation, the biggest challenge is to devise means and measures to reduce the post harvest loss of fruits and vegetables, maintain their quality during retail and ensure their safety for consumption.

82

Reducing Drudgery through User Friendly Equipment in Horticulture

C.R. Mehta*, B. Gaikwad, R.K. Tiwari and B.S. Gholap

AICRP on Farm Implements and Machinery,
ICAR-Central Institute of Agricultural Engineering,
Nabi Bagh, Berasia Road, Bhopal – 462 038, Madhya Pradesh
**E-mail: cr.mehta@icar.gov.in, crmehta65@yahoo.co.in*

The horticultural crops include fruits, vegetables, spices and condiments, plantation crops and flowers. India is the second largest producer of fruits and vegetables in the world. It produces nearly 15 per cent of world's vegetables and 12 per cent of fruits. It accounts only for 1.7 per cent of the global trade in vegetables and 0.5 per cent in fruits. During the last decade, there has been a tremendous growth in the area under horticultural crops. Horticulture accounts for 30 per cent of India's agricultural GDP from 8.5 per cent of cropped area. Similarly, the production of horticultural crops has also nearly tripled from 97.83 million tonnes in 1991-92 to 283.50 million tonnes in 2014-15. Horticultural crops present a great diversity in terms of morphology, agronomy, plant protection, nutrition, reproduction, micro-climate, and utilization.

The cultivation of horticultural crops is predominantly depended upon human labour. Besides, horticultural produces are highly perishable in nature, and due to inadequate or improper harvesting and post-harvest handling facilities, the losses are very high. The farmers as a result are not able to get adequate income from cultivation of these crops. Modern production techniques and better harvest and post-harvest practices besides increasing the productivity are likely to reduce the dependence on intensive labour, and thus reduce the cost of cultivation. Mechanization is one of the ways of increasing the production and productivity of crops by achieving timeliness, increasing cropping intensity, crop density, and through efficient use of inputs.

A large number of farm equipment have been developed in the country for cereals, pulses and oilseeds. Very little efforts have been made in mechanising fruits and vegetable production in the country. The mechanization of selective unit operations of horticultural crops in flat and hilly regions of country can save inputs, time, labour and cost of operation by 20-40 per cent and reduce drudgery of agricultural workers. A few hand tools such as spade, *kodali*, hand hoes, *khurpi*, sickle, prongs, hand rake, dibblers, budding and grafting knives, manual fruit harvester, small hand tools *etc.* are being manufactured by small scale manufacturers and are being used by the growers. The quality of these tools was not upto the mark. Recently a few manufacturers have started manufacturing good quality horticultural tools and are also exporting them to the other countries.There are various cost effective modern equipment suitable for different power sources which are capable for maintaining timeliness of operations. The light weight equipment will serve the purpose under terrace farming for horticulture. The small equipment have scope for use under greenhouses for accomplishing jobs in time with precision. The land preparation, seedling raising and spraying operations

Table 1: Mechanization Gaps in Horticultural Sector

1. Horticultural hand tools	Small hand tools for various horticultural operations like grafting, pruning, harvesting, decortication, *etc.* are needed for time and energy saving and drudgery reduction
2. Planting of fruit tree nurseries	Power tiller operated post hole digger (pit size 1×1×1 m size) for planting of seedlings of horticultural crops at hilly terrains
3. Extension of growing season under cold climatic conditions	Protected cultivation technologies for cucumber, tomato, capsicum, cauliflower, cabbage, okra, roses, gerbera and lilliums.
4. Stacking system for vegetable crops like pea, tomato, beans, cucurbits, kiwi *etc.*	Develop/adopt the stacking systems
5. Pruning, canopy management and harvesting of fruit crops	✩ Develop/modify an improved ladder for pruning and harvesting of fruit crops for hill horticulture ✩ Adoption/modification of pneumatic pruning secateurs/power secateurs ✩ Develop/adopt pluckers for apple, mango, peach, apricot, walnut, *aonla*, chilgoza *etc.*
6. Plant protection in orchards	Low cost plant protection equipment having potential to spray tall as well as medium fruit trees with mist/fog type spray nozzles
7. Planting and harvesting of ginger and turmeric	Develop/adopt planting and harvesting equipment for rhizome crops *viz.*, ginger, turmeric *etc.*
8. Efficient irrigation systems for horticultural crops	✩ Drip and sprinklers irrigation techniques needs to be propagated for judicious and timely use of available water, especially for horticultural crops ✩ Design of gravity fed water system of drip and sprinkler irrigation for hill horticulture
9. Nursery tools and devices	Development of tools/equipment for uprooting of fruit/forestry plants from nurseries
10. Weeding	Low cost weeding tools for food and horticultural crops
Polyhouse mechanisation	
11. Greenhouse designs	Develop/adopt design of polyhouses for varying slopes of hills
12. Protected cultivation for extension of growing season	Small poly houses for nurseries and cultivation of vegetables, flowers, *etc.* under controlled atmospheric conditions

have been mechanized for vegetable cultivation during the last three decades. However, attempts are required to mechanize transplanting and harvesting operations for vegetable cultivation.

Improved Horticultural Equipment

There are 24 ergonomically designed women friendly technologies for reduction of human drudgery which include manual grubber, hand ridger, Naveen dibbler, rotary dibbler, twin wheel hoe, manual serrated sickle, manual fruit harvester, manual *bhindi*plucker, manual single/two row seed drill, manual fertilizer broadcaster, *etc.* suitable for various operations in horticultural crops. A good number of horticultural equipment have been developed in the country which need to be promoted. The brief details of some of improved equipment suitable for horticultural crops are as follows.

Seedbed Preparation

The seedbed preparation/tillage operations for cultivation of different horticulture crops (fruits, vegetables and plantations) are more or less mechanized and equipment for different power sources are available. The equipment like MB plough, tine cultivator, disc harrow, ridgers, furrowers, bed formers, levellers, post hole diggers, *etc.* are commercially manufactured. Transplanting of fruit and plantation crops involves digging of pits or trenches and placement of seedlings. At present, these operations are carried out manually. Tractor, power tiller and engine operated post hole diggers have been developed and are commercially available. An auger digger is a high speed machine fitted with 1.3 kW single cylinder two stroke engine, auger and handle (Figure 1). The capacity of auger digger is 30 pits/h. It is very efficient equipment for horticultural cultivation.

Light weight tiller with rotary attachmentis a single axle power unit to till fields in dry condition as well as puddle the fields for rice crop (Figure 2). It can

Figure 1: Self-propelled Post Hole Digger.

Figure 2: Light Weight Tiller.

plough fields, make ridges and interculture in crops with different attachments. It can renovate pastures, cut grasses, and level fields, dig potatoes, spray insecticides, pump water and haul up to 1.5 t load. This implement is commercially available in the country and costs about Rs. 90,000/-.It saves 40 per cent time in ploughing as compared to traditional practice.

A power tiller operated bench terrace-cum-leveller of 1.0 m width was developed to mechanize the bench terracing work in hilly region (Figure 3). The capacity of the terracer is 0.12 ha/h. The maximum volume of soil handled at a time by the terracer is 0.6 m^3.

Figure 3: Power Tiller Operated Bench Terracer-cum-Leveller.

A tractor operated check basin former has been developed by MPKV, Rahuri with the help of M/s BhansaliTrailorsPvt. Ltd., Kopargaon (Figure 4). It scrapes, collects and distributes the collected soil uniformly to form check basins of 2 x 6 m size. The cost of the equipment is Rs. 65,000/- (approx.).

Plastic mulching has been found to be useful for moisture conservation, reduction of weeds and increase in yield for vegetable and fruit crops. A tractor operated plastic mulch laying machine has been developed at CIAE, Bhopal for mechanizing the plastic mulch laying for vegetables.

Figure 4: Tractor Operated Check Basin Former.

Nursery Mechanization

Mechanized cultivation of nursery is generally required for raising and supplying seedling for large scale transplanting. The unit operations such as preparation of soil, seed and trays/mats, spreading of soil, water application, sowing of seeds, covering the mats, handling/conveying of trays *etc.* are mechanized to ensure uniformity and quality operations. The system essentially consists of soil crusher, soil manure mixer, soil sieve, soil acidity detector, conveyor belt, sprout promoting machine, water pump, dehydrator, sowing plant, automatic seedling box, automatic seed supplier, seeding conveyor (with watering function), germination units, automatic seedling box stacker and growing chamber.

Sowing/Planting or Transplanting Operation

The vegetable crops are sown both by seeding and transplanting. Direct seeding of vegetable crops is mostly done manually except for crops like pea, okra,

Figure 5: Six Row Small Seed Planter.

beans, *etc.* having medium or large size seeds which can be easily sown with the help of mechanical/pneumatic seed drill or planters. For direct sowing of small seeded vegetable crops having very low seed rates like carrot, onion, *etc.*, mechanical/pneumatic planting systems are being adopted/developed. A six row tractor operated small seed planterfor sowing small seeds (onion) at 150 mm row spacing and adjustable plant to plant spacing was developed at PAU, Ludhiana (Figure 5). The field capacity and cost of operation are 0.16 ha/h and Rs. 5,090/ha, respectively. There is a saving of 50 per cent in cost of operation and 81 per cent in labour requirement as compared to traditional method.

The transplanting of vegetable seedlings is one of the major challenging field operation and presently is being done manually or by use of semi-automatic transplanters. Various types of vegetable seedlings (bare root, plug and pot seedlings) are being used for vegetable cultivation. Based on the type of seedlings, attempts have been made to develop mechanical transplanters to overcome the tedious, cumbersome and labour intensive manual transplanting operation. The developed transplanters are either semi-automatic or automatic (riding type or walk behind type) machines. Tractor operated two-row semi-automatic vegetable transplanter suitable for bare root seedlings have been developed at CIAE, Bhopal and PAU, Ludhiana (Figure 6). A two row semi-automatic planter having picker wheel type metering mechanism has been developed to transplant the root wash type seedlings on beds as well as on flat fields. Two persons (one for each row) are required to place the seedlings in the flappers when these open at the top position. After the seedling is dropped in the furrow, the soil is compacted around it with the help of two moving inclined wheels. Average field capacity of machine is 0.10 ha/h at forward speed of 1.0 km/h. The cost of operation with the machine is Rs. 2,200/ha as compared to Rs. 2,800/ha with manual transplanting

Tractor operated vegetable transplanters for cup type seedlings developed at TNAU, Coimbatore and IIHR, Bengaluru are under limited commercialization and need further popularization. An effort is being made for development of automatic onion seedling transplanter at CIAE, Bhopal.

Figure 6: Two Row Semi-Automatic Vegetable Transplanter.

Potato planters both semi-automatic and automatic types have been developed and are commercially available. A manual, animal and tractor operated garlic planters have been developed at PAU, Ludhiana and MPUAT, Udaipur. A tractor operated garlic planteris a 4/6 row unit with actuating spoon type metering mechanism suitable for planting of garlic (Figure 7). The effective field capacity ranges 0.18–0.21 ha/h. The cost of the machine is Rs. 1.5 lakh and cost of operation is Rs. 6,170/ha. There is saving of 82 per cent in labour and 57 per cent in cost of operation as compared to manual planting.

Figure 7: Tractor Operated Garlic Planter.

The tractor rear mounted turmeric rhizome ridger planter has been developed by TNAU, Coimbatore (Figure 8). It consists of 3 ridger bottoms and planting mechanism for planting on one side of the ridges in one pass. The effective field capacity of the unit is 0.15 ha/h and costs Rs. 45,000/- (approx.).

Figure 8: Turmeric Rhizome Planter.

Weeding/Interculture/Fertilizer Applications

Number of ergonomically designed manual hand tools, bullock drawn and power operated equipment have been developed for weeding and interculture operations for operator comfort and are commercially available. Appropriate tools and equipment needs to be adopted/popularized among the vegetable growing farmers. Power weeders, both sweep and rotary types have been developed for vegetable crops. A light weight self-propelled power weederconsists of a 4.1 kW diesel engine mounted on the power tiller chassis, power transmission system, two MS wheels, a frame and a rotary (Figure 9). The effective field capacity varies from 0.10 to 0.13 ha/h. The equipment saves 90 per cent in operating time and 30 per cent in cost of weeding as compared to hand weeding by khurpi. The machine is also useful in row crops, horticultural and vegetable crops. The average cost of weeding with power weeder is Rs. 1,000/ha.

Figure 9: Self-Propelled Power Weeder.

The preparation of basins around orchard trees for manuring and fertilizer application needs to be mechanized as the operation is carried out manually. Power tillers with rotary are useful for these operations and need to be demonstrated and adopted.

Pruning, Budding and Grafting Operations

Pruning, budding and grafting are some of essential operations in fruit crops. Several improved tools have been developed and are commercially manufactured. Power chain saws are commercially available but they are still not widely adopted. Pruning in medium and tall trees is still a problem. Hydraulic and pneumatic pruners, available abroad need to be introduced in the country.A manual tree pruner consists of semi-circular blade and handle made of aluminium pipe of 2.08 m (Figure 10). It could access up to a height of 4.0 m. It weighs 0.950 kg and is suitable for pruning of fruit branches up to 20 mm thickness.

Spraying Operation

Sprayers and mist blowers for orchard crops are commercially available. However, their adoption level is still low. The latest development in the spraying application is the aero-blast sprayers, air sleeve boom sprayer and tall tree sprayer suitable for trees up to a height of 20-25 m. The effective and precise spraying of chemicals in orchards and on tall trees in plantation crops is another area need attention of researchers.

Figure 10: Manual Tree Pruner.

The air assisted vertical boom sprayer (Figure 11a) can be used for applying insecticides and other chemicals to fruit crops to effectively control pest attack or diseases as the chemicals can penetrate in the canopy of the trees due to air blast created by the sprayer. The sprayer was operated at 540 PTO rpm with nozzle discharge of 1.42 l/min. The average effective field capacity and field efficiency of the air assisted vertical boom sprayer was 1.70 ha/h and 81.0 per cent, respectively. The cost of operation was Rs. 453/h and Rs. 266/ha.

Figure 11a: Air-assisted Vertical Boom Sprayer.

The ultrasonic sensors based pomegranate spraying systemis equipped with sensor, programmer and non-return valve for spraying on fruits trees (Figure 11b). The effective field capacity and number of plants covered are 0.88 ha/h and 1370 plants/h, respectively. The unit price of the machine is Rs. 3.5 lakh. The saving of liquid is up to 45-50 per cent with hollow cone nozzles.

Harvesting Operation

Most of the horticultural fruit and plantation crops are harvested by manual picking all over the country. Manual tools/aids for harvesting of fruit and plantation

Figure 11b: Sensor-based Spraying System.

crops have been developed (Figure 12) and need to be adopted. TNAU, Coimbatore has developed an efficient coconut climber.Power/hydraulically operated elevated platforms are being tested for harvesting of fruits from tall trees. A self-propelled (8.2 kW petrol engine) multi-purpose hydraulic system has been developed by CIAE, Bhopal (Figure 13). The self-propelled hydraulic multi-purpose system has a maximum vertical reach of 6 m, load carrying capacity of 200 kg and can be operated at maximum ground speed of 3 km/h. It is useful for harvesting, pruning and spraying of fruit crops like mango, sapota and citrus. The cost of machine is Rs. 7,50,000/- (approx.).

Figure 12: Manual Tools/Aids for Harvesting of Fruits.

Figure 13: Self-Propelled Hydraulic Platform.

The harvesting of most of vegetable crops like tomato, chilli, peas, brinjal, okra, beans, *etc.* is manual due to selective and multi-stage harvesting/picking of fruits. As such no equipment/tools are available for these crops suiting to present harvesting practices followed by the farmers.

For harvesting of root crops like potato, good design of harvesting and windrowing equipment are available. Harvester/digger needs to be adopted for harvesting of onion, carrot, rhizomes (ginger, turmeric, *etc.*) and other crops. Tractor mountedroot crop harvester-cum-elevator is suitable for digging onion and other root crops. The field capacity of the machine was 0.28, 0.24, 0.21 and 0.21 ha/h for digging of carrot, potato, garlic and onion crops, respectively. The approximate cost of the machine is Rs. 40,000/-. The saving in labour ranged from 62 to 71 per cent.

Figure 14: Banana Clump Remover.

A tractor drawn turmeric digger of 1.45 m width and rake angle of 55° has been developed andis capable of digging four rows at 300-350 mm depth (Figure 14). The effective field capacity is 0.36 ha/h in single pass. The cost of the implement is Rs. 22,000-25,000/unit. The cost of digging turmeric with the implement is Rs. 4830/ha.

Tractor operated garlic harvesterconsists of a blade with triangular point knives mounted over a MS flat for its use in heavy soil for smaller size soil clods and less draft requirement (Figure 15). The machine has an effective field capacity of 0.26 ha/h and digging losses ranges 2-3 bulbs/m^2. The cost of harvesting is Rs 3885/ha.

Figure 15: Tractor Mounted Garlic Harvester.

A tractor operated banana clump remover has been developed by adopting frame of nine-tyne cultivator (Figure 16). Two sub-soilers perform as a fork for removing the banana clump. A deflector was provided to push the soil sideways. The field capacity of machine is 0.5 ha/h. The cost of operation with the equipment is Rs 700/ha as compared to Rs. 2,000-2,500/ha during manual digging. Thus, there is a saving of Rs. 1,800/ha.

Figure 16: Tractor Mounted Turmeric Digger.

Harvesting or dehusking of plantation crops

A hand operated coconut dehusker has been developed by OUAT, Bhubaneswar (Figure 17). It has an output capacity of 212 dehusked nuts/h, against 99 nuts/h under traditional dehusking method. The dehusker costs approx. Rs. 800/- and more than 100 units have been purchased by farmers and are in use.

Figure 17: Manual Coconut Dehusker.

A pedal cum power operated arecanut dehusker has been developed for better man machine system efficiency based on ergonomical principles. For pedal operation, two workers are needed for its operation, one for pedaling the unit and the other one for feeding arecanuts. The dehusking efficiency, kernel breakage and dehusking capacity are 97 per cent, 7 per cent and 15 kg/h, respectively. The power operated unit is equipped with 0.5 kW single phase electric motor and has dehusking efficiency, kernel breakage and dehusking capacity of 98 per cent, 10 per cent and 20 kg/h, respectively. The weight of the unit is 110 kg and mounted on wheels for mobility. The unit costs Rs. 26,000/-.

A tractor mounted aerial access hoist has been designed and developed by TNAU, Coimbatore in collaboration with M/s Vanjax, Chennai for management of coconut and tall trees. The aerial access hoist can go up to 15 m height and four coconut trees can be accessed from a single position. The time required for re-locating the unit and operating stabilizers is 1 min and positioning against a tree of 10 m height is 2 min.

Future Trends in Horticultural Mechanisation

High-tech horticulture is modern, less environment dependent, capital intensive and has the capacity to improve the productivity and quality of horticultural crops. The adoption of this technology in horticultural crops is necessary to ensure the nutritional security

Table 1: Important Areas which need Immediate Attention for Horticultural Mechanisation

Strategy	*Action Plan*
☆ Mechanization of vegetable cultivation ☆ Harvesters for root and leafy vegetables ☆ Hydraulic operated post hole digger ☆ Primary processing of leguminous plants	☆ Multi-purpose vehicle for various operations in vegetable crop ☆ Tractor mounted ginger planter ☆ Three row automatic vegetable transplanter for plug type seedlings ☆ Vertical cup belt type vegetable transplanter for cell feed nursery ☆ Mechanical weeder for wide spaced deep rooted crop ☆ Automatic metering mechanism for portray seedlings ☆ Semi-automatic raised bed vegetable transplanter ☆ Design and development of semi-automatic vegetable transplanter for plastic mulch ☆ Development of harvester for carrot and cabbage ☆ Development of tractor front mounted hydraulic operated post hole digger
☆ Mechanization of grape cultivation ☆ Package of equipment for ginger cultivation	☆ Development of power tiller operated tiller-cum-fertilizer applicator for orchards ☆ Development of planter and harvester for ginger
☆ Mechanization of floriculture/vegetable culture in glass house ☆ Development of safe spraying systems for greenhouse applications ☆ Mechanisation of different operations under covered cultivation	☆ Equipment for different operations under covered cultivation ☆ Spraying system for greenhouse applications

of future generation. Hi-tech horticulture includes micropropagation, micro-irrigation, fertigation, protected cultivation (greenhouse/polyhouse cultivation), organic farming, mechanization and use of sensors, remote sensing and robotics. The futuristic approach for mechanization of harvesting of fruit crops will be to develop sensor based equipment with machine vision. Some of the important areas which need immediate attention for horticultural mechanisation are as shown in Table 1.

Conclusion

The approach given below may be followed for mechanization of horticulture in India:

i) Identifying suitable machinery and equipment available worldwide with potential for direct adoption or adoption with modifications for individual ownership and or custom hiring basis.

ii) The tremendous potential of growing fruits and vegetables in the hill agriculture calls for developing appropriate technologies for mechanizing hill agriculture.

iii) There is need for large-scale demonstration and popularization of technologies developed for sowing and planting of vegetable crops.

iv) The harvesters/diggers need to adopted for harvesting of onion, carrot, rhizomes (ginger, turmeric, *etc.*) and other crops.

v) Effective and precise spraying of chemicals in orchards and on tall trees in plantation crops is another area needing attention of researchers.

vi) There is a need to introduce hydraulic and pneumatic pruners, available abroad in India.

vii) There exists lot of potential to digitally connect progressive farmers through smartphones, reverse engineer proven farm mechanization technologies not available in India, develop sensors required in agriculture at lower costs, use of UAV's and other smart technologies in precision horticulture and therefore need special focus for development under Digital India, Make in India and National Skills Development Mission.

References

Mehta, C.R., Gaikwad, B., Chandel, N.S. and Gholap, B.S. 2016. Trends of Innovations in Agricultural Mechanization. Conference on Innovation in Agricultural Mechanization – Development of linkage among R&D Institutes – Industry - Farmers (July 7-8, 2016), Mechanization and Technology Division, Department of Agriculture, Cooperation and Farmers Welfare, Ministry of Agriculture and Farmers Welfare (GoI), New Delhi, pp 5–12.

www.data.gov.in/catalog/gross-area-under-major-crops

www.thehindubusinessline.com/opinion/columns/g-chandrashekhar/growth-with-equity-need-of-the-hour-to-develop-rural-india/article3390881.ece

www.satnetasia.org/sites/default/files/cusom_hiring_prevailing_practices_english.pdf

Horticultural Statistics at a Glance 2015. (www.nhb.gov.in/area-pro/horst_galance_2016.pdf)

www.business-standard.com/article/companies/chinese-firms-expand-dominance-in-indian-farm-equipment-market-110120400001_1.html

83

Precision Agriculture Applications in Horticulture

P.S. Tiwari*, S. Prakash, R. Sahni, V. Kumar and N.S. Chandel

Agricultural Mechanization Division,
ICAR-Central Institute of Agricultural Engineering,
Nabi Bagh, Berasia Road, Bhopal – 462 038, Madhya Pradesh
**E-mail: pstiwari60@yahoo.com*

The intensification of agricultural production over the last few decades has greatly increased food production, but at a high environmental cost. In order to manage the resources and nutrients variable rate application systems were developed to apply inputs according to the actual requirements of the soil and plants in the specified management zones. Precision agriculture (PA) refers to the management of spatial and temporal variability of the fields using mechatronics. Agricultural field and crop related required data are collected from different sources, stored to GIS data bases, analysed using geostatistical methods to develop management zones and decision support systems (DSS). The controllers, actuators and applicators use the DSS to apply the agricultural inputs as per requirement of specific management zones. Precision agriculture precisely establishes various operations, such as tillage, sowing, irrigation, application of fertilizer, pesticide spraying, harvesting *etc.*, and thus, turns traditional extensive production to intensive production according to space variable data (Shanwad *et al.*, 2004). PA benefits with increased yieldand better quality produce and increases the profit to farming community with better environment conservation and working conditions (Zarco-Tejada, 2014). The application of mechatronics in horticultural crops was developed in the last fifteen years.

Horticultural commodity production and its quality is effected by many factors like proper planting, weeding, irrigation, attack of insects and harvesting technique. The feasibility of RTK-GPS enabled seed planters or transplanters that can map the geo-position of crop seeds or transplants as they are released by the planter has been successfully demonstrated (Ehsani *et al.*, 2004; Sun *et al.*, 2010). Weeds, as one of the most significant factors in affecting yield quality and quantity, demand a holistic and interdisciplinary approach for their adequate control (Schans *et al.*, 2006). It is known that weeds not only lower the yields, but they constitute one of the most important means for spread and survival of crop pathogens. In many cases weeds have been found as symptomless carriers of vector-borne viruses. Considering the worldwide average, about 10 per cent of losses in the total yield are caused by weeds. Weeds affect crop yield about 45 per cent more as compared to insects and diseases, due to competition to acquire plant nutrients and resources (Anonymous, 2013). Site-specific herbicide applicators not only control the weeds effectively, but also reduce the quantity of herbicide used thus, reducing the carbon footprints generated during its manufacturing.

In the perspective of harvesting of horticulture produce, 'handpicking method' is common practice for harvesting the fruits on the individual fruits trees. It is well known that harvesting in the large scale is still required technological intervention to reduce the cost of operation (Li *et al.*, 2011). There is rapid development of precise agricultural technologies for harvesting, threshing and yield monitoring of fruits and vegetable crops. Introduction of robotics in agriculture have played pivotal role to overcome the problem of shortage of labour and timely harvesting of horticulture produce. The techniques like vision based control and fruit recognition system has been employed to solve the two major problems of detecting objects in tree canopies and picking objects using visual information (Zhao *et al.*, 2016). Agricultural robots are perceptive and intelligent machines which are programmed to perform a variety of agricultural tasks such as transplanting, cultivating, spraying, trimming and harvesting (Edan *et al.*, 2009). Various sensors are used to detect the fruits and vegetable for harvesting. This recognition system gathers the information of position, orientation and size of the fruits. Control system sends the information to robot manipulator arm for picking, grasping, pruning or cutting. Harvesting robots for tomato, cherry tomato, cucumber, strawberry, grapes, and watermelon have been already commercialized. It is predicted that more robotic harvesting systems will be commercialized in early 21stCentury (Umeda *et al.*, 1999).

Development of geographic information system (GIS), global positioning system (GPS), remote sensing and several intelligent devices and implements (IDIs) during 1960's to 1990's gave birth to the idea of aligning information technology (IT) with agricultural practices and methods. Although the method of collecting data and making decisions based on them has been the fundamental method in precision agriculture, but as the size of farm grew it became difficult to implement, and that brought the necessity of these techniques and tools.Precision machinery in horticulture is necessary to alleviate the above said problems. PA includes microcontroller, sensors, image processing, programming *etc.* to full fill our requirements. Thus,additional need-based research is required.

Precision Agriculture Techniques

RTK-GPS Based Vegetable Transplanter

Real-time Kinematic-Global Positioning System (RTK-GPS) mounted ontractor for geographical location mapping of planting events occurring on the tractor-drawn transplanter have been developed and successfully adopted. The mechanical hitch interface between the tractor and the transplanter is instrumented with orientation sensors (Figure 1). A ruggedised, real time, embedded control system has been used for monitoring of sensor and logging of GPS location, planting events and transplanterodometry data. The system is capable of producinghighly accurate maps of crop plant location for subsequent precision plant care tasksconducted at the centimetre scale.

Variable Rate Irrigation (VRI)

The site-specific or variablerate irrigation is considered as a necessary or essential component of precision irrigation. Most researchers expect a reduction in water use on at least parts of fields, if not areduction in the value aggregated over entire fields (Sadler *et al.*, 2005). It has been reported that variable rate irrigation (VRI) could save 10 to 15 per cent of water used in conventionalirrigation practice (Yule *et al.*, 2008). Hedley and Yule (2009) demonstrated that water savings of around 25 per cent are possible through improvements in application efficiency obtained by spatially varied irrigation applications. Osroosh *et al.* (2016) compared irrigation automation algorithms for irrigation scheduling in apple trees and preliminary results supported the use of weather- and plant-based algorithms, decision support and monitoring software, and wireless sensor network for automatic irrigation management of drip-irrigated orchid trees.

Wireless sensor network (WSN) is a network of small sensing devices known as sensor nodes or motes, arranged in a distributed manner, which collaborate with each other togather, process and communicate over wireless channel about some physical phenomena. Irrigation control systems based on WSN and real time soil moisture data are a potential solution to optimise water management by remotely accessing in-field soil water conditions and then site-specifically controlling

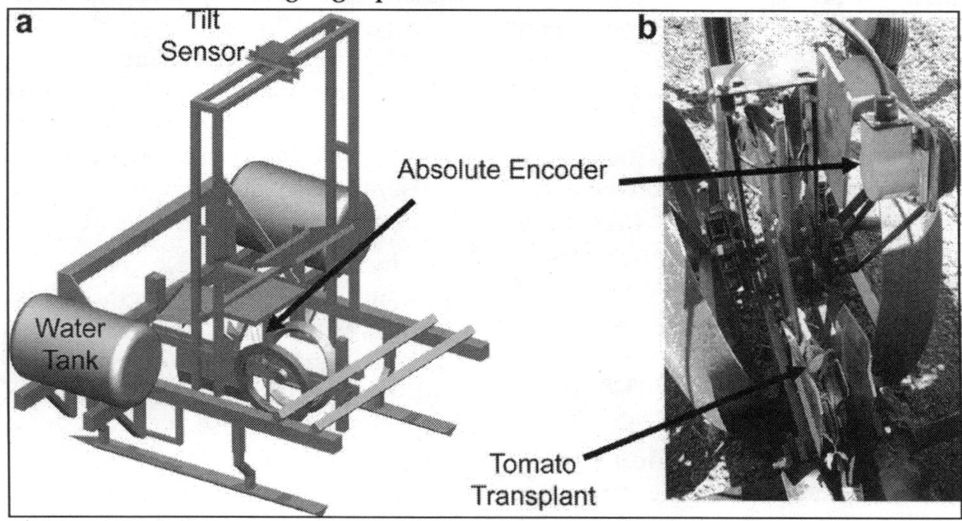

Figure 1: Vegetable Crop Transplanter, Setup for Geospatial Mapping of Crop Plants (Perez-Ruiz *et al.*, 2016).

irrigation systems (Kim and Evans, 2009; Hedley *et al.*, 2011). The system requires seamless integration of the real-time soil moisture data *via* a 3G cellular or ADSL network to control individual sprinklers on a precision irrigation system (Figure 2). Peters and Evett (2007) used a slight warming of crop leaf to indicate water stress, and measure this on a fully automated centre-pivot irrigation system, using infrared thermocouple thermometers attached to the trusses of the pivot. This canopy temperature method provides a useful indicator of the initiation of water stress, however it does not indicate when a plant is approaching moisture stress. It can only indicate when the plant is suffering from stress, which could be too late for irrigation if we are aiming to eliminate water stress impact on yield reduction.

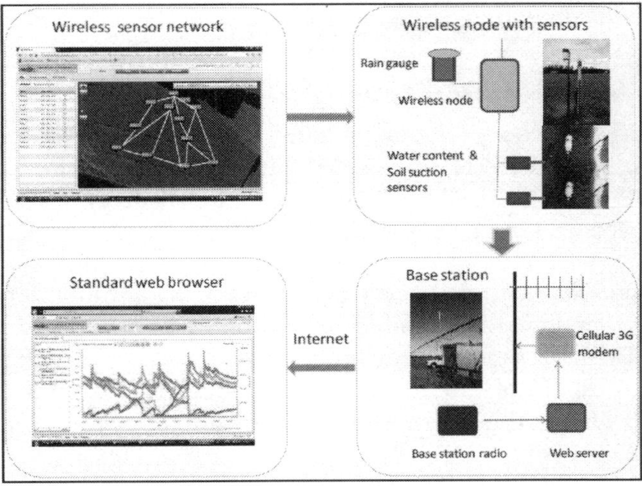

Figure 2: Schematic Flowchart of Wireless Soil Moisture Sensor Network.

Precision Mechanical Weeder

Robotic control systemsare also available for real time weed control. The direct chemical application end effector is used to directly apply chemical *via* a weed's vascular tissue by cutting the weed's stem and wiping chemical on to its cut surface. The robotic system consists of two micro-pumps, a custom-designed circular saw with a DC motor, an applicator, a chemical tank, a flow controller, and a micro-reservoir (Figure 3).

There are a numberof machines available for inter-row weeding operation but machines forintra-row weeding are still very few. Both, inter-row as well as intra-row weeds affect the yield and quality of produce. Radis mechanisation (FR) developed an intra-row weed control system, which consisted of blades mounted on a pivoting arm. A light sensor for detection of plant was used. Arm blade of weeding system is controlled by an air pressure cylinder. When no plant is detected the pivoting arm is moved in the intra-row area, thus cultivating and removing the intra-row weeds (Figure 4). Bakker (2003) expressed that if system is driven at a speed of 5 km/h then weeds are removed up to 20 mm from the plant. However, Bleeker (2005) reported that maximum speed of 3.0 km/h is suitable for weeding operation, because of more damage of plant due to mechanical transition of the intra-row hoe. The system is designed for wide spaced vegetables and the minimum intra-row spacing that the system can work is 220 mm (Bakker, 2009).The problem with system was in detection of plant in wide row crop and also speed limitation for intra-row weeding operation. Tillett (2008) reported that high accuracy can be obtained from ultrasonic guidance with accuracy of 99 per cent at the distance range of 100 mm to 10 m. It was also reported that problems are encountered with stray foliage as distance is calculated from the time taken for an ultrasonic signal to reach and reflected back from the target, thus the reflected signal may bounce back from a weed rather than the crop.

Tillett *et al.* (2008) designed and testeda weeding machine based on computer vision to detect plants for removing intra-row weeds. This machine consisted of a rotating half circle disc that rotated to avoid contact with the crop plants during weeding operation (Figure 5). A camera was mounted centrally on the implement at a

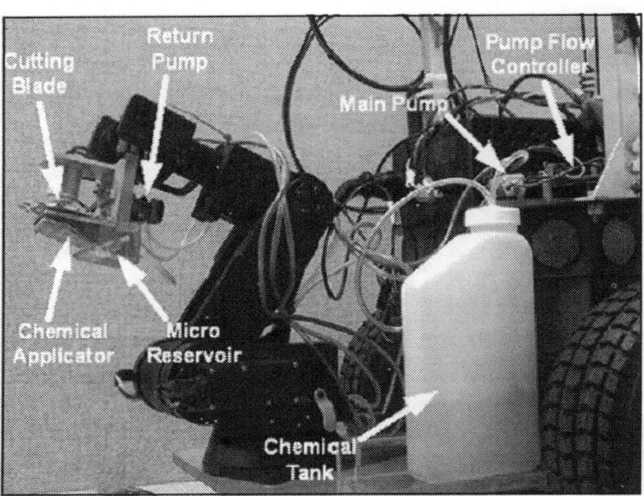

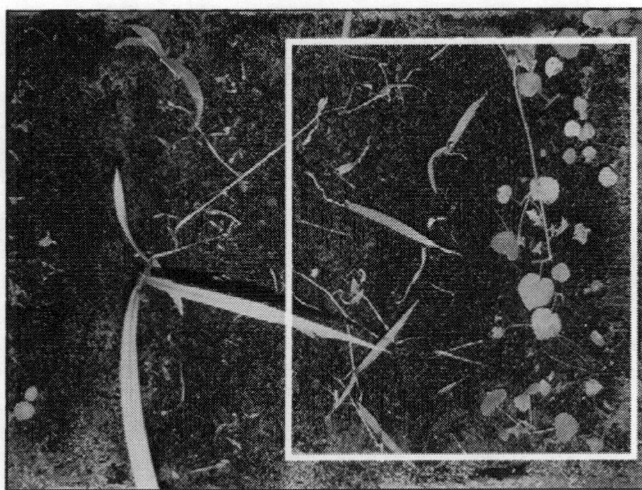

Figure 3: Robotics Control Weeder with End Effector (Jeon and Tian, 2009)

Figure 4: SarlRadis intra-row Weeder (Cloutier *et al.*, 2007).

height of 1.7 m looking ahead and down for capturing image at a length of approximately 2.5 m.

Figure 5: Automated Semi-circle Disc Weeder Machine (Tillett *et al.*, 2008).

An experiment conducted on a cabbage plot withplant spacing of 0.3 m and forward speed of 1.8 km/h at 16, 23, and 33 days after transplanting indicated that best results are obtained at 16 days with weeding efficiency 77 per cent and 23 days with 87 per cent weeding efficiency.

Figure 6: Two-row Laser Sensor Based Herbicide Applicator (Agrawal *et al.*, 2012).

Herbicide Application Techniques

A two-row tractor operated laser sensor based site-specific herbicide applicator (Figure 6) was developed at ICAR-CIAE, Bhopal to spray herbicide on inter-row weeds. The developed herbicide applicator has two components, laser sensor for sensing the inter-row weeds and applicator part, which has controller, solenoid valve and nozzles for application of herbicide. The laser sensor is trained to sense green colour with a threshold of +10 per cent (green colour for which it is trained). The missing percentage varied between 5 to 26 per cent.

A microcontroller based contact type variable rate herbicide applicator (Figure 7) developed at IIT, Kharagpur was used to control the inter-row weeds. The system automatically computes the amount of herbicides to be applied on the weed present in real time image processing. In this system the amount of herbicide applied was based on weed density. The average reduction in herbicides application was 26 per cent with weeding efficiency of 78 per cent. Contact type roller application wasalso helpful for drift reduction.

Figure 7: Microcontroller Based Variable Rate Herbicide Applicator (Tewari *et al.*, 2014).

Spectroscopy for Identification of Disease in Horticultural Crops

Sindhuja *et al.* (2013), developed and evaluated a computer vision and machine learning technique for classification of infected and healthy leaves using fluorescence image spectroscopy (Figure 8). Fluorescence images were segmented using normalized graph and texture features were extracted from the segmented images using co-occurrence matrix. The extracted features were used as an input in the classifier, support vector machine. In Florida for identification of disease of citrus, machine vision control system and artificial intelligence was used to detect the early diseases and to apply selective fungicide (Pydipati *et al.*, 2006). Indirect method of disease detection includes plant properties or stress based disease detection imaging techniques like hyper spectral and fluorescence imaging.

Figure 8: Spectroscopy and Artificial Intelligence for Detection of Disease in Orchards (Sindhuja *et al.*, 2013).

Canopy Sensor Based Pesticide Applicator

Orchard crops suffer from insect and pest attacks and indiscriminate spraying of pesticides causes environmental problems.In order to solve this problem an ultrasonic canopy sensor based pesticide applicator (Figure 9) has been developed at IIT, Kharagpur (Tewari, 2015). The sensor based automated orchard sprayer, delivers pesticide spray only when it senses the canopy of the targets.The tractor mounted sensor based system for spraying saves 45-50 per cent insecticide/pesticide.

A sprayer using ultra sound sensors (Prowave 400EP14D) with appropriate electronics to detect the canopy has been also developed abroad. Ultra sound signalsare processed by a personal computer and fed in real-time to solenoid controlled spray nozzles, which open and close in relation to the canopy structure (Figure 10). The demonstrated concept of precise application of pesticide sprays supports a decrease in the amount

Figure 9: Field Testing of Canopy Sensor Based Spraying System (Tewari, 2015).

of delivered spray, thereby reducing both costs and environmental pollution by plant protection products.

Figure 10: A Prototype Sprayer with Ultrasound Sensors (Jejèiè *et al.*, 2011).

Evar *et al.* (2010) developed a prototype of automated orchard sprayer, which can deliver pesticide spray selectively with respect to the characteristics of the targets (Figure 11). The shape of the apple tree canopy was detected by a machine vision system using an RGB camera and appropriate image analysis. Information captured by RGB camera and processed by specific software was fed in real-time to a spraying arm, with three individually controlled sections, which adapt the pesticide spray flow to the canopy shape. The system allowsvariation in the liquid flow rate and volume of chemicals by means of controlled electric valves, whereby the amount of spray depends on the shape of the tree crowns.

In order to identify the characteristics of the target canopy, in terms of its size and density, a Crop Identification System (CIS), based on ultrasonic sensors, was studied and produced (Figure 12). The results indicated that, by adapting the application rate to the characteristics of the target, the CIS enabled the amount

Figure 11: A Prototype Automated Sprayer Mounted on the Tractor (Hoèevar *et al.*, 2010).

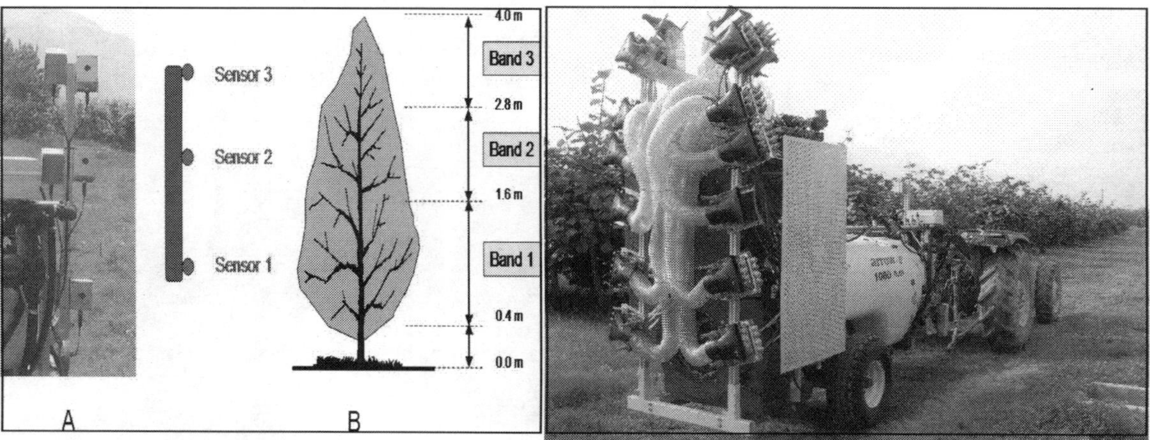

Figure 12: Concept of CIS Sprayer and Prototype ISAFRUIT (Balsari *et al.*, 2009).

of spray deposited on the leaves to be maximised with respect to lower and higher volume rates.

Automated harvesting System

Different methods have been proposed, investigated and practiced since early 1960s for mechanical harvesting of fruits. The mechanical harvesting methods consisting of limb shaking, air blasting, canopy shaking, trunk shaking, and the use of an abscission chemical agent to loosen the fruits have been used (Li *et al.*, 2011). The quality and size selection cannot be maintained by mechanical harvesting systems. On the other hand a vision control system can maintain the quality and do the size selection. The automated harvesting system for horticultural crops has been considered as an alternative solution for mechanical harvesting system (Schertz and Brown, 1968). Automated harvesting system consists of machine vision control system for recognizing the fruits in tree (Figure 13). The vision control system consists of colour camera for providing the information of the location and distance of the fruits to the control system (Li *et al.*, 2011). Automated harvesting systems are not ready to commercialization stage because of their low efficiencies, low intelligence, and high initial investment (Zhao *et al.*, 2016). Many researchers are working to design efficient harvesting robots for horticultural crops. Details of major efforts made for development of horticulture harvesting robots are given in Table 1.

Figure 13: Robotic Harvesting of Sweet Pepper and Citrus Fruits (*Source*: http://www.crops-robots.eu/and http://www.energid.com/experience/citrus-harvest).

Automated Yield Monitoring System

A yield monitoring system is vital element in the precision agriculture. Farmer can get direct feedback for variability in his field by quantifying the variability

Table 1: Details of Harvesting Robots for Horticultural Produce (Zhao *et al.*, 2016)

Crop	*Robots*	*Vision Scheme*	*Success Rate, (%)*	*Speed (s/fruit)*
Fruits	Apple harvesting	Camera-in-hand and positioning sensor	77	15
		Camera-in-hand and laser range sensor	>90	7.1
	Citrus harvesting robot	A fixed camera and a camera-in-hand	NR	<8
		A fixed camera	50	36
	Melon harvesting robot	A far-vision CCD and a near-vision CCD	>85	<22
	Strawberry harvesting robot	A stereovision system and a central camera		
		A stereo camera, a camera and a laser sensor	<41.3	11.5
			NR	7.0
	Watermelon harvesting robot	A stereo vision sensor and a camera in hand	100	12s
		A stereo vision sensor	66.7	NR
	Kiwifruit harvesting robot	Eight cameras (four stereo vision systems)	NR	NR
	Grape harvesting robot	A camera-in-hand	NR	NR
	Cherry harvesting robot	A red and infrared laser active sensor	66.7	14
Vegetables	Tomato harvesting robot	A binocular stereo vision sensor	70.0	3-5
	Cucumber harvesting robot	A fixed camera and a camera-in-hand	80.0	45
		A near-infrared camera and a camera	74.4	65.2
	Mushroom Harvesting robot	A monochrome camera in hand	>80	6.7
	Sweet-pepper harvestingrobot	Two cameras, a stereo camera and acamera	79	NR

in yield within the field (Pelletier *et al.*, 1999). It is a tool to reduce costs, increase yield, and increase efficiency of the system. An automated yield monitoring system consists of color cameras, real time kinematics-global positioning system, custom software, and laptop computer which are mounted on a Specialized Farm Motorized Vehicle (SFMV) for real-time fruit yield mapping. An automated yield monitoring system can perform well to estimate the yield of fruits and vegetable crops. A yield monitoring system was developed using optical sensors for measuring the yield of wild blueberry (Chang *et al.*, 2012). Yield monitoring of horticultural crop can be easily done on mechanized crop harvesting. Research work needs to be carried out on yield mapping of harvesting fruits and vegetables.

Conclusion

Precision agriculture in India is capable of bringing next green revolution to produce food security as well as rural wealth. Although it is in dawning stage in India, but has lot of adoption opportunities.Scarcer inputs like labour, water, fertilizer and change of weather pattern affecting majority of the farms in India because they are rainfedand seek help of engineers, scientists and agriculturists together with government's intercession to take application of PA forward. Progressive farmers in India have adopted mechanization, are identifying better crops to use the seed to be propagated further and likewise if they adopt PA, it would help Indian agriculture in increasing yields and economic returns per field with minimised harm to environment. But Indian farmers will adopt PA only if it brings more or atleast similar profit as compared to conventional practices andsupported/managed by government and private sectors to volunteer in initial costs. Also training centres to train implementation of these technologies and guidance to use them is very much required. Therefore an organised structure of researchers, engineers, industries and growers can coordinate to implement PA and achieve sustainable agricultural production.

References

Anonymous 2013. http://www.dwr.org.in/AICRP–WM per cent 20Document/AICRP–WC–ANNUAL REPORT–2012–13.pdf

Agrawal, K.N., Singh, K., Tiwari, P.S. and Chandra, M.P. 2012. Laser sensor based tractor mounted herbicide applicator. *Proceedings of AIPA*, 2012, India.

Bakker, T. 2003. *Autonomous Weeding: Literature Review.* SCO report 03–01, Wageningen University, Systems and Control group, p. 43.

Bakker, T. 2009. *An Autonomous Robot for Weed Control: Design, Navigation and Control.* Wageningen University.

Bleeker, P. 2005. First experiments with intra-row weed control with a camera. **In:** *Proceedings of the Spatial and Dynamic Weed Measurements and Innovative Weeding Technologies*, Bygholm, Denmark, European Weed Research Society.

Balsari, P., Marucco, P. and Tamagnone, M. 2009. A crop identification system (CIS) to optimise pesticide applications in orchards. *J. Hort. Sci. Biotech.*, **84**(6): 113–116.

Chang, Y.K., Zaman, Q., Farooque, A.A., Schumann, A.W. and Percival, D.C. 2012. An automated yield monitoring system II for commercial wild blueberry double-head harvester. *Comp. Elec. Agr.*, **81**: 97–103.

Cloutier, D.C., Weide, R.Y.V.D., Peruzzi, A. and Leblanc, M.L. 2007. Mechanical weed management. **In:** *Non-*

Chemical Weed Management, (Eds.) M.K. Upadhyaya and R.E. Blackshaw. CAB International, p. 111–134.

Edan, Y., Han, S. and Kondo, N, 2009.Automation in agriculture. **In**: *Springer Handbook of Automation*, pp. 1095–1128.

Ehsani, M.R., Upadhyaya, S.K. and Mattson, M.L. 2004. Seed location mapping using RTK GPS. *Transactions of the ASAE*, **47**(3): 909.

Evar, M., Sirok, B., Jeji, V., Godesa, T., Lesnik, M. and Stajnko, D. 2010. Design and testing of an automated system for targeted spraying in orchards. *J. Pl. Dis. Prot.*, **117**(2): 71–79.

Hedley, C.B. and Yule, I.J. 2009. Soil water status mapping and two variable-rate irrigation scenarios. *Prec. Agr.*, **10**(4): 342–355.

Hedley, C., Bradbury, S., Watson, E., Dalrymple, H. and Wright, J. 2011. Farm scale trials of variable rate irrigation to assess the benefits of modifying existing sprinkler systems for precision application. *Int J. Agril. Mgt.*, **1**(2): 51–55.

Hoèevar, M., Širok, B., Jejèiè, V., Godeša, T., Lešnika, M. and Stajnko, D. 2010. Design and testing of an automated system for targeted spraying in orchards. *J. Pl Dis. Prot.*, **117**(2): 71–79.

Jejèiè, V., Godeša, T., Hoèevar, M., Širok, B., Malneršiè, A., Štancar, A., Lešnik, M. and Stajnko, D. 2011. Design and testing of an ultrasound system for targeted spraying in orchards. *Strojniškivestnik–J. Mech. Engg.*, **57**(7–8): 587–598.

Jeon, H.Y. and Tian, L.F. 2009. Direct application end effector for a precise weed control robot. *Biosystems Engg.*, **104**(4): 458–464.

Kim, Y. and Evans, R.G. 2009. Software design for wireless sensor-based site-specific irrigation. *Comm. Elec. Agri.*, **66**(2): 159–165.

Li, P., Lee, S.H. and Hsu, H.Y. 2011. Review on fruit harvesting method for potential use of automatic fruit harvesting systems. *Procedia Engg.*, **23**: 351–366.

Osroosh, Y., Peters, R.T., Campbell, C.S. and Zhang, Q. 2016. Comparison of irrigation automation algorithms for drip-irrigated apple trees. *Comm. Elec. Agri.*, **128**: 87–99.

Peters, R.T. and Evett, S.R. 2007. Spatial and temporal analysis of crop conditions using multiple canopy temperature maps created with center-pivot-mounted infrared thermometers. *Trans. ASABE*, **50**(3): 919–927.

Pelletier, G. and Upadhyaya, S.K. 1999. Development of a tomato load/yield monitor. *Comm. Elec. Agri.*, **23**(2): 103–117.

Perez-Ruiz, M., Slaughter, D.C., Gliever, C. and Upadhyaya, S.K. 2012. Tractor-based real-time kinematic-global positioning system (RTK–GPS) guidance system for geospatial mapping of row crop transplant. *Biosystems Engg.*, **111**(1): 64–71.

Pydipati, R., Burks, T.F. and Lee, W.S. 2006. Identification of citrus disease using color texture features and discriminant analysis. *Comm. Elec. Agri.*, **52**: 49–59.

Sadler, E.J., Evans, R., Stone, K.C. and Camp, C.R. 2005. Opportunities for conservation with precision irrigation. *J. Soil Water Cons.*, 60(6): 371–378.

Schans, D. and Bleeker, P. 2006. *Practical Weed Control in Arable Farming and Outdoor Vegetable Cultivation without Chemicals*. Wageningen UR.

Schertz, C.E. and Brown, G.K. 1968. Basic considerations in mechanizing citrus harvest. *Trans. ASAE*, **11**(3): 343–0346.

Shanwad, U.K., Patil, V.C. and Gowda, H.H. 2004. *Precision Farming: Dreams and Realities for Indian Agriculture*. Map India.

Sindhuja, S., Joe, M.M., Sherrie, B. and Reza, E. 2013. Huanglongbing (Citrus greening) detection using visible, near infrared and thermal imaging techniques. *Sensors*, **13**: 2117–2130.

Sun, H., Slaughter, D.C., Ruiz, M.P., Gliever, C., Upadhyaya, S.K. and Smith, R.F. 2010. RTK GPS mapping of transplanted row crops. *Comm. Elec. Agri.*, **71**(1): 32–37.

Tewari, V.K., Kumar, A.A., Nare, B., Prakash, S. and Tyagi, A., 2014. Microcontroller based roller contact type herbicide applicator for weed control under row crops. *Comm. Elec. Agri.*, **104**: 40–45.

Tewari, V.K., Nare, B., Prakash, S., Chandel, A.K. and Tyagi, A., 2014. A six-row tractor mounted microprocessor based herbicide applicator for weed control in row crops. *Int. Pest Cont.*, **56**(3): 162–164.

Tewari, V.K. 2015. Progress Report, IIT, Kharagpur centre of AICRP on Farm Implements and Machinery.

Tillett, N.D., Hague, T., Grundy, A.C. and Dedousis, A.P., 2008. Mechanical within-row weed control for transplanted crops using computer vision. *Biosystems Engg.*, **99**(2): 171–178.

Umeda, M., Kubota, S. and Iida, M., 1999. Development of "STORK", a watermelon–harvesting robot. *Artificial Life and Robotics*, **3**(3): 143–147.

Yule, I.J., Hedley, C.B. and Bradbury, S., 2008. Variable-rate irrigation. **In**: *12th Annual Symposium on Precision Agril. Research and Application in Australasia*, Sydney.

Zarco-Tejada, P., Hubbard, N. and Loudjani, P., 2014. Precision Agriculture: An Opportunity for EU Farmers—Potential Support with the CAP 2014–2020. Joint Research Centre (JRC) of the European Commission.

Zhao, Y., Gong, L., Huang, Y. and Liu, C., 2016. A review of key techniques of vision-based control for harvesting robot. *Comm. Elec. Agri.*, **127**: 311–323.

84

Role of Primary Processing in Improving Income from Horticulture Crop Production

Pitam Chandra* and Ashutosh Upadhyay

National Institute of Food Technology Entrepreneurship and Management, Kundli, Sonepat – 131 028, Haryana
****E-mail: pc1952@gmail.com***

Horticultural production in India has vastly improved during the past three decades. Not only area under fruits and vegetables has increased during this period, the productivity also improved significantly. Over the last decade, the area under horticulture grew by about 2.7 per cent per annum and annual production increased by 7.0 per cent. During 2013-14, the production of horticulture crops was about 283.5 million tonne from an area of 24.2 million hectares. While horticultural production has increased, the horticultural post-harvest losses have also come down to below 20 per cent. However, the major challenge that Indian horticulture faces today is that of profitability. The profitability, in turn, is affected by input costs, climatic uncertainties, crop productivity, post-harvest losses, absence of post-harvest processing and value-addition opportunities, and linkages with markets. These constraints on supply side affect adversely the food processing sector as well.

The majority of horticultural produce is highly perishable on account of higher moisture content. Therefore, as compared to food grains, horticultural produce requires more care at the farm as well as in the subsequent supply chain. Appropriate post-harvest strategy would permit considerable reduction in quantitative as well as qualitative post-harvest losses leading to higher price realization for the farmer and better quality material reaching the consumer. Given the skills and the infrastructure in production catchments, there could be a possibility of even secondary processing, value-addition, and linkage with a nearby food processing industry to further add to the income of the farmer. Therefore, with the increased availability of horticultural produce on farm, there is need for appropriate skills, infrastructure, and entrepreneurship to permit the farmer receive due benefits from his/her efforts.

In view of small and marginal land holdings of most farms, income of farm families from production needs to be supplemented through post-harvest management and value-addition activities. ICAR has been emphasising such post production activities through the establishment of agro-processing centres (APC) in production catchments (Nanda, 2011). There have been several success stories of these APCs establishing the proof of concept (Singh *et al.*, 2007; Borkar *et al.*, 2014). Ministry of Agriculture (GoI) has been promoting the concept through its schemes. However, the concept has not had large scale acceptance by potential entrepreneurs.

Objectives

The present study has been taken to evolve a model of profitable horticulture by combining the production activity with post-harvest management and value-addition activities in a production catchment. The study incorporates hybrid energy system with the model to relax the constraint of grid power. The following is the list of deliverables from the study.

- ✰ Creating effective backward linkages at farm gate level
- ✰ Demonstrating basic processing technologies to villagers
- ✰ Organizing the collection of the horticulture produce of respective area
- ✰ Empowering farmers at village level by providing storage facilities
- ✰ Enhancing the shelf-life of the perishable produce resulting into better remunerative prices
- ✰ Creation of minimum processing facilities at village level
- ✰ Organization of marketing of the processed material

Methodology

Large scale interactions with farmers and rural youth on all India basis through NIFTEM's Village Adoption Programme (VAP) indicated that lack of awareness and technical skills has been one of the serious constraints in accepting the APC proposal. Lack of access to appropriate infrastructure including uninterrupted power supply creates another hurdle. Finally, the entrepreneurial efforts for putting in place a sound business plan including marketing strategy are required for the success of such a micro-enterprise. This line of reasoning found favour with ICAR that led to the formulation of the project on "Development of a typical village level Primary Processing Centre with hybrid power system for post-harvest loss reduction and value-addition".

The project is being implemented in one of the villages adopted under VAP where awareness efforts have led to motivating the farmers in the village to form a FPO and allocating a piece of land for setting up the primary processing centre to enhance processing in the production catchment and profitability of farmers. Necessary registrations and construction of the building for housing the PPC have been discussed with the FPO members and the team of our students working in that village is facilitating them in carrying out the defined activities.

Facilities to be planned at primary processing centre should aim at fruits and vegetables grown in the vicinity. The PPC proposal should contain layout diagram, specifications of each item for quantified produce, inter-linking of the facilities for easy flow of produce *etc.* The civil structure of varying size depending upon the produce available for the purpose is also required where facility is to be housed. Some of the components can be pre-cooler cum storage, ripening chambers, ancillary facilities like grading/sorting tables, washing *etc.* In fact, contribution of such components shall vary depending upon the local produce and the proposed level of processing.

The list of processing machinery for the PPC has been finalized in consultation with the FPO and is customized for the horticultural produce available in the village and nearby area. A solar-cum-biomass based power generation system makes the PPC independent of the grid power and imparts the advantage of long term sustainability.

Equipment for the Processing Centre

The PPC would have a cold storage facility of adequate capacity. Since the fresh produce is going to be stored only for a period of a few days for which temperature control alone is adequate. It is desirable to have a few simple instruments/gadgets for checking the quality of fresh produce such as TSS meter, a hand lens, an electronic balance, colour charts, vernier calliper and a digital camera to check and record the physical and biochemical quality attributes periodically.

Table 1: Summary of Equipment of the PPC

Equipment	*Qty*	*Cost*
Cold Room		
Specifications		
Capacity: 8 MT fresh fruits and vegetables		
No. of chambers: One		
Chamber size: 3.25 m x 3.25 m x 2.6 m		
Ambient: 40°C		
Design storage temp.: 2°C (+/-2°C)		
Relative humidity: 90% RH		
Insulation: 60 mm thick PUF panel	1	**42.00 lakh**
Fruit sorting, grading and packing machine as perscope of work, 500 kg/h suitable for fresh fruits and vegetables, Weighing scales, quality measuring instruments, etc	1 set	
Plastic crates along with washing tank	1000	
10 KW solar Hybrid PV power plant	one	
Biomass gasifier along with basic accessories for power generation and distribution	one	
Total		

The processing machinery, a cold store and the hybrid power system are estimated to cost about Rs. 42 lakh. The processing unit is planned to operate for 250 days in a year with four tonne per day processing capacity.

Business model should cover initial cost including building and construction cost, plant equipment, installation cost and fixed operational costs like repair – maintenance, electricity charges and depreciation along with wages of persons working in the PPC.

Marketing channels (front end activities) are the flow of processed and packed fruits and vegetables from PPC to final consumer. Our customers are the wholesalers, retail chains, and retailers. Marketing of primary processed fruits and vegetables attracts special attention given their highly seasonal and perishable nature. Our marketing is characterized by fragmented supply chain, concentration of market power with the wholesalers, existence of large number of intermediaries, little or no quality control, absence of standards, lack of product innovation, small volume for transactions and low inventories. The PPC would gradually work to minimize these constraints.

There are some of the innovative marketing models (Dastagiri MB and Immanuelraj TK, 2012) along with the traditional marketing models. The proposed PPC marketing model would assimilate the goodness of these models. After analysis by different means it has been observed that the marketing cost of horticultural commodities under different supply chains in India is about 10-15 per cent of total consumer price. The marketing costs were estimated in accordance with nearby states (Haryana, Punjab and Uttar Pradesh).

Analysis

Assumptions

i) The annual capacity of the primary processing centre is 1,000 tonne for fruits and vegetables with short term cold storage facility. Thus, daily capacity is 4.0 tonne.

ii) The unit will operate for 250 days in a year.

iii) Processing cost has been estimated at Rs. 1,135/tonne (without cold storage facility) and Rs 1,355/tonne with cold storage facility. Therefore, average processing cost of Rs. 1,250/tonne will be assumed.

iv) The initial cost of investment is the sum of plant and machineries, commissioning, miscellaneous expenditure, pre-operative expenses and net working capital.

v) The operation of PPC plant is assumed for 10 years.

vi) The total initial cost of investment is Rs. 42.00 lakh (Total equipment cost + installation + working capital of one years).

Estimated Running Cost

Hourly capacity: 500 kg/hr @ Daily processing capacity: 4,000 kg/day

Assumed working per year: 250 days. Yearly capacity: 1,000 MT

Table 2: Estimation of Operating Costs

Sl.No.	*Particulars*	*Numbers*	*Expenditure (Per Month)*	*Total Cost (Per Year)*
1.	**Manpower**	5		
A.	Supervisor (Diploma in Food Processing)	1	25,000	3,00,000
B.	Worker (Skilled)	2	20,000	2,40,000
C.	Worker (Unskilled)	2	14,000	1,68,000
			Total	**708,000**
2.	**Utilities**		*Running Cost/Day*	*Running Cost/Year*
A.	Power			
	For cold storage Assumed- Running 300 days/year (Production and storage)		3 kW	65,000
	For equipment Continuous/Batch Process		10 kW Rs 480/day	1,20,000
B.	Water			
	5000 L/day (pottable water), pump and other charges		Rs. 80	20,000
3.	**Packaging Cost** Assumed 50% for Bulk packaging (Crates) and 50% for Retail Packaging			60,000
4.	**Depreciation on equipment cost** @ 10%/year			2,07,500
			Total	**11,80,500**

The total running cost of post-harvest (pre-processing) management along with short cold storage facilities of different fresh fruit and vegetable has been estimated to be about Rs. 11,80,500, say, Rs. 12,00,000 if the unit runs for 250 days/year and cold room runs for 300 days/year.

Economic Parameters

1. NPV (Net Present Value)

NPV is the present value of a project's cash flows over the project period. The decision rule is to accept projects that have a positive NPV, and reject projects with a negative NPV.

$$NPV = \sum_{t=1}^{n} \frac{C_t}{(1+k)^t} - C_0$$

where,

C_t Is net cash inflow in time t.

k Is opportunity cost of capital

C_0 Is initial cost of investment.

n Is the expected life of the investment

2. IRR (Internal Rate of Return)

Internal rate of return (IRR) is the interest rate at which the net present value of all the cash flows (both positive and negative) from a project or investment equals zero.

Internal rate of return is used to evaluate the attractiveness of a project or investment. If the IRR of a new project exceeds a company's required rate of return, that project is desirable. If IRR falls below the required rate of return, the project should be rejected

$$C_0 = \sum_{t=1}^{n} \frac{C_t}{(1+r)^t} = 0$$

where,

C_0 Is initial cost of investment.

C_t Is net cash inflow in time t.

r Is the internal rate of return.

3. BCR (Benefit-cost ratio) or PI (Profitability Index)

Profitability index is the ratio of the present value of cash inflows, at the required rate of return to the initial cash outlay of the investment.

$$BCR = \frac{\text{PV of cash inflows}}{\text{Initial cash outlay}} = \frac{PV(C)_t}{C_0}$$

$$= \sum_{t=1}^{n} \frac{C_t}{(1+k)^t} \div C_0$$

The following results have been obtained from the assumptions and the above analytical tools.

4. Net Present Value

Net Present Value given in Table 3.

Table 3: Net Present Value Analysis Table

Sl.No.	*Present Value Factor PVF*	*Assumed Cash Inflow*	*NPV*
1.	10%	1,56,0000	53,84,640
2.	15%	1,56,0000	36,29,640
3.	20%	1,56,0000	23,42,640
4.	25%	1,56,0000	13,70,760
5.	30%	1,56,0000	6,21,960
6.	35%	1,56,0000	29,160
7	36%	1,56,0000	–73,800

The NPV turns negative somewhere between 35 per cent and 36 per cent.

5. IRR (Internal Rate of Return)

The IRR is calculated by Trial and Error method and found to be 35.28 per cent.

The internal rate of return is higher than the opportunity cost of investment. So we can accept the project. (r > k)

r = 35.28 per cent

k = 10-35 per cent

The net present value is zero **(NPV=0)** at IRR 35.28 per cent. Therefore, we can accept the investment project.

6. BCR (Benefit-cost ratio) or PI (Profitability Index)

Table 4: Benefit Cost Ratio Analysis

Sl.No.	*Present Value Factor (PVF)*	*Assumed Cash Inflow*	*NPV*	*Initial Cash Outlay*	*BCR*
1	10%	1,56,0000	53,84,640	42,00,000	2.28
2	15%	1,56,0000	36,29,640		1.86
3	20%	1,56,0000	23,42,640		1.56
4	25%	1,56,0000	13,70,760		1.33
5	30%	1,56,0000	6,21,960		1.149

The investment project can be accepted when PI is greater than one (PI>1).

7. Payback (PB)

Payback is the number of years required to recover the original cash outlay invested in a project. If the project generates constant annual inflow then

$$\text{Payback} = \frac{\text{Initial investment}}{\text{Annual cash inflow}} = \frac{C_0}{C}$$

$$PB = \frac{4200000}{1560000} = 2.69 = 2 \text{ years } 9 \text{ months}$$

The Payback of the investment project is 2 year and 9 months. Hence, one we can accept the proposal.

The processing cost including temporary storage and packaging is estimated to be Rs. 1,250/tonne. The payback period for such an investment is about three years with benefit-cost ratio of about 2.0. It is hoped that when the primary processing centre is operational, it brings out the intended benefits triggering large scale adoption of primary processing centres in India.

We decided to establish primary and secondary schools in every village to promote primary education in the country. There are 6.41 lakh villages in India and the total number of primary schools in India is 7.5 lakh. Then, we decided to establish primary health centres in the country to make health services available to all. India has about 1.5 lakh health centres. Agriculture is the main activity for earning livelihood in villages and food security of India. Therefore, we must decide to establish primary/secondary processing centres in each village to ensure higher income to the farmers and reduction of post-harvest losses. In fact, establishment of processing infrastructure could be the major contributor to the goal of doubling the farmers' income by 2020.

References

Borkar, P.A., Murumkar, R.P., Rajput, M.R. and Rathod, P.K. 2014. Development and economic analysis of agro processing centre in production catchment of Vidarbha region. *IJISET*, **1**(10): 379–388.

Dastagiri, M.B. and Immanuelraj, T.K. 2012. Supermarket supply chains in horticulture in India: The novel marketing models, effects and policies. *Agrotechnol.*, S11:001. doi:10.4172/2168-9881.S11-001.

Nanda, S.K. 2011. Agro processing centres: Success stories. All India coordinated research project on post-harvest technology. ICAR-Central Institute of Post-harvest Engineering and Technology, Ludhiana, 73 p.

Singh, K.P., Srivastva, A.K., Srinivas, K., Singh, S.R.K. and Gupta, H.S. 2007. Entrepreneurship development in agriculture through agro processing centre: A case study of Almora district in NW Himalaya. *Invited Overview for Agricultural Engineering International: CIGR J.*, **9**(2): 1–14.

85

Machine Vision and Sensor-based Applications for Pre- and Post-Harvest Operations in Horticulture

Nachiket Kotwaliwale*, Km. Sheetal Banga and Shekh Mukhtar Mansuri

Agro Produce Processing Division,
ICAR-Central Institute of Agricultural Engineering,
Nabi Bagh, Berasia Road, Bhopal – 462 038, Madhya Pradesh
**E-mail: nachiket.kotwaliwale@icar.gov.in*

Due tothe growing concern among the consumers for safety as well as nutritional quality of foods, there is frightful scope of modernizing existing pre and post-harvest operations of horticultural produce. Horticulture sector provides a substantial portion of fresh as well as processed foods. Over the last decade, the area under horticulture grew by about 2.7 per cent per annum and annual production increased by 7.0 per cent. However, assuring product quality is one of the most important and challenging tasks of the industries before export or local consumption of food and horticultural produce. With increased expectations for food and horticultural products with high quality and safety, there is a need for accurate, fast and objective quality determination of food and agricultural products. Journey of a quality product starts right from the production field; hence technological interventions are required for rapid, accurate and appropriate measures for quality vis-à-vis safety assurance of the product. In past years, with the rapid exploitation of information technology, the computer vision has been advancing in the theory and technology.

Vision is the most powerful sense to ascertain quality of a product at variety of stages. To emulate the strength of human vision and to go beyond its capabilities technologies like machine vision have evolved. Machine vision is a rapid, economic, consistent and objective inspection technique, which has expanded into many diverse areas. Its speed and accuracy satisfy ever increasing production and quality requirements, hence aiding in the development of totally automated processes. Applications of machine visionand sensor based technology have been developed for pre and post-harvest operations of horticultural produce, such as land-based and aerial-based remote sensing for pre-harvest operations, precision farming, post-harvest product quality and safety detection, classification and sorting and process automation. It does not only recognize size, shape, colour and texture of objects, but also provide numerical attributes of the objects or scene being imaged.

Moreover, machine vision opens up the possibility of studying these objects in regions of the electromagnetic spectrum in which the human eye is not sensitive, as is the case of the ultraviolet (UV) or infrared regions (Zude, 2008). The information received from objects in invisible light regions can be very useful in determining pre-harvest plant maturity, disease, or stress states,plant and vegetable variety, maturity, ripeness and quality, postharvest quality and safety, such as defects, composition, functional properties, diseases and contamination of plants, vegetables and fruits.

Machine Vision and Sensor System

A complete machine vision system consists of sensors mounted on appropriate device, computation device for image processing and actuation device, if any. The sensors commonly acquire reflectance, transmittance, or fluorescence images of the horticultural materials under UV, VIS, or NIR illumination (Chen *et al.*, 2002). Images could be acquired in mono, multi-spectral or hyperspectral imaging domain and also techniques like radiography (X-rays), computed tomography, ultra sonography, thermal or magnetic resonant imaging (MRI). A basic machine vision system consists of a camera, a computer equipped with an image acquisition board and a lighting system. Also, computer software is required for transmitting electronic signals to computers, acquiring images and performing storage and processing of the images.

Image Processing Operations

There could be various optional steps in image processing operations. The ultimate goal is to arrive at decision based of certain desirable or target attributes. The process involves use of appropriate software platform and series of mathematical operations to reach the desired result.

Actuation Devices`

Variety of actuation devices like servo motors, stepper motors, solenoid valves, relay gates *etc.* are normally employed to actuate/operate a device based on the decision taken based on image processing. Not all the machine vision systems have actuation devices. The systems where information is processed off-line or processed by subject matter experts do not normally employ actuation devices. While, the systems with online processing capability have some kind of actuation device. Digital data control cards installed on the computation module of the machine vision system send appropriate signals to the actuation devices.

Sensors

A sensor defined as any device that measures some type of chemical or physical characteristic and converts it into a signal that can be read by an observer or record automatically. Thermography, hyperspectral, fluorescence and chlorophyll sensors are some of the most reliable sensor. The sensors provide quantitative information to help, guide and automate the decision-making process.

Thermography Sensors

Thermal sensors have been used for estimating the number of fruits in grooves and orchards (Stajnko *et al.*, 2004; Wachs *et al.*, 2009). The fruits detection based on the assumption that there is significant temperature difference from the surroundings. The temperature of fruits is also affected by the intensity of the incident radiation, ambient temperature, relative humidity and wind speed.

Hyperspectral Sensors

Hyperspectral imaging equipment measures spectral reflectance as the amount of reflected light from a surface. It is the process by which images are taken and each pixel assigned by spectral radiance (numerical values), range of wavelengths across the electromagnetic spectrum utilizes, including visible and infrared regions.

Hyperspectral imaging technology was used for measuring fruit firmness, maturity and soluble solids content and for detecting bitter pits and bruises on mushroom and apple (Nicolaï *et al.*, 2006; Gowen *et*

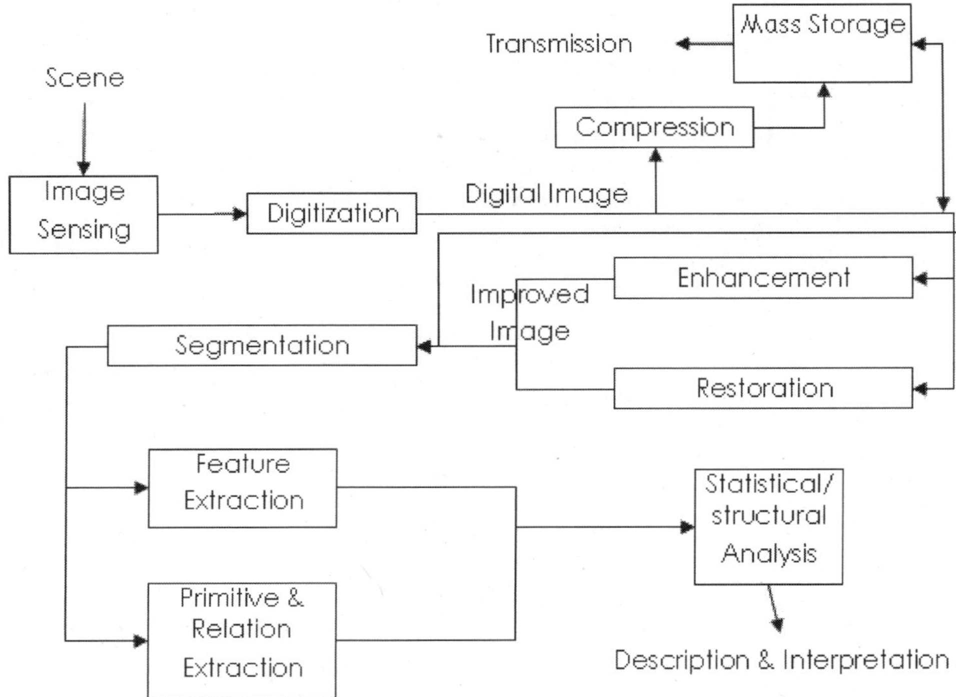

Figure 1: Different Stages of Image Processing.

al., 2008; Yang *et al.*, 2011) deterioration in mushroom (Taghizadeh *et al.*, 2010) and internal defect and chilling injury of cucumber (Cheng *et al.*, 2004; Ariana and Lu, 2010).

Chlorophyll Sensors

Water and chlorophyll are two primary visible and infrared absorbing components of plant leaves. The concentration of chlorophyll, pigments (carotenoids, xanthophylls and anthocyanins) and absorbing biochemicals in the visible (400-700 nm) wavelength region, mesophyll structure in the near infrared (700-1200 nm) region and amount of water in the middle infrared (1200-2400 nm) region are used for determining spectral properties of leaves (Knipling, 1970). Sensors used for chlorophyll measurements for plants leaves are GreenSeeker™ (Trimble Company, Westminster, USA), Soil-Plant Analysis Development (SPAD) meter, atLEAF+ (FT Green LLC., Wilmington, USA) *etc.*

Fluorescence Sensors

Fluorescence spectroscopy measures the longer wavelengths lights after it is excited with a beam of light of short wavelengths (usually in the UV region) emitted from the object. Recent decades, laser-induced fluorescence has been using for vegetative studies, such as to monitor physiological states and stress levels in plants (Belasque *et al.*, 2008). Two types of fluorescence produced by green leaves: (i) blue-green fluorescence in about 400-600 nm range and (ii) chlorophyll fluorescence in about 650-800 nm range. It has been used to monitor stress and diseases in plants, nutrient deficiencies, in laboratory or field conditions. Application of this technology to industrial processes is yet to be fully developed.

Range Sensing

Range sensors are commonly 'active' sensor systems in which illumination is supplied as part of the system. The sensing systems are more robust to variations in ambient lighting than 'passive' sensor systems comprising only cameras. Similar to the visual hull method (Shlyakhter *et al.*, 2001) using multiple camera images, range sensing provides information about the canopy bounding geometry. Range sensing is typically used for generating maps of overall canopy size at high spatial resolution in the field. Geiger (2004) used an array of infrared emitters on either side of a cotton row to measure cotton plant height on-the-go. Tumbo *et al.* (2002), Wei and Salyani (2004) and Schumann and Zaman (2005) all used laser scanning and/or ultrasound to estimate the volume of trees in a citrus orchard.

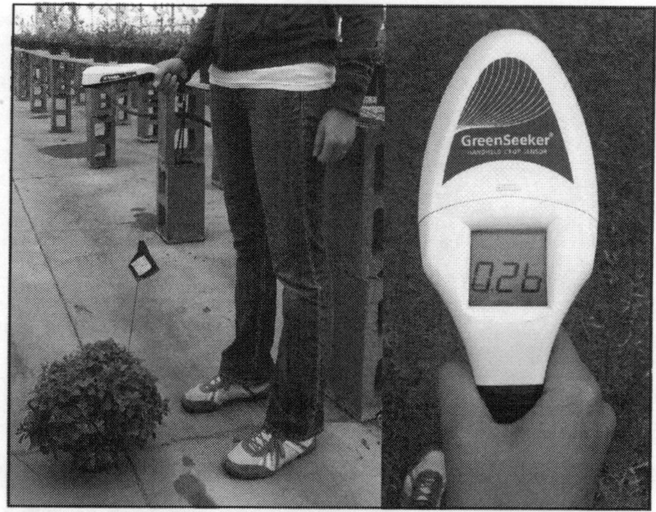

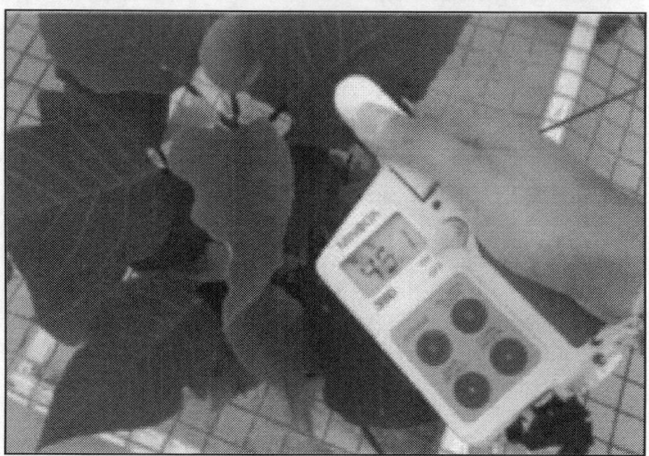

Figure 2: GreenSeeker™ Handheld Sensor, SPAD-502 Meter, atLEAF+ Meter. (Giulio Reina and Annalisa Milella, 2012).

Different Applications of Machine Vision and Sensor Systems

Machine Guidance

Vision guidance has been applied with great technical success to tractors (Billingsley and Schoenfisch, 1993). Within each received image, the software locates the rows of crop and assesses the control action needed to bring the vehicle back on course. The algorithms were developed in the early 1990s, when computing power was much less plentiful. A frame-grab from the steering software is shown in Figure 3.

Use of machine vision with GPS allows a tractor to follow a row of crops, also used for locating the plants and for guiding applications in the pre harvest

Figure 3: Frame-grab from the Video Image that is Steering a Tractor at Speed.

operations of horticultural produce. Figure 4 depicts theautonomous used in the experiments (Reina and Milella, 2012) This tractor wasequippedwitha 3Dscanninglaserrangefinder,a radar, a trinocular camera, a thermos camera and a stereo camera. Use of machine vision with GPS allows a tractor to follow a row of crops, also used for locating the plants and for guiding applications in the pre harvest operations of horticultural produce.

Use of Drones with Sensors to Increase Yields and Reduce Crop Damage

Drones can provide farmers with three types of detailed views. First, seeing a crop from the air can reveal patterns that expose everything from irrigation problems to soil variation and even pest and fungal infestations that aren't apparent at eye level. Second, airborne cameras can take multispectral images, capturing data from the infrared as well as the visual spectrum, which can be combined to create a view of the crop that highlights differences between healthy and distressed plants in a way that can't be seen with the naked eye. Finally, a drone can survey a crop every week, every day, or even every hour. Combined to create a time-series animation, that imagery can show changes in the crop, revealing trouble spots or opportunities for better crop management.

Canopy Measurement

For precise irrigation, fertilizer application, chemical applications, as well as health assessment and relates to crop yield canopy volume of horticulture crops is an important factor (Haselgrove *et al.*, 2000). For horticultural crops estimation of size, growth stage and water needs are important because most plants grow in limited water environments and need irrigation.

For orchard precision fertilizing management quick determination of mineral nutrition contents of fruit trees is essential. Yang and Lv (2011) developed a multispectral imaging system for the measurement of nitrogen content of fruit trees leaves. Some other researchers also reported based on canopy volume measurements savings in chemical application (Solanelles *et al.*, 2006; Allen *et al.*, 2007).

Sensing Plant Disease

In horticultural industry diseases cause major production and economic losses. Remote sensing and near range techniques have exhibited a high potential in monitoring crop and detecting diseases in stands for sub-areas with infected plants. The incidence of plant diseases depends on definite epidemiological and environmental factors; therefore, diseases often have

Figure 4: The Tractor Test Platform to Enable Safe Autonomous Navigation in High-Vegetated, Off-road Terrain. (a) its sensor suite (b). (Reina and Milella, 2012).

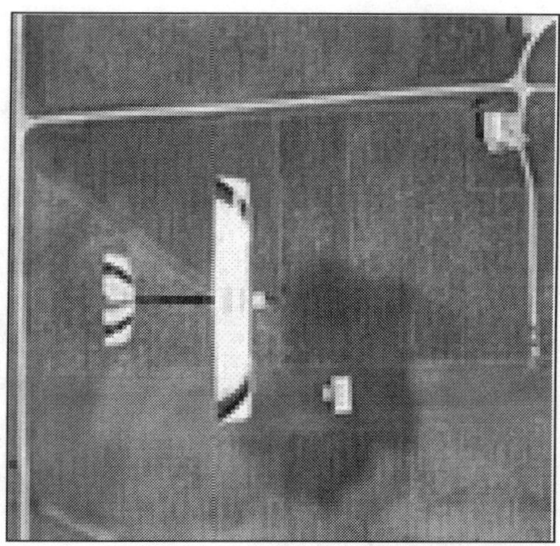

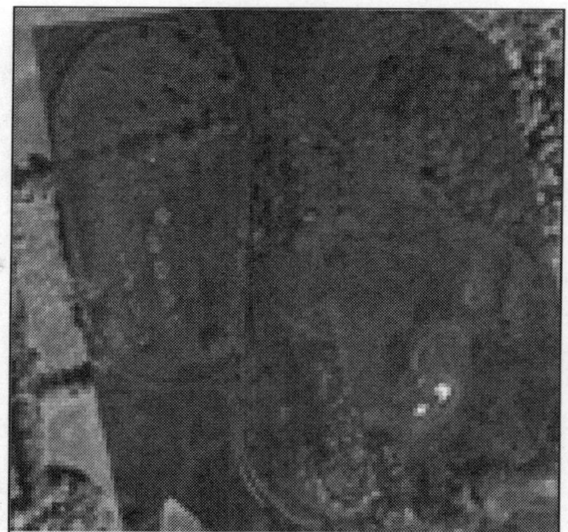

Figure 5: (a) A drone from Precision Hawk is equipped with multiple sensors to image fields. (b) Image depicts vegetation in near--infrared light to show chlorophyll levels (*Courtesy*: Chris Anderson).

the patchy distribution. For plant disease detection and monitoring, imaging systems are superior to non-imaging systems. For instance, the reflectance at visible wavelength provides the information on the leaf pigmentations while, reflectance at infrared wavelength provides the physiological condition of the plant (Huang *et al.*, 2007).

Detection of Weeds and Species Identification

Color discrimination can also be used in the field to discriminate between plants and weeds for the application of selective spraying (Åstrand and Baerveldt, 2002; Zhang *et al.*, 2008). It is possible for the color channels of the camera to include infra-red wavelengths, something that can be achieved by removing the infra-red blocking filter from a simple webcam. Figure 5 is a low-resolution frame-grab showing one stage in the detection of a grass-like weed, "panic," during trials in sugar cane. The computer has marked pixels determined to be "weed" in yellow. Komi *et al.* (2007) combined spatial information from a low-cost RGB camera with spectral data from a line scan spectral camera (400-1000 nm) to classify the detached leaves of six plant species under halogen lighting with a shaded enclosure.

Laser Sensor Based Herbicide Applicator

A two row tractor operated laser sensor based herbicide applicator has been developed to spray herbicide on weeds between rows (CIAE, Bhopal). It consists with two components – laser sensor for sensing the greenness and applicator part, which has controller, solenoid valve and applicator for application of herbicide.

Crop Yield Monitoring

For management of crops in precision agriculture yield monitoring is one of the key information. It combines the effects of various spatial features *viz.* soil properties, topography, fertilization, irrigation, plant population and pest infestations (Lee *et al.*, 2010). For

Figure 6: Tractor Operated Laser Sensor Equipped Herbicide Weeder (ICAR-CIAE, Bhopal).

yield estimation in large geographic areas satellite imaginary has been used but in precision agriculture for mapping within-field yield variability and other applications high resolution satellite imagery and airborne multispectral and hyperspectral imagery have been used.

Plant Material Identification

Applications of machine vision of plant structures include robotic harvesting of fruit in greenhouses. A differential two-waveband infrared vision system was designed and tested that made use of the spectral differences in fruit and leaves at 850 and 970 nm wavelengths to identify cucumbers on a vine (van Henten *et al.*, 2002). This spectral difference also occurs between stems and leaves (Kondo and Ting, 1998). Additional image analysis enabled selective harvesting of only ripe or mature fruit by modeling the fruit size or volume. Noordam *et al.* (2004) presented a comparison of a variety of methods for locating a cutting position on a rose stem, for the purpose of automation in the cut flowers industry. One of the methods evaluated was X-ray imaging and it was demonstrated that thin leaves

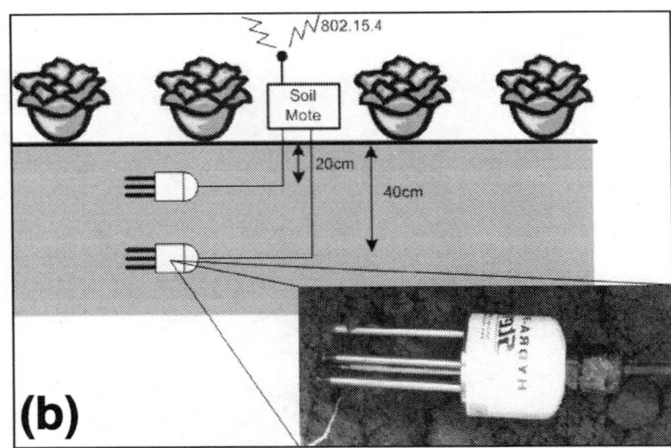

Figure 7: (a) Wireless Sensor Mote for Precision Horticulture.
(b) Sensors Buried at different Depths to Monitor the main Soil Parameter.

were completely invisible in X-ray images. However, the approach was limited by stems occluding other stems and by the severe safety regulations of X-ray usage.

Automated Harvesting

Many attempts have been made over the years to use machine vision to detect and locate fruit for automatic picking and harvesting of horticultural crops. Several machines are developed for harvesting the crops by using machine vision and sensor technology. Although positive research results have been reported, operational speeds are generally low and there has been very limited commercial success.

Sorting of Horticultural Produce by Colour

Machine vision has been used for grading and sorting of fruits and vegetables on the basis of color. For some decades, a simplified form of machine vision hasbeen used for sorting produce (Tao *et al.*, 1995). An automated system is used for grading apples, as seenin Figure 9.

Quality Assessment

True machine vision concerns shape information. Oneexample is the assessment of fodder quality. From a webcamimage of a sample handful of hay, the stem-widthscan be measured and a histogram displayed. Color can alsobe determined by comparison of the video signal againstthat from the image of a calibration card. This makes anobjective standard possible, to achieve agreement betweenvendor and purchaser (Dunn and Billingsley, 2007). Shape information can also be used to monitor thegrowth of a crop, measuring stem length between nodesautomatically (McCarthy *et al.*, 2009).

Figure 8: (a) Instrumented Glove for Fruit Quality; (b) Pea Viner; (c) Apple and Cherry Harvester; (d) Strawberry Robotic Harvesting (Errol W. Hewett).

Figure 9: (a) Apples Moving Towards Washing and Grading. (b) X-ray Imaging Based Mango Sorting System (Mehta, 2013).

Machine vision is also applicable in thermal imaging (above 2500 nm) for horticultural products; for example thermal imaging camera of mid-wavelength infrared (3500–5000 nm) is used for detection of early bruises in apples (Baranowski *et al.,* 2012).

Applications of Computer Vision

The ultimate purpose of many inspection systems based oncomputer vision is to estimate one or several features of the product of interest in a particular moment and relate themwith the quality, which is normally associated to maturity,absence of deformities and blemishes and so forth. Othersystems are aimed at determining the evolution of theproduct over time in order to determine whether a particulartreatment or process is valid or not.In most of these applications, image analysis isemployed to assess features like colour, size, shape, textureand presence of damage.

Fruits and vegetables travel at a high speed under the camerain current inspection lines in packinghouses. In order toacquire images with sufficient quality, cameras freeze themovement by using high-speed electronic shutters that arecombined with adequate illumination, because as the shutterspeed increases, the intensity of lighting must also increaseto avoid underexposure. In order to achieve real-time operation, image processing must be very fast. For this reason, some authors have used specific hardware to minimize processing time. For instance, Aleixos *et al.* (2002) developed a multispectral camera capable of acquiring two images simultaneously (one RGB image and one monochromatic NIR image) from the same scene in a citrus packing line.

Use of Hyperspectral Imaging, X-Ray Imaging and other Techniques

Hyperspectral imaging systems provide hyperspectral images consisting of numerous spatial image planes of the same object at different wavelengths. Hyperspectral imaging systems can be operated either in reflectance or transmittance modes. To acquire images in transmittance mode, thin sample sizes are usually used to allow light to travel through the sample. Thicker samples can be used in reflectance hyperspectral imaging measurements. Thus, food materials can be inspected as a whole in reflectance mode without the need to make slices. Examples include apples (Peng and Lu, 2008), cucumbers (Ariana and Lu, 2010) and mushrooms (Gown *et al.,* 2009).

X-ray imaging technique is used for identification of internal defects. Due to low penetration power andability to reveal the internal density changes soft X-rays are more suitable to be used in fruits. Commercial application ofthese techniques will be beneficial for the consumer as well as the producer to get uniform high-quality products. Shahin and Tollner (1997) used X- ray imaging for the detection of watercore in apples. Kotwaliwale *et al.* (2007a, b) developed a soft X- ray imaging digital imaging system and used for quality determination of pecans.

Computed tomography has been employed for non-destructive internal quality assessment of Peaches (Barcelon *et al.,* 1999), mango (Kotwaliwale *et al.,* 2009). Nuclear magnetic resonance or magnetic resonance imaging is sensitive to the existence of mobile water, oil and sugar, which are major components of agricultural materials therefore it has a high potential for use in various internal quality factors, such as bruises, dry regions, worm damage, stage of maturity and presence of voids, seeds and pits, internal damage due to freezing (Chen *et al.,* 1989; McCarthy, 1994; Kotwaliwale *et al.,* 2012).

Machine vision system is seen as an easy and quick way to acquire data that would be otherwise difficult to obtain manually. Since, the capabilities of digital image analysis technology to generate precise descriptive data on pictorial information have contributed to its more widespread and increased use. An ambiguity of computer vision is that its results are influenced by the quality of the captured images. Often due to the unstructured nature of typical agricultural settings and biological variation of plants within them, object identification in these applications is considerably more difficult. Also if the research or operation in being conducted in dim or night conditions artificial lighting is needed.

Advantages

The main advantages of machine vision system and sensing technology are:

- ☆ **Inspection of total production**: machine vision system can be used to inspect each part produced instead of sampling.
- ☆ **Continuous inspection:** As there no tiredness like human, the machine vision system can be used 24/7 without interruption.
- ☆ **Repeatability:** All inspections are performed in the same conditions; each property of the product acceptance is assured.
- ☆ **Consistency:** machine vision will not have any subjectivity due to fatigue and so its decisions are consistent.
- ☆ **Cost reduction:** Due to web inspection, the fault products are identified and instruct the production system to rectify immediately which reduce the cost.
- ☆ **Customer satisfaction:** Due to inspection of all products, customer receives only valid and fault-free products that makes customer happy.

Table 1: Summary of Machine Vision and Sensors Applications for Horticulture Produce

Product Type	*Product*	*Application*	*Accuracy*	*Reference*
Fruits	Apple	Defects detection	>95%	Yang *et al.* (2011)
		Classification	78%	Steinmetz *et al.* (1999)
		Grading	95%	Yang (1994)
		Bruise	95%	Baranowski *et al.* (2012)
	Oranges	Classification	93%	Ruiz *et al.* (1996)
		Quality Evaluation	87%	Kondo (1995)
	Citrus	Decay	89%	Lorente *et al.* (2013a)
			87.46–95.52%	Lorente *et al.* (2013b)
	Strawberries	Sorting	94–98%	Nagata *et al.* (1997)
			98–100%	Bato *et al.* (2000)
	Oil palm	Sorting	–	Alfantni *et al.* (2008)
	Papayas	Classification	94%	Riyadi *et al.* (2007)
	Nuts	Classification	95%	Pearson and Toyofuku (2000)
	Pomegranate	Sorting (Aril)	90%	Blasco *et al.* (2009)
		Volume estimation	–	Arendse *et al.* (2016)
	Peach	Maturity	70-80%	Herrero-Langreo *et al.* (2011)
			83%	Lleo *et al.* (2011)
	Persimmon	Maturity Detection	90.24	Mohammadi *et al.* (2015)
	Pecans	Attenuation Coefficient	9.350	Kotwaliwale *et al.* (2006)
	Mango	Inspection	–	Kalne *et al.* (2012)
		Surface micro-flora		Jha et. al.(2010)
	Kiwifruit	Bruise	87.5%	Lü *et al.* (2011)
Vegetables	Potatoes	Classification	84%	Wooten *et al.* (2000)
		Sorting	96.2%	Elmasry *et al.* (2012)
	Chilli	Classification	87% and 96.3%	Hahn and Sanchez (2000)
	Lemon	Classification and sorting	——	Lino *et al.* (2008)
	Cucumber	Insect damage	82-93%	Lu and Ariana (2013)
		Classification	86%	Ariana and Lu (2010)
	Mushroom	Bruise	100%	Esquerre *et al.* (2012)
	Tomato	Contamination	>99%	Yang *et al.* (2012)
		Defect	>99%	Cho *et al.* (2013)
	Onion	Skin disease	80-100%	Wang *et al.*, 2010

Disadvantages

While the benefits of machine vision sensors are substantial, the disadvantages of machine vision can be prohibitive. Machine vision systems are unable to cope with unforeseen circumstances and input. While ultimately cost-saving, high development costs can be expected for installation and personnel training. Constant levels of appropriate illumination can be difficult to maintain and cameras can have difficulties isolating products in congested environments.

References

Agarwal, K.N. 2016. Identification, mapping and application equipment for weed management. **In**: ICAR Short Course on *"Machine Vision Systems for Precision Agriculture and Food Safety: Sensors and Protocols"*.

Aleixos, N., Blasco, J., Navarro´n, F. and Molto, E. 2002. Multi-spectral inspection of citrus in real-time using machine vision and digital signal processors. *Com. Elec. Agri.*, **33**(2): 121–137.

Alfantni, M.S.M., Shariff, A.R.M., Shafri, H.Z.M., Saaed, O.M.B. and Eshanta, O.M. 2008. Oil palm fruit bunch grading system using red, green and blue digital number. *J. Appl. Sci.*, **8**(8): 1444–1452.

Allen, R.G., Tasumi, M. and Trezza, R. 2007. Satellite-based energy balance for mapping evapotranspiration with internalized calibration (METRIC) Model. *J. Irrig. Drain. Engg.*, **133**(4): 380–394.

Arendse, E., Fawole, O.A., Magwaza, L.S. and Opara, U.L. 2016. Estimation of the density of pomegranate fruit and their fractions using X-ray computed tomography calibrated with polymeric materials. *Biosyst. Engg.*, **148**, 148–156.

Ariana, D.P. and Lu, R. 2010. Evaluation of internal defect and surface color of whole pickles using hyperspectral imaging. *J. Food Engg.*, **96**(4): 583–590.

Ariana, D.P. and Lu, R. 2010. Hyperspectral waveband selection for internal defect detection of pickling cucumbers and whole pickles. *Comput. Electron. Agric.*, **74**: 137–144.

Åstrand, B. and Baerveldt, A.J. 2002. An agricultural mobile robot with vision-based perception for mechanical weed control. *Autonomous Robots*, **13**(1): 21–35.

Baranowski, P., Mazurek, W., Wozniak, J. and Majewska, U. 2012. Detection of early bruises in apples using hyperspectral data and thermal imaging. *J. Food Engg.*, **110**(3): 345–355.

Bato, P.M., Nagata, M., Cao, Q.X., Hiyoshi, K. and Kitahara, T. 2000. Study on sorting system for strawberry using machine vision (Part 2): Development of sorting system with direction and judgement functions for strawberry (Akihime variety). *J. Jpn. Soc. Agr. Mach.*, **62**(2): 101–110.

Blasco, J., Cubero, S., Gomez-Sanchis, J., Mira, P. and Molto, E. 2009 Development of a machine for the automatic sorting of pomegranate (*Punica granatum*) arils based on computer vision. *J. Food Engg.*, **90**: 27–34.

Chen, Y.R., Chao, K. and Kim, M.S. 2002. Machine vision technology for agricultural applications. *Comp. Elec. Agri.*, **36**(2): 173–191.

Cheng, X., Chen, Y.R., Tao, Y., Wang, C.Y., Kim, M.S. and Lefcourt, A.M. 2004. A novel integrated PCA and FLD method on hyperspectral image feature extraction for cucumber chilling damage inspection. *Transactions of the ASAE*, **47**(4): 1313.

Cho, B.K., Kim, M.S., Baek, I.S., Kim, D.Y., Lee, W.H., Kim, J. and Kim, Y.S. 2013. Detection of cuticle defects on cherry tomatoes using hyperspectral fluorescence imagery. *Postharv. Biol. Tech.*, **76**: 40–49.

Dunn, M.T. and Billingsley, J. 2007. The use of machine vision for assessment of fodder quality. **In**: *14th International Conference, Mechatronics and Machine Vision in Practice, 2007 (M2VIP 2007)*, IEEE, pp. 179–184.

ElMasry, G., Sergio, Cubero, Enrique, Moltó and José Blasco. 2012. In-line sorting of irregular potatoes by using automated computer-based machine vision system. *J. Food Engg.*, **112**: 60–68.

Errol W. Hewett. *Automation, Mechanisation and Robotics in Horticulture*. Institute of Food, Nutrition and Human Health Massey University Albany, Auckland New Zealand

Esquerre, C., Gowen, A., Downey, G. and O'Donnell, C. 2012. Wavelength selection for development of a near infrared imaging system for early detection of bruise damage in mushrooms (*Agaricus bisporus*). *J. Near Infr. Spect.*, **20**(5): 537.

Geiger, D. 2004. Extending the utility of machine based height sensors to spatially monitor cotton growth, Biological and agricultural engineering, Texas A and M University.

Gowen, A.A., O'donnell, C.P., Taghizadeh, M., Cullen, P.J., Frias, J.M. and Downey, G. 2008. Hyperspectral imaging combined with principal component analysis for bruise damage detection on white mushrooms (*Agaricus bisporus*). *J. Chemometrics Chemometrics*, **22**(34): 259–267.

Gowen, A.A., Taghizadeh, M. and O'donnell, C.P. 2009 Identification of mushrooms subjected to freeze damage using hyperspectral imaging. *J. Food Eng.*, **93**: 7–12.

Gustavsson, J., Cederberg, C., Sonesson, U., van Otterdijk, R., Meybeck, A. 2011. *Global Food Losses and Food Waste: Extent Causes and Prevention*. Rome, Food and Agriculture Organization (FAO) of the United Nations.

Hahn, F. and Sanchez, S. 2000. Carrot volume evaluation using imaging algorithms. *J. Agr. Engg. Res.*, **75**: 243–249.

Haselgrove, L., Botting, D., Heeswijck, R.V., Høj, P.B., Dry, P.R., Ford, C. and Land, P.G. I. 2000. Canopy microclimate and berry composition: The effect of bunch exposure on the phenolic composition of *Vitis vinifera* L cv. Shiraz grape berries. *Aust. J. Grape Wine Res.*, **6**(2): 141–149.

Herrero-Langreo, A., Fernández-Ahumada, E., Roger, J.M., Palagós, B. and Lleó, L. 2012. Combination of optical and non-destructive mechanical techniques for the measurement of maturity in peach. *J. Food Engg.*, **108**(1): 150–157.

Huang, W., Lamb, D.W., Niu, Z., Zhang, Y., Liu, L., Wang, J., 2007. Identification of yellow rust in wheat using *in situ* spectral reflectance measurements and airborne hyperspectral imaging. *Prec. Agri.*, **8**: 187–197.

Jha, S.N., Jaiswal, P., Narsaiah, K., Bhardwaj, R., Sharma, R., Kumar, R. and Basediya, A.L. 2010. Post-harvest micro-flora on major cultivars of Indian mangoes. *Scientia Horticulturae*, **125**(4): 617–621.

Kalne, A., Kotwaliwale, N., Singh, K. and Singh, V.K. 2012. Non-destructive inspection of mango fruit using digital radiography, computed tomography and magnetic resonance imaging. *J. Agril. Engg.*, **49**(4): 33–41.

Komi, P.J., Jackson, M.R. and Parkin, R.M. 2007. Plant classification combining colour and spectral cameras for weed control purposes. **In**: *IEEE International Symposium on Industrial Electronics*, pp. 2039–2042.

Kondo, N. 1995. Quality evaluation of orange fruit using neural networks. **In**: *Food Processing Automation IV Proceedings of the FPAC Conference*. ASAE, 2950 Niles Road, St. Joseph, MI 49085–9659, USA

Kotwaliwale, N., Singh, K. and Kalne, A. 2012. Monitoring of mango (*Mangifera indica* L.) (Cv.: Chousa) ripening using X-ray computed tomography. **In**: *Sensing Technology (ICST), 2012 Sixth International Conference*, pp. 326, 330, 18–21 December. doi: 10.1109/ICSensT.2012.6461696

Kotwaliwale, N., Subbiah, J., Weckler, P.R., Brusewitz, G.H. and Kranzler, G.A. 2007a. Calibration of a soft X-ray digital imaging system for biological materials. *Trans ASABE*, **50**(2): 661–666.

Kotwaliwale, N., Weckler, P.R., Brusewitz, G.H. and Kranzler, G.A. 2007b. Non-destructive quality determination of pecans using soft X-rays. *Postharv. Biol. Tech.*, **45**: 372–380.

Kotwaliwale, N., Weckler, P.R. and Brusewitz, G.H. 2006. X-ray attenuation coefficients using polychromatic x-ray imaging of pecan components. *Biosystems Engg.*, **94**(2): 199–206.

Kotwaliwale, Nachiket, Curtis, Evan, Othman, Shadi, Naganathan, Govindarajan, K. and Subbiah, Jeyamkondan. 2012. Magnetic resonance imaging and relaxometry to visualize internal freeze damage to pickling cucumber. *Postharv. Biol. Tech.*, **68**: 22–31.

Lee, W.S., Alchanatis, V., Yang, C., Hirafuji, M., Moshou, D. and Li, C. 2010. Sensing technologies for precision specialty crop production. *Com. Elec. Agri.*, **74**(1), 2–33.

Lino, A.C.L., Sanches, J. and Fabbro, I.M.D. 2008. Image processing techniques for lemons and tomatoes classification. *Bragantiacampinas*, **67**(3): 785–789.

Lleó, L., Roger, J.M., Herrero-Langreo, A., Diezma-Iglesias, B. and Barreiro, P. 2011. Comparison of multispectral indexes extracted from hyperspectral images for the assessment of fruit ripening. *J. Food Engg.*, **104**(4): 612–620.

Lorente, D., Aleixos, N., Gómez-Sanchis, J., Cubero, S. and Blasco, J. 2013a. Selection of optimal wavelength features for decay detection in citrus fruit using the ROC curve and neural networks. *Food Biopro. Tech.*, **6**: 530–541.

Lorente, D., Blasco, J., Serrano, A.J., Soria-Olivas, E., Aleixos, N. and Gómez-Sanchis, J. 2013b. Comparison of ROC feature selection method for the detection of decay in citrus fruit using hyperspectral images. *Food Biopro. Tech.*, http://dx.doi.org/10.1007/s11947–012–0951–1 (in press).

Lü, Q., Tang, M.J., Cai, J.R., Zhao, J.W. and Vittayapdung, S. 2011. Vis/NIR hyperspectral imaging for detection of hidden bruises on kiwi fruits. *Czech J. Food Sci.*, **29**(6): 595–602.

Lu, R. and Ariana, D.P. 2013. Detection of fruit fly infestation in pickling cucumbers using a hyperspectral reflectance/transmittance imaging system. *Postharv. Biol. Tech.*, **81**: 44–50.

McCarthy, C.L., Hancock, N.H. and Raine, S.R. 2009. Automated internode length measurement of cotton plants under field conditions. *Trans. ASABE*, **52**(6): 2093–2103.

Mehta, V.B., Haldankar, P.M., Burondkar, M.M., Jadhav, B.B., Shinde, V.V., Khandekar, R.G. and Kadam, S.G. 2013. Use of a soft x-ray imaging system for online detection of spongy tissue in 'Alphonso' mango fruit. *Acta Hortic.*, **992**: 575–578.

Mohammadi Vahid, Kamran Kheiralipour, Mahdi Ghasemi-Varnamkhast. 2015. Detecting maturity of persimmon fruit based on image processing technique. *Scientia Horticulturae*, **184**: 123–128.

Monta, M., Kondo, N. and Ting, K.C. 1998. End-effectors for tomato harvesting robot. **In**: *Artificial Intelligence for Biology and Agriculture*, Springer Netherlands, pp. 1–25.

Nagata, M., Cao, Q., Bato, P.M., Shrestha, B.P. an Kinoshita, O. 1997. Basic study on strawberry sorting system in Japan. **In**: *1997 ASAE Annual International Meeting Technical Papers*, Paper No. 973095, ASAE, 2950 Niles Road, St. Joseph, Michigan 49085–9659, USA.

Nicolaï, B.M., Beullens, K., Bobelyn, E., Hertog, M.L.A. T.M., Schenk, A., Vermeir, S. and Lammertyn, J. 2006. Systems to characterise internal quality of fruit and vegetables. *Acta Horticulturae*, **712**(1): 59.

Pearson, T. and Toyofuku, N. 2000. Automated sorting of pistachio nuts with closed shells. *Appl. Eng. Agric.*, **16**(1): 91–94.

Peng, Y. and Lu, R. 2008. Analysis of spatially resolved hyperspectral scattering images for assessing apple fruit firmness and soluble solids content. *Postharv. Biol. Tech.*, **48**: 52–62.

Reina, G. and Milella, A. 2012. Towards autonomous agriculture: Automatic ground detection using trinocular stereovision. *Sensors*, **12**(9): 12405–12423.

Riyadi, S., Rahni, A.A.A., Rahni, M., Mustafa, M.M. and Hussain, A. 2007. Shape characteristics analysis for papaya size classification. **In**: *The 5th Student Conference on Research and Development* (*SCOReD*), 11–12th December.

Ruiz, L.A., Molto, E., Juste, F., Pla, F. and Valiente, R. 1996. Location and characterization of the stem calyx area on oranges by computer vision. *J. Agril. Engg. Res.*, **64**: 165–172.

Schoenfisch, M. and Billingsley, J. 1993. A comparison of automated guidance systems for a variety of operations in the growing of cotton. **In**: *Proc. 3rd Int. Workshop IXRP*, Brisbane, pp. 8–16.

Schumann, A., and Zaman, Q. 2005. Software development for real-time ultrasonic mapping of tree canopy size. *Comp. Elec. Agr.*, **45**(1): 25–40.

Shahin, M.A. and Tollner, R.W. 1997. Detection of watercore in apples using x-ray line-scans feature extraction and classification. **In**: *Proceeding from the Sensors for Non-destructive Testing International Conference and Tour*, Orlando, 18–21st February, pp. 389–400.

Shlyakhter, I., Rozenoer, M., Dorsey, J. and Teller, S. 2001 Reconstructing 3D tree models from instrumented photographs. *IEEE Comp. Graph. Appl.*, **21**: 53–61.

Stajnko, D., Lakota, M. and Hoc¡evar, M., 2004. Estimation of number and diameter of apple fruits in an orchard during the growing season by thermal imaging. *Comp. Elec. Agri.*, **42**(1): 31–42.

Steinmetz, V., Roger, J.M., Molto, E. and Blasco, J. 1999. On-line fusion of colour camera and strawberry using machine vision (Part 2): Development of sorting system with direction and judgement functions for strawberry (Akihime variety). *J. Jpn. Soc. Agri. Machinery*, **62**(2): 101–110.

Taghizadeh, M., Gowen, A., Ward, P. and O'Donnell, C.P. 2010. Use of hyperspectral imaging for evaluation of the shelf-life of fresh white button mushrooms (*Agaricus bisporus*) stored in different packaging films. *Innov. Food Sci. Engg. Emerg. Tech.*, **11**(3): 423–431.

Tao, Y., Heinemann, P.H., Varghese, Z., Morrow, C.T. and Sommer, H.J., III, 1995. Machine vision for color inspection of potatoes and apples. *Transactions of the ASAE*, **38**(5): 1555–1561.

Tumbo, S., Salyani, M., Whitney, J., Wheaton, T. and Miller, W. 2002. Investigation of laser and ultrasonic ranging sensors for measurements of citrus canopy volume. *Appl. Engg. Agri.*, **18**(3): 367–372.

Van, H.E., Hemming, J., van Tuijl, B., Kornet, J., Meuleman, J., Bontsema, J. and van Os, E. 2002. An autonomous robot for harvesting cucumbers in greenhouses. *Autonom. Robots*, **13**(3): 241–258.

Wachs, J., Stern, H.I., Burks, T., Alchanatis, V. and Bet-Dagan, I., 2009. Apple detection I natural tree canopies from multimodal images. **In**: *Proceeding of the Joint International Agricultural Conference, JIAC*, pp. 293–302.

Wang, W., Li, C., Gitaitis, R., Tollner, E.W., Rains, G. and Yoon, S.C. 2010. Near-infrared hyperspectral reflectance imaging for early detection of sour skin disease in Vidalia sweet onions. In 2010 Pittsburgh, Pennsylvania, June 20–June 23, 2010 (p. 1). American Society of Agricultural and Biological Engineers.

Wei, J. and Salyani, M. 2004. Development of a laser scanner for measuring tree canopy characteristics: Phase 1 Prototype development. *Trans ASAE*, **47**(6): 2101–2107.

Wooten, J.R., White, J.G., Thomasson, J.A. and Thompson, P.G. 2000. Yield and quality monitor for sweetpotato with machine vision. **In**: *2000 ASAE Annual International Meeting*, Paper No. 001123, ASAE, 2950 Niles Road, St. Joseph, MI 49085–9659, USA.

Yang, C.C., Kim, M.S., Kang, S., Tao, T., Chao, K., Lefcourt, A.M. and Chan, D.E. 2011. The development of a simple multispectral algorithm for detection of fecal contamination on apples using a hyperspectral line-scan imaging system. *Sensing and Instrumentation for Food Quality and Safety*, **5**(1): 10–18.

Yang, C.C., Kim, M.S., Millner, P., Chao, K. and Chan, D.E. 2012. The development of the line-scan image recognition algorithm for the detection of frass on mature tomatoes. **In**: *SPIE Defense, Security and Sensing,* International Society for Optics and Photonics, pp. 836908–836908.

Yang, H.Q. and Lv, G. 2011. Determination of pear leaf nitrogen content based on multi-spectral imaging technology and multivariate calibration. **In**: *Key Engineering Materials*, Vol. 467, pp. 718–724, Trans Tech Publications.

Yang, Q. 1994. An approach to apple surface feature detection by machine vision. *Comp. Elec. Agri.*, **11**: 249–264.

Zhang, Z., Kodagoda, S., Ruiz, D., Katupitiya, J. and Dissanayake, G., 2008. Classification of bidens in wheat farms. **In**: *Proceedings of Mechatronics and Machine Vision in Practice*, pp. 505–510.

Zude, M. (Ed.) 2008. *Optical Monitoring of Fresh and Processed Agricultural Crops*. CRC Press.

Managing Climate Risks

86

Climate Resilience in Horticulture: A Way Forward in Improving Farmers Income

N.N. Reddy

ICAR-Central Research Institute for Dryland Agriculture, Hyderabad – 500 059, Telangana
E-mail: drreddynn@gmail.com

There has been a significant rise in the frequency of extreme weather events in the recent years affecting farm level productivity and adversely impacting stability at the global level and more so at national level. Climate change and its variability are emerging as major challenges to modern agriculture all over the world (Table 1).

Table 1: Phenological Changes in Selected Tree Species

Location	*Species*	*Time Period*	*Change in First Flowering*
North America	Elm	1956–1977	–8 days
Europe	Numerous	1960s	1.4+3 or 1.4-3 days
Canada	Aspen	1901–1997	–26 days
Canada	Cherry	1936–2000	–8 days
US	Sugar maple	1957–2004	–5 to –10 days

Derived from Walther (2002) and Cavaliere (2009).

Weather extremes, such as, drought and flood are being experienced in the same season posing serious problems. Suitable technology interventions are of paramount importance to cope with these aberrant weather conditions. The technologies available to cope with climate variability *viz.*, the use of drought (Tables 4 and 5)/submergence/heat (Tables 2–7)/cold (Tables 8 and 9)/salinity tolerant varieties during

Table 2: Symptoms of Heat and Solar Injury of Fruit and Vegetable Crops.

Crop	*Symptoms*
Apple	Skin discolouration, pigment breakdown and water-soaked areas
Avocado	Skin and flesh browning; increased decay susceptibility
Lime	Juice vesicle rupture; formation of brown spots on fruit surface
Pineapple	Flesh with scattered water-soaked areas; translucent fruit flesh
Snap dragon	Brown and reddish spots on the pod; spots can coalesce to form a water-soaked area
Cabbage	Outer leaves showing a bleached, papery appearance; damaged leaves are more susceptible to decay
Lettuce	Damaged leaves assume papery aspect; affected areas are more susceptible to decay; tipburn is a disorder normally associated with high temperatures in the field; it can cause soft rot development during post-harvest
Muskmelon	Characteristic sunburn symptoms: dry and sunken areas; green color and brown spots are also observed on rind
Bell pepper	Sunburn: yellowing and, in some cases, a slight wilting
Potato	Black heart: occur during excessively hot weather in saturated soil; symptoms usually occur in the center of the tuber as dark-gray to black discoloration
Tomato	Sunburn: disruption of lycopene synthesis; appearance of yellow areas in the affected tissues

Moretti *et al.* (2010).

Table 3: Vegetable Varieties Suitable for Cultivation under Delayed Monsoon

Crop	Variety(ies)	Suitability Zone
Ash gourd	Kashi Dhawal	North Eastern Plain
Bottle gourd	Kashi Ganga	North Eastern Plain
Brinjal	GJB-2 and GJB-3	Gujarat, NW Arid, South Saurashtra
	PKM-1	Southern
	PPI-1	High rainfall
	Kashi Sandesh, Kashi Taru	UP, Bihar, Jharkhand
Cauliflower	Pusa Sharad, Pant Gobi-4, Pant Shubra, Pusa Dipali, Daina	New Alluvial, Red and Laterite
Chillies	Kashi Anmol	NE Plains, Central South
	KKM-1, K-1	Southern
Cluster bean	RGC 936, RGC 1002, RGC-1003	NWPlains, Fatehpur (Sikar) Rajasthan NSC, RSSC, NSP
	RGC-936, RGC-1017	Arid Western Plains and Hyper arid Rajasthan RSSC, NSC
Colocasia	Panchmukhi	Bastar Plateau
Cowpea	IT-38956-1, PKB-4	Central, Eastern Dry
	RC-19, RC-101	Sikar, Rajasthan RSSC
Cucurbits	Rajmata	Coastal Saline
Elephant foot yam	Bidhan Kusum, Gajendra	New Alluvial, Red and Laterite
	NDA-9	Bastar Plateau
French beans	YCD-1	High altitude
Indian bean	GJIB-11, GJIB-2	Middle, South and N Gujarat
Okra	Arka, Anamika, Azad, Sonal, Avantika	Red and Laterite, Coastal Saline
Onion	N-53	Transitional Plain, Fatehpur Rajasthan NSC and RSSC
Tapioca	CO3, CO4	Western
	CO4	High rainfall
Tomato	JT-3	South Shurashtra Gujarat VRS
	Paiyur-1	North Western
Yam bean	RM-1	New Alluvial

Table 4: Fruit Varieties Suitable for Cultivation under Drought Stress

Crop	Variety(ies)	Suitability Zone
Apple	York Imperial	Warm Temperate
Apricot	Badami, Inzhirnyl, Rannil	Intermediate
Banana	Karpuravalli (ABB)	Southern
Ber	Seb, Gola, Umran	Hyper arid and Western Plains
	Seb, Mundia, Jogia, Gola	Sub-Tropical
	Seb, Gola	Plains of Luni Basin, Jaipur
Citrus	Mosambi	Sub-Tropical
Water Melon	Durgapura, Madhu, Durgapura Kesar	Eastern Hills and Plains
Guava	Allahabad Safeda, L-49	Sub-Tropical
Mango	Arka, Neelanchal, Kesar	Coastal regions
	Sinduri	Luni Basin, Jaipur
	Jalore seedless/Sindhuri	Arid Western Plains

aberrant weather situations (Table 3) is one of the key interventions. The impacts of climate change on fruit and vegetable production will depend not only on climate, but also on the internal dynamics of agricultural systems, including their ability to adapt to the changes.

As most of the farmers in the country are smallholders, a fewer options are available and must rely heavily on resources available in their farms or within their communities.

Table 5: Vegatable Varieties Suitable for Cultivation under Drought Stress

Crop	*Variety(ies)*	*Suitability Zone*
Ash gourd	Kashi Dhawal	North Eastern Plain
Bottle gourd	Thar Samridhi	North Western Plain
	Kashi Ganga	North Eastern Plain
Brinjal	PKM-1	Southern
	Kashi Sandesh, Kashi Taru	North Eastern Plain
Carrot	Ooty	High altitude
Cassava	Shri Sahya	Bastar Plateau
Chillies	Samrudhi	Eastern Dry
	Kashi Anmol	All s
Cluster bean	RGC-936, RGC-1017	Sub-humid southern plains and Aravali Hills
	RGC-1017, 1003, 1066	Western Plains, Thar Bhadavi NW
Colocasia	White Goriya, Haloo Kesoo	Bastar Plateau
Cowpea	IT-38956-1, TVX -944- 02E	Central Dry, Eastern Dry
	RC-19, RC-101	Fatehpur (Sikar) Rajasthan RSSC
Cumin	GC-4	Transitional Plains of Luni Basin
Drumstick	Bhagya-KDM-01	Northern, North Easternand Central Dry
Elephant Footyam	NDA-5, Gajendra	Bastar Plateau
Onion	Agrifound Dark Red, Arka Kalyan	Chhattisgarh Plains, Northern hills
	Raseedpura local	Fatehpur (Sikar) Rajasthan NSC and RSSC
Taramira	RTM-314	Southern plains and Aravali hills
Tomato	Arka Meghali, Arka Vikas	Chhattisgarh Plains, N Hills, Bastar plateau

Table 6: Fruit Varieties Suitable for Cultivation under Heat Stress

Crop	*Variety(ies)*	*Suitability Zone*
Aonla	Francis Chakkia	Semi-arid Eastern Plains, Navgaon (Alwar)
Apricot	Badami, Inzhirnyl and Rannil	Intermediate
Banana	Shrimanti and Grand Naine	TN, Kerala, A.P., Karnataka,
	Poovan, Karpuravalli	Bihar, W.B.
Peach	Flordasun and Sunlet	Intermediate
Sweet orange	Mosambi	Sub-Tropical
Water melon	Durgapura Madhu, Durgapura Kesher and Local	Transitional Inland Drains

Table 7: Vegetable Varieties Suitable for Cultivation under Heat Stress

Crop	*Variety(ies)*	*Suitability Zone*
Bottle gourd	Thar Samridhi	Northern
Brinjal	Kashi Sandesh, Kashi Taru	North Eastern Plain
Cauliflower	Sabour Agrim	North eastern plain
Clusterbean	RGC-197, RGC-936	Transitional Plain of Inland Drainage, Fatehpur (Sikar) Rajasthan RSSC
	RGC-936	Arid Western Plains Rajasthan RSSC, NSC
Cowpea	Kashi Kancha, Kashi Nidhi	North Eastern Plain
Okra	Kashi Pragati, Kashi Kranti	North Eastern Plain
Onion	NP53	Navgaon (Alwar)
	Raseedpura local	A Transitional Plain of Inland Drainage, Fatehpur (Sikar) Rajasthan NSC and RSSC
Pea	Matar Ageta-6	North Western Plain
	Azad Pea G 10	A Semi-arid Eastern Plains
Tomato	Arka Meghali, Arka Vikas	Chhattisgarh Plains, Northern Hills, Bastar Plateau
	Varkha Bahar-1 and -2	Vegetable Crops Punjab.

Table 8: Fruit Varieties Suitable for Cultivation under Cold Stress

Crop	Variety(ies)	Suitability Zone
Banana	Poovan, Karpuravalli	NE Plains, Southern
Cashew	Indira Cashew	Bastar Plateau
Sweet Orange	Indira Navin	Bastar Plateau

Table 9: Vegetable Varieties Suitable for Cultivation under Cold Stress

Crop	Variety(ies)	Suitability Zone
Carrot	Ooty-1	High altitude
Cumin	RZ-223, RZ-209	Plains of Luni basin

Environmental Constraints Limiting Production

Increased temperatures, reduced irrigation-water availability, flooding, and salinity will be the major limiting factors in sustaining and increasing vegetable productivity. Plants may respond similarly to avoid one or more stresses through morphological or biochemical mechanisms (Capiati *et al.*, 2006).

High Temperature

Winter season vegetables are more sensitive to harsh weather than the summer season vegetables. A constantly high temperature causes an array of morpho-anatomical changes affecting seed germination, plant growth, flower shedding, pollen viability, gametic fertilization, fruit setting, fruit size, fruit weight, fruit quality *etc.*

Morphological Traits for Heat Tolerance

- ✰ Long root length has a good ability to uptake water and nutrients from the soil.
- ✰ Short life-span which help to minimize the temperature effect on plant.
- ✰ Hairiness provides partial shade to cell wall and and repels sun rays.
- ✰ Small size of leaf which resists evaporation due to reduction of stomata.
- ✰ Leaf orientation enhances photosynthetic activity and tolerance to heat stress.
- ✰ Leaf glossiness and waxiness repel sunlight.

The reproductive development in tomato was more sensitive to high temperatures than vegetative development. The optimum temperatures for tomato cultivation are between 25° and 30°C during the photoperiod and 20°C during the dark period. However, only 2-4°C increase in optimal temperature adversely affected gamete development and inhibited the ability of pollinated flowers into seeded fruits and thus, reduced crops yields.

High temperatures also interfere with floral bud development due to flower abortion. High temperatures can cause significant losses in tomato productivity due to reduced fruit set, and smaller and lower quality fruits. High temperature tolerance has been genetically engineered in plants mainly by over expressing the heat shock protein genes or indirectly by altering the levels of heat shock transcription factor proteins. Apart from heat shock proteins, thermo tolerance has also been altered by elevating the levels of osmolytes, increasing the levels of cell detoxification enzymes and through altering membrane fluidity.

Chilling Tolerance in Tomato

The cultivated tomato (*Solanum lycopersicum* L.) displays limited growth and development at temperatures under 12°C (Hu *et al.*, 2006). At temperatures between 0 and 12°C, plants are damaged by the chilling stress. Wild tomato species such as *S. habrochaites*, *S. chilense* (Dunal) Reiche and *S. peruvianum* L., recover rapidly after exposure to sub-optimal temperatures. These genotypes can be grown at high elevations where temperature remains low, below 10°C. Through backcrosses and selection assisted by molecular markers, cold-resistant genes from wild species can be bred into cultivated tomato varieties.

Physiological Traits of Chilling Tolerance in Tomato

- ✰ Thin stem, short dense glandular hairs and narrow leaflets (*L. peruvianum*)
- ✰ Densely hairy stem, leaves and fruits (*L. hirsutum*), high photosynthetic rate
- ✰ *L. chilense* can survive on rock as it has a deep root system.

Drought Tolerance

In combination with elevated temperatures, decreased precipitation could cause reduction in availability of irrigation water and increase in evapotranspiration, leading to severe crop water-stress conditions.

Nearly all vegetable crops are sensitive to drought during two periods:

☆ Flowering
☆ Two-to-three weeks before harvesting.

Leafy Vegetables

Cabbage, lettuce, and spinach are generally planted at or near field capacity. Being shallow rooted, these crops benefit from frequent irrigation throughout the season. As leaf expansion relates closely to water availability, these crops, especially cabbage and lettuce, are particularly sensitive to drought-stress during the period of head formation through harvest.

Roots, Tuber and Bulb Vegetables

In sweet potatoes, potatoes, carrots, and onions, the yield depends on the production and translocation of carbohydrates from the leaf to the root or bulb. The most sensitive stage of growth generally occurs as these storage organs enlarge. Carrots require an even and abundant supply of water throughout the season. Stress causes small, woody, and poorly flavoured roots.

Fruit and Seed Vegetables

Cucumbers, melons, pumpkins and squashes, lima beans, snap beans, peas, peppers, sweet corn, and tomatoes are most sensitive to drought-stress at flowering and as fruits and seeds develop. Fruit-set on these crops can be seriously reduced if water becomes limited. An adequate supply of water during the period of fruit enlargement can reduce the incidence of fruit cracking and blossom-end rot in tomatoes. Irrigation is often reduced as fruit and seed crops mature.

Physiological Traits for Screening against Drought/ Water Stress

Drought Avoidance Characteristics

☆ High relative water content
☆ Low membrane injury index
☆ Low epidermal conductance
☆ High transpiration efficiency
☆ Early vigour/ground cover
☆ Fibrous and moderate lateral roots
☆ Deeper taproots early in the season.

Drought Tolerance Characteristics

☆ High harvest index
☆ High water use efficiency
☆ More leaf pubescence
☆ Partial closure of stomata
☆ Leaf osmotic adjustment

Salinity

About 20 per cent of cultivated lands and 33 per cent of irrigated agricultural lands worldwide are afflicted by high salinity. High salinity in the root area sternly inhibits normal plant growth and development, resulting in reduced crop productivity or total crop failure. Young seedlings and plants at anthesis appear to be more sensitive to salinity stress than at the mature stages (Lutts *et al.*, 1995).

Flooding

In general, the damage to vegetables by flooding is due to reduction of oxygen in the root zone, which inhibits aerobic processes. Flooded tomato plants accumulate endogenous ethylene that causes damage to the plants. The severity of flooding symptoms increases with rising temperatures.

Cultural Management

The use of organic and inorganic mulches is common in high-value vegetable production systems. In the areas where temperatures are high, dark-colour plastic mulch is recommended in combination with rice straw. Dark colour of plastic mulch prevents sunlight from reaching the soil surface and the rice straw insulates the plastic from direct sunlight thereby preventing the soil temperature rising too high during the day.

Grafting for Stress Management

Grafting of vegetables was used primarily to control soil-borne diseases affecting the production of vegetables such as tomato, eggplant, and cucurbits. *Solanum lycopersicum* x *S. habrochaites* rootstocks provide tolerance to low soil temperatures (10 to 13°C) for their grafted tomato scions, while eggplants can be grafted on wild brinjal (*S. integrifolium*) as rootstocks to overcome low temperatures (18 to 21°C). Until now, genetic variability for tolerance of excess soil moisture is limited or inadequate to prevent losses. Many accessions of eggplant are highly tolerant to flooding, thus, can be grafted to improve the flood tolerance of tomato using eggplant rootstocks which were identified with good grafting compatibility with tomato and high tolerance to excess soil moisture.

Use of Heat- and Cold-Tolerant Genotypes

The heat-tolerant tomato lines were developed using heat-tolerant breeding lines and landraces from the Philippines (*viz.*, VC11-3-1-8, VC 11-2-5, Divisoria-2) and the United States (*viz.*, Tamu Chico III, PI289309) ([10]Opena *et al.*, 1992). A CL5915 line is considered best combiners for percentage fruit set and total yield in hybrids developed for heat-tolerance.

Tomato varieties: Pusa Sadabahar, Pusa Sheetal and Pusa Hybrid-1

Radish variety: Pusa Chetaki

Carrot variety: Pusa Vrishti

Cauliflower variety: Pusa Meghna

Drought Tolerance

Most of the vegetables are sensitive to drought; however brinjal, cowpea, amaranth, and tomato can tolerate drought to a certain extent. Genetic variability for drought tolerance found in the cultivated tomato (*S. lycopersicum*) is limited and inadequate. *S. chilense* is five-times more tolerant to wilting than cultivated tomato. *S. pennellii* has the ability to increase its water-use efficiency under drought conditions, unlike the cultivated *S. lycopersicum.*

Salt Tolerance

A few vegetables like beet, palak, tomato, *etc.* can tolerate salt to some extent. Most commercial tomato cultivars are moderately sensitive to increased salinity and only limited variation exists in the cultivated species. The tomato genotypes, LA1579 and LA1606, (*S. pimpinellifolium*) and LA4133 (*S. lycopersicum* var. *cerasiforme*) from AVRDC, Taiwan, have shown salt tolerance. Wild tomato species, *S. cheesmani, S. peruvianum, S. pennellii, S. pimpinellifolium,* and *S. habrochaites* are the potential source of salt tolerance.

Strategies to Alleviate the Effect of Climate Change

i) Screening for stress tolerance (high/low temperature, drought/flood, salinity)
ii) Development of cold and hot set varieties of tomato
iii) Development of low chill requiring varieties in temperate vegetables like cabbage, carrot, beet root, turnip, knol-khol, Brussels sprout, *etc.*
iv) Develop leafy vegetables like palak, spinach, coriander, *etc.* with wider adaptability and greater tolerance to high temperature.
v) Development of 'N'-use and water use efficient genotypes.

Physiological Effects on Tree Crops and Potential Consequences for Pollinating Insects

The responses of flowering plants to warming are likely to affect floral visitors in a number of ways. Elevated temperatures have been found to have varying effects on flower production. Some plants grown under higher temperatures may be less likely to flower or may produce fewer flowers. Individual flower size can be affected by temperature. Experimentally warmed (23 °C) pumpkin plants *Cucurbita maxima* produced flowers of smaller diameter (Hoover *et al.*, 2012). The production of floral scent, nectar, and pollen can also be affected by temperature. Warmer temperatures might increase emissions and/or volatility of organic compounds

Table 11: Physiological and Quality Parameters of Fruit and Vegetable Crops Affected by Exposure to Increased O_3 Levels (Moretti *et al.*, 2010)

Physiological or Quality Parameter	*Effect of High CO_2*	*Produce*	*Reference(s)*
Respiration	I	Blueberry; broccoli; carrot	Song *et al.* (2001)
Dark Respiration	I	Sugar maple	Tjoelker *et al.* (1993)
Photosynthesis	D	Strawberry; conifers; hardwoods	Amthor and Cumming (1988), Reich (1987)
Visible injury	I	Black cherry	Chappelka *et al.* (1997)
Viscosity	I	Potato	Donnelly *et al.* (2001)
Starch	NC	Potato	Vorne *et al.* (2002)
Reducing sugars	D	Potato	Vorne *et al.* (2002)
Citric acid	I	Potato	Piikki *et al.* (2003)
	NC	Tomato	Tzortzakisa *et al.* (2007)
Electrolyte leakage	I	Persimmon	Salvador *et al.* (2006)
Malic acid	D	Potato	Piikki *et al.* (2003)
Ascorbic acid	D	Potato	Vorne *et al.* (2002)
	NC	Tomato	Tzortzakisa *et al.* (2007)
N, P	NC	Potato	Heagle *et al.* (2003)
K, Mg	I	Potato	Piikki *et al.* (2003)
Nitrate	I	Potato	Vorne *et al.* (2002)
Colour (L)	I	Broccoli;	Skog and Chu (2001)
	D	Cucumber	
Browning index	I	Mushroom	Skog and Chu (2001)
Firmness	I	Cucumber	Skog and Chu (2001)
Isocumarin	I	Carrot	Hildebrand *et al.* (2008)
Sucrose	NC	Potato; carrot	Hildebrand *et al.* (2008)

I: increase; D: Decrease; NC: No perceptible change in effect of O_3.

Table 10: Physiological and Quality Parameters of Fruit and Vegetable Crops Affected by Exposure to Increased CO_2 Levels (Moretti *et al.*, 2010)

Physiological or Quality Parameter	*Effect of High CO_2*	*Produce*	*Reference(s)*
Photosynthesis	I	Potato; spinach	Katnya *et al.* (2005), Jain *et al.* (2007)
Respiration	D	Asparagus; tomato; broccoli; mung blueberry; pear	Beaudry (1993), Peppelenbos and Leven (1996)
	I	Potato; lettuce, eggplant; lemon; cucumber; mango	Pal and Buescher (1993), Fonseca *et al.* (2002), Bender *et al.* (1994)
	NC	Apples	Peppelenbos and Leven (1996)
Ripening	D	Tomato	Klieber *et al.* (1996)
Stomatal conductance	D	Spinach	Leakey *et al.* (2006), Jain *et al.* (2007)
Firmness	NC	Tomato	Klieber *et al.* (1996)
	I	Strawberry; raspberry	Siriphanich *et al.* (1998), Haffner *et al.* (2002).
Colour intensity	I	Grape	Bindi *et al.* (2001)
Dry matter	I	Potato	Vorne *et al.* (2002)
Starch	I	Potatao	Vorne *et al.* (2002)
Alcohol	I	Grape, mango, pear	Bindi *et al.* (2001), Bender *et al.* (1994)
Titratable acidity	NC	Grape	Bindi *et al.* (2001)
Citric acid	D	Potato; tomato	Donnelly *et al.* (2001), Islam *et al.* (1996)
Malic acid	D	Potato; tomato	Vorne *et al.* (2002), Islam *et al.* (1996)
Ascorbic acid	I	Potato; strawberry; orange; tomato	Vorne *et al.* (2002), Wang *et al.* (2003), Idso *et al.* (2002), Islam *et al.* (1996)
Reducing sugars	I	Potato	Vorne *et al.* (2002); Högy and Fangmeier (2009)
	I	Tomato	Islam *et al.* (1996)
Total phenolics	I	Grape; strawberry	Bindi *et al.* (2001); Wang *et al.* (2003)
Flavonoids	I	Grape	Bindi *et al.* (2001)
Anthocyanins	I	Grape; strawberry	Bindi *et al.* (2001), Wang *et al.* (2003)
Glycoalkaloids	D	Potato	Vorne *et al.* (2002)
pH	NC	Grape	Bindi *et al.*, 2001
Nitrate	D	Potato; celery; lettuce; cabbage	Vorne *et al.* (2002), Jin *et al.* (2009)
Volatile compounds	D	Mango	Lalel *et al.* (2003)
Antioxidant capacity	D	Scallion; strawberry	Levinea and Paré (2009), Shin *et al.* (2008)

I: Increase; D: Decrease; NC: No perceptible change in effect of CO_2.

produced by flowers, although some evidence suggests that endogenous floral scent production decreases with increasing temperature. Pollinators, too, are susceptible to many changes as a direct result of climate change. Higher temperatures tends to produce adults that are smaller in size, possibly because development is accelerated. Warmer temperatures associated with climate change could reduce the life span of pollinating insects.

Antioxidant Activity

Antioxidants in fruit and vegetable crops can also be altered by exposure to high temperatures during the growing season. Wang and Zheng (2001) observed that 'Kent' strawberries grown in warmer nights (18–22°C) and warmer days (25°C) had a higher antioxidant activity than berries grown under cooler (12°C) days. The investigators also observed that high temperature conditions significantly increased the levels of flavonoids and, consequently, antioxidant capacity.

Physiological Disorders and Tolerance to High Temperatures

Exposure of tomato fruit to temperature extremes approaching 40°C can induce metabolic disorders and facilitate fungal and bacterial invasion. Electrolyte leakage in harvested 'Dorado' tomatoes exposed to direct sunlight (34 ± 2°C) for 5 h was 73 per cent higher than fruits held in shaded (29 ± 2°C) conditions.

Effects of Carbon Dioxide Exposure

Changes in CO_2 concentration in the atmosphere can alter plant tissues in terms of growth and physiological behavior. Increased atmospheric CO_2 alters net photosynthesis, biomass production, sugars and organic acids contents, stomatal conductance, firmness, seed yield, light, water, and nutrient use efficiency and plant water potential.

Effects of Ozone Exposure

Visible injury symptoms of exposure to low ozone concentrations include changes in pigmentation, also known as bronzing, leaf chlorosis, and premature senescence (Felzer *et al.*, 2007). Since leafy vegetable crops are often grown in the vicinity of large metropolitan areas, it can be expected that increasing concentrations of ozone will result in increased yellowing of leaves with reduction in photosynthesis and increased turnover of antioxidant systems.

Physiological Bases of Plants to Adaptability

In order to survive during the winter period, fruit trees from cold winter climates fall dormant. Dormancy is a physiological state where quite accurate factors like temperature below certain threshold and longevity (duration) are necessary. Chilling requirements are crucial for choosing the right tree cultivar for a certain climatic zone in order to have high orchard productivity. Period of dormancy enable fruit trees to tolerate freezing temperatures in their original habitats. Early blooming and harvest date for fruits around the world indicate that dormancy processes are changing.

Strategies to overcome chilling requirements under global warming includes careful selection of cultivars, dormancy avoidance, manipulation of micro-climate, use of colour sheds result in higher productivity, better storage life and higher antioxidants content.

Medicinal Plants

A range of adaptation strategies provide the opportunity to improve the sector management towards the development of climate resilient pathways. Elevated CO_2 levels (4,000 µl CO_2/litre of air) increased fresh weight and leaf and root numbers in cultures of lemon basil (*Ocimum basilicum* L.), oregano (*Origanum vulgare* L.), peppermint (*Mentha piperita*), spearmint (*M. spicata* L.) and thyme (*Thymus vulgaris* L.) shoots compared with cultures grown on the same media under ambient air.

Climate Warming vs. Secondary Metabolite Production

- Secondary metabolite production may increase in stressed plants
- Increase in volatile organic compounds
- The availability of menthol crystals was affected by heavy monsoon rainfall, which occurred earlier than usual in Northern India and reportedly damaged field mint (*Mentha arvensis*, Lamiaceae) crops in 2008.

Effect of Elevated CO_2 on Quality

- Effect of atmospheric CO_2 enrichment on plant constituents of significance to human health varied from increase, decrease or no effect on protein contents of various foods
- Tripling atmospheric CO_2 concentration increased vitamin C an antioxidant of great importance to human health in tomato modestly by 7 per cent only and also vitamin A increased significantly
- In sour orange crop, atmospheric CO_2 enrichment by 75 per cent increased vitamin C concentration by 15 per cent compared to the control. This will be of great public health significance as it may help in minimising scurvy.
- Foxglove (*Digitalis lanata*) produces cardiac glycoside digoxin, used in the treatment of cardiac insufficiency. Tripling the air CO_2 in a phytotron, increased the concentration of digoxin by 11 per cent under well-watered and 14 per cent under water-stress conditions.

The endemic plant species are considered more vulnerable to climate change and may face high risk of extinction due to confined distribution. Research

Table 12: Temperature Sensitivity in Citrus by Phenological Stage

Phenological Stage	*Optimum Range of T (ºC)*		*Characterization of Sensitivity*
Dormancy	–4	14	Low physiological activity
(a) Hardening	–4	8	Reduced losses due to freezing
(b) Pre-bloom	0	14	Required 'resting period' for bloom
Flowering	10	27	Requires daily mean over 20°C to begin
Fruit set	22	27	Range end of bloom, single day max over 38°C may cause losses. At 48°C losses of 50%
Fruit growth	20	33	Requires heat but expansion is reduced if temp. is high due to low turgor and low physiological activity
Maturation			Increased sugars and decreased acids increase with heat units
(a) Soluble solids	13	27	Breakdown of green chlorophylls hastened by cool temperatures.
(b) Colour	8	18	Heat may lead to re-greening

Based on Ben Mechlia and Carroll, 1989 and Reuther, 1973.

Table 13: Physiological Disorders of Vegetable Crops Caused by High Temperatures (Adopted from Rai and Yadav 2005)

Crop	Disorder	Aggravating Factor
Asparagus	High fibre in stalks and spheres	High temperature
Asparagus	Feathering and lateral branch growth	> 32°C, picking frequency is reduced
Bean	High fibre in pods	High temperature
Carrot	Low carotene content	Temperature < 10°C or > 20°C
Cauliflower	Blindness, buttoning, riceyness	Temperature fluctuations
Cauliflower, broccoli	Hollow stem, leafy heads, no heads and bracting	High temperature
Cole crops and lettuce	Tip burn	Drought with high temp; high respiration
Tomato, pepper and watermelon	Blossom end rot	High temperature, combined with drought and high transpiration.

on habitat range and secondary chemical production efficiency of threatened Himalayan medicinal plants under climate warming scenario is essential for developing conservation strategies as well as agro-technologies for cultivation.

Table 14: Effect of Water Stress on Morpho-Physiological Charecters Vegetable Crops

Vegetable	Symptom(s)
Brinjal	Reduced extension of main stem and No. of branches per plant
Beans	Few flowers, delayed flowering, seed yield reduction, decreased starch content, low seed protein, accumulation of free proline.
Potato	Yield loss, decreased starch content, increase in reducing sugars
Cauliflower	Ricey, leafy, loose, yellow, small and hard curds
Tomato	Blossom end rot, accumulation of free proline.
Cassava	Reduction of leaf area
Lettuce	Bitter taste, accelerated development of tip burn
Spinach beat	Quick bolting

Mango

Cloudy weather, with considerable day and night temp variations, increased humidity results in out breaks of powdery mildews, blossom blight/anthracnose, mango hopper and thrips. Increasing temperatures would enhance more rapid fruit growth in mango. In Australia, the estimated duration of mango fruit development decreased by 12-16 days (7-8 per cent) as a consequence of the 1.5°C increase of winter temperatures over the last 45 years. High temperature could also have a positive effect on fruit quality because of stress-induced synthesis of secondary compounds, some of them being of nutritional value. Higher light intensity enhances skin color and fruit size for colored cultivars.

Table 15: Adaptation to Heat and Cold Stress in some Vegetable Genotypes

Crop	Varieties	Tolerance
Tomato	Pusa Sadabahar, Pusa Sheetal, Pusa Hybrid-1	Tolerant to high and low temperatures
Radish	Pusa Chetki	Better root formation under high temperature regime, *i.e.* April-August
Carrot	Pusa Vrishti	Form root at high temperature and high humidity, *i.e.* March-August
Early Cauliflower	Pusa Meghana	Can form curd at high temperature

Tuber Crops

Sweet potato requires warm weather of at least four to five months. It needs temperature range between 21 to 27°C and a well distributed rainfall of 75 to 150 cm.

Table 16: Abiotic Stress Tolerance in some Vegetable Varieties and Advanced Lines

Tolerant	Crop	Variety	Advanced Line
Drought/train fed	Tomato	Arka Vikas	RF- 4A
	Onion	Arka kalyan	MST-42 and MST-46
	Chilli	Arka Lohit	IIHR Sel.-132
Photo insensitive	Dolichos bean	Arka Jay, Arka Vijay, Arka Sambram, Arka Amogh, Arka Soumya	IIHR Sel.-16-2
	Cowpea	Arka Garima, Arka Suman, Arka Samrudhi	
High temperature	Capsicum		IIHR Sel.-3
	French bean		IIHR-19-1
	Peas		IIHR-1 and IIHR-8
	Cauliflower		IIHR 316-1 and IIHR- 371-1

It is a tropical and sub-tropical crop. It can not tolerate heavy rainfall which induces excessive vegetative tuber development. Is does not stand frost. Cassava is a perennial woody shrub with an edible roots responds to changed CO_2 levels by adjusting evapo-transpiration rates or water use relations, indirectly determining crop yields (Ainsworth and Long, 2005).

Way Forward

Enhance resilience by focussing on adaptation to climate variability with appropriate strategies to contingent situations. Identifying climate resilient crop varieties is one of the foremost important criteria to deal with the impending peril. Improved and tolerant varieties along with the proper management practices can enhance the coping ability through risk reduction in vulnerable environments. Ensuring seed and propagule availability of the resilient varieties in various crops at the appropriate time is an important challenge to be addressed immediately.

References

Ainsworth, E.A. and Long, S.P. 2005. What have we learned from 15 years of free-air CO_2 enrichment (FACE): A meta-analytic review of the responses of photosynthesis, canopy. *New Phytologist*, **165**(2): 351–371.

BenMechlia, N. and Carroll, J.J. 1989a. Agroclimatic modeling for the simulation of phenology, yield and quality of crop production: 1. Citrus response formulation. *Int. J. Binmeteorology*, **33:** 36–51.

Capiati, D.A., Pais, S.M. and Tellez-Inon, M.T. 2006. Wounding increases salt tolerance in tomato plants: Evidence on the participation of calmodulin-like activities in cross-tolerance signaling. *J. Exp. Bot.*, **57**: 2391–2400.

Cavaliere, C., 2009. The effects of climate change on medicinal and aromatic plants. *Herbal Gram*, **81**: 44–57.

Felzer, B.S., Cronin, T., Reilly, J.M., Melillo, J.M. and Wang, X. 2007. Impacts of ozone on trees and crops. *Compters Rendus Geoscience*, **339**: 784–798.

Hoover, S.E.R., Ladley, J.J., Shchepetkina, A.A., Tisch, M. and Gieseg, S.P. 2012. Warming, CO_2 and nitrogen deposition interactively affect a plant-pollinator mutualism. *Ecol. Lett.*, **15**: 227–234.

Hu, W.H., Zhou, Y.D., Du, Y.S., Xia, X.J and Yu, J.Q. 2006. Differential response of photosynthesis in greenhouse and field ecotypes of tomato to long-term chilling under low light. *J. Plant Physiol.*, **163**: 1238–1246.

Lutts, S., Kinet, J.M. and Bouharmont J, 1995. Changes in plant response to NaCl during development of rice (*Oryza sativa* L.) varieties differing in salinity resistance. *J. Exp. Bot.*, **46**: 1843–1852.

Moretti, C.L., Mattos, L.M., Calbo, A.G. and Sargent, S.A. 2010. Climate changes and potential impacts on post-harvest quality of fruit and vegetable crops: A review. *Food Res. Int.*, **43**: 1824–1832.

Opena, R.T., Chen, J.T., Kuo, C.G. and Chen, H.M. 1992. Genetic and physiological aspects of tropical adaptation in tomato. **In**: *Adaptation of Food Crops to Temperature and Water Stress* (Eds.) C.G. Kuo, AVRDC, Shanhua, Taiwan, p. 321–334.

Rai, N. and Yadav, D.S. 2005. *Advances in Vegetable Production*. Researchco Book Centre, New Delhi, India.

Reuther, W. 1973. Climate and citrus behavior. In: *The Citrus Industry*, 2nd edn. Vol. 3, (Ed.) W. Reuter. University of California Press, Berkeley, pp. 280–337.

Walther, G.R., 2002. Ecological responses to recent climate change. *Nature*, **416**(6879): 389–395.

Wang, S.Y. and Zheng, W. 2001. Effect of plant growth temperature on antioxidant capacity in strawberry. *J. Agril. Food Chem.*, **49**: 4977–4982.

87

Development of Abiotic Stress Tolerant Climate Smart Varieties of Horticultural Crops

***Major Singh*[1]* *and J.K. Ranjan*[2]**

[1]All India Coordinated Vegetable Research Project (Vegetable Crops),
[2]Division of Crop Improvement,
ICAR-Indian Institute of Vegetable Research, Varanasi – 221 305, Uttar Pradesh
**E-mail: singhvns@gmail.com*

Although climate change would result in both challenges and opportunities for horticulture, it is well understood that factors that have led to the current food security crisis are likely to multiply under climate change. The reasons for climate change are not completely known today, but as per the available information anthropogenic activities like industrialization and mechanization may contribute up to some extent. Drought and salinity are the two important consequences of increase in temperature worsening vegetable cultivation. Among atmospheric gases CO_2 is affecting growth and development as well as pest and diseases of vegetable crops and it is the major component of greenhouse gases, which are responsible for global warming. Under changing climatic situations crop failures, shortage of yields, reduction in quality and increasing pest and disease problems are common and they render the vegetable cultivation unprofitable. Horticulture will have to adapt to climate change, but it can also help mitigate the effects of climate change. Unless measures are undertaken to adapt to the effects of climate change on vegetable production, nutritional security in developing countries will be under threat.

Climate change is a complex phenomenon and may be referred as any change in climate over time, whether due to natural variability or as a result of human activity. Climate change *per se* is not necessarily harmful but the problems arise from extreme events that are difficult to predict are the main concern (FAO, 2001). According to the Intergovernmental Panel on Climate Change (IPCC) atmospheric CO_2 will increase from the current level of 360 ppm to 400-750 ppm and sea level will rise from 15 to 95 cm by 2100, which in turn will result in an increase in global average temperature and drastic shifts in the annual precipitation. Average global air temperature has increased by about 0.8°C above pre-industrial levels and a 2001 report by the Intergovernmental Panel on Climate Change projected a rise from 1.4 to 5.8°C by the year 2100. While this prediction is extremely important, there is a significant amount of spatial variability in the air temperature. Rising temperatures will alter precipitation patterns so that wet areas become wetter, and dry areas become dryer. The temperature increase will alter the timing and amount of rainfall, availability of water, wind patterns and causes incidence of weather extremes, such as droughts, heat waves, floods or storms, changes in ocean currents, acidification, forest fires and hastens rate of ozone depletion. The current fertilizer use efficiency that ranges between 2 to 50 per cent in India is likely to be reduced further with increasing temperatures. Greater fertilizer use to boost agricultural production will in turn lead to higher emission of greenhouse gases. Small changes in temperature and rainfall will have significant impact on quality of fruits and vegetables with resultant implications in domestic and external trade. Higher average temperatures will also stimulate the emergence of new and aggressive pests including weeds which are likely to invade our crops.

Studies on climate impacts and adaptation strategies are increasingly becoming major areas of scientific concern. Tackling this immense challenge, breeding of climate smart varieties for abiotic stress would certainly pave the way.

Impact of different Abiotic Stress

Climate change impacts vegetable production both directly and indirectly. However, the type and magnitude of impact varies depending on the degree of change in climate, geographical region and type of production system. Impact of climate change may be visualised by change in productivity with reference to quantity and quality of crops, changes in agricultural practices like use of water, fertilizers, and pesticides and environmental influences particularly in relation to the frequency and intensity of soil drainage which may lead to loss of nitrogen through leaching, soil erosion and reduction of crop diversity. Climate change is basically associated with extremities of four factors, *viz.*, temperature, drought, salinity and CO_2 concentration which are causing environmental stress leading to crop losses worldwide, reducing average yields for most of the crops by more than 50 per cent (Boyer, 1982; Bray *et al.*, 2000). Vegetables are in general more sensitive to environmental extremes, and thus high temperatures and limited soil moisture becomes the major causes of low yields and poor quality produce which will be further magnified by climate change.

Impact of Increasing Temperature and Water Stress

Temperature is the main factor which limits the range and production of many crops including vegetables. In general, most of the vegetables require low temperature for their cultivation and that is why average productivity of temperate countries are much higher than tropical one. However, since higher temperature conditions are often prevalent in the tropics, with changing climate, crops in this area will be subjected to increased temperature stress because the crops are already near the limit of their temperature tolerance, even a slight increase in temperature will result in drastic fall in crop productivity.

Increasing temperature will reduced the irrigation water availability, decrease soil fertility, increase soil erosion and flooding and salinity incursion; which are major restrictive factors in sustainable vegetable production (Wheeler and von Braun, 2013; Anwar *et al.*, 2015). A warmer climate is likely to increase the frequency of high day or night temperature stress events that can negatively affect flowering, fruit set, and quality. Analysis of climate trends in tomato-growing locations suggests that temperatures are rising and the severity and frequency of above-optimal temperature episodes will increase in the coming decades (Bell *et al.*, 2000). This is may be due to the fact that vegetative and reproductive processes in tomatoes are strongly modified by temperature alone or in conjunction with other environmental factors (Abdalla and Verkerk, 1968). High temperature stress disrupts the biochemical reactions fundamental for normal cell function in plants. It primarily affects the photosynthetic functions of higher plants (Weis and Berry, 1988). In tomato, if temperatures exceed 90 F during critical flowering and pollination periods (Sato *et al.*, 2001), it cause significant losses in tomato productivity due to reduced fruit set, and smaller and lower quality fruits (Stevens and Rudich, 1978). The symptoms causing fruit set failure at high temperatures in tomato includes bud drop, abnormal flower development, poor pollen production, dehiscence, and viability, ovule abortion and poor viability, reduced carbohydrate availability, and other reproductive abnormalities (Hazra *et al.*, 2007). Higher temperature during pre-anthesis is associated with developmental changes in the anthers, particularly irregularities in the epidermis and endothesium, lack of opening of the stromium, and poor pollen formation (Sato *et al.*, 2002). However, in pepper pre-anthesis period is not sensitive to high temperature exposure as it did not affect pistil or stamen viability, but high post-pollination temperatures inhibited fruit set, suggesting that fertilization is sensitive to high temperature stress (Erickson and Markhart, 2002). High temperatures can affect photosynthesis, respiration, aqueous relations and membrane stability as well as levels of plant hormones, primary and secondary metabolites. Symptoms of high temperature and solar injury varies from crop to crop as presented in Table 1. The practical effects of climate change have already been experienced in some parts

Table 1: High Temperature Injury Symptoms in different Vegetable Crops

Crop	*Injury Symptoms*
Snap bean	Brown and reddish spots on the pod; spots can coalesce to form a water-soaked area
Cabbage	Outer leaves showing a bleached, papery appearance; damaged leaves are more susceptible to decay
Lettuce	Damaged leaves assume papery aspect; affected areas are more susceptible to decay; tipburn is a disorder normally associated with high temperatures in the eld; it can cause soft rot development during post-harvest
Muskmelon	Characteristic sunburn symptoms: dry and sunken areas; green color and brown spots are also observed on rind
Bell pepper	Sunburn: yellowing and, in some cases, a slight wilting
Potato	Black heart: occur during excessively hot weather in saturated soil; symptoms usually occur in the center of the tuber as dark-gray to black discoloration
Tomato	Sunburn: disruption of lycopene synthesis; appearance of yellow areas in the affected tissues, No fruit set after 35°C day temperature

Source: Moretti *et al.* (2010).

across the globe. For example, increased temperatures in Sambalpur, Odisha, India, have delayed the onset of winter resulting in the decline of cauliflower yields (Pani, 2008). Earlier growers used to harvest 1 kg of inflorescences head but now weight has reduced to 0.25-0.30 kg each. Such reductions in yield hiked production costs of other vegetables too like tomatoes, radishes, and other native Indian vegetable crops.

Increase in temperature increases transpiration and in drier regions it leads to water stress causing yield reduction. Vegetables, being succulent products by definition, generally consist of greater than 90 per cent water (AVRDC, 1990). Thus, water greatly influences the yield and quality of vegetables; drought conditions drastically reduce vegetable productivity. Drought stress causes an increase of solute concentration in the environment (soil), leading to an osmotic flow of water out of plant cells. This leads to an increase of the solute concentration in plant cells, thereby lowering the water potential and disrupting membranes and cell processes such as photosynthesis. The timing, intensity, and duration of drought spells determine the magnitude of the effect of drought. In India only about 41 per cent area is irrigated and remaining 59 per cent is rainfed

Impact of Salinity

Vegetable production is threatened by increasing soil salinity particularly in irrigated croplands which provide 40 per cent of the world's food (FAO, 2002). In hot and dry environments, high evapotranspiration results in substantial water loss, thus leaving salt around the plant roots, which interferes with the plant's ability to uptake water and thus reduces crop productivity. According to the United States Department of Agriculture (USDA), onions are sensitive to saline soils, while cucumber, eggplant, pepper, and tomato are moderately sensitive. Plant sensitivity to salt stress is reflected in loss of turgor, growth reduction, wilting, leaf curling and epinasty, leaf abscission, decreased photosynthesis, respiratory changes, loss of cellular integrity, tissue necrosis, and potentially death of the plant (Jones, 1986; Cheeseman, 1988). Coastal regions are highly affected by salinity due to low-quality and high-saline irrigation water thus affecting crop production. Salinity fluctuates with season, being generally high in the dry season and low during rainy season when freshwater flushing is prevalent

Impact of Elevated CO_2 Level

Carbon dioxide concentrations in the atmosphere have increased approximately 35 per cent from pre-industrial times to 2005 (IPCC, 2007). Growth and physiological behoavior of plants are highly dependent on CO_2 concentrations in the atmosphere. Increased atmospheric CO_2 is reported to modify net photosynthesis, biomass production, sugars and organic acid contents, stomatal conductance, fruit firmness, seed yield, light, water, nutrient use efficiency, and plant water potential in vegetables (Bazzaz, 1990; Cure and Acock, 1986; Idso and Idso, 1994). The magnitude of the "CO_2 fertilization effect" varies tremendously among plant species and from variety to variety (Wolfe 1994). C_3 plants, to which majority of the vegetables belong, respond positively under elevated CO_2 concentrations and ambient temperature compared to C4 plants (Patterson *et al.*, 1999), but this is less evident at higher temperatures. However, attaining maximum CO_2 benefits often requires more fertilizer (to support bigger plants), optimum temperatures, unrestricted root growth, and excellent control of weeds, insects, and disease. High CO_2 can have another direct effect on plants- reducing transpiration (water loss) per unit leaf area by causing partial closure of leaf stomata. However, often this water conserving response is minimized or not observed at the whole-plant or field (ET) level because high CO_2 grown plants have more total leaf (transpirational surface) area. In general, the CO_2 fertilization effect cannot compensate for negative effects from other environmental stresses (Luo and Mooney, 1999).

Crops grown on elevated CO_2 concentration are expected to decline in nutritional quality and greater carbon to nitrogen ratios in plant foliage that may stimulate greater feeding activity in some herbivores. The fact that host plant quality is often increased by drought-induced plant stress may result in increased feeding by pests and lead to greater crop damage (Hill and Dymock, 1989; Porter *et al.*, 1991); however, this theme remains controversial. Clark (2004), while working on tropical forests, argued that increasing atmospheric CO_2 has no or little effect on biomass production rates and stressed that the growth of tropical forests is not carbon limited, since higher temperature increase, respiration, and other metabolic processes result in increased atmospheric CO_2, which reduces forest productivity. There will also be associated changes in the movement of crop pests and diseases, thereby changing the dynamics of their interactions (Harrington and Woiwod, 1995; Bale *et al.*, 2002). It has been reported that many of the most invasive and noxious weeds respond more positively to increasing CO_2 than do most of our cash crops (Ziska, 2003) and weeds are much more difficult to control with herbicides at CO_2 levels we anticipate will occur in the coming decades (Ziska *et al.*, 1999). It is mostly speculation at this time as to which crops and regions will benefit and which will be worse off in the future with regard to weed and pest control. Some leaf-feeding insects appear to do more crop damage to plants under high CO_2 conditions (Hamilton *et al.*, 2005). It remains difficult to predict future rainfall patterns, but wetter summers would tend to favour many foliar pathogens (Coakley *et al.*, 1999). Climate change also influences the ecology and biology of insect pests. Anticipated effect of climate change on some insect pests of vegetables is presented in Table 2.

Table 2: Anticipated Effect of Climate Change on some Insect Pests of Vegetables (Yukawa, 2008; Cannon, 2008)

Pest	*Present Distribution*	*Anticipated Effect of Climate Change*
Melon fly (*Bactrocera cucurbitae*)	South of Japanese archipelago	Invasion to Japanese archipelago
Sweet potato weevil (*Cylas formicarius*)	Tropical and sub-tropical regions of world	Invasion to temperate zones due to increase
Aphids (*Aphis gossypi; Aulacorthum solani*)	Worldwide distribution	Increase in number of generations per year resulted in more infestation due to lower developmentalzero point, low thermal totals required for onegeneration
Fruit and Pod borer (*Helicoverpa armigera; Spodoptera litura*)	Worldwide distribution	Increase in number of generations per year resulted Higher infestation
Colorado potato beetle (*Leptinotarsa decemlineata*)	North America, Southern Europe, Asia and much of Pacific Ocean	Expansion in itspotential northern limit by 400 km causing more potential risk to 99% potato growing areas
Diamond back moth (*Plutella xylostella*)	European origin, cosmopolitan in distribution	More frequent overwintering and increase in pest status drastically
Cabbage butter fly (Pieris *brassicae*)	Europe, NW Africa and Asia, Introduced to America and Australia	Increase in range, abundance and diversity. Decrease in diversity is reported by Roy *et al.* (2001)
Cabbage root fly (*Delia radicum*)	Europe, N. Africa, W. Asia and N. A	Populations become active a month earlier in UKwith a mean temperature increase of 3°C

Source: Ayyogari *et al.* (2014).

Developing Climate-Resilient Smart Genotypes

Improved germplasm is the most cost-effective option for farmers to meet the challenges of a changing climate. While breeding for high yields, we have been counter selecting genotypes that are water and fertilizer responsive but at the present juncture of climate change, there is a need to develop superior genotypes adapted to a wider range of climatic conditions through the discovery of novel genetic variation for tolerance to different biotic and abiotic stresses.

Breeding for Stress Tolerance

Genotypes with improved attributes conditioned by superior combinations of alleles at multiple loci could be identified and advanced. Improved selection techniques are needed to identify these superior genotypes and associated traits, especially from wild and related species that grow in environmentsm which do not support the growth of improved high yielding varieties. Plants native to climates with marked seasonality are able to acclimatize more easily to variable environmental conditions and provide opportunities to identify genes or gene combinations which confer such resilience. Since cucurbits are mostly warm season crop, any increase in temperature will largely affect its productivity. Hence, there is urgent need to develop varieties of cucurbitaceous vegetable crops with general adaptation to hot and humid tropical environments. AVRDC - The World Vegetable Center, Taiwan has developed tomatoes and Chinese cabbage with general adaptation to hot and humid tropical environments and low-input cropping systems since the early 1970s. This has been achieved by developing heat-tolerant and disease-resistant breeding lines.

Tissue tolerance to severe dehydration is not common in crop plants but is found in species native to extremely dry environments (Ingram and Bartels, 1996). Some *Cucurbita* sp. posses some xerophytic characters essential to adopt under water scarcity condition. These gene pool should be thoroughly studied to isolate drough tolerant line in pumpkin and squash. Drought stress is a major environmental factor influencing plant growth and development. *Citrullus colocynthis* is a very drought tolerant cucurbit species with a deep root system. Differences in gene expression during drought were studied using cDNA-AFLP. Two genes, CcrbohD and CcrbohF, encoding respiratory burst oxidase proteins were cloned using RACE. Real-time RT-PCR analysis showed that expression of CcrbohD was rapidly and strongly induced by abiotic stress imposed by PEG, ABA, SA, and JA treatment. CcrbohD has great promise for improving drought tolerance of other cucurbit species.

Attempts to improve the salt tolerance of crops through conventional breeding programs have very limited success due to the genetic and physiologic complexity of this trait (Flowers, 2004). In addition, tolerance to saline conditions is a developmentally regulated, stage-specific phenomenon; tolerance at one stage of plant development does not always correlate with tolerance at other stages (Foolad, 2004). Success in breeding for salt tolerance requires effective screening methods, existence of genetic variability, and ability to transfer the genes to the species of interest. Screening for salt tolerance in the field is not a recommended practice because of the variable levels of salinity in field soils. Screening should be done in soil-less culture with nutrient solutions of known salt concentrations (Cuartero and Fernandez-Munoz, 1999).

Severity of diseases and pests are the main havoc of climate change, which needs to be managed well. Development of resistant/tolerant genotypes will be the best option to tackle this problem. Large gene pool is available in Cucurbitaceae family which can be utilized for the development of pest tolerant/resistant varieties or elite breeding lines in cucumber, muskmelon, watermelon and others (Table 3).

Table 3: Genetic Resources in different Cucurbitaceous Vegetables Resistant to different Diseases and Insect Pests

Crop	*Disease/Insect Pest*	*Resistance Source*	*References*
Cucumber	Powdery mildew	PI 197087, Poinestee, Yomaki, Sparton Salad, PI 197088, *Cucumis ficifolia, C. anguria, C. dinteri* and *C. sagittatus, C. ficifolia* accessions IVf 1801 and PI 280231, *C. anguria* PI 147065, *C. anguria* var. *anguria, C. dinteri* PI 374209 and *C. sagittatus* PI 282441	Barnes (1966), Imam and Morkes (1975), Omara (1979), Munger *et al.* (1979), Lebeda (1984)
Cucumber	Downy mildew	Chinese Long and Poinsette	Imam and Markes (1975), Seshadri (1986)
Cucumber	Anthracnose	PI 197087 and PI 175111	Barnes and Epps (1952), Abul-Hayja and Peterson (1978)
Cucumber	CMV	TMG-1, Tokyo Long Green, Chinese Long, Wisconsim and Table Green	Provvidenti (1985)
Cucumber	CGMMV	*Cucumis anguria*	Den-Nij (1982)
Cucumber	WMV	Table Green and Sarinam	Takeda and Gilbert (1975) and Provvidenti (1985)
Musk melon	powdery mildew	Edisto, PMR-45 and PMR-450), Georgia-47 and C-68), Campo and PMR- 6)), Arka Rajhans, RM-43 and Pusa Sharbati Campo, Jacumba, Levlita, PM-5 and PMR-6, PI 164323 and PI 180283	Copeland (1957), Bohn and Whitaker (1964), Takada *et al.*, (1975), Norton and Cosper (1985), Choudhury and Sivakami (1972), Khan (1973)
Musk melon	Downy mildew	Edisto, Seminole), Buduma Type -1, 2 and 3, Phoontee, Goomuk, Nakkadosa, Ex- 2, Annamalai, Edisto and Harvest Queen), *Cucumis callosus*, WMR-29, MR-I, Punjab Rasila, Cinco, DMDR-I and DMDR-2), Punjab Rasila), EC 163888), Snapmelon collections like SP- I, SP-2, SP-3, KP-2, KP-7 and KP-9	Copeland (1957), Whitner (1960; Sambandam *et al.* (1979), Zink *et al.* (1983), Nandpuri (1993); Singh (1996)
Musk melon	Fusarium wilt	Delicious-51 and *C. melo* var. *reticulatus, indorus, chito* and *flexuosus*	Munger (1954) and Zink *et al.* (1983)
Musk melon	Gummy stem blight	Line PI 140471	Norton and Cosper (1989)
Musk melon	CMV	Freeman	Karchi (1975)
Musk melon	WMV	PI 414723, B 66-5 and *C. metuliferus*	Webb and Bohn (1962), Webb, (1979), Provvidenti and Robinson, (1977)
Musk melon	Zuchini yellow mosaic virus	PI 161375	Lecoq and Pitrat (1985)
Watermelon	Powdery mildew, Downy mildew and Anthracnose	Arka Manik	Nath (1973)
Watermelon	Anthracnose	Black Stone, Charleston Gray and Cargo	Robinson and Shail (1975), Suvanjrakorn and Norton (1980)
Watermelon	*Fusarium* wilt	Citron, Calhoun Gray, Sornkylee and Summit, Dixielle, All Sweet, Crimson Sweet, Charleston Gray and Lousiana Queen	Orton (1911), Elmstrom and Hopkins (1981)
Pumpkin and squash	Powdery mildew	*C. moschata*	Sowell and Corley (1973)
Pumpkin and squash	Bacterial wilt	*C. pepo, C. maxima, C. andreana* and *C. lundellina*	Watterson *et al.* (1971)
Pumpkin and squash	Squash mosaic virus	*C. pepo, C. maxima* and *C. moschata*	Salama and Sill (1968)
Pumpkin and squash	WMV and CMV	*C. ecuadorensis* and *C. foetidissima*	Provvidenti *et al.* (1978)
Cucurbits	Two-spotted spider mite	Cucumber	Gould (1978)
Cucurbits	Melon aphid, fruit fly, leaf miner and red spider mite	Musk melon	Bohn *et al.* (1972), Khandewal and Nath (1978), Dhooria *et al.* (1987)

Climate-Proofing through Genomics and Biotechnology

Stress tolerance in plants is a complex trait and involves many genes (Wang *et al.*, 2003). Genome sequencing of some of the vegetable crops including cucumber, musk melon, water melon, potato, tomato, pepper has opened the way to genetic manipulation of genes associated with tolerance to environmental stresses. These tools promise more rapid, and potentially spectacular, returns but require high levels of investment. The use of molecular markers as a selection tool provides the potential for increasing the efficiency of breeding programs by reducing environmental variability, facilitating earlier selection, and reducing subsequent population sizes for field testing. Molecular marker analysis of stress tolerance in vegetables is limited but efforts are underway to identify QTLs underlying tolerance to stresses. QTLs for drought tolerance have been identified in tomato (Martin *et al.*, 1989). In plants, RNA and protein expression profiles change during stress. Approximately 130 drought-responsive genes have been identified using microarrays (Seki *et al.*, 2001; Reymond *et al.*, 2000). These genes are involved with transcription modulation, ion transport, transpiration control and carbohydrate metabolism. *DREB1A* and *CBF*, and *HSF* genes are transcription factors implicated in drought and heat response, respectively (Sung *et al.*, 2003; Sakuma *et al.*, 2002). The *ERECTA* gene regulates plant transpiration efficiency in *Arabidopsis thaliana* (Masle *et al.*, 2005), and the *NHX* and *AVP1* genes are associated with ion transport (Zhang and Blumwald 2001). There are many more genes implicated with stress response and the current challenge is to identify the ones that confer a tolerant phenotype in the crop of interest. Although the function of these genes has been elucidated, particularly in *A. thaliana*, only a few genes have contributed to a tolerant phenotype when over-expressed in vegetables (Zhang *et al.*, 2004). Expression of *AVP1*, a vacuolar H^+ pyrophosphatase from *A. thaliana*, in tomato resulted in enhanced performance under soil water deficit (Park *et al.*, 2005). The engineered tomato has a stronger, larger root system that allows the roots to make better use of limited water. The control plants suffered irreversible damage after five days without water as opposed to transgenic tomatoes which began to show water-stress damage only after 13 days but recovered completely as soon as water was supplied. The *CBF/DREB1* genes have been used successfully to engineer drought tolerance in tomato and other crops (Hsieh *et al.*, 2002).

Exploring Climate Resilient Vegetable Crops

One of the important steps towards combating the ill effects of climate is to look for those species which may adapt to climatic changes in a better way and make the dynamics of carbon sequestration faster. There is a need to diversify our food basket with new species which have the potential to combat the challenges of food and nutrition even under adverse climatic conditions. The crops for the future are to be identified and promoted in an emerging production scenario. Crops with either C-4 cycle or CAM physiology, like amaranth, Indian spinach, opuntia, portulaca *etc.*, may be promoted as they make carbon sequestration faster under high CO_2 level. Other potential vegetable crops which may be grown on marginal lands with low inputs should be promoted.

References

Abdalla, A.A. and Verderk, K. 1968. Growth, flowering and fruit set of tomato at high temperature. *The Netherlands J. Agric. Sci.*, **16**: 71–76.

Abul-Hayja, Z., Williams, P.H. and Peterson, C.E. 1978. Inheritance of resistance to anthracnose and target leaf spot in Cucumbers. *Plant. Dis. Rep.*, **62**: 43–45.

Anwar, M.R., Li, L.D., Farquharson, R., Macadam, I., Abadi, A., Finlayson, J. and Ramilan, T. 2015. Climate Change impacts on phenology and yields of five broad acre crops at four climatologically distinct locations in Australia. *Agric. Syst.*, **132**: 133–144.

AVRDC 2005. Annual Report. AVRDC – The World Vegetable Center, Shanhua, Taiwan.

Ayyogari, K., Palash, S. and Pandit, M.K. 2014. Impact of climate change on vegetable cultivation: A review. *Int. J. Agri. Env. Biotech.*, **7**(1): 145–155

Bale, J.S., Masters, I.D., Hodkinson, C., Awmack, T.M., Bezemer, V.K., Brown, J. *et al.*, 2002. Herbivory in global climate change research: Direct effects of rising temperature on insect herbivores. *Glob. Chan. Biol.*, **8**: 1–16.

Barnes, W.C. 1966. Development of multiple disease resistance hybrid cucumbers. *Proc ASHS*, **89**: 390–393.

Barnes, W.C. and Epps, W.M. 1952. Two types of anthracnose resistance in cucumbers. *Plant Dis. Rep.*, **36**: 479–480.

Bazzaz, F.A. 1990. The response of natural ecosystems to the rising global CO_2 levels. *Ann. Rev. Ecol. Syst.* **21**: 167–196.

Bell, G.D., Halpert, M.S., Schnell, R.C., Higgins, R.W., Lowrimore, J., Kousky, V.E., Tinker, R., Thiaw, W., Chelliah, M. and Artusa, A. 2000. Climate Assessment for 1999. Supplement June 2000 *Bull. Am. Meteorol. Soc.*, **81**.

Bohn, G.W. and Whitaker, T.W. 1964. Genetics of resistance to powdery mildew race 2 in muskmelon. *Phytopathology*, **54**: 587–591.

Bohn, G.W., Kishaba, A.N. and Toba, H.H. 1972. Mechanisms of resistance to melon aphid in a muskmelon line. *Hort. Sci.*, **7**: 281–282.

Boyer, J.S. 1982. Plant productivity and environment. *Sci.*, **218**: 443–448.

Bray, E.A., Bailey-Serres, J. and Weretilnyk, E. 2000. Responses to abiotic stresses. In: *Biochemistry and Molecular Biology of Plants*, (Eds.) W. Gruissem, B. Buchannan and R. Jones, ASPP, Rockville, MD, pp. 1158–1249.

Cannon, R.J.C. 2008. Annexure–1 In: *Climate-Related Transboundary Pests and Diseases*, technical background document from the expert consultation held on 25 to 27 February. FAO, Rome.

Cheeseman, J.M. 1988. Mechanisms of salinity tolerance in plants. *Plant Physiol.*, **87**: 57–550.

Choudhury, B. and Sivakami, N. 1972. Screening musk melon (*Cucumis melo* L) for breeding resistant to powdery mildew. *Third Inst. Symp. Subtrop. Trop. Hortic.*, **2**: 10.

Clark, D.A. 2004. Sources or sinks? The responses of tropical forests to current and future climate and atmospheric composition. *Phil. Trans. R. Soc. B*, **359**: 477–491.

Coakley, S.M., Scherm, H. and Chakraborty, S. 1999. Climate change and plant disease management. *Ann. Rev. Phytopath.*, **37**: 399–426.

Copeland, J. 1957. Downy mildew in musk melon. *Seed World*, **81**(9): 8.

Cuartero, J. and Fernandez-Munoz, R. 1999. Tomato and salinity. *Sciencia Horticulturae*, **78**: 83–125.

Cure, J.D. and Acock, B. 1986. Crop responses to carbon dioxide doubling: a literature survey. *Agric. For. Meteorol.*, **38**, 127–145.

Den Nij, A.P.M. 1982. CGMMV resistance in *Cucumis anguria. Cucurbit Genet. Coop. Rep.*, **5**: 57.

Dhooria, M.S., Singh, D., Shukla, B.S. and Bhathol, G.S. 1987. Differential response of watermelon accessions for resistance to red spider mite, *Tetranvchus cinnaharinus* (Boisduval). *J. Punjab Agric. Univ.*, **24**: 590–594.

Elmstrom, G.W. and Hopkins, D.L. 1981. Resistance of watermelon cultivar to fusarium wilt. *Plant Dis.*, **65**: 825–827.

FAO 2001. http://www.fao.org

FAO 2015. http://www.fao.org/docrep/017/i3138e/i3138e05.pdf

FAO (Food and Agricultural Organization of the United Nations) 2009. *State of Food Insecurity in the World: Economic Crises – Impacts and Lessons Learned*. FAO, Rome.

Flowers, T.J. 2004. Improving crop salt tolerance. *J. Exp. Bot.*, **55**: 307–319.

Foolad, M.R. 2004. Recent advances in genetics of salt tolerance in tomato. *Plant Cell Tissue Organ Cult.*, **76**: 101–119.

Gould, F. 1978. Resistance of cucumber varieties to *Tetranychus urticae*: Genetic and environmental determinants. *J. Econ. Entomol.*, **71**: 680–683.

Hamilton, J.G., Dermody, O., Aldea, M., Zangerl, A.R., Rogers, A., Berenbaum, M.R. and Delucia, E. 2005. Anthropogenic changes in tropospheric composition increase susceptibility of soybean to insect herbivory. *Env. Entomol.*, **34**: 479–485.

Harrington, R. and Woiwod, I.P. 1995. Insect crop pests and the changing climate. *Weather*, **50**: 200–208.

Hazra, P., Samsul, H.A., Sikder, D. and Peter, K.V. 2007. Breeding tomato (*Lycopersicon esculentum* Mill.) resistant to high temperature stress. *Int. J. Plant Breed.*, 1(1).

Hill, M.G. and Dymock, J.J. 1989. *Impact of Climate Change: Agriculture/Horticulture Systems.* DSIR Entomology Division Submission to the New Zealand Climate Change Programme. Department of Scientific and Industrial Research, Auckland, New Zealand.

Hsieh, T.H., Lee, J.T., Charng, Y.Y. and Chan, M.T. 2002. Tomato plants ectopically expressing *Arabidopsis* CBF1 show enhanced resistance to water deficit stress. *Plant Physiol.*, **130**: 618–626.

Idso, K.E. and Idso, S.B. 1994. Plant responses to atmospheric CO_2 enrichment in the face of environmental constraints: A review of the past 10 years' research. *Agric. Forest Meteorol.*, **69**: 153–203.

Imam, M.K. and Morkes, H. 1975. Downy mildew resistance in cucumber. *Egypt J. Genet. Cytol.*, **4**: 475–481.

Ingram, J. and Bartels, D. 1996. The molecular basis of dehydration tolerance in plants. *Ann. Rev. Pl. Physiol. Pl. Mol. Biol.*, **47**: 377–403.

IPCC 2007. *Climate Change*, Fourth Assessment Report, Cambridge University Press, London.

IPCC 2007. Appendix I: glossary. **In**: *Climate Change: Impacts, Adaptation and Vulnerability*, (Eds.) M.L. Parry, O.F. Canziani, Palutik of J.P., *et al.*, Contribution of working group II to the fourth assessment report of the intergovernmental panel on climate change. Cambridge: Cambridge University Press, pp. 79–131.

ITC (International Trade Center) 2007. *Organic Farming and Climate Change.* www.intracen.org/Organics/documents/Organic-Farming-and-Climate-Change.pdf.

ITC (International Trade Centre UNCTAD/WTO) and FiBL (Research Institute of Organic Agriculture)

2007. *Organic Farming and Climate Change*. ITC, Geneva.

Jones, R.A. 1986. The development of salt-tolerant tomatoes: Breeding strategies. *Acta Hort.*, **190**.

Karachi, Z. 1975. CMV resistance in musk melon. *Phytopath.*, **65**: 479.

Khan, F. 1973. Resistance to powdery mildew in muskmelon. **In**: Meeting of the melon working group of *Eucarpia*. *Avignon Montfavet*, 19–27 June.

Khandelwal, R.C. and Nath, P. 1978. Inheritance of resistance to fruit fly in water melon. *Canadian J. Genet. Cytol.*, **20**: 31–34.

Lebeda, A. 1984. Screening of wild Cucumis species for resistance to cucumber powdery mildew (*Erysiphe cichoracearum* and *Sphaerotheca fuliginea*). *Sci. Hortic.*, **24**: 241–249.

Lecoq, H. and Pitrat, M. 1985. Specificity of the helper component mediated aphid transmission of three potyviruses infecting musk melon. *Phytopath.*, **75**: 890–893.

Lee, Jung-Myung, Kubota, C., Tsao, S.J., Bie, Z., Hoyos Echevarria, P., Morra, L. and Oda, M. 2010. Current status of vegetable grafting: Diffusion, grafting techniques, automation. *Scientia Hortic.*, **127**: 93–105.

Luo, Y. and Mooney, H.A. (Eds.) 1999. *Carbon Dioxide and Environmental Stress*. Academic Press. New York.

Martin, B., Nienhuis, J., King, G. and Schaefer, A. 1989. Restriction fragment length polymorphisms associated with water use efficiency in tomato. *Science*, **243**: 1725–1728.

Masle, J., Gilmore, S.R. and Farquhar, G.D. 2005. The ERECTA gene regulates plant transpiration efficiency in *Arabidopsis*. *Nature*, **436**: 866–870.

Moretti, C.L., Mattos, L.M., Calbo, A.G. and Sargent, S.A. 2010. Climate changes and potential impacts on post-harvest quality of fruit and vegetable crops: A review. *Food Res. Int.*, **43**: 1824–1832.

Munger, H.M. 1954. Delicious 51, an early *Fusarium* resistant musk melon. *Farm Res.*, **2**(1): 8.

Munger, H.M. 1979. The influence of temperature on resistance to powdery mildew in cucumber. *Cucurbit Genet. Cooperative Rep.*, **2**: 9–10.

Nandpuri, K.S. 1993. Breeding musk melon for disease resistance. *PAU J. Res.*, **30**: 125.

Nath, P. 1973. Arka Manik a multiple disease resistant variety of water melon. *SABRAO Newsl.*, **5**: 71–79.

Norton, J.D. and Cosper, R.D. 1989. AC-70-154: A gummy stem blight-resistant muskmelon breeding line. *Hortic. Sci.*, **24**: 709–711.

Omara, S. 1979. Dominant genes for resistance to powdery mildew (*Sphaerotheca fuliginea* poll.) in cucumber (*Cucumis sativus* L.). *Cucumber*, Cornell University.

Orton, W.A. 1911. The development of disease resistant varieties of plant. **In**: *Fourth International Conference on Quantitative Genetics*, pp. 247–265.

Pani, R.K. 2008. Climate change hits vegetable crops. *Indian Express*. Available from: http://www.expressbuzz.com.

Park, S., Li, J., Pittman, J.K., Berkowitz, G.A., Yang, H., Undurraga, S., Morris, J., Hirschi, K.D. and Gaxiola, R.A. 2005. Up-regulation of a H-pyrophosphatase (H-PPase) as a strategy to engineer drought-resistant crop plants. *PNAS*, **102**: 18830–18835.

Patterson, D.T., Westbrook, J.K., Joyce, R.J.V., Lingren, P.D. and Rogasik, J., 1999. Weeds, insects and diseases. *Clim. Change*, **43**: 711–727.luo.

Porter, J.H., Parry, M.L. and Carter, T.R. 1991. The potential effects of climatic change on agricultural insect pests. *Agric. Forest Meteorol.*, **57**: 221–240.

Provvidenti, R. 1985. Sources of resistance to viruses in two accessions of *Cucumis sativus*. *Cucurbit Genet. Cooperative Rep.*, 8: 12.

Provvidenti, R. and Robinson, R.W. 1977. Inheritance of resistance to watermelon mosaic virus 1 in *Cucumis metuliferus*. *J. Hered.*, **68**: 56–57.

Provvidenti, R., Robinson, R.W. and Munger, H.G. 1978. Multiple virus resistance in *Cucurbita* species. *Cucurbit Genet. Coop. Rpt.*, **1**: 26–27.

Reymond, P., Weber, H., Damond, M. and Farmer, E.E. 2000. Differential gene expression in response to mechanical wounding and insect feeding in *Arabidopsis*. *Plant Cell*, **12**: 707–720.

Robinson, R.W. and Shail, J.W. 1981. A cucumber mutant with increased hypocotyl and internode length. *Cucurbit Genet. Coop. Rpt.*, **4**: 19–20.

Sakuma, Y., Liu, Q., Dubouzet, J.G., Abe, H., Shinozaki, K. and Yamaguchi-Shinozaki, K. 2002. DNA-binding specificity of the ERF/AP2 domain of *Arabidopsis* DREBs, transcription factors involved dehydration- and cold-inducible gene expression. *Biochem. Biophys. Res. Comm.*, **290**: 998–1009.

Sambandam, C.N., Parthasarathy, V.A. and Raj, S.A. 1979. Reaction of certain form of *Cucumis* sp. to Downey mildew. *Auara*, **7**(8): 149–150.

Sato, S., Peet, M.M. and Gardener, R.G. 2001. Formation of parthenocarpic fruit and aborted flowers in tomato under moderately elevated temperatures. *Scientia Hortic.*, 90: 243–254.

Sato, S., Peet, M.M. and, Thomas, J.F. 2002. Determining critical pre- and post-anthesis periods and physiological process in *Lycopersicon esculentum* Mill. Exposed to moderately elevated temperatures. *J. Exp. Bot.*, **53**: 1187–1195.

Seki, M., Narusaka, M., Abe, H., Kasuga, M., Yamaguchi-Shinozaki, K., Carninci, P., Hayashizaki, Y. and Shinozaki, K. 2001. Monitoring the expression pattern of 1300 Arabidopsis genes under drought and cold stresses by using a full-length cDNA microarray. *Plant Cell*, **13**: 61–72.

Seshadri, V.S. 1986. Cucurbits. **In**: *Vegetable Crops in India*, (Eds.) T.K. Bose and M.G. Som, Naya Prokash, Kolkata.

Singh, P.P. 1996. Screening musk melon for downy mildew resistance. *Indian Phytopath.*, **49**: 188–196.

Sowell, G. and Corley, W.L. 1969. Screening cucurbits for resistance to powdery mildew. *Proc. Ga. Acad. Sci.*, **27**: 82.

Stevens, M.A. and Rudich, J. 1978. Genetic potential for overcoming physiological limitations on adaptability, yield, and quality in tomato. *Hort. Sci.*, **13**: 673–678.

Sung, D.Y., Kaplan, F., Lee, K.J. and Guy C.L. 2003. Acquired tolerance to temperature extremes. *Trends Plant Sci.*, **8**: 179–187.

Suvanjrakorn, K. and Norton, J.D. 1980. Anthracnose resistance in watermelon. *J. Amer. Soc. Hort. Sci.*, **105**: 862.

Takeda, K.Y. and Gilbert, J.C. 1975. Inheritance of resistance to watermelon mosaic virus 2 in cucumber (*Cucumis sativus* L.). *Hort. Sci.*, 10: 319.

Von Westarp, S., Chieng, S. and Scheier 2004. A comparison between low. cost drip irrigation, conventional drip irrigation, and hand watering in Nepal. *Agric. Water Mgt.*, **64**: 143–160.

Wang, W.X., Vinocur, B. and Altman, A. 2003. Plant responses to drought, salinity and extreme temperatures: Towards genetic engineering for stress tolerance. *Planta*, 218: 1–14.

Watterson, J.C., Williams, P.H. and Durbin, R.D. 1971. Response of cucurbits to *Erwinia tracheiphila. Plant Dis. Rep.*, **55**: 816–819.

Webb, R.E. 1979. Inheritance of resistance to WMV in *Cucumis melo* L. *Hort. Sci.*, **14**: 265–266.

Webb, R.E. and Bohn, G.W. 1962. WMV resistance in musk melon. *Phytopath.*, **52**: 12–21.

Weis, E. and Berry, J.A. 1988. Plants and high temperature stress. *Soc. Expt. Biol.*, pp. 329–346.

Wheeler, T. and von Braun, J. 2013. Climate change impacts on global food security. *Sci.*, **341**: 508–513.

Whitner, B.F.J. 1960. Breeding for downy mildew resistance in musk melon. *Circ. Univ. Florida Agric. Excp. Stn.*, **5**: 122.

Wolfe, D.W. 1994. Physiological and growth responses to atmospheric CO_2 concentration. **In**: *Handbook of Plant and Crop Physiology*, (Ed.) M. Pessarakli. Marcel Dekker, New York.

Zhang, H.X. and Blumwald, E. 2001. Transgenic salt-tolerant tomato plants accumulate salt in foliage but not in fruit. *Nat. Biotech.*, **19**: 765–768.

Zhang, J.Z., Creelman, R.A. and Zhu, J.K. 2004. From laboratory to field: Using information from Arabidopsis to engineer salt, cold, and drought tolerance in crops. *Plant Physiol.*, **135**: 615–621.

Zink, F.W., Gubler, W.D., Grogan, R.G. 1983. Reaction of muskmelon germplasm to inoculation with *Fusarium oxysporum* f. sp. *melonis* race 2. *Plant Dis.*, **67**: 1251–1255.

Ziska, L.H. 2003. Evaluation of the growth response of six invasive species to past, present, and future atmospheric CO_2. *J. Exp. Bot.*, **54**(389): 395–406.

Ziska, L.H., Teasdale, J.R. and Bunce, J.A. 1999. Future atmospheric carbon dioxide may increase tolerance to glyphosate. *Weed Sci.*, **47**: 608–615.

88

Mango Production in the Emerging Climate Change for Ensuring Sustainability

H. Ravishankar

Ex-Director, ICAR-Central Institute for Sub-tropical Horticulture, Rehmankhera, Lucknow, Uttar Pradesh Principal Scientist (Horticulture) and in charge, IIHR-ATIC, ICAR-Indian Institute of Horticultural Research, Bengaluru – 560 089, Karnataka E-mail: ravishankar@iihr.res.in, drhravishankar@gmail.com

Mango (*Mangifera indica* L.) is one of the choicest and widely cultivated fruit crops of tropical and sub-tropical regions of the world. Mango the *'king of fruits'*; is believed to have its origin in south-east Asia most probably in North-East India, north western Myanmar, and Bangladesh, from these mentioned gene centers the mango later spread all around the Asia and other parts of the world. India, where it is acclaimed as the *'national fruit'* is the largest producer in the world, contributing 40.48 per cent of the total world mango production (Anonymous, 2015). Uttar Pradesh, Gujarat, Karnataka, Andhra Pradesh and Bihar are the major producing states and during 2014-15 the crop covered an area of about 2.22 million ha producing 18.50 million MT. Mango has been under cultivation in India for about 4,000 years and over 1,200 varieties are said to exist in the country. The cultivated mango varieties in India, over years have co-evolved in different agro-ecologies of the country, exhibiting an unusual diversity of fruit forms, flavours and tastes (Mukherjee, 1948) and have acclaimed regional acceptance, preferences and are significantly contributing to the local economies except a few *viz., Alphonso, Kesar, Banaganapalli etc.* that have made their presence in the export trade too. Some of the cultivars have acquired the Geographical Indication (GI) tag also. Names of the mango cultivars registered under GI include *Appemidi* in Karnataka, *Laxman Bhog, Khirsapati* (*Himsagar*) and *Fazli* in West Bengal, and *Malihabadi Dusseheri* in Uttar Pradesh, and *Gir Kesar* in Gujarat state. Northern, eastern, western and southern regions of India are recognized as distinct mango growing regions with choice varieties, dominating the production scene though some cultivars are grown over wider areas. Mango is well adapted between 25°N and 25°S of the equator and from sea level up to the height of 1200 m, but cannot be grown commercially in areas above 600 m. A rainfall of about 125 cm, received mostly during monsoon, is considered very suitable. The favorable temperature for mango ranges between 24-27°C during the growing season along with high humidity. It can be grown in a wide range of soils from alluvial to lateritic soils however, loamy, alluvial, and sandy loams are found ideal. Mango production and associated economic activities are providing employment to a sizeable population in these regions and helping in providing sustainable livelihoods to the local population by generating employment and incomes. Mango production dynamics in different parts of the country in the recent years is gripped with considerable uncertainties arising due to occurrence of weather extremes, inadequacies of water, and nutrient management and impacts of altered pests and diseases prevalence as witnessed in the recent past in Maharashtra, Gujarat, Odisha, West Bengal and some southern states. Notable among them include, recurrent flowering, fruit size grades heterogeneity at harvest, reduced edible quality and marketability, incidence of blight and other disorders (in Alphonso), jelly seed formation in

Dashehari and impaired post-harvest behavior *etc.*, The Inter Governmental Panel on Climate Change (IPCC) has projected a temperature rise of 1-3°C by 2020 which may adversely impact agriculture by reducing crop yields up to 10 per cent for Asia (IPCC, 2007). The major trend of weather extremes likely to impact the mango crop is temperature, a crucial environmental cue with projected increases from the present to 2°C to 4°C through 2030-2099 periods for South Asia. This is also expected to be accompanied by the CO_2 fluxes, a key component of the atmospheric greenhouse gases (GHG), which is also witnessing progressive trends. Mango production sustainability obviously, is expected to be seriously challenged by these climate change paradigms if, global warming continues on its current trajectory. CO_2 is the most important greenhouse gas for its role in the current climate change. Its atmospheric concentration shows a dramatic increase since the middle of the 19th century, from 278 ppm before the industrial era up to 391 ppm in September 2012 (World Bank, 2012). This increase is mainly caused by anthropogenic CO_2 due to burning of fossil fuels and deforestation. The present paper based on the current knowledge of mango tree functioning and the factors affecting it, appraises the future of mango production scenario as impacted/affected by emerging weather dynamics through influencing photosynthesis, phenophases, yield and quality in the Indian context by taking in to account the climate change paradigms projected by IPCC (2007) and the consequences of the emerging impacts indicated by other researchers (Normand *et al.*, 2013) elsewhere especially for South Asia.

Features of Agro-climate and Soils of Major Mango Producing States of India

Mango being a tropical fruit, but can be grown up to 1,100 m above mean sea level. The trees can tolerate a wide range of climatic conditions. The crop is found successfully cultivated under conditions which vary from very humid to cool and dry, to very hot and arid. The trees can survive in swampy conditions for an extended period of time, but will also survive in areas with an annual rainfall of less than 300 mm and temperatures as high as 45°C. The average minimum temperature during the winter should preferably be above 5°C. There should not be high humidity, rain or frost during flowering. The temperature between 24°C and 27°C is ideal for its cultivation. Low temperatures when the trees are in full bloom can cause the fruit to develop to approximately golf-ball size, turn yellow and then be aborted. High temperatures and low-humidity conditions found to result in low carbon accumulation, which lowers the tree's ability to hold heavy crop loads. Low temperature stress is necessary in sub-tropical areas while a period of water stress is found essential in the tropics for floral induction in mango.

Higher temperature during fruit development and maturity gives better quality fruits. The areas

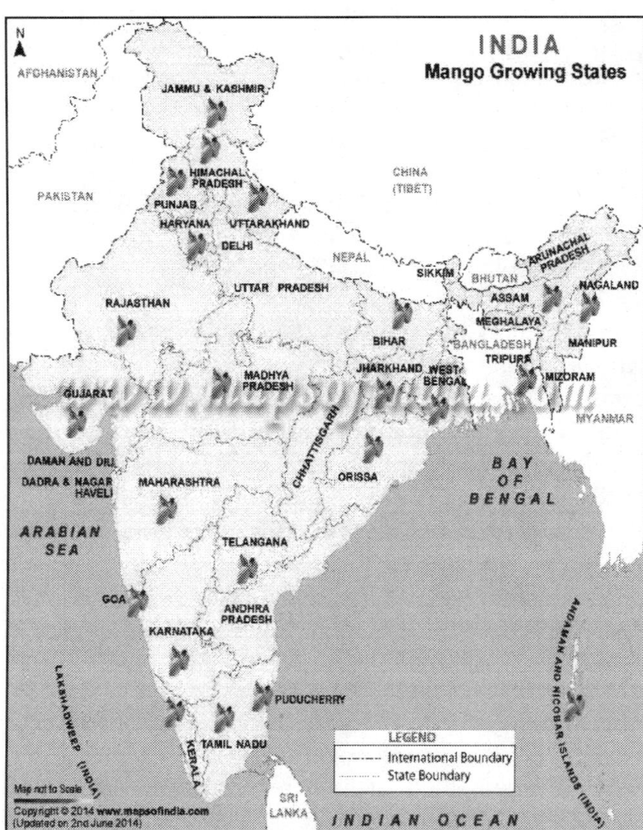

Figure 1: Mango Map of India (2014 *www. maps india.com*).

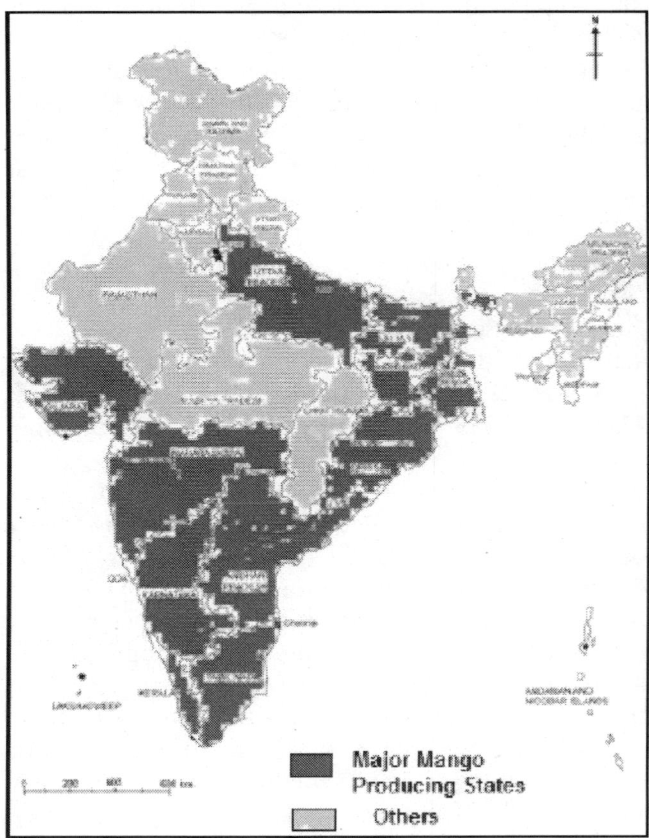

Figure 2: Major Mango Producing States (Red marked).

experiencing frequent showers and high humidity are prone to many pests and diseases. Thus it can be grown best in regions with a rainfall between 25 cm and 250cm. Rain or free moisture (high humidity, heavy dew, and fog) during the flowering and fruiting period is conducive to the development of fungal diseases that cause flower and fruit drop. Regions having bright sunny days and moderate humidity during flowering are ideal for mango growing. For maximum fruit production, a prolonged and severe dry season is necessary. The longer and more severe the dry season, the more regular is the cropping habit. Young trees will be killed outright by the mildest frosts, but larger mature trees can regenerate, although defoliation and death of some terminal buds and branches will occur as was seen in parts of Uttar Pradesh and Uttarakhand in 2012 fruiting season.

Soils

Mango in the Indian sub-continent is grown on a wide variety of soils under varied climatic conditions. It is being grown from alluvial to lateritic soils except in black cotton soil having poor drainage. It is found growing well in soils with slightly acidic pH and found not performing well in soils having pH beyond 7.5. Ideal pH range however, is from 6.5 to 7.5 but the tree is found growing outside this range, with low pH (acid) and high pH (>7.5) being the most deleterious to growth. Soils having good drainage are found ideal for mango. Ideal soil for mango is red loamy. The plant has a strong vigorous rooting system which will penetrate a large area to obtain nutrients. Production is found best on well drained sandy or gravelly soils that dry out rapidly after the wet season, forcing the trees into a dormant period, essential for heavy flowering. Mangoes do not grow or produce well in saline soils, but the rootstock variety ***'13-1'*** from Israel reportedly has some salt tolerance and has recently been planted in some parts of Gujarat.

Phenophases

Mango trees predominantly present three types of growth patterns, *viz.*, flushing or emergence of new vegetative growth (leaves upon exposure to solar energy produce carbohydrates by photosynthesis to support fruit development in the ensuing season); dormancy or quiescence (tree carbohydrate reserves levels increase as against tree N level. Thus C: N ratio of the trees increase); flowering (mature shoots upon becoming physiologically active become floriferous under inductive conditions), fruit set and development. Mango trees put forth growth in a discontinuous manner of course depending upon the variety and *'On'* and *'Off'* phases of the tree; the tree size increments accrue from the dynamics of vegetative flushes production that arise periodically. The vegetative growth restriction precludes flowering, since fruit bud differentiation event unless and otherwise not adversely impacted by factors *viz.*, unseasonal rainfall (both in tropics and subtropics) and disturbances to cold (in subtropics); incidence of frost and hailstorms determine the fate of a particular meristem in favour of vegetative phenophase. Temperatures between 20 to 30°C favor vegetative shoot development while, temperatures between 6 to 18°C incenti*vi*ze the initiation of the inflorescence development (Ravishankar *et al.*, 1979 and Davenport, 2007). The development of the terminal inflorescence occurs after a cold period associated to a degree of drought stress. In locations, where winter is not cold enough for an adequate flowering induction (minimum daily T° above 15°C), it is common to produce this induction by exposing trees to soil moisture deficiencies (drought stress) before flowering. A number of instances where the occurrence of these factors has even caused reversal of the already determined meristem fate of floral status in to vegetative growth instead of reproductive organs developing have been recorded. Further, in mango there exist growth correlations between vegetative growth, flowering intensity and the crop load.

Mango growth and quality traits are affected by pre-harvest factors such as light, temperature and carbon and water availability (Léchaudel and Joas, 2007). Boudon *et al.* (2012) described structural sub-model and functional sub-models for mango for cv. Cogshall in the context of weather dynamics with the former taking into account carbon allocation to branches, old wood, coarse roots and fine roots while the latter addressed the issues related to tree phenology and architectural development. Dambreville *et al.* (2013) observed that phenological asynchronisms in mango depended upon the interplays between structural and temporal components of mango architectural development.

Meteorological observations of 2012-13 at CISH, Rehmankhera indicated that fruit development stages were vulnerable under high pan evaporation regimes (>6 to 10 mm/day) particularly during the last week of March and onwards. High pan evaporation values indicated higher rate of evapo-transpiration resulting into drier soils with soil moisture stress developing silently within the soil and leading to abiotic stress complexes barring water and nutrient flow through xylem. Thus protective irrigations are required as an intervention strategy so as to maintain optimum soil moisture and leaf water potential in order to sustain fruit set, growth and development and minimizing pre-harvest fruit drop unless otherwise this period is intervened by unseasonal rains. Reports have indicated that mango is drought tolerant, due to its ability to maintain turgor through osmotic adjustment attributable to the existence of lacticifer system. Mango, on the other hand, exhibit some degree of flood tolerance though production of hypertrophied lenticels that facilitate increased O_2 absorption besides offering sites for excretion of toxic by-products of anaerobic metabolism of roots (Bruce Schaffer *et al.*, 1994). Perhaps flooding tolerance of mango trees in north Bihar during periods of flood could be explained to the above mentioned phenomenon. Precise understanding of these features however, is constrained by the lack of quantitative data on water relations in respect of mango tree physiology on real time scale.

The vulnerability of mango to weather dynamics also emerged at some other important mango growing parts of the country. A recurrent flowering phenomenon, which has become common in the Konkan and Bengaluru regions for the past four years appears to be a consequence of weather impacts which needs micro-analysis with at least the Konkan case coming under the influence of western disturbances occurring over the Arabian sea. Meteorological data analysis at Bengaluru indicated that T_{min} ranged between 13 to 16°C, on majority of the days during January 2013 being 14-15°C, a critical period for flowering. In the first fortnight of January 2013, T_{min} ranged between 10-16°C. Hence during January 2013, attributable to this variation, flowering occurred in more than one flush, termed 'recurrent flowering'. This kind of phenomenon was also reported in '*Kesar*' from Gujarat (Personal communication, 2013) which has led to the occurrence of different stages of fruit growth and development stages on the same tree, consequently multiple harvests of varying quality may ensue affecting markets. During February 2013, the T_{max} was 28.7°C and the T_{min} was 17.4°C with the RH of 73 per cent. The data recorded for two fortnights revealed that there was very little difference in T_{max} and T_{min}. However, large variations in the T_{min} emerged during February 2013. On some days, the T_{min} was 21°C and after a gap of 6 to 7 days, the T*min* fell to 14°C. Such variations in T_{min} could greatly affect the pollinator dynamics and their activity as was discernible during 2013 fruiting season at Lucknow and adjoining regions. This kind of T_{min} dynamics also probably triggered a second flush of flowering during 2013 at Bengaluru region, infusing disturbed source-sink relationships as fruits of different growth and developmental stages became mutually competitive that could probably impact both quantity as well as quality of harvests. February 2013, however, witnessed T_{min} fluctuating between 13 to 15°C without much change.

It has been reported that the mango production dynamics of the Konkan region is getting influenced by the global warming phenomenon leading to heating of the waters of the Pacific Ocean during the current decade although the exact dimension of the scenario remains to be captured. The cursory analysis of the data (Ravishankar *et al.*, 2011 and 2014) on mango production of the region indicated that, new vegetative flushes emerged in the second week of November (2011-2012). The first flowering flush to the tune of about 5 per cent occurred in the third week of December even though there were extended rains up to 29th November, 2011 in continuation to the normal rainfall in the monsoon. The second flowering flush was noticed during January second week which was to the tune of about 30 per cent. The total flowering up to March, 2012 in '*Alphonso*' was about 48 per cent and in other varieties it was about 25 per cent. The late showers received up to November end in that year affected the fruit set of the first flowering flush as the flowering got delayed despite hermaphrodite flowers percentage in the first flowering flush being normal up to 9.8 per cent; the panicles emerging during the second flowering flush were predominantly of mixed panicles nature having extra length (size-41 cm length) that contributed to low fruit set as was seen at Vengurle, Mulde, Deogad and Lanja in about 10.4 per cent plants in both the cases there was failure to set any fruits. Since 2009, unpredictable weather conditions have been occurring in the premier mango growing belts of the country every year, like incidence of unseasonal rains (up to Dec. 23; March-May); cloudy weather, severe cold (8 to 12°C), and extremes of T_{min} continuously for 60 to 70 days, (instead of 4 to 5 spells of 12 to 15 days cold periods) and unfavorable T_{max} prevailing during fruit growth, development and maturity. These aspects have come to sharp focus from 2009 onwards especially the Konkan region of Maharashtra. The normal rainfall period of this region is June to September with 1 to 2 dry spells of 8 to 10 days period, with mean annual rainfall of 3000 mm received over an average of 105 days. However, deviations to this normal showed, the extended rainfall (15 to 42 days) with increased intensity (122 mm to 1500mm). This situation apparently lead to unavailability of sufficient physiological stress conditions, very much crucial for satisfactory induction of flowering in '*Alphonso*' under tropical humid conditions, thereby severely impacting flowering adversely.

Normally, rainfall ceases in the Konkan region by September end followed by onset of winter from November that continues up to February. However, due to delayed and extended rains, this normal weather cycle is found disturbed and the much needed physiological stress conditions preceding fruit bud differentiation were lacking and as a consequence, uncontrolled vegetative flushes found occurring, thus jeopardizing flowering prospects. Agro-climate, further impacted by the extended cold periods (>122 days of <17°C in 2011-12), '*Alphonso*' manifested three reproductive flushes in the fruiting season despite favorable weather conditions prevailing subsequently. Similarly, continuance of severe cold for extra long periods (62 to 74 days) also prevailed during 2009 and 2010 fruiting seasons that resulted in altered phenophases dynamics in '*Alphon*so'. As a result, the inflorescences emerging during this period were found abnormal like extra long flowering panicles (42 to 45 cm); instead of the normal (29 to 32 cm), mixed panicles and changed sex ratio. During 2010-2011, severe cold period(<17°C) continued for 72 days (instead of 4 to 5 spells of 15 days period) which led to the development of extra long inflorescences with altered sex ratio (4 to 5 per cent hermaphrodite flowers) characterized by preponderance of male flowers (96 per cent) adversely impacting fruit set; the reduced number and activity of pollinators further compounding the problem. Majority of the inflorescences of this period failed to set fruits due to inadequate pollination and hence subsequently dried up. Intermittent flowering phenomenon in the peninsular region in the recent years since 2008 is posing serious problems. During January 2013, T_{min} of the region, ranged between 13 to 16°C, but on majority of the days it was between 14 to 15°C. In the first fortnight of January 2013, T_{min} recorded was between 11 to 18°C and 10 to 16°C. Hence, during 2013 fruiting season, the flowering flushes occurred more than twice attributable to this variation. During February 2013, the T_{max} was 28.7°C and the T_{min} was 17.4°C and the RH was 73 per cent. The data recorded for two fortnights revealed very little difference between T_{max} and T_{min}. However, large variations in the T_{min} were seen during February. On some days, the T_{min} was 21°C and in some days after a gap of 6 to 7 days, the same declined to 14°C. This probably triggered a wave of second flush of flowering in 2013 fruiting season (*Personal communication*, 2013). However, during February 2012, the minimum temperature always had hovered between 13 to 15°C without much change. Similar observations of *'recurrent flowering'* phenomenon and occurrence of fruits of different developmental stages on the same tree had also been reported from Gujarat and Odisha (Personal communication, 2013).

According to Normand *et al.* (2013), the expected changes in climate variables though are global; they will be with regional and seasonal variability. The major trend of weather extremes likely to impact the mango crop is temperature, a crucial environmental cue with projected increases from the present to 2°C to 4°C through 2030-2099 periods for South Asia. This is also expected to be accompanied by the CO_2 fluxes, a key component of the atmospheric greenhouse gases (GHG) which is also witnessing progressive trends. Mango production sustainability obviously, is expected to be seriously challenged by these climate change paradigms if, global warming continues on its current trajectory. The consequences of emerging weather dynamics experienced by mango production scenario in the country may broadly include, occurrence of warmer winters, wetter climates, more extremely warm and wet seasons, higher frequency and intensity of tropical cyclones. A cursory analysis of probable impacts of these weather paradigms may result in phenological asynchronysms, reduced floral induction, good fruit growth but reduced fruit quality and increased pests and diseases/disorders problems as a result of hot and wet seasons. Perceived climate changes with their potential to severely impact phenophases, production, and fruit quality certainly will have serious consequences on the prospects of mango cultivation in the sub-continent. These impacts are expected to be mediated through disturbances in such key physiological processes as photosynthesis, stomata conductance and fruit growth and quality build-up as evidenced by the thermal time models: relationships between temperature and development events deduced by some researchers elsewhere. Increased projected global warming phenomenon would consequentially impact mango production through increases in temperature, light, CO_2, vapor pressure deficit (VPD), drought and flooding (áâ) which adversely impacts photosynthesis. Extreme temperatures (> 45°C) and high levels of light intensity could possibly damage the photosynthesis machinery in mango. It is also possible, that these effects are counterbalanced by increased photosynthetic assimilation and increased respiratory losses. Some studies in mango (Normand *et al.*, 2013), elsewhere have indicated that increases in temperature will lead to more rapid growth rhythm, no growth cessation (not very conducive to mango tree vegetative phenology/physiological system where, growth is cyclic in general while, regular bearing cultivars could in all probability defy this rule,if otherwise offset by certain environmental cues), reduced fruit bud differentiation (FBD), higher atmospheric CO_2 concentration resulting in increased dry matter allocation to roots, on the contrary, drought and flooding could lead to reduced tree growth and even death. Focused studies on different parameters of photosynthetic aptitude of leaves of *'flowering'* and *'non-flowering'* terminals *viz.*, photosynthetic rate (P_n), stomata conductance (g_s), internal CO_2 (C_i) concentration, stomata behavior, chlorophyll fluorescence (F_v/F_m) and electron transport rate (*ETR*) in leaves of important commercial cultivars implied that photosynthetic aptitude of different kind of shoots in *'Dashehari'* mango perhaps may not be the sole factor deciding the status of meristems' fate *per se* in any given year but governed by other internal regulatory mechanism as well (Singh and Ravishankar, 20014).

Likely Consequences of Emerging Weather Dynamics on Key Weather Components on Mango

a) Light and Temperature

The floral induction of mango tree is mainly driven by cool temperatures as is being encountered in sub-tropical situations of northern plains of the country. Consequently, under these situations, increasing temperatures would have a negative effect on floral induction. But in regions with particularly cool temperature during flowering, increasing temperatures would have a positive effect on pollen viability and fruit set. Temperature also has a negative effect on inflorescence size (Dambreville *et al.*, 2013) and on the number of flowers per inflorescence (Sukhvibul *et al.*, 1999). Both light and temperature are important factors for photosynthesis besides CO_2 concentration. Increase in the levels of these factors would have a positive effect on photosynthesis. However some studies have indicated that, higher temperatures and CO_2 concentrations have counter balancing effects because they increase both photosynthetic assimilation and respiratory losses. Extreme temperatures (> 45°C) or high levels of light intensity provoke damages to the photosynthesis machinery. The warmer climates during flowering and the warmer and wetter climates during the season of vegetative rest will probably lead to a lower floral induction. On the opposite, the hot and wet climates during fruit growth and vegetative growth will promote good fruit growth and important vegetative growth after harvest. But fruit quality could be reduced and pests and diseases problems and other disorders could increase. Provided nutrient and water are not limiting, the increasing temperature will lead to more rapid growth rhythms (shorter time-lag between successive vegetative flushes) and development of growth units and leaves and to the delay or the avoidance of growth cessation at the beginning of winter. The effect of high temperature on growth unit and leaf size is not clear. Whiley *et al.* (1989) showed that the mean leaf size of trees grown at 30/25°C (day/night temperature) was 300 per cent larger than that of trees growing at 20/15°C. On the other hand, Dambreville *et al.* (2013) did not observe a significant effect of temperature on leaf size for leaves which developed in the mean daily temperature range of 20-28°C. A cultivar effect, observed by Whiley *et al.* (1989), possibly could explain these different results.

The experience of the author at Lucknow and earlier indicated that diurnal variations in temperature could play important role, especially the T_{min} that appeared crucial not only for flowering but also for the fruit set problems *viz.*, nubbin fruit formation. Recent reports of Indian Institute of Tropical Meteorology (IITM) and the World Meteorological Organization in Geneva and the publication in the *American Journal of Climate Change* appears significant for mango production too that points out to significant temperature rise (ranging from 0.21-0.38°C) across some metropolis from 1971-2013. Far more significant analysis in this regard from mango floral induction point is the increasing trends in T_{min} after 1971 which underscores the possible vulnerability and the authors have attributed this phenomenon to such factors as solar radiation and aerosol concentrations.

Temperature, light and CO_2 concentration are important factors affecting photosynthesis, and an expected increase of the level of these factors would have a positive effect on photosynthesis. But higher temperature and CO_2 concentration have counterbalanced effects because they increase both photosynthetic assimilation and respiratory losses. Extreme temperatures (> 45°C) or high levels of light intensity provoke damages to the photosynthesis machinery.

The prediction of the effects of climate change on mango production and cultivation requires therefore a complete mango crop model. Several partial models, simulating specific processes of the mango tree, are available: a biochemical model of photosynthesis (Urban *et al.*, 2003), a model of stomata conductance (Damour *et al.*, 2010), a model of fruit growth and quality build-up (Léchaudel *et al.*, 2005, 2007) and thermal time models (Mosqueda-Vazquez *et al.*, 1993; Normand and Léchaudel, 2006; Dambreville *et al.*, 2013). But there is no complete mango crop as of now available. Crop models have been mainly developed for annual crops. Few models at the whole-plant scale that have existed for temperate fruit trees such as apple (Costes *et al.*, 2008) or peach (Lopez *et al.*, 2010; Miras-Avalos *et al.*, 2011) could facilitate some understanding of impacts of climate variations. They are however, found not always oriented toward fruit production and there is a lack of such models for tropical fruit trees. Consequently, many researchers have averred that it is not currently possible to assess the effects of climate change paradigms on the cultivation and production of mango because the partial models available are felt not sufficient for this purpose. They however, have opined that it may be possible to assess the response of the mango trees to the predicted future climate paradigms based on the current knowledge of the mango tree functioning and of the factors affecting it. Information on the influence of climatic factors on four key processes of mango production: photosynthesis, vegetative development, flowering and fruit set, fruit growth and quality are available (Davenport, 2009; Schaffer *et al.*, 2009; Crane *et al.*, 2009). Although mango tree is adapted to dry environments experiencing water stress and high evaporative demands, studies have indicated that the projected increase of drought and vapor pressure deficit (VPD) under emerging climate change paradigms would have negative effects on photosynthesis because of the rapid stomata closure.

b) CO_2 Concentration

The increasing concentration of atmospheric CO_2, which is a main driving force of climate change, has to be considered for appraising consequences of climate change on mango production too, because CO_2 is involved in such key process, as photosynthesis besides impacting temperature dynamics. CO_2 concentration is important factor for photosynthesis and likely to have a positive effect on photosynthesis although precise details about the carbon acquisition, assimilation and carbon foot prints of an adult mango orchard is presently unavailable. Empirical studies have indicated that probably more dry matter would be allocated to the roots under increasing CO_2 concentrations to some limits.

c) Precipitation/Water Management

In the tropics of India, floral induction of mango tree is mainly driven by stress conditions during the period preceding fruit bud differentiation (Ravishankar, 1977) and any intervening precipitation would predispose growth units (GUs) to vegetative phenology. The type of irrigation adopted for mangoes, hence is an important consideration if flowering management is desired. Most growers in the tropics, who have irrigation to provide water to the trees through the typical six-month annual dry season, use furrow irrigation along the tree rows (*Personal observation*). In-row furrow irrigation has the disadvantage of providing water periodically around the base of trees. The major problem is that many roots outside the limits of the irrigation ditches never get watered during the dry season; hence, water moves from roots located in or near the irrigation ditches to not only the canopy but also out to the dry roots in response to water potential gradients in the root system (Boyer, 1985; Canny, 1995; Passioura, 1988). This backwards xylem flow towards root tips prevents upward xylem movement of shoot initiating hormones (cytokinins) that are synthesized in the root tips (Skene, 1975; Mok, 1994) and the cytokinins, instead, accumulate in the tips of the roots. Active water uptake by these roots when the first rains arrive move the accumulated cytokinins to stem buds in the canopy, thus providing the stimulus to initiate an undesired flush of vegetative growth in insufficiently physiologically mature stems, when the first rains of the rainy season arrived(Davenport, 2008). In some areas, the first rains of the rainy season stimulated flowering in those stems that had attained sufficient time in rest due to lack of flushes during the dry season.

As a consequence of the emerging climate change paradigms projected (IPCC, 2007), precipitation will in general, decrease in the tropical regions, and increase in the equatorial and temperate regions. Soil moisture will also decrease in the tropics, leading to more severe drought scenarios. As a consequence, the vapor pressure deficit (*VPD*) will increase. The tropics and subtropics will undergo more frequent scorching heat events and more frequent and intense precipitation or droughts. More frequent and intense hurricanes are also expected. Drought and flooding have both negative effects on the vegetative development of mango trees as they potentially reduce tree growth. If mango trees can survive without rain or irrigation for more than eight months, flooded trees could die within few days. Flooding is expected to increase in regions with more intense and frequent rainfalls, and decrease in regions becoming drier. In both the cases, the effect on mango tree photosynthesis would probably be negative. Mango is moderately flood-tolerant as could be seen in the flood-prone areas of Bihar during 2013 and flooding leads to rapid decrease of transpiration, stomata conductance and photosynthate assimilation. The decrease of flooding occurrence however corresponds to the increase of drought.

Besides the above, unseasonal rains especially at the time of fruit bud differentiation (FBD) could transform the already differentiated shoot meristems in favor of reproductive ones to vegetative phenology, attributable to their unfavorable intervention in the requirements of drought stress in the tropics while, low temperature in the subtropics; occurrence of frost (-0.2° to -1.2°C on 9th and 10th January, 2013) as it happened at CISH, Lucknow however, virtually killed portions of the shoots and triggered the emergence of new vegetative flushes from the lateral buds subtending the dead portion of shoots, thus contributing to transgression of tree physiology from an expected reproductive one to vegetative predominantly. Perhaps, this was also compounded by the occurrence of hailstorm on 18th and 19th January, 2013. Such a situation altered the dynamics of thrips incidence in several cases, resurgence on new vegetative flushes despite spraying of thiomethoxam (0.02 per cent) and in some instances, damages to the fruits at pea stage of development. Temperature dynamics together with soil moisture conditions (due to unseasonal rainfall that occurred on 18th and 19th January, 2013) that prevailed during 2013 witnessed however, below threshold levels incidence of mealy bugs, hoppers and powdery mildew.

d) Relative Humidity

The mean temperature of the Earth's surface is expected to rise by 2-4°C by the end of the century (IPCC, 2007), this increase is projected to be larger at the high Northern latitudes (up to more than 4°C) than in the tropics (1-3°C). South-Asia region that extends from 5 to 30° North latitude and from 64 to 100° East longitude and includes India, the leading mango producing country, and Sri Lanka. The predicted climate is for the end of the 21st century, with respect to the mean climate of the last 20 years of the 20th century, Normand *et al.* (2013) observed that it will be warmer and wetter, with a mean annual increase of 3.3°C for temperature and of 11 per cent for precipitation. The temperature and precipitation increase will not be uniform among seasons, with the largest temperature increase during the flowering season (+3.6°C) and the lowest during the harvest and vegetative growth season (+2.7°C). According to them, the precipitation is expected to be lower during flowering (-5 per cent) and higher during the other seasons, in particular during the end of vegetative growth and vegetative rest (+15 per cent). The frequency of extremely warm season and extremely wet season are reported to be high, in particular for warm season and it is expected that there would be more frequent and intense hurricanes.

Possible Effects of Predicted Weather Dynamics on Mango Production

a) Mango Phenological Stages

i) Vegetative

The optimum temperature for mango tree growth has been reported to be 24-30°C (Bruce Schaffer *et al.*, 1994); though it can tolerate air temperatures up to 48°C, cold temperatures influence crop production. Sub-zero temperatures for even a single day or few hours severely damage or even kill the trees as could be seen during the 2013 fruiting season in Lucknow (portion of twig death) and Dehradun region of Uttarakhand where nearly 10,000 trees have been reportedly found dead. Trees manifest strong vegetative bias with increasing temperatures provided water and nutrients are not limiting (Whiley *et al.*, 1989). They observed that the median (mean of the maximum and minimum daily temperatures) daily temperature for zero shoot growth of ten mango cultivars to be 15°C which, was subsequently confirmed to be critical for critical daily mean temperature for vegetative growth. Soil temperatures reportedly have a strong control on synchronization of growth phases in the tree. The same authors found that periodic shoot growth, changing between phase of activity and dormancy, occurred when soil temperatures were held at 27°C or 32°C for 120 days but at 21°C an extended dormant period developed. Temperature also influenced the number and size of leaves. An average of 7.1 leaves that developed on shoots at 20/15°C increased to 13.6 at 30/25°C. Dry matter partitioning to roots was the greatest in trees growing at 15/10°C *i.e.*, temperatures that suppressed shoot growth. With increase in temperatures, partitioning to roots declined but correspondingly increased for stem and leaves. In *'Irwin'*, there was a greater allocation of dry matter to roots at higher temperatures of 25/20°C and 30/25°C, a reflection of its reduced flushing and semi-dwarf architecture.

ii) Reproductive

Studies under controlled conditions have shown, of the different weather variables, temperature exerts an important influence on different facets of reproductive phenology starting right from FBD (Ravishankar, 1977), and flowering (Bruce Schaffer *et al.*, 1994). They opined that mango gets affected when temperature falls below 0°C, hence low temperature factor more than any others defines the range of mango cultivation. A period of cold stress generally precedes flowering, which occurs following stress relief, and temperature, particularly in the subtropics provides the strong environmental stimulus for flower induction, the threshold being cultivar-specific. On the contrary, dry period preceding flowering appears necessary for satisfactory flowering in the tropics highlighting the role of drought stress to be the key environmental cue. Studies of Pongsomboon (1991) showed that no flowering occurred in trees held at 30/20°C at either soil temperatures, while at 15/10°C air or 12.5°C soil temperature, all the trees had flowered after 16 weeks, but only 40 per cent trees flowered when soil temperatures were held at 25°C. These results suggested that *'root signals'* probably played significant role in induction of flowering in mango. Similar indications emerging from the work of Ravishankar (1987) on *'Alphonso'* (distinct alternate bearer) and *'Totapuri'* (a regular bearer) strengthens this view point. Export of cytokinins from roots impacting reproductive morphogenetic responses of shoots under the influence of temperature cannot be ruled out as high temperatures provided increased vegetative bias in mango (Whiley *et al.*, 1989). The fact that foliar gibberellins applications suppressed flowering in mango despite trees being grown under inductive conditions (Kachru *et al.*, 1971) emphasized that interpolations of promoters and inhibitors under the influence of high soil temperatures perhaps played crucial role in coordinating shoot-root communications (Ravishankar,1987) ultimately determining the flowering responses in mango which however, needs further investigation as major focus of research so far in mango converged on signals and responses generated in shoot apical meristems (SAMs) largely (Ravishankar,1987 and Chacko, 1991).

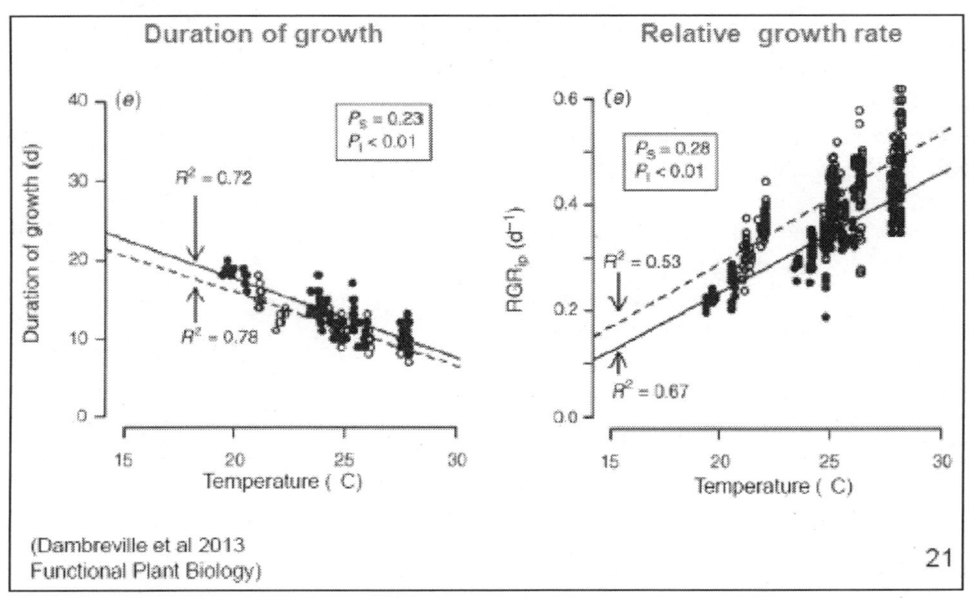

Figure 3: Effect of Temperature on the Duration of Growth and on the Relative Growth Rate of Leaves of Two Mango Cultivars.

Negative Effect of Temperature on Inflorescence Size and Number of Flowers

Studies have shown, T_{min} (<15°C) during the development of inflorescences resulted in deformation of flowers which were found smaller as compared to those that developed at about 7°C higher (Issarakraisila *et al.*, 1992) and these abnormal flowers at such low temperatures had small ovary, dark and reduced style length, dark ovule and black anther instead of the usual red one; pistil abnormalities also ensued. Issarakraisila and Considine (1994) noted, although mature pollen was tolerant to low temperatures even at 10°C, this temperature being critical during meiosis, below which the pollen viability falls to about 50 per cent. They further observed prevalence of high temperatures of 36°C and above during the pre-vacuolated-meiosis stage of microsporogenesis of pollen, drastically reduced the pollen viability. Many tropical and sub-tropical species respond to low temperature by initiating reproductive development. The temperature range is higher, often around 10–15°C, and is sometimes referred to a cool temperature response (Wilkie *et al.*, 2008). At present, because little detail exists on the molecular basis of tropical species responses, it is not known to what extent, the mechanisms are similar to vernalization. However, there is accumulating evidence that at least some components of cool temperature responses involve perception outside the SAMs, *i.e.*, in leaves, and thus, include a systemic signaling component that awaits further elucidation. A clear demonstration of the importance of leaves comes from the cool temperature induction of mango (Davenport *et al.*, 2006 and Ramirez *et al.*, 2010) where, defoliation of girdled shoots prevented flowering but as little as a single leaf on one-meter stem was found sufficient to result in substantial floral induction. It will be interesting to ascertain whether *FT* mobilization from those leaves is the causative signal, and whether altered photosynthate transport in defoliated plants had a signicant effect.

Higher amounts of *Pn* with concomitantly high leaf water potential during periods of rank vegetative growth (March and September-November) coincided with periods characterized by the prevalence of correspondingly high leaf water potential (Singh and Ravishankar, 2013) however, a period of stress precludes this phenomenon and under sub-tropical conditions, this is provided by cold. This stage could turn out to be vulnerable, if intervened by heavy rainfall so that the reproductive primordia could revert to the vegetative one. Such an unfavorable event did not occur during the study period (2012-13). Decrease in *Pn* in leaves during flowering is attributable to the decrease in stomata conductance. Reduction in this parameter could also be due to differences in the leaf water state of shoots. Thus it is possible that flowering results due to local water stress. Increase in *Pn* during fruit set could be attributed to higher sink activity demanding increased mobilization of photosynthate to the developing fruit lets. Based on the pattern of flushing, the maximum flushes were recorded during March in both the years.

Table 1: Competence of Important Phenophases in Respect of Key Vulnerable Physiological Attributes in '*Dashehari*' to Weather Dynamics (Singh and Ravishankar, 2013)

Type of Shoot/Stage	*Rate of Leaf Water Potential (ψw)*	*Photosynthesis (Pn) (μ mol m^{-2} s^{-1})*
Vegetative growth	9.25	–7.36
Reproductive growth	–	–
a) FBD	7.61	–7.95
b) Flowering	6.25	–8.46
c) Fruit set	7.27	–6.50

However, in 2011 the significantly higher flush was also recorded during June-August (16.91-38.26 per cent) in addition to March flushing which is attributed to relatively lower intensity of reproductive phase of the previous fruiting season.

Mango flowering involves hormonal regulation of shoot initiation and induction events resulting in reproductive shoot formation. A balance or ratio of endogenously regulated phytohormones, thought to be auxins from leaves and cytokinins from roots, appears to govern the initiation cycle independently from inductive influences. Induction of reproductive or vegetative shoots is thought to be governed by the ratio of a temperature-regulated florigenic promoter and an age regulated vegetative promoter at the time of shoot initiation. Management of off-season flowering in mango trees is being accomplished in the tropics by successfully synchronizing shoot initiation through tip pruning and use of nitrate sprays coupled with management of the stem age to induce flowering such that it can be accomplished during any desired week of the year (Davenport, 2006).

The floral induction of mango tree is mainly driven by cool temperatures as pointed out elsewhere. Consequently, increasing temperatures would have a negative effect on floral induction. But in regions with particularly cool temperature during flowering, increasing temperatures would have a positive effect on pollen viability and fruit set. Temperature also has a negative effect on inflorescence size (Dambreville *et al.*, 2013) and on the number of flowers per inflorescence (Sukhvibul *et al.*, 1999). Floral induction also requires the exposure of mature leaves to light, and thus higher levels of light intensity could have a positive effect on mango flowering. Drought and higher VPD would have a negative effect on fruit set and retention. Drought could also have an indirect positive effect on floral induction by promoting early growth cessation and vegetative rest required for floral induction.

Hormonal Control of Flowering in Mango

There seems to be a floral stimulus in mango that is transient, graft transmissible, and generated by the leaves. Kulkarni (2004) proposed a tri-factor hypothesis to explain the association of three factors in flowering. Vegetative growth in mango is through cyclic flushing. The flushing is found more frequent as temperature increases (Whiley *et al.*, 1989; Singh *et al.*, 1998). Episodic or recurrent flushing, common in sub-tropical and tropical trees, is where apical or axillary buds are released and the new shoots expand continuously through several nodes and then mature. After a period of dormancy, the cycle begins again with further bud release. The timing of flush development is important for successful flowering because, bud release for vegetative or reproductive growth, can only occur from mature flushes (Nunez-Elisea and Davenport, 1995). Pruning treatments to manipulate the timing of flush development and synchronize canopy flushing have been successful in increasing flowering intensity (Yeshitela, 2005; Balamohan and Devi, 2014; Anonymous, 2013).

Under tropical conditions, in which cold inductive temperatures may be brief, erratic or non-existent in some seasons, it is found unclear what exactly leads to floral initiation; however, some factors improve the likelihood of flowering. First, some cultivars flowered at higher temperatures more reliably than others (Pandey, 1989). For example '*Florigon*' was the only cultivar that flowered at the higher temperature treatment of 20/15°C and ten cultivars tested by Whiley *et al.* (1989). Secondly, the treatments which reduced vegetative vigor such as paclobutrazol (GA biosynthesis inhibitor), and manipulated the timing of flush development, such as water stress (Davenport, 2003), might help in bud release around the time of inductive temperatures, although this requires more testing. However, water stress has not been shown to induce flowering or decrease the cold requirement of mango (Chaikiattiyos *et al.*, 1994). Third, potassium nitrate is thought to induce flowering through ethylene (Bondad and Apostol, 1979), but seems to be ineffective in many environments (Davenport and Nunez-Elisea, 1997; Singh, 1998).

Root Dynamism and Root-Shoot Correlations

The perennial crops system like mango, present extreme difficulties in carrying out real time studies on root growth, replenishment, sink activities including assimilate transformations and other aspects of root dynamics non-destructively. Although in recent years, rhizotron studies have significantly enhanced our knowledge on root functions, more focused studies on root-shoot communication in regulation of shoot morphogenesis contributing to productivity of mango are flagged in order to understand the efficiency of genotypes based on dynamic root systems. This appears to be particularly significant in view of the fact that in mango tap root had the highest DMC (dry matter content) of (53.4 per cent) with starch grains being

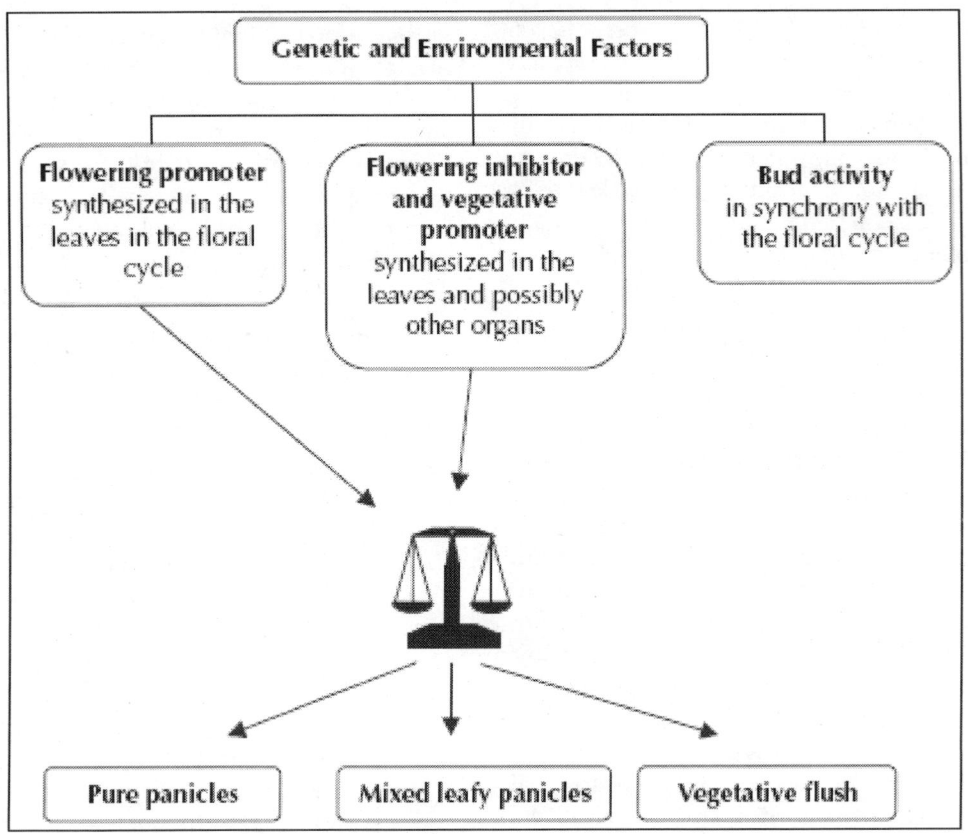

Figure 4: Hormone-Regulated Flowering Model in Mango Based on Tri-factor Hypothesis (Kulkarni, 2004).

predominantly located in xylem parenchyma with longitudinal gradient established in tap root, coarse roots, stump and branches (Damour and Normand, 2003; Normand *et al.*, 2003). Further studies however are needed to understand the phenological relationships of adult trees with starch content and whether the longitudinal gradient is constant during phenological cycle? It also becomes imperative to study these aspects in alternate and regular bearing genotypes in order pyramid the genes governing such dynamism for articulating crop improvement initiatives accordingly.

Teskey *et al.* (1987) opined that net carbon gain of a tree can be considered a function of the rate of photosynthesis per unit area of foliage, the respiration rate of the photosynthetic tissue, leaf area, and the surface area and respiration rate of non-photosynthetic tissue (*e.g.*, root, stem, branch). These components according to them can be related by the following formula:

$$\text{Carbon gain} = \sum_{i=1}^{n} L_i (A_i - R_i) - B_x(R_X)$$

Where, L_i is the leaf area of an age class of foliage, A_i and R_i are the photosynthetic rate and respiration rate of that age class, respectively, B, is the surface area of non-photosynthetic tissue of type x (*e.g., root*, stem, branch) and R, is the respiration rate of that tissue. Their study pointed out though water and temperature are important factors regulating the productivity, they alone did not control productivity. They surmised that, productive potential of a tree or forest is defined by a combination of factors and interactions, including water, temperature, nutrients, light, disease, insects, as well as less understood factors such as allelopathic and *root growth* patterns.

Fruit Set and Development

Increasing temperatures would have a positive effect on mango fruit growth which would be more rapid. In Australia, the estimated duration of mango fruit development decreased by 12-16 days (7-8 per cent) as a consequence of the 1.5°C increase of winter temperatures over the last 45 years (Olesen, 2011). High temperature could also have a positive effect on fruit quality because of stress-induced synthesis of secondary compounds, some of them being of nutritional value.

Studies on the stages of fruit growth and development in *'Dashehari'* and their relative vulnerability to weather variables indicated that *initiation* (up to 20 days-30 days)- February- March; vulnerable to temperature and moisture dynamics; *Lag phase* -30-40 days – March-April- vulnerable to high temperatures; low RH; moisture dynamics and under hormonal interpolations; fruit borer; pre-harvest fruit drop; *Grand growth phase*-40-70 days – April-May – vulnerable to hormonal interpolations; nutrients and water dynamics; fruit fly; pre-harvest fruit drop; shoulder browning if exposed to rainfall; consolidation of fruit quality principles; *Climacteric* -70-80 days - Mid June- Mid July; high temperature effects on fruit size and quality; harvest maturity; jelly seed formation, exposure to rainfall; *Senescence*- 80 days onwards- high temperature effects; concentrated maturity; market gluts; exposure to rain fall; reduced profitability.

Fruits approaching maturity are greatly affected by high temperatures exceeding 35°C as it happened in the Konkan region (day temperatures peaking to 40°C and beyond in May) during 2010 and 2015 fruiting seasons where crores worth *'Alphonso'* mangoes was lost, while in 2011 fruiting season in Malihabad area of Lucknow witnessed sharp focused maturity and market gluts during the third week of June when the day temperatures peaked to 43°C, a sudden increase of 2°C over the previous week in a short span of a week's time.

Fruit Quality

The pre-harvest factors affect both fruit growth during its development by changing the accumulation of water and dry matter, including biochemical and mineral compounds, and fruit behaviour during its storage. These pre-harvest factors affected both fruit growth during its development by changing the accumulation of water and dry matter, including biochemical and mineral compounds, and fruit behaviour during its storage. Studies have indicated that final mango quality at the consumer level depends not only on the maturity stage at harvest and post-harvest conditions during storage and marketing, but on environmental factors as well. These factors can be controlled by various cultural practices, *i.e.*, light and temperature through tree pruning, carbon availability through fruit thinning, or water availability through irrigation management (Léchaudel and Joas, 2007).

Increasing temperatures are shown to have a positive effect on mango fruit growth which would be more rapid. In Australia, the estimated duration of mango fruit development decreased by 12-16 days (7-8 per cent) as a consequence of the 1.5°C increase of winter temperatures over the last 45 years (Olesen, 2011). High temperature could also have a positive effect on fruit quality because of stress-induced synthesis of secondary compounds, some of them being of nutritional value. Higher light intensity enhanced pericarp color for colored cultivars (Léchaudel *et al.*, 2013). Since excessive light could be a stress factor, it could, as high temperature, enhance fruit quality. It could also have a positive effect on fruit size through improved photosynthesis (Urban *et al.*, 2003). Higher CO_2 concentration enhanced photosynthesis and could therefore have a positive effect on fruit quality. Fruits grown at higher CO_2 concentrations had a larger fruit dry mass, mainly due to an increase of the pulp dry mass.

The effects of drought on fruit quality are positive and negative and well-known in non-irrigated orchards. Drought reduced fruit size (Spreer *et al.*, 2009) and increased fruit quality, in particular the dry matter content and sugar concentration (Léchaudel *et al.*, 2005).

Physiological Attributes

i) Photosynthesis-RUBISCO

Assimilate Partitioning Priorities

Ravishankar (1987) studied the pattern of ^{14}C-Sucrose translocation and assimilation in distinctly alternate (*'Alphonso'*) and *regular* (*'Totapuri'*) bearing mango varieties during pre, FBD and post-FBD stages. Based on ^{14}C-Sucrose translocation, assimilation and metabolic labeling pattern, he provided evidence on the possible nature of interrelationships between the shoot apex and other parts of the mango tree *viz.*, stem portion and leaves and the presumptive role of root system in the regulation of translocation and assimilation in flowering of mango cultivars studied. His studies also pointed out to the strong possibilities of regular bearing feature of *'Totapuri'* attributable to higher metabolic turnover, energy transducing mechanism probably under up/down regulation of genes at specific stages of fruit bud differentiation in response to environmental cues *via* hormonal route since, paclobutrazol mediated flower induction in mango have been widely and successfully practiced globally.

ii) Photosynthesis and Stomata Conductance

Mango being a C_3 plant, the photosynthetic performance among cultivars is important for productivity under variable climatic conditions. Higher photosynthetic efficiency during reproductive phenophases especially flowering and fruit growth was identified as the important traits contributing to higher productivity. Under Lucknow conditions photosynthetic fluxes under different temperature regimes in 25 cultivars representing different agro-climatic regions of the country was evaluated (Singh and Rajan, 2009) where, in the commercial cultivars of Lucknow region were found more efficient as compared to others. Precise understanding of effects of light interception patterns in mango canopies on productivity is seriously constrained by lack of sufficient quantitative data.

A key enzyme associated with abiotic stress responses of crops, *Rubisco* limited rate of photosynthesis is reported less sensitive to temperature as the decrease in carboxylation at low temperatures is partially compensated by suppression of photorespiration. *RUBP*-regeneration limited rate of photosynthesis which depends upon much less on oxygenation however, is fairly sensitive to temperature. Observations at CISH, Lucknow on different varieties of regular (*Amrapali, Totapuri*) and biennial (*Langra, Chausa, Dashehari*) during different phenophases which experienced range of temperatures in 2012-13, exhibited different levels of sensitivity to low temperatures for their photochemical reaction (Personal observation; Singh *et al.*, 2012). Further studies in this regard is underscored as the behavior of this enzyme under emerging weather paradigms appears to hold key for the resilience of mango cultivars.

Incidence of Pests and Diseases

Weather dynamics is shown to impact the incidence of wide range of pests and disease problems through impacting phenophases. Key pests (stem borer, hoppers, mealy bugs, leaf Webbers, red banded caterpillar, shoot gall psylla, inflorescence midge *etc.*,), diseases (anthracnose, powdery mildew, blight, bacterial canker, tear stain disease, die-back *etc.*,) are known to severely affect sustainability of mango production in the Indian sub-continent for quite some time. As, based on the sporadic studies available in different cultivars, it

Table 2: Photosynthesis Rate (*Pn*) Stomata Conductance (*gs*), Inter Cellular CO_2 (*Ci*) and Mesophyll Efficiency (*Pn/Ci*) in Leaves of Non-flowering, Flowering Shoots and Leaves Proximal to Panicles Bearing Fruits

Shoot	P_n (μ mol m^{-2} s^{-1})	g_s (m mol CO_2 m^{-2} s^{-1})	C_i (ppm)	P_n/C_i
Non-flowering	9.25±0.38c	95.32±5.45	193±11.50a	0.047±0.01a
Flowering	6.25±0.46a	69.45±6.85a	198±9.85c	0.032±0.008a
Leaves proximal to panicles bearing fruits set	7.27±0.55ab	79.89±9.4b	195±9.68b	0.037±0.11a

Means are presented ±SE; n= 10. Within each row, values followed by different letters differ significantly (P <0.05).

Singh and Ravishankar, 2013.

Table 3: Chlorophyll Fluorescence (*Fv/Fm*), Electron Transport Rate (*ETR*) in Leaves of Non-Flowering, Flowering Shoots and Leaves Close to Panicles Bearing Fruits

Shoot	*Fv/Fm*	*ETR*
Non-flowering	0.79 ±0.15 c	265.89 ± 11.56 c
Flowering	0.27 ± 0.08 a	160.87 ± 9.75 a
Leaves proximal to panicle bearing fruits	0.53 ± 0.12 b	198.43 ± 6.86 b

Means are presented ±SE; n=10. Bars represent standard error. Means followed by different letters differ significantly at P<0.05.

Singh and Ravishankar, 2013.

appears that the emerging weather change paradigms could have far reaching implications to the life cycle, etiology/epidemiology and population dynamics of key pests and diseases and the region and cultivar-specific quantified data are unavailable to develop sustainable management strategies. Anthracnose, an important pre and post-harvest factor threatening sustainability needs due attention. Systematic studies on biotic stresses, therefore focusing especially, in relation to weather change paradigms and need of effective forecasting and growers' advisory systems is underscored.

Incidence of Disorders

Spongy Tissue in 'Alphonso'

Shivashankara and Mathai (1999) reported significant and negative relationship between the fruit transpiration rate and the spongy tissue suggesting that the lower fruit transpiration rates in cv. *Alphonso* are a varietal specific trait, which results in slower movement of water and minerals to the fruits from soil leading to the development of spongy tissue and opined that technologies that increase fruit transpiration could help in reducing the incidence of spongy tissue. Similar findings of high humidity (>85.0 per cent) associated with the rains reducing the transpiration rate of fruit; increased the incidence of softening of tissue has been reported in Dashehari.

Nubbin Fruit Formation

'Dashehari' cultivar, being highly sensitive to temperature fluxes in respect of key reproductive phenophases, *viz*., flowering and fruit set, also gets highly impacted by the abnormal weather conditions especially, the spatio-temporal relationships between T_{max} and T_{min}. The clustering (nubbin) of fruit lets phenomenon may be due to prevalence of high fluctuations in temperatures during February and March especially, the T_{min} of cold winters which coincided with the period of flowering and fruit development during 2013 fruiting season probably impacted by hormonal interpolations; this type of weather extremism despite good intensity of flowering and set, often ends up in clustering of fruit lets at the distal end of the inflorescence where, higher percentage of perfect flowers normally occurred in the panicle configuration. Such fruits invariably with aborted embryos prematurely drop off and are not carried to maturity, a direct loss to total crop outputs. Hormonal relationships also explained that, the normal fruits contained higher concentrations of gibberellins and cytokinins and lower concentrations of auxins and abscissic acid as compared to the nubbin fruits. These findings perhaps explained the reasons of occurrence nubbin fruits that have a slower growth rate than the seeded one and small in size, majority of these fruits dropped and failed to reach full size (Shaban and Ibrahim, 2009). Variety differences, effects of cold winter temperatures in this regard though have been implicated; exact reasons for this phenomenon however are not clearly understood (Campbell and Campbell, 1991).

Jelly Seed Formation

'Jelly seed' formation, a physiological disorder widely occurring in *'Dashehari'* affects the market prospects, appears to be related to reduced internal calcium allocation to fruits under the influence of high temperature holds ground in light of the observations of Batten *et al.* (1988) as its incidence increased in late harvested fruits exposed to high temperatures (T_{max} > 40°C). This view is also substantiated by the results of Singh *et al.* (2006) that highlighted the factors of late harvesting after a spell of rains, and association of fruit transpiration with internal break down in 'Dashehari'. Singh (2007) found spraying of dehydrated Calcium chloride (2.0 per cent) in combination with potassium sulphate (1.0 per cent) along with soil application of borax (150-250 g per tree) in October-November, a month before harvesting of fruits controlled this disorder.

Conclusion and Future Thrusts

The 2001-2010 decade reportedly was the warmest since 1880 - the year when enough temperature records became available worldwide to calculate a global average. Over the last 50 years, the number of cold days and record low temperatures in various locations has declined, while the number of hot days and heat waves has risen in majority of places worldwide. The best projections show that average global temperatures are likely to increase by 3.1-7.2°F (1.8-4.0°C) by the end of the century depending on the amount of carbon emissions. Now, a new study suggests that natural cycles in the Pacific Ocean are the culprit. Since the end of last *El Niño* warming event of 1997 to 1998, the tropical Pacific Ocean has been in a relatively cool phase-strong enough to offset the warming created by greenhouse gas emissions. But, according to climate researchers, this is just a temporary balm: When the switch flips and the waters turn warm again, the researchers say, Earth will likely continue warming.

The precise understanding of the impacts of different weather components either individually or conjunctively influencing individual phenophases and their contribution to over all crop outputs presently suffers from lack of availability of complete crop model for mango; as it is felt that partial models are not sufficient to predict the effects of emerging weather dynamics on mango and it also needs to integrate the effects on soil and pests also which are going to be influenced eventually. At present, what is perhaps possible would be to assess the response (s) of the region-specific commercial mango cultivar (s) to the predicted future paradigms of weather variables, based on the current knowledge and resources available in the country and also relate to the work carried out in other mango producing countries based on which, future lines of works could be defined.

In light of this and work done so far, the following future lines of work are suggested:

- ✰ There exist considerable knowledge gap in different tropical fruit tree crop models including mango especially with regard to the limited and incomplete database of good quantitative data on carbon acquisition, photo-assimilates production and transformations occurring in response to a range of environmental cues in commercial cultivars.
- ✰ Some researchers have prominently flagged for consolidation, important data and knowledge bases generally *viz.*, phenology; leaf photosynthesis (light and temperature responses); assimilate transfer and transformations; shoot growth and leaf area development and fruit growth and respiration patterns.
- ✰ Among the perennial fruit crops, despite the best data being available in apple and citrus, variations do prevail while, tropical and sub-tropical fruits, like mango in specific suffer from dearth of critical knowledge perhaps due to low number of scientists pursuing the above lines of work on commercial cultivars across mango producing agro-ecologies of the country in order to understand cultivar-environment interactions in response to weather dynamics. Adopting regional climate models for such studies is helpful. Use of model of Gavin *et al.* (2015) which has the advantage of straightforward calculation of an estimate of transient climate response based on the model and an estimate of equilibrium climate sensitivity appears best.
- ✰ Based on the information available so far, major gaps in perennial fruit crops like mango, appear to be, deciphering root growth patterns, respiration, root turnover rates, rhizosphere competence and its implications; respiration rates especially responses to temperature fluxes on long-term; role of *rubisco* and its behavior; seasonal demands for carbon by different tree organs, and crop load relations.
- ✰ Systematic research work on developing mango crop model merits primary consideration. Participation of genetic engineers, physiologists and horticulturists to pursue symphonic approach is highlighted.
- ✰ Need to engineer efficient, resilient and metabolically dynamic root systems and harmonize shoot-root communications in mango under the emerging weather paradigms is underscored.

References

Anonymous. 2015. *Horticultural Statistics at a Glance 2015*. Horticulture Statistics Division, Department of Agriculture, Cooperation and Farmers Welfare, Ministry of Agriculture and Farmers Welfare, Government of India, Oxford University Press, p. 437.

Boudon *et al.*, 2012. L-Py: An L-system simulation framework for modeling plant architecture development based on a dynamic language. *Front. Pl. Sci.*, **3**(76).

Campbell, R.J. and Campbell, C.W., 1991. The Parvin mango. *Proc. Fla. State Hort. Soc.*, **104**: 47–48.

Cawley, Gavin C., Cowtan, Kevin, Way, Robert G., Jacobs, Peter and Jokimäki, Ari. 2015. On a minimal model for estimating climate sensitivity. *Ecol. Model.*, **297**: 20–25.

Chacko, E.K. 1991. Mango flowering-still an enigma. *Acta Hort.*, **291**: 12.

Dambreville *et al.*, 2013a. Deciphering structural and temporal interplays during the architectural development of mango trees. *J. Exp. Bot.*, **64**: 2467–2480.

Dambreville, A., Normand, F. and Lauri, P.E. 2013. Plant growth co-ordination in natura: A unique temperature-controlled law among vegetative and reproductive organs in mango. *Funct. Pl. Biol.*, **40**: 280–291.

Damour, G. and Normand, F. 2003. Distribution and variability of dry matter in mango tree cv. *Cogshall*. Cirad-Flhor, Reunion Island, France, ISHS ppt.

Davenport, T.L. 2007. Reproductive physiology of mango. *Braz. J. Pl. Physiol.*, **19**(4): 363–376.

Davenport, T.L. 2009. Reproductive physiology. p.97–169. **In**: *The Mango: Botany, Production and Uses*, 2nd edn, (Ed.) R.E. Litz, CABI, Wallingford.

Davenport, T.L., Ying, Z., Kulkarni, V. and White T.L. 2006. Evidence for a translocatable florigenic promoter in mango. *Scientia Horticulturae*, **110**: 150–159.

Grechi, L., Léchaudel, M. Normand, F. and Boudon, F. 2015. A crop simulation model to predict fruit yield and quality on mango tree: Recent progress and future steps. Paper presented at the *X International Mango Symposium*, 28th September–2nd October, Darwin, Australia.

Kachru, R.B., Singh, R.N., and Chacko, E.K. 1971. Inhibition of flowering in mango (*Mangifera indica* L.) by gibberellic acid. *Hort. Sci.*, **2**: 140.

Léchaudel *et al.*, 2005. Modeling effects of weather and source–sink relationships on mango fruit growth. *Tree Physiol.*, **25**: 583–597.

Léchaudel *et al.*, 2007. An analysis of elastic and plastic fruit growth of mango in response to various assimilates supplies. *Tree Physiol.*, **27**: 219–230.

Léchaudel, M. and Joas, J. 2007. An overview of pre–harvest factors influencing mango fruit growth, quality and post–harvest behaviour. *Braz. J. Pl. Physiol.*, **19**: 287–298.

Litz, Richard E. (Ed.) 2009. *The Mango: Botany, Production and Uses*, 2nd edn, CABI.

Meehl, G.A., Stocker, T.F., Collins, W.D., Friedlingstein, P., Gaye, A.T., Gregory, J.M., Kitoh, A., Knutti, R., Murphy, J.M., Noda, A., Raper, S.C.B., Watterson, I.G., Weaver, A.J. and Zhao, Z.C. 2007. Global climate projections. **In**: *Climate Change 2007: The Physical Science Basis*, (Eds.) S. Solomon, D. Qin, M. Manning, Z. Chen,M. Marquis, M.B. Averyt, M. Tignor and H.L. Miller. Contribution of Working Group I to the Fourth Assessment Report of the Intergovernmental Panel on Climate Change. Cambridge University Press, Cambridge, and New York, NY.

Morton, J. 1987. "Mango": Fruits of Warm Climates. New CROP, New Crop Resource Online Program, Center for New Crops and Plant Products, Purdue University. pp. 221–239.

Normand, F. Lagier, S., Escoutes, J., Verdiel, J.L. and Mialet–Serra, I. 2003. *Starch Localization in Mango Tree: Histological Observations*. Cirad–Flhor, Reunion Island, France, and Cirad–Amis Montpellier, France, ISHS ppt.

Normand, F., Lauri, Pierre–Eric and Legave, Jean–Michel 2013. Climate change and its probable impacts on mango production and cultivation. **In**: *Mango: Opportunities and Challenges in the 21st Century. X International Mango Symposium*, 3–7 June, Punta Cana, Dominican Republic.

Normand, F., Lauri, Pierre–Eric and Legave, Jean–Michel, 2013. Climate change and its probable impacts on mango production and cultivation. **In**: *Mango: Opportunities and Challenges in the 21st Century. X International Mango Symposium*, 3–7 June, Punta Cana, Dominican Republic.

Olesen, T. 2011. Late 20th century warming in a coastal horticultural region and its effects on tree phenology. *New Zealand J. Crop Hort. Sci.*, **39**(2): 119–129.

Pongsomboon, W. 1991. Effects of temperature and water stress on tree growth, flowering, fruit growth and retention of mango (*Mangifera indica* L.).*Ph.D Thesis*, Kasetstart University, Bangkok, Thailand.

Rajan, Shailendra, Ravishankar, H., Tiwari, Divya, Singh, V.K., Saxena, Pooja, Singh, Shilpi, Reddy, Y.T.N. Reddy, Upreti, K.K., Burondkar, Bhagwan, A, and Kennedy, R. 2012. Harmonious phenological data: A basic need for understanding the impact of climate change on mango. **In**: *National Dialogue on Climate Resilient Horticulture: Adaptation and Mitigation Strategies for Climate Resilient Horticulture* held at IIHR, Bengaluru from 28–29th January, pp. 80–89.

Ramirez, F., Davenport, T.L. and Fischer, G. 2010. The number of leaves required for floral induction and translocation of the florigenic promoter in mango (*Mangifera indica* L.) in a tropical climate. *Scientia Horticulturae*, **123**: 443–453.

Ravishankar, H. 1987. Studies on physiological and biochemical aspects in to the causes and control of alternate bearing in mango (*Mangifera indica* L.). *Ph.D. Thesis*, University of Agricultural Sciences, Dharwad, p. 310.

Ravishankar, H. and Kailashkumar, 2011. Appraisal of some critical abiotic factors in mango (*Mangifera indica* L.) culture. Paper presented in *Global Conference on Augmenting Production and Utilization of Mango: Biotic and Abiotic Stresses* organized by The Society for Development of Sub–tropical Horticulture, Lucknow in collaboration with ISHS, Belgium from 21st–24th June, *Souvenir* pp. 87–92.

Ravishankar, H. and Rajan, S. 2011. Possible impacts of climate change on mango and guava productivity. **In**: *Impact of Climate Change on Fruit Crops*, (Eds.) W.S. Dhillon and P.S. Aulakh, Narendra Publishing House, Delhi, pp. 151–156

Ravishankar, H., Adak, T., Singh, V.K., Pandey, B.K., Singh, A.K. and Salvi, B.R. 2014. Empirical appraisal of some weather parameters dynamics for their possible implications on mango production in some important mango growing regions with special reference to Lucknow region of Uttar Pradesh. **In**: *Climate Dynamics in Horticultural Sciences, Vol. 1: Principles and Applications*, (Eds.) M.L. Choudhary, V.B. Patel, M.W. Siddiqui and S.S. Mahdi, CRC Press (Apple Academic Press), Taylor and Francis Group, USA, pp. 61–84.

Ravishankar, H., Rajan, S. and Salvi, B.R. 2011. Analysis of Alphonso mango production in the Konkan region of Maharashtra: Impacts of weather dynamics. Lead Paper Presented in *International Conference on Issues for Climate change, Land Use Diversification and Biotechnological Tools for Livelihood Security* (*ICLBDT–2011*) held at Sardar Vallabhbhai Patel University of Agriculture and Technology, Meerut, U.P., India, *Souvenir*, pp. 22–28.

Ravishankar, H., Rajan, S., Singh, V.K., Pandey, B.K., Achal Singh and Verma, Ajay. 2012. Status of mango production and its appraisal for improving sustainability of mango industry in India. **In**: Year Book 2012, *Agriculture Today*, pp. 73–78

Ravishankar, H., Rao, M.M. and Bojappa, K.M. 1979. Fruit–bud differentiation in mango '*Alphonso*' and '*Totapuri*' under mild tropical rainy conditions. *Scientia Horticulturae*, **10**: 95–99.

Schaffer, Bruce, Anthony, W. and Crane, J.H. 1994. Mango. **In**: *Handbook of Environmental Physiology of Fruit Crops*, (Eds.) Bruce Schaffer and P.C. Anderson. CRC Press Inc., Boca Raton, Florida, pp. 165–196.

Shaban, A.E.A. and Ibrahim, A.S.A. 2009. Comparative study on normal and nubbin fruits of some mango cultivars. *Australian J. Bas. Appl. Sci.*, **3**(3): 2166.

Singh, V.K. and Ravishankar, H. 2013. Photosynthetic aptitude of leaves on flowering and non–flowering terminals in mango (*Mangifera indica* L.) cv. Dashehari. *Paper presented in International Conference on Issues for Climate change, Land Use Diversification and Biotechnological Tools for Livelihood Security (ICLBDT–2011)* held at Sardar Vallabhbhai Patel University of Agriculture and Technology, Meerut, U.P., India, *Souvenir*, pp. 174–179.

Singh, V.K., Singh, D.K. and Pathak, S.M. 2006. Relationship of leaf and fruit transpiration rates to the incidence of softening of tissue in mango (*Mangifera indica* L.) cultivars. *American J. Pl. Physiol.*, **1**(1): 28–33.

Sukhvibul, N., Whiley, A.W., Smith, M.K., Hetherington, S.E. and Vithanage, V. 1999. Effect of temperature on inflorescence development and sex expression of mono– and poly–embryonic mango (*Mangifera indica* L.) cultivars. *J. Hort. Science and Biotechn.*, **74**: 64–68.

Teskey, R.O. Bongarten, B.C., Cregg, B.M., Dougherty, P.M. and Hennessey, T.C. 1987. Physiology and genetics of tree growth response to moisture and temperature stress: An examination of the characteristics of loblolly pine (*Pinus taeda* L.). *Tree Physiol.*, **3**: 41–61.

Wilkie, J.D., Sedgley, M. and Olesen, T. 2008. Regulation of floral initiation in horticultural trees. *J. Exp. Bot.*, **59**: 3215–3228.

World Bank, 2012. Turn down the heat: why a 4°C warmer world must be avoided. Washington DC. World Bank. *http://documents.worldbank.org/curated/en/2012/11/17097815/*turn–down–heat–4 per cent C2 per cent B0c–warmer–world–must–avoided.

89

Climate Change: Management Strategies in Potato Production

V.K. Dua, Tanvi Kapoor, Radhika Pathania and S.K. Chakrabarti*

ICAR-Central Potato Research Institute,
Shimla – 171 001, Himachal Pradesh
**E-mail: vkdua65@yahoo.co.in*

Climate change is posing serious threat to human survival on the earth in one or the other way. The disturbances in earth's climate have been greatly accelerated during the past century. It's a well-established fact that greenhouse gas (GHG) emission is responsible for climate change and is increasing rapidly at an alarming rate.

Potato is the third major food commodity of India after rice and wheat in terms of production. India has made tremendous progress in potato production and the per capita availability of potato has increased from 4.37 kg in 1950 to 21.52 kg in 2012. However, the climate change is likely to have a negative effect on potato growth in India, including productivity, production and profitability. The increase in CO_2 is expected to bring on increase in productivity of potato as reported by many workers. However, increase in temperature and atmospheric CO_2, both are interlinked and occur simultaneously and the CO_2 enrichment does not appear to compensate for the detrimental effects of higher temperature on tuber yield. Worldwide potato crop losses due to late blight have been estimated at €12 billion (Haverkort *et al.*, 2009). In India, severity of late blight varies from region to region, being more severe in temperate highlands than in the sub-tropical plains with an average of 15 per cent crop losses annually (Collins, 2000). In case of late blight, the effect of climate change on the most dreaded disease of potato in India also, could be mixed. Late blight caused by *Phytophthora infestans* is often regarded as the most important disease of potato globally. In potato cultivation, potato seed is most expensive input accounting for 40 to 50 per cent of the production cost. Since the eastern, northeastern, Deccan and south western parts of the country are not suitable for quality seed production barring few locations, the farmers of these areas have no option but to buy the seed potato from northern India. The high hills of Himachal Pradesh, Indo-Gangetic plains of Punjab, Haryana, north-western parts of Uttar Pradesh, and Bihar are suitable for nucleus and breeder seed production. However, Punjab and western Uttar Pradesh are two major states, which supply quality potato seed to rest of the country. In case of climate change scenario, the temperature change is likely to affect the late blight outbreak in Punjab and western Uttar Pradesh, thus affecting the potato seed production.

Bacterial wilt is causing huge losses in warmer area of Madhya Pradesh and plateau region. Presently, in North Western Indo-Gangetic region, the disease severity is very less but due to global warming situation, the disease may enter the new areas, where it is not present. The present studies were conducted to assess the yield losses due to climate change, to develop suitable adaptation strategies and to assess effect of climate change on late blight and bacterial wilt prevalence.

Impact on Potato Productivity

Selection of Model

In the present study, WOFOST (**WO**rld **FO**od **ST**udies) crop growth simulation was used. WOFOST, a mechanistic model which simulates the growth of a crop based upon eco-physiological processes is widely used to assess the effect of climate change on the growth and yield of many crops which includes wheat, rice, maize potato, barley, soybean and sugar beet throughout the world (Boogard *et al.*, 1998; Wolf *et al.*, 2010) and is one of three most widely used crop growth model for climate change studies (Diepen *et al.*, 1987). WOFOST has been calibrated for Indian potato cultivars using the time course data on potato growth and development (Dua *et al.*, 2014). Hence, WOFOST model was used for impact assessment of climate change on potato productivity and scheduling planting date and selection of suitable cultivar to minimize climate impact in Uttar Pradesh, Punjab, Madhya Pradesh, Gujarat, Bihar and West Bengal.

Selection of Potato Cultivars

The simulation studies were carried out for four potato cultivars, Kufri Badshah, Kufri Bahar, Kufri Pukhraj and Kufri Jyoti. The model was run for different planting dates depending upon the locations and the cultivar selected. The simulation study was carried out for potential yields of potato cultivars for all the scenarios.

Weather Data

In the present study, IMD district normals of 1971-2000 of 61, 13, 29, 13, 20 and 38 districts of Uttar Pradesh, West Bengal, Bihar, Gujarat and Madhya Pradesh, respectively were used for baseline scenario (year 2000). The simulation studies were carried out for A1FI high emission scenario. For generation of scenario for 2020 and 2055, projected changes in surface air temperature for sub-regions of the Asia under SRES A1FI pathway, based on the Fourth Assessment Report (AR4) Atmosphere-Ocean General Circulation Models (AOGCMs) were added on the baseline data. Projected atmospheric CO_2 concentration based on the Bern-CC model for A1FI scenario was used for incorporating the effect of change in CO_2 concentration in WOFOST model. The figures used in this study for atmospheric CO_2 concentration were 367 ppm (for baseline), 415 ppm (for 2020) and 590 ppm (for 2055).

Incorporation of CO_2 Impact in the Model

For incorporating the impact of CO_2 in the WOFOST model, changes were made in the parameters light-use efficiency of single leaf and maximum leaf CO_2 assimilation rate. A 25 to 40 per cent (mean 32.5 per cent) increase in yield in C3 plants due to doubling of CO_2 from 355 ppm to 710 ppm has been reported. Based on literature reviews, Wolf *et al.* (2010), have made changes in initial angle (+11 per cent) and in maximum of the CO_2 assimilation-light response curve response curve (+60 per cent) parameters of WOFOST model for doubling CO_2 concentration from 355 to 710 ppm. Under the experimental conditions with non-limiting supply of water and nutrients, and where temperatures are kept near the optimum for crop growth, the yield increase for C_3 crops with a doubling of CO_2 has been estimated at 30 per cent by various workers, therefore we have taken these figures as +10 per cent (30/32.5 × 11) and +55 per cent (30/32.5 × 60) for doubling CO_2 concentration for potato, assuming roughly linear relationship between the CO_2 increase and the growth processes.

Impact on Late Blight

JHULSACAST model was used to work out the effect of climate change on potato late blight outbreak, duration of favourable period and disease severity in Uttar Pradesh, Punjab and West Bengal. The model requires hourly temperature data (°C), relative humidity (RH per cent) and daily rain (mm) data as input. The model was run for baseline scenario (year 2000) and future climate scenarios (years 2020 and 2055). IMD district normal of 1971-2000 were used for baseline scenario (year 2000) and for future climate scenarios of the years 2020 and 2055 A1FI scenario of temperature (SRES A1FI pathway) was used. It is not possible to predict the daily humidity in years 2020 and 2055 during cropping period, therefore, we assumed that favourable RH would start from a week of emergence of the crop in western Uttar Pradesh and Punjab respectively and proceed with 10 days intervals after the setting of required RH. Thus, the model results are the outcome of the effect of temperature input. To estimate the number of sprays required in seed crop, the daily severity values were calculated on the basis of decision support system developed for western Uttar Pradesh and Punjab (http://cpri.ernet.in). The number of sprays needed in seed crop was calculated by dividing total accumulated severity by 180 for contact fungicides. The value of divisor 180 was derived empirically.

Study Areas

The study was conducted for entire Punjab state, 23 contiguous districts of western Uttar Pradesh and 11 districts of West Bengal. The JHULSACAST model was run on available weather data for 13 districts of Punjab, 21 districts of western Uttar Pradesh and 11 districts of West Bengal and the results were extended to the entire study area employing Kriging technique.

Impact on Survival of *R. solanacearum*

For studying the survival of *Ralstonia solanacearum* in soils of plateau region of India, results of the experiment conducted under controlled condition of various temperatures on the survival of *R. solanacearum* in soil were used to develop. A logical mathematical model giving the time (days) in which the bacterium will not survive in the soil. The model gives the results on the basis of daily mean soil temperature. For baseline survival, IMD normal were used and for future climate, temperature rises for A1F1 scenario of the climate change for 2020 and 2055 were added on the baseline temperature. The presence of *R. solani* in the soil was presumed on 1st of January and the days it will survive in the soil of 60 potato growing districts of eight states in plateau region of India were worked out, based upon soil temperature.

Change in Potato Yield

Punjab

The potential productivity of Kufri Badshah, Kufri Pukhraj and Kufri Jyoti was 52.6, 51.2 and 49.0 t/ha under baseline scenario. Under future climate scenario, when the effect of temperature and CO_2 fertilization was considered together, productivity of Kufri Badshah is projected to remain unchanged in 2020 (-0.08 per cent) and will reduced by 2.62 per cent in 2055. However, in case of Kufri Jyoti an insignificant 0.4 per cent increase is expected in 2020 (from 48.9 to 49.0 t/ha), which will decline by 4.59 per cent to 46.6 t/ha) in 2055 as compared to baseline yield. The maximum increase is likely in productivity of Kufri Pukhraj from 50.8 to 51.2 t/ha (0.69 per cent) in 2020 and maximum decline (to 48.1 t/ha; 5.33 per cent) in 2055.

Uttar Pradesh

The productivity of Kufri Badshah, Kufri Bahar and Kufri Pukhraj was 46.0, 46.8 and 45.3 t/ha, respectively under baseline scenario. The productivity under climate change scenario is expected to decrease by 5.5, 6.1 and 7.0 per cent, respectively for Kufri Badshah, Kufri Bahar and Kufri Pukhraj during 2020 and by 9.4, 10.9, 13.4 percent, respectively in 2055 as compared to the baseline scenario (Table 1).

Bihar

In the baseline scenario, the potential productivity of potato in Bihar was 40.7, 38.7 and 40.8 t/ha for Kufri Badshah, Kufri Jyoti and Kufri Pukhraj, respectively. A decrease of 5.1, 6.2 and 6.9 percent in productivity was recorded during 2020, while productivity is likely to decrease by 8.7, 10.8 and 12.7 per cent during 2055 for the three cultivars, respectively.

West Bengal

In West Bengal under the baseline scenario, the potato productivity was 38.9, 37.2 and 39.2 t/ha for Kufri Badshah, Kufri Jyoti and Kufri Pukhraj, respectively. A decrease of 5.1, 5.9 and 6.1 per cent in productivity was recorded during 2020, while productivity decreased by 8.8, 10.5 and 12.0 percent during 2055 for the three cultivars, respectively.

Table 1: Change in Potato Yield over Baseline Scenario

			Change in Yield in 2020		*Change in Yield in 2055*	
State	*Cultivar*	*Baseline Yield (t/ha)*	*Without Adaptation (t/ha)*	*With Adaptation (t/ha)*	*Without Adaptation (t/ha)*	*With Adaptation (t/ha)*
Bihar	Kufri Badshah	40.7	–5.1	–1.91	–8.7	–6.6
	Kufri Jyoti	38.7	–6.2		–10.8	
	Kufri Pukhraj	40.8	–6.9		–12.7	
West Bengal	Kufri Badshah	38.9	–5.1	–4.1	–8.8	–8.1
	Kufri Jyoti	37.2	–5.9		–10.5	
	Kufri Pukhraj	39.2	–6.1		–12.0	
UP	Kufri Badshah	46.0	–5.5	–5.2	–9.4	–9.2
	Kufri Bahar	46.8	–6.1		–10.9	
	Kufri Pukhraj	45.3	–7.0		–13.4	
MP	Kufri Badshah	43.6	–6.4	1.7	–10.9	–3.6
	Kufri Jyoti	40	–7.3		–12.8	
	Kufri Pukhraj	42.2	–7.6		–14.3	
Gujarat	Kufri Badshah	40.3	–7.6	–0.8	–19.5	–10.4
	Kufri Jyoti	36.1	–9.7		–22.1	
	Kufri Pukhraj	37.5	–10.9		–25.2	
Punjab	Kufri Badshah	52.6				
	Kufri Jyoti	49.0				
	Kufri Pukhraj	51.2				

Gujarat

The potential productivity of Kufri Badshah, Kufri Jyoti and Kufri Pukhraj was 40.3, 36.1 and 37.5 t/ha respectively under baseline scenario 2000. The productivity under climate change scenario was found to decrease by 7.6, 9.7 and 10.9 per cent respectively for Kufri Badshah, Kufri Jyoti and Kufri Pukhraj during 2020 (Table 1). The productivity was found to increase by 19.5 per cent for Kufri Badshah, while it decreased by 22.1 and 25.5, respectively for Kufri Jyoti and Kufri Pukhraj during 2055 as compared to baseline scenario.

Madhya Pradesh

Under baseline scenario of 2000, the potential productivity of Kufri Badshah, Kufri Jyoti and Kufri Pukhraj was 43.6, 40.0 and 42.2 t/ha, respectively. The productivity under climate change scenario was found to decrease by 6.4, 7.3, 7.6 per cent, respectively for Kufri Badshah, Kufri Jyoti and Kufri Pukhraj during 2020 and it decreased by 10.9, 12.8 and 14.3, respectively for the three cultivars during 2055 as compared to baseline scenario.

Overall, the impact analysis using WOFOST crop growth model has shown that under A1FI scenario of climate change although there will be no change in potato productivity in Punjab in 2020, in other states the productivity is likely to decline in 2020 (Uttar Pradesh = 5.5 to 7.1 per cent, Bihar = 5.1 to 6.9 per cent, West Bengal = 5.1 to 6.1 per cent, Madhya Pradesh = 6.4 to 7.6 per cent and Gujarat = 7.6 to 10.9 per cent) and 2055 (Uttar Pradesh = 9.4 to 13.4 per cent, Bihar = 8.7 to 12.7 per cent, West Bengal – 8.8 to 12.0 per cent, Madhya Pradesh – 10.9 to 14.3 per cent and Gujarat = 19.5 to 25.2 per cent) (Table 1). With the simple and practical adaptation measures like change in date of planting and selection of suitable variety, the decline in productivity can be brought down to 5.2, 1.9, 4.1, +1.7 and 0.8 per cent in 2020 and 9.2, 6.6, 8.1, 3.6 and 10.4 per cent in 2055 in Uttar Pradesh, Bihar, West Bengal, Madhya Pradesh and Gujarat, respectively.

Impact on Late Blight

Appearance of late blight is expected to be delayed in both 2020 and 2055 climatic scenarios in all the three states based on Jhulsacast Model. The number of days favorable for the disease will increase in Punjab. The earliest Late blight appearance, which is 13-15 October, in 2000, may be delayed by 0-6 days in 2020 and 12-14 days in 2055. However, due to rise in temperature in future, there will not be mid-season breaks in late blight favourable days. The late blight favorable days during 2000 were 105 and these increased by 30 and 35 days by 2020 and 2055 respectively. Hence, more number of fungicide sprays will be required for the control of this disease. However, in the remaining two states the number of favorable days for late blight may decrease and hence less number of fungicides sprays will be required.

In western Uttar Pradesh, earliest late blight appearance during the potato crop season is predicted during 13th October to 1st November during baseline year 2000 and is expected to be delayed by 0 to 8 days in 2020 and 10-21 days during 2055. The delay in late blight appearance in 2020 and 2055 is due to expected increase in temperature by 1.08 and 2.98°C respectively over the baseline year 2000. In western Uttar Pradesh, potato growing season would be warmer, which would decrease late blight favorable days by 7 and 27 in 2020 and 2055, respectively. The maximum number of sprays required to control late blight in potato seed crop thus would be 7.3 and 8.0 in future scenario (2020 and 2055) in comparison to 6.5 in baseline (2000) in Punjab. In contrast, there would be no change in number of sprays in year 2020 (over baseline year 2000) in western Uttar Pradesh, however, due to further increase in temperature in year 2055, it could be reduced by 2.

In West Bengal, earliest late blight appearance during the potato crop season was predicted during 4th to 11th October in baseline year 2000 and is expected to be delayed by 7.6 days in 2020. The delay in late blight appearance in 2020 is expected due to increase in temperature by 1.08°C over the baseline scenario. However, due to delay in appearance of late blight in the season, the total late blight favourable period is likely to be reduced by 17.7 days in 2020, respectively. The number of sprays required to control late blight in potato seed crop would be 4.2 in future scenario of 2020 in comparison to 5.5.5 in baseline (2000) in West Bengal. Hence, less number of fungicides sprays will be required.

Impact of Climate Change on *R. solanacearum*

Tamil Nadu and Kerala

Under the baseline climate scenario (2000), the bacterium would survive till 18th March in soils of Idukki, Kannur, Malappuram, Pathanammitha and Wayanad districts of Kerala and Dharamapuri and Virudinagar districts of Tamil Nadu and no changes are expected in 2020. However, in 2055, the bacterium is likely to survive in soil of these districts till 03 March only. The bacterium is likely to disappear 0-05 days earlier in 2020 and 05 to 15 days in 2055.

Karnataka

The bacterium could survive in soil till 18 March in Gadag, 7 April in Dharwad, Bidar, Chitradurga, Karwar, Udupi and 27 April in Rural and Urban Bengaluru districts of Karnataka in the baseline scenario. In 2020, the bacterium is likely to survival in soil till 18th March in some areas of Dharwad and Chitradurg, while in rest of districts, no change is expected. The bacterium is likely to disappear 0-08 day earlier in 2020 and 11-15 days in 2055.

Andhra Pradesh, Maharashtra, Odisha and Chhattisgarh

The bacterium may survive in soil from 7 to 27 April in different districts and is likely to disappear 3-8 and 11-25 days earlier in Andhra Pradesh and Maharashtra, 3-8 and 11-20 days in Odisha and 3-5 and 11-15 days in Jharkhand in 2020 and 2055, respectively.

Jharkhand

Under the baseline (2000), the bacterium could survive in soil maximum till 27 April in Gumla and 17 May in Ranchi and Singhbhum. No change is expected in Gumla in 2020, however, in Ranchi and Singhbhum, the survival is likely to be up to 6th May only.

Madhya Pradesh and Gujarat

The bacterium may survive in soil from 27 April to 17 May in different plateau districts of Madhya Pradesh and Gujarat in 2000. The bacterium is likely to disappear 3-8 days earlier in 2020 and 5-15 days in 2055 in Madhya Pradesh and 3-5 days earlier in 2020 and 11-26 days in 2055 in Gujarat.

Rajasthan

Under the baseline climate scenario (2000), the bacterium survived in soil till 17 May. The period is likely to reduce by 0-5 days in 2020 and 5-15 days in 2055.

Adaptation and Input Strategies

The various adaptation strategies to combat the impact of climate change on potato productivity may include breeding short duration and heat tolerant cultivars, developing potato cultivars that tube rises at higher night temperatures. Mining for biodiversity to heat tolerance should be given priority. Breeding drought and salinity tolerant cultivars would be effective to face the future challenges of climate change. Use of wind breaks around fields and crop residue mulches for some period after planting, using drip and sprinkler irrigation in place of furrow and basin methods and altering cultural management in potato based cropping systems are few examples of agronomic management practices to reduce the impact of climate change. Besides, conservation tillage and on farm crop residue management are required to increase input use efficiency. Advance planning for possible relocation and identifying new areas for potato cultivation is needed. Improvement and augmentation of cold storage facilities and air conditioned transportation from producing to consumption centers will be required for storage and transportation of this semi-perishable commodity. Strengthening education, research and development in warm climate production technology for ware and seed potato crop is also required to meet the production targets in future climates.

References

Boogaard, H.L., Van Diepen, C.A., Rötter, R.P., Cabrera, J.M.C.A. and Van Laar, H.H. 1991. WOFOST 7.1: User's guide for the WOFOST 7.1 crop growth simulation model and WOFOST Control Center 1.5. Wageningen (Netherlands), DLO Win and Staring Centre. Technical Document.

Collins, W.W. 2000. The global initiative on late blightalliance for the future. **In**: *Potato, Global Research and Development*, Vol. 1, (Eds.) S.M.P. Khurana *et al.*, Indian Potato Association, CPRI, Shimla, India.

Diepen, V.C.A., Van Keulen, H., Penning de Vries, F.W.T., Noij, I.G.A.M. and Goudriaan, J. 1987. *Simulated variability of wheat and rice in current weather conditions and in future weather when ambient CO_2 has doubled.*, Simulation Reports CABO-TT 14. CABO-DLO, WAU-TPE, Wageningen.

Dua, V.K., Govindakrishnan, P.M. and Singh, B.P. 2014. Calibration of WOFOST model for potato in India. *Potato J.*, **41**(2): 105–112.

Haverkort, A.J., Struik, P.C., Visser, R.G.F. and Jacobsen, E. 2009. Applied biotechnology to control late blight in potato caused by *Phytophthora infestans*. *Potato Res.*, **52:** 249–264.

Wolf, J., Mandryk, M., Kanellopoulos, A., van Oort, P., Schaap, B., Reidsma, P. and van Ittersum, M. 2000. Methodologies for analyzing future farming systems and climate change impacts in Flevoland as applied within the AgriAdapt project. AgriAdapt Project Report no. 1.Wageningen University, Groups Plant Production Systems and Centre for Crop Systems Analysis, The Netherlands, pp. 108.

90

Carbon Sequestration Potential in Perennial Horticultural Trees for Managing Global Warming Issues

Debashis Mandal, D.M. Kadam and Rajesh Kaushal

ICAR-Indian Institute of Soil and Water Conservation
Dehradun – 248 195, Uttarakhand
E-mail: dmandalcswcrti@gmail.com

Current concern over climate change requires that carbon (C) sources and sinks within managed landscapes are quantified so that they can be actively managed to help reduce net GHG emissions. By 2050, it is estimated that global potential of biological mitigation would be about 100 Gt C with largest potential is in the tropical and sub-tropical regions. Fruit based cropping systems are an important land use types in Indian agriculture with 6.53 million ha plantation, in spite of their potential to sequester atmospheric C and to mitigate climate change, relatively little is known regarding the influx and outflux of C in these perennial horticultural systems while a wide data exist for C storage in grassland and forests. In contrast to annual crops, perennial Horticulture trees are able to sequester C for longer periods with smaller annual ûuctuations, moreover, structural characteristic of these crops allowing them to sequester significant quantities of atmospheric carbon.

As a general rule, plantation of horticultural or forestry stands in agricultural or degraded lands would tend to increase soil carbon. It is evident from research studies across the world that perennial fruit trees like cashew, mango and guava have similar potential like forest trees to sink atmospheric carbon. The capacity of perennial horticultural trees to sequester carbon is a function of the productivity of the site and the potential size of the various pools such as soil litter, down woody materials, standing dead wood, live steams, branches and foliages. A 20 year study in temperate region of Asia revealed that an apple orchard has the potentiality to sequester about 20 Tg C as net carbon sink while carbon storage in biomass varied between 230 and 475 Tg C (Wu *et al.*, 2012). Perennial horticultural plantation of 25 year lifespan in New Zealand has shown the potentiality to sequester 70 t C ha^{-1}. In north eastern hilly region, it was reported that agri-horti practices substantially reduce soil erosion and loss of carbon (Deurer *et al.*, 2008). Based on 6 years observation, it was established that agri-horti system could enhance 86 per cent more soil carbon than sole cropping. Similarly, in semi-arid ecosystem of Maharashtra, citrus orchard showed C-sequestration potentiality of about 745 kg ha^{-1} yr^{-1}. Evergreen species such as sweet orange (*Citrus sinensis* L.) having an area of 3.9 million ha globally could create a sustainable ecosystem with favourable C sequestration potential, due to their year-round activity and tree longevity (>30 years), vigorous growth, large root system, high leaf area index (LAI) and slow leaf turnover. In Nagpur mandarin the carbon balance was almost four times higher in the intensive system (1.8 t C ha^{-1} yr^{-1}) than in the traditional one (0.5 t C ha^{-1} yr^{-1}), due to large differences in soil respiration. aonla-green gram based agri-horticulture system have indicated 1.60 times (8.35 to 13.42 t ha^{-1}) enhancement in soil organic carbon and 3.5 times (8.7

t ha^{-1} to 30.12 t ha^{-1}) carbon sequestration in different carbon pools, respectively, over a period of 25 years.

A study conducted across three states in southern India (Kerala, Karnataka and Andhra Pradesh) under tropical monsoon climate with; Figure 1), long-standing history of diverse agroforestry-based land-use practices showed that deforestation for agriculture resulted in a strong decrease in SOC stocks across all climate zones, creating a loss of 50 to 61 per cent of the original SOC stock. The resulting SOC stocks of agro-forestry systems (AFSs) however were dependent on AFS type (Hombegowda *et al.*, 2016). While SOC stocks in home garden and coffee AFSs rebounded to near forest levels, the SOC stocks in both coconut and mango AFSs increased only marginally compared to the agricultural reference (Figure 1). Of the AFSs studied, SOC stocks in the top meter of soil were highest in coffee (156±10 Mg C ha^{-1}) followed by home garden (151±5 Mg C ha^{-1}), coconut (98±7 Mg C ha^{-1}) and lowest in mango AFSs (76±3 Mg C ha^{-1}). AFS establishment on agricultural land caused SOC stocks to increase significantly in all AFSs, with SOC gains ranging from ~ 45 per cent in coconut in dry sub-humid zones to ~ 103 per cent in home garden in humid zones (Figure 1). Furthermore, significant SOC stock gains were measured at all soil depths (with the exception of coffee at 60–100 cm). Home gardens exhibited the highest overall SOC stock gains, with relative changes within each depth layer being relatively uniform throughout the soil profile. This was followed by coffee, which in contrast had highest SOC stock gains at the soil surface and decreased with depth. Lastly, SOC stock gains in mango and coconut AFSs were comparatively low but constant throughout the soil profile (Figure 1). When expressed in terms of overall changes throughout the whole soil profile, most SOC gains were concentrated at the soil surface (Figure 1). Nevertheless, when considering the whole soil profile (which has a much bigger volume), there are large SOC gains below 30 cm. For home gardens, 58 per cent of the gains occurred below 30 cm, for coffee 26 per cent, for coconut 59 per cent, and for mango 50 per cent.

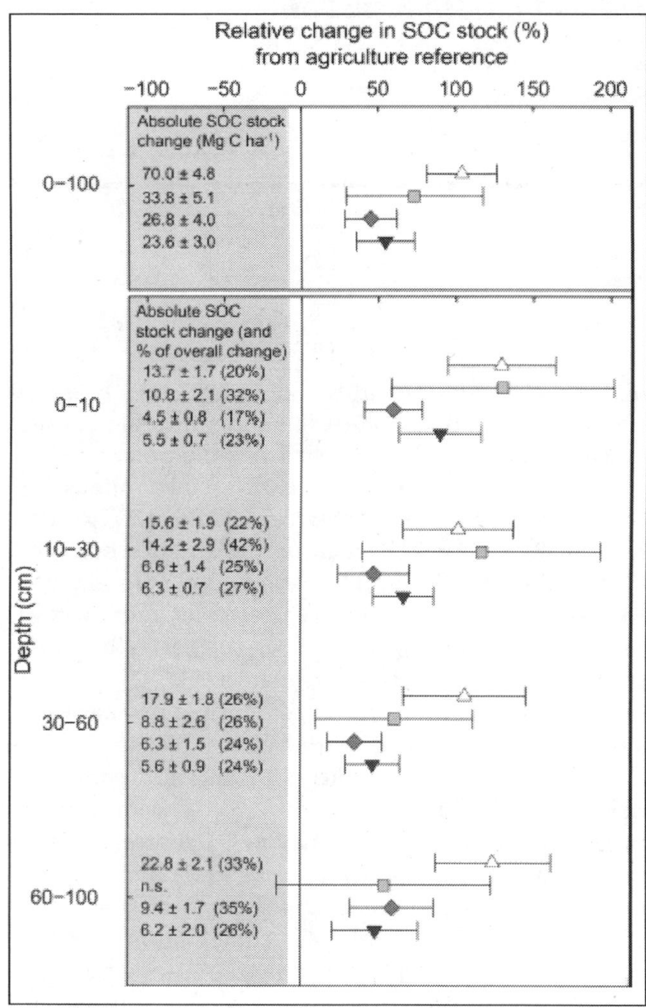

Figure 1: Relative Changes in SOC Stock in 1 m Soil Profile from Agriculture to Alternative Land Use Systems: Home Garden (Δ; n= 27), coffee (■; n= 9), coconut (♦; n= 28), and mango (▼; n= 28).

Error bars indicate the 95 per cent confidence intervals based on the Student's T distribution. The figures in the shaded area show the absolute changes in SOC stocks (MG C ha^{-1}). The values in the parentheses are the corresponding carbon change in percent of the overall change in the soil profile. (ns = not significant).

In Tamil Nadu, different ages of different types of horticultural tress such as mango, sapotta and coconut were studied for determining carbon sequestration potential. The organic carbon content at 0-20 cm soil depth was higher in mango (0.64 to 0.85 per cent followed by sapotta (0.36 to 1.07 per cent and coconut (0.07 to 0.81 per cent. the equivalent CO_2 and carbon stock was derived from the bulk density and organic carbon content of the soil. The equivalent CO_2 was also found highest in mango followed by sapotta and coconut (Selvavaraj *et al.*, 2016).

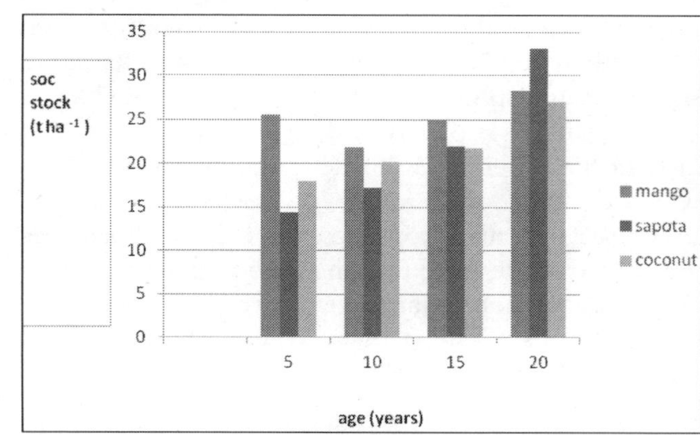

Figure 2: Soil Carbon Stock in different Horticultural Trees at different Ages.

C accumulation under different types of shade trees is given in Table 1. The large difference in different trees can probably be explained by the density of shade trees and climatic factors. Annual accumulation of C was highest in temperate region of Himachal Pradesh in India followed by Rubber plantation in Eastern Ghana.

Table 1: Carbon Sequestration Potential (Mg c ha^{-1}yr^{-1}) of Various Tree Based Agro-forestry Systems

Location	*Agroforestry System*	*Tree species*	*No of Trees/ha*	*Age (Year)*	*CSP (Mg C ha^{-1}yr^{-1})*	*References*
Himachal Pradesh	Agri-silviculture	Fruit trees	69	-	12.15	Goswami *et al.* (2014)
Jhansi, Uttar Pradesh	Silviculture	E. Officinalis	312	10	1.55	Rai *et al.* (2000)
Ghana	Orchard	Orange	NA	25	3.1	Kongsager *et al.* (2013)
Ghana	Plantation	Rubber	NA	12	5.1	Kongsager *et al.* (2013)
Ghana	Plantation	Rubber	NA	44	4.9	Kongsager *et al.* (2013)

The 213.6 tC/ha from the 44-year-old rubber plantation seems high, but the result was assessed as reasonable, given that the average DBH in the 1967 plantation was twice the average in a primary forest, where 130.2 tC/ha.

Conversion of long-term arable crop land to agro horticulture resulted in a significant increase in SOC (Table 2) under a different system of intercropped fruit trees, cultivation of coconut intercropped with guava enhanced the SOC concentration from 6.8 to 19.6 Mg ha^{-1} and 4.8 to 13.4 Mg ha^{-1} in two different plantation sites. Whereas in northeast hill states of india, alternative land use systems, *viz.* agro-forestry, agro-horticulture and agi-silviculture are more effective for SOC restoration as compared to sole cropping (Table 2). In a 6 year period the organic carbon content was about double in agro-horticulture and agroforestry systems as compared to shifting cultication. In addition these systems enhanced the biological activities approximately 2-fold in 38 years.

Strategies

There are many management options for enhanced carbon sequestration in soils these include one or combination of the options given in Table 3. In tropics, potential carbon sequestration rates for smallholders, sustainable agro-forestry systems range from 1.5 to 3.5 Mg ha^{-1} yr^{-1}.

Table 2: Relative Changes of SOC due to Land Use Change to Horticulture based Production Systems

	Field Crop	*Original Level*	*Carbon Latest*	*Changed Carbon*	*Relative Change (%)*	*References*
Site 1	Coconut	6.8	9	2.2	32	Manna *et al.* (2003)
	Coconut+sapota	6.8	18.9	12.1	178	
	Coconut+guava	6.8	19.6	12.8	188	
Site 2	Coconut	4.8	6.4	1.6	33.3	Manna *et al.* (2003)
	Coconut+guava	4.8	13.9	9.1	189	
	Coconut+banana	4.8	15.4	10.6	220	
	coconut+custard apple	4.8	12.6	7.8	162	
	Coconut+Sapota	4.8	16.7	11.9	248	
	Coconut+litchi	4.8	16.1	11.3	235	
Site 3						Das and Itnal (1994)
	Agri-forestry	0.42	0.71	0.29	69	
	Agri-horticulture	0.42	0.73	0.31	64	
	Agri-silviculture	0.42	0.56	0.14	33	

Table 3: Management Strategies for better C-sequestration in Horticultural Systems

System	*Management*
Land use	Bringing degraded and culturable waste lands under perennial horticulture crops like mango, ber, anona, amla, litchi, pomegranate, apple, peach depending upon the location of these lands. Super imposing with cover crops, erosion control measures and fertilization, life saving irrigation through water harvesting.
Farming system	Avoid mono-cropping/monoculture. A mixed orchard along with cover corps. These will enhance biodiversity and minimize expenditure on nutrients.
Tillage	Following conservation tillage practices and mulching.
Fertility maintenance	Use of fertilizers and organic manures judiciously, enhancing nutrient use efficiency (NUE) nutrient cycling through cover crops use of legume, VAM and biofertilizers.
Biochar incorporation	Convert all pruned materials and crop residues available under horti-system into biochar and incorporate in the soil to enhance C-sequestration

Source: Ganeshamurthy (2013).

Conclusion

Enhance protection and sustainable management in perennial horticultural tree crops can play critical role in climate change mitigation and strategy. Moist tropical fruit trees have evolved adaptive mechanisms such as deep root system (>1.0 m) for getting access to water in deep soil during dry season resulting in higher carbon uptake. Perennial horticultural tree based system could be a more feasible mitigation solution in many parts of developing countries including India compared to pure afforestation and reforestation projects, since tree crop plantations also provide work, income, food and nutritional security, especially when established in degraded smallholder systems where local people have control over production.

The C sequestration potential in plantations can be achieved, but it requires plantations to be established on land with modest C content, such as degraded forest or agricultural land. For example, the transformation of agricultural areas to grassland ecosystems is a common problem (Cotter *et al.*, 2009), and converting these land-use types into plantations could be beneficial from both C and economic perspectives (Li *et al.*, 2008). The consequences of converting old-growth forest to plantations have shown to result in large C debts except for rubber plantations, where the C payback time will be around 40 years. In our study, it was also found that rubber plantations have the highest C content and can sequester C at an equal or higher level than natural forests in the area. The potential for conserving C is so high that it may be considered a good practice to be included in future mitigation agreements under a revised Clean Development Mechanism. Rubber plantations also have the dual economic potential of both latex and timber production (Rahaman and Sivakumaran, 1998), and the sequestrated C can go into long-term storage if it is converted into permanent wood products, or decays over a moderate period of time if the wood is used for disposable products (Yang *et al.*, 2005). Wood from cocoa, orange, and rubber plantations can also serve as fuelwood and thereby reduce the pressure on forest areas.

The general allometric equations available for estimation of C sequestration in forest species are based on measurement of DBH, which cannot be used for fruit species as horticultural trees are mostly grafted, resulting in profuse branching just above grafted union. Development of new allometric equations applying innovative methods for non-destructive estimation of carbon sequestration in perennial horticultural trees is very much needed. The growth rates and hence carbon sequestration potential diminish as trees approach maturity owing to altered balance between photosynthetic and respiring tissues, decreasing soil nutrient availability and Increasing stomata limitations.

References

Cotter, M., Martin, K. and Sauerborn, J. 2009. How do "renewable products" impact biodiversity and ecosystem services: The example of natural rubber in China. *J. Agr. Rural Dev. Trop.*, **110**: 9–22.

Das, S.K. and Itnal, C.J. 1994. Capability based land use systems: Role in diversifying dryland agriculture. **In**: *Soil Management for Sustainable Agriculture in Dryland Areas*. Bull. Indian Society of Soil Science, **16**: 92–100.

Deurer, M., Sivakumaran, S., Ralle, S., Vogeler, I., McIvor, I., Clothier, B., Green S. and Bachmann, J. 2008. A new method to quantify the impact of soil carbon management on biophysical soil properties: The example of two apple orchard systems in New Zealand. *J. Env. Qual.*, **37**(3): 915–924. doi: 10.2134/jeq2007.0508.

Ganeshamurth, A.N. 2013. Strategies for soil carbon sequestration through horticultural crops. **In**: *Climate-Resilient Horticulture: Adaptation and Mitigation Strategies*, (Eds.) H.C.P. Singh, N. Krishnamurthy, S. Rao and K.S. Shivshankar. Springer Science and Business Media, pp. 221–235.

Goswami, S., Verma, K.S. and Kaushal, R. 2014. Biomass and carbon sequestration in different agroforestry systems of a Western Himalayan watershed. *Biol. Agr. Hort.*, **30**(2): 88–96.

Hombegowda, H.C., Straaten O. Van, Kohler, M. and Holscher, D. 2016. On the rebound: Soil organic carbon stocks can bounce back to near forest levels when agroforests replace agriculture in southern India. *Soil*, **2**: 13–23.

Kongsager, R., Jonas Napier, J. and Mertz, O. 2013. The carbon sequestration potential of tree crop plantations. *Mitigation and Adaptation Strategies for Global Change*, **18**: 1197–1213 DOI 10.1007/s11027–012–9417–z.

Manna, M.C., Ghosh, P.K. and Acharya, C.L. 2003. Sustainable crop production through management of soil organic carbon in semi-arid and tropical India. *J. Sust. Agri.*, **21**(3): 85–114, DOI: 10.1300/J064v21n03_07.

Rahaman, W.A. and Sivakumaran, S. 1998. Studies of carbon sequestration in rubber. UNCTAD/IRSG Rubber Forum, Bali.

Rai, P., Solanki, K.R. and Singh, U.P. 2000. Growth and biomass production of multipurpose tree species in natural grassland under semi-arid condition. *Ind. J. Agrofor.*, **2**: 101–103.

Selvaraj, A., jayachandran, S., Thirunavukkarasu, D.P., Jayaraman, A. and Karuppan, P. 2016. Carbon sequestration potential physioco-chemical and microbiological properties of selected trees *Mangifera indica*, *Manilkara zapota*, *Cocos nucifera* and *Tetona grandis*. *Bioscience Discovery*, **7**(2): 131–139.

Wu, T., Wang, Y., Yu, C., Chiarawipa, R. and Zhang, X., *et al.*, 2012. Carbon sequestration by fruit trees: Chinese apple orchards as an example. *PLoS ONE*, **7**(6): e38883. doi: 10.1371/journal.pone.0038883.

Yang, J., Huang, J. and Tang, J. *et al.*, 2005. Carbon sequestration in rubber tree plantations established on former arable lands in Xishuangbanna, SW china. *Chin. J. Pl. Ecol.*, **29**: 296–303.

91

Impact of Climate Change on Apple Production and Strategies to Mitigate its Effects

J.I. Mir, D.B. Singh, N. Ahmed, O.C. Sharma, A. Sharma, S. Lal, W. Hassan Raja and Aarifa Jan*

ICAR-Central Institute of Temperate Horticulture, Rangreth, Srinagar – 190 007, J&K
**E-mail: javidiqbal1234@gmail.com*

The North Western Himalayas have a unique and fragile eco-system, where people are heavily dependent on their natural environment for their sustenance and livelihood and draws about 60 per cent of Gross Domestic Product (GDP) from agri-horticultural system. The climate in this region is temperate type mainly characterized by extreme cool winters and mild summers. It offers tremendous opportunity to produce high quality horticulture crops like apple, pear, peach, plum, almond, apricot, walnut, off-season vegetables, and ornamental crops. These crops cover an area of more than 500 thousand hectares and produces fruits of approximately 31 lakh tonnes. After independence there has been seen marked growth in area and production of these crops but on the other hand productivity has left far behind as compared to advanced countries. The low productivity is mainly attributed to several factors including environmental, physiological and biological. But over the years, environmental changes playing a significant role like occurrence of erratic rain and snowfall, droughts, increase in temperature, fast receding of glaciers *etc.* Decrease in productivity of apple crop in the recent years due to change in climatic conditions have become a serious concern of the growers in India. Apple trees develop their vegetative and fruiting buds in the summer and as winter approaches the developed buds undergo dormancy due to shorter day lengths as well as cooler temperatures. The apple buds remain dormant until they have accumulated sufficient chilling units (CU) during the winter season. When required chilling units accumulated, the buds grow in response to warm temperatures. Different combinations of the temperatures during the dormant phase and the duration of the chilling period fulfill the plants chilling requirements. But due to lack of sufficient chilling temperatures during the winter apple plant will develop physiological anomalies like delayed foliation, reduced fruit set, increased buttoning and reduced fruit quality. A significant change in climate at global and national level is certainly impacting horticulture and affecting our fruit production and quality. But understanding of impact of climate change on perennial horticultural production system and the potential effects on fruit quality have drawn a little attention of researchers. The current and future changes in climate patterns and presence of higher concentrations of anthropogenic gases in the atmosphere can have dramatic effects on yield, flavour and nutritional quality of fruit. These effects are likely to affect both growers and consumers. The projected increase in surface-air temperature can have both positive and negative effects on eating quality of fruit. The yield and quality may be inferior in fruit produced under water-deficit or high rainfall conditions. The staggered flowering due to inadequate chilling in temperate fruit crops can lead to low fruit set, non-uniform fruit quality and wider harvest windows. Fruit coloration may be severely affected because the biosynthesis of colouring pigments is strongly influenced by the temperature. The nutritional value and antioxidant potential of fruit may be affected due to decrease in skin pigments. The severity of certain physiological disorders in fruits is likely to increase, contributing economic losses to the growers. The projected increase in the incidence of insect–pests and diseases would further affect the fruit quality and consequently low pack–outs for farmers.

Air temperature near the earth surface rose by 0.74°C from 1906 to 2005 and it could be increased as much as 6.4°C on average during 21st century, while CO_2 concentration is projected to increase in the range from 550 to 850 ppm (IPCC, 2009). Increase in temperature and greenhouse gases are among most important issues associated with climate change. Increase in temperature affects photosynthesis directly causing alterations in sugars, organic acids, firmness and antioxidant activity. The global warming caused loss of vigour, fruit bearing ability, reduction in size of fruits, less juice content, low colour, reduced shelf-life and increased attack of pests resulting in low production and poor quality apple crop (Jangra and Sharma, 2013). Cool temperature in winter is essential for fulfilling chilling requirements to ensure homogenous flowering and fruit set and generate economically sufficient yields. In order to escape the damage of sensitive tissues from winter the temperate trees undergo dormancy. After certain duration of cold conditions dormancy is broken and trees resume growth in following spring (Luedelling *et al.*, 2011). Apple trees require 1000-1500 h of chilling below 7°C. Chilling hours less than 1000 lead to low yield of poor quality. For breaking dormancy, trees must not only fulfill their chilling but also heat requirement at proper time (Campoy *et al.*, 2011). The delay in blooming dates of trees indicate that proper chilling is not met and dormancy breaking process is changing which is most likely in response to climate change (Lagave *et al.*, 2013). In Himachal Pradesh climate change had greatly affected apple cultivation. Due to increasing trends in temperature and decreased precipitation the apple cultivation is shifting towards

Table 1: Changes in Temperature and Precipitation *vis-a-vis* Apple Acreage in Himalayan States during 1980–2010

State	*Period (°C)*	*Avg. Annual Temp.*	*Rise in Temperature (°C) (1980-2008)*	*Precipitation (mm)*	*Area (000 ha)*	*Approx. New Area Covered under Higher Elevations*
J&K	1980-85	13.01	1.45	726	63.09	–
Ladakh	1986-90	13.58		817	66.85	–
	1991-95	13.12		784	71.33	–
	1996-2000	13.91		585	82.18	6510
	2001-2005	14.46		682	96.34	8496
	2006-2008	13.32		763	138.19	25110
	2001-2002	–		Reduced	0.609	–
	2009-2010	–		(1973-2008)	0.836	227
Uttarakhand	1980-85	12.4	1.51	1394	–	–
	1986-90	11.45		1430	–	–
	1991-95	13.69		1104	52.7	–
	1996-2000	13.90		1067	51.8	–
	2001-2005	13.84		935	55.98	4180
	2006-2009	13.91		1245	31.66	–24320
Himachal Pradesh	1980-1985	13.03	1.37	1323	46.8	–
Solan/Kangra	1991-2000	13.77		1270	83.2	36400
Mandi/Chamba/Sirmaur	2001-2007	14.40		1023	90.2	7000
Shimla/Kullu	1980-1985	–		–	953	–404
Kinaur/Lahul-Spiti	2005-2007	–		–	549	16804
	1980-1985	–		–	12368	20516
	2005-2007	–		–	29172	6433
	1980-1985	–		–	30975	
	2005-2007	–		–	51491	
	1980-1985	–		–	2532	
	2005-2007	–		–	8965	

Source: IMD, Srinagar.

higher altitudes. In Uttarakhand the area under apple cultivation is decreasing drastically which might be due to less rainfall and higher temperature in winter. The chilling requirement of apple trees is not met which is important for producing good quantity and quality apples (Sharma, 2015). It has been reported that rise in average temperature, long spells of drought in summer, delay in start of winter and reduced snowfall have reduced large area under apple cultivation. Keeping in view the concern of climate the review paper focuses on impact of climate change on apple production and its remedies to reduce the risk and take advantage of potential opportunities. Chand *et al.* (2015) provide the details on assessment of chill units accumulation and apple productivity under changing climatic conditions in high hill temperate zone of NW Himalayas. They have shown decreasing trend of the order of 11.40 chill units per year with chill units for 1986-1995, 1996-2005 and 2006-2015 revealing a decreasing trend of the order of 119.30 CU decrease per decade. Among different months maximum decrease of 6.24 CU has been observed in the month of November. Trend analysis of last five years 2011-2015 showed that there was decrease of 2.92 CU per year (Chand *et al.*, 2015). Due to decreasing trend of chill units accumulation due to climate change indicates that climatic conditions might become unfavorable in coming years for cultivation of apple and thus may lead to shift in apple belt to higher altitude in the region so as to accumulate the required chilling units. The area under apple crop has increased significantly in the higher altitude of Ladakh district of JK and Kinnaur and Lahaul Spiti district of H.P. (Table 1).

Problems Associated with Climate Change

There are four factors of plant environment *viz.*, moisture, soil, light and temperature of which fruit grower can modify two considerably (Gunduz *et al.*, 2011). The orchardist can irrigate or drain the orchard and can fertilize or modify soil structure to some extent if necessary, but light and temperature has to be taken as such. Amongst all the climatic components, temperature seems to be the most crucial factor in apple production, however, the role of spring frosts, hails, summer droughts and unseasonal spring rain in lowering the productivity and quality of apple cannot be overlooked (Hussein and Shaban, 2012).

i) Lack of Sufficient Chilling Hours

In temperate fruits, the requirement of certain hours of winter chill (below 7°C) to break dormancy in spring and then flowering in spring is termed as chilling hours. Warming of winter due to climate change in last few decades have been adversely affecting the winter chilling requirements of temperate fruits. Apple produced in India has greatly been influenced by warming over years. Inadequate chilling hours make buds to break slow in spring or not at all particularly in mid hills and marginal areas. Chilling temperatures are important in apple and other fruit production because they are needed to break full dormancy, which is a prerequisite for effective and synchronous bud-break and flowering. Temperature increase related to climate change could lead to inadequate chilling in certain areas, which could affect the suitability for some species or cultivars to survive or yield in that location. Sufficient chilling hours are not being met with as there is lack of snowfall and rainfall during the winter months especially in the mid hill areas of Himachal Pradesh (Jindal *et al.*, 2001). During the second half of 20th Century apple cultivation failed in some districts of Himachal Pradesh due to insufficient of chilling hours in these belts due to rising temperatures. Climate change is also potentially threatening current production in major apple growing region of Jammu and Kashmir and Uttrakhand. It has also been a general experience that in years when the snowfall is less in temperate fruit growing regions and the winters are comparatively warm, chilling requirement of the apple trees is not met as a result of which bud break is erratic and delayed and the per cent flowering is also reduced and as a consequence the production is considerably reduced (Jindal *et al.*, 2004). Temperate fruit and nut trees require adequate winter chill to produce economically viable yields. Global warming has the potential to reduce available winter chill and greatly impact crop yields (Luedeling *et al.*, 2011). Aditya *et al.* (2013) reported that decrease in chill units due to increase in temperature adversely affected apple production. Trees in temperate regions need a chilling period so they can grow and flower in spring. Rising temperatures pose a special problem for temperate but comparatively warm areas where adequate winter chill is already in short supply.

ii) Lack of Sufficient Soil Moisture

Water stress is one of the most important factors limiting apple production. Perennial crops use more water simply by virtue of being in place longer. The most critical period of water requirement is during active vegetative growth and fruit development in summer months. Climate change has reduced winter rains and snow and water supply from glaciers has already been depleted which has aggravated the problem of water stress in summer. Decreased volume of snow in winter and rain during winter, spring and summer has resulted in drought like conditions, which creates moisture stress and trees do not flower normally (Chand *et al.*, 2015). Water stress affects apple trees in many ways like, mid and late summer drought triggers June as well as pre harvest fruit drop, the incidence of fruit cracking increases, water stress leads to boron deficiency; which affects flowering, fruit development and causes apple corky spot.

iii) Spring Frosts

Due to climate change the winters remain warm and dry but during spring time the inclement weather prevails and temperatures go low and spring frost is experienced resulting in frost damage to flowers and poor fruit set which in turn results in poor yields (Awasthi *et al.*, 2001). The spring frost adversely affects the flowers and fruit set is poor, which results in low fruit retention and poor yields. Apples bloom in the spring and are threatened every spring by freezing temperatures during bloom.

iv) Poor Pollination

The lack of early cold in December and January adversely affect the chilling requirements. Late cold during April month delay blossoming and reduce the pollination activity of bees (Sharma, 2012). The occurrence of spring frost and low temperature at the time of flowering and fruit setting adversely affects apple production. Weather at blossom time plays an important role in pollination. In apple low temperature (below 41°F) can cause the pollen not to germinate and pollen tube growth is slow below 51°F (Sharma, 2015). Weather plays a vital role in spring blooming fruit trees due to the combined effects on bee activity, flower opening, pollen germination and fertilization. If cold weather prevails from green tip to half inch green, while the pollen is in the process of production the quality of pollen may be seriously reduced. Hot and dry weather will reduce the stigmatic receptivity. Wind also plays an important role in cross-pollination of apple trees. It has an adverse effect on insect pollination that is required for cross-pollination. The various effects of climate change on pollination include; irregular flowering, low temperature causes injury to flowers, pollens are washed away due to rainfall, slow pollen germination and seed abortion due to low temperature and finally no or poor fruit set.

v) Yield and Quality

The production of apple has gradually increased but productivity has fallen from 10.8 to 5.8 t/ha (Awasthi *et al.*, 2001). The reasons attributed to it are climate variability, poor soil health, less progress in crop improvement *etc.* Among all the productivity reducing factors, climate is difficult to manage. According to Vedwan and Rhoades (2001) climatic changes alter the pattern of blossoming, bearing and, therefore, fruit yield, and the quality of apple under Western Himalayan condition of India. In Himachal Pradesh apple growing areas are being shifted to higher elevations and larger area has been reduced to unfit for cultivation. Productivity of apple has shown a declining trend of 40-50 per cent in Himachal Pradesh upto 1,500 m above mean sea level due to warmer climate and lack of chilling hours during winter and adequate growth during warmer summers (Gautam *et al.*, 2004). Increase in CO_2 will cause positive impacts, which may be nullified by increased temperature and less water availability ultimately decreased production (Singh, 2010). Chilling affects flowering and subsequent fruit setting qualitatively and quantitatively (Sharma *et al.*, 2015). In Kashmir valley, from 1974 to 2009, areas devoted to apple production rose from 49,190 ha to 133, 2810 ha. However, unprecedented storms as well as rain and snowfall in the region are now threatening to limit production. Fruit growers of Kashmir valley reported record apple production of over 2.2 million tonnes in 2014. Climate changes have brought alteration in the pattern of blooming, bearing and ultimately the fruit yield (Singh *et al.*, 2016).

Regarding the quality of fruits, increase in temperature should have more pronounced effect than the elevated CO_2 (DaMatta *et al.*, 2010). High temperature and moisture stress has resulted in increased incidence of sun burn and cracking in apples which has markedly decreased the fruit quality (Rai *et al.*, 2015). Increase in temperature and greenhouse gases are among most important issues associated with climate change. Increase in temperature affects photosynthesis directly causing alterations in sugars, organic acids, firmness and antioxidant activity (Moretti *et al.*, 2010). In Kashmir valley, apples fail to change into their specific red shades or an increase of apple sunburn due to climate change. If the temperatures are enough low most of the apples fail to turn into their specific shades. The Ladakh province becomes a potential area for apple cultivation due to climate change (Lal and Ahmed, 2014). In past few decades, yield and quality of apple have been adversely affected due to climate change which further may pose risk of substantial economic losses for apple industry worth an estimate of Rs. 10,000 crores annually (Sharma, 2015).

Other quality parameter in apple is fruit flavor which is affected by high temperatures. Apple fruits exposed to direct sunlight have higher sugar content compared to those fruits grown on shaded sides. Increase in 10°C temperature caused a 50 per cent reduction in tartaric acid content. Fruit rmness is also affected by high temperature conditions during growth. Changes in cell wall composition, cell number, and cell turgor properties were postulated as being associated with the observed phenomenon. Antioxidant in fruit crops can also be altered by exposure to high temperatures during the growing season. The high temperature conditions signicantly increased the levels of flavonoids, antioxidant capacity and fruit quality (Wang and Zheng 2001; McKeon *et al.*, 2006).

Higher CO_2 and GHG significantly effect the fruit quality and yield. Carbon dioxide is important because carbon atoms form the structural skeleton of the plant. A doubling of carbon dioxide levels may increase plant growth by 40-50 per cent though continuous high levels saturate the plant's ability to use carbon dioxide and the benefits decrease with time. If other factors remain favorable, increased carbon dioxide concentrations will lead to greater rates of photosynthesis in plants. Current carbon dioxide concentrations limit plant photosynthesis. Growers of protected horticultural crops have already aware from so many years that artificially raising the concentration of carbon dioxide up to certain stage in greenhouses can substantially increase crop growth and yield. Effect on timing of bud burst, cessation of growth, altered concentrations of carbohydrates and plant hormones in turn altered the dormancy status of trees thereby changing the timing of bud burst and the length of the active growing period. Flowering and fruiting of apple trees are likely to be hastened under conditions of elevated carbon dioxide. Leaves are able to detect and respond rapidly to carbon dioxide concentration. Stomata opening decreases in response to increased carbon dioxide concentration.

Ozone exposure also effects apple fruit quality and yield. Higher concentrations of atmospheric ozone are found during summer due to increase in nitrogen species and emission of volatile organic compounds (Mauzerall and Wang, 2001). Concentrations are at maximum values in the late afternoon and at minimum values in the early morning hours, notably in industrialized cities and vicinities. The opposite phenomenon occurs at high latitude sites. Another potential source for increased levels of ozone in a certain region is *via* the movement by local winds or downdrafts from the stratosphere.

vi) Physiological Disorders and Tolerance to High Temperatures

Frequent exposure of apple fruit to high temperatures, such as 40°C, can result in sunburn, development of watercore and loss of texture. Moreover, exposure to high temperatures on the tree, notably close to or at harvest, may induce tolerance to low-temperatures in post-harvest storage (Buescher, 1979) (Figure 1).

vii) Effect on Incidence of Insect, Pest and Disease

Erratic changes in temperature and precipitation leads to more incidences of insect, pest and disease. In the last few years, the attack of apple scab, powdery mildew in apple, flee beetle in almost all the temperate fruit crops has been increased (Figure 2).

Figure 1: Incidence of Physiological Disorders in Apple to Climate Change.

Strategies for Mitigating Effects of Climate Change

i) Adaptation

Climate change is now known to be a universal phenomenon and is to be continued. Farmers of different zones of Himalayas should be made aware of the adaptive horticultural production system which is less vulnerable to climate change.

ii) Diversification in Fruit Cultivation

Diversification necessarily involves its own costs and risks, but the idea is to avoid larger losses and larger risks. Climate change has adversely affected apple production and farmers are steadily shifting towards other crop options. Farmers have shifted to cultivation of pomegranate, kiwi fruit and other vegetable crops (Randev, 2009). From past experience appropriate strategies are needed to develop in respect of diversification for sustained and consistent fruit production (Belsare *et al.*, 2015). In areas where chilling requirement of commercial apple trees is not met

Figure 2: Powdery Mildew of Apple (R) at Dry and Warmer Summers and Flea Beetle (L) at Warm and Humid Climate.

diversification of low chill cultivars is recommended. Reevaluation of fruit varieties as per the indicated climate change is imperative for planning a new orchard (Rai *et al.*, 2015).

iii) Breeding Low Chill and Late Blooming Varieties

'This is the most feasible solution to the problem of insufficient chilling, however, it is very difficult to breed low chill cultivars. Introduction and adaptation of low chilling cultivars of apple in certain areas of lower hills and North Indian plains where they could be grown commercially (Rai *et al.*, 2015). Scientists need to develop apple varieties which have low chilling requirements. Farmers should select low chilling varieties in each ecological niche of temperate fruits (Sharma, 2015). Some of the low chilling apple cultivars are; Anna, Mayan, Tamma, Vared, Neomi, Tropical Beauty, Michael, Schlomit *etc.* The early flowering cultivars get damaged as winters remain warm and dry but during spring time the inclement weather prevails and temperatures go low and spring frost is experienced resulting in frost damage to flowers. To escape from spring frost growing of late blooming apple cultivars would be an appropriate method to adopt to such a climate change. Lesser damages to buds due to spring frost has been found in late blooming apple cultivars like, Golden Delicious, Royal Gala, Rome Beauty, Northern Spy *etc.*

iv) Water: Accommodating Flood and Drought

For proper growth and development in apple trees the supply of water at critical stages is important. In recent years there has been a wide range of proposed novel approaches to irrigation scheduling which are based on sensing the plant response to water deficit directly, as opposed to sensing the soil moisture status (Jones, 2004). In plants two types of water stresses are experienced *viz.*, excess water (floods) and deficit water (drought). Drought is one of the most important environmental stresses which affect most growth and physiological aspects of plants and causes huge damages to agriculture products every year. There are several reports of osmolyte increase in apple under drought stress. Jie *et al.* (2008) reported that under drought stress proline at high levels was accumulated in different organs of apple. The selection of apple cultivars and rootstocks that are able to maintain productivity at low water potential is of utmost importance. In apple drought tolerant rootstocks like MM 111, M 7 and EMLA 9 (Fernandez *et al.*, 1997) should be used. Flooding, drought, and premature defoliation impair tree defense mechanisms and trigger biochemical responses that release carbohydrates, sugars, and other nutrients which seem to invite insect and fungal pathogen attack. Northern Spy and Malling Merton series of apple rootstocks are tolerant to collar rot. Yao and Neilson (2001) studied the effects of water stress on growth and mineral composition of 'Gala' apple fruit and found that water stress treatment affected fruit diameter and fruit weight. The accumulation of Ca per fruit increased continuously throughout the growing season, peaking two weeks prior to harvest, and then declining at harvest. Excessive soil moisture also provides an ideal environment for crown and collar rots, while insufficient moisture results in drought, stress and reduces fruit quality (Black *et al.*, 2008). Makeredza *et al.* (2013) reported that sunburn necrosis increased with increasing water stress and concluded that water stress aggravates sunburn development under conditions conducive for its development by increasing fruit surface temperature.

v) Soil Building

Apple trees can grow in a wide range of soils but poor soils will produce poor results and the best crops are found on fertile sandy soils and loams. Soil building is increasing organic matter in soil which gives protection from both water stresses flood as well as drought. The soils with high organic matter holds higher water contents. A high organic matter soil is better at buffering temperatures as these soils heat and cool slowly reducing plant stress. Other benefits of building organic matter include improving fertility, improving fertilizer efficiency and resisting erosion. Bacterial activity hinders in case of unexpected cold temperatures and decomposing activities reduced in orchards therefore reducing organic matter content not sufficient to support apple production. A permanent organic mulch is often the choice of small scale organic apple growers as it provides many soil building and water conserving benefits. Asghar *et al.*, 2012 reported that orchard soils showed a mean value of 74 per cent moisture content fairly good to support high productivity in apple. The deposition of pollutants in the soil, altered chemical characteristics such as pH value affecting availability of nutrients and increasing the leaching of minerals from the soil (Oren, 1996). It may also result in deficiencies of the nutrients (mainly Mg, Ca and K) in plants along with severe nutritional imbalances.

vi) Technology, Information and Policy

Most farmers adopt best affordable technology to improve farming. The apples trees in Washington are sprayed with Surround™, a pulverized kaolin clay product that reduces heat stress in plants and inhibits sunburn on the fruit. The technology is promising in times of climate change and improved frost prevention systems. Functioning together on farm weather stations linked in a network is able to predict specific weather events. The future agriculture will involve more better computer based information based technologies to guide irrigation, planting, cultivar choices *etc.* On June 30, 2008, Former Prime Minister, Dr. Manmohan Singh released India's first National Action Plan on Climate Change (NAPCC) outlining existing and future policies and programs addressing climate mitigation and adaptation.

The plan identifies eight core "National Missions" running through 2017. National mission for sustainable agriculture aims to support climate adaptation in agriculture through the development of climate-resilient crops, expansion of weather insurance mechanisms, and agricultural practices. Apple is known in North-West states as a most significant commercial crop. The farmers are unaware about scientific agro commercial practices, horticulture schemes and agri inputs. The present changes in climate are posing serious threats on apple production. It is important to understand the variations in patterns of climate change and also to identify management practices and alternatives for farmers in order to cope up with vagaries of changing climate.

vii) Agronomic Management Strategies

- ✰ Assessment of the vulnerability and climate risks associated with apple fruit production system in temperate region.
- ✰ Development of apple based cropping systems under various agro-climatic conditions.
- ✰ Microirrigation for enhancing WUE and drainage to avoid water logging.
- ✰ Development of appropriate tillage and intercultural operations.
- ✰ Integrated approaches for management of pests, diseases, weeds and nutrients.
- ✰ Development of water harvesting techniques.

viii) Biotechnological Innovative Strategies

- ✰ Development of molecular markers for traits related to biotic and abiotic stress resistance and utilization of molecular tools for assisting breeding of apple for development of new varieties with traits like disease resistance, drought tolerance, late flowering etc
- ✰ Development of tissue culture techniques like DH to speed up the breeding programmes.
- ✰ Development of molecular diagnostic and prognostic tools for diseases and pests.

ix) Mitigation Strategies for Higher CO_2/GHG

- ✰ Assessment the carbon sequestration potential of various apple cultivars and apple based cropping systems.
- ✰ Waste processing and recycling.

Conclusion

Climate change impacts are to be looked not in isolation but in conjunction with all the aspect of agriculture and allied sectors. The effects of climate change on horticulture sector are still uncertain but a significant overhaul in pre-harvest and post-harvest practices may be required as an adaptation measure to maintain productivity and fruit quality under changing climatic conditions. In the light of possible global warming, researchers should give more emphasis on development of heat- and drought-resistance varieties. Research is needed to define the current limits to these resistances and the feasibility of manipulation through modern genetic techniques. In apple identification and development of varieties with traits like low chill requirement, resistance to abiotic and biotic stress, drought tolerance, longer shelf-life, late flowering *etc.* need to be done and such varieties should be improved further for quality and yield. Both crop architecture and physiology may be genetically altered to adapt to warmer environmental conditions. The continuation of current and new initiatives to research potential minimize the effects of climate change at farm, regional, national and international level and will help to provide a more detailed picture of how world horticulture and agriculture could change. Only then may we see the implementation of policies and other adaptations in agricultural systems that would minimize the negative effects of climate change and exploit the beneficial effects.

References

Aditya, Rana, R.S., Chauhan, R.C. and Sen, V. 201. Farmers' perception in relation to climate variability in apple growing regions of Kullu district of Himachal Pradesh. *J. Agril. Phys.*, **13**(1): 48–54

Asghar, A., Ali, S.M. and Yasmin A. 2012. Effect of climate change on apple (*Malus domestica* var. Ambri) production: A case study in Kotli Satian, Rawalpindi, Pakistan. *Pak. J. Bot.*, **44**(6): 1913–1918.

Awasthi, R.P., Verma, H.S., Sharma, R.D., Bhardwaj, S.P. and Bhardwaj, S.V. 2001. Causes of low productivity in apple orchards and suggested remedial measures. **In**: *Productivity of Temperate Fruits*, (Eds.) K.K. Jindal and D.R. Gautam, pp. 1–8.

Belsare, C.R., Sharma, N., Sharma, D.P. and Moankotia, M.S. 2015. Impact of different location and altitudes on fruit quality parameters of apple cv. 'Starking Delicious' in Himachal Pradesh. *Green Fmg.*, **6**(5): 1056–1060.

Black, B., Hill, R. and Cardon, G. 2008. Orchard irrigation: Apple. Horticulture/Fruit/2008-01pr. US Department of Agriculture, Utah State University.

Brady, N.C. and Weil, R.R. 2007. *The Nature and Properties of Soils*, 14th edn. Pearson Prentice Hall.

Buescher, R.W. 1979. Influence of high temperature on physiological and compositional characteristics tomato fruits. *Lebensmittel-Wissenenshaft*, **26**: 237–268.

Campoy, J.A., Ruiz, D. and Egea, J. 2011. Dormancy in temperate fruit trees in a global warming context: A review. *Scientia Horticulturae*, **130**: 357–372.

Chand, H., Verma, S.C., Bhardwaj, S.K., Sharma, S.D., Mahajan, P.K. and Sharma, G. 2015. Assessment of chill units accumulation and apple productivity under changing climatic condition in high hill temperate wet zone of North–Western Himalayas. *J. Hill Agri.*, **6**(2): 229–233.

DaMatta, F.M., Grandis, A., Arenque, B.C. and Buckeridge, M.S. 2010. Impacts of climate changes on crop physiology and food quality. *Food Res. Int.*, **43**: 1814–1823.

Fernandez, R.T., Perry, R.L. and Flore, J.A. 1997. Drought response of young apple trees on three rootstocks: Growth and development. *J. Amer. Soc. Hort. Sci.*, **122**: 14–19.

Gautam, D.R., Sharma, G. and Jindal, K.K. 2004. Fruit setting problem of apple changing climate scenario of North-Western Himalayas of India. *Acta Horticulturae*, **662**: 435–442.

Gunduz, O., Ceyhan V. and Bayramoglu Z. 2011. Influence of climatic factors on apricot (*Prunus armeniaca* L.) yield in the Malatya province of Turkey. *Asian J. Agril. Sci.*, **3**(2): 150–155.

Houghton, J., Ding, Y., Griggs, D., Noguer, M. and Van der Linden, P. 2001. *Climate Change 2001: The Scientific Basis*. Published for the Intergovernmental Panel on Climate Change. Cambridge University Press, Cambridge, UK and New York. 881p.

Hussein, A. and Shaban, D. 2012. Climate change and its impact on the productivity and quality of vegetable crops. *J. Appl. Sci. Res.*, **8**(8): 4359–4383.

IPCC, 2013. *Climate Change, 2009: Mitigation of Climate Change*. Cambridge University Press, Cambridge.

Jangra, M.S. and Sharma, J.P. 2013. Climate resilient apple production in Kullu valley of Himachal Pradesh. *Int. J. Farm Sci.*, **3**: 91–98.

Jie, Y., Yang, H., Zhao, H. and Li, D. 2008. Promotion of proline accumulation in apple leaves by bioregulayors. *Acta Hort.* **774**: 237–242.

Jindal, K.K. and Mankotia, M.S. 2004. Impact of changing climatic conditions on chilling units, physiological attributes and productivity of apple in western Himalayas. *Acta Hort.*, **662**: 111–118.

Jindal, K.K., Chauhan, P.S. and Mankotia, M.S. 2001. Apple productivity in relations to environmental components. **In**: *Productivity of Temperate Fruits*, (Eds.) K.K. Jindal and D.R. Gautam, pp. 12–20.

Jones, H.G. 2004. Irrigation scheduling: advantages and pitfalls of plant-based methods. *J. Exp. Bot.*, **55**: 2427–2436.

Lal, S. and Ahmed, N. 2014. Strategies to minimize impact of climate change on fruit production. *Biotech Articles*.

Luedeling, E., Girvetz, E.H., Semenov, M.A. and Brown, P.H. 2011. Climate change affects winter chill for temperate fruit and nut trees. *PLoS One*, **6**: e20155.

Makeredza, B., Schmeisser, M. and Lotze, E. 2013. Water stress increases sunburn in 'Cripps' Pink' apple. *Hort. Sci.*, **48**(4): 444–447.

Mauzerall, D.L., and Wang, X. 2001. Protecting agricultural crops from the effects of tropospheric ozone exposure: Reconciling science and standard setting in the United States, Europe, and Asia. *Ann. Rev. En. Env. Tech.*, **12**: 162–164.

McKeon, A.W., Warland, J., and McDonald, M.R. 2006. Long-term climate and weather patterns in reation to crop yield: A mini review. *Can. J. Bot.*, **84**: 1031–1036.

Moretti, C.L., Mattose, L.M., Calbo, A. and Sargent, S.A. 2010. Climate changes and potential impacts on post-harvest quality of fruit and vegetable crops: A review. *Food Res. Int.*, **43**: 1824–1832.

Oren, R. 1996. Nutritional disharmony in plants: soil and weather effects on source-sink interactions. *Pl. Resp. Air. Poll.*, **34**: 75–97.

Rai, R., Joshi, S., Roy, S., Singh, O., Samir, S. and Chandra, A. 2015. Implications of changing climate on productivity of temperate fruit crops with special reference to apple. *J. Hort.*, **2**(2): 1–6.

Randev, A.K. 2009. Impact of climate change on apple productivity in Himachal Pradesh, India. In: *60th International Executive Council Meeting and 5th Asian Regional Conference*, 6–11, December, New Delhi, India.

Sharma, I.M. 2012. **In**: *Proceedings of the National Symposium on Blending Conventional and Modern Plant Pathology for Sustainable Agriculture*. Indian Institute of Horticultural Research, Bengaluru.

Sharma, N., 2015. Impact of climate change on temperate fruit and nut production. **In**: *Temperate Fruits and Nuts*, (Eds.) K.L. Chadha, N. Ahmed, S.K. Singh and P. Kalia, Astral International Pvt. Ltd., New Delhi, pp. 117–133.

Singh, H.P. 2010. Impact of climate change on horticultural crops. **In**: *Challenges of Climate Change: Indian Horticulture*. Westville Publishing House, New Delhi, pp. 1–8

Singh, N., Sharma, D.P. and Chand, H. 2016 Impact of climate change on apple production in India: A review. *Curr. World Env.*, **11**(1): 251–259.

Vedwan, N. and Rhoades, R.E. 2001. Climate change in the Western Himalayas of India: A study of local perception and response. *Clim. Res.*, **19**: 109–117.

Wang, S.Y. and Zheng, W. 2001. Effect of plant growth temperature on antioxidant capacity in strawberry. *J. Agril. Food Chem.*, **49**: 4977–4982.

Yao, S., Neilsen, G.H. and Neilsen, D. 2001 Effects of water stress on growth and mineral composition of 'Gala' apple fruit. *Acta Hort.*, **564:** 449–456.

Creating Employment Opportunities for Youth

92

Horticulture Information System Network (HORTNET) 2.0 for Increasing Farmers' Incomes through Digital India Programme

M. Moni

Centre for Agricultural Informatics and e-Governance Research Studies, Shobhit University, Delhi

E-mail: moni@shobhituniversity.ac.in; moni@nic.in

Rural India has been witnessing "technology-led development of Horticulture" to sustain its economy. In India, digital economy (Informatisation in economic activities) is intensifying at faster pace, through the identified "Nine Pillars of Growth" of **Digital India Program 2014.** The Indian Horticulture, which consists of (a) Fruits including nuts (b) Vegetables (c) Flowers including ornamental plants, (d) Spices, (e) Plantation crops (arecanut, cashew nut and cocoa), (f) Coconut, (g) Potato, Tuber Crops and Mushrooms, and (h) Medicinal and Aromatic plants, has become a key driver for economic development in many Indian States.

The EXIM Bank Report 2013 has drawn attention that (a) there is a wide gap between "yield obtained" and "potential yield", (b) the quality of produce has been relatively lower than the International Standards, and (c) on the supply-side, the wastage level is very high, significantly, in the post-production stage resulting in low market surplus.

There is a need to develop a platform for farmers to interact with (Figure 1): Input Suppliers (Seeds, fertilizers, pesticides, Machinery and credit), Marketing Channels (Wholesale Markets, Rural Markets, Commission agents/traders, processing industries *etc.*), Agricultural laborers and Agricultural extension workers. Farmers face distress problems during their entire crop cycle: high input cost, unpredictable weather conditions during production, and distress sale of output. Farm-Gate Price, Wholesale Price and Consumer Price of farmer produces are to be inter-related for farmers to overcome "distress sale".

Wider adoption of ICT (Information and Communication Technologies) in farming life cycle system is of strategic importance to agricultural stakeholder groups: Farmers, Researchers, Entrepreneurial Startups, Government, Industries, Consumers and Civil Society Organisations. Digitalization of micro-level farming system has its associated challenges and diversified

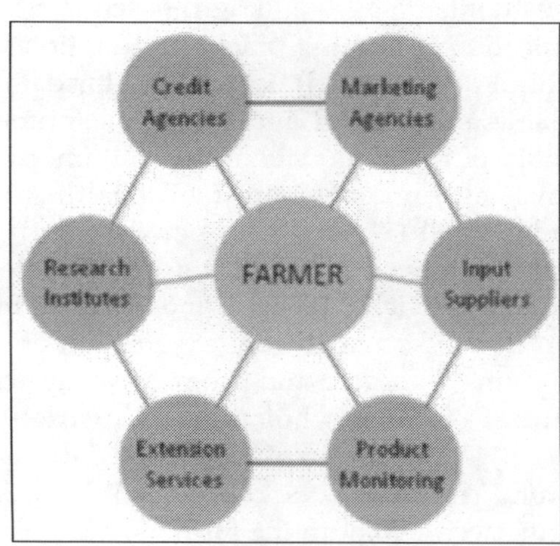

Figure 1: Farmer's Linkage with Various Stakeholders.

opportunities. Conversion of farming community (Small and Marginal Farmers, Medium Farmers and Large Farmers) as "e-Farmer" community through JAM (Jandhan, Aadhaar and Mobile) Number Trinity 2015, is a formidable challenge. Digitizing farms is therefore of importance for increasing income.

This paper discusses the means and methods through which the Horticulture Information System Network (HORTNET) requires to emerge as an Integrated ICT enabled value Chain (HORTNET 2.0) for Horticulture Farmers who have been categorized as: Fruit Crop Farmer, Vegetable Crop Farmer, Flower Crop Farmer, Medicinal Plant Crop Farmer, Plantation Crop Farmer, Coconut Crop Farmer, Bamboo Crop Farmer, Apiculture Farmer *etc.*, so as to provide progressive measures for increasing Farmers' Income through Digitalization.

1. Digital India Programme 2014: Targeted Objectives for Facilitating ICT for Farming Sector

The Digital India Program 2014 aims at transforming India into "digitally empowered society and knowledge economy" and providing thrust to nine Pillars of Growth. With the Internet of Thing (IoT) – where Smart Phones, Tablets and PCs are not included as part of IoT – Digital India Programme will usher in a new feature in e-Governance, through a seamless connected network of embedded objects/devices, with identifiers, in which M2M communication will reign supreme.This will facilitate "precision farming" in the Horticulture sector immensely. Digital Identification will be continuous to be challenging. The Digital India 2014 Programme in Horticulture Sector has the potential to transform the existing public service delivery system, improve productivity, create jobs and induce economic activity in areas which are not yet digitally connected or digitised.

1.1 Priority Areas for ICT Intervention

There has been an emphasize on: (a) development through good governance, (b) Improving supply-side constraints on agro and agro-based products, (c) modernization of farming practices (adoption of good agricultural practices), (d) national land use policy for sustainable agricultural and rural development, (e) availability of real-time data on agricultural products, (f) farm health management (soil health card), (g) "lab-to-land", (h) "more crop per drop", (i) multi-skill development, (j) Youth led development, (k) Internet connectivity to villages, and (l) digital India. In addition these, discussions on (m) creation of a talent pool of young, educated and progressive farmers, and agricultural research scholars in all districts of the country, (n) Development of a digitized database of all agricultural research in the country, (o) Reaching out agriculture technology to the farmers, have also been making inroads. These 15 areas are, in fact, national projects to be undertaken under the Digital India Programme in the Farming Sector. Appropriate Detailed Project Reports (DPRs) for these areas will facilitate Non-Governmental Organisations (NGOs) for their involvement to implement these priority projects at the grassroots level.

e-Governance and Informatics is the pathway for "Development 2.0" in Food and Agriculture, which requires an Institutional approach by bridging the development gaps in Human Resources: (a) Professionals through M.Tech/B.Tech courses in Agricultural Informatics and MBA in Agri-Business Management and Digital Marketing, and (b) e-Farmers through Competency Development Programme on "e-Governance in Farming Sector" for Agricultural Stakeholders. India requires about 2,00,000 Agricultural Informatics professionals by 2022, which is a great opportunities for Rural Youths. B.Tech in Agricultural Informatics Programme, for 10+2 students with Biology/ Mathematics subjects, will be the game changer (multiple skill development and youth-led development).

1.2 Benefitting from Digital Revolution

According to Mckinsey Global Research Institute's Report "12 Technologies to empower India" (2014), the estimated collective impact of technology interventions (mainly Information technology) in Framing sector is about "US $45 Billion to US $80 Billion" by 2025. The COST (www.cost.eu) Foresight 2030 Workshop (2009) on Food Security on the theme: "Benefitting from the Digital Revolution", has identified about 20 Technologies needed to enhance food security.

OpenData, OpenGov, Open Standards based Open Technologies, Computational Informatics, Indian Languages Computing, JAM Number Trinity, BharatNet, RailNet, Digital Cable TV Network, Wi-Fi Network, Cloud Computing, Localisation (L10N) and Internationalization (I18N) pave the way for bridging the development gaps in this sector. Drone advocates have cited precision agriculture-crop management that uses GPS and big data-as a way to boost crop yields and profits while resolving water and food crises [2].

Social Media, Mobile Computing, Analytics and Business Intelligenceand Cloud Computing (SMAC) technology, facilitate e-Markets and Value-Chain Analysis providing technological advantages to farmers. In addition, Geospatial Technologies safeguard the efforts of farmers by reducing cultivation uncertainties, and sowing seeds of prosperity.

1.3 Horticulture Production System: Challenges and Opportunities

As per the published Reports and Documents, the challenges faced in India are across the value-chain from supply-side (including production and post-harvest management) to demand-side (including non-tariff barriers). At the production level, the challenge is low productivity and at the post-production level, the challenge is high wastage rate due to technological gap (slow rate of development of post-harvest technologies and its dissemination). The suggested Imperatives to enhance Indian Horticulture Produces exports are as follows:

a. Efficient F and V supply-chain so as to reduce wastages to minimal level (w.r.t. both quantatively and qualitatively).

b. Value-addition (i) Primary: farm level and Post-harvest handling mainly by Farmers; (ii) Secondary : Primary processing into concentrates and juices, and (iii) Tertiary: Hi-end processing advanced technology; for all the three levels of value-addition, the following measures are to be complied with, to penetrate into international markets:

   i) Requirements of Good Agricultural Practices (GAP).

   ii) Grading/Sorting/Cleaning for uniform quality.

   iii) Minimizing Pesticide residues (as per international regulations).

   iv) Pest/fungal infections management.

   v) Taste and size preferences.

   vi) Packaging as per consumer preferences.

c. Networking of 27,479 Agricultural and Food Processing Units out of which only 4000 are involved in processing of Fruit and Vegetables.

d. Diversification of markets.

e. Ability to tap the potentials Small farmers to generate surpluses for export to overcome the supply-chain problems of:

   ☆ Scattered Smallholders production in isolated places.

   ☆ Limited availability of road and market infrastructure at the point of sourcing/ production.

f. Formation of Farmers' Cooperatives and Cluster Approach to undertake PH Management activities - cleaning, sorting, grading and scientific storage, reefer vehicle, skill development *etc*..

g. Providing Infrastructure and Collaborative Planning, Forecasting and Replenishment (CPFR) : PH Management Infrastructure, Cold Chain Infrastructure and Foreign Direct Investment.

h. Information dissemination on Weather updates, Prices and Farming techniques.

i. Market Intelligence : Introduction of Standard Operating Procedures (SOPs) and Electronic Data Capture to get correct and real time market information.

j. Promotion of Use of Plastics in Horticulture.

k. Precision Horticulture – GIS and RS data analysis - Soil data mapping, landscape features analysis and crop performance (yield maps).

l. Water and Irrigation Management.

m. Integrated Nutrient Management.

n. Integrated Pest Management - Disease and Pest Management.

o. Marketing Channels of Fruits and Vegetables.

p. Mapping of Cold-Storages in India – 6488 in Nos.

q. ICT enabled Extension services.

The EXIM Bank Report (2013) [1] suggestedto have an operational end-to-end ICT System for making the "Technology-led Horticulture development" process in the Country, especially in the states which are involved in exporting various horticulture produces:

A. States: Andhra Pradesh, Bihar, Gujarat, Haryana, Himachal Pradesh, Jammu and Kashmir, Kerala, Madhya Pradesh, Maharashtra, Odisha, Tamil Nadu, Uttar Pradesh and West Bengal.

B. Major Fruits: Banana, mango, citrus, papaya, guava, apple, pineapple, sapota, grapes, pomegranate and litchi.

C. Major vegetables: Potato, tomato, onion, brinjal, tapioca, cabbage, cauliflower, okra, peas, pumpkin and sweet potato.

The EXIM Bank Paper (2013) has also suggested: (a) undertaking steps in improvement of market intelligence through "internet enabled market information system" at horticulture markets to help in improving market integration and market efficiency - domestic demand and supply chain and export demand and supply chain, (b) ensuring consistency in supply through customized products, (c) complying with International standards and specifications including recorded and demonstrated traceability of products; and (d) generating India's capability to generate surpluses for export ability to tap the potential of small farmers to the large map of the horticulture sector.

A Centrally Sponsored Scheme of Mission for Integrated development of Horticulture (MIDH) (http://midh.gov.in) has been launched for the holistic development of horticulture in the country during XII Plan. The scheme, which has taken off from 2014-15, integrates the ongoing schemes of National Horticulture Mission, Horticulture Mission for North East and Himalayan States, National Bamboo Mission, National Horticulture Board, Coconut Development Board and Central Institute for Horticulture, Nagaland. The http://midh.gov.in website has about 14 related websites, linking to:

1. http://agricoop.nic.in
2. http://nhm.nic.in
3. http://tmnehs.gov.in
4. http://nbm.nic.in
5. http://nccd.gov.in
6. http://nhb.gov.in
7. http://coconutboard.nic.in
8. http://dccd.gov.in
9. http://cihner.gov.in
10. http://apeda.gov.in
11. http://mofpi.nic.in
12. http://agmarknet.nic.in
13. http://dial.gov.in
14. http://ncpahindia.com
15. http://farmer.gov.in
16. http://mkisan.gov.in

The Operational back-end Computerised Database System on the Horticulture Development in Central and State Government departments requires to be"interoperable". Horticulture stakeholders need to be facilitated to use one "Intelligent" Portal based on SMAC technology (Social Media, Mobile, Big data Analytics and Cloud Computing) for horticulture produce value-chains.

A comprehensive "Farmer-Centric" and "Farm-centric" Database for decision support is also lacking in the Horticulture sector, even though there are more than 15 websites (non-interoperable) operational for monitoring various horticultural schemes of the Government. The Ongoing Digital Networks for Farmers (DNFs) - AGMARKNET, HORTNET and AGRISNET- are required to be given focused thrust for "Market-led Extension" and "Value-chain" Analytics through M and E system.

1.4 Irrigation Access to Every Farm: SMART Water Technology Applications and Digitized District Irrigation Plan

The World Economic Forum (2014) highlighted the water crisis as the third most significant risk to economic growth. Sustainable management of the natural resource base supporting agriculture is one of the three major strategic objectives of the CGIAR Research Programme Water, Land and Ecosystems [3].

The Slogan "**More Crop per Drop**" pronounces Water Use Efficiency (WUE) measures very effectively. Remote sensing (RS) and Geographic Information Systems (GIS) technologies have huge potential to improve WUE in Irrigation commands through estimating: irrigated areas, cropping pattern, cropping intensity, soil moisture availability, evapotranspiration, crop water stress, land and water productivity, prospective yields, and extent of land degradation due to salinization, water logging, flooding and soil erosion.

The Pradhan Mantri Krishi Sinchayee Yojana (PMKSY) 2014 – irrigation - envisages creating and strengthening water distribution network and enhancing on–farm Water Use Efficiency (WUE) and management, through converging all ongoing irrigation related schemes *viz.*, Accelerated Irrigation Benefit Programme (AIBP), Integrated Watershed Development Programme (IWDP), Mahatma Gandhi National Rural Employment Guarantee Scheme (MGNRES), National Mission on Sustainable Agriculture (NMSA), Swarnajayanthi Gram Swarozgar Yojana (SGSY), Backward Regions Grant Fund (BGRF) and Rashtriya Krishi Vikas Yojana (RKVY) *etc.* The PMKSY irrigation scheme envisages connecting the irrigation system's three crucial components - water sources, distribution network and field applications - for optimal usage, and also developing a detailed "District Irrigation Plan".

Implementation of PMKSY scheme needs deployment of Remote Sensing and GIS technology and Database Technology for mapping water resources and distribution metwork, and on-Farm Internet of Things (IoTs) sensor based "SMART Water" for water budgeting. It requires Interoperability standards among various water management ICT systems for achieving higher WUE. To quote, Tomás Robles *et al.* (2014) proposed a Smart Water Management Model integrating Internet of Things (IoT) technologies for decoupling Decision Support Systems and Monitoring (DSS and M) from business processes coordination and subsystem implementation.

The Water Management ICT System (on-Farm) has the ability to adjust watering run times based on weather data or soil moisture data on a daily basis (which will amount to large scale water savings in a year), and also **to view** and make changes to an IOT-based controller from computer, tablet or smart phone.

To ensure that 'irrigation access to every farm" requires to get established, a District level irrigation portal having "water requirement and water released to every farm" with details linked to Computerised land records database and Aadhaar number of the owner of farmland (farmer) is essential. This means "creation of one web page of water budgeting and accounting" for

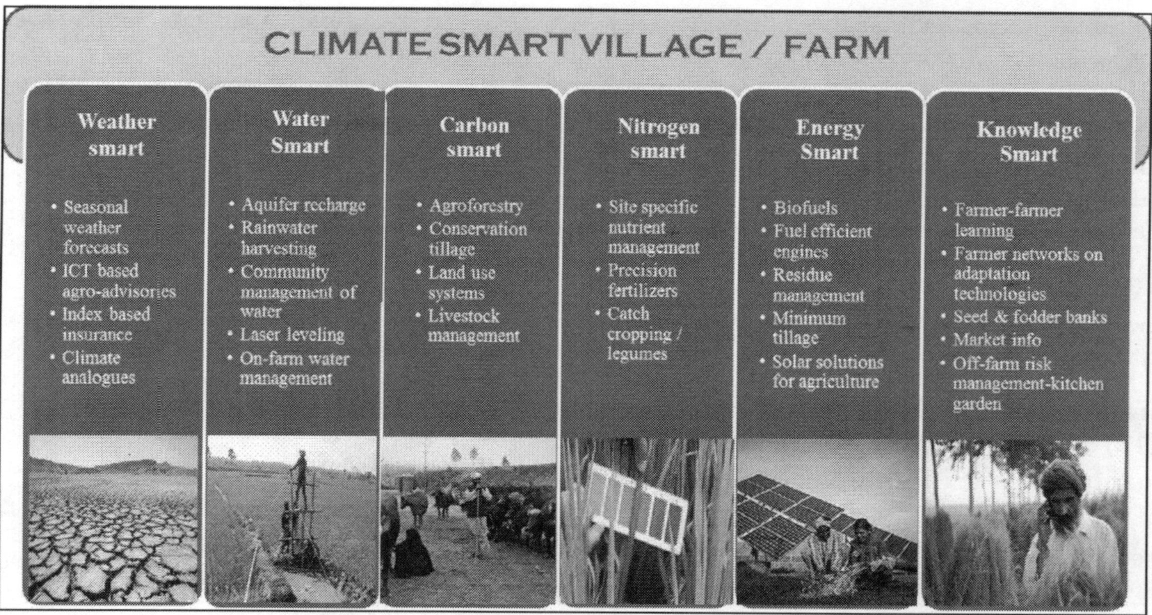

farmers of all categories (marginal, small, semi-medium, medium, and large).

Pilot Projects on SMART Water (http://csa2015.cirad.fr) are required to be undertaken in villages to calculate, through empirical evidences, the savings accrued on expenditure ofFarmer for raising a particular crop, as the step towards achieving "Digitized District Irrigation Plan" which will be inclusive of all categories of Farmers.

1.5 Warehousing, Logistics, Pledging, Clearing and Settlement–Digitisation to Remove Distress Sale by Farmers

There is no centralized database giving details of location and capacity of these storage structures and its associated road network, rail network, seaport network and airport network. There is a requirement for providing information to farmers on pre-harvest and post-harvest storage structures "any-time any-where" basis and establishment of on-line national warehousing receipt, for overcoming "distress sale" and achieving farmers' income rise.

There is a need to create a nationwide ICT enabled integrated warehousing management system by introducing "electronic warehousing receipt" (Electronic Warehousing Receipt Management System – EWRMS) for agricultural commodities. This EWRMS will be a Web-and-PKI based Application available 24x7. This is over-due to overcome the farmers' distress.The Private sector initiatives such as http://www.staragri.com and http://www.marketyard.com (announced) on ICT enabled warehousing management systems are encouraging. Under the Digital India Programme, this is expected to have its priority.

1.6 Agricultural Statistical System: A Quintessential Base for Decision Support

The Agricultural e-Governance System requires that, at least, the Block Level GovernmentInfrastructures associated with agricultural development sector (agriculture, livestock and fisheries) and rural development should be equipped with databases, workflow systems, portals and decision support systems (DSSs) based on spatial and non-spatial datasets on "farm-based" and "house-hold based". This means the Operationalisation of an Agricultural Resources Information (AgRIS) at the Block level. It requires application of "divide, build and integrate" algorithm to establish about 6,500 block level systems to realize "Agricultural e-Governance System" in the country.

The National Statistical Commission Sub-Group-V Report 2011 has recommended for overhauling existing Indian Agricultural Statistical System through ICT enabled process from "Farm-household" and "Farm-level" to Panchayat, block, district, state and centre, by bringing the National Agricultural Research System (NARS) Institutions and about 500 Departments of Life Sciences (Botany and Zoology) and about 250 Departments of Geography and Spatial Informatics in to the system, to facilitate implementation of the World Bank-FAO's Global Strategy to improve agricultural and rural statistics.

The establishment of a comprehensive computerised horticulture database management system is crucial for ensuring effective planning to facilitate a systematic development of the horticulture sector in the country. This database will help in exploiting the huge export potential for the sector. The area under horticulture crops which was 12.77 million hectares during 1991-1992 has increased to 23.69 million hectares during 2012-13;

the total production has increased by nearly 2.8 times and corresponding productivity has increased 1.5 times.

This data Book also draws data on domestic market prices from http://agmarknet.nic.inand Data onImport and export from http://apeda.gov.in, whereas the area and production statistics from the state government departments. There has to be huge manual intervention is required to cull out information from various data sources: Computerised databases, digitized but non-interoperable databases, non-digitized formats *etc*. It is still a Challenge.

1.7 Competency Development Programme (CDP) for Horticulture Farmer Producers Organisations (FPOs): e-Learning Portals

The Competency Development Programme (CDP) needs to be customized with respect to major Horticulture Export Products and issues related to:

i. agro-climatic, socio-economic technological, infrastructural, institutional and policy factors.
ii. economic aspects including production, profitability, equity and viability of small and marginal farms.

During this programme, efforts are to be undertaken as to how to establish a dynamic and enabling environment where ICT for Horticulture Sector can be grow with respect to:

A. ICTs and Value Chains: Integrated ICT solutions

- ✰ Mobile applications and ICT solutions
- ✰ Changing architecture of agricultural extension and rural advisory services
- ✰ To enhance and monitor agricultural processes
- ✰ To boost access to markets and facilitate agri-business
- ✰ Well-organised farm records - Farm database and Farm-house database
- ✰ Production, transaction and market data need to be fully integrated into market information services for driving economic growth and individual farmer's benefits.

B. Advocacy and Policy development

- ✰ Agricultural policies
- ✰ ICT policies
- ✰ E-horticulture policies

The CDP modules need to be designed "to generate India's capability to generate surpluses for export ability to tap the potential of small farmers, through Farmer Producer rganisations (FPOs), to the large map of the Horticulture Sector". Each CDP will be a 5-day programme (residential), excluding the dates of arrival and departure, with the following characteristics:

I. Arrival of participants on Sunday evening and registration
II. Day-1 Programme - Monday
   a. Horticulture trade and commerce issues - commodity-wise
   b. Promoting opportunities for youth in horticulture and ICTs
III. Day-2 Programme - Tuesday
   a. Horticulture productivity and production issues
   b. Field visit and demonstration
IV. Day-3 Programme - Wednesday
   a. Horticulture commodities – research and technology availability issues
   b. Field visit and demonstration
V. Day-4 Programme - Thursday
   a. Horticulture produce processing issues
   b. Post-harvest management and marketing issues
   c. Field visit and demonstration
VI. Day-5 Programme - Friday
   a. Value-chain and ICT strategies
   b. ICT enabled extension services
VII. Departure of participants on Saturday

The potential farmer producer organisations/ farmers club for capacity building can be identified through NABARD and SHM Directorates in States.

The CDP modules will have:

i) Case studies (both national and International) – film based, to be presented and discussed to workout "best practices to be adopted" for the benefit of all stakeholders;
ii) Group discussions to be arranged with ICAR/ SAUs experts, NABARD technocrats, KVKs, progressive farmers, practitioners, industries, banking and financial institutions, traders, and inspection agencies, *etc.;*
   a. ICT Device for integration with Commerce and Trade - Tablet PC with local language interface with Horticultural Commodity Portal to be provided to the Horticulture farmer;
iii) Portal for buyer-seller e-platform for horticultural produces.

Expected Outputs

i) Capacity Building of FPOs and Horticulture sub-district level officials, on Good Agricultural Practices (GAP) for Horticultural produces;
ii) Training materials for undertaking stakeholders programmes through on-line;

iii) ICT enabled interactive hubs (portals) for gathering and organizing information that will help research and development organizations and farmers to seize new opportunities and solve problems constraining horticulture systems;

iv) Market prices information access from AGMARKNET Portal, NHB Website, NHM website;

v) Portal for Buyer-seller e-platform for horticultural produces;

vi) Horticultural crop suitability map of India based on agro-climatic conditions, depicting most suitable areas for optimum productivity of a particular crop;

vii) Value-chain and supply-chain for major horticulture export commodities (in the similar lines of GrapeNet of APEDA).

Expected Outcome

i) Effective supply-chain management for export commodities so as to ensure consistency in supply of horticulture commodities for export;

ii) Recorded and demonstrated traceability of horticulture products;

iii) Increased country's capability to generate surpluses for exports by tapping potentials of small farmers;

iv) Reduction of quantitative and qualitative losses between the time of harvest and consumption;

v) Dissemination of accurate and efficient information towards effective supply-chain management

vi) Value-addition at primary, secondary and tertiary levels in marketability of Fruits and Vegetables

vii) Overcoming production imperatives through adopting good agricultural practices

viii) ICT enabled extension and advisory services to horticulture farmer producers organisations (FPOs) with respect to: Supply constraints, yield gaps and huge logistic costs affect India's competitive and comparative advantage in world trade market;

This CDP will eventually become an operational e-Learning portal.

2. Agri-Business Management: Emerging as an ICT enabled Value Chain and Potential Scope for large Scale Private Investment through Entrepreneurial Startups

India has a huge potential for growth, development and business in the agribusiness and food processing sector. It is estimated that by 2030, the agriculture and agribusiness industry is projected to be 3 trillion US$ in Africa and almost 4.5 trillion US$ in Asia. India emerges as an investment destination in the agribusiness and food processing sector, in view of the sound factors *viz.*, supply-side advantages, supply demand growth, conducive policy environment, increasing investments, and strategic location". ICT enabled Rural Business Hubs (RHBs) can facilitate to achieve this collective impact in Rural India. Table 1 exhibits entrepreneurial startups in agribusiness management in India. Human resources programmes such as MBA and PG Diploma in agribusiness management courses and MBA/ PG Diploma/Diploma in digital marketing and e-entrepreneurship course are required to be supported to facilitate this initiative on a largescale.

3. *e*-Marketing of Horticultural Produce

Internet has effectively facilitated e-markets which is expected to be more transparent and competitive than physical markets. The national market, including global markets, is accessible to the farming community through their websites. As horticulture produces are perishable in nature, proper handling and packaging, timely delivery, storage conditions, and knowledge of Internet operations and marketing in local languages are crucial. An ICT enabled value-chain (input industry-farmer-output industry-consumer) is required for each produce and as such, India requires about 400 agricultural commodity ICT value-chains. Concepts and principles of *e-commerce, e-business, e-auctioning, and e-marketing,* and their applications to agribusiness need to be understood. Skill development – Competency development – is essential for various stakeholders in the value-chains.

During the period of 2000-10, there have been some path-breaking initiatives (*e.g.* AGMARKNET, e-*Krishi Vipnan*, e-*Mandi*, e-*Choupal*, Hariyali Kisan Bazar), besides strengthening traditional information sources such as individual State Agricultural Marketing Boards, Commodity boards, Commodity Exchanges, National Federations, and Agri-Business organizations. However, the absence of inter-linkages (interoperability) among various boards demanded establishment of a comprehensive information networking system that straddles across commodity groups and markets, while delivering real-time, authentic information in a user-friendly format through localization. Table 1 gives a partial list of Websites (Government and Non-Government) on agricultural marketing information and on-line trading in India.

At the national scene, both AGMARKNET (a Government Initiative) and e-*Choupal* (a Private sector initiative) systems, had emerged as operational systems with scaling up in a sustainable manner, and received national and international appreciation and awards, for their impact at grassroots level.

Table 1: Websites (Government and non-government) on Agricultural Marketing Information and Online Trading in India – A Partial List

Sl.No.	*Websites URL*	*Characteristics*
1.	http://www.agmarket.nic.in of the Ministry of Agriculture, Government of India, and operationalized by NIC.	☆ Networked about 3297 Agricultural Wholesale Produce Markets (AWPMs), covering 29 States and 6 UTs/NCT with details about market prices and arrivals for about 400 commodities/2000 Varieties, and disseminating information in about 12 Indian languages apart English; ☆ Launching of Voice based Personalized Multi-modal and Multi Lingual Agro Commodity Pricing System on AGMARKNET developed by IIT Kanpur, Under the BSNL-TCOE Cooperation Programme on 29th August 2011; ☆ Mandi – Automatic Speech Recognition (ASR) for Commodity Price Teller, over AGMARKNET, in Six Indian Languages – Assamese, Bengali, Hindi, Marathi, Tamil and Telugu- by a consortium of seven institutions (IIT-M, IIT-K, IIT-B, IIT-G, IIIT-Hyd, TIFR and CDAC-Kol) funded by TDIL of DeitY, India; ☆ **AgriMarket mobile APP :** to get the market price of crops in the markets within 50 km of the device's location of CDAC Mumbai - launched by Hon'ble Agriculture Minister in December 2015; ☆ **Kissan Suvidha mobile app** (an Omnibus Mobile app developed by DAC) launched by Hon'ble PM having Market prices component data sourced from AGMARKNET portal in March 2016; and ☆ **IFFCO (IKSL) mobile APP** : IFFCO Kisan ☆ **Digital mandi India APP**, ☆ StartUp Mandi Prices and **mandi trade APP** (http://www.farmobi.com and www.manditrades.com) draws data from this Portal (as published in http://www.opengov.in) for marketing advisories; ☆ **Gramseva APP** - Gramseva is a project aimed at helping villagers tackle their everyday problems. This app focuses on helping farmers and wholesalers keep track of the commodity market prices of their farm produce. The app uses reliable data, directly from government servers: http://data.gov.in ☆ **Market watch APP**.
2.	EKVI (e-Krishi Vipanan) project (http://www.worldaginfo.org/ http://www.mapit.gov.in) – **not Operational**	☆ This e-Agriculture Marketing project of Government of Madhya Pradesh was operationalized through BOO basis with a consortium of vendors on PPP model, through MAP-IT Agency, with the aim to achieve "grainless mandis" in 2001; ☆ Vendors were paid a percentage of fees collected by the Agricultural Marketing Yards; ☆ Automation of Mandi Board Head Office, 7 Regional Offices, and 231 Mandis and their associated Sub-market yards and Nakas (Interstate barriers) across the State of Madhya Pradesh; ☆ Most of the Mandis and Sub Mandis are located in villages having 6 million Farmers with 70,000 licensed traders; ☆ Possibilities of e-Commerce, where in farmer could sell his produce at the village level to a license holder trader who could receive e-copy of essential Transit Permit "anugya patrak" from the requisite Mandi enabling him t0 transport the material directly to the processor/ godown *etc.* ☆ Operational during 2003 – 2008 and afterwards, the project suffered due to Non-charging of Service Tax Litigation case.
3.	ITC's e-CHOUPAL (http://www.echoupal.com)	☆ Covering about 4 million farmers in 40,000 Villages linked to about 6500 e-Choupal Kiosks in 14 States with facilities to transact in 8 Indian languages; ☆ Transactions with respect to soya, wheat, pulses, groundnuts, sesame, paddy, coffee, black pepper and marine products *etc.*
4.	http://www.dcmshriram.com/ haryali-kisaan-bazaar and http://www.shriramfarmsolutions.com/	☆ **Haryali Kissan Bazaar** - 37 Retail outlets in 6 States **to provide farmers with agricultural services** - improved crop inputs, agronomic support, fuel, banking, and consumer goods, **and buy back** some of the farmers' production at harvest time; ☆ Shriram Farm Solutions has been serving farmers for over four decades through an extensive network of around 3000 channel partners and 30000 retailers.
5.	http://www.agriwatch.com	☆ Information Service provided by Indian Agribusiness Systems Pvt. Ltd. (IASL) which has an information collection network through the country giving us access to over 5 lakh farmers and 25,000 of traders based in all States; ☆ Planning to introduce e-Commerce;
6.	http://www.kisanmandi.com	☆ **Online Agri-Market** to Buy or Sell or Advertise fruits and vegetables, agri produce or any agri machinery or Tools or Tractors *etc.* ☆ **Retail** – B2C just Like Dedicated Online Vegetable store in Main cities of India. ☆ **Online Farmers (wholesale)** – B2B: Online Portal Where farmers can submit their Agri Produce for Sale. ☆ **Vendors (Channel Partners)** – Online Portal where Manufacturer/Big Distributor can sell Agri Machinery and products to Farmers.

Contd...

Table 1–*Contd...*

Sl.No.	*Websites URL*	*Characteristics*
7.	http://www.efreshglobal.com	☆ Support to branding and marketing of Farm produce and other Food products directly to the consumer market; ☆ Dissemination of Good Agricultural Practices to farmers and connecting them to the market place and consumers. ☆ Building online linkages and Trade relationships between producers, marketers and consumers by ushering in a "Prosumer" (Producer-Consumer) society.
8.	http://www.Indiagriline.com	☆ to catalyze e-commerce in agricultural and non-farm products (*e.g.* Sugarcane procurement) by M/s EID Parry Ltd.
9.	Safal National Exchange of India Limited (SNX), Bengaluru – Not operational	☆ A joint venture between Financial Technologies, MCX, and Mother Dairy Foods Processing Ltd (MDFPL) of NDDB, with a target of doing business worth of INR 6 billion; ☆ Online trading in Fruits and Vegetables (Kinnow, banana, mango, onion, potato and coconut *etc.*) and milk; ☆ to act as aggregator for the lakhs of fresh produce farmers who may not be able to reach the nearest big market or be able to store their produce for a viable length of time; ☆ to allow traders, warehouse managers, bankers, retailers, wholesalers and exporters in the chain to do business with the same pool of farmers;
10.	www.ikisan.com	☆ Ikisan Limited, Nagarjuna Group's new initiative on 'farm management services and IT enabled services in agri-sector', to address Information, Knowledge and Business requirements of various players in the Agri arena *viz.,* Farmers, Trade channel partners and Agri Input/Output companies;
11.	http://www.enam.gov.in	☆ National on-line trading Platform of Ministry of Agriculture and Farmers Welfare, based on e-mandi model of the Government of Karnataka; developed by M/s. Nagarjuna Fertilizers and Chemicals Ltd., who is the Strategic Partner (SP) responsible for development, operation and maintenance of the platform.
12.	http://www.krushakbazar.com	☆ **e-agri trading portal** which promotes entrepreneurial, Industry oriented farming community and skilled Persons and organizes Technology transfer through effective extension methods catalyzing Agro-Industrial growth and prepare diversification of Agriculture and Agro based Industries to boost Productivity across the entire value chain;
13.	http://www.akashganga.in/	☆ **Automatic Milk Collection System** ensuring transparency, mutual faith and error-free operations of Milk Collection by integrating Electronics Weigh Scale, Quality testing equipment (EMT or Milk Analyzer) with computer system; ☆ **GPRS based Data Transfer** from Collection Point to Chilling Plants/Bulk Milk Coolers/Dairy Plants;
14.	http://www.fishtra.in	☆ A fish trading portal related to Fish (marine and inland), Farm fish, hatcheries and ornamental fish.
15.	http://www.pescafresh.com	☆ An **ICT enabled Supply-Chain Portal** on Fish Distribution : from 50 fishing centres – from "sea/Pond" to "Plate";
16.	https://www.ekgaon.com	☆ ekgaon Platform facilitates "Buy Direct from Farm";
17.	http://karshika.com	☆ Platform for local producers, manufactures, farmers and service providers to reach locally and globally; ☆ Avail rare and valuable Indian products and services globally; ☆ Karshika provide small scale producers "The power to sell";
18.	e-PashudhanHaat (announced in the 2016-17 Budget) of the Ministry of Agriculture and Farmers Welfare	☆ **e-Market Portal** for trading bovine germplasm by connecting breeders and farmers; ☆ **needs** to get integrated with Animal Health Card and emerge as a Virtual Livestock Market; ☆ http://www.agmarknet.nic.in has incorporated commodity Groups for Livestock and Poultry Markets;
19.	Vegetable and Fruit Promotion Council, Keralam (http://www.vfpck.org)	☆ VFPCK Swasraya Karshaka Samithi (Farmers Markets) facilitates trading between farmers and traders improving their bargaining power through better access with markets and traders through physical infrastructures. ☆ The primary objective of the Council is to improve the livelihood of Vegetables and Fruit farmers by empowering them to carry on vegetables and fruit production, value-addition and marketing as a profitable venture in a sustainable way. ☆ e-marketing is not yet introduced;
20.	http://www.agribazaar.co	☆ **Online Agriculture Market in India**
21.	http://www.indianagri.in	☆ **A B2B Platform for Indian Agricultural exporters, suppliers of agricultural products and manufacturers - Post a BUY Request, Get Competitive quotes, Select the Best options at a very low cost;** ☆ **Acts as a knowledge base for all issues related to export and import of agricultural products in India;**

Contd...

Table 1–*Contd...*

Sl.No.	*Websites URL*	*Characteristics*
22.	http://www.agricart.com	☆ Online shop agriculture and gardening needs of India
23.	http://www.IndiaMart.com – agribusiness Section	☆ India's largest B2B marketplace, offering comprehensive business solutions to the global EXIM community through its wide array of online services, directory services and facilitation of trade promotional events.
24.	http://www.TradeIndia.com - Agribusiness Section	☆ India's Largest B2B Marketplace, Connecting Buyers and Sellers.
25.	http://www.shopping.merakisan.com	☆ A "Farm to Folk" Portal. ☆ Unlike other online grocery shopping portals, Merakisan aims at supporting farmers to sell their products at fair price.
26.	http://www.krishidoot.in	☆ A leading information and market place platform for the agricultural community in India, trading with about 400 FPOs across 20 States. ☆ Krishidoot is an ICT based aggregation and transaction platform for Agri Communities, launched jointly by the Small Farmers Agri-Business Consortium (SFAC) of the Ministry of Agriculture and Farmers Welfare, and RML Information Services Pvt. Ltd (RML) in 2013.
27.	http://delhikisanmandi.in/[similar to http://vfpck.org of the Kerala State]	☆ To provide a platform for FPOs (producers) to directly sell to the Wholesale and Retail buyers in NCR of Delhi for which the Physical market is likely to come up in Alipur (Delhi) supported by SFAC – Not yet a digital mandi.
28.	http://www.farmpik.com	☆ Adani Agri Fresh Limited has instituted India's largest integrated apple supply chain initiative with ultra-modern storage infrastructure to provide fresh farm products to the consumers and improve the livelihood of about 15,000 apple farmers of Himachal Pradesh; Payment to Producer (Farmer) through Bank account;
29.	http://www.agriclinics.net	☆ This Government of India Programme facilitates to set up Agri Business Centres.
30.	http://upajj.com	☆ Farmer-Supplier-Buyer-Advisor Link; ☆ Comprehensive Mobile Platform for selling Crop and Farm products; managing entire lifecycle of farmer needs; providing e-extension services.
31.	http://www.halyaagro.com	☆ An agricultural platform to connect farmers to direct market, both domestically and globally;
32.	http://www.remsl.ine-Mandi and http://krishimaratavahini.kar.nic.in	☆ Trading of commodities in Karnataka APMCs operationalized during 2009; ☆ Provides United Market Platform (electronic Platform); ☆ Rashtriya eMarket Services Private Limited (ReMS) - a joint venture company with Government of Karnataka and the NCDEX Spot Exchange Limited; ☆ Market prices are available through http://krishimaratavahini.nic.in – a Kannada language version of http://www.agmarknet.nic.in; ☆ Data transfer between krishimaratavahini and agmarknet periodically;
33.	http://www.ekutirsb.com	☆ Developing micro-entrepreneurs to generate income streams, link markets, and create opportunities; ☆ Facilitating increase in income, productivity, and better livelihoods and providing avenues for growth and development; ☆ Working through a decentralized network of micro-entrepreneurs, who engage with the farmers and rural communities.
34.	http://www.farmobi.in	☆ Providing a "Farm to Shop" Trading Platform ☆ Facilitating a farmer to post details of his/her crops directly from their smart phone; ☆ Enabling Buyers and Traders to access the crops available locally within their location; ☆ Facilitating to make decision easy by accessing prices of farm commodities as provided by the Open Data Portal (http://www.data.gov.in which sources structured data through APIs from http://agmarknet.nic.in); ☆ Auto tagging of Farmer's location; ☆ Map based view of available farm products; ☆ Multi-Lingual Support.
35.	http://www.ficci-b2b.com	☆ Online promotion of agricultural products;
36.	http://www.msmemart.com	☆ Online promotion of agricultural products;
37.	http://www.agricultureinformation.com	☆ Connecting growers, buyers and experts;
38.	http://agriexchange.apeda.gov.in	☆ Promoting online trading platform in agri business for buyers and sellers to offer negotiate and transact a deal;

Contd...

Table 1–*Contd...*

Sl.No.	Websites URL	Characteristics
39.	https://www.ciitrade.in	☆ Gateway to Global Supply (buy, sell) Chains
40.	http://www.ahimsaact.com	☆ Supply Chain Marketing Outlets of Vegetables : "Farm to Home" in Chennai only, reducing considerably the middle agents between farmers and customers; ☆ Yet to introduce e-Commerce;
41.	http://www.kovaipazhamudir.com	☆ Supply Chain Marketing Outlets of Fruits and Vegetables through 44 outlets in South India; ☆ Yet to introduce e-Commerce;
42.	https://gem.gov.in	☆ Government e-Market Place (GeM): After comprehensive study through a consultant followed by engaging a Managed Service Provider (MSP), a full-fledged version of GeM is likely to be positioned by March, 2017 by DGS and D (GOI). The purchases through GeM by Government users have been also authorized by Ministry of Finance by adding a new Rule No. 141-A in the General Financial Rules, 2005.
43.	http://www.ebiz.gov.in	☆ Offers on-line access to core services needed to obtain necessary clearances, licenses and complete mandatory tax registrations and also on-line access to a number of regulatory filing services that required for operating one's business/industrial unit, within the country;
44.	http://www.sewarudi.com	☆ RUDI Bazaar – Rural Distribution Network for all villages in 14 Districts of Gujarat; ☆ Enables direct promotion and marketing of the agro commodities procured through Farmer's Association in the rural retail network of farmers and agriculture labourers in villages;
45.	http://www.marketyard.com (announced)	☆ An e-auction Platform for auctioning Warehouse Receipts (WR) issued to farmers and traders against agricultural produces (goods) stored in Warehouses of National Collateral Management Services Ltd (NCML); ☆ NCML expects 29,000 clients who avail its services to participate in auction platform on the buy side. ☆ Farmers who wish to avail of funding, have to store their produce at one of 10,000 NCML Warehouses located throughout India;
46.	http://www.levelacommodities.com	☆ A global strategy, knowledge management and consulting firm that offers advisory on agriculture value chains, trade and investment, international trade and analytics to clients worldwide.
47.	http://www.staragri.com	☆ Star Agriwarehousing and Collateral Management Limited (StarAgri) aims to improve farmer access to markets by building modern, organized agri trade and marketing infrastructure across India; ☆ Establishing electronic trading platforms and modern rural retail and wholesale markets that are efficient, professionally managed and responsive to farmer's needs; ☆ Modern Mandi: Aims to create 50 agri trade markets licensed by the Agricultural Produce Marketing Committees (APMCs) that serve as information, innovation and trade centres for agri produce (*e.g.* Kota in Rajasthan) ☆ With clean, hygienic and green environments, to provide one-stop services with the quality and transparency of international standards; ☆ With the help of India's leading financial institutions and banks, StarAgri currently holds commodities worth INR 75 billion across 200 collateral management locations.
48.	http://my-rml.android.informer.com	☆ RML Farmer - Krishi Mitr- An app where farmers can keep up with commodity and mandi prices, farm and farmer related news, weather information and advisory. ☆ Empowers those involved in agriculture and allied occupations in taking informed decisions by providing easy access to personalized and unbiased agriculture insights as per their location and in their preferred language. ☆ Users can choose from over 450 crop varieties, 1300 mandis, 3500 weather locations across 50,000 villages and 17 states of India.
49.	www.kisanmarket.com	☆ A platform for farmers and consumers to directly interact; ☆ An India's Largest online Agriculture Market(kisan) place to connect agriculture communities in India and provides online platform to connect Suppliers and farmers to facilitate hassle-free and convenient market to buy and sell all agriculture related products.
50.	i. www.nhb.gov.in ii. www.cmfri.org.in iii. www.nhrdf.org iv. www.indiacoffee.org v. www.indianspices.com vi. www.trifed.in vii. www.siffs.org, www.commodityindia.com	☆ Prominent Websites provide marketing intelligence information on agricultural commodities.

The Government of India, under its Digital India Programme, has launched its pan-India online National Agriculture Market Portal (http://www.enam.gov.in) on 14 April 2016 [10].In the Country, there are more than 22000 rural (primary and periodic) markets, being the first contact points of farmers with the market economy, both for selling and buying. As there have been high price differentials, many times between the Wholesale Markets and the Rural Markets, there has been room for arbitrage which is being exploited by the traders to their advantage.Rural Markets can become "Rural Virtual markets" (Rural Business Hubs) through private investment and entrepreneurial Startups in the agricultural sector (*e.g.* farmobi.com, eKutrisb.com, Kisanmandi.com, efreshglobal.com, ekgoan.com, vfpck.org, ahimsact.com, sewarudi.com, marketyard.com, *etc.*).

Therefore, it is imperative to make Wholesale Markets as the price discovery point and Rural Markets as the price takers with due consideration for transport and other costs. It is necessary for complete integration of e-NAM 2016 (a Pan-India electronic Trade Platform for APMCs) of NDA-II with AGMARKNET 2000 (a Pan-India with 3297 Nodes for Agricultural Produce Marketing Information System) of NDA-I, for the benefit of Indian Farmers to successfully operationalize about 400 Agricultural Commodities Value Chains as the bye-product by 2018.

In India, the various Trade regulatory and facilitating agencies, have established electronic interfaces (*e.g.* e-Trade, e-PQIS, GrapeNet, AnarNet and e-Biz) between them as well as with the trading community to allow electronic delivery of services facilitating enhancements in both exports and imports. ICT enabled Value Chains such as GrapeNet and AnarNet are required about 400 agricultural commodities to reap the benefits of Globalisation.

4. Horticulture Information System Network (HORTNET 2.0): Restructured ICT Intervention Programme for Increasing Farmers' Income

The Government of India has envisioned, during 1999-2004,"Doubling Food Productions" in India, and now has outlined an achievable 7-point strategy (with a good Strategy, well-designed Programmes, adequate Resources and good governance in implementation), to double Farmer's Income by 2022:

1. Big focus on irrigation with large budgets, with the aim of "per drop, more crop.";
2. Provision of quality seeds and nutrients based on soil health of each field;
3. Large investments in warehousing and cold chains to prevent post-harvest crop losses;
4. Promotion of value-addition through food processing;
5. Creation of a national farm market, removing distortions and e-platform;
6. Introduction of a new crop insurance scheme to mitigate risks at affordable cost;
7. Promotion of ancillary activities like poultry, beekeeping and fisheries.

Horticultural Production system is well suited to site-specific data techniques and information management systems which, interalia, includes: (a) Agricultural Resources Information System (AgRIS) for Micro-level Planning, (b) GIS based Network System of infrastructure facilities *viz.*, cold-storages, roads and distribution and management of horticulture products, (c) Agricultural Inputs System Network – Technology, Credit, Farm Equipments, Water, Seeds, Pesticides and Fertilizers *etc.* (d) Integrated ICT Enabled Certification Process for Organic Farming – TraceNet of APEDA (for Export Purposes) and PGS of NCOF (for Domestic consumption) are to be made interoperable in conjunction with GAPs of ICAR, (e) SeedNet (http://seednet.gov.in) and Soil Health Card Database System, as an interoperable Database system, (f) ICT Enabled Ubiquitous RFID Tag System/2-D Barcode for traceability of Packaging and marketing of Horticulture products, (g) adoption of SMART VILLAGE Programme to operationalize Climate-Resilient-SMART-Agriculture, and (h) Operationalisation of Competency Development Programme (CDP) for Horticulture farming stakeholders, through e-learning methods in local languages. These measures need to be made operational as a "GAPS - Citizen Charter".

The operational Horticulture Informatics System Network,HORTNET1.0(http://hortnet.gov.in), needs to be revamped from "Government-centric" to "Farmer-centric" and "Farm-centric" (HORTNET 2.0)with sub-informatics network components as:

1. Fruits Informatics Network System (FruitNet),
2. Vegetables Informatics Network System (VegNet),
3. Flowers Informatics Network System (FlowerNet),
4. Aromatics and Medicinal Plants Informatics Network System (AroMedNet),
5. Plantation Crops Informatics Network System (PlantCropNet),
6. Coconut Informatics Network System (CoconutNet),
7. Spices Informatics Network System (SpicesNet),
8. Bamboo Farm Produces Informatics Network (BambooNet),

9. Apiculture Informatics Network (ApiNET),
10. Marketing Information System Network (AGMARKNET) integrated with the proposed Virtual NAM System (AGMARKNET 2.0), and
11. Farm Health Management Information System.

The revamped HORTNET (HORTNET 2.0) will provide many opportunities for strengthening and establishing entrepreneurial Startups (*e.g.* e-Markets, eKutirsb.com, manditrades.com, ekgaon.com, vfpck.org, halyagro.com *etc.*), at grassroots level, having Agri-Business Value-Chains and Architecture as given in Figure 2.

5. Actionable Points: Way Forward

Digital Network for Farmers (DNF) is a strength, wealth and prosperity for farming households in India. It is required "to leverage" on a mix of emerging and existing technologies for effective and inexpensive ICT penetration in farming sector, which includes – Input system, production system, output system and (non-farm) rural development. Digital inclusion is the step towards reduction of farmers' distress. In order to boost farmer's income, India requires strategic intervention of ICT in farming system life cycle. To begin with, the following measures are certainly the way forward:

1. Operationalisation of Digital Network for Farmers (DNF) – HORTNET 2.0 :
    a. Fruits Informatics Network System (FruitNet),
    b. Vegetables Informatics Network System (VegNet),
    c. Flowers Informatics Network System (FlowerNet),
    d. Aromatics and Medicinal Plants Informatics Network System (AroMedNet),
    e. Plantation Crops Informatics Network System (PlantCropNet),
    f. Coconut Informatics Network System (CoconutNet),
    g. Spices Informatics Network System (SpicesNet),
    h. Bamboo Farm Produces Informatics Network (BambooNet),
    i. Apiculture Informatics Network (ApiNET),
    j. Marketing Information System Network (AGMARKNET) integrated with the proposed Virtual NAM System (AGMARKNET 2.0),
    k. Farm Health Management Information System.
2. Operationalisation of farm health management information system integrating plant health, soil health, water health – one health/eco-health – at farm level; the National Academy of Agricultural Sciences has already published an approach paper in this regard;
3. Operationalisation of agricultural resources information system (AgRIS) and micro level planning at village level;
4. Operationalisation of SMART Village Project;
5. Design and development of ICT enabled value-chain for all horticulture commodities (*e.g.* **Fruits**: Mango, banana, citrus, papaya, guava, apple, pineapple, sapota, grapes, pomegranate and litchi; **Vegetables**: Potato, tomato, onion, brinjal, tapioca, cabbage, cauliflower, okra, peas, pumpkin, sweet potato)
6. ICT enabled warehousing receipt management system;
7. Design and development of farmer centric service delivery (e-Farmer) system in 22 languages;
8. ICT enabled management of climate risk in horticulture sector;
9. Bridging the development gaps in human resources for digitalization of farming sector:
    a. Promotion of M.Tech. in agricultural informatics for B.Sc (Ag and allied sciences) graduates programme operational in the Shobhit University, Meerut.
    b. Promotion of B.Tech. in agricultural informatics for 10+2 students of rural India having agricultural background, with PCB/PCM subjects.

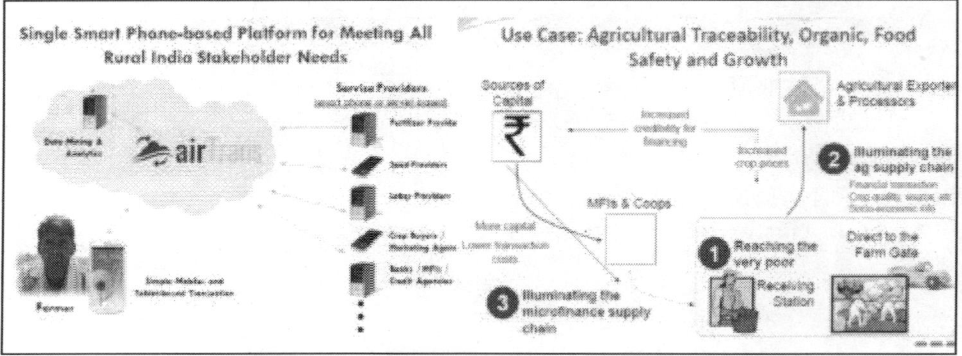

Figure 2: Agri-business Value Chain and Architecture (*Source*: http://www.shuvgray.com).

c. Promotion of MBA in Agri-business management to strengthen "market-led extension" in villages
d. Promotion of one-month programme on "e-governance in horticulture sector" for stakeholders horticulture farmers, FPOs, YOUTHs, government officers, extension officials *etc.* This programme is being offered by the Shobhit University Meerut (www.shobhituniversity.ac.in);
e. Promotion of One Week Programme on "Capacity Building for Horticulture Farmers" through e-learning Portals;

10. Networking FPOs (Farmer Producer Organizations)in every village and Farmer Producer Companies (FPCs) operational in every district, to strengthen Agricultural Marketing and promoting of *"Good Agricultural Practices"* in production of agricultural produces; Competency building of members of FPOs and FPCs in "Good Agricultural Practices" on priority basis;

This suggested strategy through ICT Intervention in Horticulture Sector can act as catalyst on Farm factors (Price and Non-Price) to maximising Value Creation by minimising costs in a commodity from farm-to-fork, which will necessarily boost farmers' income in the desired way. The Integrated ICT enabled value Chain (HORTNET 2.0), based on the features given in Figure-2, for horticulture farmers who have been categorized as: Fruit crop farmer, vegetable crop farmer, flower crop farmer, medicinal plant crop farmer, plantation crop farmer, coconut crop farmer, bamboo crop farmer, apiculture farmer *etc.*, is recommended so as to provide progressive measures for increasing farmers' income through digitalization.

India requires, however, an operational Integrated Rural Informatics System for sustaining productivity and income, so as to facilitate the Government of India's ongoing interventions in the farm and non-farm sectors to double the income of the farmers by 2022. The centre for Agricultural Informatics and e-Governance Research Studies (CAIRS) of the Shobhit University has initiated research project activities to evaluate this strategy.

6. References

Dillow, Clay. 2015. Why 2015 is the year agriculture drones take off. in http://fortune.com/2015/05/18/drone–agriculture/; May 18.

eWorld Forum. 2012. "HORTNET : National Horticulture Mission and All State Horticulture Missions" – Best ICT-enabled Agriculture Initiative of the Year, http://ewf.eletsonline.com/2012/05/hortnetnational-horticulture-mission-all-state-horticulture-missions, held at new Delhi on 15–16 June.

EXIM Bank. 2013. *Indian Horticulture – Imperatives to enhance Trade from India*. Occasional Paper No. 162 of the EXIM Bank of India, October.

FICCI. 2014. The report *"Unlocking the Food Belts of Asia and Africa"*, released by FICCI during the Asia Africa Agribusiness Summit 2014 held in February, New Delhi.

Indian Horticulture Statistics Book – 2014.

IWMI. 2010. CGIAR Research Programme on Water, Land and Ecosystems (WLE) led by IWMI – www.iwmi.cgiar.org.

Moni, M. 2010. Agriculture e-governance, logistics and warehousing facilities. Invited talk presented in the *National Conference on Public Private Partnership Warehousing in Semi Urban and Rural India: Issues and Solutions*, November 15th, ASSOCHAM House, New Delhi.

Moni, M. 2010. HORTNET: A horticulture informatics network for productivity increase and farmers' income Rise in India. Invited speaker in the Technical session: *New Initiatives in Technology Transfer and Adoption* of the 4th Indian Horticulture Congress 2010, organized by Horticultural Society of India, New Delhi; Published in the Proceedings of the 4th Indian Horticulture Congress.

Moni, M. 2011. Agricultural e–governance system for Indian farmers. *Good Governance Magazine*, September.

Moni, M., Mishra, Hare Krishna and Roy, Siddhartha. 2011. Report of the sub-Group-5 of the Committee on Statistics of Agriculture and Allied Sector: IT and IT usage. **In**: *Agriculture Data, Agro-Climatical Level and Small Area Farm Business*, April, National Statistical Commission, Ministry of Statistics and Programme Implementation, Government of India, New Delhi.

NAM. 2015. *National Agriculture Market: An introduction*. Pamphlet, Small farmers Agribusiness Consortium (SFAC), Ministry of Agriculture, Government of India, New Delhi, June.

Pradhan Mantri Jan–Dhan Yojana progress report (http://www.pmjdy.gov.in/account–statistics–country.aspx)

93

Fruit Nursery Production: An Attractive Entrepreneurship Venture for Youth

M.I.S Gill* and N.K. Arora

Department of Fruit Science,
Punjab Agricultural University, Ludhiana – 141 004, Punjab
****E-mail: misgill@pau.edu***

The backbone of quality fruit production is the elite planting material. The planting material is considered elite if it meets all the requirements prescribed by the authority to verifyauthenticity of the nurseries at national level.National Horticulture Board (NHB) with its head office at Gurgaon is the prime body having responsibility for accreditation of the fruit plant nurseries in India. As per the prescribed guidelines for accreditation of nurseries, a particular nursery (public or private) is rated upto a maximum of 5 star, which signifies the highest authenticity of quality planting material produced in a particular nursery.Mango, citrus, banana, apple, guava, pomegranate, papaya and grapes are important fruit crops of India, which constitute more than 95 percent of the total area under fruits in the country. Of the total fruit plant nurseries accredited in India, there are only about 140 fruit nurseries having NHB rating 3 or above, in other words, which produce quality planting material of important fruit plants. These nurseries have a production capacity of around 90 lakhs of major fruit plants (Apple, mango, citrus, grapes, guava, and pomegranate) per annum, which is much lower than the actual yearly requirement of quality planting materialof these fruit plants in our country (Table 1). Based on the data of previous years, an area expansion under fruit cultivation in India in next decade is likely to be 1.4 percent per annum among important fruit crops. To meet the target of 1.4 per cent growth per annum by 2026 in area of fruits, about 98 crore quality fruit plants are required. Thus there exists immense demand of quality planting material of important fruits (Table 2).

In the present scenario of inadequate availability of quality planting material, non-accredited nurseries are proliferating at an exponential rate. Moreover, most of these non-accredited nurseries are being run by uneducated/illiterate or minimally educated persons (*malis*), having least knowledge of quality planting material and about its significance in the field. In other words, there are specific requirements for production of quality nursery plants, the foremost being the provision of land for raising mother plants of stock and scion. The nursery cannot be principally accredited if there is no provision of mother plants each of stock and scion. There is a mismatch between demand and availability of quality planting material of fruit plants, which needs immediate attention, as fruit plants have long productive life.

Keeping in view the huge demand of quality planting material against the few accredited nurseries of fruit plants providing such material, as well as in the scenario where more than 2/3rd nurseries are unaccredited, more so, being run by illiterate persons, there is a huge scope for raising quality nursery fruits plants. To understand the basics of nursery production and to raise its potential upto entrepreneurship level, literate youth can play an important role. Nursery production if taken up on a mission mode, is a profitable

Table 1: Production Capacity of NHB Star Rated (3 to 5 star) Fruit Plant Nurseries in India

Sl.No.	*Fruit Plant*	*Production Capacity (in lakhs)*			
		Star Rating			
		3	*4*	*5*	*Total*
1.	Almond	0.00044	0.068	–	0.068
2.	*Aonla*	3.04	0.385	–	3.42
3.	**Apple**	**3.01**	**2.29**	–	**5.3**
4.	Apricot	0.07	0.32	–	0.39
5.	Arecanut	0.01	2.00	–	2.01
6.	Assam lemon	0.15	–	–	0.15
7.	*Bael*	0.23	0.065	–	0.29
8.	Cashewnut	1.00	2.15	–	3.15
9.	Cherry	–	0.052		0.052
10.	**Citrus**	**12.34**	**4.98**	**3.20**	**20.52**
11.	Coconut	0.395	3.08	–	3.47
12.	Custred apple	0.255	0.04	–	0.259
13.	Fig	0.05	–	–	0.05
14.	**Grape**	**1.26**	**–**		**1.26**
15.	**Guava**	**7.55**	**4.4**	**–**	**11.95**
16.	Jackfruit	0.15	–	–	0.15
17.	Jamun	0.12	–	–	0.12
18.	Khasi mandarin	1.0	–	–	1.0
19.	Kiwi	0.131	–	–	0.131
20.	Lemon	0.18	0.02	–	0.20
21.	Lime	1.17	–	–	1.17
22.	Litchi	2.55	–	–	2.55
23.	**Mango**	**44.73**	**2.59**	–	**47.32**
24.	Olive	20.0	–	–	20.0
25.	Orange	2.0	–	–	2.0
26.	Papaya	0.15	–	–	0.15
27.	Peach	0.03	0.493	–	0.52
28.	Pear	0.38	0.43	–	0.81
29.	Plum	0.18	0.37	–	0.55
30.	**Pomegranate**	**2.82**	**2.63**	–	**5.45**
31.	Sapota	2.8	1.68	–	4.48
32.	Walnut	0.25	0.041	–	0.29
	Total	**108.0**	**28.0**	**3.2**	**139.2**

Source: NHB database.

venture and can economically surpass the existing traditional crop cultivation system in the country. In other words, by running nursery production as a entrepreneurship venture, the rural as well as urban youth having specialised training/diploma in nursery production after even 10+2 grade and having limited land can make more profit by selling quality plants as compared to conventional crops/orchards from same piece of land.

The unemployed graduates can become self-sufficient by producing quality nursery plants.Theuse of specialized techniques like tissue culture or by using protected conditions farmers can produce plants all around the year, which becomes good source of income for an average family throughout the year. As compared to illiterate persons in nursery business, the educated youth, has high marketing IQ, which can help him earn remunerable returns even through distant marketing of produced quality nursery plants. Central government in the recent past has given emphasis on the growth of horticulture in the country, as a result the demand of fruit plants has also risen significantly. Moreover, to get away from the traditional system of cropping, diversification particularly through horticulture is a most viable option. Diversification in real sense can be achieved, only if accredited nurseries are promoted and strict restrictions are imposed on the private/non-accredited nurseries especially selling fruit plants. National Horticulture Board (NHB) provides financial help for establishment of quality nurseries. An educated youth can thus adopt fruit nursery production as an entrepreneurship venture and can become financially independent and self-sufficient to meet family needs.

Raising Nursery Plants

a. Establishment of Mother Block

The most important requirement for production of quality fruit plant nursery, is the availability of sufficient space for establishment of mother plants each of rootstock and scion. The mother plant refers to the elite plants selected on the basis of quality traits, which differ w.r.t. rootstock and scion. Until the mother plants are there in a particular nursery, the assurance of quality cannot be made to the customer/farmer having demand of nursery plants. To establish, mother blocks we need to first understand the kind of fruit plants to be propagated and the method of propagation most adequate for commercial propagation.

Rootstock play an important role, to maintain productive long life of fruit plants and it plays an important role with respect to tolerance/resistance of fruit plants to biotic and abiotic stresses. However, most of nurseries in and around cities are being run on a limited land area, especially non-accredited nurseries do not have the mother plants of rootstock and are also not aware what type of rootstock they are using. The use of rootstock of unknown origin creates lot of problems, related to biotic or abiotic stresses in fruit plants. As fruit plants have a long life span in the field, as compared to annual crop/vegetable/flowering plants, the use of rootstock as well as scion of unknown pedigree/source hamper the growth and yield potential of fruit plants, year after year in succession. Different fruit plants are budded or grafted on specific rootstocks as detailed in Table 3.

Table 2: Projected Demand of Nursery Plants of Major Fruit Crops in India

Fruit Plant	Area 000 ha		Increase per annum (%)	Projected area under Fruits after			Projected Demand of Nursery Plants	
				5 year	10 years	15 years		
	2006-07	2015-16		20-21	25-26	30-31	2025-26 (Crore)	2030-31 (Crore)
Apple	252	314	2.5	355	400	452	3.46	5.53
Citrus	798	970	2.2	1079	1201	1336	6.34	10.06
Grapes	65	122	8.8	186	283	430	35.69	68.48
Guava	176	248	4.1	303	370	453	3.36	5.62
Mango	2164	2243	0.4	2284	2326	2369	1.04	1.58
Papaya	72	124	7.2	176	249	353	38.59	70.65
Pomegranate	117	190	6.2	257	348	471	9.88	17.56

Table 3: Propagation Method and Rootstock of Important Fruit Crops grown in India

Fruit Crop	Method of Propagation	Rootstocks Used
Apple	Grafting	Crab apple (seedling rootstock); M 27 (ultra-dwarf),M 9, M 26 (dwarf), MM106 (semi-dwarfing), MM111 (vigorous)
Citrus	T-budding	Rough lemon, Rangpur lemon, Pectinifera, Cleopatra mandarin, Sour orange, *Poncirus trifoliata*, Carrizo citrange, Dragon Fly, *Severinia buxifolia*.
Grapes	Cutting, budding	Dogridge, 110 R, St. George, 1616, 1613, SO4, Teleki
Guava	Inarching, patch budding, wedge grafting	L-49, Pusa Srijan (dwarfing)
Mango	Inarching, wedge grafting, side grafting, veener grafting	*Desi* mango seedling, Vellaicollumban, Olour, Pahutan, Goa, Kurrukan (polyembryonic)
Peach	Grafting	Sharbati, Florda Guard
Pear	Grafting	Kainth (*Pyrus pashia*), Sucker (*Pyrus serotine*), Quince

B. Ideal Rootstock

- ✰ It should be compatible with the scion, which in turn will ensure long productive life of a fruit plant.
- ✰ It should impart tolerance/resistance to soil borne diseases or physiological disorders. For instance nematode tolerant rootstock should be preferred for the fruit plants like peaches having susceptibility to nematodes. The other requirements are salt tolerance, tolerance to *Phyophthora* and other diseases or rootstock should have ability to impart biotic and abiotic stress tolerance to the fruit plant.
- ✰ An ideal rootstock should impart precocity to the scion. In other words, in fruit plants like pear which have long juvenile period and come into bearing after at least 5-6 years, should be grafted on rootstock which can impart early bearing in such fruit plants.
- ✰ The rootstock should impart fruit quality over the years of its productive life.
- ✰ It should impart dwarfness to the plant, without compromising the yield potential and quality attributes.

C. Ideal Scion

- ✰ The scion primarily should be a proven variety having known pedigree for high quality fruit production.
- ✰ It should have capacity to tolerate incidence of insect-pest and diseases to a reasonable extent.
- ✰ It should have compatibility with ideal rootstock.

Thus the quality fruit plant is one which is a combination of an ideal/quality rootstock and a scion. The nursery plants such developed have a long productive life and produce quality fruits.

To establish a mother block, apart from the knowledge of ideal stock and scion, we need to have sufficient space for raising these plants under field conditions. The other basic requirements are trained technical staff for maintenance of mother plants as well as for budding and grafting of fruit plants; provision of basic amenities like regular water supply, connection of electricity, nearness to the city and ease of transportation facilities.

Tips for Successful Nursery Production

Apart from the selection of suitable rootstock, scion, open or protected structures, some important technical and general points must be kept in mind for success in nursery production of fruit plants:

a. Technical Points

- ☆ Rootstock and scion mother blocks of known pedigree, is the foremost requirement for success in nursery production *vis-à-vis* to get remunerative returns from nursery production venture.
- ☆ Scion should be of a known variety having commercial significance in local and distant markets and as per consumer requirement/ demand.
- ☆ Rootstock and scion should be compatible with each other for ensuring long life of fruit plants at grower's field and long credibility of a nursery.
- ☆ Grafting in deciduous fruit plants like apple, pear, peach and plum should be done when plant is dormant.
- ☆ In grapes under north Indian conditions, unlike other deciduous fruit plants like pear, peach, plum, the grafting success is maximum just a few days before natural sprouting (end February) of plants.
- ☆ In deciduous fruit plants, after grafting, waxing of joints and terminal ends of stock using bee wax/ordinary wax helps to enhance successful grafts.
- ☆ Semi-hard wood scion has a better success rate in case of grafting, as compared to soft wood and hard wood scions.
- ☆ Simultaneous grafting of fresh stock and scion as in plum and peach, followed by planting in nursery beds or directly in field for rooting saves one year, however, can only be successful with the use of suitable rooting hormones like IBA or IAA.
- ☆ The clean cut between stock and scion ensures better grafting success in survival of plants. In other words there should be no air passage through the graft union.
- ☆ In wedge grafting, maintaining humidity by use of paper bags helps to improve the grafting success in plants such as guava and mango.
- ☆ For inarching to be successful in mango, the scion should be prepared 7-10 days in advance by removal of leaves along with petioles from the scion. This helps in formation of plump buds in leaf axils, which when grafted on rootstock ensures maximum success of grafted plants.
- ☆ Budding particularly patch and T budding should be done, when sap flow is maximum and there is ease in removal of bud from the scion. For instance in citrus and guava, T budding and patch budding is done in Feb-March, Aug-Sep and May-June, respectively.
- ☆ For successful budding in citrus, the rootstock should preferably be round and of pencil thickness. The angular stock do not support the bud and thus fails to produce successful plants.
- ☆ To have maximum success rate, patch budding in guava is done in the months of May-June, while wedge grafting can be done in Feb-March.
- ☆ While doing *in situ* budding, particularly in evergreen fruit plants like citrus and guava, there must be sufficient soil moisture in the field.
- ☆ Bending and tying of scion, particularly in citrus after few days of budding improves the success rate of successful plants, as well as hastens the time for production of nursery plants. Bending helps to directs the free flow of metabolites towards bud union.
- ☆ Constant removal of shoots emerging from rootstock after budding, beneath the bud union helps to improve the success of budded plants.
- ☆ For better success in propagation during off season, *in situ* budding or grafting ensures better success of plant propagation. For instance, grapes which do not go dormant under south Indian conditions, can be successfully established under North Indian conditions, by *in situ* grafting of a particular variety on a rootstock already established in the field.

b. General Points

- ☆ Nursery men should have the idea of demand of nursery plants in nearby locality and also in distant markets. For distant marketing E-marketing links should be established.
- ☆ Always select the well-connected site for nursery for ease in sale and transport of nursery plants.
- ☆ Always maintain the advance booking register in the nursery and timely inform the buyer on start of season. This will add to the credibility of a particular nursery.
- ☆ Near the planting season advertisement in local newspaper and media should be given to promote the sale.
- ☆ Nursery staff should be well trained, particularly having knowledge of budding and grafting. A diploma holder of Garden Training Course (GTC) or Horticultural Supervisor Training course (HSTC) can be an asset to the nursery.
- ☆ Keep nursery free from weeds, insect-pest and disease infestation. The sale point particularly should be attractive and clean.
- ☆ At sale point along with fruit tree seasonal annual/seeds/pots plants may also be kept for additional income.

- ✰ Proper record of sale should be maintained by issuing receipt to buyer and mentioning name of plant and variety.
- ✰ Record of inputs used in nursery is also important along with sale record to know the profit and loss in quantitative terms and for sustaining nursery business over the years.
- ✰ Each and every plant in nursery should be tagged, mentioning name of scion and rootstock.
- ✰ Disease free sale of quality nursery plants having known pedigree should be the sole motto for success in nursery production business and successful entrepreneurship.

The fruit plant nursery can be raised under open field conditions or under protected structures. The choice of raising fruit plants under open or protected conditions depends upon several factors:

i) Kind of fruit plant to be propagated. The protected structures are preferred for fruit plants which require extra care and can be sold off at premium rates. The fruit plants which can be easily propagated under open field conditions should not be opted for propagation under protected structures.

ii) Economic status or risk taking ability of the entrepreneur. If the nursery is to be established only on the loan money and with low economic buffering, easy to propagate fruit plants under open field conditions should be selected at first instance.

iii) The nursery fruit plants in which quality is affected particularly due to viruses, should preferably be produced under protected conditions. Although the plants produced, are not virus resistant/tolerant, but healthy plant at initial stage have better performance under field conditions, as compared to disease infested plants.

iv) Cost benefit ratio is the final criteria to decide, whether fruit plants should be produced under open or protected conditions.

Protected Nursery Raising of Fruit Plants

To produce fruit plants under protected conditions, different structures are in commercial use and each of them has its pros and cons. The different protected structures are discussed as under:

1. Shade Nethouse

To protect the nursery fruit plants especially during summer months shade net house is a viable option. During summer, the tender nursery plants are severely affected due to sun scorching thereby adversely affecting its growth. Shade net house can be of any dimension, as per the requirement of the plants and financial status of the entrepreneur. In the market, different shade nets are available which can partially or prevent the solar radiations and even UV safe nets are also available. Ideally a shade net which can allow a reasonable photosynthesis as well as protect the plants from sun rays is most adequate. A shade net which can prevent upto 60 percent of the sun rays is generally considered ideal for growth of nursery plants. During summer, if shade net has the provision of misting accompanied with proper exhaust fans it can further add to its value.

2. Polyhouse/Glasshouse

During winter, polyhouse/glass house is a good option to protect the plants from frost and cold waves. In general, under open field conditions plant growth is restricted during severe winters. Preferably a transparent material should be used as it allows solar radiations to enter the protected structure. By keeping the plants under poly/glasshouse during winter, a continuous growth can be obtained, thus can shorter the time span to produce nursery plants. In other words the plants become buddable early, as compared to under open field conditions. As compared to net house, this structure is more durable, but expensive. This cannot be used throughout the year, as during summer months, the temperature in such protected structures is not bearable for tender plants.

3. Mist Chamber

For propagation of more tender plants, specialized protected structures having a facility of regular misting is a basic requirement. Under mist chambers, the plants need to be kept under regular misting, which allows rooting of cuttings and thus tender plants especially those in which soft wood cuttings are the method of propagation can be propagated on commercial scale. The need for regular misting makes this method of propagation more expensive than net house.

4. Screen House

This kind of protected structure is preferred in those plants which are affected by insect vectors and thus suffer from viruses. Citrus for an example is affected by several viruses such as tristeza, exocortis and ringspot. The nursery plants raised under screen house are initially healthy and are known to have early establishment and better growth under field conditions. It is pertinent to mention over here that such plants are not virus resistant, but initially free from viruses.

5. Tissue Culture Propagation

The plants which are difficult to propagate under open field conditions can also be propagated through tissue culture. However, the cost of production of plants through tissue culture or via micropropagation can only be justified if the propagation rate exceeds the propagation under open field or under other protected structures. Propagation of plants through tissue culture

Table 4: Estimated Cost of Commercial Fruit Plant Nursery Production under Protected Structures

A	**Cost of structure (Rs.)**	
i.	Polyhouse for Primary Nursery: Size 32' x 96' (one) for 50,000 seeds	2.5 lakhs
ii.	Shade nethouses for secondary nursery: Size 32' x 96' (four)-30,000 seeds	36 lakhs
iii	Screen houses for maintaining mother plants: Size 32' x 96' (Two) Required for approximately 100 plants for obtaining 300 buds per plant	26 lakhs
	Total cost of structure (approx.)	65 lakhs
	Annual depreciation (5% per annum suppose 25 years average life)	3.25 lakhs
	Interest over fixed capital (7% rate of interest)	4.55 Lakh
	Annual depreciation/cost on fixed structure (approx.)	**7.8** Lakh
B	**Variable cost**	
	Cost of seed, containers, fertilizers, labour, repair of implements, irrigations *etc.*	3.37 Lakh
	Interest on variable costs (@ 7% for 9 months)	0.23 Lakh
	Approximate annual variable cost	3.60 Lakh
	Approximate total cost per annum = **Cost of Structure + Variable cost (7.8 + 3.6)**	11.4 Lakh
	Cost of one plant (approx.)	**Rs. 45.6***

* *Note*: Cost will vary with quality of the components used and from one place to another.

have advantage such as around the year propagation of plants, as no environmental barriers which hamper plant propagation exists. The true to type plants can be produced, where normally sexual method (seed propagation) is adopted for commercial propagation of plants as is the case of papaya.

Cost of Raising Nursery Plants

The cost of raising nursery fruit plants depends upon several factors such as scale of nursery production/ number of plants to be produced per annum, nursery production under open or protected structures, cost of components-which vary from place to place, mechanization in nursery production, irrigation and fertilization with or without drip system, rooting media and quality of components used.

The estimated cost (Table 4) for production of disease free citrus plants on commercial scale (25,000 plants per annum) under protected structures is detailed below:

As per rough cost estimate, under open field conditions, the nursery plants can be produced at half the rate (Rs. 20 per plant), as compared to under protected structures. The plant propagation through tissue culture being the most expensive. However, it is important to note that choice of structures for nursery production depends upon kind of fruit plant and its value in the market. For instance, if fruit plant is difficult to propagate and can be sold at high rates in the market, it can be propagated even using tissue culture facility. On the contrary if the fruit plant can be easily propagated through cuttings and by seeds), propagation under open field conditions can be most viable option.

Source of Funding for Nursery Enterprise

A. Assistance from NABARD

As per Model Scheme of National Bank for Agricultural and Rural Development (NABARD) for establishment of fruit plant nursery a financial support of Rs 17,75,500/- (Table 5) over period of three years is provided. The youth, to start nursery production as an entrepreneurship venture may obtain the loan from banks up to 75 percent of total cost. The bank loan is phased over period three years.

Table 5: Cost of Establishment of Fruit Tree Nursery as per Model Scheme of NABARD

Source of Funds	*Year 1*	*Year 2*	*Year 3*	*Total*
Total cost (Rs.)	12,59,300	2,08,400	3,07,800	17,75,500
Margin (upto 25%)	3,15,300	52,400	76,800	4,44,500
Bank Loan (upto 75%)	9,44,000	1,56,000	2,31,000	13,31,000

Table 6: Assumptions of Financial Viability of Project

Parameters	*Year 2*	*Year 3*	*Year 4*	*Year 5 Onwards*
No. of plants grafted (on basis of availability of scion and rootstocks)	12000	22500	37500	37500
Saleable graft (survival @ 80% success)		9600	18000	30000
Sale price per graft		35	35	40

Source: Model scheme of NABARD.

If scion is available from reliable source, nursery infrastructure may be utilized for production of fruit plants from first year itself or nursery men may utilize infrastructure for production of vegetable nursery/ flower nursery for additional income.

B. Assistance from National Horticulture Mission (NHM)

Other government agencies like NHM may also help (vary from state to state) for production of nursery plants or provide the subsidies for increasing area under fruits which indirectly enhance the requirement of nursery of fruit plants. Youth may also use facilities provided by government agencies for establishment of entrepreneurship from time to time.

- ✰ **Establishment of Hi-tech Nurseries**: The assistance is provided by NHM for establishment of hi-tech nurseries with an area of 1 to 4 ha and with capacity to produce 50,000 plants per ha of mandated perennial fruit crops per year. The plants produced are duly certified for their quality. National Horticulture Mission provide the support for fencing of nursery, Scion/ Mother block of improved varieties, Rootstocks block, Net house, irrigation facilities, Hi-tech greenhouse having insect proof netting on sides and fogging and misting systems, hardening/ maintenance in insect proof net house with light screening properties and sprinkler irrigation system, pump house to provide sufficient irrigation to the plants and water storage tank to meet atleast 2 days requirement, Soil solarisation – steam sterilization system with boilers.
- ✰ **Establishment of Small Nurseries:** National Horticulture also provide assistantship to small nurseries with an area of up to 1.00 ha. These nurseries have the provision for naturally ventilated greenhouses and net houses. Small nurseries will produce 25,000 plants of the mandated perennial vegetatively propagated fruit plants per year, duly certified for its quality.
- ✰ **Establishment/strengthening of tissue culture units**: For quality production of planting materials, new tissue culture units with a capacity to produce 25 lakh plants of the mandated crop for which protocols are available for commercial use, except Date palm. In the event of production of lesser number of plants, cost of lab will be reduced on a pro data basis (subject to its viability). Each tissue culture unit would get itself accredited as per standards and norms of Department of Bio Technology within eighteen months from the date of receiving funds, failing which the assistance provided under the scheme will have to be returned.

Conclusion

To conclude, fruit nursery production is a profitable venture and can be adopted by youth as an attractive entrepreneurship. However, to have remunerative returns over longer periods of time, production of quality nursery plants needs to be the key motto of a nurserymen. The nursery plants for sale must be accredited from NHM, and preferably should have high star rating. The quality nursery plants are must for establishment of healthy orchards in the country. In a parallel move, to encourage young entrepreneurs to adopt nursery production as a business on commercial scale, the nursery act must be enforced in letter and spirit. Those states where the nursery act is not there, it must be enacted. The sale of fruit plants from non-accredited government and private nurseries should be strictly banned in the country.

References

NABARD: Model Scheme for establishment of Modern nursery unit for production of quality planting material for perennial horticultural crops. www.nabard.org

NHB: Operational guidelines (2015) – National Horticulture Board. www.nhb.gov.in

NHM: Operational guidelines (2014) - Mission for Integrated Development of Horticulture. www.midh.gov.in

94

Vegetable Seed Production: A Remunerative Entrepreneurial Venture

Sudipta Basu[1]*, Malavika Dadlani[2], B.S. Tomar[3] and Rashmi Singh[4]

[1]Division of Seed Science and Technology,
[2]Former Joint Director (Research),
[3]Division of Vegetable Sciences,
[4]Division of Agricultural Extension,
ICAR-Indian Agricultural Research Institute, New Delhi – 110 012
E-mail: bst_spu_iari@rediffmail.com

India is the second largest producer of vegetable crops next to China covering an area of about 94,00,000 ha with a production of about 162.9 million tonnes and average productivity of 17.3 t/ha (NHB, 2014). India contributesnearly 15 per cent of the world's vegetable production from 2.8 per cent of cropped area. Even though intensive research and development in vegetable crops has given a substantial boost to its productivity, there is still a vast gap between average and potential productivity. Thus to bridge the gap, adoption of hybrids or high yielding OPVs along with best production technologies and integrated pest management is required. This necessitates timely availability of high quality seeds/seedlings of suitable hybrids or HYVs to the farmers.

Globally, seed industry is worth US$ 45 billion and Indian seed industry ranked 5th in the world with a net worth of US$ 2.2 million with major contribution from cotton, rice and vegetable seeds (Dubey, 2013). Among vegetable seeds, public sector has major share in seed production and marketing of OPVs and private sector with high value low volume hybrid seeds. Vegetables seed segment has a significant share in the Indian Seed Market. Though the actual contribution made by the public and private seed companies in vegetable seed market is difficult to assess, as per the industry estimates, the hybrid vegetable seed market in India is of about Rs.1500 crore which grew at a rate of 10-15 per cent per annum.

The public funded research institutes played a key role in establishing the vegetable improvement and seed production programmes in the country. However, the R&D on vegetables is also very active in the big and medium private seed companies. Out of about 110 vegetable hybrids released by the All India Coordinated Project on Vegetables, nearly 60 percent are developed by the private sector. With the increased availability of quality seeds of high yielding OPVs and hybrids, the SRR in most of the vegetables, which was ~ 20 percent in the early 80s, rose to 60-90 percent by early 2000.

India is endowed with diverse agro-climatic conditions which facilitates seed production of a wide range of vegetable crops. Availability of cheap labour and skilled technical manpower supported by liberal import and export government policies since the adoption of New Seeds Policy (1988), facilitated a great potential for vegetable seed production both for domestic and export markets.The strengths and weaknesses of Indian vegetable seed sector are:

Strengths

a) Availability of varied agro-climatic conditions to suit seed production of large number of hybrids and OPVs of different crops.

b) Well developed vegetable seed production chain under public sectorand high quality maintenance and good marketing in private sector.

c) A network of seed certification agencies for maintaining high seed quality during seed production.

d) Well established big and medium seed companies to bridge gap between demand and supply of vegetable seeds.

e) Availability of cheap trained man power and technical expertise for vegetable seed production.

Weaknesses

a) No restriction for seed production of any crop in any region, thus maintaining isolation becomes difficult.

b) Non availability of parental lines of private sector hybrids for seed production

c) Involvement of large number of small private seed companies with poor R&D facilities in seed production.

d) Localised areas of vegetable seed production.

e) Lack of awareness, resources and technical know-how about vegetable seed production technology to small and marginal farmers.

f) Unorganisedseed production, marketing and distribution of quality seeds.

g) Inability of public sector seed agencies to produce sufficient quality seed of improved varieties/ hybrids.

h) Higher seed cost of OPVs and hybrids sold by private seed companies.

i) Lack of efforts to popularize public sector hybrids besides development of viable seed production technology.

Naik *et al.* (2012), estimated the vegetable seed requirement, production and deficit in different vegetables in the country (Table 1) and estimated that the F_1 hybrid seed requirement is around 3,000 MT (Table 2). It was also estimated that an area of 25,000 acre is under hybrid seed production exclusively for domestic market (Table 3). To meet the requirement of quality seed of vegetables the following measures can be adopted:

a) Assessment of area under vegetables to estimate the actual requirement of quality seeds

b) Popularisation of public sector hybrids through frontline demonstrations.

c) Development of seed villages for intensive and systemic seed production.

Table 1: Estimated Vegetable Seed Requirement and Production in India

Crop	*Total Requirement (tonnes)*	*Production from Organized Sector (tonnes)*	*Production from Un-organized Sector (tonnes)*	*Shortage (tonnes)*
Tomato	360	190	80	90
Brinjal	465	85	50	330
Chilli	200	68	80	52
Cauliflower	260	130	35	115
Cucumber	70	37	20	13
Muskmelon	110	37	40	33
Watermelon	320	135	100	85
Bottle gourd	615	105	100	410
Onion	1200	200	400	600
Okra	4250	1350	800	2100
Radish	650	300	200	150
Carrot	700	200	250	250
Beat root	400	100	120	180
Peas	6000	1500	2000	2500
Cabbage	200	80	60	60

Source: State of Indian Agriculture, NAAS, 2009, pp. 108.

Table 2: Estimated Vegetable F_1 Hybrid Seed Requirement

Vegetable	*F_1 Hybrid Seed Requirement (MT)*
Okra	1700
Gourds	800
Watermelon	130
Cole crops	110
Tomato	90
Chilli	80
Brinjal	75
Cucumber	30
Capsicum	4
Other melons	20

Source: Reddy (2011).

Table 3: Estimated Production and Acerage under F_1 Vegetable Seed Production (Domestic market)

Vegetable	*Domestic (t)*	*Area (acre)*
Tomato	67	1787
Watermelon	120	2400
Chilli	70	933
Okra	1500	15000
Melon	15	250
Brinjal	50	625
Cucumber	40	667
Ridge Gourd	50	500
Bitter gourd	100	1000
Bottle gourd	120	640

Source: Anand (2011).

d) Financial assistance to both public and private sector seed companies to strengthen quality seed production.

e) New untapped areas should be explored indigenously for production of seeds of tropical as well as temperate vegetable crops.

f) Rural folk should be encouraged and trained in seed production of varieties and hybrids.

g) Encouraging public, private and farmers' partnership for vegetable seed production.

In India, Karnataka, Maharashtra and Gujarat are the leading states in quality seed production of vegetables. Among which nearly 90 per cent of the total hybrid vegetable seeds are produced in Rannebenneur area of Karnataka due to congenial weather conditions, availability of skilled and trained labour, guaranteed returns from seed production and incentives for quality. This has also helped in improving the socio-economic scenario in this region, including overall prosperity, narrowing down of rural/urban divide and employment generation especially for village women and youth. This success story has to be simulated in other potential seed production regions of the country.

An effort for diversification of seed production area to northern states is also need of the hour. Repeated seed production of a crop in a given location not only increases pest problem, it also adversely affects the soil health. As majority of hybrid vegetable seed production is undertaken in Karnataka and seed is transported to north India for cultivation which increases the seed cost. Thus, to ensure timely availability of quality seeds at affordable cost,diversification of seed production areas in north India is an alternative, where vegetable seed production could be a viable option for entrepreneurship development. Given adequate training, enterprising and progressive farmers and rural youth could undertake seed production for income generation and create a seed hub for quality seed supply of vegetables to the farmers of the same/neighboring villages.However, it has not yet become popular amongst farmers and rural youth due to the following major constraints:

a. Unorganised production, marketing and distribution of quality seeds of hybrids and high yielding OPVs.

b. Lack of awareness, resources and technical know-how of vegetable seed production as an entrepreneurship venture.

c. Non availability of the parental lines and package of practices for quality seed production.

d. Lack/low incentives for seed production and poor buy back facilities for quality seed produced by the farmer.

e. Lack of information on cost benefit ratio of seed production as compared to crop production to facilitate adoption of seed production.

f. Localised seed production areas in Karnataka and Maharashtra restricting adoption of large scale seed production of hybridsin other potential areas.

Vegetable seed production in India is undertaken asdirect contract or custom production basis by both public and private sector.

A. Contract Seed Production

The seed production of vegetable crops is being undertaken by progressive farmers on contract basis, mostly in compact area approach. In compact area approach, a group of farmers (50-150, or more) are identified in a favourable locationand treated as one unit to undertake seed production of one variety/hybrid of one crop. The advantages of compact area approach of seed production are:

i. A large number of farmers/farm-labours, particularly women can be trained on location.

ii. Integrated approach can be adopted for input management and seed production technologies.

iii. Need for isolation distance to prevent contamination could be avoided/reduced.

iv. Genetic purity of the hybrids or OPVs is better ensured.

v. One trained supervisor can efficiently monitor seed production in a village or a village cluster.

vi. Ease of accessibility to new technologies to the farmers, including farm machineries and irrigation facilities.

vii. Ease of harvesting, processing, storage and transport.

viii. Ease of field inspection and seed certification.

ix. Plant protection measures can be taken up more effectively.

x. Overall reduction of cost of seed production

B. Custom Seed Production

Seed is produced by one company, either for its other global markets, other companies for solely for foreign markets. This has gained popularity due to economic advantage in terms of skilled labour, availability of conducive production climate and technical expertise and production of international standard hybrid seeds in the open or contained conditions.

The economic growth and development of any nation depends primarily on the role played by their entrepreneurs. An entrepreneur is a person, who strives to maximize his profits by adoption of innovations. Entrepreneur are not simply innovators, but have a will to act, take risk and bring a change through organization of human efforts. Seed production can be adopted for entrepreneurship development which could shift farmers from subsistence farming to commercially and economically viable agriculture. Different entrepreneurial venturesavailable for seed production of vegetables are:

a. Hybrid Seed Production

Hybrid seed production of vegetable crops provides great opportunities for fulfilling the seed demand and generating employment opportunities. Seed production could be undertaken either by progressive farmers or small and marginal farmers on contract basis in compact area approach. Due to high value of hybrids, seed production is undertaken either on contract basis through farmers or by the private seed producers, which provides a more profitable option of employment generation from small holdings of land for small and marginal farmers employing rural youth. To the loyal and skilled small farmers, several seed companies also provides 30 to 50 percent of their estimated net income in advance enabling them to meet the initial investment required for hybrid seed production in vegetable crops. The assured procurement price, buy back arrangement, technical guidance and fund advances to ensure the financial security of the small and marginal farmers engaged in hybrid seed production

As per an estimate made a decade back by Singh and Dutta (2005), India produced 994 tonnes of hybrid seeds of vegetables from nearly 8,000 acres, generating employment of 2.71 million man-days per year, of which solanaceous crops contributed for more than 56.46 per cent towards employment generation, followed by cucumber (28.08 per cent) and okra (15.46 per cent) (Table 4). Approximately 994 tonnes of hybrid seed is being produced in tropical vegetables from an estimated area of 7,957 acres leading to an employment generation of 2.71 million man days per year. Among vegetables, hybrid seed production of sweet pepper was most remunerative, followed by hot pepper, tomato and water melon. It is estimated that the total employment generation from vegetable seed sector was estimated as over 7,00,000 man days per year. Though no such data in the current scenario is available, the scale of economic advantage of hybrid vegetable seed production over that of vegetable crops remain unchanged.

b. Seed Production of Open-pollinated Varieties

Production of seeds of improved open-pollinated varieties could be taken up as a profitable entrepreneurial venture, in crops like okra, pea, bittergourd, tomato, brinjal and chillies, which are relatively short duration crops and do not require seed production in two stages, as in case of cole crops. Presently, certified/truthfully labeled seeds of public and private bred varieties are being produced by public agencies like NSC, SFCI and State Seed Corporations as well as private seed companies by progressive and trained farmers on contractual basis

As per estimates of Kapur (2011), out of total OPVs sold in India, 70 per cent consists of beans, coriander, onion and peas by value. Around 40,000 MT seeds of OPVs of 25 vegetables are sold with an estimated value of US$ 200 million.

c. Organic Seed Production

Organically produced common, traditional and indigenous vegetables have gained importance among elite populations in the metro cities. Thus organic seed production of these vegetables for use in vegetable production can be a viable option for entrepreneur development especially in peri-urban areas.

Table 4: Estimated Employment Generation through Hybrid Seed Production in Vegetable Crops

Crop (Hybrid)	*Hybrid Seeds Produced (tones)*	*Area Covered (Acres)*	*Man-Days Employed per acre*	*Crop Duration (Days)*	*Total Man-days Employed (million)*	*% Contribution*
Tomato	38	1,583	460	120	0.75984	27.97
Hot pepper	30	375	1,800	150	0.67500	24.85
Okra	700	2,333	180	120	0.41994	15.46
Watermelon	65	1,083	165	110	0.17870	6.58
Cucumber	15	180	450	100	0.08100	2.98
Ridge gourd	20	166	472	110	0.07835	2.88
Melons	10	200	350	100	0.07000	2.58
Eggplant	15	115	600	150	0.06900	2.54
Bitter gourd	20	200	150	150	0.03000	1.10
Bottle gourd	20	166	150	120	0.02490	0.92
Sweet pepper	1	56	534	150	0.02990	1.10
Squash	60	1,500	200	120	0.3000	11.04
TOTAL	**994**	**7,957**	**5,531**		**2,71663**	**100**

Source: Singh and Dutta, 2005.

d. Seed Production under Protected Cultivation

The seed production under open field conditions is often affected by various biotic and abiotic stresses which affects seed production in terms of quality and quantity. Seed production could be undertaken under protected conditions namely insect proof net house, naturally ventilated greenhouse, climatic controlled greenhouse, walk in tunnels which gives higher seed yield, better seed quality (virus free) with lesser use of chemicals and also creates higher employment generation. Large number of farmers, in peri urban areas have successfully diversified by adopting different protected cultivation technologies for seed production of vegetables depending on availability of area, resources, manpower, mechanisation and demand of vegetables seeds. In general, the cost of construction of protected structure varied from Rs. 200-250 to 500 per m^2 and with adequate training the farmers themselves can construct these structures at village level by using rural artisans. To encourage its adoption, subsidies are also available for construction of these structures from different state governments are:

i) Higher seed yield (2-3 times) and better seed quality as compared to open field.
ii) Lesser problems of maintenance of isolation.
iii) Better seed and fruit setting as compared to open field.
iv) Longer stigma receptivity and pollen viability, extended crop duration favouring higher yields.
v) Better pest management due to reduced usage of pesticides.
vi) Feasibility of seed production under adverse climatic conditions.

Presently, seed production of sweet pepper, hot pepper, indeterminate tomato, melon, cucumbers are undertaken under protected conditions. The major advantages of vegetable seed production under protected conditionshas been reviewed by Singh (2010) and Singh and Tomar (2015). They reported better vegetative growth, longer pollen viability,stigma receptivity, prolonged duration of pollination, higher fruit and seed set and better seed quality in solanaceous and cucurbitaceous vegetables under net house,naturally and artificially ventilated polyhouses, as compared to open fields.The protected cultivation has proved highly profitable with limited resources and manpower to produce quality seeds of vegetables.

Seed/Seedling Production under Protected Conditions

a. Plug Tray Technology

Seedlings raised under open conditions during rainy and post rainy season are prone to get infected with soil borne fungi, nematodes and diseases especially viruses transmitted by insects. In addition, high cost of hybrid seeds (Solanaceous vegetables) warrants the farmers to improve nursery raising method to ensure establishment of each seedling and protect them from various biotic and abiotic stresses. Raising of healthy vegetable seedlings in plug trays in artificial potting mixture under protected conditions is an innovative technology to raise virus-free vigorous seedlings with good root development. Theses seedlings could to sold to farmers for raising an commercial or a seed crop. Plug tray nursery raising technology has been standardized for different vegetables especially solanaceous and cucurbitaceous ones under greenhouse conditions (Singh and Tomar, 2015).

b. Plastic Low Tunnels

Seed production of cucurbits under open field conditions may get frost and cold spell in northern plains during early winter or spring seasons. Raising initial crop in plastic low tunnel protects the crop from frost and advances it by 20-30 days over their traditional system of cultivation. This ensured prolonged duration of flowering, higher fruit and seed set and seed yield.

c. Shade Net-House

Shade nets are plastic nets which reduced solar radiation and temperature during summer months of April-August as the scorching sun desiccates the leaves, flowers and pollen grains which severely affected the seed production. These shade nets are commonly used for seed production in summer season which enhanced crop growth and prolonged flowering and improved seed set as compared to un shaded control. Among nets, black coloured shade nets were most effective due to maximum absorbance of heat as compared green, white or silver coloured shade nets. Seed production of leafy vegetables, cucurbits and early cauliflower could be undertaken under these nets.

d. Insect Proof Net House

Insect proof net house is a dome shaped protected structure constructed with galvanized iron frame covered with insect proof net (35-40 mesh/inch^2) which prevents entry of insects, hoppers, aphids and restricts transmission of diseases. Hybrid seed production under net house is a profitable and environment friendly technology wherein the crop is more vigorous, insect free, exhibits higher fruit and seed yield and better seed quality as compared to open field condition. Thus it helped in reducing cost of seed production and indiscriminate use on pesticides for insect and pest control. But for growing these crops under insect proof net houses, it is pre-requisite to raise virus free healthy seedlings of these crops either in the greenhouse or by covering the nursery beds with insect proof net.

e. Low Cost Polyhouse

Low cost polyhouse is a GI pipe structure, which is covered with 200-800 micron plastic and the sidewalls are covered with insect proof net of 40 or 50 mesh up to 5-6 feet height followed by plastic sheet. During winter season polythene is rolled down which increases the temperature inside the polyhouse and in summers (March-April) all side rolls of plastic is rolled up for cross ventilation. These structures can be used for hybrid seed production of tomato, chilli and other solanaceous vegetables and year round virus free healthy nursery raising of vegetables.

The villagers and women folk in tribal and rural areas, can also initiate nursery production of vegetables under open or protected conditions on communitybasis for growing low cost seedlings/plantlets for crop production.

Entrepreneurship Development: IARI Experience

Hybrid seed production in vegetables like tomato, brinjal, bottle gourd, bittergourd, pumpkin, summer squash, cauliflower *etc.* are highly remunerative especially in the peri urban areas for farmers and small entrepreneurs. ICAR-Indian Agricultural Research Institute (IARI) has released improved varieties and hybrids of above vegetables and developed their cost effective hybrid seed production technology under open field and protected conditions for adoption by farmers.

I. Hybrid Seed Production of Vegetables under Open Field

a. Cauliflower: Pusa Sankar 1 (Early Group): An average yield of 15 kg hybrid seed can be achieved from 500 sq m area (Cost of hybrid seed Rs. 4,000/kg).

b. Brinjal: Pusa hybrid 5, Pusa hybrid 6 Pusa hybrid 9. An average seed yield of 180-220 kg can be achieved from one ha area (Cost of hybrid seed Rs. 16,000/kg).

c. Bottle gourd: Pusa hybrid 3 and Pumpkin: Pusa hybrid 1. An average seed yield of 170 kg can be achieved from one ha area (Cost of hybrid seed Rs. 6,000/kg).

II. Hybrid Seed Production under Protected Cultivation

1. *Hybrid Seed Production under Net House*

a. Brinjal: An average yield of 2.0-2.5 kg of hybrid seed/100 sq m area of net house could be achieved per season. A farmer can earn Rs. 32-40,000 (Rs. 16,000/kg of seed). The net house can be retained for 3-4 years with minor repair, thus the returns on seed production increases in the following years. The benefit cost ratio is 1.9:1.

b. Bitter gourd: An average yield of 2.0-2.5 kg of hybrid seed/100 sq m area of net house can be achieved per season. A farmer can earn Rs. 12,500 [5 kg seed/year (2 seasons) @ Rs. 2,500/kg of seed]. The benefit cost ratio is 2.3:1.

2. *Hybrid Seed Production under Polyhouse*

A. Tomato: An average yield of 3.5 kg of hybrid seed/100 sq m area of polyhouse can be achieved per season from which a farmer can earn Rs. 70,000 per year (Cost of hybrid seed : Rs. 20,000/kg of seed). The benefit:cost ratio is 2.87:1.

Entrepreneurshipdevelopment through vegetable seed production was undertaken in a project on *"Developing Agricultural Entrepreneurs for Vegetable Seed Production"*, funded by NABARD. The project was launched with creation of awareness and mobilization of farmers for vegetable seed production by on farm and institutional trainings. Interactive meetings were organized with district agricultural/horticultural officers, subject matter specialists, *sarpanch*, BDOs, NABARD officials and farmers. Hands-on training on emasculation and pollination techniques were provided to the farmers in adopted villages to promote seed production of vegetables.Farmers were also provided technical input, planting material and four net houses were constructed in adopted villages; *Tigipur* (Alipur Block), *Badarpur Said* (Faridabad Dist.), *Jhatikara* (Najafgarh) and *Kansala* (Rohtak Dist). Bitter gourd hybrid seed production technology under net house was adopted in three villages; *Tigipur*, *Badarpur Said* and *Kansala*. A trained farmer from *Badarpur Said* produced 1.8 kg of hybrid seed from 100 sq m net house. The crop was jointly inspected by scientific personnel from IARI, New Delhi and National Seeds Corporation, New Delhi. The seed was procured from the farmers by IARI, New Delhi and bought by National Seeds Corporation, New Delhi for marketing under Farmer NSC buy back agreement facilitated by IARI, New Delhi.

References

Anand, N. 2011. Growth of hybrid vegetable seed industry in India: Progress towards a bright future. *Indian Seed and Planting Material*, **4**(4): 7–11.

Anonymous. 2014. *Indian Horticulture Database*. National Horticulture Board, GOI.

Anonymouus. 2009. *State of Indian Agriculture*, NAAS, pp. 108.

Dubey, Bhupen. 2013. Vegetable seeds export: India and world. *Seed Times*, p. 23–24.

Kapur, Arvind. 2011. Onion tears brought back the focus on vegetable seed industry. *Indian Seed and Planting Material*, **4**(1): 3–6.

Naik, S.S., Singh, P.N. and Singh, B. 2012. Hybrid revolution in vegetables. *Seed Times*, p. 73–84.

Reddy, A.S.N. 2011. Indian seed Industry: Status, challenges and opportunities in vegetable crops. *Indian Seed and Planting Material*, **4**(3): 7–12.

Singh, Balraj and Tomar, B.S. 2015. Vegetable seed production under protected and open conditions in India: A review. *Indian J. Agril. Sci.*, **85**(10): 1251–1259.

Singh, Balraj. 2010. Protected cultivation of vegetables. **In**: *Farmers Training Manual*, Centre for Protected Cultivation Technology, Indian Agricultural Research Institute, New Delhi, 15th–23rd June, pp. 121–127.

Singh, U. and Dutta, O.P. 2005. Seed industry's contribution in improving farmer's income through contract hybrid seed production. *Indian Seed Industry*, N.S.A.I, pp. 19–24.

95

Mushroom Production: A Viable and Profitable Avocation for Educated Youth

Manjit Singh* and Mahantesh Shirur

ICAR-Directorate of Mushroom Research,
Solan – 173 213, Himachal Pradesh
**E-mail: manjitbhandal122@gmail.com*

There are thousands of edible species of mushrooms, but only a few hundred have been cultivated. Only six species (*Agaricus, Pleurotus, Lentinula, Flammulina, Auricularia* and *Volveriella* species) grown commercially constitute 90 per cent of total world mushroom production (Royse, 2014) (Figure 1). In the beginning of the 20th Century, almost whole of world mushroom production comprised of button mushroom. Over the decades the production of other species has been on increase and as per the latest estimates, shiitake and oyster seem to have overtaken button mushroom as far as contribution to global mushroom production is concerned. The reason for this is the exponential increase in production of oyster, shiitake, wood ear and *Flammulina* in the last decade in China (*Zhang et al.*, 2014).

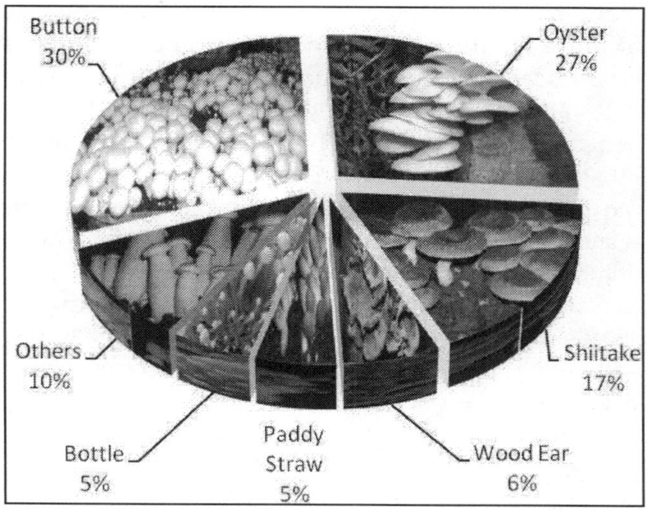

Figure 1: The Six Species Contributing 90 per cent of Global Mushroom Production (Royse, 2014).

Even today few species are collected from nature, but there is a growing interest in cultivation to supplement or replace wild harvest. It has been well understood that mushroom cultivation has high income generating potential and there is increased recognition of the nutritional value of mushrooms. Mushrooms can play an important role contributing to the livelihood of youth both in rural and urban areas through food security and income generation. The demand for quality and non-conventional foods is going to increase in the current century. There is a rapid change in culinary habits led by increasing literacy level and economic development. People wish to have foods that not only meet calorie needs but also have nutritional and medicinal benefits. Mushrooms represent such quality food and their demand is on increase in India and across the globe. Globally mushroom industry has grown more than 25-fold during the last 35 years (from about 1 billion kg in 1978 to 27 billion kg production in 2012 (Royse, 2014).

Mushrooms are a health food. The consumption of mushrooms can make a valuable addition to the often unbalanced diets of people in developing countries. Mushrooms are a good source of vitamin B, C and D, including niacin, riboflavin, thiamine, and folate, and various minerals such as potassium, phosphorus, calcium, magnesium, iron and copper. They provide carbohydrates, but are low in fat and fibre, and contain no starch. Furthermore, edible mushrooms are an excellent source of high quality protein. Mushrooms can make a valuable dietary addition through protein and various micronutrients and, coupled with their medicinal properties, mushroom cultivation can represent a valuable small-scale enterprise option. Mushrooms can be successfully grown without access to land, and can provide a regular income throughout the year. Growing mushrooms also helps avoid some of the challenges facing collectors of wild fungi, including species identification. Cultivation is also independent of weather, and can recycle agricultural by-products as composted substrate which, in turn, can be used as organic mulch in growing other horticultural crops, including vegetables. Mushroom cultivation is highly combinable with a variety of other traditional agricultural and domestic activities (Marshall and Nair, 2009).

Small scale cultivation of some mushroom species as a seasonal activity does not require much investment. Mushrooms are grown on abundantly available raw materials - agro wastes; to harvest protein rich quality food stuff. The water required to produce one kg mushroom is about 25 litre, possibly the lowest required by any commercial crop. Ecologically too, cultivating mushrooms is having great relevance as after producing wealth from the waste, the remaining substrate gets recycled as manure in fields thereby improving soil health, reducing use of chemicals and thus promoting organic farming. Substrates like rice straw, wheat straw, sugarcane bagasse, saw dust, corn cobs, pearl millet, sorghum, mustard stalks, cotton waste, banana waste *etc.* along with other materials like rice or wheat bran, molasses, various cakes, chicken manure, *etc.* can be used for cultivation of oyster, paddy straw mushroom, button, shiitake, wood-ear, milky mushroom, winter mushroom and many others. Each of these has a specific requirement in terms of temperatures ranging from 10 to 40 degree Celsius. India produces around 700 million tonnes of crop residues of which 400 million tonne are cereal residues and even if 10 percent of these cereal residues are used for mushroom cultivation, we can produce over 20 million tonne of mushrooms- about half of world production and 400 times more than what we are producing today. At present our production is around 0.12 million tonnes only.

In 1978 mushroom production in China was 0.06 million tonnes which is half of our current mushroom production. But the comparison ends there as in last four decades there has been more than 500 times increase in mushroom production (Figure 2). China is producing ¾[th] of world's mushrooms (about 32 million tonnes) providing jobs to 20 million of its people and mushroom cultivation there is 5th important crop (Wang *et al.*, 2015).

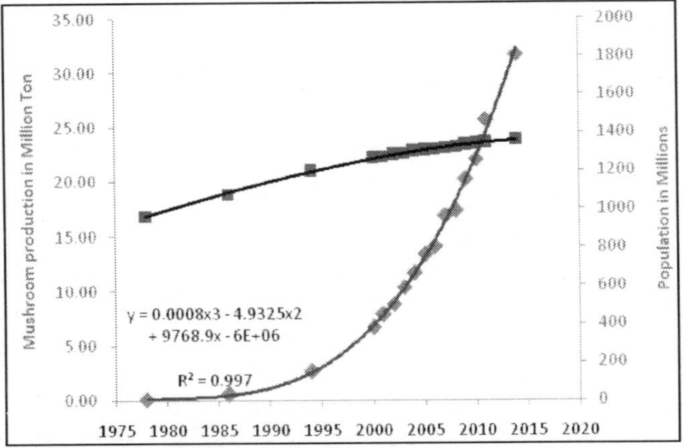

Figure 2L Mushroom Production and Population Growth in China. (The Edible Fungi Association of China, 2009; Wu *et al.*, 2013; Wang *et al.*, 2015).

As we can see from the graph, in the initial years the growth in mushroom production was slow and from 1978 to 1994 only 0.16 million ton mushrooms were added every year. There is a dramatic shift from 2000 onwards where 1.76 MT mushrooms were added per year. The primary reason for this shift was change in policies, higher investments on scientific inputs and better market systems.

In China, the mushroom availability in 1978 per person per year was 60 g. Despite the export and growth in population, the availability of mushroom per person per year at present is about 20 kg. Going by the Chinese production statistics, the present world production is over 40 million tonnes implying about 6 kg mushroom availability per person per year. In India, the mushroom consumption is less than 100 g per person per year. Educated people constitute the bulk of mushroom consumers in recent years (Shirur and Shivalingegowda, 2015). Hence, there is lot of potential in this venture.

If India has to catch up with the increasing demand for mushrooms from domestic and international markets, the educated youth must be promoted to undertake the mushroom entrepreneurship in a larger scale. Because several aspects of mushroom cultivation demands a better education status and hence, an educated entrepreneur can fare better in mushroom entrepreneurship. Mushroom cultivation requires better understanding of technical aspects related to insulation, air conditioning, air flow, control of humidity, temperature, carbon dioxide, hygiene, *etc.* and also the production of quality spawn. The road to success in mushroom cultivation is in understanding and following the good cultivation practices in totality as mistake at any one stage renders the whole mushroom unit a loss making one.

The educated youth represent the right section of society to understand the principles of mushroom cultivation, and execute these in totality and handle the issues related with proper packaging and marketing. There is need for better training set up in the country as the research institutes are not fully equipped and mandated to train people across the country. The institutions at lower level are not competent to provide answers to all the issues involved in commercial mushroom cultivation. In fact, we lack the machinery, technical know and proper consultancy services. Over time, it has been observed that most of the growers maintain secrecy and this lack of sharing results in copying of mistakes from one unit to another. All new units require trained manpower. But we do not have any mechanism in the country where we train people in quality compost production, cultivation or processing both technically and professionally through on farm trainings. Modern agriculture is different than traditional agriculture requiring much better technical support and understanding of market dynamics. Similarly, there is ambiguity that whether cultivating mushrooms is agriculture or not and the parameters for providing electricity and other facilities for transport and marketing vary across states. The interpretation also varies over ministries as income tax department does not consider mushroom cultivation and many other agricultural vocations performed above the ground as agricultural activities. Such anomalies unless addressed holistically, will discourage the educated youngsters from trying their hand in becoming new generation agri-preneurs.

The best way to demonstrate that mushroom Production is a viable and profitable avocation for educated youth is to use real life examples. In the recently held national mushroom mela (10 September, 2016) at ICAR-DMR Solan, five young mushroom growers, two of which were girls, were awarded with the progressive mushroom grower award (Table 1).

Ms. Divya Rawat is a progressive mushroom grower from Dehradun district of Uttrakhand. After training on mushroom production technology for entrepreneurs from ICAR-DMR, Solan in 2014, she started cultivating white button, oyster and milky mushrooms round the year. Her mushroom unit Soumya Foods Pvt. Ltd. presently produces 300-350 kg fresh mushroom a day. She started mushroom production in abandoned buildings of Garhwal and Dehradun district and finally set up a successful business model in Dehradun for production and practical training. Ms. Divya has taken up mushroom cultivation as a social entrepreneurship giving hope for migrating youth in the Uttarkhand State.

M/s Kulkarni Farm Fresh is an agri-business enterprise promoted by Kulkarni family in their native village Tigadi in Belagavi district of Karnataka. The new generation brothers of the Kulkarni family; Mr. Satish, Mr. Shrikar and Mr. Jayatheerth ventured into mushroom cultivation to set the example for an ideal agri-business entrepreneurship. The most meticulously designed environment controlled unit of the Kulkarni farm fresh is the largest mushroom production unit in Karnataka producing nearly 500 tonnes of fresh mushroom per annum. The unit has set an example with their efficient production system, spent mushroom substrate disposal and marketing chain management. The three brothers share among them three post graduate degrees with different experience and expertise to be used to achieve success in mushroom entrepreneurship.

M/s VG Mushroom is one of the most popular manufacturer and supplier of wide array of mushroom products in Southern India. The proud owner of the VG mushroom unit Mr. C Vijayan started VG Mushroom culture farm at Kamalapuram, Salem with initial production of 10 kg fresh oyster mushroom per day. Subsequently he added milky mushroom production facility and spawn production to VG mushroom unit. At present VG Mushroom production unit produces more than 250 tonnes of fresh mushroom per annum. Besides his successful mushroom cultivation entrepreneurship, he has guided many small mushroom growers to make their livelihood. He also supplies quality spawn to the growers from the spawn production facility of the VG mushroom unit. He has made use of his professional

Table 1: List of Persons Conferred with the 'Progressive Mushroom Grower' Award during the National Mushroom Mela-2016 at ICAR-DMR, Solan (HP)

Name and State	*Educational Qualification*	*Age*	*Experience in Mushrooms*	*Production/day*					*Estimated Annual Income*
				Button Mush.	*Oyster Mush.*	*Milky Mush.*	*Paddy Straw*	*Spawn (kg)*	
Ms Divya Rawat (Uttarakhand)	MSW (Masters in Social Work)	27 year	3 years	100 kg	100 kg	100 kg	–	–	25 lakh
Mr Shrikar Kulkarni (Karnataka)	MA. LLB	36 year	7 years	1200 kg	–	–	–	–	150 lakh
Mr C. Vijayan (Tamil Nadu)	M.E (refrigeration)	37 year	4 years	–	90 kg	900 kg		300 kg	100 lakh
Ms Hiresha Verma (Uttarakhand)	MBA	49 year	6 years	300 kg	150 kg	–	–	–	40 lakh
Mr Ajay Kumar Prusty (Odisha)	12th Class (Science)	49 year	25 years	–	100 kg		100 kg	200 kg	60 lakh

expertise to the best advantage to set up his mushroom unit.

Ms. Hiresha Verma of Dehradun is one of the successful mushroom growers in North India who has carved a niche for herself in a very short span of time. What started as a small scale seasonal cultivation has turned into a state of the art environment controlled mushroom unit. At present, her mushroom production unit 'Hanzen International farm' is producing about 500 kg of fresh mushrooms per day. She has realized the importance of emphasizing more on marketing than the production itself. Her MBA degree has often given her the edge in succeeding in the mushroom enterprise.

Mr. Ajaya Kumar Prusty has been a successful mushroom grower since 1991. Though, a science student in his intermediate, he dropped education to become a trailblazer in mushroom entrepreneurship in his native district. At present, his mushroom unit is producing 100 kg each of fresh oyster mushroom and paddy straw mushroom per day. The mushroom production unit of Mr. Prusty owns a well maintained state of the art spawn unit and mushroom production facility at village Indipur of Dhenkanal in the State of Odisha. At present, his marketing approaches of usina a mobile van with cooling facilities has enabled him to deliver his mushrooms to consumers across the State. Besides, he has popularized many post-harvest products on mushrooms through value-addition.

All the five progressive mushroom grower awardees epitomize the importance of education as key profile character to succeed in mushroom cultivation.

There are numerous such examples where youth have ventured into mushroom cultivation and related activities. Every year since 1997, DMR has been awarding the mushroom growers and majority of these have been educated youth. Our experience in training programmes held every year at DMR Solan is that most of the participants represent educated youth with nearly 93 per cent of the trainees possessing graduate or post graduate degrees. (Shirur *et al.*, 2014)

There is hardly any other agricultural activity that can provide such high degree of income per unit time. Mushrooms are considered highest protein producers per unit area and time. In Odisha, paddy straw mushroom is cultivated almost throughout the year in coastal areas. The total time taken by the paddy straw mushroom from a crop cycle is 20-22 days, implying that even if there is 20 percent profit over investment and if the farmers harvest same crop after every 22 days, his income will be more than 3 times that of his initial investment. There may be hardly any venture that can theoretically lead to such benefits. Then where is the problem?

The major limiting factor for mushroom cultivation is the non-availability of quality spawn. The other limiting factors are lack of assured electricity, limited availability of standard infrastructural designs, rural markets, lack of trained man power, non-availability of replicable demonstration models, hands-on training on site and literature in local language on both cultivation and consumption.

Some of the other factors for this are lack of awareness, myths and misinformation about mushrooms, lack of proper scientific training, problems in availability of raw materials, non availability of mushroom specific inputs like poly propylene (PP) bags, pp rings, non absorbent cotton, machinery and equipments specifically for mushroom cultivation, non availability of technology for mushroom value-added products, lack of organized market, *etc.* (Pandey and Senthil, 2016). Some solutions to bridge the gaps suggested by these workers include Increasing awareness about the excellent nutritional and vegetable status of mushrooms; units to produce Ready to fruit (RTF) bags; enhancing the awareness for the usage of dry mushroom in the form of dry mushroom powder Inclusion of mushroom technology in national schemes like MGNREGA to create rural employment among women and rural youth; establishment of dedicated scientific training centres with state of the art facilities; more spawn production labs with stringent quality control; organized collection and transportation of agricultural residues; multidisciplinary research for complete package of practices; awareness about the use of mushroom fungi for high quality manure production to maintain the soil health; and encouraging co-operative or satellite mushroom farming. Union is strength. Cooperatives increase the confidence of people, promote social harmony, enhance the ability to get loans, help in exchange of knowledge and the division of labour in work, processing, marketing and other activities results in a win-*win situ*ation. Hence it is important to form mushroom associations of growers.

On the consumption front, Shirur *et al.* (2014 b) found that, non-availability of mushrooms in the local market and lack of fresh mushrooms in the market are two prime reasons of non-purchase of mushrooms by the consumers. Of all the respondents, more than half said that, non-availability of mushrooms in the market hinders their purchasing frequency. Nearly 40 per cent said that mushrooms available in market are not fresh. The third reason in the order is about the respondents' unawareness about making different recipes from the mushrooms. Non availability of different varieties of mushrooms is another important cause of their non-purchasing. Hence, it can be said that, the mushroom consumption in India presents a story of asymmetry between the supply and demand.

Small scale cultivation may be possible in huts or available covered space. Commercial cultivation requires specific and specialized facilities. Small scale growers have to purchase spawn, though the commercial growers can start making their own spawn. Own spawn production normally makes economic sense when the average mushroom production is more than a ton per day. Similarly, for sales, normally it helps to sell fresh produce and one should think of canning or processing only if production levels are really high. however, there is an exception. In areas where marketing is a problem, growers may dry or pickle mushrooms or make local mushroom products.

When we say mushroom cultivation as a profitable avocation for youth, the implication is not just growing mushrooms. It has sub components like producing mushroom spawn (=seed), preparing quality compost and selling ready to fruit bags, growing, processing and marketing mushrooms. One can adopt one or more of these sub-components.

Spawn availability is a major constraint because its production requires technical knowhow and good knowledge of scientific aspects related to microbes. This is possibly one of the best avocation for educated youth. It is a totally indoor activity requiring sterile environment and highest degree of hygiene. Of course all the farmers cannot make the spawn and depend on other sources. At present there are no standards for spawn, some people are importing it, others are multiplying it after importing few bags, while few others are providing their own labels and names for the existing strains. Hence, on the policy side there is need to develop procedure to release of varieties of mushrooms, need for developing spawn standards and detailed information about their carbon dioxide, temperature, water requirements and pattern of flushes. There are strains that produce high yield only initially whereas others give gradual yields. Some strains can start pinning even at higher levels of carbon dioxide. Hence, a strong policy and scientific back up is required. Many times the reasons for non adoption of an avocation by the youth is that nationally the right technology and environment is not available.

The second input for mushroom production is the compost. Preparing good quality compost which selectively allows the growth of mushroom alone but other fungi and competitor moulds is an art as well as science. One of the secret of success in mushroom cultivation is to know how to prepare the compost. In mushrooms like button, compost below particular quantity will not ferment and produce its own heat, requires infrastructure like bunkers and tunnels and also machinery to turn the compost and fill it in tunnels. Hence it may not be possible for every grower to make the compost on his own. The short cut advocated till date was to prepare compost outside by long method that takes one month, involving the use of chemicals like Furadan and others, is prone to diseases and hence less productive. With increasing awareness about the pesticide residues, it is time that we shift to short method of compost preparation. Making compost and selling it after spawning/spawn running is another important activity by the youth. In fact, in The Netherlands only 2-3 companies are making compost and supplying it to farmers at their door steps and are even exporting compost to various countries. In India too there are persons who are preparing only compost of button mushroom and selling it.

Cultivation of mushrooms like king oyster, shiitake, wood year mushroom and many others requires autoclaving of substrate for 1-2 h for which facilities may not be available with every grower. Even for cultivation of oyster, sterilisation/pasteurisation is required. In the west, ready to fruit bags were available since long. As a part of training, many institutions have been providing spawn-run bags to trainees for hands on experience of mushroom cultivation. It was a good development that some youngsters have now come forward to prepare such bags and one recent example is the launch of one firm 'Pluck Fresh, with the help of IIHR Banglore.

One of the primary limiting factor in mushroom cultivation is its short shelf-life. Consumers are also vary of purchasing mushrooms which lose their sheen but are still very much edible. This implies need and scope for value-addition. There are hardly any products available in Indian market clearly specifying mushrooms as one component. Mushroom pickle is one simple alternative. Some growers in South India came out with mushroom milk rusk. A few tried to market mushroom noodles, mushroom pasta, *etc.* The market is still wide open and there is scope of large number of products. In addition to the fact that mushroom is a health food, it is also a low calorie high fibre food and has lot of scope in slimming diets. There is however, need for standards of labelling any product as mushroom product.

Majority of population is deficient in Vitamin D due to the changed life style. Mushrooms are the only vegetable food that contain Vitamin D. It has been demonstrated that exposure to UV light for about 15 minutes increases the Vitamin D content by 100 times. Such vitamin D enriched mushrooms are marketed in some developed countries. The educated youth can market such mushrooms. The traceability and actual laboratory analysis report has to be part of such product.

The possibilities are enormous. There is need for better interaction of policy makers, scientists, marketers and mushroom growers for a uniform policy to promote this unique venture. Doubling farmers income requires proper emphasis on recycling agro-wastes and secondary agriculture. Using these materials for mushroom cultivation not only lead to income from waste, but it also means ecological benefits as these wastes get recycled. The materials left after cultivation of mushrooms, commonly referred as spent mushroom substrate, can be easily converted into compost and put in the fields. In the words of Paul Stamets - "Mushroom growing isn't just a rapidly expanding agribusiness; it's also a significant tool for the restoration, replenishment and remediation of earth's overburdened ecosphere" (Stamets, 2002).

References

Marshall, E. and Nair, N.G. 2009. Make money by growing mushroom. FAO diversification booklet number 7. Rural Infrastructure and Agro-industries division, FAO Rome, pp. 1-64.

Pandey, Meera and Santhil Kumaran, G. 2016. Indian mushroom industry: Constraints and solution for sustainable development. **In**: *Mushrooms in India: Diversity, Improvement, Cultivation, Medicinal uses and Value-addition for Rural Development*, (Eds.) Sharma, S., Singh, M., Prasad, S. and Rathore, H. Indian institute of Technology Delhi, Hauz Khas, New Delhi, pp. 206–217.

Royse, D.J. 2014. A global perspective on the high five: Agaricus, Pleurotus, Lentinula, Auricularia and Flamulina. **In**: Proceedings of the 8th International Conference on Mushroom Biology and Mushroom Products (ICMBMP8), pp. 1–6

Shirur, M. and Shivalingegowda, N.S. 2015. Mushroom marketing channels and consumer behaviour: A critical analysis. *Mysore J. Agric. Sci.*, **49**(2) 390–393.

Shirur, M., Ahlawat, O.P. and Manikandan, K. 2014a. Profile characteristics and entrepreneurial attributes of trainees of national training programme on mushroom cultivation. *Mushroom Res.*, **23**(1): 99–104.

Shirur, M., Ahlawat, O.P. and Manikandan, K., 2014b. Mushroom consumption and purchasing behaviour in India: A study among selected respondents. *Mushroom Res.*, **23***(2): 225–231.*

Stamets, P., 2002. *Growing Gourmet and Medicinal Mushrooms*. Ten Speed Press, California.

The Edible Fungi Association of China (China Mushroom Association) 2009. China Mushroom Statistical Yearbook. China's Statistic Publishing House, Beijing, China.

Wang Jianan, Li Jing, Wei Shuxin and Liu Lina 2015. China mushroom processing industry: Current situation and development trends. *Mushroom Times*, **5**: 4–7.

Wu, S.R., Zhao, C.Y., Hou, B., Tai, L.M. and Gui, M.Y. 2013. Analysis of Chinese fungus production area layout nearly five years. *Edible Fungi China*, **1**: 51–53.

Zhang, Y., Geng Wei, Shen, Y., Wang, Y. and Dai, Y.C. 2014. Edible mushroom cultivation for food security and rural development in China: Bio-innovation technological dissemination and marketing. *Sustainability*, **6**: 2961–2973.

96

Beekeeping as an Attractive Agro-based Enterprise for Youths, Especially in Rural Areas: Perspectives and Opportunities

B.L. Sarswat

Director(C/C) and Executive Director,
National Bee Board, Department of Agriculture, Cooperation and Farmers Welfare,
Ministry of Agriculture and Farmers Welfare, Govt. of India, Krishi Bhawan, New Delhi – 110 001
E-mail: bl.sarswat@nic.in

Beekeeping is an agro-based activity which is being undertaken by farmers/landless labours in rural area as an integrated framing practice. Beekeeping supplements income generation and nutritional intake of rural population. It is a very fascinating occupation. Though the honeybees are best known for the honey they produce, their economic role in nature is to pollinate hundreds and thousands of flowering plants and assure setting of seed or fruit. Honeybees have been offering services to the society through ensured pollination in cross-pollinated crops as well as by providing honey and a variety of beehive products. Honey Bees have their vital role in sustaining plants bio-diversity resulting in environmental stability.

Major Honeybee Species

Some of the major honey bee reared in India is shown in Table 1.

Importance of Beekeeping

Till mid-20th century, honeybees were equated with the production of honey and beeswax only. But since past 3-4 decades, utilizing honeybees to pollinate large number of agricultural and horticultural crops to increase per acre yield has become a routine practice in many developed countries. According to agricultural scientists value of additional yield obtained by pollination service rendered by honeybees alone is about 15-20 times more than the value of all the hive products put together (Kaloo, 2004).

Table 1: The Details of Major Honey Bee Species found in India

Species	*Nature*	*Foraging Range (km)*	*Avg. Honey Yield (kg/yr)*
Indian bee (*Apis cerana, A. himalaya, A. indica*)	Domesticated	0.8-1.0	10-12/ hive
Western bee (*Apis mellifera, A. ligustica, A. carnica*)	Domesticated	1.0-2.0	50-60/ hive
Apis dorsata	Wild	–	50/hive
Little bee (*Apis florea*)	Wild	–	0.25/hive
Trigona (can survive in high temp.)	Stingless		Few grams/ colony

Honey bees play very important role in pollinating various agricultural and horticultural crops and increase their yield and improve the quality of produce. The potential benefits, due to bee pollination, in the form of increase in yields of various crops including fruits and vegetables, oilseeds, pulses and others, varies from 5 to 33,150 per cent. It is being increasingly realized that bees could be less expensive input for promoting sustainable and eco-friendly agriculture and enhancing crop productivity.

Table 2: The Crops-wise details of Increase in Yield of Various Crops Due to Bee Pollination

Oilseed	*% Increase in Yield*	*Legume/Pulses*	*% Increase in Yield*
Mustard	128.1 to 159.8	Alfalfa	23.4 to 19,733.3
Rai	18.4	Berseem and other clovers	23.4 to 33,150
Rapeseed	12.8 to 139.3	Vetches	39 to 20,000
Toria	66 to 220	Broad Beans	6.8 to 90.1
Sarson	222	Dwarf beans	2.8 to 20.7
Safflower	4.2 to 114.3	Kidney beans	500 to 600
Linseed	1.7 to 40	Runner beans	20.6 to 1,100
Niger	260.7	*Arahar*	21 to 30
Sunflower	20 to 3,400	Other pulses	27-30 (RAU)
Fruit Crop	*% Increase in Yields*	*Vegetables for Seed/Fruits*	*% Increase in Yield*
Apple	180 to 6,950	Radish	22 to 100
Pear	240 to 6,014	Cabbage	100 to 300
Plum	6.7 to 2,739	Turnip	100 to 125
Cherry	56.1 to 1,000	Carrot	9.1 to 135.4
Strawberry	17.4 to 91.9	Onion	353.5 to 9,878
Raspberry	291.3 to 462.5	Brinjal	35-67
Persimmon	20.8	Cucumbers	21.1 to 411
Litchi	4,538 to 10,246	**Miscellaneous crop**	
Citrus	7 to 233.3	American cotton	5 to 20
Grape	756.4 to 6,700	Egyptian cotton	16 to 24
Squash	771.4 to 800	Buckwheat	62.5
Guava	70-140	Coffee	16.7 to 39. 8
Papaya	22.4-88.9		
Mosambi	36-750	This increase in yield is in addition to the value of honey and other hive products.	
Sweet orange	471-900	Bee pollination also improves the quality of seed/fruit.	

The increase in yield of different crops (Table 2) due to honey bee pollination is as shown in Table 2.

Beekeeping industry has quadruple benefits, *viz.*, i) providing self-employment to rural and forest based population; ii) production of honey, pollen and beeswax, venom, royal jelly, *etc.*; iii) providing employment to rural educated youths in collecting, processing and marketing of bee/beehive-products and the most important; and iv) cross-pollination of various agricultural and horticultural crops and improving their quality and increasing yields. This way, beekeeping plays a vital role in sustainable agriculture and rural development.

Government Initiatives

After independence, Govt. of India took policy decision to revive various traditional village industries and an All India Khadi and Village Industries Board (KVIB) was formed in 1954. Through coordinated efforts of well-knit organizations like Khadi and Village Industries Commissions (KVIC), State KVIBs, Beekeepers' Co-operatives, Public Institutions *etc.* the beekeeping industry came on the map of village industries of India within two decades. In 1981, an All India Coordinated Research Project on Honeybee Research and Training was launched by Indian Council of Agricultural Research (ICAR) involving State Agricultural Universities (SAUs).

Figure 1: Shri Narendera Modi, Hon'ble Prime Minister of India Understanding the Process of Honey Extraction.

In 1994-95, the Ministry of Agriculture, renamed as Ministry of Agriculture and Farmers Welfare, took initiative of launching a Central Sector Scheme entitled *"Development of Beekeeping for Improving Crop Productivity"*, during the VIII Plan. A Beekeeping Development Board also functioned under the Chairpersonship of Secretary (A and C) to coordinate the beekeeping activities. The Scheme was approved for continuation during the IX Plan. However, the scheme got sub-sumed under the Macro Management Scheme, with effect from October, 2000. The focus on beekeeping also got diminished under such an arrangement. The Department facilitated efforts by forming the National Bee Board in 2000. With effect from May, 2005, beekeeping has been included as an activity under National Horticulture Mission (NHM), which has now been merged in Mission for Integrated Development of Horticulture (MIDH), for promoting cross-pollination of Horticultural Crops in all parts of the country. Under MIDH, among others, assistance for following components for promoting Scientific Beekeeping under the component of *'Pollination Support through Beekeeping'* is available and Mission/Scheme is being implemented by the State Departments of Horticulture/Agriculture in the field.

The different components of the scheme are enlisted in Table 3.

National Bee Board (NBB) and its Role

The National Bee Board (NBB) was reconstituted in June, 2006 by the Department of Agriculture, Cooperation and Farmers Welfare. Secretary (AC and FW), DAC and FW, Ministry of Agriculture and Farmers Welfare, Government of India is the Ex-Officio Chairman of the Board (NBB). Executive Director is the Principal Executive Officer of the Board. The main objective of the National Bee Board (NBB) is overall development of Beekeeping by promoting scientific beekeeping in the country to increase the productivity of crops through pollination and increase honey production for increasing income of the beekeepers/farmers.

The National Bee Board act as a nodal agency for formulation, implementation, execution, supervision and monitoring of projects/programmes in the field of beekeeping and also an appraisal agency for appraising the Annual Action Plan/Schemes/Projects to be implemented by the State Governments/UTs, international agencies, private sector and other agencies/organizations for development of scientific beekeeping in the country. Some of the activities/components are also implemented by the Board (NBB) directly in the field/states through its Expert Member Beekeeping Societies/Associations/Federations/other Members including State Bee Boards/State Beekeeping and Honey Missions, State Khadi and Village Industries Boards (SKVIBs)/SDAs, State Department of Horticulture, Agriculture *etc.* spread all over country.

Beekeeping Industry in India

At present there are about 30 lakhs bee colonies in India, with estimated annual production of around 89,000 metric tonnes of honey (2015-16) including honey from wild honey bees. India is one of the honey exporting countries. The major markets for Indian honey

Table 3: The Major Feature of the Scheme *Pollination Support through Beekeeping*

Sl.No.	Component	Rates of Assistance Approved under MIDH (NHM/HMNEH)
1.	Development and multiplication of bee stock	
	i. Production of nucleus (Pedigree) stock	Rs. 20.00 lakhs/project
	ii. Production of bee colonies by bee breeders	40% of cost or Rs. 4.00 lakhs/project (whichever is less)
2.	Distribution of 8 frame bee colonies (50 bee colonies per beneficiary)	40% of cost or Rs. 800/per set of bee colony (whichever is less)
3.	Distribution of beehives, supers, *etc.* (50 beehives, supers, *etc.* per beneficiary).	40% of cost or Rs.800/per set of beehive, supers, *etc.* (which ever is less)
4.	Distribution of bee equipment [a set of one honey extractor of SS (4 frames) and 10 containers (30 kg each) of FGP/SS, 1 net and a set of other tools]/unit of 50 bee colonies/beneficiary	40% of cost or Rs. 8000/- per set/per beneficiary (whichever is less)
5.	Human Resource Dev.(HRD) activities	
6.	Conf./Seminar/Workshop	
	i. International level	Rs. 10.00 lakh/event
	ii. National level	Rs. 5.00 lakh/event
	iii. State level	Rs. 3.00 lakh/event
	iv. District level	Rs. 2.00 lakh/event
7.	**Trainings** – (i) Within state (WST)	@Rs. 1000/- participant/day
	(ii) Out of state (OST)	Project based (OST)
8.	Exposure visits – Outside states and outside India	Project based

are Germany, USA, UK, Japan, France, Italy, Spain *etc.* In India two types of honey, *viz.*, apiary honey (of domesticated bees) and squeezed honey (of wild bees) is produced. *Apis cerana* and *Apis mellifera* are two types of bees, which are being domesticated and kept in hives. In India honey is not used in the form of food as its per capita per year consumption is about 10.00 g.

World Scenario of Beekeeping

Millions of honey bee colonies, mostly, *Apis mellifera*, are maintained all over the world. The world production of honey has bee ranging between 14 to 15 lakh M tonnes per year. There are 15 countries in the world which account for the 90 per cent of the world honey production. China is the only Asian country producing nearly 2.5 lakh metric tonnes of honey. The per capita consumption of honey in world, on an average, is 250-300 g. Whereas in Asia, Japan has the highest per capita honey consumption, *i.e.* about 600 g. The honey consumption in some of the countries where it is considered as a food, for example, in Germany, is > 2,000 g per capita per annum.

Beekeeping as an Enterprise

The main benefits of beekeeping are summarized as under:

- Bees help in cross-pollination thus they increase the productivity of crops
- Honeybee produces:
  - **Honey:** Nutritious whole food with medicinal property.
  - **Bees wax:** More valuable, used in candle, pharmaceutical and cosmetic industry.
  - **Bee venom:** Used for treatment of arthritis, rheumatic and pains (Apitherapy).
  - **Royal jelly:** Nutritious, increase vigour and vitality/fertility.
  - **Propolis:** Resinous substance uses as gum.
  - **Honey and hive products:** Source of income to the rural people and also generate employment.
- Unemployed youth can start this business with minimal funds
- Generates 3.75 lakhs man-days to maintain 10,000 Bee colonies in Bee hives
- Proper utilization of natural resources
- Different sectors and trades benefit from a strong beekeeping industry
- Beekeeping encourages ecological awareness
- Beekeeping helps in increasing National income
- Income from 100 Bee colonies is around Rs. 2.50-3.00 lakhs per annum
- It helps in rural development and promotes small village industry
- Beekeeping is benign: Beekeeping generates income without destroying habitat
- Encouraging beekeeping encourages biodiversity.

Keeping in view the above, beekeeping may be adopted as an enterprise by anyone with minimum resources after getting training on the subject.

Constraints in Beekeeping

The major constraints confronting the development of beekeeping are given as under:

- Lack of scientific data on the choice of honey bee species for commercial beekeeping and for promoting cross-pollination
- Lack of quality nucleus stock of bees
- Lack of infrastructure for producing large volumes of genetically superior queen bees for supply to beekeepers
- Lack of technical knowledge for efficient management of bee colonies for higher honey yield
- Lack of infrastructure at grass root level and national level for promoting beekeeping
- Poor quality control for production of honey and other beehive products
- More emphasis for production of honey instead of other bee products such as bee wax, pollen, propolis, bee venom and royal jelly
- Lack of adequate laboratories for disease prevention, control, analysis and testing of quality of bee hive products
- Lack of institutional support for beekeeping in terms of bank loans, *etc.*
- Lack of proper pricing policy for honey and those engaged in packaging, processing and storing honey
- Lack of consumer awareness of honey and its products
- This is a non-traditional newly introduced industry
- Tribals and illiterates from forests and remote rural areas generally practice this industry
- Flowering of plants and secretion of nectar and production of pollen – sole food of honeybees, is influenced by climatic conditions
- The behaviour and life cycle of honeybees depend completely on climatic and floristic conditions, which vary from place to place
- De-forestation
- Indiscriminate use of insecticides, pesticides, weedicides, *etc.*
- Wild fires
- Mono-Cropping culture

☆ Pollution of water and air

☆ Global warming and unforeseen changes in climatic conditions.

Figure 2: Officers of NE States Learning Procedure of Honey Extraction while Participating in Training in Beekeeping Conducted by NBB.

Opportunities

India has vast potential for Beekeeping. The diversity in flora and fauna provides more opportunities for the development of beekeeping industry. The National Commission on Agriculture had visualized the need for deploying about 150 million bee colonies for pollinating the agricultural crops in the country. This industry does not need any sophisticated technology, high capital investment or infrastructure. Compared to the potential, not even the fringe of it, in terms of number of honeybee colonies is achieved. There is thus great potential and opportunities for the development of beekeeping industry in India.

Beekeeping industry has great self-help potential for the rural people, tribals, marginal and small farmers, land-less labourers *etc.* The great potential and opportunities in beekeeping are given as under:

a) As per the cropped area under the major insects pollinated crops, about 200 million bee colonies are required in the country to enhance the yield levels of these crops at par with the yield levels of developed countries. It will provide jobs to about 215 lakh persons.

b) Honey has great food value and provides cash income.

c) Beeswax which is twice as much costly as honey is in great demand.

d) Other products, *viz.*, bee-collected pollen, propolis, bee-venom and royal jelly are several times costlier than honey and beeswax.

e) Providing bee pollination service to farmers for increasing crop production and quality and productivity of honeybees – A double benefit service.

f) Maintenance of biodiversity by pollination of flowering plants.

g) Apitherapy medicine using bees' products.

h) Processing and value-added products of bee-hive products; *etc.*

i) Few resources are needed.

j) Land ownership not essential.

k) Nectar and pollen are otherwise not harvested.

l) Everybody can be a beekeeper.

Figure 3: Regional Seminar on Beekeeping at Barabanki (U.P.) for Creating Awareness in Farm Women on Apiculture.

Beekeeping as a Source of Livelihood

The beekeeping may be a source of livelihood for rural and tribal population. The activity may be adopted by anybody and helpful in generating income and employment as summarized under:

i) ***Employment generation***: Beekeeping has vast potential in employment generation and according to estimates 3,00,000 man-days are generated for maintaining 10,000 bee colonies. Besides, it also creates employment opportunities in appliances and equipment manufacturing sector. It is estimated that around 75,000 man-days are created for manufacturing enough appliances for 10,000 bee colonies.

ii) ***Enhanced income generation***: Production and employment are the basic factors to bring enlistment and prosperity. As per economics worked out for beekeeping, the net income from 100 bee colonies varies from Rs. 2,50,000/- to 3,00,000/- annually, which is directly linked with the prevailing market prices of honey and other beehive products and Beekeeping equipments.

Table 4

Sl.No.	*Item*	*Rate/per Unit Cost (in Rs.)*	*Total Cost (in Rs.)*
A.	**One time cost for establishment/Fixed cost**		
1	50 Beehives with supers and tools, *etc.*	2000/-per set	1,00,000
2	50 bee colonies each of 8 frames @ Rs. 250/- frame.	2000/-colony	1,00,000
3	50 iron stands	100/-each	5,000
4.	Honey Extractor (SS) and other equipment, FGP containers, honey extraction net *etc.*	–	20,000
	Sub total of A.	–	**2,25,000**
B.	**Recurring cost/working capital per year**		
1	Comb foundation sheets (wax sheets) 100 kg for one unit	250 per kg	25,000
2	250 kg sugar for feeding in dearth period	40 per kg	10,000
3	Interest on fixed capital	12%/annum	27,000
4	Depreciation on fixed capital	20%/annum	45,000
5	Misc. exp. including labour, migration cost, *etc.* per annum	–	1,50,000
	Sub-total of B.		2,57,000
C.	**Per year income from 50 colonies**		
1	Honey production @ 40 kg/colony Total = 2000 kg.	80 per kg	1,60,000
2	Sale price of BCs. of 8 frames each multiplied in a year, *i.e.* 50 colonies (50 x 250 x 8).	2000/- colony	1,00,000
3	Production of bee pollen (250 kg) @ 5 kg/colony	400 per kg	1,00,000
4	Income from production of other bee hive products like propolis, bees wax *etc.*		20,000
	Total Income (C)		3,80,000
D.	**Net income per year (C-B)**		**1,23,000**

Note: Economics is directly linked with market prices of honey and other beehive products and price of bee colonies, bee hive, equipment *etc.*, which fluctuate from time to time.

It may be remembered that all the bee products are used either as food or in pharmaceutical and cosmetic industries. For this reason, hygienic collection, handling, processing, storage, *etc.* and maintaining National and International purity standards are of prime importance.

Economics of Beekeeping

For rearing honey bee *Apis mellifera*, one man power required on regular basis along with two casual labour. As per the present prevailing market prices of honey and other beehive products and price of bee colonies are the other expendiutures (Table 4).

Conclusion

Beekeeping is an agro-based activity, which is being undertaken by farmers/fruit growers/landless labours in rural areas as an integrated farming practice. It supplements income and create employment generation and also high quality nutritional produce, ensure adequate pollination in different crops, thus improving yield, quality and additionally provide honey and a variety of other bee hive products, which have high value. With a short duration skill development and training any youth can start this enterprise and can make handsome income to sustain his expenses and also make a decent living and also creating jobs in rural areas. India has vast potential for beekeeping. Beekeeping industry has great self-help potential for the rural people, tribals, marginal and small farmers, land-less labourers *etc.* As per the cropped area under the major insect-pollinated crops, about 200 million bee colonies are required in the country to enhance the yield levels of these crops at par with the yield levels of developed countries. It will provide jobs to about 215 lakh persons. Beekeeping may be a source of livelihood for rural and tribal population.

97

Custom Hiring of Farm Machinery/Equipment/Implements: A Way for Increasing Farm Efficiency and Enhancing Profitability

Indra Mani

Division of Agricultural Engineering,
ICAR-Indian Agricultural Research Institute, New Delhi – 110 012
E-mail: head_engg@iari.res.in

There are seven vital inputs to agricultural production system - seed, fertilizer, irrigation water, plant protection chemicals, agro machinery, transfer of knowledge and credit. The efficiency of all non-engineering inputs depends on efficiency, accuracy and precision of engineering input, *i.e.* agricultural machines. In fact, mechanization alone enhances productivity of crop by 15 per cent and reduces cost of crop production by 20 per cent (Shyam, 2012). However, farm mechanization in India is still in its nascent stages and during the last two decades has been able to achieve a meager growth of less than 5 percent. The sector faces critical challenges in terms of large share of small and marginal farmers, declining land holding sizes, high cost of farm machinery and equipment, inappropriate technologies, undeveloped markets, complex operations, maze of legislation and insufficient policy framework. Unlike other agricultural sectors, farm mechanization sector in India has a far more complex structural composition. With continued decrease in average farm size, more farms will fall into the adverse category thereby making individual ownership of agricultural machinery increasingly uneconomical. Custom Hiring envisages promoting establishment of farm machinery banks for hiring by way of providing financial assistance to self-help groups or farmers' co-operatives since the prohibitive cost of hi-tech and high productive equipment renders it difficult for individual ownership. The Custom Hiring Centres (CHCs) offer farm equipment and machineries on rental basis to farmers who cannot afford to purchase highend agricultural machineries and equipment. The CHCs play a pivotal role in introducing high technology agricultural machinery to even small farmers with the objective to boost crop production, improve quality, timeliness and efficiency of agriculture operations. As the small and marginal holdings constitute 80 per cent of total land holdings in India, the potential for CHCs which will cater to the farm machinery requirement of such a vast area, is quite huge. Government of India, in recognition of this potential had envisaged increase of farm power availability from the present level of 0.93 to 2 kW/ha during the XII Plan period (2012-2017). 'Sub-Mission on Agricultural Mechanization (SMAM)' is one such initiative towards achieving this objective. Custom Hiring facilities for agricultural machinery is one of the major components of this mission. State governments of Karnataka, Andhra Pradesh, Madhya Pradesh and Punjab have been promoting Custom Hiring on Public Private Partnership (PPP) basis through training, demonstration and financial incentives. Private sector participation in promoting farm mechanization in India through establishment of CHCs is gradually evolving with unique business models. As majority of Indian farmers belong to small and marginal category, purchase of farm equipment is a significant investment for them. Reasonable financing norms are a must for making farm equipment and machineries available at affordable price and enhance farm mechanization. Custom Hiring in India faces constraints like high initial cost of equipments, lack of knowledge in the aspects of operation, maintenance and repair of equipment, repair and maintenance under individual ownership coupled with lack of space for shelter, orientation towards the use of tractors and allied equipment, suboptimal asset capacity utilization on account of crop specific requirements. To overcome this, virtual or real consolidation of the widely fragmented and scattered land holdings in many parts of the country, extension of benefits of mechanization to all cropping systems including horticultural crops, enhancement of the average farm power availability to minimum 2.5 kW/ha to assure timeliness and quality in field operations and use of precision and efficient equipments to improve the quality of operations is required. The Custom Hiring model holds the potential to be the best way to introduce capital intensive, high quality and efficient farm mechanization to the small farming structures prevalent in India. The Custom Hiring model enables new machines to be used at their maximum capacity and enables farmers to gain access to latest technology they would otherwise not be able to afford. Custom Hiring can significantly facilitate diversification in agriculture specifically from wheat and paddy to other crops. Custom Hiring is the evolving concept in India and holds an immense potential to change the farm mechanization landscape of India. There is a need to study and replicate successful business models along with incentivisation and policy support for the adoption, capacity building and skill enhancement, development and promotion of farm mechanization technologies. In this paper the various components of farm mechanization and the role of custom hiring in agricultural production and productivity has been discussed.

Status of Farm Mechanization in India

India witnessed unprecedented growth in agriculture that helped the country to graduate from import dependence to self-sufficiency in food grains by increasing the food grain productivity from 0.64 MT/ha in year 1965-66 to 2.07 MT/ha in 2015-16 resulting in export. This growth was mainly attributed to adoption of the agricultural technology during green revolution supported by positive governmental policies, liberal public funding for agricultural research and development and untiring work of farmers and manufacturers of agricultural machinery.

The different sources of power available on the Indian farm for doing various mobile and stationary operations are mobile power, *viz.*, human, draught animals, tractors, power tillers and self-propelled machines (combines, dozers, reapers, sprayers *etc.*) and stationary power, *i.e.* diesel/oil engines (for pump sets, threshers, sprayers and other stationary operations) and electric motors (for pump sets, threshers, sprayers and other stationary operations). The time series data (1960-61 to 2013-14) of population of farm power sources and power availability from various farm power sources shows that animate power is being replaced by inanimate power (Figure 1). This is an indicative of the fact that slowly and steadily India has been progressing towards higher farm mechanization.

Status of Farm Implement Manufacturering in India

The adoption of mechanization technology depends upon the local manufacture and after sales-services besides credit and financial incentive provided by the government. The manufacture of agricultural machinery in India is quite complex comprising from village artisans, tiny units, small scale industries to State Agro Industrial Development Corporations and organized tractor, engine and processing equipment industries. As per Agricultural Machinery Manufacturers Association (AMMA) of India, there are approximately 250 medium

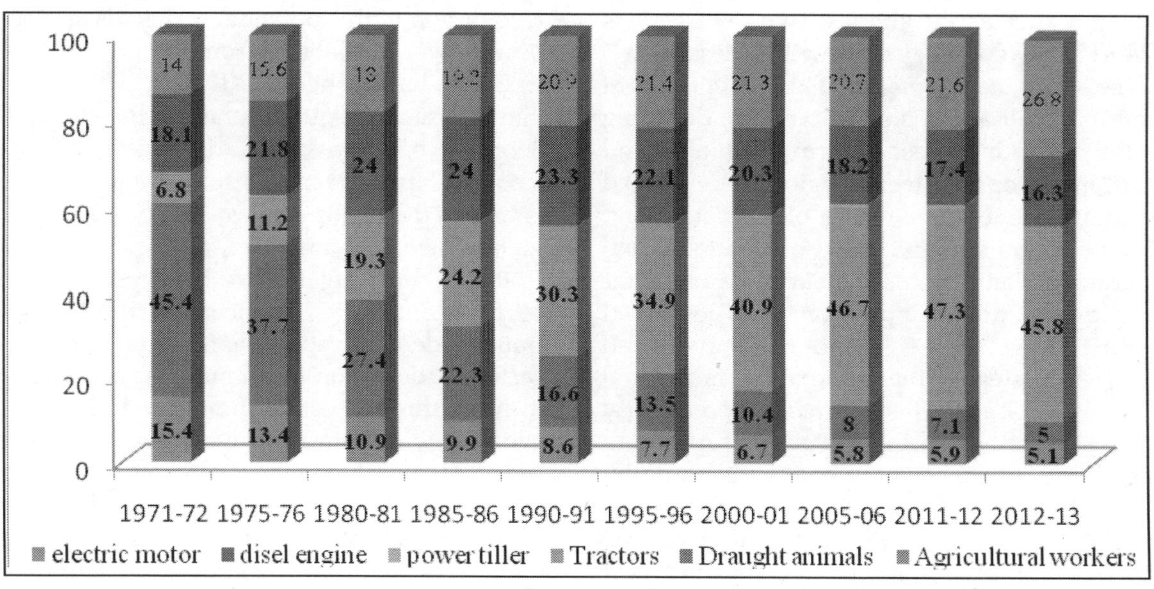

Figure 1: Relative Share of different Sources to Power Availability at Indian Farms.
***Source*: Trends of Agricultural Mechanisation in India CSAM Policy Brief, June 2014.**

to large scale units, 2,500 small scale industries, 15,000 tiny industries and 1,00,000 village level artisans (Table 1). Major farm machinery used in India includes tractors, threshers and power tillers. Among these, the biggest market in terms of annual sales is that of tractors (around 6 lakh units annually), threshers (around 1 lakh units annually) and power tillers (around 56,000 units annually) (Table 2). Among farm machinery, tractors are most widely used by the domestic farmers with the total market size estimated at around Rs. 34,000 crores annually.

Table 1: Status of Farm Mechanization Industry in India

Equipment Manufacturer	*No. of units*
Agricultural tractors	22
Power tillers	5
Irrigation pumps	600
Plant protection equipment	300
Combine harvesters	48
Reapers	60
Threshers	6,000
Seed drills and planters	2,500
Diesel oil engines	200
Plough, cultivators and harrows	5,000
Chaff cutters	50
Rural artisans	>1 million

Source: Trends of Agriculture Mechanization in India, CSAM Policy Brief, June 2014

Table 2: Major Farm Machinery Used in India

Machinery	*Market Size Annually (units)*	*Annual Industry Size (Rs. crores)*
Tractors	6,00,000	34,200
Power tillers	56,000	706
Combined harvesters	4,000–5,000	770
Threshers	1,00,000	1,230
Rotavators	60,000–80,000	693
Rice transplanters	1,500–1,600	62
Self-propelled reapers	4000–5,000	45
Zero till seed drills	25,000–30,000	132
Multi-crop planters	1,000–2,000	8
Laser land levellers	3,000–4,000	129
Weeders	25,000	1,275

Source: Trends of Agriculture Mechanization in India, CSAM Policy Brief, June 2014.

Today, India is recognized as a leading country in the world for the development and manufacture of agricultural implements and equipments. India is exporting increasing volumes of these to various countries including USA, Africa and Asia.

Role of Farm Mechanization in Agricultural Production/Productivity

Besides its paramount contribution to the multiple cropping and diversification of agriculture, mechanization also enables timeliness of operations, a very important aspect of agricultural production system. Farm mechanization is key for agricultural productivity as it has positive correlation with level of farm mechanization (Figure 2).

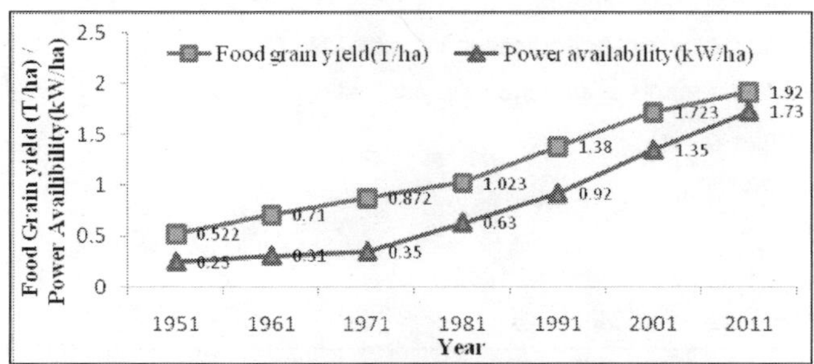

Figure 2: Relation between Farm Power Availability and Agricultural Productivity.

Cropping intensity also increased with an increase in per unit power availability (Mehta *et al.*, 2014). It was 120 per cent with power availability of 0.48 kW/ha during 1975-76 and increased to 139 per cent with increase in power availability to 1.71 kW/ha in 2009-10 (ICAR, 2013).

Farm Mechanization Potential

Farm mechanization has immense potential in India. As a general trend, farm operations requiring high power inputs and low control (tillage, transport, water pumping, milling, threshing, *etc.*) are mechanized first and those requiring medium levels of power and control (seeding, spraying, intercultural operations, *etc.*) are mechanized next. The operations requiring high degree of control and low power inputs are mechanized last (transplanting, planting of vegetables, harvesting of fruits and vegetables *etc.*). This happens so because any work, which is power intensive, can be done faster mechanically and at a lower cost. Whereas, converting human knowledge into machine knowledge is difficult and costly. Growth of the mechanization in India has also followed the same general pattern found worldwide. As a result of this most of the precision requiring operations, *i.e.*, planting and transplanting of vegetables and harvesting of fruits and vegetables are yet to explore. There is a great scope to reduce the input cost and increase the productivity of farmer through mechanization of vegetable and horticultural crops (Table 3).

Role of Custom Hiring in Agricultural Mechanization

Custom hiring plays an important role in mechanizing small farms as well as allows timeliness

Table 3: Opportunities in Farm Mechanization in India

Operation	Plains			Hilly Region			Precision Farming Machines
Prime Mover	Manual	Animal Drawn	Tractor/ Power Tiller/ Self Propelled/ Power Operated	Manual	Animal Drawn	Tractor/ Power Tiller/ Self Propelled/ Power Operated	
Seed bed preparation	x	x	x	x	x	x	x
Sowing/planting/transplanting/ fertilizer application	p	x	p	p	y	y	s
Irrigation and drainage	y	s	p	x	y	y	s
Weeding/herbicide application/intercultural	x	y	p	x	y	y	s
Fertilizer/manure/agro-chemical application	x	y	p	x	y	y	s
Spraying	y	s	p	x	y	y	s
Harvesting	y	y	p	y	y	y	s
Threshing	y	y	P	y	y	y	s
Special package of equipment	y	s	s	y	s	s	y
Other machines/equipment	s	s	s	s	s	s	s

Notes: x: Available, possibility on improvement, needs to be demonstrated; y: Normally, the operation is done using conventional tools or the power source is not used for this operation/system or not application; p: Available for some crop or in some parts of country, needs refinement, feasibility trials in other parts/crops; s: Needs development/refinement.

of operations in different farm holdings. Custom hiring helps in reducing the cost of operation by spreading variable costs over more number of acres. The custom hiring of agricultural machinery is supposed to emerge in the form of a big business model and/or employment source to in the future. Bansal and Mukesh (2014) studied the impact of custom hiring on farm mechanization in Haryana. Results revealed that the dissemination of improved technology in Haryana through custom hiring has been found to be very effective and productive as far as mechanization is concerned. There was encouraging adoption of improved machinery which resulted in timely operation, increase in production and additional source of income generated. This ultimately benefited the farmers of Haryana to the tune of Rs. 643 crores by adopting improved farm implements.

Key Drivers of Agricultural Mechanization

Labour shortage is being experienced at peak seasons due to the enactment of the National Rural Employment Guarantee Act and huge demand from the construction sector in cities. Labour is available at a higher cost per hectare and this would increase the demand for mechanization. It has been observed that the percentage of agricultural workers to total workers in India has been gradually declining from 59.1 per cent in 1990-1991 to 54.6 per cent in 2010-2011, which is further expected to decline to 25.7 per cent by 2050 leading to severe farm labour shortage.As per the Vision 2030 document by Indian Council of Agricultural Research, domestic demand for food grains is expected to increase at around 2 per cent CAGR in 2030 against that of year 2000. Food grains demand is expected to reach 355 MT in 2030 *vis-à-vis* 192 MT in year 2010. Fruits and Vegetables demand is expected to reach 290 MT in 2030 *vis-à-vis* 136 MT in year 2010. However, given the limitations in land use and in increasing cropping intensity over a certain period, increasing the yield from the same land is an urgent requirement to meet the needs of a growing domestic population. With limited land and increased demand for food, the cropping intensity needs to be increased through timeliness of operations. Also, increased participation of corporates through corporate farming has become very popular. Companies are entering into agreements with farmers through contract farming thereby requiring mechanization.

Key Challenges to Agricultural Mechanization

The key challenges faced by the farm mechanization in India are as follows:

i) The average farm size in India is small (less than 2 ha) as compared to the European Union (14 ha) and the United States (170 ha). Therefore, there will be little mechanization unless machines appropriate for small holdings are made available or substantial farm amalgamation takes place. Due to small size of land holdings, it is difficult for the farmers to own machinery. As a result, the benefits of mechanization are enjoyed by only a section of the farmers who have large farm holdings.

ii) Mechanizing small and non-contiguous group of small farms is against economies of scale especially in operations like land preparation and harvesting. With continued shrinkage in average farm size, more farms will fall into the adverse category thereby making individual ownership of agricultural machinery progressively more uneconomical.

iii) The major constraint of increasing agricultural production and productivity is the inadequacy of farm power and machinery with the farmers. The average farm power availability needs to be increased to minimum 2 kW/ha to assure timeliness and quality in field operations, undertake heavy field operations like sub-soiling, chiseling, deep ploughing and summer ploughing.

iv) There is an urgent need to design, develop and adopt machinery especially suitable to Indian farming conditions such as dry land farming, paddy transplanting, sugarcane harvesting, potato combining and cotton picking.

v) The quality and after-sales service of farm machinery are the other concerns in India as the majority of farmers are cost conscious. There are inadequate service centres for proper up-keep of the machinery. In addition, the inability of local low cost manufacturers to come up to the levels of standard designs of equipment also poses a big challenge to farm mechanization.

vi) Almost 90 per cent of tractors are sold in India with the assistance of some financial institutions. Sale of farm machinery is driven by factors like financial support, limit of funding (in terms of percentage of the cost), funding/financing institution and the applicant's profile (deciding the credibility of the loanee).

vii) The high cost and energy efficient farm machinery are capital intensive and majority of Indian farmers are not able to acquire these assets due to shortage of capital with them. Therefore, an arrangement to provide custom hiring service facility for these farm machineries to the farmers by engaging unemployed rural youth will go a long way in meeting the requirements.

viii) Matching equipment for tractors, power tillers and other prime movers are either not available or farmers make inappropriate selection in the absence of proper guidance, resulting in fuel wastage and high cost of production.

ix) Hill agriculture, which covers about 20 per cent of cultivated land, has little access to mechanization. This situation has to be improved by developing and promoting package of technology for mechanization of hill agriculture to achieve higher productivity

Role of Custom Hiring in Providing Machinery and Farm Equipment

Custom Hiring envisages promoting establishment of farm machinery banks for hiring by way of providing financial assistance to self-help groups or farmers' co-operatives since the prohibitive cost of hi-tech and high productive equipments renders it difficult for individual ownership. The CHCs give farm machinery on rental basis to farmers who cannot afford to purchase high-end agricultural machinery and equipments apart from servicing old machinery. The centres play a pivotal role in introducing high technology agriculture machinery to even ordinary farmers with the objective to boost crop production and improve quality of agriculture operations. The total land under agriculture for an average Indian farmer is shrinking, which is a mere 1.5 acre per farmer at present. With such small land holdings, a farmer faces difficulty in justifying ownership of any kind of agri-machinery. Hence, renting out farm machines becomes more viable. Out of 120 million strong farmer population in India, majority of them are incapable of buying machines. Moreover, the growing reality of labour shortage is making farmers more inclined towards mechanization to ensure faster work at a lower cost.

Government Initiatives in Promotion of Custom Hiring

Government of India has kept special provision for custom hiring in the project 'Sub-mission of Mechanization on Agriculture' which is running during the current Five Year Plan (2012-17). In addition different State governments have taken measures to scale up the custom hiring services in their respective states. In July 2014, Karnataka government appointed two private entities to run 178 CHCs around the state for six years. Shri Kshethra Dharmasthala Rural Development Project, popularly known as SKDRDP, a charitable trust runs 161 centres. The Indian Society of Agribusiness Professionals (ISAP), New Delhi has been assigned to run the remaining 17 centres. Under the PPP model, each centre has been given a budget for 50 lakhs in the first year and 25 lakhs in the second year. Similar initiatives have been taken by the Government of Punjab, Haryana and Madya Pradesh to promote custom hiring of agricultural machinery in a big way.

Private Initiatives in Custom Hiring

A number of private partners like Zamindara Farm Solutions (ZFS), OLAM India, Tata Trusts and John Deere are entering in the custom hiring business and playing a big role in agricultural mechanization, rural employment, resource conservation and in improving the economic status of the farmer. Zamindara Farm Solutions (ZFS) Fazilka Punjab, has been one of the most successful custom hiring enterprenuer of the county. The company is associated with more than 5,000 families for tractors and farm implements on rent. The farmers have also started high tech equipment like Reversible Mould Board (RMB) ploughs and sub soilers. In association with Bill and Melinda Gates Foundation, the company trained farmers for baling operations which not only increased their incomes but also helped in protecting the environment. Moreover, 800 village youth got employment on Custom Hiring of residue collection machines. ZFS uses a combination of library model and radio taxi model. In the library model, all the machines are parked in the same place, farmers take the machine, use it and return it; while the radio taxi model works through a call centre. The farmers can reach them at their call centres and the company sends the requested services. The company also provide extensions services. The company rents out a wide range of machines including tractor, trencher, fodder, direct paddy seeder, laser leveler, turbo happy seeder, sub-soiler, rice transplanter, bed maker, gen set, front loader and back hoe, pit digger, precision and many others. Village based staffs and rural entrepreneurs provide all the services. The company's junior partners organize the delivery to farmers. A pilot project has been launched in Punjab by Zamindara Farming Solutions to allow companies and individuals rent out their tractors and equipment to a registered user who can expect the services at rates cheaper than those offered by traditional companies. The pilot project has received an initial funding from NABARD as a loan of Rs. 22 and Rs. 3 lakh as grant to carry out this innovative project called "*Chaalak Bane Malik*" based on popular taxi service like Ola. This initiative of "*Chaalak Bane Malik*" will provide services to about 3,000 farmers in the area under which tillers, farmers will be given loan at low rates. The amount of Equated Monthly Installments (EMI) will be deducted from the business they generate and eventually a farmer becomes the tractor/implement owner.

Conclusion

Though mechanization has improved the state of agriculture in certain parts of the country, it is still a bottom of the pyramid story and it will remain so unless concrete measures are taken to propel farmers towards adoption of efficient farm mechanization practices. The Custom Hiring is the only practical way to introduce capital intensive, high quality mechanization to the small farming structures prevalent in India. Custom Hiring can significantly facilitate diversification in agriculture specifically from wheat and paddy to other crops. Machinery required for operations like sowing, planting, transplanting, plant protection, harvesting and product recovery is highly crop specific. Thus, diversification would require use of a vast variety of additional equipment for these operations on limited area in the initial stages, making it highly uneconomical on ownership basis. However, Custom Hiring through private entrepreneurs or co-operatives will help to increase annual use of these equipment thereby making them viable.

Following are the key interventions required across the value chain to enable the wide spread adoption of custom hiring model

i) With increased participation of stakeholders across the agri supply chain and handholding farmers by supplying all equipments for entire life cycle of a crop sequentially, Custom Hiring concept can be made successful.

ii) Concept of Custom Hiring holds good potential provided there is an integration of all the operations *viz.*, provision of agri inputs like seeds, fertilizers, implements *etc.* through partnerships with various companies.

iii) There is a need for incentives and policy support for the adoption, development and promotion of farm mechanization technologies particularly suitable for dry land farming, horticulture and orchards, hill agriculture, sugarcane harvesting, cotton picking, rice production *etc.*

iv) Commercial banks and financial institutions need to develop hassle free loan origination and disbursement process for tractors and farm machinery on individual ownership basis or Custom Hiring basis.

v) Training and farmer workshops to show and identify the benefits of new technologies as well as intensive communication with the value chain is the key.

vi) KVKs and extension wings of universities need to play a very key role in sensitizing the farmers on Custom Hiring adoption.

vii) Manufacturing units that are set-up in areas with lower mechanization needs to be supported by extending tax and duty sops. This would result in easier reach of the equipment to farmers in those areas.

viii) There is a need to innovate Custom Hiring model by institutionalization for high cost farm machinery such as combine harvester, sugarcane harvester, potato combine, paddy transplanter, laser guided land leveler, rotavator *etc.*

ix) Indigenization of many specialized machineries by the private players which are otherwise imported thereby bringing down the cost.

References

Bansal, N.K. and Mukesh, S. 2010. Impact of custom hiring on farm mechanization in Haryana. *A Report of All India Coordinated Research Project on Farm Implements and Machinery*, Department of Farm Machinery and Power Engg., CCS HAU, Hisar, pp: 1–64.

Farm Mechanization in India: The Custom Hiring Perspective. Available at https://www.yesbank.in/pdf/farm_mechanization_in_india_per cent E2 per cent 80 per cent 93_the_custom_hiring_perspective.pdf

ICAR. 2013. *Hand book of Agricultural Engineering*. ICAR Publications, New Delhi

Indian Council of Agricultural Research – Vision 2030.

Mehta, C.R., Chandel, N.S. and Senthil Kumar, T. 2014. Status, challenges and strategies for farm mechanization in India. **In**: *Agricultural Mechanization in Asia, Africa and Latin America*, **45**(4): 43–50.

Mehta, C.R., Chandel, N.S., Senthilkumar, T. and Singh, K.K. 2014. *Trends of Agricultural Mechanization in India*.CSAM Policy Brief, June.

Proceedings of 2nd Regional Forum on Sustainable Agricultural Mechanization at Serpong, Indonesia, 9th–11th September, 2014.

Shyam. 2012. Developement of custom hiring model for rice-wheat cropping system for irrigated areas. Unpublished *M.Tech. thesis*, Division of Agril. Engg, IARI, New Delhi.

SMAM. 2014. Submission on Mechanization in Agriculture. *Policy Report*. Mechanization and Technology Division, Department of Agriculture, Cooperation and Farmers Welfare Ministry of Agriculture and Farmers Welfare, MoAFW.

Trends of Agriculture Mechanization in India, CSAM Policy Brief, June 2014.

Vision 2050. Document of Central Institute of Agricultural Engineering, Bhopal, 2015.

98

Horti-Tourism: A Venture in India for Creating Employment Opportunities for Youth

Adavi Rao Desai and Narendra Pratap Singh*

ICAR-Central Coastal Agricultural Research Institute,
Old Goa – 403 402, Goa
**E-mail: adavirao.desai@icar.gov.in desaiavars@gmail.com*

Horticulture crops and commodities are unique not only in terms of their nutritive/ health values but also for their exemplary aesthetic beauty, thereby signifying their importance in everyday life almost on every occasion. These features of them envisage enormous economic derivatives.

Subsistence farming of earlier times has no relevance in the present context, especially due to radical changes in social life style, rapid urbanization and their consequent onslaught on ecological balance. Therefore, Subsistence farming has suitably undergone diversification for sustainability reasons through Agri-business. Horticulture with its multiple components, has been playing a greater role in this transformation thereby imparting considerable opportunities in Agri-busniness sector by way of supporting and catering to the needs of Tourism Industry. This is how Horti-Tourism concept has emerged as an opportunity for many farm families to diversify their enterprises through Agro-eco tourism arena in which new generation youth is offered with tremendous opportunities for showcasing their potential to earn a very lucrative livelihood.

Incredible India has vast country side Tourist destinations showcasing multi cultural heritage in respect of varied life styles, traditional food and crops, paraphernalia of regional people and natural environment around them. Traditional or ethnic horticultural crops and their products of the region form the great fascination for the tourists. This brings farmers and visitors closer for effective social and economic interactions for deriving mutual benefits of aesthetic pleasure. Thus, the concept of Agro-eco tourism comprising of Horti-tourism as well, is envisaged as a new dimension to the farming sector with a view to ensure and enhance farm income in a sustainable way. Launching of "Rainbow Revolution" has been a major pep and fillip imparting multiple dimensions to diversify Indian horticulture industry in general. Horti-Tourism, indeed, adds a new dimension to galloping horticulture industry especially, to the country side and hinterland horticulture in the country. Due to the hectic life style in urban areas, there is a growing tendency of people to take a break from the urban fatigue and visit the serene country sides for enjoying natural beauty and aesthetic pleasure. Hinterland Horticulture is a niche area for accomplishing this and therefore form great fascination for tourism industry.

Conceptual Domain

Like Agro-eco-tourism, "Horti-Tourism" is the functional association of farming sector, tourism industry and farm business. It is indeed, a vibrant activity that occurs when people link travel with agricultural/ horticultural products, services or experiences to cater to the needs of mutual demands. Here, horticulture components play a vital role In Horti-Eco Tourism to secure a multifunctional, sustainable and competitive enterprise, maintain the landscape and the countryside, contribute to the vitality of rural communities, respond to consumer concerns and demands regarding food quality and safety, environmental protection, aesthetic pleasures, *etc*.

The societal demands include contribution to enhanced employment opportunities, viability of the rural country side and regional development, Horticultural landscape and cultural heritage values, good farming technologies and practices that respect ethnic and social welfare, environment protection and food safety and security.

Key drivers for Horti-tourism

For successful Horti-tourism in an area, the following key drivers play functional role and therefore need to be kept in view while venturing:

- ☆ Harmonious integration of ethnic horticulture and its diversified products with tourism for sustainable economic returns.
- ☆ Local employment generation.
- ☆ Delivering Aesthetic goods and services
  - I. Environmental protection and management;
  - II. Eco-tourism industry development;
  - III. Infrastructure development and
  - IV. Community development leading to socio-economic transformation

Key Elements of Horti-Ecotourism

- ☆ Ethnic crops and cultural Products: Specialty crops and their products are the fascination for the visitors.
- ☆ People : Horti-tourism includes mutual interaction between local farmers and people paying visit and/or staying on farms to experience rural life and or learn about different farming activities. It involves display of traditional skills by the local artisans.
- ☆ Natural resources and education: This is nature-based tourism that involves education and interpretation of the natural environment and is managed to be ecologically sustainable. It is an unobtrusive development reflecting environmental values while seeking pleasant sustainable economic derivatives through diligent management of natural resources. Therefore, Natural Resources, like water bodies, water streams, riversides, hillocks, *etc*., available in and around the farm have to be accordingly managed to amalgamate with crop production function.
- ☆ Social interactions: It is a relationship between tourism, the natural environment and the local community for exchange of culture, experience, commodities, *etc*., for mutual needs and benefits.
- ☆ Fiscal components : It is labour and knowledge intensive and encourages dispersion of income directly into regional areas rather than to larger urban areas. and provides the **opportunity for economic growth** across other community based service providers. This has the potential to contribute significantly to the local, state and regional economies on a sustainable ecologically sound approach.
- ☆ Multifarious range of tourism products: Agro-eco tourism has the potential of being an active participant and contributor to the tourism product range.
- ☆ Complementary functions: It can complement existing tourism products and link with accommodation, transport and other tourism related activities and service providers.

Why Horti-Ecotourism ?

The constant instability of net farm income and the loss of jobs in rural areas have led to a human and capital drain from many rural areas, leaving many farming families and business under economic stress. As has been realized already, agriculture alone can no longer be dependent upon to provide the economic stability for rural villages/ country side communities. But, their strength lies in serene and aesthetic environment, abundant natural resources, traditional crops and culture, local skills and innocence of the local communities. Thus, integration of all these commodities with tourism will open up a new horizon and can play the key role as new employment partners for rural communities thereby improve the economic status. Besides this, Horti-eco tourism forms a potential "Alternative as value added product to routine tourism" which is "Farm based and Harmonious with nature.

Opportunities for Youth

Horti-tourism based farms can be powerful centres for creating awareness among the people about conservation of regional biodiversity, rural artisans, and their role in nation building in pursuit of rural employment opportunities, besides providing aesthetic pleasure, peace and tranquility to the visitors. These Centres are the platforms which open up several livelihood opportunities of varied levels for youths of the region. Firstly, an enterprising youth can contemplate converting one's own existing horticulture farm into Horti-tourism unit by linking the activities with tourism operators.

These centres in the hinterlands can be hubs for showcasing the potential of rural talent, once the rural artisans, artists, local skills, *etc*. are linked for providing pleasant services and goods to the visitors or tourists. Thus, Horti-tourism endeavours pave the red carpets for the youth for a number of agribusiness activities (like on Farm marketing skills for regional handicrafts, products and goods and services, *etc*.), agro-based industries (processing of spices, fruits, flower products, virgin coconut oil, health drinks/ beverages, *etc.*) and indigenous medical therapy centres for which a number of horticulture plants, produce and products are required. In this way, Horti-tourism hubs will provide effective linkage between young farmers for taking up cultivation of need based horticulture crops in an enterprising manner and cottage industries in which entire rural folk in surrounding areas will seek a number of employment opportunities in local domain itself instead of migrating to the urban sectors.

Youth can take up secondary horticultural activities thereby linking themselves with Horti-tourism units for effective marketing right at their door steps, without involving any uscrupulous middle agencies.

Nursery ventures within and/or around Horti-tourism units for production and supply of quality planting material and seeds of regional specialty crops or high value plants of aesthetic importance will also open great opportunities for entrepreneurial youth.

Youth also can also get engaged themselves in Short time Training options for visitors, amateurs, school children and other enthusiasts about various propagation techniques, gardening practices, kitchen/ nutrition gardening, organic gardening, *etc.*, which also open up multiple employment opportunities.

Youth can also get opportunities to interact with people of different social strata from among national and international tourists for exchange of social, cultural and economic endeavours right in their home places and even these hubs provide opportunities to the youth to earn foreign currencies right in their places.

Horti-tourism domain as such can impart to the youth generation a total managerial skill for dealing with influx of visitors of different kind and thus become experts in the novel approach of additional dimension to the horticulture.

Spice Crops-based Tourism in Goa and Some Case Studies

Some case studies are presented here which are classical examples of enterprising youth who have taken up horticulture based farm tourism programmes and have earned 2-3 fold higher income in the range of Rs. 15 - 40 lakhs annually.

Paradigm shift from "on shore tourism" to "off shore tourism" is a key turning point in Goan Tourism industry that has envisaged a tremendous potential for horticultural based farm tourism. Predominantly, coconut/arecanut based spice gardens are playing major role farm based tourism.

There are several spice based farms in the range of small to commercial scale in the state. "Sahakari Spice Farm" at Curti, Ponda, "PausCal Spice Village" at Khandepar, "Tropical Spice Plantations" at Keri, Ponda and "Savoi Spice plantations" at Savoi Verem are among the prominent ones. Organic farming or natural farming is the principle of all these gardens.

Organic production of Vanilla, pepper, *etc.*, use of dung slurry for plastering the dining arena, Coconut and areca nut thatches are the special features of Sahakari Spice Farm, while abundant tropical orchids, Bimla (*Averrhoa carambola*) avenues, picturesque riverside location, rafter water sports, fishing pleasure, availability of quality planting material of orchids, anthriums and other ornamental plants, bird watching opportunity, swing hangs on river bank, *etc.* offer real eco-farm pleasure at "Spice Village". On the other hand, "Tropical Spice Plantations" has such special features as display of rural artisan and coconut handicrafts, majestic entry on bamboo bridge across the rivulet, elephant shower bath and elephant riding, organic spices, *etc.* for the visitors. One can enjoy the hill side location of "Savoi Plantations" adjoining the traditional rice fields in valleys and Crocodiles ponds in the farm besides cool hill side breeze under palm thatchet. Traditional welcome is a pleasant moment in all these farms and makes a total departure from urban style.

Spice crops are the destinations in these Agro-eco farms for the visitors especially the Western or European tourists in Goa. Visitors include school children as well, for whom these gardens offer opportunity to learn and enrich the knowledge about spices and their importance. Tropical spice crops as Kokum, black pepper, clove, nutmeg, cinnamon, cardamom, allspice, vanilla, *etc.* coupled with native fruit crops like banana, jack fruit, bread fruit, mango, cashew, *etc.*, are incorporated in palm based farming systems or agroforestry systems with specific orientation towards Agro-eco tourism. These gardens provide the tourists or the visitors single window opportunity to see, observe, and aesthetically enjoy the wealth of spices in their natural habitat in the back drop of lush- green-forest covers with natural scenic beauty. Here visitors can breathe not only fresh air but also feel the fresh aroma of spices in their fresh forms and can get the fresh, pure, organically grown spices and spice products. These gardens also function as health gardens for which visitors attach utmost importance, especially, those who come from the hustle and bustle of the metropolitan/cosmopolitan mechanical urban life. Curative function to overcome the mental melancholy is another feature of such Agro-eco tourism gardens for which also visitors throng to these destinations. Cottages admixtured with natural habitats of crops, fishes, live-stocks and herbal health care units coupled with Goan cuisine, provide the visitors comfortable stay in the midst of aroma gardens – a predominant philosophy of the Agro-eco tourism, where in the spice components play key role. Agro-eco tourism manifests multiple functions by way of bringing eco-friendly (natural) farming for sustainable production and tourism industry together for deriving dual flow of economic benefits. This is very much required in states like Goa where

Agrarian practices have to withstand the onslaught of Tourism industry. This year, Goa government has cleared 14 new projects worth Rs 717 crore, which will help generate employment to 2,600 persons.

Home Stay in Coffee Land: Agro-ecotourism in Kodagu (Coorg) District, Karnataka

Coffee based cropping systems comprising of spices like cardamom and pepper are envisaged as potential Horti-eco tourism destinations in Kodagu district of Karnataka. A number of coffee, tea and other plantations have been providing "Stay Home" facilities to the nature loving city dwellers by offering fresh agricultural products with local delicious cuisines. This is fast catching up in Coorg (Kodagu), Karnataka. Coorgese are known for their hospitality and home stay system which is common in foreign countries, is being seen everywhere in Coorg. Treating tourists as guests with affection and courtesy (Royal treatment) is making an annual business of 5-10 crores in a tiny district. Affectionate Hospitality that makes visitors feel "True Home stay away from Home" is a unique feature of Home Stays, besides, preference for local dishes, limited availability of drinks (Beverages). Based on tourist demand round in the coffee plantations, swimming in the river, catching of fish, walking in rice field bunds, visit to coffee processing will be arranged. These practices have provided enormous opportunities to the women folk for a lucrative earning annually. The Tariff rates for Home Stays are between Rs. 500/- to Rs. 3000/- depending on facilities. This fosters friendship through farm tourism. In few places cultural programmes depicting Coorg's life and culture will be presented by local youth folk to tourists. Home stay activities are managed mostly by ladies earning a sizeable income of Rs. 25,000/- to Rs. 50,000/- per month thereby serving as alternative sources of income during free period or distressed price for coffee beans.

Experience of Farm-based Tourism in Maharashtra State

In the state of Maharashtra as many as 90 agro-tourism centres have been registered according to Maharashtra Agriculture and Rural Tourism (MART), though many more are running unregistered. All these have horticulture components to play major role. For instance, vineyards in Nasik have set up villas for tourists to enjoy tasting various wines and experience the pleasure of living on a vineyard. Some vineyards use this to educate tourists about wines. Here, youth have opportunities for earning through resin industry, Wine industry, *etc.*

A farmer, Shri Ganpat Parthe, from Bhilar district in Maharashtra, has cultivated strawberries, mulberries and raspberries, is earning profitable income through Horti-tourism as well. He takes tourists around the farm to give them a glimpse into a farmer's life, and offers them fruits and food devoid of extravagance. He has two cottages on his farm, for renting out to groups/ families at Rs 700 per head for a day, including food. This way, his income has gone up by 20-30 per cent in the last couple of years due to agri tourism, at times even compensating losses incurred sometimes in farming due to unexpected situations.

In another case, Sunil Mankikar, a farmer in Wardha district of Maharashtra, earned Rs 27 lakh last year

Figure 1

only through agri tourism. Farmers are increasingly taking up agri tourism, along with farming, to increase their income. They are now building cottages on their farms and keeping these open for guests who want to experience rural life. (http://www.business-standard.com/article/economy-policy/agri-tourism-gets-farmers-extra-dough-111040300035_1.html)

Vednandini Agro-Tourism premises of Shri Praksh Pohre, near Akola in Vidarbha region is another success story which started with idea to effectively deal with infertile or uncultivable land, unreliability of electricity, misuse of water, moody behavior of nature the reasons that are primarily responsible for farmer suicides in Vidarbha (Nagpur Todau, 2015)

Farm Tourism in Kerala

Of the total of 32.22 lakh hectares (accounting for 56.78 per cent of state's geographical area), 1/3rd of this is mainly covered by coconut and another 1/3rd covered by plantations of cashewnut, Tea, Coffee, Rubber, pepper, cardamom and ginger, *etc*. Thus, it is reported that the Government has decided to allow 5 per cent of the farm area for farm based tourism purpose to encouraging the tourism industry for farmers' benefits. several schemes underway are intended to prepare the Farms/Plantations of Kerala to receive tourists by presenting a positive image of the **farm** and of agriculture as a whole, in view of revitalising the agriculture sector of Kerala through tourism. (http://www.keralatravels.com/stateresources.php?Topic=Kerala-Farm-Tourism and id=97 and sid=14)

'Wayanad County' (Priyadarshini Tea Plantation Resort) in Kerala is another example of tea plantation based tourism centre. A cooperative agency founded primarily to rehabilitate indigenous tribes has linked farm tourism with rehabilitation of several hundreds of tribes by initiating Horti-tourism activities. This has provided job opportunities to the tribal youths in Wayanad, Kerala. This tea plantation based tourism cnetre has become sought-after Horti-tourism centre. Spread over in an area of 400 acres, this unit has a Bunglow for the comfortable sty of visitors, trained youths as guides, trained youths as chefs, a tea factory employing tribal youth, Trekking route map and view point called "Vishwas" as a symbol of belief and hope for the people. The welfare of the tribal community is the driving force of this centre which achieved through farm based tourism.

Hill Regions and Horti-tourism

Himachal Pradesh

Horticulture is a vital sector in the state's economy as it generates more than Rs.3,200 crore ($525 million) annual income. In Himachal Pradesh, events like "Apple Festival" being organized jointly by Himachal Pradesh Tourism Development Corporation (HPTDC) and State Horticulture Department facilitate the linking of tourism with Orchard farming and opening up many opportunities for apple farmers and youth for additional income generation.

The exciting sights of blooming and fruiting in the extensive plantations of apple and cherries on higher hills of Shimla, Kullu, Mandi and the slopes of Kinnaur are the real paradise for tourists from other places. The Apple cultivation has made inroads into the high altitude cold desert valley of Spiti. Fruit maturing on the fields surrounding the over 1000 year old Tabo monastery in Spiti valley are some of the most delicious ones produced in Himachal Pradesh. Like the famed Japanese Cherry Blossom season, a walk through a cherry orchard in Kotkhai, Kumarsain, Baghi or Kotgarh is a very blissful experience. Some home stays are being made available for the visitors. Youth folk can engage themselves for driving the visitors through this countryside and for their stay in Farm rest houses thereby linking and realizing the potential of Horticulture based tourism. The rural homestay scheme in the state started in 2008. At present, over 550 homestay units with a room capacity of around 1,500 have been registered with

Figure 2: A View of Umlakro Farm at Jirang, Ri-bhoi District.

Figure 3: Shri Ganesh Chetri with Cardamom Plantation.

the state. Out of these, over 187 alone are in the Kullu-Manali region. The Shimla and Dharamsala areas have 77 and 69 units respectively. There is a local tradition to welcome the guests with a garland of dry fruits and a round Kinnauri cap with green flap. It's also known for legally brewing two popular drinks, 'angoori' and 'ghanti. While the former is made from distinctively flavoured black grapes grown in high altitudes, the latter is made from locally grown apples and apricots. These impart ample opportunities for horti-tourism activities.

Meghalya

With the aim of bringing development closer to the people in the villages, creating basic infrastructure to provide gainful employment opportunities for the people and empower them so that they may have the liberty to work as employers rather than employees, the Directorate of Horticulture released fund of Rs. 6.25 lakhs for developing basic infrastructure construction of shade house, overhead water tank and land development to develop Umlakro as a model horticulture farm. (Shabong, 2016)

Sikkim

Sikkim's economy has strong support from orchids and other flower crops and spices like large cardamom and ginger and thereby horticulture

Shri Ganesh Kumar Chettri, an young enthusiastic farmer from West district Sikkim, has given a new direction in the cardamom farming by merging it with tourism. While majority cardamom farmers in Sikkim bemoan that the income from the cardamom is not sufficient to run the house, Shri Chettri came up with an innovative idea and merged farming business with tourism and this brought additional income and helped him to sustain tourism in the west part of Sikkim. He recounts the importance of eco-tourism that helps to retain the cultural life and migration of youths from village to cities for job. He started a 'homestay' and named it 'Dhungay Homestay'. The tourists experience a 'rural stay' and live with the people of different mountain communities of Sikkim and get to witness the entire process of cardamom cultivation 'live'.

Orchids are another important resource for Horti-tourism in Sikkim with more than 523 species spread over 134 genera. The orchid diversity in Sikkim is so diverse and rich that its total number of species is yet to be assessed properly and this study does not claim to have surveyed and sampled all the forest and woodland areas of Sikkim. There are still good chunks of virgin lands in Sikkim yet to be explored to unfurl the richness of its biodiversity. Youth can establish orchid Sanctuaries and attract tourists for lucrative livelihood earnings. (http://www.bayultour.com/sikkim/orchid_diversity.html). Deorali Orchid Sanctuary, Gangtok (East Sikkim) is a classical example for the youth of the state. This venture not only helps conserve the native orchid resources for the next generation but also provide

Figure 4: A Beautiful View of Deorali Orchid Sanctuary, Gangtok (East Sikkim).

villagers around several opportunities when linked with Horti-tourism activities.

Conclusion

Potential of Horti-Tourism Venture is already realized by some enthusiastic farmers. But, comprehensive planning by the stake holders for this venture opens up vast and multiple employment opportunities to the youth. Multiple linkages within the niche domain make this a reality through the empowerment of villages and the villagers.

In this article, some successful cases of Agro-ecotourism Centres of Goa, Maharashtra, Karnataka and Kerala, and scope for such endeavours in other places like hill regions and North East India, are discussed with objective examples. Horti-tourism centres can be the effective learning centres for school children where, children can see crops practically, understand their role in human health, enjoy the local skills and products displayed there and develop, right from that age, the awareness and interest in the area of sound and safe food production. Thus, this article is an effort made to envisage importance of Horti-tourism in the current perspectives and its need and constraints in the relevant areas for encashing the potential opportunities for livelihood security for youth. Practicing Horti-Tourism imparts aesthetic pleasure and tranquility not only to the urban mass or any visitors but even to the entire local community involved in its domain.

References

Business Standard, 2016. Press Trust of India, Panaji June 20, 2016 Last Updated at 20:32 IST

http://martindia.org/english/solapur.aspx

http://www.bayultour.com/sikkim/orchid_diversity.html, S. Z. Lucksom, Director Himalayan Zoological Park, RVP and EPC, Department of Forest, Env and Wildlife, Management, Government of Sikkim.

http://www.business-standard.com/article/economy-policy/agri-tourism-gets-farmers-extra-dough-111040300035_1.html

https://in.news.yahoo.com/thoughtful-ias-officer-used-tourism-105901994.html

Nagpur Today, 2015. Vednandini Agro Tourism venture proving a boon to farmers in many ways, april.30, 2015.

Shabong C.S., 2016. UMLAKRO FARM – Model for Sustainable Horticulture Tourism Development in Meghalaya.

99

Quality Seed Production and Profitability of Horticultural Crops in Northern Karnataka

D.R. Patil[1]*, Shivayogi Ryavalad[1], Bapurayagouda Patil[1], Deelep Masuti[2] and Manjunath Tattimani[3]

[1]Seed Unit, University of Horticultural Sciences, Bagalkot – 587 104, Karnataka
[2]KRC College of Horticulture, Arabhavi, UHS, Bagalkot – 587 104, Karnataka
[3]MHREC, University of Horticultural Sciences, Bagalkot – 587 104, Karnataka
*E-mail: adre.uhsbagalkot@gmail.com

Indian farming system is characteristically low land holding in nature and seed can play pivotal role in achieving higher productivity as it is small holder friendly and in principle equally applicable on small or large farms. Seed can play critical role in achieving higher productivity, use of quality seeds alone could increase productivity by 20-25 per cent highlights the important role of seed in agriculture (Anon, 2013).

India is blessed with assorted agro-climatic conditions ranging from tropical to temperate which make possible for cultivation and seed production of horticulture crops belonging to different temperature regimes. Karnataka state is also endowed with varied agro-climatic conditions making it suitable for successful cultivation of different horticulture crops like fruits, vegetables, spices, plantations crops, flowers, medicinal and aromatic plants on commercial scale. It produces about 17.80 million MT of horticulture produce from an area of 1.87 million ha accounting for 7.40 per cent of horticultural production in the country (Anon., 2014). Major share of production is from vegetables (50.88 per cent), fruits (35.25 per cent) and plantation crops (10.07 per cent).

India is the second largest producer of vegetable in the world constituting around 11 to 12 per cent, next only to China. But, the productivity of vegetables in India (15.3 t/ha) and Karnataka (16.07 t/ha) is low as compared to other vegetable producing countries (Figure 1). Now, the area under vegetable cultivation is expanding enormously year by year due to high profits in vegetable cultivation. This creates huge demand for vegetable seed in the market. The current estimated requirement of vegetable seeds at the country's present production level is to be around 25,000 tonnes for tropical and sub-tropical vegetables and 250 tonnes in the temperate kind.

Karnataka is considered as hub of seed production and have been identified in the global seed map. Vegetable seeds produced here are exported to every corner of the world by the private seed industry. As vegetable seed industry is dominated by the private industry. Public sector industry is less identified in vegetable seed industry. Presently, majority of the vegetables grown in the state are dominated by private hybrids rather than the public hybrids. However, in crops like onion, chilli, beans *etc.* still varieties are ruling as there are no popular hybrids in this crop. Hence,

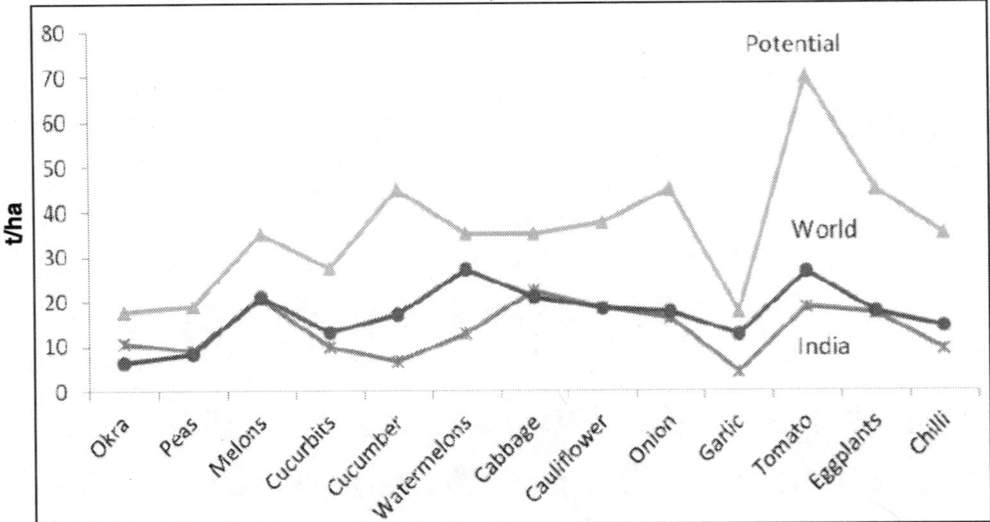

Figure 1: Vegetable Productivity Scenario.

there is a lot of scope to public sector organizations to compete with private seed companies in production of quality seeds of this type of vegetable crops. Public sectors can produce seeds in crops like onion, chilli, *bhendi*, leafy vegetables, seed spices and papaya. The list of crops and approximate seed requirement in the state is presented in Table 1.

There is a lack of quality seed availability in the state every year. Further, Public sectors alone cannot produce seeds and meet the demands. Hence, the farmers need to be involved as a seed producer. In view of the above, to supply quality seeds of vegetable crops to farming community at affordable price, a programme was initiated with alternate method of seed production and marketing by involving scientists and farmers under participatory approach with buy back system to produce large quantity of quality seeds (Vishwanath and Gangadhariah, 2012).

Material and Methods

In order to enhance the availability of quality seeds of newly released public varieties, the Seed unit, University of Horticultural Sciences, Bagalkot, Karnataka implemented Scientists-Farmers participatory mode seed production programme with buy back system to produce quality seeds of vegetable crops like onion, chilli, drumstick, beans, *bhendi etc.* in large quantity. In which, every year university is selecting interested farmer or group of farmers in three districts of northern Karnataka, which comes under university jurisdiction based on the criteria which are required for quality seed production with economic yield (Basave Gowda *et al.*, 2014).

The selected farmers were supplied with the good quality foundation seeds for production of certified seed for one acre each. The seed production specialists of the university organize 2-3 trainings programmes at village level in the selected beneficiary group on various aspects of quality seed production. In addition to the training programmes on-farm demonstrations and field days were also conducted about different practices of quality seed production.

Further, seed production was offered for certification from Seed Certification Officers (SCO) of Karnataka State Seed Certification Agency (KSSCA), custom seed processing and seed testing facilities were also provided after processing. After the results of seed testing, those seeds produced are meeting minimum standards of seed certification were procured by the university, after procuring the seeds were treated, bagged, labeled and sold to the end user at reasonable price .

Table 1: Major Vegetables Production Area and Seed Requirement in Karnataka State

Crop	*Area (in lakh ha)*	*Seed Requirement (in tonnes)*	*Popular Varieties*
Onion	1.64	820	Arka Kalyan, Bellary Red, Rose onion, Nasik Red
Beans	0.15	600	Arka Komal, Suvidha
Chilli (green + dry	1.35	337	Pusa Jwala, Byadagi Kaddi and Dabbi
Bhendi	0.88	132	Arka Anamika, Parbhani Kranthi
Coriander	0.765	918	Arka Isha and Local varieties
Fenugreek	0.0687	275	Co-1, Local varieties
Leafy vegetables	0.154	30	Local types
Papaya	0.633	1.58	Solo, Surya

Responsibilities of University of Horticultural Sciences, Bagalkot

- ✰ Identification and selection of seed growing farmers, preference should be given to progressive farmers as seed production need technical skill.
- ✰ Supply of basic seed materials to confirm the authentication source of seeds used for sowing.
- ✰ Technical guidance to seed growers in maintaining the field and seed standards.
- ✰ Field inspection for seed quality control and assurance.
- ✰ Seed Processing and upgrading the seed quality.
- ✰ Seed Testing for quality management.
- ✰ Seed Packing, labeling storage and marketing.

Responsibilities of Farmers

- ✰ Production of seeds under the guidance of scientists of the University and also certification officials.
- ✰ Follow instructions given by the scientists and carryout all cultural and special operations like isolation and rouging to meet the field and seed standards.
- ✰ Submit raw seeds having moisture not more than 8 per cent to the university.

Result and Discussion

Under the novel programme called "Scientist Farmers Participatory Mode Seed Production Programme" the university has produced quality seeds of new and popular varieties of onion (Var. Arka Kalyan) Chilli (Byadagi Dabbi and Byadagi Kaddi) and Drumsick (Bhagya and KDM-1) involving the farmers/ seed growers of more than 10 villages in three districts which comes under university jurisdiction (Table 2).

Table 2: Details of seeds produced at UHS, Bagalkot under SFPMSP

Year	Crop	No. of Farmers	No. of Villages Covered	Quantity of Seeds Produced (kg)
2014-15	Onion	12	04	1196
	Chilli	05	02	619
	Drumstick	10	10	217
Total	27	16	2037	
2015-16	Onion	47	15	6972
	Chilli	23	12	1237
	Drumstick	15	10	310
Total	**85**	**37**	**8240**	

With implementation of this programme, more than 5000 farmers are in direct contact with the university in getting the benefit of new technologies, varieties, seeds and know how. Due to this programme, there will linkage between scientist and the farmers in identifying the problems faced by the farmers and which helped the scientists in formulation of technical programme and research activities. The availability of skilled human resources at the grass root level can be used for quality seed production and promotion of horticultural technologies both by public and private (Jhani Rani *et al.*, 2013).

Impact of SFPMSP Programme on Quality Seed Production and Enhancing the Horticulture Production

- ✰ Bridging the gap between supply and demand of the quality seeds of public sector.
- ✰ Seed production potential of the university is increased for timely supply of quality seeds to the farmers.
- ✰ University can also supply basic seeds to other seed producing agencies like KSSC, NSC, KOF and other private agencies for large-scale multiplication of quality seeds.
- ✰ Dissemination of seed production skills to farmers through trainings, demonstrations and field days.
- ✰ Employment generation through engaging the unemployed youths in the seed production programme.
- ✰ Develop intimate Scientists-Farmers relationship, resulting in efficient transfer and implementation of new technologies including varieties/hybrids which help in overall development of farming community.
- ✰ Improves socio economic status of the farmers.

References

Anonymous. 2013. *Indian Agriculture: Performance and Challenges, State of Indian Agriculture 2012-13.* Directorate of Economics and Statistics, Department of Agriculture and Cooperation, Ministry of Agriculture, Government of India, New Delhi, pp. 1–22.

Anonymous. 2014. *Indian Horticulture Database-2014.* National Horticulture Board, Ministry of Agriculture, Government of India, New Delhi.

APEDA. 2013. Exports from India of fruit and vegetable seeds.

Basave Gowda, Lokesh, G.Y., Rakesh, C. Mathad, Ganiger, B.S., Lokesh, K., Prasanth, S.M., Bellad, S.B., Radha, K. and Patil, S.B., 2014, UAS, Raichur-Quality seeds for food security and sustainable agricultural productivity in North Eastern Karnataka. Regional

Science Conference 4-5 January, Bengaluru, pp. 127–128.

Jhani Rani, K., Keshavalu, K., Razia Sultana and Brahmiah, M., 2013. Seed village programme in Andhra Pradesh: Problems and perspectives. *13th National Seed Seminar*, 8-10 June, Benguluru, pp. 177.

Viswanath, K. and Gangadharai, S., 2012. Impact of scientists farmers participatory seed production on socio-economic status of farmers in Cauvery command area. *National Seed Congress*, 23rd-25th January, Chandigarh, pp. 220.

100

Production of Bioagents: A Potential Enterprise for Youth

A.N. Shylesha

ICAR-National Bureau of Agriculturally Important Insect Resources, Hebbal, Bengaluru – 560 024, Karnataka
E-mail: anshylesha@gmail.com

Biological control is a naturally occurring phenomenon acting on the crop pests and maintain them at certain densities. We have to supplement the activity of the naturally occurring biocontrol agents *viz.*, parasitoids, predators and entomopathogens with introduction of appropriate agents. There are two important aspects, *e.g.*, conservation of natural enemies and enhancement of their activities to keep the pests under check and is use of biocontrol agents by way of mass culturing and release of selective natural enemies. This can be by inoculative release wherein the released biocontrol agent establishes and multiplies generation after generation, persists for longer periods and brings about control of pest in a shorter period.

Advantages of Biological Control

Biocontrol agent like pathogens, parasitoids, predators and antagonistic organisms have become invaluable components in agricultural IPM system in view of high level of specificity, safety and sustainability. Their role has expanded in crops and forest protection with the discovery of new agents, genetic improvement, application options and compatibility with other interactions. Biopesticides find extensive use where the value of a crop is too low and pesticides usage will be a costly affair and overuse of insecticides has created pesticides resistance and in cropping systems where biological methods are not currently part of an already adequate pest management programme.

Entomophagous Insects

Work on biological control using entomopahges in India commenced with the introduction of the coccinellid, *Cryptolaemus montrouzieri* from Australia in coffee in the Lower Pulney hills for the control of coffee green scale *Coccus viridis* (Mayne, 1953). Though not successful against *C. viridis*, it established as a common predator of mealy bugs (Rao *et al.*, 1971). Since then, exploitation of exotic entomophagous insects has been encouraged as a form of biological control.

Conservation of Natural Enemies

The rate of success of natural enemies has been primary consideration in pest management programmes. Some of the important conservation measures adopted were selective use of chemicals that do not kill the natural enemies rather they create of favourable environment for survival and multiplication of natural enemies and helped judicious selection of crop mix that facilitate colonization and providing refugia and alternate food (Hokkanen, 1993).

The use of laboratory bred natural enemies to increase the numbers and species composition in crop habitat has gained momentum due to our successes in mass breeding programmes. In our country, successful experimentation with entomophagous arthropods for crop pests control has been demonstrated in different crops. The establishment of biocontrol agents production facilities both under governmental and private efforts has enabled year round production and availability of species for field augmentation (Jayaraj *et al.*, 1994). In addition, economics involved in the rearing of major natural enemies is receiving wider attention. It may be more cost – effective to rear and release a natural enemy that is less effective in terms of field performance by inundation, if it can be reared for a much lower cost per individual, than to release a very effective natural enemy that is expensive to rear. This adds a dimension in manipulation of rate of release, species composition and numbers.

The science and art of insect rearing has increased in complexity and sophistication as the need for insects of known traits has grown phenomenally. Today scientists are able to modify characters in the laboratory principally by selection. The development of artificial diets, assurance of diet ingredient quality and new container designs had major impact on the development of rearing procedures.

The basic requirements in a production system will depend on the types of insect species being reared in the system. There is always a commonality in the use of materials in multiple species rearing systems. To achieve this end the desired characteristics must be known. The development in the technology of insect rearing has moved forward from the natural diet to artificial diet containerization. Currently, metal, fibreglass and thermoplastics are in use in many of the lepidopterous rearing systems.

Artificial Diets

Artificial diets have relieved the researchers from the maintenance of expensive greenhouse facilities and host plants for phytophagous insects. The continual advancements in artificial diets have made it possible for the maintenance of many additional species also resulted in the consistent and economical production of vigorous competitive insects. Standardization of the insect diet is important and it is essential to apply quality standards to the formulation of standard diets. The diet shall include many variables: protein, carbohydrate, lipid minerals, vitamins, microbial inhibitors and secondary substances. Also the pH, moisture content, texture, consistency are considered important.

Sanitation and Decontamination

Due to inefficient protocols the rearing environment frequently gets gross microbial contamination. Therefore, strict sanitation measures are needed to minimize or eliminate microorganisms in the rearing facility. Sensible use of various sterilization procedures *viz.*, heat, steam and chemical methods has to be followed.

Laboratory Colony

The success of the rearing programme depends upon obtaining a disease-free and a genetically efficient and uniform parental stock. The preferred stage from the field for starting the colony is gravid female. A starter colony of *H. armigera* can be established by field collection of grown-up larvae from ladies finger chickpea, pigeonpea, cotton, groundnut, sunflower, bhendi, *etc.* so that a continuous supply of adult moths of both sexes will be available for the breeder stock. The colonies established from collections made from different crops and areas have to be kept isolated and examined for the incidence of protozoan pathogens.

Production of Host insects

During the last two decades much effort has been expended on developing suitable diets and containers to house larval stages of *H. armigera.* Many of the research laboratories depend on the semi-synthetic diet developed by Shorey and Hale (1965).

Maintenance of Colony

The pupae from primary parental colony are kept in a 30 x 30 cm adult emergence cage for eclosion. Daily an adult feed solution containing 10 per cent sucrose fortified with vitamins is provided in the cage for adults. Ten pairs of healthy adults are transferred to oviposition chamber consisting of a 30 x 20 cm plastic bucket. The mouth of the unit is covered with muslin cloth which serves as a oviposition substrate. It is daily removed and replaced with fresh ones. Adult feed is also changed daily. To improve sanitation, adults which are weak are removed.

The muslin cloth are collected and labeled properly. They are kept in an atmosphere of higher saturation keeping them inside a humidifying chamber. As inexpensive chamber consists of a plastic bucket with a lid. Sodium hypochlorite 0.25 per cent (v/v) is poured inside the bucket upto 2 cm height level. The egg laden muslin cloths are kept inside a clear plastic container (20 x 10 cm) and allowed to float on the solution. The lid is firmly placed over the bucket. 24 h after the collection of egg-cloth, and incubation, the eggs are watched for development of embryo. Eggs showing distinct germ band are suitable for surface sterilization, which is done to eliminate the contaminant microbes. Surface sterilisation is achieved by treatment in formaldehyde. The egg-laden clothing is submerged in a 10 per cent formaldehyde solution for 10 minutes. After the expiry of time, the clothes are washed in tap water for 20 min to eliminate the traces of formaldehyde and are shade dried and kept in sterile plastic cups inside a sterile humidifying chamber for eclosion.

The larvae are supported by a semi-synthetic diet. It is prepared by blending and cooling with the ingredients given in Table 1. Chickpea seeds (previously soaked for 6-8 h in water) are boiled in water (300 ml) and transferred to a blender. Wesson's salt mixture, sorbic acid and methyl-para-hydroxy benzoate and yeast, are added while blending. Simultaneously, agar is boiled in the remaining portion of water (360 ml), cooled to

60°C and then finally added to the above mix. After thorough blending for 2-3 min. the remaining fractions of the chemicals have to be added one after another. The hot liquid depending upon the need is dispensed into rearing trays and glass vials for rearing early and older stages of the larvae respectively. After solidification the containers and vials are kept in insect proof area for subsequent use the next day.

Table 1: Semi-Synthetic Diet for Rearing Larvae

Chickpea seeds: 100 g	Wesson salt mix: 7.2 g
Agar agar: 12.8 g	Streptomycin sulphate: 40 mg
Yeast: 30 g	Vitamin supplement: 2 ml
Methyl-para-hydroxy benzoate: 2 g	Formaldehyde 40%: 1ml
Sorbic acid: 1 g	Carbendazim: 500mg
Ascorbic acid: 3.2 g	Water: 720 ml

Newly hatched larvae are transferred and confined properly with the help of orgami tissues reinforcement beneath the lid. The trays in the rearing tray are kept inverted to facilitate the positively photropic and negatively geotaxic larvae to feed unhindered on the diet. When the larvae reach second instar premoult stage they are transferred to glass vials (5ml) and plugged with cotton.

The prepupae of uniform size are collected and transferred to diet trays filled up to 1/3rd level with cotton wool for pupation. Pupae are removed from the culture and washed in 0.25 per cent solution sodium hypochlorite and kept in adult emergence cage. The production cycle is repeated. Culturing is a labour intensive process and involves handling of live stages. Development of fungus, bacteria in the diet used for mass and culture can be overcome by proper sanitation, restriction of personnel inflow into the rearing area.

Guidelines for Production Flow Maintenance

The following important guidelines should be observed during the production process (Sathiah and Jayaraj, 1996).

- ☆ Continuously examine each step in production for ways to reduce the time and labour input.
- ☆ Maintain good hygiene and cleanliness and monitor microbial contamination.
- ☆ Keep biological records of the insects.
- ☆ Check quality of the insect frequently.

Preparation of Schedules

It is often important to draw schedules for daily routine work. The following sequence of work has to be attended in order to maximise the productivity:

- ☆ Prepare the facility for use.
- ☆ Sanitize the rearing areas.
- ☆ Prepare adult feed.
- ☆ Prepare semi-synthetic diet and cool.
- ☆ Release the freshly hatched larvae into the trays.
- ☆ Sanitize the implantation area.
- ☆ Move the rearing trays to holding rooms.
- ☆ Harvest, surface sterilize and incubate eggs.
- ☆ Provide adult feed.
- ☆ Collect adults, sex, pair and allow for oviposition,
- ☆ Harvest pupae, surface sterilize and keep in adult emergence cages.
- ☆ Sanitize the adult holding room.
- ☆ Sort out trays for host culture and subsequent recycling and for virus production.
- ☆ Prepare materials for wash up operation.
- ☆ Surface sterilize the materials appropriately.
- ☆ Record production, quality *etc.*
- ☆ Close up operations

Mass Production of some Entomophagous Insects

Corcyra cephalonica

The green lacewing, *Chrysoperla carnea*, Stephens is a successful predator on many species of aphids, caterpillar eggs and several soft bodied sap feeders, mites and eggs of lepidopterans. For successful mass production a continuous supply of the eggs of rice moth *Corcyra cephalonica* is required. The rice moth *C. cephalonica* is cultured in broken grains of pearl millet or sorghum or maize in plastic round basins. Heat sterilized broken grain is taken at 2.5 kg per plastic basin (30 cm dia) to which groundnut kernel powder is added at 100 g and yeast powder added at 5 g (16 tablets). To prevent bacterial infection in the food medium streptomycin sulphate 0.05 per cent spray is given at 10-20 ml/basin using a hand sprayer or atomizer. Sulphur W.P. is added at 5 g/basin to prevent storage mite. Then nucleus *Corcyra* eggs are added at 0.5 l/basin containing 2.5 kg of grain medium (Wajnberg and Hassar, 1994). After uniform mixing of the contents of the basin, it is covered with *kada* cloth (0.25 mt) and secured by rubber band. The young *Corcyra* larvae that hatch out from the eggs in 3-4 days feed on the medium, constructing webbing. The adult *Corcyra* moths start emerging from the medium from 30-35 days onwards and continued to emerge upto 90 days after inoculation of eggs due to staggered development of larvae in the medium.

The adult moths rest on the inner surface of the cloth cover. They are collected in morning hours using glass specimen tubes (14 x 2.5 cm) or a specially designed modified vacuum aspirator (TNAU model). The moth collection is effectively done by keeping the basins inside a mosquito net house so that escape of moths is prevented. The adult moths are then transferred into a specially designed mating drum made of G.I. sheets with wire mesh bottom. Adult moths are provided with honey solution (50 per cent) added with Vitamin E as feed (one capsule/20 ml of 50 per cent honey) to boost the vigour of the adults and to get higher quantity of healthy eggs. The adult food is given by dipping cotton swab and allowing them to hang inside the drum with a thread.

Daily, fresh moths are collected and allowed into a fresh mating drum is cleaned and dried under sun earlier. *Corcyra* eggs are loosely laid by moths and are collected through the wire mesh bottom on a receiving container with funnel set up or enamel tray. Eggs are collected continuously for 4 days from each drum. On 5th day it is vacated and cleaned. A sheet of blotting paper is spread of the tray or in the funnel set up. It retains most of the moth scales and body fragments while the eggs easily roll out during cleaning.

The eggs thus obtained are cleaned with the help of plastic sieves of different meshes. One cc (or) ml of *Corccyra* eggs approximately contains 18,000 eggs. About 100 pairs of *Corcyra* moths (50 per cent female) produce 1.5 cc of eggs during its egg laying period 4 days. From each *Corcyra* rearing basin an average of 2500 moths emerge. Hence from each basic 18-20 cc of eggs are obtained during the period of 90 days. After 90 days the contents of the basin are discarded and the basic is washed, disinfected with 2 per cent formalin solution and dried thoroughly before reusing.

Materials Required for *Corcyra* rearing (King and Leppla, 1984)

- ☆ Broken bajra (pearl millet) grain
- ☆ Mosquito net (6′ x 6′ x ′6′)
- ☆ Plastic basins (30 cm dia)
- ☆ Specimen tubes (glass 15 x 2.5 cm)
- ☆ Kada cloth, aspirator (TNAU model), Yeast tablets
- ☆ Rubber bands and twines
- ☆ Moth scale separator (TNAU model)
- ☆ Groundnut kernel, sieves and filters (plastic)
- ☆ Sulphur W.P. streptomycin sulphate
- ☆ Home milling machine, vacuum pump
- ☆ Exhaust fan/Ceiling fan
- ☆ Measuring cylinders 10, 50, 100 ml
- ☆ G.I. mating drums (25 x 25 cm)
- ☆ Formaldehyde 40 per cent, enamel tray
- ☆ Hand sprayer or hand atomizer
- ☆ Blotting paper sheets, honey
- ☆ Camel hair brush, Vitamin E capsules
- ☆ Shoe brush, absorbing cotton

Chrysoperla Mass Culture

It is done in GI round basins (28 cm dia) at 250 larvae/basin covered with *kada* cloth. The eggs of *Corcyra cephalonica* are given as feeding material for the larvae in the laboratory. For rearing 500 *Chrysoperla* larvae the total quantity of *Corcyra* eggs required is 25.0 c.c. at the rate of 5.0 c.c./feeding for 5 feedings in alternate days. The *Chrysoperla* larvae pupate into round white coloured silken cocoons in 10 days. The cocoons are collected with fine brush and transferred into 1 lit plastic containers with wire mesh window for emergence of adults. From the coccons, pale green coloured adults with transparent lace like wings emerge in 9-10 days (Singh, 1995).

The adults are collected daily and transferred to pneumatic glass troughs or G.I. round troughs (30 cm x 12 cm). Before allowing the adults, the rearing troughs are wrapped inside with brown sheet which act as egg receiving card. About 250 adults (60 per cent females) are allowed into each trough and covered with white nylon or georgette cloth secured by rubber band. On the cloth outside three bits of foam sponge (2 sq.in) dipped in water are kept. Besides an artificial protein rich diet is provided in semisolid paste form in three spots on the cloth outside. This diet consisted of one part of yeast powder, one part of fructose, one part of honey and one part of Proteinex. Water is mixed to make it as paste. The adults fed the food and lay eggs on the brown sheet. The adults are collected daily and allowed into fresh rearing troughs with fresh food. From the old troughs, the brown paper sheets along with *Chrysoperla* eggs is removed.

The brown paper sheet kept inside the adult rearing troughs contain large number of eggs each laid on a stalk or pedicel. As such the sheet is stored at 10°C in B.O.D. incubator or refrigerator for about 21 days. When the eggs are required for culturing or for field release the egg sheet is kept at room temperature for a day and the eggs turned into brown and hatched on 3rd day. The first instar larvae are either taken for culture tray for recycling or for field release.

The first instar larvae of *Chrysoperla* are released in horticultural crops at 50,000 to 1,00,000/ha for 3-5 times at 10 days interval to control aphids, whitefly, *Spodoptera*, *Helicoverpa*, pink bollworm, thrips and mites. The larvae are taken in plastic containers containing a small quantity (1-2 c.c) of *Corcyra* eggs and loose paper strips. The paper strips along with larvae sticking on them are dropped in the field at random while walking across the fields (Ridgeway and Vinson, 1977).

Storage is done by keeping destalked eggs after sodium hypochloride egg wash (at 0.1 per cent for 1 minute) at 10°C in BOD incubator adjusted for constant temperature. Daily a sample of 10 destalked eggs can be taken and kept outside and the per cent hatchability can be checked. This method is best for storage.

Cryptolaemus montrouzieri Mulsant (Coleoptera: Coccinellidae) Rearing

The predator, *Cryptolaemus montrouzieri* is native of Australia. In 1892 it was introduced into California for the control of citrus mealybugs. Following the success the beetle it was introduced into India in 1898 by New Port. It has given effective control of mealybugs in fruit crop like citrus, grapes, guava *etc.*, To release the predator in large number in the field, mass culture of *Cryptolaemus* is a pre-requisite. *Cryptolaemus* is easily cultured on a large scale on the meal bugs in the laboratory.

Mealy Bug Production

In the large scale production of mealy bugs, potato sprouts or ripe pumpkins are utilised. The mealybug *Planococcus citri* (Risso) is better adapted to the insectary programme than the other mealy bug species by reason of its shorter life cycle, higher fecundity, and suitability as host for *Chryptolaemus*.

The use of potato sprouts as an insectary host for mealybug culture has undergone several modifications since reported first. Planting trays made of red wood (45x15x10 cm) are used for this purpose. Tiny tubers of potato, three months after harvest or when sprouts begin to appear, are ready for planting. Whole potatoes are used and 25 to 30 tubers are placed about 1/2" layer of sandy silt soil in the tray and covered with slightly moist soil. These trays are kept in racks in the production at a temperature of 70-74°F room and watered for facilitating sprout growth. The time from planting until infesting with the mealybugs is usually 21 days in summer and 10 days in winter. Stock from one mealybug tray is sufficient to infest 20-25 trays of sprouts.

The culture of mealybugs on pumpkin is another common method. The pumpkins are selected with ridges and grooves with a small stalk which makes handling very easy. Pumpkins are cleaned with water to get rid of any dust on them. To prevent rotting, pumpkins are treated with 0.1 per cent benlate (1 g/lit.). The wounds, if any are plugged with paraffin wax. Ovisacs of the mealybug are placed over the pumpkin for about 48 hrs. The infested pumpkins are kept on a plastic stand in wooden cages, with glass sliding in the front and cloth on other sides. In due course, crawlers emerge from ovisacs, settle on all sides of pumpkin and develop into fully mature mealybugs in 30-40 days.

Cryptolaemus Production

When mealybugs are 8 days old, a stock of 15-20 females of *Cryptolaemus* is placed in each tray in the dark room at 74-76 F. Their eggs are deposited liberally on the sprouts and wooden trays. Twelve days later, the ovipositing adults are removed from the tray. Burlap strips are attached to the fronts of racks to accommodate pupating *Cryptolaemus* and the room is against darkened. Eighteen days after removal of sting stock, the emerging adults are collected into plastic vials by means of a broad flat scoop funnel which is attached to the open top of the vial. Each plastic vial contains 100 beetles.

In about 20-25 days after the infestation with the mealybug, *Cryptolaemus* adults are released into the cage through its sleeve. The adult beetles, besides feeding on the mealybugs, lay their eggs singly or in groups of 4-12 in the ovisacs of female mealybugs. If the females are not fully developed, eggs are laid anywhere near the mealybugs. The grubs are visible in about a week's time. Initially, these feed on the eggs of mealybugs or smaller nymphs. As they grow, they start feeding on all stages of the mealybugs. Cannibalism is observed when the mealybug population is low. The older grubs feed upon young grubs. The fully developed grubs pupate on the pumpkin or anywhere inside the breeding cage.

The first beetle emerges in 30 days time from the date of exposure of the mealybugs to the beetles. The beetles continue to emerge for another 5-10 days and are collected in glass vials using an aspirator. Each breeding cage yields 100-200 beetles. They are fed with honey solution (50 per cent) or honey agar in the laboratory. In about 10-15 days, when the adult beetles complete the mating and preoviposition, they are ready for field release. The production cost of 100 beetles has been estimated to be about Rs.150/-

The beetles are also fed on diet containing agar powder (1 g), sugar (20 g), honey (40 cc) and water (100 cc). The adult beetle diet is prepared by boiling sugar in 70 cc of water, adding 1 gm agar, diluting 40 cc honey in 30 cc of water and adding to the sugar and agar mixture when it comes to boiling point. The hot liquid diet is kept on small while plastic cards in the form of droplets which get solidified on cooling. Such cards containing adult diet can be fed not only to *C. montrouzieri* but also to many other species of coccinellids. From each cage about 175 beetles are obtained. The emergence of the beetles is completed within 10 days. The cost of production of 100 larvae at 1977 price index was Rs. 6/- and at 1992 Rs. 60/-. The cost of production of 100 beetles at 1977 price index was Rs. 10/- and at 1999 price index Rs. 150/-

Beetles can also be reared on *Corcyra cephalonica* eggs but empty ovisacs of *P. citri* are to be kept for inducing egg laying by the beetles. The beetles are also multiplied on semisynthetic diet which is still in the process of further refinement. All due precautions should be taken to avoid scarcity of food for the grubs to avoid cannibalism by grubs. All the pumpkins showing sign of rotting should be properly incinerated.

Egg Parasitoids

The members of the family Trichogrammatidae are minute 0.2 – 1.0 mm long, widely distributed and are characterised by 3-segmented tarsi. They constitute an economically important group of hymenopteran parasitoids that attack eggs of various groups of insect pests mostly lepidopteran and hemipteran, few on coleopteran, dipteran and thyasopteran (Yousuf and Shafee, 1987). Genus *Trichogramma* was erected by Westwood (1840), whereas Genus *Trichogrammatoidea* was erected by Foerster (1856). Trichogrammatids are one of the important components of bio-intensive pest management (BIPM) and has proved to be an excellent alternate to chemical pesticides in many ecosystems. In the field, both abiotic factors and natural enemies are responsible in reducing the pest population by up to 80 per cent. However, these parasitoids are active in the field under natural conditions only after the damage to crop has already been done. The folly of sole dependence on pesticides has been realised the world over and the concern for preserving a cleaner world for posterity has increased. Still, it is a challenge to make biological control work in the field for the benefit of the small and marginal farmers.

Tomato

Trichogramma chilonis, *T. brasiliense* and *T. pretiosum* have been extensively tested against *H. armigera* in tomato fields (Krishnamoorthy, 1987; Krishnamoorthy and Mani, 1987; Singh, 1991; Rawat and Pawar, 1993). In Gujarat, *H. armigera* infestations on tomato were checked by releasing *T. chilonis* @ 2.5 lakhs parasitoids/ha at weekly intervals during December-February (Yadav *et al.*, 1985). Releases of *T. brasiliense* @ 50,000/ha/week/ release (6 releases) were made which resulted in egg parasitism up to 78.4 per cent as compared to 12.3 per cent in control areas. Fruits bored in parasitoid released area were 19.0 per cent in comparison with 42.0 per cent in unreleased areas in Karnataka (Singh, 1991). Krishnamoorthy and Mani (1996a) had evaluated *T. brasiliense* and *T. pretiosum* @ 2.5 lakhs/ha. The borer damage was 8.9 and 7.2 per cent respectively as against 23.1 per cent and 13.7 per cent in the control when *T. brasiliense* and *T. pretiosum* were released in Karnataka.

Cabbage/Cauliflower

On cabbage, cauliflower and other cole crops Diamondback moth (DBM), *Plutella xylostella* is major pest. *Trichogrammatoidea bactrae* Nagaraja and Nagarkatti was recorded on *Bactra venosana* Zeller attacking the nutgrass (*Cyprus rotundus* L.) in India. It was found quite effective against DBM in places of cooler temperatures ranging 15 to 25°C. In India, preliminary trials conducted by Singh and Jalali (1993) also indicated that *T. bractrae* would be useful for trials against DBM. During 1998-99, *T. bactrae* was mass multiplied on eggs of *Corcyra cephalonica* and evaluated by releasing @ 2.5 lakhs/ha (five weekly releases starting 20 days after transplanting) in Pune, Hyderabad and Bengaluru. There was about 30 per cent reduction in larval population and the marginal increase in the yield in plots released with *T. bactrae* (Krishnamoorthy and Mani, 1999). Jalali *et al.* (2001) reported about 55 per cent reduction in DBM larval population in net houses where *T. bactrae* was released, indicating its effectiveness for large-scale evaluation in the field.

Predatory Mites

Phytophagous or plant feeding mites infest almost all cultivated and ornamental crops and wild plants. They reduce the yield of cultivated crops besides serving as vectors for several plant diseases. Yield losses in different crops due to mite damages are: 50-80 per cent in mango, 30-40 per cent in litchi, 20-25 per cent in paddy, 20-30 per cent in sugarcane, 15-30 per cent in red gram, 20-30 per cent in cotton, 10-15 per cent in vegetables, 5-11 per cent in tea and 25-30 per cent in *ber*. Besides, instances of total failure of crop due to mite infestation are not uncommon.

Mites are gaining increasing importance in the recent years as pests. Mites of the family phytoseiidae have been recognized as potential predators of phytophagous mites since the beginning of this century. Phytoseiidae, Stigmaeidae, Anystidae, Cheyletidae Tydeidae, Cunaxidae and Erythraeidae are important predatory mite groups. The Phytoseiid mites are worldwide in distribution and they feed specially on mites of the families Tetranychidae, Eriophyidae, Tarsonemidae and Tenuipalpidae. In addition, reports are also available on their feeding upon small phytophagous insects like aphids, coccids and thrips (Bonde,1989).

Life Styles of Predatory Mites

The general pattern of life cycle of phytoseiids consists of egg, larva, protonymph, deutonymph, in some cases tritonymph and adult and the egg to adult emergence takes about or less than 6 days and hence the generation is much quicker than most of the phytophagous mites like Tetranychidae, Tenuipalpidae and Eriophyidae. *Amblyseius longispinosus* completes the life cycle in 3.98 to 4.13 days (Mallik and Channabasavanna, 1983).

Specialized Predator

They are genus specific in regards to prey preference and is represented only by *Phytoseiulus* species. These species are mainly associated with spider mites of the genus *Tetranychus* having highly dense, complicated, and irregular web. Morphological adaptation by having long setae in the mid-dorsal and lateral part of the body shield might have made them more efficient to move through the thick stranded webs of spider mites (Sabelis and Baker, 1992). Comparatively large size, high searching capacity, high fecundity and very high prey consumption capacity of both nymphs and adults have made this group of phytoseiids a highly specialised and promising predator of tetranychid mites (Sabelis, 1981)

It is reported that *Phytoseiulus* species only thrive under high humidity and temperatures < 38°C (McMurtry, 1997). So, restricting the ability to crops grown under greenhouse condition and a very rare chance to be effective in open field cultivation. Another problem is that when the tetranychid preys are fully consumed, the predators disperse due to lack of food. Thus, specialised predatory mites only provide short-term control of spider mites.

Selective Predator

This group represented by the genus, *Galendromus* and *Neoseiulus* and readily prey on the tetranychid which produce "Little Webbing" in a leaf depression or along the midrib or main leaf veins (Saito, 1983). Pollen and plant exudes may act as a supplementary food but reproductive capacities and developmental rate will be very low as compared to prey mites (McMurtry, 1992). They also have long setae on the median and lateral surface of the dorsal shield. Members of this group are smaller in size and require less number of preys for completing development, in comparison to specialised predators.

Generalist Predator

Preying occurs on various genera of tetranychid and even on other insect prey like thrips, white flies, mealy bugs, scales and psocids. Pollen and various plant exudes can also be a supplementary food but reproductive potential is high in the presence of prey (McMurtry, 1992). Cannibalism and interspecific predation is very common in this group of predatory mites (Croft and Croft, 1996). Body size may vary from large to very small and they need comparatively more prey to complete their development (Zhang and Croft, 1994). Morphological adaptation is unfavorable to move in the "dense web" of spider mite and the plants having smooth leaves are the habitat of this group of predators (de Moraes *et al.*, 1986).

The generalist group of predaceous mites are omnivorous. They can survive and reproduce on pollen, leaf sap, spider mites, thrips and other small insects. This group of mites can regulate the pest population over a longer period of time. In addition, generalist predators show wider range of tolerance for temperature and humidity.

Mass Culturing

Phytoseiids can be divided into two main groups.

- ✰ Those species such as *P. persimilis* or *Metaseiulus* (= *Typhlodromus*) *occidentalis* that must be reared on the natural hosts, the tetranychidae and
- ✰ Those polyphagous species of *Typhlodromus*, *Amblyseius* and *Eusieus* that can be reared on pollens, alternate prey or even artificial diets.

Strategy in Selecting of Predatory Mites for Biological Control

Based on the morphological adaptation and food preference of phytoseiid mites, specialised and selective groups are suitable for augmentation and generalist and pollen feeder are suitable for conservation strategies of biological control of phytophagous mites. The inundative releases of *Phytoseiulus persimilis* and *Amblyseius cucumeris* have given successful control of *T. urticae* and thrips respectively on various crops grown under greenhouse condition. But large-scale application under open field conditions have not yet been shown to be feasible, probably because of the greater expense involved in rearing the host prey. It would be highly practical as well as economical if we can exploit those phytoseiid species which can easily be reared on pollen/ artificial diet (McMurthy and Scriven, 1966) and can build up the predator population by dusting the pollen as a supplementary food on the prey mite infested plant.

Phytoseiid Mites as Biocontrol Agents

The strategies made to utilize different phytoseiid predators as biocontrol agent are conservation, augmentation and importation. The developed countries have made major contributions in biological control of phytophagous mites by using phytoseiid predators.

A large number of phytoseiid mites belonging to the genera *Amblyseius*, *Phytoseius*, *Phytoseiulus*, *Typhlodromus* and few mites from Ascids, Tydeids, Cheyletids, Anystids, Cunaxids, Erythraeids are recorded as potential predators in India

Studies have been carried out in India mainly on the incidence, biology, laboratory multiplication and effect of pesticides on a promising predatory mites *Amblyseius tetranychivorus* (Gupta) by Puttaswamy and Channa Basavanna (1981), Nangia and Channa Basavanna (1984, 1995), Jagadish and Channa Basavanna (1983) and Krishnamoorthy (1982a, 1982b, 1983). Efforts have also been made to establish *P. persimilis* in India by importing it in 1994 under the all India Co-ordinate Research Project on Biological Control of Crop Pests and Weeds, (Krishnamoorthy, 1988). Although, the potentiality of this predator has been demonstrated under Indian climatic conditions but it has failed to establish under field conditions.

Cultivation of roses under polyhouse conditions is gaining importance in India. Prevailing high humidity and temperature in the polyhouse is very favourable for the rapid population build up of spider mites, which have developed resistance to commonly used acaricides. Release of laboratory reared colonies of the indigenous obligate predatory mite, *A. longispinosus* has been found to be effective in controlling *T. urticae* on roses (Mallik *et al.*, 1998). The predatory potential of *A. tetranychivorus* against *T. urticae* under both laboratory and polyhouse conditions was demonstrated by Ghosh *et al.* (2000).

Microbial Pathogens

Among the different microbial agents developed and tested, bacteria, viruses, protozoans and fungi are considered promising agents to control insect pests.

Information on these pathogens in pest control is available in surplus in India and few candidate agents are commercially exploited.

Insect Viruses

Work on insect viruses in India was initiated as early as 1968 with the report of nuclear polyhedrosis virus (NPV) from *H. armigera* (Patel *et al.*, 1968) a pest of national importance and *S. litura* (Ramakrishnan and Tiwari 1969) a polyphagous pest attacking several crops. Since then studies on insect viruses have progressed rapidly and several viruses were reported to occur in insect pests, most of them from the order lepidoptera. These comprise of NPV, granulosis viruses (GV) and cytoplasmic polyhedrosis viruses (CPV). Recently, a GV from the diamond back moth *Plutella xylostella* (Rabindra *et al.*, 1996) was reported.

Attempts made to utilize the baculovirus for the suppression of the coconut rhinoceros beetle *O. rhinoceros* has given encouraging results. Release of infected beetles spread the disease to subsequent generation of beetles and larvae in breeding sites (Mohan *et al.*, 1983).

Bacillus thuringiensis

The spore forming, crystalliferous and ubiquitous bacterium *B. thuringiensis* (*Bt*) is one of the earliest microbial insecticides to be commercially produced world wide. In India, this bacterial insecticide has been used in the management of several insect pests notably, *P. xylostella* larvae on cruciferous vegetables (Asokan *et al.*, 1996). Other insect pests controlled by *Bt* are *H. armigera* on pigeonpea (Putambekar *et al.*, 1997), *S. litura* on cabbage (Datta and Sharma, 1997) and fruit borers of okra (Satpathy and Panda, 1997). There are several reports however, indicating the ineffectiveness of *Bt* preparations particularly against the noctuids. Delfin was ineffective in controlling *H. armigera* on chickpea (Srinivasan *et al.*, 1994).Therefore, it is important that new strains of *Bt* with increased host spectrum and efficacy are developed. A few indigenously prepared *Bt* formulations are not as effective as the imported preparations. There is an urgent need to isolate indigenous *Bt* strains with high pesticidal activity, develop techniques of mass production and stable formulations and patent them. NBAIR, Hebbal, Bengaluru has come up with local efficient strains and has started mass production of the same.

Fungal Pathogens

Fungal pathogens particularly *Beavaria bassiana*, *Melarhizium anisopliae*, *Verticilium lecanii* and *Nomureae rileyi* have been found to be promising in the control of several agricultural pests. *B. bassiana* and *M. anisopliae* were found effective against *S. litura* (Gopalakrishnan and Narayanan, 1989), the sweet potato weevil *Cylas formicarius* (Khader Khan *et al.*, 1990), the termite *Odontotermis brunneus* and *O. obesus* (Khader Khan *et al.*, 1993). *M. anisopliae* when applied to manure pit suppressed the population of *O. rhinoceros* larvae (Sundara Babu *et al.*, 1983). *V. lecanii* effectively controlled the coffee green bug *C. viridis* in coffee plantations and addition of glycerol as a humectant increased the efficacy of the fungus (Easwaramoorthy and Jayaraj, 1978).

N. rileyi active against several lepidopteran insects was found to occur in epizootic form on *H. armigera* (Gopalakrishnan and Narayanan, 1989), *S. litura* (Gopalakrishnan and Narayanan, 1988) and *S. exigua* (Phadke *et al.*, 1978) suppressing the pests under field conditions. While *N. rileyi* killed *A. janata* larvae, it was not pathogenic to *Telenomus proditor* an egg parasitoid of *A. janata* (Phadke and Rao, 1978). These reports clearly indicate the scope for entomopathogenic fungus for pest control in India.

Insect Virus

The use of viral pathogens of insects in most agricultural crops is inundative and does not utilize their full epizootic potential, but takes advantage of their virulence and specificity (Payne, 1982). Their efficacy, specificity and production of secondary inoculum make baculoviruses attractive alternatives to broad spectrum insecticides and ideal components of IPM systems. The most successful commercial baculovirus in terms of total area covered is the Nuclear Polyhedrosis Virus (NPV)

In India, *Helicoverpa armigera* NPV, *Spodoptera litura* NPV and *Chilo infuscatellus* GV are being used in management of these pests. The use of baculoviruses for insect control within the IPM context is expected to increase in the coming years, particularly in developing countries where the cost of imported insecticides is high and that of labour is lower, *in vivo* production could provide both a viable means of producing large quantities of virus and a source of employment (Lacey *et al.*, 2001).

I. Mass Production of Baculoviruses

Helicoverpa armigera NPV (HaNPV)

Baculoviruses are obligate pathogens having a high degree of host specificity and hence can be multiplied only in their respective host insects. The viruses can also be multiplied *in vitro* on insect cell lines but the media used for cell production are costly and the production systems are out of our reach (Evans and Shapiro, 1997; Kennedy and Sathiah, 2001). Research on large-scale use of cell lines that could reduce cost of production is warranted. Hence HaNPV is mass multiplied *in vivo* on larva of *H. armigera*. The following are the steps involved in *in vivo* mass production of HaNPV.

Host Insect Multiplication

- First the host insect *H. armigera* has to be multiplied in the laboratory on artificial diet following standard procedures available for this purpose (Sathiah *et al.*, 2001; Rabindra, 1973) and a disease-free laboratory culture of the host insect must be established.

- ☆ Ninety percent of this laboratory-reared insects can be used for mass production of virus while 10 per cent of insects may be retained for continuing the disease-free host insect production.
- ☆ Standard semi-synthetic diet (minus formalin) may be half filled in 5 ml vials.

Virus Inoculation and Harvest

- ☆ Nucleus HaNPV viral preparation may be obtained from standard laboratories and diluted to required concentration.
- ☆ Viral suspension having 5 x 10^7 POB/ml (Sathiah, 2001) may be prepared.
- ☆ Ten µl of the viral suspension may be dropped on the surface of the diet in 5 ml vials and spread with the blunt end of a glass rod.
- ☆ Early fifth instar larvae (8-9 day-old larvae -head capsule measurement 1.6 mm) may be transferred using a fine forceps to the virus-contaminated diet and plugged tightly with cotton. The stage of larvae is highly crucial for maximizing virus production, as the larvae of lesser age will yield lower POB/larvae, while very late stage larvae when released may escape from disease
- ☆ The larvae are allowed to feed on the contaminated diet by incubating the vials at 25°C.
- ☆ Infected larvae when moribund or dead are collected from 5th day after inoculation and frozen.

Extraction and Purification of Virus

- ☆ For extraction, the larvae are macerated in ten times volume of distilled water
- ☆ The homogenate is then filtered through double layer muslin cloth.
- ☆ The filtrate is centrifuged at 500 rpm for half a minute to remove larger contaminants (discard the pellet)
- ☆ The supernatant is spun at 5000 rpm for 15 min. to pellet the virus
- ☆ The supernatant is discarded and the pellet containing virus particles is re-suspended in small volume of water and stored at refrigerated condition (4°C).
- ☆ One virus killed larva can yield 2 x 10^9 POB
- ☆ Hence approximately 750 infected caterpillars are required to get 1.5 x 10^{12} POB, which is the dose needed to cover one hectare
- ☆ Enumeration of the POB can be done following procedure that will be explained in later part of this paper

Spodoptera litura NPV

The procedure explained for the mass multiplication of HaNPV may as well be used for multiplying SlNPV also.

- ☆ The host insect (*Spodoptera litura*) may be mass multiplied either on natural host (castor leaves) or on artificial diet.
- ☆ The virus may be propagated by contaminating the artificial diet in 5 ml vials and allowing early fifth instar larvae to feed individually on contaminated diet
- ☆ Alternatively, the virus can be propagated by providing castor (*Ricinus communis*) leaves dipped in viral suspension.
- ☆ In the latter method, the leaves are dipped in viral suspension (5 x 10^7 POB/ml) and shade-dried.
- ☆ The starved early fifth instar *S. litura* larvae (7 - 9 days old larvae with head capsule measurement of around 1.4mm) are provided with the contaminated leaves.
- ☆ On the second day also virus treated leaves are provided
- ☆ Subsequently untreated leaves may be provided.
- ☆ The infected larvae showing characteristic symptom after 4-5 days may be collected and frozen.
- ☆ Extraction and purification procedures are same as in case of HaNPV.

Amsacta albistriga NPV

Mass culturing of the host insect, Red hairy caterpillar of groundnut, *Amsacta albistriga* has been difficult due to diapause in pupal stage of this insect. Hence field collection of 10 day old larvae is recommended. The following steps may be followed in propagation of the virus.

- ☆ Ten day old larvae (4th instar) measuring 2.5 - 3 cm may be field collected and starved overnight
- ☆ Virus suspension of *Amsacta albistriga* may be prepared at 1 x 10^6 POB/ml concentration.
- ☆ *Calotropis* leaves are collected from field and the wax coating is removed with cotton wool and the leaves are dipped in viral suspension and shade-dried.
- ☆ The starved larvae are provided with the contaminated *Calotropis* leaves.
- ☆ On the second day also virus treated leaves are provided

- ☆ Subsequently untreated leaves may be provided.
- ☆ Larvae start dying after day 5 of inoculation. Infected larvae show pinkish white colouration on the ventral surface of the body due to accumulation of polyhedra. After death the skin ruptures and the liquified body content oozes out.
- ☆ Collect the dead (virosed) caterpillars and freeze.
- ☆ Follow the steps mentioned under HaNPV for extraction of virus.

Chilo infuscatellus Granulosis Virus (GV)

i) Propagation of Virus

- ☆ Third or 4th instar field collected larvae are suitable for multiplication of the virus.
- ☆ A virus suspension containing 10^7 - 10^8 Inclusion bodies (IB)/ml of water is prepared
- ☆ The larvae are fed with a drop of virus suspension through a pin head or the head of larvae dipped in virus suspension.
- ☆ The virus fed larvae are reared on sugarcane shoot bits at the rate of 3-5 per plastic container (7.7 x 6.4 cm dia). The plastic containers are provided with filter paper for absorption of excess moisture and three pieces of sugarcane shoot bits split open at one end. The shoot bits and filter paper are changed on alternate days.
- ☆ The infected larvae begin to show symptoms in 5-8 days.
- ☆ The larvae start dying from 8 days upto 20 days and stored in refrigerator at minus 20°C.

ii) Extraction and Purification of Virus

- ☆ Larvae are macerated in distilled water and filtered through double layered muslin cloth to remove debris.
- ☆ The filtrate is spun at 500 rpm for 3 min. and the sediment containing larger debris are removed.
- ☆ The supernatant is centrifuged at 10,000 rpm for 30 min to sediment the virus.
- ☆ The supernatant is discarded and pellet re-suspended in small volume of water which gives a fairly pure virus preparation.
- ☆ The virus may be stored in a refrigerator at 4-5°C in distilled water in a amber coloured bottle. The virus can be stored for three or more years without apparent loss in infectivity.

II. Enumeration and Standardization of NPV and GV

Use of appropriate dose of the virus is highly essential for successful control of pests. Hence it is necessary to estimate the concentration of polyhedral occlusion bodies in the virus suspension before it is applied on the crop. Most of the crop pests are successfully managed with a dose of 1.5 to 3 x 10^{12} POB/ha. It has been found that a full grown larva of *H. armigera* or *S. litura* yields on an average 2 x 10^9 POB (Rabindra, 2001).

Enumeration of inclusion bodies (IB) of GV can be done by using a counting chamber having a depth of only 0.02 mm instead of 0.1mm. The inclusion bodies of GV can be observed and counted under a dark field microscope.

P. xylostella GV (PxGV)

The granulosis virus of *P. xylostella* is mass multiplied on laboratory reared larvae, which are raised on cauliflower leaves. Early third instar larvae fed with a virus dosage of 350. 0 OB/mm² has been found to be optimum for the mass production of P x GV. Larvae of the optimum size are allowed to feed on leaf discs (3.3cm dia) treated with the GV suspension of 24 µl (12µl on either side of the lamina) containing 1 x 10^8 OB/ml. The inoculated larvae are transferred to untreated leaves after 24h and they are incubated at 25°C. Virosed larvae are collected from 5th day after treatment and stored at -20°C.

S. litura GV (SlGV)

Mass production of granulosis virus of *S. litura* is carried out on laboratory reared larvae maintained on semi-synthetic diet. The optimum stage of larvae and dose of the inoculum are late fourth instar and 1.25 x 10^5 OB/mm² respectively. Treatment of the larvae is carried out by diet surface contamination method. Semi-synthetic diet taken in 5 ml glass vial is contaminated with the GV by dispensing 10µl of virus suspension containing (3.0 x10^9 OB/ml) and spread uniformly over the entire surface. Late fourth instar larvae are allowed to feed on the contaminated diet. Treated larvae are incubated at 25°C and cadavers are collected from the 12th day onwards and stored at -20°C.

H. armigera GV (HaGV)

Mass production of *H. armigera* GV is similar to the *S. litura* GV. The optimum stage of larvae and the dose of inoculum are fourth instar and 1 x 10^7 OB/larvae. Treatment of the larva is carried out by diet surface contamination method. Virosed larvae are collected from 10 – 12 days after treatment and stored at -20°C.

Formulations

Biopesticide formulations used in crop pest control are usually a physical mixture of one or more biologically active chemicals with an inert carrier or active biocide with a secondary material. In commercial development of biopesticides the four main functions of formulations are the stabilization of microorganisms during production, distribution and storage; improvement in handling and application of the product; protection from the harmful effect of environmental factors at the target site, and enhancement in the activity of the pathogen and its contact and interaction with the target pest.

Formulation of biopesticides are of two *viz.*, basic comprising of liquids, wettable powders, dusts, baits and granules, and tank mix formulations prepared from basic formulation with additives (Couch and Ignoffo, 1981; Rhodes, 1993).

Mass Production of Fungi

Most of the entomopathogenic fungi are facultative pathogens and can be mass produced in synthetic, semisynthetic or natural media containing suitable nutrient source. Selection of strains of fungi having high virulence, good growth and sporulation is considered important in mass culturing. The spore yield depends upon the type of medium used to culture the fungus (Roberts and Yendol, 1971). A number of successful commercial products based on conidia/spores of fungal pathogens have already been developed and sold (Jenkins *et al.*, 1998).

1. Production System

The method of production varies according to the use either for laboratory testing or large scale field use. For bioassays, isolate screening and maintenance of fungal pathogens can be cultured in agar slants in glass tubes or 250 ml Erlenmayer flask. Autoclavable polypropylene plastic bags or wide mouthed jars or fermenters are widely used for large scale field application or commercial use. Most of them are grown as surface culture on solid or liquid (broth) media. Some of the entopathogenic fungi are dimorphic and when grown in submerged shake flask cultures produce yeast like cells instead of mycelium. These cells are called blastospores and can be mass produced quickly. However, blastospores are short lived than conidia. In submerged culture a few fungal pathogen produce mycelia or conidia.

2. Media Used Production and Maintenance of Fungal Pathogens

Culturing can be done in variety of media containing sugar and salts as sources of carbon and nitrogen respectively. Specific media have been identified for culturing different types of fungi (Table 3).

2.1 Natural Media

Different types of cereals, pulses, brans and vegetables have been reported to be suitable for mass production of entomopathogenic fungi (Table 3).

3. Isolation of Fungal Pathogens

3.1 Collection of Cadavers

Mycosed insects are collected individually using fine forceps in glass tubes or screw cap vials and brought to laboratory. If the fungus has sporulated cadavers can be stored in the fridge till the isolation of the pathogen. The cadavers without external growth of fungus can be kept at 20 - 25°C in an environment with high humidity (*e.g.* moistened filter paper in sterile petridish). This will induce the growth of the fungus and sporulation.

3.2 Isolation

The spores/condia on the surface of the cadavers can be transferred aseptically using sterile inoculation needle to appropriate agar medium containing antibiotics in a laminar hood. Another isolation method is cutting of cadavers into small bits after surface sterilization in 70 per cent alcohol or 0.1 per cent mercuric chloride and transferring bits to agar medium in a laminar hood. The cadavers can also be homogenized in sterile distilled water and plated on appropriate agar medium.

3.2. Purification

It has to be ensured that after isolation, fungus is free from contaminant microorganisms and represent a single genotype. Purification is done by single hyphal isolation or single spore/conidia isolation method.

Table 2: Synthetic/Semi Synthetic Media Used for Culturing Entomopathogenic Fungi

Medium	*Composition*	*Fungal Pathogen(s)*
Sabouraud's dextrose agar with yeast extract (SDAY)	Dextrose 40 g; peptone 10 g; yeast extract 2 g; agar 15 g; distilled water 1000 ml	*Beauveria bassiana, B. brongniartii Verticillum lecanii*
Subouraud's maltose agar with yeast extract (SMAY)	Maltose 40 g; peptone 10 g; yeast extract 5 g; agar 15 g; distilled water 1000 ml	*Nomuraea rileyi, Zoophthora radicans, Paecilomyces farinosus*
Potato dextrose agar (PDA)	Peeled potato 200 g; dextrose 20 g; agar 15 g; distilled water 1000 ml	*B. bassiana, V. lecanii*
Emerson's YpSs agar	Soluble starch 15 g; K_2HPO_4 1 g, Mg SO_4. $7H_2O$ 0.5 g, yeast extract 4 g, agar 15 g, distilled water 1000 ml	*Metarhizium anisopliae, M. flavoviride*
Nutrient agar	Beef extract 3.0 g; peptone 5.0 g, NaCl 5 g; agar 15 g; distilled water 1000 ml	*Paecilomyces fumosoroseus*
Oat meal agar	Oats 100 g, agar 15 g, ditilled water 1000 ml	*P. fumosoroseus*
Czapek - Dox agar (CDA)	Sodium nitrate 2 g; K_2HPO_4 1 g, $MgSO_4$ 0.5 g; KCl 0.5 g, $FeSO_4$ 0.01 g; sucrose 30 g; agar 15 g; distilled water 1000 ml	*B. bassiana*

Table 3: Natural Media for Culturing Entomopathogenic Fungi

Medium	*Fungal Pathogens*
Sorghum grain	*Verticillium lecanii* (Easwaramoorthy and Jayaraj, 1977); *Paecilomyces farinosus* (Gopalakrishnan *et al.*, 1999); *Beauveria bassiana* (Devaprasad, 1989)
Rice	Nomuraea rileyi (Silva and Loch, 1987)
Carrot	*B. bassiana* (Sivasankaran *et al.*, 1990); *Metarhizium anisopliae* (Sundarababu, 1980)
Crushed sorghum grains with yeast extract	*N. rileyi* (Vimala Devi, 1994)
Jack seeds	*V. lecanii* (Easwaramoorthy and Jayaraj, 1977)
Cooked rice	*B. bassiana; M. anisopliae* (Jenkins *et al.*, 1998)
Wheat bran	*B. bassiana* (Rombach *et al.*, 1988)
Oats	*M. anisopliae* (Villacorta, 1976)
Coconut water	*M. anisopliae* (Chandrika Mohan, 2001)
Cassava chips + rice bran + fish meal	*M. anisopliae* (Chandrika Mohan, 2001)

Culturing of Fungal Pathogens

Preparation of Synthetic Media

For preparing solid agar media like SDAY, SMAY, Emerson's YpSs agar, Nutrient agar and CDA half the quantity of water is used to dissolve the chemicals by slight heating/boiling. The agar is melted in the other half of water by slow heating and constant stirring. The melted agar is added to the other half of medium containing chemicals and mixed thoroughly. Usually pH is adjusted to 6.5 – 6.8 before mixing with agar. The medium is then dispensed in test tubes (5 ml/tube) or 250 ml Erlenmeyor flask (50 – 100 ml/flask). The mouth of the test tube or flask is plugged with non absorbent cotton, covered with paper and autoclaved at 15 psi for 20 min.

The tubes are allowed to cool in slanted position for getting agar slants. The flask is allowed to stand until they can be held by hand and the medium is poured asceptically into sterile petridishes to produce agar plates.

Preparation of Semi-Synthetic Medium

Potato Dextrose Agar

Washed and peeled potato is made into thin chips, boiled in 500 ml of water and filtered through muslin cloth. To the extract, weighed quantity of glucose is added and mixed. The agar is melted in another half of water and mixed with potato extract containing glucose. The volume of the medium is made upto to one litre and dispensed in test tubes or Erlenmeyer flasks and autoclaved as described previously

Oat Meal Agar

The oats is first cooked in 500 ml of water and filtered through muslin. The agar is melted in another 500 ml of water, added to the extract of oats and final volume is made upto one litre. Dispensing and autoclaving are done as described earlier.

Some of the synthetic or semisynthetic media are commercially available. The required quantity of the media is added to distilled water, if necessary boiled to dissolve the ingrediants and autoclaved according to manufacturer's instructions.

Preparation of Natural Medium

Sorghum Medium

A quantity of 100 g of sorghum grains is thoroughly washed, cooked in 500 ml of distilled water and extract is strained through muslin. Extract is dispensed in Erlenmeyer flasks at the rate of 100 ml/flask and autoclaved. If solid medium is needed agar can be added.

Carrot Medium I

A quantity of 100 g of carrot is washed well in tap water, cut into small bits and cooked in 500 ml distilled water. The extract was filtered through muslin cloth dispensed in 250 ml Erlenmeyer flask at the rate 100 ml/flask and autoclaved. If solid medium is needed of agar can be added.

Carrot Medium II

The cut pieces of carrot are transferred to 250 ml Erlenmeyer flasks containing 100 ml of water at the rate of 40 g per flask. Autoclaving is done as described elsewhere.

Inoculation

Inoculation should always be carried out in a laminar flow hood. The working area inside-well the hood must be disinfected with 70 per cent alcohol before starting the inoculation. A Bunsen burner is kept inside the hood for flaming the loop of inoculation needle.

i. Remove the cotton plug of test tube or Erlenmeyer flask containing culture medium/fungal culture and heat the mouth over the flame in Bunsen burner.

ii. Sterilize the inoculation needle by heating over the flame till it becomes red rot
iii. Cool the needle and introduce it into the tube/ flask containing mother fungal cultures. Pick up the spores and hyphal bodies and inoculate by drawing it lightly over the surface of fresh medium in slants/flasks/Petridishes.
iv. Flame the mouth of the flasks or tubes again and recap
v. Incubate the culture at 25-27°C in an incubator. It is desirable to keep the flask or tube inside a humidified chamber.

Culture Maintenance

Normally sporulation will occur within 7-10 days and subculturing can be done at this stage. Continuous subculturing on artificial media can result in attenuation of virulence and affect sporulation. Virulence can be restored after a single or several passages through the insect host.

Storage

The conidia, blastospores and mycelia can be stored at 4°C upto several weeks or even months depending on the species, strain and media used. Slant cultures can be stored using mineral oil or distilled water (Humber, 1997). For long term storage the spores/condia can be stored on sterile anhydrous silica gel crystals at -20°C. Fungal spores stored in slica gel may remain viable for more than ten years.

Large Scale Production

Efficient mass production technologies have been developed for a few fungal pathogens. Production efficiency was highest in the case of *B. bassiana* (Wraight and Carruthers, 1999). In South and Central America, Europe and Asia *M. anisopliae* and *B. bassiana* were produced commercially using cooked rice or other grains in trays or autoclavable plastic bags or glass jars. The average yield was 1-5 x 10^9 spores g of substrate (Alves and Pereira, 1989; Aregger, 1992).

Industrial production of fungal pathogens is generally based on diphasic or two stage system in which fungal mycelium or hyphal bodies (blastospores) are produced in liquid culture either in shake flasks or fermenters and transferred to solid substrate for production of conidia (Jenkins *et al.*, 1998).

Mass Production of Entomopathogenic Nematodes

Entomopathogenic nematodes (EPN) in the families Steinernematidae (*Neosteinernema longicurvicauda* and *Steinernema* spp.) and Heterorhabditidae (*Heterorhabditis* spp.) are the potential candidates for biological control of insects among beneficial nematodes. The nematodes are symbiotically associated with the bacteria *viz.*, *Xenorhabdus* or *Photorhabdus* species. Moreover these nematodes possess attributes of both insect parasitoids or predators and microbial pathogens. They are motile, highly virulent, killing their hosts quickly (24-48 h), can be cultured easily *in vitro* and *in vivo* and have a high reproductive potential. They have a broad host range, safe to non-target organisms, free from environmental hazards, easily applied using standard spray equipments and compatible with many chemical pesticides and other methods of insect management.

Entomopathogenic nematodes are not effective at temperatures above 36°C, sensitive to ultraviolet light, require adequate moisture and high relative humidity for survival for reasonable length of time and to be effective. These negative attributes can be overcome to a large extent by adjusting the time of application and using a right formulation against a suitable pest.

Mass Production

Amenability for mass production of EPN is an important attribute which is considered suitable for field release. Mass production of EPN has evolved from the first large-scale *in vivo* solid media on potato mash (McCoy and Glaser, 1936) and dog food (House *et al.*, 1965) to the *in vivo* production (Dutky *et al.*, 1964; Poinar, 1979; Lindegren *et al.*, 1993) to the three dimensional solid media (Bedding, 1984) and the solid media process has successfully produced pathogenic Steinernematids and Heterorhabditids.

A liquid culture technique using 15,000–18,000 l capacity fermenters developed for *S. carpocapsae* and *S. feltiae* yields more than 10,000 infective Juveniles (IJ) per ml (Georgis and Hauge, 1991). Pilot – scale liquid culture technology is also available for *H. bacteriophora* (Surrey and Davies, 1996). The liquid fermentation process is highly efficient for Steinernematids but not for Heterohabditids (Gaugler and Georgis,1991).

A. *In vivo* Production

Corcyra cephalonica and *Galleria mellonella* are often used as hosts because they are widely available, easily reared, highly susceptible and are excellent hosts for nematode production. Average production of IJ is in the order of 30,000 to 50,000 per insect. The basic *in vivo* production method developed by Poinar (1979) is outlined below.

1. Infecting *Galleria* or *Corcyra*

Let the IJ suspension warm to room temperature. Dilute 1ml suspension in an appropriate quantity of sterile distilled water (sdw) to yield a suspension of 200 IJ per ml. Evenly distribute 1 ml of the suspension on 9.0 cm dia Whatman No. 1 filter paper in the lid of a 100 x 14 mm petri dish. Add 10 *Galleria* or *Corcyra* larvae. Cover the lid with inverted petri dish bottom and store them in a plastic bag (to conserve moisture) at room temperature. Place dead larvae into White's traps (White, 1927) after 5- 7 days of infection.

2. Harvesting

To make White's trap, place a 9.0 cm dia Whatman No.1. filter paper in a concave side up watch glass in a large petri dish. Pour about 50 ml sdw or 0.1 per cent formalin into the petri dish. Drape the filter paper over the watch glass so that it comes into contact with the liquid. Place infected larvae on the filter paper over the edge of the watch glass. IJ will start to exit 10 –12 days after infection. Once nematodes begin to appear, they should be harvested daily until production drops (3-4 days). To harvest, remove the watch glass with dead insect larvae (cadavers), pour the IJ into a beaker, add 50 ml of sdw or 0.1 per cent formalin and replace the watch glass.

Rinse IJ by allowing them to settle in a beaker. Then decant the supernatent and add more sdw until the suspension is clean (3-4 times). Finally transfer the nematodes to a storage container (beaker, conical flask, tissue culture flask) and store them at a temperature of 4 - 10°C.

B. *In vitro* Production

Bedding (1984) has developed a technique whereby huge number of IJ may be economically produced using a chicken, duck or turkey offal medium on a porous polyether polyurethane foam substrate. Glass flasks or autoclavable bags are used as rearing containers. This system produces on an average of over one billion IJ per bag or flask of 500 ml capacity (100 g medium). The technique outlined below is for 500 ml wide month Erlenmeyer flasks.

Preparation of Rearing Flasks

Impregnate small foam pieces with chicken, duck or turkey offal homogenate at the rate of 12.5 parts medium to 1 part foam by weight. Fill the glass conical flasks with foam homogenate mixture to the 250 – 300 ml mark (about 100 g). Wipe the mouth of the flasks, plug with cotton wrapped with cheese muslin cloth and autoclave for 20 mts at 121°C.

Instead of chicken, duck or turkey offal homogenate, a medium containing nutrient broth – 0.44, Yeast extract – 0.16 g, Soya flour – 7.20 g, Corn oil – 5.20 g and Water – 27.00 g (Wouts,1981) or Wheat flour – 150 g, Bengal gram flour – 50 g, Coconut oil - 12 g, Agar (1 per cent) – 400 ml, Yeast extract – 6 g and Beef extract – 6 g (Ponniah, 1997) or supplementing Groundnut cake and Palm oil for Soy flour and corn oil (Subramaninan, 2000) in Wouts (1981) medium coated onto polyether polyurethane foam produces over 10, 1 and 6.4 million IJ, respectively from a 250 ml conical flask (50 g medium) one month after inoculation of IJ.

1. Inoculation with Bacteria and Nematodes

Cells of the appropriate *Xenorhabdus* or *Photorhabdus* bacterium should be aseptically transferred to 5 ml of nutrient broth in a test tube and kept overnight on a shaker.

Allow the autoclaved flasks to cool to room temperature and inoculate the bacterium by pouring the contents of one tube of bacterial culture into each flask. Shake well and store for 2–3 days at 25°C to allow the bacterial population to build up. Add surface sterilized 500 – 1000 IJ in 5 ml sdw into each flask and incubate at 25°C.

2. Harvesting

The flasks will be ready for harvest in about 3 – 4 weeks. The foam pieces may be piled 5 cm deep on a 20 mesh sieve. Place sieves in a pan of tap water with water level adjusted so that the foam pieces are just submerged. Within two hours IJ will migrate into the water. The nematodes may be sedimented and rinsed to remove particulate matter and inactive or dead IJ. The nematodes should be rinsed several times until the water suspension appears clear.

3. Trouble Shooting

A drop in production of IJ could indicate contamination, a reversion to the secondary phase of the bacterium, unsuitable incubation temperatures and improper moisture content or many other problems.

4. Storage

After mass production, nematodes may be stored in sdw with a drop of Triton X – 100 or Tween 20 (wetting agents that prevent nematodes from sticking to the sides of the storage containers) or 0.1 per cent formalin. If stored without proper aeration, the nematodes should be concentrated at the rate of 10,000 – 20,000 per ml and the water depth should be less than 1 cm. Canted tissue culture flasks are ideal for this type of storage since there is a large surface - to - volume ratio (Woodring and Kaya, 1988). Steinernematids store well for 6 -12 months while Heterorhabditids for 2 - 4 months at a temperature of 4 - 10°C.

Control of Papaya Mealybug *Paracoccus marginatus*

Procedure for the Rearing of the Mealybug *Paracoccus marginatus*

Classical biological control involves the use of co-evolved natural enemies of the alien pest. Therefore, for the management of the papaya mealybug National Bureau of Agriculturally important Insects (NBAII) imported three encyrtid parasitoids *viz., Acerophagus papayae, Pseudleptomastix mexicana* and *Anagyrus loecki.* In order to carry out the mandatory tests of specificity and safety, these parasitoids were multiplied in the qurantine facility of NBAII at Bengaluru on the papaya mealybugs.

Although many methods and host plants were tried for the mass production of the host mealybug *P.*

Figure 1: Potato Sprouts with Mealy Bugs.

marginatus, rearing the mealybug on potato sprouts was found to be easy and cost effective. Potatoes of about 2 months old are procured from the market washed with clean water were surface sterilized with common bleach or 5 per cent sodium hypochlorite solution. The potatoes were then treated with 1 ppm solution of gibberlic acid by making a small cut on the potatoes and immersing the same in the solution for half an hour. The potatoes after air drying are placed over a wet cloth and covered by black cloth and kept in a dark humid place for the development of sprouts. The sprouts developed within a week's time. Or you can also make potato sprouts by keeping potatoes in moist sand in dark place for a week.

Each sprout was infested with five mature ovisacs of *P. marginatus* and kept in a cage @ 20 potato sprouts per cage of 30 x 30 x 30 cm. After 20 days of release of the ovisacs the crawlers settlement is stabilized and they can be used for parasitization (Figure 1). The other method can also be followed where we can infest the sprouts with crawlers. The sprouted potatoes are kept spread over an area in a dark room surrounded by potato sprouts with mature mealybugs which are producing the crawlers. A table lamp or some source of light is directed towards the fresh potato sprouts and kept for a day. The crawlers are attracted to light and colonize on the fresh sprouts.

These infested sprouts are taken out and maintained in labeled cages for use in parasitoid production and are covered with black (or dark) cloth. The dark fabric cover protects them from the light, and curtails the movement of crawlers away from the buds, a temperature between 27° to 28°C and a relative humidity between 40 to 60 per cent is maintained in the rearing facility.

Water is sprayed on the sprouts to reduce the moisture loss of potatoes. Lower turgidity in the sprouts was found to initiate crawler migration hence one spray of freshwater in the morning and one spray in the evening was found to reduce crawler movement and good settlement of crawlers and their establishments. Between 14 and 20 days, infested potatoes can be removed and set-up in cages to be used for the rearing of the parasitoids with proper labeling for keeping of records. Potato plants in pots can also be used for mass production of parasitoids (Figure 3).

Rearing of the Parasitoids

After choosing the potatoes infested with the adequate life stages of the mealybug (Figure 2), cages were set up cages for the different parasitoids of *P. marginatus* and held in a room with controlled temperature and humidity as mentioned above. These cages holding the infested potatoes were placed on shelves by date; new cages being set-up weekly. For rearing *Acerophagus papayae* and *Pseudleptomastix mexicana* 2nd stage mealybugs are used. For parasitization by *Anagyrus loecki*, 3rd stage or last instar nymphs are provided. Each cage containing 20 potato sprouts are infested with 30 adult parasitoids and are provided with

Figure 2: Different Stages of *P. marginatus*.

Figure 3: Potato Plants with Mealybugs.

honey strips or honey streaked on the glass surface as food and left in dark for oviposition. After 10-12 days of oviposition, the potatoes containing parasitized nymphs are moved to emergence cages where the emergence is monitored and the ensuing adults are collected by aspirators and used for further culture or releasing in field conditions. Mealybugs grown on papaya plants as well as on alternate host plants like Parthenium, *Sida acuta, Plumaria alba, Hibiscus rosa sinensis,* and okra can also be used for the multiplication of the parasitoids. The parasitoids eclose from host at different times: *A. papayae* – (14-16 days), *A. loecki* – (14-16 days) and *P. mexicana* – (20-24 days). The collected adults can be stored in vials of 25 x 100 mm with honey streakings for about 5-12 days depending up on the species. *Acerophagus papyae* leaves for 5-6 days, *Anagyrus loecki* 8-9 days and *Pseudleptomastix mexicana* for about 10-12 days.

Large Scale Production of the Parasitoids

Large scale production of the papaya mealy bug parasitoids can be done on the mealybugs cultured on alternate hosts. The most preferred and extensively available host of papaya mealybug is Parthenium (*Parthenium histerophorus*). Parthenium plants are collected from field and planted in polythene covers or pots and are maintained for a week till nicely established. These plants are kept at a place with good light source and are infested with papaya mealy bug from culture brought from field. Colonization of papaya mealybug on the parthenium takes about a weeks time and once the different stages crawlers are settled the group of plants are covered with nylon mesh forming a cage. About 100 parasitoids of *A. papayae* are released inside the cage and allowed for oviposition. After 12-14 days the parasitization can be observed (in case of *A. papayae and A. loecki*) and after 18-20 days incase of *P. mexicana* in the form of mummies of mealybugs. Either they may be allowed to emerge in the same cage or the shoots can be cut and placed in an emergence cage of 30 x 30 x 30 cm and emerging adults can be procured. Other commonly available hosts like well grown potato sprouts, *Plumeria, Acalypha and Cassia* species can also be used for mass production of the parasitoids.

Transport and Release of Parasitoids

The adult parasitoids can be collected using asperator and are trasfered to plastic or glass vials (4 x 1.5 cm) for further use and transport using muslin cloth cover on the mouth of the vial. providing honey for survival in transit. Inoculative releases of the parasitoids are done at the site of severe infestation by mealybugs. @ 100-200 per infested plant/tree. Severely affected plants may be covered with nylon/muslin net cage and parasitoids released inside. Provide paper strips streeaked with 10 per cent honey as adult feed. After 4 days the nets can be removed. After about 14 days the parasized mealybugs can be seen as mummified nymphs. Emergence of adults will happen in 4-5 days. The area of release is monitored after every 18 days to see the emergence of parasitoids incase of *A. papayae* and *Anagyrus loecki*, and 25 days incase of *P. mexicana*. The survey for parasitization is followed up after 3 months and 6 months after initial release of the parasitoids. Each time data on the parasitization by all the three parasitoids, different host plants on which observed and emergence of parasitoids are recorded and analysed for assesing the impact of released parasitoids.

Cost effective methods of mass production of all the three parasitoids was achieved by rearing the mealy bug on potato sprouts About 20 potato sprouts are kept in a cage of 30 x 30 x 30 cm and the mealy bugs yielded about 10,000 parasitoids in a span of 20-25 days at temperature between 27° to 28°C and a relative humidity between 40 to 60 per cent is maintained in the rearing facility. The parasitoids of *P. marginatus* eclose from host at different times: *Anagyrus loecki* – (14-16 days), *Acerophagus papayae* – (14-16 days) and *Pseudleptomastix mexicana* – (20-24 days).

NBAII in collaboration with entomologists from CSR and TI, TNAU, and other Agricultural universities and Krishivigyana Kendras conducted the impact assesment of the released parasitoids. Continuous supply of the parasite is ensured by mass productions through various KVKs, Dept of Entomology of Universities and State biocontrol station. Trainings on mass production of the parasioids have been conducted and altogether 250 scientists have been trained so far who in turn will conduct and monitor production of these parasitoids in their respective areas.

References

Alves, S.B. and Pereira, R.M. 1989. Production of *Metarhizium anisopliae* (Metsch) Sorok and *Beauveria bassiana* (Bals.). Vaill in plastic trays. *Ecossistema*, **14**: 188–192.

Aregger, E. 1992. Conidia production of the fungus *Beauveria brogniartii* on barley and quality evaluation during storage at 2°C. *J. Invertebr. Pathol.*, **59**: 2–10.

Bedding, R.A. 1984. Large-scale production, storage and transport of the insect-parasitic nematodes, *Neoaplectana* spp. and *Heterorhabditis. Ann. Appl. Biol.*, **101**: 117–120

Chandrika Mohan. 2001. Techniques in microbial control of the Rhinoceros beetle *Oryctes rhinoceros*. **In**: *Microbial Control of Crop Pests*, (Eds.) R.J. Rabindra, J.S. Kennedy, S. Sathiah, B. Rajasekaran and M.R. Srinivasan. Tamil Nadu Agricultural University, Coimbatore, India, pp. 271–274.

Crook, N.E. 1994. Baculoviruses: Granulosis virus. **In**: *Encyclopedia of Virology*, Vol. 1, (Eds.) R.G. Webster and A. Granoff, pp. 127–130. Academic Press, New York.

Dutky, S.R., Thompson, J.V. and Cantwell, G.E. 1964. A technique for the mass propagation of the DD-136 nematode. *J. Insect Pathol.*, **6**: 417–422.

Geetha, B. and Swamiappan, M. 1998. Improved adult rearing cages for the predator *Chrysoperla carnea*. *Madras Agril. J.*, **85**: 333–334.

Hagen, K.S. 1987. Nutritional ecology of terrestrial insect predators. **In**: *Nutritional Ecology of Insects, Mites, Spiders and Related Invertebrates*, (Eds.) F. Slansky and J.G. Rodriquez. John Wiley and Sons, New York, pp. 533–577.

Hassan, S.A. 1992. Integration of biological and chemical methods to control pests in greenhouses. *Arab J. Pl. Prot.*, **10**: 57–54.

Kind, *e.g.* and Leppla, N.C. 1984. Advances and challenges in insect rearing. *Agric. Res. Service*, U.S.D.A., New Orieans, La., p. 306.

Laemmli, U.K. 1970. Cleavage of structural proteins during the assembly of the head of bacteriophage T4. *Nature*, **227**: 680–685.

Mallik, B., Onkarappa, S. and Harish Kumar, H. 1998. Management of the Spider mite *Tetranychus urticae* Koch, on rose using phytoseiid predators, *Amblyseius longispinosus* in polyhouse. *Pest Mgt. Hortl. Ecos.*, **4:** 46 – 48.

Momen, F.M. 1996. Effect of prey density on reproduction, prey consumption and sex ratio of *Amblyseius barkeri* (Acari: Phytoseiidae). *Acarologia*, **37**: 3–5.

Ridgway, R.L. and Vinson, S.B. 1977. *Biological Control by Augmentation of Natural Enemies*. Plenum Press, New York, p. 480.

Santharam, G. 1986. Studies on the nuclear polyhedrosis virus of tobacco cutworm *Spodoptera litura* (Fabricius) (Notuidae: Lepidoptera). *Ph.D. Thesis*, Tamil Nadu Agricultural University, Coimbatore, India.

Sathiah, N. and Jayaraj, S. 1996. Technology for Mass Production of *Helicoverpa armigera* and its NPV: A Constraint Analysis. Paper Presented at the National Seminar on *Advancement in the Production and Use of Biocontrol Agents* held on 11th October, Entomology Research Institute, Loyola College, Madras.

Shorey, H.H. and Hale, R.L. 1965. Mass rearing of the larvae noctuid species on simple artificial medium. *J. Econ. Entomol.*, **58**: 522–524.

Singh, S.P., 1995. *Technology for Production of Natural Enemies*. Project Directorate of Biological Control, Bengaluru, India, p. 221.

Singh, S.P. and Jalali, S.K. 1993. Evaluation of Trichogrammatids against *Plutella xylostella*. *Trichogramma News*, **7**: 27.

Singh, S.P. and Jalali, S.K. 1994. Trichogrammatids. Technical Bulletin No. 7, 93 pp. Project Directorate of Biological Control, Bengaluru.

Stehr, F.W. and Haynes, D.L. 1972. Establishment in the United States of *Diaparsis carnifer*, a larval parasite of the cereal leaf beetle. *J. Econ. Entomol.*, **65**: 405–417.

Subramanian, S. 1998. Studies on the nuclear polyhedrosis virus of *Spodoptera litura* (Fabricius). *M.Sc. (Ag.) Thesis*, Tamil Nadu Agricultural University, Coimbatore, 112 p.

Travers, R.S., Martin, P.A.W. and Reichelderfer, C.F. 1987. Selective process for efficient isolation of soil bacterium *Bacillus* sp. *Appl. Environ. Microbiol.*, **53**: 1263–1266.

Varma, G.C., Rataul, H.S., Shenmar, M., Singh, S.P. and Jalali, S.K. 1991. Role of inundative releases of *Trichogramma chilonis* Ishii in the control of *Chilo auricilius* Dudgeon on sugarcane. *J. Insect Sci.*, **4**: 165–166.

Wajnberg, E. and Hassan, S.A., 1994. *Biological Control with Egg Parasitoids*. CAB International, Wallingford, UK, p. 265.

101

Value-addition in Horticultural Crops as an Entrepreneurial Option for Creating Employment Opportunities for Youth

Vimal Pant[1] *and Prachi Pathak*[2]

[1]*National Institute of Food Technology, Entrepreneurship and Management, Kundli – 131 028, Haryana*
[2]*Doon University, Banjarawala, Kedarpur, Dehradun – 248 001, Uttarakhand*

Blessed with diverse soil and climate types, India has been at the forefront of fruits and vegetable production in the world. It is considered as the second largest producer of fruits after China. The horticulture sector is poised to be the key driver of the agriculture produce in the near future. As per latest estimates available, the country witnessed a total production of around 89 million metric tonnes of fruits and 163 million metric tonnes of vegetables till the year end 2014 (NHB, 2015). The last five years have also seen an impressive growth of 30 per cent in production with banana, mango, citrus, papaya, guava and grape being the major fruit crops while Potato, tomato, onion, brinjal, cabbage, cauliflower and tapioca account for maximum share in vegetable production.

However, of all this only 2 per cent of vegetables and 4 per cent of fruits gets processed while more than 20 per cent produce gets wasted. Various studies have indicated that 30-40 per cent of farm produce gets

Table 1: Production of Various Horticultural Crops over the Years

Year	*Fruits*	*Vege-tables*	*Flowers and Aromatic*	*Plantation Crops*	*Species*	*Total*
1991-92	28632	58532	0	7498	1900	96562
2001-02	43001	88622	535	9697	3765	145785
2011-12	76424	156325	2218	16359	5951	257277
2012-13	81285	162187	2647	16985	5744	268848
2013-14	88977	162897	3192	16301	5908	277352
2014-15	88819	168300	3233	17131	5908	283468

Figures in thousand tonnes.

Source: Department of Agriculture, Cooperation and Farmers Welfare, Ministry of Agriculture and Farmers Welfare, Government of India.

Table 2: Leading Fruit-producing States of 2013-14

State	*% Share*
Maharashtra	15
Andhra Pradesh	12
Gujarat	9
Tamil Nadu	8
Uttar Pradesh	8
Karnataka	8
Madhya Pradesh	6
Telangana	5
Bihar	5
West Bengal	3
Others	21

Source: Horticultural Statistics published by Dept of Agriculture, Cooperation and Farmers Welfare, Ministry of Agriculture and Farmers Welfare, Government of India.

wasted after leaving the farm gates thereby reducing the return for the farmer and increasing the price for the consumer. Although meager in share, processed fruits and vegetables in India find their way as exports to United States, Saudi Arabia, United Kingdom, Netherland, United Arab Emirates and Japan. Apart from processed products dried and preserved vegetables have also find markets in countries like Germany, Russia, United Kingdom, United States, France and Brazil for the worth of Rs. 742.74 crores during the year 2013-15. Looking at these nascent levels of processing, India holds a tremendous potential in increasing the processed output for domestic as well as international consumption. Huge varieties of fruits are grown across the nation with Maharashtra and Andhra Pradesh accounting for more than one-fourth of the total produce.

Despite increasing production of fruits and vegetables, there are various challenges faced by the sector in terms of processing the produce for limiting wastage. The horticulture sector is characterized by small size farms with low productivity per hectare and high post-harvest losses.

Processed fruits can take different forms- prepared fresh (adding ingredients like flavours, spices, dressings *etc.* to fruits), frozen or canned fruits, dried fruits and juice extraction. The processing can either be done at packing level by cleaning, grading and sorting in an assembly line or extracting produce in the form of pulp, slices, juice *etc.* At secondary level, the fruit produce is heated, fermented or dehydrated for a longer shelf-life like production of jams, pickles, juices *etc.* All the processing procedures aim to stem the decay and damage of the F and V produce on the one hand and increasing value and thereby revenue generation on the other. The food processing sector has been growing at an Average Annual Growth Rate (AAGR) of 8.4 per cent as compared to 3.3 per cent in Agriculture and 6.6 per cent in manufacturing. (Fourth Report by Committee on Agriculture, Sixteenth Lok Sabha, December 2014). Growth in this sector could create immense opportunities of increasing employment for the rural youth as opportunities in agriculture are shrinking.

Review of Literature

Nain *et al.* (2013) conducted a research on development of efficient marketing network, organisation of fruit growers, development of fruit mandies or markets in main growing areas, development of entrepreneurial abilities and skills to establish small fruit processing industries in Doha and Baramulla districts of Jammu and Kashmir. The study suggested that awareness, motivation, technical skill, the right assistance and support from family at extension level and government and other organizations can help the horticultural entrepreneurs to strengthen their capacities besides adding to the family income and national productivity. Hrangao and Sorokhaibam (2015) studied the opportunities and challenges of small scale farmer entrepreneurs and their contribution to the Manipur's state economy where in it was concluded that state government and society's proper intervention may lead to a shift from negative to positive mindset. It was also suggested by the authors that unless entrepreneurs were well equipped with technical and business skills, they may not able to flourish amidst various problems they encounter at every stage of their business life.

D'Silva *et al.* (2011) attempted to study the various factors which influence the level of acceptance of sustainable agriculture among contract farming entrepreneurs. It opined that entrepreneurship was an important mechanism to develop novel opportunities for farmers to strengthen their role in the farming industry by way of enhancing their productivity and marketing strategies and enable them to enhance income levels. Hanan (2015) carried out a research based on previous survey reports indicating that the Horticulture sector warrants intense structural changes through diversification, value-addition, harmonious integration of modern and indigenous know-how, organized marketing strategies and infrastructure development for the sustainable production of fruits and their processed products. Mega Food Parks Scheme (MFPS) from the Government in India is one such scheme which holds a great opportunity for the farmer-entrepreneur in food processing sector.

Pannekoek *et al.* (2005) focused their research on studying the critical success factors in entrepreneurial innovation, and the role of cooperation with chain partners, knowledge institutions and colleague firms in the Dutch glasshouse industry. The results concluded that characteristics demonstrated by innovative entrepreneurs were essential for success. The innovative entrepreneur had a strategic vision of how to serve the market and create value. Lanset al (2004) studied work-related lifelong learning for entrepreneurs in the agri-food sector where learning needs, learning preferences, learning motivation and conditions in the context of lifelong learning were identified. Results indicated that technology, IT and entrepreneurial competencies will become of increasing importance in the future. Mulder *et al.* (2007) studied the competence development of entrepreneurs in innovative horticulture where strengths and weakness of entrepreneurial competencies and their learning methods were critically analysed. Pathak (2016) attempted to study the rural development through agri-preneurship through entrepreneurs experience in agriculture identifying the problems and schemes for the rural development. The results showed with a strong agro-based programme, nonfarm activities could also be initiated when the entrepreneurs became more experienced and capable of taking risks.

Patel and Chavda (2013) attempted to find out the problems and challenges for the potentiality of rural entrepreneurship focusing on the major problems faced by rural entrepreneurs especially in marketing of agri-products, financial amenities and other primary amenities. Paucity of funds, lack of infrastructure facilities, financial risk bearing ability, market competition, middlemen, lack of IT knowledge, legal formalities, poor quality of products, low skills of workers and negative attitude of farmers were identified as the core issues. Kumar (2015) conducted a research on entrepreneurship in agriculture development with special reference to Indian agricultural sector where a way of approach was analysed in which agricultural sector in India could get a boost through entrepreneurship.

Objectives

The Indian food processing industry contributes to about 32 per cent per cent of the total food market and contributes around 14 per cent of manufacturing Gross Domestic Product (GDP), 13 per cent of India's exports and six per cent of total industrial investment. *[ibef.org]*. The country has exported 3,20,732.58 MT of Other processed fruits and vegetables to the world for the worth of Rs. 2,900.33 crores during the year 2015-16. *[apeda.gov.in.]*. Since the horticulture processing sector could provide a new fillip to the employment generation initiatives for rural India, this study aims to address the following objectives:

i) To highlight the importance of horticulture processing in creating employment opportunities for the rural youth in the country.
ii) To critically evaluate the factors affecting the growth of horticulture processing in India
iii) To suggest possible remedies for addressing the constraints faced by the industry.

Methodology

This study utilizes descriptive research for analyzing the issues governing the Indian horticulture processing sector; explore the scope of increasing employment by way of investment growth in the industry and thus finding out measures to address the impediments. A review of contemporary literature was done as well as various resources in the print and digital space were utilized along with research databases like Ebsco, Wiley, J Host. International and National Journals, Government publications, discussions with Industry experts are part of the exhaustive information and assessment exercise undertaken by the writers.

Entrepreneurship and Employment in Horticulture Processing

As per the results published in the 68th round of the Employment and Unemployment Situation Among Major Religious Groups In India report released by the Government of India, the unemployment rate in rural India has increased. With a large part of this population engaged in agriculture, there is unemployment for at least six months a year as they are engaged seasonally in their occupation. The most critical reasons for unemployment are lack of job opportunities, no access to affordable capital as well as inefficient utilization of natural resources. The problem is further compounded by poverty, lack of formal education and poor health of the inhabitants. Apart from the usual unemployed people, there are several with concealed unemployment. With per capita farm holdings shrinking in size, it is becoming more than evident. Professional education facilities in and around rural areas are scarce making it difficult for youth with modest family means to take up employment in a demanding industry environment in urban areas.

As is evident from the above figure, horticulture has outpaced the food grain production over the last 15 years. In this prospect, horticulture processing initiatives in rural areas can prove to be very useful in addressing the employment challenges for the youth population. Investments in horticulture processing would not only

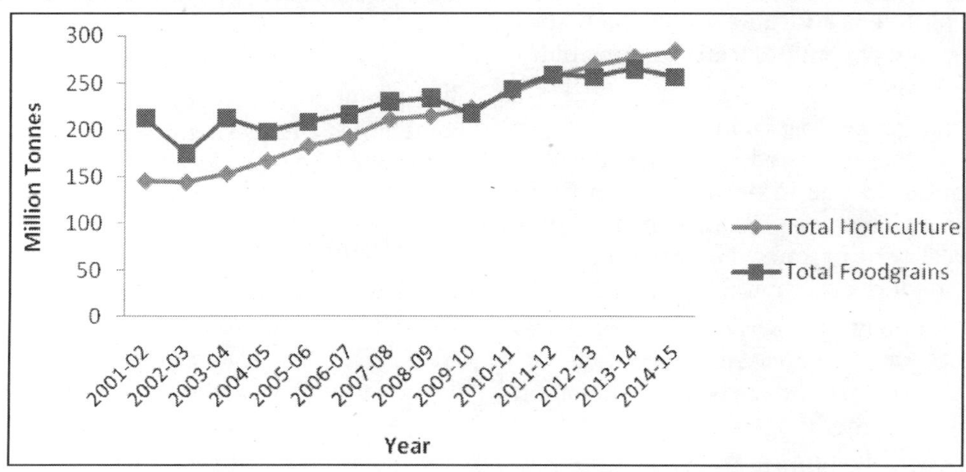

Figure 1: Production of Horticulture *vis-à-vis* Foodgrains Data.
***Source*: Horticultural Statistics published by Dept of Agriculture, Cooperation and Farmers Welfare, Ministry of Agriculture and Farmers Welfare, Government of India.**

help reduce wastage but also increase opportunities for technical innovation, further opportunities in value-addition and enhance competitiveness in manufacturing. Stimulus in value-addition activities for horticulture produce would have the following advantages for addressing the issue of employment:

i) It can open new vistas of opportunities for the rural youth as they would not have to migrate to far off places in search of employment. This would lower the cost of living due to which it would be a win-win situation for the employer as well as employee.

ii) As the rural population is well familiar with the raising, harvesting and other issues related to the fruits and vegetable crops, training the workforce in the sector would be easier and impactful. Skill set required in this sector is relatively lower than other manufacturing industries. Also, since the trainees would not have to travel far off for training courses, the enrolment would be higher.

iii) The burden of creating livelihood on Agriculture can be greatly reduced with members of the family engaged in other activities than farming which anyway is a seasonal occupation.

iv) Wastelands across rural areas can be utilized for establishment of processing units of different scales and capacity. This would reduce capital cost of the projects as well as allow for closer proximity of processing centre to the farm.

v) Farmers after being exposed to value-addition/ processing methods could be encouraged to take upon entrepreneurial initiatives by themselves and operate their own units thereby increasing the return on their produce.

vi) Investors in horticulture processing units can reduce the operational costs of their businesses by having access to cheap labour, reduction in transport costs of raw material to the factory.

vii) Establishment of such units in rural areas leads to rural upliftment and thus various activities pertaining to Corporate Social Responsibility can be undertaken.

viii) The vast unorganized sector in this industry can be linked to the organized sector wherein the former could indulge in basic value-addition (mostly through manual practices) and then pass the produce to the organized sector which uses relatively modern and capital intensive.

ix) Skill Development and R&D sector could also get a boost by way of creating opportunities in training of vast workforce as well as innovation in processing methodologies.

x) Expansion of Horticulture processing units across the rural belts would also bring in infrastructure

Challenges for Horticulture Processing in India

The horticulture processing sector holds lot of promise for more fruitful engagement of all stakeholders- the farmers, rural unemployed workforce, entrepreneurs as well as backward integration benefits for large and established big players but for this to succeed, some of the impediments need to be addressed. These have marred the sector for long and restricted the potential of this sector. The most critical of these are as under:

Inadequate Infrastructure

Horticulture produce being a highly perishable commodity demands strong infrastructural support at each level. There is an urgent need to establish multi-purpose cold storages across India as about 75 per cent of the current capacity is suitable only for potato storage and only 23 per cent is meant for multi-purpose. [*Study titled 'Opportunities in Cold Chain: Emerging Trends and Market Challenges' jointly conducted by The Associated Chambers of Commerce and Industry of India (ASSOCHAM) and consultancy firm TechSci Research, 2014*]. Although it requires heavy initial investment but the long term returns can be rewarding. Inadequate post-harvest support in the form of warehouses (cold or controlled atmospheric) near the farm is one of the key ailments of the sector. On the other hand, many established warehouse storage facilities are running with under-utilized capacity due to unavailable supplies.

The prime reason for this is unaffordable pricing and locational issues due to which growers do not have access to such facilities. The above mentioned study also states that out of the 130MT of fruits and vegetables produced in India, only 4MT is transported in refrigerated containers. Similarly, there is dearth of primary processing facilities accessible to the growers. Hilly terrains and lack of road connectivity with proper transport facility is another big challenge. This either leads to post-harvest losses or distress sale by the produce owners. Add to this, the lack of power facility at most of the remote areas and the prohibited cost for operating a CA store. All these challenges make it difficult for development of a thriving fruits and vegetables processing industry. Providing suitable infrastructure is the key towards attracting small scale entrepreneurs in the sector.

Low Productivity Level

For the processing industry to deliver economies of scale, it is essential to increase productivity of horticulture crops as a quick view of the productivity levels around the world reveals that India is still behind a lot of leading fruits and vegetable producers in the world. Lower yields coupled with alarming post-harvest losses create substantial deficit for processing scales. Increased output could be an incentive to invest in a micro-processing unit for a farmer.

Table 3: Leading Countries–Productivity (mt/ha) in Fruits and Vegetables

Country	*Fruit Productivity*	*Country*	*Vegetable Productivity*
USA	23.3	USA	32.5
Indonesia	22.3	Iran	26.8
Brazil	16.5	Egypt	25.7
Turkey	13.6	Turkey	25.0
Phillipines	13.2	China	23.4
India	11.6	India	17.6

Source: FAO and NHB India (http://mospi.nic.in/Mospi_New/upload/SYB2015/CH-9-HORTICULTURE/horticulture.pdf).

Although productivity has increased in the last few years (As per Ministry of Agriculture and Farmers Welfare, the area under horticulture decreased from 24,198 thousand hectares in 2013-14 to 23,216 thousand hectares in 2014-15 but the production increased from 277,352 to 277,743 MT) but it is still to come on par with international leaders.

Lack of Technical Skills

The rural youth is not aware about the technical know-how on post-harvest management and continue to engage in age-old traditional practices. Right from selection of quality seed to primary processing techniques, there is vast knowledge gap at the village level. Some government departments have made efforts to impart knowledge and information on cluster level but it has largely remained ineffective. Since the youth lack the technical skills to start a processing enterprise, they normally migrate towards other occupations. Their ignorance is commercially exploited by middlemen and traders. Also, there is a gap at creating linkages between R&D labs and grassroots level functionaries.

Restricted Credit Availability

The credit accessibility at rural areas is mostly in the form of small public sector bank branches and private money lenders. While the latter's cost of funds is prohibitively exorbitant, the former's risk perception about the sector is high due to which they refrain from sanctioning credit facilities in the sector. Banks in semi-urban or rural locations are hesitant to finance small F and V processing units as they doubt on technical and marketing skills of the applicants, which to a great extent is justified. Although many government schemes have been floated with subsidy element but the discretion for lending is still decided by the rural banker. Working Capital Limits are decided without proper assessment and there are no consulting services for the same. FICCI Survey on challenges in Food Processing Sector has revealed that credit access is still limited to 'top section of the farmers, thus choking credit for marginal farmers at the bottom of the pyramid.

Lower Quality of Produce

Quality of the produce is dependent on various pre and post-harvest interventions. Irrigation, use of pesticides and fertilizers, climatic conditions, state of the tree are some of the pre-harvest factors affecting quality while time, methods and tools of harvesting, storage facilities, transportation, quality of packing material *etc.* are some of the post-harvest factors which decide the ultimate product quality. Non-adherence to many of these standards due to ignorance and want of facility results in poor quality of the raw material thereby decreasing its market value and return.

Suggestions

The irony is that in spite of India being the world's second largest fruits and vegetables processor, it has a miniscule share in F and V processing. This has to be dealt with a multi-pronged strategy. This is the most viable and suitable alternative to create employment opportunities for the youth. As agriculture growth is stunted and provides little scope for engaging the ever growing population, horticulture with its low crop area and high output can provide immense opportunities for entrepreneurship and thereby increase in employment. With just an estimated 2 per cent of output being processed at present, the scope seems to be huge. The processing industry presently is in its nascent stage and is unorganized. Some of the interventions proposed in this regard are as under:

Infrastructure Overhaul

Efficient road network by way of connectivity as well as construction quality, cold storage warehouses in village areas, availability of grading machinery, ensuring uninterrupted power supply, availability of testing labs are the key areas in infrastructure which need to be addressed. Many of these are areas which are required for development of any industry and are thus on the top of the agenda. However, road connectivity to farms and establishment and access to block-level cold storage facility at major horticulture regions of the country will go a long way in promoting entrepreneurship in value-addition activities. This will shorten the time taken by the produce to reach preservation stage as well as reduce operational costs. A pilot scheme for a cluster of districts could be taken up for the purpose.

Micro Units Establishment for Primary Processing

The small entrepreneurs or farmers are not expected to invest heavily in large plant and machinery for setting up a processing unit. This is primarily due to risk averse attitude, lack of capital and technical know-how to run. However with ample produce at farm gates, primary processing centres (grading, peeling, cutting, drying, sorting etc) with manual labour as the major input can be undertaken easily. However, this activity needs to be incentivized at village level with standardized schemes to suit the needs of a particular geography. Employment generation in these units should be rewarded by way of providing interest subsidy, exemption in taxes and infrastructure support. These micro units will later expand into bigger units and inspire others to take up entrepreneurship and create a backward integration model at the rural level.

Strong Marketing Linkages and Support

The village level entrepreneur can put all his efforts into standardizing processes at the processing unit, arranging raw material, assuring quality, supervising human resources and effective administration but the biggest hurdle facing such micro-enterprises is the marketing of produce. Getting market information, accesses to channels of distribution, contract manufacturing or supply to a tertiary producer, brand equity are areas which are outside the scope of the small rural player. It cannot be imparted as a training module in the initial stages. It would be critical to provide supply and marketing linkages by either creating co-operatives, providing incentives to secondary processors who source supplies through primary processing centres, procurement by government departments for official supplies, creating awareness among consumers about availability of local products. However for these to be effective an agency needs to be created for certifying product and process quality of the producer.

Better Credit Access

One of the key requirements for any entrepreneurial intervention to succeed is the easy access to credit facilities. Although Institutions like NABARD have formulated schemes to provide concessional credit to food processing units at Mega Food Parks and designated food parks but ease of availing a credit facility in rural areas is still a distant dream. Credit schemes have to be customized at district level for area specific products and administrative agencies should constantly monitor their progress. The lengthy processes needs to be cut short, credit assessment to be done through standardized software, documentation and other formalities to be simplified and interest rates differentiated according to stage of the enterprise. Public Sector Banks could form a department with Lead Bank Office for continuous improvement in quality of credit services to the prospective entrepreneurs.

Promotion of New Technologies

Small scale plants running on obsolete technologies lead to lower efficiency realizations and compromise on product quality. Hence it is paramount to introduce newer efficient technologies in the sector as entrepreneurs would need maximum possible returns. This could be initiated by way of promoting new technologies through boosting R&D activities, providing incentives for development of scalable and economically viable technology as well as effective promotion of such technologies at grassroots level. Government can also tie-up with major technology companies nationally or internationally for developing efficient and low cost technologies for crops with bulk production.

Relevant and Practical Training near farms

Customers and Companies involved in tertiary processing demand quality products through standard processes. In order to supply to these processors as well as compete in the market with established businesses, an entrepreneur needs skilled manpower with training in modern methods. In order to turn farmers into entrepreneurs it is necessary to provide training facility in the vicinity of village/farms. The workforce especially would not like to move far for training, the benefits of which are relatively unknown to them. Simpler and interactive methodologies need to be evolved and success stories well documented and publicized to motivate people. One of the effective ways would be to develop trainers in the village itself so that the villagers have resource person available at all times to cater to their needs.

Public-Private Partnership

The various interventions especially in the areas of technology and infrastructure development can be undertaken under public private partnership mode. The private sector should be taken on board to develop viable models of investment in developing cold chains, controlled atmospheric warehouses, logistics support like refrigerated vehicles *etc.* which after due expert analysis should be financed by the Government. Long term break even targets can be given to large and critical projects. Similar approach may be useful in innovation for efficient and affordable technologies in preservation and processing. Schemes like Startup India are ideal launch-pads for these but here the initiative and funding has to be taken by the Government for a better response from the private sector.

Outreach Programmes from Institutions

Various educational Institutions as part of social outreach programs can provide strong interventions at rural level for educating farmers, motivating and helping entrepreneurs start micro business units, help in spreading information about latest technical advancements, assist in following regulatory provisions and link them with various governmental initiatives. Educational courses should have inbuilt credits for such extension programs. Such initiatives should also be undertaken by other Government Institutions specially those in the policy making domains.

Conclusion

Entrepreneurship is critical to creating job opportunities for the ever growing agri-dependent population of India in the near future. The need of the hour is to stop them from migrating to urban areas paying for higher living costs or struggling for jobs. Instead they should get employment opportunities in their domicile thereby creating sustainable rural economic prosperity. Being an agrarian population, horticulture processing in India offers immense scope as the present levels of processing are miniscule. The expansion of processing Industry would help the farmers increase their income, youth getting jobs, consumers getting fruit and vegetable based products at lower price and the government increase its revenue. However, it requires sustained all round meaningful interventions and investment so that India emerges as one of the top destinations for horticulture processing industry.

References

Breen, J.P. 2004. Enterprise, entrepreneurship and small business: where are the boundaries. *Int. J. Entrep. Sm. Buss.*, **1**: 21–34.

Cope, J. and Watts, G. 2000. Learning by doing: An exploration of experience, critical incidents and reflection in entrepreneurial learning. *Int. J. Entrep. Beh. Res.*, **6**(3): 104–124

D'Silva, Jeffrey Lawrence, Samah, Bahaman Abu, Uli, Jegak and Shaffril, Hayrol Azril Mohamed. 2011. Towards developing a framework on acceptance of sustainable agriculture among contract farming entrepreneurs. *Afr. J. Buss. Mgt.*, **5**(20): 8110–8116.

De, Wolf, P. and Schoorlemmer, H. 2007. *Exploring the Significance of Entrepreneurship in Agriculture.* Research Institute of Organic Agriculture FiBL, Frick, Switzerland.

Desai, Vasant. 2011. *Entrepreneurship Development: A New Venture Creation.* Himalaya Publishing House Pvt. Ltd., Mumbai, India.

Gartner, W.B. 1989. Who is an entrepreneur? is the wrong question. *Entrep. Theo. Pract.*, **13**(4): 47–68.

Gatewood, E. and Hoy, F. Superior. 1989. Functionalist vs. conflict theories: Entrepreneurship disrupts the power structure in a small southern community. *J. Comm. Dev. Soc.*, **20**(1): 70–83.

Gerard, McElwee. 2013. *Working Paper–The Enterprising Farmer: A Review of Entrepreneurship in Agriculture.* Lincoln Business School, University of Lincoln.

Ghani, Ejaz, Kerr, William R. and Stephen, D. O'Connell. 2014. Political reservations and women's entrepreneurship in India. *J. Dev. Econ.*, **2**(1): 55–61.

Gielen, P.M., Hoeve, A. and Nieuwenhuis, A.F.M. 2003. Learning entrepreneurs as experts. *J. Agril. Edu. Ext.*, **9**(3).

Hanan, Ebtisam. 2015. Entrepreneurship perspective for trade and management of horticulture sector in Kashmir Himalayan valley. *Int. J. Soc. Sci. Mgt.*, **2**(3): 284–289.

Hrangao, Blesson R.R. and Sorokhaibam, Robita. 2015. Enterprising the farmer entrepreneur: A case study of Manipur, India. *Int. J. Human. Social Sci. Inv.*, **4**(4): 28–64

Krishna, K.V.S.M. and Singh, K. Anand. 1994. Agricultural entrepreneurship: The concept and evidence. *J. Entrep.*, **3**(1): 97–111.

Kumar, Ashish. 2015. Entrepreneurship in agriculture development with special reference to Indian agricultural sector. *Research Front.*, **3**(1): 85–94

Lans, Thomas, Wesselink, Renate, Biemans, Harm, J.A and Mulder, Martin. 2004. Work-related lifelong learning for entrepreneurs in the agri-food sector. *In. J. Train. Dev.*, **8**(1): 73–89.

Mulder, Martin, Lans, Thomas, Verstegen, Jos, Biemans, Harm and Meijer, Ypie. 2007. Competence development of entrepreneurs in innovative horticulture. *J. Workplace Learning*, **19**(1): 32–44.

Nain, M.S., Singh, Rashmi, Sangeetha, V., Chandel1, S.S., Kumari, Parveen and Peer, Javeed Ahmad. 2013. Strategies for entrepreneurship development through fruit Production in Jammu and Kashmir state. *Agr. Sci. Dig.*, **33**(3): 165 – 171.

Nandanwar Kalpana, P. 2011. Role of rural entrepreneurship in rural development. *Int. Ref. Res. J.*, **2**(26).

Negi, Saurav and Anand, Neeraj. 2015. Issues and challenges in the supply chain of fruits and vegetables sector in India: A review. *Int. J. Mang. Value Supply Chains (IJMVSC)*, **6**(2): 47–60.

Omkar, Pathak. 2016. Rural development through agri-preneurship: Entrepreneurs experience in agriculture. *Int. J. Mark. Fin. Serv. Mgt. Res.*, **1**(6): 143: 149

Pannekoek, L., Kooten, Van O., Kemp, R and Omta, S.W.F. 2005. Entrepreneurial innovation in chains and networks in Dutch greenhouse. *J. Chain Net. Sci.*

Patel, Brijesh and Chavda, Kirit. 2013. Rural entrepreneurship in India. *Int. J. Adv. Res. Comp. Sci. Mgt. Stud.*, **1**(2): 28–37.

Santhi, N. and Rajesh Kumar, S. 2011. Entrepreneurship challenges and opportunities in India, Bonfrin. *Int. J. Ind. Engg. Mgt. Sci.*, **1** (special Issue).

Saxena, Sandeep. 2012. Problems faced by rural entrepreneurs and remedies to solve it. *J. Buss. Mgt.*, **3**(1).

Shane, S. 2003. *A General Theory of Entrepreneurship: The Individual Opportunity Nexus*. Edward Elgar, Cheltenham.

Singh, Surendra P., Tegegne, Fisseha and Ekenem, Enefiok. 2012. The food processing industry in India: Challenges and opportunities. *J. Food Dist. Res.*, **43**(1): 81–88.

Smith, N.R. and Miner, J.B. 1983. Type of entrepreneur, type of firm and managerial motivation: Implications for life cycle theory. *Strat. Mgt. J.*, **4**: 325–340.

Developing Horticulture in North East India

102

Horticulture in North Eastern Region of India

Lallan Ram

Central Institute of Horticulture, Ministry of Agriculture and Farmers Welfare, Govt. of India, Medziphema – 797 106, Nagaland

E-mail: lallanram09@gmail.com

The horticulture which includes a wide variety of crops such as fruits, vegetables, spices, plantation crops, floriculture, medicinal and aromatic plants, *etc.* is recognized as an important product for potential diversification and value-addition in agriculture. With contribution of 18.8 per cent in total agriculture production and 52 per cent in total agriculture export, horticulture has emerged as a prominent sector offering wide scope for diversification in agriculture. It has a vital scope in foreign exchange earnings and employment generation. The area under horticulture crops which was 12.77 million ha during 1991 – 1992 has increased to 23.69 million ha. The total production during this period has increased by nearly 2.8 times and corresponding productivity has increased 1.5 times. As compared to 257.1 million tonnes of food grain production during 2012-13, the total horticulture production was 268.9 million tonnes. The annual growth rates for area and production of horticultural crops during 2012-13 over 2011-12 were 1.9 per cent and 4.5 per cent, respectively. The percentage share of vegetable production in the total horticulture production was highest (60.3 per cent during 2012-13) as compared to other horticulture crops (fruits – 30.20 per cent). The productivity level of most of the horticultural crops in the country is still low as compared to developed countries.

1. Horticulture Status of NE States

North east India consists of 8 states, namely, Arunachal Pradesh, Assam, Manipur, Meghalaya, Mizoram, Nagaland, Sikkim and Tripura. Nearly 10 per cent of the total geographical (Table 1) area in India comes under the NEH region. Horticulture in North east is a significant and upcoming sector in India and has proved to be the best diversification option for agriculture land use, because of assured and remunerative returns to the farmers.

Table 1: State-wise Geographical Area of NE Region

State	*Geographical Area (mha)*
Arunachal Pradesh	8.374
Assam	7.844
Manipur	2.233
Meghalaya	2.243
Mizoram	2.108
Nagaland	1.658
Sikkim	1.049
Tripura	0.71
Total NER	26.219
All India	328.726
% share to NER	8.0

Source: Basic statistics of North Eastern Region 2011, North Eastern Council, Shillong, Ministry of Home affairs, GoI.

The diversity for horticultural crops of this region has mainly been managed by local farmers, often women. Considerable diversity exists among the regional horticultural species including variation in plant type, morphological and physiological characteristics, reactions to diseases and pests, adaptability and distribution. Apart from the nutritional value, many

Table 2: Major Horticultural Crops of NE Region

State	*Fruit*	*Vegetables*	*Spices*	*Others*
Arunachal Pradesh	Apple, orange, kiwi, walnut	Potato, Tomato, Cabbage	Ginger, Turmeric Cardamom (large)	Medicinal plants and herbs, Flowers/Orchid
Assam	Banana, orange, pineapple, arecanut, lemon	Potato, cauliflower, cabbage and other summer vegetables	Ginger, turmeric	Medicinal plants and herbs, flowers
Manipur	Pineapple, papaya, banana, lemon, orange, passion fruit	Potato, cauliflower, pea, tomato	Chili, turmeric	Medicinal plants and herbs, flowers
Meghalaya	Pineapple, banana, orange, passion fruit, strawberry	Cabbage, chow-chow	Ginger, chili	Medicinal plants and herbs, flowers
Mizoram	Orange, banana, papaya, grapes, passion fruit	Potato, squash, cabbage, broccoli	Ginger, chili	Flowers/orchid, medicinal plants and herbs
Nagaland	Pineapple, banana, orange, passion fruit	Tapioca, cucumber	Chili, ginger, large cardamom	Medicinal plants and herbs
Sikkim	Orange, passion fruit	Cabbage, potato, off-season vegetables	Large cardamom, ginger	Medicinal plants and herbs, flowers/orchids
Tripura	Pineapple, banana, litchi, jackfruit, mango	Potato, off-season vegetables	Ginger	Medicinal plants and herbs, flowers

regional horticultural crops are used for medicinal purposes and income generating source in the rural areas. The region offers scope for cultivation of a wide variety of horticultural crops (Table 2) such as fruits, vegetables, flowers, tuber and rhizomatous crops and spices because of its diversities in topography, altitude and climatic conditions. A range of fruit crops varying from highly temperate types like walnut, apple, *etc.*, to sub-tropical as well as tropical fruits are coming up well in this region. Similarly, wide and diverse types of vegetables including indigenous ones are cultivated in the region. Despite of these favourable factors and the scope for cultivation of horticultural crops, the development of horticulture has not picked up momentum as desired. State-wise area and production in North-Eastern states for fruits (Table 3), vegetables (Table 4), flowers and spices (Tables 5 and 6) is presented.

Cold storage and processing industries in North-Eastern states are given in Table 5.

Table 3: State-wise Area and Production of Fruit Crops in NE Region (2013-14)

States	*Area (000 ha)*	*Production (000 t)*	*Productivity (t/ha)*
Arunachal Pradesh	86.9	312.2	3.59
Assam	150.7	2073.8	13.76
Manipur	51.9	440.6	8.48
Meghalaya	33.2	316.6	9.53
Mizoram	49.7	293.0	5.89
Nagaland	37.2	276.0	7.41
Sikkim	14.7	24.0	1.63
Tripura	60.1	697.9	11.61
NEH Region	484.4	4434.1	9.15
India	6982	81285	11.64

Source: Indian Horticulture Database, 2014, NHB.

Table 4: State-wise Area and Production of Vegetable Crops in NE Region (2013-14)

States	*Area (000 ha)*	*Production (000 t)*	*Productivity (t/ha)*
Arunachal Pradesh	1.5	37.6	25.06
Assam	278.7	3415.1	12.25
Manipur	21.7	219.8	10.12
Meghalaya	40.5	403.4	9.96
Mizoram	39.3	236.7	6.02
Nagaland	26.0	207.7	7.98
Sikkim	25.6	132.5	5.17
Tripura	45.1	754.1	16.72
NEH Region	478.4	5406.9	11.30
India	9205	162187	17.61

Source: Indian Horticulture Database, 2014, NHB.

Table 5: State-wise Area and Production of Flower Crops in NE Region (2013-14)

States	*Area (000 ha)*	*Production*	
		Loose (000 t)	*Cut (Lakh nos.)*
Arunachal Pradesh	1.2	0.0	297
Assam	–	–	–
Manipur	–	–	–
Meghalaya	–	–	–
Mizoram	0.1	166.8	605
Nagaland	0.1	0.0	97
Sikkim	0.2	26.5	214
Tripura	–	–	–
NEH Region	1.6	193.3	1213
India	232.7	1729.2	76732

Source: Indian Horticulture Database, 2014, NHB.

Table 6: State-wise Area and Production of Major spices Crops in NE Region (2013-14)

States	Area (000 ha)	Production (000 t)	Productivity (t/ha)
Arunachal Pradesh	10.2	64.3	6.3
Assam	96.7	287.5	3.0
Manipur	10.5	24.1	2.3
Meghalaya	16.9	74.8	4.4
Mizoram	22.5	59.6	2.7
Nagaland	9.8	39.2	4.0
Sikkim	26.6	60.1	2.3
Tripura	5.7	18.0	3.2
NEH Region	198.9	627.6	3.15
India	3075.9	5743.5	1.9

Source: Indian Horticulture Database, 2014, NHB.

Table 7: Cold Storage and Processing industries in NE Region

States	No. of Cold Storage	No. of Horticultural Crops Processing Industries
Assam	30	58
Arunachal Pradesh	1	2
Manipur	1	16
Mizoram	1	3
Meghalaya	3	12
Nagaland	2	8
Sikkim	1	6
Tripura	14	7
NER Total	53	112

Source: www.agmarknet.nic.in and www.kiran.nic.in.

2. Biodiversity of Horticultural Crops

The North Eastern region is considered to be the richest reservoir of genetic variability of large number of horticultural and plantation crops. The enormous diversity makes the region a gene pool for the varietal improvement but in spite of potentiality no worth mentioning development in the field of horticulture has taken place. It may be mentioned that in hill area particularly horticultural crop cultivation as an alternative to jhuming may prove to be a boon in the regional economy. In NEH region farming being the main stay of the people, development of horticulture will markedly improve the economy of the people. Establishment of orchards and planting of plantation crops on hill slopes will prevent soil erosion which may solve the problem of shifting cultivation and out migration of people to towns.

a) Temperate Fruits

Large number of wild and semi wild species of temperate fruits particularly of the Rosaceae family occurs in this region. Mukherjee *et al.* (1969,1970) reported the distribution of wild relatives of pome and stone fruits in the north eastern region. *Malus baccata* widly used as rootstock of apple and *Pyrus pashia* common rootstock of pear could be located in Khasi hills. *Pyrus pyrifolia* is also another temperate fruit grown considerably in this area and is often used as rootstock of pear in Assam. Besides there are some others native/ endemic species such as *Sorbus verrucosa* and *Pyrus polycarpa* found in Khasi hills.

b) Tropical/Sub-tropical Fruits

Banana, pineapple and citrus constitute the major fruit crops followed by papaya, guava, litchi and jackfruit. The region has been described as one of the major centres of diversity for citrus, banana and mango *etc.* Cultivation of mandarin is distributed in all across the North-East with Meghalaya leading the area. Assam lemon, a seedless lemon, is under cultivation to a considerable extent in the foot hills. Systematic exploratory survey of mandarin orange showed that *Citrus indica* is one of the most primitive species of citrus available in the region. Other promising natives so for identified and commercialised are citron (*C. medica*), sweet lime (*C. lamittoides*), sour orange (*C. aurantium*), sweet pumalo (*C. grandis*), sour pumelo (*C. megaloxycarpa*), Khasi paida (*C. latipes*) and rough lemon (*C. jambhiri*) *etc.*

Mango is found growing wild in many parts of the region. Two species, *i.e., M. sylvatica* and *M. khassaina* are found in Arunachal Pradesh. The main problem of mango cultivation in the region is the attack of mango root weevil (*Sternochestus mangiferae*). One dwarf type mango called March mango found in Tripura can be used for evolving dwarf and resistant varieties to stone weevil.

Pineapple cultivation in all the states of the region is done on slopes under rainfed conditions with two cultivars namely Giant Kew and Queen dominating the production. Tripura leads the area under Queen.

Banana another important fruit of the region is found growing wild at varying altitudes. The species like *Musa balbiciana* and *M. acuminata* growing abundantly in NE region are highly seedy and lack in good taste. Besides cultivated and wild species there exist a number of strains which requires improvement. One such variety locally known as 'Bhim Kol' (*Musa balbiciana*), though seed, has very soft pulp and is used as infant food in Assam. Another type grown in Sikkiim, called 'Gheo Kela' (Butter banana) is orange or yellow coloured; its soft buttery pulp is highly productive and preferred by local people.

Besides the three important fruits as discussed above a number of other tropical and sub-tropical fruits belonging to genus *Artocarpous, Phyllanthus, Anona, Averrhoa, Persia, Aegle, Carrisa, Passiflora, Psidium,* Litchi, *etc.*, are found wild. The guava has already shown high promise in Meghalaya, Manipur and Nagaland.

The genetic resources of temperate fruits represent even more better picture in the north eastern region of India. A number of species belonging to genus *Malus, Pyrus, Prunus, Rubus, Ribes, etc.*, are growing wild. *M. baccata* is found growing in Meghalaya, Manipur and Nagaland and *M. sikkimensis* in Sikkim. *Prunus nepalensis* and *Prunus cerasoides*, which flower during November at Shillong are non deciduous and need improvement in their quality.

c) Vegetable and Tuber Crops

NE region is well known for its rich genetic resources and variabilities for edible and non-edible types of cucurbits. Rare edible cultivars include *Momordica cochinchinensis, Momordica dioica,* summer squash (*Cucurbita pepo*), *etc.* An array of vegetable and tuber crops are available in this region. The region abounds in cucurbitaceous vegetables like pumpkin, bottle gourd, ridge gourd, bitter gourd and cucumber, *etc.* The introduced cucurbit chow-chow (*Sechium edule*) locally known as squash needs a special mention that after having introduced into this region the crop has acclimatised so well that every house in the city of Shillong besides Nagaland, Sikkim, *etc.*, has atleast a single plant of chow-chow.

Among the leguminous vegetables, rich diversity is available in *Dolichos, Vigna, Psophocarpus* and *Phaseolus vulgaris* (French bean). In the entire hill region of the north eastern India a wide variability is available. Variability of *Dolichos lablab* is present in whole of Tripura. *Psophocarpus tetragonolobus* (Winged bean) has been observed to be growing all along Indo-Burma region of Mizoram and Manipur. One underutilized legume-cum-tuber vegetable crop, known as Sohphlong (*Vigna vexillata*), is a wonderful vegetable crop, which possesses edible pods as well as underground tubers. The vegetable offers great promise as 'Soybean' rival to combat the nutritional deficiencies in Indian sub-continent.

Among the solanaceae, the *S. melongena* (brinjal) is widely distributed in Bengal and Assam region extending upto Manipur. The botanical varieties namely, *S. depressum* and *S. serpentinum* are found widely in Manipur. In addition to cultivated brinjal an unidentified Solanum species related to brinjal and similar to *S. gilo* Raddi having red tomato like fruits and another kind of brinjal with enlarged calyx almost covering the round fruits and intermediate in appearance between tomato and brinjal are found in the region. Besides brinjal, tomato, chillies and capsicum also hold great promise for cultivation and improvement. Among the tuber crops, an array of variability is present in colocasia and yams.

d) Spice Crops

Among the different spices crops that are grown in the region are ginger, turmeric, chillies and bay leaf. Though recently introduced, the region has a potential for commercial cultivation of black pepper, cumin, large cardamom, and saffron. Three commercial crops need mention in this respect *viz.*, ginger, turmeric and large cardamom. A number of local cultivars exist in north eastern region. In case of turmeric the local variety 'Lakadong' grown mainly in Jawai area of Meghalaya has shown high curcumin content (7.45 per cent) as compared to 6.7 and 7.2 per cent in high yielding varieties like G.L. Puran and Daghi.

The large cardamom (*Amomum subulatum* Roxb.) is an important spice crop growing abundantly in Sikkim and in some parts of Arunachal Pradesh. The total annual production of dry capsules is to the tune of 4,000 tonnes from these states. Some other species like *A. delabatum* and *A. aromaticum* are also exist. A wild type of Ammomum known as 'Belak' in Arunachal Pradesh has got very small sized seeds, although the capsules are large. If the astringency of its seeds could be reduced, it will find scope for cultivation.

e) Ornamental Crops

A vast treasure of ornamental plants and orchids exists in NE India. The important ornamental species that have now been adopted for cultivation include *Bauhenia, Cassia, Calestemon, Erythrina, Jacatrenda, Magnolia, Rhodedendron, Myria, etc.* Some of the shrubs and climbers like *Azalia, Achenia, Baugainvellea, Camilio, Gardenia, Hibiscus, Jatropa, Narium* and *Thunbergia* are colourful ornamentals. However among the flowering plants, special mention may be made about the orchids, which have both ornamental and medicinal value. Out of 1300 orchid species reported, about 600 species occur in north eastern region alone. Plants of the epiphytes origin have great opportunities for development and growth of industries. Mention may be made of *Vanda coerules* (Blue Vanda), *Renanthera imschootina, Paphiopedilum hirsutissimum, Dendrobium falconerii* and *Paphiopedilum fairicanum* (The lady's slipper) and Cymbidium, *etc.*

f) Medicinal and Aromatic Plants

The NE Indian forest is an important repository for a large number of naturally occurring medicinal and aromatic plants with distinct photochemical, pharmaceutical, therapeutic and industrial properties. Six important genera *viz., Coetus, Coptis, Dioscorea, Epecal, Rauroltia* and *Solanum* have been identified for this purpose. Two important aromatic plants namely, agarwood (*Aqualaria agallocha*) and Java citronella (*Cympogon winteri anus*) have been exploited for commercial cultivation for extraction of essential oils by RRL, Jorhat. Similarly, three endemic medicinal plant species namely *Dioscorea floribunda, D. prazeri* and *Solanum khasianum* have been recommended for cultivation for extraction of steroids. The four important orchid species, namely, *Dendrobium paciflorum, D. nobile, Diplomeris hirsute* and *Paphiopedilum* with potential for use in traditional medicines are now facing extinction in this region. The promising oil-yielding aromatic plants of NE India are *Citronella*, Lemon Grass, agar wood, turpentine, cinnamon, mentha and *Eucalyptus*. There is a great scope for commercial cultivation of these aromatic plantation crops to derive a sizeble production of precious essential oils for industrial use. Some of the plant species that grow usually wild in the region and may be used as a potent of agarbati and other related perfumery products are *Juneperus macropoda, J. recurva, Rhododendrium anthopagon, etc.* Several medicinal plants species with insecticidal properties have also been identified.

g) Post-harvest Management in NEH Region

To minimize post-harvest loss the correct stage of maturity of fruit is very important at harvesting time. This maturity is indicated by several parameters based on visual, physical and chemical indices. These not only help in deciding the date of harvest but also to sort out the harvested fruits into lots of varying maturity for their optimum use. Studies conducted at ICAR Research Complex for NEH region, Umiam shows that Khasi mandarin harvested during November can successfully be stored for 45 days. In north-eastern region of India, the harvesting of fruits is still done manually. Oranges are normally harvested by hand-picking or by climbing on the trees directly or with the help of ladder or sometimes directly by beating the branches with long bamboo stick. Pineapple and banana are generally harvested by a sharp sickle-shaped hand tool locally known as 'Dao', whereas tuber and rhizomatous crops are harvested mainly with the help of spade. The vegetables like tomato, capsicum, cucurbits and cole crops are harvested manually. The techniques for harvesting of banana, citrus, grape, papaya and pineapple are available elsewhere in the country. But no such techniques have been standardized for harvesting of Khasi mandarin, indigenous varieties of banana and pineapple in this region.

3. Schemes for Promotion of Horticulture in NER

Mission for Integrated Development of Horticulture (MIDH), a Centrally Sponsored Scheme has been launched by Ministry of Agriculture, Government of India for the holistic growth of the horticulture sector covering fruits, vegetables, root and tuber crops, mushrooms, spices, flowers, aromatic plants, coconut, cashew, cocoa and bamboo. The scheme, which has taken take off from 2014-15, integrates the ongoing schemes of National Horticulture Mission (launched 2005-06), Technology Mission for North East (TMNE) (launched 2001-02) later renamed as Horticulture Mission for North East and Himalayan states (HMNEH) with addition of Himalayan states (launched 2002-03) and National Bamboo Mission (NBM) (launched 2006-07).

The Central Sector Schemes included in MIDH are National Horticulture Board (NHB), Coconut Development Board (CDB) and Central Institute for Horticulture (CIH), Nagaland. MIDH also works closely with National Mission on Sustainable Agriculture (NMSA) now Pradhan Mantri Krishi Sinchai Yojna (PMKSY) towards development of Micro-Irrigation for all horticulture crops and protected cultivation on farmers' field. MIDH also provides technical advice and administrative support to State Governments/ State Horticulture Missions (SHMs) for the Saffron Mission and other horticulture related activities like Vegetable Initiative for Urban Clusters (VIUC), funded by Rashtriya Krishi Vikas Yojana (RKVY).

4. Major Constraints in Horticulture Sector in NE Region

Though the NE region has high potential for the development of horticultural crops, efforts have not been made to develop it as a commercial venture. Factors inhibiting horticultural development in the region are as follows:

a) Shifting Cultivation

Shifting cultivation also known as jhuming is widely prevalent in North Eastern states of India. This jhuming cycle which extended to 15-20 years earlier has now been shortened to 2-3 years because of increased population pressure on land, decrease in productivity leading to utilization of more area under jhuming. At present about 0.5 million hectare area is under shifting cultivation in whole NE region. This system has caused large-scale deforestation, soil degradation/erosion (removes nutrient rich top soil) and depletion of resource base.

b) Poor Cultivation Practices and Low Yield

General neglect and non-adoption of scientific cultivation practices are the major constraints for poor return from most of the horticultural crops in this region. Despite conducive environment, the rate of production and growth of all horticultural crops are far below the all India average.

c) Lack of Desirable Planting Material

The disease free, true-to-type genuine planting material is absolutely lacking in a number of horticultural crops. It is imperative to produce disease free propagules. Screening of planting materials before its distribution is of utmost importance.

d) Lack of Marketing Facilities

Due to lack of organized marketing structure in this region, farmers are getting low return compared to the other parts of India, whereas the middleman gets the profit at their expenses. Except the organized tea industry, almost all the commodities including specialized products like citronella oil the producers face considerable marketing problems. Due to thin primary markets and perishable nature of the products the farmers sell their produce at a throw away prices to the middleman without even getting the opportunity to display them. Transportation of perishable produces is perhaps the most serious constraints in the horticultural development of this region.

e) Scarcity of Trained Manpower and Extension Support

Dearth of trained manpower and low priority to horticulture in the development plans of states despite high potential are some of the factors responsible for ineffective extension programme. Unlike other states of India like Punjab, Himachal Pradesh, Haryana *etc.*, where the extension services are very efficient, the NE region on the other hand is lagging far behind in this aspect. To strengthen this wing not only trained manpower but determined extension activities with full government support are most urgently required.

f) Land Tenure System or Land Ownership System

Land tenure or land ownership system and laws are somewhat peculiar in the NE region. The whole system fall under the following three broad categories;

- ✰ Lands owned by the villagers collectively
- ✰ Lands owned by the chiefs, who allot land among the individual households for shifting cultivation purposes, and
- ✰ Lands owned by individual families.

Since horticultural crops have long gestation period and initial cost of establishment of orchard or plantation is high, it becomes almost impossible for the marginal farmers to go for such ventures without long-term credits from financial institutions like banks. Nationalized banks, do not find it a very favourable investment and are not sure about the recovery of loans because the land tenure system particularly in the tribal belts does not permit land mortgages in favour of lending banks. As the tribal farmers cannot ask for loans individually, intermediaries like District Councils or Village Council can take active part in securing loans for them. Apart from these difficulties, the farmers are not tuned to the idea of considering agriculture as a business proposition and are not accustomed to bank loans. From the available information the share of agriculture in most of the states is less than 10 per cent of outstanding credit of the banks with exception to Assam and Tripura. Thus, until and unless this system is changed, the financial investment will not increase to the national level.

g) Problems of Processing

For a region like this the success of fruits and vegetable growing is closely linked with the success of fruit processing units, because of poor marketing and transport facilities. The processing industry can help in sorting out the problem of proper disposal of perishable commodities. Till today, there are hardly any cold storage facilities available; few processing units exist but are not functioning up to the desired capacity. Use of appropriate pre and post-harvest practices for horticultural crops is vital for the success of the crops and to provide good return to the growers. Unfortunately this is the weakest spot in the entire region. Value-addition should be given top priority for the crops like ginger and turmeric. Production of oleoresin from ginger, turmeric and chilli using improved techniques as developed by CFTRI, Mysore needs to be tested in the region.

h) Financial Constraints

The high capital cost involved in establishing orchard/plantation and setting up of required infrastructure is a serious constraint in the expansion of area under horticultural crops. The situation becomes all the more difficult in view of the large number of small holdings.

i) Less Expenditure on Research Work

Investments for research on horticulture have always remained low when compared to the large number of crops it covers. This results in poor technological support. The extension system is also weak. The Department of Horticulture has been created in many states but, do not have adequate manpower and infrastructure to address the entire problem of horticulture.

j) Absence of Crop Insurance Facility

Risk management in horticultural crops is non-existent although crops like onion and potato are covered under the National Agriculture Insurance Scheme. There is a need to cover the risk in case of other horticultural crops in a different manner, perhaps on the basis of potential production coverage instead of average yield. This would encourage higher investment to achieve high productivity.

5. Future Strategies for Improvement of Horticulture Industry in NE Region

a) Collection, Characterization, Conservation and Utilization of Germplasm

There is a need for extensive survey of the region and collection of all the germplasm available in the region. These germplasm should be characterized on morphological and molecular basis and conserved at one place. These germplasm should be utilized in strategic breeding programme involving high yielding national/ exotic varieties. The systematic characterization of high value molecules present in the national/exotic germplasm and their utilization.

b) Identification of Area-specific Major Crops

There is need to identify important horticultural crops for different areas of the region. The infrastructure facilities for commercial cultivation, marketing, export and processing/value-addition for identified crop should be developed in that part of the region, for example, Khasi mandarin and ginger for Meghalaya, arecanut for Assam, Litchi and pineapple for Tripura, orchids for Sikkim *etc.* There is also need to identify the high yielding varieties and hybrids available in the country in selected crops for different agro-climatic zones of the region.

c) Hi-tech Horticulture

High-tech horticulture is the deployment of any technology, which is modern, less environment dependent, capital intensive and has the capacity to improve the productivity and quality of horticultural crops. Adoption of this technology in horticultural crops is necessary to ensure the nutritional security of future generation. Hi-tech horticulture includes micropropagation, micro-irrigation, fertigation, protected cultivation (greenhouse/polyhouse cultivation), organic farming, mechanization and use of remote sensing.

d) Infrastructure for Horticulture

The basic infrastructure facilities like pre-cooling units, packing and grading shed, short and long term cold storage facilities, refrigerated containers, storage and phytosanitary facilities at *mandi* are lacking in the region. Therefore, there is urgent need to create basic infrastructure facilities to boost the horticulture in the region.

e) Establishment of Agricultural Technology Information Centre (ATIC)

There is need to establish ATIC in different states of the region. Information through internet connection, should be provided regarding demand, supply, price market outlook, knowledge of consumers' preference, marketing channels and practices.

f) Conduction of On-farm Trials/Frontline Demonstration (FLD)

The farmers of the region are not aware with the recent technologies of horticulture. Therefore, there is need to conduct demonstration/FLD as much as possible at farmers' field in the identified crops to convince the farmers about the efficacy of measures in enhancing the productivity of identified crops. Apart from this the extension personnel should try to bring the maximum number of farmers to demonstration plots, organize farmers day, fair and yield competition, distribute leaflets/bulletins to the farmers.

g) Post-harvest Management and Processing

The region is lacking in trained personnel with sound knowledge of post-harvest management of produce. There is also a need for integrated research for post-harvest handling, packaging, transportation, storage and quality control of perishable commodities. There is also a need to establish processing units to formulate value-addition products of excess produce.

h) Strengthening of Horticultural Farms and Nurseries

The different horticultural farms and nurseries of the region which are meant for supply of disease free, healthy, true to type planting materials of fruits, vegetables and ornamental crops should be strengthened so as to meet the increasing demand for planting materials. The important rootstocks/mother tree stock should also be maintained at these horticultural farms/nurseries.

i) Training to Farmers/Extension Functionaries

Non-availability of trained manpower is one of the major problems of the region. As horticulture requires highly skilled personnel for grafting, pruning, orchard management and also for vegetables and ornamental plants. The farmer as well as extension functionaries should be given training from time to time regarding recent advances in horticulture. The entrepreneurship should also be generated by providing training to entrepreneurs for overall development of horticulture in the region.

j) Protected Cultivation

The NE states are dominated by hilly areas thus they are highly suitable for protected/off season cultivation of high value vegetable crops like tomato and capsicum and flowers/ornamentals crops.

k) Emphasis on Organic Farming

There is need for research on standardization of doses of different organic fertilizers like FYM, green manure, vermi-compost, neem cake, biofertilizers (*Rhizobium, Azotobacter*, VAM), *etc.*, in different crops. The emphasis should be given on the use of locally available organic materials.

l) Research on Underutilized Crops

There is need for research on underutilized/lesser known horticultural crops for their commercialization, because these crops are grown at very large scale, in the region. For example passion fruit which is grown in Mizoram at very large scale is becoming popular in other states also. Passion fruit is having export potential due to pleasant flavour and attractive natural colour of the juice.

m) Crop Diversification

While giving emphasis on one or two crops, the other crops should also be taken into account to make agriculture sustainable. For example, after ginger and turmeric, which are heavy nutrient feeder, leguminous vegetables like cowpea, pea, *etc.*, should be grown to maintain the fertility of the soil. Similarly, agri-horti-silvi-pastoral system or multi-storey system and agro-forestry will be more successful in hilly areas.

6. Conclusion

Development of location specific high yielding varieties/hybrids, mass multiplication of quality planting materials through participatory approach by use of biotechnology, cultivation of off-season vegetable and high-value low-volume horticultural crops under protected condition *etc.* can increase the productivity to meet the future demand of the region.

Apart from HMNEH interventions, the Central Institute of Horticulture (CIH) has been imparting various training programmes both for officials, farmers entrepreneurs, SHGs, NGOs or allied as a part of capacity building promoting production of quality planting materials, protected cultivation, organic farming, nursery accreditation and certification, post-harvest management activities and marketing aspects. Besides, introduction of four certificate courses for unemployed youth, the institute also has been organizing exposure tours for potential entrepreneurs, officials and progressive farmers. The institute has been actively co-ordinating with various reputed institute, different organization and stake holders of horticulture in the region and the state government departments of North east region in an effort to achieve its objectives.

7. References

Annonymous. 2011. *Basic Statistics of North Eastern Region 2011*. North Eastern Council, Ministry of Home Affair, GoI, Shillong.

Annonymous. 2013. *Agril. Research Data Book*, ICAR, 2013.

Annonymous. 2013. *National Horticulture Board, Yearbook-2013*. Associated Publishing Company, New Delhi,

Annonymous. 2014. *Proceedings: Conference of Agriculture and Horticulture Ministers of North-eastern States*. HMNEH, DAC Krishi Bhawan, New Delhi.

Ghosh, S.P. 1985. Horticulture in North Eastern India. *Indian J. Hill Fmg.*, **7**: 11-21.

Sarkar, A.N. 1994. Resource potential and bio-diversity of North Eastern Region.

Shadeque, 1989. Genetic resources diversity in horticulture crops of the North Eastern Region. *Proc. Indian national Sci. Acad.* B55 Nos. 5 and 6, pp. 473-476

www.agmarknet.nic.in/asnew.htm browse on 06.10.2014.

Yadav, R.K., Yadav, D.S., Rai, N. and Patel, K.K. 2003. Prospects of horticulture in north eastern region. *ENVIS Bulletin : Himalayan Ecology*, **11**(2).

103

Horticulture Based Farming Systems: The Engine for Prosperity in Northeast India

***Bidyut C. Deka*[1]*, *H. Rymbai*[2], *N.A. Deshmukh*[2] *and H.D. Talang*[2]**

[1]ICAR-Agricultural Technology Application Research Institute (ATARI),
[2]ICAR Research Complex for NEH Region,
Umroi Road, Umiam – 793 103, Meghalaya
**E-mail: bidyutdeka@yahoo.com*

The North East of India is a region of mystic splendors with rich cultural heritage comprising of eight states, *viz.*, Assam, Arunachal Pradesh, Meghalaya, Manipur, Mizoram, Nagaland, Tripura and Sikkim. It is stretched between 89.46° to 97.30° East longitude and 21.57° to 29.30° North latitude. The region is characterised by diverse agro-climatic and geographical situations. It has total geographic area of 18.4 m ha which is 5.6 per cent of total geographic area of India. About 54.1 per cent of the total geographical area is under forests, 16.6 per cent under crops, and the rest either under non-agricultural uses or uncultivated land. Shifting cultivation, also known as *jhum* cultivation is the most traditional and dominant land use system and predominantly followed in all North-eastern states. Presently, around 367,000 tribal families are involving with *jhum* cultivation, which affects 3,85,400 ha annually (Patiram and Verma, 2001).

The *jhum* has sustainability when the cycle last for 10-15 years. However, increasing food demand with the increase of population and shrinkage of resources has shortened the *jhum* cycle into 3-4 years, which resulted in low crop productivity. In addition, the shortening of cycle gradually depletes the soil nutrients, and flora and fauna, which leads the land into fallow and slowly land become degraded wasteland. Continuous dilution of the forest cover in the region due to shifting cultivation, firewood, and timber collection is posing the most crucial problem resulting in poor soil health and environmental degradation in the hills.

Land degradation in the North-eastern region is 36.64 per cent of the total geographical area, which is almost double than the national average of 20.17 per cent (Anon, 2000). The states like Manipur, Nagaland, and Sikkim are facing major threat where more than 50 per cent of total geographical area is under wastelands. Soil erosion under shifting cultivation is highly erratic varying with years depending on rainfall patterns. The soil loss from hill slopes (60–79 per cent) under first year, second year, and abandoned *jhum* was estimated to be 147, 170, and 30 t ha^{-1} y^{-1} (Singh and Singh, 1981). Of various degradation types, water erosion, reduced infiltration, acidification, nutrient leaching, burning of vegetation, decline in vegetative cover, and biodiversity are important in context to the NE region and these are factors mainly responsible for degradation of land and water resources.

Under the scenario of climate change and gradual degradation of natural resources including lesser per capita availability of land, there has been an urgent need for location-specific measures to conserve, utilize and manage these resources for optimizing production on sustainable basis without adversely affecting its quality. The age old practices of *Bari* (Backyard farming) system in Assam, *Zabo* system in Nagaland and *Apatani* system in Arunachal Pradesh are some of the glaring examples of effective utilization of the bio-resources in farming systems mode. The wide agro-climatic variation from sub-tropical to alpine provide enough scope for growing different horticultural crops with little modification in abandoned *jhum* lands and existing *jhum* fields based on the ecosystem prevailing. However, development of horticultural based farming system in varying topography, soil and environmental conditions *etc.* will be a very challenging task. While devising any such programmes, social, economic and cultural traditions of the *jhumias* have to be introspected. Moreover, priority should be given to techniques and practices that can enhance production and productivity with regular income at farm level. Notions and conceptions have been integrated in a more systematic way; there is no reason that Northeast India be lagging behind. Keeping all these in mind, an attempt has been made to popularize the idea of Horti-based farming system for doubling the farm income in Northeast India.

Table 1: Potential Horticultural Crops for NEH Region

Crop	*Species*
Fruits	Pineapple, Khasi mandarin, banana, passion fruit, Assam lemon, peach, guava, kiwi, strawberry and apple
Vegetables	Colocasia, chow-chow, tapioca, potato, pea, French bean, cabbage, cauliflower, tomato, brinjal, onion, garlic, onion, leafy vegetables *etc.*
Flowers	Gladioli, roses, lilliums, carnation, gerbera, orchids, chrysanthemum, marigold, anthurium and foliage plants
Spices	Ginger, cardamom, turmeric, king chilli and black pepper
Medicinal and aromatic plants	Patchouli, neem, agave, alovera, garcinia and ginseng *etc.*
Plantation crops	Arecanut, coconut and cashew

Selection of Crops across the Elevation

The different horticultural crops can be grown in different altitudes based on their climatic requirements in *jhum* or degraded lands.

Table 2

Altitude (above msl)	*Suitable Crops*
High hills (900–2000 m)	Apple, peach, pear, plum, apricot, kiwifruit, strawberry, potato, colocasia, cabbage, cauliflower, radish, beans, *etc.*
Mid hills (below 800 m)	Citrus, banana, pineapple, papaya, guava, ginger, turmeric, chilli, brinjal, tomato, bean, sweet potato, tapioca, colocasia, *etc.*
Foot hills	Jackfruit, arecanut, cashew nut, coconut, black pepper *etc.*

Profitbale Cropping Systems

The following cropping models/techniques may be followed in the entire region based on the availability of lands suitable for the system.

A. Mixed Horticultural Land Use for Hilly Slopes

- ✰ 2/3rd areas from top towards lower hillside are converted into contour using logs/bamboo.
- ✰ 3-4 fruit blocks may be developed *i.e.* mandarin block, peach block, guava block, lemon block.
- ✰ The lower 3/4 contour may be utilized for cultivation of rhizomatous crops. After that 3-4 terraces may be completely utilized for the sole vegetable cultivation.
- ✰ The lower 1 or 2 contours may be utilized for strawberry under plastic mulch.
- ✰ The contour bunds can be utilized for planting of pineapple.
- ✰ Alternate contour bunds may be utilized for marigold, gerbera *etc.*

B. Multi-tier Horticultural Land Use System

Horti-horti three-tier system: Areca nut/coconut + black pepper + ginger/turmeric/pineapple/Assam lemon

- ✰ Silvi-horti-three tier system: MPT + black pepper + ginger/turmeric/pineapple
- ✰ Silvi-horti-two tier system: Tree bean and pineapple
- ✰ Alder-based farming system: Alder and vegetables like potato, cole crops or alder and cereals like maize, rice *etc.*
- ✰ Alder-based large cardamom system
- ✰ MPT + Assam lemon

C. Horticulture-Based Farming System

Horticulture based farming system approach may be a sustainable option for abandoned/existing *jhum*. ICAR Research Complex for NEH region, Barapani has developed a number of models and in one such model comprising of 1 hectare area, the following components were tried to get income throughout the year.

- ✰ Fruit crops: *Khasi* mandarin/sweet orange, pineapple, Assam lemon and guava
- ✰ Vegetable crops: As inter-crops
- ✰ Protected cultivation of vegetable in polyhouse
- ✰ Flower crops: Gerbera and marigold
- ✰ Vermicompost: 1 unit: 3 chambers
- ✰ Vermiwash: 2 units
- ✰ Water harvesting structure cum fish pond

The following are some of the other models developed by ICAR Complex, Barapani.

1. Tuber Crop-based Farming System

The *tilla* (moderate hillock upland) land is suitable for cultivation of tuber crops in Dhalai District of

Table 3: Income Generation from Horticulture Based Farming System within One Year

Crop(s)	Area (m²)	Planting Season	Gross Income (Rs.)
Sweet orange	2000	July	
Pineapple + Assam lemon	1200	April	2,000 (Pineapple)
Guava	500	July	
Polyhouse	100	Whole year	16,000
Vermicompost unit	1 unit	October	Utilized for cultivation
Vermiwash unit	2 nos.		
Turmeric, colocasia *etc.*	200	April	9,900
Water harvesting pond	1000		3,000 (fish)
Flowers (Gerbera and marigold)	Intercrop		3,450
Cabbage, capcicum, brinjal	intercrop		2,330
Pea, tomato, French bean	Intercrop		1,748
Broccoli, cowpea, okra	Intercrop		2,750
Total income	**Nov, 08 – Dec, 09**		**41,178**

Source: Annual Report (2009-10, ICAR Research Complex, Umiam).

Tripura. In this model a *tilla* was taken up for the study as most of the landscape in Tripura is *tilla* only. A *tilla* having no source of irrigation was divided into plots having the size of 800 m^2 each for the individual tuber crops. In these plots tuber and rhizomatous crops like diascorea, elephant foot yam, tapioca, ginger, sweet potato *etc.* were cultivated. Boundary of each plot was planted with banana suckers. One unit of The total cost of production was Rs. 18,150.00 per ha and farmers earned a gross return of Rs. 39,100/-. with a B:C ratio of 2.15 (Datta *et al.*, 2012).

2. Agri-Hort-Silvi-Pastoral System

The system was standardized in 0.80 ha area in ICAR Research Complex for NEH Region, Umiam, Meghalaya. In this system 0.10 ha of foothills was used for agricultural crop, 0.25 ha for horticulture and 0.44 ha for silvi-pastoral crops. The Agri-horti-silvi-pastoral system produced 8,344 kg REY (rice equivalent yield). The highest REY of 3,000 kg was estimated with cow milk followed by Capsicum-turmeric (2,006.5 kg REY). Vegetable component registered a net income of Rs. 18,365, while the fruit orchard (guava and pineapple) gave a net income of Rs. 5,071 amounting to a total net income of Rs. 45,092 from the system (Annual Report 2015-16, ICAR Complex, Umiam).

3. Silvi-Horticultural System

The total area of silvi-horticultural system was 3.13 ha with a forest land of 2.17 ha and planned land use of 0.96 ha of which 0.50 ha area was kept for system study. The average slope of the area was 53.18 per cent. Lower terraces covering an area of 490 m^2 was utilized for growing of spices and vegetables in combinations like turmeric + bottle gourd, turmeric + pumpkin and turmeric alone. The middle portion of the system was utilized for fruit crops such as guava. Upper portion of the system was covered with the forest tree spp. *Alnus nepalensis*. A gross income of Rs. 34,400.00 was recorded from an area of 490 m^2 during 2015-16 (Annual Report 2015-16, ICAR Complex, Umiam).

Nutrient Recycling in Integrated Farming System

Soil fertility can be easily maintained when the different components like crop-livestock included in cultivation within a farm. Soil-plant-animal continuum recycles all the resources available in it. The integrated farming system has the potential to recycle a large portion of nutrients used by the crop back to the soil, because 75 per cent of animal feed is excreted in manure or urine. It is illustrated in the Figure 1. Singh and Dhyani (1995) from ICAR Research Complex, Umiam, Meghalaya studied the effect of fruit based integrated farming system on soil nutrients recycling. They stated that organic carbon, available P, and exchangeable cations contents in surface soil ranged in between 2.0–2.5 per cent, 10.4–13.2 ppm, and 5.9–8.4 cmol (p$^+$) kg^{-1}, respectively, under jackfruit-based AFS, while 1.5–1.8 per cent organic carbon, 3.8–6.7 ppm available P, and 3.9–5.9 cmol (p$^+$) kg^{-1} total cations were found under arecanut/Khasi mandarin-based AFS.

Majumdar *et al.* (2004) evaluated the long term effect of agri-horticulture (comprising Khasi mandarin + agricultural crops, and Assam lemon + agricultural crops), agri-silviculture (multipurpose tree species + annual agricultural crops), silvi-horti-pastoral (alder + pine apple + fodder grasses), and multi-storeyed cropping system (alder + tea + black pepper + annual agricultural crops between the tree rows) on soil properties and fertility status acid Alfisol of Meghalaya compared with natural forest as a control. The study reveald that organic carbon was significant increased in all the models when compared to initial status. The

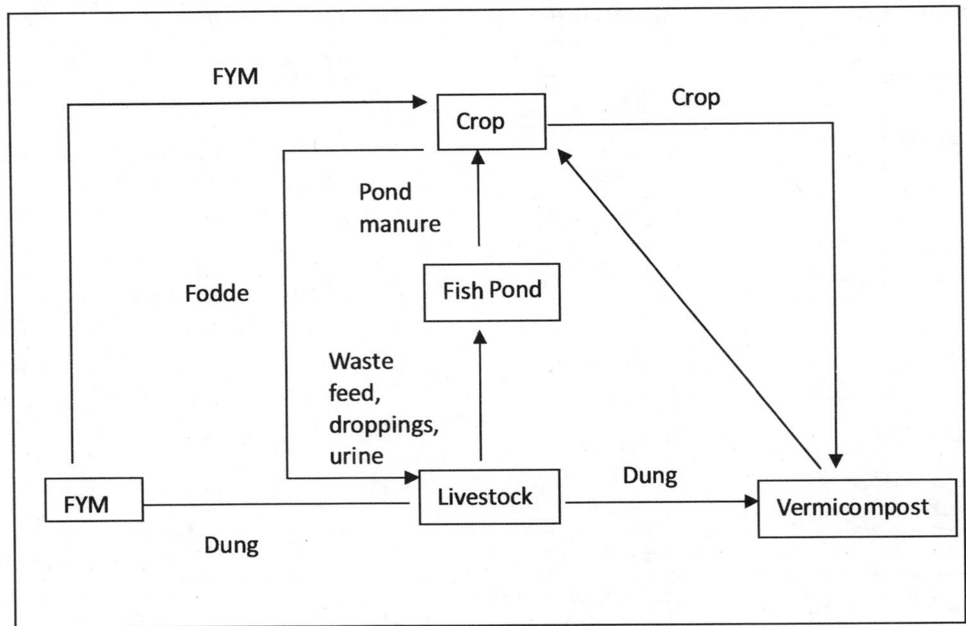

Figure 1: Nutrient Recycling in Integrated Farming System.

exchangeable Ca, Mg, Na, K, and Al and available N, P, and K content were higher in all the systems compared to natural forest and the content of these nutrients decreased with increasing soil depth (Table 3).

Enhancing Water Productivity through Integrated Farming S(IFS)

The water is the most critical resources in any system of crop production. The horticulture based integrated farming system offers a great scope for efficient utilization of water. Kijne *et al.* (2003) described the three dimensions of water productivity include: physical productivity, it is expressed in kg per unit of water consumed; combined physical and economic productivity, it is expressed in terms of net return per unit of water consumed, and economic productivity, it is expressed in terms of net income returns from a given amount of water consumed against the opportunity cost of using the same amount of water.

North east India receives high precipitation during the monsoon season and it faces severe water scarcity during post monsoon season between November to April. The maximum precipitation received during monsoon is lost via surface runoff and evaporation due to wide variations in slope, altitude, rocky layers in the soil and lack of knowledge on water harvesting methods.

Table 3: Effect of Integrated Farming System on Soil Properties

Property	*Agri-silviculture*	*Agri-horticulture (Khasi mandarin + crops)*	*Agri-horticulture (Assam lemon + crops)*	*Silvihorti-pastoral (Alder + pine apple + fodder grass)*	*Multistoried AFS (Alder + tea + blackpepper + crops)*	*Natural Forest*
pH	4.65	4.62	4.80	4.25	4.61	4.62
Organic C (%)	1.62	1.55	2.02	2.19	1.91	1.92
Exchangeable Ca^{++} [cmol (p^+) kg^{-1}]	0.40	0.86	0.74	0.31	0.65	0.26
Exchangeable Mg^{++} [cmol (p^+) kg^{-1}]	0.75	0.51	0.33	0.48	0.71	0.16
Exchangeable K^+ [cmol (p^+) kg^{-1}]	0.232	0.244	0.249	0.238	0.201	0.169
Exchangeable Na^+ [cmol (p^+) kg^{-1}]	0.201	0.220	0.194	0.195	0.197	0.196
Exchangeable Al^{3+} [cmol (p^+) kg^{-1}]	2.65	2.70	2.20	3.15	2.05	2.20
Available N (ppm)	190.1	180.8	203.6	199.4	216.9	167.2
Available P (ppm)	2.75	4.10	5.36	0.94	3.36	0.63
Available Fe (ppm)	8.9	10.4	12.8	10.9	13.9	7.3
Available Mn (ppm)	0.58	0.92	0.79	0.83	1.04	0.04
Available Zn (ppm)	0.08	0.05	0.07	0.006	0.08	0.025
Available Cu (ppm)	0.21	0.23	0.37	0.30	0.27	0.10

Rain water harvesting has very good potential in this region to conserve the water for agriculture purpose in the post monsoon season as well as for domestic use. The *Jalkund*, spring water harvesting, surface runoff harvesting and rooftop rainwater harvesting are the best ways to harvest and store the water for off season cultivation of crops. In possible areas, utilization of pressurised irrigation system like drip and sprinkler irrigation systems will enhance the crop productivity and also it saves irrigation water by around 60 per cent. The multiple use of water is illustrated in Figure 2.

Water can be effectively utilized for several purposes simultaneously at unit piece of land to integrate the different components like crop, fish, poultry, piggery, duckery *etc.* to get more income and to create more employment opportunities that leads to economic upliftment of resource poor small and marginal farmers. Different components under integration have synergistic effect on each other, which results in better utilization of natural resources thereby it increases the crop and water productivity. It also generates employment round the year in a farm and also it conserves the ecosystem.

Many experiments have been carried out to use water for integration of different components within a farm like crop – fish, crop – poultry – fish, crop – livestock – fish – duck, crop – fish – vegetables on bund, *etc.* (Rangaswamy *et al.*, 1992, 1996; Mahapatra, 1992; Behera and Mahapatra, 1998; Samra *et al.*, 2003; Rautaray *et al.*, 2005). Maximum productivity of land was obtained by crop – livestock components when vegetables were included as one of the components in the system due to its high yield at short period of time and less water requirement (Samra *et al.*, 2003). The economic analysis of this system revealed that the net return and employment opportunities could be increased with minimum inputs in a defined land area. Similarly, the study conducted under Eastern Indian situations revealed that integrated farming system having fish-vegetable-rice based on pond water gave the highest productivity.

Behera and Mahapatra (1999) developed an integrated farming system using the pond water based on the resource availability of a farm for small farmers of Odisha. Further, they stated that the diversified integrated farming system developed for small and marginal farmers gave maximum output and more employment opportunities (Table 4).

Table 4: Integrated Farming System Models and its Economics for Small Farmers

Component	*Employment Generation (Man days)*	*Total Expenditure (Rs.)*	*Net Returns (Rs.)*	*Returns/Rs. Invested (Rs.)*
Field crops	98.2	3315.00	5638.00	2.70
Multi-storeyed cropping	87.0	3831.00	9089.00	3.37
Pomology	18.4	900.00	1466.00	2.63
Olericulture	96.4	3812.00	8302.00	3.18
Floriculture	4.0	125.00	100.00	1.80
Pisciculture	31.0	3722.00	16603.00	5.46
Poultry	23.0	9240.00	981.00	1.11
Duckery	23.0	5387.00	713.00	1.13
Mushroom	180.0	18181.00	12856.00	1.70
Apiary	1.0	170.00	1180.00	7.94
Biogas	11.0	600.00	1431.00	3.38
Total	573.0	49286.00	58360.00	2.18

The pond plays a central role in rain-fed systems, as it is essentially the only source of water supply in majority of farms. Households in rain-fed systems regularly irrigate fruits and vegetables using pond water

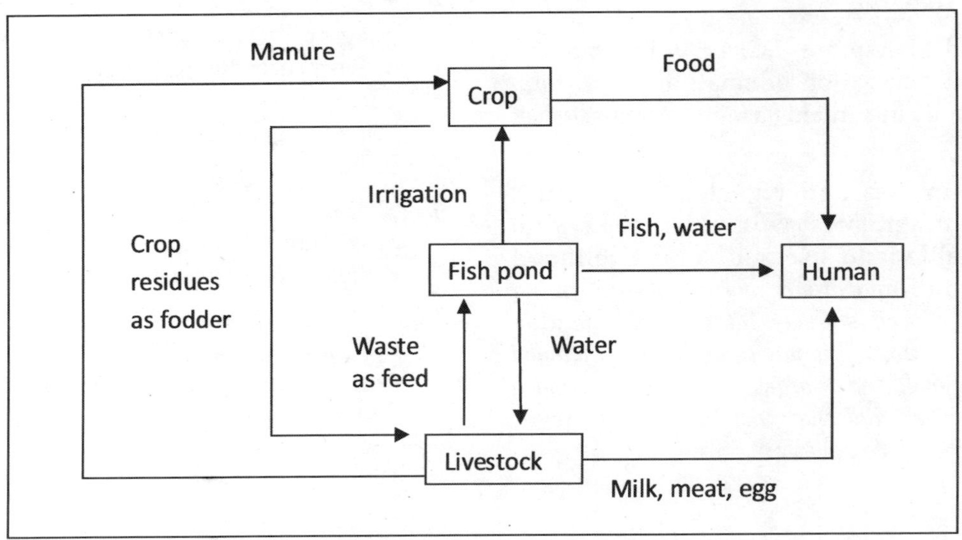

(For human, water should be properly treated)

Figure 2: Multiple Use of Water in Integrated Farming System.

and occasionally rice, particularly during transplanting stage when rainwater is inadequate to puddle the field (Mohanty *et al.*, 2004).

The success in enhancing the water productivity is mainly depends on the choice of the crop. The crop selected for integration should be less water demand crops. The integration of components based on pond water not only increase the crop and water productivity, it also enhances the food and nutritional security of the poor farmers by supplying food grains, fish, meat, fruits and vegetables and more income from a unit of land.

Conclusion

Horticulture-based Integrated Farming System (IFS) has emerged as a well accepted, single window and sound strategy for harmonization and simultaneous management of land, water, vegetation, livestock and human resources, The horticulture based integrated farming system developed for hilly areas would create sustainable environment by reducing the risk of soil and other environmental degradation, while enhancing the productive potential of soil. Besides, the interventions through horticulture based farming system having a tree crop would improve the physicochemical conditions of the soil due to its high quality of leaf litter and root binding ability. Furthermore, this combination would create employment besides increasing the income of the farmers with sustainable socio-economic prosperity.

References

Anonymous 2000. *Wastelands Atlas of India*, Ministry of Rural Development, Government of India and National Remote Sensing Agency, Hyderabad, India, pp. 2–6.

Behera, U.K. and Mahapatra, I.C. 1998. Income and employment generation of small and marginal farmers through integrated farming systems. *Indian Farm*, **48**(3): 16–28

Behera, U.K and Mahapatra, I.C. 1999. Income and employment generation of small and marginal farmers through integrated farming systems. *Indian J. Agron.* **44**(3): 431–439

Datta, M., Anup Das and Ngachan, S.V. 2012. Innovations in agroforestry-based farming system for sustainable land use and rural livelihood in Tripura. **In**: *Innovations in Integrated Farming System for Livelihood Improvement.* Book of Extended Summaries of the *National Seminar on Livelihood Options for Small and Marginal Farmers Livelihood Options for Small and Marginal Farmers in Fragile Ecosystems*, 9–10th August, pp. 58–64.

Kijne, J., Barker, R., and Molden, D. 2003. Improving water productivity in agriculture. **In**: *Water Productivity in Agriculture: Limits and Opportunities for Improvement* (*Comprehensive Assessment of Water Management in Agriculture*), (Eds.) J. Kijne *et al.*, CABI, International Water Management Institute, UK, pp. 11–19.

Mahapatra, I.C. 1992. Farming systems research challenges and opportunities. *East Indian Farm Syst. Res. Extens. Newslett.*, **6**: 3–10

Majumdar, B., Kumar, K., Venkatesh, M.S., Patiram and Bhatt, B.P. 2004. Effect of different agroforestry systems on soil properties in acid Alfisols of Meghalaya. *J. Hill Res.*, **17**(1): 1–5.

Mohanty, R.K., Verma, H.N. and Brahmanand, P.S. 2004. Performance evaluation of rich-fish integration in rainfed medium land ecosystem. *Aquaculture*, **230**: 125–135

Patiram and Verma, N.D. 2001. Reformed system to increase the productivity of shifting cultivation in north-eastern hills. *J. North Eastern Council*, **21**: 24–32.

Rangaswamy, A., Venkataswamy, R., Premsekhar, M. and Palaniappan, S.P. 1992. Sustainable agriculture for rice (*Oryza sativa*) based ecosystem. *Indian J. Agron.*, **37**(2): 215–219.

Rangaswamy, A., Venkitaswamy, R., Purushothaman, M. and Palaniappan, S.P. 1996. Rice-Poultry-Fish-mushroom integrated farming systems for lowlands of Tamil Nadu. *Indian J. Agron.*, **41**(3): 344–348.

Rautaray, S.K., Das, P.C. and Sinhababu, D.P. 2005. Increasing farm income through rice-fish based integrated farming system in rainfed lowlands of Assam. *Indian J. Agric. Sci.*, **75**(2): 79–82.

Samra, J.S., Sahoo, N., Roy Chowdhury, S., Mohanty, R.K., Jena, S.K. and Verma, H.N. 2003. Sustainable integrated farming system for waterlogged areas of Eastern India. Research Bulletin No. 14. Water Technology Centre for eastern Region (ICAR) Chandrasekharpur, Bhubaneswar, Odisha, 751023, India.

Singh, A and Singh, M.D. 1981. Effect of various stages of shifting cultivation on soil erosion from steep hill slopes. *Indian Forester*, **106**(2): 115–121.

Singh, B.P and Dhyani, S.K. 1995. Significance of jackfruit in restoration of soil fertility. *Annual Report*, ICAR Research Complex, Umiam, Meghalaya.

104

Floriculture in North Eastern States of India: Prospects and Challenges

D.R. Singh[1]* *and Ram Pal*[2]

[1]*National Research Centre for Orchids, Pakyong – 737 106, East Sikkim, Sikkim*

[2]*National Research Centre for Orchids, Darjeeling Campus, Darjeeling – 734 101, West Bengal*

**E-mail: nrcorchids@rediffmail.com, drsinghhort66@gmail.com*

The north eastern region of India include eight states *i.e.* Arunachal Pradesh, Assam, Manipur, Meghalaya, Mizoram, Nagaland, Tripura and Sikkim. This region lies between 21°57′ and 29°28′ north latitude and 89°40′ to 97°50′ east longitude. This region has 2.55 lakh km^2, which is about 8 per cent of the total geographical area of the country. The physiographically this region is divided into three divisions namely Meghalaya plateau, the northeastern hills and the Brahmaputra valley. The NE hills alone accounts for 65 per cent of the total land area whereas the Brahmaputra valley and the Meghalaya Plateau constitute 22 per cent and 13 per cent of the area, respectively. More than 64 percent of total area in north-eastern states is covered by thick and deciduous forests. Prof. MS Swaminathan, renowned Indian Scientist described the region as cultural and genetic paradise and granary of megadiversity in terms flora and fauna. Despite of richness natural resources and climatic diversity, the economic development in north-eastern states of India is poorest from the rest of country. The change in climate, decreasing crop productivity, poverty and unemployment are creating pressure on natural resources. In result, the natural resources in the Region are jeopardising very fast. Floriculture is an very important sector of horticulture industry, which is labour intensive and gives highest per unit profitability. It also creates job opportunities ranging from cultivation to post-harvest handling, marketing, value-addition and sale of products by retailers. Among floriculture crops, high value and low volume crops hold good promise for these areas. The orchids are one of them. The region has abundant orchid genetic diversity resource and nearly 900 species are concentrated in this region. Orchids in this region can be grown on unused/wastelands since these are grown in containers. A few success stories of commercialization of orchids has already been written in states like Sikkim, Arunachal Pradesh, Tripura, Assam and Meghalaya. However, the concerted efforts are required in planning, promotion and mitigating the problems of orchid growers for sustainable development of orchid industry in the region.

Indian Floriculture: Past, Present and Future

India has a long tradition of growing ornamental plants. The history of gardening in India is as old as civilization. The evidence found in the excavation of Indus and Mohan jo Daro civilization (2400 BC and 1750 BC) indicates the importance of ornamental plants. Aryans (1600 BC) called India as *Aryavrta* meaning the country of Lotus and abundant sunshine. They saw the lakes were studded with lotus flowers and there were wide-open spaces. They appreciated the beauty of ornamental plants, mountains, lakes and flowers. The association of Lord Buddha with the trees is well known. He was born under the Ashoka (*Saraca indica*) tree, attained his enlightenment under Pipal (*Ficus religiosa*), spread his new teachings under banyan (*Ficus benghalenlesis*) and mango (*Mangifera indica*) tree and breadth last in Sal (*Shorea robusta*) grove. The Aryan Epics Ramayana and Mahabharata (500 BC) mention trees and creepers like *Anthocephalus cadamba, Ficus bengalensis, F. religiosa, Saraca indica, Michelia champaca, Terminalia arjuna, Butea monosperma etc.* The great Emperor Ashoka (264-227 BC) adopted arboriculture as one of his state policies. The sculptures of Kushan period (70-200 AD) have been inscribed with *Saraca indica, Anthocephalus cadamba,* and *Ixora parviflora*. The writings of Sudraka (100 BC), Asvaghosha, and Kalidasa have mentioned several trees and climbers. Vatsayana (300-400 AD) mentioned four types of gardens, Promodyan, Udyan, Brikshvatika and Nandavana in his book Kamsutra. Feroz Shah Tughlaq (1351-1388 AD) developed 1300 garden in Delhi. The mughals had great love for flowers and gardens they established several gardens in their capitals and other places as well. The Nawabs and Rajputs also established several garden during their rule. The Sikandar Bagh developed by Nawab Wajid Ali (1847-1856) Shah is now National Botanical Research Institute, Lucknow. During the British rule in India there were many activities related gardening by British and Indian kings. King Hyder Ali established most famous Lal Bag Garden at Mysore. In North India, Maharaja Ranjit Singh made the garden at Amritsar. The British introduced annual flowers that were not known in gardening of India. They also introduced several trees and flowering plants in India. They established Royal Agri-Horticultural Societies and Botanical Gardens in India. Among them Royal Agri-Horticultural Society Garden, Calcutta; Lloyd Botanical Garden, Darjeeling; Botanical Garden, Saharanpur; National Botanical Garden, Lucknow; Botanical Garden, Ootacmund, *etc.*

Floriculture gained the status of farming much late, and the recognition as an industry is only after the liberalization of economy in 1991-92 and modified EXIM policies of 1995-96 and 1999-2002. After liberalization, the Government of India identified floriculture industry as a sunrise industry and accorded 100 per cent export-oriented status. Opening of the world market has encouraged Indian entrepreneurs for establishing world class export oriented floriculture units. The boost in export oriented floriculture in India was triggered by the access to contemporary exotic genetic materials along with world class technology. The Indian floriculture industry is mostly characterized by sale of loose flowers (rose, chrysanthemum, jasmine, marigold, crossandra, tuberose *etc.*) cut flowers (rose, carnation, chrysanthemum, gladiolus, gerbera, orchids, anthuriums, liliums, alstroemeria, tulip *etc.*), Cut greens and nursery materials (seeds of annuals, propagating materials of foliage and landscaping plants). In the recent year (2013-14) the country has produced 5,43,000 MT of cut flowers and 1,75,4000 MT of loose flowers. Traditional loose flowers form a major section of the internal flower market. In current years India has occupied the top position in terms of area under floriculture crops but modern cultivation under greenhouses and export from the country is still in its nascent stage. According to a study conducted by ASSOCHAM, The Associated Chamber of Commerce and Industry of India Indian floriculture sector values at Rs. 3,700 crore (US $ 68.6 million), expanding at a compounded annual growth rate (CAGR) of about 30 per cent. At this growth rate, the industry is likely to cross the Rs. 8,000 crore (US $ 148.4 million) mark by 2015. The study further says that Floriculture sector in India has meagre share (0.61 per cent) in the global floriculture industry which is likely to reach 0.89 per cent by 2015. Floriculture as an industry has a tremendous potential in our country for generation of employment and earning of valuable foreign exchange, through achieving consistent growth in production and export. The hills of India has greater potential for exploiting this increasing demand of ornamental crops with the virtue of its wide climatic variation, unique niche areas and traditional resources management practices.

In order to harness the opportunities and overcoming the hurdles of setting up floriculture in the north east India one has to understand the composition of the silvi-horti- pastural farming system of the hills and integrate modern technologies and infrastructure in the existing frame work. To achieve this, specialized policies are required for developing floriculture and horticulture in the hills. The implementation Horticulture Mission for the North East and Himalayan States (HMNEH) as a part of the Mission for Integrated Development of Horticulture (MIDH) scheme is such an initiative by the Government of India. The Mission looks forward in developing technologies suitable for the hilly terrain and also addresses the entire spectrum of horticulture from production to consumption through forward and backward linkages. The Mission covers all the North Eastern States including Sikkim and three Himalayan states of Himachal Pradesh, Uttarakhand and Jammu and Kashmir and has given a fillip in development of floriculture through area expansion, supply of planting material and financial support to the growers. Emphasis should also be given to conservation of the rich genetic wealth and sustainable exploitation of the same in developing improved cultivars and introduction of new species in the world of ornamentals. These hills also offer several niche commodities and there is ample scope to reap this uncommon opportunities. But the linkage between the market and the producers is missing. Hence, to develop floriculture industry in the north east India devolution of appropriate strategies is required.

Floriculture Research and Development in India

There are 10 ICAR institutes (with 24 regional stations) 6 Directorates and 7 National Research Centres (on major crops) engaged in horticulture research in the country. Several of these institutes carry out research in floriculture. Area specific, multi-disciplinary research is also being conducted under All India Co-ordinated Research Projects each on Floriculture located at various research institutes and State Agricultural Universities. Research and education on floriculture is also undertaken in multi-disciplinary institutes, departments of horticulture in 35 Agricultural Universities, two Deemed Universities and 3 Universities of Horticulture and Forestry. Horticulture Division of ICAR plans and promotes major research programs in relation to horticultural crops including floriculture. The main focus of program is to safeguard the genetic resources, develop suitable cultivar of different crops, augment the production and protection technologies and also technologies for enhanced shelf-life. The systematic research in floriculture began with the establishment All India Coordinated Research Project (AICRP) on Floriculture during 1970-71. AICRP on floriculture was an effort to link ICAR Institutes with the State Agricultural Universities (SAUs) to carry out nation-wide interdisciplinary research in floriculture crops. The AICRP on Floriculture is mandated to coordinate research on genetic resource management and utilization, standardization of production technology, resource utilization such as efficient use of water, repository of data bank *etc.* Currently, AICRP on Floriculture has 22 co-ordinated centres (16 budgetary, four institutional and two voluntary) working on 13 ornamental crops. Realizing the importance of the floriculture sector, the AICRP on floriculture has been upgraded to a full-fledged Directorate of Floriculture to cater the research needs of the sector. The floriculture research in the country now is carried out at Directorate of Floriculture Research, Pune, Maharashtra and National Research Centre for Orchids, Sikkim. Area specific, multi-disciplinary research is also being conducted under All India Co-ordinated Research Projects on Floriculture at centres located in various agroclimatic conditions with major work on orchids at NRC Orchids, IIHR, Bengaluru, Kerala Agricultural University, Thrissur, TNAU, Coimbatore, *etc.* Several other institutions engaged in various aspects of research development of orchids in the country are Orchid Laboratory, Chandigarh, IHBT, Palampur, Botanical Survey of India, SFRI, Arunachal Pradesh, Departments of Botany in several universities, TBGRI, Trivandrum.

Floriculture as Industry: The Global Scenario

Floriculture is established as a dynamic and fast-growing global industry, today. In 2013, the global exports of cut flowers and ornamental plants accounted for US$ 20.6 billion (World Floriculture Map 2015). Floriculture gained the status of industry only in 1800s when large-scale plantings were set up in the United Kingdom. Presently, more than 140 countries are involved in the flower business with a total area of 3, 05,105 ha under production (Current Status of Floriculture, www. farmrest.com). Europe has always remained the centre of floral industry producing and consuming nearly half of World's cut flower production. The Netherlands is the dominant player in world flower trade with 52 per cent share (2015). The other European flower producing countries are the Germany, Italy, UK, Spain and France. In recent years, however, there is a paradigm shift from the conventional production areas to the other parts of the world endowed with more suitable environment. Now, entrepreneurs are more interested in investing in developing countries because of the huge cost of production in the traditional European countries, where considerable amount is spent on energy cost for heating of greenhouses and labour, and stringent environmental consciousness. On the contrary, the tropical developing countries have more congenial natural condition, cheap labor and lax environmental laws. This is the reason behind the speedy growth of flower industry in the Latin American, African and Asian countries like Columbia, Kenya, Israel, Ecuador, Ethiopia, Brazil, South Africa, Australia, Thailand and Malaysia. India is sure to reap the benefit of this shifting trend with its climatic and other advantages.

Floriculture as Industry: National Scenario

Area and production of flowers under both greenhouse and open field cultivation have shown increasing trends in India. It is interesting to look at the figures of area and production in India over the period of time. The area under flower crops has increased from 106 thousand hectares to 255 thousand hectares over a period of last fifteen years. Similarly, the production of loose flowers increased from 735 thousand MT in 2001-02 to 1,754 thousand to 1,754 in 2013-14. The cut flower production has risen from 8040 to 76,732 lakhs in 2012-2013. The value of export of flowers reached from Rs. 105.2 crores to 460.75 crores in 2014-15. The production of loose and cut flowers showed increasing trends over a period of time. However, as compare to loose flowers, it shows that, cut flower production increasing more rapidly. In last few years, there can be seen changes in area and production in various states of India. During 2012-13, Tamil Nadu held first place in production of loose flowers (312.97 thousand MT) accounting for 18.10 per cent of total cut flower production. It is followed by Andhra Pradesh (224.41 thousand MT) and Karnataka (207.50 thousand MT). During the same year the cut flower production was highest in West Bengal (25,429 lakhs) followed by Karnataka (9,441.8 Lakhs) and Maharashtra (7,914 lakhs). It can be seen that there would be increasing trends in production of both the loose and cut flowers in India. The Government of India has identified floriculture as high export potential. It was proposed that export of floriculture should be increased Rs.100 crore per annum in VIII Plan period. For this purpose, National Commission on Agriculture had set a target to bring five lakh hectares of land to be under floriculture up to 2000 AD. With the help of supporting agencies, export of floriculture has been increasing year by year. At the beginning of current decade, India's flower export to world market was of about US$ 50 billion crore per annum which less than 0.1 per cent, it was definitely negligible.

Policy Changes for the Development of Floriculture

After liberalization of economy and announcement of National Seed Policy that paved the way for the import of planting material of elite, exotic varieties from different parts of the world. The liberalization of Indian economy also favoured Foreign Direct Investment (FDI) in floriculture, infrastructure and input supply industries. This helped access to technology for state of the art greenhouse production of flowers during early 90's. A large number of public and private companies were established for 100 per cent EOUs in different parts of the country. Both public and private sector played a proactive role in development of infrastructure facilities and logistic network which are the backbone of Indian flower industry. The cold storage facilities at airports, fleet of refrigerated vans for effective transportation were developed. The state of the art international flower auction centres were commissioned at Bengaluru and NOIDA. The floriculture sector in India is growing much above 7 per cent annually, in terms of value and creating job opportunities and livelihood. The sister organizations of Govt. of India provide for setting up commercial horticulture units, development of post harvest infrastructure and development of technology for horticultural crops including floriculture. There are more than 300 export-oriented units in India. More than 50 per cent of the floriculture units are based in Karnataka, Andhra Pradesh and Tamil Nadu. APEDA provides assistance for development of infrastructure facilities, quality development and transport assistant to floriculture industries engaged in production and export of floriculture products.

World Flower Growing Environment: Opportunities North East India

A review of the flower growing environment of the world reveals that the floriculture has developed in three different horizons in the world- i) The European countries which was the center of economy and knowledge. Here floriculture got a boost as a consequence of the acquired knowledge and genetic materials gathered from all over the world and backed by the strong economic uplift. ii) Latin America and Africa (Columbia, Kenya, Ecuador) - the upcoming countries where the industry was initialized in recognition of the comparative natural advantage and government policies to combat the high production cost of the Europe or North America. Most of the flower industry of these countries are owned by the investors of the developed countries and production is dependent on the technology, genetic materials and market of those countries iii) The South East Asian countries of (Thailand, Malaysia). Here the flower industry is mostly based on the indigenous genetic base and technology and specialized on specific crops, mostly orchids, of which the region is a niche area. Country like India has an advantage of both the second and the third group of countries, specially, in its hilly region. The other advantage of the country is the huge and growing internal market; this gives more stability to the market comparative to the other developing exporting countries.

However, if we take an account of Columbia, Kenya and Ecuador, the leading countries of the world flower export, we can see that the crops cultivated are mainly rose, carnation and chrysanthemum and foliage and filler ornamentals. The cultivation areas of Kenya and Columbia mostly lay at higher altitudes of 1500m msl. This offers greater diurnal variations and more bright sunshine hours which are congenial for rose, carnation (standard and spray) and chrysanthemums. Similar situations are available in the hills of India, mainly in the Himalayas. Exploitation of these areas with proper infrastructure development, technology support and strong marketing strategy can lead world class production of roses, carnations and chrysanthemums.

The roses require a light intensity of more than 8000 lux, temperature between 15–21°C and wintering for its proper growth and flowering. Most of Ecuador's roses are found in the Pichincha in the Andes at an altitude between 2,800 and 3,000m msl which offers more natural light all through the year and cooler nights. Roses grown at this height has much longer growing cycle (15 weeks) than those cultivated at the sea level (8 weeks) and is perfect for long stemmed varieties with big heads. The cold night encourages the bicolour varieties which are sought after in the international market. The average light intensity and the temperature regime of the hills in our country are optimum for rose cultivation. The crop can easily be grown in open condition if protection is given from harsh weather and disease pests. The carnations are more sensitive to diurnal variation and light intensity. The hilly region of northern India, due to its topography, shows a wider variation of day and night temperature than the plains. This can be exploited for the cultivation of carnations. On the other hand, chrysanthemum is more sensitive to day length. These plants can be tailored to flower by manipulating the night duration. Adaptation of simple technologies enables year round production of these flowers in the mid and low hills. Other crops like gladiolus, gerbera, orchids like *Cymbidium* can also be cultivated in the middle and lower altitudes. The higher altitudes are suitable for temperate crops like liliums, alstromerias, tulips, narcissus, hyacinths *etc.* The western Himalayas are especially suitable for cultivation for lilies and tulips, which are difficult to grow in the warmer climate. The Jammu and Kashmir and Himachal Pradesh are endowed with the temperate climate with cold winter and bright cool summer which are required for the growing these crops. Apart from cut flowers the propagating materials of lilies and tulips are in high demand in the world market and can be exported to earn the foreign exchange.

The lower altitudes of Shiwalik range and Southern hills, which are characterized by mild warm temperature and heavy rainfall can be effectively utilized for cultivation of high value crop like orchids, gerbera and anthuriums under shade and tuberose, marigold, rose and gladiolus in open. Orchids and anthuriums require a humid condition with mild temperatures. Crops like *Dendrobium, Pahlaenopsis, Oncidium* and *Vanda* and Mokara hybrids are very suitable for growing in the foot hills. Hills of Meghalaya, Manipur, Arunachal Pradesh, Mizoram, Sikkim and Darjeeling are very apt for growing these crops.

Potted plants and nursery stocks of *Azalea, Hedera, Glaxonia, Cyclamen, etc.* can be raised in the higher altitude over 1500 m msl whereas, seed production of annuals can be better option in the mid hills. The foot hills are also favorable for growing of foliage plants like *Dracaena*, cordyline, palms, cycads, crotons, *aliums*, ferns, cacti, succulents and *Juniperus etc.*

Apart from the climatic suitability, India has a wealth of large genetic diversity of indigenous plants and flowers. The biodiversity hotspots of the country are mostly spread on the hills and mountainous regions, examples the Indo-Burma mega-diversity region covering the Himalayas and biodiversity hotspot of Western Ghats. These regions offer an amazing diversity of orchid flora along with the assortment of palms, cycads and flowering plants of other families. This genetic diversity can be improved and introduced to the international trade, the strategy adopted by the south-east Asian countries for orchids.

Challenges of Floriculture Industry in North East India

In spite high export potential, the export of India Floriculture has not been very encouraging. The export of floriculture remained stagnant since last few years. The Indian floriculture stands 0.19 per cent of total world floriculture exports. The low export of flower products is attributed to high import tariff *vis-a-vis* African countries. The high freight cost availability of logistics for export of perishables, the small size of farms, cold chain management are major constraints. The Indian floriculture industry is facing several challenges at production level like availability of quality seed and planting materials, farm inputs, quality irrigation water and skilled manpower. At marketing level product diversification, quality and environmental issues are of major concern.

In addition to above challenges floriculture in north east India, flower growing areas are to be affected by the large-scale deforestation, soil erosion, (46.0 t/ha of top soils washes away annually), dry water bodies, soil acidity, high intensity of rainfall, snow and frost, lack of infrastructure and transportation facilities in hostile topography. The agriculture and horticulture are the dominant activity in the hill economy, but it faces multiple risks and challenges like:

i) The staggered land holding makes it challenging to use modern technologies and equipment.
ii) The ethnic practice of *jhum* or shifting cultivation is also a limiting factor for setting up of high-value floriculture units.
iii) The single most limiting factor is the inaccessibility due to the fragile and difficult terrain of the regions.
iv) Remoteness from the market as well as inputs.
v) Lack of availability of modern technologies customized for use in the hilly terrain.
vi) Lack of access to genetic materials or planting materials.
vii) The fast depletion of biodiversity is also a major concern for the area.

Strategies for Promotion of Floriculture in North East India

The domestic floriculture market is estimated to be about Rs. 5,000 million and growing by 25 per cent annually. The preference of domestic market is shifting from traditional cut flowers to exotic flowers which is good sign for orchids and other exotic flowers. The shift from traditional to exotic flowers offers good scope for orchids due to their variety in colour and long shelf-life. The Delhi market alone is estimated to be about of Rs. 1,000 million.

(i) Determining Product Specific Zones

India is endowed with diverse ecological zones, making it feasible to cultivate almost all kinds of orchids grown for commercial purpose. The modern cultivars are combination of several species and thus suitable climatic conditions needs to be ascertained before taking up large scale production. There is need to earmark the production zones for each product of cultivated genera like *Cymbidium, Dendrobium, Oncidium, Paphiopedilum etc.* The work done in past may be useful in assessing potential areas. There is need to find out nontraditional areas for growing orchids *e.g.* there is possibility of producing good quality of *Cymbidium* and other exotic flowers in Uttarakhand and Himachal Pradesh with slight manipulation of greenhouse conditions.

(ii) Human Resource Development

In India cheap skilled and non skilled manpower is the major advantage in floriculture and orchids in particular. The human resource could be developed through training of entrepreneurs in selected institutions of the country and a backward linkage with the institutes specializing in the region needs to be created for technical support and advice of entrepreneurs. However there is need to start a diploma course in cultivation of orchids as well as tissue culture so that these people may become future orchids entrepreneurs or contribute significantly by getting employed in companies engage in production of orchids.

(iii) Infrastructural Support

The infrastructure is prime requirement for the development of commercial cultivation perishable meant shipping to distant places. Both, on farm or off farm infrastructure facilities would require to be created for safe delivery of quality products. The on farm facilities like pack house, refrigerated van, and cool store may be owned by individual or a group of growers. The off farm facilities are being created by APEDA. A substantial progress has been made to develop off farm infrastructure facilities like construction of cool store at major airport of the country, development of flower auction centers, and an international flower auction centre at Bengaluru. An Agri export zone for promotion of orchids is being setup in Sikkim with an investment of Rs. 32.21 crores.

(iv) Floriculture Education and Skill Development

The floriculture is fast growing segment of Indian horticulture. A key challenge is to harness and sustain the strength and achievements made in this sector. The educated youth have inclination towards floriculture as it provides higher income and social status. The avenues of employment are expanding in the fields of farm/greenhouse managers, supervisors, consultants in landscape architecture, town and city planning, parks, public gardens, floral designers, and horticultural therapists *etc.* Apart from these employment opportunities one can have his own enterprise. The capacity building programme needs to reorient using modern information technology. The Indian floriculture sector moving towards Hi-tech horticulture the institutions and learning centres would be in actual fact required to serve as knowledge bank and grow as an interactive learning centres. These should be backed by complete research and reliable hands on training. These learning centres should provide superior learning opportunity along with perfect blend technical skill to demonstrate their ability at workplace. The manpower and technology development through education and research institutions should be as per demand of the market in both quality and quantity. The change in technology, preferences of consumers, national and international completions affect on the profitability and sustainability of the enterprise. Accordingly, the course curriculums for diploma, graduate and post-graduate courses have to be redesigned.

(v) Research and Development for Hill Floriculture

To develop floriculture industry in the hilly terrains, strong R&D support is the most essential factor. Prioritization of research on developing and refinement of appropriate technologies and production of suitable genetic material is required. It is also essential to assess the impact of such technologies and address constrains of adaptation. Dedicated institutes for research and development of hill horticulture can serve the need. The National Research Centre for Orchids with its mandate to develop the orchid cultivation in the country is situated in the hilly terrain of Sikkim. The institute is devoted in addressing the various problems and issues associated to orchid cultivation, especially in the hilly regions and have provided location specific strategies and techniques for the hills of Sikkim and Darjeeling and other parts of the country. The centre is also involved in conserving the orchid wealth of the country, improvising the genetic stock and development of market driven cultivars. Similar institutes, with sole objective of developing floriculture in hills can manifold the efforts and lead to a bright floral future in the hills of India.

Strategies for Growth of Floriculture

Rapid technological agri-business, international economic integration, saturated markets and free market mechanism have provided opportunity, but also the challenges. Retailing on markets will be more complex. Service, quality and reliability would be an essential factor for securing position in international market. Producers have to organize the production so as to supply the necessary quantities according to the required quality standards. Any parties in the chain, which do not contribute to higher added value, will disappear. The advantages of large-scale market could be found for efficient purchasing process and also in terms of logistics and sue of information technologies. Accordingly, our efforts have to be directed to harness the potential through strategic promotion of market. Strategies could be for policy support, infrastructural development, professionalism in market management, networking of markets and quality assurance.

All these developments provide opportunity for production and marketing. This would need strategic marketing approach having backward and forward linkages coupled with horizontal and vertical integration. By providing sufficient attention and support, attaining the goal of reliable production of high-quality product consistent in quantities could be attained. Resultantly, Asian flower sector would soon become a major player in the region as well as in European flower market. Moreover, given the rapidly increasing rate of spending among Asian consumers for cut flowers, it is also reasonable to expect that the Asian flower industry will soon surpass consumption rates for cut flowers compared to other regions.

Cooperation and commitment, in terms of education, research, funding and communication in Asia would be a driving force to become a leader of commercial floriculture worldwide, in years to come. The strategies have been chalked out to meet the challenges and to make floriculture a most viable activity in Asia to ensure employment with enhanced farm income. The challenges are to capture emerging trend in marketing through innovation and skilled professional management. Therefore, strategies to promote effective marketing should include, quality assurance, transportation, hub development for effective delivery, institutional support for information and training, specialty production, reducing cost and widening products, developing domestic market, promoting indigenous plants and flowers, developing professional skill and knowledge management, promoting uses of flowers and providing policy support product and delivery. Therefore, there is a need for quality products and delivery, and to develop quality certification system, as developed in Holland, which ensures the quality for the brand.

Conclusion

India has conducive climatic conditions, richness of genetic resources, basic infrastructure, favorable government policies, and large scientific and technical expertise in the field of orchids. A substantial progress has been made in basic and applied research on flower crops including orchids. Our genetic resources are depleting very fast, they need to be conserved and protected from biopiracy, at the same time they need to be judiciously utilized for commercial gains. In globalised economy easing of trade barriers, it is obvious that orchids and exotic flowers from other countries would dominate our markets and it is required to protect the interests of our growers. We must devise ways and means to hedge our markets with competitive countries and harness orchid potential for export markets. New consumption centres are also emerging in South-east Asia, Middle-East and Eastern Europe. The scenario provides opportunity to capitalize on the strength and convert weaknesses into opportunity However, in the consumer-driven market, quality of service and delivery system will play a significant role. Marketing is no more a meeting of buyers and sellers, but it is complex, which is driven by quantity of products and reliability of delivery. New distribution structure with the aid of IT is likely to be in place, which would be a virtual market, demanding, branding, cataloguing and quality assurance emphasizes.

105

Marketing and Market Infrastructure Development for Improving Farmer's Income

S. Bhattacharjee

NERAMAC Ltd.,
A Govt. of India Enterprise, Guwahati – 781 005, Assam
E-mail: edfmd.niramac@gmail.com

NE Region is blessed with vast natural resources. It has immense opportunities for trade and commerce both in fresh and processed form. NE region being home to diverse exotic variety of fruits, vegetables, medicinal plants and other crops could immerge as major centre of food processing industry. Enhancing income of farmers through value-addition, product diversification, and entrepreneurship development must be the thrust areas of development Departments of the region. In the north east, large losses from farm to plate are attributed to poor handling, distribution, storage, and purchase/ consumption behavior. Huge resources that could otherwise be spent on more productive activities go into producing and transporting goods that only go to waste. Losses at almost every stage of the food chain can be reduced by using appropriate processing and packaging. Packaging being an essential part of a long-term incremental development process to reduce losses will have to employ a blend of both technologies and processors to yield maximum result.

In a difficult terrain like north east, marketing of processed products is more remunerative than raw produces where logistics, seasonal crops and infrastructure matters most. Farmer and processor linkages are needed to add value as per the requirements and demands of consumers. Storage and handling technologies and infrastructure requirements needs to be assessed and designed in a manner that losses are minimized and the produce retains its original quality. Very often value-addition and product diversification are easier said than done because of numerous constraints in micro scale rural enterprises. Lack of technological know-how, high cost of machinery, insufficient capital, and absence of hassle-free credit, taxation rules and food safety concerns create a fear psychosis.

The New Generation Cooperatives (NGC) model and FPOs are ideally suited since it binds both farmers and processors to honour commitments and agreements. Collective farming in clusters involving small farmers needs to be promoted including small cooperative processing/Self Help Group units in rural areas to undertake primary processing, grading and cleaning of produce for value-addition Rural processing and value-addition group needs to be encouraged so as to allow small and marginal farmers to be part-time farmers and work in rural enterprises for enhanced income and returns on their produce.

Infrastructural development is the most crucial factor for tapping the potential for industrialization. The State Governments in the region has to supplement effort by development of infrastructure and offer its participation in the form of Government share or partner in the PPP models of projects which are of prime importance considering the location of the state. In terms of FDI intention, the region has been given low priority by the foreign investors and there is a need to encourage foreign participation through simplification of its business environment. Region holds enormous potential to emerge not only as the main production and distribution hub but can also develop into a centre of cross-border trade. Broadly, the region needs to create dynamic institutional capacities for leveraging on its resource endowments and location advantage.

Economy of North Eastern Region being largely rural and agrarian, Agriculture is the principal occupation of majority of the rural population in the state in terms of employment and livelihood. Agriculture sector continues to support more than 75 percent population directly or indirectly providing employment to about 50 percent of the total workforce.

North Eastern Region is traditionally rich in horticultural production due to its diverse and unique agro-climatic condition which is conducive for growing wide range of horticultural crops like various fruits, vegetables, flowers, spices, nuts, tuber crops, medicinal and aromatic plants. To make the horticulture sector more viable and sustainable, due emphasis have been given in raising productivity of various commercially potential crops through the use of better technology, management and planting materials. Similarly due importance has also been given on various aspects of facility like pre-harvest treatment, proper harvest, setting up of collection centers with grading, transport, marketing and processing facilities *etc.* Despite significant growth registered through different scheme of Govt. of India region is faced with numerous problems such as good governance, unemployment, inadequate basic infrastructure and above all lack of agricultural and industrial infrastructure, sickness in the SME sector, environmental degradation issues *etc.*

Marketing of agro-horticulture produces is a serious problem for the farming community in the region. Lack of organized marketing facility, lack of storage infrastructure near the field, poverty and need of money after harvest for repayment of agricultural debt forced the farmers to dispose off their hard earned produces at the earliest and that too at low prices. Keeping in view of the interest of farmers there is a need to play proactive role to develop adequate facility for farmers by initiating the following on priority:

- ☆ Setting up of high quality infrastructure in PPP mode for horticulture promotion. Undertake widespread promotion of horticulture and floriculture, as well as medicinal and aromatic plants and herbs including organic farming, to capture highly remunerative niche markets abroad.
- ☆ Creations of new agri-markets, certification/ standardization centres and other post-harvest operations including food processing in PPP basis near the production centres may be given priority.
- ☆ Establishing a network of cold storage facilities and information centres and organizing marketing and financial support through SHG will have to be initiated on a large scale.
- ☆ Proper infrastructure for handling the post-harvest operations with limited wastage should be attached to production centres in PPP mode.
- ☆ Government should invest in developing regulated markets and trading centres.
- ☆ State should provide adequate storage facilities such as space where farmers, traders, and in a limited way even retail buyers can interact in the price discovery process
- ☆ Set up an Agricultural and Horticultural Marketing Board as a nodal agency for marketing, value-addition and processing for North-east.

NE Region has been witnessing a strong demand for employment as large portion of population coming into the working age group of population. An ASSOCHAM study estimate made for Assam, the gate way state of north east, it is seen that 30.8 percent of population fall in the age group of 15 to 30 years. Assuming the current level of workforce participation rate, the demand for employment would account for 3.7 million. To meet the potential demand of employment, it needs to grow at a double digit rate. At the investment front, Assam has recorded 11.1 compound annual growth rates during 2006-07 to 2015-16. Almost similar is the case with other states in the region, which needs to be looked on priority basis so as to accelerate investment activities in the region as well as encourage private sector to invest in the state.

NERAMAC is taking initiative to organize Food Processing Investors meet in all the eight states HQ in the region as a prime step in this regard besides taking up seminars on Organic Mission for north east, awareness on GI initiatives, buyers sellers meet *etc.*

Govt. of India is making efforts to promote startup companies and develop entrepreneurship particularly in the north east to make a favourable change in the scenario. Assam government has recently introduced a single window system to clear all required permissions for an industrialist/entrepreneur within 30 days to set up a unit in the state. Last week, this entire process has been made online. Now an entrepreneur, sitting anywhere in the country, can log on to the website www. easeofdoingbusinessinassam.in and fill up a simple form providing details about him, the unit, its location, contact address to apply for permission of starting business/ industrial units. In order to spur industrial growth and make Assam an investor-friendly state, this initiative of the Govt. of Assam will be a boon for the economical development in the state. Assam being the gateway of north east, it will allow other north eastern states also take the cue for a series of possibilities.

In the north east, Mission on Integrated Development of Horticulture (MIDH) is in practice in different form for long resulting to increase in production and productivity manifold. But it's estimated that more than 40 per cent of the produce are wasted due to lack of post-harvest facility and that a very negligible amount of it is processed in the region. Loss of nutritional value and monitory loss on that count is huge. In order to boost farmers income, Govt. of India must review the provisions of National Mission on Food Processing (NMFP) for the development of food processing sector in the north east, which in turn would help in increase in farm productivity thereby increase in farmers' income as it will curbing the post-harvest loss of fruits and agricultural produce to a great extent.

It is imperative that centrally, an Agricultural and Horticultural Marketing Board as a nodal agency for marketing, value-addition and processing for north east is created soon to address the situation in the north east. This marketing Board once set up can run a Market Intervention scheme where the cost of procurement and transportation from farm gate level to the terminal markets could be supported by formalizing minimum support price (MSP) scheme for horticulture produces in the north east so that it would be in a position to pass on the benefit of higher price to the farmers and play a pivotal role in bringing horticulture prosperity in the region. It can focus on creating infrastructure for post-harvest and market linkages between producers and markets by establishing forward and backward linkages with farmers to provide assured market to their produce by generating rural employment and prosperity in the region through enhancement of farmers' income. Region would also be able to carve a niche in the organic food market once the naturally grown organic production, certification and marketing can be done in an institutionalized manner with better post-harvest technology, supply chain, branding and certification of produce with appropriate aggregation model for marketing in the north east.

106

Nutritional Security of Tribals of NEH Region of India through Underutilized Vegetables

A.K. Pandey, Siddhartha Singh and S.D. Warade*

College of Horticulture and Forestry, Central Agricultural University, Pasighat – 791 102, Arunachal Pradesh

**E-mail: pandey.ajai1@gmail.com*

About 600 species constitute the global diversity in vegetable crops. However, presently only one fourth is utilized as a major vegetable crops and rest are named as minor, underutilized, rare vegetables, wild edible vegetables or so on. The distinction between underutilized and life support species is by no means strict. The life support species help to sustain life in stress prone habitats under abiotic stresses like temperature, soil and water and in emergency situations (Paroda *et al.*, 1988). Underutilized crops, like life support species mostly tropical and sub-tropical are not limited to specific environments. Their distinguishing feature is a potential for expansion of what is currently a limited area of cultivation. Some had been more widely used in the past but have given away to more remunerative alternatives. The main impetus for revivals or promotions of underutilized crops is to diversify and stabilize agricultural production. For this there are several agronomic and economic reasons. Alternative and additional crops should make system more sustainable through modifying and diversifying crop rotations, improving the soil structures and the nutrient status (*e.g.* by the inclusion of nitrogen assimilating crops) and through reducing the buildup of pests and parasites which occur in mono-crop or short rotation agricultural systems. The economic advantage of diversifications is the stabilization of farm incomes that depend on major commodities subject to price fluctuations.

1. Nutritional Potential of Underutilized Vegetables

Underutilized vegetable have immense potential for contribution to a particular pocket's of food production because they are well adapted to existing as well as adverse environmental conditions and are generally resistant to pests and pathogens. *Euryale ferox* grown in ponds and other water bodies of Madhubani, Bihar is considered one of the most viable sources of income of the local people. Underutilized legumes may be a cheap, alternate source of protein and can alleviate protein malnutrition among preschool children in rural areas. Nutritious pods of *Parkia roxburghii* are consumed as staple legume vegetable in the NEH region of the country. Similarly, *Mucuna pruriens*is considered one of the most preferred legume vegetables in tribal people of Uttar Pradesh, Bihar and Jharkhand. A vast reservoir of leafy vegetables belongs to group of underutilized vegetables, which are rich source of vitamins, minerals, fibre and diversity in the diet. Furthermore, they have been a traditional part of cropping systems, especially in home gardens. Their cultivation, utilization and acceptability should not be a problem (Ghosh and Kalloo, 2001). Underutilized flower and fruit vegetables have longer shelf-life than leafy vegetables. Underutilized vegetables as a whole can, therefore, make an impact on the nutritional status of population, yet among food crops, they are neglected. They are generally low in energy and dry matter content, but immensely important as source of protective nutrients, especially vitamins, mineral and phytochemicals. Vegetables are the most important source of vitamin A, which is deficient in most part of the world; where rice based diets predominate, blinding thousands of children, annually.

(a) Folic Acid Content of Underutilized Vegetables

Underutilized vegetables are rich source of folic acid, which is commonly referred to as folate. Folic acid is involved in the multiplication and maturation of cells, and its deficiency results in certain types of anemia especially in infants and in pregnant women. Recent researchers have associated it with the reduced risk of cardiovascular disease. Earlier studies revealed that diets with higher folic acid reduced the risk of colon cancer. Folic acid is crucial in foetal development in the early weeks of pregnancy, hence, nutritionists recommend that woman of child bearing age should consume the recommended amount (400 µg/day) of folic acid by taking a healthy balanced diet that includes vegetables. Folic acid content of underutilized vegetables is given in Table 1.

Table 1: Folic Acid Content in some of Underutilized Vegetables

Vegetable	*Folic Acid (µg per 100 g edible portion)*	
	Free	*Total*
Amaranth	41.0	149.0
Cluster bean	50.0	144.0
Coccinia	18.0	55.0
Colocasia	16.0	94.0
Curry leaf	23.5	93.0
Mint	9.7	114.0
Plantain green	1.6	16.4
Snake gourd	7.5	15.5
Sorrel	40.0	125.0
Spinach	51.0	123.0
Yam	0.9	17.5

(b) Mineral Contents of Underutilized Vegetables

On an average a man excretes daily about 20 to 30 g of mineral salts, consisting mostly of chlorides, sulphate and phosphates of sodium, potassium, magnesium and calcium and this output must be recuperated by the intake of food stuffs. In case of the growing body, provision must be made for additional amounts of many of the elements to ensure adequate growth of the tissues. Underutilized vegetables are considered to be a reservoir of various mineral nutrients. Looking into the prevalence of high level of micronutrient malnutrition among the vulnerable sections in the developing countries and the increasing prevalence of chronic degenerative diseases globally, the need for exploration of underutilized foods is significant to overcome the nutritional disorders. The diet and food based approach in combating micronutrient malnutrition is essential for its role in increasing the availability and consumption of micronutrient rich foods (FAO, 1997). Increasing the utilization of green leafy vegetables (GLV) in our diet, known to be rich sources of micronutrients as well as dietary fiber can be a food-based approach for ensuring the intake of these nutrients. It is essential that the locally available GLV, which are inexpensive and easy-to-cook, be used in the diets to eradicate micronutrient malnutrition and also to prevent the degenerative diseases.

Table 2: Mineral Contents of Underutilized Leafy Greens (per 100 g of edible portion)

Leafy Vegetable	*Ash (g)*	*Ca (mg)*	*P (mg)*	*K (mg)*	*Na (mg)*	*Mg (mg)*
Trianthema portulacastrum	2.29	52	22	317	16.0	153
Celosia argentea	2.65	188	35	476	240.6	233
Polygala erioptera	0.77	41	43	125	9.8	57
Boerhaavia diffusa	2.91	330	27	381	—	167
Centella asiatica	2.06	174	17	345	107.8	87
Coleus aromaticus	1.06	158	16	138	4.7	88
Digera arvensis	3.54	506	63	604	—	232
Cocculus hirsutus	1.44	126	18	343	9.4	35
Commelina benghalensis	1.85	113	19	473	20.7	77
Amaranthus tricolor	2.51	239	33	433	84.0	253
Gynandropsis pentaphylla	1.80	151	38	360	15.7	77
Cucurbita maxima	3.04	302	34	368	12.0	150
Delonixelata	1.36	112	29	365	10.8	59

Source: Gupta *et al.* (2005).

Table 3: Trace Element Content of Underutilized Leafy Greens (mg per 100 g of edible portion)

Leafy Vegetable	*Iron*	*Zinc*	*Cop*	*Chro.*	*Mang.*
Trianthema portulacastrum	4.16	0.46	0.12	0.200	0.43
Celosia argentea	13.15	0.49	0.15	0.153	0.27
Polygala erioptera	4.76	0.68	0.15	0.029	0.18
Boerhaavia diffusa	7.83	0.44	0.22	0.040	—
Centella asiatica	14.86	0.97	0.24	0.046	—
Coleus aromaticus	2.62	0.33	0.12	0.022	—
Digera arvensis	17.72	0.57	0.16	0.243	0.23
Cocculus hirsutus	9.86	0.55	0.22	0.059	—
Commelina benghalensis	7.11	0.63	0.09	0.115	0.08
Amaranthus tricolor	15.01	0.60	0.09	0.140	—
Gynandropsis pentaphylla	4.84	0.57	0.13	0.078	—
Cucurbita maxima	4.38	0.62	0.19	0.049	—
Delonix elata	6.20	0.76	0.27	0.068	—

Source: Gupta *et al.* (2005).

(c) Antioxidants

Natural antioxidants that are present in herbs and spices are responsible for inhibiting or preventing the deleterious consequences of oxidative stress. Underutilized vegetables have a long list of antioxidants some of which are discussed in Table 4. Several underutilized vegetables contain free radical scavengers like polyphenols, flavonoids and phenolic compounds. Minerals like selenium, copper, manganese, zinc, *etc.* also act as antioxidants. However, vitamins like C, A and E are most popular antioxidants that plays crucial role in preventing peroxidation damage to the biological system by interacting directly with radicals like O_2^- and $OH^\bullet$. Antioxidant value of some of the underutilized vegetables is given in Table 4.

Table 4: Antioxidant Content of Underutilized Vegetables (mg/100g of the edible portion)

Underutilized Vegetable	*Ascorbic Acid*	*Tocopherol*	*Total Carotenoids*
Clerodendrum colebrookianum	89.6	25.34	13.42
Diplezium esculentum	106.2	26.2	14.0
Sida acuta	78.2	22.6	10.1
Sauropus androgynus	105.2	24.9	21.03
Solanum nigrum	92.0	21.0	6.83
Ceratopteris thalictroides	551.8	3.5	—
Monochoria vaginalis	438.1	1.1	—
Limnophila aromaticoides	844.6	14.8	—
Gynura cusimbua	87.6	18.89	15.23

The most important chain breaking antioxidant is α-tocopherol present in human membranes. Ascorbic acid and α-tocopherol both help to minimize the consequences of lipid peroxidation in membranes. Carotenoids are natural pigments which are synthesized by plants and are responsible for the bright colors of various fruits and vegetables. There are several dozen carotenoids in the foods th and most of these carotenoids have antioxidant activity. Carotenoids play a major role in the protection of plants against photo-oxidative processes. They are efficient antioxidants scavenging singlet molecular oxygen and peroxyl radicals. In the human organism, carotenoids are part of the antioxidant defense system. They interact synergistically with other antioxidants.

(d) Anti-nutritional Component

Vegetables are known to be important sources of protective foods. However, they may contain anti-nutritional factors that may affect the availability of nutrients. Anti-nutritional components of some of the underutilized vegetables are given in Table 5.

Anti-nutritional factors are known to interfere with metabolic processes such that growth and bioavailability of nutrients are negatively influenced. Phytate and oxalates have the ability to form chelates with di-and trivalent metallic ions such as Cd, Mg, Zn and Fe to form poorly soluble compounds that are not readily absorbed from the gastrointestinal tract thus decreasing their bioavailability. Phytate inhibits the functions of some digestive enzymes. High level of hydrogen cyanide has been implicated for cerebral damage and lethargy in man and animals.

Table 5: Anti-nutritional Component of Underutilized Vegetables (μg/100 g)

Underutilized Vegetable	*Phytic Acid*	*Cyanogens*
Clerodendrum colebrookianum	27.0	70.1
Diplezium esculentum	48.0	44.36
Sida acuta	66.00	54.16
Sauropus androgynus	48.0	10.7
Solanum nigrum	29.0	117.8
Gynura cusimbua	30.0	47.8
Solanum spirale	48.0	44.36
Zanthoxylum rhetsa	28.0	39.4

2. Medicinal Properties

A number of underutilized vegetables possess several desired medicinal properties. Drumstick (*Moringa oleifera*) is known for its medicinal properties since time immemorial and its leaves are used by physicians of traditional medicine for the treatment of hypertension. Bhalla *et al.* (1983) have reported the hypotensive action of alcoholic extract obtained from the dried drumstick leaves in patients with moderate to severe hypertension. Coleonol, a diterpenoid isolated at CSIR-Central Drug Research Institute, Lucknow from *Coleus forskohlii* may become a good antihypertensive agent (Rastogi and Dhavan, 1982). The anti-parkinsonian and prolactin reducing effects of an underutilized legume vegetable, *Mucuna pruriens* have been reported by Vaidya *et al.* (1978 a, b). Eating leaves of *Polygonum plebeium* as vegetable improves lactation. Sinha *et al.* (1984) have carried out a detailed macro and micro-scopical studies on the underutilized leafy vegetable *Alternanthera sessilis* and have given the inflorescence analysis, ash and extractive values of the leaf. Lal (1980) has investigated in detail the macro and microscopical features of the underutilized leafy vegetable *Portulaca quandrifida* and the inflorescence analysis of its powder, quantitative microscopy and micro chemical tests. The study is helpful to identify the drug when it is adulterated or substituted. Lal and Khan (1982) have studied the pharmacognosy of the stem of *Portulaca quandrifida* and *P. oleracea* and reported their distinguishable features. Trivedi *et al.* (1980) have studied the pharmacognosy of the starches from the rhizome of *Canna indica* and *Nelumbo nucifera*. Antifibrionolytic activity of the roots of an under-utilized leafy vegetable *Boerhaavia diffusa* has been found to be due to a phenol glucoside, punarnvoside. It was found effective in controlling IUD induced bleeding in monkeys (Srivastava *et al.*, 1980). List of under-utilized vegetables reported to have significant pharmacological activity is given in Table 6.

Table 6: Vegetables Possessing Medicinal Properties

Vegetable	*Family*	*Activity*
Boerhaavia diffusa L.	Nyctaginaceae	IUCD
Carica papaya L.	Caricaceae	Anti- androgenic
Cissus quandragularis L.	Vitaceae	Analgesic
Coleus forskohlii (Willd.) Briq.	Labiatae	CNS depressant, hypotensive spasmolytic
Costus speciosus (Koen. ex Retz.) Sm	Costaceae	Hypotensive
Cyamopsis tetragonoloba (L.) Taub	Papilionaceae	Hypoglycaemic, hypolipidaemic
Gymnema sylvestre (Retz.) Schult	Asclepiadaceae	Hypoglycaemic
Mollugo cerviana (L.) Ser.	Molluginaceae	Cordiostimulant
Momordica charantia L.	Cucurbitacea	Hypoglycaemic, hypolipidaemic
Trianthema portulacastrum L.	Aizoaceae	Analgesic, antipyretic, CNS depressant

3. Adaptive Ability under Adverse Abiotic Conditions

Underutilized vegetables possess unprecedented ability to stand against several stress conditions like vulnerable soil, moisture stress, and unfavorable temperature. Cassavas are adaptable to soils of low fertility and are able to tolerate low Ca, N and K. The edaphic tolerances give cassava an advantage over many other crops in acidic infertile soils. Sweet potato is fairly tolerant to Al and Mn and less tolerant to soil acidity. Yams are also less tolerant to soil acidity. In general, commonly grown vegetables are very prone to excessive moisture conditions where as some of underutilized vegetables are fully adapted to excess moisture; water logged or flood prone conditions. The underutilized vegetables which are catering the needs of flood affected people are given in Table 7.

The cold desert and sub-desert lies in the north of great Himalayas. Ladakh, the land between earth and sky, is known for its remoteness, difficult terrain and a short agriculture season. Nearly 50 per cent of its total land lies between 1,65,000 and 1,25,000 ft above mean sea level (msl). It is virtually unfit for human life and vegetation. The local people of Ladakh use these leafy vegetables afresh during April-June when there is no fresh vegetable. In addition, some leaves are solar dried during July-September for winter consumption. Besides above mentioned leafy vegetables, leaves of other species like *Allium sphaerocephalum*, *Allium* spp., *Aconitum heterophyllum*, *Polygonum viviparum*, *Sedum rhodiola*, *S. tibeticum* and *Scorzonera mollis* are eaten by the natives of Ladakh and Lahaul both in raw and cooked form (Singh and Gupta, 1988).

Table 7: Underutilized Vegetables for Flood Prone and Swampy Areas Conditions

Underutilized Vegetable	*Family*	*Parts Used*
Achyranthes aquatic	Amaranthaceae	Leaves
Alternanthera philoxeroides	Amaranthaceae	Leaves
Aponogeton chinatum	Aponogetonaceae	Leaves
Aponogeton crispum	Aponogetonaceae	Leaves
Aponogeton natans	Aponogetonaceae	Leaves
Colocasia sp.	Araceae	Corms
Enhydra fluctuans	Asteroceae	Potherb
Euryale ferox	Nymphaeaceae	Rhizome, young leaves
Ipomoea aquatic	Convolvulaceae	Leaves
Nelumbo nucifera	Nyphaeaceae	Rhizome
Nymphaea nouchali	Nyphaeaceae	Petioles and pedicels
Nymphoides cristata	Menyanthaceae	Stem, fruits and leaves
Otteliaa lismoides	Hydrocharitaceae	Stem, leaves
Sonneratiaca seolaris	Sonneratiaceae	Young edible fruits
Trapabi spinosa	Trapaceae	Immature kernel
Typha angustata	Typhaceae	Rhizome

4. Use as Rootstocks

Grafting is a simple method of propagation in which desired rootstocks are obtained to induce vigour, precocity, enhanced yield and quality, better survival under biotic and abiotic stress conditions (Pandey and Rai, 2003). Technique of grafting is in vogue in fruit crops since time immemorial, however, the raising of grafted plants of vegetables first started in Japan and Korea in the late 1920s with watermelon [*Citrullus lanatus* (Thunb.) Matsum. and Nakai] onto gourd rootstock (Lee, 1994). Eggplant was grafted onto scarlet eggplant (*Solanum integrifolium* Poir.) in the 1950s. Since, then the area of grafted vegetables has increased multifold. The proportion of the area in Japan producing grafted watermelon, cucumber, melon, tomato and eggplant reached 57 per cent of the total production area in 1960 and 90 per cent in 1990 (Oda, 1993). Presently, more than 95 per cent of the watermelon and oriental melon in Japan, Korea and Taiwan are grafted on squash and gourd rootstocks before transplanting. The number of vegetables being grafted is estimated to 540 million seedlings per year in Korea and 750 million in Japan (Lee *et al.*, 1998). A number of underutilized cucurbits and solanaceous vegetable possess immense potential towards the resistance for biotic and abiotic stresses and can be utilized as rootstocks. A list of the promising rootstocks is given in Table 8.

Table 8: Underutilized Vegetables as Source for Grafting of Vegetables

Scion	*Rootstock(s)*
Cucumber	*Cucurbita moschata*
	Cucurbita ficifolia
	Cucurbita maxima
	Sicyos angulata
Melons (for open field)	*Cucurbita* spp.
	C. moschata × *C. maxima*
	Cucumis melo
Melons (for greenhouse)	*Cucumis melo*
	Benincasa hispida
	Cucurbita spp.
	C. moshata × *C. maxima*
Watermelon	*Citrulus lanatus*
	Cucurbita maxima
	C. moschata
	C. moschata × *C. maxima*
	Lagenaria siceraria
Bitter gourd	*Cucurbita moschata*
	Lagenaria siceraria
	Luffa aegyptica
Eggplant	*Solanum torvum*
	Solanum melongena
	Solanum nigrum

Source: Pandey and Pandita (2002).

Conclusion

Underutilized vegetables embedded with rich nutrient potentials along with ability to stand against adverse climatic conditions may prove boon to all concerns - growers, consumers and environmentalists, provided that they are tamed properly. The possible reasons for the low utilization of underutilized vegetables in spite of their recognized importance are the lack of seeds, lack of information about their performance and input requirements, lack of information on how they can fit into production systems and non-viability of indigenous vegetable production like the major cultivated species of vegetables such as tomato, pepper, eggplant, cauliflowers, cabbage, *etc.*, whose improvement and seed production are taken care of by the private sector as well as government institutions, whereas the underutilized vegetables are a neglected lot. Most research on production covers only a few economically important vegetable crop species. Furthermore, the potential role of underutilized vegetables in sustainable agriculture through diversification of the agricultural environment has not been fully exploited (Ghosh and Kalloo, 2001).

References

Bhalla, I.N., Singh, S.P., Singh, N. and Bhargava, K.P. 1983. A clinical trial of an indigenous drug *Moringa oleifera* (Sahjan) on hypertensive patients. Abstr. II. *World Conference on Clinical Pharmacology and Therapeutics*, Washington, p. 95.

Ghosh, S.P. and Kalloo, G. 2001. Genetic resources of indigenous vegetables and their uses in South Asia. *Technical Bulletin* 4, Indian Institute of Vegetable. Research, Varanasi.

Gupta, S.A. Jyothi, Lakshmi, Manjunath, M.N. and Prakash, J. 2005. Analysis of nutrient and anti-nutrient content of underutilized green leafy vegetables. *Food Sci. Tech.*, **38**(4): 339–345.

Lal, J. 1980. Foliar pharmacognosy of *Portulaca quadrifida* L. *Bull. Med. Ethnobot. Res.*, **1**: 46–54.

Lal, J. and Khan, A.M. 1982. Pharmacognosy of the stems of *Portulaca quadrifida* and *P. olerocea*. *Proc. Indian Academy of Science.* (*Plant Sci.*), **91**: 235–240.

Lee, Jung Myung, Bang, HaeJeen and Ham, Hyunsook 1998. Grafting of vegetables. *J. Japanese Soc. Hort. Sci.*, **67**(6): 1098–1104.

Lee, J.M. 1994. Cultivation of grafted vegetables. I. Current status, grafting methods and benefits. *Hort. Sci.*, **29**: 235–239.

Oda, M. 1993. Present state of vegetable production using grafted plants in Japan. *Agr. Hort.*, **68**: 442–446.

Pandey, A.K. and Rai, M. 2003. Prospects of grafting in vegetables: An appraisal. *Veg. Sci.*, **30**(2): 101–109.

Pandey, A.K. and Pandita, M.L. 2002. Hi-Tech transplant production in vegetables. *Proceed. Int. Conf. Veg.*, Bengaluru, pp. 155–206.

Paroda, R.S., Kapoor, P., Arora, R.K. and Bhagmal 1988. *Life support Plant Species: Diversity and Conservation*, National Bureau of Plant Genetic Resources, New Delhi.

Rastogi, R.P. and Dhawan, B.N. 1982. Research on medicinal plants at the Central Drug Research Institute, Lucknow (India). *Indian J. Med. Res.*, **76**(Suppl. 27–45).

Singh, P. and Gupta, J.N. 1988. Life support species used by rural people under extreme environmental conditions. **In**: *Proceedings: Life support Plant Species*, (Eds.) R.S. Paroda, P. Kapoor, R.K. Arora and Bhag Mal. NBPGR, pp. 165–172.

Sinha, P., Arora, V.K., Ansari, M.S. and Wahi, S.P. 1984. Pharmacognostical studies of the leaf of *Alternathera sessilis* R.Br. *J. Sci. Res. Plant Med.*, **5**: 41–47.

Srivastava, K., Jain, G.K. Sarin, J.P.S., Khanna, N.M., Liedholm, P., Asteolt, B. and Bhadu, B. 1980. Effect of *Boerhaavia diffusa* and d-amino caproic acid on fibrinolytic activity. In: *IUD Pathology and Management*, Vol. 3, (Eds.) E.S.E. Hafez and W.A.A. Van O.S. M.T.P. Press Ltd., Lancaster, England, p. 121.

Trivedi, A, Mishra, P. and Srivastava, M.P. 1980. Pharmacognostical investigation of the starches from the rhizomes of *Canna indica* and *Nelumbo nucifera* Gaertn. *East Pharm.*, **23**(272): 185–186.

Vaidya, A.B., Rajagopalan, T.G., Mankodi, N.A., Antarkar, D.S., Tathed, P.S., Purohit. A.V. and Wadia, N.H. 1978a. Treatment of Parkinson's disease with the cowhage plant (*Mucuna pruriens* Bak.). *Neurol.*, **26**: 171–175.

Vaidya, R.A., Sheth, A.R., Aloorkar, S.D. Rege, N.R., Bagadia, V.N., Devi, P.K. and Shah, L.P. 1978b. The inhibitory effect of the cowhage plant (*Mucuna pruriens* Bak.) on chlorpromazine–induced hyperprolactinaemia in man. *Neurol.*, **26**: 177–178.

107

Kiwi: A Potential Fruit Crop, in North East India

K.K. Jindal[1] and Laxman Sharma[2]

[1]*Former Director Research,*
Dr YS Parmar Horticulture and Forestry Nauni, Solan, Himachal Pradesh
and Central Agricultural University Imphal, Manipur
[2]*Department of Horticulture, School of Life Sciences, Sikkim University, Gangtok*

North Eastern Hills have rich resources of soil and agro-climate making them an ideal region for horticulture industry. North East is sixth mega-centre of rich plant-biodiversity in the world and therefore a natural home of origin of many important horticulture crops. The wild relatives of apple, kiwi temperate fruits and nuts are found growing in natural form in NE forests, reflecting scope of wide adaptability of various temperate fruit crops. Fruits namely apple, pear, plum, peach are grown semi commercially in the NE-Hill states particularly in Arunachal Pradesh, Sikkim, Nagaland and Meghalaya. Kiwi fruit, walnut, and strawberry are also extended to new areas.

Commercial Temperate Fruits of North-East States

Among the temperate fruits, cultivation of apple, pear, kiwi, peach and plum is done on commercial scale. The best suited tract for apple and kiwi cultivation is in Arunachal Pradesh. Cultivation of pear is promising in Sikkim, while stone fruits like peaches and plums are mostly concentrated around Maram and Ukhrul area of Manipur and around Shillog in Meghalaya, which constitute sub-temperate areas. In the sub-tropical hills low chilling varieties of apple, peach, plum, apricot and pear are getting popularity as crop diversification.

The lack of quality planting material, package of practices, modern technology and trained manpower are the major constrains in enhancing the productivity of temperate fruits in general and stone fruits and kiwi in particular. The technologies generated in Dr Y S Parmar University of Horticulture and Forestry, Solan have been transferred with the co-operation of scientists, State Departments of Horticulture and farmer to farmer interaction facilitated by national agencies like National Horticulture Board and APEDA. The technologies available for the farming communities in NE, particularly in high hills of Arunachal Pradesh and Sikkim for low chilling fruits such as Kiwi fruit are discussed here in.

It is pertinent to mention that recently kiwifruit has caught imagination of producers in Arunachal, Sikkim, Nagaland, Manipur and Meghalaya. However, productivity of temperate fruits is low. Causes of low productivity visualized are selection of areas for expansion in high hills and valleys having spring frost and temperature fluctuations during flowering time adversely affected fruits yield. Trees hardly receive desired training and pruning resulting in poor fruiting. Pollinizing varieties are insufficient and disproportionate resulting in the lack of pollination and fruit set. Lack of quality planting material and technical knowhow have added serious dimensions to productivity problems.

Present Scenario of Temperate Fruits in North East

In North East Region, large parts of Arunachal Pradesh, Sikkim and some part of Nagaland and Meghalaya fall under sub-temperate zone. Whereas, low hills of Assam, Meghalaya, lower hills of Arunachal

Pradesh and Sikkim fall under sub-tropical zone, while part of Tripura and Mizoram under coastal/tropical zone. Areas suitable for temperate fruit growing in N.E. states are presented in Table 1 and area and production in Table 2.

ICAR's Roving Team Road Map for Temperate Horticulture in the North East Hills

1. Asses the present status and future scopes of temperate horticultural crops, fruits and nuts in particular, in the high hills and mountains of North Eastern States.
2. Analyze climate change parameters and suggest R&D intervention areas for successful commercial cultivation of apple, nut crops, kiwi and wine grapes in suitable pockets.
3. Suggest immediate actions under MM I and MM II of Horticulture Technology Mission for Hill States.

Temperate fruits, namely, apple, pear, plum, peach are grown semi-commercially in the NE Hills states, particularly in Arunachal Pradesh, Sikkim, Nagaland and Meghalaya. Walnut, strawberry and kiwi also experienced area expansion. Apple in rain shadow belts in Arunachal Pradesh with high density planting system has been reported to enhance productivity in a big way. Similarly, Kiwi has caught imagination of producers in Arunachal Pradesh, where not only productivity is good, but also the quality of the produce is reported to compare favourably with Australian kiwi fruits that are in good demand in the international market. The road map for the establishment of a sound temperate fruit industry in Arunachal Pradesh and Sikkim suggested below:

Thrust Areas in Apple Production

A) Arunachal Pradesh

Arunachal Pradesh is the only leading state in the NEH suitable for apple production. In more recent times Kiwi and Walnut are showing promise in the temperate- alpine zone, particularly in the rain shadow belts. High hills of West Kameng, Tawang, Ziro, Anini, Upper reaches of Anjaw, Upper Subansiri and upper Siang are suitable areas for apple.

B) Sikkim

Sikkim had a meagre area of 200 ha in apple with limited potential for expansion. The existing orchards were sick, infected with scab and cankers and had no systematic training and pruning of trees. They needed rejuvenation, proper training and technological interventions for management of pests and diseases. The other temperate crops which have potential are kiwi fruit, pear and walnut. Apple has not been included as a focused crop under Horticulture Technology

Table 1: Areas Suitable for Temperate Fruits in different States of North-Eastern India

State	*Location(s)*	*Approximate Range of Elevation (amls)*	*Fruit Crops that can be Successfully Tried*
Arunachal Pradesh	Kameng,Tawang, Siang, Trirap and Lohit district	900-4000 m	Apple, peach, plum, pear, apricot, walnut, chestnut, cherry
Meghalaya	Central plateau-Khasi and Jaintia hills districts	950-1900 m	Plum, pear, peach, apricot,persimmon
Manipur	Sekmai, Maram, Tadubi, Mao,Ukhrul, Tengnoupal	900-2000 m	Peach, plum, pear, apricot,persimmon
Nagaland	Mokokchung, Wokha, Tuensang, Kohima. Phek districts	1000-3000 m	Peach, plum, pear, apricot

Table 2: Area and Production of Temperate Fruits in North-Eastern States of India

Fruits	*Arunachal Pradesh*	*Manipur*	*Meghalaya*	*Mizoram*	*Nagaland*	*Sikkim*
Apple	A 12,800 P 10,000	– –	– –	– –	A 750 P 150	A 50 P <50
Pear	A 125 P 681	A 160 P 1000	A 30 P 185	A 60 P 210	–	
Kiwi fruit	A 4,052* P 4,956*	A 20*			A 2000* P – NA	A 98 P 78**
Low chill plum, peach and others	A 430 P 961	– –	– –	A 36 P 99	A 327 P 592	

Based on ICAR Roving Team (2012-13).

A=Area in hectares; P=Production in tonnes.

* Arunachal and Manipur Kiwi data from MIDH Database 2014-15. * Nagaland Feb 2016 – Personal communications.

* Sikkim Organic Mission (2013).

Mission and State Govt. to come forward with specific programme for temperate fruits, Asiatic Pear and Kiwi in particular.

Thrust Areas in Kiwi Fruit Production

1. Arunachal Pradesh

The Kiwi fruit is a berry, oblong in shape with brown skin similar to Sapota (Chikoo). The fruit has a refreshing, delicate flavour, pleasing aroma and high nutritive and medicinal values with low calories. The centre of origin of Kiwifruit lies in China and hence is crown as Chinese gooseberry. It is grown well in the mid hills under moderate to high rainfall conditions with a wide range of adaptability having no serious pests and diseases. The fruit ripens during October/ November which is a lean period for other fruits in the market. During this period rains stop and temperature slides down providing less chance of road blockade and spoilage thereby ensuring possibility of supply for long periods without creating glut and fetching high prices.

Although the Kiwifruit is of recent introduction in the state it has assumed tremendous popularity and preference among the growers as well as consumers due to its favourable properties for easy maintenance and marketing besides having high nutritional and medicinal values. The Kiwifruit of the state has already attained commercial identity not only in the local markets but also in the national as well as international markets. As the Kiwifruit originated in China on the northern front of the eastern Sub Himalayan ranges, the inherent potentiality for its commercial cultivation on the southern Sub Himalaya ranges is obviously evident. This is the future crop of the state which could provide sustenance to the economy of the rural masses and the state.

Diversification in any farming system provides sustainability to the major occupation. Cultivation of minor temperate fruit crops can impart sustainability to the promotion of temperate horticulture venture.

2. Sikkim

West Sikkim is taking lead in kiwi fruit production. More and more farmers in Richenpong, Deythang and Soren areas are growing the fruit with support from State Department of Horticulture under Horticulture Technology Mission guidance. There are now many progressive kiwi fruit growers of Dehap, Singyang, Narboo, Yuksam, Neshe, Kaluk, Uttarey, Yangum area of west Sikkim. It is interesting is note that wherever Kiwi is grown, large cardamom is there. This means that large cardamom can be inter-planted in new kiwi orchard.

Precision Farming in Kiwi Fruit

In general,the precision farming of Kiwi fruit in Arunachal and Sikkim require scientific manual of training and pruning, quality improvement of fruits besides proper post-harvest management practices as major attributes to sustain its cultivation to meet the international standards. The broad outlines are given below:

Training and Pruning of Kiwi Plants for Quality Production

The exotic kiwifruit has shown tremendous potential in the recent years and has been assessed as a model fruit for diversification of mid hill horticulture in West Sikkim with scope in Upper Gyathang, Dzongu areas of North Sikkim Ranipul, Rhenok, Pakheyang and in East and Kewzong and Zaubari in South under Horticulture Technology Mission. The transfer of technology on the improved cultivation in production with main emphasis on training and pruning will further enhance the scope of kiwifruit in mid hills of other districts of Sikkim where stone fruits have shown declining trend in the recent past due to their perishable nature and poor marketability. In order to have high yield of quality fruit of Kiwi a training manual is required for proper training and pruning of Kiwi fruit plants to boost the productivity of quality fruits with international marketing standards and grades.

Pollination

Wind and insects play a significant role in pollination, however, introduction of honeybees in the orchards further increase fruit-set and size of the fruits. Palmer Jones and Clinch (1974) recommended 9 colonies per hectare for better pollination. Male flowers produce viable pollen for 2-3 days after opening, after which they become senescent and die. Female flowers in contrast are receptive for 7-9 days after opening even when the petals have started falling. Besides, insect and wind pollination, hand pollination is essential to get fruits of better size and quality.

Canopy Management

Training

Kiwifruit is a vine like grape, thus require similar training structure but more stronger than grape. In T-bar trellis system, the pillars of iron or concrete about 1.8 m in height above the ground level are erected at a distance of 6 m from each other in a row in straight line. A cross arm (1.5 m) is fixed on each pole, which carries 5 outrigger wires at a distance of 45 cm each. Vine is trained upto the wire as single stem, than two leaders in opposite direction along the center wire are selected or developed. From these permanent leaders, temporary fruiting arms 25-30 cm apart are selected at right angle along both sides of each leader.

Training of vine on pergola system is similar to that of T-bar. A flat topped network of criss-cross wires is prepared on the erected pillar. This system is costly to prepare but gives higher yield.

Pruning

The fruit is developed on current season's growth arising from one year old shoot. Only the basal buds of nodes 4 to 12 on current season's growth are productive. Vine grows 2-3 m every year, whichbecome over-crowded if not controlled by both summer and winter pruning. The shoots developed on older wood by heading back will not fruit normally in the first season. Vine pruning is carried out in such a way that the fruiting areas are available every year requiring the wood to be young. This is achieved by 3 to 4 years lateral replacement system.

In dormant pruning, the fruiting lateral is cut back to 2 vegetative buds beyond the last fruit. In the second year these vegetative buds produce the fruiting shoots, which is pruned again. The arm on lateral shoots are pruned and allowed to fruit for 3 to 4 years. After this the lateral is removed from the main branch and other laterals are selected and pruned accordingly so that the balance between vegetative and reproductive growth is maintained for the continuity in the fruit production. In summer pruning, a fruiting shoot is headed back beyond 6-8 buds from the last fruiting bud during June-July.

Crop and Quality Regulations

All the cultivars of kiwifruit except Hayward bears heavily every year. This heavy crop create a severe competition between the fruits for water, nutrients and photosynthates, which leads to production of small sized fruits. Therefore, to harvest quality crop of good size, hand thinning is essential, as chemical thinning is ineffective. In a study on thinning in kiwifruit, flowers or fruit thinning (20 per cent thinning) to the extent of retaining 5 to 6 fruits/flowering shoot produced more fruits of A grade without any adverse effect on total yield. In hand thinning only lateral flowers or fruits are removed.

Maturity, Harvesting and Marketing

In kiwifruit, determination of optimum fruit maturity is difficult because no change take place into the skin or flesh colour or size of fruit due to maturity. A maturity index of 6.2°B TSS is considered satisfactory for fruit harvest. Days from full bloom to harvest in lower elevations ranges between 190+4 and 205+4 to 223+4 at higher elevation. Beside this at maturity the hairs present on the fruit skin are removed very easily and can be used to judge the harvesting or maturity.

After harvesting, fruits can be stored for 3-4 weeks at room temperature and for 4-6 months in cold storage at 0°C with 80-90 per cent relative humidity. Fruits can be processed into jam, juice and squash.

Strategies and Priority Areas in North East

In the North East kiwi fruit is experiencing expansion among the temperate fruits. However it has specific production problems which need to be addressed.

In kiwi fruit major production constraints are:

(a) Lack of genetically superior planting material.

(b) High cost of plantation whole establishment.

(c) Lack of standardized techniques of cultivation like pruning cum training.

(d) Since Sikkim is an organic state, package of practice of organic cultivation is *adhoc.*

(e) Lack of use of proper proportion of pollinizers, practice of hand pollination.

(f) Concept of hand thinning and use of biostimulants to increase fruit size is non-existent.

The priority areas and activities suggested for precision farming in kiwi aimed at crop improvement and improved agro techniques:

A. Introduction, Evaluation and Mass Multiplication of Suitable Varieties which can bear Biotic and Abiotic Stress

The orchards in West Sikkim has mainly, Monty, Allison and Bruno. Hayward variety which has less area is now being introduced widely under Technology Mission.

B. Training and Pruning

The technique of trellis on T-bar is very scientific and needs precise practice to be performed by farmers so as to have maximum fruiting wood. The farmers are to be educated practically.

C. Crop and Quality Regulation

All cultivars of kiwi fruit except Hayward bear heavy crop every year. The heavy crop creates a severe competition between fruits for nutrients and water which leads to production of small size fruits. Poor management practices, improper ratio of pollinizers, their placement in plantations and lack of hand pollination also contribute to small fruits.Therefore to increase quality kiwi crop of good size hand thinning is essential. The studies conducted in thinning of fruits to 4-6 fruits per bunch immediately after fruits set and treatment with CPPU bio-stimulant/plant growth promoter standardized by Dr YS Parmar University of Horticulture and Forestry Nauni, Solan need to be replicated in kiwi orchards of Sikkim.

D. Improvement for Propagation and Quality using Biotechnological Tools

Quality production of planting material of kiwifruit through improved propagation technologies and improvement of fruit growth by plant hormones.

E. Guidelines for Crop and Quality Regulation (Hand Thining)

All the cultivars of kiwi fruit except Hayward bear heavily every year. This heavy crop creates a severe competition between the fruits for water and nutrient material Which leads to production of small sized fruits. Therefore, to harvest quality fruits hand thinning of small fruits is essential to the extent of retaining 5-6 fruits per bunch or flowering shoot will produce more 'A' grade fruits without having any adverse effect on total quality yield.

Table 3: Estimated Economics of Kiwi Cultivation (based on a case study, HP):**

A.) Estimated cost of establishment of 100 plant Kiwi block including iron T-bar supporting structure, cost of plants and plantation, say Rs 3.5 Lakh. Detailed below:

1)	Cost of leveling of the land: 10 Hours – JCB @ Rs. 800 per hour + Fuel	= Rs. 10,000
2)	Digging of pits: 20 man days X Rs. 300 per man day	= Rs. 6,000
3)	Training structure and erection: Rs. 3000 X 100	= Rs. 3,00,000
4)	Cost of plants: Rs. 50 X 100	= Rs. 5,000
5)	Bamboo staking for support: to young plants:	= Rs. 5,000
6)	Miscellaneous expenses	= Rs. 15,000
TOTAL:		= Rs. 3,41,00, say 3.5 Lakh
Cost of Fertigation unit and installation:		= Rs. 1.5 Lakh
GRAND TOTAL FOR INITIAL ESTABLISHMENT OF BLOCK :		= Rs. 5 Lakh
Annual maintenance expenditure including FYM, fertilizer, labour [60 man days @ 300]		= Rs. 1 Lakh*

* Therefore the cost per plant shall be around 1000 rupees initially, with annual maintenance @ 20%.

** Input cost are subsidized in the North East States under Horticulture Technology Mission

B) Yearly production and gross income per plant

Age	*Production*	*Amount [@ Rs 60/Kg, A Grade Fruit at UHF Nauni Farm Gate]*
1st and 2nd year	Nil	Nil
3rd year	1 kg	Rs. 60
4th year	3 kg	Rs. 180
5th year	20 kg	Rs. 1,200
6th year	30 kg	Rs. 1,800
7th year	40 kg	Rs. 2,400
8th year	50 kg	Rs. 3,000 [productive phase]
9th year	60 kg	Rs. 3,600
10th year	70 – 80 kg	Rs. 4,200–4,800
11th to 25th year	70 – 80 kg	Rs. 4,200–4,800
25th to 30th year	60 - 70 kg	Rs. 3,600–4,200
30th to 35th year	50 – 60 kg [Mostly B and C grade]	Rs. 2,500–3,000
35th to 40th year	Decline in productivity	<Rs. 2,000
After 40th year	Needs total replacement	Unproductive

Note: The benefit to cost ration (BC) is based on empirical figures, provided by HP kiwifruit growers.

Prospects

Kiwifruit has a bright prospect in our country in general and North East in particular. It has been assessed as one of the important future commercial fruits. It provides high return per unit area and the farmers can earn about Rs. 4 to 5 lakhs per hectare annually. Kiwifruit bears heavily every year with no crop failure. The fruit is highly acclaimed for the nutritive and medicinal value. It holds a wealth of health giving property, thus recommended for patients suffering from diabetes and heart diseases. In kiwifruit, no serious pests and diseases attack have been observed, thus it has a better scope to become commercial eco-friendly fruit crop of the country. Due to hard nature of the fruit with hairy skin surface, the fruits are not damaged by any bird and even not by the monkeys. It has a longer shelf-life and can be stored for one month in open at room temperature and for 4-6 months in cold storage.

Conclusion

It is strongly recommended that kiwi fruit production be taken up in commercial scale in all the states of North East except Assam and Tripura, if the basic requirements of development like appropriate self-reliant technologies, infrastructure and techno-physical marketing facilities are addressed in a mission mode with long term vision with the active participation of North East Horticulture Technology Mission of different states supported by Departments of Agriculture, Cooperation and Farmers Welfare Ministry of Agriculture Govt. of India, Central Institute of Horticulture, Medziphema, and Assam Agriculture University.

The temperate fruits of the North Eastern Hill states, particularly kiwi fruit, could be developed into a vital industry for sustainable economy of hilly region rural people in particular and North-east region in general for their livelihoods and export to neighbouring countries.

References

Chandel, J.S. and Rana, Ramesh. 2004. Kiwi fruit. **In**: *Recent Trends in Horticulture in the Himalayas*, (Eds.) K.K. Jindal and R.C. Sharma. Indus Publication Co., New Delhi, pp. 147–154.

Jindal, K.K. 2003. Impact of changing climate on productivity of temperate fruits. **In**: Proceedings of *"The 7th International Symposium on Temperate Zone Fruits in Tropics and Sub-tropics"* (*ISHS TZFTS Working Group Acta Horticulture*) held at UHF, Nauni, Solan, October, pp. 692 and 696.

Jindal, K.K. 2015. Potential and strategies for development of temperate fruits and nuts in the North-Eastern region. **In**: *Temperate Fruits and Nuts: A Way Forward for Enhancing Productivity and Quality*, (Eds.) K.L. Chadha, Nazeer Ahmed, S.K. Singh and P. Kalia. Daya Publishing House, pp. 43–65.

Jindal, K.K. 2016. Prospects and strategies for production of higher quality and productivity of temperate fruits with special reference to Arunachal Pradesh and Sikkim states of North-East. **In**: Proceedings of *"National Conference on Horticulture in North Eastern Region"* (Souvenir), (Ed.) A.K. Pandey *et al.*, held at CAU, Pasighat, January, pp. 56–58.

Jindal, K.K. and Sharma, Laxuman. 2016. Emerging horticultural crops of North Eastern states of India suitable for precision farming to improve livelihoods of farming community with special reference to kiwi fruit. *Life Sci. Int. J.*, **3**(2): 1–9.

Jindal, K.K. and Singh, R.K. Dileep. 2016. Climate change and adaptive technologies: Low chilling temperate fruits for sub-tropical and tropical zones of North-East Indian states to improve livelihoods of farming community: An overview. **In**: Proceedings of *"National Seminar on Integrating Agri–Horticlture and Allied Research for Food and Nutritional Security in the Era of Global Climate Disruption"* (Souvenir) (Ed.) N. Prakash *et al.*, held at ICAR Research Complex, Imphal, Manipur, March, pp. 71–75.

Jindal, K.K. and Singh, R.K. Dilip. 2015. Low chilling temperate fruits for sub-tropical and tropical zones of North-East states. **In**: Proceedings of *"International Symposium on Next Generation Approaches for Sustainable Development of Hill and Upland Horticulture"*, held at Sikkim University, Gangtok, November, pp. 43–46.

Jindal, K.K., 2015. Impact of changing climate on the productivity of economically important plant species with special reference to apple, ways and means to overcome weather vagaries to improve livelihoods of farming community. **In**: Proceedings of *"International Symposium on Next Generation Approaches for Sustainable Development of Hill and Upland Horticulture"*, held at Sikkim University, Gangtok, November, pp. 16–21.

Jindal, K.K., 2015. Potential and strategies for production of higher productivity and quality temperate fruits and nuts in North-Eastern regions. **In**: Proceedings of *"National Conference on Temperate Fruits and Nuts– A Way Forward for Enhancing Productivity and Quality"*, held at CITH Srinagar, November, pp. 43–66.

Jindal, K.K., 2016. Possibilities and strategies in temperate fruits production with special reference to precision farming in kiwi fruit production in North-East states: Case study of Sikkim and Arunachal. **In**: Proceedings of *"National Seminar on Integrated Development of Horticulture in Sub-Tropical and Hill Region"*, held at HRS Kahikuchi Guwahati organized by CIH, Medziphema, Nagaland and Assam Agriculture University, February, pp. 120.

Jindal, K.K., Chauhan, P.S., Mankotia, M.S. 2002. Apple production in the changing climate scenario: Issues and constraints. **In**: *Proceeding Enhancement of Temperate Fruit Production in the Changing Climate*, (Eds.) K.K. Jindal and D.R. Gautam. ICAR Publications, pp. 46–55.

Rathore, D.S. 2001. Kiwi fruit production: Present status and future prospects. **In**: *Productivity of Temperate Fruits: Issues and Strategies*, (Eds.) K.K. Jindal, D.R. Gautam. ICAR–UHF Publications, pp. 355–364.

108

Revisiting Citrus Genetic Resources of Northeast India

S.K. Malik[1]*, I.P. Singh[2], Susheel Kumar[1,3] and Rekha Chaudhury[1]

[1]ICAR-National Bureau of Plant Genetic Resources,
Pusa Campus, New Delhi – 110 012
[2]ICAR-Central Citrus Research Institute,
Amaravati Road, Nagpur – 440 033, Maharashtra
[3]Rubber Board Office, Parlakhemundi, Gajapati,
Odisha – 761 200
E-mail: skm1909@gmail.com

The domestication and distribution of edible citrus types began several thousand years ago in Southeast Asia and spread globally, following ancient land and sea routes (Albert Wu *et al.*, 2015). South-east Asia, Australia, the intervening islands between Asia and Australia as well as central Africa are recognised as important centres of origin of *Citrus* and related genera (Tanaka, 1958; Swingle and Reece, 1967; Mabberley, 2004; Jena *et al.*, 2009). Despite being extensively studied because of its commercial relevance, the origin of cultivated *Citrus* species and the history of its domestication still remain an open question (Carbonell-Caballero *et al.*, 2015). The domestication of *Citrus*, is poorly understood (Albert Wu *et al.*, 2015). Cultivated types are selections from, or hybrids of, wild progenitor species, whose identities and contributions remain controversial.

In India, north-eastern Himalayan region and foothills of the central and western Himalayan tracts are rich sources of *Citrus* genetic diversity (Malik *et al.*, 2006; 2013). The north-eastern region of India includes the states Assam, Arunachal Pradesh, Meghalaya, Manipur, Mizoram, Nagaland, Tripura and Sikkim and this area is considered as home and reservoir of various cultivated and wild *Citrus* species (Tanaka, 1958; Singh and Chadha 1993; Chadha, 1995). In this region, many *Citrus* species originated and later dispersed to various parts of the world where substantial improvement of citrus industry has been taken place. Due to its unique combination of diverse soil-physiographic and climatic set up, the region offers a number of *Citrus* species with extensive variability (Hazarika, 2012). In north-east India, Meghalaya is the leading state in both area and production, followed by Manipur, Assam, Tripura, Mizoram, Nagaland, Arunachal Pradesh and Sikkim (Singh, 2001).

Citrus genetic diversity collection from northeastern part of India was taken up intensively by Tanaka (1928, 1937) and Bhattacharya and Dutta (1956) during 3rd to 5th decade of the last century. Most of the new *Citrus* species were identified, collected and documented from north-eastern parts of India during this period. Since then, no significant information has been added to our knowledge about the occurrence of any additional species of *Citrus* from this part of India. However, recent extensive surveys and explorations during last 2-3 decades by ICAR-National Bureau of Plant Genetic Resources, New Delhi and ICAR-Central Citrus Research Institute, Nagpur enumerated loss of some species/cultivars from various pockets where these were earlier reported to occur and also several new local types and new locations of already existing species and unknown citrus types were reported from north-east India (Singh and Singh, 2003; Malik, 2006; Malik *et al.*, 2012; Malik *et al.*, 2013).

In this context there is an urgent need to revisit present status of *Citrus* genetic wealth of north-east India using modern tools of survey, collection, characterization, evaluation, conservation and utilization. Citrus gene sanctuary designated in the year 1981 in the buffer zone of Nokrek Biosphere Reserve in the Meghalaya has not been taken up seriously for protection and management of existing Citrus Germplasm. Vast genetic variability of various *Citrus* species existing in this area has been vanishing due to increasing human intervention and slow regeneration capacity of some of the species. Increasing global emphasis on *in situ* and *in situ* on farm conservation of plant genetic resources due to its dynamism and support to ecosystem services should be taken as guidance and direction to revive this gene sanctuary which is truly the treasure house of citrus diversity in north-east India. In the present paper current status of citrus genetic resources in north-east India and urgent emphasis to revisit this indigenous diversity and need for reviving and strengthening the Citrus Gene Sanctuary have been highlighted.

Present Status of Germplasm in NE India

Northeast India has an enormous diversity of *Citrus* genetic resources, both in cultivated and wild types, comprising 24 species, one subspecies and 78 varieties, including wild, endangered and endemic species (Sharma *et al.*, 2004; Malik *et al.*, 2013). Some of the wild species, namely *C. indica*, *C. ichangensis*, *C. latipes*, *C. macroptera*, *C. assamensis*, *C. megaloxycarpa* and *C. medica* have special importance in Indian germplasm being of use in socio-economy and health of local people (Malik *et al.*, 2012), besides utility in citrus industry and improvement programmes. In a systematic account on Indian *Citrus*, Nair and Nayar (1997) included 18 species, which comprised of eight species under Subgenus *Citrus*, three species under Subgenus *Papeda* and seven other indigenous species of probable hybrid origin and uncertain taxonomic affinities.

ICAR-National Bureau of Plant Genetic Resources in collaboration with ICAR-Central Institute of Citrus Research has taken up a comprehensive programme on collection and conservation of Indian Citrus germplasm. During last several years systematic explorations have been taken up to collect, characterize and conserve the Citrus genetic diversity from different states of north-east India. This structured programme on revisiting the Citrus genetic resources of India especially north-east India have been undertaken after more than six decades of publication of monograph by Bhattacharya and Datta (1956). During these explorations vast diversity of various *Citrus* species representing important rootstocks, mandarins, acid lime, pummelo, grapefruit and sweet oranges along with the indigenous wild species were collected (Singh and Singh, 2003; Malik *et al.*, 2006; Malik *et al.*, 2012). During these explorations major emphasis was on collecting diversity of wild, semi-wild and endangered *Citrus* species such as *C. indica*, *C. assemensis*, *C. ichangensis*, *C. megaloxycarpa*, *C. macroptera* and *C. latipes* from entire northeastern India and for which specific explorations were undertaken. Since last four decades germplasm representing sizable diversity of cultivated, wild, rootstocks and several intermediate Citrus types have been collected during various explorations and national identity for more than 1400 accessions has been allotted by NBPGR, out of which 678 are collected from the north-eastern states (Figures 1A, B).

Revisit Citrus Genetic Diversity in NE India

During the extensive survey and collections in northeast India several unidentified Citrus types

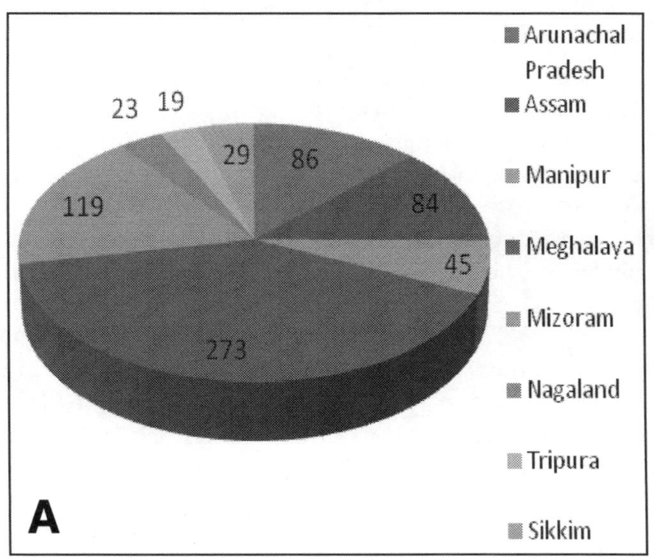

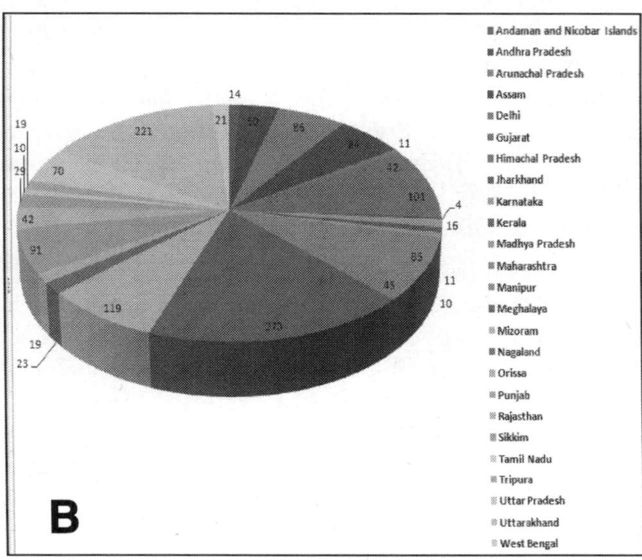

Figure 1: A. Citrus collection made for India; B: Citrus collection made from northeast India.

were reported with uncertain identity especially in Khasi and Garo hills of Meghalaya the area, which falls under the Citrus gene sanctury. Several such unidentified variants or probable natural hybrids of *Citrus* still exist in northeastern India with no taxonomic identity. Several of these variants like Memang athur, Themachhi, Chimbal, Chamba, Seranga, Salonga, Atol, Bamsim, Kagzi-wak-ma, Soronga, Kampil, Kim, *etc.* are established local cultivars and used for various economic purposes (Figure 2). Based on the morphological characterization, identification of these unknown variants of *Citrus* is very difficult due to plasticity in the morphological characters. In such cases, molecular markers, particularly DNA-based markers, *i.e.* Random Amplified Polymorphic DNA (RAPD), Inter Simple Sequence Repeat (ISSR), Simple Sequence Repeat or microsatellites (SSR), Directed Amplification of Minisatellite DNA (DAMD), Amplified Fragment Length Polymorphism (AFLP), *etc.* are highly useful for understanding level of genetic diversity, identification of closely related varieties/hybrids/cultivars, inferring systematics, phylogeny, management and improvement of crops, conservation and management of endangered species.

The taxonomy of *Citrus* is controversial, mainly due to the sexual compatibility between *Citrus* species and its related genera, apomixis (adventive nucellar polyembryony), high frequency of bud mutations, long history of cultivation, and wide dispersion (Frost and Soost, 1968). Consequently there has been no consensus among the taxonomists as to the actual number of species that constitute the genus. Swingle's system of *Citrus* classification (Swingle and Reece, 1967) recognized 16 species (10 sub species in subgenus *Citrus* and 6 subspecies in subgenus *Papeda*) on the basis of biological species concept. Tanaka (1954, 1977), based on horticultural species concept, recognized up to 162 species in two subgenera: *Archicitrus* and *Metacitrus*. Advanced studies based on the biochemical and morphological characterization, suggested that there are only three basic species, *i.e.* citron (*C. medica* L.), mandarin (*C. reticulata* Blanco), and pummelo [*C. maxima* (Burm.) Merr.] within the subgenus *Citrus*, whereas the other edible citrus, *e.g.* lemon, lime, sour orange, sweet orange, grapefruit, *etc.* have been considered as apomictically perpetuated biotypes with probable hybrid origin (Scora, 1975).

As far as germplasm management and utilization is concerned effective mapping of citrus genetic resources of northeast India needs to be undertaken. Presently, available modern tools of Geographic Information System (GIS) technology may not be much proficient in the effective mapping of genetic resources of citrus. Worldwide for mapping of biodiversity, mostly background data in the form of passport data or herbarium data has been used. Effective mapping in diverse genus like citrus with several species and genotypes available in the northeast and northwest part of India need correct identification of germplasm which is still lacking in many cases. Looking to the diverse issues faced in management of citrus genetic diversity and large area still unexplored in the remote parts of north-

Figure 2: Unidentified and Ambiguous Taxa: Need to Revisit and Ascertain Taxonomic Identity of Northeast Indian *Citrus*.

east India, there is need to have well planned targeted explorations for collection of germplasm. Recent report of existence of *C. indica* in the Siang valley of Arunachal Pradesh and Manipur is a relevant example of need for extensive explorations and collection missions in the northeast part of India. There is also need to have extensive ground surveys of potential areas where citrus diversity is expected to occur to have effective mapping of citrus genetic resources. Once the ground surveys are completed the modern tools using available or newly developed models can be effectively utilized for mapping of existing diversity. The use of Weighted Value Scheme in assessing the level of population size and frequency of species or genotypes would be of high significance especially in case of citrus diversity mapping of northeast India where concentrated species and genotypic diversity is available. It is, therefore, desired to have a consolidation of available information at various levels and about the diverse species at single platform to synthesize and use this data for mapping the Indian diversity using modern tools and models.

There is need to study taxonomic status of ambiguous taxa of NE Indian *Citrus* to arrive at well-founded conclusions on their taxonomy, origin and genetic diversity, which would facilitate its utilization for citrus improvement programmes. An integrated approach involving morphological, micro-morphological and molecular characterization of these taxonomically uncertain local cultivars, natural wild types and taxa with ambiguous identity at *Citrus* species or sub-species level need to be undertaken. Following are the taxa which require further study and revisiting.

i) **Memang athur** is one of the important natural hybrid and also known as Memang narang big in South Garo hills, showing very close affinity with *C. indica*. Its fruit has high medicinal value, used for several alimentary disorders, small pox, antidote, typhoid, *etc.* The natural populations of Memang athur are found only in Daribokgre (East Garo hills) and Ronwak (South Garo hills) in Meghalaya. However, numbers of individuals are very less due to low regeneration capacity. Kumar *et al.* (2010) have studied Inter simple sequence repeats (ISSR) polymorphism in *Citrus indica* Tanaka (Rutaceae), an endemic and threatened wild species, was examined along with three other closely related wild taxa (*C. medica* L., *C. latipes* (Swingle) Tanaka, and Citrus sp. 'Memang athur') by analyzing 53 representative accessions sampled from North-east India. The study confirmed the "Memang athur" being intermediate of *C. indica* and *C. medica*.

ii) **Themachhi** is found naturally in Daribokgre Sasatgre Chandgre, Mandalgre, Rogsak, and asanangre of East Garo hills in Meghalaya. It is considered as a wild citron, which is morphologically very much similar to *C. medica* except presence of articulation in leaf and shape of lamina. Recently, Kumar *et al.* (2014) based on results derived from morphology and molecular characterization indicated that 'Themachhi' is very close to *C. medica* and considered as natural cultivar of typical citron, which is now being domesticated and cultivated by the tribal people in Garo hills. Sequence analysis of psbM-trnD spacer region of cpDNA of Indian taxa of *Citrus* was successfully utilized for resolving taxonomic identity of 'Themachhi'.

iii) **Chambal** is found in East Garo hills especially Rengsangre and Did are adjoining villages. Frequency of individuals of this natural variant is very low. Chambal is a medium size tree and very much similar to *C. macroptera* based on broadly winged petiole and fruit characters.

iv) **Chinara**, a natural variant of *Citrus* found in Daribokgre, Mandalgre, Rongsangre and sasatgre of East Garo hills. 2 to 4 individuals have been recorded in these localities during the trip. Chinara is morphologically very close to lemon group.

v) **Soronga** is natural variant of *Citrus*, found in Nokatgre of West Garo hills and Mandalgre of east Garo hills. It is a large tree having broadly lanceolate leaf with narrowly winged petiole. Tree habit, leaf and fruit characters of Soronga are very much similar to that of Jamir-tenga, a variety of *C. megaloxycarpa* (sour pummelo) found in Garo hills of Meghalaya.

vi) **Chamba**, a natural variant of *Citrus*, morphologically more close to *C. paradisi* (grapefruit). It was collected from Simsangre, East Garo hills. Individuals of this variant are decreasing at alarming rate in East Garo hills, although it was earlier reported to grow in other localities like Chandgre and Sasatgre of East Garo hills.

vii) **Bamsim**, a new variant of *Citrus* was collected from Rengsangre, East Garo hills. It is morphologically much similar to sour orange.

viii) **Narang juskil (Atol)** is also one of the important natural hybrids and found in Rongsak, East Garo hills. Very few individuals of this variant were growing in that region under semi-domestication. It is morphologically very much similar to *C. indica* except fruit characters as its small sized yellowish fruits have no seeds and very less amount of juice.

The factors like inter-specific hybridization, apomixes, bud mutations and diverse eco-geographical distribution worked together to produce enormous variability in *Citrus*, resulting in presence of diverse cultivars with in a species. Presence of these ambiguous

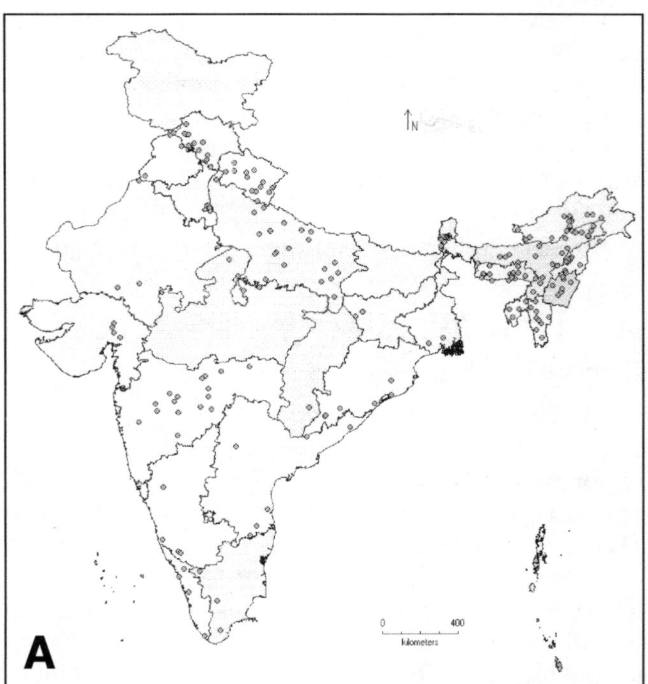

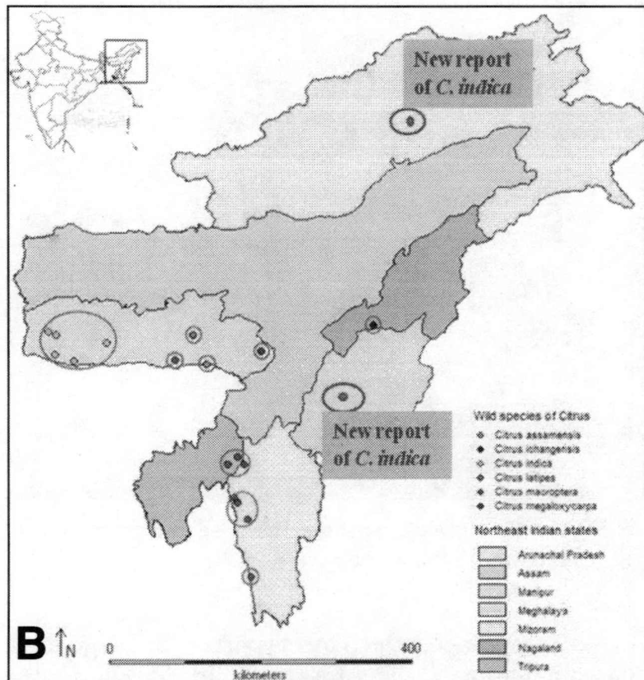

Figure 3: A. Collection Sites of Citrus Germplasm from Various States of India. B. New Sites of *C. indica* Collection.

taxa and several other unreported genotypes in the vast genetic diversity of citrus in northeast India need in depth investigation and surveys and explorations.

New Locations of Species

Recent reports of presence of *C. indica* in the East Siang and Upper Siang valleys of Arunachal Pradesh and Dailong village in Manipur's Tamenglong district are exciting reports for the researchers of plant genetic resources especially explorers. This clearly indicates the need for more extensive surveys and explorations in the northeast India for indigenous plant species. New locations of the presence of these species not only enrich existing diversity and new populations with inherent variations due to genetic and edapho-climatic factors but also provide very interesting information on diverse and localized socio-economic, cultural and religious use of these wild species in different states of northeast India. In case of *C. indica* which is known as "Memong Narang" (Ghost Fruit) in the Garo hiils of Meghalaya, is interestingly known as "Bango Shinkin" in Siang Valley and "Biurengthai" in the of Tamenglong district of Manipur. In Siang valley trees of this species were small to medium sized with very heavy fruiting. Fruits were deep red, variable in shape and size and were less hardy, comparatively large, with shining skin and full of large shiny seeds. This species is known as "Bango Shinkin" by the tribals of this area where "Bango" means small monkey and "Shinkin" means fruits. It was interesting to note that this fruit was not being used by the local tribals for any economic or social or religious purposes as was being used in Garo hills. Only use for this fruit in this area is that it is consumed raw occasionally by tribals, when moving in the forest and is not being harvested for use or sale as fruit being very sour and not of any liking and importance. Due fruit being non-edible and of no socio-economic important, plants of this species are being removed very fast by farmers in Siang Valley while clearing forest land for agriculture. Tribal farmers were not even aware of any use and importance of this species, which has got a Geographical Indicator (GI) tag for Meghalaya state and is known as "Indian Wild orange" with lot of scientific importance in Citrus taxonomy. This species is present in forest of almost all the parts of East Siang and Upper Siang district of Arunachal Pradesh but never reported earlier in literature. Similar case is with *C. indica* ("Biurengthai") in the parts of Tamenglong district of Manipur, where there is still need to thoroughly study the morphology, variability, socio-economy and traditional importance of this fruit species.

Citrus Gene Sanctuary

The World's only "Citrus Gene Sanctuary" is at Nokrek Biosphere Reserve (NBR) in Garo Hills of Meghalaya and was established in 1981 covering an area of about 10, 266 hectares (Singh, 1981). This unique Citrus gene sanctuary probably first of its kind in the World is endowed with highly specified microclimate with a combination of tropical and mild temperate climate and experiencing very high humidity and rainfall. In 2009, the Nokrek Biosphere Reserve was designated as UNESCO World Heritage Site. The Nokrek is the highest peak of the East and West Garo hills located

Figure 4: Nokrek Biosphere Reserve in Meghalaya (*Source*: Borah, 2009).

on the Tura Range, which forms part of the Meghalaya Plateau (Figure 4). The reserve holds a remarkable variety of endemic *Citrus* spp., especially *Citrus indica* (Indian wild orange). Very low genetic diversity and destruction of its natural habitat pose serious threat to *C. indica* even in the Citrus Gene Sanctuary in Nokrek Biosphere Reserve (Kumar *et al.*, 2010; Malik *et al.*, 2013). Narrow distributional range, high habitat specificity, low natural regeneration and extraction of fruits from wild by Garos for local use and sale in local markets are the other constraints associated with *in situ* germplasm conservation and management of *C. indica* and other *Citrus* species growing in gene sanctuary area. Recently, *Citrus indica* (Memong narang) has been awarded the Geographical Indicator tag for the state of Meghalaya in India being endemic to northeast India. This highly rich area of citrus and other plant genetic diversity is also facing lot of damage due to human interventions and slow regeneration capacity of some of the species.

There is an urgent need to study the existing genetic diversity of entire Citrus Gene Sanctuary and to revive this area by marking the habitats for indigenous species and types along with other forest species for sustainable and dynamic habitat conservation. This "Citrus Gene Sanctuary" amply deserves the status of "Biodiversity Heritage Site" under the provisions existing in "The Biological Diversity Act, 2002" of India (Anonymous, 2004). Complementary *ex situ* conservation, propagation and cultivation methods need to be urgently undertaken for protection and maintenance of genetic diversity in this rapidly declining genetic resource. Establishment of *ex situ* germplasm conservatories in the form of a field gene bank and associated propagation centers in the vicinity of Nokrek Biosphere Reserve will be an alternate strategy for augmentation of germplasm materials of *C. indica* and other wild and semi-wild species, which can serve as important resources for recovery and rehabilitation of these threatened plant species.

There is an urgent need to develop "Citrus parks" or "Citrus Biodiversity Gardens" in other states of Northeast India where indigenous germplasm of diverse *Citrus* species can be maintained for conservation and utilization. It would help in exchange, conservation and sustainable use of citrus germplasm for genetic improvement as well as knowledge site for students, researchers and teachers and tourism site for national and international visitors.

Need for Dynamic *In situ* On-farm Conservation

India in general and tribal regions in particular still have workable on-farm conservation systems, but in the present context they are also facing threats. *In situ* on-farm conservation is a common practice in citrus diversity rich areas of north-eastern Himalayas. In citrus several "local selections" or "farmers varieties" have been selected or identified since time immemorial in India. Due to their local socio-economic importance, special liking as a food or additive, cultural or religious purposes, on-farm conservation for these semi-wild or semi-domesticated, local natural selections/cultivars/ farmer's varieties have been attempted. These local cultivars are being grown as isolated plants or in small numbers in the homestead gardens, farmers fields, backyards or in the common Panchayat lands in villages. Such selections need urgent attention for further characterization, evaluation, taxonomical identity and on-farm conservation. *In situ* on farm conservation of Kachai lemon in Kachai village of Okhrul district of Manipur is a model of exclusive farmers participation in on farm conservation and sustainable horticulture providing livelihood support to tribal farmers (Malik *et al.*, 2013). *In situ* conservation areas for wild and semi-wild species and several local Citrus types have been proposed (Figure 3A) based on detailed survey in the north-east India which needs to be urgently taken up with the cooperation of local farmers (Malik *et al.*, 2013). Recent concept and role of custodian farmers in on farm conservation may be further strengthen especially for citrus genetic resources of India being highly diverse in nature (Sthapit *et al.*, 2013).

Revised targets for 2020 of Global Strategy for Plant Conservation (GSPC) of CBD and Strategic Plan for Biodiversity (2011-2020) and the Aichi Targets for Biodiversity under their Strategic goals and objectives amply suggest to the member countries to assess the conservation status and effectively conserve at least 75 per cent of known threatened plant species using *in situ* and *ex situ* methods (Sharrock, 2012; CBD, 2011). In case of Indian Citrus plant genetic resources concerted efforts are required to conserve existing diversity using both *in situ* and *ex situ* conservation. Species specific conservation strategy is to be adopted depending upon biological status, population size, diversity and endemism in the northeast India. Some of the species namely *C. indica*, *C. latipes* and *C. assamensis* are endemic to this area requiring to grow in the specific microclimate. In contrarast, *C. macroptera* and *C. indica* due to their increasing commercial, religious and cultural importance are now being prefered for cultivation by the farmers and tribal people in some parts of the Northeast india. *In situ* and *in situ* on-farm conservation strategies are proposed for species like *C. indica*, *C. macroptera*, *C. latipes*, *C. ichangensis*, *C. assamensis* and *C. megaloxycarpa* (Figure 5). *Ex situ* conservation in the field genebanks have been tried for wild and semi-wild species of *Citrus* but it has not been of much success due to the requirement of specific microclimate for better survival and overcoming/breaking juvenility. Therefore, conservation of wild and semi-wild species of *Citrus* may be attempted using dynamic *in situ* approach. Farmers and tribal communities of these areas may be encouraged to maintain and grow few plants of these wild species in thier fields, backyards or in community lands for reasonable incentive as per the provisions already made in the "Protection of Plant Varieties and Farmer's Right Act, 2001" of India (Anonymous, 2001). Farmers and farmers communities may be further supported for their efforts for the conservation and sustainable use of genetic resources including *in situ* and *ex situ* conservation especailly in the agro-biodiversity hotspot under the "National Gene Fund".

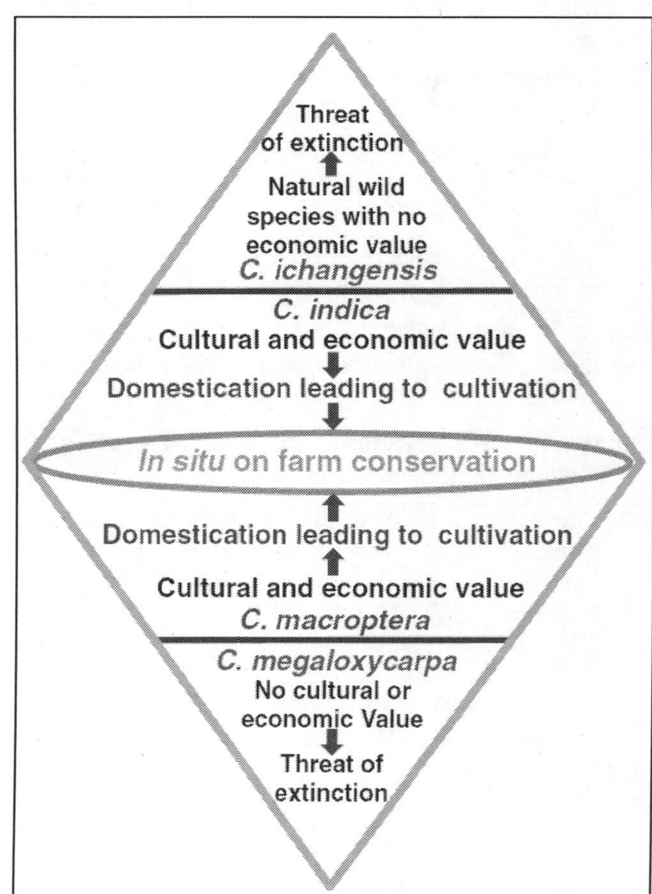

Figure 5: Domestication *vis-à-vis* on-farm Conservation Model for Four Wild *Citrus* Species of North-East India (*Source:* Malik *et al.*, 2013).

Conclusion

In conclusion efficient management of citrus genetic resources in India needs further impetus. Northeastern citrus genetic resources need to be revisited in the light of vast diversity and distribution. Explorations in northeast and northwest India especially remote and difficult areas are to be targeted for collection of rare germplasm from new sites. Characterization, evaluation, taxonomic studies and domestication trend and existing bottle necks in indigenous citrus plant genetic resources need to be addressed using biotechnological tools and techniques. This would help in realizing the real wealth of available genetic diversity which can be further exploited for citrus improvement programmes. As far as conservation is concerned, both *in situ* and *ex situ* approaches are strongly recommended, while *ex situ* conservation (*i.e.* field genebanks, clonal repositories and cryobanking) would address the immediate need, the dynamic *in situ* conservation (*in situ* on-farm conservation, gene sanctuary and national parks), especially for wild and semi-wild species is inevitable for long-term conservation and evolution of these species.

References

Anonymous. 2001. *The Protection of Plant Varieties and Farmers' Rights Act, 2001* (No. 53 of 2001). Universal Law Publishing Co. Pvt. Ltd., New Delhi, India.

Anonymous. 2004. *The Biological Diversity Act, 2002 and Biological Diversity Rules, 2004.* Triplicane, Chennai, India: National Biodiversity Authority. Frontline Offset Printers.

Bhattacharya, S.C. and Dutta, S. 1956. Classification of citrus fruits of Assam. *Sc. Monogr.*, ICAR, New Delhi, **20**: 110.

Borah, A. 2009 Local citrus goes global. *Down to Earth*, January.

Carbonell-Caballero *et al.*, 2015. A phylogenetic analysis of 34 chloroplast genomes elucidates the relationships between wild and domestic species within the genus Citrus. *Mol. Bio. Evol.* Doi: 10.1093/molbev/msv082.

CBD. 2011. https://www.cbd.int/doc/strategic–plan/2011./Aichi–Targets–EN.pdf

Chadha, K.L. 1995. Status report of tropical fruit species in South Asia. **In**: *Proceedings Expert Consultation on Tropical Fruit Specially Asia*, (Eds.) R.K. Arora and V. Ramanatha Rao, MARDI Serdang Kuala Lumpur, IPGRI Office for South Asia, New Delhi, pp. 45–60.

Frost, H.B. and Soost, R.K. 1968. Seed reproduction: Development of gametes and embryos. In: *The Citrus Industry, Vol. 2*, (Eds.) W. Reuther, L.D. Batchelor and H.B. Webber. Univ. California Press, pp. 290–324.

Hazarika, T.K. 2012. Citrus genetic diversity of north-east India, their distribution, ecogeography and ecobiology. *Genet. Resour. Crop Evol.*, **59**: 1267–1280.

Jena, S.N., Kumar, S. and Nair, N.K. 2009. Molecular phylogeny in Indian *Citrus* L. (Rutaceae) inferred through PCR-RFLP and trnL-trnF sequence data of chloroplast DNA. *Sci. Hort.*, **119**: 403–416.

Kumar, S., Jena, S.N. and Nair, N.K. 2010. ISSR polymorphism in Indian wild orange (*Citrus indica* Tanaka Rutaceae) and related wild species in North-east India. *Sci. Hort.*, **123**: 350–359.

Kumar, S., Malik, S.K., Chaudhury, R. and Bhat, S.R. 2014. A new wild type of citron (*Citrus medica* L., Rutaceae) identified through morphology and psbM-trnD spacer region of chloroplast DNA. *Tree Genet. Genomes*, 10.1007/s00468-014-1022-1.

Mabberley, D.J., 2004. *Citrus* (Rutaceae): A review of recent advances in etymology, systematics and medical applications. *Blumea*, **49**: 481–498.

Malik, S.K., Chaudhury, R., Dhariwal, O.P. and Kalia, R.K. 2006. Collection and characterization of *Citrus indica* Tanaka and *C. macroptera* Montr.: Wild endangered species of northeastern India. *Genet. Res. Crop Evol.*, **53**: 1485–1493. http://dx.doi.org/10.1007/s10722-005-7468-7.

Malik, S.K., Chaudhury, R., Kumar, S., Dhariwal, O.P. and Bhandari, D.C. 2012. *Citrus Genetic Resources in India: Present Status and Management.* National Bureau of Plant Genetic Resources, New Delhi.

Malik, S.K., Rohini, M.R., Kumar, S., Choudhary, R., Malik, S.K., Kumar, S., Singh, I.P., Dhariwal, O.P. and Chaudhury, R. 2013. Socio-economic importance, domestication trends and *in situ* conservation of wild Citrus species of Northeast India. *Genet. Res. Crop Evol.*, **60**: 1655–1671 http://dx.doi.org/10.1007/s10722-012-9948-x.

Nair, K.N. and Nayar, M.P. 1997. Rutaceae. **In**: *Flora of India, Vol. IV*, (Eds.) P.K. Hajra, V.J. Nair and P. Daniel, Botanical Survey of India, Calcutta, India, p. 229-407.

Scora, R.W., 1975. On the history and origin of *Citrus*. *Bull. Torrey Bot. Club*, **102**: 369–375.

Sharma, B.D., Hore, D.K. and S.G. Gupta 2004. Genetic resources of Citrus of north-eastern India and their potential use. *Genet. Res. Crop Evol.*, **51**: 411–418.

Sharrock, S. 2012. *A Guide to the GSPC: All the Targets, Objectives and Facts*. Botanic Gardens Conservation International Descanso House, 199 Kew Road, Richmond, TW9 3BW, UK, p 36.

Singh, B. 1981. *Establishment of First Gene Sanctuary for Citrus in Garo Hills*. Concept Publ. Co., New Delhi, p. 182.

Singh, H.P and Chadha, K.L. 1993. Genetic resources of citrus. In: *Advances in Horticulture, Vol. 1: Fruit Crops, Part 1*, (Eds.) K.L. Chadha and O.P. Pareek. Malhotra Publishing House, New Delhi, India, pp. 95–121.

Singh, I.P. and Singh, S. 2003. *Exploration, Collection and Mapping of Citrus Genetic Diversity in India*. Technical Bulletin No. 7, National Research Centre for Citrus, Nagpur.

Singh, S. 2001. Citrus industry in India. **In**: *Citrus*, (Eds.) Shyam Singh and S.A.M.H. Naqvi. International Book Distributing Co, Lucknow, pp. 3–41.

Sthapit, B., Lamers, H., and Rao, R. 2013. Custodian farmers of agricultural biodiversity: Selected profiles from South and South East Asia. Paper presented at: *Workshop on Custodian Farmers of Agricultural Biodiversity*, Biodiversity International, New Delhi, India.

Swingle, W.T. and Reece, P.C. 1967. The botany of Citrus and its wild relatives. In: *The Citrus Industry, Vol. 1*, (Eds.) W. Reuther, H.J. Webber and L.D. Batchelor, University of California, Berkeley, pp. 190–430.

Tanaka, T. 1928. On certain new species of *Citrus*. *Stud. Citrol.*, **2**: 155–164 (Japanese with English and Latin resume).

Tanaka, T. 1937. Further revision of Rutaceae-Auranticeae of India and Ceylon. *Indian Bot. Soc.*, **16**: 227–240.

Tanaka, T. 1958. The origin and dispersal of *Citrus* fruits having their centre of origin in India. *Indian J. Hort.*, **15**: 101–115.

Tanaka, T. 1977. Fundamental discussion of *Citrus* classification. *Stud. Citrol.*, **14**: 1–6.

Wu, Albert *et al.*, 2015. Sequencing of diverse mandarin, pummelo and orange genomes reveals complex history of admixture during citrus domestication. *Nat. Biot.*, **32**(7) 656–662.